IFMBE Proceedings

Volume 41

Series Editor

Ratko Magjarevic

Deputy Editors

Fatimah Binti Ibrahim
Igor Lacković
Piotr Ładyżyński
Emilio Sacristan Rock

For further volumes:
http://www.springer.com/series/7403

The International Federation for Medical and Biological Engineering, IFMBE, is a federation of national and transnational organizations representing internationally the interests of medical and biological engineering and sciences. The IFMBE is a non-profit organization fostering the creation, dissemination and application of medical and biological engineering knowledge and the management of technology for improved health and quality of life. Its activities include participation in the formulation of public policy and the dissemination of information through publications and forums. Within the field of medical, clinical, and biological engineering, IFMBE's aims are to encourage research and the application of knowledge, and to disseminate information and promote collaboration. The objectives of the IFMBE are scientific, technological, literary, and educational.

The IFMBE is a WHO accredited NGO covering the full range of biomedical and clinical engineering, healthcare, healthcare technology and management. It is representing through its 60 member societies some 120.000 professionals involved in the various issues of improved health and health care delivery.

IFMBE Officers
President: Ratko Magjarevic, Vice-President: James Goh
Past-President: Herbert Voigt
Treasurer: Marc Nyssen, Secretary-General: Shankhar M. Krishnan
http://www.ifmbe.org

Laura M. Roa Romero
Editor

XIII Mediterranean Conference on Medical and Biological Engineering and Computing 2013

MEDICON 2013, 25–28 September 2013, Seville, Spain

Volume 2

Editor
Laura M. Roa Romero
Biomedical Engineering Group
Engineering School of the University of Seville
Sevilla
Spain

ISSN 1680-0737 ISSN 1433-9277 (electronic)
ISBN 978-3-319-00845-5 ISBN 978-3-319-00846-2 (eBook)
Printed in 2 Volumes
DOI 10.1007/978-3-319-00846-2
Springer Cham Heidelberg New York Dordrecht London

Library of Congress Control Number: 2013949465

© Springer International Publishing Switzerland 2014

This work is subject to copyright. All rights are reserved by the Publisher, whether the whole or part of the material is concerned, specifically the rights of translation, reprinting, reuse of illustrations, recitation, broadcasting, reproduction on microfilms or in any other physical way, and transmission or information storage and retrieval, electronic adaptation, computer software, or by similar or dissimilar methodology now known or hereafter developed. Exempted from this legal reservation are brief excerpts in connection with reviews or scholarly analysis or material supplied specifically for the purpose of being entered and executed on a computer system, for exclusive use by the purchaser of the work. Duplication of this publication or parts thereof is permitted only under the provisions of the Copyright Law of the Publisher's location, in its current version, and permission for use must always be obtained from Springer. Permissions for use may be obtained through RightsLink at the Copyright Clearance Center. Violations are liable to prosecution under the respective Copyright Law.

The use of general descriptive names, registered names, trademarks, service marks, etc. in this publication does not imply, even in the absence of a specific statement, that such names are exempt from the relevant protective laws and regulations and therefore free for general use.

While the advice and information in this book are believed to be true and accurate at the date of publication, neither the authors nor the editors nor the publisher can accept any legal responsibility for any errors or omissions that may be made. The publisher makes no warranty, express or implied, with respect to the material contained herein.

The IFMBE Proceedings is an Official Publication of the International Federation for Medical and Biological Engineering (IFMBE)

Printed on acid-free paper

Springer is part of Springer Science+Business Media (www.springer.com)

Welcome Message

It is a great pleasure and honor to welcome all participants in the XIII Mediterranean Conference on Medical and Biological Engineering and Computing (MEDICON 2013), which will be held in Seville, the capital of the Spanish region of Andalusia, from September 25 to 28, 2013.

MEDICON is a regional conference with a long tradition and high scientific level, which is organized every three years in a Mediterranean country under the umbrella of the International Federation of Medical and Biological Engineering (IFMBE).

The general theme of MEDICON 2013 is "Research and Development of Technology for Sustainable Healthcare". This decade is being characterized by the appearance and use of emergent technologies under development. This situation has produced a tremendous impact on Medicine and Biology from which it is expected an unparalleled evolution in these disciplines towards novel concept and practices. The consequence will be a significant improvement in health care and well-fare, i.e. the shift from a reactive medicine to a preventive medicine. This shift implies that the citizen will play an important role in the healthcare delivery process, what requires a comprehensive and personalized assistance. In this context, society will meet emerging media, incorporated to all objects, capable of providing a seamless, adaptive, anticipatory, unobtrusive and pervasive assistance. The challenge will be to remove current barriers related to the lack of knowledge required to produce new opportunities for all the society, while new paradigms are created for this inclusive society to be socially and economically sustainable, and respectful with the environment. In this way, the conference program will be focused on the convergence of biomedical engineering topics ranging from formalized theory through experimental science and technological development to practical clinical applications.

In spite of all the problems that many Mediterranean countries are currently facing due to the financial and political crisis, we are proud of the great diffusion of this Conference, with participants coming from all the continents. This participation is a sign that the Biomedical Engineering Community can contribute to overcome the challenges that are affecting society. I would like to share this success with all the members of the committees, institutions, participants, and individual persons and that have made possible the celebration of this event.

Seville is a millenary town with a rich artistic heritage as a consequence of the fusion of the three cultures. The conjunction of the Spanish weather, history and natural resources have shaped it as a modern city with an outstanding projection at the levels of technological research and industry, and biomedical research. All the improvements made on the city during the last decades, have allowed Seville to become a cosmopolitan and easy-to-get destination.

I am sure that you will enjoy a wonderful experience of visiting Seville during your stay.

Laura M. Roa　　　　　　　　　　　　　　　　　　　　　　　　　　　　　　　　　　　　Conferencechair, MEDICON 2013

Preface

We are pleased to introduce the Proceedings of the 13th Mediterranean Conference on Medical and Biological Engineering and Computing (MEDICON 2013), that are published as the 39th volume of the IFMBE Proceedings, published by Springer under the sponsorship of the International Federation for Medical and Biological Engineering (IFMBE).

Over 500 papers have been submitted for review by the International Program Committee/International Scientific Committee, and the program is structured under ten themes covering the latest advances on Medical and Biological Engineering and Computing. These Proceedings are structured as follows: the first section includes the conference committees, themes and tracks, and parallel activities comprising special sessions, workshops and round tables; next, keynote speakers with their plenary lectures and invited papers; finally, the papers selected for presentation at the conference. The indexing of the Proceedings are eased by two registers for authors and themes and tracks.

We are grateful to Grupo Pacifico Technical Secretary and to Springer for their help and support for the edition of these Proceedings.

We wish that this volume of IFMBE Proceedings will serve as a valuable source of information for professionals, scholars, researchers and students as well as a reference for the state-of-the-art in this period of rapid development and changes in our fields of Medical and Biological Engineering and Computing.

Laura M. Roa Romero
Igor Lackovic
Kang-Ping Lin
Javier Reina-Tosina
Publication Committee

Conference Committee

Conference Chair

Laura M. Roa — Universidad de Sevilla (Spain)

Local Organizing Committee Chair

Javier Reina-Tosina — Universidad de Sevilla (Spain)

Publication Committee

Laura M. Roa — Universidad de Sevilla (Spain)
Igor Lackovic — University of Zagreb (Croatia)
Karg-Ping Lin — Chung Yuan Christian University (Taiwan)
Javier Reina-Tosina — Universidad de Sevilla (Spain)

Young Investigator Competition Committee

Marcello Bracale — Università degli Studi di Napoli Federico II (Italy)
Fumihiko Kajiya — Kawasaki University of Medical Welfare (Japan)
Nicolas Pallikarakis — University of Patras (Greece)
Joe Barbenel — University of Strathclyde (UK)
Enrique J. Gómez — Universidad Politécnica de Madrid (Spain)

Local Organizing Committee

José A. Milán — Hospital Universitario Virgen Macarena (Spain)
Isabel Román — Universidad de Sevilla (Spain)
Jorge Calvillo-Arbizu — Universidad de Sevilla (Spain)
David Naranjo-Hernández — Universidad de Sevilla, CIBER BBN (Spain)
Amparo Callejón — Universidad de Sevilla (Spain)
Miguel Ángel García — Universidad de Sevilla (Spain)
Alejandro Talaminos — Universidad de Sevilla (Spain)
Gerardo Barbarov — Universidad de Sevilla (Spain)
Alfonso Lara — Hospital Universitario Virgen Macarena (Spain)

Regional Conference Organizing Committee

Associazione Italiana di Ingegnaria Medica e Biologica

Leandro Pecchia — University of Nottingham (UK)
Paolo Melillo — Università degli Studi di Napoli Federico II (Italy)

Croatian Medical and Biological Engineering Society

Igor Lackovic	University of Zagreb (Croatia)
Vedran Bilas	University of Zagreb (Croatia)

Cyprus Association of Medical Physics and Biomedical Engineering

Prodromos A. Kaplanis	University of Cyprus (Cyprus)
Constantinos Pattichis	University of Cyprus (Cyprus)

European Alliance for Medical and Biological Engineering and Science

Birgit Glasmacher · Leibniz Universitaet Hannover (Germany)

Greek Society for Biomedical Engineering

Nicolas Pallikarakis	University of Patras (Greece)
Panagiotis Bamidis	Aristotle University of Thessaloniki (Greece)

IEEE-Engineering in Medicine and Biology Society

Donna Hudson · UCSF Fresno (USA)

Israel Society for Medical and Biological Engineering

Haim Azhari	Technion-IIT (Israel)
Idit Avrahami	Ariel University (Israel)

Italian Association of Clinical Engineers

Paola Freda	Paola Freda, AReSSPiemonte - Agenzia Regionale per i Servizi Sanitari (Itlay)
Paolo Lago	Fondazione IRCCS - Policlinico San Matteo Pavia (Italy)

Société Française de Génie Biologique et Médical

Catherine Marque	Université de Technologie de Compiègne (France)
Véronique Migonney	Université Paris 13 (France)

Sociedade Portuguesa de Egenharia Biomedica

Mario F. Secca	Universidade Nova de Lisboa (Portugal)
Pedro M.C. Vieira	Universidade Nova de Lisboa (Portugal)

Slovene Society for Medical and Biological Engineering

Damijan Miklavcic	University of Ljubljana (Slovenia)
Tomaz Vrtovec	University of Ljubljana (Slovenia)

Spanish Society of Biomedical Engineering

Enrique J. Gómez	Universidad Politécnica de Madrid (Spain)
Raimon Jané	Universitat Politècnica de Catalunya (Spain)

International Program Committee

Zulfiqur Ali	Teesside University (England)
Metin Akay	University of Houston (USA)

Conference Committee

Ahmad Taher Azar	Misr University for Science and Technology (Egypt)
Haim Azhari	Technion-IIT Haifa (Israel)
Panagiotis Bamidis	Aristotle University of Thessaloniki (Greece)
José López Barneo	Universidad de Sevilla (Spain)
José Becerra	Universidad de Málaga (Spain)
Anna M. Bianchi	Politecnico di Milano (Italy)
Paolo Bonato	Harvard Medical School (USA)
Lodewijk Bos	ICMCC (Holland)
Marcello Bracale	Università degli Studi di Napoli Federico II (Italy)
Saide Calil	State University of Campinas (Brazil)
Pere Caminal	Universidad Politécnica de Cataluña (Spain)
Alicia Casals	Universidad Politécnica de Cataluña (Spain)
Bernardo Celda	Universidad de Valencia (Spain)
Sergio Cerutti	Politecnico di Milano (Italy)
Walter H. Chang	Chung Yuan Christian University (Taiwan)
Febo Cincotti	Sapienza Università di Roma (Italy)
Jean Louis Coatrieux	Université de Rennes 1 (France)
Yadin David	Engineering Consultants LLC (USA)
José M. Delgado-García	Universidad Pablo de Olavide (Spain)
Manuel Desco	Universidad Carlos III (Spain)
Manuel Doblaré	Universidad de Zaragoza (Spain)
José M. Ferrero-Jr	Universidad Politécnica de Valencia (Spain)
Mario Forjaz Secca	Universidade Nova de Lisboa (Portugal)
Dimitris Fotiadis	University of Ioannina (Greece)
Alejandro Frangi	Universitat Pompeu Fabra (Spain)
Enrique Gómez	Universidad Politécnica de Madrid (Spain)
Tomás Gómez-Cía	Hospital Universitario Virgen del Rocío (Spain)
Juan Guerrero	Universidad de Valencia (Spain)
Carlos H. Salvador	Instituto de Salud Carlos III (Spain)
Elena Hernando	Universidad Politécnica de Madrid (Spain)
Roberto Hornero	Universidad de Valladolid (Spain)
Gerhard Holzapfel	Technische Universität Graz (Austria)
Donna Hudson	UCSF Fresno (USA)
Helmut Hutten	Technische Universität Graz (Austria)
Antonio Fernando C. Infantosi	Federal University of Rio de Janeiro (Brazil)
Robert Istepanian	Kingston University (England)
Christopher James	University of Warwick (England)
Raimon Jané	Universidad Politécnica de Cataluña (Spain)
Akos Jobbagy	Budapest University of Technology and Economics (Hungary)
Fumihiko Kajiya	Kawasaki University of Medical Welfare (Japan)
Roger Kamm	Massachusetts Institute of Technology (USA)
Gyeon-Man Kim	CEIT (Spain)
Pablo Laguna	Universidad de Zaragoza (Spain)
Thomas Lango	SINTEF Health Research Trondheim (Norway)
Alberto Leiva	Hospital Sant Pau (Spain)
Nigel Lovell	University of New South Wales (Australia)
Ratko Magjarevich	University of Zagreb (Croatia)
Nicos Maglaveras	Aristotle University of Thessaloniki (Greece)
Luca Mainardi	Politecnico di Milano (Italy)

Jaakko Malmivuo	Aalto University (Finland)
Meyya Meyyapan	NASA (USA)
Véronique Migonney	Université Paris 13 (France)
Damijan Miklavcic	University of Ljubljana
José Luis Monteagudo	Instituto de Salud Carlos III (Spain)
Joachim Nagel	University of Stuttgart (Germany)
Robert Newcomb	University of Maryland (USA)
Konstatina Nikita	National Technical University of Athens (Greece)
Marc Nyssen	Vrije Universiteit Brussel (Belgium)
Aníbal Ollero	Universidad de Sevilla (Spain)
Ramón Pallás	Universidad Politécnica de Cataluña (Spain)
Nicolas Pallikarakis	University of Patras (Greece)
Rosario Pásaro	Universidad de Sevilla (Spain)
Constantinos Pattichis	University of Cyprus (Cyprus)
Josep A. Planell	IBEC Barcelona (Spain)
Leandro Pecchia	University of Nottingham (UK)
Antonio Pedotti	Politecnico di Milano (Italy)
Barbara Perscionek	Kingston University (England)
Ronald Phlypo	University of Maryland (USA)
José Luis Pons Rovira	CSIC (Spain)
Francisco del Pozo	Universidad Politécnica de Madrid (Spain)
Jose Principe	University of Florida (USA)
Richard Reilly	Trinity College Dublin (Ireland)
Juan Ribas	Universidad de Sevilla (Spain)
Charles Robinson	Clarkson University (USA)
Montserrat Robles	Universidad Politécnica de Valencia (Spain)
Joaquin Roca	Universidad Politécnica de Cartagena (Spain)
Christian Roux	Télécom Bretagne (France)
Carmelina Ruggiero	Università degli Studi di Genova (Italy)
Juan Sabaté	Universidad de Sevilla (Spain)
Josep Samitier	Universitat de Barcelona (Spain)
Andrés Santos	Universidad Politécnica de Madrid (Spain)
Javier Saiz	Universidad Politécnica de Valencia (Spain)
Francisco Sánchez-Margallo	Centro de Mínima Invasión Jesús Usón (Spain)
Niilo Saranummi	VTT Finland (Finland)
Jos Spaan	Universiteit van Amsterdam (Holland)
Christos Schizas	University of Cyprus (Cyprus)
Leif Sornmo	Lunds Universitet (Sweden)
Ron Summers	Loughborough University (England)
Toshiyo Tamura	Chiba University (Japan)
Heikki Terio	Karolinska University Hospital (Sweden)
José María Tormos	InstitutGuttmann (Spain)
Shoogo Ueno	Kyushu University (Japan)
Gozde Unal	Sabancı Üniversitesi (Turkey)
Max Valentinuzzi	Universidad de Favaloro (Argentina)
Juan José Vaquero	Universidad Carlos III de Madrid (Spain)
Jaume Veciana	Institut de Ciencia de Materials de Barcelona (CSIC) (Spain)
Herbert Voigt	Boston University (USA)
Y.T. Zhang	Chinese University of Hong Kong (Hong Kong)

International Scientific Committee

Begoña Acha	Universidad de Sevilla (Spain)
Carlos Alberola	Universidad de Valladolid (Spain)
Mariano Alcañiz	Universidad Politécnica de Valencia (Spain)
Ricardo L. Armentano	Universidad Favaloro (Argentina)
Sergio Arana	CEIT (Spain)
Joe Barbenel	University of Strathclyde (UK)
Giuseppe Battaglia	University of Sheffield (UK)
Anastasios Bezerianos	National University of Singapore (Singapore)
Ewart Carson	City University London (UK)
Mario Cifrek	University of Zagreb (Croatia)
Rui Cortesao	Universidade de Coimbra (Portugal)
Hariton Costin. Grigore T	Popa University of Medicine and Pharmacy, Iasi (Romania)
Gianaurelio Cuniberti	Technische Universität Dresden (Germany)
Stefanie Demirci	Munich Technical University (Germany)
André Dittmar	CNRS (France)
Olaf Doessel	Karlsruhe Institute of Technology (Germany)
Jaime Domínguez Abascal	Universidad de Sevilla (Spain)
Kim Dremstrup	Aalborg University (Denmark)
Christohper Druzgalski	California State University. Long Beach (USA)
Miguel Ángel Estudillo	CIBER-BBN (Spain)
José Carlos Fernández de Aldecoa	Sociedad Española de Electromedicina e Ingeniería Clínica (Spain)
Monique Frize	University of Ottawa (Canada)
Javier García-Casado	Universidad Politécnica de Valencia (Spain)
Amit Gefen	Tel Aviv University (Israel)
Beatriz Giraldo	Universitat Politècnica de Catalunya (Spain)
Balwant Godara	Institut Superieur d'Electronique de Paris (Francia)
Alberto del Guerra	Università de Pisa (Italy)
Enrique Guijarro	Universidad Politécnica de Valencia (Spain)
James C.H. Goh	National University of Singapore (Singapore)
Hiie Hinrikus	Tallinn University of Technology (Estonia)
Ernst Hofer	Medizinische Universität Graz (Austria)
Roman Hovorka	University of Cambridge (UK)
Ernesto Iadanza	UniversitàdegliStudi di Firenze (Italy)
David Izquierdo-Garcia	Harvard University (USA)
Luis Kun	NationalDefenseUniversity (USA)
Damien Lacroix	University of Sheffield (UK)
Rajaram Lakshminarayan	University of South Florida (USA)
André C. Linnenbank	Academisch Medisch Centrum (Netherlands)
Miguel López Coronado	Universidad de Valladolid (Spain)
José Alberto Maldonado Segura	Universidad Politécnica de Valencia (Spain)
José D. Martín	Universidad de Valencia (Spain)
Ruben Martín	Universidad de Sevilla (Spain)
José Luis Martínez de Juan	Universidad Politécnica de Valencia (Spain)
Marco Masseroli	Politecnico di Milano (Italy)
Olivier Meste	Université Nice Sophia Antipolis (France)
José Millet Roig	Universidad Politécnica de Valencia (Spain)
Umberto Morbiducci	Politecnico di Torino (Italy)
Elena De Momi	Politecnico di Milano (Italy)

Julio Moreno	Hospital Universitario Virgen Macarena, Universidad de Sevilla (Spain)
Víctor F. Muñoz	Universidad de Málaga (Spain)
Klaus Neuder	VBE-DKE (Germany)
Quentin Noirhomme	Université de Liège (Belgium)
Tadeusz Palko	Warsaw University of Technology (Poland)
Alfonso Palma	Universidad de Sevilla (Spain)
Gernot Plank	Medizinische Universität Graz (Austria)
Carmen Poon	Chinese University of Hong Kong (Hong Kong)
Jesús Poza	Universidad de Valladolid (Spain)
Andriana Prentza	University of Piraeus (Greece)
Petia Radeva	Universitat de Barcelona (Spain)
Jose Joaquin Rieta	Universidad Politécnica de Valencia (Spain)
José Luis Rojo	Universidad Rey Juan Carlos (Spain)
Isabel Román	Universidad de Sevilla (Spain)
Xavier Rosell Ferrer	Universitat Politècnica de Catalunya (Spain)
Mercedes Salgueira	Hospital Universitario Virgen Macarena (Spain)
Francisco Sánchez Doblado	Universidad de Sevilla. Sociedad Española de Física Médica (Spain)
Daniel Sánchez Morillo	Universidad de Cádiz (Spain)
Roberto Sassi	Università degli Studi di Milano (Italy)
Francisco Sepulveda	University of Essex (UK)
Carmen Serrano	Universidad de Sevilla (Spain)
Javier Serrano	Universidad Politécnica de Madrid (Spain)
Maria Siebes	University of Amsterdam (Netherlands)
Slavik Tabakov	King's College London (UK)
Edward Vigmond	Université Bourdeax 1 (France)
Karin Wardell	Linköping University (Sweden)
Jan Wojcicki	Institute of Biocybernetics and Biomedical Engineering (Poland)
Vicente Zarzoso	Université Nice Sophia Antipolis (France)

Reviewer's List

Alberto del Guerra	Alicia Casals	Helmut Hutten
Juan Guerrero	Bernardo Celda	Antonio Fernando C. Infantosi
Elena Hernando	Sergio Cerutti	Raimon Jané
Meyya Meyyapan	Walter H. Chang	Akos Jobbagy
Gernot Plank	Jean Louis Coatrieux	Fumihiko Kajiya
Carmen Poon	Yadin David	Gyeon-Man Kim
Luis Javier Reina Tosina	José M. Delgado-García	Pablo Laguna
Zulfiqur Ali	Manuel Desco	Thomas Lango
Ahmad Taher Azar	Manuel Doblaré	Alberto Leiva
Haim Azhari	José M. Ferrero-Jr	Nicos Maglaveras
Panagiotis Bamidis	Mario Forjaz Secca	Luca Mainardi
José López Barneo	Dimitris Fotiadis	Jaakko Malmivuo
José Becerra	Enrique J. Gómez	Véronique Migonney
Anna. M. Bianchi	Tomás Gómez-Cía	Damijan Miklavcic
Lodewijk Bos	Carlos H. Salvador	José Luis Monteagudo
Marcello Bracale	Roberto Hornero	Joachim Nagel
Saide Calil	Gerhard Holzapfel	Robert Newcomb
Pere Caminal	Donna Hudson	Konstantina Nikita

Conference Committee

- Marc Nyssen
- Aníbal Ollero
- Ramón Pallás
- Nicolas Pallikarakis
- Rosario Pásaro
- Constantinos Pattichis
- Leandro Pecchia
- Antonio Pedotti
- Ronald Phlypo
- José Luis Pons Rovira
- Jose Príncipe
- Richard Reilly
- Juan Ribas
- Charles Robinson
- Montserrat Robles
- Joaquin Roca
- Christian Roux
- Carmelina Ruggiero
- Josep Samitier
- Andrés Santos
- Javier Saiz
- Francisco Sánchez-Margallo
- Niilo Saranummi
- Christos Schizas
- Leif Sornmo
- Ron Summers
- Toshiyo Tamura
- Heikki Terio
- Shoogo Ueno
- Gozde Unal
- Juan José Vaquero
- Jaume Veciana
- Herbert Voigt
- Begoña Acha
- Carlos Alberola
- Mariano Alcañiz
- Ricardo L. Armentano
- Sergio Arana
- Joe Barbenel
- Anastasios Bezerianos
- Ewart Carson
- Mario Cifrek
- Rui Cortesao
- Hariton Costin
- Gianaurelio Cuniberti
- Stefanie Demirci
- Olaf Doessel
- Kim Dremstrup
- Cristopher Druzgalski
- Miguel Ángel Estudillo
- José Carlos Fernández de Aldecoa
- Monique Frize
- Javier García-Casado
- Amit Gefen
- Beatriz Giraldo
- Balwant Godara
- Enrique Guijarro
- James C.H. Goh
- Hiie Hinrikus
- Ernst Hofer
- Ernesto Iadanza
- David Izquierdo-García
- Damien Lacroix
- Rajaram Lakshminarayan
- André C. Linnenbank
- Miguel López Coronado
- José Alberto Maldonado Segura
- José D. Martín
- Rubén Martín
- José Luis Martínez de Juan
- Marco Masseroli
- Olivier Meste
- Umberto Morbiducci
- Elena De Momi
- Víctor F. Muñoz
- Quentin Noirhomme
- Tadeusz Palko
- Alfonso Palma
- Jesús Poza
- Andriana Prentza
- Petia Radeva
- Jose Joaquin Rieta
- José Luis Rojo
- Isabel Román
- Xavier Rosell Ferrer
- Mercedes Salgueira
- Francisco Sánchez Doblado
- Daniel Sánchez Morillo
- Roberto Sassi
- Francisco Sepulveda
- Carmen Serrano
- Javier Serrano
- María Siebes
- Slavik Tabakov
- Edward Vigmond
- Karin Wardell
- Jan Wojcicki
- Vicente Zarzoso
- Jaime Domínguez Abascal
- Jorge Calvillo Arbizu
- Nada Philip
- Barbara Pierscionek
- Robert Istepanian
- José Millet Roig

Table of Contents

Volume 1

Part I: Keynote Speakers

Design, Development, Training and Use of Medical Devices; with Practical Examples from Cardiovascular Medicine and Surgery 3
Alan Murray

Part II: Invited Presentations

David Dewhurst – Biomedical Engineer and IFMBE Pioneer 9
R.L.G. Kirsner, J.S. McKenzie

Did Jan Swammerdam Do the First Electric Stimulation over 100 Years before Luigi Galvani? 13
J.A. Malmivuo, J. Honkonen, K.E. Wendel

Part III: Biomechanics, Robotics and Minimal Invasive Surgery

Shear Stress Rapidly Alters the Physical Properties of Vascular Endothelial Cell Membranes by Decreasing Their Lipid Order and Increasing Their Fluidity 19
K. Yamamoto, Akira Kamiya, Joji Ando

An In Vitro Analysis of the Influence of the Arterial Stiffness on the Aortic Flow Using Three-Dimensional Particle Tracking Velocimetry 23
U. Gülan, B. Lüthi, M. Holzner, A. Liberzon, A. Tsinober, W. Kinzelbach

A New Method for Coronary Artery Mechanical Properties 27
Omer Shalev, Dar Weiss, Yigal Kassif, Rami Haj-Ali, Shmuel Einav

Numerical Analysis of a Novel External Support Device for Vein Bypass Grafts 29
T. Meirson, E. Orion, G. Bolotin, M. Brand, I. Avrahami

Hemodynamical Aspects of Endovascular Repair for Aortic Arch Aneurysms 33
A. Nardi, M. Brand, M. Halak, M. Ratan, D. Silverberg, I. Avrahami

Optical Tracking System Integration into IORT Treatment Planning System 37
E. Marinetto, V. García-Vázquez, J.A. Santos-Miranda, F. Calvo, M. Valdivieso, C. Illana, M. Desco, J. Pascau

Comparative Study of Two Laparoscopic Instrument Tracker Designs for Motion Analysis and Image-Guided Surgery: A Technical Evaluation 41
J.A. Sánchez-Margallo, F.M. Sánchez-Margallo, I. Oropesa, M. Lucas, J. Moreno, E.J. Gómez

A Decision Support System Applied to Lipodystrophy Based on Virtual Reality and Rapid Prototyping 45
G. Gómez Ciriza, C. Suárez Mejías, C. Parra Calderón, T. Gómez Cía, R. López García

Comparison between Optical and MRI Trajectories in Stereotactic Neurosurgery 49
K. Wårdell, N. Haj-Hosseini, S. Hemm

EVA: Endoscopic Video Analysis of the Surgical Scene for the Assessment of MIS Psychomotor Skills 52
I. Oropesa, P. Sánchez-González, J.A. Sánchez-Margallo, J. García-Novoa, F.M. Sánchez-Margallo, E.J. Gómez

Quantitative Evaluation of a Real-Time Non-rigid Registration of a Parametric Model of the Aorta for a VR-Based Catheterization Guidance System 57
P. Fontanilla-Arranz, B. Rodriguez-Vila, H. Fontenelle, J. Tarjuelo-Gutiérrez, O.J. Elle, E.J. Gómez

Transtibial Amputee Gait: Kinematics and Temporal-Spatial Analysis 61
A.E.K. Ferreira, E.B. Neves, A.G. Melanda, A.C. Pauleto, D.D. Iucksch, L.A.M. Knaut, R.M. da Silva, R.F. M da Cunha

Development and User Assessment of a Body-Machine Interface for a Hybrid-Controlled 6-Degree of Freedom Robotic Arm (MERCURY) 65
N. Moustakas, A. Athanasiou, P. Kartsidis, P.D. Bamidis, A. Astaras

EMG and Kinematics Assessment of Postural Responses during Balance Perturbation on a 3D Robotic Platform: Preliminary Results in Children with Hemiplegia 69
C. De Marchis, F. Patané, M. Petrarca, S. Carniel, M. Schmid, S. Conforto, E. Castelli, P. Cappa, T. D'Alessio

Simulation-Based Planification Tool for an Assistance-as-Needed Upper Limb Neurorehabilitation Robotic Orthosis 73
R. Pérez-Rodríguez, C. Rodríguez, F. Molina, C. Gómez, E. Opisso, J.M. Tormos, J. Medina, E.J. Gómez

Total Joint Replacement: Biomaterials for Application in the Temporomandibular Joint 77
N. Fakih-Gomez, L.M. Gonzalez-Perez, B. Gonzalez Perez-Somarriba

Kinematic Indexes' Reproducibility of Horizontal Reaching Movements 81
G. D'Addio, L. Iuppariello, M. Romano, F. Lullo, N. Pappone, M. Cesarelli

Handling Disturbances on Planned Trajectories in Robotic Rehabilitation Therapies 85
V. Rajasekaran, J. Aranda, A. Casals

Computational Modelling of the Shape Deviations of the Sphere Surfaces of Ceramic Heads of Hip Joint Replacement 89
V. Fuis, Koukal, Z. Florian, P. Janicek

An Innovative Multisensor Controlled Prosthetic Hand 93
N. Carbonaro, G. Anania, M. Bacchereti, G. Donati, L. Ferretti, G. Pellicci, G. Parrini, N. Vitetta, D. De Rossi, A. Tognetti

Robot-Assisted Surgical Platform for Controlled Bone Drilling: Experiments on Temperature Monitoring for Assessment of Thermal Bone Necrosis 97
M.A. Landeira Freire, J.C. Ramos González, E. Sánchez Tapia

A Kinematic Analysis of the Hand Function 101
J. Martin-Martin, A.I. Cuesta-Vargas

Single Incision Laparoscopic Surgery Using a Miniature Robotic System 105
I. Rivas-Blanco, M. Cuevas-Rodriguez, E. Bauzano, J. Gomez-deGabriel, V.F. Muñoz

Characterization of Anastomosis Techniques for Robot Assisted Surgery 109
Jordi Campos, Enric Laporte, Gabriel Gili, Carlos Peñas, Alicia Casals, Josep Amat

Considering Civil Liability as a Safety Criteria for Cognitive Surgical Robots 113
   E. Berges, A. Casals

Kinematic Quantification of Gait Asymmetry in Patients with Elastic Ankle Wrap Based on Cyclograms 117
   P. Kutilek, J. Hejda, Z. Svoboda

Fat Percentage Equation for Children with Cerebral Palsy: A Novel Approach 121
   E.B. Neves, E. Krueger, B.R. Rosário, M.C.N. Oliveira, S. Pol, W.L. Ripka

Functional and Structural of the Erector Spinae Muscle during Isometric Lumbar Extension 125
   M. González-Sánchez, A.I. Cuesta-Vargas

On Evaluation of Shoulder Roundness by Use of Single Camera Photogrammetry 129
   A. Katashev, E. Shishlova, V. Vendina

Effects of Superimposed Electrical Stimulation Training on Vertical Jump Performance: A Comparison Study between Men and Women 133
   D.C. Costa, M.N. Souza, A.V. Pino

Attentional Focus and Functional Connectivity in Cycling: An EEG Case Study 137
   S. Comani, S. Di Fronso, E. Filho, A.M. Castronovo, M. Schmid, L. Bortoli, S. Conforto, C. Robazza, M. Bertollo

ERD/ERS Patterns of Shooting Performance within the Multi-Action Plan Model 141
   S. Comani, L. Bortoli, S. Di Fronso, E. Fiho, C. De Marchis, M. Schmid, S. Conforto, C. Robazza, M. Bertollo

Upper Limb Joint Torque Distribution Resulting from the Flat Tennis Serve Impact Force 145
   W.D. Masinghe, T. Nanayakkara, G. Collier

The Fatigue Vector: A New Bi-dimensional Parameter for Muscular Fatigue Analysis 149
   S. Conforto, A.M. Castronovo, C. De Marchis, M. Schmid, M. Bertollo, C. Robazza, S. Comani, T. D'Alessio

Kinetic Analysis of Manual Wheelchair Propulsion in Athletes and Users with Spinal Cord Injury 153
   M. Solís-Mozos, A. del Ama-Espinosa, B. Crespo-Ruiz, E. Pérez-Rizo, J.F. Jimenez-Díaz, A. Gil-Agudo

Multimodal MRI Evaluation of Physiological Changes on Leg Muscles due to Fatigue after Intense Exercise 157
   Mario Forjaz Secca, Sergio S. Alves, Ana Rita Pereira, José Nuno Alves, Filipa Joao, Antonio P. Veloso, Michael Noseworthy, Nuno Jalles Tavares, Cristina Meneses

Part IV: Biomedical Imaging and Processing

Design and Implementation of a Bipolar Current Source for MREIT Applications 161
   H.H. Eroglu, B.M. Eyüboglu, C. Göksu

Automatic Detection of Heart Center in Late Gadolinium Enhanced MRI 165
   K. Engan, V. Naranjo, T. Eftestøl, L. Woie, A. Schuchter, S. Ørn

The Application of Highly Accelerated MR Acquisition Techniques to Imaging the Peripheral Vasculature 169
   Stephen J. Riederer

A New Label Fusion Method Using Graph Cuts: Application to Hippocampus Segmentation 174
   C. Platero, M.C. Tobar, J. Sanguino, O. Velasco

Computer Aided Decision Support Tool for Rectal Cancer TNM Staging Using MRI 178
   A. Torrado-Carvajal, T. Martin Fernandez-Gallardo, N. Malpica

Volume of Tissue Activated in Patients with Parkinson's Disease from West-Center Region of Colombia Treated with
Deep Brain Stimulation.. 182
   Hernán Darío Vargas Cardona, Mauricio A. Álvarez, Genaro Daza, Enrique Guijarro, Álvaro Orozco

Computer-Aided Diagnosis of Abdominal Aortic Aneurysm after Endovascular Repair Using Active Learning
Segmentation and Texture Analysis.. 186
   G. García, J. Maiora, A. Tapia, M. Graña, M. De Blas

Assessment of Magnetic Field in the Surroundings of Magnetic Resonance Systems: Risks for Professional Staff 190
   V.M. Febles Santana, J.A. Hernández Armas, M.A. Martín Díaz, S. de Miguel Bilbao, J.C. Fernández de Aldecoa, V. Ramos González

Morphological Characterization of the Human Calvarium in Relation to the Diploic and Cranial Thickness Utilizing
X-Ray Computed Microtomography... 194
   E. Larsson, F. Brun, G. Tromba, P. Cataldi, K. Uvdal, A. Accardo

A Programmable Current Source for MRCDI and MREIT Applications... 198
   C. Göksu, B.M. Eyüboğlu, H.H. Eroğlu

How Does Compressed Sensing Affect Activation Maps in Rat fMRI?... 202
   C. Chavarrias, J.F.P.J. Abascal, P. Montesinos, M. Desco

Resting State Functional Connectivity Analysis of Multiple Sclerosis and Neuromyelitis Optica Using Graph
Theory.. 206
   E. Eqlimi, N. Riyahi Alam, M.A. Sahraian, A. Eshaghi, S. Riyahi Alam, H. Ghanaati, K. Firouznia, E. Karami

Quantitative Evaluation of Patient-Specific Conforming Hexahedral Meshes of Abdominal Aortic Aneurysms and
Intraluminal Thrombus Generated from MRI... 210
   J. Tarjuelo-Gutierrez, B. Rodriguez-Vila, D.M. Pierce, T. Fastl, E.J. Gomez

Compressed Sensing for Cardiac MRI Cine Sequences: A Real Implementation on a Small-Animal Scanner.......... 214
   P. Montesinos, J.F.P.J. Abascal, C. Chavarrías, J.J. Vaquero, M. Desco

The Importance of a Valid Reference Region for Intensity Normalization of Perfusion MR Studies in Early Alzheimer's
Disease... 218
   M. Lacalle-Aurioles, Y. Alemán-Gómez, J. Guzmán-De-Villoria, J. Olazarán, I. Cruz, J.M. Mateos-Pérez, M.E. Martino, M. Desco

A Full Automatic Method for the Soft tissues Sarcoma Treatment Response Based on Fuzzy Logic 221
   E. Montin, A. Messina, L.T. Mainardi

Reconstruction of DSC-MRI Data from Sparse Data Exploiting Temporal Redundancy and Contrast Localization 225
   D. Boschetto, M. Castellaro, P. Di Prima, A. Bertoldo, E. Grisan

The Influence of Slice Orientation in 3D CBCT Images on Measurements of Anatomical Structures 229
   W. Jacquet, E. Nyssen, C. Politis, B. Vande Vannet, P. Bottenberg

A Prior-Based Image Variation (PRIVA) Approach Applied to Motion-Based Compressed Sensing Cardiac Cine
MRI.. 233
   J.F.P.J. Abascal, P. Montesinos, E. Marinetto, J. Pascau, J.J. Vaquero, M. Desco

Symmetry Based Computer Aided Segmentation of Occluded Cerebral Arteries on CT Angiography 237
   Emilie M.M. Santos, Henk A. Marquering, Olvert A. Berkhemer, Wim van der Zwam, Aad van der Lugt, Charles B. Majoie, Wiro J. Niessen

Automatic Segmentation of Gray Matter Multiple Sclerosis Lesions on DIR Images 241
  E. Veronese, M. Calabrese, A. Favaretto, P. Gallo, A. Bertoldo, E. Grisan

Calibration of a C-arm X-Ray System for Its Use in Tomography 245
  C. de Molina, J. Pascau, M. Desco, M. Abella

Quantitative Assessment of Prenatal Aortic Wall Thickness in Gestational Diabetes 249
  Elisa Veronese, Silvia Visentin, Marius Linguraru, Erich Cosmi, Enrico Grisan

Fast Anisotropic Speckle Filter for Ultrasound Medical Images 253
  G. Ramos-Llordén, G. Vegas-Sánchez-Ferrero, S. Aja-Fernández, M. Martín-Fernández, C. Alberola-López

Parallel Implementation of a X-Ray Tomography Reconstruction Algorithm for High-Resolution Studies 257
  J. Garcia, M. Abella, C. de Molina, E. Liria, F. Isaila, J. Carretero, M. Desco

Stacked Models for Efficient Annotation of Brain Tissues in MR Volumes 261
  Fabio Aiolli, Michele Donini, Enea Poletti, Enrico Grisan

Mumford-Shah Based Unsupervised Segmentation of Brain Tissue on MR Images 265
  A. Cevik, B.M. Eyuboglu

Validation of a Computer Aided Segmentation System for Retinography 269
  M. Baroni, P. Fortunato, L. Pollazzi, A. La Torre

Computer-Assisted System for Hypertensive Risk Determination through Fundus Image Processing 273
  S. Morales, V. Naranjo, F. López-Mir, A. Navea, M. Alcañiz

Interventional 2D-3D Registration in the Presence of Occlusion 277
  S. Demirci, F. Manstad-Hulaas, N. Navab

Marker-Controlled Watershed for Volume Countouring in PET images 281
  V. Naranjo, F. López-Mir, C. Marín, S. Morales, J.J. Fuertes, E. Villanueva

Predictive and Populational Model for Alzheimer's Disease Using Structural Neuroimaging 285
  D. López-Rodríguez, A. García-Linares

Accurate Cortical Bone Detection in Peripheral Quantitative Computed Tomography Images 289
  T. Cervinka, J. Hyttinen, H. Sievänen

Spatial Aliasing and EMG Amplitude in Time and Space: Simulated Action Potential Maps 293
  B. Afsharipour, K. Ullah, R. Merletti

Lacunarity-Based Inherent Texture Correlation Approach for Wireless Capsule Endoscopy Image Analysis .. 297
  V.S. Charisis, L.J. Hadjileontiadis, G.D. Sergiadis

Tracer Kinetic Modeling with R for Batch Processing of Dynamic PET Studies 301
  J.M. Mateos-Pérez, M. Desco, J.J. Vaquero

Tensor Radial Lengths for Mammographic Image Enhancement 305
  S.E. Chatzistergos, I.I. Andreadis, K.S. Nikita

Magnetic Resonance Texture Analysis: Optimal Feature Selection in Classifying Child Brain Tumors 309
   Suchada Tantisatirapong, Nigel P. Davies, Daniel Rodriguez, Laurence Abernethy, Dorothee P. Auer, C.A. Clark,
   Richard Grundy, Tim Jaspan, Darren Hargrave, Lesley MacPherson, Martin O. Leach, Geoffrey S. Payne,
   Barry L. Pizer, Andrew C. Peet, Theodoros N. Arvanitis

Image Segmentation for Treatment Planning of Electroporation-Based Treatments of Tumors 313
   M. Marcan, D. Pavliha, R. Magjarevic, D. Miklavcic

Detection of Retinal Vessel Bifurcations by Means of Multiple Orientation Estimation Based on Regularized
Morphological Openings 317
   Álvar-Ginés Legaz-Aparicio, Rafael Verdú-Monedero, Juan Morales-Sánchez, Jorge Larrey-Ruiz, Jesús Angulo

Segmentation of Basal Nuclei and Anatomical Brain Structures Using Support Vector Machines 321
   A. Bosnjak, R. Villegas, G. Montilla, I. Jara

Color Analysis in Retinography: Glaucoma Image Detection 325
   Roberto Román Morán, Rafael Barea Navarro, Luciano Boquete Vázquez, Elena López Guillén,
   Jaime Campos Pavón, Lucía de Pablo Gómez de Liaño, David Escot Bocanegra, Luis de Santiago,
   Miguel Ortiz

An Image Analysis System for the Assessment of Retinal Microcirculation in Hypertension and Its Clinical
Evaluation 330
   X. Zabulis, A. Triantafyllou, P. Karamaounas, C. Zamboulis, S. Douma

Automatic Selection of CT Perfusion Datasets Unsuitable for CTP Analysis due to Head Movement 336
   F. Fahmi, H.A. Marquering, G.J. Streekstra, L.F.M. Beenen, N.Y. Janssen, A. Riordan, H. De Jong, C.B.L. Majoie,
   E. van Bavel

An In-Vitro Model of Cardiac Fibrillation with Different Degrees of Complexity 340
   A.M. Climent, M.S. Guillem, P. Lee, C. Bollensdorff, F. Atienza, M.E. Fernández-Santos, R. Sanz-Ruiz, P.L. Sánchez,
   F. Fernández-Avilés

Semi-automatic Segmentation of Sacrum in Computer Tomography Studies for Intraoperative Radiation Therapy 344
   E. Marinetto, I. Balsa-Lozano, J. Lansdown, J.A. Santos-Miranda, M. Valdivieso, M. Desco, J. Pascau

Anatomical Discovery: Finding Organs in the Neighborhood of the Liver 348
   C. Oyarzun Laura, K. Drechsler, S. Wesarg

A Comparative Study of Different Methods for Pigmented Lesion Classification Based on Color and Texture
Features 352
   J.A. Pérez-Carrasco, B. Acha, C. Serrano

Automatic Burn Depth Estimation from Psychophysical Experiment Data 356
   Begoña Acha, Tomás Gómez-Cía, Irene Fondón, Carmen Serrano

Segmentation of Retroperitoneal Tumors Using Fast Continuous Max-Flow Algorithm 360
   J.A. Pérez-Carrasco, C. Suárez-Mejías, C. Serrano, J.L. López-Guerra, B. Acha

Evaluation of the Visibility of the Color and Monochrome Road Map by the Searching Distance and Time of a Locus of
Eye Movement 364
   Takakazu Kobayashi, Tatsuya Oizumi, Kazushige Kimura, Katsumi Sugiyama

3D Segmentation of MRI of the Liver Using Support Vector Machine 368
    J.L. Moyano-Cuevas, A. Plaza, I. Dopido, J.B. Pagador, J.A. Sanchez-Margallo, L.F. Sánchez,
    F.M. Sánchez-Margallo

Interpolation Based Deformation Model for Minimally Invasive Beating Heart Surgery 372
    A.I. Aviles, A. Casals

A Framework for Automatic Detection of Lumen-Endothelium Border in Intracoronary OCT Image Sequences 376
    G. Cheimariotis, V. Koutkias, I. Chouvarda, K. Toutouzas, Y.S. Chatzizisis, A. Giannopoulos, M. Riga,
    A. Antoniadis, C. Doulaverakis, I. Tsampoulatidis, I. Kompatsiaris, C. Stefanadis, G. Giannoglou, N. Maglaveras

Learning Optimal Matched Filters for Retinal Vessel Segmentation with ADA-Boost 380
    E. Poletti, E. Grisan

Using Optical Flow in Motion Analysis for Evaluation of Active Music Therapy 384
    M. Suzuki, S. Kataoka, E. Shimokawa

Application of Gaussian Mixture Models with Expectation Maximization in Bacterial Colonies Image Segmentation
for Automated Counting and Identification 388
    I. Silva Maretić, I. Lacković

Need of Multimodal SPECT/MRI for Tracking of ^{111}In-Labeled Human Mesenchymal Stem Cells in Neuroblastoma
Tumor-Bearing Mice 392
    L. Cussó, I. Mirones, S. Peña-Zalbidea, L. Lopez-Sánchez, V. García-Vázquez, J. García-Castro, M. Desco

First Steps towards a USE System for Non-invasive Thyroid Exploration 395
    J. Rodriguez, O. Moreno-Perez, J. Sabater-Navarro

Registration of Small-Animal SPECT/MRI Studies for Tracking Human Mesenchymal Stem Cells 399
    V. García-Vázquez, L. Cussó, J. Chamorro-Servent, I. Mirones, J. García-Castro, L. López-Sánchez,
    S. Peña-Zalbidea, P. Montesinos, C. Chavarrías, J. Pascau, M. Desco

Segregation of Emotional Function in Subcortical Structures: MEG Evidence 403
    C. Styliadis, A.A. Ioannides, P.D. Bamidis, C. Papadelis

AFM Multimode Imaging and Nanoindetation Method for Assessing Collagen Nanoscale Thin Films
Heterogeneity 407
    A. Stylianou, S.V. Kontomaris, D. Yova, G. Balogiannis

Segmentation of Corneal Endothelial Cells Contour through Classification of Individual Component Signatures 411
    E. Poletti, A. Ruggeri

Split Bregman-Singular Value Analysis Approach to Solve the Compressed Sensing Problem of Fluorescence Diffuse
Optical Tomography 415
    J. Chamorro-Servent, J.F.P.J. Abascal, J. Ripoll, J.J. Vaquero, M. Desco

Development of an Optical Coherence Tomograph for Small Animal Retinal Imaging 419
    Susana F. Silva, José P. Domingues, José Agnelo, António Miguel Morgado, Rui Bernardes

Inhomogeneous Modification of Cardiac Electrophysiological Properties Due to Flecainide Administration 423
    A.M. Climent, M.S. Guillem, P. Lee, C. Bollensdorff, F. Atienza, M.E. Fernández-Santos, R. Sanz-Ruiz, P.L. Sánchez,
    F. Fernández-Avilés

Predictive Analysis of Photoacoustic Tomography Images in Dermatological Liposarcoma 427
   F. Fanjul-Vélez, D. Martín-Ruiz, N. Ortega-Quijano, I. Salas-García, J.L. Arce-Diego

Monte Carlo Simulation of Radiation through the Human Retina Using Geant4 431
   G. Lopes, D. Tendeiro, J.P. Santos, P. Vieira

Semiautomatic Evaluation of Crypt Architecture and Vessel Morphology in Confocal Microendoscopy: Application to Ulcerative Colitis 435
   E. Veronese, E. Poletti, A. Buda, G. Hatem, S. Facchin, G.C. Strurniolo, E. Grisan

3D Reconstruction of Skin Surface Using an Improved Shape-from-Shading Technique 439
   G. Balogiannis, D. Yova, K. Politopoulos

QuantiDOPA: A Quantification Software for Dopaminergic Neurotransmission SPECT 443
   A. Niñerola, B. Marti, O. Esteban, X. Planes, A.F. Frangi, M.J. Ledesma-Carbayo, A. Santos, A. Cot, F. Lomeña, J. Pavia, D. Ros

Spatial Normalization in Voxel-Wise Analysis of FDG-PET Brain Images 447
   M.E. Martino, V. García-Vázquez, M. Lacalle-Aurioles, J. Olazarán, J. Guzmán de Villoria, I. Cruz, J.L. Carreras, M. Desco

EndoTOFPET-US a High Resolution Endoscopic PET-US Scanner Used for Pancreatic and Prostatic Clinical Exams 451
   C. Zorraquino

Light Emission Efficiency of $Gd_3Al_2Ga_3O_{12}$:Ce (GAGG:Ce) Single Crystal under X-Ray Radiographic Conditions 455
   I.E. Seferis, C.M. Michail, S.L. David, A. Bakas, N.I. Kalivas, G.P. Fountos, G.S. Panayiotakis, K. Kourkoutas, I.S. Kandarakis, I.G. Valais

Calcification Detection Optimization in Dual Energy Mammography: Influence of the X-Ray Spectra 459
   V. Koukou, N. Martini, G. Fountos, P. Sotiropoulou, C. Michail, I. Valais, I. Kandarakis, G. Nikiforidis

X-Ray Spectra for Bone Quality Assessment Using Energy Dispersive Counting and Imaging Detectors with Dual Energy Method 463
   P. Sotiropoulou, G. Fountos, N. Martini, V. Koukou, C. Michail, I. Valais, I. Kandarakis, G. Nikiforidis

CDMAM Phantom Optimized for Digital Mammography Quality Control by Automatic Image Readout 467
   M.J. Floor-Westerdijk, W.N.J.M. Colier, R.J.M. van der Burght

Experimental Evaluation of a High Resolution CMOS Digital Imaging Detector Coupled to Structured CsI Scintillators for Medical Imaging Applications 471
   C.M. Michail, I.G. Valais, I.E. Seferis, F. Stromatia, E. Kounadi, G.P. Fountos, I.S. Kandarakis

Estimation of Contrast Agent Motion from Cerebral Angiograms for Assessing Hemodynamic Alterations Following Aneurysm Treatment by Flow Diversion 475
   T. Benz, M. Kowarschik, J. Endres, P. Maday, T. Redel, N. Navab

Graphical User Interface for Breast Tomosynthesis Reconstructions: An Application Using Anisotropic Diffusion Filtering 479
   A. Malliori, K. Bliznakova, A. Daskalaki, N. Pallikarakis

Image Analysis of Breast Reconstruction with Silicone Gel Implant 483
   A. Daskalaki, K. Bliznakova, N. Pallikarakis

Quantitative Analysis of Marker Segmentation for C-Arm Pose Based Navigation 487
    T. Steger, S. Wesarg

Iterative Dual-Energy Material Decomposition for Slow kVp Switching: A Compressed Sensing Approach 491
    A. Sisniega, J. Abascal, M. Abella, J. Chamorro, M. Desco, J.J. Vaquero

Brain Gray – White Matter Discrimination in Dual Energy CT Imaging: A Simulation Feasibility Study 495
    A. Dermitzakis, G. Gatzounis, K. Bliznakova

Part V: Biomedical Signal Processing

Removal of Low Frequency Interferences from Electrocardio Signal Based on Transform of TP Segment Samples 501
    O.V. Melnik, A.A. Mikheev

Noise Reduction Using 2D Anisotropic Diffusion Filter in Inverse Electrocardiography 505
    A. Mazloumi Gavgani, Y. Serinagaoglu Dogrusoz

Detection and Identification of S1 and S2 Heart Sounds Using Wavelet Decomposition and Reconstruction 509
    K. Hassani, K. Bajelani, M. Navidbakhsh, John Doyle

Estimation of the Hemoglobin Glycation Rate Constant Based on the Mean Glycemia in Patients with Diabetes 515
    P. Ładyżyński, P. Foltyński, J.M. Wójcicki, M.I. Bąk, S. Sabalińska, J. Krzymień, J. Kawiak, W. Karnafel

Spatial Dynamics of the Topographic Representation of Electroencephalogram Spectral Features during General
Anesthesia 519
    B. Direito, C. Teixeira, B. Ribeiro, A. Dourado, M.P. Santos, M.C. Loureiro

The Properties of the Missing Fundamental of Complex Tones 523
    T. Matsuoka, Y. Iitomi

Synchrony Analysis of Unipolar Cardiac Mapping during Ventricular Fibrillation 527
    J. Caravaca, E. Soria-Olivas, A.J. Serrano-López, M. Bataller, A. Rosado, L. Such-Belenguer, J.F. Guerrero

Symbolic Analysis of Heart Period and QT Interval Variabilities in LQT1 Patients 531
    V. Bari, T. Bassani, A. Marchi, G. Girardengo, L. Calvillo, S. Cerutti, P.A. Brink, L. Crotti, P.J. Schwartz, A. Porta

Mechanical Stimulation and Cardiovascular Control in Parkinson Disease 535
    T. Bassani, V. Bari, A. Marchi, S. Tassin, L. Dalla Vecchia, M. Canesi, F. Barbic, R. Furlan, A. Porta

Variability of EEG Theta Power Modulation in Type 1 Diabetics Increases during Hypo-glycaemia 539
    A. Goljahani, A.S. Sejling, G. Sparacino, C. Fabris, J. Duun-Henriksen, L.S. Remvig, C. Cobelli, C.B. Juhl

Analysis of Systematic Inter-channel Time Offsets in FECG from Maternal Abdominal Sensing 543
    P. Melillo, D. Santoro, M. Vadursi

Spatiotemporal Brain Activities on Recalling Names of Body Parts 547
    T. Yamanoi, Y. Tanaka, M. Otsuki, H. Toyoshima, T. Yamazaki

Elimination of ECG Artefacts in Foetal EEG Using Ensemble Average Subtraction and Wavelet Denoising Methods:
A Simulation 551
    F. Abtahi, F. Seoane, K. Lindecrantz, N. Löfgren

External Uterine Contractions Signal Analysis in Relation to Labor Progression and Dystocia 555
    H. Gonçalves, P. Pinto, D. Ayres-de-Campos, J. Bernardes

Quality Based Adaptation of Signal Analysis Software in Pregnancy Home Care System 559
   J. Wrobel, K. Horoba, J. Jezewski, T. Kupka, M. Jezewski, T. Przybyla

A Recovery of FHR Signal in the Embedded Space 563
   T. Przybyła, T. Pander, J. Wróbel, R. Czabański, D. Roj, A. Matonia

Online Drawings for Dementia Diagnose: In-Air and Pressure Information Analysis 567
   Marcos Faundez-Zanuy, Enric Sesa-Nogueras, Josep Roure-Alcobé, Josep Garré-Olmo, Karmele Lopez-de-Ipiña,
   Jordi Solé-Casals

Spontaneous Speech and Emotional Response Modeling Based on One-Class Classifier Oriented to Alzheimer Disease
Diagnosis.. 571
   K. Lopez-de-Ipiña, J.B. Alonso, N. Barroso, J. Solé-Casals, M. Ecay-Torres, P. Martinez-Lage, F. Zelarain,
   H. Egiraun, C.M. Travieso

Complexity of Epileptiform Activity in a Neuronal Network and Pharmacological Intervention 575
   D. Abásolo, L. González, Y. Chen

Evaluation of Different Handwriting Teaching Methods by Kinematic Analysis 579
   A. Accardo, M. Genna, I. Perrone, P. Ceschia, C. Mandarino

Analysis of MEG Activity across the Life Span Using Statistical Complexity 583
   J. Poza, C. Gómez, M. García, A. Bachiller, A. Fernández, R. Hornero

Impact of Device Settings and Spontaneous Breathing during IPV in CF Patients 587
   E. Fornasa, A. Accardo, M. Ajcevic, R. Sartori, F. Poli

Estimation of Respiratory Mechanics Parameters during HFPV 591
   M. Ajcevic, A. Accardo, E. Fornasa, U. Lucangelo

Signal Source Estimation Inside Brain Using Switching Voltage Divider 595
   Yusuke Sakaue, Shima Okada, Masaaki Makikawa

A Phase-Space Based Algorithm for Detecting Different Types of Artefacts 599
   A. Brignol, T. Al-ani

Automatic Classification of Respiratory Sounds Phenotypes in COPD Exacerbations 603
   Daniel S. Morillo, M.A. Fernández Granero, A. León, L.F. Crespo

Response Detection in Narrow-Band EEG Using Signal-Driven Non-Periodic Stimulation 607
   M. Cagy, A.F.C. Infantosi

EMG-Based Analysis of Treadmill and Ground Walking in Distal Leg Muscles 611
   F. Di Nardo, S. Fioretti

Temporal Variation of Local Fluorescence Sources in the Photodynamic Process 615
   I. Salas-García, F. Fanjul-Vélez, N. Ortega-Quijano, J.L. Arce-Diego

Statistical Analysis of EMG Signal Acquired from Tibialis Anterior during Gait 619
   F. Di Nardo, A. Mengarelli, G. Ghetti, S. Fioretti

Auto-Mutual Information Function for Predicting Pain Responses in EEG Signals during Sedation 623
   U. Melia, M. Vallverdú, M. Jospin, E.W. Jensen, J.F. Valencia, F. Clariá, P.L. Gambus, P. Caminal

Prony's Method for the Analysis of mfVEP Signals 627
  A.J. Fernández-Rodríguez, L. de Santiago, R. Blanco, C. Amo, R. Barea, J.M. Miguel-Jiménez,
  J.M. Rodríguez-Ascariz, E.M. Sánchez-Morla, M. Ortiz, L. Boquete

EEG Denoising Based on Empirical Mode Decomposition and Mutual Information 631
  A. Mert, A. Akan

Decomposition Analysis of Digital Volume Pulse Signal Using Multi-Model Fitting 635
  Sheng-Cheng Huang, Hao-Yu Jan, Geng-Hong Lin, Wen-Chen Lin, Kang-Ping Lin

A Neural Minimum Input Model to Reconstruct the Electrical Cortical Activity 639
  S. Conforto, I. Bernabucci, N. Accornero, M. Bertollo, C. Robazza, S. Comani, M. Schmid, T. D'Alessio

Improved Splines Fitting of Intervertebral Motion by Local Smoothing Variation 643
  P. Bifulco, M. Cesarelli, G. D'Addio, M. Romano

Effects of Wavelets Analysis on Power Spectral Distributions in Laser Doppler Flowmetry Time Series 647
  G. D'Addio, M. Cesarelli, P. Bifulco, L. Iuppariello, G. Faiella, D. Lapi, A. Colantuoni

Outliers Detection and Processing in CTG Monitoring 651
  M. Romano, G. Faiella, P. Bifulco, G. D'Addio, F. Clemente, M. Cesarelli

Changes in Heart Rate Variability Associated with Moderate Alcohol Consumption 655
  A. Fratini, P. Bifulco, F. Clemente, M. Sansone, M. Cesarelli

EEG Rhythm Analysis Using Stochastic Relevance 658
  L. Duque-Muñoz, C.A. Aguirre-Echeverry, G. Castellanos-Domínguez

Optimize ncRNA Targeting: A Signal Analysis Based Approach 662
  N. Maggi, P. Arrigo, C. Ruggiero

Mathematical Modelling of Melanoma Collective Cell Migration 666
  J.V. Gallinaro, C.M.G. Marques, F.M. Azevedo, D.O.H. Suzuki

Synthetic Atrial Electrogram Generator 670
  M.W. Rivolta, L.T. Mainardi, R. Sassi, V.D.A. Corino

EMG Topographic Image Enhancement Using Multi Scale Filtering 674
  K. Ullah, B. Afsharipour, R. Merletti

Model Based Estimates of Gain between Systolic Blood Pressure and Heart-Rate Obtained from Only Inspiratory or Expiratory Periods 678
  D.S. Fonseca, A. Beda, A.M.F.L. Miranda de Sá, D.M. Simpson

Dielectric Properties of Dentin between 100 Hz and 1 MHz Compared to Electrically Similar Body Tissues .. 682
  T. Marjanović, I. Lacković

Hypoglycaemia-Related EEG Changes Assessed by Approximate Entropy 686
  C. Fabris, A.S. Sejling, G. Sparacino, A. Goljahani, J. Duun-Henriksen, L.S. Remvig, C. Cobelli, C.B. Juhl

Non-linear Indices of Heart Rate Variability in Heart Failure Patients during Sleep 690
  R. Cabiddu, S. Mariani, J. Henriques, S. Cerutti, A.M. Bianchi

Detrended Fluctuation Analysis of EEG in Depression 694
  M. Bachmann, A. Suhhova, J. Lass, K. Aadamsoo, Ü. Võhma, H. Hinrikus

Can Distance Measures Based on Lempel-Ziv Complexity Help in the Detection of Alzheimer's Disease from
Electroencephalograms? 698
  S. Simons, D. Abásolo

Predictive Value on Neurological Outcome of Early EEG Signal Analysis in Brain Injured Patients 702
  A. Accardo, M. Cusenza, L. Prisco, F. Monti, A. Draisci, W. Calligaris

Lempel-Ziv Complexity Analysis of Local Field Potentials in Different Vigilance States with Different Coarse-Graining
Techniques 706
  D. Abásolo, R. Morgado da Silva, S. Simons, G. Tononi, C. Cirelli, V.V. Vyazovskiy

Heart Rate Variability in Pregnant Women before Programmed Cesarean Intervention 710
  Juan Bolea, Raquel Bailón, Eva Rovira, Jose María Remartínez, Pablo Laguna, Augusto Navarro

Fractal Changes in the Long-Range Correlation and Loss of Signal Complexity in Infant's Heart Rate Variability with
Clinical Sepsis 714
  E. Godoy, J. López, L. Bermúdez, A. Ferrer, N. García, C. García Vicent, E.F. Lurbe, J. Saiz

Estimation of Coupling and Directionality between Signals Applied to Physiological Uterine EMG Model and Real
EHG Signals 718
  A. Diab, M. Hassan, J. Laforêt, B. Karlsson, C. Marque

Dynalets: A New Tool for Biological Signal Processing 722
  J. Demongeot, O. Hansen, A. Hamie

Cyclostationarity-Based Estimation of the Foetus Subspace Dimension from ECG Recordings 726
  M. Haritopoulos, J. Roussel, C. Capdessus, A.K. Nandi

Feature Extraction Based on Discriminative Alternating Regression.. 730
  C.O. Sakar, O. Kursun, F. Gurgen

Influence of Signal Preprocessing on ICA-Based EEG Decomposition 734
  Z. Zakeri, S. Assecondi, A.P. Bagshaw, T.N. Arvanitis

Muscle Synergies Underlying Voluntary Anteroposterior Sway Movements 738
  S. Piazza, D. Torricelli, I.M. Alguacil Diego, R. Cano De La Cuerda, F. Molina Rueda, F.M. Rivas Montero,
  F. Barroso, J.L. Pons

Recognition of Brain Structures from MER-Signals Using Dynamic MFCC Analysis and a HMC Classifier 742
  Mauricio Holguin, German A. Holguin, Hernán Darío Vargas Cardona, Genaro Daza, Enrique Guijarro,
  Alvaro Orozco

Feed-Forward Neural Network Architectures Based on Extreme Learning Machine for Parkinson's Disease
Diagnosis 746
  P.J. García-Laencina, G. Rodríguez-Bermúdez, J. Roca-Dorda

Breast Tissue Microarray Classification Based on Texture and Frequential Features 750
  M.M. Fernández-Carrobles, G. Bueno, O. Déniz, M. García-Rojo

Fuzzy System for Retrospective Evaluation of the Fetal State 754
  R. Czabanski, J. Wrobel, K. Horoba, J. Jezewski, A. Matonia

Advanced Processing of sEMG Signals for User Independent Gesture Recognition	758
A. Doswald, F. Carrino, F. Ringeval	
Adaptive Classification Framework for Multiclass Motor Imagery-Based BCI	762
L.F. Nicolas-Alonso, R. Corralejo, D. Álvarez, R. Hornero	
Computer Program for Automatic Identification of Artifacts in Impedance Cardiography Signals Recorded during Ambulatory Hemodynamic Monitoring	766
P. Piskulak, G. Cybulski, W. Niewiadomski, T. Pałko	
Comparison between Artificial Neural Networks and Discriminant Functions for Automatic Detection of Epileptiform Discharges	770
C.F. Boos, G.R. Scolaro, F.M. Azevedo	
Classification of Early Autism Based on HPLC Data	774
T. Kristensen	
Feature Selection Techniques in Uterine Electrohysterography Signal	779
D. Alamedine, M. Khalil, C. Marque	
Studying Functional Brain Networks to Understand Mathematical Thinking: A Graph-Theoretical Approach	783
Georgios Bamparopoulos, Manousos A. Klados, Nikolaos Papathanasiou, Ioannis Antoniou, Sifis Micheloyannis, Panagiotis D. Bamidis	
A Short Review on Emotional Recognition Based on Biosignal Pattern Analysis	787
Manousos A. Klados, Charalampos Styliadis, Panagiotis D. Bamidis	
Matching Pursuit with Asymmetric Functions for Signal Decomposition and Parameterization	791
K.J. Blinowska, W.W. Jedrzejczak, K. Kwaskiewicz	
Analysis of Intracranial Pressure Signals Using the Spectral Turbulence	795
M. García, J. Poza, D. Santamarta, D. Abásolo, R. Hornero	
Graph-Theoretical Analysis in Schizophrenia Performing an Auditory Oddball Task	799
A. Bachiller, J. Poza, C. Gómez, V. Molina, V. Suazo, A. Díez, R. Hornero	
Spectral Parameters from Pressure Bed Sensor Respiratory Signal to Discriminate Sleep Epochs with Respiratory Events	803
Giulia Tacchino, Guillermina Guerrero, Juha M. Kortelainen, Anna M. Bianchi	
Wavelet Energy and Wavelet Entropy as a New Analysis Approach in Spontaneous Fluctuations of Pupil Size Study – Preliminary Research	807
W. Nowak, E. Szul-Pietrzak, A. Hachol	
Rhythm Extraction Using Spectral–Splitting for Epileptic Seizure Detection	811
L.M. Sepúlveda, J.D. Martínez, L. Duque, C.D. Acosta, G. Castellanos	
EEG Metrics Evaluation in Simultaneous EEG-fMRI Olfactory Experiment	815
Eva Manzanedo, Ana Beatriz Solana, Elena Molina, Ricardo Bruña, Susana Borromeo, Juan Antonio Hernández-Tamames, Francisco del Pozo	
An Emboli Detection System Based on Dual Tree Complex Wavelet Transform	819
G. Serbes, B.E. Sakar, N. Aydin, H.O. Gulcur	

New Indices Extracted from Fetal Heart Rate Signal for the Assessment of Fetal Well-Being 823
A. Fanelli, G. Magenes, M.G. Signorini

A Novel Method of Measuring Autonomic Nervous System Pupil Dynamics Using Wavelet Analysis 827
G. Leal, C. Neves, P. Vieira

Part VI: Bio-micro and Bio-nano Technologies

Analysis Sensitivity by Novel Needle-Type GMR Sensor Used in Biomedical Investigation 833
H. Shirzadfar, J. Claudel, M. Nadi, D. Kourtiche, S. Yamada

Focusing of Electromagnetic Waves by Non-Spherical, Au-Si Nano-particles 837
A.P. Moneda, D.P. Chrissoulidis

Hybrid Microfluidic Biosensor for Single Cell Flow Impedance Spectroscopy: Theoretical Approach and First Validations 841
J. Claudel, M. Ibrahim, H. Shirzadfar, M. Nadi, O. Elmazria, D. Kourtiche

Integrating an Electronic Health Record Graphical User Interface into Nanoelectronic-Based Biosensor Technology 845
Ana María Quintero, Carlos Cavero Barca, Carlos Marcos Lagunar, César Mediavilla, José Luis Conesa, Miguel Roncalés, Yaiza Belacortu, Alejandro Juez, Luis Fernandez, Alexandra Martin, Renzo Dal Molin

Synthesis of Gadolinium-doped Fluorescent Au/Ag Nanoclusters as Bimodal MRI Contrast Agents 849
Walter H. Chang, Cheng-An J. Lin, Ching-Yi Chang, Wen-Fu. Lai

Microchannel Modification to Enhance the Sensitivity in Biosensors 852
M. Gomez-Aranzadi, M. Mujika, S. Arana, D. Hansford

Towards Point-of-Use Dielectrophoretic Methods: A New Portable Multiphase Generator for Bacteria Concentration 856
B. del Moral Zamora, J.M. Álvarez Azpeitia, J. Colomer Farrarons, P.Ll. Miribel Català, A. Homs Corbera, A. Juárez, J. Samitier

Flagella – Templates for the Synthesis of Metallic Nanowires 860
L. Deutscher, L.D. Renner, G. Cuniberti

A Versatile Microfabricated Platform for Single-Cell Studies 864
A. Benavente-Babace, D. Gallego-Pérez, D.J. Hansford, S. Arana, E. Pérez-Lorenzo, M. Mujika

Photoacoustics of Gold Nanorods under Low Frequency Laser Pulses in Optical Hyperthermia 868
C. Sánchez, J.A. Ramos, T. Fernández, F. del Pozo, J.J. Serrano

Demonstration of an On-Chip Real-Time PCR for the Detection of *Trypanosoma Cruzi* 872
R.C.P. Rampazzo, M. Cereda, A. Cocci, M. De Fazio, M.A. Bianchessi, A.C. Graziani, M.A. Krieger, A.D.T. Costa

A Stand-Alone Platform for Prolonged Parallel Recordings of Neuronal Activity 876
G. Regalia, E. Biffi, A. Menegon, G. Ferrigno, A. Pedrocchi

Interdigitated Biosensor for Multiparametric Monitoring of Bacterial Biofilm Development 880
S. Becerro, J. Paredes, S. Arana

Human Splenon-on-a-chip: Design and Validation of a Microfluidic Model Resembling the Interstitial Slits and the Close/Fast and Open/Slow Microcirculations... 884
   L.G. Rigat-Brugarolas, M. Bernabeu, A. Elizalde, M. de Niz, L. Martin-Jaular, C. Fernandez-Becerra, A. Homs-Corbera, H.A. del Portillo, J. Samitier

Microbioreactor Integrated with a Sensor for Monitoring Intracellular Green Fluorescence Protein (GFP)............. 888
   Godfrey Pasirayi, Meez Islam, Simon M. Scott, Liam O'Hare, Zulfiqur Ali

Simultaneous Electrochemical Detection of Dopamine, Catechol and Ascorbic Acid at a Poly(acriflavine) Modified Electrode.. 892
   M. Rashid, V. Auger, Z. Ali

Induced Transmembrane Voltage during Cell Electrofusion Using Nanosecond Electric Pulses 896
   L. Rems, D. Miklavčič, G. Pucihar

Gene Transfer to Adherent Cells by *in situ* Electroporation with a Spiral Microelectrode Assembly.................. 900
   Tomás García Sánchez, Maria Guitart, Javier Rosell-Ferrer, Anna M. Gomez-Foix, Ramon Bragós

Part VII: Cardiovascular, Respiratory and Endocrine Systems Engineering

Low Contents of Polyunsaturated Fatty Acids in Cultured Rat Cardiomyocytes... 907
   D. Sato, T. Karimata, T. Wakatsuki, Z. Feng, A. Nishina, M. Kusunoki, T. Nakamura

Classification of Chronic Obstructive Pulmonary Disease (COPD) Using Integrated Software Suite 911
   Almir Badnjevic, Mario Cifrek, Dragan Koruga

Mathematical Model of Apico Aortic Conduit in Presence of Steno-Insufficiency.. 915
   G. Fragomeni, M. Rossi, F. Condemi, R. Mazzitelli, G.F. Serraino, A. Renzulli

Computational Hemodynamic Model of the Human Cardiovascular System 919
   A. Talaminos, L.M. Roa, A. Álvarez, I. Valverde, J. Reina

Object-Oriented Modeling and Simulation of the Arterial Pressure Control System by Using MODELICA........... 923
   J. Fernandez de Canete, J. Luque, J. Barbancho, V. Muñoz

Electrical Remodeling in the Epicardial Border Zone of the Human Infarcted Ventricle 927
   J.V. Visconti, L. Romero, J.M. Ferrero, J. Sáiz, B. Trenor

Comparing Hodgkin-Huxley and Markovian Formulations for the Rapid Potassium Current in Cardiac Myocytes: A Simulation Study... 931
   E. Godoy, L. Romero, J.M. Ferrero

Mechanical Properties of Different Airway Stents.. 935
   A. Ratnovsky, N. Mell, S. Wald, M.R. Kramer, S. Naftali

Site-Specific Mechanical Properties of Aortic Bifurcation 939
   J. Kronek, L Horny, T. Adamek, H. Chlup, R. Zitny

Cardiac Autonomic Nervous System Activity on Breathing Training 943
   H. Nakamura, H. Saito, M. Yoshida

Context Aware Contribution in Ambulatory Electrocardiogram Measurements Using Wearable Devices 947
   F.J. Martinez-Tabares, G. Castellanos-Dominguez

A Versatile Synchronization System for Biomedical Sensor Development 951
   M. Zakrzewski, A. Joutsen, J. Hännikäinen, K. Palovuori

ECG Acquisition Using Fluid-Repellent, Wearable Electrodes 955
   F.J. Martinez-Tabares, G.P. Cardona-Cuervo, G. Castellanos-Dominguez

Accuracy of the Oscillometric Fixed-Ratio Blood Pressure Measurement Using Different Methods of Characterization of Oscillometric Pulses 958
   J. Talts, R. Raamat, K. Jagomägi, J. Kivastik

Differences in QRS Locations due to ECG Lead: Relationship with Breathing 962
   M.A. García-González, A. Argelagós, M. Fernández-Chimeno, J. Ramos-Castro

Synchrosqueezing Index for Detecting Drowsiness Based on the Respiratory Effort Signal 965
   N. Rodríguez-Ibáñez, M.A. García-González, M. Fernández-Chimeno, Member IEEE, H. De Rosario, J. Ramos-Castro, Member IEEE

Applying Variable Ranking to Oximetric Recordings in Sleep Apnea Diagnosis 969
   D. Álvarez, G.C. Gutiérrez-Tobal, J. Gómez-Pilar, F. del Campo, M. López, R. Hornero

Paroxysmal Atrial Fibrillation Termination Prognosis through the Application of Generalized Hurst Exponents 973
   M. Julián, R. Alcaraz, J.J. Rieta

Evaluation of Laplacian Diaphragm Electromyographic Recordings in a Static Inspiratory Maneuver 977
   L. Estrada, A. Torres, J. Garcia-Casado, Y. Ye-Lin, R. Jané

Analysis of Normal and Continuous Adventitious Sounds for the Assessment of Asthma 981
   M. Lozano, J.A. Fiz, R. Jané

Automatic Extrasystole Detection Using Photoplethysmographic Signals 985
   A. Solosenko, V. Marozas

ECG Signal Reconstruction Based on Stochastic Joint-Modeling of the ECG and the PPG Signals 989
   D. Martín-Martínez, P. Casaseca-de-la-Higuera, M. Martín-Fernández, C. Alberola-López

Application of Impedance Cardiography for Haemodynamic Monitoring in Patients with Ischaemic Stroke 993
   A. Zielińska, H. Dudek, G. Cybulski

AR versus ARX Modeling of Heart Rate Sequences Recorded during Stress-Tests 996
   J. Holcik, T. Hodasova, P. Jahn, P. Melkova, J. Hanak

Cancellation of Cardiac Interference in Diaphragm EMG Signals Using an Estimate of ECG Reference Signal 1000
   A. Torres, J.A. Fiz, R. Jané

Relevance of the Atrial Substrate Remodeling during Follow-Up to Predict Preoperatively Atrial Fibrillation Cox-Maze Surgery Outcome 1005
   A. Hernández, R. Alcaraz, F. Hornero, J.J. Rieta

A New Method to Estimate Atrial Fibrillation Temporal Organization from the Surface Fibrillatory Waves Repetitiveness 1009
   R. Alcaraz, F. Hornero, J.J. Rieta

Study on Atrial Arrhythmias Optimal Organization Assessment with Generalized Hurst Exponents 1013
   M. Julián, R. Alcaraz, J.J. Rieta

New Sleep Transition Indexes for Describing Altered Sleep in SAHS 1017
   O. Urra, R. Jané

Blood Pressure Variability Analysis in Supine and Sitting Position of Healthy Subjects 1021
   B.F. Giraldo, A. Calvo, B. Martínez, A. Arcentales, R. Jané, S. Benito

Noninvasive Interdependence Estimation between Atrial and Ventricular Activities in Atrial Fibrillation 1025
   R. Alcaraz, J.J. Rieta

Prototype Development of a Computerized System for Interpretation of Heart Sounds Using Wavelet 1029
   I.S.G. Brites, N. Oki

Oxygen Dynamics in Microcirculation of Skeletal Muscle 1033
   M. Shibata, S. Hamashima, S. Ichioka, A. Kamiya

Computational Study on Aneurysm Hemodynamics Affected by a Deformed Parent Vessel after Stenting 1037
   W. Jeong, M.H. Han, K. Rhee

Inflation Tests of Vena Saphena Mangna for Different Loading Rates 1041
   J. Veselý, L. Horný, H. Chlup, R. Žitný

Modeling of Stent Implantation in a Human Stenotic Artery 1045
   G.S. Karanasiou, A.I. Sakellarios, E.E. Tripoliti, E.G.M. Petrakis, M.E. Zervakis, Francesco Migliavacca,
   Gabriele Dubini, Elena Dordoni, L.K. Michalis, D.I. Fotiadis

In Vitro High Resolution Ultrasonography Measurements of Arterial Bifurcations with and without Stenosis as Inputs
for In-Silico CFD Simulations 1049
   D. Suárez-Bagnasco, G. Balay, L. Cymberknop, R.L. Armentano, C. Negreira

Study on the Dynamic Behavior of Arterial End-to-End Anastomosis 1053
   P.C. Roussis, A.E. Giannakopoulos, H.P. Charalambous

Volume 2

Part VIII: Clinical Engineering

A New Approach for Preventive Maintenance Prioritization of Medical Equipment 1059
   N. Saleh, A. Sharawi, M. Abd Elwahed, A. Petti, D. Puppato, G. Balestra

2D and 3D Photogrammetric Models for Respiratory Analysis in Adolescents 1063
   W.L. Ripka, L. Ulbricht, E.B. Neves, P.M. Gewehr

Evaluation of Gastroesophageal Reflux in Infants Treated with Osteopathy Using the I-GERQ-R Questionnaire 1067
   M. Gemelli, L. Ulbricht, E.F.R. Romaneli

A Comparison of Three Techniques for Wound Area Measurement 1071
   P. Foltynski, J.M. Wojcicki, P. Ladyzynski, S. Sabalinska

Monitoring Changes of the Tibialis Anterior during Dorsiflexion with Electromyography, Sonomyography,
Dynamometry and Kinematic Signals 1075
   M. Ruiz-Muñoz, J. Martín-Martín, M. González-Sánchez, A.I. Cuesta-Vargas

Automation of Analytical Processes in Immunohematology: Hospital Based-HTA Approach 1079
   P. Lago, C. Lombardi, B. Milanesi, P. Isernia, L. Raffaele, A. Berzuini, P. Giussani, C.E. Battista, M. Angeleri,
   R. Santini

Total Laboratory Automation and Clinical Engineering 1083
  P. Lago, I. Vallone, A.M. Fenili

Development of Vertebral Metrics – An Instrument to Analyse the Spinal Column 1088
  C. Quaresma, A. Gabriel, M. Forjaz Secca, P. Vieira

Global Program for Certification of Local Clinical Engineers: Back to the Future 1092
  M. Medvedec

Emerging of Human Factor in Risk Analysis of Home Care Service 1096
  F. Clemente, G. Faiella, G. Rutoli, M. Romano, P. Bifulco, A. Fratini, M. Cesarelli

Selection and Deployment of a Business Intelligence System (BI) at a Hospital's Clinical Engineering Department 1100
  C. Pérez-Martín, J.C. Fernández-Aldecoa, J. Hernández-Armas, R. Cánovas-Paradell

Health Technology Assessment Applied to Health Technology Management through Clinical Engineering 1104
  F.A. Santos, A.E. Margotti, F.B. Ferreira, R. Garcia

Biomedical Engineering at the Centre for Preclinical Research and Technology 1108
  T. Pałko, N. Golnik, K. Pęczalski

'E-HCM' Project: Cost Calculation Model for Surgical Procedures 1111
  P. Perger, M. Buccioli, V. Agnoletti, E. Padovani, G. Gambale

Analysis of Combined Ultrasound Therapy and Medium Frequency Current in Abdominal Adiposity 1115
  P.A.G. Gripp, A.M.W. Stadnik, A.A. Bona, E.B. Neves

Cooperative Experiential Learning Models for Biomedical Engineering Students 1119
  S.M. Krishnan

Biomedical Engineering Research Improves the Health Care Industry 1124
  O.A. Lindahl, B.M. Andersson, R. Lundström, K. Ramser

Effect of Different Visual Feedback Conditions on Maximal Grip-Strength Assessment 1127
  A. Chkeir, R. Jaber, D.J. Hewson, J.-Y. Hogrel, J. Duchêne

Health Technology Assessment of Home Monitoring for Patients Suffering from Heart Failure 1132
  L. Pecchia, P. Melillo, M. Attanasio, A. Orrico, E. Pacifici, E. Iadanza

Monitoring and Evaluation of Non-pharmacological Anti-smoking Treatment by Monoximetry: The First
Three Months 1136
  A.M.W. Stadnik, L. Ulbricht, M.F. Sarturi, D. Meirelles, M. Maldaner

Patient Satisfaction Evaluation of Telemedicine Applications Is Not Satisfactory 1140
  Shuai Zhang, Sally I. McClean, Duncan E. Jackson, Chris Nugent, Ian Cleland

Hospital Based Economic Assessment of Robotic Surgery 1144
  R. Miniati, F. Dori, G. Cecconi, F. Frosini, F. Saccà, G. Biffi Gentili, F. Petrucci, S. Franchi, R. Gusinu

Evaluating Care Coordination and Telehealth Services across Europe through the ACT Programme 1147
  C. Maramis, D. Filos, I. Chouvarda, N. Maglaveras, S. Pauws, H. Schonenberg, C. Westerteicher, C. Bescos

Early HTA to Inform Medical Device Development Decisions – The Headroom Method 1151
  A.M. Chapman, C.A. Taylor, A.J. Girling

Analysis of Exposure of Social Alarm Devices in Near Field Conditions 1155
    S. de Miguel-Bilbao, F. Solano, J. García, O.J. Suárez, D. Rubio, V. Ramos

Assessment of Statistical Distribution of Exposure to Electromagnetic Fields from Social Alarm Devices 1159
    S. de Miguel-Bilbao, J. García, E. Aguirre, L. Azpilicueta, F. Falcone, V. Ramos

Increased Complexity of Medical Technology and the Need for Human Factors Informed Design and Training 1163
    P.L. Trbovich, S. Pinkney, C. Colvin, A.C. Easty

Effects of Combined Use of a Patient-Tracking System and a Smart Drug-Dispenser on the Overall Risk of the
Diagnostic-Therapeutic Process 1166
    E. Iadanza, S. Bargellini, G. Biffi Gentili

Risk Management Process in a Microwave Thermal Ablation System for CE Marking 1170
    E. Iadanza, C. Ignesti, G. Biffi Gentili

Medical Devices Recalls Analysis Focusing on Software Failures during the Last Decade 1174
    Z. Bliznakov, K. Stavrianou, N. Pallikarakis

Evaluation of Physical Environment Parameters in Healthcare 1178
    G.A. Elias, S.J. Calil

Manufacturer Training Impact in Heuristic Analysis: Usability Evaluation Applied on Health Devices 1182
    V.H.B. Tsukahara, F.N. Bezerra, R.H. Almeida, F.S. Borges, E. Pantarotto, R.V. Martins, M. Iglesias, P. Trbovich,
    S. Pinkney, S.J. Calil

Heuristic Analysis: A Tool to Improve Medical Device Safety and Usability 1186
    F.N. Bezerra, V.H.B. Tsukahara, R.H. Almeida, F.S. Borges, E. Pantarotto, R.V. Martins, M. Iglesias, P. Trbovich,
    S. Pinkney, S.J. Calil

Measurement and Analysis of the Impact of the Environment on the Spread of Ionizing Radiation in Medical
Facilities 1190
    P. Gabriel, M. Penhaker

Part IX: Health Informatics, E-Health, Telemedicine and Information Technology in Medicine, Bioinformatics

Comparing Two Coaching Systems for Improving Physical Activity of Older Adults 1197
    H. Similä, J. Merilahti, M. Ylikauppila, S. Muuraiskangas, J. Perälä, S. Kivikunnas

Detecting and Analyzing Activity Levels for the Wrist Wearable Unit in the USEFIL Project 1201
    James D. Amor, Vijayalakshmi Ahanathapillai, Christopher J. James

Home Monitoring of Elderly for Early Detection of Changes in Activity and Behavior Patterns 1205
    S. de Miguel-Bilbao, J. García, F. López, P. García-Sagredo, M. Pascual, B. Montero, A. Cruz Jentoft, C.H. Salvador

Automatic Detection of Health Emergency States at Home 1209
    M. Vela, D. Ruiz-Fernández

An Adaptive Home Environment Supporting People with Balance Disorders 1213
    Iliana Bakola, Christos Bellos, Evanthia E. Tripoliti, Athanasios Bibas, Dimitrios Koutsouris, Dimitrios I. Fotiadis

AALUMO: A User Model Ontology for Ambient Assisted Living Services Supported in Next-Generation Networks 1217
P.A. Moreno, M.E. Hernando, E.J. Gómez

Internet of Things for Wellbeing – Pilot Case of a Smart Health Cardio Belt 1221
E. Kovatcheva, R. Nikolov, M. Madjarova, A. Chikalanov

Advances in Modelling of Epithelial to Mesenchymal Transition 1225
R. Summers, T. Abdulla, J.-M. Schleich

Glucose-Level Interpolation for Determining Glucose Distribution Delay 1229
Tomas Koutny

A Supervised SOM Approach to Stratify Cardiovascular Risk in Dialysis Patients 1233
J. Ion Titapiccolo, M. Ferrario, S. Cerutti, C. Barbieri, F. Mari, E. Gatti, M.G. Signorini

Logistic Regression Models for Predicting Resistance to HIV Protease Inhibitor Nelfinavir 1237
L.M. Raposo, M.B. Arruda, R.M. Brindeiro, F.F. Nobre

Application of Special Parametric Methods to Model Survival Data 1241
J. Holčík, K. Opršalová

Protein Function Prediction Based on Protein-Protein Interactions: A Comparative Study 1245
K.S. Ahmed, S.M. El-Metwally

Inhibitory Regulation by microRNAs and Circular RNAs 1250
J. Demongeot, H. Hazgui, J. Escoffier, C. Arnoult

Public Electronic Health Record Platform Compliant with the ISO EN13606 Standard as Support to Research Groups 1254
R. Sánchez de Madariaga, J. Cáceres Tello, A. Muñoz Carrero, O. Moreno Gil, I. Velázquez Aza, A. Castro Serrano, R. Somolinos Cristóbal

Development of a Visual Editor for the Definition of HL7 CDA Archetypes 1258
David Moner, José Alberto Maldonado, Diego Boscá, Alejandro Mañas, Montserrat Robles

homeRuleML Version 2.1: A Revised and Extended Version of the homeRuleML Concept 1262
H.A. McDonald, C.D. Nugent, J. Hallberg, D.D. Finlay, G. Moore

Detailed Clinical Models Governance System in a Regional EHR Project 1266
D. Bosca, L. Marco, D. Moner, J.A. Maldonado, L. Insa, M. Robles

An Extensible Free Software Platform for Managing Image-Based Clinical Trials 1270
M.A. Laguna, N. Malpica, J.A. Hernández-Tamames

Reuse of Clinical Information: Integrating Primary Care and Clinical Research through a Bidirectional Standard Interface 1274
Paolo Fraccaro, Valeria Pupella, Roberta Gazzarata, Mauro Giacomini

Archetype-Based Solution to Tele-Monitor Patients with Chronic Diseases 1278
Juan Mario Rodríguez, Carlos Cavero Barca, Paolo Emilio Puddu, John Gialelis, Petros Chondros, Dimitris Karadimas, Kevin Keene, Jan-Marc Verlinden

Table of Contents

Connecting HL7 with Software Analysis: A Model-Based Approach 1282
    A. Martínez-García, M.J. Escalona, C.L. Parra-Calderón

Integrating the EN/ISO 13606 Standards into an EN/ISO 12967 Based Architecture 1286
    J. Calvillo, I. Román, L.M. Roa

Design of a Semantic Service for Management of Multi-domain Health Alarms 1290
    J. Calvillo, I. Román, L.M. Roa

SILAM: Integrating Laboratory Information System within the Liguria Region Electronic Health Record 1294
    Alessandro Tagliati, Valeria Pupella, Roberta Gazzarata, Mauro Giacomini

A Model for Measuring Open Access Adoption and Usage Behaviour of Health Sciences Faculty Members 1298
    E.T. Lwoga, F. Questier

EHR Anonymising System Based on the ISO/EN 13606 Norm 1302
    R. Somolinos, A. Muñoz, M.E. Hernando, M. Pascual, J. Cáceres, R. Sánchez-de-Madariaga, J.A. Fragua,
    M. Carmona, A.L. Castro, O. Moreno, C.H. Salvador

The Status of Information Systems in the Hospitals of the Greek National Health System 1306
    George Aggelinos, Sokratis Katsikas

Operating Room Efficiency Improving through Data Management 1310
    P. Perger, M. Buccioli, V. Agnoletti, E. Padovani, G. Gambale

A Semantically Enriched Architecture for an Italian Laboratory Terminology System 1314
    Silvia Canepa, Sabrina Roggerone, Valeria Pupella, Roberta Gazzarata, Mauro Giacomini

A National Electronic Health Record System for Cyprus 1318
    K.C. Neokleous, E.C. Schiza, E. Salameh, K. Palazis, C.N. Schizas

Information Driven Care Pathways and Procedures 1322
    R.J. Dickinson, R.I. Kitney

Decision Making in Screening Diagnostics E-Medicine 1326
    G. Balodis, I. Markovica, Z. Markovics, D. Matisone

Using Social Network Apps as Social Sensors for Health Monitoring 1330
    I. Pagkalos, L. Petrou

Advanced Medical Expert Support Tool (A-MEST): EHR-Based Integration of Multiple Risk Assessment Solutions for
Congestive Heart Failure Patients 1334
    Carlos Cavero Barca, Juan Mario Rodríguez, Paolo Emilio Puddu, Mitja Luštrek, Božidara Cvetković,
    Maurizio Bordone, Eduardo Soudah, Aitor Moreno, Pedro de la Peña, Alberto Rugnone, Francesco Foresti,
    Elena Tamburini

Modeling and Implementing a Signal Persistence Manager for Shared Biosignal Storage and Processing 1338
    S. Pirola, E. Opri, A.M. Bianchi, S. Marceglia

Heart Rate Variability for Automatic Assessment of Congestive Heart Failure Severity 1342
    P. Melillo, E. Pacifici, A. Orrico, E. Iadanza, L. Pecchia

Support System for the Evaluation of the Posterior Capsule Opacificaction Degree 1346
    D. Ruiz, L. González, A. Soriano

A Custom Decision-Support Information System for Structural and Technological Analysis in Healthcare............ 1350
  A. Luschi, L. Marzi, R. Miniati, E. Iadanza

Performance Assessment of a Clinical Decision Support System for Analysis of Heart Failure..................... 1354
  G. Guidi, P. Melillo, M.C. Pettenati, M. Milli, E. Iadanza

Self-reporting for Bipolar Patients through Smartphone 1358
  P. Berg Andersen, A. Babic

Comorbidities Modeling for Supporting Integrated Care in Chronic Cardiorenal Disease 1362
  E. Kaldoudi, N. Dovrolis

Overall Survival Prediction for Women Breast Cancer Using Ensemble Methods and Incomplete Clinical Data 1366
  *Pedro Henriques Abreu, Hugo Amaro, Daniel Castro Silva, Penousal Machado, Miguel Henriques Abreu,
  Noémia Afonso, António Dourado*

Automatic Blood Glucose Classification for Gestational Diabetes with Feature Selection: Decision Trees vs. Neural
Networks... 1370
  E. Caballero-Ruiz, G. García-Sáez, M. Rigla, M. Balsells, B. Pons, M. Morillo, E.J. Gómez, M.E. Hernando

On the Global Optimization of the Beam Angle Optimization Problem in Intensity-Modulated Radiation Therapy 1374
  H. Rocha, J.M. Dias, B.C. Ferreira, M.C. Lopes

Effective Supervised Knowledge Extraction for an mHealth System for Fall Detection 1378
  G. Sannino, I. De Falco, G. De Pietro

Chemoprophylaxis Application for Meningococcal Disease for Android Devices 1382
  *M. Parejo-Bellido, E. Dorronzoro-Zubiete, M. Zurbarán, F.J. Sánchez-Laguna, L. Fernández-Luque,
  A.A. Muñoz-Macho*

Automation of Evaluation Protocols of Stand-to-Sit Activity in Expert System 1386
  M.J. Cunha, G. Cardozo, F. de Azevedo

Digital Diary for Persons with Psychological Disorders Using Interaction Design 1390
  H. Sørheim, A. Babic

A Novel Data-Mining Platform to Monitor the Outcomes of Erlontinib (Tarceva) Using Social Media................ 1394
  A. Akay, A. Dragomir, B.E. Erlandsson

Advanced Networked Modular Personal Dosimetry System 1398
  *R. Chil, L.M. Fraile, J. Vaquero, E. Picado, A. Rodriguez-Moreno, M.C. Rodriguez-Sanchez, S. Borromeo,
  M. Desco, J.M. Udías, J.J. Vaquero*

Methods for Personalized Diagnostics 1402
  D.L. Hudson, M.E. Cohen

The Prediction of Blood Pressure Changes by the Habit of Walking 1406
  Toshiyo Tammura, Soichi Maeno, Yutaka Kimira, Yuichi Kimura, Takumu Hattori, Kotaro Minato

Parallel Workflows to Personalize Clinical Guidelines Recommendations: Application to Gestational Diabetes
Mellitus.. 1409
  G. García-Sáez, M. Rigla, E. Shalom, M. Peleg, E. Caballero, E.J. Gómez, M.E. Hernando

Case Based Reasoning in a Web Based Decision Support System for Thoracic Surgery ... 1413
A. Babic, B. Peterzen, U. Lönn, H.C. Ahn

Data Mining in Cancer Registries: A Case for Design Studies ... 1417
G. Kanza, A. Babic

Software Prototype for Triage and Instructional for Homecare Patients ... 1421
L. Boom, N. Escobar, C. Ruales, L. López

Analysis and Impact of Breast and Colorectal Cancer Groups on Social Networks ... 1425
I. De la Torre, B. Martínez, M. López-Coronado

Health Apps for the Most Prevalent Conditions ... 1430
B. Martínez-Pérez, I. de la Torre-Díez, M. López-Coronado

A Mobile Remote Monitoring Service for Measuring Fetal Heart Rate ... 1435
G. Lanzola, I. Secci, S. Scarpellini, A. Fanelli, G. Magenes, M.G. Signorini

Investigating Methods for Increasing the Adoption of Social Media amongst Carers for the Elderly ... 1439
Kyle Boyd, Chris Nugent, Mark Donnelly, Raymond Bond, Roy Sterritt, Lorraine Gibson

AMELIE: Authoring Multimedia-Enhanced Learning Interactive Environment for e-Health Contents ... 1443
P. Sánchez-González, I. Oropesa, P. Moreno-Sánchez, J.M. Martínez-Moreno, J. García-Novoa, E.J. Gómez

mHealth: Cognitive Telerehabilitation of Patients with Acquired Brain Damage ... 1447
C. Suárez-Mejías, M. Parejo, M.J. Zarco, A. Naranjo, C. Echevarría, J. Barros, R. Díez, G. Escobar, M. Elena, C.L. Parra

Mobile Telemedicine Screening Complex ... 1451
J. Lauznis, Z. Markovics, I. Markovica

Radial-Basis-Function Based Prediction of COPD Exacerbations ... 1455
M.A. Fernández Granero, Daniel S. Morillo, A. León, M.A. López Gordo, L.F. Crespo

Research Benefits of Using Interoperability Standards in Remote Command and Control to Implement Personal Health Devices ... 1459
H.G. Barrón-González, Student Member, IEEE, M. Martínez-Espronceda, Member, IEEE, S. Led, L. Serrano, Senior Member, IEEE

Low Cost, Modular and Scalable Remote Monitoring Healthcare Platform Using 8-Bit Microcontrollers ... 1464
F.B. Cosentino, J. Marino-Neto, F.M. Azevedo

Adaptive Healthcare Pathway for Diabetes Disease Management ... 1468
Giuseppe Fico, Alessio Fioravanti, Maria Teresa Arredondo, Chiara Diazzi, Giovanni Arcuri, Claudio Conti, Giampiero Pirini

Real Time Health Remote Monitoring Systems in Rural Areas ... 1473
M.G. Sánchez, R. Nocelo López, J.A. Gay-Fernández

Detecting Accelerometer Placement to Improve Activity Classification ... 1477
Ian Cleland, Chris D. Nugent, Dewar D. Finlay, Roger Armitage

Incorporating the Rehabilitation of Parkinson's Disease in the Play for Health Platform Using a Body
Area Network 1481
   F. Tous, P. Ferriol, M.A. Alcalde, M. Melià, B. Milosevic, M. Hardegger, D. Roggen

Activity Classification Using 3-Axis Accelerometer Wearing on Wrist for the Elderly 1485
   D.I. Shin, S.K. Joo, J.H. Song, S.J. Huh

Intelligent Chair Sensor – Classification and Correction of Sitting Posture 1489
   L. Martins, R. Lucena, J. Belo, R. Almeida, C. Quaresma, A.P. Jesus, P. Vieira

Non-invasive System for Mechanical Arterial Pulse Wave Measurements 1493
   Mikko Peltokangas, Jarmo Verho, Timo Salpavaara, Antti Vehkaoja

Automatic Identification of Sensor Localization on the Upper Extremity 1497
   S. Lambrecht, J.L. Pons

Part X: Medical Devices and Sensors

A New Device to Assess Gait Velocity at Home 1503
   R. Jaber, A. Chkeir, D.J. Hewson, J. Duchêne

Magnetic Induction-Based Sensor for Vital Sign Detection 1507
   H. Mahdavi, J. Rosell-Ferrer

Unconstrained Night-Time Heart Rate Monitoring with Capacitive Electrodes 1511
   A. Vehkaoja, A. Salo, M. Peltokangas, J. Verho, T. Salpavaara, J. Lekkala

Wrist-Worn Accelerometer to Detect Postural Transitions and Walking Patterns 1515
   V. Ahanathapillai, J.D. Amor, M. Tadeusiak, C.J. James

Comparative Measurement of the Head Orientation Using Camera System and Gyroscope System 1519
   P. Kutilek, O. Cakrt, J. Hejda, R. Cerny

A Noninvasive Method of Measuring Force-Frequency Relations to Evaluate Cardiac Contractile State of Patients
during Exercise for Cardiac Rehabilitation 1523
   M. Tanaka, M. Sugawara, Y. Ogasawara, I. Suminoe, T. Izumi, K. Niki, F. Kajiya

A Phase Lock Loop (PLL) System for Frequency Variation Tracking during General Anesthesia 1527
   C.A. Teixeira, B. Direito, A. Dourado, M.P. Santos, M.C. Loureiro

Evaluation of Arterial Properties through Acceleration Photoplethysmogram 1531
   R. Gonzalez, A. Manzo, E. Cardenas, J. Herrera, F. Martinez, J. Gomis, J. Saiz

Comparing over Ground Turning and Walking on Rotating Treadmill 1535
   A. Olenšek, J. Pavčič, Z. Matjačić

Displacement Measurement of a Medical Instrument Inside the Human Body 1539
   D.A. Fotiadis, A. Astaras, A. Kalfas, K. Papathanasiou, P. Bamidis

"Algorithmically Smart" Continuous Glucose Sensor Concept for Diabetes Monitoring 1543
   G. Sparacino, A. Facchinetti, C. Zecchin, C. Cobelli

Piezoresistive Goniometer Network for Sensing Gloves 1547
   G. Dalle Mura, F. Lorussi, A. Tognetti, G. Anania, N. Carbonaro, M. Pacelli, R. Paradiso, D. De Rossi

Preliminary Study of Pressure Distribution in Diabetic Subjects, in Early Stages 1551
    M.L. Zequera, L. Garavito, W. Sandham, Á Rodríguez, J.A. Alvarado, C.A. Wilches

System for Precise Measurement of Head and Shoulders Position 1555
    J. Hejda, P. Kutílek, J. Hozman, R. Černý

An Online Program for the Diagnosis and Rehabilitation of Patients with Cochlear Implants 1559
    T. Lopez-Soto, A. Castillo-Armero

Analysis of Anomalies in Bioimpedance Models for the Estimation of Body Composition 1563
    D. Naranjo, L.M. Roa, L.J. Reina, M.A. Estudillo, N. Aresté, A. Lara, J.A. Milán

Risk Analysis and Measurement Uncertainty in the Manufacturing Process of Medical Devices 1567
    P.E. Baru, N.M. Roman, C. Rusu

Development of an Equipment to Detect and Quantify Muscular Spasticity: SpastiMed – A New Solution 1571
    V. Fernandes, I. Clemente, C. Quaresma, P. Vieira

Comparison between Two Exercise Systems for Rodents 1575
    T. Ödman, N. Ödman, E. Rabotchi, S. Åkervall, M. Lindén

Tripolar Flexible Concentric Ring Electrode Printed with Inkjet Technology for ECG Recording 1579
    Y. Ye-Lin, E. Senent, G. Prats-Boluda, E. Garcia-Breijo, J.V. Lidon, J. Garcia-Casado

Measurement System for Pupil Size Variability Study 1583
    W. Nowak, A. Żarowska, E. Szul-Pietrzak, A. Hachoł

Patellar Reflex Measurement System with Tapping Force Controlled 1587
    Y. Salazar-Muñoz, L.A. Ruano-Calderón, J.A. Leyva, O.S. Martínez, O. García-Cano, B. García-Caballero

A Bioelectric Model of pH in Wound Healing 1591
    M. Amparo Callejón, Laura M. Roa, Javier Reina-Tosina

Part XI: Molecular, Cellular and Tissue Engineering and Biomaterials

Competitive Adsorption of Albumin, Fibronectin and Collagen Type I on Different Biomaterial Surfaces:
A QCM-D Study 1597
    H. Felgueiras, V. Migonney, S. Sommerfeld, N.S. Murthy, J. Kohn

Biomimetic Poly(NaSS) Grafted on Ti6Al4V: Effect of Pre-adsorbed Selected Proteins on the MC3T3-E1 Osteoblastic
Development 1601
    H. Felgueiras, V. Migonney

Bioactive Intraocular Lens – A Strategy to Control Secondary Cataract 1605
    Yi-Shiang Huang, Virginie Bertrand, Dimitriya Bozukova, Christophe Pagnoulle, Edwin De Pauw,
    Marie-Claire De Pauw-Gillet, Marie-Christine Durrieu

Ultrathin Films by LbL Self-assembly for Biomimetic Coatings of Implants 1609
    M. Giulianelli, L. Pastorino, R. Ferretti, C. Ruggiero

Nanohelical Shape and Periodicity Dictate Stem Cell Fate in a Synthetic Matrix 1613
    R.K. Das, O.F. Zouani, R. Oda, M.-C. Durrieu

The Role of Adult Stem Cells on Microvascular Tube Stabilization 1617
    O.F. Zouani, Y. Lei, M.-C. Durrieu

Processes of Gamma Radiolysis of Soluble Collagen in the Aspect of the Scaffolds Preparation 1622
    K. Pietrucha

Building Up and Characterization of Multi-component Collagen-Based Scaffolds...................................... 1626
    K. Pietrucha

Fabrication of Gelatin/Bioactive Glass Hybrid Scaffolds for Bone Tissue-Engineering 1630
    S. Borrego-González, J. Becerra, A. Díaz-Cuenca

Collagen-Targeted BMPs for Bone Healing 1634
    P.M. Arrabal, R. Visser, L. Santos-Ruiz, J. Becerra, M. Cifuentes

Layer by Layer: Designing Scaffolds for Cardiovascular Tissues 1638
    B. Glasmacher, A. Repanas A., O. Gryshkov, F. AL Halabi, T. Rittinghaus, R. Kortlepel, S. Wienecke,
    M. Müller, H. Zernetsch

Part XII: Neural and Rehabilitation Engineering

Voice Activity Detection from Electrocorticographic Signals 1643
    V.G. Kanas, I. Mporas, H.L. Benz, N. Huang, N.V. Thakor, K. Sgarbas, A. Bezerianos, N.E. Crone

Assessment of an Assistive P300–Based Brain Computer Interface by Users with Severe Disabilities 1647
    R. Corralejo, L.F. Nicolás-Alonso, D. Álvarez, R. Hornero

Single-Trial Detection of the Event-Related Desynchronization to Locate with Temporal Precision the Onset of
Voluntary Movements in Stroke Patients 1651
    J. Ibáñez, M.D. del Castillo, J.I. Serrano, F. Molina Rueda, E. Monge Pereira, F.M. Rivas Montero,
    J.C. Miangolarra Page, J.L. Pons

Hybrid Brain Computer Interface Based on Gaming Technology: An Approach with Emotiv EEG
and Microsoft Kinect 1655
    Nuno André da Silva, Ricardo Maximiano, Hugo Alexandre Ferreira

Rehabilitation Using a Brain Computer Interface Based on Movement Related Cortical Potentials -A Review 1659
    K. Dremstrup, I.K. Niazi, M. Jochumsen, N. Jiang, N. Mrachacz-Kersting, D. Farina

Numerical Simulation of Deep Transcranial Magnetic Stimulation by Multiple Circular Coils 1663
    M. Lu, S. Ueno

Volitional Intention and Proprioceptive Feedback in Healthy and Stroke Subjects 1667
    M. Gandolla, S. Ferrante, F. Molteni, E. Guanziroli, T. Frattini, A. Martegani, G. Ferrigno, A. Pedrocchi, N.S. Ward

Nonlinear Relationship between Perception of Deep Pain and Medial Prefrontal Cortex Response Is Related to
Sympathovagal Balance 1671
    R. Sclocco, M.L. Loggia, R.G. Garcia, R. Edwards, J. Kim, S. Cerutti, A.M. Bianchi, V. Napadow, R. Barbieri

Dynamics of Learning in the Open Loop VOR 1675
    P. Colagiorgio, G. Bertolini, C. Bockisch, D. Straumann, S. Ramat

Numerical Modeling of Optical Radiation Propagation in a Realistic Model of Adult Human Head 1679
    N. Ortega-Quijano, F. Fanjul-Vélez, I. Salas-García, J.L. Arce-Diego

Individual Noise Responses and Task Performance and Their Mutual Relationships 1683
    T. Niioka, S. Ohnuki

Is the Software Package Embedding Dynamic Causal Modeling Robust? 1686
    P. Tayaranian Hosseini, S. Wang, S.L. Bell, J. Brinton, D.M. Simpson

The Focuses of Pathology of Electrical Activity of Brain in Assessment of Origin of Syncope 1690
    K. Peczalski, T. Palko, D. Wojciechowski, W. Jernajczyk, N. Golnik, Z. Dunajski

Neuroanatomic-Based Detection Algorithm for Automatic Labeling of Brain Structures in Brain Injury 1694
    M. Luna, F. Gayá, A. García-Molina, L.M. González, C. Cáceres, M. Bernabeu, T. Roig, A. Pascual-Leone,
    J.M. Tormos, E.J. Gómez

Brain Activity Characterization Induced by Alcoholic Addiction. Spectral and Causality Analysis of Brain Areas
Related to Control and Reinforcement of Impulsivity 1698
    J. Guerrero, A. Rosado, M. Bataller, J.V. Francés, T. Iakymchuk, A. Luque-García, V. Teruel-Martí, J. Martínez-Ricós

A Novel RMS Method for Presenting the Difference between Evoked Responses in MEG/EEG – Theory
and Simulation 1702
    I. Nemoto, M. Kawakatsu

The Influence of Neuronal Density on Network Activity: A Methodological Study 1706
    E. Biffi, G. Regalia, A. Menegon, G. Ferrigno, A. Pedrocchi

Effective Connectivity Patterns Associated with P300 Unmask Differences in the Level of Attention/Cognition between
Normal and Disabled Subjects 1710
    S.I. Dimitriadis, Yu Sun, N.A. Laskaris, N. Thakor, A. Bezerianos

A General Purpose Approach to BCI Feature Computation Based on a Genetic Algorithm: Preliminary Results 1714
    S. Ramat, N. Caramia

Modular Control of Crouch Gait in Spastic Cerebral Palsy 1718
    D. Torricelli, M. Pajaro, S. Lerma, E. Marquez, I. Martinez, F. Barroso, J.L. Pons

Time-Frequency Analysis of Error Related Activity in Anterior Cingulate Cortex 1722
    G. Mijatovic, T. Loncar Turukalo, E. Procyk, D. Bajic

Depth-Sensitive Algorithm to Localize Sources Using Minimum Norm Estimations 1726
    B. Pinto, A.C. Sousa, C. Quintão

Measurement of Gait Movements of a Hemiplegic Subject with Wireless Inertial Sensor System before and after
Robotic-Assisted Gait Training in a Day 1730
    Takashi Watanabe, Jun Shibasaki

Glenohumeral Kinetics in Gait with Crutches during Reciprocal and Swing-Through Gait: A Case Study 1734
    E. Perez-Rizo, R. Casado-Lopez, V. Lozano-Berrio, M. Solis-Mozos, M. Nieto-Diaz, S. Martin-Majarres, J.L. Pons,
    A. Gil-Agudo

A Feasibility Study to Elicit Tactile Sensations by Electrical Stimulation 1738
    S.H. Hwang, J. Ara, T. Song, G. Khang

Dystonia: Altered Sensorimotor Control and Vibro-tactile EMG-Based Biofeedback Effects 1742
   C. Casellato, S. Maggioni, F. Lunardini, M. Bertucco, A. Pedrocchi, T.D. Sanger

Closed-Loop Modulation of a Notch-Filter Stimulation Strategy for Tremor Management with a Neuroprosthesis 1747
   J.A. Gallego, E. Rocon, J.M. Belda-Lois, J.L. Pons

Objective Metrics for Functional Evaluation of Upper Limb during the ADL of Drinking: Application in SCI 1751
   A. de los Reyes-Guzmán, I. Dimbwadyo-Terrer, S. Pérez-Nombela, F. Trincado, D. Torricelli, A. Gil-Agudo

Efficacy of TtB-Based Visual Biofeedback in Upright Stance Trials 1755
   C. D'Anna, D. Bibbo, M. Goffredo, M. Schmid, S. Conforto

A Data-Globe and Immersive Virtual Reality Environment for Upper Limb Rehabilitation after Spinal Cord Injury 1759
   Ana de los Reyes-Guzman, Iris Dimbwadyo-Terrer, Fernando Trincado-Alonso, Miguel A. Aznar, Cesar Alcubilla,
   Soraya Pérez-Nombela, Antonio del Ama-Espinosa, Begoña Polonio-López, Ángel Gil-Agudo

Wearable Navigation Aids for Visually Impaired People Based on Vibrotactile Skin Stimuli 1763
   M. Reyes Adame, K. Möller, E. Seemann

Brain Injury MRI Simulator Based on Theoretical Models of Neuroanatomic Damage 1767
   L.M. González, M. Luna, A. García-Molina, C. Cáceres, J.M. Tormos, E.J. Gómez

Haptic Feedback Affects Movement Regularity of Upper Extremity Movements in Elderly Adults 1771
   M. Schmid, I. Bernabucci, S. Comani, S. Conforto, B. D'Elia, B. Fida, T. D'Alessio

Dysfunctional Profile for Patients in Physical Neurorehabilitation of Upper Limb 1775
   M.A. Villán-Villán, R. Pérez-Rodríguez, C. Gómez, E. Opisso, J.M. Tormos, J. Medina, E.J. Gómez Aguilera

Video-Based Tasks for Emotional Processing Rehabilitation in Schizophrenia 1779
   R. Caballero-Hernández, A. Vila-Forcén, S. Fernandez-Gonzalo, J.M. Martínez-Moreno, M. Turon,
   R. Sánchez-Carrión, E.J. Gómez

Preliminary Experiment with a Neglect Test 1783
   C. Lassfolk, M. Linnavuo, S. Talvitie, M. Hietanen, R. Sepponen

A Subject-Driven Arm Exoskeleton to Support Daily Life Activities 1787
   E. Ambrosini, S. Ferrante, M. Rossini, F. Molteni, G. Ferrigno, A. Pedrocchi

Evaluation of a Novel Modular Upper Limb Neuroprosthesis for Daily Life Support 1791
   S. Ferrante, E. Ambrosini, M. Rossini, F. Molteni, M. Bulgheroni, G. Ferrigno, A. Pedrocchi

Evaluating Spatial Characteristics of Upper-Limb Movements from EMG Signals 1795
   O. Urra, A. Casals, R. Jané

An Adaptive Rreal-Time Algorithm to Detect Gait Events Using Inertial Sensors 1799
   N. Chia Bejarano, E. Ambrosini, A. Pedrocchi, G. Ferrigno, M. Monticone, S. Ferrante

A Graphical Tool for Designing Interactive Video Cognitive Rehabilitation Therapies 1803
   J.M. Martínez-Moreno, P. Sánchez-González, A. García, S. González, C. Cáceres, R. Sánchez-Carrión, T. Roig,
   J.M. Tormos, E.J. Gómez

Part XIII: Young Investigator Competition

Motion-Related VEPs Elicited by Dynamic Virtual Stimulation 1809
    P.J.G. Da Silva, B.P. Rosa, M. Cagy, A.F.C. Infantosi

A Distributed Middleware for the Assistance on the Prevention of Peritonitis in CKD 1813
    M.A. Estudillo-Valderrama, A. Talaminos-Barroso, L.M. Roa, D. Naranjo-Hernández, L.J. Reina-Tosina

Bioactive Nanoimprint Lithography: A Study of Human Mesenchymal Stem Cell Behavior and Fate 1817
    Z.A. Cheng, O.F. Zouani, K. Glinel, A.M. Jonas, M.-C. Durrieu

Automated Normalized Cut Segmentation of Aortic Root in CT Angiography 1821
    Mustafa Elattar, Esther Wiegerinck, Nils Planken, Ed vanbavel, Hans van Assen, Jan Baan Jr., Henk Marquering

Classification Methods from Heart Rate Variability to Assist in SAHS Diagnosis 1825
    J. Gómez-Pilar, G.C. Gutiérrez-Tobal, D. Álvarez, F. del Campo, R. Hornero

AdaBoost Classification to Detect Sleep Apnea from Airflow Recordings 1829
    G.C. Gutiérrez-Tobal, D. Álvarez, J. Gómez-Pilar, F. del Campo, R. Hornero

Effect of Electric Field and Temperature in E.Coli Viability 1833
    A.M. Oliva, A. Homs, E. Torrents, A. Juarez, J. Samitier

3D Shape Landmark Correspondence by Minimum Description Length and Local Linear Regularization 1837
    M. Valenti, C. Chen, E. De Momi, G. Ferrigno, G. Zheng

β-Band Peak in Local Field Potentials as a Marker of Clinical Improvement in Parkinson's Disease after Deep Brain Stimulation 1841
    P.D. Frangou, K.P. Michmizos, P. Stathis, D. Sakas, K.S. Nikita

Ergonomics during the Use of LESS Instruments in Basic Tasks: 2 Articulated vs. 1 Straight and 1 Articulated Graspers 1845
    M. Lucas-Hernández, F.J. Pérez-Duarte, A.M. Matos-Azevedo, J.B. Pagador, F.M. Sánchez-Margallo

Network-Based Modular Markers of Aging Across Different Tissues 1849
    Aristidis G. Vrahatis, Konstantina Dimitrakopoulou, Georgios N. Dimitrakopoulos, Kyriakos N. Sgarbas, Athanasios K. Tsakalidis, Anastasios Bezerianos

Experimental Characterization of Active Antennas for Body Sensor Networks 1853
    D. Naranjo, L.M. Roa, L.J. Reina, G. Barbarov, A. Callejón

Part XIV: Special Sessions

Models of Arrhythmogenesis in Myocardial Infarction 1859
    Natalia A. Trayanova

Multiscale-Multiphysics Models of Ventricular Electromechanics – Computational Modeling, Parametrization and Experimental Validation 1864
    G. Plank, A.J. Prassl, R. Arnold, Y. Rezk, T.E. Fastl, E. Hofer, C.M. Augustin

Effect of Purkinje-Myocyte Junctions on Transmural Action Potential Duration Profiles 1868
    R. Walton, O. Bernus, E.J. Vigmond

Morphometry and Characterization of Electrograms in the Cavotricuspid Isthmus in Rabbit Hearts during Autonomic Sinus Rhythm 1871
 E. Hofer, D. Sanchez-Quintana, R. Arnold

Cavotricuspid Isthmus: Anatomy and Electrophysiology Features: Its Evaluation before Radiofrequency Ablation 1875
 D. Sánchez-Quintana, J.A. Cabrera

Preoperative Prognosis of Atrial Fibrillation Concomitant Surgery Outcome after the Blanking Period 1879
 A. Hernández, R. Alcaraz, F. Hornero, J.J. Rieta

A Decision Support System for Operating Theatre in Hospitals: Improving Safety, Efficiency and Clinical Continuity 1883
 R. Miniati, G. Cecconi, F. Frosini, F. Dori, G. Biffi Gentili, F. Petrucci, S. Franchi, R. Gusinu

LICENSE: Web Application for Monitoring and Controlling Hospitals' Status with Respect to Legislative Standards 1887
 E. Iadanza, L. Ottaviani, G. Guidi, A. Luschi, F. Terzaghi

Relation among Breathing Pattern, Sleep Posture and BMI during Sleep Detected by Body Motion Wave 1891
 Junya Wada, Tadashi Yajima, Takenori Imamatsu, Hiroaki Okawai

Sophisticated Rate Control of Respiration and Pulse during Sleep Studied by Body Motion Wave 1895
 H. Okawai, T. Yajima, T. Imamatsu, J. Wada

A Tool for Patient Data Recovering Aimed to Machine Learning Supervised Training 1899
 G. Guidi, M.C. Pettenati, M. Milli, E. Iadanza

The Potential of Machine-to-Machine Communications for Developing the Next Generation of Mobile Device Monitoring Systems 1903
 R.S.H. Istepanian, B. Woodward, D.J. Mulvaney, S. Datta, P. Harvey, A.L. Vyas, O. Farooq

A Study on Perception of Managing Diabetes Mellitus through Social Networking in the Kingdom of Saudi Arabia 1907
 T.M. Alanzi, R.S.H. Istepanian, N. Philip, A. Sungoor

Validity of Smartphone Accelerometers for Assessing Energy Expenditure during Fast Running 1911
 C. Easton, N. Philip, A. Aleksandravicius, J. Pawlak, D.J. Muggeridge, P.A. Domene, R.S.H. Istepanian

Medical Quality of Service Analysis of Ultrasound Video Streaming over LTE Networks 1915
 Ali Alinejad, R.S.H. Istepanian, N. Philip

Smart Social Robotics for 4G-Health Applications 1919
 R.S.H. Istepanian, A. Good, N. Philip

Regional Cooperation in the Development of Biomedical Engineering for Developing Countries 1923
 K.I. Nkuma-Udah, G.I. Ndubuka, E.E.C. Agoha

Author Index 1927

Keyword Index 1939

Part VIII
Clinical Engineering

A New Approach for Preventive Maintenance Prioritization of Medical Equipment

N. Saleh[1], A. Sharawi[2], M. Abd Elwahed[2], A. Petti[3], D. Puppato[4], and G. Balestra[1]

[1] Politecnico di Torino/Electronics and Telecommunications department, Turin, Italy
[2] Faculty of Engineering/Systems and Biomedical Engineering department, Cairo University, Giza, Egypt
[3] ASL BI, Biella, Italy, [4] AReSS, Turin, Italy

Abstract—Efficient maintenance of medical equipment is crucial phase in medical equipment management. Preventive maintenance is a core function of clinical engineering and it is essential to guarantee the correct functioning of the equipment. The aim of this paper is to develop a new model for preventive maintenance priority of medical equipment using the quality function deployment (QFD) as a new concept in maintenance of medical equipment. We developed a 3 domain framework consisting of requirements, function, and concepts. The requirements domain is the house of quality matrix (HOQ) or planning matrix. The second domain is the design matrix. Finally, the concept domain contains the critical criteria for preventive maintenance prioritization with its weights. According to the final scores of the criteria, the prioritization of medical equipment is performed. The data set includes 200 medical equipment belonging to 17 different departments of 2 hospitals in Piemonte; Italy. It includes 70 different types of equipment. Our model proposes 5 levels of priority for preventive maintenance. The results show a high correlation between risk - based criteria and prioritization.

Keywords—Medical Equipment, Preventive Maintenance, QFD, prioritization, Healthcare Technology Management.

I. INTRODUCTION

Medical equipment is an asset that directly affects human lives. It is important, therefore, to have well planned and managed maintenance programs that are able to keep the medical equipment in healthcare organizations reliable, safe, and available for use when it is needed. Maintenance activity no longer involves only corrective actions as in the past [1], but aims towards a more complete risk management approach. Preventive maintenance is mainly a risk based approach and considered as a core function of clinical engineering department.

Hospitals should be compliant to the most stringent maintenance requirements. In the late of 2011, centers for Medicare & Medicaid Services (CMS) issued a clarification of hospital equipment maintenance requirements conflicts with The Joint Commission (TJC) standards for preventive maintenance frequency [2]. This fact reflects the open challenge for preventive maintenance management.

Maintenance prioritization is a crucial task in management systems, especially when there are more maintenance work orders than available people or resources that can handle those [3]. In preventive maintenance management field, several models have been proposed for this purpose. *Josegh, J. et al* [4] have been developed a model for preventive maintenance index considering Risk Level Coefficient (RLC) of the instrument. Risk level coefficient was calculated through five different classified factors related to the medical equipment electrical risk.

In [1], the authors developed System of Information Technology and Support System for Maintenance Actions (SISMA) system. The system considers two aspects for evaluation; technical and economic needs to assess PM plan for equipment.

Quality function deployment (QFD) is one of the total quality management (TQM) quantitative tools and techniques that could be used to translate customer requirements and specifications into appropriate technical or service requirements. QFD uses visual matrices that link customer requirements, design requirements, target values, and competitive performance into one chart. This technique has been applied in manufacturing and production area, also it utilized to develop performance measurement systems in maintenance in 2001 [5].

One form of QFD is a four-phase model, which includes the house of quality (HOQ), parts deployment (design matrix), process planning, and production planning [6].

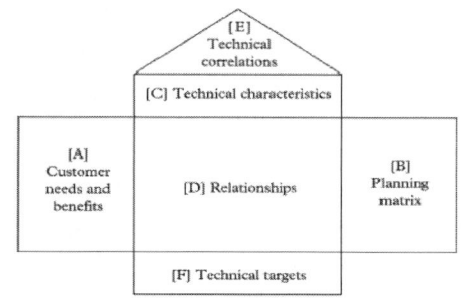

Fig. 1 The house of quality of function deployment [7]

Among the various matrices, the house of quality, HOQ shown in Fig. 1 is commonly used. The matrix displays the voice of customers (VOC) or customer requirements that are known as WHATs against the technical requirements or voice of engineers (VOE) that are known as HOWs [5, 6]. The order suggested by letters A to F is normally followed during the process. Room A contains a list of customer needs, each of which is assessed against competitors and the results, which are absolute and relative weights for customer needs prioritization are reported in to room B. Room C has the information necessary to transform the customer expectations into technical characteristics and the correlation between each customer requirements and technical response is put into D. The roof, room E, considers the extent to which the technical responses support each other. The prioritization of the technical characteristics, information on the competition and technical targets weights all go into F [7].

II. METHODOLOGY

By using the quality function deployment, we proposed a 3 domains framework for preventive maintenance priority, as illustrated in Fig.2.

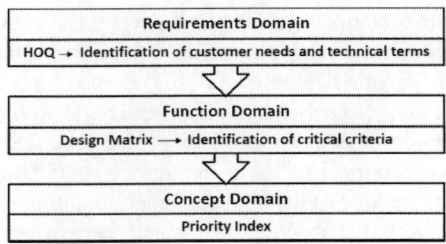

Fig. 2 A 3 domains framework for preventive maintenance of medical equipment prioritization

The first domain considers the customer requirements and the technical characteristics that meet the customer requirements, i.e. the house of quality matrix. The second domain is the function domain or design matrix which translates the top technical specifications into critical criteria for prioritization. The last stage of this model is the generation of priority equation regarding the critical criteria with weights to produce the priority index of medical equipment preventive maintenance.

A. Requirements Domain

Requirements domain is the house of quality matrix of preventive maintenance prioritization. For our case, the customers are the patients and the clinical staff, since they have direct interface with medical devices (WHATs). The clinical engineering department is responsible to satisfy the customer requirements (HOWs). For patient's requirements, no doubt that safety and availability of medical equipment are essential. Clinical staff requirements were chosen based upon literature [8] and experience. Table 1 depicts customer's requirements and technical characteristics.

Table 1 customer requirements and technical characteristics

Customer Requirements	Technical terms	Technical category
Safety of medical equipment	Physical risk	Risk
Efficiency	Function	
Durability	Maintenance requirements	
Quick response of technical team	Mission criticality	performance assurance
Back up availability	Functional verification	
Check the device after maintenance	Age	
Regular monitoring of the devices	Labeling	
Priority based on importance	Electrical safety	
Obvious operating instructions	Parts replacement	
Knowledge of maintained devices	Regular inspection	
Existence of a contact person 24h	Qualification of technicians	User competence
Avoiding suspension of services	Complexity of devices	
	Equipped workshop	
	Test equipment availability	
	Service manual availability	
	Activities recording	
	Updating or loan	The costs
	Spare parts availability	
	Service provider type	
	Meet standards	standards

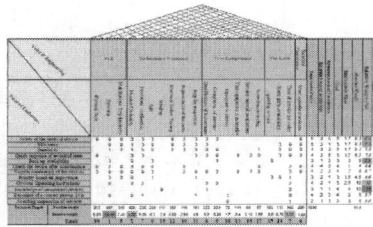

Fig. 3 The house of quality of preventive maintenance prioritization

Prioritization of customer's requirements is performed considering a comparison between Italian hospital and Jordanian hospital [8]. The relationships between WHATs and HOWs are indicated by scores 9 for strong relation, 3 for medium relation, 1 for low relation, and nothing for no relation [6]. Fig.3 illustrates the house of quality of developed model; the single parts are described in detail in the text.

C. Function Domain

The next stage of our proposed model is to identify the critical criteria among the technical criteria for preventive maintenance priority. We selected the top 11 criteria of technical terms based upon their weights and importance to become the inputs (WHATs) of the second matrix. The criteria are function, mission criticality, service provider type, standards meet, maintenance requirements, age, functional verifications, team qualification, device complexity, physical risk, and regular inspection.

We classified the critical criteria into 3 categories for HOWs of the second matrix; risk-based criteria including function, physical risk, and maintenance requirements; mission-based criteria including utilization level, area criticality, and device criticality; and finally maintenance-based criteria incorporating, failure rate, useful life ratio, device complexity, number of missed maintenance, and downtime ratio as illustrated in Fig. 4. The figure presents the relationships among the parts that are described in detail in the text.

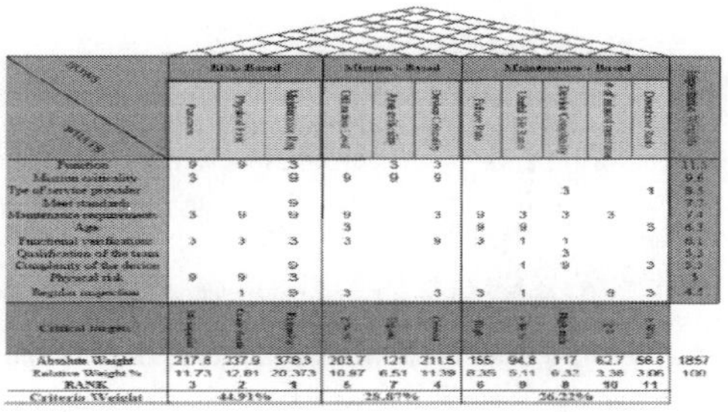

Fig. 4 The design matrix of preventive maintenance prioritization

C. Concept Domain

The output of the design matrix is prioritization equation considering the most critical criteria with weights. Equation (1) shows the priority index of preventive maintenance, and Table 2, presents the proposed factors with brief description and their proposed scores.

$$PS = 11.7(FN) + 12.8(PR) + 20.4(MR) + 11(UL) + 6.5(AC) + 11.4(DC) + 8.3(FR) + 5.1(LR) + 6.3(CM) + 3.4(MM) + 3.1(DR) \quad (1)$$

PS: priority score
FN: function of equipment
PR: physical risk
MR: maintenance requirements
UL: utilization level
AC: area criticality
DC: device criticality
FR: failure rate
LR: useful life ratio
CM: device complexity
MM: missed maintenance
DR: downtime ratio

Table 2 brief description of critical parameters and their scores

Parameter	Description	Thresholds	Scores
Function	Device function	Life support	5
		Therapeutic	4
		Diagnostic / monitoring	3
		Analytical	2
		Miscellaneous	1
Physical risk	Probable harms caused by equipment failure	Death	5
		Injury	4
		Misdiagnosis	3
		Equipment damage	2
		No risk	1
Maintenance requirements	Maintenance activities depending on equipment type	Extensive	5
		A above average	4
		Average	3
		Below average	2
		Minimal	1
Utilization level	Number of working days a week	>4 days	3
		3-4 days	2
		<3 days	1
Area criticality	Assessment of area criticality for patients	Urgent	5
		Intensity care units	4
		Diagnostic area	3
		Law intensity area	2
		Non clinical area	1
Device criticality	The importance level of equipment in serviced area	Critical	3
		Important	2
		Necessary	1
Failure Rate	Number of failures a year based on device criticality level	≥2 for critical, ≥4 for important, ≥5 for necessary	3
		1 for critical, 2-3 for important, 3-4 for necessary	2
		0 for critical, ≤1 for important, ≤2 for necessary	1
Useful life ratio	Ratio between age to expected life time of a device	Ratio > 80 %	3
		50% < Ratio ≤80%	2
		Ratio ≤ 50 %	1
Device complexity	Technical complexity based on a model	Score 6 - 8	3
		Score 3 - 5	2
		Score 0 - 2	1
Missed maintenance	Number of missed maintenance a year	≥ 2	3
		1	2
		0	1
Downtime ratio	Ratio between the duration of downtime in days to days a year	Ratio ≥ 20 %	3
		10% ≤ Ratio<20%	2
		Ratio < 10 %	1

The complexity model [9] assesses the technical complexity of a device considering four factors; equipment maintainability, installation requirements, repair, and connectivity. The scores are given for each factor range from 0 to 2 for evaluation.

III. RESULTS

For model verification, we utilized a data set of two hundreds medical equipment of two hospitals in Piemonte; Italy. Seventy different types of equipment belonging to 17 departments were analyzed. Table 3, shows sample data of various types of investigated equipment along with priority scores.

Table 3 Sample data of investigated equipment for priority

Equipment	FN	PR	MR	UL	AC	DC	FR	LR	CM	MM	DR	PS	PS %
Ventilator	5	5	5	2	5	3	2	3	3	3	1	377	93
C-arm	3	3	4	2	5	3	3	3	3	2	2	316	78
Monitor	3	3	3	3	3	2	3	3	2	1	1	269	66
Centrifuge	2	3	3	3	2	2	1	3	1	1	1	228	56
Scale	1	1	1	1	3	1	1	2	1	2	1	121	30

By using the result priority scores percentages, the preventive maintenance priority is classified into five categories. The first class is very high priority class and includes equipment with priority score percentage equal or greater than 80. In second class, preventive maintenance should be performed within one month if priority percentage in range 70 to 80. Class 3 is medium priority, contains all equipment should be considered for preventive maintenance within 2 months in case of priority percentage in range 60 to 70. Class 4 is low priority, includes all equipment with priority percentage of 50 to 60, and in this case preventive maintenance should be performed within 3 months. Finally all equipment with priority percentage less than 50 could be visually inspected and considered for next preventive maintenance as minimal preventive maintenance.

In our data set of equipment and according to the proposed model, 15 % needs very high priority preventive maintenance, 19.5 % should be included as high priority, 29.5 % should be considered for medium priority, 27 % for low priority .and finally 9 % should be with minimal priority preventive maintenance.

By analyzing the results, very high priority class incorporates all equipment with high risk criteria, relatively high mission based criteria, also with high complexity. High priority class contains relatively high risk criteria and mission based criteria in addition to high missed maintenance. Medium priority class is considered for equipment with relatively high utilization level, area criticality, and old equipment. Low priority class contains old not risky devices. Relatively stable equipment doesn't need preventive maintenance. The results are consistent with the classification given by an experienced clinical engineer.

IV. CONCLUSION

In this study, quality function deployment is presented for first time to solve the problem of preventive maintenance prioritization of medical equipment. The proposed model has proven its validity in real environment correctly separating equipment that needs preventive maintenance from those that do not need it. It is important to note that the classification is based on the requirements of patients and clinical staff. By analyzing the results, we can state that the risk-based criteria have a great impact on preventive maintenance prioritization decision in addition to criticality and age of medical equipment.

REFERENCES

1. Miniati R, Dori F, and Gentili G (2012) Design of a decision support system for preventive maintenance planning in health structures. Technology and Healthcare 20 (2012) , pp 205-214 DOI 10.3233/THC-2012-0670, Italy
2. Wang B, Rui T, Balar S (2013) An estimate of patient incidents caused by medical equipment maintenance omissions. Biomed Inst & Tec.2013 Jan-Feb; 47(1):84-91. DOI 10.2345/0899-8205-47.1.84.
3. Ni J, Jin X (2012) Design support systems for effective maintenance CIRP Annuals- Manufacturing technology 61 (2012) 411-414
4. Justin J, Madhukumar S (2010) A novel approach to data driven preventive maintenance scheduling of medical instruments. International conference on systems in medicine and biology, India. IEEE Eng. Med. Biol. Soc.,2010; pp 193-197
5. Duffuaa S, Al Ghamdi A, Al Amer A (2002) Quality function Deployment in maintenance work planning process. 6th Saudi engineering conference, KFUPM,2002,V4, pp 503-5012
6. Shen X, Tan K, Xie M (2000) An integrated approach to innovative product development using kano's model and QFD. European journal of innovation management, V3, N2,2000, pp 91-99
7. Deglado D, Aspinwall E et al (2007) quality function deployment in construction. Construction management and economics, 2007, 25, pp 597-609
8. Adnan A et al (2012) Building medical devices maintenance system through quality function deployment. Jordan journal of mechanical and industrial engineering , V6, N1, 2012, pp 25-36
9. Nataly Y, Hyman W (2009) A medical device complexity model: A new approach to medical equipment management. Journal of clinical engineering, 03/2009; 34(2), pp 94-98 DOI :10.1097/JCE.0b013e31819fd711

Author: Neven Saleh
Institute: Department of Electronics and Telecommunications, Politecnico di Torino
Street: Duca degli Abruzzi, 24
City: Turin
Country: Italy
Email: nevensaleh@hotmail.com

2D and 3D Photogrammetric Models for Respiratory Analysis in Adolescents

W.L. Ripka[1], L. Ulbricht[2], E.B. Neves[2], and P.M. Gewehr[1]

[1] Federal University of Technology-Paraná, Graduate Program in Electrical Engineering and Computer Science, Curitiba, Brazil
[2] Federal University of Technology-Paraná, Graduate Program in Biomedical Engineering, Curitiba, Brazil

Abstract—The aim of this study was to test the correlation between a two dimensional (2D) and three dimensional (3D) biomechanical photogrammetric model, with inspiratory and expiratory lung capacities in adolescents. This was a cross sectional study, in which 40 adolescents of both genders were assessed, aged between 14 and 17 years old. Anthropometric data were collected (body mass, height and thorax length). Then, three tests were performed for forced vital capacity (FVC) in the supine position, using a spirometer. For FVC proof assessments, the volunteers were subjected to a mapping of body structures with adhesive markers and, after the test, one was selected with better respiratory performance for computational analysis of photogrammetric thoracoabdominal mobility. The results showed average values of 15.40, 61.29, 1.66 and 25.54, for age (years), body mass (kg), height (m) and thorax length (cm), respectively. The analysis found respiratory values of: 3:22 l for forced expiratory volume in one second, 3.87 l for forced vital capacity, 6.50 l/s for peak expiratory flow and 3.59 l for deep inspiration. The correlation showed better values for 3D photogrammetry as compared to 2D. The inclusion of the third dimension has been applied in order to adequate photogrammetric methodology to existing kinematics analysis tools as videogrammetry computed and plethysmography.

Keywords—Photogrammetry, Respiratory Analysis, Child.

I. INTRODUCTION

Human systems are essential for the maintenance of life [1]. Among them, the respiratory system, which comprises a complex set of organs, muscle and bone structures that work in synchronism to generate the ventilation phenomenon is vital for the human being [2]. When the respiratory system does not work in perfect condition, obstructive or restrictive diseases may be suspected or low fitness [3].

On this issue, it was highlighted the importance of assessments to measure and evaluate breathing performance, since they represent a procedure for professionals decision making in healthcare practice.

In the instrumentation field, there are several validated methods for respiratory assessment, among them: plethysmography [4], nitrogen washout [5] e three-dimensional kinematic analysis [6]. These methods evaluate effectively and non-invasively volumes, capacities and breathing patterns, but using high cost equipment and of difficult transportability. Also these equipment usually only exist in major laboratories for the study of the respiratory system.

As a low-cost strategy, there is a known technique for respiratory analysis (photogrammetry), dating from the late nineteenth century, which is based on the relationship between surface movement with thoracoabdominal lung volumes [7, 8]. Incorporated in the respiratory evaluation, the photogrammetry showed good results in two dimensional respiratory analysis, proved to be a reliable method, reproducible and affordable [9, 10].

However, studies in adolescents such as the application in ventilatory three dimensional simulation represent gaps to be filled. Thus, the objective of this study was to test the correlation between a two dimensional (2D) and three dimensional (3D) biomechanical photogrammetric model, with inspiratory and expiratory lung capacities in adolescents.

II. MATERIALS AND METHODS

The sample was chosen intentionally and was composed by 40 subjects of both genders aged between 14 and 17 years old.

The evaluations were divided into: anthropometric and photogrammetric analysis associated with spirometry. All participants in this study signed a consent form. This study was approved by the ethical committee of the Federal University of Technology-Paraná.

For anthropometric measurements, several parameters were collected: body mass (kg), height (m) and thorax length (cm). Body mass was measured by using a 100g resolution mechanical scale (Filizola). The height was measured by using a stadiometer with 0.1cm resolution. The thorax length was measured with the subject standing with his arms away from the body. A anthropometric caliper (WCS model), with precision of 0.1mm, was positioned in line with the xiphoid process of each individual, who was asked to perform a deep breath for obtaining the measure.

For photogrammetric analysis, adhesive markers, with known diameter (13 mm), were placed on the participants on the projections of the umbilicus (PU), the inferior angle of the 10th rib (AR), the manubrium sterni (MS), the xiphoid process (XP) and the right anterior superior iliac spine (SI)

(Fig. 1a e 1b). Secondarily, subjects were positioned supine and were instructed to a forced vital capacity (FVC) maneuver, with a bidirectional digital spirometer (CareFusion - Microloop) with accuracy of 10 ml for volume and of 0.03 l/s for flow. The use of the FVC maneuver allowed the determination of the values: FVC, peak expiratory flow (PEF), forced expiratory volume in one second (FEV1), inspiratory volume in deep maneuver (IP). The procedures followed the recommendations of the Brazilian Thoracic Association.

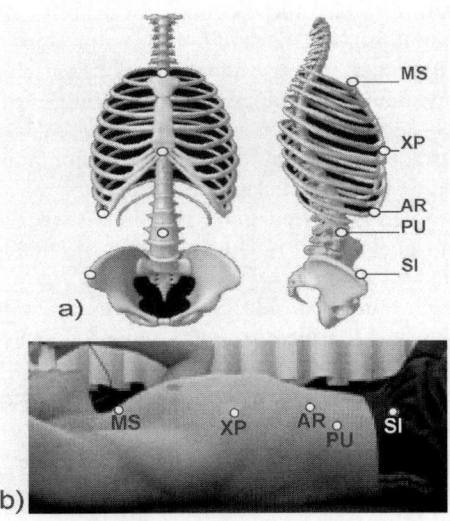

Fig. 1 a) Theorical representation of the adhesive markers on the thoracoabdominal region. (MS) manubrium sterni; (XP) xiphoid process; (AR) inferior angle of the 10th rib; (PU) projection of the umbilicus; (SI) anterior superior iliac spine; b) Practical representation with subjects in dorsal decubitus position.

The tests were performed and filmed three times, being considered for analysis, only the exam with best result of FVC. Movies with the best evidence were imported into Windows Movie Maker software since it extracts the frames related to moments of maximal inspiration and expiration. These images were transferred to the AutoCad software (version 2012) for calculations of thoracoabdominal areas (Figure 2a) and three dimensional extrapolation (Figure 2b) to obtain volumes of thoracic and thoracoabdominal contribution for breath.

All calculations were made after correcting the scale factor for each image based on the marker adhesive with known diameter, so each image received individual treatment. The 3D extrapolation was made from the insertion length measurement in thoracic 2D image, and for this is used the AutoCad extrusion tool. In expiratory and inspiratory frames, compartments were drawn over the thorax and abdomen to examine in more detail the different respiratory behavior of Thorax (TX), Abdomen (AB) and Total Area (TA) (Figure 2b).

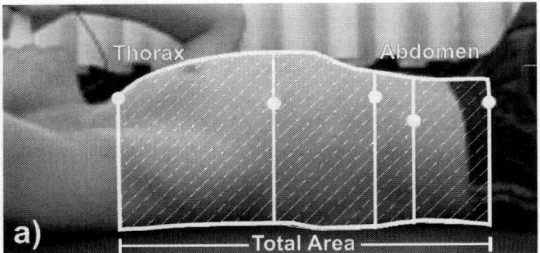

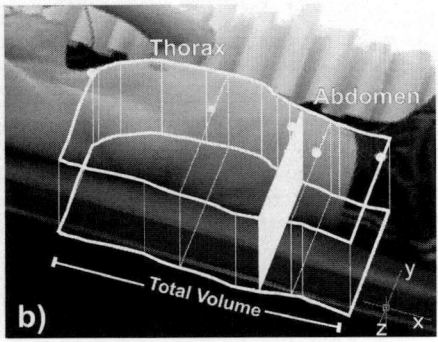

Fig. 2 a) Representation of two dimension photogrammetric model for maximal inspiration moment; b) Photogrammetric model of three dimensional extrapolation from two dimensional image.

For quantitative analysis, it was used descriptive statistics and the Pearson correlation test to check the relationship between spirometric volumes and the offset of thoracoabdominal surface. As statistical significance, it was adopted $p < 0.05$.

III. RESULTS

Table 1 shows the descriptive values for anthropometric variables: age, weight, height and thorax length, as average values obtained by the FVC test.

Table 1 Mean values, standard deviation, minimum and maximum values for anthropometric variables, forced expiratory volume in the first second (l), forced vital capacity (l), peak expiratory flow (l / s) and deep inspiration (l).

Variables	Mean	S.D.	Minimum	Maximum
Anthropometric Values				
Age (years)	15.40	0.98	14.00	17.00
Body Mass (kg)	61.29	12.27	43.50	87.00
Height (m)	1.66	0.09	1.52	1.87
Thorax (cm)	25.54	1.87	21.5	29.50
Pulmonary Function				
FEV1 (l)	3.22	0.82	1.77	5.16
FVC (l)	3.87	0.91	2.39	5.81
PEF (l/s)	6.50	2.41	2.84	12.15
DI (l)	3.59	0.91	2.05	5.40

Table 2 Correlations among respiratory spirometric values and photogrammetric analysis (2D) areas.

	TX	AB	TA	TX	AB	TA
	Expiration			Inspiration		
FEV_1	0,645*	0,531*	0,631*	0,727*	0,564*	0,695*
FVC	0,655*	0,561*	0,650*	0,764*	0,626*	0,744*
PFE	0,667*	0,580*	0,668*	0,762*	0,619*	0,739*
DI	0,611*	0,554*	0,617*	0,724*	0,621*	0,715*

* $p<0,05$

Table 3 Correlations among respiratory spirometric values and photogrammetric analysis (3D) areas.

	TX	AB	TV	TX	AB	TV
	Expiration			Inspiration		
FEV_1	0,668*	0,594*	0,659*	0,728*	0,616*	0,708*
FVC	0,698*	0,627*	0,691*	0,780*	0,664*	0,761*
PFE	0,697*	0,658*	0,701*	0,773*	0,665*	0,756*
DI	0,654*	0,609*	0,655*	0,739*	0,651*	0,728*

* $p<0,05$

Strong correlation was noted for the protocol 2D and 3D. The inclusion of the thorax length measure for 3D study improved the correlation for all points analyzed.

IV. DISCUSSION

The breath comprises a dynamic act that is based on the difference in pressure inside and outside of the rib cage to capture the atmospheric air [2].

It was decided to use the spirometry test since it is considered an important tool for assessment, quantification and monitoring of the respiratory function [11]. This test enables monitoring and detection of airway narrowing because it analyzes the airflow in the airways [11].

During respiratory mechanics, the phases of inspiration and expiration cause deformation in the surface thoracoabdominal perceived in three different directions: laterolateral, craniocaudal and anteroposterior [2, 12, 13] (Fig. 3).

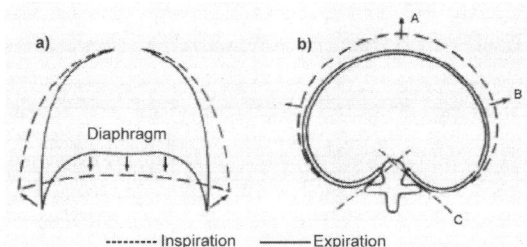

Fig. 3: Increase in thorax diameters during inspiration and expiration. a) It illustrates a front view of the rib cage; b) It shows a cross section of the same region, where the letters A, B and C refer to the anteroposterior, lateral-lateral and cranial-caudal, respectively.

Authors describe that the most significant changes in the inspiration and expiration are perceived in the anteroposterior view [12-14]. The findings in this study indicate a positive contribution of the thorax length on estimating respiratory volumes, as observed in the correlations presented in Table 3. Thus, the thorax length expands the possibility of analyzing respiratory capacities by photogrammetry.

In a previous study, developed with 25 adolescents with mean age of 15.5 years old, in which the thoracic length was also considered, the authors found similar results to our study regarding the correlation between areas and FVC values [15].

Quantifying the properties of the respiratory system has been the focus in several studies using methods based on the relationship between inspired volume and mobility of thoracoabdominal surface [3, 16]. The technological evolution has made digital imaging being progressively incorporated as an option for diagnostic methods in health.

Photogrammetric studies has been shown to be a tool of wide applicability and reproducibility, as well as having advantages such as low cost and easy transportability [10, 15, 17, 18]. Another advantage of photogrammetry is the reduced examination time, which on average demands 15 minutes (for the respiratory test, video capture and photogrammetric analysis). In traditional laboratory systems, an examination lasts about 180 minutes [3]. The limitation of this method is the use of straight measurements of the segments. It is believed that with the development of algorithms to estimate the body curves, it is possible to improve the determination of respiratory volumes. The correlation between the 3D photogrammetry method and the value of FVC was 0.780. For laboratory measurements, values can be above 0.900[3].

V. CONCLUSION

Photogrammetry (with two dimensional and three dimensional measurements) showed a useful and reproducible technique for estimating respiratory volumes. It was obtained strong correlation between areas/volumes measured by photogrammetry and by the spirometer.

Since it is a relatively simple method, quick, and of easy portability, the photogrammetry for respiratory analysis enables the use of pulmonary function tests in various environments such as: schools, gyms and clubs. Furthermore, photogrammetric studies in adolescents represent an advance in literature, since up to a short time ago, the image analysis of respiratory mechanics was restricted to children and adults.

ACKNOWLEDGMENT

We would like to thank CNPq, CAPES (for granting a master's scholarship) and Araucaria Foundation for financial support.

REFERENCES

1. A. C. Guyton and J. E. Hall, *Tratado de fisiologia médica* vol. 11. Rio de Janeiro: Elsevier, 2006.
2. J. B. West, *Fisiologia Respiratória*, 8 ed. São Paulo: Artmed, 2010.
3. C. M. A. Loula, A. L. Pachêco, K. J. Sarro, and R. M. L. Barros, "Análise de volumes parciais do tronco durante a respiração por videogrametria," *Rev. Bras. Biomec.*, vol. 9, pp. 21-27, 2004.
4. A. E. F. Da Gama, "Eletromiografia e pletismografia óptico-eletrônica na avaliação respiratória," Mestrado Dissertação, Departamento de Fisioterapia, Universidade Federal de Pernambuco, Recife, 2011.
5. J. G. Webster, *Medical instrumentation: application and design*. New Delhi: Wiley-India, 2009.
6. G. Ferrigno, P. Carnevali, A. Aliverti, F. Molteni, G. Beulcke, and A. Pedotti, "Three-dimensional optical analysis of chest wall motion," *J. Appl. Physiol.*, vol. 77, pp. 1224-1231, 1994.
7. T. Kotani, S. Minami, K. Takahashi, K. Isobe, Y. Nakata, M. Takaso, M. Inoue, T. Maruta, T. Akazawa, and T. Ueda, "An analysis of chest wall and diaphragm motions in patients with idiopathic scoliosis using dynamic breathing MRI," *Spine*, vol. 29, p. 298, 2004.
8. L. Coelho and J. N. Brito, *Fotogrametria digital*. Rio de Janeiro: EDUERJ, 2007.
9. D. V. Ricieri and N. A. Rosário Filho, "Efetividade de um modelo fotogramétrico para a análise da mecânica respiratória toracoabdominal na avaliação de manobras de isovolume em crianças," *J. Bras. Pneumol.*, vol. 35, pp. 144-150, 2009.
10. W. L. Ripka, D. d. V. Ricieri, L. Ulbricht, E. B. Neves, I. A. Guimarães, E. Romaneli, and A. M. W. Stadnik, "Biophotogrammetry Model of Respiratory Motion Analysis applied to child," presented at the 34rd Annual International Conference of the IEEE Engineering in Medicine and Biology Society, San Diego, 2012.
11. C. A. C. Pereira, J. M. Jansen, S. S. M. Barreto, J. Marinho, N. Sulmonett, and R. M. Dias, "Espirometria," *J. Pneumol.*, vol. 28, pp. 1-82, 2002.
12. A. De Groote, M. Wantier, G. Cheron, M. Estenne, and M. Paiva, "Chest wall motion during tidal breathing," *J. Appl. Physiol.*, vol. 83, pp. 1531-1537, 1997.
13. T. A. Wilson, K. Rehder, S. Krayer, E. A. Hoffman, C. G. Whitney, and J. R. Rodarte, "Geometry and respiratory displacement of human ribs," *J. Appl. Physiol.*, vol. 62, pp. 1872-1877, 1987.
14. A. I. Kapandji, *Fisiologia articular: tronco e coluna vertebral* vol. 6. Rio de Janeiro: Guanabara Koogan, 2008.
15. W. L. Ripka, E. B. Neves, D. V. Ricieri, L. Ulbricht, and P. M. Gewehr, "Análise Fotogramétrica para Predição de Volume e Comportamento Respiratório em Adolescentes.," presented at the XXIII Congresso Brasileiro de Engenharia Biomédica, Porto de Galinhas, 2012.
16. K. Konno and J. Mead, "Measurement of the separate volume changes of rib cage and abdomen during breathing," *J. Appl. Physiol.*, vol. 22, pp. 407-422, 1967.
17. D. V. Ricieri, "Princípios processuais da Biofotogrametria e sua adaptação para medidas em estudos sobre movimentos respiratórios toracoabdominais," Doutorado em Saúde da Criança e do Adolescente Dissertação, Departamento de Ciências da Saúde, Universidade Federal do Paraná, Curitiba, 2008.
18. A. Perin, L. Ulbricht, D. d. V. Ricieri, and E. B. Neves, "Use of biophotogrammetry for assessment of trunk flexibility," *Revista Brasileira de Medicina do Esporte*, vol. 18, pp. 176-180, 2012.

Corresponding author:

Author: Leandra Ulbricht
Institute: Federal University of Technology-Paraná (DAEFI)
Street: Av. Sete de Setembro, 3165
City: Curitiba -PR
Country: Brazil
Email: prof.leandra@gmail.com

Evaluation of Gastroesophageal Reflux in Infants Treated with Osteopathy Using the I-GERQ-R Questionnaire

M. Gemelli[1], L. Ulbricht[2], and E.F.R. Romaneli[3]

[1] Federal University of Technology – Paraná/PPGEB, Curitiba, Brazil
[2] Federal University of Technology – Paraná/PPGEB, Curitiba, Brazil
[3] Federal University of Technology – Paraná/DAELT, Curitiba, Brazil

Abstract - Gastroesophageal reflux (GER) in infants has important clinical relevance by high incidence and risk of complications, besides family inconvenience and discomfort for babies. Today, treatment is based on medication for relief of symptoms. The goal of this study is to analyze the impact of osteopathic treatment in the course of GER in babies from zero to one year of age in the city of Curitiba. This is a retrospective pilot study of a database where four clinical files of individuals diagnosed with GERD in clinical treatment were analyzed. The data used for analysis were taken from questionnaire I-GERQ-R starting and ending at the osteopathic treatment. It was observed that in 28 ± 5.72 days scores fell from an average of 15.5 ± 4.45 to 0.50 ± 0.50, determining reduction of all symptoms of reflux. A child presenting respiratory disorders has presented total relief of symptoms. Osteopathy has been effective to eliminate reflux symptoms suggesting that control group studies must be carried out on a statistically significant sample to assess the real relationship between osteopathy and acceleration of improvement of babies with GER. In addition, the questionnaire I-GERQ-R has proved efficient to demonstrate this evolution.

Keywords: Osteopathy, Gastroesophageal reflux, infant, I-GERQ-R

I - Introduction

There are several known causes of reflux and other still not well clarified. Studies suggest that the vagus nerve (X) influences directly the tonus (relaxation) of the lower esophageal sphincter (LES) in individuals with GER [1-3]. Hiatal hernia [4] and the immaturity of the gastrointestinal system (GIS) are pointed out as responsible for reflux. Other studies demonstrate the GER inducing milk protein intolerance (MPI) in lactating children with less than 1 year of age [5].

From those with reflux symptoms, 50% of children with until three months may have vomiting that tend to improve until 12 months of age [1]. Studies indicate a close relationship between reflux and asthma symptoms by up to 75% of cases [2-4] and other respiratory problems of lower airway like cough, pneumonia [5, 6], increased inspiratory resistance in spirometry [7] in rhinitis and upper airways [8]. Several of these respiratory symptoms seem to have correlation with vagal reflexes [9]. Reflux may cause heartburn which justifies the fact that some children do not feed well [10], cry after breastfeeding and bend to backside (hyperextension of trunk)[11, 12]. As a complication, the GER in children may cause liquid aspiration and pneumonia in addition to the chronic presence of acid in the distal third of the esophagus that may predispose to Barrett's esophagus and some types of cancer [13].

After using Osteopathic Maneuvers Treatment (OMT), studies indicate positive influences on the GIS symptoms in adults and children [14-16], which suggests that the tensions that have an impact on the vagal system may be causing the main symptoms of GER and that it is possible to accelerate the process of improvement and elimination of its causes.

This pilot study is to analyze if there is this impact of osteopathic treatment in the course of GERD in infants and quantify it using the questionnaire I-GERQ-R GER maneuvers [17-19] in order to determine protocol Osteopathic studies that might lead to the reduction of exposure to medicines and the complications of GER.

II - Methodology

This is a retrospective study. Information has been obtained from the database of a clinic in the city of Curitiba, in the month of March, aimed at individuals from 0 to 1 year of age who were under treatment for gastroesophageal reflux disease and that met the proposed criteria for inclusion and exclusion.

The questionnaire I-GERQ-R, clinically used to diagnose the GER, presents the 0 to 25 score, which indices greater than seven indicate possible reflux and greater than nine probable diagnosis of GER. Higher scores determine greater severity of symptoms [18].

The selection of four clinical files of four patients has been randomly taken and, after analysis of the anamnesis and the patient's history, were included patients who: had been clinically diagnosed GER for a pediatrician, were conducting drug based treatment for reflux, use the assessment questionnaire I-GERQ-R to measure the evaluation.

It was excluded from the study: all individuals who do not have medical diagnosis of GER, who were making use of medication that was not related to the diagnosis of GER, who had any associated pathology that could produce symptoms similar to GER, individuals who have not completed the treatment with medical discharge by Osteopath e those who had left the drug treatment without doctor's prescription.

Data from anamnesis and the questionnaire I-GERQ-R have been compiled and analyzed the scores for the two phases of the evaluation: one at the beginning of the osteopathic treatment and the other at the end of it.

Descriptive statistics with measures of location and dispersion were used for the treatment of the data.

III - Results

The sample composed of four subjects presented average age of 4 ± 1.41 months. The number of sessions was 4.5 (± 0.58), where the average interval of days between the first and last sessions was 28 (±5.72) days. The average of the initial and final score of the questionnaire I-GERQ-R (Fig. 1) was 15.5 (5.45 ±), and the final average was 0.25 (±0.50).

The patient 1 had an evolution of indexes I-GERQ-R from 23 to 1. In the initial assessment, the patient was crying 1-3 hours per day, vomiting more than 5 times per day, cried during or after breastfeeding and stopped breathing showing cyanosis. In the final evaluation, the patient did not show any symptom that could be reported on I-GERQ-R except crying about 1 hour per day. Pre-treatment medication was Domperidone. In the end, no medicine was used.

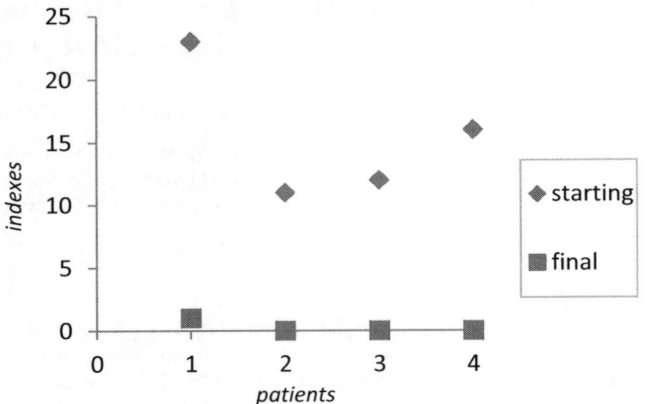

Figure 1. Index evolution I-GERQ-R starting and final

The patient 2 had an evolution of indices I-GERQ-R from 11 to 0 (zero). In the initial assessment, the patient had vomiting more than five times a day, crying 1-2 hours per day, did not cry during breastfeeding and had no respiratory signs. In final assessment the patient did not show any symptoms that could be related to I-GERQ-R. Pre-treatment medication was Domperidone. In the end, no medicine was used.

The patient 3 had an evolution of indices I-GERQ-R from 12 to 0 (zero). In the initial assessment, the patient had vomiting 1-3 times a day in small quantities, referred to much discomfort and presented no respiratory symptoms. In the final evaluation, the patient had a zero score without indication of reflux on I-GERQ-R questionnaire. Pre-treatment medication was Ranitidine Hydrochloride. In the end, no medicine was used.

The patient 4 had an evolution of indices I-GERQ-R from 16 to 0 (zero). In the initial assessment, the patient had vomiting 3-5 times a day, reported crying to breastfeed, followed by frequent back arching and with absence of respiratory symptoms. In the final assessment, the patient did not show any symptom related to the I-GERQ-R. Pre-treatment medication was Domperidone and Omeprazole. In the end, no medicine was used.

IV - Discussion

Studies describe reductions in the number of regurgitation when thickening formulas are added to milk, however, they are not statistically significant [20]. In this study, it was observed total reduction of vomiting and regurgitations after an average period of 28 ± 5.72 days, associating the osteopathic treatment to thickening diets, overcoming the results of literature in terms of reducing the number of episodes.

Reflux may cause heartburn which justifies the fact that some children do not eat [10], cry after breastfeeding and frequent back arching [11, 12, 21]. These signs and symptoms were eliminated in all four cases in this study in 28 ±5.72 days after the osteopathic treatment, suggesting that the OMT corrected mechanical processes that originated these symptoms. No child in this study continued to present symptoms that suggested discomfort to breastfeed or after feeding.

The use of medications doesn't seem to interfere in the course of the reflux. A good effectiveness of antacids such as Omeprazole and Cimetidine is demonstrated for the symptom of heartburn and esophagitis [22]. However, the incorrect indication in cases without specific diagnosis of esophagitis reduces its efficiency in the treatment of symptoms of GER[23]. The use of Metoclopramide was not effective and there are still signs of the occurrence of side effects such as cramping in your use [24], where the literature suggests the use of these medicines only in last case when other treatments have not been effective.

The results found in this study are justified by the good effects of OMT on the gastrointestinal system, in particular on the vagus nerve as described in the literature [25, 26]. In this study, it was found an average time of 28 ± 5.72 days for the ceasing of symptoms and withdrawal of all medicinal products for GER.

Finally, the vomiting, the most evident symptom of reflux in baby, was also eliminated in the average time of 28 ± 5.72 days corroborating with osteopathic studies that describe the average time acceleration of improvement [26] compared to that described in the medical literature, that happen around 12 months of age or later [20].

Even if it is described that the GER has a benign and spontaneous course [6], this time is long (after 12 months of age) and points to the susceptibility of food problems in young children [12]. The literature mention that the improvement is subject to age, generally reducing the symptoms of vomiting in 4% of individuals after the first year of life [11, 20]. However, there are indications that the low symptomatic untreated reflux in baby can appear close to 9 years or in adolescence as reflux [32, 33] or respiratory manifestations [9].

Therefore, reflux not properly treated in childhood suggests that consequences as the reflux and breathing problems can manifest in other stages of development [11, 13]. Reflux diagnosis both overrated (unnecessary prescription medication) or underestimated can bring important consequences in the life of an individual. Appropriate and effective treatments should be studied and prescribed by competent professionals who identify in your routine this type of pathology.

V - CONCLUSION

Osteopathy seems to accelerate the improvement of GER in babies when compared with the spontaneous evolution described in the medical literature or drug treatment.

The I-GERQ-R questionnaire has proved efficient to measure the evolution of the osteopathic treatment in the reflux in babies.

With these results, randomized prospective studies case-control are suggested to be carried out on larger samples in order to assess whether the same evolution is evidenced.

ACKNOWLEDGMENT

We would like to thank CNPq, CAPES and Araucaria Foundation for financial support.

REFERENCES

1. Cox, M.R., et al., *Effect of general anaesthesia on transient lower oesophageal sphincter relaxations in the dog.* Australian and New Zealand Journal of Surgery, 1988. **58**(10): p. 825-830.
2. Ricard, F. and E.M. Loza, *Osteopatía y pediatría*2005: Medica Panamencana.
3. Cobeta, I., A. Pacheco, and E. Mora, *The role of the larynx in chronic cough.* Acta otorrinolaringologica espanola, 2013.
4. Vandenplas, Y. and E. Hassall, *Mechanisms of gastroesophageal reflux and gastroesophageal reflux disease.* Journal of pediatric gastroenterology and nutrition, 2002. **35**(2): p. 119-136.
5. Salvatore, S. and Y. Vandenplas, *Gastroesophageal reflux and cow milk allergy: is there a link?* Pediatrics, 2002. **110**(5): p. 972-984.
6. Nelson, S.P., et al., *Prevalence of symptoms of gastroesophageal reflux during infancy: a pediatric practice-based survey.* Archives of pediatrics & adolescent medicine, 1997. **151**(6): p. 569.
7. Blake, K. and W.G. Teague, *Gastroesophageal reflux disease and childhood asthma.* Current Opinion in Pulmonary Medicine, 2013. **19**(1): p. 24-29.
8. Liang, B., et al., *Association of gastroesophageal reflux disease risk with exacerbations of chronic obstructive pulmonary disease.* Diseases of the Esophagus, 2013.
9. Komatsu, Y., T. Hoppo, and B.A. Jobe, *Proximal Reflux as a Cause of Adult-Onset AsthmaThe Case for Hypopharyngeal Impedance Testing to Improve the Sensitivity of DiagnosisProximal Reflux as a Cause of Adult-Onset Asthma.* JAMA surgery, 2013. **148**(1): p. 50-58.
10. Euler, A.R., et al., *Recurrent pulmonary disease in children: a complication of gastroesophageal reflux.* Pediatrics, 1979. **63**(1): p. 47-51.
11. Leggett, J.J., et al., *Prevalence of Gastroesophageal Reflux in Difficult AsthmaRelationship to Asthma Outcome.* CHEST Journal, 2005. **127**(4): p. 1227-1231.
12. Eidani, E., et al., *A Comparison of Impulse Oscillometry and Spirometry Values in Patients with Gastroesophageal Reflux Disease.* Middle East Journal of Digestive Diseases (MEJDD), 2013. **5**(1).

13. Prokopakis, E., et al., *Chronic rhinosinusitis: observation or treatment.* Current Opinion in Allergy and Clinical Immunology, 2013. **13**(1): p. 31-36.
14. Gurski, R.R., et al., *Manifestações extra-esofágicas da doença do refluxo gastroesofágico.* Jornal Brasileiro de Pneumologia, 2006. **32**: p. 150-160.
15. Hyman, P.E., *Gastroesophageal reflux: one reason why baby won't eat.* The Journal of pediatrics, 1994. **125**(6): p. S103-S109.
16. Czinn, S. and S. Blanchard, *Gastroesophageal Reflux Disease in Neonates and Infants: When and How to Treat.* Paediatric drugs, 2013.
17. Nelson, S.P., et al., *One-year follow-up of symptoms of gastroesophageal reflux during infancy.* Pediatrics, 1998. **102**(6): p. e67-e67.
18. Forssell, L., *Gestational age and size at birth and risk of esophageal inflammation and cancer.* At https://publications.ki.se/xmlui/bitstream/handle/10616/41350/Thesis_Lina_Forssell.pdf?sequence=1 , viewed in April, 20th, 2013
19. Pizzolorusso, G., et al., *Effect of osteopathic manipulative treatment on gastrointestinal function and length of stay of preterm infants: an exploratory study.* Chiropractic & manual therapies, 2011. **19**(1): p. 15.
20. Miller, J.E. and K. Benfield, *Adverse effects of spinal manipulative therapy in children younger than 3 years: a retrospective study in a chiropractic teaching clinic.* Journal of manipulative and physiological therapeutics, 2008. **31**(6): p. 419-423.
21. da Silva, R., et al., *Increase of lower esophageal sphincter pressure after osteopathic intervention on the diaphragm in patients with gastroesophageal reflux.* Diseases of the Esophagus, 2012.
22. Orenstein, S.R., *Symptoms and reflux in infants: Infant gastroesophageal reflux questionnaire revised (I-GERQ-R)—utility for symptom tracking and diagnosis.* Current gastroenterology reports, 2010. **12**(6): p. 431-436.
23. Orenstein, S.R., T.M. Shalaby, and J.F. Cohn, *Reflux symptoms in 100 normal infants: diagnostic validity of the infant gastroesophageal reflux questionnaire.* Clinical pediatrics, 1996. **35**(12): p. 607-614.
24. Kleinman, L., et al., *The infant gastroesophageal reflux questionnaire revised: development and validation as an evaluative instrument.* Clinical gastroenterology and hepatology: the official clinical practice journal of the American Gastroenterological Association, 2006. **4**(5): p. 588.
25. Hegar, B., et al., *Natural evolution of infantile regurgitation versus the efficacy of thickened formula.* Journal of pediatric gastroenterology and nutrition, 2008. **47**(1): p. 26-30.
26. Salvatore, S., et al., *Gastroesophageal reflux disease in infants: how much is predictable with questionnaires, pH-metry, endoscopy and histology?* Journal of pediatric gastroenterology and nutrition, 2005. **40**(2): p. 210.
27. Rudolph, C.D., et al., *Guidelines for evaluation and treatment of gastroesophageal reflux in infants and children: recommendations of the North American Society for Pediatric Gastroenterology and Nutrition.* Journal of pediatric gastroenterology and nutrition, 2001. **32**: p. S1-S31.
28. Khoshoo, V. and D. Edell, *Overprescription of antireflux medications for infants with regurgitation: in reply.* Pediatrics, 2008. **121**(5): p. 1070-1071.
29. Jordan, B., et al., *Effect of antireflux medication, placebo and infant mental health intervention on persistent crying: a randomized clinical trial.* Journal of paediatrics and child health, 2006. **42**(1-2): p. 49-58.
30. Mirocha, N.J. and J.D. Parker, *Successful treatment of refractory functional dyspepsia with osteopathic manipulative treatment: a case report.* Osteopathic Family Physician, 2012.
31. Nemett, D.R., et al., *A randomized controlled trial of the effectiveness of osteopathy-based manual physical therapy in treating pediatric dysfunctional voiding.* Journal of Pediatric Urology, 2008. **4**(2): p. 100-106.
32. Martigne, L., et al., *Prévalence du reflux gastro-œsophagien (RGO) chez l'enfant et l'adolescent en France: résultats d'une étude observationnelle transversale.* Gastroentérologie Clinique et Biologique, 2009. **33**(3S1): p. A40.
33. Martin, A.J., et al., *Natural history and familial relationships of infant spilling to 9 years of age.* Pediatrics, 2002. **109**(6): p. 1061-1067.

Corresponding author:

Author: Leandra Ulbricht
Institute: Federal University of Technology-Paraná (DAEFI)
Street: Av. Sete de Setembro, 3165
City: Curitiba -PR
Country: Brazil
Email: prof.leandra@gmail.com

A Comparison of Three Techniques for Wound Area Measurement

P. Foltynski, J.M. Wojcicki, P. Ladyzynski, and S. Sabalinska

Nalecz Institute of Biocybernetics and Biomedical Engineering, Polish Academy of Sciences, Warsaw, Poland

Abstract—Reduction of wound surface area has been identified as one of the important parameter of the applied therapy assessment in diabetic foot ulceration. There are some methods of wound area measurement and this paper presents the comparison of the Visitrak device, the SilhouetteMobile device and the TeleDiaFoS system. All devices were designed for such a measurements and are based on digital planimetry. Sixteen wound shapes made of adhesive film and prepared based on pictures of actual diabetic foot wounds were placed on feet of 10 volunteers and measured with all devices. The Pearson's r correlation coefficient and linear regression equation were calculated between each pair of methods. All methods are highly correlated with r equal to at least 0.97 ($p<0.0001$). The linear regression equations were calculated for all data (0 – 17 cm^2) and in two sub-ranges: 0-2 cm^2 and 2-17 cm^2. These equations may be used for recalculation of results from one method into another. The use of appropriate equation according to range of value to be recalculated will provide more accurate result.

Keywords—wound area measurement, diabetic foot syndrome, diabetic ulceration.

I. INTRODUCTION

Diabetic foot syndrome (DFS) is one of the late complications of diabetes which may lead to risk of developing foot ulcers and wounds. Those injuries are slow healing because of angiopathy causing decreased oxygen supply to the tissues. Long lasting wounds are also prone for bacterial infections and therefore the proper medical care is very important otherwise they can lead to foot amputation. Wound area reduction over time is an important factor in wound treatment evaluation [1-4]. One of the guidelines of DFS treatment concerns the progress of wound area change after 4 weeks from the start of treatment [5]: the therapy should be reevaluated if the wound area reduction is not greater than 40%. Medical teams may use different methods for area measurement and their interchangeability was the subject of a research. In two studies [6, 7] four measurement methods were used: the area estimation from length and width of the wound measured with a ruler, the Visitrak device (Smith & Nephew, England), the SilhouetteMobile device (Aranz Medical, New Zeland), and the TeleDiaFoS system (Nalecz Institute of Biocybernetics and Biomedical Engineering, Polish Academy of Sciences, Poland). The methods were compared using correlation coefficient (Pearson's r) and difference plot. The comparison of the methods may be performed using actual wounds, as in two mentioned above studies, but the wound border uncertainty may influence wound area determination and therefore adversely affect the comparison of methods. Therefore in other comparative studies the equivalents of actual wounds were used: Langemo et al. [8] used wound models made of plastic, whereas Haghpanah et al. [9] used wound shapes cut out of paper. The limitation of the former study was the use of only three wound shapes of areas 28.7, 34.0 and 35.0 cm^2. In the later study the digitizing tablet (Visitrak, Smith & Nephew) and the software (VeV MD, Canada) using the images from a digital camera were used, but the wound shapes were not similar to the actual wounds. Our new approach is to use wound shapes of actual DFS wounds, which are cut out of self-adhesive foil and can be sticked on the foot skin. Such an approach eliminates the wound border uncertainty and enables the wound shapes measurement in almost natural conditions i.e. when they are present on naturally curved skin.

The aim of the present study was to compare three methods (the Visitrak and the SilhouetteMobile devices and the TeleDiaFoS system) of wound area measurements in DFS wounds using wound shape stickers placed on the foot skin.

II. MATERIALS AND METHODS

A. Wound Shapes Preparation

In the first step of the work 16 pictures with foot wounds registered in the TeleDiaFoS system during monitoring of DFS patients [10, 11] were selected. The chosen wounds had the areas from about 0.2 to 17 cm^2 (mean±SD = 5.1±4.7 cm^2) and came from different parts of the foot sole. The wound pictures were imported into CorelDraw (Corel Corp., Ottawa, Canada), where the shapes of wounds were redrawn in real dimensions (i.e. in scale 1:1) as closed curves in order to be appropriate for a digital plotter machine. The wound shapes were cut out of self-adhesive vinyl film (Avery 500 Event Film, Avery Dennison Graphics Corp.) of thickness of 0.07 mm, using a plotter. Finally, 16 sets of wound shapes (Fig. 1 A) were ready to be applied onto volunteers' foot skin. For the hygienic reasons once the wound shape was applied onto the skin, it was not used again.

The wound shapes were placed on the sole of left and right foot (Fig. 1 B) of 10 healthy volunteers (3 women and 7 men). The mean age of volunteers was 55.2±15.6 years (range 30 - 72).

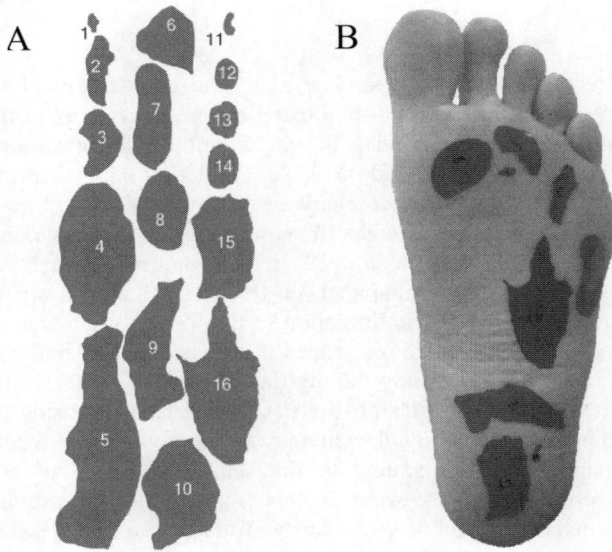

Fig. 1 (A) The wound shapes showed with IDs used in the study; (B) The photograph of nine wound shapes at a volunteer's foot.

B. Methods of Wound Area Measurement

The area of each wound shape was measured using Visitrak device (Smith & Nephew, UK), SilhouetteMobile device (Aranz, New Zealand) and the Patient's Module of the TeleDiaFoS system (Nalecz Institute of Biocybernetics and Biomedical Engineering, Warsaw, Poland).

Visitrak device. The portable device called Visitrak calculates the area of a closed shape after tracing it with an enclosed pen onto a sensitive area of the device. When the area of wound is to measure, the wound outline is traced with a permanent water-resistant marker on a double-layer film placed onto the wound. The film layer which was in contact with the wound is disposed of and the clean layer is placed onto the device and the wound shape is retraced by operator. The area is automatically calculated and displayed when the tracing is completed.

SilhouetteMobile device. The SilhouetteMobile device is a PDA computer equipped with a head allowing at taking a picture of the wound and emitting two laser beams. The software calculates the wound area considering the skin curvature and the wound plane skewness determined from the laser beam shapes displayed at the skin nearby the wound. The wound edges identification is not automatic and the operator needs to trace an outline of the wound. The SilhouetteMobile device was calibrated before measurements according to the procedure described in the manual.

TeleDiaFoS system. The patient's module of TeleDiaFoS system takes a picture of foot with wound and sends to the clinical server. The TeleDiaFoS system software uses the taken picture and calculates the wound area after the operator draw the wound contour at the foot picture using a computer mouse [11].

C. Statistics

The data were analyzed using linear regression analysis with Pearson's *r* correlation coefficient. The slopes of two regression lines b_1 and b_2 were compared using *t* statistics [12]:

$$t = \frac{b_1 - b_2}{s_{b_1 - b_2}}, \qquad (1)$$

where $s_{b_1-b_2}$ is the standard error of the difference between the slopes.

A p-value lower than 0.05 was considered statistically significant.

III. RESULTS

Each method used in the present study was compared with each other and the results are presented in Table 1.

Table 1 Regression equations and correlation coefficients between the used methods.

Pair of methods	Area range 0-17 cm² (N=160)	Area range 0-2 cm² (N=60)	Area range 2-17 cm² (N=100)
TeleDiaFoS vs. Visitrak	T=0.947·V+0.073 V=1.048·T-0.038 r = 0.996	T=1.004·V+0.034 V=0.972·T-0.011 r=0.988	T=0.949·V+0.047 V=1.038·T+0.065 r=0.993
Silhouette-Mobile vs. Visitrak	S=0.982·V+0.199 V=1.003·S-0.120 r = 0.992	S=1.098·V+0.039 V=0.858·S+0.019 r = 0.970	S=0.968·V+0.348 V=1.003·S-0.113 r = 0.985
TeleDiaFoS vs. Silhouette-Mobile	T=0.954·S-0.065 S=1.014·T+0.134 r = 0.994	T=0.882·S-0.032 S=1.093·T+0.002 r = 0.982	T=0.961·S-0.134 S=1.014·T+0.338 r = 0.987

T – TeleDiaFoS; V – Visitrak; S – SilhouetteMobile

The dynamics of data were almost 100, i.e. from 0.19 cm² to 17 cm² and next the data were divided into 2 ranges: 0-2 cm² and 2-17 cm² of almost equal dynamics of about 10. The initial comparison was performed using all area data (range 0-17 cm²). The division of data seems to be a good solution when the data have wide range, what is the case here. There were calculated 2 regression equations for each compared pair of methods in each range. If, for instance, the

TeleDiaFoS (T) system is compared with the Visitrak (V) device, there are two functions: T(V) and V(T). The correlation coefficients are equal in both cases.

Fig. 2 A B and C presents the graphs of data from the TeleDiaFoS system against the data from the Visitrak device for the range 0-17 cm² and sub-ranges 0-2 cm² and 2-17 cm², respectively. The other graphs from comparison of other methods are not presented due to limited space.

All correlation coefficients were statistically significant with $p<0.0001$ as well as all slopes of the regression equations ($p<0.0001$), whereas almost all (except of 2) y-intercepts were nonsignificant.

The comparison of the regression slopes revealed that for each compared pair of methods the slope for all data were nonsignificantly different from the slope for the 2-17 cm² sub-range and significantly different from the slope for the 0-2 cm² sub-range. Additionally, the slope for each 0-2 cm² sub-range were significantly different from the slope for the sub-range 2-17 cm² for each pair of compared methods.

IV. DISCUSSION

All methods used in the current study are highly correlated with each other with correlation coefficients from 0.970 to 0.996 (Table 1). These correlation coefficient are as high as earlier were reported by Molik et al. [6] and by Ladyzynski et al. [7] who used the same methods as in the present study and the measurement methods were compared based on the data gathered during patients' examinations in the outpatient clinic.

The regression equations (see Table 1) obtained for the sub-range 2-17 cm² are similar to the equations for the range 0-17 cm² and all the slopes of regression lines are nonsignificantly different for these ranges. Such a results suggest that the data from the sub-ranges 0-2 cm² have little effect on regression equations in the range 0-17 cm². By contrary, the regression equations from the sub-range 0-2 cm² are different from the equation for the whole data range i.e. 0-17 cm², and also the slopes of regression lines are significantly different. We have checked that a value from the range 0-2 cm² obtained during recalculation from one method to the other using the regression equation for the range 0-17 cm² may differ even more than 10% from the value obtained from the equation from the sub-range 0-2 cm². It is therefore clear that the user should use proper equation for the recalculation of the area values from one method into another.

The slopes of regression lines are generally close to 1, however in the sub-range 0-2 cm² for the comparison of the SilhouetteMobile device with the Visitrak device and the TeleDiaFoS system with the SilhouetteMobile device the slopes are 1.098 and 0.882 respectivelly. Such a result

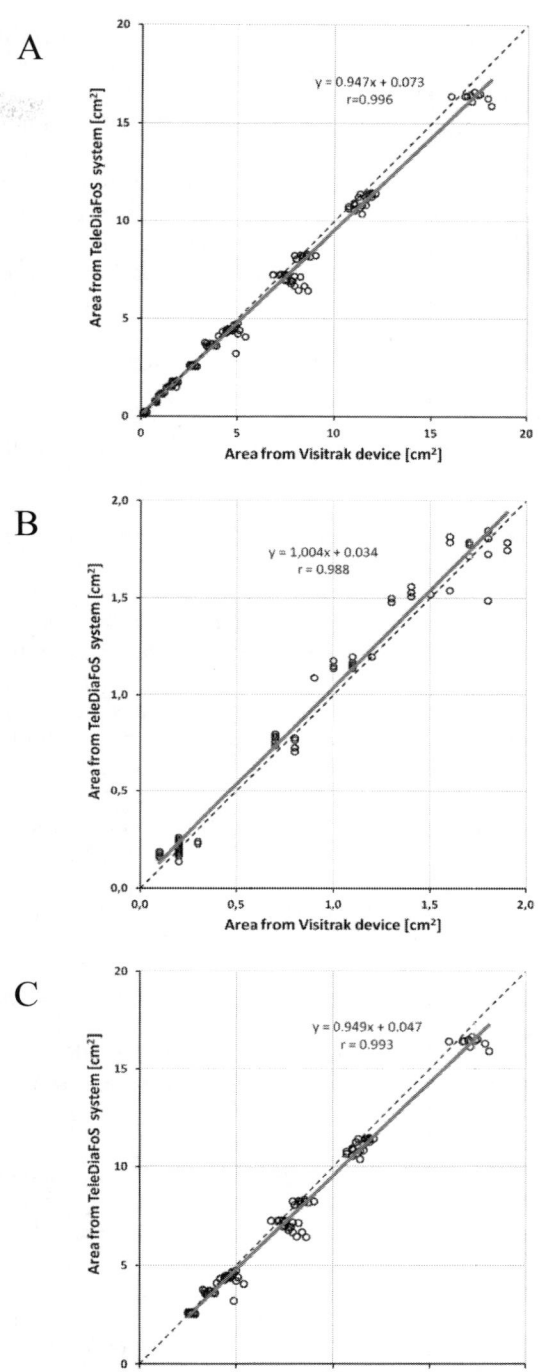

Fig. 2 Scatterplots of areas from the TeleDiaFoS system against values from the Visitrak device for the data range 0-17 cm² (A), sub-range 0-2 cm² (B) and sub-range 2-17 cm² (C). The grey line is a regression line for the data and the dashed line is a line of equity. Regression equation and Pearson's corelation coefficient r are presented in upper part of each graph.

suggests that either the Visitrak device underestimates measured area in this range or the SilhouetteMobile overestimates. As it was shown earlier [13], the Visitrak device underestimates the actual area in such an area range, therefore the former hypothesis is more probable. The analysis of second pair of devices (i.e. TeleDiaFoS and SilhouetteMobile) suggests that the TeleDiaFoS system underestimates measured area rather than the SilhouetteMobile device overestimates. Such a hypothesis may be supported by the fact that former does not consider (in contrast to the later) the skin curvature in the area calculation, what causes the area underestimation.

Although the slopes of regression lines in the current study are close to 1 for the comparison of the TeleDiaFoS system and the two other devices, but in other studies by Molik [6] and Ladyzynski [7] the slopes were close to 0.8. The possible reason for such a difference is underestimation of large wound areas located at plantar part of foot and extending to its lateral part. Using the data from the study of Molik et al. [6] and removing the 4 outlier points from the data, the slopes of 0.946 and 0.962 were obtained between the TeleDiaFoS system vs. the Visitrak device and TeleDiaFoS system vs. the SilhouetteMobile device, respectively, which is almost the same as in the current study.

The measurement of wound shapes stuck on the skin is superior over measurement of actual wounds not only due to elimination of wound boarder uncertainty. The repeated measurements over the same wound shape are possible, what is difficult to do over the actual wound in outpatient clinic environment due to limited time when foot wound is without a dressing. The important issue both in the case of wound shape stickers and actual wounds is deformation of skin in different foot position which influence the wound area. It is therefore important to ask the patient to keep the foot in the same position and with the same muscle tension during each measurement. That is easier to achieve when the patient takes a comfortable position for his legs. The skin on the foot sole is very liable to deformations therefore a care must be taken during the placement of wound shapes and during the area measurements.

The successful interchangeability of the measurement data from different devices for wound area measurement may be achieved by using regression equation obtained for the related sub-range. The use of regression equations obtained for whole data range of large dynamics may generate large errors for the recalculated small values. It seems that the dynamics of the data used for the calculation of regression equation should not be greater than 10.

REFERENCES

1. Sheehan P, Jones P, Caselli A et al. (2003) Percent change in wound area of diabetic foot ulcers over a 4-week period is a robust predictor of complete healing in a 12-week prospective trial. Diabetes Care 26:1879–82
2. Cardinal M, Eisenbud DE, Phillips T, Harding K. (2008) Early healing rates and wound area measurements are reliable predictors of later complete wound closure. Wound Repair Regen 16(1):19-22
3. Lavery LA, Barnes SA, Keith MS et al. (2008) Prediction of healing for postoperative diabetic foot wounds based on early wound area progression. Diabetes Care 31(1):26-9
4. Coerper S, Beckert S, Kuper MA, Jekov M, Konigsrainer A. (2009) Fifty percent area reduction after 4 weeks of treatment is a reliable indicator for healing - analysis of a single-center cohort of 704 diabetic patients. J Diabetes Complications 23(1):49-53
5. Steed DL, Attinger C, Colaizzi T, Crossland M, Franz M, Harkless L et al. (2006) Guidelines for the treatment of diabetic ulcers. Wound Repair Regen 14:680–92
6. Molik M, Foltynski P, Ladyzynski P, Tarwacka J, Migalska-Musial K, Ciechanowska A, Sabalinska S, Mlynarczuk M, Wojcicki JM, Krzymien J, Karnafel W (2010) Comparison of the wound area assessment methods in the diabetic foot syndrome. Biocybernet Biomed Eng 30:3–15
7. Ladyzynski P, Foltynski P, Molik M et al. (2011) Area of the Diabetic Ulcers Estimated Applying a Foot Scanner–Based Home Telecare System and Three Reference Methods, Diabetes Technology & Therapeutics 13(11): 1101-7
8. Langemo D, Anderson J, Hanson D, Hunter S, Thompson P. (2008) Measuring wound length, width, and area: which technique? Adv Skin Wound Care 21(1):42-5
9. Haghpanah S, Bogie K, Wang X, Banks PG, Ho CH. (2006) Reliability of electronic versus manual wound measurement techniques. Arch Phys Med Rehabil. 87(10):1396-402
10. Foltynski P, Wojcicki JM, Ladyzynski P, Migalska-Musial K, Rosinski G, Krzymien J, Karnafel W. (2011) Monitoring of diabetic foot syndrome treatment: some new perspectives. Artif Organs 35:176–82
11. Foltynski P, Ladyzynski P, Migalska-Musial K, Sabalinska S, Ciechanowska A, Wojcicki JM. (2011) A New Imaging and Data Transmitting Device for Telemonitoring of Diabetic Foot Syndrome Patients, Diabetes Technology & Therapeutics 13(8): 861-867
12. Kleinbaum DG, Kupper LL, Muller KE. (1978) Applied Regression Analysis and Other Multivariable Methods, Duxbury, Boston
13. Foltynski P, Ladyzynski P, Sabalinska S, Wojcicki JM. (2013) Accuracy and Precision of Selected Wound Area Measurement Methods in Diabetic Foot Ulceration. Diabetes Technology & Therapeutics "in press"

Corresponding author

Author: Piotr Foltynski
Institute: Nalecz Institute of Biocybernetics and Biomedical Engineering, Polish Academy of Sciences
Street: 4, Trojdena Str.
City: Warsaw
Country: Poland
Email: pfoltynski@ibib.waw.pl

Monitoring Changes of the Tibialis Anterior during Dorsiflexion with Electromyography, Sonomyography, Dynamometry and Kinematic Signals

M. Ruiz-Muñoz[1], J. Martín-Martín[1], M. González-Sánchez[1], and A.I. Cuesta-Vargas[1,2]

[1] Faculty of Health Sciences, University of Málaga, Málaga, Spain
[2] School of Clinical Sciences, Faculty of Health, Queensland University Technology, Queensland, Australia

Abstract—Dorsiflexion (DF) of the foot plays an essential role in both controlling balance and human gait. Electromyography and Sonomyography can provide information on several aspects of muscle function. The aim was to describe a new method for real-time monitoring of muscular activity, as measured using EMG, muscular architecture, as measured using SMG, force, as measured using dynamometry, and kinematic parameters, as measured using IS during isometric and isotonic contractions of the foot DF. The present methodology may be clinically relevant because it involves a reproducible procedure which allows the function and structure of the foot DF to be monitored.

Keywords—electromyography, sonomyography, dynamometry, kinematic, tibialis anterior, dorsal flexion.

I. INTRODUCTION

By parameterising foot DF during isometric and isotonic contractions and synchronising EMG, SMG, dynamometry and IS, new variables can be studied to facilitate the monitoring of key aspects of this foot gesture and therefore gait.

The main aim of this study was to describe a new method for real-time monitoring of muscular activity, as measured using EMG, muscular architecture, as measured using SMG, force, as measured using dynamometry, and kinematic parameters, as measured using IS during isometric and isotonic contractions of the foot DF. The second aim was to establish a descriptive analysis of each of the variables of interest.

II. METHODS

A. Participants

6 healthy young adults (3 men and 3 women) aged 28.17 ± 6.55 years, 1.68 ± 0.11 metres tall and 68.5 ± 10.85 kg in weight were brought in for this study. Each of the participants gave informed consent in writing prior to the study. Ethical approval for the study was granted by the Ethics Committee of the Faculty of Health Sciences at the University of Malaga.

B. Experimental Procedure

The subjects sat in a chair which had been specially adapted in line with their size, as previously established. The hip and knee were positioned at 90°. Attached to the chair was a specially designed height-adjustable device comprising two platforms, one vertical and one horizontal, which the load cell was connected to. The platforms formed an angle of 90°, therefore allowing maximum DF whilst preventing plantar flexion of the foot (Figure 1 and 2). The sole of the right foot was placed on the horizontal platform, whilst the posterior lower half of the leg was in contact with the vertical platform, forming a maximum angle of 90° between foot and leg. The foot and the lower half of the leg were attached to the device with Velcro straps in order to prevent any changes of position during the test. The bisection of the knee joint and the centre of the rotation axis of the load cell had an angle of 0° in the frontal plane and in the sagittal plane measured with a dual-axis goniometer.

The electrodes were positioned and the skin prepared in accordance with European Recommendations for Surface Electromyography (sEMG) [1].

A free area was left in the TA muscle belly in order to position the ultrasound probe without affecting the position and connection of the electrodes. The probe stayed fixed in the chosen position thanks to a mechanical articulated arm system in which the probe head was placed, thus allowing its height and angle to be adjusted.

The load cell was positioned between the ground and the horizontal platform, and secured to both using ring clamps. A series of non-extendable links allowed the distance to the ground to be adjusted in accordance with the subject's leg length (Figure 1).

The inertial sensor was placed and fixed on the distal end of the horizontal platform so as not to disturb foot placement or the development of the test (Figure 2).

After several contractions for the purpose of familiarisation, each subject was asked to use the right foot and the tests described below were carried out in the same order:

Maximal Voluntary Contraction (MVC). The maximal isometric DF of the foot or the maximal voluntary contraction (MVC) used for normalisation of the study variables was recorded for each subject. Three maximal isometric DFs of the right foot were carried out for 5 seconds, with a 90-second rest between each one. An artificial horn was sounded to mark the start of each contraction. All subjects received the same initial instructions with regards to the gesture and the same verbal stimuli were given as feedback during each contraction. Ultrasound signals, electromyography signals and the force generated by the resistance offered by the load cell secured to the ground and to the horizontal platform were collected during this test.

MVC submaximal contractions. Subjects performed 75%, 50% and 25% of their MCV. These values were calculated from the maximum peak recorded during the MCV. The selected protocol (Trainer Dyn. MEGA) showed feedback (vertical bar) of electrical activity of the TA muscle on the computer screen. A visual reference was placed on the computer screen for the subject to know how far to lift in each contraction. The rest of the test development was the same as the MCV.

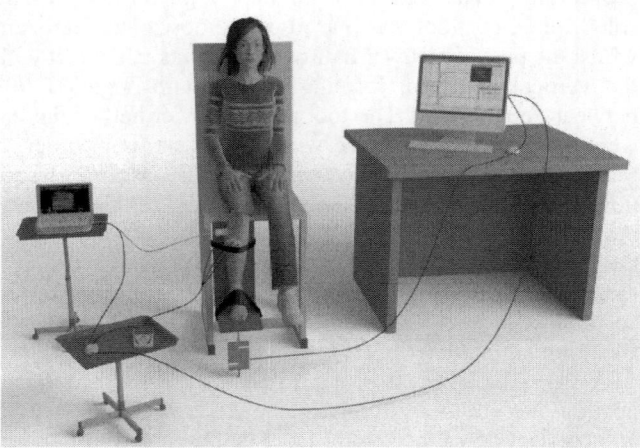

Fig. 1 Experiment Setup (Isometric test)

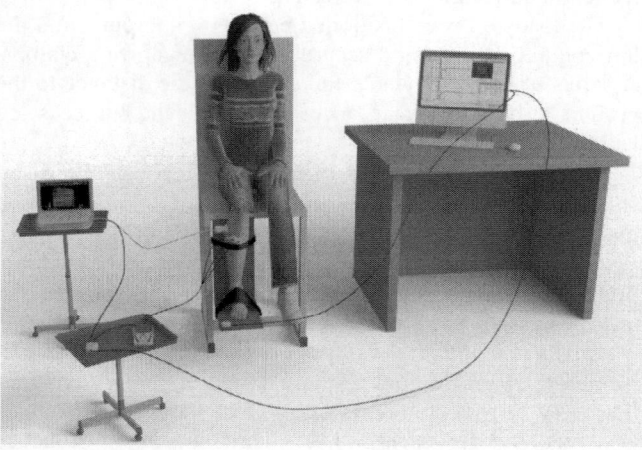

Fig. 2 Experiment Setup (Isotonic test)

Isotonic DF. Isotonic foot DF without any resistance consisted of a dynamic test in which the foot started from a position of 90° with respect to the leg, with the participating subject having to reach the maximal foot DF range as quickly as possible. This test also included three consecutive contractions with a 90-second rest between each one. The rest of the test development was the same as the MCV. Electromyographic, kinematic and ultrasound records were taken during the test.

Before the test protocol, each subject performed as many repetitions of the gesture as deemed necessary in order to become familiar with it.

C. Data Acquisition

All data were recorded continuously and synchronously during each test using the Biomonitor ME6000 [2] console with Megawin 3.0.1 software (Mega Electronics Ltd, Kuopio, Finland), which each of the devices were connected to. Image acquisition was carried out using a duly adapted image capture device and a software add-on (Video EMG Option). This allowed offline searching for the ultrasound image and the electromyographic data for the selected instant. For this study, the maximum muscular activation peak was located and a snapshot was taken in order to subsequently measure the muscular architecture variables (thickness and pennation angle) of the TA.

The start and end of the synchronising of all systems during each test were marked by an activation device or trigger (DV Trigger Mega Electronics Ltd). The recording started before the first contraction with the foot in the starting position and stopped when the subject had finished the last contraction and returned to the starting position.

Electromyographic acquisition. The electrical activation of the TA muscle was measured using the Biomonitor ME6000 electromyography with a sampling frequency of 1000 Hz. Raw data were recorded and processed by MegaWin 3.0.1 software and filtered using a bidirectional fourth-order, 20 HZ low pass Butterworth filter to remove high-frequency noise from the sample.

Ultrasound acquisition. Ultrasound images were obtained using the Esaote MyLab25 Gold scanner with a model LA523 probe [3].

Dynamometric acquisition. Force parameters were obtained using the load cell from ME6000 additional accessories [2].

Kinematic acquisition. Inertial sensor InertiaCube3™ [4] was used to obtain kinematic data.

Data Analysis

EMG Data Analysis. The electromyography variables, maximum peak and area under the curve (AUC) were extracted from the basic results for the selected area of interest for all subjects. This area includes the maximum activation peak of the electromyographic register of the TA and the two seconds around it (one second before and one second after).

SMG Data Analysis. The muscular architecture variables muscle thickness and pennation angle from the ultrasound images were taken following the procedure described by Hodges, 2003 [5]. All muscular architecture variables were obtained from the photo extracted from the video captured during synchronous measurement. F.205.0.0 AutoCAD 2012-English SP2 software was used to extract these parameters.

Kinematic Data Analysis. Kinematic data used as study variables were the maximum peaks of angular velocity and acceleration.

The values extracted from the electromyographic records and the muscular architecture obtained in the submaximal contractions and isotonic DF were normalised for each subject with regards to the maximum values obtained in the MVC.

E. Statistical Analysis

Descriptive statistical analysis was carried out with mean and standard deviation. Statistical analysis was carried out with SPSS 15.0 statistical package for Windows. The confidence level was established with a statistically significant p-value of less than 0.05.

Table 1 Descriptive data from Isometric test

Isometric	100%	75%	50%	25%
	Mean (Minimum - Maximum)			
Thickness TA (mm)	23,41 (19,03-30,62)	21,1 (18,13-22,68)	20,16 (17,31-21,58)	18,17 (14,31-20,27)
Pennation Angle TA (°)	11,40 (6-17)	10,8 (7-15)	8,60 (4-13)	5,60 (2-8)
Maximum Peak EMG TA (uV)	598,2 (403-873)	506,8 (302-722)	360,60 (213-602)	174,6 (106-233)
Area EMG TA (uV)	813,6 (559-1113)	658,6 (397-1085)	488,40 (347-635)	262 (159-381)
Torque DF Ankle (N/m)	1469,83 (474-2591)	1165,67 (334-2126)	852 (221-1417)	336,6 (56-951)

III. RESULTS

Descriptive data included Mean and Standard Deviation (SD) from EMG, US images and Torque to Isometric test (submaximal contractions and MVC) and, on the other hand, EMG, US and kinematic parameters to Isotonic test as shown in Table 1 and 2, respectively.

Table 2 Descriptive data from Isotonic test

Isotonic	Minimum	Maximum	Mean (SD)
Thickness TA (mm)	17,73	24,43	20,12 (24,46)
Pennation angle TA (°)	8	12	9,67 (1,36)
Maximum Peak EMG TA (uV)	177	663	441 (119,94)
Area EMG TA (uV)	130	711	283,17 (218,78)
Maximum Peak Angular Velocity (°/s)	18,29	272,4	120,82 (88,49)
Maximum Peak Acceleration (m/s^2)	4926	17500	11562,67 (5388,85)

IV. DISSCUSION

Many quantitative methods have contributed to research into the mechanisms of normal and pathological gait and its evaluation (footprint analysis, pressure-sensitive conductive walkway, electrogoniometers, computerised video analysis with joint marker, surface and fine wire electrodes, force plate in walkway or treadmill, oxygen consumption by respirometer, etc.) [6]. The contribution of this work is the synchronization of several of these methods.

Comparative study of muscular architecture with EMG is becoming ever more widespread since it could provide a safe, non-invasive way of determining the muscular function of the superficial muscles [5,7]. This paper aims to contribute to this methodology by offering the analysis of quantitative values which can provide changes during disease or treatment processes.

There are studies which have developed new variables for assessing and treating musculoskeletal function in different parts of the body. Some authors have focused their studies on the trunk, whilst others have specified abdominal musculature [5,8,9] or back musculature [10–14]. Other authors have focused on the role of ADL in the upper limb [5,15,16]. The lower limb is also widely studied, mainly due to its relevance in human gait [17,18]. Research has been carried out into the TA muscle during isometric contractions at different intensities using EMG and SNM [5,17].

The method used may be of interest to monitor neuromuscular activity measured using sEMG, muscular architecture measured using SMG, force measured using dynamometry and kinematic parameters measured using IS during isometric and isotonic contractions of the foot DF.

Although they present some methodological differences, there are other studies based on the synchronisation of instruments, mainly sEMG and SMG, focused on the TA during foot dorsiflexion in healthy subjects. Descriptive data shown by these authors are consistent with those obtained in this study. Hodges and collaborators [5] found an increase in thickness and pennation angle of the TA muscle during isometric contractions. Manal and collaborators [17] also found an increase in pennation angle from rest to MVC in both males and females. Maganaris and Baltzopoulos [19] obtained a decrease in pennation angle in a foot plantarflexion test; however, no significant changes were shown in TA thickness, which may be because, unlike other studies and as previously mentioned, the test was carried out from -15° of ankle dorsiflexion (rest or initial condition) to +30° of plantar flexion (MVC or final condition). Ilse M.P. Arts and collaborators recruited 95 healthy volunteers and provided normative muscle ultrasonography data for muscle thickness which are consistent with our results [20].

V. CONCLUSIONS

The present methodology may be clinically relevant because it involves a reproducible procedure which allows the function and structure of the foot DF to be monitored. The use of this synchronised recording method may be extended to diagnosis and to evaluation of therapies [8,21–24].

REFERENCES

1. SENIAM at http://www.seniam.org/
2. Mega Electronics Ltd Pioneers in Biosignal Monitoring Technology at http://www.megaemg.com/
3. Esaote España at http://www.esaote.es/modules/core/page.asp?p=MYLAB25 GOLD&t=OVE
4. InterSense Precision Motion Tracking Solutions InertiaCube3™ at http://www.intersense.com/pages/18/11/
5. Hodges PW, Pengel LHM, Herbert RD, Gandevia SC (2003) Measurement of muscle contraction with ultrasound imaging. Muscle Nerve 27(6):682-92
6. Dobkin BH. (2003) The Clinical Science of Neurologic Rehabilitation. Oxford University Press
7. ISEK-2012-Conference-Proceedings.pdf at http://isekconference2012.com/wp-content/uploads/2012/07/ISEK-2012-Conference Proceedings.
8. Whittaker JL, Warner MB, Stokes M. (2013) Comparison of the sonographic features of the abdominal wall muscles and connective tissues in individuals with and without lumbopelvic pain. J. Orthop. Sports Phys. Ther. 43(1):11-9
9. Brown SHM, McGill SM. (2010) A comparison of ultrasound and electromyography measures of force and activation to examine the mechanics of abdominal wall contraction. Clin. Biomech. Bristol Avon. 25(2):115-23
10. Stokes M, Hides J, Elliott J, Kiesel K, Hodges P. (2007) Rehabilitative ultrasound imaging of the posterior paraspinal muscles. J. Orthop. Sports Phys. Ther. 37(10):581-95
11. Masuda T, Miyamoto K, Oguri K, Matsuoka T, Shimizu K. (2005) Relationship between the thickness and hemodynamics of the erector spinae muscles in various lumbar curvatures. Clin. Biomech. Bristol Avon. 20(3):247-53
12. Watanabe K, Miyamoto K, Masuda T, Shimizu K. (2004) Use of ultrasonography to evaluate thickness of the erector spinae muscle in maximum flexion and extension of the lumbar spine. Spine. 29(13):1472-7
13. Masuda T, Miyamoto K, Shimizu K. (2006) Intramuscular hemodynamics in bilateral erector spinae muscles in symmetrical and asymmetrical postures with and without loading. Clin. Biomech. Bristol Avon. 21(3):245-53.
14. Cuesta-Vargas AI, González-Sánchez M. (2011) Arquitectura muscular y momento de fuerza: relación en el erector espinal durante extensión isométrica máxima. Rev. Iberoam. Fisioter. Kinesiol. 14(2):77-82
15. Huang QH, Zheng YP, Chena X, He JF, Shi J. (2007) A system for the synchronized recording of sonomyography, electromyography and joint angle. Open Biomed. Eng. J. 1:77-84
16. Shi J, Zheng Y-P, Huang Q-H, Chen X. (2008) Continuous monitoring of sonomyography, electromyography and torque generated by normal upper arm muscles during isometric contraction: sonomyography assessment for arm muscles. IEEE Trans. Biomed. Eng. 55(3):1191-8
17. Manal K, Roberts DP, Buchanan TS. (2008) Can pennation angles be predicted from EMGs for the primary ankle plantar and dorsiflexors during isometric contractions? J. Biomech. 41(11):2492-7
18. Guo J-Y, Zheng Y-P, Xie H-B, Chen X. (2010) Continuous monitoring of electromyography (EMG), mechanomyography (MMG), sonomyography (SMG) and torque output during ramp and step isometric contractions. Med. Eng. Phys. 32(9):1032-42
19. Maganaris CN, Baltzopoulos V. (1999) Predictability of in vivo changes in pennation angle of human tibialis anterior muscle from rest to maximum isometric dorsiflexion. Eur. J. Appl. Physiol. 79(3):294-7
20. Arts IMP, Pillen S, Schelhaas HJ, Overeem S, Zwarts MJ. (2010) Normal values for quantitative muscle ultrasonography in adults. Muscle Nerve. 41(1):32-41
21. Guo J-Y, Zheng Y-P, Huang Q-H, Chen X. (2008) Dynamic monitoring of forearm muscles using one-dimensional sonomyography system. J. Rehabil. Res. Dev. 45(1):187-95
22. Zheng YP, Chan MMF, Shi J, Chen X, Huang QH. (2006) Sonomyography: monitoring morphological changes of forearm muscles in actions with the feasibility for the control of powered prosthesis. Med. Eng. Phys. 28(5):405-15
23. Guo J-Y, Zheng Y-P, Kenney LP, Xie H-B. (2009) Evaluation of sonomyography (SMG) for control compared with electromyography (EMG) in a discrete target tracking task. Conf. Proc. Annu. Int. Conf. Ieee Eng. Med. Biol. Soc. Ieee Eng. Med. Biol. Soc. Conf. 2009:1549-52
24. Shi J, Chang Q, Zheng Y-P. (2010) Feasibility of controlling prosthetic hand using sonomyography signal in real time: preliminary study. J. Rehabil. Res. Dev. 47(2):87-98

Author: María Ruiz-Muñoz
Institute: University of Malaga
Street: Paseo de Martiricos, S/N
City: Málaga
Country: Spain
Email: marumu@uma.es

Automation of Analytical Processes in Immunohematology: Hospital Based-HTA Approach

P. Lago[1], C. Lombardi[1], B. Milanesi[2], P. Isernia[3], L. Raffaele[4], A. Berzuini[4], P. Giussani[5], C.E. Battista[5], M. Angeleri[5], and R. Santini[5]

[1] Department of Clinical Engineering, Hospital of San Matteo, Pavia, Italy
[2] Department of Clinical Chemistry, Hospital of Desenzano del Garda, Desenzano del Garda, Italy
[3] Department of Immunohematology and Transfusion, Hospital of San Matteo, Pavia, Italy
[4] Department of Immunohematology and Transfusion, Hospital of Lecco, Pavia, Italy
[5] Diamed, Italy

Abstract—Problems related to the interpretation of results and clinical consequences of analysis in Immunohematology laboratory are discussed. On the basis of an Hospital-Based HTA (HB-HTA) approach, most common errors of the analytical processes have been highlighted, divided in pre-analytical, analytical and post analytical phase. For each phase, the adoption of automated instruments has enabled a strong reduction of errors, monitored by using indicators related to the analytical phases.

Keywords—Hospital-Based HTA, Immunohematology and Transfusion, Automation, Error, telemedicine.

I. INTRODUCTION

HTA multidisciplinary approach to the healthcare context is represented by Hospital-Based HTA, supported by the assessments at the hospital level [1]. HB-HTA is founded on four steps, that are: needs assessment; recovery of available technical and economical evidences; selection of purchasing procedures, with the comparison of different technologies; analysis of the impact given by the adopted solution, using adequate indicators. Diagnostic Immunohematology is one of the most critical laboratory activities, which must take into account all the overall cycle of investigations pertaining to transfusion, starting from the appropriateness of the request to the correct use and interpretation of the diagnostic information in the management patient records. Each phase of the immune-hematologic laboratory activities (pre-analytical, analytical and post-analytical phase) can be characterized by errors that may compromise the quality of the examinations, and the resulting therapeutic course of the patient [2]. Adverse reactions due to transfusion errors represented about 70% of all adverse events. Among these, about 20% are transfusion reactions from ABO incompatibility [3].

II. MATERIAL AND METHODS

The activity of laboratory in the area of Immunohematology, traditionally divided in pre-analytical, analytical and post-analytical phases, can be affected by mistakes, with negative results on the care of the patients. It's important to investigate every possible lack that it is checked in the analytical process.

Complying with the HB-HTA method, we have decided to create an assessment, to individuate the most relevant problems about laboratory activities, that are the request from the clinicians, the correct evaluation of the exam, and the consequent action on the therapeutic decision.

A. Pre-analytical Phase

Literature evidence in the pre-analytical phase shows the great number of: human errors, for incomplete, incorrect or illegible identification of a patient; duplicated or not pertinent requests for laboratory tests; the presence of not acceptable blood samples for the analysis; the execution of incorrect blood taking or wrongly labelled samples; the wrong identification of patients. However, there are some procedures not performed directly in the laboratory, and therefore not subjected to direct control of laboratory personnel, such as requesting the examination, collection and management of the samples, transport of the samples.

B. Analytical Phase

In blood assignment phase (analysis of ABO/Rh, irregular antibody screening...), purpose of laboratory clinicians is to use the most suitable test on the appropriate sample, achieving a correct result, aimed at the delivering the right blood product to the right patient. Errors in the analytical phase may therefore be due to the imprecision of diagnostic tests for low sensitivity and specificity and / or incorrect identification of samples. Particular attention must be given to the correct operation of the instrumentation, to the reagent preparation, to the calibration tasks of analytical systems, and to the activities of Internal Quality Control for all tests.

C. Post-analytical Phase

The post-analytical phase encloses the overall quality of all the analytical process. The context of the patient and the donor and the ability of the clinician to use the analysis information properly are very important. Procedures include the transfer of analytical data from the equipment to the information management system, the check of the results and their communication. Therefore a report is created, able to point out possible mistakes, causing alarming results. Similarly to the pre-analytical phase, there are external procedures to the laboratory, that are the post-transfusion follow-up of the patient, the donor look back and the haemovigilance. Due to the inadequate experience about transfusion field, showed by any clinician, errors in the communication and in the interpretation of reports may adversely affect treatment decisions adopted by clinicians.

III. RESULTS

Protocols and strategies can help reduce the number of errors committed during laboratory activities.

In the Department of Immunohematology and Transfusion Medicine in San Matteo Hospital, it was decided to adopt a technological solution, that is an automated analytical system (Fig. 1, IH1000, Bio Rad, Berkeley, California). The analytical process is based on the determination of the antigen/antibody, that is detectable in media called "ID-Cards", consisting of tubes filled with Sephacryl gel. The gel is composed of microspheres with a variable diameter, mixed with antibodies or specific buffers, depending on the reaction that must be observed. For each phase of the analytical process, this system allows to reduce the number of errors, improving safety and outcome of the patient and the donor, and protecting the operator at the same time.

Laboratory activity was analyzed in a period between April and December 2012. In the first 4 months, laboratory operators learned the utilization of IH-1000, maintaining the old laboratory equipment. From August to December 2012, the activity passed completely to IH-1000 equipment, showing a strong growth in the number of tests on donors and patients. In Table 1 the results of analysis are shown.

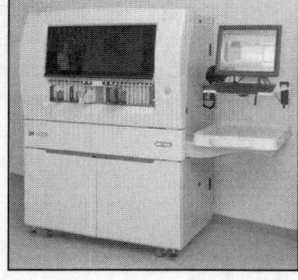

Fig. 1 Automation equipment IH1000, Bio Rad

Table 1 Comparison between two periods of utilization of IH-1000

Quantity (#)	Old equipment (from April to July 2012)	IH-1000 equipment (from August to December 2012)
# patients tested	3000	7500
# donors tested	500	600
# AB0 tests on patients	500	2500
# AB0 tests on donors	500	4000
# AB0 tests on indirect reverse group	3500	8000
# Type and Screen tests	500	6000
# Trio Coombs tests	200	500
# Crossmatch C tests	1000	3500
# Crossmatch BR tests	200	3000

A. Automation in Pre-analytical Phase

The automated equipment identifies the resources required to start the analytical process, by setting a reading of bar codes of the items inside the instrument, and comparing them with the resources needed to run the selected test. The unambiguous identification of the input sample is given by the link with the information management system of the Department and of the Hospital, that collects all the requests. The equipment also controls the lot number and the expiration date of the used devices (reagents, ID-Cards, thinners), the level of liquid reagents and solutions, and the presence of clots in the hydraulic circuit during the process of dispensing. Problems related to the use of expired products and the failures of the instrument caused by the obstructions (clots, fibrin) are therefore solved.

The instrument has an optical sensor for the identification of test tubes accidentally loaded with the stopper, and acceptance options of test tubes of different sizes can be configured, such as pediatric samples.

Logs of events and display areas for loading samples and reagents are available for the operator, ensuring a view of the current status of each loaded rack. The warehouse of ID-Cards is displayed on the monitor, with information on the amount, location and type, ensuring the operator to have the possibility to know the availability of ID-Cards for tests. By software management IH-Com, the operator can combine several different types of tests, creating customized job profiles. The software module contains the Reflex Test, that ensures the possibility to perform more tests on the basis of previous results or requests defined by the user, with reference to patient parameters such as age and sex.

B. Automation in Analytical Phase

The analytical cycle (dispensation of ID-Cards, incubation, centrifugation, reading and interpretation) is completely inside the instrument, protected by a front panel whose opening is detected by the instrumental control software. In this way, no external element can affect the process, and there are not risks of contamination to the outside. The analytical phase begins with the identification, by the carrier arm, of the ID-Card type loaded on each support, while samples and reagents are identified during the loading rack. More precisely, the uploading of the rack of samples and reagents takes place frontally and in dedicated positions, to minimize possible errors of insertion and optimize the activity and the continuity of the work. Tubes are equipped with a bar code, for the recognition of the positive samples, avoiding errors due to incorrect matching between sample and patient, and between patient and donor. Expired reagents and solvents are identified and discarded by the system, by reading the bar code on the package.

The logic random access allows the operator to process the samples in the order of insertion, ensuring optimal management of urgency. In fact, it is possible to automatically load some rack, equipped with specific barcode, which allow to discriminate samples priority, enabling the instrument to handle them immediately. Priorities are performed automatically by optimizing the cycles of centrifugation and incubation that could be in progress. The instrument is equipped with automated accessories (3 centrifuges, 2 arm / needle dispensers, double wash equipment of tanks and exhaust) that allow the optimization of the working flow, with the ability to load a large number of samples, and the continuity of the work even in the presence of an hardware failure (auto-backup).

The reading of the ID-cards is via color camera high definition, and it is based on the detection of antigen-antibody complexes. The camera provides a picture of the reaction and evaluates the results for the corresponding ID-Card. The result of the reaction for each microtube can be positive for 4 degrees (+ + + +, + + +, + +, +), negative (-), undefined (?), or with double erythrocyte population (Dp) (Fig. 2).

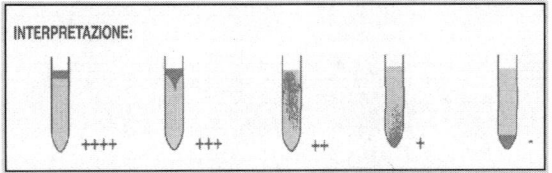

Fig. 2 Kinds of possible reactions

The sequence of steps followed by a test (dispensation, incubation, centrifugation, reading) is represented by a screen, that shows the status and timing of the test. The screen showing relevant maintenance gives to the operator the options for the weekly maintenance and system initialization. Some of the maintenance activities have a daily periodicity (the check of the outer surface of the tool and of work plan), others a weekly one (washing water circuit with dedicated solution and cleaning device for removing the ID-cards). The initialization and the control of the modules and of the work plan are performed, during the start and the stop of the equipment.

The system is equipped with a quality control module that allows the verification of the used reagents. The strategy for the controls can be set by the operator, and controls validity time is monitored by the system. Then the instrument can detect the presence of clots in the needle: the user is guided by instrument itself in the cleaning procedure of the needle, while the second needle ends the dispensation in progress, ensuring work continuity.

C. Automation in Post-analytical Phase

The system allows the processing of reports as shown in the current guidelines, considering all the necessary information for the correct interpretation of the results:

- The name of Department which writes the report;
- The patient/donor identification;
- The date of the sample and of the report;
- The indication about the exam done;
- The results of the exam;
- The reporting of any discrepancies
- The information about significant events that may interact with the results;
- Any notes
- The pictures of the wells with their score of reaction

Through the management software of the system, users can validate results. The results that contain inconsistencies (discrepant results, doubts) or that need deepening tests are shown (antibody screening tests or cross-posi-tive, subgroups, RhD weak). It's 'also possible to set a double validation.

The system proposes an elaboration and a clear display of the results, allowing the operators to enlarge the selected wells, to modify and then confirm the results interpreted by the analyzer.

The system is composed of basic elements (a module of reading images of ID-Cards, a module for user authentication, an user interface module, a module for the supervision, a module for automatic interpretation) that communicate as HTTP traffic. Clinical physicians may at any time remotely connect to server computers in the blood transfusion service, for viewing and validating the results. In addition to an increased safety, the use of telemedicine ensures a better organization in the blood assignment, with consequent optimization and reduction of costs for the hospital.

IV. DISCUSSIONS

Transfusion Medicine regards collection, processing and distribution of blood components and blood products. Problems shall be continuously identified, corrected and monitored, implementing any interventions in an attempt to improve the performance and the patient and donor health [4].

As defined in the last crucial step of Hospital-Based HTA, any appropriate performance indicators must be used for monitoring automation.

Monitoring indicators, grouped in relation to the phases of laboratory work, are adopted in order to check the main error situations that could happen.

The indicators are shown below:

1. Pre-analytical phase: appropriateness in the application of examinations; patient and donor satisfaction; indicators on the transport of specimens; percentage of misidentification; optimization of the technician time for the management of the specimens;

2. Analytical phase: validity of the tests; use of human and material resources; organizational changes; treatment times of the samples;

3. Post-analytical phase: time required for the availability of results; accuracy of laboratory reports; clinicians satisfaction for the service offered by laboratory work; appropriateness of the request for further analytical tests.

V. CONCLUSIONS

Health Technology Assessment is a useful tool to provide as objective as possible criteria, in order to decide on future investments of resources in the healthcare sector. In this context it was decided to evaluate the immunodiagnostics field, especially critical sector of transfusion medicine, according to a perspective of Hospital-Based HTA.

The major problems that may arise during the phases of acquisition requests, execution of tests, interpretation, reporting, counseling and eventual assignment of the blood were identified, on the basis of available information sources.

The help of automation ensures high activity through continuous loading, work autonomy (walk away), a chance for uninterrupted monitoring of the state of the instrument, a guaranteed traceability for all requests, a safe management and interpretation of the results.

Last but not least, the telemedicine module assures user-friendly work processes, with increases in the efficiency and effectiveness of validation tests, and rationalization of available resources.

The introduction of automation equipment must still be followed by a monitoring of its performance and of the organizational changes, through the identification and the analysis of previously evaluated and defined indicators.

REFERENCES

1. Francesconi A (2007), Innovazione organizzativa e tecnologica in sanità. Il ruolo dell'Health Technology Assessment
2. Plebani M (2006), Error in clinical laboratories or errors in laboratoriy medicine?,Clinical Chemistry Lab Med,44:750-9
3. Dipartimento della Qualità, Direzione Generale della Programmazione Sanitaria, dei Livelli di Assistenza e dei Principi Etici di Sistema, (2008), Raccomandazione per la prevenzione della reazione trasfusionale da incompatibilità ABO, 5
4. Anyaegbu C. C. (2011), Quality indicators in Transfusion Medicine: the building blocks, ISBT Science Series; 6, 35–45
5. Vladislav Chizhevsky, Diane Esagui, Eduardo Delaflor–Weiss, (2005), Evaluation of an Automated System for /D Typing and RBC Antibody Detection System in a Hospital Transfusion Service, Science; Volume 36 n. 1
6. Faber JC (2004), Worldwide overview of existing haemovigilance systems, Transfus Apher Sci; 31:99–110
7. Kaplan HS (2005), Getting the right blood to the right patient: the contribution of near-miss event reporting and barrier analysis, Transfusion Clin Biol; 12:380–4
8. Favaretti C (2009), Health Technology Assessment in Italy, International Journal of Technology Assessment in Health Care, 127-133
9. Goodman S (2004), Introduction to Health Technology Assessment, The Lewin Group
10. Bjerner J, Nustad K, Norum LF, Olsen KH, Bormer OP (2002), Immunometric assay interference: incidence and prevention, Clinical Chemistry; 48:613–21

Author: Paolo Lago, Cesare Lombardi
Institute: San Matteo Polyclinic Hospital
Clinical Engineering Department
Street: Viale Golgi, 19
City: Pavia 27100
Country: Italy
Email: p.lago@smatteo.pv.it, c.lombardi@smatteo.pv.it

Total Laboratory Automation and Clinical Engineering

P. Lago, I. Vallone, and A.M. Fenili

San Matteo Hospital, Clinical Engineering Department, Viale Golgi, 19-Pavia 27100, Italy
p.lago@smatteo.pv.it, i.vallone@smatteo.pv.it, fenili.alessandra@gmail.com

Abstract—The introduction of an automation system in a clinical laboratory represents an innovative issue as for IVD technology is concerned. It requires changes in internal organization, review of workflows and design of laboratory re-engineering. The final aim is to improve services and outcomes for patients.

In San Matteo Hospital, Clinical Engineering Department (CED) has an important coordination role in laboratory technology-related issues. Through the application of Hospital Based - Health Technology Assessment (HB-HTA) methodology, CED can support high-level management decisions, for rationalizing processes and evaluating technological solutions. Moreover, for the peculiar reality of Clinical Chemistry Laboratory we apply also Lean principles and tools in order to translate clinical needs into technical data.

Keywords—Total Laboratory Automation, Lean, Hospital Based-Health Technology Assessment, Clinical Engineering.

I. INTRODUCTION

Lean and *Hospital Based-HTA* methodologies represent useful engineering tools in order to improve the organization of Clinical Laboratory.

Lean approach allows to remove all useless activities for the process. This method is aimed at reducing costs, optimizing resources and increasing efficiency and productivity [1, 2]. In view of the introduction of an automation system in the Clinical Chemistry Laboratory of San Matteo Hospital, a suitable preliminary analysis has been carried out by Lean method.

After studing how the system now worked, the market analysis of available automation solutions and best technology choice in our contest will be evaluated through Hospital Based-HTA approach.

The present paper is organized as follows. The analysis carried out on the current and the future work organization is described in Section II. The workload analysis, layout and times analysis, clinical risk analysis and cost analysis are introduced. Some conclusions end the paper.

II. MATERIAL AND METHODS

San Matteo Hospital will purchase in full service a Total Laboratory Automation (TLA) system [3], specifically composed by automatic equipment to carry out analysis of Clinical Chemistry and Immunochemistry. It will perform pre-analytical, analytical and post-analytical tasks including reagents, suitable equipment and related technical support.

According to many authors [4, 5, 6], the main goals are:
- maintaining high levels of analytical quality;
- consolidating the major clinical areas of Laboratory;
- rationalizing procedures, reduce repetitive and manual processing steps, unify the management processes of routine samples and urgency;
- ensuring samples traceability, by monitoring each step in the sample processing;
- minimizing risk and number of human errors, and increase safety for health care operators' work with biohazardous samples;
- reducing amount of primary tubes and therefore amount of biological sample collected for each patient;
- reducing and standardize time of reporting;
- ensuring samples storage conservation in an optimal way and its availability for any repetition and extension of one or more required tests;
- rationalizing number and type of available analytical instrumentation.

A. Workload Analysis

Clinical Chemistry Laboratory gives services to all Departments displaced in the hospital and provides diagnostic activity every day of the year round the clock.

Each year, about 950.000 samples reach the Laboratory with an average of 3.200 samples on workdays and 1.100 on weekends. The workload is not evenly distributed throughout the day (Fig. 1), and, in particular, about 90% of samples comes during the morning. This distribution, which reaches the peak of more than 900 samples between 10 am and 11 am, is due to the time of blood collection. Of course, on Saturdays and Sundays the load of tubes significantly decreases, without vanishing completely for urgent samples.

Since from 6 am to about 6 pm routine tubes represent the majority of the samples, at night there are almost only samples processed in case of emergency. While the distribution of the first ones presents peaks in the morning, the distribution of the tubes of urgency is rather homogeneous throughout the whole day (Fig. 2).

The distribution of tubes in relation to the type of required exams is derived using Laboratory Information System (LIS),: Clinical Chemistry (23%), Blood Count (22%), Coagulation

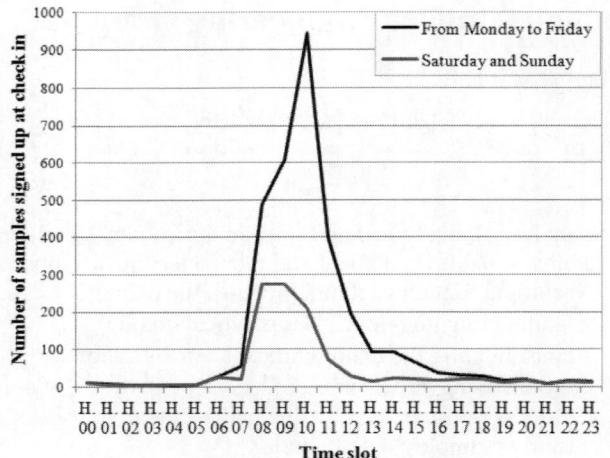

Fig. 1 Daily samples distribution on workdays (solid black line), and on weekends (solid red line).

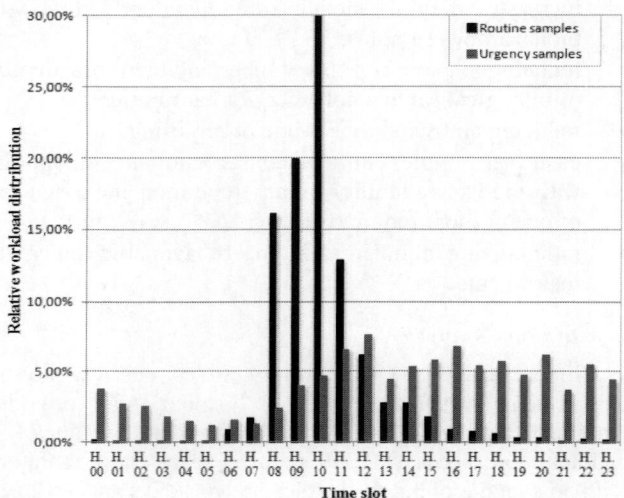

Fig. 2 Relative workload distribution of urgency samples (red bars), and routine samples (black bars).

(13%), Specific Proteins (8%), Immunochemistry and Tumor Marker (7%), and then the other exams (17%).

The automation impact depends on analytical sectors. Functional modular blocks (sorting, preparation, analysis, and storage) build different processes included in the design [8, 9], as shown in Fig. 3. First, Urine samples (4%) will not be loaded on the automation system; tubes of Blood Count and Glycated Mutations (27%) will be sorted but their path will end before centrifugation; the preparation of samples for Coagulation, Autoimmunity, Cardiac Markers, Electrophoresis, PCR and Specific Protein (39%) will automatically happen, but arranged in suitable racks, tubes will be dedicated to analysis in related sector. Clinical Chemistry and Immunochemistry analyzers, including Endocrinology and Tumor Marker (30%), will be integrated in the automation system; all samples processed in Laboratory (with the exception of samples of Urine, Coagulation, NH4, Homocysteine and PTH) will be ultimately stored in a refrigerated system, placed at the end of automation system (81%, of which 30% comes from analysis on board and 51% from other areas of Laboratory).

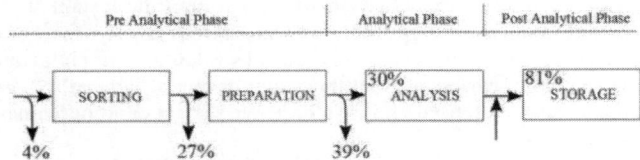

Fig. 3 Load distribution of test tubes on analytical process.

To correctly combine hypotheses indicated by clinicians and to evaluate volume actually engaged by the samples in the refrigerated store, the performance of content storage has been simulated (Fig. 4). After the first week, which is necessary to go full speed, the filling follows a periodic behaviour for 7 days, with maximum load on Wednesdays, Thursdays and Fridays. The result is reasonable if we consider the reduction of sampling, except emergencies, planned for the weekend. By estimating 5 tubes for each patient, it is required a refrigerated system characterized by the minimum capacity of 10.500 samples.

The automation system should be designed on the basis of the results obtained through this workload analysis.

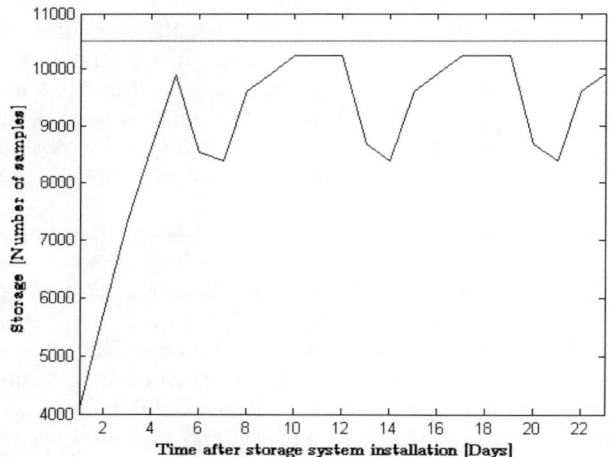

Fig. 4 The filling of the storage (solid black line) and minimum required capacity (solid red line).

B. Layout and Time Analysis

The description of layout includes distribution of locations and machines, operators' movements and provision of materials. For this purpose a *Spaghetti Diagram* is used: this instrument consists in the graphical representation of the movements of the sample and the corresponding operators' movements within the Laboratory.

The architectural features currently represent the major significant constraint: the disposition on different floors does not make the exchange of information easy, and does not simplify the work processes. The operators can not follow a single stream of samples but they are forced to simultaneously devote attention to different process steps, and therefore to samples stored in different places (Fig. 5).

The automation system should be positioned in an open space, with all the equipment, from check in to storage (Fig. 6). Free spaces will be available for the reorganization of the other analytical sectors and for a possible extension of the diagnostic panel. An introduction of a total automation system allows the improvement in the use of space to minimize internal transports.

As regards necessary times, it is not possible to exactly obtain measurements because of lacking traceability of samples in present organization of work and places.

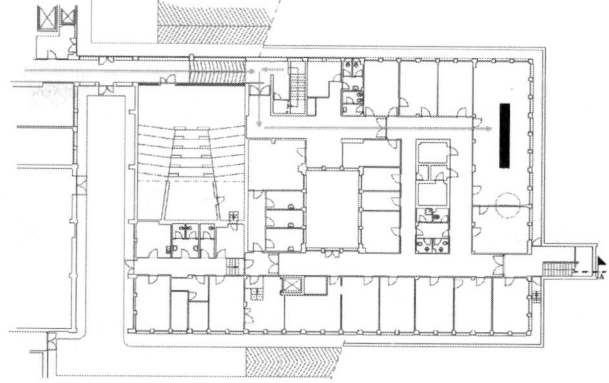

Fig. 6 Future Spaghetti Diagram (green line) and location of automation system.

However, we can say that the major part of routine samples is processed within the first hours in the afternoon.

At present a routine blood sample's journey involves 44 steps, among which 14 are dead times and process delays. It is reasonable to imagine a significant reduction of these times. According to clinicians' experience, it is expected that the automation solution can ensure the execution of at least 90% of Clinical Chemistry tests within 60 minutes from check in, and 90% of Immunochemistry tests within 120 minutes from check in. As regards tubes of urgent Clinical Chemistry, it necessary that the process is completed in 45 minutes from check in.

C. Clinical Risk Analysis

In order to examine clinical risks in the biological sample processing [7], *Failure Mode and Effect Criticality Analysis (FMECA)* is used. We have identified 25 events that directly affect the work of the Laboratory. These events consider all the possible direct or indirect consequences related to the errors in which Laboratory is intrinsically involved. They have been sorted out according to the characteristic phase of the process in which they occur or that they predominantly affect.

Tables of assessment are realized, as progressive scales depending on the accuracy required in the evaluation, experience and available data. Tables 1, 2 and 3 report severity (S), detection (D) and probability to occur (P) of the effects of the possible error/failure.

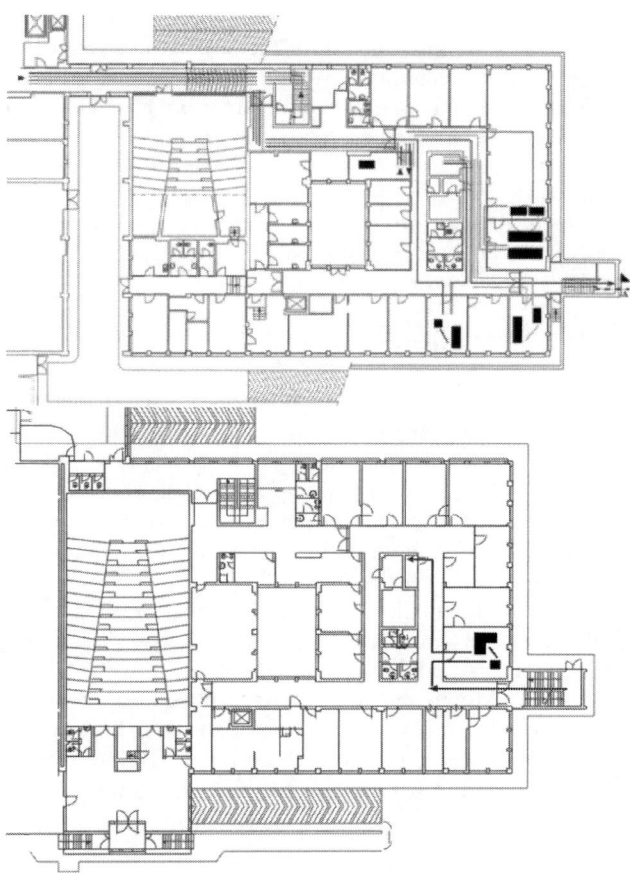

Fig. 5 Present Spaghetti Diagram. Clinical Chemistry (green line), Immunochemistry (blu line), Tumor Marker (light blu line), and urgency samples (red line).

Table 1. Scale for assessment of severity.

Seriousness	Criteria	Values
Null	No relevant consequences for patient health	1
Low	Need of operator's action, ligth delay	2
Moderate	Need of test ripetition, significant delay	3
High	Need of new blood collection, serious delay	4
Dangerous	Relevant consequences for patient health	5

Table 2 Scale for assessment of detection.

Detection	Criteria	Values
Very easy	Detectable through alarm systems	1
Easy	Promptly detectable by technician	2
Moderate	Slowly detectable by clinician	3
Difficult	Hard to detect	4
Very difficult	Very hard to detect	5

Table 3 Scale for assessment of probability.

Probability	Criteria	Values
Very low	Remote probability, totally automatic procedure	1
Low	Few probable	2
Moderate	Random	3
High	Probable	4
Very high	Frequent, totally manual procedure	5

Each event can be associated to a quantitative assessment, defined as the product of the values of three considered parameters, called risk priority index (RPI). Then, the whole process can be described by a table, ordered by decreasing values of RPI and by identifying critical ones with RPI greater than the threshold fixed to 25.

The process steps identified by a high value of RPI require the development of preventive measures and protection, aimed to reduce the level of risk and, in general, to improve the process. Considering the identified 25 events, 11 of them have a value of RPI greater than or equal to 25. The most serious events concern the Pre Analytical phase and are external to Laboratory procedure, while no event involved in Analytical phase, presents a higher value than the critical threshold. These results comply with existing literature [10, 11]. The introduction of automation contributes to reduction of risk, by replacing practical and repetitive operator's skills, such as the sorting and transport of samples between the areas of the Laboratory. Automation system does not fully solve the problem but contributes to decrease risk level, by reducing the probability of occurrence and by simplifying the detection of the error. Therefore, automation can lead to significant indirect effects, thanks to better organization of work, workers' workload and stress reduction.

D. Cost Analysis

Current supplies include analytical equipment, with its installation and technical support, against the payment of reagents, calibrators, solvents and controls, consumables and anything else needed for tests execution. The total spending of Laboratory, limited to analytical sectors involved in the project of reorganization, has a regular trend and an average of about € 960.000 for year, of which € 640.000 allocated to the sector of Immunochemistry and € 320.000 allocated to the sector of Clinical Chemistry.

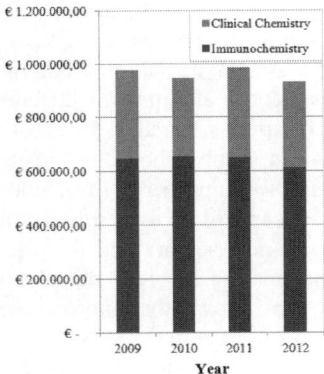

Fig. 7 Total costs incurred by analytical areas of key concern of automation project.

Following the introduction of the automation system, we will introduce the method of remuneration defined *price for test* in order to make the whole system more efficient and to guarantee a better service to the community. Automation supply will be awarded on the basis of cost indicated for each test, including reagent, consumer products, instruments, hardware, and customer service. The supplier will bill the number of tests actually carried out, according to this defined cost.

III. CONCLUSIONS

Workflow analysis shows the strengths and weaknesses of the existing process and allows to fix quality scores according to which it is possible to evaluate competing products and, further, reinforce the potential benefits of automation. Indeed, the results of this analysis provide important information to define technical specifications and the future review of the project.

We have shown that Clinical Engineering Department (CED) acts like a bridge between technologies, Laboratory clinical needs and economic-administrative hospital constrains. Moreover, the close cooperation between Laboratory, Administration, Chemist's, Information Systems and CED could lead to an effective improvement of processes and technological resources in health care.

ACKNOWLEDGMENT

The authors gratefully acknowledge the contribution of all members of Clinical Laboratory and Clinical Engineering Department of San Matteo Hospital.

REFERENCES

1. D. Jones and A. Mitchell, *Lean thinking for the NHS*. NHS Confederation 2006. Lean Enterprise Academy UK,.
2. P. G. D. Role, *Lean for manifacturing*, September 2011 in Six Sigma Master Black Belt.
3. R. A. Felder, *Automation: Survival tools for the hospital laboratory,* in Second International Bayer Diagnostics Laboratory Testing Symposium, July 1998.
4. C. W. Lam and E. Jacob, Implementing laboratory automation system: experience of a large clinical laboratory, JALA, vol. 40, pp. 1–9, 2010.
5. M. Zaninotto and M. Plebani, *The hospital central laboratory. automation, integration and clinical usefulness*, Clinical Chemistry Lab Med, vol. 48(7), pp. 911–917, 2010
6. M. S. Zaleski, Automation, the workforce, and the future of the laboratory. MLO, July 2011
7. C. Signori, F. Ceriotti, A. Sanna, M. Plebani, G. Messeri, C. Ottomano, F. D. Serio, and P. Bonini, *Process and risk analysis to reduce errors in clinical laboratories*, Clinical Chemistry Lab Med, vol. 45(6), pp. 742–748, 2007.
8. P. Najmabadi, A. A. Goldenberg, and A. Emili, Hardware flexibility of laboratory automation systems: Analysis and new flexible automation architectures, JALA, vol. 203, pp. 203–216, August 2006.
9. S. E. F. Melanson, N. I. Lindeman, and P. Jarolim, *Selecting automation for the clinical chemistry laboratory*, Arch Pathol Lab Med, vol. 131, pp. 1063–1069, July 2007.
10. M. Plebani, Errors in clinical laboratories or errors in laboratory medicine?, Clinical Chemistry, vol. 44, pp. 750–759, 2006.
11. M. Plebani, *Exploring the iceberg of errors in laboratory medicine*, Clinica Chimica Acta, vol. 404, pp. 16–23, 2009.

Development of Vertebral Metrics – An Instrument to Analyse the Spinal Column

C. Quaresma[1,2], A. Gabriel[1], M. Forjaz Secca[1,4], and P. Vieira[2,5]

[1] Cefitec, Departamento de Física, Faculdade de Ciências e Tecnologia, Universidade Nova de Lisboa,
Quinta da Torre, 2829-516 Caparica, Portugal
[2] Departamento de Física, Faculdade de Ciências e Tecnologia, Universidade Nova de Lisboa,
Quinta da Torre, 2829-516 Caparica, Portugal
[3] Departamento de Saúde, Instituto Politécnico de Beja, Portugal
[4] Ressonância Magnética – Casellas, Portugal
[5] Centro de Física Atómica, Departamento de Física,
Faculdade de Ciências e Tecnologia – Universidade Nova de Lisboa – Monte da Caparica, Portugal

Abstract—**In order to outline prevention and intervention strategies in the PublicHealth area it is crucial to determine the quantitative characteristics of biomechanical changes in the spinal column. Most researchers links rachialgiae aetiology to these biomechanical modifications, but these studies are limited due to the invasive nature of the technics available for quantification of the spinal column anatomic parameters. . For this reasons it is important to develop non-invasive methodologies that can be applied to general population. Most of existing non-invansive techniques have the handicap of not providing an overall view of all biomechanical parameters of the spine**

In order to fill this gap, it was built the Vertebral Metrics, a mechanical instrument designed to identify the X, Y, and Z positions of each spine apophyses (vertebrae), in a standing position. After its validation, the need has arisen to further develop this equipment in order to automate the data acquisition reducing the acquisition time and increasing its reprodubility. Therefore, a semi-automatic and non-invasive instrument was designed to detect the position of the spine apophyses (vertebrae). The aim of this paper is to describe the automation process concerning the collection of data with Vertebral Metrics.

Keywords—**Spine, Evaluation, Medical Device, Automatic.**

I. INTRODUCTION

Currently, back pain is considered a major problem in modern society [1].The incidence of handicaps related to this symptomatology is so common and usual that it should be studied as an epidemic and as a social disease [2]. Most researchers attribute its aetiology to the biomechanical changes of the spinal column, thus it is important to obtain a quantitative characterization of the spinal column in order to delineate prevention and intervention strategies in the area of Public Health. However, further analysis has been held back due to the fact that the standard comprehensive analytical procedures are invasive in nature and cannot, as such, be applied to the general population [3,4]. On the other hand, the non-invasive diagnostic tools that exist in the market only allow for partial scans of the spinal column or are extremely expensive, and none allowed for a three-dimensional analysis of the vertex of the spinous apophysis and did not, for this reason, provide a comprehensive overview.

Thus, in order to fill this gap, we built *Vertebral Metrics* (registered with the Portuguese name of "*Métrica Vertebral*") a mechanical device that permits us to identify the positions of X, Y and Z of the vertex of each spinous apophysis (vertebra) in the standing position [5].

The mechanical instrument *Vertebral Metrics* (Fig. 1) consists of two parts: one we call the *Body* and another which we define as the *Support* [6, 7]. The *Body* consists of one vertical piece and 18 horizontal pieces that we call *2D Positioners* [5, 6, 7].

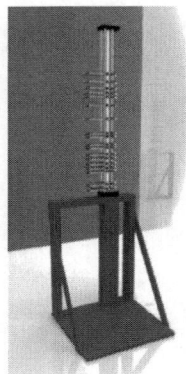

Fig. 1 Image of the *Vertebral Metrics*

Each *2D Positioner* is adjustable in order to identify the position of X, Y and Z of each vertex of the spinous apophyses, from the first cervical vertebra to the first sacral vertebra. The contact is made through the cone-shaped end of the horizontal piece, which we call the point of contact

Immediately afterwards, the need to develop this device arose, in order to automate and reduce the time for data acquisition. Therefore we constructed a semi-automatic and non-invasive instrument to detect the position of the vertices of the spinous apophyses [6, 7,8].

The purpose of this paper is to describe the automation process concerning the collection of data with Vertebral Metrics.

II. METHODOLOGY

1.1 Construction of an Optoelectronics Device

The automation of the vertebral metric was based on the use of an optical camera aligned along the Y's axis together with a laser diode bearing a specific angle of said axis. The whole assemblage was mounted on a positioner with degrees of freedom on the other two coordinates (Fig. 2).

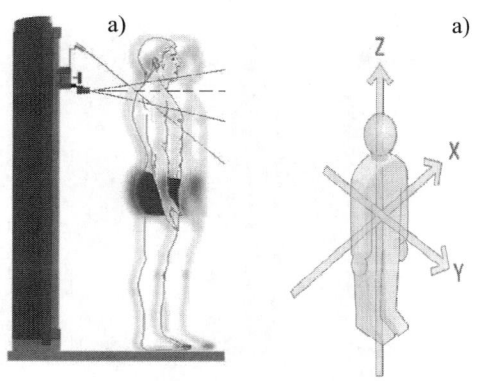

Fig. 2 Schematic representation of the optical *Vertebral Metrics*); b) Coordinate system used

The laser is focused on the patient, being its spot visible by the CCD camera. The coordinate on the CCD where the image of the laser is observed has a proportionality factor in relation to the Y coordinate.

At the beginning of each measurement the camera will distinguish one or more marks of the spinous apophysis that had been previously marked by the expert. The X and Z coordinates of these marks are obtained directly from the information contained in the image on the CCD camera. There is a known scale relationship between each pixel of the CCD and the actual coordinate of X and Z. Information on the Y coordinate (depth) can not be obtained in this manner due to the fact that the CCD is not, in itself, capable of determining the depth of an object. Thus to obtain the Y coordinate, it was used a laser in a different axis of the camera [9].

The spatial extent of the entire spinous apophysis is obtained in the following manner.

- The mechanical positioners are placed in such a way that the camera projects an image of the first apophysis to be measured

- Then the vertical positioner begins a constant motion in order to find the marks of other apophysis. An image processing algorithm was developed for this specific use.
- The x, y and z coordinates of each apophysis are found sequentially from the coordinates of the positioners and the coordinates of the laser spot viewed by the camera.

Next, Details are given regarding the manner in which the X, Y and Z coordinates are obtained, as well as the image processing involved in identifying the apophysis and the laser diode mark.

1.1.1. Calculation of the Anterior-Posterior Distance (Y's)

In order to calculate the Anterior-Posterior (Y's) distance, the physical restrictions of the system are established (Fig. 3):

- D - distance from the camera to the reference plane, represented in orange in Fig. 3.
- α - the angle of the laser in relation to the vertical
- θ – the field of view of the camera
- m – real dimension of each camera pixel in the reference plane. Obtained from the knowledge of D, θ, e R.
- R – Resolution, of pixels in the camera

For safety reasons, the initial distance between the reference plane (D) and de camera cannot be too small. This could place the individual being studied too close to the equipment and because the same is in motion, this could cause physical damage. That distance can also not be too big because the study area that is represented in the sensor increases, and as the number of pixels in the camera remains, there will be a smaller number of pixels to represent each region, ie, a smaller resolution. Bearing this in mind, and considering that distance of 20 cm was the minimum for safety, it was decide to use 30 cm as the set distance of the reference plane[9].

Having defined the D distance, all the other parameters can be found in order to allow accurate measurements of the position of the apophysis. Taking into account published information, it was considered that if the reference plane lies 30 cm from the camera, the system should be able to take measurements from 23cm to 42 cm. Using only the geometric considerations of the system and bearing in mind that the camera's angle of vision is 10°, then the laser of the diode laser must be placed at a height of 9 cm in relation to the camera and at an angle of 72° in relation to the vertical.

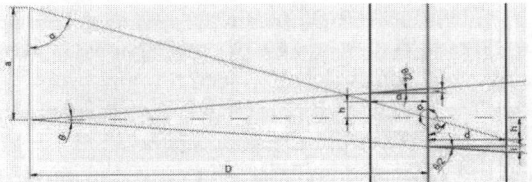

Fig. 3 Diagram to demonstrate the calculation of the anterior-posterior distance

1.1.2. Detection Algorithm

In order to build the automatic *Vertebral Metrics* a process that is capable of automatically determining the location of the marks on the spinous apophysis, as well as the laser spot, is essential. In order for the detection algorithm to be as simple as possible and, consequently, be able to process a large number of images per unit time, it is crucial to select the correct choice of laser colour and marks on the apophysis. When using a colour camera, it is known that the image is comprised of a combination of three components of primary colours, red, green and blue. Because human skin has a strong red component, the contrast of the markers will be higher if the green and blue components are employed. For these reasons was used, a green laser of 533nm; and, a pure blue pigment with a maximum reflectance at 480 for the marker of the spinous apophysis.

Geometric and algebraic principles are involved in the definition of the plan. The process starts with the identification of the regions and the coordinates that are going to be used. In practice, we are defining the coordinates of three points. Then, we use the previous points to get two vectors and, by doing the vector product, the result is an orthogonal vector to the plan. With an orthogonal vector and a point of the plan, its equation can be calculated:

$$z(x,y) = \frac{-1080(m3 - m2)(x - 100) + 850(m2 - m1)(y - 100)}{918000} + m1$$

where:
- $z(x,y)$ is the intensity of each point of the plan;
- $m1$, $m2$ and $m3$ are the means of intensity of all of the pixels of each region.

An automatic threshold algorithm was defined to perform the image binarization. An automatic threshold algorithm was defined to realize the image binarization. This algorithm is based on a common property of the histogram of all the images captured by the video camera: the appearance of a peak with a high ordinate value. This means that the majority pixels of the images have an approximate intensity.

The threshold algorithm identifies the x coordinate of the peak and, from this one, it searches the first point with y coordinate equal or smaller than 400, because it is much lower than peakLastly, after the point has been found, its x coordinate is converted to the binary scale and as a result we have the value of the threshold to do the binarization.

As the binarized image has many artifacts, it is necessary to minimize them. Consequently, after the binarization process is completed, morphologic functions of erosion and dilation of the image are applied. The number of functions implemented in the points detection algorithm is not equal to the laser detection algorithm; however, the mask used is always the same: it has a disk form with two pixels of dimension. In Fig. 4, it is presented an example of the binarization plus the morphologic functions.

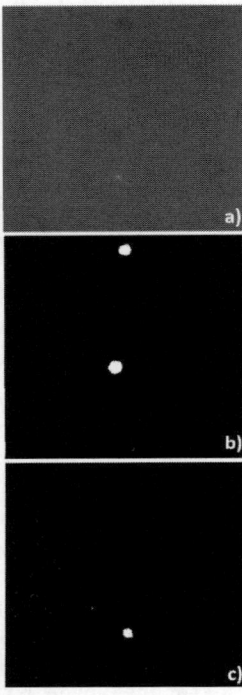

Fig. 4 a) Aspect of the blue and laser mark on the skin; b) Binarization plus morphologic functions of the blue mark; c) Binarization plus morphologic functions of the laser mark.

The last step of both algorithms is the assortment of the objects. This stage is based on the shape and size of the objects that each one of the algorithms must detect. Therefore, the conditions defined in the algorithms are slightly different too.

The validation process, the efficacy of the methodology and algorithms developed, tests were performed with success in ten volunteers with no associated pathology.

III. CONCLUSIONS

The automatic Vertebral Metrics presents a sufficient accuracy for the global analysis of the spinal column, i.e., it is a totally functional prototype that is capable of registering the x, y and z coordinates on the vertex of each vertebrae.

In light of the aforementioned, we confirmed that we successfully carried out the objective to automate the data collection process of *Vertebral Metrics* and to develop software that allows for the real-time 3D representation of the spinal column and the calculation of the inter-vertebral angles.

By building *Vertebral Metrics* we hope to contribute towards an improved identification of the dysfunctions and/or pathologies of the spinal column, in the standing position. By means of an improved diagnosis, it will be possible to develop and implement intervention programmes that are more directed to the specific problems of each individual. This device allows us to conduct dynamic studies of the spinal column and is easily applied in areas such as orthopaedics, neurosurgery, paediatrics and rehabilitation[7,8,9].

It is worth highlighting the fact that *Vertebral Metrics* represents an innovation in the field of prevention due to the fact that it can be applied repeatedly without causing damage to the individual and general population. Moreover, it is relatively inexpensive, easy to transport and presents few logistic requirements. Consequently, it can be used in different contexts, such as ambulatory: public and private (health centre / doctor's office), hospital and as a potential screening device for back pain in the populating.

ACKNOWLEDGMENT

The authors express gratitude for the precious collaboration of the following elements:

- Pedro Duque and António Jordão, for the construction of a optoelectronic device that allows for the measurement of the spatial position of the spinous apophysis
- NGSS, for the funding and know-how provided and that allowed for the completion of this project

REFERENCES

1. Moraes A. Modelo de avaliação fisio funcional da coluna vertebral. Rev. Latino-am Enfermagem 2001; 9(2): 67-75.
2. Knoplich J. Enfermidades da coluna vertebral. Robe Editorial. São Paulo, 3ªed. 2003.
3. Ashton-Miller, J. A., Schultz A. B. (1968). Biomechanics of Human Spine. Em: *Basic Orthopaedic Biomechanics*. 2nd Edition. Lippincott-Raven Publishers. Philadelphia, 1997. 353-393.
4. Descarreaux, M., Blouin, J. S., Teasdale, N. (2003). A non-invasive technique for measurement of cervical vertebral angle: report of a preliminary study. Spine J. 12: 314-319.
5. Secca, M.; Quaresma, C; Santos, F. (2008) "A Mechanical Instrument to Evaluate Posture of the Spinal Column in Pregnant Women". In IFMBE Proceedings, 4th European Conference of the International Federation for Medical and Biological Engineering. 23-27 November 2008, Antwerp, Belgium.
6. Quaresma, C.; Forjaz Secca, M.; Goyri O'Neill, J.; Branco, J.(2009a). "Development of a mechanical instrument to evaluate biomechanically the spinal column in pregnant women". Proc. International Conference Biodevice, 310-113.
7. Quaresma, C.; João, F.; Fonseca, M.; Forjaz Secca, M.; Veloso, A.; Goyri O'Neill, J.; Branco, J.(2009b) "Validation of Vertebral Metrics: a mechanical instrument to evaluate posture of the spinal column". O. Dössel and W.C. Schlegel (Eds.): WC IFMBE Proceedings 25/VII, 711–713.
8. Quaresma, C.; João, F.; Fonseca, M.; Forjaz Secca, M.; Veloso, A.; Goyri O'Neill, J.; Branco, J.(2010) "Comparative evaluation of the tridimensional spine position measured with a new instrument (Vertebral Metrics) and an Optoelectronic System of Stereophotogrammetry". Medical, Biological Engineering and Computing.
9. Jordão, A; Duque, P.; Quaresma, C; Vieira, P. (2011). Development of Vertebral Metrics - An Instrument to Study the Vertebral Column. Proc. Internacional Conference Biodevices2011: 224-229.

The corresponding author:

Author: Cláudia Quaresma
Institute: Cefitec, Departamento de Física, Faculdade de Ciências e Tecnologia, Universidade Nova de Lisboa,
Street: Quinta da Torre, 2829-516 Caparica
City: Caparica
Country: Portugal
Email: q.claudia@fct.unl.pt

Global Program for Certification of Local Clinical Engineers: Back to the Future

M. Medvedec

Department of Nuclear Medicine and Radiation Protection, University Hospital Centre Zagreb, Zagreb Croatia

Abstract—For some, Clinical Engineering (CE) profession is relatively a new profession but, according to the recent data, it is part of the fastest growing community of expanding professions that make up healthcare workforce in current decade. The Clinical Engineering Division of the International Federation for Medical and Biological Engineering (CED/ IFMBE) has initiated a project that included a global survey of CE certification programs. The purpose of this project was to gather, analyze and synthesize available information to determine the need and the model for establishing an international program for certification in CE. In simple words, certification is issuing the document of completion or qualification, accreditation is like certification of the certification body, registration is recording or registering the certificates, and licensing is issuing a permission to do something that otherwise is forbidden.

In that context, an e-letter on behalf of the CED/ IFMBE was sent to (1) affiliated National Societies of the IFMBE, (2) members of the Yahoo! Group - CED Global, Clinical Engineering Division, and (3) all CED/ IFMBE Board members. It has been estimated that survey requests were sent to approximately 200 e-mail addresses in about 50 countries.

Only a few e-mails were reported by e-mail server software as undeliverable. So far, information has been received from 18 countries, resulting in a survey's response rate of more than one-third. The certification in CE appears to exist in 6 out of those 18 countries. In 2 countries there is an explicit legal framework i.e. acts on biomedical/clinical engineering, including the scheme for mandatory professional certification. In additional 4 countries the certification in CE exists on a voluntary basis.

Global CE community in the 21st century needs and deserves an international CE certification programme to come finally into life.

Keywords—certification, clinical engineering, health worker, legislation, occupation.

I. INTRODUCTION

Clinical biomedical engineering profession i.e. clinical engineer (CE) is for many people still relatively new profession, but it is projected as one of the fastest growing occupations in the current decade [1]. According to the latest International Standard Classification of Occupations of the International Labour Organization, a number of such 'upcoming' professions are clearly considered to be a part of the health work force. Apart from 'usual' health professionals (medical doctors, nursing and midwifery professionals, traditional and complementary medicine professionals, paramedical practitioners, veterinarians, etc.), such occupations include but are not restricted to biomedical engineers and medical physicists as well [2]. Similar considerations of clinical qualified professionals for physical and technical jobs have been or are supposed to be explicitly promoted through publications of the World Health Organization, International Atomic Energy Agency, European Union and other international authorities [3-5]. Thus, it stands to reason that formal career paths of clinical engineers or medical physicists should be demanding similar, if not the same, as career paths of other health professionals with commensurate duration or level of basic (university) education, as for example of medical doctors, medical biochemists, pharmacists, etc. Despite this recent encouraging news, it appears that there is a slow embracement by national healthcare systems in many countries to fully recognize, regulate and integrate clinical engineers. Although the world of sophisticated healthcare technologies substantially depends on the application of engineering within the clinical environment, in many ILO-, WHO-, IAEA- and EU-member states there is actually no legal framework for clinical engineering profession. So, clinical engineers and medical physicists, like brothers in arms in everyday practice and during the joint world congresses of medical physics and biomedical engineering, continue struggling for better perception, recognition, regulation and integration of their professions [6-14]. Well-defined specific qualifications and experience in clinical practice are necessary requirements for those interested to actively participate in the field of clinical engineering. Such qualifications and experience should be subject to certification, otherwise anybody could claim to be a CE or a clinical engineering practitioner and take part in responsibility for the healthcare technology and patients in a hospital without the need for certification, i.e. without a proof of the necessary qualifications and skills. Therefore, one of the professional milestones is certainly the establishment of certification program(s), because the purpose of certification is to promote healthcare delivery improvement nation- and world-wide through the certification and continuing assessment of competency of professionals who support and advance patient care by applying engineering and management skills to healthcare technology.

The Clinical Engineering Division of the International Federation for Medical and Biological Engineering (CED/IFMBE) is dedicated, among other goals, to the advancement of international guidelines for professional education, professional development and certification in CE, and to the advancement of the CE role in the institutional frameworks of healthcare policy, strategy, planning, and management worldwide. Therefore, in 2010 the CED/IFMBE has initiated a project to examine global status of CE professional recognition that included a global survey of CE certification programs. The main purpose of this effort was to gather, analyze and synthesize all available information in order to determine the need and the model for establishing an international program for certification in CE. The main goal of the project is to provide recognition for meeting benchmarks, a certificate – as a formal document issued by a global awarding body, which records the achievements of candidate following an assessment and validation against a predefined standard.

The objective of this paper is to present current status of the CED/IFMBE activities towards international program for clinical engineering certification (CEC).

II. MATERIALS AND METHODS

Since people may confuse the terms like accreditation, certification, licensing or registration in the professional context, it may be worth knowing that there are differences and what those differences are. For example, a certification is a 'third-party attestation related to...persons', as defined by ISO/IEC (International Organization for Standardization/International Electrotechnical Commission). Thus, certification is the process of issuing a certificate, diploma or title formally attesting that a knowledge, know-how, skills and/or competences acquired by an individual have been assessed and validated by a competent body against a predefined standard. Accreditation is a 'third-party attestation related to a conformity assessment body conveying formal demonstration of its competence to carry out specific conformity assessment tasks' as defined by ISO/IEC. Accreditation is a process of quality assurance through which accredited status is granted, showing it has been approved by the relevant legislative or professional authorities by having met predetermined standards. In simple words, certification is issuing the document of completion, qualifications or experience, accreditation is like certification of the certification body, registration is recording or registering certificates, and licensing is issuing a permission to do something that otherwise is forbidden.

In that context, an e-letter on behalf of the CED/IFMBE was sent to Affiliated National Societies of the IFMBE (using contact data at http://ifmbe.org/members/members_directory/ and at http://who.ceb.unicamp.br/), to the members of the Yahoo! Group CEDGlobal - CED Global, Clinical Engineering Division (http:// health.groups.yahoo.com/group/CEDGlobal/), and to all CED/IFMBE Board Members in June/October 2010. The e-letter was kindly asking recipients to try to provide as much as possible information (criteria, procedures, titles, legislation, documents, links...) on any attempted or established national or international program for certification in clinical engineering. Further information on certification in other clinical sciences or for other professionals within healthcare system was also asked for and expected to be very useful. The final reminders were sent again in November 2010. It has appeared that a survey request was sent to approximately 200 e-mail addresses in about 50 countries.

III. RESULTS

Only a few e-mails were reported by e-mail server software as undeliverable. So far, the information has been received from 18 countries (Argentina, Australia, Austria, Brazil, Canada, Chinese Taipei/Taiwan, Croatia, Czech Republic, Greece, Italy, Japan, Netherlands, Nigeria, Singapore, South Africa, Sweden, Ukraine and United States), resulting in a survey's response rate of 36%. The certification in clinical engineering appears to exists in 6 (Canada, Chinese Taipei/Taiwan Czech Republic, Japan, Sweden, USA) out of 18 countries. In 2 countries (Czech Republic, Japan) there is an explicit legal framework i.e. acts on biomedical/clinical engineering, including the scheme for mandatory professional certification. In other 4 countries (Canada, Chinese Taipei/Taiwan, Sweden, USA) the certification in clinical engineering exists on a voluntary basis.

IV. DISCUSSION

Currently, CED/IFMBE is still exploring and discussing all aspects of the whole process of CEC: eligibility requirements (education, practice, job content), administration, application procedure (resume, references, transcripts, fees), grandfathering (national/regional Board of Examiners), preparation (study materials, literature availability/accessibility), examination (written, oral, language), levels of certification (clinical engineer, clinical engineering expert), certificate attainment, renewal and revocation, etc. International programs for CEC are likely to be administered regionally by a regional Board of Examiners and certainly have to be adjusted to acknowledge and respect cultural and professional differences in the distribution of the information in the clinical engineering body of knowledge required by regional groups of clinical engineers to function in their day to day job. So far, the certification in clinical engineering by the Healthcare Technology Certification Commission Program sponsored by the Healthcare

Technology Foundation and with the examination conducted by the US Board of Examiners for Clinical Engineering Certification appears to be probably the most advanced and elaborated program having already an international capacity [15].

Within the excellent European BIOMEDEA project, protocols for the Europe-wide harmonized accreditation of biomedical and clinical engineering programs, training of clinical engineers, continuing education, and the certification of clinical engineers have been developed. Also, a management structure for the European Clinical Engineering Certification Protocols has been worked out which is also applicable to the global development of clinical engineering as a regulated healthcare profession. In a next step the various BIOMEDEA protocols have been expected to be presented to the European national IFMBE member societies for adoption. However, practical real-life outcomes and impact on CE regulation Europe-wide are still unclear [6,16].

Finally, it is noteworthy to take a look more than thirty years back and to remind on the document entitled 'Agreement on Mutual Recognition of Qualifications for Clinical Engineers', where 22 affiliated National Societies of the IFMBE (Austria, Australia, Belgium, Canada, Denmark, Federal Republic of Germany, Finland, France, German Democratic Republic, Hungary, Israel, Italy, Japan, Mexico, Netherlands, Norway, Spain, South Africa, Sweden, United Kingdom, USA, Yugoslavia) mutually agreed in 1981 to recognize any holder of the IFMBE's certificate of registration as a clinical engineer. Formally, the agreement seems to be still in place, but since certification and registration have never been made mandatory by national legislations in most of the countries, the agreement has been neglected [17].

V. CONCLUSIONS

We all need to make every effort to demonstrate more clearly and timely to policy makers and the public, but in particular to national governments, all the benefits derived from well trained health professionals who achieved and sustain mastering of the defined body of knowledge. A formal career path and opportunities in clinical engineering should be similar to those of other university-degree health professionals. Clinical engineers around the globe desperately need and deserve international professional certification program to come into life as soon as possible, in order to help to tread the path towards full perception, recognition, regulation and integration of clinical engineering professionals in healthcare systems of many countries.

Primarily for the benefit of patients, WHO, IAEA, IFMBE, IOMP, IUPESM and other international organizations and authorities should work more intensively on developing the most efficient mechanisms of urging, pleading and reminding all national governments of the member-states to adopt systems of clinical quality assurance, control and safety and to make certifications within the whole health workforce mandatory.

ACKNOWLEDGMENT

The author is grateful to Yadin David, Tom Judd, Kang-Ping Lin, Frank Painter, Nils-Erik Pettersson, Calil Saide, James Wear and others for their help, support and cooperation.

REFERENCES

1. US Department of Labor, Bureau of Labor Statistics at http://www.bls.gov/ooh/About/Projections-Overview.htm
2. International Standard Classification of Occupations - ISCO-08 Volume I (2012) at http://www.ilo.org/wcmsp5/groups/public/---dgreports/---dcomm/--publ/documents/publication/wcms_172572.pdf
3. International Atomic Energy Agency (2011). Radiation Protection and Safety of Radiation Sources: International Basic Safety Standards: General Safety Requirements - Interim Edition. IAEA, Vienna at http://www-pub.iaea.org/MTCD/Publications/PDF/p1531interim_web.pdf.
4. Council directive 96/29/Euratom of 13 May 1996, laying down basic safety standards for the protection of the health of workers and the general public against the dangers arising from ionising radiation. Official Journal of the European Communities; 1996; 39: No L-159, p1.
5. Directive 97/43/Euratom of 30 June 1997 on health protection of individuals against the dangers of ionising radiation in relation to medical exposure. Official Journal of the European Communities; 1997; 22: No L-180, p.5
6. Nagel JH (2010) The regulation of the clinical engineering profession as an important contribution to quality assurance in health care. IFMBE Proc. vol. 25, World Congress on Med. Phys. & Biomed. Eng., Munich, Germany, 2009, pp 376-378
7. Lhotská L, Cmíral J (2010). Accreditation and certification in biomedical engineering in the Czech Republic. IFMBE Proc. vol. 25, World Congress on Med. Phys. & Biomed. Eng., Munich, Germany, 2009, pp 372-375
8. Ask P, Pettersson NE, Andersson K (2010). Certification of clinical engineers in Sweden. IFMBE Proc. vol. 25, World Congress on Med. Phys. & Biomed. Eng., Munich, Germany, 2009, pp 430-431
9. Khambete ND (2010) Biomedical engineering education and training in India - a need for new approach. IFMBE Proc. vol. 25, World Congress on Med. Phys. & Biomed. Eng., Munich, Germany, 2009, pp3 68–371
10. Medvedec M (2012). Current legal framework for biomedical engineering profession in Croatia. IFMBE Proc. vol. 37, 5th European IFMBE Conference, Budapest, Hungary, 2011, pp 1404-1407

11. Geoghegan DS, Stefanoyiannis AP, Christofides S et al. (2013) The international context of education, training and certification for medical physicists in Europe, North America and Australasia. IFMBE Proc. vol. 39, World Congress on Med. Phys. & Biomed. Eng., Beijing, China, 2012, pp 2260–2263
12. Oliver LD (2013) International medical physics certification: but what else is needed for international acceptance?. IFMBE Proc. vol. 39, World Congress on Med. Phys. & Biomed. Eng., Beijing, China, 2012, pp 1679–1682
13. Ravindran PB (2013) Medical Physics Certification in India. IFMBE Proc. vol. 39, World Congress on Med. Phys. & Biomed. Eng., Beijing, China, 2012, pp 1676–1678
14. Wu RK (2010) Creating an independent international medical physics board. IFMBE Proc. vol. 25, World Congress on Med. Phys. & Biomed. Eng., Munich, Germany, 2009, pp 68–70
15. Healthcare Technology Foundation - HTC at http://www.thehtf.org/certification.asp
16. BIOMEDEA at http://www.biomedea.org
17. International Federation for Medical and Biological Engineering (1981) The IFMBE international register of clinical engineers - Agreement on mutual recognition of qualifications for clinical engineers at http://www.biomedea.org/Documents/IFMBE%20Int%20Reg%20Stuttgart.pdf

Author: Mario Medvedec
Institute: Department of Nuclear Medicine and Radiation Protection, University Hospital Centre Zagreb
Street: Kispaticeva 12
City: Zagreb
Country: Croatia
Email: mario.medvedec@kbc-zagreb.hr

Emerging of Human Factor in Risk Analysis of Home Care Service

F. Clemente[1], G. Faiella[2], G. Rutoli[3], M. Romano[2], P. Bifulco[2], A. Fratini[2], and M. Cesarelli[2]

[1] IsIB CNR - Institute of Biomedical Engineering of National Research Council, Rome, Italy
[2] D.I.E.T.I., University of Naples Federico II, Naples, Italy
[3] Biotechnology Unit, ASL NA1, Naples, Italy

Abstract—Risk management in healthcare represents a group of various complex actions, implemented to improve the quality of healthcare services and guarantee the patients safety. Risks cannot be eliminated, but it can be controlled with different risk assessment methods derived from industrial applications and among these the Failure Mode Effect and Criticality Analysis (FMECA) is a largely used methodology. The main purpose of this work is the analysis of failure modes of the Home Care (HC) service provided by local healthcare unit of Naples (ASL NA1) to focus attention on human and non human factors according to the organization framework selected by WHO.

Keywords—Risk Management, Failure Mode, Effect and Criticality Analysis (FMECA), Home Care, Human factors.

I. INTRODUCTION

In recent years, continuous improvement approach received an increasing attention in healthcare system and in its component. To this, professionals and managers focus their attention to methods and techniques widely adopted in industry to improve quality [1]. The Institute of Medicine in USA has identified in the medical domain six dimension of quality. Among these, safety is the first point [2]. Therefore analysis of adverse events and the implementation risk assessment solicited different stakeholders in adopt quality concepts and methodologies procedure in the clinical environment [3]. One of the most used methods for risk assessment is Failure Mode, Effect and Criticality Analysis (FMECA), which, in 2001, has been adopted as the leadership standard for healthcare risk assessment by the Joint Commission on the Accreditation of Healthcare Organization (JCAHO) [4].

Furthermore, in risk identification and analysis, some authors [5;6;7] suggest to classify risks sources into areas of healthcare delivery system in which they originate. A good approach to reduce risk, according to Reason model [6] for complex systems (as healthcare system), is the definition of the contributing factors in all the areas [6].

In the healthcare system, the human factors play a key role in failure, errors and clinical outcome [3]. In the interest of simplification of the improvement actions designed after a risk analysis, it could be useful to focus attention on adverse events that are related to human and non human factors [7].

Between the services of healthcare system, the Home Care (HC) is showing a huge growth in European countries [8]. HC satisfies the necessity to assist the chronic patients at home that receive continuous assistance provided mainly by human resources (doctors, paramedics, technicians, care givers, etc.) and material ones (including medical devices) [8]. As a matter of fact, home is not a place destined and built to guarantees medical cures then new safety measures must be designed to move the global environment to comply to requirements of a complex assistance system [9].

In this work a risk assessment for a Home Care service has been applied using FMECA methodology. The failure modes resulted by the analysis, have been classified in human and non human factors according to the classification proposed by the World Health Organization (WHO) [7].

In the following, section II reports the description of FMECA methodologies and its main steps, section III describes its application to the HC service, section IV reports the methodology followed to classify failures and section V shows the results of such classification.

II. METHODS

The FMECA is one of the most extensively adopted risk analysis method in the healthcare sector [10]. Actually, it has been proposed in reducing risk in different fields of healthcare such as in blood transfusions [11], in intravenous drug infusions [12], for drug distribution systems [13] and drug prescription in hospitals [14].

The technique is a logic process and it is structured in the following steps [15] :

A. Selection of a process/service/product.
B. Collect information about process/service/product.
C. Definition of multidisciplinary team.
D. Identification of risk areas and main activities.
E. Identification of failure(s), effect(s), cause(s) and the control(s), for each activity.
F. Failure Modes scoring.
G. Definition of corrective actions.
H. Re-evaluation of residual risk.

Here a preliminary study including points A to F is considered.

III. MATERIALS

The FMECA Analysis is applied to the Home Care service following the steps previously described.

A. Selection of the Service

The analyzed Home Care service is provided by the local healthcare unit of Naples ASL NA1. ASL NA1 covers a population of one million of citizens and assists continuously in average 70 patients, mainly affected by Amyotrophic Lateral Sclerosis (ALS). The service is modeled according to regional level standards [16]. The FMECA analysis is limited to study the tasks assigned to the Biotechnology unit of ASL NA1 which are better described in point D.

B. Collection of Information

The information necessary to perform FMECA is collected performing an internal audit [17] involving experts and operators. Moreover, 73 technical reports compiled after an inspection at patient's houses are examined.

C. Definition of the Multidisciplinary Team

The team is composed by six operators with different levels and types of expertise. This aspect is very important to avoid bias errors in identifying vulnerabilities that could be overlooked by experts because considered too obvious. [18].

Team of work is composed by: the team leader, three experts of the target service and two inexperienced people. The team leader is the main expert of the home care service and he guides the analysis.

D. Identification of Risk Areas and Main Activities

The identification of risks is done aaccording to the main task of the Biotechnology Unit [16] and two principle *risk areas* are defined:

- Risks related to the *use of medical devices*, mainly life support devices (lungs ventilators, oxygen cylinders, mucus aspirators, etc.) [19;20];
- Risks related to *fire and electrical hazards*, because of the constant presence of oxygen and other inflammable products [19; 20].

E. Identification of the Failure Modes, Causes, Effects and Controls

Each activity is analyzed through a brainstorming session. The main questions that conduct the session are: "what could interfere with the execution of the examined activity?", "why?" and "what could happen? [21]. The answers were used to defined the failure modes, the causes and the effects. Each activity is correlated to a set of failure modes.

F. Failure Modes Scoring

The FMECA analysis assigns three parameters to each failure: *detectability* (D), *occurrence* (O) and *severity* (S).

Detectability (D): it measures the attitude of a failure mode to be identified by controls or inspections.
Occurrence (O): it indicates the probability of a failure occurring.
Severity (S): it indicates the gravity of the effects of a failure which affect the system or consumer.

Each parameter is scored according to specific scales, proposed by the Joint Commission [10] and adapted to the context examined and patients conditions [22].

The *detectability* rating scale:

- *impossible (10-9);*
- *low (8-7);*
- *high (6-4);*
- *sure (3-19).*

The *occurrence* rating scale:

- *frequent* (10-9), if it is expected to occur daily;
- *possible* (8-7), if it is expected to occur monthly;
- *unlikely* (6-4), if it is expected to occur annually;
- *rare* (3-1), if it is expected to occur for years.

The *severity* rating scale:

- *catastrophic (10-9);*
- *occasional (8-7);*
- *uncommon (6-4);*
- *remote (3-1).*

After the assignment of the three scores, the Risk Priority Number (RPN) is calculated.

$$RPN = D \times O \times S$$

This because the single indexes do not have meaning, whereas their product is informative [15].

The RPN attributes a weight to each failure mode under consideration. The higher is the RPN of a failure mode, greater is the risk and consequently the need to re-examine the correspondent activity.

Finally, the Master List represents the result of the FMECA analysis and it is obtained sorting in a descending order the RPNs previously calculated.

IV. FAILURE MODE CLASSIFICATION

Usually the FMECA results are classified in order to obtain extra information and defined better solutions to solve the problematic situations resulted from the analysis. In this work the failure modes are classified in two steps.

In the first step the failure modes are filtered according to a cut-off threshold for the identification of the failure modes that need to be treated immediately because their RPN score is higher than the security threshold and so need a prior intervention [23]. The threshold is a RPN value chosen according to the classification of the risks in criticality areas (Low, Medium, High and Extreme) [22]. The critical RPN is 343 that is equal to 7x7x7 and corresponding to the risks having extreme and high entity.

In the second step, the critical failure modes are classified in risk related to human factor (H) and risk related to non human factor (N-H). The latter are related to technologies' malfunctioning.

V. RESULTS

The results of FMECA analysis are presented separated for the risk areas considered: *use of medical devices* and *fire and electrical hazards*.

As previously said, the attention is focused on the failure mode with high priority.

A. Risks Related to Medical Devices

The identification step of FMECA analysis counts a total amount of 56 failure modes, 64 % are related human factors and 36% are not related to human factors (Fig. 2). High priority risks are reported in Tab. 1 where the classification in H and NH is reported as well.

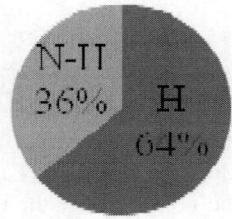

Fig. 1 Percentages of Human factors and Non Human factors related to the use of medical device

Table 1 High priority FMs related to the use of medical device

Failure Modes	RPN	Classification
Electro-Magnetic field exposure	720	H
Incorrect storage of devices	720	H
Lack of communication	640	H
Autonomous/ incorrect reparations	576	H
Reserve lung ventilator not operable	540	H
Reserve aspirator not operable	512	H
Malfunctioning of Mucus aspirator	450	N-H
Obsolete and damaged UPS	432	N-H
Liquid substances above the devices	405	H

B. Risks Related to Fire and Electrical Hazards

The identification step of FMECA analysis counts a total amount of 36 failure modes, 67% are related to human factors, while just the 33% are not related to human factors (Fig. 3). High priority risks are reported in Tab. 2 where the classification in H and N-H is reported as well.

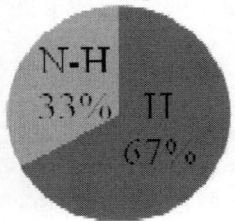

Fig.2 Percentages of Human factors and Non Human factors related to fire and electrical hazards

Table 2 High risk FMs related to fire and electrical hazards

Failure Modes	RPN	Risk areas
Improper execution of evacuation procedures	720	H
No communication of difficult situations	640	H
Autonomous and incorrect reparations	576	H
Malfunctioning of fire extinguisher	560	N-H
No safety training against electrical risks.	504	H
Absence of fire extinguishers	450	H
No knowledge about emergency procedures	448	H
Incorrect storage of oxygen cylinders	432	H
Improper usage of plugs, extension cables, adapters	432	H
Malfunctioning of fire exit emergency light	392	N-H
Extra voltage caused by metallic objects	378	H

VI. CONCLUSIONS

Figures 1 and 2, and respective tables 1 and 2, show that the most of the critical failure modes are related to human factors.

To activate suitable improvement it will be of course necessary to study how single failure is generated in the process and which and how stakeholders (i.e. medical staff, engineer, care giver, etc.) are involved.

ACKNOWLEDGMENT

This work was partially supported by the DRIVE IN^2 project, founded by the Italian P.O.N. 2007/13 and by the QUAM project, founded by Italian Ministry of Economic Development.

REFERENCES

1. Clemente F., D'Arco M., .D'Avino E., (2013) The use of a conceptual model and related indicators to evaluate quality of healthcare in ICUs, Quality Engineering, in press.
2. Barach P., (2002) Lessons from the USA, Improving Patient Safety: Insights from American, Australian and British healthcare. ECRI and Department of Health, London, UK.
3. Verbano C., Turra F., (2010) A human factors and reliability approach to clinical risk management: Evidence from Italian cases, Safety Science, 48:625-639.
4. Trucco P. and Cavallin M., (2006) A quantitative approach to clinical risk assessment: The CREA method, Safety Science, 44:491–513.
5. Cagliano A. C., Grimaldi S.,C. Rafele (2011) A systemic methodology for risk management in healthcare sector, Safety Science, 49: 695–708.
6. Reason J., (1995) Understanding adverse events: human factors., Quality in health care, 4:80–89.
7. World Health Organization (WHO), (2009) Human Factors in Patient Safety, Review of Topics and Tools.
8. Matta A., Chahed S., Sahin E., and Dallery Y., (2011) Modelling home care organizations from an operations management perspective, Flexible Services and Manufacturing Journal, 24.
9. Mestas C.A., Calil S.J., Risk Involved with medical device use in Home Care Program, IFBME Proc., vol. 25/7, World Congress on Med. Phys. & Biomed. Eng, Munich, Germany, 2009, pp. 519-522.
10. Mummolo G., Pilolli R., Pugliese D., Ranieri L., Cesare P., FMECA analysis of healthcare services: the evaluation of the efficiency of a hospital department, Proc. of the International Workshop on Applied Modeling & Simulation, 2010, pp. 93-100.
11. Burgmeier J., (2002) Failure mode and effect analysis: an application in reducing risk in blood transfusion. Joint Commission Journal on Quality Improvement 28: 331–339.
12. Apkon M., Leonard J., Probst L., DeLizio L., Vitale R., (2004) Design of a safer approach to intravenous drug infusions: failure mode effects analysis. Quality and Safety in Healthcare, 13:265–271.
13. Lyons M., Adams S., Woloshynowych M., Vincent C., (2004) Human reliability analysis in healthcare: a review of techniques. International Journal of Risk and Safety in Medicine 16 (4), pp. 223–237.
14. Saizy-Callaert S., Causse R., Thebault A., Chouaid C., (2001) Analysis of mode of failure, their effects and criticality: improving of the hospital drug prescribing process. Therapie 56: 525–531.
15. Institute for healthcare improvement, Failure modes and effects analysis (FMEA) at http://www.ihi.org/IHI/Topics/PatientSafety/MedicationSystems/Tools/ 2002
16. Regione Campania, (2011) The system of Home Care services in Campania Region (in italian) Official journal of Campania Region (BURC), 35: 77-85.
17. Emslie S.,Oliver C., Bruce J., (2006) Getting governance right at board level: the Policy Governance approach to building better National Health Service boards, Clinician in management, 14:69–77.
18. Apollo Program, (1966), Procedure for Failure Mode, Effects and Criticality Analysis (FMECA), National Aeronautic and Space Administration (NASA).
19. Radunovic A., Mitsumoto H., Leigh P.N., (2007), Clinical Care of patient with amyotrophic lateral sclerosis, Lancet Neurol, 6: 913-25.
20. Krivickas L. S., Shockley L., Mitsumoto H., (1997) Home Care of patient with amyotrophic lateral sclerosis (ALS), Journal of the Neurological Sciences, 152: S82-S89.
21. Australian/New Zealand Standard AS/NZS 4360:2004 (2004) Risk management.
22. Bowles J. B., (2012) Fundamentals of Failure Modes and Effects Analysis, Tutorial Notes Annual Reliability and Maintainability Symposium.
23. Cupryk M., (2011) Standardizing Patient Safety Risk Management, Pharmaceutical Engineering, 31:1-10.

Author: Antonio FRATINI
Institute: D.I.E.T.I., University "Federico II"
Street: via Claudio, 21
City: Naples
Country: Italy
Email: a.fratini@unina.it

Selection and Deployment of a Business Intelligence System (BI) at a Hospital's Clinical Engineering Department

C. Pérez-Martín[1], J.C. Fernández-Aldecoa[1], J. Hernández-Armas[2], and R. Cánovas-Paradell[3]

[1] Hospital Universitario de Canarias / Biomedical Engineering Department, La Laguna, Tenerife, Spain
[2] Hospital Universitario de Canarias / Medical Physics Department, La Laguna, Tenerife, Spain
[3] Hospital Universitario Vall D'Hebron / Clinical Engineering Unit, Barcelona, Spain

Abstract—This project aims to provide guidelines for the implementation of a business management process within the Clinical Engineering department of a large university hospital by using Business Intelligent (BI) platforms. The platform will be selected among leading market and free/open-source software developers following four phases: Initiate, Select-Plan, Deploy and Start-up. Among all options reviewed, the two platforms preselected for further analysis, were SAP Business Objects and Pentaho CE. The latter one was finally selected due to economic reasons. In the Deployment phase there is an analysis of the present situation, expected results and benefits of the BI implementation as well as the identification of main key and risk factors. The implementation will take place in the Biomedical Engineering Department (BED) of the Hospital Universitario de Canarias (HUC), whose mission is the management, installation and maintenance of medical equipment. In order to accomplish the BI tools development and the project implementation, there has been an analysis and redesign of procedures within the BED based on real cases.

Keywords—Business Intelligence, Enterprise Resource Planning, SAP-NetWeaver, SAP-BusinessObjects, Pentaho-CE.

I. INTRODUCTION

The term Business Intelligence (BI) groups the set of strategies, processes and technology tools that support the decision making in an organization.

BI systems combine very different software applications that provide:

- Real-time access to the company activity data analysis.
- Information about the actual state of the company, its performance and scope.
- Ability to provide information on future circumstances of the company or entity.

II. OBJECTIVES

1. To perform a comparative market analysis of BI options and select the best platform to be incorporated to the Maintenance Management Module of SAP NetWeaver, also called LOPM. SAP has been the Enterprise Resource Planning (ERP) at the Hospital Universitario de Canarias (HUC) since 2000.

2. To implement the selected system in the Biomedical Engineering Department (BED) of HUC in order to improve efficiency in decision-making and to optimize key factors such as cost reduction, risk identification and medical equipment availability rate.

By achieving these objectives, this project aims to:

- Provide precise, real time and reliable information analysis tools that increase knowledge, help decision-making and save time and expenses.
- Easily create reports in different formats (Web, PDF, Microsoft Office, etc...).

Parameters to be monitored among others are:

- Number of repair and preventive maintenance, technical labour hours.
- Number of internal and outsourced technical interventions.
- Response Rate, Uptime and Downtime.
- Spare Parts replaced.
- Medical Device Tracking – Alerts.
- Technicians involved in each intervention.
- Unitary maintenance cost.

III. METHODOLOGY

The Hospital Universitario de Canarias, located in Laguna, on the island of Tenerife, is a public institution, which depends on the Ministry of Health of the Autonomous Government of the Canary Islands. It provides medical care on the island through the hospital itself, the Mental Health External Area and the Specialized Care Centres (CAE). It is also a referral hospital for the neighbour island of La Palma. With 767 beds, 19 operating rooms and a staff of 3,520 workers, it serves a population of over 450,000 people [1].

The BED is attached to the Engineering Department included in the Hospital Management Department. Its main functions are maintenance, installation and management of medical devices. The inventory of biomedical equipment includes over 6,500 items with different levels of technical complexity. It staffs 3 engineers, 8 technicians and a secretary and in 2012 received 4,400 repair recalls and planned 450 preventive maintenance actions [2].

A. Platform Selection

Of the four phases that are considered in the methodology of a BI system [3], as shown in Figure 1, this project is focused on the selection and deployment phases.

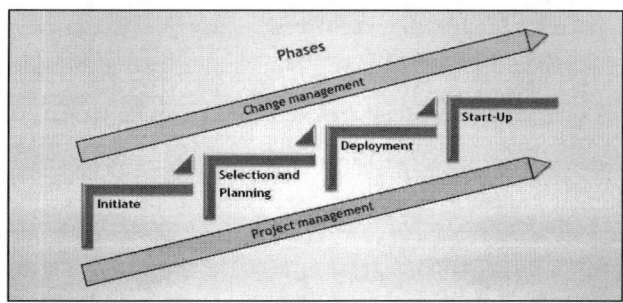

Fig.1 Methodology

In order to ensure the correct choice of platform during select-plan, milestones are set taking into account the platform functionalities and technicalities and using an iterative and filtering process.

The *"Wisdom of Crowds Business Intelligence Market Study"* (2012) by Howard Dresner [4] helps us to understand the state of the art of Business Intelligence. According to this report, main suppliers (and their platforms) are: Microsoft (SQL Server OLAP, DataMarket), IBM (Cognos 8, Cognos 10), SAP (BusinessObjects), Oracle (Oracle Business Intelligence), SAS, MicroStrategy, QlikTech (QlikView) and Information Builders (WebFOCUS) whereas the emerging suppliers are: Yellowfin, Dimensional Insight, Arcplan, Pentaho and Jaspersoft.

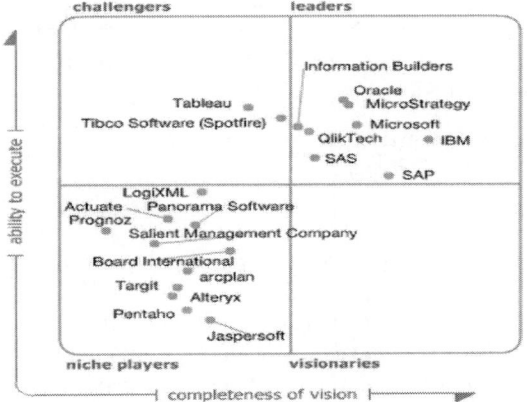

Fig 2 Magic Quadrant for Business Intelligence Platforms [5]

The selection process will help to propose the BI tools to be deployed in SED at HUC. As starting point for the platform evaluation and analysis, the *"Magic Quadrant for Business Intelligence Platforms* (2012)" report by Gartner Inc.Consulting [5] has been reviewed. Figure 2 shows the quadrant position of the different platforms according to this report.

A.1 Candidate Search and Preselection

The first selection, based on the following arguments, provided two platforms candidates:

SAP: Extensively covers the proposed requirements at this stage. Since SAP NetWeaver is running as ERP at HUC an optimal integration is expected with its applications [6]. Moreover, the Hospital is satisfied with the system, it has trained technicians for administration and development and SAP is a consolidated company that is a worldwide leader in ERP platform installations.

Pentaho: Covers the proposed requirements at this stage. It is a complete open source platform, which has a very popular free Community Edition (EC).Users expressed faith in the future of the developer. However product support is below average for most of its tools and it has shortcomings of some functionalites. Being a free platform has been a decision factor.

Other platforms were rejected [5] based on:

- *Microsoft* presents deficiencies in its functionalities, customer support and client expertise and has a complex integration.
- *IBM* presents functionality and performance problems and high licensing costs.
- *Oracle* has difficulties with the functionality and quality of support.
- *SAS* is a singular product but it is focused on predictive analysis that lacks interest for our project.
- *MicroStrategy* and *QlikTech* present high licensing and software costs.
- *Jaspersoft* presents low support evaluations and works only with low data volume.
- *Information Builders* does not offer skilled staff.

A.2 Second Selection

The platforms of the two candidates, SAP BusinessObjets and Pentaho CE, meet the requirements to achieve the objectives set in this work.

It is to remark that the user's opinion, reflected in the Gartner's report [5] it's very similar to that of the HUC's projects participants.

A.3 Final selection

The usual scoring method in public tenders for public hospitals will be used for the final platform selection.

In this method the cost represent 50% of the total score as showed in Table 1.

Table 1 Items considered and % distribution for the platforms evaluation

Functionalities	20%
Technicalities	15%
Market positioning	15%
Cost	50%

The result of the above mentioned items analysis (shown in Table 2) concludes that SAP BusinessObjects scored 9 points above Penthao CE without taking cost into account. However the economic factor clearly favours Penthao CE since it has no software /license cost.

Table 2 Different platforms scoring

Items	%	SAP	Pentaho
Functionalities	20	15,7	13,9
Technicalities	15	11,9	8,8
Market positioning	15	12,2	8,2
Total (without cost):	50	**39,9**	**30,9**

Since the two platforms meet the basic requirements and due to the current economic scenario with limited financial resources, the Pentaho CE solution is finally selected. Nevertheless it is clear that license and software cost saving results in the lowering the level of excellence in both technical and functionality. The selected applications within this platform are:

- Pentaho Reporting [7].
- Pentaho Data Integration (Kettle) [8].
- Pentaho BI Server.
- Pentaho Data Mining (Weka) [9].

IV. RESULTS AND DISCUSSION

A functional analysis of the Biomedical Engineering Department (BED) is preformed in order to describe present situation and to set up the expectations for BI system deployment.

Nowadays analysis and decision making tools are a series of Excel files extracted from the Hospital ERP (e.g. work orders, equipment records, entry and exit of equipment...).

The monitoring of technical collaboration contracts with outsourced companies and corrective/preventive maintenance are the two main processes that make it crucial for BED to have a BI system. Other activities, such as medical device management, are currently supported by hospital information system and therefore BI tools are neither essential nor a priority.

A. Present Situation

A.1 Technical Collaboration Contracts

In HUC contracts for high-tech medical equipment are not at all-risk. Maintenance contracts signed with the manufacturers are in a collaborative contract formula, where the workload is shared between technicians of BED and outsource companies with different responsibilities and intervention levels. Nowadays BED technician's activity and technical data are managed with in ERP (SAP NetWeaver) whereas outsource company's interventions are registered in Excel files and paper support. Evaluation data for monitoring the contracts need to be prepared in an Excel file manually.

A.2 Corrective Maintenance

It is the core business of the BED and therefore it requires a fluid and updated information system for control and monitoring. LOPM (the Maintenance Management Module of SAP NetWeaver current ERP) allows many information extractions to excel files filtered by different technical parameters such as: devices, technicians and users.

A.3 Preventive Maintenance

Also one of the central processes in the activity of BED, it ensures the quality of medical devices operability and helps to reduce the number of corrective interventions. The scheduled maintenance is planned through the ERP (SAP NetWeaver). To monitor this activity, it is necessary to search and filter in SAP and export to excel files.

B. Expected Results after BI Implementation

The aim of the whole project is to have a single datawarehouse that provides updated and reliable information by integrating current information sources and by using dynamic tools for advanced data analysis. In addition its goal it is to avoid time-consuming processes for the generation of reports (either pdf, excel or text files) and help monitoring and decision making.

C. Analysis of Key and Risk Factors and BI Implementation's Benefits

C.1 Key Factors

- Higher efficiency and cost reduction in collaboration contracts.
- Better control and management of preventive and corrective maintenance associated with downtime reduction.
- Better knowledge of the service's activity and quick detection of technically conflictive devices.
- Increase the profitability due to on time and continuous, decision making.

C.2 Risk Factors

- Poor adaptation to the new system due to cultural impact and change resistance.
- Technical incidents due to integration or compatibility problems.
- Cost Increase resulting from a poor implementation or a wrong platform choice.

C.3 Identification of Benefits

- Easier and safer decision making.
- Reduction in information dispersal.
- Better treatment and monitoring of medical device incidents and alerts.
- Improvement of down/response/repair time for the benefit of patient care.
- Control and supervision of outsourced companies.
- Cost reduction of outsourced contracts.
- Risk control by predictive maintenance.
- Maintenance costs reduction (a 2% reduction of the total annual cost is to be expected).
- Cost reduction associated with patient care activities due to breakdowns.
- Satisfaction increase of staff and patients.
- SEB's efficiency improvement.
- HUC managers' improved perception of quality and image of BED.

D. Project Deployment

The project's success requires a methodology that takes into account the business know-how and its infrastructure by relying on the experienced staff. Therefore interdisciplinary workgroups with IT and BED staff will be implemented. In order to involve all stakeholders, a change management plan has been developed that includes soft-type actions such as communication strategies and training but does not consider hard-type actions related to jobs or reimbursement.

Performance tests with all integrated applications have taken place. Proratio supplied the plug-in [10] to test the connection to SAP for 30 days, which was installed on Kettle for conversion work. Further test with its own tools were conducted for other applications with excellent results.

V. CONCLUSIONS

This project is framed within the Biomedical Engineering Department (BED) of the Hospital Universitario de Canarias (HUC). The implementation of a BI platform aims to cover the requirements of knowledge for decision-makers in order to improve the efficiency, quality and safety of processes carried out by this department.

Pentaho CE covers the proposed objectives. Its functionality and results are really satisfactory since it generates useful knowledge that allows the evaluation of SEB's processes.

On the wrong side, it is worth mentioning that certain graphic presentations and designs of the application suffer from an excess of simplicity. Therefore the IT department released an in-house development upgrade, using standard design tools, to improve this deficiency. Notice that the ETL processes are somehow complex and require long development and testing.

Nevertheless the result of this study is that Pentaho CE is a fine solution that expands the present utilities of ERP (SAP NetWeaver) by using a full BI platform that generates knowledge. It provides an instant, reliable and unambiguous query platform in a useful way (e.g. automatic reports, alarms and flags). Concerning its integration capacity it fully integrates data sources from current applications (SAP and several applications from outsourced companies) and it is foreseen that its future developments might include other applications in which BED might also be interested.

REFERENCES

1. Servicio Canario de Salud. Gobierno de Canarias. *Memoria Asistencial y Científica 2011. Complejo Hospitalario Universitario de Canarias* (2012). http://www2.gobiernodecanarias.org/sanidad/scs/content/92ea95a1-49bd-11e2-a5ba-c162420e469e/MemoriaAC2011.pdf
2. Hospital Universitario de Canarias. Datos de actividad del Servicio de Ingeniería Biomédica. *Informe de revisión por la Dirección 2012 para la Certificación de Calidad ISO 9001:2008* (2012).
3. Guill H, Guitart I, Joana JM, Rodríguez JR. *Fundamentos de sistemas de información* (2011). http://blog.interweb-consulting.es/biblioteca/fundamentos-sistemas-informacion.pdf
4. Dresner H. *Wisdom of Crowds Business Intelligence Market Study* (2012). http://www.yellowfinbi.com/YFCommunityNews-Download-Dresner-s-2012-Wisdom-of-Crowds-Business-Intelligence-Market-Study-114810
5. Hagerty J, Sallam RL, Richardson J. *Magic Quadrant for Business Intelligence Platforms* (2012). http://www.gartner.com/technology/reprints.do?id=1-196VVFJ &ct=120207
6. SAP. *Soluciones de BI de SAP BusinessObjects* (2012). http://www.sap.com/spain/solutions/sapbusinessobjects/large/business-intelligence/index.epx
7. Pentaho Community. *Pentaho Reporting Community Documentation* (2011). http://wiki.pentaho.com/display/Reporting/Report+Designer
8. Pentaho Community. *Latest Pentaho Data Integration Documentation* (2012). http://wiki.pentaho.com/display/EAI/Latest+Pentaho+Data+Integration+%28aka+Kettle%29+Documentation
9. Weka. *Data Mining Software in Java*. http://www.cs.waikato.ac.nz/ml/weka/
10. Proratio. *ProERPconn Quick Start and Reference Guide* (2012). http://www.proratio.de/systeme-und-produkte/proerpconn/

Health Technology Assessment Applied to Health Technology Management through Clinical Engineering

F.A. Santos[1], A.E. Margotti[1], F.B. Ferreira[2], and R. Garcia[1]

[1] Institute of Biomedical Engineering IEB-UFSC, Federal University of Santa Catarina, Florianopolis, Brazil
[2] Faculty of Science and Technology FCT-UNL, Universidade Nova de Lisboa, Lisbon, Portugal

Abstract— Clinical Engineering promotes actions to strengthen health technology management (HTM), considering the specificities of each phase of the technologies' life-cycle. This study presents the HTM model associated with other tools and methodologies, in particular Health Technology Assessment (HTA). The model is supported by a base of three domains infrastructure, technology, and human resources to deliver consistent parameters for the evaluation of each phase of the medical equipment life-cycle. From this evaluation, some key impacts can be analyzed and then used to support the health managers in decision-making process. A case study was conducted with the application of this model. The application focused on the incorporation process of the robotic surgery system and this contributed to recommendations to health managers, allowing more quality to the technological processes in health.

Keywords—Clinical Engineering, Health Technology Management, Health Technology Assessment, Medical Equipment life-cycle.

I. INTRODUCTION

According to World Health Organization (WHO), medical devices are essential components within the context of current healthcare systems. To be suitably used, particularly in the case of medical equipment, it is necessary that the equipment matches the health problem to maximize its effectiveness. However, systems and providers of health services are not reaching the entire panorama. Some factors that contribute to this situation are lack of needs and resources assessment, lack of information for development and maintenance, as well as untrained health professionals. These issues are embedded in a broader problem, which is still found in some countries - the lack of a management system for medical equipment [1].

In this sense, Clinical Engineering seeks to expand its strengthening, considering the perspective that technology management is the essence of its performance through assessment activities, strategic planning, acquisition, use, maintenance, control, planning, and quality assurance replacement towards medical equipment [2].

The model of Health Technology Management (HTM), which states that all Clinical Engineering actions are based on three domains, namely infrastructure, human resources, and technology, has supported many activities related to equipment management developed by IEB-UFSC (Brazil).

This model allows clinical engineers to improve the quality of the technological process in health, considering that it is not only an analysis related to technology (maintenance, acquisition, disposal, among other activities), but also an analysis of the aspects related to infrastructure and human resources required for the proper use of medical equipment. Thus, Clinical Engineering acts in a holistic and integrated way with the needs of health systems.

This study presents the context of the Clinical Engineering performance, specifically, its HTM model. In this model tools have been included to serve the needs of the technological process, as HTA.

II. MATERIALS AND METHODS

Through academic researches and experiences acquired by the action of the IEB-UFSC Clinical Engineering in healthcare establishments, Clinical Engineering strengthens and promotes healthcare technology management, focusing on medical equipment.

With respect to the potentials and challenges associated with the management process of technologies, it is necessary to systematize the structures, methodologies, and observed impacts. For this purpose, we have elaborated a model of Clinical Engineering action, focused on healthcare technologies management, as shown in Fig 1.

The proposed model is supported by a base constituted by three domains: infrastructure, technology, and human resources. All actions motivated by HTM should be undertaken to consider the domains. This base, in association with the study of medical equipment's life-cycle, determines the entire HTM model. Thus, it is ensured that the equipment, regardless of its life-cycle phase, is analyzed by means of technological parameters, by the quantity and quality of human resources, and also by the whole group of prerequisites about the infrastructure necessary for proper equipment use. This broad view allows equipment to be available for

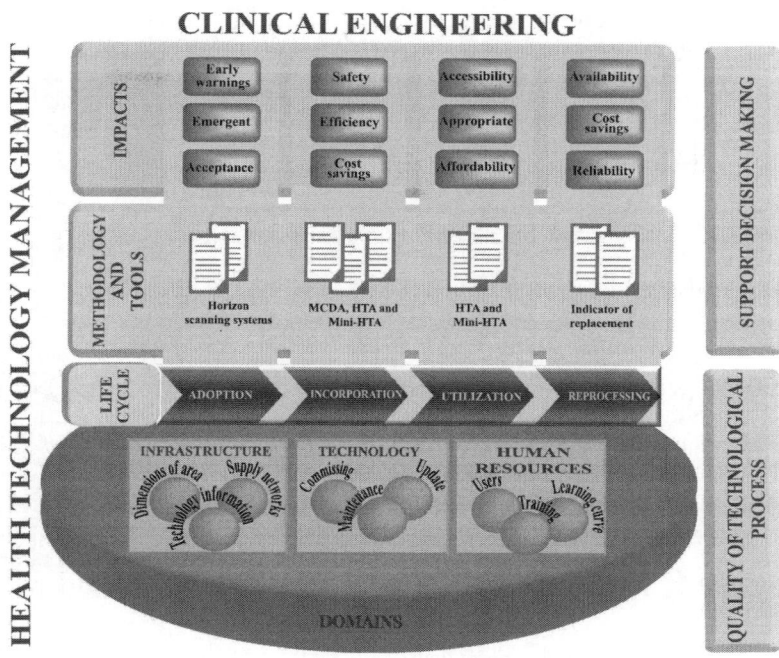

Fig. 1 Model of clinical engineering performance based on HTM, where three domains are considered: infrastructure, technology, and human resources. These domains support all actions in the life-cycle as well as the methodologies and tools to obtain the projected impacts.

operation in reliable, safe, and efficient conditions, characterizing the technological process quality.

Several requirements of analysis about equipment come from its current life-cycle phase, from adoption to reprocessing. Thus, within the HTM, the analysis of medical equipment life-cycle is crucial for a proper contribution to health managers. This allows mapping of the real conditions of each device inserted in the technology park. In this way, actions can be planned for the appropriate use in short and medium term.

To allow the implementation of integrated actions with the medical equipment life-cycle, methodologies and tools from other areas of knowledge should be applied according to the specificities of each life-cycle phase. These methodologies, consolidated into other areas, are identified in the literature and transferred to the Clinical Engineering area, according to the necessity. Thus, new approaches are developed, as the HTA approach, widely applied to drugs and adapted to equipment.

HTA is characterized as a comprehensive assessment, where clinical, technical, operational, and economic impacts, among others, of health technologies are measured [3]. This tool is intended to assist the decision-making process in healthcare. Therefore, Clinical Engineering incorporated the HTA as a tool in its field of performance. In other terms, the HTA rationality and systematization is being used for discerning evaluation of medical equipment, considering the inherent characteristics of each life-cycle phase.

However, as the study of Clinical Engineering is based on the medical equipment life-cycle, the HTA tool was adapted for its utilization in each stage, because the criteria that should be investigated are different. Particularly, in the incorporation phase, it is essential to assess the impacts on safety, efficacy, and accuracy, on the installation parameters, functional characteristics, and performance tests before the equipment is available for use. These impacts should be investigated in a specific HTA model for this phase. Regarding the equipment assessment in the utilization phase, the essential parameters are maintenance, effectiveness, attendance of technical and clinical needs, among others. Thus, some approaches such as the Multi-Criteria Decision Aid (MCDA), in conjunction with the HTA and its byproducts such as MINI-HTA, generated new methodologies for Clinical Engineering.

According to this HTM model, the final answer is the accomplishment of quality in the technological process in healthcare. Consequently, the application of different methods and tools has significant impacts that are transferred to the generation of security, reliability, and efficiency.

The proposed model was applied to the assessment of a robotic surgery system used to perform prostatectomy. Thus, to verify a possible incorporation of the equipment in a network of hospitals, we opted for the assessment of some impacts through the MINI-HTA methodology.

III. RESULTS

Initially, we checked the registration of equipment with the Brazilian Regulatory Agency (ANVISA). It was found that only one device was registered-da Vinci Robotic Surgical System.

In the assessment of the technology domain, the operation of the da Vinci robotic surgery must also be checked. This consists of three main components: the surgeon console, a patient-side cart, and a vision system. The surgeon's console is the control center that promotes an interface between the surgeon and surgical robotic arms. The patient-side cart, where the instruments and endoscope are stored, supports the robotic arms. The vision system allocates the equipment to process the images, perfecting the 3D images of the operative field [4].

In the context of assessing equipment incorporation, it was observed that the equipment could be added to the technology park already existing in the hospitals network; hence, it may not replace other technology.

Another aspect of the technology domain is related to the needs of cleaning, maintenance, and disposal of equipment and accessories. It was found that the system requires preventive maintenance every 3 months, and this maintenance, as well as the corrective maintenance, must be performed by a specialist supplier. There is no protocol for different disposals of the system and its accessories.

The assessment of human resource domain revealed that a team of specialized professionals is required for the proper use of the equipment. These professionals are surgeons who move the system through the console, auxiliary nurses who operate the patient-side cart, and technicians of the clinical engineering team who can operate the vision system. The team should be widely knowledgeable. Training should be conducted in international centers linked to the manufacturer. A plan for continuous training is needed, during which the team performs training *in loco*, in hospitals, to upgrade their learning skills. In the da Vinci robotic surgery system utilization for prostatectomy procedures, the learning curve is established between the cases 40th and 60th [5].

For this equipment to be incorporated into a hospital, it must respect the standards regarding infrastructure. Initially, a special room is needed to accommodate all system elements, whose size can range from 37 to $70m^2$. In the room, there must be a network of computers with an internet connection, and four outlets with an independent circuit.

Table 1 Capital and Operating Costs of the da Vinci Surgical System, in Canadian Dollars [7]

Item	Year 1	Year 2	.....	Year 7
da Vinci Si Surgical System	2,643,680.00	-		-
Start-up reusable equipment and accessories	203,360.00	-		-
Disposables and consumables	330,460.00	330,460.00		330,460.00
Training of surgeons	-	6,101.00		6,101.00
Annual maintenance	-	177,940.00		177,940.00
Total annual cost	3,177,500.00	514,501.00		514,501.00
Cumulative annual cost	3,177,500.00	3,692,001.00		6,264,506.00

Furthermore, in this room, certain environmental conditions must be secured and guaranteed by an air conditioning system (Temperature:10 to 35^oC, Humidity: $10 - 85\%$ non-condensing).

According to the available evidence related to the economic assessment, the costs of the system were quantified in its first year of use, and from the second to the seventh year, as the equipment's useful life was estimated to be 7 years [6]. In Table 1, it can be seen that the capital and operating costs of the system, over 7 years, exceed $C\$ 6$ million.

In the economic domain, it was possible to evaluate the effectiveness and quality-adjusted life years (QALYs). Table 2 provides a comparative study of economic evaluation among prostatectomy performed through open surgery and robotic system.

The studies compiled in Table 2 show a residual advantage for radical prostatectomy assisted by robotic surgery system, in terms of effectiveness and QALYs, when compared with open radical prostatectomy. The costs per unit of QALYs are higher in prostatectomy assisted by the equipment. It is important to note that the studies found in the literature are of poor quality, characterized by heterogeneous populations, biases, and other confounders. Furthermore, the authors have not clearly described the methods adopted for estimation.

From the above-mentioned points, the emergence of new evidence provided by studies with methodological rigor in formulating recommendations can be noted.

According to the evidence assessed, the benefits of equipment incorporation do not outweigh the difficulties, when used for prostectomy; i.e., the general recommendation is low for the equipment. However, in studies to be undertaken in the future, this recommendation can be considered with the recommendations for the use of robotic surgery in other surgeries.

Table 2 Economic evaluation studies comparing radical prostatectomy through robotic surgery system (RPRS) and open radical prostatectomy (ORP).

Study	Population	Results
[8]	ORP (n=100) RPRS (n=500)	RPRS: 0.093 of QUALY increment and $24,475.43/QUALY
[9]	ORP (42 studies; n=132,402) Laparoscopic (34 studies; n=19,324) RPRS (22 studies; n=6,819)	RPRS: 0.16 of QUALY increment and $1,740/QUALY
[10]	n=231 (50-69 years-old)	RPRS: 64.34 of QUALY increment, efectiveness of 7%

IV. CONCLUSION

The model developed aims to strengthen HTM through its integration with other methodologies such as HTA to generate impacts at each stage of the medical equipment life-cycle.

The characterization of each phase of the life-cycle (adoption, incorporation, utilization, and reprocessing/ obsolescence) through the use of evidence to support the decision-making process provides clinical engineers and health managers a better position to face the challenges raised by proper technologies management.

Through the case study conducted, a recommendation was generated to support the decision-making process regarding the incorporation of robotic surgical system for prostatectomies. This was possible through the assessment of relevant multicriteria about the medical equipment incorporation process. The assessment was intended to prevent unsafe and inefficient technologies with low growth in cost-utility terms from being available in healthcare services. Thus, the selection of technological alternatives could be carried out according to the clinical and technical needs given the budget constraints.

REFERENCES

1. WHO. *Medical Devices: Managing the Mismatch. An outcome of Priority Medical Devices Project*. Switzerland: WHO Library Cataloguinin- Publication Data 2010.
2. Dyro J. *Clinical Engineering Handbook*. USA: Elsevier Academic Press 2004.
3. Brasil Ministerio da Saude. *Directrizes Metodologicas: Elaboracao de estudos para avaliacao de equipamentos medico-assistenciais*. Brazil: Editora MS 2013.
4. Intuitive surgical at http://www.intuitivesurgical.com/
5. F V Novara et al. Retropubic, laparoscopic, and robot-assisted radical prostectomy: A systematic review and cumulative analysis of comparative studies *Eur Urol*. 2009;9;55:1037-1063.
6. AHA. *Estimated Usefuk Lives of Depreciable Hospital Assets*. Washington: AHA Health Data Management Group 2008.
7. C Ho et al. *Robot-Assisted Surgery Compared with Open surgery and Laparoscopic Surgery: Clinical Efectiveness and Economic Analyses*. Ottawa: Canadian Agency for Drugs and Technology in Health 2011.
8. S P O'Malley E Jordan. Review of a decision by the medical services advisory comitte based on health technology assessment of an emerging technology: the case for remotely assisted radical prostectomy *Int J Technol Acess Health Care*. 2007;23;2:286-91.
9. D A Ollendorf et al. Active surveillance and radical prostectomy for the management of low-risk, clinically localized prostate cancer *Institute for Clinical and Economic Review*. 2009.
10. L Hohwu et al. A short-term cost-effectiveness study comparing robot-assisted laparoscopic and open retropubic radical prostectomy *J Med Econ*. 2011;23;14;4:403-9.

Author: Francisco de Assis Souza dos Santos
Institute: Institute of Biomedical Engineering IEB-UFSC
Street: Campus Universitario
City: Florianopolis
Country: Brazil
Email: [franciscoassis, ana.morgotti, renato]@ieb.ufsc.br ;
fb.ferreira@campus.fct.unl.pt

Biomedical Engineering at the Centre for Preclinical Research and Technology

T. Pałko, N. Golnik, and K. Pęczalski

Warsaw University of Technology/Institute of Metrology and Biomedical Engineering, Warsaw, Poland

Abstract—In the article, the research investment and scientific areas at the new Centre for Preclinical Research and Technology (CePT), located in Warsaw, Poland are presented. Also the instrumentation and research program related to biomedical engineering field has been described.

Keywords—preclinical research, medical technology biomedical engineering, medical physics.

I. Introduction

Contemporary medicine is challenged to seek methods for delaying and alleviating undesirable and pathological disturbances of the ageing processes, which translate into researching structural and functional properties of proteins and developing possible pharmacological therapies. The existing research centres are not always capable of meeting new challenges in this area

Research groups collaborating within the Centre for Preclinical Research and Technology (CePT), which was formed at Research Campus Ochota in Warsaw, will strengthen this area of scientific exploration in Poland.

II. Centre for Preclinical Research and Technology (CEPT) in Warsaw

The Centre for Preclinical Research and Technology (CePT) in Warsaw is a scientific investment project undertaken by consortium of the three Polish Universities and seven research institutes with support of European Commission. These are the following universities: Medical University of Warsaw (WUM) being the coordinator, University of Warsaw (UW) and Warsaw University of Technology (WUT). The research institutes of Polish Academy o Sciences are: Nencki Institute of Experimental Biology, Institute of Biochemistry and Biophysics, Mirosław Mossakowski Research Centre, International Institute of Molecular and Cell Biology, Institute of Fundamental Technological Research, Institute of High Pressure Physics and Nałęcz Institute of Biocybernetics and Biomedical Engineering.

The CePT consists of 10 smaller specialized centres for: preclinical research, experimental medicine, neurobiology, functional and structural analysis of proteins, molecular biotechnology, research on physico – chemical properties of biomaterials and systems, particle acceleration and positron emission tomography, large-scale computing and biomedical data processing, bionanomaterials, biomedical technology and medical physics. In each Centre there are several specialized laboratories.

The Warsaw University of Technology (coordinator - prof. Tadeusz Pałko) is responsible for biomedical engineering and technological problems in the CePT.

Eight Faculties of WUT take part in the development of these problems: Chemical and Process Engineering, Chemistry, Electrical Engineering, Electronics and Information Technology, Mechatronics, Material Science and Engineering, Power and Aeronautical Engineering and Production Engineering. The program is realized in two centers: Centre for Bionanomaterials and Centre for Biomedical Technology and Medical Physics.

III. Research areas in the CEPT

The large part of the resources will be devoted to preclinical research in therapy of neurodegenerative diseases, which is of particular importance because of societal and economic impact. The cost of treatment for such diseases accounts for 30% of all health protection expenses in EU countries and increases due to ageing of the society and new possibilities offered by modern medicine.

Another important and prestigious area of research is exploration of interdependencies between the human genome, development of diseases and oncogenesis of newgrowths constitutes. Centre for Preclinical Research will serve to improve methods of diagnosis and experimental treatment in this area.

The centre will also focus on research of the cardiovascular system, an area, in which institutions from the Ochota Research Campus play a leading role on the national level. Research of new medications will be extended to link activities of the Ochota Research Campus with market activities of innovative pharmaceutical companies. Studies of proteins in silico (bioinformatics and computer modelling) will facilitate a systemic research approach within this area of study.

Improving the technology of preparation of bionanomaterials determines the basic direction in development of nanotechnology and its medical applications (e.g. peptides and proteins, which possess a richer structure and are cheaper than DNA).

The Centre will also carry out research focusing on designing complex structure nanomolecules with a functional layer of peptides, which serve as vehicles for medicines directed to specific areas of the organism, while minimizing their harmful effect on other organs.

Research areas in the CePT were divided into three parts:

Area 1: Fundamental research on pathogenesis of the most important diseases associated with the progress of civilization.
Problems:
a) Research on the nervous system diseases (neurodegenerative diseases, plasticity of the nervous system and neuro-regeneration);
b) Research on cardiovascular diseases;
c) Cancer;
d) New technologies supporting scientific discoveries and their application to the study of physiology.

Area 2: New diagnostic methods.
Problems:
a) Chemical, physical and biological markers for biomedical diagnostics of cardiovascular diseases, diseases of the nervous system, cancer and tissue damage;
b) New methods for studying and imaging the properties of cells, tissues and organs in cardiovascular diseases, diseases of the nervous system, cancer and tissue damage ;
c) New methods for programming and modelling biological processes within the cardiovascular system, nervous system and tissue damage.

Area 3: New therapeutic methods.
Problems:
a) Research on new biologically active compounds in the context of new treatment methods for cardiovascular diseases, diseases of the nervous system, cancer and tissue damage;
b) New biotechnologies based on genomics, molecular biology and microbiology for medicine and environmental protection;
c) Design, synthesis and characterization of new functional materials applied in medicine to achieve therapeutic change of the properties of tissues within organs of the cardiovascular system, the nervous system and other systems.

The structure of the programme bases on common research fields, in which scientists from all the units will carry out their projects. The potential danger of fragmenting the research programme and dividing it into ten fields assigned to individual CePT research units was thus avoided.

The Research Programme is therefore a tool building and strengthening the CePT project synergy.

IV. LABORATORIES, INSTRUMENTATION AND RESEARCH PROGRAM ON BIOMEDICAL ENGINEERING

The main biomedical engineering program of CePT is realized in nine laboratories of the Centre for Biomedical Technology and Medical Physics (BIOFIM) and in three laboratories of the Centre for Bionanomaterials.

The first Centre is organized by Warsaw University of Technology and consists of the following laboratories:
Biomaterials Design and Production Techniques,

- Artificial Locomotion Organs,
- Biomechanics and Rehabilitation Engineering,
- Artificial Organs and Modelling of Metabolic Processes,
- Computer-Aided Medical Diagnostics,
- Hybrid Modelling of the Circulatory and Respiratory Systems,
- Sensors, Ultrasonic Methods and Weak Biological Signals,
- Miniature Analytical Systems,
- Electronics, Medical Informatics and Bioinformatics,
- Radiological Equipment, Dosimetry and Nanodosimetry.

BIOFIM Centre forms a unique complex of core laboratories facilitating effective cooperation of research teams in medical engineering and physics and their close cooperation with leading medical clinics. The Centre is equipped with instrumentation unique in European scale and consolidates research in the following areas:

- New materials for medicine, including nanomaterials,
- Electronic medical equipment and testing,
- Medical signal and image analysis,
- Medical imaging techniques,
- Hybrid modelling of circulatory and respiratory systems and functions of other organs,
- Biomechanics and equipment for implants,
- Medical sensors,
- Micro analysis systems „Lab-on-a-Chip" for medical diagnosis.

Centre for Bionanomaterials consists of three laboratories:

- Biomaterials Production and Characterization, Laboratory organized by Faculty of Materials Science and Engineering of WUT,
- Laboratory of Modelling and Imaging in Biomechanics located in Institute of Fundamental Technological Research PAS (Polish Academy of Sciences)

Laboratory of the Nanostructures for Photonics and Medicine located in Institute of High Pressure Physics PAS.

In biomedical engineering the above mentioned laboratories of instrumentation have equipment for 10 groups of apparatus arrangements:

1. Instrumentation for CVD and plasma technological processing,
2. Stands for biomechanical studies of human joints and simulation of the car accidents,
3. Fluorescent microscopy and capillar electrophoresis,
4. MRI tomograph,
5. Stand for examining the quality of ionizing radiation applications,
6. Shielding room and instrumentation for measurement of weak electrophysiological signals,
7. Special ultrasonograph for investigation of tissue properties,
8. Hybrid simulators for modeling of circulatory and respiratory systems,
9. Computer systems for signal processing and imaging,
10. Equipment for material characterizations and tissue engineering.

The research program for biomedical engineering and medical physics includes four main topic areas:

- materials used in implants and tissue regeneration,
- methods to examine the properties of tissues,
- instrument – aided methods for altering the properties and state of tissues,
- biological process modeling and the production and evaluation of artificial tissues and organs.

The Centre for Biomedical Technology and Medical Physics will comprise a network of open laboratories forming the base for research and deployment projects in biomedical engineering and biophysics. The center will conduct research on new nanomaterials for medicine, devices used in medical examination, methods of signal processing, medical imaging methods and techniques dosimetry, nanodosimetry and quality review of radiological devices, new sensors and measurement methods, biomechanics and prostheses and miniature "lab-on-a-chip" analytical systems. The center will also examine artificial organs and create models of the neural, cardiovascular and respiratory systems.

The Centre for Bio-Nanomaterials will bring together laboratories working on the medical applications of advanced materials and nanotechnology, especially in the area of tissue regeneration. It will research the applications of state-of-the-art ultrasonography, develop models of interaction between tissues, biomaterials and physical signals and methods of microflux research and modeling, seeking to use achievements in nanotechnology to support research, development and deployment projects. The center will also develop biomaterials and nanomedicine in line with the Strategic Research Areas of the European Technology Platform on Nanomedicine and the Polish Nanotechnology Platform.

The Biomechanical Modeling and Imaging Laboratory will study applications of ultrasounds in medical imaging. Research will include therapeutic contrast and thermal effects connected with heat production and cavitation. The laboratory will also model and study processes occurring in tissues as a result of interaction with biomaterials and physical/mechanical factors. Moreover, the lab will study the use of nanofibers in localized drug delivery systems.

The Laboratory of Nanostructures for Photonics and Medical Diagnostics will produce third-generation nanoparticles, or nanostructures that enable the identification and neutralization of pathogens. The lab will also produce functional implants and nanoparticles with antibacterial properties.

The Biomaterials Production and Characterization Laboratory will produce growth media for tissues, in particular bone tissue.

V. CONCLUSIONS

A unique feature in the development strategy of the Centre for Preclinical Research and Technology, nationally and regionally, is its interdisciplinary approach to scientific and technological problems in the areas important to society and economy. The field of activities includes diseases of the central nervous system, civilization- and ageing-related diseases, new growth tumours, degenerative changes within the cardiovascular system as well as arterial hypertension. Biomedical research is combined with research of physicochemical properties and bio-nano-materials, supported by state of the art information technologies.

We can expect that in the nearest future CePT will become a modern platform for medicine oriented research.

Results of technological research will be applied in preclinical studies of instrumental methods for:

- early detection of structural or functional changes on cell and organ level,
- treatment;
- support of organ functions.

CEPT is also market oriented. There is a well recognized need to have a harmonized vision and appropriate instruments to support a continuum of actions from "idea-to-market" for research labs collaborating within the Centre, in order to increase the success of innovative products.

ACKNOWLEDGMENT

The work was partially supported by the National Centre for Research and Development within the project WND-POIG. 02.02.00 – 14 – 024/08.

REFERENCE

1. www.cept.wum.edu.pl

Author: Tadeusz Pałko
Institute: Institute of Metrology and Biomedical Engineering, Warsaw University of Technology
Street: Sw A. Boboli 8
City: 02-525 Warsaw
Country: Poland
Email: T.Palko@mchtr.pw.edu.pl

'E-HCM' Project: Cost Calculation Model for Surgical Procedures

P. Perger[1], M. Buccioli[2], V. Agnoletti[3], E. Padovani[4], and G. Gambale[5]

[1] Research Assistant, University of Bologna
[2] Data Manager, Forlì Hospital
[3] Anesthetists, Forlì Hospital
[4] Professor of Business Administration, University of Bologna
[5] Head of Emergency Department and Anesthesia and ICU, Forlì Hospital

Abstract—The economic and financial crisis has also had an important impact on the healthcare sector. Available resources have decreased, while at the same time costs as well as demand for healthcare services are on the rise (Mladovsky et al., 2012; Oduncu, 2012). This coalescing negative impact on availability of healthcare resources is exacerbated even further by a widespread ignorance of management accounting matters. Little knowledge about costs is a strong source of costs augmentation. Although it is broadly recognized that cost accounting has a positive impact on healthcare organizations, it is not widespread adopted (Kaplan/Porter, 2011). Hospitals are essential components in providing overall healthcare (McKee/Healy, 2002). Operating rooms are critical hospital units not only in patient safety terms but also in expenditure terms (Guerriero/Guido, 2011; Maryamaa/Kirvela 2007). Understanding OR procedures in the hospital provides important information about how health care resources are used (Elixhauser/Roxanne, 2007). There have been several scientific studies on management accounting in healthcare environments (see for example van Rensburg/Jassat 2011; Lievens/Van den Bogaert/Kesteloot 2003; Chan 1993) and more than ever there is a need for innovation, particularly by connecting business administration research findings to modern IT tools. IT adoption constitutes one of the most important innovation fields within the healthcare sector, with beneficial effects on the decision making processes (Tan, 2005). The e-HCM (e-Healthcare Cost Management) project consists of a cost calculation model which is applicable to Business Intelligence. The cost calculation approach comprises elements from both traditional cost accounting and activity-based costing. Direct costs for all surgical procedures can be calculated through a seven step implementation process.

Keywords—Cost Accounting, Cost Control, Information Technology, Hospital Management.

I. INTRODUCTION

Healthcare Sector Financial Difficulties. Healthcare organizations are facing a paradoxical situation where resources are decreasing just as and costs and demand are increasing. A decrease in the availability of resources can be attributed primarily to negative effects of the economic and financial crisis (De Belvis et al., 2012). There has also been an increase in demand for healthcare services due to demographic ageing and a rise in public expectations (demand side changes). Also costs for purchased goods and services within the healthcare sector have increased (supply side changes) mainly due to introduction of new technology (McKee et al., 2002; Rechel et al., 2009). 'The importance of hospitals in overall healthcare budgets makes them the obvious targets for governments trying to cap public expenditure or to slow the rate of growth' (McKee et al., 2002). Achieving financial balance and sustainability is increasingly challenging.

Cost Accounting in Healthcare Organizazions. 'Cost is increasingly recognized as an important dimension in providing clinical care. In the current climate requiring fiscal restraint, it is becoming more important for hospital managers to assess the costs of services and clinical procedures provided' (van Rensburg/Jassat, 2009). Little knowledge about costs is a strong source of costs augmentation. Although it is broadly recognized that cost accounting has a positive impact on healthcare organizations, it is not widespread adopted (Kaplan/Porter, 2011). Accurate Management Accounting information have the power to introduce the following advantages to healthcare organizations: Improve pricing decisions or if reimbursement is set on basis of agreements with other organizations or institutions important negotiation advantages; better profitability assessments and resource allocation; possibility to encourage/discourage use of services; comparative analyses; income and asset management for external parties; budgeting and resource control improvements, positive changes of the decision making process, healthcare professional's education toward cost matters; department level cost determination and the introduction of responsibility accounting (Baker, 1998; Finkler et al., 2007; Young 2008; Macario 2010).

E-Healthcare: Information Technology in Healthcare. IT provides interesting possibilities to improve healthcare services processes and overall management. Hospitals have been forced to down-size their activity, decreasing personnel and reducing hospital beds. This is due to the government's decreased capabilities and willingness to spend enormous quantities of money for health care service and delivery. Healthcare providers had to search for new alliances and modalities to be able to cope with this difficult situation. IT adoption has played and continues to play a key role within these changes, offering interesting opportunities to save costs by increasing efficiency and efficacy and maintaining a high level of overall service quality. Especially as important tools to sustain decision making processes, IT has an fundamental role (Tan, 2005). The e-HCM approach wants to combine the positive aspects of Management Accounting with modern IT by implementation of a Cost Accounting Model on Business Intelligence.

II. MATERIALS AND METHODS

The Cost Accounting Model. E-HCM (e-HCM) project has the aim to calculate direct costs of all surgical procedures of the operating room block (ORB). The cost accounting model was developed for implementation on Business Intelligence. The model is a "hybrid" cost accounting model taking elements from activity-based costing (when resource consumption is related to the process) and traditional cost accounting methodology (where source consumption is not related to the process). The seven implementation steps were closely modeled according to Upda (1999) but modified in line with the research contexts' requirements.

Step 1: Building of a cross-functional steering committee

An interdisciplinary steering committee was build. Meetings discussed development and matters related to practical application. The participation of professionals from various disciplines (anesthesiologist, physicians, nurses, clinical engineers) was fundamental for the creation of a successful and effective cost calculation model. This was particularly the case when the operational team was trying to understand if the developed cost calculation design fitted with the clinical reality.

Step 2: Identification of the cost object

The team decided to choose single surgical procedures as cost objects. Direct costs referred to this will be calculated.

Step 3: Activity mapping of the surgical path

One of the first steps of the cost calculation process was the illustration of all activities of the surgical path as a flowchart where the whole process was broken down into its single path steps. This flowchart must take into consideration all potential steps of a patient's path within the ORB. Any surgical path mapping analysis should take into consideration at least the following four resource categories: Personnel workforce (e.g. Induction/Positioning stage: Anesthesiologist and Anesthesia nurse); High or/and low cost materials (e.g. Induction/Positioning stage: gloves of the personnel, patient anti-thrombosis stockings, antibiotic and anesthetic, syringe and needles); Depreciation of building and devices (e.g. Induction/Positioning stage: Induction room and anesthesiological devices) and; Purchased services from other internal units or the market (e.g. Induction/Positioning stage: Induction room cleaning).

Step 4: Investigation of basic cost information

It is important to investigate the availability of basic cost information. Data received from the management accounting department or the financial accounting department of the hospital must be thoroughly checked, verified and comprehended in terms of composition and structure.

Step 5: Surgical procedure input information

It is important to note that the hospital is not registering input for single surgical procedures (for instance, materials used for each surgical procedure). If surgical input data are not registered automatically, this type of information can be gained by a specific surgeon questionnaire. This information can be connected to cost per unit information (gathered at step 4). Following information should be asked within the questionnaire: Name of the surgical procedure; average time of surgical procedure (in minutes); DRG assigned to the procedure (Procedure Code and Description of the procedure); number of surgeons; anesthesia type (local or general); high cost materials (Description, Quantity and Number of Utilizations); number of Surgical Instrument Kits; Robot Usage (Yes or No); Open Procedure (Yes or No)

Step 6: Data processing model

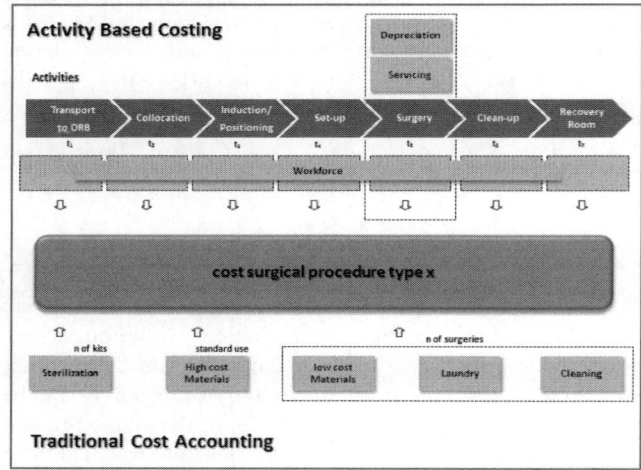

Step 7: Data analysis output

The basic idea of e-HCM is to integrate the cost calculation analysis on Hospitals' Business Intelligence and process data to simple and understandable performance pie charts and tables. 'Every profile includes a few subcategories where operators can access more detailed data analyses. The fundamental concept is that the first data output screen shows general information and guides the user towards more detailed data analysis. The hierarchy inside the software enables the user to have a complete insight of data regarding his/her profile' (Agnoletti/Buccioli/Padovani et al. 2012). The economic profile entrance screen (E1) shows the following data output: Total number of procedures; Total time spitted as regular and overtime;Total number of Scheduled/Unscheduled procedures; Total costs in Mio; Total costs in Mio as a pie chart as Workforce, Services, Depreciation and Materials costs; List of the 5 most expensive procedures in terms of total cost for the whole ORB (cost of surgical procedure*number of procedures performed). The subcategory E2 displays a list of all wards with their corresponding total costs, direct costs and allocated costs. There an illustrated a bar chart provides a quick comparison overview of costs of the different wards. The third subsection E3 separately illustrates wards analyses: Total number of procedures; Total time split into regular and overtime; Total number of Scheduled/Unscheduled procedures; Total costs in Mio; Total costs in Mio as a pie chart as Workforce, Services, Depreciation and Materials costs; List of the 5 most expensive procedures in terms of total costs for the ward. The subcategory E4 deals with data analysis referred to process step costs of all surgical procedures (procedures the horizontal line and process step costs the vertical line of the table). Data output on quantity of procedures of a certain procedure type as well as connected process step costs (average) is specified. At the end of the horizontal table line total average procedure costs are indicated. E5 is planned as a cost comparison simulation section where regular time and overtime differences as well as cost variations if times, material costs and number of surgeons are varied.

III. RESULTS

A prototype on Excel shows interesting cost accounting results. Final implementation on Business Intelligence is not already happened. The prototype shows that average cost of tonsillectomy surgical procedure in regular working time is 1.382 € with a minimum cost of 938 € and a maximum cost of 2.743 €. The standard deviation is 281 €. 91 % of resources are consumed during the surgical procedure itself and only the remaining 9% are constituted of collateral activities (transport, induction ect.). With regard to the cost type distribution workforce requires 52%, materials 37%, depreciation 10% and services 1%. If the same procedure is performed in overtime its average cost is 1.858 € with a minimum cost of 1.177 € and a maximum cost of 4.004 €. The standard deviation in overtime is of 432 €. The cost type distribution differs significantly. Workforce, because of different unit costs, consumes 65%, materials 27% and depreciation 8%.

IV. CONCLUSION

The e-HCM project provides hospital managers with important cost information. This is important for decision making processes, especially with regard to financial sustainability.

REFERENCES

1. AGNOLETTI, Vanni/ BUCCIOLI, Matteo/ PADOVANI, Emanuele et.al. (2012): *Operating room data management: Improving efficiency and safety in a surgical block*. In: BMC Surgery. In Review.
2. BAKER, Judith (1998): *Activity-based costing and activity-based management for health care*. Gaithersburg: Aspen Publishers.
3. CHAN, Yee-Ching Lilian: *Improving hospital cost accounting with activity-based costing*. Healthcare Management Review 1993 (18): 71-78.
4. DE BELVIS, Giulio Antonio/ FERRÈ, Francesca/ SPECCHIA, Maria Lucia et.al.: *The financial crisis in Italy: Implications for the healthcare sector*. In: Health Policy 2012, (106): 10-16.
5. ELIXHAUSER, Anne/ ANDREWS, Roxanne: *Profile of Inpatient Operating Room Procedures in US Hospitals*. In: Arch Surg 2010, (12):1201-1208.
6. FINKLER, Steven A./ WARD, David M./ BAKER, Judith (2007): *Essentials of Cost Accounting for Healthcare Organizazions*. Third Edition. Sudbury: Jones & Barnett.
7. GUERRIERO, Francesca/ GUIDO, Rosita: *Operational research in the management of the operating theatre: a survey*. In: Health Care Manag Sci 2011, 14:89–114.
8. KAPLAN, Robert/ PORTER, Michael: *The Big Idea: How to Solve the Cost Crisis in Health Care*. In: Havard Business Review 2011, (89): 46-64.
9. LIEVENS, Yolande/ VAN DEN BOGAERT, Walter/ KESTELOOT, Katrien: *Activity-based costing: A practical model for cost calculation in radiotherapy*. Int J Radiat Oncol Biol Phys 2003 (2):522-535.
10. MACARIO, Alex: *What does one minute of operating room time cost?* In: Journal of Clinical Anesthesia 2010: 233-236.
11. MARYAMAA, Riitta/ KIRVELA, Olli: *Who is responsible for operating room management and how do we measure how well we do it?* In: Acta Anaesthesiol Scand 2007, 7:809–814.

12. MCKEE, Martin/ HEALY, Judith (Eds) (2002): *The significance of Hospitals: An introduction.* In: Hospitals in a Changing Europe. European Observatory on Health Care Systems Series. Page: 3-14.
13. MLADOVSKY, Philippa/ SRIVASTAVA, Divya/ CYLUS, Jonathan et.al (2012): *Health policy responses to financial crisis in Europe.* Edited by: WHO Regional Office for Europe and European Observatory on Health Systems and Policies. Copenhagen: WHO Regional Office for Europe.
14. ODUNCU, Fuat S.: *Priority-setting, rationing and cost-effectiveness in the German healthcare system.* Med Health Care Philos 2012. Published online: Jun 13
15. RECHEL, Bernd/ WRIGHT, Stephen/ EDWARDS, Nigel et.al. (2009): *Introduction: hospitals within a changing context.* In: Investing in Hospitals of the future. European Observatory on Health Systems and Policies. 3-27.\
16. TAN, Joseph (Edt.) (2005): *E-Health: The Next Health Care Frontier.* In: E-Healthcare Information Systems. An Introduction for Students and Professionals. San Francisco: Jossey-Bass. 3-27.
17. UDPA, Suneel (1999): *Activity-Based Costing for Hospitals.* In: Cost Accounting for Healthcare Organizazions. Concepts and Applications. Second Edition. Gaithersburg: Aspen Publishers. 723-739.
18. VAN RENSBURG, Janse A.B./ JASSAT, Waasila: *Acute mental health care according to recent mental health legislation Part II.* Activity-based costing. In: Afr J Psychiatry 2011; (14):23-29.
19. YOUNG, David W. (2008): *Management Accounting in Health Care Organizazions.* Second Edition. San Francisco: Jossey-Bass.

Analysis of Combined Ultrasound Therapy and Medium Frequency Current in Abdominal Adiposity

P.A.G. Gripp[1], A.M.W. Stadnik[2], A.A. Bona[3], and E.B. Neves[4]

[1] UTFPR/Graduate Program in Biomedical Engineering, Student, Curitiba, Brazil
[2] UTFPR/ Graduate Program in Biomedical Engineering, Professor, Curitiba, Brazil
[3] UTFPR/Department of Electrical Engineering and Industrial Informatics, Student, Curitiba, Brail
[4] UTFPR/ Graduate Program in Biomedical Engineering, Professor, Curitiba, Brazil

Abstract—Brazil is the second country in the world in aesthetic surgical procedures, with 905,000 procedures, being behind only the U.S., according to information corresponding to the year 2011 from the International Society of Aesthetic Plastic Surgery (ISAPS) and the Brazilian Society of Plastic Surgery (BSPS). Given this scenario, we chose to study the adipose cells, which are a organ with intense metabolic and endocrine activity. Accumulated in excess adipocytes causes the disadvantages metabolic and endocrine in the organism, leading to obesity in general and located. This research referred to obesity marked by deposition in the abdominal area, that is denominated central or androide obesity, in the case of a localized adiposity. The objective of the research was to analyze the effects of therapeutic ultrasound and medium frequency current in this particular type of adiposity. The therapeutic protocol consisted of the utilization therapy of therapeutic ultrasound 3 MHz and medium frequency current (1 kHz, with Bursts of 2 ms), performed in ten sessions of 40-minute each. The sample consisted of 16 women, of whom 14 have finalized treatment. The assessment instruments used were: percentage of total body fat by bioimpedance, skinfold measurement in the region where localized adiposity was. We observed the reduction in measures measured in abdominal circumference and percentage decrease in total body fat measured on a digital scale bipolar bioimpedance, as well as the decrease measured by skinfold plicometer in all patients submitted to treatment. The results of submission to the treatment of "Combined Therapy" can be considered satisfactory, due the record in reducing the parameters analyzed, despite being necessary to a more thorough study on the subject and scientific substantiation in a larger sample. Based on this study has observed a dearth of studies on the subject so that the same password be a non-invasive alternative in the treatment of localized adiposity and dermatological alterations such as gynoid lipodystrophy.

Keywords—combined therapy, ultrasound, medium frequency current, adiposity, abdominal.

I. INTRODUCTION

The International Society of Plastic and Aesthetic Surgery (ISAPS) [1] and the Brazilian Society of Plastic Surgery (SBPC) [2] disclosed data relating to aesthetic surgical procedures performed in Brazil, showing that the country is the second in the world to accomplish that type of surgery, with 905,000 procedures investigated in the year 2011, behind only the U.S. which recorded 1.1 million.

Corroborating the data from ISAPS, accordance the World Health Organization (WHO) [3] most performed procedure was the reduction of localized fat, about 23% of the total, justifying to IBGE [4] counting on partnership Ministry of Health and are presented in the Household Budget Survey (HBS), approximately 50% of Brazilian adults are overweight (IBGE Household Budget Survey 2002-2003).

The localized adiposity is defined [5] as regional distribution of fat. In this case, the present adipocytes are increased in specific regions of the tissue and irregularly wavy appearance, as affirmed by Cardoso [6].

In 1947, Vague described two types of body fat distribution: the androide or type masculine and gynecoid exter-nalizing characteristics somatic feminine [7]

The Obesity marked by deposition in the abdominal area is denominated central obesity or androide. When excessive deposition are located in the hips and thighs, obesity is denominated peripheral or gynoid [8].

Among the physiotherapy functional dermato most used in Brazil, is located therapeutic ultrasound which has demonstrated good results when combined with other techniques such as medium frequency current. This utilization in conjunction also known as "Combined Therapy" [9].

The fibroblasts when subjected to sound waves of ultrasound subjected to pressure of one atmosphere, found an increase of collagen synthesis [10]. A similar result had already been described by Guirro & Guirro [11], reporting that ultrasonic irradiation could enhance the synthesis of fibroblasts, these vittaly cell are important in the process of repair of tissues

This study aimed to analyze the effects of combined therapy of ultrasound 3 Mhz and medium frequency alternating current in the treatment of localized adiposity, thus being improving body contour and aspect of the skin of patients submitted to treatment.

II. MATERIALS AND METHODS

For this study, was selected a random group of 16 volunteers, where 14 completed the treatment.

The inclusion criteria of were: age group not determined, white race, healthy or not practitioners of physical activities being allowed to perform it more than twice a week and can not change this frequency during treatment, do not use controlled drugs, layer thickness adipose at least 1.5 cm in the treated area and not belonging to the group of people with contraindications to treatment.

The exclusion criteria of were: initiating alimentary diet during treatment; make use of pharmaceuticals and / or other alternative therapies (except contraceptives), having performed in less than one year of treatment to reduce localized adiposity; present Body Mass Index (BMI) equal to or greater than 35, do not attend at any session, including assessment, reassessment or any session for application of the treatment, treatment withdrawal during the research time.

Were used in the research: medical records constituted of anamnesis and personal data (name, age, occupation, address, phone number, city and living habits), ballpoint pen, tape measure Ibramed measured in mm and cm 120cm in length, balance digital bipolar bioimpedance Britânia, adipometer branded by Ibramed pencils, Heccus combined Therapy device (frequency of 1 kHz combined with Bursts of duration equal to 2 ms), Ibramed® constituted of a headstock with total area of 18 cm^2 and composed of three ultrasound generator 3 MHz with effective radiation area of 6 cm^2 and a power of 18W each one totaling a potency of 54W/cm^2 accordance with the manufacturer, 12.1 megapixel photo camera branded by Fugifilm® , conductive gel with nine active brand by Mega® Gel.

The gel composition: Arnica montana antiinflammatory action, healing and skin irritations; Centella asiatica -increases the peripheral vasculature; Algae of sea -antiinflammatory; Hamamelis Virinianas stimulates the reabsorption of fat, Indian Chestnut increased vascularity and skin tonicity; Cavalinha stimulates lymphatic drainage, Yerba Mate -cicatrizant and protective HDL; Hedera helix -flexibility, tonicity and elasticity of the skin and Rosmarinus officinalis -cicatrizant.

Possibility were to exist that during the application choose to use three diferent modes of utilization of the headstock being only emission ultrasound of 3.0 MHz, only emission so three-dimensional medium frequency current or association the ultrasound of 3.0 MHz with a current media frequently emitted three dimensional way , this being the therapy chosen for conducting the study.

The collection of data, evaluation and sessions have been held in the medical office of the principal investigator, in the period from July to December 2012, being held sessions twice a week, totaling ten sessions of 40 minutes each, this constituted by the physical therapy evaluation that was performed in first attendance where it happened the explanation of the proposed treatment, submitted to acceptance of research for inclusion in the aforementioned requirements and signing of free and informed consent form and authorization form for image.

It was also performed a report submitted by anamnesis and completing a form records of assessment which was noted the reaction of the patient in all sessions held, the patient submitted triage with measurement of skinfolds, perimetry in predetermined regions, weigh, photograph the area to be treated, and the first session of the application of treatment.

This first session also included the parameters of low use of the device aiming adherence and knowledge of the submitted to treatment. Were held to the whole ten sessions, which were described parameters of each patient according to pain threshold and acceptance of the respectives. The reassessment and measurement of measurements were made at the end of all the sessions and noted the medical records.

In the cases the weeks that have holidays and / or lack inevitable attendance was transferred and performed on another day of the same week, previously agreed between the researcher and the participant, without interference of the final result.

The measurements were taken with standard previously established by the researcher, which would be the largest measured abdominal diameter, the smaller abdominal diameter, five centimeters above the umbilical scar and five centimeters below the umbilical scar. All demarcated with pencil in patients. Repeating also on fifth and tenth session before each session, was performed to asepsis skin of participants with cotton and alcohol 70.

In the tenth session the patient was reevaluated, photographed again and presented the results obtained, photography, initial and final, was performed with the participant in orthostatic position in one of the walls of the medical office of the researcher, with undergarments, light on, the machine without flash, with a distance of 40 cm from the area to be treated. Four photos were taken of each participant: front, rear and side, the evaluation and reevalua-tion. Results were recorded by the researcher on the medical records at the beginning and end of all sessions.

Risks and or discomforts: pain or discomfort during the session due to the discomfort of medium frequency current generated during the application of the device can reach the pain threshold of each patient, increased local temperature skin burns may occur, which would be caused by poor application of the device or lack of knowledge of the operator, discomfort in individuals with sensitivity to ultrasound, temperature increase intra-abdominal harm to the fetus can occur abortion if the woman is pregnant and omits and or is unaware this information. The conductive gel used in the therapy is based cosmetic H2O, and standards antiallergics however not prevent rejection and dermatitis in the skin of volunteers.

Benefits: ultrasound associated with the medium frequency current besides reducing the abdominal circumference, performs fluidization of fat by mechanical effects of the ultrasonic waves, as already mentioned in the literature, improves the appearance of abdominal fat and can in this way raise the self-estimates of voluntary.

III. RESULTS

Of the total 16 patients, two abandoned treatment and other three did not achieve satisfactory results. With the achievement of treatment, there was a reduction in measures initially measured abdominal circumference (AC) and a decrease in the percentage of total body fat (% TBF) measured on a digital scale bipolar bioimpedance and as a consequence a reduction in total body mass (TBM), two volunteers obtained a reduction of less than 1% in the percentage of total body fat, five of 11 volunteers who were revalued by bioimpedance, although their results were presented to be trustworthy due to measurement have been made during the menstrual period or earlier at the same by stress and hormonal changes may be interfered with the change in total body mass and hydric retention, data described in table 1.

It was noted, reducing the localized adiposity, even those who did not complete treatment or which submitted failed to comply with the basic requirements for admission, keeping the body fat percentage referred to in the final total bioimpedance and / or increasing the total body mass, compounding the positive results the final measurements of skinfold measurements, performed by plicometer, decreased compared with the initial measurement.

Five of the submitted obtained a reduction ten centimeters or more in diameter abdominal referring to the measures summed (AC 5 cm above the umbilical scar (ACAUS) + AC in the umbilical scar (ACUS) + AC 5 cm below the umbilical scar (ACBUS) described in Table 2.

Table 1 Description of individual patients with the results of the reduction of AC, % TBF and TBM in kilograms.

	Age	AC (cm)	% TBF	TBM (kg)
1	26	-----	----------	------------
2	40	------	----------	------------
3	38	1,0	0,9	+1,3
4	44	2,3	+1,3	maintained
5	32	0,9	+1,6	+1,2
6	28	8,6	3,0	3,4
7	41	5,3	1,9	3,6
8	24	3,5	1,9	4,5
9	27	1,4	1,4	2,4
10	35	3,9	0,4	1,2
11	47	6,7	0,9	3,2
12	56	4,2	4,9	2,1
13	47	5,6	2,7	1,7
14	32	6,1	4,1	2,5
15	35	2,2	1,5	1,6
16	54	4,6	2,3	4,8

Sum of t he measures reduced waist circumference:

Table 2 Description of five patients with a reduction greater than the sum of 10 cm in the AC measurements given in centimeters

Age	ACAUS (cm)	ACUS (cm)	ACBUS (cm)	Total (cm)
24	5,9	3,5	2,6	12,0
47	4,2	6,7	4,9	15,8
47	4,8	2,7	3,1	10,6
56	4,3	4,9	5,7	16,7
28	4,4	8,3	3,1	15,8

IV. DISCUSSION

Societal factors, sociocultural influences, pressures from the media and the relentless quest for ideal body associated with achievements and happiness are among the causes of changes in body image perception, generating dissatisfaction especially for female individuals [12].

Although the results present reductions in abdominal circumference and reducing the percentage of total fat in some voluntary notice to maintaining total body mass and or small reduction this may be due to possible hormonal changes and / or stressful situations during treatment by preventing the result trustworthy for this type of patients.

For the relevance to the treatment needs to be further investigation into the topic, since from those submitted three of the patients reported not realize the positive results due to past situations of stress during treatment, changing the psychological and increasing food compulsiveness, still that the results were visible in images taken and also by measuring measured during the reassessment, we observed the influence of changes in body image.

Through the utilization of combined therapy found the reduction of localized fat, supposing occurred lipolysis and lipoclasis cellular due to the effects of ultrasound [6]. According to the literature, there is release of triglycerides, being done by reusing energetic by the cells. When supplied the needs of the body, these cells can be stimulated by current medium frequency that degrades fatty acids across the electro stimulation.

V. CONCLUSIONS

The results of treatment compliance of "Combined Therapy" can be considered satisfactory, since in all patients were recorded decrease in the measured parameters and the photographic comparisons prior to treatment.

This study demonstrated that the combination of ultrasound and medium frequency current treatment is tolerated by patients submitted especially presented fast and satisfying results for the large majority. So with the new market technologies can reduce small adipocytes measurements without the submission of invasive therapies such as liposuction.

Due to the scarcity of studies that prove the real effectivity of electrical stimulation medium frequency associated with the application of 3 MHz ultrasound in the treatment of localized adiposity, knowing that treatment consists in the stimulus for the improvement of local circulation in adipocyte lipolysis and lipoclasis and reducing edema regionalized, it became necessary to for this study to verify the action of combined therapy in the treatment of localized adiposity.

ACKNOWLEDGMENT

We thank those responsible for the Laboratory of Ergonomics from Federal Technological University of Paraná for the support given to the study.

REFERENCES

1. Procedures Performed in 2011 (2011). ISAPS International Survey on Aesthetic/Cosmetic.
2. Sociedade Brasileira de Cirurgia Plástica (2011). Demografia Médica no Brasil, Vol 1.
3. Organização Mundial de Saúde(OMS).(2003).Relatório da IOTF-Internacional Obesity Task Force.
4. Pesquisa de Orçamentos Familiares (2002-2003). IBGE-Instituto Brasileiro de Geografia e Estatística.
5. Spiegelman, BM., Flier JS.(2001) Obesity and the regulations of energy balance. Cell Press, v.104, p.531.
6. Cardoso E (2002). A evolução e as novas técnicas utilizadas na estética corporal. Revista Vida Estética, n.104.
7. Vague J (1956). The degree of masculine differentiation of obesities: a factor determining predisposition to diabetes, atherosclerosis, gout, and uric calculous disease. American Journal of Clinical Nutrition, v. 4, p. 20-34.
8. Guirro ECO. Guirro, RRJ (2004). Fisioterapia Dermato Funcional: Fundamentos, Recursos e Patologias. 3ª ed. Revisada e ampliada por Manole. São Paulo.
9. ContI BZ, Pereira TD (2003). Ultra-som terapêutico na redução da lipodistrofia ginóide.
10. Harvey W, Dyson M, Pond JB, Graharme R (1975). The in vitro stimulations of protein synthesis in human fibroblastos by therapeutic levels of ultrasound. Proceeding of the Second European Congresso n Ultrasonics in Medicine. Expecta Medica, Amsterdan, p. 10-21.
11. Guirro E, Guirro R. (1996), Fisioterapia Dermato Funcional: Funda-mentos, Recursos e Patologias. -Ed Manole -2ª Ed,116-129; 222-223.
12. Conti MA; Frutuoso MFP, Gambardella AMD. Excesso de peso e insatisfação corporal em adolescentes. *Rev Nutr*2005; 18 (4): 491-7.

Author: Paula Alessandra Garcia Gripp
Institute: Universidade Tecnológica Federal do Paraná (UTFPR).
Street: Av. Silva Jardim 1364, ap1104.
City: Curitiba.
Country: Brazil
Email: paulagripp_fisio@hotmail.com

Cooperative Experiential Learning Models for Biomedical Engineering Students

S.M. Krishnan

Department of Biomedical Engineering, WIT, Boston

Abstract—Undergraduate biomedical engineering (BME) programs have increased considerably and the drives for continuous improvement of the programs persist. While covering a comprehensive base of multiple disciplines required for BME studies on campus has been a big challenge, the Industry Professional Advisory Committee members usually recommend incorporating experiential learning modules of co-op or internship in the curriculum. Embedding cooperative modules within the undergraduate BME educational program is aimed to assist students in gaining the highly valuable real-life experience. The cooperative work modules facilitate the students in exploring different realistic aspects of the complex work processes in the biomedical engineering field. The values of the cooperative learning modules are recognized by the academicians. However, the developers of the BME curriculum find the inclusion of cooperative work experience or internship with a heavy course load in the program a formidable challenge. The main objective of the present work is to develop few cooperative experiential learning models for BME undergraduate students. Some variations in the models can be applied to graduate students though the constraints are different. In this paper, designs of a few co-op/internship models embedded in the undergraduate BME curriculum are described. A BME Co-op model with three phases at a teaching hospital is illustrated. Other models of one semester at a medical device company or at an academic research lab are mentioned. The results obtained clearly support the proposed co-op/internship scheme. In conclusion, integrating the cooperative work experience will be of significant value in biomedical engineering education by giving opportunities for real-life work experience to the students. For sustained success, it is essential that a suitable model must be selected to blend with the mission of the overall training program at the academic institution.

Keywords—BME Cooperative Education, Experiential Learning Models, Hospital Internship.

I. INTRODUCTION

It is generally agreed that professional concepts and skills could not be learned effectively in the classroom, but required practical experience for their understanding and mastery. Co-op programs are routinely adopted in professional studies such as medicine, nursing, pharmacy, law, accounting, etc. Literature on engineering education even pays scant attention to how the co-op is a jewel in education. A main goal in including co-op learning modules as an integral component of the BME undergraduate educational program is to assist students to gain real-life practical experience in the chosen discipline. Many programs make participation in such activities available to students, but in general it is up to the students to take advantage of this experiential learning [1]. The Industry Professional Advisory Committee members frequently make recommendations to increase practical experiences for students through feasible schemes. The main challenging factors are the time/durations required for comprehensive training in multi-disciplinary fields such as BME, and non-availability suitable of collaborators to provide cooperative work experience to the BME students.

For earning a baccalaureate degree in BME, students are required to take a multitude of courses in different basic sciences and engineering disciplines as well as in biomedical engineering. This requirement places a very severe constraint on the available and assignable time for practical work experience training since the trend is to reasonably manage the total number of academic credits. In Clinical Engineering, which can be considered a sub-specialization within the BME domain, it seems that there are not that many programs offering cooperative work experience training [2]. It is difficult to design courses within the available spread of semesters to train the students to be adequately competent for meeting the tough demands in a biomedical industrial or clinical setting.

The cooperative work modules facilitate the students in exploring different realistic aspects of work processes in the field. On-the-job training truly complements what the students learn in classrooms and labs. During the cooperative work modules, the students apply their engineering expertise and medical equipment knowledge in a complex clinical setting or in a device manufacturing facility setting to seek solutions to real and practical problems. A well-designed cooperative work experience program provides a very good on-the-job training to complement in-class and in-lab learning in the academic environment [3]. Limited resources and unpredictable business climates make setting up co-ops/internships with biomedical device manufacturers more difficult. This problem extends to small city locations as well as to hospitals. Despite these constraints, the overall benefits and track records prove that incorporating cooperative internships in the BME curricula gives the best results for graduates. The challenge rests mainly with selection of a suitable model to ensure a good fit for the co-op student, the academic institution and the host organization.

The objective of the present project is to formulate different models for experiential learning modules in the form of co-op or internships and to develop collaborators or host organizations to accept students. This paper describes general BME curriculum requirements and different models for co-op modules as well as potential collaborating host organizations. The paper also reflects on the three phases and corresponding work assignments for clinical engineering co-operative training at a very large hospital affiliated with a medical school.

II. BME ACADEMIC CURRICULUM

The number of BME programs has steadily increased over the past two decades. Clear guidelines for the BME curricula have been proposed by VaNTH [4]. Wide variations in course requirements in different programs can be observed based on literature review [5]. It is indeed a challenge to develop a comprehensive BME programs at the undergraduate level [6, 7]. The required courses may comprise of the following clusters: BME core courses including biomedical electronics, biomechanics and biomaterials, BME capstone projects, design and BME elective courses as well as courses in Electrical Engineering & Computer Science, Mechanical Engineering and Design, Math and Physical Sciences, Biology, Biochemistry, Life Sciences, Anatomy and Physiology, complemented with an array of courses in Communications, Humanities and Social Sciences.

Depending on the emphasis to be placed in a BME program, the students may be offered a choice of tracks such as biomechanics, bioelectronics, biomaterials, bioinformatics, etc. The selected specialization will require the student to take appropriate advanced courses to gain relevant expertise. There has been a general trend of increased components in laboratory, design and project work, in order to ensure a well-rounded training [8-10]. A sample cluster of courses designed for an undergraduate BME program is developed by the author [13].

Practical experience, observed evidence and standard practices in several professional education programs, lead support to embedding it is internship or cooperative work modules will constitute a comprehensive training for the biomedical students.

Coupled with the time constraints, the availability of suitable sites and positions for placing the students with funding is a difficult problem. The following sections different co-op models, potential collaborations and a sample clinical engineering co-op at a hospital are described.

III. BME COOPERATIVE EXPERIENTIAL LEARNING MODELS

From the websites of institutes offering BME program and other references, it is observed that co-operative work modules vary and there can be a few models extending for different durations [11-13]. While internship or co-op experience is encouraged, it is not a compulsory component to meeting graduation requirements in many institutions. The durations can vary from one full semester, six months full time, one year, two to six quarters, part-time or even one day or week.

Four models of cooperative work/internship are described in the following section. Model 1 is designed to have two or three semester cooperative work modules interspersed with the curriculum, as shown in Table 1. It is worth noting that industries prefer this model since they could have co-op students or interns from an institution throughout the year. This model, developed by the author, facilitates funding and planning of activities by the co-op/intern student as well as supervision and mentoring by the host organization.

Table 1 Two or Three semester Co-op model [Ref [13]]

	Fall	Spring	Summer
Freshman	Class	Class	Off
Sophomore	Class	Class	Co-op (optional)
Junior	Class	Co-op	Class
Senior	Co-op	Class	Graduation

Models 2 and 3 apply to quarter systems as offered in some universities. A model with two quarter co-op op can be achieved in a four year program, as shown in Table 2.

Table 2 Two Quarter Co-op Option in a Four year Program [Ref [14]]

	Fall	Winter	Spring	Summer
Freshman	Class	Class	Class	Off
Sophomore	Class	Class	Class	Class
Junior	Co-op	Co-op	Class	Class
Senior	Class	Class	Class	Graduation

Table 3 Six Quarter Co-op Option in a Five Year Program [Ref [14]]

	Fall	Winter	Spring	Summer
Freshman	Class	Class	Class	Off
Sophomore	Co-op	Co-op	Class	Class
Pre-Junior	Co-op	Co-op	Class	Class
Junior	Co-op	Co-op	Class	Class
Senior	Class	Class	Class	Graduation

It can be observed in the model shown in Table 3, there are six quarters of co-op modules. This results in the duration of the overall program to be five years. It must be emphasized that the students from this program would have about 18 months of industrial experience, which greatly enhances their potential for employment immediately after graduation. It is to be noted that the Two sequential quarter blocks of internship can be typically carried out at one organization, whereas, the students will have the option to choose different sites for co-op or internship as depicted in Figure 1.

Model 4 is a commonly used scheme with co-op/internship in the summer when several summer research programs and scholarships are made available to undergraduate students on a competitive basis at the national level. This model does not disrupt the standard four year curricula. However, with respect to the host organization, especially a medical device company, the students are typically available only during a third of the year.

Obtaining partnerships with medical device companies or hospitals with a goal to enhance student training is not easy. A scheme for forming collaboration to act as host organizations to provide practical co-operative experiential training to BME students is proposed [Fig 1]. The design and implementation of the scheme pose significant challenges and necessitate diligent assessment of the student's academic requirements, professional interests, prerequisite training, and available resources at the training institution prior to discussions with the like-minded and willing partners. With the globalization trend, the potentials for collaboration at international levels exist at various companies, hospitals, research institutions, and regulatory agencies. To ensure the program's success, careful selection of students to match the plans and the goals of collaborating organizations is absolutely essential.

From the illustration in figure 1, it can be observed that a wide array of collaborators can act as host organizations providing suitable co-op/internship training to BME students.

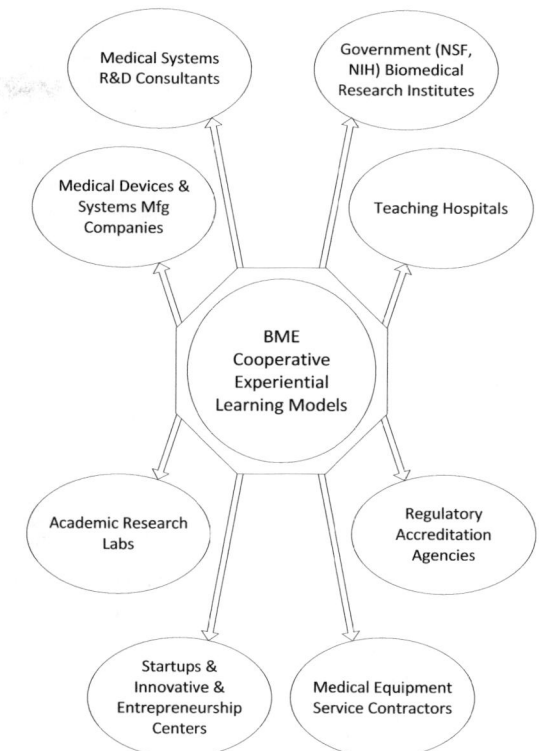

Fig. 1 Potential host organizations for BME Co-op/Internship

Having a diverse distribution of potential collaborators with varied interests and opportunities will relieve the stress on the co-op placement administrators and will offer the students a broader spectrum of BME activities and avenues for gaining familiarity and some expertise. The proposed scheme of having multiple collaborators at the local, regional and national levels will enhance the intended comprehensiveness of student training.

Enthusiastic and productive performance by the participating students will pave the way for continued renewals and expansions of the program by the hosts.

IV. Results

Data from different sites and various collaborative efforts have linked cooperative and intern experiences with positive results. The students usually worked their co-ops or internships at medical device manufacturing sites or in the BME departments of teaching hospitals affiliated with medical schools or with service contractors or consultants. The skill set gained and experiences of co-op students at medical companies depended on the division they were placed at and on the host company's products, manufacturing processes and, protocols. The exposures and assignments covered a spectrum of divisions from R&D to production, test engineering, quality assurance and service.

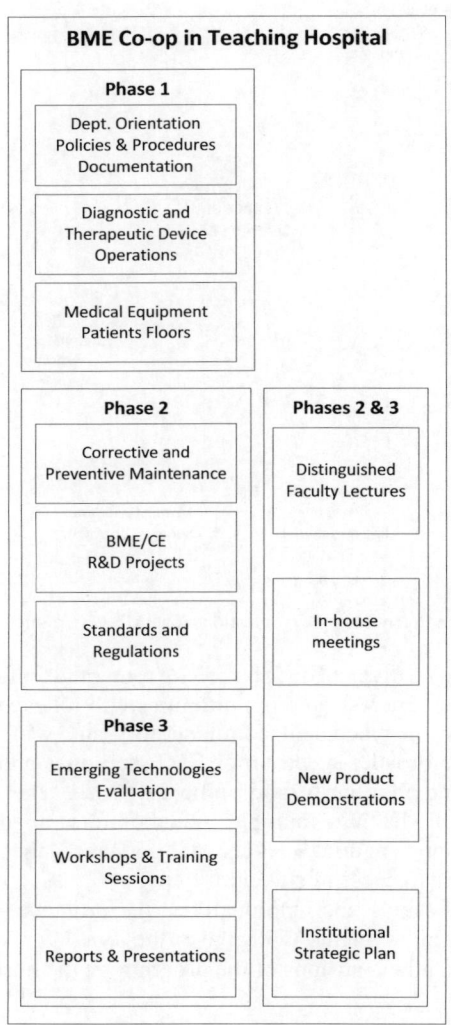

Fig. 2 BME Co-op at a teaching hospital in three phases.

Students placed at hospitals gained knowledge of the multiple functions of clinical engineering covered with an understanding of overall operations of the hospital. A co-op program with three phases and multiple training/activity segments at a large teaching hospital is illustrated in Fig.2.

Some students did their co-op at a research center or an academic lab, or with service contractors, or with project teams on installation and commissioning of specialized equipment. Generally, the co-op students prepared a written report with appropriate approval from employers.

The placement of BME co-op/internship was done over several years at various institutions mainly in Illinois, Massachusetts, Missouri and Georgia and in Singapore. Based on the practical real-life experience and skill sets gained during the co-op/internship, with an exception of less than 10%, the programs were considered to be very beneficial to students and achieved the intended goals. The exception with mixed results corresponded to less than 10% of co-op placements. Feedback from students as well as co-op hosting organizations confirmed the added value of cooperative experiential learning based on the students' real-life experiences coupled with benefits to the hosts.

V. DISCUSSION

Experiential learning with co-op is a superior way for students to apply what they have learned in the classroom and labs. Not only do students gain insight into their career field, they also make contacts with professionals, clarify their career goals, increase the potential of getting hired and give themselves an edge over the competition. During co-op, the strengths and weaknesses become more visible to the students. Some students may choose to transfer to other fields as a direct result of these realizations [16]. In most cooperative placements, the collaborating organizations provide funding to the students undergoing training. The cooperative/intern students are treated as temporary employees without medical or other benefits. The interns get an opportunity to learn a wide variety of unplanned and non-anticipated activities in the BME field. In most cases, within about a six-month training period, the interns could learn various functions in the BME department and can then develop and sharpen their professional skills. Co-op length usually has a direct correlation with outcomes and benefits from the experience [16]. In addition to technical work, the importance of the overall processes of experimentation and proper documentation, as well as learning the protocols of non-disclosure, confidentiality and deadlines is appreciated. The host organization also benefit by training a possible future staff. Students benefit mainly from familiarizing themselves with complex and expensive biomedical systems used for diagnosis and therapy practices. International cooperative placements could be highly beneficial and could help boost self-confidence, despite encountering a number of logistical difficulties such as long delays in securing work permits and adequate funding [16]. The student is also exposed to their possible area of interest and in the event that they feel uncomfortable or realize they do not like the work, they have an opportunity to change fields while still a student so as to save on both time and money [17].

It is to be pointed out some constraints and hurdles are inevitable in carrying out the CE internship programs and efforts are to be made to circumvent the difficulties. It has been reported that there are some early warning signs of lack of adjustment that are related to both motivational anxiety and lack of initial social contacts. However, mid-semester reports of proactive behavior by the student can have a significant impact on both learning outcomes and well-being [18].

Future studies plan to promote the co-op and internship models to the academic institutions which do not presently include them in their undergraduate curriculum. Input will be sought from BME program directors interested in considering experiential learning models for undergraduate BME students and corresponding pros and cons will be discussed.

VI. Conclusion

In conclusion, a few models for integrating co-op modules with undergraduate BME curricula and industries are formatted. An array of potential co-op hosting organizations is presented. The results of cooperative work/internship experiences of students fully support cooperative experiential learning in BME. For sustained success, a suitable co-op model has to be selected to blend well with the interests of students and hosting organizations as well as with the academic institution's overall training mission.

Acknowledgment

The author wishes to acknowledge the participation and support by various hospitals and medical device companies in Illinois, Massachusetts, Missouri, and Singapore for hosting co-op/internships. The author expresses his thanks to H. C. Lord Fund at WIT, Adam Paczuski and Adityen Sudhakaran, and numerous student participants.

References

1. Luzzi, David E. "Beyond the Classroom." *ASEE PRISM*. ProQuest Central, 20 Dec. 2010. Web. 28 July 2013.
2. E. S. Sovilla, "Co-op's 90-Year Odyssey," *ASEE Prism.7.5,1998, 20.*
3. T.R. Harris, "Recent Advances and Directions in Biomedical Engineering Education." IEEE Engineering in Medicine and Biology Magazine 22: pp 30-31, 2003.
4. The Whitaker Foundation. (2006). BME Curriculum Database [Online]. <http://www.bmes.seas.wustl.edu/Whitaker> June 29, 2010.
5. T.R. Harris and S.P. Brophy, "Challenge-based instruction in Biomedical Engineering: A scalable method to increase the efficiency and effectiveness of teaching and learning." Medical Engineering and Physics 27: pp 617-624, 2005.
6. C.B. Paschal, "The Need for Effective Biomedical Engineering Education." IEEE Engineering in Medicine and Biology Magazine 22: 88-91, 2003.
7. P. King and R. Fries, "Designing Biomedical Engineering Design Courses." International Journal of Engineering Education 19: pp346-353, 2003.
8. R. Fries, "An Industry Perspective on Senior Biomedical Engineering Design Courses." IEEE Engineering in Medicine and Biology Magazine 22: pp 111-113, 2003.
9. T. Martin, S.R. Revale, and K.R. Diller, "Comparison of student learning in challenged-based and traditional instruction in biomedical engineering." Annals of Biomedical Engineering 35, 2007.
10. The Biomedical Engineering Graduate Program Handbook, University of Conn, 2007 [Online]. http://www.bme.uconn.edu/pdf/
11. J. Dyro, "Clinical Engineering Handbook." pp 297-298, Clinical Engineering Internship, 2007.
12. A. Lozano-Nieto, "Internship Experiences in Biomedical Engineering Technology: An Overview of Students and Prospective Employers Perception," ASEE Conference, Session #1149. Seattle, WA 1998.
13. "Academic Catalog." *Biomedical Engineering Department: Wentworth Institute of Technology Catalog*. Wentworth Institute of Technology, Web. 28 July 2013. <http://www.wit.edu/catalog/2012-Catalog/academic-programs/eng-tech/biomedical.html>.
14. "Drexel University." Steinbright Career Development Center. Drexel University, n.d. Web. 28 July 2013. <http://www.drexel.edu/scdc/co-op/undergraduate/>.
15. Coll, Richard K., and Richard Chapman. "Advantages and Disadvantages of International Co-op Placements: The Students' Perspective." *Journal of Cooperative Education* 35.2/3 (2000): 95-105. Web. 29 July 2013
16. Loftus, Margaret. "A Head Start." *ASEE*. ASEE Prism, 21 Nov. 2011. Web.28July 2013<http://www.prism-magazine.org/nov11/tt_01.cfm>
17. Lynch, Daniel R., and Jeffrey S. Russell. "Experiential Learning in Engineering Practice." *Journal of Professional Issues in Engineering Education and Practice, ASCE* 135.1 (2009): 31-39. *Gale: Academic OneFile*. Web. 28 July 2013
18. C. K. Parsons, E. Caylor, and H. S. Simmons. "Cooperative Education Work Assignments: The Role of Organizational and Individual Factors in Enhancing ABET Competencies and Co-op Workplace Well-Being." *Journal of Engineering Education* 94.3 (2005): 309. Web. 29 July 2013

Author: S.M. Krishnan
Institute: Department of BME, WIT
Street: 550 Huntington Avenue
City: Boston
Country: United States of America
Email: smkrishnan@gmail.com

Biomedical Engineering Research Improves the Health Care Industry

O.A. Lindahl[1,2,3], B.M. Andersson[1,4], R. Lundström[1,3], and K. Ramser[1,2]

[1] Centre for Biomedical Engineering and Physics, Umeå University, Umeå, Sweden
[2] Department of Computer Science, Electrical and Space Engineering,
Luleå University of Technology, Luleå, Sweden
[3] Department of Radiation Sciences, Biomedical Engineering, Umeå University, Umeå, Sweden
[4] Department of Applied Physics and Electronics, Umeå University, Umeå, Sweden

Abstract—The health care industry is dependent on new innovations for its survival and expansion. Health care innovations are also important for improving patient care. Through activities at the centre for biomedical engineering and physics (CMTF) we have generated growth both in academia at the universities and in the industry in northern Sweden. Fruitful cooperation was generated between 26 research projects and about 15 established companies in the field of biomedical engineering. The established researcher-owned company for business development of the research results from the CMTF, CMTF Business Development Co Ltd, has so far launched three spin-off companies and has 10 new business leads to develop. The activities have also increased the interest for commercialization and entrepreneurship among the scientists in the centre. So far a total of nine spin-off companies have resulted from the CMTF-research since the year 2000 that has improved the health care market in northern Sweden.

Keywords—Health care, Biomedical engineering, business development, innovation, science centre.

I. INTRODUCTION

The health care industry is dependent on new innovations for its survival and expansion. Health care innovations are also important for improving patient care. Triple-Helix methods, i.e. cooperation between scientific research, industry and health care, with the aim to commercialise scientific research results are well established in northern Sweden through the existing innovation system. The Centre for Biomedical Engineering and Physics (CMTF) formed an organisation for triple-helix cooperation between scientific research, biomedical industry and health care. The aims were to establish intense co-operation with the health care industry and create a good milieu for creating and growing new innovations and start spin-off companies to the benefit of the health care industry and the patients.

CMTF was funded through local support from regional foundations and the EU structural foundation, Objective 1 and Objective 2. In total the CMTF turned over 6 million Euro during the years 2000-2007, and has a budget of 12 million Euro for 2008-2014.

The two northernmost universities in Sweden, Umeå University (UmU) and Luleå University of Technology (LTU) have collaborated since 2007 combining the strong technical research at LTU with the strong medical/biomedical research at UmU.

The practical work at CMTF is organized in 26 research projects and one project management (Figure 1).

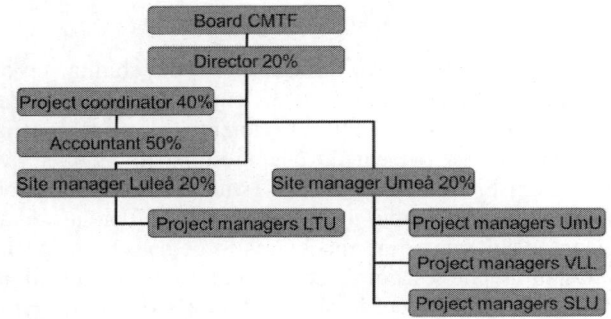

Fig. 1 Organization of CMTF. LTU is Luleå University of Technology, UmU is Umeå University, VLL is county council of Västerbotten, SLU is Swedish University of Agricultural Sciences.

A development company CMTF Business Development (CMTF BD) Co. Ltd., owned by the scientific leaders and the local innovation system represented by Uminova Innovation Co. Ltd. and LTU Holding Co. Ltd., was inaugurated in 2007. The purpose was to facilitate with the business development of the scientific research results from the centre as well as establishing business contacts with the health care industry.

The goal for the CMTF establishment is to create a strong, sustainable and virtual organisation for scientific research resulting in business development in northern Sweden. A further aim is to form a model for how to develop new biomedical viable spin-off companies from the research results through triple-helix activities where the health care industry plays an important role.

II. METHODS

The CMTF is organized with a board of directors assigned by the universities. The board is chosen to give the

CMTF a stable leadership and to reinsure a good cooperation in-between the two counties and the two universities as well as with the health care industry in the region. The broad professional expertise of the board members guarantees high competence for decision making on industrial as well as scientific matters.

Before joining the CMTF, all original 26 research projects were evaluated by the board concerning three criteria; scientific excellence, clinical and industrial relevance and scientific research management. The evaluated and approved projects could make use of the CMTF logotype and refer to CMTF as their research milieu. About 200 researchers and supporting staff were engaged within CMTF at the start of 2012.

A majority of the projects in CMTF has both scientific and industrial cooperation with international partners both outside EU, e.g. Japan and USA, and within EU, e.g. Norway, Finland and Italy.

An important step was the establishment of a research company CMTF Business Development Co., Ltd. (CMTF BD), in 2007. The owners were the scientists/project managers from CMTF as well as representatives from the regional innovation system in both Umeå and Luleå. This was in order to form an organization that could be a part of the existing innovation system but with a special emphasis to launch the biomedical engineering research results on the health care market. In 2011 the CMTF BD was expanded with two business councils, one in Umeå and one in Luleå. The project leaders signed over the IPR to the CMTF BD through an agreement. For identified research innovations, a contract was signed with the scientists about the sharing of future profit from the innovation, a so called incentive agreement giving a stream of money to the researchers and one stream of money to the company at success.

The CMTF research and development was funded by several local and global organisations. The biggest founder was the EU structural foundation, objective 2.

A majority of CMTF BD was funded with support from private means from the scientific leaders, Uminova Innovation Co Ltd, LTU Holding Co. Ltd. and the County councils in Västerbotten and Norrbotten.

III. RESULTS

Nine new health care companies were established from research results from CMTF (Table 1). Eight of them were based on patented innovations and one is the CMTF Business Development Co. Ltd. (CMTF BD). CMTF BD has started three new companies that are currently performing multi-centre studies and developing quality systems for the CE marking process. More than 11 patents have been filed from the research in CMTF and 10 innovations are currently under business verification through CMTF BD and the innovation systems in Umeå and Luleå.

Table 1. Results from the CMTF, the amount of spin off companies and related activities. The figures show the amount.

Year/Activity	2000-2012
Spin-off companies	9
Senior scientists	25
Scientific publications	300
PhD/Lic exams	30
Graduation works	120
Projects	26
Patents	11
Workshops/Conferences	24
New employment	35
Industrial co-operation	15

Twenty-four workshops have been arranged together with the industry and established spin-off companies, with a mean of 60 participants. In addition CMTF was the local organizer of the Nordic Baltic Conference on Biomedical Engineering and Physics in the year 2005 (NBC2005), with about 200 participants and the Swedish Biomedical Engineering conference 2011 with 350 participants.

CMTF has built up an industrial network for triple helix cooperation and currently about 15 national and international companies are involved in the CMTF projects. The co-operations with investment companies and other financers are intense in order to finance new company start-ups.

IV. DISCUSSION

The CMTF scientists have been successful with their research (Table 1). In parallel, the CMTF BD is putting effort into being successful with the implementation of the research results on the health care market. During 2011, two business advisory groups have been established to facilitate the evaluation process of the best projects for commercialisation. One group is located in Umeå and one in Luleå. The groups give advice to the board of directors of the CMTF BD that finally decides about further investments. Nine new product companies have been established since the year

2000 and several new innovations are under development at CMTF BD. The CMTF BD was established in cooperation with the innovation system in Luleå and Umeå and the strategy was to be a branch and market oriented complement to the existing regional innovation system, e.g. Uminova Innovation Co Ltd[1] and LTU Innovation[2] (CENTEC). The aim was to strengthen the health care industry and gather all positive forces for the business development to minimize the so called "time to market" for the products developed.

The CMTF BD[3] has also become a place for the CMTF scientific leaders to meet and discuss and debate business questions and innovations. Thus the company has contributed to the encouragement of entrepreneurship in the field of health care.

CMTF[4] has also been successful in attracting funding from national scientific foundations, support from the EU structural foundation as well as from local funds. This funding was necessary to build up and support the centre organisation and thus for establishing of new spin-off companies that has improved the health care industry.

V. CONCLUSION

The CMTF research network and CMTF BD form today an established complementing innovation organisation for the development of research into spin-off companies in the area of biomedical engineering and health care in northern Sweden. This has given positive results in increased growth of the biomedical engineering activities both in academia and in the health care industry in Northern Sweden.

ACKNOWLEDGMENT

The study was supported by the EU structural foundation objective 2 North Sweden.

REFERENCES

1. Uminova Innovation at http://www.uminova.se
2. LTU Innovation at http://www.ltu.se
3. www.cmtfab.se
4. www.cmtf.umu.se

Corresponding author:
Author: Olof Lindahl
Institute: Centre for Biomedical Engineering and Physics
Street: Umeå University
City: 901 87 Umeå
Country: Sweden
Email: olof.lindahl@umu.se

Effect of Different Visual Feedback Conditions on Maximal Grip-Strength Assessment

A. Chkeir[1], R. Jaber[1], D.J. Hewson[1], J.-Y. Hogrel[2], and J. Duchêne[1]

[1] University of Technology of Troyes, UMR CNRS 6279 STMR, Troyes, France
[2] Institute of Myology, Paris, France

Abstract—Grip strength was assessed using the Grip-ball dynamometer for twenty-one subjects (11 men and 10 women). Five different verbal feedback conditions were tested: no visual feedback (NVF), non-quantified visual feedback (IVF), accurate visual feedback (CVF), and both over-estimated (OVF) and under-estimated (OVF) visual feedback. In the latter three conditions, subjects were presented with a target force to strive for. Significantly greater grip strength was observed for the CVF condition in comparison with the NVF and IVF conditions (6% increase for both comparisons, $p<0.05$). No other significant differences were observed. The magnitude of the differences suggests that visual feedback could be of clinical relevance when maximal grip strength testing is performed.

Keywords—Grip strength, visual feedback.

I. INTRODUCTION

Grip strength assessment is a commonly used tool in a range of healthcare evaluation protocols. Grip-strength has been used to predict post-operative outcomes after surgery [1] as well as the effect of a rehabilitation program [2]. Grip strength is also related to overall strength, and can therefore be used to assess work capacity [3, 4]. Grip strength has also been shown to be a reliable measure of patients' nutritional status [5] and the level of malnutrition [1]. Many clinical evaluations also include grip-strength assessment, for a range of conditions such as Huntington's disease [6] and Duchenne muscular dystrophy[7]. Grip strength is also an element of the most commonly used frailty index proposed by Fried and colleagues [8].

In respect to the measurement of grip-strength, the most commonly used device is the Jamar dynamometer (Sammons and Preston, Bolingbrook, IL, USA), which is considered to be the gold-standard, according to the American Society of Hand Therapists (ASHT) [9]. The Jamar is an isometric dynamometer with an adjustable handle offering five positions from 34.9 to 85.7 mm in width, thus ensuring that grip strength can be measured for various hand sizes. Although the Jamar is in widespread clinical use, the isometric nature of the device can lead to discomfort among users [10].

To solve this problem of comfort, a new grip-strength dynamometer, the Grip-ball was developed (Fig. 1). A detailed description of the Grip-ball can be found in [11]. In short, the device consists of a supple ball in which pressure and temperature sensors and data acquisition and communication systems have been placed. Communication in real-time is performed via Bluetooth, thus ensuring interoperability with local devices that could store or transfer the data (typically a computer, tablet PC, or smartphone).

In addition, the internal pressure of the Grip-ball can be varied by inflating or deflating the ball, thus the stiffness of the device can be modified depending on use of the ball. For instance, a lower pressure could be used for exercises performed as part of a rehabilitation program, while a firmer ball could be used for fatigue or maximal grip-strength evaluation.

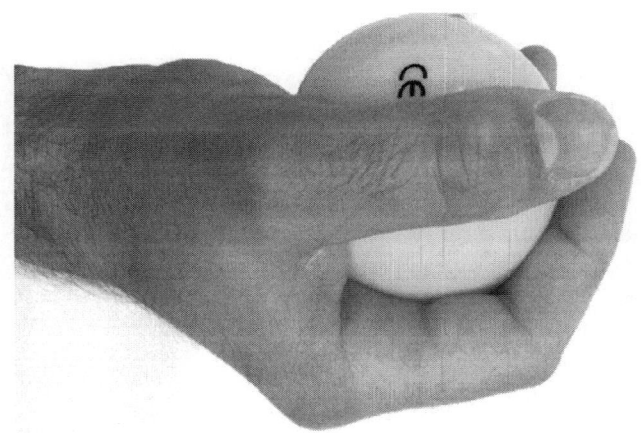

Fig. 1 The Grip-ball

In respect to an evaluation of maximum force, it can be difficult to ensure that subjects have genuinely exerted a maximal effort, either involuntarily or voluntarily. In the first case, some subjects recovering from an upper extremity injury, or who have a fear of pain might not produce a maximal effort [12-14]. In the latter example, some subjects might not want to produce a sincere maximal effort. In case of the lack of a sincerity of effort during a strength test, the results could be interpreted as a lack of response to treatment, leading to potentially unnecessary therapeutic procedures and an increase in health care duration and associated costs [15]. Such a concept has been studied for hand-grip strength [15-18] with a range of testing protocols used such as the five-handle position test [16], the coefficient of variation [17], and the rapid exchange grip test [18].

Another technique commonly used by therapists during both evaluation and rehabilitation is verbal encouragement in the belief that it motivates patients to perform a maximal contraction [19]. Verbal encouragement has been shown to increase maximum force by 5% [20, 21]. The addition of visual feedback can also increase maximal force by 3% [22].

The provision of visual feedback is a straightforward procedure for digital dynamometers such as the Grip-ball due to the presence of a Bluetooth connection to a tablet PC, which can provide a display of the force exerted by the subject. Visual feedback is of particular importance in the case of a remote assessment where it is difficult to provide any verbal encouragement. The presence of visual feedback during self-measurements by a user of the Grip-ball could provide a reliable alternative to an evaluation of grip strength for which the patient would otherwise have to be transported to the clinic.

The aim of this study was to investigate a range of different visual feedback techniques, including the absence of any feedback, on the maximal force produced during grip-strength measurement.

II. METHODOLOGY

A. Subjects

Twenty-one subjects (11 men and 10 women) with no prior history of trauma in either their forearms or their hands and no prior or existing neurological or musculoskeletal disorders were tested (Table 1). Subjects were given a detailed description of the objectives and requirements of the study prior to the experiment, while an informed consent was read and signed by each subject before testing.

Table 1 Subject Characteristics

	Men (n=11)	Women (n=10)
Age (y)	33.1 ± 9.2	25.5 ± 0.7
Height (cm)	177.6 ± 6.8	164.3 ± 5.7
Weight (kg)	82.0 ± 9.2	58.0 ± 10.5

B. Equipment

All testing in the present study was performed using a hybrid device, the Vigoriball, which consisted of a modified Vigorimeter [23], including the Grip-ball electronics (Fig. 2). The Vigoriball has a pneumatic connection enabling the bulb to be inflated at different pressures. For the present study only atmospheric pressure was used. A validity study for the Vigoriball can be found in [24].

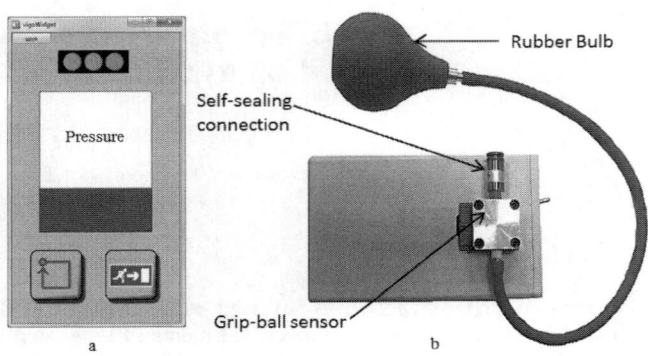

Fig. 2 The computer interface (left) and the Vigoriball (right) used for testing.

C. Procedure

All testing was performed with subjects seated on a chair facing the evaluator, with their feet flat on the floor and their back straight and placed against the back of the chair. Subjects' shoulders were adducted, with the arm tested close to their body, with a 180° elbow extension and a slight wrist extension [25]. Prior to testing, subjects were allowed to practice gripping the device to familiarize them with the task. Subjects were asked to use their dominant hand for all tests, with maximal grip strength evaluated three times per session, with a two-minute rest between trials. Each grip force contraction lasted for three seconds, with the instruction to stop given by the investigator. All tests with the Vigoriball were performed with the largest of the three Vigorimeter bulbs.

Five different visual feedback conditions were evaluated:

- No Visual Feedback (NVF). The pressure exerted in this trial was retained as the maximal pressure (P_{MAX}) used in all of the other feedback conditions.
- Instantaneous Visual Feedback (IVF). The feedback was provided by a vertical green bar of which the level varyied dynamically as the pressure exerted varied (Fig. 2).
- Controlled Visual Feedback (CVF). A target (red bar) corresponding to P_{MAX} was displayed (as a red bar) for the subject to strive for. The pressure each subject exerted during the trial was displayed as a green bar superimposed on the target.
- Overestimated Visual Feedback (OVF). A target corresponding to 110% of P_{MAX} was displayed as an objective for the subject to strive for. The pressure each subject exerted during the trial was displayed as a green bar superimposed on the target.
- Underestimated Visual Feedback (UVF). A target corresponding to 90% of P_{MAX} was displayed as an objective for the subject to strive for. The pressure each subject exerted during the trial was displayed as a green bar superimposed on the target.

D. Data Analysis

The first value of each condition was taken as the reference level and used to calculate the target for the feedback condition provided. Thus, for each subject n, an average pressure value $\overline{P}_j(n)$ was calculated for each feedback condition j from the results of the second and the third trial. All statistical tests were applied on the relative differences for each pair of conditions:

$$D_{j,k}(n) = 200 * \frac{[\overline{P}_j(n) - \overline{P}_k(n)]}{[\overline{P}_j(n) + \overline{P}_k(n)]}$$

Where j and k are the indices of the pair of feedback conditions and n is the subject number.

All statistical analysis was performed using the Statistical Package for Social Sciences (SPSS Inc., Chicago, IL, USA; version 20). Repeated measures analysis of variance (RM-ANOVA) was used, with sex as a between factor. A Newman-Keuls test was used for post-hoc comparisons. As the number of subjects was too small to demonstrate data normality, a non-parametric sign-test was applied in addition to the RM-ANOVA. An alpha of less than 0.05 was considered as a significant difference.

III. RESULTS

There was a significant difference between men and women for all conditions, as shown by the Newman-Keuls test, which gave a mean difference of 29.9 kPa compared to the Critical Difference value of 17.5 kPa (p<0.05). This difference was consistent across all feedback conditions, which corresponded to a difference of 34% (Fig 3.). A significant was also observed between men and women for all feedback conditions (Fig. 4)

There was a significant effect of feedback condition (p<0.05). The results for the non-parametric sign-test were in agreement with those of the RM-ANOVA. Results of a post-hoc analysis with the Newman-Keuls test can be seen in Fig. 5. Significantly higher grip strength was observed for the CVF condition in comparison to both the NVF and IVF conditions. No other effect of feedback condition was observed. The absolute mean differences compared to the CVF condition were 4.9 and 4.8 kPa for NVF and IVF, respectively, which corresponded to 5.9% and 5.8% decreases in grip strength for NVF and IVF, respectively.

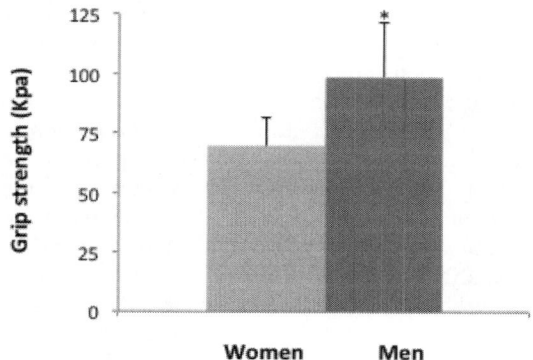

Fig. 3 Grip strength for men and women across all conditions. Data are means ± SD. *Significant difference (p<0.05).

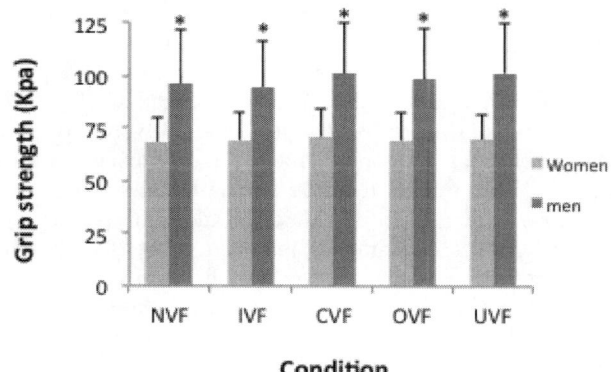

Fig. 4 Grip strength for men and women across all conditions. Data are means ± SD. *Significant difference (p<0.05).

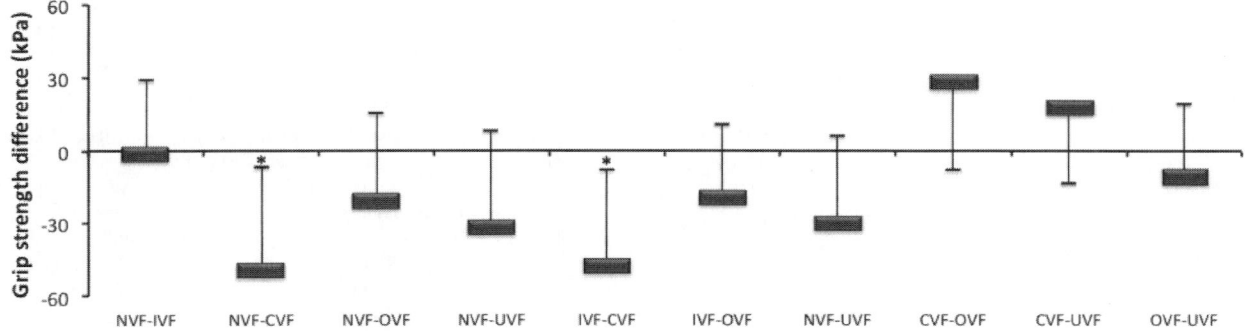

Fig. 5 Post-hoc comparison of grip strength between all conditions. Mean differences, with critical differences displayed as errors bars. *Significant difference (p<0.05)

IV. DISCUSSION

This study proposed a new approach to determine maximum grip strength based on the presence of visual feedback. A range of different conditions was compared, ranging from no visual feedback though to visual feedback with both positive and negative erroneous information.

There was a marked difference in the pressure applied by men and women, with women having significantly lower pressure than men. Such a result was expected, given the results of previous studies of grip strength that have reported higher strength in men [26]. However, the magnitude of the difference was far greater than that reported previously for grip strength using the Martin Vigorimeter. Thorngren and Werner reported a 15% increase in grip strength for men compared to the 34% observed in the present study. The large difference in the present study could have been due to the limited size of the population studied. In addition, most of the female subjects tested were students unaccustomed to physical activity.

In respect to the effect of the type of feedback provided, few differences were observed, with only two significant differences out of 10 experimental comparisons. Both of the differences observed between conditions were for the controlled visual feedback (CVF) condition, where subjects were presented with their actual maximal force from previous trials as a target. The two conditions for which significantly lower pressures were exerted were those with no feedback at all, or instantaneous feedback, meaning subjects had no target to strive for but could see the results of their contraction in real time. In addition, grip strength from CVF was also greater than the other two feedback conditions, OVF and UVF, in which erroneous feedback information was provided, although these differences were not significant. Nevertheless, these differences corresponded to effect sizes of 0.57 and 0.79 for UVF and OVF, respectively.

It must be stressed that the present study had certain limitations, particularly in relation to the subject sample size. As it was an initial study, only 21 subjects participated. In addition, subjects were aged from 24-54 years, meaning that no children or elderly were included. It is possible that the effect of visual feedback might be different between adult subjects and children or the elderly.

A second point that is worthy of further investigation is that the erroneous targets provided were relatively close to the real maximal force, with only a 10% change displayed. It would be interesting to compare different values such as 15 or 20% of variation from true maximum force.

Despite these possible enhancements to the methodology the effect of visual feedback in respect to clinical recommendations on grip strength should be noted. The increase in strength of 5% could be clinically relevant, particularly if the test is used for diagnosis. These differences correspond to 6%, double that reported previously for visual feedback [22] and similar to that observed for verbal encouragement. Only in the presence of appropriate visual feedback did subjects exerting true maximal force

REFERENCES

1. P. Sultan, M. A. Hamilton, and G. L. Ackland, "Preoperative muscle weakness as defined by handgrip strength and postoperative outcomes: a systematic review," BMC Anesthesiol, vol. 12, pp. 1-10, 2012.
2. P. Petersen, M. Petrick, H. Connor, and D. Conklin, "Grip strength and hand dominance: challenging the 10% rule," Am J Occup Ther, vol. 43, pp. 444-7, 1989.
3. L. Wolf, L. Matheson, D. Ford, and A. Kwak, "Relationships among grip strength, work capacity, and recovery," J Occup Rehabil, vol. 6, pp. 57-70, 1996.
4. T. Tietjen-Smith, S. W. Smith, M. Martin, R. Henry, S. Weeks, and A. Bryant, "Grip strength in relation to overall strength and functional capacity in very old and oldest old females," J Phys Occup Ther Geriatr, vol. 24, pp. 63-78, 2006.
5. K. Norman, N. Stobaus, M. C. Gonzalez, J. D. Schulzke, and M. Pirlich, "Hand grip strength: outcome predictor and marker of nutritional status," J Clin Nutr, vol. 30, pp. 135-42, 2011.
6. M. E. Busse, G. Hughes, C. M. Wiles, and A. E. Rosser, "Use of hand-held dynamometry in the evaluation of lower limb muscle strength in people with Huntington's disease," J Neurol, vol. 255, pp. 1534-40, 2008.
7. F. L. Mattar and C. Sobreira, "Hand weakness in Duchenne muscular dystrophy and its relation to physical disability," J Neuromuscul Disord, vol. 18, pp. 193-8, 2008.
8. L. P. Fried, C. M. Tangen, J. Walston, A. B. Newman, C. Hirsch, J. Gottdiener, T. Seeman, R. Tracy, W. J. Kop, G. Burke, and M. A. McBurnie, "Frailty in older adults: evidence for a phenotype," J Gerontol A Biol Sci Med Sci, vol. 56, pp. 146-157, 2001.
9. E. E. Fess, Grip strength. in: Casanova JS, editor. Clinical assessment recommendations vol. 2nd ed. Chicago, 1992.
10. L. Richards and P. Palmiter-Thomas, "Grip strength measurement: A critical review of tools, methods, and clinical utility," J Crit Rev Phys Rehabil Med, vol. 8, pp. 87-109, 1996.
11. D. Hewson, K. Li, A. Frerejean, J. Y. Hogrel, and J. Duchene, "Domo-Grip: functional evaluation and rehabilitation using grip force," in 32nd Annual International Conference of the IEEE EMBS, Buenos Aires, Argentina, 2010, pp. 1308-11.
12. A. A. Czitrom and G. D. Lister, "Measurement of grip strength in the diagnosis of wrist pain," J Hand Surg Am, vol. 13, pp. 16-19, 1988.
13. C. P. Van Wilgen, L. Akkerman, J. Wieringa, and P. U. Dijkstra, "Muscle strength in patients with chronic pain," J Clin Rehabil, vol. 17, pp. 885-889, 2003.

14. T. Pienimaki, T. Tarvainen, P. Siira, A. Malmivaara, and H. Vanharanta, "Associations between pain, grip strength, and manual tests in the treatment evaluation of chronic tennis elbow," Clin J Pain, vol. 18, pp. 164-70, 2002.
15. J. C. Gilbert and R. G. Knowlton, "Simple method to determine sincerity of effort during a maximal isometric test of grip strength," Am J Phys Med Rehabil, vol. 62, pp. 135-44, 1983.
16. S. Goldman, T. D. Cahalan, and K. N. An, "The injured upper extremity and the JAMAR five-handle position grip test," Am J Phys Med Rehabil, vol. 70, pp. 306-308, 1991.
17. O. Shechtman, "Using the coefficient of variation to detect sincerity of effort of grip strength: a literature review," J Hand Ther, vol. 13, pp. 25-32, 2000.
18. D. H. Hildreth, W. C. Breidenbach, G. D. Lister, and A. D. Hodges, "Detection of submaximal effort by use of the rapid exchange grip, " J Hand Surg, vol. 14, pp. 742-745, 1989.
19. M. J. Bickers, "Does verbal encouragement work? The effect of verbal encouragement on a muscular endurance task," J Clin Rehabil, vol. 7, pp. 196-200, 1993.
20. M. S. Hallbeck, "Quantification of the effects of instruction type, verbal encouragement, and visual feedback on static and peak handgrip strength," Int J Ind Ergon, vol. 34, pp. 3-8, 2004.
21. M.-C. Jung and M. S. Hallbeck, "The effects of instruction, verbal encouragement, and visual feedback on static handgrip strength," in Proceedings of the Human Factors and Ergonomics Society 43rd Annual Meeting 1999, pp. 703-707.
22. G. Weinstock-Zlotnick, J. Bear-Lehman, and T.-Y. Yu, "A test case: does the availability of visual feedback impact grip strength scores when using a digital dynamometer?," J Hand Ther, vol. 24, pp. 266-276, 2011.
23. A. Chkeir, R. Jaber, D. J. Hewson, and J. Duchêne, "Reliability and validity of the Grip-Ball dynamometer for grip-strength measurement," presented at the 34th Annual International Conference of the IEEE EMBS, San Diego, California USA, 28 August - 1 September 2012.
24. R. Jaber, D. J. Hewson, and J. Duchêne, "Design and validation of the Grip-ball for measurement of hand grip strength," presented at the Med Eng Phys, 2012.
25. K. Li, D. J. Hewson, and J.-Y. Hogrel, "Influence of elbow position and handle size on maximal grip strength," J Hand Surg (Eur Vol), vol. 34, pp. 692-694, 2009.
26. K.-G. Thorngren and C. O. Werner, "Normal grip strength," Acta Orthop, vol. 50, pp. 255-259, 1979.

Author: David Hewson
Institute: University of Technology of Troyes
Street: 12 rue Marie Curie
City: Troyes
Country: France
Email: david.hewson@utt.fr

Health Technology Assessment of Home Monitoring for Patients Suffering from Heart Failure

L. Pecchia[1], P. Melillo[2,3], M. Attanasio[2], A. Orrico[2], E. Pacifici[4], and E. Iadanza[4]

[1] Department of Electrical & Electronic Engineering, University of Nottingham, Nottingham, United Kingdom
[2] Dipartimento multidisciplinare di specialità medico-chirurgiche, Second University of Naples, Naples, Italy
[3] Dipartimento di Ingegneria dell'Energia Elettrica e dell'Informazione "Guglielmo Marconi", University of Bologna, Bologna, Italy
[4] University of Florence, Florence, Italy

Abstract—Congestive Heart Failure (CHF) is a major problem for developed countries. In this study we present the results of a meta-analysis comparing the effectiveness of the Home Monitoring (HM) versus the usual care.

The home monitoring seems to be more effective than the UC in reducing readmission, mortality and hospital staying, for all-causes and HF-related. Nonetheless, this result is not supported by statistical significance. The relations between HM features and some relevant study design features seems to suggest that more complex HM models are more effective and that the HM could be more beneficial for more severe patients.

Keywords—Congestive Heart Failure (CHF), Home monitoring, telemonitoring, telemedicine, Health technology Assessment (HTA).

I. INTRODUCTION

Congestive Heart Failure (CHF) is the leading cause of hospitalization among the elderly in developed countries [1,2]. Moreover, these patients are often readmitted, up to 50% within three months of initial discharge from hospital [3,4]. Luis and al.[5] in their review report that 30% of patients with a discharge diagnosis of heart failure are readmitted at least once within 90 days and readmission rates range from 25 to 54% within 3–6 months. Current mortality of HF is related to its severity, ranging from 5% to 10% in patients with mild symptoms, to 30% to 40% in severe cases [6]. All this implies that in industrialized countries, the direct treatment costs of HF represent 2–3% of the total healthcare budget and 10% of the cardiovascular health care expenditure[7]. In Europe the estimated total cost associated with heart failure in 2000 was 1.9% of healthcare costs [8].

Therefore, the identification of the correct management strategy for the management of patient suffering from CHF is a difficult and timely challenge. The models that have been proposed in literature for the management of CHF can be organized in three main categories: ambulatory follow-up (usual care), disease management programs and home monitoring.*

Usual care (UC), as international guideline recommends, consist on long-term ambulatory follow-up, based on outpatient controls under the responsibility of a General Practitioner (GP). It could be said that in this model the patient, in coordination with his GP, goes to the hospital for specialized medical effort. Disease Management Programs (DMP) [9,10] are models of care which consist in coordinated and multidisciplinary healthcare interventions and communications for populations with conditions in which patient self-care efforts are significant [11]. DMPs generally include telephone calls which enable patients to interact with trained nursing professionals, and require an extended series of interactions, including a strong educational element. Patients and their relative are expected to play an active role in managing their diseases, also assuming a correct style of life. Differently from UC, in a DMP, the specialized medical effort is brought to the patient house. **Home Monitoring (HM)** services are models in which the disease management is integrated with remote monitoring of signs, symptoms, physiological parameters and biomedical signals. This implies that HM is, generally, more complex then DMP, requiring more resources, and so far is more expensive.

Healthcare professionals are committed to follow only the best available evidence according to well-designed trials [12], meta-analyses [13] or network meta-analyses [14]. Therefore, in this study, assuming the UC being the benchmark, we performed two independent meta-analyses of RCTs studies comparing DMP with UC and HM with UC. Then the relative effectiveness of HM programs was correlated with relevant characteristics of the different programs (i.e. frequency interventions, severity of cases etc.). This paper presents the preliminary result on the HM.

II. METHODS AND MATERIALS

A. Study Design

In planning this study, we considered the PRISMA statement [15], so we decided to limit our search to only RCTs and we assessed a checklist to select studies.

B. Search Strategy

We performed an electronic search using the MedLine's and the Cochrane's Databases and we also considered most relevant citations. Keywords adopted in our search considered those used by Inglis [16] and included all the MESH and all the useful Boolean combinations of the following terms: heart failure, telemedicine, case management, comprehensive health care, disease management programs, home care services, patient care planning, randomized controlled trial. Three investigators independently analysed the full texts of each paper to assess the coherence with the objective of the Study. We restricted our research to the studies published from the 2008 to be complementary to the meta-analysis of Inglis [16].

C. Trials Selection

To assess the RCTs, we prepared a checklist with a series of questions: is the paper coherent with the objective of our Study? Are the treatment group and the control group responding to the requirements of our Study? Are the outcomes of interest described in the paper with enough statistical data?

Then we considered the following exclusion criteria to refine our research: not heart failure; editorial or review; not an RCT; other heart failure intervention or research; only study design; other languages; short follow-up.

D. Statistical Analysis

In both meta-analyses, continuous outcomes were expressed by Mean Differences, with the relative 95% Confidence Interval; binaries outcomes were expressed by Relative Risks, with the relative 95% Confidence Interval. We assessed the heterogeneity by using the Chi-square Test; a P Value less than 0.05 was considered statistically significant. When the heterogeneity was statistically significant, we used the Random Effect Model, otherwise, the Fixed Effect Model [17]. We used the software Meta-Analyst to process data [18]. The results were represented by forest plot and potential bias for publication was explored by funnel plot. To determine the variation between studies, we used the statistic Chi-squared as the difference of the overall heterogeneity and the measures of the heterogeneity in the two primaries meta-analyses [19].

E. Outcomes

The following six outcomes were meta-analyzed: mortality for CHF and for all causes, bed days for CHF and for all causes, readmissions for CHF and for all causes.

F. HM Effectiveness and Patient Severity

The majority of the papers identified described the target population giving the mean age of the patients, the NYHA class and the maximum ejection fraction to be enrolled. According to these values, we classified the severity of the target population in the tridimensional space described in the Figure 1, and then we investigated the correlation.

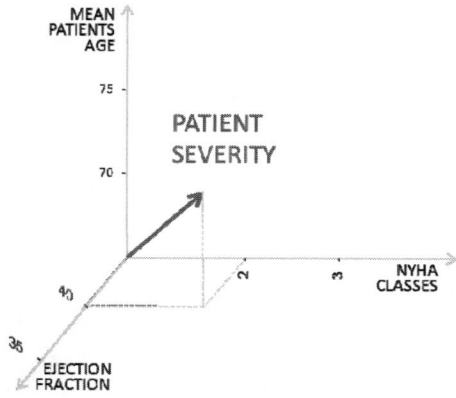

Fig. 1 Patient severity

G. HM Complexity and Its Effectiveness

The selected papers described study design of different complexity. There were differences in: the periodicity of the monitoring: from more one per day to weekly; the data acquired: only symptoms, symptoms + parameters, symptoms + parameters + bio-signals (mainly ECG).

According to these, we classified the studies according to their complexity and investigated the correlation between the complexity and their effectiveness.

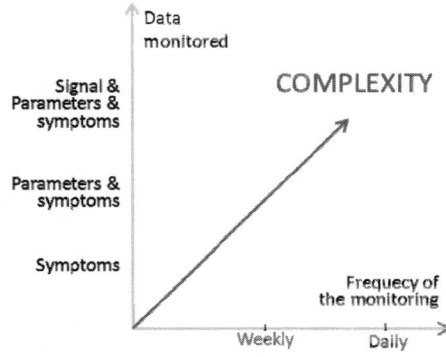

Fig. 2 Intervention complexity

III. RESULTS

Following the search strategy described above, 314 journal papers were identified on the different database analysed, after repetition removal. Frome these, only 32 papers (about 10%) met our inclusion criteria (Figure 1).

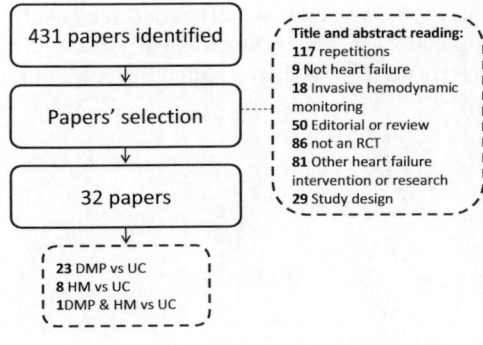

Fig. 3

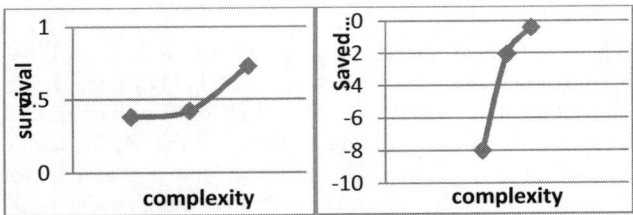

Fig. 4 Patient survival, bad days saved and complexity

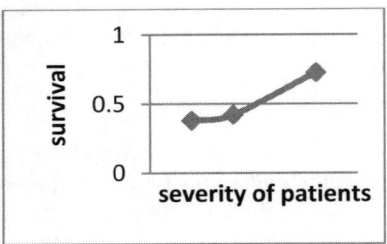

Fig. 5 Patient severity and survival

Tables 1, 2 and 3 show the effectiveness of the HM to reduce respectively patient readmission, bed of days reduction and mortality, for all causes and HF-related. The results do not achieve statistical significance (p value >0.05). These results demonstrated that the HM seems to be more effective than the UC, but more studies are needed to prove this result with statistical significance. One of the reasons why these results resulted not significant is the high level of heterogeneity among the different study design.

Table 1 HM readmission

Readmission	RR (95% CI;p)
All-causes	0.970 (0.811; 1.159; 0.377)
HF-related	0.816 (0.559; 1.192; 0.230)

Table 2 HM bed days

Bed Days	MD (95% CI;p)
All-causes	1.109 (-1.100; 3.317; 0.311)
HF-related	1.937 (-0.792; 4.666; 0.124)

Table 3 HM mortality

Mortality	RR (95% CI;p)
All-causes	0.723 (0.402; 1.302; 0.223)
HF-related	0.602 (0.362; 1.001; 0.059)

Figure 4 shows respectively the correlation between HM complexity and patients' survival (calculated as 1-RR) and saved hospital days. This suggests that a higher level of complexity results in a more affective HM program.

Figure 5 shows the relation between the patient severity and the survival, suggesting that the HM is more beneficial for more severe patients.

This paper presents some limits as the limited number of RCT studies on HM, which may be the main cause of low significance of the achieved results.

IV. CONCLUSIONS

The HM seems to be more effective than the UC in reducing readmission, mortality and hospital staying. Nonetheless, this result is not supported by statistical significance. The relations between HM features and some relevant study design features seem to suggest that more complex HM models are more effective and that the HM could be more beneficial for more severe patients.

ACKNOWLEDGMENT

LP acknowledges support of this work through the MATCH Programme (EPSRC Grant EP/F063822/1) although the views expressed are entirely his own. PM, MA and AO acknowledge support of this work through the project Smart Health and Artificial intelligence for Risk Estimation (grant PON04a3_00139 to PM; 2007-2013 National Operational Programme for Research and Competitiveness).

REFERENCES

1. Cowie MR, Mosterd A, Wood DA, Deckers JW, PooleWilson PA, Sutton GC, Grobbee DE (1997) The epidemiology of heart failure. European Heart Journal 18 (2):208-225
2. McMurray JJ, Stewart S (2000) Epidemiology, aetiology, and prognosis of heart failure. Heart 83 (5):596-602
3. Vinson JM, Rich MW, Sperry JC, Shah AS, McNamara T (1990) Early readmission of elderly patients with congestive heart failure. J Am Geriatr Soc 38 (12):1290-1295
4. Harjai KJ, Thompson HW, Turgut T, Shah M (2001) Simple clinical variables are markers of the propensity for readmission in patients

hospitalized with heart failure. Am J Cardiol 87 (2):234-237, A239. doi:S0002-9149(00)01328-X [pii]
5. Louis AA, Turner T, Gretton M, Baksh A, Cleland JGE (2003) A systematic review of telemonitoring for the management of heart failure. European Journal of Heart Failure 5 (5):583-590. doi:Doi 10.1016/S1388-9842(03)00160-0
6. Massie BM, Shah NB (1997) Evolving trends in the epidemiologic factors of heart failure: rationale for preventive strategies and comprehensive disease management. Am Heart J 133 (6):703-712. doi:S0002870397001981 [pii]
7. Berry C, Murdoch DR, McMurray JJ (2001) Economics of chronic heart failure. Eur J Heart Fail 3 (3):283-291. doi:S1388-9842(01)00123-4 [pii]
8. Hunt SA, Abraham WT, Chin MH, Feldman AM et al. (2005) ACC/AHA 2005 Guideline Update for the Diagnosis and Management of Chronic Heart Failure in the Adult: a report of the American College of Cardiology/American Heart Association Task Force on Practice Guidelines (Writing Committee to Update the 2001 Guidelines for the Evaluation and Management of Heart Failure): developed in collaboration with the American College of Chest Physicians and the International Society for Heart and Lung Transplantation: endorsed by the Heart Rhythm Society. Circulation 112 (12):e154-235. doi:10.1161/CIRCULATIONAHA.105.167586
9. Gonseth J, Guallar-Castillon P, Banegas JR, Rodriguez-Artalejo F (2004) The effectiveness of disease management programmes in reducing hospital re-admission in older patients with heart failure: a systematic review and meta-analysis of published reports. European Heart Journal 25 (18):1570-1595. doi:DOI 10.1016/j.ehj.2004.04.022
10. Holland R, Battersby J, Harvey I, Lenaghan E, Smith J, Hay L (2005) Systematic review of multidisciplinary interventions in heart failure. Heart 91 (7):899-906. doi:DOI 10.1136/hrt.2004.048389
11. Pecchia L, Schiraldi F, Verde S, Mirante E, Bath PA, Bracale M (2010) Evaluation of short-term effectiveness of the disease management program "DI.PRO.DI." on continuity of care of patients with congestive heart failure. Journal of the American Geriatrics Society 58 (8):1603-1604. doi:10.1111/j.1532-5415.2010.02985.x
12. Bracale U, Rovani M, Picardo A, Merola G, Pignata G, Sodo M, Di Salvo E, Ratto EL, Noceti A, Melillo P, Pecchia L (2012) Beneficial effects of fibrin glue (Quixil) versus Lichtenstein conventional technique in inguinal hernia repair: a randomized clinical trial. Hernia. doi:10.1007/s10029-012-1020-4
13. Bracale U, Rovani M, Bracale M, Pignata G, Corcione F, Pecchia L (2011) Totally laparoscopic gastrectomy for gastric cancer: Meta-analysis of short-term outcomes. Minimally invasive therapy & allied technologies : MITAT : official journal of the Society for Minimally Invasive Therapy. doi:10.3109/13645706.2011.588712
14. Bracale U, Rovani M, Melillo P, Merola G, Pecchia L (2012) Which is the best laparoscopic approach for inguinal hernia repair: TEP or TAPP? A network meta-analysis Surgical endoscopy Epub ahead of print
15. Knobloch K, Yoon U, Vogt PM (2011) Preferred reporting items for systematic reviews and meta-analyses (PRISMA) statement and publication bias. Journal of cranio-maxillo-facial surgery : official publication of the European Association for Cranio-Maxillo-Facial Surgery 39 (2):91-92. doi:10.1016/j.jcms.2010.11.001
16. Inglis SC, Clark RA, McAlister FA, Ball J, Lewinter C, Cullington D, Stewart S, Cleland JG (2010) Structured telephone support or telemonitoring programmes for patients with chronic heart failure. Cochrane Database Syst Rev (8):CD007228. doi:10.1002/14651858.CD007228.pub2
17. Sutton AJ (2000) Methods for meta-analysis in medical research. Wiley series in probability and mathematical statistics. J. Wiley, Chichester ; New York
18. Wallace BC, Schmid CH, Lau J, Trikalinos TA (2009) Meta-Analyst: software for meta-analysis of binary, continuous and diagnostic data. BMC Med Res Methodol 9:80. doi:10.1186/1471-2288-9-80 1471-2288-9-80 [pii]
19. Bucher HC, Guyatt GH, Griffith LE, Walter SD (1997) The results of direct and indirect treatment comparisons in meta-analysis of randomized controlled trials. J Clin Epidemiol 50 (6):683-691. doi:S0895-4356(97)00049-8 [pii]

Author: Leandro Pecchia
Institute: Department of Electrical & Electronic Engineering, University of Nottingham, Nottingham, United Kingdom
Street: University Park
City: Nottingham
Country: United Kingdom
Email: Leandro.Pecchia@nottingham.ac.uk

Monitoring and Evaluation of Non-pharmacological Anti-smoking Treatment by Monoximetry: The First Three Months

A.M.W. Stadnik[1], L. Ulbricht[2], M.F. Sarturi[3], D. Meirelles[4], and M. Maldaner[5]

[1] UTFPR/Graduate Program in Biomedical Engineering, professor, Curitiba, Brazil
[2] UTFPR/Graduate Program in Biomedical Engineering, professor, Curitiba, Brazil
[3] PMC/Department of Public Health, nurse, Curitiba, Brazil
[4] UTFPR/Department of Physical Education, student, Curitiba, Brazil
[5] UTFPR/Department of Mechanical Engineering, professor, Curitiba, Brazil

Abstract— **This paper presents the process and results of a research of monitoring and evaluation of the attempt to quit smoking in a group of smokers. The technique that smokers used was Neural Stimulation (NS), taught and applied to smokers participants by a group of volunteers from a non-Governmental Organization (NGO). The objective of the research was to determine the effectiveness of this technique. The researchers chose to perform the measurement of the rate of expired carbon monoxide through a monoximeter. The methodology included to track all phases of the treatment carried out by the volunteers of the NGO and to evaluate results after the third month of treatment. 27 workers started the treatment but only seven of these made all stages of data collection. One of them showed a considerable smoking decrease and claimed stopped smoking, but also only one claimed to have carried out self-treatment of NS, as recommended by the technique and it was not the same person who stopped. We conclude that monitoring by volunteers was an important incentive in the beginning, considerably reducing the rates of expired carbon monoxide, but in this study the most part of participants themselves were unable to leave their addiction.**

Keywords— **non-pharmacological, anti-smoking, treatment, monoximetry, NGO.**

I. Introduction

Cigarette smoking is associated with the mortality of about six million people each year and continues to be the leading cause of preventable death in the world and it is considered a worldwide epidemic [1].

The World Health Organization (WHO) recognizes as a basic requirement for the reduction of tobacco use that each person be made aware of this use represents a damage to health: its addictive nature and potential to cause widespread and various diseases, including premature deaths both for consumers and for those directly exposed to their smoke [2].

Additionally, the population in general, has the fundamental right of assistance to quit, in the case of a commitment to humanitarian and a legal obligation of member countries of the WHO. The WHO also encourages countries to develop and conduct data collection on the use and implementation of policies for tobacco control, which can be used in general health surveys or as part of research in the area. Brazil was one of the countries that signed the Framework Convention on Tobacco Control (FCTC) [3].

The agreement provides, among other things, the control of tobacco use by the public and therefore the country that signed the agreement is committed to establishing effective measures to promote cessation of use of this substance and the appropriate treatment for dependence. Also, preparing and implementing effective programs with these objectives are supposed to be done [1].

It is considered that the university, through concrete actions of their students, professors and researchers, has the moral and social mission to contribute to the advancement of research in the area of combating smoking, helping Brazil in compliance with their commitments, not only made by the WHO, but also made to the health of the population as a whole. In this sense, it is essential to assess possible methods of combating smoking in the country.

In this direction, the American Cancer Society [4], states that Brazil has developed strong and comprehensive actions in relation to tobacco control, giving the country recognizing international leadership in this area, despite being the second largest exporter and producer of tobacco in the world.

Therefore, the research group Quality of Life, Health and Labor at the Federal Technological University of Paraná (QVSAT-UTFPR) decided to monitor and evaluate a non-pharmacological anti-smoking initiative proposed by a NGO, which obtained from a Brazilian public system of safety, permission to work with their workers.

It is necessary to clarify that we do not seek to describe the technique used by the NGO and its application, but only address the issue of effectiveness and evidence that give support or not to the technique.

II. MATERIALS AND METHODS

The said NGO offers a comprehensive program to improve the quality of life (QL) for workers involved in their project. The program was initially made up of a series of courses, repeated every week for six months, aiming to create the opportunity for all workers access. A part of the course was focused on the "treatment for tobacco dependence" - according NS methodology, and workers who smoked were invited to participate in treatment.

To participate in the research, participants were presented to their goals, advantages, disadvantages, the form of participation and informed that they could leave the research at any time without any personal injury. In case of acceptance, participants signed the Instrument of Consent (IC) before data collection begins.

In this sense, one group of participants was investigated, monitored and tested. The application of these tests included: completing an instrument to collect data about the participant's profile, an individual field journal and measuring the amount of carbon monoxide exhaled by participant, through monoximeter BabyCO Meter (Cat No. 36-BC01-STK) until the end of their treatment proposed by NGO (two weeks).

Moreover, considering the Protocol Minimum Approach Smoker, participants also were evaluated at the end of the third month from the date of initiation of treatment. It is considered this period as relevant because it represents the months with the greatest possibility of relapse and also, being estimated as a period of improved pulmonary function [5].

Monitoring and evaluation will be terminated at the end of the sixth month after initiation of treatment, verifying the final situation in which the participant is at that time. This time at least five to six months to verify the effectiveness of treatment was determined by comparative studies of antismoking treatments from Boletim Brasileiro de Avaliação de Tecnologias em Saúde [6] and the wording of the Cochrane Library, that, according Presman et al. [7], adopts as a criterion for successful abstinence period at least six months.

Among the possible outcomes are three options to be investigated: the completely stop smoking; reduce consumption or not quit smoking, and the proposed approach will clarify the main difficulties for obtaining the first result.

Data collection using the monoximeter (a three seconds breath on the device, using individual disposable nozzles):

1st At the beginning of the second day of QL course, taught by the NGO, after signing the IC;

2nd On the same day, at the end of the course, when the participant is supposed to smoke its last cigarette, after non-pharmacological treatment of NS performed by volunteers of the NGO in smokers;

3rd After the first week of non-pharmacological treatment of NS performed by participant;

4th After the second week of non-pharmacological treatment of NS performed by participant;

5th At least three months after starting treatment;

6th At least six months after starting treatment.

In each of these six moments of data collection the research team used an individual field diary in which were recorded all participants complaints, perceived changes, explanations about what happened during the week or the months that followed, to help us to elucidate from the point of view of qualitative data, numerical results found by the use of monoximeter.

Additionally, there was a specific space in this field journal to record daily during the two first weeks of treatment if the participant performed or not the self-treatment of NS.

III. RESULTS

Between the months of July and October 2012, 10 courses of QL were administered, with the participation of 500 workers. Of the total workers, 27 reported being smokers wishing to quit smoking and decided to perform the non-pharmacological treatment proposed by the NGO, agreeing to join this study through their signatures in IC. There were 25 men and two women.

Until the release of this study, equivalent to the fifth data collection - after three months, we have 20 dropouts (19 men and one women), 12 of them have already given up on the first day, and they did not return any time.

They have an average age of 39 years (31 years old the youngest and 50 years old the oldest) and started smoking on average to 16 years. In Brazil, smoking is considered a pediatric disease, 90% of adult smokers started smoking before 19 years of age [5].

The group showed variations in the level of education: 13 were graduates, 11 completed high school, two had only primary education and one was a postgraduate degree.

The main reason that led them to smoking was accompany friends (16), followed by curiosity (eight) and for fun (seven).

On average they smoke 16 cigarettes per day and the first one is consumed before the first hour after waking up (56 minutes on average).

Only five (18.5%) had never tried to quit smoking and the average of the group is 4.3 attempts to quit smoking. Different studies show differences in the average number of times that a smoker presents relapses, Reichert et al. [8], for example, report three to ten attempts.

In research conducted in the Brazilian city where the study presented here was made, 84.82% of smokers want to stop smoking [5].

The main symptoms they feel when stop smoking are restlessness (23) and nervousness (18) and they back to smoke mainly due to stress (13) or seeing someone else smoking (11).

Eight report having problems due to smoking as shortness of breath, cough, weight loss, pain throat and laryngitis and 24 of them (89%) believe that smoking is definitely harmful.

On average the first measure with monoximeter gave 9.3, the second after they smoke their, supposedly, last cigarette, gave 11.1, the third after the first week of self-treatment 3.3, the fourth after the second week of self-treatment, 3, and the fifth after three months, 6.6. This result showed a tendency to decrease or cessation of smoking during the two weeks in which participants were accompanied, first, by the researchers who collected the data and after, by the volunteers who accompanied the last day of the first week and the last day of the second week, talking with participants and making them the technique of NS.

Figure 1 shows the results obtained with the workers who participated in all stages of data collection (seven smokers) using the monoximeter.

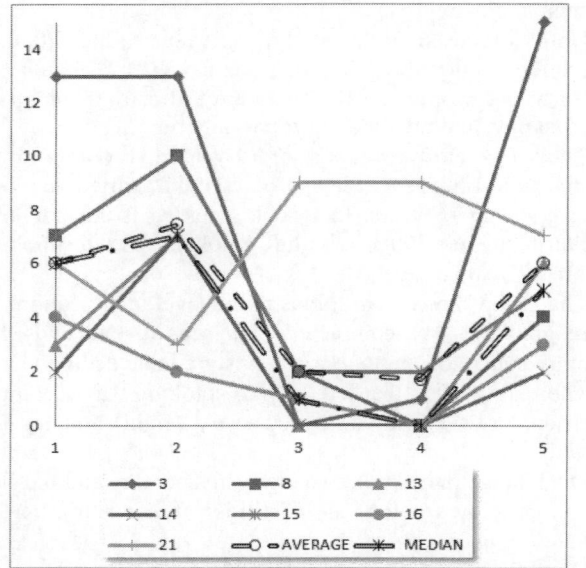

Fig. 1: Tendency to improve in the measurements 3 and 4.

IV. DISCUSSION

Each colored line in the graph (Fig. 1) is one of the seven participants who reached the measure n° 5 - after at least three months of treatment. They were numbered according to the order of entry into the process of data collection. And those black and dotted lines represent the average and median. The graph shows a tendency of reduced values at measurements 3 and 4. The average got somewhat higher than the median due to one participant that had higher results during those two weeks (measurements 3 and 4) - that received the tracking of both: the research team, who performed measurements via monoximeter as the volunteers who carry out the treatment by the technique of NS. This participant (participant no. 21 in the Fig. 1) increased cigarette consumption during this time. The other ones decreased at the same time.

We observed that this participant was one of the women. According to Perkins [9], who performed a specific study on smoking cessation in women, the main symptoms of withdrawal syndrome are more frequent in this social group. However, this study presented here did not reveal more data, so we believe that we could not corroborate this information.

In a field journal entry, we observe that this participant, particularly was going through a change in her work sector and regarded her new role very stressful. Also, referring to other field journal notes from other participants, we found that this was a very common complaint in the group studied. Workers were actually performing a function normally very stressful, public safety. In this regard, Shiffman et al. [10] described that the majority of relapses related to a period of abstinence from smoking are related, especially at situations of psychological stress and the presence of smokers in the environment. In this study presented here we received reports of smokers who wanted to quit but have for example a spouse who is a smoker or friends who are smokers too. According to the participants so far hindered its success.

As the universe of the study consisted of workers of a Brazilian public system of safety we considered absolutely normal that the sample had been composed by 93.6% of men (25) and 7.4% of women (two). This situation only reflected the reality found in place.

The non-pharmacological technique of tobacco control used by the NGO cited here is the NS and we believe that this type of treatment offered approaches itself from other activities recognized by the Brazilian Public Health System (Sistema Único de Saúde - SUS) as Integrative Practices and Complementary Practices in health [11].

These practices comprise the universe of approaches called by WHO Traditional Medicine and Complementary / Alternative - TM / CAM, such as Acupuncture, Homeopathy, Herbal Medicine, Hydrotherapy and Crenotherapy. These therapies involve approaches which stimulate the natural mechanisms of prevention and restoration of health with emphasis on receptive listening, developing the therapeutic relationship and the integration of humans with the environment and society. Other points shared by several approaches included in this field are the enlarged view of the health / disease process and the global promotion of human care, especially self-care [12].

According to studies found in the Cochrane Database [13, 14] there is still no scientific evidence that these types of methods increase the rate of smoking cessation. Are cited as methods Acupuncture, Hypnotherapy, Laser Therapy, Electro-stimulation and Biomedical Risk Assessment (measurement of expired CO and spirometry).

V. CONCLUSIONS

We chose to investigate a non-pharmacological technique that began to be used in Brazil in the past ten years, Neural Stimulation. We conclude that we have not studied sufficiently to perform validation of such treatment. None of the participants truly stopped smoking after to the first three months of treatment. Of the seven participants who were present at all stages of data collection, only one showed a considerable decrease and the participant claimed to have stopped smoking (participant no. 8 in the Fig. 1). However, only one of the participants performed the self-treatment regularly as indicated by the technique. Additionally, the participant (a woman) who was more disciplined in relation to treatment was one that had the highest measurements by monoximeter during the two weeks of following-up (measurements 3 and 4).

On the other hand, experience reports coming from the NGO group of volunteers present successful cases regarding smoking cessation, hence our interest in researching the technique. Although it seems promising still requires further studies for their understanding and validation.

We consider an important aspect to be pointed out the issue of psychosocial support; we observed that during the time in which participants were followed directly by the volunteers of the NGO the results were better.

Also, it is important to clarify that the use of monoximeter may have caused some positive or negative influence. There were no clinical studies proving its effectiveness in the treatment of smokers [7].

In this research the monoximeter was only used as a tool of measurement to monitor and evaluate the treatment carried out by the volunteers of the NGO, however, we believe that this tool can also influence the participants, some participants were motivated to check your measurements, others were intimidated or annoyed by the results or with the number of times they were searched for this action: five times to the presentation of this article and six times in total.

ACKNOWLEDGMENT

To UTFPR's Laboratory of Ergonomics (LAERG); Cardiomed; and Instituto Internacional de Estimulación Neural y Terapias Naturales.

REFERENCES

1. OMS, Organización Mundial de la Salud (2011). Informe OMS sobre la epidemia mundial de tabaquismo, 2011: advertencia sobre los peligros del tabaco. Resumen. Ginebra, Servicio de Producción de Documentos de la OMS.
2. WHO, World Health Organization (2011). WHO report on the global tobacco epidemic, 2011: warning about the dangers of tobacco. Geneva, WHO Press.
3. OMS, Organización Mundial de la Salud (2005). Convenio Marco de la OMS para el Control del Tabaco. Geneva, The WHO Document Production Services.
4. American Cancer Society (2003). Luther Terry Awards Leadership on Tobacco Control. Helsinki.
5. PMC/SMS, Prefeitura Municipal de Curitiba (2007). Como ajudar seu paciente a deixar de fumar: Protocolo de orientações para o profissional de saúde na abordagem mínima do fumante. Curitiba, Secretaria Municipal de Comunicação Social.
6. BRATS, Boletim Brasileiro de Avaliação de Tecnologias em Saúde (2005). Estratégias clínicas para cessação do tabagismo. Brasília-DF, Agência Nacional de Vigilância Sanitária - Anvisa.
7. Presman S, Carneiro E, Gigliotti A (2005). Tratamentos não-farmacológicos para o tabagismo. Rev Psiq Clín 32(5):267-275.
8. Reichert J, Araújo AJ, Cantarino MC et al. (2008). Diretrizes para cessação do tabagismo. J Bras Pneumol 34(10):845-880.
9. Perkins KA (2001). Smoking cessation in women. Special considerations. CNS Drugs 15(5):391-411.
10. Shifman S, Read L, Matese J et al. (1993). Prevenção de recaída em ex-fumantes: uma abordagem de automanejo. In: Marllart GA, Gordon JR. Prevenção de Recaída: estratégia de manutenção no tratamento de comportamentos Adictivos. Artmed, Porto Alegre.
11. Brasil (2006). Portaria n° 971, de 03 de maio de 2006. Aprova a Política Nacuonal de Práticas Integrativas e Complementares (PNPIC) no Sistema Único de Saúde.
12. Brasil (2006). Anexo Portaria n° 971, de 03 de maio de 2006. Aprova a Política Nacional de Práticas Integrativas e Complementares (PNPIC) no Sistema Único de Saúde.
13. White AR, Rampes H, Campbell JL. (2006) Acupuncture and related interventions for smoking cessation. Cochrane Database Syst Rev (1):CD000009. 104.
14. Bise R, Burnand B, Mueller Y, Cornuz J. Biomedical risk assessment as an aid for smoking cessation. Cochrane Database Syst Rev (4):CD004705.

Author: Adriana Maria Wan Stadnik
Institute: UTFPR/ Graduate Program in Biomedical Engineering
Street: Desembargador Motta, 2350 apto.1903
City: Curitiba
Country: Brazil
Email: stadnik@utfpr.edu.br

Patient Satisfaction Evaluation of Telemedicine Applications Is Not Satisfactory

Shuai Zhang, Sally I. McClean, Duncan E. Jackson, Chris Nugent, and Ian Cleland

Faculty of Computing and Engineering, University of Ulster, Northern Ireland

Abstract—Patient satisfaction is an important metric for the evaluation of telemedicine applications to be able to provide patient-centred care. This paper analyses trends in patient satisfaction reporting of telemedicine applications, specifically comparing three telemedicine classes (store-and-forward, real-time consultation, telecare) and the two study stages of pilot and routine delivery. We also discuss the methods used to acquire satisfaction. This paper aims to challenge current approaches and provide recommendations on improving satisfaction evaluation of new telemedicine applications towards provision of patient-centred care. Literature searches were conducted for selected telemedicine studies which report measures of satisfaction from sources of Web of Knowledge and PubMed. Evaluations of patient satisfaction were analysed quantitatively and qualitatively. Our analysis reports that patient satisfaction evaluation is insufficient. However, an increased trend has been detected, which was most evident in the major growth area of telecare. Evaluations are not as actively performed in the routine delivery stage as the pilot stage. Our qualitative analysis found that measuring methods used are usually unsophisticated, poorly described, and often fails to allow patients to communicate their experience in a useful manner. Great attention must be paid to address overall deficiencies in the important area of patient satisfaction evaluation. Satisfaction evaluations are necessarily telemedicine class-specific with an emphasis on telecare applications. There is an obvious need to adopt standard methodologies for measuring satisfaction and ways of incorporating this into economic evaluation, to ensure comparability of data and applications, and not least to compel researchers to adopt an agreement on the dimensions of satisfaction to be evaluated.

Keywords—Patient satisfaction, Statistical methods, Health technology assessment and Patient-centred care.

I. INTRODUCTION

Telemedicine refers to the application of telecommunications technology in healthcare provision, which encompasses most specialties and application domains [1]. Adequate evaluation of telemedicine is a vital requirement to support the vast investment in this technology worldwide, which promises improved quality in healthcare provision, greater access to care, prompt service and crucially costs savings [2]. The evaluation of programmes will vary under different healthcare regimes but it is widely recognised that the fundamental considerations for any new healthcare application must be clinical effectiveness and safety, cost effectiveness [3] and of patient satisfaction [4]. Patients' perspective of services is a valuable information and powerful motivator when designing, development and upgrading telemedicine applications for better patient experience and greater engagement for potential economic savings. Patient satisfaction is an important metric for the evaluation of telemedicine studies to be able to provide patient-centred care to take into consideration of patients' circumstance, preference, lifestyle etc [5]. Active involvement of patients is essential in the development of telemedicine applications. Though, patient satisfaction is often viewed as a commonly reported metric in telemedicine studies [6], the actual frequency of reporting patient satisfaction in telemedicine studies remains un-quantified. The oft-cited major review of satisfaction with telemedicine conducted was a systematic review which did not address the frequency of this measure's evaluation, but instead analysed 32 high quality studies with randomised controlled study designs [7].

In this paper, we investigate, in details, patient satisfaction evaluation for telemedicine applications. For the first time this study quantifies the frequency of patient satisfaction reporting in the telemedicine research. Such quantification reveals the deficiency of patient satisfaction evaluation in certain areas of telemedicine applications, identifies any problem areas and obstacles that require attention and solutions. We analyse recent trends in reporting of patient satisfaction in telemedicine applications both qualitatively and quantitatively. Relationships have been captured on the patient satisfaction evaluations with the telemedicine classes and study stages. Understanding current trends in satisfaction reporting will facilitate better design of future evaluations. We also review methods used to acquire satisfaction evaluations, and the utility of these measures in the ongoing development of telemedicine applications. Based on our analysis, we provide recommendations on the design and development of new telemedicine applications and directions of future work to improve the satisfaction evaluations.

II. METHODS

A. Literature Search

Automated literature searches were conducted for published studies on telemedicine applications which reported evaluations of satisfaction (or acceptability) using two online resources, Web of Knowledge (WoK) and PubMed. Separate searches were conducted for the terms telemedicine, telemonitoring, telecare and e-health publications using the criteria detailed in our search strategy, shown as follows {(1) Telemedicine; (2) Telecare; (3) e-health or ehealth; (4) Telemonitor*; (5) OR 1 – 4; (6) Evaluat* AND 5; (7) Validat* AND (5); (8) Feasib* AND (5); (9) Assess* AND (5); (10) Value AND (5); (11) Satisfaction AND (5); (12) Accept* AND (5)} where * indicates the use of wildcard character(s). The literature search was conducted together with our general review of telemedicine applications evaluation, as an aspect of evaluations along with other aspects such as economic analysis. Full details of this general review appear in Jackson and McClean [8]. The filtered abstracts were manually analysed to identify which studies explicitly reported measures of patient satisfaction (or acceptability). Studies were identified with telemedicine application class and study stage detailed below.

B. Telemedicine Class and Study Stage

Technological advances have enabled telemedicine to evolve from simple consultations via standard telephone to internet-based image transfer, video consultation and remote monitoring for care. Studies are categorised into three telemedicine classes:

- Store-and-forward (SAF): is the most widely used form of telemedicine involves pre-recording of images which are then relayed for evaluation by a clinician at a later date [9]. The diagnostic accuracy of store-and-forward telemedicine has been well established in areas such as dermatology and ophthalmology screening.
- Real time consultation (RTC): is an interactive process where the patient is located at one site consulting with a specialist located at another via a telecommunications network, increasingly a video conference facility.
- Telecare (CARE): this class of telemedicine describes care, support or monitoring delivered in the home, which enables rapid response to emergencies, or the provision of ongoing medical advice [10, 11].

Studies are also classified into one of the two evaluation stages in a clinical environment:

- Pilot or feasibility study: early stage assessment of a new telemedicine application in a clinical environment.
- Routine delivery: the evaluation in everyday use, typically after a minimum of one year in service.

C. Data Analysis

Evaluations of patient satisfaction in telemedicine applications are analysed quantitatively and qualitatively. We analysed for any trends in satisfaction reporting, specifically comparing studies among the three telemedicine classes and in pilot and routine delivery stages. Literature search results have been tabulated by years into contingency tables for telemedicine classes and study stages separately. Chi-square tests or Fisher's exact tests have been employed to identify any significant differences in the reporting patient satisfaction assessment and types of applications including the telemedicine classes and study stages. The conventional 0.05 for p-value has been used for statistically significance. Tests were performed using IBM® SPSS® Statistics 19.0.0. Then the quality of the patient satisfaction assessment for telemedicine applications are analysed and evaluated.

III. RESULTS

A. Patient Satisfaction Evaluation Frequency

Our search returned 478 published studies which used quantitative methods to evaluate telemedicine in the survey period. 158 studies (33.1%) reports contained measurements of patient satisfaction. For purposes of statistical analysis we pooled reports into two years sample intervals, because of low sample numbers in the early stages of our reported survey period. Data collected from part of 2009 have been combined with the interval of 2007 and 2008. Table 1 illustrates the increase in frequency of studies in telemedicine applications during the survey period and suggests a general but unspectacular upwards trend in the proportion which evaluated patient satisfaction, from 21.7% of 60 studies in the period of 1999 and 2000 to 33.3% out of 183 studies in years of 2007and 2009. Satisfaction was measured more frequently in pilot studies (38.3% of 222 studies), compared to applications in routine delivery (28.5% of 256 studies).

B. Satisfaction Evaluation by Telemedicine Class

Statistical test result on the overall literature data reveals an extremely strong relationship between reporting of patient satisfaction and the telemedicine classes (First row in Table 1). In SAF class, patient satisfaction evaluations have been reported in only 19 studies (14.3%), which are less than 58 (31.0%) evaluations in RTC class and CARE class with 81(51.3%) evaluations. The relationship of patient satisfaction assessment and telemedicine classes has been further investigated in two-year intervals during the search period. Results show that in all of the study periods, this relationship between patient satisfaction evaluation and the telemedicine classes remains significant (marked *).

For each telemedicine class, we investigate trends in the reporting of patient satisfaction during the literature search period in the intervals of two years. The figures reveal very different trends in the three classes. Telecare was the only application class where an increase in reported studies was evident, and this was coupled with a trend towards an increased evaluation of patient satisfaction (Figure 1).

Table 1 Two yearly patient satisfaction evaluations of telemedicine applications by three telemedicine classes

Employment of patient satisfaction evaluation		Number of applications by telemedicine class			Total
		SAF	RTC	CARE	
All*	Yes	19	58	81	158
	No	114	129	77	320
1999 and 2000*	Yes	0	11	2	13
	No	18	22	7	47
2001 and 2002*	Yes	0	9	5	14
	No	18	11	7	36
2003 and 2004*	Yes	7	14	15	36
	No	28	16	10	64
2005 and 2006*	Yes	4	7	23	34
	No	15	23	13	51
2007 and 2009*	Yes	8	17	36	61
	No	35	47	40	122

Asterisk * indicates a statistical significance with $p<0.05$

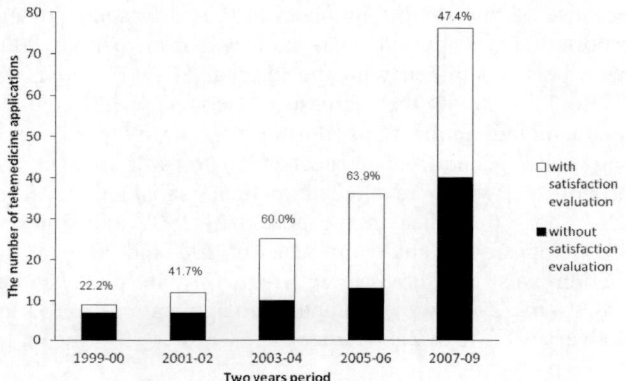

Fig. 1 Number of telemedicine applications in the telecare (CARE) class with (white area of bar) and without (black area of bar) reported patient satisfaction evaluation, with percentage values at the top of the bars

C. Satisfaction Evaluation by Study Stage

Literature searches results have also been tabulated into a contingency table for the two study stages of applications reporting patient satisfaction. The statistical test result reveals an overall relationship between the telemedicine application study stage and the reporting of patient satisfaction assessment where 85 telemedicine applications in the pilot study stage (38.3%) have performed satisfaction assessment in application pilot study, which is significantly more than the 73 evaluations (28.5%) carried out in the routine delivery stage of everyday service.

D. Quality of Satisfaction Evaluation Methods

In terms of qualitative analysis, we found that satisfaction questionnaires typically lack clear methodological description and fail to address potential sources of bias in experimental design [7]. Most studies simply report that a satisfaction questionnaire was used without qualifying this statement with the contents of the questionnaire, or details regarding the mode of its delivery. Also, usually the evaluations are conducted using simple patient survey technique and most often solely after experiencing the novel telemedicine application. Surveys of attitudes prior to experiencing the new intervention would clearly be of utility in establishing a comparative baseline.

Typically patient satisfaction was reported on the overall service, which was measured by "how satisfied" the patient is with the telemedicine service experience. Often a Likert scale was employed in questionnaires, ranging from "very dissatisfied" to "very satisfied". The goal of measuring satisfaction is not simply reporting that a service approaches an adequate level but it should aim to identify areas requiring improvement. The scale approach fails to allow the patient to communicate their experience in a useful manner. Aspects of the service that could be improved or replaced must be considered in a carefully constructed questionnaire.

Where methodological descriptions were reported in detail, we found frequently that satisfaction measurements incorporated questions to evaluate impacts of the service on patient costs and access to care [12]. These parameters must be considered separately in other evaluation aspects [1], though they can be satisfaction determinants.

IV. DISCUSSION

For the first time this study quantifies the frequency of patient satisfaction evaluations in the telemedicine research area. Previous satisfaction surveys, taken the form of systematic reviews, prescribe strict fitness characters so as to select the best studies available for useful comparisons when evaluating new healthcare interventions. However, for most interventions the decision on adopting a new technology or treatment is context-specific and made locally based on trial data. Therefore, for many researchers and decision-makers it is useful to know the frequency with which satisfactions are reported in studies and the methods used to acquire such data. Here we found that contrary to common opinion, often cited in published media e.g. in [6], patient satisfaction is not widely reported, despite the general but unspectacular upwards trend in the evaluation proportion. The frequency of satisfaction reporting varied considerably among the three distinct telemedicine classes and the two study stages. We detected a trend to increased reporting of satisfaction, which was most evident in the major growth area in telecare [13]. On the other hand, the proportions of satisfaction evaluation in the SAF and RTC classes are relatively low and they do not show as clear an increasing trend as for the telecare class. Telecare class applications transform health care delivery into homes to support both patients and carers [14]. Telecare encompasses prevention as well as treatment of diseases and aims to discover deterioration at an early stage with regular monitoring and interventions. With the request of such close monitoring, often in patients' living environment, the emphasis of telecare class applications on measuring patient satisfaction is particularly important and therefore should be addressed accordingly. For the evaluation trend in the study stages, satisfaction evaluation for applications in the 'routine delivery' stage is not paid as much emphasis as the evaluation for the 'pilot' stage applications. Evaluation in routine delivery investigates the sustainability of telemedicine services and is particularly beneficial to the service maintenance and upgrade. This deficiencies need to be addressed.

Our qualitative analysis found that methods used for satisfaction evaluation are usually unsophisticated and poorly described. The primary objective of the evaluation must be to identify any shortcomings in a service so as to contribute to enhanced service provision. Instead, satisfaction ratings in telemedicine are universally skewed in reporting very high satisfaction ratings [7], often exceeding a rate of 90% with little evidence that the process is to be repeated to ensure continued monitoring and sustained quality improvement. There is an obvious need to adopt standard methodologies for measuring satisfaction, to ensure comparability of data and not least to compel researchers to adopt a common agreement on the dimensions of satisfaction to be evaluated [15]. We recommend that assessment of the key criteria of 'satisfaction' must begin with a clearly stated hypothesis allied to measurement using clearly defined, established methodologies to evaluate the requisite dimensions of patient satisfaction. The crucial components identified for assessing patient satisfaction are the interpersonal relationship between the patient and the provider; the effects of the method of service delivery on the patient-provider relationship; identifying communication issue in healthcare delivery; and patient perceptions of limitations of the service received [7].

To summarise, patient's perspective is a powerful driver in the design and improvement of the telemedicine applications. A properly conducted satisfaction evaluation can increase potential patient compliance with provided service and thereby improve cost-effectiveness of the application. Our work reveals that patient satisfaction evaluation is not routinely collected during evaluations of telemedicine and varies under different healthcare regimes. A growing trend is discovered in both the number of telecare class applications and their employment of patient satisfaction evaluations which is of particular importance for the possibly long term employment of the telecare class services. In addition to an overall deficiency in patient satisfaction assessment, such evaluation is particularly lacking in the telemedicine applications in the routine delivery stage when compared to pilot studies. The address of such deficiency is important in both stages for the further improvement of the telemedicine application services. Current methods of collecting patient satisfaction data need to be standardised for both satisfaction definition and design of the satisfaction evaluation. Qualitative research methods can be considered to allow greater patient input and feedback to support patient-centred care and its continuous service improvement. Despite the challenges of patient satisfaction evaluation, we strongly advocate the collection of important patient satisfaction evaluation metrics, to facilitate take-up and improvement in telemedicine services. Such an approach is essential if patient-centred care is to become a reality.

ACKNOWLEDGMENT

The authors wish to acknowledge support from the EPSRC through the MATCH programme (EP/F063822/1 and EP/G012393/1).

REFERENCES

1. Roine R, Ohinmaa A, Hailey D. Assessing telemedicine: a systematic review of the literature. Can Med Assoc J 2001; 165: 765-771
2. Singh SN, Wachter RM. Perspectives on medical Ooutsourcing and telemedicine – rough edges in a flat world? N Engl J Med 2008; 358: 1622-1627
3. Hailey D. The need for cost-effectiveness studies in telemedicine. J Telemed Telecare 2005; 11: 379-383
4. Darzi A. High quality care for all: NHS next stage review final report. London: Department of Health, 2008
5. Maizes V, Rakel D, Niemiec C. Integrative medicine and client-centered care. Explore (NY) 2009; 5: 277-289
6. Coughlan J, Eatock J, Eldabi T. Evaluating telemedicine: a focus on patient pathways. International Journal of Technology Assessment in Health Care 2006; 22: 136-142
7. Mair F, Whitten P. Systematic review of studies of patient satisfaction with telemedicine. BMJ 2000; 320: 1517-1520
8. Jackson DE, McClean SI. Trends in telemedicine assessment indicate neglect of key criteria for predicting success. Journal of Health Organization and Management 2012; 26: 508-523
9. Loane MA, Bloomer SE, Corbett R, Eedy DJ, Hicks N, Lotery HE, Mathews C, Paisley J, Steeles K, Wootton R. A comparison of real-time and store-and-forward teledermatology: a cost-benefit study. Br J Dermatol 2000; 143: 1241-1247
10. Hermens HJ, Vollenbroek-Hutten MM. Towards remote monitoring and remotely supervised training. J Electromyogr Kinesiol 2008; 18: 908-919
11. Tang P, Venables T. 'Smart' homes and telecare for independent living. J Telemed Telecare 2000; 6: 8-14
12. Smith KB, Humphreys JS, Jones JA. Essential tips for measuring levels of consumer satisfaction with rural health service quality. Rural Remote Health 2006; 6: 594
13. Barlow J, Singh D, Bayer S, Curry R. A systematic review of the benefits of home telecare for frail elderly people and those with long-term conditions. J Telemed Telecare 2007; 13: 172-179
14. Heinzelmann PJ, Lugn NE, Kvedar JC. Telemedicine in the future. J Telemed Telecare 2005; 11: 384-390
15. Hall JA, Dornan MC. Meta-analysis of satisfaction with medical care: description of research domain and analysis of overall satisfaction levels. Social Science and Medicine 1988; 27: 637-644

Hospital Based Economic Assessment of Robotic Surgery

R. Miniati[1], F. Dori[1], G. Cecconi[1], F. Frosini[1], F. Saccà[1],
G. Biffi Gentili[1], F. Petrucci[2], S. Franchi[2], and R. Gusinu[2]

[1] Information Engineering Department, University of Florence, Florence, Italy
[2] HTA Department, Teaching Hospital AOU Careggi, Florence, Italy

Abstract—The aim of the following paper is reporting the hospital based assessment of robotic surgery at the teaching hospital AOU Careggi in Florence (Italy). The analysis was mainly focused on the economic assessment of robotic surgery at the hospital context by evaluating different acquisition simulated scenarios and by comparing two different robotic solutions: the daVinci© and the Kymerax©.

The methodology is based on the calculation of specific economic indicators which takes into consideration the number of interventions and different clinical uses. The indicators are based on scientific evidence coming from scientific literature, simulated clinical and consumption data and expert opinions.

DaVinci system usability is really high especially for those clinical application needing complex movements while Kymerax could represent a valid alternative to laparoscopy for less complicated surgery.

Keywords—Hospital Based Assessment, Robotic Surgery.

I. INTRODUCTION

The paper is organized as follows: a first part introducing the robotic systems, then the methodology description including the working hypothesis and explaining the simulated data, afterwards the results and finally the conclusions.

The daVinci ©, owned by the US company Intuitive Surgical, is a robotic system for surgical application developed in the 1999. The system is composed as follows:

- Surgical console with 3D vision technology (InSite®Vision) and the controls system;
- Patient side equipment, with three robotic arms (plus Endowrist® tools) guided by the surgeon;
- A vision system equipped with monitors, lights and the CO_2 insufflator;
- Alerts and alarms system.

Further features of the system regard the scaling 5:1 of the surgical movement by the surgeon, the filtering of the surgeon hands vibrations and the tactile feedback.

KYMERAX© is a robotic system by the Japanese Terumo Corporation developed in 2011. The system is technologically less complex than the daVinci, as described in figure 1:

- A portable console;
- The vision system;
- Four terminal tools with 6 dof capable of 360° movement (180° clockwise and 180° anti-clockwise).

The system includes 6:1 scaling (6 mm displacement made by the surgeon hand corresponds to 1 mm displacement made by the system) and tactile feedback.

Finally, for both systems, users' training is included before the use.

II. METHODOLOGY

A. Comparing Acquisition Scenarios

The analysis of scientific literature [1-4] on hospital based assessment with the daVinci robot leads to define reliable economic and clinical data and indicators. For the economic part, costs are moderately high compared to those of traditional laparoscopic and "open" surgeries. Acquisition cost amounts to € 2.5 million, maintenance costs are estimated in 9-10% of the purchase price. Consumable costs vary according to the clinical specialty: 3.000 € for gynecology and 5.000 € for urology per intervention, while the national health system economic refund is about 5.000 € per surgery (eg. urology). Once the surgical team is expert with the robot application, Operating Theatre (OT) costs (time plus personnel) can be assumed similar to the ones evaluated for traditional laparoscopy.

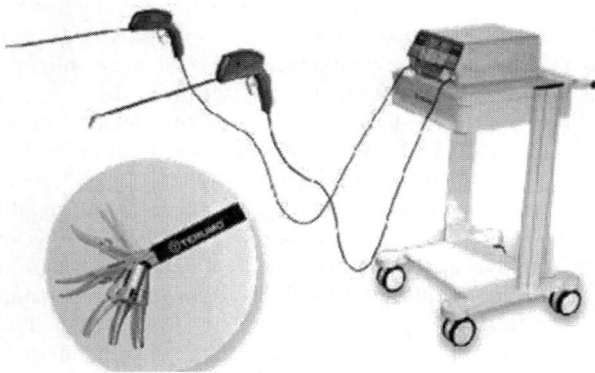

Fig. 1 The robotic system Kymerax©

For clinical aspects in urology, benefits coming from the use of robotic system than traditional laparoscopy have been numerously reported, with the advantage of faster patient recovery and shorter hospitalization length and less number of relapses and re-intervention.

On the other hand, in gynecology the use of daVinci is not very convenient for both the economic (confirming the general literature data) and the clinical dimensions as it is not more effective than laparoscopic procedure, especially for benign tumors [5].

As reported in table 1, for the simulated scenarios, only urology was considered in 200 annual interventions composed as follows:

- 112 interventions on pelvis without complications.
- 84 kidney and ureter cancer;
- 11 interventions on pelvis with complications.

The choice of these types of intervention depends on the fact that they represent the most clinical treated diseases and the ones with the higher costs.

The acquisition options considered in the analysis were service, leasing and purchase. Moreover, the purchase cost is estimated in 2.400.000€ while for correspondent maintenance is quantified as 9% of the purchase cost and 10 years life cycle.

B. DaVinci and Kimerax Evaluation

For a better results comparison between daVinci and Kymerax systems, the analysis involved for both technologies, the same economic refunds, hospital and personnel costs, and volume of activity.

Different values were considered for cost of accessories and consumables according to the national distribution company simulation (Terumo-Italy®) which defined a purchase cost of 150.000 €, disposables cost of 2.750 € per intervention (considering a 20 intervention offer) and cost of the accessories around a € 91.000.

Only the acquisition option was considered for the economic comparison with daVinci robot.

III. RESULTS

As reported in figure 2, the most convenient acquisition option may be direct purchase of the robot, as the annual cost of the device (by considering 7 years of expected life cycle) is lower than annual service or leasing. For a further simulation, table 1 reports all the simulated data included in the assessment and in the comparison.

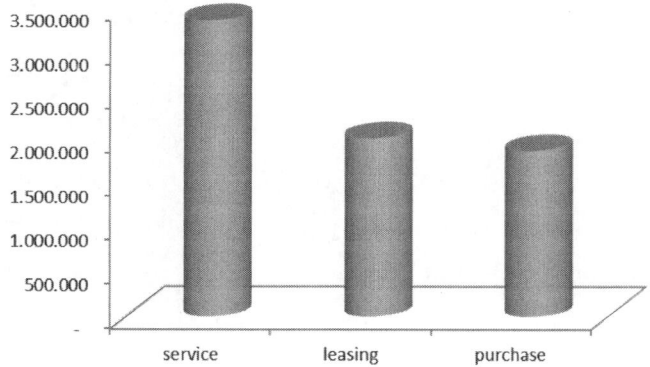

Fig. 2 Economic analysis on acquisition options for the daVinci system.

According to figure 3, it would be more convenient remaining under 500 interventions as the costs for each procedure gets more expensive over it (as showed by an higher angular coefficient).

Figure 4 shows the economic comparison between daVinci and Kimerax systems. Kymerax shows lower costs coming from less expensive disposable and purchasing costs.

It is interesting to underline how, also for Kymerax, 500 interventions represent the specific threshold for the increase of the single intervention cost, even if its relative cost increase (over 500 interventions) is more moderate than in the daVinci system.

Clinical comparison cannot be carried out as there are not high quality clinical evidence yet on KYMERAX.

Table 1 Simulation of costs and profits for daVinci systems in urology.

	profits (€)	hospitalization_day (€)	hospitalization (day)	OT (€)	disposable (€)	personnel (€)	Intervention (n°)
pelvis	4.234	235	7,8	1.457	5.123	1.366	112
kidney	5.948	235	7,1	1.339	4.471	1.366	84
pelvis (complication)	5.418	235	9,7	1.457	5.123	1.366	11

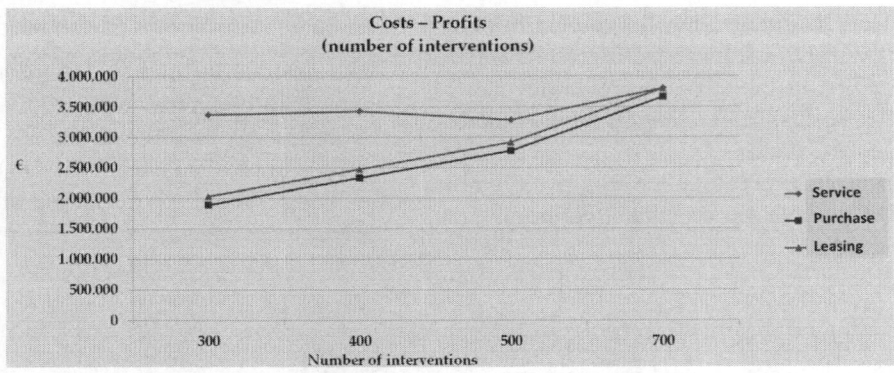

Fig. 3 Annual daVinci© cost-profits according to the number of interventions (simulated data).

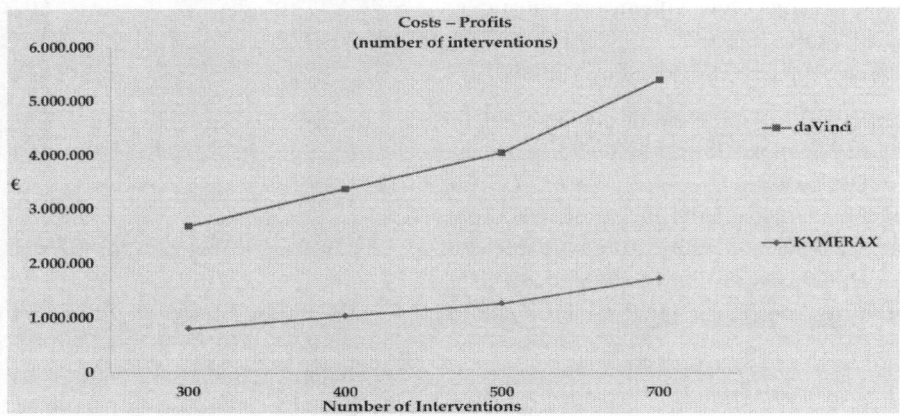

Fig. 4 daVinci© - Kymerax© economic annual comparison according to the number of interventions (simulated data).

IV. CONCLUSIONS

The study confirmed the strong economic impact of the surgical robot on the hospital budgets. The direct purchase is the most convenient option for robot acquisition, costs of the robot are very high and of great impact, especially those related to the consumable costs.

Although the clinical comparison between the two systems is still an open problem, as the KYMERAX is an emerging technology without enough medical evidence, it could probably be a valid alternative to laparoscopy procedures for less complicated surgery.

ACKNOWLEDGMENT

The authors sincerely thank the essential and fundamental contribution to the study provided by Dr. Nerattini Marco, Dr. Mercatelli Andrea and Dr. Presicce Giorgio from the teaching hospital AOU Careggi in Florence (Italy).

REFERENCES

1. G. Gautam, A. L. Shalhow, Section of Urology, Department of Surgery, University of Chicago Medical Center, Chicago, Illinois, Vol.183, 858-861, March 2010
2. G. Turchetti, I. Palla, F. Pierotti, A. Cuschieri, "Economic evaluation of da Vinci-assisted robotic surgery: a systematic review"
3. Bernard J. Park, MD, FACS "Cost concerns for robotic thoracic surgery", Division of Thoracic Surgery, Hackensack University Medical Center.
4. Robotic Assisted Surgery, Health Technology Assessment Program, DRAFT EVIDENCE REPORT, March 22nd, 2012, Health Technology Assessment Program (HTA) Washington State Health Care Authority, http://www.hta.hca.wa.gov
5. J. D. Wright, C. V. Ananth, S. N. Lewin, W. M. Burke, Yu-Shiang Lu, A. I. Neugut, T. J. Herzog, D. L. Hershman, "Robotically assisted vs laparoscopic hysterectomy among women with benign gynecologic disease", JAMA, Vol. 309, No. 7, 2013.

Author: Roberto Miniati
Institute: Dept. of Information Engineering
Street: Via Santa Marta 3
City: Florence
Country: Italy
Email: roberto.miniati@unifi.it

Evaluating Care Coordination and Telehealth Services across Europe through the ACT Programme

C. Maramis[1], D. Filos[1], I. Chouvarda[1], N. Maglaveras[1], S. Pauws[2], H. Schonenberg[2], C. Westerteicher[3], and C. Bescos[3]

[1] Lab of Medical Informatics, Medical School, Aristotle University of Thessaloniki, Greece
[2] Philips Research Laboratories, Eindhoven, The Netherlands
[3] Philips Telehealth Europe, Home Healthcare Solutions, Philips Healthcare, Boeblingen, Germany

Abstract—This paper presents the main concepts and ideas of the research program ACT (Advancing Care Coordination & TeleHealth Deployment). ACT focuses on studying Coordinated Care (CC) and TeleHealth (TH) services for chronically ill patients, such as heart failure, COPD, diabetes, as well as co-morbid patients. The potential of these services to improve quality of care within European healthcare systems can be fully released only when certain barriers are overcome and organizational changes progress. Therefore, the purpose of ACT, as described in this work, is to study in a qualitative and quantitative manner, with a solid and reproducible research methodology, the organizational and structural procedures supporting effective implementation of CC&TH services in the routine management of chronic patients. The consortium will investigate key organizational and structural drivers in 5 European healthcare regions, and eventually produce a toolkit for use across Europe.

Keywords—telehealth, care coordination, chronic illness, regional evaluation, best practices

I. INTRODUCTION

Chronic illness is defined as a condition of persistent or long lasting effects. In EU 10 million people suffer from heart failure (HF) [1], 20 million have chronic obstructive pulmonary disease (COPD) [2] and 60 million live with diabetes [3]. Chronic disease poses a huge social and economic burden. Each year, the aforementioned chronic diseases cost EU healthcare around €125 billion. The direct healthcare costs, such as hospitalization, for HF alone is over €10 billion per year, and mortality rates range from 5% to 30% [1].

Coordinated Care (CC) and TeleHealth (TH) services have the potential to effectively aid chronically ill people in their home through remote management systems and integrated networks of caregivers. These systems can both reduce the economic burden of chronic care and improve the delivery of clinical support – despite the shortage of skilled professionals within Europe – by increasing patients' independence and control over their health and lifestyle [4]. The value of TH services on reducing hospital admissions, days in hospital and mortality rates, has been highlighted by [5] in the Cochrane Review 2010 on Telemonitoring in HF, and by [6] in the COPD Cochrane Review 2011, as well as in the Whole System Demonstrator (WSD) findings [7].

Challenges in translating the positive effects of CC&TH into routine clinical practice are currently primarily organizational and structural in nature. To date, healthcare systems have mainly superimposed CC&TH onto their existing care delivery structures. To generate maximum impact however, care ecosystems need to be restructured, and care providers educated as to how to take full advantage of CC&TH and see its impact in health outcomes, administrative efficiency, cost effectiveness and user (patient and health professional) experience. Translating evidence into practice is complex and requires significant organizational change.

The first EU-wide attempt to tackle the aforementioned challenges in CC&TH is the ACT Programme. ACT aims to undertake the translational tasks that have been described in the previous paragraphs in order to produce and distribute a set of guidelines for successful exploitation of CC&TH services across Europe. This paper presents the vision of ACT and it is structured as follows. Section II describes the ACT Programme focusing on two of its key components. Section III presents the expected outcomes of the programme, while Section IV draws the conclusion of this work.

II. PROGRAMME DESCRIPTION

The "Advancing Care Coordination and Telehealth Deployment" (ACT) Programme is a recently initiated (Feb. 2013) EU Health Project that aims to overcome current organizational and structural barriers in order to transfer the advantages of CC&TH services into routine clinical practice. This will be attempted by a powerful European consortium of companies, universities, hospitals and healthcare authorities. The consortium will investigate 3 core questions: 1) *How do CC&TH services work around Europe?*, 2) *What needs to be done to make them work better?*, and 3) *How to deploy high-quality CC&TH services in new European regions?*

The programme has two main objectives. The first is to *understand the working mechanism* of existing CC&TH services. Mastering this mechanism will allow ACT to predict whether a new CC&TH service is going to be a 'success' or not, to fine-tune a running service in order to optimize its outcomes, and – most importantly – to 'compose' optimal CC&TH services using the best components of existing services as building blocks. When the first objective is achieved, the programme will aim towards its second objective, which is to *compile its conclusions into a set of guidelines for sharing* with other European healthcare authorities. This shall facilitate the deployment of high-quality CC&TH services in other European regions as well.

In order to achieve its first objective ACT needs to make use of a set of appropriately selected *Key Drivers & Indicators*. The drivers are generic aspects of a healthcare programme that are considered when determining the quality of CC&TH, while the indicators are measurable or inferred healthcare outcomes used for assessing the drivers. The main task of the first objective is to reveal and eventually describe the relation between the drivers and the indicators. This is undertaken by the *Evaluation Engine*, i.e., the computer system infrastructure of the programme. The Key Drivers & Indicators and the Evaluation Engine components of ACT are described in detail in the following subsections.

ACT addresses CC&TH services for the management of chronic patients in the area of HF, COPD, diabetes and also the – more complex – case of patients with multiple conditions (co-morbid patients). By developing key drivers and indicators that best apply to the aforementioned medical conditions, the programme aims to longitudinally capture the impact of CC&TH services in the integrated care. Contrary to previous efforts for assessing integrated care services or technology interventions, ACT is focusing on not only covering the high level factors that facilitate or hinder care-coordination and telehealth but also evaluating their variation over time and – ultimately – understanding the reasons of success or failure of the different regional interventions.

Five European regions, namely Groningen (the Netherlands), Scotland (United Kingdom), Lombardy (Italy), Basque Country (Spain), and Catalonia (Spain) with long experience in CC&TH programmes have been brought together in ACT's consortium. These regions have been committed to deploy and operate their preferred CC&TH solutions in at least 3000 patients each, sharing their experiences and providing the data required for the implementation of the programme.

In brief, the timeplan of ACT is as follows: Starting in Feb. 2013, an initial six-month study is going to determine the baseline status of CC&TH in the participating regions (*phase 1*). The care structures and procedures are then going to be optimized during an 18-month iterative evaluation procedure (*phase 2*). Finally, the optimized procedures are going to be compiled into a 'cookbook' of best practices for large-scale CC&TH deployment and the findings are going to be replicated in other EU health regions (*phase 3*). The implementation phases of ACT are summarized in Table 1, whereas the overall perspective of the programme is illustrated in Fig. 1.

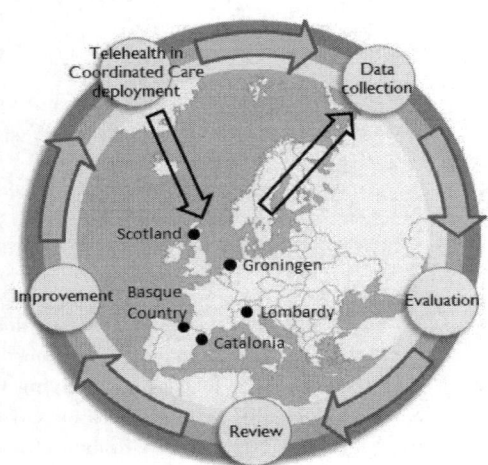

Fig. 1 The ACT perspective.

Table 1 Implementation phases of the ACT Programme.

Phase		Primary Results
1	baseline assessment	evaluate key results at baseline within regions, compare across regions
2	iterative refinement	modify key drivers to optimize outcomes, assess key results on an ongoing basis (3 iterations)
3	dissemination	compile cookbook, replicate findings in new regions

Since the evaluation process will be performed in existing healthcare ecosystems that support large cohorts of patients, the 'best practice' results will be verified with the necessary 'real-world' rigour. This shall ensure the reproducibility of the results in other regions, taking into account regional/national boundary conditions and other diversities.

ACT is fully aligned with the EC European Innovation Partnership on Active and Healthy Ageing (`https://webgate.ec.europa.eu/eipaha`) objectives to deploy integrated care for chronically ill patients. For more information on ACT the reader may visit the programme's website at `http://act-programme.eu/`.

A. Key Drivers and Indicators

Within the framework of ACT, a key driver has been defined as *an essential aspect of a healthcare programme that can cause an intended change in a healthcare-related outcome*. Thus, the key drivers characterize the quality of a healthcare programme and can be viewed as a set of objectives that need to be optimized in order to improve its quality.

Given that ACT is looking for 'best practices' in CC&TH programmes, it becomes evident that it should start by developing a set of strategically-selected key drivers. The members of this set, which must be minimal, will be sought within four significant aspects of healthcare:

1. **Patient risk stratification** – to provide services that match patient needs,
2. **Patient and professional engagement** – in the integrated healthcare services,
3. **Organization and workflow structures optimization** – to ensure the alignment of the organizational structure and the service delivery, and
4. **Programme efficacy and efficiency** – evaluating outcomes versus employed resources.

In order to assess the – abstract – key drivers one must define a set of healthcare outcomes that can be either measured directly or inferred/estimated from other measurable entities. In the framework of ACT, these outcomes are called key indicators. For instance, with respect to the 'patient stratification' key driver, a list of possible key indicators could include outcomes related to: 1) clinical effectiveness (mortality, health care resources utilization), 2) the patients (disease severity, exacerbation rate, frailty), and 3) the stratification program itself (number of patients who are eligible, stratified, received care, resource utilization per patient sub-group).

The – quantitative – key indicators are of critical importance in ACT's effort to understand and model the working mechanism of CC&TH services. Therefore, the set of indicators to be utilized by ACT needs to be selected very carefully so as to ensure that 1) all the key drivers can be assessed, 2) the set is minimal – to avoid unneeded over-complication of the problem, and 3) the data needed for computing the indicators is available in all the participating regions.

In order to ensure the aforementioned requirements the ACT programme has carefully designed a *key indicator selection process*. Using the MAST framework [8] as a starting point, ACT compiled a long list of over 300 candidate key indicators (classified per associated key driver). Then, the candidate indicators were ranked with respect to their significance according to the driver-specific authorities of the consortium and the participating regions via a voting process.

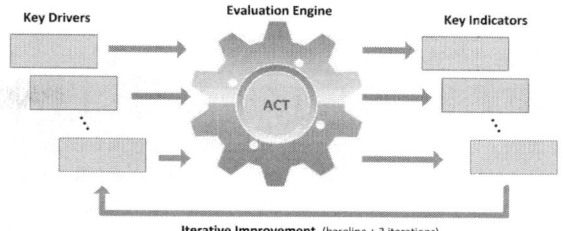

Fig. 2 The Evaluation Engine of ACT.

Also, the availability of the necessary data for each candidate indicator in the five regions was checked. Finally, the driver-specific authorities and the participating regions selected the final, minimal set of key indicators via a consensus process.

B. Evaluation Engine

The Evaluation Engine is the organizational and technological infrastructure of ACT that is employed to *collect, analyse and report* CC&TH related data. The Evaluation Engine supports monitoring of key indicators for assessing key drivers in all participating regions. The engine makes sure that the necessary data on CC&TH deployment is transparent, interrogative and actionable to all regions at several predetermined moments in time (see Table 1). Fig. 2 visualizes the relation of the Evaluation Engine with the programme's key drivers and indicators.

The Evaluation Engine will perform the following tasks:

- **Key Indicator Definition.** A minimal set of key indicators that are supported by all regions is going to be defined in a concrete, unambiguous and computer processable manner (as described in the previous subsection), including their target values and limits.
- **Data Collection.** The data that is required to compute the key indicators is going to be collected from the regions. The format, frequency and granularity of the data will be defined, according also to the needs of the following task.
- **Analysis and Reporting.** The collected data will be analysed using statistical processing and/or data mining techniques and custom reports for the regions will be prepared. This also involves the computation of inferred or aggregated indicators. Foreseen analysis will include comparison between and within regions, as well as generation of reports and interactive web-based views.

The Evaluation Engine will be developed on the basis of 'off-the-shelf' data analytics technologies (e.g., the BI suites of Jaspersoft or Pentaho) supporting data transformation, statistical processing, data mining and reporting functionalities.

III. EXPECTED OUTCOMES

Two principal outcomes are expected upon the completion of ACT. The first concerns the participating regions. Within these 5 regions, ACT is going to systematically evaluate the key drivers. The conclusions/findings of this evaluation shall allow the effective deployment of CC&TH services in the regions: Existing services will be optimized, unsuccessful services will be abandoned, and new services will be composed from the 'best' components of existing ones.

The second expected outcome of ACT has to do with healthcare regions outside the consortium. The knowledge and experience gained from the systematic evaluation of the key drivers are going to be incorporated in a best practice manual, the so called *ACT Cookbook*; this is going to specify how the drivers can be leveraged to expedite deployment of CC&TH services in other European healthcare regions.

The cookbook will provide explicit recommendations and examples regarding each of the key drivers. These recommendations will take into account the regional/national boundary conditions, and, if appropriate, other diversity aspects. Best practice recommendations regarding the following CC&TH drivers will be summarised in the cookbook:

- **Optimising CC&TH organization and workflow structures** – Providing effective structures, including examples of how the care provider ecosystems can be organized, as well as IT tools and solutions that support patient-centric management.
- **Effective patient stratification** – Moving from disease-oriented stratification models to more holistic approaches. This involves patient stratification concepts with demonstrated practical and sufficient specificity to ensure that patients are clinically best served.
- **Staff engagement and education** – Prescribing a landscape of tools and incentive programmes that motivate staff to engage with CC&TH.
- **Improving patient adherence** – Presenting strategies to improve patient adherence (to disease management regimes, etc.) and empower patients to more effectively self-manage their condition via CC&TH delivery.
- **Improving care provider efficacy and efficiency** – Including guidelines to assist CC&TH in delivering adequate health outcomes to the eligible population in a cost efficient manner.

By the end of the project, 10 to 15 European healthcare regions are expected to become affiliated with ACT. The ACT Cookbook is going to help these regions avoid beginners' mistakes and save considerable time in deploying their own CC&TH programmes in an optimized manner.

IV. CONCLUSION

In this paper, the vision of the recently-initiated EU ACT programme was presented. Starting from 5 European regions, ACT will evaluate a variety of existing CC&TH services for HF, COPD, diabetes and co-morbid patients. The evaluation will revolve around patient stratification, patient and professional engagement, organization and workflow structure optimization, programme efficacy and efficiency.

The main objectives of the programme is to discover the underlying mechanism that allows CC&TH services to actually work in the investigated regions, and, based on these discoveries, to identify 'best-in-class' practices within the services. The conclusions of this effort will be compiled into a 'manual' (the ACT Cookbook), aiming to assist other European regions to increase their capacity in CC&TH services.

ACKNOWLEDGEMENTS

This work is partially supported by the HEALTH-20121209 project, entitled "Advancing Care Coordination and Telehealth Deployment" (ACT), funded by the Commission of the European Community (CEC).

REFERENCES

1. Braunschweig F, Cowie MR, Auricchio A. What are the costs of heart failure? *Europace*. 2011;13:ii13–ii17.
2. European COPD Coalition at http://www.copdcoalition.eu/about-copd/prevalence.
3. Diabetes Atlas of the International Diabetes Federation (5th ed.) at http://www.idf.org/diabetesatlas/europe.
4. Maglaveras N, Reiter H. Towards closed-loop personal health systems in cardiology: The HeartCycle approach in *Conf Proc IEEE Eng Med Biol Soc*:892–895 2011.
5. Inglis S, Clark R, McAlister F, et al. Structured telephone support or telemonitoring programs for patients with chronic heart failure *Cochrane Database Syst Rev*. 2010;2010:1–138.
6. Inglis SC, Clark RA, McAlister FA, Stewart S, Cleland JGF. Which components of heart failure programmes are effective? A systematic review and meta-analysis of the outcomes of structured telephone support or telemonitoring as the primary component of chronic heart failure management in 8323 patients: Abridged Cochrane Review *Eur J Heart Fail*. 2011;13:1028–40.
7. Bower P, Cartwright M, Hirani SP, et al. A comprehensive evaluation of the impact of telemonitoring in patients with long-term conditions and social care needs: protocol for the whole systems demonstrator cluster randomised trial *BMC Health Serv Res*. 2011;11:184.
8. Kidholm K, Pedersen CD, Rasmussen J, et al. A new model for assessment of telemedicine – MAST *Int J Integr Care*. 2011;11.

Early HTA to Inform Medical Device Development Decisions – The Headroom Method

A.M. Chapman, C.A. Taylor, and A.J. Girling

University of Birmingham, Multidisciplinary Assessment of Technology Centre for Healthcare (MATCH), Birmingham, UK

Abstract—The headroom method offers medical device developers a simple way to integrate health economics into the decision of whether or not to develop a medical device. By estimating the maximum reimbursable price (MRP) for a new device idea, and comparing this reimbursement opportunity with a developer's expected costs, the method offers a way to ensure developers invest only in devices that are commercially viable. This paper explains the headroom method, and describes a study whose aim was to evaluate the method by applying it retrospectively to a large and diverse set of case studies. The method was applied systematically to 20 devices / diagnostics that were invented in the past (identified from the UK National Horizon Scanning Centre (NHSC)'s 2000 to 2009 database). Predicted 'headroom' was then compared with later UK National Health Service (NHS) uptake, in order to assess the performance of the method as a predictive tool. The headroom method predicted uptake with a sensitivity of 92% and a negative predictive value of 67%. When numerical headroom assessments were considered alongside the more qualitative factors identified (which generally reflected the clinical and market context), the method offered a good indication of commercial opportunity.

Keywords—Health Technology Assessment, Early economic evaluation, Headroom, Investment decisions.

I. INTRODUCTION

In a healthcare context, an economic evaluation is simply the comparison of two clinical interventions (generally current practice versus its proposed replacement) in terms of their costs and consequences [1]. This forms a key part of a health technology assessment (HTA), which guides decision-makers in their consideration of medical technology uptake. Methods of economic evaluation at the demand-side are well established. Less well developed, however, is the use of economic evaluation to inform *supply-side* decision-making in the early stages of development. Hartz and John [2], who searched for both methodological and empirical studies using early data, found limited research into or application of methods of early economic evaluation. With companies keen to direct resources away from unsuccessful products as early and accurately as possible, there is a clear need for such a tool.

The headroom method, which has provided a key research focus for the Multidisciplinary Assessment of Technology Centre for Healthcare (MATCH), proposes to act as an early warning signal for industry, so that only commercially viable innovations are pursued [3]. However, this, and other methods of early economic evaluation, have only been applied to a small number of examples, and their predictive capacity has not been tested [4]. The study outlined in this paper therefore assesses the possible implications of using the headroom method to guide development decisions.

II. METHODS

A. The Headroom Method

The headroom method allows device developers to consider the commercial viability of a product idea, before any resources are directed toward its development. By considering the potential value of a device to the health service, the price it might fetch in the future can be calculated. The earlier this value proposition can be investigated, the better.

At such an early stage the developer will have limited or no data on the performance of a medical device. However, the developer should have an idea of its intended impact on health service costs (SC) (e.g. staff time, hospital bed occupancy, etc.) and/or its potential impact on a patient's health (reduced pain, faster recovery, reduced mortality, etc.). With these estimations, a simple re-arrangement of the standard incremental cost-effectiveness ratio (ICER) equation can identify the maximum reimbursable price (MRP) for the new device. That is to say, based on early expectations of service cost and patient health impact (as compared with the current gold standard in clinical practice), we can calculate how much the health service should be willing to pay for the innovation.

Health benefit is expressed in quality adjusted life years (QALYs), which provides a standard measure of health benefit which can reflect changes in quality of life (QoL: a weighting between 0 and 1) as well as the time over which this improvement is sustained. In the UK (the perspective from which this study has been undertaken) the National Institute of Health and Care Excellence (NICE) explicitly provides an ICER threshold of £20,000 to £30,000 [5]. This represents the decision-maker's willingness to pay (WTP) for one unit of health benefit (1 QALY). For the purposes of this study, we take this WTP to be £20,000 per QALY.[1]

By considering the decision problem of the healthcare provider, we can calculate the net benefit of a proposed intervention, and use this analysis to calculate the MRP:

$$MRP = (£20,000 * \Delta QALY) - \Delta SC \quad (1)$$

where Δ denotes incremental costs or QALYs (as compared with current practice). The MRP is therefore equal to the difference between: (a) the value of the potential health benefit, and (b) the net impact on costs to the health service. The latter includes any change in service costs resulting from the implementation of the new device (which add to the MRP if it is cost saving or detract from it if it will be more expensive to implement) and also the price of any equipment currently used which would be displaced if the new technology were implemented. The only variable unaccounted for in this health service net benefit (the right-hand-side of Equation (1)), therefore, is the price of the new device. It is this price that we wish to identify the 'headroom' for: the MRP.

The MRP represents the potential unit value of a new device, which should be compared with likely costs of production in order to judge commercial viability. Although not discussed in this paper it should be noted that, for business decisions, the MRP can be used in conjunction with market size to inform an assessment of recoverable research and development (R&D) costs.

B. The Process

A structured process and template of resources for the headroom method's application was followed. In the first instance, an analysis of the relevant disease or condition and the relevant patient population is required, as well as an understanding of the proposed function of the new device. Consideration of the current clinical care pathway should follow: how would the new device alter this, and what deficiencies in current clinical practice would it address? Costs are then investigated, by considering plausible health service resource impact and using national unit cost data to quantify these. Potential patient health impact must subsequently be considered, which similarly should be quantified using relevant literature resources, e.g. cost-effectiveness studies addressing related topics. Peripheral developments in the clinical or market context should be noted throughout the analysis, and articulated into research questions. These research questions should help to guide R&D if a positive development decision is indicated.

---

[1] The WTP threshold will differ by country, based on healthcare budgets and the opportunity cost of spending. In many countries this figure is not explicit, though could be inferred from past decisions (see for example Simeons [6]).

C. A Validation of the Headroom Method

In this paper I present the results of a retrospective case study analysis. The method was applied systematically (using a structured pro forma) to 20 devices/diagnostics invented in the past, retrieved from the UK National Horizon Scanning Centre (NHSC)'s 2000 to 2009 database. A UK NHS perspective was taken. Literature searches used to inform headroom analyses were date-restricted to end 3 years before the publication of the NHSC briefing (or 3 years prior to launch date if this was known to predate the briefing's publication), in order to mimic the availability of information at 'concept' stage of the device. The method was applied to each case study under the premise of 'headroom in a day'. The medical technologies were then followed-up to the present day, to observe actual market success (and price where relevant). This was then compared with the headroom decision outcome, to assess the method's performance.

III. RESULTS

A. Case Study Example: Stapled Hemorrhoidectomy

Hemorrhoids are swollen blood vessels in or around the anus which can cause itching, mucus discharge, pain and bleeding. A 1995 time perspective was taken for this case study, at which time conventional practice for prolapsed third and fourth degree hemorrhoids was surgical removal using the Milligan-Morgan technique. Stapled Hemorrhoidectomy proposes to offer a less painful and less invasive alternative to surgical removal; an intraluminal circular stapling device is introduced into the anal canal, the staples applied, and the redundant tissue removed [7].

Conventional surgical removal usually involved three days in hospital [7]. According to hospital episode statistic data, around 10% of procedures were carried out as day-cases [8]. The main cost saving proposed by the Stapled Hemorrhoidectomy derives from its minimally invasive nature, meaning that all procedures should be carried out as day-cases. Therefore, it can be estimated to save 2 days worth of hospital costs (valued at £300 per day) in 90% of cases. This is worth an average of £540 per patient in service cost savings.

The proposed new procedure is also expected to be less painful and allow for faster recovery. It was considered that QALY impact could be reasonably represented as two *fewer* weeks (0.038 of a year) in a health state involving 'some problems performing usual activities' and 'moderate pain or discomfort' (QoL weighting 0.76). This was valued, using the EQ-5D UK value sets [9], and is worth 0.00912 QALYs per person (0.038 * [1 − 0.76]). This results in the following

MRP, which represents the maximum the NHS should be willing to pay for Stapled Hemorrhoidectomy per patient, given the assumptions made:

$$MRP = (£20{,}000 * 0.00912) + £540$$
$$= £722 \qquad (2)$$

This is represented graphically on a cost-effectiveness plane in Figure 1.

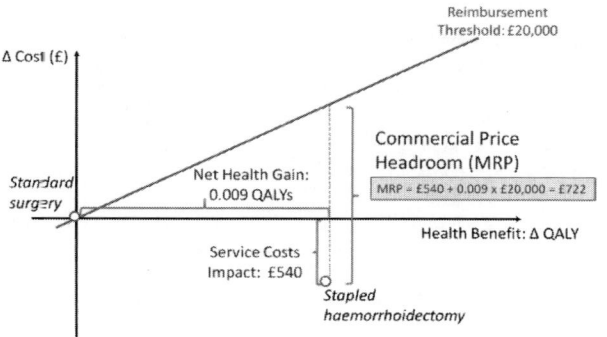

Fig. 1 The MRP for Stapled Hemorrhoidectomy

Upon follow-up, the price to the NHS per Stapled Hemorrhoidectomy procedure is £420 [10]; the headroom analysis would therefore have indicated ample room for development. NICE now recommends the procedure, finding it to be both clinically and cost-effective. Trial results indicate that Stapled Hemorrhoidectomy does indeed involve less pain and a shorter hospital stay compared with standard surgical removal, though not the extent modeled in the headroom analysis. The two factors that cause the procedure to be less cost-effective than the situation modeled in the headroom analysis were both highlighted qualitatively as potential threats during the write-up (that not all are carried out as day-cases, and subsequent prolapse and re-intervention is increased).

B. Relating Headroom to Subsequent Market Uptake: Performance of the Headroom Method

Headroom analyses (pre follow-up) took on average 9 hours. Results presented in Tables 1 and 2 relate the development decision indicated by the headroom analysis to subsequent NHS uptake. The postulated headroom decision relates to the numerical headroom assessment only (excluding consideration of the qualitative threats or advantages that were identified and articulated during the headroom analysis). One case study was excluded from these results, as it transpired to be a patient purchase.

Table 1 Headroom and NHS uptake contingency table

		FOLLOW-UP		
		NHS Uptake	No NHS Uptake	
HEADROOM	Favorable	12	2	14
	Unfavorable	1	4	5
		13	6	19

Table 2 Summary headroom performance statistics

Performance statistic	Value	Description
Sensitivity	0.92	Of those that achieved NHS uptake, the proportion for which headroom was favorable
Specificity	0.67	Of those that did not achieve NHS uptake, the proportion for which headroom was unfavorable
Positive predictive value	0.86	Of those for which headroom was favorable, the proportion that went on to achieve NHS uptake
Negative predictive value	0.80	Of those for which headroom was unfavorable, the proportion that were unsuccessful in achieving NHS uptake
Accuracy	0.84	'Getting it right': the proportion of case studies whose subsequent NHS uptake was correctly predicted

The low specificity reaffirms the fact that a favorable headroom result does not guarantee future uptake, meaning that false positives are not uncommon. The use of optimistic early assumptions means that the headroom method should most appropriately be seen as a 'rule-out' tool, to avoid investment in devices that could *never* be cost-effective. More importantly, therefore, the method should not rule-out technologies that have some chance of future uptake (which can be observed in its sensitivity). Sensitivity was not found to be 100%, due to one case (a cell therapy product for knee cartilage defects). Although, as the headroom analysis had predicted, NICE did not deem the product to be cost-effective in its own assessment, it has achieved reimbursement on the NHS through pass-through payments (a special reimbursement scheme for expensive innovative technologies). The negative predictive value (which represents the performance of the headroom method as a rule-out tool) illustrates that 80% of those for which headroom appeared unfavorable did not achieve NHS uptake (4 out of 5). This, again, was driven by the cell therapy case described above.

IV. DISCUSSION

The two most important caveats to the interpretation of the result statistics outlined are: (a) the close-to-market nature of the sample (the NHSC provides briefings within three months of launch, thus biasing the sample toward successful products); and (b) that the headline 'result' in terms of headroom favorability (to which subsequent uptake was compared) was based only on the numbers that were generated by the analyses, rather than the information provided by the exercise as a whole. Considering the full analysis of each case study, it was demonstrated that even at a very early stage of development, through the process of assessing headroom, the relevant threats to uptake can also be brought to light, as well as the additional opportunities for a product that have not been characterized by a numerical headroom assessment. By considering these factors alongside the monetary estimates provided, the commercial opportunity was generally well captured.

The study identified a number of challenges associated with the application of the headroom method as well as opportunities. It was found that, often, an MRP could not be fully quantified. This was sometimes due to the scarcity of clinical and cost data that could be identified in the literature on the relevant condition. Where devices would be used for multiple patients, a value *per patient* rather than per device (MRP) was sometimes given, as patient through-put was unknown. However, 'partial' headroom estimates were often straightforward to quantify, and on their own may be sufficient to support a development decision.

An important difficulty identified was in applying the headroom method to evaluate devices that could cause a significant shake-up to the care pathway of a patient (e.g. diagnostic devices). The direct impact of such devices is difficult to predict, and therefore headroom estimates must be based on particularly speculative assumptions.

Where a device would be used for multiple clinical applications, the headroom method was generally used to value the device in just one of these contexts. However, the method could plausibly be used to compare alternative market entry opportunities. Additionally, the value provided by a headroom estimation could also be used to inform product design.

V. CONCLUSION

There is an important role for the early integration of health economics into decision-making by medical device developers. The headroom method could offer a practical way to guide investment, by grounding development choices in the potential healthcare value of a product. The headroom method 'works' best when a flexible and inclusive stance is adopted, in which case the method can help to identify essential research questions that should be addressed through the development process.

ACKNOWLEDGMENT

The authors acknowledge funding from the EPSRC under the MATCH Programme (grant EP/F063822/1). The views expressed, however, are entirely those of the authors.

REFERENCES

1. M. F. Drummond, B. O'Brien, G. L. Stoddart, and G. W. Torrance (1997) Methods for the Economic Evaluation of Health Care Programmes, 2nd ed. Oxford: Oxford University Press, 1997.
2. S. Hartz and J. John (2008) Contribution of economic evaluation to decision making in early phases of product development: A methodological and empirical review, Int J Technol Assess Health Care, vol. 24, no. 04, pp. 465-472.
3. E. Cosh, A. Girling, R. Lilford, H. McAteer, and T. Young, (2007) Investing in new medical technologies: A decision framework, Journal of Commercial Biotechnology, vol. 13, no. 4, p. 263
4. M. Bartelmes, U. Neumann, D. Lühmann, M. Schönermark, and A. Hagen (2009) Methods for assessment of innovative medical technologies during early stages of development, GMS Health Technology Assessment, vol. 5, p. 15
5. NICE (2008) Guide to the methods of technology appraisal, London. Available at: http://www.nice.org.uk/media/B52/A7/TAMethodsGuideUpdatedJune2008.pdf
6. S. Simoens (2009) Health economic assessment: a methodological primer, International Journal Of Environmental Research And Public Health, vol. 6, no. 12, pp. 2950-2966
7. NHSC (2001) Stapled haemorrhoidectomy [New & Emerging Technology Briefing]. Available at: http://www.hsc.nihr.ac.uk/outputs/
8. HES (2000) HES online. Total procedures and interventions [Freely available data, Inpatient data]. Available at: http://www.hscic.gov.uk/hes
9. A. Williams (1995) A measurement and validation of health: a chronicle (Discussion paper 136), University of York: Centre for Health Economics
10. NICE (2003) IPG 34. Circular stapled haemorrhoidectomy. Available at: http://www.nice.org.uk/nicemedia/live/11075/30891/30891.pdf

Author:	Chapman, A.M.
Institute:	School of Health and Population Sciences, University of Birmingham
City:	Birmingham
Country:	UK
Email:	a.m.chapman@bham.ac.uk

Analysis of Exposure of Social Alarm Devices in Near Field Conditions

S. de Miguel-Bilbao[1], F. Solano[2], J. García[1], O.J. Suárez[3], D. Rubio[3], and V. Ramos[1]

[1] Telemedicine and e-Health Research Unit, Health Institute Carlos III, Madrid, Spain
[2] Audiovisual Engineering and Communications Unit, EUIT Telecommunication, Polytechnic University of Madrid
[3] RF Laboratory, General Direction of Telecommunications and Information Technologies,
Ministry of Industry, Energy, and Tourism

Abstract—Despite of the rapidly increasing number of short range wireless applications and devices in healthcare environments, very little is know about the radiofrequency (RF) exposure due to such devices.

The most widespread model of social alarm devices was identified from among the most used in telecare monitoring activities. The levels of electromagnetic (EM) field in near field conditions were measured inside an anechoic chamber with a specific absorption rate (SAR) fully automated test system, DASY5PRO. The obtained results were compared with the levels set by international regulations to analyze the exposure to EM fields.

In very close areas to the device, the obtained electric field levels show that local exposure can reach the most restrictive value of 3 V/m that is established in the International Electrotechnical Commission Standard of Electromedical Devices.

Keywords—social alarm devices, electric field, exposure threshold, near field exposure.

I. INTRODUCTION

A variety of short range wireless communication technologies and devices has become widespread in recent years [1]. Wireless technologies are increasingly used in environments focused on providing personal care settings with the primary intent of supporting patients and elderly people. Social alarm devices are wireless devices that work in the frequency range of 869.2-869.25 MHz, and are used to assist patients, elderly and disabled people in distress situations.

The use of such devices based on short range wireless technologies involves the increase of levels of electromagnetic (EM) fields in residential environments. Given the increasing use of domiciliary telealarm devices, and the nonexistence of previous studies about the working conditions and the emission levels, the electric field (E-field) levels were measured to compare them with the existing standards.

The proposed study provides a global, immediate and accurate vision that can help to characterize the E-field levels, and monitor the exposure to EM fields of people using and in the proximity of social alarm devices in home environments.

New health solutions based on any kind of short range technology must consider the issues of electromagnetic compatibility and regulatory compliance. Currently, the degree and type of EMF exposure need to be characterized in household settings, in order to ensure that applications operate properly and exposure guidelines are not exceeded.

II. METHOD AND MATERIALS

The measurements were performed in an anechoic chamber. The instrument of measurement was DASY5PRO that is usually used to measure the specific absorption rate (SAR). This instrument is provided with an E-field probe that allows near-field measurement of any device embedded in any casing. Fig. 1 shows the measurement device DAISY5PRO inside the anechoic chamber.

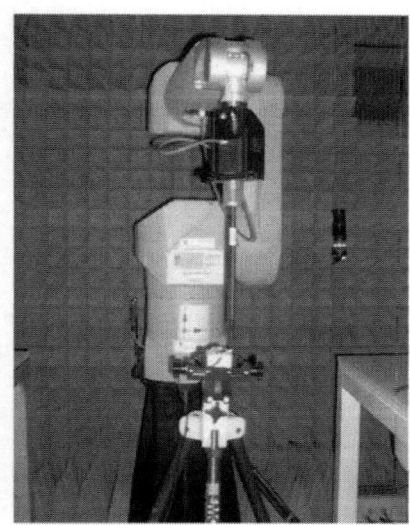

Fig. 1 The measuring instrument, DASY5PRO, and a tripod holding the device under test

The systems of social alarm devices consist of two operational units: the buttons that are worn by the users, typically hung over the neck or attached at the wrist, and the fixed unit that is connected to the home phone. When the user is in a distress situation, he can push the button, a radio frequency signal will be transmitted to the fixed unit and an emergency phone call will be made to the monitor centre. The buttons transmit a signal that typically consists of three pulses (depending on the model) at the frequency range of 869.2-869.25 MHz. This involves that effective information is transmitted only during a slot of time.

For this work, several models of social alarm devices were chosen from among the most used in telecare monitoring activities. Precisely the tested model was Tunstall AMIE+, which is shown in Fig. 2, whose working frequency has been fixed to 869.25 MHz.

Fig. 2 Selected model of social alarms devices: Tunstall AMIE+.

DASY5PRO is usually used with a unit that simulates the behavior of a base station in order to establish a communication with the tested device. This unit supports several communication protocols (CDMA, Bluetooth, GSM and UMTS), so it was necessary to modify several parameters of the measuring system to adapt it to the pulse transmission that characterize the emissions from the device under test (DUT).

A Rhode & Shwarz HE300 log periodic antenna and a Rhode & Shwarz FSH18 spectrum analyzer have been used to detect the pulsed signal generated by the DUT and extract the features of the signal:

- Period: 5.6 s.
- Duration of the effective transmission: 0.6 s.

Therefore the duty factor is 10.71% that is defined as the relation between the duration of the effective transmission and the period.

To measure of the E-field generated by a signal whose duty factor is less than 100%, it is necessary to obtain the peak value of the E-field during an interval of time greater than the period of the signal.

The DASY5PRO system calculates the average values of electric field by default, and provides the possibility of measuring peak values by introducing a factor called Peak to Average Ratio (PAR). This means that once the mean value of the E-field has been measured, it is possible to obtained the peak value with a correction factor that is defined as the relation between the mean power of a continuous signal (with a duty factor of 100%), and the mean power of a pulsed signal that does not transmit effective information for the entire duration of the transmission.

The correction factor PAR needed to measure the E-field of a pulsed signal is defined according to the following equation:

$$PAR = 10\log\left(\frac{P_{peak}}{P_{mean}}\right) \quad (1)$$

Where P_{mean} is the mean power of a pulsed signal, and the P_{peak} is the peak power.

To obtain this correction factor the first step is to validate the measurement system. The objective of the validation process is to know the relation between the mean power of a signal with duty factor of 100% (without modulation), and the mean power of a modulated signal with a duty factor of 10.71%.

This validation process has been carried out with a simulated signal of the same features than the signal transmitted by the DUT, therefore the validation signal is a pulse at the frequency of 869.25 MHz that modulates a square wave whose period is 5.6 s, and duty factor of 10.71%. This signal was generated by a Wavetek 395 signal generator and the radiation source was a dipole.

After measuring the E-field of the modulated and non modulated signals the obtained factor that relates the mean power of both types of signals is 10.

In order to check this calculation, the PAR has also been obtained theoretically. The mean power of the modulated signal that does not transmit effective information for the entire duration of the transmission is defined as follows:

$$P_{mean} = \frac{1}{T}\int_{-T/2}^{T/2} x^2(t)dt \quad (2)$$

Considering the features of the signal (power: 1 W, period: 5.6 s, and duration of the effective transmission: 0.6 s), the mean power can be defined as the relation between the duration of the effective transmission and the period of the signal, that is, the duty factor:

$$P_{mean} = \frac{0.6}{5.6} = 0.1071 \quad (3)$$

Substituting this result in the expression of the PAR (1):

$$PAR = 10\log\left(\frac{1}{0.1071}\right) = 10\log(9.35) \approx 10 \quad (4)$$

The calculated value is the same than that obtained experimentally.

The near field measurements were realized in the interpolated points belonging to a predefined grid whose dimensions are 9cmx6cmx0.6cm. Precisely, the differential distances of each axis between the points of measurement are dx=1.5mm, dy=1.5mm, and dz=0.4mm.

The obtained E-field values were compared with the thresholds of the recommended exposure levels (ICNIRP-98) [2], and the thresholds for the safety and basic performance of the electromedical equipment (IEC 60601-1-2) [3].

III. RESULTS

In this section the results of the E-field around the tested device are provided. Fig. 3 shows the E-field values in the three axes: (x,z), (y,z), and (x,y).

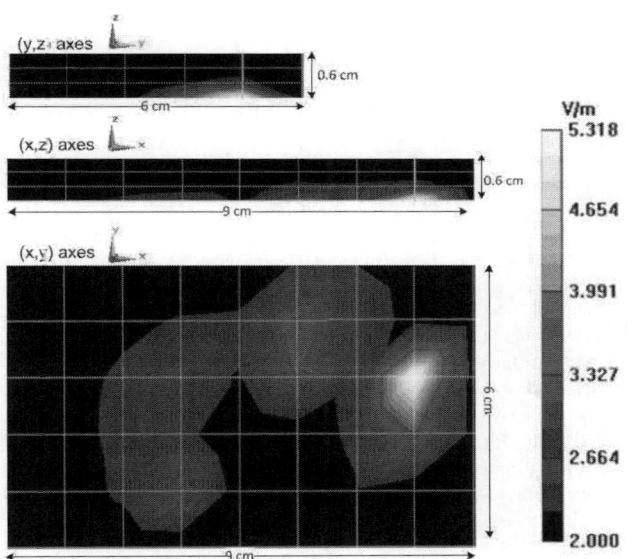

Fig. 3 E-field around the tested device in the (x,z), (y,z) and (x,y) axes

A relevant disadvantage of the measurement system is the lower value of the sensibility range of the measurement system that is 2 V/m. The measured values lower than 2 V/m are set to the value of the detection threshold (2 V/m).

The highest measured level of E-field is 5.318 V/m. Fig. 4 shows the marked area belonging to the tested device that matches with this maximum value of E-field:

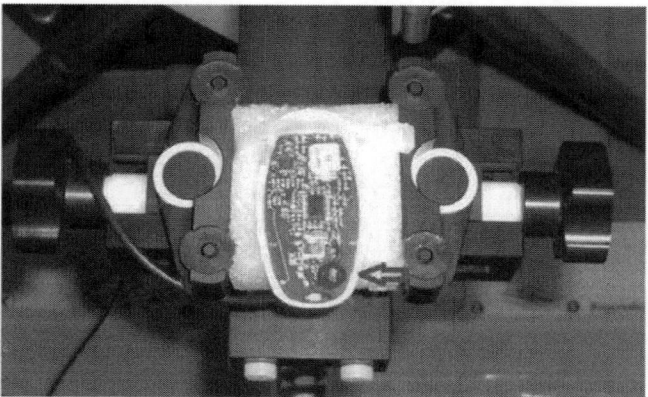

Fig. 4 The point of the tested device that presents the maximum E-field level

IV. DISCUSSION

The objective of this study is to quantify the exposure in near field conditions due to the increasing use of social alarm devices, and to analyze the compatibility between equipment and networks in assisted environments at home.

The greater value of the E-field is 5.318 V/m, this involves that the more restrictive threshold of 3 V/m, established in the International Electrotechnical Commission Standard of Electromedical Devices [3], is exceeded.

There are studies that advice about possible interferences between electromedical devices themselves, and these devices with wireless networks in healthcare environments [4]. The increase of the E-field strength is localized in a concrete point of small dimensions and immediately decreases with the distance, so it is nearly impossible the existence of possible risks because the recommended thresholds have been exceeded.

All the field strengths recorded in this study are well below the corresponding ICNIRP reference level of 40 V/m defined for the general public at the working frequency (869.25 MHz) [4]. It means that electric field strength levels in healthcare home environments are apparently safe according to the health and safety requirements of the exposure of patients, professionals and the general public for protection against possible health effects from nonionizing radiation.

This work provides a useful insight into the exposure levels caused by the working conditions of the social alarm devices, so it is also important to consider the absorption of radiofrequency energy by the body of the user that wears the device. ICNIRP guidelines are also expressed in terms of specific absorption rate (SAR), measured in W/Kg, in the body tissues [2].

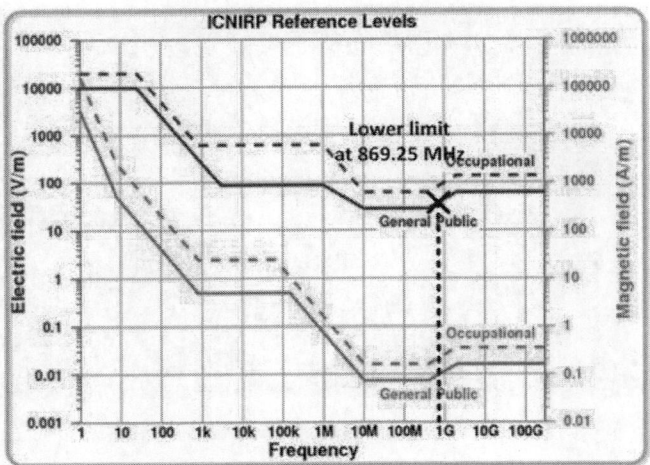

Fig. 5 ICNIRP reference levels and the lower limit at working frequency of social alarm devices (869.25 MHz)

After analyzing the results that are shown in Fig. 3, it is important to note that the measured values of the E-field are under the detection limit of the measure systems (2 V/m) at points very close to the emitter. This is due to the low level of power (10 mW) that characterized the emission of this type of devices [5]. Therefore, if the procedure of the SAR measurements was carried out, the obtained results would not be significant because the level of the E-field around the devices is registered as a constant value of 2 V/m.

V. CONCLUSION

The emissions of one of the most widespread model of social alarm devices have been characterized in near field conditions to analyze the influence on other electromedical devices and the possible over-exposure of people in the proximities.

Measurements in near field conditions have been realized in an anechoic chamber with the measurement system DAISY5PRO that is provided with the possibility of measure the E-field in predefined and programmed positions.

Under usual operating conditions the levels of E-field caused by the tested social alarm device do not exceed the limits of personal exposure according to ICNIRP 1998 [2].

In the cases of very small distances (2mm) from the tested device, the highest level of E-field strength exceeds the more restrictive threshold of 3 V/m that is established in the International Electrotechnical Commission Standard of Electromedical Devices [3]. But the E-field strength decreases rapidly in a very short distance, so it can be concluded that the risk of overexposure is nonexistent.

ACKNOWLEDGMENT

This work has been funded by the grant CA12/00038 from the Health Strategic Action, belonging to the National Scientific Research, Development and Technological Innovation Plan (National R&D&i Plan 2008-2011)

This work has been realized thanks to the valuable cooperation and help offered by the staff of the Radio Frequency Laboratory El Casar of the General Direction of Telecommunications and Information (Spanish Ministry of Industry, Energy, and Tourism).

REFERENCES

1. Schmid G, Lager D, Preiner P, et al. (2007) Exposure caused by wireless technologies used for short-range indoor communication in homes and offices. Radiat Prot Dosim 124(1): 58–62.
2. [ICNIRP (1998) Guidelines for limiting exposure to protection time-varying electric, magnetic and electromagnetic fields (up to 300 GHz). International Commission on Non-Ionizing Radiation.
3. International Electrothecnical Commission (IEC) Standard IEC 60601-1-2 Electromedical devices. 2007
4. Miguel-Bilbao S, Martín MA, Pozo A, et al. (2013) Analysis of exposure to electromagnetic fields in a healthcare environment: simulation and experimental study. Health Phys, accepted for publication.
5. 2011/829/EU: Commission Implementing Decision of 8 December 2011 amending Decision 2006/771/EC on harmonisation of the radio spectrum for use by short-range devices.

Author: Silvia de Miguel Bilbao
Institute: Health Institute Carlos III
Street: Monforte the Lemos, 5
City: Madrid
Country: Spain
Email: sdemiguel@isciii.es

Assessment of Statistical Distribution of Exposure to Electromagnetic Fields from Social Alarm Devices

S. de Miguel-Bilbao[1], J. García1[1], E. Aguirre[2], L. Azpilicueta[2], F. Falcone[2], and V. Ramos[1]

[1] Telemedicine and e-Health Research Unit, Health Institute Carlos III, Madrid, Spain
[2] Electric Engineering and Electronic Department, Public University of Navarra, Pamplona, Spain

Abstract—Social alarm devices are used to assist patients, elderly and disabled people in distress situations. Environments equipped with such wireless, sensitive and responsive devices are referred to as being Ambient Assisted Living (AAL). The use of such devices and transmission sources based on short range wireless technologies involves the increase of levels of electromagnetic (EM) fields in residential environments.

Laboratory measurements have been carried out to characterize and analyze the radiofrequency emissions of one of the more extended models of social alarm devices. The electric field (E-field) strength has been measured to examine the compliance with exposure guidelines. The maximum value obtained is much lower than the 3 V/m that is established in the International Electrotechnical Commission Standard of Electromedical Devices. Results show a high correlation in terms of E-field cumulative distribution function (CDF) with the most typical statistical distribution in indoor environments.

Keywords—social alarm devices, electric field, exposure threshold, near field exposure.

I. INTRODUCTION

The increasing average age of people and the subsequent rise of chronic diseases will result in a dramatic growth of the need for assistance and healthcare within the years to come. There is an increasing demand for outpatient care accessibility, for maintaining and restoring health, as well as for maximizing the independence of patients [1].

Social alarm devices provide direct benefits when applied in healthcare environments. The special implication of these devices with welfare and safety requirements involves a special interest in its operating conditions as well as in promoting habits of usage.

For some years now the short range technology has been considered a very promising option to cope with healthcare monitoring challenges. Some models of social alarm devices are based on a wide range of wireless technologies (RFID, UWB, NFC, WLAN, etc...). Specific architectures of short range technologies have been adapted for medical applications (ZigBee Health Care and Bluetooth Health Device Profile). Other type of social alarm devices operate in its own working frequency, from 869.2 to 860.25 MHz.

The buttons transmit a signal that typically consists of three pulses (depending on the model) at the frequency of 869.21 MHz. The systems of social alarm devices consist of two operational units: the buttons that are worn by the users typically hung over the neck or attached at the wrist, and the fixed unit that is connected to the home phone. When the user is in a distress situation, he or she can push the button, a radio frequency signal will be transmitted to the fixed unit and an emergency phone call will be made to the monitor centre.

These are emissions in domestic settings that can affect the electromagnetic environments and can involve the increase in the exposure to electromagnetic fields of users, patients, medical workers and people in general.

Laboratory measurements have been carried out to characterize and analyze the emissions of a very widely spread model of social alarm device that operates at 869.21 MHz.

II. METHOD AND MATERIALS

For this work, several models of social alarm devices were chosen from among the most used in telecare monitoring activities. Precisely the tested model was Tunstall AMIE+, which is shown in Fig. 1.

Fig. 1 Selected model of social alarms devices: Tunstall AMIE+.

One of the objectives of the laboratory measurements was to establish the radiation pattern (i.e. angular distribution of electric field (E-field) strength around each type of device) and identify the orientation at which the field is the maximum.

The measurements to obtain the radiation pattern were performed in a semianechoic chamber, shown in Fig. 2. The room has dimensions of 9,76 m x 6,71 m x 6,10 m, the walls are lined with a foam based radiofrequency absorber material (RANTEC Ferrosorb300) specified to have a reflection/absorption coefficient of -18 dB at the frequency of 869.21 MHz.

The device under test was mounted on a manual positioning device, allowing the device to be rotated in two orthogonal planes and permitting the measuring antenna to sample the radiation pattern at any angle. The distance between the device under test and the measuring antenna was 3 m, and the distance to the floor was 1.5 m.

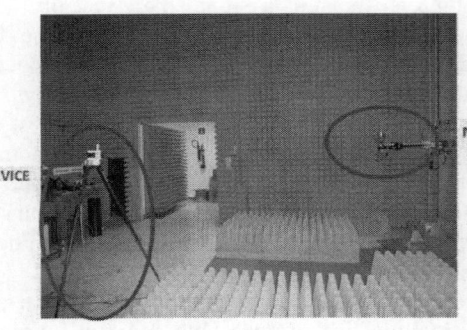

Fig. 2 Measuring antenna and positioners required for the radiation patterns measurements inside the anechoic chamber

A positioner with an EMCO 1051 motor allows the changes of the measuring antenna between the horizontal and vertical position. The measurements were carried out with an EMI Test Receiver ESIB26, Rhode & Schwartz with a frequency range of 20 Hz - 26.5 GHz, and the measuring antenna is a VBAA-9144 Schwarzbeck biconical antenna with a frequency range of 80 MHz - 1 GHz.

After obtaining the radiation pattern, the position of each tested device at which the electric field strength is maximum was fixed. In that position the electric field strength was measured as a function of the distance in near field conditions to check the influence on other wearable electrical devices. The chosen environment to carry out the measurements was the EMF laboratory of the Telemedicine Research Unit.

A handmade positioning device was used to determine the distances between the social alarm device and the measuring antenna. A handmade goniometer was used allowing the device to be rotated in two orthogonal planes to orientate the tested device in the position of maximum radiation. The handmade positioning device and the goniometer are shown in Fig. 3.

At 869.21 MHz, the wavelength is about 34 cm, which means the reactive near field extends to around 5.5 cm from the source (based on the usual $\lambda/2\pi$ criterion, where λ is the wavelength).

Fig. 3 Handmade goniometer to orientate the tested device according to the position of maximum radiation in polar coordinates

The emission levels of tested devices were measured in near field conditions, as it is shown in Fig.3. In these conditions the measurements were carried out with a Rhode & Schwartz FSH6 spectrum analyzer, a near field probe R&S HZ-11, and Model 907B preamplifier is used to increase the sensitivity of the test system. The measurements were made in 0.5 cm steps. The electric field was calculated as in equation (2):

$$E = V + AF + ATT - G \qquad (2)$$

where E is the electric field strength (dBuV/m), V is the measured voltage (dBuV), AF is the antenna factor (dBm−1), and ATT is the cable attenuation (dB), and G is the gain of the amplifier (dB).

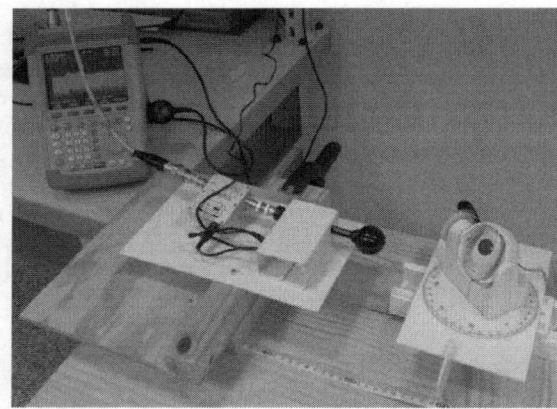

Fig. 4 R&S FSH6 Spectrum analyzer, near field probe, preamplifier, and tested device in the near field measurements

The field strength recorded from the tested devices was compared with the thresholds of the recommended exposure levels (ICNIRP-98) [3], and the thresholds for the safety and basic performance of the electromedical equipment (IEC 60601-1-2) [4].

The measurements have been set with typical statistical distributions in indoor environments (Lognormal, Nakagami, Rayleigh, Ricean and Weibull) [5]. The KS test allows for the probability that the CDF of the E-field values follows a specific statistical distribution.

III. RESULTS

Fig. 5 shows the radiation pattern of the selected model of social alarm device.

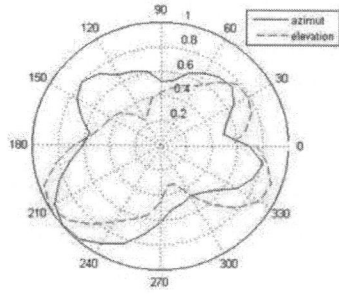

Fig. 5 Radiation pattern of the selected social alarm device.

Fig. 6 is a graph of the E-field strength in the orientation of maximum radiation from the tested device in near field conditions in function of the distance.

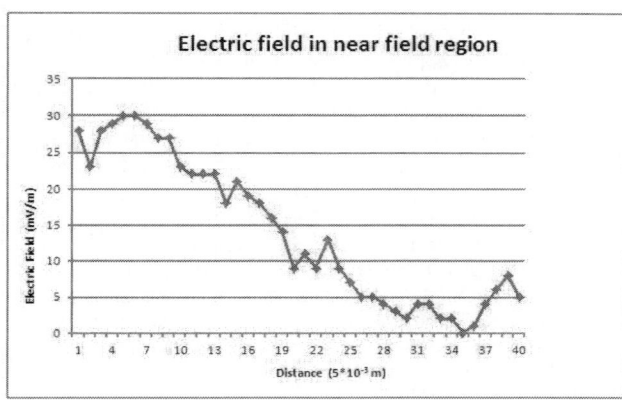

Fig.6 Variation of the E-field strength as a function of distance for the selected social alarm device in near field conditions.

One of the purposes of this work is to demonstrate that the CDF of the calculated E-field in indoor conditions can be characterized with one of the most typical statistical distributions in indoor conditions.

The Kolmogorov-Smirnov (KS) test has been employed to determine the statistical distribution that adapts more easily to the cumulative distribution function (CDF) of the obtained results of the E-field in near field conditions.

The KS test is an indicator of the probability that the E-field CDF follows a specific statistical distribution. The bigger the p-value the more the E-field CDF fits a given statistical distribution. Table 1 shows the p-values provided by the KS-test between the E-field CDF and the most typical statistical distribution in indoor environments.

Table 1 The p-values provided by the KS-test between the E-field CDF and the most typical statistical distribution in indoor environments

Distribution	P-value
Lognormal	0.1975
Nakagami	0.4515
Normal	0.2084
Rayleigh	0.0142
Rician	0.0142
Weibull	0.3565

P-value results show a high correlation between the Nakagami distribution and the E-field CDF.

IV. DISCUSSION

After selecting the most widely used model device, the emission levels were analyzed with regard to potential risks and operational disturbances in accordance with the existing standards.

In this work the EM conditions have been analyzed and the radiation pattern of one of the most extended social alarm device has been obtained. Given the increasing use of domiciliary telealarm devices, and the non-existence of previous studies of the working conditions and the emission levels, this paper analyzes two of the aspects that have to be considered to ensure a proper, reliable and safe usage of these systems. The first is the compatibility with other communication networks and implanted electric devices. The second is the compliance with exposure levels threshold, to quantify and analyze the risk of exposure caused by the use of these devices. Data from the analysis have been saved and processed, in order to compare them with the International Electrotechnical Commission (IEC) Standard 601-1-2 (IEC, 2002) and the ICNIRP-98 standard (ICNIRP, 1998)

All the field strengths recorded in this study are well bellow the corresponding ICNIRP reference level of 40 V/m defined for the general public at the working frequency [3]. It means E field level in healthcare environments at home is apparently safe according with the health and safety requirements regarding the exposure of patients, professionals and general public for protecting against possible health effects from non ionizing radiation.

One prominent concern has involved possible interference with medical devices. The International Electrotechnical Commission (IEC) Standard IEC 60601-1-2 [4], sets a minimum immunity level of 3 V/m for non-life supporting devices. Examining the near field results, the maximum value of the electric field is much lower than the 3 V/m.

It has been shown that the CDF of the dataset in near field conditions, seem to follow one of the most typical statistical distributions in indoor environments. P-value results show a high correlation between the Nakagami distribution and the E-field CDF generated by the analyzed device in conditions of near field.

Fig. 7 shows a comparison between the E-field CDF and the typical statistical distributions in indoor environments, which present a considerable similarity.

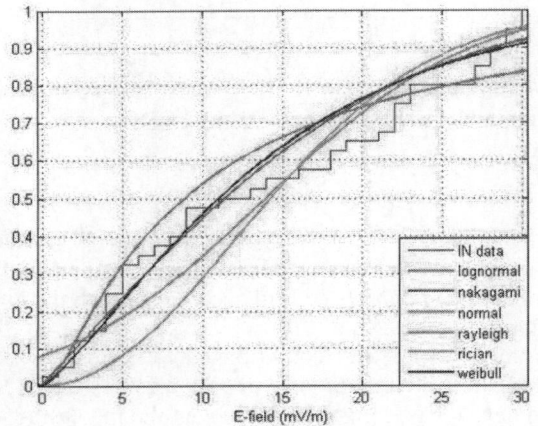

Fig. 6 Set result of the E-field levels using common statistical distributions

Analyzing Fig. 7, it should be noted the distribution that better fits with the E-field CDF is Nakagami, but there is a great similarity with other distributions, Weibull and Log-normal, so any of them could characterize the propagation of the E-field in indoor environments [5].

V. CONCLUSION

This work presents an analysis of the exposure of the E-field in near field condition due to the use of social alarm devices. This paper provides a useful insight into the statistical distributions associated with the propagation and exposure of the E-fields in indoor environments.

The performed environmental study of the working conditions of the alarm devices helps to quantify the exposure of assisted people, and to analyze the compatibility between equipment and networks that operate in the surroundings

The proposed study provides a global, immediate and accurate vision that can help to avoid EM interferences on mobile telemedicine systems, and monitor the exposure to EM fields of people using and in the proximity of social alarm devices in home environments.

New telemedicine solutions based on the applications of social alarm devices must consider the issues of electromagnetic compatibility and regulatory compliance. Currently, the degree and type of EMF exposure need to be characterized in domestic settings, in order to ensure that applications operate properly and exposure guidelines are not exceeded.

ACKNOWLEDGMENT

This work has been realized thanks to the valuable cooperation and help offered by the staff of the Radio Frequency Laboratory El Casar of the General Direction of Telecommunications and Information (Spanish Ministry of Industry, Energy, and Tourism)

REFERENCES

1. 2011/829/EU: Commission Implementing Decision of 8 December 2011 amending Decision 2006/771/EC on harmonisation of the radio spectrum for use by short-range devices.
2. Carranza, N., Ramos V., Lizana F.G., et al., "A Literature Review of Transmission Effectiveness and Electromagnetic Compatibility in Home Telemedicine Environments to Evaluate Safety and Security," Telemed J e-Health 16(7):530-541; 2010.
3. NPL. A guide to power flux density and field strength measurement. Middlesex London: National Physical laboratory, UK; The Institute of Measurement and Control; 2004.
4. ICNIRP (1998) Guidelines for limiting exposure to protection time-varying electric, magnetic and electromagnetic fields (up to 300 GHz). International Commission on Non-Ionizing Radiation.
5. Miguel-Bilbao S, Martín MA, Pozo A, et al. (2013) Analysis of exposure to electromagnetic fields in a healthcare environment: simulation and experimental study. Health Phys, accepted for publication

Author: Silvia de Miguel Bilbao
Institute: Health Institute Carlos III
Street: Monforte the Lemos, 5
City: Madrid
Country: Spain
Email: sdemiguel@isciii.es

Increased Complexity of Medical Technology and the Need for Human Factors Informed Design and Training

P.L. Trbovich[1,2], S. Pinkney[1], C. Colvin[1], and A.C. Easty[1,2]

[1] University Health Network/Techna Institute, HumanEra, Toronto, Canada
[2] University of Toronto/Faculty of Medicine, Institute of Biomaterials and Biomedical Engineering, Toronto, Canada

Abstract—Numerous studies of errors in health care identify medication-related errors as at or near the top of the list of concerns, both in terms of frequency and severity of impact. Despite the development of automated technologies, such as smart drug infusion pumps, medication errors are still prevalent and typically occur in more than half of all intravenous (IV) infusions. Incorrect set-up, programming and management of IV systems, particularly complex secondary and multi-line infusions, are often caused by knowledge and performance deficits. Furthermore, current clinician training regarding infusions does not consistently cover fundamental infusion principles. Thus, there is a great need for (a) medical device design that optimizes capability and ease of use, and (b) new and effective means of educating clinicians to improve basic knowledge and critical thought necessary for decreasing medical errors such as IV infusion errors.

We have undertaken a series of studies to determine the underlying cause of medication-related errors and, to compare types of errors that were best mitigated through improved device design to those that benefited most from improved user training.

Methods used in our studies included literature reviews, mining of incident databases, ethnographic observations in clinical environments, and laboratory-based simulations designed to uncover performance limitations, and to test proposed benefits of improved technology designs and clinician training.

Results showed that safety-based improvements to known shortcomings in technology design for IV infusions were beneficial in reducing, and even eliminating, errors of omission. Conversely, training-based interventions were beneficial in reducing errors where actions required higher-level clinical decision-making.

These studies highlight the need for mitigation strategies aimed at reducing medication errors that include both technological solutions that can achieve higher accuracy and reliability than human processes as well as user-centric based training solutions that enhance clinicians' problem solving skills.

Keywords—complex medical devices, technology design, user training, patient safety.

I. INTRODUCTION

Adverse events and medical errors pose a serious problem to health care systems [1]. In 1999, the Institute of Medicine (IOM) report, "To Err is Human: Building a Safer Healthcare System" [2], concluded that medication errors account for 7,000 deaths annually, while total preventable medical errors cause between 44,000 and 98,000 annual deaths in the United States alone. Even when considering the lower estimate, deaths in hospitals due to preventable adverse events exceed the deaths attributable to motor vehicle accidents, breast cancer, and AIDS combined [2]. The IOM estimated that the annual cost associated with preventable medical errors is as much as $29 billion US annually. Similarly, research on incidence rates of adverse events in Canadian hospitals indicated that approximately 185,000 admissions per year are linked to an adverse event, and that nearly 70,000 of these may be avoidable [3]. Medication errors are the most frequent cause of medical injuries, representing 19.4% of all adverse events [4]. Furthermore, a subsequent report by the IOM on medication errors [5] estimates that a minimum of 1.5 million people are harmed per year due to preventable adverse drug events, suggesting that every hospital patient may be subjected to as much as one medication error per day. As a result of these landmark reports, there has been an increased awareness of patient safety issues over the past decade, particularly in the context of medication safety.

Clinicians' user errors and improper use of medical devices have been linked to multiple patient injuries [6]. Adverse events involving medical devices have led to serious problems, including incorrect or delayed diagnosis and treatment or patient injuries and deaths. ECRI Institute [7], estimates that approximately 75 percent of the reported problems they receive are related to user error. Specifically, the majority of problems reported concern users who do not fully understand the devices and systems they are being asked to use. When errors involving medical devices recur repeatedly, people typically blame the users rather than investigate whether a poorly designed interface between the medical device and the user may be a factor. Furthermore, the Medicines and Healthcare product Regulatory Agency (MHRA) has identified inadequate staff training as a primary cause of incidents with medical devices [8]. There is a need to understand how to better ensure that (a) medical devices are designed for optimum capability and ease of use, and (b) clinician training is thorough and effective.

In health care, the objective of human factors is to improve human performance with medical devices and systems, and to reduce the likelihood of error or injury, thereby improving patient and workplace safety [9].

II. OBJECTIVES

In the present paper, we will discuss a series of Human Factors studies that we have undertaken to determine the underlying cause of medication-related errors and compare types of errors that were best mitigated through improved device design to those that benefited most from improved user training.

III. BACKGROUND

The studies discussed in this paper focused on the safe delivery of intravenous (IV) medication therapies as these procedures have become more complex over the years due to the introduction of technologies such as smart large volumetric infusion pumps. That is, to address the high incidence of infusion errors, manufacturers have developed pumps that have dose error reduction systems (DERS), which include hospital-defined drug libraries with dosing limits and clinical advisories (i.e., smart pumps). While traditional general-purpose infusion pumps have a wide range of acceptable programming settings/parameters, smart pumps are designed with drug-specific safety software to help nurses' avoid programming errors. Smart pumps provide either a "soft" limit warning (allows nurse to override the limit and continue infusing) or "hard" limit warning (requires nurse to reprogram the pump within acceptable parameters).

Although the purported benefits of these technologies is that medications can be given accurately and reliably at all times, clinicians often encounter difficulties using the devices, which can increase the risk of patient injury.

IV. METHODS

Methods used in our studies included literature reviews, mining of incident databases, ethnographic observations in clinical environments, and laboratory-based simulations designed to uncover performance limitations, and to test proposed benefits of improved technology designs and clinician training.

V. RESULTS/DISCUSSION

In our own simulation study of traditional vs. smart pump infusion systems, we observed that although current smart pumps are designed to reduce pump programming errors, they did not significantly improve nurses' ability to successfully operate secondary infusions in comparison with traditional large-volume pumps. Secondary (also referred to as "piggyback") infusion is a convenient set-up that allows clinicians to administer two or more medications intermittently to patients through a single channel in the infusion pump. The error rates for nurses to complete all secondary infusion tasks scenarios were as high as 50% [10]. Moreover, we identified that many secondary infusion issues, such as misalignment of infusion bags, errors in tubing setup, and the failure to open the roller clamp on secondary IV tubing, cannot be detected or intercepted by the commercially available smart infusion technologies that are commonly used in the clinical setting. Our experimental data demonstrated that although smart pumps contain safeguards against dosing errors, they do not prevent errors related to the remaining five rights of medication administration (i.e., the right drug, patient, route, and time).

Our research assessing the extent to which fundamental infusion principles are taught in baccalaureate and postgraduate nursing programs revealed that the administration and management of IV infusions, particularly multiple IV infusions to a single patient, is not reliably covered in detail at either stage of training [11]. That is, education around key theoretical concepts (e.g., hydrostatics, fluid mechanics) required for an understanding of patient safety risks associated with administration and management of IV infusions is lacking from the curriculum. Furthermore, the scarce educational content that is covered is presented in a decontextualized, didactic lecture style format.

We recently conducted a study in which we empirically compared the effectiveness of technology-based interventions to training-based interventions on nurses' ability to safely administer IV infusions. Results showed that safety-based improvements to known shortcomings in technology design for intravenous infusions (e.g., automatic clamp detector that alarm users when a roller clamp is closed at the start of a secondary infusion) were beneficial in reducing errors of omission (e.g., user knows that s/he must open the secondary clamp but forgets to perform the action). Conversely, training-based interventions (e.g., education module that addresses information on basic infusion principles and known failure modes) were beneficial in reducing errors where actions required higher-level clinical decision-making (e.g., user must integrate various pieces of information to arrive at a decision).

In sum, our research findings suggest that IV infusion errors are attributable to both (1) the way the technology to deliver infusions is designed and (2) clinicians' knowledge of key infusion principles.

VI. CONCLUSION

It has been well established that technological systems can achieve a much higher accuracy and reliability than any human processes [12]. Consequently, considerable attention has been directed towards automating these mechanistic tasks of the IV infusion process through design of smart pumps with bar code readers. Less attention, however, has been paid to other (non-mechanistic) components of the IV infusion system. Although automated technological innovations are effective at assisting in the performance of mechanistic tasks, they are not as effective at assisting in the performing of tasks requiring critical thought.

Automated technical systems are limited in that they are designed to deal with the foreseen. That is, technological solutions are based on tasks, actions, or procedures that can be anticipated and therefore built into the design. The reality of health care environments, however, is that they are complex, dynamic, and often unpredictable. Consequently, clinicians must be adaptive problem solvers. Thus, mitigation strategies aimed at reducing medication administration errors must include both technological solutions that can achieve higher accuracy and reliability than human processes as well as user-centric training solutions that enhance clinicians' problem solving skills.

ACKNOWLEDGMENT

The authors gratefully acknowledge the support from Health Quality Ontario, the Institute for Safe Medication Practices Canada, and the Multiple IV Infusions Expert Panel.

REFERENCES

1. Schwappach DL, Wernli M. (2010) Medication errors in chemotherapy: incidence, types and involvement of patients in prevention. A review of the literature. European Journal of Cancer Care 19: 285–292
2. Kohn LT, Corrigan JM, Donaldson MS. (2000) To Err is Human: Building a Safer Healthcare System. Institute of Medicine.: National. Academy Press, Washington, D.C
3. Baker GR, Norton PG, Flintoft V, Blais R, Brown A, Cox J, Etchells E, Ghali WA, Hébert P, Majumdar SR, O'Beirne M, Palacios-Derflingher L, Reid RJ, Sheps S, Tamblyn R. (2004) The Canadian Adverse Events Study: the incidence of adverse events among hospital patients in Canada. CMAJ 170 (11):1678-1686
4. Leape LL, Brennan TA, Laird N, Lawthers AG, Localio AR, Barnes BA, Hebert L, Newhouse JP, Weiler PC, Hiatt H. (1999) The nature of adverse events in hospitalized patients: Results from the Harvard Medical Practice Study II. N Engl J Med 324:377-384
5. Aspden P, Wolcott J, Bootman JL, Cronenwett LR. (2007) Preventing Medication Errors. Institute of Medicine. National Academy Press, Washington, D.C
6. Cooper JB, Newbower RS, Kitz RJ. (1984) An analysis of major errors and equipment failures in anesthesia management: considerations for prevention and detection. Anesthesiology 60(1):34-42.
7. ECRI at www.ecri.org/PatientSafety/ReportAProblem/Pages/default.aspx
8. MHRA: Device Bulletin Managing Medical Devices- Guidance for healthcare and social services organisations at http://www.mhra.gov.uk/home/groups/dtsbs/documents/publication/con2025143.pdf
9. Gosbee JW & Lin L (2001). The role of human factors engineering in medical device and medical system errors. In Charles Vincent (ed.) Clinical Risk Management: Enhancing Patient Safety, 2nd Edition. London: BMJ Press. Pp. 301-317
10. Trbovich PL, Pinkney S, Cafazzo JA, Easty AC. (2010) The impact of traditional and smart pump infusion technology on nurse medication administration performance in a simulated inpatient unit. Qual Saf Health Care 19(5):430-434
11. Cassano-Piché a, Fan M, Sabovitch S, Masino , Easty AC. (2012). Multiple Intravenous Infusions Phase 1b: Practice and Training Scan, Ontario Health Technology Assessment Series 12:1-116
12. Drews, F.A. (2008). Patient Monitors in Critical Care: Lessons for Improvement. In: Advances in Patient Safety: From Research to Implementation. Volume 5, AHRQ Publication Nos. 050021 (1-5). Agency for Healthcare Research and Quality, Rockville, MD.

Patricia L. Trbovich
University Health Network
R.Fraser Elliott Bldg, 4th Floor, 190 Elizabeth Street
Toronto, Ontario
Canada
patricia.trbovich@uhn.ca

Effects of Combined Use of a Patient-Tracking System and a Smart Drug-Dispenser on the Overall Risk of the Diagnostic-Therapeutic Process

E. Iadanza, S. Bargellini, and G. Biffi Gentili

Department of Information Engineering, University of Florence, Florence, Italy

Abstract—Clinical risk and the incidence of adverse event are relevant problems, that have been subject of many studies during last years. The objective of this study is to assess the potential clinical risk reduction obtained with the combined implementation in the diagnostic-therapeutic process workflows of a patient-tracking system and a smart drug-dispenser, both based on RFId technology. The study will be driven by a multidimensional FMECA, considering in the risk analysis all the subjects involved in the diagnostic-therapeutic process, not only the dimension of patient safety indeed, but also medical staff dimension and structure (hospital facility) dimension. Mitigation interventions are found and they often result in a further development and enhancement of the functionality of the two studied RFId systems.

Keywords—adverse event, risk management, FMEA, diagnostic-therapeutic process, RFId.

I. INTRODUCTION

The clinical risk has been defined in [1] as the probability for a patient of being subject to an adverse event.

During the last years many studies had determined a relevant incidence of adverse events in hospitalized patients, showing the need for more attention in the whole diagnostic-therapeutic process. [2][3]

Clinical Risk Management has the objective of ensuring patient safety and improving health-care services. Reducing the incidence of adverse events is consequently one of the primary goals of clinical risk management.

For implementing technological solutions, it has to be guaranteed that technologies and new devices implemented are safe for all subjects involved in the process and that the benefits and the advantages that are gained with new devices are relevant and indisputable.

II. MATERIALS AND METHODS

A brief description of the two analyzed system is given.

- Patient Tracking: it is a hardware/software system that provides a method for patient identification and tracking, based on an RFId system of active tags and a software for the visualization and management of the information [4][5][6][7][8];

- Drug-Tin: a smart-drug dispenser based on RFId standard ISO/IEC 15963, that allows access to the drug only in case of correct patient-drug binding. [9][10][11]

The first necessary step for the risk assessment is the definition of the objective of the evaluation: the diagnostic-therapeutic process is broken down into 9 sub-process, organized in two macro-phases based on the position of the patient in the hospital.

The ward chosen for the analysis is a typical Internal Medicine ward, mainly because such ward can be considered paradigmatic also for others wards.

The workflow analysis is developed according to Standard ANSI/PMI 99/001/2008 and it's based for the most part of the processes on a previous work from the Department of Information Engineering of the University of Florence [12].

Processes that were not studied in [12], like "Surgical intervention", has been broken down using documents of international institutions, like the Safe Surgery Manual of World Health Organization. Sub-process of macro-phases A and B are reported respectively in table 1 and table 2.

Table 1 Sub-process of macro-phase A

A. Inside the ward
1. Hospital admission
2. Medication administration
3. Hospital Stay
4. Diagnostic intervention at patient's bed

Table 2 Sub-process of macro-phase B

B. Outside the ward
1. Diagnostic examination
2. Surgical intervention
3. Therapeutic or rehabilitation activity
4. Patient moved in a new ward
5. Patient Discharge

Every sub-process is indeed univocally broken down into a series of coded tasks, indicating the responsible for the activity and comparing the *a priori* situation, without using the two system described and neither similar systems, and the implementation situation, using the systems in the workflows of the processes and named *Patient Tracking + Drug Tin*. The entire diagnostic-therapeutic process is broken down into a total of 80 tasks; these tasks constitute the base for the risk analysis.

For the second phase of the analysis, risk identification and assessment, it was decided to use the Failure Mode, Criticality and Effect Analysis (FMECA).

In particular, in this study, the method chosen for the risk management has been recently proposed by Italian Ministry of Health: a multidimensional FMECA characterized by different scales for assessing the severity of the effects of the FM, depending on the area of interest or the subject involved by the FM (patient safety, medical staff safety, patient satisfaction, clinical effectiveness, economical expenses). [13]

The value selected are, for occurrence scale, detection scale and severity scales: 1,3,5,8,10 (see Table 3).

Table 3: Severity Scale Rank for the three dimensions

Severity Rank	Patient	Medical staff	Structure
1	No consequences.	No consequences.	No consequences.
3	Delay and discomfort.	Intervention required.	Workflow delay; Low expenses increase.
5	Minor Complications.	Error determining a bad reputation.	Mild expenses increase; continuity of service interruption.
8	Major Complications.	Serious error determining a very bad reputation; Possible health complications.	High expenses increase; Structure's trust loss.
10	Serious complications till death.	Serious complications till death.	Very high expenses increase; Serious structure's trust loss.

Fig. 1 Table FMECA for the identifying of failure modes

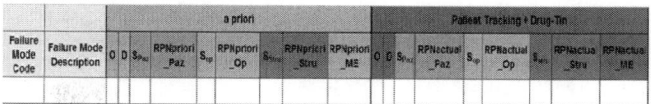

Fig. 2 Table FMECA for the calculation of RPNs

289 Failure modes have been identified in the entire diagnostic-therapeutic process; Risk Priority Numbers (RPNs) have been calculated for every failure mode according to the IEC 60812 Standard, for the three dimensions and comparing the two situation *a priori* and *Patient Tracking + Drug Tin*.

The two parts constituting the FMECA table are reported in Fig.1, while in Fig. 2 is shown the RPNs calculation of the identified failure modes, where Paz, Op and Stru are respectively the abbreviation for the three dimensions Patient, Medical Staff (Operator) and Structure.

III. RESULTS

Total RPN is calculated as the sum of all the RPNs for a dimension in the entire diagnostic therapeutic process.

The result of RPNs calculation shows how clinical risk is reduced for all the three dimensions considered in this analysis, as indicated in table 3.

Table 3. Total RPNs of the diagnostic-therapeutic process

RPN_Paz *a priori*	RPN_Paz *Pat+Drug*	RPN_Op *a priori*	RPN_Op *Pat+Drug*	RPN_Stru *a priori*	RPN_Stru *Pat+Drug*
9979	4429	8589	4808	9538	5389

Pat+Drug = *Patient Tracking + Drug Tin*, situation using the described RFId systems

The implementation of the two described systems reduces clinical risk versus *a priori* situation of 55% for the Patient dimension, (4429 vs. 9979), 44% for the Medical staff dimension (4808 vs. 8589) and 43% for the Structure dimension (5389 vs. 9538)

The total RPN is however not sufficient for ensuring the safety for the patient and the other dimensions; in fact there can be risks that are too high and unacceptable and they are "hidden" by the information of total RPN. Hence, for every dimension, it is given the RPN distribution, by a pie chart.

In Fig.3 it is reported the distribution of the Structure dimension, giving the comparison between *a priori* situation and *Patient Tracking + Drug Tin* situation.

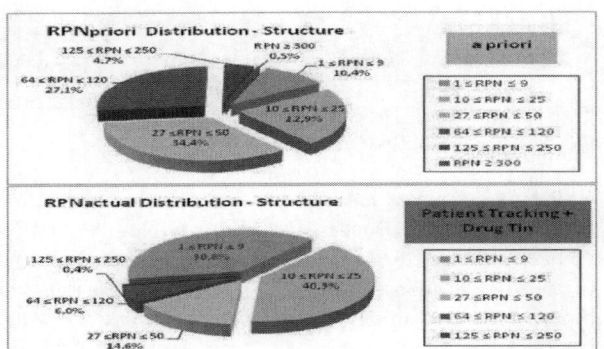

Fig. 3 Comparison of the RPN distribution for structure dimensions

Fig. 3 shows how there are high RPNs also implementing RFId systems in the workflows of the diagnostic-therapeutic process. Similar graphs has been calculated also for Patient dimension and Medical staff dimension, but they are not reported; it is important to underline that there are high RPNs also in the other two dimensions. The next step of the FMECA is indeed finding a RPN threshold and suggesting mitigation intervention for all those failure modes that have RPN equal or greater than such threshold.

An interesting method for detecting the RPN threshold, named "scree plot" is given in [14]. Scree plot settings require preliminary ordering of the RPNs, by value, from lowest to highest and then these values are plotted on a graph. According to [14], in a "good FMEA" the scree plot appears like a cliff, descending to base level of ground. There can be noticed two well distinguished trend lines, one is characterized by a gradual increase step by step and the other is steeper. At the intersection of these two lines, it is located the threshold for risk acceptability; RPNs under this threshold can be considered "information noise" while RPNs greater than this threshold, defined RPN Jumps, must be considered for mitigation interventions.

This method is applied to the three dimensions considered in this analysis: Patient, Medical staff and Structure and the three threshold detected are RPN=27 for patient, RPN=27 for medical staff and RPN=27 for structure.

From Fig.3 it is visible that 21% of the failure modes have inacceptable RPNs for Structure dimension; the other distributions, calculated but not reported, show for Patient dimension and Medical staff dimension respectively 12% and 16,8% of the failure modes associated with inacceptable risk.

The three detected thresholds are independent and the criterion adopted for the selection can be represented by the "OR" of the Boolean Algebra. So for every failure mode it is sufficient that only one dimension have a RPN value greater than the threshold, to highlight the need of a mitigation intervention. The typology of the interventions depends also on which and how many dimensions have RPN greater than respective threshold in the considered failure mode.

In order to prioritize the need of mitigation intervention it's used the descending order of RPNactual_ME parameter, shown in table 2; such parameter is given for every failure mode from the sum of RPNs of the three dimension patient, medical staff and structure.

When a mitigation intervention is indicated, additional risks that can be generated by such intervention are considered and the associated RPNs are calculated, as in the previous phase, for the three dimensions, to ensure that mitigation interventions do not generate inacceptable risks for any subject involved in the diagnostic-therapeutic process.

It can be useful to report some examples of risk control methods; it is important to underline that most of the times the mitigation interventions result in development and enhancement of functionality of the two RFId system, as the following:

- Supply Drug-Tin with an alarm-clock to remind patient, and also medical staff, administrating drug at the right time;
- Interfacing the Patient Tracking with the HIS of the ward for highlighting only the right patient and his medical record and for avoiding errors and mistakes, e.g. in case of movement in a new ward or patient discharge;

The actuation of interventions determines the iteration of RPN calculation; the new simulated situation after such actions is named *After risk mitigation* and will be compared with *Patient Tracking + Drug Tin* situation.

In Fig. 4 it's reported the scree plot of the entire process for the Patient dimension. RPN Jumps that need interventions are highlighted by the red oval and the RPN threshold is plotted as a red line.

The situation *After risk mitigation* is represented in blue and it's visible only one trend, a gradual and almost linear increase, step by step. This means that no further mitigation intervention is required and it is easy to note that all failure modes are below the highlighted threshold.

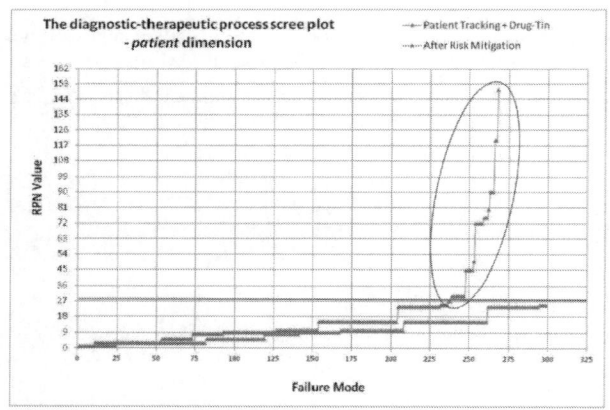

Fig. 4 Scree plot of the diagnostic-therapeutic process pre and post mitigation intervention for the patient dimension

Fig. 4 represents the Patient dimension; Medical staff dimension and Structure dimension have been calculated and graphed but they are not reported. The trend is similar to Patient one and all the failure modes are below threshold also for the other two dimensions.

It is interesting to underline that the number of failure modes is greater in the *After risk mitigation* situation; this is due to the additional risks considered, as described before. No one of these risks is however greater than the threshold (for any dimension), therefore all the mitigation interventions can be realized.

IV. CONCLUSIONS

It has been showed that after mitigation interventions all the failure modes are below the detected RPN thresholds. So it is now possible to consider the total RPN (equal, as the previous step, to the sum of all RPNs of the process for every dimension) for the last situation analyzed, *After risk mitigation.*

The first aim is comparing the values with the values of Total RPN for *a priori* situation, reported in table 3.

We note a final risk reduction of 70% for Patient dimension (9979 vs. 2974), 63% for Medical staff dimension (8589 vs. 3154) and 63% for Structure dimension (5389 vs. 3812).

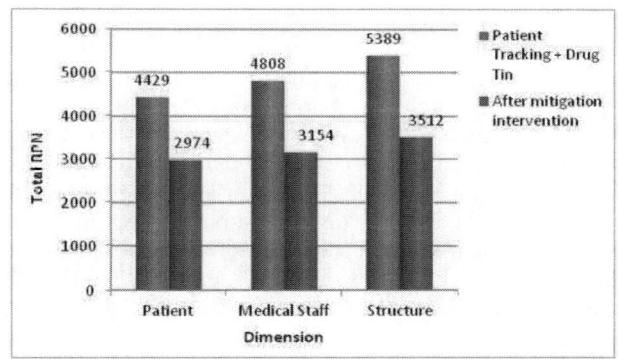

Fig. 5 Total RPN of the entire diagnostic-therapeutic process; comparison of pre e post mitigation intervention

Then, referring to Fig. 5, the efficiency of mitigation interventions can be represented by the further risk reduction respect to the *Patient Tracking + Drug Tin* situation; the efficiency is so 35% for structure dimension, 34% for medical staff dimension and 33% for patient dimension.

The innovative multidimensional approach has allowed to detect an highest number of critical failures, that could not be detected in a one-dimensional approach, involving all the subjects of the diagnostic-therapeutic process in the improvement of the process and of the two systems.

The analysis of the RFId systems, based on the breaking down of the entire diagnostic-therapeutic process, provides in fact, a greater number of better and more efficient indications for the next development of the systems, ensuring the safety and the importance of them, not only for the Patient but also for the other figures involved in the process, like Medical staff and Structure.

REFERENCES

1. Kohn L, Corrigan J and Donaldson M et al. (1999) To err is human: Building a safer health system. Institute of Medicine, National Academy Press, Washington
2. Baker GR, Norton PG, Flintoft V et al. (2004) The Canadian adverse event study: the incidence of adverse events among hospital patient in Canada. CMAJ 170: 1678-86.
3. Brennan TA, Leape LL, Liard NM et al. (1991) Incidence of adverse events and negligence in hospitalized patients; results from the Harvard medical practice study. N Eng J Med 324: 370-6.
4. Iadanza, E., & Dori, F. (2009). Custom active RFId solution for children tracking and identifying in a resuscitation ward. Conference Proceedings : 31th Annual International Conference of the IEEE Engineering in Medicine and Biology Society. IEEE Engineering in Medicine and Biology Society Conference, 2009, 5223-5226.
5. Biffi Gentili, G., Dori, F., & Iadanza, E. (2010). Dual-frequency active RFID solution for tracking patients in a children's hospital. design method, test procedure, risk analysis, and technical solution. *Proceedings of the IEEE, 98*(9), 1656-1662.
6. Iadanza, E., Dori, F., Miniati, R., & Corrado, E. (2010). Electromagnetic interferences (EMI) from active RFId on critical care equipment. Paper presented at the *IFMBE Proceedings, , 29* 991-994.
7. Iadanza, E., Dori, F., Miniati, R., & Bonaiuti, R. (2008). Patients tracking and identifying inside hospital: A multilayer method to plan an RFId solution. Paper presented at the *Proceedings of the 30th Annual International Conference of the IEEE Engineering in Medicine and Biology Society, EMBS'08 - "Personalized Healthcare through Technology",* 1462-1465.
8. Iadanza, E., Chini, M., & Marini, F. (2013). Electromagnetic compatibility: RFID and medical equipment in hospitals. Paper presented at the *IFMBE Proceedings, , 39 IFMBE* 732-735.
9. Iadanza, E., Baroncelli, L., Manetti, A., Dori, F., Miniati, R., & Gentili, G. B. (2011). An rFId smart container to perform drugs administration reducing adverse drug events. Paper presented at the *IFMBE Proceedings, , 37* 679-682.
10. Iadanza, E., Burchietti, G., Miniati, R., & Biffi Gentili, G. (2013). Drugs administration: How to reduce risks to patients. Paper presented at the *IFMBE Proceedings, , 39 IFMBE* 743-745.
11. Iadanza, E., Pettenati, M. C., Bianchi, L., Turchi, S., Ciofi, L., Pirri, F., Giuli, D. (2012). Telematics integrated system to perform drugs prescription and administration reducing adverse drug events. Paper presented at the *Proceedings of the Annual International Conference of the IEEE Engineering in Medicine and Biology Society, EMBS,* 6082-6085.
12. E. Iadanza, F. Gaudio, F. Marini (2013) The diagnostic-therapeutic process. Workflow analysis and risk management with IT tools. 35th Annual International Conference of the IEEE EMBS, Osaka, Japan, July 3 – 7 no. In Press.
13. Ministero della Salute (2011) Sviluppo di una metodologia per la valutazione delle tecnologie finalizzate alla sicurezza dei pazienti. Dipartimento della qualità
14. Bluvband Z, Grabov P, Nakar O (2004) Expanded FMEA (EFMEA). 2004 Annual Reliability and Maintainability Symposium Proceedings, 2004, pp.31–36

Risk Management Process in a Microwave Thermal Ablation System for CE Marking

E. Iadanza, C. Ignesti, and G. Biffi Gentili

Department of Information Engineering, University of Florence, Florence, Italy

Abstract—Microwave thermal ablation (MWA) is a promising technique capable of achieving larger ablation zones quicker than other energy forms. The aim of this work is the risk management of a brand new microwave thermal ablation system, for the purpose of CE marking. The risk analysis will be driven by a multidimensional Failure Mode, Criticality and Effect analysis, which will take into account the effects on the patient, the operator and the hospital facility. All the principal mitigation interventions found will be described.

Keywords—Microwave, thermal ablation, CE marking, FMEA, risk management.

I. INTRODUCTION

Thermal ablation, in its deeper meaning, can be defined as an extreme hyperthermia procedure aiming to remove unwanted tissue, as cancer masses. Thermal ablation techniques have become more and more popular and exploited as a safe, efficient and economic option for the treatment of primitive and secondary unresectable malignancies. Microwave thermal ablation (MWA), is a new, promising, technique capable of achieving larger ablation zones quicker than other energy forms [1][2][3]. The operation is applicable with excellent results, during short sessions and not necessarily with general anesthesia, usually with low risks and minimal invasiveness [4]: the short postoperative course enhances patient's quality of life and alleviates health facilities expenses. Nevertheless, MWA has not reached broad diffusion in clinical practice, mainly due to the difficulties in realizing applicator antennas able to assure minimal invasiveness together with optimal control of the ablation pattern, in order to guarantee safety and predictability of the treatment.

II. MATERIALS AND METHODS

The main characteristics of the project studied in this paper are:

- A single microwave generator working at 2.45 GHz, able to manage up to four antennas between 17 and 14 G, to achieve asynchronous microwave thermal ablation;
- An automatic power distribution unit, which supervise the synergic operation of the antennas, guaranteeing high result controllability and predictability.
- An active gas cooling system of the antenna shaft, obtained by means of carbon dioxide and Joule-Thomson effect, which has been demonstrated since the very first tests more practical and efficient than water cooling;
- A brand new low loss applicator technology (patent pending) that allows to eliminate cooling systems for a wide variety of operations, ensuring higher safety, reliability and usability for the system.

This project is a great challenge from the safety standpoint, mainly because of the active gas cooling system. The first steps that have to be taken into account are a deeper risk analysis and assessment, for the purpose of CE marking, mainly conducted with international standard ISO 14971. The lack of particular standards for the basic safety and essential performance of microwave surgical equipment, forces risk management process to be based only on general standard IEC 60601-1.

The chosen method for the risk management, according to ISO 14971, is the one recently proposed by Italian Ministry of Health: a multidimensional Failure Mode, Criticality and Effect Analysis (FMECA), characterized by a single evaluation scale for *occurrence* and *detection* of the Failure Mode (FM), and by various scales for assessing the severity of the effects of the FM, depending on the area of interest involved (patient safety, operator safety, patient satisfaction, clinical effectiveness, economical expenses) [5]. This instrument has been adapted and utilized during the development of the present work and for its goals. It has been decided to use the Failure Mode, Criticality and Effect Analysis (FMECA) mainly because of its flexibility in the definition and calibration of risk parameters (i.e. hazard likelihood, severity and detectability). Indeed the use of the FMECA framework led to the definition of a multi-attribute analysis where all the dimensions of service effectiveness were taken into account. The FMECA approach requires to identify all the potential failure modes (FMs) of the process – including technical failures, human errors and organizational vulnerabilities – and to estimate, for each one of the identified FMs, three risk parameters:

- likelihood of FM occurrence (Occurrence Scale Rank);
- detectability of the FM (Detection Scale Rank);
- severity of the FM for patient, operator and structure (Severity Scale Rank).

Those three scales have been calibrated based on both available data on thermal ablation adverse events and international quality standards (Good Manufacturing Practices, GMP). It has been decided not to considerate the entire range of values proposed by the international standard IEC 60812 about FMEA, not only to draw a parallel between the Ministerial document, but also due to a better analysis repeatability and an easier use and result interpretation.

Table 1 Occurrence Scale Rank

Occurrence Scale	Rank
Remote	1
Low	3
Moderate	5
High	8
Very High	10

Table 2 Detection Scale Rank

Detection Scale	Rank
Very High	1
High	3
Moderate	5
Low	8
Remote	10

Table 3 Severity Scale Rank

Severity Scale	Rank	Patient	Operator	Structure
Insignificant	1	No consequences	No consequences	No consequences
Marginal	3	Need of additional diagnostic investigations	Workflow delay	Workflow delay
Moderate	5	Minor Complications	Need of therapeutic treatments	Mild expenses increase
Critical	8	Major Complications	Need of hospital admission	Activity interruption
Catastrophic	10	Death	Death	Structure trust loss

Three dimensions have been taken into account: *patient* (mainly based on the effects on patient's health), *operator* (mainly based on the effects on operator's health and his professionalism) and health *structure* (mainly based on an economical profile and on patient trust in the facility itself).

To identify all the possible FMs, the first step is devoted to the detailed description of the process in terms of activities, physical and information flows, tools and resources, operators and different professional profiles involved in the entire process, by means of a workflow analysis according to ANSI Standard ANSI/PMI 99/001/2008. The correct level of detail in the process description is crucial for an effective risk analysis and also allows to anticipate the identification of both critical points and opportunities of improvement in the process.

There have been identified 49 different tasks for the MWA activity. For each one of those, all possible FMs have been investigated, identifying 273 overall FMs.

In Fig.1 is showed the header of the FMECA table used during this analysis.

Fig.1 FMECA table header

III. RESULTS

Proceeding with risk analysis, *Risk Priority Numbers* (RPNs) have been calculated according to international standard IEC 60812. In Fig.2, Fig.3 and Fig.4 the distributions of RPN for the patient, operator and structure dimension, respectively, are reported.

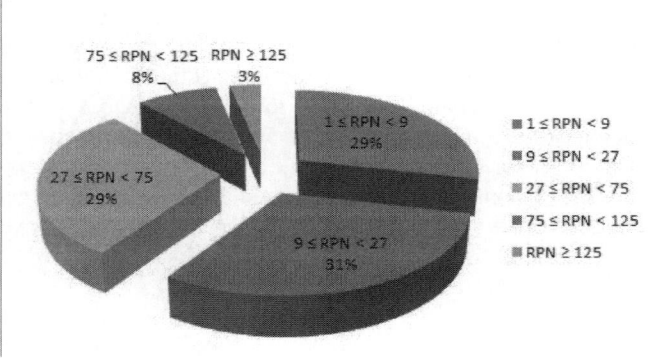

Fig.2 RPN distribution for patient dimension

Concerning risk assessment, a threshold has been graphically identified by means of a scree plot, similar to the one utilized in the Principal Components Analysis (PCA) [6]. Scree plot settings require preliminary ordering of the RPN values by size, from smallest to largest. These values are then plotted, by size, across the graph. Normally, when

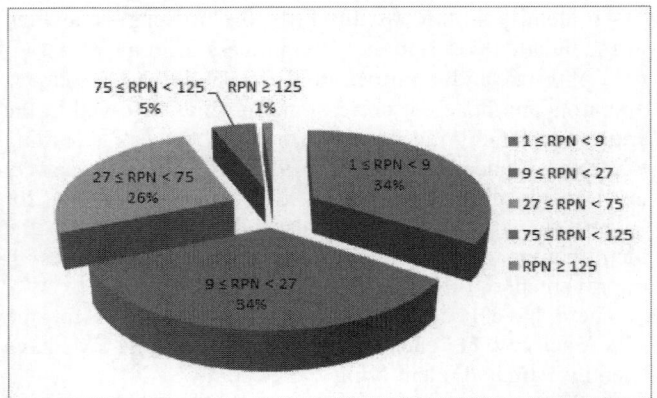

Fig.3 RPN distribution for operator dimension

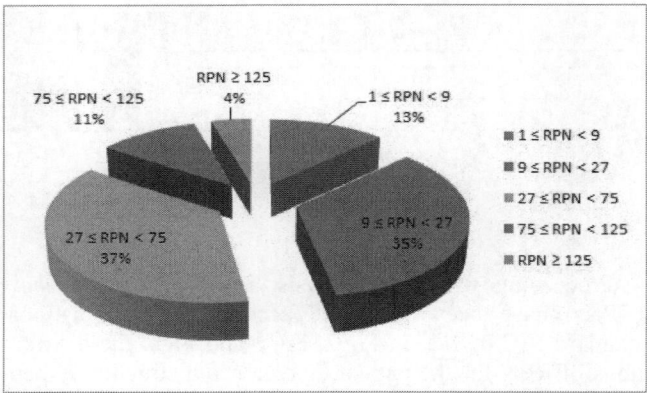

Fig. 4 RPN distribution for structure dimension

observing from the right, scree plot appears like a cliff, descending towards ground level. Two well distinguished trendlines can be noticed. At the intersection of these two lines, it's located the threshold for risk acceptability. The identified thresholds are RPN = 50 for patient dimension, RPN = 40 for operator dimension and RPN = 50 for structure dimension. RPN distribution and Table 4, clearly show the need of a wide mitigation maneuver: the elements above this threshold are quite relevant indeed, and about 30% of FMs has an unacceptable risk value.

Table 3 Failure Modes with risk exceeding threshold

	RPNp		RPNo		RPNs	
	FMs	%	FMs	%	FMs	%
FM w/RPN> threshold	67	24.5%	69	25.3%	91	33.3%

To prioritize the mitigation interventions, it has been used RPN_{tot} value, representing the sum of the RPNs of the three dimension considered during the analysis.

The four most compelling intervention involve the following failure modes:

- Impossibility to cool the antenna due to the depletion of cooling gas;
- Attempt to reutilize and/or re-sterilize the disposable applicator;
- Breakage of the antenna shaft due to excessive force applied by the operator;;
- Choose of dangerous parameters for the health of the patient.

For each one of those, it has been reported the proposed risk control solution:

- Providing an internal thermocouple, to continuously monitor the temperature inside the ablation zone, in combination with an ultrasound sensor outside the gas tank which informs the user in case the gas level would become inadequate for the correct completion (with wide margin) of the longest possible operation;
- Providing a control and authentication system for the applicators through the internal microcontroller, which should be able to interrupt the distribution of microwave energy in case of test failure;
- Providing a pressure sensor near the radiating tip which informs the user about the risk of breaking the applicator; furthermore it would be useful to design the applicator with reasonable flexibility for clinical purposes;
- Providing for a popup window which shows duration and distributed power values, that have to be necessarily validated by the user before the beginning of the treatment.

Being MWA an inherently destructive operation, even after the mitigation interventions, it will be necessary that the user's guide will inform the radiologist about the various potential risks still present: therefore it will be essential to perform a series of interventions to drive formation and training to the user himself.

IV. CONCLUSIONS

The mitigation interventions proposed here have ensured an RPN reduction near to 70% for each one of the three dimension considered, as seen in Fig.5.

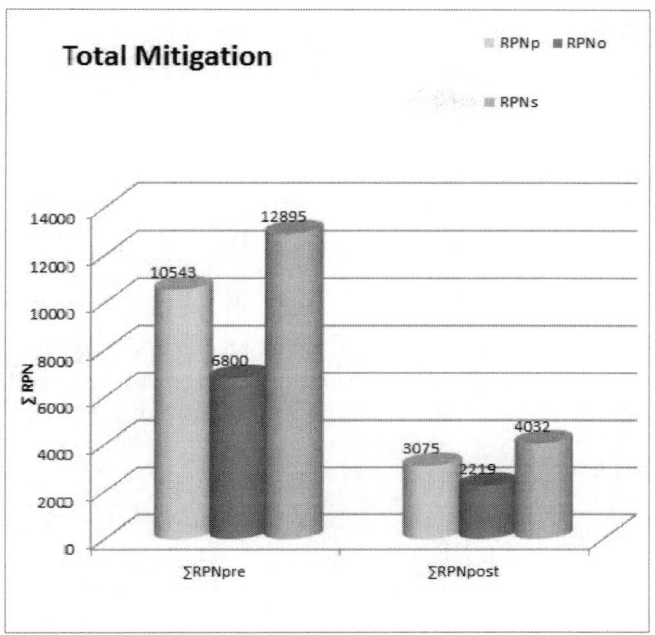

Fig.5 Mitigation effects

The multidimensionality has allowed the identification of some criticalities not otherwise detectable and has included in the improvement process all the principal subjects involved in the microwave thermal ablation activity.

Every identified risk, including the additional ones closely related to the mitigation interventions themselves, has been brought back under the acceptability threshold: the system object of the paper is ready to go further, towards a clinical evaluation and all the next steps needed for CE marking.

REFERENCES

1. Brace C L (2009) Radiofrequency and microwave ablation of the liver, lung, kidney, and bone: what are the differences ?. Curr Probl Diagn Radiol 38:135-143
2. Brace C L (2011) Thermal tumor ablation in clinical use. IEEE Pulse 2(5):28-38
3. Tabuse K (1998) Basic Knowledge of a microwave tissue coagulator and its clinical applications. J Hep Bil Pancr Surg 5:165-172
4. Liang P, Wang Y, Yu X et al. (2009) Malignant liver tumors: treatment with percutaneous microwave ablation - Complications among cohort of 1136 patients. Radiology 251(3) 933–940
5. Ministero della Salute (2011) Sviluppo di una metodologia per la valutazione delle tecnologie finalizzate alla sicurezza dei pazienti. Dipartimento della qualità - Direzione generale della programmazione sanitaria, dei livelli di assistenza e dei principi etici di sistema
6. Bluvband Z, Grabov P, Nakar O (2004) Expanded FMEA (EFMEA). 2004 Annual Reliability and Maintainability Symposium Proceedings, 2004, pp.31–36
7. Bambi F, Spitaleri I, Gianassi S, Verdolini G, Perri A, Dori F, Iadanza E (2009) Analysis and management of the risks related to the collection, processing and distribution of peripheral blood haematopoietic stem cells. Blood Transfus. 2009 January; 7(1): 3–17. doi: 10.2450/2008.0006-08
8. Dori F, Iadanza E, Miniati R, Mattei S (2010) Risk management process and CE marking of software as MD. IFMBE Proceedings MEDICON and HEALTH TELEMATICS 2010 "XII Mediterranean Conference on Medical and Biological Engineering", Vol. 26, 2010, Chalkidiki, Grecia. ISBN: 978 3 642 13038 0

Author: Ernesto Iadanza
Institute: Department of Information Engineering, Florence
Street: via di Santa Marta, 3
City: Florence
Country: Italy
Email: ernesto.iadanza@unifi.it

Medical Devices Recalls Analysis Focusing on Software Failures during the Last Decade

Z. Bliznakov, K. Stavrianou, and N. Pallikarakis

Biomedical Technology Unit, Department of Medical Physics, School of Health Sciences,
University of Patras, Rio–Patras, Greece

Abstract—Medical devices play a vital role in the delivery of high quality healthcare. Although recent technological advancements have led to much more reliable and safer medical devices, potential risks of failure and the associated adverse incidents cannot be neglected. Medical device (MD) recalls by manufacturers contribute to the safe function of the devices, in order to avoid incidents that could lead to injuries and deaths.

The purpose of the present work is to present up-to-date information concerning MD recalls due to software failures. For this purpose, data from the United States Food and Drug Administration (FDA) Enforcement Reports for the period 1999-2010 have been used. The outcomes from data collection and analysis are presented through the use of ratio indicators and their distribution over the time. Furthermore, classifications of the MD recalls according to reasons of failure and the level of health hazard have been performed.

The results reveal that almost half of the medical devices being recalled make use of software for their operation, indicating the growing role of the software in the domain of medical equipment. Furthermore, four out of every ten medical devices incorporating software have failed due to a problem in the software itself, while compared to the total FDA MD recalls this reaches 18.3% of software failures during this period. The present recalls analysis has demonstrated significant increase of MD software failures during the last decade, compared to previous studies.

Keywords—Medical Devices, Adverse events, Recalls, Software failures, FDA Enforcement reports.

I. Introduction

Biomedical technology has contributed decisively to the impressive progress of modern healthcare over the past fifty years. Medical devices (MD) play a vital role in the delivery of high quality health care. Innovative medical technologies, providing new preventive, diagnostic and therapeutic means, enter continuously into the market. Although recent technological advancements have led to much more reliable and safer MD, potential risks of failure and the associated adverse incidents cannot be neglected. In fact adverse incidents, due to MD, have recently increased in absolute terms, due to the exponential increase of the MD used today.

Medical devices nowadays multiply rapidly, and today there are hundreds of thousands of different products available on the worldwide markets. The impact of informatics on the healthcare sector has also increased enormously over the last twenty years. The evolution of computer technology in combination with the new advances of software engineering sustains the high level of progress in this domain. Today, software is either embedded in a wide range of medical equipment, used as an accessory, or as a standalone medical device on its own right. It becomes critical to face the issue of quality design of software products, in order to ensure safe and reliable performance and to bring the benefits from their usage right to the patient with minimum risks.

Software dramatically differs from hardware in majority of the aspects. It is very easy to modify and update software without precise control, whereas producing copies might be accomplished without supervised management. This makes the traceability very difficult. Additionally, due to its high complexity, it is generally almost impossible to entirely check and control software for quality and safety [1]. During the last two decades there were indications of a trend for an increasing share of MD failures due to software, reflecting the growing importance of software in these products. Fatal accidents have occurred due to software errors that have caused among others wrong patient blood groups identification or misleading electronic prescriptions [2].

The purpose of the present work is to present up-to-date information concerning medical device recalls due to software failures. Specifically, the objectives are to: (1) collect data available from major medical device vigilance systems concerning recalls due to software problems; (2) perform analysis, using different methods and classification schemes; (3) present the results in a comprehensive way; and (4) make conclusions and identify future trends.

II. Materials and Methods

Medical device recalls by manufacturers contribute to the safe and qualitative function, in order to avoid incidents which could lead to injuries and deaths. The current study uses data from the United States Food and Drug Administration (FDA) Enforcement Reports [3] for the period

1999-2010. The most significant information from the FDA MD recalls used in the present work includes:

- Recalling Firm - The firm that initiates a recall or has primary responsibility for the product manufacture.
- Class - Numerical designation (I, II, or III) that is assigned by FDA to a particular product recall that indicates the relative degree of health hazard.
- Product Description - Brief description of the product being recalled.
- Reason for Recall - Information describing how the product is defective.

The medical devices that are referenced in the FDA recalls enforcement reports can be as simple as single-use devices, such as surgeon gloves and syringes, or complicated systems, such as computed tomography scanners and magnetic resonance imaging systems. For the purposes of this study, emphasis has been given to the medical devices that contains software and have been recalled due to software problems. The total number of the recalls studied exceeds 7000 over the period 1999-2010. Data were processed and presented through the use of relative indicators and their distribution over the time. Further on, classification of the recalls in the collected data has been performed according to the failure symptoms. Four major categories of failure reasons have been distinguished and relative classification has been performed.

A. Data Collection and Presentation

Collection of FDA recalls data has been performed over a twelve-year period of time 1999-2010. Three absolute values have been measured and presented analytically for each year:

- total number of medical device recalls (MDR);
- number of medical device recalls involving medical devices incorporating software (MDRSW);
- number of medical device recalls involving medical devices incorporating software and due to software failures (MDRSWF).

Further on, three ratio indicators have been derived and used for presentation and analysis:

- percentage ratio of recalls of medical device incorporating software vs. total number of medical device recalls (MDRSW / MDR);
- percentage ratio of recalls involving medical devices incorporating software and due to software failures vs. recalls involving medical devices incorporating software (MDRSWF / MDRSW);
- percentage ratio of recalls involving medical devices incorporating software and due to software failures vs. total number of medical device recalls (MDRSWF / MDR).

B. Classification of Medical Device Recalls According to Reason of Failure.

The medical devices recalls have been classified into four major categories according to the primary reason of failure of the recalled medical device for the 2008 - 2010 time period. Medical devices recalls have been grouped into the following four categories:

- Hardware failures
- Software failures
- User interface problems
- Others

A hardware failure is considered to be a malfunction in the mechanical components, electromechanical components or within the electronic circuits of a medical equipment system. Recovery from a hardware failure requires repair or replacement of the defective part before the equipment can be placed back into operation. A software failure is an error, bug, or fault into the software computer program or system that produces an incorrect or unexpected result, or causes it to behave in unintended ways. Most faults arise from mistakes and errors made by people (the software developers) in either the program's source code or its design, and a few are caused by compilers producing incorrect code. A failure associated with the user interface is considered to be any problem directly related to the interface, operation controls and communication between the user and the medical device. The "Others" general group contains any medical devices recalls which do no fall into any of the above three groups.

C. Classification of Medical Device Recalls According to Level of Hazard.

The FDA medical devices recalls have been classified into three different categories according to the level of hazard involved:

- Class I: Dangerous or defective products that predictably could cause serious health problems or death;
- Class II: Products that might cause a temporary health problem, or pose only a slight threat of a serious nature;
- Class III: Products that are unlikely to cause any adverse health reaction, but that violate FDA labelling or manufacturing laws.

III. RESULTS

Analytical data of the MD recalls from the FDA enforcement reports for each year separately are presented in table 1. Overall, 7004 recalls of medical devices have been studied, covering the period from January 1999 to December 2010. The number of MD recalls using software for their function is found to be 3224, representing 46,0% of the total number. The MD recalls due to software failures are found to be 1283, representing 18,3% of the total number of MD recalls. The percentage of MD recalls due to software failures vs. the MD recalls incorporating software is 39,8%.

The percentage ratio of recalls of MD incorporating software vs. total number of MD recalls (MDRSW / MDR) is given analytically for each year. This indicator reflects the range of software introduced in MD, resulting from the computer technology progress. At the beginning of the period, MDRSW / MDR is about 25%; while after 2005 this ratio increases and exceeds 50% until the end of the period.

The percentage ratio of recalls involving MD incorporating software and due to software failures vs. recalls involving MD incorporating software (MDRSWF / MDRSW) is given analytically for each year. The data indicate the software involvement in MD failures. The values vary from 20% to over 50% for the whole period 1999-2010 with a mean value of 39,8%.

The percentage ratio of recalls involving MD incorporating software and due to software failures vs. total number of MD recalls (MDRSWF / MDR) is given analytically for each year. The values vary from about 10% at the beginning of the period to over 30%, with a mean value of 18,3% showing more intensive presence of the software failures in the MD failures.

Figure 1 shows graphically the percentage ratios of MD recalls (MDR); the number of MD recalls involving medical devices incorporating software (MDRSW); and the number of MD recalls involving medical devices incorporating software and due to software failures (MDRSWF) analytically for each year within the study period 1999-2010.

Classification of MD recalls according to the reason of failure for the period 2008-2010 have been accomplished. The results reveal that almost half of the medical devices

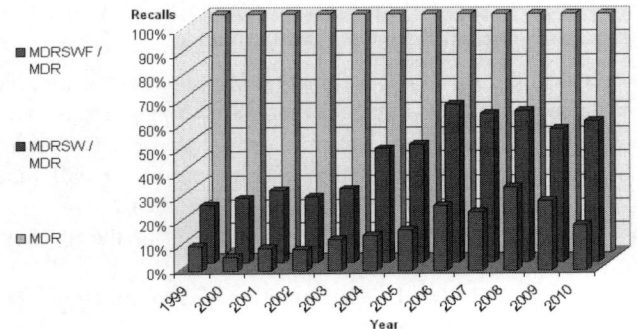

Fig. 1 Graphical presentation of percentage ratios of MDR, MDRSW / MDR, MDRSWF / MDR vs. time.

have been recalled due to failure in the hardware itself. In percentage ratios, 45,7% of the total FDA recalls are attributable to hardware failures, 39,9% to software failures, 9,0% to user-interface issues and the rest 5,4% to none of the above (Fig. 2). Comparison with previous studies demonstrates that the number of software failure recalls has significantly increased during the last decade [2]. The increase in the number of medical devices being recalled due to software problems in the latest years is a direct consequence of the increased MD number incorporating software.

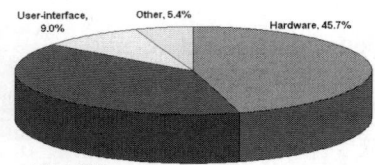

Fig. 2 Percentage classification of medical devices recalls according to the reason of failure for the period 2008-2010.

Table 1 Overall recalls data from the FDA enforcement reports for the period 1999-2010.

FDA recalls	1999	2000	2001	2002	2003	2004	2005	2006	2007	2008	2009	2010	Total
MDR	498	460	593	674	525	468	553	585	603	663	598	784	**7004**
MDRSW	118	123	179	185	161	222	273	384	372	417	331	459	**3224**
MDRSW / MDR	23,7%	26,7%	30,2%	27,4%	30,7%	47,4%	49,4%	65,6%	61,7%	62,9%	55,4%	58,5%	**46,0%**
MDRSWF	51	27	57	60	69	68	93	158	147	231	173	149	**1283**
MDRSWF / MDRSW	43,2%	22,0%	31,8%	32,4%	42,9%	30,6%	34,1%	41,1%	39,5%	55,4%	52,3%	32,5%	**39,8%**
MDRSWF / MDR	10,2%	5,9%	9,6%	8,9%	13,1%	14,5%	16,8%	27,0%	24,4%	34,8%	28,9%	19,0%	**18,3%**

Classification of MD recalls according to the level of the health hazard has been also accomplished for the period 2008-2010. The results reveal that 6,7% of the MD recalls have been considered to be of a high level risk for health hazard. The majority of the cases 91,3% has been classified as medium risk and only 2% have been found to be low risk.

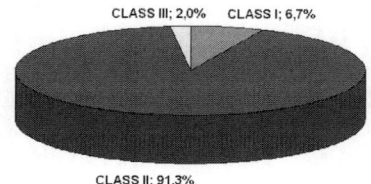

Fig. 3 Percentage classification of medical devices recalls according to the FDA Class, representing the level of the hazard for the period 2008-2010.

IV. DISCUSSION

The increased percentage of the MD recalls incorporating software (demonstrated in figure 1) reveals the extended introduction of software in medical equipment. Nowadays, software is found on a large scale of even simple devices or inevitably appears in new sophisticated medical equipment. Almost half of the MD (46%) being recalled make use of software for their operation. The increased number of MD being recalled due to software problems in the latest years is a direct consequence of the increased number of MD with embedded software. The percentage of software failure recalls vs. the recalls of MD using software for their operation has an average value almost 40% over the whole period studied. This means that two out of five medical devices using software have failed due to a problem in the software itself.

The percentage of software failure recalls compared to the MD total number recalls exceeded 30% in 2008. Concerning the next two years, although there is a small decrease, this value is still significantly high. This decrease shows a tendency of the manufacturer to pay more attention on safety issues concerning software failures.

Regarding the reason of failure, the results over the period 2008-2010 reveal that almost half of the MD have been recalled due to failure in the hardware itself. In percentage ratios, 45,7% of the total FDA recalls are attributable to hardware failures, 39,9% to software failures, 9,0% to user-interface issues and the rest 5,4% to none of the them.

Concerning the FDA Class, representing the level of the hazard, the results over the period 2008-2010 reveal that the majority of the MD (over 91%) being recalled are considered to be medium risk products that may cause a temporary health problem. The percentage of the high risk medical devices is relatively low with less than 7% from the total number and 2% of low risk medical devices.

V. CONCLUSIONS

The present study on MD recalls has demonstrated significant increase of recalls due to software failures during the last ten years. Results have shown that approximately almost half of the medical devices being recalled make use of software for their operation, indicating the growing role of the software in the domain of medical equipment. However, this leads to an increase of the number of MD being recalled due to software problems.

Comparison with previous studies demonstrates that the number of software failure recalls has significantly increased during the last decade. This increase in the number of medical devices being recalled due to software problems is a direct consequence of the increased number of medical devices using embedded software. This reveals the tendencies and the areas to focus for the achievement of highest level of safety and quality in biomedical technology.

ACKNOWLEDGMENT

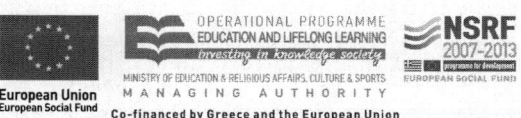

This research has been co-financed by the European Union (European Social Fund – ESF) and Greek national funds through the Operational Program "Education and Lifelong Learning" of the National Strategic Reference Framework (NSRF) - Research Funding Program: Thalis. Investing in knowledge society through the European Social Fund.

REFERENCES

1. Andersen P (2003). A suggestion for guidance to the Medical Devices Community on the use of software validation and approval, In Proceedings of the Medical Device Software Workshop, EC Joint Research Centre, Ispra, Italy.
2. Pallikarakis N (2003) Medical Devices Software. In Proceedings of the Medical Device Software Workshop, EC Joint Research Centre, Ispra, Italy.
3. U.S. Food and Drug Administration Enforcement Reports. http://www.fda.gov/Safety/Recalls/EnforcementReports/default.htm

Address of the corresponding author:

Author: Zhivko Bliznakov
Institute: University of Patras
Street: Department of Medical Physics, School of Health Sciences
City: Rio - Patras
Country: Greece
Email: jivko@upatras.gr

Evaluation of Physical Environment Parameters in Healthcare

G.A. Elias[1] and S.J. Calil[2]

[1] CEFET-MG/Department of Technological and Professional Education, DEII, Belo Horizonte, Brazil
[2] Unicamp/Department of Biomedical Engineering, FEEC, Campinas, Brazil

Abstract—The physical environment in hospitals should provide adequate conditions in terms of lighting, thermal comfort, air quality, and noise level. If such conditions are not appropriate, both workers and patients may be negatively affected. In this work is was developed a human factors and ergonomics based methodology to enable the evaluation of the physical environment in patient care areas. The methodology was developed in six steps. First, literature research was performed to determine the parameters to be evaluated. Second; three methods to evaluate the selected parameters were defined: measurement, observation, and written survey. In the third step two charts were created to aid in the parameters measurement and observations. The fourth step involved the development of a written survey in the form of a questionnaire to be applied to healthcare staff. In the fifth step it was developed a method to process the collected data (measurements, observations, and written survey). Finally, in the sixth step, dashboards were developed to report the collected data. The methodology was applied in the emergency department observation room of a public hospital. The analysis of the reports showed that the temperature and noise were not always in accordance with the established limits. Moreover, the fact that some workers were negatively affected by noise could be verified through survey answers. In addition, there were complaints regarding the risk of slip, trip or fall.

Keywords—Ergonomics, human factors, physical environment, environmental risk.

I. INTRODUCTION

Hospital activities are characterized by intensive labor requiring high productivity in a limited period of time, often under inadequate working conditions and with possible problems related to the environment, equipment and processes. To allow workers to appropriately perform their tasks, the physical environment should provide adequate conditions in terms of lighting, thermal comfort, air quality, and noise level. The healthcare workers' ability to perform their tasks accordingly is linked to the prevailing environmental conditions within the workplace. If such conditions are not appropriate, those workers may be affected by health problems, dissatisfaction, fatigue, and low productivity [1,2]. The patient, on the other hand, can be affected both directly and indirectly by the same environmental conditions that influence the workers and by the poor quality of the service they may receive.

Although it is known that the physical environment significantly affects healthcare workers and patients, in general, hospital design is not human-centered. In fact, most construction codes do not take into account ergonomics criteria [3,4]. As a result, workplace problems related to inadequate design do affect workers' as well as patients' health and safety. Moreover, even when an environment is ergonomically designed but not properly maintained, problems will appear sooner or later.

II. METHODOLOGY

A human factors and ergonomics (HF/E) based methodology was developed to evaluate the physical environment in patient care areas. After an extensive review of the literature, the parameters to be analyzed were selected and grouped as following: work area, noise, lighting, power outlets, and medical gas outlets. These parameters were analyzed by physical measurement, observation, or written survey (questionnaire). The application of the methodology comprised six steps as shown in figure 1.

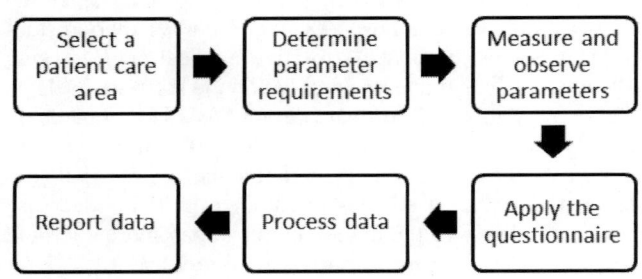

Fig. 1 Methodology application steps

The description of each step is given hereafter:
1. Since the methodology is intended to be applied in any patient care area (e.g. operating room, intensive care unit, neonatal intensive care unit, emergency room, etc.), it is necessary to select one area to be evaluated at a time, in order to determine the specific parameter requirements. So, the measured values can be properly compared to the reference values for this area;
2. With the aid of a form, specific parameter (e.g. noise, temperature, relative humidity, Carbon Dioxide

concentration, etc.) requirements for the selected area should be determined. These requirements may be the parameter's minimum value, the parameter's maximum value, or the parameter's range. For example, a determined patient care area may require as the temperature limits the range of 22 to 24 degrees Celsius, while the maximum noise level at that environment be 45 dB. It is recommended that standards be consulted to determine these levels;

3. It was defined to measure parameters regarding work area, noise, lighting, power outlets, and medical gas outlets to verify compliance with previously established requirements. However, some elements, such as liquids on the floor, objects in passageways, and lamp flickering, cannot be objectively measured and it is necessary to use observation to verify whether they are present in an environment or not. A form was created to guide the applicant during this stage;

4. A written survey using self-administered questionnaire composed of 31 closed-ended questions was developed aiming to discover how the workers perceive certain parameters in the environment (e.g. noise, temperature, air quality) and also to determine the physical, mental, and emotional symptoms these parameters may be causing on workers. Three types of scales for the answers were chosen: verbal frequency (e.g. never, rarely, sometimes, often, always), semantic differential (e.g. dry/humid, low/loud, small/big) and multiple responses;

5. Data regarding measurement and questionnaires need further processing in order to be properly reported, which is done in this stage;

6. Finally, all the data processed and collected so far is visualized with the aid of dashboards created in Microsoft Excel™ 2010.

In summary, to apply the methodology, a person should select the patient care area that will be analyzed. The parameter requirements such as temperature, relative humidity, and others to be determined. Then, the parameters should be measured and observed, followed by the application of a questionnaire to the healthcare workers. After all data is collected, it should be processed to be properly analyzed.

III. DISCUSSION

The methodology was applied in the emergency department observation room (ED-OU) in a public teaching hospital in São Paulo state, Brazil from 9:38 am to 1:19 pm and from 4:05 pm to 7:23 pm, performing a total of 7 hours and 19 minutes. All the steps described earlier were followed. It will be discussed some results regarding noise and environmental parameters. The graphs show the measurements of noise and temperature, while the horizontal and vertical bar charts displays questionnaire answers.

The measured noise level in the environment and the users' perception about it is shown in figure 2 shows. The measured average noise was 67.8 dB, 22.8 dB higher than the recommended level of 45.0 dB. It can be seen that about 93% of the workers considered the noise level in the environment to be loud or very loud.

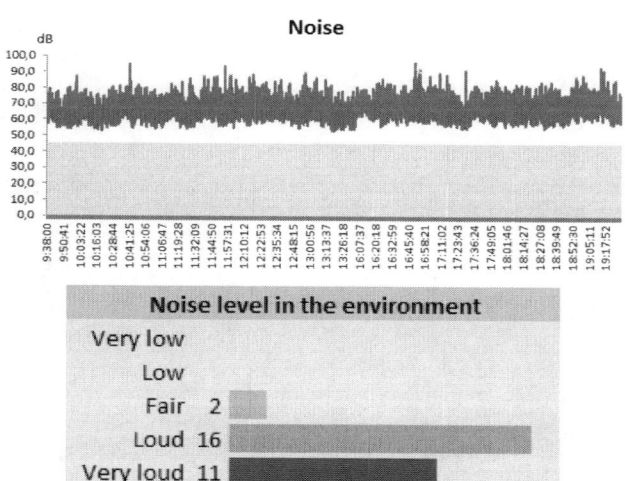

Fig. 2 Noise level in the ED-OU

Figure 3 displays the noise symptoms, gathered with the aid of the written survey, according to the workers, showing that only one worker did not present any noise related symptom. It is possible to see that 80% of respondents complained about the need to speak up due to the noise what, in turn, contribute to increase the noise level in the environment, creating a vicious cycle. About 63% of the respondents complained about the difficulty in hearing during conversations and mental fatigue which could compromise the quality of care.

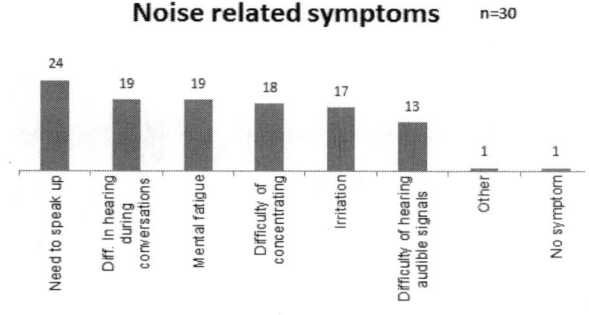

Fig. 3 Noise related symptoms

In figure 4, it is possible to see that the temperature in the ED-OU was above the upper limit of 26.0°C during all the measurement period, reaching 29.0°C in some moments. About 83% of the staff felt uncomfortable of very uncomfortable regarding temperature.

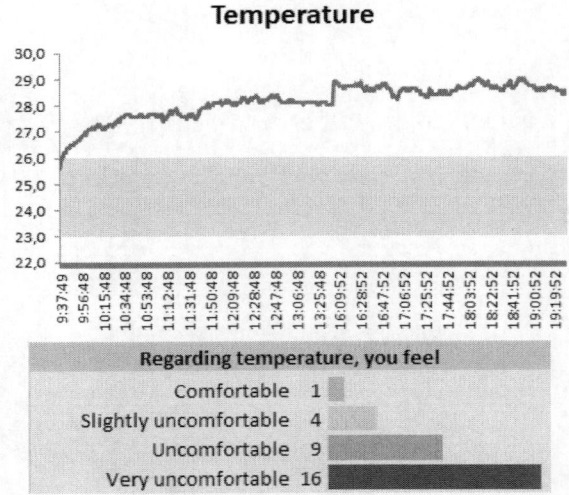

Fig. 4 Temperature in ED-OU

In figure 5, the observations and questionnaire regarding the floor and the slip, trip, and fall (STF) risks are shown. It is possible to see that the floor is slippery and there are objects in passageways, what increase the STF risk. In fact, 55% of workers considered the environment STF risk as being medium or even high, with half of them suffered at least one slip, trip, or fall in the 15 days prior to the research.

Figure 6 below presents the workers opinion related to air quality parameters originated from questionnaire analysis. It can be seen that about 90% of the respondents consider the environment as often or always stuffy, while 100% agreed that it has unpleasant odors and 85% stated that air quality is poor or very poor. In general, it can be said that this environments is not provided adequate working conditions regarding the air quality.

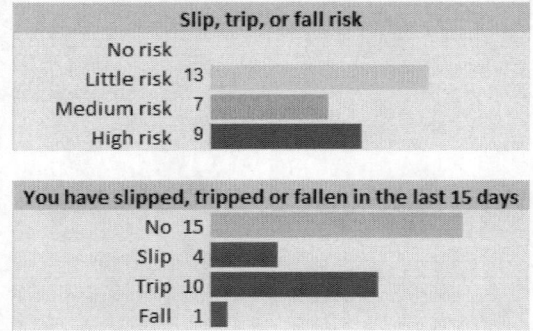

Fig. 5 Slip, trip, and fall risk

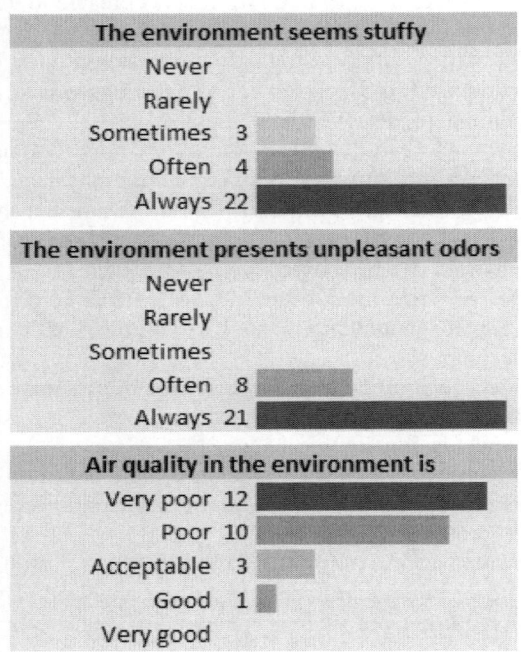

Fig. 6 Air quality related parameters

IV. CONCLUSIONS

It was possible to identify environmental parameters, such as noise, temperature and air quality negatively affecting workers. The analysis of the problems found with the purpose of developing and implementing solutions could bring benefits to both workers and patients. These actions could improve both workers and patients well-being, creating better work conditions, and eventually improving the quality of health care.

The structure of the methodology is more important than the parameters analyzed by it. The way the methodology was developed allows the inclusion or exclusion of parameters to be analyzed as well as the inclusion of other groups of parameters such as biological contaminants and volatile organic compounds, alarms, vibration, and arrangement of components. It is still possible to perform a more thorough analysis of the parameters depending on the available physical, human and financial resources. This should be done by defining the parameters to be evaluated as well as how to evaluate them; writing instructions in the forms; adding questions to the questionnaire; adding information on how to process the collected data; and designing a dashboard to report the collected data.

ACKNOWLEDGMENT

Capes/Setec by the financial support.

REFERENCES

1. Hedge A. Environmental methods. In: Stanton, N. et al. (2005) Handbook of Human Factors and Ergonomics Methods. CRC Press, Boca Raton.
2. Rosa MTL. O desenvolvilmento técnico-científico da engermagem: uma aproximação com instrumentos de trabalho, Congresso Brasileiro de Engermagem, Brasil, 1999, pp 97-126.
3. Reiling J, Chernos S. Human factors in hospital safety design. In: Carayon, P (ed) (2007) Handbook of Human Factors and Ergonomics in Health Care and Patient Safety. CRC Press, Boca Raton.
4. Villeneuve J et al. Ergonomic intervention in hospital architecture. In: Pikaar, R et al (2007) Meeting diversity in ergonomics. Elsevier, Oxford.

Author: Gustavo Alcântara Elias
Institute: CEFET-MG
Street: Av. Amazonas 5253
City: Belo Horizonte
Country: Brazil
Email: gustavo@deii.cefetmg.br

Manufacturer Training Impact in Heuristic Analysis: Usability Evaluation Applied on Health Devices

V.H.B. Tsukahara, F.N. Bezerra, R.H. Almeida, F.S. Borges, E. Pantarotto, R.V. Martins,
M. Iglesias, P. Trbovich, S. Pinkney, and S.J. Calil

[1] Unicamp/Biomedical Engineering Department (DEB), Campinas, São Paulo, Brazil
[2] HCor, São Paulo, São Paulo, Brazil
[3] University Health Network, Toronto, Ontario, Canada

Abstract—The adoption of new technologies into hospitals has been improved diagnosis and patient care, and it has been also increased the effectiveness of the services concerned. Despite all the benefits of using medical devices, however, this new reality has brought a new kind of problem: errors caused by misuse. Errors which may result in adverse events and usually are credited to health professionals' mistakes, often solved through actions like training and institute protocols that sometimes are inefficient. The study of interaction between man and machine may help to understand this new misuse issue and the causes that may lead to use errors. Aiming to improve equipment usability, the use of tools from Human Factors Engineering like Heuristic Analysis can contribute to usability problems identification. The objective of this paper is to verify the influence of equipment training on its usability problems by means of comparing the results from applying Heuristic Analysis technique on a traditional infusion pump before and after manufacturer training.

Keywords—**Human Factors Engineering, Usability, Heuristic Analysis, Infusion Pump, Usability Problem.**

I. INTRODUCTION

The introduction of new technologies into the healthcare system significantly improved diagnosis and diseases treatments, as well the effectiveness of the services concerned [1]. However, even with all the benefits derived from this improvement, such as better work conditions for health professionals and better patient assistance, the use of medical devices introduced a new kind of problem: device use errors. Operating errors or device misuses represent 60-80% percent of the total hospital adverse events in Brazilian's Health System (SUS) [2]. ECRI Top Ten Hazards 2012 [3] states that errors on the administration of medications using infusion pumps are the third most important concern related with use of health technologies by professionals. The same publication [4] moved up its concern to the second position in 2013.

Related to use errors, it is verified that lack of operator training, stress and equipment interface project are factors that must be considered on this scenario [5].

One way to analyze this human-machine interaction could be through device usability evaluation, which allows the identification of problems that may lead to human errors [6]. A practical tool that can be used for error identification is Heuristic Analysis.

This paper presents the outcome of the application of Heuristic Analysis technique, by a multidisciplinary team on a commonly used infusion pump model. The technique was applied twice: once before the team received training on the pump by the manufacturer, and once after the training.

II. BACKGROUND

Initially developed in Usability Engineering (related to Computer Science, Ergonomics area), Heuristic Analysis was later adapted to medical devices evaluation [6].

This tool consists in a practical technique, in which at least three evaluators apply a set of heuristic rules to identify usability problems in a medical device.

To apply the technique, a total of 14 rules are used as a reference by the evaluator to better explain the usability problems identified during device's Heuristic Analysis, the Nielsen-Schneiderman heuristics[7]. After identifying a problem, the heuristics violations are associated, and then the severity levels of each violation are defined, that represents the gravity level associated to the usability problem that can lead to an adverse event.

To apply the Heuristic Analysis technique, the evaluator doesn't necessarily need to be an expert on heuristics (the minimum requirement is to have a basic knowledge on Heuristic Analysis technique), and neither to have a deep knowledge about the studied device, like a medical device for example. The adoption of a multidisciplinary evaluation team also influences positively the analysis. The observation of a medical device by professionals with different backgrounds allows the identification of a larger number of problems, because of the different viewpoints applied in the evaluation.

III. Methodology

To apply Heuristic Analysis it is necessary for the evaluator to explore the object under evaluation, purposing to investigate its usability. With this experience the evaluator could point out usability problems and determine corresponding heuristic violations and severity level.

A "heuristic violation" means that some characteristic (physical, software, use, hardware) violates one of the fourteen heuristic rules described below.

The study was based on the 14 Nielsen-Schneiderman heuristic rules [7]. Although rules definitions were presented in this literature, the team had to discuss them to ensure evaluators the same understanding over all the fourteen rules, because it was the first team contact with the technique. The following interpretation was achieved from this discussion:

1. Consistency and patterns: aspect that must be followed and observed on product interface, through layout and positioning, color patterns, action sequences, language;
2. Visibility of system state: the system must present information to user in order to provide a clear current state;
3. Match between system and world: interface must be intuitive. The operator perception about the system must correlate to user vision about the system;
4. Minimalist: information must be summarized. However, caution is necessary to not transform it in abstract information for the user;
5. Memory: memory load must be minimized. User doesn't have to memorize a lot of information to use the product;
6. Feedback: user needs confirmation of their system actions;
7. Flexibility and efficiency: operators have different modes of interaction with the system. Therefore, the system must provide a flexible interface in such a way as to consider users preferences and variability to ensure maximum efficiency;
8. Messages: during system problems, the user must be informed to understand, learn and solve the presented error;
9. Error preventions: system interface must be developed to prevent misuse;
10. Closure: executed tasks must have well defined starting and ending points;
11. Reversible actions: the operator must be capable to recover from errors;
12. Language: the language used must be clear to the user profile;
13. Control: the system must allow at least minimum control over the user actions;
14. Help and documentation: operator must have easy access to help when necessary.

Still according to [7], severity levels for equipment problems can be assessed using a severity scale:

0, when it does not cause a misuse problem, but it's an opportunity for future improvements;
1, only when it is a cosmetic problem, and correction must be done only when there is free time;
2, small usability problem. It may have a small priority to correct;
3, major usability problem. High priority must be given to solve the problem, because it is important to be fixed;
4, catastrophic error. It is essential to solve. If the object under evaluation has not been released, it is mandatory previous correction.

The Heuristic Analysis of the device is realized individually by evaluators, to avoid bias. Each heuristic violation is associated to one of the fourteen rules and subjectively assigned a severity level.

After identification and rating processes, individual evaluations are compiled and discussed with team, to generate a combined analysis. The comparison of results contributes to reach consensus on the heuristic violations as well as on the severity levels of each problem.

IV. Applied Methodology

A traditional infusion pump model was selected for the case study, because of its wide use in healthcare and medium complexity of operation.

In this study, evaluators executed an initial Heuristic Analysis without any training on that specific infusion pump model. After receiving training from the manufacturer (company instructor providing details about programming, cassette priming and troubleshooting), evaluators performed the technique again.

Heuristic Analyses were done by a multidisciplinary team, composed of: one pharmacist, two nurses, six engineers and two psychologists specialized in cognitive analysis.

Each Heuristic Analysis (before and after manufacturer training) was developed simulating the processes of preparation and programming of intravenous medication infusion. The usability problems identified were analyzed according to the fourteen Nielsen-Schneiderman heuristic rules and then the severity level for each of those violations was assigned through team discussion. Based on discussion, a chart with problems and respective heuristic violations and severities was elaborated.

V. Results

A. Heuristic Analysis Before Manufacturer Training

The team found 22 usability problems, resulting in 88 heuristic violations, separated in 13 heuristic rules. Seven of the problems were rated as severity level 3 – it means a major usability problem, as described above – and one problem was rated as severity level 4 – catastrophic problem. Most of the problems identified were related with aspects about patterns adopted for infusion pump (i.e. programming sequence, alarm), difficulties for user to understand the current state of system (i.e. discerning whether primary or secondary infusion is running), icons and messages, and user memory load. The final histogram of the Heuristic Analysis is presented in Figure 1. The number of problems for each severity level is showed in Figure 3.

B. Heuristic Analysis Done After Manufacturer Training

The team found 26 usability problems (18 of them were new), resulting in 94 heuristic violations, separated in 13 heuristic rules. Eight of the problems were rated as severity level 3 and one problem was rated as severity level 4. In this analysis, the main rules violated remain similar to first Heuristic Analysis with problems regarding to patterns adopted, visibility of system state, icons and messages, and memory load. For better visualization of violated heuristic rules, the resulting histogram of Heuristic Analysis is presented in Figure 2 and the number of problems for each severity level is showed in Figure 3.

C. Comparative Results from Heuristic Analyses

From both analyses, some information can be extracted. The number of total problems identified was 40, with 14 exclusively identified in the first evaluation, 18 discovered only during the second analysis, and 8 identified in both Heuristic Analyses (four of them were rated as severity level 3). Figure 2 show percentages representing those data.

After manufacturer training, some problems identified on first analysis were redefined, and some of them were not considered a usability problem. For example, the activities to turn the pump on and the configuration of the parameter "Keep Vein Open" (KVO) were considered a usability problem on first analysis, but after training they were not considered a problem anymore. Also three of the eight problems found in both analyses were modified. The reason to change them was due some unclear activities during first analysis that became more intuitive, or because of the interpretation after training that brought up new usability issues.

The first analysis showed mainly problems related with intuitive aspects, because none of the evaluators had experience in handling the device. The second analysis allowed the evaluation according to manufacturer guide. So it made possible to observe a slight difference about violated heuristics (some of them were less violated and contrariwise) and the number of problems (more grasp of equipment revealed new problems).

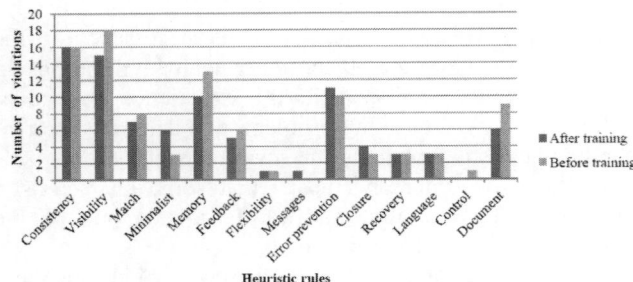

Fig. 1 Heuristic Analysis results on traditional infusion pump after and before manufacturer training

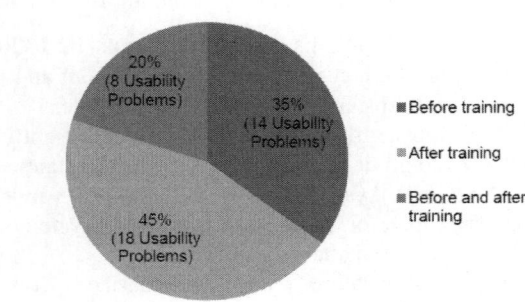

Fig. 2 Percentages of usability problems identified on Heuristic Analyses

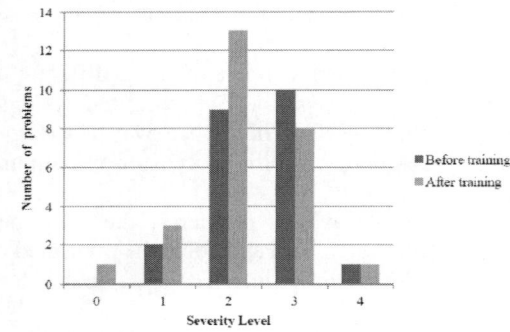

Fig. 3 Number of usability problems related with their severity levels for Heuristic Analysis before and after manufacturer training

VI. DISCUSSION

Observing the histogram on Figure 1 (Heuristic Analysis before training), it's possible to conclude that "Consistency" rule have the larger incidence among all of the problems identified (16 violations), followed by "Visibility", "Error Prevention" and "Memory" respectively. Considering only the two more incident rule violations, they represent 35% of all violations, and changing this analysis up to 4 most incident heuristic violations, the percentage represents 59%, an expressive result. Nine usability problems were identified as severity level 3 or 4, which mean that approximately 41% of all pump problems are considered a serious or catastrophic, requiring urgent solutions. It is a expressive and worrying data.

On Figure 1 about Heuristic Analysis results after training, it is possible to verify that "Visibility" rule is the most violated heuristic rule (18 violations), followed by "Consistency", "Memory" and "Error Prevention" respectively. Comparing the results obtained from first Heuristic Analysis, the same top 4 rule violations represent 60% of total in this second analysis. This suggests consistency between analyses. Considering problems severity levels, it's verified that 11 of 22 represent a serious or catastrophic issue. In other words, 42% of all usability problems need urgency to solve. Thus, this data represents consistency between analysis and it is a expressive and worrying data.

Another important issue observed after applying again the technique, refers that 8 problems remained on results (Figure 2), which represents 20% of total pump problems identified. Thinking about training aspect, even acquiring more knowledge about the infusion pump, it wasn't enough to finish up the problem. It suggests that it might be a device usability problem that could only be solved performing infusion pump redesigning. Four of those eight problems were rated as severity level 3. It means that 50% of them are major usability problems that need high priority to solve. Sometimes, when a near miss or an adverse event occurs during a drug administration process, in which health professionals are using technology, there's a common reasoning that misuse is caused usually due to the user's lack of training. This result could contribute to indicate that usability problems that exist in product design may lead to operator misuses.

Furthermore, there was better understanding about product usability after training. It changed three of the eight problems identified during both analyses. The problem about how to reset an alarm may clarify it: without training, evaluators didn't discover what the right procedure was, they were turning the pump off and switching on again to reset the alarm. This was pointed out as a usability problem. After the manufacturer training, the team learned how to proceed in this activity. However it did not completely solve the problem, and from the right way to execute the procedure new problems were revealed, like icons that weren't intuitive about their functions, and the lack of a user guide sequence in the pump. This changed the focus about analyzed problems, influencing on violated heuristics and severity level on second Heuristic Analysis, indicated by the identification of 45% percent of all pump usability issues and a higher number of problems with severity level 3.

VII. CONCLUSIONS

The use of Heuristic Analysis proved to be useful in identifying usability problems. By applying this technique before and after manufacturer training, it was possible to recognize a higher number of usability problems that represent serious or catastrophic issues. In addition it was identified problems that cannot be solved through training, suggesting that they could be eliminated only by infusion pump redesigning. The multidisciplinary approach allowed a better understanding about user interaction with device, providing a comprehensive analysis. Ultimately, as the fact that Heuristic Analysis is a practical technique, it could be a good alternative for health devices usability inspections.

ACKNOWLEDGMENT

This project has financial support from São Paulo State Research and Support Funds (FAPESP).

REFERENCES

1. C. Koutsojannis, J. Prentzas, I. Hatzilygeroudis (2001), A web-based intelligent tutoring system teaching nursing students fundamental aspects of biomedical technology, Proceedings of the 23th Annual EMBS International Conference, Istanbul, Turkey, 2001, pp 4024-4027.
2. M.V. Lucatelli (2002), Proposta de Aplicação da Manutenção Centrada em Confiabilidade em Equipamentos Médico-Hospitalares. Thesis – (Doctorate), Biomedical Engineering Institute, Santa Catarina Federal University – UFSC. Florianópolis, Brazil, 286p.
3. ECRI Institute, Top 10 technology hazards for 2012: the risks that should be at the top of your prevention list (2011), Health Devices 40: 358-369.
4. ECRI Institute, Top 10 health technology hazards for 2013: key patient safety risks, and how to keep them in check [guidance article] (2012), Health Devices 41: 342-365.
5. L. Lin, R. Isla, K. Doniz et al. (1998), Applying human factors to the design of medical equipment: patient-controlled analgesia, Journal of Clinical and Monitoring Computing 14: 253—263.
6. M. J. Graham, T.K. Kubose, D. Jordan et al. (2004), Heuristic evaluation of infusion pumps: implications for patient safety in Intensive Care Units, International Journal of Medical Informatics 73: 771-779.
7. J. Zhang, T.R. Johnson, V.L. Patel et al. (2003), Using usability heuristics to evaluate patient safety of medical devices, Journal of Biomedical Informatics 36: 23-30.

Author: Victor Hugo Batista Tsukahara
Institute: Campinas State University (Unicamp)
Street: Cidade Universitária Zeferino Vaz
City: Campinas, São Paulo
Country: Brazil
e-mail: vhbtsukahara@gmail.com

Heuristic Analysis: A Tool to Improve Medical Device Safety and Usability

F.N. Bezerra[1], V.H.B. Tsukahara[1], R.H. Almeida[2], F.S. Borges[2], E. Pantarotto[2], R.V. Martins[2], M. Iglesias[2], P. Trbovich[3], S. Pinkney[3], and S.J. Calil[1]

[1] Unicamp/Biomedical Engineering Department (DEB), Campinas, Brazil
[2] HCor, São Paulo, Brazil
[3] University Health Network, Toronto, Canada

Abstract—Although there are efforts from regulatory agencies, academic studies and end-users demonstrating the need to consider human factors during the manufacturing process and acquisition of new medical devices, our study showed that there are still widely-used products with usability problems in the market. Deficiencies in the design of a device, coupled with complex activities (e.g., task is rarely performed by a single operator) and high stress environments, may contribute to usability errors and lead to adverse events. This article shows how the application of the Heuristic analysis to health technology can help to reduce problems of use pre and post-market.

Keywords—**Human Factors Engineering, Usability, Heuristic Analysis, Design, Infusion Pump.**

I. INTRODUCTION

Over time, many use errors have occurred due to pushing the wrong button, mistaking a number or neglecting an error message when using medical equipment. This is largely because of the culture that requires people to adapt to the demands of technology, instead of designing technologies and systems that are more intuitive and easier to operate [1].

In the hospital environment, there are safety concerns due to the danger of failure of medical devices that can lead to death or irreparable consequences to the patient.

Studies focusing on patient safety caution about the high rate of preventable adverse events that occur in North America. According to the Institute of Medicine (IOM) estimates show that 44,000 to 98,000 people die each year because of errors in care processes [2]. Errors could be avoided if more efficient protection barriers were in place. In Canada, studies show that around 9,000 to 23,000 deaths annually are due to preventable adverse events, with errors related to medication administration as the second leading cause [3].

It is estimated that over 90% of hospital patients receive intravenous therapy. In analyzing data related to volumetric infusion pump, the Food and Drug Administration (FDA) indicates that 56,000 adverse event cases have already been reported, resulting in 500 deaths until 2010. As a result, between 2005 and 2009, 87 recalls of infusion pump were conducted in order to respond to the safety concerns identified, according to FDA data [4].

Despite the introduction of new technologies to improve processes such as therapy and diagnosis in hospitals, this situation has brought new kind of errors related to misuse. These errors have led some technologies to be abandoned or recalled due to problems related to design shape, impact on workflow during its operation and end-user dissatisfaction.

A way to address this concern, which is identifying risks and developing systems that are more usable by end users, is through the use of Human Factors Engineering (HFE). HFE focuses on the understanding of human capabilities and human-machine interaction [5].

When analyzing the context of misuse problems, the human-machine interaction is an important aspect to consider as it can contribute to the reduction of failures in the handling of medical equipment and increase patient safety. One of the HFE tools that has proven to be very useful for evaluating the usability of devices is the Heuristic analysis [6].

The use of Heuristic analysis to identify usability problems in 3 infusion pump models, currently in use in the health market, is presented here.

II. METODOLOGY

The Heuristic analysis technique was adopted to evaluate the usability of infusion pumps. It consists of a systematic analysis of all aspects of the design, to identify usability problems that do not comply with a set of predefined heuristic rules and assesses the severity of each of these violations [7].

A multidisciplinary team of evaluators consisting of five clinical engineers, two nurses, and a psychologist was composed to identify problems from different points of view. Furthermore, according to studies conducted by Nielsen, the use of more than three participants is required to identify a range of 60 to 75% of usability problems [7].

In this project, the following 14 Nielsen-Schneiderman rules [8] were taken as reference: consistency, visibility, match, minimalist, memory, feedback, flexibility, error message, error prevention, closure, undo, language, control and documentation.

One of the limitations of this technique is that the focus is on a single medical device considering an ideal condition without environmental factors such as noise and interruptions that can affect its use. However, the technique is satisfactory in this study since the purpose of the analysis is to identify usability problems related to the equipment interface which could be troublesome for users and lead to errors and frustration.

For the development of practical activity, all evaluators analyzed the three infusion pumps in the following sequence: traditional pump - brand A (Pump 1), traditional pump - brand B (Pump 2) and smart pump, again brand A (Pump 3). The process of Heuristic analysis was performed as follows:

1. It was ensured the evaluators' consensus on the knowledge of 14 heuristic rules of Nielsen-Schneiderman.
2. Each evaluator operated the pump model to explore the device. Tasks were performed individually to assure impartiality in the activity.
3. A specifically designed chart, where the evaluator could describe the usability problems, was filled to determine the severity level of the violations and corresponding heuristics rules. Severity was ranked on a scale from 0 to 4 [7]:

- 0 - When it does not cause misuse problem, but it is an issue with opportunity for future improvement;
- 1 - When it is a cosmetic problem and correction must be done with a very low priority;
- 2 - Minor usability problem. Correction of issue has a low priority;
- 3 - Major usability problem. High priority must be given to solve the problem, because it is important to be fixed;
- 4 - Catastrophic error. If the evaluated product was not released, it is mandatory for the issue to be corrected.

4. After each evaluator completed the charts, a meeting was held to discuss and analyze the results. The purpose of this meeting was to achieve group consensus on the problems, to determine the level of impact on usability of the equipment, and to identify risks that could potentially lead to the occurrence of medical errors.
5. Results were presented in a histogram to facilitate the visualization of the equipment analysis. The histogram shows the number of occurrences of each violation and the frequency of heuristic usability issues for each severity level. This histogram, as shown in Figures 2 and 3, allows identification of major usability problems and levels of severity.

III. RESULT

Results are presented in three perspectives correlating the results from 3 pumps analyzed. Figure 1 shows the total number of identified problems, violations and the average critical severity ratings.

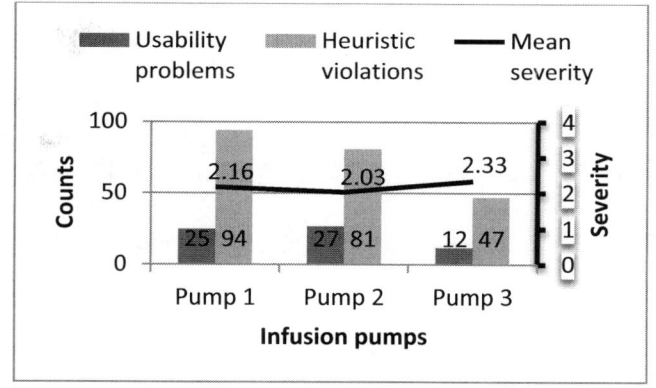

Fig. 1 Amount of usability issues by pump

As shown in Fig. 1, Heuristic analysis of Pump 2 generated the highest number of usability problems: 27 in all, totaling the recognition of 81 heuristic violations with 2.03 severity mean value. The existence of a higher number of heuristic violations than identified problems is explained based on evaluator's choice. For example, "Memory" and "Consistency" could be pointed out at the same time if root problem violates more than one heuristic rule.

Pump 1 was characterized with the highest number of violated heuristics (94), along with 25 usability problems which generate a 2.16 severity mean value.

Regarding the analysis of usability problems with high severity level, Pump 3 stands out with 2.33 mean for 12 problems identified that violated 47 heuristic rules.

In Figure 2, a graphic showing total violations for each of 14 Nielsen-Schneiderman heuristics [8]. A noteworthy observation is that, in Pump 1, the most violated heuristic rules were Visibility" (18 violations - 19%), "Consistency" (16 violations - 17%), "Memory" (13 violations - 14%) and "Error Prevention" (10 violations or 11%). It was also observed that the above-mentioned four major heuristic rules represent approximately 61% of all violations, suggesting that those four points are the main references to determine the usability problems for this infusion pump model. For example, it could be used as a guide to design the equipment, in order to improve device usability and security.

Regarding Pump 2, the most violated heuristic rule was "Match" (14 violations - 17%), followed by "Consistency", "Memory" and "Error Prevention" (13 violations - 16%) rules. For this analysis, the four major device heuristic violations represent 65% of total.

On Pump 3, the highest number of violations fall on the rules "Memory" and "Visibility" (7 violations - 15%). In second place, the same numbers of violations were attributed to the rules "Match" and "Feedback" (5 violations

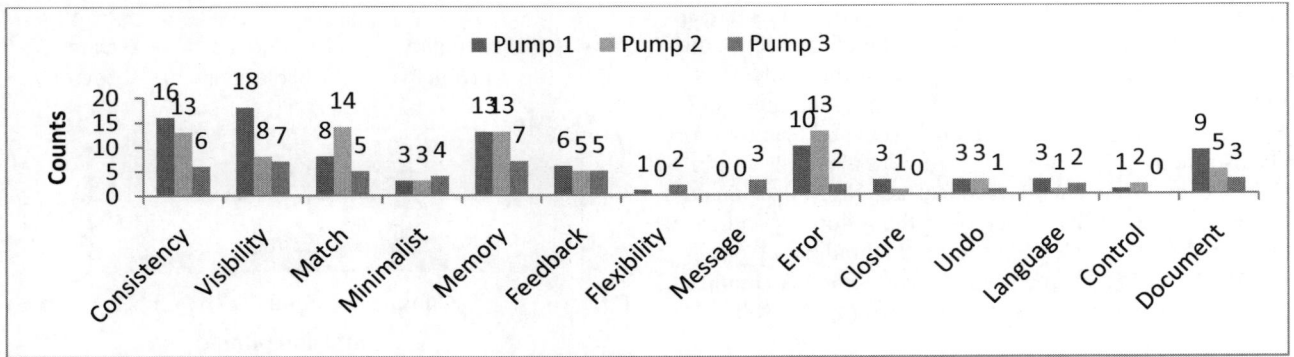

Fig. 2 Frequencies of usability problems by heuristic violations

- 13%). In this case, four rules were responsible for 56% of violated heuristics.

According to the graphic on Fig. 2, the three pumps together resulted in recognition of 222 heuristic violations total. It was also found that all pumps showed a high number of usability problems that violated "Consistency", "Memory", "Visibility" and "Match" rules, aspects that directly impact on operators' ability to understand how to use the equipment interface. It may represent an important risk when considering infusion pumps, as users must understand how to select different functions and set parameters.

In Figure 3, a graphic showing total problems identified separately according to their severity level on a scale ranging from 0 to 4 (4 is the most severe level).

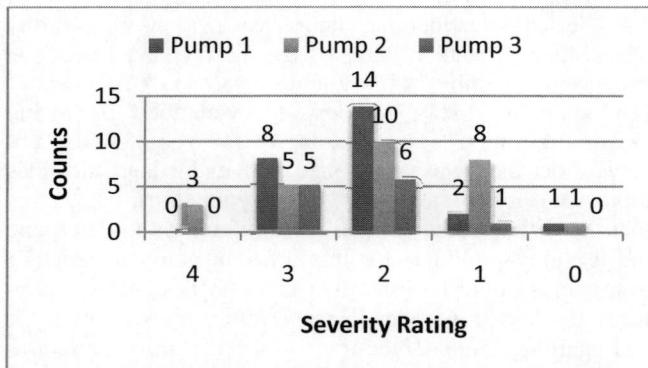

Fig. 3 Frequencies of usability problems by severity

In Fig. 3, a noteworthy observation is that Pump 1 had 8 problems (31%) rated as severity level 3. Though not catastrophically affecting the use of the equipment, they require high priority to be solved with potential to lead to an operational error, e.g. display does not show a clear indication that the infused volume represents the total value of both channels, so it is confusing to write down infused values for each drug afterwards by nurses or nursing assistants.

Pump 2 had 5 problems attributed as level 3 and other 3 serious problems with severity level 4 – these represent usability issues that could potentially affect the use of the device with high probability of patient harm. For example, on Pump 2 it is possible to put cassette line in the wrong position, resulting in reverse flow, i.e., removing fluids instead of injecting drugs to the patient. Those problems account together for 8 usability problems (30%) with severity 4 and 3 that require users' attention.

Pump 3 has the least amount of problems with greater severity. Specifically, 5 problems were identified with level 3 severity, e.g. task for removing air isn't obvious what results in users forgetting about such sequence for air removing. However considering the total number of identified usability problems, this represents 42%, i.e., a high risk level was attributed to almost half of the problems.

IV. DISCUSSION

Usability problems were identified in all three pump models. Traditional pumps 1 and 2 showed a higher number of usability problems when compared to Pump 3. However, Pump 3 demonstrates the importance of considering usability improvement with technology evolution, because many smart pump functions resulted in the highest severity level in comparison with other identified problems.

Altogether, 64 usability problems were identified across the 3 pumps, with severity average ranging from 2 to 2.5, showing the need for modifications on equipment design, in order to improve usability.

Despite considering Heuristic analysis a quick and practical technique to evaluate usability interface issues, it does not use real users to confirm risks in process context. So it is recommended to use other human factors tools to perform complete usability analysis, such as usability test, tool that allows considering environmental factors integrated into overall workflow, consequently, identifying the real severity and occurrence probability of risks associated with usability problems.

One possible explanation for the lack of usability of commercially available medical devices is the quick release of new technologies by manufacturers [9], who make an effort to evaluate health risk to patients, while product usability represents a small part of product development.

As a result, sometimes manufactures provide insufficient evidence of usability study. It is a hazardous issue, because of this approach goes against ensuring the quality and safety goals during device development.

Another point of view is that although regulatory agencies are starting to require an analysis of human factors, during the approval process for new medical devices following international standards such as ANSI/AAMI 75, IEC 62366 and IEC 60601-1-6 [10], the analyses are not always independent or accessible to public. So this impartiality or inaccessibility contributes to develop a superficial usability testing during pre-market approval.

Though human factors analyses add expenses to the development of equipment, this kind of study allows the reduction of risks, improves use of the equipment and lowers the possibility of legal exposure and a judicial or regulatory action. Furthermore, equipment recall to perform hardware changes or constant software updates result in high costs and a negative impact to the company [11].

Another benefit about focusing of human factors issues analysis is that more intuitive equipment leads to better user acceptance and decreases post-market support [12]. In addition, features such as comfort, work load and quality are used by hospitals as criteria to make procurement decisions, so improvement of device usability can lead to increased sales [13].

Thus, manufacturers' awareness as well the adoption of strict policies for equipment pre-market approval concerning usability would contribute to reduce operation errors in long term. However, it's necessary also emphasize that many hospitals still use medical devices developed before human factors standards definitions for manufacturers, which need more study to generate recommendations in short term based on database about usability problems till now registered, as a way to prevent adverse events with current equipment.

V. CONCLUSIONS

Heuristic analysis allowed the usability evaluation of three models of infusion pumps (i.e., two traditional pumps and one smart pump), which resulted in the quick identification of problems and risks associated with each one of them.

All pumps that could be used at the hospital showed serious usability problems that may lead to a device misuse. It demonstrated that even with the existence of medical device design standards, pumps with usability problems can be found on market.

This work illustrates the advantages of the application of human factors to the development cycle of medical equipment as well as to inform hospital procurement decisions. Human factors methods, such as a Heuristic analysis, enable designing technologies to meet the needs, cognitive processes and environments of health professionals.

REFERENCES

1. Vicente K (2006) The Human Factor, Taylor & Francis Group.
2. Kohn L, Corrigan J, Donaldson M (2000) To Err is Human: Building a Safer Healthcare System. Washington, D.C.: Institute of Medicine. National Academy Press.
3. Baker G, Norton P, Flintoft V, et al (2004) The Canadian Adverse Events Study: the incidence of adverse events among hospital patients in Canada, 170(11), pp 1678-1686.
4. FDA news & events (April - 2013) http://www.fda.gov/NewsEvents/Newsroom/PressAnnouncements.
5. Gosbee J, Gosbee L (2005) Using Human Factors Engineering to Improve Patient Safety, USA: Join Commission Resources.
6. Graham M, Kubose T, Zhang J et al (2004) Heuristic evaluation of infusion pumps: implications for patient safety in Intensive Care Units, International Journal of Medical Informatics, 73, pp 771-779.
7. Nielsen J (2000) Usability Engineering. San Diego: Kaufmann, 2000.
8. Zhang J, Johnson T, Patel V, Paige D, KuboseT (2003) Using usability heuristics to evaluate patient safety of medical devices, Journal of Biomedical Informatics, 36, pp 23-30.
9. Cafazzo J, Trbovich P, Chagpar A, Rossos P et al (2009) Human Factors Perspectives on a Systemic Approach to Ensuring a Safer Medication Delivery Process. Healthcare Quarterly, 12, pp 70-74.
10. Fairbanks R, Wears R (2008) Hazards With Medical Devices: The Role of Design, Annals of Emergency Medicine, Volume 52, No. 5.
11. Chagpar A, Cafazzo J (2010) Human Factors Recommendations and Testing Considerations for FDA Premarket Medical Device, Human Factors Horizon, ProQuest Central, pp 33.
12. Wiklund M, Kendler J, Strochlic A (2011) Usuability Testing of Medical Devices, CRC Press, London: Taylor & Francis Group.
13. Ginsburg G (2005) Human factors engineering: A tool for medical device evaluation in hospital procurement decision-making, Journal of Biomedical Informatics, 38, pp 213–219.

Author: Fagner Nakahata Bezerra
Institute: Campinas State University (UNICAMP)
Street: Cidade Universitária Zeferino Vaz
City: Campinas, São Paulo
Country: Brazil
Email: fagnek@hotmail.com

Measurement and Analysis of the Impact of the Environment on the Spread of Ionizing Radiation in Medical Facilities

P. Gabriel and M. Penhaker

VSB – Technical University Ostrava, Faculty of Electrical Engineering and Computer Science,
Department of Cybernetics and Biomedical Engineering, Ostrava, Czech Republic

Abstract—This paper discusses the measurement of the spread of X-rays, and the detection and evaluation of their effects on people. Part of the aim of this work is to create an application in Matlab to calculate the distribution of X-ray radiation and its inhibition in different environments based on the measured reference points. Realization of practical measurements using two kinds of conventional X-ray equipment were measured using a precision disimeters. The measurement results are compared with theoretical calculations and evaluated primarily in terms of the impact on human health in the context of safety standards.

Keywords—Gamma Rays, Radiation, Roentgen, Roentgen Rays, Dosimeter.

I. Introduction

Over the last decade, a trend has developed, as more and more private practices of all specialties have begun to move away from hospital equipment and buildings known as clinics, in areas that were not primarily intended for medical treatment. The main reason for some doctors, was to improve their services and create a more pleasant environment for both patients and medical staff. Another reason was because it was often impossible to reach a reasonable agreement with the other co-owners in matters of medical devices.

These efforts however, raised issues that were caused by the fact that the building or commercial space was not primarily designed for medical use. The issues were mainly in meeting the different standards of hygiene, the need for the presence of different installation options and wheelchair access. [1 - 3]

Another important issue that must be addressed is the possibility of using X-ray equipment in areas where they were not originally intended for. Dental clinics are already using basic equipment including intraoral x-rays or OPG scanning machines. Furthermore, doctor's offices or veterinarians often use conventional X-ray machines. It is a small source of ionizing radiation, but in this case it is a potentially dangerous device for which it is necessary to pay attention when operating, especially in the long term.

The aim of this paper is to describe the spread of X-ray radiation in space. The result is the determination of the total attenuation of ionizing radiation, depending on the distance from the source (in the direction of the emission) and after passing through the patient's tissue, air and walls that surround the treatment space itself. The result will be compared with hygiene standards, evaluating the level of risk for people who are both in the clinic itself and beyond the wall in the next office or clinic. To perform this work real measured values have been taken at the two devices used for the X-ray work at the University Hospital in Ostrava. Part of this work is the application of Matlab, which models the spread of X-ray radiation, depending on the parameters.

II. Standards for Exposure to Ionising Radiation

National legislation sets limits on the exposure of individuals to protect the health of people who come into contact with sources of ionizing radiation. In setting the exposure limits, exposure to natural sources of radiation outside the natural resources that are knowingly and deliberately exploited are not included. [9]

Decree No. 307/2002 Coll. divides the limits for the exposure of individuals into main categories:

A. General Limits

General limits apply to the total radiation exposure from all activities, in addition to occupational exposure, medical exposure and accidental exposure. The limits for the sum of effective doses are:

- 1 mSv per calendar year;
- 5 mSv for a period of 5 consecutive years;
- The value of an equivalent dose of 15 mSv in the crystalline lens in a calendar year;
- The average equivalent dose in 1 cm2 of skin 50 mSv per calendar year. [9]

B. Limits for Radiation Workers

These limits apply to occupational exposure, i.e. radiation that workers are exposed to in direct relation to the work performed. The limits for the sum of effective doses for radiation workers are:

- 50 mSv per calendar year;
- 100 mSv for a period of 5 consecutive years;
- The value of an equivalent dose of 150 mSv in the crystalline lens in a calendar year;
- The average equivalent dose in 1 cm2 of skin 500 mSv per calendar year;
- The average equivalent dose to the hands of the forearm and toes from the feet up to the ankles of 500 mSv per calendar year. [9]

Table 1 Table of radiation from various sources

Sources of ionizing radiation	The absorbed dose
X-ray arm	1 µGy
X-ray examination of the chest	20 µGy
CT scan of the chest	5.8 mGy
X-rays of the teeth	5 µGy
CT scans of the whole body	10 – 30 mGy
mammographic examination	0.5 mGy
The average daily dose of natural radiation	10 µGy/24 hours
The average daily dose on board aircraft at an altitude of 10,000 m	66 µGy/24 hours
Radioisotopes in the body	7.7 µGy/24 hours
X-ray body scanner	250 µGy
Sources of ionizing radiation	50 Gy/10 mins.

III. CALCULATING THE ENERGY OF X-RAY RADIATION, DEPENDING ON THE DISTANCE

Any radiation emitted from an X-ray passes through various kinds of environments, where it interacts with the atoms of substances and loses its energy. The attenuation of X-rays depends on the density of matter and their atomic weights and the initial energy of X-rays.

The basis of knowledge for the absorption half-thickness for various types of radiant energy and materials. [4]

Based on the knowledge of the energy the tissue absorbs, and by knowing the weight of the tissue, it is now possible to calculate the radiation dose absorbed using a formula:

$$D = \frac{dE}{dm} = \frac{\Delta E}{m} = \frac{1}{m} \cdot \left[E_A - \frac{E_A}{2^{\left(\frac{L_{MO}}{X_{\frac{1}{2}MO}}\right)}} \right] \quad [Gy], \quad (1)$$

where the target object (e.g., metre or tissue irradiated persons), ΔE is the portion of energy transferred from photons to charged particles in a unit mass. EA is the energy of X-ray radiation that penetrates into the tissue, m is the mass of irradiated tissue, LMO width (normal person 30 cm) and the absorption half-thickness of human tissue. [5] - [7]

IV. MEASURING RADIATION

The measurements used standard X-ray equipment and it is possible to compare the results generalised and used for the safety evaluation of X-ray sources, which have similar characteristics. Measurements were performed partly within the normal operation of the radiodiagnostic workplace and so a real predictive value were obtained. Finally, a comparison was made of measurement results with theoretical calculations.

Measurements were carried out at the workplace radiodiagnostic University Hospital Ostrava, the cardiovascular department. A total of 2 units were used. The first device type Carestream DRX-Evolution allows the imaging of patients in both standing and lying down positions. It can thus measure the intensity of ionizing radiation for both of these examples. The other device, the X-ray imaging OPG machine, was a Kodak 9000 extraoral IMG system. [8]

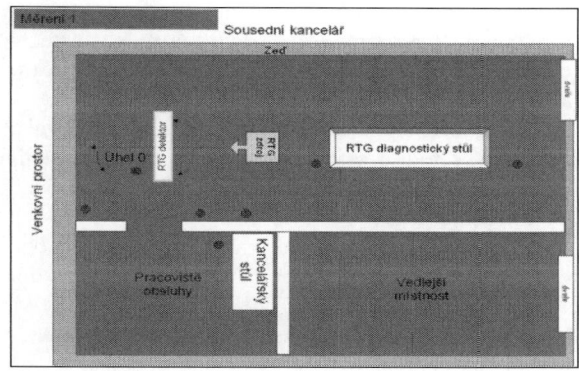

Fig. 1 A stationary X-ray in position I - plan

A. Measurement in Position I

The x-ray source emits energy in a horizontal direction toward the windows of the room. A total of 10 measurements were carried out in seven positions as shown in Fig. 3. Some measurements were performed on a phantom and some on the patients (see the table of measurements). Measuring sensors are mostly located at a height of 1.15 m above the ground. This is the same height, where the source of X-rays was.

B. Measurement in Position II

The x-ray source is located above the X-ray examination table at a height of 160 cm and emits energy perpendicular to the ground down to the x-ray table, where the phantom was placed. In the case of measuring position No.2 the dose was measured in two different positions of the beam aperture. The dose measured from 5-7 depending on the height.

V. EVALUATION OF THE MEASUREMENT

In the case of measurements I and II, which were carried out on stationary X-ray apparatus, it was found that if the dose of the conventional x-ray that passes through the patient (phantom) is 4.4, then the sensory system of the X-ray equipment is very low (26.7 nGy). A louver system and collimator that can focus the beam precisely on the areas examined and the X-ray sensor is very precise and emits a direct beam of ionising radiation that does not penetrate into the surrounding area. The reason for this fact is the possibility to precisely adjust the level of light intensity and time of emission of radiation, depending on the size of the irradiated body.

However, it was found that part of the energy of the ionising radiation was not absorbed by the patient (phantom), but was instead dispersed into the surroundings by the mechanism of Compton (quantum) scattering. In this case, there was a direct correlation - the larger the patient, the greater the emitted dose, the greater the intensity of radiation in space. In a room with X-rays a dose was detected per one frame ranging from 68 nGy in the farthest corners to values of over 200 nGy near the tube and the patient.

In terms of height distribution the highest intensity of radiation was found at the level of the patient (phantom), which was a major source of the particles produced by Compton scattering.

Two main factors impacted on the spatial intensity distribution of the radiation effect. The first was the distance from the patient (phantom). The intensity of radiation decreased with the square of distance. Another influence was the gradually decreasing energy quanta of radiation (Compton scattering losing a substantial part of energy) and thus decreasing the absorption half-thickness of the air that despite the relatively low density, can significantly suppress residual ionising radiation.

Comparison of Simulated Values with Measurement

To compare the simulated values the measured values which were measured for the devices Carestream DRX-Evolution and the values measured in the test of the long-term stability of the device Satelec X-Mind DC were used. Due to the variables that have been identified, the second version was used.

Satelec X-Mind DC - The measured value of the absorbed dose of ionizing radiation after passing through the phantom (6 mm aluminum) according to the protocol stability tests was 0.072 mGy. The calculated value was 0.095 mGy.

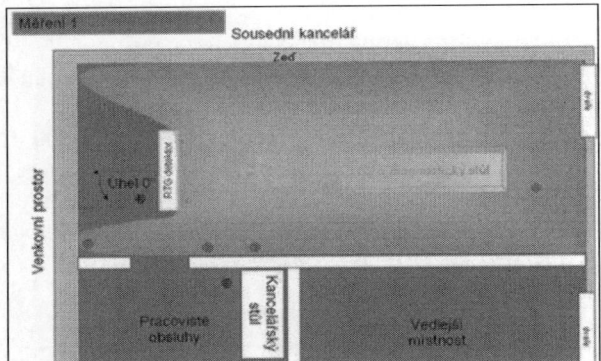

Fig. 2 Calculation of the intensity distribution of ionizing radiation

Carestream DRX-Evolution - The measured value of the absorbed dose of ionizing radiation after passing through the phantom (20 cm water phantom) was 0.0043 mGy. The calculated value was 0.0055 mGy.

The difference in the values of the absorbed dose is due to the imperfection of the environment in which the actual radiation passes. Simulated calculations are set to the constant and coherent properties of the environment, but they are not in fact identical. Another reason is the fact that there is only a limited list of slice thickness absorption. If the source has a different value on the X-ray tube voltage, it is necessary to estimate the absorption half thickness according to certain laws, which is implied in the table. Another source of uncertainty is limited to the materials, for which the half-thickness is calculated. If the material is not in the list, it is necessary to estimate the half-thickness according to its density – an analogy to one of the materials listed in the table. The programme is tuned to account for reserves – i.e. that the calculated value was slightly higher in reality. It is for this reason that the purpose of this program is to determine the possibilities of risk. If the programme proves the risk it is assumed to have taken the measurement.

VI. EVALUATION OF THE SAFETY AND POTENTIAL RISKS OF THE USE OF X-RAY EQUIPMENT

Given the measured values of absorbed doses of ionising radiation, any possibility of acute radiation exposure for the operator and patient can be clearly ruled out. These values are at the level of tens to hundreds nGy and are not so dangerous.

But in a day there are about 20 such procedures over a full year, with 259 working days in 2013. If the operator remained at the edge of the examination room, exposed to the lower dose of reflected radiation (count 100 nGy), the staff receive a total dose per year of a value of $5180 \cdot 100 nGy = 518 mGy$. Over five years, the staff receive a dose of $518 mGy \cdot 5 = 2,59 Gy$.

Under current legislation, the present value of the annual limit of gamma absorbed dose to workers is 50 mGy, 100 mGy over a five-year limit. The value of the absorbed dose would thus be exceeded by more than ten times, in the case of the five-year limit by 26 times even. [10]

If we compare the dose with Table 1, it can be seen that the dose exceeded 400 mGy over the already observed symptoms of radiation sickness. Everything is of course very individual and everybody can cope differently with absorbed dose. However, in the event of such an ionic load there is a significant risk of tumour development.

VII. CONCLUSIONS

It can thus clearly be concluded that even if the diagnostic X-ray equipment for small energy ionizing radiation exists for non-compliance of safeguards serious risk to the operator. Given the size of the received dose will represent the main risk of tumor development is most exposed tissues, which will no longer suffice repair mechanisms.

To protect the operating X-ray equipment it is essential to ensure compliance with the principles of protection against ionizing radiation. It is all about the location of the control devices in the separate rooms of the building, where it will remain in operation when the X-rays are emitted.

In the case of X-rays located in areas that are primarily built as a radiology department, we can say that the current levels of energy quanta up to 140 keV is fully adequate protection for ordinary walls made of brick, concrete or other "heavy" construction material. In the case of plasterboard walls, this protection is insufficient and it is necessary to add a layer of material that has a small absorption half thickness (lead, iron).

ACKNOWLEDGMENT

The work and the contributions were supported by the project SP2013/35 "Biomedical engineering systems IX" and TACR TA01010632 "SCADA system for control and measurement of the process in real time". The paper has been written within the framework of the IT4 Innovations Centre of Excellence project, reg. no. CZ.1.05/1.1.00/02.0070 supported by the Operational Programme 'Research and Development for Innovations' funded by Structural Funds of the European Union and the state budget of the Czech Republic.

REFERENCES

1. Ullmann, Vojtěch. Nuclear and Radiation Physics. First issue. University of Ostrava, Faculty of Health Studies, 2009. 173 pp. ISBN 978-80-7368-669-7.
2. Gray, Josef. Dosimetry of ionising radiation. First issue. Prague, SNTL, 1983. 420 s, currently L11-C3-IV-41f/17832.
3. Bazant, Zdenek and Šimerda, George. Dosimetry. First issue. SNTL, by 1960. Type No. 274 s L 11-C2-13-II/1321
4. NAVRÁTIL, Leos and Jozef Rosina et al. Medical Biophysics. First issue. Praha: Grada, 2005. 524 p 352-353. ISBN 80-247-1152-4.
5. Jakes, Dušan and Cecil final. Nuclear chemical tables. First issue. Prague: SNTL, 1964th ISBN 04-647-64.
6. Siman, Čestmír and Petr OTČENÁŠEK. Screening of radioactive sources and nuclear facilities. First issue. Prague: CTU, 1974. ISBN 404-1680.
7. Munroe, Randall. Radiation Chart. Xkcd: The Blag of the webcomic [online]. 2011 04.19 [Cit. 2013-05-01]. Available from: http://blog.xkcd.com/2011/03/19/radiation-chart/
8. Czech Republic. The law on the peaceful use of nuclear energy and ionising radiation (the Atomic Act) and on amendments to certain laws, as amended. In: 1997. 1997 n 18 Available from: http://www.sujb.cz/legislativa/
9. Czech Republic. Decree of the State Office for Nuclear Safety of Radiation Protection. In: 2002. 2002, No. 307 Available from: http://www.sujb.cz/legislativa/
10. SONS - State Office for Nuclear Safety [online]. 2013 April 29, 2013 [cit. 2013-05-01]. Available from: www.sujb.cz

Author: Marek Penhaker
Institute: VSB - Technical University of Ostrava
Street: 17. listopadu 15
City: Ostrava
Country: Czech Republic
Email: marek.penhaker@vsb.cz

Part IX
Health Informatics, E-Health, Telemedicine and Information Technology in Medicine, Bioinformatics

Comparing Two Coaching Systems for Improving Physical Activity of Older Adults

H. Similä[1], J. Merilahti[2], M. Ylikauppila[2], S. Muuraiskangas[1], J. Perälä[2], and S. Kivikunnas[1]

[1] VTT Technical Research Centre of Finland, Oulu, Finland
[2] VTT Technical Research Centre of Finland, Tampere, Finland

Abstract—Physical activity can prevent diseases and disablement in older adults. Two touch-screen based coaching systems developed for motivating older adults to increase their physical activity are presented. The systems had similar content, but they utilized a) text-based and b) animated virtual coach approaches to interact with the user. Both systems included health related tips and information, questionnaires, physical activity prompts and video guided exercises. Integrated sensors enabled self-monitoring of physical activities and health related parameters. Preliminary experiences of a short intervention study with four older users highlighted some important issues for the future work and the most advantageous features from both systems should be taken into account. Specifically the content of the activities and messages, and the visualization of the self-monitoring items were emphasized. Further studies are required to evaluate whether the proposed systems would actually increase physical activity of a person and the costs effects they pose.

Keywords—motivation, physical exercising, virtual coach, self-monitoring

I. INTRODUCTION

According to [1], "Active ageing is the process of optimizing opportunities for health, participation and security in order to enhance quality of life as people age." Demographic ageing increases also the prevalence of chronic diseases such as diabetes and cardiovascular diseases, which are the leading causes of death and disability in the world [2]. Physical activity has been recommended as one of the highest priorities for preventing and treating disease and disablement in older adults [3, 4]. In addition, as the research show that people spend most of their waking hours sedentary [5], there is a need for motivational tools to provoke more physically active behavior.

There are a few papers that report new types of concepts and designs that use technology in encouraging older adults to be physically more active. For example, in [6] a design of a persuasive virtual coach that aims at motivating older adults to walk more is presented. Their concept incorporates a pedometer wirelessly connected to a touch-screen photo frame that shows the general overview with a flower image and weekly and monthly overviews of pedometer data. A bit similar approach is UbiFit Garden. It is a mobile application for increasing physical activity [7]. It includes sensors and interactive application that automatically detects activities and the activity status is visible on the mobile phone wallpaper in the form of flowers (activity types) and butterflies (reached goals).

This paper presents preliminary user experiences from two technological solutions that aim at motivating older adults (over 65 years old) to be physically more active utilizing e.g. suggestions, reminders and information sharing. The aim was to compare a more traditional text-based user interface with an animated virtual coach in delivering the messages to the user.

II. METHODS

The design of the developed solution was driven by the aim of motivating older adults to be physically more active. An integral part of the solution is to remove environmental barriers such as difficult access to exercise facilities [8] and provide home exercising videos [9,10] via an easy-to-use user interface. A 22-inch touch screen PC with Windows operating system and wireless internet connection was used as a home terminal. Such a screen enabled large enough fonts and images also for potential users with vision deficits and a touch-based user interface was anticipated easy to use.

The home technology can also be used as a channel for providing appropriate knowledge about benefits of exercising [8,11] and delivering prompts and suggestions [8]. The developed software applications enabled scheduling of pop-up reminders, questions and wellness related tips to the user.

As older adults often believe themselves to be too old or frail for physical activity [8], boosting their self-efficacy is important e.g. by demonstrating the achieved goals [11]. Besides health information, sensors can provide objective data about achievements and compliance to the plans. The sensors integrated into the system were Bluetooth blood pressure monitor and weight scales (A&D Medical UA-767PBT-C and UC-321PBT-C, www.aandd.jp), and an USB accelerometer (X6-2mini, www.gcdataconcepts.com) for detecting the number of steps and activity level. The sensor data was automatically transferred to the server. Integration of the blood pressure monitor and weight scales followed the guidelines by Continua Health Alliance (www.continuaalliance.org). The collected sensor data was visualized to the users by giving historical views on user's data as daily/weekly summaries and trends. In addition, an actigraph (www.actigraphcorp.com) was used for collecting reference data.

Two types of user interfaces were tested. System 1 used text-based coaching, and it was based on the software developed in the AmIE project [12]. In the default state the system appears as a simple desktop view that offers selected activities e.g. calendar functionality, video exercises, internet browser and games as large buttons to the end user (on the left in Figure 1). The messages, e.g. tips and suggestions, appear as full screen pop-ups with answer alternative buttons at scheduled times.

The System 2 interacted with the user via an animated virtual coach that also spoke up the scheduled messages that appear on the screen for the user using text-to-speech converter. The coach is almost always visually present on the screen for the user. In addition, the system has a widget view for retrieving calendar and sensor information. A typical visual presentation of the virtual coach is presented on the right hand side in Figure 1.

The virtual coach based system deployed only a push type approach for suggesting the exercising events, whereas in the text-based system the user was also able to start the exercising sessions by himself. The sensor data visualizations slightly differed between the systems. In addition to trend views, the text-based system also showed performed measurements as icons, e.g. scales depicting a weight measurement, on a month calendar view.

A. Procedure

The two systems were tested in a short health coaching intervention study. The intervention content, i.e. wellness related information, suggestions for exercise, reminders for health measurements and questionnaires, was organized into an individual task calendar so that they integrate into the participant's normal routines. The content was based on the health living plan [13], physical activity pie [14] and personally perceived life values in order to internalize extrinsically motivated behaviors [15].

The study was targeted for 65 year-olds or older, who were able to move alone outside, do easy home exercises and willing to participate in the study. The participants were recruited from the guided exercise groups arranged by city of Tampere (in southern Finland) and via city's home care service integrator. Info lectures were organized to explain the trial procedure and give information about healthy lifestyle. Finally four volunteers participated in the study. They were all female (aged 63-84y, mean 78y). Due to difficulties in recruiting the participants, also a person younger than 65 years was accepted as a test user. The participants were asked to sign a written consent and the study was approved by the Ethical Committee of the local Hospital District.

The field trial was divided into three periods (one to two weeks each). 1.) The first period worked as a baseline period when the user filled only a paper diary and wore the actigraph. 2.) During the second period the System 1 was introduced to the users and 3.) System 2 was introduced during the third period. Before starting to use the systems, the participants got a brief, about 1.5 hours, usage guidance.

B. Data Collection

User experience data related to the two systems were collected via a) in-depth, open- and closed-ended questionnaires/interviews, b) direct observation, and c) other written documents, such as diaries [16]

- *Baseline questionnaire and interview:* Baseline questionnaire asked about e.g. demographics, current use of health care services, health status, physical activity, use of technical devices and personal values. The questionnaire guided the baseline interview, which purpose was to personalize the task calendar for the participant.
- *Intermediate interview:* After the periods 2 and 3, the users were inquired for their experiences with the

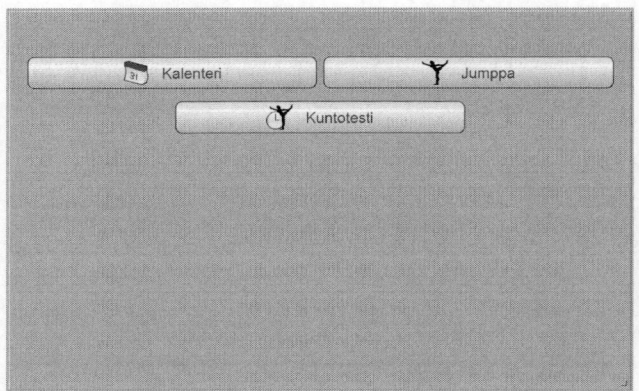

Fig. 1 Left: Desktop view of text-based system (System 1). Right: Visual presentation of the virtual coach (System 2).

Systems 1 and 2, respectively. The questions covered e.g. frequency of use, usefulness, ease of use, feelings towards usage and the system.

- *Final questionnaire and interview:* The users were inquired for their experiences with the systems using a questionnaire. A semi-structured interview tackled with the system usage, comparison between the two systems, evaluation of the concept and motivational issues.
- *Observations:* During the introduction of the systems the researchers observed the usage focusing on the barriers of use.
- *Written documents:* Users kept diary during the trial, in which they filled in a sleep log, activities and exercises, experience with the system, and any other open comments.

C. Data-Analysis

The collected data from questionnaires, interviews and observations were first analyzed using qualitative content analysis by two researchers who were not directly in contact with the participants. The results were then discussed and agreed on via discussions between the authors. The focus was on user experiences related physical activity coaching, how the experiences differed between the two systems and what lessons were learnt. In-depth sensor data analysis is not in the scope of this paper.

III. RESULTS

A. User Experience on System Functionalities and Motivational Aspects

In general the users perceived the intervention trial as a positive experience. Even if the participants were not familiar with touch screen technology, all the users learned to use both systems relatively quickly.

One user commented that the timed events gave rhythm to the day's activities and the possibility of fitting the exercise events to the user's personal schedule was valued. One user said the guidance and instructions to exercises are important, since she is not motivated to do the exercises on her own and needs a nudge to do them.

Personalization and variation of exercise videos is important. The provided exercise videos were too easy for one person and, on the other hand, one person with arthritis had difficulties with the exercise movements that needed an elastic band. Two of the participants remarked that the current content got boring after some time and one user was irritated by the instructions between the exercise movements.

The content of the delivered information has to be carefully drawn up. None of the users found the tips and information delivered by the system relevant particularly for them, but one user found the idea of using the system as an information channel interesting.

Self-monitoring with sensors was perceived easy to use and interesting by all the participants. Especially counting the steps was appreciated. There were differing opinions regarding sharing information with others. One person commented that it could help communicating about the health issues with one's family and close ones, if the health information is shared online.

The network breaks caused a lot of the problems in the trial. During a network break the sensor measurements were not synchronized between the home device and the server database and the user application of the System 2 could not be started at all without an internet connection. USB activity monitor was malfunctioning with two users and the other one was given a traditional pedometer instead for monitoring the step count. In addition, all the programs had to be reinstalled for one user in the middle of the trial.

B. Comparing the Two Systems

Both systems had their benefits according to the participants and they said it was not easy to compare the two. The sensor data visualization of the System 1 was evaluated better by two of the users due to clearer visibility of the health measurements as icons on the monthly calendar view. On the other hand, the System 2 was liked, because of its speech functionality and it was seen more versatile. For one participant the System 2 was more difficult to grasp because its operation logic; the system is deploying push-type events and the user has no control over the system, i.e. she cannot do much without the scheduled events. This was found inconvenient by another user as well. Two of the users saw advantages in both approaches.

IV. DISCUSSION

The area is very potential for new innovations as portion of older people is increasing leading to a higher demand. Customer and user-centered solutions are important and might make the acceptance of the solutions easier in the field. The content in the trial was very limited and in real life the personalization has to be designed and implemented carefully. The service provider has to take into account very versatile and complex subject group that are of different ages and functioning status. It would be beneficial to have tools for automated personalization. Cross-disciplinary knowledge from e.g. coordination, physiotherapy, healthcare, medical, and technological perspective is needed for providing the services. In addition, it should be noted that people who would really benefit from this kind of solutions, i.e. inactive persons mostly staying at home, are difficult to reach. This became evident during recruiting the test users, as people that participated in the info lectures were quite active already and not the best candidates to take part in the trial.

Based on our study, the prompts and suggestions for exercising are useful motivators for some people, which were also stated in [8]. Similarly to the feedback in [17], the flexibility in setting up an individual schedule for the exercises was appreciated. The anticipated wow effect of virtual coach was not noticed. One reason was the technical problems and the similar content in both of the systems. The push type realization of System 2 decreased user's control over the system and the user becomes a passive receiver who either can accept to start activity or reject it.

The technical problems caused a lot of work which should be considered in the resource planning before such a study. When exact timing of events is important, wired broadband internet connection should be used. The system should also contain local data storage that allows triggering of timed events even without internet connection.

In the future work, the major technical issues must be resolved before testing the solutions on the field. In addition, long term studies are required to evaluate whether using this kind of systems actually increase physical activity of a person. In cost-effectiveness analysis even years-long trials might be needed.

V. CONCLUSIONS

This study was conducted to reveal early user experiences regarding two types of coaching systems aiming at improving physical activity of older adults. The most advantageous features from both systems should be taken into account in the future work. Not to mention the importance of resolving the major technical issues before bringing the solutions on the field, the actual content of the messages delivered to the end users should be carefully planned. The information must be relevant to the target group in question. In addition, there should be enough variety in suggested activities. Self-monitoring and visualizations of performed activities and health parameters were interesting for the test group. Furthermore, longer term studies are required to evaluate whether the proposed systems would actually increase physical activity of a person and the costs effects they pose.

ACKNOWLEDGMENT

This work was part of the Adaptive Ambient Empowerment of the Elderly (A2E2) project funded by the Tekes – the Finnish Funding Agency for Technology and Innovation under the Ambient Assisted Living (AAL) Joint Programme. The authors would like to thank Mawell Care Ltd for valuable assistance and feedback during the field trials and the other project partners: Center for Advanced Media Research at the Vrije Universiteit Amsterdam, Amsta and Hospital IT AS. Furthermore, authors would like to thank SalWe Ltd (Strategic Centre for Science, Technology and Innovation in Health and Well-being) for financial support.

REFERENCES

1. http://www.who.int/ageing/active_ageing/en/index.html, Retrieved 29.1.2013
2. Noncommunicable diseases country profiles 2011, World Health Organization, 2011
3. Nelson ME, Rejeski WJ, Blair SN et al. (2007) Physical activity and public health in older adults: Recommendation from the American College of Sports Medicine and the American Heart Association, Med Sci Sports Exerc, 39:1435-1445
4. Global atlas on cardiovascular disease prevention and control Policies, strategies and interventions, (2011) WHO; World Heart Federation; World Stroke Organization
5. Owen N, Bauman A, and Brown W (2009) Too much sitting: a novel and important predictor of chronic disease risk?, Br J Sports Med, 43:80–81
6. Albaina IM, Visser T, van der Mast CAPG et al. (2009) Flowie: A persuasive virtual coach to motivate elderly individuals to walk, IEEE Proc. 3rd Int Conf on Pervasive Computing for Healthcare, London, UK, 2009, pp. 1-7
7. Consolvo S, McDonald D, Landay J (2009) Theory-Driven Design Strategies for Technologies that Support Behavior Change in Everyday Life. CHI 2009 Proc. Boston, USA, pp. 405-414
8. Schutzer KA, and Sue Graves B (2004) Barriers and motivations to exercise in older adults, Preventive Medicine, 39:1056–1061
9. LIKES, Kunnossa kaiken ikää 75+ senioreiden tuolijumppa ja seniorijumppa istuen ja seisten (Fit for life chair exercise, and sitting and standing exercise for seniors 75+ years), 2007, Video material, In Finnish.
10. Ikäinstituutti, Voimisteluohjelma iäkkäille 1, 2, 3. Voima- ja tasapainoharjoittelu arjen apuna (The Age Institute, Exercise program for older adults 1, 2, 3. Strength and balance exercising helping in everyday life), 2009, Video material, In Finnish.
11. Bandura A (2004) Health promotion by social cognitive means, Health Education & Behavior, 31:143-164
12. AmIE. ITEA2 Call 1 project 06002, Ambient Intelligence for the Elderly, 2007-2010.
13. Kelley K, and Abraham C (2004) RCT of a theory-based intervention promoting healthy eating and physical activity amongst out-patients older than 65 years, Social Science and Medicine, 59:787-797
14. Physical Activity Pie, http://www.ukkinstituutti.fi/en/products/physical_activity_pie, Retrieved 29.1.2013
15. Ryan RM and Deci EL (2000) Intrinsic and extrinsic motivations: Classic definitions and new directions, Contemporary Educational Psychology, 25:54-67
16. Patton MQ (1990) Qualitative evaluation and research methods (2nd ed.). Newbury Park, Sage Publications
17. Romero N, Sturm J, Bekker T et al. (2010) Playful persuasion to support older adults' social and physical activities, Interacting with Computers, 22. 485-495

Author: Heidi Similä
Institute: VTT Technical Research Centre of Finland
Street: Kaitoväylä 1
City: FI-90571 Oulu
Country: Finland
Email: heidi.simila@vtt.fi

Detecting and Analyzing Activity Levels for the Wrist Wearable Unit in the USEFIL Project

James D. Amor, Vijayalakshmi Ahanathapillai, and Christopher J. James

Institute of Digital Healthcare, WMG, University of Warwick, Coventry, CV4 7AL

Abstract—Aging populations present a new set of health challenges. In particular the need to provide evermore healthcare with relatively fewer resources is becoming an increasing issue. This particular problem is highly suited to a technological approach and in this paper we present the USEFIL project and examine how wearable devices can be used to enhance independent living for older people. We present the work that has been undertaken to date on the development of a wrist wearable unit that can capture the user's physical activity. Initial steps in the processing of this accelerometry data to determine a score for the user's level of physical activity in one minute increments are described and the results from this processing are shown.

Keywords—Independent living, wearable devices, ambulatory monitoring, activity monitoring, digital healthcare.

I. INTRODUCTION

There is a growing worldwide trend for people to live longer, with an increasing population of older people, aged 60 and over [4], which in turn leads to two important observations. Firstly, as the number of older people in the population increases, the number of people with age related health conditions increases. Secondly, as the proportion of older people increases, the proportion of people able to provide care decreases. This impacts the number of available carers and the relative amount of money with which care can be provided [3].

These effects combine to create a pressure on society to provide healthcare for its older population using relatively fewer resources per person. However, in parallel with this pressure, the last decade has seen a rapidly expanding boom in technology. As the price of technology falls and the performance increases, technology is being used to solve a number of emerging challenges.

The emerging healthcare need presents an opportunity to leverage technology to provide affordable, scalable healthcare [2]. Through using appropriate technologies, healthcare systems can be delivered that provide effective healthcare within emerging resource constraints.

The Unobtrusive Smart Environments for Independent Living (USEFIL) project [5] is aiming to utilize established and emerging technology to develop a system to assist independent living for older people. The USEFIL project will combine off the shelf devices to create an independent living system within which specific services can be delivered. Examples include fall detection and remote GP consultation. The USEFIL project will develop the system; an initial set of services; and a framework to enable other parties to develop additional services.

The USEFIL project is a collaboration between several universities and companies and within this the University of Warwick is developing a wrist wearable unit (WWU). The WWU will provide a wearable platform for monitoring older people. Data gathered by the WWU will be analyzed and used by the USEFIL system.

In this paper we present the device that is being used in the USEFIL project, explain how it is integrated into the USEFIL system and present an early analysis of the accelerometer data that has been gathered.

II. THE WRIST WEARABLE UNIT

There are a number of devices available on the market that are either specifically designed for health based monitoring, or that can be adapted to this use. Of these devices, the types that best suit the needs of the USEFIL project are smart-watches and watch-phones [1]. These devices combine some integrated sensors with the computational power and communication capabilities that are required for the USEFIL project.

The specific device that has been chosen for the USEFIL project is the Z1 Watch Phone, shown in Fig 1, which is a fully featured Android 2.2 smartphone in a wrist wearable form factor. The Z1 has both an accelerometer and GPS receiver as well as full suite of processor, memory and communications options (including GSM and WiFi) that are standard for smart-phones. The processor on the Z1 is sufficient to allow for multi-tasking, enabling the Z1 to perform background data gathering tasks at the same time as foreground user interaction tasks.

A. Integration with USEFIL

The USEFIL system will provide a number of services and applications to assist independent living. Within the USEFIL system, the WWU will serve as both a data gathering device and a service delivery platform.

Fig 1. The Z1 Watch-Phone that is used as the base unit for the USEFIL WWU.

Data gathering is one of the core functions of the WWU as it enables the collection of information relating to the elder. The WWU will principally use the accelerometer to gather data about the physical activity of the user. This can be converted into estimates of activity level and physical exertion. This information, when combined with other information in the DSS, can be used to make decisions about the user's healthcare.

In addition to the accelerometer, the WWU will be able to gather data from the GPS to provide the user's location if required. This is likely to be used most in an emergency situation when the location of the user needs to be determined to enable assistance to be properly directed.

The Z1 will maintain communication with the USEFIL system primarily through WiFi when the user is in the home. When the user is outside the home GSM can be used for high priority communication. Data gathered and processed by the WWU can be sent back to the main USEFIL system and the USEFIL system can make use of the WWU to provide services to the user.

Messaging and alerting are two key services that USEFIL will provide. Messaging allows the system to send messages to the WWU, which is vital for some WWU functions and allows communication with the elder. This enables many different functions such as reminders and social connection. Alerts allow the system to generate various levels of emergency alert in response to any number of anomalous or pre-set conditions.

B. Activity Detection

As discussed above, the WWU gathers data from its accelerometer and uses this to perform activity level detection, which is the detection of a person's physical activity. This can range from a basic measure of activity level to the estimation of calorific expenditure and can also include elements of the detection of specific activities such as walking and postural transitions. Depending on the method applied, activity detection can be used in a variety of situations.

In a system such as USEFIL, activity level detection or estimation of energy expenditure can be used to help determine activity patterns and daily rhythms. Posture and walking detection can also be used at to determine behavior patterns but can also be used directly as a health indicator and may be useful to detect problems with movement.

Furthermore, through the detection of physical activity it is possible to extrapolate patterns in behavior, such as sleep-wake cycles. Changes in behavioral pattern can be key indicators to health status. A person who is ill might spend longer in bed for example.

III. METHODOLOGY

To enable the Z1 to function as the WWU in the USEFIL system we have written a bespoke app for the device to perform on-going data collection. The app can be configured in a number of ways to adjust sampling frequency and sampling periods so that battery life can be managed. For this paper the device was worn by one of our researchers for a section of a normal working day. The WWU was set to sample continuously and data were broken up into consecutive one minute long segments.

Data from the WWU are analyzed to determine the level of activity during each minute of sample time. Data from the accelerometer in three axis of acceleration are represented as

$$\boldsymbol{x} = \{x_1, x_2, \ldots, x_n\}, \quad (1)$$

$$\boldsymbol{y} = \{y_1, y_2, \ldots, y_n\}, \quad (2)$$

$$\boldsymbol{z} = \{z_1, z_2, \ldots, z_n\}. \quad (3)$$

Data, in all three axis were initially processed to remove noise caused by the accelerometer accuracy by adopting a baseline smoothing method, shown here for the x-axis. Initially the baseline is set as

$$\boldsymbol{b} = x_1. \quad (4)$$

Subsequently, we use the following update rules

$$x'_i = \begin{cases} x_i & \text{if } |x_i - b| < T, \\ x'_{i-1} & \text{otherwise,} \end{cases} \quad (5)$$

$$b = \begin{cases} x_i & \text{if } |x_i - b| < T, \\ b & \text{otherwise,} \end{cases} \quad (6)$$

where T is the threshold value and x' the values after smoothing.

The data are then interpolated to a constant sample rate and filtered to remove the gravity component from all three axes. The root mean squared (RMS) value is calculated as

$$r = \sqrt{\frac{1}{n}\sum_{i=1}^{n} x_i^2 + y_i^2 + z_i^2}. \qquad (7)$$

The calculation of r is carried out for each minute long segment and allows us to calculate activity scores over an entire day, or multiple days.

The final step of processing, which is performed to aid data visualization, is to classify the data into one of five activity categories from very low to very high with an additional category for 'not worn'. An initial threshold is calculated to classify data with a very low ($r < 0.05$) score as 'not worn'. To calculate the activity level thresholds the range of r values is calculated and thresholds set to give five equally spaced categories in this range.

This provides a simple way of visualizing the amount of activity the person is undertaken and we display this as an activity bar that shows the level of activity over time.

IV. RESULTS

Fig 2 shows an activity bar for the time the researcher wore the WWU. There are clearly periods of greater intensity activity, around 11:30-12:00 for example, and periods of lower intensity activity, 14:00-15:00 for example.

The periods of activity in this plot match well to the known activity of the participant over the day. Coffee breaks at 11:30 and 15:30 result in higher levels of activity being registered. Lunch at around 13:00 results in two periods of high intensity activity with a low intensity period in the middle. This is consistent with walking, sitting and eating, and walking again.

Fig 3 shows examples of accelerometry recorded from the WWU and classified according to our methodology. Acceleration is shown in three axes prior to the removal of the gravity component.

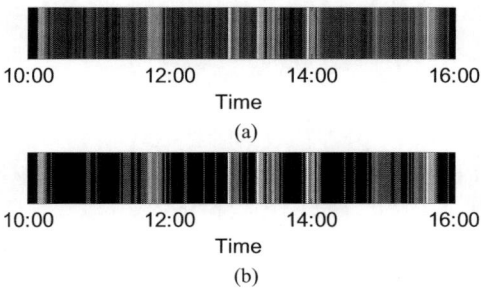

Fig 2. The activity bar showing activity levels before threhsolding (a) and after thresholding (b). In both cases, brighter colours indicate more intense activity.

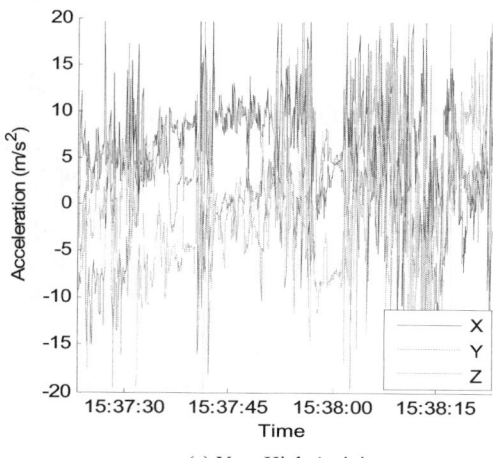

(a) Very High Activity

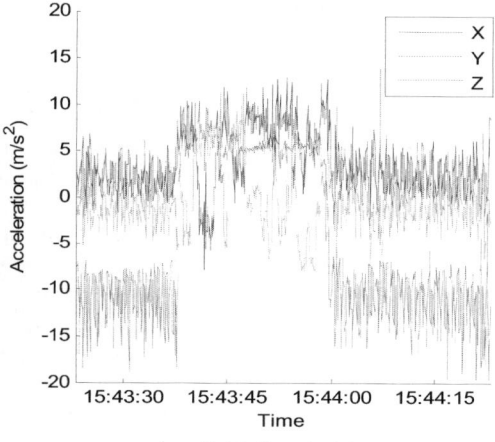

(b) Medium Activity

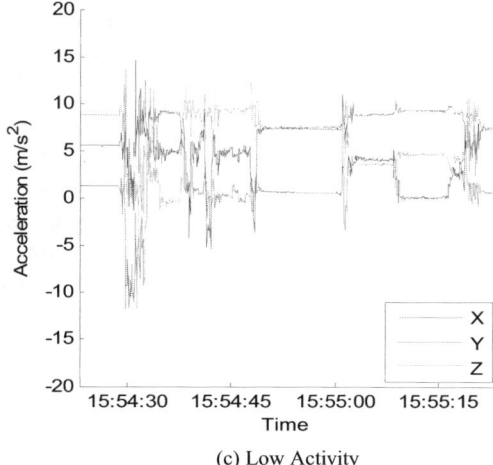

(c) Low Activity

Fig 3. Activity traces from the WWU showing minute recordings of different activities and the classifications derived by our system.

V. DISCUSSION

A visual analysis of the graphs in Fig 3 shows that there are varying degrees of activity across the graphs and is the expected result given that we are classifying from RMS values. Fig 3(a) shows very high activity and a trace that is consistent with a brisk walk with orientation changes in the arm (evidenced by the altering positions of the three axes) with a pause around 15:38. Fig 3(b) shows medium activity and is consistent with a slow walk (evidenced by the very rhythmic patterns in much of the trace) broken with a short task. Fig 3(c) shows low activity and is consistent with little movement punctuated by posture changes.

High intensity periods correlate to greater amounts of movement, such as walking, whereas lower intensity periods correlate to more sedate activities, such as desk work. Very short duration high intensity activity in a minute of otherwise low intensity activity tends to result in a medium activity score. A good example of this is retrieving a document from the printer where the participant is seated, gets up and walks to the printer and then returns to their desk.

There is one limitation on the WWU that affects the work presented in this paper; the capabilities of the WWU are limited by battery life. There is a trade off in the WWU between data gathering and battery life. More of the former shortens the latter. In order to maximize the length of time that data can be recorded for, sacrifices must be made either in sampling frequency or in sample interval. In this paper we have shown an activity bar collected under constant sampling. This however limits the battery life and for the USEFIL system a configuration leaning towards less dense sampling and a longer operation time will likely be required.

VI. CONCLUSIONS AND FUTURE WORK

This paper has presented a wearable device for independent living and shown how it will be used in the USEFIL project. The WWU will be used to gather data on the activity level of the user and this data can be processed to estimate their activity level. The activity level estimation in turn can feed into a number of different health indicators. Initial results have been shown that demonstrate how accelerometry data can be processed to extract a measure of activity level and that these can be categorized easily for display purposes. Furthermore, it has been shown, particularly in Fig 1 that these category values map sensibly onto different intensities of activity.

We have discussed the battery life limitations of the WWU and how a trade-off must be made between data gathering and battery life. We do not consider this to be a major issue going forward and are taking steps to improve the battery life. An example of this work includes changing the sampling interval or frequency on the fly based on incoming data so that periods of little activity receive little monitoring and more battery resources are used to monitor higher activity behaviors. Furthermore, the ever advancing state and miniaturization of technology continue to present increasingly better devices to which all the techniques described in this paper could be easily adapted.

The work presented in this paper is our initial progress towards achieving our vision for the WWU in the USEFIL system. As such there are numerous avenues of extension for the work presented here. A principal area of work is the improvement of activity level estimation in a number of directions including calculating step counts, estimating calorific expenditure and detecting the user's posture and postural transitions. Through developing these areas of activity monitoring, and addressing the battery life concerns we will develop a platform capable of performing a broad range of activity monitoring tasks suitable for the USEFIL system and indeed in any situation where activity monitoring is required.

ACKNOWLEDGEMENT

The authors would like to acknowledge the input from the rest of the USEFIL team and EU funding under FP7, project reference 288532.

REFERENCES

1. Amor, J., James, C., *Wearable devices for independent living: Analysing suitability for the USEFIL project*. Internal technical report.
2. Grguric, A., 2012. *ICT towards elderly independent living*. Research and Development Centre, Ericsson Nikola Tesla.
3. Humphereys, G., 2012. The Health-Care Challenges Posed by Population Ageing. *Bulletin of the World Health Organization*, 90, 2.
4. World Health Organization, 2012. Health Topics Ageing [Online]. Available: http://www.who.int/topics/ageing/en/ [Accessed 12th September 2012].
5. USEFIL, 2012. [Online] http://www.usefil.eu/ [Accessed 12th September 2012]

Author: Dr. James D. Amor
Institute: Institute of Digital Healthcare, WMG, University of Warwick
Street: International Digital Laboratory, University of Warwick
City: Coventry
Country: UK
Email: J.D.Amor@warwick.ac.uk

Home Monitoring of Elderly for Early Detection of Changes in Activity and Behavior Patterns

S. de Miguel-Bilbao[1], J. García[1], F. López[1], P. García-Sagredo[1], M. Pascual[1], B. Montero[2], A. Cruz Jentoft[2], and C.H. Salvador[1]

[1] Telemedicine and e-Health Research Unit, Health Institute Carlos III, Madrid, Spain
[2] Geriatric Service, University Hospital Ramón y Cajal, Madrid, Spain

Abstract—28 real homes of elderly have been monitored during three weeks by a platform based on the BUSing protocol that consists of 5 sensors of presence and 5 sensors of action.

The objective of this study is to implement a monitoring platform in order to obtain the user profile characterized by the level of activity and periodic habits.

This short term monitoring provides the configuration parameters, indicators, and thresholds that characterize the behavior of the user independently of the topology of the monitored home. The extracted information will constitute the inputs and the configuration parameters of future long term monitoring in order to early detect physical and mental dysfunctions and the objectification of geriatric assessment.

Keywords—home monitoring, elderly, daily living activity, geriatric assessment.

I. INTRODUCTION

Nowadays, there is an increasing demand for outpatient care accessibility, for maintaining and restoring health, as well as for maximizing the independence of patients [1]. Comprehensive care of the elderly at home is an important need that requires the implementation of systematic tools that provide the detection and prevention of hazardous situations and the detection of physical and cognitive dysfunctions [2], especially for people that live alone.

In this context, our group is working on two related research areas:

- The monitoring of daily living activity of elderly patient at home for early detection of changes in the activity and behavior habits.
- The searching for mappings between indicators derived from the process of monitoring and subjective indicators from standardized geriatric assessment scales in order to advance in the objectification of comprehensive geriatric assessment process.

This work is based on the following assumptions:

- It is possible to define a personal profile of activity and behaviour of the user at home. This profile consists of activities, level of activity, and habits that the user carries out periodically.
- Temporal variations of this personal profile during long term monitoring, may be an indicator of the change of the biopsycosocial condition of the user. If so, such variations may involve the anticipatory detection of physical and mental dysfunctions.

The objective of this work is to define a pattern that characterizes the activity and the behaviour of elderly at home. This pattern consists of a set of indicators that are representative of the following functions: biomedical, physical, mental and social. These indicators constitute the support of the geriatric evaluation and they might be characterized by the information extracted from the monitoring of the daily activity at home.

This work presents the obtained results of the development, implementation, and subsequent stage of information processing of a home monitoring platform that provides the profiles that characterize the activity and behaviour of elderly according to the proposed pattern.

II. METHOD AND MATERIALS

A. Specification of the Activity and Behaviour Pattern

The geriatric assessment is based on four areas of evaluation that provide a global vision of the elderly state:

- Biomedical function is the classical history and physical examination.
- Physical function is an index of the mobility, the personal care, and interaction with the environment.
- Mental function involves the psychological, intellectual, cognitive and affective activities.
- Social Function includes social activities and human relationship.

In this framework, it was established a set of ten constructors that describe the complete state of the user. Fig. 1 enumerates these constructors, where each of them is the connecting joint between two types of components:

- Subjective component that is extracted from the standard questionnaires of geriatric evaluation.
- Objective component that is provided with the result of the monitoring process of the daily living activities and whose components constitute the activity and behaviour pattern.

Function	Geriatric Assessment Tests (subjective) ←							Constructs	→ Monitoring inf. (objective)
	Barthel	Lawton	MNA	MMSE	GDS	Morisky Green	SPPB		
Biomedical		7		*		*		Medication	Pill (N)(T)
Physical	8 & 10	4			*		*	Mobility within home	ChL (N)+ A/B (H)
	2, 4 & 7			*	*			Grooming: hygiene	Bath (N) + Bath (H)
	5 & 7		*	*				Grooming: toilet use	
	3			*	*			Get Dressed	Bedr (N)+Bedr (H)
		3, 5	*	*	*		*	Houseworks	Kitc (N)+Kitc (H)+ Frid(N)
Mental			*	*	*			Emotional/ Affective	Out (N)+Out (H)+ A/B (H)
			*	*	*			Cognitive	Rr(N)
Social	1	3	*	*	*			Nutrition	Kitc (N)+Kitc(H)+ Frid(N)
		1, 3 & 8		*	*		*	Social activity	Acc(N)+Acc(H)+Out(H)

[Test questions related to constructs (* full test)]	Pill: Open/Close pillbox	Out: Outdoor stays
	A/B: Armchair/Bed use	Rr: Repeated routes
	ChL: Changes of location (room)	Acc: Accompanied
	Bath: Bath visits	(N): Number of events
	Bedr: Bedroom visits	(H): Accumulated time (hours)
	Kitc: Kitchen visits	(T): Time (hours)
	Frid: Open/Close fridge	

Fig. 1 Geriatric evaluation areas, geriatric assessment tests involved (subjective component), constructors, and pattern (objective component).

The activity and behaviour pattern is defined by a set of explicit and implicit parameters extracted during a 7 days monitoring process.

A. Home Monitoring Platform

The platform used to monitor each tested home consists of the following components that are shown in Fig. 2 [3] [4]:

- A wireless sensor network and a gateway (Ingenium S.L.), based on the BUSing®-radio protocol that operates at the frequency of 2.4 GHz and 868MHz [3]. Concretely, ten sensors were installed; five are presence sensors that detect the presence in each of the five rooms of each monitored home: livingroom, kitchen, bath, bedroom and hall. The rest are action sensors, three sensors that detects the opening and closing of doors, concretely the front door, the door of the fridge, the door of the pillbox, and the other two are sensors of pressure of the armchair and the bed to detect when the user sits on the arm chair and lies down on the bed. The gateway collects the information from the sensors and establishes a TCP-IP socket to transmit the presence information to the router.
- The router, 3G Huawei B970, transmits the information collected by the gateway to the central server.
- The central server receives, stores and processes the data from monitored homes. The central server is located in the Data Processing Center of the Telemedicine and eHealth Research Unit.

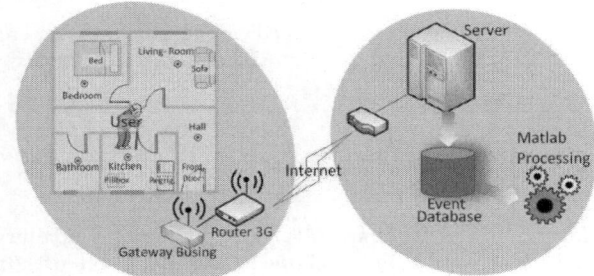

Fig. 2 Components of the monitoring platform located at homes.

B. Processing Stage

The aim of this stage is to obtain the activity and behavior pattern (explicit and implicit components) by off-line processing of monitored information.

The explicit parameters are mainly based on the quantification of the presence events in each of the five rooms, the time spent in each room, the number of changes of location and repeated routes, the action events of opening and closing doors (pillbox, fridge and front door) and the events of pressure of the armchair and the bed.

After obtaining all the information from the monitoring platform and before starting the quantification of explicit parameters, it is necessary to assure the consistency of the detected events. This pre-processing stage provides several advantages, being possible the treatment of certain inconsistencies. After reviewing the presence events, it was detected the existence of unreal alternations between pairs of presence events of very short duration. This is due to the type, the number, and the location of the sensors, the topology of the home, and the algorithm of capture of the presence events.

To detect these inconsistencies it was necessary to calculate a threshold per home and room that allow the distinction between presence events of short and long duration. The presence events, whose duration is under the calculated thresholds, are defined as short presence events, and the events whose duration is above the corresponding threshold, are called long presence events.

These presence thresholds provide another advantage in the quantification of the number of presence events in each of the rooms. It has been checked that the duration of short presence events are not significant compared with the duration of the long presence events. The calculation of these presence thresholds justifies the importance of a short monitoring to fix the parameters that characterize and configure the behavior of the user independently of the topological features of the home.

The implicit parameters (corresponding to the emotional/affective and social activity constructs) cannot be extracted from the information that is directly provided by the sensors, and it is needed additional processing to obtain them. They are defined by the periods of time during which the user is either absent or accompanied, and they are obtained by contrasting the detected events provided by the combination of the presence and the action sensors (doors) and activity level thresholds.

The periods of absence are defined by periods of time without presence events between two front door events, the first being the output of the home and the second the arrival.

In the case that the user is provided with the help of a support person, there are periods of time characterized by multiple presences at home, therefore the level of activity is greater than when the user is alone. It is considered that the user is accompanied when there is a period of time characterized by a number of presence events above a calculated presence threshold, and when either of these two conditions is met: there are hours of help registered or there is a front door isolated event due to another person that enters or leaves the home.

III. RESULTS

This work presents the results of the processing carried out after a three weeks monitoring of 28 homes where elderly live alone. First week is required to adapt the working functions of the sensors; second week is required for the user to become accustomed to the monitored environment. And the information that is taken into account is the information collected during the last week, when the operating conditions are stable.

The users that have participated in this project are patients of the Geriatric Service of the University Hospital Ramón y Cajal (Madrid), and they are attendees of the elderly centre Queen Sofia (Cruz Roja Española). In total, 28 patients were monitored; of them, 15 were male and 13 female, they were more than 65 years old, and had at least one chronic disabling disease. Patients also had to comply with the following requirements: no pets at home, and if the patient is provided with a support person, the estimated visitation schedule must be registered.

Fig. 3 shows the activity of one of the monitored homes, and consists of seven graphs, one for each day of the week. Concretely, Fig. 3 provides the implicit information about the detected periods of absence, the periods when the user is accompanied, and also shows the information needed to detect these periods: front door events, activity level and the presence threshold.

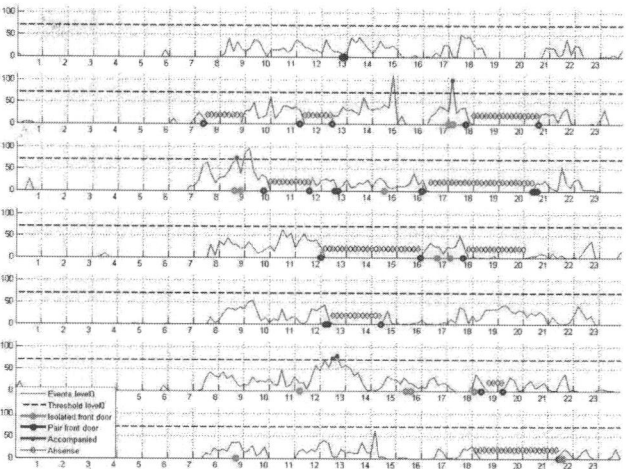

Fig. 3 Periods of time when the user is accompanied and outside home. Front door events, activity level and threshold of presence events.

For each patient, it has been obtained his/her activity and behaviour pattern of the stable week. Table 1 shows the constructors, and the pattern of activity and behaviour of one of the monitored homes. Each parameter is indicated with the mean and the standard deviation. The cognitive constructor is not indicated by a quantification parameter; otherwise it is expressed as "most common route" and "route with more intermediate rooms".

Table 1 Pattern of activity and behavior of a patient.

Constructor	Parameters	Mean±SD
1.- Medication	N° Opening/closing pillbox	2,29±0,95
2.- Mobility within home	N° Changes of location	21,71±12,91
	Time spent on bed and armchair	5,30±2,13
Grooming:	N° Bath visits	12,43±5,72
3.- Toilet use	Time spent in bath	0,9±0,45
4.- Hygiene		
5.- Get dressed	N° bedroom visits	8,14±5,72
	Time spent in bedroom	1,98±2,99
6.- Houseworks &	N° kitchen visits	19,14±8,89
7.- Nutrition	Time spent in kitchen	3,65±1,85
	N° Opening/closing fridge	9,00±3,00
8.- Emocional/effective	N° outdoor stays	1,57±0,98
	Time spent outside	3,10±2,39
	Time spent on bed	8,4±1,84
	Time spent on armchair	3,61±1,29
9.- Cognitive	Most common route	hall - bedroom
	Route with more intermediate rooms	bath-kitchen
10.- Social activity	N° times accompanied	0,43±0,53
	Time accompanied	0,10±0,13
	N° way outs	1,57±0,98
	Time spent way out	3,10±2,39

Patients also completed the geriatric assessment tests that are the basis of the subjective component. The comparison between objective components (pattern) and subjective components (tests) has not been carried out yet.

IV. DISCUSSION

It is important to emphasize the real context that frames the main objective of this project. The platform was installed in real homes to monitor the daily activity of elderly during three weeks. The number of sensors was quite low: five presence sensors and five actions sensors. This contrasts with other studies that employ a large number of invasive sensors.

This work presents the advantages of the off-line processing of the information gathered by a real monitoring platform of elderly. The processing stage provides the calculation of configuration parameters that characterize the behaviour of the user independently of the topology of the monitored home.

By crossing the information collected by presence and action sensors, the absence periods have been determined. The detected absence involves the detection of the periods of time during which the user is not at home, being able to consider these periods as the stay in another room that is called as "outside". The inclusion of the external presence as another room provides reliable and complete information about the activity of the user, and the possibility of inclusion the sixth room in future longer monitoring.

The processing stage also includes the calculation of a threshold per home and room. These thresholds provide the possibility of eliminating inconsistencies of the collected information and also it has been checked that more reliable information can be provided in the quantification of the number of events.

After analyzing the results of the processing stage it can be said that this monitoring carried out during three weeks can be described as a previous short monitoring that is necessary to fix intrinsic configuration. This type of short monitoring provides fix thresholds and parameters that are characteristic of the behaviour of the user and also are characteristic of the monitored environment.

In order to contrast the information obtained from the monitoring platform (objective results) the patients completed several questionnaires of comprehensive geriatric assessment (subjective results) that will provide more reliable conclusions, nevertheless this analysis has not been performed yet.

V. CONCLUSION

This work describes a home monitoring platform of the daily life activities. This is a non invasive platform that consists of action sensors and presence sensors. A useful insight of the advantages of the post processing stage is presented, providing the independence of the monitoring process from the home topology.

This study expects to be not only a starting point in the monitoring of elderly based in real environments, but also a necessary stage to configure the monitored homes and users. The extracted parameters can be considered as inputs of long term monitoring whose objective is the detection and variation of the daily activities that enables the early detection of physical and cognitive dysfunctions [5].

The monitoring of household activity can help to improve global geriatric evaluation and to make possible a better follow-up and control of elderly people in their homes.

ACKNOWLEDGMENT

Authors want to thank the valuable information transferred from the project "Definition, design, development and evaluation of monitoring services, tracking and control of elderly in assisted environments for independent living" FIS-PI080435 (DGPY 1368/08).

REFERENCES

1. Action Plan for EIP AHA: B3 Integrated Care. Brussels: European Commission; 2012. Available at: http://ec.europa.eu/research/innovation-union/pdf/active-healthy-ageing/b3_action_plan.pdf
2. Eklund K, Wilhelmson K. (2009) Outcomes of coordinated and integrated interventions targeting frail elderly people: a systematic review of randomized controlled trials. Health Soc Care Community, 17:447-58
3. García J, Pozo A, Pascual M, et al. (2010) Seguimiento de Personas Mayores en Espacios Asistidos para la Vida Independiente," Proc INFORMED 2010, San Sebastián, Spain, 2010, pp: 45-50.
4. Pozo A, Pascual M, et al. (2010). Plataforma pública para el despliegue y evaluación de nuevos servicios asistidos basados en telemedicina. INFORSALUD 2010, Madrid, Spain.
5. Rashidi P, Cook DJ, Holder LB, Schmitter-Edgecombe M (2011) Discovering Activities to Recognize and Track in a Smart Environment. IEEE T Knowl data en, 23(4):527-539.

Author: Silvia de Miguel Bilbao
Institute: Health Institute Carlos III
Street: Monforte the Lemos, 5
City: Madrid
Country: Spain
Email: sdemiguel@isciii.es

Automatic Detection of Health Emergency States at Home

M. Vela and D. Ruiz-Fernández

Departament of Computing Science and Technology, University of Alicante, Alicante, Spain

Abstract—New technologies have lead to a better quality of life and a rise in life expectancy. The elderly face a number of needs and problems which must be resolved in order to maintain such quality of life. Some of these problems are related to muscle weakness, a lack of accuracy in sensory information or loss of equilibrium. These become a major issue at elder age making the elderly become dependent and susceptible to home accidents. The aim of this research is to increase the security, confidence and independence of the elderly as well as their relatives. To achieve such objective, this research proposes using a network of multiple Kinect sensors installed in different areas of a house monitoring the individual with the ability to interact with the person by voice or video, if required. The Kinect sensors analyze the everyday life and routines of the person. This analysis allows the individual to be traced, and means an automatic detection of risks in real time, decreasing the response time required in the event of a health emergency.

Keywords—Health emergencies, Kinect sensor, computer vision, dependent people.

I. INTRODUCTION

As a result of the improvement in quality of life, the average life expectancy in Europe for 2010 was 79.9 years (82.6 for females and 76.7 for males), reaching its highest in Spain with an average of 85.4 years [1]. It is expected that population at age 65 and above will rise from 17.4% in 2010 to 20% in 2040, and 30% by 2060 [1], meaning the ratio of population age 80 and above will become significantly important.

Elderly and dependant people suffer more home accidents than other sector of the population. The main causes of home accidents are due to moving and handling objects; slips, trips and falls; loss of equilibrium, motion sickness, and falling out of bed [2] [3]. These accidents may either have no further consequences or may lead the elderly to a health emergency situation involving anxiety, serious injuries, unconsciousness or even death. In such situations, medical care is critical. The time elapse from the moment the accident happens to the moment assistance is provided must be minimal, as it can become vital for the individual.

The overall objective is therefore to identify the human states of the monitored person. Following the state recognition and analysis, the system will define whether or not we are facing a health emergency situation, and subsequently medical care will be arranged and relatives will be contacted. The definition of the health emergency is based on behavior patterns, background information and the surroundings of the individual. Identifying the state and analyzing the additional background and surrounding information allow a significant reduction of false alarms [4]. Equally it supposes a less number of devices needed in comparison to other similar systems, whilst providing the person with more independence. It also reduces the direct interaction of the individual with the system as the person is completely independent from the system and does not need to wear or carry any portable device.

In order to meet this objective a system based on computer vision and voice recognition techniques has been designed. This system uses technologies like Microsoft Kinect for Windows [5] [6], Arduino Uno microcontrollers [7] and Dynamixel AX-12 robot actuators [8].

II. STATE OF THE ART

For the purpose of analyzing similar research, three categories have been established [9]: Systems whose results are solely based on the data received from infrared sensors and intelligent cameras (Microsoft Kinect, Asus Xtion [10] and other high cost specialized devices). Secondly, other systems that require a complex technical installation of equipment with different kinds of sensors placed in strategic locations to compile information about the surroundings. As a third category, the systems that involve wearing portable devices with sensors have also been considered [11] [12] (medical alert bracelets, panic button alarms, specialized smartphones). Such latter systems essentially link the individual with the device and require the direct involvement of the person. These do not provide solutions in the event of the individual being unable to interact with the device – as a result of the state, for instance, unconsciousness. Therefore, this research will focus on the two previous categories: Systems based on gathering surroundings information and those based on computer vision and image processing.

Research such as [13], involves a big investment in professional devices of different nature as well as a complex technical installation which are not scalable and require professional experts for adaptations and ongoing maintenance.

Others [14] [15] also use the Kinect sensor. Typically two human states are analyzed [14]: on one hand, falling on the floor in a specific position that is known by the system at a distance previously calculated; on the other hand, the inactivity state. The research [15] studies the point of the person's centroid in the final position after falling in order to specify the distance with the floor and define if it means an actual fall. These systems are limited as they cannot offer a solution for blind spots or in the event of the person being partially hidden when falling.

There are also some more sophisticated techniques. These place the individual inside a 3D Bounding Box [16], which presents its height reduced and its width increased when a fall happens. Similarly, [17] creates a knee-hip-thorax-neck axis which works similarly as the bounding box. As it is the case for [14], these systems require the sensor to recognize the person being unconscious or inactive, which does not pose a solution if the person is partially disabled, asleep, seated or in a blind spot. Systems using fuzzy logic [17] [18] to specify if a fall has or has not happened have also been considered. However, these latter systems require the individual carrying a portable device. Therefore, these represent the same limitations as those previously mentioned above.

None of the previously analyzed systems consider using multiple sensors, thus reducing the space for analysis to a lab-compartment space far from being a real scenario. Nor they use the voice recognition functionality.

Projects based on Kinect technology are focus on fall recognition. These projects are limited as they rely on the individual falling in a particular sequence ending in the floor at a specified programmed position, which is quite unlikely to happen in reality. This research treats falls as one of the many cases considered within what it is hereby referred to as "health emergency state".

III. SYSTEM ARCHITECTURE

The system architecture is based on a central server computer that has the software installed to analyze the data compiled by the network of Kinect sensors. The sensors are distributed in all the rooms of the house. The central computer stores the system settings data as well as all the events detected for processing. The system will send a sms, mms or an email including a picture/video of the accident via a cloud service or GSM/GPRS modem with a SIM card (to allow operation in the event of lack of Internet connection), hence it will alert the relevant contacts (previously stored) in the event of the system recognizing the individual in a non desired state.

For setting the communication in between the sensors and the server, several possibilities were considered: long USB cables, wireless USB devices, RJ45 to USB adapters to cover bigger distances than those with USB, and the implementation of a client-server system with sensors connected to Raspberri Pi [19] devices establishing a wireless socket with the server. For security and reliability purposes, it was decided to use long USB cables with repeaters for scenarios requiring more space and WUSB for smaller scenarios. The system architecture is displayed in Figure 1.

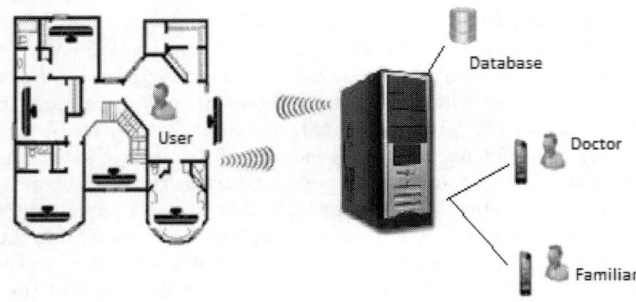

Fig. 1 System architecture

A. System Settings

The initial settings of the system are simple and intuitive. The sensors are identified by the rooms they monitor and these are set up depending on the type of room (kitchen, bedroom…). Thus allows an early detection of the behavior expected to be performed in such room, and guides the heuristics and algorithms when making decisions. The relation of each room with other sensors is also established, hence setting a logic home network, as it is illustrated in Figure 2.

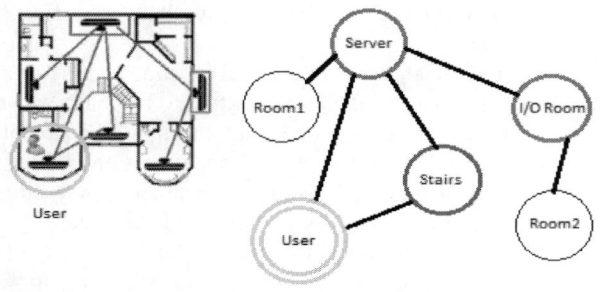

Fig. 2 Example of the setup of a logic network

The end user can set up his own network by plugging or unplugging the Kinect sensors at any time (plug&play). The system will detect at any time if a sensor has been plugged, and the user will be able to update it with the settings of the given scenario.

The logic network of sensors allows the identification of the rooms where the individual could be found in the event of going missing, as the system knows the immediate surroundings of the last room where the individual was last seen. The sensors scan the room as they have the ability to 360° degrees rotate thanks to the swivel stand where they are positioned. If the individual still cannot be found, the system will consequently take the appropriate next step. In addition to the cameras, the microphones built in the sensor allow the possibility of tracing the individual immediately if the person goes missing through the voice and sound recognition feature. This also can guide the sensors when scanning the room. To enable this feature, the system uses grammars that have been programmed to deploy a language combining key words which have been previously entered by the user, thus enabling a simple and intuitive communication with the system, fairly close to natural human language.

Simultaneously, the individual common behavior patterns are defined by the system by tracking the person movements in the house, the human states, and the duration of the movements as well as some other additional data that is analyzed by the system. Some of these patterns could be the following: The person suffering from insomnia if often waking up during the night; the person going to the bathroom more often than normal; the person's time in seated position or even the detection of unexpected and suspicious individuals. All this allows the creation of reports from the logs and history stored in the system that could potentially be used for different purposes like medical diagnosis, informative updates for the relatives or security purposes.

For fall detection, the system is based on the recognition of the person position (seated, lying down, standing up and others), and the subsequent analysis of the position: Whether or not it is known by the system. It also analyses the angles created by the different straight lines that can be aligned with the joints of the body (head-hip, hip-knee...) when intersecting with the floor plan, as well as uses other heuristics similar to those described in the projects previously mentioned above.

B. Skeletons Confidence Level

The detection of skeletons by the SDK of MS Kinect [20] allows tracking up to 20 joints. In order to establish the confidence level of a skeleton in a given state, the algorithm used gives a confidence value to each joint. Such confidence value is multiplied by the following values depending on the reported joints being tracked (x10), inferred (x3) or not tracked (x0). If the skeleton does not reach a set confidence level, the state can be discarded, hence reducing the number of false positives.

C. Swivel Stand

The purpose of having each Kinect sensor placed in a swivel stand is to widen the field of view covered by each sensor. Because of the swivel stand the number of Kinect sensors required to cover big spaces can be reduced, hence reducing the final cost of the system. Moreover, the rotary movements of the stand guarantee an accurate location of the individual, overcome the problem of blind spots and allow a more in depth scanning of the rooms when searching for the individual. The stand makes movements in accordance with the logic set by the system for the purpose of monitoring the person.

The stand works with one Arduino Uno microcontroller board and two Dynamixel AX-12A robot actuators. The Arduino controller allows setting the necessary movements required to position the cameras in the direction set by the system logic to scan the room or track the person in a particular space. The two actuators are perpendicularly interconnected and also connected to the Arduino board, thus allowing the Kinect sensor to swivel through both X-axis and Y-axis. The data collected by the Kinect sensor is processed, and, if approved by the sensor, certain commands will be sent to the Arduino microcontroller so that the actuators perform the relevant movements. The communication in between the Arduino board and the application can be established through either USB cables or wireless by acquiring additional hardware.

The purpose of adding other technologies to the information reported by Kinect is to increase the level of accuracy and intelligence of the system by collecting data of the surroundings being tracked and monitored. Such objective can be achieved thanks to the swivel stand hardware, the camera settings, and the context (time, period of the year, geolocation, etc) and person background information previously stored.

IV. TEST SCENARIO

Several different types of tests were performed in a house with three Kinect sensors installed communicating three rooms. The scenarios covered the following: Search and location of the person, tracking the person through the rooms, identification of different states and home accident simulation considering different falls. For the assessment of the falls, a total of 25 fall tests simulating home accidents were carried out. In particular, 5 users of different sizes simulated 5 different falls. The falls did not follow a specific pattern, and the tests aimed to represent different fall sequences (particularly at the start and end positions of the fall). For the tracking, and search and location scenarios, a total of 20 tests in 2 different versions were carried out. The first version consisted of 10 tests where the person was moving naturally through the rooms. The second version consisted of 10 tests where the person was trying to hide from the cameras and create confusion while moving through the rooms. The first version of the tests was carried out without losing track of the individual. For the second version, the system had no problem at all finding the individual in less than a minute every time the person disappeared from the system. Overall, the tests results were 100% score.

V. CONCLUSION

Using a network of Kinect sensors enabled to swivel 360º degrees facilitates a constant location and tracking of the person through the different rooms of the house, which allows using different heuristic methods to identify the state of the person in accordance with the individual and surroundings information. The benefits of the system can be translated into more independence for the user, independence for the relatives who can limit their dedication to the dependent person, economic savings (reduced spend on home care assistants), and an added security feeling for both the family and user.

It is important to highlight that the number of false positives is reduced thanks to the context data collected, for example, a false unconsciousness state when the individual is actually lying down in bed in a bedroom. The context data also allows putting more focus on non typical behavior patterns in locations like the staircase where the risk of home accident is bigger.

As the system is not a customized solution for certain houses, it therefore does not involve a complex technical installation and guarantees covering a wider variety of final scenarios.

Last, the estimated cost of the system is reasonably affordable when compared to other similar systems as the main data source is based on the Kinect sensor, which is easily found and available in the market at a lower cost than other more specialized ones.

REFERENCES

1. Mortality and life expectancy statistics and demography report (2010), European Comission Eurostat (October 2012)
2. Centers for Disease Control and Prevention, National Center for Injury Prevention and Control. Web–based Injury Statistics Query and Reporting System (WISQARS). November 30, 2010
3. Stephen N Robinovitch, Fabio Feldman, Yijian Yang, Rebecca Schonnop, Pet Ming Leung, Thiago Sarraf, Joanie Sims-Gould, Marie Loughin, Video capture of the circumstances of falls in elderly people residing in long-term care: an observational study, The Lancet, Volume 381, Issue 9860, 5–11 January 2013, Pages 47-54, ISSN 0140-6736
4. M. Kangas, I. Vikman, L. Nyberg, R. Korpelainen, J. Lindblom, T. Jamsa: Comparison of real-life accidental falls in older people with experimental falls in middle-aged test subjects, Gait & Posture article (2011)
5. Microsoft Kinect for Windows at http://www.microsoft.com/en-us/kinectforwindows/ (April 2013)
6. Stone, E.E.; Skubic, M., "Evaluation of an inexpensive depth camera for passive in-home fall risk assessment," Pervasive Computing Technologies for Healthcare (PervasiveHealth), 2011 5th International Conference on , vol., no., pp.71,77, 23-26 May 2011 at IEEE Xplore
7. Arduino Uno specs (April 2013) at http://arduino.cc/en/Main/arduinoBoardUno
8. Dynamixel AX-12A specs (April 2013) at http://www.robotis.com/xe/dynamixel_en
9. Mubashir, M., Shao, L., Seed, L.: A Survey on FallDetection: Principlesand Approaches. Neurocomputing100, pp. 144 -- 152. (2012).
10. Asus Xtion specs (April 2013): http://www.asus.com/Multimedia/Xtion_PRO_LIVE/
11. Lombardi, A.; Ferri, M.; Rescio, G.; Grassi, M.; Malcovati, P., "Wearable wireless accelerometer with embedded fall-detection logic for multi-sensor ambient assisted living applications," Sensors, 2009 IEEE , vol., no., pp.1967,1970, 25-28 Oct. 2009
12. Alert1, http://www.alert-1.com/ (April 2013)
13. Liang Liu; Popescu, M.; Skubic, M.; Rantz, M.; Yardibi, T.; Cuddihy, P., "Automatic fall detection based on Doppler radar motion signature," Pervasive Computing Technologies for Healthcare (PervasiveHealth), 2011 5th International Conference on , vol., no., pp.222,225, 23-26 May 2011
14. Ong Chin Ann, Lau Bee Theng, Hamid Bagha; An injury recognition model using single rotatable infrared sensor (2013). The Second International Conference on e-Technologies and Networks for Development (ICeND2013) – Malaysia
15. Rougier. C, Auvinet. E, Rousseau. J, Fall Detection from Depth Map Video Sequences, 9th International Conference on Smart Homes and Health Telematics, ICOST 2011, Montreal, Canada, June 20-22, 2011. pp 121-128 (2011) ISBN: 978-3-642-21534-6
16. Mastorakis G, Makris D, Fall detection system using Kinect's infrared sensor (2012) , Journal of Real-Time Image Processing; pp 1-12, Springer-Verlag.
17. Planinc. R, Kampel. M, Introducing the use of depth data for fall detection, Personal and Ubiquitous Computing, Springer-Verlag 1617-4917 May 2012
18. M. Kepski, B. Kwolek, and I. Austvoll, "Fuzzy Inference-Based Reliable Fall Detection Using Kinect and Accelerometer", ;in Proc. ICAISC (1), 2012, pp.266-273.
19. Raspberry Pi specs at http://www.raspberrypi.org/quick-start-guide
20. Zhengyou Zhang, "Microsoft Kinect Sensor and Its Effect," Multi-Media, IEEE , vol.19, no.2, pp.4,10, Feb. 2012

An Adaptive Home Environment Supporting People with Balance Disorders

Iliana Bakola[1], Christos Bellos[2], Evanthia E. Tripoliti[1], Athanasios Bibas[3],
Dimitrios Koutsouris[2], and Dimitrios I. Fotiadis[1]

[1] Unit of Medical Technology and Intelligent Information Systems, Dept of Materials Science and Engineering,
University of Ioannina, Ioannina, Greece
[2] Biomedical Engineering Laboratory, Institute of Communications and Computer Systems (ICCS)-National
Technical University of Athens (NTUA), Greece
[3] 1st Department of Otolaryngology – Head & Neck Surgery, University of Athens, Greece

Abstract—This paper presents an architecture allowing simple access to Ambient Intelligent Systems deployed in domestic environments, particularly focused on people with balance disorders. Our aim is to develop a systemic life-space solution, for indoor environments, by integrating heterogeneous physiological and environmental real-time data that will result in a continuous monitoring and early warning system of people with mobility problems due to loss of balance. The designed control system is adaptive, and it can accommodate to changing conditions of inhabitants. The system recognizes the state of the environment by integrating different contextual data from different devices and sensors. It processes both video and audio signals to detect dangerous events and trigger automatic warnings. The proposed unobtrusive system permits elder people to augment their perception of independence and safeness at home.

Keywords—Ambient Intelligence, Balance Disorders, smart environments.

I. INTRODUCTION

Loss of balance is the inability to maintain an upright posture during standing and walking. The human postural control is a complex skill which requires the contribution of vision, vestibular sense, proprioception, muscle strength and reaction time. With increased age, there is a progressive loss of function of these systems which can contribute to balance perturbations [1]. Loss of balance should not be dismissed as an unavoidable consequence of aging, as most of these pathologies can be handled properly as long as there exist early detection and appropriate rehabilitation.

Due to the demographic and social changes the percentage of population aged 65 and above in the developed countries will rise from 7.5% in 2009 to 16% in 2050 [2]. In addition, about one-third of the elderly population reports difficulty with balance or walking. Balance disorders occur across the entire age span in up to 30–40% of the population by 60 years of age and are associated with substantial occupational and healthcare costs [3]. In adults over the age of 65, balance problems may lead to falls which is a pervasive but under investigated problem, partly due to the difficulties in determining the cause of the fall. Falls are highly prevalent with increasing age, with 30% of people over age 65 falling every year, and more than half of the people over age 75 in nursing homes or care facilities being unable to live independently because of falls, but with only about 20% of fallers get medical attention and appropriate remediation for the problem. Falls that take place indoors are expected to result in hip fracture [4]. The 85% of fractures occur at home and 25% of them attributed to environmental risks within the house. In parallel, the majority of elder people lives in private households and along with increasing age and declining health, tend to spend more and more of their time indoors or in its close surroundings. Furthermore, fear of falling often leads to social isolation, inactivity and sometimes depression. In addition, more recent studies provide information on the predictive value of balance and most of these conclude that elderly people with decreased balance abilities have a higher risk of developing Activities of Daily Living (ADL) disability [5]. Without receiving sufficient care, elderly are at risk of losing their independence. Thus, a system permitting elderly to live safely and independently at home is of profound importance.

In recent years special attention has been given to applications of ambient intelligent aimed at increasing the security of elderly people in their living environment, prolonging the autonomous living at home, while assuring a high level of assistance when needed. These kinds of systems are usually based on deploying different kinds of sensors and actuators over the environment in order to capture relevant information and to act accordingly inside the ambient. Several methods have been developed focusing on motion capture and gait measurement since research has shown that the parameters which describe locomotion are indispensible in the diagnosis of frailty and fall risk [6] and that the measurement of a person's gait is crucial to a variety of health conditions [7]. Clinical research has pointed out that variability in stride parameters may be prognostic of future falls in elder people [8-9]. It should be noted that there aren't any integrated software systems or standalone tools towards the early diagnosis and management of the balance disorders.

In addition, due to the diversity of the patients and their impairments, as well as, the multifactorial nature of loss of balance, the research led to the development of a number of tools which assess balance and more specifically, the functioning of the above mentioned systems. The results of the tests are either discrete or graded scores or continuous measurements that need further analysis. However, no patterns have been extracted expressing the alteration of the function of these systems. Toward this direction many studies focus on gait analysis aiming to understand age-related declines in gait performance, screen balance impairments in the elderly and to early identify of at-risk gait for falls prevention in the older population. Gait analysis is based on measurements including basic spatial-temporal data, kinematic or displacement/velocity/acceleration-time data, kinetic or force/moment time data, electromyography or electrical activity of lower limb muscles, generates parameters that reflect gait degeneration and through the utilization of machine learning techniques reveals biomechanical abnormalities [17, 18].

Recent research in activity monitoring of older adults has focused on the use of passive infrared (PIR) motion sensor in the home. These sensors generate information about the daily activity of monitored subjects, and arrays of such sensors have been used to obtain velocity measurements on a continuous basis in domestic environments [10, 11]. In addition wearable devices for measuring gait, based on accelerometers are an area that has received significant attention. The acceleration data is often augmented with data from sensors such as gyroscopes [12], magnetometers, ambient light [13] or ambient and skin temperature [14], aiming to extract a more detailed environmental user context. Another widely researched area that has been applied to gait assessment is vision-based monitoring systems. Human motion analysis using vision technology addresses the need for passive, environmentally mounted hardware that does not require those being monitored to wear any devices. Moreover, several mobility aids have been deployed to assist walking or otherwise enhance the mobility of people with mobility impairment. There are numerous walking aids such as wheelchairs and mobility scooters which can help with impaired ability to walk or for more severe kinematic disabilities. Walking aids comprise assistive canes (commonly referred to as walking sticks), crutches and walkers. These devices allow maintaining an upright posture control by providing any or all of improved stability, reduced lower-limb loading and generating movement. Technical advances increased the scope of these devices considerably, for example by use of sensors and audio or tactile feedback.

The goal of this paper is to present a provisional architecture of an intelligent and highly automated home environment, specifically targeted for elderly people susceptible to loss of balance disorders. The underlying core of this environment is based on the Ambient Intelligent (AmI) concept, therefore resulting in a threefold contribution: (i) significant independent mobility enhancement, (ii) seamless alignment in everyday life with minimal effort and involvement of the patient, (iii) remote monitoring and alarming in critical situations. In the sections that follow, we present the design of this supportive environment, describe the technical details and discuss its components in terms of functionality.

II. SYSTEM OVERVIEW

Based on current findings there are varied factors that affect mobility and balance in the elderly as well as additional factors that are manifested by balance pathology itself. For instance, the feeling of insecurity in people with balance problems may trigger a response from the autonomic nervous system that may lead to an increase in blood pressure and heart rates as well as changes in skin conductance. In addition, reduced or excessive lighting, unsymmetrical or extremely high or narrow steps, humidity, slippery surfaces and unmarked edges are some of the factors that can pose risks.

Although many aspects of balance and gait can influence risk of falling, a critical factor is the ability to react effectively to balance disturbances. These perturbations can arise from: (a) slips and missteps; (b) collisions or other physical interactions with objects (animate or inanimate) in the environment; or (c) the destabilizing effects of volitional movement [15, 16]. A "hostile" living environment can increase potential perturbations. In order to accomplish this, inhabitant's actions would have to be monitored continuously, because the individual environment may change due to a change in the patient's functional or cognitive activity, or an alteration in their routines.

The intelligent ambient proposed in this paper is composed by several audio and video sensors deployed in the environment combined with a wearable sensor attached to the wrist and a wireless pressure insole sensor. A first prototype scenario has been formed, in which a user has placed an ambient system at home that is able to monitor his or her activity to detect incidents related to loss of balance. The prototype apartment has a bedroom, a hall, a toilet, a kitchen and a living room.

The current measurements will be gathered through the sensors in order to obtain information about the environment and vital signs of the elderly people. The selection of the sensors will be based on criteria such us small size, light weight, unobtrusiveness, potential usefulness for our target audience, aesthetics, safety, privacy, operational capacity, ease of integration and battery-life information. The system is open to receive additional information from wearable devices such as the wristband and the wireless pressure insole sensor which measure triaxial accelerometer signals, physiological signs including heart rate, blood pressure and body temperature as well as foot pressure signals, respectively.

More specifically, the proposed system is divided to three sensors units, and the system architecture is shown in Fig. 1:

- Balance Sensor Unit: This Unit incorporates sensors for measuring angular velocity, acceleration, sense the translation movement and provide information on tilt during static periods. Passive infrared motion sensors, also, are installed at each location and, in addition, the bedroom has a pressure sensor positioned in bed and a magnetic sensor has been settled to the doorframe to detect the opening and closing of the entrance door.
- Physiological Sensor Unit: The wristband, mentioned before, is included in this category as well as personal wearable sensors that measure sweat, blood pressure and heart rate.
- Environmental Sensor Unit: All sensor boards have a supplementary temperature and humidity sensor, while ambient light can be measured as well.

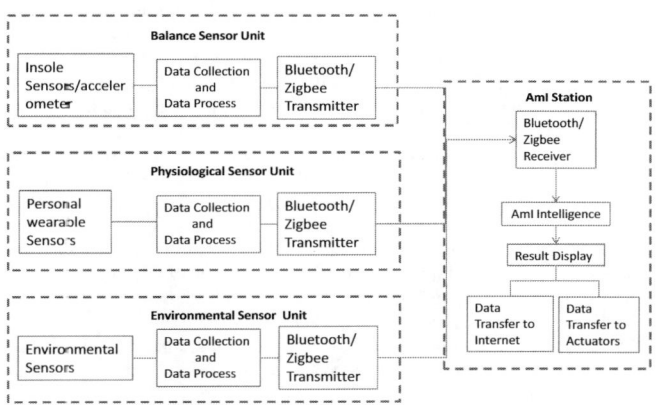

Fig 1 Architecture of the proposed system

Each sensor unit incorporates a data collection, processing and transmission module. These modules act as a facilitator between the sensors and the AmI Station. They manage and coordinate the multi-sensor system, which acts in a dynamic and uncertain environment, in order to improve the performance of the System Ambient Intelligence and ultimately the performance of the system's perception of patient's balance.

Moreover, these modules undertake the wireless communication either with Bluetooth or Zigbee protocol between the Sensor Units and the AmI Station enhancing the efficient data transmission and facilitating battery savings.

The smart home platform, called AmI Station consists of a back-end and network infrastructure, and a home gateway that controls a wireless sensor network, network communications and the delivery of a range of services including: health, security and smart energy. The AmI Station integrates three (3) main modules:

- Bluetooth/Zigbee Receiver: Data streams from the sensors are collected and processed and only when an event is identified it is transmitted through the wireless network to a PC-base station (i.e. AmI Station) in order to decide what actions to take, enhancing, thus, the battery and resources savings.
- AmI Intelligence: The system predicts the inhabitants' needs using patterns in the inhabitants' behavior. The system will initially be trained by feeding in data from individual users, thus making it personalised, extracting individualized patterns and building implicit and explicit profiles for each user in ordert o enhance and personalize the final output of the AmI Intelligence module.
- Result Display: The system will be able to provide efficient feedback to elderly users as well as augmented and personalized alarming and recommendations for an immediate medical assistance.
- Data Transfer: The processed data are published to online Repositories, in order to be managed by clinicians, while specific information is tranferred to in-house actuators, triggering them in order to assist the elderly people on a specific balance disorder situation.

III. RESULTS AND DISCUSSION

As described earlier, in our Ambient Assisted Living (AAL) smart home environment a variable number of heterogeneous wireless devices and sensors provide context information in diverse forms. In order to provide flexibility and scalability, integrating context data from different sources to comprehend the overall environmental state demands a significant programming effort because devices' source code must be updated whenever a new device is added. Since the connection between devices is not fixed, the role of middleware architecture is of profound importance. We need an adaptable and light-weight operating system not only for connecting sensors and actuators, but for providing seamless connection from devices to users.

A wireless monitoring system able to passively and unobtrusively monitor in-home mobility perturbations in real time is presented in this paper. The proposed system consists of three sensor units: the balance, the physiological and the environment sensor unit. The three sensor units have a small size due to the use of microsensors and they are integrated with Bluetooth transmitters. The Bluetooth data transmission allows the monitoring system to be mobile and wireless so that the user can use it freely. The PC connected to the internet as the user interface terminal provides telemedicine capabilities. A series of monitoring tests will be conducted using the proposed system in order to evaluate the accuracy of the method and verify the feasibility of the monitoring system.

IV. CONCLUSIONS

Assisted living home environments, monitor a wide variety of factors and although there are many examples of ambient home environments only a small number of them employ mobility and autonomic nervous system monitoring as part of their functions. The reason is that it is quite difficult to measure mobility only by using ambient sensors. Monitoring the mobility of an elderly person in a smart home is comparatively simple if the person is living alone, because all the detected activity can be attributed to that person. The ambient smart home must have the ability to identify the monitored subject, and distinguish between their location and the location of others if the monitored person is living with others or regularly receives visitors. This can be achieved using video and audio recognition, passive infrared sensors located in doorframes wearable accelerometers or footstep analysis. Wearable accelerometers and are the most common sensors applied to measure mobility. A fundamental aspect of the design of an AAL system is energy consumption.

However, ambient home care systems have several disadvantages including the requirement to identify the monitored subject from others in the home and the inability to monitor the person's mobility outside of the home environment. Therefore smart home systems are not suited to monitoring the mobility levels of active persons who are frequently and irregularly, outside of the home.

In spite of the huge advances that have been made in the last ten years the challenge to design and develop stable and general systems still persists, as most systems only solve specific problems in very particular environments. Advances at the field of AAL systems, are of profound importance as personal autonomy and quality of life for elderly can be improved enormously, while at the same time care costs can be reduced remarkably.

REFERENCES

1. L Sturnieks, D., R. St George, and S. R Lord, Balance disorders in the elderly. Neurophysiologie Clinique/Clinical Neurophysiology, 2008. **38**(6): p. 467-478.
2. Nations, U., World Population Ageing. 2009.
3. Royal College of Physicians, Hearing and Balance Disorders, Report of a working party. London: RCP,2007.
4. Nordell, E., Jarnio, G., Jetsen, C., Nordstrom, L., Thorngren, K., Accidental falls and related fractures in 65-74 year olds: a retrospective study of 332 patients. Acta Orthop Scand, 2000..
5. Onder, G., Pennix, B., Measures of physical performance and risk for progressive and catastrophic disability: results from the Women's Health and Aging Study. The Journals of Gerontology Series A: Biological Sciences and Medical Sciences, 2005. **60**(1): p. 74-79.
6. Runge, M. and G. Hunter, Determinants of musculoskeletal frailty and the risk of falls in old age. Journal of Musculoskeletal and Neuronal Interactions, 2006. **6**(2): p. 167.
7. Hodgins, D., The Importance of Measuring Human Gait, Medical Device Technology, vol. 19, pp. 42-47, Sep. 2008
8. Hausdorff, J., Rios, D., Edelberg, H., Gait variability and fall risk in
community-living older adults: a 1-year prospective study, Arch Phys Med Rehabil, 2001;82:1050–6.
9. Barak, Y., Wagenaar, R., Holt, K., Gait Characteristics of Elderly People with a History of Falls: A Dynamic Approach, Physical Therapy, 86, 11. 1501-1510 (2006).
10. Austin, D., Haves, t., Kave, J., Mattek N., Pavel, M., On the disambiguation of passively measured in-home gait velocities from multi-person smart homes. Journal of ambient intelligence and smart environments, 2011. **3**(2): p. 165-174.
11. Hagler, S., Austin, D., Hayes, T., Kaye, J., Pavel, M., Unobtrusive and Ubiquitous In-Home Monitoring: A Methodology for Continuous Assessment of Gait Velocity in Elders, IEEE Transactions on Biomedical Engineering, 2010. **57**(4): p. 813-820.
12. Altun, K. and B. Barshan, Human activity recognition using inertial/magnetic sensor units, in Human Behavior Understanding. 2010, Springer. p. 38-51.
13. Borazio, M. and K. Van Laerhoven. Combining wearable and environmental sensing into an unobtrusive tool for long-term sleep studies. Proceedings of the 2nd ACM SIGHIT International Health Informatics Symposium. 2012: ACM.
14. Krause, A., Siewiorek, D., Smailagic, A.,Farringdon, J., Unsupervised, dynamic identification of physiological and activity context in wearable computing. in Proceedings of the 7th IEEE International Symposium on Wearable Computers. 2003.
15. Maki, B.E., Mcilroy, W., and Fernie, G., Change-in-support reactions for balance recovery. Engineering in Medicine and Biology Magazine, IEEE, 2003. **22**(2): p. 20-26.
16. Maki, B.E., Sibley, KM., Jaglal, SB., Reducing fall risk by improving balance control: Development, evaluation and knowledge-translation of new approaches. Journal of safety research, 2011. **42**(6): p. 473-485.
17. Lozano-Ortiz C, Muniz A, Nadal J, "Human Gait Classification after Lower Limb Fracture using Artificial Neural Networks and Principal Component Analysis". In Proc. 32nd IEEE EMBS Ann Internat Conf, Buenos Aires, Argentina, pp. 1413-1416, 2010.
18. Alaqtash M, Sarkodie-Gyan T, Yu H, A, "Automatic Classification of Pathological Gait Patterns using Ground Reaction Forces and Machine Learning Algorithms". In Proc. 33rd IEEE EMBS, 2011.

Author:	I. Bakola
Institute:	University of Ioannina
Street:	University Campus of Ioannina
City:	Ioannina
Country:	Greece, GR 45100
Email:	iliana_bakola@hotmail.co.uk

AALUMO: A User Model Ontology for Ambient Assisted Living Services Supported in Next-Generation Networks

P.A. Moreno[1,2], M.E. Hernando[1,2], and E.J. Gómez[1,2]

[1] Bioengineering and Telemedicine Group, ETSI Telecomunicacion- Technical University of Madrid, Madrid, Spain
[2] Biomedical Research Networking center in Bioengineering, Biomaterials and Nanomedicine (CIBER-BBN), Zaragoza, Spain

Abstract—Ambient Assisted Living (AAL) services are emerging as context-awareness solutions to support elderly people's autonomy. The context-aware paradigm makes applications more user-adaptive. In this way, context and user models expressed in ontologies are employed by applications to describe user and environment characteristics. The rapid advance of technology allows creating context server to relieve applications of context reasoning techniques. Specifically, the Next Generation Networks (NGN) provides by means of the presence service a framework to manage the current user's state as well as the user's profile information extracted from Internet and mobile context. This paper propose a user modeling ontology for AAL services which can be deployed in a NGN environment with the aim at adapting their functionalities to the elderly's context information and state.

Keywords—context-aware applications, ambient assisted living, next-generation network, user model, ontology.

I. INTRODUCTION

Nowadays, the world population is experiencing a progressive ageing due to medical and technological advances which improves the life expectancy. Therefore, an enormous interest is growing to use the information and communication technologies (ICT) to support elderly people to live independently for longer period in their home. Presently, Ambient Assisted Living (AAL) [1] is one of the significant ICT initiatives for elderly people with special needs. The AAL services belong to the context-aware paradigm in which computing technology becomes closely invisible embedded into everyday objects making services more personalized, adaptable, interactive and therefore useful.

Since context can be defined as "any information that can be used to characterize situation and specifies the elements that must be observed to model situation" [2], context-aware applications employ sensors and context sources to drive their performance in specific user's conditions or preferences. Thus, context modeling explains contexts and the relationships among the distributed heterogeneous contexts. Context and user models describe through formal and conceptual language user's aspects (profile, preferences, interest, education, profession or living conditions) as well as environment information to improve services' features.

Ontologies are used to represent context data of models, so systems and devices can interpret the semantics of processing information allowing interoperability in an AAL environment. Ontology-based languages and tools enable a formal analysis of the domain knowledge and promote contextual knowledge sharing and reuse in context-aware systems [3]. Ontology Web Language (OWL) allows to construct complex, graph-like hierarchies of user model concepts, which is especially important for ontology integration. Several ontologies as UserML, OntobUM or GUMO allow modeling user's context. The User Modeling Mark-up Language (UserML) models user in ubiquitous computing; and OntobUM integrates three ontologies: one for user, another for defining relations between applications and other to define the user-application interaction semantics. The General User Modeling Ontology (GUMO) is used for a uniform interpretation of distributed user models in intelligent Semantic Web enhanced environments [4].

With the rapid advance in technology, users can take advantage of the devices and services enriched with context information. The evolution of ICT toward an horizontal service platform, supporting anytime and anywhere access to information and knowledge-driven and context-aware decision making, can act as a driver toward the delivery of AAL services. A good amount of research has been conducted lately to develop integrated context management frameworks aimed at acting as context servers, relieving the final context-aware application of all context-management related operation [3].

With the appearance of Next-Generation Networks (NGN) telecom operators can leverage All-IP networks to design external service interfaces that integrate a diverse set of sources and context inference processes that are easily scalable, extendable, and robust at the same time [5]. NGN is based on the IP Multimedia Subsystem (IMS) architecture which supports context inference that enables multimedia application services to acquire rich user information from both Internet and mobile spheres regardless of the access network. The horizontality of IMS architecture provides a group of functions called "service enablers" that are common to many third-party applications and therefore should be built once and reuse many times. Therefore, context management can take advantage of user provisioning, management and security features offered by the NGN infrastructure. Despite users are reluctant to give personal information to third parties, telco operators have traditionally been considered reliable and trustable. Thus, a telco is particularly suitable to offer context-aware services because it is an intermediary between the user, who gets access to the communication network, and software developers, who use the NGN to provide their enriched applications.

Some related context-aware works employ ontological approaches to manage information in smart homes and health services [6] or to support ambient intelligence mobile services [7]. Some context-aware services use NGN infrastructure to build a map of user meaningful context status [8]; or a NGN service enabler to handle context information of device, network, user, and position to decide the appropriate services for a user [9]. Finally,[10] propose the creation of a context knowledge layer in the IMS layered architecture with the aim at helping 3rd party service providers to create personalized services considering user's context (preferences, device, networks or service).

The AAL user model ontology proposed in this work is aimed at being employed by an IMS architecture in order to provide third party AAL applications with personalization and adapting functionalities considering the elderly's context and state. Therefore, the ontology will be described as well as the implementation of user model classes in the IMS architecture.

II. MATERIALS AND METHODS

A. GUMO Ontology

GUMO has been the ontology selected to develop the user model as it is the most comprehensive user modeling ontology represented by modern semantic web languages which eases the user model exchange between different applications.

GUMO divides the user model dimensions into three parts: auxiliary, predicate and range. Therefore if one wants to say something about "the user's interest in football", one could divide this into the auxiliary (has interest), the predicate (football) and the range (low-medium-high).

GUMO exposes three main user's dimensions (Basic, Context and Domain) that are modeled within user-adaptive systems and [11]. The *Basic User Dimension* entails the information related to the physical and psychological user conditions. The classes which are contained in this dimension are: Contact Information, Demographics, Abilities, Personality, Characteristics, Emotional State, Physiological State, Mental State, Motion, Role, Mood, Nutrition, Facial Expression, Relationships and Basic Human Needs. The *Context Dimension* defines classes regarding the user's environment or product used as Location, Physical Environment, Social Environment, Sensor Dimensions, Product Information and Travel Contexts. Finally, the *Domain Dependent Dimension* reflects classes as Interest, Knowledge, and Preference.

The auxiliaries employed by GUMO are: hasBelief, hasDone, hasInterest, hasKnowledge, hasLocation, hasPlan, hasPreference, hasRated, hasExperience, hasRegularity, and hasGoal.

B. NGN and IP Multimedia Subsystem Presence Service Enabler

The use of IMS in this work is justified due to its presence service enabler that allows managing user information in order to personalize others services provided. The core of IMS architecture is composed of several servers which handle registered users' information and signaling functions to allow multimedia session management through the Session Initiation Protocol (SIP) [12]. The Presence Service is deployed in IMS architecture through an added value application server, named Presence Server. The SIP/SIMPLE protocol is employed for publishing, subscription and notifying information through events definition by using PUBLISH, SUBSCRIBE and NOTIFY SIP methods. This information is described by means of XML document with PIDF (Presence Information Data Format) format [13] which has been extended with others formats (e.g. RPID, GEOPRIV) with the aim at adding more user information. The presence server can request or update user's profile data in a XML Document Management Server (XDMS) by using the XML Configuration Access Protocol (XCAP). This XDMS server stores user's information which is adjusted to a defined format, known as *appusage*, which refers to the structure to be followed by user's profile data. The server allows storing as much appusages as type of information needed (contact list, user's images, etc.). Thus, the presence server could offer to 3rd party applications user's current state and user's profile with for inferring the user context.

III. RESULTS

A. AALUMO Ontology Specification

Considering the scenarios of AAL services (homecare, health and wellness, supply with goods and chores; safety, security and privacy; social interaction; information and learning; working life; mobility and hobbies), the proposed ontology includes specific classes that characterizes users of AAL services. The AALUMO ontology includes some classes and subclasses that belong to GUMO ontology's dimensions, but other classes have been defined to cover as wide as possible the older people conditions and environments.

The *Basic Human Needs* class is inherited directly from GUMO and expresses the data related to physiological and psychological needs of elderly people. The *Preferences* class refers to user's preferences in several areas, where their daily activities are developed, and is composed of seven subclasses: interest, nutrition (both from GUMO), privacy, devices, leisure activities, places entertainment and social communities. The *Profile* class stores user's relevant data which are regular as abilities, contact information, demographics, knowledge, contact persons (all of them extracted from GUMO basic dimension) and chronic diseases. Since elderly people usually suffers any chronic disease, is important to note that we created the subclass chronic disease, not covered by GUMO, that is composed of others subclasses as: disease, limitations (both physical and psychological), medications and threshold medical data parameters. The created *Service* class is aimed at informing about some characteristics (devices employed, access information like password, and a description) of different AAL services which are usually used by the older person. Considering that our proposed user model is intended to be used in context-aware AAL services, the knowledge about the user's state is essential to infer the context and consequently provide him with the suitable service. Thus, the *Current State* and *Previous State* classes refer to aspects which are likely to change in a certain period of time, as emotional state, location, mental state, mood, motion, physical environment, which are extracted from Basic and Context GUMO dimension; and reminders, current device, current medications and current service (developed in this work). Other subclasses as date and time are set in previous state class to note the instant in which the user had that specific state. In addition, the proposed ontology inherits the auxiliaries from GUMO.

B. Overall Architecture

The AAL user model ontology has been defined to be employed by means of the presence service of an IMS architecture. As Fig. 1 shows, the AAL User Model Ontology is deployed in the XDMS server by means of creating appusages according to classes and subclasses defined in the ontology. Any user's device capable of gathering context information is allowed to publish this information using the IMS presence service. Thus, the user's context information will be stored in the presence server. On the one hand, 3rd party applications can subscribe to user's presence information through a SIP SUBSCRIBE message sent to the presence server. Once the user's presence information has arrived, the presence server will notify about the user's state to the application sending a XML document in a SIP NOTIFY message. Moreover, the presence server will update with the user's presence information the ontology classes related to the user's state (i.e. CurrentState class) using the XCAP PUT method. On the other hand, the applications can access to the user model's information using the XCAP GET method. As the ontology has been deployed through XDMS server's appusages, the user model's information will have a known structure which will be useful to 3rd party application's developer. Thus, the application will be able to apply specific reasoning techniques with the aim at inferring a user's contextual situation and adapting its functions to a certain user when the application has both AAL user model information and presence information.

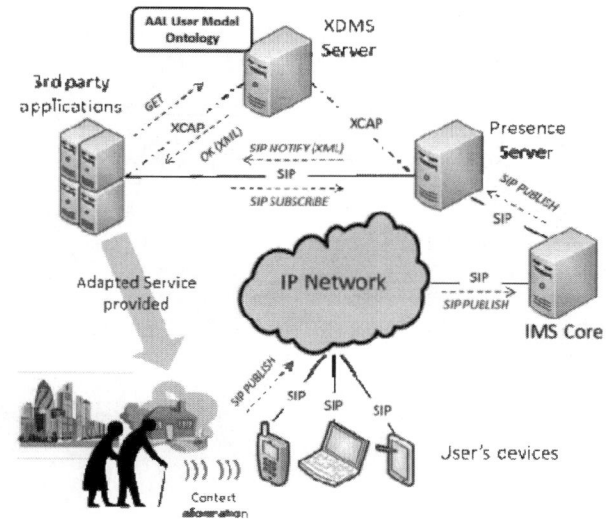

Fig. 1 Overall architecture

C. AAL User Model Appusages

Regarding the AAL User Model Ontology, the XDMS server will store the user model information which will be based on two appusages: *AALUserState* and *AALUserProfile*. The AALUserState appusage refers to the current user's information which is likely to change in a short period of time. The class CurrentState and Service of the AAL User Model Ontology will provide the labels to the AALUserState appusage. An example of a XML document compliant with this appusage format is shown below:

```
<ns:UserState
xmlns:ns='http://AALUMO.org/schema/AALUserState'
<ns:EmotionalState>  <ns:happiness>  </ns:… >
<ns:Location>
    <ns:coordinates>  <ns:latitude>40:27:11.04N</ns:latitude>
     <ns:longitude>3:43:36.85W</ns:longitude>  </ns:…>
    <ns:spatiallocation>  <ns:home>  </ns:…>
    <ns:virtuallocation>www.myweb.com  </ns:…></ns:Location>
<ns:MentalState><ns:nervousness></ns:…>
```

```xml
<ns:Mood><ns:excited></ns:…>
<ns:Motion> <ns:sitting> </ns:…>
<ns:PhysicalEnvironment> <ns:weather>sunnny</ns:…></ns:…>
<ns:PhysiologicalState> <ns:temperature>36.5</ns:…></ns:…>
<ns:Device>mobilePhone001</ns:…>
<ns:Medication>none</ns:…>
<ns:Service>id0034</ns:…>
```

The ontology classes BasicHumanNeeds, and Prefer-ences and Profile are used to create the AALUserProfile appusage that includes the user information which is not going to change in a short period of time. An example of a XML document which shows this appusage is shown as following:

```xml
<ns:UserProfile
xmlns:ns='http://AALUMO.org/schema/AALUserProfile>
<ns:BasicHumanNeeds> <ns:PsychologicalNeeds>
  <ns:SocialNeeds><ns:Family></ns:…></ns:…></ns:…>
<ns:Preferences>
  <ns:Interest><ns:Sports><ns:football></ns:…></ns:…>
  <ns:Nutrition><ns:Diabetic></ns:…>
<ns:Profile >
  <ns:ChronicDiseases>
    <ns:Diseases><ns:diabetesmellitus></ns:…>
    <ns:Medications><ns:IDMedication>med002</ns:…>
    <ns:DescriptionMedication>insuline</ns:…>
    <ns:ScheduleMedication>every8hours</ns:…>
  <ns:ContactInformation>
    <ns:familyname>Mateo</ns:…>
    <ns:fullname>Higueras</ns:…> </ns:…>
  <ns:ContactPersons >
    <ns:familyname>Julia</ns:…>
    <ns:fullname>Higueras</ns:…> </ns:…>
<ns:PreviousStates > id002845</ns:…>
```

IV. DISCUSSION AND CONCLUSIONS

Context-awareness services are becoming essential to support the autonomy and independence of elderly people due to the heterogeneity of situations or environment they usually experience. User model and ontologies are very useful to represent user's characteristics and their relationships with environment and applications. On the other hand, the Next-Generation Networks provides enriched multimedia services that allow the context inference considering information from both Internet and mobile spheres. Thus, this work has introduced a user modeling ontology to adapt and personalize Ambient Assisted Living services provided through NGN networks. By using the Presence service, 3rd party applications can access to user's presence and user profile information with the aim at applying context reasoning techniques. Therefore, two appusages has been defined formatting the information contained in the AAL user model ontology presented.

As future work, the ontology could be extended including as much requirements (i.e. extended contact information, or more user's devices and services characteristics) as AAL services may meet in order to support the elderly people's daily life. Furthermore, the AALUMO ontology could also support the definition of a context service enabler with context management capabilities that could relieve 3rd party applications of context managing tasks.

ACKNOWLEDGMENTS

We would like to acknowledge for contributing to this work to Liss Hernandez.

REFERENCES

[1] Ambient Assisted Living at http://www.aal-europe.eu/
[2] Y. Cao, L. Tao, and G. Xu, "An event-driven context model in elderly health monitoring," in Ubiquitous, Autonomic and Trusted Computing, 2009. UIC-ATC'09. Symposia and Workshops on, 2009, pp. 120–124.
[3] F. Paganelli and D. Giuli, "An Ontology-Based System for Context-Aware and Configurable Services to Support Home-Based Continuous Care," IEEE Transactions on Information Technology in Biomedicine, vol. 15, no. 2, pp. 324–333, Mar. 2011.
[4] D. Heckmann, T. Schwartz, B. Brandherm, M. Schmitz, and M. von Wilamowitz-Moellendorff, "Gumo–the general user model ontology," in User modeling 2005, Springer, 2005, pp. 428–432.
[5] P. Gutheim, "An ontology-based context inference service for mobile applications in next-generation networks," Communications Magazine, IEEE, vol. 49, no. 1, pp. 60–66, 2011.
[6] Y. Evchina, A. Dvoryanchikova, and J. L. M. Lastra, "Ontological framework of context-aware and reasoning middleware for smart homes with health and social services," in Systems, Man, and Cybernetics (SMC), 2012 IEEE International Conference on, 2012, pp. 985–990.
[7] D. Preuveneers, J. Van den Bergh, D. Wagelaar, A. Georges, P. Rigole, T. Clerckx, Y. Berbers, K. Coninx, V. Jonckers, and K. De Bosschere, "Towards an extensible context ontology for ambient intelligence," in Ambient intelligence, Springer, 2004, pp. 148–159.
[8] C. Baladron, J. M. Aguiar, B. Carro, L. Calavia, A. Cadenas, and A. Sanchez-Esguevillas, "Framework for intelligent service adaptation to user's context in next generation networks," Communications Magazine, IEEE, vol. 50, no. 3, pp. 18–25, 2012.
[9] J. Kim, J. Jeong, S. Nam, and O. Song, "Intelligent Service Enabler Based on Context-Aware in Next Generation Networks," 2008, pp. 802–806.
[10] A. Moon, Y. Park, and S. Kim, "Higher order knowledge management platform for the personalized services in next generation networks," in Networked Digital Technologies, 2009. NDT'09. First International Conference on, 2009, pp. 452–457.
[11] UbisWorld web page: http://ubisworld.ai.cs.uni-sb.de/index.php
[12] Y. ITU-T, General overview of NGN. December, 2004.
[13] H. Sugano et al. "Presence Information Data Format (PDIF)" IETF RFC 3863, 2004.

Author: Pedro Antonio Moreno Sánchez
Institute: Bioengineering and Telemedicine Group- Technical University of Madrid
Street: E.T.S.I. Telecomunicación. Avda Complutense 30,
City: Madrid
Country: Spain
Email: pmoreno@gbt.tfo.upm.es

Internet of Things for Wellbeing – Pilot Case of a Smart Health Cardio Belt

E. Kovatcheva[1], R. Nikolov[2], M. Madjarova[1], A. Chikalanov[2]

[1] Sofia University "St. Kl. Ohridski", Sofia, Bulgaria
[2] VIRTECH Ltd, Sofia, Bulgaria

Abstract—This paper deals with the opportunities that the advanced telemedicine and information technologies can give for preventing patients with cardiovascular diseases – the leading cause of deaths worldwide. The selected pilot case presents a Smart Health Cardio Belt (SHCB) system developed by the Bulgarian team on the basis of the existing TEMEO prototype. Through the SHCB system, the patient is linked to the medical centre which monitors the patient 24 hours per day while in the same time, the patient is capable in accomplishing all activities of his normal day. The medical centre reacts adequately when a critical event appears. The SHCB system is based on the existing and new developed Internet services. They are piloted, fine tuned and implemented under the FP7 ELLIOT Project. The impact of the Internet of Things (IoT) embedded in it is measured and carefully analysed through a set of knowledge-social-business (KSB) experience models. Their functionality and effectiveness are analysed, exploring the integrity of the social, intellectual, cognitive, economical, legal and ethical aspects of its usage.

Keywords—Telemedicine, Cardio belt, Internet of things (IoT), KSB models

I. INTRODUCTION

The Cardiovascular Diseases (CVDs) are globally number one among those causing death: more people die annually from CVDs than from any other disease. In 2008, an estimated 17.3 million people died from CVDs which represents 30% of all global deaths. Each year 9.4 million deaths or 16.5% of all deaths can be attributed to the high blood pressure. This includes 51% of deaths due to strokes and 45% of deaths due to coronary heart diseases [1].

The problem for prevention against CVDs is a hot topic nowadays. During the day everyone has diversity of activities and emotions which can influence the heart itself and the entire cardiovascular system. Its continuous monitoring could prevent the negative events. If the patient is linked to an observing medical centre, it can react when critical events appear (Fig. 1).

These were the main considerations which gave the impetus for development of the Bulgarian Smart Health Cardio Belt (SHCB). This pilot is developed and improved under the implementation of the FP7 ELLIOT (**E**xperiential **L**iving **L**abs for the **I**nternet **O**f **T**hings) Project.

The technology for monitoring the cardiovascular system is not new. The cardiac "Holter" sensor exists since early

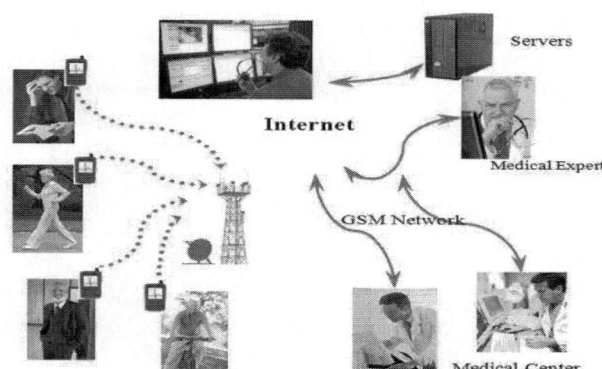

Fig. 1 Usage of Smart Health Cardio-Belt

1960s. But it worked offline, the patient data were collected in it and Medical Doctor had access to them after certain amount of time.

The first Bulgarian prototype of the cardio belt monitored online 24 hours per day was developed three years ago by TEMEO Company. This prototype became the basis for further development of the SHSB system.

The SHCB pilot had been selected after thorough analyses considering the public needs not only in Bulgaria but in Europe wide, the level of the necessary technologies' development and the possible impact of the product for overcoming the current gaps and lacks of the existing TEMEO prototype.

II. BACKGROUND

A. Cardio Belt – New Functionalities Based on IoT

The Smart Health Cardio Belt system is based on the usage of the already existing TEMEO system (www.temeo.org) - an innovative telemedicine system with a potential to save the life of thousands of people with cardio-vascular diseases. The SHCB extended substantially the functions of the system that had been already run. Based on TEMEO, algorithms for ECG (Electrocardiography) and HRV (Heart rate variability) analyses were developed. TEMEO had been expanded also with tools for blood pressure measurement and analyses.

The origin TEMEO was designed only to observe the patient and was not able to record the data received. The new function for archiving the patient's profile unfolds new opportunities for full value observation of the patient and thus, to give more precise diagnosis and more adequate programme for its medical treatment. Next step in the development of this function is to enrich this archive with exhaustive records of the entire state of health of the patient, linked to other accompanying deceases or simply observations linked to other systems of the patient's constitution. This would lead to a new approach in health care where health institutions can deliver Global support through sharing information between doctors from different hospitals for consulting the complex state of health of the patient.

Medically oriented business intelligence is another new function which gives the opportunity to identify events in risk, appearing during the observation of the patient. Based on this early warning mechanism, the necessary measures for treatment can be undertaken in a shorter period of time.

The integrity established between the embedded Android system and the Google maps made possible the Geo observation of the patient. Through it, the position of the patient can be localised, as well as his movement.

When the system defines an event in risk for the observed patient, it informs the medical staff sending visual indications both – on the map showing the localisation of the patient and in the data base where the record with patient's evidence are saved. In the first case the name of the patient blinks in red colour on the map and in the second one the evidence record containing "risky" data is coloured in red.

The achievements on the SHCB improvements raised the partners' interest and lead to the idea for developing the multilingual function of the TEMEO platform. This will open the prospective for a monitoring performed by several hospitals and will give new impetus to the further development of the system and its wider utilisation.

B. ELLIOT Project

ELLIOT aims at developing IoT technologies and Ambient Intelligence (AmI) services designed for users and developed with their essential contribution [2]. The early involvement of users/citizens in the research and innovation phases of IoTs' development is conducted according to the precepts of the Open User-Centred Innovation paradigm and through the co-creation and experimentation mechanisms of the Living Lab (LL) approach. The functionality and effectiveness of the IoT is measured through a set of Knowledge-Social-Business (KSB) experience models implemented in an innovative ELLIOT Experiential Platform operating as an environment for knowledge and experience gathering.

C. The KSB Model

The KSB Experiential Model integrates social, intellectual cognitive, economic, legal and ethical aspects that enables data collection from user behaviour and usage analysis. One of the project tasks is to produce validated methods, techniques, and tools and an experiential platform for user-driven innovation on IoT. These efforts should lead to vastly increase of the IoT adoption and to enhance the potential of collaborative innovation for the discovery of innovative IoT application/service opportunities in bridging the technological distance with users/citizens. The methods, techniques, and tools could be used in other LL settings as well.

III. DESCRIPTION

The Bulgarian pilot is based on the implementation of a Smart Health Cardio-Belt (SHCB) designated to facilitate the medication of people with cardio-vascular diseases. This technology supports the Doctors' efforts to observe patients and undertake the most appropriate treatment. SHCB allows the observation to be carried out at a distance and permanently for the period in which the patient is equipped with SHCB while in the same time, the patient is free to conduct a normal style of life.

A. General Description

The prototype development of the SHCB was split into three phases. The first one was dedicated to the database establishment and connection with a sensor. Through the second phase the visual data were maintained and interpreted. The third phase is devoted to the improvement of the patient's Android device performance. Some individual interviews with the involved patients and doctors were organised. Their suggestions were used for further specifications.

The further steps will be focused on the quality assurance of the SHCB and on the implementation of the business intelligence module (Fig. 2).

The implementation of the Smart Health Cardio-Belt gives a new opportunity for detecting the small aberrations in the behaviour of the patient. This contributes to better observation of the health care principle to prevent from disease, and when the patient is sick, to treat him in the most adequate way. The Smart Health Cardio-Belt was designed and the new health cloud was established in the virtual space. The most significant outcome is that 89% of the observed patients feel themselves safer with the SHCB.

The SHCB system had been developed in the "IoT for Wellbeing" Living Lab (LL). It is based on the Smart Sensors Medical System for Mobile Patients (SSMSMP) platform and brings together all stakeholders (such as end-users, researchers, developers, industrialists, policy makers, etc.) in an open innovation process aiming to create and validate digital products and services in real life settings. The process consists of iterative design and evaluation cycles and involves end-users (both patients and doctors) in

Fig. 2 Functionality of SHCB

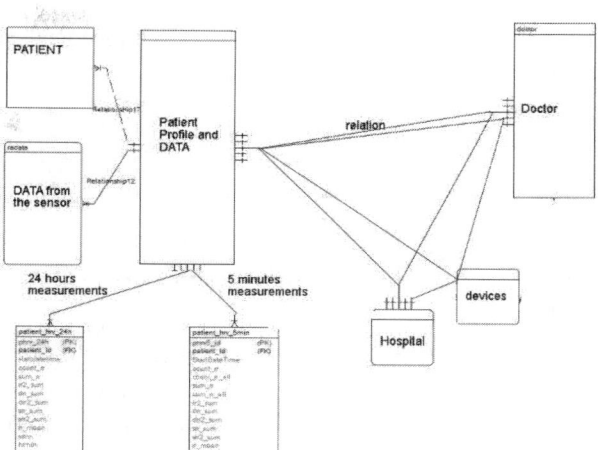

Fig. 3 The Data-base Created for New Human-Things Interaction

the respective phases (i.e. design, evaluation) of the product/service life cycle. A national focus group of end-users (at least 25) are the main assessors. Their responses and reactions are recorded, consequently depict the corresponding values to the associated user acceptance indicators, calculate the final measurement per indicator; and consolidate and analyse the findings. Through this process, project participants receive a significant amount of feedback, in order to have a broader insights into the system's user acceptance, and consequently into its perspectives for success.

B. Indicators Measured

The SHCB sensor transfers the data each 5 minutes:

- Absolute arrhythmia (atrial fibrillation);
- Missed heartbeats;
- Bradycardia;
- Tachycardia;
- Atrial fibrillation,

and the corresponding ECG graphic is drawn and provided to the end-users. These indicators help to be defined the initial diagnostics for the patient.

C. Development

The developed SHCB system is based on the database relational scheme (Fig. 3). The central information entity in the DB is the Patient table. The patient's profile contains:

- Personal health related information;
- Row sensor data like ECG, pulse measurement etc.;
- Processed data which is readable by medical staff (it is generated in intervals of 5 minutes and later aggregated in intervals of 24 hours);

- Devices attached to the patient;
- Physicians observing the patient;
- Hospitals where the patient is situated.

D. Global Positioning

An Android device indicates the patient's global position (Fig. 4). In case of emergency it sequentially dials to 3 phone numbers. For the improvement of the Android device design and functions, the end-users are also involved following the methodology of the Living Lab.

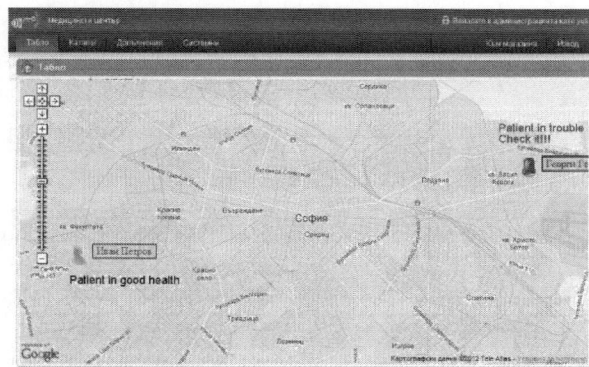

Fig. 4 The patient's position and status

The initial idea was that when a patient is in a trouble, (s)he should press a red button for help. Later on, more useful solution was found – simply to leave the device on the floor. The same effect would be achieved if the patient is not able to react. In case of an accident, a message "Are you OK?" appears on the device's screen. If the patient is able to reply on the message – (s)he is in a relatively stable condition. Otherwise, the device starts dialling to the Medical Centre, sends a message to the Doctor's screen and shows a red lamp on the map in order to catch medical staff attention.

E. SHCB Data

The data received from the belt are structured in daily reports on each 5 minutes and in monthly reports (Fig. 5) where an average value (per day) is shown. The system compares the measured results with standardised frontier values [3] and if they are out of the relevant rate, then the system reacts.

For one epoch (5 minutes -first note starting time; ending time) the system reports patient's data based on the collected data by the sensor. They are very important for identification of patients' conditions. Some of these data are:

- Hrmin-minimal pulse rate/ Hravg-average pulse rate/ Hrmax-maximum pulse rate
- rMSDD - average square between differences in two neighbor RR intervals in [ms];
- APC - number of supraventricular extrasistols per 24 hours; APC/h - number supraventricular extrasistols per 1 hour;

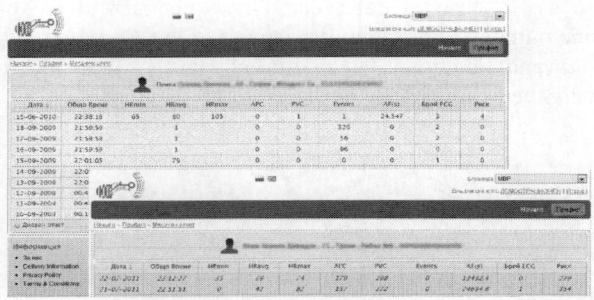

Fig. 5 Monthly report

- PVC - number of ventricular extrasistols per 24 hours; PVC/h - number of ventricular extrasistols per 1 hour;
- Events - how many times the Android device is fallen down;
- AF(s) - the time from one epoch when the patient was in absolute arrhythmia;
- AFHRmax - maximum pulse for the moments with arhythm

IV. FURTHER WORKS AND CONCLUSIONS

The Bulgarian team is strongly motivated to continue with the further improvement of the SHCB system and enrichment of its functionality. The efforts of the next steps will be oriented towards:

- Extension of the capability and the level of intelligence of the devices to gather different types of data;
- Application of Business Intelligence in order to allow the system to support the physician by giving him possible hypothesis for diagnostics;
- Application of Expert systems;
- Open interface to external systems by applying web services.

The adoption of IoT and enhancement of the potential of collaborative innovation are the contemporary way for satisfying essential needs of the society and improving its wellbeing.

REFERENCES

1. World Health Organization. (2013, March). *Cardiovascular diseases (CVDs), Fact sheet N°317, Updated March 2013*. Retrieved May 9, 2013, from Media centre: http://www.who.int/mediacentre/factsheets/fs317/en/
2. FP7 ELLIOT project. (2010). *ELLIOT*. Retrieved 5 9, 2013, from Experiential Living Lab for the Internet of Things: http://www.elliot-project.eu/
3. Carola, R., & Harley, J. P. (1990). Human Anatomy and Physiology. McGraw-Hill.

Author: Eugenia Kovatcheva
Institute: Sofia University "St. Kl. Ohridski"
Street: 5 James Bourchier, Str.
City: Sofia
Country: Bulgaria
Email: epk@fmi.uni-sofia.bg

Advances in Modelling of Epithelial to Mesenchymal Transition

R. Summers[1], T. Abdulla[1], and J.-M. Schleich[2]

[1] School of Electronic, Electrical and Systems Engineering, Loughborough University, Loughborough, UK
[2] LTSI, University of Rennes 1, Rennes, France

Abstract—This paper presents simulations of in vitro Epithelial to Mesenchymal Transition (EMT). The conditions of 2D migration on the surface, and 3D invasion into a collagen gel are represented as a cellular Potts model. The model demonstrates that a loss of endocardial adhesion is a sufficient condition for 2D migration behaviour, while a simultaneous loss of endocardial cohesion and gain in endocardial to collagen gel adhesion is a sufficient condition for 3D invasion. The 3D model captures the hierarchy effective surface tensions that correspond to the three experimental conditions of stable monolayer, 2D migration, and 3D invasion. A 2D cellular Potts model is used to investigate the relationship between cell shape changes, motility and adhesion during the condition of 2D migration.

Keywords—Cellular Potts, Development, EMT, Agent Based, Adhesion.

I. INTRODUCTION

Epithelial to mesenchymal transition (EMT) is the process by which epithelial cells lose their cohesion, migrate into the extracellular matrix (ECM) and adopt a mesenchymal phenotype. It is a fundamental mechanism of general embryonic development, and is also reactivated in adult physiology e.g. during tissue regeneration and cancer progression. Improved understanding of EMT and the conditions under which it takes place has the potential to be leveraged in regenerative medicine and targeted drug delivery. The importance of EMT has led to a wide array of *in vivo* and *in vitro* research efforts to develop a caricature of the protein and genetic interactions involved.

A recent *in vitro* study of endocardial cells cultured on a collagen gel demonstrated that 2D scattering of cells on the surface could be induced independently of 3D invasion into the gel [1]. This was achieved by constitutively activating Notch1 in the ventricular explants, without treatment with TGFβ2 or BMP2. Treatment with TGFβ led to similar 2D scattering and anti-TGFβ2 both counteracted this and maintained the monolayer in Notch1 activated cells.

Treatment with BMP2 induced both 2D and 3D invasiveness of wildtype cells (Fig. 1). This suggests that the actions of both TGFβ and Notch1 in reducing endocardial adhesion are independent of factors that induce 3D invasion (including increased endocardial-matrix adhesion).

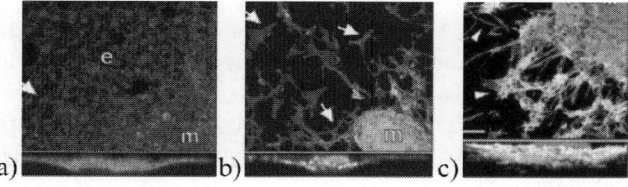

Fig. 1 In vitro endocardial explants. a) Wildtype tissue remains in a monolayer b) Notch activated cells scatter on the surface c) BMP2 treatment causes wildtype cells to both scatter on the surface and invade the gel [1].

In endocardial cells the Notch1 protein is known to induce EMT by up-regulating Snail proteins, which in turn acts to down-regulate VE-Cadherin [1]. Cell doublet experiments have been used to measure the strength of cell adhesion as a function of contact time. In these experiments, two E-cadherin expressing cells are held in contact and the force required to separate them measured after given durations [2]. The adhesion force required to separate two E-cadherin expressing cells is initially of the order of a few nano Newtons (nN). However, this adhesion strength increases rapidly when neighbouring cells remain in contact between 30 seconds and 30 minutes, followed by a slower increase when contact duration is between 30 minutes and an hour, reaching a peak force over 200 nN [2]. The initial E-cadherin mediated contact adhesion does not require connection to the actin cytoskeleton. The stronger junctional adhesion forms over a longer period of contact, by connection to the actin cytoskeleton between the two cells.

II. CELLULAR POTTS MODELING

Cellular Potts models (CPM) simulate cell behaviours as terms within a generalised Hamiltonian energy, H. Cells occupy multiple lattice sites, and thus have size, shape and surfaces that may be adjacent with other cells. During a simulation step, lattice copy attempts at cell surfaces will occur with a probability so as to reduce H. This includes interactions between cells; and between cells and the ECM. CPM can also include constraints on cell volumes, surface areas, chemotaxis and mitosis. In the 3D simulations presented, H is given by the boundary energy between two neighbouring cells per unit area, and the volume constraints:

$$H = H_{Boundary} + H_{Volume}$$

$$H = \sum_{i,j} J\left[\tau\{\sigma(\bar{\imath})\}, \tau\{\sigma(\bar{\jmath})\}\right]\left[1 - \delta\{\sigma(\bar{\imath}), \sigma(\bar{\jmath})\}\right]$$
$$+ \sum_{\sigma} \lambda_{vol}(\sigma)[v(\sigma) - V_t(\sigma)]^2$$

Where for cell σ, λ_{vol} is the volume constraint, V_t is the target volume, and for neighbouring lattice sites $\bar{\imath}$ and $\bar{\jmath}$, J is the boundary coefficient between two cells (σ, σ') of given types $\tau(\sigma), \tau(\sigma')$, and the boundary energy coefficients are symmetric: $J[\tau(\sigma), \tau(\sigma')] = J[\tau(\sigma'), \tau(\sigma)]$, and the Kronecker delta is $\delta_{x,y} = \{1, x = y; 0, x \neq y\}$.

As the 2D simulations additionally investigate the role of cell morphology, they add to the Hamiltonian a surface and elongation constraint:

$$E_{Surface} = \sum_{cells} \lambda_{Surface} (s - S)^2$$
$$E_{Elogation} = \sum_{cells} \lambda_{Length} (l - L)^2$$

Where $\lambda_{Surface}$ and λ_{Length} denote the strength surface and length constraints; s and l denote the current surface and major axis length; and S and L denote the target surface and length.

Additionally, the 2D simulations investigate the different roles played by the loss of labile adhesion and plastic coupling in cell migration. Labile adhesion is represented by the normal contact energy parameters of cellular Potts models, while plastic coupling represents the stronger forces that link cells across their cytoskeletons, which form over a longer period of contact. Plastic coupling is represented by a breakable spring force between cell centres, and this term is added to the Hamiltonian equation governing the simulations:

$$E_{Plastic} = \sum_{i,j - cell\ neighbours} \lambda_{ij} (p_{ij} - P_{ij})^2$$

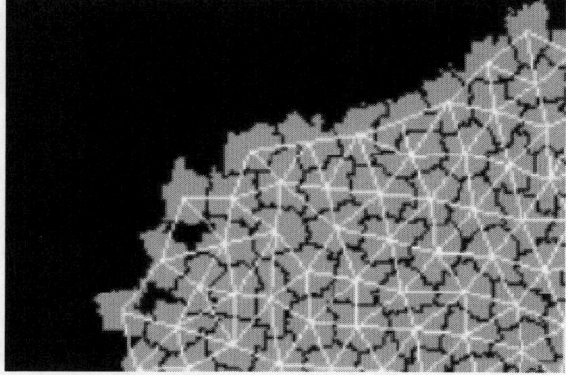

Fig. 2 Plastic coupling in a 2D simulation, represented by breakable spring forces (white lines) that connect cells.

Where p_{ij} is the distance between the centre of masses of cells i and j; P_{ij} is the target distance corresponding to p_{ij}, and λ_{ij} is a constraint representing the strength of the plastic coupling. Additionally, a maximum distance is set, which determines the distance between the centre of mass of neighbouring cells when the link between them breaks, and also the distance at which links can be established between unconnected cells. Plastic coupling is illustrated in Fig. 2.

Cellular Potts models provide a good representation for any mechanism where cell rearrangement is principally determined by differences in adhesion. This is because surface energy is a good (inverse) analogue of the overall adhesive force between cells. CPM has been widely used for modelling developmental mechanisms. It is thus an appropriate formalism for modelling EMT. Compucell3D [3] is the most widely used modelling environment for developing CPM. It is open source and extensible, enabling the sharing of results. Compucell3D was used for all simulations described in this paper.

III. 3D MODELS OF MIGRATION AND INVASION

Compucell3D includes a type 'medium' by default, which is often treated as the ECM. In the 3D models, this is treated as the space above the culture, with no intrinsic surface energy. The 3D simulations also include endocardial cells (ECs) and ECM. An assumption is that EC-EC adhesion is stronger, in the wild type situation, than EC-ECM adhesion, which is stronger than ECM-ECM adhesion. The contact energy with the surrounding space is taken to be higher between EC-medium than ECM-medium, due to the lower deformability of cell membranes compared to ECM. Therefore, to simulate a wild type ventricular explant on collagen gel 'ECM' the following energy hierarchy is assumed:

$$J_{EC,medium} > J_{ECM,medium} > J_{ECM,ECM} > J_{EC,EC} > J_{medium,medium} = 0$$

An initial layout of 100 cells in a circular monolayer was generated and simulated for 1000 Monte Carlo Steps (MCS). By giving these initial volumes of 100 voxels, and target volumes of 400 voxels, a randomised arrangement of regular sized cells was generated. From parameter searching, it was found that the parameters in set 1 (Table 1) ensured this system remained in equilibrium, and cells did not scatter or invade the ECM. This suggests the energy hierarchy assumption above is reasonable. Set 2 corresponds to a loss of endocardial cohesion (increase in $J_{EC,EC}$). Set 3 corresponds to a gain in EC-ECM adhesion (reduction in $J_{EC,ECM}$). Set 4 corresponds to both effects simultaneously.

Table 1 Surface energy parameters J, in $10^{-15}kg^1s^{-2}$
KEY: EC=Endocardial Cell; ECM=Extra Cellular Matrix

Surface Energy J	EC, Medium	ECM, Medium	ECM, ECM	EC, ECM	EC, EC	Medium, Medium
Set 1	16	14	8	4	2	0
Set 2	16	14	8	4	10	0
Set 3	16	14	8	1	2	0
Set 4	16	14	8	1	10	0

The base case scenario (Fig 3a) using parameters from set 1 (Table 1) was perturbed by adopting the parameters in sets 2-4 and running the simulation for a further 1000 MCS in separate experiments.

With set 2, ECs scattered on the surface of the matrix without invading it (Fig. 3b). With set 3, the ECs invaded the ECM, but without delaminating from each other (results not shown). With set 4, all ECs delaminated from each other, and some invaded the matrix (Fig. 3c).

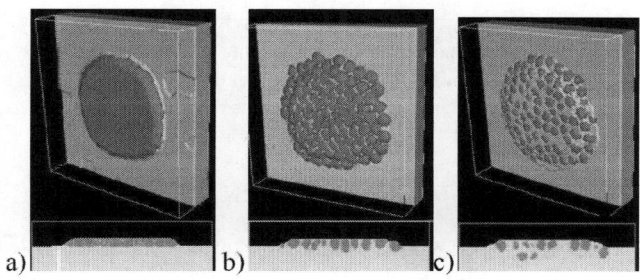

Fig. 3 Simulations of in vitro EMT. a) Endothelial monolayer on the surface of collagen gel. b) With reduced endocardial cohesion, cells scatter on the surface, but do not invade the gel. c) With reduced endocardial cohesion and increased endocardial-matrix adhesion, some cells invade the matrix. Simulations run in Compucell3D.

IV. 2D MODELS OF CELL MORPHOLOGY, MOTILITY AND ADHESION

The time and space scales of the 2D models were fit in the following way. The length scale was set to 1 μm per pixel, and model cells given a width of 15 μm, based on the dimensions of cultured murine endocardial cells [1]. Experimentally, *in vitro* epithelial cells such as MCF-7 cells move at a rate of about 0.28 μm/min, whereas mesenchymal cells move at about 0.4 μm/min [4]. For typical parameter settings, simulated epithelial cells move at about 0.01 pixels/MCS (e.g. 0.01 μm/MCS). Equating the experimental and simulated cell speed implies 0.28 MCS = 0.01 min, or 1 MCS = 0.036 min (about 2 seconds). Each multicellular simulation was run for 5×10^4 MCS, which equates to 27.8 hours. Cellular migration and morphology was found to be quite stable after this time.

There are a few measures that can be used for quantifying the shape of cells. The main criteria for selecting suitable metrics is that they should enable a comparison to be made between the *in vitro* results [1] and the simulation results, that they should describe shape property relevant to the changes which take place during EMT, and preferably that they should be widely used in other experimental work, to allow for direct comparisons. Based on these criteria, the two metrics selected were aspect ratio (length/width) and circularity ($4\pi*$area/perimeter^2). Aspect ratio indicates the elongation of a cell. Circularity gives a number between 0 and 1, where a score of 1 indicates a perfect circle, and smaller values indicating a less rounded shape. These metrics could be calculated during a simulation run, from basic cell attributes, and also calculated from the confocal images of endocardial explants with image processing. This allowed a direct comparison to be made between simulated representations and *in vitro* cells.

The results of the image processing indicated that there is a significant difference in terms of circularity and aspect ratio between wild-type and Notch1 activated ventricular endocardial cells. As shown in Fig. 3, the N1ICD cells have both a significantly less circular morphology, but are not significantly elongated compared to the wild-type endocardial cells.

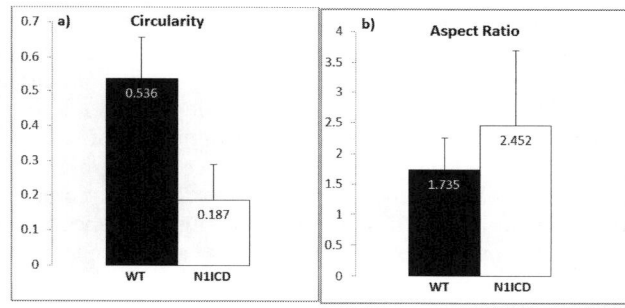

Fig. 4 Shape metric comparison of wild-type and N1ICD murine ventricular endocardial cells. a) Wild-type cells have a mean circularity of 0.536, and N1ICD cells 0.187 (p<0.001). b) Wild-type cells have a mean aspect ratio of 1.735 and N1ICD cells 2.452, however this difference is not significant. Error bars show standard deviation.

A. Larger Surface Area Induces Greater Motility

In simulations, an increase in surface area, relative to volume, can be induced by increasing the target surface parameter, while keeping other parameters constant. This constrains cells into adopting a fibroblastic (spindle-shaped) morphology. However this change in morphology is also accompanied by an increase in motility, as shown in Fig. 5.

This can be explained by the greater number of interactions between cell surfaces and medium leading to a greater number of pixel copies being attempted and accepted. This has the biological equivalent of a cell with a more fibroblastic morphology having a greater surface area over which to interact with and adhere to the matrix.

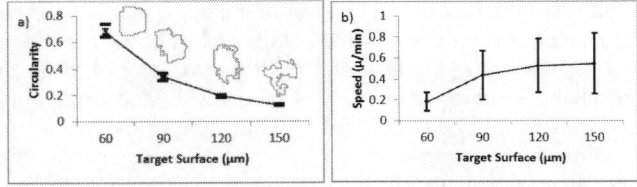

Fig. 5 a) Circularity of simulated cells falls with increasing Target Surface. b) This is accompanied by increased speed (and hence motility). Error bars show standard deviation, and caps show the range, from 10 simulation replicas

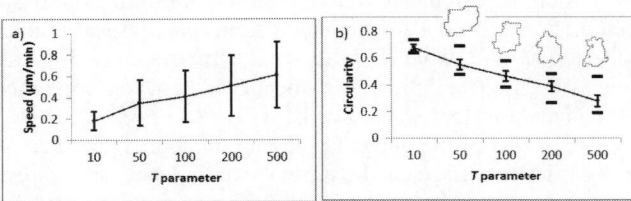

Fig. 6 Speed of simulated cells increases as a result of increasing the T parameter b) This is accompanied by a reduction in circularity. Error bars show standard deviation, and caps show the range, from 10 simulation replicas.

B. Greater Motility Induces Fibroblastic Morphology

Increasing the value of the T parameter increases the speed of movement of the cells. However, this increase in cell motility is accompanied by the adoption of a fibroblastic morphology. Furthermore, the magnitudes of reduced circularity associated with increased speeds are similar to those that result from increasing the target surface (compare Figure 5 with Figure 6). In both cases, migration speeds are consistent with those of *in vitro* epithelial and mesenchymal cells [4].

There are two plausible explanations for this. Either the increase in the T parameter has the side of effect of giving the cells the flexibility to deviate from a rounded morphology; or the easiest (lowest entropy) way for a cell to be more motile is for it to adopt an elongated or fibroblastic morphology. This has the biological equivalent of a cell changing shape in order to permeate the cell matrix.

C. Rounded Cells Migrate Only in the Absence of Plastic Coupling

Based on the results of the single cell simulations, it was postulated that an increase in cellular motility (via the T parameter), coupled with a loss of epithelial adhesion, might be sufficient to induce an EMT process in a simulated multicellular tissue. This was not the case, as the epithelial integrity and morphology were preserved even under large increases in the T parameter value (from 20 to 500). However, when this motility increase was coupled with a moderate increase in cell-matrix adhesion, an intermediate phenotype is observed, whereby a proportion of cells migrate, while maintaining a rounded morphology. This intermediate phenotype only occurred in the cases with a loss of plastic coupling, illustrating that this has a greater effect in maintaining epithelial morphology in this case (Fig. 7).

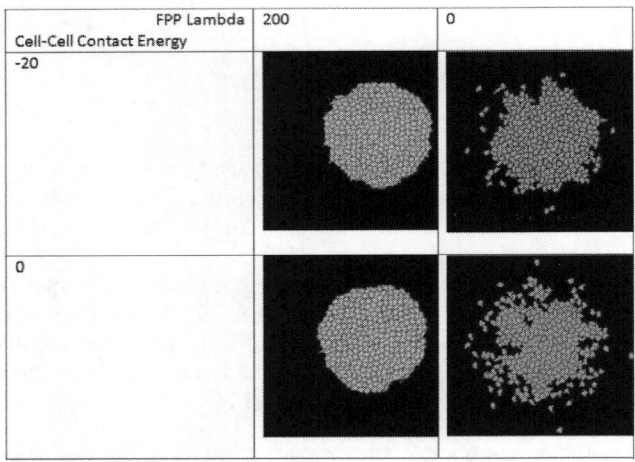

Fig. 7 Under conditions of increased motility (T=500) and moderate cell-matrix adhesion (-20), cells scatter under conditions with a loss of junctional coupling, but not with a loss of labile adhesion alone. In all cases cells maintain a rounded morphology.

V. Conclusion

EMT is a diverse process and is context dependent. EMTs in different contexts are regulated by different signaling pathways and thus different roles may be played by cell morphology, motility and adhesion in each case. Computational modeling provides the flexibility to simulate cells with different combinations of physical parameters. This allows investigation of the conditions under which cells remain epithelial, undergo EMT or exhibit an intermediate phenotype.

References

1. Luna-zurita L, Prados B, Grego-bessa et al. (2010) Integration of a Notch-dependent mesenchymal gene program and Bmp2-driven cell invasiveness regulates murine cardiac valve formation. The Journal of Clinical Investigation 120:3493-3507
2. Chu Y –S, Thomas W A , Eder O et al. (2004) Force measurements in E-cadherin-mediated cell doublets reveal rapid adhesion strengthened by actin cytoskeleton remodeling through Rac and Cdc42. The Journal of Cell Biology, 167:1183-1194.
3. Swat M H, Hester S D, Heiland R W et al. (2009) Multi-Cell Simulations of Development and Disease Using the CompuCell3D Simulation Environment. Methods Mol. Biol 500:361-428.
4. Mendez M G., SKojima S I, Goldman R D (2010) Vimentin induces changes in cell shape, motility, and adhesion during the epithelial to mesenchymal transition, FASEB Journal, 24:1838-1851.

Ron Summers, Loughborough University, UK,
R.Summers@lboro.ac.uk

Glucose-Level Interpolation for Determining Glucose Distribution Delay

Tomas Koutny

Department of Computer Science and Engineering, Faculty of Applied Sciences,
University of West Bohemia, Univerzitni 8, Plzen 30614, Czech Republic
txkoutny@kiv.zcu.cz

Abstract—As a part of research on glucose transporters, I already proposed a hypothesis that the change of the blood glucose level includes information about the estimated rate with which the hypothalamus expects the blood glucose level to return to normal range, by means of regulatory mechanisms of glucose homeostasis. As the interstitial glucose level change reflects the blood glucose level change, I proposed a method to estimate the blood-to-interstitial glucose level delay prior formulating the hypothesis. For the estimation, measured glucose levels can be either approximated or interpolated. Each method has its pros and cons. However, the estimated delay converges into narrower bounds, as more glucose levels are measured, if the measured glucose levels are interpolated. In this paper, I present further details on the interpolation method which were not presented in previously published papers.

Keywords—approximation, interpolation, glucose level.

I. INTRODUCTION

Glucose homeostasis is maintenance of normal blood glucose level that is accomplished by a network of hormones, neural signals, and substrate effects that regulate endogenous glucose production and glucose utilization by tissues other than the brain [1]. The pancreas, liver and hypothalamus have a regulatory function in the glucose homeostasis. As the blood glucose level changes, these compartments react and the level changes again. A portion of blood glucose passes through the capillary wall into the interstitial fluid. Therefore, we can observe that the level of interstitial-fluid glucose level changes with a delay, and in dependency on blood glucose level. The blood capillaries are present in subcutaneous tissue, skeletal muscle tissue and visceral fat. The glucose level for these compartments is measured in their interstitial fluid [2]. The interstitial fluid is found in the intercellular spaces between tissue cells [3]. It supplies the cells with nutrients, i.e. with glucose among others.

The pancreas produces two important hormones – insulin and glucagon. Pancreas production is driven by sensing the glucose level in the blood. Additionally, the hypothalamus exerts a degree of control over the pancreas production [4]. The hypothalamus is a neural network, whose output affects the pancreas production. The input is glucose-sensing in the blood that crossed the blood-brain barrier. With a neural network, the output may change even without a corresponding change in the input. Determining the blood-to-interstitial glucose level delay could be a complex, daunting task.

In a previously published paper [5], I proposed a method to estimate the delay. Using it, I proposed a hypothesis that hypothalamus set-points could possibly encode parameters of regulatory processes of glucose homeostasis [5].

To estimate the delay, the measured glucose levels of blood and interstitial fluid must be either approximated or interpolated. As more glucose levels are measured, accuracy of the estimation improves. Therefore, convergence is an important factor. Consecutively estimated delays should converge into a range that bounds the real reaction delay. As the blood glucose level may change rapidly, when compared to changes of the interstitial glucose level, the approximation-based analysis converges faster. Therefore, it does not need as many measured values as the interpolation-based analysis. On the other hand, the interpolation-based analysis seems to converge into narrower bounds. However, if the period between two measuring would grow too long, the approximation would not be applicable anymore and the interpolation stays as the only choice. For example, it could be a diabetic patient that self-measures blood glucose level.

In another previously published paper [6], I presented the approximation method. In this paper, I present further details on the interpolation method which were not presented in previously published papers. These details describe efficient finding of the interpolation control points, solving the sub-sampling problem, comparison with cubic and spline interpolations and evaluation with glucose levels measured in the blood, cerebrospinal fluid (glucose supply of the brain), subcutaneous tissue, skeletal muscle tissue and abdominal subcutaneous tissue – i.e. the visceral fat.

II. INTERPOLATION

Let us briefly recapitulate the approximation method [5] first. The method approximates the time course of a measured quantity from a given time series $\{y([t(i)])\}$, where y denotes the measured glucose levels, t denotes the time of measuring and i=0..n-1. n is the number of glucose levels.

Let us use exponential functions in the $y=Ce^{kx}$ form. In a generic [t(i+1), t(i+2)] interval, the approximation is obtained by averaging the two following exponential functions: the first functions fits {y[t(i), t(i+2)]} and the second function fits {y[t(i+1), t(i+3)]}. In [t(0), t(1)] interval, the approximation is obtained with an exponential function that fits {y[t(0), t(1)]}. In [t(n-2), t(n-1)] interval, the approximation is obtained with an exponential function that fits {y[t(n-2), t(n-1)]}. Then, the method approximates entire interval [t(0), t(n-1)].

To smooth the approximation, the method adds a midpoint between each two adjacent times. Then, it generates a new time series where the glucose levels are obtained with the approximation of the original time series. This is called an approximation pass. The degree of smoothness is given by the number of approximation passes, which are applied recursively. As each new time series replaces the original glucose level with an approximated glucose level, this reduces the measurement noise automatically.

The measured glucose levels are the control points for the approximation curve. Therefore, to interpolate the measured glucose levels, the interpolation method has to find such control points for which the approximation curve will fit the measured glucose levels.

Let the given time series {y[t(i)]} be the test vector. Then, let us construct a control vector as a copy of the test vector. In subsequent iterations, the glucose levels of the control vector are adjusted until the approximation curve fits the test vector with a desired accuracy. In the iteration, the approximation curve is calculated for the current control vector. Then, a glucose level difference for each element of the control vector is added to that element. The glucose level difference is the glucose level given by the respective test-vector's element less the approximated glucose level. Once the control vector is adjusted, the calculation proceeds with next iteration, or it stops on a custom condition. A pseudo-code follows.

```
Construct the test and control vectors;
do {
  Calculate the approximation curve
  for the current control vector;
  for (i=1; i<control_vector.size()-1; i++) {
    difference = test_vector[i].glucose_level
                 – approximated glucose level
                   at test_vector[i].time;
    control_vector[i].glucose_level += difference;

  }
} while (the stop condition is not met);
```

The test vector does not change with iterations. Also, the first and last glucose levels of the control vector do not change. Their difference is always zero because of the approximation method. The other glucose levels of the control vector change, as they control the shape of the curve.

The interpolation method presented offers flexibility when designing the custom condition. For example, let us consider the following conditions:

• The number of iterations is fixed. Table 1 relates interpolation error to the number of iterations.
• The number of iterations is given by a global epsilon. The global epsilon is calculated as maximum absolute difference between all measured glucose levels and their respective approximated glucose levels.
• The interpolation stops, if average absolute difference between measured and respective approximated glucose levels falls under a given threshold, for a desired number of measured glucose levels.

Furthermore, the method has two important properties:

1. The measured glucose levels do not need to be equidistant on the time axis. This is important for such a case, when a particular measuring fails e.g. due to a technical difficulties.
2. Per a particular glucose level, except the first and the last measured glucose levels, it is possible to set whether the level should be ignored, interpolated or approximated, while calculating a single curve. Furthermore, it is possible to set a custom accuracy of the interpolation for each measured glucose level separately. This is important, if the measuring reads such a level that differs considerably from the preceding and succeeding measured levels.

III. SUB-SAMPLING

Thanks to the similarity of sugar and insulin physiology between humans and rats, experimenters may carry out the experiments on rats. The results presented were obtained using hereditary hypertriglyceridemic rats.

When the laboratory rat is given an intravenous bolus of glucose, a steep increase of the blood glucose level follows immediately. Before most of the glucose bolus undergoes the glycogen synthesis, the blood glucose level can be more than twice greater. A possible effect of the intravenously-given glucose bolus is a sub-sampled glucose-level change, if the sampling frequency is not increased. The same sub-sampling issue may arise in such a case, when the rat is given a bolus of short-action insulin (and the glucose-infusion is stopped, if there is any).

In the experiments performed, the sampling limit was five minutes – it was the shortest measuring interval with the continuous glucose monitoring system used. In some experiments, possibly sub-sampled glucose levels were observed with the intravenous bolus of glucose.

The approximation is a sequence of non-weighted averages of the overlying exponential functions. If we consider interpolating each measured glucose level, the interpolation curve could diverge because of the sub-sampled glucose level changes. With the increasing number of iterations, the global epsilon would not be monotonically decreasing in such a case.

To solve the sub-sampling issue, it is necessary to identify the possibly sub-sampled glucose level changes. In a generic $[t(i), t(i+1)]$ interval, let us construct $y=kx+q$ line that fits $\{y[t(i), t(i+1)]\}$. The glucose-level change is possibly sub-sampled between such two measured glucose levels where $k<-489.6$ or $k>489.0$. The threshold values of k were determined experimentally, and they are subject to the time scale where 24 hours = 1.0.

To prevent the interpolation curve from diverging, let us construct an artificial glucose level in each interval with a possibly sub-sampled glucose level change. The artificial glucose level's time is obtained by averaging $t(i)$ and $t(i+1)$. Using $y=Ce^{kx}$, the artificial glucose level is obtained with an exponential function that fits the $\{y[t(i), t(i+1)]\}$.

IV. NUMERICAL SPEEDUP

Let us consider an interpolation with at least 10 approximation passes and more than 100 iterations. To reduce the computational complexity, it is possible to trade off the first iterations for a quick estimation of the final control vector.

The first 20 iterations can use 3 approximation passes only. After the first 20 iterations, and while more than 75 iterations remain, 5 approximation passes are sufficient to calculate the interpolation curve. Finally, the last iterations are calculated with the desired number of approximation passes. As each subsequent approximation pass adds n-1 glucose levels to the current n glucose levels, this effectively reduces the number of floating-point operations a processor has to do. The threshold values of iterations' counts were determined experimentally.

Finally, the interpolation curve should be re-sampled, before it is used in a subsequent calculation. To resample the curve, the curve is stepped with the desired numerical stepping, while constructing a new approximation curve. As the newly constructed curve is populated from the interpolation curve, it will fit the measured glucose levels as well.

V. EXPERIMENT SETUP

The experimenter installed cannulas into the vena jugularis and arteria carotis. In the subcutis, there was a sensor of a continuous glucose monitoring system (CGMS). Rats were given insulin and glucose infusions into the vena jugularis. As an anesthetic, the experimenter administered a combination of urethane, xylazin and ketamine.

The insulin infusion rate was 50mUI/kg/min. Such a rate is considered to suppress endogenous insulin secretion. As a result, the glucose utilization rate is considered constant as well. Reference [7] gives a related study on the suppression.

After 15 min of a steady state, since the CGMS was calibrated, the experimenter administered a bolus of glucose, 0.5 g/kg. Then, the experimenter tried to achieve a new steady state, with a blood glucose level at 12 mmol/l, by adjusting the glucose infusion rate as needed. After 60 min, the experimenter administered a bolus of short-action insulin, 0.5 UI/kg, and stopped the glucose infusion. Then, the experimenter continued to monitor the glucose levels for another 80 min. Insulin infusion was kept at a constant rate.

In total, 57 measurements were done in four different compartments. The blood glucose level was measured in the arterial blood. CGMS was used for other compartments: cerebrospinal fluid, subcutaneous tissue, skeletal muscle tissue and abdominal subcutaneous tissue – i.e. the visceral fat. The interval between readings was 5 min. The measuring tolerance for the blood glucose level was ±0.2 mmol/l and ±0.4 mmol/l for the interstitial and cerebrospinal fluids. For the skeletal muscle tissue and the visceral fat, the measuring tolerance was 15%.

VI. RESULTS

The method is being used with work such as reference [5]. The performance is about the same as with three particular rats, which I chose to demonstrate particular results. Rats 0904, 0906 and 0405 demonstrate the interpolation error testing. For Table 1, rat 0906 provided the blood and the subcutaneous tissue glucose levels. Rat 0904 provided the cerebrospinal fluid glucose level. Rat 0405 provided the skeletal muscle tissue and the visceral fat glucose levels. For Table 2, rat 0405 demonstrates the glucose level reconstruction. Rat 0904 was female and weighed 243 g. Rat 0906 was male and weighed 410 g. Rat 0405 was male and weighed 420 g.

Table 1 Interpolation Error per Compartment and Number of Iterations

Number of Iterations	0	10	100	1000	10000
Blood	0.622	0.129	0.029	0.013	0.000
[mmol/l]	5.784	0.386	0.072	0.022	0.001
Subcutaneous Tissue	0.094	0.031	0.019	0.004	0.000
[mmol/l]	0.400	0.143	0.051	0.008	0.000
Cerebrospinal Fluid	0.100	0.049	0.040	0.010	0.000
[mmol/l]	0.462	0.262	0.103	0.018	0.001
Skeletal Muscle	0.062	0.012	0.002	0.000	0.000
[mmol/l]	0.371	0.036	0.005	0.000	0.000
Visceral Fat	0.080	0.014	0.010	0.009	0.001
[mmol/l]	0.708	0.111	0.039	0.018	0.002

Table 1 gives the results of interpolation error testing. The average and the maximum absolute differences are given in [mmol/l]. Average difference is given in the upper line; maximum difference is given below. I ran the method with 10 passes and variable number of iterations over the measured glucose levels.

Interpolation constructs new data points within the range of known data points. In this case, the interpolation reconstructs glucose level from a set of measured glucose levels. To test the glucose level reconstruction, I incrementally constructed several subsets of the measured glucose levels by removing subsequent levels with an increasing stepping. For the first subset, I removed every second measured glucose level. For the second subset, I removed every second and third measured glucose levels from the original set of measured levels. For the third subset, I removed every second, third and fourth measured glucose levels from the original set. In total, I constructed four subsets, thus omitting up to 80% of the measured glucose levels. The glucose levels of the original set were measured with five-minute interval between each two measured glucose levels.

Having the subsets, I interpolated each of them. Then, I calculated average absolute difference using each of the original measured glucose levels. I did such a glucose-level reconstruction test with the blood and the subcutaneous-tissue interstitial-fluid glucose levels. Furthermore, I compared the interpolation method presented with cubic and spline interpolation methods [8].

Table 2 gives the comparison on rat 0405. In the Method column, 'cubic' is the cubic interpolation from the four nearest neighbors, 'spline' is the cubic spline interpolation-smooth first and second derivatives throughout the curve, 'exponential' is the interpolation method presented.

VII. CONCLUSION

I presented a numerical method for interpolation of measured glucose levels. The method was tested on measured glucose levels of blood and cerebrospinal fluids, interstitial-fluid glucose levels measured in subcutaneous tissue, skeletal muscle tissue and the visceral fat.

As Table 1 shows, the precision of the interpolation method presented improves with the increasing number of iterations. This demonstrates the numerical stability of the method, even with a sub-sampled glucose-level change. In accordance with the physiology, we can see that the interstitial glucose level interpolation converges faster than the blood glucose level interpolation.

The reconstructed glucose level does not differ to an important degree, when compared with the cubic and spline interpolation methods. However, the method presented addresses specific issues of the experiment setup described.

Table 2 Average Absolute Differences for the Blood and Subcutaneous Tissue Interstitial Fluid Glucose Level Reconstruction

Interleaving	Method	Blood [mmol/l]	Subcutaneous Tissue [mmol/l]
1	Cubic	0.365	0.035
	Spline	0.403	0.037
	Exponential	0.357	0.035
2	Cubic	0.657	0.063
	Spline	0.708	0.062
	Exponential	0.611	0.088
3	Cubic	0.842	0.161
	Spline	0.916	0.165
	Exponential	0.769	0.168
4	Cubic	1.180	0.170
	Spline	1.393	0.197
	Exponential	1.173	0.177

The calculation times do not grow linearly with the number of iterations due to the numerical speedup. It effectively reduces the number of required floating-point operations.

The method presented was used successfully in proposing a hypothesis about hypothalamus set-points which are related to sustaining the glucose homeostasis [5].

ACKNOWLEDGMENT

Experimental data were provided by Diabetology Center, University Hospital in Pilsen, Charles University in Prague.

REFERENCES

1. Longo D, Fauci A, Kasper D, Hauser S, Jameson J, Loscalzo J (2011) Harrison's principles of internal medicine, 18th ed. Mc Graw-Hill, New York
2. Cengiz E, Tamborlane WV (2009) A tale of two compartments: interstitial versus blood glucose monitoring. Diabetes Technol Ther 11:11-17
3. Guyton AC, Hall JE (2006) Medical textbook of physiology. Elsevier Inc., Philadelphia
4. Grassiolli S, Bonfleur ML, Scomparin DX, de Freitas Mathias PC (2006) Pancreatic islets from hypothalamic obese rats maintain K+ATP channel-dependent but not -independent pathways on glucose-induced insulin release process. Endocrine 30:191-197
5. Koutny T (2011) Estimating reaction delay for glucose level prediction. Med Hypotheses. 77:1034-1037.
6. Koutny T (2010) Modeling of compartment reaction delay and glucose travel time through interstitial fluid in reaction to a change of glucose concentration, IFMBE Proc. vol. 10, IEEE International Conference on Information Technology and Applications in Biomedicine, Corfu, Greece, 2010
7. Pretty CG, Docherty PD, Lin J, Pfeifer L, Jamaludin U, Shaw UM, Le Compte AJ, Chasel JG (2011) Endogenous insulin secretion and suppression during and after sepsis in critically ill patients: implications for tight glycemic control protocols. Crit Care 15:389
8. Burden RL, Faires JF (2010), Numerical analysis, 9th ed., Brooks/Cole, Boston

A Supervised SOM Approach to Stratify Cardiovascular Risk in Dialysis Patients

J. Ion Titapiccolo[1], M. Ferrario[1], S. Cerutti[1], C. Barbieri[2], F. Mari[2], E. Gatti[2,3], and M.G. Signorini[1]

[1] Department of Electronics, Informatics and Bioengineering, Politecnico di Milano, Milano, Italy
[2] Fresenius Medical Care, Bad Homburg, Germany
[3] Center for Biomedical Technology at the Danube, University of Krems, Austria

Abstract—Chronic renal failure (CRF) patients experience a 30% higher risk of cardiovascular (CV) death compared to the general population. In this study data of 3581 incident hemodialysis (HD) patients were considered, i.e. patients who started for the first time the HD treatment. In this work supervised SOM were used with an innovative strategy to built a predictive model to estimate the probability that incident CRF patients experience a CV event in the second semester after a semester of HD treatment. A feature selection approach based on the minimum redundancy maximum relevance (mRMR) algorithm, was wrapped on the self-organizing maps (SOMs) model and a subset of 17 physiological variables with higher performance capability than the complete set of 39 variables was identified. AUC of the ROC curve of the shrunk model was 67±4%. The obtained model permits to investigate non-linear relationships among features related to an increased CV risk condition.

Keywords—Hemodialysis, cardiovascular events, self-organizing maps, feature selection, wrapper.

I. INTRODUCTION

Chronic renal failure (CRF) has been defined as a "vasculopathic state" [1] because of the high-risk condition of the cardiovascular system of these chronic patients. This is the reason why the understanding and proper management of the crucial reasons of cardiovascular disease have recently become a clinical target in nephrology care. Cardiovascular risk in end stage renal disease (ESRD) patients receiving hemodialysis (HD) treatment three times per week is enhanced by the growing proportion of elderly individuals, by the high incidence of co-morbidities (diabetes, hypertension, congestive heart failure, multiple organ failure...), by the metabolic derangement caused by renal failure and because they are affected by an hemodynamic instability condition [2]. This is mainly due to the hemodynamic overload due to plasma volume expansion, the presence of the arterio-venous fistula, anaemia, hyperparathyroidism, electrolyte imbalance and increased oxidant factors [2].

Despite great recent advances in nephrology care, cardiovascular events (CVE) remain the primary cause of death in HD patients, with an incidence 30% higher than in the general population [3]. In 2012 it has been reported that more than 42% of all deaths in incident dialysis patients is due to cardiac causes [4]. During dialysis therapy administration a large amount of treatment and patient data can be collected. The exploration of huge quantity of data employing machine learning tools can open the possibility to discover hidden patterns and relations in the data useful to suggest clinical interventions and provide a support for clinical decisions [5]. Self-organizing map (SOM) [6] is a well-known method for data clustering, i.e. to search for the presence of similarities and particular patterns in the data. For clustering problems, SOM is the most commonly used network, because many visualization tools can be used to analyze the resulting clusters in order to get useful insights in large data sets [7]. In this work a large HD database was analyzed by means of a feature selection wrapper approach and by a clustering method such as supervised self-organizing maps (SOMs) in order to identify physiological patterns associated to high risk of cardiovascular events in ESRD patients in their first year of HD treatment.

II. MATERIALS AND METHODS

A. Dataset Structure

Data used in this study was collected in Fresenius Medical Care clinics through the Clinical Management System EuCliD, a system used to handle hemodialysis treatments and related information in 626 clinics belonging to 25 countries all over the world since 1999 [8]. In our study the data comes from patients in clinics located in Portugal and Spain and were collected from 2006 to 2010.

Cardiovascular condition of the patients at the beginning of renal replacement therapy (RRT) strongly influences patients' outcome and it needs to be taken into account in the estimate of hemodialysis cardiovascular risk [9]. Moreover, a correct estimate of cardiovascular risk in incident HD patients can lead to early clinical interventions in order to prevent sudden deaths or sudden derangement in cardiovascular system during the next months or years. The first six months of HD treatment were considered and the following variables

Table 1 Variable description and relevant information

VARIABLE DESCRIPTION	Missing data %	mRMR ranking
Mean Systolic pressure (pre HD) (mmHg)	0,0%	37
Mean Systolic pressure (post HD) (mmHg)	0,0%	2
Mean Delta Systolic(HD post-HD pre) (mmHg)	0,0%	38
Mean Diastolic pressure (pre HD) (mmHg)	0,0%	33
Mean Diastolic pressure (post HD) (mmHg)	0,0%	36
Mean Delta Diastolic(HD post-HD pre) (mmHg)	0,0%	28
Mean Pulse pressure (pre HD) (mmHg)	0,0%	34
Mean Pulse pressure (post HD) (mmHg)	0,0%	39
Mean Delta pulse pressure (post - pre) (mmHg)	0,0%	32
Mean heart rate (pre HD) (bpm)	0,0%	30
Mean heart rate (post HD) (bpm)	0,0%	35
Mean Delta heart rate (HD post - HD pre) (bmp)	0,0%	25
Mean Delta weight (HD post - HD pre) (Kg)	0,3%	3
Weight percentage loss in six months (%)	4,8%	4
Modality (0=HDF, 1=HD)	0,0%	31
Mean sodium dialysate concentration (mEq/l)	0,7%	5
Mean bicarbonate dialysate concentration (mEq/l)	0,0%	29
Mean total fluid lost per HD session (ml)	0,0%	26
Dialysate temperature (°C)	0,0%	23
Mean dialyzer blood flow (ml/min) (from 50th day)	1,3%	8
Mean Urea Reduction Rate - blood test (%)	6,6%	9
Mean potassium - blood test (mEq/l)	6,8%	10
Mean sodium - blood test (mEq/l)	7,3%	11
Mean calcium phosphate - blood test (mg/dl)	5,5%	12
Mean phosphate - blood test (mg/dl)	4,3%	13
Mean PTH value - blood test (ng/l)	11,5%	14
Mean calcium - blood test (mg/dl)	4,6%	16
Mean haematocrit - blood test (%)	6,5%	17
Mean haemoglobin - blood test (g/dl)	4,7%	18
Mean totalprotein content - blood test (g/dl)	26,3%	19
Mean albumin content - blood test (g/dl)	16,5%	20
Mean albumin percentage-blood test (%)	33,9%	21
Mean Creatinine (Pre HD) - blood test (mg/dl)	12,0%	22
Mean C Reactive Protein - blood test (mg/dl)	25,8%	24
Age (years)	0,0%	7
Diabetes	0,0%	15
Heart Disease	0,0%	27
Angina	0,0%	1
Peripheral Vascular Disease	0,0%	6

were extracted from EuCliD database: time series of the physiological variables measured at each HD session and of monthly blood test values; dialysis settings; dialysis modality (hemodialysis or hemodiafiltration) and anamnestic data of the patient. The series of values were averaged and the mean value enter the set of variables considered for this study. The complete list of extracted features is reported in Tab. 1. Patients were divided in two groups accordingly to the cardiovascular (CV) outcome in the six months following the first semester of HD. CVEs encompass CV hospitalization, insurgence of a new CV comorbidity or CV death, which occurred in months 7-12 of treatment. Patients, who underwent kidney transplantation or other events of different origin than CV, were excluded from the analyses. Patients did not experience CVEs were gathered in the control group. Overall 3581 incident hemodialysis patients were considered. 375 patients out of the 3581, i.e. 10.5% of enrolled patients experienced a CVE in the next 6 months.

B. Preprocessing

Data were normalized to have unitary variance and zero mean and randomly divided in training (80% of data) and testing sets. As the two groups of patients are unbalanced, the CV group of patients selected for the training set was randomly oversampled with replacement to get two equal size classes and thus an equal space representation in the SOM. Some feature values may be missing, but SOM algorithm computes distance between data using just the complete fields [10]. Percentage of missing values in each variable is reported in Tab. 1.

C. Self Organizing Maps

A SOM is a mathematical model based on artificial neural networks able to project data from a high dimensional input space onto a two-dimensional plane of neurons [6]. SOM is a topology-preserving map: two data points (patients in this case) with similar features will be closed in the resulting map [11]. SOM implementation was carried out in Matlab® environment using SOM toolbox.

A supervised SOM approach was chosen, in order to understand the relationships among variables in patients affected by CV disorders and derangement, in the second six months of HD treatment. Class information, i.e. the information about the belonging class of each patient (CV or control group), was concatenated to the input vector in order to get two distinct areas in the resulting map: an area encompassing the similar characteristics in patients of the CV group and an area classifying with a higher probability patients belonging to the control group.

Once the map was built on training set instances, it was used to estimate the performance on the testing set patients. First of all a label was assigned to each neuron of the map: if that neuron resulted to be "best matching unit" for mostly CV patients, then a CV label was assigned to that map unit. Next a k-means algorithm was used to identify the best number of clusters of similar neurons in the map. Finally a label was assigned to each cluster of neurons (label 1: CV; label 0: control) on the basis of the predominant label of the neurons in that cluster. In the testing phase, each patient of testing set was presented to the network and the distance between that instance and each centroid of the identified clusters was computed. A weighted sum of the distances was used to compute the probability for the ith patient to belong to a CV cluster according to the following formula:

$$p_i = \frac{\sum_{j_{CV}} \lambda_{CV} \frac{1}{d_{ij_{CV}}} + \sum_{j_0} \lambda_0 \frac{1}{d_{ij_0}}}{\sum_{j_{CV}} \frac{1}{d_{ij_{CV}}} + \sum_{j_0} \frac{1}{d_{ij_0}}} \quad (1)$$

where λ_{CV} and λ_0 are the labels of CV group and control group (1 and 0 respectively), j_{CV} is the index of CV clusters, j_0 is the index of control clusters and d_{ij} is the Euclidian distance between the ith patient and the jth cluster. Using the obtained pseudo-probability values and the actual labels, a ROC curve is built and AUC is computed to assess the performance of each obtained map as a classifier.

A Supervised SOM Approach to Stratify Cardiovascular Risk in Dialysis Patients

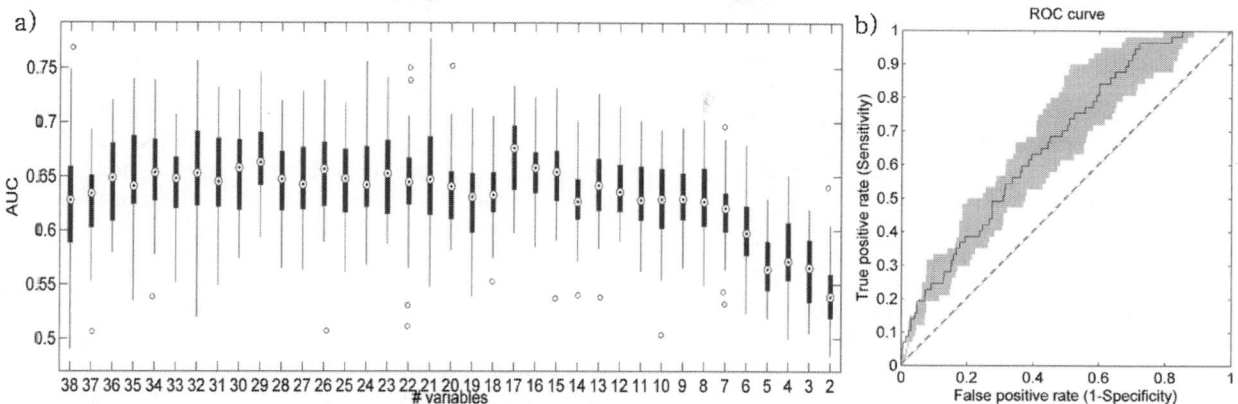

Fig. 1 a) Boxplots of the AUC values obtained at each step of mRMR wrapper procedure: the least promising variable is removed and the number of variables left in the dataset to build the supervised SOM model is reduced at every run up to leave 2 variables only. Notice that the optimal model (maximum median value of AUC curve) used 17 variables only. b) ROC curve of the optimal model (17 variables), confidence interval of the curve is shown (AUC = 67±4%). Dotted gray curve refers to a random classifier.

D. Feature Selection Approach

SOM permits an easy-readability and an easy-interpretability of the result. Nevertheless the high number of variables can reduce the interpretability of the map. A feature selection approach can solve the problem [12]. In particular, a backwards wrapper approach based on the minimum redundancy maximum relevance (mRMR) score of variables was adopted. The mRMR algorithm computes the mutual information among variables and tries to maximize the ratio of relevance and redundancy. It provides also a ranking of variables [13]. The meaning of the obtained ranking, reported in Tab. 1, is that the best combination of just N variables to explain the target class, is represented by the first N variables in the ranking. In the backward wrapper, variables are eliminated one by one and a SOM classifier was built and tested

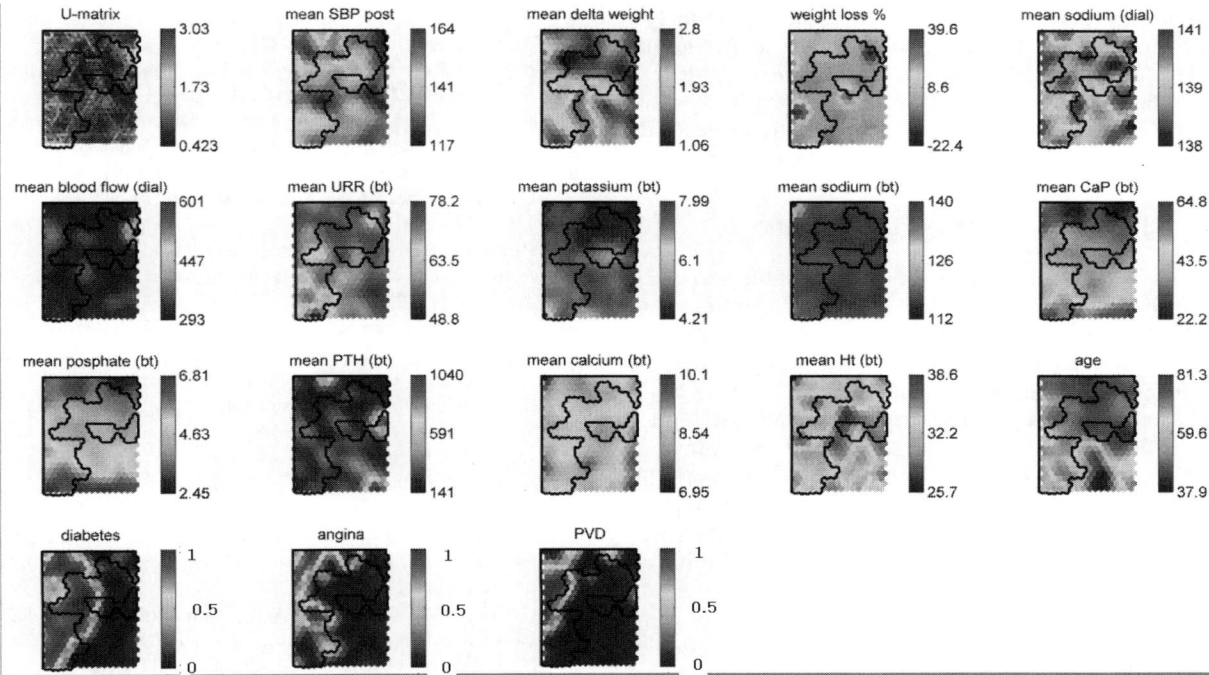

Fig. 2 U-matrix and components planes of the 17 variables supervised SOM model. For categorical variables 0 means absence and 1 means presence of the disease. Black surrounded regions collect patients belonging to CV group. Note: bt=blood test; dial=dialysate concentration.

for the subset of remaining features. The order of variable removal is the mRMR ranking: the least important one (ranking score 39) is removed first and so on. Each wrapper step was repeated 40 times randomly selecting training and testing sets to strengthen the statistical significance of the result. To identify the best nested subset of features, the median AUC maximization criteria was used [14].

III. RESULTS

The supervised SOM model built on the entire set of variables showed an AUC equal to 62.8±5.4% with sensitivity, specificity and accuracy at best cut-off point of 81.8%, 51.3% and 54.6%, respectively.

The maximum of the median value of AUC_{ROC} corresponds to a subset of 17 variables and according to this criterion it can be considered the optimal features set. This model showed best performance than the model built on the entire set of 39 variables as AUC_{ROC} is 67±4%. Fig. 1a shows the boxplots of AUC_{ROC} values at each wrapper step. The selected features have ranking 1 to 17 in Tab 1.

U-matrix and component planes of one of the SOM models built just on the 17 selected variables are reported in Fig. 2. U-matrix shows the distance values among neurons of the map and components planes show the values of the features in the different points of the map. On each plane, regions of neurons collecting CV patients are outlined using a black contour. The number of identified clusters on the net used for probability estimation was 18.7±1.5 and approximately half of them voted for CV group and half for control group. The ROC curve of the optimal model (17 variables) is shown in Fig. 1b. The AUC of this model was 67.2% (s.e. 4.1%) and accuracy, sensitivity and specificity at best cut-off point were 62.1%, 69.1% and 58.5%, respectively.

IV. DISCUSSION AND CONCLUSIONS

A predictive model built on selected features showed a better performance than the model built on the entire set of variables. Thus mRMR approach succeeded in selecting the information from the collected data for a classifying method based on SOMs. The performance of the optimal SOM model is comparable to the performance of a logistic regression classifier built on a similar set of data in a previous work [15]. Nevertheless the SOM model gives the opportunity to further and efficiently investigate the relationships between the variables involved in the prediction. Moreover using the supervised approach it is possible to identify on the final map the region of neurons voting for CV group. In this way features patterns of CV patients can be easily identified. Patients belonging to the CV group showed higher values of SBP measured after HD treatment, a prevalence of the three considered comorbidities, lower values of Urea Reduction Ratio (URR) and are older than patients in the control group. Moreover the regions delimited by a black contour in Fig.2, i.e. the region associated to CVE, include patients with a high loss of weight in the first six months and patients with low values calcium, phosphate and hematocrit, as expected. A low value of urea reduction rate (URR) outlines a higher cardiovascular risk: a not sufficiently efficient treatment in the first six months of HD places patients at a higher risk of CV system derangement.

Concluding, a new classification method based on SOM approach has been presented. An mRMR feature selection approach was wrapped on the classification method and a more effective model for the prediction of CV events was built. The investigation of the final model allowed the identification of high risky patterns in the physiological variables, able to stratify the patients and to help the development of new clinical strategy in order to prevent the insurgence of CVE in HD patients.

REFERENCES

1. Luke RG, "Chronic renal failure - a vasculopathic state", New Engl J Med, vol. 339, Sep. 1998, pp. 841–843
2. Locatelli F, Marcelli D, Conte F et al (2000) Cardiovascular disease in chronic renal failure: the challenge continues. Nephrol Dial Transplant 15(5): 69-80
3. U.S. Renal Data System, USRDS 2012 Annual Data Report: Atlas of Chronic Kidney Disease and End-Stage
4. Brunner FP, Selwood NH (1992) On behalf of the EDTA registry committee. Profile of patients on RRT in Europe and death rates due to major causes of death groups. Kidney Int 42(38): S4-S15
5. Savage N. (2012) Better medicine through machine learning. Communications of the ACM 55(1): 17-19
6. Kohonen T (1995) Self organizing maps. Springer series in information sciences, 30, Springer Berlin Heidelberg
7. Berkhin P (2006) Survey of clustering data mining techniques. Springer Berlin Heidelberg, Grouping multidimensional data: 25-71
8. Marcelli D, Kirchgessner J, Amato C et al (2001) EuCliD (European Clinical Database): a database comparing different realities. J Nephrol 14(4): S94–100
9. Pafrey PS, Foley RN, Harnett JD et al (1996) The outcome and risk factors for left ventricular disorders in chronic uraemia. Nephrol Dial Trasplant 11: 1277-1285
10. Samad T and Harp SA (1992) Self organization with partial data. Network 3: 205-212
11. Hagenbuchner M, Tsoi AC (2005) A supervised training algorithm for self-organizing maps for structures. Pattern Recogn Lett 26: 1874-1884
12. Guyon I, Elisseeff A (2003) An introduction to variable and feature selection. J   Mach Learn Res 3: 1157-1182
13. Peng H, Long F, Ding C (2005) Feature selection based on mutual information: criteria of max-dependency, max-relevance, and min-redundancy. IEEE Transactions on Pattern Analysis and Machine Intelligence, vol. 27(8): 1226-1238
14. C. Cortes, M. Mohri (2003) AUC Optimization vs. Error Rate Minimization. Advances in Neural Information Processing Systems (NIPS 2003), vol. 16, Vancouver, Canada, 2004, MIT Press
15. Ion Titapiccolo J, Ferrario M, Cerutti S et al (2013) Artificial intelligence models to stratify cardiovascular risk in incident hemodialysis patients. Expert Syst Appl 40: 4679-4686

Logistic Regression Models for Predicting Resistance to HIV Protease Inhibitor Nelfinavir

L.M. Raposo[1], M.B. Arruda[2], R.M. Brindeiro[2], and F.F. Nobre[1]

[1] Department of Electronics, Informatics and Bioengineering, Politecnico di Milano, Milano, Italy
[2] Fresenius Medical Care, Bad Homburg, Germany
[3] Center for Biomedical Technology at the Danube, University of Krems, Austria

Abstract — The development of models to predict the resistance to the antiretroviral drugs can be useful in making a decision regarding the best therapy for HIV+ individuals. This study developed predictive models of resistance to the protease inhibitor Nelfinavir using logistic regression. The data comprises a total of 625 patients for which HIV-1 genotype was available, with 130 resistants to Nelfinavir in the last regimen. Feature selection was carried out using a combination of bootstrap resampling procedure with the stepwise selection technique. Additionally, due to the unbalanced nature of the dataset, we develop four balanced final models. The accuracies of the models ranged from 70.40 to 76.80% and areas under the ROC curve (AUC) ranged from 0.657 to 0.687. The best model had AUC equal to 0.687, accuracy of 76.80%, specificity of 84.21% and sensitivity of 53.33%. The agreement between this model and the known resistance level was fair, Kappa index of 0.3712.

Keywords— HIV, Logistic Models, Drug Resistance

Introduction

In Brazil, since 1996 the antiretroviral therapy has been offered freely and universally [1], with a reduction in morbidity and mortality as well as an increased quality of life of patients [2]. In some cases, the therapy did not show durable clinical benefit due to several factors such as poor adherence to the treatment, loss of one or more clinic visits, high viral load, low count of CD4 + T cells and resistance to antiretroviral drugs [3-5], which has been a major obstacle in getting a lasting treatment.

HIV drug susceptibility can be evaluated either by genotyping or phenotyping. Genotyping identifies genetic mutations associated with resistance to antiretroviral drugs. It is faster, has a lower cost and is more accessible [6]. Phenotyping provides a direct measure of susceptibility to HIV antiretrovirals, but is very expensive, and requires special laboratories [6,7]. Several studies have proposed models to predict HIV resistance using genotyping. The range of methods varies from basic statistical methods [8,9] to modern machine learning algorithms [6,10,11]. Here, we develop models to predict resistance to the HIV protease inhibitor Nelfinavir using logistic regression, a generalized linear model widely used to make predictions or to explain the occurrence of a specific event when the response variable is binary, and a combination of bootstrap with stepwise procedure to select the variables for the model.

Material and Methods

The data used in this study are from 625 infected individual, and was made available by the Laboratory of Molecular Virology from the Center of Health Sciences, Federal University of Rio de Janeiro, Brazil. It consists of the amino acid sequences of the protease enzyme of the pol gene from HIV-1 subtype B, counting of CD4+ T cells, and viral load in the last period of treatment as markers of treatment failure for Nelfinavir.

For patients that were not using or did not show resistance to the drug in the last therapeutic regimen, the response variable was coded as 0 and, for those who showed resistance to the therapy, the variable was coded as 1.The set of explanatory variables were the mutations described by the International AIDS Society (IAS) [12] associated with resistance to Nelfinavir: L10, D30, M36, M46, A71, V77, V82, I84, N88 and L90. The sequence is represented by a letter corresponding to one of the 20 amino acids and a number, indicating the wild type amino acid residue and its position in the sequence. The amino acids were coded according to the hydrophobicity scale of Eisenberg [13] and used as input in the logistic model. Additionally, we used the count of CD4 + T cells and the viral load in the last period of drug use.

The sample was initially split into two subsets. The first, with 500 subjects; 100 with therapeutic failure, was used for the selection of variables and to obtain four models. The second subset, with 125 patients; 30 with therapeutic failure, was used as a test set to evaluate the models. The best set of features was selected using the bootstrap technique [14] such that the logistic regression is applied 1000 times to bootstrap samples obtained by resampling with repetition from the 100 samples with therapeutic failure in the training set, and combining with 100 patients randomly selected from the 400 non-resistant samples, generating a balanced set for model estimation. The variables at each bootstrap were selected by the stepwise method, using the Akaike Information Criterion (AIC) [15].

The most frequent (> 60%) variables in the models were D30 (90.7%), M36 (77.0%), V77 (78.7%) and L90 (91.1%). From four subsets obtained from the training set (100 resistant individuals and 100 non-resistant randomly selected from a total of 400 non-resistant), we created four models using as explanatory variables the most frequently variables.

The performance of the four models was evaluated at the cutoff of 0.5, according to accuracy, sensitivity, specificity and Kappa index. In addition we obtained the area under the receiver operating characteristic (ROC) curve (AUC). The accuracy (A) is defined as the proportion of correct classification by the model over the total sample. This metric is given by the following formula:

$$A = (TP + TN) / (TP + FP + VN + FN) \quad (1)$$

where TP, FP, TN and FN are true positives, false positives, true negatives and false negatives, respectively.

The sensitivity (S) is defined as the proportion of true positives as compared to the total positive class, whereas specificity (E) comprises the proportion of true negatives in relation to the total negative class.

$$S = TP / (TP + FN) \quad (2)$$

$$E = TN / (TN + FP) \quad (3)$$

The ROC curve is obtained by drawing pairs of sensitivity and false positive rate (1 – specificity) at different cutoff points. It shows the relationship between the sensitivity and specificity of a test and can be used in deciding the best cutoff [16].

The Kappa index measures the degree of agreement between different techniques beyond what would be expected by chance. It is calculated by dividing the difference between the expected and observed agreement and the difference between the absolute and expected agreement. The Kappa index is evaluated according to the scale of Landis, where a value equal to 1 indicates perfect agreement, between 1 and 0.80 indicates almost perfect agreement, between 0.60 and 0.79 represents considerable agreement, between 0.40 and 0,59 indicates moderate agreement, between 0.20 and 0.39 indicates reasonable agreement, between 0 and 0.19 represents low concordance and values less than 0 indicate no agreement [17].

All analyzes were performed using the software R version 2.15.

RESULTS

The average age of patients was 38.46 years (± 11.55), with a predominance of males (67.36%). About clinical characteristics, viral load had an average of 4.57 million copies of HIV/ml of blood, with inter-quartile range equal to (4.09 to 5.00), and the count of CD4 + T cells showed an average from 300.50 cells/mm³ of blood, with inter-quartile range equal to (127.0 to 420.0).

The four models obtained using logistic regression showed the following equations and AICs:

Model 1: Resistance = 18.83 + 23.73D30 − 1.59L90 + 0.87M36 + 2.21V77, AIC equal to 231.26.

Model 2: Resistance = 3.09 + 6.22D30 − 2.15L90 + 1.58M36 + 2.36V77, AIC equal to 259.55.

Model 3: Resistance = 17.15 + 23.91D30 − 1.75L90 + 0.99M36 + 3.79V77, AIC equal to 226.27.

Model 4: Resistance = 11.44 + 17.72D30 − 2.41L90 + 1.53M36 + 4.07V77, AIC equal to 231.87.

Table 1 shows the confusion matrices indicating the distribution of patients according to the classification given by the models and the actual outcome and table 2 shows the performance of the models.

Table 1. Distribution of patients in the test set according to the models and the actual outcome (n = 125).

	Resistant	Non-resistant	Total
Model 1			
Resistant	17	23	40
Non-resistant	13	72	85
Model 2			
Resistant	18	24	42
Non-resistant	12	71	83
Model 3			
Resistant	16	15	31
Non-resistant	14	80	94
Model 4			
Resistant	17	24	42
Non-resistant	13	71	83
Total	30	95	125
Total	30	95	125

Model 3 showed, in general, better performance compared to the other models with higher values to accuracy (76.80%), specificity (84.21%), AUC (0.6877) and Kappa coefficient (0.3712).

Table 2 Performance of the logistic regression models.

	Model 1 [CI95%]	Model 2 [CI95%]	Model 3 [CI95%]	Model 4 [CI95%]
AUC	0.6623	0.6737	0.6877	0.657
Accuracy	71.20 [62.72 – 78.41]	71.20 [62.72 – 78.41]	76.80 [68.67 – 83.33]	70.40 [61.89 – 77.70]
Sensitivity	56.67 [39.20 – 72.62]	60.00 [42.32 – 75.41]	53.33 [36.14 – 69.77]	56.67 [39.20 – 72.62]
Specificity	75.79 [66.28 – 83.29]	74.74 [65.16 – 82.40]	84.21 [75.57 – 90.19]	74.74 [65.16 – 82.40]
Kappa Index	0.2913	0.3056	0.3712	0.2790

Figure 1 shows the ROC curves of the four models. The circles marked on the graph indicate the sensitivity and specificity of the models when a cutoff equal to 0.5 was assumed.

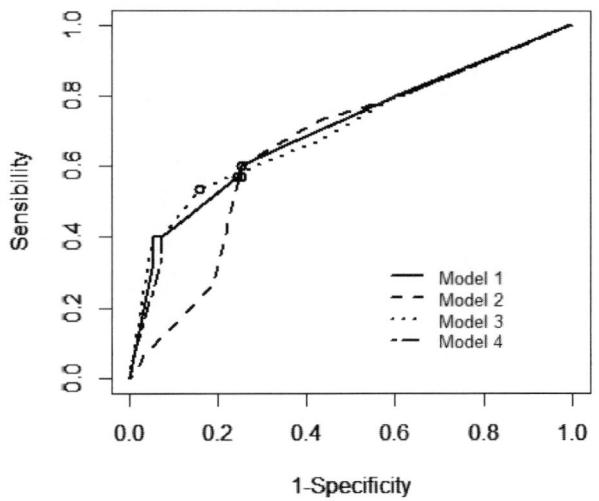

DISCUSSION

In the present study, we developed logistic regression models to predict the resistance to the antiretroviral Nelfinavir. We used data from the National Network of Genotyping (RENAGENO). This is the first study where these data were used for developing resistance models.

Our modeling approach differs from most others that used logistic regression by using a bootstrap technique combined with the classical stepwise method for feature selection. A particular advantage of this approach is that it allows to better assess the relative importance of each variable for the final logistic regression. The use of balanced data was also considered in our study. In classification problems it is not adequate to use imbalanced data sets, since there is a tendency of the model to be biased towards the majority class over the minority one.

The accuracies and areas under the ROC curves ranged from 70.40 to 76.80% and from 0.657 to 0.6877, respectively. The four logistic regression models showed performances very similar, with no significant difference between them at the significance level of 5%.

The best model was model 3, showing better AUC (0.6877), accuracy (76.80%) and specificity (84.21%), but a reduction in sensitivity (53.33%). This model also had the highest Kappa coefficient (0.3712), indicating reasonable agreement.

When selecting a model, it is important to know different performance parameters such as sensitivity and specificity. Many studies do not present these parameters, reducing the interpretation of their results. In the study developed by Prosperi (2009) [8], the values of accuracy (68.27 to 73.32%) and AUC (0.619 to 0.773) were very close to those of this study.

Logistic regression has some limitations since it exploits a linear combination of the variables. However, it is simple to implement and is available in several statistical packages, easy to interpret and a well-known model approach in different areas. Additionally, there are still some controversies regarding the use of more sophisticated and non-linear models in predicting antiretroviral drug resistance [8].

The models proposed in this study may contribute for the classification of new individuals in relation to the development of resistance to Nelfinavir and is a simple tool that can help in each HIV+ individual.

Nelfinavir was used for being a drug extensively used in the therapy and so, a rich database of information is available. However, this first generation protease inhibitor is currently has little clinical use. Nevertheless, the proof of the the proposed mathematical model will enable its use to uncover other complex patterns of resistance due to new antiretroviral drugs, which still lack extensive data on associated resistance mutations associated, especially considering the genetic variability of HIV-1 and its several subtypes.

REFERENCES

1. Dourado I, Veras M A D S M, Barreira D et al. (2006) Tendências da epidemia de Aids no Brasil após a terapia anti-retroviral. Revista de Saúde Pública 40:9-17

2. Bushman F, Landau N R, Emini E A. (1998) New developments in the biology and treatment of HIV. Proc Natl Acad Sci USA 95(19):11041-11042
3. Fätkenheuer G, Theisen A, Rockstroh J et al. (1997) Virological treatment failure of protease inhibitor therapy in an unselected cohort of HIV-infected patients. Aids 11(14):F113-F116
4. Zolopa A R, Shafer R W, Warford A et al. (1999) HIV-1 genotypic resistance patterns predict response to saquinavir-ritonavir therapy in patients in whom previous protease inhibitor therapy had failed. Ann Intern Med 131(11):813-821
5. Robbins G K, Daniels B, Zheng H et al. (1999) Predictors of antiretroviral treatment failure in an urban HIV clinic. Journal of Acquired Immune Deficiency Syndromes 44(1):30
6. Wang D, Larder B. (2003) Enhanced prediction of lopinavir resistance from genotype by use of artificial neural networks. J Infect Dis 188(5):653-660
7. Vermeiren, H., Van Craenenbroeck, E., Alen, P., et al. (2007) Prediction of HIV-1 drug susceptibility phenotype from the viral genotype using linear regression modeling. J Virol Methods 145(1):47-55
8. Prosperi M C F, Altmann A, Rosen-Zvi M et al. (2009) Investigation of expert rule bases, logistic regression, and non-linear machine learning techniques for predicting response to antiretroviral treatment. Antivir Ther 14:433-442
9. Van Der Borght K, Verheyen A, Feyaerts M et al. (2013) Quantitative prediction of integrase inhibitor resistance from genotype through consensus linear regression modeling. Virology journal 10(8):1-12
10. Draghici S, Potter R B. (2003) Predicting HIV drug resistance with neural networks. Bioinformatics 19(1):98-107
11. Beerenwinkel N, Däumer M, Oette M et al. (2003) Geno2pheno: Estimating phenotypic drug resistance from HIV-1 genotypes. Nucleic Acids Res 31:3850-3855
12. Johnson V A, Calvez V, Günthard H F et al. (2011) 2011 update of the drug resistance mutations in HIV-1. Top Antivir Med 19(4):156-164
13. Eisenberg D, Weiss R M, Terwilliger T C. (1984) The hydrophobic moment detects periodicity in protein hydrophobicity (protein structure/a helix/, sheet/31, helix/secondary structure). Proc. Nadl. Acad. Sci. USA 81:140-144
14. Efron B. (1979) Bootstrap methods: another look at the jackknife. The annals of Statistics 1-26
15. Akaike H. (1974) A new look at the statistical model identification", IEEE Trans. Automatic Control AC-19:716-723
16. Zweig M H, Campbell G. (1993) Receiver-operating characteristic (ROC) plots: a fundamental evaluation tool in clinical medicine. Clin Chem 39(4):561-577
17. Landis J R, Koch G G. (1977) The measurement of observer agreement for categorical data. Biometrics 33(1):159-174

Application of Special Parametric Methods to Model Survival Data

J. Holčík[1,2] and K. Opršalová[1,3]

[1] Institute of Biostatistics and Analyses, Masaryk University, Brno, Czech Republic
[2] Institute of Measurement Science, Slovak Academy of Sciences, Bratislava, Slovak Republic
[3] Department of Mathematics and Statistics, Faculty of Science, Masaryk University, Brno, Czech Republic

Abstract—This paper deals with some new forms of probability distributions that can be used for parametric survival analysis. A special transformation is introduced to transform standard distributions to more flexible ones that allow us to model bathtub hazard functions and prevent inaccuracies in estimates of survival and hazard functions in longer survival times. Some commonly used distributions are transformed to the new forms. Basic characteristics of these transformed distributions are described in details. The standard distributions and the new ones are used to model the survival function of the patients suffering from breast cancer. The results are compared according to Akaike's information criterion and examples of the use of the new derived distributions are presented, as well.

Keywords—Parametric methods, Survival analysis, Transformed distributions.

I. Introduction

Many various types of methods can be used to analyze survival data. Nonparametric methods are represented mainly by the Kaplan-Meier estimate of the survival function and are used to explore a basic character of given data. Unfortunately, this type of methods is not suitable for more detailed analysis, because of its simplicity and due to the fact that it is not able to capture the impact of various explanatory variables.

Semiparametric Cox proportional hazard model is the most widely used method in survival analysis. It can be used to handle with explanatory variables, but there are many situations where such a model is not very appropriate. Especially when the assumption of the hazard proportionality is not met and the hazard is time dependent. In such a case the parametric methods look like the best choice to handle the data.

There is no need to assume any proportionality when using the parametric methods. A possibility to extrapolate the estimates to the future that can be done neither by nonparametric nor by semiparametric methods is another their great advantage.

The greatest disadvantage of the parametric methods is the need to assume a particular probability distribution of the data. Particular distribution for a given situation is usually chosen based on past observations and a desired shape of the hazard function. Exponential, Weibull, lognormal or gamma distributions are the most often used for the purpose. It is possible to model constant, monotone or unimodal hazard rates by these distributions. Problems arise when it is necessary to model more complex shapes of the hazard function, such as a bathtub shaped hazard function where the hazard declines at the beginning of observed time, remains almost constant in the middle and rises at the end. Bathtub hazard function occurs very often in survival data. There are some more complex distributions that can be used to handle the problem, such as three parametric generalized Weibull or generalized gamma distribution. These are not so popular, because they are not implemented in any basic statistical software.

All of the above mentioned probability distributions are defined on infinite interval $\langle 0, \infty)$, that may cause survival overestimation, especially at the longer survival times. Our aim was to develop some new distributions defined in finite interval that will be flexible enough to follow various shapes of hazard function and will avoid that overestimation.

II. Data and Methods

A. Transformation Function

We applied a sigmoidal transformation function defined as

$$y = \frac{2a}{1 + e^{-kt}} - a, \quad (1)$$

where $k > 0$ is a shape parameter, $t \in \langle 0; a)$ is realization of a random variable T that represents the survival time and $a > 0$ can be interpreted as a maximum value of y.

The inverse function used for transformation can be expressed as

$$y^{-1} = \frac{1}{k} \ln \frac{a-t}{a+t}. \quad (2)$$

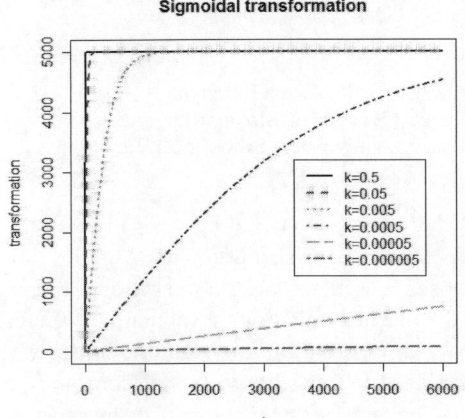

Fig. 1 Sigmoidal transformation for various values of k

B. Maximum Likelihood Function

Parameters of the new distributions were estimated by the maximum likelihood method. A special form of the method had to be used due to particular properties of survival data, namely censoring [1]

$$L((t,c),\beta) = \prod_{i=1}^{n} \{[f(t_i,\beta)]^{c_i}[S(t_i,\beta)]^{1-c_i}\}. \quad (3)$$

The function is more often applied in its logarithmic form

$$l((t,c),\beta) = \sum_{i=1}^{n} \{c_i \, ln[f(t_i,\beta)] + (1-c_i) \, ln[S(t_i,\beta)]\}, \quad (4)$$

where $f(t)$ is a probability density function and $S(t)$ a survival function, β is a vector of estimated parameters and c is an indicator of censoring

$$c = \begin{cases} 1, & \text{if } t \text{ observed} \\ 0, & \text{if } t \text{ censored}. \end{cases}$$

C. Used Data

The maximum likelihood method was applied to the data of 333 women who suffered from a breast cancer diagnosed in the 4th stage of the disease in Czech Republic in 1990. The study ended in 2007. The sample contained 8 right censored observations. Maximum observed uncensored survival time was 5753 days, which was approximately 16 years.

III. RESULTS

We decided to transform Weibull, gamma and generalized gamma distributions because of their potential to fit the hazard function based on the data properly. Exponential and log-normal distributions were not suitable for our data. Hazard function of the exponential distribution is constant over the whole interval of survival times and hazard for log-normal distribution increases at the beginning of the interval and decreases for longer survival times [2]. Hazard function for our data had the bathtub shape.

We transformed every single distribution, estimated the parameters and compared the survival functions of the patients made by new distribution and the original one. We used a nonparametric Kaplan-Meier estimate of the survival function as a reference.

A. Transformed Weibull Distribution

Weibull distribution is one of the most popular in survival analysis. It is simple to handle and flexible enough to fit a large range of various shapes.

We obtained transformed Weibull distribution applying the above mentioned sigmoidal transformation function for the standard two-parametric Weibull distribution. Basic properties of the new distribution are as follows.

Probability density function

$$f(t) = \frac{2ab}{l(a^2-t^2)}\left(-\frac{1}{l}ln\frac{a-t}{a+t}\right)^{b-1} e^{-\left(-\frac{1}{l}ln\frac{a-t}{a+t}\right)^b}. \quad (5)$$

Survival function

$$S(t) = 1 - F(t) = e^{-\left(-\frac{1}{l}ln\frac{a-t}{a+t}\right)^b}. \quad (6)$$

Hazard function

$$h(t) = \frac{f(t)}{S(t)} = \frac{2ab}{l(a^2-t^2)}\left(-\frac{1}{l}ln\frac{a-t}{a+t}\right)^{b-1}. \quad (7)$$

where $l = \lambda k > 0$ and $\lambda, b > 0$ are parameters of the Weibull distribution.

Maximum likelihood estimates of the parameters obtained for the experimental data are

$$\begin{aligned}\hat{b} &= 0{,}613 \\ \hat{l} &= 0{,}148 \\ \hat{a} &= 8357 \text{ days}\end{aligned} \quad (8)$$

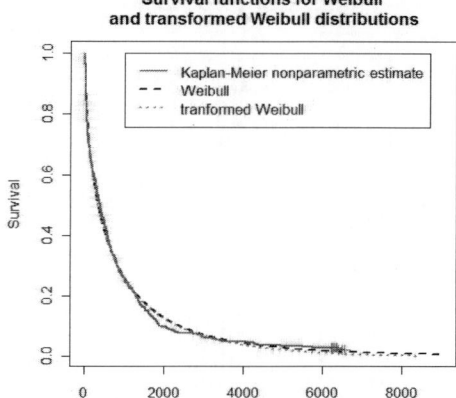

Fig. 2 Comparison of the survival functions made by Weibull and transformed Weibull distributions

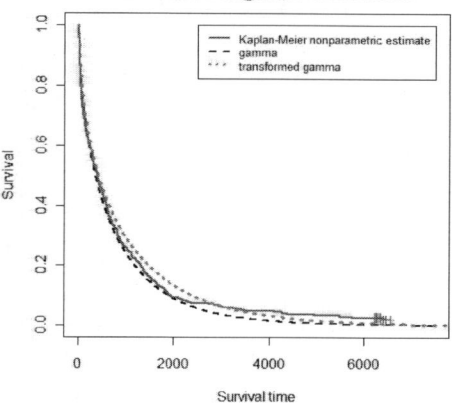

Fig. 3 Comparison of the survival functions made by gamma and transformed gamma distributions

B. Transformed Gamma Distribution

Gamma distribution has very similar properties as Weibull distribution, but it is not used so often due to its greater complexity and the need to work with gamma functions.

The transformed gamma distribution can be described by the following functions.

Probability density function

$$f(t) = \frac{2a}{\Gamma(p)q^p}\left(-\ln\frac{a-t}{a+t}\right)^{p-1} e^{-\left(\frac{-\ln\frac{a-t}{a+t}}{q}\right)}. \quad (9)$$

Survival function

$$S(t) = 1 - \frac{\gamma\left(p, \frac{-\ln\frac{a-t}{a+t}}{q}\right)}{\Gamma(p)}. \quad (10)$$

Hazard function

$$h(t) = \frac{2a}{q^p(a^2-t^2)\left(\Gamma(p)-\gamma\left(p,\frac{-\ln\frac{a-t}{a+t}}{q}\right)\right)}\left(-\ln\frac{a-t}{a+t}\right)^{p-1} e^{-\left(\frac{-\ln\frac{a-t}{a+t}}{q}\right)}, \quad (11)$$

where $= rk > 0$, $r, p > 0$ are parameters of the gamma distribution, $\Gamma(p) = \int_0^\infty x^{p-1}e^{-x}dx$ is a standard gamma function and $\gamma\left(p, \frac{-\ln\frac{a-t}{a+t}}{q}\right) = \int_0^{\frac{-\ln\frac{a-t}{a+t}}{q}} x^{p-1}e^{-x}dx$ is lower incomplete gamma function.

Maximum likelihood estimates of the parameters

$$\begin{aligned}\hat{p} &= 0{,}486 \\ \hat{q} &= 0{,}366 \\ \hat{a} &= 10492 \text{ days}\end{aligned} \quad (12)$$

C. Transformed Generalized Gamma Distribution

Generalized gamma distribution is the only one that is able to fit a bathtub hazard function. It is not so known, because it is usually not implemented in basic statistic software packages.

Transformation can be described as

Probability density function

$$f(t) = \frac{2ap}{\Gamma(p)v^d(a^2-t^2)}\left(-\ln\frac{a-t}{a+t}\right)^{d-1} e^{-\left(\frac{-\ln\frac{a-t}{a+t}}{v}\right)^p}. \quad (13)$$

Survival function

$$S(t) = 1 - \frac{\gamma\left(\frac{d}{p}, \left(\frac{-\ln\frac{a-t}{a+t}}{v}\right)^p\right)}{\Gamma\left(\frac{d}{p}\right)}. \quad (14)$$

Hazard function

$$h(t) = \frac{2ap}{v^d(a^2-t^2)\left(\Gamma\left(\frac{d}{p}\right)-\gamma\left(\frac{d}{p},\left(\frac{-\ln\frac{a-t}{a+t}}{v}\right)^p\right)\right)}\left(-\ln\frac{a-t}{a+t}\right)^{d-1} e^{-\left(\frac{-\ln\frac{a-t}{a+t}}{v}\right)^p}, \quad (15)$$

where $= ok > 0$, $o, p, d > 0$ are the parameters of the generalized gamma distribution.

Maximum likelihood estimates of the parameters

$$\begin{aligned}\hat{p} &= 0{,}369 \\ \hat{d} &= 0{,}831 \\ \hat{v} &= 0{,}014 \\ \hat{a} &= 7500 \text{ days}\end{aligned} \quad (16)$$

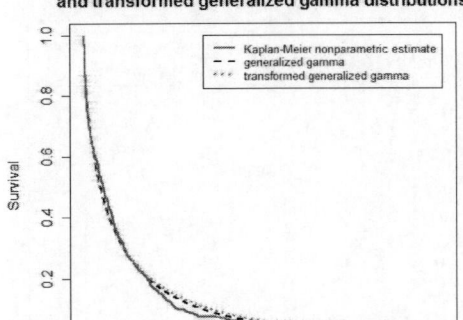

Fig. 4 Comparison of the survival functions made by generalized gamma and transformed generalized gamma distributions

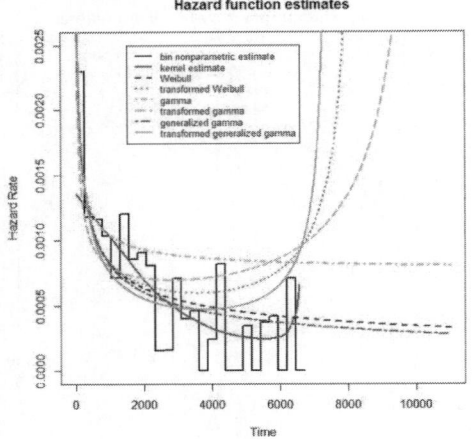

Fig. 5 Comparison of the hazard functions made by all above mentioned distributions

IV. CONCLUSIONS

According to Akaike's information criterion [3] we can say that the best results for our data can be obtained when applying Weibull or generalized gamma distributions. It seems that the standard distributions are more suitable than the transformed ones. But the differences are very small. The biggest difference makes only 5‰. It is largely caused by a greater complexity and by higher number of parameters. But the new distributions have some special properties that can make them more suitable in some situations. The great benefit of them is the ability to fit the bathtub hazard function. They are defined in finite interval that gives us a basic estimate of a maximum survival time for particular disease. We can also obtain more precise life time estimates for patients who are still alive at the end of the study.

Table 1 Akaike's information criterion

Distribution	Log-likelihood	AIC
Weibull	-2446,821	4897,642
Transformed Weibull	-2450,613	4907,226
Gamma	-2453,238	4910,476
Transformed gamma	-2457,237	4920,474
Generalized gamma	-2446,013	4898,026
Transformed generalized gamma	-2448,596	4905,192

ACKNOWLEDGMENT

This research was partially supported by the ESF project No. CZ.1.07/2.2.00/28.0043 "Interdisciplinary Development of the Study Programme in Mathematical Biology.

REFERENCES

1. Hosmer D W, Lemeshow S (1998) Applied Survival Analysis: Regression Modeling of Time to Event Data. John Wiley& Sons, NY
2. Lee E T, Wang J W (2003) Statistical Methods for Survival Data Analysis. John Wiley& Sons, New Jersey
3. Bradburn M J , Clark T G, LOVE S B, ALTMAN D G (2003) Survival Analysis Part III: Multivariate data analysis – choosing a model and assessing its adequacy and fit. British Journal of Cancer 89: 605-611

Author:	Jiří Holčík
Institute:	Institute of Biostatistics and Analyses, Masaryk University
Street:	Kamenice 126/3
City:	Brno
Country:	Czech Republic
Email:	holcik@iba.muni.cz

Protein Function Prediction Based on Protein-Protein Interactions: A Comparative Study

K.S. Ahmed[1] and S.M. El-Metwally[2]

[1] Department of Bio-Electronics, Modern University for Technology and Information Cairo, Egypt
[2] Department of Systems and Biomedical Engineering, Faculty of Engineering, Cairo University, Giza, Egypt

Abstract—Since interactions among proteins are of central importance for virtually every process in living cells, their information improves our understanding of diseases and can provide the basis for new therapeutic approaches. The study of protein function prediction based on protein-protein interactions (PPI) is one the most important issues in the field of bioinformatics since it is crucial for the understanding of cell activities. In this paper, a comparative study of different methods for protein function prediction based on PPI is presented. Five selected methods including the neighbor counting, Chi-square, Markov random field, Prodistin, and weighted interactions methods are applied to yeast proteome and their prediction performance is compared. Results revealed variant differences in the sensitivity and specificity of prediction for the functional category: cellular role. The weighted interactions method showed the best sensitivity values, while the Chi-square method resulted in the worst sensitivity/specificity values.

Keywords—Protein function, Computational method, PPI

I. INTRODUCTION

One of the most important challenges of the postgenomic era is determining protein functions. For this reason, automated function prediction became currently an extremely active research field. In the past, biologists tried to determine protein functions from the structure of the studied protein and other similar proteins. Possible roles of similarity between the protein and its homologies; from other organisms; were investigated to predict protein functions. Because of the diverse groups of homologous, these methods were found to be exhaustive and non-certain. Other techniques were suggested for the protein functions prediction as analyzing gene expression patterns [1,2], phylogenetic profiles [3,4,5], protein sequences [6, 7], protein domains [8], and integrated multi sources [9], protein structure [10, 11], and methods relying on collecting relations between the known functions [12, 13, 14] . However, these technologies often suffer from high error rates because of their inherent limitations. The computational approach has been adopted to solve these problems by using information gained from physical and genetic interaction maps to predict protein functions. Recently, researchers introduced different techniques to determine the probability of protein function prediction using the information extracted from PPI. A PPI network is often described as a complex system of proteins linked by interactions. The simplest representation takes the form of a network graph consisting of nodes and edges [15]. Proteins are represented as nodes in the graph and each two physically interacting proteins are represented as adjacent nodes connected by an edge [16]. Based on this graphical representation, various computational approaches, such as data mining, machine learning, and statistical approaches can be performed to reveal the PPI networks at different levels. In general, the computational analysis of PPI networks is challenged with some major problems. The first problem is the unreliability of protein interactions which comes from large-scale experiments that have yielded numerous false positives (Y2H). A second problem is that a protein may have more than one function and may be considered in one or more functional groups which lead to overlapping function clusters. The third problem is that proteins with different functions may interact. And this means that PPI has connections between proteins in different functional groups which expand the topological complexity of the PPI networks.

Currently, the lack of a systematic comparison of the different methods for function prediction is the major issue faced in the automated functional inference field, thereby hindering bench scientists to choose the most appropriate tools to predict the protein function for their interest [17]. Since different methods have been previously proposed to perform well for different functional schemas, a comparison of prediction methods with gold standard datasets may provide insights about the appropriate choice of methods for the specific problem of interest.

This paper presents a systematic comparison of some methods including the neighbor counting, Chi-square, Markov random field, Prodistin, and weighted interactions methods using the yeast proteome gold standard dataset. The performance of each of these methods is evaluated and compared to the others. The paper is organized as follows. The applied techniques are explained in section II with a description of the used dataset. Section III presents the obtained results. Finally, the paper ends with a discussion of the comparison results.

II. MATERIALS AND METHODS

The used dataset comprises 2559 protein-protein interactions between 6416 proteins collected from the Munich Information Center of Protein Sequences for yeast Saccharomyces cerevisiae [18]. Before applying the selected algorithms, a basic procedure is done as:-1) Specifying an ID for every protein that should not be duplicated. 2) Collecting the basic functions of every protein in the studied function category: cellular role. 3) Determining the interactions for every protein (level-1, and level-2 interactions). For example, protein ID-1 named (AAC1/ANC1/YMR056C) has basic function 'Small molecule transport function' in cellular role category. Also, this protein has three levels interactions: (MRS11/ TIM10/ YHRS01C), (SNF5/ TYE4/ HAF4/ SWI10/ YBR2036), and (TAF25/ TAF23/ TAFII25/ YD9489.0).

Then, for the given set of N proteins and E pair-wise interactions, each protein is labeled depending on whether the corresponding protein performs or does not perform a particular function. The labeling is encoded in N-dimensional binary vector x, i.e. $x_i = 1$ if the ith protein performs a particular function, $x_i = 0$, if it does not. The main objective is to assign each un-annotated protein to one of the two possible states (protein has function or not). The applied methods are described as follows.

A. Neighbor Counting Method

Neighbor counting is a method proposed by Schwikowski et al. in [19] to infer the functions of an un-annotated protein from the PPI. This method finds the neighbor proteins and gets their assigned functions and the frequencies of functions occurrence. Then, these functions are arranged in descending order according to their frequencies. The first k functions are considered and assigned to the un-annotated protein. In [20], k was set to 3. Figure 1 shows four annotated proteins (p1, p2, p3, p5) and their functions (f8, f9), (f1, f7), (f1, f3), and (f1-f6), respectively.

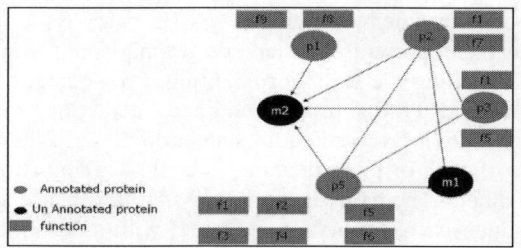

Fig.1 Physical interactions of annotated and un-annotated proteins

m2 is an un-annotated protein. It is predicted to have the most frequent functions: f1 is found three times from (p2, p3, p5), f5 is found two times from (p3, p5), while the rest of functions are seen only once.

B. CHI- Square Method

In this method, developed by Hishigaki [21], the problem of confidence level which was considered one of the drawbacks of neighbor counting methods, has been addressed. The Chi-square statistics (χ^2- statistics) as used to calculate significance value for the probability of the predicted functions. For a protein P_i, let $n_i(j)$ be the number of proteins interacting with P_i and having the function f_j. $e_i(j)$ is the result of multiplication between #Nei(i), the number of all proteins interacting with P_i, and π_j, the expected number of proteins in Nei(i) having function f_j. A score value $s_i(j)$ is calculated as in equation (1) to assess the strength of function confidence relating to the protein.

$$s_i(j) = [n_i(j) - e_i(j)]^2 / e_i(j) \quad (1)$$

C. Markov Random Field

A number of probabilistic approaches to the annotation problem have been suggested, all relying on a Markovian assumption: the function of a protein is independent on all other proteins given the functions of its immediate neighbors. This assumption gives rise to a Markov random field (MRF) model. The framework for protein function prediction based on MRF was originally proposed by Deng et.al. [22]. In MRF method, the probability that a protein performs a particular function depends on two numbers, namely the number of its direct neighbors in the network that perform the function and the number of those that do not. A model is adopted in which the probability that a protein v is assigned with a certain function that occurs with frequency f represents a logistic function as shown as equation (2).

$$\log(f/1-f) + \beta N(v,1) + \alpha(N(v,1) - N(v,0)) - N(v,0), \quad (2)$$

Where α and β are model parameters and N(v,1) and N(v,0) are the numbers of neighbors of v that are assigned or not assigned with the function, respectively. The parameters of the model are first learned from a training set by logistic regression [23] using these numbers as predictors. Then, Gibbs sampling [24] is employed for functional inference of the proteins with unknown function (un-annotated proteins).

D. PRODISTIN

Unlike the above mentioned methods that do not assign any weights for the edges of interaction network and consider them with same strength, PRODISTIN [25] assigns a weight to every pair of proteins by using a graph theory measure known as the Czekanowski-Dice distance (CD-Distance) or a distance metric. The approach uses a simple first-order model of the sampling variances and covariance of evolutionary distance estimates. This model leads to a simple expression of the minimum variance reduction, which is fully consistent with the agglomerative approach. These elements are combined to form BIONJ algorithm. And the minimum results will be selected. The formula of this technique is shown in equation (3)

$$D(u,v) = \frac{|N_u \triangle N_v|}{|N_u \cup N_v| + |N_u \cap N_v|} \quad (3)$$

where:
N_u refers to the set that contains u and its level-1 neighbors, $N_u \triangle N_v$ refers to the symmetric difference between two sets u and v, and $D(u, v)$ is the distance between u and v:

$D(u, v) < 1$, if u and v are level-1 neighbors,
$D(u, v) = 0$, if $N_u = N_v$, and
$D(u, v) = 1$, if $N_u \cap N_v = \emptyset$

Every pair of proteins that share at least one interaction neighbor will have a CD-distance less than 1. The distance will be large for protein pairs that share few neighbors, with the maximum distance being 1 when there are no common neighbors. The minimum distance between two proteins is 0 when both have identical set of neighbors. PRODISTIN defines a functional class for every function k as the largest possible sub-tree in the classification tree that contains at least three proteins annotated with k and has at least 50% of its annotated members annotated with k. Then, un-annotated proteins in the functional class are predicted to have this function.

E. Weighted Interactions Method

In this method, a new technique based on assigning weights to interactions is presented [26]. The weights are related to the network topology (local and global) and the number of identified methods (experimental methods with high significance). The method has introduced a comparative study between the proteins in PPI networks to the connected routers in the same autonomous number of networking. The protein acts as a router, and edge acts as the connection between two routers. The method has calculated the average between three different weights. After considering the number of experimental methods as first weight; interactions can be identified by more than one experimental method as Y2H [18, 27], mass spectrometry of co-immuno-precipitated protein complexes (Co-IP) [28], gene co-expression, TAP purification cross link, co-purification and biochemical methods.

In yeast, approximately ten experimental methods can be used to identify protein-protein interactions. For example, (Edge between Proteins (YBR0904) and (YDR356W) can be identified by ten experimental methods while edge between (AAC1) & (YHR005C-A) can be identified by just one method. The Technique has considered weight equal 1 for the interactions identified by more than one method and 0.5 for interactions identified by one. The second weight is the IG1 concept (Interaction Generality 1) for assessing the reliability of protein-protein interactions (as local topology). It is assumed that interactions with an IG1 value less than four (as threshold) have high confidence (100%) and those with more than four have low confidence. The third weight is the IG2 (Interaction Generality 2). This algorithm explores the five major sub-graphs of a network to obtain information concerning the global topology of the network. It is assumed that IG2 values less than the threshold are more accurate than those above the threshold. Regarding the three previous weights, high confidence interactions can be collected. After collecting the previous weights, a new weighted strategy can be created using an average of the three values as indicated in equation (4). The neighbor counting method was applied to the average weights and new sensitivity and specificity are calculated.

$$W = 1/3 \, [W_1 + W_2 + W_3] \quad (4)$$

where, W_1 is the number of experimental methods score ($W_1 =1$ if $n > 3$, otherwise $W_1 = 0$), W_2 is the IG1 (Interaction generality 1) that calculates the number direct leafs for interaction pair, W_3 is the IG2 (Interaction generality 2) that calculates the average between the most common five figures inside the protein network using principal component analysis.

III. PERFORMANCE EVALUATION

The prediction performance of the different applied methods is measured by the leave-one-out method. For each annotated protein with at least one annotated interaction partner, it is assumed to be un-annotated and its functions are predicted by all the above methods. The estimated functions are then compared with the actual annotations of the protein. The leave-one-out experiment is repeated for all such proteins. Let n_i be the number of functions for protein P_i, m_i the number of predicted functions for protein Pi, and k_i the overlap between the set of observed functions and the set of predicted functions. The specificity (SP) and the sensitivity (SN) are calculated as:

$$SN = \frac{\sum_i^K k_i}{\sum_i^K n_i}, \qquad SP = \frac{\sum_i^K k_i}{\sum_i^K m_i} \qquad (5)$$

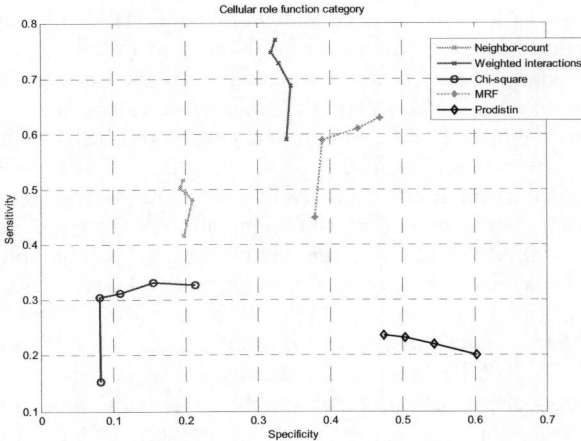

Fig. 2 The specificity & sensitivity for all applied methods.

IV. RESULTS

The obtained results based on the various methods are analyzed by applying the leave-one-out for all proteins having at least one interaction partner. Figure 2 illustrates the specificities and sensitivities calculated for all the applied methods. It can be seen that the weighted method shows the highest sensitivity values followed by the MRF method, the neighbor-counting, the Chi- square method then Prodistin. However, Prodistin reveals the best specificity values among all methods. Also, for a given specificity, it can be seen that the sensitivity increases with the number of interaction partners. As expected, the more interaction partners a protein has, the more accurate are the obtained predictions.

V. DISCUSSION

All the selected approaches were applied to yeast proteins network collected from MIPS and the protein function annotations based on YPD. The sensitivity and specificity of each method was studied by the leave-one-out approach and the obtained results were compared. By examining the calculated sensitivity and specificity values for the applied methods, the obtained results can be attributed due to following:

Although neighbour counting method exploits information from the protein neighbours, it has some drawbacks as it considers the interactions to be of equal weights, for example if the interaction between the proteins p2 and m2 is more reliable than the interaction between p3 and m2, this suggests that m2 may have function f7 instead of f5. Another drawback is that it does not consider the nature of the function and whether it is dominant or not. Also, it does not provide a confidence level for assigning a function to the protein. This method also couldn't predict functions of protein found in un-annotated proteins group.

Although the Chi-square method overcomes one of the neighbour counting method limitations, i.e., the confidence/significance value of the prediction, as it depends on statistical measurements (χ^2) computing the deviation of the observed occurrence of function in the neighbours, it introduced poor results compared to the neighbour counting method its self. It has been found that the total number of overlapping functions, i.e., between basic & estimated functions, is less than that computed for the other methods, which reflects the least sensitivity and specificity values of the Chi-square method.

The MRF method allows for the sampling of the state of every un-annotated protein knowing the parameters and the states of its neighbours. In the parameter estimation step, some or all neighbours may be with an unknown state. The main problem in MRF method is that it disregards the neighbourhood uncertainty in the parameter estimation step, but takes it into account during the labelling step. By disregarding un-annotated proteins in the first task, they are pruned compared to the full network. Although it is expected that this strategy will work worse as the proportion of un-annotated proteins becomes larger, it significantly outperforms the neighbour counting and Chi-Square methods.

PRODISTIN has some advantages over neighbour counting and the Chi-Square methods. Firstly, function prediction of a certain protein is not limited to its immediate neighbours, e.g., two proteins that share some neighbours can have a weight less than 1 even if they do not interact. Moreover, during the construction of the tree, proteins that do not share any neighbours may also be clustered together. Secondly, the algorithm assigns a weight to each protein pair using common interaction neighbours, while the other methods simply consider protein pairs that interact but do not distinguish between them in any way.

As discussed earlier in this paper, protein-protein interaction data, especially those derived from high-throughput experimental assays tend to have a lot of false positives. The likelihood that two proteins are falsely detected as interacting is much higher than the likelihood they are falsely sharing a number of interacting neighbours; especially if a large number. The assignment of the CD-distance helps to reduce the effect of false positives on the prediction results. It was shown that the weighted interactions method revealed great improvement in the sensitivity and specificity of prediction in terms of cellular role function category. This is due to its dependence on the number of experimental methods identifying the interaction and, furthermore, on the interaction network topology which is considered either locally for direct adjacent neighbours or globally for the common graphical figures through the network.

VI. CONCLUSIONS

By applying the previous mentioned techniques to yeast proteome network, the integrated weighted method has revealed the highest sensitivity while the highest specificity values were shown for the Prodistin method.

REFERENCES

1. Zhao M, Aihara K (2008) Gene function prediction using labeled and unlabeled data. BMC Bioinformatics 9: 57-71
2. Zhao H, Wu B (2003) DNA-Protein binding and gene expression patterns. Lecture Notes-Monograph Series, Statistics and Science: A Festschrift for Terry Speed 40: 259-274
3. Morin M (2007) Phylogenetic networks: simulation, characterization, and reconstruction. New Mexico
4. Sun J, Zhao Z (2007) Construction of phylogenetic profiles based on the genetic distance of hundreds of genomes. Biochem Biophys Res Commun 335(3): 849-853
5. Pellegrini M, Marcotte E M, Thompson M J et al. (1999) Assigning protein functions by comparative genome analysis: protein phylogenetic profiles, Proc Natl Acad Sci, vol. 96, no. 8, USA, pp 4285-4288
6. Harrington E D, Singh A H, Doerks T et al. (2007) Quantitative assessment of protein function prediction from metagenomics shotgun sequences, Proc Natl Acad Sci, vol. 104, USA, August, 2007, pp 13913-8
7. Spriggs R V, Murakami Y, Jones S (2009) Protein function annotation from sequence: prediction of residues interacting with RNA. Bioinformatics 25: 1492-7
8. Nariai N, Kolaczyk E D, Kasif S (2007) Probabilistic protein function prediction from heterogeneous genome-wide data. PLoS One 2: 337-344
9. Liu Y, Zhao H (2008) Protein interaction predictions from diverse sources. Drug Discov Today 13: 409-16
10. Whisstock J C, Lesk A M (2003) Prediction of protein function from protein sequence and structure. Q Rev Biophys 36: 307-40
11. Friedberg I (2006) Automated protein function prediction--the genomic challenge. Brief Bioinform 7: 225-42
12. Sayed K, Soloma N, Kadah Y (2010) Estimation of the correlation between protein sub-function categories based on overlapping proteins, Proc. 27th National Radio Science Conference, Menouf, Egypt, March, 2010.
13. Sayed K, Soloma N, Kadah Y (2011) Exploring Protein Functions Correlation Based On Overlapping Proteins and Cluster Interactions, 1st middle east conference for biomedical engineering, 21-24 Feb, Sharqah, 2011.
14. Sayed K, Soloma N, Kadah Y (2011) Determining The Relations Between Protein Sub-Function Categories Based On Overlapping Proteins. JCC 2011.
15. Wagner A (2003) How the global structure of protein interaction networks evolves, Proc Biol Sci, vol. 270 (1514), pp. 457-466, 2003.
16. Aidong Z (2009) Protein interaction networks: Computational Analysis. New York, Press.
17. Janga S, Dı́az-Mejı́a J, Moreno-Hagelsieb G (2011) Network-based function prediction and interactomics: The case for metabolic enzymes. Metabolic Engineering 13: 1–10
18. Ito T, Chiba T, Ozawa R et al. (2001) A comprehensive two-hybrid analysis to explore the yeast protein interactome, Proc Natl Acad Sci, vol. 98, no. 8, USA, 2001, pp. 4569-4574
19. Schwikowski B, Uetz P, Fields S (2000) A network of protein-protein interactions in yeast. Nat Biotechnol 18(12): 1257-1261
20. Sharan R (2006) Analysis of biological networks: Protein-protein interaction networks – functional. Annotation.
21. Hishigaki H, Nakai K, Ono T et al. (2001) Assessment of prediction accuracy of protein function from protein--protein interaction data. Yeast 18(6): 523-531
22. Deng M, Zhang K, Mehta S et al. (2003) Prediction of protein function using protein-protein interaction data. Journal of Computational Biology 10: 947–960
23. McCullagh P, Nelder J. (1989) Generalized linear models (Monographs on statistics and applied probability). London: Chapman Hall.
24. Geman S, Geman D (1984) Stochastic relaxation, gibbs distributions, and the bayesian restoration of images. IEEE Transactions on Pattern Analysis and Machine Intelligence PAMI 6: 721–741
25. Brun C, Chevenet F, Martin D et al. (2003) Functional classification of proteins for the prediction of cellular function from a protein-protein interaction network. Genome Biol, vol. 5.
26. Ahmed K, Soloma N, Kadah Y (2011) Improving the prediction of yeast protein function using weighted protein-protein interactions. Theoretical Biology and Medical Modelling, vol. 8.
27. Tong A, Drees B, Nardelli G et al. (2002) A combined experimental and computational strategy to define protein interaction networks for peptide recognition modules. Science 295: 321-324
28. Gavin A, Bosche M, Krause R et al. (2002) Functional organization of the yeast proteome by systematic analysis of protein complexes. Nature 415(6868): 141-147

Author:	Khaled Sayed Ahmed
Institute:	Modern University for Technology and Information
Street:	Katameia
City:	Cairo
Country:	Egypt
Email:	khaled.sayed@k-space.org

Inhibitory Regulation by microRNAs and Circular RNAs

J. Demongeot, H. Hazgui, J. Escoffier, and C. Arnoult

I AGIM CNRS FRE 3405, Faculty of Medicine, University J. Fourier of Grenoble, 38B700 La Tronche, France

Abstract—**The microRNAs are responsible of a general post-transcriptional inhibitory activity partly unspecific due firstly to a possible direct negative action during translation by hybridizing tRNAs, and secondly to the large number of their putative targets. Recently a second regulation layer has been discovered involving both proteins as microRNA transcription factors and circular RNAs as microRNA inhibitors ("miRs sponges"). We will show the existence of circuits inside the role of miR-7 in the regulation of the spermatozoa morphogenesis.**

Keywords—**MicroRNAs, Genetic regulatory networks, Cyclic RNAs inhibition, spermatozoa morphogenesis regulation.**

I. INTRODUCTION

The genetic regulatory networks involved in the control of morphogenesis have only a few number of asymptotic dynamical behaviors, called attractors. This small number of attractors is directly linked with the possibilities of differentiation of the cells in the tissue in construction, *e.g.* the spermatozoa, and this number is controlled inside the interaction graph of the regulatory network by the circuits of its strongly connected components [1-4], giving to the network the possibility to have if necessary, when the tissue has more than one function, more than one attractors (due to positive circuits, having inside an even number of inhibitions) and the ability to be stable (due to negative circuits, having inside an odd number of inhibitions). We will describe successively a recently discovered regulation due to circular RNAs, inhibiting microRNAs (miRNAs) responsible of a partly unspecific inhibition in genetic networks, then the network regulating the spermatozoa morphogenesis in its controlling environment, and eventually, a method to calculate the reduction of the attractor numbers due to interactions between intersecting circuits and apply it to the control of the gene Dpy9l2, responsible for the morphology of the acrosome in spermatids, involved in fecundation.

II. CIRCULAR RNAS

Mature messenger and transport RNAs are inhibited by miRNAs in a partly unspecific way due to the great number of their putative targets [5,6] and other RNAs play also a regulatory role by hybridizing miRNAs, the circular RNA sequences (ciRS) discovered firstly in plants and now unambiguously validated in animals: in [7,8] authors found that a human ciRS, ciRS-7, antisense to the cerebellum degeneration-related protein CDR1, was binding sites of the miRNA, miR-7, perfectly conserved from annelids to humans, with practically the same role in protein translation regulation, despite of the fact that for example it regulates in *Drosophila* some proteins not regulated in mammals [9]. miR-7 functions in several interlocking feedback and feed-forward loops, and we have chosen its role in a negative circuit involving the couple of genes Egfr/RAS, increasing the stability of the MAPK network, *i.e.* buffering them against external perturbations. Conserved miRNAs like miR-7 may increase robustness of genetic regulatory networks. Hence a second layer of regulation could serve to prevent an over-inhibition by such critical miRNAs, leading to pathologic down-regulations. On Fig. 1, we can see good matches between a sequence called AGAT homologous from 13 species to a human CDR1 sequence using BLAT and the UCSC Genome browser [7]. This sequence AGAT matches the half part of the ABEL sequence, proposed as an archetypal sequence for the ancient genetic code [10-12]. The main characteristic of ABEL is that it contains from start codon AUG until end codon UGG one and only one codon per synonymy class of amino-acids [10]. ABEL is close to another sequence AL, which can be considered as a the barycentre of the miRNA set for Hamming distance, and AL shares with any miRNA subsequence of lengh 5 (like UUCCA) more than expected by chance [11], and is close to sequences coming from the UTR viral genomes [12].

ABEL sequence	5'-AUGGCACUGAAUGUUCCAAGAU-3'
AGAT sequence	5'-GTCTTCCAAGAT-3'
hsa-miR-7	5'-AACAAAAUCACUAGUCUUCCA-3'

Fig. 1 Matches between the ABEL, AGAT and human miR-7 sequences

The antisense sequence 3'-5' of AGAT anti-matches the corresponding sequence 5'-3' of miR-7, which explains the inhibitory power of ciRS-7. By searching in miRNAs bases [6], we found the sequence UCUUCCA in about 25 of among 500 miRNAs, significantly more than expected (about 2±4). In [5], are reconstructed 16S and 23S rRNA data sets and trees spanning all sequenced type strains of classified species of Archaea and Bacteria, and we found inside these RNAs 249 times the sequence AUCUUCCA, significantly more than expected frequency (about 160±40).

L.M. Roa Romero (ed.), *XIII Mediterranean Conference on Medical and Biological Engineering and Computing 2013*, IFMBE Proceedings 41,
DOI: 10.1007/978-3-319-00846-2_309, © Springer International Publishing Switzerland 2014

III. THE SPERMATOZOA MORPHOGENESIS REGULATION

The Dpy19l2 protein is expressed in spermatids with a specific localization on the inner nuclear membrane facing the acrosomal vesicle [15,16]. The absence of Dpy19l2 leads to the destabilization of both the the junction between the acroplaxome and the nuclear envelope. Consequently, the acrosome fails to be linked to the nucleus leading to the failure of sperm nuclear shaping and eventually to the elimination of the unbound acrosomal vesicle rendering impossible the fecundation. Hence, the control of Dpy19l2 has to be very reliable and Figs. 2 and 3 show the corresponding regulatory network in its genomic environment (after [17]) and its possible dynamics (after [2]).

IV. NUMBER OF ATTRACTORS

The different components of the sub-networks of the Fig. 3 have been extracted from the recent literature, in particular those concerning the inhibitory action of the miRNAs [19-26]. Let us remark that the gene Dpy19la is inhibited by miR-146a, whose dynamics is dependent on two negative circuits, one of size 12 coming from the miR-146a and passing through the couple of genes Egfr/RAS (counted only for one node) and the gene p21, and the other of size 4 coming from the miR-7 passing through the node Egfr/RAS, the gene RAF and the couple of genes MEK/ERK (counted only for one node). These two negative circuits are tangent on the node Egfr/RAS.

The number of attractors of the sub-networks of the Fig. 3 can be calculated by using an algorithm presented in [2], by supposing that each gene can have only two states, 1 if the gene is expressed and 0 if not. The central elements of the network architecture fixing the number of possible asymptotic dynamical behavior of the gene states (called attractor), when the updating schedule is supposed to be synchronous, are the circuits of the strongly connected components of the network. Using the results of [2], it is for example possible to calculate the number of attractors of any sub-network of the Fig. 3. For example, at the intersection of the line 12 and of the column 4, we meet the integer 2, which is the number of the attractors of the couple of two tangential negative circuits of size respectively 12 and 4. One of the two attractors corresponds to both miR-146a and ciRS-7 expressed. Then miR-7 is not expressed, but thegenes Egfr/RAS, RAF and MEK/ERK of the negative circuit of length 4 as well as the gene Dpy19la are not expressed. This attractor corresponds to a pathologic spermatozoa morphogenesis. On the contrary, when ciRS-7 is expressed, but miR-146a is inhibited by c-Myc, then the genes of the negative circuit of length 4 are expressed as well as the gene Dpy19la, and the corresponding attractor leads to a physiologic spermatozoa morphogenesis.

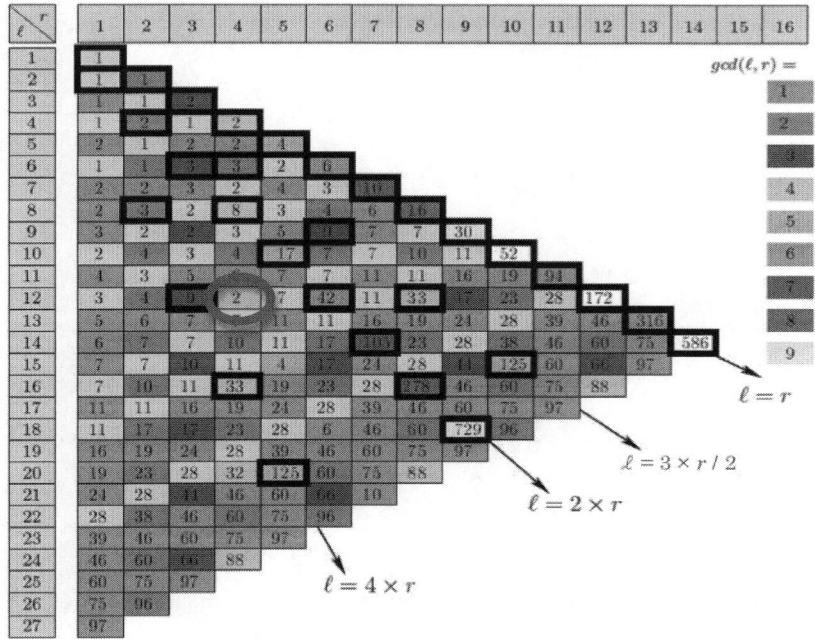

Fig. 2 Attractors number of tangential negative circuits of size l and r. Colors indicate the different values of the greatest common divisor (gcd).

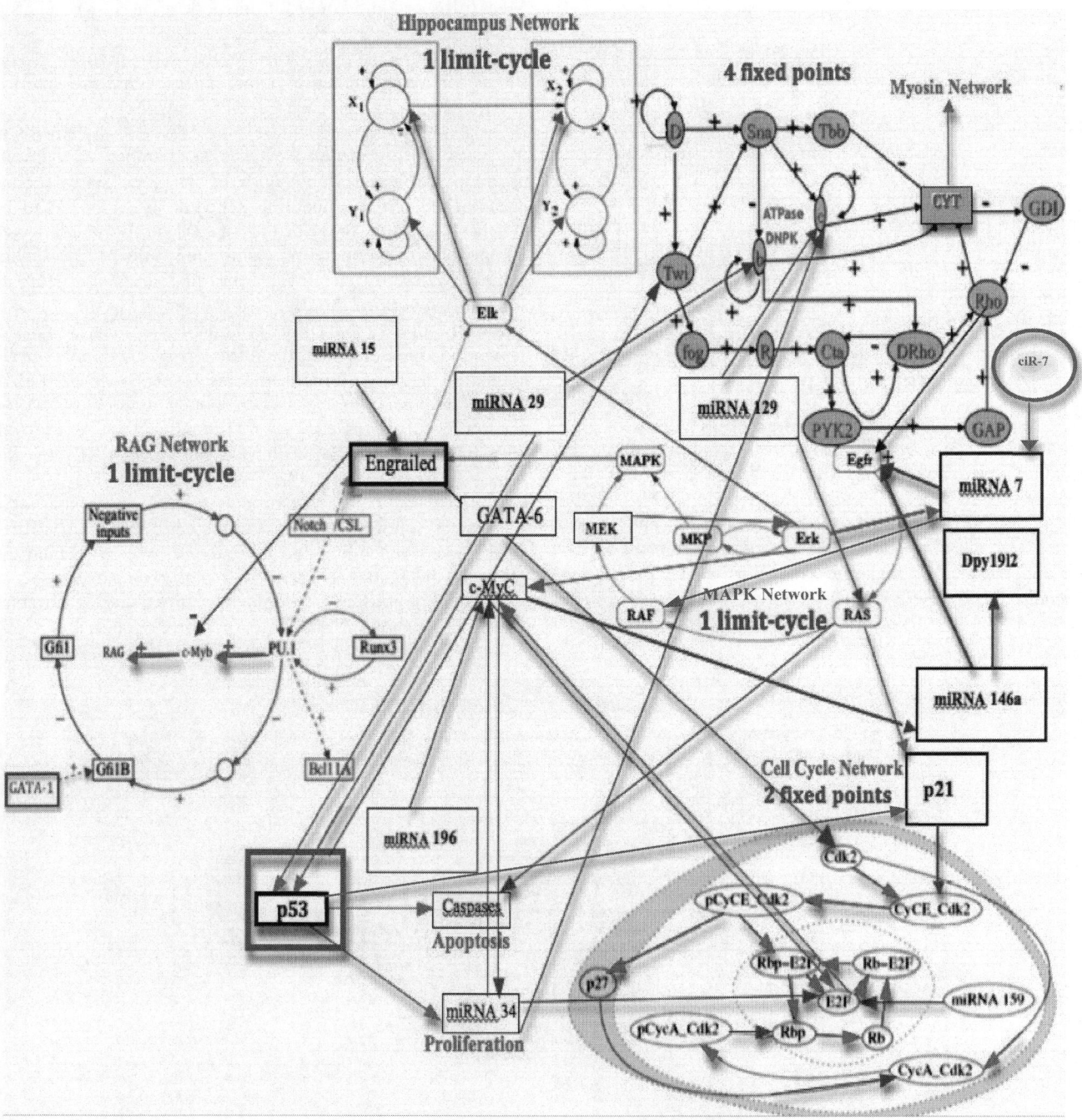

Fig. 3 Genetic network controlling the gene Dpy19la with its environment centred on the gene Engrailed with its up and down-sub-networks. The number and nature of attractors of these sub-networks are indicated in blue. Bottom left: interaction signed digraph modelling the genetic regulation network controlling the cell cycle in mammals [18]. Black (resp. white) arrows represent activations (resp. inhibitions).

The size of the genetic regulatory networks is increasing and the complexity of their architecture leads to use recent mathematical approaches on the dynamics of the Boolean networks, in order to identify their attractors (their number and their nature). The recent discovery of two layers of gene inhibition, due respectively to the circular RNAs and to the micro-RNAs, partly unspecific, renders this mathematical framework still more necessary. Surprisingly, the addition of a great number of such inhibitory RNAs is in general reducing the number of possible asymptotic configurations of the genetic regulatory networks and increases their stability and robustness [2-4], leading to biological functions with a few number of physiologic and pathologic phenotypes, hence focusing a tissue on a specific well controlled role inside the whole organism, resisting to environmental external perturbations, like here the spermatozoa morphogenesis.

Acknowledgment

The present work has been supported by the ANR Project REGENER.

References

1. Cinquin O, Demongeot J (2002) Positive and negative feedback: striking a balance between necessary antagonists. J Theor Biol 216:229-241
2. Demongeot J, Noual M, Sené S (2012) Combinatorics of Boolean automata circuits dynamics. Discrete Appl Math 160:398-415
3. Demongeot J, Cohen O, Doncescu A, Henrion-Caude A (2013) MitomiRs and energetic regulation, IEEE Proc., AINA' 13, pp 1501-1508
4. Demongeot J, Goles E, Morvan M, Noual M, Sené S (2010) Attraction Basins as Gauges of Environmental Robustness in Biological Complex Systems. PloS ONE 5:e11793
5. http://mamsap.it.deakin.edu.au/~amitkuma/mirna_targetsnew/sequence.html
6. http://mirdb.org/miRDB/
7. Memczak S, Jens M, Elefsinioti A, Torti F, Krueger J, Rybak A, Maier L, Mackowiak S D, Gregersen L H, Munschauer M, Loewer A, Ziebold U, Landthaler M, Kocks C, le Noble F, Rajewsky N (2013) Circular RNAs are a large class of animal RNAs with regulatory potency. Nature 495:333-338
8. Hansen T B, Jensen T I, Clausen B H, Bramsen J B, Finsen B, Damgaard C K, Kjems J (2013) Natural RNA circles function as efficient microRNA sponges. Nature 495:384-388
9. Li X., Cassidy J.J., Reinke C.A., Fischboeck S., Carthew R.W. (2009) A microRNA Imparts Robustness Against Environmental Fluctuation During Development. Cell 137: 273–282
10. Weil G, Heus K, Faraut T, Demongeot J (2004) An archetypal basic code for the primitive genome. Theoret Comp Sc 322:313-334
11. Demongeot J, Moreira A (2007) A circular RNA at the origin of life. J Theor Biol 249:314-324
12. Demongeot J, Drouet E, Moreira A, Rechoum Y, Sené S (2009) MicroRNAs: viral genome and robustness of the genes expression in host. Phil Trans Royal Soc A 367:4941-4965
13. Kumar A, Wong A K, Tizard M L, Moore R J, Lefèvre C (2012) miRNA_Targets: a database for miRNA target predictions in coding and non-coding regions of mRNAs. Genomics 100:352-356
14. http://www.arb-silva.de/fileadmin/silva_databases/living_tree/LTP_release_93/type_strains_LTP_s93_unaligned.fasta
15. Pierre V, Martinez G, Coutton C, Delaroche J, Yassine S, Novella C, Pernet-Gallay K, Hennebicq S, Ray PF, Arnoult C (2012) Absence of Dpy19l2, a new inner nuclear membrane protein, causes globozoospermia in mice by preventing the anchoring of the acrosome to the nucleus. Development 139:2955-2965
16. Escoffier J, Boisseau S, Serres C, Cheng C C, Kim D, Stamboulian S, Shin H S, Campbell K P, De Waard M, Arnoult C (2007) Expression, localization and functions in acrosome reaction and sperm motility of $Ca_V3.1$ and $Ca_V3.2$ channels in sperm cells: An evaluation from $Ca_V3.1$ and $Ca_V3.2$ deficient mice. J Cell Physiol 212: 753-763
17. Demongeot J, Pempelfort H, Martinez J M, Vallejos R, M. Barria M, Taramasco C (2013) Information design of biological networks: application to genetic, immunologic, metabolic and social networks., IEEE Proc., AINA' 13, pp 1533-1540
18. Kohn K W (1999). Molecular Interaction Map of the Mammalian Cell Cycle Control and DNA Repair Systems. Molecular Biology of the Cell 10:2703–2734.
19. Maurel M, Jalvy S, Lareiro Y, Combe C, Vachet L, Sagliocco F, Bioulac-Sage P, Pitard V, Jacquemin-Sablon H, Zucman-Rossi J, Laloo B, Grosset C F (2013) A functional screening identifies five miRNAs controlling glypican-3: Role of miR-1271 down-regulation in hepatocellular carcinoma. Hepatology 57:195-204
20. Huang N, Lin J, Ruan J, Su N, Qing R, Liu F, He B, Lv C, Zheng D, Luo R (2012) MiR-219-5p inhibits hepatocellular carcinoma cell proliferation by targeting glypican-3. FEBS Letters 586:884-891
21. Boominathan L (2010) The Tumor Suppressors p53, p63, and p73 Are Regulators of MicroRNA Processing Complex. PLoS ONE 5:e10615
22. Boominathan L (2010) The guardians of the genome (p53, TA-p73, and TA-p63) are regulators of tumor suppressor miRNAs network. Cancer Metastasis Rev 29:613-639
23. Segura M F, Greenwald H S, Hanniford D, Osman I, Hernando E (2012) MicroRNA and cutaneous melanoma: from discovery to prognosis and therapy. Carcinogenesis 33:1823-1832
24. Tennant D A, Dur R V, Gottlieb E (2010) Targeting metabolic transformation for cancer therapy. Nature Reviews Cancer 10:267-277
25. Georgescu C, Longabaugh W J R, Scripture-Adams D D, David-Fung E S, Yui M A, Zarnegar M A, Bolouri H, Rothenberg E V (2008) A gene regulatory network armature for T lymphocyte specification. Proceedings of the National Academy of Sciences USA 105:20100–20105
26. Elkon R, Linhart C, Halperin Y, Shiloh Y, Shamir R (2007) Functional genomic delineation of TLR-induced transcriptional networks. BMC Genomics 8:394
27. Jiang L, Liu X, Chen Z, Jin Y, Heidbreder C E, Kolokythas A, Wang A, Dai Y, Zhou X (2010) MicroRNA-7 targets IGF1R (insulin-like growth factor 1 receptor) in tongue squamous cell carcinoma cells. Biochem J 432:199–205

Public Electronic Health Record Platform Compliant with the ISO EN13606 Standard as Support to Research Groups

R. Sánchez de Madariaga[1], J. Cáceres Tello[1], A. Muñoz Carrero[1], O. Moreno Gil[1],
I. Velázquez Aza[1], A. Castro Serrano[1], and R. Somolinos Cristóbal[2]

[1] Unidad de Investigación en Telemedicina y e-Salud,
Instituto de Salud Carlos III. 28029-Madrid, España
[2] Unidad de Bioingeniería y Telemedicina,
Hospital Universitario Puerta de Hierro. 28222, Majadahonda-Madrid, España

Abstract—This paper presents the design and implementation of a new platform compliant with the ISO EN-13606 standard which facilitates semantic interoperability between clinical platforms. The presented work forms a point of convergence of clinical information originating in distinct institutions such as hospitals, primary attention centers, etc. It enables clinical care continuity, Electronic Health Record (EHR) realistic and practical edition, clinical decision support systems development and secondary research development using medical data mining. The Telemedicine and e-Health Research Unit at the Instituto de Salud Carlos III, as a public institution, has developed this platform allowing its use as support to other research groups and taking one step closer to everyday generalized use of EHRs in hospitals, healthcare and research centers.

Keywords—Electronic Health Record, semantic interoperability, ISO EN13606 standard, clinical care continuity, secondary research development.

I. INTRODUCTION

It is an accepted fact these days that medical information systems need to be semantically interoperable in order to achieve continuity of care. One way to achieve semantic interoperability is to normalize Electronic Health Record (EHR) transfer. In doing so, not only can continuity of care be obtained [1, 2], but also clinical decision support systems can be developed and adopted [3] and normalized information can also be used for secondary research purposes [4].

A. Related Work

The amount of published work presenting developments of platforms using the EN-13606 standard to facilitate semantic interoperability between clinical platforms is growing fast. Early work dealt with conversion between different standards [5, 6]. Works on the central paradigm of the EN-13606 standard as facilitating semantic interoperability between distinct clinical platforms from different perspectives is becoming more important [7, 9].

II. MATERIALS AND METHODS

EHR normalization means to incorporate Information and Communications Technologies (ICT) into healthcare activity as part of an integrated clinical information system [10]. Generally speaking, an EHR system is a complex structure incorporating many different information components. The ISO EN-13606 standard suggests a dual model for the development of healthcare information systems [11]. This dual model separates information from knowledge managed by the information systems. Knowledge may change substantially over time for strictly medical reasons. By separating information from knowledge the system is able to evolve and it can be adapted easily and automatically to changes.

The EN-13606 standard defines two complementary concepts: on one side a Reference Model (RM) for the representation of clinical information; and on the other side an archetypes model representing clinical concepts with a higher degree of semantics. In this context a framework to put into practice the structures suggested by the standard and to carry out a number of proofs like the degree of interoperability with other platforms or storage and information retrieval optimization becomes necessary. The present work presents the development of an ISO EN-13606 standard compliant platform for the management of EHR and also clinical archetypes.

III. RESULTS

The platform architecture follows a modular design that can be used by the platform itself, but also by other clinical and research centers (see Figure 1). In order to meet these goals web service oriented architecture was used. This architecture uses web protocols (*http*, *https*) avoiding complexity in its access. These services are accessible by any platform, including the platform itself, and thus are independent of the programming language used by the client.

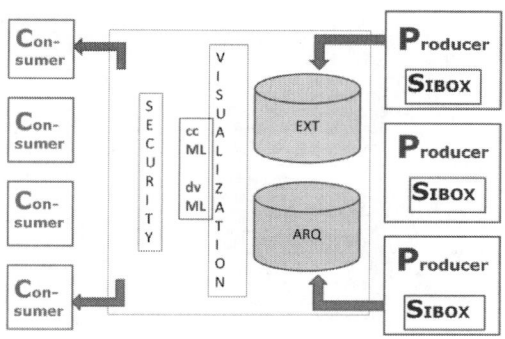

Fig. 1 Interoperability platform architecture.

Actual functioning of the platform is based upon the concept of Contributor Organization. This can be defined as any entity or public organization wishing to participate in a project to translate and manage its patient's EHR in compliance with the ISO EN-13606 standard.

The platform functioning architecture consists of three different levels of users.

The first level gives information to anyone publicly connecting to the platform [12]. Anyone connecting to the public web page of the platform can obtain examples of information data like those stored in it.

The second level is formed by any person or organization wishing to sign in and to upload, store and retrieve information from the platform. This person or organization may see and retrieve the information, and only this information, that he or she has formerly uploaded to the platform.

The Contributor Organization constitutes the third level in the platform's architecture. At the moment this Contributor Organization concept includes the *Hospital Universitario de Fuenlabrada*, *Hospital Puerta de Hierro*, *Hospital Clínic de Barcelona* and the *Universidad de Zaragoza*. In the case of the *Hospital Universitario de Fuenlabrada*, a software component called *SIBOX* (*Semantic Interoperability Box*) was installed on its servers (Figure 1). This component makes the EHRs anonymous [13] and sends them to the platform for ulterior processing. The platform has extracts and archetypes storage, and a visualization layer that uses languages ccML and dvML (see below) and a security layer. In this fashion a user assigned to a given project or healthcare institution will only be able to access those EHRs coming from that origin. This makes possible visualization of all the EHRs from a given center or project enabling subsequent Data Mining processing [14].

A. Security

Web services use free and open standards such as *http*, *XML* or *SOAP* for reading and interpretation of queries. This poses a serious security problem since these technologies are opaque to traditional security systems such as IP firewalls. In order to deal with these problems a threefold security policy was established. This policy was designed to improve three important aspects: a role or profile-based access control, communications and access to clinical information. Regarding access control a hierarchical structure of roles was designed in which roles assume functionalities of lower level roles. This security policy is carried out using the ISO22600 [15] model where privilege management establishes authoring restrictions; these restrictions are set using the *OASIS XACML* format [16] which allows definition and interchange of access policies between systems using the *XML* standard.

Communications between the Web Service and the client platform are encrypted. The protocol SSL was used. This protocol is not safe [17] and several measures have been adopted, namely SSL-MITM and anti-sniffy techniques [18].

IV. DISCUSSION

One important aspect of EHR management is that of EHR extracts edition and visualization. Regarding this issue the platform includes a visual extracts editor. This tool allows extracts edition by a practitioner from a practical and realistic standpoint. Since EHR extracts include 70% of their data from the so-called medico-legal context a new mark-up programming language called *ccML* (*context completion Markup Language*) has been designed, developed and tested [17, 18] in order to make the edition process automatic by the system and thus relieving the user from doing it manually and making extracts edition a realistic and practical task.

A new markup programming language *dvML* (*dual visualization Markup Language*) is currently under development in order to deal with the EHR visualization aspects related to the underlying dual software management model. This new language tackles the visualization aspects related to the RM used by the EHR management system and by those related to the archetypes model encoding the medical domain knowledge. The new *dvML* enables a medical practitioner to write a configuration file indicating the system how he or she wants to visualize the EHRs (Figures 2 and 3). For instance the user can tell the system which elements of the EHR he or she wants to be visualized

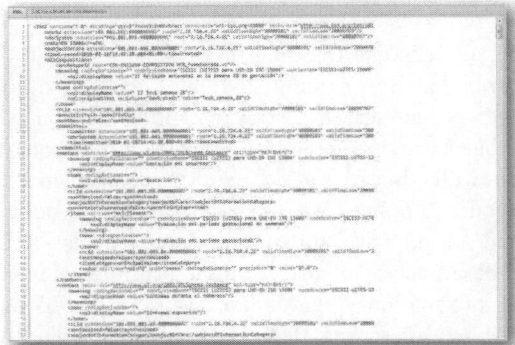

Fig. 2 Visualization of an EHR extract as XML.

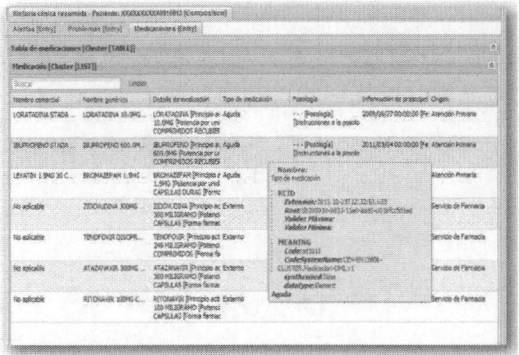

Fig. 3 Visualization of an EHR extract in a user-friendly manner.

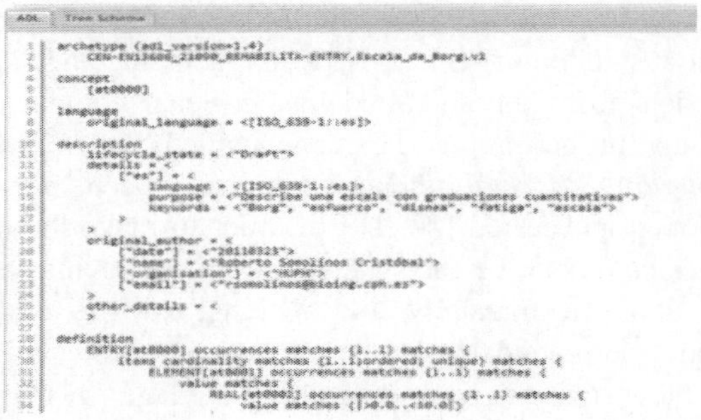

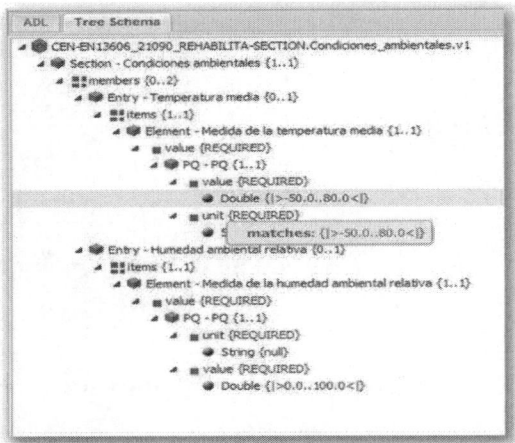

Fig. 5 Visualization of an archetype in the form of a data tree.

Regarding archetypes visualization the platform allows it in two different ways: in ADL (*Archetype Description Language*), the from in which they are first defined using the ISO EN13606 standard (Figure 4) and in the form of a data tree (Figure 5). This last visualization mode enables the user a clearer conceptualization of the knowledge stored in the archetype.

V. CONCLUSIONS

The platform developed by the Telemedicine and e-Health research Unit at the Instituto de Salud Carlos III constitutes a major milestone in the development and practice of systems based upon the ISO EN-13606 standard. It manages real data coming from many different origins and opens up the possibility to innovative research paradigms like medical knowledge discovery from databases, medical data mining and clinical decision support systems. It also makes possible the practical edition and visualization of EHRs and archetypes using newly designed markup programming languages and the practical interoperable communication of EHR systems moving one step closer to the everyday and generalized use of EHRs in hospitals and health centers.

Fig. 4 Visualization of an archetype in the form of an ADL file.

or hidden or in which format or color they will be displayed. This visualization is independent from the information stored, i.e. it should be transparent to any other user visualizing the same data.

The platform also includes an archetypes management subsystem. This subsystem includes a number of functionalities including insertion, validation and visualization of archetypes.

ACKNOWLEDGEMENTS

This work has been partially financed by projects PI09-90110 - PITES and PI08-1148 - CAMAMA from the "Fondo de Investigación Sanitaria (FIS)", "Plan Nacional de I-D-i" and by project CEN-20091043 - REHABILITA from the Plan Nacional de Investigación Científica, Desarrollo e Innovación Tecnológica (CENIT-E).

REFERENCES

1. Commission of the European Communities – COM. 356: e-Health – making health care better for European citizens: An action plan for a European e-Health Area. European Union, Publications Office. Brussels, 2004-04-30 2004.
2. US Department of Health and Human Services. Development and Adoption of a National Health Information Network (NHIN); Request for Information, Nov. 09, 2004, p.2.
3. Sim, I., Gorman, P., Greenes, R A., Haynes, R B., Kaplan, B., Lehmann, H., Tang, P C. Clinical decision support systems for the practice of evidence-based medicine. J Am Med Inform Assoc; 8: 527-34. 2001.
4. Giannopoulo, EG, ed. Data Mining in Medical and Biological Research. 2008. In-Teh.
5. Muñoz A, Somolinos R, Pascual M, Fragua JA, González MA, Monteagudo JL, et al (2007). "Proof-of-concept design and development of an EN13606-based electronic health care record service". J.Am.Med.Inform.Assoc. 14(1):118-129.
6. Muñoz, A. *"Interoperabilidad semántica entre los modelos de historia clínica electrónica de CEN y HL7. Propuesta de un modelo de armonización"*. Tesis Doctoral. Universidad Politécnica de Madrid. 2007
7. Muñoz Carrero A., del Pozo Guerrero F., Hernández Salvador C., Tous Llull F., Ferriol Monserrat P. *"Diseño de una pasarela HL7/EN 13606 para el intercambio de información de Telemonitorización domiciliaria de pacientes"*. XXIV Congreso Anual de la Sociedad Española de Ingeniería Biomédica. 2006.
8. Schloeffel, P., Beale, T., Hayworth, G., Heard, S. and Leslie, H. (2006). *"The relationship between CEN 13606, HL7 and openEHR"*. HIC 2006, Sydney, Australia, 2006.
9. Caceres Tello J., Muñoz Carrero A., Sánchez de Madariaga R., Castro Serrano A.L., Velázquez Aza I., Somolinos Cristóbal R. *"Plataforma de Visualización de Extractos conforme a la norma UNE-EN ISO 13606. Caso práctico"*. OpenHealth Spain 2011.
10. Rector, A., Qamar, R., Marley, T. Binding ontologies and coding systems to Electronic Health Records and messages. Proceedings KR-MED, 11-19. 2006.
11. Beale, T. Archetypes: constraints-based domain models for future-proof information systems. 2002. http://www.openehr.org/files/publications/archetypes/archetypes_beale_oopsla_2002.pdf (accessed April 2013).
12. http://amaltea.telemedicina.isciii.es/interServer/
13. R. Somolinos Cristóbal, A. Muñoz Carrero, M. Pascual Carrasco, M. Carmona Rodríguez, J.A. Fragua Méndez, M.A. González de Mingo, J. Cáceres Tello, R. Sánchez de Madariaga, A.L. Castro Serrano, I. Velázquez Aza, M.E. Hernando Pérez. "Sistema anonimizador conforme a la norma UNE-EN ISO 13606". CASEIB 2012.
14. Witten, I.H., Frank, E., Mark, A.H. Data Mining Practical Machine Learning Tools and Techniques, Third Edition. 2011. Elsevier.
15. ISO/DIS 22600-3: Health Informatics Privilege management and access control. Implementations.
16. eXtensible Access Control Markup Language (XACML). https://www.oasis-open.org/committees/tc_home.php?wg_abbrev=xacml (accessed April 2013)
17. Chomsiri, T. HTTPS Hacking Protection. 21st International Conference on Advanced Infromation Networking and Applications Workshops, 2007, AINAW'07. Pages 590-594.
18. LIN Ting-pi. *"The Analysis of Advantages and Disadvantages of the Network Sniffer Behavior"*. Journal of Sanming University 2008-02
19. Sánchez de Madariaga, R., Muñoz, A., Cáceres, J., Somolinos, R., Pascual, M., Martínez, I., Salvador, C.H., Monteagudo, J.L. ccML, a new markup language to improve ISO/EN 13606-based Electronic Health Record extracts practical edition. J Am Med Inform Assoc (JAMIA) 2013;**20**:298-304. doi:10.1136/amiajnl-2011-000722
20. Sánchez de Madariaga, R., Cáceres, J., Somolinos, R., Castro, A., Merino, LM., Velázquez, I., Muñoz, A. 2012. miML, un nuevo lenguaje de marcado diseñado para la edición práctica de Historia Clínica Electrónica según la norma UNE-EN ISO 13606. XV Congreso Nacional de Informática de la Salud, Inforsalud 2012. Madrid, 20-22 de marzo de 2012.

Development of a Visual Editor for the Definition of HL7 CDA Archetypes

David Moner, José Alberto Maldonado, Diego Boscá, Alejandro Mañas, and Montserrat Robles

Grupo de Informática Biomédica, Instituto ITACA, Universitat Politècnica de València, Spain

Abstract—This paper describes the design and implementation of a clinical-oriented tool for the definition of HL7 CDA templates based on the archetype methodology proposed by CEN/ISO 13606 and openEHR. The use of archetypes together with HL7 CDA brings along new possibilities for clinical model definitions based on CDA. It allows taking profit of the formal nature of archetypes as well as their ability of binding medical knowledge and technical requirements into the same artefact. To achieve this objective, LinkEHR archetype editor has been adapted to support and facilitate the use of the HL7 CDA reference model by clinical users. This paper shows the problems faced and the solutions that have been developed in order to implement a tool that can ease the way HL7 CDA templates are currently defined.

Keywords—Archetype, HL7 CDA, Clinical information model, Electronic health record, Semantic interoperability.

I. INTRODUCTION

The definition of clinical information models, also called detailed clinical models, is a relevant topic for faithful health information management. A clinical information model defines a discrete set of precise clinical knowledge and data that documents a specific health topic. The new generation of standards for the representation of Electronic Health Record (EHR) data, including CEN/ISO 13606 and HL7 CDA, recognize the need of a separate formal definition of those models, in the form of archetypes or templates respectively.

Our proposed approach combines both standards, by using archetypes to define template definitions of HL7 CDA documents, thus solving some limitations of the CDA template model. For example, templates provide limited reuse capabilities. They do not include an inheritance mechanism and only support the inclusion of references to external components by explicit naming them. They also provide a much less elaborated definition of the semantics of the clinical model, only by constraining an attribute code to a particular coded value, and do not support multilingualism either. Finally, the constraining expressivity of templates is limited in comparison with archetypes [1].

To achieve our objective, appropriate tools must be developed, that hide the inherent complexity of the standards and bring closer the archetype/template definition process healthcare professionals.

II. METHODS

A. HL7 CDA

HL7 Clinical Document Architecture (CDA) [2] is currently one of the most widely adopted standard for the representation of clinical information in the form of persistent XML documents.

A CDA document can contain any kind of clinical information in a narrative form (in XHTML format) and at the same time, structure it using predefined coded entries such as Observation, Procedure, Organizer, Supply, Encounter, Substance, Act, etc.

Given that CDA has a generic structure, a mechanism is needed to define specific schemas or document types. This role is covered by templates. A template is a definition (usually in the form of Schematron or implementation guides) that describes a configuration or combination of the CDA classes, and constraints on the attribute values. It gives instructions and validation rules for the creation of specific CDA instances for a particular clinical domain.

B. Archetypes

The archetype approach or dual model architecture is currently part of CEN/ISO 13606 norm [3] and openEHR specifications (www.openehr.org). Archetypes are definitions of the documentation associated to domain-level information models in a formal syntax, which provide a powerful, reusable and interoperable mechanism for managing the creation, description, validation, and query of EHRs. For each model, an archetype can be developed in terms of constraints on structure, types and values of classes of an object-oriented reference model (RM). A RM provides the basic structure to represent clinical data and its context information (dates, authorship, etc.) Examples of archetypes include prescriptions, health problems, pregnancy reports or blood pressure observations. The underlying idea behind the archetype approach is to provide a means for health professionals to express their domain knowledge separately from the technical representation of clinical data instances and without the intervention of technical specialists.

Other benefits of archetypes include being reusable (they can be specialized to better fit the requirements or be aggregated to create more complex archetypes), being multilingual (they can be defined in different languages to improve human readability) and serving as a link between an information structure and the terminologies or ontologies that semantically describe that information.

C. LinkEHR Editor

The work reported in this paper has been carried out by extending the functionality of LinkEHR Editor (www.linkehr.com). LinkEHR Editor is a reference model-independent archetype editor that offers a generic interface that can be used for the definition of archetypes based on any RM or data standard that has been previously imported into the tool. Additionally to this generic interface, specific, clinical-oriented interfaces can be developed to ease the use of the tool by archetype designers. Before this work, a specific editor existed only for CEN/ISO 13606.

A second functionality of LinkEHR is to help to the transformation and normalization of existing legacy data into standardized data that are compliant with both the reference model and the archetype definitions. Detailed information about LinkEHR can be found in [4] and [5].

III. RESULTS

In order to support a new RM, it must be first imported into LinkEHR. Among other formats, LinkEHR supports XML Schema as a format for the RM definition. Since CDA is normatively defined as an XML Schema, the process of importing it is nearly straightforward. During this import process we have to select which classes of the reference model will be available as root entities of archetypes. In the case of CDA, it was decided to support the ClinicalDocument and Section classes, together with all specializations of the coded entries (Observation, Act, etc.). The import process analyzes and stores internally all the data structure, fixed constraints or values and data types definitions of the RM.

Once imported, we could use directly the generic interface to define CDA archetypes, but the main objective of this work was to simplify the usage of the tool specifically with that standard. It is a complex standard where many different data structures coexist in order to support the representation of any type of clinical information. This requires to create deeply nested structures to represent the clinical data, both in structured and un-structured form, together with context data needed to safely interpret that information.

For example, if we want to represent the drug used in a SubstanceAdministration act, we would need to create an instance of the ClinicalDocument, which includes an instance of a medication Section, with a SubstanceAdministration entry, with a ManufacturedProduct, which finally includes the LabeledDrug. Additionally, each instance will need to define the value of several internal attributes. The complete structure can be seen at Figure 1.

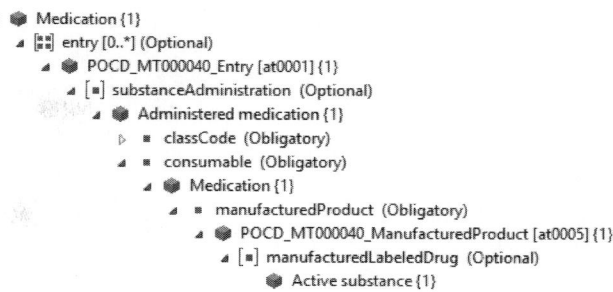

Fig. 1 Example of the generic editor view of a CDA archetype

LinkEHR was modified to hide all that complexity for a clinical user, so that he or she can be focused in defining the clinical model instead of dealing with the technical complexity of the standard. Following with the previous example, the simplified editor only shows those classes and attributes that are clinically relevant (Figure 2).

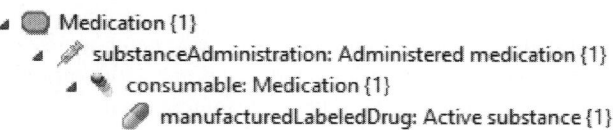

Fig. 2 Example of the simplified editor view of a CDA archetype

Specific classes are created to encapsulate the management of each component of the complete CDA structure. The selection of classes has been made taking into account existing implementation guides, in order to learn which structures are usually constrained in CDA documents. Once an instance of these container classes is created, they can automatically generate the equivalent CDA structures compliant with the complete RM. This simplified representation has been developed for the ClinicalDocument class, header classes (Author, Custodian, RecordTarget, Title, Code, ConfidentialityCode, TemplateId), body classes (StructuredBody, NonXMLBody, Section, Act, Organizer, Encounter, Procedure, Supply, SubstanceAdministration, ObservationMedia, Observation), and a selection of most relevant data types (Boolean, Integer, Real, SimpleText, TimeStamp, PhysicalQuantity, several coded values and multimedia data).

A key problem also faced was to simplify the archetype tree as seen in Figure 2 without losing the access to the most relevant properties of the model, which could be of interest for the archetype designer to be constrained. This was achieved by creating forms associated to each node, where those properties were properly displayed to be easily identified and modified by the user (Figure 3).

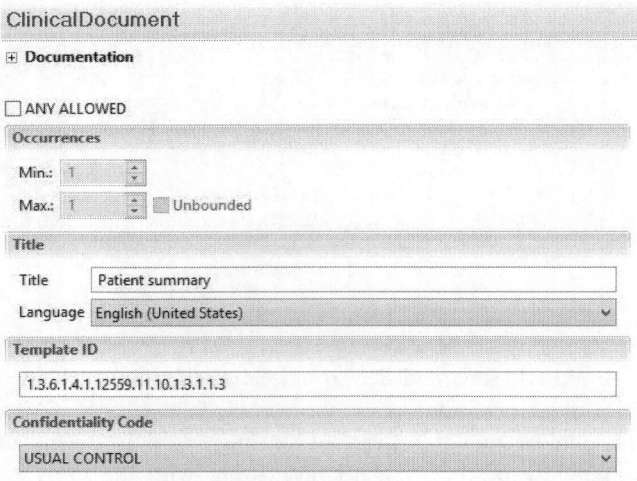

Fig. 3 Some properties of the ClinicalDocument class

The access and use of terminologies was also a relevant topic. Some CDA attributes must use a fixed terminology or predefined value sets. For example, CDA recommends the use of LOINC for describing the meaning of the different sections of the document. Other attributes, such as ConfidentialityCode, ClassCode or MoodCode, have a fixed set of possible codes. To ease the inclusion of those codes, the internal terminology of CDA and some external terminologies such as LOINC have been imported into LinkEHR. The tool provides a simple access to them and makes possible for a clinical user to select the codes he needs in an integrated environment, as can be seen in Figure 4.

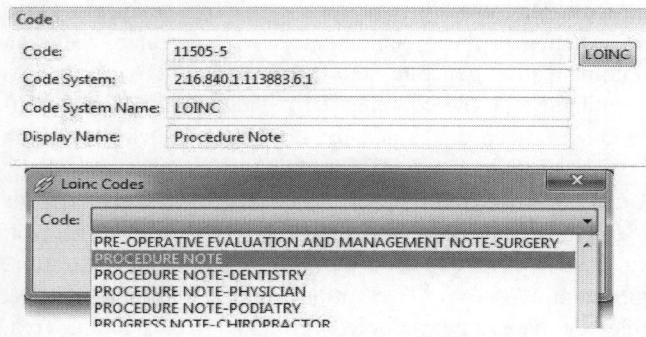

Fig. 4 Inclusion of LOINC codes through the editor interface

IV. DISCUSSION

The resulting HL7 CDA archetype editor can be evaluated from two different perspectives. On the one hand, we can evaluate if the approach of using archetypes for defining CDA templates is technically feasible and covers all the needs of clinical modelers. On the other hand, we can evaluate the benefits of CDA archetypes and developed tools in a real use case, in order to confirm the practical advantages of this technology.

A. HL7 CDA Reference Model Support in LinkEHR

As previously said, LinkEHR Editor can import any XML Schema as the definition of a RM. But while archetypes define constraints over an object-oriented RM, HL7 CDA XML Schema is a document-oriented model. It is necessary to make some decisions on how to adapt the XML Schema specifications to a pure object-oriented model during its importation process into the tool.

For example, the "Choice" structure in an XML Schema allows the definition of alternatives of elements even if both are also defined as mandatory (e.g. a choice between structuredBody and nonXMLBody). In an object-oriented model, these elements become represented as optional attributes of a class, so that it is a responsibility of the archetype designer to decide which one will be the mandatory.

Another problem was to preserve the namespace information of the CDA schema. This information will be needed to generate valid XML instances of the RM by the transformation and normalization functionality included in LinkEHR. In this case, the namespace information is stored in a documentation file that is generated in parallel to the importation process. This information will be recovered and used afterwards, during the generation of normalized instances.

A third problem exists, which is related to the support of "mixed nodes". In XML Schema, a complex type can be defined as "mixed". This means that it can contain both plain text and other XML elements or complex types (for example, embedded markup). This is a common case in CDA, for example in any narrative section, where plain text and XHTML constructions are mixed. This behavior cannot be directly represented as an object-oriented class, where all information must be stored in explicit properties. This fact made necessary to generate a virtual attribute of type String that we called "linkEHR_mixed_content", to avoid collisions with other existing attributes. Other attributes representing the inner elements are also generated. This allows to define plain text content inside this new attribute and to use the other structural attributes to add more data.

B. Reduction of the Clinical Model Development Cost

A question to be answered is if this dedicated editor simplifies the task of creating HL7 CDA archetypes. We can measure the cost of creating a clinical model archetype by analyzing the number of steps or interactions of the user needed to define a constraint. We have measured the number of clicks needed to create a Section, a CustodianOrganization and a LabeledDrug with and without the simplified editor (Table 1). Study cases were selected as representatives of general use cases (the Section is a mandatory node at any clinical document, the CustodianOrganization is representative of context information and LabeledDrug is representative of detailed information of clinical acts).

Table 1 Interactions needed to create a node constraint

Constrained node	Number of steps (generic editor)	Number of steps (simplified editor)
Section	16	4
CustodianOrganization	11	6
LabeledDrug	36	10

C. Use Case: HL7 CDA Genetic Test Report

The HL7 CDA implementation guide for the Genetic Test Report (GTR) [6] is a 107 pages document that defines the template to be followed by any CDA report about genetic information. LinkEHR Editor was used to build CDA archetypes following the indications of the GTR implementation guide. Over 50 different data items (archetype nodes) were defined and constrained following the definitions of the guide, including the name of the node, its textual description, terminology binding (semantic description), occurrences, identifiers (templateId) and data types. Several individual archetypes were created to maximize the reusability of those definitions. To mention a few, Genetic report, Cytogenetics, Genetic variation and Specimen archetypes were created. The genetic report, which defines the global document that includes all the other archetypes, can be seen in Figure 5.

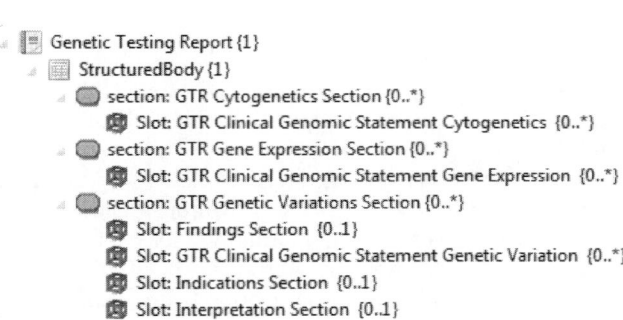

Fig. 5 Partial view of the archetype for the genetic test report

The experience demonstrated that archetypes can represent faithfully all constraints and conditions that are part of the definition of a CDA clinical information model.

As a second use case, a more detailed example of the application of the HL7 CDA archetype editor for defining archetypes for nephrology information was presented in [7].

V. CONCLUSIONS

The developed work has demonstrated that it is feasible to apply the archetype methodology to a reference model such as HL7 CDA by using LinkEHR Editor. To the author's knowledge, it is the first existing tool that allows the definition of archetypes for different standards or reference models. It is clear that a complex RM requires specific implementations in order to create a clinically-friendly tools. Having archetypes as a basic semantic resource will bring new possibilities to automatically generate other useful resources such as Schematron or other validation scripts. Regarding the use of this approach in real deployments, it showed as an effective solution to define archetypes by clinicians in an integrated and controlled environment that hides the unnecessary complexity of the CDA model.

ACKNOWLEDGMENT

This work has been partially funded by the Spanish Ministry Science and Innovation and the FEDER program through grant TIN2010-21388-C02-01.

REFERENCES

1. Bointer K. Duftschmid G. (2009) HL7 Template Model and EN/ISO 13606 Archetype Object Model – A Comparison. Stud Health Technol Inform. IOS Press; 150:249
2. Dolin B.H., Alschuler L., Boyer S., Beebe C., Behlen F.M., Biron P.V., Shabo A. (2006) HL7 Clinical Document Architecture, Release 2. J Am Med Inform Assoc; 13:30-39 (doi:10.1197/jamia.M1888)
3. ISO 13606-2:2008. Health informatics -- Electronic health record communication -- Part 2: Archetype interchange specification
4. Maldonado J.A., Moner D., Boscá D., Fernández J.T., Angulo C., Robles M. (2009) LinkEHR-Ed: A multireference model archetype editor based on formal semantics. International Journal of Medical Informatics 78(8), pp. 559-570.
5. Maldonado JA et al. (2011) Using the ResearchEHR platform to facilitate the practical application of the EHR standards. J Biomed Inform (2011), doi:10.1016/j.jbi.2011.11.004
6. CDA Implementation Guide for Genetic Testing Report (GTR) (September 2011 Draft) at http://www.hl7.org/documentcenter
7. Moner M, Moreno A, Maldonado JA, Robles M, Parra C. (2012) Using Archetypes for Defining CDA Templates. XXIV Int. Conf. of the EFMI (MIE 2012). Quality of life through quality of information, pp. 53-57, IOS Press BV, Amsterdam. ISBN 978-1-61499-100-7

homeRuleML Version 2.1:
A Revised and Extended Version of the homeRuleML Concept

H.A. McDonald[1,*], C.D. Nugent[1,2], J. Hallberg[2], D.D. Finlay[1], and G. Moore[1]

[1] School of Computing and Mathematics, University of Ulster, Jordanstown, Northern Ireland
swan-h2@email.ulster.ac.uk, {cd.nugent,d.finlay,g.moore}@ulster.com
[2] Department of Computer Science, Electrical and Space Engineering, Luleå University of Technology, Sweden
josef.hallberg@ltu.se

Abstract—As a direct result of the changes in global demographics, a significant amount of research has been undertaken in the area of home based support and healthcare provision, particularly in the direction of smart home environments. When applied to data generated within a smart home environment, decision support rules have the potential to recognise an inhabitant's behaviour and provide suitable support and assistance when required. homeRuleML is an XML-based format for the storage and exchange of decision support rules generated within smart home environments. In our current work we have extended upon the concepts of homeRuleML and have subsequently developed an improved format. The evolution of homeRuleML from version 1.0 to version 2.1 has been documented within this paper.

Keywords—Smart Home Environments, Decision Support Rules, homeML and XML.

I. Introduction

Studies have shown that the global population is experiencing a transition from a predominantly younger population to one with a much higher percentage of older adults [1]. Coupled with an increase in the number of people suffering from chronic diseases such as heart failure, diabetes and dementia [1], there is a growing demand for the development and implementation of smart home environments to facilitate the independent living of its inhabitants [2]. Monitoring an inhabitant within such an environment has the potential to support and assist said inhabitant in a non-intrusive way [2].

Monitoring an inhabitant within a smart home environment can produce an abundance of data which lacks context and meaning. Through the application of decision support rules to this data it is possible to recognise an inhabitant's behaviour, including what activity they are performing, how they are performing it and what stage of the activity the inhabitant has reached [3].

As a direct result of the large amount of research being performed within the smart environments research domain, an abundance of decision support rules have been produced, with a largely heterogeneous nature. As a result, the opportunity to share and re-use decision support rules is becoming limited. Similarly to the generation of data within a smart home environment, decision support rules can be time consuming and expensive to both test and validate. Therefore, to enable the sharing of tested decision support rules, a common standard is required. To the best of the authors' knowledge no standard format currently exists within this research domain specifying how decision support rules generated within a smart home environment should be formatted and stored. The work presented within this paper extends a concept, homeRuleML [4], proposed as a means of addressing the need for a standard format to store decision support rules generated within a smart home environment.

The remainder of this paper is organised as follows: Section 2 provides an overview of the notion of decision support rules within smart environments. Section 3 presents the methods used in evaluating and extending homeRuleML and Section 4 summaries the study and provides an overview of the future research direction for homeRuleML version 2.1.

II. Background

Decision support rules are used to express algorithms and reactions to an event typically consisting of a set of conditions and an ordered list of actions, which are executed when all conditions are met [5]. Decision support rules when applied to data generated within a smart home environment create an opportunity for the reliable recognition of inhabitants' Activities of Daily Living (ADL). They also create the opportunity to identify what stage of an ADL an inhabitant has reached and render adequate assistance to support the inhabitant as they complete specific tasks [6]. In addition to being able to recognise what stage of an ADL an inhabitant has reached; decision support rules can potentially recognise dangerous situations which require some form of alarm to be raised or intervention to be delivered.

One possible structure for decision support rules is in the form of the ECA (Event Decision Action) notion; were an event can be viewed as a signal that triggers a rule, the condition is a logical test that if met starts an action and an action updates or changes data. Figure 1 depicts an example of a decision support rule conforming to the notion of ECA. Although ECA rules allow a person to reason more precisely in terms of the relationship between the occurrence and the actions that should follow as a response to them, they are limited in regard to the representation of complex event detection [7]. In light of this there is a recognised practical need to compliment the ECA rule representation with the representation of other important aspects of knowledge [8].

> **ON** Medication dispenser not activated
> **IF** Patient is at home
> **DO** Send reminder to inhabitant
> **AND** Inform Care Provider

Fig. 1 Representation of a rule written in EAC format illustrating when an inhabitant is required to take their medication once a day. If this rule is not met an alarm is trigged informing the care provider and a reminder is sent to the inhabitant reminding them to take their medication.

III. HOMERULEML

homeML is an XML-based open format for the storage and exchange of data generated within a smart home environment [10]. As technology has evolved it is now possible to monitor a person both inside and outside of their home environment. It was therefore essential that the homeML format also evolve in order to accommodate these advances. Subsequently, the homeML format has evolved from version 1.0 in 2007 [10] to version 2.2 in 2012. At the time of its original proposal, homeRuleML was designed to complement the use of homeML. Therefore, as homeML has evolved there was a necessity for the evolution of homeRuleML, in order for both formats to continue to complement each other when deployed within a smart home environment.

homeRuleML [4] is an XML-based open format for the storage and exchange of decision support rules generated within a smart home environment. It was proposed as a means of solving the problems caused by the heterogeneous nature of decision support rules generated within a smart home environment. It was also proposed as an attempt to reduce the duplication of rules which have already been tested and evaluated within the research domain. The main benefits of using XML are its application, vendor and platform independence; whilst following a straight forward hierarchical data structure [9]. homeRuleML has evolved from version 1.0 in 2007 [4] to version 2.1 in 2012. This evaluation process is documented within the remainder of this section.

The original homeRuleML format [4] was proposed in 2007 as a series of two hierarchical data tree structures, called Rule and Operation. Initially the homeRuleML version 1.0 format was compared to the homeML version 2.2 format. As a result of this review process discrepancies were identified and a number of amendments made to the original homeRuleML format. A list of the amendments made to the format can be viewed in Table 1.

Table 1 Amendments made to homeRuleML version 1.0 Rule Tree and Operation Tree following the initial review process.

Rule Tree		
Element	Alteration	Rational
Description	Name changed to: ruleDescription	Aligns element name with the homeML naming convention.
Device	Device element removed and locationDevice element and mobileDevice element added which can both occur multiple times.	Aligns format with the the homeML format.
Operation Tree		
Element	Alteration	Rational
Description	Name changed to: ruleDescription	Aligns element name with the homeML naming convention.
Device	Device element removed and locationDevice element and mobileDevice element added which can both occur multiple times.	Aligns format with the homeML format.
deviceType	Element removed.	As the homeRuleML schema is designed to complement the homeML format, this element is not required.
Unit	Element removed.	As the homeRuleML schema is designed to complement the homeML format, this element is not required.
Description	Name changed to: ruleDescription	Aligns element name with the homeML naming convention.

Within a smart home environment it is recognised that data from sensors can be unreliable, noisy, imprecise and prone to delay [11]. Taking this into consideration the addition of an 'Uncertainty' tag to both the Rule and Operation trees was proposed as an addition. Within the homeRuleML format uncertainty refers to either the reliability of a sensor or interference rule uncertainty [11]. As the amendments made during this review process were considered to be significant, the versioning of the format increased incrementally from version 1.0 to 2.0.

IV. OPEN SOURCE DECISION SUPPORT RULES

At the time of its original development homeRuleML was not validated or reviewed. Subsequently a second review process was performed and its purpose was to ensure that the homeRuleML format was accurate and capable of storing a range of rules that could potentially be used within a smart home environment. A number of open-source decision support rules, listed in Table 2, were accessed; in addition to the two decision support rules outlined within the original homeRuleML publication [4].

Table 2 Table showing the source of the decision support rules used to review the homeRuleML version 2.0 schema.

Institution	Project	Rule	Reference
University of Ulster	COGKNOW	Preparing Tea Preparing Dinner Reading	[12]
National Institute for Health and Clinical Excellence (NICE)	PH16	Mental Health and Older People	[13]
	CG127	Hypertension: Clinical Management of Primary Hypertension in Adults	[14]

The decision support rules used during this process were selected as they have all been published within the research domain and have therefore previously been critically reviewed and deemed acceptable and correct. The authors also felt it was important to cover a wide range of decision support rules that could potentially be deployed within a real smart home environment. Each of the decision support rules were manually converted into the homeRuleML version 2.0 format in order to identify any unsupported data and any unnecessary elements.

Table 3 Amendments made to homeRuleML version 2.0 Rule Tree and Operation Tree following the second review process.

Rule Tree		
Element	Alteration	Rational
ruleDescription	Element now optional rather than essential.	Not every rule recorded a description.
dateRecordCreated	Element now optional rather than essential.	Not every rule recorded the date it was created.
Uncertainty	Element now optional rather than essential.	Not every rule recorded uncertainty.
Operation Tree		
Element	Alteration	Rational
Uncertainty	Element now optional rather than essential.	Not every rule recorded uncertainty.

The review process identified a small number of instances where the homeRuleML version 2.0 format was unable to encapsulate the rules. The amendments and rational for each adjustment are listed in Table 3. As minor amendments were made to the homeRuleML version 2.0 format the version number increased incrementally from version 2.0 to version 2.1. The hierarchical data trees representing the latest version of homeRuleML can be viewed in Figure 2.

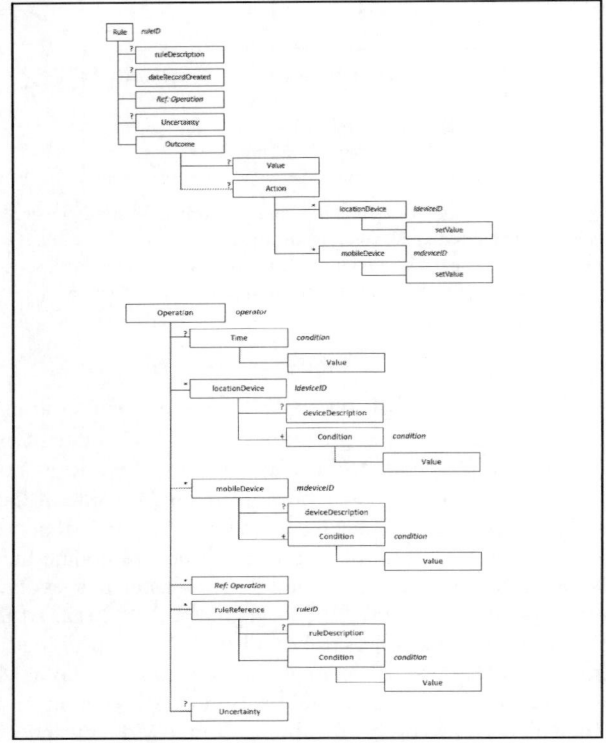

Fig. 2 Overview of the homeRuleML version 2.1 tree structure.

V. DISCUSSION AND FUTURE WORK

It has been the aim of this study to develop a solution to the heterogeneous nature of decision support rules generated within a smart home environment. This study extends upon the previously proposed solution referred to as homeRuleML [4]. homeRuleML [4] is an XML-based open format for the storage and exchange of decision support rules generated within a smart home environment. It has been proposed as a means of addressing the need for a standard format to store decision support rules generated within a smart home environment.

Over the duration of the study the homeRuleML format was iteratively redesigned, evolving from version 1.0 in

2007 [4] to version 2.1 in 2012. The authors hypothesise that the proposed format has the potential to improve the exchange and reuse of decision support rules generated within a smart home environment amongst researchers and between systems.

The homeML Suite[1] is an online resource existing of a repository and corresponding suite of tools to facilitate and support the use of homeML [9].

In our current work we have extended the homeML Suite in order to incorporate homeRuleML. Users have access to a tool that enables them to design decision support rules using the homeRuleML version 2.1. Figure 3 is a screenshot showing the web form available via the homeML Suite enabling users to design a decision support rule and an excerpt of an XML file created using the tool.

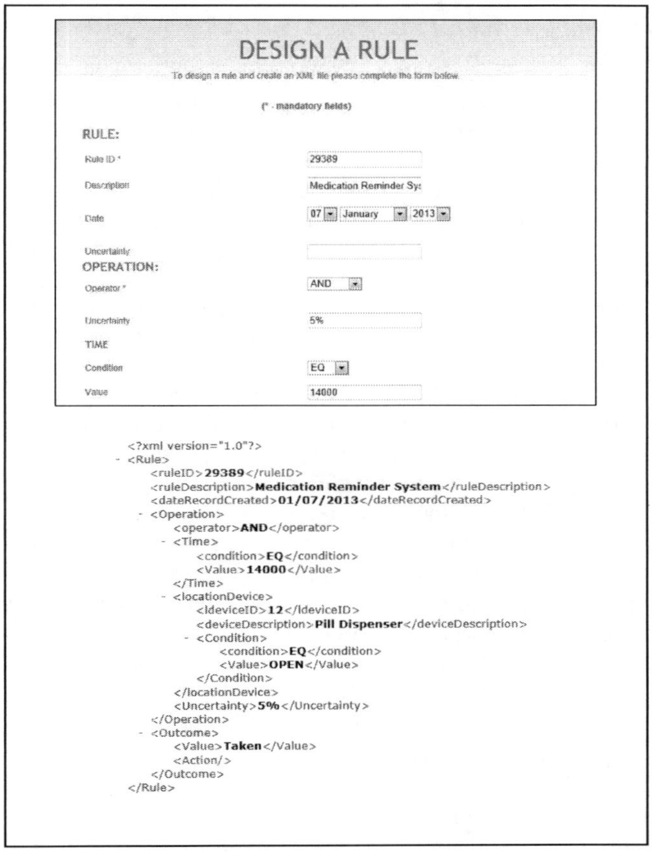

In addition to this a repository is also available which will provide researchers with the facilities to share and compare decision support rules generated within a smart home environment.

Although each version of homeRuleML underwent a rigorous review process and the results from preliminary evaluations have been positive, additional validation tests are necessary in order to ensure the latest version of homeRuleML is complete and ready for use within the research domain. Currently a number of domain experts are performing evaluations of both the format and the homeML Suite in order to gain additional validation of the homeRuleML version 2.1 format.

References

1. Coughlin J. F and Pope J. (2008) Innovations in health, wellness, and aging-in-place. IEEE Eng. Med. Biol 27: 47-52
2. Chan M, Campo E, Estève D and Fourniols J.Y. (2009) Smart homes—Current features and future perspectives. Maturitas 64:90-97
3. Philipose M, Fishkin K.P, Perkowitz et al. (2004) Inferring activities from interactions with objects. IEEE Pervasive Computing 3:50-57
4. Hallberg J, Nugent C.D, Davies R.J et al. (2007) HomeRuleML-A model for the exchange of decision support rules within smart environments. Automation Science and Engineering 513-520
5. Terfloth K, Wittenburg G and Schiller J. (2006) FACTS–A rule-based middleware architecture for wireless sensor networks. Communication System Software and Middleware 1-8
6. Storf H, Becker M and Riedl M. (2009) Rule-based activity recognition framework: Challenges, technique and learning. Pervasive Computing Technologies for Healthcare 1-7
7. Augusto J.C, and Nugent C.D. (2004) The use of temporal reasoning and management of complex events in smart homes. ECAI 778–782
8. Papamarkos G, Poulovassilis A and Wood P.T. (2003) Event-condition-action rule languages for the semantic web. Workshop on Semantic Web and Databases
9. Feki M, Abdulrazak B, and Mokhtari M. (2001) XML modelisation of smart home environment. Smart Homes and Health Telematics 55-60
10. Nugent C.D, Finlay D, Davies R et al. (2007) homeML–an open standard for the exchange of data within smart environments. Pervasive Computing for Quality of Life Enhancement 121-129
11. McKeever S, Ye J, Coyle L and Dobson S. (2009) Using dempster-shafer theory of evidence for situation inference. Smart Sensing and Context 149-162
12. Mokhtari M, Aloulou H, Tiberghien T et al. (2012) New trends to support ageing people with mild dementia : A mini review. Gerontology 58:554–563
13. NICE, "Occupational therapy interventions and physical activity interventions to promote the mental wellbeing of older people in primary care and residential care," vol. 2013, October 2008.
14. NICE, "Hypertension Clinical management of primary hypertension in adults," vol. 2013, August 2011.

Fig. 3 Screenshot of the 'Design a Rule' form, used to design a decision support rule that adheres to the homeRuleML version 2.1 format and an excerpt of XML code designed using the tool.

---

[1] Access to the homeML Suite can be obtained using the link: `http://www.home-ml.org/Browser/index.php`

Detailed Clinical Models Governance System in a Regional EHR Project

D. Bosca[1], L. Marco[1], D. Moner[1], J.A. Maldonado[1], L. Insa[2], and M. Robles[1]

[1] Grupo IBIME, Instituto Universitario de Aplicaciones de las Tecnologías de la Información y de las Comunicaciones Avanzadas (ITACA), Universitat Politècnica de Valencia, Spain
[2] Agència Valenciana de Salut, Valencia, Spain

Abstract—**In this work we present the Concept Oriented Repository (ROC), a system developed for the management of clinical information models, also known as detailed clinical models (DCM). It has been developed to be used in the Electronic Health Record project of the Valencia regional health agency (AVS). The system uses DCMs as a way to define clinical models independently of the healthcare standard chosen by the organization. These definitions create a framework where different actors can come to agreements on which information has to be represented and managed in the project. These concepts can be used later for the definition of technical artifacts (archetypes, templates, forms or message definitions) to be used by AVS information systems.**

Keywords—**Detailed Clinical Models, Archetype, Knowledge management, Semantic interoperability, Standards, Electronic Health Records.**

I. INTRODUCTION

Putting the knowledge in the center of Health Information Systems is one of the most important trends all around the world. But currently, different terminologies and standards are used at different stages of care process, which makes difficult knowledge and information sharing between healthcare providers. Several organizations and committees are working in different alternatives to define community agreed reusable clinical concepts. In fact, the use of formal concept definitions is already a reality in countries such as UK, Australia, Brazil, Netherlands, or Sweden. There are more than 15 clinical concepts repositories in the world that use different formats for concept definitions. The NEHTA Clinical Knowledge Manager [1] (a national version of the openEHR Clinical Knowledge Manager [2]), Clinical Element Models (CEM) by Intermountain Healthcare [3] or HL7 CDA and DICOM IHE profiles [4] are examples of the importance of this approach.

The main problem is that each repository represents the concept in a particular standard or format. The relationships between similar concepts represented in different standards are unknown, which at the end hinders interoperability. Detailed Clinical Models (DCMs) were created due to the need of making compatible and connect those concept definitions. DCMs provide an abstraction layer independent of the chosen standard. DCMs group available definitions under one single clinical model (i.e. discharge report, patient summary, etc.). This grouping is made from a semantic perspective, not forcing syntactic transformations of existing data.

The objective is to have a unique semantic reference for each clinical model where we can also find associated technical resources and documentation that implement those clinical models in deployed systems.

II. METHODOLOGY

A. Detailed Clinical Models

DCMs can be defined in a conceptual and structural way [5]. Conceptually, DCMs are information models of a discrete set of precise clinical knowledge which can be used in a variety of contexts. Structurally DCM are descriptions of items of clinical information that include the clinical knowledge on the concept, the data specification, a model and where possible, technical implementation specifications. Provide data elements and attributes, including possible values and attribute types, needed to express clinical reality in an understandable way for both clinical domain experts and modelers.

If we want to introduce the DCM workflow and strategy into our systems, it is needed to coordinate the development, maintenance, and evolution of DCM avoiding the overlap of concepts over different healthcare domains. It is also needed to promote the creation of a validated reference set of DCM based on clinical evidence (when possible) and to allow the persistence of DCM independently of their format or standard. Finally, we will need mechanisms for accessing and locating existing DCMs.

B. Concept Oriented Repository

The Electronic Health Record (HSE) project of the Valencia regional Health Agency (AVS) is a centralized system whose main purpose is to guarantee the access to relevant standardized clinical information to both patients and health professionals. This system, currently in pre-production is based the EN 13606 standard and archetypes, but also support for HL7 CDA templates and local data structures. Thus, the main requirement was to develop a mechanism for the formal management of all those different models.

The Concept Oriented Repository (ROC) has been developed in order to provide a methodology and a clinical-oriented tool to manage these models and ease its use by the HSE project. The implementation of ROC has been guided by the philosophy of DCMs, by managing high level generic and standard-independent models that are afterwards associated to technical resources (archetypes, templates, and documentation). The development was a joint effort with the AVS functional group.

III. RESULTS

ROC is a web application developed in Java using Vaadin framework [6]. It is deployed as a central service of the AVS, together with the EHR data repository, the terminology server and the identification service.

The designed system architecture (Figure 1) provides support for the management, discovery, and publication of concepts through a user interface and web services. It also provides an audit system for administrators.

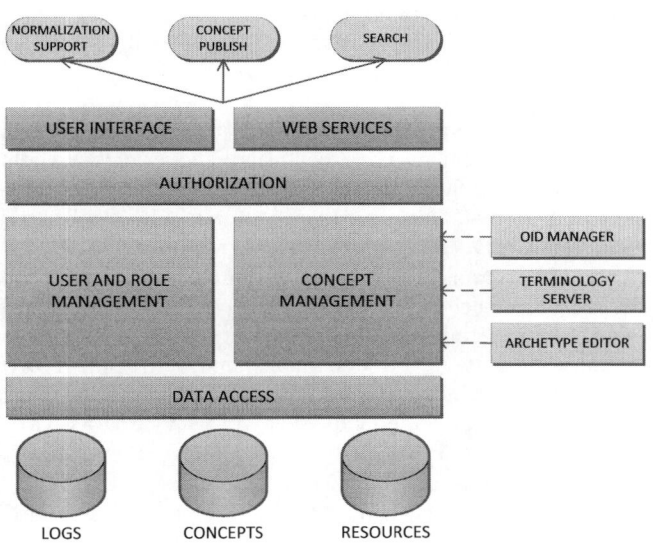

Fig. 1 ROC architecture

The main characteristics of ROC are described in the following sections.

A. Independency of Standards, Specification Models and Formats

The repository defines a DCM as a set of metadata. Associated to it, specifications in any native format can be attached, from text documents, spreadsheets, or pdf documents to computable definitions such as ADL archetypes or references to the executable programs used to give support to that concept (such as Java or XQuery programs).

B. Management of DCM Relationships

ROC provides mechanisms to link DCMs between them. In addition to versioning and specialization, the system allows the definition of semantic links between DCMs in the form of inclusions, exclusions, and associations. Inclusion can be seen as the containment of one concept inside of another one. Exclusion marks the prohibition of containment (the concept is not part of the other). Association is a weaker relationship meaning that two concepts are somehow related. Figure 2 shows a simulated example of DCM relationships supported on the repository.

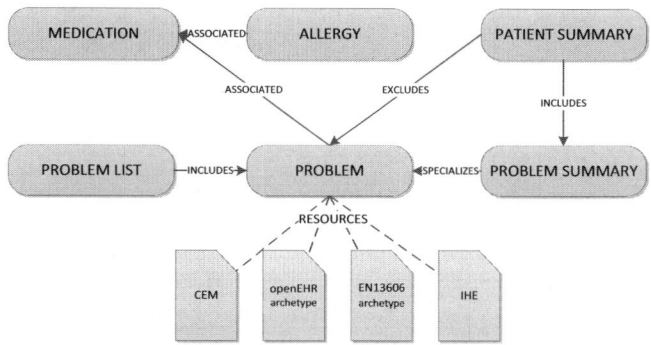

Fig. 2 Example of DCM in ROC

In this example, problem summary is a specialization of problem. Both medication and allergy are associated with problem DCM. Patient summary only contains a problem summary, but not the full problem, so problem is excluded and problem summary is included into patient summary. Problem is included into the DCM problem list (we could also have excluded problem summary from problem list). The problem DCM has also different associated resources: three structured resources (two archetypes, openEHR-EHR-EVALUATION.problem.v1 and CEN-EN13606-ENTRY.problem.v1, and the CEM definition for problem called HealthIssue) and the IHE profile of Concern Entry. All these metadata, relationships, and associated resources are what we call a DCM on the repository. It is worth noticing that ROC should not be considered as an ontology, since DCMs rely on these three parts to be fully defined.

All these kinds of relationships between concepts are shown in the tool as mindmaps, which makes them easier to understand and maintain.

C. Management of DCM Lifecycle

ROC implements a model to manage the lifecycle of DCMs from their creation as drafts to being published or marked as obsolete. This, in addition to the version history makes possible to retrieve a present or past DCM definition when desired. In particular, concepts can be in one of the following stages: Draft, Team review, Public, Validated, Obsolete, and Revoked. 'Draft' status is used for the first upload or creation of a concept which its validity is yet unknown. 'Team review' status is used for a concept which is being iteratively revised by a team to reach a consensus on it. 'Public' status is used to mark a concept which is mature enough to be used but has yet to be validated on a live system. The latter is marked with 'Validated' status. From there a concept could be marked as 'Obsolete', marking that concept is not valid on its current form and probably would need a new version to be correct again or 'Revoked' that implies that the concept itself is no more valid and should not be versioned or specialized.

D. Collaborative Edition

One of the basic aspects of the tool is to ease the concurrent and collaborative edition of concepts. The tool allows the creation of specific working groups for domains or even specific concepts by defining roles and permissions both at concept and application level. The tool also provides the user with more mechanisms to ease collaborative edition of concepts, like being able to subscribe to concepts to be informed of their changes or providing a notification wall with the latest changes on the subscribed concepts. The repository also includes a comment system to let users suggest improvements during concept lifecycle.

E. Search DCM by Its Structure, Content, Metadata, or Ontology

Repository includes the definitions of a set of metadata associated to each DCM. Metadata include fields such as unique name, description, DCM Type, version, original language and translations, lifecycle state, authors, managers, authoring and revision dates, etc. To obtain this minimum metadata subset we analyzed the proposed metadata in ISO 13972 draft "Quality processes regarding detailed clinical model development, governance, publishing and maintenance" [7], CEN TS 15699 [8], which is an extension of Dublin Core norm [9] for healthcare domain, and metadata set defined in CEN EN13606 part 2 [10] for archetype definition. Additionally, resources related to each DCM have also associated metadata (such as author, description, language, format, or organization) to ease resource identification and discovery. Finally, if the associated resource is defined in a computable format (such as an ADL archetype) system can search for specific information inside it.

F. Multilingual Support

Both concepts definition and the repository User Interface allow the use of multiple languages. A subset of the concept metadata has been considered to be language dependent and can be translated within the tool.

G. User Management

ROC is connected to the AVS user authentication system in order to be easily included on the AVS workflow.

Additionally, different roles with different permission set are defined depending on the role of the user in the system (concept creators, reviewers, translators, technical staff, etc.). The repository provides different functionality depending of the effective role of each user.

H. Connection to Other Systems

The repository is also related with other tools available in the AVS, such as the OID manager (GOID) to assign unique identifiers to DCMs, archetypes and implementation guides, access to the terminology service for the mapping of the concepts to available medical terminologies, and archetype editors that will be used for the definition and validation of the concepts stored in ROC.

IV. DISCUSSION

There are several clinical information model repositories around the world. However, they usually cover only a single format to define those models. Some define their own representation format, such as Intermountain CEML [3], the Netherlands DCM project [11] or Japan MedXML MML[12]. In fact the Netherlands approach was one of the first to introduce the use of DCM (in a sense of standard independent definitions), but they were defined in a non-computable way (word documents and UML diagrams) which makes difficult their automatic use on the systems. Also, neither the Netherlands DCM project nor MedXML MML provide a repository where manage the concepts lifecycle, versions, or relationships.

Some countries such as Australia [1] and Russia [13] have adopted the Clinical Knowledge Manager approach [14], which is a collaborative repository for defining and managing openEHR archetypes. Other national projects dealing with archetypes define their own public repository, like Minas Gerais regional EHR system in Brazil [15], which has developed and made public a set of ISO EN13606 archetypes.

The main difference between ROC and all discussed repositories is that ROC supports definition of DCMs independently of any representation or standard. Concepts defined in ROC can reference any of the former repositories instead of keeping a copy of the resource. Metadata defined for ROC also includes the subset of metadata that is used to describe the archetypes. As archetypes carry their own metadata, they can be used as basis to create new DCM.

V. CONCLUSIONS

The definition of clinical information models independently of the standard of the health information systems is currently one of the main trends around the world. The work currently being developed at Clinical Information Modeling Initiative (CIMI) working group [16] tries to create a model for the concept definition independently of chosen clinical records representation standards and propose it as a standard to the OMG. CIMI participants include IHTSDO, HL7, CDISC, EN13606 association, openEHR, Mayo Clinic, National Health Service (UK), or National Institute of Health (US). ROC is totally aligned with this philosophy and could support this future standard.

One of the proposed improvements is to integrate the repository in the concept development process. Connecting ROC with a tool like LinkEHR archetype editor [17] assures that archetypes included in ROC would be valid syntactically and semantically according to a chosen reference model.

Both DCM definition process and lifecycle management require the implication and communication of all the different actors of the project. Success in this project would come from the understanding and cooperation among the members of the multidisciplinary team in charge of the creation and management of DCMs. Finally success will also come from the effective use of the information models as reference definitions for the development of AVS information systems.

ACKNOWLEDGMENT

This work has been funded by Electronic Health Record project from Valencia Health Agency (HSEAVS).

REFERENCES

1. National eHealth Transition Authority Clinical Knowledge Manager (http://dcm.nehta.org.au/ckm/)
2. openEHR Clinical Knowledge Manager (http://www.openehr.org/ckm/)
3. Intermountain Healthcare CEM Browser (http://www.clinicalelement.com)
4. Integrating the Heathcare Enterprise, IHE (www.ihe.net)
5. HL7 Detailed Clinical Models Wiki (http://wiki.hl7.org/index.php?title=Detailed_Clinical_Models) (Accessed on 24-04-2013)
6. Vaadin, the open source Web application framework for rich client applications (https://vaadin.com/)
7. ISO 13972 Standard: Health Informatics - Detailed Clinical Models. Part 1: Quality processes regarding detailed clinical model development, governance, publishing and maintenance'. Committee draft (2011)
8. CEN TS 15699 Standard: Health Informatics – Clinical Knowledge Resources - Metadata (2009)
9. ISO 15836:2009 Information and documentation -- The Dublin Core metadata element set
10. CEN EN 13606 Standard: Health informatics – Electronic health record communication - Part 2: Archetype interchange specification (2008)
11. Goossen WT, Goossen-Baremans A. Bridging the HL7 template: 13606 archetype gap with detailed clinical models. Stud Health Technol Inform. 2010;160:932–936
12. MedXML consortium Medical Markup Language (MML) specifications (http://www.medxml.net/E_mml30/MMLV3Spec.pdf)
13. емиас Clinical Knowledge Manager (http://simickm.ru/ckm/)
14. Garde S, Chen R, Leslie H, Beale T, McNicoll I, Heard S. Archetype-based Knowledge Management for Semantic Interoperability of Electronic Health Records. Medical Informatics Europe 2009; 2009 September; Sarajevo, Bosnia & Herzegovina.
15. Santos, M. R.; Bax, M. P; Kalra, D. Dealing with the Archetypes Development Process for a Regional EHR System. Applied Clinical Informatics, v. 3, p. 258-275, 2012.
16. Clinical Information Modeling Initiative (http://informatics.mayo.edu/CIMI)
17. Maldonado JA, Moner D, Boscá D, Fernández-Breis JT, Angulo C, Robles M.. LinkEHR-Ed: A multi-reference model archetype editor based on formal semantics. Int. Journal of Medical Informatics, vol 78, Aug 2009, pp. 559-570.

Corresponding author:

Author: Diego Boscá Tomás
Institute: IBIME Group, ITACA, UPV
Street: Camino de Vera s/n
City: Valencia
Country: Spain
Email: diebosto@upv.es

An Extensible Free Software Platform for Managing Image-Based Clinical Trials

M.A. Laguna[1], N. Malpica[2], and J.A. Hernández-Tamames[2]

[1] Hospital General Universitario de Ciudad Real, Ciudad Real, Spain
[2] Universidad Rey Juan Carlos, Móstoles, Spain / Fundación CIEN - Fundación Reina Sofía, Madrid, Spain

Abstract—The Multiclinical Trial Web PACS (MCTWP) is a free software platform to support image management in clinical trials. This means not only long-term image storage, but also trial protocol support, post-processing, task management and result registration. The application makes it possible to ensure traceability and replicability in experiments and measurements.

MCTWP has been developed in Java. It follows the MVC pattern and uses several free software projects. It has been designed to be image-format independent, so it can manage DICOM, Analyze and Nifti, and can also be extended to include other standards.

The resulting application is currently being used in Fundación CIEN in Madrid, for performing studies on neurodegenerative diseases.

The application increases quality and consistency of trial processes and results, and allows to revisit a certain trial in the future, as new techniques become available, or to confront results with new clinical findings.

Keywords— Clinical Trial, Image Management, Data Provenance.

I. INTRODUCTION

Medical imaging-based clinical trials are a growing approach in drug testing and clinical research. In some cases, new image processing and analysis software is even developed specifically for a clinical trial. Despite that, many clinical trials are still being managed without a proper information system [1].

Clinical trial management gets more complicated when using images to obtain biomarkers. There are several factors that limit the role of imaging in clinical trials, such as: management of data collections, standardization of imaging protocols, standardized image review and assessment [2].

These factors can be controlled using an adequate information system that addresses the right functionalities to support image management in clinical trials. We have detected the need for the following functionalities:

- Establishing an image review process. This can be implemented as a manual quality control process, to review all images received before they are accepted in order to check that every image fulfills all requirements.

- Storing and classifying images. Once an image is accepted, it has to be stored and identified in the clinical trial. A typical trial stores images into patient studies, and divides patients into groups.

- Supporting trial protocol. Images of a clinical trial are processed in order to extract data or obtain new graphical representations according to the clinical trial protocol.

- Ensuring traceability and replicability in the experiments and measurements. Input images cannot be deleted once task results are added to them. It should also be possible to retrieve the source images and the processes associated to every result.

When no information system is applied to manage clinical trials, it is hard to support the above functionalities. In such a scenario, images are dispersed in several media formats such as external hard disks, DVD's, laptop computers and so on.

We propose a free software application which allows not only to manage images, but to control the clinical trial workflow in a centralized way, avoiding dispersion of images and ensuring compliance with the trial protocol.

Our solution supports the trial protocol by means of processes, tasks and image types that make it possible to define rules that create new tasks automatically, track task changes, record results, and so on.

II. BACKGROUND

Several previous works have faced these problems, using PACS [3, 4] as starting point. Some shortcomings of these solutions are the following:

- PACS are DICOM based. Thus it is not possible to store other image file formats, such as Nifti or Analyze.
- The data model of PACS is patient focused.
- PACS are designed for the health care process.

PACS need new functionalities to address some clinical trial needs [5]. For example:

- The Medical Image Resource Center (MIRC) [6], is an Electronic Teaching Files (ETF) Server software created by RSNA(Radiological Society of North America.), that is based on PACS concepts.

- IHE (Integrating the Healthcare Enterprise) includes a technical framework for radiology which includes a profile for teaching files and clinical trial export [7].
- MIDAS [8], is another open-source web-based digital archiving system that handles large collections of scientific data, that it is not tied to DICOM.

Lately, some projects have applied grid software to clinical trials. A good example of this is caGRID which is a project of the National Cancer Institute of the United States, which aims to develop a federation of interoperable research information systems.

The National Biomedical Imaging Archive (NBIA) [9], is a part of this project and is in charge of image management. NBIA is also DICOM based and does not support clinical trial workflow, as it is managed by other caGRID components.

XNAT [10], is a recent solution for clinical image management and clinical trial workflow. It is a software platform developed by a free software community and supported by BIRN (Biomedical Informatics Research Network, http://www.birncommunity.org/), HHMI (Howard Hughes Medical Institute, http://www.hhmi.org/), Harvard University and Washington University.

It gives support to neuroimaging laboratory needs, and it is somehow similar to this work, although there are some differences:

- Trial protocol support. Our solution allows researchers to define protocols that guide the workflow of image analysis pipelines, tracking every step and its results, ensuring traceability and replicability in experiments and measurements.
- Interface: Our solution has a less broad field of application, which results in a simpler and more usable interface for managing centralized image-based clinical trials.

MCTWP has been developed at Laboratorio de Análisis de Imagen Médica y Biometría of Universidad Rey Juan Carlos and Fundación CIEN (Fundación Centro de Investigación en Enfermerdades Neurológicas, http://www.fundacioncien.es/). We are evaluating MCTWP at Fundación CIEN, in Madrid.

Our group has been working for several years on this subject. We have already proposed a first model for image-based clinical trials [11]. The model divides the clinical trial into seven different steps, ranging from image acquisition to the visualization of results. Image processing and statistical analysis are core steps of the process.

Figure 1 shows the main data entities and the relationship among them. We want to ensure efficiency and data sturdiness, implementing task, protocol and image management together.

III. METHOD

Free software plays a very important role in this project. Several free software projects have been used in order to build our system. Some of them implement image support, while others deal with data persistence, communications or view components. We also wanted to contribute to this effort, so we have licensed this project as free software, using the GPL v3 license. Source code is stored in Google code at http://code.google.com/p/mctwp/.

A. Architectural Principles

This software is a web application that follows the Model View Controller pattern. The business layer implements data persistence, workflow, image management and storage, DICOM communications, and so on. This layer has been divided into sublayers, each fulfilling its own responsibilities.

Model entities, workflow and Image management are explained in the following sections. Data persistence was implemented using the Hibernate project. DICOM communications are implemented with dcm4che2 (as suggested in [12]) and it is used to allow the application to send and receive studies through DICOM C-STORE.

Command pattern [13], is used to link the view and business layers. Every function of the business layer is encapsulated into a command. Commands ease the implementation of several transversal concerns as authorization, audit trail and transactions configuration.

B. Data Model and Workflow

Figure 1 depicts the main entities in clinical trials. Briefly, the application manages trials that contain groups. Every group contains patients, and these contain studies. Images are stored into studies, and every image is assigned one or several tasks. Researchers perform tasks, obtaining statistical data or new images that are recorded as task results.

The trial protocol must be defined at the beginning of the trial. The protocol is defined by means of processes and image types. The application has an entity called ProcessDef, which defines a kind of image processing technique that could be applied to any type of image, for example: inhomogeneity correction, tissue segmentation, normalization, etc.

Images acquired and result images must have only one image type. The application has an entity called ImageType

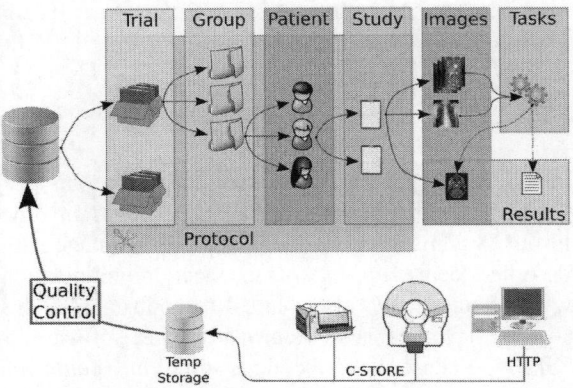

Fig. 1 Workflow and main entities of the application

that defines a kind of image, for example: structural images, functional images, diffusion images, etc. It is important to note that image type is not related to the image format, but to the functionality of the image inside the trial. Actually, the image type of each image is manually selected, but we are working on defining rules that select the right image type using meta-data.

A task is an entity of the application that relates one or more images with a process. All the images of a task have the same image type. Every task is assigned to a researcher that must complete it and record the obtained results, such as quantitative measures on the individual images, population statistics or even new images. All tasks are traced, so that every change is properly recorded.

As stated before, the protocol defines the task that must be created automatically. Every `protocolable` entity is an abstract class which can hold a protocol definition. For example, `Trials`, `Groups` or `Patients` are `protocolable` entities. These entities can also hold images, so when new images are associated with any `protocolable` entity, the application creates a new task for every image if the image type matches that of the entity.

The application allows any authorized user to delete or modify images if there are no results registered to any task of the images. When results are registered, the application controls that source images remain unmodified. This feature and task traceability, allow the application to ensure replicability of experiments.

C. Image Management

The application needs to manipulate images using specific standard operations, such as loading images, obtaining demographic data, converting between formats, and so on. To decouple these operations from the application, the interface `ImagePlugin` interface is used. This interface, declares all operation that must be implemented for every specific standard. We denote by image plugin every class implementing this interface.

Image identifiers management are taken into account. This application uses its own image identifiers instead of general ones, ensuring consistency, by means of an identifier mapper that knows how to match specific image IDs with our own internal Ids.

Every image is stored as is, although the application is able to transform non DICOM result images to DICOM in order to send them as secondary capture type images to DICOM workstations or PACS.

D. Exporting Image Results as Secondary Capture

Another important feature of our solution is that it is able to export functional analysis results using DICOM Secondary Capture. We used a similar approach to Maldjian,[14]. There are other ways to do this, as explained in CAD-PACS,[15]. They implement IHE recommendations, but we don't need such a complex scenario to fulfil our needs.

E. Storage Management

This application has to serve several purposes that require more than just a relational database. On the one hand, it is necessary to store all relational information about model entities, such as users, trials, patients, tasks, results, etc. On the other hand, it is necessary to store the original images.

Every image format has its own meta-information about modality, acquisition parameters, facility, demographics, etc. This information can be helpful, and it is worth not only storing it, but allowing the user to search into this information. As this information is not suitable for relational data bases (because it is not well structured information, and it is not the same between standards), it is necessary to use a different storage technology.

For these reasons we chose three different storage systems:

- Image collection storage. Here images remain unmodified, and they are grouped into collections. A collection represents a single clinical trial. This has no implication on image storage location; it simply means that images must be retrieved by their application ID.
- Meta information collection storage. XML databases are good at storing semi-structured data sets. It does not matter what XML database is selected, but it must support xPath queries in order to search for images. Here, every image must be indexed at least by its application ID.

- Relational database. All information about trials, groups, tasks, protocols, patients, results, etc. are stored into a relational database management system. Images are not contained in the relational model, but their application ID is. In this way, it is possible to coordinate all three parts.

The main advantage of this idea is its versatility. It is possible to request from the application images satisfying any meta-data criteria, and to obtain the original images at any time. As this requires maintaining all three parts synchronized, the business logic implements controls to avoid data inconsistencies.

IV. RESULTS

The resulting application is currently being used at Fundación CIEN in Madrid, for performing several studies on neurodegenerative diseases. The application can be easily tested, using the virtual machine that we have created for this purpose. This virtual machine and a brief manual are hosted in http://code.google.com/p/mctwp/.

The application is currently used to manage four clinical trials. All trials are focused on neurodegenerative diseases, and have a similar complexity. Each of them involves around a hundred patients divided into four groups and different types of image analysis processes are applied to all of the images (segmentation, volumetry, etc.) or to groups of images (i.e., morphometrical group analyses).

The application can manage several trials simultaneously. It has an administration area to manage trials, roles and users, and a research area where researchers can work on the trials, performing tasks on the images.

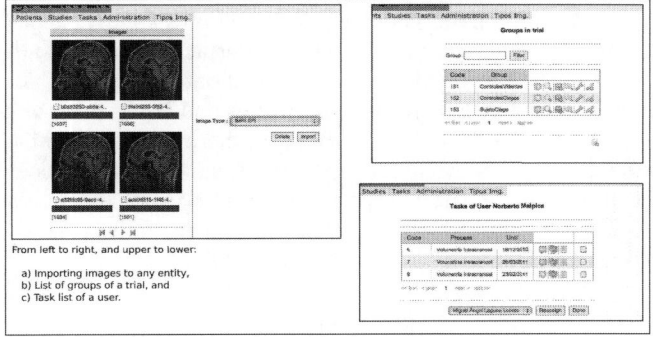

Fig. 2 Screenshot composition

Figure 2 shows a screenshot composition of the application.

A. Conclusion

We have developed a platform that allows the storage and management of all the images in a clinical trial. The management includes image storage according to the different groups in the study, the possibility of adding intermediate processed series of images, as well as incorporating the final results of the study.

The different image analysis steps required in the study can also be managed, assigning individual or groups of tasks to the different researchers working on the study. This allows tracking of the state of the different tasks at any moment, as well as a fluid distribution of different analysis steps among the different researchers. The tool will not only allow a more efficient and error-free study management, but will also allow revisiting a certain study when new image analysis or statistical tools are available. At the time of writing this article, the application does not allow tasks interfacing with third party applications, but this will be the subject of our future work.

Source code and further information about the project is available at http://code.google.com/p/mctwp/.

REFERENCES

1. Sharib A. K., Philip R. O., Payne M. P., Jhonson J. T., Kukafka R.. Modeling Clinical Trials Workflow in Community Practice Settings AMIA 2006 Symposium Proceedings. 2006:419-423.
2. Erickson B. J., Buckner J. C.. Imaging in clinical trials. Cancer Inform. 2007;4:13-8.
3. Huang H. K.. PACS is only in the beginning of being used as a clinical research tool Biomed Imaging Interv. 2007;3:e12-41.
4. Langer S., Bartholmai B.. Imaging informatics: challenges in multi-site imaging trials. J Digit Imaging. 2011;24:151–159.
5. Huang H. K.. PACS and Imaging Informatics: Basic Principles and Applications. Wiley-Liss 2004.
6. Lim C. C., Yang G. L., Nowinski W. L., Hui F.. Medical Image Resource Center–making electronic teaching files from PACS. J Digit Imaging. 2003;16:331–336.
7. IHE International, Inc. . IHE Radiology Technical Framework . 2011.
8. Jomier J., Aylward S., Marion C., Lee J., Styner M.. A digital archiving system and distributed server-side processing of large datasets Proc. SPIE. 2009;7264.
9. Wiley A., Meenan C.. National Biomedical Imaging Archive (NBIA) web 2012.
10. Marcus D. S., Olsen T. R., Ramaratnam M., Buckner R. L.. The Extensible Neuroimaging Archive Toolkit: an informatics platform for managing, exploring, and sharing neuroimaging data. Neuroinformatics. 2007;5:11–34.
11. Hernandez J. A., Acuna C. J., Castro Ma, Marcos E., Lopez M., Malpica N.. Web-PACS for Multicenter Clinical Trials Information Technology in Biomedicine, IEEE Transactions on. 2007;11:87-93.
12. Vazquez A., Bohn S., Gessat M.. Evaluation of open source DICOM frameworks MCIM Workshop 2007 Proceedings. 2007:C-10.
13. Gamma Erich. Design Patterns. Boston: Addison-Wesley 1995.

Reuse of Clinical Information: Integrating Primary Care and Clinical Research through a Bidirectional Standard Interface

Paolo Fraccaro, Valeria Pupella, Roberta Gazzarata, and Mauro Giacomini[*]

Department of Informatics, Bioengineering, Robotics and System Engineering; University of Genoa, Genoa, Italy

Abstract—Nowadays, most healthcare providers archive and manage patients' information in electronic format. The reuse of such Clinical Information is essential from both a scientific and financial perspective, and for the integration of Primary Care and Clinical Research; which has become a principal goal. The solution proposed aims at integrating data between a web platform, developed for research purposes, and Hospital Information Systems. Information is shared through a Bidirectional Standard Interface, which adopts biomedical controlled vocabularies and ontologies (LOINC, ICD and SNOMED), and HL7 messaging (CDA V3 R2) for data exchanging. Such an interface allows the importation of data from Hospital Information Systems to a web platform, and the exportation of information from the platform to Hospital Information Systems. Currently, this project has been implemented in the Infectious Diseases context, and complete data integration and semantic interoperability has been obtained between the "Ligurian HIV Clinical Network" and the Infectious Diseases Department of the San Paolo Hospital (Savona, Italy).

Keywords—Reuse of Clinical Information; Standard Interface; HL7; Clinical Document Architecture.

I. INTRODUCTION

Recent surveys have highlighted the importance of Information Technology in health care [1] and the positive aspects (notwithstanding some limitations) related to the adoption of Electronic Health Record Systems [2]. Despite encouraging results, the utilization of Health Information Technology is still insufficient and, also in industrialized countries, the scenario is heterogeneous [3]. However, even though integrated Electronic Health Records have still not been developed in many countries, most health care providers store and manage health information in an electronic format. Consequently, a large amount of data is produced during Primary Care and is already available in Hospital Information Systems. The reuse of such information for Clinical Research has become a primary goal. The economic and scientific benefits of integration between Primary Care and Clinical Research have already been shown [4,5] and many workgroups have been focusing on this goal [6,7]. The final objective is to achieve the Single-Source Data-Entry which would permit the complete reutilization of health data. Therefore, the automatic transmission of information between Hospital Information Systems and Clinical Data Management Systems, which are widely used to manage data in the Clinical Studies context, would increase the quantity and quality of available clinical information for Clinical Research. Consequently, this would generate an improvement in patients' and public health.

Previous works [8–11] have shown that semantic interoperability is the essential requirement for obtaining an integration between Primary Care and Clinical Research. Unfortunately, this requisite is not easily achieved, and to obtain scalable and general applications seems to be a complex task. For the implementation of such systems, standards of communication and codification have to be used within Hospital Information Systems and Clinical Data Management Systems. Unfortunately in many countries Hospital Information Systems providers are still not adequately receptive to these themes. Furthermore, recent surveys proved there is a substantial heterogeneity concerning Clinical Data Management Systems in Europe [12]. Therefore, most of both commercial and open source applications often do not have a suitable architecture for the achievement of semantic interoperability.

The goal of this paper is to prove the possibility to obtain an effective and correct semantic interoperability between Primary Care and Clinical Research by the realization of a working prototype. The objective of the solution proposed is to realize data integration between a web platform, which has a general and scalable structure and is suitable for performing research within several medical contexts, and Hospital Information Systems. Information is exchanged through a Bidirectional Standard Interface, which adopts biomedical controlled vocabularies and ontologies (LOINC, ICD and SNOMED) for tagging values and HL7 messaging (CDA V3 R2) for data exchange. Furthermore, the bidirectionality also allows the importation of data which has been produced for research purposes and to reuse such data within routine clinical practice. Subsequently, both Primary Care and Clinical Research receive benefits.

II. METHODS

A highly structured information framework was necessary on both sides of the interface, in order to preserve the semantic meaning of the recorded values through standard codes and biomedical ontologies. Concerning the research side, a general prototype, appropriate for creating and managing Multicenter Clinical Trials in several Medical

---

[*] Corresponding author.

Domains, has already been developed by the authors [13]. The prototype was based on a relational database with a highly normalized structure supported by a large set of standard codes (e.g. LOINC, ICD and SNOMED) and a complete collection of information regarding the partner's work environment (e.g. units and normality ranges) [13]. These types of features permit semantic data sharing and comparison within Clinical Trials. From the Hospital Information Systems' perspective, a commercial Medical Record System has been used with similar characteristics of semantic enrichment.

From an architectural point of view, a Service Oriented Architecture was selected. In particular, a bidirectional Web Service and a Desktop Client Application was adopted to share information between the Research System and the Hospital Information System based on XML HL7-V3 messages. Specifically, the Version 3 of the standard was implemented because it has more complete and structured characteristics from a semantic point of view, than Version 2. Subsequently, within the Version 3, Common Message Element Types and the Clinical Document Architecture Release 2 (CDA-R2) [14] were chosen to perform actual information exchange. The HL7 CDA-R2 infrastructure is used to communicate any type of Clinical Information, independently from the Domain, and has high data structuring characteristics. Accordingly, the CDA-R2 was the ideal option to achieve the required aims and to exploit all the characteristics which were introduced in the previous step. Since, how described above, the Research database contained structured and coded information, it was possible to crate CDA-R2 documents using an iterative algorithm which dynamically filled a CDA-R2 template according to the Client Application's requests.

Finally, to avoid legal and privacy problems, the last challenge was to preserve the real identity of patients. To resolve such issues a HASH encryption algorithm was adopted. A similar type of algorithm has already been utilized within the web-platform [13] with a manual data insertion in the previous version of the system. In particular, during the registration process in the platform, the HASH encryption algorithm is applied to the national ID code of each patient to obtain a unique anonymous identifier. This operation is performed within the Hospital Information Systems' framework through a Desktop Client Application. Consequently, from one side, such an application has the rights to access all the information concerning patients and can manage the ID codes without any legal problems. From the other side, only HASH codes and limited registry information, from which it is impossible to trace to the patients' real identity, are sent in standard format across network, by using the HL7 "*PRPA_MT201301UV02Patient*" class, and stored through the Web Service. Therefore, for these reasons, the HASH encryption algorithm is both the optimal solution for simultaneously preserving privacy and identifying patients unequivocally.

III. RESULTS

Firstly, the proposed structure and architecture to integrate Primary Care and Clinical Research is scalable and repeatable. In fact, if the Hospital Information Systems have the right characteristics of Structured Clinical Information in relation to biomedical controlled vocabularies and ontologies, which have been previously described, such a structure can be reported in all the Hospitals of the partners who are participating in a research project independently from the medical domain. These results are possible due to the bidirectional HL7 standard interface of the Web Service. Accordingly, the only requirement in order to exploit the standard interface is to develop a Desktop Client Application which can access the Structured Clinical Information of Hospital Information Systems and then utilize the interface's web services through HL7 messaging. The list of the HL7 interaction between the Web Service and the Desktop Client Application is reported below:

a) **Clinical Studies list Pull:** The Web Service exposes the list of all ongoing Clinical Studies and some information regarding research purposes. The XML message adopts *ST* and *CE* HL7 V3 Data types.

b) **Clinical Study info Pull:** As a result of a Client request, the Web Service exposes the detailed information concerning a particular Clinical Trial. The request and the response utilize respectively *CE* and *POCD_MT000040Section* HL7 V3 Data types.

c) **Patient Push:** The Client sends to the Web Service the registry information of a new patient who requires registration and the list of the Clinical Studies in which the patient participates. The message adopts *PRPA_MT201301UV02Patient* and *CE* HL7 V3 Data Types and the unique identifier of the patient is the described HASH code.

d) **Patient info Pull:** As a result of a Client request, the Web Service exposes all the studies in which a patient participates and his/her registry information. The request and the response use respectively *PRPA_MT201301UV02Patient* HL7 V3 class and *CE* HL7 V3 Data types. The unique identifier of the patient is the HASH code.

e) **Clinical Information Pull:** As a result of a Client request, the Web Service exposes particular patient's Clinical Information. The request uses *PRPA_MT201301UV02Patient* HL7 V3 class and

CE HL7 V3 Data type while the response uses the *HL7 V3 CDA R2*. The unique identifier of the patient is the HASH code.

f) **Clinical Information Push:** The Client sends the patient's Clinical Information to the Web Service. The *HL7 V3 CDA R2* is the adopted infrastructure and the HASH code is the unique identifier.

```
<Patient>
  <Person>
    <id root="2.16.840.1.113883.2.9.3.17" extension="/F0w0UY3wByNd97cQDQL+g=="></id>
    <name text="MV"></name>
    <birthTime value="19880207000000"></birthTime>
    <administrativeGenderCode code="F"></administrativeGenderCode>
  </Person>
</Patient>
```

Fig 1: Example of Patient Push

Currently, the proposed system has been developed to integrate Primary Care and Clinical Research in the context of Infectious Diseases, particularly in the field of HIV infection. In the last years, HAART therapies have increased the survival rates of HIV+ patients and has led the infection to be considered a chronic disease. Therefore, HIV+ patients need a complex longitudinal follow-up. In addition, HIV+ patients are widely involved in Clinical Research to optimize the HAART therapies, which often cause side effects, with the goal of ultimately discovering a definitive cure. Consequently, the integration between Primary Care and Clinical Research is particularly suitable. A similar type of integration was developed between "The Ligurian HIV Clinical Network" and the Infectious Diseases Department of the San Paolo Hospital (Savona, Italy). "The Ligurian HIV Clinical Network", which has been previously developed by authors [13], is a web-platform adopted to conduct Multicenter Clinical Trials within the HIV context in North Italy (Ligurian and Piedmont region).The web-platform has the appropriate characteristics of data structuring to obtain semantic interoperability and it widely adopts biomedical ontologies. Furthermore, the effectiveness of its methods has been proved by the works which have been published through the utilization of this instrument [15]. Concerning the San Paolo Hospital, the Infectious Diseases Department is participating in the "Ligurian HIV Clinical Network" and the "Whale" information system, provided by El. Co. (Cairo Montenotte, Italy),which has the suitable characteristics to fulfill the required objectives. Examples of interactions between the two Systems already introduced are reported below. Figure 1 shows a part of an example of a *Patient Push*: the Web Services receives this message which contains the registry patient information e.g. initials, birth time and gender) and the patient HASH code (highlighted in the figure). The second example, shown in Figure 2, effectively explains the level of semantic interoperability attained. In the figure, the observations of a CDA sample which contains the Clinical Information of one patient are shown. The LOINC codes related to the monitored parameters (highlighted in green) and the information concerning values (highlighted in orange) permit a correct and effective comprehension without any errors.

IV. CONCLUSIONS

The methods and results which have been previously described permit a complete semantic interoperability between Primary Care and Clinical Research. Furthermore, one of the fundamental characteristics is the bidirectionality of the Web Service which has been developed. On the one hand, these features allow the exploitation of the large amount of Clinical Information which is already available in Hospital Information Systems for research purposes. This aspect leads to an increase in data quantity and quality within Clinical Research and above all avoids copy operations, time wasting and errors carriers, by physicians. On the other hand, the additional data which has been produced within Research activities can be used in Primary Care through the Web Service's HL7 Standard Interface and the Client Desktop Application. Consequently, physicians can exploit such information to improve patients' health during the routine Primary Care. Furthermore, since both methods and structure which have been presented are of a standard and general, such infrastructures are also appropriate for other medical domains. In conclusion, in the near future, developments are planned to exploit the Web Service's HL7 Standard Interface for data sharing between "The Ligurian HIV Clinical Network" and other research databases. In particular, work in this direction is already being carried out within a project which provides a complete data synchronization between "The Ligurian HIV Clinical Network" and "ARCA" which contains the largest quantity of HIV+ patients' virological data in Italy.

```xml
<observation classCode="OBS" moodCode="EVN">
    <code code="8122-4" codeSystem="2.16.840.1.113883.6.1" codeSystemName="LOINC" displayName="CD3 cells [#/volume] in Blood"/>
    <!-- Code for type of information within the observation-->
    <statusCode code="completed"/>
    <effectiveTime value="20120130110000"/>
    <!-- Date and time of observation -->
    <!--Observed value-->
    <value xsi:type="RTO_PQ_PQ">
        <numerator value="1690" unit="N"/>
        <denominator value="1" unit="mmc"/>
    </value>
    <!--Observed value-->
</observation>
<observation classCode="OBS" moodCode="EVN">
    <code code="27011-6" codeSystem="2.16.840.1.113883.6.1" codeSystemName="LOINC" displayName="CD14 cells [#/volume] in Blood"/>
    <!-- Code for type of information within the observation-->
    <statusCode code="completed"/>
    <effectiveTime value="20120130110000"/>
    <!-- Date and time of observation -->
    <!--Observed value-->
    <value xsi:type="RTO_PQ_PQ">
        <numerator value="0.01935" unit="N"/>
        <denominator value="1" unit="mmc"/>
    </value>
    <!--Observed value-->
</observation>
```

Fig 2: Observations of a CDA sample

ACKNOWLEDGMENT

The authors would like to thank: El.Co. S.R.L. and specifically Manuela Bagnasco for the collaboration and partial support of this reseach; Marco Anselmo, Director of the Infectious Diseases Department of San Paolo Hospital (Savona, Italy) and Doct. Pasqualina De Leo for their collaboration; Moyra Watson for the English review.

REFERENCES

[1] M. B. Buntin, M. F. Burke, M. C. Hoaglin, and D. Blumenthal, "The benefits of health information technology: a review of the recent literature shows predominantly positive results.," *Health affairs (Project Hope)*, vol. 30, no. 3, pp. 464–71, Mar. 2011.

[2] J. M. Holroyd-Leduc, D. Lorenzetti, S. E. Straus, L. Sykes, and H. Quan, "The impact of the electronic medical record on structure, process, and outcomes within primary care: a systematic review of the evidence.," *Journal of the American Medical Informatics Association: JAMIA*, vol. 18, no. 6, pp. 732–7, Jan. 2011.

[3] A. K. Jha, D. Doolan, D. Grandt, T. Scott, and D. W. Bates, "The use of health information technology in seven nations.," *International journal of medical informatics*, vol. 77, no. 12, pp. 848–54, Dec. 2008.

[4] J. G. Williams, W. Y. Cheung, D. R. Cohen, H. A. Hutchings, M. F. Longo, and I. T. Russell, "Can randomised trials rely on existing electronic data? A feasibility study to explore the value of routine data in health technology assessment.," *Health technology assessment (Winchester, England)*, vol. 7, no. 26, pp. iii, v–x, 1–117, Jan. 2003.

[5] M. Dugas, M. Lange, C. Müller-Tidow, P. Kirchhof, and H.-U. Prokosch, "Routine data from hospital information systems can support patient recruitment for clinical studies.," *Clinical trials (London, England)*, vol. 7, no. 2, pp. 183–9, Apr. 2010.

[6] C. Ohmann and W. Kuchinke, "Future developments of medical informatics from the viewpoint of networked clinical research. Interoperability and integration.," *Methods of information in medicine*, vol. 48, no. 1, pp. 45–54, Jan. 2009.

[7] H. U. Prokosch and T. Ganslandt, "Perspectives for medical informatics. Reusing the electronic medical record for clinical research.," *Methods of information in medicine*, vol. 48, no. 1, pp. 38–44, Jan. 2009.

[8] F. Gerdsen, S. Müeller, S. Jablonski, and H.-U. Prokosch, "Standardized exchange of medical data between a research database, an electronic patient record and an electronic health record using CDA/SCIPHOX.," *AMIA ... Annual Symposium proceedings / AMIA Symposium. AMIA Symposium*, p. 963, Jan. 2005.

[9] R. Kush, L. Alschuler, R. Ruggeri, S. Cassells, N. Gupta, L. Bain, K. Claise, M. Shah, and M. Nahm, "Implementing Single Source: the STARBRITE proof-of-concept study.," *Journal of the American Medical Informatics Association: JAMIA*, vol. 14, no. 5, pp. 662–73, 2007.

[10] A. El Fadly, B. Rance, N. Lucas, C. Mead, G. Chatellier, P.-Y. Lastic, M.-C. Jaulent, and C. Daniel, "Integrating clinical research with the Healthcare Enterprise: from the RE-USE project to the EHR4CR platform.," *Journal of biomedical informatics*, vol. 44 Suppl 1, no. null, pp. S94–102, Dec. 2011.

[11] F. Fritz, P. Bruland, S. Balhorn, and M. Dugas, "Quality of Information for Pruritus Research through Single Source," in *Medical Informatics Europe 2012*, Pisa (Italy), August 2012; available at: http://person.hst.aau.dk/ska/mie2012/CD/Interface MIE2012/MIE_2012_Content/MIE_2012_Content/SCO/268_CD_SC_Oral_ID_360.pdf.

[12] W. Kuchinke, C. Ohmann, Q. Yang, N. Salas, J. Lauritsen, F. Gueyffier, A. Leizorovicz, C. Schade-Brittinger, M. Wittenberg, Z. Voko, S. Gaynor, M. Cooney, P. Doran, A. Maggioni, A. Lorimer, F. Torres, G. McPherson, J. Charwill, M. Hellström, and S. Lejeune, "Heterogeneity prevails: the state of clinical trial data management in Europe - results of a survey of ECRIN centres.," *Trials*, vol. 11, p. 79, Jan. 2010.

[13] P. Fraccaro and M. Giacomini, "A Web-based Tool for Patients Cohorts and Clinical Trials management.," *Studies in Health Technology and Informatics*, pp. 554–558, 2012.

[14] R. Dolin and L. Alschuler, "HL7 clinical document architecture, release 2," *Journal of the American Medical Informatics Association: JAMIA*, pp. 30–39, 2006.

[15] C. Dentone, P. Fraccaro, D. Fenoglio, E. Firpo, G. Cenderello, R. Piscopo, G. Cassola, V. Bartolacci, G. Casalino Finocchio, P. De Leo, M. Guerra, G. Orofino, E. Mantia, M. Zoppi, G. Filaci, A. Parodi, A. De Maria, F. Bozzano, F. Marras, M. Sormani, A. Signori, B. Bruzzone, N. Nigro, G. Ferrea, M. Giacomini, C. Viscoli, and A. Di Biagio, "Use of maraviroc in clinical practice: a multicenter observational study.," *Journal of the International AIDS Society*, vol. 15, no. 6, p. 18265, Jan. 2012.

Archetype-Based Solution to Tele-Monitor Patients with Chronic Diseases

Juan Mario Rodríguez[1], Carlos Cavero Barca[1], Paolo Emilio Puddu[2], John Gialelis[3], Petros Chondros[3], Dimitris Karadimas[3], Kevin Keene[4], and Jan-Marc Verlinden[4]

[1] ATOS, ATOS Research and Innovation (ARI), Spain
[2] Sapienza University of Rome, Department of Cardiological Sciences, Rome, Italy
[3] Industrial Systems Institute/RC Athena, Platani Patras, Greece
[4] ZorgGemak BV, Voorschoten, Netherlands

Abstract— Health tele-monitoring systems can be applied to improve chronic diseases treatment and reduce cost of care delivery. Behind this innovative and promising philosophy in the care of people with chronic diseases several benefits can be found: hospitalizations may be reduced, improvement in the patients' quality of life and clinical evaluations more precise. The tele-monitoring includes measuring and collecting health information about individual patients, thus the evolving concept of Electronic Health Record (EHR) is crucial. Getting a shareable and universally accessible EHR is a challenge whose importance is considered by organizations that establish and manage standards. In the context of the CHIRON project[1], Congestive Heart Failure (CHF) patients are enrolled in an observational study to be tele-monitored by experienced doctors. Technical solutions have been designed to deal with EHR desirable features and visual requirements for remote visualization and study. An EN13606/openEHR compliant kernel is the core component dealing with multisource patient data making up a complete EHR system assuring semantic interoperability. EN13606 [1] and more concretely openEHR [2] follows the two-level modelling approach describing specifications of a reference model and archetypes to store, retrieve, exchange and manage health data in EHRs. Dynamic components to access, visualise and insert data into patients' records are shown through a doctor-friendly user interface called Slim MEST (light-weight Medical Expert Support Tool) and combines high flexibility and adaptability as it is built upon the same archetypes defined in openEHR-kernel. Functionality is extended by means of an ECG signal viewer application, which proofs versatility of data collected by sensors used in the tele-monitoring process providing the clinicians with a practical tool for their diagnosis.

Keywords—EHR, archetypes, tele-monitoring, EN13606, openEHR.

I. INTRODUCTION

The use of communication technology to monitor patients and their health status is a focal point for improving chronic disease management. Hospitalizations can be reduced, patients' self-care is enhanced and outcomes are meant to be more valuable. Risk identification, particularly short- and very short-term, and multiple interactions among old and newly investigated risk indexes [3] may be an effective tool for patient-centric tele-monitoring [4] with the potential of fostering the continuum of care [5] in patients with chronic diseases.

To detect diseases in an early stage, before symptoms have occurred, the CHIRON project investigates patients with Congestive Heart Failure (CHF) both at home and in ambulatory conditions. Newly envisaged risk functions are applied to go beyond what has been done recently [6].

Provision of Electronic Health Record (EHR), as well as demographic information and data integration, are assessed following openEHR open standard specification [1], which relies on archetypes as the standardized methodology to get reusability and scalability. Patient data obtained from heterogeneous sources can be stored using the openEHR reference model [7].

The project equips the patients with unobtrusive sensors to collect multiple physiological parameters such as ECG, skin temperature, sweating index and activity patterns, as long as means to store other parameters, getting a complete list of 67 parameters which were found relevant for CHF patients, after a survey between cardiologists experts. Raw data collected by the sensors is processed to obtain relevant statistics, which are incorporated to the EHR. The system also includes other EHR-based information coming from the Hospital Information System (HIS) such as medications or interventions. Furthermore, raw data are made available for visualisation purposes, in the case of ECG, using the ECG Viewer tool. This division of tasks creates a conceptually coherent and functional model, taking into account not only medical standards compliance but actual professionals needs. Slim MEST tool has been conceived to take advantage of the openEHR flexible dual model concept which separates out the reference model and the clinical knowledge (archetypes) modelled by doctors [9]. It encapsulates archetypes, data storage and retrieval and offers a doctor-oriented interface which gives a complete overview of parameter trends and the health profile of the patient by accessing the EHR. Figure 1 shows the components involved in the solution and their communication flows.

---
[1] www.chiron-project.eu

Archetype-Based Solution to Tele-Monitor Patients with Chronic Diseases

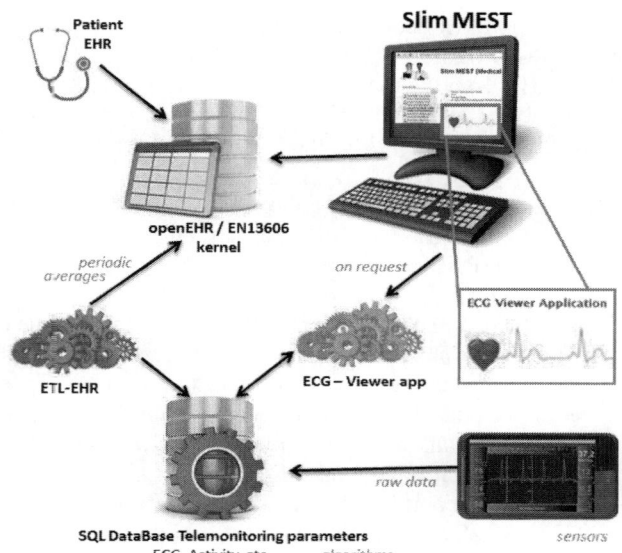

Fig. 1 Components diagram.

The achievements of the medical objectives are assessed through a six months observational study started in May. Although number of patients and people involved is limited, it attempts to provide a proof of concept and useful conclusions about the potential benefits of the exposed solution.

II. EHR INTEROPERABILITY

A. Archetypes

The use of EN13606 in eHealth systems is a proven solution for standardization [10]. To store (medical) parameters for the CHIRON project, the EN13606 standard has been used, and more specific the openEHR reference model. As the EN13606 is a subset of openEHR classes only shareable elements has been used (compositions and observations) to be compliant with both reference models. The key concept of EN13606/openEHR is the so-called archetypes to model the medical knowledge providing interoperability, standardisation and computability of EHR between systems. Each medical concept has been modelled by means of archetypes created ad-hoc or searched for in the openEHR clinical knowledge manager [11] which is a repository of archetypes already stored, validated and shared by community members. For the project these archetypes were designed and clustered following the cardiologists' suggestions. Concepts like NYHA classification, daily activity, daily weight change and CHF aetiology were covered, but also other archetypes already available in the openEHR knowledge manager about blood pressure, ECG and lab tests.

Finally the archetypes selected were approved by the cardiologists, which also validated the medical workflow.

B. Transforming and Storing Raw Data into Archetypes

Heterogeneous medical information comes from multiple systems. The openEHR kernel and middleware permits the connection with the data contained in the patients' EHRs transparently in a standardized way. The openEHR kernel complies with the EU standard for medical data transport using the approved and authorized CKM archetypes. Postgres database is used to store and retrieve the information to / from the archetypes via the openEHR kernel middleware REST-based API (https + JSON) using the ADL path/value approach. Further research is being done to use XML database (eXist-db) using XQuery as the query language.

The raw sensor data are stored in a MySQL database and has to be available in the EHR to provide the doctors a general overview of the patient's health. Therefore during night the daily averages of the parameters are extracted out and calculated from the MySQL database and published to the kernel using the archetypes EN13606 compliant included in the configuration provided in the openEHR configurator (summaries of compositions / entries). For instance, to store the relevant ECG information, instead of storing every single point, a calculation was done with an algorithm to extract the useful parameters from the ECG concerning CHF.

III. SLIM MEST

A. openEHR Configurator

Clinical concepts modelled as archetypes in openEHR are written in Archetype Definition Language (ADL) [7]. Due to the path/value nature of archetypes, a component to configure the clinical concepts and the corresponding archetypes was found useful and consequently developed.

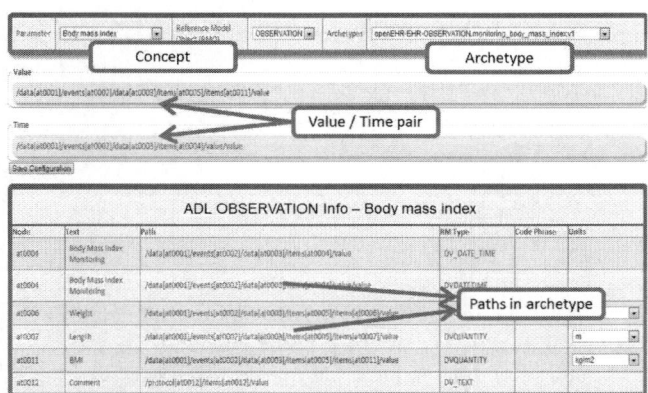

Fig. 2 openEHR configurator.

This openEHR-configurator simplifies the associations between the parameters needed by the doctors and the archetypes stored in the openEHR kernel. Scalability and flexibility are key factors to easily include new concepts and archetypes in the future, so the Graphical User Interface is dynamically built. Figure 2 shows an example of concepts and archetypes configuration.

Some initiatives have already been deployed with full archetype-based development [9-14] demonstrating at a vendor level the lack of dynamicity to constraint the archetype nodes, so the stakeholders continued to use their own interfaces linking them to the dual model solution. Archetypes are defined as maximum datasets, therefore there is a need of selecting the information to be shown.

OpenEHR-configurator plays an important role in the automation of EHR accesses. It sets the properties which allow the correct communication between the client application and the openEHR-kernel. As a result, heterogeneous medical parameters can be treated uniformly through its API and client applications maintenance gets deeply simplified.

B. Graphical User Interface

Interface for medical experts is provided through Slim MEST. This tool is a web based application that allows ubiquitous secure access to doctors, where they can visualise, update and analyse the data of their patients. The application accesses the EHR, located in openEHR-kernel, remotely and shows up data collected in the observation period plus some data coming from the EHR.

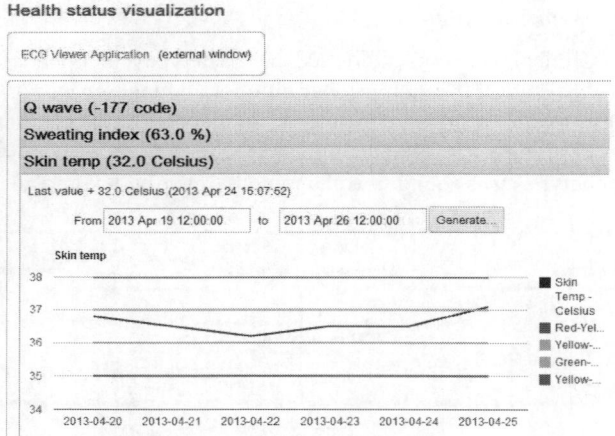

Fig. 3 Slim MEST: Visualisation.

Slim MEST relies on archetypes that are included in the EHR. Following this strategy makes the tool specially robust and flexible to the variations introduced in the openEHR-kernel it accesses to. This allows also a simpler adaptation to another supposed openEHR system in any other context. A communication layer, configured by openEHR-configurator (Section III. A) makes remote calls independent from the tool and relates archetypes in the EHR with concepts that will let the GUI being built dynamically.

The Slim MEST provides the following features:

- *Visualise parameters health status*: the last patient health status (last sensor measurements and heterogeneous EHR values, as medications or previous diagnosis) is shown.

    A self-expandable table is provided showing parameter name beside its last value and allows generating graphs between certain periods of time of the numeric parameters illustrating the variation together with the stipulated thresholds. Diseases and medications are displayed indicating start and end of diagnoses and treatment. A printable report is available with this information.

- *Insert parameters health status*: this option brings out a paginated list of all available parameters in the system and allows introducing new measured or calculated values. As a reference, last value stored is shown beside the insertion area, accompanied by the date when the measurement was taken. Doctors guidelines have been widely taken into account in the design of the data update mechanism. As a result, new values are stored with the current date and time. A printable document that summarizes the actions taken can be obtained.

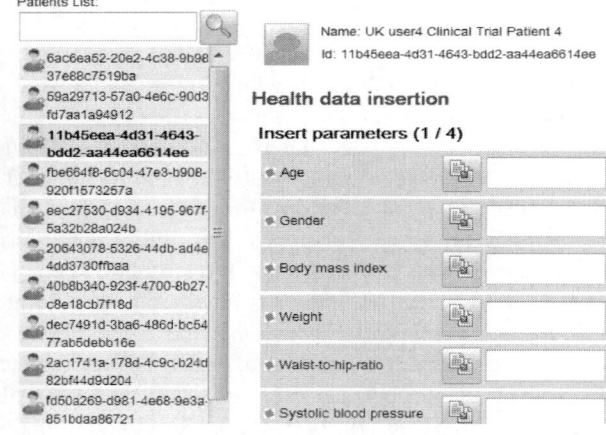

Fig. 4 Slim MEST: Insertion

IV. ECG VIEWER

As described in the previous section the overall objective of the proposed solution is the effective management of the patient data coming from various sources coming from the EHR; but the medical experts need to analyse the raw data collected during the observational study protocol.

Under this concept the present research also includes an ECG (Electrocardiogram) Viewer application aiming at real electrocardiogram signal visualization. The ECG Viewer allows the doctors to visualise a specific part of the real ECG signal when such a demand rises. This means that when medical experts receive a critical alarm for a specific patient from the Slim MEST they are able to have access in the original ECG signal, in an appropriately visualized depiction, for that specific patient. The ECG Viewer has been set to provide a 30-secs, 60-secs or 90-secs ECG signal after the time of interest has been selected, at which the critical situation has been observed in the Slim MEST. The expert is able then to move back and forth in time, observing the ECG signal and make up his conclusions based on his knowledge. The ECG Viewer provides a user-friendly, easy and responsive environment with unlimited time-axis zoom capabilities, signal coordinates tracking and even exportation of the ECG signal in various formats.

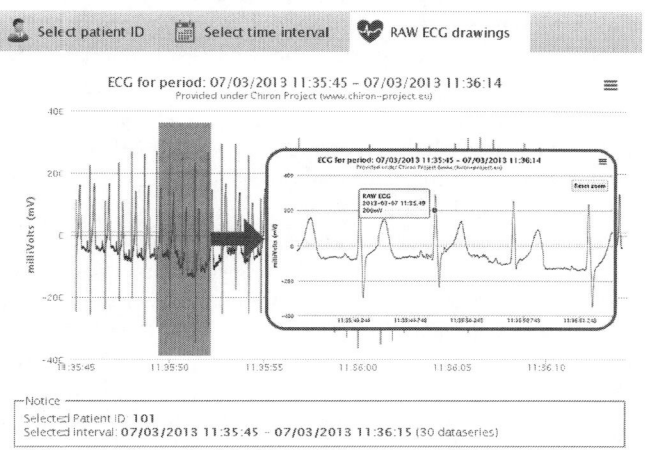

Fig. 5 ECG Viewer and zoom feature demonstration.

V. CONCLUSIONS

Health is a topic that concerns everybody. The conjunction between health care and technology has always been a source of hope to improve quality of life. This paper studies in depth an EHR-based tele-monitoring implantation for CHF patients and presents the satisfactory solution that has been reached through a collaborative work. EN13606 / openEHR model provides many benefits as an EHR standard and creates an effective synergy with tools designed to ease the doctor participation and intervention in the care of the patients with chronic diseases converting the archetypes in the basis of the overall system. What makes this approach especially significant is that tele-monitored patients and doctors get benefit from this progress due to its dynamicity and simplicity. Additional parameters do not require new releases; just archetype modelling and concepts assignment.

Future success in eHealth, tele-monitoring and standardisation resides in collaboration, intercommunication and investigation that becomes exciting when day by day idealistic forecasts become closer.

ACKNOWLEDGMENT

This research was carried out in the CHIRON project (Cyclic and person-centric Health management: Integrated appRoach for hOme, mobile and clinical eNvironments) co-funded by the ARTEMIS Joint Undertaking (grant agreement #2009-1-100228) and by national authorities.

REFERENCES

1. CEN/TC251-ISO/TC215 (2010) Electronic Healthcare Record (EHR) Communication. Parts 1: Reference Model, Part 2: Archetype Model, Part 3: Reference Archetypes and Term lists, Part 4:Security and Part 5: Interface Specification.
2. Beale T., Heard S. (2008) Architecture overview. [Online] openEHR Foundation. [Cited: 24th April 2013] www.openehr.org/releases/1.0.2/architecture/overview.pdf.
3. Puddu PE, Morgan JM, Torromeo C et al. (2012) A clinical observational study in the Chiron Project: rationale and expected results. In Proceedings of ICOST, 2012. pp 74-82
4. Chaudhry SI, Mattera JA, Curtis JP et al. (2010) Telemonitoring in patients with heart failure. N Engl J Med 363: 2301-2309
5. Bonfiglio S. (2012) Fostering a continumum of care. In Proceedings of ICOST, 2012., pp 35-41
6. Klersy C, De Silvestri A, Gabutti G et al. (2009) A meta-analysis of remote monitoring of heart failure patients. J Am Coll Cardiol (2009) 54: 1683-1694
7. Beale T., Heard S., Kalra D., Lloyd D. (2007) The openEHR Reference Model - EHR Information Model. [Online] openEHR foundation. [Cited: 24th April 2013] www.openehr.org/releases/1.0.1/architecture/rm/ehr_im.pdf.
8. Beale, T (2013) Archetype Definition Language ADL 1.5, openEHR Foundation. [Online] [Cited 24th April 2013]: http://www.openehr.org/releases/trunk/architecture/am/adl1.5.pdf.
9. Beale T. (2002) Archetypes constraint-based domain models for future-proof in-formation systems. s.l. : OOPSLA-2002 Workshop on Behavioural Semantics, 2002.
10. Roberta Gazzarata, Jan-Marc Verlinden et al. (2012) The integration of e-health into the clinical workflow – Electronic Health Record and standardization efforts. Proceedings of ICOST, 2012. pp 107-115.
11. Clinical Knowledge Manager. (2010) [Online] OpenEHR Foundation. [Cited: 24th April 2013] http://www.openehr.org/ckm/.
12. Stroetman K. A., Artmann J., Stroetman V. N. et al. (2011) European countries on their journey towards national eHealth infrastructures. s.l. : eHealth Strategies Report,
13. Chen, R., G. Klein, et al. (2009) Archetype-based conversion of EHR content models: pilot experience with a regional EHR system. s.l. : BMC Medical Informatics and Decision Making, 9(1): 33..
14. Bernstein, K., I. Tvede, et al. (2009) Can openEHR archetypes be used in a national context? The Danish archetype proof-of-concept project. s.l. : Stud Health Technol Inform. 150: 147-151.

Connecting HL7 with Software Analysis: A Model-Based Approach

A. Martínez-García[1], M.J. Escalona[2], and C.L. Parra-Calderón[1]

[1] Technological Innovation Group, "Virgen del Rocío" University Hospital, Seville, Spain
[2] Web Engineering and Early Testing research group, University of Seville, Seville, Spain

Abstract—HL7 is an international organization that defines standards on health information systems. Most HL7 domain information models are designed according to an own graphic language. All this domain models are based in a unique metamodel. In the last years, many researchers have considered the possibility of using HL7 in the Model Driven Engineering (MDE) context. However, we have identified a weakness: most MDE tools do not support HL7 own model language, but all of them support UML standard model language. The present research area aims to connect HL7 with software analysis through a model-based approach.

Keywords—HL7, Model Driven Engineering, Metamodel, UML, Model Interchange Format.

I. INTRODUCTION

HL7 (Health Level 7) [1] is an international non-profit organization that promotes and defines standards associated with health information systems. HL7 members develop standards related to exchange and modeling health information, with the objective of supporting clinical practice, management, development and evaluation in health services.

A domain model is a conceptual model that describes concepts related to the problem domain [2,3]. It copes with concepts linked to the problem itself, instead of describing software system concepts.

HL7 defines different domain models to explain each working problem or scenario that has been identified throughout the process. These conceptual schemes cover all areas that range from the necessary information to define system messages to clinical documents themselves.

Model Driven Engineering (MDE) [2,3] is a new paradigm that centers on models creation and exploitation. Using MDE, we increase productivity since we maximize compatibility among systems (thanks to reutilization), simplifying the design process. Models act as systems basis. This way, we can separate applications' conceptual definition from technology where they are executed.

For this purpose, metamodel is a fundamental concept [2,3] as it describes concepts used in a specific model. There are many accepted notations to represent metamodels. In this case, we use UML class diagrams because they are the notation used both in HL7 and UML.

HL7 has a metamodel, called Model Interchange Format (MIF) [4], from which we can model all HL7 domain models. HL7 has developed its own graphic language to model its models elements. Considering the wide range of entities that MIF needs to cover to collect all the necessary concepts in a general health system, we must argue that MIF is very extensive and is presented in such an abstract way that, although it results very interesting from the conceptual point of view, it can be difficult to manage.

Keeping in mind that HL7 models are modeled in their own graphic language and the extension they present to cover all the necessary entities in a health system, we think that designing a software solution that should fulfill a HL7 standard is not an easy task for a software engineer. By contrast, software engineers generally know UML and they can design solutions by means of this standard. Besides, there are many MDE tools that, through a UML model, carry out a series of actions automatically, such as generating code or documentation.

Therefore, working to connect HL7 with software analysis has been relevant for us. Our long-term objective is that software engineers can design their solutions using the UML metamodel and, in an automatic way, the HL7 metamodel. Consequently, we would offer the capability to use standard MDE tools that need the problem to be modeled with UML modeling, apart from simplifying solutions design.

This article, lays the foundation for this research area that we have just begun and is motivated by our previous experiences, such as the Diraya Specialized Attention project [5,6]. In that project, we carried out a practical experience in the MDE context, consisting in applying NDT (Navigational Development Techniques) Web Engineering methodology [7,8] when performing the requirements and the analysis phases in a large-scale Web system that aimed to support the health information systems in Andalusia. This experience concluded that MDE could reduce the development time, as well as detects possible errors or inconsistencies in the early phases.

The main target of the research presented in this paper is to use the HL7 metamodel in the MDE context.

Figure 1 illustrates the general process that we aim to reach through this research.

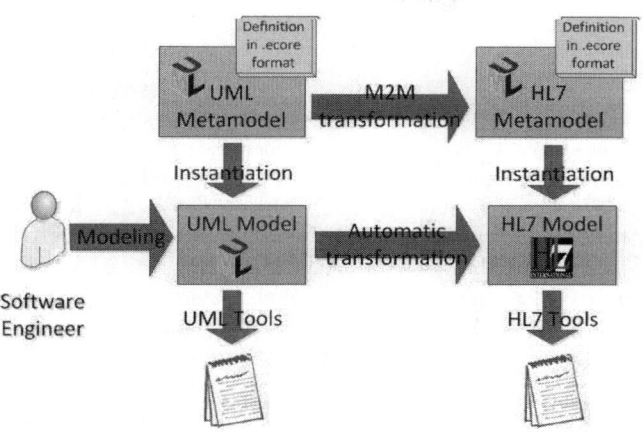

Fig. 1: Solution using the HL7 metamodel in the MDE context

Secondary goals:

- To propose a solution using the benefits of UML general proposed standard, the standards proposed by HL7, and MDE existing tools, to the software engineers working in the healthcare area.
- To acquire a more complete knowledge both of HL7 and UML metamodels.
- To take advantage of the potential of the existing tools that work with the new domain models exploitation paradigm: MDE.

This paper is structured in the following way: After this introduction, Section II reviews and presents previous experiences. Then, Sections III and IV explain the method and results, respectively. Finally, Section V provides a discussion and Section VI states final conclusions.

II. PREVIOUS EXPERIENCES

Some members of the HL7 community have considered the need of using a modeling standard instead of the modeling language that defines the domain models generated from MIF.

Previous experiences have studied the connection between HL7 v2.X and UML structures [9]. One of the first steps to use HL7 in MDE context consists in implementing MIF in a computer-workable language. There are experiences of implementation in computer-workable languages of specific domain model, for example HL7 v3, but they do not cover the HL7 metamodel completely [10].

A research team of the Polytechnic University of Cataluña has conducted an experience in this same line. They identified difficulties as a result of the use of HL7 modeling language, and proposed translating HL7 domain models into the UML nomenclature. They even implemented HL7 v3 domain model transformations to UML models [11]. The obstacle they may face is that the global HL7 community may not find it convenient to abandon the original MIF and, consequently, they may reject the proposal.

No previous studies that intend to use HL7 in the MDE context making a correspondence between its metamodel elements and UML metamodel elements have been found, with the aim that the software engineer can use UML directly (being able to get help from all existing UML tools in the MDE context) and work automatically on the HL7 metamodel.

III. MATERIALS AND METHODS

We are working to reach a solution in the MDE context evaluating the possibility of associating both metamodels, UML and HL7.

To get this goal, firstly we have conceptually analyzed UML and HL7 metamodels and secondly, we have studied theoretically the correspondence between their elements.

Below, table 1 shows the result of this correspondence study.

As expected, there are many elements in the HL7 metamodel that we could not identify with UML (approximately 65% against 35% that we could identify).

This is mainly because UML is a general standard and HL7 is a specific standard (healthcare) that defines specific elements for health needs.

It is necessary to seek a solution using all HL7 metamodel conceptual richness in our MDE solution. For this purpose, we suggest the use of UML extensibility mechanisms.

These extensibility mechanisms enable UML to be presented as an open standard that can model aspects not previously covered in its principal and general specification, allowing us to broaden the notation and semantics of this standard.

Once theoretical foundations have been laid, we will use model transformation techniques in the MDE context, by ensuring that the results of these transformations are consistent with source models. This fact will reduce effort and errors, for these techniques allow automating models construction and text generation.

Table 1 Correspondence between both metamodels

HL7 Metamodel Diagram	HL7 Metamodel Element	UML Metamodel Element
Information Model	Class	Class
	Attribute	Property
	Structural_attribute	Direct correspondence not found
	Association	Association
	Composite_aggregation	Direct correspondence not found
	Generalization_relationship	Generalization
	State	State
Data types and Vocabulary domain	Data_type	Type
	Data_type_category	Direct correspondence not found
	Attribute_type	TypedElement
	Vocabulary_concept	Direct correspondence not found
	Domain_version	Direct correspondence not found
	Code_system	Direct correspondence not found
	Coded_term	Direct correspondence not found
	Concept_relationship	Direct correspondence not found
Use cases and Interaction	Actor	Actor
	Use_case	UseCase
	Use_case_relationship	DirectedRelationship
	Storyboard	Direct correspondence not found
	Interaction	Interaction
	Application_role	Direct correspondence not found
Messages design	Design_information_model	Direct correspondence not found
	DIM_class_row	Direct correspondence not found
	DIM_attribute_row	Direct correspondence not found
	DIM_relationship_row	Direct correspondence not found
	DIM_state_row	Direct correspondence not found
	Hierarchical_message_description	Direct correspondence not found
	HMD_class_row	Direct correspondence not found
	HMD_attribute_row	Direct correspondence not found
	HMD_relationship_row	Direct correspondence not found
	Message_type	Direct correspondence not found

IV. RESULTS

This research area is currently being addressed on a PhD thesis work. Even though the present work is still in its early stages, it has already previous experiences, such as the Diraya Specialized Attention project mentioned afore, which constitutes the first previous experience on this research topic.

The short-term objective deals with designing a solution in the MDE context to connect both metamodels, UML and HL7, but focusing on requirements and analysis levels.

We would also like to test these results in the NDT methodology context [7,8]. This methodology, developed mainly in the software projects context, has been applied in biosanitary fields related to HL7 and offers a suitable framework to focus the practical assessment of our results. In addition, we will execute a proof of concept in this line in the context of OncoInves project (code PI-0116-2012) by designing its scenario based on the EHR functional model, clinical research profile for HL7 [12], and therefore according to the HL7 metamodel underlying to any model.

V. DISCUSSION

HL7 involves many standards development lines and working groups focused on different objectives. Each HL7 subgroup conducts the work based on a domain model. All these domain models are based on the same metamodel.

We find it interesting to lead our work to these lines below, once the correspondence between UML and HL7 metamodels has been studied and the transformation between models has been implemented:

- To design the software systems models by means of the UML standard and obtain the HL7 correspondence automatically. The learning curve of software engineers, who would like to design these systems in accordance with HL7, would be shorter as they would model directly in UML. Most software engineers know UML, since it is the most commonly used conceptual model language.
- To use UML-based tools. There are lots of UML-based tools, both in open-source and private markets that allow designing a system with UML. Additionally, there are lots of tools that can use UML diagrams to make easier software engineers work, such as NDT [7,8].
- To align, in a simple way, the concepts used in HL7 models with a system of concepts in a health scenario: ISO 13940. Studying the correspondence between concepts used in the HL7 metamodel is the only requisite to align all HL7 models with a standard system of concepts, since all HL7 models are based on the same metamodel.

- To certify the compliance of HL7 domain models with UML information systems designed models. Once UML models have been automatically transformed into HL7 models, a tool that reports on errors found will be designed. Thanks to Model-to-Model transformation (M2M), we will be able to validate models, as we will specify the metamodel with which the source model must comply. It may be interesting for HL7 to have a tool that can validate a UML requirement model, which should include the system requirements of HL7 functional model, in order to initiate a validation process of the existing systems.

Moreover, our research must face up a challenge: We expect that software engineers can design systems using UML and align them automatically with HL7 metamodel. Nevertheless, it raises the following question: What would happen, if we wanted them to be aligned with a specific HL7 model as, for example, RIM? We will have to find a solution to this challenge, and we may get it by aligning with HL7 Development Framework (HDF).

VI. CONCLUSIONS

Domain models proposed by HL7 standards are very interesting from a conceptual point of view despite they are very complex. For this reason, software engineers who try to design systems based on these standards may find some difficulties when handling them, together with a long learning curve.

Using HL7 standard in the MDE context remains an unexplored area, from which many benefits can be obtained as well as many research areas providing software engineers, who try to design health information systems with a solid support.

ACKNOWLEDGMENT

This research project has been supported by the Tempos project (code TIN2010-20057-C03-02), the NDTQ-Framework (code TIC-5789), the "Carlos III" Health Institute within the call Strategic Help in Health (PITeS project, code PI09/90518) and RETICS Innovation in Healthcare Technology call (code RD09/0077/00025).

REFERENCES

1. HL7 official web page at http://www.hl7.org/
2. Schmidt DC (2006) Model-Driven Engineering. IEEE Computer 39(2):25-31 DOI 10.1109/MC.2006.58
3. Van Der Straeten R, Mens T, Van Baelen S (2009) Challenges in Model-Driven Software Engineering. Models in Software Engineering 5421:35-47 DOI 10.1007/978-3-642-01648-6_4
4. Spronk R, Ringholm C (2010) The HL7 MIF - Model Interchange Format. At http://www.ringholm.com/docs/03060_en_HL7_MIF.htm
5. Escalona MJ, Parra CL, Nieto J (2007)Diraya Project. The power of metamodels in real experiences with Web Engineering, Proceedings of XI International Congress on Project Engineering (AEIPRO), Lugo, Spain, 2007, pp 167
6. Escalona MJ, Parra CL, Martín FM, Nieto J, Llergó A, Pérez F (2008) A Practical Example From Model-Driven Web Requeriments. Information System Development. Challenges in Practice, Theory and Education 1:157-168
7. Escalona MJ, Aragón G (2008) NDT a Model-Driven Approach for Web Requirements. IEEE Transactions on Software Engineering 34:377-390 DOI 10.1109/TSE.2008.27
8. IWT2 Research Group at www.iwt2.org
9. Oemig F, Blobel B (2011) A formal analysis of HL7 version 2.x. Stud Health Technol Inform 169:704-708 DOI 10.3233/978-1-60750-806-9-704
10. Bánfai B, Ulrich B, Török Z, Natarajan R, Ireland T (2009) Implementing an HL7 version 3 modeling tool from an Ecore model. Stud Health Technol Inform 150:157-161 DOI 10.3233/978-1-60750-044-5-157
11. Ortiz D, Villegas A, Sancho MR, Olive A, Vilalta J (2011) Automatic transformation of HL7 v3 information models into equivalent UML models. HL7 Spain Technical Report.
12. HL7 EHR Clinical Research Functional Profile (CRFP), Release 1 at http://www.hl7.org/implement/standards/product_brief.cfm?product_id=16

Author: Alicia Martínez-García
Institute: "Virgen del Rocío" University Hospital
Street: Manuel Siurot Avenue
City: Seville
Country: Spain
Email: alicia.martinez.exts@juntadeandalucia.es

Integrating the EN/ISO 13606 Standards into an EN/ISO 12967 Based Architecture

J. Calvillo[1,3], I. Román[1,2], and L.M. Roa[1,3]

[1] Grupo de Ingeniería Biomédica, Universidad de Sevilla, Sevilla, Spain
[2] Área de Telemática, Dpto. de Ingeniería de Sistemas y Automática, Universidad de Sevilla, Sevilla, Spain
[3] Centro de Investigación Biomédica en Red en Bioingeniería, Biomateriales y Nanomedicina (CIBER-BBN), Spain

Abstract—Electronic health record (EHR) systems have acquired an outstanding role on the evolution of healthcare. The wide spectrum of efforts (of implementation and standardization) is a sample of such importance. EHR systems by itself are often considered as the ultimate goal of IT policies in health, by disregarding the potential of acting as supporting building blocks of advanced systems (e.g., knowledge generators). Thus, to promote the cooperation and interoperability among EHR systems and others is a major challenge on the road ahead of health IT systems. The use of service architectures could ease the interoperability, but the integration of EHR systems (even standardized ones) within an ecosystem or service architecture is a not deeply explored issue. This work presents a formalization approach for one of the most relevant international initiatives in EHR systems, the family of standards EN/ISO 13606. This formalization is made according to the methodological principles of the ISO 10746 standard (RM-ODP) and integrated in a health service architecture conformed to the EN/ISO 12967 standard (HISA).

Keywords—Health informatics, EHR, EN/ISO 12967, RM-ODP, HISA.

I. INTRODUCTION

Electronic health record (EHR) systems have been the focus of numerous efforts in last decades due to they have been pointed as one of the essential keys for the electronic revolution of the healthcare domain [1]. A sample of such acknowledgement is the wide spectrum of existing standardization initiatives. Standards address different features of EHR systems (requirements, functional and information models, etc.) with the objective of easing the interoperability of final solutions and building a federation of cooperating EHR systems instead of close ad-hoc solutions [2].

Because of the benefits that can be obtained from the deployment of normalized EHR systems, these systems are often considered as the ultimate goal of IT policies in health. Indeed EHR systems are an essential key of the evolution of healthcare because they support the electronic storage and communication of health information. However there is a strong need of tools using capabilities of EHR systems for advanced purposes such as the transmission of information anywhere at any time, the discovering of risk events and tendencies, etc., in general, generation of knowledge. Those high level services exploiting the potential of EHR systems are also required for the advancement of healthcare. Thus, EHR systems should not be considered as isolated and final components but as another integrated element within the health domain. The interactions among EHR systems and systems covering other areas (administrative and demographic information, CPOE, security, terminology, communications, advance capabilities, etc.) should be analyzed and promoted.

The vision of a healthcare environment as a cohesive whole will allow future solutions to cooperate in order to reuse capabilities and accomplish more complex goals. Each particular solution will be a building block of a bigger system (the healthcare domain) that will be evolvable and more and more complex, sophisticated and efficient.

In such a cohesive environment, any system (even EHR ones) is part of an architecture seen as a set of components operating in a distributed way. By following the fundamental principles of that architecture boost the interoperability and cooperation of systems working upon it. There are several standardization efforts for the formalization of distributed system architectures (Reference Model – Open Distributed Processing [3], Model-Driven Architecture [4], etc.), and one of the few initiatives focused on the health domain is the ISO/EN 12967 standard, Health Information Service Architecture (HISA) [5].

Due to the development in parallel (often disconnected) between efforts in EHR systems and health architectures, the current work aims to formalize one of the most relevant international initiatives in the domain of EHR systems (the family of standards EN/ISO 13606 [6]) according to the methodological principles of ISO 10746 (RM-ODP), and integrate the formalization within a health service architecture conforming the HISA standard.

The main contribution of this work is to identify the difficulties of harmonization between the standards and to serve as a first approach of integration of standards under the paradigm of normalized health service architectures. The EN/ISO 13606 family of standards has been chosen due to its international relevance and the maturity of its approach although other initiatives equally relevant (such as HL7 [7] or openEHR [8]) could have been selected. The harmonization of all them will be the objective of future work.

II. MATERIALS AND METHODS

A. EN/ISO 13606 - Health Informatics – Electronic Health Record Communication

The EN/ISO 13606 family of standards, whose first part was released as European standard in 2007, is one of the most popular international efforts in the domain of EHR systems [6]. Its objective is to define an information architecture to communicate all (or part) the EHR of a single subject of care. An important contribution of these standards is a two-level modeling approach, which distinguishes a reference model to represent the generic properties of health record information, and archetypes (conforming to an archetype model) which are constraints on the underlying information model used to define patterns for specific characteristics of the clinical data.

The EN/ISO 13606 is composed of five standards, each one focused on specific features of EHR communication (the information architecture needed to achieve interoperable communications, archetypes linking elements from the reference model for the knowledge abstraction and the boost of semantic interoperability, the specification methodology of privileges needed to access to EHR data, and the interfaces to be provided by the EHR systems for the communication of EHR data).

B. EN/ISO 12967 – Health Informatics – Service Architecture (HISA)

The ISO 10746 standard (RM-ODP) [3], released in 1997, has the aim of developing standards easing the exploitation of distributed information systems in heterogeneous environments and across administrative domains. The reference model defines five viewpoints as separate projections of the whole description of the system. Such viewpoints are:

- Enterprise viewpoint: identifies activities and responsibilities of the system, its environment, and the communication between them;
- Information viewpoint: determines logic information entities, their content, repositories and objects responsible of the information flow;
- Computational viewpoint: formalizes the system behavior;
- Engineering viewpoint: identifies infrastructure services needed for the system functioning; and
- Technology viewpoint: describes the implementation of system services and other components by means their hardware and software configuration.

UML modeling tools can be used for the formalization of architectures according to RM-ODP. In particular, there is a standardized UML profile for RM-ODP, the ISO 19793 standard (UML4ODP) [9].

The HISA standard defines an architecture for health information systems according to the RM-ODP, i.e. decomposed in viewpoints [5]. Trying to stand generic, HISA only provides common architectural features of the three technology independent viewpoints (i.e., Enterprise, Information, and Computational viewpoints). The two viewpoints dependent of technology have to be addressed by the architect. In [10], the authors identified divergences between HISA and RM-ODP and presented a formalization approach of the HISA normative principles according to RM-ODP and the profile UML4ODP. That work is the base upon the integration of normative principles of EN/ISO 13606 standards is going to be done.

III. RESULTS

The requirements and normative principles of the EN/ISO 13606 standards have been formalized according to the three technology independent viewpoints of HISA, as is showed below, and integrated with the fundamental features of the normalized architecture.

A. Enterprise Viewpoint

This viewpoint corresponds to the business architectural model with no connection with technological features of the implementation supporting it. The scope of the EN/ISO 13606 standards is the communication of health records between systems, thus three processes are added to the Enterprise Viewpoint of the architecture:

- Request EHR (process 'Request EHR_EXTRACT' of EN/ISO 13606): allows to request a particular electronic health record.
- Request Archetype (process 'Request ARCHETYPE' of EN/ISO 13606): for recovering of archetypes.
- Request Audit Record (process 'Request EHR_AUDIT _LOG_EXTRACT' of EN/ISO 13606): allows to request a audit record.

Basic processes such as add, delete or update instances of HISA objects are underlying methods for the management of EHR, archetypes and audit records. Also HISA identifies several higher processes relevant to the domain of EHR systems, e.g., processes for registering, deleting or validating clinical information.

From an organizational viewpoint, HISA establishes processes for the local management of information objects meanwhile the EN/ISO 13606 provides those for the request of records and archetypes between systems. Both standards are complementary in this perspective.

B. Information Viewpoint

The ISO/EN 12967-2 standard determines two sets of normative elements: the information objects, and the attributes common to all of them. The attributes are classified in two mandatory groups (i.e., *SystemAttributes* and

VersionAttributes) and four optional groups (i.e., *ExtendedAttributes*, *StateChanges*, *BusinessRule* and *ClassificationCriteria*). Thus, all the information objects identified from the EN/ISO 13606 standard must adopt, at least, the two mandatory set of attributes.

The information model of the EN/ISO 13606-2 standard is stated as an RM-ODP information viewpoint of the EHR Extract. Then, for interoperability's sake, harmonization between both information model (EN/ISO 13606-2 and HISA) must be done. Table 1 shows a sample of the correspondence between HISA information objects and elements from the EN/ISO 13606 information model. The divergences between the models are discussed in Section IV.

In addition, the messaging objects cluster from the HISA information model must accommodate the structured communication of EN/ISO 13606 in order to ease the information exchange between healthcare information systems.

C. Computational Viewpoint

The third part of the EN/ISO 12967 standard describes three kind of computational objects: basic, general-purpose, and complex. The EN/ISO 13606-5 standard determines one interface for the communication of health records (*REQUEST_EHR_EXTRACT*), one for archetypes (*REQUEST_ARCHETYPES*), and one more for audit records (*REQUEST_EHR_AUDIT_LOG_EXTRACT*). Each interface exposes three methods: request element, return element value, and reject request. According to the EN/ISO 13606-5 standard, these interfaces are specializations of basic methods *Detail* defined by the EN/ISO 12967-3 standard, that allow recover instances of HISA objects.

Then, after the integration of both standards, three kind of computational objects are formalized:

- Basic computational objects: include those supporting the interfaces determined by the ISO/EN 13606-5 standard and, at least, one per each information object defined in the information viewpoint and supporting the basic methods.
- General-purpose objects: an object from HISA is relevant for EHR systems, including the methods *login*, *logout*, *setParameter* and *getParameter*;
- Complex computational objects: defined by HISA and reusable for EHR systems. An object related to the subject of care workflow (methods *registerSOC*, *getContact* and *getSOC*, among others), and another for clinical information (methods *annul*, *registerCI*, *validateCI* and *generateAggregation*).

Table 1 Relations between HISA and EN/ISO 13606 information objects

HISA	ISO/EN 13606
Clinical Information / CI Complex	RECORD_COMPONENT
Id	rc_id
value	ELEMENT.data_value
unitMeasure	CEN Datatype of ELEMENT.data_value
status	AUDIT_INFO.version_status
Subject of Care	EHR_EXTRACT.subject_of_care
(0..*) Agent	IDENTIFIED_ENTITY
(0..*) Activity	links
TypeOfClinicalInformation	meaning | archetype_id
Organization element	ORGANISATION
name	name
address	POSTAL_ADDRESS
telcom	TELECOM
Person	PERSON
id	id
name	ENTITY_NAME
birthTime	SUBJECT_OF_CARE_PERSON_IDENTIFICATION.birthTime
deceasedTime	SUBJECT_OF_CARE[…].deceasedTime
gender	SUBJECT_OF_CARE[…].administrativeGenderCode
address	POSTAL_ADDRESS
telcom	TELECOM

IV. DISCUSSION

The work of formalization of the EN/ISO 13606 standards according to RM-ODP and UML4ODP standards, and its integration within a HISA-based architecture, can be publicly accessed in [11]. From this work, some points can be discussed.

One of the main features of the Enterprise Viewpoint is the business processes. The EN/ISO 13606 determines only those related to the communication of health and audit records and archetypes. It is out of the scope of these standards to cover local processes of EHR systems such as creation, validation, or deletion of health records. Although HISA covers these basic processes (add, update, validate…), more EHR-related efforts should be considered. There are several initiatives (even standards) that tackle the functional model of EHR systems and could be integrated in the Enterprise Viewpoint [12][13].

For the Information Viewpoint, the provision of information models of both standards eases the integration of them. However the different levels of granularity of concepts make difficult a one-to-one mapping. For example, Table 1 related the class Person of both standards. In HISA, any person has determined its birth and deceased time, and gender; meanwhile in EN/ISO 13606-2 only subjects of care have such attributes. A lot of similar discrepancies deserve further harmonization efforts.

From the computational perspective, a greater effort to normalize interfaces of EHR systems may be done in order to ease the standard-compliant implementations. Infrastructure and high level services could be defined supporting the management of the complexity of information and structures of EHR systems. HISA provides some guidelines such as the methods *generateAggregation* (for creating clinical information structures with specific purpose, for example, reports or arquetypes) or *validateCI* (for validating clinical information by an authorized user) but a wider set of services could be useful. For example, all services related to the management of archetypes (beyond the request through the interface REQUEST_ARCHETYPES from the EN/ISO 13606-5 standard) or high-level services using information from EHR system to provide end-user capabilities (*viewMedication*, *recordAllergies*, etc.).

Finally, it is worthy to be mentioned that there are a lot of efforts and standards addressing features of the EHR domain from a diversity of perspectives (EHR system functional model [12], requirements [14], security issues in EHR systems [15][16], etc.). All relevant standards must be considered in order to establish a normalized formalization of EHR systems within health service architectures, and that work will be the objective of future efforts.

V. CONCLUSIONS

EHR systems have acquired great relevance in last decades as basic building blocks for the improvement and evolution of healthcare, but to exploit their greatest potential, they should be integrated in the health information architectures as any component. The standards EN/ISO 13606 and EN/ISO 12967 are reference initiatives in EHR communication and health service architectures respectively. Both approaches aim to ease the development and deployment of a normalized and cohesive health information environment.

The integration of the EN/ISO 13606 normative principles in a health architecture based on HISA has been analyzed in this work. HISA features such as enterprise processes or computational methods can be used by EHR systems beyond the communication principles defined by EN/ISO 13606 although an important effort to define others is needed. The information models from these standards can be easily integrated by paying special attention to different granularity levels or approaches for some concepts (e.g., Person or abstract class Record_Component).

This work is a first step in the issue of integration of normalized EHR efforts within health information service architectures. The authors consider that a wider study is needed including international initiatives in the EHR system domain by different organizations (e.g., openEHR or HL7).

ACKNOWLEDGMENT

This work has been partially supported by the CIBER-BBN, the Biomedical Engineering Group at University of Sevilla, an Excellence Project of the Andalusian Council (P10-TIC-6214) and a grant from the Fondo de Investigación Sanitaria inside project PI11/00111. CIBER-BBN is an initiative funded by the VI National R&D&i Plan 2008-2011, Iniciativa Ingenio 2010, Consolider Program, CIBER Actions and financed by the Instituto de Salud Carlos III with assistance from the European Regional Development Fund.

REFERENCES

1. Brender J, Nøhr C, McNair P. (2000) Research needs and priorities in health informatics. Int J Med Inform 58: 257-289.
2. Muñoz P, Trigo JD, Martínez I, et al. (2011) The ISO/EN 13606 standard for the interoperable exchange of electronic health records. Journal of Healthcare Engineering 2(1):1-24.
3. ISO 10746: Reference Model – Open Distributed Processing, 1998.
4. Model-Driven Architecture (MDA) at http://www.omg.org/mda
5. EN/ISO 12967: Health informatics – Service architecture, 2008.
6. EN/ISO 13606: Health informatics - Electronic health record communication, 2008.
7. HL7 Group. HL7 Reference Information Model, 2007.
8. OpenEHR: future proof and flexible EHR specifications at http://www.openehr.org/ home.html.
9. ISO 19793: Information technology – Open Distributed Processing - Use of UML for ODP system specifications, 2008.
10. Calvillo J, Román I, Rivas S, Roa LM. (2013) Easing the development of healthcare architectures following RM-ODP principles and healthcare standards. Comput Stand Inter 35:329-337. DOI 10.1016/j.csi.2011.12.002
11. HISA - EN/ISO 13606 formalization at http://gibserv.us.es/ wiki/doku.php
12. 2011.HL7/ISO 10781: Electronic Health Record-System Functional Model, Release 1.1, 2009.
13. ISO 18308: Health informatics -- Requirements for an electronic health record architecture, 2011.
14. EuroRec Institute. EuroRec EHR Quality Seals at http://www.eurorec.org/services/seal/index.cfm
15. ISO/TS 21547: Health informatics -- Security requirements for archiving of electronic health records, 2010.
16. ISO/DTS 14441: Health informatics -- Security and privacy requirements of EHR Systems for use in conformity assessment, 2012.

Author: J. Calvillo
Institute: Grupo de Ingeniería Biomédica, Universidad de Sevilla
Street: Camino de los Descubrimientos S/N 41092
City: Sevilla
Country: Spain
Email: jcalvillo@gib.us.es

Design of a Semantic Service for Management of Multi-domain Health Alarms

J. Calvillo[1,3], I. Román[1,2], and L.M. Roa[1,3]

[1] Grupo de Ingeniería Biomédica, Universidad de Sevilla, Sevilla, Spain
[2] Área de Telemática, Dpto. de Ingeniería de Sistemas y Automática, Universidad de Sevilla, Sevilla, Spain
[3] Centro de Investigación Biomédica en Red en Bioingeniería, Biomateriales y Nanomedicina (CIBER-BBN), Spain

Abstract—Alarm management is an essential capability of health information systems, but an inefficient dissemination of alarms produces the unawareness and fatigue of health professionals. Most initiatives in this field focus on particular scenarios or systems disregarding the whole health environment, and more sophisticated alarm systems are needed to address the complexity of such domain. The adoption of personalization and context awareness concepts in this field can ease the dissemination of alarm notifications to the proper users in the most suitable manner. In this work a service for notification of health alarms, capable of managing and combining profiles of health professional, alarm characteristics, and context conditions, is designed. It is based on semantic technologies to promote the interoperability of systems and establish a multi-domain framework. In particular, characteristics of professional users, alarms, and context are covered in an OWL ontology, and the user preferences and profiles are codified through SWRL rules. When an event happens, the service reasons about the user profiles and context conditions to disseminate the notification to those users (if there is any) willing to receive it, by choosing the most suitable channel for each one, and by adapting the notification to the reception device/s in each case. The design approached is independent of the underlying technology and can be translated to different technologies, what will be focus of future work.

Keywords—Health alarm systems, User preferences, Semantic technologies

I. INTRODUCTION

The health domain is heterogeneous and complex, and healthcare professionals need more and more, mainly in critical settings, to be informed of events at the time they happen. Due to this, alarm and notification management in real-time should be a basic and transversal component of health information systems. Alarms can be categorized according to a wide set of features such as the kind of event that they report (technical, security, domain-specific...), priority level, source of the event (information system, person, device...), repeatability, content, diffusion channel, etc.

In last decades, a lot of efforts focused on the field of alarm management have been developed. In the health domain, two main obstacles have been identified in alarm notification: the unawareness of alarms by professionals (because of a wrong transmission to interested users), and the fatigue or insensitivity of alarms [1]. Often users are overwhelmed with alarms interrupting their natural workflow, and the cost of a notification depends on the grade of interruption for the user, what also can put in danger to subjects of care [2].

One of the possible causes of these problems is that two essential features are often disregarded: human factors and efficient transmission. By human factors we refer to requirements and preferences of users to be notified and that influence greatly the alarm diffusion. Users usually require different levels of reliability and timeliness for different categories of alarms, but most alarm notification systems do not provide fine-grained personalization capabilities. The concept of context-awareness is crucial in this kind of systems, i.e., the user context influences the value of receiving a notification. Among the factors involved are: the importance of the event, the activity the user is performing at that moment, his social activity (if the user is interacting with others), and his location. Besides that, due to the wide variety of devices and communication systems existent (instant messaging, telephony, e-mail...), alarm data can be transported and displayed to the user in different formats. Thus, to provide with discrimination capabilities among types of content, devices, and user preferences is essential to achieve the most efficient alarm diffusion possible.

To sum up, the efficient application of alarm and notification systems to the health domain requires fundamental capabilities of adaptation and personalization to: the end-user profiles (preferences, disabilities, organizational role...), the characteristics of reception devices (screen size, content format supported...), user and device context (presence, location, availability....), and the own features of the notification (priority, timeliness, criticality...).

In the current work a semantic service of health alarm notification is designed. This service is capable of managing and combining user profiles, alarm characteristics, and context conditions, in order to notify alarms to proper users at the proper time. It is based on semantic technologies to promote the interoperability of systems and establish a multi-domain framework. In particular, the characteristics of users, alarms, and context are covered in an OWL ontology, and the user preferences and profiles are codified through SWRL rules. When a event happens, the service, by means its inference engine, reasons about the user preferences and context conditions to disseminate the notification to those users willing to receive it (if there is any), by choosing the most suitable channel for each one, and by adapting the notification to the reception device in each case. The design approached is independent of the underlying technology and can be translated to different technologies, what will be focus of future work.

A. Motivating Scenario

Mr. Abad works as cardiologist in the Hospital Virgen del Rocío. His working activities are operating on surgery, seeing hospitalized patients, and performing administrative tasks in his office. Few times in a month, he has to be 24-hour on call, i.e., he can leave the hospital but must be available to attend emergencies. Among his preferences are:

- If he is seeing patients or in his office, he is available to receive any kind of alarm related to his patients or other issues. As exception, if he is in the room of a patient, he does not want to receive non-emergency alarms that could divert his attention from the patient he is seeing at that moment.
- If he is in the operating room, and he does not desire receive any alarm or warning that is not emergency.
- In out-of-duty periods he wants all the events be registered to consult them when he comes back to work. The urgent ones will be referred to an available doctor.
- If he is on call, he is only available for emergencies, and alarms will be sent to his smartphone.

These preferences, which could be modified in real-time by the user to adapt them to his desires, should be grouped in profiles and always one profile should be active. The transitions among profiles will be performed by the own user or automatically by verifying the context conditions (duty periods, his location, etc).

II. MATERIALS AND METHODS

A. Alarms and Service Personalization in the Health Domain

There are a lot of efforts centred on alarm and notification management in the health domain for specific systems such as monitoring devices [3] or e-Prescribing [4], and some approaches introduce the context-awareness concept as key for alarm triggering (e.g., [5]). Out of the health domain, alarm systems have been applied to different fields such as academic community [6] or security networks [7].

Most of these approaches do not consider the personalization of alarm systems, although in last decades the field of service personalization has acquired outstanding relevance due to the consideration of the end user as an individual active party (with particular desires, preferences and limitations) instead of as a uniform passive population. The tendency is to group the preferences and characteristics of each user in profiles that will be consulted by access systems personalizing the services according to them. One of the most active normalization organizations in user profiles is the European Telecommunications Standards Institute (ETSI) [8]. It has released three publications of user profile management of general-purpose covering the definition of user profile and information, and an architecture supporting the sharing and common understanding of user profiles among systems [9-11]. In addition, these principles have been particularized for the health domain in the standard 'Personalization of eHealth Systems' [12].

B. Semantic Technologies: Ontologies and Inference Engines

During the last decade, semantic technologies have been developed enormously because of the benefits which they provide to distributed systems. One of the most popular semantic tools is the Web Ontology Language (OWL) [13]; a knowledge representation language based on description logic and the Resource Description Framework (RDF) representation. OWL allows the specification of domain knowledge by using classes in ontologies. Reasoners or inference engines work over instances of these classes and allow inferring implicit information about the instances according to the domain ontology. Although simple inferences can be realized on OWL, a limitation in the reasoning process exists because OWL does not allow using more complex rules than the inheritance of classes. A special rule language is required in order to write rules composed of OWL concepts and to reason over ontology instances. A promising approach is the Semantic Web Rule Language (SWRL) [14], which allows establishing complex relations among properties by extending the OWL expressivity. SWRL supports the construction of "Horn-like" rules expressed over OWL concepts.

In this work, OWL is used to design and develop an ontology of alarms, context characteristics, and user profiles for the event management and dissemination. Notification policies and user preferences are expressed as SWRL rules. There is a wide range of inference engine to reason upon SWRL rules in an OWL knowledge base. Jess [15] is used in this work to perform the inference process due to its compatibility with the Protégé-OWL platform [16] which allowed develop the knowledge base (i.e., OWL ontology and SWRL rules). These semantic technologies have been used to build a mechanism of health alarm management enhancing the interoperability for multi-domain settings.

III. RESULTS

A. The Ontology of Alarms and User Profiles

The first contribution of this work is an OWL ontology to describe alarms, user profiles, context characteristics and all the features needed to perform the notification management.

This ontology is the foundational component supporting the service of alarm management because inference processes will allow conjugate user preferences, features of the alarm and of the context to disseminate the event in the most suitable and efficient way possible.

Figure 1 shows the core of the developed ontology. Each person in the system will have one or more user profiles that, in this first approach, only contain preferences about alarm reception. A user profile could include also information for other kind of personalization of services such as configuration parameters of user interface. Each user profile identifies a set of reception parameters (*ReceiverProfile*) to apply to a set of alarms (*AlarmProfile*). Among the reception parameters are the characteristics of reception device/s (coverage, reachability, battery, etc.), filters and rules of content information to discriminate between similar alarms (*AttendanceRule* and *InformationConstraint*) as well as features of the receptor context (*ReceiverContextFeatures*) such as his location, the activity that he is performing, his availability, disabilities, etc. The user will determine a reception profile for one or more alarms.

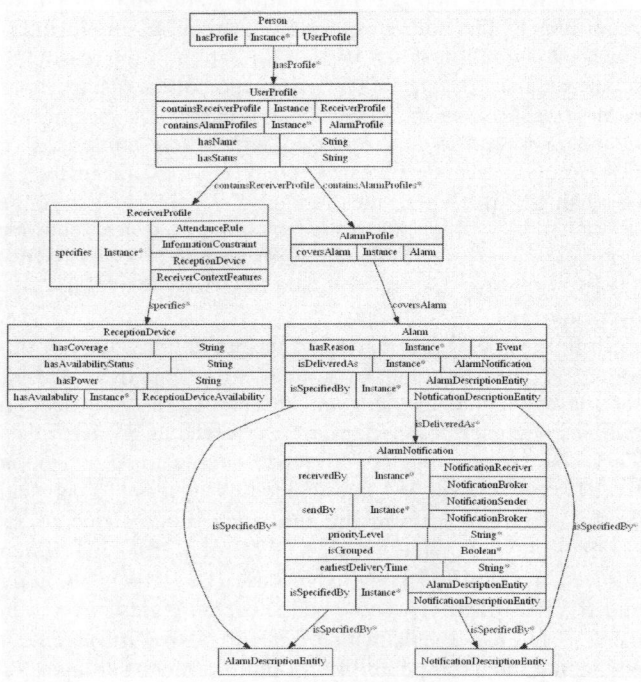

Fig. 1: Ontology core for user profile and alarm management

Each alarm has one (or several) event as source, is carried by a notification (*AlarmNotification*), and can be described by a set of parameters (subclasses of *AlarmDescriptionEntity* and *Notification DescriptionEntity*). Some characteristics describing an alarm are status (*cleared, overriden, posponed,* *rejected, triggered...*), gravity level (*critical, major, minor, informational, warning, indeterminate...*), repeatability, required response action (if exists), and validity period. The description parameters of the notification cover features such as its reliability, validity period of the notification, the communication channel, content format, notification mode, etc.

B. The Semantic Service of Alarm Management

User profiles, alarms, and characteristics are introduced in the system as individuals of the ontology classes, and user preferences as SWRL rules. A modular design is approached to allow several event listeners, reasoners, and notifiers to work in parallel, even in separate domains. When a listener detects an event, informs to its domain reasoner. The inference engine will combine the OWL ontology and its individuals with the SWRL rules and will execute the inference process in order to verify if there are users with active profiles willing to be notified of that event. For each user to be notified, several queries to the knowledge base will be performed to extract the reception preferences. The queries to the knowledge base will be made by means of the SQWRL language.

As illustrative example, an event of dropping pressure of a monitored patient is detected. The reasoner composes and applies the following SWRL rule to the knowledge base to infer if there are healthcare professionals with an active profile covering that alarm:

PressureAlarm(?alarm) ^ HealthcareProfessional (?doc) ^ hasProfile (?doc, ?profile) ^ hasStatus(?profile, active) ^ containsAlarmProfile (?profile,?Aprofile) ^ coversAlarm (?Aprofile, ?alarm) -> AlarmNotification(?not) ^ isDeliveredAs(?alarm, ?not) ^priorityLevel(?not, critical)

The instance of the notification is created in the knowledge base (with, e.g., id Notif32) and next, the inference engine will recover the reception preferences of identified users and verify if there are conditions to be satisfied. For example, a doctor (with id HProf45) can have determined that he only desire to receive that kind of alarms if he is in his office. The following rule specifies that if the current location of the user is equal to the location determined in the active profile, the notification is send.

hasProfile (HProf45, ?profile) ^ hasStatus(?profile, active) ^ containsReceiverProfile (?profile, ?reception) ^ specifies (?reception,?location) ^ currentLocation(HProf45,?currentloc) ^ equals (?location,?currentloc) -> isSendTo(Notif32,HProf45)

The rest of user preferences (content, device, format, etc.) must be also verified and, according to all of them, the service will compose the proper notification for each user.

From a technology perspective, the use of semantic technologies (in particular, ontologies) boosts the interoperability between domains and the independence of this solution from the underlying technology to support it. A lot of technologies and patterns of notification could be used (for example, publisher/subscriber paradigm) even simultaneously in different domains without losing the common understanding and capabilities of personalized alarm management.

IV. CONCLUSIONS

Alarm management is a crucial process in the health domain. A lot of initiatives of alarm systems exist but most of them are centred on particular scenarios, devices or purposes. To consider the wide variety of users, alarms and settings in the complex health domain requires sophisticated and advanced methods for addressing alarm notification. In addition, alarm management services can adopt the concept of personalization and context awareness in order to reduce alarm fatigue by disseminating alarms to the proper users in the most suitable way (format, content, repeatability...) according to their preferences.

Semantic technologies can ease the specification of user preferences and the alarm management at high level across systems or domains. The use of ontologies allows establishing a common understanding between systems and the independence of alarm management from the underlying technologies that could transport the notifications.

ACKNOWLEDGMENT

This work has been partially supported by the CIBER-BBN, the Biomedical Engineering Group at University of Sevilla, an Excellence Project of the Andalusian Council (P10-TIC-6214) and a grant from the Fondo de Investigación Sanitaria inside project PI11/00111. CIBER-BBN is an initiative funded by the VI National R&D&i Plan 2008-2011, Iniciativa Ingenio 2010, Consolider Program, CIBER Actions and financed by the Instituto de Salud Carlos III with assistance from the European Regional Development Fund.

REFERENCES

1. Gee T, Moorman BA. (2011) Reducing Alarm Hazards: Selection and Implementation of Alarm Notification Systems. Patient Safety and Quality Healthcare, vol. March/April 2011.
2. Monegain B. (2013) 'Alarm fatigue' endangers patients at http://www.healthcareitnews.com/news/%E2%80%98alarm-fatigue%E2%80%99-endangers-patients?topic=06,19.
3. Fensli R, Gunnarson E, Hejlesen O. (2004) A wireless ECG system for continuous event recording and communication to a clinical alarm station. Proc. Int Conf IEEE Eng in Medicine and Biology 26 III: 2208-2211.
4. Weingart SN, Massagli M, Cyrulik A, Isaac T, Morway L, Sands DZ, Weissman JS. (2009) Assessing the value of electronic prescribing in ambulatory care: A focus group study. Int J Med Inform 78(9): 571-578.
5. Paganelli F, Giuli D. (2007) An ontology-based context model for home health monitoring and alerting in chronic patient care networks. Proc 21st Int Conf Advanced Information Networking and Applications, AINAW'07, art. no. 4224210.
6. Leonidis A, Baryannis G, Fafoutis X, Korozi M, Gazoni N, et al. (2009) AlertMe: A Semantics-based Context-Aware Notification System. Proc Int Conf Computer Software and Applications, art. no. 5254125.
7. Tsang TMF, Yeung TMW, Chiu DKW, Hu H, Zhuang Y. (2010) Security alert management system for Internet Data Center based on ISO/IEC 27001 ontology. Proc IEEE Int Conf E-Business Engineering, ICEBE 2010, art. no. 5704314.
8. European Telecommunications Standards Institute (ETSI) at http://www.etsi.org/
9. ETSI Guide 202 325 V1.1.1 (2005) - Human Factors (HF); User Profile Management.
10. ETSI Standard 202 746 V1.1.1 (2010) - Human Factors (HF); Personalization and User Profile Management; User Profile Preferences and Information.
11. ETSI Technical Specification 102 747 V1.1.1 (2009) Human Factors (HF); Personalization and User Profile Management; Architectural Framework.
12. ETSI Standard 202 642 V1.1.1 (2010) – Human Factors (HF); Personalization of eHealth systems by using eHealth user profiles (eHealth).
13. Patel-Schneider P, Hayes P, Horrocks I. (2004) OWL Web Ontology Language Semantics and Abstract Syntax at http://www.w3.org/TR/owl-semantics/.
14. Horrocks I, Patel-Schneider P, Boley H, Tabet S, Grosof B, Dean M. SWRL: a semantic web rule language combining OWL and RuleML at http://www.w3.org/Submission/SWRL/.
15. Jess rule engine at http://www.jessrules.com/jess/index.shtml.
16. Knublauch H, Fergerson R, Noy N, Musen M. (2004) The Protégé OWL Plugin: an open development environment for semantic web applications. Proc 3rd Int Conf Semantic Web.

Author: J. Calvillo
Institute: Grupo de Ingeniería Biomédica, Universidad de Sevilla
Street: Camino de los Descubrimientos S/N 41092
City: Sevilla
Country: Spain
Email: jcalvillo@gib.us.es

SILAM: Integrating Laboratory Information System within the Liguria Region Electronic Health Record

Alessandro Tagliati[1], Valeria Pupella[1], Roberta Gazzarata[1], and Mauro Giacomini[1,2]

[1] Department of Informatics, Bioengineering, Robotics and Systems Engineering and [2] Center of Excellence for Biomedical Research
University of Genoa, Liguria, Italy

Abstract—In the last few years a significant challenge has been faced in the healthcare world, after the development of the LOINC (Logical Observation Identifiers Names And Codes) and the HL7 (Health level 7) standards; in order to provide systems of interoperable platforms, interconnected and utilizing standard language for the processing, communication and treatment of data. The aim of this paper is to demonstrate how a small local laboratory can easily communicate with a central data repository using these standards. This was realized by creating: a database with laboratory observations codified in LOINC, a web server that represents the general repository of data and by using two clients that directly link to the Web Service (WS) and which also allows clinicians and laboratory biologists to communicate with each other. The results clearly demonstrate that these standards are able to realize interoperability among heterogeneous systems. A future development should consist in testing this reality at a national level, with a national data repository.

Keywords—HL7, LOINC, Interoperability, Standard Language, Electronic Health Record.

I. INTRODUCTION

During recent years a great challenge has been posed in the healthcare world to develop standards for interoperability that could improve medical care, optimize workflow and reduce ambiguity; thus improving the transfer of knowledge among the actors including clinicians, agency government and patients. Processes must be rapid, as a diagnosis is often the result of multiple evaluations of a number of clinicians. Presently, in Italy no legislation exists on these issues, which could regulate the entire process of an Electronic Health Record (EHR) [1]. Each hospital and laboratory generates and processes data created with their own language, with the result of data incompatibility and very poor interoperability. For example, each laboratory can have its own database of exams coded in its own language: thus, a simple exam such as glucose measurement could be codified in several ways.

Additionally, there is no type of EHR system, which is able to trace the clinical history of a patient. Consequently, it is not possible nowadays for clinicians to do any type of historical analysis and analysis of correlation of a disease among patient's relatives.

The aim is to demonstrate that it is possible to realize an interoperable system that could link a small laboratory, clinicians and a standard repository of data, centralized in a localized area.

This project, called SILAM (Medical Analysis Laboratory Information System), currently involves the University of Genoa - DIBRIS department and the Infinity Technology Solutions S.P.A. Company and it is developed within the Regional Operative Programme of the Ligurian Region.

II. MATERIALS AND METHODS

The first goal was to create a standard dictionary of coded laboratory tests, which was achieved by establishing contact with a small hospital in Pietra Ligure, in Liguria. Clinicians sent a list of exams executed in their structure. This list was first coded in LOINC terms and then an infrastructure based on the paradigm of a Service Oriented Architecture (SOA) was created; afterwards the environment of a General Practioner(GP) was created, where a list of exams can be prescribed and directly sent to the laboratory involved. Finally, a client for the laboratory, receiving the patient and generating the final report, was created.

The instruments utilized for such a purpose were:

A. HL7 Standard

HL7 standard has been chosen as the background of the entire system. Following the natural workflow of a GP request generation, once the exams of interest are selected, he/she sends the laboratory request directly to the laboratory involved by a HL7 CDA (Clinical Document Architecture) document. The choice of adopting the CDA structure was made for two reasons: it was suggested by the "Tavolo Sanità Elettronica" (an Italian committee about e-Health) and there are suitable implementation guides localized into the Italian reality for GP prescriptions [2] and for laboratory reports [3]. Moreover it maintains the clinician – patient structure. Subsequently, the creation of a client for the laboratory was necessary to read the CDA document and to generate the CDA document containing the final report for the exam. Finally, once this report has been sent, it is automatically stored in the central database of the Ligurian

Region and it is made available for clinicians, for reference and further studies.

B. LOINC Database

In order to realize data exchange efficiently, two SQL databases were included into the framework. The first one contains 35 tables and it is used to create and store laboratory requests. The second one is a partial view of the laboratory environment and contains only 3 tables. It is the only part meant to be made public by the laboratory managers for these kinds of applications and it is used for managing a series of events, such as:

- Patient acceptance;
- Final confirmation of the prescription;
- Execution of laboratory observations;
- Generation and sending of the reports;

At the moment, the above mentioned system is a highly developed prototype. In order not to interfere with the laboratory routine, only two tests were chosen to verify the effectiveness of the system itself: a simple one (*Glucose in serum or plasma*) that returns a unique value and a complex one (*Hemoglobin*), identified by a single name, but composed of different types of values.

C. Clinician Interface

In the Visual Studio 2012 environment, an interface based on the paradigm of SOA has been developed; a GP makes real-time laboratory requests, which are sent to the laboratory.

In order to define how to design this interface, the HL7 classes were implemented, starting from the choice of the storyboard.

For example, the storyboard selected is called PRPA_ST201307UV02 "Patient registry get demographics query" included in the Patient Administration domain. Such a storyboard provides two interactions, one for questioning the database (PRPA_IN201307UV02), and another one for sending the answer to the system (PRPA_IN201308UV02). First, the storyboard verifies the authenticity of the "Sender" through a control on the patient and clinician's identifiers. If these are correct, the system executes the extraction of the patient data, which is then returned to the sender as HL7 objects.

In detail, the following classes were developed:

- A class PRPA_MT201307UV02 populated with a patient's identifier that is sent to the server;
- A class QUQI_MT021001UV01ControlActProcess completed with the clinician's identifier sent to the server;
- A class MCCI_MT000100UV01 containing the response of the authentication;
- A class containing patient's data coming from the server, in case of positive authentication;

D. Laboratory Interface

In the Visual Studio 2012 environment, an interface, allowing users to welcome patients and directly visualize the steps to be followed, was developed for the laboratory. Once the operator confirms the request sent by a clinician, exams can be executed. Afterwards, biologists make the reports and finally send all the information to the central repository. This workflow has been implemented with these HL7 objects:

- Patient authentication, with the class PRPA_MT201307UV02;
- Data extraction (finding the CDA stored in the repository);
- Request confirmation, with HL7 POLB_TE004100 and POLB_TE004200;
- Report creation;
- Sending of the final report (Interaction with Web Server, CDA Transmission and storage);

III. RESULTS

As far as the process of standardization is concerned, the vocabulary of the tests was completed with only two complete encodings of observations. Consequently, this has led to two general examples, which were useful for testing the functionalities of the project. Nowadays in the Ligurian Region, there are 25 private/covered by insurance laboratories, which generate an account of nearly 10000 Hemoglobine tests per year and 5000 Glucose in serum or plasma per year.

As indicated in the previous section, the first goal of SILAM was to translate a complete dictionary of tests into LOINC. A semi-automated procedure, which will be described in a future paper, was designed to help in executing the translation [4].

Nowadays, the infrastructure is installed in the DIBRIS department of the University of Genoa. Therefore, in order to avoid any problem linked to privacy violation and problems related to authorization in using sensitive data, tests have been completed using anonymized personal data coming from the laboratory structure "Laboratorio Albaro".

The experimentation, up to now, has concerned one week's daily activity, following the workflow already described.

As an example, we have indicated the scenario of report generation and sending to the general repository in figure1:

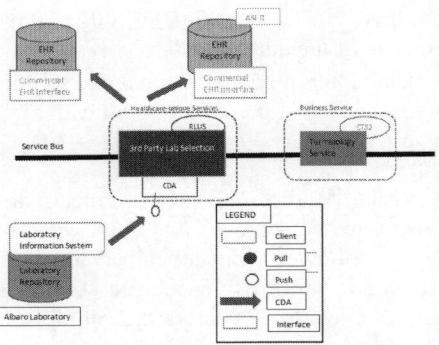

Fig. 1 Overall architecture of the system

The Laboratory Information System (LIS) pushes a CDA to the 3rd Party Lab Selection, which automatically sends it to the appropriate commercial EHR interfaces. Before sending such CDA documents to the interfaces, the appropriate terminology is completed by the Web service.

Subsequently, a clinician may be interested in visualizing these results; this action is depicted in figure 2, where: The general practitioner asks for the patient's data. So the GP client asks for different CDA documents through a Pull function to the 3rd Party Lab Selection, which sends a request for such documents to the Commercial EHR interfaces, thanks to a Pull function. These interfaces send the CDA documents requested to the 3rd Party Lab Selection which then answers to the initial request made by the GP client with all the CDA documents gathered as indicated in figure2.

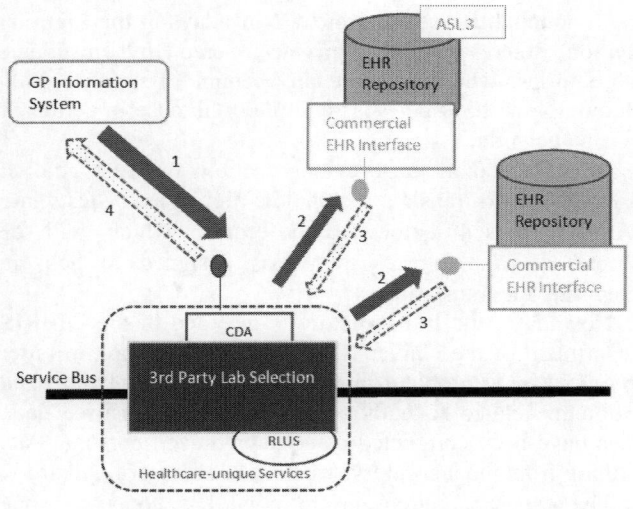

Fig. 2 Interactions for result deployment.

The results obtained clearly demonstrate a full integration with the HL7 and LOINC standards at both software production and graphical levels. It is provided at the first level by the structure of several HL7 fields, which provides LOINC parameters, while at the second level, it is due to the choice of modularity distributed in three levels; provided by the SOA paradigm and in particular, by the object oriented programming.

IV. DISCUSSION

One problem to deal with, is the fact that laboratories manage their LIS using different forms of database, coded in different ways; almost impossible to manipulate properly, causing a lacking of flexibility and reciprocal binding, although there is a transversal support infrastructure. As discussed in [5], the need for a mature and efficient framework, accepted at a national level that could improve the use of data is well known. In fact, it could lead to advanced clinical studies, better care, following the new phenomenon of big data, thanks to the new statistical analysis instruments. National guidelines written for LIS implementation follow the accepted and shared standard-de-facto HL7 and LOINC; hence perfectly fitting for the context. The proposed solution is based on a middleware acting as a bus, able to receive and transfer all the operations needed by a LIS.

V. CONCLUSIONS

The SILAM project was started in March 2011 and concerns the design and the realization of a complete interface, fully implementable in the Ligurian regional healthcare system over the next five years. The first two years were dedicated to the feasibility study of the project and to the realization of demonstrative prototypes of the development.

At present, this phase has been concluded successfully, continuing with the extension of the dictionary of exams provided by the laboratory and mapped in LOINC. The subsequent objective is to realize a complete matching system, fully integrated with the service bus and that offers an automated system for coding heterogeneous data, coming from multiple sources.

The next step will be to provide real patient data. Such a step will pose new challenges in privacy management, linked to the discretion of personal data and the management of the security systems of the platform; which are currently in a study phase.

The ultimate goal of SILAM is to offer a fully standardized automated and user-friendly system, which should be further developed within the context of an Italian legislation background, when it will be clearly defined. Currently, no precise regulations have been established for the EHR and the general process of e-Health in Italy.

However, this tool could be able to help small laboratories in following the future guidelines providing data for the Italian Personal Health Record (FSE).

SILAM will be able to generate structured clinical data, easily interpretable all over the world, also considering that the combination HL7-LOINC is shared in many countries.

SILAM should make the healthcare workflow centralized, generating less problems and allowing all clinicians to actualize evaluations based on the patients' clinical history.

The prototype proposes the standardization of medical processes at a global level as a central topic, demonstrating that nowadays this is feasible and the importance of following this perspective.

ACKNOWLEDGMENTS

The authors would like to thank: Simone Naso, Viola Parodi and Dario Passi for the collaboration in this research and Moyra Watson for the English review. This research was partially supported by the Ligurian "Programma Operativo Regionale POR-FESR(2007-2013).

REFERENCES

1. http://www.governo.it/Presidente/Comunicati/testo_int.asp?d=69362, 2012.
2. Valeria Burchielli, Marisa Soprano, Implementation Guide Clinical Document Architecture (CDA) Rel. 2 Prescrizione, 2009.
3. Valeria Burchielli, Marisa Soprano, Implementation Guide Clinical Document Architecture (CDA) Rel. 2 Rapporto di Medicina di Laboratorio, 2009.
4. Swapna Abhyankar, Dina Demner-Fushman, Clement J. McDonald Standardazing clinical laboratory data for secondary use. In: Journal of Biomedical Informatics. Elsevier, 2012; pp. 642-650.
5. A. Berler, S. Pavlopoulos, D. Koutsouris Design of an Interoperability Framework in a Regional Healthcare System, Proceedings of the 26th Annual International Conference of the IEEE EMBS San Francisco, CA, USA, September 1-5, 2004.

Corresponding author:

Author: Alessandro Tagliati
Institute: DIBRIS - Università di Genova
Street: Via Opera Pia 13
City: Genova
Country: Italy
Email: tagliati.alessandro@gmail.com

A Model for Measuring Open Access Adoption and Usage Behavior of Health Sciences Faculty Members

E.T. Lwoga[1] and F. Questier[2]

[1] Department of Informatics, Bioengineering, Robotics and Systems Engineering and [2] Center of Excellence for Biomedical Research
University of Genoa, Liguria, Italy

Abstract - The study sought to investigate factors that affect the adoption and use of open access in Tanzanian health sciences universities. Based on the social exchange theory (SET), and the unified theory of acceptance and use of technology (UTAUT), the study developed a model suitable for assessing open access adoption and usage in academic institutions. The validated model demonstrated that facilitating conditions, extrinsic benefits (professional recognition), behavioural intention and individual characteristics (professional rank, technical skills and number of publications) influenced actual usage of open access. Other factors related to contextual factors (attitude, and open access culture), and extrinsic benefits (academic reward, accessibility and preservation) influenced behavioural intention to use open access. Fear to violate publisher's copyright policies and effort expectancy however de-motivated faculty to adopt open access, while copyright concerns inhibited faculty's actual usage of open access.

Keywords - open access, social exchange theory, technology acceptance model, faculty, Tanzania, Africa

I. INTRODUCTION

Health scientists in developing countries have a critical role to play in disseminating their research results to optimize access and use of such scholarly information and to reduce global disparities in health. The Open Access (OA) movement has changed how researchers conduct and share research, primarily by increasing the reach of scholarly communication across the world, including Africa. Several studies however, report that the adoption and participation of faculty in OA publishing and self-archiving practices is low across the world [1,2], including Africa and Tanzania [3,4]. Access to health literature is an essential component in strengthening local teaching and research, improving local medical practices, empowering local experts to find solutions to local health issues, and supporting government officials to make informed decisions and formulate sound policies [5].

II. CONCEPTUAL FRAMEWORK

This study used a combination of the social exchange theory (SET) [6,7], and the Unified Theory of Acceptance and Use of Technology (UTAUT) [8]. UTAUT considers four constructs to play a significant role as direct determinants of user acceptance and usage behavior: performance expectancy, effort expectancy, social influence, and facilitating conditions. The effect of independent variables on dependent variables is moderated by four moderating variables: gender, age, experience, and voluntariness of use. SET is based on the premise that people interact to help each other for profit or by expecting that they will gain something in return. Based on SET, Kim proposed the following variables, influencing self-archiving behaviour: (1) costs (additional time and effort; copyright concerns, fear of plagiarism); (2) extrinsic benefits (academic reward; professional recognition; accessibility; publicity; trustworthiness); (3) intrinsic benefits (altruism, self-interest) (4) contextual factors (self-archiving culture; influence of external actors) and (5) individual traits (number of publications; professional rank; age, technical skills). On the whole, factors from SET [2] were added to UTAUT.

III. METHODOLOGY

The study was conducted in all eight health sciences universities in Tanzania. A stratified random sampling procedure was used to select a sample of faculty (researchers and library staff, n=415) from a total population of 679, with a response rate of 71.1% (see Table 1). A structured questionnaire was used to collect data, where survey questions were developed based on existing, tested and verified instruments [3,9]. The questionnaire was pretested and optimized with a small pilot group. Binary logistic regression analyses were performed to examine the ability of each factor to predict actual usage and behavioural intention to use open access. The items were also tested for validity using factor analysis with principal components analysis and varimax rotation. All factors showed alphas greater than 0.70.

IV. FINDINGS

Factors affecting adoption and usage of open access
Statistical evidence of the model's fitness to the collected data was found with Omnibus Test of Model coefficients significant ($p < 0.001$) for both behavioural intention and actual usage of OA. The model predicted 90.1%

of the responses correctly for the intention, and 78.2% for the usage.

Table 1: Demographic details (N=295)

		No	%
Gender	Male	189	64.1
	Female	106	35.9
Age	30 years and below	21	7.1
	31-40	101	34.2
	41-50	112	38.0
	51-60	54	18.3
	61 and above	7	2.4
Academic Qualification	PhD	93	31.5
	Master	141	47.8
	Postgraduate Diploma	18	6.1
	Bachelor	43	14.6
Professional Rank	Professor	15	5.1
	Associate Professor	26	8.8
	Senior Lecturer	83	28.1
	Lecturer	68	23.1
	Assistant Lecturer	60	20.3
	Tutorial Assistant	43	14.6
Discipline	Medicine	137	46.4
	Nursing	40	13.6
	Biological sciences	36	12.2
	Pharmacy	30	10.2
	Public health & allied sciences	33	11.2
	Allied health sciences	9	3.1
	Dentistry	10	
Number of Publications per year	0-1 publications	98	33.2
	2 to 3 publications	57	19.3
	More than 3 publications	140	47.5

The results indicated that seven independent variables were significantly related to the behavioural intention and four independent variables to the actual usage (see Table 2).

V. IMPLICATIONS FOR THEORY

This study advances theoretical development in the open access research. This study conceptualized and developed a model for open access adoption and use among scientists based on social exchange theory, and UTAUT, and the validated model consists of seven dimensions: facilitating conditions, cost, extrinsic, contextual, individual traits, behavioural intention on open access usage, and actual use of open access. The study findings showed that facilitating conditions, extrinsic benefits (professional recognition), behavioural intention on OA usage and individual characteristics (professional rank, technical skills and number of publications) influenced the usage of open access. Contextual factors (attitude and culture), and extrinsic benefits (academic reward, accessibility and preservation) also had positive effects on behavioural intention on open access usage. Copyright concerns and effort expectancy however had negative effects on behavioural intention to use open access, while copyright

Table 2: Results of binary logistic regression analysis regarding faculty's behavioural intention and actual usage of open access

Independent variables		Behavioural intention on open access usage		Actual usage of open access	
		p-value	Odds Ratio	p-value	Odds Ratio
Cost	Copyright concerns	**0.002**	**0.249**	**0.045**	**0.565**
	Fear of plagiarism	0.506	0.832	0.178	0.768
	Additional time & effort	0.436	0.787	0.953	1.013
	Effort expectancy	**0.045**	**0.391**	-	-
Intrinsic	Altruism	0.330	0.706	0.874	0.964
Extrinsic	Academic reward	**0.017**	**2.698**	0.083	1.633
	Accessibility	**0.009**	**2.615**	0.303	1.301
	Publicity	0.481	1.278	0.056	0.611
	Professional recognition	0.969	0.985	**0.016**	**1.950**
	Trustworthiness	0.117	1.933	0.084	1.666
	Preservation	**0.013**	**2.619**	0.683	0.903
Contextual	Culture	**0.031**	**1.932**	0.924	1.023
	Influence of other actors	0.172	0.464	0.233	0.720
	Attitude	**0.011**	**3.359**	-	-
Facilitating conditions		0.068	1.941	**0.033**	**1.687**
Individual Characteristics	Professional rank (ref Tutorial Assistant)	0.540		**0.017**	
	Assistant Lecturer	0.742	0.777	**0.005**	**4.485**
	Lecturers	0.903	1.100	**0.001**	**6.540**
	Senior Lecturers	0.750	0.746	**0.003**	**5.326**
	Associate Professors	0.236	4.209	**0.007**	**8.805**
	Professors	0.298	4.104	**0.036**	**7.106**
	Age	0.254	0.674	0.806	0.945
	Technical skills	0.886	1.039	**0.006**	**1.636**
	Number of publications	0.898	0.963	**0.008**	**1.731**
Behavioural intention to use		-	-	**0.050**	**1.522**

concerns had negative effects on actual usage of open access. Based on the study findings, the original model was refined, and a validated research model is proposed that can better explain the adoption and usage of open access by health sciences faculty (see Figure 1). The factors examined provide a strong basis for understanding the open access adoption and usage.

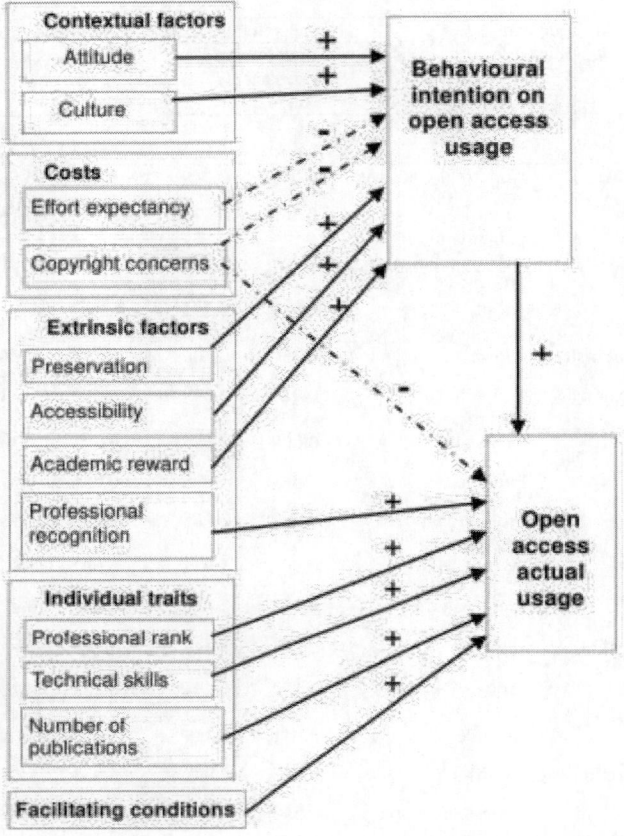

Figure 1: Validated model on open access adoption and usage

The study further contributes to theory by revealing factors that did not have effects on open access adoption and usage, which include trustworthiness, publicity, altruism, additional time, fear of plagiarism, and influence of other actors. Although these factors were found to significantly influence open access usage behaviour in previous studies, these factors were not significant in this study. These findings suggest that future research should examine the effects of these factors on open access adoption and usage. The study has also contributed to the body of knowledge on the open access adoption and usage because little empirical findings exist in the developing world context although much of the literature exists in other developed countries. Thus, the validated open access adoption and usage model can be adapted to test the adoption and usage of open access in other institutions with similar conditions. The results can provide more understanding into how to plan and implement successful open access projects in academic institutions.

VI. IMPLICATION FOR PRACTICE

This study has several implications for the successful adoption and usage of open access. These study findings offer suggestions to academic and research institutions management and librarians on how to plan, manage and promote open access usage among scientists. Firstly, institutions should develop institutional repositories and policies to encourage and motivate faculty to use open access. The study findings showed that faculty support the OA initiative and are willing to contribute their research findings. However, the survey results revealed that only one university had established an institutional repository. Secondly, institutions should improve the ICT infrastructure, ensure adequate technical support and increase internet bandwidth to enable faculty to adopt and use open access. The study findings revealed that the facilitating conditions were among the significant factors that influenced faculty to use open access. With adequate technical support and infrastructure, OA initiatives can be successful, especially in institutions with limited resources. Thirdly, the academic and research institutions should review their academic reward policies to recognize new forms of scholarly communication in order to motivate faculty to self-archive their research outputs in institutional repositories. The study findings indicated that faculty perceived that tenure and promotion criteria had no effect on their use of open access. Generally, faculty members were using the open access journal articles for promotion. However, none of the academic reward policies in the surveyed universities formally recognized the open access journal articles in their promotion criteria. This implies that universities need to formally recognize open access publications in the academic reward policies. Fourthly, the librarians in academic and research institutions should create awareness and conduct information literacy programmes to enable faculty understand OA issues and benefits, and eventually enhance their capacity to self-archive and publish in open access web avenues. These programmes will also enable faculty to understand copyright issues and their right to self-archive in repositories. Lastly, libraries should provide information services that focus on open access issues, such as copyright management in order to assist researchers to understand the legal implications of self-archiving their research outputs.

VII. CONCLUSIONS

The study findings indicated that despite the low usage of open access in terms of self-archiving and publishing in OA web avenues, health sciences faculty are positive about adopting and using open access. Tanzania faculty members who do deposit their research outputs in repositories and publish in OA journals are primarily motivated by perceived extrinsic OA benefits from user perspective (i.e. professional recognition), behavioural intention to use OA and facilitating conditions such as availability of reliable infrastructure and technical assistance. On the adoption of open access, the faculty's attitude toward open access had the greatest effects. Other factors that motivated faculty to adopt open access include extrinsic benefits (academic reward, accessibility and preservation), and contextual factor (attitude, and open access culture). The findings further showed that faculty rank, technical skills and number of publications influenced positively OA actual usage. The study findings also demonstrated that open access culture was common across all faculty members' disciplines; however there were differences among the surveyed institutions, which may greatly influence adoption of open access in the future. The study findings also specified that faculty were de-motivated to use open access due to the fear of violating publisher's copyright policies, and perceived difficulties in using the system (effort expectancy). Overall, this study reveals findings that are useful for planning and implementing open access initiatives in research and academic institutions in Tanzania and beyond. The study also developed a model that can be adapted to test the adoption and usage of open access in other institutions with similar conditions. Future research can also extend the revised theoretical model (see Figure 1) to account for the remaining unexplained variance in open access adoption and usage by scientists in research and academic institutions.

ACKNOWLEDGMENT

The authors would like to acknowledge the Swedish International Development Cooperation Agency (SIDA) for funding the data collection and analysis for this study; and the Flemish Interuniversity Council – University Development Cooperation (VLIR-UOS) for funding the writing of the manuscript and international collaboration possibilities through the Short Research Stay Programme.

REFERENCES

1. Creaser C, Fry J, Greenwood H, Oppenheim C, Probets S, Spezi V, et al. Authors' awareness and attitudes toward open access repositories. New Review of Academic Librarianship [Internet]. 2010 [cited 2013 Jan 8];16(S1):145–61. Available from: http://www.tandfonline.com/doi/abs/10.1080/13614533.2010.518851
2. Kim J. Faculty self-archiving: Motivations and barriers. Journal of the American Society for Information Science and Technology [Internet]. 2010 [cited 2012 Jul 13];61(9):1909–22. Available from: http://onlinelibrary.wiley.com/doi/10.1002/asi.21336/full
3. Dulle F, Minishi-Majanja M. The suitability of the Unified Theory of Acceptance and Use of Technology (UTAUT) model in open access adoption studies. Information Development [Internet]. 2011 Feb 9 [cited 2013 Jan 6];27(1):32–45. Available from: http://idv.sagepub.com/cgi/doi/10.1177/0266666910385375
4. Southern African Regional Universities Association. Open access to knowledge in Southern African Universities [Internet]. 2008. Available from: http://www.sarua.org/files/publications/ST.
5. World Health Organization. Access to the world's leading journals [Internet]. 2006. Available from: http://www.who.int/hinari/Hinari Agora 8pp English leaflet FINAL.PDF
6. Homans G. Social behavior: its elementary forms. Taylor & Francis; 1961.
7. Homans G. Social behavior as exchange. American Journal of Sociology. 1958;63(6):597–606.
8. Venkatesh V, Morris M, Davis G, Davis F. User acceptance of information technology: toward a unified view. MIS quarterly [Internet]. 2003 [cited 2013 Jan 28];27(3):425–78. Available from: http://www.jstor.org/stable/10.2307/30036540
9. Kim J. Motivations of faculty self-archiving in institutional repositories. The Journal of Academic Librarianship [Internet]. Elsevier B.V.; 2011 May [cited 2012 Nov 20];37(3):246–54. Available from: http://linkinghub.elsevier.com/retrieve/pii/S0099133311000310

Author: Frederik Questier
Institute: Vrije Universiteit Brussel
Street: Pleinlaan 2
City: 1050 Brussels
Country: Belgium
Email: fquestie@vub.ac.be

EHR Anonymising System Based on the ISO/EN 13606 Norm

R. Somolinos[1], A. Muñoz[2], M.E. Hernando[3], M. Pascual[2], J. Cáceres[2], R. Sánchez-de-Madariaga[2],
J.A. Fragua[1], M. Carmona[1], A.L. Castro[2], O. Moreno[2], and C.H. Salvador[2]

[1] Bioengineering and Telemedicine Laboratory, University Hospital 'Puerta de Hierro' Majadahonda, Madrid, Spain
[2] Telemedicine and Information Society Department, Health Institute 'Carlos III' (ISCIII), Madrid, Spain
3 Bioengineering and Telemedicine Group, Polytechnic University of Madrid, Madrid, Spain

Abstract—This paper presents a service for the anonymisation of electronic health care record (EHR) extracts for secondary use based on the ISO/EN 13606 norm. The sending of clinical data for secondary use, in accordance with current legislation, must be carried out using anonymised data. The ISO/EN 13606 standard has characteristics which favour the development of an anonymisation service, thanks to a design that separates clinical information from demographic information and allows a semantic interoperability to be achieved in the exchange of information. The developed system, based on ISO/EN 13606, consists of two modules: demographic server and an anonymising module. El demographic server is able to work independently, while the anonymising module must always work with an associated demographic server. The anonymisation is the process through which it is no longer possible to establish the link between the data and the subject to whom it refers. The demographic server is responsible for the permanent storage of the demographic entities. The anonymising module is responsible for eliminating everything linked to the demographic data of a given extract. The anonymisation process consists of four phases: storage of the demographic information included in the extract, substitution of identifiers, elimination of the demographic information of the extract and final validation. The anonymising system has been integrated into Telemedicine projects with favourable results. The sending of anonymised data for a secondary use allows the generation of large clinical databases from which knowledge can be deduced using data-mining techniques.

Keywords—ISO/EN 13606, interoperability, anonymisation, electronic health care record (EHR), secondary use.

I. INTRODUCTION

Biomedical and health science research have changed both methodologically and conceptually thanks to the appearance of new analytical data tools. These changes have given rise to the need to update the regulations, at both the national and international level, as regards access to and the use of personal data. To this general end, Spain has developed the Constitutional Law for the Protection of Personal Data (LOPD) and the Spanish Law 14/2007 on Biomedical Research for the handling of clinical data. In accordance with these laws, if you want to use clinical information for a secondary use, this information must be anonymised, that is, all links that might connect the information with the subject to whom it refers must be eliminated.

Information and Communications Technologies (ICT) applied to Medicine facilitate the use of clinical information for secondary uses such as research and teaching. But, in many cases, the clinical information, collected from very different sources, is accompanied by demographic information, and as has already been mentioned, the Electronic Health Care Record (EHR) extracts, set aside for secondary use must be anonymised prior to being sent.

In the transmission of EHR extracts, sensitive information is sent, of both a clinical and demographic nature, therefore guaranteeing its security is fundamental. There are different standards that regulate the transfer of EHR. These standards include protection mechanisms for the transmitting of data. However, one way of increasing the security of the information transmitted is to separate the clinical information from the demographic information. In this way, if the message is intercepted by agents external to the transmission or in the system for receiving the information, its clinical information cannot be associated with any specific entity.

For this reason an anonymising system has been designed and developed in accordance with the ISO/EN 13606 norm. The anonymisation of EHR extracts basically consists of the elimination of all demographic information from the extract and the substitution of all means of identification present within them which might be associated with specific demographic entities.

The developed anonymising system has already been deployed in an ongoing project (PI 08/1148) of our research group. In this project an uploading of anonymised information from different sources to an EHR server for its later secondary use takes place

II. ISO/EN 13606

The European Committee for Standardization (CEN) [1] has been put in charge of drawing up the ISO/EN 13606 norm, whose main objective is to standardise the transfer of EHR, or part of them. This regulation is based on a double model that separates information from knowledge [2] from the design stage, Thanks to this separation a semantic interoperability has been achieved between the systems that exchange information. Based on a similar philosophy, there are other initiatives and standards such as HL7v3 [3] and openEHR [4]. There are also projects to reach common interoperability solutions from among the different solutions, as has occurred in the CIMI [5].

The ISO/EN 13606 standard consists of five parts, of which part 1 describes the reference model (information model) of the regulation. This model provides the classes necessary to represent the clinical information and context, and includes a separate package for the demographic information of all those actors that intervene in the EHR (patients, health staff, organizations, devices, etc). This separation is very useful at the time of anonymising the information, since it allows the entities in all of the register to be represented by means of just one code, which may be private, with no relationship to the rest of the identifiers that they might have. In this way the information sent retains its anonymity. The classes of the demographic package and the relationships between them can be seen in the UML diagram in figure 1.

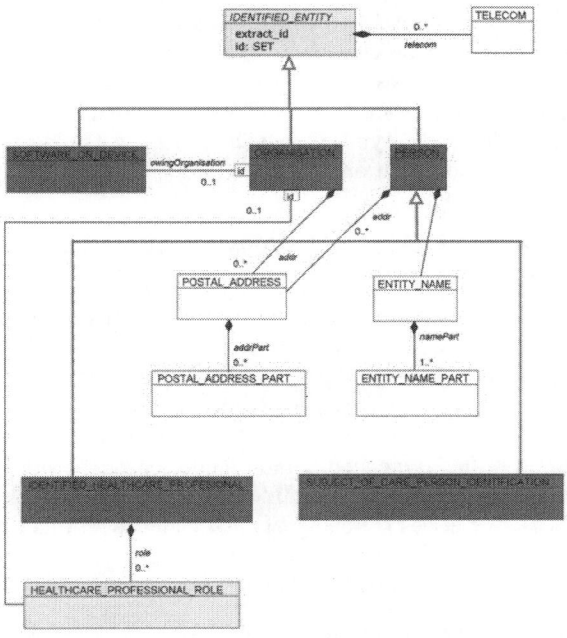

Fig. 1 Demographic package of the ISO/EN 13606 norm

A. IDENTIFIED_ENTITY Class

The main class of the demographic package is the *IDENTIFIED_ENTITY* class. This is an abstract class which is used to group together all of the types of demographic entities. The rest of the classes that represent the different types of entity inherit from it: *SOFTWARE_OR_DEVICE, ORGANISATION, PERSON, IDENTIFIED_HEALTHCARE_PROFESSIONAL* and *SUBJECT_OF_CARE_PERSON_IDENTIFICATION*.

The *IDENTIFIED_ENTITY* class includes the following fields common to all of the demographic entities:

- *extract_id*: type *II* field, is the single identifier used to represent this demographic entity within the extract
- *id:* is a series of identifiers (of type *II*) from which this demographic identifier can be referenced

B. Type II (Instance Identifier)

The type *II* data (*Instance Identifier*) is used to represent identifying objects. The most important fields of this class are *root* and *extension*, since two type *II* objects are considered the same if, and only if, its values for the *root* and *extension* fields are the same.

The *root* field is a single identifier that guarantees the overall unicity of the type *II* objects. It is a species of "name space" which ensures that all of the type *II* objects that are generated under this value of the *root* field will be single.

The *extension* field is a chain of characters which make up a unique identifier.

III. ANONYMISING SYSTEM

The work presented has as its basis an EHR server in accordance with the EN/ISO 13606 norm [6] developed by our research group in recent years. The anonymising system uses the libraries of the reference model and the types of data of the EHR server. The system has been developed using Java as the programming language, MySQL [7] for the databases, XML as the markup language, XML Schemas to define the structures of the data, JPA libraries [8] for the permanent storage, JAXB libraries [9] for the automatic generation of Java classes and Web Services have been used as communication technologies implemented by means of the Axis2 tool and deployed on an Apache Tomcat applications server [10].

The anonymiser is an independent system that offers its services to clinical information systems. The anonymiser must be integrated into the information emitting systems in such a way that it avoids information being sent without anonymising it to the exterior of the system. Another possibility, more ambitious but currently less realistic would be that the anonymisation were a service provided by a centralised organisation in which the anonymiser system is installed, and gives the service to multiple information systems. In this way, the associated demographic server would register the overall data, not just those related to a specific emitting system. However, it should guarantee the security of the communications between the central service and information systems.

The system consists of two main modules (figure 2): a demographic server [11] and an anonymising module. Both modules use Web Services to offer access to its clients by means of a series of public functions. The demographic server is able to work in a totally independent way (without interacting with other modules) with clients who wish to store or recover the demographic information of certain entities. However, the anonymiser always works collaboratively with an associated demographic server, in such a way that the anonymiser accesses the functions that the demographic server offers as a client.

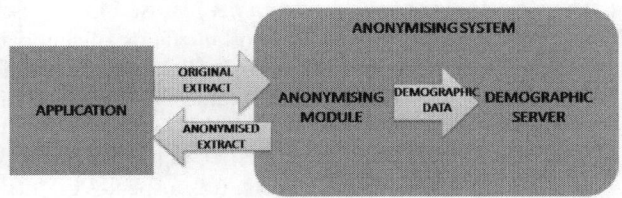

Fig. 2 Anonymising process

A. Demographic Server

The main functions of the demographic server are the permanent storage of the demographic entities and the management of their identifiers. For this reason it has functions to store and recover demographic entities by means of its identifiers. It also has functions for the management of the identifiers of the demographic entities, especially useful for the "anonymiser" clients. These functions are as follows:

- *registerIdentifiedEntity*: this function allows data on an IDENTIFIED_ENTITY object to be stored in the databases of the demographic server which is set as an input parameter.
- *recoverIdentifiedEntity*: this function returns, in the case of success, an IDENTIFIED_ENTITY object that contains the demographic data corresponding to the entity referenced by the type *II* identifier made up of the *root* and *extension* values passes as input parameters in the function.
- *existII*: it has two textual *root* and *extension* fields as input parameters that define the type *II* identifiers. This function looks to see if there is a type *II* object stored in the demographic server with the same values in the same field. If one is found, it returns the *true* value and if not it returns a *false* value.
- *equivalentExtension*: this function looks in the demographic server to see if there is a type *II* object equivalent (that identifies the same entity) to another type *II* identifier passed by input parameters (*root* and *extension*) and whose *root* field is the same as the input parameter *equivalentRoot*. If it finds the said identifier it returns the value of its *extension* field, but returns *null*.
- *updateSetId*: this function updates the demographic server to store a new *II* parameter (input parameter) in its databases and makes it refer to the same demographic entity as the other *II* identifier (input parameter) already registered in the server.

B. Anonymising Module

The anonymising module is in charge of anonymising a given extract from a determined name space (*rootProject*). The anonymiser sends all of the demographic information of the extract to the associated demographic server for its storage and eliminates it from the extract. It also manages all of the relevant type *II* identifiers that appear in the extract to be substituted by others, if necessary, whose *root* field value is *rootProject*. It is also responsible for updating the demographic server with the new identifiers and their links to the demographic entities. The anonymisation generates a new extract with the same clinical information as the initial extract, but with no explicit or implicit demographic information.

This module has a single public function accessible to clients known as *anonymiseExtract* and with which the *anonymization* process of the extracts starts. It has two input parameters: the extract to be anonymised (*extract*) and a textual chain (*rootProject*). The result that this function returns is the anonymised extract.

The anonymisation of EHR extracts consists of the following phases:

a) Storage of the demographic information included in the extract:

13606 extract contains a field known as *demographic_extract* which contains all of the demographic information known about the participating entities in the extract. For each of these entities included in the *demographic_extract* it is checked to see whether it is already stored in the associated demographic server (by means of a series of identifiers). If it is already stored, it goes on to update the series of identifiers associated with this entity in the demographic server. If it is not stored the complete entity is sent to the demographic server for its storage

b) Substitution of the entity identifiers of the extract:

In the clinical part of the extract there are several type *II* fields that contain identifiers that refer to demographic entities that intervene in the extract. Although the identifiers, by themselves, do not contain demographic information, they must be substituted as external agents might already know to which entity each identifier refers. The main identifier that must be substituted is *subject_of_care* of the EHR_EXTRACT class, which refers to the subject of the attention. But other identifiers must be substituted such as the *performer* field of the FUNCTIONAL_ROLE class and the *party* field of the RELATED_PARTY class. The new type *II* identifiers to be used will have *rootProject* as the value of its *root* field. The value of its *extension* field will be awarded in such a way as to ensure that there are no identifiers replicated and that the demographic entities stored in the server with the new assigned identifiers are updated.

c) Elimination of the demographic information included in an extract:

All data included in the *demographic_extract* field of the EHR extract is eliminated. Some data (sex, birth date) of interest for secondary use may be retained in the *demographic_extract* field, as it is not possible to access the rest of the demographic entity data from it.

d) Final validation:

On using the anonymised system with real EHR extracts a problem is detected which gives rise to the creation of this final phase. Some extracts contain data that allow demographic entities to be linked to locations in the reference model not intended for this use. For this reason, on finalising the anonymisation, the entire extract is gone through in search of this type of data. If any are found, they are substitutes by its equivalent in the anonymised extract (such as identifiers) or they are eliminated if substitution is not possible.

IV. CONCLUSIONS

The developed system is one of the first implementations of a service for the anonymisation of EHR based on the ISO/EN 13606 norm. The ISO/EN 13606 standard was chosen for its development because it has characteristics which favour the creation of an anonymisation service: its reference model clearly separates the clinical information from the demographic, and the demographic entities may have more than one identifier that might reference them as they are grouped together in the same field in accordance with the regulation. In this way the management of the identifiers and anonymisation of the extracts are facilitated.

The anonymisation of the EHR extracts avoids the identification of the owners of the clinical data sent, thus compliance with current law on the protection of data is facilitated.

The demographic information stored in all of the participating entities in a demographic server will always be available, both for the project that generated it and for any other use. Thanks to this, the use of the databases for secondary uses is expanded substantially, making it possible to establish mechanisms in them to infer new knowledge through data-mining techniques [12].

The anonymising system has been used in several Telemedicine projects of our research group with satisfactory results. Its use in the CAMAMA project stands out. Still active, it aims to achieve 200,000 extracts of summarised EHR exchanged.

ACKNOWLEDGMENT

This research has been partially supported by projects PI08/1148 (CAMAMA), PI09/90094 (coord PI09/90110) (PITES), PI12/01305 (coord PI12/00508) (PITES-ISA) and CEN-20091043 (REHABILITA).

REFERENCES

1. CEN. European Committee for Standardization. http://www.cen.eu (accessed 2013 May)
2. Beale T. Archetypes: constraints-based domain models for future-proof information systems. http://www.openehr.org/files/publications/archetypes/archetypes_beale_oopsla_2002.pdf (accessed 2013 May)
3. Health Level Seven International. http://www.hl7.org (accessed 2013 May)
4. OpenEHR. http://www.openehr.org (accessed 2013 May)
5. CIMI Wiki. Clinical Information Modeling Initiative. http://informatics.mayo.edu/CIMI (accessed 2013 May)
6. Muñoz A, Somolinos R, Pascual M et al. Proof-of-concept design and development of an EN13606-based electronic health care record service. J Am Med Inform Assoc 2007 Jan;14(1):118-29
7. MySQL. http://www.mysql.com (accessed 2013 May)
8. Java Persistence API. http://en.wikipedia.org/wiki/Java_Persistence_API (accessed 2013 May)
9. JAXB. http://jaxb.java.net (accessed 2013 May)
10. Apache Tomcat. http://tomcat.apache.org (accessed 2013 May)
11. Somolinos R, Muñoz A, González MA et al Demographic server according to UNE-EN ISO 13606. Annual Congress of Biomedical Engineering Spanish Society (CASEIB 2010)
12. Witten IH, Frank E, Mark AH. Data Mining. Practical Machine Learning Tools and Techniques. Third ed. Elsevier; 2011

The Status of Information Systems in the Hospitals of the Greek National Health System

George Aggelinos and Sokratis Katsikas

University of Piraeus, Dept. of Digital Systems,
Piraeus GR-18532, Greece
gaggelinos@yahoo.com, ska@unipi.gr

Abstract—**Hospitals increasingly rely on their information systems to efficiently and effectively carry out both their administrative and medical function. In Greek National Health System hospitals, the situation with HISs has started to change in recent years. In this paper, we present the results of a survey, performed among all public Greek hospitals regarding their information systems. The results show that the general strategy of developing smaller ISs for each medical department has changed towards developing holistic systems; however, considerable room for further improvement still exists.**

Keywords—**Survey, Hospital Information Systems, Information systems infrastructure.**

I. INTRODUCTION

Hospital Information Systems have been around since the 1960's. Initially they provided mostly financial, billing and hotel services but they gradually moved into providing a growing range of services, including medical and clinical services. On the other hand, Information and Communication Technologies found their way into a very broad range of applications in healthcare. All these complexes or systems of processing data, information and knowledge in healthcare environments are commonly referred to as Health Information Systems; Hospital Information Systems thus constitute one instance of Health Information Systems, with a hospital as the healthcare environment [1] According to [2], Hospital Information Systems (henceforth HIS) are "*integrated, computer-assisted systems designed to store, manipulate, and retrieve information concerned with the administrative and clinical aspects of providing medical services within the hospital*".

Originally, the medical part of HISs was mostly focused on small and functionally limited applications in special departments of a hospital (e.g. laboratory, radiology, cardiology etc); thus the HIS was mostly a collection of departmental information systems [1]. Later on, these departmental systems started being integrated into clusters and then on into a single HIS. However, this process was not easy, as making interoperable systems that have been designed to function independently is a difficult –and sometimes impossible– task.

This line of development was followed in many countries, irrespective of their economic development status (see e.g. [3], [4], [5]). Greek National Health System (GNHS) hospitals have followed, for many years, the same route of HIS development. The hospitals were acquiring (more seldom developing), separate departmental information systems. This was mainly due to the limited funds made available for investment in ICT at the hospital level. As a result, HIS in GNHS hospitals constituted a fragmented picture, making the operation and maintenance of the HIS much more difficult than if a holistic system were in place.

Recently, HIS development strategy was changed towards designing and implementing "holistic" HIS. In a holistic HIS, the software for all the system components in the logical architecture model comes from one manufacturer and is based on a common data model [6]. At the same time, many hospitals started providing their physicians with portable machines in order to be able to access patients' information from remote locations. In order to obtain a clear picture of what the current status of HIS in GNHS hospitals is, so that an assessment of further needs and recommendations for further actions are made, a survey was performed. This paper reports the results of this survey.

The remaining of the paper is structured as follows: In section 2 we describe the survey methods and process. In section 3 we discuss the survey findings. Finally, section 4 summarizes our conclusions and offers some recommendations for future development.

II. METHODS AND PROCESS

The survey was directed to the 139 National Health System hospitals in Greece; these are all state-owned institutions.

The survey was performed by means of a structured questionnaire; possible answers were either of the (yes/no) type or of a Likert scale type. Space for free (unstructured) comments on the answers was also provided. The questions were aimed at the structure and organization of the hospital information system (30 questions) and at the coverage of hospital functions provided by the information system (33 questions). As in [3], the questionnaire was sent by email to the IT Director of each hospital and the follow-up of the survey was done by telephone contact with them. The Directors were asked to assign the duty of filling in the questionnaire to the competent personnel. Any remarks or questions were answered either by phone or e-mail communication. Completed questionnaires were received by mail. The responses were processed with SPSS v.19.

III. RESULTS

A. Response Rates

Out of the 139 questionnaires sent, 100 fully completed responses were received (71.94%); 14 hospitals (10.06%) stated that they were unable to respond, either because they do not have an IT Department or because they simply did not want to take part in the survey. 25 (18%) hospitals did not respond at all. Further, 14 hospitals responded with partially filled questionnaires; taking these into account, the response rate rises to 82%. Similar response rates have been reported in [7] and [8], while lower rates have been reported in [3], [9]. The GNHS is administratively organized into seven regions; the response rate per region ranges between 61.29% (Peloponnese, Ionian Islands, Epirus and Western Greece) and 85% (Piraeus and Aegean Islands). Questions on the structure and organization of the information system were answered in 99.58% of the received response, whereas the respective percentage for the questions on the hospital functions supported by the HIS is 99.45%.

B. Hospital Characteristics

Responses were received from General Hospitals (63%), University General Hospitals (13%), Regional General Hospitals (7%), whereas a significant percentage of responses (17%) were received from healthcare establishments self-identified as Health Centers or Clinics.

Size, expressed in number of beds, ranges from 50 or less (7%) to more than 900 (3%), with the majority (36%) being in the 51-200 beds range. Size, expressed in number of clinics, ranges from 0 (1%) to more than 50 (3%), with the majority (44%) having 10 or less clinics.

C. HIS Characteristics

The size of the HIS, expressed in number of workstations, ranges from less than 50 (17%) to more than 500 (13%), with the majority (23%) being in the 101-150 range (Fig. 1). In fact, 60% of all cases have less than 150 workstations. The number of users ranges from less than 100 (34%) to more than 600 (11%), with the majority being in the less than 100 region; 64% of the cases report 200 users or less (Fig. 2). Only 26% of the hospitals report mobile computing and teleworking ability, of which 16 have integrated systems. This finding, correlated with the number of users, leads to the conclusion that the smaller hospitals tend to provide these options more often than the larger ones.

The strategy followed by GNHS hospitals in the past was the acquisition or development of small information systems tailored to the needs and requirements of each department (e.g. LIS). The interconnection of these small systems so as to appear as one system has been troublesome. This strategy proved to be inefficient, in fact, and the survey revealed that it has started to change. 51% of the hospitals answered that their system is "one and only integrated system" across the hospital. 12% of those systems are two years old and the remaining 39% is a little older than three years. These findings indicate that the latest strategy for the deployment of HIS is quite different than the one dominant in the past. Client/server architecture has been used almost throughout (91.9%).

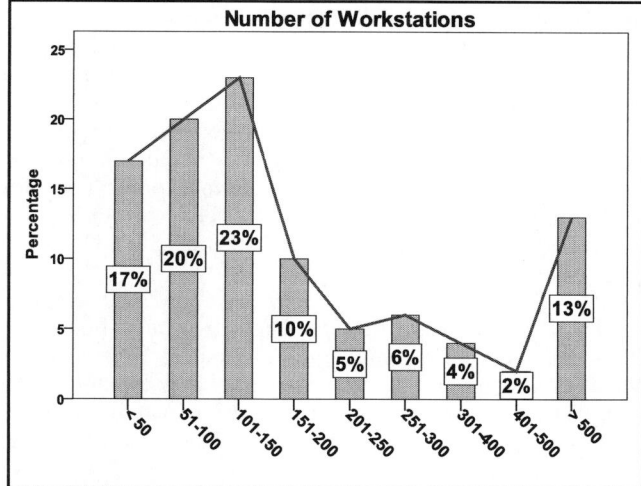

Fig. 1 IS workstations

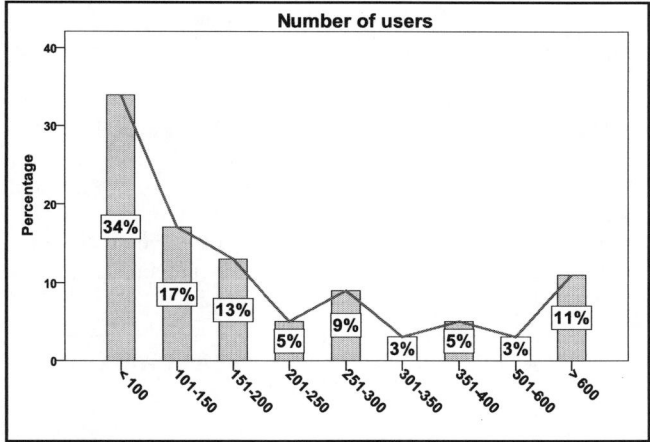

Fig. 2 IS users

Redundancy of system units is necessary for any system that needs to operate 24/7. The situation with respect to this aspect of the HIS it does not seem to follow a specific pattern. Out of the 50 hospitals that have such redundancy, only 34 have the redundant equipment online, so that the time to respond to a failure is minimized.

D. HIS Development and Maintenance

31% of the respondents believe that their IT Department is capable of developing information systems. Similarly,

35% of the respondents believe that their IT department is capable of developing software for processing medical information. These rather low percentages most likely stem from the scarcity of IT specialised personnel in GNHS hospitals. On the other hand, 53% of the hospitals outsource software development. One can conclude that system maintenance is largely done within the context of maintenance contracts with software vendors.

Health legislation changes frequently in Greece, particularly in the past few years. This means that HIS have to be frequently updated, to reflect these changes; this in turn means that formal procedures for software change control and system or update acceptance should be in place. Unfortunately, only 33% applies formal procedures for software change control, with the respective percentage for system or update acceptance procedures being at 44%.

E. HIS Functional Components

GNHS hospitals are structured into two large functional units, namely the Administrative-Financial unit and the Medical-Nursing unit. Further, they are often structured into wings and departments. HIS are found to span and support both functional units only in 36% of the cases, whereas wings are covered in 59.2% of the cases, with 31% of those fully covering their needs and 27.6% covering their needs at a level between 76% and 90%. 61% of the hospitals answered that their departments are covered by applications of the HIS, with 26 of them reporting coverage at the 76-90% level, and 13 reporting full coverage. These large percentages are achieved in hospitals with holistic information systems.

The Medical and Nursing unit of a hospital constitutes its core business. The coverage of their services requires extremely specialized applications, respective of the different medical specializations and the nursing requirements. 41.3% of the hospitals have responded that they cover their needs with only one application; the next lower percentage of 25% covers their needs with five applications (Fig. 3).

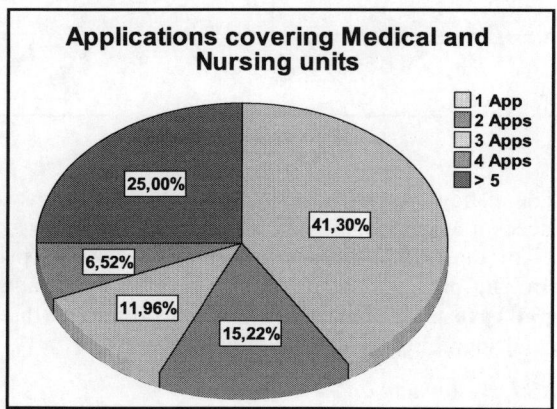

Fig. 3 Applications covering Medical and Nursing needs

Operating theatres and Intensive Care units are covered in 37.1% and 43.3% of the cases, respectively. Both of these departments have limited margin time to react and the need for accurate and direct information is more pronounced. Due to the fact that these departments are operationally connected, we surveyed the Intensive Care units' coverage level. The responses revealed that 14 hospitals have a 51-75% coverage level of their IC units' needs, whereas about the same number of hospitals (13) support less than 50% of their IC units' needs (Fig. 4).

The hospital Laboratories present enough room for improvement. Even though 68% of the hospitals' HIS cover their labs, a large number have no coverage at all. 37.1% of the hospitals that cover their labs report full coverage, while the next level of 76-90% coverage accumulates to 31.4%. It is apparent that there is a significant gap between the hospitals that cover the services of their labs and those that do not. This is one of the most important gaps, because the labs serve most of the medical departments.

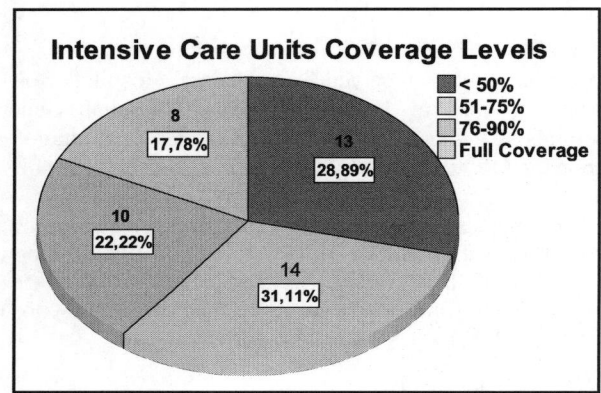

Fig. 4 Coverage Levels of Intensive Care units

The coverage of the outpatient departments and of the physicians' offices in the hospital was also surveyed. 56% of the hospitals responded positively about the coverage of their outpatient departments, the majority of which (17) support them fully or at 76-90% coverage level (14). The physicians' offices in the hospital are covered in 43% of the hospitals, with only 8 hospitals responding positively for full service coverage.

The Administrative and Financial unit is fully covered by the HIS in 87% of the cases. Although there is still a 13% not covering these needs, this is confined to smaller hospitals with HIS of up to 200 workstations. Human resources needs are covered by 80% of the hospitals with most of them (31) reporting full coverage.

The hospital, being an organization that interacts with other similar or governmental bodies needs to continuously process a significant volume of documents.

The requirements of document management are covered in 84% of the cases. Those that do not, tend to be the smaller hospitals with HIS of up to 100 users.

Repair and Maintenance of the medical equipment is of paramount interest for the purposes of a hospital. Only 30% responded affirmatively to the question whether their HIS cover the process of equipment repair or maintenance; this is considered to be quite low. The quality coverage falls into the same low levels, with only six out of thirty hospitals reporting full coverage of repair and maintenance needs, while eight more are in the 76-90% level.

The procurement of pharmaceutical and health supplies affects the core business services of the hospital. Both of these two categories of hospital materials are covered in 83% of the cases. As before, the majority of negative answers come from hospitals (10) with HIS of up to one hundred users. A similar situation (80%) exists with the coverage of general supplies (e.g. blankets, towels). These three categories of procurement were investigated separately, due to the fact that the acquisition process is totally different for each of them.

Auxiliary functions (e.g. depositories, cook-rooms etc) were investigated with the aim to identify the coverage of auxiliary processes that take place in a hospital (e.g. dieticians' offices). It turns out that 70% of the HIS support such functions.

The financial management of a hospital, including accounting tends to be universally covered (98%), with only smaller hospitals with HIS up to one hundred users reporting lack of such coverage. The coverage of the accounting department services is at a satisfactory level, with 38.78% of the cases fully covering the department's processes, and another 41.84% reporting coverage at the 76-90% level.

Finally, we considered the Business Intelligence capability of the systems. Only 42% of the HIS have this capability. It is interesting to note that 10% of the smaller hospitals, with HIS of less than one hundred users do have BI capability –this is the largest percentage-, while the next higher percentage (9%) is found among the larger hospitals with HIS of over six hundred of users.

IV. Conclusions

The past practice of acquiring and installing departmental information systems in GNHS hospitals led to fragmented HIS and has deprived the hospital management and staff of the ability to access unified information. The survey showed that this strategy appears to have changed in the last five years, with a turn towards developing holistic systems.

Older HIS tend not to cover the whole hospital structure and functions; on the contrary, newer HIS tend to cover more hospital functions. Medical and nursing functions are mostly covered by one or five software applications, thus raising the costs of keeping the systems alive. This weakness is reduced in the new holistic systems.

The coverage of hospital wings, departments and laboratories is at a very good level, both quantitatively and qualitatively. In contrast, low levels of coverage are observed for the operating theatres and the IC units; the respective percentages are improved in hospitals with holistic systems. Middle coverage levels have been reported for the outpatient departments, whereas the hospital physicians' offices are covered at levels between low and medium, either quantitatively or qualitatively.

The Management and Financial units are covered at very high coverage levels; this is certainly related to the fact that the introduction of ICT in hospitals started from these units.

There is much room for improvement with regard to HIS in GNHS hospitals. The new generation of HIS cover much more hospital services; this, in conjunction with the experience that the IT personnel has acquired through their involvement with older HIS, are encouraging factors.

References

[1] R. Haux, "Health information systems - past, present, future," *International Journal of Medical Informatics*, vol. 75, pp. 268-281, 2006.

[2] "Hospital Information Systems," Reference.MD, 6 June 2012. [Online]. Available: http://www.reference.md/files/D006/mD006751.html. [Accessed 9 May 2013].

[3] L. V. Lapao, "Survey on the Status of the Hospital Information Systems in Portugal," *Methods Inf Med 4*, pp. 493-499, 2007.

[4] M. F. Collen, "A brief historical overview of hospital information system (HIS) evolution in the United States," *Int J Biomed Comput*, vol. 29, pp. 169-189, 1991.

[5] "The White Paper on China's Hospital Information Systems," China Hospital Information Management Association, 2008.

[6] P. Haas and K.A. Kuhn, "Hospital Information Systems," in *Springer Handbook of Medical Technology*, Springer, 2012, pp. 1095-1118.

[7] E. Smith et al., "Managing health information during disasters: a survey of current specialised health information systems in Victorian hospitals," *Health Information Management Journal*, vol. 36, pp. 23-29, 2007.

[8] S. Landolt et al., "Assessing and comparing information security in Swiss hospitals," *Interactive Journal of Medical Research (i-JMR)*, vol. 1, p. e11, 2012.

[9] J. Kim, M. Piao and J. Wu"The current status of Hospital Information Systems in Yanbian, China," *J. Kor. Med. Informatics*, vol. 15, no. 1, pp. 133-140, 2009.

Operating Room Efficiency Improving through Data Management

P. Perger[1], M. Buccioli[2], V. Agnoletti[3], E. Padovani[4], and G. Gambale[5]

[1] Research Assistant, University of Bologna
[2] Data Manager, Forlì Hospital
[3] Anesthetists, Forlì Hospital
[4] Professor of Business Administration, University of Bologna
[5] Head of Emergency Department and Anesthesia and ICU, Forlì Hospital

Abstract—Data recording has a long tradition in healthcare (Murphy, 2012; O'Connor/Neumann, 2006). Due to evidence-based medicine influences and opportunities provided by modern information technology the number and overall importance of clinical data measurement has increased (Walshe/Rundall, 2001). Unfortunately the huge amount of collected data has not always high quality standards, neither their analysis is based on scientific results (Wager/Wickham Lee/Glaser 2005). Operating rooms (OR) are critical hospital units not only in patient safety terms but also in expenditure terms (Guerriero,/Guido, 2011; Maryamaa/Kirvela 2007). There is the urgent need for new tools which sustain Hospitals decision makers work (Agnoletti/Buccioli/Padovani et al., 2013). The primary goal of the ORMS (Operating Room Data Management) project was to provide the OR decision makers with an information basis to increase OR efficiency and patient safety. Data analysis is based on scientific literature (Macario, 2006; Williams/DeRiso/Figallo et al., 1998; McIntosh/Dexter/Epstein, 2006; Dexter/Dexter/Masursky et al., 2009; Wachtel/Dexter, 2009; Dexter/Epstein/Marcon 2005; Wachtel/Dexter 2009) and the project teams' experience with tracked data. The system login is layered and different users have access to different data outputs depending on their professional needs (Manager, Anesthesiologist and Surgeon). Every profile includes subcategories where operators can access more detailed data analyses. The total number of surgical procedures has increased from 4892 in 2009 to 5616 in 2010 and decreased to 5120 in 2011. In 2012 there is to state a small increase to 5180 surgical procedures. Due to the introduction of ORMS managers optimized significantly performance indicators of the surgical theater over the years and enabled in this way to find an output number which is in line with given resources.

Keywords—**Operating Room, Surgical Path, Management, Indicators, Outcomes, Efficiency, Safety, Sustainability.**

I. Introduction

Data Management in Healthcare. Data recording has a long tradition in healthcare (Murphy, 2012; O'Connor/Neumann, 2006). Due to evidence-based medicine influences and opportunities provided by modern information technology the number and overall importance of clinical data measurement has increased (Walshe/Rundall 2001). Unfortunately the huge amount of collected data has not always high quality standards, neither their analysis is based on scientific results (Wager/Wickham Lee/Glaser 2005). Operating rooms (OR) are critical hospital units not only in patient safety terms but also in expenditure terms (Guerriero,/Guido, 2011; Maryamaa/Kirvela 2007). Therefore, there is the urgent need for new tools which sustain OR decision makers work (Agnoletti/Buccioli/Padovani et al., 2013). Healthcare researchers and managers are under pressure to create and implement tools which sustain decision-making activities and in this way ensure efficiency and guarantee at the same time improvement of patient safety matters. Moreover, the collection of information through performance measurement can assist these organizations to move toward an improved allocation of resources through management control systems (Padovani/Young 2012). Professor Bruce Keogh (Medical Director of the Britain NHS) expresses such concept with the words: 'If you do not know what you are doing and how well you are doing it, you have no right to be doing it at all.'

Research Purposes. The aim of this project is to render the operating room process efficient and safe for patients in terms of clinical risk management. The operating theatre represents one of the most critical hospital units, both in patient safety and financial terms (Guerriero,/Guido, 2011; Maryamaa/Kirvela 2007). The team has chosen the topic of operating room management because of an urgent need to deliver high quality care with limited resources and the correct management of operating theaters represents an important step towards achieving this. We wanted a system able to elaborate data in line with literature in order to identify each phase of patient flow.

Research Context Characteristics: National contracts for healthcare workers and nursing staff (not to mention doctors) foresee payment for a fixed amount of hours. Any extra hours which do not derive from overtime (hours worked immediately after the official end of a shift) or oncall hours, go unpaid. This lack of flexibility compromises the optimization of human resources. Therefore, unlike in the USA or other countries, in Italy it would be impossible to ask a nurse scheduled for an afternoon shift, to work additional hours in the morning, or to call extra staff to clean operating rooms when scheduled staff are struggling to maintain a rapid turnover time. In Italy this flexibility stems from a lack of financial resources (it is not possible to pay workers more or hire extra temporary staff for a few hours every week) as well as from a probable "lacuna legis".

II. MATERIALS AND METHODS

Applied Research Approach. The so called "Operating Room Management System" (ORMS) can be regarded as practical analysis tool embedded in a Oracle Business Intelligence Environment, which processes data to simple and understandable performance tachometers and tables. The analysis of data is based on scientific literature on ORB efficiency (Macario, 2006; Williams/DeRiso/Figallo et al., 1998; McIntosh/Dexter/Epstein, 2006; Dexter/Dexter/Masursky et al., 2009; Wachtel/Dexter, 2009; Dexter/Epstein/Marcon 2005; Wachtel/Dexter 2009) and the research teams' experience and analysis of tracked data. The data recording system is able to read every step of the surgical path (12–16) and all the delta-times between every step and the next. It is possible to obtain a maximum of 25 deltatimes. Data recorded by the data recording system (DRS) is sent immediately via wifi connection to a central hospital server which functions as interim storage. At the end of every week data is sent to the ORMS system where they are processed and added to previous data analyses.

Interdisciplinary Research. Interdisciplinary research has multiple advantages. Joint work of scientists of different disciplines creates a format for connection and interaction that lead to new knowledge. Particularly in the healthcare sector, where experts of different scientific disciplines have to work side by side, an interdisciplinary research approach would be ideal. Only when cross boundary working environments are created, problem comprehension, analysis and solving can be efficient and adequate. The one-dimensional views of different scientific fields are no longer able to cope with the complex dynamics of stratified health sector. Hence, Karl Popper (1963) states: 'We are not students of some subject matter, but students of problems. And problems may cut right across the borders of any subject matter or discipline.'

Ergonomics. IT (Information Technology) was introduced to ensure an appropriate and accurate gathering of the large quantity of data needed by the project. The underlying logic used for IT implementation was not to interfere with the daily activities in the operating rooms. Rather, the project team set up a "system ergonomics development" as its primary strategy of implementation of the new information system for the operating room. "Ergonomics" means that methodologies, equipment and devices were designed in a way that fit the human body, its movements, and its cognitive abilities (McCauley Bush, 2012). In essence, the basic thrust of the project was to simplify the procedures of tracking the surgical path in a way that they would be simple and easy to use in other contexts.

Design of Data Output. The ORMS login (with password) is layered and every user has access to data depending on his/her professional needs. The system is divided in tree main profile types (manager), A (anesthesiologist) or S (surgeon); each profile type can access required information in the profile content. Every profile includes a few subcategories where operators can access more detailed data analyses. The first data output screen shows general information and guides the user towards more detailed data analysis as precise surgical procedure time of every single surgical units. The hierarchy inside the software enables the user to have a complete insight of data regarding his/her profile in a very simple and clear way. The manager's profile is aimed at hospital managers and presents data concerning the entity of operations. Within the surgeons profile the business intelligence software works out data which is important for surgeons and anesthesiologists alike.

Manager Profile (M): The manager's profile comprises 5 different data analysis subcategories. The first output screen (M1) is a global vision of the entire surgical activity in terms of total number of procedures, number of scheduled / unscheduled procedures, raw utilization (total hours of cases performed ÷ total hours of OR time allocated) [22], and a description of all surgical units' workload. M2 is a comparison of the productivity of each surgical unit. Variables used to describe the workload are: number of surgical procedures, number of procedures together with duration, and logistic pathway (induction area, ward, recovery room or ICU) M3 gives a view on surgical units in terms of number of procedures, surgical time average and logistic patient flow analysis (ward, RR or ICU admission). M4 displays the efficiency indicators and expressed as KPIs (6 dashboards with red, yellow and green color schemes). M5 represents the Transport-Induction-Surgery-Awakening (TISA) graph. This graph maps the time it takes to bring the patient from the ward to ORB, the induction time, the surgery procedure time and the awakening time. Each time interval is referred to the surgical procedure chosen by the operator, so the TISA graph represents the total amount of time, expressed as average time and standard deviation required to perform a specific procedure.

Anesthesiologist Profile (A): The anesthesiologist profile includes 4 different data analysis levels. A1 shows the total surgical activity in terms of number of anesthesiological procedures and the average anesthesia time (per year and expressed in 12 months). A2 deals with ORB logistics in term of patient flows. This analysis shows how many patients changed their scheduled pathway and which pathway the patients follow after the surgical procedure (ward, RR, ICU). A3 displays an Induction and Awakening graph (IA) where anesthesia times are mapped; much like the TISA graph, the average time and the standard deviation is related only to the surgical procedure chosen. A4 illustrates statistical description (mean, SD, median, min, max) of the all phases of the entire surgical patient pathway. At this level, recorded data is divided into three groups: surgical time, recovery room time and anesthesia time.

Surgeon Profile (S): The surgeon profile consists of 4 subdivisions. S1 represents a general description of the surgical activity. Data displayed includes: the number of procedures, raw utilization, the efficiency indicators and the five most performed surgical procedures (expressed in terms of quantity, average time and standard deviation). S2 displays a performance comparison between different years/months/weeks. The variables used are: the number of surgical procedures, scheduling analysis (scheduled/unscheduled), logistic patient flow analysis (ward, RR or ICU admission and the number of procedures with a duration of more / less than 120 minutes. S3 displays an Induction-Surgery-Awakening graph (ISA), similar to the TISA graph, but without the time. S4 creates a link between the Diagnosis Related Group (DRG) classification and the surgical procedures of a specific surgical unit. The chart presents a quantitative analysis in terms of numbers of surgical procedures per DRG.

III. RESULTS

The DRS enabled the registration of 20.808 surgical procedures performed over 4 years (from 2009 to 2012). Scheduled procedures increased from a starting percentage of 74% in 2009 to 85% in 2012. The total number of surgical procedures has increased from 4892 in 2009 to 5616 in 2010 and decreased to 5120 in 2011. In 2012 there is to state a small increase to 5180 surgical procedures. Due to the introduction of ORMS managers optimized significantly performance of the surgical theater. Especially the fact that the tool enabled to find an output number in line with analyzed performance indicators has to be underlined. Despite a higher number of performed surgical procedures in 2010 compared to the two following years (2011 and 2012) start-time tardiness decreased (from 84% in 2010 to 80% in 2012). Also the number of surgical procedures performed in overtime decreased (from 27% in 2009 to 21% in 2012). Reassuming this concept we can state that ORMS enabled OR managers to plan and perform more efficiently, with an output which is in line with given resources.

IV. CONCLUSION

This project represents a successful experiment of the introduction of managerial innovation in a public hospital a data management solution. It is interesting to note that although the project was developed by healthcare professionals, it aims to align managerial and professional goals. This is an important step forward, when compared to solutions typically based on a "trade-off" between efficiency (managerial side) and effectiveness (professional side).

REFERENCES

1. AGNOLETTI V, BUCCIOLI M, PADOVANI E et.al. (2012): *Operating room data management: Improving efficiency and safety in a surgical block.* BMC Surgery 2013, 13:7
2. DEXTER F, Epstein RH, MARCON E, et al: *Estimating the influence of prolonged turnover times and delays by time of day.* Anesthesiology 2005, 102:1242–1248.
3. DEXTER EU, DEXTER F, MASURSKY D, et al: *Both bias and lack of knowledge influence organizational focus on first case of the day starts.* Anesth Analg 2009, 108:1257–61.
4. GUERRIERO F, GUIDO R: *Operational research in the management of the operating theatre: a survey.* Health Care Manag Sci 2011, 14:89–114.
5. MACARIO A: *Are Your Hospital Operating Rooms "Efficient"?: A Scoring System with Eight Performance Indicators.* Anesthesiology 2006, 105:237–240.
6. MARYAMAA RA, KIRVELA OA: *Who is responsible for operating room management and how do we measure how well we do it?* Acta Anaesthesiol Scand 2007, 7:809–814.
7. MCCAULE BUSCH P (2012): *Ergonomics: Foundational Principles, Applications and Technologies.* Boca Raton: Taylor&Francis
8. MCINTOSH C, DEXTER F, EPSTEIN R: *The Impact of Service-Specific Staffing, Case Scheduling, Turnovers, and First-Case Starts on Anesthesia Group and Operating Room Productivity: A Tutorial Using Data from an Australian Hospital.* Anesth Analg 2006, 103:1499–1516.
9. MCKEE M, HEALY J (Eds) (2002): *The significance of Hospitals: An introduction.* In: Hospitals in a Changing Europe. European Observatory on Health Care Systems Series. Page: 3-14.
10. MURPHY PJ.: *Measuring and recording outcome.* British Journal of Anaesthesia 2012 109: 92–8
11. O'CONNER RJ, NEUMANN VC.: *Payment by results or payment by outcome? The history of measuring medicine.* J Roy Soc Med 2001; 99: 226–31
12. PADOVANI E, YOUNG D: *Managing Local Governments: Designing Management ControlSystems That Deliver Value.* Milton Park: Routledge; 2012

13. RECHEL B, WRIGHT S, EDWARDS N et.al. (2009): *Introduction: hospitals within a changing context.* In: Investing in Hospitals of the future. European Observatory on Health Systems and Policies. 3-27.
14. WACHTEL RE, DEXTER F: *Influence of the operating room schedule on tardiness from scheduled start times.* Economics, Education, and Policy 2009, 108:1889–1901.
15. WACHTEL RE, DEXTER F: *Reducing tardiness from scheduled start times by making adjustments to the operating room schedule.* Anesth Analg 2009, 108:1902–1909.
16. WALSHE K, RUNDALL TG: *Evidence-based management: from theory to practice in health care.* Milbank Q. 2001, 79:429-57
17. WILLIAMS BA, DERISO BM, FIGALLO CM, et al: *Benchmarking the perioperative process: III. Effects of regional anesthesia clinical pathway techniques on process efficiency and recovery profiles in ambulatory orthopedic surgery.* J Clin Anesth 1998, 7:570–8.

A Semantically Enriched Architecture for an Italian Laboratory Terminology System

Silvia Canepa[1], Sabrina Roggerone[1], Valeria Pupella[1], Roberta Gazzarata[1], and Mauro Giacomini[1,2]

[1] Department of Informatics, Bioengineering, Robotics and Systems Engineering, and
[2] Center of Excellence for Biomedical Research University of Genoa, Liguria, Italy

Abstract—This paper presents the architectural solutions and the first implementation results of a terminology service that aims to support Italian health institutions in the deployment of their clinical data in a semantically standardized format while maintaining their internal coding habits unchanged. As a first example of this general idea, an implementation structure of a translation system in LOINC for laboratory tests is presented. The first prototype of this system is at a testing stage within some clinical institutions of the Ligurian region. In this paper, the workflow of collaboration between the staff of a medical informatics academic laboratory and some hospital analysis laboratories is presented. With this system more than 400 records relative to clinical tests were coded with the appropriate LOINC code and these translations were inserted into a Common Terminology Services 2 (CTS2) based tool to support future cooperative maintenance of the coding system. A comparison with similar implementations in other English and non-English speaking countries is present. The authors thinks that the present example could be easily adopted both at a regional and at a national level in order to form an interconnected laboratory network towards real semantic interoperability.

Keywords—Semantic Interoperability, Laboratory LOINC based coding system, Common Terminology Services 2 (CTS2) framework.

I. INTRODUCTION

Man has always felt obliged to leave written evidence of the ills that have afflicted him and their care. Hippocrates observed that to achieve successful of treatments it was necessary to observe patients by noting the symptoms and diagnostic inferences [1]. From these early insights, in the 800s, the concept of medical records was born: "The medical record is the who, what, why, when and how of patient care during hospitalization". From the second half of the last century from empirical medicine was succeeded by evidence based medicine, which aims to explain the morbid phenomena through controlled clinical trials and guidelines for clinical practice. It is therefore obvious that this type of medicine can be only managed through the efficient sharing of information through computer systems. The creation of systems for sharing clinical data for purposes of care, research and administration is the focus of much development all over the world. Also in Italy, this need has been acknowledged by the legislature in the decree of 12/11/2012 [2]. Nevertheless, the Italian health facilities are not yet equipped for this type of management, as when present, the sharing systems are nor analytical neither standardized. One of the key elements for correct semantically significant data sharing is the use of a standardized terminology; at least in the data that will be shared. Various studies show that a medical department works more efficiently if it maintains an internal common terminology that rarely reconciles with the needs of more standardized vocabularies [3, 4, 5]. Our proposal aims to create a terminology service that supports Italian health institutions in the deployment of their clinical data into a semantically standardized format while maintaining their internal coding habits unchanged. As a first example of this general idea, an implementation structure of a translation system in LOINC [6] for laboratory tests is presented. The first prototype of this system is at a testing stage within some clinical institutions of the Ligurian region.

II. MATERIALS AND METHOD

In Italy, the only effort to standardize the nomenclature carried out has been for purely economic purposes and it was produced by the Italian national healthcare system; which reported outpatient specialist care and, in this respect, defined the essential level of assistance in this system of service. The nomenclature currently in force was established by the Ministerial Decree 332 of 08/27/1999 [7]. In Nomenclature, each service is identified by a specific DM code number derived from the Italian translation of the International Classification of surgical, diagnostic and therapeutic procedures (ICD-9-CM) [8]. This nomenclature suffers from its purely economic nature and it is excluded from the rapid evolution that has been occurring in the world of clinical care. This limit led to the creation of many different local terminologies, which then need to be harmonized. An effort to create uniformity is taking place in various Italian regions, for example in Liguria, where a permanent committee of the various laboratory directors has produced an Excel file; where for each type of performed test in Ligurian laboratories, some common fields have been defined. Among these fields, the following are present: Exam name, evaluated material, DM code, medical definition and administrative details of the exam (see Table 1).

Table 1 Example of EXCEL file

EXAM	MATERIAL	NATIONAL CODE DM	SANITARY DESCRIPTION	ADMINISTRATIVE DESCRIPTION
OMOCISTEINA	Siero	90.22.3	DOSAGGIO OMOCISTEINA	FERRITINA [P/(Sg)Er]
ALBUMINA	Siero, Liquidi Vari	90.05.1	ALBUMINA	ALBUMINA [S/U/dU]
ALDOSTERONE	Siero	90.05.3	ALDOSTERONE	ALDOSTERONE [S/U]

This translation project has been developed in two aspects. From one side the interaction was directed towards the medical and technical personnel in laboratories, to define the exact correspondences among the couples of nomenclatures. Whilst from the other the realization of a data processing infrastructure to deploy the translated terms toward registered application and to help the maintenance of the translating system was developed.

To realize the exam encoding, a comparative analysis was carried out between the hospital's Excel file and the LOINC database. This work was performed with the cooperation of laboratory technicians by several cycles of translation proposals, review and approval on small set of codes. Every document was managed using a commercial versioning system (Sharepoint). This tool allows automatic numbering of the successive iterations of each document. The process was also supported by a content approval system: by which partners, who have approval rights, control of the final publication of each content. Finally, a Check-out and Check-in tool was used to manage an overall control of the life cycle of each document within the system [9]. Every exam was encoded in a series of steps:

1. Italian-English laboratory's exam translation.
2. Go to the LOINC website (www.loinc.org) and search the exam's codes; usually the research produces several records.
3. Compare the codes through the used material to reduce the number of selected codes (timing is 5 minutes for each code);
4. Create a new Excel's file where the selected codes are inserted;
5. Send the laboratory's technicians requests for clarification through the messaging service;
6. Laboratory technicians, through specific manuals, analyze the records and fix the result in a revised Excel file with the request of a final check out control point;
7. The final reviewer examines the technicians' results and gives his/her approvals, adds modifications to the file and sends it to the data processing technicians.
8. The coding is now inserted into a standard repository

To develop the terminology service the Common Terminology Services 2 (CTS2) [10] framework was referred to; a database was developed by the authors to record the defined relations and to deliver the required translations. The goal of the CTS2 specification stack is to provide a standardized interface for the usage and the management of terminologies. In a shared semantic environment, CTS2 provides a modular, common and universally deployable set of behaviors, which can be used to deal with a set of terminologies chosen by the users of the service in their deployment environment. CTS2 provides the terminology community with a defined set of standards interfaces that can be used to evaluate the structure, the source content and the tools of terminology.

III. RESULTS

The laboratory's exam coding has been developed using 400 records. The personnel effort used was equivalent to 4 men/months performed on a part time basis. Specifically 2 men/months from two people from the informatics- technical side (points 1, to 5 and 8), 1 man/month of a laboratory technician (point 6) and 1 man/month of the reviewer for the final medical content approval (point 7). Out of these 400 records, 74.2% were codified using an exact LOINC code. However, 20.4% were codified but with difficulty due to the ambiguity of the translated names in English and to an inaccurate definition of the method used in the analysis, obtaining 379 codified exams (94.4%) with respect to the initial 400 (see Table 2). This data has been inserted into the database of the terminological service, which will be described afterwards. It represents an initial prototype, which is in an advanced phase of development, planned according to the model described by CTS2. The proposed use can be described as the following: the laboratory analysis of a hospital presents referrals with its local terminology to the CTS2 Service (of course, the used local terminology must be previously recorded in the system). The terminology repository completes a comparison with its standard, obtaining a uniquely standardized code for every exposed exam and communicates it, through the service bus, to an Enabled Terminological Application (See Figure 1). For a more detailed description, the logical structure of a "repository standard" is enclosed. This is the core of the system, where all the relations, necessary to perform the translations previously described, are memorized. A database consisting of seven tables, where the root node is "Code_System" (See Figure 2), was created. Special attention was focused on the definition and on the description of the considered analytes in the laboratories (table Analyte) so that they can have a common comparison element. Each Analyte can be measured with different methods; this will result in a cluster of various equivalent codes for different specific methods. Each of these specific local codes has a unique standard code in the system, but at a higher level they can be unified with specific links.

Table 2 Results of encoding

	Number of record	% of Total
Exat encoding	297	74.2
Ambiguous but likely encoding	82	20.4
-Ambiguous name	53	13.2
-Ambiguous method	17	4.3
-Ambiguous materials	12	2.9
Unable to encode	21	5.2
-Ambiguous name	15	3.6
-No LOINC code availble	6	1.6

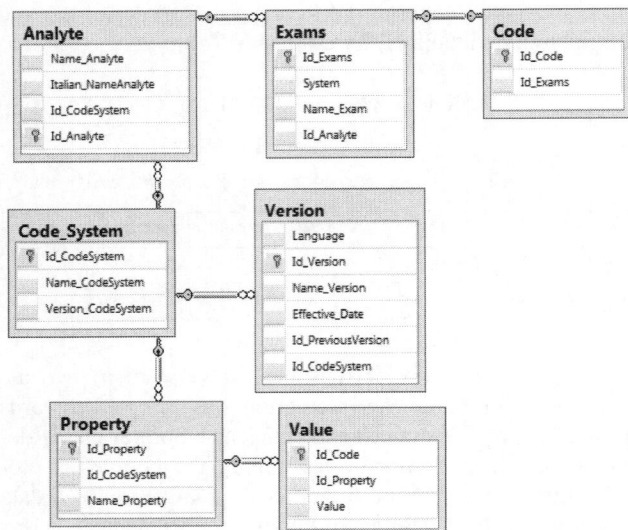

Fig. 2 Logical structure of a Repository Standard

IV. DISCUSSION

In the world, there are some terminological infrastructure examples already available. One of the most popular is the "LexGrid project" developed in USA. This project aims to develop a common terminology model, whose key aspect is to accommodate multiple vocabularies and terminology distribution formats and to support multiple data storages for federated vocabulary distribution. Consequently, the LexGrid project builds upon a set of common tools, data formats, and read/update mechanisms for storing, representing and querying biomedical terminologies and vocabularies. This was a useful example that inspired the planning of the presented terminological service applied to the Italian reality. What differentiates the approach is in the first place the presence of the linguistic barrier, which forced the insertion of the translation step in the workflow. Moreover, the close cooperation between the developers of this service and applications will make the information exchange more efficient even within the standard scheme. There are also some other examples of terminological tools which tackle the problem of the management of linguistic differences, as the one developed in Korea [11]. This work, as far as the authors know, has been limited to the creation of a translation vocabulary and not of an automatic service. A similar translation work, in Italy, is in an advanced development stage at the University of Catanzaro [12].

Similar problems, as the ones faced in the presented work, related to the impossibility of an automatic translation

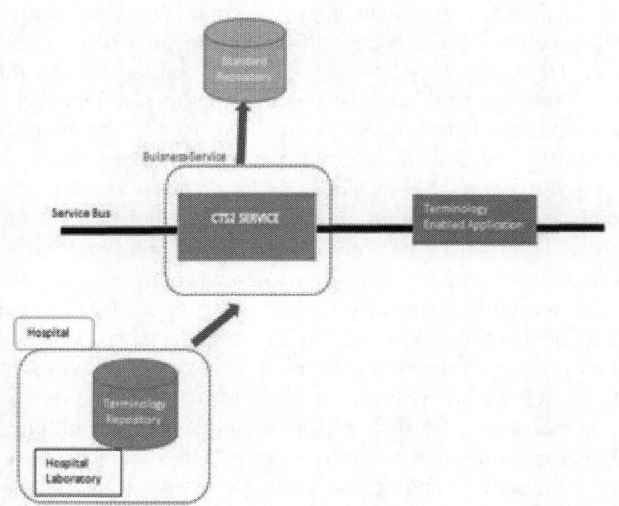

Fig. 1 Architecture of CTS2 Service

from local codes to standard codes in laboratory referral was also illustrated by Lin et al [13], but still limited to a completely English-speaking environment.

Our objective was to provide both an integrated terminological service and a linguistic support to share clinical data in a transparent way and consequently to improve care continuity and efficiency in clinical research.

V. CONCLUSIONS

The purpose of the presented work is to build a service that allows a local reality to interface with international environments through a standardized coding system, keeping the internal coding habits unchanged. Cooperation with the hospital laboratory technicians was very profitable. The automatic versioning system allows the interaction with small numbers of records, so that supplementary work within the laboratory environment was easily accepted; 400 records were translated in a month. It should be emphasized that the present example could be easily adopted both at a regional and at a national level in order to form an interconnected laboratory network towards real semantic interoperability.

ACKNOWLEDGMENTS

The authors would like to thank: Rita Scarso and Flavia Lillo for the collaboration in this research and Moyra Watson for the English review.

REFERENCES

1. Christos yapijakis, Hippocrates of Kos, the Father of Clinical Medicine, and Asclepiades of Bithynia, the Father of Molecular Medicine, In Vivo July-August 2009 vol. 23 no. 4 507-514.
2. Decreto 14 novembre 2012, Gazzetta Ufficiale N. 266 of 14 Novembre 2012
3. C G Fraser, P H Petersen, Desirable standards for laboratory tests if they are to fulfill medical needs, Clinical Chemistry, July 1993, vol. 39 no. 7, 1447-1453
4. Arthur S Elstein, professor and Alan Schwarz, assistant professor of clinical decision making, Clinical problem solving and diagnostic decision making: selective review of the cognitive literature, BMJ. 2002 March 23; 324(7339): 729–732
5. Vaishali R. Choksi, Charles S. Marn, Yvonne Bell and Ruth Carlos, Efficiency of a Semiautomated Coding and Review Process for Notification of Critical Findings in Diagnostic Imaging, April 2006, Volume 186, Number 4
6. Agha N.Khan, Dorothy Russell, Catherine Moore, Arnulfo C. Rosario, Jr. Stanley P. Griffith, and Jeanne Bertolli, AMIA Annu Symp Proc.2003;2003: 890.
7. Decreto del 27 agosto 1999, Gazzetta Ufficiale del 27 settembre 1999 n. 227
8. Cherkin, Daniel C. PhD; Deyo, Richard A. MD, MPH; Volinn, Ernest PhD; Loeser, John D. MD,Use of the International Classification of Diseases (ICD-9-CM) to Identify Hospitalizations for Mechanical Low Back Problems in Administrative Databases July 1992 - Volume 17 - Issue 7
9. Website: http://technet.microsoft.com/en-us/library/cc262378.aspx
10. Organisation HL7,"HL7 Version 3 Standard: Common Terminology Services HL7 Draft Standard for Trial Use DSTU Release 2", Ottobre 2009 HL7, Inc. HL7 Draft Standard for Trial Use
11. Kap No Lee, Jong-Hyun Yoon, Won Ki Min, Hwan Sub Lim, Junghan Song, Seok Lae Chae, Seongsoo Jang, Chang-Seok Ki,¶Sook Young Bae, Jang Su Kim, Jung-Ah Kwon, Chang Kyu Lee, and Soo-Young Yoon, Standardization of Terminology in Laboratory Medicine II, J Korean Med Sci.2008 August;23(4): 711–713
12. Chiaravalloti M. T. , Vreeman D. J. , " LOINC Learns to Speak Italian: Translation and Mapping of LOINC in Italy". Atti del convegno "AMIA 2011 Annual Symposium", Washington, 22-26 Ottobre, 2011, 2011, pp. 1719-1719.
13. A Characterization of Local LOINC Mapping for Laboratory Tests in Three Large Institutions, M. C. Lin; D. J. Vreeman; C. J. McDonald; S. M. Huffl, Methods Inf Med. 2011;50(2):105-14. doi: 10.3414/ME09-01-0072. Epub 2010 Aug 20.

A National Electronic Health Record System for Cyprus

K.C. Neokleous[1,2], E.C. Schiza[1], E. Salameh[1], K. Palazis[1], and C.N. Schizas[1]

[1] Department of Computer Science, University of Cyprus, Nicosia, Cyprus
[2] IBM Italia, S.p.A., Cyprus Branch, Nicosia, Cyprus

Abstract—This paper presents part of the work done for the "Electronic Health Record (EHR) at National Level project" under the Framework Program for Research, Technological Development and Innovation 2009-2010 of the Research Promotion Foundation. A summary of the limitations and threats of the local infrastructure that compromise the future deployment of a National EHR in Cyprus are analyzed. Some of the derived solutions in respect to a number of difficulties that seem to prevent the adaptation of a national Electronic Health Record System for Cyprus are presented. Lessons learned from this exercise can be very beneficial when other countries attempt to do the same exercise.

Keywords—Electronic Health Record, Security, Modular System, Cyprus.

I. INTRODUCTION

Electronic Health Record is defined as a collection of continuously updated health related information and medical data related to a patient. The EHR is a longitudinal electronic record for each person that stores data from nine months before birth to death. EHR systems enable hospitals to store and retrieve detailed patient information, facilitate physicians to provide safer and more effective care through embedded clinical decision support and intelligent diagnostic systems and can provide useful information through the collection of data for medical research purposes. In addition, EHRs can help hospitals monitor, improve, and report data on health care quality and safety.

Indeed, EHRs are the cornerstone of an eHealth strategy for every government that wishes to deliver faster and more effective treatment and follow-up to the citizens. Especially for Cyprus which is currently under the serious impact of the global financial crisis, the adoption of a national EHR system is a one way road leading to it. With a national EHR system, there will be an improvement in the availability of medical information, improving in the protection of medical data and development productivity (less time for recording and updating general information of the patient). Important economic benefits in this difficult time for Cyprus will be derived by saving time and money from the reduction of bureaucracy, minimizing errors, reducing paper consumption, duplication of tests and misdiagnosis incidents. Furthermore Cypriot citizens, can have direct access to their EHR maintain in Cyprus, from medical centers outside Cyprus, in cases where something medical occurs and action should be taken immediately. Similarly European citizens visiting Cyprus will be able to access their EHR held in their country at any time, thus making Cyprus an even better tourist destination. Cyprus economy is largely dependent on the tourist and services economy. The permanent population of Cyprus is less than a million and at the same time over than three millions tourists and business people are visiting Cyprus every year.

To this end, the Ministry of Health has decided to proceed with a Health Care Information Support (HCIS) System in all government hospitals, outpatient departments and rural health centers [1].

Still, electronic patient data storage is not yet very common in Cyprus. According to the Empirica study: Pilot on eHealth Indicators, 2007 in comparison to the other EU Member States, Cyprus has to be regarded as one of the laggards as only around one half of the GP practices store at least one sort of individual medical patient data [2].

II. CYPRUS SURVEY

In an effort to find out what is the current situation in Cyprus and inform the local medical community with issues related to electronic health records, an ongoing survey is in progress, mainly through a questionnaire that has been circulated in the whole island and administered by the members of the Cyprus Medical Association and the Cyprus Society of Medical Informatics. The questionnaire is consisted of three parts and a total of 47 questions that examine mainly the level of usage of EHR systems in the private sector, regain the requirements from the doctor's perspective and register the willingness of the doctors to accept important issues such as a patient centric philosophy. The questionnaires were already completed by 58 physicians, and although more data is expected to arrive shortly, important results emerged and are presented here.

71% of the physicians claim to use an electronic patient management and from those the 69% are not satisfied with their system, mainly because it does not support interoperability for communicate with other existing systems. Of those who use some sort of EHR systems only 50% provided all the requested data for forming the patient summary as this is defined by the European eHealth project "epSOS" (http://www.epsos.eu/). In fact, according to the so far analysis, 66% of the doctors who use a kind electronic patient record are using the telephone to share patient information.

As for the remaining 29% of the physicians who do not use any kind of EHR system, the main reasons why they chose not to, are mostly due to the lack of computer knowledge and the limited time that they have to dedicate for becoming familiar with the EHR philosophy. Yet, the 75% of all the doctors believe that an EHR system will improve the quality of their services.

III. IDENTIFYING THE PROBLEMS

Creating an EHR system is not something unfeasible for IT programmers. However, it requires lot of effort and input from many management services. The main difficulties that have to be overcome can be characterized as technical, engineering, political, financial and legal.

Until now no authority, committee or organization has developed a worldwide accepted set of standards and rules that can define an EHR. This fact allowed many governments and institutions to develop their own proprietary systems. Those systems are structured based on their needs: relational databases and data formats structured using common or private technologies which may not necessarily satisfy others to use them. Some standards that have been developed so far are Health Level 7 (HL7) Clinical Document Architecture (CDA)[3], CEN EN 13606 EHRcom [4], and openEHR [5]. Heterogeneous systems are being used worldwide having their distinct usability and customization according to their needs in each case. For these reasons some international standards are established from the World Health Organization (WHO), which is the directing and coordinating authority for health within the United Nations network [6].

Moving away from standards, another issue raised is the security of data and the safety of such systems. Most hospitals and healthcare centers do not have adequate secure network infrastructures. This is a basic requirement for an EHR system and it gets more serious when users access it remotely, from external networks. Also, questions like where EHR data is stored and who will be responsible for maintenance are needed to be clarified before any development. Data protection must be enforced into the systems since they deal with sensitive patient information.

IV. PROPOSED SOLUTIONS

Any proposed solutions for a national EHR system must be in agreement with the overall objectives of the eHealth Network (http://ec.europa.eu/health/ehealth/policy/network/index_en.htm) which aim to enhance continuity of care through interoperable and secure eHealth services, ensure a high level of trust and results in high quality healthcare, in accordance with the Directive on patients' rights in cross border healthcare (2011/24/EU).

In addition, it is believed that it is crucial to follow a patient-centric philosophy in any EHR solution in order to establish a level of trust among the patients, being one of the main drawbacks for not moving forward in adopting a national EHR system in Cyprus. The patient-centric philosophy mainly relies on the concepts that: a) the patient is the owner and responsible for the management of his/her record and b) physicians can have access to medical records only after permission of the EHR owner or the person authorized by the owner.

In accordance to this, a very important area that is strongly linked with the trust of the patients towards any EHR system corresponds to the security issues.

Our proposal is that policies should be re-examined and evaluated at different levels, i.e. both at the organisational and national levels in terms of security issues since the right use of an EHR system and the fact that the corresponding data are protected, may be the ultimate factor for a successfully national EHR system. Some of the most important security challenges that may arise from the adoption of a national EHR system in Cyprus correspond to the implementation architecture of a national EHR system which determines the place of storage of the patient data. Therefore the architecture affects the security of the whole system. Most of the EHR systems are either centralized or decentralized.

A centralized EHR system sends the patient's data to a central repository and a role-based access is granted to each person accessing the system. Organizations, health centres and institutions need to have continuous communication with the main repository, follow policies and standards so as to be able to ensure the interoperability of their system with the central repository; this is also necessary to protect the patient data that is retrieved from the repository and used in their system. The main problem with the centralized scheme is that the data repository belongs physically to one single organization. Holding the data in a central server has troubled both the experts and the health care providers, as attackers can have access to the whole data by attacking once the repository [7].

Alternatively, patient data in decentralized systems are distributed in many locations, and hence the full record of a patient is built by retrieving the individual partial data from every location; in this case, each system participating in this scheme is responsible to maintain its part of the patient data and allow controlled access to this data.

In Cyprus the population is approximately 780.000 people therefore it is believed that a combination of the above solutions will be the most appropriate approach. For example, group of hospitals can assign to private companies the storage in a centralized manner of their patient data, while following interoperability protocols for communication purposes with centralized systems of other groups of clinics. Hence, it will result into a global decentralized system that will be controlled privately and independently for each healthcare institution. Furthermore, the small population of Cyprus allows to additionally providing for each patient a memory device (e.g. USB stick) that can be updated for back up purposes each time the patient visits his/her physician.

Besides the implementation approach, software itself may play an important role in the security, thus software criteria should also be considered.

A list of security criteria can be used to evaluate the security of EHR software. For example, each organization must be able to assign rights from a restrictive set of privileges to groups and users. A strict privileged based-scheme can control the access to the patient data; both system users and relatives should be granted the right to see patient data only after the patient's consent. Thus, the patient is the one that assigns privileges to others. For example, a patient can allow a doctor, e.g. a dentist, to access their health data; however, this access may be limited only to the dental medical data. With the use of a profile, patients could be able to easily assign roles to service providers and physicians. Authorized system administrators should also be able to assign restrictions or/and privileges to users or/and groups.

General audit procedures should also be considered for events that happened out of the authority of the EHR system. Access to audit trail should be granted only to authorized personnel, e.g. high level administrators, and the system shall prohibit modifications to audit records.

Furthermore, multiple levels of authentication can minimize security risks. For example, system users should be authenticated, especially before accessing sensitive data, such as the patient data, and the system should force users of inactive sessions to authenticate before using the system again.

Clinicians can mark some of the patient data as blinded for prohibiting access to the data by other users; according to the legislation in Cyprus and other European countries, a doctor has the right to hide some information from the patient if such data can harm (physically or mentally) the patient. Access to hidden data can be granted if it is proved to be crucial for an emergency condition.

The system must support backup of the system and patient data. In addition, it should be capable of preventing corruption of the data and users should be able to restore the data and the state of the system in only a short time. This is most essential when the data are centralized as in centralized systems there are many more safety issues; in the UK for example, a hospital with centralized EHR system, lost a day transactions after a power failure [7].

Finally, documentation of the functions of the system is part of a proper design/measure that can minimize the security risk. EHR software should be fully documented for allowing users to use effectively and completely its security functions. Adequate documentation for errors and security events should be provided so as to allow the users to take proper actions.

V. A MODULAR SYSTEM SOLUTION

In respect to the implementation of the system we propose a patient oriented modular system that it's able to deal with the complexity and the separate needs from a diversity of medical specialties and all the users (patients, doctors, nursing staff, pharmacist, administration stuff, etc.). Since standards are yet to be fully approved and no regulations have been enforced, the need for a platform that will be able to be evolved, optimized, maintained or modified in the future is strongly recommended. A platform like this is ideal for the case of Cyprus, since from the survey (section II) it was shown that most medical personnel are using proprietary outdated systems with limited capabilities and interconnectivity.

Modules can use or expand other modules to retrieve information or to undertake a specific task. Patients can view and authorize any transaction with his/her profile. At the same time, security functions will ensure the patient privacy and integrity of data. The system is structured in Model-View-Controller (MVC) architecture and will allow developers to build and integrate modules rapidly and easily.

The system is designed from a patient prospectus, which allows patients to access the system, view their history and any other medical details, authorize the users that will access their profile and transfer their medical record to any place. With this way the EHR becomes a nationwide and worldwide accessed tool and the patient is contributing and educated for his/her health.

In the diagram below (**Figure 1**) we provide an overview of the proposed system. The system will consist from a backend primary module which will play the role of the regulator and many sub-modules that will complete the system. Under the primary module there are three categories (or set) of modules: **Main Module**: contains all the necessary modules required from the system to operate. **Secondary Modules**: are modules that support the main modules enriching the system characteristics. **Medical Modules**: includes specific and specialized functionalities needed by the users of the system.

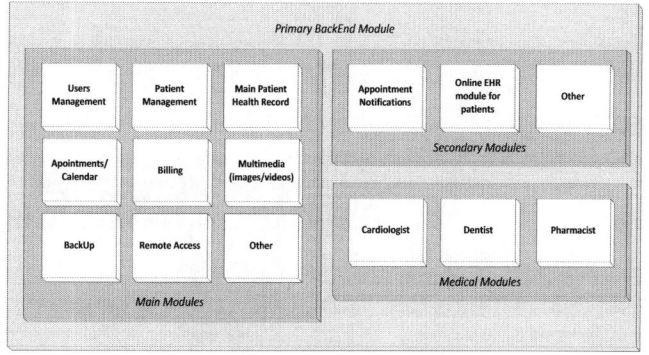

Fig. 1 Proposed system architecture

Our proposal suggests the usage of open source technologies which is proven to be efficient, fast and reliable. Open source technologies can reduce the development cost, eliminate the bugs and mistakes and provide a long term support from the developer's community.

VI. CONCLUSIONS

Patient-centric healthcare philosophy is promoted as the future dictates. Consequently all healthcare services should be built around the patient preferences, convenience, needs, and values, and ensure that patient is well informed about any clinical actions and decisions. In essence, the EHR belongs to the patient and not to the healthcare system. To minimize the security risks, HC providers must invest in the software functionality and specifically in features related to security, auditing, control of transactions, authorization of users and data monitoring.

Finally, one should take into consideration that the healthcare environment is a rapidly growing field with many needs to be fulfilled. It requires a state of the art management system so it can fulfill the needs of all the users (patients, doctors, nursing staff, pharmacist, administration stuff, etc.) while ensuring reliability and confidentiality. To be able to build such a system it should break it down in independent modules with inter-modular connections. The main benefit of a modular system is the scalability that will offer through the isolation of different functionalities. A modular system will allow developers to expand the system easily without interfering with the existing functionality.

ACKNOWLEDGMENT

This research was financially supported by the research grant ΥΓΕΙΑ/ΔΥΓΕΙΑ/0311(ΒΙΕ)/47 from the Cyprus Research Promotion Foundation

REFERENCES

1. Angelidis P., Giest, S., Dumortier, J., Artmann J., (2010). Report on eHealth strategies – country brief Cyprus. Available at http://ehealth-strategies.eu/database/documents/Cyprus_CountryBrief_eHStrategies.pdf
2. Empirica study: Pilot on eHealth Indicators, Country profile: Cyprus (2007).
3. American National Standards Institute, "ANSI. ANSI/HL7 CDA-R2 (2005). HL7 Clinical Document Architecture, Release 2.0," 2005.
4. Technical report, European Committee, "CEN prEN 13606-1. (2004) Health Informatics—Electronic Health Record Communication—Part 1: Reference Model."
5. T. Beale and S. Heard, (2007) "OpenEHR Architecture: Architecture Overview"
6. National Library of Medicine, National Institutes of Health, U.S. Department of Health and Human Services. Mapping SNOMED-CT to ICD-10-CM final Release notes 2012.
7. Ross Anderson (2008).Patient confidentiality and central databases, British Journal of General Practice
8. Terry K.J. Doctors' 10 Biggest Mistakes When Using EHRs. Medscape. May 01, 2013. Available at: http://www.medscape.com/viewarticle/803188 Accessed May 10, 2013.

Information Driven Care Pathways and Procedures

R.J. Dickinson[1] and R.I. Kitney[2]

[1] Imperial College, Department of BioEngineering, London, UK
[2] Imperial College, Department of BioEngineering, London, UK

Abstract—The paper addresses the issue of the implementation of care pathways in electronic form. Within the National Health Service (NHS) of England, Care Pathways are becoming increasingly important. These are typically provided by the Department of Health. The Pathways provided are in the form of paper-based schema. They either have to be implemented via paper forms or, as presented here, in electronic form. In addition, care pathways must be seen in the context of the T-Model of health care which comprises the care continuum and the biological continuum. The two care pathways which had been chosen as exemplars are myocardial infarction and stroke. However, the objective of the paper is not to discuss the specific care pathways in detail, but, rather, to describe technology which has been developed for their electronic implementation. The result of this implementation is that all the data and information acquired from the implementation of the care pathway is stored in a single clinical information system (CIS), which has incorporated in it the SQL database. Another important element of the system which has been developed is the ability to display data and information in terms of two dashboards (i.e. single screens which show the most important information). The two dashboards display clinical information (the point of care dashboard) and management information (the management dashboard).

Keywords—CIS (clinical information system), clinical pathways, clinical dashboard, management dashboard.

I. INTRODUCTION

The aim of the project is to optimize care pathways using information- based techniques. Two examples of care pathways, namely chest pain and stroke, have been chosen for this purpose, although it is intended that the results of the project will be used over a much broader clinical spectrum. The core of the project is the electronic implementation of care pathways (CPs) for chest pain and stroke (both of which have been designated by the UK Government as important clinical areas). Throughout the project there has been emphasis at all stages on clinicians working with engineers to achieve a fully working pilot system within two hospitals, St Mary's Paddington and The Chelsea and Westminster (both in London, UK). The work has centered on the core concept of the T model of the care and the biological continuums (Fig 1) [1].

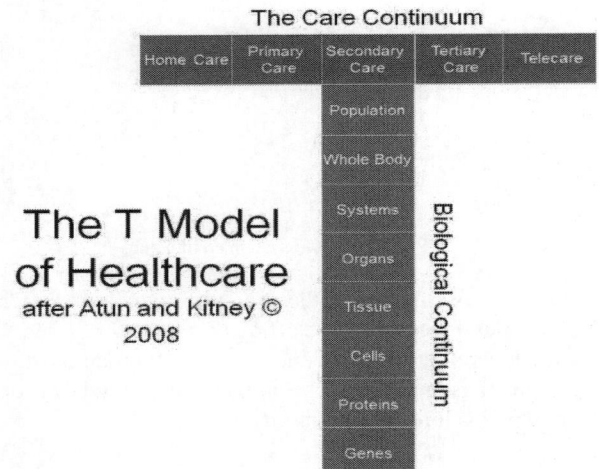

Fig. 1 The T Model of Healthcare

II. METHODS

A. Clinical Information

A web-based clinical information system (CIS) has been developed and implemented which provides access to both the patient record and image studies. The top layer of the system comprises a web-based implementation of the care pathways with a timeline. Taking chest pain or stroke, the care pathway is divided into a number of steps; these steps and the timeline are represented on an electronic dashboard, where each step is represented as part of a flow diagram.

This comprises a web-based IT system which incorporates a full DICOM library and HL7 functionality. The clinical information system (CIS) was originally developed in Kitney's laboratory. The CIS comprises both full EPR and PACS functionality, can accommodate and display data across the biological continuum [2][3], and is fully operational. It was important to use the system as the base of the IT system for the project because it has taken many years for it to be developed and successfully implemented[4][5]. The system is capable of being easily linked to a wide range of scanners, databases and laboratories by means of interface technology.

A principal objective of the project has been to develop and implement web-based versions of the CPs and to link them to the common CIS. Part of this work has been to develop a clinical dashboard. This presents the CP in electronic form and enables the user to observe the progress of the patient along the CP, which test/scans etc. are outstanding, and whether they are late. It also monitors data entry, to ensure that critical data are properly entered, as well as highlighting data errors (thus improving patient safety). Finally, the dashboard is designed so that decision support systems, based on the analysis of patient data contained in the EPR, can be easily integrated into the new CIS.

Pathway-Centric Data Models: A data management mechanism has been developed to provide a uniform view of information along an entire clinical pathway. Within such a model, each element of an intervention for each patient from an underlying EPR is stored in a consistent format and organized by a data schema - which can be dynamically generated, according to the underlying pathway structure. Thus, all the information along a pathway can be dynamically captured, integrated and then analyzed for the calculation of overall cost and outcome metrics.

Value Focused Analysis: Based on the data models, analysis can be performed to provide reports and dashboards that deliver value based metrics across the pathway from senior management to clinical practitioners. Such an analysis is a key component of the iHealth clinical decision support system. The system is integrated into the physician's work process to assist clinical decisions.

For the effectiveness of the decisions, such a decision support system requires the association of clinical pathway data with the proper patient profiles (cohorts and their associated therapeutic and biological information).

B. Management Information

A key challenge in developing a system for the integration of a wide range of information and data in the clinical context, is the need to store a wide range of information in a common archive. And associated problem is the need to be able to access this information in forms which are compatible with the requirements of different users. Hence, in addition to clinical staff, information and data are required by senior hospital management and ward managers. This means that the information and data needs to be presented in a form which is compatible with its use for management purposes. In the system which has been developed this is achieved by storing the data required from the CIS in an SQL database data - and information are entered into the CIS via electronic versions of standard NHS forms. For the purposes of hospital management, data and information are transferred from the database using SQL Server. This is achieved by using HL7 data transfer and HL7 messaging,

coupled to a data miner. This scheme enables management level data to be extracted and presented.

C. Dashboards

On the basis of the clinical and management information procedures described above, we have also developed dashboard technology to enable clinicians and managers to see important information at a glance on a single screen. In practice, the dashboard displays time information, such as when a particular test was ordered, when it was due for completion etc. A lot of this information can be color coded in terms of "traffic lights", with the ability to "drill down" for more detailed information. In the case of the management dashboard, the work has focused on building a system for computing the pathway oriented performance metrics based on the integrative analysis of operational data along clinical pathway practices of NHS as in, for example, the Map of Medicine (www.mapofmedicine.com). These metrics aim to provide quantitative understanding of the value of healthcare by quantifying both the clinical outcomes and costs of interventions across all points of a clinical pathway. Referring to figure 3, the full range of patient and associated data are stored in the SQL database of CIS. The schema for the data can be envisaged in terms of a two-dimensional matrix, where the rows of the matrix contain all the data and information relating to an individual patient. The columns of the matrix contain different types of data and information across the patient population (a single column containing a particular type of information, for example MR scans, for the whole patient population). Taking the example of the management dashboard (although this equally applies to the point of care dashboard), management data are drawn down from the CIS SQL database by means of an SQL Server. The SQL Server communicates with a data miner via HL7 messaging. Data and information are then fed to the management dashboard from the data miner.

III. RESULTS

D. The Care Pathway

In its original form, as shown in Fig 2, the CP does comprise a series of episodes along the care pathway (or patient journey). Referring to the figure, episode one of the patient journey comprises clinician intervention. This takes the form of a verbal discussion, the results of which are included in the patient's paper record. In the second episode, again there is clinical intervention in the form of an examination. As shown in the figure, results of the examination are again recorded on paper – but, where appropriate, there may, for example, be a digital image resulting from the examination. In the schema this is stored in the first of two clinical Information Systems. It can be seen from the figure that de-

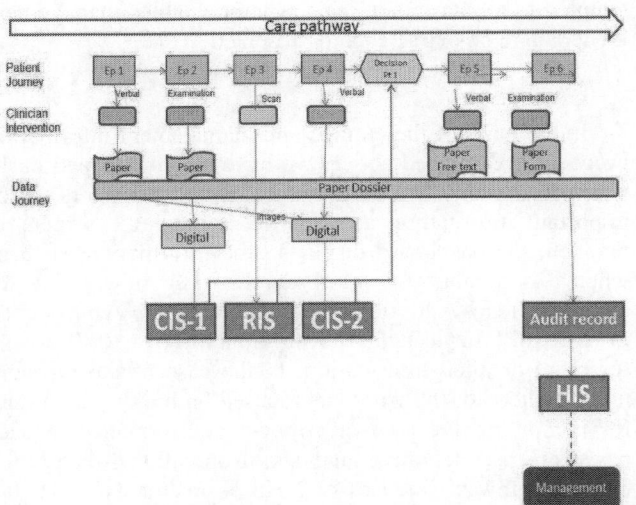

Fig. 2 Traditional Care Pathway Schema

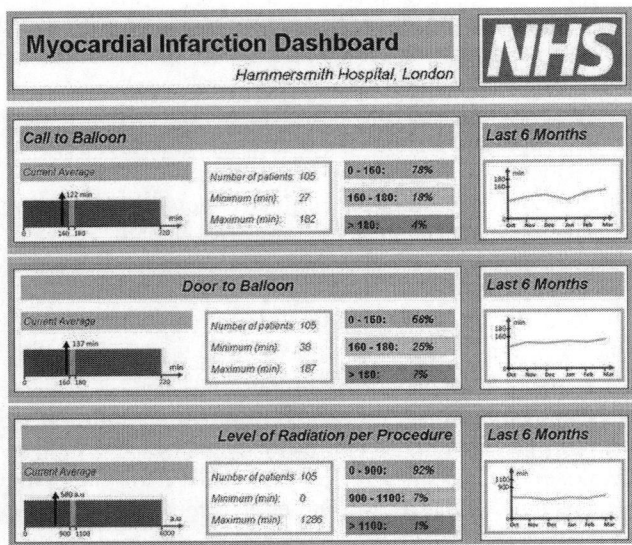

Fig. 4 Example of a Management Dashboard

pending on the nature of the examination there may be associated scans and in the case shown (Episode 3) information eventually ends up in the radiological information system (RIS). Hence, it can be seen that data and information stored by means of a number of different types of systems: paper dossier, electronic clinical information systems - and in the example given, in a RIS. By the end of the patient journey information is, additionally, stored in paper free text and other paper forms. Information in the paper dossier is ultimately transferred to an audit record and the hospital information system (HIS). The hospital management can then get access to the patient information via the HIS. Fundamentally, the problem with this type of patient system is that it is impossible for other clinicians to access the full range of information on the patient without going to a range of paper and electronic systems.

Figure 3 illustrates the implementation of the fully electronic care pathway. Comparing figures 2 and 3, it can be seen that there are a number of key differences. Whilst the structure of the patient journey is similar in both cases, in figure 3 electronic systems have now all been incorporated into a single clinical information system. The patient journey is still based on a series of episodes with clinician intervention. However, if the Data Journey section in both figures is compared, it can be seen that in figure 3 (i.e. the fully electronic implementation of the care pathway) that for each episode the patient information and data acquired through clinical intervention is stored via an EPR compatible electronic form. In the case of our implementation of the care pathways, these forms are exact electronic versions of the original NHS paper forms (i.e. as far as the clinician is concerned, the forms look identical to their paper versions). What underlies this approach is the ability to integrate a number of different types of medical equipment into a common information environment. This is achieved by the use of the DICOM and HL7 standards, together with data format converters, where necessary.

D. The Dashboards

As previously stated, we have developed two dashboards – a clinical dashboard and a management dashboard. Both dashboards are designed to provide the most important information for the clinician or manager (depending on which dashboard is being viewed) on a single screen. The full range of patient and associated data are stored in the SQL database of CIS. Taking the example of the management dashboard, Figure 4 illustrates the dashboard for myocardial infarction. In this case the information displayed represent figures derived from the Hammersmith Hospital in London. The graphs, bar charts and data shown all corresponds to UK Department of Health guidelines and indicators for myocardial infarction.

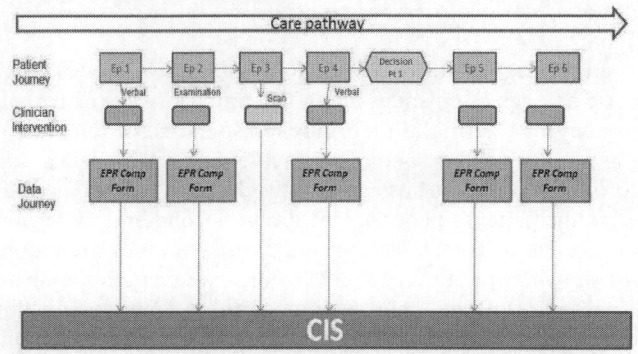

Fig. 3 Integrated Electronic Care Pathway Schema

IV. DISCUSSION

The display of detailed clinical and management information at numerous points across the hospital is becoming increasingly important. This paper has primarily focused on an implementation of care pathways within the NHS in England. Two care pathways, one for myocardial infarction and the other for stroke, are the basis of the project. Details of the actual care pathways are not the subject of this paper. The subject of this paper is, rather, the underlying technology which allows, for example, the full range of patient information to be accessed in real-time at the clinical decision point. The systems which have been developed based on the concept of the T-model for healthcare (illustrated in figure 1). As illustrated in figure 2, current implementations of care pathways within the NHS store data and information in multiple systems. These comprise both paper records and electronic systems such as the EPR, CIS and RIS. The technology described in the paper allows the full integration of patient data into a common CIS with its associated SQL database. Data entry is via specially designed electronic forms, which correspond exactly to the original, equivalent NHS paper forms. A key element of the system which has been developed is the use of dashboards to provide clinicians and management with all the key information they require at a glance on a single screen. This is achieved by creating a database structure which effectively comprises a two-dimensional matrix – where the rows contain the full range of data and information for a single patient, and the columns contain one type of data (e.g. MR scans) for the whole patient population. By using a data miner it is possible to construct dynamic dashboards showing point of care information and management information, respectively. An example of a management dashboard for myocardial infarction is illustrated in figure 4. This contains a range of key information and data as defined by the Department of Health indicators and the clinicians involved in the hospital. Whilst this type of data per has been available for some time in paper form, there is usually quite significant delay (sometimes weeks) before the data are available. In the system described in the paper the dashboards are updated automatically in real-time, providing minute by minute access to a wide range of key information.

V. CONCLUSION

The system described in the paper is now being implemented in the hospital. Clinicians have been involved at every stage of its design and implementation and the pilot results which are being obtained look very encouraging. Two key features which the clinicians and hospital managers find particularly attractive are that the data presented are comprehensive - and can be accessed anywhere in the hospital in real-time as this is a web-based implementation.

ACKNOWLEDGMENTS

We would like to acknowledge the support of the UK Engineering and Physical Sciences Council and the Medical Research Council in providing the funding for the project.

We would particularly like to acknowledge the technical assistance of Dr Matthieu Bultelle and Dr Stefan Claesen in the execution of the project.

REFERENCES

1. Poh, C.L.; Kitney, R.I.; Geometric Framework linking different levels of the Biological Continuum. (2006). Engineering in Medicine and Biology Society, IEEE-EMBS 27th Annual International Conference. pp 4068 – 4071
2. Poh C L, Kitney RI and Shrestha RB. (2007) Addressing the Future of Clinical Information Systems—Web-Based Multilayer Visualization. IEEE Trans on information technology in biomedicine. Vol 11, Issue 2, pp 127 -140
3. Kitney, R., Poh, C.L., Claesen, S. and Dewey, C. F. Accommodating Multiple Data Inputs across the Biological Continuum. Proceedings of the World Congress on Medical Physics and Biomedical Engineering, 2006, Seoul, Korea, [DVD-ROM] ISBN 3-54036839-6, Paper No. 2204, pp 4142
4. Poh, C.L.; Kitney, R.I.; Cartilage Thickness Visualization using 2D WearMaps and TrackBack. Engineering in Medicine and Biology Society, 2007. EMBS 2007. 29th Annual International Conference of the IEEE 22-26 Aug. 2007 Page(s): 2883 – 2886
5. Poh CL, Kitney RI and Akhtar S (2009). Web-based Multi-Layer Viewing Interface for Knee Cartilage. IEEE Transactions on Information Technology in Biomedicine, 13(4) pp 546-553

Decision Making in Screening Diagnostics E-Medicine

G. Balodis[1], I. Markovica[1], Z. Markovics[1], and D. Matisone[2]

[1] Riga Technical University/ Faculty of Computer Science and Information Technology, Riga, Latvia
[2] University of Latvia / Institute of Cardiology, Riga, Latvia

Abstract—The paper discusses development of the e-medicine system for preventive examinations and screening and its computer implementation. System is based on a mobile telemedicine screening complex (MTSC), which includes a laptop, a questionnaire about the state of organism's 12 systems and 14 measurement modules for objective measurement acquisition.

Computer's functions include measurement management, questionnaire management, measurement value calculation, digitization of the results, wireless transmission to the analysis center, indication of the measurement results and decision-making. Decision making unit includes processing of the interactive questionnaire, findings on subjective complaints, discoveries of objective abnormalities and decisions on further actions.

The paper focuses on information processing and decision-making methodology. Analysis and classification of possible final conclusion has been done. Analysis of the questionnaire responses and the results of measurements and their connection to the possible final conclusion has been done. Information flow graph has been developed. Configuration of decision-making tree has been developed. The law set of production rules has been developed as the basis for decision-making.

Keywords—Telemedicine, screening, diagnostics, e-health, decision support system.

I. INTRODUCTION

Public health is one of key priorities in the world. And so, the European Union Charter of Fundamental Rights Section 35 provides that every person of any country, including Latvia, has a right to preventive health care and medical treatment. This means not only high quality and high levels of treatment of diseases, but also timely diagnosis and prevention. An important role here is to regular preventive examinations. Currently investigations and analysis are usually carried out by health care institutions because only in very rare cases family doctor has the necessary diagnostic equipment and skills of evaluation of the information obtained in the investigations.

To organize and facilitate the process of screening at health care institutions but even more so at consulting rooms of family doctors, information technologies, including decision support systems [1, 2], are used. To improve this process we have developed Mobile Telemedicine Screening Complex (MTSC) [3], consisting of 2 parts. First part include 14 measurement modules for objective data acquisition for 9 systems of the organism and detection of 4 risk factors (RF) – increased blood pressure, determination of increased level of glucose and cholesterol in blood, obesity. The second part is interactive questionnaire that can be considered as the fifteenth module. It is composed of person identification data, subjective complaints about the organism's 12 systems and information about 7 potential risk factors – smoking, high blood pressure, increased blood glucose and cholesterol levels, unfavorable heredity, sedentary lifestyle and obesity. A separate chapter in the questionnaire is devoted to chronic disease detection. Full list of examined systems of organism and list of measurement modules can be seen in table 1.

Table 1 Examined systems and measurement modules

Systems	Measurement modules
S1 Cardiovascular	1. Electrocardiography (ECG)
S2 Respiratory	2. Blood pressure measurment
S3 Kidneys and urinary tract	3. Pulse oximetry
S4 Possible tumors	4. Digital phonendoscopy
S5 Bones and joints	5. Cholesterol test strips
S6 Endocrine	6. Spirometry
S7 Vision	7. Digital thermometry
S8 Hearing	8. Urine test strips
S9 Digestive	9. Dermascopy
S10 Neurological	10. Anthropometry
S11 Psychoemotional	11. Weight measurement
S12 Reproductive	12. Glucose test strips
	13. Acuity of vision
	14. Audiometry

II. THE SYSTEM

The central node of the MTSC is mobile computer with monitor whose tasks [4] include:

- individual measurements,
- calculating measurement values,
- displaying of measurement results for on-site medical staff and patients in an intelligible form (in decimal system),
- conversion of measurements into digital format and wireless transformation to the analysis center,

- processing of the results of interactive questionnaire,
- decision making about consultations needed on findings of the screening results for the organism's systems based on questionnaire and objective measurements,
- decision making on further actions (referrals for more precise investigation, taking into account of risk factors, strategic actions and recommendations).

Interactive questionnaire and measurement modules provide subjective and objective information about the person being investigated allowing to make 5 types of conclusions:

S Type conclusion comprises information from interactive questionnaire (subjective complaints from 12 organ systems) and objective measurements showing deviations in 9 organ systems.

The conclusion in this case consists of 3 components:

- consulting part in which subjective complaints and objective findings are indicated in encoded form,
- decision part in which further steps are specified,
- explanation part where the above listed fields are explained in a simple language understandable to the patient.

R type conclusion comprises information from interactive questionnaire about 7 risk factors and 7 objective measurements attributable to the risk factors. Conclusion comprises information whether or not the investigated person has any risk factors, what kinds of risk factors and how many risk factors the investigated person has. If the investigated person has risk factors, the recommendation to attend family doctor is obligatory.

H type conclusion shows whether or not the investigated person has any chronic disease, based on questionnaire data.

P type conclusion is based on questionnaire data and shows whether or not the investigated person is having regular health checks.

N type conclusion is formulated if the investigated person has no complaints and objective measurements are normal. The conclusion may consist of one or both of the findings:

- no subjective complaints,
- no objective findings.

III. ORGANIZATION OF INFORMATION FLOW

Conclusion making and decision making in screening system is based on two sources of information: an interactive questionnaire data and objective measurements and test results.

In the interactive questionnaire blocks are arranged in sections corresponding to the above mentioned conclusion types S, R, H, P.

In the S section there are blocks about all 12 systems of the organism, such as:

S2. The respiratory system.
S 2.1 Do you often have a cough that cannot be associated with colds or viral infection? Yes □ No □
S 2.2 Have there been any noises or squeaking while breathing in the last year? Yes □ No □
S 2.3 Have you had any dyspnoea attacks during or after physical exertion in the last year? Yes □ No □
S 2.4 Have you woken up at night from shortness of breath in the last year? Yes □ No □
S 2.5 Have you woken up at night from coughing in the last year? Yes □ No □
S 2.6 Does your morning usually starts with a cough? Yes □ No □
S 2.7 Does your morning usually starts with expectoration? Yes □ No □
S 2.8 Do you have/had an allergic reaction to any medication, insect bites, allergic rhinitis, hay fever? Yes □ No □
S 2.9 Have you had any causeless temperature rising episodes? Yes □ No □ I cannot tell □

Section R contains question blocks about 6 risk factors such as:

R1. Unfavorable heredity
Do your blood relatives (parents, sisters, brothers) have or have been
R 1.1 high blood pressure Yes □ No □ Cannot tell □
R 1.2 myocardial infarction and / or stroke
Yes □ No □ Cannot tell □
R 1.3 diabetes Yes □ No □ Cannot tell □
R 1.4 oncological diseases Yes □ No □ Cannot tell □
R 1.5 glaucoma or cataracts Yes □ No □ Cannot tell □

Section H contains question blocks about 14 chronic diseases.

Section P has only one question about the existence of regular health checks.

There is no section N in the interactive questionnaire because this conclusion is formed only in case when the all the questionnaire questions are answered negatively and all the objective measurements show no abnormalities.

The 14 measurement modules and tests forming the complex together can measure more than 60 values, for example blood pressure monitor gives systolic and diastolic blood pressure value, pulse rate and calculates average pressure. Urine test strips provide answers for the 11 values.

The resulting values are compared with norms and conclusions are formed according to symptoms for screening diagnostic. For example:

№ 3. Pulse oximetry has 2 symptoms
3.0 - normal
3.1 - hypoxemia exists

№ 11. Body mass index (BMI) has three symptoms
11.0 - normal
11.1 - increased
11.2 - reduced
№ 4. Phonendoscopy has four symptoms
4.0 - normal heart and breath sounds
4.1 - a heart murmur
4.2 - noises on blood vessels
4.3 - lung rales

IV. OVERALL SCHEME OF THE DECISION DEVELOPMENT

It is intended that the screening system forms conclusions on all 12 body systems S1 ÷ S12, which are included in the interactive questionnaire. 9 of them also have objective measurement information from measurement modules. 3 systems (neurological, psychoemotional and reproductive) have only information from the questionnaire. So conclusions about 9 systems should be based on two kinds of information – the subjective questionnaire data and objective measurement results.

The situation is made even more complicated by the fact that the decision about abnormalities of functioning in particular system may be affected not only by direct measurements and direct answers to questions of this system, but also by questions and measurements from several other systems as well. For example monitoring the status of respiratory system S2 should take into account not only the answers to direct questions in S2, but also the answers to questions such as the 1.8 S – "Do you have shortness of breath during light physical activity" from the heart and blood vessels block and S 4.1 - "Is there secretion coughing" of possible neoplasms block. Indications of respiratory system status from objective measurements may be pulse oximetry (3), phonendoscopy (4), spirometry (6) and thermometry (7). Thus, the links to the final decision from responses to the questionnaire and from the objective measurements of symptoms may be complex (Figure 1).

V. DECISION TREES

Because of the complicated structure of the decision-making it is useful to develop decision making trees, which should include specific symptoms from the questionnaire and also from the measurement part.

Decision making tree for respiratory system's S2 case is shown in figure 2 where S 2.1 ÷ S 2.9 are questionnaire questions from the respiratory system block, S 1.8 and S 4.1 are explained above, 3.1 – a symptom of abnormality in pulse oximetry measurements, 4.3 is a symptom of lung sounds, 6.1, 6.3, 6.5 are symptoms of abnormality in spirometry measurements and 7.1 are symptoms of abnormality in measurements of thermometry.

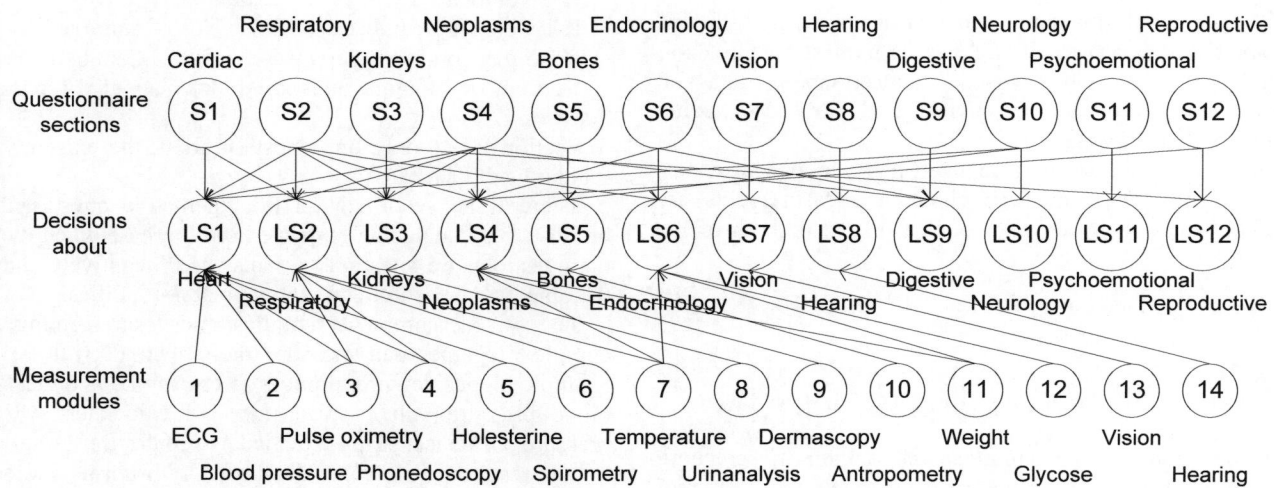

Fig. 1 Connection between symptoms and decisions

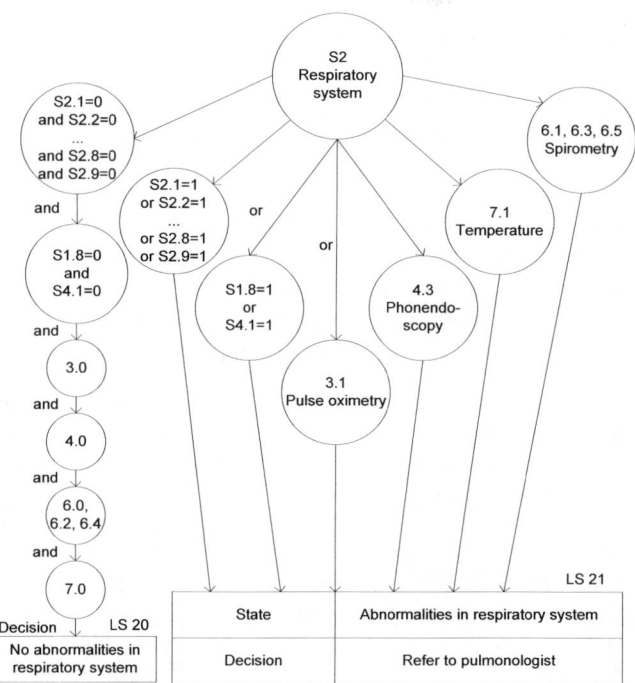

Fig. 2 Decision tree about respiratory system

For computer implementation the information contained in the decision-making trees can be expressed as production rules in IF...THEN form. For practical convenience the format used in the decision making tree is left intact in the condition part, example is shown in table 2:

Table 2 Respiratory system

Condition part	Conclusion part
IF S 2.1=1 or S 2.2=1 or S 2.3=1 or S 2.4=1 or S 2.5=1 or S 2.6=1 or S 2.7=1 or S 2.8=1 or S 2.9=1	THEN decision L S 21 "Respiratory system has abnormalities." "Send to pulmonologist"
IF S 1.8=1 or S.4.1=1	
IF exists S 1	
IF exists 4.3	
IF exists 6.1 or 6.3 or 6.5	
IF exists 7.1	

Similarly decision making trees and production rules are made for all 12 systems, for 7 risk factors and for sections H, P and N.

It may be noted that the observed subjective complaints, the existence of chronic diseases, objective findings and the calculated values, as well as all computer decisions are indicated on site on the monitor screen of MTSC for the knowledge of medical staff and the patient. It is possible to print out the results of screening diagnostics. Information in digital form is wirelessly transmitted to the analysis center where there are specialists that can make diagnosis of ECG and dermascopy images and approve or correct the computer made decision. Also at the analysis center data is stored in a database for further reference.

VI. CONCLUSIONS

The developed computer system allows efficient use of MTSC capabilities and provides computer based decision making about the state of organism's 12 systems and 7 risk factors. Collaboration with analysis centre provides confirmation or correction of the decision in cases when decision can be made by a suitably qualified specialist which in our case if ECG and dermascopy.

Potential of the system can also be used in medical practices for patient diagnosis because measurement modules used in the complex are some of the most advanced in its class, such as ECG, digital phonendoscope, pulse oximeter, anthropometric measurements complex and others. The complex can also be used by a practitioner to provide additional services to their patients.

ACKNOWLEDGMENT

Project is supported and financed thanks to European Regional Development Fund, agreement Nr: 2011/0007/2DP/ 2.1.1.1.0/10/APIA/VIAA/008

REFERENCES

1. Bright TJ, Wong A, Dhurjati R et al. (2012) Effect of clinical decision-support systems: a systematic review. Annals of Internal Medicine 157: 29–43
2. Main C, Moxham T et al. (2010) Computerised decision support systems in order communication for diagnostic, screening or monitoring test ordering: Systematic reviews of the effects and cost-effectiveness of systems. Health Technology Assessment 14(48):1-227 DOI 10.3310/hta14480
3. Katasevs A, Markovics Z, Markovica I, Balodis G, Lauznis J. Development of New Mobile Telemedicine Screening Complex, International Symposium on Biomedical Engineering and Medical Physics, IFMBE Proceedings, Riga, Latvia, 2012, pp. 31–34
4. Lauznis J, Markovics Z, Balodis G, Strelcs V. On Resource Distribution in Mobile Telemedicine Screening Complex. Scientific Journal of Riga Technical University, Computer Science, vol. 13, Riga, Latvia, 2012, pp 28–31

Author: Gunars Balodis
Institute: Riga Technical University, Faculty of Computer Science and Information Technology
Street: Meza str. 1/3
City: Riga
Country: Latvia
Email: gunars.balodis@rtu.lv

Using Social Network Apps as Social Sensors for Health Monitoring

I. Pagkalos and L. Petrou

Department of Electrical and Computer Engineering, Aristotle University of Thessaloniki, Thessaloniki, Greece

Abstract—A wealth of data that can be useful to health monitoring is often available in Social Networking Sites, such as Facebook. This includes a person's social data (friends, connections, group memberships) which is typically hard to obtain by traditional means, as well as user-generated content (Facebook wall posts, tweets). Due to their unstructured nature, gleaning meaning from this user-generated content typically requires text mining or natural language processing techniques which need specialized vocabularies and may prove inaccurate. We propose using Social Network applications, such as those provided by the Facebook Developer platform to get the best of both worlds – structured user-submitted content as well as social data and other useful information present on the Social Networking platform. We employ Semantic Web techniques to convert an application's output to sensor observations in order to a) homogenize and integrate data from multiple applications and b) treat applications as Social Sensors, efficiently integrating physical and social sensors where needed. With the appropriate application design, a social network that is application-capable can become an infinite repository of human sensor observations and their accompanying social data that can be queried and used for a variety of feature-rich health applications. We present the design of our prototype monitoring framework, SENHANCE, which includes a Facebook app and show how it can used for health monitoring.

Keywords—Social Network, Sensors, Social Sensor, Semantic Web.

I. Introduction

With the explosion of Web 2.0 marking a new era for the Web, many people now have a "digital life" where they produce digital data streams as well as digital footprints on various web services and increasingly so in Social Networking Sites (SNS). Facebook[1] alone, the current "champion" of SNSs, apart from being the second most visited website in the world [1], is fast approaching 1 billion users [2]. Even if Facebook is dethroned from this position, other SNSs will quickly fill the gap, as what was once a trend of specific age groups is now a staple of the Web. Therefore, information about people, their social circles and their activities is available on SNSs of various forms, and in increasing volume.

Over the years, SNSs have implemented a wide variety of technical features, but their backbone consists of visible profiles that display an articulated list of Friends who are also users of the system [3]. Many of the SNSs, including Facebook, allow the user to create user-generated content, in the form of written text, web, audio, video, or picture links. This has served as a platform of expression for a whole generation of internet-savvy users, aided by the proliferation of internet-connected mobile devices. It is there that humans act implicitly as sensors, making observations that are extremely varied – from their opinion on what the weather is like, to what news is trending at the moment to how many miles they ran today during their daily workout routine. In health care, this translates to people tracking their workout routines, posting reviews of their medical treatments, and raising awareness about certain health conditions [4]. These observations are bundled with a wealth of metadata such as social data (friends, connections, group memberships etc.) that can play multiple roles - aid the research itself in reaching conclusions about the user and his/her environment, establish context and even aid in calculating user trust [5].

Nevertheless, this data is mostly unstructured and usually provided for different purposes. We propose using Social Network Applications (SNApps), such as those provided by the Facebook Developer platform to get the best of both worlds – structured user-submitted content that is fit-for-purpose as well as social data and other useful information present on the Social Networking platform. In addition, SNApps have the benefit of being able to 'traverse' a social graph by advertising themselves or being inadvertently advertised by users, thus expanding the potential user base of a monitoring scenario.

We employ Semantic Web technologies in order to provide a framework to systematically include data from SNSs as a sensor (i.e. a semantic social sensor) into a generic health monitoring scenario in a way that is implementation-agnostic. We can, thus, treat a SNS as a repository of unstructured human sensor observations, where the sensor is the human entity that operates the SNS account. These observations are then available to be integrated with true sensor data, and exploited along with users' social data by health professionals who place high value in self-reported data.

We present our prototype monitoring framework, SENHANCE, which includes a Facebook app and show how it can be used for health monitoring alongside body sensors that measure physical activity and energy expenditure.

---

[1] Facebook.com: a popular (at the time of writing) SNS site.

II. SNS APPLICATIONS AS SOCIAL SENSORS

Data on SNSs, and thus data that can be collected via a SNApp can be seen as a twofold structure: Data describing the social network structure (*Social Data*) and data describing the user-generated content (*UGC*) [6]. We make a further categorization of UGC by distinguishing between *active UGC* (status updates, tweets and data from SNApps) and *passive UGC*, data that "lives" on the platform. Passive UGC can be defined as information the user keeps on the SNS platform, usually in the forms of variables on a user profile, such as his Age, Sex and Location and possible implementation-specific variables such as height and weight. An interesting characteristic of passive UGC is that it can exist in the platform independent of a user's active UGC activity. A user that creates a profile on a SNS only for networking purposes and never posts or interacts with other members may still keep useful information in his profile, especially if that forms part of a sign-up procedure.

A. Modeling Social Data

Let $U = \{u_1, u_2, ..., u_n\}$ be a group of n SNS users. In Social Network theory, a graph $G = (V, E)$ where V a finite set of vertices and E a finite set of edges such that $E \subseteq V \times V$ can be used to model the social network that describes U. Furthermore, the matrix $M := (m_{i,j})_{n*n}$ where

$$m_{i,j} = \begin{cases} 1 & (u_i, u_j) \in E \\ 0 & otherwise \end{cases}$$

and $n = |V|$, associated with that graph, is used to describe the relationships between individuals (reciprocal or not) [7].

To describe this social space semantically, we employ FOAF [8], one of the most popular and tried Semantic Web Ontologies. By modeling social connections via FOAF, we can leverage the extensibility of the RDF/OWL language, to define more precise notions of relationships where needed. Extensions to FOAF that more clearly define relationships and tie strengths can be custom-built or found as published ontologies (e.g. [9],[10],[11]) which further proves the power of FOAF as a starting point for modeling social data. In addition, FOAF profiles can be generated and gathered from other services, be they SNApps or SNSs in general in order to integrate social data for monitored users from various sources (some people already keep a personal FOAF profile on their website or on a shared hosting space and there are also SNSs that export FOAF [12]).

We model a tie ($m_{i,j} = 1$) as a foaf:knows relationship between u_i and u_j, which are both instances of foaf:person. A social group SG_i (a community, group of interest etc.) that consists of users $u_i \in U$ is modeled as a foaf:group that has as foaf:member one or more foaf:person. This provides an abstract approach towards a user's social data that can be further refined if needed.

B. Modeling Active and Passive UGC

For the purposes of this framework we chose to model a SNS application as a sensor, i.e. a Social Sensor. This way we can homogenize and integrate data from multiple SNApps as well as true hardware sensors. The term Social Sensor has been used before in literature ([13],[14],[15]) but we use the term herein to indicate the gathering of observations on a SNS from a human entity acting as a sensor. For example, in a scenario where users are submitting physical activity data via a Facebook application, they are acting as human sensors in the sense that they self-report the activity they performed as an observation. The value of a user profile variable is defined in a similar way. The Facebook application acts as a Social Sensor in the sense that it is responsible for detecting, annotating and aggregating these human observations. We believe it is necessary to make this distinction between human social sensors because of *accountability*; it is the human who produces (senses) the event and whose credibility is in question in regards to data quality.

We use the SSN (Semantic Sensor Network) ontology [16] to describe the sensors, what they measure and the result of these observations. SSN provides sensors with semantic interoperability and a basis for semantic reasoning. SSN can be seen as complimentary to OGC's Sensor Web Enablement (SWE) activities [17] which is a long-standing community standard, in the sense that SWE is intended to provide standardization at the syntactic and service levels and does not explicitly address semantic level interoperability [18]. Thus, by using SSN as the base for sensor observations, we address the common issue of heterogeneity between health sensors by creating reusable, shareable semantic web components that can exploit the benefits of OWL. This approach requires a semantic adapter that converts a sensor's (hardware or social) output to SSN-compliant semantics, which is discussed in the chapter that follows.

Let $P = \{p_1, p_2, ..., p_k\}$ represent the set of k properties that describe a user $u_i \in U$, such as Age, Weight and Height. Then $(u_i, p_j)_t$ denotes a time-stamped by t sequence of observations in a particular time period of T time units where $t \in [0, T]$. We do not differentiate as to the origin of these observations – they may come from the user profile or the SNApp, depending on the implementation. We map the above concepts to SSN OWL constructs as follows: a user u_i, apart from a foaf:person, is also a ssn:Sensor (as a human sensor) and a

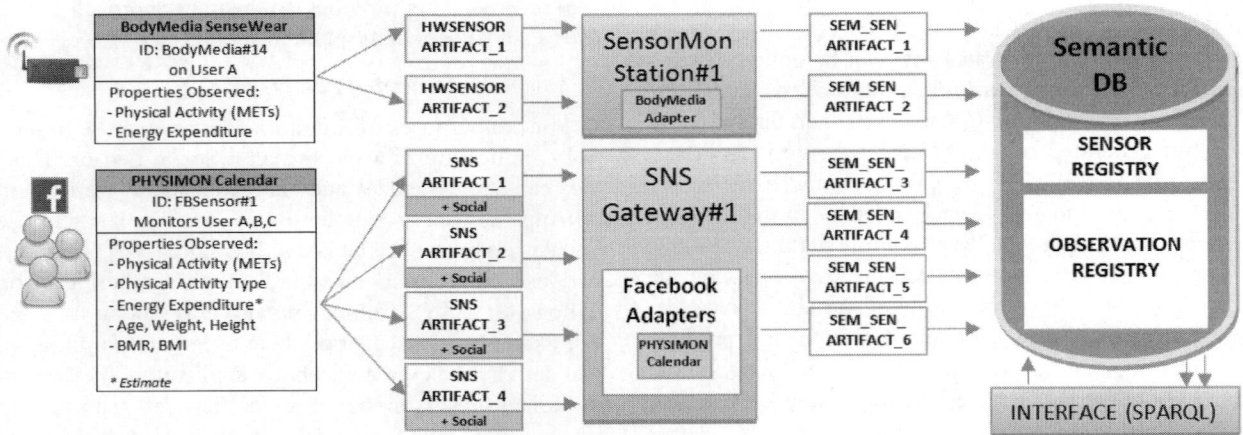

Fig. 1 An overview of the SENHANCE monitoring framework, adapted to a Physical Activity Monitoring motivational scenario

ssn:FeatureOfInterest (as a target of sensing) with instances $u_i \in U$. A property p_j is a ssn:Property with instances $p_j \in P$, and an observation $(u_i, p_j)_t$ is an ssn:Observation. Because SSN does not dictate how the result of an observation (ssn:SensorOutput) is defined, the semantic representation of the set of acceptable values is implementation specific (e.g. using the SWEET ontology or custom-built domain ontologies), as is the implementation of a time-reference mechanism for t. This flexible mechanism of representation allows for the creation of implementation-agnostic and reusable components, which agrees with the chosen representation of a user's social data and the Semantic Web vision in general.

III. SYSTEM ARCHITECTURE

Figure 1 presents an overview of the SENHANCE framework, a semantic platform that systematically integrates SNApps as Social Sensors in an implementation-agnostic health monitoring scenario. The figure is adapted to a use case from the PHYSIMON project, an ongoing inter-disciplinary collaboration between the Department of Elec. & Comp. Engineering of AUTH, Greece and the Department of Nutrition of TEITHE, Greece that aims to explore the feasibility of gathering user-submitted data from a SNS for the purposes of discovering patterns associated with Physical Activity. In this use case, a pre-determined group of users are being monitored by a dietician. Some users are equipped with a BodyMedia SenseWear wearable body sensor that detects physical activity and energy expenditure. Every user has a profile on Facebook from where basic information about the user, his/her interests and social ties can be extracted to gather useful research data. In addition to hardware sensing, the users disclose information about their activities and other variables (Age, Sex, Weight, Height and automatically calculated values such as Body Mass Index (BMI) and Basal Metabolic Rate (BMR)) on Facebook via the PHYSIMON Calendar app[2] that can be collected and analyzed [19].

The SENHANCE architecture is comprised of a number of hardware and/or social sensors that create output in the form of *artifacts,* an abstracted notion of a typical sensor output. These artifacts are forwarded to a semantic adapter entity which is responsible for converting them into *semantic artifacts,* expressed in SSN semantics (OWL/N3). The semantic adapter follows the mappings presented in chapter 2 for the social and UGC data of users and generates the appropriate RDF with the help of external ontologies such as owl-time[3] and implementation-specific ontologies such as the physact Physical Activity ontology (custom-built). These are subsequently stored in a Semantic (triples) Database that can be queried via SPARQL to provide "smart" complex queries that are agnostic of sensor type such as the queries presented in Figure 2.

The presented prototype system uses HTML, PHP and JavaScript for the Facebook app and Python scripts for converting and forwarding data to a Sesame Triplestore. The triplestore runs on Apache Tomcat and uses OWLIM[4] for reasoning, which we have found is an excellent and speedy OWL reasoner, at least in comparison to other open-source/free choices. The scalability of Sesame+OWLIM in regards to large scale datasets remains to be seen but for our small dataset (<1M triples) the performance is acceptable.

---

[2] http://apps.facebook.com/physimoncal
[3] http://www.w3.org/TR/owl-time/
[4] http://www.ontotext.com/owlim

Example Query 1: "Sensors that measure Physical Activity and their observations" (agnostic of hardware or social sensor type)

```
SELECT ?sensor ?obs
WHERE {
    ?sensor ssn:observes physact:PhysAct.
    ?obs ssn:observedBy ?sensor.
}
```

Example Query 2: "Show me the physical activity of "John Smith""

```
SELECT ?obs
WHERE {
  ?user foaf:name "John Smith" .
  ?obs ssn:featureOfInterest ?user;
       ssn:observedProperty physact:PhysAct.
}
```

Example Query 3: "Physical activity of "John Smith" and his friends"

```
SELECT ?obs
WHERE {
  ?user foaf:name "John Smith";
        foaf:knows ?friend .
  { ?obs ssn:featureOfInterest ?user;
         ssn:observedProperty physact:PhysAct. }
  UNION
  { ?obs ssn:featureOfInterest ?friend;
         ssn:observedProperty physact:PhysAct. }
}
```

Fig. 2 Example SENHANCE SPARQL queries

IV. CONCLUSIONS & FUTURE WORK

We discussed the use, modeling and implementation of SNApps for the purposes of Health monitoring as well as proposed a system architecture for their integration with true sensor data, the SENHANCE platform. With the variety of properly-annotated and semantically-rich data available in such a platform and the powerful query capabilities of SPARQL, we believe that this innovative approach can serve as the basis for a multitude of feature-rich health applications.

With our architecture and proper design considerations, such as proper user incentives, a SNApp can provide useful health monitoring data and essentially act as a social sensor for health monitoring and public health research. Nevertheless, human self-reported data always carries with it the risk of uncertain data quality (DQ). Therefore, we're currently exploring how to best implement DQ assessment techniques such as treating sensor data as ground truth where available and maintaining user reputation tables according to the self-reported data's accordance to sensor data. In addition, even though the expressiveness of SPARQL introduces many benefits, it can also act as a limiting factor for health applications where a more natural-language-style of querying may be of more use. More work is needed towards that area in order to make these feature-rich health applications as user-friendly as possible.

REFERENCES

1. 'Alexa.com - Facebook.com Site Info' at http://www.alexa.com/siteinfo/facebook.com
2. 'Facebook Statistics and Metrics by Continents' at http://www.socialbakers.com/countries/continents/
3. Boyd DM, Ellison NB (2007) Social Network Sites: Definition, History, and Scholarship. J Comput-Mediat Commun 13(1):210–230
4. 'The Social Life of Health Information' (2011) at http://pewinternet.org/Reports/2011/Social-Life-of-Health-Info.aspx
5. Golbeck J (2008) Weaving a web of trust. Science 321(5896):1640–1641
6. Erétéo G, Limpens F, Gandon F, Corby O, Buffa M, Leitzelman M, Sander P (2011) Semantic Social Network Analysis: A Concrete Case. Handb. Res. Methods Tech. Stud. Virtual Communities Paradig. Phenom. 1:122–156
7. Mika P (2007) Social Networks and the Semantic Web. Springer Berlin
8. 'FOAF Vocabulary Specification' at http://xmlns.com/foaf/spec/
9. 'RELATIONSHIP: A vocabulary for describing relationships between people' at http://vocab.org/relationship/
10. 'Descriptions of Social Relations' at http://www.cs.vu.nl/~pmika/research/foaf-ws/foaf-x.html
11. 'FoafExtensions - FOAF Wiki' at http://wiki.foaf-project.org/w/FoafExtensions
12. Golbeck J, Rothstein M (2008) Linking social networks on the web with FOAF: a semantic web case study, Proc. vol. 2, 23rd Nat. Conf. on Art. Int., 2008, pp 1138–1143
13. Rosi A, Mamei M, Zambonelli F, Dobson S, Stevenson G, Juan Y. (2011) Social sensors and pervasive services: Approaches and perspectives, IEEE Int. Conf. on Perva. Comp. and Comm. Workshops, pp. 525–530.
14. Sakaki T, Okazaki M, Matsuo Y (2010) Earthquake shakes Twitter users: real-time event detection by social sensors, Proc. of the 19th Int Conf on WWW, New York, NY, USA, 2010, pp. 851–860
15. Christakis NA, Fowler JH (2010) Social Network Sensors for Early Detection of Contagious Outbreaks. PLoS ONE 5(9): e12948. doi:10.1371/journal.pone.0012948
16. Compton M, Barnaghi P, Bermudez L, Castro RG, Corcho O, Cox S, Graybeal J, Hauswirth M, Henson C, Herzog A (2011) The SSN Ontology of the Semantic Sensor Networks Incubator Group. J. Web Semant. 17:25-32 DOI 10.1016/j.websem.2012.05.003.
17. Botts M, Percivall G, Reed C, Davidson J (2008) OGC® Sensor Web Enablement: Overview and High Level Architecture, in GeoSensor Networks, Springer Berlin
18. Sheth A, Henson C, Sahoo SS (2008) Semantic sensor web. Internet Comput. Ieee 12(4):78–83
19. Pagkalos I, Rossiou D, Papadopoulou S (2012) Monitoring physical activity through social networks: a Facebook case study, Presented at the 6th DIETS/EFAD Conference, Portoroz, Slovenia, 2012 (Proceedings pending)

Author: Ioannis Pagkalos
Institute: Aristotle University of Thessaloniki
Street: Egnatia Str., University Campus, Dept. of Electrical & Computer Engineering, 54124, Thessaloniki, Greece.
City: Thessaloniki
Country: Greece
Email: ipagkalo@auth.gr

Advanced Medical Expert Support Tool (A-MEST): EHR-Based Integration of Multiple Risk Assessment Solutions for Congestive Heart Failure Patients

Carlos Cavero Barca[1], Juan Mario Rodríguez[1], Paolo Emilio Puddu[2], Mitja Luštrek[3], Božidara Cvetković[3], Maurizio Bordone[4], Eduardo Soudah[4], Aitor Moreno[5], Pedro de la Peña[5], Alberto Rugnone[6], Francesco Foresti[6], and Elena Tamburini[6]

[1] ATOS, ATOS Research and Innovation (ARI), Madrid, Spain
[2] Sapienza University of Rome, Department of Cardiological Sciences, Rome, Italy
[3] Jožef Stefan Institute, Department of Intelligent Systems, Ljubljana, Slovenia
[4] CIMNE, Barcelona, Spain
[5] Ibermatica, Donostia, Spain
[6] I+ s.r.l., Florence, Italy

Abstract—More and more the continuum of care is replacing the traditional way of treating the subjects of care putting people in the centre of the healthcare process. Currently clinicians start treatment after a problem occurs due to the low adoption of Clinical Decision Support Systems (CDSS) integrated with standardised Electronic Health Record (EHR) systems; The volume to value revolution in the healthcare (from stakeholder-centric to patient-centric) will allow doctors to follow the evolution of the individual before a medical episode happens, treating the patient based on statistical trends to forecast the future. The CDSS techniques applied on telemonitoring tools permit the doctors to predict forthcoming events, improve the diagnosis and avoid continuous visits to the hospital, therefore saving costs. Advanced Medical Expert Support Tool is a step towards achieving the patient-centric approach by incorporating the health information into the EHR using European standards (ISO/EN 13606) to provide semantic interoperability by means of the dual model approach (reference model and archetypes). Three different CDSS modules have been implemented and contextualised publications are provided to the cardiologist to facilitate their daily work. A person-centric Graphical User Interface (GUI) facilitates the visualization of the health status of the patients providing meaningful information to the cardiologists. The use of archetypes allows scalability, transparency and efficiency to the hospital environment.

Keywords—Risk assessment, EHR, Congestive Heart Failure, Clinical Decision Support System, Person-centric.

I. INTRODUCTION

A. Clinical Problem

High interest and intense research can be observed on developing effective tools to assess impending risks in Congestive Heart Failure (CHF) patients, including home and/or ambulatory tele-monitoring [1]. However, apart from few studies with implanted materials and less than optimal results [2][3], most work up to date concentrated on ECG and blood pressure changes [4]. Multi-parametric time-domain integration was not considered, nor were parameters such as ambient or skin temperature and humidity or the amount of exercise [5]. Risk modeling was classically evaluated at relatively medium- or long-term so that predictability was frequently diluted into a long list of relatively unspecific elements without a constant relevance consensus among experts. By contrast, short- and very short-term risk parameters and their multiple interactions were not deeply assessed [6].

B. Technical solution

EHR standards must be considered in CDSS in order to facilitate the exchange of information in a machine-readable format [7]. The adoption of standardised EHR allows processing the information more effectively across multiple Hospital Information Systems (HIS) and care settings. Standards need to clearly represent (i) Terminologies and information models and (ii) Clinical knowledge [8].

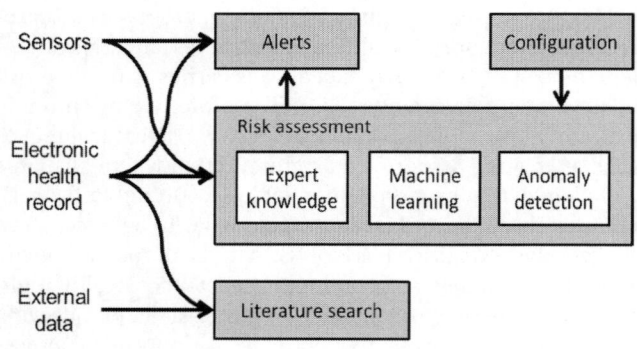

Fig. 1 The architecture of the CDSS

Three different CDSS modules have been implemented and an intelligent search tool to recommend external content from the literature, contextualised to the specific patients. All the information is stored using archetypes and each component can operate independently from each other.

The modules are seamlessly integrated in the Advanced Medical Expert Support Tool (A-MEST), which also has the following features:

- **Scalability and transparency**
- **Openness**, by means of standards
- **Robustness**, integrity and confidentiality
- **User-friendliness**, through rich visualization

The use of archetypes permits the EHR integration fully compliant with the European standard EN13606 [9].

II. ARCHETYPES FOR EHR INTEROPERABILITY

The semantic interoperability (SIOp) addresses the issue of how to seamlessly exchange the information between health services providers and patients [10]. The openEHR and ISO/EN 13606 standard follow the two-level modeling approach which separates out the clinical knowledge from the information model. The former are the clinical concepts which are modeled using archetypes and the latter remains static and it is composed by a few immutable classes leaving the technical solution unaltered.

For the presented research several archetypes have been created or reused to (i) store the parameters considered relevant for CHF and (ii) store the risk assessment data using evaluation archetypes. The CHIRON data are collected from sensors and stored using the archetypes included as entries inside one composition of the patient EHR.

III. RISK ASSESSMENT

A. Configuration of the System

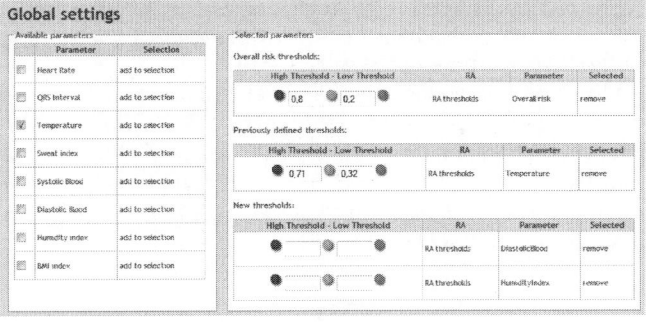

Fig. 2 Global Monitoring Settings

Monitored parameters used by the risk assessment module are set for each patient, since the risk to different patients' health may depend on different parameters. Furthermore, green/yellow/red risk areas are defined by terms of customizable thresholds doctors may modify if they judge different values to suit their patients better.

Expert System

The first module for risk assessment relies on expert knowledge. It is thus able to leverage the existing experience in the field of cardiology. The construction of the expert system started with an extensive search of the medical literature, which yielded over 60 parameters considered potential risk factors for CHF. To estimate the importance, a survey was conducted among the European opinion leaders in cardiology. Based on 32 responses, each parameter was assigned low, medium or high importance [6].

Three risk assessment models were constructed for different time horizons: long-, medium- and short-term. The following information was required on each parameter:

- The relation between the parameter value and the risk.
- The minimum and the maximum parameter value.
- Two parameter values representing thresholds.
- The importance: low, medium or high.
- The frequency at which the parameter changes.

Each parameter value was first transformed into a risk value by linearly scaling the parameter value to the (0, 1) interval. The same transformation was used to obtain the thresholds between the low-, medium-, and high-risk areas. The risk values of individual parameters were combined into an overall risk value by a weighted sum. Each weight in the overall-risk was the product of the importance and a model-specific weight related to the time horizon.

C. Machine Learning

The second module is an artificial neural network (ANN) which is a tool able to give response almost in real time, in order to support clinical decision in the detection of CHF class 2 to 3 alert problems. ANN consists of simple elements massively interconnected in parallel, denoted as neurons, with a hierarchical organization similar to biological nervous systems. Drawing an analogy between the synaptic activity and the ANN, we can set the following concepts [11]:

- Signals arriving at the synapses are the neuron entries.
- Entries are weighted through a parameter denoted weight, which is associated with a specific synapse.
- Signals can excite or inhibit the neurons.
- The effect on the neuron results from the addition of all the entries.

Structure of ANN consists of three levels or neuron layers: input level, hidden layers level, with 16 hidden neurons, and the output level. The network adapts the different weights during its learning process. After building an ANN for the risk assessment, the validation of the results is crucial [12]. Pre-processing transformations were applied to the input data using simple linear rescaling of the data. Mean and standard deviation of the training set were normalized. This process forced input variables to have similar ranges for easier training. Due to the amount of available data, the input was reduced to 12 variables giving 3 levels of risk (low, medium and high). In order to obtain the best performance on new data, 35 different ANNs were compared. The Levenberg-Marquardt back-propagation algorithm was chosen. Performance validation [12,13] of the network was performed with cross-validation, on a data set that is independent of the training data; practically mean error value of 0.003503% was obtained.

D. Anomaly Detection

The final module for risk assessment requires neither knowledge nor data weighted with the risk, but only some data considered normal. The parameters describing the patient's normal condition tend to follow recurrent patterns. Such patterns can be learned, and when a new potentially anomalous pattern is detected, the module raises an alert that medical professionals can confirm or not.

Local Outlier Factor (LOF) algorithm [14] is used to detect anomalies. The algorithm compares the density of data instances around a given instance X with the density around X's neighbors. If the former is low compared to the latter, it means that X is relatively isolated – that it is an outlier and thus considered anomalous. The LOF algorithm assigns a value to each instance that indicates the degree of its anomalousness, which we consider to correspond to the risk. To also compute the degree of anomalousness of individual parameters, which corresponds to their contributions to the risk, an extension was made to the algorithm.

The anomaly detection was tested on a dataset of five healthy test subjects performing a range of activities, and a subset of the parameters relevant to CHF [15]. These data were used to set the parameters of the LOF algorithm and thresholds between the low- medium- and high-risk areas. To set the thresholds, anomalous data was generated by randomly replacing some parameter values during one activity with those during another.

IV. LITERATURE SEARCH

Questions and Answers (Q&A) systems attempt to extract direct answers from large data sets, regardless of the complexity or ambiguity of the question and size and amount of data sources. To date, few Q&A systems have focused on designing effective interfaces and avoiding long lists of retrieved documents.

Clinical Q&A system approach is founded on Evidence-Based Medicine (EBM). It works on plain text, extracting the relevance of certain medical paper using Medical Subject Heading (MeSH) codes and summarizes the publication's content into a few annotated sentences using Natural Language Processing (NLP) techniques. Firstly, the system is implemented creating rules manually to discover new semantic information, in order to enrich the user's questions with inferred new conclusions. The module is contextualised with information coming from the EHR using an ontological semantic system[1], which permits to run rules over the patient information and extract new diagnoses to enrich the queries.

The system automatically reads, identifies and extracts binary relations from whole publications texts extracting the relevant information by means of triples (subject, object, predicate). The CHF ontology used is constantly updated and filled automatically with the information received from the medical sources. This brings two response levels: direct answers from the knowledge database and complex responses using reasoning. The first one includes as knowledge data a system with a SVM classifier (Support Vector Machine) over a tokenized input (clinical terms) coming from the user questions. If the confidence is not enough (85%) ontology classifier looks for inferences over questions terms also included in the working memory referenced to the query to provide complex responses. Finally, the conclusions are presented to the user in a natural language sentence.

V. GRAPHICAL USER INTERFACE

A-MEST provides a full support to clinical risk management process with a simple set of GUIs, whose design followed a person-centric approach, based on interviewing medical professionals. The process, driven by medical experts, starts from risk evaluation; it goes iteratively exploring its causes and finishes with actions whose purpose is to mitigate the current values of risk [16].

The first component, the Monitoring Settings, provides a mean to pre-configure the system (Fig. 2. Section III. A); the second component is the Risk Assessment (Fig. 3. Section III. B, C, D), where the medical experts are allowed to review the clinical status of their patients. Risk level source is set by a toggle selector.

---

[1] Ontology used was developed by the Laboratory for Information Systems; RBI Zagreb Croatia, Dragan Gamberger, Rudjer Boskovic Institute.

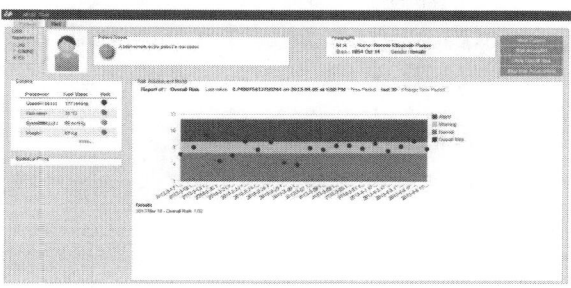

Fig. 3 A-MEST Graphical User Interface – Risk Assessment activity

Upon patient selection, the system shows a more detailed view including recent-past evaluations of the patient's risk presented in a graphical way that highlights the overall trend. If the clinician starts a new Risk Assessment procedure, he/she can access a broad spectrum of information regarding the patient together with literature search module. The relevant publications (Section IV) are shown below the parameters. The system also raises alerts if the risk values exceed the thresholds defined.

VI. CONCLUSIONS

In this paper the Advanced Medical Expert Support Tool has been presented, which permits the doctors to evaluate the health risk of a CHF patient and get the relevant publications contextualised to specific cases.

Three different Decision Support Systems are used to assess the risk to the patient from different viewpoints. The seamless integration is fulfilled by means of standards, concretely the European norm EN13606. Health data and the automatic risk assessment calculation are incorporated in the Electronic Health Record providing a common solution to store and use the data.

Predicting the future has been pursued during centuries of history, but right now the technology makes it possible to do this accurately in the healthcare environment. The proposed tools will facilitate the daily monitoring of the CHF patients, and reduce the visits to hospitals because the doctor will know in advance the possible problems instead of treating them when they have already happened.

ACKNOWLEDGMENT

This research was carried out in the CHIRON project (Cyclic and person-centric Health management: Integrated appRoach for hOme, mobile and clinical eNvironments) co-funded by the ARTEMIS Joint Undertaking (grant agreement #2009-1-100228) and by national authorities.

REFERENCES

1. Chaudhry SI, Mattera JA, Curtis JP et al. (2010) Telemonitoring in patients with heart failure. N Engl J Med 363: 2301-2309
2. Conraads VM, Tavazzi L, Santini M et al. (2011) Sensitivity and positive predictive value of implantable intrathoracic impedance monitoring as a predictor of heart failure hospitalizations: the SENSE-HF trial. Eur Heart J 32: 2266-2273
3. Abraham WT, Adamson PB, Bourge RC et al. (2011) Wireless pulmonary artery haemodynamic monitoring in chronic heart failure: a randomised controlled trial. Lancet 377: 658-666
4. Klersy C, De Silvestri A, Gabutti G et al. (2009) A meta-analysis of remote monitoring of heart failure patients. J Am Coll Cardiol 54: 1683-1694
5. Bonfiglio S. (2012) Fostering a contimuum of care. In Proceedings of ICOST, 2012., pp 35-41.
6. Puddu PE, Morgan JM, Torromeo C et al. (2012) A clinical observational study in the Chiron Project: rationale and expected results. In Proceedings of ICOST, 2012. pp 74-82.
7. Kashfi, H. (2011). The Intersection of Clinical Decision Support and Electronic Health Record: A Literature Review. Proceedings of the Federated Conference on Computer Science and Information Systems.
8. Kawamoto, K., G. Del Fiol, et al. (2010). "Standards for Scalable Clinical Decision Support: Need, Current and Emerging Standards, Gaps, and Proposal for Progress." The Open Medical Informatics Journal 4: 235.
9. CEN/TC251-ISO/TC215 (2010) Electronic Healthcare Record (EHR) Communication. Parts 1: Reference Model, Part 2: Archetype Model, Part 3: Reference Archetypes and Term lists, Part 4: Security and Part 5: Interface Specification.
10. Semantic HEALTH report. (2009) Semantic Interoperability for better health and safer healthcare: European Commission. Information Society and Media Directorate-General.
11. Dadvand P, Lopez R. and Oñate E. (2006). Artificial Neural Networks for the Solution of Optimal Control Problems. Proceedings of the International Conference ERCOFTAC.
12. Lopez R., Balsa-Canto E. and Oñate E (2005). Artificial Neural Networks for the Solution of Optimal Control Problems. Proceedings of the Sixth Conference on Evolutionary and Deterministic Methods for Design, Optimisation and Control with Applications to Industrial and Societal Problems.
13. Michel A, Farrell J.A, Porod W. (1989). Qualitative analysis of neural networks IEEE Trans. Circuits Systems, 36, pp. 229−243.
14. Breunig MM, Kriegel H-P, Ng RT, Sander J (2000) LOF: Identifying density-based local outliers. Proc. ACM SIGMOD International Conference on Management of Data, pp 93–104
15. Luštrek M, Cvetković B, Kozina S (2012) Energy expenditure estimation with wearable accelerometers. Proc. IEEE International Symposium on Circuits and Systems, pp 5–8
16. Barca C. C., Rodríguez J.M., Rugnone A. et al. (2012) Medical Expert Support Tool (MEST): A Person-Centric Approach for Healthcare Management. In Proceedings of ICOST, 2012. 99-106.

Modeling and Implementing a Signal Persistence Manager for Shared Biosignal Storage and Processing

S. Pirola*, E. Opri*, A.M. Bianchi, and S. Marceglia

Dipartimento di Elettronica, Informazione e Bioingegneria, Politecnico di Milano, Milan, Italy

Abstract—Biosignal recording is gaining growing attention in several biomedical application, for example Brain Computer Interface (BCI).

The availability of systems providing shared management and processing of signals is still lacking.

A centralized system for signal persistence management and processing, with the ability to provide predefined analysis algorithms, would hence be an useful asset.

In this work, we designed and implemented a platform to support the management and analysis of signals for different neurophysiological applications, including BCI, that could be also expanded to other applications.

It is composed mainly by a client framework (Matlab compliant), and a web interface to manage the platform itself, priviliging a centralized approach.

To validate the platform, and to test it, we performed an experiment during which, in normal subjects, we recorded signals during real movements (right hand, left hand, and feet) and during motor imagery. Following the literature on BCI, we developed specific algorithms for signal analysis and uploaded them on the platform.

Recorded signals were uploaded to the platform and the analysis packages available were used to extract Event related desynchronization (ERD)/Event related synchronization (ERS) on μ and ß bands.

The validation showed that our platform can be easily adopted to share signals and analysis chains to design BCI applications.

Keywords—Databank, Signal Processing, biosignals.

I. INTRODUCTION

The recording and processing of signals (EEG, EMG and others) is a common practice for different purposes, and the increase in the amount and complexity of data needs an improvement in processing power and data management.

Despite the general increment in processing power, there is a lack of availability of a large centralized Signal Management System, especially for Brain Computer Interface (BCI).

Research labs usually have their own signal archives, and data are often not shared, not only for privacy issues, but also because the system does not support interoperability even across different departments.

For example, BCI research requires many subjects to be analyzed to obtain reliable results and, because different labs are executing similar or complementary experiments, a centralized and a shared approach, adding an Open Source and an Open Data Approach, would improve BCI research [1].

Hence, we implemented a platform, called Signal Persistence Manager (SPManager), to provide tools to store and process signals, especially for BCI applications, also making available a set of utility to work with BCI.

To validate the platform, to test its robustness, and to provide feedback for further implementations, we performed an experiment during which, in normal subjects, we recorded signals during real movements (right hand, left hand, and feet) and during motor imagery.

We used SPManager to store and analyze the recorded signal. We analyzed movement-related event related desynchronizations (ERD) and event related synchronizations (ERS) on μ (8-13 Hz) and ß (14-25 Hz) rhythms [2], considering also improvement like Electrooculography (EOG) artifact realtime removal. After offline ERD/ERS recognition, we obtained parameters to be used into an online classifier for pattern recognition.

The developed system, thanks to its flexible and modular structure, is not limited to BCI application, but could host different kind of signal processing.

II. SYSTEM OVERVIEW

The proposed system is represented in Fig.1. It aims to manage, process and archive signals acquired focusing on BCI approach. After acquisition, signals can be stored on a database management system (DBMS) through a webinterface and can be processed thanks to the framework available on the client side that provides tools for new analyses or visualizations.

The three layered system is composed by:

- Client Layer: it contains the client side framework, that is used to process the data acquired, and the webclient, that gives a user friendly Graphical User Interface (GUI) to manage the platform itself

---

* The first two authors equally contributed to this work.

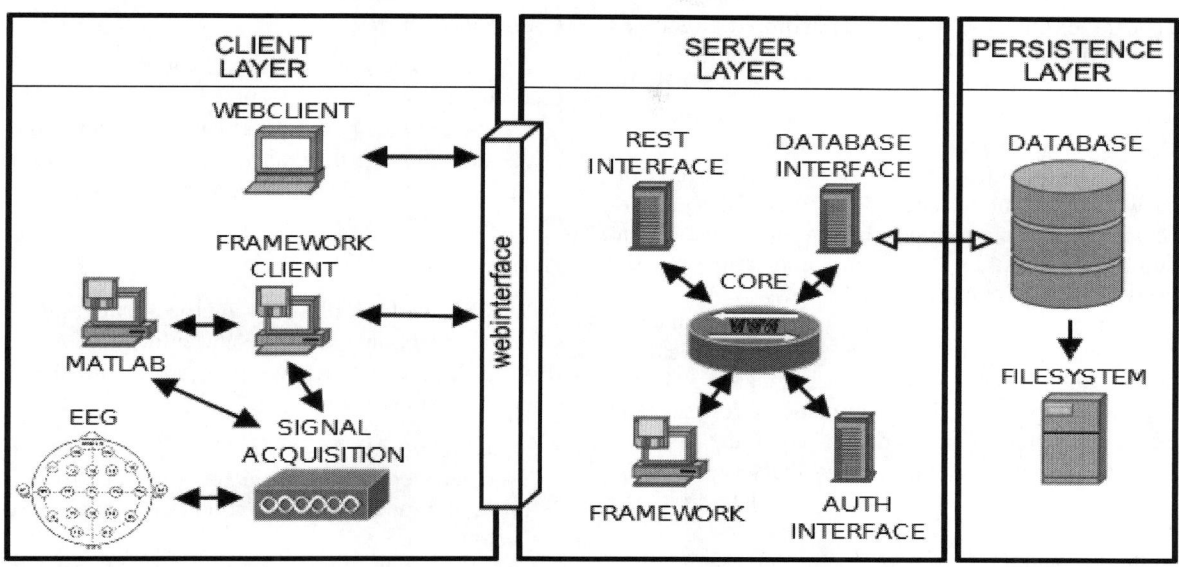

Fig. 1 The modeled platform. The platform is composed by a Client Layer (left panel), representing data acquisition and user operations, that communicates through a webinterface to the Server Layer (central panel). The Server Layer can manage data processing and storage thanks to the Persistence Layer (right panel), through the database interface.

- Server Layer: it contains all the webservices and server side routines. It also exposes an interface to the client layer and can process data.
- Persistence Layer: this layer groups all the subsystems dedicated to data persistence, as the database and the file system, whom role is to store big binary data (as signals) and files.

A. Persistence Layer

In signal processing, one of the most important entities is the signal per se, therefore we placed this entity as the cornerstone of our system.

This entity was designed to be auto-referenceable, allowing an efficient management of the results of consecutive analyses (referenced to a parent signal), by linking each parent (raw) signal to its children (processed signals).

The persistence of large binary data is managed in the filesystem.

We chose to use a relational DBMS (r-DBMS) to simplify the management of the relations between the different entities.

Specifically MySQL was chosen because it is commonly used in the IT departments, being stable and supporting recursive foreign keys. Also it does not have licence costs and restrictions like other proprietary alternatives.

It was implemented using InnoDb engine, with UTF8 as charset.

The actual persistence referenced with Data_store is managed by using a dedicated part of the file system, and linking each record to their correlated physical part. One directory was used to store all the data using a flat approach.

B. Server Layer

The key of the centralized approach is the Server Layer performing the actual management of every part of the system itself. It provides an interface between the framework, the database and the other submodules.

It also provides a set of Application Programming Interface (API) to work with and to do data manipulation.

It was developed in Python, based on a Linux Apache MySQL Python (LAMPy) platform that could provide both stability and versatility, without having license limitations and being free of charge.

Python was chosen because it allows to use the efficient SciPy and NumPy[4] libraries to execute some signal processing directly on the server side.

The Server side can simply retrieve and save the data or execute complex queries on the data themselves (when possible).

C. Client Layer

To interact with data, the user can use a Web based GUI or directly access to the application programming interface (API) that is provided by the Server Layer.

Within this layer it is possible to visualize and process data (thanks to the Matlab Framework), but also to control the server configuration and services.

It is developed mainly in HTML and javascript, based on a LAMPy platform.

Signals can be imported, after being acquired directly as binary data (and without an import processing) or as ASCII, European Data Format (.EDF) format, Micromed datafile format (.TRC), by parsing and subdividing them in their components (Recording and linked Channels).

An important submodule of the Client Layer (strictly intertwined with the webinterface) is the WebClient, implemented completely as a webapp to be platform independent (being html and javascript compliant) and managing the transactions with other modules with JSON.

This submodule provides a user-friendly GUI to manage the Web Services Submodules and to give a graphical representation of the evolution of the signal processing chain.

The framework gives, on the client side, the routines and functions to manipulate signals and data acquired.

It also serves as an interface between the server and the client side, allowing to store and retrieve data from the database.

The system is mainly developed for a Matlab compliant system and it was designed to manage BCI data and motor-imagery.

The modular structure allows expanding the analysis library. Specifically, the main modules developed are those for signal import/export, filtering functions and wrappers, utility for statistical analysis, utility for informative content extraction (rhythm, variance), database interface.

III. SYSTEM VALIDATION: EXPERIMENTAL DESIGN

To validate our system, we performed a data acquisition and analysis session, simulating a classical BCI experiment.

A. Subjects

Our study involved 12 right-handed, healthy subjects (6 males and 6 females), aged from 20 to 25.

All of them were involved in a real-movement session and in a motor-imagery session.

B. Experimental Session

We asked each subject to perform or imagine the following movements:

- closing the right hand
- closing the left hand
- lifting the right foot

Two different trials were performed: in the first one the movement, clearly visualized on the screen, was effectively performed, in the second one, the movement had to be imagined.

Each trial, of about 12 minutes, was made of 60 repetitions: 20 for each movement, executed in random sequence with a 6s elapsing between two consecutive movements.

C. Signals Recorded

Among the derivations given from the 10-20 system, we focused on those of C3, C4 and Cz.

Hands and feet of every subject were monitored with an EMG derivation. In the real-movement session we used it as a trigger, whereas in the motor-imagery session, when no movement had to be performed, as a control.

D. Data Analysis

We followed the classical approach suggested by Pfurtscheller and colleagues [5]. We extracted ERS and ERD from each subject, then we proceeded with the following elaboration steps:

1. Qualitative analysis of ERS and ERD. Trials from each subject were analyzed in order to identify an average behavior and exclude outliers or signals affected by adaptation.
2. Evaluation of the amplitude of ERD end ERS considered as the maximum percentage variation, positive or negative, from baseline.
3. Estimation of the plateau length.
4. Creation of an identification threshold for ERD and ERS, defined as M (amplitude mean) ±2,5*SD. This threshold allowed an automatic identification of synchronization and desynchronization.

IV. VALIDATION RESULTS

Signals acquired through the acquiring system and saved locally as EDF files were imported in the platform through the functions available in the <export> package.

We chose to import raw signals (1024 Hz) from the EDF files. The function "exportFromFile" was used to downsampled the signals to 128 Hz and to apply a notch filter. We used the "Pwelch" function, available in the Matlab Signal Processing Toolbox, to analyze the power spectrum.

"exportAll" gave us the possibility to filter our signals in order to extract mu rhythm trends.

After these preliminary steps, a statistical analysis and a preselection of the signals (with visual interaction) were executed by using the <statistics> package: signals with a trend far from a Gaussian distribution were discarded.

The <extraction> package allowed us to identify and extract intervals correlated to real-movement or motor-imagery and to divide these into the three states (foot, right hand, left hand movement or motor-imagery). In particular, for the real-movement analysis, the function "extract_rhythm_mot", provided for an automatic selection and "extract_rhythm_mot_visualSelection" gave us the possibility to discard outliers and to correct contingent classification errors.

In motor imagery trials, because the triggers were more difficult to be identified, we adopted a manual classification strategy implemented by the "extract_thythm_imag_visualSelection" function, to extract intervals. We obtained a ".txt" output file.

With the <ErsErd> package function "getErsErs_from_Extract_rhythm", we used the output of the previous phase to extract from the filtered signals a percentage change variation to detect an ERD or ERS correlated to hand/ foot movement or motor-imagery. This function gave as output the average ERD/ERS for foot, right hand and left hand, subdivided into C3, C4 and Cz derivations, as shown in Fig. 2 (only a subset of the results).

By using the ".txt" file as input, "extract_template" function generated the ERD/ERS template: in this case we chose a classification method based on plateau length and amplitude. The template, passed as input to the function "analyze_signal_ErsErd", allowed us to test our classification method: in particular we studied simulated signals to determine the parameters for classification method evaluation (e.g. confusion matrix).

Thanks to "filterLMS_offline_with_pow" function (<filter> package) we started a preliminary study about the correlation between the EOG signals and signals of interest for BCI (mu rhythm). We also worked on the beta rhythms (thanks to "exportAll" function from <export> package) in order to choose between mu and beta ranges the most reliable frequency range for BCI.

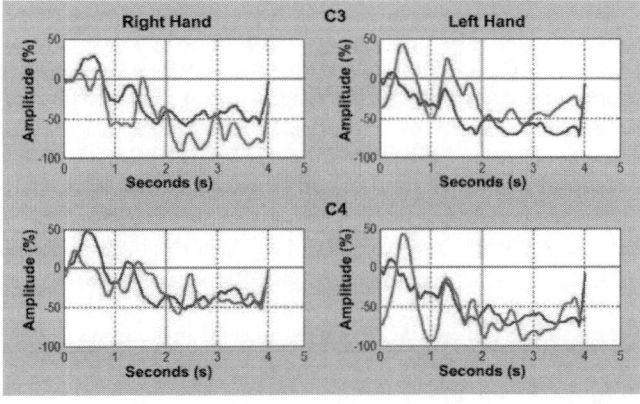

Fig. 2 ERD-ERS identification on C3 (top panels) and C4 (bottom panels) channels. The plots show the activity related to real hand movements (blue lines) and motor-imagery (magenta lines). ERD-ERS are obtained averaging all the available epochs.

V. CONCLUSIONS

In this work, we proposed a system to support the signals management and processing (mainly EEG), oriented to BCI research.

The SPM system was implemented and tested with the BCI experiment described, giving interesting results and resulting in a quite robust system, open to be expanded and improved, given its modular structure.

We are currently working on other modules as a realtime server for signal acquisition and an offline middleware to simulate a realtime acquisition.

Regarding the experiment, we are collecting more data to verify the quality of the classifier and improve the lateralization study.

We are considering the introduction of an Error Potential evaluation (ErrP) [6] to improve the quality of the classifier, by introducing the capability to discriminate when an incorrect classification occurred.

However the experiment was only executed in one center, thus not really taking advantage of the sharing capability.

Hence, doing multi center experiments could let us test the possibility of adding new analyses.

REFERENCES

1. Zander TO, Ihme K, Gärtner M, Rötting M (2011 Apr) A public data hub for benchmarking common brain-computer interface algorithms. J Neural Eng DOI 10.1088/1741-2560/8/2/025021
2. Pfurtscheller G, Brunner C, Schlögl A, Lopes da Silva FH (2006 May 15) Mu rhythm (de)synchronization and EEG single-trial classification of different motor imagery tasks. Neuroimage 31(1):153-9
3. Masse M. (2011 Oct) REST API Design Rulebook. O'Reilly Media, Sebastopol
4. SciPy and NumPy at http://www.scipy.org
5. Pfurtscheller G, Lopes da Silva FH (1999 Nov) Event-related EEG/MEG synchronization and desynchronization: basic principles. Clin Neurophysiol 110(11):1842-57
6. Buttfield A, Ferrez PW, Millán Jdel R (2006 Jun) Towards a robust BCI: error potentials and online learning. IEEE Trans Neural Syst Rehabil Eng 14(2):164-8

Corresponding Author:

Author: Selene Pirola
Institute: Politecnico di Milano
Street: 32, Piazza Leonardo Da Vinci
City: Milan
Country: Italy
Email: selene.pirola@mail.polimi.it

Heart Rate Variability for Automatic Assessment of Congestive Heart Failure Severity

P. Melillo[1,2], E. Pacifici[3], A. Orrico[1], E. Iadanza[3], and L. Pecchia[4]

[1] Dipartimento multidisciplinare di specialità medico-chirurgiche – Second University of Naples, Naples, Italy
[2] Dipartimento di Ingegneria dell'Energia Elettrica e dell'Informazione
"Guglielmo Marconi",
University of Bologna, Bologna, Italy
[3] University of Florence, Florence, Italy
[4] Department of Electrical & Electronic Engineering,
University of Nottingham, Nottingham, United Kingdom

Abstract—The aim of this paper is to describe an automatic classifier to assess the severity of congestive heart failure (CHF) patients. Disease severity is defined according to the New York Heart Association classification (NYHA). The proposed classified aims to distinguish very mild CHF (NYHA I) from mild (NYHA II) and severe CHF patients (NYHA III), using long-term nonlinear Heart Rate Variability (HRV) measures. 24h Holter ECG recording from 2 public databases was performed, including 44 patients suffering from CHF.

One non-linear HRV feature was effective in distinguishing very-mild CHF from mild CHF, by achieving a sensibility and specificity rate of 75% and 100% respectively. Moreover, we combine the results obtained by LDA in a classification tree (previously described) in order to obtain an automatic classifier for CHF severity assessment.

Keywords—Heart Rate Variability (HRV), Congestive Heart Failure (CHF), linear discriminant analysis.

I. INTRODUCTION

Chronic Heart Failure (CHF) is a major health problem with high prevalence (≈ 15 million individuals in greater Europe) [1], high mortality (25% patients die after one year) [2] and as well as high cost for National Health Systems (accounts for 2% of total healthcare expenditure).

CHF diagnosis and severity assessment are important because CHF is a chronic, degenerative condition with continuous worsening of the symptoms, which may bring to more complex and more costly interventions if not early detected.

CHF diagnosis remain a challenging problem because international guidelines[3] suggests that for an objective assessment of CHF the most effective diagnostic test is the comprehensive 2-D echocardiogram coupled with Doppler flow studies to determine whether abnormalities of myocardium, heart valves, or pericardium are present and in which chamber. As far as ECG is concerned, the same guidelines evidenced that conventional 12-lead ECG should not form the primary basis for determining the specific cardiac abnormality responsible for the development of CHF, because of low sensitivity and specificity.

Furthermore one of the most diffused measurements of the severity of CHF is the New York Heart Association (NYHA) classification[4], which is a symptomatic functional scale. Nonetheless, this scale is criticized because, being based on subjective evaluation, it is amenable to inter-observer variability.

Since CHF is characterized by striking abnormalities of the autonomic nervous system (ANS) [5] and Heart Rate Variability (HRV) analysis can be used as an noninvasive and reliable measure to investigate autonomic nervous system[6], HRV has been widely studied in patients suffering from CHF, in particular, for disease detection[7-9]. Many studies demonstrated that HRV is an effective means for the risk assessment of mortality. A number of studies [10-14] demonstrated the relationship of HRV measures and the NYHA classification scale. In our previous papers we demonstrated that HRV might be used to detect CHF using short-term[15] or long-term measures [16]. Moreover, we proposed a classifier based on short-term HRV measures to individuate severity of CHF[17]. In a recent report, Melillo et al. [18] studied the discrimination power of HRV linear parameters in order to distinguish automatically mild CHF patients from severe CHF patients basing on Classification And Regression Tree (CART). The authors obtained sensitivity and specific rates of 93.3 and 63.6% respectively. Nonetheless, this study failed to automatic classify NYHA I versus NYHA II, while this is paramount important because very mild patients may have a good quality of life for a long time, if appropriately monitored.

In this study, we describe an automatic classifier for CHF severity assessment, in particular, to distinguish very mild CHF patients from mild and severe CHF and based on HRV features. The automatic classifier developed achieved a sensitivity and a specificity rate of 75% and 97%, respectively, although the number of patients in NYHA I class was very limited.

II. METHODS AND MATERIALS

A. Data

We analyzed 44 nominal 24-hour records, from 4 NYHA I patients, 8 NYHA II patients and 34 NYHA III patients. Only patients with a satisfactory signal quality as described in the section II.B were selected as eligible. The data were retrieved from the Congestive Heart Failure RR Interval Database[19] and from the BIDMC Congestive Heart Failure Database [19]. The former database includes RR intervals extracted from 24-hour ECG-Holter recordings of 8 men, 2 women, and 19 unknown-gender subjects, aged 34 to 79 (55±11 years). The original ECG records were digitized at 128 samples per second and the beat annotations were obtained by automated analysis with manual review and correction. The latter database includes long-term ECG recordings from 11 men and 4 women, aged 22 to 71 (56±11 years). The records were sampled at 250 samples per second and were automatically annotated.

B. Heart Rate Variability Analysis

We performed standard long-term HRV analysis on nominal 24-hour recordings according to International Guidelines[20]. The series of normal to normal (NN) beat intervals were obtained from the beat annotation files of the selected databases and the NN/RR ratio was computed as the fraction of total RR intervals classified as NN. This ratio has been used as a measure of data reliability, excluding records with a ratio less than a threshold. We chose the threshold of 80% [19] as it was a satisfactory trade-off between numbers of included subjects and quality of NN signals. Using this technique 3 records were excluded (1 NYHA II patients and 2 NYHA III patients) and the final dataset consisted of 41 subjects.

Basic time- and frequency-domain HRV measures have been analyzed elsewhere[18]. Nonlinear properties of HRV were analyzed by the following methods: Poincaré Plot [21], Approximate and Sample Entropy[22], Correlation Dimension[23], Detrended Fluctuation Analysis[24,25], and Recurrence Plot[26-28].

The Poincaré Plot (PP) is a common graphical representation of the correlation between successive RR intervals, for instance the plot of RR_{i+1} versus RR_i. A widely used approach to analyze the Poincaré plot of RR series consists in fitting an ellipse oriented according to the line-of-identity and computing the standard deviation of the points perpendicular to and along the line-of-identity referred as *SD1* and *SD2*, respectively[21].

Approximate entropy, *sample entropy* and *correlation dimension* D_2 measure the complexity or irregularity of the RR series. Detrended Fluctuation Analysis measures the correlation within the signal and two parameters are defined: short-term fluctuations ($\alpha1$) and long-term fluctuations ($\alpha2$)[24,25]. Further details about the computation of the selected parameters can be found in [29].

C. Feature Selection and Classification

We tested each nonlinear HRV feature and we adopted Linear Discriminant Analysis (LDA) as classification method. LDA aims to find linear combinations of the input features that can provide an adequate separation between the classes. LDA uses an empirical approach to define linear decision plans in the feature space. Further details about LDA can be found in Krzanowski[30]. In order to evaluate the classifier, we computed the common measures for binary classification performance measurement[31] using the formulae reported in Table 1. To estimate the performance measures we adopted the leave-one-out cross-validation scheme[32].

Table 1 Performance measurement

Measures	Abbr.	Formula
Accuracy	ACC	$\frac{TP + TN}{TP + TN + FP + FN}$
Sensitivity	SEN	$\frac{TP}{TP + FN}$
Specificity	SPE	$\frac{TN}{FP + TN}$

III. RESULTS

The discrimination power of all the computed nonlinear HRV features was reported in Table 2.

Table 2 Discrimination power of nonlinear HRV measures

Measures	ACC	SEN	SPE
SD1	78%	50%	81%
SD2	95%	75%	97%
Approximate Entropy	71%	75%	70%
Sample Entropy	66%	75%	65%
D_2	85%	75%	86%
α_1	73%	75%	73%
α_2	56%	50%	57%
l_{mean}	68%	50%	70%
l_{max}	54%	25%	57%
REC	59%	50%	59%
DET	39%	50%	38%
Shannon entropy	56%	0%	62%

Finally, based also on our previous research, we propose an automatic classifier based on if-then rules for CHF assessment (see Figure 1). This classifier achieved an overall accuracy rate of 95% (resubstitution estimate).

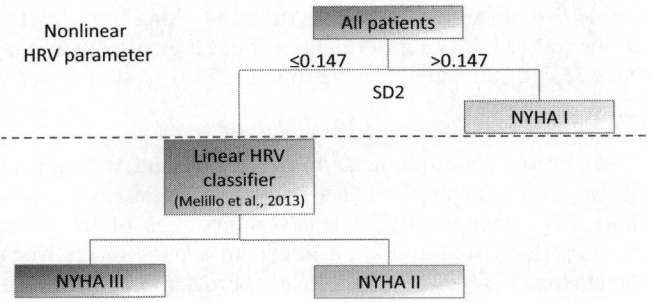

Fig. 1 The proposed classifier for CHF severity assessment

IV. DISCUSSIONS

In this study, we propose a classification tree based on long-term HRV for severity assessment in patients suffering from CHF.

As regards the other classifier proposed in literature for CHF assessment, Guidi [33] compared different algorithms to automatically classify CHF patients in three groups (mild, moderate and severe) and achieved an accuracy of 86% (independent set estimate; sensitivity and sensibility are not reported) by using neural network. Guiqiu [34] proposed a classifier based on support vector machine, which achieved an accuracy of 74% (10-fold-cross-validation estimate) in discriminating between mild CHF (NYHA I) and moderate/severe CHF patients (NYHA II and III). We underline that the classifier proposed by Guidi [33] was based on anamnestic and instrumental data (not including HRV measures), and the one by Guiqiu [34] was based on twelve parameters including LF/HF and other parameters from clinical tests (blood test, echocardiography test, electrocardiography test, chest radiography test, six minute walk distance test). For that reason some parameters needed by the automatic classifier proposed by Guidi [33] or Guiqui [34] should be entered by physicians, while the adoption of only HRV measures, as in the current study, enables a completely automatic assessment.

Our study had the following limitations related to the employed Holter databases: a small and unbalanced dataset, the differences in the sampling frequency of ECG recordings and the different extraction procedures of NN intervals.

In future research we are planning to develop a similar study on a larger (although non-public) dataset of ECG recordings digitalized at the same sampling frequency and annotated with a stated procedure, i.e. the database of Regione Campania Network [35,36]. Moreover, new HRV indexes[37,38], point process time-frequency[39] analysis and/or introduction of other noninvasive measurement (i.e. pupillometry[40]) could provide additional useful measures for automatic classification of cardiac rhythm. Finally, the classification algorithms could be integrated in portable sensing devices[41].

ACKNOWLEDGMENT

LP acknowledges support of this work through the MATCH Programme (EPSRC Grant EP/F063822/1) although the views expressed are entirely his own. PM and AO acknowledge support of this work through the project Smart Health and Artificial intelligence for Risk Estimation (grant PON04a3_00139; 2007-2013 National Operational Programme for Research and Competitiveness).

REFERENCES

1. Agren S, Evangelista LS, Hjelm C, Stromberg A (2012) Dyads affected by chronic heart failure: a randomized study evaluating effects of education and psychosocial support to patients with heart failure and their partners. J Card Fail 18 (5):359-366. doi:10.1016/j.cardfail.2012.01.014S1071-9164(12)00037-1 [pii]
2. Pocock SJ, Wang D, Pfeffer MA, Yusuf S, McMurray JJ, Swedberg KB, Ostergren J, Michelson EL, Pieper KS, Granger CB (2006) Predictors of mortality and morbidity in patients with chronic heart failure. Eur Heart J 27 (1):65-75. doi:ehi555 [pii]10.1093/eurheartj/ehi555
3. Jessup M, Abraham WT, Casey DE, Feldman AM, Francis GS, Ganiats TG, Konstam MA, Mancini DM, Rahko PS, Silver MA, Stevenson LW, Yancy CW, Hunt SA, Chin MH, Comm HFW, Members WC (2009) 2009 Focused Update: ACCF/AHA Guidelines for the Diagnosis and Management of Heart Failure in Adults A Report of the American College of Cardiology Foundation/American Heart Association Task Force on Practice Guidelines. Circulation 119 (14):1977-2016. doi:Doi 10.1161/Circulationaha.109.192064
4. New York Heart Association. Criteria Committee., New York Heart Association. (1979) Nomenclature and criteria for diagnosis of diseases of the heart and great vessels. 8th edn. Little, Brown, Boston
5. Francis GS, Cohn JN (1986) The autonomic nervous system in congestive heart failure. Annu Rev Med 37:235-247. doi:10.1146/annurev.me.37.020186.001315
6. Pumprla J, Howorka K, Groves D, Chester M, Nolan J (2002) Functional assessment of heart rate variability: physiological basis and practical applications. Int J Cardiol 84 (1):1-14. doi:S0167527302000578 [pii]
7. Jovic A, Bogunovic N (2011) Electrocardiogram analysis using a combination of statistical, geometric, and nonlinear heart rate variability features. Artif Intell Med 51 (3):175-186. doi:10.1016/j.artmed.2010.09.005S0933-3657(10)00119-3 [pii]
8. Isler Y, Kuntalp M (2007) Combining classical HRV indices with wavelet entropy measures improves to performance in diagnosing congestive heart failure. Computers in Biology and Medicine 37 (10):1502-1510. doi:DOI 10.1016/j.compbiomed.2007.01.012
9. Jovic A, Bogunovic N (2012) Evaluating and comparing performance of feature combinations of heart rate variability measures for cardiac rhythm classification. Biomedical Signal Processing and Control 7 (3):245-255. doi:http://dx.doi.org/10.1016/j.bspc.2011.10.001
10. Musialik-Lydka A, Sredniawa B, Pasyk S (2003) Heart rate variability in heart failure. Kardiol Pol 58 (1):10-16

11. Panina G, Khot UN, Nunziata E, Cody RJ, Binkley PF (1996) Role of spectral measures of heart rate variability as markers of disease progression in patients with chronic congestive heart failure not treated with angiotensin-converting enzyme inhibitors. American Heart Journal 131 (1):153-157
12. Arbolishvili GN, Mareev VY, Orlova YA, Belenkov YN (2006) Heart rate variability in chronic heart failure and its role in prognosis of the disease. Kardiologiya 46 (12):4-11
13. Casolo GC, Stroder P, Sulla A, Chelucci A, Freni A, Zerauschek M (1995) Heart-Rate-Variability and Functional Severity of Congestive-Heart-Failure Secondary to Coronary-Artery Disease. European Heart Journal 16 (3):360-367
14. Mietus JE, Peng CK, Henry I, Goldsmith RL, Goldberger AL (2002) The pNNx files: re-examining a widely used heart rate variability measure. Heart 88 (4):378-380
15. Pecchia L, Melillo P, Sansone M, Bracale M (2011) Discrimination power of short-term heart rate variability measures for CHF assessment. IEEE Trans Inf Technol Biomed 15 (1):40-46. doi:10.1109/TITB.2010.2091647
16. Melillo P, Fusco R, Sansone M, Bracale M, Pecchia L (2011) Discrimination power of long-term heart rate variability measures for chronic heart failure detection. Med Bio Eng Comput 49 (1):67-74. doi:10.1007/s11517-010-0728-5
17. Pecchia L, Melillo P, Bracale M (2011) Remote health monitoring of heart failure with data mining via CART method on HRV features. IEEE Trans Bio Med Eng 58 (3):800-804. doi:10.1109/TBME.2010.2092776
18. Melillo P, De Luca N, Bracale M, Pecchia L (2013) Classification Tree for Risk Assessment in Patients Suffering From Congestive Heart Failure via Long-Term Heart Rate Variability. Biomedical and Health Informatics, IEEE Journal of PP (99):1-1. doi:10.1109/jbhi.2013.2244902
19. Goldberger AL, Amaral LAN, Glass L, Hausdorff JM, Ivanov PC, Mark RG, Mietus JE, Moody GB, Peng C-K, Stanley HE (2000) PhysioBank, PhysioToolkit, and PhysioNet : Components of a New Research Resource for Complex Physiologic Signals. Circulation 101 (23):e215-220
20. Malik M, Bigger JT, Camm AJ, Kleiger RE, Malliani A, Moss AJ, Schwartz PJ (1996) Heart rate variability: Standards of measurement, physiological interpretation, and clinical use. Eur Heart J 17 (3):354-381
21. Brennan M, Palaniswami M, Kamen P (2001) Do existing measures of Poincare plot geometry reflect nonlinear features of heart rate variability? IEEE Trans Bio Med Eng 48 (11):1342-1347
22. Richman JS, Moorman JR (2000) Physiological time-series analysis using approximate entropy and sample entropy. American Journal of Physiology-Heart and Circulatory Physiology 278 (6):H2039-H2049
23. Carvajal R, Wessel N, Vallverdú M, Caminal P, Voss A (2005) Correlation dimension analysis of heart rate variability in patients with dilated cardiomyopathy. Computer Methods and Programs in Biomedicine 78 (2):133-140. doi:10.1016/j.cmpb.2005.01.004
24. Penzel T, Kantelhardt JW, Grote L, Peter JH, Bunde A (2003) Comparison of detrended fluctuation analysis and spectral analysis for heart rate variability in sleep and sleep apnea. IEEE Trans Bio Med Eng 50 (10):1143-1151. doi:Doi 10.1109/Tbme.2003.817636
25. Peng CK, Havlin S, Stanley HE, Goldberger AL (1995) Quantification of Scaling Exponents and Crossover Phenomena in Nonstationary Heartbeat Time-Series. Chaos 5 (1):82-87
26. Trulla LL, Giuliani A, Zbilut JP, Webber CL (1996) Recurrence quantification analysis of the logistic equation with transients. Phys Lett A 223 (4):255-260
27. Webber CL, Zbilut JP (1994) Dynamical Assessment of Physiological Systems and States Using Recurrence Plot Strategies. Journal of Applied Physiology 76 (2):965-973
28. Zbilut JP, Thomasson N, Webber CL (2002) Recurrence quantification analysis as a tool for nonlinear exploration of nonstationary cardiac signals. Medical Engineering & Physics 24 (1):53-60. doi:Pii S1350-4533(01)00112-6
29. Melillo P, Bracale M, Pecchia L (2011) Nonlinear Heart Rate Variability features for real-life stress detection. Case study: students under stress due to university examination. Biomed Eng Online 10 (1):96
30. Krzanowski WJ (2000) Principles of multivariate analysis : a user's perspective. Oxford statistical science series, vol 22, Rev. edn. Oxford University Press, Oxford Oxfordshire ; New York
31. Sokolova M, Lapalme G (2009) A systematic analysis of performance measures for classification tasks. Inf Process Manage 45 (4):427-437
32. Jain AK, Duin RPW, Jianchang M (2000) Statistical pattern recognition: a review. IEEE Trans Pattern Anal Mach Intel 22 (1):4-37
33. Guidi G, Iadanza E, Pettenati M, Milli M, Pavone F, Biffi Gentili G (2012) Heart Failure Artificial Intelligence-Based Computer Aided Diagnosis Telecare System Impact Analysis of Solutions for Chronic Disease Prevention and Management. In: Donnelly M, Paggetti C, Nugent C, Mokhtari M (eds), vol 7251. Lecture Notes in Computer Science. Springer Berlin / Heidelberg, pp 278-281. doi:10.1007/978-3-642-30779-9_44
34. Guiqiu Y, Yinzi R, Qing P, Gangmin N, Shijin G, Guolong C, Zhaocai Z, Li L, Jing Y A heart failure diagnosis model based on support vector machine. In: Biomedical Engineering and Informatics (BMEI), 2010 3rd International Conference on, 16-18 Oct. 2010 2010. pp 1105-1108. doi:10.1109/bmei.2010.5639619
35. Melillo P, Izzo R, Luca N, Pecchia L (2012) Heart rate variability and target organ damage in hypertensive patients. BMC Cardiovasc Disord 12 (1):105. doi:1471-2261-12-105 [pii] 10.1186/1471-2261-12-105
36. Melillo P, Izzo R, De Luca N, Pecchia L Heart Rate Variability and renal organ damage in hypertensive patients. In: Engineering in Medicine and Biology Society (EMBC), 2012 Annual International Conference of the IEEE, 2012. IEEE, pp 3825-3828
37. Piskorski J, Guzik P (2011) Asymmetric properties of long-term and total heart rate variability. Med Biol Eng Comput 49 (11):1289-1297. doi:DOI 10.1007/s11517-011-0834-z
38. Garcia-Gonzalez MA, Fernandez-Chimeno M, Ferrer J, Escorihuela RM, Parrado E, Capdevila L, Benitez A, Angulo R, Rodriguez FA, Iglesias X, Bescos R, Marina M, Padulles JM, Ramos-Castro J (2011) New indices for quantification of the power spectrum of heart rate variability time series without the need of any frequency band definition. Physiol Meas 32 (8):995-1009. doi:Doi 10.1088/0967-3334/32/8/001
39. Kodituwakku S, Lazar S, Indic P, Chen Z, Brown E, Barbieri R (2012) Point process time–frequency analysis of dynamic respiratory patterns during meditation practice. Med Biol Eng Comput 50 (3):261-275. doi:10.1007/s11517-012-0866-z
40. Melillo P, Pecchia L, Testa F, Rossi S, Bennett J, Simonelli F (2012) Pupillometric analysis for assessment of gene therapy in Leber Congenital Amaurosis patients. Biomed Eng Online 11 (1):40
41. Mougiakakou SG, Kyriacou E, Perakis K, Papadopoulos H, Androulidakis A, Konnis G, Tranfaglia R, Pecchia L, Bracale U, Pattichis C, Koutsouris D (2011) A feasibility study for the provision of electronic healthcare tools and services in areas of Greece, Cyprus and Italy. Biomed Eng Online 10. doi:10.1186/1475-925X-10-49

Author: Paolo Melillo
Institute: Dipartimento multidisciplinare di specialità medico-chirurgiche – Second University of Naples
Street: Via Del Grecchio
City: Naples
Country: Italy
Email: paolo.melillo@unina2.it / paolo.melillo2@unibo.it

Support System for the Evaluation of the Posterior Capsule Opacificaction Degree

D. Ruiz[1], L. González[2], and A. Soriano[1]

[1] Dpt. Tecnología Informática y Computación. University of Alicante, Alicante, Spain
[2] IBIS Group, University of Alicante, Alicante, Spain

Abstract—Patients with intraocular lens can suffer problems related with a progressive opacification of the lens. A human expert observes the state of the capsule located behind the lens and evaluates the level of vision described by the patient. In this article we described a method to detect the cells that can be grew between the posterior capsule and a lens inserted. Cells detection method integrates a series of techniques as thresholding techniques and edge detection.

Keywords—Edge Detection, IOL, Opacification, Posterior Capsule, Thresholding.

I. INTRODUCTION

Patients who suffer of cataract [1] have a progressive decrease of vision. Depending on the degree of development of the cataract, the patient can see only shadows, stains or lights.

The procedure used to diagnose if a patient is suffering from ocular opacity in the crystalline [2] is based on fundus examination. For this examination it is necessary dilating pupils. If the ophthalmologist diagnoses cataract in the eye, the only way to treat this condition is surgery.

Cataract surgery involves removal of the crystalline. One of the most disseminated techniques for the lens extraction is the extracapsular surgery. In this type of surgery, the surgeon removes the opaque portion of the cataract crystalline. Later, the surgeon places an intraocular lens in the same place where the crystalline was. The rear part of the capsule is not extracted because it serves to support the intraocular lens.

There are adverse reactions that can appear in some patients after cataract surgery. Among these undesirable effects, there is the decline in visual acuity after the surgery due to a posterior capsule opacification, which is in contact with the lens. In this case, the rear of the capsule reacts on contact lens and cells are formed between the surface of the capsule and the intraocular lens.

It is important quantifying the level of opacity to decide if it is necessary to change the lens. For ophthalmologists is also important to know which part of the lens is the most affected. Thus they can assess whether the most affected part is the most critical, usually the center. The posterior capsule opacification is a problem today that needs to be treated objectively. For this reason, it arises the idea of the quantification of posterior capsule opacification.

Conducting a system capable of quantifying the posterior capsule opacification entails a number of significant drawbacks. The intervention of the ophthalmologist to determine the degree of opacity is another factor that connotes some subjectivity. The reason for this is that the technique used to make the diagnosis is based on observation by the expert of an image captured by the medical system used. The ophthalmologist can determine whether or not there is opacity. Then, this specialist will qualify opacity with indeterminate terms as much, little, etc.

Another drawback involved in determining the degree of opacity is the instrumentation used for the images. The capture of the image is usually performed with cameras coupled to zoom objectives machines. This capture method introduces undesirable external components in the image, as the same camera flash, making it difficult to calculate the degree of opacity.

The remaining of the paper is organized as follows: section 2 discusses some related work. Section 3 describes the proposed method and the intermediate results, which are represented by images. Section 4 concludes the done work and presents the possible future work.

II. STATE OF ART

The problem of quantifying posterior capsule opacification has been approached in recent years using different computational techniques. Among these studies, there are solutions such as the one proposed in [3], a commercial software that calculates the degree of opacification. This software takes into account the different levels of existing opacification. Grade zero indicates no opacification or that the opacification is practically inappreciable. The greatest degree specified in this work is the degree three, which corresponds to a severe opacification.

Another system similar used to obtain the degree of opacification is presented in [4]. In these two works cited an image is acquired, typically through a photographic machine. The user who interacts with the software is an expert ophthalmologist, who gets the result of posterior capsule opacification following a few steps in the program. The program asks the user to indicate where the intraocular lens is and the areas of opacification. The difference between the two systems is that in the first, ophthalmologist or the expert has to be noted exactly the opacification area which will be classified according to different degrees of opacity. However, the second one detects automatically the opening of the anterior capsule extracted in the previous operation. Then, the total area is divided into regions asking the user to specify which of those regions opacification is greater than 50%. Both systems have in common the feature that the user has to manually indicate in which areas there are opacification. The result of specifying the areas where there are opacification is that subjectivity contributes to the final result. These systems help to the ophthalmologist to catalog the degree of opacification. Besides, these types of systems are used for studies of the posterior capsule opacification. One of the reasons for its use is to ensure a sufficiently acceptable standard deviation. Works like [5] or [6] use these systems to perform such studies.

In addition to the aforementioned type of software, there have been several studies that have used different computer technologies associated with this type of problem. In [7], the work shows a method based on unsupervised learning using the Hopfield neural network to divide the analyzed space into regions according to the color value in each of the pixels of the image. The degree of opacity is determined from the number of regions found. The number of regions is directly proportional to the degree of opacification. This is calculated dividing the image into groups of pixels depending on the roughness generated on the surface of the intraocular lens. Extraction of the number of regions involves an image pre-processing where it computes a circular portion in the center of the intraocular lens, which is the area to be analyzed. It implements an algorithm to classify the pixels depending on its color. Afterwards, clusters are formed, from which it calculates the number of these regions. The method used to group the pixels is presented in [8] (K-means technique), where the partition is performed into clusters to the area to analyze. For this, it takes into account the pixels with similar color values. Hopfield neural network [9], which is implemented in this work, has the number of entries as a result of multiplying the number of pixels and the number of clusters. The number of iterations of the neural network depends on the time at which the output becomes constant. As it is explained in the same article, this always occurs. Other techniques that are used to implement the neural network are Mahalanobis distance [10] and the Euler approximation. Additionally, a median filter [11] 5 x 5 is used to obtain better results.

The work introduced in [12] presents the quantization resolution of the problem of the posterior capsule opacification using Contourlet transform [13]. To determine the region of interest, that is, the area belonging to the posterior capsule, it applies to the image a color space transformation: from the RGB system to the YCbCr system. The resulting image is converted to a binary image using Otsu's method and taking into account the Cr component. In this way, it obtains the area of the posterior capsule where the lens is. One of the interesting aspects of this work is the reconstruction of the areas where the image information is lost due to the reflection of the flash when the picture is taken. The neighboring pixel information is used to rebuild theses areas. Contourlet Transform is used to find the contour and feature extraction. The image is divided into 64 equal regions where the feature vector is obtained for each. This feature vector is used to classify the image according to four levels of opacification. Each of these vectors is compared with a previously trained database calculating the Euclidean distance. To establish the final degree of opacification in the entire area it is assigned a coefficient to each of the 64 regions indicated above by performing a weighted average.

The exposed methods do not calculate the percentage of the opacification in the capsule. Furthermore, results in some of these methods are the grade of opacification, which are represented with qualitative adjectives only. We propose a method that calculates the percentage of opacification divided by areas.

III. METHOD PROPOSED

The main goal consists of obtaining the percentage of quantification of posterior capsule opacification. To do this, we proceed to the application of different mathematical methods or transformations to the sample image of the posterior capsule. The methods we use are the Otsu's method, Sobel operator, Hough transform and morphological transformations with different structuring elements.

One of the objectives of this work is to minimize user interaction to get the final result. To do this, first, it is detected the lens edge automatically. Subsequently, over the area where the lens is, different transformations are applied to calculate the affected area.

We use Fast Hough Transform for circumferences to determine the area occupied by the intraocular lens. This transform searches geometric shapes for the entire image, circumferences in this case, by finding the parameters of the shape that contain more points on it. Therefore, the curve is defined parametrically describing the circumference, as shown in (1).

$$(x - c_1)^2 + (y - c_2)^2 = c_3^2 \qquad (1)$$

For a point belonging to an edge, (c_1, c_2), the space of possible circumferences is given varying c_3 from 0 to infinity.

However, for the particular problem of the opacification quantification, c_3 parameter variation is delimited taking into account the smaller side of the sample image. We obtain two values: the total length and half the total, previously calculated. This transform provides good results when the flashlight is minimal in the image. The flash is considered an undesired artifact, which has occurred due to the instrumentation equipment used. This kind of noise cannot be eliminated. Therefore, we must analyze the remaining sample ignoring those pixels that are part of the flash.

To detect the pixels that are part of the flash, it is transformed the given image in RGB format to gray scale. Subsequently, three techniques are use to detect the flash zone:

- One of the techniques consists in using Otsu's method taking into account the values of the gray scale image that have been previously calculated.
- The second technique also uses the Otsu method. Nevertheless, in this case, we transform the original image to HSV color model [14]. Subsequently, the threshold value is determined taking into account the component H (hue).
- The latter technique that it is used to locate the pixels that belong to areas of flash consists of determining which pixels are below a fixed threshold. The value of these pixels will be given by the components S (saturation) and V (value) of the HSV color model.

These three techniques are combined by weighted average. If the value is larger than 0.5, the pixel is considered a flashlight pixel. The result is shown in Figure 1.

The detected flash area is discarded to process the rest of the image and apply the different transformations. Thus, the flash will not affect the processing of thresholding, as the value of the pixels is not taken into account.

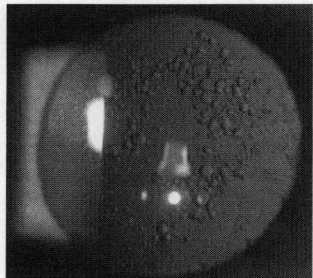 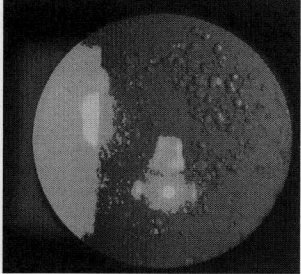

Fig. 1 Left: original image. Right: Flash detected combining the three techniques.

The next step consists in detecting the contour of the cells that give rise to the opacification. For this purpose, there are different techniques for edge detection, which seek those points where there is a variation of the intensity. The detection of such intensity variation can be performed by applying the first or second derivative, gradient or Laplacian respectively.

For this particular case, a gradient technique is applied: Sobel operator (Figure 3). This operator is based on the First-Order Derivate. Therefore, applying Sobel operator is applying de gradient of function f at coordinates (x, y) in the image, as is defined in (2).

$$\nabla f \equiv grad(f) \equiv \begin{bmatrix} g_x \\ g_y \end{bmatrix} = \begin{bmatrix} \frac{\partial f}{\partial x} \\ \frac{\partial f}{\partial y} \end{bmatrix} \quad (2)$$

It is more suitable computationally to approximate the calculation and defining a discrete approximation to the preceding equation and from there obtain the filter mask. For Sobel, it is denoted the intensities of image points in a 3x3 region (figure 2).

-1	-2	-1
0	0	0
1	2	1

g_x

-1	0	1
-2	0	2
-1	0	1

g_y

Fig. 2 A 3x3 region of an image: Sobel operators.

Applying Sobel operators to the image gives the result shown in the figure 3.

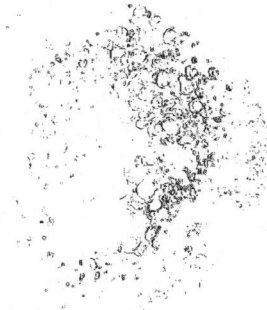

Fig. 3 Binary image whith the detected edges of the cells by Sobel operator

Finally, morphological operators are applied to fill the cells detected. Closing is used to smooth contour detected portions and merging small cracks thereof. Dilation operators are used to fill holes.

Figure 4 shows the final result of applying the process described. This result is divided into concentric regions calculating the percentage of opacity in each of the areas. This allows a more objective evaluation by the expert. These calculations help the ophthalmologist to perform diagnosis taking into account the importance of each of the areas: the central area is the most critical of all because it is the area where more information is collected from the outside world that includes the patient.

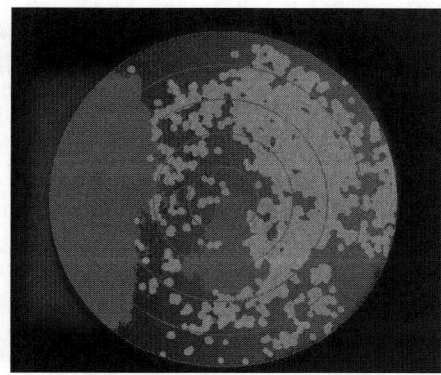

Fig. 4 Splitted image into concentric regions with detected cells.

IV. CONCLUSIONS

The system implemented calculates an objective quantification of the grade of opacification automatically (without active participation of the specialist). Furthermore, this quantification is presented by sectors, which facilitates the specialist interpretation.

According to the obtained results, the location of the opacification due to the cells generated with lens intraocular depends on the quality of the image, as well as external factors such as the flash of the camera.

Likewise, the application of morphological operators is more effective when the cells are small, because if the outline of the image becomes too large, it cannot completely fill inside the portion in the image.

As future works we will apply different techniques such as detection of blobs or image segmentation, combining those with the described transformations to improve performance and extend the possibilities of analyzing different types of intraocular lenses. Furthermore, a combination of the Hough transform is applied to detect cells, combining this technique with first and second order operators.

ACKNOWLEDGMENT

Finally, thank those who have provided different images to order to perform the work done, particularly, the company *Vissum Coorporación Oftalmológica*.

REFERENCES

1. Roger F S (2009) Cataract surgery, in The patology of Cataracts. Expert Consult. Saunders Elsevier, 2009, pp 3–10.
2. Jay B L (2001) The Eye Care Sourcebook, in Anatomy of the Eye. McGraw-Hill, Contemporary Books, 2001, pp 9–20.
3. Tetz M R, Auffarth G U, Sperker M, Blum M, Völcker H E (1997) Photographic image analysis system of posterior capsule opacification. Journal of cataract and refractive surgery, Vol 23, Edition 10, 1997, pp 1515–1520.
4. Bender L, Spalton D J, Uyanonvara B, Boyce J, Heatley C, Jose R, Khan J (2004) POCOman: New system for quantifying posterior capsule opacification. Journal of Cataract and Refractive Surgery, Vol 30, Edition 10, 2004, pp 2058 – 2063.
5. Nixon D R, Woodcock M G (2010) Pattern of posterior capsule opacification models 2 years postoperatively with 2 single-piece acrylic intraocular lenses. Journal of cataract and refractive surgery, Num. 6, Vol 36, 2010, pp 929–934.
6. Jose R M J, Bender L E, Boyce J F, Heatley C (2005) Correlation between the measurement of posterior capsule opacification severity and visual function testing. Journal of cataract and refractive surgery, Num. 3, Vol 31, 2005, pp 534–542.
7. Naoufel W, Rachid S, Fatma A (2010) An unsupervised learning approach based on a hopfield-like network for assessing posterior capsule opacification. Pattern Analysis and Applications, Vol 13, 2010, pp 383–396.
8. Aristidis L, Nikos V, Jakob J V (2003) The global k-means clustering algorithm. Pattern Recognition, Vol 36, Num 2, 2003, pp 451 – 461.
9. John J H (1998) Artificial neural networks: theoretical Concepts, in Neurons with graded response have collective computational properties like those of two-state neurons, IEEE Computer Society Press, 1998, pp 82–86.
10. De Maesschalck R, Jouan-Rimbaud D, Massart D L (2000) The mahalanobis distance. Chemometrics and Intelligent Laboratory Systems, Vol 50 Num 1, 2000, pp 1–18.
11. David R K B (1984) The weighted median filter. Communications of the ACM, Vol 27, Num 8, 1984, pp 807–818.
12. Muhammad P, Hammid P, Oliver F, Ramin D, Wolf B (2009) CPCO: Contourlet Based PCO Quantification System. IEEE International Conference of Soft Computing and Pattern Recognition, Malacca, 2009, pp 409–413.
13. Minh N D, Martin V (2005) The Contourlet Transform: An Efficient Directional Multiresolution Image Representation. IEEE Transactions on Image Processing, vol 14, Num 12, 2005, pp 2091–2106.
14. Pradeep K B (2008) Computer Graphics, in Color Models, I.K. International Publishing House, 2008, pp. 318-325.

A Custom Decision-Support Information System for Structural and Technological Analysis in Healthcare

A. Luschi[1], L. Marzi[2], R. Miniati[1], and E. Iadanza[1]

[1] Department of Information Engineering, University of Florence, Italy
[2] Department of Architecture, University of Florence, Italy

Abstract—This paper presents a custom informative system called SACS, a software that drives Autocad to manage and analyze digital plans of hospital buildings coded on specific layers. The software maps Departments, Operative Units, Destinations of Use, healthcare technologies and environmental comforts grouping info by single rooms and homogeneous areas, giving quantitative and qualitative results. System outputs can be used by top-management as a decision-support aid to assess parameters to improve the hospital's structure and organization.

Keywords—healthcare, workplace, management, monitoring, support.

I. INTRODUCTION

The modern hospital organization is based upon an enormous amount of data and people and a major problem is to organize and make functional relations among them. In fact hospitals must now undergo a numerous quantity of requirements set by national and international institutions in order to have sufficient hygienic, qualitative and organizational standards granted.

Following this, some technical tools have been developed to monitor the hospital structure by measuring quantitative architectonical, technological and people-relational parameters which give a detailed "picture" of the effectiveness and the efficiency of the hospital itself.

Many of these are based upon special hospital dedicated DBMS (Database Management Systems), BIM (Building Information Models) or GIS (Geographic Information Systems). By using databases all sorts of data can be stored and then aggregated in different ways to answer to a wide range of queries. Numeric indexes (Key Performance Indicators – KPI) have been developed by the scientific community throughout the recent past years to monitor the performance of buildings so that quantitative analysis can be made and more structures can be compared.

GIS-related software usually perform typical measurements like path-analysis and people/units finding and unfortunately there are no quantitative comparable data [1] [2].

Facility Management (FM) systems are decision-support tools based upon Integrated Healthcare Facility Management Model (IHFMM) which outputs KPIs on the process that can affect the performance of the facilities [3] [4] resulting very useful for the top-management for maintenance, performance and risk evaluation, business management and development [5].

Workplace Management Systems (WMS) are solutions designed to manage real estate facilities, allowing users to assess, analyze and reorganize company assets in order to preserve their value, improve their effectiveness and respond to multiple needs. They provide access to stored information regardless of the workplace: data and plans can be acquired through web services, using a common browser over an intranet network [6].

Our main scope is to develop a tool which monitors the whole hospital and gives unbiased data to assess the building/technological estate assigning priorities to necessary interventions by whomever it is responsible. The developed system is an integrated hospital Workflow Management System (WMS) tool which can also output KPIs and quantitative parameters typical of FM systems. The software [7] [9] is used in the university hospital campus of Careggi (Florence, Italy) to assess the status-quo of the buildings in terms of beds, square meters, destination of use, functional areas and many other features for every room by driving DWG maps in Autocad. This software, named SACS, has been developed by the Monitoring Laboratory, an autonomous office made by University of Florence's researchers (Department of Information Engineering together with Department of Architecture).

II. METHODS

A. Careggi Organization

Careggi polyclinic is a pavilion hospital and it is hence made by different buildings spread over a 74 hectares area. Its inner organization has been reviewing over the years and it is now structured in Departments, Activities Area and Operative Units. The Departments (or Integrated Activity Department – IAD) are 20 functional macrostructures working among more buildings, using different structures, technology and rooms throughout the whole hospital area.

For each IAD there are functional substructures with a lower level of aggregation: Activity Areas (AA) and

Operative Units (OU). The first ones join together physical spaces in relation to the carried out activity. The OU join together medical staff in relation to the medical activity they are assigned to. Therefore it is possible to have many OUs operating in different rooms associated with different AAs: a single room is assigned to only one Activity Area but it can be used by many medical OUs (for example ambulatories and surgery rooms are used in many cross-specialties, but assigned to only one AA).

The 20 IADs, together with the 276 Activity Areas and 181 OUs make a three-dimensional matrix where each cell identifies a single medical center inside the hospital.

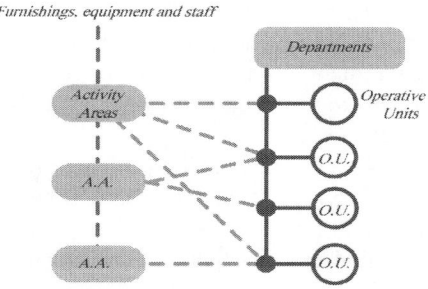

Fig. 1 Careggi Polyclinic organization

B. Critical Design Review

The idea behind SACS is to bring the knowledge-sharing to its maximum, allowing hospital staff to potentially know everything about the building estate (spreading of Operative Units, Activity Areas and Departments, beds, detailed destination of use of every single room). The informative system itself is basically composed by the core-engine software which drives CAD maps, makes automations on them, and the MSSQL Server 2008 database that contains all the tables used to decode the amount of data stored inside the DWG files. This is the great potential of the system: all data are stored inside the maps (and also backed-up on the database for emergency recovery) so that the drawings are self-sufficient and the info can be always rebuilt using nothing but the CAD files.

The current release of the software (SACS v4.0) electronically analyzes the plan of the hospital and allows to obtain information about dimensions and aggregation of rooms and spaces, subdivided into 42 Main Destination of Use (MDU) and 246 Secondary Destination of Use (SDU).

Data is collected using techniques which have reference to Post Occupational Evaluation (POE) through on-site surveys and personnel interviews. Survey information is then data-entered into SACS and linked to the DWG AcadPolyline objects; each one of them outlines the perimeter of a room. Polyline color has been used as a parameter to identify MDU and it is associated with the intensive of care of each destination. This so-called "everything inside DWG" in-map storing allows to rebuild anytime the whole information having nothing but the DWG files and the translation tab used to link every color to an exclusive destination of use.

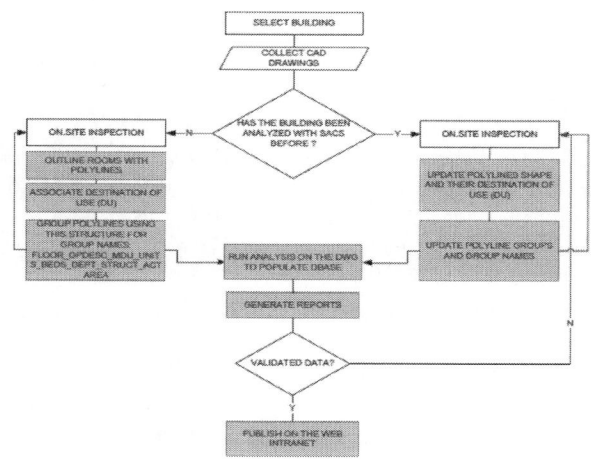

Fig. 2 SACS flowchart

This detailed set of destination of use enhance the performance of the software by storing a whole new set of information in order to assess accreditation issues in terms of technological, structural and plant requirements. Furthermore the in-data allows to compute a set of structural Key Performance Indicators (KPI) [3] [4].

SACS also manages the following listed data in order to give users more information about the analyzed spaces:

Room code (BLD_LEVELROOM): an unique alphanumerical code where BLD is the code of the building, LEVEL is the number of the floor and ROOM is a 3-digit formatted increasing number.

ATU: a code of the Air Treatment Unit that feeds the room.

EG: Electrical Group for the medical rooms as described by the Italian Electrical Committee, that can be 0, 1 or 2 (X is used for the non-medical rooms).

NGAS: number of medical gasses connections detailed for air, oxygen, nitrogen monoxide and void terminals.

H: the height of the room.

SLR: the square-light ratio.

In order to maintain the original idea of "everything inside DWG" approach, all this info is grouped in a data array, which is written directly to the DWG file, attached to the room-polyline proprieties known as XData (Autocad Extended-Data) that allow to store up to 64bit data as structured as the user wants.

Structural KPIs are calculated and outputted as additional information about the hospital rooms, giving an even more useful tool for the top management (for example the "air volumes per hour" can be evaluated knowing area, height and feeding ATU parameters).

The information in each file is extracted by the software itself and inserted in a MSSQL Server 2008 support table linked via ActiveX Data Objects (ADO) in order to make reports and to assure redundancy backup functionality. The database also provides several tables used as translation keys for the codes embedded in the DWG file.

C. Intranet and Reporting

Autocad hyperlink property of the polyline is used to associate every single entity to a hypertext. Because of the available large amount of data, it is clear the it can be useful to access them even from the intranet consulting. Therefore SACS publishes all the gathered attributes together in a DWF file, embedded in a HTML page: this allows everyone to access the inner information recorded in the maps even without having the SACS software installed but only with a common web-browser.

Before publishing the DWF, SACS is able to make some automation on the CAD map like generating hatches colored according to the MDU's RGB color or texts of the attributes of the single room, automatically centered in the polyline that outlines it.

The structure of the reports can be arranged by buildings or by departments (useful to study the scattering of an IAD in the hospital estate and to plan possible improvements).

III. RESULTS AND DISCUSSION

SACS is applied to Careggi polyclinic in Florence, Italy. It maps 14.062 rooms, 317.850m^2 and 1.796 beds in 52 medical and non-medical buildings. DWF files were produced for every floor and published to the local Hospital Intranet.

The tool is used every day by different hospital offices, especially by the Clinical Engineering Service and the Technical Department and it is frequently updated by destined personnel. The average of logs on the intranet SACS site is about 100 per day.

The software outputs both numerical and graphical reports. Hence it is very useful as support tool for the healthcare planning. A list of few examples grouped for users follows:

- Directors or Nurse Coordinators use SACS to know the spreading of their units/departments and which ones they eventually have to coexist with.
- Firemen query SACS to know the escaping pathways along the buildings, where the fire-escapes and fire-stairs are and which places are more sensible and thus ask for more attention (rooms with combustive agents like ward or ICUs, super-magnet room of MRIs, etc.).
- Technical staff uses SACS almost every day to retrieve parameters used for managing the hospital and to calculate indicators for quality of service or accreditation requirements like number of beds for activity-area/department, number of beds for square-meter, total number of ward rooms with one, two, three, four or more-than-four beds, etc.
- Everybody inside the hospital can query SACS to know building's code and name or any medical or non-medical activities done inside a given room by knowing only its number. This function can be more expanded by introducing info about the personnel so that it may become an useful tool also for people-finding.

The software is used for many purposes and in really different scenarios like transfers management, assessment of accreditation requirements, electro-medical devices management (through the interface with the Clinical Engineering database) and general designing and remodeling.

A. Transfer Management

The first problem which SACS solved was about the transfer management. Careggi has been being remodeled for about the last 10 years: during this process a lot of transfers have been made following the demolition and the new building of many areas of the hospital. SACS has been able to answer the main question in a transfer process: does the target room meet the requirements of the new use it has to fulfill? By consulting the data stored along the room-representing polyline it is possible to know that and then easily answer the previous question. For example if a 2-beds wage room needs to be transferred to another area, it can be simply known if the square-light ratio is enough or if the number of medical gas connections are sufficient or not. Otherwise if an Operating Room must be moved, parameters like air volumes per hour, and electrical continuity groups have to be verified. Thus SACS gives a prevision of which transfers are more or less complex to manage or generally not allowed at all.

B. Accreditation Requirements Assessment

Recently Careggi is undergoing a process of accreditation which plans to certify the whole 25 medical buildings. By using SACS and its reporting features it is possible to determinate which structural requirements are already satisfied and which are not: KPIs give a relevant, strong and very useful information about that. Obviously not all the requirements can be assessed by using the software, so on-site surveys need to be done and other Hospital Information System (HIS) databases must be consulted. Plus, SACS is recorded in the inner organization paper as the official hospital tool used to verify the possession of the accreditation requirements.

C. Space Management

SACS is used in hospital facility management and governance activities like destinations of use verification and cost and space analysis of OUs. SACS offers a complete database with multifunctional data which allows many typologies of aggregation and enquiries by different categories of users, with the possibility to make more complex studies like cost-benefit analysis or comparative analysis.

IV. CONCLUSIONS AND FUTURE WORKS

The idea behind SACS is to empower the knowledge sharing among different typologies of users. Medical staff like nursing coordinators, IAD or OU directors can use SACS to know where their "frameworks" are spread on, which rooms are involved and, in case of OUs, whom they have to coexist with. Instead technical staff may be interested from more technological and procedural ways, but this info is resident inside the system so that it can be known by anyone who wants to accede to it (with access privileges provide by system administrator). This is also very useful for the continuous updating of SACS, with updating/changing claims coming from nurses, doctors, top-clinicians, technicians, engineers, managers and any other professional figure that may works in a hospital.

As said above, the main resource of SACS is the "everything inside DWG" approach: all the info is carried on with the CAD maps themselves in addition to the database. This offers a double degree of backing up and safety: if database (or map) crashes or is accidentally deleted, all data can be easily recovered by the map (or database). Besides the system is very independent by the technology of DBMS used: data is read from maps, parsed by the software, and then stored in any type of database. The system itself provides very useful and automated features like hatches drawing, texts writing and centering, hyperlinks producing, database feeding and html and pdf reporting, which considerably decreases the time spent to publish the whole data up to the hospital server.

A new software release is about to be developed with even more data stored in. Information about the entrance typologies to map and manage the people path along the hospital (particularly useful to remove the architectural barrier for the disabled users), names of the occupants of the rooms, typologies of the Cleaning Facility Macro-Area to evaluate the cost of every space in terms of cleaning operations, automatic door plate production for the Mobility and Surveying Service, electrical and data systems management and an intranet search engine are just a part of the future available features.

REFERENCES

1. Naves Givisiez GH (2001) Hospital Demand: Using GIS and spatial analysis for estimation. Pan American Health: 1–33
2. Muresan F, Tirt DP, Haindu I (2006) Specific features of GIS database for hospital. Database 1: 133–138
3. Rodriguez E, et al. (2003) A new proposal of quality indicators for clinical engineering. 25th Annual International Conference of the IEEE Engineering in Medicine and Biology Society Proc. (IEEE Cat. No.03CH37439): 3598–3601
4. Iadanza E, Dori F, Biffi Gentili G, Calani G, Marini E, Sladoievich E, Surace A (2007) A hospital structural and technological performance indicators set. IFMBE Proceedings MEDICON and HEALTH TELEMATICS 2007 "XI Mediterranean Conference on Medical and Biological Engineering", Vol. 16, 2007, Ljubljiana, Slovenia. ISBN: 978 3 540 73043 9
5. Lavy S, Shohet IM (2007) Computer-aided healthcare facility management. Journal of Computing in Civil Engineering 21: 363–372
6. InfoCAD at http://www.infocad.fm
7. Iadanza E (2009) An unconventional approach to healthcare (geographic) information systems using a custom VB interface to AutoCAD. In: Biomedical Engineering Systems and Technologies: International Joint Conference, Biostec 2009 Porto, Portugal, January 15-17, 2009 INSTICC PRESS. ISBN: 978-989-8111-78-4
8. Iadanza E, Dori F, Biffi Gentili G (2006) The role of bioengineer in hospital upkeep and development. IFMBE Proceedings Aug, Vol. 14, 2006, Seoul, Korea. ISBN: 3/540 36839 6
9. Iadanza E, Marzi L, Dori F, Biffi Gentili G, Torricelli MC (2006) Hospital health care offer. A monitoring multidisciplinar approach. IFMBE Proceedings WC 2006 "World Congress on Medical Physics and Biomedical Engineering", Vol. 14, 2006, Seoul, Korea. ISBN: 3 540 36839 6

Author: Ing. Alessio Luschi
Institute: Dept. of Information Engineering (University of Florence)
Street: Via S. Marta, 3
City: Florence
Country: Italy
Email: alessio.luschi@unifi.it

Performance Assessment of a Clinical Decision Support System for Analysis of Heart Failure

G. Guidi[1], P. Melillo[2], M.C. Pettenati[3], M. Milli[4], and E. Iadanza[1]

[1] University of Florence, Dept. of Information Engineering, Italy
[2] University of Bologna, Dept of Electrical Energy and Information Engineering, Italy
[3] ICON Foundation, Florence, Italy
[4] Azienda Sanitaria Fiorentina Dept. of Cardiology, Italy

Abstract—In this paper we compare five machine learning techniques in dealing with typical Heart Failure (HF) data. We developed a Clinical Decision Support System (CDSS) for the analysis of Heart Failure patient that provides various outputs such as an HF severity evaluation, an HF type prediction, as well as a management interface that compares the various patient's follow-ups. To realize these smart functions we used machine learning techniques and in this paper we compare the performance of a neural network, a support vector machine, a system with fuzzy rules genetically produced, a Classification and regression tree and its direct evolution which is the Random Forest, in analyzing our database. Best performances (intended as accuracy and less critical errors committed) in both HF severity evaluation and HF type prediction functions are obtained by using the Random Forest algorithm.

Keywords—Neural network, SVM, CART, Random Forest, Fuzzy, Genetic, machine learning, Heart Failure.

I. INTRODUCTION

In our work we are developing a Heart Failure (HF) Clinical Decision Support System (CDSS) that, combined with a hand-held device for the automatic acquisition of a set of clinical parameters, enables to support telemonitoring functions. The system provides an HF severity/type assessment function using five machine learning techniques: a Neural Network [1], [2], a Support Vector Machine [3], a Fuzzy Expert System whose rules are produced by a Genetic Algorithm [4], a Classification And Regression Tree (CART) [5] and a Random Forest [6]. The whole system is accurately described in [7–9]. In this paper we compare the five techniques to determine which is the one that better behaves with the typical data of HF. Each technique has strengths and weaknesses, and in this paper they will be highlighted. The system we developed produces many outputs, coordinated by a management interface designed together with physicians. These outputs have been denominated: HF Severity assessment, HF Type prediction, Short term prediction, Chronological follow-up comparison and Score-based prognosis. Not for all these functions the involvement of a machine learning technique is required. The function called Chronological follow-up comparison creates a graphical trend of all the data of the various patient's follow-ups and points out to the doctor if the patient has worsened or improved, in comparison with the previous check. The Score-based prognosis function is the computer implementation of prognosis models known in literature. The functions called Severity Assessment, HF type prediction and Short term prediction instead need machine learning techniques and provide, respectively: an assessment of the state of the patient's HF severity at the time of the parameters measurement, a predictive evaluation of the type of HF (defined as stable HF, or HF with rare or frequent exacerbations) that the patient will undergo, and a prediction of the imminence of an acute episode. Currently the latter output is still at work in progress stage because, for a proper training of the machine learning techniques, we need data coming from a real scenario of daily telemonitoring of patient's parameters (operations to be performed automatically, using wearable devices or setting up a semi-automatic measurement desk at patient's home), that is still not been activated. In this paper we therefore evaluate the effectiveness of the five mentioned techniques in providing the outputs Severity assessment and HF type that are both on three levels, respectively mild - moderate - severe and stable - rare exacerbations (<= 2/year) - frequent exacerbations (> 2/year).

II. MATERIALS AND METHODS

A. Database

The two above explained functions are trained using a de-identified database of HF patients, with varying severity degrees, all treated by the Cardiology Department at the St. Maria Nuova Hospital in Florence, Italy in the period 2001-2008. The database consists of a total of 136 records of 90 patients, including baseline and follow-up data (if available). At the time of the data collection, the specialist physician provided an HF severity assessment in the desired three levels: 1-Mild, 2-Moderate and 3-Severe, which was stored in the database. Note that in determining this severity level the physician evaluates the patient in his general condition and not just relying on the levels of the best known HF severity markers such as the Brain Natriuretic Peptide (BNP), Ejection Fraction (EF) or New York Heart Association (NYHA) class that are often at odds each other in patients with complex situations. In this way the system will train itself to find a model for assessing general HF condition and not to set simple thresholds in individual parameters. Moreover, after 12-24 months from the data collection, the status of each patient in terms of HF type was assessed and associated with the correspondent record. In this manner the system, after a training phase, can recognize typical parameter patterns (e.g. a patient that may develop rare exacerbations). Being the output relative to future time we can classify this function as a prediction. Thanks to these target-output assignment operations, performed by a specialist cardiologist, we can perform supervised machine learning. [10]

Variables in database that are used as input of the Machine Learning Techniques are the following twelve:

- Anamnestic data: Age, Gender, NYHA class
- Instrumental data: Weight, Systolic Blood Pressure, Diastolic Blood Pressure, Ejection Fraction (EF), Brain Natriuretic Peptide (BNP), Heart Rate, ECG-Parameter (atrial fibrillation true/false, left bundle branch block true/false, ventricular tachycardia true/false

In the database the classes mild-moderate-severe and stable-rare-frequent, respectively for the functions HF severity assessment and HF type prediction, are distributed as shown in Table 1.

Table 1 Class distribution

Severity assessment	Mild	Moderate	Severe
N° of patients	51	37	48
Type prediction	Stable	Rare	Frequent
N° of patients	110	14	12

B. Machine Learning Methods Details

For each machine learning technique we have changed the internal parameters in order to obtain a good compromise between learning ability and generalization capability. In Table 2 are summarized the parameters that were varied. Note that SVM is a binary classificatory, so we have to combine two SVM to obtain a three level output, building an "SVM tree". Best performances are obtained by setting the investigated parameters shown in Table 2 as follows:

Table 2 Parameters that were varied to obtain best results

Method	Investigated parameter
Neural Network	N° of hidden Neurons (by automatic cycle)
SVM	Combination order of the two SVM
Fuzzy-genetic	N° of fuzzy rules, N° of Generation
CART	Level of Pruning (by automatic cycle)
Random Forest	N° of features (m) to be used for each tree, N° of trees, Class cut-off levels

- **Neural Network:** We cyclically retrained the Neural Network (NN) varying Hidden Neurons from 2 to 8. The best configurations are: 5 hidden neurons for HF severity assessment and 8 hidden neurons for HF Type prediction.
- **SVM:** We tried all the possible permutations of SVM tree and we obtained the best results with the combination that first detects the non-severe vs. severe status then going to recognize the mild and moderate.
- **Fuzzy-Genetic:** The best results are achieved with a population of 30 individuals, each composed by 45 rules. The algorithm evolves for 600 generations.
- **CART:** The system automatically tests the CART with various levels of pruning. Best results in assessing severity are obtained with a prune level = 2.
- **Random Forest:** We performed various tests obtaining the best performances with a number of features (*m*) to be used for each tree equals to 4. Analysing the OOB (Out Of Bag) error rate, related to the increasing of number of trees, we chose the value of 2000 trees, because above that value the error rate is sufficiently stabilized. We performed several tests, cycling the thresholds for the three classes and establishing the combination which generated better accuracy, and the resulting cut offs are: stable class, 50; rare class, 20; frequent class, 30.

C. Performance Measurement

In order to measure and compare the performance of the machine learning methods, we adopted a ten-fold cross validation. To better exploit all the available data, we made the assumption of considering each record of the database, ie "follow-up information," as if it were a patient. In this way we have considered to have a database composed of 136 different patients each with a single follow-up (instead having 90 patient with a total of 136 follow-ups). This assumption is justified by the fact that the follow ups are spread on a large period of time (1-2-3 years) and the parametric situation and health of the patient was changed so as to justify the approximation described. During the cross validation process we have taken precautions so that follow-ups of the same patient are grouped within the same fold, thus our assumption does not affect the independence of the folds.

In Results section we reported Test set Accuracy (Generalization Capability) for each machine learning technology. The Multiclass Accuracy formula we used, according to [11], is shown below (TP: True Positive, TN: True Negative, FN: False Negative, FP: False Positive).

$$Accuracy = \frac{\sum_{i=1}^{N° class} \frac{TP_i+TN_i}{TP_i+TN_i+FP_i+FN_i}}{N} \quad (1)$$

In addition we reported for each method the number of "critical errors" committed, meaning e.g. the classification of a severe HF patient as mild and vice versa.

III. RESULTS

10-folds cross validation results of various machine learning techniques are summarized in Tables below.

Table 3 Performance in severity assessment

Machine learning method	Accuracy %	STD	Critical Errors
Neural Network	77.8	7.4	0
SVM	80.3	9.4	3
Fuzzy-genetic	69.9	9.9	1
CART	81.8	8.9	2
Random Forest	83.3	7.5	1

Table 4 Performance in type prediction

Machine learning method	Accuracy %	STD	Critical Errors
Neural Network	84.73	10.9	0
SVM	85.2	11.7	8
Fuzzy-genetic	85.9	11.5	6
CART	87.6	11.2	9
Random Forest	85.6	11.1	5

IV. DISCUSSION

A. Choice of Parameters

As explained in Materials and Methods section, in each machine learning method we varied some parameters in order to contain overtraining and obtain a good generalization capability. Best results are obtained with parameters configuration summarized in Table 2 and in this section we justified our choices and results. For the NN, we selected 5-8 hidden neurons as the best compromise between learning ability and generalization capability; this is conceivable by using a pattern of 12 inputs. The distribution of the three states of HF Type, shown in Table 1, requires a strong learning ability (and so 8 hidden neurons) to prevent the system from training itself to always say "Stable". Regarding SVM we selected the solution of splitting first between "Severe" vs "Others" class, because this choice showed the best results. This means that there is a greater separation of parameters from severe state against others. Regarding the CART, in the cross validation process are produced trees with different split levels that use multiple variables up to a maximum of 5, but BNP, EF and Weight are always present as main variables. This confirms that these three variables are the most important in describing the severity of HF. In predicting HF Type, the CART selects the following split variables: Systolic Blood Pressure, Weight and Age. Not pruning the trees gives an optimal learning ability but very poor results on the test-set. Fuzzy-genetic technique is the one which is more affected by having relatively few patients in the database. Indeed, while the fact of having 12 inputs would force to have a larger number of fuzzy rules, having few patients in the training set requires keeping down the number of rules in order not to compromise the generalization capability. We consider 45 rules and 600 generations as a good compromise. Adding rules or further evolving the algorithm produces over fitting, while too few rules or generations are not sufficient for a correct system training.

Regarding Random Forest, as showed in Materials and Methods section, we obtained better results with m=4; as we have 12 inputs, this figure is in line with the literature that states that a well-balanced value for m is the square root of the number of input features [12]. Class cut off values were set in order to rebalance the database. This is particularly important in determining the HF type because, as shown in Table 1, the database is very unbalanced in these data. The fact that the cut offs with which we obtain best results are "stable-rare-frequent:50-20-30" confirm the known rule that lowering the cutoff makes a class an easy-winner.

B. Choice of Performance Measurement Methods

Cross validation is useful to assess whether results were bound to a special train/test set patients or they are patient-independent. Hold-out method instead permits the evaluation of the various machine learning techniques operating on the same train/test patient subset but results are linked to these special sets, so we chose Cross Validation. Folds are not completely randomly built because, in order to maintain folds independency, we need to not separate follow-ups of the same patient between different folds.

C. Discussion of Results

Random Forest and CART produce good results in severity assessment if compared with other studies that assess HF severity such as [13] that combined two Support Vector Machines (SVM) to classify HF patients in three groups (74.4% Global Accuracy, 78.8% - 87.5% - 65.6% Accuracy to classify Healthy - HF prone - HF respectively). In [14] are used decision tree techniques to classify patients in three groups of severity (Healthy, Moderate, Severe) using Heart Rate Variability measurements (HF vs Normal Subject: 96% Accuracy - Severe vs Moderate: 79.3% Accuracy). The performances of our system are not directly comparable with some HF binary classifiers that distinguish Healthy from HF patients (for example [15]); in this case patient parameters are much more correlated with the desired output (for example BNP levels of a HF Patient is much more high than those of a Healthy patient). As shown in cross validation tables the Standard Deviation in assessing the severity is very high. This means that there are some lucky folds where accuracy is 92% or 100% and some folds where accuracy is <50%. These unlucky folds have an high percentage of "Moderate" patients and this fact cause worse results revealing the difficulty that the system has in classifying patients whose parameters are in a "gray zone". So the system performance is quite fold-dependent, because the system fails in detecting Moderate status. For critical errors the Neural Network is the best, committing no critical error; the Random Forest also performed very well, with only one critical error. For this we consider the Random Forest the winner technique in determining HF severity using as input the data at our disposal. About HF type prediction results are generally better than in severity assessment, but they are quite distorted because of the high asymmetry between the number of patients with "Chronic Stable HF" and those with Frequent or Rare Exacerbations, as shown in Table I. A correct HF prediction function operation would require a more homogeneous database.

D. Advantages and Disadvantages of the Used Algorithms

- NN pros: quick and easy to train, relatively easy to control over-fitting. It shows best performances in classifying severe status. NN cons: does not provide an easily understood representation of the learned knowledge.
- SVM pros: all the advantages of SVM (optimal hyperplane etc.). SVM cons: these are binary classifiers, and combining two SVM for three levels output is an improper use of this type of machine learning technique.
- CART pros: humanly understandable threshold type "IF-THEN" rules are obtained. It shows the best overall performance. CART cons: high risk of over-fitting, in our case a 2 level pruning has improved the generalization ability
- Random Forest pros: possibility of internal control for re-symmetrising the database, by using both the bootstrap technique and setting cut off on classes. Cons: none, in our application.
- Fuzzy-Genetic pros: the knowledge is in the form of humanly understandable "IF-THEN" rules. Fuzzy-Genetic cons: It needs a long and computationally complex genetic evolution: note that performing a cross validation to the fuzzy-genetic method was very time-consuming because a single genetic evolution takes about 3-4 hours to properly produce rules. It requires a very high number of training samples to work properly.

REFERENCES

1. Rosenblatt, Frank (1957), The Perceptron--a perceiving and recognizing automaton. Report 85-460-1, Cornell Aeronautical Laboratory.
2. C. M. Bishop (1995), Neural Networks for Pattern Recognition., vol. Oxford Uni.
3. C. Cortes and V. Vapnik (1995), Support-vector networks, *Machine Learning*, vol. 20, no. 3, pp. 273–297.
4. S. F. Smith (1980), A Learning System Based on Genetic Adaptive Algorithms. PhD dissertation, Department of Computer Science, University of Pittsburgh.
5. L. Breiman, J. H. Friedman, R. A. Olshen, and C. J. Stone (1984), Classification and Regression Trees. Belmont, CA:Wadsworth.
6. L. Breiman (2001), Random Forests, *Machine Learning*, vol. 45, no. 1, pp. 5–32.
7. G. Guidi, E. Iadanza, and M. C. Pettenati (2012), Heart Failure Artificial Computer Aided Diagnosis Telecare System Using Various Artificial Intelligence Techniques. *Congresso Nazionale Di Bioingegneria 2012*. Roma, ISSN: 978-88-555.
8. G. Guidi, E. Iadanza, M. C. Pettenati, M. Milli, F. Pavone, and G. Biffi Gentili (2012), Heart Failure Artificial Intelligence-based Computer Aided Diagnosis Telecare System, *ICOST 2012, Lecture Notes in Computer Science*, vol. 7251, pp. 278–281.
9. G. Guidi, M. C. Pettenati, R. Miniati, and E. Iadanza (2012), Heart Failure analysis Dashboard for patient's remote monitoring combining multiple artificial intelligence technologies, in *2012 Annual International Conference of the IEEE Engineering in Medicine and Biology Society*, pp. 2210–2213.
10. G. Guidi, M. C. Pettenati, R. Miniati, and E. Iadanza (2013), Random Forest For Automatic Assessment Of Heart Failure Severity In A Telemonitoring Scenario, *EMBC2013, in press*.
11. M. Sokolova and G. Lapalme (2009), A systematic analysis of performance measures for classification tasks, *Information Processing and Management*, vol. 45, no. 4, pp. 427–437.
12. L. Breiman at: http://oz.berkeley.edu/users/breiman/Using_random_forests_V3.1.pdf
13. G. Yang, Y. Ren, Q. Pan, and G. Ning (2010), A heart failure diagnosis model based on support vector machine, *IEEE International Conference on Biomedical Engineering and Informatics*, no. Bmei, pp. 1105–1108,.
14. L. Pecchia, P. Melillo, and M. Bracale (2011), Remote health monitoring of heart failure with data mining via CART method on HRV features., *IEEE transactions on bio-medical engineering*, vol. 58, no. 3, pp. 800–4, Mar.
15. F. S. Gharehchopoghi and Z. A. Khalifelu (2011), Neural Network Application in Diagnosis of Patient: A Case Study, *Computer Networks and Information Technology (ICCNIT), 2011 International Conference on*, pp. 245–249.

Self-reporting for Bipolar Patients through Smartphone

P. Berg Andersen[1] and A. Babic[1,2]

[1] University of Bergen/Department of information and media science, Bergen, Norway
[2] Department of Biomedical Engineering, Linkoping University, Sweden

Abstract—Self-reporting of symptoms is widely used and validated in the field of psychiatry, also in the context of bipolar disorder. This paper presents work on a self-reporting system for bipolar patients using a smartphone to gather data from the patient, which is communicated to a server via a secure connection. The data is presented in a web application to a patient for his/hers self-monitoring, and to medical personnel associated with the treatment of the patient. The work described here is part of an ongoing system development and gives insights into the field research and motivation for choosing Life Charting Methodology as a structural element. Leaning on such well accepted and validated therapeutic tools should secure validity and feasibility of the final system that would appear to patients as familiar and easy to use. Consequently, the application is expected to be directly understandable to everyone involved in the treatment. Programming solutions will capture the essence, but will be adjusted to the electronic environment which will be validated for its correctness and user-friendliness.

Keywords—Self-reporting, psychiatry, bipolar disorder, information system development, application, life charting.

I. INTRODUCTION

Bipolar disorder, formerly known as manic-depressive illness, can be described in the following way: "…a brain disorder that causes unusual shifts in mood, energy, activity levels, and the ability to carry out day-to-day tasks." [1]

However, it has been said that people suffering from bipolar disorders can become their own best therapist [2] and the Norwegian guide for treatment of bipolar disorders [3] claims that mood diaries can be of help in treatment. And it has been indicated through experiences of patients that ongoing monitoring is one of several factors of importance for well-being and recovery [4].

There are several methods that are currently making their way into clinical practice and being recommended to patients for use in order to better understand the condition of mood swings. Typically they are introduced in paper form and they become important for the consultation. It seems obvious that there are several problems with charting and self-reporting of the mood in the form of a diary by using a paper format – it is easy to forget it, it takes up space and it becomes incomprehensible to analyze and interpret the data over time. It has actually been shown that utilizing an online web solution named MoodChart.org leads to more days rated and more complete acquisition of data [5].

Even though there are several solutions for smartphone and other solutions available on the web (such as Mood-Chart.org), it is believed that there is a strong need to do research in this area to provide the patients with even better solutions in the context of user interaction and readability of reports. Also, the study aims at investigating the question whether this is of aid to doctor-patient conversation, and whether the process of self-reporting is a useful activity in itself in the process of helping the patient become more aware of his own state. A prototype will be developed to test this hypothesis.

The paper is structured in the following way: firstly, the research framework and the methods used in the mood monitoring process are described, secondly, recent work in the field is presented,, thirdly, an overview of existing solutions are reviewed, and finally the prototype is presented with a discussion and conclusions regarding future work.

II. RESEARCH FRAMEWORK AND METHODS

A solid theoretical foundation and well-tuned software are the main guiding motives of the development, following Nunamaker et. al [6] who define Information Systems (IS) research and describes systems development as a research methodology in IS research. The framework describes how one can use a mixed-methodological approach, combining several research methods with the aid of systems development in order to do research. A development cycle starts with a theoretical foundation and experimenting in real environments, then performing case studies which allows for a competent and comprehensive approach. Repeating cycles could result in refinement of theoretical understanding, and yet more theory may be helpful in setting experimental conditions and evaluation. Different areas for which the system is being built will add their own specifics and demands. This can be applied in the area of psychiatry and in the context of clinical treatment of bipolar disorder.

The intended approach will be to conduct research in the clinical field and come up with the prototype that will be evaluated by both patients and health care givers, applying field and survey studies. This, in turn, may provide answers to our research questions and generate new challenges. One question is how new applications and tools will influence old well established methods and move practitioners to use and interpret results coming from new tools such as smartphones.

A. Applications on the Internet

There are already several tools developed by various vendors, and some of them are free to download or open for use on dedicated web-sites. None seem to be developed in Norway. Patients keep using Internet resources to help themselves and often choose an application that they find useful and are comfortable with. A quick review shows a variety of tools which differ in the amount of information they accept or provide in return (basic mood and summary of the patient reports) to quite elaborate inquires and rather complex graphical summaries. Some of them are based on the well-established methodology (Life Chart Methodology) and the others are more generic in their domain profiling.

A preliminary study of existing systems is being conducted and this is a work in progress. The study will be performed using a user satisfaction questionnaire [7] asking the patients to evaluate existing solutions using an analogue scale to address different dimensions listed here as following:

- Visibility of system status
- Relation between the real world and the system
- Control and freedom for the user
- Consistency and Standards
- Prevention of errors
- Familiarity (something the user recognizes)
- Flexibility and efficiency in use
- Aesthetic and minimalistic design
- Help for users to recognize, diagnose and recover from errors
- Help and documentation

The results will be analyzed to gain insights into patient experiences with the tools/ applications which they have chosen to use.

B. System Development

Currently, scientific articles and knowledge of recent research work are guiding the systems development process. Once the prototype is ready for user testing, it will be tested by patients over a shorter period of time. Before this, the prototype will be tested by a control group. The testing on patients will be done between two therapy sessions, and the aim is to interview the patients and their therapist after this period to find out:

- In what ways is the solution implemented within the prototype perceived to have a therapeutic effect in itself?
- Was the solution of use to guide and make the therapy session more effective?

This will be done in a questionnaire and in semi-structured interviews.

It could be argued that there are several possibilities to analyze the data that is being gathered, but it will certainly have two dimensions: one concerning the information system development and the other concerning the clinical value and potentials for patients regarding self-monitoring.

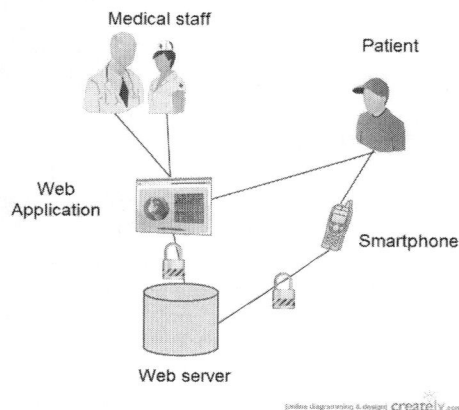

Fig. 1 An overview of the systems architecture.

The overall architecture of the solution is based on:

- An Android smart phone app to collect the relevant data from the patient using notifications to request input from the user at appropriate, configurable times. For instance, at some specific time in the morning, at 10 a.m., the patient is asked to fill in how many hours he/she had slept during the night; the user could reply later during the day, but the remainder will be sent out at some predefined time. For instance, if he gets out of bed at 11 a.m., he can still report his hours of sleep.
- A web based API which the smart phone app communicates with over a secure connection.
- A web application where both the patient and the medical professional can view reports and visualizations of the collected data.

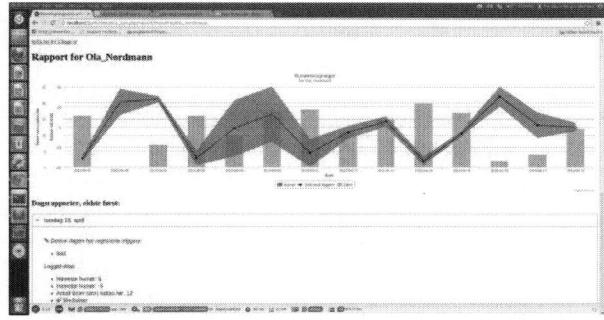

Fig. 2 A screenshot of the web application at an early stage

This outline is inspired by Life Chart Methodology (LCM-p) [8] but it will still take the user evaluation into consideration before the decision will be made to fully implement this methodology. Other software and paper forms use different values; while the patients in the solution proposed in this paper fills in "trigger words", ChronoRecord[9] do not.

A "trigger word" is a word that the patient is logging to define a feeling or event that may trigger the mood swings. It is logged in the prototype on a daily basis and could be aggregated and analyzed. It is expected that by knowing the triggers influencing the mood would help the patient to adjust the behavior and avoid stressors. Triggers are therefore a mechanism of the right potential [10].

There are however other variables in this software that are not included in the prototype in this study. For example, ChronoRecord keeps track of other patient data such as menstrual data [11].

There are also many user interface aspects to consider before the tests, but that could not be addressed in this article. It will however be considered for the prototype later. Besides, it has not yet been decided what data will be gathered and how it will be rated. For instance, the use of a visual analogue scale from -50 to 50 is currently used in reports.

III. RECENT WORK

There are clinical findings that will influence the system development and this overview will brief on the research in the area of life-charting in general, self-reporting and digital solutions for self-reporting and its outcome.

An early example of charting of the mood of bipolar patients was done at the Psychiatric Clinic of Heidelberg, led by Emil Kraepelin between 1891 and 1903 [12]. It had been shown that charting of bipolar patient's mood has been used to aid clinicians in diagnosis and treatment for quite a while.

More recent studies of self-reporting of symptoms have been done using software. Most of the academic field work has been done based on ChronoSheet, using the software ChronoRecord, and Life Chart Methodology. These studies validated the methods, studied different aspects of the disease and even used statistics to predict episodes [13]. But it is unknown what the therapeutic effect was. It is also not very well documented whether it could support the therapeutic consultation between the therapist and the patient.

Work has also been done concerning the effect of keeping a diary to assess patient experiences [14]. For unipolar depression, a wristwatch was utilized to gather actigraphic data from patients in addition to self-reported data [15].

It has been very hard to find articles concerning the therapeutic effect of keeping a diary in itself – as a tool for gaining awareness.

IV. AN OVERVIEW OF SOME EXISTING SOLUTIONS

Table 1 A few similar applications

Name of software	Platform	Description
eMoods Bipolar Tracker	Android	Tracks mood and other variables. Uses e-mail to send reports
Mood Journal Plus	iPhone and BlackBerry	Not reviewed yet
T2 Mood Tracker	iPhone and Android	Not reviewed yet
Moodchart.org	Web and e-mail	Sends a rating mail every day
ChronoRecords	Desktop (probably Windows)	Used in research

Table 1 provides a list of applications available on the market which seems to be often used in research and downloaded by patients. A quick search in the iOS app store shows numerous results. The same happens when searching the Android Market. The three first apps in the table were mentioned in [16].

eMoods have been reviewed to some extent by the authors, and it was found that it does not display the data in a good, understandable way – the graphs are hard to read. Besides, the data is exchanged through e-mail, which could be an unsecure way of transferring the data.

Moodchart.org has been tested by the authors and delivers quite well. It sends a rating mail every day and has been validated in previously mentioned work. It is however not designed for the cell phone, and the rating mail is only sent once a day asking what happened yesterday. The prototype in progress will notify the user at different times.

V. DISCUSSION

System development presented in this paper aims at building a user friendly application that is based on the field knowledge which in practical term meant relying on tools used by medical professionals. That is where validity and continuity of care could be transferred into the new media.

Comfort, ease of use and availability of mobile phones have become central to everyday functioning of the majority of the population and have shown development of reminders, alarms and informative, record-keeping functions. Specialized diaries like the one discussed here will also have their clinical evaluation. It will have to be justified whether a smartphone could be efficient enough, or even more efficient than existing input methods, to spot mood swings and meaningful changes in behavior to alert patients to act in prevention.

Further studies must be done to assess whether there are better ways of providing data. It would also be interesting to consider data from other sources. ESM (Experience Sampling Methodology (ESM) [13] is for instance a nice example of combining patient data - a wristwatch was used to gather self-reported data and sensor data. And an interesting question for further research might be whether it would be better to use a smartphone as the major point of data input or not? Such a solution could be more cost- efficient. In addition, what role could a Smartphone have in the patient-doctor communication? Self-reporting in the paper form has already been accepted and is often summarized before the treatment. System development will be carried out considering these aspects and implement the system out accordingly.

VI. CONCLUSION

A fully developed prototype will enter the clinical and patient evaluation. The results will be used to assess the mobile application and its potential to be used in the therapy. The prototype will be built on a strong theoretical groundwork and fetch inspiration from existing solutions taking into account information regarding their scope, performance, and user satisfaction, but the prototype will also differ where problems arise in other systems. Our research intends to integrate feasible, effective solutions and avoid shortcomings seen in already exiting solutions. The final system will be evaluated by patients and a control group before its release.

REFERENCES

1. NIMH - what is bipolar disorder? (n.d.). Retrieved April 12, 2013, from http://www.nimh.nih.gov/health/publications/bipolar-disorder/what-is-bipolar-disorder.shtml
2. Haver, B. 1942-, Ødegaard, K. J., & Andreassen, O. A. 1949-. (2012). Bipolare lidelser . Bergen: Fagbokforl. Page 16
3. *Nasjonal faglig retningslinje for utgreiing og behandling av bipolare lidingar*. (n.d.). Retrieved from http://helsedirektoratet.no/publikasjoner/nasjonal-retningslinje-for-utgreiing-og-behandling-av-bipolare-lidingar/Sider/default.aspx
4. Suto, M., Murray, G., Hale, S., Amari, E., & Michalak, E. E. (2010). What works for people with bipolar disorder? Tips from the experts. *Journal of affective disorders*, *124*(1-2), 76–84. doi:10.1016/j.jad.2009.11.004
5. Lieberman, D. Z., Kelly, T. F., Douglas, L., & Goodwin, F. K. (2010). A randomized comparison of online and paper mood charts for people with bipolar disorder. *Journal of Affective Disorders*, *124*(1–2), 85–89.
6. Nunamaker, J. F. J. (1990). Systems Development in Information Systems Research. *Journal of management information systems*, *7*(3), 89–106. Livianos-Aldana & Moreno (2006) Life-Chart Methodology: a long past and a short history. Bipolar Disorders Volume 8 Issue 2 p 200-202
7. Sharp, H., Rogers, Y., & Preece, J. (2007). Interaction design: beyond human-computer interaction . Chichester: John Wiley. Page 242-243
8. Prospective, S., Leverich, G. S., & Post, R. M. (2002). The NIMH L ife C hart M anual for R ecurrent A ffective I llness : The LCM - S / P, (June 1997).
9. Bauer, M., Wilson, T., Neuhaus, K., Sasse, J., Pfennig, A., Lewitzka, U., Grof, P., et al. (2008). Self-reporting software for bipolar disorder: validation of ChronoRecord by patients with mania. *Psychiatry research*, *159*(3), 359–66. doi:10.1016/j.psychres.2007.04.013
10. Proudfoot, J., Whitton, A., Parker, G., Doran, J., Manicavasagar, V., & Delmas, K. (2012). Triggers of mania and depression in young adults with bipolar disorder. *Journal of Affective Disorders*, *143*(1–3), 196–202. doi:http://dx.doi.org/10.1016/j.jad.2012.05.052
11. Bauer, M., Grof, P., Gyulai, L., Rasgon, N., Glenn, T., & Whybrow, P. C. (2004). Using technology to improve longitudinal studies: self-reporting with ChronoRecord in bipolar disorder. Bipolar disorders, 6(1), 67–74. Retrieved from http://www.ncbi.nlm.nih.gov/pubmed/14996143
12. Livianos-Aldana, L., & Rojo-Moreno, L. (2006). Life-Chart Methodology: a long past and a short history. Bipolar Disorders, 8(2), 200–202. Retrieved from http://dx.doi.org/10.1111/j.1399-5618.2006.00301.x
13. Glenn, T., Whybrow, P. C., Rasgon, N., Grof, P., Alda, M., Baethge, C., & Bauer, M. (2006). Approximate entropy of self-reported mood prior to episodes in bipolar disorder. *Bipolar disorders*, *8*(5 Pt 1), 424–9. doi:10.1111/j.1399-5618.2006.00373.x
14. Campbell, J. M. (1992). ADULTS: USE OF DIARIES, (1988), 19–29.
15. Peeters, F., Berkhof, J., Delespaul, P., Rottenberg, J., & Nicolson, N. a. (2006). Diurnal mood variation in major depressive disorder. *Emotion (Washington, D.C.)*, *6*(3), 383–91. doi:10.1037/1528-3542.6.3.383
16. Prociow, P., Wac, K., & Crowe, J. (2012). Mobile psychiatry: towards improving the care for bipolar disorder. *International journal of mental health systems*, *6*(1), 5. doi:10.1186/1752-4458-6-5

Comorbidities Modeling for Supporting Integrated Care in Chronic Cardiorenal Disease

E. Kaldoudi and N. Dovrolis

School of Medicine, Democritus University of Thrace, Alexandroupoli, Greece

Abstract—This paper presents work towards constructing a generic information model for comorbidities management. This involves the following conceptual steps: (a) develop an information model and ontology of the management of disease and comorbidities based on ground medical knowledge; (b) enrich the generic model to reflect current-state-of-the-art medical evidence, which will determine in detail the connections and conditions for such connections for comorbid progression pathways; (c) instantiate the enriched model for each specific patient, coupling patient personal information of a variety of sources. The model is based on the UMLS semantic network. Our specific aim is to address the medical domain of cardio-renal disease and comorbidities. The ultimate goal is to provide the means for patients with comorbidities to take an active role in care processes, including self-care and shared decision making, and also to support medical professionals in understanding and treating comorbidities via an integrative approach.

Keywords—comorbidity management, information modeling, medical ontologies.

I. INTRODUCTION

Comorbidity refers to the presence of one or more disorders in addition to a primary disease or disorder (either independently, or as a consequence of the primary condition or otherwise related) [1]. As approximately half of all patients with chronic conditions, even in a nonelderly population, have comorbidities [2], comorbidity management is a hot topic in current medical literature [3,4]. When addressing disease in the presence of comorbidities, each different medical condition the patient presents should not be viewed independently, but a "patient as a whole" view approach should be followed [5]. This places an emphasis on and extra burden of dealing successfully with all associations, interactions, co-dependencies, implications, adverse events, etc. that occur between different conditions co-presenting at the same patient at the same time, as well as between the different treatment regimens these conditions involve.

Our work aims at developing the technological infrastructure for understanding and managing disease progression pathways and comorbidities trajectories and their dynamics, enriched with up-to-date medical evidence and personalized for the individual patient. This involves the following conceptual steps: (a) develop an information model and ontology of the management of disease and comorbidities based on ground medical knowledge; (b) enrich the generic model to reflect current-state-of-the-art medical evidence, which will determine in detail the connections and conditions for such connections for comorbid progression pathways; (c) instantiate the enriched model for each specific patient, coupling patient personal information of a variety of sources, including medical patient data demographics, general medical status, physiological and activity related real-time signals, information on personal intention, mood, preferences, lifestyle, travel planning, etc. The ultimate goal is to employ the personalized model of comorbidities for shared decision support services targeting personalized education, complex risk calculation for disease progression and comorbidity trajectories, alerts for adverse events of multiple co-existing treatments and personalized planning for monitoring

In this paper we present initial work towards constructing a generic information model for comorbidities. Our specific aim is to address the medical domain of cardio-renal disease and comorbidities as this is a very common, life threatening and costly condition and because it also presents a number of challenges and opportunities for the demonstration of comorbidities patient empowerment and management.

II. BACKGROUND

Chronic cardiorenal disease is the condition characterized by simultaneous kidney and heart disease while the primarily failing organ may be either the heart or the kidney. Very often the dysfunction occurs when the failing organ precipitates the failure of the other. The cardio-renal patient (or the person at risk of this condition) presents an interesting case example for addressing and demonstrating novel patient empowerment services for personalized disease & comorbidities management and prevention for a number of reasons as chronic cardiorenal disease has an increasing incidence and a number of serious (and of increasing incidence) comorbidities.

One of the most important aspects of cardiorenal disease and comorbidities diagnosis and treatment is early detection and aggressive management of underlying causes. Preventing progression to end stage renal and cardiac deficiency may improve quality of life and help save health care costs.

Prevention of the disease includes: lifestyle modification (controlling obesity, diabetes and hypertension), public-health education for reduction of excessive bodyweight, regular exercise, and dietary approaches, control of hypertension, dietary protein restriction and blood-pressure control, proteinuria management, dyslipidaemia management and smoking cessation. Delaying disease progression is crucial and must include patient education and aggressive treatment and management of chronic cardiorenal disease and its comorbidities [6]. However, effective implementation of such strategies will only come when both the general public and the renal community work together towards public awareness and lifestyle management on a personal basis and following an integrated care approach. The notion of integrated care is a central issue in the domain of healthcare. Integrated care has many meanings [7]; the notion of "re-unite parts of a whole" is an underlying commonality, whether integration refers to integration within different healthcare settings or in terms of integration of healthcare, social care, long-term and self-care or even integration of patient management for different conditions. In any case, a patient-centered bottom-up approach is favored [7].

The etiology of comorbidity in relation to the risk factors associated with individual diseases led clinical researchers to propose 13 clinical models of comorbidity (including risk factors) [8,9]. These models formalize the process of risk factors affecting one or more comorbid diseases, the diseases themselves affecting each other in various degrees and even signifying the presence of another previously unrelated condition that plays a significant role. A simplified approach summarizes comorbidity in 5 etiological models [10]:

1. No etiological association between coexisting diseases (the "luck" factor)
2. Direct causation: one of the diseases may cause the others;
3. Associated risk factors: the risk factors for each disease are correlated;
4. Heterogeneity: the risk factors for each disease are not correlated but each one of them can cause either disease;
5. Independence: the presence of the diagnostic features of each disease is actually due to a different distinct disease.

Although a lot of work has been conducted towards a common understanding and integration within the healthcare enterprise, as presented by a recent thorough healthcare standards review [11] a "semantic gap" is revealed. This gap mainly refers to (a) the still missing semantic integration of the personal environment of the patient; (a) the lack of integration amongst clinical guidelines addressing individual medical conditions; and (c) the lack of modeling medical context in an integrative approach.

Many approaches have been proposed for the modeling and management of information in healthcare. One of the common approaches relies on traditional database modeling principles, explicitly representing information that is required to be in the relevant biomedical domain [12]. The other approach [13] is based on ontology modeling, which provides semantic descriptions of the concepts used in the healthcare field under study. The semantic descriptions facilitate the integration and interoperation of independent datasets and applications. Data sources are represented using additional (mediator) layers that can expose the sources in terms of domain ontology, following a local-as-view approach to data integration [14]. The scalability of this viewpoint largely depends on the effort required to create the domain ontology. Some initiatives build it anew [13], while others try to reuse as much as possible [14].

The aim of the work presented here is to use the clinical models of comorbidity in order to create an information model that connects comorbid diseases with their respective risk factors and symptoms, weighing their influence on each other and on the patient's health.

III. TOWARDS A MODEL FOR ADDRESSING COMORBIDITIES

The health environment for the patient and/or the healthy citizen comprises of various coexisting and strongly interlinked entities: (a) individuals, including patients, healthy citizens and healthcare professionals; (b) organizations, including any institutional or organizational entity involved in any way in the healthcare process, e.g. healthcare providers, social services, health insurances, medical research institutions, research projects, pharmaceutical companies, well-being and fitness clubs, etc.; (c) health conditions, i.e. any health or medical condition; and (d) health interventions, including interventions on diet, life-style, therapy and drugs, supporting devices, etc.

The UMLS (The Unified Medical Language System, http://www.nlm.nih.gov/research/umls/) semantic network [15] concepts and relationships have been used a basis for developing the proposed model. This network has 135 semantic types and 54 relationships, and covers with these effectively all concepts and connections in the healthcare domain. The Unified Modeling Language (UML) version 2.4.1 has been used for this work (http://www.uml.org).

The proposed model presented here focuses on health conditions and their inter-relations in the case of comorbidities. In terms of modeling, risk factors and diseases in the case of comorbidity can be viewed as subclasses of the same superclass named 'condition'.

Table 1 Building blocks of the proposed model

Model Entity	UMLS SN Term
Disease	Disease or Syndrome
Risk Factor	Qualitative Concept
Symptom	Sign or Symptom

A condition is a UMLS semantic type 'event'. It is mainly characterized by a start and end times (or groups of such when more than one episode occur). A condition may can be managed (i.e. diagnosed and/or treated) via clinical protocols, that contain a series of diagnostic and therapeutic procedures.

Finally, a condition may be caused by and may cause other conditions. This casual association is subject to one or more factors (referred to as Evidence Reference Values) as described by dynamically changing medical evidence. Such an evidence reference value may have one more value ranges and the likelihood attached to them of causing one or more conditions. The class diagram picturing the above is given in Fig 1.

Table 2 Relationships between model entities

Entities Involved	UMLS SN Relationship
Disease - Diseases	Precedes, co-occurs with, result of, affects, associated with, temporally related to, causes, degree of
Risk Factor – Risk Factor	Co-occurs with, complicates, Evaluation of, issue in, result of, affects, causes, degree of
Risk Factor - Disease	

The generic entity "condition" refers both to a disease as well as a risk factor and is modeled as a parent class. The main entities 'disease' and 'risk factor' and their respective UMLS semantic network types are shown in Table 1. The main relationships between these entities and their corresponding UMLS semantic network associations are listed in Table 2.

Risk factors can be of a general nature and can be traced for every possible disease/disorder. Genetics, demography, life style, social and physical environment all affect a person's health in specific ways. Even more specific ones like biological or medical factors contribute to the onset and course of diseases. Thus, in terms of modeling risk factors correspond to a variety of UMLS semantic network entities as shown in Table 3.

Table 3 Risk factors and their respective UMLS SN Terms

Risk Factor	UMLS SN Term
Genetics	Genetic Function, Family Group
Demography	Age Group, Geographic Area
Lifestyle (e.g. smoking, drinking, nutrition, activity)	Social Behavior, Daily or Recreational Activity, Occupational Activity
Environmental (e.g. air pollution)	Environmental Effect of Humans, Hazardous or Poisonous Substance
Biological risk factor (e.g. gender, cholesterol, obesity)	Organism Attribute, Immunologic Factor,
Health care	Health Care Activity

Based on the above, the five different clinical models of comorbidity can be described in the activity diagram shown in Fig 2.

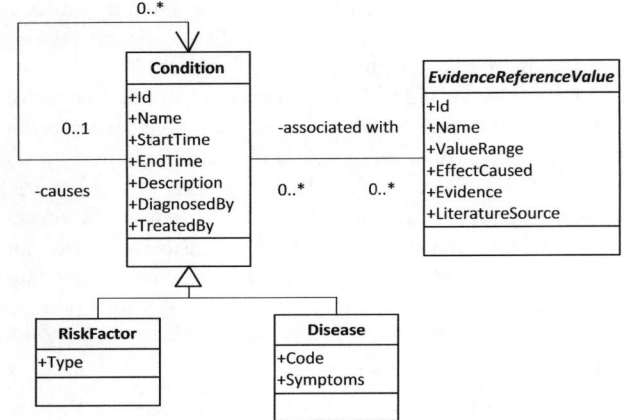

Fig. 1 A simplified class diagram of condition and sub-classes.

IV. CONCLUSIONS

The work presented here addresses comorbidity management via an information model based on standardized technologies in healthcare. The model is expected to allow for semantic interlinking of three types of data (a) medical ground knowledge (b) up-to-date medical evidence and (c) personal patient data in order to create a personalized model of the disease and comorbidities progression pathways and trajectories. Finally, the personalized model of comorbidities will be used for shared decision support services targeting personalized education, complex risk calculation for disease progression and comorbidity trajectories, alerts for adverse events of multiple co-existing treatments and personalized planning for monitoring.

The ultimate goal is to use this personalized model of comorbidities in order to provide the means for patients with comorbidities to take an active role in care processes, including self-care and shared decision making, and also to support medical professionals in understanding and treating comorbidities via an integrative approach.

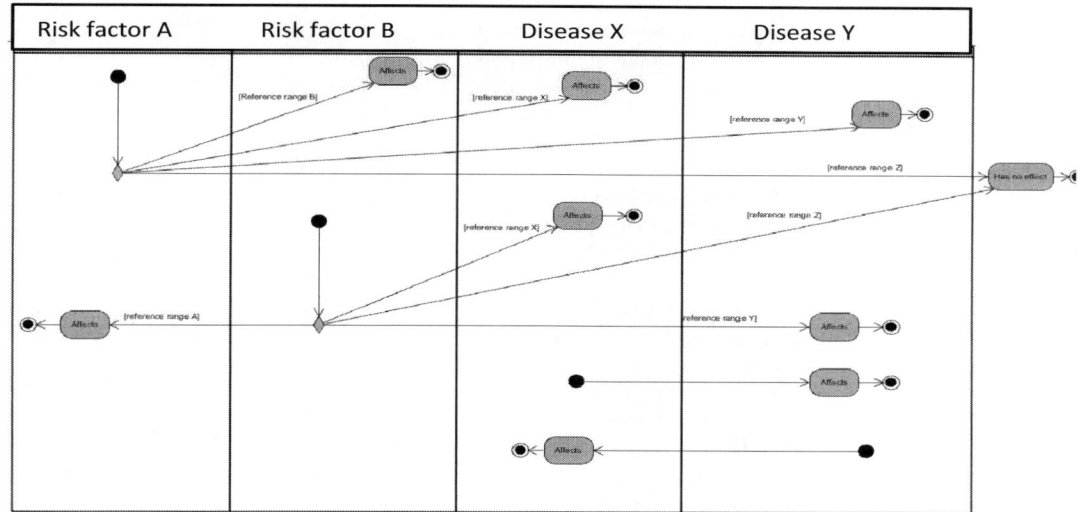

Fig. 2 An activity diagram of the 5 clinical models of comorbidity.

ACKNOWLEDGMENT

This work has been partly funded by the project HELCOHOP, 2012-2015, MIS 375876 (EU and National co-funding).

REFERENCES

1. Akker M van den, Buntix F, Knottnerus JA (1996) Comorbidity or multimorbidity: what's in a name? Eur J Gen Pract, 2:65-70
2. Starfield B, Lemke KW, Bernhardt T et al. (2003) Comorbidity: implications for the importance of primary care in 'case' management. Ann Fam Med. 1:8-14
3. Barnett K, Mercer SW, Norbury M et al. (2012) Epidemiology of multimorbidity and implications for health care, research, and medical education: a cross-sectional study. Lancet 380:37-43
4. Tinetti ME, Fried TR, Boyd CM (2012) Designing healthcare for the most common chronic condition: multimorbidity. JAMA 307:2493
5. Nagarajan V, Tang WH (2012) Management of comorbid conditions in heart failure: a review. Med Clin North Am 96: 975-85
6. Khwaja A, Kossi M El, Floege J et al (2007) The management of CKD: a look into the future. Kidney Int 72:1316-23
7. Kodner DL, Spreeuwenberg C (2002): Integrated care: meaning, logic, applications, and implications – a discussion paper. Int J Integrated Care 2:e12
8. Neale, M.C., Kendler, K.S. (1995) Models of comorbidity for multifactorial disorders. American journal of human genetics 57, 935.
9. Klein, D. N., & Riso, L. P. (1993). Psychiatric disorders: problems of boundaries and comorbidity.
10. Valderas, J.M., Starfield, B., Sibbald, B., Salisbury, C., Roland, M.(2009) Defining Comorbidity: Implications for Understanding Health and Health Services. Ann Fam Med 7, 357–363.
11. Lenz R, Beyer M, Kuhn KA (2007) Semantic integration in healthcare networks, Int J Med Inform 76:201-207
12. Elger BS, Ianvindrasana J, Iacono LL et al (2010) Strategies for health data exchange for secondary, cross-institutional clinical research. Comput Methods Programs Biomed 99:230-51
13. Tsiknakis M, Brochhausen M, Nabrzyski J et al (2008) A semantic grid infrastructure enabling integrated access and analysis of multilevel biomedical data in support of post-genomic clinical trials on Cancer. IEEE Trans Inf Tech Biomed 12:205-17
14. Lenzerini M. (2002) Data integration: a theoretical perspective. ACM SIGMOD-SIGACT-SIGART p. 233-246
15. National Library of Medicine (2009) Chapter 5 - Semantic Networks. UMLS Reference Manual. Bethesda, MD: U.S. National Library of Medicine, National Institutes of Health.

Corresponding Author:

Eleni Kaldoudi
School of Medicine, Democritus University of Thrace
University Campus, Dragana
Alexandroupoli 68100, Greece
kaldoudi@med.duth.gr

# Overall Survival Prediction for Women Breast Cancer Using Ensemble Methods and Incomplete Clinical Data

Pedro Henriques Abreu[1], Hugo Amaro[1], Daniel Castro Silva[1], Penousal Machado[1], Miguel Henriques Abreu[2], Noémia Afonso[2], and António Dourado[1]

[1] Department of Informatics Engineering,
University of Coimbra, Portugal
[2] Portuguese Institute of Oncology of Porto,
Porto, Portugal

Abstract—Breast Cancer is the most common type of cancer in women worldwide. In spite of this fact, there are insufficient studies that, using data mining techniques, are capable of helping medical doctors in their daily practice.

This paper presents a comparative study of three ensemble methods (TreeBagger, LPBoost and Subspace) using a clinical dataset with 25% missing values to predict the overall survival of women with breast cancer. To complete the absent values, the k-nearest neighbor (k-NN) algorithm was used with four distinct neighbor values, trying to determine the best one for this particular scenario. Tests were performed for each of the three ensemble methods and each k-NN configuration, and their performance compared using a Friedman test. Despite the complexity of this challenge, the produced results are promising and the best algorithm configuration (TreeBagger using 3 neighbors) presents a prediction accuracy of 73%.

Keywords—Ensemble Methods, Overall Survival Prediction, Classification, Women Breast Cancer Dataset.

I. Introduction

Nowadays, cancer is one of the leading causes of death worldwide. According to Siegel [1], more than 1 million new cancer cases will be diagnosed and more than 580 thousand cancer deaths will occur in 2013 in the United States alone. Breast cancer is the most common cancer in women, and accounts for 29% of all cancer cases.

In the literature, some prognostic factors were described that influenced clinic decision. Patients with large tumors, not well differentiated, with no expression of hormonal receptors are expected to have worse prognosis and were treated more aggressively. These tumors are particularly prevalent in young patients (women less than 35 years old) and in spite of many developments in this area, this group is still a special one [2]. The research mark in the past two decades was the discovery of HER2, which showed that patients' treatment must be supported by a molecular understanding of breast tumors. This new marker was only detected in almost 20% of the cases but predicts a bad survival. The work of Slamon et al. [3] was the paradigm of this, demonstrating a survival benefit of HER2 blockage (with a drug called trastuzumab) associated with a classical chemotherapy regimen. Despite the early enthusiasm with this discovery, there have been few new prognostic markers in breast cancer after that. The gene signatures, as Mamaprint [4] [5], try to identify patients at high risk of distant recurrence following surgery, based on the analysis of many genes; however, the majority of these gene signatures is not validated for clinical practice nor cost-effective [6], and clinicians still decide based on a set of variables (patient- and tumor-dependent). Over the last 30 years, more than 3 million studies regarding cancer were conducted (values obtained using the ISI web of knowledge). However, in 2007, and according to Cruz and Wishart [7], less than 120 articles were related to cancer prediction/prognosis using soft-computing techniques. In this work, a survival model is presented based on 15 variables that are available in clinical practice. The challenge of this research is to understand if ensemble methods and k-NN algorithm can be used to create accurate predictors, in an oncological center using 847 patient files where 25% of the values are missing. The percentage of missing values found in the patient files reflects the reality of an oncological center that still uses physical patient files and constitutes by itself a good research challenge. The results are based in three different ensemble methods with different strategies in the classification process, and not only they show to be promising, but also open new perspectives for future works in the area.

The remainder of this paper is organized as follows: Section II presents a brief review of the literature, while section III outlines the methodological steps used in this project and section IV presents the collected results. Finally, in section V, the conclusions and some proposals for further studies are presented.

II. Literature Review

Over the years many studies have been developed in the area of cancer. Following the classification proposed by Cruz and Wishart [7], the cancer research area can be divided into cancer prediction and prognosis or cancer detection and diagnosis. As the work presented in this article is based on data collected from an oncological center, the second research area (cancer detection and diagnosis) will not be subject to analysis. In the area of cancer prediction and prognosis, many studies appeared over the last two decades. In spite of the fact that these studies are difficult to compare, mainly because they present different characteristics, such as number of cases to be examined or type of abnormalities, among others, we decided to divide the studies into classification of the tumor based in different types of clinical techniques (X-ray [8], microarray techniques [9]) or prediction, including cancer risk or susceptibility [10], cancer survivability [11] and cancer recurrence [12]. Having the main goal of this project in mind, in the breast cancer area there are still few works that used data mining techniques to predict patient survival [13]. In this work, the authors presented a comparison study that tried to predict patient survival using more than 200 thousand files and three different algorithms: Naive Bayes, Artifical Neural Networks and C4.5. However, this work presented some important drawbacks: authors eliminated incomplete data from the database, which substantially decreased the size of the original database; patient files included patients from different countries and some have more that 40 years, which means that many of the 16 variables used would already be outdated. At the end of the process, none of the algorithms proved to be better than the other. Similar issues are presented in the work presented by Endo et al. [14]. In this work, authors used the same database (provided by SEER - Surveillance Epidemiology and End Results[1]) and performed a comparison between Naive Bayes, Decision Trees (ID3 and J48) and a combination between Naive Bayes and Decision Trees. Also, the authors used only 10 variables in order to characterize the patient (most of then not clinical) and the range of the accuracy results revolves around 80%, which is far from ideal given that they eliminated noise from data at the beginning of the process. Finally, and following the same research line, Wang et al. [15] proposed a new method to predict breast cancer patients' survival using the SEER dataset. Doing a comparison with the other two analyzed studies, Wang improved the results to 90% of accuracy, but with the other detected issues still remaining.

In conclusion, and in spite the fact that some research studies addressed the problem of predicting breast cancer patient survival, none of the studies used only data from one oncology center and updated patient data; none of the studies used exclusively clinical variables; and none of the studies used incomplete data and ensemble methods in the prediction process, which constitutes the main contributions of this project.

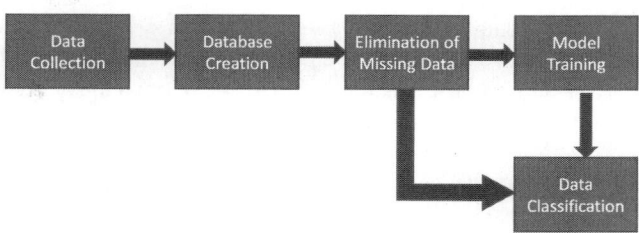

Fig. 1 Project Architecture

III. METHODOLOGY

The goal of this work is to predict the overall survival of women with breast cancer having as a base a dataset composed by more than 840 patient files. Generically, the goal is to identify which is the ensemble method that presents better performance classifying a breast cancer dataset with incomplete data. Having this goal in mind, 5 steps were defined (Figure 1):

1. **Data Collection:** The data was collected by a team composed by 4 medical doctors that collected information from 847 patient files with breast cancer over 2 months from the same oncological center. Also, it is important to state that two other medical doctors performed a cross validation in the collected data in order to minimize the error in this process. Each patient was characterized by 15 variables, including age, tumor site and topography, contralateral breast involvement, tumor stage (according to [16]), variables included in TNM classification (T: tumor size, N: nodes involved, M: metastasis), histological type, degree of differentiation, expression of hormonal receptors, expression of HER2 and type of treatment (including type of surgery, chemotherapy regimen, type of hormonotherapy, if applied).

2. **Database Creation:** After selecting and processing the patient files, a dataset was created to store patient data. Also in this step, a team of two medical doctors performed the cross validation in the stored data.

3. **Elimination of Missing Data:** As often happens in clinical environments, some processes did not contain all patient information. This can be a result of many factors as explained in [17]. Analyzing more deeply the dataset produced in the previous step, we observe that 25%

---

[1] Available at http://www.seercancer.gov

of the data was missing. To solve this problem, many strategies can be used, e.g omitting the instances with missing values, which is far from ideal, or using an algorithm in order to complete such data. In this project, the k-NN algorithm with four distinct values for neighbor (3, 5, 10 and 20) was used to complete the missing data by detecting similarity between data. The choice of the algorithm is based in its implementation simplicity [18] and its good performance in such contexts [17]. At the end of this process, the neighbor value that minimizes classifier error will be detected. Finally, it is important to state that by the end of this step the dataset was split into two groups: the first group, composed by 240 randomly selected patients, will be used as new instances in the classification process; the second group (the remaining 607 patients) will be used in the next step (model training).

4. **Model Training:** To construct a classifier model, three ensemble methods were tested. Over the last decade, ensemble methods have proven themselves to be very effective and extremely versatile in a broad spectrum of problem domains and real-world applications [19]. For this project, three distinct methods from distinct data mining families were used: TreeBagger, LPBoost and Subspace.

5. **Data Classification:** After the construction of the ensemble models, the dataset with 240 patients (produced at the end of step 3) was used to analyze the classification performance of each ensemble method.

IV. EXPERIMENTAL RESULTS

To produce the experimental setup of this work, the dataset produced in step 5 (explained in the previous section) was used. Due to the high percentage of missing data, the k-NN algorithm was used with four distinct values (3, 5, 10 and 20) for k, which produces not one but four distinct datasets to test. As 3 ensemble methods were used, that resulted in 12 distinct comparisons. To compare the 3 ensemble methods, the Friedman rank test was used. The averages of the results of each of the four configurations per algorithm were compared. The 240 patients in each of the four produced datasets were divided into 12 distinct groups (each group containing 20 randomly selected patients). The obtained ranks are shown in Table 1, where 'Number of NN' means number of nearest neighbors used and the number of the ranks varies between 1 (highest accuracy) and 12 (lowest one). Finally, in the case of a draw, average ranks are assigned [20]. Following the work presented by Demsar [20], and as $N > 10$ (number of split groups – 12) and $k > 5$ (number of classifiers used – 12) the proposed F_f value was calculated (7.34) and compared to the F distribution $F(0.05) = 2.69$. As a consequence, the null hypothesis of equivalence between the twelve predictors is rejected. Comparing the twelve configurations (four for each ensemble) for a 5% significance level using the Nemenyi test [20], it was possible to obtain CD = 4.810366. The CD is the critical value for the difference of mean ranks between the twelve predictors. It was proved that TreeBagger (NN–3) and LPBoost (NN–20) performed better than TreeBagger (NN–5) and Subspace (NN–5 and 10). Also, Subspace (NN–20) presented better performance comparing to TreeBagger (NN–5) and Subspace (NN–10). Finally, TreeBagger (NN–20) presented better performance than Subspace (NN–10).

Regarding to the classification performance and attending exclusively to the two algorithms that presented the highest mean in the Friedman Table (Table 1), TreeBagger (NN–3) presented 73% and LPBoost (NN–20) presented 70% of median concerning to hit rate in classification process, which constitutes a good and promising result, attending to the high percentage of missing data (25% of the values). Albeit previous results [14] [15] attained higher accuracy ratings, in these previous studies entries with missing values were eliminated, which obviously leads to better performance. However, it is important to stress that in the real world most of the patient files will be incomplete and therefore those studies do not reflect what can actually be achieved in practice, but rather what can be achieved in an idealistic scenario. This study reflects what can be achieved in the real world and presents a valid solution for overcoming the problem of missing data.

V. CONCLUSIONS AND FUTURE WORK

In this research work, an overall survival prediction approach for the most common cancer pathology in women (breast cancer) was presented. Based on 847 patients files collected at the same oncological center, the performance of three ensemble methods was compared. The results showed that, even with a high percentage of missing values (around 25%), it is possible to obtain good results in the prediction of overall survival.

Further developments in this research project shall focus in several distinct areas: increase the number of patients or expand the study to predict the disease free survival. The first identified direction will have huge similarity with the presented project concerning to the collection of the data and it will be very interesting if the new study focuses exclusively in a group of patients, e.g. younger ones rather than englobing patients with a wide range of ages, allowing the identification of features that most influence survival in those groups. The second future direction consists in expanding this study to predict the disease free survival. Nowadays, and fortunately

Table 1 Ranks of the Friedman test for the three ensemble method and k-NN configuration (Number of NN) for each of the 12 groups of patients. The last column presents the mean ranking accross the 12 groups

Ensemble Method	Number of NN	Groups												MEAN
		1	2	3	4	5	6	7	8	9	10	11	12	
TreeBagger	3	2.5	1	3	5.5	1.5	6.5	5.5	3	2	5.5	1	3.5	3.375
	5	12	12	11.5	12	10.5	4.5	7.5	11.5	8	2.5	11	11	9.500
	10	2.5	5.5	1	10.5	10.5	1	12	10	2	2.5	5	12	6.208
	20	5.5	4	3	2.5	5.5	12	1	2	4.5	5.5	11	3,5	5
LPBoost	3	2.5	5.5	6	8.5	1.5	6.5	9	4.5	4.5	9	6.5	10	6.167
	5	7.5	2.5	6	5.5	9	9.5	10	8	8	9	11	6	7.667
	10	7.5	11	9	1	4	8	3	7	2	5.5	2	8	5.667
	20	2.5	2.5	3	2.5	7	3	2	6	6	5.5	3.5	5	4.042
Subspace	3	9	8	6	7	5.5	4.5	7.5	4.5	11	9	6.5	1.5	6.667
	5	11	10	10	8.5	8	9.5	5.5	9	8	11	8	8	8.875
	10	10	9	11.5	10.5	12	11	11	11.5	12	12	9	8	10.625
	20	5.5	7	8	4	3	2	4	1	10	1	3.5	1.5	4.208

in some cases, the survival of a cancer patient is very good; however, some tumors recur over time. Because of that, it is important to know what the disease free survival of a patient would be. At the end of this study, a new study can emerge in the area of optimization problems. Combining overall survival and the free survival, the goal is to find the treatment that optimizes both previously defined target functions, supporting the clinician in his treatment decision.

ACKNOWLEDGEMENTS

This work is partially supported by iCIS project (CENTRO-07-ST24-FEDER-002003) which is co-financed by QREN, in the scope of the Mais Centro Program and FEDER.

REFERENCES

1. Siegel R, Naishadham D, Jemal A. Cancer statistics, 2013 *A Cancer J. for Clinicians*. 2013;63:11–30.
2. Banz-Jansen C, Heinrichs A, Hedderich M, et al. Are there changes in characteristics and therapy of young patients with early-onset breast cancer in Germany over the last decade? *Archives of Gynecology and Obstetrics*. 2013:1-5.
3. Slamon D, Eiermann W, Robert N, et al. Adjuvant Trastuzumab in HER2-Positive Breast Cancer *The new England J. of Medicine*. 2011:1273–1283.
4. Mesquita JM Bueno, Harten WH, Retel VP, et al. Use of 70-gene signature to predict prognosis of patients with node-negative breast cancer: a prospective community-based feasibility study (RASTER) *Lancet Oncology*. 2007;8:1079–87.
5. Slodkowska EA, Ross JS. MammaPrint 70-gene signature: another milestone in personalized medical care for breast cancer patients *J. of Expert Review of Molecular Diagnostics*. 2009;9:417–422.
6. Williams C, Brunskill S, Altman D, et al. Cost-effectiveness of using prognostic information to select women with breast cancer for adjuvant systemic therapy *J. of Health Technology Assessment*. 2006;10:1–204.
7. Cruz J A, Wishart D S. Applications of Machine Learning in Cancer Prediction and Prognosis *J. of Clinical Informatics*. 2017;2:59–77.
8. Vasantha M, Bharathi V, Dhamodharan S. Medical Image Feature, Extraction, Selection And Classification *International J. of Eng. Science and Technology*. 2010;2:2071–2076.
9. X Ruan J Wang, Li H, Li Xiaoming. A Method for Cancer Classification Using Ensemble Neural Networks with Gene Expression Profile in *Conference on Bioinformatics and Biomedical Eng.*:342–346 2010.
10. Dumitrescu R, Cotaria I. Understanding breast cancer risk where do we stand in 2005? *J. of Cellular and Molecular Medicine*. 2005;9:208–211.
11. Futschik M E, Kasabov N, Reeve A, Sullivan M. Prediction of clinical behavior and treatment for cancers *J. of Applied Bioinformatics*. 2003;2:53–58.
12. Fan Q, Zhu C-J, Yin L. Predicting breast cancer recurrence using data mining techniques in *International Conference on Bioinformatics and Biomedical Technology (ICBBT)*:310–311 2010.
13. Sarvestani A S, Shiraz I, Safavi A A, Parandeh N M, Salehi M. Predicting breast cancer survivability using data mining techniques in *2nd International Conference on Software Technology and Engineering (ICSTE)*:227–231 2010.
14. Endo A, Shibata T, Tanaka H. Comparison of Seven Algorithms to Predict Breast Cancer Survival *Biomedical Soft Computing and Human Sciences*. 2008;13:11–16.
15. Wang K-M, Makond B, Wu W-L, Wang K-J, Lin Y S. Optimal data mining method for predicting breast cancer survivability *International J. of Innovative Management,Information*. 2012;3:28–33.
16. Edge S B, Byrd D R, Carducci M A, et al. , eds.*AJCC Cancer Staging Handbook*. Springer-Verlag New York Inc. 2009.
17. Twala B, Cartwright M, Shepperd M. Comparison of various methods for handling incomplete data in software engineering databases in *International Symp. on Empirical Software Eng.*, 2005:10 pp 2005.
18. Jain A. Data clustering: 50 years beyond K-means *Pattern Recognition Letters*. 2010;31:651–666.
19. Zhang C, Ma Y. , eds.*Ensemble Machine Learning*. Springer-Verlag New York Inc. 2012.
20. Demšar J. Statistical Comparisons of Classifiers over Multiple Data Sets *J. of Machine Learning Research*. 2006;7:1–30.

Author: Pedro Henriques Abreu
Institute: Center for Informatic and Systems (CISUC)
Street: Pólo II, Pinhal de Marrocos
City: Coimbra
Country: Portugal
Email: pha@dei.uc.pt

Automatic Blood Glucose Classification for Gestational Diabetes with Feature Selection: Decision Trees vs. Neural Networks

E. Caballero-Ruiz[1], G. García-Sáez[1], M. Rigla[2], M. Balsells[3], B. Pons[2], M. Morillo[2], E.J. Gómez[1], and M.E. Hernando[1]

[1] Polytechnic University of Madrid, Bioengineering and Telemedicine Centre, Madrid, Spain
[2] Hospital de Sabadell, Endocrinology and Nutrition Dept. CSPT, Sabadell, Spain
[3] Hospital Mutua de Terrassa, Endocrinology and Nutrition Dept., Terrassa, Spain

Abstract—Automatic blood glucose classification may help specialists to provide a better interpretation of blood glucose data, downloaded directly from patients glucose meter and will contribute in the development of decision support systems for gestational diabetes. This paper presents an automatic blood glucose classifier for gestational diabetes that compares 6 different feature selection methods for two machine learning algorithms: neural networks and decision trees. Three searching algorithms, Greedy, Best First and Genetic, were combined with two different evaluators, CSF and Wrapper, for the feature selection. The study has been made with 6080 blood glucose measurements from 25 patients. Decision trees with a feature set selected with the Wrapper evaluator and the Best first search algorithm obtained the best accuracy: 95.92%.

Keywords—Classification, decision support, diabetes, decision trees, neural networks.

I. INTRODUCTION

Pregnancy is associated with changes in insulin sensitivity which may lead to changes in plasma glucose levels [1]. Gestational Diabetes Mellitus (GDM) is defined as glucose intolerance with onset or first recognition during pregnancy. Approximately 7% of all pregnancies (ranging from 1 to 14%, depending on the population studied and the diagnostic tests employed) are complicated by GDM [2]. Several adverse outcomes are associated with it, as preeclampsia, fetal macrosomia, perinatal mortality or neonatal respiratory problems and metabolic complications. Although most cases resolve with delivery, the woman maintains a more elevated risk of developing type 2 diabetes in the future, and this chronic hyperglycemia is associated with long-term damage, dysfunction, and failure of different organs, especially the eyes, kidneys, nerves, heart, and blood vessels [2].

Improving maternal glycemic control can reduce the risk of GDM complications, so patients should self-monitor their blood glucose (BG) levels with a glucose meter, and write their measurements down in a control book, along with information about intakes for a clinician to check it over once a week. The specialist determines the best treatment which consists in nutritional prescription, recommendation to practice physical activity and, if it is necessary, insulin administration.

Telemedicine in combination with Decision Support tools (DST) can improve GDM outcomes [3,4] without increasing clinician's workload [5]. Our final research goal is to develop intelligent tools integrated in a telemedicine system that allows control of GDM automatically, guarantying glucose control objectives consecutions and unnecessary in person visits to the health care center. DST can improve GDM treatment by helping the specialist in the control book inspection. These tools, following the expert indications, can preprocess the monitoring data contained in the control book, and determine which patient is evolving satisfactorily and which one needs a deeper examination by the specialist. They can also be integrated into a telemedicine system since the current glucose meters allow data download. Patients can send their BG levels directly to the system to be analyzed, and according to this information, the specialist will decide the corresponding treatment.

Automatic analysis of glucose meter files have to deal with the problem of lack of intake information associated to the measurements. Any DST requires to know whether the measurement was taken in breakfast, lunch or dinner time and if it is a pre-prandial or a postprandial measurement. However, most of the available glucose meters do not allow registering this information, or even if they do, patients forget to introduce these data. This information is essential to the specialist in order to evaluate the state of the patient so an automatic blood glucose classifier should be developed.

This paper presents a comparison of two well known supervised machine learning algorithms [6,7]: Decision trees and Neural networks (NN) for automatic BG classification. Different feature selection (FS) methods have also been compared in order to select the optimum feature set.

II. MATERIALS AND METHODS

Both learning algorithms, decision trees and NN, have been combined with 9 different feature sets obtained applying 6 different FS methods to our data set. A total of 18 classifiers have been built, which performance has been evaluated testing their accuracy.

A. Glycemic Data

The data set (DS) consists of 6080 BG measurements from 25 patients, who were told to measure their BG with a glucose meter at least 4 times a day, in a fasting state and after the 3 main meals: breakfast, lunch and dinner. However some patients control their BG levels more often: before or/and between meals, at night or repeat some of them. They also wrote down the results in their control book, during a period ranging from the diagnosis date until the delivery date.

B. Data Preprocessing

a) Inputs

In order to procure the best classifier more features were obtained from the ones available in the glucose meter memory file, since in a previous study [8] we observed that classifiers accuracy improved with a large number of features. Directly from the memory file we acquire three features from each measurement: *"date"* and *"time"* when the measurement was taken and the *"bg"* concentration in mg/dL. We calculated another 17 features explained below.

From the *"date"* feature we calculate 5 more features: *"day"*, *"month"*, *"doy"*, *"dow"* and *"workable"*, related respectively to the day and month of the date, the day of the year, day of the week and if the day is workable or not, as schedules and eating habits may change on weekends.

In a previous study [9], an expert determined that the insulin bolus administered close to a BG measurement was an important input to decide the measurement mealtime, which is our objective. We obtained 7 attributes related to insulin from the clinical history: *"insulin"*, a boolean indicating if the patient has insulin treatment or not, *"insulin_type"*, the type of the insulin treatment, *"rib"*, *"ril"*, *"rid"*, *"si1"* and *"si2"* representing respectively the breakfast, lunch and dinner rapid insulin dose and the night and morning slow insulin dose. According to the same study, the time difference with the previous measurement was important too, so we calculated these 2 features related to that: *"interval_prev"*, and *"interval_post"*, which are the time difference with the previous and subsequent measurement. Counting on consecutive measurements BG values can provide information of whether an intake has taken place, so features *"bg_prev"* and *"bg_post"* have been calculated.

Finally we calculate the feature *"intake"*, which represents the most probable intake according to the patient schedules. It may have three possible values: breakfast, lunch or dinner and is obtained grouping each patient measurement in 3 subgroups according to *"bg"* and *"time"* attributes. This is done implementing a cluster in Octave [10] using the K-means algorithm [11].

In summary, we have a total of 20 features: *"date"*, *"day"*, *"month"*, *"doy"*, *"dow"*, *"workable"*, *"time"*, *"interval_prev"*, *"interval_post"*, *"bg_prev"*, *"bg"*, *"bg_post"*, *"insulin"*, *"insulin_type"*, *"rib"*, *"ril"*, *"rid"*, *"si1"*, *"si2"* and *"intake"*. We applied the "Remove useless" Weka filter to our DS to remove features that do not vary at all or that vary too much. The attribute *"slow ins 1"* is removed because none of our patients has that kind of treatment, leaving a total of 19 features.

b) Outputs

Ten different output classes have been used for the measurements classification: *"break-prep"*, *"break-post"*, *"lunch-prep"*, *"lunch-post"*, *"dinner-prep"* and *"dinner-post"*, corresponding to main meals pre-prandial and post-prandial measurements; *"morning"*, *"afternoon"*, *"night"* and *"repeated"* corresponding to other measurements patients can make. The DS was labeled according to patients annotations contained in their control book.

C. Machine Learning Algorithms

The two learning algorithms used are the C4.5 Quinlan decision tree [12] and a Multilayer Perceptron (MLP) neural network [13]. The first one is characterized by good accuracy in a wide range of problems in addition to producing a comprehensible structure summarizing the knowledge it induces. It is also robust and fast, and it may degrade significantly its performance when dealing with many irrelevant features [14]. NN present lower accuracy than decision trees but are more robust. Another disadvantage of NN is that they work as a black box system where inputs and outputs are known but the output function is unknown. The architecture chosen has been MLP with 3 layers, input, hidden and output layers.

D. Feature Selection

In order to select potentially relevant features from the ones we calculated, we have tested 6 different FS methods, combining three searching algorithms with two evaluators.

a) Evaluators

There are two main approaches for FS evaluation: wrappers and filters. We tested one of each approach, a wrapper and the Correlation based Feature Selection (CFS) which uses the filter approach:

Wrapper: Evaluates attribute sets by using a learning scheme. Cross validation is used to estimate the accuracy of the learning scheme for a set of attributes [14]. It is very computationally intensive.

CFS: Uses the Evaluates the worth of a subset of attributes by considering the individual predictive ability of each feature along with the degree of redundancy between them. Subsets of features that are highly correlated with the class while having low intercorrelation are preferred [15].

b) Searching Algorithms

We tested three searching algorithms:

Greedy: Performs a greedy forward search through the space of attribute subsets. It starts with no attributes and stops when the addition of any remaining attributes results in a decrease in evaluation. It can also produce a ranked list of attributes [16].

Best First: Searches the space of attribute subsets by greedy hillclimbing augmented with a backtracking facility. Setting the number of consecutive non-improving nodes allows controlling the level of backtracking done. It considers all possible single attribute additions [16].

Genetic: Performs a search using the simple genetic algorithm described by Goldberg [17].

E. Evaluation Method

FS and classifiers performance evaluation have been executed in an Intel(R) Core(TM) i7-2600 CPU @ 3.40GHz using the Weka 3.6.9 [16] tool, because it provides the algorithms implementation we needed. Cross validation evaluation method [7] has been used for both tasks, with 3 folds for FS due to the wrapper execution time, and 10 folds for classifiers evaluation.

III. RESULTS

A. Classifiers Accuracy

Table 1 shows classifiers accuracy with each feature subset. The first three rows show results with the 3 initial features available in the glucose meter, with the features selected by the expert and with all the features we calculated in the preprocessing. In the first column appears the code to identify the FS method used, in the second column the FS evaluator, in the third column the FS search algorithm and in the last columns the C4.5 and MLP accuracy with the features selected by each FS method.

Table 1 Classifiers accuracy

Code	FS Eval.	FS Search Alg.	Learning Algorithm C4.5	Learning Algorithm MLP
GlucoMeter.	-	-	90,905%	86,875%
Expert	-	-	92,007%	87,823%
All	-	-	94,885%	93,273%
I	CFS	Greedy	95,395%	93,470%
II	CFS	Best First	95,395%	93,470%
III	CFS	Genetic	95,395%	93,470%
IV	Wrapper	Greedy	**95,921%**	94,340%
V	Wrapper	Best First	**95,921%**	**94,408%**
VI	Wrapper	Genetic	95,839%	94,079

Table 2 shows the number of features contained in each feature sets obtained with the feature selection methods showed above. Last columns shows which features are the ones selected.

Table 2 Feature selection (date[1], day[2], month[3], doy[4], dow[5], workable[6], time[7], interval_prev[8], interval_pos[9], bg_prev[10], bg[11], bg_post[12], insulin[13], insulin_type[14], rib[15], ril[16], rid[17], si2[18], intake[19])

Code	Nº of Features C4.5	Nº of Features MLP	Features selected C4.5	Features selected MLP
GlucoMeter	3	3	1,7,11	1,7,11
Expert	4	4	7,8,11,13	7,8,11,13
All	19	19	DS	DS
I	6	6	1,7-9,11,19	1,7-9,11,19
II	6	6	1,7-9,11,19	1,7-9,11,19
III	6	6	1,7-9,11,19	1,7-9,11,19
IV	**12**	11	**1,2,7-9,11,13-17,19**	1,7-11,14,15,18,19
V	**12**	13	**1,2,7-9,11,13-17,19**	**1,7-15,17-19**
VI	10	13	1,7-9,11,13-16-19,19	1,7-16,18,19

IV. DISCUSSION

We observed that C4.5 achieves higher accuracy than MLP in all cases. Adding features to the initial three available in the glucose meter increases accuracy in both learning algorithms, though in MLP the improvement is higher. Adding 1 feature, we observed an improvement of 1.1% in C4.5 and 1% in MLP which rises to 4% in C4.5 and 6.4% in MLP when adding the rest of features we calculated in the preprocessing. Applying feature selection (Wrapper + Best First) to these features we achieved a total improvement of 5% in C4.5 and 7.5 % in MLP.

CFS evaluator selects the same features regardless of the search algorithm and achieves less accuracy than wrapper (95.3% vs. 95.7% for C4.5 and 93.5% vs. 94.2% for MLP, in average). Wrapper execution time for MLP is very high,

72hours, while the execution of different alternatives for the C4.5 or the CFS takes less than 10 minutes.

BestFirst and Greedy achieved same results for C4.5 because they behave similarly, though the first one is a bit more thorough search technique.

The best accuracy results in both learning algorithms have been obtained with the feature selection method consisting in the Wrapper evaluator and the BestFirst search algorithm. In our case, increasing the feature search effort improved classifiers performance, though it is not always like that because of the bias-variance tradeoff [14].

V. CONCLUSION

C4.5 achieves higher accuracy than MLP, it is much faster to train and for feature selection, in addition of being more understandable for clinicians. The FS method consisting in wrapper evaluator and Best first search algorithm has proved to find the optimum feature set for the C4.5 and to achieve the best accuracy for our data set: 95,92%.

Automatic blood glucose classification is essential for automatic glucose meter file inspection. This not only will save pregnant women from problematic and unnecessary displacements, as they can download their measurements and send them to the system at any time, but also avoid the risk of making mistakes or oversights when transcribing the glucose meter results. Automatic BG classification will contribute to the development of DSS that can help in BG data interpretation counting with more exact and more available data.

ACKNOWLEDGMENT

This work has been funded by the Spanish grant "SINEDiE" (PI10/01125), co-funded by FEDER and the FP7 project "MobiGuide" (FP7-287811). We would like to thank Hospital de Sabadell and Hospital Mutua de Terrasa for their collaboration.

REFERENCES

1. IDF Clinical Guidelines Task Force. (2009). Global guideline on pregnancy and diabetes. Brussels: International Diabetes Federation.
2. American Diabetes Association (2012) Diagnosis and classification of Diabetes Mellitus. Diabetes Care, 35(1):64–71 DOI:10.2337/dc12-s064.
3. Dalfrà, M. G., Nicolucci, A., & Lapolla, A. (2009). The effect of telemedicine on outcome and quality of life in pregnant women with diabetes. J Telemed Telecare, 15, 5:238–42. DOI:10.1258/jtt.2009.081213.
4. Ferrara, A., Hedderson, M. M., Ching, J., Kim, C., Peng, T., & Crites, Y. M. (2012). Referral to telephonic nurse management improves outcomes in women with gestational diabetes. Am. J. Obstet. Gynecol. 6, 206:491. DOI:10.1016/j.ajog.2012.04.019
5. Klonoff, D. C., & True, M. W. (2009). The missing element of telemedicine for diabetes: decision support software. J Diabetes Sci Technol,5,3: 996–1001
6. Kotsiantis, S. B., Zaharakis, I. D., & Pintelas, P. E. (2007). Machine learning: a review of classification and combining techniques. Artif Intell Rev, 3, 26:159–190. doi:10.1007/s10462-007-9052-3
7. Geurts, P., Irrthum, A., & Wehenkel, L. (2009). Supervised learning with decision tree-based methods in computational and systems biology. Mol Biosyst, 1–18.
8. Caballero-Ruiz E, Garcia-Sáez G, Rigla Cros M, Balsells M, Pons B, Gómez Aguilera EJ, Hernando Pérez ME (2012). Clasificación de medidas de glucemia en función de ingestas en diabetes gestacional. CASEIB Proc. XXX Congreso Anual de la Sociedad Española de Ingeniería Biomédica, San Sebastián, Spain.
9. García-Sáez G, Alonso JM, Molero J, Rigla M, Martínez-Sarriegui I, de Leiva A, Gómez E, Hernando M (2009) Mealtime Blood Glucose Classifier Vased on Fuzzy Logic for the DIABTel Telemedicine System. Artif Intell Med, Lecture Notes in Computer Science 5651:295-304 DOI 10.1007/978-3-642-02976-9_42
10. GNU Octave at http://gnu.org/software/octave
11. Hartigan JA, Wong MA. (1979). Algorithm AS: A k-means clustering algorithm. Appl Stat-J Roy St C 28: 100-108
12. Quinlan, J. R. (1993). C4.5: Programs for Machine Learning. Morgan Kaufmann.
13. Hykin S (1994). Neural networks: a comprehensive foundation, Macmillan College Publishing Company Inc, New York.
14. Ron Kohavi, George H. John (1997). Wrappers for feature subset selection. Artificial Intelligence. 97(1-2):273-324 DOI 10.1016/S0004-3702(97)00043-X
15. M. A. Hall (1998). Correlation-based Feature Subset Selection for Machine Learning. Hamilton, New Zealand
16. Witten H, Frank E, (2005). Data Mining: Practical machine learning tools and techniques. (2nd ed.)Morgan Kaufmann Publishers, San Francisco.
17. David E. Goldberg (1989). Genetic algorithms in search, optimization and machine learning. Addison-Wesley.

Author: Estefanía Caballero Ruiz
Institute: Polytechnic University of Madrid
Street: Avd. Complutense nº 30, Ciudad Universitaria. 28040
City: Madrid
Country: Spain
Email: ecaballero@gbt.tfo.upm.es

On the Global Optimization of the Beam Angle Optimization Problem in Intensity-Modulated Radiation Therapy

H. Rocha[1], J.M. Dias[1,2], B.C. Ferreira[3,4], and M.C. Lopes[3,4]

[1] INESCC, Coimbra, Portugal
[2] FEUC, Coimbra, Portugal
[3] IPOC-FG, Coimbra, Portugal
[4] I3N, Aveiro, Portugal

Abstract—The beam angle optimization (BAO) problem remains an important and challenging problem in intensity-modulated radiation therapy (IMRT) treatment planning. BAO consists on the selection of appropriate radiation incidence directions and may influence the quality of the IMRT plans, both to enhance organs sparing and to improve tumor coverage. This is a very difficult global optimization problem since it is a highly non-convex continuous optimization problem with many local minima. Many conventional BAO approaches are based on single-beam metrics to solve a relaxed combinatorial formulation of the BAO problem. Typically, the quality of the solutions obtained is not simply related to the final value of an objective function but rather judged by dose-volume histograms or considering a set of physical dose metrics. For that reason, and also due to the fact that the global optimum value is unknown, it is difficult to perceive, in medical physics point of view, how good a solution is or how much could it be improved. In a mathematical point of view, it is difficult to acknowledge how far a solution is from the global optimum. The objective of this paper is to present the difficulties in obtaining near global optimum solutions for the BAO problem, particularly when using single-beam approaches considering discrete subsets of all possible beam angles. The benefits of using a derivative-free approach for a continuous formulation of the BAO problem are discussed using a retrospective treated case of head-and-neck tumor at the Portuguese Institute of Oncology of Coimbra.

Keywords— Intensity-modulated Radiation Therapy, Beam Angle Optimization, Global Optimization, Single-beam Metrics, Derivative-free Optimization.

I. INTRODUCTION

The goal of radiation therapy is to deliver a dose of radiation to the tumor volume to sterilize all cancer cells minimizing the damages on the surrounding healthy organs and tissues. An important type of radiation therapy is intensity-modulated radiation therapy (IMRT), a modern technique where the radiation beam is modulated by a multileaf collimator allowing the irradiation of the patient using non-uniform radiation fields from selected angles aiming to deliver a dose of radiation to the tumor minimizing the damages on the surrounding healthy organs and tissues. The IMRT treatment planning is usually a sequential process where initially a given number of beam directions are selected followed by the fluence map optimization (FMO) at those beam directions. Beam angle optimization (BAO) consists on the selection of appropriate radiation incidence directions and may influence the quality of the IMRT plans, both to enhance better organs sparing and to improve tumor coverage. The BAO problem is quite difficult since it is a highly non-convex optimization problem with many local minima – see Fig. 1. Regardless the evidence presented in the literature that appropriate radiation beam incidence directions can lead to a plan's quality improvement [1], in clinical practice, most of the time, the number of beam angles is assumed to be defined a priori by the treatment planner and the beam directions are still manually selected by the treatment planner in a time-consuming trial and error iterative process.

In most of the previous works on beam angle optimization, the entire range $[0°, 360°]$ of gantry angles is discretized into equally spaced beam directions with a given angle increment, such as 5 or 10 degrees, where exhaustive searches are performed directly or guided by a variety of different heuristics including simulated annealing [2], genetic algorithms [3] or other heuristics incorporating a priori knowledge of the problem [4]. On the other hand, the use of single-beam metrics has been a popular approach to address the BAO problem as well, e.g., the concept of beam's-eye-view [5]. Despite the computational time efficiency of these approaches, the optimality of the solutions proposed cannot be guaranteed since the interplay between the selected beam directions is ignored.

It is well known that, when the BAO problem is not based on the optimal FMO solutions, the resulting beam angle set has no guarantee of optimality and has questionable reliability since it has been extensively reported that optimal beam angles for IMRT are often non-intuitive [6]. Therefore, the optimal FMO solutions will be used both to drive and compare our BAO experiments. Typically, the quality of the

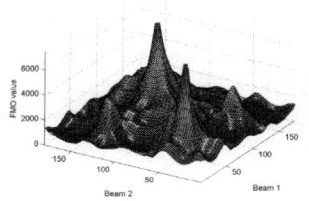 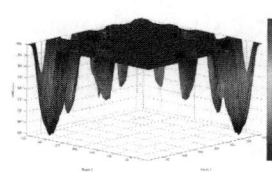

Fig. 1 2-beam BAO surface (left) and truncated surface (right) to highlight the many local minima.

solutions obtained is not simply related to the final value of an objective function but rather judged by dose-volume histograms or considering a set of physical dose metrics. For that reason, and also due to the fact that the global optimum value is unknown, it is difficult to perceive, in a medical physics point of view, how good a solution is or how much could it be improved. In a mathematical point of view, it is difficult to acknowledge how far a solution is from the global optimum.

The objective of this paper is to present the difficulties in obtaining near global optimum solutions for the BAO problem, particularly when using single-beam approaches considering discrete subsets of all possible beam angles. The benefits of using a derivative-free approach for a continuous formulation of the BAO problem are discussed using a retrospective treated case of head-and-neck tumor at the Portuguese Institute of Oncology of Coimbra. The paper is organized as follows. In the next section we describe the BAO problem and the FMO problem formulation used. Section 3 briefly presents the single-beam approaches tested and our derivative-free method proposed. Section 4 presents the obtained results. In the last section we have the conclusion.

II. BEAM ANGLE OPTIMIZATION PROBLEM

A quantitative measure is required to compare the quality of different sets of beam angles. For the reasons presented before, we will use the optimal solution of the FMO problem as measure of the quality of a given beam angle set. A convex penalty function voxel-based nonlinear model is used for the FMO problem [7]. In this model, each voxel is penalized according to the square difference of the amount of dose received by the voxel and the amount of dose desired/allowed for the voxel. This nonlinear formulation implies that a very small amount of underdose or overdose may be accepted in clinical decision making, but larger deviations from the desired/allowed doses are decreasingly tolerated.

The FMO optimal value is used to compare the solutions obtained by single-beam approaches and to drive the derivative-free approach we propose [8, 9, 10]. Our formulation of the BAO problem is briefly presented. Let us consider n to be the fixed number of (coplanar) beam directions, i.e., n beam angles are chosen on a circle around the computed tomography (CT)-slice of the body that contains the isocenter (usually the center of mass of the tumor). In our formulation we consider all continuous $[0°, 360°]$ gantry angles instead of a discretized sample. A basic formulation for the BAO problem is obtained by selecting an objective function such that the best set of beam angles is obtained for the function's minimum:

$$\min \quad f(\theta_1, \ldots, \theta_n)$$
$$s.t. \quad \theta_1, \ldots, \theta_n \in \mathbb{R}^n.$$

Here, the objective $f(\theta_1, \ldots, \theta_n)$ that measures the quality of the set of beam directions $\theta_1, \ldots, \theta_n$ is the optimal value of the FMO problem for each fixed set of beam directions.

III. BEAM ANGLE OPTIMIZATION APPROACHES

A. Single-Beam Approaches

Two different single-beam approaches will be tested. One related to the beam's-eye-view concept and the other similar to the successful strategy used in [11].

The beam's-eye-view concept uses topographic criteria to rank the candidate beam directions. For IMRT, the geometrical considerations are not as important. Some variations of the beam's-eye-view concept consider dosimetric criteria to rank the candidate beam directions selecting those with higher scores [12]. Unlike conventional beams-eye-view (BEV) tools that considers only geometric criteria, beams-eye-view dose metrics (BEVD) evaluate each possible beam direction using a score function that accounts for beam modulation. In IMRT, beam directions are often non-intuitive and have to go through sensitive organs to achieve an optimal compromise between target coverage and organs sparing, which makes the geometrical criteria used by BEV limited. An intensity-modulated beam can intercept a large volume of an organ at risk (OAR) or normal tissue and may not be necessarily a bad beam direction. The dose tolerances of the involved structures should be considered also when constructing a metric for measuring the quality of incident beam directions. Therefore, in IMRT, it is more appropriate to measure the quality of a radiation beam direction using a score function based on dosimetric criteria. A technique based on sensitive structures tolerance dose as a determinant factor for deliverable target dose [5, 12], denoted BEVD, was used to find a set of beams that are not too close to each other and have the largest scores. We should emphasize that a computationally intelligent algorithm should balance the BEVD scores and the beam interplay as a result of the overlap of radiation fields.

The other single-beam strategy tested is similar to the strategy used in [11] and can be described as follows:

1. Consider the 72 candidate beam angles $\{0, 5, \ldots, 355\}$.
2. For $n = 1$ beam direction:
   Find the best beam irradiation direction, i.e., for the 72 beam angles in the candidate beam angle set, find the beam direction with lowest optimal FMO value.
3. For $n = 2$ to $n = 9$ beam directions:
   Find the best set of n directions considering fixed the $n - 1$ beam directions determined in the previous iteration, i.e., find direction n among the remaining $72 - (n - 1)$ beam directions similarly to procedure for $n = 1$.

In the end of this sequential procedure, here denoted by SEQ, we will obtain "optimal" beam angle sets for $n = 1$ to $n = 9$ beam directions. This is a clear advantage of this strategy because it allows simultaneously an implicit optimization of the number of beam angles.

B. Derivative-Free Approach

Our derivative-free approach is based on a pattern search methods (PSM) framework. PSM are derivative-free optimization algorithms that require few function evaluations to progress and converge and have the ability to better avoid local entrapment making them a suitable approach for the resolution of the highly non-convex BAO problem [8, 9, 10].

PSM are directional search methods that use positive bases to move in a direction that produces a decrease in the objective function. The main feature of a positive basis, that motivates PSM, is that for any given vector, in particular for the gradient vector, there is a vector of the positive basis that forms an acute angle with the gradient vector which means that it is a descent direction.

PSM are organized around two phases at every iteration: one that assures convergence to a local minimum (poll), and the other (search) where flexibility is conferred to the method allowing searches away from the neighborhood of the current iterate. Within the search step we use beams-eye-view dose metrics so that directions with larger dose metric scores are tested first improving results and computational time. This pattern search approach [8], denoted PSM-BEVD, was tested along with the single-beam strategies.

IV. NUMERICAL TESTS AND DISCUSSION

A clinical example of a retrospective treated case of head-and-neck tumor at the Portuguese Institute of Oncology of Coimbra is used to highlight the difficulties in obtaining near global optimum solutions for the BAO problem. The patients' CT set and delineated structures were exported via Dicom RT to a freeware computational environment for radiotherapy research. In general, the head-and-neck region is a complex area to treat with radiotherapy due to the large number of sensitive organs in this region (e.g. eyes, mandible, larynx, oral cavity, etc.). For simplicity, in this study, the OARs used for treatment optimization were limited to the spinal cord, the brainstem and the parotid glands. The tumor to be treated plus some safety margins is called planning target volume (PTV).

Our tests were performed on a 2.66Ghz Intel Core Duo PC with 3 GB RAM. In order to facilitate convenient access, visualization and analysis of patient treatment planning data, the computational tools developed within MATLAB and CERR [13] (computational environment for radiotherapy research) were used as the main software platform to embody our optimization research and provide the necessary dosimetry data to perform optimization in IMRT. The dose was computed using CERR's pencil beam algorithm (QIB). To address the convex nonlinear formulation of the FMO problem we used a trust-region-reflective algorithm (*fmincon*) of MATLAB 7.4.0 (R2007a) Optimization Toolbox. We choose to implement the incorporation of BEVD into the pattern search methods framework taking advantage of the availability of an existing pattern search methods framework implementation used successfully by us to tackle the BAO problem [8, 9, 10] – the last version of SID-PSM [14, 15].

Typically, in head-and-neck cancer cases, patients are treated with 5, 7 or 9 equispaced beams in a coplanar arrangement. Therefore, results for the equispaced solution, denoted EQUI, for the BEVD solution, for the SEQ solution and for the PSM-BEVD solution are presented in Table 1 for 5, 7 and 9 beams. For this number of beams we can see that results obtained by PSM-BEVD are slightly better than SEQ solutions and way better than BEVD that struggles to be competitive with the traditional equispaced solutions. For larger numbers of beams the advantage of PSM-BEVD over SEQ is residual, indication that for larger number of beams the optimization process becomes harder. On the other hand, for single-beam approaches, particularly for SEQ, results tend to improve by increasing the number of beams. PSM-BEVD is also very competitive in terms of number of functions evaluations.

Solutions for 2 beams are also presented since the global optimum for the candidate beams $\{0, 5, \ldots, 355\}$ was computed through exhaustive search (see Fig. 1) and we aim to perceive how close solutions are to the global optimum. For 2 beams, the global optimum found in the candidate set $\{0, 5, \ldots, 355\}$ was 591.4. Two interesting conclusions can be withdrawn: the PSM-BEVD solution is better than the

Table 1 BAO results for $n = 2, 5, 7$ and 9 beams.

n	Equi Fval	BEVD Fval	SEQ Fval	SEQ Fevals	PSM-BEVD Fval	PSM-BEVD Fevals
2	1278.6	936.5	624.5	143	588.1	84
5	186.8	189.3	187.5	350	175.3	127
7	173.5	174.8	168.1	483	162.4	158
9	169.2	168.0	155.9	612	154.7	197

global global optimum found in the candidate beam set and the single-beam approaches behave poorly for few beams. It is not straightforward to extrapolate conclusions for more beams but it is expected that differences between the solutions obtained and the global optimum increase since the optimization problem becomes harder with a larger search space to be explored.

V. CONCLUSION

The BAO problem is a continuous global highly nonconvex optimization problem known to be extremely challenging and yet to be solved satisfactorily. Many conventional BAO approaches are based on single-beam metrics to solve a relaxed combinatorial formulation of the BAO problem. We have shown, using an head-and-neck cancer case, that single-beam strategies behave better for larger number of beams. On the other hand, PSM-BEVD has shown to yield solution of superior quality, in particular for smaller number of beams. This feature might be important for prostate or breast cancer cases where few beams are typically used. PSM-BEVD has shown ability to avoid local entrapment and efficiency by converging faster which is of the utmost importance in a busy clinical practice.

ACKNOWLEDGEMENTS

This work was supported by QREN under Mais Centro (CENTRO-07-0224-FEDER-002003) and FEDER funds through the COMPETE program and Portuguese funds through FCT under project grant PTDC/EIA-CCO/121450/2010. This work has also been partially supported by FCT under project grant PEst-C/EEI/UI0308/2011. The work of H. Rocha was supported by the European social fund and Portuguese funds from MCTES.

REFERENCES

1. Das SK, Marks LB. Selection of coplanar or non coplanar beams using three-dimensional optimization based on maximum beam separation and minimized nontarget irradiation *Int J Radiat Oncol Biol Phys.* 1997;38:643–655.
2. Bortfeld T, Schlegel W. Optimization of beam orientations in radiation therapy: some theoretical considerations *Phys Med Biol.* 1993;38:291–304.
3. Dias JM, Rocha H, Ferreira BC and Lopes MC. A genetic algorithm with neural network fitness function evaluation for IMRT beam angle optimization *Cent Eur J Oper Res.* "in press"
4. D'Souza WD, Meyer RR, Shi L. Selection of beam orientations in intensity-modulated radiation therapy using single beam indices and integer programming *Phys Med Biol.* 2004;49:3465–3481.
5. Pugachev A, Xing L. Computer-assisted selection of coplanar beam orientations in intensity-modulated radiation therapy *Phys Med Biol.* 2001;46:2467–2476.
6. Stein J, Mohan R, Wang XH, Bortfeld T, Wu Q, Preiser K, Ling CC, Schlegel W. Number and orientation of beams in intensity-modulated radiation treatments *Med Phys.* 1997;24:149–160.
7. Aleman DM, Kumar A, Ahuja RK, Romeijn HE, Dempsey JF. Neighborhood search approaches to beam orientation optimization in intensity modulated radiation therapy treatment planning *J Global Optim.* 2008;42:587–607.
8. Rocha H, Dias JM, Ferreira BC and Lopes MC. Beam angle optimization for intensity-modulated radiation therapy using a guided pattern search method *Phys Med Biol.* 2013;58:2939–2953.
9. Rocha H, Dias JM, Ferreira BC, Lopes MC. Selection of intensity modulated radiation therapy treatment beam directions using radial basis functions within a pattern search methods framework *J Glob Optim.* "in press"
10. Rocha H, Dias JM, Ferreira BC, Lopes MC. Pattern search methods framework for beam angle optimization in radiotherapy design *Appl Math Comput.* "in press"
11. Breedveld S, Storchi PRM, Keijzer M, Heemink AW, Heijmen BJM. A novel approach to multi-criteria inverse planning for IMRT *Phys Med Biol.* 2007;52:6339–6353.
12. Pugachev A, Xing L. Pseudo beam's-eye-view as applied to beam orientation selection in intensity-modulated radiation therapy *Int J Radiat Oncol Biol Phys.* 2001;51:1361–1370.
13. Deasy JO, Blanco AI, Clark VH. CERR: A Computational Environment for Radiotherapy Research *Med Phys.* 2003;30:979–985.
14. Custódio AL, Vicente LN. Using sampling and simplex derivatives in pattern search methods *SIAM J Optim.* 2007;18:537–555.
15. Custódio AL, Rocha H, Vicente LN. Incorporating minimum Frobenius norm models in direct search *Comput Optim Appl.* 2010;46:265–278.

Author: Humberto Rocha
Institute: INESC-Coimbra
Street: Rua Antero de Quental 199
City: Coimbra
Country: Portugal
Email: hrocha@mat.uc.pt

Effective Supervised Knowledge Extraction for an mHealth System for Fall Detection

G. Sannino[1,2], I. De Falco[1], and G. De Pietro[1]

[1] Institute of High Performance Computing and Networking, CNR, Naples, Italy
[2] University of Naples "Parthenope", Department of Technology, Naples, Italy

Abstract—Fall detection is an important task in telemedicine. In this paper an approach based on supervised knowledge extraction is presented. A fall recordings database is analyzed offline and a set of IF...THEN rules is obtained. This way, also selection of the most relevant features for fall assessment is automatically carried out. The approach is embedded within a real-time mobile monitoring system, and is used to discriminate in real time normal daily activities from falls. If the data collected in real time by wearable sensors of the system allow recognizing a fall, suitable alarms are automatically generated.

Keywords—Knowledge extraction, IF...THEN rules, fall recording, real-time monitoring system, wearable sensors.

I. INTRODUCTION

In this paper we introduce an innovative approach to tackle the problem of fall detection by making use of wearable sensors, so as to understand in real time whether or not they have fallen. In this latter case, an immediate alarm should be sent. This can be very useful in speeding up first aid, should a fall take place.

The main problem here is to accurately discriminate falls from normal daily activities which could somehow resemble falls, such as for instance quickly sitting or lying on a bed, or running, jumping and other similar movements.

In general, to fulfill any given real-time detection task, knowledge should be provided to the intelligent core of the mobile monitoring system about how to discriminate among possible events. For many medical problems knowledge is explicitly available either as a set of rules provided by doctors or as agreed guidelines. Unfortunately, for fall detection none of these kinds of knowledge is available.

Therefore, to efficiently tackle this problem, in our approach knowledge is extracted from raw acceleration data by using a supervised learning methodology.

With respect to existing literature, our approach is innovative in many ways. Firstly, most existing systems are based on investigation of the total acceleration only (for example [1][2][3]). Our approach, instead, takes into account both the three acceleration components along the sensor axes and the total acceleration.

Secondly, a set of parameters is extracted from these four signals to obtain a database onto which artificial intelligence techniques can be applied to perform classification.

Thirdly, and more importantly, our system automatically extracts from that database a set of IF...THEN rules, by means of a supervised learning methodology. This set of rules is extracted offline, and can then be inserted into the knowledge base of a reasoner being part of a control system, so that they can be used in real time. IF...THEN rules are user-friendly and can always explain the reasons why a given decision is taken. This is an enormous advantage with respect to the other existing methods, which behave as black boxes, without providing experts with any explanation about their behavior. It should be remarked that some other papers in literature try to find useful rules, as for example [1], [2]. [4], yet in all of them the rules are provided by the expert who examines the fall recordings, hypothesizes that some parameters are more important, and imagines some rules and some thresholds for them. Moreover, in those papers a very small number of parameters is considered, for example in [2] just three (the average acceleration magnitude variation, the free fall interval, and the free fall average acceleration magnitude) whereas in [1] the total acceleration and its components along the three axes are considered.

A quite similar approach is in [5], yet they lack the third part of our methodology: they do not extract IF...THEN rules over all the database attributes, so they do not get knowledge about how to discriminate falls from non-falls.

Finally, the approach proposed here is embedded within a real-time mobile monitoring system [6] developed at iHealthLab [7] (ICAR-CNR). This system continuously monitors a wide set of vital parameters for patients through the use of wearable sensors, and can help in checking patient's health with respect to many issues at the same time.

II. THE DATABASE

The database exploited in this paper has been provided by Prof. Sebastian Fudickar of the Department of Computer

---

* This work has been partly supported by the project "Sistema avanzato per l'interpretazione e la condivisione della conoscenza in ambito sanitario A.S.K. – Health" (PON01_00850).

Science, University of Potsdam, Germany, and has been built within his activity of developing a fall-detection simulator for accelerometers with in-hardware preprocessing [8].

This data set consists of 86 recordings with simulated falls performed by 3 people, and of nine recordings containing normal activity such as sitting, walking, or dancing.

The data were acquired at a frequency of 800 Hz by using an ADXL 345 sensor. Each recording contains the accelerations recorded for each axis, and the total acceleration.

Starting from this set of four signals, we have extracted a database suitable to undergo classification with the following procedure based on three steps.

A. Annotation

Firstly, for each fall example, the start and the end times of the fall action have been annotated. Similarly, for each normal activity example, suitable start and end times have been chosen so that the delimited time span contains activities very similar to falls, i.e. with high-varying signals.

B. Windowing

Secondly, we have processed raw acceleration data by means of windowing [9]. Windowing technique is typically used to divide the sensor signal into smaller time segments, called windows, and each window will be a separate sample for the database under construction.

The windowing technique used here is that based on sliding windows. Therefore, each portion of signal selected in the previous step has been divided into windows of fixed length and with a fixed overlap size with previous and next windows [10]. Good values for both window size and overlap have been found through preliminary experiments as 800 samples and 400 samples, which correspond to one second and half second respectively. Fig. 2 shows an example of a fall acceleration signal with sliding windows.

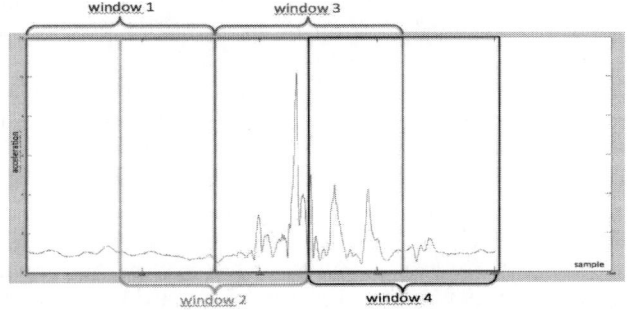

Fig. 1 An example of a fall acceleration signal with windows.

C. Computation

Thirdly, in each window we have computed some statistic values: for each of the four signals (the three acceleration components and the total acceleration) we have extracted:

- the average value in the window
- the standard deviation
- the highest value
- the lowest value
- the range width, i.e. the difference between the highest and the lowest value.

This results in a total of 20 parameters.

As a consequence of these three steps, we have obtained a database which contains a total od 249 samples, 121 of which represent falls and 128 non-falls. Each sample consists of 20 real values, one for each extracted attribute, plus the class, encoded as 1 for non-falls and 2 for falls.

III. KNOWLEDGE EXTRACTION

We wish to automatically extract from the database described above a set of explicit IF-THEN rules that can be profitably used to detect falls. To fulfil this goal, the DEREx tool [11] has been used. It is based on Differential Evolution [12], a fast and effective evolutionary algorithm specifically devised to tackle real-valued multivariable optimization problems, and automatically extracts a set of IF-THEN rules from a database. More specifically, DEREx uses a 10-fold cross-validation to select the set of rules that maximize the correct classification rate over unseen examples. Furthermore, DEREX is run 25 times because it is not deterministic, rather its execution depends on an initial random seed. It is impossible to describe here the way the tool works, readers can refer to [11]. The best set of rules found in the 25 executions consists of five rules and is:

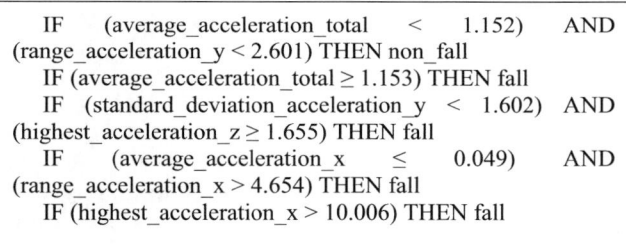

IF (average_acceleration_total < 1.152) AND (range_acceleration_y < 2.601) THEN non_fall
IF (average_acceleration_total ≥ 1.153) THEN fall
IF (standard_deviation_acceleration_y < 1.602) AND (highest_acceleration_z ≥ 1.655) THEN fall
IF (average_acceleration_x ≤ 0.049) AND (range_acceleration_x > 4.654) THEN fall
IF (highest_acceleration_x > 10.006) THEN fall

It is worth noting that the first two rules imply average_acceleration_total as a very discriminating attribute: in fact, if its value is higher than or equal to 1.153 then the movement is a fall, whereas if it is lower than 1.152 then the movement is not a fall, provided that another condition on range_acceleration_y is verified.

In total, the rules contain seven out of the 20 parameters, so the classifier has also automatically carried out the selection of the most relevant ones.

Among the parameters used, the group related to the average accelerations is the more frequently present (three times), followed by ranges and maximal values (twice each), and by the standard deviations (once). The minimal values, instead, do not appear in this best set of rules.

The above set of rules has a discriminating ability reported in Table 1 in terms of the typical medical-based metrics. It is worth recalling here that sensitivity relates to the test's ability to identify positive results, i.e. falls, and the higher the number of false negatives the lower the sensitivity. Specificity, instead, measures the ability of the classifier to identify negative results, i.e. non-falls, and for it too the higher the number of false positives the lower the specificity. All parameters range in our tests from a minimum of 0.0 (completely wrong predictive ability) to a maximum of 100.00 (perfect predictive ability with no errors).

Table 1 Discriminating ability of the best set of rules found

	Training Set	*Testing Set*	*Whole Database*
Accuracy	92.44	100.00	93.17
Sensitivity	85.32	100.00	86.77
Specificity	99.13	100.00	99.21
ROC area	98.12	100.00	98.39

This set of rules classifies with no errors on the Testing Set, so there are neither falls incorrectly classified as non falls nor non falls that are regarded as falls. They correctly classify 93% of the instances in the whole database.

Table 2 Average results for the testing set over the 25 runs

	Accuracy	*Sensitivity*	*Specificity*	*ROC area*
Average	91.88	91.26	92.09	91.67
Variance	0.53	0.23	0.21	0.21

Table 2 shows the average values for accuracy, sensitivity, specificity, and ROC area, achieved for the testing set over the 25 runs. Average values are quite high, and variance is very low, which suggests independence of the results on the initial random seed.

A comparison with results achieved by other classifiers has been made. Due to lack of space, it cannot be reported here. However, results have shown the superiority of our approach in terms of accuracy, sensitivity and specificity.

IV. WORK IN PROGRESS

This set of rules should now be tested in a real-world case. Therefore, it has been included in our mobile health monitoring system developed at iHealthLab, and experiments are being carried out with volunteers from ICAR staff. They have the tasks of wearing sensors of the system and of simulating both falls and daily activities which, due to their accelerometer-related features, could cause false positives. Very preliminary tests on just three people seem to confirm the goodness of the approach, yet preliminary numerical results coming from a wider set of volunteers will be provided in a future paper.

It is worth providing readers with both software and hardware details on our mobile health monitoring system.

A. Software Details

The mHealth monitoring system has been developed at iHealthLab [1]. The overall software architecture is outlined in Fig. 1. It is devised to simply accommodate technological or functional changes, as the introduction of new devices or software modules. It is divided into three different layers: Data Layer, Decisional Layer, and Action Layer. All layers and modules are implemented for resource-limited mobile devices, by using Java, but the system could be used also to build desktop applications, except for the user interfaces.

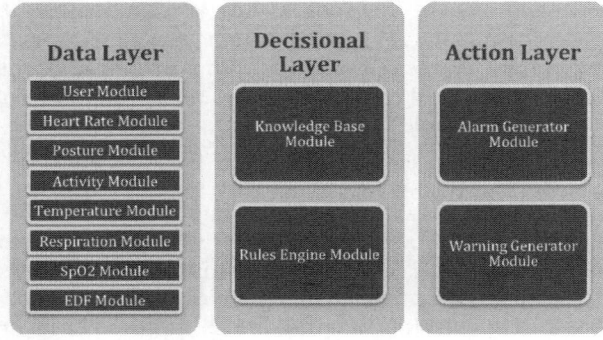

Fig. 2 The System Architecture.

The Data Layer provides user interfaces and mechanisms to manage sensors data and patient information that will be processed by the Decisional Layer. The Data Layer is in charge of collecting information about the patient monitored, such as name, gender, age, etc.. and is also responsible for collecting data from wearable medical devices, like sensors for ECG, SpO2, and acceleration. In this layer there are also modules for calculating more parameters such as physiological values, like Heart Rate or Heart Rate Variability, or activity values, like motor velocity of the patient, or other parameters like the average value in a temporal window or the standard deviation.

Finally, this layer provides a module to store the acquired data in an EDF file, which is a standard format designed for exchange and storage of medical time series.

The Decisional Layer represents the intelligent core of the system and includes the rule engine described in [13]. In this layer, data coming from the Data Layer are elaborated by the rule engine on the basis of the knowledge formalized through IF…THEN rules, so recognizing in real-time the possible critical or dangerous situations and determining the appropriate actions to be performed by the Action Layer.

Finally, the Action Layer executes the actions inferred by the Decisional Layer by means of mechanisms to generate alarms and/or warning messages.

B. Hardware Details

To deal with the specific fall detection use case, it could be possible to use just one wearable sensor in which an accelerometer is embedded. For this reason, the system has been developed and tested by using just the wearable wireless medical sensor, the Zephyr BioHarness BH3, that is an advanced physiological monitoring device. Of course more sensors could be added, as for instance one for SpO2, so as to gather as much data as possible related to fall situations.

The monitor could be used with the BioHarness™ strap that is a lightweight elasticized component which incorporates Zephyr Smart Fabric ECG and Breathing Rate sensors.

The BioHarness BH3 module contains also a 3-axis accelerometer for monitoring attitude and activity. The device permits to recover axis mapping information for the accelerometer. Any accelerometer axis can be mapped to any other axis as well as being inverted to allow the device to be used in a number of different orientations (e.g. worn on the front, worn on the side, upside down, etc.). This feature is very important, because it has allowed us to utilize a fall database gathered with a different sensor to formalize the knowledge, without worrying about orientation.

All monitored data is transmitted by bluetooth and can be monitored using any suitably-configured mobile device with bluetooth technology, such as a laptop, phone, or PDA.

V. CONCLUSIONS

Fall detection is an important task in telemedicine. In this paper an approach based on supervised knowledge extraction has been presented. A fall recordings database has been analyzed offline and a set of IF…THEN rules has been obtained. This way, also selection of the most relevant features for fall assessment has been automatically carried out. The approach, embedded within our developed mobile monitoring system, can be used to discriminate in real time normal daily activities from falls.

To further validate these results on the field, the approach is being tested through a set of simulated falls performed in our iHealthLab, and by examining the fall discrimination ability of the proposed set of rules. Very preliminary tests on just three people seem to confirm the goodness of the approach, yet preliminary numerical results coming from a wider set of volunteers will be provided in a future paper.

ACKNOWLEDGEMENT

The authors wish to thank their colleagues Dr. M. Esposito and Dr. A. Minutolo for their useful contribution in permitting us to use their developed DSS and Prof. S. Fudickar of the Department of Computer Science, University of Potsdam, Germany, for the fall recordings database.

REFERENCES

1. Lan C-C, Hsueh Y-H, Hu R-Y (2012) Real-Time Fall detecting System Using a Tri-axial Accelerometer for Home Care, Proc. International Conference on Biomedical Engineering and Biotechnology, IEEE Computer Society, 2012, pp. 1077-1080.
2. Abate S, Avvenuti M, Cola G et al. (2011) Recognition of false alarms in fall detection systems, Proc. IEEE Consumer Communications and Networking Conference (CCNC), 2011, pp. 23-28.
3. Erdogan S Z, Bilgin T T (2012) A data mining approach for fall detection by using k-nearest neighbor algorithm on wireless sensor network data, IET Communications, vol. 6, issue 18, pp. 3281-3287.
4. Mirchevska V, Lustrek M, Gams M (2011) Towards robust fall detection, Proc. 14th International Multiconference Information Society (IS2011), 2011, pp. 75-78.
5. Kerdegari H, Samsudin K, Rahli A R et al. (2012) Evaluation of Fall Detection Classification Approaches, Proc. Fourth International Conference on Intelligent and Advanced Systems, pp. 131-136.
6. Sannino G, De Pietro G (2010) An Intelligent Mobile System For Cardiac Monitoring, Proc. IEEE Healthcom, Lyon, France, pp: 52-57.
7. iHealthLab - http://ihealthlab.icar.cnr.it/
8. Fudickar S, Karth C, Mahr P et al. (2012) Fall-detection simulator for accelerometers with in-hardware preprocessing, Proc. 5th International Conference on Pervasive Technologies Related to Assistive Environments (PETRA'12), 2012, p. 41.
9. Preece S J, Goulermas J Y, Kennedy L P J et al. (2009) Activity identification using body-mounted sensors - a review of classification techniques, Physiological Measurement, vol. 30, no. 4, pp. R1-R33.
10. Estudillo-Valderrama M A, Roa L M, Reina-Tosina J et al. (2009) Design and Implementation of a Distributed Fall Detection SystemPersonal Server, Information Technology in Biomedicine, IEEE Transactions on , vol.13, no.6, 2009, pp.874-881.
11. De Falco I (2013) Differential Evolution for automatic rule extraction from medical databases, Applied Soft Computing, vol. 13, Elsevier, 2013, pp. 1265-1283.
12. Price K, Storn R, Lampinen J (2005) Differential Evolution: A Practical Approach to Global Optimization, Springer, 2005.
13. Minutolo A, Esposito M, De Pietro G (2011) A Mobile Reasoning System for Supporting the Monitoring of Chronic Diseases, Proc. 2nd International ICST Conference on Wireless Mobile Communication and Healthcare MobiHealth 2011, pp. 225-232.

Author: Giovanna Sannino
Institute: Institute of High Performance Computing and Networking
Street: Via Pietro Castellino 111
City: Naples
Country: Italy
Email: giovanna.sannino@na.icar.cnr.it

Chemoprophylaxis Application for Meningococcal Disease for Android Devices

M. Parejo-Bellido[1,2], E. Dorronzoro-Zubiete[3], M. Zurbarán[4], F.J. Sánchez-Laguna[4],
L. Fernández-Luque[3], and A.A. Muñoz-Macho[2]

[1] Open University of Catalonia, Telemedicine Master, Catalonia, Spain
[2] Thotalmed, Seville, Spain
[3] Salumedia, Seville, Spain
[4] University Hospital Virgen del Rocío, Seville, Spain

Abstract—**This paper presents a feasibility study about a Clinical Decision Support System (CDSS) application for Android devices, Chemopro, for meningococcal disease using antimicrobial chemoprophylaxis of close contacts. The application implements an algorithm provided by the Regional Ministry of Health of Andalusia, Spain, and it has been extended to consider pregnant women sensible to ceftriaxone. Having the application on a smartphone or a table device allows physicians to have access to the protocol and could reduce errors when it is applied. All the data introduced on the application is sent and stored in a web, which can be consulted to control use of antibiotics on the study of contacts. The implementation was done using ODK, a free and open-source set of tools for building data collection forms or surveys. The forms have been created using ODK and enriched to support constraints and rules to control the branches of the algorithm.**

Keywords—**CDSS, ODK, chemoprophylaxis, meningococcal disease.**

I. INTRODUCTION

Each year, an estimated 1,400--2,800 cases of meningococcal disease occur in the United States (CDC, unpublished data, 2004) [1]. Meningococcal disease is caused by the bacterium Neiserria meningitidis. The infection produced by this bacterium can cause meningitis, widespread blood infection (sepsis) or a combination on both.

One method of prevention for meningococcal disease is antimicrobial chemoprophylaxis of close contacts of a patient with invasive meningococcal disease. Chemoprophylaxis refers to the administration of a preemptive medication. It must be applied only when the benefits overweigh the risks produced by the side effects of the medication.

In order to support physicians to make medication recommendations, we have designed an application for Android devices that implements an algorithm provided by the Regional Ministry of Health of Andalusia, Spain. The application is presented as a CDSS. CDSS is a computerized system that, responding to inputs or generating alarms, provides useful information to support the physician in decision-making [2].

This paper contains the following sections: 1) Introduction, 2) Objective, 3) Materials and methods, 4) Results, 5) Discussion and 6) Conclusions. In the objective section the challenges of this project are detailed. In the materials and methods section, the preliminary study will be presented with the description of the application developed. In the Results section, results of this preliminary study are showed. The state of Chemopro application and the next steps to perform to evaluate its effectiveness are explained in the discussion section. Finally, considerations of the Chemoprophylaxis application are reported in the conclusion section.

This paper presents a Knowledge-Based CDSS application that implements a protocol to determine which antibiotic should be administered according to the characteristics of the person who has been in contact with a case of meningococcal disease.

II. OBJECTIVE

The objective of Chemopro is to make a feasibility study of the implementation of a CDSS for meningococcal disease using ODK, a free and open-source set of tools for building data collection forms and surveys.

III. MATERIALS AND METHOD

For developing the application, we have used ODK, a free and open source suit of tools, explained below in detail. It provides the tools to implement a fast and effective CDSS to allow physicians to apply decision algorithms.

A. Decision Support Systems

A CDSS is a computerized system that helps the physicians to make a diagnosis, medication recommendations, provide alarms, etc. The CDSS can be active and provide alarms based on a particular situation or it can be passive and respond to the physician inputs providing a response. It must be the physician who has to decide if the information provided by the CDSS is valid or not.

At the beginning CDSS, even they were used at healthcare information systems, were focused on assisting in administrative and financial tasks in a retrospective way. As they became more useful they were applied at the clinic domain but keeping its retrospective feature. They were useful for planning (treatment guides, critical pathways). Actual CDSS are able to provide an output to before, during and after the clinical decision has to be made [3].

CDSS include differing areas of care, preventive care, diagnosis, planning or implementing treatment, follow up management, cost reductions and improved patient convenience, etc.

They have proved to be useful. The meta-analyses that have been done on CDSS using RCT (Randomized Controlled Trials) reflects that they can alter the decision of the physician, reduce the medication errors, use evidence-based recommendations on medical prescription [4].

There are different types of CDSS. As it has been previously commented, it is important "when" the output is provided or if the output is triggered by an alarm or as a response of an input made by a physician. But the two main categories of the CDSS are: Nonknowledge-Based and Knowledge-Based.

• Nonknowledge-Based CDSS: This type of CDSS is designed to learn by itself. It can learn form past experiences or recognize patterns in the clinical data. The approach to these systems has been made using AI (Artificial Intelligence) techniques such as neural networks, fuzzy logic, multi-agent based systems.

• Knowledge-Based CDSS: Medical experts generate the knowledge base. Their inference engine uses the knowledge base and/or external input to generate the outputs. This output is displayed to the physicians.

B. Open Data Kit

ODK is a free and open-source suite of tools that provides an out-of-the-box solution for users to build a data collection form or survey, collect the data on a mobile device, send it to a server, and aggregate the collected data on a server and extract it in useful formats [5]. These tools can be used independently or with each other. ODK use XForm [6] standard, an XML-based form description standard designed by the W3C for the next generation of web forms, implementing the OpenRose [7] subset of XForms.

Some tools provided by ODK [8]:

• ODK Collect (Smart Phone Client) is a client on Google's open source Android platform written in Java.

• ODK Aggregate (Server Storage) is a ready-to-deploy server that hosts forms and submitted results.

• ODK Build is a form designer with a drag-and-drop user interface. It is an HTML5 web application and works best for designing simple forms.

The algorithm for a CDSS can be built following XForm standards and setting restrictions and conditions that must be satisfied to automatically take a branch or another through the decision tree in order to get the correct result depending of the introduced parameters.

C. Chemopro

The knowledge base is the algorithm to decide which antibiotic should be prescribed. The algorithm is taken from the Monitoring and Meningococcal Disease Alert Protocol of the Epidemiological Surveillance Service of Andalusia of the Regional Ministry of Health, updated in June 2011. The protocol has been modified to consider pregnant women sensible to ceftriaxone. In these cases, ciprofloxacin is recommended by the NHS guidelines.

The protocol is presented in the Fig.1. Based on the inputs of the physician to the different questions the protocol presents the recommended medication. Following these guidelines for a female, pregnant and allergic to penicillin patient the algorithm will recommend Ciprofloxacin 500mg.

Implementing the protocol as an application for Android devices eliminates the need to carry the algorithm printed.

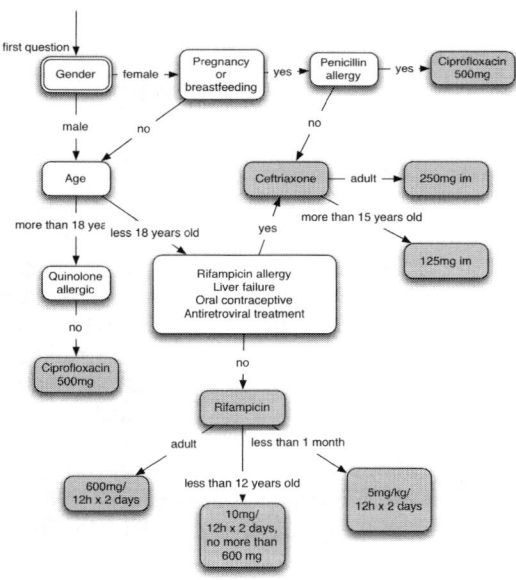

Fig. 1 Chemoprophylaxis Algorithm

The physicians can evaluate a scenario just using his smartphone or tablet in an easy and fast way. It also reduces the possibility of making input error as the system inputs are controlled by constraints.

The application can send the collected data to a server. With the collected data saved in this server the physician has control about the use of antibiotics on the study of contacts.

D. Implementation

Chemopro has a forward chaining inference engine, also called data-directed inference because inference is triggered by the arrival of new data. When a user answers a question the introduced data triggers a rule whose conditions are matched. These rules perform their actions, asking for more input data to the user. The new data trigger new rules and so on until get a final result.

Chemopro has been designed graphically using ODK Build. As ODK Build does not support complex constraints and rules the form created has been modified following XForm standards adding this new feature.

Range constraints are used to validate the data introduced by the user. Using constraints input errors are avoided.

Relevant attribute is used to control which branch of the algorithm will be taken in each step to reach the correct result. The value of the relevant attribute is an expression that references the values of previously answered questions. If this evaluates to true, the question will be shown; if it evaluates to false, it will be skipped. The code shown in Fig.2 makes that the question "Pregnancy or Breastfeeding?" was only asked if the gender question was answered 'Female'.

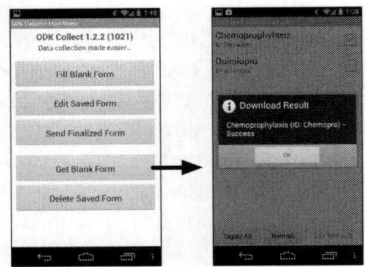

Fig. 3 Get chemoprophylaxis form

```
<xf:bind nodeset="/data/HistoryPregnancy"
required="true()" type="string"
relevant="/data/Gender = 'Female'"/>
```

Fig. 2 Code for Checking Gender

Once Chemopro XForm is designed, it can be uploaded to a server with ODK Aggregate installed. We have used Google Application Engine server (GAE) in which we have configured an application and installed ODK Aggregate. Then, we have uploaded the Chemoprophylaxis Algorithm XForm. From this moment the form is available to be downloaded for Android clients with ODK Collect installed. It is possible to configure the server so that only privileged users can connect and download the form.

The application is configured so that the server URL and other configuration parameters are pre-established and no accessible. In this way, users do not have to worry about settings.

The application presents an intuitive selection screen where different actions can be performed.

The first step when launching the application for the first time is to get the form that implements the Chemoprophylaxis Algorithm. "Get the blank form" button connects to the server and returns all the available forms. The form can be selected and downloaded to the application as shown in Fig. 3. As soon as the download process is finished the form is ready to be used.

Once the form is in the device there is no need to download it again. It can be used for as many times as it is required.

"Fill blank form" starts the form of the CDSS as shown in Fig.4. To answer a question is as easy as introducing the required information and sliding with the finger to the next question.

This new question is presented based on the input of the physician and following the previously presented algorithm.

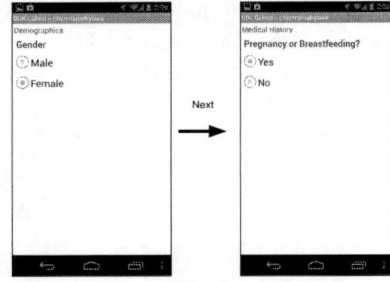

Fig. 4 Chemopro algorithm

As the last question is answered the application shows on screen the recommended medication. The form can be saved and the application returns to the main screen.

Back to the main menu there is an option to send the completed forms to the server, edit previously completed forms and remove them.

Physicians can access to the server using a web navigator and study and evaluate the values of all the Chemopro forms sent.

These values and results (comparative use of antibiotics) can also be presented as pie charts, bar graphs or maps. It is possible to filter them by any of the parameters defined in Chemo-pro as well, as shown in Fig.5.

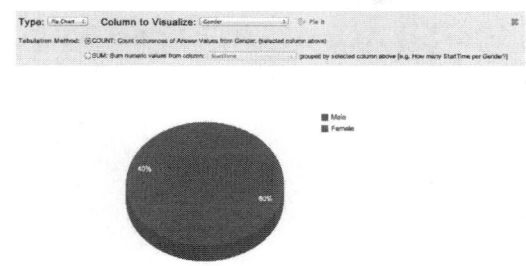

Fig 5 Pie chart: % gender in Chemopro forms

ODK Aggregate allows exporting all the data saved in different formats such as CVS, KML and JSON. This feature allows exporting the collected data in order to be used in other medical applications.

IV. RESULTS

We have performed a complete battery of tests to the Chemopro application checking all the possible combinations without patients involved. Physicians from the University Hospital Virgen del Rocio have verified that Chemopro works according to the algorithm taken from the Monitoring and Meningococcal Disease Alert Protocol of the Epidemiological Surveillance Service of Andalusia of the Regional Ministry of Health.

V. DISCUSSION

Chemopro could help physicians in making-decisions and reduce medication errors. The data is sent to a centralized server so that it is available to do statistical studies and to have control about the use of antibiotics on the study of contacts.

Testing Chemopro on real patients could have endangered them by wrong advise to physicians, therefore the application has been tested through laboratory trials. We have verified that Chemopro works according to the algorithm implemented. It is necessary that the application complaints with the medical devices regulations for safety and security, and implements interoperability standards such as those defined in HL7 (Health Level 7) [9] for its integration with the EHR (Electronic Medical Record). It is also very important a pilot controlled phase in order to evaluate the usability and effectiveness of Chemopro.

VI. CONCLUSIONS

In this paper a feasibility study of the implementation of a CDSS for meningococcal disease using ODK is presented. The algorithm that has been used is taken from the Monitoring and Meningococcal Disease Alert Protocol of the Epidemiological Surveillance Service of Andalusia of the Regional Ministry of Health. It has been expanded to consider pregnant women sensible to ceftriaxone.

The application has been built using ODK, which provides a suit of tools that make possible to implement similar CDSS in a fast and effective way. The form has been created using ODK Builder and it has been enriched to control which one of the different branches of the algorithm to take according to the input data and to provide constraints to the inputs that reduces errors.

The application could be useful to support physicians in decision-making and for reducing medication errors. Moreover, all the data is sent to a server were it can be presented in tables and charts to be studied. The data can also be exported in different formats such as CVS, KML and JSON.

REFERENCES

1. Bilukha, O., & Rosenstein, N. E. (2005). Prevention and Control of Meningococcal Disease.
2. Yaw Anokwa. Improving Clinical Decision Support In Low-Income Regions (Slides). Dissertation at University of Washington. 2012.
3. Berner ES. Clinical decision support systems: State of the Art. AHRQ Publication No. 09-0069-EF. Rockville, Maryland: Agency for Healthcare Research and Quality. June 2009.
4. Berner, E. S., & Lande, T. J. (2007). Clinical Decision Support Systems (pp. 3–22). New York, NY: Springer New York. Medical Records, Medical Education and Patient Care. 2nd ed. Cleveland: Case Western Reserve University Press, 2007.
5. Carl Hartung, Yaw Anokwa, Waylon Brunette, Adam Lerer, Clint Tseng, Gaetano Borriello. Open Data Kit: Tools to Build Information Services for Developing Regions. Information and Communication Technologies and Development (ICTD). 2010.
6. XForms at http://www.w3.org/MarkUp/Forms/
7. OpenRosa Consortium at http://openrosa.org/
8. Open Data Kit at http://opendatakit.org
9. Health Level 7 at http://www.hl7.org/

The address of the corresponding author:

Author: Mónica Parejo Bellido
Institute: Universitat Oberta de Catalunya
Street: Finlandia, 2, H, 1º D C.P. 41012
City: Seville
Country: Seville
Email: monica.parejo@thotalmed.es

Automation of Evaluation Protocols of Stand-to-Sit Activity in Expert System

M.J. Cunha[1], G. Cardozo[2], and F. de Azevedo[2]

[1] Medical Sciences Postgraduate Program - University Federal of Santa Catarina, Brasil
[2] Biomedical Engineering Institute, Federal University of Santa Catarina, Florianópolis, Santa Catarina, Brazil

Abstract—The evaluation of sit-to-stand and stand-to-sit activities is often used by physiotherapists in patients with neurological and musculoskeletal disorders and it is essential in identifying movement problems. There are different methodologies used to describe the stand-to-sit activity and its evaluation is not standardized, which makes difficult the practical application of resources on clinical observation. The use of Expert Systems able to aid in systematizing the decision making process can contribute significantly to the professional in their practice. Two Expert Systems were implemented on this study through the Expert Sinta Shell. Each system generates a number of questions, functioning as a guide for the physiotherapist and supporting to standardize the evaluation. The ES´s were assessed by 12 physiotherapists experts in the area through questionnaires with items related to their characteristics and were considered useful in clinical practice.

Keywords—Expert system, Physiotherapy, Activities of Daily Living, Disability Evaluation.

I. INTRODUCTION

The evaluation of sit-to-stand and stand-to-sit activities is often used by physiotherapists in patients with neurological and musculoskeletal disorders [1]. The observation of the way these activities are executed is essential to identify problems on movements.

There are different methodologies used to describe the stand-to-sit activity and there is not standardization on its evaluation [2-6], which makes difficult the practical application of resources on clinical observation.

Kralj et al. [7] developed a study aiming to establish normative data for the STS, and described it in 4 phases: *initial phase* (anterior tilt of trunk), *descending* (vertical displacement), *seat loading* (weight transfer to the seat) and *stabilization* (trunk and balance adjustment). Perracini et al. [8] consider this activity to be performed in sequential order of *anterior tilt* (phase I), *vertical displacement* (phase II), *angular displacement of knee* (phase III) and *stabilization* (phase IV).

Cunha et al. [9] developed two protocols to support the observation of the movements during the stand-to-sit activity to sit on anterior and lateral views, and described it in 4 consecutive phases: "Initial Position", "Pre-squat", "Squat" and "Stabilization" and for each phase the protocol gives 3 options of answers. The proposal of utilization of these instruments is to facilitate the analysis of the stand-to-sit activity in clinical practice [9], since there is not a gold standard method for identifying the positions of body segments during its execution [10].

However, these instruments can used only manually, taking a long time and doesn´t seem to be practical for clinical use. Perhaps this instrument should be implemented in to a computerized system to supply better conditions for the physiotherapist execute the evaluation.

Currently, physiotherapists and other specialists in health area have had their most routine tasks aided by computerized systems. Activities involving decision making, in particular, have been the object of study and implementation of software.

Decision support systems have been implemented, initially using traditional computational techniques. However, some problems in health area are extremely complex to be structured in algorithm form, which implies on the need to use techniques of Artificial Intelligence (AI), also known as Computational Intelligence (CI).

Among these, the Expert Systems (ES's) are the most known. ES´s have been used to solve problems trying to simulate human behavior of experts in a particular domain.

An Expert System (ES) deals with complex problems of the real world which require the interpretation of an expert, solve these problems by using a computational model of reasoning of a human expert, reaching the same conclusions that the human expert would reach if faced with a similar problem [10].

For the creation of SE's, specific programming languages may be used, such as Prolog or the Shells. In the first case, it requires a great knowledge of both programming language such as the AI techniques, in the second case, the shells enable specialists in other areas, other than those related to AI can develop their systems with relative ease. For this, the domain expert must create a knowledge base in which all the information is structured. This knowledge base is composed by variables, values, truth, rules, issues and goals.

The stand-to-sit activity has been less investigated due it lower functional impact compared to the sit-to-stand activity [11]. However, the transfer form stand to sit is also performed several times a day and may indicate musculoskeletal abnormalities.

According to the exposed, the aim of this study was to automating the protocols as expert systems to support the evaluation of stand-to-sit activity on anterior (AV) and lateral views (LV).

II. MATERIALS AND METHODS

This study was approved by the Ethics Committee on Human Research of Federal University of Santa Catarina with the protocol number 1093.

The Shell chosen was Expert Sinta, which was developed at the Federal University of Ceará [12] and is released for use. It uses AI techniques for automatic generation of ES's.

This symbolic ES could be implemented using the Prolog language exploring all the flexibility and resources made available by it or a Shell could be used to facilitate the process of constructing the system by a non-expert on computer systems. Considering the need for fast prototyping, the last option was chosen in order to allow the domain expert (physiotherapist and researcher) could implement the system and the expert on AI acted only as a consultant.

The Shell has an user friendly interface and its architecture consists on:

- **Knowledge base:** the information that the system uses, formed by the rules and facts, which are appointed by the user while browsing ES.
- **Editor bases:** it is the resource which the system allows the implementation of the bases.
- **Machine inference:** is the part of the ES which is responsible for deductions about the bases.
- **General database:** are the facts mentioned by the user while browsing the ES.

The Expert Sinta allows the implementation of different knowledge bases. These are implemented thorough an editor which is linked to the system. During a base execution, the Shell uses a run time system, where the base charged is processed by the inference engine. In order to determine a sequence analysis of knowledge, Expert Sinta uses a procedural methodology known as "backward chaining." This is the method most commonly used in an ES. In it, the developer must include in the definition of the base which assignments must be found, ie the goals of the ES. The inference machine takes care of finding the desired attribute in the rules conclusions (after THEN...). For the rule be approved, it assumptions must be satisfied, forcing the machine to find the attributes of the assumptions, so they can be judged, setting a recursive chain. If the assignment searched is not found in any conclusion rule, a direct question is made to the user.

A. Implementation of the Expert System

This step was the definition of subjective and objective variables and their values (Figure 1), elaboration of rules, truth values, questions that the system should make to the user, in this case the physiotherapist, with their respective explanations.

The system was implemented according to clinical needs of the physiotherapist, and the knowledge bases were elaborated according to clinical protocols to support the evaluation the stand-to-sit activity on anterior and lateral views.

Thus, the final model is as follows:

IF...

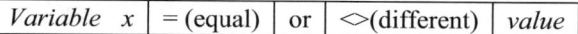

THEN...

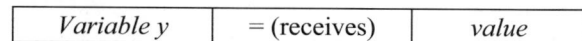

The system allows to create rules formed by a set of conditions. Likewise, there is the possibility to include more than one assignment to each result for conclusion

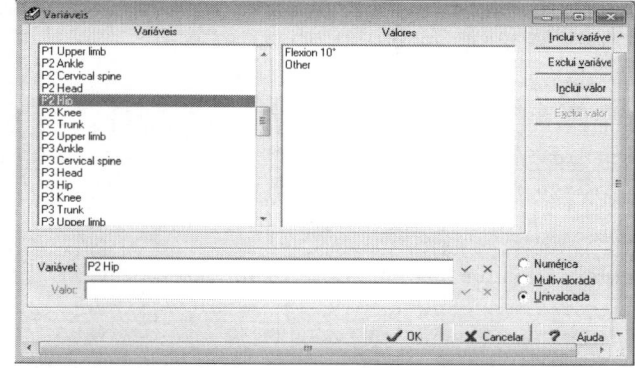

Fig. 1 Definition of variables and it respective vaues

III. USING THE EXPERT SYSTEM

The use of a SE for evaluation of stand-to-sit activity is necessary in order to standardize and make the kinesiological analysis easier, since the computerization of this protocol has the purpose to support the physiotherapist through make the procedure faster and serve as a reference guide.

A. Purpose of the Project

Considering that in practical and clinical applications for analysis and functional performance of an activity it is necessary to know the movement patterns considered appropriate to identify possible abnormalities and that the stand-to-sit activity does not have their established standards, the objective of this study was to implement the protocol of evaluation in order to supply a tool to the physiotherapist which holds the description of this activity with the movements of reference, so it's possible to identify abnormalities during the evaluation.

The technical purpose of the project was to implement the information contained in the protocol to support the stand-to-sit activity in a computational tool.

B. ES's Assessment

The ES's were assessed by 12 physiotherapists experts in the area, who belong to the service of the Central Institute of Physiotherapy – Hospital of Clinics in Medicine Faculty of São Paulo University.

The assessment of ES were made through questionnaires with items related to the characteristics of each one, as "utility for clinical practice of the physiotherapist" and "easy to use" with dichotomous options of answers "yes" or "no".

IV. RESULTS

Two ES's were implemented, one for the AV and other for LV. Each one generates a number of questions, which serve as a guide for the physiotherapist, supporting to standardize the evaluation.

A. Evaluation of the Patient

To evaluate the stand-to-sit activity, the patient should be with the proper attire as swimsuit or gym clothes so that body segments are properly seen during the execution of the movements.

The physiotherapist may ask the execution of stand-to-sit activity to the patient and observe the movements of each body segment according to the 4 phases of the activity.

From the beginning of ES query, it starts the release of sequences questions regarding the position of each body segment in a given phase (Figure 2). Thus the ES serves as a guide able to guide sequence for the analysis of body segments.

It is recommended that the physiotherapist runs the ES after the first visualization of the movements of his patient. This reduces the possibility of some body segments be unnoticed.

B. Operation of the Program

The SE performs the sum of the situations regarding the movements according to each phase, and suggests the condition related to each segment according to the legend of the points (Figure 3).

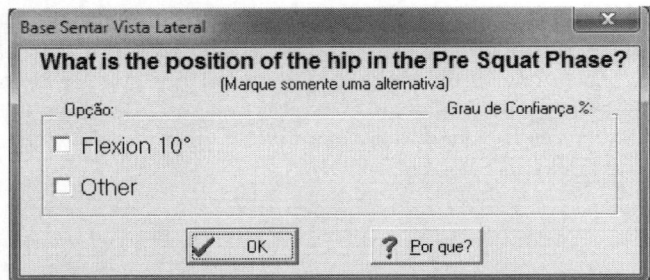

Fig. 2 Model of a question made to the user by the ES

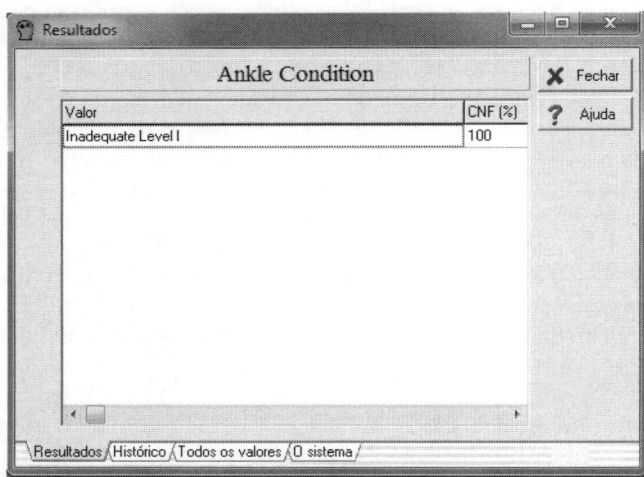

Fig. 3 Model of a result suggested by the ES

From the classification of conditions for each body segment with their respective values (Table 1), there are possibilities for conclusion.

Table 1 Conclusion of the results about the segments during the activity.

General Conclusion of Body Segments in the Activity
Adequate Condition (0 points):
Inadequate   Condition Level I (1 point):
Inadequate   Condition Level II (2 points):
Inadequate   Condition Level III (3 points):
Inadequate   Condition Level IV (4 points):

C. Interpretation of the Results in the Evaluation

From the use of the ES, the physiotherapist can understand the conclusion of the results about the body segments through a classification related to the adequacy of the movements in each phase of activity.

The degree of inadequacy correspond the number of phases in which the segment showed a different situation than what it is described in the ES.

The use of ES to evaluate these activities allows a quantitative conclusion, in order to indicate the condition of the body segment through the "degree of inadequacy".

D. ES's Assessment

The question about the "utility of the system in clinical practice of the physiotherapist", the majority (67%) of the evaluators answered "Yes" and 33% answered "No" for both systems.

The "ease of using" was considered 100% positive for system in LV. For the system in LV, the majority (92%) answered "Yes" and only 8% considered that the system is not easy to use.

V. CONCLUSIONS

The implementation of protocols to support the evaluation of stand-to-sit activity in to ES's allowed the utilization of a new evaluation resource for physiotherapists.

The use of ES for the evaluation of these activities allows a quantitative conclusion, to suggest the condition of the body segment through the "degree of adequacy".

The automated process was considered useful in clinical practice by the evaluators. This will allow the identification of inappropriate movements to orientate the therapeutic decision-making, considering that kinesiological abnormalities compromise the functionality of the human beings.

Considering the practicality regarding the use of these protocols should be prioritized because they must attend the physictherapist daily requirements in the clinical practice, the ES's elaborated on this study are promising tools for these professionals.

REFERENCES

1. Soni SR, Khunteta A, Gupta M. A Review on Intelligent Methods Used in Medicine and Life Science (2011). International Conference and Workshop on Emerging Trends in Technology (ICWET) – TCET, Mumbai, India.
2. Mourey F, Pozzo T, Rouhier-Marcer I, Didier JP (1998). A kinematic comparison between elderly and young subjects standing up from and sitting down in a chair. Age and Ageing. 27:137-146.
3. Goulart FRDP, Valls-Sole J (1999). Patterned electromyographic activity in the sit-to-stand movement. Clinical Neurophysiology. 110:1634±1640.
4. Kim MH, Yi CH, Yoo WG, Choi BR (2011). EMG and kinematics analysis of the trunk and lower extremity during the sit-to-stand task while wearing shoes with different heel heights in healthy young women. Human Movement Science.
5. Mohammad R. Fotoohabadi, Elizabeth A. Tully, Mary P. Galea (2010). Kinematics of Rising From a Chair: Image-Based Analysis of the Sagittal Hip-Spine Movement Pattern in Elderly People Who Are Healthy. Physical Therapy. 90(4)
6. Kerr A, Durward B, Kerr KM (2004). Defining phases for the sit-to-walk movement. Clinical Biomechanics. 19:385–390.
7. Kralj A, Jaeger R J, Munih M. Analysis of Standing up and Sitting down in humans: definitions and normative data presentation. *J. Biomechanics*. 1990;23:1123-1138.
8. Perracini MR, Fló CM. Funcionalidade e envelhecimento. Rio de Janeiro, Guanabara Koogan, 2009. 557 p. Fisioterapia : teoria e prática clínica.
9. Cunha MJ (2011). Bases de conhecimento para sistemas especialistas de suporte na avaliação das atividades de levantar e sentar – protocolos clínicos / Maíra Junkes Cunha. [master dissertation]. Florianópolis: Universidade Federal de Santa Catarina.
10. Roorda LD, Roebroeck ME, Lankhorst GJ, Van Tilburg T, Bouter LM (1996). Measuring functional limitations in rising and sitting down: development of a questionnaire. Arch Phys Med Rehabil.77:663-9.
11. Dubost V, Beauchet O, Manckoundia P, Herrmann F, Mourey F. Decreased Trunk Angular Displacement During Sitting Down: An Early Feature of Aging. *Physical Therapy*. 2005 May; 85(5): 404-412.
12. EXsinta; Tutorial do software (1997), LIA (Laboratório de Inteligência Artificial), São Paulo.

Use macro [author address] to enter the address of the corresponding author:

Author: Fernando Mendes de Azevedo
Institute: Biomedical Engineering Institute (Federal University of Santa Catarina)
Street: Campus Universitário - Trindade
City: Florianópolis
Country: Brazil
Email: azevedo@ieb.ufsc.br

Digital Diary for Persons with Psychological Disorders Using Interaction Design

H. Sørheim[1], and A. Babic[2]

[1] Institute for Information and Media Science, University of Bergen, Bergen, Norway
[2] Department of Biomedical Engineering, Linkoping University, Sweden

Abstract—The main goal is to facilitate information technology to support patients by providing them a digital tool for daily logging of psychiatric symptoms and the integration of a crisis plan. This is presented in a project which places focus on a design of a mobile diary application. Interaction design and evaluation methodologies will be used throughout this project to design and test the concept and implemented application. The focus will be on the active user role in all phases. The first design has been done in close sessions with users one of whom had a long experience with the acute and sever disease episodes (schizophrenia). The initial evaluation has shown that a high degree of security could be expected from using this application. Additional value could be seen in therapy when the self-entered data becomes a part of the medical assessement. The evaluation has also indicated that the current design was easy to understand and interperate.

Keywords—Interaction design, digital diary, design research, schizophrenia, sustainable design.

I. INTRODUCTION

The purpose of this paper is to present how interaction design can be integrated in a mobile digital diary which is personalized for people with psychological disorder. Functionality of the diary includes logging of psychiatric symptoms and the digitalization and integration of an individual crisis plan. Here the purpose includes strengthening persons control over their psychiatric symptoms from a pre-psychiatric stage. This is through means of enhancing the usability of already existing principles from paper diaries and a crisis plan by using a customized and sustainable design. The idea introduced is based on principles of renewal and reuse from sustainable interaction design [1].

The question of whether or not persons can exert control over their own psychiatric symptoms is introduced by Breier and Strauss (1983) [2]. Here it is presented that this can be done through the usage of a three phase set of self-regulation. This includes a process of becoming aware of pre-psychotic behavior by self-monitoring, executing self-evaluation and then executing mechanisms of self-control. This opens for possibilities of self-monitoring within the usage of both paper and electronic diaries.

The reason why the design of a digital diary have been chosen is based on research concerning the usability of electronic diaries contra paper diaries, as introduced in Stone et al.'s (2002, 2003) studies [3, 4]. Here they present electronic diaries as more effective and used within logging psychiatric symptoms compared to paper diaries. Mutschler et al's [5] studies also highlights advantages in the usage of electronic diaries such as the availability of collecting real-time data, reminding patients to take medication and self-observing symptoms in a longitudinal assessment. However criticism to a unilateral diary design in Stone et al.'s [4] study as well as limitations to possible alternatives has been presented [6, 7].

To achieve an effective and user friendly design to a mobile digital diary, design research guidelines as introduced by Hevner [8], principles of sustainable design by Blevis (2007) [1] and a model specified for HCI design collaboration between experienced anthropologists and designers by Zimmerman, Forlizzi, and Evenson (2007) [9] will be involved. This is based on advantages which interaction design has from gaining knowledge through the use of design research techniques such as observation, analysis, modeling and the evaluation of the needs of a target user group [9]. The achievement gained is based on a closer sense of cooperation and collaboration between the design and the target users for the purpose of a stronger usable design.

The expected results of this project revolve around an increased effectiveness and user friendliness of an interface in a mobile diary application. It is to be expected that the system has a higher integrated set of functionalities from existing sustainable design, which is relevant according to user needs.

II. MATERIAL

The application will be implemented for a wide patient group which has stabile periods and is capable of managing their own lives. Prior focus will however be on patients with schizophrenia and bipolar disorder. The support network (parents, spouses, siblings, friends, and the health network) is to be considered as a special user group.

Data which is designed in the system is the contact data, basic clinic profiles such as treatment and crisis plan, user registered data regarding sleep and medication. Technology solutions are provided by the mobile and Internet technology.

III. METHOD

The guidelines which Hevner (2004) [8] introduces includes a process of the creation of a unique artifact which has an evaluated and purpose based function, as well as a representation of a well communicated and formally presented research. To implement these guidelines; interaction design principles will be used throughout the project for the creation of a design and the execution of the testing of concept and implemented application. For the integration of a collaborative based model introduced by Zimmerman, Forlizzi, and Evenson (2007) [9], a focus on an active user role will be placed in all phases throughout the interaction design process. This is for gaining qualitative user input based on user experience, through the enabling of closer observation and probing of the target user group and the values which is desired [10].

The execution of this study will be done through a process of discovering sustainable design [1] from given user enactments and the creation of an artifact [8]. This is done through analyzing and comparing different designs and interpretations of functionalities with the user experiences [9]. The advantage of placing this focus is to gain first-hand experience on the results of the given values and thus increasing the quality of the explored design concepts.

The design will be focused on medium fidelity prototypes through the usage of advanced sketching tools. This is to easier communicate the dialogue with the user of which kind of design can best fulfill the set requirements [11].

The design of the medium fidelity prototypes will also have a focus on Krug's (2005) [12] approach to web usability. Here it is important to highlight a certain level of simplicity on the web interface, as well as the impact of how the design is presented. This is especially important in a mobile device where the available space on the interface is limited.

IV. DESIGN OUTLINES

By using the guidelines which Hevner introduces as well as the collaboration based model presented by Zimmerman, Forlizzi, and Evenson, we have integrated a focus on user driven development. Here collaboration has been done closely in sessions with representatives of the support network, one whom had a long experience with the acute and severe disease episodes (schizophrenia). This has been to gain an effective analysis from user experience based on observation on patient needs.

Based on the input gained from the executed sessions, we have made several medium fidelity prototypes through sketches which illustrate user needs and how the design of an electronic diary can log symptoms and use an integrated crisis plan. The design of the sketches will implement a collection of several sustainable functionalities which originates from own unique and respective sustained principles and functionalities. The principle of a crisis plan is implemented through both a digitalization and integration for the purpose of active usage. The different unique logging functionalities have been collected and reused in simple interfaces with the purpose of easier availability of a broader context of self-monitoring.

The central requirements which have been addressed revolve around the needs of communication, self-monitoring and addressing necessary security. Below; the design solutions based on the requirements are presented.

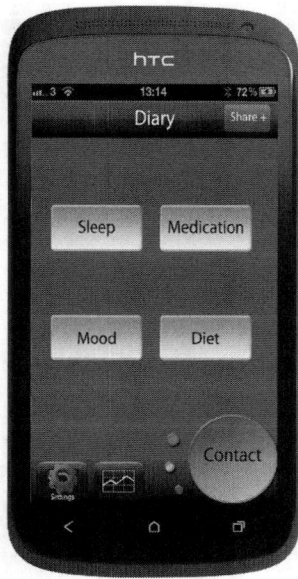

Fig. 1 Main menu with the exposed contact button

The main site design is simple and user friendly as it does not require lots of interaction from the users nor background understanding. The design has a distinguished contact button and four (4) main functionalities: sleep, medication, mood, and diet routine (Figure 1). There an option for data import and export and as well as both a general and more detailed graphical overviews. There is an option for sharing logged- and crisis plan data with represented guardians from the support network. Lastly there is also an option for interface settings for the purpose of personalization of user logging needs. However to address the need of security in the system, a log-in function is implemented before the enabling of the main site screen.

Sleep patterns (Figure 2) can easily be inputted using well established features such as calendar, an analogue scale from 1 to 10 to input quality of sleep, and rotating button for hours.

Fig. 2 Sleep data entry

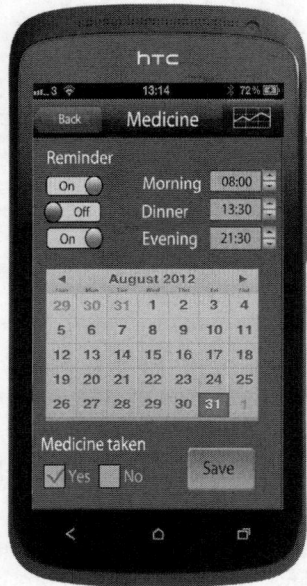

Fig. 3 Medicine Reminder Data

The intention is that the patient does this in a straight forward manner with no pressure to write the same information in his own words.

The design of a medicine reminder is again following well established features such as calendar and 'moving scales' (Figure 3). Entries' are easy to make, but the demand some

Fig. 4 Graphical overview over sleep patterns for day and month periods

interaction such as check buttons as reminders that the medication is taken. This is a very important function that will remind the user daily to keep with the mediation regime.

The value of a summarization of data will give the patient a better undersanding and alertness to act in prevention (Figure 4). Long term data could be used to reveal the patterns of medication, sleep and food routines.

The vital contacts including close relatives and personal network are implemented according to the patient's crisis plan (Figure 5). The colors used to highlight each contact represents a function of the availability of the target contact. On each contact row is a function to different types of communication (message and phone call) for quick and easy contact with the target person. The message function will send the target contact an automatically generated message which is based on the users crisis plan, for easier communicating needs of communication with the user and how problem areas should be handled.

V. RESULT

The first design was done in close sessions with users one of whom had a long experience with the acute and sever disease episodes (schizophrenia) which is reflected in the design that is logistically conscious and lean. The most important features are based on the documents used in clinical treatment and therefore containing sensitive and the most essential data.

Fig. 5 Menu with the vital network contacts: green currently available, red momentarily not accessible

The results of the executed sessions shows possibilities of reuse on functions and principles from different existing design. Relevant existing designs includes the functionality of a diary and a crisis plan, as well as several existing logging application and functions.

The initial evaluation has shown that a high degree of security could be expected from using this application. Additional value could be seen in therapy when the self-entered data becomes a part of the evaluation. The current design has advantages to easily be understood and interpreted as it is displayed.

VI. DISCUSSION AND CONCLUSION

The project is still in an early stage. This means that upgrades, refinement and building up on the functionality to address all the expectations are a central issue in this project. Upgrades can include more integrated functionality within the system, such as adding more control to the reminder buttons involved when taking or not taking medicine as well as possibilities of reporting different logging activities to the guardian and the function of triggering an alarm. Refinement on the different designs should better be presented to support a finished application.

There are additional work concerning the design of the paper versions of the important documentation and interactions based on that. An example of this concern is how to implement the entirety of the patient's crisis plan, including measures and warning signs of upcoming disease activity.

Testing in a broader user group by integrating controlled experiments [13] based on quality and quantity based data will be necessary.

REFERENCES

1. Blevis E. Sustainable interaction design: invention and disposal, renewal and reuse. in *Proceedings of the SIGCHI Conference on Human Factors in Computing Systems CHI '07*.(New York, NY: ACM Press) 2007.
2. Breier Strauss J. S.. Self-control in Psychotic Disorders *Archives of General Psychiatry*. 1983;40:1141–1145.
3. Stone Shiffman S. Schwartz J. E. Broderick J. E. Hufford M. R.. Patient non-compliance with paper diaries. *British Medical Journal*. 2002;324:1193–1194.
4. Stone Shiffman S. Schwartz J. E. Broderick J. E. Hufford M. R.. Patient compliance with paper and electronic diaries. *Control Clin Trials*. 2003;24:182–99.
5. Mutschler J Rössler W Grosshans M.. Application of electronic diaries in patients with schizophrenia and bipolar disorders. *Psychiatria Danubina*. 2012;24:206–10.
6. Green Rafaeli E. Bolger N. Shrout P. E. Reis H. T.. Paper or plastic? Data equiva-lence in paper and electronic diaries. *Psychol Methods*. 2006;11:87–105.
7. Tennen Affleck G. Coyne J. C. Larsen R. J. DeLongis A.. Paper and plastic in daily diary research: Comment on Green, Rafaeli, Bolger, Shrout, and Reis. doi: 10.1037/1082-989X.11.1.112 in *Psychological Methods*.;11:112–118 2006.
8. Hevner A. Park J. Ram S.. Design science in information systems research. *MIS Quarterly*. 2004;28:75–105.
9. Zimmerman Forlizzi J., Evenson S.. Research through design as a method for interaction design research in HCI in CHI07 Proceedings(New York, NY: ACM Press):493–502 2007.
10. Odom Zimmerman J. Davidoff S. Forlizzi J. Dey A. K. Lee M. K.. A Fieldwork of the Future with User Enactments in *Proceedings of the Designing Interactive Systems Conference*:338–347 2012.
11. Fallman D.. Design-Oriented Human-Computer Interaction in *Proceedings of Human Factors in Computing Systems Conference*:225-132 2003.
12. Krug S.. *Don't Make Me Think: A Common Sense Approach toWeb Usability*. 2.Ed. Berkeley, California USA: New Riders Publishing 2005.
13. Blandford Cox A. L. Cairns P. A. Controlled Experiments in *In Cairns, P.A., Cox, A.L. (eds.) Research Methods for Human Computer Interaction. CUP*.(New York, NY: ACM Press):1–16 2008.

Helen Sørheim, BSc:
Institute for Information and Media Science, University of Bergen
Fosswinckelsgt. 6, postbox 7802, 5020 BERGEN
Bergen
Norway
helen.soerheim@gmail.com

A Novel Data-Mining Platform to Monitor the Outcomes of Erlontinib (Tarceva) Using Social Media

A. Akay[1], A. Dragomir[2], and B.E. Erlandsson[1]

[1] Swedish Royal Institute of Technology/School of Technology and Health/IEEE Member/Huddinge/Sweden
[2] University of Houston/Department of Biomedical Engineering/IEEE/Houston/Texas

Abstract—A novel data-mining method was developed to gauge the experiences of the oncology drug Tarceva. Self-organizing maps were used to analyze forum posts numerically to infer user opinion of drug Tarceva. The result is a word list compilation correlating positive and negative word cluster groups and a web of influential users on Tarceva. The implica-tions could open new research avenues into rapid data collec-tion, feedback, and analysis that would enable improved solu-tions for public health.

Keywords—**Social Media, Self-Organizing Map, Data-Mining, TF-IDF, Graph.**

I. INTRODUCTION

Social media media is providing limitless opportunities for patients to discuss their experiences with drugs and devices and for companies to receive feedback on their products and services [1]. Pharmaceutical companies are prioritizing social network monitoring for their IT departments, poten-tially creating an opportunity for rapid dissemination and feedback of products and services to optimize and enhance delivery and reduce costs [2].

Existing network model and computational tools (i.e., graph theory) can be used to extract knowledge and trends from the information 'cloud.' Under this paradigm, a social network is a structure composed of interconnected nodes in various relationships such as interests, friendship, kinship, etc. A graphical representation is a common and useful method for visualizing and representing the information.

Network modeling can offer a deeper understanding of social network dynamics. A network model could be used for simulating many network properties such as understand-ing how users disseminate information among themselves. Another example is studying the enhancement of certain edges of networks (and how certain information affects the enhancements (e.g. how certain user communities evolve based on common interests about specific diseases).

A sociomatrix (or adjacency matrix) is a matrix repre-senting information extracted from social media. It can help construct the network representation. Social networks, though sparse, are leveragable for performing efficient analysis of the constructed networks. Node degree and other large-scale parameters can derive information about the importance of certain entities within the network (drug brands, healthcare providers or pharmaceutical companies). Such communities are *clusters*, or *modules*.

Specific algorithms can perform network-clustering, one of the fundamental tasks in network analysis. Finding a community in a social network means identifying nodes that interact with each other more frequently than nodes outside of the group. Community detection can facilitate the extrac-tion of valuable information for the healthcare industry. Pharmaceutical companies could benefit from this for better targeting their marketing spending. Healthcare providers could better understand the level of satisfaction in their services among patients. Doctors could collect important feedback (stored in the labels characterizing these network modules) from other doctors and patients that would help them in their treatment recommendations. Lastly, patients could evaluate and leverage other consumers' knowledge before making healthcare decisions.

Social networks are heterogeneous, multi-relational, and semi-structured, hindering easy data collection. One potential method is link (relationship) mining: it combines social networks, link analysis, hypertext and Web mining, graph mining, relational learning, and inductive logic pro-gramming. Researching links involve several steps: link-based object classification (categorizes objects based on links and attributes), object type prediction (predicts object types based on attributes, links, and objects linked to it), link type prediction (predicts the purpose of the link based on the objects involved), link existence prediction (predicts the existence of a link), link cardinality estimation (predict-ing the number of links (and objects reached) to an object), object reconciliation (determining whether two objects are the same based on their links), group detection (predicting if an object set belong to together), sub-graph detection (dis-covering sub-graphs within networks), and metadata mining (mining for data about data). [3]

Other methods include link prediction, which uses the in-trinsic features of the social network's current model to develop future connection models within the network. Viral marketing uses word-of-mouth effect by measuring the interactions among customers and to carefully market to individuals with the most social connections. Newsgroups discussions exploit 'response' relationships based on how often people respond to messages they agree (or disagree) with using graph-partitioning algorithms. Relation selection and extraction of a multi-relational network measures and ranks different relations based on user information (ac-quired through queries). [4]

Traditional social sciences use surveys and involve sub-jects in the data collection process, resulting in limited raw data (typically hundreds of subjects per study). However, thousands of users of social media produce inordinate amounts of data with rich user interactions. There are two ways to extract this information: 1) crawling using site provided APIs, or 2) scraping needed information from rendered html pages. Many social media sites provide APIs: Twitter, Facebook, YouTube, Flickr, etc. We can also fol-low how its properties change over time, which would greatly interest public health studies.

We used the self-organizing map (SOM) because of its visual benefits and high-level capabilities that greatly facili-tated the high-dimensional data analysis. Bonato et al. has shown how vector quantization algorithms reduce the fea-ture space's size without losing information for identifying clusters in the classification space. [5]

II. METHODS

A. Forum Search

We used the forum The Cancer Forums (http://www.cancerforums.net) based on its popularity. We compiled a list of oncology drugs and separated them by pharmaceutical manufacturer. We then performed a search in the forum for threads discussing oncology drugs, and separated them based on manufacturer. We chose Tarceva based on the large number of posts relating to it compared to the limited number of posts on other drugs and medical devices.

B. Text Mining

We used Rapidminer (www.rapidminer.com) to convert the search results of the drug Tarceva ('positive,' and 'nega-tive') into a network-ranking system reflecting the degree to which the respective network is involved in the opinion formation, and the degree of influence specific users have over the opinion formation, of Tarceva.

All Tarceva-related posts were put into an Excel spread-sheet and fed into a modified Rapidminer operator decision-tree that removed unwanted characters (HTML tags, punc-tuation, numbers) and common stop words (e.g. a, the, it, etc.). The process further broke down words into tokens and roots (e.g. working->work; lost, lose->los), resulting in a list with the term-frequency-inverse document frequency (TF-IDF) scores of selected words.

We based this model on how a vector space model can represent text retrieved from the forum: One document (forum post) represents one vector. Each vector component is associated with a particular word. A vector component is assigned a weight, denoting its importance in the document [7]. The weighted components of each vector uses the term-frequency-inverse document frequency (TF-IDF) scheme:

$$weight_{t,d} = \begin{cases} \log(tf_{t,d} + 1) \log \frac{n}{x_t} & \text{if } tf_{t,d} \geq 1 \\ 0 & \text{otherwise} \end{cases}$$

where $tf_{t,d}$ is the frequency of word t in document d, n, number of documents in collection, and x_t, number of documents where word t occurs. [8]

We measured the TF-IDF scores of the initial wordlist before splitting it into three word lists (positive words, and negative words).

C. Wordlists and Self-Organizing Maps

We fed each word list into the SOM toolbox (www.cis.hut.fi/projects/somtoolbox/) in Matlab to see which vectors clustered together based on the specified words. Each feed resulted in different vector groups clus-tered together. The clusters were checked for vector similar-ities. Cluster groups containing less than three posts, and no words of interest, were eliminated. The remaining words were counted in the remaining cluster groups.

The SOM is an artificial neural network that produces low-dimensional representation of high-dimensional data. It is a network where a neural layer (projecting the input data) represents the output space, with each neuron corresponding to a cluster with an attached weight vector. The weighted vectors' values reflect the cluster content they are attached to. The SOM presents the available data to the network, linking similar data vectors to the same neurons.

The training process presents new input data to the net-work that determines the closest weight vector and assigns the data vector to the matching neuron: such neurons (and its neighbors) change to reflect their new value. The neu-rons farther from the changed neurons rarely change. The process repeats for all input vectors until all convergence criteria are met, resulting in a two-dimensional map. We then visually identified subgroups within the map ('positive words' and 'negative words'), and ascertained which posts were gravitating towards which words and whether the map reflected consumer disposition towards Tarceva.

D. Modeling Forum Postings Using Graph Theory

Graphs were then built from forum posts and their replies. Graphs consist of nodes and connections, and are either non-directional (a connection between two points without a direction) or directional (a connection with a point of origin to an end). A non-directional nodal degree measures the number of connections of a node whilst a directional nodal degree measures the number of connec-tions from an original node and its destination(s). Wasser-mann et al. (1994) identified four different nodes within a network: Isolated (connects to no other nodes), Transmitter (connects to other nodes but does not receive them), Recep-tor (does not connect to other nodes but receives them), and Carrier (connects and receives connections). The network's density measures the current number of connections. [6] Measuring the density depends on the network: non-directional networks divide the maximum number of con-nections with the number of connections as shown below:

$$\Delta = \frac{L}{f(f-1)/2} = \frac{2L}{f(f-1)}$$

where L is the number of connections and f is the total num-ber of nodes.

Directional networks divide the maximum number of connections with the number of arrowed connections as shown below:

$$\Delta = \frac{L}{f(f-1)}$$

where L is the number of connections and f is the total num-ber of nodes.

We used Graph Theory because of its widespread use in social network analysis, and the ease with which to study, and model, user interactions and relationships. We used the directional nodal degree graph because of the nature of the forum and its internal dynamics among the members, as shown in Figure 1.

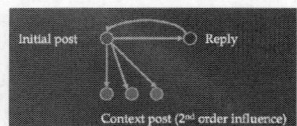

Fig. 1 The nodes represent users/posts and the edges represent information among users.

The most important posts contained the most replies. As shown in Figure 1, we added additional edges to the subse-quent posts (coded in green) in addition to the replier (coded in blue) because the subsequent posts continued to discuss the topic thread (initial post).

The forum posts were modeled as initial posts-reply (a pair of edges with opposite directions) and context posts-reply (one edge flowing to one direction). We only needed the post and replier indexes as shown in Figure 2:

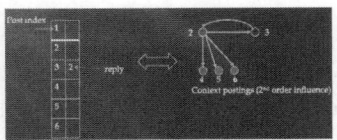

Fig. 2 The post and replier indexes without the wordlist vectors as this point.

E. Identifying Sub-Graphs

Our modeling framework has converted the forum posts into a directional graph containing a number of densely connected units (or sub-graphs) and unconnected nodes shown in the Figure 3 below:

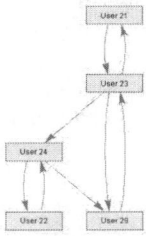

Fig. 3 The pruned sub-graph and influential users

F. Module Average Opinion and User Average Opinion

Using the wordlists and TF-IDF scores, we then defined the module average opinion (MAO) within each module by examining the TF-IDF scores of all postings corresponding to the nodes within the current module:

$$MAO = \frac{Sum_+ - Sum_-}{Sum_{all}}$$

Where: $Sum_+ = \sum \sum x_{ij}$ is the sum of all TF-IDF scores corresponding to positive words for all users within the current module. The unit i is the node/post index. The unit j is the wordlist index (numbered 1 through 15).

$Sum_- = \sum \sum x_{ij}$ is the sum of all TF-IDF scores corresponding to negative words for all users within the current module. The unit i is the node/post index. The unit j is the wordlist index (numbered 16 through 30).

$Sum_{all} = \sum_{i=1}^{N} \sum_{k=1}^{M} x_{ik}$ is the total of both sums. The unit k represents the index of the whole wordlist.

Rank users within each module via their edge degrees (within each module compute for each user the UAO score (using the TF-IDF scores from wordlists))

Similarly, we defined the user average opinion (UAO) by examining the TF-IDF scores of the post corresponding to the specific node within the current module:

$$UAO = \frac{Sum_{i+} - Sum_{i-}}{Sum_{iall}}$$

Where: $Sum_{i+} = \sum_{x_i \neq 0} x_i$ is the sum of all TF-IDF scores corresponding to positive words for all users within the current module. The unit *i* represent the indexes corresponding to the positive words in our wordlist.

Where: $Sum_{i-} = \sum_{x_j \neq 0} x_j$ is the sum of all TF-IDF scores corresponding to negative words for all users within the current module. The unit *i* represent the indexes corresponding to the positive words in our wordlist.

$Sum_{iall\_} = \sum_{k=1}^{M} x_k$ is the total of both sums. The unit k represents the index of the whole wordlist.

III. CONCLUSIONS

Figure 4 is the graphical representation of the SOM of the positive words group. We used a 13 x 13 map size with thirty variables (the most frequently used positive and nega-tive words from the forum) present to ascertain the weight of the words corresponded to the opinion of the drug Tarce-va. A sizable number of positive words and negative words converged in certain points of the map (the upper and lower portions, respectively). The U-matrix shows three high-concentration areas (with the highest concentration located at the upper right-hand corner), revealing a slightly favora-ble view of the drug Tarceva. The negative opinion stems from either the drug's side effects, drug performance issues, and concern over family members who have cancer.

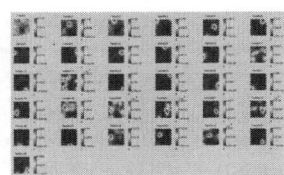

Fig. 4 Positive and Negative Words.

We next identified specific, influential users within the forum. One user out of the 568 posts was identified using the algorithm as information brokers. This user was identi-fied based on the quality of his posts, which other forum members used to search for more information on Tarceva.

IV. DISCUSSION

The goal of this study was to transform the posts in an oncology forum into vectors to scan for patterns in consum-er disposition towards Tarceva. The results open new ave-nues of research into the development of a more thorough web of user influence (based on quality of posts and ranking within the forum) algorithms and how that influence (based on quality of posts and ranking within the forum) affects interactions with other users (replies, friendships, the timing and quality of posts in threads). [9] The new research ave-nues will counter the limitations in this study (focus on one drug, focus on one forum, in-depth interaction analysis). Social media represents a goal mine for companies seeking to optimize health delivery and reduce costs.

REFERENCES

1. Alberto Ochoa, Arturo Hernández, Laura Cruz, Julio Ponce, Fernando Montes, Liang Li and Lenka Janacek (2010). Artificial Societies and Social Simulation Using Ant Colony, Particle Swarm Optimization and Cultural Algorithms, New Achievements in Evolutionary Computation, Peter Korosec (Ed.), ISBN: 978-953-307-053-7, InTech (http://www.intechopen.com/articles/show/title/artificial-societies-and-social-simulation-using-ant-colony-particle-swarm-optimization-and-cultural)
2. "Pharma 2.0 – Social Media and Pharmaceutical Sales and Market-ing"
3. Jiawei Han and Micheline Kamber "Data Mining: Concepts and Techniques" 2nd ed., Morgan Kaufmann, 2006
4. Gundecha, P., Liu, H. "Mining Social Media: A Brief Introduction" P. Mirchandani, ed. INFORMS TutORials in Operations Research, Vol. 9. INFORMS, Hanover, MD, pp.1-17
5. Bonato P, Mork PJ, Sherrill DM, Westgaard RH., "Data mining of motor patterns recordedwith wearable technology," *IEEE Eng Med Biol Mag.*," vol. 22. No. 3, pp. 110-119, May-June 2003] Bonato P, Mork PJ, Sherrill DM, Westgaard RH., "Data mining of motor pat-terns recordedwith wearable technology," *IEEE Eng Med Biol Mag.*," vol. 22. No. 3, pp. 110-119, May-June 2003
6. Passmore, D., "Social Network Analysis: Theory and Applications"
7. Lynn A. Dunbrack. Feb. 23 2010.
8. Beyond TFIDF Weighting For Text Categorization in the Vector Space Model, P. Soucy, G. Mineau, IJCAI 2005
9. Identifying influential users in an online healthcare social network, X. Tang, C.C Yang, ISI 2010

Author: Akay, Altug
Institute: Swedish Royal Institute of Technology
Street: Alfred Nobels Alle 10, SE-141 52
City: HuddingeCountry: Sweden
Email: altu@kth.se

Advanced Networked Modular Personal Dosimetry System

R. Chil[1], L.M. Fraile[2], J. Vaquero[3], E. Picado[2], A. Rodriguez-Moreno[3], M.C. Rodriguez-Sanchez[3], S. Borromeo[3], M. Desco[1,4], J.M. Udías[2], and J.J. Vaquero[4]

[1] Instituto de Investigación Sanitaria Gregorio Marañón, Madrid, Spain
[2] Grupo de Física Nuclear, Univ. Complutense de Madrid, Madrid, Spain
[3] Departamento de Tecnología Electrónica, Univ. Rey Juan Carlos, Móstoles, Spain
[4] Departamento de Bioingeniería e Ingeniería Aeroespacial. Univ. Carlos III de Madrid, Leganés, Madrid

Abstract—A personal radiation level monitoring system has been developed based on an autonomous, advanced personal dosimeter. This dosimeter uses a scintillator-SiPM combination for the radiation detection, it integrates spectroscopy and isotope identification capabilities. In order to choose the optimum scintillator crystal and photosensor for a portable gamma dosimeter, several tests on energy resolution and internal activity were performed. We have chosen a 9x9x20mm^3 CsI:Tl crystal and a 50 μM SiPM. All measurements are time and location stamped, even for indoor locations. The system implements Wireless Sensor Networks access in such a way that several networked personal dosimeters can be considered mobile sensors providing a high sampling rate over an extended area. The system developed is modular, it consists of three elements: the Personal Dosimeters, a Wireless Gateway deployment and a web-based remote storage, monitoring and management system, named Central System.

Keywords—Radiation level monitoring, personal dosimetry SiPM, gamma-spectroscopy, mobile wireless sensor network, intelligent sensing.

I. INTRODUCTION

Personal dosimeters are commonly used in sectors were occupational exposure is to be expected, such as nuclear power plants, nuclear science researchers, and hospital personnel working in nuclear medicine or radiation-therapy.

To this end it is common practice to use film-badges and thermo luminescence detectors to measure the dose over a certain period of time. Spectroscopic dosimetry gives a more accurate measurement of the dose rate, allowing us to better estimate the biological damage due to radiation exposure. Also electronic dosimeters offer the advantage of real time read-out, which is not possible with passive compact radiation sensors like radiographic films [1].

Some commercial applications can be found offering dose rate measurement and isotope identification [2, 3], but none of them include the possibility to monitor large areas simultaneously using real time monitoring systems. Although the subject of radiation level monitoring has been treated before, focusing mainly on Homeland Security [4], not many details about detector implementation were given. Similar systems that relayed on pulse counting [5] could not be used for isotope identification.

The relevance of Wireless Sensor Networks (WSNs) is commonly accepted [6]. They are based on modules (nodes or motes) equipped sensors, smart data processing, data storage capabilities, autonomous power supply and a radio chipset. Usually, these are low power devices oriented to perform distributed sensing and actuation tasks [7], including features such as localization and synchronization among different sensors. A Personal Dosimeter (PD) can be considered as a sensor, and any facility in which PD are mandatory can be treated as a WSN based system if the PD are mobile sensors providing a high sampling rate.

II. SYSTEM ARCHITECTURE DESCRIPTION

The architecture of the complete system consists of three elements; the previously mentioned PD, for radiation detection and data pre-processing, storage and transmission; second a Wireless Gateway (WG) deployment, for communication between the PD and the third element, a web-based storage, monitoring and data management system, named Central System (CS), for remote data storage, processing and PD management. The number of WG deployed depends on the application scenario necessities and restrictions, in order to obtain the required coverage for the PD.

A. Personal Dosimeter

The dosimeter consists of a radiation detector with its corresponding conditioning electronics, an analog-to-digital converter (ADC) and a microprocessor-based system with data pre-processing (radiation histograms generation of single events), data storage, conditions or alarms detection, and wireless communications capabilities. It also has a positioning system and a real time clock, so every histogram is time and place stamped. Finally, each dosimeter has a unique identification.

The PD stores and wirelessly sends an instantaneous and a cumulative histogram every two seconds, to any of the available WG. If there is no WG coverage, the PD keeps storing the data (histogram, time and position) in an internal memory until a WG is found, automatically restarting the communication.

B. Wireless Gateway

A WG is a communication bridge between PD and CS. It consists of a wireless router that connects with the existing PD in its coverage area. It collects the data from different PDs and sends it to the CS, via Internet, using the available connection (Ethernet, Wi-Fi or 3G). The collected histograms are stored in a remote database on the CS. When GPS signal is not available (indoors), localization information about the PD can be collected using beacons and Received Signal Strength Intensity (RSSI) techniques.

C. Central System

The CS implements a database where histograms and all information related to PD identification is stored. A multiplatform web-based software interface enables access to this database for histogram processing, isotope identification and PD management.

It also allows remote near real-time PD monitoring, including localization and alarms detection, and generation of dynamic radiation maps. Although not yet implemented in the current version of the PD, the CS could also send commands or messages to one or all (broadcasting) PDs in order to inform or warn the user.

III. SYSTEM IMPLEMENTATION

A prototype of the architecture described has been physically implemented as shown in figure 1.

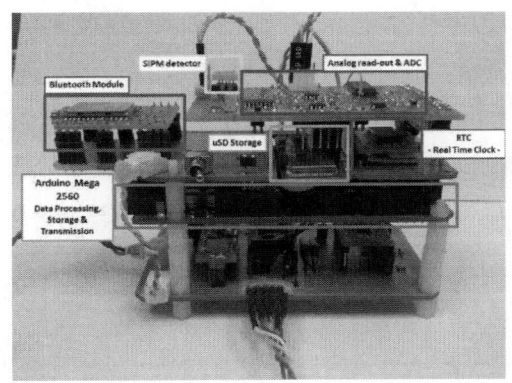

Fig. 1 System prototype

A. Detector and Crystal Selection

Crystals and photosensors: We tested five alkaline scintillators ($LaBr_3$:Ce, $CeBr_3$, NaI:Tl, KI:Tl, CsI:Tl) and seven non-alkaline ones (BaF_2, GSO, BGO, MLS, LYSO, LuAG:Pr, LFS). The crystals were coupled to four different types of photosensors. Two of them were photomultiplier tubes (PMT, XP2020Q and XP20D0, Photonis) and the others were silicon photomultipliers (SiPM, S10985-50C Hamamatsu, and SPM Micro3035x13, SensL). The SiPMs are multipixel arrays of 2x2 (Hamamatsu) or 4x4 (SensL) pixels. Specific read out circuits were developed for each SiPM. *Intrinsic activity* for each crystal was measured with the help of an HPGe detector, and gamma sources (^{22}Na and ^{137}Cs) were employed to measure energy resolution.

B. Personal Dosimeter

Starting with a standard photodetector front end, we simplified the design in order to reduce both dimensions and power consumption. For this fast detector the maximum expected count rate is around 40 kcps so a base-line restorer probed not to be needed. We used a combination of a latch and a comparator to drive the integration of the detector signal that was set to 10ns for a CsI, which is enough to get a good energy resolution.

The typical pulse width obtained with a CsI crystal is around 2 μs, which allows us to use a 16 bit microcontroller to control the conversion timing. The system is depicted in Fig. 2

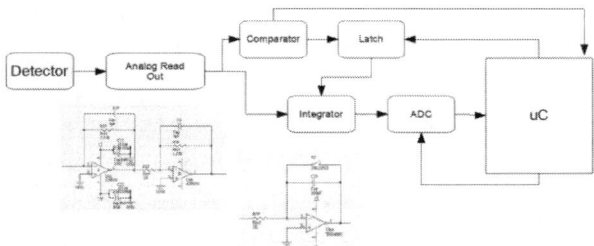

Fig. 2 Electronics diagram

The data from the ADC is read by a microcontroller, (ATmega 2560, Atmel) and from its 12-bit resolution only the 10 most significant bits are used in order to achieve the desired linearity. The maximum data rate is 40ksps, so every data is read and processed in less than 10μs. Up to 65,535 pulses could be stored using two bytes of memory per channel. For the cumulative histogram a total of four bytes per channel has been reserved.

Both histograms (instantaneous and cumulative) are stored locally in a *microSD* non-volatile memory card of 2GB, allowing more than one million instantaneous histograms plus the cumulative one. Therefore, the PD could permanently store histograms corresponding to 24 days of continuous operation, batteries allowing. This memory also stores the unique identification of each PD, the Media

Access Control (MAC) address of the radio chipset used for wireless communication in each PD, associated with the user in the CS database.

For the time stamp of each histogram, a Real Time Clock (RTC, DS3234, Maxim Integrated) has been used. Communications of both the microSD memory and the RTC with the microcontroller is done via Serial Peripheral Interface (SPI) protocol.

Because of the required data rate of 2kB (instantaneous histogram) plus 4kB (cumulative histogram) and 15 bytes of maximum overhead every two seconds, the bandwidth of typical WSN communication protocols, such as ZigBee is not enough, so Bluetooth (BT) was chosen for PD wireless communication. The radio chipset (WT12, Bluegiga) sends the information packet in less than 300ms. A Wi-Fi chipset could be used, but at the expense of a much higher power consumption. The communications with the microcontroller is done via Universal Asynchronous Receiver-Transmitter protocol (UART) with a maximum data rate of 2Mbps, enough for the application.

Outdoors localization is implemented with a GPS chipset (XT65, Cinterion) that communicates via UART with the microcontroller. Although not yet implemented, the access to Internet, and so to the CS, could be done directly via 3G, without the WG bridge.

Finally, alarms and user indications have been implemented, giving visual information to the user about the PD on-off state, wireless communications active (on coverage of WG), or pre-fixed radiation levels reached. These events are also sent with their associated histograms to the CS.

The PD is powered from a rechargeable Li-Polymer battery (3.7V, 2.000mAh, Varta). The estimated maximum total consumption is 250mA, so a minimum of 10 hours of continuous operation is expected.

A step-up DC/DC converter (NCP1450, On-Semiconductor) powered from the battery provides a regulated +5V for the digital electronics. A ±5V DC/DC converter (NMA5050, Murata) powers the analog electronics from the regulated +5V. Finally, a programmable DC/DC high voltage converter (SIP90, Emco) supplies the bias voltage for the SiPM detector with nominal +70V. This voltage must be controlled, with a precision of 100mV, according to the detector temperature, so a temperature sensor with Inter-Integrated Circuit (I^2C) communication protocol (LM73, Texas Instruments), is read by a microcontroller (ATmega 328, Atmel). The programmable value is set via a digital-to-analog converter (DAC, AD5231, Analog Devices). A voltage feedback loop dynamically adjusts the +70V output.

C. Wireless Gateway

The WG has been implemented using a Raspberry Pi platform, which includes an ARM11 processor at 700MHz

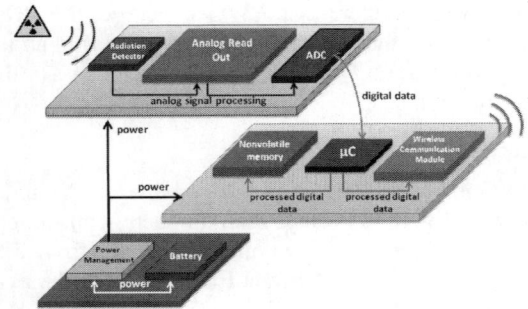

Fig. 3 Personal dosimeter layout

running a Linux Operating System (OS). It has two USB 2.0 ports where Bluetooth-USB adapters (F8T017, Belkin) are connected for PD coverage. They implement BT 2.1+EDR version with a data rate up to 3Mbps and 100 meter coverage (line of sight, LOS). It also has an Ethernet port for Internet access to connect with the CS database. A USB-Wi-Fi adapter (MN-WD152B, MicroNEXT) has been used where Ethernet connection is not available. Finally a GPS/3G module, Raspberry Pi compatible, based on the SIM5218 module from SIMcom, could be used for 3G-network coverage.

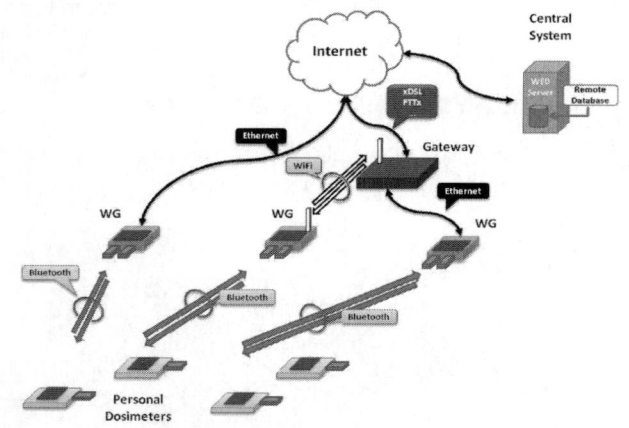

Fig. 4 System Architecture

D. Central System

The CS is an Apache Tomcat web server on a Linux, Ubuntu distribution, and a MySQL database. A web interface connects to a Java application running on the server for database accessing. This provides secure remote PD monitoring and data storage and processing. The architecture is scalable, so several CS could also be monitored from another CS.

The web application allows access to the web server database only to authorized users. Although not yet implemented, the application may allow the CS to send control and query commands to the PD. Finally, it also includes PD management features such as user registration.

IV. RESULTS AND CONCLUSIONS

Crystals with important internal activity, mostly coming from β–transitions of ^{176}Lu and ^{138}La, were deemed useless for low-level gamma measurements, which are one of the main purposes of the dosimeter. From the remaining crystals, the 9x9x20 mm3 of CsI:Tl has the best energy resolution, acceptable sensitivity and it's emission wavelength suits well the photodetector spectral sensitivity. Therefore it was selected as our best detector choice.

A full characterization of the dosimeter is underway. Spectra from ^{22}Na and ^{137}Cs sources have already been obtained allowing us to expect an energy range between 50 keV and 2 MeV, with a maximum count rate of 40 kcps.

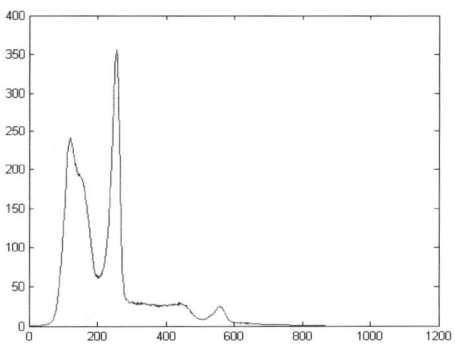

Fig. 5 ^{22}Na spectrum.

We have chosen as photosensor a SiPM because of the size limitations for this application. They are easy to use, require simple electronics read-out, have high gain, small size, and they are competitive in price compared to PMTs. Their limitations are their temperature dependency and their saturation at high energies. For the application considered here, saturation at high energies is of no concern, and linearizing the response in that range is feasible. The temperature dependence was compensated adjusting the bias supply.

Although the current demonstrator is made of a series of independent PCBs, once integrated the personal dosimeter will have a size similar to that of a mobile phone so that users can easily carry it. The expected battery life is 10 hours. It would allow real time radiation level monitoring over wide areas, including the creation of radioactivity maps and the detection of hotspots. It also has isotope detection capabilities whose reliability has been tested. The dosimeter will also have built in alarm functions to alert the user if needed.

ACKNOWLEDGMENT

This work was supported in part by projects PR13/09-16797 (Consejo de Seguridad Nuclear, CSN), CENIT-AMIT, CEN-20101014 (Ministerio de Ciencia e Innovación), ARTEMIS S2009/DPI-1802 (Comunidad de Madrid, Fondo Social Europeo).

REFERENCES

1. Knoll G.F.. Radiation detection and measurement. Wiley 2000.
2. Harvey Richard. Radiation Detector Module C12137 Hamamatsu News. 2012;1:21.
3. Angel Bennett FLIR Systems Inc. FLIR Systems Introduces New Personal Radiation Detector, nanoRaider FLIR Company News. 2011.
4. Neuer Marcus J., Ruhnau Kai, Ruhnau Arne, et al. Surveillance of nuclear threats using multiple, autonomous detection units in Nuclear Science Symposium Conference Record, 2008. NSS '08. IEEE: 3324-3329 2008.
5. Waguespack R., Pellegrin S., Millet B., Wilson C.G.. Integrated System for Wireless Radiation Detection and Tracking in Region 5 Technical Conference, 2007 IEEE:29-31 2007
6. J. Yick B. Mukherjee, Ghosal. D. Wireless sensor network Survey Computer Networks. 2008;52:12.
7. Kasimoglu I. F. AkyildizandI. H.. Wireless sensor and actor networks: Research challenges, Ad Hoc Networks Computer Networks. 2004;vol. 2, no. 4:351367

Methods for Personalized Diagnostics

D.L. Hudson and M.E. Cohen

University of California, San Francisco, Graduate Group in Biomedical Engineering, San Francisco, CA, USA

Abstract—Historically medical diagnosis has relied on observation of signs and symptom followed by comparison of these signs and symptoms to profiles of known instances of the disease. While this approach works well in straight-forward diagnosis in commonly found diseases it may not work well for rare diseases or for a common disease that does not present with the normal symptom pattern or when symptoms are the result of the presence of more than one disease. New technologies can assist in the diagnostic process through a number of paradigms, including examination of the life-long patient history, analysis of genetics components, and the inclusion of multidimensional medical decision making that makes full use of automated scanning and analysis. In this article specific approaches that can contribute to personal diagnosis are coupled with a structure for maintaining life-long patient and family history data.

Keywords—Personal health records, multidimensional decision making, disease models, special data types, individual healthcare.

I. INTRODUCTION

The practice of medicine has developed rapidly in the last century based on numerous scientific discoveries and advances in technology [1]. In spite of these advances some traditional components of diagnosis and treatment have remained. The increasingly rapid advance of technology has had a major impact on diagnosis, particularly through discoveries of pertaining to heart monitoring including ECG analysis [2] and Holter monitoring [3] for the detection of cardiac disorders and the invention of computed tomography [4] and magnetic resonance imaging that have yielded insight [5] into the internal functioning of the human body. Because of recent discoveries a much larger quantity of variables are available for diagnosis. The drawback is that human decision making is only highly functional when fewer than five or six variables are present. The first step is to define and implement automated methods for decision making that will make use of all relevant variables [6]. Secondly, the previous history of each patient can lend insight into the current state of the patient [7]. In order to take full advantage of the patient history a lifetime history should be available. This history also presents a challenge to human decision making due to the increasing content as the patient ages. Again, automated methods must be developed to help in the integration of the patient's personal history with the current presenting signs and symptoms. In this article methods will be presented for the structure, maintenance, and access to the life-long medical record along with a basis for providing automated analysis of the patient record in the context of the current patient encounter.

II. METHODS

A. Numerical Data

Traditionally numerical data is the most easily handled by automated systems. Many common medical tests are recorded in this format. An example is the basic blood panel that is often ordered in a regular check-up. Table I shows examples of the components of this record. While this test gives a snapshot of the current condition of the patient, its usefulness is improved when compared to previous tests. If this information is stored in the personal health record, automated analysis can compare current values to the previous test as well as do an analysis of the history of the values over any specified timeframe. The same can easily be implemented for any numeric variable as long as a personal health record is available and also portable for

Table 1 Medical Blood Test Results

Test	Result	Normal Range
Glucose	90	70-110
LDL	140	< 100
HDL	50	> 30
Systolic BP	128	115-135
Diastolic BP	78	60-80
BUN	17	8-24
Creatinine	1.1	0.6-1.2
Potassium	3.0	3.5-5.4
Sodium	141	133-146
RBC	4.0	4.2-6.9
WBC	9.1	4.3-10.8
Albumin	4.1	3.4-5.4
Bilirubin	.33	< 0.4
Calcium	7.8	8.2-10.6
CO_2	36	35-45

access from other physician offices or hospitals. Solutions for portability are discussed later in this article.

Neural networks are a reliable decision making method as long as the following conditions hold: all variables are numeric and sufficient data are available for definition of both a training set and a test set to develop weighting factors for each of the components. In the applications described here, the Hypernet neural network developed by the authors is employed [8].

A. Symbolic Data

Symbolic data are common in medical applications. The traditional method for handling symbolic information is the expert system in which premises and conclusions are utilized to analyze data items. In work by the authors [9] several extensions are included to permit the use of uncertain reasoning. In this approach premises are weighted to indicate their significance in the decision and symptoms are evaluated based on the degree of presence. Evidence is then aggregated to determine the strength of the evidence. The condition or finding is substantiated if the strength exceeds the pre-specified threshold level. In the traditional use of the method the physician enters the degree of presence of each symptom. The automated version searches the patient record to determine the degree and automatically invokes the rule [10]. The rule format is given below:

Antecedent	Importance	Degree of Presence
Finding 1	a_1	d_1
Finding 2	a_2	d_2
.		
.		
.		
Finding n	a_n	d_n

The condition is confirmed if S > Threshold T where

$$S = \sum_{i=1,n} a_i d_i q \tag{1}$$

C. Signal Analysis

Signal data is an increasingly important aspect of medical diagnosis and treatment. It encompasses a number of technologies including electrocardiograms (ECG) [11] and Holter monitoring [12] for diagnosis of cardiac disorders, electroencephalograms (EEG) for diagnosis of neurological conditions, as well as a number of other signals such as the electromyogram (EMG) [13] that measures electrical activity of the muscles and is a measure how well and how fast the nerves can send electrical signals. It is used to find diseases that damage muscle tissue, nerves, or the junctions between nerve and muscle and to find the cause of weakness, paralysis, or muscle twitching. Automated analysis in the patient record requires the availability of prior results along with the current result. In general, the information is stored in the record in one of two formats: 1) as a textual evaluation of the results or 2) as a numerical evaluation. The latter is usually only applied to long time series such as the Holter monitor that consists of a 24-hour recording of the time between heartbeats, known as the R-R interval. A recording device that hangs from the patient's shoulder is used for the recording done while the patient goes about daily activities. The result is a time series in excess of 100,000 points. A method based on chaos theory is used to summarize this information. It uses the central tendency measure (CTM), computed by selecting a circular region around the origin of radius r, counting the number of points that fall within the radius, and dividing by the total number of points. Let t = total number of points, and r = radius of central area [14]. Then

$$CTM = [\sum_{i=1}^{t-2} \delta(d_i)]/(t-2) \tag{2}$$

where

$$\delta(d_i) = \begin{cases} 1 & if\ [(a_{i+2}-a_{i+1})^2 + (a_{i+1}-a_i)^2]^{.5} < r \\ 0 & otherwise \end{cases}$$

In previous studies [15] there were no patients with a CTM value over 0.90 who had congestive heart failure and no patients with a CTM below 0.50 who did not have CHF. In the range between these two values additional medical parameters were required to determine disease status. In automated analysis of the patient record it is vital to tract the CTM to determine if changes are occurring.

D. Automated Comparison within the Patient Record

The flowchart in Figure 1 illustrates the overall procedure using available data in the patient's personal health record to take full advantage of current and historical information. The current visit in general will only record information that is relevant to a small part of the life-long patient record. The process can take a number of forms:

- Compare current values to previous value
- Compare and current values with data for 1 year
- Compare current values with data for 3 years
- Compare current values with all available data.

Results can be displayed as text or as graphical trends. An additional step to compare findings with disease profiles.

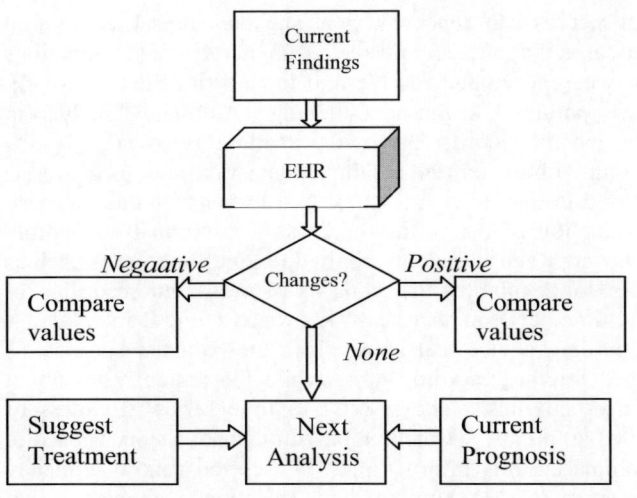

Fig 1 Automated Analysis

A trend analysis algorithm can be used to give immediate automated feedback to the physician [16].

For all confirmed conditions
   If condition i is present with $\delta(C)=a$
     If condition i was previously present with $\delta(C) = b$
     and if $|a-b| > x$ then send alert
     If condition i was previous not present then send alert
For all previously-confirmed conditions
   If condition is not currently present send notification
   If condition i is present with $\delta(C)=a$
     If condition i was previously present with $\delta(C) = b$
     and if $|a-b| > x$ then send notification

The value for x is determined for each condition based on expert input and statistical evaluation of previous cases. Note that this process simply alerts the physician who decides if action is necessary. It should also be noted that this process will send a notification if a condition improves or no longer exists which may result in re-evaluation of treatments.

III. RESULTS

The methods described above have been used both alone and together as a hybrid system to develop disease models for a variety of conditions including myocardial infarction, lung cancer, melanoma, congestive heart failure, and Alzheimer's disease, among others. Each disease model has its own characteristics which may change over time as more information in gathered. Some examples follow.

Disease Model 1: Task: Rule out possible myocardial infarction

The symbolic model, EMERGE, is used. A sample rule is given below. The evidence s aggregated using eqn (1). If the threshold is exceeded, the condition is verified.

Table 2 EMERGE Rule

Rule for Chest Pain/MI Threshold T = 0.6	Weighting Factor	Degree of Presence
BP < 100/60	0.5	0.6
Abnormal mental status	0.1	0.1
Cold, clammy skin	0.1	0.1
Gray, cyanotic skin	0.1	0.1
Weak peripheral pulses	0.1	0.3
Urinary output < 30 cc/hr.	0.1	0.6

Disease Model 2: Task: Rule out possible coronary artery disease

Methods: Combines two methods: hybrid neural network and CTM model

In this model, the value for component x_1 is determined from the CTM value determined through the Holter tape evaluation. Components x_2 through x_7 are numerical values taken from the patient's health record.
  x_1 Abnormal Holter ECG
  x_2 Dyspnea
  x_3 Orthopnea
  x_4 Edema
  x_5 Functional impairment
  x_6 PND
  x_7 BUN

Values are normalized to the interval [0,1]. The pre-trained neural network:

$$D(x) = \sum_{i=1}^{7} w_i x_i + \sum_{i=1}^{7} \sum_{\substack{j=1 \\ i \neq j}}^{7} w_{i,j} x_i x_j \qquad (7)$$

IV. DISCUSSION

A number of barriers exist which must be overcome to achieve both a life-long personal health record and sophisticated and complete disease models.

Barrier 1: The Comprehensive Personal Health Record
The goal is a life-long electronic portable personal health record [PHR] that would permit diagnosis and treatment based on the individual rather than on population statistics.

Portability can be achieved by either a standardized record format or through the use of HL7-type standards cam be properly identified].

Barrier 2: Definition of Disease Models

Automated analysis requires the definition of a broad range of disease models that must teams that include physicians and other health care professionals with deep knowledge of the specific domain working with bioengineers and other IT professionals. De-identified databases for specific diseases can assist model development [17].

Barrier 3: Model updating

Disease models are not static. When more evidence becomes available a seamless method must be established for updating existing models and creating new models when necessary.

Barrier 4: Acceptance

Both healthcare professionals and patients must have confidence in the automated analysis process.

V. CONCLUSION

The journey to comprehensive life-long medical records will require collaboration among many research groups as well as the introduction of new methods for definition of the life-long health record. Once established the current approach of diagnosing and treating disease can shift from knowledge-based population statistics to true personalized medicine. Both new methods of analysis and the introduction of efficient search engines will be required to bring all evidence to bear to bring about the goal of true personalized medicine. The methods described here will need to expand to encompass new knowledge regarding specific diseases as well as new disease paradigms. In the future keeping models up-to-date will require large dedicated groups of researchers.

REFERENCES

1. Viangteeravat, T, Anyanwu, MN, Nagisetty, V, Kuscu, E, Sakauye, ME, Wu,,D, Clinical data integration of distributed data sources using Health Level Seven (HL7) v3-RIM mapping, Journal of Clinical Bioinformatics, 1:32, doi:10.1186/2043-9113-1-32, 2011.
2. Rowley, DW, Glagov, S,Heart Attacks, Heart Rate, and Gel Electrodes: The Invention of Ambulatory Cardiology, Perspective in Biology and Medicine, 49, 346-356, 2006.
3. Cohen, ME, Hudson, DL, Deedwania, PC, Measurement of variability in Holter tape R-R intervals for patients with congestive heart failure, IEEE Engineering in Medicine and Biology, 16: 127-128, 1994.
4. Moon SH, Cho SK, Kim WS, Kim SJ, Chan Ahn Y, Choe YS, Lee KH, Kim BT, Choi JY, The Role of 18F-FDG PET/CT for Initial Staging of Nasal Type Natural Killer/T-Cell Lymphoma: A Comparison with Conventional Staging Methods, J Nucl Med, 2013.
5. Cheriff, AD, Kapur,AG, Qiu, MC, Cole, CL,Neurovisualization of the dynamics of real and simulation biofeedback: functional magnetic resonance imaging study, International Journal of Medical Informatics, 79(7), 492-500, 2010
6. Hudson, DL, Cohen, ME, distributed Medical Decision Support Systems, ISCA Computer Applications in Industry and Engineering, 23:6-11, 2011.
7. Hudson, DL, Cohen, ME, Uncertainty and Complexity in Personal Health Records, IEEE Engineering in Medicine and Biology, 32:6773-6776, 2010.
8. Faust O, Acharya UR, Tamura T.Formal design methods for reliable computer-aided diagnosis: a review, IEEE Rev Biomed Eng. 2012;5:15-28.
9. Hudson, DL, Cohen, ME, (2010) Diagnostic Models Based on Personalized Analysis of Trends (PAT), IEEE Transactions on Information Technology in Biomedicine, 14(4): 941-948
10. Hudson, DL, Cohen, ME, (2010) Automated Signal Processing for Biomedical Decision Making, ISCA Computers and Their Applications, 25:87-92, 2010
11. Kramer, MA, Chang, FL, Cohen, ME, Hudson, D, Szeri, AJ. (2007) Synchronization measures of the scalp electroencephalogram can discriminate healthy from Alzheimer's subjects, International Journal of Neural Systems, 17(2):61-69
12. Cohen, ME, Hudson, DL. (2007) Combining evidence in hybrid medical decision support systems, IEEE Engineering in Medicine and Biology, 29:5114-5147
13. Cukic M, Oommen J, Mutavdzic D, Jorgovanovic N, Ljubisavljevic M.The effect of single-pulse transcranial magnetic stimulation and peripheral nerve stimulation on complexity of EMG signal: fractal analysis,Exp Brain Res. 2013 May
14. Cohen, ME, Hudson, DL, Deedwania, PC. The use of continuous chaotic modeling in differentiation of categories of heart disease, Inf. Proc. Manage. of Uncertainty in KB Systems, 7:548-554, 1998.
15. Hudson, DL, Cohen, ME. Temporal Trend Analysis in Personal Health Records, IEEE Engineering in Medicine and Biology, 30:3811-3814, 2008.
16. Hudson, DL Cohen, ME, Merging Medical Informatics and Automated Diagnostic Methods, IEEE Engineering in Medicine and Biology, 35, 2013, in pres
17. Hudson, DL, Cohen, ME. Temporal Trend Analysis in Personal Health Records, IEEE Engineering in Medicine and Biology, 30:3811-3814, 2008

Author: Donna L.Hudson, Ph.D.
Institute: University of California, San Francisco
Street: 155 North Fresno Street
City: Fresno, CA 93701
Country: USA
Email: dhudson@fresno.ucsf.edu

The Prediction of Blood Pressure Changes by the Habit of Walking

Toshiyo Tammura[1], Soichi Maeno[2], Yutaka Kimira[3], Yuichi Kimura[2], Takumu Hattori[4], and Kotaro Minato[4]

[1] Department of Biomedical Engineering, Osaka Electro-communication University, Neyagawa, Osaka, Japan
[2] Department of Biomedical Engineering Chiba University, Chiba, Japan
[3] Health Science Centre, Kansai Medical University, Hirakata, Osaka
[4] Graduate School of Information Science, Nara Institute of Science and Technology

Abstract—A number of walking step per day is an important parameter to evaluate daily activities as well as blood pressure control. In this study, we used blood pressure and walking step data, obtained from 48 patients in a home healthcare system, to investigate the effectiveness of the number of walking steps in a clinical setting. Changes in blood pressure and walking steps per day were compared. Our results indicate that walking, as a regular form of exercise, contributed to lowering of blood pressure. Thus the daily walking is useful for improving the quality of life of subjects in the home healthcare setting.

Keywords—**Blood pressure, walking step, Clusgter Clinical evaluation.**

I. INTRODUCTION

The application of information communication technology (ICT) to healthcare is now very common. There are two ways; one is general healthcare, the other is healthcare for post-operated patients. There are many researches as well as commercial system for both applications. The post-operative healthcare was effective to reduce the cost both medical insurance and medical treatment [1]. In contrast, the home healthcare for the elderly and middle aged people who need to care and prevention and prediction of disease could not find any good evidence of the ICT system. Last ten year, many researchers investigated the best ICT system to healthcare. However, home healthcare is not popularly distributed. The main disadvantage is to the motivation of the clients.

However, personal health care is very important issue to reduce the medical insurance cost.

We have developed and evaluated the home healthcare system. The main system consists of automatic monitoring, data transfer and data storage. The most commercially available devices are the same structure. The automatic monitoring for physiological parameters such as systolic and diastolic blood pressures, body weight and walking steps is common for the market. The Continua interface, G3 transformation as well as general infrared wireless communication and low power-wireless transmission are applied to the system. As an engineering point of view, the wireless and standard protocol of interface are useful and motivated. However, participant ratio of home healthcare system is not so high. Some literatures showed the non-effectiveness of ICT systems [2]

In this paper, we investigated the effectiveness and usage of ICT system. Then from the result of usage in home healthcare system, the possibility of the prediction in the blood pressure changes by the walking step records.

II. METHOD

A. Home Healthcare System

The home healthcare system monitored blood pressure, body weight, and the number of steps taken per day. These parameters were measured using home healthcare devices that met the Continua Health Alliance standards, as shown in Figure 1. Systolic and diastolic blood pressures were measured by a blood pressure monitor (BP-301, Tanita, Tokyo, Japan). Body weight was assessed using a body-composition monitor (BC-503, Tanita). The data of blood presure and body weight were transmiited by Bluetooth. The number of steps walked per day was determined using a pedometer (FB-723, Tanita). The data obtained from the pedometer were transmitted via infrared wireless communication to a Web server using a receiver (MY101, Tanita, Tokyo, Japan). These data were then uploaded and stored on the home healthcare server, where it was made available to medical personnel for monitoring purposes.

All paramters are sutomatically uploaded without any special handlings. The client switched on the devices and the measrued data were automatically uploaded and transmitted to the Web server.

The system configuation and protocol were based on our previous study [3] and coniua health alliance [4].

Using this system, our main priority was lowering of blood pressure via walking exercise. Personality assessments were performed to taylor goals to each patient. Data from the initial physical examination (*i.e.*, body weight, waist circumference, and blood pressure) were recorded. Goals were set for each patient on the basis of these data, and methods of losing weight through changes in food intake and/or exercise were recommended.

Such commercial home healthcare system are available and everyone undestood the effectiveness, but there were no reports for usability and effectiveness. Ina first stage we investgated the effectiveness of the system.

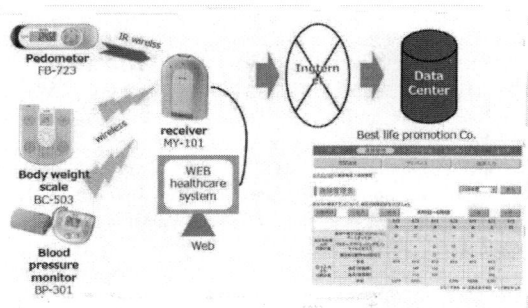

Fig. 1 Automated health monitoring and management system

B. Subjects

A total of 223 participants (average age: 43.0 ± 8.1 years) were enrolled initially: 205 male (average age: 43.7 ± 7.6 years) and 18 female (average age: 37.1 ± 7.9 years). During the 6-month study period, the participants did not monitor their walking and blood pressure at all times, even though the recording was automatic. Participants who recorded at least 50% of their walking steps and 20% of their systolic blood pressure readings over the 6-month period were included in the analysis. Participants taking antihypertension medications were excluded. Forty-eight participants, all male (average age: 47.5 ± 5.2 years), fulfilled the criteria cited above.

C. Analysis

a) Constancy of walking

For verification of the effectiveness of walking as an exercise, the relationship between changes in blood pressure and number of walking steps was evaluated over a 6-month period. Changes in blood pressure (mmHg) were defined as follows:

Changes in blood pressure (mmHg) = final 1-month average systolic blood pressure − initial average 1-month systolic blood pressure

Assuming a continuous walking pattern, we defined constancy and continuity with respect to the average number of walking steps as follows:

Averaged walking steps = average number of walking steps during the 6-month study period

Constancy = Average of total weeks of coefficient of variation in each week

$$A \cdots \cdots_{6-month} [\frac{Standard\ deviation\ of\ walking\ steps\ per\ week}{Average\ numner\ of\ walking\ steps\ per\ week}]$$

Continuity = Coefficient of variation of the average number of walking steps per week:

$$\frac{Standard\ deviation\ of\ 6-month\ walking\ steps}{Average\ of\ 6-month\ walking\ steps} \times 100\ \%$$

The averaged number of walking steps indicates the average number of walking steps taken per day. Lower constancy means the constant walk per week. Lower continuity means constant walk per measurement period.

Linear and multiple regression analyses were applied to the walking step data, followed by Student's *t*-test to evaluate the relationship between walking steps and blood pressure, and the predictor variables. The **Akaike** information criterion (AIC) and coefficient of determination (R) were used as measures of the relative goodness-of-fits of the statistical ***models***.

b) Clustering

From the habit of walking, then we classified the one-week walking style versus blood pressure changes. Our assumption is if the subject constantly walking per week, the blood pressure will be lower.

For clustering, we use K-means for clustering which is one of the simplest unsupervised learning algorithms [5]. The number of cluster in the data set was estimated by gap statistics [6]. The constancy of walking was classified into three; 5 mmHg evaluation, no changes and 5 mmHg decline. Then liner discriminant analysis was applied each cluster to calculate the threshold value.

III. RESULTS AND DISCUSSION

Each cluster of walking constancy in 12 weeks is shown in Fig.1 and Table 1. The subject corresponded to cluster A is only one, and the constancy is very high. The subjects corresponded to clusters B and C are 26 and 21, respectively. The constancy in clusters B and C were 13.9 and 41.6%, respectively. The average constancies in Cluster B and C were 23.95 %and 41.6 %, respectively. In Cluster B no subjects elevate the blood pressure. While only 4 out of 21 was declined the blood pressure in cluster C. This provides the low constancy of walking per week is more effective to lower the blood pressure.

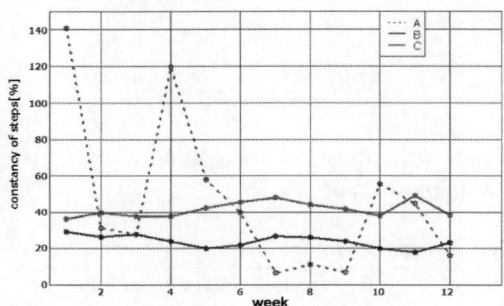

Fig. 2 Average trends in constancy of walking steps in 12 weeks at each cluster

Table 1 Cross tabulation of changes in blood pressure and clustering analysis results

		Cluster A	Cluster B	Cluster C
Changes in blood pressure	Elevation	0	0	5
	No changed	0	12	12
	Decline	1	14	4

The average constancy is shown in Fig. 3. The Clusters B and C were significantly different. The discriminant variable D is D=17.6constancy -5.8. When D=0, the constancy was 32.8 %. Figure 3 shows cluster B is lower and cluster C is higher than discriminant variable. The lower constancy shows the lower blood pressure. Thus the subject needs to walk constantly. If subjects used to walk around 8000 steps and average variation was ±2624. This suggested constant daily walking need to lower the blood pressure.

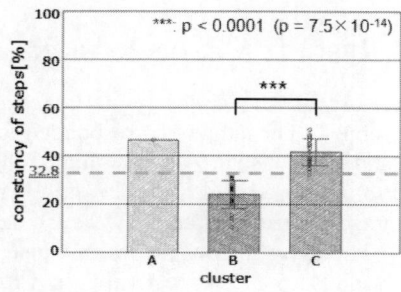

Fig. 3 Average constancy of walking steps at each cluster.

IV. CONCLUSIONS

For prediction of blood pressure changes, the habit of walking steps per week was evaluated. If the constancy of walking step per week is lower, i.e. constantly walking per week, the changes in blood pressure was significant.

REFERENCES

1. A.Giordano, S.Scalvini, E.Zanelli, U Corrà, GL Longobardi, VA Ricci, P Baiardi, F Glisenti (2008) Multicenter randomised trial on home-based telemanagement to prevent hospital readmission of patients with chronic heart failure.. Int J Cardiol. 131(2):192-9
2. Woolhandler S. Campbell T and Himmelstein D.U. (2003) Costs of health care administration in the United State and Canada. New Engl J Med 349: 768-775
3. Mizukura I, Tamura T, Kimura Y. and Yu W.(2009) New application of IEEE 11073 to home health care. The Open Med Infor J 3:44-53
4. Continua health alliance http://www.continuaalliance.org/ access on June 30
5. J. B. MacQueen (1967): Some Methods for classification and Analysis of Multivariate Observations, *Proceedings of 5-th Berkeley Symposium on Mathematical Statistics and Probability"*, Berkeley, University of California Press, 1:281-297
6. R. Tibshirani, G Walther and T. Hastie (2001); Estimating the number of clusters in a data set via the gap statics.. J. R. Statis. Soc. B 63 Part2 pp411-423

The address of the corresponding author:

Author: Toshiyo Tamura
Institute: Department of Biomedical Engineering, Osaka Electro-communication University
Street: 18-8 Hatsu-cho,
City: Neyagawasi
Country: Japan
Email: tamurat@isc.osakac.ac.jp

Parallel Workflows to Personalize Clinical Guidelines Recommendations: Application to Gestational Diabetes Mellitus

G. García-Sáez[1,2], M. Rigla[3], E. Shalom[4], M. Peleg[5], E. Caballero[2], E.J. Gómez[2,1], M.E. Hernando[2,1]

[1] Networking Research Centre for Bioengineering, Biomaterials and Nanomedicine, CIBER-BBN, Zaragoza, Spain
[2] Bioengineering and Telemedicine Centre, Politechnic University of Madrid, Spain
[3] Endocrinology and Nutrition Dept., CSPT, Hospital de Sabadell, Sabadell, Spain
[4] Department of Information Systems Engineering, Ben Gurion University of the Negev, Israel
[5] Department of Information Systems, University of Haifa, Israel

Abstract—The MobiGuide system provides patients with personalized decision support tools, based on computerized clinical guidelines, in a mobile environment. The generic capabilities of the system will be demonstrated applied to the clinical domain of Gestational Diabetes (GD). This paper presents a methodology to identify personalized recommendations, obtained from the analysis of the GD guideline. We added a conceptual parallel part to the formalization of the GD guideline called "parallel workflow" that allows considering patient's personal context and preferences. As a result of analysing the GD guideline and eliciting medical knowledge, we identified three different types of personalized advices (therapy, measurements and upcoming events) that will be implemented to perform patients' guiding at home, supported by the MobiGuide system. These results will be essential to determine the distribution of functionalities between mobile and server decision support capabilities.

Keywords—m-health, clinical guidelines, personalized care, decision support system.

I. INTRODUCTION

Clinical Practical Guidelines (CPGs) collect recommendations on the appropriate treatment and care of people with a specific disease and conditions, and they are based on a systematic review of clinical evidence. Care providers can use CPGs for treatment, but time constraints hamper the access to the knowledge accumulated during patient encounters.

Computer-interpretable Guidelines (CIGs) are usually developed based on the text of CPGs [1]. They can be used to support clinicians' decision making processes, by automatically generating recommendations about what medical procedures to perform tailored for each individual patient [2].

A significant amount of knowledge is still left as implicit in CPGs and should be turned into explicit knowledge during the Knowledge Acquisition (KA) process. Some of the implicit knowledge is specific to the local implementing institution, and includes organizational workflow and constraints, goals, and scope. The KA process is usually performed by knowledge engineers (KEs) in collaboration with Expert Physicians (EPs) from the institution at which the CIG is to be implemented.

It is difficult to tailor care workflows to the needs of the individual patient. Implementing CPGs in a decision support system with an interface to an electronic health record (EHR) makes the application of guidelines more personal and more likely to be accepted during clinical care [3].

The EU FP7 funded project MobiGuide (MG) [4] provides patient guidance services based on CIGs supported by an intelligent decision-support system. The system helps patients manage their illness by monitoring disease parameters and providing the appropriate feedback personalized to patients' preferences and context in a mobile environment. The system is based on the best available clinical evidence personalized to the patient's personal circumstances and technological context.

The main components of the MG System are: 1) A backend Decision Support System (DSS) devoted to the representation and execution of CIGs, which is complemented by a mobile DSS that supports the distribution of guideline parts; 2) A Body Area Network (BAN) that provides real-time monitoring of biosignals and communicates with the backend server and; 3) A Personal Health Record (PHR) that is ubiquitously and securely accessible and collects patients' personal and monitoring data. The MG system and its clinical effectiveness will be evaluated in patients with Gestational Diabetes (GD) as one of the application areas.

This paper presents a methodology to identify personalized patient recommendations obtained from the analysis of a specific CPG. The GD local guideline has been elicited to formalize the knowledge representation and to allow creating the ontology specific consensus [5], which describes schematically the interpretation of the GL agreed by both the EPs and the KEs. In this work, we added a conceptual parallel part to the process of formalization that focuses on patient behaviour, called "parallel workflow". The aim was to consider patient context and patient preferences so that we can generate recommendations, warnings or reminders able to guide patients at home supported by the MG system.

II. MATERIALS AND METHODS

In order to elicit the Gestational Diabetes CPG, we implemented the KA methodology that was previously developed and evaluated for the specification of CPGs by EPs and KEs [5, 6].

This methodology starts with choosing the specification language and continues with instructing the EPs about the KA process. Then, the appropriate CPGs for formalization are selected. The EPs of 'Hospital de Sabadell' created the local GD CPG implementation [7]. The process continues with making an ontology specific consensus. This phase is important because although CPGs are evidence based, some actions do not appear in the CPG but are still considered correct by some physicians, and because not all physicians will necessarily consider everything in the CPG as relevant and applicable to the patient. We focused on creating the local consensus and elaborated it with patient-customization, personalization and technological–context aspects. We used a CPG graphical representation iteratively to produce this local consensus.

In order to extract patient-tailored parallel workflows from the CPG, we identified those recommendations that might imply an advice for the patient, typically to be carried out at home (e.g. to measure blood glucose (BG)). Recommendations that imply advice for the patient are also the starting point of a process of care (workflow) that is parallel to the CPG and completely patient-centric. The result of the process is a customized CIG including the recommendations for care professionals and the corresponding recommendations to be addressed to patients (see Figure 1).

Figure 2 shows an example of the parallel workflow applied to GD monitoring plan. In "traditional" CIGs, recommendations 1-5 are directed toward the care professionals, and the care professionals explain them to their patients, but they are not actually applied in a functional automatic manner by a DSS. For example, after enhancing the plan for BG monitoring with a parallel workflow, it now contains recommendations for the patient and can be further

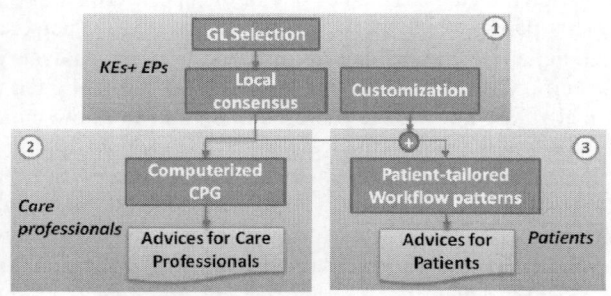

Fig. 1 Methodology for GL formalization and patient-tailored workflow patterns identification that produces a customized CIG

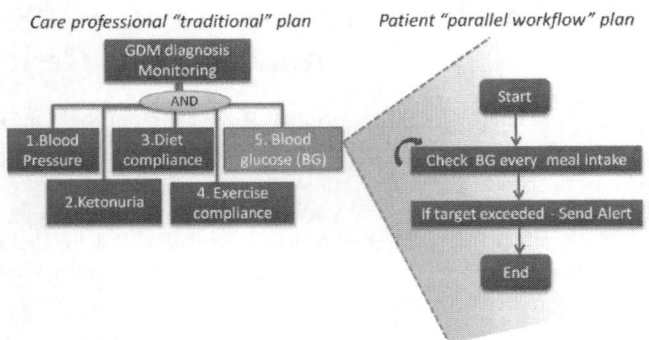

Fig. 2 An example of parallel workflow for BG monitoring

decomposed into a formal monitoring plan which can be executed by the patient's mobile (Smartphone).

III. RESULTS

The analysis of the GD guideline, done by KEs in consultation with EPs of 'Hospital de Sabadell', took a total time duration of 4 months along 10 iterations. It led to the extraction of a set of workflow patterns, related to three types of advices:

A. Therapy Advisors, to Help the Patient to Comply with Her Treatment

Patients with GD are responsible for following therapy prescriptions at home. Therapy prescription in GD is mainly related to three types of recommendations: a) Nutritional therapy; b) Exercise; and c) Insulin therapy.

For some women, showing a good compliance to nutritional and exercise recommendations can avoid to administer insulin. Due to this reason, mechanisms to reinforce the patient's compliance to diet and physical activity could contribute significantly to keep a good metabolic control. Patients with GD are trained to estimate the amount of carbohydrates for each meal so that they can follow nutritional prescription correctly. Also they are instructed to write down in their personal's logbook whether they are not following nutritional prescription or eating forbidden nutrients and the motivation for not following recommendations.

Clinicians check the patient's compliance to diet and exercise during clinical encounters, where they can assess the importance of not following compliance. MG will detect possible situations of non-compliance related to diet and physical activity, and ask the patient for confirmation.

In MG, physical activity practicing and its intensity is detected automatically with a physical activity detector. In order to estimate compliance to diet, the system will need to trust the subjective estimation of carbohydrates for each meal intake.

The parallel workflow pattern for diet and exercise compliance can control the patient's adherence to recommendations. When not following recommendations repeatedly, the patient will receive an alert to show the importance of following compliance such as: (e.g. *"Try to avoid food with high glucose content such as cakes"* or *"Try to increase the duration of physical activity to improve glycemic control"*). Also, patients will receive feedback from the system when they follow recommendations, so that they can feel that their efforts while managing the disease are recognized.

It is especially important to know if the patient is not following nutritional prescription when there are high BG levels, as BG measurements are associated to diet intakes along the day. Patients could receive specific recommendations when one of these situations is detected (e.g. *"Your BG has been elevated associated to high glucose intake, you should follow the recommendations"*).

Depending on the side effect of non-compliance, related to exercise and/or diet prescription, the clinician in charge of the patient should be informed.

As in the example in Table 1, patients will be asked specifically about non-compliance related to the presence of high BG levels. Additionally, the patient will be able to write down the reason why she is not following diet recommendations (as in first data entry). When the importance of non-compliance is medium or high and repeated in different days/meal-intervals (as in third and fourth data entries), a clinical decision about therapy might be necessary.

For patients that require insulin administration, insulin doses will be registered to assess compliance. The patient can receive specific reminders about when to administer insulin, following her personalized schedule of meals (considering working days and weekends).

Insulin compliance needs to be monitored to determine the relationship cause-effect when analyzing BG levels. The system will ask the patient about insulin compliance, especially when anomalous BG levels are detected.

Table 2 shows an example of non-compliance related to insulin administration. Patient P_1 presents hypoglycemia levels associated to the administration of an insulin dose

Table 1 Example of non-compliance to nutritional therapy and high BG
(* repeated anomalous BG due to non-compliance)

Abs. Time	BG (mg/dL)	Intake	BG moment	Diet	Type of food	Reason	Importance
16/03/2013 22:30	125	Dinner	Postprandial	+	Cake	Party	Low
16/03/2013 15:30	165	Lunch	Postprandial	++	Ice cream	-	High
17/03/2013 21:30	148	Dinner	Postprandial	+	-	Birthday	Medium*
18/03/2013 21:30	145	Dinner	Postprandial	++	Double carbs	Skip prev. meal	High*

Table 2 Example related to non-compliance to insulin administration

Abs. Time	ID	BG value	Intake	BG moment	Insulin	Reason	Importance
16/03/2013 13:30	P_1	70	Lunch	Postprandial	+1 U	Eat more carbs.	Medium
16/03/2013 15:30	P_2	165	Breakfast	Postprandial	-3 U	Unknown	High
18/03/2013 14:13	P_2	148	Lunch	Postprandial	0 U	Forgot	High
18/03/2013 14:30	P_1	65	Dinner	Postprandial	+2 U	Eat more carbs.	High

higher than recommended in two situations, so she would receive specific recommendations to reinforce the correct therapy use (e.g. *"Administering more insulin than recommended is producing hypoglycemia events"*). Patient P_2, instead, seems to be skipping insulin doses and presents repeated hyperglycemia, so she will receive recommendations to reinforce the importance of following prescription.

B. Measurements Advisors, to Remind the Patient to Take Specific Measurements

GD management requires monitoring several variables daily and assessing the results periodically. Usually clinicians perform this process during encounters at the hospital, which happens every week or every two weeks. MG will monitor variables with specific sensors or manually in the Smartphone's logbook. For some patients, MG can be used to remind when each measurement should be performed. Only patients interested in this option will receive reminders about specific measurements such as BG or ketonuria. It is important to take into account that patients should not be overloaded with excessive reminders, as this might have the opposite result than expected. Patients could finish leaving pass messages if they are too demanding for them. Other parameters required to monitor compliance (diet, insulin or exercise) will not be part of this type of reminders.

Also, non-compliance in the process of measurement will be considered for the main clinical parameters (BG, ketonuria, blood pressure). When the patient does not perform measurements frequently enough or when she is not downloading data to the system so that data can be analyzed according to periodic requirements, the patient will be reminded. If the same situation is kept, the patient could be cited to attend an on-site visit, after warning the clinician.

C. Upcoming Event Advisors, for Dealing with Personal Situations

There are several significant personal events that can lead to delivery of GD guideline recommendations in the MobiGuide system. These can be related to situations where the diet, physical activity and schedule of the patient may not be routine, as in an upcoming holiday or party. Respective advisors could be addressed to the patients using the Smartphone interface, regarding changes to therapy and measurement schedule associated to different types of personalized events.

We have identified several parameters to be personalized by each patient with GD, depending on the specific schedule or the presence of special events (see the example in Table 3). When required, personalization parameters will be supervised by clinicians.

For example, patients could personalize their preferred time to perform physical activity, the intensity level or duration, depending on working days or holidays. The MG Smartphone will allow configuring schedules and registering special events. As a result, the system will adapt reminders to reinforce compliance to physical activity.

Also therapy related recommendations will be modified to consider parameters depending on the patient's personal context. Following the personalized parameters in Table 3, the system will not alert about practicing low intensity physical activity while the patient is in a work travel. When the patient is going to a dancing party, she is increasing her routine level of physical activity and she is eating not recommended nutrients. In this situation, the MG system will not insist on un-compliance but it will account for un-compliance events in case of repetitions.

Table 3 Example: Preferred configuration of upcoming events

	Working days	Holidays	Special event (e.g. dancing party)	Special event (e.g. work travel)
#Required daily BG measurements	4	2-3	Same as holidays	Same as working days
#Required daily ketonuria levels	1	0	1	3
Max. BG ranges (mg/dL)			Same as holidays	Same as holidays
Preprandial	<95	<105		
Postprandial	<130	<145		
Preferred time to perform physical activity	18:30	11:00	--	Afternoon
Intensity level of physical activity	Medium	Medium	High	Low
Physical activity duration	1 hour	30 min.	2 hours	1 hour
Wake-up time	07:00	10:00	--	08:00
Preferred number of meals	5	6	6	5
High sugar food	--	--	Yes	Yes
Reminders before BG measurements	No	Yes	Yes	Yes

IV. CONCLUSIONS

This work presents three different types of advices (therapy, measurements, upcoming events) extracted from the Gestational Diabetes local consensus guideline, which are parallel to recommendations for clinicians included in the CPG. The process has permitted to define the necessary steps to create a customized CIG that includes personalization to each individual patient including specific preferences and context. The results will help to determine the level at which decisions should be placed in terms of MG DSS (at mobile-side DSS or backend server-side DSS). Those decisions which could be taken by patients self-management will be part of the mobile DSS while decisions that need to be supported by clinicians will be placed in the server DSS. The distribution of DSS capabilities could be also modified depending on patients' customization or on the technological context, such as the absence of connectivity.

ACKNOWLEDGMENT

This work was supported by the European project Mobi-Guide under the European Commission FP7 (287811).

REFERENCES

1. Peleg M, Tu, S, Bury J et al (2003). Comparing Computer-interpretable Guideline Models: A Case-study Approach. J Am Med Inform Assoc 10:52-68 DOI 10.1197/jamia.M1135
2. González-Ferrer A, Ten Teije A, Fernández-Olivares J et al (2013) Automated generation of patient-tailored electronic care pathways by translating computer-interpretable guidelines into hierarchical task networks. Artif Intell Med 57(2):91-109 DOI 10.1016/j.artmed.2012.08.008
3. Gooch P, Roudsari A (2011) Computerization of workflows, guidelines, and care pathways: a review of implementation challenges for process-oriented health information systems. J Am Med Inform Assoc 18:738-748 doi:10.1136/amiajnl-2010-000033
4. MobiGuide "Guiding patients anytime everywhere", EU FP7 project at http://www.mobiguide-project.eu/
5. Shalom E, Shahar Y, Taieb-Maimon M et al (2008) A quantitative assessment of a methodology for collaborative specification and evaluation of clinical guidelines. J Biomed Inform, 41(6): 889-903 DOI 10.1016/j.jbi.2008.04.009
6. Shalom E, Shahar Y, Taieb-Maimon M et al (2009) Ability of expert physicians to structure clinical guidelines: Reality versus perception. J Eval Clin Pract 15(6):1043-1053 DOI
7. Rigla M, Tirado R, Caixàs A et al (2013). Gestational Diabetes Guideline CSPT. Internal document MobiGuide Project (FP7-287811).

Author: Gema García-Sáez
Institute: Polytechnic University of Madrid
Street: Avda. Complutense, 30, Ciudad Universitaria, 28040
City: Madrid
Country: Spain
Email: ggarcia@gbt.tfo.upm.es

Case Based Reasoning in a Web Based Decision Support System for Thoracic Surgery

A. Babic[1,2], B. Peterzen[3], U. Lönn[3], and H.C. Ahn[3]

[1] Department of Information Science and Media Studies, University of Bergen, Norway
[2] Department of Biomedical Engineering, Linkoping University, Sweden
[3] Linköping Heart Centre, Linköping University Hospital, Linköping, Sweden

Abstract—Case Based Reasoning (CBR) methodology provides means of collecting patients cases and retrieving them following the clinical criteria. By studying previously treated patients with similar backgrounds, the physician can get a better base for deciding on treatment for a current patient and be better prepared for complications that might occur during and after surgery. This could be taken advantage of when there is not enough data for a statistical analysis, but electronic patient records that provide all the relevant information to assure a timely and accurate clinical insight into a patient particular situation.

We have developed and implemented a CBR engine using the Nearest Neighbor algorithm. A patient case is represented as a combination of perioperative variable values and operation reports. Physicians could review a selected number of cases by browsing through the electronic patient record and operational narratives which provides an exhaustive insight into the previously treated cases. An evaluation of the search algorithm suggests a very good functionality.

Keywords—Case Based Reasoning, Nearest Neighbor algorithm, heart failure, web- based system, thoracic surgery.

I. INTRODUCTION

Linköping Heart Centre at the Linköping University Hospital performs about 850 open-heart operations every year. In approximately 1-3% of these, there is need for a Left Ventricular Assist Device (LVAD) due to heart failure when weaning from cardiopulmonary bypass (CPB) [1]. The use of a LVAD indicates a critically ill patient and hence a very high mortality rate. However, different studies report varying mortality rates, ranging from 7% up to 86% [2]. It is clear that to optimize the chance for survival for the patient it is important to choose the right treatment according to the situation.

We have developed a web-based Clinical Decision Support System (CDSS) which uses Case-Based Reasoning (CBR) to enable the physicians to find previously treated patients with similar preoperative backgrounds. Information about these patients are presented with variable values as well as operation reports, giving the physicians full insight in the course of events until the patient leaves the operation theatre. By studying previously treated patients with similar backgrounds, the physician can get a better base for deciding on treatment for a current patient and be better prepared for complications that might occur during the surgery.

The CBR engine is implemented as a part of a system under development called AssistMe [3]. An additional feature of this system is information for patients and relatives about cardiac diseases and open-heart surgery, patient questionnaires and a Rule-Based Reasoning (RBR) engine. The RBR and CBR engines will work parallel to each other giving decision support to the physicians.

CBR as a method has been considered being advantageous in this domain due to the complex nature of the information about patient undergoing operations with LVADs. The information consists of a large number of variables with numerous interrelations and many missing values. CBR is not as dependant of well structured data as many other methods [4].

II. CASEBASE AND METHODOLOGY

The case-base used during the development of the CBR engine was a pilot research database consisting of 24 patients that had undergone open cardiac surgery where the Hemopump™ (HP) LVAD [5] was used postoperative. Between September 1991 and November 1994, the HP was used in 24 patients. During this period, 2,603 open-heart surgical procedures were performed at the Department of Cardiothoracic Surgery. During the same period, the intra-aortic-balloon-pump (IABP), another mechanical assist device was used in 20 patients. The frequency of mechanical assist device during this time was 1.7%. 58% (14) of the HP patients survived the surgery. The cases in the case-base were described by 25 variables.

A. Algorithm

The algorithm developed to find similar cases is based on the Nearest Neighbour (NN) algorithm [6] that uses Euclidean distances and includes Some modifications have been done to as rescale the variables to standardize them.

Based on their professional experiences and preferences, the system users are defining their own sets of variables to search for similar cases using a clinical database. The number of variables could be chosen freely and is not limited by the CBR algorithm. Each variable is assigned a significance value, corresponding to the importance of the variable for the search. The significance values range from 1 representing the most important variable to either a fixed number or infinity, depending on how they are connected to the weights used in the algorithm. More than one variable could have the same significance value.

All values are then squared and the new variable values are after that summed for each case in the case-base. The sum of the cases is normalized with respect to the sum of squares of the weights used in the algorithm. Finally, the square root is calculated for each case, giving the Euclidean distance in the rescaled and normalized data space. For each case, we now have a similarity score:

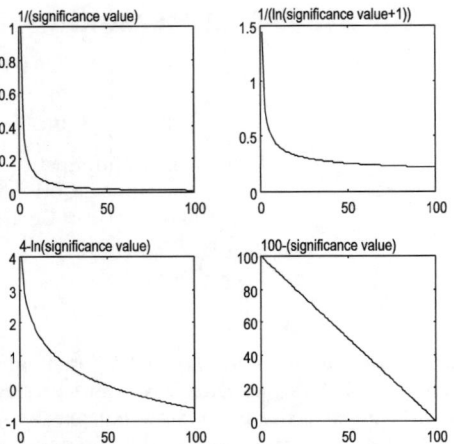

Fig. 2 Four different types of weight functions.

$$\sqrt{\frac{\sum_{f \in \mathbf{F}} (w_f d(x_f, y_f))^2}{\sum_{f \in \mathbf{F}} w_f^2}}$$

Fig.1 Euclidian distance used in the NN algorithm.

where x_f and y_f are the values of the variable f of the evaluated case and the compared case respectively, $w(f)$ is the weight connected to the variable and $d(x_f, y_f)$ is the distance between the variable values in the rescaled and logarithmic value is calculated. **F** is the data space of the chosen variables.

Since the algorithm cannot handle certain types of variables, such as categorical variables with more than two categories, a separate list consisting of the variable names that can be used in the algorithm has been created. In addition, as with other NN algorithms, redundant, irrelevant and noisy values have as much impact as any other value [6].

The significance values can be connected to weights in a number of different ways. The only restraint on the weights is that they should be derived from a monotonically decreasing positive function of the significance values. Four different types of weight functions had been considered during the development of the CBR engine.

The Figure 2 shows that the bottom left function does not fulfill the demands on weight functions since it becomes negative. The same would be true for the bottom right function for significance values greater than 100. For these weight functions to be applied, the choice of significance values must be constrained to the part of the function that is positive.

B. Implementation

The system has been implemented using a web-based environment called HAHTsite IDE 4.0 (former HAHT Commerce Inc, now part of Global exchanges Services) [7]. The software provides a possibility to develop web pages, include Java code and to connect to databases via ODBC.

Upon a start of the CBR engine, a physician is faced with two choices of actions, editing a patient or running the CBR algorithm to retrieve similar cases to a patient at hand.

Upon a start of the CBR engine, a physician is faced with two choices of actions, editing a patient or running the CBR algorithm to retrieve similar cases to a patient at hand.

Automatic Case Retrieval: Editing and Retrieving

In order to edit or add new patients, a physician can enter a page connected to the database of patients currently being treated. Entering a legal identification number determines the choice of patient; a list of all variables appears together with a list of variables that have already been assigned values. New variable values can be entered and existing values can be altered. The physician can also choose to store the patient only into the computer memory.

Retrieval of similar patients begins with by choosing or entering variables on which a search is to be performed. To achieve a better precision, significance is assigned to each of these variables. In addition, the physician can choose how many cases to present. The search can also be restricted to only give results for a certain type of operation.

When the variables have been chosen and given their significance values, the algorithm is started. The result is presented as a list of cases organized in descending order in terms of their similarity scores. If a case is chosen to be further studied, all available information about it is presented as in Figure 3.

At the top of the screen is information about the patient that is a set of 'key words' for searching similar cases. The bottom part of the screen contains the short version of the operation report. This allows for the physician to decide whether the case is interesting or not in a quick glance. Should the case be interesting, physician can choose to read the full operation report using a button at the bottom of the page.

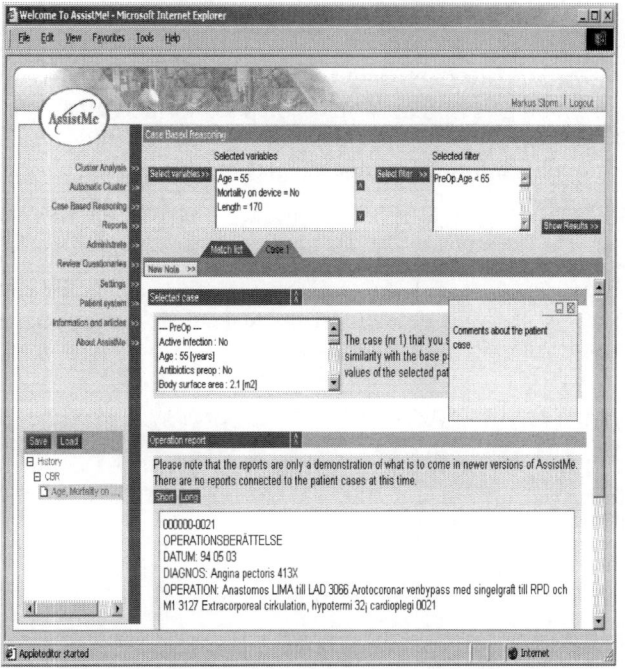

Fig. 3 A patient case with the operation report.

Evaluation

The focus of the evaluation was the case retrieval and similarity of the results between the KK algorithm and the treating physicians. As the choice of search variables was left to the surgeons we hypothesized that the outcome would depend on the professional experience, as well as on their user preferences concerning computer aided decision making. We assumed that the acceptance of the system would be favorable; the system design is using features well known to the physicians such as their own clinical databases and discharge narratives. The three surgeons with different levels of professional experience were handed out printouts of the case-base and for the first three (3) patients to find out the most similar cases among remaining twenty-one cases (21). They were also asked to rank the chosen case in order of similarity. Finally, they were to describe which variables they had been studying when deciding on which cases were most similar and how they handled missing values. Physician no. 1 was interviewed while he was looking for similar cases in order to get a clearer picture of how he was reasoning when matching cases.

Physician no. 1 had been a heart surgeon for three years and had previously been an anesthesiologist. Physician no. 2 had ten years of experience as a heart surgeon, and physician no. 3 was a professor in the domain.

The variables that the physicians reported having studied were entered into the CBR engine and were used in the algorithm to find the most similar cases to the 3 studied cases. The results were then compared with what the physicians considered being the most similar cases.

A comparison of cases that the algorithm and the physicians considered being similar gave varying results. While the physicians no. 1 and 3 matches were of good or very good similarity with the matches given by the CBR engine, physician no. 2 ordered the cases in a way that did not correspond with the CBR engines ranking at all.

The physicians reported that they had tried to replace a missing value with a guessed value based on other most appropriate variables for the case. However, physician no. 1 explained that he had problems figuring out if cases with many missing values had matched the evaluated cases or not. Furthermore, the choice of variables varied. All physicians used two variables, with different levels of importance, and between three and five additional variables. As for the usability of the CBR engine, the three physicians were demonstrated the system's main features and asked to test the 'find' and 'edit' patient functionalities. . They all claimed they could use the CBR engine relatively easily, without much prior training; they could move across the pages keeping a strong sense of orientation. The possibility of combining the clinical records and operation narratives one screen was seen as an advantageous and skilful tool not commonly offered in hospital settings. A quick retrieval of similar cases and its presentation was found useful, natural, and self documenting. The process of evaluation was also a learning experience in which a manual review of the cases contributed to a good understanding of the CBR methodology.

III. DISCUSSION

Case-based reasoning is an artificial intelligence methodology that solves new problems by retrieving stored records of prior problem-solving cases and adapting their solutions to fit new cases [8]. The AssistMe system does not adapt, but rather present selected cases information to physicians when deciding on treatment for a patient with heart failure. The functionality of the CBR engine is however dependant on the users backgrounds, their experience influences choice of clinical variables that the initiate retrieval. The method of finding similar cases is very similar to that of a physician, choosing only a few out of many variables and assigning the different levels of importance. It has however been shown that physicians with different backgrounds tend to concentrate on different variables when evaluating a case.

From a comparison of what the CBR engine considered being similar and physicians opinions it is obvious that the functionality of the CBR engine is depending on which variables are being used. A study of the choice of variables of physician no. 2 reveals that he has considered a variable called "ASATM" being most important. Studying this variable more closely reveals that the values range from 1 to 9 except for 3 cases that have values of 12, 16 and 21. Cases 2 and 3 had values of 8 and 9 respectively for this variable. The most similar cases according to physician no. 2 had values of 16 and 21 respectively. It is plausible that the physician has considered 8 and 9 as "high" values, and hence also considered 16 and 21 as "high" and therefore similar. This shows that the mathematical similarity that is defined by the algorithm does not necessarily represent a medical similarity.

Substituting a missing value for a mean value has not shown to be an appropriate method. Many of the cases that were ranked higher by the CBR engine than by the physicians were patients that did not survive and hence have many missing values. Though it is important to present these cases in order to make the physicians aware of risks connected to their evaluated patient, missing values should be substituted with a value more close to the truth. Other methods such as substitution by a previous value, a random value or a median value would probably lead to the same problem. One method that would resemble the physicians way of reasoning about these values would be to let the CBR engine evaluate each case in the case-base with respect to the others using all variables that are available for this case. Missing values could then be replaced by values taken from the most similar cases among the rest of the case-base, hence giving a substitution value that would fit the case.

The evaluation has shown usefulness of the CBR engine and made the users confident into the value of the methodology for receiving relevant cases on which they could base their estimates of the outcomes (surgical success, morbidity, and mortality).

IV. CONCLUSION

The CBR engine retrieves patient cases up to users' expectations and satisfaction. A set of the closes cases is presented using variable values as well as operation reports, giving useful information to physicians for deciding about the best treatment and risks that could happen as in the previous similar cases. The KK nearest neighbours algorithm used resembles the way some physicians compare cases, while others tend to look at features not detectable by the algorithm.

Further work regards developing a method for imputing missing values when feasible and appropriate and developing the functionality for adapting cases. The latter would help to enhance the system's learning capacity which would enlarge the user base to those less experienced or coming from other professional groups that treat patients.

The operation reports that are presented contain much information that can be only viewed, but could be used if a text mining reasoning would be added to the CBR engine. These should be studied to find an automatically way of extracting information that then could be entered into the case-base.

The current version of the CBR engine does not make any attempt of giving automatically advice to the user. For the engine to be able to give automatically advice, the solutions of previous cases must be adapted by making use of sets of rules.

ACKNOWLEDGEMNTS

This research has been funded by the Centre for Industrial Information Technology (CENIIT) of Linköping University, Sweden.

REFERENCES

1. B Peterzén, U Lönn, A Babic, H Granfeldt, H Casimir-Ahn, H Rutberg. Postoperative Management of Patients With Hemopump Support After Coronary Artery Bypass Grafting. Ann. Thorac. Surg. 1996;62: 495-500.
2. Frazier O H. New Technologies in the Treatment of Severe Cardiac Failure. The Texas Heart Institute Experience. Ann Thorac Surg 1995;59:pp 39-45.
3. Antonsson J, Granfeldt H, Kircher A, Babic A, Lönn U, Casimir-Ahn H. Design of a Clinical Decision Support System for Assist Support Devices in Thoracic Surgery. Proc. AMIA Symp 2000:p 954.
4. Deng P S. Using Case-based Reasoning for Decision Support. IEEE Proceedings of the Twenty-Seventh Annual Hawaii International Conference on System Sciences. 1999; 4:pp 552-561.
5. Casimir-Ahn H, Lönn U, Peterzén B. Clinical use of the Hemopump Cardiac Assist System for Circulatory Support. Ann Thorac Surg 1995;59:pp 39-45.
6. Wettschereck D, Aha D W, Mohri T. A review and Empirical Evaluation of Feature Weighting Methods for a Class of Lazy Learning Algorithms. AI Review 1997;11:pp 273-314.
7. http://www.haht.com/
8. Holt A et al., The Knowledge Engineering Review / Volume 20 / Issue 03 / September 2005, pp 289-292

Data Mining in Cancer Registries: A Case for Design Studies

G. Kanza[1] and A. Babic[1,2]

[1] University of Bergen/Department of Information Science and Media Studies, Bergen, Norway
[2] Department of Biomedical Engineering, Linkoping University, Sweden

Abstract—Cancer registries are created, managed and data mined to gain knowledge about long term patient outcomes, effects of medication, clinical factors influencing patients' well-being. Equally important is the insight into the cost effectiveness of cancer treatments, and securing data input from different medical centers and enable competent data analysis and meaningful results. Interest among different user groups (physicians, researchers, health care administrators, policy makers) cerates expectations regarding the results and active role in the development and in interactive use of the information. This paper discusses several design cases in which data mining could be implemented to enable efficient and user friendly knowledge extraction. Three important design cases have been identified following the pathways that the users typically make: 1. ensemble data mining from long term national registries; 2. ensemble data mining form the dedicated clinical web-databases; 3. ensemble distributed data mining and analysis.

Keywords—HCI design, big data, data visualization, data mining, cancer registries

I. INTRODUCTION

Long term studies with a high number of subjects are trusted to give statistically meaningful results and a reliable insight into the patient outcomes. Clinical trials have established very strict study conditions and collect and process a wide range of clinical variables to provide answers to the study hypotheses. Registries have been developed using fewer data items, limiting the number of primary and secondary end points, but they gather data from several centers, both nationally and internationally. They allow several periodic analyses and some flexibility in generating reports. However, for many years researchers and physicians had to deal with more than one tool to obtain, analyze and visualize data from the registry. Reports have been published off line: once the gathered data was verified, data sets are sent to bio-statisticians. Reports resulting from those analyses typically come back in a form of a file which could be tedious to follow: having to look at never ending rows and columns of numbers is cognitively a demanding task that could be eased by human computer interaction methodology. A hypothesis could be made to explore how a smart design for visualizing could benefit mining the data from what is usually called big data [1]. The notion of big data stands for the combination of various clinical data, images, and textual information, all of which is the reality of the patient management, clinical research and medical education. We will look at the example of Swedish Cancer Registry and the report based on it (Design Case 1).

In addition, we will look at specialized web-sites offering information regarding registries covering a great number of clinical indications and outcomes, study duration, data types, clinical centers involved, study status. At best those sites offer some information retrieval function and no possibility for data mining. A good example of a very well established web-site is *ClinicalTriasl.gov* that is visited by clinicians, researches, industries, and patients interested in the clinical trials (Design Case 2).

Cancer registries, and web-sites containing information about cancer related studies, are subjects of change as new data brings new information and holds insights into the patient population, disease treatment, and changing patterns. It is one of the challenges of data mining to design a distributed system in which users would contribute their data and track those changes (Design Case 3).

II. RESEARCH FRAMEWORK AND METHODS

There is a large amount of clinical and administrative data that healthcare information systems produce for each patient in different care processes. However sufficient knowledge discovery methods can be developed to retrieve information and implicit knowledge about evidence-based care processes and patient treatments. When statistical methods are combined with data mining techniques, reuse of previous experiences can be utilized in order to improve the clinical guidelines for treatments and services [2].

Table 1 presents two main approaches to the data mining offered by the statistics and artificial intelligence.

Table 1 Overview of the data methodology.

MULTIVARIATE STATISTICS	Descriptive Statistics, Cluster Analysis
	Discriminant Analysis, Regression Analysis
ARTIFICIAL INTLLIGENCE	Decision Trees, Neural Networks, Bayesian Rules, Fuzzy Logics, Rough Sets

An excellent example of the data mining systems implemented in Java, the WEKA system [3] is a collection of machine learning algorithms useful for extracting information from large databases. It is an open source system executable on different platforms. It could provide data in two main ways. The first is by allowing the possibility of loading data from databases, files and Universal Resource Locators (URLs) with the help of supported formats. The second is the possibility to generate data from artificial data sources where the generated data can be edited manually with a dataset editor.

Applications of data mining in a specific domain such as cancer have special requirements, users expect a certain kind of result such as optimal treatment or a survival rate. Methods that are capable of extracting knowledge in forms closer to human perception are those that induce decision trees, classification rules. However, often and more used are statistical methods due to the robustness and validity which are not transparent. Clinical areas with successful applications are numerous and growing [4,5], however more research should be done to aid users by designing the interactions that would provide transparency and clarity of data mining steps. Therefore, the research need to consider work flow and in particular the tasks, procedural steps, organisations and people involved in managing and analysing cancer registries. Resulting patterns such as resource, data, and interaction patterns will be assessed for their potentials to influence the design process.

An overview of the research described in this paper is illustrated in the Figure 1. The main goal was to explore the clinical data and indication for analysis, as well as to understand clinical and all other related work practices. By further looking at how data mining is conducted currently a set of design features could be identified (Figure 1).

The research has been focused on the system design (objective, elements, services, workflow) and interactions (functions, user behaviours, workflow) [6]. The resulting, new automatic, user-friendly system for data mining from the clinical registries is to be tested and evaluated by various expected user groups such physicians, health care administrators, and potentially medical students. The design science knowledge has been instrumental in this study, and In particular the seven design guidelines presented by Hevner et al. (2004) [7]: *design as an artifact, problem relevance, design evaluation, research contributions, research rigor, design as a search process, and communication of research.*

From the various evaluation methods proposed by Rogers, Sharp, and Preece (2011) [8], this research employs the *analytical evaluation* where experts will be involved in the evaluation process.

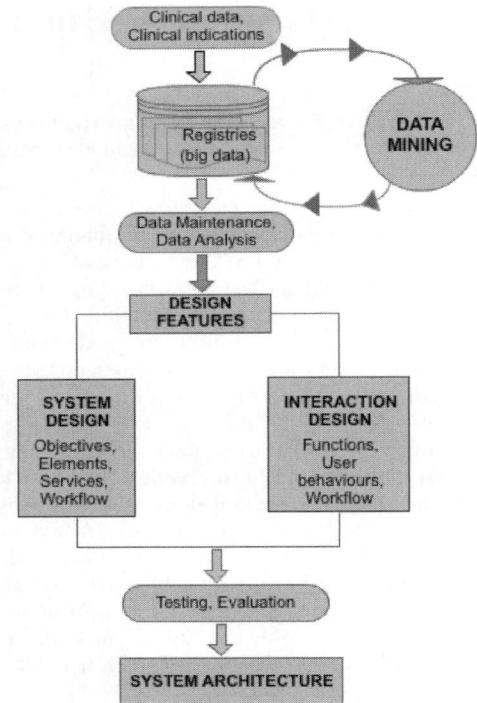

Fig.1 Design steps in data mining.

The design should also look at the number of quality issues and how they could resonate with the users, both medical and administrative. Table 2 lists the quality aspects such as appropriateness of the data, way the results and knowledge could be presented and validation issues.

Table 2 Aspects of quality assurance of importance for data mining.

DATA	Is the data representative, reproducible, of good quality?
PATTERNS	Are the patterns presentable, understandable, easy to interpret and reason about?
KNOWLEDGE	What is feasible/ which is the best way to extract and present the knowledge: Regression and/or classifications formulae, decision tree, neural networks, statistical tables?
VALIDATION	Can we validate methods? Are results comparable to the published ones?

Design Studies

There are three major cases of design to be considered. They are resulting from a pre-study in which we looked at three most typical ways of working with the cancer registries: 1. obtaining in depth analysis, 2. conducting meta analysis of web-based resources, 3. working with the distributed databases in different locations.

Design Case 1: Data mining from long term national registries

The Swedish Cancer Registry [9] has been a data source for epidemiologists and clinicians for over fifty years now. This data is used both at national and international levels to provide an overview of the country's cancer incidences. Researchers, physicians and other medical staff continuously use this data to study the disease in order to improve the treatments that already exist while aiming at finding its cure [9, 10]. Currently, extracting data from the registry is done by sending a request to the cancer registry (statisticians) who retrieves the data, analyses it and sends back tables, graphs back to researchers and physicians who analyze, draw conclusions, and suggest additional analyses. This process is time consuming as it involves several stages starting with often repeated requests and corresponding, resulting sets of results. Each iteration could take up to few weeks. In spite of already well established routines, there are times when doctors may want to get an original solution for a problem at hand. *For example: A research and administrative group wants to do a series of investigations focused on the intersection of aging and cancer, a study should discover information about treatment efficacy, non-compliances, tolerance and effects of co-morbidities in the treatments of elderly colorectal cancer patients.* In this case, data mining should consider different age groups, but that may cause problems since the age is measured purely chronological in most contexts. Another challenge might be to conduct a long term follow up where other populations registries needs to be followed- up and their data combined.

Design Case 2: Data mining form the dedicated web-databases to make their resources available to the users

ClinicalTrials.org is an international registry with 185 countries participating with their own data in the studies carried out for a number of disease indications. In this online registry, there is currently a total of 716 cancer studies registered from Sweden, which involve international collaborators from different institutions. Of these, 230 studies are still open and 486 closed.

Institutional commitment to so many national and international studies shows a strong intention to do research and assistance in making bigger databases in which population representative and meaningful patterns could be extracted.

The sense of community, collaborative work forms are important design factors to explore. The Table 3 shows just a few registries with a long duration and a high enrollment.

Table 3 Examples of currently running colorectal cancer registries.

STUDY TITLE	TO ENROLL	COLLABORATORS	TIME
Epidural Versus Patient-controlled Analgesia for Reduction in Long-term Mortality Following Colorectal Cancer Surgery	300	-Örebro University, Sweden - University Hospital, Linköping	Mar 2011 to Dec 2018
The Northern-European Initiative on Colorectal Cancer	66000	Norwegian Department of Health and Social Affairs - Maria Sklodowska-Curie Memorial Cancer Center, Institute of Oncology - Erasmus Medical Centre, Dep. Of Gastroenterology, Rotterdam, The Netherlands - Landspitali University Hospital - Uppsala University Hospital, Dep. Of Surgery, Uppsala, Sweden - Karolinska Institutet - Riga Eastern Clinical University Hospital, Riga, Latvia - Memorial Sloan-Kettering Cancer Center - Harvard School of Public Health	May 2009 to Jun 2026
Rectal Cancer And Pre-operative Induction Therapy Followed by Dedicated Operation. The RAPIDO Trial	885	- University Medical Centre Groningen - Karolinska University Hospital - Leiden University Med. Center -Uppsala University Hospital -Dutch Cancer Society	Jun 2011 to Jun 2016

Design Case 3: Distributed data mining data

Clinically based information systems, patient records and registers are designed for recording data more than reporting data. In normal clinical settings, there are several sources of information that could be eligible and useful for exploring clinical hypotheses (Figure 3).

Data mining technique and machine learning methods can bring the facility to discover patterns and connections (hidden knowledge) within the medical databases. Therefore, it is important will utilize the data by:

- integrating information coming from various database covering clinical, research, economic, and other aspects of patient management;
- outlining recommendations for users on how to move across platforms and systems.

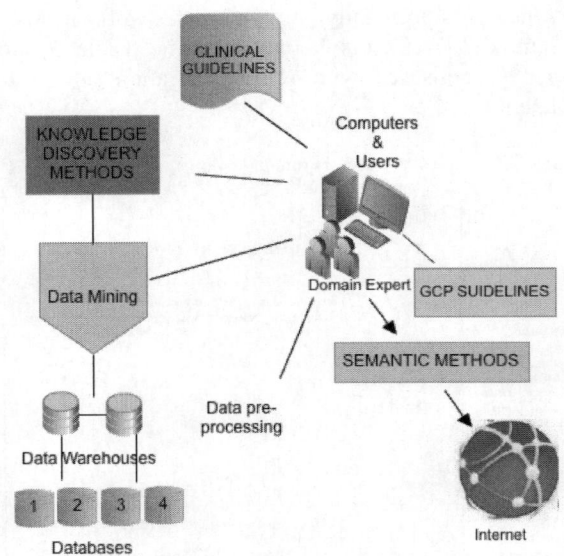

Fig. 3 Design points in distributed data mining.

III. DISCUSSION AND CONCLUSION

Here presented design research aims at transforming the current state of interacting with cancer registries to a better and user friendlier stage. User experiences in working with the data sources are different, but demanding in several ways. The most detailed clinical knowledge is being extracted *off line* and with a help of a statistician. A quality checks performed throughout, do guarantee a reliable facts representative of the patient population. However, there is little flexibility in extracting knowledge, for example, it may not be simple to learn about *just* the mortality in a certain age group given a selected treatment. Automated systems would allow to get several interesting comparisons among selected patient groups. Communication in real time would allow testing hypotheses, identifying special cases which are invisible within the whole population. Ad-hoc searches could be performed to confirm findings reported in the literature in a cost efficient manner.

Retrieving information from the dedicated web-sites results in the meta-knowledge of the studies or conditions that physicians are interested in; some sites offer final and periodic reports, but this could be furthered by semantic methodology that would look deeper into the content. In the cases when there are several database covering different clinical areas (*cancer and palliative care*, for example), user is in even bigger need of a good data mining system that would overcome geographical and electronic barriers. We plan on developing several design solutions for the design cases identified and reported in this paper.conclusion.

Human computer interaction design methodology provides meaningful means to make data mining easier for users [11]. It needs to be seen what factors will be key to the acceptance of new means of working with the clinical data. Clinical experts will need to continuously evaluate the results of knowledge extraction and to provide suggestions for improvements of the whole process [12].

REFERENCES

1. Eaton, Chris, Deroos, Dirk, Deutsch, Tom, Lapis, George, & Zikopoulos, Paul. (2012). *Understanding Big Data: Analytics for Enterprise Class Hadoop and Streaming Data* Retrieved from http://public.dhe.ibm.com/common/ssi/ecm/en/iml14296usen/IML14296USEN.PDF
2. Bifet, Albert, Holmes, Geoff, Pfahringer, Bernhard, Kranen, Philipp, Kremer, Hardy, Jansen, Timm, & Seidl, Thomas. (2010). MOA: Massive Online Analysis, a Framework for Stream Classification and Clustering (pp. (3-16)): University of Waikato, Hamilton, New Zealand, Aachen University, Germany.
3. Hall, Mark, Frank, Eibe, Holmes, Geoffrey, Pfahringer, Bernhard, Reutemann, Peter, & Witten, Ian H. (2009). The WEKA Data Mining Software: An Update. *SIGKDD Explorations, 11*(1), (10-18).
4. Delen, D., Walker, G., & Kadam, A. (2005). Predicting breast cancer survivability: a comparison of three data mining methods. *Artif Intell Med, 34*(2), 113-127. doi: 10.1016/j.artmed.2004.07.002
5. Wilson, Andrew M., Thabane, Lehana, & Holbrook, Anne. (2003). Application of data mining techniques in pharmacovigilance. *British Journal of Clinical Pharmacology, 57*(2), (127-134). doi: 10.1046/j.1365-2125.2003.01968
6. Shaw, Michael J., Subramaniam, Chandrasekar, Tan, Gek Woo, & Welge, Michael E. (2001). Knowledge management and data mining for marketing. *Decision support systems, 31*(1), (127-137).
7. Hevner, Alan R., March, Salvatore T., Park, Jinsoo, & Ram, Sudha. (2004). Design Science in Information Systems Research. *MIS Quarterly, 28*(1), 75-105
8. Rogers, Yvonne, Sharp, Helen, & Preece, Jenny. (2011). *Interaction design: beyond human-computer interaction* (3rd ed.). UK: John Wiley & Sons Ltd.
9. Åberg, Anders, Ericsson, Jan, Holmberg, Lars, Rozell, Barbro Lundh, Ayoubi, Shiva, Khan, Staffan, & Klint, Åsa. (2011). Cancer Incidence in Sweden 2010 *Statistics – Health and Medical Care* (pp. 1-100).
10. NSD, Norsk samfunnsvitenskapelig datatjeneste. (2012). Personvernombudet for forskning . Retrieved 30 March, 2013, from http://www.nsd.uib.no/personvern/forskningstemaer/registerstudier.html
11. MacLean, Allan, Young, Richard M., & Moran, Thomas P. (1989). Design Rationale: The argument behind the artifact. *SIGCHI Bull., 20*, 247-252. doi: 10.1145/67450.67497
12. Nielsen, Jacob, & Molich, Rolf. (1990). Heuristic evaluation of user interfaces. *CHI '90*, 249-256. doi: 10.1145/97243.9728

Software Prototype for Triage and Instructional for Homecare Patients

L. Boom, N. Escobar, C. Ruales, and L. López

Universidad Pontificia Bolivariana, Medellín, Colombia
lauraboom.md@gmail.com

Abstract—Home healthcare services were created with the aim of bringing medical services to patients; in this scenario the developments in telemedicine offer significant support, providing or facilitating care and treatments at patient's homes or remote sites. This paper describes the development of a software prototype that enables homecare users, classify their symptoms at home and then apply for the appropriate type of care as indicated by the software. For the development of the software prototype, there were selected the 10 most common reasons of consultation at Empresa de Medicina Integral (EMI) and algorithms were constructed according to the pathophysiology of each symptom associated. Being the doctor present, patients tested prototype before being assisted. For validation, it was calculated kappa coeffiecient. The information given by the software has a aggrement of 92%. According prototype indication 72.5% of patients required medical attention and 27,4% should make phone attention. The most common reasons of consultation were diarrhea and upper respiratory tract symptoms.

Keywords—Homecare, software, telemedicine, prototyping, triage.

I. INTRODUCTION

The origins of telemedicine are chronologically close to home health care (USA, 1950). These two models share the objective to provide, facilitate and optimize the delivery of medical services outside the hospital, with major advances in both directions [1,19,43].

Worldwide homecare activities include the full range of health services, from the promotion and protection of health, primary care, clinical prevention, hospitalization at home (with the use of equipment traditionally hospitable house), high-cost chronic diseases and rehabilitation procedures [4] [5] [7] [8].

The application of telemedicine and / or telehealth for home care, defined as telehomecare or more recently as home telehealth or eHealth home [6] [7] [8], provides solutions to administrative and welfare activities involved in the process of home healthcare [4], [5]. Information systems, teleconsultation, rehabilitation, control, software for healthcare institutions, virtual medical records, monitoring and alerting equipment, systems analysis, processing and transfer of biological signals, educational tools, smart phone applications.

Developments go hand in hand with the changing roles of health [9]. The idea of treating short clinical symptoms at home, prompted the establishment of private entities dedicated exclusively to immediate medical consultation, emergency and ambulance transportation [10] [11].

This research was conducted at EMI, which has its own system of consultation and emergency care. It provides users with homecare and services faster and easier than using traditional consultations, because the visit is scheduled within the next two hours of receiving the call [12], which is why patients prefer more often home visits for mild symptoms. These low complexity consultations may delay patient care more seriously and tend to overwhelm the system. For this project we develop a prototype software that enables patients with mild symptoms (emergency triage V), rated their clinical status through a questionnaire that indicates the type of service to ask EMI, resulting in the prototype testing: make telephone consultation, home consultation or assistance to medical services institution provider if necessary.

II. METHODOLOGY

This work was performed in two phases, each of which had in turn with different stages as described in Figure 1.

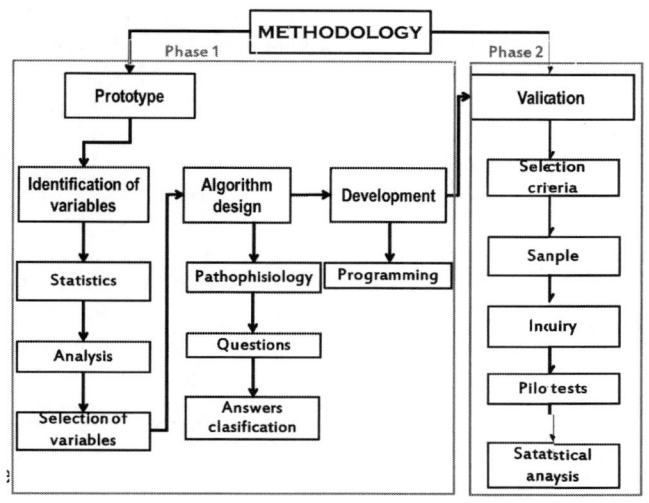

Fig. 1 Methodology

As indicated by Figure 1, the first phase prototype was developed; in the second, software and validation was performed using pilot tests. Described below activities performed in each of the phases.

1. Development

Development of prototype software was conducted in three stages:

A. Identification of variables. Statistical reports were consulted for EMI in 2012 *focusing the search for reasons OF consultation with priority level V of the run (EMI* cataloged as banal queries). Selecting 10 of them considering their frequency and banality.

B. Algorithm design. For each of the selected symptoms was determined duration, causation underlying disease, treatment management, improvement and severity. Pathophysiology correlated with each response and classified the clinical picture, and then assign the type of care appropriate to the patient's condition.

C. Programming. The tools used for development are freeware to be more accessible in the health sector. We worked with PHP programming language in version 5.0 that uses an Apache web server version 2.4. The database used version 5.5 of the MySQL DBMS.

2. Validation

For validation of the prototype the methodological design selected was agreement study. It was calculated kappa coefficient.

A. Sample. Depending on the type of study selection criteria was defined. Patients between fifteen (15) and eighty (80) years with no medical history of physical or mental condition that would affect participation, those who consult for symptoms triage V (trivial queries).

The method for sample selection was convenience sampling, according the inclusion and exclusion criteria. For the type of study sample size was not calculated.

B. Data Collection. Data was collected through inquiry to assess user acceptance of this technology and implementing pilot tests by EMI in medical patients' homes before making medical care. Interrogation was made according to the algorithm for the symptoms reported by the end user and the decision indicated by the software was annotated as the type of consultation that the patient actually received.

C. Statistical Analysis.

According to the chosen study design, analysis of the results obtained from the pilot test was conducted by calculating the kappa coefficient to determine correlation between the results obtained by the software and healthcare. The data were analyzed in Epidat 4.0.

III. RESULTS

The most frequent reasons for consultation are respiratory symptoms in 29.4% of cases (cough, sore throat) and diarrhea (23%), both related to high prevalence of viral diseases in our environment causing frequent medical consultations both hospital and residential. So the implementation of strategies for classification and clinical care of these may lead to significant improvements in medical services.

According to prototype, 72.5% of patients who sought care at home requiring treatment (medication management) and the remaining 27.4% mild symptoms that did not require medical procedures and could be assisted by telephone consultation.

Table 1 Prototype vs. Medical indication

		Medical decision		
		Homecare	Phone call	Total
Prototype	Homecare	35	2	37
	Phone call	2	12	14
	Total	37	14	51

A prior classification effective of mild symptoms and use of other types of attention (phone calls) could result in a 27.4% reduction in the number of queries, reducing waiting times and costs, while providing timely care patients with more complex symptoms.

Kappa coefficient was 0,8031 with an observed agreement of 0.9216 (92%), expected agreement 0,6017 (60%). Minimum kappa, -0,0408 and a maximum kappa of 0,8441. Z-test: 5,7352 and p value: 0,0000.

The level of agreement found (92%) indicates that the algorithms are reliable for sorting at the home of the symptoms, as it has high correlation with the type of care recommended by the doctor.

Table 2 Kappa coefficient

Kappa coefficient	Standar error	Confidence interval
0,8031	0,0941	0,6186 - 0,9875

The initiative for the classification in the home through this type of applications was accepted among the users. The main factor of dissatisfaction was the perception of being evaluated by the computer and not the doctor. But patients showed intent to use and conform to the instructions provided by the tool.

IV. DISCUSSION

Home telemedicine developments aimed at strengthening and supply healthcare needs of a patient group that typically includes: elderly, chronically ill, young people with some degree of disability, living in remote sites that require all acute hospital care and people with terminal illness whose condition THEY favor remain in place of residence. [10] [11] [13] [14] Some categories are useful to describe the attributes and benefits of technology in these tools:

A. Active devices that provide users therapy [15].
B. Non-active [15]
C. General Assistance and control devices [14] [15].
D. Modifying property [15] [16] [17].
E. Telemedicine. [14] [15] [18] [19].

iTriage and Sana are similar telemedicine applications. With iTriage user load symptoms, then software make diagnosis about the signs as described, find possible causes to guide treatment or simply recommend a visit to a doctor [18]. With Sana mobile (Android ®) decision support for nurses and health workers (non-medical), is possible, with remote consultations between specialists in the care of health and community health workers in remote areas [19].

This study highlighted as differentiators, which developed software is designed to run for patients whose health status is severely compromised, but with short clinical symptoms and mild. No offers diagnosis or treatment, and it replaces medical care. But to manage the rational use of homecare thanks to the triage provided to patients before requesting service; besides deliver practical and reliable medical information with instructions for users. The homecare attention process is optimized with the active participation of the patient from a computer or mobile phone; leading home care beyond the concept of having a doctor in the house.

V. CONCLUSIONS

We developed a software prototype for clinical classification of patients using home medicine, reasons for consultation is based triage V as EMI statistical reports. After development, tests were performed to assess agreement and determine the reliability of the results obtained by the algorithm to classify. The validation process was not brought to its final stage by limitations in permits to have several evaluators, a significant sample, and access to the patient record.

This tool helps to provide care faster and make appropriate use of available resources could also be used in hospitals and remote sites for triage of patients and as a medical guide for users when care is not available.

The high concordance of the prototype to classify patients, be reliable and safe for the classification process. Besides the results obtained so far, aims to integrate additional applications, such as teleconsultation and telemetry.

ACKNOWLEDGMENT

For planning and implementing the project was necessary to have a real home health care scenario. Some phases of research were conducted in EMI, and we appreciate the support and involvement of its administrative and assistance workers during research.

REFERENCES

[1] Badia JG. Does the hospital, must compete, dominate or share with primary care?. Primary Care.1998;21(4):16-7.
[2] Onder G, Liperoti R, Soldato M, Carpenter I, Steel K, Bernabei R, et al. Case management and risk of nursing home admission for older adults in home care: Results of the aged in home care study. J Am Geriatr Soc 2007, 55:439-44.
[3] Field M, Telemedicine: A Guide to Assessing Telecommunications for Health Care. Washington DC, 1996.
[4] Sicurello F. Some aspects on telemedicine and health networks. Proceedings of the 2nd International Symposium on Image and Signal processing and analyisis. Pula Croatia 2001. IEEE. Pg 651-54.
[5] Roncancio D, et al. Prototype mobile telemedicine and remote home healthcare.Eighth LACCEI Latin American and Caribbean Conference for Engineering and Technology (LACCEI'2010) "Innovation and Development for the Americas", Peru, June, 2010. [citado febrero 2011]. Disponible en: http://www.laccei.org.

[6] Koch S. Home telehealth−Current state and future trends. International Journal of Medical Informatics. 2006:8: 565−576.
[7] Wootton R. Recent advances: Telemedicine. BMJ. 2001 Sep 8; 323(7312):557-60..
[8] OMS. A. Health telematics policy (Document DGO/98.1) Ginebra, 1997.
[9] De Rouck S. Jacobs A., Leys M. A methodology for shifting the focus of e-health support design onto user needs: A case in the homecare field. International Journal of Medical Informatics 2008:9:589−601.
[10] Espinosa J y Cols. Organizational models of home care in primary care. Available: http://www.semfyc.es/es/actividades/publicaciones/documentossemfyc/docum015.html.
[11] Garberi I. Reflections after ten years of work in an interdisciplinary team of home care in the social health. Rev Mult Gerontol. 2003;13.(2):114-16.
[12] Servicio médico EMI. Available: www.grupoemi.com.
[13] First Argentine Congress Home Hospitalization. 2005,35:34-37 medical journal. Available at: www.revistamedicos.com.ar
[14] Le Bihan B, Martin C. A comparative case study of care systems for frail elderly people: Germany, Spain, France, Italy, United Kingdom and Sweden. Social Policy and Administration. 2006;40(1):26-46.
[15] Tarricone R, Tsouros A. The solid Facts. Home Care in Europe. World Health Organization Europe, 2008.
[16] Warren I, Weerasinghea T, Maddison R, Wang Y OdinTelehealth: A Mobile Service Platform for Telehealth. Procedia Computer Science 5 (2011) 681-688. The 8th International Conference on Mobile Web Information Systems
[17] Michaud F, Boissy P, Labonte D, Briere S, Perreault K,Corriveau H et al. Exploratory design and evaluation of a homecare teleassistive mobile robotic system. Mechatronics 201020(7):751-66.
[18] Sana Mobile. Software. Avalilable en www.sana.mit.edu.
[19] iTriage. Available en   www.itriagehealth.com/Mobile.

Analysis and Impact of Breast and Colorectal Cancer Groups on Social Networks

I. De la Torre, B. Martínez, and M. López-Coronado

Department of Signal Theory and Communications, and Telematics Engineering, University of Valladolid, Valladolid, Spain

Abstract—There are few researches about the use of social networks for medical purposes. Starting from De la Torre and her colleagues research "A content analysis of chronic diseases social groups on Facebook and Twitter" (2012) and due to the presence of both breast and colorectal cancer over the social networks; the status and impact of these cancer-related groups will be analyzed. We searched on Facebook (www.facebook.com) and Twitter (www.twitter.com) using the term 'colorectal cancer' and 'breast cancer'. The following data has been extracted in base to analyze the real repercussion of these groups: name of the networks, member number, creator, interests (based on messages off the site Wall, plus relevant information), and Website. We found 241 breast cancer and 173 colorectal cancer groups. The main aim of the groups about these kinds of cancers is the awareness-raising (37%). Added to the 21% of groups dedicated to fight the disease are the main target of the networks towards these issues. This target is consciousness-raising as a way to eradicate and prevent the cancers. Groups dedicated to raise funds also need to be mentioned; they gather the 46% of users, due to the fact that only by belonging to the groups means more resources for mammograms, treatments and investigations.

Keywords—Breast Cancer, Colorectal Cancer, Facebook, Social Network, Twitter.

I. INTRODUCTION

Nowadays, social networks are a very important piece of the mass media matrix and also in terms of information interchange. The speed and easy-access to loads of information make of them a relevant way of communication. Inside social networks, we can find all kinds of profiles, from personal ones to institutional or corporation profiles [1]. Medicine has also gone through this land of social networks. It allows us to access easily to health related issues, offering us information, advice and even support [2].

Research over online groups oriented to health has been mainly focused on the use and effects of forums and mailboxes between ill people [3]. These groups create a space for members where they can treat fragile issues, reducing loneliness and incertitude over the prognostic and painful symptoms; this way, people can be more informed and better prepared for their interaction with their own health system [4]. Recent studies have shown that these community groups reduce depression, stress and cancer related trauma. Community groups also produce the increase of social support [5-7]. However, little is known concerning the use of social networks owning to health reasons [8]. In two of his investigations [9-10], Keelan, tested the use of YouTube videos, and Myspace blogs as an informational source about the immunity. As a result he found a sub-community of critics, the different points of view over the use of injections. The research made by Scanfeld [11] shows that Twitter has been used to share information about the use of antibiotics and their secondary effects. As we can see, there are not many studies on the use of social networks as a way of treating the health; there is one [12] where De la Torre and her colleagues analyzed chronic diseases social groups on social networks. Specifically they studied how diabetes, colorectal and breast cancer-related groups are present on Facebook and Twitter.

Next, this research tries to clarify the presence of colorectal and breast cancer in social networks. It also tries to show the repercussions of the groups over the networks, its aim and creator; so it gives an idea about the current state of these diseases, being a starting point for future researches.

The remainder of this paper is structured as follows. Firstly, after justifying why breast and colorectal cancer have been chosen, we describe the method used. In the following section, we show the results obtained and a discussion. Finally we extract the main conclusions of this study.

II. MOTIVATION

Starting from the analysis made by De la Torre [12], and according to the fear factor that cancers provoke on the society; we can see how its presence over the social networks has been growing up to date [12]. Consequently, this research tries to study in depth the status and impact of both breast and colorectal cancer. In addition, breast cancer is the most common cancer among women all over the world [13], with 1.38 new million cases every year, what means the 23% of the whole cancer diagnostics [14]. But thanks to the improvements in detection and treatment, its mortality rate has been reduced significantly. According to International Agency for Cancer Research sources, today the numbers are about 19 dead for every 100,000 inhabitants [14].

Colorectal cancer is the second most common cancer between women, with 571,000 diagnosed cases over the world every year, representing 9.4% of the whole cancer diagnostics. It is the third most common cancer in men with 663,000 cases over the world every year, a 10% of total [14-15]. This supposes 1.23 new million cases every year.

According to the United States Society of Oncology, colorectal cancer is one of the most important dead reasons between the cancer victims. Mortality induced by this kind of cancer goes to 10 dead every 100,000 inhabitants, with an increasing trend in the developed countries [16].

The survival rate is very high in both diseases. However, there is a problem: the period after the treatment goes along with several physical and emotional needs that professional medics do not concern about [17].

III. METHODOLOGY

On November 2012, we made a deep search in Facebook and Twitter websites using the search engine provided by these platforms. Search key words used were: 'colorectal cancer' and 'breast cancer'. Our searching incorporates all kinds of public groups of Facebook and Twitter. Besides, pages of individual members, organizations, events, and applications were included.

In order to get the statistics, the program used has been SPSS version 15 to classify and compare the size, referring to the amount of members and the aim of the group. Due to the asymmetric distribution of the data, a median will be made along with the common mistake range, instead of making the common average, a measure more adequate for symmetric distribution.

Starting from the classification made by Bender JL and his colleagues [18] about the aim of every Facebook group that treats breast cancer, our several groups were analyzed. The conclusion reached was that these groups were quite close to those targets. On the other hand these groups needed a small change due to the fact that there are two diseases to analyze now. According to their purpose, the groups can be classified in:

- Fundraising groups. Created to raise funds (financial resources) and then directed to investigation, free treatment, etc., through an event, product or service.
- Awareness-raising groups. Created to attract the attention spot over the importance of these diseases and divulgating knowledge to confront them better and get to know their symptoms.
- Support groups. Created to satisfy the informational and emotional needs. It is achieved through the interchange of experiences and testimony of the sick, the patients who have overcome the disease, their close friends or family.

- "Promote-a-site" groups. Created to raise the protagonist of an external fund raising website. Also to raise awareness of society through the selling of products or services to advertise collaborating enterprises with foundations.
- Disease fighting groups. With the purpose of eradicating the diseases, they encourage the science to find a cure. They promote a bigger investigation and several campaigns to fight and defeat the diseases and their origin.

IV. RESULTS

In this section, we try to test the impact of both most relevant cancers in the social networks. A total of 414 colorectal and breast cancer groups was found. The size of the simple goes up to 8,156,264 users distributed between 200 groups. The median of the members per group is 822.5 with a mistake range of 23,562.5.

It has been tried to get the co-relation between the aim of the group and its creator. Results show that there is not co-relation, and we can assure that there is no conduct pattern that defines the relations between both variables.

Fig. 1 shows the difference between breast cancer and colorectal cancer members. This huge contrast is due to two reasons: breast cancer is the most common one, represented by 23% of cancer diagnosed cancer [15], and that mainly focuses in an only sex, female. Data show the higher risk of having this kind of cancer, what gives explanation to the awareness raising and worry about breast cancer prevention on the women side.

Twitter and Facebook searching engines do not allow identifying the sex of every member of a group, but it is sure, within a mistake range, that almost 2/3 of breast cancer group members are women.

On the other hand, the few members of colorectal cancer groups find explanation in the less frequency of this kind of cancer (10% of the whole cancer cases). As well, it finds answer in the low probability of having this cancer (it affects both to women and men). As a result the consciousness-raising and worry towards colorectal cancer is far lower.

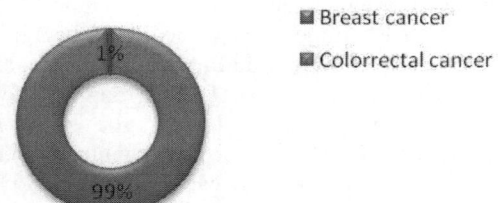

Fig. 1. Percentage of users according to the type of disease

Table 1 General purpose and size of groups of cancer

Main Issue	Total members. Target	Members median	Mistake range. Member numbers.	Minimum of members	Maximum members
Fundraising	3,763,06	1,324	409,633	61	2,264,45
Awareness-raising	3,220,76	701.5	26,7545.	60	2,270,47
Support	250,739	688	20,374.1	61	106,202
"Promote-a-site"	34,287	1,008	3,125.68	76	14,198
Disease fighting	887,406	971.5	85,955.9	66	472,997

As it can be seen in Fig. 2, a 37% of colorectal and breast cancer groups have as their aim the awareness-raising, followed by a 21% dedicated to fight the disease. This means that almost 50% of the groups want the society to get sensible about both kinds of cancer; so they follow prevention advice and fight to destroy breast and colorectal cancer. A 16% of groups are directed to raise money and 16.5% offer support. And only a 9.5% of groups are "promote-a-site" ones.

In order to see the real insight of this kind of targets within the social networks; we must focus in the amount of members of each group rather than focusing on the number of groups for each target.

Table 1 classifies users owning to the aim of the group they belong to. Groups which aim is fund-raising count with 3.7 million users. The biggest group concerns about breast cancer. This group is formed by 2.2 million people. Consciousness-raising groups also have a lot of members, more than 3 million. As fund-raising groups, the biggest one has 2.2 million users and concerns about breast cancer. These groups get special attention because both of them belong to the same non-lucrative organization: "the breast cancer site". It looks like this organization is doing a great job because it embraces almost 50% of users. Its program 'fund mammograms' casted this site; social networks did the rest.

Fig. 3 shows the percentage of members according to the group's purposes for the different groups of colorectal and breast cancer.

Another proof of the groups oriented to awareness-raising and disease fighting getting their targets is the 51% of members they own. It could be seen as a promising spot of viral marketing towards breast and colorectal cancer, due to the spreading effect of the networks.

As it can be seen in Fig. 3, a 46% belong to fund-raising groups. This high number owns its origin to the bigger number of users of breast cancer groups, also because only by belonging to this group results in financing research and mammograms. It seems poor for support groups to represent only a 3% of users, if we compare to the rest. But as it can be seen in Table 1; support groups count with more than 250,000 users, that is, 250,000 disease victims or their relatives who are in touch, helping each other. And increasing the social support to get over the disease too [6]. Fig. 4 shows like creators of the groups are fairly distributed.

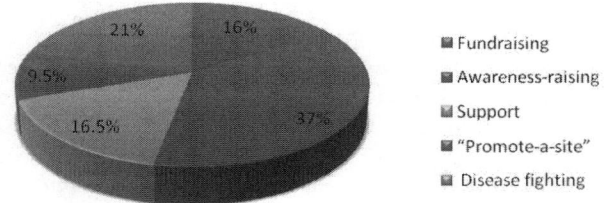

Fig. 2 Main issues of the Facebook groups and Twitter for colorectal and breast cancer

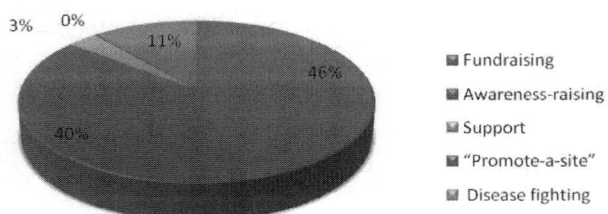

Fig. 3 Percentage of users according at group purpose for breast cancer and colorectal cancer

Table 2 Size of groups according to social media and type of disease

Social network	Total breast cancer's groups	Total colorectal cancer's groups	Total members. Target	Members median	Mistake range. Member number
Facebook	96	26	8,045,928	1,359	298,958
Twitter	4	74	110,336	217.5	4,616.97

This data is not very representative because a large part of the profiles were private. In the graphic they appear as "others" and it was impossible to determine their role. If we take it as representative, the creator of a 24% of the groups is a cancer affected person, a close relative or a survivor. Due to their experiences they promote groups embracing all the aims, from 'awareness-raising' to 'support'. A 20.5% of creators are non-lucrative organizations; they work as an alternative to increase the knowledge about the disease and fight it. An 18% of groups have been created by an institution or medicine Enterprise; so they get advertised, also they do 'awareness-raising' and 'fundraising'.

Only a 5% of the identified creators are medicine experts, but still reduced (only 10 out of 200 analyzed groups), which does not mean they cannot get involved in groups with other role. In addition, a higher number of professionals could be a good way to get patient-medic communication, better advice, and a correct treatment tracking, all around the Medicine 2.0.

Fig. 5 and Table 2 show how colorectal and breast cancer groups are distributed on the social networks. A 61% of breast and colorectal cancer groups are on Facebook, against the 39% Twitter profiles. These results are due to the higher insight of Facebook against Twitter, by two reasons: Twitter is newer than Facebook and its use is not as common Facebook can be. Also it must be mentioned that colorectal cancer is not a very treated issue on Facebook. Only 26 groups are between the most popular versus the 96 analyzed about breast cancer. At the same time, we can appreciate the difference in the Table 2, watching the highest number of members: more than 2 million users in the biggest Facebook group, versus 32,000 users in the Twitter ones.

V. DISCUSSION

Social Networks are a useful tool for certain health-related situations as: giving support to the disease affected or the awareness-raising towards breast or colorectal cancer. Groups formed about these issues have also their aims: fund-raising, awareness raising, support, advertise medical institutions and fighting the disease.

The research has been focused on breast and colorectal cancer; two of the most common cancers of the XXI century. These two cancers share some aspects as they differ in others in terms of the treatment they are given in social networks. As it can be seen on the Fig. 1, breast cancer is far more present than colorectal cancer on the networks. This presence is shown according to the existing groups and the number of users that compose them. Breast cancer has 241 groups and more than 8 million users between the 100 most important groups. On the other hand, colorectal cancer that counts with 173 groups and only 54,330 users between the 100 most important groups. This large difference is a mirror of reality and that society is more aware with breast cancer. This is mainly due to two reasons: It represents 23% of all cancer diagnoses [15] and is focused primarily on women. Causing more fear of suffering it and therefore a greater awareness and prevention against it. In terms of the aim of the groups, as seen on Fig. 2, it is shown that the main aim of the groups in both of the diseases is 'awareness-raising'. 'Awareness-raising' represents more than a 37%; this highlights the idea that the main work of the social networks relating the cancer is to raise the population awareness towards the danger and importance of it.

In the same way, on Fig. 2 it can be seen the importance of groups which aim is to support (16.5%) due to the 18% of breast cancer and 15% of colorectal cancer groups. Although they have a low amount of users (only 3% total users) because the public is limited: sick people, family, survivors and medics, it shows hope. The fund-raising also is one of the main targets of these networks. In groups of breast cancer, this aim represents a 46% of the groups, while colorectal cancer represents little more than 20%. This situation finds its reason in the next: only by belonging to the group, an increment is gotten on the funds for research, mammograms, treatments and centers. As Table 2 shows, 96% of the breast cancer groups are present on Facebook due to the fact that Twitter is a newer network and has far less members; only the biggest groups were chosen.

On the other hand, and as seen on the Table 1 too, a 74% of groups are present on Twitter. It is provoked by the lower social insight of colorectal cancer and the higher awareness-raising through the same on Twitter.

VI. CONCLUSION

In this paper, we analyze the status and impact of the (breast and colorectal) cancer-related groups on Facebook and Twitter.

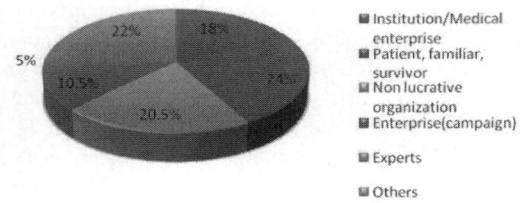

Fig. 4 Creator of cancer groups

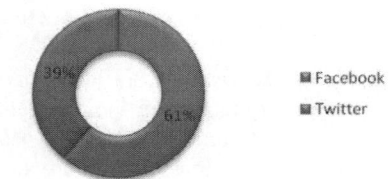

Fig. 5 Use of Facebook and Twitter for cancer groups

We found 173 colorectal cancer groups and 241 breast cancer groups. The main aim of the groups about these kinds of cancers is the awareness-raising. 21% of groups dedicated to fight the disease are the main target of the networks towards these issues. Other important groups are dedicated to raise funds.

Several researches confirm that social networks improve the prospects towards the disease and also decrease cases of depression and other injuries. This is thanked to the interchange of information, experiences and feelings between the disease victims [5-7].

Future researches could focus on the contents produced by users who belong to any group related to the diseases, so their involvement and its profit could be tested.

ACKNOWLEDGMENT

This research has been partially supported by the Spanish Social Security Administration Body (IMSERSO) under the project 85/2010.

REFERENCES

1. Boyd DM, Ellison NB, Social Network Sites: Definition, History, and Scholarship (2007) J Comput Mediat Commun 13: 210–230.
2. Giustini D (2006) How web 2.0 is changing medicine. BMJ 333:1283
3. Davison KP. Pennebaker JW, Dickerson SS (2000) Who talks? The social psychology of illness support groups. Am Psychol 55:205-217.
4. Shaw BR, McTavish F, Hawkins R, Gustafson DH, Pingree S (2000) Experiences of women with breast cancer: exchanging social support over the CHESS computer network. J Health Commun 5:135-159.
5. Gustafson DH, Hawkins R, Pingree S, McTavish R, Arora NK, Mendenhall J, et al (2001) Effect of computer support on younger women with breast cancer. J Gen Intern Med 16:435-445.
6. Gustafson DH, McTavish FM, Stengle W, Ballard D, et al (2005) Use and impact of eHealth system by low-income women with breast cancer. J Health Commun 10:195-218.
7. Winzelberg AJ, Classen C, Alpers GW, Roberts H, Koopman C, Adams RE, et al (2003) Evaluation of an internet support group for women with primary breast cancer. Cancer 97:1164-1173.
8. Preece J, Maloney-Krichmar D (2005) Online communities: design, theory and practice. J Comput Mediat Commun 10.
9. Keelan J, Pavri-Garcia V, Tomlinson G, Wilson K (2007) YouTube as a source of information on immunization: a content analysis. JAMA 298:2482-2484
10. Keelan J, Pavri V, Balakrishnan R, Wilson K, An analysis of the human papilloma virus vaccine debate on MySpace blogs. Vaccine 2010; 28(6):1535-1540.
11. Scanfeld D, Scanfeld V, Larson EL (2010) Dissemination of health information through social networks: Twitter and antibiotics. Am J Infect Control 38:182-188
12. De la Torre-Díez I, Díaz-Pernas FJ, Antón-Rodriguez M (2012) A content analysis of chronic diseases social groups on Facebook and Twitter. Telemed J E-health 18:404-408.
13. Reis N, Rodrigues JJPC, Moutinho JAF, De la Torre I (2012) Breast Alert: an On-line Tool for Predicting the Lifetime Risk of Women Breast Cancer. J Med Syst 36: 1417-1424.
14. GLOBOCAN. World Health Organization. Breast and Colorectal Cancer Incidence, Mortality and Prevalence Worldwide at http://globocan.iarc.fr/factsheet.asp
15. De la Torre I, Díaz FJ, Antón M, Barragán E, Rodrigues JJPC, Pires RA (2012) Telematic Tool to Predict the Risk of Colorectal Cancer in White Men and Women: ColoRectal Cancer Alert (CRCA). J Med Syst 36: 2557-2564.
16. Smith RA, Cokkinides V, Brooks D, Saslow D, Brawley OW (2010) Cancer screening in the United States, 2010: a review of current American Cancer Society guidelines and issues in cancer screening. CA Cancer J Clin 60:99-119.
17. Surbone A, Peccatori FA (2006) Unmet needs of cancer survivors: supportive care's new challenge. Support Care Cancer 14:397-399.
18. Bender JL, Jimenez-Marroquin MC, Jadad AR (2011) Seeking Support on Facebook: A Content Analysis of Breast Cancer Groups. J Med Internet Res 13(1):e16

Author: I. De la Torre
Institute: Department of Signal Theory and Communications, and Telematics Engineering. University of Valladolid
Street: Paseo de Belén, 15
City: Valladolid
Country: Spain
Email: isator@tel.uva.es

Health Apps for the Most Prevalent Conditions

B. Martínez-Pérez, I. de la Torre-Díez, and M. López-Coronado

Department of Signal Theory and Communications, and Telematics Engineering,
University of Valladolid, Valladolid, Spain

Abstract—In the last years, significant improvements in communications and technology have boosted the m-health to the point of being available thousands of health apps for smartphones and tablets. The main aim of this paper is to develop a study of the existing apps in the most important commercial stores for the 5 most prevalent health conditions issued by the last update of the *Global Burden of Disease* by the World Health Organization, understanding prevalence as the number of individuals who have the disease or condition at any moment. These conditions are iron-deficiency anemia, hearing loss, migraine, low vision and asthma. For this objective, it has been carried out a review of the apps for these conditions on the most important apps stores and an in-depth analysis of a sample of reference apps related to them. 356 relevant apps were found on the stores Google play, iTunes, BlackBerry World and Windows Phone Apps+Games in February 2013. From these apps, the most relevant of each condition by specific selection criteria were selected for being analyzed. Some interesting findings could be deduced from the results obtained. There is an alarming low number of apps for BlackBerry compared to the rest of systems. Despite anemia is the most prevalent condition, there are very few apps for it and no one aimed for patients. There are more apps for migraine and asthma than for the rest and they are divided into informative and monitoring apps whereas the majority of those for hearing loss and low vision are assistive.

Keywords—apps, m-health, prevalent health conditions, smartphones' stores, World Health Organization.

I. INTRODUCTION

A. Background

The health is the most important matter in the life of an individual. If one is not healthy other issues lacks importance, hence humans have been always studying aspects of health and medicine. In the last decades, the irruption of the Internet, and the important advances in wireless communications have propitiated the creation and the expansion of m-health [1]. The more than 6 billion mobile phone subscriptions worldwide out of which 1.08 are smartphones [2-4] (not to mention the tablets) make these mobile devices an exceptional opportunity for providing health care everywhere and at every moment by using apps designed for health care [5].

But even with the last advances in these fields, there are still many diseases affecting people worldwide. According to the last Global Burden of Disease [6] by the World Health Organization (WHO), 58.8 million deaths occurred globally in 2004 and, out of every 10 deaths, only one was caused by injuries. In this year, there were also 18.6 million people severely disabled and 79.7 million moderately disabled. It has especial importance the prevalence of a disease or health condition, understood as the number of individuals who have the disease or condition at any moment.

The main aim of this paper is to study the commercial apps of the most important apps stores exclusively dedicated to the 5 most prevalent conditions by [6] and to carry out an analysis of features of a representative selection of the apps for each condition. The remainder of the paper is structured as follows. Next subsection is about the 5 most prevalent conditions. The following section describes the methodology used. After this, there are shown the results obtained from the review of the stores and the analysis of the mobile applications. Finally the discussions and conclusions are presented.

B. The Most Prevalent Conditions

The 5 most prevalent conditions by [6] are iron-deficiency anemia, hearing loss, migraine, low vision, and asthma. Figure 1 shows the prevalence in million individuals of each condition and below there is a brief description of these conditions.

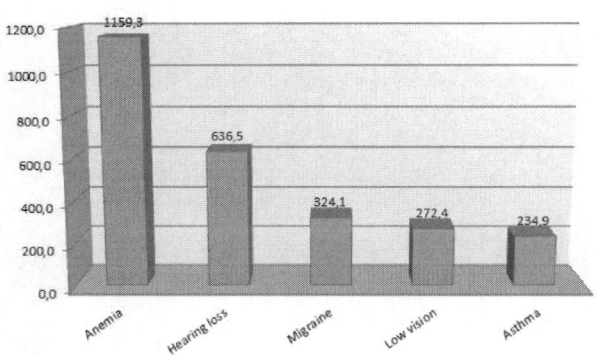

Fig. 1 Prevalence of the most prevalent conditions by the WHO

Incorrectly, it is common to use the concepts anemia and iron-deficiency anemia (IDA) indistinctively, but they are not the same: anemia has its source in several causes together or isolated, but the most important contributor is iron deficiency [7]. IDA has also multifactorial sources, but all of them result in the fact that there is not enough absorption of iron by the person, no matter the cause. IDA's consequences are diminished exercise capacity, immune dysfunction, impaired thermoregulation and gastrointestinal disturbances, among others. [7-8].

Hearing loss or hearing impairment is defined by the WHO as *"the inability to hear as well as someone with normal hearing"*. This includes minor hearing loss to deafness [9]. There are various types of hearing loss and some of the most common causes are congenital, infectious diseases, excessive noise, ototoxic medications, head or ear injuries and presbyacusis. The effects of this condition have social and economic consequences, not only in individuals, but also in communities [10-11].

Although the pathophysiology of migraine is not totally known, it has its origins in an *"activation of a mechanism deep in the brain which causes release of pain - producing inflammatory substances around the nerves and blood vessels of the head"* [12]. It is not known why these attacks occur periodically and what process finishes spontaneously these attacks. Some common characteristics of migraine are strong headache, nausea and intolerance for some light and noise conditions. These episodes entail important social and economic negative aspects [12-14].

Understood as visual acuity less than 6/18 and equal or better than 3/60 due to refractive errors, cataracts, glaucoma or macular degeneration, low vision cause difficulties in almost every aspect of the life of the person: things we see normal such as dressing, walking, communicating and interacting can be very hard for people with low vision, not to mention blind [15-17].

The WHO defines asthma as *"a chronic disease characterized by recurrent attacks of breathlessness and wheezing, which vary in severity and frequency from person to person"* [18]. The causes are not fully understood, but it is known that the genes and the environment are very important in its development. 75-80% of all asthmatic cases are caused by a response to an allergy and the triggers of asthma attacks are inhaled particles [18-20].

II. METHODS

The research of the apps were carried out in the most important application stores [21-22] which are, in descending order of market share of operative systems for smartphones, Google play of Google Android [23], iTunes of Apple [24], BlackBerry World of BlackBerry (previously known as RIM or Research In Motion Limited) [25] and Windows Phone Apps+Games Store of Microsoft [26].

The name of the condition was searched but if the search returned few or no results, it was tried other name, as happened with iron-deficiency anemia, searching just "anemia". The criteria of selection of apps were the following: applications had to be in English or in the language of the country where the search was executed (Spanish), focused on the sought condition, not included in the categories of entertainment, games or music, designed for human care (not veterinarian) and not being magazines or journals.

During the search we faced two issues. The first was that, on iTunes, apps for iPod and iPhone were separated from the ones for iPad, therefore only apps for the first one were searched. The second difficult was a problem with the web page of Google Play [23]. In some searches, the site said that a certain number of results have been found but while exploring the pages of the results the last pages were blank, being the number returned different from the number of apps showed. In these cases it was considered the last one.

For the in-depth analysis an application was chosen for each condition. It was decided to select apps for Android as it is the most extended software in smartphones [21-22]. For each condition, the first relevant free app with a rating by users of 3 or more stars which Google play showed when searching by the condition and sorted by popularity was selected. Furthermore, another requisite was that the app had to be developed for patients, not for health care professionals. So, summing up, the app chosen for analysis was the most popular free one associated to the condition with an evaluation over the mean and designed for patients. Nevertheless, all the results with a rating of 3 stars or more in the case of anemia were created for caregivers, so just in this case the app chosen did not fulfill the last requisite of being aimed for patients. The apps were tested on a Samsung Galaxy S SCL GT-I9003.

III. RESULTS

A. Review of Commercial Apps

Table 1 includes the results of the review, showing the number of significant apps out of the total number of applications found in each store for each condition. The last row shows the addition of all the apps found for store and the last column contains the addition of all the apps found for condition. It is important to state that the last does not mean the total number of apps for a specific condition because there are apps designed for several systems. For example, one app found on iTunes can also be found on Google play.

Table 1 Results of the apps review

	Google play	iTunes	BlackBerry	Windows	Total
Anemia	7/74	7/21	0/0	0/0	**14**
Hearing loss	17/42	32/37	0/0	3/5	**52**
Migraine	57/201	46/102	5/6	4/8	**112**
Low vision	33/43	30/46	0/0	4/8	**64**
Asthma	44/226	57/124	6/7	4/14	**112**
Total	158	172	11	15	

In light of the results, it is obvious that iTunes and Google play are the stores with more apps, followed distantly by Windows Phone Apps+Games and being BlackBerry World the last. Comparing the conditions by their numbers of apps, migraine and asthma are the ones with more apps, followed by low vision and hearing loss. Finally, anemia is the last.

B. Analysis of Representative Apps

The commercial apps that fulfill the requisites explained in the Methods section and, therefore, selected for the indepth analysis were: MD Series: Anemia – Free [27]; Hearing Tests [28]; My Headache Log Pro [29]; A.I.type EZReader Theme Pack [30] and SIGN Asthma Patient Guide [31].

MD Series: Anemia – Free is an app designed for caregivers which provides several tools for the diagnosis and treatment of different types of anemia in adult patients. These tools have an educational purpose [27]. Hearing Tests is basically a hearing examination which uses sounds of diverse frequencies to provide to the user an auto-diagnosis for his hearing. [28]. My Headache Log Pro is a diary where the user can note the headache attacks suffered, with its triggers, symptoms and medications used, permitting also to email these records to the doctor [29]. A.I.type EZReader Theme Pack is a keyboard design for Android smartphones with bigger keys than the ones of normal keyboards, higher contrast, helpful colors and even audio aid which reproduces what the user have written, and it is specially developed for people with visual problems, [30]. SIGN Asthma Patient Guide is a guide for both asthmatic patients and relatives in order to help families to know and take control over this condition. For this purpose, it has two sections, one dedicated to patients and other dedicated to carers of asthmatic children [31]. Figure 2 shows some snapshots of the apps analyzed.

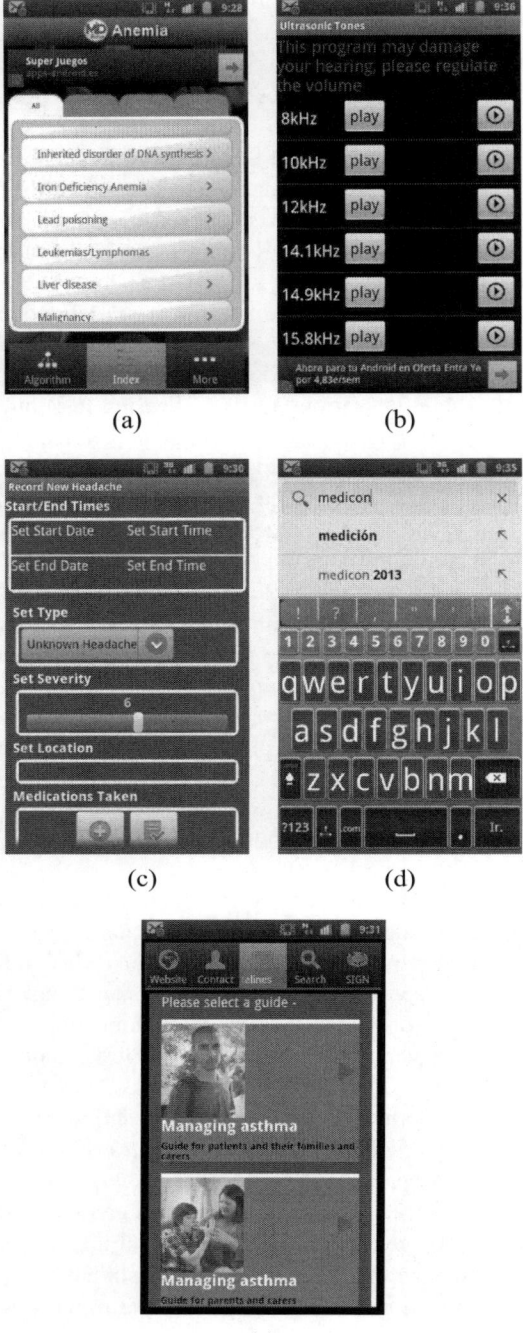

Fig. 2 Snapshots of (a) MD Series: Anemia – Free, (b) Hearing Tests, (c) My Headache Log Pro, (d) A.I.type EZReader Theme Pack and SIGN Asthma Patient Guide.

In Table 2 there are shown the results of the analysis of some characteristics of the selected applications.

Table 2 Analysis of features of the representative apps

	MD Series: Anemia	Hearing Tests	My Headache Log Pro	A.I.type EZ-Reader	SIGN Asthma
Rating	4.9	3	4.1	4.4	4.9
Class	Educational	Diagnosis	Monitoring	Assistive	Informative
Internet requirement	No	No	Only for mails	No	Some functions
Clinical/Non-clinical	Both	Non-clinical	Both	Non-clinical	Non-clinical
Data visualization	Text	Text	Text, graphs	Text, audio	Text, pictures
Context awareness	No	No	Preferences, location	Language	No
Therapist intervention	-	No	Possible	No	No
Frequency of use	Occasional	Occasional	Frequency of attacks	Continuous	Occasional
Interface	Complex	Basic	Complex	Basic	Basic
Public	Specialists	General	General	General	General

IV. DISCUSSION AND CONCLUSIONS

Several interesting findings can be extracted from the results. The most striking is that the most prevalent condition (IDA) is the one with less apps while the less prevalent (asthma) is the one with the highest number of apps (with migraine). This can be explained by two facts. The first is that the majority of the cases of IDA are located in underdeveloped or developing countries [6, 32] where smartphones or tablets are not spread so it is not worthwhile developing apps for IDA. However, this fact alone does not explain the situation because IDA is even more extended than asthma in developed countries and it is necessary a second fact: there is not a social conscience of IDA as strong as the one formed for asthma and, therefore, it is under-investigated. For these reasons and in light of the numbers, it can be profitable to create apps for IDA or anemia in general, filling this gap in the market.

Excluding the case of IDA, the conditions with more apps are migraine and asthma whereas hearing loss and low vision have less apps, despite hearing loss is more prevalent than migraine and asthma, and low vision is more prevalent than asthma. These results can be explained by the sort of treatment these conditions have. On one hand, hearing loss and low vision are conditions that affect human senses and can be easily treated in most cases with hearing aid and glasses respectively, or even with surgical operations. On the other hand, migraine and asthma have not cure and, in fact, the origin of these conditions is unknown [12, 14, 18-19], therefore the treatment consists of monitoring the attacks of the conditions and the use of some medication for a better management and control [13, 20]. Developers may have seen this point and used it for developing apps for monitoring these conditions, whereas this type of apps cannot be used for hearing loss or low vision, explaining the difference in the numbers of apps for these conditions.

Comparing the number of apps available in each store, Android and iOS are the systems with more apps, followed distantly by Windows Phone, whereas BlackBerry is the one with less apps. This classification corresponds with the market share except the case of BlackBerry, which is the third smartphone platform in market share over Windows Phone [21-22]. It seems that developers are not spending their time in this platform and this can be both cause and consequence of the loss of market share of BlackBerry.

From the review and the analysis of the apps it can be said that, in general, apps for anemia are aimed for aiding in its diagnosis whereas apps for hearing loss are divided in two types: hearing checks and informative apps. As it was mentioned previously, apps for migraine and asthma are designed for monitoring and there are also informative apps. Low vision has principally assistive apps.

Usually, Internet is not required or just needed for some functions such as sending emails, which is an advantage in situations where the Internet connection is not available. Normally, the apps are not designed for clinical purposes, at least not only for them. In addition to this, if an app is for non-clinical use then there is no therapist intervention but if it can have both clinical and non-clinical use, the therapist intervention can be possible.

The preferred method of data visualization is text, which is used in every case and can be combined in some occasions with graphs, common in monitoring apps to show data in a more comfortable way, and pictures, normally used in apps with informational or educational aims. Audio is usually used in low vision apps and in hearing loss apps for hearing tests. The interface is connected to data visualization in the way that apps with only text or with text and pictures have a simple interface whereas apps with graphs have a complex one. The exception occurs with the app for anemia, which has a complex use, even not instinctive at first sight.

As it can be seen, there is no relation between the type of apps and the context awareness. Focusing on frequency of use, monitoring and assistive apps have a continuous or very frequent use and in some cases, such as migraine, it depends on the frequency of the attacks. On the other hand, apps for diagnosis and educational or informative apps are used more occasionally. Finally, all the analyzed apps are aimed for general public, excluding the one for anemia which is designed for specialists. This is normal because developers want the most number of users for its applications, not focusing only on a specific type of users.

For future work, various lines can be followed. It was seen that there are no apps for anemia aimed for patients so it can be a good idea to design and create one, which can be educational and informative and, this way, fill this gap in the market. Another way of action can be developing applications for low vision or hearing loss, since there are few apps compared with other health conditions such as asthma and migraine.

ACKNOWLEDGMENT

This research has been partially supported by *Ministerio de Economía y Competitividad*, Spain.

REFERENCES

1. El Khaddar MA, Harroud H, Boulmalf M, Elkoutbi M, Habbani A (2012) Emerging wireless technologies in e-health Trends, challenges, and framework design issues. 2012 International Conference on Multimedia Computing and Systems (ICMCS); 2012 Oct 10-12; pp 440-445
2. RT Lester, van der Krop M, Taylor D, Alasaly K, Coleman J, Marra F (2011) M-health: Connecting patients to improve population and public health. BCMJ 53(5):218
3. International Telecommunication Union (2011) The World in 2011 ICT Facts and Figures. The Union, Switzerland
4. GO-Gulf.com Web Design Company. Smartphone Users Around the World – Statistics and Facts (Infographic) at http://www.gogulf.com/blog/smartphone/
5. Suleymanova A (2013). Empowering patients: How mobile apps are influencing the future of health plan customer loyalty. Health Manag Technol 34(2):10-1
6. World Health Organization (2008) The global burden of disease: 2004 update. The Organization, Switzerland
7. World Health Organization. de Benoist B, McLean E, Egli I, Cogswell M (2008) Worldwide prevalence of anaemia 1993–2005: WHO Global Database on Anaemia. The Organization, Switzerland
8. Clark SF (2008) Iron deficiency anemia. Nutr Clin Pract 23(2):128-141
9. World Health Organization. Health topics: Deafness and hearing loss at http://www.who.int/topics/deafness/en/
10. World Health Organization. Deafness and hearing loss Fact Sheet N° 300 at http://www.who.int/mediacentre/factsheets/fs300/en/index.html
11. Hearing Loss Association of America. Basic Facts About Hearing Loss at http://www.hearingloss.org/content/basic-facts-about-hearing-loss
12. World Health Organization (2011) Atlas of headache disorders and resources in the world. The Organization, Italy
13. Donnet A, Becker H, Allaf B, Lantéri-Minet M (2010) Migraine and migraines of specialists: perceptions and management. Headache 50(7):1115-25
14. Steiner TJ, Stovner LJ, Birbeck GL (2013) Migraine: the seventh disabler. Headache 53(2):227-9
15. World Health Organization. Visual impairment and blindness Fact Sheet N°282 at http://www.who.int/mediacentre/factsheets/fs282/en/index.html
16. Chung ST, Bailey IL, Dagnelie G, Jackson JA, Legge GE, Rubin GS, Wood J (2012) New challenges in low-vision research. Optom Vis Sci 89(9):1244-5
17. Pascolini D, Mariotti SP (2012) Global estimates of visual impairment: 2010. Br J Ophthalmol 96(5):614-8
18. World Health Organization. Asthma Fact sheet N°307 at http://www.who.int/mediacentre/factsheets/fs307/en/
19. Madore AM, Laprise C (2010) Immunological and genetic aspects of asthma and allergy. J Asthma Allergy 20(3):107-21
20. Shahidi N, Fitzgerald JM (2010) Current recommendations for the treatment of mild asthma. J Asthma Allergy 8(3):169-76
21. IDC. Press release - Android and iOS Combine for 91.1% of the Worldwide Smartphone OS Market in 4Q12 and 87.6% for the Year at http://www.idc.com/getdoc.jsp?containerId=prUS23946013#.UWfEsKIqyCI
22. Gartner. Press release - Gartner Says Worldwide Mobile Phone Sales Declined 1.7 Percent in 2012 at http://www.gartner.com/newsroom/id/2335616
23. Google. Google play at https://play.google.com/store
24. Apple. iTunes at http://www.apple.com/itunes/
25. BlackBerry. BlackBerry World at http://appworld.blackberry.com/webstore/product/1/
26. 68.   Microsoft. Windows Phone Apps+Games at http://www.windowsphone.com/en-us/store
27. Beach-Rak Medicine LLC (Google play). MD Series: Anemia – Free at https://play.google.com/store/apps/details?id=com.enifyservices.MDS.Anemia.Free&feature=search_result#?t=W251bGwsMSwxLDEsImNvbS5lbmlmeXNlcnZpY2VzLk1EUy5BbmVtaWEuRnJlZSJd
28. mikecaroline2008 (Google play). Hearing Tests at https://play.google.com/store/apps/details?id=com.kk.tones&feature=search_result&hl=en
29. Solar Embedded (Google play). My Headache Log Pro at https://play.google.com/store/apps/details?id=com.dontek.myheadachelog&feature=search_result
30. A.I.type (Google play). A.I.type EZReader Theme Pack at https://play.google.com/store/apps/details?id=com.aitype.android.theme.ezreader&feature=search_result
31. SIGN Executive (Google play). SIGN Asthma Patient Guide at https://play.google.com/store/apps/details?id=com.rootcreative.asthma&feature=search_result
32. Shaw JG, Friedman JF (2011) Iron Deficiency Anemia: Focus on Infectious Diseases in Lesser Developed Countries. Anemia 2011:260380

Author: Borja Martínez Pérez
Institute: University of Valladolid
Street: Paseo de Belén 15
City: Valladolid
Country: Spain
Email: borja.martinez@uva.es

A Mobile Remote Monitoring Service for Measuring Fetal Heart Rate

G. Lanzola[1], I. Secci[1], S. Scarpellini[1], A. Fanelli[2], G. Magenes[1], and M.G. Signorini[2]

[1] Department of Electrical, Computer, and Biomedical Engineering, University of Pavia, Pavia, Italy
[2] Department of Electronics, Information Technology and Bioengineering, Politecnico di Milano, Milano, Italy

Abstract—This paper illustrates an architecture and an implementation of a tele-monitoring service measuring fetal heart rate. The service helps pregnant women in self-accomplishing the exam at their homes wearing a textile garment connected to an 8-channel ECG monitoring device which delivers the signal to a mobile device such as a smartphone or tablet. The mobile device then sends the acquired signal over the air to a clinic server where further processing is performed and a medical report is eventually issued. A web application is available on the clinic server in order to enable the medical staff to retrieve and classify the acquired signals.

Keywords—Biosignal processing, Wearable technologies, Tele-monitoring, e-Health, Point-of-care systems.

I. INTRODUCTION

Tele-monitoring is that branch of telemedicine dealing with the exploitation of Information and Communication Technologies *(ICT)* in support for diagnosis, treatment delivery and patient care, occurring at a distance from a clinical setting [1]. Until the past decade only few parameters could be collected and sent in real-time over the network mainly because of technical and connection-related limitations. However, the recent achievements in ICT, together with the appearance of new biosensor families made available through advances in miniaturization and nanotechnologies [2], are deeply reshaping the way in which real-time surveillance of physiological parameters is accomplished.

Following this paradigm shift, tele-monitoring is becoming prominent in the management of chronic diseases on the basis that home care delivery is considered a key issue to enforce a better control on the disease, delay the onset of its complications, reduce hospitalization episodes and save any related cost [3]. Nevertheless, the aim of improving patient comfort, along with the requirement of optimizing resources usage in health care, is targeting remote monitoring also at different patient classes. The Carelink™ Network, for example, has been conceived by Medtronic for managing heart failure patients treated with Implantable Cardioverter Devices (*ICD*) [4]. Case studies are starting to appear in the literature demonstrating the effectiveness of continuous remote monitoring over ICDs on a large scale either for improving the early detection of problems with those devices [5] or for optimizing the therapy on each patient [6].

A further impulse to remote monitoring is given by the introduction of low cost miniaturized wearable sensors and devices that can be embedded in standard garments. Those allow to extend the use of tele-monitoring to almost all healthy population, in most every-day life activities with the aim of checking and preventing unphysiological conditions. Moreover, intelligent garments and tele-monitoring systems may be employed to monitor highly risky human activities such as firefighting [7] and can also be used to generate alarms and prevent the onset of pathological conditions in healthy subjects when they are defenseless, as it happens for fetuses during pregnancy. Previous studies have shown that pregnant women represent a patient category for which wearable remote monitoring is particularly effective, potentially leading to high savings [8-9] through the reduction of in- or out- stays for those only undergoing routine examinations, thereby yielding a substantial decrease both in bed shortages and in the overall costs borne by health services.

On this basis we are carrying out a project for remote antepartum fetal monitoring accomplished through wearable sensors and in this paper we propose the architecture and the specific implementation of the fetal tele-monitoring service. Besides plain data transmission, the service will be complemented by modules performing an automatic analysis of the signals sent, helping the clinical staff in promptly recognizing any possible misbehavior.

II. THE CLINICAL CONTEXT

An important issue for pregnant women is given by the analysis of fetal conditions in order to anticipate as much as possible the detection of problematic situations calling for an action. A good predictor of the unborn child health status is considered to be the Fetal Heart Rate (FHR) which also helps in assessing the maturity level of the fetus circulatory system [10]. In the last three decades cardiotocography (CTG) has become the preferred examination for measuring FHR [11], albeit it can only be accomplished at specialized clinics since it exploits the ultrasound technology which requires trained people for its management.

However, current research is positive about the possibility of measuring FHR also through non-invasive means, such as through ECG recordings performed on the mother's abdomen [12]. Despite the fact that an ECG signal acquired in that way is corrupted by several sources of noise, a dedicated device has been already developed at the Politecnico di Milano [13] and proved to be a valuable alternative to the traditional CTG approach. That system is based on a garment which every pregnant woman could wear in order to autonomously accomplish the FHR measurement at her own domicile without the supervision of health care staff.

Relying on ECG signals for FHR measurement may therefore represent a concrete opportunity for cost reduction. In fact, during her pregnancy a woman may undergo one to three CTG examinations which are presently accomplished at a clinic. In situations requiring an enhanced monitoring during the last quarter a pregnant woman may undergo weekly or even daily examinations. The requirement of accomplishing CTG at a clinic besides being the cause of discomfort for patients and their caregivers, severely limits the possibility of performing repeated examinations. On the contrary, the possibility of acquiring ECG signals at the patient's home and sending them to a centralized server for FHR measurement and diagnostic classification may improve both the clinical outcome and the comfort for the patient while at the same time help in curbing the expenses.

III. THE FUNCTIONAL ARCHITECTURE

The main goal of this work encompasses the implementation of a telemedicine service for the remote acquisition of a CTG signal exploiting electrical activity in pregnant women. In Figure 1 we illustrate the functional architecture of the system, pointing out all the components involved, which are better described in the following.

The acquisition of the cardiac electrical signals takes place through a dedicated hardware device which is connected to the patient through a sensorized garment. That garment consists of an underwear t-shirt made of an elastic fabric with a matrix of 9 textile electrodes interwoven in it which may be donned by the patient whenever an acquisition is to be performed. In order to prepare for the acquisition it is only required that the patient connects the leads coming out of the acquisition device with the contacts available on the garment. To reduce the costs connected with its manufactoring, the hardware device has no display for interfacing with the user and only consists of an 8-channel differential amplifier paired with a Bluetooth^{TM} wireless communication module. In fact, smartphones or tablets available nowadays are endowed with a high resolution color screen whose capabilities outpace those of any other rendering device and parallel those of the computers available in the past decade. Furthermore, they are also running a multitasking operating system which translates in a full-fledged windowing system that enables their exploitation through a simple and high-level programming.

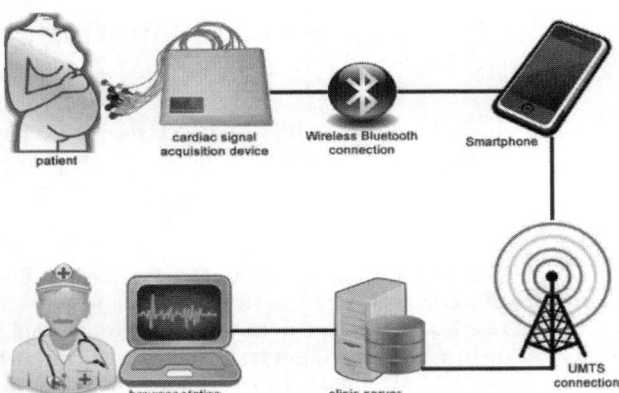

Fig. 1 The functional architecture of the service.

In order to start signal capturing a mobile terminal is required which accomplishes both tasks of interfacing with the hardware device to control it and sending the acquired signals across the network to the clinic server. On the hardware device side the connection is accomplished using the Bluetooth^{TM} wireless technology which has become a *de facto* standard and is now available on every mobile terminal, while a GRPRS/UMTS link is used for remotely sending the signal through the mobile carrier.

Once the mobile terminal is successfully connected through the Bluetooth^{TM} link to the hardware device, the patient may start the acquisition and visualize in real time on its screen the signals being acquired. This is particularly useful since it gives an immediate feedback to the patient about the proper operation of the system. Thus, if she experiences any problem with one of the channels, possibly due to a noisy or wrongly positioned lead, the acquisition may be stopped, the lead may be repositioned and the acquisition may be started again.

The signal is acquired by the hardware device over 8 channels which are sampled 250 times/sec, and each sample is converted into 2 bytes. This means that the data production rate is 32 kbit per second, which lies in-between the upload throughput rates of GPRS/EDGE and UMTS without applying any compression. From a diagnostic perspective there was no point in pursuing a real-time operation, which had the inherent risk of saturating the bandwidth, causing a massive packet drop to recover, thereby spoiling the recorded signal quality. Thus we opted for a deferred transmission privileging instead error detection and retries which are more valuable features for our project.

With this choice the mobile terminal acts according to a *store-and-forward* paradigm saving a track of variable length growing at a rate of 250 kBytes for every minute of sampling. When, according to the directions given by the health care staff, the expected length is achieved, the patient may stop the acquisition and select whether to enable its transmission to the center, keep it locally, or discard it altogether.

A Mobile Remote Monitoring Service for Measuring Fetal Heart Rate

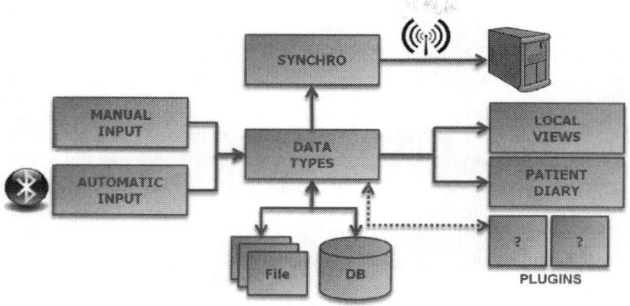

Fig. 2 The computational architecture of the mobile device application.

IV. THE MOBILE APPLICATION

A key element in our architecture is represented by the mobile terminal which is meant to act as a matching peer for the hardware capturing device. The combination of the mobile terminal and the hardware device gives rise to a remote station at a patient's home capable of acquiring electric signals on her and sending those to the center. A mobile terminal was selected in order to provide a high degree of comfort for the patient and allow data acquisition virtually anytime and anywhere, posing no other constraint than the availability of network coverage. Furthermore, given that the mobile acts according to a *store-and-forward* paradigm, in case of poor or no network coverage any unsent data will be cached until the suitable conditions for data transfer are restored.

The Android™ operating system has been selected for the implementation of the mobile platform due to its reliability combined with a wide availability of low cost devices and the open source nature of its developing environment. Figure 2 illustrates the computational architecture of the mobile platform which has been loosely based on a multi agent paradigm [14]. That architecture emerged as a result of several other projects previously developed by us and exploiting mobile technologies for the remote monitoring of patient data [15]. More specifically we designed a modular framework for the implementation of telemedicine services which favors the reusability and configuration of its components therefore speeding up its porting to different medical contexts.

The main components of the architecture are represented by an extensible *Data Types* core which can be configured either to use the *Sqlite3* database available with each Android implementation or even plain text files. In our case, descriptive data concerning patient information, such as demographic data or textual notes introduced by the patient herself for further describing the acquisition, are better saved exploiting the relational structure provided by the Sqlite3 implementation, while the sequence of values, representing the actual signal acquired, better fit the file

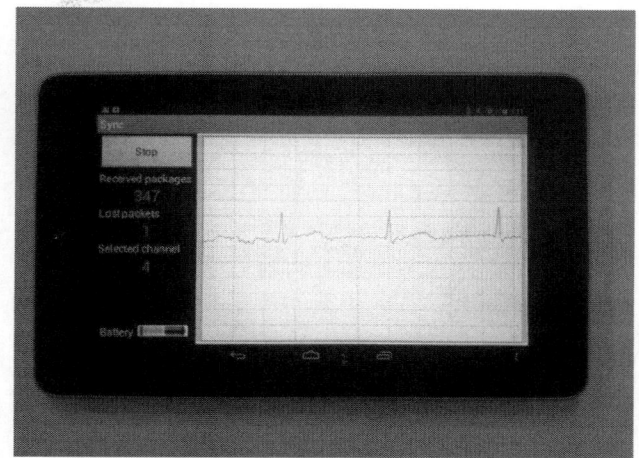

Fig. 3 A signal being acquired on a tablet.

paradigm. The *Data Types* structure directly interfaces with the *Synchronization* module responsible for implementing a bidirectional data exchange with the clinic server. This module is also a configurable one designed to fit a wide set of data types and implementing functionalities such as retries and error recovery, which has been validated and used in several projects concerning remote monitoring in different areas [16-17]. The architecture also encompasses a module for the *Manual Input* of data which uses the smartphone touch-screen and another one exploiting Bluetooth™ for the *Automatic Input* which is used in our case for interfacing with the capturing hardware. The architecture also foresees some *Plugins* encapsulating knowledge which is specific for a given application and allows several views on the data stored (i.e. *Patient Diary* or *Local Views*) which are used for displaying data as shown in Figure 3.

V. THE SERVER

The server mainly consists of a synchronization engine storing the signals sent by the remote stations, a set of signal processing modules for extracting the FHR from the acquired signals and a web application for analyzing the results. All the 8 tracks sent after a successful recording will be archived in the patient's Personal Health Record (PHR) based on a set of credentials associated with the remote station and indexed with the date-time of their acquisition. Then a filtering pipeline is applied. The first component is responsible to suppress both any superimposed noise as well as the mother ECG signal, while emphasizing at the same time the fetus ECG component. Then the subsequent element in the pipeline is responsible for selecting the most interesting track upon which FHR should be measured, which is then accomplished by the third component [18].

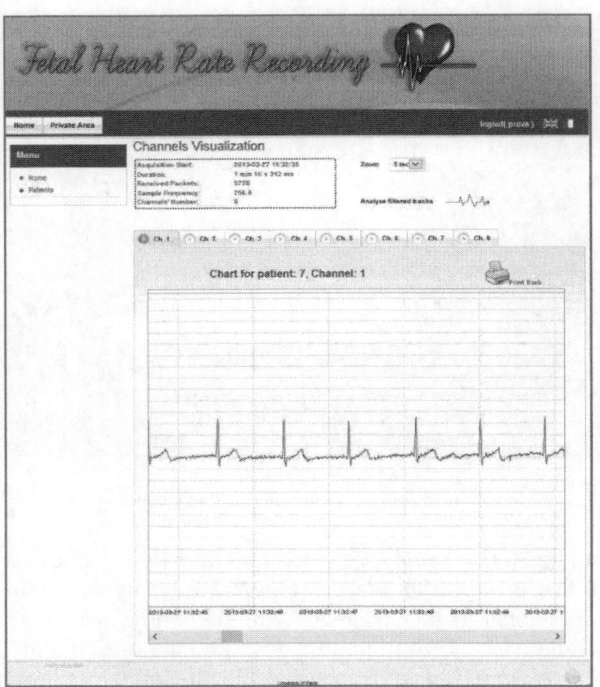

Fig. 4 A screenshot of the web application running on the clinic server.

The server also hosts a web application whose intended user is the physician involved with overseeing the service. To this aim the web application will have facilities for selecting patients and navigating across the past sessions of each one. For each session the application may display each of the 8 tracks as well as the FHR for visual investigation. All tracks may be exported and downloaded for further analysis by the physician with specialized software, and to this aim the web application also makes available for download some reference software tools.

VI. CONCLUSIONS

This paper reports on a remote monitoring service supporting an innovative way of acquiring CTG signals based on electrical signals which doesn't require echo-graphic techniques and the trained staff for accomplishing them. The remote monitoring service is at an advanced state of implementation and will be tested as soon as it will be completed.

REFERENCES

1. Norris A (2002) Essentials of Telemedicine and Telecare. Wiley.
2. Luong J, Male K, and Glennon J (2008) Biosensor technology: Technology push versus market pull. *Biotechnology Advances*, 26(5):492–500.
3. Gaikwad R and Warren J (2009) The role of home-based information and communications technology interventions in chronic disease management: A systematic literature review. *Health Informatics Journal*, 15(2):122–146.
4. Schoenfeld MH, Compton SJ, Mead RH, et al. (2004) Remote monitoring of implantable cardioverter defibrillators: A prospective analysis. *Pacing and Clinical Electrophysiology*, 27(6):757-763.
5. Varma N, Michalski J, Epstein AE et al. (2010) Automatic Remote Monitoring of Implantable Cardioverter-Defibrillator Lead and Generator Performance. *Circulation: Arrhythmia and Electrophysiology*, 2: 428-436.
6. Varma N, Stambler B and Chun S (2005) Detection of atrial fibrillation by implanted devices with wireless data transmission capability. *Pacing and Clinical Electrophysiology*, 28(1):S133-S136.
7. Curone D, Secco EL, Tognetti A, et al. (2010) Smart garments for emergency operators: the ProeTEX project. *IEEE Trans Inf Technol Biomed* 14(3):694-701. doi: 10.1109/TITB.2010.2045003.
8. Di Lieto A, De Falco M, Campanile M, et al. (2006) Four years' experience with antepartum cardiotocography using telemedicine. *Journal of Telemedicine and Telecare*, 12(5):228-33.
9. Buysse H, De Moor G, Van Maele G et al. (2008) Cost-effectiveness of telemonitoring for high-risk pregnant women. *International Journal of Medical Informatics*, 77(7):470-476.
10. Van Geijn HP (1996) Developments in CTG analysis. *Bailliere's Clinical Obstetrics and Gynaecology* 10(2):185-209.
11. Daly N, Brennan D, Foley M, et al. (2011) Cardiotocography as a predictor of fetal outcome in women presenting with reduced fetal movement. *European Journal of Obstetrics & Gynecology and Reproductive Biology*, 159(1):57-61.
12. Peters C, Vullings R, Bergmans J et al. (2006) Heart Rate Detection in Low Amplitude Non-Invasive Fetal ECG Recordings. *Proc. of the 28th IEEE EMBS Conf. 2006*, pp. 6092-6094.
13. Fanelli A, Signorini MG, Ferrario M et al. (2011) Telefetalcare: a first prototype of a wearable fetal electrocardiograph. *Proc. of the 33rd IEEE EMBS Conf. 2011*, pp. 6899-6902.
14. Bui T (2000) Building agent-based corporate information systems: an application to telemedicine. *European Journal of Operational Research*, 122(2):242-257.
15. Lanzola G, Capozzi D, D'Annunzio G et al. (2007) Going mobile with a multiaccess service for the management of diabetic patients. *Journal of Diabetes Science and Technology*, 1(5):730-737.
16. Capozzi D and. Lanzola G (2013). A generic telemedicine infrastructure for monitoring an artificial pancreas trial. *Computer Methods and Programs in Biomedicine*, doi:10.1016/j.cmpb.2013.01.011 (in press).
17. Ginardi MG and Lanzola G (2013) A Mobile Platform for Administering Questionnaires and Synchronizing their Answers. *Proc. of the IADIS Int.. Conf. on Mobile Learning*, Lisbon, Portugal.
18. Fanelli A, Signorini MG, Heldt T (2012) Extraction of Fetal Heart Rate from Maternal Surface ECG with Provisions for Multiple Pregnancies. *Proc. of the 34th IEEE EMBS Conf.* 2012, pp. 6165-6168.

Investigating Methods for Increasing the Adoption of Social Media amongst Carers for the Elderly

Kyle Boyd[1], Chris Nugent[1], Mark Donnelly[1], Raymond Bond[1], Roy Sterritt[1], and Lorraine Gibson[2]

[1] Computer Science Research Institute University of Ulster, Jordanstown, Northern Ireland
[2] Belfast Health and Social Care Trust, Belfast, Northern Ireland

Abstract—Social Media has become one of the biggest successes on the World Wide Web. The Online Social Network (OSN) Facebook is by far the most popular with over 1 billion users. OSNs have the potential to minimise the effects of social isolation in older people which could alleviate some of the responsibility placed upon older carers. Nonetheless, a number of barriers still exist to the usage of Social Media. This study has investigated how to increase adoption of Social Media usage among older users by providing personalised training to elderly carers relating to how OSNs work and how they could be used in their lives. This paper presents the quantitative and qualitative findings from a training day that observed ten carers being introduced to Social Media. The paper documents the issues that were encountered and discusses how these can be addressed to help aid adoption of Social Media, in the future.

Keywords—Carers, Social Isolation, Web 2.0, Social Media, Online Social Networks, Facebook, Adoption, Training.

I. INTRODUCTION

The World Wide Web has made significant gains over the last decade in terms of enhancing people's knowledge, entertainment, collaboration and social communication. By the middle of the last decade, 'Web 2.0' was being used to describe the modern Internet era [1]. It went from being a static Web to a dynamic and collaborative platform, a growing phenomenon that was a collection of technologies, business strategies and social trends [2]. It has allowed users to be more collaborative, allowing for social interaction and allowing developers to easily and quickly create applications that draw on data, information or services from the Internet that can be easily shared [3]. In our current work, we explore the suitability of particular Web technologies for use by older users with a focus on leveraging these technologies to support older age carers to maintain a positive social position in society..

II. SOCIAL MEDIA ON THE WORLD WIDE WEB

Social Media refers to Internet based applications built on the ideological and technological foundations of Web 2.0. Social Media allows for the creation of words, pictures, video and audio, which can be shared with others through Social Media technologies. Various Social Media solutions exist such as Blogs, Wikis, Social tagging, Video and Web Mashups. Nonetheless, the most popular form of Social Media is the Online Social Network (OSN). The most popular OSN is Facebook [4] with over one billion users, a figure that equates to one in six people on earth [5]. Facebook's core aim is to encourage people to feel more connected to each other by allowing them to share the details of their private life as a means of socialising with others. Underlying this functionality are various features and applications such as a timeline to display information relating to education, relationship status, work background, newsfeeds, photos, event calendars and video. In addition, users can also communicate via chat, personal messages, wall posts and status update services [6].

III. USING ONLINE SOCIAL NETWORKS TO COMBAT SOCIAL ISOLATION

By 2050, 2 billion people in the world will be 60 years old [7]. With such a rise in the elderly population there will be an inevitable increase in long-term conditions such as chronic pain, cardiovascular diseases and most notably dementia with estimating 65.5 million people will be diagnosed with the condition by 2030 [8]. Such conditions place a significant strain on health and social care and more recently have led to strategies being developed to reduce institutional care provision by utilising technology to provide remote care to people within their own home environment (known as Connected Health solutions) [9].

These are positive steps, however, this impact means that there will bean increase in the number of required carers for persons with dementia (PwD) and other long term chronic diseases. Caring for someone can be an overwhelming experience, which brings changes to lives and relationships that are more often than not irreversible. In the UK there are 1.25 million people who provide unpaid care of over 50 hours per week and 670,000 of those are acting as primary carers for PwD [7]. Over 52% of those are not receiving sufficient support financially, training, information about the PwD needs and access to peer support and respite care. Brownie and Horstmanhof state a lack or loss of companionship and an inability to integrate into the social environment are critical correlates of loneliness and social isolation. Steptoe et al. state that there is a higher risk of mortality in older men and women because of social isolation [10].

Approaches to support improved communication and interaction could reduce, for example, levels of depression and feelings of isolation in older carers [11]. Shapira et al. [12] reported that that older people have a greater sense of empowerment through online interpersonal interactions. They help them maintain cognitive functioning and have a greater independence in their lives. Goswami et al. [13] reported that the characteristics of OSNs makes it easier and cheaper for the elderly to keep social ties active and therefore enhances feelings of social connectedness, reducing loneliness and improved psychological well-being.

IV. SOLUTIONS TO THE BARRIERS

There are a number of inherent challenges, not least, that barriers remain in an older person's usage of the Web and in particular OSNs. These have been summarised in Table 1. To address these barriers it is hypothesised that specialist training could aid in the adoption of new technologies amongst older people. Warburton, Cowan and Bathgate [14] state that "Specialist training, appropriate programs and IT support are all needed to address the key concerns of rural, older Australians, who are disengaged or excluded from ICTs". Heidrum Mollenkopf of the Universal Accessibility and Universal Design Expert Group at the COFACE e-Health Seminar, April 2012 said "In the beginning older people can show distance towards innovative technology but they are happy to use the technologies if they address their needs, if they are accompanied with social attention and support and if they receive the necessary training".

Table 1 Barriers to use of World Wide Web in older users.

Barrier	Source
Disinterest, prohibitive cost to purchase and maintain, poor access to Internet service	Ewing and Thomas 2010 [15]
Inability to grasp the purpose or benefits of technology like Online Social Networks	Gibson et al. 2010, Sundar et al. 2011, Sayago, Blat 2011, Bennett 2011[16-19]
Daunted by the technology because of a complete lack of awareness of digital products	Norval et al. 2011, Bennett 2011, Xie et al. 2012 [16-18]
They were worried about breaking it	Lehtinen, Naassanen&Sarvas 2009, Wen-Hui Chou, Yu-Ting Lai &Kuang-Hsia Liu 2010 [19, 20]
They were worried about privacy and security concerns	Mellor, Firth & Moore 2008, Norval et al. 2011 [11, 16]
Language was a problem, and terminology used	Gibson et al. 2010 [21]

The current research aims to consider the training methods required for older carers to aid their adoption of Social Media technology and particularly OSNs to increase uptake and usage.

V. METHOD

An introduction to a Social Media training course was conducted during January 2013 at [details removed for review]. Participants were invited to attend this course prior to the study. Questionnaire data was collected and participants were asked to attend a presentation, which included videos of Social Media technology. Verbal discussion was also encouraged throughout the session. Two researchers were present to act as note takers and deliver demonstrations. Ten participants (2 Male and 8 Female) were recruited to take part in the study. Each participant gave consent to be included in the study in accordance with the ethics approval granted by the authors' University ethics committee. One participant had a visual impairment hence their data was not used in the results. Ages were between 45 and 85, with 5participants being above the age of 60. English was the native speaking language. Four participants said they used Social Media on a daily basis and one used email and Skype weekly. The remainder had no experience in using the Internet. The study was conducted in the activity room of [details removed for review] (Fig 1). The room was divided into two parts a presentation area and a live demo area. The presentation area was at the front of the room with a projector screen and a seated area. This also acted as an open discussion area. The live demo area had two circular tables with a PC, keyboard and mouse and an Internet connection. Each table had an arced seated area around each computer to encourage discussion.

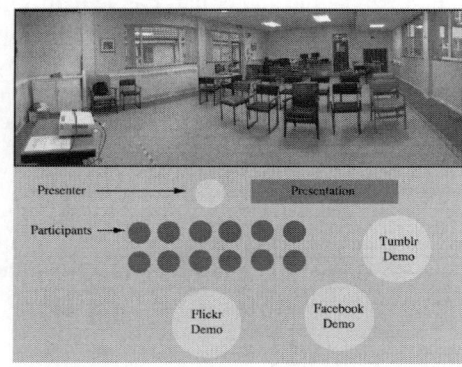

Fig. 1 Room setup and layout

The procedure was initially explained to the participants. The first questionnaire aimed to gather personal information relating to Age, Gender, Social Media usage and understanding along with other demographics. The participants then watched a presentation on Social Media. This presentation explained Social Media and its potential uses. Social Media was introduced using three of its main technologies, OSNs, Blogs and Photo repositories; to aid understanding, videos were used. After each technology was presented the participants were encouraged to discuss what they had watched and to understand if they could use this in their own lives. Consequently, after the presentation, each participant was provided with a live demo of the available technology and was encouraged to verbally discuss what had been presented as well as have a go at using the technology. Following the demo, each participant completed a second questionnaire relating to how they found the introduction to Social Media, i.e. how easy or difficult they found to understand the technology and terminology, were they more confident in their understanding of the technology, would they use Facebook in the future to communicate with friends and family or be more motivated to do so. They were also asked whether they would like to receive additional training.

VI. RESULTS

These results are derived from nine of the participants as one participant has a visual impairment. All participants felt that they could be regarded as a 'novice' in terms of computer literacy. Nonetheless, six of the nine felt they had an understanding of what Social Media was. Four indicated that Social Media was difficult to use and they wouldn't use the Internet to support real world connections during the session. Initially during the presentation stage participants were quite hostile to Social Media because of anxiety of privacy and security as well as the knowledge required to use it. One effort was to present Facebook as an analogy called Social Delivery, a hypothetical service whereby a company would organize the person's photos from the previous ten years and deliver this on a DVD for the participant to watch. Only three participants said they would sign up for the service and on average it would not be important or valuable for them to use. The participants were again asked for their views on Social Media to determine whether perceptions had changed. From all the participants, eight said they now had a better understanding of Social Media with five stating that it met their expectations in usefulness and that they were now more confident in using the technologies. They were also now more likely to use it to communicate with friends and family. The participants 'opinions on the training were positive with responses either stating that it was good or very good. Interestingly, seven said that they would be willing to attend additional training in Social Media technologies. One participant stated that they would sign up for Facebook, privately, following the training.

VII. DISCUSSION

It was decided to introduce Social Media technology to a group of carers who looked after people with a range of chronic diseases such as dementia and multiple sclerosis to try and increase their adoption of Social Media technology. The benefits of which have already been discussed in this paper. The session was split into two parts: a presentation and a set of live demos. During the presentation, the existence of Social Media technology was discussed alongside three of the most common Social Media technologies, OSNs, blogs and online photo repositories. Subsequently, the participants were shown live demonstrations of the three technologies. The group discussion sessions were more beneficial than the initial engagement during the presentation stage and this helped deal with hostilities toward the technology. This was expected due to previous research [16, 18, 22] . Introducing three different technologies at the one time was confusing for participants. Many Social Media technologies have similar functionality, for example, Facebook allows for sharing of photos, however, as does Flickr. If this study were to be repeated one technology would be shown per session. The videos showing each technology and how it worked are regarded as having been successful with four of the participants agreeing that "The videos make it very easy to understand". To introduce Facebook we used an analogy to try and explain why and how they could use the technology. Nonetheless this confused participants, only three out of the nine said they would use it. During live demonstrations the participants were broken into two smaller discussion groups that they became a lot more engaged. Most of the problems revolved around not having a computer, no Internet connection or just disinterested in using technology. One participant stated, "What's wrong with using the telephone, you can feel connected then?" It was explained to them that Social Media allowed them to have many conversations at once, even if the other person(s) there were connecting too weren't there. "I still wouldn't use it, it's not for me". A woman in her 70's used Skype and email to contact her son in Canada on a weekly basis was particularly interested. It was explained that she could keep up to date with them without having to schedule a time. Nonetheless she stated, "It sounds really useful but I just wouldn't have the time to get started and to use it, I wouldn't know how too!" This was a common theme and probably why seven of the participants would be willing to take more free training courses in other specific Social Media technologies to try and learn more about them. Most elderly, 65 and over are described as old. Nonetheless it is inadequate to fit all older persons into this category due to the growing number of older adults in the oldest ages. The research community has divided elderly into the young old 65-74 years old, the old-old, 75-85 years old and the oldest old 85 years and older [23] . When we look at our participants and break these into the various categories over half of our participants are between 41-60 with the average age of all participant's being 60. The findings for the study shows that adoption of Social Media in the elderly is significantly lower in comparison to usage with other assistive technologies. Moving forward this study has informed us that the design of current social media does not meet the expectations or usefulness of older persons. Many technologies contain features that they do not require. Simple services like messaging, commenting and viewing photographs would be useful according to feedback provided through the open verbal discussions. This will be considered in further studies, focusing on what they want, not what they might need.

VIII. LIMITATIONS

The limitations of this study include the small participant numbers reported. Generally, a participant numbers can be up to 30, however on the day of our study only 10 arrived. This potentially shows that elderly carers are not interested in Internet technologies and this contributed to the low turnout, justifying previous points that older people may be afraid of technology.

IX. CONCLUSIONS

To conclude, in this paper we have shown that the problem of social isolation causing increased mortality in older carers, which needs to be addressed [10]. Web 2.0 technology and OSNs could alleviate this problem. We need a methodology to increase the adoption of Social Media among elderly carers exploring the possibility that the age of the older old is significantly lower when addressing technology adoption. Didactics doesn't engage and educate carers as well as discussion groups. Collaborative learning can be a useful method for improving older adults' e-health literacy [24]. Based on feedback future work is planned to run a sustained training course on one Social Media technology, teaching them how to use the technology's features while communicating safely and efficiently.

REFERENCES

[1] T. OReilly, "What is Web 2.0: Design patterns and business models for the next generation of software," 2007.
[2] S. Murugesan, "Understanding Web 2.0". IT Professional," vol. 9, pp. 34-41, 2007.
[3] D. L. Hoffman and T. P. Novak, "Why Do People Use Social Media? Empirical Findings and a New Theoretical Framework for Social Media Goal Pursuit," *SSRN eLibrary*, 2012.
[4] Facebook.24/01/2012.Facebook Home Page.http://www.facebook.com.
[5] Facebook Statistics. 24 January2012 http://tinyurl.com/mwf76j.
[6] S. Campbell. How Does Facebook Work? 2012.
[7] Alzheimers Association. Facts and Figures. 2012.
[8] Alzheimer's Disease International, "World alzheimer report." 2010.
[9] S. Lauriks, A. Reinersmann, H. G. Van der Roest, F. J. Meiland, R. J. Davies, F. Moelaert, M. D. Mulvenna, C. D. Nugent and R. M. Droes, "Review of ICT-based services for identified unmet needs in people with dementia," *Ageing Res. Rev.*, vol. 6, pp. 223-246, Oct, 2007.
[10] A. Steptoe, A. Shankar, P. Demakakos and J. Wardle, "Social isolation, loneliness, and all-cause mortality in older men and women," *Proceedings of the National Academy of Sciences*, vol. 110, pp. 5797-5801, April 09, 2013.
[11] D. Mellor, L. Firth and K. Moore, "Can the Internet Improve the Well-being of the Elderly?" *Ageing Int.*, vol. 32, pp. 25-42, 2008.
[12] Shapira, N., Barak, A., Gal, I., "Promoting older adults' well-being through Internet training and use." *Aging & Mental Health*, vol. 11, pp. 477-484, 2007.
[13] S. Goswami, F. Köbler, J. Leimeister M and H. and Krcmar, ""Using online social networking to enhance social connectedness and social support for the elderly", 2010.
[14] J. Warburton, S. Cowan and T. Bathgate, "Building social capital among rural, older Australians through information and communication technologies: A review article," Australasian *Journal on Ageing*, 2012.
[15] S. Ewing and J. Thomas, "The Internet in Australia," *ARC Centre of Excellence for Creative Industries and Innovation*, 2010.
[16] C. Norval, J. L. Arnott, N. A. Hine and V. L. Hanson, "Purposeful social media as support platform: Communication frameworks for older adults requiring care", pp. 492-494, 2011.
[17] J. Bennett, "Online Communities and the Activation, Motivation, and Integration of Persons Aged 60 and Older – a Literature Review," 2011.
[18] B. Xie, I. Watkins, J. Golbeck and M. Huang, "Understanding and Changing Older Adults' Perceptions and Learning of Social Media," *Educational Gerontology*, vol. 38, pp. 282-296, 2012.
[19] V. Lehtinen, J. Naassanen and R. Sarvas, "A little silly and empty-headed: Older adults' understandings of social networking sites," in *Proceedings of the 23rd British HCI Group Annual Conference on People and Computers: Celebrating People and Technology*, Cambridge, United Kingdom, pp. 45-54, 2009.
[20] Wen-Hui Chou, Yu-Ting Lai and Kuang-Hsia Liu, "Decent digital social media for senior life: A practical design approach," in *Computer Science and Information Technology (ICCSIT)*, pp. 249-253, *2010*.
[21] L. Gibson, W. Moncur, P. Forbes, J. Arnott, C. Martin and A. S. Bhachu, "Designing social networking sites for older adults," *BSC HCI*, 2010.
[22] S. Sayago and J. Blat,"An ethnographical study of the accessibility barriers in the everyday interactions of older people with the web," *Universal Access in the Information Society*, vol. 10, pp. 359-371, 2011.
[23] B. Lee, Y. Chen and L. Hewitt, "Age differences in constraints encountered by seniors in their use of computers and the internet," *Comput. Hum. Behav.*, vol. 27, pp. 1231-1237, 2011.
[24] B. Xie, "Older adults, e-health literacy, and collaborative learning: An experimental study," *J.Am.Soc.Inf.Sci.Technol*, vol. 62, pp. 933-946, 2011

AMELIE: Authoring Multimedia-Enhanced Learning Interactive Environment for e-Health Contents

P. Sánchez-González[1,2], I. Oropesa[1,2], P. Moreno-Sánchez[1,2],
J.M. Martínez-Moreno[1,2], J. García-Novoa[1,2], and E.J. Gómez[1,2]

[1] Biomedical Engineering and Telemedicine Centre, ETSI Telecomunicación,
Universidad Politécnica de Madrid, Madrid, Spain
[2] Biomedical Research Networking center in Bioengineering, Biomaterials and Nanomedicine, Madrid, Spain

Abstract—This paper presents the AMELIE Authoring Tool for e-health applications. AMELIE provides the means for creating video-based contents with a focus on e-learning and telerehabilitation processes. The main core of AMELIE lies in the efficient exploitation of raw multimedia resources, which may be already available at clinical centers or recorded ad hoc for learning purposes by health professionals. Three real use cases scenarios involving different target users are presented: (1) cognitive skills' training of surgeons in minimally invasive surgery (medical professionals), (2) training of informal carers for elderly home assistance and (3) cognitive rehabilitation of patients with acquired brain injury. Preliminary validation in the field of surgery hints at the potential of AMELIE; and its versatility in different medical applications is patent from the use cases described. Regardless, new validation studies are planned in the three main application areas identified in this work.

Keywords—Medical education, e-learning, telerehabilitation, AMELIE, authoring tool.

I. INTRODUCTION

The evolution of technology-enhanced learning (TEL) has brought forth a new education paradigm where cost, time and geographical barriers are no longer a constraining issue. Due to its nature, it provides the necessary learning conditions anytime and anywhere, as well as personalize the learner's experience and formative paths without compromising the consistency of training programs [1].

All advantages considered, TEL poses an opportunity to ease formative requirements in health applications at all levels: from the training and assessment of new professionals (doctors, nurses) to the instruction of informal carers or the rehabilitation of neural injuries.

A critical aspect in the definition of TEL environments resides in the quality and appeal of the provided didactic contents. It has been pointed out that these contents should promote interactivity with the learner rather than be simple passive units [2]. Multimedia resources, especially in the form of procedural videos, are often available at hospitals and clinical centers. While they may be used for classroom tuition, more often than not they are not fully exploited.

Moreover, the creation of these contents can be tedious, counterintuitive and difficult [3].

In this paper, we present AMELIE (Authoring Multimedia-Enhanced Learning Interactive Environment), an authoring tool for the creation of multimedia didactic contents in clinical applications by health professionals. The tool contemplates the definition of complete, personalized didactic units focused on the exploitation of augmented video resources, with an emphasis on reusability and sharing of contents. The work also presents three real medical use cases where the authoring tool is currently being applied, addressing the different actors involved in the clinical system (medical professionals, informal carers and patients).

II. AUTHORING TOOL

The main core of AMELIE lies in the efficient exploitation of raw multimedia resources for video-based learning and telerehabilitation, which may be already available at hospitals (e.g.: surgical interventions, 4D image studies, etc.) or recorded ad hoc for learning purposes by health professionals. Didactic units may be created, structured in an indeterminate number of sections (determined by the author), including text, image and enhanced videos.

For every section the content creator wishes to include, he/she may add a new tab to the didactic unit. Two main content tabs can be added: (1) regular and (2) video-based. In the first, space is provided to include a main text description, along with images and any additional relevant files. The latter allows the inclusion of an augmented video resource, and provides direct access to an editor where the author may process the raw video (Fig. 1).

The editor offers intuitive methods to modify a video resource by: (1) adding text, audio or image overlays at different points of the recording; (2) crop the video; (3) modify its visual properties (contrast, brightness, etc.); (4) create clips; (5) document the videos; (6) include assessment breakpoints, where the student must answer questions on the video before moving on; and (7) tag and track objects on the screen (by means of segmentation and tracking algorithms specifically designed to that end).

Finally, the possibility of creating an assessment tab is provided. Available functionalities allow for the definition, edition and deletion of evaluation questions (multiple choice/sorting test-type questions). Additionally, the creator may choose where to place said questions, whether it be within a specific section or on a whole new one at the end of the didactic unit.

A special video player is provided to account for augmented features of the video resources. It allows playback of the different text/audio/assessment/segmentation overlays, as well as the possibility of toggling them on/off according to users' preferences.

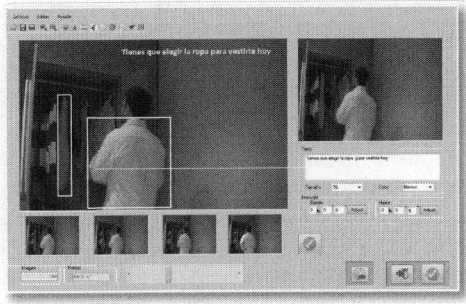

Fig. 1 AMELIE video editor

The current version of the AMELIE tool and video player are implemented in C++ and C# respectively, using Microsoft's (Redmon, WA) .Net Framework. OpenCV library is used to handle all video-based objects, as well as for the implementation of video-processing algorithms.

III. MEDICAL USE CASES

A. Minimally Invasive Surgery

Minimally invasive surgery (MIS) has become a procedural standard for many surgical sub-specialties. These techniques forgo the traditional approach of open interventions, allowing surgeons to perform surgeries through minimal incisions in the patient's body. MIS interventions are less painful for the patient; have fewer post-operative complications associated, and can shorten hospital stays [4].

Surgeons approaching MIS must become proficient on a series of cognitive and technical skills particular to these surgeries. Moreover, educational processes are gradually being adapted from mentor-apprentice-based approaches towards structured, objective training and assessment programs, where direct involvement of residents in real surgeries is delayed until becoming proficient in the required skills. Several motivators can be identified behind this: social awareness on medical errors, the need to reduce costs in hospitals, or the overloaded schedules of surgeons [4].

Adoption of TEL for cognitive online training can be a useful way of optimizing education programs, allowing for anytime/anywhere training. Combined with the strengths provided by laparoscopic videos [5], e-learning in surgery is an effective way of breaking time, space, and cost barriers by offering online or blended education alternatives that are potentially more viable and feasible than on-site courses.

In this context, the AMELIE Authoring Tool provides the means to exploit the numerous raw OR-video repositories stored in hospitals to create augmented video-based structured didactic contents (Fig. 2). Contents may then be used as class material, or most importantly, uploaded to a TEL environment for online training. In general, the authoring tool enables content sharing, task reproducibility and adaptive and ubiquitous learning; thus, it contributes to the shortening of MIS cognitive skills' learning curves and optimizes the efficiency of training.

B. Informal Carers

Nowadays, the increasing population aging index makes customized care necessary to keep the quality of life of older adults at home. New models of care must be deployed, in which informal carers play a crucial role as one of the most important sources of care for older adults, suffering stress episodes, over-work and depression. The carer supervises the patient's daily activities, spends a lot of time with him and assists in the care process. For this reason, it is vital to provide the necessary support in order to help carers in this task and reduce their workload [6].

One of the main informal carers' requirements is the need of knowing how to manage particular elderly's assistance and their consequent problems [7]. The use of TEL can provide new approaches for carers to satisfy the growing demand for attendance and support. This learning approach will allow the informal carer to enhance the assistance provided, adapting and personalizing contents addressing cognitive, functional, behavioral and personality changes of their elderly relatives [8]. Video-based enhanced learning technologies provide informal carers with an interesting opportunity for the acquisition of assistance skills through an interactive and use-case based experience.

Employing AMELIE, several issues are being addressed: (1) creating personalized video-based contents considering carer's preferences and learning requirements as well as the assisted person's condition; (2) enhancing user experience by showing real use-case videos; (3) focusing on particular aspects of older people assistance by means of detailed videos with added information (Fig. 2); and (4) allowing carers to navigate through several assistance pathways thanks to the interactivity provided by the tool.

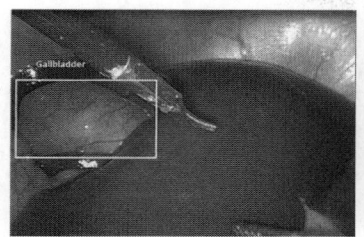 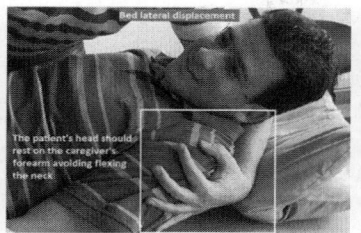 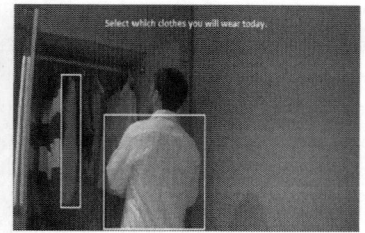

Fig. 2 Examples of augmented videos: Left: minimally invasive surgery; Middle: Informal carers; Right: Cognitive neurorehabilitation

Cognitive Neurorehabilitation

Acquired brain injury (ABI) is defined as a brain damage that suddenly and unexpectedly erupts in people's life. An ABI can be produced by different causes, such as stroke and traumatic brain injury. Associated cognitive impairments are the main cause of disability in developed societies [9]. Cognitive neurorehabilitation is the process whereby people with brain injury work together with health service professionals and others to remediate or alleviate said impairments. New technologies have led to higher intensity rehabilitation processes, extending therapies in an economically and sustainable way.

Some studies have proved that activity daily living (ADL) observation treatment is a good rehabilitative approach in stroke patients [10]. Additionally, the effectiveness of interactive videos (IV) has been verified on learning applications, where interactive dynamic visualization allows users to adapt the learning process to their individual cognitive skills [2][11]. IV refers to any video whose sequences and displayed information depend on the users' responses. Interactivity is provided by associating an interaction with any element tagged in the video scenes, modifying the video flow according to the way users interact with it.

Current research in this field is trying to ascertain whether IV is an appropriate technology to sustain personalized telerehabilitation processes. More specifically, the contemplated goal calls for patients to deal with real situations of daily living based on enhanced video-based environments [12]. In this context, AMELIE is used to define, tag and track the hot zones of the IV that will enable interactivity [13]. Using the enhanced video, the goal is to guide patients along every single scene (Fig. 2). Depending on their cognitive sequels, additional text, audio or virtual objects are inserted in order to help them reach the final goal. In this context, current work is focusing on adapting AMELIE's breakpoint inclusion functionality to set up the interactive videos and enable joining clips according to interactions and hot spots.

IV. USER VALIDATION

The AMELIE tool has been validated in the context of project TELMA, which resulted in the creation of a TEL environment for MIS cognitive skills training [3]. The environment provided surgical trainees ubiquitous access to multimedia didactic units. These were created and uploaded by expert surgeons by means of AMELIE, based on enhanced laparoscopic videos.

Five surgeons (4 experts with teaching experience and 1 novice) performed a two-phase validation using AMELIE. The alpha test was performed under supervision of a developer. In the beta test supervision was avoided, and surgeons were allowed to freely use AMELIE for 30-60 minutes. A likert-type questionnaire was filled to obtain information regarding operational, functional and usability data.

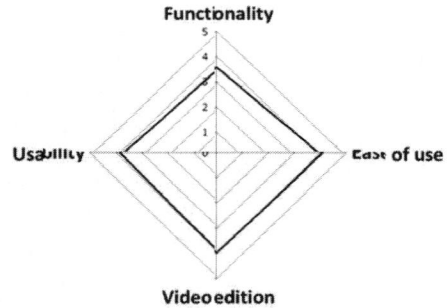

Fig. 3 Global results. (1: Low performance; 5: High performance)

Validation global results are shown in Fig. 3. Functionality of the tool, video edition tasks, ease of use and learning and usability were analyzed, in many cases showing good preliminary results. For a complete analysis of results the reader is referred to [3].

V. DISCUSSION

In this work we have presented AMELIE, an authoring tool for video-based didactic content creation in e-health applications. Three real application scenarios have been described, each in a different stage of implementation and with a different target user.

A key advantage of AMELIE lies in its versatility. Creation of didactic contents is adaptable to any kind of clinical application; mostly related with training processes (medical professionals, informal carers), but also adjustable to telerehabilitation scenarios (patients). Moreover, the concept could be easily extended beyond the clinical world.

While AMELIE provides means to create complete contents combining multimedia information and assessment capabilities, its true kernel lies on its video editor. Proprietary solutions such as Adobe Premiere are usually aimed at professionals with background knowledge on video edition, and thus present a steep learning curve for non-experienced users. Additionally, the focus of their functionalities does not fall on edition for didactic purposes. AMELIE covers the gap by providing health professionals ways to easily enhance any video with didactic information.

Moreover, the tool can be adapted to different environments independently of their content presentation standard. Contents may be uploaded/downloaded directly to/from the e-learning platform of choice. For storage and retrieval purposes, multi-language documentation is handled via metadata fields generated in a relational data model through a predefined thesaurus. Integration of the video-player in the platform guarantees the correct content playback.

Depending on the final application, however, different approaches to its use may be contemplated. Deviations from the default scenario (creation of didactic units for e-learning applications) may be considered, for example with the definition of contents for classroom teaching, or the preparation of an IV for a telerehabilitation session.

Preliminary validation in MIS has yielded positive reception on its functionality and usability. While the rest of case uses described remain to show valid proof, these results encourage the usefulness of such tools in e-health applications. In order to corroborate them in diverse areas of medical specialties, new and larger validation studies are planned in the three application areas described in this work.

VI. CONCLUSIONS

The AMELIE tool can be a valuable asset for the creation of medical video-based contents in e-learning applications. The authoring tool's greatest strengths fall on its effective enhancement of videos from any source with didactic purposes and its versatility in any number of e-health applications. The tool is intuitive, easy to use and robust, and is accessible to any medical professional regardless of their computing knowledge.

ACKNOWLEDGMENT

Authors participate under the following research projects: MISTELA (LLP-Leonardo da Vinci-528125-LLP-1-2012-1-UK), CARECLOUD (INNPACTO IPT-2012-0599-300000) and REHABILITA (CENIT-E CEN-20091043).

REFERENCES

1. Bloomfield JG, Jones A (2013) Using e-learning to support clinical skills acquisition: Exploring the experiences and perceptions of graduate first-year pre-registration nursing students - A mixed method study. Nurs Educ Today [Epub ahead of print].
2. Merkt M, Weigand S, Heier A et al (2011) Learning with videos vs. learning with print: The role of interactive features. Learn Instr, 21: 687-704.
3. Sánchez-González P, Burgos D, Oropesa I et al. TELMA: Technology enhanced learning environment for minimally invasive surgery (2012), J Surg Res, 182(1): 21-29.
4. Oropesa I, Sánchez-González P, Lamata P et al (2011) Methods and Tools for Objective Assessment of Psychomotor Skills in Laparoscopic Surgery. J Surg Res, 171: e81-e95.
5. Sánchez-González P, Cano AM, Oropesa I et al (2011) Laparoscopic Video Analysis for Training and Image Guided Surgery, Minim Invasive Ther Allied Technol, 20: 311-320.
6. Ducharme FC, Levesque LL, Lachance LM et al (2011) Learning to Become a Family Caregiver' Efficacy of an Intervention Program for Caregivers Following Diagnosis of Dementia in a Relative. Gerontologist 51: 484-494.
7. Powell J, Chiu T, Eysenbach G (2008) A systematic review of networked technologies supporting carers of people with dementia, J Telemed Telecare 14: 154-156.
8. Powell J, Gunn L, Lowe P, et al (2010) New networked technologies and carers of people with dementia: an interview study. Ageing Soc 30: 1073-1088.
9. World Health Organization. Burden of Disease Statistics; Available online: http://www.who.org/ (last access: april 2013).
10. Franceschini M, Agosti M, Cantagallo A et al (2010) Mirror neurons: actions observation treatment as a tool in stroke rehabilitation. Eur J Phys Rehab Med 46:517-523.
11. Martin M, Weigand S, Heier A et al (2011) Learning with videos vs. learning with print: The role of interactive features. Learn Instr 21: 687-704.
12. Martínez-Moreno JM, Solana J, Sánchez R, et al (2013) Monitoring visual attention on a neurorehabilitation environment based on Interactive Video. International Conference on Recent Advances in Neurorehabilitation: 182-185.
13. Luna M, Sánchez-González P, Bonilla E et al (2012) Detección y seguimiento de objetos en vídeos de actividades de vida diaria para rehabilitación de pacientes con daño cerebral adquirido. XXIX Annual Congress of the Spanish Society for Biomedical Engineering.

Corresponding author:

Author: Patricia Sánchez González
Institute: Biomedical Engineering and Telemedicine Centre, ETSI Telecomunicación, Universidad Politécnica de Madrid.
Street: Avda. Complutense, 30
City: Madrid
Country: Spain
Email: psanchez@gbt.tfo.upm.es

mHealth: Cognitive Telerehabilitation of Patients with Acquired Brain Damage

C. Suárez-Mejías[1], M. Parejo[2], M.J. Zarco[3], A. Naranjo[1], C. Echevarría[3], J. Barros[3], R. Díez[3],
G. Escobar[1], M. Elena[4], and C.L. Parra[1]

[1] Grupo de Innovación Tecnológica, Hospital Universitario Virgen del Rocío, Seville, Spain
[2] El Sendero, Thotalmed, Sevilla, Spain
[3] Unidad de Gestión de Rehabilitación, Hospital Universitario Virgen del RocíoInstitution y Distrito de Atención Primaria, Seville, Spain
[4] Departamento Ingeniería Electrónica de Universidad de Sevilla, Spain

Abstract—In this paper, we present an eHealth platform-called mHealth that involves a professional methodology for cognitive rehabilitation of patients with acquired brain damage. mHealth provides patients with Cerebral Vascular Accident (CVA) and Traumatic Brain Injury (TBI) a system for training and contacting with professionals responsible for their rehabilitation without either patients or professionals having to leave home. The platform has a custom-tracking system that allows videoconferencing. With mHealth, health professionals can carry out common clinical assessments and monitor neurocognitive exercises made by patients. Also, with mHealth, patients can consult and resolve issues relating to self-training and/or technical issues by sending voice messages or during a videoconference. The platform has been designed with requirements defined by doctors and engineers from Virgen del Rocío University Hospital and Seville District. Preliminary results of the pilot are presented.

Keywords—Telerehabilitation, Cognitive, Acquired brain damage, Traumatic Brain Injury and Cerebral Vascular Accident.

I. INTRODUCTION

The WHO reports that acquired brain damage is the second leading cause of death and the leading cause of disability in adults in recent years. The data show that Cerebral Vascular Accident (CVA) or stroke has the highest incidence, followed by Traumatic Brain Injury (TBI). 16 million people suffer acquired brain damage each year, of which if is estimated that 5.7 million die and another 5 million are permanently disabled. In the elderly population of the 15-country Europe, estimates show 2.700.000 prevalent cases and 536.000 incidence cases yearly [1, 2]. The consequences of the suffering caused by these diseases are often very diverse and very severe. They affect physical, cognitive, emotional, and behavioral levels of the patient and family and other relationships. It is therefore of great importance to perform a neurorehabilitador treatment so as to minimize the impact of this effect. Health care requires emergency services but later will require the services of General Hospital and Rehabilitation. Moreover, these facts become even more relevant if we assess the economic impact of dependency caused by head injury and / or stroke. From a social standpoint a high dependence on the family develops. Productive capacity is considerably decreased, so eventually the family of the person must seek rehabilitation through social reintegration [2].

In this sense, not only pathologies in the earlier months but also damage during subsequent years require health care, so this means patients after hospital discharge, should continue their cognitive treatment at home. Today, thanks to the development of telemedicine it is possible to convey the rehabilitation services they need to those who are geographically or financially limited. This also allows filling the shortage of human resources, prevents movement of patients, and reduces the costs involved in rehabilitation processes. Studies indicate that telerehabilitation services are a valuable practice as a method of care when resources are limited, when there are long distances and for patients with reduced mobility [3].

There are experiences of the application of telemedicine in all these conditions and the results are very favorable. Clinical validation of cognitive rehabilitation therapy was conducted by a telemedicine platform called PREVIRNEC. It worked with attention, memory and executive functions in 80 patients with TBI. 40 were extra-hospital treated another 40 intra-hospital. After the pilot they concluded that the extra-hospital group had improved their functional capacities and in their daily responsibilities [4]. Forducey et al. [5] highlights the findings of 2 pilot studies on the use of telecommunications technologies in promoting self-care management and enhancing health care outcomes in persons with brain injuries and their family caregivers. 15 patient with were recruited over an 18 month period in the first pilot and 32 patient during 12 month in the second pilot. The obtained results were significant improvements in the level of dependence for patient, as well as it provides a cost-effective alternative to standard home care services.[6]R. Rietdijk et al done a systematic review about the effectiveness of using telehealth programs to provide training of patient with TBI. In this paper, 14 studies involved only participants with TBI, 9 studies focused on children and 4 studies included long-term follow-up after intervention. The variety of intervention programs in the studies reviewed in this paper demonstrates the potential for using telehealth and the effectiveness. In Hong Kong [7], trained clinicians conducted videoconferences with older people who had suffered stroke, checking improvements in areas such as balance, strength and social support, self-esteem, etc. Therefore, one can conclude that the use of tele-neurorehabilitation has a high potential for the neurological rehabilitation of traumatic brain injury and cerebral stroke [8-10].

In this paper, we present the mHealth platform which was designed by doctors and engineers from our hospital. mHealth is an platform that involves a professional methodology for cognitive rehabilitation in TBI and CVA patients.

II. OBJETIVE

The main objective of mHealth is to design and to implement a platform for telerehabilitation of patient with acquired brain damage. The mHealth intends to fill the gap in the evolution and rehabilitation of these people via tele-monitoring, tele-rehabilitation and putting in contact all agents: patients, relatives and healthcare professionals. Specifically, the specific objectives that mHealth pursues are:

- Improve accessibility of patients with acquired brain injury to neurocognitive treatments.
- Optimize time patients and caregivers dedicate to training and monitoring neurorehabilitation by reducing the trips to the hospital.
- Provide a mechanism enabling the clinician monitoring by the conditions and rehabilitation exercises results using mHealth and digital records.
- Conduct a personalizing treatment teleneurorehabilitation where cognitive abilities are trained without travel by the patient or the caregiver.

III. MATERIALS AND METHOD

A. Medical Case

The mHealth was designed with requirements identified by engineers and health care professional from Virgen del Rocío University Hospital. For this purpose, different qualitative analyses based on individual and group techniques were made. Specifically, structured interviews were used as individual techniques and expert panel and person method as group techniques. The different functionalities were prioritized using the Horizon model. Also, a Business Process Modelling (BPM) was done to model the relationship between patients and health care professionals in the experimental group. The functionalities for physicians and physiotherapists in the mHealth are the following:

- Professional management: A user with Administrator role can add a new user with a professional role (physician or therapist) or patient role, modify his data or remove it.
- Patient management: A user with a professional role is allowed to add a new patient, see the data for the patients he has assigned, assign one of his patient to another professional, modify or remove a patient.
- Appointment management: Give an appointment to the patient for videoconferencing with one of the physicians or therapists he has assigned, modify or delete the appointment.
- Patient Voicemail: physicians and therapists can listen to voice messages sent by patients they have assigned in order to solve the problems they could have.
- Videoconferencing between professionals and patients. Physicians can evaluate patients with an implementation of the Mini-Mental State Examination (MEC) and using videoconferencing. The MEC, including the exercises, the punctuation given by the physicians and the videoconference, are stored in the server and can be visualized by physicians and therapists in order to monitor the progress of the patients.

The functionality for the patients in the mHealth application is the following:

- Android Application includes brain training exercises for the patients. Once the patient finalizes an exercise, the application sends it to the server so that the physicians and therapists can evaluate it at any time.
- Patients can consult when they have a videoconferencing appointment with their physician or therapist.
- Patients can send voice mail to the server to ask for some help. These voice mails can be listened to by the assigned professional in order to solve the problem.
- Patients can videoconference with the professionals they have assigned in established appointments.

The pilot was designed as quasi-experimental composed by experimental and control group. On the one hand, patients with TBI in the control group were trained with a neurocognitive treatment at the Virgen del Rocio University Hospital. On the other hand, patients with TBI belonging to the experimental group were trained at home using mHealth application. It allows us to elucidate the result obtained after processing through teleneurorehabilitation.

B. mHealth Arquitecture

mHealth is a client / server platform in which many web and Android clients can request and receive services from a centralized server. It has been built completely with open source software. mHealth is built with a videoconferencing server, a web application for physicians and physiotherapists and an Android client application for the patients. The platform allows sending information and videoconferencing between the web and Android. The web application is developed using Spring's Web MVC 3.0 framework so that the business logic or model, the view and the controller are completely separated improving scalability. We have used Hibernate 3.3.1 GA framework for object-relational mapping to achieve high performance, reliability and scalability.

Executing automatized tasks for the web application is possible thanks to the integration of Quartz 1.6.3. scheduling services. In addition, we have integrated a logging utility, log4j library, to write information messages with detailed context for application failures. We have also used log4j to write messages about who and when is accessing which page in order to satisfy the Spanish data protection law, L.O.P.D. [11]. The web application works with HTTP over SSL and implements Apache Axis web services that are used by the Android Application in order to send and receive information from the server.

The platform has a Red5 Media Server 0.9.1 for videoconferencing. Red5 is a Java Server application that allows streaming video, audio and recording client streams using RTMPT, RTMPS and RTMPE protocols. In order to get Red5 working properly in the mHealth platform. It is possible to configure the path where the videoconferences are stored, adding lines in the red5-web.xml file. The web application and Red5 have been developed in an Apache Tomcat servlet engine, allocated in a centralized server. In this server, we have also installed a MySQL database to manage all the data, as it is shown in the following Fig. 1:

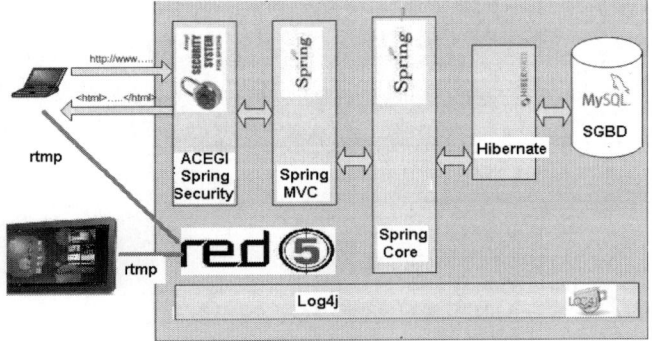

Fig.1 mHealth Platform Architecture

You can see the mHealth application on the Fig. 2 and the implemented videoconference on Fig. 3.

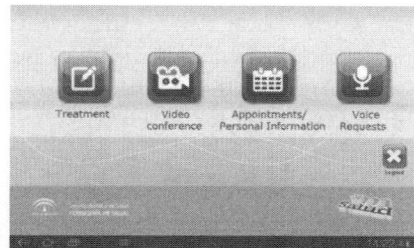

Fig. 2 Android App Screenshot - Samsung Galaxy Tab 10.1

Fig. 3 Android Application Screenshot: videoconferencing. Samsung Galaxy Tab 10.1

IV. EXPERIMENTAL RESULTS

A quasi-experimental study was performed to demonstrate the benefits with mHealth application. Ten patients with moderate and severe TBI have been included in the pilot. All of them satisfy inclusion and exclusion criteria. Their cognitive function the "Ranchos de los amigos" scale is greater than 4, they have speaking capacity that allowed them to communicate with health care professionals and collaborate in the pilot. The patients are divided into experimental group and control group. The duration of the training has been 4 months. All the patients were evaluated at the beginning and they were evaluated at the end of the pilot with the following tests: Minimental Test, Digit SPAM, the Trail Making Test and the Brief Neuropsychologist scale. Epidemiological variables of patients are measured and shown on Table 1. We can prove that there were no significant differences between both groups.

Table 1 Epidemiological variables of patients those were included in the pilot.

Variables	Experimental group	Control group
Men / (woman)	75% / (25%)	75% / (25%)
Injury agent: motor vehicular	75%	75%
Injury agent: fall	25%	25%
Severy based on Glasgow Come Score	7.2±3.9	7.7±1
Moderate TBI	25%	25%
Severe TBI	75%	75%

As you can see in Table 2, the patients using mHealth compared with control patients, improved in the control of digital spasms, in working memory, sustained attention and global cognitive flexibility. Furthermore, they improved in abstraction, as well as in the interaction field due to obtaining positive results in the clock test. Also, a test was administered to patients and caregivers for a qualitative evaluation of the application. The application was rated at 4.5 out of 5.

Table 2 Results of pilot in experimental and control group.

Variables	Experimental group		Control group	
	Baseline	After treatment	Baseline	After treatment
LCFS level V	25%	0%	100%	0%
LCFS level VI	50%	0%	0%	25%
LCFS level VII	25%	25%	0%	75%
LCFS level VIII Score	0%	75%	0%	0%
Minimental Test	26.2	32.2	22.7	29.5
Digit SPAM	4.75	7	3.7	4.5
Trail Making Test	77.5	25.4	77.5	45
Brief Neuropsychologist scale memory	9.5	14.25	10.2	16.75
Brief Neuropsychologist scale abstraction	4.5	5.5	3	3.2
Brief Neuropsychologist scale clock test	6.5	8	4.7	5.7

V. CONCLUSIONS

We have designed and developed a telerehabilitation platform called mHealth. mHealth provides patients with neurocognitive training from in the homes or anywhere thank to use of mobile devices such as a tablet PC with internet connection. mHealth allows doctors to establish an individual and structural cognitive treatment from an evaluated neurocognitive symptomatology. mHealth also allows doctors to collect information and monitor the activity and results of the treatment performed by the patients. Furthermore, it provides doctors a mechanism for exchanging information in real time with patients using videoconferencing. Doctors can evaluate patients without having to move to the hospital due to videoconferencing. Ten patients with moderate and severe TBI were involved in the pilot. Patient with similar features were divided into experimental and control group. Patients belonging the experimental group were trained with mHealth. On the other hand, patients belonging the control group were trained with conventional treatment in the hospital. The pilot proves that mHealth allows patients to improve in the control of digital spasms, working memory, sustained attention, global cognitive flexibility, abstraction, as well as in the field of personal interaction. Furthermore, as an added value, mHealth allows extending the rehabilitation treatment while reducing the movement of patients, caregivers and their associate costs.

ACKNOWLEDGMENT

mHealth was designed by Innovation Technologies Group and Rehabilitation Department from Virgen del Rocio University Hospital and Home Care from Seville District with the collaboration of El Sendero enterprise. This project was cofinanced by Vodafone and RETICS project. RETICS project was financed by Carlos III Institute. (RD09/0077/00025), 2010.

REFERENCES

1. WHO. World Health Organization (2007). Health Statistics and Health information Systems.
2. Nichols M, Townsend N, Luengo-Fernandez R, et al (2012). European CardiovascularDisease Statistics 2012. European Heart Network, Brussels, European Society of Cardiology, Sophia Antipolis.
3. Schwamm LH, Audebert HJ, Amarenco P et al (2009) Recommendations for the implementation of telemedicine within stroke systems of care: A policy statement from the american heart association. Stroke, 40: 2635-2660.
4. García-Molina A, Gómez A, et. al. (2010) Programa clínico de tele-neurorehabilitación cognitiva en el traumatismo craneoencefálico. Trauma Fund Mapfre 21 1:58-63.
5. Forducey P, Glueckauf R, Berquist T et al (2012) Telehealth for Persons with severe functional disabilities and their caregivers: Faciliting self-care management in the home setting, Psychological Services, 9:2:144-162, DOI 10.1037/a0028112.
6. Rietdijk R, App B, Toguer L et al (2012) Supporting family members of people with traumatic brain injury using telehealth: a system review. J rehabil Med 44:913-921.
7. Lai J, Woo J et al. (2004) Telerehabilitation a new model for community-based stroke rehabilitation. J Telemed Telecare, 10: 199–205.
8. Audebert H, Schwamm L (2009). Telestroke: Scientific Results. Cerebrovasc Dis. 28(4): 323-330.
9. Taylor DM, Cameron JI, Walsh L, et al (2009) Exploring the Feasibility of Videoconference Delivery of a Self- Management Program to Rural Participants with Stroke. Telemed J E Health., 15 (7): 646-654.
10. Turkstra L, Quinn-Padron M, Johnson J (2012) In Person Versus Telehealth assessment of discourse abiliy in adults with traumatic brain injury, J Head Trauma Rehabil, 27:6: 424-432

The address of the corresponding author:

Author: Cristina Suárez Mejías
Institute: Virgen del Rocío University Hospital
Street: Avda. Manuel Siurot s/n C.P. 41013
City: Seville
Country: Seville
Email: cristina.suarez.exts@juntadeandalucia.es

Mobile Telemedicine Screening Complex

J. Lauznis, Z. Markovics, and I. Markovica

Riga Technical University, Faculty of Computer Science and Information Technology, Riga, Latvia

Annotation—Article is dedicated to research work in telemedicine field including statements required for e-medicine and based on preliminary development results of computerized screening complex, with wireless transmission of measurements to analysis center, computerized decision-making realization with certified expert conformation and feedback of conclusion to user.

The developed Computerized Screening Complex is based on two main information sources: Interactive questionnaire to register subjective symptoms in about 12 human organism subsystems; and Modular electronic device for 14 physiological test registration with additional 3 test programs for objective parameter measurement and system testing.

System hardware part of Mobile Telemedicine Screening Complex (MTSC) is based on set of specially designed measurement modules. Overall MTSC modules can acquire more than 60 parameter values - numeric, nonnumeric, number strings, or curves, measurement values and calculated values. Computer included in MTSC provides more additional features, result visualization and data flow management. Measuring modules in the unit include separate measuring operations management and recalculation of raw parameter values. Information transmission unit send the results of digitization and preprocessed data to analysis center. Its task is to check the information received, provide detailed analysis, including computerized and expert evaluation of individual tests (like ECG, dermascopy), correction and approval, confirmed by certified specialist at analysis center and storage of tests and results in the database. Consequently link to analysis center is duplex, the indication is provided on both ends of the system.

Keywords—Telemedicine, mobile, screening, diagnostics, e-health, emergency, consulting, wireless transmission.

I. INTRODUCTION

It has been proved that not early diagnostics of diseases and early patient medicamentose or surgical treatment and rehabilitation cost for person itself, his employer and, generally for country, significantly more than prevention, reduce of risk factors, and early diagnosis [1]. It is based on necessity to check population with aim to determine so called risk groups for regular screening to find out individuals with health abnormalities before they need serious and costly medical treatment. If screening procedure founds abnormality it guides potential patients for further diagnosis to certain specialists - cardiologists, oncologists and other. Authors offer their solution to screening by creating a new portable telemedicine screening complex (MTSC) and analysis methods used. It consists of a set of up to 14 physiological parameter measuring modules, computerized interactive questionnaire (CIQ) of subjective symptoms on 12 body subsystems, portable computer with monitor, wireless data transmission means to remote analysis and consultancy center with database [2].

One of basic statements is that CIQ should be done before real measurements to reduce or extend their amount according of results of CIQ, in most cases it may significantly lower costs of necessary investigations and analysis performed.

MTSC has a number of functions. Measuring function unit includes a separate measurement functions, their management and data preprocessing. Transmission of information includes the results of digitization and preprocessing, and, wireless transmission to the analysis center. Decision-making is based on questionnaire data, objective measurement and test results. Data display function provides access to information in measuring people familiar decimal system or verbal form, including those, received as feedback from analysis center.

II. TARGET FUNCTIONS OF SCREENING

Decision making functionality of MTSC provide target screening functions formulated as follows:

- Discover abnormal deviations of patient health status before feeling discomfort or problems;
- Provide recommendations to patient for necessary actions, stating under which professionals turn to more accurately identify possible problem;
- Identify potential risk factors;
- Provide recommendations for risk reduction;
- Create patient database to compare data dynamics in following exams.

Information sources of system are:

- The patient's history and complaints, acquired from interactive questionnaire;
- Objective measurement data provided from existing 14 measurement modules of MTSC;
- Complex of medical service personnel evaluations of individual measurements;

- Primary results of formal data processing and analysis of measurements;
- Analysis center specialist assessments and adjustments of the primary data.

Based on this information screening results are established and followed by recommendations for further action.

III. TECHNOLOGY OF MEASUREMENTS

MTSC includes a portable computer (PC) with a touch screen monitor and at least rear WEB camera (for anthropometric measurements using photogrammetry) [3] and separate modules for physiological parameter measure-ment connected to PC using wireless (WLAN or Bluetooth) or USB connection. Modules provide more than 60 physiological parameters measured, each of the modules can perform a number of functions, like ECG module provides 12 standard Leads, Blood pressure module measures 3 values, Urine test strips give 11 values, Hearing test measures 14 parameters and so on. System is designed using distributed resource principle [4], meaning that each module processes acquired data to appropriate level and format, stated by specilly developed Transmission Protocol. The measuring side, partially in modules, additionally in PC, calculates also valuable additional parameters (see Table 1.).

Table 1 Measured and Calculated Parameters

Nr.	Module	Number of Measured Parameters	Parameter Description	Calculated Parameters
1.	ECG device	8	At least 500Hz sampled potentials at RA; LA; C1-C6, 10 second segment	12 Lead ECG, other depends from Analysis Software if used
2.	Blood pressure (NiBP)	3	P_{SYS}, P_{DIA}, Pulse Rate (PR)	Mean Arterial Pressure (P_{MAP})
3.	Pulse Oxsimeter (SpO_2)	3	Pletismogram curve, SpO_2 level in % (H_6O_2 level)., Pulse Rate (PR)	Compare to normal
4.	Digital Phonendoscope	3	Sound in 3 frequency ranges (Bell, Diaphragm, Extended)	Depends from PC Software for sound interpretation
5.	Cholesterol Strip Test	1	Cholesterol level in blood	Compare to normal
6.	Spirometrer	1	Flow velocity in time	FVC, FEV_1, Tifno index (FEV_1/FVC in %), relative to calculated normal
7.	Digital Thermometer.	1	Body temperature in °C	
8.	Urine Strip Test	11	Standard parameters of urine analysis	Compare to normal
9.	Dermascope	Individual number	Photos of suspicious formations in visible and infrared light	
10.	Anthropometric measurement module	6	Distance (in relative units) between predetermined anatomical points on human body; Determined constitutional type,	Body length (A), waist circumference (V), hip circumference (G), Scoliosis index, Kyphosis index, Relationship V/G
11.	Weight and fat/muscle ratio meter	2	Weight, fat/muscle ratio	Body mass index (BMI)
12.	Glucose Strip Test	1	Glucose level in blood	Compare to normal
13.	Computer based Visual Acuity Meter	4	Visual acuity in each eye, color vision in each eye	Acuity index for each eye
14.	Audiometer	14	Hearing threshold for each ear at 7 frequency values	Graphic hearing level presentation in db relative to hearing threshold (0db)

Measurement module technical capabilities are higher. Table summarizes the parameters used only for screening diagnostics.

The information obtained from measurement modules can be in several forms:

1. Measurements when human physiological parameter can be expressed as a single number, such as height, weight, temperature, frequency, etc. Parameter value in digital form is for wireless data transmission to a remote data analysis centers. Parameter value in decimal form is used in on-site measurement at local indication side. This form is used in 10 modules and partially in the interactive questionnaire (CIQ), as a special number.

2. Measurements, formed as a series of numerical values, usually presented and analyzed as curves, functional relationships, such as an ECG, audiogram, etc.Transmitted are only digital values obtained by fixing parameters at certain time intervals.

3. Presentation of information in nonnumeric form, such as a sound from phonendoscopes evaluating noise found (or not found) in certain locales: the heart, lungs, and major blood vessels. The results are transmitted in coded form to analysis center and may be indicated at the measurement side. Sound file record, transmission and analysis are not used in screening.

Something similar occurs with ECG evaluation at measurement side, when medical personnel visually analyses and states only 3 situations – normal, some abnormalities, probably dangerous situation. Detailed measurements and deeper evaluation is done in the analysis center by certified cardiologist.

Interactive questionnaire responses are verbal form (yes, no, can't answer), and are entered in the system in encrypted form. This group includes also the person's constitutional type found in anthropometric data (Weakness, normastenic or hyperastenic type).

4. The measurements being transmitted to analysis center without a local evaluation. This is one module - dermascope who photos abnormal area of skin and images are assessed and evaluated by expert in analysis center.

5. Information obtained by a calculation of the individual parameter measurements, like mean blood pressure; body mass index, etc. (See Table 1). The calculation process is organized in the measurement side of the complex by software in portable computer. The results are indicated on site and sent to analysis center.

Every measurement module is characterized by several values, for illustration is given short description of non-invasive blood pressure (NiBP) module:

Measurement Principles

Classic method with compression cuff and osilometric detection of blood pressure border values is used by module electronics and embedded software [5]. Obtained results in digital form are used locally and presented in decimal form and also transmitted to analysis center.

Parameters Measured by Module

- Systolic blood pressure P_{SYS} in mmHg;
- Diastolic blood pressure P_{DIA} in mmHg;
- Pulse rate PR in bpm (beats per minute);

Mean arterial pressure MAP is calculated value.

$P_{MAP} = P_{SYS} + ⅓ (P_{SYS} – P_{DIA})$ in mmHg (Fig.1.)

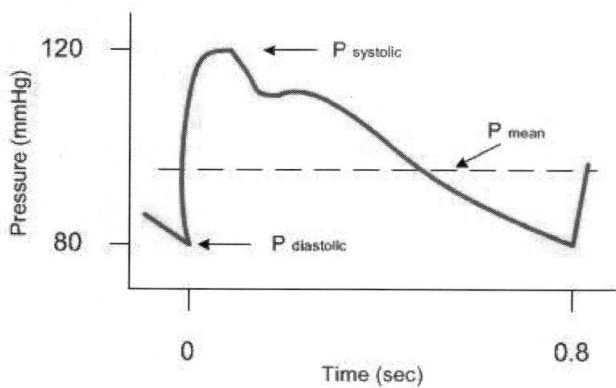

Fig. 1 Presentation in time of normal Systolic, Diastolic and Mean blood pressure at pulse rate 75 bpm.

Format of Results, Border Values:

Format for all measured and calculated parameters P_{SYS}; P_{DIA}; PR; P_{MAP} is equal and is presented as 3 decimal numbers. Boarder values for P_{SYS}, P_{DIA} and P_{MAP} are 50 – 260 mmHg, for PR 30 – 230 bpm.

Normal Values

For screening diagnostics classic normal blood pressure and pulse rate values are used as follows:

P_{SYS} = 120 – 140 mmHg,
P_{DIA} = 80 – 90 mmHg,
PR = 60 – 80 bpm.

Symptomatic

Decision making in screening uses three symptoms and their code numbers:

2.0 – normal or 0-class,
2.1 – increased blood pressure or A, B and C class.
2.2 – lowered blood pressure

Technical Requirements for Module

Required connection with computer using interface RS 232 TTL or virtual COM using USB or Wireless connection mode (as serial port or, medical device profile) with data exchange rate at least of 9.6 Kbit/s. Software must provide possibility of automated periodic measurements and display of results, including calculated and decision.

Measurement use in screening diagnostics

Blood pressure values are basic parameters for arterial hypertension diagnostics and together with results from interactive questionnaire may determine necessity of consulting at professional cardiologist.

IV. ACQUIRED INFORMATION TRANSMISSION AND PRESENTATION

MTSC has two information flows:

1. From sensors and measurement modules to portable computer;
2. From computer to final establishment called Analysis Center

As Analysis Center may act:

- University Clinics;
- Existing Medical Centers, having necessary experts;
- Specialized Analysis Centers, formed in frames to support this project;
- Family doctors and other private praxes;

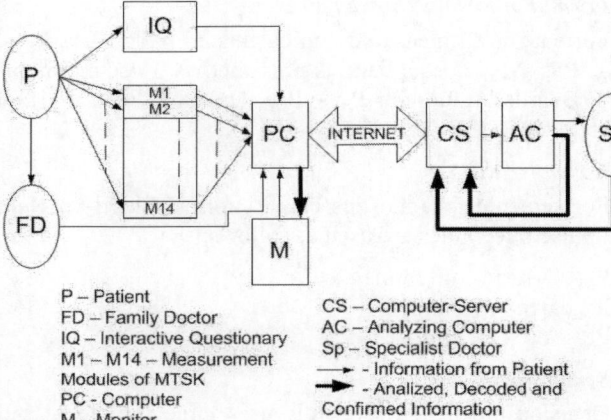

Fig. 2 Information flow in MTSC and Analysis Center side

First information flow is realized on standard wired connections (RS232 or USB) with computer or, using Wireless (WLAN, Bluetooth) for separate modules. Second flow is realized from computer to Analysis Center (Includes also Server and Database) using wireless Internet connection.

Transmission to Analysis Center include encoded initial patient data, digital values of measurements, calculated results, medical staff comments, interactive questionary data, pictures. The aim of transmission is:

- Provide necessary information processing by means of advanced and powerful analysis software, not available on site, like for multiprofile ECG evaluation;
- To obtain high qualification specialist conclusions, when formalized processing is not possible on site, like dermascopy picture analysis.

Information flows are presented in Fig.2. Computer included in set of MTSC provides decision only about total screening results and recommendations for future actions.

Decisions are based on decision making theory and are realized by means of production laws.

V. CONCLUSION

Nevertheless MTSC is targeted to screening needs; its functionality may be increased by additional modules and software. This may find additional applications of complex as full featured diagnostic tool in family doctor's praxis and, as more rugged version, for emergency, rescue and army use.

ACKNOWLEDGMENT

Project is supported and financed thanks to European Regional Development Fund, agreement Nr: 2011/0007/2DP/2.1.1.1.0/10/APIA/VIAA/008

REFERENCES

1. Wolf S, Husten C, Levin L, Marks J, Fielding J, Sanches E. The Economic Argument for Disease Prevention: Distinguishing Between Value and Savings. Partnership for Prevention, 2009
2. Katasevs A., Markovics Z., Markovica I., Balodis G., Lauznis J. Development of New Mobile Telemedicine Screening Complex. International Symposium on Biomedical Engineering and Medical Physics: IFMBE Proceedings, Latvia, Riga, 2012., pp. 31 – 34
3. Celinskis D, Katashev A. On criteria for wide-angle lens distortion correction for photogrammetric applications. International Symposium on Biomedical Engineering and Medical Physics: IFMBE Proceedings, Latvia, Riga, 2012., pp.153 – 158
4. Lauznis J, Markovics Z, Balodis G, Streļcs V. On Resource Distribution in Mobile Telemedicine Screening Complex. Scientific Journal of Riga Technical University, Computer science, vol. 13., Riga, Latvia: RTU, 2012, pp. 28-31
5. Balodis G, Markovitch Z, Lauznis J, Development of online blood pressure monitoring system using wireless mobile technologies. Proceedings of conf. "Biomedical Engineering", Kaunas University of Technology, Kaunas, Lithuania, 2009, pp. 190 – 193.

Author:	Juris Lauznis
Institute:	Riga Technical University, Faculty of Computer Science and Information Technology
Street:	1/3 Meža str.
City:	Riga
Country:	Latvia
Email:	Juris.Lauznis@rtu.lv

Radial-Basis-Function Based Prediction of COPD Exacerbations

M.A. Fernández Granero[1], Daniel S. Morillo[1], A. León[2],
M.A. López Gordo[1], and L.F. Crespo[1]

[1] Biomedical Engineering and Telemedicine Lab, University of Cádiz, Cádiz, Spain
[2] Pulmonology and Allergy Unit, Puerta del Mar University Hospital, Cádiz, Spain

Abstract—**Chronic Obstructive Pulmonary Disease (COPD) is a very serious progressive lung disease with a high socio-economic impact and prevalence levels worldwide. Admissions for acute exacerbation of respiratory symptoms (AECOPD) have the highest proportion of economic and human cost. During a 6-months field trial in a group of 16 patients, a novel electronic questionnaire for the early detection of COPD exacerbations was evaluated. Data mining strategies were applied. A Radial Basis Function (RBF) network classifier was trained and validated and its accuracy in detecting AECOPD was assessed. 94% (31 out of 33) AECOPD were early detected. Sensitivity and specificity were 73.8% and 87.0% respectively and area under the ROC curve was 0.82. The system was able to early detect AECOPD with 5.3 ± 2.1 days prior to the day in which the patients required medical attention.**

Keywords—**COPD, exacerbation, telehealth, symptoms, questionnaire, early detection, RBF, telemonitoring.**

I. INTRODUCTION

COPD is a leading cause of death in the world [1]. COPD is a very serious progressive lung disease that is regarded as the respiratory disease with the greatest socio-economic impact and prevalence levels in our society [2]. The admissions for AECOPD have the highest proportion of economic and human cost [3]. Furthermore, exacerbations increase morbidity and cause an impaired quality of life [4].

Home telemonitoring is a sustainable alternative for the follow-up of patients with chronic conditions. Patients who present for therapy before the onset of symptoms have shown better outcomes [5]. Therefore, the human and economic burden of the disease could be reduced by early detection and prompt treatment of AECOPD [6].

The Whole System Demonstrator (WSD), that is the world's largest randomized control trial of telehealth and telecare in the world, has reported a reduction in mortality and in hospital admissions on long-term diseases including COPD, diabetes and heart failure [7].

Although home telemonitoring has been lately applied to COPD, only a few studies have attempted to achieve early detection of exacerbations on a daily basis [5,8,9].

In this study a mobile health system was designed for the early detection of AECOPD on a day-to-day basis. During a 6-months field trial an electronic questionnaire (AQCE) [10] was evaluated. Data mining techniques were applied and accuracy in early prediction of AECOPD was assessed.

II. METHODS

A. Participants

A sample of 16 COPD patients recruited in the University Hospital Puerta del Mar of Cádiz (Spain) from the Pneumology and Allergy Department participated during 6 months in this study. All the patients were equipped with a home base station to daily respond to a questionnaire.

Participants were all aged over 60 years. They had a diagnosis of COPD confirmed by spirometry, with COPD severity II, III or IV (GOLD guidelines). Patients had cumulative tobacco consumption greater than 20 Packs-Year and at least one hospital admission for exacerbation or two exacerbations treated with oral antibiotics or corticosteroids in the past year.

The Hospital's research ethics committee approved the study and signed informed consent was obtained from all patients.

B. Components of the Telemonitoring System

The system emulated a medical consultation including a medical interview and auscultation [11]. The telemedical system included a dedicated mobile device (*DmD*) with speech and touch interfaces, a telemedical server located at the Hospital, a sensor delivered to patients for recording the respiratory sounds and a communication network. The remote server was connected to each *DmD* through a secure private IP-based wireless telemedicine network.

A multimodal interface application was developed to allow the patients to perform the required daily tasks [12]. It was specially designed to compensate possible sensory decline and physical and cognitive deficits in elderly users [13].

A XML file with the information obtained from the patient was sent daily to an electroni patient record (EPR). EPR was ad-hoc created for this study and could be accessed by physicians.

C. Data Preparation and Prediction Model

The multimodal interface guided the patients to record their symptoms. All symptom-related questions were presented in agreement to the developed AQCE, based on COPD exacerbation symptoms and prodromes. Data were sent to the remote server after recording symptoms. A variety of definitions of a COPD exacerbation has been proposed [14]. In this study, exacerbation was defined as a non-programmed medical visit to the primary care center, to the specialist or to the emergency unit because of respiratory symptoms.

Yes/no answers were assigned Boolean outputs (1 for present symptom). Likert-scale was used to assign the possible answers from 0 (much better/much more) to 4 (much worse/much less). Twelve predictor variables were processed. Missing data were processed using backward and forward imputation [15]. Four additional parameters were calculated: 3-days moving average applied to the total score and average scores for symptoms associated with major, minor and complementary symptoms.

The increase of symptoms in the prodromal stage of 7 days prior to the onset of an exacerbation is well-known [16]. Therefore, the target was defined as a categorical variable with two levels: a label: "1" was assigned for the exacerbation onset (the day in which the patient required medical care) and the previous 7 days and "0" otherwise. Recovery periods of two weeks after the AECOPD were discarded [17].

A supervised radial basis function neural network (RBF) was used for classification. RBF was introduced by Powell [18] in 1987 for the purpose of exact interpolation. A RBF network has three layers: input, hidden and output layer (Figure 1). Neurons in the hidden layer have Gaussians functions. The activation level is expressed as [19]:

$$h_i(X) = exp\left(\frac{\|X-u_i\|^2}{2\sigma_i^2}\right) \quad (1)$$

where X represents the input vector and u_i is a vector with the same dimension as X that denotes the center vector. Activation level of the radial basis function is maximum when X is at the center u_i. The final output of the RBF network can be evaluated as the weighted sum of the outputs:

$$y_i = \sum_{i=0}^{N} \omega_i h_i(X) \quad (2)$$

Where N is the number of neurons in the hidden layer, ω_i is the weight of neuron i in the linear output neuron. The proposed binary classifier has two neurons in the output layer.

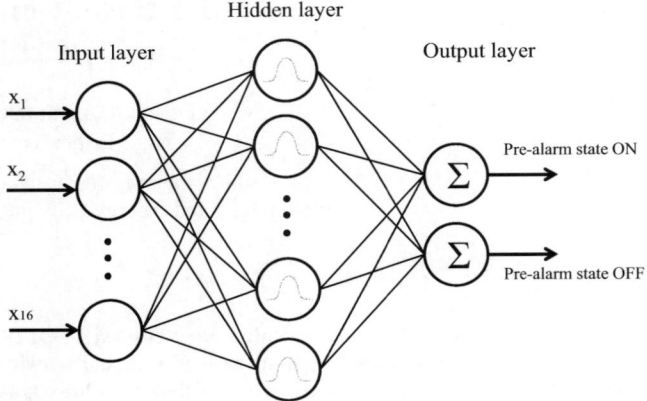

Fig. 1 Architecture of the proposed Radial Basis Function Network

With the aim of reducing the false positive rate, the classifier output was forwarded to a simple decision rule. Alarm state was set after two consecutive days with a positive output in the classifier.

10-Cross-validation was used to gauge the generalizability of the classifier and to ensure stability of the results.

Evaluating the performance of the monitoring system to early detect AECOPD is the primary objective of this study.

The area under the receiver operating characteristic (ROC) curve (AUC) was used for a total estimation of the accuracy of the classifier. ROC curve relates the sensitivity and specificity of a dichotomous classifier [20]. Two ROC curves were estimated; one for the training set and another one for the validation set. Performance was assessed according to sensitivity, specificity, accuracy, confusion matrix, positive predictive value (PPV) and negative predictive value (NPV).

The software package used for graphical representation and signal processing was MathWorks MATLAB®. Statistical analysis was completed using the SPSS® statistical package.

III. RESULTS

15 patients were included in the final study (one patient was excluded). The Table 1 shows the clinical and demographic characteristics of the participants. An average of 2.1±1.7 AECOPD episodes was confirmed. 10 patients needed 51 medical attentions (33 AECOPD and 18 events associated to non-recovered exacerbations).

789 days, each one characterized by 16 input features, was used to train the RBF network classifier. Figure 2 shows the ROC curves for both the train and validation stages. AUC was 0.94 and 0.82 for the training and for the validation set respectively.

Table 1 Clinical and demographic characteristics of the study group

Variable	N(%)
Age (years)	70.2±6.6
Male (%)	93% (14)
COPD Stage (%)	
Stage I (>80%)	0% (0)
Stage II (50%-79%)	13.3% (2)
Stage III (30-49%)	46.7% (7)
Stage IV (<30%)	40.0% (6)
AECOPD (%)	
None	33.3% (5)
1 to 2	13.3% (2)
3 or more	53.4% (8)
Days hospitalized (%)	
0	60.0% (9)
1 to 10	13.3% (2)
11 or more	26.7% (4)

Table 2 details some performance parameters of the resulting classifier.

The system was able to early detect AECOPD with 5.3 ± 2.1 days prior to the day in which the patients required medical attention. Fig. 3 shows the box plot with prediction margins. 94% (31 out of 33) AECOPD were early detected.

Table 2 Classifier performance evaluation

Metric	Value
Total records	789
True Positives (TP)	186 (23.6%)
False Positives (FP)	70 (8.9%)
True Negatives (TN)	467 (59.2%)
False Negatives (FN)	66 (8.3%)
Accuracy	82.8%
Sensitivity	73.8%
Specificity	87%
Positive Predictive Value (PPV)	72.7%
Negative Predictive Value (NPV)	87.6%

IV. DISCUSSION

An automatic classifier for the prediction of AECOPD was trained and validated using the daily responses to an electronic questionnaire designed for the remote monitoring of patients with COPD.

In this work, all results achieved were produced using an approach applied in many recent studies, patient-dependent 10-fold cross-validation. The analysis would ideally include a larger stratified sample of patients. Such approach would enable using leave-one-subject-out cross validation (LOSO-CV) to avoid optimistic bias that we could had with possible inclusion of some samples (although from different days) for the same subject in both training and test datasets.

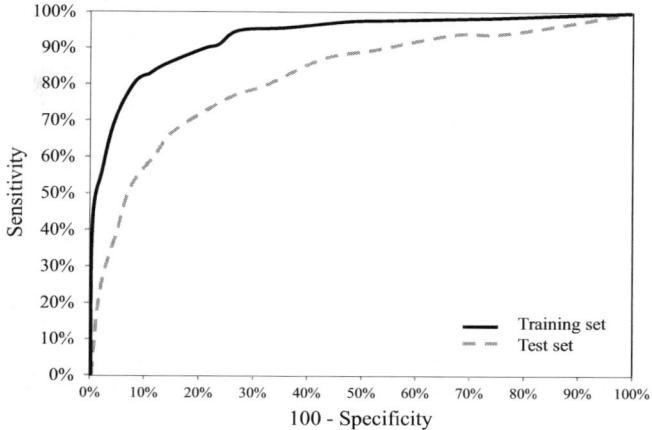

Fig. 2 ROC curves for the trained and validated classifier

The six-month duration of the trial period was selected to include the more detrimental seasons for patients in order to optimize the number of exacerbations. The prediction algorithm early detected 31 of 33 exacerbations. Therefore, the proposed method could be used to forecast an incoming AECOPD. One of the two undetected exacerbation had a pattern without prodrome characterized by a sudden onset. This pattern has been recently described and it is difficult to early detect [21].

Regarding false positives, 7 false positives were detected during the week after recovery, in which symptoms could probably be not stabilized. 45 cases could be predictive episodes of AECOPD beyond the defined prodromal phase that were detected in days 7 to 14 prior to onset.

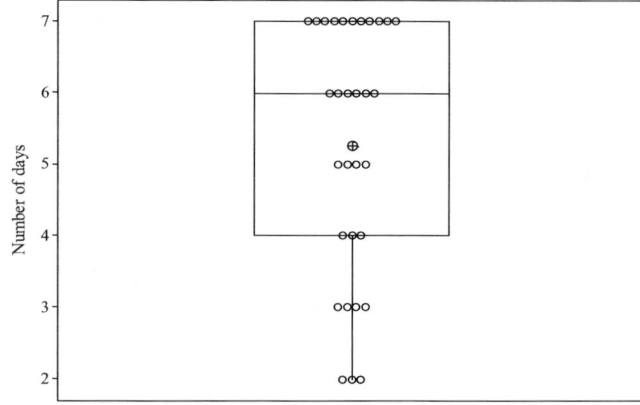

Fig. 3 Box plot of prediction margins.

On the other hand, there are non-informed exacerbations characterized by great symptoms scores variability without requesting medical assistance. In these cases, the training algorithm was supplied with misleading targets that could reduce its accuracy.

Finally, a larger field trial sample would be needed in order to corroborate the results.

V. CONCLUSIONS

This study was aimed to early prediction of COPD exacerbations using a daily remote questionnaire. COPD exacerbations are major events that affect patients' mortality, morbidity and quality of life and that present a major health burden.

The proposed system was able to early predict AECOPD with a margin of 5.3±2.1 days. Detection accuracy was 94%. 31 out of 33 exacerbations were early detected and a reduce rate of false positive was achieved.

The proposed electronic questionnaire and the designed classifier could help to early detect COPD exacerbations and therefore could provide support both to physicians and patients.

ACKNOWLEDGMENT

This work was supported in part by the Ambient Assisted Living (AAL) E.U. Joint Programme, by grants from Ministerio de Educación y Ciencia of Spain and Instituto de Salud Carlos III under Projects PI08/90946 and PI08/90947.

REFERENCES

1. Divo M, Cote C, de Torres JP, Casanova C, Marin JM, Pinto-Plata V, Zulueta J, Cabrera C, Zagaceta J, Hunninghake G, Celli B (2012) Comorbidities and risk of mortality in patients with chronic obstructive pulmonary disease. Am J Respir Crit Care Med 186(2):155-161.
2. Rennard S, Decramer M, Calverley PMA, Pride NB, Soriano JB, Vermeire PA, Vestbo J. (2002) Impact of COPD in North America and Europe in 2000: subjects' perspective of Confronting COPD International Survey, Eur Respir J, vol. 20, pp. 799-805.
3. Ornek T, Tor M, Altın R, Atalay F, Geredeli E, Soylu O, Erboy F. (2012) Clinical factors affecting the direct cost of patients hospitalized with acute exacerbation of chronic obstructive pulmonary disease. Int J Med Sci. 9(4):285-90.
4. Terzano C, Conti V, Di Stefano F, Petroianni A, Ceccarelli D, Graziani E, Mariotta S, Ricci A, Vitarelli A, Puglisi G, De Vito C, Villari P, Allegra L (2010) Comorbidity, hospitalization, and mortality in COPD: results from a longitudinal study. Lung 188(4):321-329.
5. Walters E, Walters J, Wills K, Robinson A, Wood-Baker R (2012) Clinical diaries in COPD: compliance and utility in predicting acute exacerbations. International Journal of Chronic Obstructive Pulmonary Disease 7:427-435.
6. Global Initiative for Chronic Obstructive Lung Disease (GOLD). Global strategy for diagnosis, management, and prevention of COPD. http://www.goldcopd.org/uploads/users/files/GOLD_Report_2013_Feb20.pdf Accessed May 1, 2013.
7. Steventon A, Bardsley M, Billings J, Dixon J, Doll H, Hirani S, et al. (2012) Effect of telehealth on use of secondary care and mortality: findings from the Whole System Demonstrator cluster randomised trial. BMJ 344:e387.
8. Jensen MH, Cichosz SL, Dinesen B, Hejlesen OK (2012) Moving prediction of exacerbation in chronic obstructive pulmonary disease for patients in telecare. J Telemed Telecare 18(2):99-103.
9. Sund ZM, Powell T, Greenwood R, Jarad NA (2009) Remote daily real-time monitoring in patients with COPD - A feasibility study using a novel device. Resp Med 103(9):1320-1328
10. León A, Astorga S, Crespo M, Morillo DS, Failde I, Crespo LF (2011) Development of an automated questionnaire for the early detection of COPD exacerbations (AQCE). Annual Congress ERS 2011, Eur Respir J, 38(55), 82s.
11. Foix LC, Morillo DS, Crespo M, Gross N, Kunze C, Giokas K, Jimenez JA (2009) AMICA telemedicine platform: a design for management of elderly people with COPD. Proceedings of the 9th International Conference of the IEEE Information Technology and Applications in Biomedicine, Larnaca, Cyprus, IEEE Conference Publications, pp 1-4.
12. Crespo M, Morillo DS, Crespo F, León A (2010) Collaborative Dialogue Agent for COPD Self-management in AMICA: A First Insight. In: Demazeau Y et al (eds) Advances in Practical Applications of Agents and Multiagent Systems, Advances in Intelligent and Soft Computing, Springer Berlin Heidelberg, pp 75-80.
13. Schieber F (2003) Human factors and aging: identifying and compensating for age-related deficits in sensory and cognitive function. In: Charness N, Schaie KW (eds). Impact of Technology on Successful Aging. Springer, New York, pp 42-84.
14. Pauwels R, Calverley P, Buist AS, et al. (2004) COPD exacerbations: the importance of a standard definition. Respir Med. 98(2):99–107.
15. Trappenburg JCA, Monninkhof EM, Bourbeau J, Troosters T, Schrijvers AJP, Verheij THJ, Lammers (2011) Effect of an action plan on recovery and health status impact of exacerbations in COPD patients; a multicenter randomized controlled trial. Thorax 66(11):977-84.
16. Seemungal TA, Donaldson GC, Bhowmik A, Jeffries DJ, Wedzicha JA (2000) Time course and recovery of exacerbations in patients with chronic obstructive pulmonary disease. Am J Respir Crit Care Med. 161(5):1608-13.
17. Jadwiga WA, Donaldson GC (2003) Exacerbations of chronic obstructive pulmonary disease. Respir Care 48:1204–15
18. Powell MJD. (1987) Radial Basis Functions for Multivariable Interpolation: A Review. In Mason and Cox: Algorithms for Approximation, Clarendon Press, Oxford, pp 143-167.
19. Begg R, Kamruzzaman J, Sarker R. (2006) Neural Networks in Healthcare. Potential and Challenges. Idea Group Publishing.
20. Metz CE. (1978) Basic principles of ROC analysis. Semin Nucl Med 8:283-298.
21. Aaron SD, Donaldson GC, Whitmore GA, Hurst JR, Ramsay T, Wedzicha JA (2012) Time course and pattern of COPD exacerbation onset. Thorax, 67(3):238-243

Corresponding author:
Author: Miguel Ángel Fernández Granero
Institute: University of Cádiz. School of Engineering
Street: C/Chile,1
City: Cádiz
Country: Spain
Email: ma.fernandez@uca.es

Research Benefits of Using Interoperability Standards in Remote Command and Control to Implement Personal Health Devices

H.G. Barrón-González[1], Student Member, IEEE, M. Martínez-Espronceda[2], Member, IEEE,
S. Led[2], and L. Serrano[2], Senior Member, IEEE

[1] FIME/ Universidad Autónoma de Nuevo León, San Nicolas De Los Garza, – México
[2] Department of Electrical and Electronic Engineering, Public University of Navarre, Pamplona, Spain

Abstract—The initial heterogeneity in the development of Personal Health Devices (PHDs) highlighted the lack of interoperability between agents and managers. The standardization work of the PHD Working Group (PHD-WG) has recently adapted the ISO/IEEE 11073 interoperability standard (X73), initially oriented to clinical environments, to include Personal Health Devices (X73PHD) communications. This new evolution allows the development of interoperable personal health ecosystems. At the moment of this writing, more than 11 specializations have been successfully published by the Personal Health Device (PHD) Working Group (PHD-WG). This new standard brings benefits to both technology producers (design cost reduction, experience sharing, marketing, etc.) and technology consumers (plug-and-play, accessibility, ease to integration, prices, etc.). Nevertheless, some recent specializations at draft stage show the need for a procedure to command and control configuration parameters and settings, which could provide additional benefits such as configurability from the cloud. As a solution, some *ad-hoc* methods have been elaborated to deal with, although the aim of the PHD-WG is to standardize a general procedure, valid for longer term. Therefore, it is needed to identify use cases requiring remote configuration services. This work identifies and studies new use cases that employ remote configuration services. The resulting use cases, discussed within the PHD-WG to get the maximum consensus, are related to the Basic Electrocardiograph (X73-10406), the Sleep Apnea Breathing Therapy Equipment (X73-10424), and the Medication Monitor (X73-10472) specializations. In addition, a classification of the findings is proposed for each use case. These findings could be the basis for the new remote configuration extension package.

Keywords—extension package, interoperability, ISO/IEEE 11073, eHealth.

I. INTRODUCTION

The research work on standardization provides benefits which may accelerate the development of technological advances which integrate market needs as well as the experience of device manufacturers and researchers in the field. In general, these are developed mainly as a global solution with high impact for a particular problem providing multiple solutions that are reused by many independent groups. The home telemonitoring standardization efforts are recently being one of the most active which was initially prompted by the constant development of new medical devices for use in intensive care units [1]. The growing number of people that requests healthcare at home, especially in adulthood, is opening the way for the medical devices to personal health environments. In addition, a Personal Health Device (PHD) market promoted by a number of manufacturers is propitiating the development of standards that enable interoperability between medical devices and monitoring systems [2].

This effort got its first results in the development of the ISO/IEEE11073 (X73) standard [3-5]. Subsequently, technological advances induced an adaptation to this standard to a new branch known as X73 for Personal Health Devices (X73PHD) [6,7]. Developed to standardize PHD communications, this new branch is mainly defined within the Optimized Exchange Protocol (11073-20601) and device specializations (11073-104xx). Nowadays, it is necessary to increase the functionality provided in some use cases within specializations, in order to optimize solutions and services centered in the patient [8]. For example, medical devices may require personalizing parameters in order to obtain a correct measurement of the data sampled by the agent.

Now, one of the main objectives of PHD-WG is focused on standardize a general procedure that allows the inclusion of remote configuration in any specialization (an extension package within the 11073-20601 toolbox), in response to new specializations being developed at this time, which require this operational feature to control configuration parameters [9].

As a solution, some *ad-hoc* methods have been elaborated so far, which provide the desired functionality. Nevertheless this functionality cannot be extended to other specializations easily. In order to define this procedure, the first step is to gather and to analyze as many use cases as possible on remote control, including both real ones and hypothetical ones.

In view of the above, this work presents the benefits that the interoperability standard X73PHD provides to both manufacturers and consumers, allowing its integration within personal health environments. In addition, this work also proposes the study of new use cases within the PHD-WG, which could benefit from the extension of remote control functionality within standard.

This paper is divided as follows. Section II shows the methodology used to get the use cases, analyzed in this research. The benefits of X73PHD Standard, the use cases, specializations that generated and the technical characteristics of each are discussed in Section III. Future lines are discussed in Section IV. Finally conclusions are drawn in section V.

II. NEW USE CASES WITH REMOTE CONTROL IN X73PHD

The PHD-WG defines standards in collaboration with the IEEE-SA. It has more than 298 members, organized into companies and universities around the world. Therefore, the group mainly develops its work by its mailing list. In addition, the group holds online weekly meetings and face-to-face (F2F) working sessions (2 to 4 calls per year).

PHD-WG has developed the optimized exchange protocol (2008, 2010 versions) and more than 10 specializations. It is working on the development of new specializations and the improvement of existing ones, including a new version of the optimized exchange protocol.

This paper gathers results of the PHD-WG in previous research to show the benefits of interoperability, and analyze them to show the requirements for the addition of an extension of the programming and remote control interoperability standard 20601.

The information provided by the agents follow a methodology based on grouping the configuration parameters on the operational functions of which they are part, which yields the following 3 functions.

- Operational modes: This classification groups the different operating modes that can be implement in an agent. It is important to know it because the ability to change parameters could not be present in all operation modes or may have different operating ranges in similar operating modes. Currently the use case Basic Electrocardiograph has two operational modes, Real-Time and Store and forward. Sleep Apnea Breathing Therapy Equipment has four operational modes, standby, Therapy, Drying and Mask Test. Be able to modify this feature by remote control, can increase the functional performance of the agent.
- Equipment, user data and settings: These are the data that identify and characterize both the device and the user. These can be classified into several categories such as device data (e.g. device type, serial number, software version), device state (e.g. stand-by, drying, therapy), maintenance data (e.g. hours of flow generation, hours of filter use), technical alerts (e.g. delivered pressure out of specification, empty humidifier), and patient preferences (e.g. patient user interface language) among others. Not only the use cases raised at this research can benefit from remote control extension that allows modify these parameters. Enable this function could enrich different specializations of the standard X73PHD.
- Therapy data and operational settings: This group represents two types of data, which provide service to the most important functional characteristics of the case studies. First, the configuration of the therapy data, include all values to set ranges to parameters necessary to implement personalized therapies in the event that the agent has enabled this special mode. This operational mode is present in one of the new specializations presented in this analysis. (e.g. CPAP/hPa, Min APAP/hPa, Max APAP/hPa). And second, operational settings are related to the correct acquisition of medical information. These settings are present in therapy devices but are more related to monitoring devices that though lack the ability to apply some therapy. These parameters are required to develop its main functions (e.g. Taqui-Lim, Asist-Lim, Alarm-duration).In the same way as in the previous classification, different specializations could benefit from this function.

The following specializations are associated to the use cases proposed by PHD-WG:

1. **Part 10406 – Basic Electrocardiograph (1 to 3 lead ECG)_[10, 11].** This specialization is intended for ECG monitoring devices acquiring and recording 1 to 3 channel (leads) electrocardiographic waveforms or analyzing the acquired signals to measure heart rate. The use case analyzed provides additional features such as event generation starting automatic form of algorithms that analyze the qrs complex, which generates instant transmission of information acquired by the agent.
2. **Part 10424 – Sleep Apnea Breathing Therapy Equipment [12].** This specialization is currently in drafting state. This specialization is intended for sleep apnea therapy devices that gather multiple parameters during sleep.
3. **Part 10472 – Medication Monitor [13].** This specialization is intended for simple *medication monitor that provides a record of the people usage of medication.* It does not include the composition of medication.

The new use case proposes to increase the function of this device by modifying the parameters for the dosage of the medication, and providing a feedback about the response by the user on the correct use of the device.

III. BENEFITS AND DRAWBACKS

The benefits of interoperability can be analyzed if it be considered the initial stage of home telemonitoring, which is based in one red PAN (Personal Area Network). Where all members can communicate with each other correctly. However, this did not happen because the initial heterogeneity between the devices that were involved did not allow interoperability between devices from different manufacturers, when lacking of a protocol for interoperability to enable integration of different devices to an existing network. This occurred because most medical devices were based on ad-hoc solutions. This situation was the origin for one of the most significant problems in home telemonitoring because they were required a larger number of medical devices to adapt to the specific needs of each user, which increases the costs of implementing services of e-Health.

The ISO/IEEE11073 (X73) standard was developed to allow the interoperability of medical devices at the point of care (Point-Of-Care, X73-PoC), and was used as an initial platform for the development of so-called X73PHD which aims to allow interoperability of medical devices wearable in personal health environments.

A standard that allows interoperability is itself one of the most significant benefits to both users and manufacturers, as it allows the development of medical devices based on 13 specializations (11073-104xx) that can monitor the most important physiological variables (respiration, heart rate, pulse, blood presure, temperature, weight, glucose, body composition). Up to this point the PHD-WG is developing seven specializations more. Within which not only monitoring functions will be added if not also mechanisms to apply therapies for various medical conditions, increasing the services provided by the e-health.

The development of the standard of interoperability ISO/11073-20601 has countless advantages. Only the Continua Health Aliance organization brings together 200 companies in the world that use these specializations for medical device development, which generates a reduction of costs for the end user to the have different possibilities to implement telemonitoring system according to their needs.

An additional benefit obtained from this work together is the reduction of production costs, to be shared the experience in the design of medical devices, also reduces development time and therefore the final cost of personal health devices.

The reusability of the devices and plug and play ability is another aditional benefit of interoperability that allows users to share and integrate different medical devices in different PAN supported by the standard.

On the other hand, the benefits to be gained from the implementation of an extension of remote control and programming within interoperability standard 20601 increase their functionality. It enables the ability to modify monitoring parameters based on patient data feedback, according to the dynamics of the disease. With this all devices used only for monitoring, may adjust the ranges of monitoring parameters.

In the case of devices that implement therapies, where using the data obtained in real-time, feedback about the evolution of the patient, the physician could define new values using the remote control to implement most adequate therapy, With this, medical devices would have the ability to enable changes generated from medical center online or even in real-time. A likely initial challenge is to have to increase the security mechanisms to prevent the intrusion of external agents, and that should also add mechanisms to avoid human error when dealing with devices that deal with the life of the patient.

Another benefit that has emerged recently is to use smartphones as a gateway to communicate data acquired by wearable devices. In the BAN network medical device communicates with the Smartphone via Bluetooth, and then to redirect to remote support points can be used gprs service. This is becoming more feasible due to the HDP profile inclusion in the operating systems of these devices, also reducing load data processing devices with limited resources, and can also reduce the use of a device extra by the user, using only their smarthphone usual.

IV. ANALYSIS OF REQUIRIMENTS

The results obtained are summarized in Tables 1 and 2. The use cases are shown in Table 1. Within this table, the operation modes proposed for these new use cases are shown in the first column. The use case A is based on a monitoring device with two operational modes. The use case B is a Medication Monitor with a single operational mode. These two agents have configuration parameters, which are shown in the third column. The use case C has several operation modes implicitly within the capabilities of the application of therapies available to the agent, which are shown in the third column. This table shows that the use cases share the requirement of a configuration service for its operational parameters.

Table 1 Use Cases

Use Case	Operational modes	Settings Parameters
(A) ECG Recorder	Store-and-forward Realtime	Operational-Mode, Taqui-Lim, Brady-Lim Asist-Lim QRS-Number T-Wave-Time Search-Back
(B) Medication Monitor		Operational-Mode, Time-Dose-Available Alarm-duration Event-Dosage-Number
(C) Sleep Apnea Breathing Therapy Equipment.	Standby, Therapy, Drying, Mask-Test	Therapy-Selector, CPAP/hPa, IPAP/hPa, EPAP/hPa, Inspiration-Trigger-Sensitivity/%, Expiration-Trigger-Sensitivity, Minimum-Respiratory-Frequency/bpm, Ramp-Start-Pressure /hPa, Ramp-Duration/min

The settings found in these use cases are summarized in Table 2. The name of the setting and the use cases where it is needed are show in the first and the second columns, respectively. The third column shows examples of data types associated to that setting.

Table 2 Setting Classification.

Parameters	Use Case	Example
Operational Mode	A B C	Real Time, Store-and-forward
Therapy Selector	C	CPAP, Auto-CPAP, Bi-Level-PAP Auto-Bi-Level-PAP
Configuration Parameter	A B C	Taqui-Lim Asist-Lim Store-Time-Auto Dosage-Number Daily-Alarm-Times CPAP IPAP

V. DISCUSIONS

In this paper we have analyzed the advantages and disadvantages of interoperability and a remote extension. But many more advantages could arise not provided in this work, could be analyzed also the interoperability of these devices with smartphones, pdas, microcontrollers, wireless routers among others to enable remote control features and interoperability in personal and body networks (Personal/Body Area Network, PAN/BAN), due to continually emerge to market new devices that implement the HDP profile, thus speeding up the development of these technologies. Will likewise be analyzed mechanisms to be implemented using any wireless communication technology like Bluetooth, Zigbee, Wi-Fi or Gprs the remote control extension for the Interoperability Standard 11073-20601.

It is also known that medical devices implemented some therapy so far are stationary in nature, making them physically feasible to have more electronic resources and energy that allow a superior performance compared to wearable devices. However the remote control extension must use a mechanism that can be implemented in the same way on both devices wearable and stationary.

It is necessary to specify data for monitoring and configuration functions, due to the different types and amount of data required for these functions.

It is noteworthy that there are more case studies within the standard specs, but mentioned here have similar characteristics to the other, so that the data can be obtained an overview of the new features and functionality that should be included within the standard.

VI. DISCUSSIONS

The aim of this work was to analyze the advantages and to detect the challenges of interoperable remote control support. In the case of interoperability, the advantages to be gained are innumerable and advances in wearable devices meet the requirements for its implementation. The benefits for manufacturers also are diverse, this shared experience in the development of devices, which reduces production time and production costs, and as a final result is devices capable of interacting with devices from other manufacturers. In the case of consumers the ability to select the best device at low cost from a wide range of possibilities and the ability to integrate a personal network with different devices at a lower cost is the main advantage. For remote control the results are useful to define a generalized configuration and control service over the existing Optimized Exchange Protocol (ISO/IEEE 11073-20601). Some of the use cases shared some common settings and modes. Therefore, future lines of this research includes to extend this results to model a configuration and control service within the context of ISO/IEEE 11073 standard to meet the requirements of the new use cases presented in this paper.

ACKNOWLEDGMENT

The authors would like to thank to Personal Health Devices Working Group (PHD WG) members for their guidance.

REFERENCES

1. J. L. Monteagudo, O. Moreno, eHealth for Patient Empowerment in Europe, [Online]. Available: http://ec.europa.eu/informationsociety/newsroom/cf/itemdetail.cfm?item_id=3448. Last access: 09/12
2. L. Kun, "Interoperability: The cure for what ails us (government affairs)" *IEEE Engineering in Medicine and Biology Magazine*, vol. 26, no. 1, pp. 87-90, 2007.
3. "Health Informatics-Point of care, Part 10201: Domain Information Model" IEEE Std 11073-10201, 2004.
4. "Health Informatics-Point of care, Part 20301: MDAP optional package remote control" IEEE Draft 11073-20301.
5. Continua Health Alliance. http://continuaalliance.org, Last visit: 9/12.
6. IEEE Std 11073-20601™-2010. Application profile—Optimized Exchange Protocol. IEEE Std 11073-104xx. Part 104xx: Device specializations.
7. M. Clarke, "Using IEEE 11073 Standards to Support Biomedical Engineering Research," in *34th Annual Int Conf of the IEEE Engineering in Medicine and Biology Society (EMBC2012)*, San Diego, 2012, ISBN: 978-1-4244-4120-4
8. M. Martínez-Espronceda, I. Martínez, L. Serrano, S. Led, J. Trigo, A. Marzo, J. Escayola, and J. García, "Implementation methodology for interoperable personal health devices with low-voltage low-power constrains" *IEEE Transactions on Information Technology in Biomedicine*, vol. 15, no. 3, pp. 398-408, 2011.
9. Lasierra, N., Alesanco, Á., García, J. "An SNMP-based solution to enable remote ISO/IEEE 11073 technical management", *IEEE Transactions on Information Technology in Biomedicine*, 16 (4), art. no. 6178799, pp. 709-719. 2012.
10. M. Martínez-Espronceda, S. Led, J. Redondo, A. Baquero, L. Serrano, "Adoption of ISO/IEEE 11073 in Electrocardiogram (ECG) Recorders: Lessons Learned," in *34th Annual Int Conf of the IEEE Engineering in Medicine and Biology Society (EMBC2012)*, San Diego, 2012, ISBN: 978-1-4244-4120-4
11. IEEE Std 11073-10406™-2011. Part 10406: Device specialization—Basic electrocardiograph (ECG) (1- to 3-lead ECG).
12. P11073-10424 - Standard for Health informatics - Personal health device communication - Device specialization - Sleep apnoea breathing therapy equipment [Online]. Available: http://standards.ieee.org/develop/project/11073-10424.html. Last access: 01/13.
13. "Health Informatics-Personal health device communication Part 10421: Device specialization- peak expiratory flow monitor (peak flow)," IEEE Std 11073-10421-2010, pp. 1-63, 19, 2010.

Author: Hector Gilberto Barron Gonzalez
Institute: FIME-UANL
Street: Pedro de Alba SN
City: San Nicolas de los Garza
Country: Mexico
Email: mcgilbertobarron@hotmail.com

Low Cost, Modular and Scalable Remote Monitoring Healthcare Platform Using 8-Bit Microcontrollers

F.B. Cosentino, J. Marino-Neto, and F.M. Azevedo

Federal University of Santa Catarina, Florianópolis, Brazil

Abstract—Medical telemetry is an attempt to reduce healthcare costs while providing ubiquitous features. Several solutions include desktop computers operated by the patient or health care professional, while others use wearable embedded devices, often with ARM or more complex processors, and an online server is a very common choice. In this context, this paper presents a system for vital signs monitoring, strongly focused on low-cost and flexibility over different scenarios. Instead of a complex central processor, the system is build entirely using 8-bit microcontrollers in a distributed, modular and scalable topology. Acquisition modules can measure signals that can be represented as a single updateable number (such as heart rate, body temperature and blood pressure) and can be in any number up to 100. Transmission modules carry this information over radio or physical links, one module for each acquisition module. One main module reads data from all sources and displays in an embedded webserver. All the information can be monitored using any device running a javascript-enabled browser, and no computers are required. The system was verified with reference signals, and processing times were measured. Results show how the data rates vary with each scenario, and respective equations have been found. The scalable modular flexibility might be useful in a systematic approach, with sources being constantly inserted and removed from the system.

Keywords— Telemedicine, u-health, monitoring, vital signs, 8 bit.

I. Introduction

The increasing prevalence of long-term conditions is a challenge to healthcare systems.[1] "Ubiquitous Health" can be defined as medical services over wired and wireless technologies, available anytime anywere, and might be an answer to this challenge.[2] U-Health devices can not only be used in a healthcare facility, but also to provide online home-care services. Among other features, remote monitoring of patients is a common implementation. Such a service can reduce hospitalization and increase quality of life.[3]

Radio transmission provides more freedom to the patient, but freedom should also be considered for the health care professional. By implementing a webserver, data can be accessed from anywhere using common web browsers, thus eliminating the need for custom client softwares.

Among existing solutions, many use desktop computers, and ARM or more complex processors. Often, a computer is used as a server to present data over internet. Having an embedded server in a microcontroller can reduce costs by not having a computer in the system at all.

A *modular* system can be defined as having parts of its function taking place in separate parts (modules) with defined tasks, regardless of the ways each module uses to accomplish its task. Being able to change modules without affecting the task outputs allows the system to be easily adapted (like chosing a radio technology instead of another). Also allows a system to be *scalable*, by adding or removing modules as needed.

II. Materials and Methods

For the proposed system, data types and three different stages were defined: data *acquisition*, *transmission* and *display*. Modules were designed to each stage, as well as data protocols between them.

A. Data Types

In order to design message protocols, three data types were defined. Most signals of interest will fall into one of these. They are related to handling requirements of the numerical values, and not to the physical nature of the source.

State: the information is of the state type when a new sample replaces any previous values. Usually, the information is an intensity, or an amount. An example would be heart rate.

Sequence: the data represent a sequence when a new sample is added to a stream of values. Usually, the information is a waveform itself, or the variability of a state over time. An example would be electrocardiogram waveshapes.

Event: the information is of the event type when the interest is on the moment in which something ocurred, not on its value. This might be requesting an intervention, or just a record to guide further actions. One example would be time of day in which medicine was administered.

In this system, state and event data types were implemented. Also, room was left in code for handling the sequence data type.

B. Data Acquisition Stage

The first stage is responsible for retrieving data and converting it into a format that is independent of the data type or source. Devices in this stage were called *acquisition modules* and take care of all calculation required to convert measurement values into desired format (such as frequencies calculated from discrete events). They are also responsible for providing identification for the data source, with two labels (a name and an optional group). Data in this stage are sent to the next stage, without any buffers, through an asynchronous serial port at 19200bps.

This stage is scalable, one module for each data source, so it was kept as simple as possible to keep costs low. The PIC 16F628A microcontroller was chosen, due to it's high availability and low cost in Brazil (where this work took place). The acquisition modules can run on a battery or powered by the next stage, depending on transmission technology.

C. Data Transmission Stage

The second stage is responsible for receiving messages from the acquisition modules, transmitting them over a technology of choice (wired or wireless), recovering data from the messages, storing in a buffer and providing them to the next stage, in a synchronous master-slave bus.

Devices in this stage are called *transmission modules*, and perform no data changes. This stage includes any radio modules of choice and a microcontroller in the receiver side. This stage is scalable, being one transmission module for each acquisition module, and the output messages from all transmission modules are sent over the same bus.

Two radio options were tested: bluetooth serial modules based on CSR BlueCore4 chips, running HC-05 firmwares, and MaxStream XBee serial modules. In both cases, modules were paired to provide transparent serial links. Wired transmission was tested as well.

The PIC 16F73 was chosen for the microcontroller in the receiving side, having similar advantages of the one chosen for the first stage, while having synchronous serial capabilities. In case of wireless transmission, the transmitting side is powered by the first stage, while the receiving side is powered by the third stage. In case of wired transmission, all parts (including the respective acquisition module) are powered by the third stage.

D. Data Display Stage

The third stage is responsible for reading the buffers in each transmission module, and displaying it in a webpage via an embedded webserver. There is only one device in this stage, called *main module*.

An extra microcontroller was added to the serial bus, acting as a RTC (real time clock) to provide time and date reference for the event data type. A custom firmware was written instead of a commercially available RTC chip, to take advantage of the already implemented serial bus, shared with the transmission modules. It also has a dedicated battery.

The PIC 18F87J60 was chosen for the main module, due to its embedded ethernet controller and flash memory size. TCP/IP libraries provided by the microcontroller manufacturer were used. For the RTC, the microcontroller chosen was the same used for the transmission modules.

E. Messages

A simple message protocol was defined for the first two stages to communicate. One byte long commands are sent from the transmission module to the acquisition module, being able to control data flux, request or write to acquisition module's identification labels, reset the device and check status. Variable length replies are sent back from the acquisition module, notifying status, sending identification labels, or streaming measurement data. Every message sent from the acquisition module starts with a 5-bit header identifying the message purpose, in such a way that the transmission module can always correctly process the message regardless of the commands sent before (that it, messages are context-independent). When sending data, two bits are sent to identify data type (state, sequence or event), and 16 bits containing the measured value itself. The whole data message comprises three bytes.

A second protocol, also simple, was defined for the second and third stages to communicate. One byte long commands are sent from the main to the transmission module, capable of requesting information regarding the respective acquisition module (data type, labels and connection status), reading data from the buffer, and resetting devices. A variable length reply is read for each command. Since this communication is synchronous, the replies are context-dependent. The main module scan transmission modules sequentially to retrieve their respective buffers. Devices inserted and removed from the bus can be detected while the system is running.

The messages to the RTC are limited to reading and updating the current timestamp.

F. Verification

In this work, *verification* is considered as a confirmation that the system is performing as supposed to (in contrast to a *validation*, being a confirmation that the system would satisfy customers). Verification was accomplished by presenting a known information as input, and comparing to the respective output shown in the embedded webpage. This was done for both state and event acquisition modules, and for wired, Bluetooth and XBee transmission modules. For the state data type, constant frequencies were generated for a set of values, as a hypothetical heart rate, and inputted into an acquisition module performing period-to-frequency conversion. For the event data type, the value read from the RTC was shown on screen through the embedded webserver, used by an observer to press a button in specific moments, inputted to an acquisition module sending simple event notifications.

G. Evaluation

In order to find the transmission capabilities, the theoretical maximum was calculated from chosen baudrate for messages sent by the acquisition modules. A microcontroller output was toggled in the beginning of each communication session in order to measure loop time including bus activity between main and transmission modules. All time measurements were taken during two minutes each, and the considered value was the average duration of all loops during this time. A transmission module was programmed to transmit fake data in all data requests, therefore keeping data reading times fixed, in order to measure average data transmission duration for each buffer length.

Equations were found relating number of modules and total loop time and number of samples in the buffer. Since the system is scalable – and therefore might work in different module quantities, data types and rates – these equations are expected to provide means of calculating the system limits for each specific scenario.

III. RESULTS

A. Verification

Constant reference frequencies, respective equivalent heart rates and values presented by the system using a Bluetooth transmission module are shown in Table 1. These values are representative, since wired and XBee modules had very similar results.

Table 1 Verification of state acquisiton over Bluetooth transmission

Frequency (Hz)	Rate (bpm)	Presented value
0.458 0,005	27.48	27
1.000 0,005	60.00	60
2.000 0,002	120.00	120
3.007 0,001	180.42	180
3.653 0,001	219.18	219

Table 2 shows time values in which the observer has pressed the button, and respective values presented by the system, using a bluetooth transmission module. These values are representative, since wired and XBee transmission modules had very similar results.

Table 2 Verification of event acquisiton over Bluetooth transmission

Observed time	Presented time
22:49:05	22:49:05
22:52:50	22:52:50
22:53:13	22:53:13
22:53:32	22:53:33
22:55:02	22:55:02

B. Evaluation

Since each data sample is sent from the acquisition module in a message consisting of three bytes, the serial channel running at 19200bps means a maximum of 640 messages (samples) per seecond.

Loop durations were captured in four scenarios regarding transmission modules and in five scenarios regarding data samples in buffer. Table 3 shows average time spent in a loop scanning empty transmission modules, for different transmission module quantities connected, and the difference between the value and the previous one. Table 4 shows loop time retrieving different (constant) number of samples in the buffer of one single module, and the average time associated with the last sample (time difference divided by sample difference, related to previous values). In both cases, the loop time includes the time spent reading the RTC timestamp and running internal tasks.

Table 3 Loop time scanning empty transmission modules

Number of modules	Scanning time (ms)	Difference (ms)
0	5.88	-
1	7.36	1.48
2	8.84	1.48
3	10.32	1.48

The total loop time for empty transmission modules can be found using Equation 1, while the loop time for one transmission module can be found using Equation 2

$$T_{Empty} = 5.88 + 1.48 \times N_{TM} \quad (1)$$

Table 4 Loop time for fixed number of samples in one module

Number of samples in buffer	Average scanning time (ms)	Last sample time (ms)
0	7.348	-
1	7.544	0.196
2	7.864	0.320
3	8.184	0.320
10	10.44	0.322

$$T_{1TM} = 7.224 + 0.322 \times S \quad (2)$$

T_{Empty} is the loop time for empty buffers, N_{TM} is the number of transmission modules, T_{1TM} is the loop time involving one transmission module, and S is the number of samples in module's buffer.

IV. DISCUSSION

Based on Tables 1 and 2, the authors considered the system verification as successful. In the state test, the numbers are rounded. However, the exact rounding point could not be determined, since the corresponding rates for all frequencies in Table 1 have decimal parts below 0.5. It can be anywhere between 0.48 and truncation. In the event test, time is resolved to seconds, and the greatest error found (in Table 2 and in all other tests) was 1 second. Therefore, systematic error can be anywhere between 1 second above or below. Also, the human observer pressing the button is a source of error, not measurable.

As for system data capacity, for one module, Equation 1 results in 7.36ms, while Equation 2 contains 7.224ms, obtained from Table 4 values. This sugests an extra time spent in processing an empty buffer. If all modules contain data, the loop time for any scenario will be given by Equation 3:

$$T_{Total} = 5.88 + \sum_{i=1}^{n}(1.344 + 0.322 \times S_{TM_i}) \quad (3)$$

where T_{Total} is the total loop time, n is the number of transmission modules, and S_{TM_i} is the number of samples buffered in transmission module i. Determining the number of samples is not trivial, but can be found for predictable sources, being the number of measurements during T_{Total}.

For a system having one transmission module holding 6 samples on each read operation, from Equation 2 the loop time will be 9.156ms, corresponding to a reading frequency of 109.218 reads/s, which means more than 655 samples/s. The acquisition module can send a maximum of 640 samples/s, being therefore the maximum data rate for a single source.

V. CONCLUSION

Having independent parallel processors, the system can work under an *isolated problem* approach: in case some external event (such as physical damage) affects a module, all the rest of the system keeps properly running. Being scalable, customers only have to pay for the modules they really need, and in health care facilities, the system can be instantly adapted to patient quantity changes.

This flexibility suggests systematic applications instead of one-person equipments. Since sources can be inserted or removed from the system at any time, patients entering and leaving a building might be monitored in real time at a facility level. The protocols can handle up to one hundred sources, being many data sources from one patient or from many.

A health care professional's analysis and test is still required for proper validation and to conform the system to specific local needs.

ACKNOWLEDGEMENTS

We thank the National Council for Scientific and Technological Development (CNPq) and the Biomedical Engineering Institute – Federal University of Santa Catarina, for their support.

REFERENCES

1. Murray E et al. Why is it difficult to implement e-health initiatives? A qualitative study *Implement Sci.* 2011;6:6.
2. Kim J et al. Development of Implementation Strategies for u-Health Services Based on the Healthcare Professionals Experiences *Telemed J E Health.* 2011;17:80-87.
3. Park D K et al. Telecare System for Cardiac Surgery Patients: Implementation and Effectiveness *Healthc Inform Res.* 2011;17:93–100.

Author: Fernando Mendes de Azevedo
Institute: Biomedical Engineering Institute
 (Federal University of Santa Catarina)
Street: Campus Universitário - Trindade
City: Florianópolis
Country: Brazil
Email: azevedo@ieb.ufsc.br

Adaptive Healthcare Pathway for Diabetes Disease Management

Giuseppe Fico[1], Alessio Fioravanti[1], Maria Teresa Arredondo[1], Chiara Diazzi[2], Giovanni Arcuri[2], Claudio Conti[2], and Giampiero Pirini[2]

[1] Life Supporting Technologies, Technical University of Madrid, Spain
[2] Azienda Unitá Sanitaria Locale di Modena, Italy
gfico@lst.tfo.upm.es

Abstract—The current clinical pathway in health facilities is affected by two strong limitations in care delivery: the absence of ICT tools that may let the physician monitor continuously patient's clinical evolution and the impossibility of managing in an effective way the complex parameters that could lead to a patient's care personalization. The Adaptive Healthcare Pathway (AHP) concept developed and integrated in a technological platform for diabetes disease management, implements a shift from a healthcare organization-centred care to patient-centred care. The system was tested in a small-scale exploratory study. Results, focused on how the AHP may have influenced the system usage and outcomes, are provided.

I. INTRODUCTION

The exponential growth of the healthcare costs in the last decades is leading to short, medium and long term changes in terms of resources restraints, improvement in efficiency and disruptive changes respectively.

This is more evident in chronic diseases where health promotion, prevention of onset, complications and hospitalization is needed. The Chronic Care Model foresees the adoption of technologies in the entire six dimensions. eHealth solutions are thus facilitators to create efficient health care system in which health is not only an outcome but a continuous process [1]. In this process, the role that the patient has in each stage (from risk to deep complicate) and the actions needed for each stage needs to be identified.

Diabetes management is a special and representative case of chronic disease, because of the pandemic dimension that is reaching, the complication associated to the disease, patients are not timely diagnosed are not fully aware of the daily actions they need to perform, especially in the first stage, when they do not "perceive" the disease. There is not a unique care solution: a variety of strategies and techniques should be used to provide patients with adequate treatment, education and development of problem-solving skills in the various aspects around diabetes management.

From the above, it appears that improving the situation of all types of patients requires the provision of a system that:
- Monitors factors having an impact on blood glucose, such as lifestyle and the surrounding environment.
- Takes into account individual characteristics for individualizing alerts.
- Integrates with individual personal healthcare program to adjust therapy for an optimal treatment.

This paper focuses on the description of a technological platform conceived as a response to the above issues, and more in detail focuses on the Personalized Healthcare Pathway concept developed within it. The platform has been deployed in four hospitals and tested in a clinical exploratory study aiming at assessing its potential.

II. MATERIALS AND METHODS

A. The Diabetes Management Platform

The platform is the result of research and development work involving industries working in Information Technologies and healthcare sectors, diabetes clinical centres and healthcare organizations, biomedical engineers and human computer interaction experts. User-centred design techniques, clinical guidelines and state-of-the-art monitoring technologies have been combined, resulting in a holistic technological platform to support health professionals and empower patients taking responsibility over the disease. The platform is the result of an iterative process during which requirements, mock-ups and prototypes were defined and validated putting the users (patients and doctors) as co-producers of the final product. Insights and results were extracted through focus groups, face-to-face interviews and usability tests as described in [2].

Further than making easier communications between patient and doctor, our approach aimed at getting diabetes follow-up effective and avoiding intrusiveness in the patient's life.

Two main interfaces were developed: patient (Patient Monitoring Devices, PMDs) and doctor applications (Control Panel, CP). As resulted from the first iterative cycle, specific applications were developed for type I and type II patients. Two interaction modalities, Smartphone and Tablet PC were provided to both patients; the first one for inserting lifestyle and disease related activities and the second more focused on visualization functionalities and as a hub for integrating different monitoring devices.

PMDs are used to gather both physiological and lifestyle information of the patient. Data is automatically synchronized with the CP.

The Control Panel is a frame and a tool of work to be shared by the Treating Professionals (TP) involved in the follow up of patients. The CP allows them performing classification and triage, structuring the clinical processes, analyzing meaningful presentation of aggregated data, and prescribe treatments that are based on therapy, education, lifestyle, goals and recommendations.to

In between, the Central System (CS), developed as a Service Oriented Architecture (SOA), let the different components interact each other, over the Internet, through SOAP messages, using HTTPs and XML serialization and other Web-related standards and interoperability health profiles (HL7, IHE profiles and Continua IEEE X73 device standards). The SOA solution let the technological infrastructure to adopt new services and applications needed in the future for further improvements in the management of patients, in or out the HC system.

The way the different components are used by the involved actors, how the services are enabled and configured and how the care process is determined by the adaptive healthcare pathway concept is explained in the following sections.

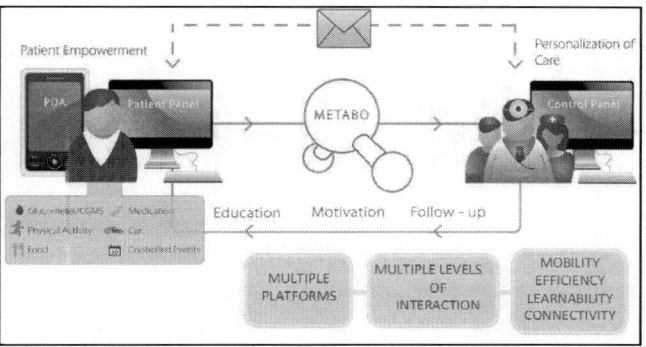

Fig. 1 Service Oriented Architecture Platform, view of the main modules

B. Personalization of the Healthcare Process

The personalization of the healthcare process is based on the the following elements:

Current Clinical Pathway: The "as-is" pathway, which represents current practice and introduces the pathway concept, highlighting its constraints.

Personalized Clinical Pathway: The current clinical pathway adapted and configured according to the current scenario and phase.

Scenario: It is the combination of the type of diabetes (type 1, type 2 not insulin treated and type 2 insulin treated) and the segmentation in six main areas of focus (constituting a total of 18 scenarios), depicted as a driver to design personalized strategies of care and solutions tailored to patient needs: 1. *Changes in the Environment*, recognizing changes that are clinically significant (workings shifts, frequent travelling, etc.); 2. *Physical Activity*, focusing on the effect of physical activity on glucose metabolism; 3. *Sudden Hypoglycemic Events*, for the early detection of hypoglycemia, especially in risky situations; 4. *Lack of Motivation*, when motivation may be the main cause of failure; 5. *Unstable Diabetes Control*, in relation to cases of sustained hyperglycaemia or unstable glucose concentrations with sudden alternations between hyperglycemia and hypoglycemic crises; and 6. *Commorbidity Disease Management*, when disease may affect progression and response to treatment of another disease.

Process Types and Switching Rules: The processes that are stored inside the platform are represented by all the steps (say events, appointments) patients can undergo within their treatment. There are three types of process:

Dynamic Process Type: Dynamically created using the data inserted by the patient representing the "past reality".

Static process type: expected personalized description of behavior according to condition and treatment.

Guidelines: Static prescriptions that are valid for a long period of time, thus providing coverage of events that may happen and serve as a starting base for further steps.

Clinical pathways are tools used to ensure high quality of standardization, aiming at reducing variability and improving clinical outcomes. One of the most recent and reliable publication is "Diabetes Clinical Pathway guidelines [3][4] developed by the British National Health Service and based on the 2008 American Diabetes Association guidelines. This model has been taken as reference for our work. Even if focused on T2DM, the approach can be generalized and represent a suitable framework. Five milestones are foreseen, we added a pre-diagnostic phase:

M0 - Prediagnostic Phase: The general practitioner (GP) prescribes a visit to a patient diagnosed with diabetes.

M1 - Diagnostic Phase: Patient has the first contact with an anti-diabetic hub. He undergoes to examinations, lab-tests and professional advice: the makes final diagnosis is done.

M2 - Educative Phase: According to patient's lifestyle, the medical staff teaches patient the way of behaving about glucose level maintenance, exercise, nutrition, medication, complications and the importance of regular reviews.

M3 - Treatment Management Phase: Patients are requested to manage their disease through food intake, drug intake, glycaemic levels and physical activity.

M4 - Complication Risk Management: To prevent or manage complications, specialist visits are planned.

M5 - Maintenance Phase: This is the routine pathway for diabetic patient follow up. Endocrinologist can request medical advices, modify the treatment and plan visits.

Based on the above milestones and the 18 scenarios, the following work was done:

1. Relationships between scenarios and milestones of were defined, merging user requirement research, evidence-based recommendations and clinical guidelines:

- For each scenario, a clear condition of belonging was set.
- Actions for each scenario-milestone were identified.
2. A deep analysis and synthesis of results, in order to build framework and rules to design the clinical pathway:
- a bottom-up analysis of all the relationships, divided into specific ones (SPECIFIC PATHWAY) and common to the belonging milestone (STANDARD PATHWAY).
- The classification criteria and the parameters through which a patient is assigned to one of the scenarios.

C. The Adaptive Clinical Pathway

The personalization of care is based on the following: during the first control (**M1 - Diagnostic Phase**), the endocrinologist, according to tests results and patient's lifestyle, can place the patient in Changes in the environment, Physical exercise and Comorbidity disease management segments since, at a very starting phase, he cannot say that the patient belongs to one of the other segments (Sudden Hypoglycemia, Lack of motivation, Unstable diabetes). For each scenario it is possible to define a specific clinical pathway, which consists in a dedicated sequence of steps (**M2, M3, M4, and M5**). The assignment of the clinical pathway automatically generates the necessary changes in the CP and on the PMD.

The five milestones have been transformed in tabs that the care professional has to go through when using the CP.

Let one consider a patient prescribed by GMP to an anti-diabetic hub visit, after the "pre-diagnostic phase". Patient makes physical activity in a semi-agonistic context: 5 days a week, duration time 60 minutes, with an intensity of 55-69% HR Max or above 5 METS. These conditions suggest GMP to classify him in "T2DM- IT- Physical activity" scenario, for which a specific Pathway is applied. After some months, the patient is diagnosed with a peripheral neuropathy. An alert message communicates the result to the endocrinologist who updates patient's pathway: he refines M4 phase - Complication\risk management: improvement of glucose control, prescription of new drugs for treatment, changes in physical activities and next review visit in 1 month.

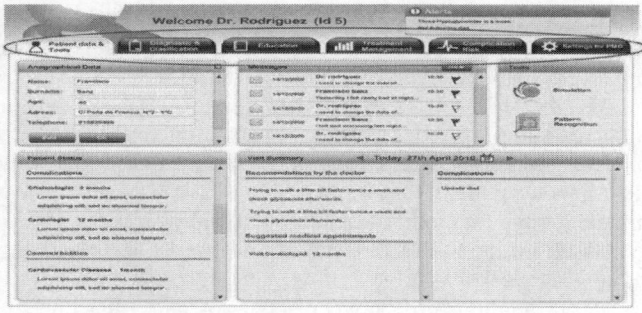

Fig. 2 the follow-up process structure is intrinsically implemented in the Control Panel. The tabs in the upper part are based on the 5 Milestones.

D. Care Plan Evaluation

Glycosylated hemoglobin (HbA1c) is the diagnostic criterion to assess the clinical status of a diabetic patient. It is typically measured every three or six months.

Based on this, we defined HbA1c significant ranges with minimum goal performance, and defined negative performance related expected action, to be suggested to caregiver for reference. For instance, lifestyle re-training and drug doses adjustment are applied if a fair poor control (Hba1c between 6.1 and 7.0) is the intermediate result.

Once defined the current clinical practice expected outcome, a set of evaluation sub-steps has been defined and integrated in the CS, the CP and the PMDs, to provide the TP with an overview of the current behavioral and clinical status and support more effective decision.

Evaluation 1 - Data Analysis: TP visualizes patient's current status, by comparing clinical and behavioral parameters against each other. Based on this, the same window allows care plan retuning (Figure 2).

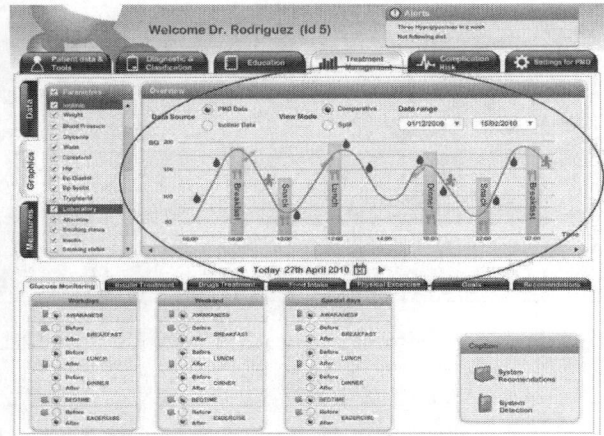

Fig. 3 CP data visualization module. Continuous glucose progressions aggregated with Food, Drugs Intakes and Physical Activity

In addition to traditional values:

1. Automatic and continuous delivery of clinical values (BG, drug intake, food intake, physical activity), through the PMDs 2. Information about patient-PMD interaction and treatment compliance and 3. Information about training activities.

Evaluation 2 - Patient Compliance: PMD interacts with patients by providing automatic messages delivered in case patients are not adhering to care plan and if they are not annotating measurements and intakes, according. Clinicians can access to these information through tables providing an overview of patient compliance (Figure 5). Thanks to this, doctors are able to understand if patients have been inserting measurements, if the adherence to the treatment is high and the most recurrent triggered alert event. During focus groups many doctors were stating: "most of patients lie during visits": with this tool we aim at making it harder for patients to do and easier for doctors to spot.

Evaluation 3 - Learning Module: It is composed of educational material and quizzes, organized in about 30 topics. The content, whose themes encompass the main information related to diabetes and self-management, is organized in web pages, stored in both CP and PMDs. Physicians are able to customize and update the contents for each patient. Quizzes are composed of questions related to assigned topics, each one including three levels (beginner, medium, advanced).

Evaluation 4 - Communication Module: Doctors can enable a tool that offers bilateral communication with patients through text messages, helping the physicians in "real-time" cases and enabling collaborative decision-making.

Evaluation 5 - Patient Self-assessment: Upon CP parametrization, PMD support patients self-evaluations through a Learning Module, Graphic/Diary (giving an overview of the inserted values compared with the prescribed ones), Feedback (messages are triggered depending on what they are doing in terms of a) compliance and b) usage) and Goals (if patients and doctors agree on some goal to meet, i.e.: weight loss, increase education level).

III. RESULTS AND DISCUSSION

An intervention study was carried out between February and September 2012 to validate the platform in terms of acceptance, usability and fit with current health care processes, to have a preliminary quantification of the impact on clinically and laboratory relevant parameters as well as to support a future translation to full scale regulatory-compliant studies. It was multi-center, multi-national, open-label, randomized, two-parallel-group exploratory study comparing the effect of the continuous use of the system vs standard care without the use of the system for four consecutive weeks on quality of life and clinically relevant parameters associated with glucose control. The intervention group was provided with the PMDs and followed up with the CP. The centres involved were public hospitals in Parma, Prague, Modena and Madrid.

22 physicians, 3 nutritionists and 1 education nurse have been involved. The results described herein are focused on the analysis of the CP usage, extracted from the study of the logfiles of CPs, the server and the entries in the DBs coming from the PMDs and on questionnaires to be filled before and at the end of the study. The intervention group was composed of 27 patients: 9 of them were explicitly assigned to a scenario by the TPs, the other 17 have been assigned a posteriori, in order to correlate, for all the patients, the clinical pathway recommended by the system, with the real treatment established by the doctor. Table 4 shows the areas of focus associated to each patient T1DM and T2DM.

Table 1 Areas of focus associated to the patients during the trials study.

	Sudden hypoglycemic events	Unstable diabetes control	Physical activity	Comorbidities disease	Changes environment	Lack of motivation
T1DM	1+(2)	4	1+(1)		(3)	(2)
T2DM	(2)	1	1+(3)	(3)	(2)	1

Results on CP-user experience have shown a very positive (above 4 on a 5 scale) trend on system usage and comprehension (70% of the total), identification with user expectations (86.7%), user stimulation on knowing all the applications skills (82%), attractiveness with regards to the system design (85%).

The average time per patient was quite high, during the first phase (about 10 minutes). After a learning phase, this decreased to 5 minutes for patient. This corresponds to the "desired" period defined by the doctors.

The section/tab most used was "Prescriptions and Clinical Data", about 90% with respect to the other sections, "Patient data & tools" (3%), patient lifestyle (5%), "Educational Messages" and "Complications" (2%).

Correlations between the scenario, the way it was treated and the corresponding meta-outcomes at the end of the trial (since the study was for about one month per patient we excluded the Hba1c related evaluations) have been performed.

Compliance: The way patients are adherent to prescription has been assessed, by comparing the first two weeks with the last two. Results show good follow-up in physical exercises and diet with an increase of 7% to the beginning of the trial, but also a progressive reduction, from 15 to 24% in glycemic control and medication intake (this may also be related with the fact that glucose measurements were sent to the CP through the devices and some patients reduced their insertions.

We compared the overall adherence vs. the focus areas, the more compliant was the Lack of motivation with an average of 93%, as showed in the table 2.

Table 2 Adherence vs areas of focus

	Sudden hypoglycemic events	Unstable diabetes control	Physical activity	Comorbidities disease	Changes environment	Lack of motivation
CMPL. AVR.	87.5	84.7	84.3	89.2	86.5	93.1

Educational level assigned and quizzes: 65% of educational content assigned by the physicians "Educational Message" section coincided with the topics suggested by the system, based on the scenario. Patients' level of knowledge increased: 62% of quizzes were filled in by T2DM, more stimulated on improving their level. Furthermore, the scores of T2DM were double with respect to T1DM, being good at changing from "beginner" level to "advanced".

Glucose monitoring System Recommendations validations: the correlation between the automatic prescriptions suggested by the system, according to the pathway associated to the patient, and the real treatment prescribed by the doctor for glucose monitoring was observed: the 82.4% of the daily prescriptions made for glucose monitoring corresponded to the ones suggested from the platform.

Messages: T2DM patients were more active on chatting with their TP (60.4%) with respect to T1DM (39.6%). Sudden Hypoglycemic event were the most active on sending message.

Table 3 Messages sent from intervention group according to the associated focus area.

	Sudden hypoglycemic events	Unstable diabetes control	Physical activity	Comorbidities disease	Changes environment	Lack of motivation
Exchanged messages	35	12	21	15	16	17

IV. CONCLUSION

The METABO project was created to support the health maintenance of diabetic patients in their living environments. The preliminary results reported here have shown the feasibility, the acceptance and potential effectiveness of the platform. The adaptive healthcare pathway had an influence on how the CP was designed and on the rules to create the different DSS components. Personalization, customization and continuum of the care process have been assessed. Further research need to be done on the information collected, correlating the technical data, with the usability results and the clinical outcomes.

ACKNOWLEDGMENT

The authors wish to acknowledge the METABO consortium for their valuable contributions to this work. This project [5] is partially funded by the European Commission under the 7th Framework Program, Theme ICT-2007.5.1, Personal Health Systems for Monitoring and Point-of-Care Diagnostics – Personalized Monitoring, grant agreement number 216270.

The authors declare no conflict of interest.

REFERENCES

[1] "Executive summary: Standards of medical care in diabetes—2010". Diabetes Care 33 (Suppl 1): S4–10. January 2010. doi:10.2337/dc10-S004. PMC 2797389. PMID 20042774.
[2] A user centered design approach for patient interfaces to a diabetes IT platform. Fico et al, Conf Proc IEEE Eng Med Biol Soc. 2011;2011:1169-72.
[3] Diabetes Care Pathway, version 3, British National Health Service, April 2008
[4] Diabetes Care Pathway, Enfield NHS, Version 3, April 2008.
[5] European Commission. Information Society Technologies Program. METABO project. Chronic diseases related to metabolic disorders. ICT-26270. www.metabo-eu.org

Real Time Health Remote Monitoring Systems in Rural Areas

M.G. Sánchez, R. Nocelo López, and J.A. Gay-Fernández

Universidad de Vigo, Departamento de Teoría de la Señal y Comunicaciones, Vigo (Spain)

Abstract—Real time health remote monitoring systems require the use of radio networks to transmit data from patients to a health center where that data are analyzed. However, the required radio networks may not have a good coverage, (or even not exist) in rural areas, where often an important number of elderly people with chronic diseases live. The performance of two remote monitoring systems based on two different radio networks (digital cellular network for mobile phones (GSM/UMTS) and Digital Mobile Radio (DMR)) to provide remote monitoring services in rural areas are compared and results and conclusions of the comparison are presented in this paper.

Keywords—Aging, Patient monitoring, Home Monitoring, Rural/Frontier Counties, Remote monitoring, radio coverage.

I. INTRODUCTION

During the last years there has been a growing interest on the development of health remote monitoring systems for patients with chronic diseases [1-6]. These systems would present several advantages for the patients that could be under real-time continuous survey while remaining at their homes, but also for the Health Authorities, that could optimize the health resources. Elderly people, with increasing chronic medical condition, are often the candidates for the use of these systems.

Thanks to the radio communication networks the information regarding patients health parameters can be transmitted continuously in real time to a remote center for analysis. With radio networks this can be done even while the patient develops daily life moving at home or around home.

However, radio coverage may not be adequate or even may not exist in rural areas. This would be a problem particularly if an important percentage of population in rural areas is aging population living alone and requires the use of these systems.

Two health remote monitoring systems are compared in this paper. The first one is based on digital cellular networks for mobile phones: second (GSM) and third generation (UMTS) systems, but it is GSM the network that should exhibit better coverage, mainly due to the lower operating frequency.

The second is based on Digital Mobile Radio (DMR) an open digital radio standard for professional mobile radio (PMR). Two frequency bands were considered for this DMR network, VHF (Very High Frequency) and UHF (Ultra High Frequency). Theoretically the lower frequency network (VHF) should provide better coverage.

Several remote monitoring systems were built to test and compare the radio networks, but the sensors used to gather information were the same: a wearable hearth rate meter, a portable blood pressure meter, and a GPS (Global Positioning System) receiver. In this way the difference between the systems is the radio technology used, but not the type or amount of information.

II. SYSTEM DESCRIPTION

A. Digital Cellular Network for Mobile Phone

The system to test the coverage was built around a 3G mobile phone, a Samsung Galaxy S-II, with integrated GPS and Bluetooth. Heart rate meter was a wearable Zephyr model HxM BT connected to the mobile phone through the wireless Bluetooth link (see figure 1).

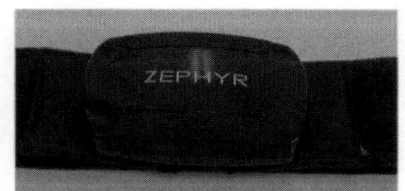

Fig. 1 Heart rate meter

The portable blood pressure sensor was an Omron model 708-BT (see figure 2), also connected to the mobile through the Bluetooth link.

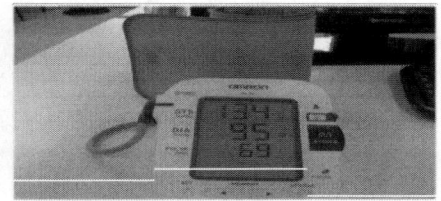

Fig. 2 Portable blood pressure sensor

An Android application was developed to gather the information from the sensors and the GPS and transmit the information through the GSM/UMTS network to a remote server where the information is stored.

B. Digital Mobile Radio

Digital radio terminals MOTOTRBO from Motorola with Bluetooth and GPS were used to build this system (see figure 3). The same sensors for hearth rate and blood pressure were connected to the DMR terminals through the Bluetooth wireless link. The information from the sensors and the terminal was transmitted to the remote database through the DMR network.

Fig. 3 DMR terminals

As DMR is a Private Mobile Radio system, and there are not commercial public networks available, a DMR base station was built both for VHF and for UHF. The corresponding radio licenses to get the privative use of the portion of the radio spectrum needed for the experiment were obtained from the Spanish Radio Spectrum Management Authority. The base station, which is shown in figures 4a and 4b, was installed at University of Vigo.

Finally the system was completed with another application to consult the data base from mobile phone, and to receive alarms in case the monitored patient leaves a predetermined area, or in case a health parameter falls out of predetermined limits. Figure 5 shows the mobile phone application with the location of the patient and the heart rate information.

Fig. 4a DMR base station antennas

Fig. 4b DMR base station

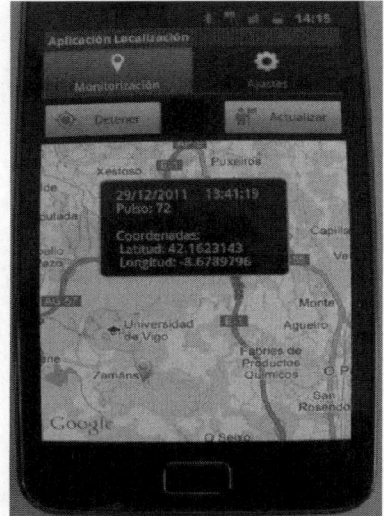

Fig. 5 Application to access the database

A complete diagram of the whole system is shown in figure 6.

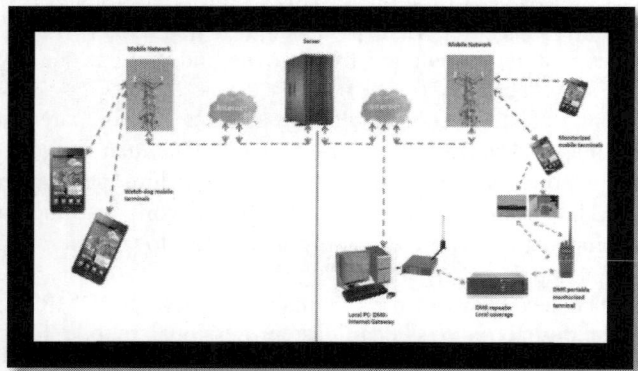

Fig. 6 Diagram of the health remote monitoring system

III. MEASUREMENT SETUP AND RESULTS

To compare the coverage area of the systems under study, the system terminals were carried in a car that was driven around the DMR base station at University of Vigo. The surroundings of the campus correspond to a mountainous rural area.

For this experiment, together with the sensors and GPS information, also the Receive Signal Strength Indication (RSSI) at the terminal was measured and transmitted to the database. Figure 7 shows an example of the path followed by the car with the received power in the DMR-VHF terminal. Received power by the GSM, DMR-VHF and DMR-UHF terminals are compared in figure 8.

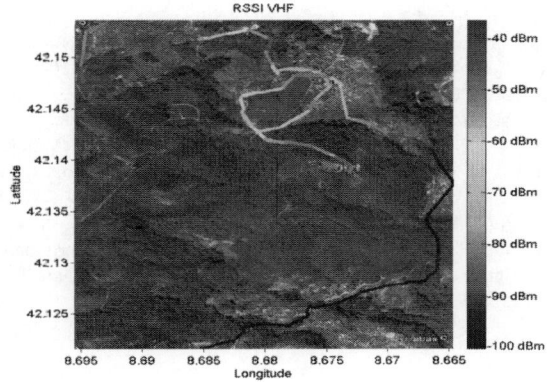

Fig. 7 An example of coverage measurement

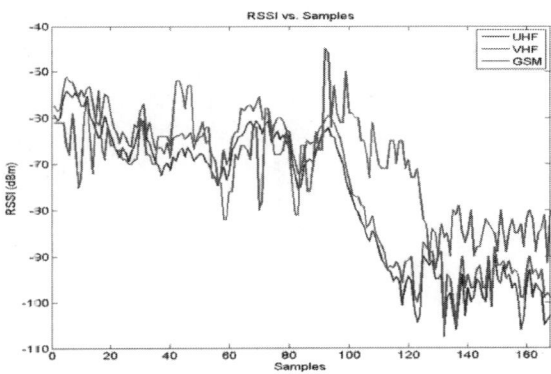

Fig. 8 Received power by the terminals

With this information an analysis was carried to calculate the corresponding Bit Error Rate (BER) of the radio link and the probability of receiving a wrong packet of information. It should be noted that, despite the signal levels are similar, GSM requires more signal level than DMR to achieve the same BER (see figure 9). This is mainly due to the different modulation schemes of both systems. While GSM uses a modulation known as GMSK (Gaussian Minimum Shift Keying) and a bit rate of 271 Kbps DMR uses 4-FSK (4 levels Frequency Shift Keying) and a bit rate of 9.6 Kbps.

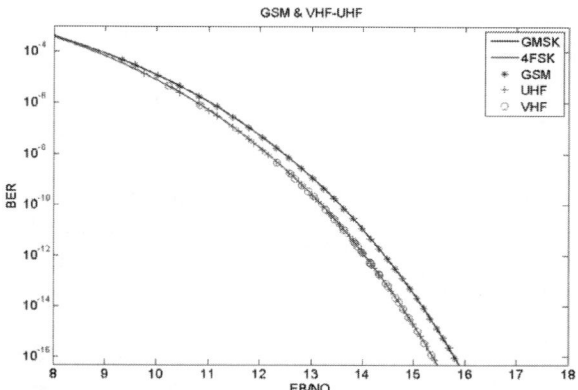

Fig. 9 Bit Error Rate of the GSM and DMR systems

Results of this analysis are plotted in figure 10, where the probability of having a BER below certain value is given for the GSM, DMR-UHF y DRM-VHF systems.

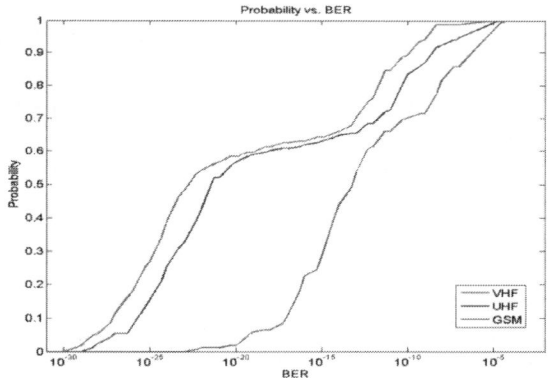

Fig. 10 Comparison of the BER probability

As can be seen in the table 1, the three systems have a high probability of having a low BER below 10^{-5}, so all three will be adequate for health remote monitoring. However, the DRM-VHF shows the better performance. As the information packet from the heart rate sensor is made of 480 bit and is transmitted at a rate of 1 packet/s, this means that we could have an erroneous packet every 30 s for GSM, every 102 s for DMR-UHF and every 266 s for DMR-VHF.

Table 1 Parameters of the three systems

	GSM	UHF	VHF
Modulation	GMSK	4-FSK	4-FSK
Bit Rate	271 Kbps	9.6 Kbps	9.6 Kbps
BER	$6.01 \cdot 10^{-5}$	$1.03 \cdot 10^{-5}$	$4.84 \cdot 10^{-6}$
Erroneous packet time	30 s	102 s	266 s

IV. CONCLUSIONS

For health remote monitoring in rural areas where no commercial mobile phone networks are available Digital Mobile Radio may be an alternative radio network that would guarantee at least as good performance as mobile phone networks.

A DMR network, however, would require the deployment of a specific base station but this may be compensated by the large DMR base station coverage areas. Depending on the base station power and the environment, the DMR coverage area could be extended tens of kilometers away from the base station.

ACKNOWLEDGMENT

Research supported by the European Regional Development Fund (ERDF) and the Galician Regional Government under project CN 2012/260 "Consolidation of Research Units: AtlantTIC" and project 10SEC322021PR.

REFERENCES

1. Weeg S (2004) Home Health and Home Monitoring in Rural and Frontier Counties: Human Factors in Implementation. Engineering in Medicine and Biology Society. IEMBS '04. 26th Annual International Conference of the IEEE, vol.2, no., pp. 3264- 3265, 1-5 Sept. 2004. DOI 10.1109/IEMBS.2004.1403918
2. Mitra M, Mitra S, Bera J N, Gupta R, Chaudhuri B B (2007) Preliminary Level Cardiac Abnormality Detection Using Wireless Telecardiology System. Digital Society. ICDS '07. First International Conference on the, vol., no., pp.14, 2-6 Jan. 2007. DOI 10.1109/ICDS.2007.34
3. Fong B, Pecht M G (2010) Prognostics in wireless telecare networks: A perspective on serving the rural Chinese population. Prognostics and Health Management Conference. PHM '10. , vol., no., pp.1-6, 12-14 Jan. 2010. DOI 10.1109/PHM.2010.5413506
4. Kappiarukudil K J, Ramesh M V (2010) Real-Time Monitoring and Detection of "Heart Attack" Using Wireless Sensor Networks. Sensor Technologies and Applications (SENSORCOMM), 2010 Fourth International Conference on , vol., no., pp.632-636, 18-25 July 2010. DOI 10.1109/SENSORCOMM.2010.99
5. Ramesh M V, Anand S, Rekha P (2012) Mobile Software Platform for Rural Health Enhancement. Advances in Mobile Network, Communication and its Applications (MNCAPPS), 2012 International Conference on, vol., no., pp.131-134, 1-2 Aug. 2012. DOI 10.1109/MNCApps.2012.34
6. Rajasekaran M P, Radhakrishnan S, Subbaraj P (2009) Elderly patient monitoring system using a wireless sensor network. Telemed J E Health, 15(1):73-9. DOI 10.1089/tmj.2008.0056

Use macro [author address] to enter the address of the corresponding author:

Author: Rubén Nocelo López
Institute: Universidad de Vigo, Departamento de Teoría de la Señal y Comunicaciones
Street: Campus Lagoas-Marcosende
City: Vigo (Pontevedra)
Country: Spain
Email: rubennocelo@uvigo.es

Detecting Accelerometer Placement to Improve Activity Classification

Ian Cleland[1], Chris D. Nugent[1], Dewar D. Finlay[1], and Roger Armitage[2]

[1] Computer Science Research Institute, School of Computing and Mathematics,
University of Ulster, Newtownabbey, UK
[2] Adidas Wearable Sports Electronics, a Trading Division of Adidas (US), Pennsylvania, USA

Abstract—This paper describes a method to improve the classification of everyday activities through detection of the location of an accelerometer device on the body. The detection of the device location allows an activity classification model, produced using a C4.5 decision tree and specifically tailored for that location, to be applied. Eight male subjects participated within the study. Participants wore six tri-axial accelerometers, positioned at various locations, whilst performing a number of everyday activities. A C4.5 decision tree was also used to detect the location of the accelerometer on the body which achieved an F-measure of 0.63. Based on this approach and applying the appropriate activity recognition model for the detected location improved activity recognition performance from an F-measure of 0.36 to 0.62, for the worst case, when using an activity model trained only one location.

Keywords—Activity recognition, accelerometry, sensor placement.

I. INTRODUCTION

Accelerometers are widely integrated into wearable systems in order to identify various activities. Previous studies have reported accuracy levels of 85% to 95% for recognition rates during ambulation, posture and activities of daily living (ADL) [1-2]. The majority of these studies have incorporated multiple accelerometers attached to different locations on the body. Whilst this provides sufficient contextual information, placing accelerometers in multiple locations can become cumbersome for the wearer and may also increase the complexity of the classification problem. For these reasons, a number of studies have opted to use a single accelerometer. Generally however, using only one accelerometer decreases the number of activities that can be accurately recognized [3].

Incorporation of accelerometer technology is becoming more common in everyday mobile devices such as mobile phones, gaming consoles and digital music players. Due to this, interest in mobile device based activity recognition is increasing.

Bieber *et al.* [4] presented a mobile phone application for identifying physical activities and estimating how many calories were expended. The majority of work on activity recognition from mobile devices assumes that the device is fixed in one location. The classifier is generally both trained and tested in this location. The phone can, however, change location on a day-to-day or much more frequent basis [5]. In such applications, changes in the location of the device may be detrimental to the performance of the classifier. This is due to the classifier no longer being able to accurately classify data from one location when it was trained on data from another [6]. Approaches for dealing with this may include the use of features which are independent of device location or the use of distinct models depending on where the device is located [7].

This paper describes a method of detecting the location of an accelerometer device on the body whilst carrying out a number of everyday activities. The appropriate activity classification model for the detected location is then used for the purposes of activity recognition.

II. METHODS

Eight male subjects volunteered to participate in the study. Subjects were members of staff and students of the University of Ulster. Subjects ranged in age from 24 to 33 (mean 26.25, sd ±2.86). All subjects provided written informed consent to participate in the study. Subjects completed a physical activity readiness questionnaire (PAR-Q) to assess their suitability to take part in the study. The study was approved by the Faculty of Computing and Engineering Research Governance Filter Committee at the University of Ulster. Subjects wore six accelerometers at various locations on the body as shown in Figure 1. Accelerometers were fixed to the body, over clothing, using elasticized strapping and holsters. This is a common method of attachment in activity recognition studies [8].

A. Data Collection

Acceleration data was collected using six Shimmer wireless sensor platforms (Shimmer 2R, Realtime Technologies, Dublin, Ireland). These tri-axial accelerometers had a range of ±6 g and sampled data at 50Hz. This sampling frequency is viewed as being sufficient for the assessment of daily physical activity [8].

Data were transmitted via Bluetooth to a notebook computer where it was saved for offline analysis. In order to achieve synchronization, data was recorded using Shimmer

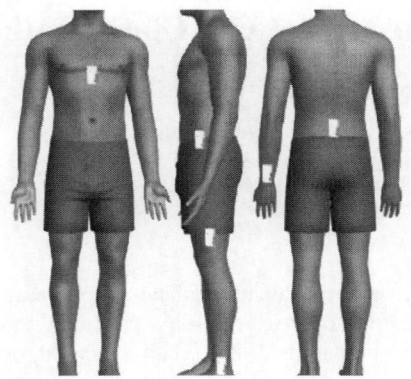

Fig. 1 Illustration showing the selected locations for the accelerometers. These include the chest, lower back, hip, thigh, wrist and foot. Accelerometers were fixed on top of clothing using elasticized strapping and holsters.

sync software (Shimmer sync Version 1.0). This synchronizes time stamp data from each of the six accelerometers. Prior to beginning the study, devices were calibrated using standard calibration techniques as described in [9].

Eight activities were studied. These consisted of whole body activities and postures including walking over ground, walking and jogging on a motorized treadmill, sitting, lying, standing and walking up and down stairs. All activities were maintained for a duration of two minutes with the exception of walking over ground and climbing stairs. These activities were carried out over approximately 60 meters and 10 flights of stairs (80 steps). These tasks were repeated in order to capture sufficient data for analysis. For treadmill based activities, users walked and jogged at a self selected comfortable speed. The maximum jogging speed was restricted to 10 km/h given that speeds above this are considered as running [10]. Data were manually labeled offline by a human observer.

B. Feature Extraction

Features were extracted from acceleration data using a window size of 256 samples with 128 samples overlapping between consecutive windows. Feature extraction on 5.12 second windows with a 50% overlap has demonstrated reasonable results in previous works [1]. This window size is capable of capturing complete cycles in repetitive action activities such as walking, running and climbing stairs.

Mean, root mean square (RMS), periodicity (energy), variance and correlation features were extracted from the x, y and z axis signal within each window. This provided a total of 15 features for each window from each accelerometer. These features have been commonly used in activity recognition studies and have been shown to provide reasonable accuracies [1, 11]. The mean acceleration value was calculated by summing the acceleration values within the window and then dividing this by the number samples within the window. The mean was also calculated in a similar manner for both the y yand z zaxis [1].

Periodicity within a signal is reflected in the frequency domain. To calculate the periodicity, the energy feature was calculated [11]. The energy feature, is the sum of the squared discrete FFT component magnitudes of the signal. Normalization was achieved by dividing the sum by the length of the window. The energy feature has been used previously for recognition of certain postures and activities [12].

Correlation is particularly useful for discriminating activities that involve movement in just one dimension [11]. For example, differentiating walking or running from stair climbing. Walking and running involves movement in one dimension whereas climbing involves movement in more than one dimension. Correlation is calculated as the ratio of the covariance between each pair of axes and the product of the standard deviations [14].

C. Classification

Activity recognition on features was performed using a decision tree (DT) based on the C4.5 rule induction algorithm (C4.5 DT) available in the Weka Machine Learning Algorithms Toolkit (Version 3.6.7). The C4.5 DT has been shown to perform well for activity recognition in previous works [1].

The classifier was trained and tested using a leave-one-subject-out protocol. In this method the classifier is trained using features from all but one subject. The classifier is then tested on the features obtained from the subject who was excluded from the training set. The leave-one-subject-out validation was repeated for all eight subjects. Population based training methods have been previously used to classify a number of activities [1]. Having a population trained activity recognition approach is beneficial as it removes the need to train the classifier on a specific individual.

The balanced F-measure was used as the performance index to evaluate the experimental results. For the test results, the F-measure is calculated for each activity at each position. The overall F-measure for the classifier is computed by averaging the F-measures for all subjects.

In order to evaluate the discriminatory power of each location, the F-measure was computed using data obtained from each accelerometer separately. The performance of the classifier at each location is presented in Table 1. Results show that the accelerometer placed at the hip was the most powerful for recognizing the eight activities studied.

Table 1 F-measure obtained using the leave-one-subject-out validation for each location. Figures presented are average F-measures for all subjects ± standard deviation

Location	Classifier F-measure (sd)
Chest	0.59 (±0.11)
Foot	0.63 (±0.23)
Hip	0.72 (±0.22)
Lower back	0.45 (±0.13)
Thigh	0.55 (±0.10)
Wrist	0.67 (±0.19)

In order to investigate the effects of training a classifier using data from one location and then the device being moved to another location, a classifier model was built using the C4.5 DT trained on data from the hip. This model was then tested on data from the foot, thigh and wrist. The performance of the classifier was then tested using the leave-one-subject-out validation method. As expected the performance of the classifier decreased with an average decrease in F-measure of 0.47, 0.29 and 0.34 for the foot, thigh and wrist, respectively (Table 2).

Table 2 F-measure for each classifier when trained on the hip and tested on the other three locations; foot, thigh, wrist.

Tested on data from:	F-measure
Hip	0.72 (±0.22)
Foot	0.25 (±0.12)
Thigh	0.43 (±0.08)
Wrist	0.38 (±0.17)

As previously discussed, the position of the accelerometer can change throughout the day. In an attempt to alleviate this problem, the current approach uses the C4.5 DT to identify the location of the accelerometer on the body. The activity recognition model for that detected location is then applied, on an instance by instance basis, in order to improve the classification accuracy. The same 15 features from the activity recognition study were used as inputs to the classifier. Again, leave-one-subject-out validation was applied.

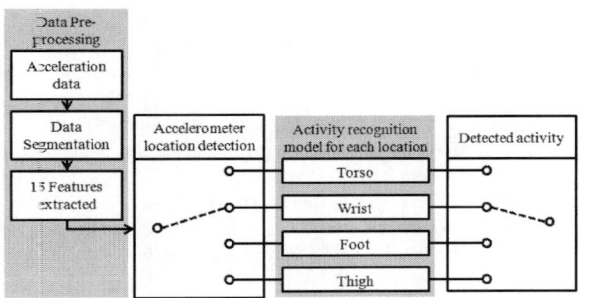

Fig. 2 Flow diagram illustrating the process used to produce and select the appropriate activity classification model.

To test this technique the model from the DT was used to detect the location of the accelerometer from 1440 instances of features from each of the 8 subjects (30 from each location). The activity recognition model from the detected location was then applied to the data on an instance-by-instance basis. For example, if the location DT detected that features were from the hip, then the activity recognition model from the hip was applied to that instance. This process was carried out manually. Figure 2 presents a summary of the approach.

III. RESULTS

This Section presents the results of the C4.5 DT to detect the location of the accelerometer. Following this, results demonstrating the effect of applying the activity recognition model for specifically detected locations will be presented.

A. Detecting Accelerometer Location

The C4.5 DT produced an average F-measure of 0.57 for detecting the location of the accelerometer on the body (Table 3). The confusion matrix, indicates that the classifier confused data from the lower back with other locations such as the hip and chest (Table 4). This may be due to similarities in body acceleration obtained from these locations as they are all located close to the body's centre of mass.

Table 3 Average F-measure of the C4.5 classifier to detect the location of the accelerometer for all six locations studied.

Accelerometer location	F-measure (sd)
Chest	0.64 (±0.27)
Foot	0.67 (±0.23)
Hip	0.66 (±0.14)
Thigh	0.61 (±0.14)
Wrist	0.52 (±0.27)
Lower back	0.33 (±0.18)
Average	0.57 (±0.23)

Table 4 Confusion matrix from the C4.5 decision tree for classifying the location of the accelerometer. All six locations are used as classes.

Classified as →	Chest	Foot	Hip	Thigh	Wrist	Lower back
Chest	1223	36	224	158	138	181
Foot	53	1312	87	95	97	316
Hip	108	52	1452	58	113	177
Thigh	74	131	58	1166	339	192
Wrist	100	74	287	161	1090	248
Lower back	191	337	244	225	302	661

By amalgamating data from the hip, chest and lower back into one class known as the Torso, the accuracy of the classifier was improved with an average F-measure of 0.63. Therefore, the subsequent activity recognition experiments were carried out using data from four locations; Torso, Foot, Thigh and Wrist, with the Chest, Hip and Lower back data combined under the single location of Torso.

Table 5 F-measure for the C4.5 decision tree in classifying the location of the accelerometer. Hip, chest and lower back classes are combined into one class referred to as torso.

Accelerometer Location	Average F-measure (sd)	
Torso	0.76	(±0.06)
Foot	0.69	(±0.24)
Thigh	0.60	(±0.12)
Wrist	0.48	(±0.28)
Average	0.63	(±0.21)

B. Activity Recognition

For the 11,520 instances tested, the detected location was the same as the actual location 67.96% of the time. Table 6 presents a summary of the classifier F-measures obtained using the activity classification model from the detected and actual locations, as well as that from each of the four investigated locations.

The F-measure obtained using the detected location was comparable to that obtained using the actual activity recognition model for that location. Using the DT to detect the location of the accelerometer improved the activity classification in comparison to always using the model from the same location (i.e. always using a model built with data from the torso, foot, thigh or wrist). The classifier F-measure improved from 0.36 when using only the thigh activity model to 0.63 when using the model for the detected location.

Table 6 Average F-measure of the activity recognition using the model for the detected location, the model for the actual location and the model for each of the four locations.

	Average F-measure					
	Detected	Actual	Hip	Foot	Thigh	Wrist
Stand	0.74 (±0.10)	0.72 (±0.12)	0.65 (±0.11)	0.59 (±0.19)	0.72 (±0.06)	0.81 (±0.08)
Walk free	0.43 (±0.13)	0.46 (±0.11)	0.29 (±0.11)	0.21 (±0.12)	0.18 (±0.11)	0.13 (±0.12)
Walk tread	0.40 (±0.19)	0.42 (±0.16)	0.29 (±0.16)	0.33 (±0.15)	0.16 (±0.16)	0.19 (±0.18)
Stairs Up	0.47 (±0.10)	0.52 (±0.09)	0.38 (±0.09)	0.30 (±0.12)	0.33 (±0.12)	0.31 (±0.12)
Stairs down	0.56 (±0.14)	0.58 (±0.14)	0.47 (±0.16)	0.35 (±0.14)	0.23 (±0.14)	0.36 (±0.10)
Run	0.93 (±0.09)	0.93 (±0.07)	0.70 (±0.04)	0.30 (±0.16)	0.69 (±0.05)	0.35 (±0.13)
Sit	0.59 (±0.20)	0.60 (±0.12)	0.43 (±0.21)	0.42 (±0.25)	0.18 (±0.15)	0.36 (±0.09)
Lying	0.81 (±0.10)	0.79 (±0.14)	0.79 (±0.09)	0.82 (±0.06)	0.37 (±0.21)	0.50 (±0.25)
Avg	0.62 (±0.19)	0.63 (±0.17)	0.50 (±0.19)	0.41 (±0.20)	0.36 (±0.23)	0.38 (±0.21)

IV. CONCLUSION

This work investigated the use of the C4.5 DT algorithm to detect the location of an accelerometer on the body. The aim of this was to improve the performance of activity recognition by applying the appropriate activity classification model for a device placed in that location. Results showed the performance of the C4.5 DT in correctly identifying the accelerometer position (F-measure 0.63). This improved the activity classification, also using a C4.5 DT, when compared to using a model from only one location. It must be noted, however, that in this case the orientation of the accelerometer is fixed. When the accelerometer is housed within a mobile device, it can change orientation in addition to location. This further complicates the ability to detect the location of the accelerometer. One solution may be to examine the use of features which are not affected by device orientation such as those associated with the magnitude of acceleration. Results within this paper are of particular interest for activity recognition using accelerometers within mobile devices taking into consideration that for mobile applications, the position of the accelerometer can change throughout the day.

REFERENCES

1. L. Bao and S. S. Intille, "Activity recognition from user-annotated acceleration data," *Lecture Notes in Computer Science,* pp. 1-17, 2004.
2. L. Atallah, B. Lo, R. King and G. Z. Yang, "Sensor positioning for activity recognition using wearable accelerometers," *Biomedical Circuits and Systems, IEEE Transactions on,* vol. 5, pp. 320-329, 2011.
3. A. G. Bonomi, G. Plasqui, A. H. C. Goris and K. R. Westerterp, "Improving assessment of daily energy expenditure by identifying types of physical activity with a single accelerometer," *J. Appl. Physiol.,* vol. 107, pp. 655-661, 2009.
4. G. Bieber, J. Voskamp and B. Urban, "Activity recognition for everyday life on mobile phones," *Universal Access in Human-Computer Interaction Intelligent and Ubiquitous Interaction Environments,* pp. 289-296, 2009.
5. F. Ichikawa, J. Chipchase, R. Grignani and N. Design, "Where's the phone?: A study of mobile phone placement in public spaces", in 2005, pp. 1-8.
6. K. Kunze, P. Lukowicz, H. Junker and G. Tröster, "Where am i: Recognizing on-body positions of wearable sensors," *Location-and Context-Awareness,* pp. 257-268, 2005.
7. A. Henpraserttae, S. Thiemjarus and S. Marukatat, "Accurate activity recognition using a mobile phone regardless of device orientation and location," in *Body Sensor Networks (BSN), 2011 International Conference on,* 2011, pp. 41-46.
8. C. V. C. Bouten, K. T. M. Koekkoek, M. Verduin, R. Kodde and J. D. Janssen, "A triaxial accelerometer and portable data processing unit for the assessment of daily physical activity," *Biomedical Engineering, IEEE Transactions on,* vol. 44, pp. 136-147, 2002.
9. F. Ferraris, U. Grimaldi and M. Parvis, "Procedure for effortless infield calibration of three-axis rate gyros and accelerometers," *Sensors and Materials,* vol. 7, pp. 311-311, 1995.
10. T. Keller, A. Weisberger, J. Ray, S. Hasan, R. Shiavi and D. Spengler, "Relationship between vertical ground reaction force and speed during walking, slow jogging, and running," *Clin. Biomech.,* vol. 11, pp. 253-259, 1996.
11. N. Ravi, N. Dandekar, P. Mysore and M. L. Littman, "Activity recognition from accelerometer data," in *Proceedings of the National Conference on Artificial Intelligence,* 2005, pp. 1541.
12. D. Figo, P. C. Diniz, D. R. Ferreira and J. M. P. Cardoso, "Preprocessing techniques for context recognition from accelerometer data," *Personal and Ubiquitous Computing,* vol. 14, pp. 645-662, 2010.

Incorporating the Rehabilitation of Parkinson's Disease in the Play for Health Platform Using a Body Area Network

F. Tous[1], P. Ferriol[1], M.A. Alcalde[1], M. Melià[1], B. Milosevic[2], M. Hardegger[3], and D. Roggen[4]

[1] Fundació BIT/Health Department, Palma, Spain
[2] University of Bologna/DEI, Bologna, Italy
[3] Swiss Federal Institute of Technology (ETH), Zürich, Switzerland
[4] Culture Lab, Newcastle University, Newcastle Upon Tyne, United Kingdom

Abstract—Play for Health (P4H) is an open and extensible telerehabilitation platform. It allows patients to perform rehabilitation exercises through a variety of videogames and interaction methods. Here, we extend P4H to address the training needs of patients with Parkinson's disease. We integrate a network of on-body inertial measurement units. This allows new motion-based interactions in serious games at home, and it is a cornerstone extension allowing P4H to now also assess the patient's movements in daily life. We introduce three new serious games in a virtual world on this platform. The game objectives are designed to be engaging, with automatically adapting challenge levels, and they realize motor-cognitive exercises according to specific therapeutic goals. We present the architecture and technical implementation of this system, including specifics in handling a large number of wireless Bluetooth sensors. We outline the scalability of the platform architecture to eventually handle other pathologies and discuss future technical improvements.

Keywords—Telerehabilitation, serious gaming, wearable computing, posture processing, context awareness.

I. INTRODUCTION

Play for Health (P4H) is an integrated telerehabilitation system offering motor and cognitive training through serious games and multiple interaction devices [1]. It is designed to be deployed at home or in clinical centers, and offers remote access to clinicians to deploy customized therapy programmes, assess progress and adjust training parameters. We aim for P4H to be a broad solution to support rehabilitation needs that we initially identify in our care centers in the Balearic Island (Spain), but with a view towards general applicability in other countries. Our philosophy is to rely mostly on commonly available commercial off-the shelf interaction devices (e.g. Kinect, WiiMote, dance mats) and largely open-source software. This allows to realize a cost-effective solution that is also easily scalable (i.e. no long lead times or production costs associated to dedicated hardware). This is advantageous over other state of the art systems, such as robotic-based rehabilitation approaches, and is competitive with other systems (e.g. GestureTeK Health IREX). P4H is appealing in cost-constrained health services, yet our initial deployments show that the approach does not compromise efficacy. So far, P4H has been used in several health centers in Mallorca (Spain), with three videogames (e.g. puzzle, maze navigation) and 6 interaction methods, mostly with stroke patients to improve cognitive and physical deficits [2].

We now expand P4H to support the rehabilitation of patients with Parkinson's disease, within the context of the European-funded FP7 project CuPiD (Closed-loop system for personalized and at-home rehabilitation of people with Parkinson's disease). Our work is based on evidence that sustained intense physical activity has a positive effect on disease severity [3], and that often physiotherapy is not accessible due to costs or geographical reasons (e.g. only 54% of the patients of Parkinson's in UK were seen by a physiotherapist in 2008).

This paper reports on two novel P4H extensions. First, we developed dedicated serious games for Parkinson's patients to increase physical activity, with a specific view to train balance, sit-stand transition, and gait and arm swing amplitude. Dual-tasking exercises allow the patient to train motor and cognitive functions. This is offered as complementary to traditional physiotherapy with the possibility to remotely assess outcomes and adapt the therapy. Through gamification, we aim to increase compliance with a more enjoyable training.

The second extension brings new interaction capabilities to P4H. We designed a module to fuse the data of on-body inertial measurement units (IMUs) into an integrated anatomical model of the human body allowing us to analyze the user's posture. The approach is scalable to integrate any number of IMUs according to their configured position on the body, while "emulating" limb movements that are not instrumented. IMUs have no range limitations compared to computer vision approaches, and they can capture movements that could not be sensed by other modalities. For instance, a rotation of the arm along its axis is easily sensed by an IMU but not by e.g. a Kinect. This second extension is a cornerstone of P4H that paves the way to daily life activity monitoring within the P4H framework. This is important for Parkinson's rehabilitation, as the intensity of physical activity can be assessed in daily life and the serious games can be adapted to encourage the movements that were less commonly performed during the day. Currently, our implementation uses custom IMUs to ensure high quality data. However, keeping to the P4H philosophy, our architecture allows us to easily swap those for off-the-shelf commercial sensors. This enables us to capitalize on the rapidly growing market of wearables (e.g. Shimmer low-cost IMUs, Amiigo wrist band, Shine sensor from Misfit Wearables, rumored smart watch from Apple, etc.). Thanks to the P4H architecture, the wearable sensors offer new interaction capabilities also to former P4H games.

II. THE P4H ARCHITECTURE

The P4H architecture is based on plugins to be easily extensible to new games and interaction devices with a message-oriented bus for communication across plugins. The architecture of P4H is composed of:

Core: This is the main program, responsible for managing the fundamental procedures of the system and organizing the execution of plugins through services.

Content Plugins: These implement multi-media videogames developed jointly by clinical an technical partners, following the "Serious Gaming" [4] philosophy. This approach focuses on the motor-cognitive component of the rehabilitation process and on patient motivation.

Interaction Method Plugins: These allow the patients to interact with the videogames and to work on the physical component of the rehabilitation process. There are 2 sub-types of interaction method plugins: *Device plugins* and *Communication plugins*. The first ones allow the system to interact with external devices such as Microsoft's Kinect or a Body Area Network (BAN) of wearables. Communication plugins allow the system to establish low-level communication channels such us USB, Bluetooth or TCP/IP. Communication plugins provide services to the Device plugins and manage the communication channels for them.

The P4H platform provides a Software Development Kit (SDK), which includes an Application Program Interface (API) for the development of new plugins. It runs on the Ubuntu 12.04 LTS operating system. According to the Linux philosophy, some of the implemented components use libraries maintained by external programming communities. In particular P4H makes use of the object-oriented graphics rendering engine Ogre3D, which provides the functionality required to render 3D scenes. This enables the abstraction from the low-level API for this purpose (OpenGL/DirectX), and the Boost libraries which extend the capabilities of the C++ programming language.

III. EXTENSIONS FOR PARKINSON'S REHABILITATION

We developed 3 videogames (Content plugins) and the Posture Processing plugin (a class of Device plugin). Fig. 1 shows the extended P4H architecture.

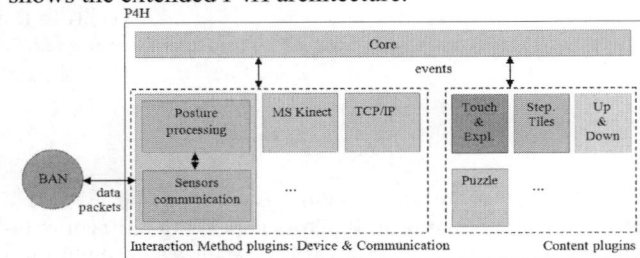

Fig. 1 P4H Extensions for IMU-based Parkinson's patient training

A. Videogames for PD Patients

Videogames are concatenations and/or repetitions of various motor-cognitive exercises designed according to predetermined therapeutic goals. In terms of motor rehabilitation, the purpose of a videogame is to achieve improvements in the execution of movements which resemble daily life motor tasks. This should be done by motivating the patient and involving cognitive challenges. The assumption regarding this cognitive aspect of videogames is that goal-oriented movements are a combined task involving both motor and cognitive aspects such as information processing, planning, strategic choices and execution of movement.

In this work, 3 typical motor exercises used for Parkinson's disease rehabilitation were implemented as videogames. In all these videogames, a 3D avatar representing the patient is moved in a virtual environment in response to the person's motions. The tasks for the patients in the 3 videogames are:

Touch'n'Explode: Pop objects (balloons) in the environment by virtually touching them.

Stepping Tiles: Step on the tiles of a kitchen floor when they light up, or follow a pre-established step sequence.

Up'n'Down: Perform a "Strength test" game by changing the position from sitting to standing and vice-versa in the correct way, and with maximum vertical acceleration.

Every videogame has configurable parameters to adjust the difficulty level, matching the characteristics of the patient and providing different game modes.

B. Wearable Sensors and Communication

Patients interact with the videogames through a BAN composed of wearable sensor nodes developed for this project. This *EXLs1* device is a compact (30x50x15mm) wearable sensor node equipped with inertial (accelerometer and gyroscope) and magnetic sensors, a Bluetooth wireless transceiver, $1GB$ Flash memory for sensor data storage when outside of communication range of P4H, and an ARM Cortex-M3 microcontroller unit (MCU). It uses state of the art digital MEMS sensors for high accuracy and incorporates an 180mAh battery, which ensures 5 hours of continuous data streaming at 100Hz, given the node current consumption of 40.9mA.

Besides serious gaming, the sensor nodes are also used in the daily life of Parkinson's patients, e.g. to quantify physical movements, detect freezing of gait incidents, and in a biofeedback system for situated training. The internal sensor parameters, such as the sampling frequency and the type of output are dynamically configured for the current service.

The sensor data is locally processed by the MCU to reconstruct the node's relative orientation in space, based on a recently developed complementary filter [5]. Thanks to its performance-computational cost tradeoff, it is the optimal choice for the resource-constrained embedded MCU. The sensor nodes are capable of streaming the sampled sensor data or the computed orientation of the device. For P4H, the computed orientation data in quaternions is streamed from each device to balance the workload among the sensors and the Posture Processing plugin.

Given the nature of quasi-static movements and postures that need to be recognized, the sensors send data packets at 15Hz, although the raw sensor data and the orientation computation are internally updated and filtered at 60 Hz to improve the orientation estimation. The low sending frequency simplifies the problem of the synchronization of the data streams from the different sensor nodes. The nodes start streaming at the same time, after receiving a command broadcasted from P4H. The data is received from the application in real-time and a packet counter, sent with the data, is used to detect the loss of packets.

The Posture Processing plugin receives the sensor data through a Bluetooth connection as a virtual serial port according to the Bluetooth's Serial Port Profile (SPP). The Linux Bluetooth protocol stack BlueZ was used to implement this profile. It is a mature and capable open-source Bluetooth stack and it is part of the official Linux kernel since the version 2.4.6. It is included in numerous Linux distributions.

At the lowest level of this integration 3 main difficulties were faced:

Trash Packets: During the connection process the reception of spurious data or corrupted packets causes a delay between the movement captured by a sensor and the movement of the avatar. This problem was solved by discarding all the received data until the detection of a valid packet to start the data stream.

Loss of Packets: Once the connection was established, a high rate of lost packets was observed (between 15% and 20%). According to a detailed analysis, the packet loss was due to an oversubscription problem. This problem arises when multiple threads are used to receive data and share the available logical CPU cores. Although the data loss rate was within an acceptable margin (it still allows a correct representation of the avatar movement on the screen), it was avoided by managing all the sensor communication in a single thread.

Connection Lost: In some cases the connection with a sensor is lost, causing the closure of the communication channel. This also causes the interruption in the flow of information of the actions undertaken by the patient, and therefore the game interaction. To make the connection more robust, P4H now constantly checks that all the devices are connected, and it automatically tries to reconnect with the devices to which it lost the connection.

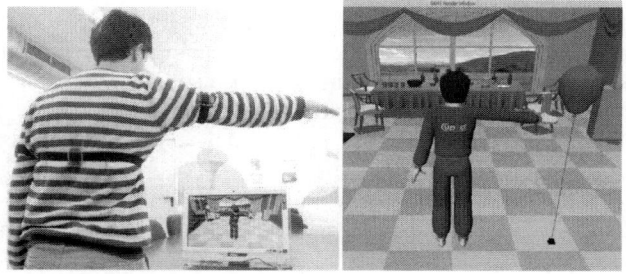

Fig. 2 User wearing 3 IMUs (torso and upper left/right arm) for the Touch'n'Explode game (left) and the corresponding virtual world (right).

C. Posture Processing

Postures in our work are sets of body segment orientations. The orientation of a single segment (e.g. upper arm, thorax, etc.) is computed from the orientation of a sensor attached to this body part. To account for alignment errors, we require the patients to perform a body-segment alignment calibration at the beginning of each exercise. In this phase, the patient has to stand upright with arms hanging down and the feet at shoulder distance. Since we know the actual orientation of body segments at this state, we can derive a set of rotation matrices describing the misalignment. These rotations are subsequently applied to the raw sensor orientation data (see [6]).

To play all videogames without the repositioning of the sensor nodes, they must be attached to both upper arms, upper legs and to the back of the patient. The total number of sensors (in that case 5) is a trade-off between the quality in the reconstruction of the body posture, and the obtrusiveness of the system (fewer sensors to wear). All videogames can also be played with 3 body-mounted sensors only, but then sensors must be repositioned between different videogames. The avatar movements in the virtual environment combine real body segment orientation data with derived orientation data for segments which are not equipped with sensors. For example, in exercises performed while sitting, the legs of the avatar are fixed, and in reaching videogames the elbow joint is assumed to be fully extended throughout the videogame.

The actual videogames are made up of a series of posture tasks. To fulfill such a task, patients must reach and hold the required postures for a minimum time. E.g. in the Up'n'Down game, the user first has to sit upright, then lean forward, lift off with the upper body, and finally go to the standing position. From a rehabilitation point of view, it is important that the patient carefully performs the intermediate steps. The Posture Processing plugin detects when a patient reaches a certain posture, based on a threshold applied to the orientation difference in degrees between the current posture and a pre-learned target posture. Reaching or losing a target posture results in an event submitted through a message-oriented event bus to the videogame. The videogame keeps track of the sequencing and timing of the postures performed by the user. It also displays animations in response to user actions (e.g. when a user makes the avatar touch a balloon, this balloon explodes), adapts the task difficulties if patients have troubles reaching the goals (to not demotivate them) based on initially estimated maximum body extension capabilities, and computes basic statistics.

By separating the posture processing from the videogames, we facilitate the development of new videogames, and we relieve the main program from lower-level tasks. Decoupling through an Event-Driven Architecture (EDA) removes dependencies, reduces module complexity and improves the system scalability. A limitation to the approach is the potentially high traffic overhead if large data volumes need to be exchanged. In our case, this is not a problem as messages are usually made up of just a few numbers (posture identifiers, threshold settings, etc.) and sent at low frequency.

IV. CONCLUSIONS

We have included the rehabilitation of Parkinson's disease patients into the P4H platform through new interaction methods based on a body area network of inertial measurement units, and dedicated videogames. We have developed algorithms for posture processing and recognition for that purpose. These new developments followed the modular structure of P4H. By separating the detection of postures from the game logic, through an event-driven architecture, these implementations are scalable and may be extended to the treatment of other pathologies. In particular, the use of an SDK has allowed external programmers to develop code and integrate it into P4H, demonstrating the extensibility of the platform.

In the near future we plan to evaluate the system performance and usability in experiments with patients and to complete the integration with the CuPiD telemedicine infrastructure. To increase usability we are working on an inductive charging station and custom color- and shape-coded wearable casings that help patients to attach the sensors at the right body location. We also plan to combine Microsoft's Kinect and the inertial sensors to capture the patients' movements and take advantage of the modularity of the platform. Thus we dynamically exploit the available sensors following opportunistic sensing principles. Furthermore, this may help to automatically identify the on-body position of sensors through data correlation analysis.

ACKNOWLEDGMENTS

The research leading to these results has been partially founded by the European Union - Seventh Framework Programme (FP7/2007-2013) under grant agreement n°288516 (CuPiD project).

REFERENCES

1. Tous F et al. (2011) Play for Health: Videogame Platform for Motor and Cognitive Telerehabilitation of Patients, eTELEMED 2011, The Third International Conference on eHealth, Telemedicine and Social Medicine, Gosier, Guadeloupe, France, 2011, pp 59-63
2. Farreny MA et al. (2012) Play for Health (P4H): a new tool for telerehabilitation. Rehabilitación 46(2):135-140
3. Schenkman M et al. (2012) Exercise for people in early- or mid-stage Parkinson disease: a 16-month randomized controlled trial. Physical Therapy 92(11):1395-1410
4. Zyda M (2005) From Visual Simulation to Virtual Reality to Games. Computer 38:25-32
5. Madgwick SOH et al. (2011) Estimation of IMU and MARG orientation using a gradient descend algorithm. IEEE Intl. Conf. on rehabilitation Robotics
6. Roetenberg D et al. (2009) Xsens MVN: full 6DOF human motion tracking using miniature inertial sensors. Xsens Motion Technologies BV, Technical Report

Author: Francisco Tous Llull
Institute: Fundació Bit
Street: Centre Empresarial Son Espanyol, C/Laura Bassi s/n. 07121
City: Palma
Country: Spain
Email: xtous@ibit.org

Activity Classification Using 3-Axis Accelerometer Wearing on Wrist for the Elderly

D.I. Shin[1], S.K. Joo[2], J.H. Song[3], and S.J. Huh[2]

[1] Biomedical Engineering, Asan Medical Center, Seoul, Korea
[2] Biomedical Engineering, University of Ulsan College of Medicine, Seoul, Korea
[3] Research Center, SooEee Electronics, SeongNam, Korea

Abstract—The monitoring of single elderly is more important due to rapid transition to aging society. There are many bio-signals to monitor the emergent state of elderly. These vital signals including ECG, PPG, blood pressure signal spend heavy processing resource and costs. We have been developed the monitoring device for the elderly emergency monitoring. In this paper we propose new criteria to classify daily life activities. We categorized activities with the motility of real action. The upper most criteria are normal and abnormal activity. The lower criteria are 'small or large movement', 'periodic or random movement', 'no movement or shock'. Then we derive some parameters to get thresholds to classify these activities according to our new criteria. The main parameters are entropy, energy and autocorrelation. Some experiments were carried out to determine classifying thresholds. Finally we got results of classified activities such as 'no movements', 'small movements', 'large movements', 'periodic movements' and 'falls'. We got nearly 100% of classifying result for falls and no movements. In this case of 'quasi-emergency state' our developing device will investigate further status of elderly by measuring of heart rate and oxygen saturate using pulse oxymeter. Finally the device decides in emergence, it sends a short message to server and then connected to the u-Healthcare center or emergency center and one's family.

Keywords—Activity, Classification, Accelerometer, Wrist, Elderly.

I. INTRODUCTION

Rapid transition to aging society becomes very important problem day by day. Especially for the single elderly, it is critical problem that whose vital situation.

According to the data from Statistics Korea, the aging index will increase rapidly from 9.5%(2006) to 14.3%(2018) and 20.8%(2026). With this trend, the number of single elderly is increases too. Knowing the emergency status of these single elderly is a critical problem in the emergency monitoring system. So we have been developed a monitoring device which can be easily worn on elderly body. The wearing position is very important because it must be very convenient for the elderly. And in the case of emergent, the reaction of elderly is also important for the decision whether he or she is in serious. There were many researches for monitoring devices(Boo-Ho Yang, Sokwoo Rhee, 2000, P. Mendoza, P. Gonzalez, B. Villanueva, E. Haltiwanger, H. Nazeran, 2004, Giuseppe Anastasi, Marco Conti, Mario Di Francesco, Andrea Passarell, 2009, Francis E.H. Tay, D.G. Guo, L. Xu, M.N. Nyan, K.L. Yap, 2009, Prajakta Kulkarni, Yusuf Ozlurk, 2010, Amr Amin Hafez, Mohamed Amin Dessouky, Hani Fikri Ragai, 2011). In these researches, there are many considerations about monitoring devices and systems in the respect of u-Healthcare Monitoring. After all, we conclude that the ideal wearing position is wrist for now. With the progress of technology, the device may be the shape of hearing aid in the future.

In this research, we classified the activity type of elderly in daily life. Recent researches classified the activity type with the real action such as walking, standing, sitting, lying etc(Arunkumar Pennathur, Rohini Magham, Luis Rene Contreras, Winifred Dowling, 2003, A. Mannini, A.M. Sabatini, 2009, G.M. Lyons, K.M. Culhane, D. Hilton, P.A. Grace, D. Lyons, 2005, Marcia Finlayson, Trudy Mallinson, Vanessa M. Barbosa, 2005, Angela L. Jefferson, Robert H. Paul, Al Ozonoff, Ronald A. Cohen, 2006, A. Godfrey, A.K. Bourke, G.M. Ólaighin, P. van de Ven, J. Nelson, 2011). But actually this kind of classification does not helpful for the decision of emergent status of an elderly. So we suggest new concept of classification criteria. We categorized activities with the motility of real action. The upper most criteria will be normal and abnormal activity. The lower criteria may be 'small or large movement', 'periodic or random movement', 'no movement or shock'.

Once we classify the elderly activity to abnormal we can further investigates the accurate status with the reaction button or pulse oximeter which will be adopted our monitoring device. The clinical importance of oxygen saturation of blood(SpO_2) is mentions on many articles(Barker SJ, Morgan S., 2004, Anna Letterstål, Fredrik Larsson, 2007, Gülendam Hakverdioğlu Yönt, Esra Akin Korhan, Leyla Khorshid, 2010, Elif Derya Ubeyli, Dean Cvetkovic, Irena Cosic, 2010)

If we can classify a person's status to normal or abnormal, we can make more powerful investment in case of abnormal status. As a result, we may reduce processing resource, power and finally physical size of the sensor. The more compact size and reduced processing power will be helpful of wearing it.

II. MATERIALS AND METHODS

A. System Overview

We extracted acceleration data from our monitoring device in developing. The acceleration data was moved from memory of monitoring device to PC via USB port. Data sampling rate is 10ms/sample and converted by 12bit depth.

Fig. 1 illustrates our processing system. Personal computer (Pentium V) is used to process and analyze activities. The LabView software from National Instruments is used to acquire and display the acceleration data from monitoring device. The Matlab^{TM} software is used to process and analyze the acceleration data.

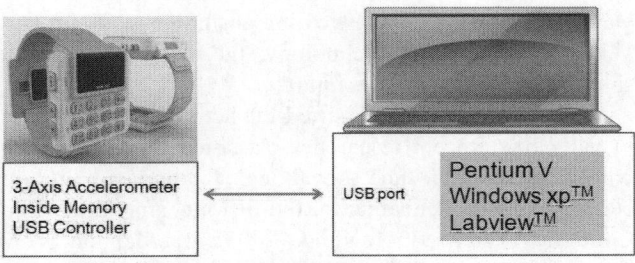

Fig. 1 The overall system configuration

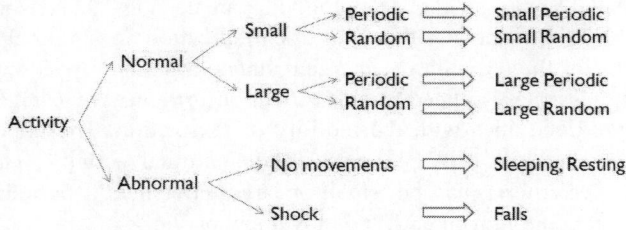

Fig. 2 Classification criteria of activities

B. Activity Classification

In this research, we classified the activity type of elderly in daily life. Recent researches classified the activity type with the real action such as walking, standing, sitting, lying etc. But actually this kind of classification does not helpful for the decision of emergent status of an elderly. So we suggest new concept of classification criteria. We categorized activities with the motility of real action. The upper most criteria will be normal and abnormal activity. The lower criteria may be 'small or large movement', 'periodic or random movement', 'no movement or shock'. Figure 3 shows the classification criteria of our new concept.

The 3-axis acceleration data were pre-processed like below.

$$A_o = \sqrt{a_x^2 + a_y^2 + a_z^2} \quad (1)$$

$$A_{os} = (Low\_pass\ Filter)A_o \quad (2)$$

A_{os} is low pass filtered data with 5Hz cutoff frequency.

To classify activities, we calculate some parameters and define threshold of classification. First, the entropy is measured like below,

$$Entropy = \nabla_{Aos}/\nabla_t \quad (3)$$

Entropy is the ratio of acceleration change per unit time. And the energy is defined like this,

$$Energy = \Sigma A_{os} \quad (4)$$

Normal Activity Classification

Normal activity is classified with two categories. The first is the magnitude of movements. This is judged by the threshold of entropy and energy. The judge function is described like this.

$$J_{mov} = a*Entropy + b*Energy \quad (5)$$

The second category is periodicity and judge function for this is,

$$J_{per} = c*Entropy + d*Autocorrelation \quad (6)$$

Abnormal Activity Classification

Abnormal state is categorized two classes. One is 'no movements', the other is 'falls'. When in 'no movements' there might be two situation which are in sleep and in emergency. In these situations our monitoring device will checks heart rate and O2 saturation in blood using pulse oxymeter.

To determine whether falls or not, we use the entropy for the threshold function.

$$J_{fall} = e*Entropy \quad (7)$$

To determine whether no movements or not, we use the entropy and the energy for the threshold function.

$$J_{nmov} = f*Entropy + g*Energy \qquad (8)$$

Classifying Algorithm

Fig. 3 shows the flowchart of activity classifying algorithm.

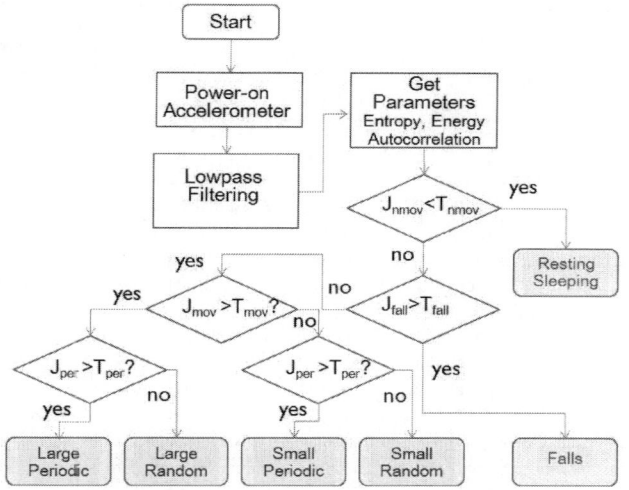

Fig. 3 Flowchart of classifying algorithm

Once we start the algorithm, the accelerometer in the device is powered on. The acceleration data are acquired with 100 samples/sec. And then the acceleration data are lowpass filtered with the 5Hz cut-off frequency. We call these procedures 'Pre-processing' and equation (1) and (2) show these procedures. Next, we calculate parameters on equations (3) to (4). Using the entropy and energy we can calculate the parameter J_{nmov}. If J_{nmov} is less than the threshold T_{nmov}, we can judge there are no movements such as resting of sleeping state. If J_{nmov} is great than the threshold T_{nmov}, we can judge that there are some movements include some kind of falls.

The next stage we investigate the parameter J_{fall} according to the equation (7). If this is greater than T_{fall}, a kind of falls must have been arisen. Once the state is classified to normal movement, we can classify to lower categories as shown in fig. 2.

In real world, situations are more complex and ambiguous. So, the classification algorithm is difficult. But as refine more accurately the algorithm, the result will be more realistic.

III. RESULTS

Fig. 4 shows the low-pass filtered acceleration data. It includes various activities. Small movements can be showed in Fig. 4(a). These movements include grapping a pen, writing, moving a paper, scratching one's skin, removing glasses etc. In small movements, they shows small accelerations under $2g(1g=9.8m/s^2)$. One the other hand Fig. 4(b) shows large movements such as stretching, doing gymnastics, standing up suddenly etc. It shows large acceleration 5g or more. Sometimes it exceeds 10g but it's slope is rather than gradual. Fig. 4(c) shows a typical periodic movement which is walking. There are two levels of valley, the upper valley represents backward peak position of hand and the lower valley represents forward peak position of hand.

Fig. 5 shows classified results for successive various activities according to our algorithm. The colored bar denotes the class of activity. Data from monitoring device are transmitted to personal computer and are processed with Labview[TM] and Matlab[TM] software to verify our algorithm.

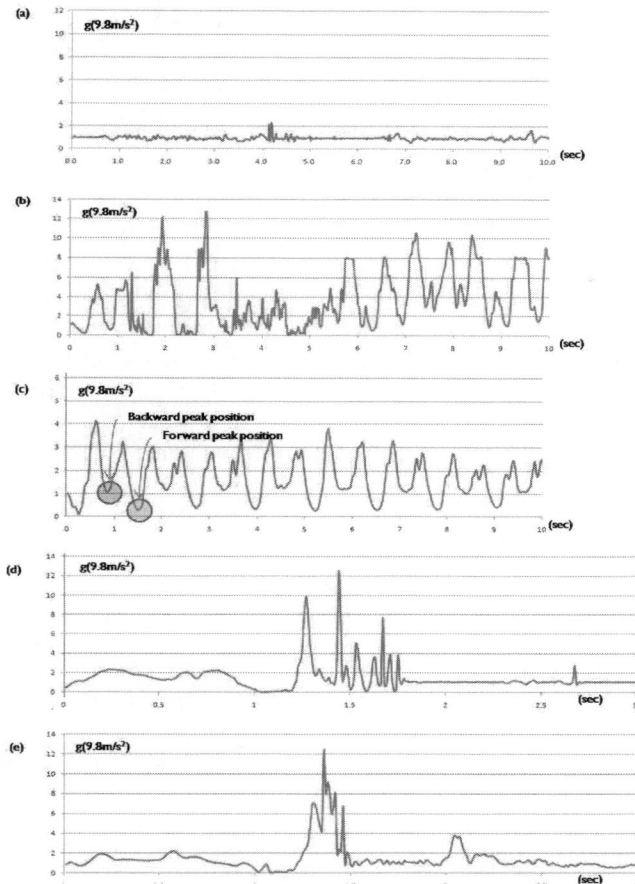

Fig. 4 Acceleration data from various activities

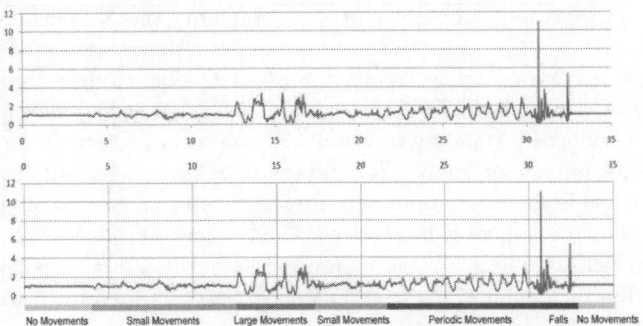

Fig. 5 Classified results for successive various activities according to our algorithm

IV. CONCLUSIONS

Knowing the emergency status of these single elderly is a critical problem in the emergency monitoring system. So we have been developed a monitoring device which can be easily worn on elderly body. The wearing position is very important because it must be very convenient for the elderly. And in the case of emergent, the reaction of elderly is also important for the decision whether he or she is in serious. After all, we conclude that the ideal wearing position is wrist for now. With the progress of technology, the device may be the shape of hearing aid in the future.

In this research, we classified the activity type of elderly in daily life. Recent researches classified the activity type with the real action such as walking, standing, sitting, lying etc. But actually this kind of classification does not helpful for the decision of emergent status of an elderly. So we suggest new concept of classification criteria. We categorized activities with the motility of real action. The upper most criteria will be normal and abnormal activity. The lower criteria may be 'small or large movement', 'periodic or random movement', 'no movement or shock'.

Once we classify the elderly activity to abnormal we can further investigates the accurate status with the reaction button or pulse oximeter which will be adopted our monitoring device.

If we can classify a person's status to normal or abnormal, we can make more powerful investment in case of abnormal status. As a result, we may reduce processing resource, power and finally physical size of the sensor. The more compact size and reduced processing power will be helpful of wearing it.

REFERENCES

1. Boo-Ho Yang, Sokwoo Rhee, Development of the ring sensor for healthcare automation, Robotics and Autonomous Systems 30 (2000) 273–281
2. P. Mendoza, P. Gonzalez, B. Villanueva, E. Haltiwanger, H. Nazeran, A Web-based Vital Sign Telemonitor and Recorder for Telemedicine Applications, in Proceedings of the 26th Annual International Conference of the IEEE EMBS San Francisco, CA, USA, September 1-5, 2004, pp. 2196-2199.
3. Giuseppe Anastasi, Marco Conti, Mario Di Francesco, Andrea Passarell, Energy conservation in wireless sensor networks: A survey, Ad Hoc Networks 7 (2009) 537-568
4. Francis E.H. Tay, D.G. Guo, L. Xu, M.N. Nyan, K.L. Yap, MEMS-Wear-biomonitoring system for remote vital signs monitoring, Journal of the Franklin Institute 346 (2009) 531–542
5. Prajakta Kulkarni, Yusuf Ozlurk, mPHASiS: Mobile patient healthcare and sensor information system, Journal of Network and Computer Applications, 2010
6. Amr Amin Hafez, Mohamed Amin Dessouky, Hani Fikri Ragai, Design of a low-power ZigBee receiver front-end for wireless sensors, Micro Electronics Journal 40 (2011) 1561-1568
7. Arunkumar Pennathur, Rohini Magham, Luis Rene Contreras, Winifred Dowling, Daily living activities in older adults:Part II—effect of age on physical activity patterns in older Mexican American adults, International Journal of Industrial Ergonomics 32 (2003) 405–418
8. G.M. Lyons, K.M. Culhane, D. Hilton, P.A. Grace, D. Lyons, A description of an accelerometer-based mobility monitoring technique, Medical Engineering & Physics 27 (2005) 497–504
9. Marcia Finlayson, Trudy Mallinson, Vanessa M. Barbosa, Activities of daily living (ADL) and instrumental activities of daily living (IADL) items were stable over time in a longitudinal study on aging, Journal of Clinical Epidemiology 58 (2005) 338–349
10. A. Mannini, A.M. Sabatini, Computational methods for the automatic classification of postures and movements from acceleration data, *Gait & Posture 30S (2009) S26–S74*
11. A. Godfrey, A.K. Bourke, G.M. Ólaighin, P. van de Ven, J. Nelson, Activity classification using a single chest mounted tri-axial accelerometer, Medical Engineering & Physics, in Press (2011)
12. Angela L. Jefferson, Robert H. Paul, Al Ozonoff, Ronald A. Cohen, Evaluating elements of executive functioning as predictors of instrumental activities of daily living (IADLs), Archives of Clinical Neuropsychology 21 (2006) 311–320
13. Barker SJ, Morgan S., A Laboratory Comparison of the Newest "Motion-Resistant" Pulse Oximeters During Motion and Hypoxemia, *Anesthesia and Analgesia* 2004;98(55),S2:A6
14. Anna Letterstål, Fredrik Larsson, Assessment of vital signs on admission to short time emergency wards improves patient safety and cost-effectiveness, Australasian Emergency Nursing Journal, Volume 10, Issue 4, November 2007, Page 191
15. Gülendam Hakverdioğlu Yönt, Esra Akin Korhan, Leyla Khorshid, Comparison of oxygen saturation values and measurement times by pulse oximetry in various parts of the body, Applied Nursing Research, 2010

Intelligent Chair Sensor – Classification and Correction of Sitting Posture

L. Martins[1], R. Lucena[1], J. Belo[1], R. Almeida[1], C. Quaresma[2,3], A.P. Jesus[1], and P. Vieira[1]

[1] Departamento de Física, Faculdade de Ciências e Tecnologia, Universidade Nova de Lisboa, Caparica, Portugal
[2] CEFITEC, Departamento de Física, Faculdade de Ciências e Tecnologia,
Universidade Nova de Lisboa, Caparica, Portugal
[3] Departamento de Saúde, Instituto Politécnico de Beja, Beja, Portugal

Abstract—In order to build an intelligent chair capable of posture detection and correction we developed a prototype that measures a pressure map of the chair's seat pad and backrest and classifies the user posture. The posture classification was done using neural networks that were trained for 5 standardized postures achieving an overall classification of around 98%. Those neural networks were exported to a mobile application in order to do real-time classification of those postures. Using the same mobile application we devised two correction algorithms that were implemented in order to create an intelligent chair capable of posture detection and correction. The posture correction is forced through the change of the conformation of the chair's seat and backrest by changing the pressure of eight pneumatic bladders.

Keywords—Sensing chair, Pressure-distribution sensors, Sitting posture, Posture Classification, Posture Correction.

I. INTRODUCTION AND RELATED WORK

Changes in the society workforce in the last decades has forced the adult population to spend long periods of time in a sitting position in the workplace that coupled with a sedentary lifestyle at home is associated to health problems, such as back and neck injuries [1]. While a person is seated, most of their bodyweight is transferred to the ischial tuberosities, to the thigh and the gluteal muscles. The rest of the bodyweight is distributed to the ground through the feet and to the backrest and armrest of the chair when they are available [2] The adoption of a lumbar flexion position for long periods of time, can lead to a decrease of the lumbar lordosis [3], causing anatomical changes to the spine and degenerate the intervertebral discs and joints, disorders that have been linked to back and neck pain.

There are a wide number of clinical views of 'correct' or 'incorrect' postures, but until recently there were little quantitative studies to define those postures. Recent studies have shown that not only adopting a seating position over long periods cause health problems, but also the adoption of an incorrect posture can also worsen the health problems [4, 5]. Other groups are trying to determine whether the so called 'good' postures actually provide a clinical advantage [6].

Several investigation groups have been working with pressure sensors placed in chairs in order to solve the problem of an incorrect posture adoption for long periods of time in the seated position. These pressure sensors were able to detect the user posture, using the acquired pressure maps and various Classification Algorithms. Various studies equipped with the sensor sheets with 2016 sensors (one for the seat pad and one for the backrest) were able to distinguish various postures [7, 8, 9]. Slivovsky et al. (2000) and Tan et al. (2001) used Principal Component Analyses (PCA) for posture detection for human-machine interactions obtaining an overall classification accuracy of 96% and 79% for familiar and unfamiliar users, respectively [7, 8]. The same data acquisition as the previous studies were used by Zhu et al. (2003) to investigate which classification algorithms would be work the best for static posture classification. The authors found that among k-Nearest Neighbor, PCA, Linear Discriminant Analysis and Sliced Inverse Regression (SIR), both PCA and SIR outperformed the other two methods [9]. One group, also using the same sensor sheets, studied the relationship between patterns of postural behaviors and interest states in children. [10] They were able to classify nine postures in real time achieving an overall accuracy of 87.6% with new subjects, while using Hidden Markov Models to study the interest levels [10].

Mutlu et al (2007) and Zheng and Morrell (2010) even reduced more drastically the number of pressure sensors for posture identification. The first group determined the near optimal placement of 19 FSR (Force Sensitive Resistors) sensors and were able to obtain an overall classification of 78%, improving the classification to 87% when the number of sensors was increased to 31 [11]. The second group adapted a chair with just 7 FSR and 6 vibrotactile actuators, in order to direct the subject towards or away from a certain position through haptic feedback. They were able to obtain an overall classification of 86.4% on those same ten postures using the mean squared error between the pressure measurements and their reference, showing also the effectiveness of haptic feedback on posture guidance [12].

The long term goal of this project is to build an intelligent chair capable of detecting the sitting posture and effectively correct an incorrect posture adoption over long periods of time in order to minimize the previously described health issues associated with the anatomical changes to the spine. In order to correct an incorrect posture we a first prototype with 8 air bladders (4 in the seat pad and 4 in the backrest), which are able to change their conformation by inflation or deflation and can increase the user discomfort when a bad posture is adopted, encouraging the user to change to a correct position. We are also able to produce slight changes in the chair conformation over a period of time, which can help to evenly distribute the applied pressure on contact zones, reducing the fatigue and discomfort of the user due to the pressure relief on compressed tissues. The pressure values inside each bladder can be used as an input to our classifier in order to analyze and detect the seating posture of a user in real-time.

This papers focus on the classification of five different postures and the correction algorithms implemented to correct the adoption of incorrect postures for long time periods.

II. METHODS AND MATERIALS

A. Equipment

This prototype was built with the aim of producing an office chair capable of detecting the user posture and also correcting an incorrect posture adoption over long periods of time, but also considering a low cost and commercially available solution that could fit the office furniture market. To accomplish this we placed a low resolution matrix of 8 air bladders inside the chair (4 in the seat pad and 4 in the backrest) coupled with pressure sensors that measure the pressure inside each bladder. The system is also comprised by an air pump and a vacuum pump for each bladder to change their conformation by inflation and deflation.

Strategically bladder placement was required in order to achieve good performance results. Previous literature identified two types of strategies: a pure mathematical and statistical approach [11] and an anatomical approach [12]. We placed the bladders, based on the second method, in order to cover the most important and distinguishable areas of the body for detecting a seated posture, such as the ischial tuberosities, the thigh region, the lumbar region of the spine and the scapula. These are also the areas where most of the bodyweight is distributed while a user is seated [2]. The distribution of pressure cells is illustrated in figure 1-A, where both the seat pad and backrest were divided into a matrix of 2-by-2 pressure sensing bladders. We used the original padding foam of the chair, placing it above the pressure cells to maintain the anatomical cut of the seat pad and backrest as shown in figures 1-B and 1-C. Honeywell 24PC Series piezoelectric gauge pressure sensor were used to measure the pressure inside each bladder. Since most of the bodyweight while seated is transferred to the ischial tuberosities, the thigh and the gluteal muscles, meaning that the seat pad will sense more pressure than the backrest, so the pressure sensors in the seat pad where chosen to be rated up to 15 psi, with a sensitivity of 15mV/psi while they were rated to 5 psi with a sensitivity of 21 mV/psi for the sensors in the backrest.

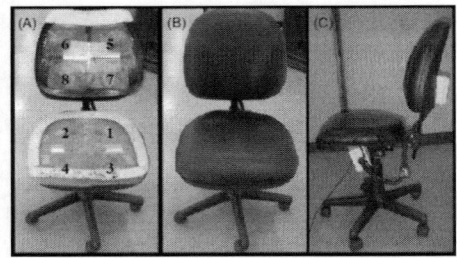

Fig. 1 (A) Distribution of the pressure cells in the chair. In the seat pad we accounted for the ischial tuberosities (Sensors 1 and 2) and the thigh region (Sensors 3 and 4). For the backrest we accounted for the scapular region (Sensors 5 and 6) and finally for the lumbar region (Sensors 7 and 8). Frontal (B) and lateral (C) view of the chair with the padding foam.

The bladder size was chosen in order to minimize the gaps between cells (large gaps would be uncomfortable for the users), while also covering the areas described above but also to be incorporated in standard office chair.

B. Experiments

The tests were done on 30 subjects (15 male and 15 female) with an average age of 20.9 years, average weight of 67.8 Kg and an average Height of 172.0 cm.

Before conducting the experiments, we needed to define the specific time of inflation for each pressure cell, in order for them to have enough air to sense the pressure of the subject in the sitting position, but not enough to cause discomfort to the users. After some tests (data not shown), we used a value of 4 seconds for inflating pressure cells represented by 1, 2, 3, 4, 7, 8 and 5 seconds to inflate pressure bladder number 5 and 6 for every subject during both experiments, which is the time chosen for the calibration process.

Before undergoing any experiment, subjects were asked to empty their pockets and to adjust the stool height so that the knee angle was at 90º (angle between the thigh and the leg) and to keep their hands on their thighs. The five postures for the classification algorithms are represented in figure 2 and were based on previous works [7, 11, 12], since they include the most common postures in literature and are classified as P1, P2, P3, P4 and P5 respectively. To obtain the experimental data for the classification algorithms, we carried out two tests, first we showed a presentation of the postures P1 to P5, each for a duration of 20 seconds, asking the subject to mimic those postures without leaving the chair. The second consisted in showing the same presentation, but every posture was repeated three times, but after every 20 seconds the subject was asked to walk out of the chair, take a few steps and sit back.

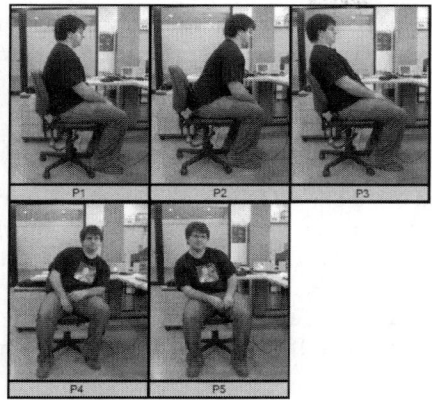

Fig. 2 Seated postures used in the experiments: (P1) Seated upright, (P2) Leaning forward, (P3) Leaning back, (P4) Leaning left, (P5) Leaning right.

Not all of the data acquired was used for the classification, because when a user changes his posture, the pressure maps will oscillate until they stabilize. In this study we focus on the Stable zone of the pressure maps and therefore, approximately 13 out of the 20 seconds were used, and since our sampling rate is 18.4 Hz, we were able to extract 240 data-points out of the 13 seconds, which were then divided into 6 pressure maps of 40 data points, giving a total of 720 maps for each posture (30 subjects * 4 repetitions * 6 pressure maps) and a total of 3600 maps (720 * 5 postures).

For each subject, we acquired an additional 12 seconds of data points in posture P1 in order to define a baseline pressure. All the 3600 maps were normalized to an input interval of [-1; 1] to create the Artificial Neural Networks (ANNs) using the MATLAB® Neural Network Toolbox™.

III. RESULTS

We tested various parameter combinations such as number of neurons, number of layers, transfer function and network training function for the classification algorithms based on ANN. In table 1 we present all considered combinations of parameters and the specific ANN returned the best overall classification. We also tried different combinations of transfer functions depending on the number of layers. For this test we used a "leave-1-person-out" program that would use 29 subjects to train the ANN, and then the last subject was used to test the network. This process was done 30 times, in order to calculate the average classification of each ANN, in order to choose the best parameters. A simple feedforward network with one-way connections from input to the output layers was able to fit our multidimensional mapping problem with a good overall classification score and can be very simply implemented in other systems without needing the MATLAB® NN Toolbox™, since we can export the weights and bias of the ANN to other systems, such as mobile devices, capable of classification in real-time.

Table 1 Parameter combination for the Neural Networks and the parameters that gave the best overall classification scores. Here, LM corresponds to Levenberg-Marquardt algorithm, SCG to Scaled Conjugate Gradient algorithm and RP to Resilient Backpropagation algorithm.

Parameters	Combination	Best
Nº of Neurons	10, 15, 20, 25, 30, 35	15
Nº of Layers	1,2,3	1
Transfer function	Tansig and Logsig	Only Tansig
Network training function	LM, SCG, RP	RP

In Table 2 we present a confusion matrix for both the test and training data of the neural network that gave the best overall classification of the eleven postures.

Table 2 Confusion Matrix for posture classification. Rows indicate the Output Class and Columns indicates the Target Class. The Target Class labels correspond to the respective postures from figure 1. The gray boxes give the percentages of correct classification in relation to the respective class. The blue box represents the overall classification score.

	P1	P2	P3	P4	P5	(%)	
P1	713	0	11	7	1	97.5%	Output Class
P2	7	710	10	2	3	97.0%	
P3	0	10	689	0	0	98.6%	
P4	0	0	10	713	10	97.3%	
P5	0	0	0	0	706	100%	
(%)	99.0%	98.6%	95.7%	99.0%	98.1	98.1%	
			Target Class				

Using the same mobile application we devised two correction algorithms in order to create a full application capable of posture detection and correction. This process starts with the seating of a user and its identification. If it is a known user the calibration profile is applied, otherwise a new calibration process is done with the protocol previously described and then we start the classification process, distinguishing between the five standardized postures. At the moment we implemented 2 correction algorithms.

The first tests if the user has been seated for more than X minutes. The variable X can vary for different persons, but should be between 90 and 180 minutes. The algorithm changes the conformation of the bladders to cause a discomfort to the user, alerting them to stand up in order to prevent that users stay too seated for long periods of time. The second algorithm was devised in order to prevent that a user stay seated in incorrect postures for periods of time that may cause health problems. This algorithm acts after Y minutes, a variable that should vary between 5 and 15 minutes, based on the ISO standards for working standards [13]. Specific bladders are inflated for correcting different posture adoption, as for example to correct posture P2 we just need to inflate bladders 1 to 4 (see figure 1) while to correct posture P3 we will need to also inflate bladders 5 to 8, but change less the conformation of bladders 1 to 4. The full process workflow is presented in figure 3.

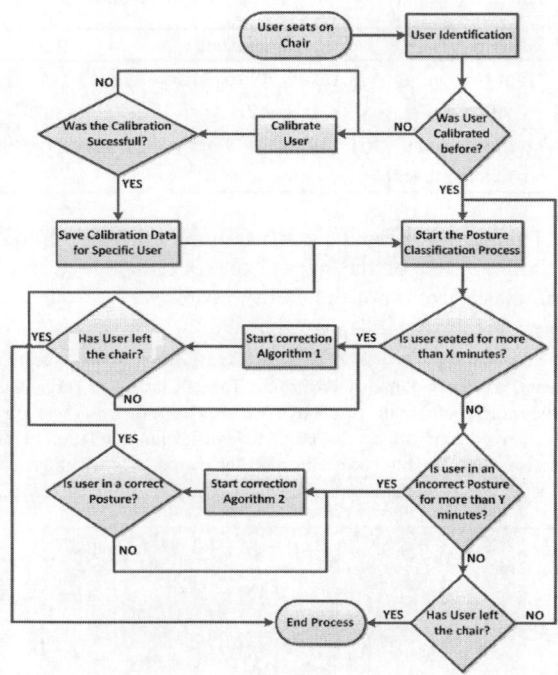

Fig. 3 Workflow of the both the Classification and Correction Processes

IV. CONCLUSIONS AND FUTURE WORK

A chair prototype with air bladders in the seat pad and backrest was developed to detect the posture and correct bad posture adoption over long periods of time. Pressure maps of five postures were gathered in order to classify each posture using ANNs. First we studied the best parameters of the ANNs for the posture classification and then we used them to create an ANN and export it to a mobile application. Results showed that for the five postures, the overall classification of each posture was around 98%. Two correction algorithms were integrated in the mobile applications in order to test if the user is seated for long periods of time and also if is seated in an incorrect posture. In any of the situations, the conformation of the chair automatically changes, so the user can adopt a correct posture.

Our next aim is to continue studying the classification algorithms in order to start classifying more postures and the correction algorithms to understand if they are being effective or if we need other correction processes. We will do clinical trials to evaluate the correction models but also to validate our classification algorithms, in order to build an intelligent chair capable of posture correction, reducing the health problems related to back and neck pain.

ACKNOWLEDGMENT

The authors wish to thank Eng. Pedro Duque for the help provided in the construction of the prototype and Eng. Marcelo Santos for the participation in the Classification studies. This project is supported by FEDER, QREN – Quadro de Referência Estratégico Nacional, Portugal 07/13 and PORLisboa – Programa Operacional Regional de Lisboa.

REFERENCES

1. Hartvigsen J, Leboeyf-Yde C, Lings S, Corder E (2000) Is sitting-while-at-work associated with low back pain? A systematic, critical literature review. Scand J Public 28(3):230-239
2. Pynt J, Higgs J, Mackey M (2001) Seeking the optimal posture of the seated lumbar spine. Physiother Theory Pract 17(1):5-21
3. van DieËn J, De Looze M, Hermans V (2001) Effects of dynamic office chairs on trunk kinematics, trunk extensor EMG and spinal shrinkage. Ergonomics 44(7):739-750
4. Cagnie B, Danneels L, Van Tiggelen D et al. (2007) Individual and work related risk factors for neck pain among office workers: a cross sectional study. Eur Spine J 16(5):679-686
5. Lis M, Black K, Korn H, Nordin M (2007) Association between sitting and occupational LBP. Eur Spine J 16:283-298
6. Claus A, Hides J, Moseley G, Hodges P (2009) Is 'ideal' sitting posture real?: Measurement of spinal curves in four sitting postures. Manual Therapy 14(4):404-408
7. Slivosky L, Tan H (2000) A Real-Time Static Posture Classification System. In ASME Dynamic Systems and Control Division
8. Tan H, Slivovsky L, Pentland A (2001) A sensing chair using pressure distribution sensors. Transactions on Mechatronics 6(3):261-268
9. Zhu M, Martinez A, Tan H (2003) Template-Based Recognitions of Static Sitting Postures. In Workshop on Computer Vision and Pattern Recognition for Human-Computer Interaction. CVPRW '03
10. Mota S, Picard R (2003) Automated Posture Analysis For Detecting Learner's Interest Level. In Workshop on Computer Vision and Pattern Recognition for Human-Computer Interaction. CVPRW '03
11. Mutlu B, Krause A, Forlizzi J et al. (2007) Robust low-cost, non-intrusive sensing and recognition of seated postures. In 20th annual ACM symposium on User interface software and technology
12. Zheng Y, Morrell J (2010) A vibrotactile feedback approach to posture guidance. In IEEE Haptics Symposium
13. ISO 11226:2000(E) Ergonomics — Evaluation of static working postures

Author:	Pedro Vieira
Institute:	Faculdade de Ciências e Tecnologia, Universidade Nova de Lisboa
Street:	Quinta da Torre P-2829-516
City:	Caparica
Country:	Portugal
Email:	pmv@fct.unl.pt

Non-invasive System for Mechanical Arterial Pulse Wave Measurements

Mikko Peltokangas, Jarmo Verho, Timo Salpavaara, and Antti Vehkaoja

Department of Automation Science and Engineering, Tampere University of Technology, Tampere, Finland

Abstract—A non-invasive multichannel measurement system for arterial pulse wave (PW) recording is presented. The system uses sensors made of EMFi (ElectroMechanical Film) material for measuring mechanical PW signals simultaneously from various arterial sites all over the body. In addition to mechanical PW, also electrocardiographic, respiration and photoplethysmographic signals can be recorded with the system. By extracting parameters from the PW contour, valuable information on the subject's vascular health can be obtained.

The data is collected by a wireless body sensor network which enables the measurement of physiological signals synchronously from eight sensor nodes in maximum. In this study, the pulse waves measured from different locations and postures are illustrated and compared as well as correlations between pulse transition times (PTT) are calculated. The result show that the PTTs obtained from the corresponding arterial sites of the left and right arm correlate strongly in most cases. We also investigated the effects of supporting bias force of EMFi sensors, and our results indicate a strong dependence between the PW amplitude and the supporting bias force.

Keywords— EMFi, EMFi pulse wave sensor, pulse transition time (PTT), pulse wave recording, wireless body sensor network.

I. INTRODUCTION

A pulse wave (PW) contour and pulse wave velocity (PWV) as a result of a left ventricular contraction provide valuable information on the arterial elasticity, which is an important indicator of the vascular health [1, 2]. The arteries tend to stiffen due to ageing [1], but also unhealthy lifestyle stiffens the arteries, increasing the risk for cardiovascular diseases such as arteriosclerosis and atherosclerosis that often lead to premature deaths. To diagnose latent disorders, there is a need for a simple clinical research method to evaluate the vascular condition in order to motivate people suffering from such disorders to change their living towards healthier lifestyle or to provide them a proper medication.

An important parameter related to arterial elasticity is a pulse transit time (PTT) which is the time delay between the PW observed at two different arterial sites. The PTT can also be measured as, for example, the time delay between the left ventricular contraction and an observed PW at a peripherial arterial site [3]. Mathematically, the PTT is related to PWV as a quotient of travelled distance and velocity. The PWV is proportional to the square root of the Young's modulus of the arterial wall and the arterial wall thickness, but the dependence can also be expressed by using Bramwell-Hill equation as

$$\mathrm{PWV} = \sqrt{\frac{A \cdot \Delta p}{\Delta A \cdot \rho}} \quad (1)$$

where A is the arterial lumen area (a cross-sectional area), ΔA and Δp are the changes in the arterial lumen area and the pressure, respectively, and ρ is the density of the blood [4]. Stiffened arteries decrease the pulsatile change in arterial lumen area ΔA, causing increased PWV as can be seen from Bramwell-Hill equation (1). In addition to arterial elasticity information, PTT provides information about inspiratory efforts and changes in systolic blood pressure. These may be helpful e.g. in sleep apnoea detection as they indicate the activity level of human autonomous nervous system [3].

Besides the PTT analysis, one method to process the PW data is to analyze the main components of a single PW [5]. First component, called a percussion wave, is originated from the blood ejection from the left ventricle to the aorta. The percussion wave is followed by a tidal wave, which is a result of the reflected pulse from the bifurcation points of the arteries at the lower body. Next, a dicrotic notch indicates the closure of the aortic valve before a dicrotic wave as a result of transient increase in arterial pressure. The remaining two main components are assumed to be reflections of earlier components, but they are not always clearly visible in the PW signal. [6, 5].

A traditional non-invasive measurement method to collect data for PW analysis is to use an electro-optical photoplethysmographic (PPG) sensor, as e.g. in [3]. The PPG determines the absorption of the light by measuring the intensity of transmitted or reflected light, which is affected by the varying arterial pressure and temporary blood vessel volume change caused by a pulsatile blood perfusion as a result of left ventricular contractions. The PPG is usually measured from a fingertip, toetip or earlobe.

An alternative non-invasive method for the arterial PW recordings is to place a force sensor, such as ElectroMechanical Film (EMFi), on the pulsating artery. EMFi is a low-cost flexible electret polymer based on polypropylene, having extremely high sensitivity to dynamic forces but not to static forces [7]. EMFi sensors can also be used for PWV measurements when two EMFi sensors with a known distance between them are placed on the same arterial branch and the PTT between these two sites is determined.

In this study, we present a wireless measurement system which is capable of measuring multiple channels of PW data synchronously from different locations on the body. Also ECG and respiration signals are simultaneously recorded by the system. A benefit of the multichannel measurement setup is that a more comprehensive view of cardiovascular health can be obtained as the PW contour differs between different arterial sites [2, 1] due to local arterial elasticity.

II. RELATED WORK

Many of the reported non-invasive pulse waveform measurement and analysis studies are based on the use of PPG sensors as e.g. in [3, 5] as it is widely available for clinical studies or the use tonometric sensors as e.g. in [8]. Related to EMFi sensors, Sorvoja et al. [9] have developed a non-invasive blood pressure transducer based on the PW signals recorded with EMFi sensors. Alametsä et al. measured PW signal with EMFi sensor stripes having a size of 15 cm × 2 cm and placing the sensors on the ankles, wrists and carotid artery [6, 2]. They concluded that the arterial elasticity information can be extracted from any of the studied arterial sites, but the differences in the derived parameters are so significant that the measurement point and the subject's position must be known when interpreting the results.

Huotari et al. reported, in addition to the analysis of PPG signal, also the analysis of mechanical PW recorded with an EMFi sensor placed on the radial artery [5]. The goal in their studies has been to develop methods for calculating parameters from a single pulse wave to describe arterial stiffness. Liu et al. in [10] presented a method to model carotid and radial pulse waves, recorded with piezoresistive force sensors, using a linear combination of three Gaussian functions. Zhang et al. developed a wavelet transform based method for extracting parameters from pressure sensor based wrist pulse signals [11].

III. MATERIALS AND METHODS

A. Measurement hardware

Our PW measurement system consists of wireless sensor nodes and a network coordinator which communicates with a computer through the USB. The synchronous sensor network is partially based on the IEEE 802.15.4 standard and it uses time division multiple access (TDMA) as the main mechanism for bandwidth management. The maximum number of sensor nodes in one measurement network is eight.

In our system, there are three different types of sensor nodes. One of these sensor nodes measures PW by using force sensors and the other one measures PW by using a PPG sensor, whereas the third type of sensor node measures ECG and respiration signals. Each force sensor node measures two dynamic force sensors and one capacitive static force sensor. The dynamic force sensors are used for PW measurements, whereas the static force sensors can be utilized to study and adjust the proper supporting bias force.

The sample rate for the dynamic force sensors is 250 Hz and for the static force sensors 10 Hz. The PPG sensor nodes measure one channel with 500 Hz sample rate and ECG/respiration nodes one ECG and one respiration channel with 250 Hz each. The recorded signals are quantized by using a 16-bit analog-to-digital converter.

Instrumentation for dynamic force sensors: The preamplifier stage of dynamic force sensors has a voltage gain of 6.6. The second-stage amplifier consists of a non-inverting first-order high-pass filter with the cut-off frequency of 0.05 Hz and pass-band gain of 2. Before the analog-to-digital conversion, the third amplification stage is implemented with a second order unity-gain low-pass filter having a cut-off frequency of 100Hz.

ECG instrumentation: Bipolar ECG is recorded by using two disposable Ag/AgCl electrodes in order to calculate the heart rate and to help PW detection for further analysis. The ECG signal is recorded by using a pass-band of 0.2 Hz–40 Hz with a gain of 123.

Respiration sensor instrumentation: Respiration rate is determined by using a thermistor integrated into a respiration mask. Thermistor is connected into a voltage divider and the respiration signal is simply recorded by measuring the voltage across the thermistor. The signal is bandpass-filtered with the cut-off frequencies of 3.4 mHz and 40 Hz.

PPG instrumentation: The pass-band for the PPG signal is 0.15 Hz–160 Hz and the signal is recorded by sensing the transmitted light with a photodiode. In our test measurements, the PPG signal is recorded from the tip of the left index finger.

B. Dynamic Force Sensor and Recording Sites

The mechanical PW signals are recorded by using sensors made of a material called EMFi. EMFi as a material consists of multiple polypropylene layers separated by air voids [7]. When an external dynamic force is applied on a piece of EMFi, the dimensions of air voids will change, modifying the charge distribution which is measured by the electrodes covering the surfaces of EMFi. Theoretically, the charge Δq generated by a dynamic force ΔF exerted on EMFi sensor is expressed as

$$\Delta q = S \Delta F \quad (2)$$

where S is the sensitivity [7]. Because of the capacitive nature of EMFi sensor, this is turned into voltage ΔV as

$$\Delta V = \frac{S \Delta F}{C} \quad (3)$$

where C is the capacitance of the EMFi sensor.

The circular EMFi sensors used in our system have a diameter of 20 mm. A rounded plastic button is glued on each sensor to improve the force transmission from pulsating artery to the sensor and to enable comfortable skin contact by decreasing the required supporting force. Eight EMFi sensors, manufactured by Emfit Ltd., are placed on different locations: left and right wrists to sense radial arteries, inside of both elbows to sense brachial arteries, neck to sense left and right common carotid arteries (CA), and both ankles to sense posterior tibial arteries. These measurement sites were selected due to their accessibility for sensor placement.

We have also successfully tested placing the EMFi sensors on the popliteal fossa above the knee for measuring the popliteal artery and on the temples for sensing the superficial temporal arteries. However, we noted that the signal amplitude from the popliteal fossa depends widely on the bending of the lower limb. In addition, we moved the sensors from the temples to the neck to measure carotid pulses which are especially interesting as they locate close to the aorta, providing information on aortic PW and elasticity.

EMFi sensors were placed on each measurement point (except neck) by using flexible textile bands around the limb. The sensors measuring carotid arteries were fixed by using minimally flexible velcro tape around the neck, but naturally ensuring that the band is not constricting the subject.

C. Static Force Sensor

The contour and the amplitude of the PW is dependent on the supporting bias force as seen in Figs. 2 and 3. To study this dependence, a capacitive static force sensor was placed between the housing of the electronics and the textile band whereas the EMFi sensor was placed between the housing of the electronics and the radial artery.

D. Test Measurements

We have tested our system with 4 test subjects S1–S4, aged 24–33, without previously known cardiovascular disorders. The actual recordings were made in the supine posture, but we also evaluated the effect of the posture by making comparative measurements in the sitting posture.

As it was obvious that the supporting bias force affects to the PW signal amplitude, more information about the effects of bias force were needed. This was studied by keeping the position of the dynamic force sensor constant and gradually decreasing the tightness of the wristband by removing thin plastic plates placed between the wristband and outer side of the wrist.

IV. RESULTS AND DISCUSSION

As expected based on several studies (e.g. [1, 6, 2]), the recorded PW contours vary widely not only between subjects, but also between different arterial sites and different measurement postures. Examples of the mechanical PW contours and the PPG signal, recorded in the supine and sitting postures from different arterial sites, are shown in Fig. 1. The signals are dependent on person, but in our experiments the signal measured from the carotid artery differ most from the signals measured from the other sites (Fig. 1). Even though it is possible to record PW signal in the supine and sitting posture, we noticed that the measurements performed in the sitting posture are more sensitive to artifacts caused by natural movements.

To evaluate the consistency of the data produced by the EMFi sensors at different locations, two different PTTs are extracted: T_p as the time delay between the R peak of the ECG and the peak of the percussion wave of the PW and T_f as the time delay between the R peak of the ECG and the foot point of the PW (a local minimum before the rise of the PW). Pearson product-moment correlation coefficients are calculated into Table 1 for both parameters T_p and T_f extracted from the EMFi signals recorded from left and right arms. The high or at least moderate correlation coefficients indicate that the EMFi sensor signals are valid. We also calculated correlation coefficients between T_p and T_f parameters determined from PPG and all EMFi signals, but the results were not consistent even in the case where the signals were recorded from the same arm: for some test subjects, both of the parameters T_p and T_f correlated strongly or at least moderately, whereas for other test subjects only one PTT parameter, either T_p or T_f correlated well. One possible reason for the differences might be in the peak shape of the PW contour: in PPG signal, the peak of a single PW is often more rounded than in the mechanical PW recorded with an EMFi sensor. From this point of view, the EMFi sensors could be more suitable selection

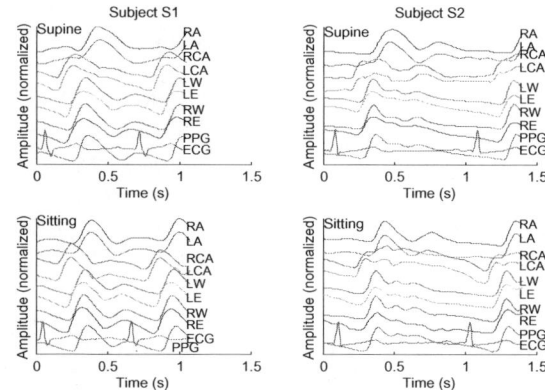

Fig. 1 The PWs with different heart rates from subjects S1 and S2, recorded with the EMFi sensors placed on different locations in the supine and sitting positions. RA=right ankle, LA=left ankle, RCA=right carotid artery, LCA=left carotid artery, LW=left wrist, LE=(inside of) left elbow, RW=right wrist, RE=(inside of) right elbow. Also ECG and PPG are presented for comparison.

Table 1 Pearson product-moment correlation coefficients between PTTs obtained from EMFi signals from left and right arm for test subjects S1–S4.

PTT Parameter	S1	S2	S3	S4
Radial T_f	0.86	0.80	0.61	0.97
Radial T_p	0.98	0.98	0.74	0.91
Brachial T_f	0.67	0.80	0.61	0.93
Brachial T_p	0.58	0.60	0.42	0.37

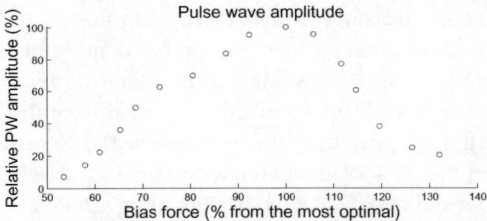

Fig. 2 The mean PW amplitude dependence on the supporting bias force for the test subject S4.

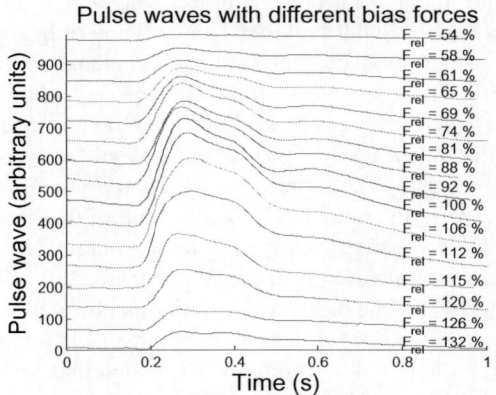

Fig. 3 Examples of the PWs at different relative bias forces (F_{rel}).

for PW recordings, although its amplitude strongly depends on the bias force, (Figs. 2 and 3) which may vary widely by a small limb movement.

In another experiment, we studied the effects of applied bias force on the radial artery PW amplitude with test subject S4. As the most optimal bias force depends on the area of the measurement device touching the skin, the bias forces presented in Figs. 2 and 3 are normalized by the bias force of the most optimal case. From Fig. 2 can be seen that the PW amplitude may decrease almost 80% if the supporting bias forces changes roughly ±40% from the most optimal case. Based on Fig. 2, we can also see that the PW contour is affected by the supporting bias force, along with the measurement site (Fig. 1).

V. CONCLUSION

The presented multichannel system could be a powerful tool for recording different physiological signals and especially mechanical arterial pulse waves. Although the pulse wave contours vary between different measurement sites, providing information on local arterial elasticity, further research is required to fully understand which parameters play the most important role in the vascular health estimation. In the future, we will study in more detail the effects of the supporting bias force applied to the EMFi sensors, as well as search and develop efficient analysis methods for the signals recorded with the presented system.

ACKNOWLEDGMENT

The authors would like to thank all the volunteer test subjects for their contribution to this study. The work was funded by the Finnish Funding Agency for Technology and Innovation (TEKES) as a part of the project Human body embedded physiological monitoring system (Healthsens).

REFERENCES

1. O'Rourke M. F., Hashimoto J.. Mechanical Factors in Arterial Aging: A Clinical Perspective *Journal of the American College of Cardiology.* 2007;50:1 – 13.
2. Alametsä J., Palomäki A.. Comparison of local pulse wave velocity values acquired with EMFi sensor *Finnish Journal of eHealth and eWelfare.* 2012;4:89 – 98.
3. Pitson D.J., Sandell A., Hout R., Stradling J.R.. Use of pulse transit time as a measure of inspiratory effort in patients with obstructive sleep apnoea *European Respiratory Journal.* 1995;8:1669–1674.
4. Westerhof N., Stergiopulos N., Noble M.I.M.. *Snapshots of Hemodynamics: An Aid for Clinical Research and Graduate Education.* SpringerLink : BücherSpringer US 2010.
5. Huotari M., Vehkaoja A., Määttä K., Kostamovaara J.. Photoplethysmography and its detailed pulse waveform analysis for arterial stiffness *Rakenteiden Mekaniikka (Journal of Structural Mechanics).* 2011;44:345–362.
6. Alametsä J., Palomäki A., Viik J.. Carotid and radial pulse feature analysis with EMFi sensor *Finnish Journal of eHealth and eWelfare.* 2012;4:80 – 88.
7. Lekkala J., Paajanen M.. EMFi-New electret material for sensors and actuators in *Electrets, 1999. ISE 10. Proceedings. 10th International Symposium on*:743–746 1999.
8. Chirinos J.A, Kips J.G., Roman M.J., et al. Ethnic Differences in Arterial Wave Reflections and Normative Equations for Augmentation Index *Hypertension.* 2011;57:1108–1116.
9. Sorvoja H., Kokko V.-M., Myllyla R., Miettinen J.. Use of EMFi as a blood pressure pulse transducer *Instrumentation and Measurement, IEEE Transactions on.* 2005;54:2505–2512.
10. Liu C., Zheng D., Murray A., Liu C.. Modeling carotid and radial artery pulse pressure waveforms by curve fitting with Gaussian functions *Biomedical Signal Processing and Control.* 2013;8:449 - 454.
11. Zhang P.-Y., Wang H.-Y.. A framework for automatic time-domain characteristic parameters extraction of human pulse signals *EURASIP J. Adv. Signal Process.* 2008;2008:1–9.

Author: Mikko Peltokangas
Institute: Tampere University of Technology / ASE
Street: Korkeakoulunkatu 3
City: FI-33720 TAMPERE
Country: FINLAND
Email: mikko.peltokangas@tut.fi

Automatic Identification of Sensor Localization on the Upper Extremity

S. Lambrecht and J.L. Pons

Bioengineering Group of CSIC, Madrid, Spain

Abstract—In this paper, we use 3D data from accelerometers and gyroscopes to identify the segment on which each inertial sensor is attached. An automatic procedure will reduce the time and skills needed to acquire motion capture data with inertial sensors, as well as reduce potential error.

Our objective was to make a computationally low-cost algorithm that can be applied on a variety of models. We propose an algorithm based on features extracted from 3D accelerometer and gyroscope data. Our algorithm does not rely on training data or standard classifiers, and is based only on basic mathematical operations allowing for a fast (0.2s on average) and light procedure that is executable on nearly all platforms.

We first tested our code on a wooden mockup of the upper extremity and subsequently applied it on data from three healthy subjects. Twelve features were selected for analysis on four different movements. Each motion was performed at three different speeds to assess the potential use of the algorithm on patient populations.

The results indicate that this algorithm can be used to identify the segments on which the inertial sensors are located. The only information required prior to execution of the algorithm is the number of segments that are involved in the model.

Keywords—Inertial sensors, motion capture, upper limb.

I. INTRODUCTION

The use of inertial sensors in biomechanics has a long history; nonetheless they have so far failed to gain popularity for use in 3D motion capture applications. Standard motion capture equipment exists of various cameras (minimum three in passive marker systems (e.g. VICON, BTS); minimum two in active marker systems (e.g. CODA, Optotrak)) that define a limited capture volume. Recent advances, such as the combination of cameras and infrared sensors have presented themselves as an easy to use and inexpensive alternative. Despite their obvious advantages these systems still have a limited capture volume and have limited precision, especially for fast and complex movements.

Current inertial sensor based motion capture systems transfer data from individual sensors to a body-worn hub. The hub facilitates synchronization and transfer to a personal computer or laptop. The sensors, either attached to the body by wearing a special suit [1] or Velcro-straps [1, 2] are manually labeled by the user. This limits the use of these systems for certain populations, such as elderly with cognitive disorders or people recovering from a stroke. Furthermore, this process is time consuming, taking up to 10 minutes.

The objective of this paper is to develop an algorithm that requires minimal prior knowledge to use an inertial motion capture system, with a focus on facilitating donning and doffing. The algorithm should not rely on specific sensor placement or order, nor demand the presence or assistance of a person with certain technical knowledge. The goal is to reduce the barriers of the use of inertial sensors for motion capture, through facilitating the donning and doffing and calibration process, to the extent that this technology can be used by clinicians or patients in a home-environment.

II. BACKGROUND

Previous studies tackling the problem of automating the identification of sensor location have all relied on computationally taxing classifications algorithms such as support vector machines [3] or decision trees based on a collection of machine learning algorithms for data mining tasks [4]. Most of these studies specific target the lower limb. The study by Weenk et al. [4] includes a full body model with up to 17 sensors. However, no low-cost method has been developed specifically for modular models. A modular model is a model that exists of several segments that can be added or discarded from the model. For example if only the wrist range of motion is of interest, then the model consists of hand and forearm (2 segments, 1 joint). A similar setup can be used to track elbow motion (forearm and humerus), whereas a more complex model is needed to identify coordination of the upper limb in reaching (e.g. 4 segments (hand, forearm, humerus, thorax) and 3 joints (wrist, elbow, shoulder)). The novelties of the proposed algorithm are that it is designed for the upper extremity, that it does not require training data or computations that are demanding for both processor and battery, and that no ideal sensor placement is required.

III. METHODS

A. Measurements

Three healthy subjects were equipped with four inertial sensors each. The order in which the sensors were attached to segments was randomized. Two inertial sensors were attached with Velcro straps, one at the forearm and one on the humerus of the right arm. Two more sensors were attached to the thorax, by a harness, and the right hand, using a neoprene glove. Each subject performed four predefined movements at three different velocities (slow, normal, and fast) in order to test the potential usability in patient populations. The subjects were instructed to imagine a patient's velocity of movement when performing the slow trials. Subjects were asked to slow down or speed up if deemed necessary. Each trial started with the subject standing upright with the arms extended and parallel to the trunk. The four movements were: abduction-adduction of the humerus without flexing the elbow, circular motion in the transversal plane with an extended arm, follow-through of push phase and an elbow flexion-extension.

The shoulder abduction-adduction was chosen because it resembled closely the motion performed on the wooden mockup used to verify the initial hypothesis leading up to this algorithm.

The circular motion represents a random movement without any clear constraints, other than to lead the motion from the shoulder.

The "follow-through" mimics the return phase that occurs in wheelchair propulsion after terminating push-off on the rims. This task has the potential to enable online verification and identification for wheelchair propulsion assessments. This movement also combines action at two joints simultaneously (shoulder and elbow).

The elbow flexion extension in turn offers the potential to use the same data for both determining the location of each sensor and the identification of the functional axis of the elbow joint [5].

Each motion trial consisted of 5 cycles of the defined movement. Both the raw sensor data (3D accelerometer data, 3D gyroscope data, and 3D magnetometer data) and orientation data (in rotation matrix format) expressed in the sensor frame were collected. For this paper, only the acceleration and angular velocity data was analyzed.

The features selected from the acceleration and angular velocity are based on the magnitude of the data are: maximum value of the full-wave rectified signal, sum of the raw data, integral of the data, and the ranges covered by the data both filtered (low-pass Butterworth filter, cutoff of 3Hz, 8th order) and non-filtered.

The algorithm extracts these features into a vector. These vectors are then ranked and the indexes obtained after ranking were compared to the known sensor placement (mockup or test subjects). Ranking was used because it is computationally not demanding, intuitive, and can be performed in the background to verify the actual calibration.

IV. RESULTS

The confusion matrices corresponding to the algorithm's performance are listed below. The real sensor number is displayed in the top row (in bold), the predicted sensor value in the columns. The represented data is pooled over subjects and velocities. The diagonal thus represents the number of sensors that have been identified correctly. The tables correspond to the follow through movement (Table 1), the humerus abduction-adduction (Table 2), the elbow flexion-extension (Table 3), and the circular motion in the transversal plane (Table 4).

Table 1 Confusion matrix of the follow through trials

Follow T	1	2	3	4
1	108	0	0	0
2	0	104	16	1
3	0	1	69	38
4	0	3	23	69

Table 2 Confusion matrix of the humerus abduction-adduction trials

AbAd H	1	2	3	4
1	108	0	0	0
2	0	59	25	22
3	0	38	65	18
4	0	11	18	68

Table 3 Confusion matrix of the elbow flexion-extension trials

Elb. FE	1	2	3	4
1	76	32	0	0
2	32	76	24	20
3	0	0	61	43
4	0	0	23	45

Table 4 Confusion matrix of the circular motion trials

Circle	1	2	3	4
1	92	12	4	0
2	16	78	12	2
3	0	17	74	26
4	0	0	18	80

The influence of the speed at which the movement is performed is illustrated by the following tables (5 and 6). The false positives are the values of the combined trials for all subjects, and represent the sum of the values that are below the diagonal of each subject's confusion matrix for that movement and execution speed.

Table 5 Influence of velocity on follow through identification

Follow T	Fast	Normal	Slow
False pos	2	36	11
False neg	25	14	16
total	27	50	27

Table 6 Influence of velocity on elbow flexion-extension identification

Elb FE	Fast	Normal	Slow
False pos	15	14	26
False neg	33	44	42
total	48	58	68

Of the features selected based on our initial test on the mockup, few remained stable when applied to the human upper extremity. In the first column are the three tasks, in the following columns are the feature(s) that for that task result in respectively zero, two, or less than 6 incorrect predictions. Given that no features could be found that accurately predicted the sensor location during the humerus abduction-adduction task, this task is not included.

Table 7 Features with respectively 0, 2, less than 6 sensors (rows) predicted incorrectly according to task (columns)

	0 sensors	2 sensors	<6 sensors
Circle	Int(gyr), sum(gyr)		Combined max (gyr)
Elb FE	Max(acc), max(acc.LPF), combined max of (gyr)	Range(gyr)	Combined max(acc)
Follow T	Max(acc), max(acc;LPF), combined max of (acc. LPF)	Combined max (acc)	Range(gyr), combined max (gyr)

V. DISCUSSION

The procedure did not perform as well as expected from the mockup trials (excluded for brevity) that were used to identify potential features. The first sensor (Tables 1-4) is most easily identified. This sensor corresponds to the thorax and was the only non-moving segment. There are trials were there is supposed to be no movement of the humerus (elbow flexion-extension) but as mentioned in literature [6] there is significant soft-tissue artifact (moving of the skin and underlying soft tissue without actual movement of the bone) to be expected in the humerus during flexion-extension tasks. No training was involved for this identification, and thus no "known" movement needed to be performed. Out of the motions performed by the subjects, the elbow flexion extension and the follow through motion performed the best.

The follow through movement further appears to be the most stable for execution at low speeds (Table 5). The elbow flexion-extension task performs better at fast speeds but degrades quickly with diminishing velocities.

Since no training is required and no specific input on the model is needed, the proposed method works equally well for modular models.

This study did not require precise sensors placement. The good performance of the elbow flexion-extension task suggests that it can be used by healthy subjects. This entails that this motion can be used to both identify and calibrate the model (e.g. thorax – humerus –forearm), although at this time the latter does require the sensors to be placed in the optimal position on each segment. The longitudinal components of the segments can be obtained at the start of the trial (accelerometer data), and the average of the full wave rectified angular velocity data from the motion will provide the functional elbow flexion axis.

In previous studies classifiers were used that required significant training and computational power [2-4, 7]. This study suggests an alternative that is based on very basic mathematical operations.

This study is only the second study to use both acceleration and angular velocity data to extract features from. The first study to do so was recently published by Weenk et al. [4].

This study is the first to use a flexible model approach, where no steps need to be taken when downsizing (reducing the number of segments and joints involved) the model.

VI. LIMITATIONS

The number of subjects used, and therefore the number of sensors that needed to be predicted, is too low to draw definite conclusions.

The low number of subjects may question the robustness when applied to a bigger population, or a patient population. However, the fact that for the movements in Table 7 several features are available that did not produce false positives or negatives, entails that these could be combined if needed to increase robustness under more challenging conditions. Being not computationally expensive, the algorithm can be performed on the background to verify the initial identification.

No sensor was placed on the scapula because this placement can only be done by another person. Our procedure is meant to facilitate the use of inertial motion capture systems outside of the constrained laboratory environment and without the need of assistance. Most current upper limb models to not include scapula markers or sensors and use a regression equation instead. We therefore feel that this simplification, although excluding some scenarios, is warranted.

VII. CONCLUSIONS

We proposed a procedure to automatically identify sensor placement on the upper limb. Our method does not require high computational power, nor is prior knowledge or training on a specific model needed. Several features from both accelerometers and gyroscopes were found that can identify the sensor placement with 100% accuracy. The follow-through motion has been identified as the most robust for changes in execution speed.

Further testing on more subjects and on a true patient population is needed to confirm these results, and to verify the suitability of the follow through task by more frail subjects.

ACKNOWLEDGMENT

This work is part of HYPER project, funded by the Spanish government and coordinated by the bioengineering group of CSIC. The goal of the HYPER project is to design and develop neuro-robotics and neuro-prostetic interventions for rehabilitation and functional compensation of motor function at upper and lower extremities suffering from stroke or SCI.
CSD2009-00067

REFERENCES

1. Xsesns Technologies B.V. website 2013 at http://www.xsens.com
2. Technaid S.L. website 2013 at http://www.technaid.com
3. Amini N, Sarrafzadeh M, Vahdatpour A, Xu W (2011) Accelerometer based on-body sensor localization for health and medical monitoring applications. Pervasive mobile computing 7(6):746-760.
4. Weenk, D., Van Beijnum, B.-J. F., Baten, C. T., Hermens, H. J., & Veltink, P. H. (2013). Automatic identification of inertial sensor placement on human body segments during walking. *Journal of NeuroEngineering and Rehabilitation*, *10*(1), 31. doi:10.1186/1743-0003-10-31
5. Luinge, H., Veltink, P. H., & Baten, C. T. M. (2007). Ambulatory measurement of arm orientation. *Journal of biomechanics*, *40*(1), 78–85. doi:10.1016/j.jbiomech.2005.11.011
6. Cutti, A. G., Paolini, G., Troncossi, M., Cappello, A., & Davalli, A. (2005). Soft tissue artefact assessment in humeral axial rotation. *Gait & posture*, *21*(3), 341–9. doi:10.1016/j.gaitpost.2004.04.001
7. Kunze K, Lukowicz P: Using acceleration signatures from everyday activities for on-body de- vice location. InWearable Computers, 2007 11th IEEE International Symposium on. Boston: IEEE; 2007: 115–116.

Author: Stefan Lambrecht
Institute: GBIO CSIC
Street: Ctra Campo Real km 0,2
City: Madrid
Country: Spain
Email: s.lambrecht@csic.es

Part X
Medical Devices and Sensors

A New Device to Assess Gait Velocity at Home

R. Jaber, A. Chkeir, D.J. Hewson, and J. Duchêne

Institut Charles Delaunay, UMR CNRS 6279, Université de Technologie de Troyes, Troyes, France

Abstract—This paper presents a technological device that is able to estimate gait velocity in non-controlled conditions, specifically at home. Based on a Doppler sensor, it provides an instantaneous signal of velocity, hence an estimation of the average velocity on a short distance (3 m). Comparison with the results produced by usual clinical tests (time measured manually on 4.5 m or 10 m) shows an excellent correlation ($R^2 > 0.93$), even though signal processing is achieved in a rudimentary manner. In addition, instantaneous recordings open the way towards extraction of other gait features like gait regularity and quality. Such a device can be used to assess frailty and risk of fall in the elderly in ecological conditions.

Keywords—Gait velocity, elderly, risk of falls.

I. Introduction

The elderly population is significantly increasing in most European countries. The majority of elderly would prefer to live at home in good conditions of security and well-being, taking advantage of the benefits of a familiar environment and the possibility of maintaining their daily life activities and social links. However, even when living independently, elderly are subject to increasingly stressful events (lower physical capacities, pathologies, etc.), which could lead to frailty and risk of falls.

Focusing on physical frailty, Fried et al [1] proposed a classification of elderly into three categories (not frail, pre-frail and frail) based on five indicators, one of them being walking speed. Theou et al [2] showed that walking speed at usual pace was the parameter with the greatest correlation with the frailty index. Montero-Odasso et al [3] assert that gait velocity can be a single predictor of adverse events in healthy seniors. This has been confirmed later by the work of Abellan van Kan et al [4]. In respect to the risk of falls, walking speed again, among other gait features, shows its ability as a relevant predictor [5] [6] [7].

Therefore, walking speed measurement seems to be essential for frailty detection and follow-up as well as for assessment of risk of falls. Typically, walking speed is measured in a clinical setting by measuring the time spent to walk 15 feet [1] or 10 m [3] distance. The tests can be performed at either usual or maximum pace. Such tests need to dispose of a sufficient distance to start, walk steadily and stop in secure conditions. When the aim is to assess walking speed in non-controlled conditions, e.g. at home within the scope of frailty follow-up, there is a need for a device that does not require a standard predefined distance, i.e. a device that directly measures gait velocity without requiring a specific distance to be walked. The aim of this paper is to describe a device that fulfills these characteristics and that also correlates well with standard clinical tests.

II. Device Description

The device is based on a commercially available Doppler sensor (X-Band Doppler Motion Detector MDU 1130, Microwave Solutions LTD., Marlowes, United Kingdom) producing a square signal, the frequency of which corresponds to the difference in frequency Δf between the emitted and reflected waves. The gait velocity is in turn proportional to Δf by the relationship:

$$v = c \frac{\Delta f}{2 f_e}$$

where c is the speed of light in space, f_e the emitted frequency (9.9 GHz) and v the gait velocity.

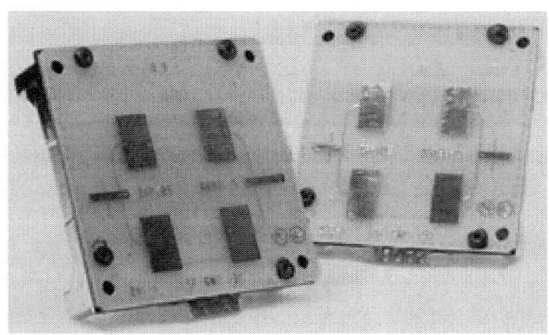

Fig. 1 The Doppler sensor (source: MDU1130 technical note)

The analog part of the electronics includes an amplifier, a 2nd order Butterworth low-pass filter and a hysteresis threshold amplifier. The whole analog chain produces a normalized square signal that is connected to an input of a microprocessor (PIC 18LF13K2, Microchip Technology, Chandler, Arizona, USA). The embedded software computes the successive inter-pulse intervals from an internal counter. Eventually the successive values are sent to a receiver (typically a tablet PC) through a Bluetooth module. An overall view of the electronics is shown in figure 2.

Fig. 2 Electronic circuit layout

The sensor, all electronic parts, batteries, power switch and LED are included in a plastic box that has no shield effect with respect to the considered waves. In the future, the box will be replaced by a more friendly and decorative object in order to improve the device acceptability.

III. EXPERIMENTATION

A. Protocol

The aim of the experiment was to compare the velocities calculated by the new device with those obtained by classical tests of time measurement on distances usually encountered in clinical practice (4.5 m and 10 m).

Twenty three subjects, all of whom were members of the laboratory, volunteered to participate in the study. All subjects gave their informed consent. All subjects walked 10 times down a corridor over a maximum distance of 10m at three self-determined speeds (slow, usual, fast).

Three investigators were present to i) run the software associated to the device that calculated the instantaneous velocities, ii) measure the time to reach 4.5 m, iii) measure the time to reach 10 m.

The device was placed on a table (height 72 cm), facing the direction down which the subject walked. Subjects were tested walking away from the device. After software initialization was performed (establishing the connection between the device and the tablet PC), the first investigator invited subjects to start walking at a steady self-determined pace (slow, usual, or fast). The time measurements started when the subject crossed the start line that was marked on the floor and stopped when the subjects crossed the 4.5m and 10m lines, respectively.

B. Data Processing

A rudimentary signal processing method was applied to the recorded signals in order to limit the effects of the quick motions of the superior members (mostly arm motions), which produce peaks of velocity in the recordings. Theoretically, classical filtering is not applicable as the sampling is irregular by nature, with each value corresponding to an interval length. However, in order to produce preliminary results, a moving average was applied on a 10-point window, and the maximum of the smoothed signal then computed. More sophisticated signal processing methods are currently under development.

IV. RESULTS AND DISCUSSION

An example of the signals produced by the device is given in figure 3, where the successive steps are clearly recognizable. In the initial version of the sensor, the maximum distance over which velocity can be estimated does not exceed 3 m, owing to the threshold used on the amplifier in order to maximize the signal to noise ratio. Tests are currently in progress to try and increase the distance up to five or six meters.

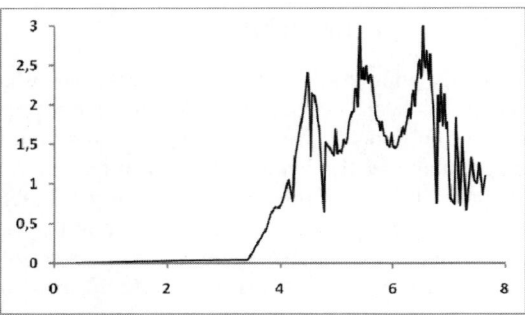

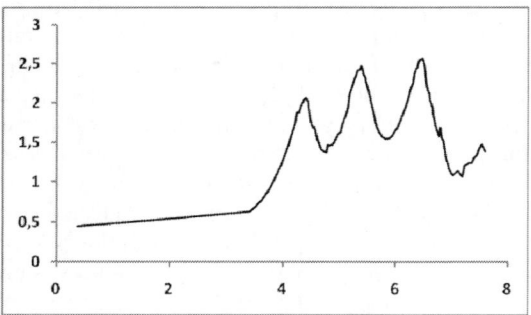

Fig. 3 An example of a raw recording (upper tracing) and the corresponding filtered signal (lower tracing). X axis in seconds, Y axis in km/h

The correlations between the velocities produced by the device and those produced by the manual time measurement of 4.5 m and 10 m, respectively, were then calculated. Results are shown in figure 4 for 4 conditions (4.5 m or 10 m, intercept computed by the modeling or forced to 0).

The best correlation obtained between the device and a stop watch measurement was observed for the short distance (4.5 m), which corresponds to the maximum distance over which the sensor is able to function. Furthermore, there are some discrepancies between the measurements on the two distances, especially for the higher velocities which correspond to the shortest times (figure 5).

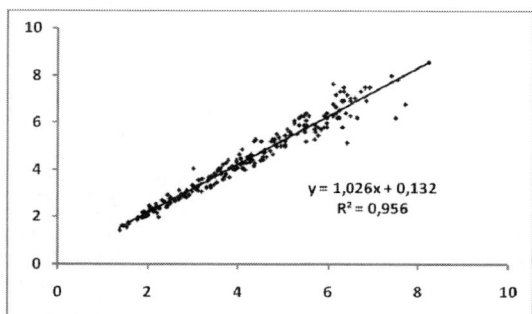

Fig. 5 Correlation between velocities computed on 4.5 m (X axis) and 10 m (Y axis). Both axes are in km/h.

The results presented here are very preliminary. No optimization has been achieved either for electronics or for data processing, and the maximum distance reached by the sensor is still very short. However, the system clearly enables instantaneous velocities to be observed (see for instance figure 3b). When applying a more adapted processing to the raw signal delivered by the sensor, it would be expected that a velocity profile that could describe the walk more precisely would be obtained.

V. CONCLUSION

A new device has been developed to measure walking speed in any environmental conditions and without the need of the presence of a health professional. Preliminary results are very promising. There are many ways to explore in order to improve the results and to increase the distance the sensor is able to reach. Such a device would be very useful to contribute to the detection and follow-up of frailty in the elderly in their usual environment.

ACKNOWLEDGMENT

This work was supported by the Champagne-Ardenne Regional Council (CRCA), the European Regional Development Fund under the Collaborative Research Program, and the Regional Nord-East Pension Fund (ARPEGE Project).

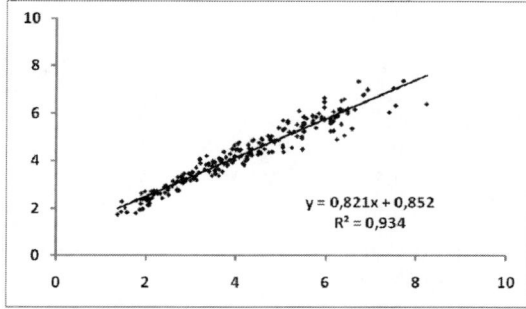
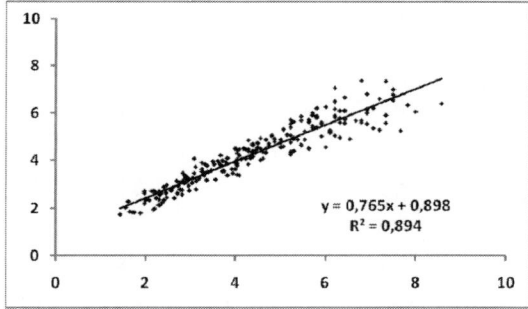
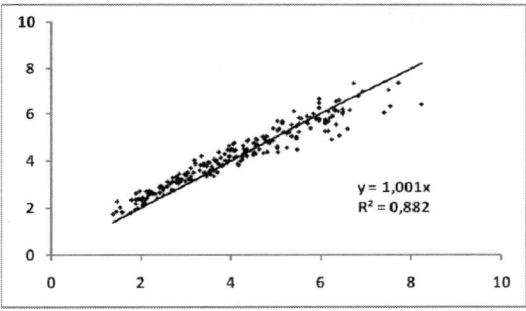
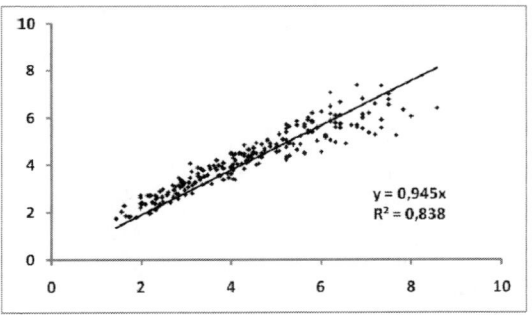

Fig. 4 Correlation between velocities obtained by hand-held stopwatch (X axis) and those obtained by the new device (Y axis). From top to bottom: 4.5 m, intercept computed by the model – 10 m, intercept computed by the model – 4.5 m, intercept fixed to 0 – 10 m, intercept fixed to 0. All axes are in km/h.

We would like to thank warmly the whole technical team of our laboratory for their major contribution to hardware and software developments.numbering.

REFERENCES

1. L. P. Fried, C. M. Tangen, J. Walston, A. B. Newman, C. Hirsch, J. Gottdiener, *et al.*, "Frailty in Older Adults: Evidence for a Phenotype," *J Gerontol A Biol Sci Med Sci,* vol. 56, pp. M146-157, March 1, 2001 2001.
2. O. Theou, G. R. Jones, J. M. Jakobi, A. Mitnitski, and A. A. Vandervoort, "A comparison of the relationship of 14 performance-based measures with frailty in older women," *Appl. Physiol. Nutr. Metab.,* vol. 36, pp. 928-938, 2011.
3. M. Montero-Odasso, M. Schapira, E. R. Soriano, M. Varela, R. Kaplan, L. A. Camera, *et al.*, "Gait Velocity as a Single Predictor of Adverse Events in Healthy Seniors Aged 75 Years and Older," *J Gerontol A Biol Sci Med Sci* vol. 60, pp. 1304-1309, 2005.
4. G. Abellan van Kan, Y. Rolland, S. Andrieu, J. Bauer, O. Beauchet, M. Bonnefoy, *et al.*, "Gait speed at usual pace as a predictor of adverse outcomes in community-dwelling older people an International Academy on Nutrition and Aging (IANA) Task Force," *J Nutr Health Aging,* vol. 13, pp. 881-9, Dec 2009.
5. H. Shimada, S. Obuchi, T. Furuna, and T. Susuki, "New Intervention Program for Preventing Falls Among Frail Elderly People - The Effects of Perturbed Walking Exercise Using a Bilateral Separated Treadmill," *Am J Phys Med Rehabil,* vol. 83, pp. 493-499, 2004.
6. L. Wolfson, R. Whipple, P. Amerman, and J. N. Tobin, "Gait Assessment in the Elderly: A Gait Abnormality Rating Scale and Its Relation to Falls," *Journal of Gerontology,* vol. 45, pp. M12-M19, January 1, 1990 1990.
7. J. Verghese, R. Holtzer, R. B. Lipton, and C. Wang, "Quantitative Gait Markers and Incident Fall Risk in Older Adults," *J Gerontol A Biol Sci Med Sci,* vol. 64A, pp. 896-901, 2009.

Corresponding author:

Author: R. Jaber
Institute: Université de Technologie de Troyes
Street: 12 rue Marie Curie
City: 10000 Troyes
Country: France
Email: rana.jaber@utt.fr

Magnetic Induction-Based Sensor for Vital Sign Detection

H. Mahdavi and J. Rosell-Ferrer

Group of Biomedical and Electronic Instrumentation, Department of Electronic Engineering,
Universitat Politècnica de Catalunya (UPC), Barcelona, Spain

Abstract—Nowadays, various medical applications demand for less-obtrusive and continuous respiratory and cardiac activity monitoring methods. Sleep monitoring and home health care are some examples of these applications. Magnetic induction could provide such a contactless unobtrusive method for continuous monitoring of vital signs. In this paper, we present the preliminary results of a developed magnetic induction sensor. The system consists of one coil for excitation and a specific shaped coil for detection. For data acquisition we have used a low noise amplifier plus a phase sensitive detection circuit. The measured phase angle and magnitude are acquired simultaneously with cardiac and respiratory signals from standard sensors of a BIOPAC system. Considering the objective of having a system to be integrated in the bed, the coils were designed in a printed circuit board (PCB). Supine and prone positions, apnea and breathing periods were parts of the experiments to investigate different parameters which affect the signal. Comparing the breathing and apnea periods, respiration and heart activity is clearly reflected in the detected signal and the results show that the system could detect these vital signs.

Keywords— Magnetic induction, Unobtrusive monitoring, Vital sign, Continuous measurements

I. INTRODUCTION

Unobtrusive non-contact methods for monitoring different physiological activities have been given a lot of attention recently. Many medical applications need continuous monitoring to be done at home environments with less complicated setup requirements and as a result minimized costs. Monitoring neonates, elderly people at home, patients with conditions that can be perturbed or worsened by contact sensors (like burn victims) and sleep monitoring are some of these applications.

Because of the stated reasons a growing interest has emerged for developing unobtrusive techniques enabling monitoring vital signs with less complexity. One approach is to measure the mechanical effects of cardiac or pulmonary activity. This method has been widely used and it is based on mechanical coupling between the patient and the sensor. Load and pressure sensors, strain gauges, piezoelectric or resistive sensors, pneumatic mattresses, etc. have been integrated in to the beds, chairs and mattresses to record the vital signs.

Photoplethysmography, infrared thermography, Doppler radars, capacitive ECG and magnetic induction are some other techniques that have been investigated and used for unobtrusive contactless monitoring. A review of these techniques could be found in [1] and [2].

This paper is focused on a magnetic induction based sensor for monitoring vital signs. The theory on which this method is based is that, when a conductive object or body is placed in a time varying magnetic field (B), eddy currents are induced in it. These currents produce a secondary magnetic field that can be detected by a properly designed receiver system. This measured signal is a function of the conductivity of the tissues and the geometry of the excitation and detection antennas. The concept has been used extensively for non-destructive testing of metals but since the conductivities of the tissues are very low comparing with metals, the perturbations in the sensed magnetic field are very small which make the measurement process difficult. For a sample of material between an excitation coil and a sensing coil [3]:

$$\Delta B / B \propto \omega(\omega \varepsilon_0 \varepsilon_r - j\sigma) \qquad (1)$$

where σ is the conductivity of the sample, ε_r is its relative permittivity, ε_0 is the permittivity of free space, ω is the angular frequency of the excitation and $j = \sqrt{-1}$. For biological tissues, the conductivity (imaginary) component will normally be dominant [4]. From (1) it is seen that the eddy currents produce a component of the secondary signal, which is proportional to σ and ω. The total detected field $(\Delta B + B)$ lags the primary field by an angle φ [5].

In addition to the small secondary signal, there is another difficulty with the MI systems. In an ideal magnetic induction system, the coupling between the coils and the body (sample) has to be magnetic coupling. In a real system, capacitive coupling is also exists, which may cause some problems in interpreting the secondary received signal. Several groups

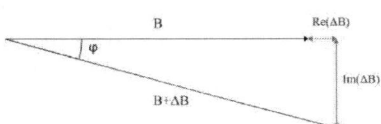

Fig. 1 Phasor diagram of the primary B and secondary (ΔB) magnetic fields

have been working on this area and the effects of capacitive coupling are well investigated [6].

Different shielding techniques and tests [7] such as individual coil screening, electric field shielding (Star pattern shielding, conductive adhesive metallic tape applied to the coil former, etc.) have been applied [5] and the true MI signal contamination from the capacitive effect were studied.

II. MATERIALS AND METHODS

A. Simulations

A simplified model of the human trunk were simulated with a finite element software. The conductivity of the lungs was considered as a variable parameter to simulate the breathing. The phase changes in the secondary magnetic field due to the inflation and deflation in the lungs at a distance of 15 cm from the body were measured. Figure 2 shows the designed model. Also, since the system is based on the exposure of an electromagnetic field, evaluation of the specific absorption rate (SAR) is of interest to be in accordance with international standards and assure the safety. Safety studies of the developed system could be found in [6] and [8].

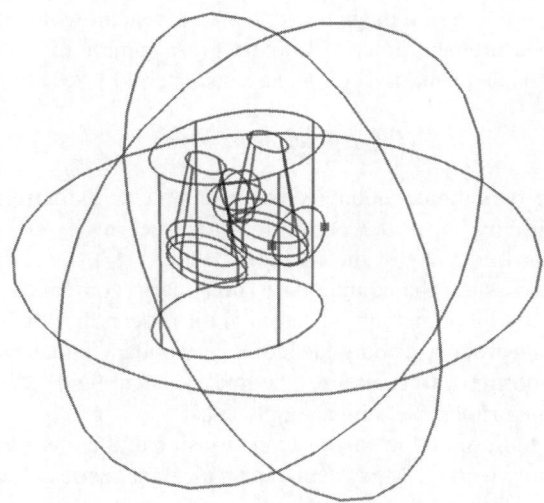

Fig. 2 General view of the designed model

B. Hardware

The system is mainly based on two coils as excitation and detection coils and a phase sensitive detector. The excitation signal is provided by a signal generator at a frequency of 10 MHz. The coils have been designed in a planar configuration

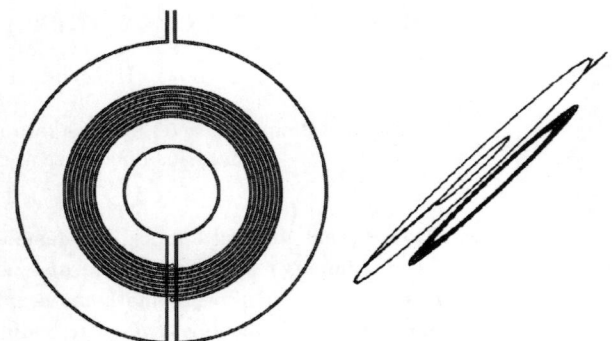

Fig. 3 Excitation-Detection coil configuration

and implemented in printed circuit board to be placed under the mattress.

As stated before, the secondary magnetic signals are very weak and since the secondary coil will receive the superposition of the primary and secondary fields, the coil structure is very important in order to suppress the large primary signal. Different coil configurations have been tested and the best results obtained by a semi concentric and symmetric coil structure. Figure 3 shows the used structure for Excitation-Detection coils.

The excitation coil is an 8-turn spiral coil with a diameter of 5 cm and the detection coil is a one turn antenna with a shape that is adapted to cancel out the primary field. The coil configuration design is based on the arrangement design explained in the patent application, " Coil arrangement for a magnetic induction impedance measurement apparatus comprising a partly compensated magnetic excitation field in the detection coil ". [9].

Both coils were implemented in PCB and placed symmetrically with a 5 mm distance. For the experiments, the coils were placed under the bed with a 15 cm separation from the body.

The detected signal at the second coil is first passed through the AD8432 to be amplified and then introduced to a low pass filter and a AD8302 for demodulation and phase detection.

The amplifier is a dual channel, low power, low noise amplifier with an input voltage noise of $0.85\ nV/\sqrt{Hz}$ and a gain of 64. Ad8302 is a RF/IF gain and phase detector with an accurate phase measurement scaling of $10 mVolt/deg$.

As a reference of the experiments, for breathing and cardiac activity signals, pulse plethysmogram (PPG) and respiratory effort transducer of a BIOPAC-MP36 system has been used together with the two channels of data (phase and magnitude) acquired from the developed system. A block diagram of the hardware is shown in figure 4.

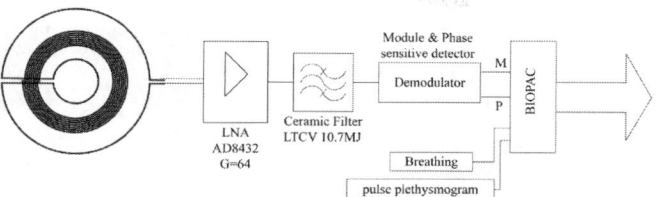

Fig. 4 Block diagram of the system

C. Experiments

As mentioned before, the coils were placed under the bed at a distance of 15 cm from the body and the BIOPAC transducers were used as reference. The test procedure started with breathing normally, followed by a period of apnea and finally deep inhalations. Experiments were done in supine and prone positions by an adult volunteer lying on the bed.

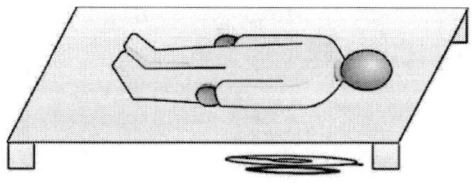

Fig. 5 Measurement setup

III. RESULTS

The differences between deep inhalation and normal breathing are clearly reflected in both magnitude and phase of the detected signal and it is in accordance with the reference breathing signal from the BIOPAC. The maximum phase change detected during breathing is about 1.06 degree (10.6 mvolts).

The periodical waveform which could be seen during apnea period is related to the heart activity. As it could be compared with the pulse plethysmogram signal, the phase detected during apnea period is in accordance with the real heart activity. The maximum phase change detected in apnea period -due to heart activity- is about 130 mdegree (1.3 mvolts) Comparing supine and prone positions, the signal obtained in prone position was less noisy and more clear than the supine. The changes due to this difference in the detected signal could be a result of the distance between the organs and the sensor and/or the fat and bone layer.

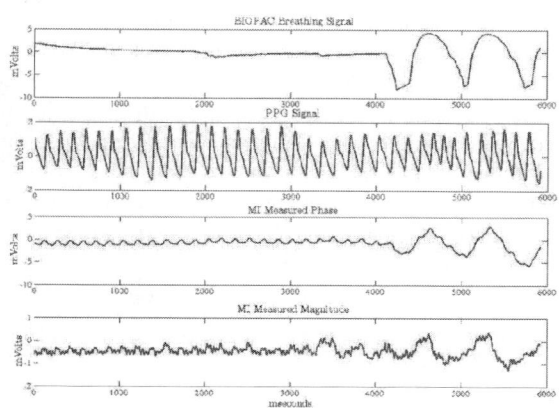

Fig. 6 Measured signals from BIOPAC and the developed MI system, including breathing and apnea periods

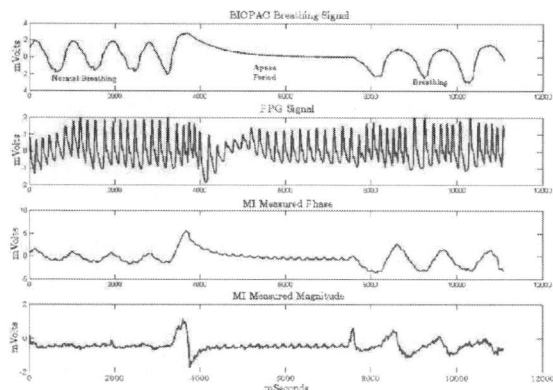

Fig. 7 Apnea vs. Breathing, Heart activity is clearly visible in the measured signal from MI system during apnea and breathing

IV. CONCLUSION

In this paper, a magnetic induction sensor for vital sign monitoring has been described. According to the objectives, the system was developed using less complex electronics to be unobtrusive and cheaper in comparison with previous works.

The first measurements with a volunteer lying on the bed demonstrate that the system is able to detect breathing and cardiac signal. The signals were obtained simultaneously from the developed MI system and BIOPAC as a reference.

The preliminary results show a good correlation between the spectrum of the detected signals from both systems. In addition, presence of the subject on the bed (going to the bed and coming out) has been detected. The results show that the received signal is highly depend on the position (supine/ prone) and the body shapes.

Future studies will focus on hardware improvement and signal processing in order to separate heart activity and breathing. Moreover the sensitivity of the system to the displacements, movements and different body shapes will be studied and examined.

ACKNOWLEDGEMENTS

This work has been funded in part by the Help4Mood project (Contract Number: ICT- 248765) from the European Union's Seventh Framework Program.

REFERENCES

1. Teichmann Daniel, Br Christoph, Eilebrecht Benjamin, Abbas Abbas, Blanik Nikolai, Leonhardt Steffen. Non-contact Monitoring Techniques - Principles and Applications in *34th Annual International Conference of IEEE EMBS*:1302–1305 2012.
2. Mahdavi Hadiseh, Ramos-Castro Juan, Giovinazzo Giuseppe, García-González Miguel A, Rosell-Ferrer Javier. A Wireless under-matress sensor system for sleep monitoring in people with major depression in *The Ninth IASTED International Conference on Biomedical Engineering* 2012.
3. Griffiths H, Stewart W R, Gough W. Magnetic induction tomography. A measuring system for biological tissues. *Annals Of The New York Academy Of Sciences*. 1999;873:335–345.
4. Griffiths H. Magnetic induction tomography *Measurement Science and Technology*. 2001;12:1126–1131.
5. Barai A, Watson S, Griffiths H, Patz R. Magnetic induction spectroscopy: non-contact measurement of the electrical conductivity spectra of biological samples *Measurement Science and Technology*. 2012;23:085501.
6. Mahdavi Hadiseh, Rosell-Ferrer Javier. A magnetic induction measurement system for adult vital sign monitoring : evaluation of capacitive and inductive effects in *XV Int. Conf. on Electrical Bio-Impedance & XIV Conf. on Electrical Impedance Tomography*;012085 2013.
7. Goss D., Mackin R. O., Crescenzo E., Tapp H. S., Peyton A. J.. Understanding the coupling mechanism in high frequency EMT in *3rd World congress on industrial process tomography*:364–369 2003.
8. Mahdavi Hadiseh, Rosell-Ferrer Javier. Simulation of a Magnetic Induction Method for Determining Passive Electrical Property Changes of Human Trunk Due to Vital Activities in *COMSOL conference* 2012.
9. Rosell-Ferrer Javier, Igney Claudia Hannelore, Hamsch Matthias. Planar coil arrangement for a magnetic induction impedance measurement apparatus 2012.

Author: Hadiseh Mahdavi
Institute: Universitat Politècnica de Catalunya (UPC)
Street: C.Jordi Girona
City: Barcelona
Country: Spain
Email: hadiseh.mahdavi@upc.edu

Unconstrained Night-Time Heart Rate Monitoring with Capacitive Electrodes

A. Vehkaoja, A. Salo, M. Peltokangas, J. Verho, T. Salpavaara, and J. Lekkala

Department of Automation Science and Engineering, Tampere University of Technology, Tampere, Finland

Abstract—An unobtrusive measurement system for night-time heart rate (HR) monitoring is presented. The system uses capacitive electrodes that do not require galvanic skin contact, thus allowing user to wear clothes while being monitored. The electrodes are located transversely on the bed and the electrocardiographic signals are measured between each electrode and a common reference. The signals are then combined in order to remove the common mode move-ment artifacts and the best one is selected for HR detection.

The average RR-interval detection coverage with five test subjects in one hour long measurements was 81.0 %. The average mean absolute error compared to the reference ECG signal was 1.78 ms in average.

Keywords—**Night-time heart rate, capacitive electrode, physiological monitoring.**

I. INTRODUCTION AND MOTIVATION

Night-time physiological information, especially the heart rate, the respiration rate and the movements can be used for example in screening of medical disorders like sleep apnea [1] and evaluation of sleeping quality [2] or detecting psychophysiological stress [3]. Many potential applications would benefit or even require continuous monitoring for detecting long term changes in sleeping pattern and sleep parameters. In these cases, it is important that the measurement equipment is unobtrusive and requires little effort from the user. Comfort is increased if the user does not need to wear the measurement equipment but the sensors are e.g. integrated into the bed.

There are several sensor options for such an unconstrained monitoring system. Heart rate and respiration can be extracted from ballistographic signals that are recorded using force sensors e.g. [4] or from the electrocardiogram (ECG) that can be measured with contact electrodes attached to a shirt or to the bed sheet as e.g. in [5]. The drawback of the ballistographic technique is that it is very sensitive to all kinds of movements. For example, movements caused by another person sleeping in the same bed can disturb the measurement and cause erroneous results. In addition, the ballistographic signal does not have as distinguishable and consistent features as the ECG's R-peak, which increases the uncertainty to the beat-to-beat interval detection. Contact ECG recorded e.g. with textile electrodes attached to the bed sheet enables monitoring of heart rate and heart rate interval generally with better coverage and accuracy than the ballistographic technique but its drawback is the requirement for direct skin contact which, in practice, prevents the user from wearing a shirt during the measurement. Using capacitive non-contact electrodes for measuring the ECG signal combines the good features of these two techniques by measuring the signal that enables accurate detection of the heart rate, yet allowing the user to wear clothes while being monitored.

II. RELATED WORK

ECG recording using capacitive bed-integrated electrodes has been studied by several researchers. Wu and Zhang used wide electrode strips made of conductive textile and placed them on the bed in transversal direction. They measured the ECG between two strips while the third strip acted as an active ground electrode [6]. They received good results when making the test measurements in an electrically shielded room. More than 98 % of the recorded data was artifact free by visual inspection. They however reported the heart rates as a mean value for one minute so it is not possible to say the amount of correctly recognized individual R-peaks. An interesting detail in their approach was that they used a high-pass cut-off frequency as high as 8 Hz when filtering the measurement signal. Usually the high-pass filter is set below 1 Hz frequency in ECG recordings.

Similar electrode setup was also used by Ishida *et al* in [7]. They also tested the effect of a layer of electrically shielding material under the electrodes. Despite that the shielding decreased the coupling of electrical interference, they were able to detect the heart rate only 27 % of the time during one night measurement.

Lim *et al.* developed a capacitive ECG measurement system that uses several active electrodes placed transversally on top of the mattress and a large sized common reference electrode located under the lower body [8]. They stated that the capacitive measurement is very sensitive to motion artifacts and that the main sources for these are the changes of the electrode and the grounding capacitances and the triboelectrical effect caused by friction of the clothes. The morphology of the ECG signal measured with a similar electrode setup has also been successfully used for detecting sleeping posture [9].

III. MATERIALS AND METHODS

A. The Measurement Setup

Our ECG monitoring system consists of four capacitive active electrodes and a large common reference electrode. The active electrodes are located transversally so that they are approximately at the location of the chest. The inter-electrode distance is set to 5 cm. The whole width of the bed cannot be covered when using such small inter-electrode distance and only four measurement electrodes, but this has not been a major problem in the laboratory measurements, where the position of the test person can be controlled. More electrodes are, however, needed if the system is used at home in overnight measurements. The capacitive reference electrode is made of a conductive fabric and it is placed under the bed sheet thus covering the whole area of the bed.

In previous studies we have measured ECG with contact electrodes attached transversally to the bed sheet by using bipolar measurement between the electrodes [5]. When using capacitive electrodes, the high-pass cut-off frequency of each electrode depends on the skin-electrode capacitance and the input resistance of the active electrode. The skin-electrode capacitances may vary during the measurement and are not necessarily the same in all electrodes. Having different cut-off frequencies in the electrodes would decrease e.g. the common mode rejection ratio of a bipolar measurement and therefore we decided to use the unipolar measurement between each active electrode and a common reference electrode.

The reference electrode is made as large as possible in order to maximize its capacitance with the body, thus making the capacitive grounding as strong as possible. If the body is in galvanic contact with the electrode, the grounding is further improved and the measurement noise decreases. Fig. 1 shows the measurement setup used.

Fig. 1 Capacitive ECG system and a close-up view of one electrode. The wireless contact ECG reference device is also seen on the bed.

B. The Capacitive Active Electrodes

An important aspect, when designing a measurement system that uses capacitive electrodes, is to provide high input resistance with respect to the electrode capacitance. We chose OPA129 from Texas Instruments as the operational amplifier for the active electrodes because it has extremely low input bias current (±30 fA typical) thus enabling the use of extremely high resistance in the bias current path and therefore also high input impedance in the capacitive electrode.

In order to increase the input impedance of the amplifier connection, the bias current is taken from the amplifier's output using so called bootstrap connection. This effectively multiplies the value of the bias resistor R1 in the Fig. 2 by the ratio of resistors R5 and R4:

$$R_{in} = (\frac{R5}{R4}+1) \cdot R1 + R3 \quad (1)$$

The benefit of the bootstrap connection is that the effects of the leakage currents on the circuit board are minimized because the value of the actual resistor can be set smaller and the voltage across the resistor is smaller. The drawback of the bootstrap circuit on the other hand is that it increases the noise and the offset voltage of the amplifier. However, the frequency bandwidth required in ECG monitoring is so small that the increased noise density is seldom a critical issue.

The diameter of our active electrodes is 25 mm. When the person being measured is wearing 0.5 mm thick cotton shirt, assuming relative permittivity of 2, the capacitance of the electrode becomes approximately 20 pF and the high-pass cut-off frequency then about 0.1 Hz.

Because of the extremely high input impedance, it is important to carefully protect the amplifier input from leakage currents that might cause the input voltage to wander uncontrolled. A guard trace is therefore placed everywhere around the electrode surface and the op amp on the layout. Special pinout of the OPA129 also supports efficient guarding.

The guard also decreases the parasitic capacitance of the circuit board layout, but the input capacitance of the amplifier component is still present. In order to minimize the remaining parasitic capacitance we used an active positive feedback for compensating it. The gain of the feedback is adjusted by the trimmer resistor R2 so that when a square wave signal is fed capacitively to the electrode, its waveform should not be distorted, besides the distortion caused by the high-pass filter in the electrode input. Similarly, the amplitude should match the gain of the active electrode, 0.9 in our case.

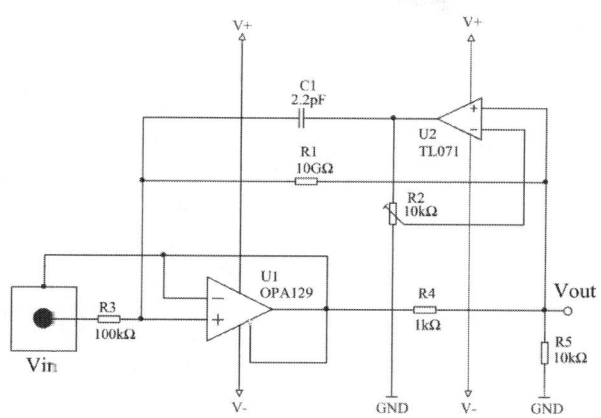

Fig. 2 The schematic of the capacitive active electrodes.

C. Data Acquisition

The four capacitive active electrodes, along with the common reference electrode, are connected to a custom made data acquisition (DAQ) device. The input voltage range of the DAQ is ± 0.436 V with a resolution of 13.4 µV when the DAQ's internal amplifier, that has a gain of 5, is used. The RMS noise level of the DAQ is less than 1 digital unit.

The DAQ is capable of transmitting the measurement data to the PC either through a USB or by a wireless Bluetooth link. We used the Bluetooth mode in the tests. The maximum sampling rate is 1 kHz per channel but 250 Hz was found high enough for the capacitive ECG recording.

C. Signal Processing

The capacitive ECG measurement is very sensitive to movements. Large movements, e.g. changing the sleeping posture or even a movement of a limb causes so strong disturbance to the measured signal that recognition of R-peaks is impossible. Even the ballistographic movements caused by heart beats and breathing may cause interference that is higher in amplitude than the electrical ECG signal as seen in the upper figure of Fig. 3. Luckily, often the movement artifact is seen similarly in multiple channels while the ECG component in the signals is different because of the different electrode location on the body. We therefore combine the measurement channels by subtracting their signals and let a channel selection algorithm to select the best signal for R-peak detection. Also the original signals are used as possible candidates in the selection phase.

We used same signal processing algorithm for finding the R-peaks, selecting the best channel, and calculating the heart rate as we have earlier used for processing ECG signal recorded with contact electrodes attached on a bed sheet. A detailed description of the algorithm can be found from [5].

All digital signal processing was done off-line using MATLAB®.

E. Reference ECG Measurement

When testing sensitive capacitively coupled ECG measurement systems, care must be taken that the reference measurement does not affect the capacitive measurement. For example, the reference device should not be connected in such a way that galvanic ground current path would be provided through its electrodes. Therefore we used a wireless ECG measurement system for recording the reference data. Disposable Ag-AgCl electrodes were attached below left and right clavicles and the reference ECG was measured between them. The reference and the capacitive ECG data were synchronized manually. The reference recording system has been presented in [10].

F. Test Measurements

We evaluated the operation of the system with five test subjects in one hour long recordings. All subjects were normal weight healthy male adults (25-33 years) with no history of cardiovascular diseases. Subjects spent 15 minutes in each main sleeping posture (supine, left side, prone, and right side) and they were wearing t-shirt made of cotton. Any galvanic contact to the measurement device was prevented.

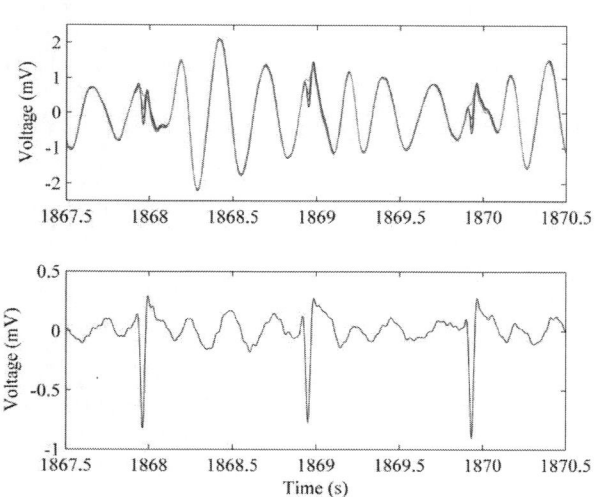

Fig. 3 Four signals recorded with the capacitive electrodes (upper figure). Large common mode artifacts are caused by the ballistographic movements caused to the heartbeats. The lower figure shows how the R-peaks are emphasized when subtracting two signals (blue and cyan).

IV. RESULTS

Table 1 shows the results calculated from the five test measurements. The RR-interval detection coverage varied between 62 % and 96 % between the subjects during the whole one hour measurement period. The average coverage was 81 %.

The average mean absolute error (MAE) of the detected RR-intervals was 1.78 ms, which can be considered a good result. Recordings of the test subjects 3 and 5 both contained 3 false positive R-peak detections, which increase the MAE slightly. The error is also partially explained by the small temporal drift between the signals of the capacitive and reference recording devices. This results from the different sampling clocks between the two devices. The drift was only approximately 0.1 seconds during the one hour measurements so we decided not to correct it.

A. The Effect of the Sleeping Posture

The Table 1 also presents the detection coverage figures for each sleeping posture separately. As seen from the results, the capacitive heart rate monitoring system performs well in all other sleeping postures but is not as reliable when the person is lying on the right side. For the subject number three, the RR-intervals were detected only 8 % of the time spent on the right side and for the subject number five, the algorithm was not able to find any R-peaks. A visual inspection of the data of the subject number five showed that the ballistographic movement component was seen in the signal but the ECG component wasn't different enough between the channels to produce usable results. A possible explanation for the missing electrical difference component is that because the heart is located on the left side of the body, the difference signal between any two electrodes at the right side was not strong enough for the R-peak detection.

The rightmost column in the table shows the average detection coverage of all postures. The time period of changing the posture is not considered in this data.

Table 1 Performance figures of the capacitive ECG monitoring system.

Subject number	Coverage	MAE (ms)	Cov. supine	Cov. left	Cov. prone	Cov. right	Mean Cov.
1	84.77	1.88	90.77	72.53	93.02	91.69	87.0
2	89.82	1.62	97.85	91.48	98.47	85.91	93.4
3	62.66	1.96	92.49	87.20	70.85	8.42	64.7
4	95.96	0.92	99.28	99.87	99.82	96.63	98.9
5	71.90	2.51	99.88	99.88	97.58	0.00	74.3
Average	81.0	1.78	96.1	90.2	91.9	56.5	83.7

V. CONCLUSIONS

A bed integrated system for capacitive monitoring of the ECG signal was proposed. The system performs well in other sleeping postures but has sometimes problems in detecting R-peaks from the ECG signal recorded when the subject is lying on the right side. Testing the monitoring system in overnight measurements is a part of the future work, as well as finding a solution for the R-peak detection in right side sleeping posture.

ACKNOWLEDGMENT

The authors would like to thank the test subjects for their participation.

REFERENCES

1. Roche F, Gaspoz J-M, Court-Fortune I et al. (1999) Screening of obstructive sleep apnea syndrome by heart rate variability analysis. Circulation, 100(3):1411–1415
2. Bulckaert A, Exadaktylos V, De Bruyne G et al (2010) Heart rate-based nighttime awakening detection. Eur J. Appl. Physiol. 109(2):317–322
3. Hall M, Vasko R, Buysse D et al (2004) Acute stress affects heart rate variability during sleep. Psychosom Med 66(1):56–62
4. Brüser C, Stadlthanner K, de Waele S, Leonhardt S (2011) Adaptive beat-to-beat heart rate estimation in ballistocardiograms. IEEE Trans. on Inf. Tech. Biom. 15(5):778–786
5. Peltokangas M, Verho J, Vehkaoja A (2012) Night-Time EKG and HRV Monitoring With Bed Sheet Integrated Textile Electrodes. IEEE Trans. on Inf. Tech. Biom. 16(5):935–942
6. Wu K-f, Zhang Y-t (2008) Contactless and continuous monitoring of heart electric activities through clothes on a sleeping bed. Proc. vol. 5, Int. Conf. on Inf. Tech. Appl. Biom. ITAB, Shenzhen, China, 2008, pp 282–285
7. Ishida S, Shiozawa N, Fujiwara Y, Makikawa M (2007) Electrocardiogram measurement during sleep with wearing clothes using capacitively coupled electrodes. Proc. vol. 29 Ann. Int. Conf. IEEE Eng. Med. Biol. Soc. in Engineering in Medicine and Biology Society, Lyon, France, 2007, pp 2647 –2650
8. Lim Y G, Kim K K, Park K S (2007) ECG recording on a bed during sleep without direct skin-contact. IEEE Trans. on Biomed. Eng. 54(4):718–725
9. Lee H, Hwang S, Lee S et al. (2013) Estimation of Body Postures on Bed Using Unconstrained ECG Measurements. IEEE J. Biomed. and Health Inf. (in press)
10. Vehkaoja A, Verho J, Cömert A et al. (2012) Wearable System for EKG Monitoring - Evaluation of Night-Time Performance. Revised Selected Papers from Mobihealth 2011, Lecture Notes of the Institute for Computer Sciences, Social Informatics and Telecommunications Engineering, 83:119-126

Author: Antti Vehkaoja
Institute: Tampere University of Technology
Street: Korkeakoulunkatu 3
City: Tampere
Country: Finland
Email: antti.vehkaoja@tut.fi

Wrist-Worn Accelerometer to Detect Postural Transitions and Walking Patterns

V. Ahanathapillai, J.D. Amor, M. Tadeusiak, and C.J. James

Institute of Digital Healthcare -WMG, University of Warwick, Coventry, CV4 7AL, UK

Abstract—The identification of postural transitions and walking patterns are important in the recognition of activities of daily living. This paper presents a feasibility study to see if postural transitions and walking patterns can be identified using a wrist-worn accelerometer-based device. Firstly, the postural transition (i.e. sit-to-stand and stand-to-sit) templates are extracted and template matching is used to identify the transitions between daily living activities such as walking within a house. Secondly, the various walking patterns such as walking on a level surface, walking up the stairs and walking down the stairs are classified using statistical features and nearest neighbor classifier. A comparison between the activity pattern from the dominant wrist and the non-dominant wrist is also presented.

Keywords—Accelerometer, Wrist-worn, Postural transitions, Activity classification, Activities of daily living.

I. INTRODUCTION

Activity monitoring to assist in maintaining a healthy daily life can be an important aide for the well-being of any person. Activity monitoring typically includes postural transition analysis, for its applications in monitoring older person or stroke patients. [1] suggests that many falls in stroke patients occur during postural transition activities, such as standing up, sitting down, or initiating walking. The automatic classification of the activities of daily living (ADL) is an active research area, as it has numerous applications. Most especially in the case of an elderly or physically challenged person, it is beneficial to identify the ability of a person to perform their ADL and also identify incidents such as falls; this can be used to assist in the care giving for such a person.

Accelerometers are widely used for the purpose of monitoring activity in humans. Activity classification studies was often been performed by placing a tri-axial accelerometer on the chest [2, 3], waist [4, 5] and multiple locations [6, 7]. Though, the placement of the accelerometers is usually chosen based on the application, for example, whole body movements (chest, sternum, under arm or waist), leg movement (shin, ankle) and Parkinsonian tremor (wrist), a simple system using only one sensor attached to the wrist is generally preferred, especially for older person [8].

As part of the USEFIL (Unobtrusive Smart Environment for Independent Living) project, we aim to develop a wrist wearable device to continuously monitor the activity of an older person to assist in their independent living, through monitoring their ADL and against falls.

The aim of this preliminary study is to determine if the wrist worn tri-axial accelerometer can be used to accurately detect postural transitions (i.e. sit-to-stand or stand-to-sit) and walking patterns. Furthermore, we would like to see if the data obtained from the accelerometer placed on the dominant / non-dominant wrist has any significant advantage in terms of data analysis. The remainder of this paper is organized as follows: the experimental setup, data collection and detailed description of the various stages involved is tri-axial accelerometer processing is discussed in section II. The results and discussions are presented in section III and section IV contains the concluding remarks.

II. METHODOLOGY

A. Device Specification

Fig.1 shows the Z1 Android Watch-Phone [9] used in this study. The inbuilt accelerometer sensor in the Z1 watch placed on the wrist measures the activity of the wearer. This device runs on an Android 2.2 operating system and contains a MediaTek MT6516 CPU (416 MHz with 256 MB RAM). The device consists of an 8 GB internal memory, which is used to store the recorded data, and has Bluetooth and Wi-Fi connections to transfer data to a computer for further analysis.

Fig.1 Z1 Android Watch-Phone used in the USEFIL project to monitor activity

B. Experimental Setup and Data Collection

The accelerometers in the watch were used to record the activity as measured at the wrist. The sampling rate is set to the maximum for this sensor at 50 Hz, which is sufficient to pick up the short activity signatures corresponding to the sit-to-stand and stand-to-sit postural transitions. Repeated sit/stand activity data was first collected. Approximately 10 sitting and standing (sitting for 5 seconds and standing for 5 seconds) data was timed via a tone feedback to the subject, as shown in Fig. 2. Next, the subjects were also asked to sit randomly while walking within a room. The activity data representing the various walking pattern of the person walking - including walking within a room, up and down the stairs was collected.

The data was collected from 12 healthy subjects, which includes 9 male and 3 female subjects. The subject's age range was between 22 to 37 years. Their height varied from 156 to 187 cm, their weight varied from 51 to 100Kg and their Body Mass Index range in the 19.59 to 28.59 kg/m^2. The dominant hand of the person was also noted during the trials. During all the trials, two identical devices were used - one on the right and one on the left wrist to assess if the measurements taken from the dominant and the non-dominant wrist were significantly different.

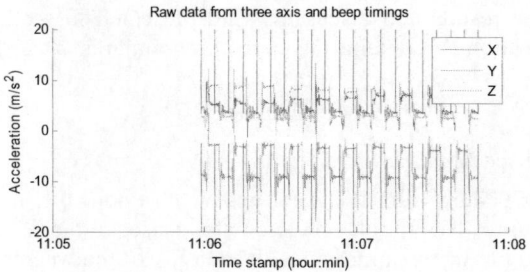

Fig.2 Sitting/Standing trial data obtained synchronized using a tone (dotted lines) feedback to the subject

C. Accelerometer Data Processing

The high level overview of the postural transition and walking pattern detection algorithm developed in this project is presented in Fig. 3. Firstly, Stand-to-Sit (St-S) and Sit-to-Stand (S-St) transition template are extracted from the Sitting/Standing trial data for each person. This template is then used to identify the S-St or St-S transition occurrences in the data that contains random walk and sit patterns. Once the transitions are identified the sit pattern and walk pattern (on level surface within the room) are segmented into sitting and standing data. An attempt was also made to further classify the walk patterns into normal walk, walk down the stairs or walk up the stairs. These stages are detailed in the following sections.

a) Postural Transition Template

The tone was used to extract the S-St and St-S transitions. These transition signals are aligned based on the dominant peak (or cross-correlation maximum). The S-St and the St-S transition template, which is the mean of the 10 aligned postural signals is extracted from the

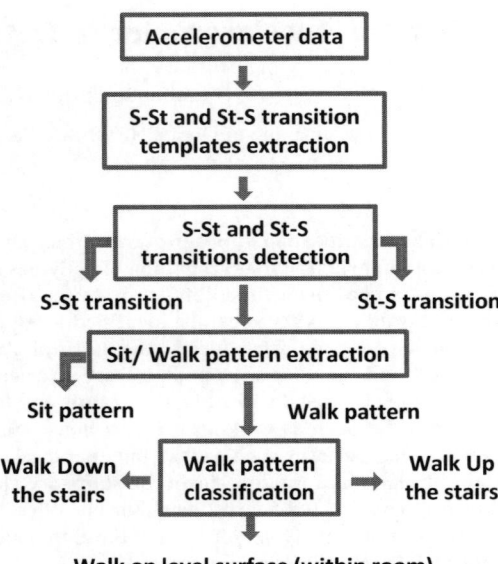

Fig. 3re 1nsactions on Biomedical Engineeringfitions using a miniature Gyroscope and its application in fall risk evaluation in th High level overview of the posture transition and walking pattern detection algorithm

Sitting/Standing trial data for each person. Aligned transitions patterns corresponding to St-S and S-St transitions are shown in Fig. 4 (a) and (b) respectively and the extracted St-S and S-St template is presented in Fig. 4 (c) and (d), respectively.

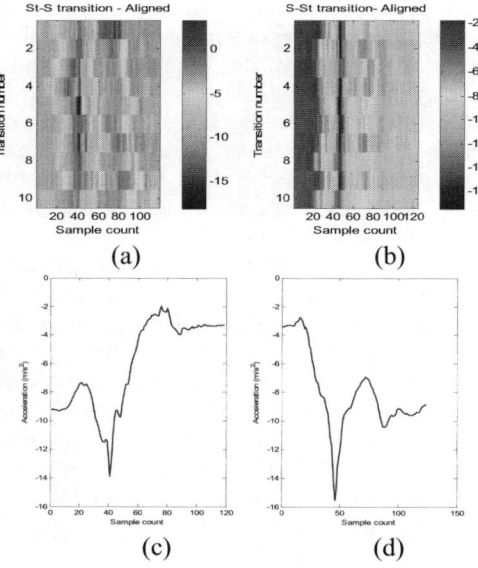

Fig. 4 Aligned transition data and template for St-S (a,c) and S-St (b,d) in a single subject

b) S-St and St-S Transition Detection

The covariance between the postural transition templates and the mixed trial data containing combined walk and sit pattern is calculated and a 50% threshold is used to extract the location of the S-St and St-S transition peaks. Fig. 5 shows y-axis accelerometer data and the extracted postural transition peak for a single subject.

c) Sit/Walk Pattern Extraction

Once the locations of the postural transitions are identified, the sit pattern can be extracted from a mixed trial experiment for each subject. The sit pattern is the data between the identified St-S transitions pattern and the S-St transition pattern. This helps in determining the time spent sitting. Similarly the walk pattern between S-St and St-S transitions is also extracted from the mixed data. Fig 6 presents the three different walking patterns for a single subject.

d) Walk Pattern Classification

Once the walking pattern is extracted, the walking pattern classification is performed to determine if the pattern denotes walking up the stairs, down the stairs or in a level surface. A set of parameters (such as mean, median, mode, maximum, minimum, range, standard deviation, variance, skewness, energy, moments and kurtosis) are determined for the walking data. These statistical features are used to classify walking into normal, up and downstairs using a nearest neighbour (KNN) classifier. The Euclidean distance metric and the nearest rule are the parameter used here for the classification.

III. RESULTS AND DISCUSSION

KNN classification results from this preliminary study are presented in Table 1. Classification results are promising given the fact that a limited data set has been collected so far for this study. It is also apparent the classification results obtained from the dominant and non-dominant hands of all the subjects were comparable.

Table 1 Classification results of various activities

Activity pattern	Left hand	Right hand
S-St transition	94.29%	97.06%
St-S transition	97.14%	94.12%
Sitting	95.71%	95.19%
Walking - Normal	83.33%	83.33%
Walking - Down stairs	91.67%	91.67%
Walking – Up stairs	75%	58.33%

The least discrimination results were obtained while classifying the normal and walking upstairs data. This is understandable, as the parameter sets overlap due to the close similarities between these two activities. Performing the experiments with a larger dataset should improve the obtained classification results. Please note that for all these trials, the subjects were asked to walk the stairs without holding the support rails. In most situations this will not be the case and would make the classification task even more challenging.

Fig. 7 shows a sample walking accelerometer data recorded on the left hand of a subject, who was holding the support rails using his left hand whilst walking down the stairs. It can be seen clearly from the recorded data that these signals are very different to those captured whilst walking without a support (as shown in Fig. 6). Also, all the subjects in this preliminary study were healthy adults. So, this walking pattern will be significantly different from more complex patterns, such as the activity of an older adult walking with or without support and for people with a disability.

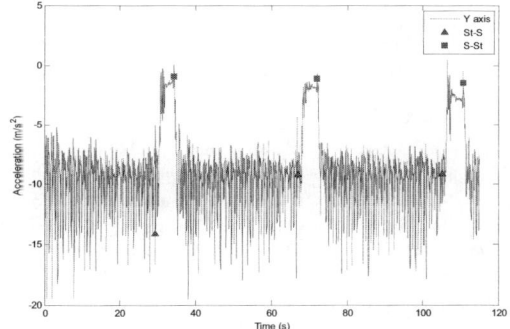

Fig. 5 St-S (triangle) and S-St (square) postural transitions identified in mixed trial data from a subject

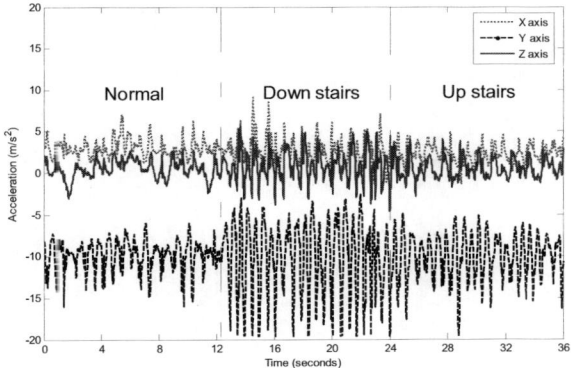

Fig. 6 Walking data extracted from a mixed trial of a single subject

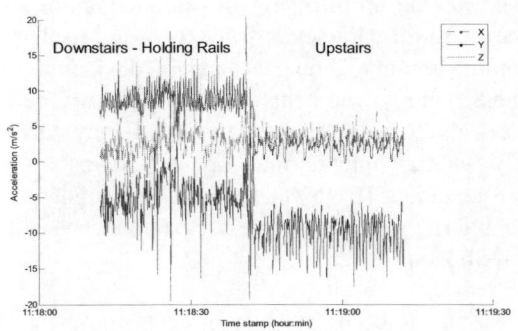

Fig. 7 Activity data recorded from the left hand whilst walking down the stairs holding the support rails (left) and walking up stairs *without* holding the rails

IV. CONCLUSIONS

A preliminary investigation into identification of postural transitions and walking patterns within activities of daily living such as walking on a level floor, using a wrist-worn accelerometer device was presented in this paper. The results to date look promising and could lay the foundation to developing a complex software algorithm that would help identify and classify activity levels.

The results from this feasibility study indicate that the placement of the accelerometer on the dominant or non-dominant wrist didn't make significant difference to the classification results. However, it should be noted that the data set used in this study consists entirely of healthy individuals. The postural transitions and walking could vary significantly in older and disabled people.

We plan to improve the present activity monitoring algorithm to include other typical activities and postural transitions, for both young and older subjects. The ultimate aim is to create a robust software algorithm that could be run on a simple wrist worn device to monitor activity levels and falls, optimized especially for older people.

ACKNOWLEDGMENT

This work is partially included in the activities related to the USEFIL project, funded by the European Commission within the 7th Framework Program (Grant agreement no: 288532).

REFERENCES

1. Nyberg L, Gustafson (1995) Patient falls in stroke rehabilitation. A challenge to rehabilitation strategies. Stroke, Vol. 26, pp. 838-842
2. Godfrey A, Bourke AK, Olaighin GM, Van de Ven P, Nelson J (2011) Activity classification using a single chest mounted tri-axial accelerometer. Medical Engineering & Physics 33:1127-1135
3. Song W, Ade C, Broxterman R, Nelson T, Warren S (2012) Activity recognition in planetary navigation field tests using classification algorithms applied to accelerometer data. International conference of the IEEE Engineering in Medicine and Biology Society. 1586-1589
4. Oshima Y, Kawaguchi K, Tanaka S, Ohkawara K, Hikihara Y, Ishikawa-Takata K, Tabata I (2010) Classifying household and locomotive activities using triaxial acceleration. Gait & Posture 31:370-374
5. Xu M, Iyengar S, Goldfain A, Roy Chowdhury A, DelloStritto J (2011) A two-stage real time activity monitoring system. International conference on Body Sensor Networks, 191-193
6. Banos O, Damas M, Pomares H, Prieto A, Rojas I (2012) Daily living activity recognition based on statistical feature quality group selection. Expert Systems with Applications 39:8013-8021
7. Chien C, Pottie G J, (2012) A universal hybrid decision tree classifier design for human activity classification. International conference of the IEEE Engineering in Medicine and Biology Society.1065-1068
8. Najafi B, Aminian K, Loew F, Blanc Y, Robert PA (2002) Measurement of Stand-Sit and Sit-Stand Transitions using a miniature Gyroscope and its application in fall risk evaluation in the elderly. IEEE transactions on Biomedical Engineering. 49(8):843-851
9. http://en.wikipedia.org/wiki/Z1_Android_Watch-Phone

Address of the corresponding author:

Author: V Ahanathapillai
Institute: Institute of Digital Healthcare,
Warwick Manufacturing Group, University of Warwick
City: Coventry - CV4 7AL
Country: UK
Email: ahanat_v@wmg.warwick.ac.uk

Comparative Measurement of the Head Orientation Using Camera System and Gyroscope System

P. Kutilek[1], O. Cakrt[2], J. Hejda[1], and R. Cerny[2]

[1] Faculty of Biomedical Engineering, Czech Technical University in Prague, Czech Republic
[2] 2nd Faculty of Medicine - University Hospital Motol, Charles University in Prague, Czech Republic

Abstract—The aim of article is to compare two fundamentally different MoCap systems used to measure head orientation. We used IR medical camera and one 3DOF orientation tracker placed on subject's head and measured 3-D data - inclination (roll), flexion (pitch) and rotation (yaw). The volumes of 3D confidence ellipsoids were used to compare the two systems. Using the volumes of confidence ellipsoids, we were able to model the distribution of the measured 3-D data. Ten healthy subjects in the study were measured and statistical analysis was performed. The results of the comparative method based on the confidence ellipsoid show that the volumes related to the 3DOF orientation tracker are significantly larger than the volumes related to the IR medical camera. Although the results are different when using different MoCap systems, the MoCap systems identified the same subject's behavior and deterioration or improvement of stability. The comparative measurement based on concept of confidence ellipsoid volume has not been used before to study the MoCap systems used to measure head orientation. The methods can be used to compare other types of MoCap systems and can also be used to study other segments of human body.

Keywords—head orientation, camera system, orientation tracker, confidence ellipsoid.

I. INTRODUCTION

The position of the head can be negatively influenced by many diseases of the nervous system [1]. Patients with vestibular deficits often show instability during stance tasks. Making the stance task more difficult by removing visual inputs has been claimed as a mean to identify a vestibular deficit, [2], [3]. The MoCap systems were used for 3-D high-accuracy measurement of human body segments instead of commonly used force platforms to study the centre of pressure. The sensing units (in the case of inertial systems) or markers (in the case of camera system) were used to measure the pitch, roll and yaw of the segments. The techniques were introduced to quantify segments movements in both anterior–posterior and medial–lateral directions during stance, [3], [4]. Measurement of head angular movements during stance can detect changes in postural stability [5].

Thus, this study is aimed to compare and evaluate the reliability and validity of 3-D data measured by two fundamentally different MoCap systems - medical camera with markers and tri-axial inertial measurement unit (i.e. 3DOF orientation tracker). Both of these fundamentally different systems are basic MoCap systems for the study of the postural stability in 3D space. Traditional method for assessing the postural instability is method based on 2-D confidence ellipse, [6], [7]. This method is based only on the description of the behavior of only two variables in twoplanes/axis. However, we need to model the distribution of the measured 3-D data (pitch, roll, yaw) because we have used MoCap systems for 3-D high-accuracy orientation measurement instead of commonly used force platforms and similar equipment. The method based on statistical technique called Principal Component Analysis (PCA) can be used [8]. By this method, we can model the distribution of the measured data (pitch, roll, yaw) by 3-D ellipsoid. The main question is whether the measurement of 3D data will lead to the same results and if the results are interchangeable or not.

II. METHODS AND MATERIALS

Ten healthy subjects (age of 22.2 (SD 1.4) years) were recruited from the students at The Czech Technical University in Prague. In brief, head orientation was measured during stance. The tasks were standing on both legs on a firm surface for 60 seconds with eyes open (EO) and closed (EC). The subject's feet were positioned next to each other splayed at 30º, arms always in hanging position, [9].

A. Motion Capture Equipment

We can use several MoCap systems for measuring movements in 2D/3D space. We used a medical camera system with active markers and an inertial measurement unit. The used camera system, Lukotronic AS 200 (Lutz Mechatronic Technology e.U.) system, is widely-used medical camera system with active markers. The markers are placed in accordance with the Frankfort horizontal, see Fig.1, for measuring angular movements, [3]. Active markers were placed on the following anatomical points: left and right tragus and left and right outer eye canthus, Fig.1. The MatLab (The MathWorks Inc.) software was used to identify the angles (pitch, roll, yaw) of the head in 3D space, [10].

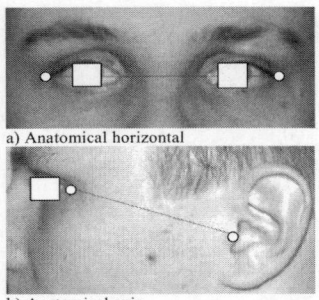

Fig. 1 Anatomical Frankfort horizontal and axis [3]

The camera system was calibrated before the experiments, and the origin of the world coordinate system was set up so that the first axis is along the symmetry axis (coinciding with the anterior posterior axis) of the platform on which the subjects stand and the other two axes are perpendicular to the symmetry axis of the platform.

For head sway measurements was also used the motion capture system Xbus Master which is a lightweight device uses MTx units for orientation measurement of segments [11], see Fig.2. The MTx unit is accurate 3DOF tracker. The one MTx unit was placed on patient's head in accordance with Raya at al. [12] and Casolo [13], see Fig.2. The head sway in 3-D space was measured in three planes and the sample frequency was 100 Hz.

Each simultaneous measurement of one subject with EO or EC took one minute. It is not necessary to normalize the data because the standard ranges of angles are the same for the three planes of the body and all adult persons if the 3DOF orientation tracker and/or markers placed on the same segment (i.e. points). Using these techniques, we can record and study the movement in 3D space.

B. Method of Quantification of Postural Stability

The used method to analyze the reliability and validity of 3-D data measured is based on Principal Component Analysis. PCA is a common technique for finding patterns in data of high dimensions [14]. We can model the distribution of the measured data (pitch, roll, yaw) by 3-D ellipsoid. The reason for the use of this method is that we can study the 3D movement by using one variable characterizing the change in all three angles.

PCA applied to a measured data forms a symmetric 3x3 covariance matrix, [8], [14]. The volume of an ellipsoid is given by an ellipsoid matrix and the ellipsoid matrix is composed of entries from the covariance matrix, [15], [16]. The eigensystem of the covariance matrix constitutes a quadratic surface which is used for visualization, Fig.3. We use 95% confidence ellipsoid (CE), [17] - [19], see Fig.3. The 95% CE (with 0.95 probability) bounds the 95% of the measured data, i.e. sizes of angles (angular displacements in the roll, yaw and pitch planes). The MatLab was used to determine the symmetric 3x3 covariance matrix of the data for each subject. The covariance matrices were used to calculate the volumes of the 95% CE:

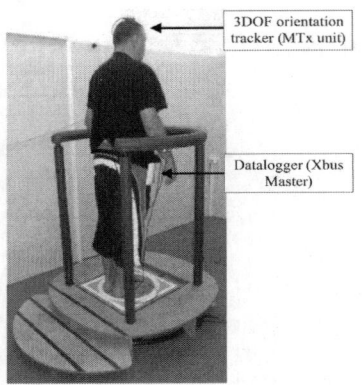

Fig. 2 Illustration of the Xsens system with gyro sensor MTx placed on patient's head

$$V = \left(\frac{4}{3} \cdot \pi \cdot \sqrt{\det(C)}\right) \cdot P_{conf} \qquad (1)$$

where the C is covariance matrix and P_{conf} is quantile for the desired confidence interval.

C. Statistical Analysis

After calculating the volume of each 95% CE of each subject with EO or EC, measured by both systems, the statistical analysis was performed with the use of the MatLab. First, we used the measured data to illustrate the relationship between the CE volumes measured by medical camera with active markers and the CE volumes measured by one 3DOF orientation tracker. We calculated the Pearson product-moment correlation.

Second, we calculated the minimum (Min), maximum (Max), median, first quartile (Q1), third quartile (Q3), mean and standard deviation (SD) for the volumes of 95% CEs. We used the descriptive statistic to illustrate the relationship between the subjects with EO and EC. The Jarque–Bera test was used to test the normal distribution of all parameters (significance level was 5%).

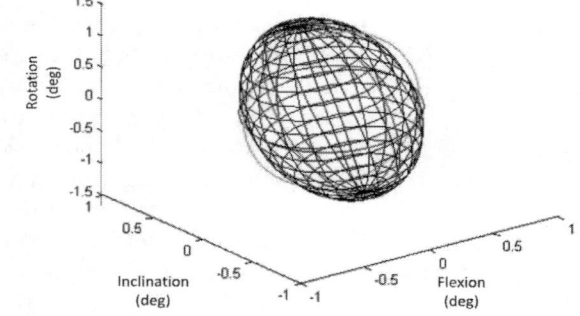

Fig. 3 Example of a three-dimensional confidence ellipsoid of the head angular displacement in the roll, yaw and pitch planes

The Wilcoxon signed rank test was used to assess the significance of the differences between CE volumes of the measured subjects, with EO and EC, measured by Lukotronic system and Xsens system. The data about the CE volumes of the subjects measured by Lukotronic system were compared with data about the CE volumes of the subjects measured by Xsens system. The significance level was set at p<0.05.

III. RESULTS

After calculating the volumes of 95% CEs, the statistical analysis was performed, Table 1. The following chart (Fig.4) shows the relationship between the CEs of the subjects with EO and EC. The Pearson product-moment correlation coefficient is 0.93 in the case of the relationship between the CE volumes of the measured subjects with EO measured by Lukotronic system and the CE volumes of the measured subjects with EO measured by Xsens system. The Pearson product-moment correlation coefficient is 0.77 in the case of the relationship between the CE volumes of the measured subjects with EC measured by Lukotronic system and the CE volumes of the measured subjects with EC measured by Xsens system. Correlation coefficients (0.93 and 0.77) are significantly different from zero and indicates a strong correlation between the CE volumes measured by Lukotronic system and the CE volumes measured by Xsens system.

The Jarque–Bera test returns h=1 in two cases (subjects with EO and measured by Lukotronic system, and subjects with EC and measured by Xsens system) and h=0 in two cases (subjects with EO and measured by Xsens system, and subjects with EC and measured by Lukotronic system). Since the some data did not have a normal distribution, the Wilcoxon test was used to analyze this data.

In the case of volume of CE of the subjects with EO and EC measured by Lukotronic system, the results did not show significant difference in volume of CEs (p=1.00). In the case of volume of CE of the subjects with EO and EC measured by Xsens system, the results also did not show

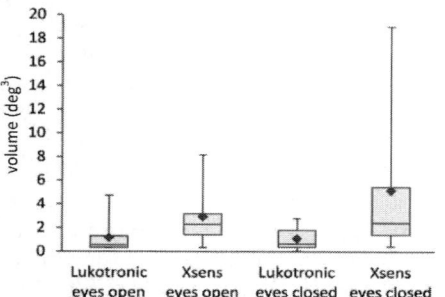

Fig. 4 Comparison of the volume of 95% confidence ellipsoids of the CG with eyes open and closed. Measured by Lukotronic and Xsens system

significant difference in volume of CEs (p=0.56). All calculated p-values were greater than the significance level (p<0.05). Therefore, we do not reject the null hypothesis, and there is no significant difference between the volume of CE of the subjects with EO and EC. The measured data of the volume of CE of the subjects with EC indicates the slight increase of the median of volume of CE. It shows deterioration in of stability and posture, but it is negligible (according to the Wilcoxon signed rank test).

Subjects with EO demonstrate significant difference in volume of CEs (p=0.002) measured by Xsens system and Lukotronic system. Subjects with EC also demonstrate significant difference in volume of CE (p=0.010) measured by Xsens system and Lukotronic system. All calculated p-values were lower than the significance level (p<0.05). Therefore, we reject the null hypothesis, and there is significant difference between the data measured by Lukotronic system and the data measured by Xsens system. The median of the volume of CEs related to the subjects with EO and measured by Xsens system is 4.3 times larger than the median of the volume of CEs related to the subjects with EO and measured by Lukotronic system. The median of the volume of CEs related to the subjects with EC and measured by Xsens system is 3.8 times larger than the median of the volume of CEs related to the subjects with EC and measured by Lukotronic system.

IV. DISCUSSION

The results show that the median of the volume of CEs related to the subjects measured by Xsens system is significantly larger than the then median of the volume of the CEs related to the subjects measured by Lukotronic system. The reason of different results is gyro sensor drift and skin artifacts which significantly affect the accuracy.

A very interesting finding is that the correlation coefficients indicate a strong correlation between the CE volumes measured by Lukotronic system and the CE volumes measured by Xsens system.

Table 1 Descriptive statistics of the volume of 95% confidence ellipsoids of the CG with eyes open (EO) and closed (EC)

		Lukotronic CG EO	Xsens CG EO	Lukotronic CG EC	Xsens CG EC
Min	(deg^3)	0.28	0.34	0.03	0.42
Max	(deg^3)	4.70	8.18	2.78	19.01
Mean	(deg^3)	1.16	2.94	1.08	5.14
SD	(deg^3)	1.40	2.44	0.97	6.42
Median	(deg^3)	0.53	2.29	0.64	2.40
Q1	(deg^3)	0.32	1.41	0.36	1.39
Q3	(deg^3)	1.28	3.18	1.78	5.43

It is also evident that statistical analysis of the subjects with EO and EC did not show significant difference in volumes of CEs. Although we expected some deterioration in postural stability of the subjects with EC, the results are not conclusive.

Although the results are different when using different MoCap systems to measure the inclination, flexion and rotation of the head, the MoCap systems identified the same subject's behavior and deterioration or improvement of stability. Therefore, the systems are not interchangeable and the same type of system must be used each time.

V. CONCLUSIONS

We found that measurement of the head orientation and use of the 3-D confidence ellipsoid volumes based on the data measured by Lukotronic system and/or the 3-D confidence ellipsoid volumes based on the data measured by Xsens system is suitable for study of the postural balance problems. The findings also showed that a one 3DOF orientation tracker placed on patient's head could replace the complex and more expensive camera system, but the results of the measurements are not interchangeable.

Although the number of articles describing different ways of studying the head angular movements, [3], the method based on concept of confidence ellipsoid volume, although known to the biomechanics community, has not been used before to compare the MoCap systems based on different measurement principles used to measure head orientation. The methods can also be used to compare other types of MoCap systems and can also be used to study other segments of human body. The future work can be focused on the comparison of the other systems and identify in detail the reasons of the different volumes of confidence ellipsoids measured by different MoCap systems.

ACKNOWLEDGMENT

This research has been supported by project SGS 13/091/OHK4/1T/17 CTU Prague.

REFERENCES

1. Cerny R, Strohm K, Hozman J et al (2006) Head in Space - Noninvasive Measurement of Head Posture, Proc. 11th Danube Symposium - International Otorhinolaryngological Congress, Bled, Slovenia, 2006, pp 39-42
2. Ochi F, Abe K, Ishigami S et al (1997) Trunk motion analysis in walking using gyro sensors, Proc. 19th International Conference of the IEEE/EMBS, Chicago, USA, 1997, pp 1824 - 1825
3. Kutilek P, Hozman J, Cerny R et al (2012) Methods of measurement and evaluation of eye, head and shoulders position in neurological practice, In: Advanced topics in neurological disorders. InTech, Rijeka, pp 25-44 DOI 10.5772/31378
4. Allum J, Adkin A, Carpenter M et al (2001) Trunk sway measures of postural stability during clinical balance tests: effects of a unilateral vestibular deficit, Gait Posture 14: 227-237 DOI 10.1016/S0966-6362(01)00132-1
5. Hozman J, Zanchi V, Cerny R et al (2007) Precise Advanced Head Posture Measurement. Proc. 3rd WSEAS Int. Conf. on Remote Sensing, Venice, Italy, 2007, pp 18-26
6. Odenrick P, Sandstedt P (1984) Development of postural sway in the normal child. Hum Neurobiol 3:241–244
7. Oliveira L, Simpson D, Nadal J (1996) Calculation of area of stabilometric signals using principal component analysis. Physiol Meas 17:305-312 DOI 10.1088/0967-3334/17/4/008
8. Jolliffe I. (2002) Principal Component Analysis. Springer-Verlag, New York
9. Cakrt O, Vyhnalek M, Slaby K (2012) Balance rehabilitation therapy by tongue electrotactile biofeedback in patients with degenerative cerebellar disease. NeuroRehabilitation 31:429-434 DOI 10.3233/NRE-2012-00813
10. Hejda J, Kutilek P, Hozman J et al (2012) Motion Capture Camera System for Measurement of Head and Shoulders Position. Biomed Tech 57:472-475 DOI 10.1515/bmt-2012-4123
11. Micera S, Sabatini A, Genovese V et al (2010) Assessment technologies for the analysis of the efficacy of a vestibular neural prosthesis. Biomed Tech 10:55-94 DOI 10.1515/bmt-2010-491
12. Raya R, Roa J, Rocon E, et al (2010) Wearable inertial mouse for children with physical and cognitive impairments. Sens Actuat A Phys 162:248–259 DOI 10.1016/j.sna.2010.04.019
13. Casolo F (2010) Elbow prosthesis for partial or total upper limb replacements, In: Motion control. InTech, Rijeka, DOI 10.5772/6972
14. Krzanowski W (1988) Principles of multivariate analysis: a user's perspective. Oxford University Press, New York
15. Nash S., Sofer A. (1996) Linear and nonlinear programming. The McGraw-Hill, New York
16. Wolf P, Ghilani CH (1997) Adjustment computations. John Wiley and Sons, New York
17. Rocchi M, Sisti D, Ditroilo M, et al (2005) The misuse of the confidence ellipse in evaluating statokinesigram. Ital J Sport Sci 12:169-72
18. Swanenburg J, De Bruin E, Favero K, et al (2008) The reliability of postural balance measures in single and dual tasking in elderly fallers and non-fallers. BMC Musculoskelet Disord 9:1-10 DOI 10.1186/1471-2474-9-162
19. Moghadam M, Ashayeri H, Salavati M et al (2011) Reliability of center of pressure measures of postural stability in healthy older adults: effects of postural task difficulty and cognitive load, Gait Posture 33:651-655 DOI 10.1016/j.gaitpost.2011.02.016

Author: Patrik Kutilek
Institute: CTU FBME in Prague
Street: Sq. Sitna 3105
City: Kladno
Country: Czech Republic
Email: kutilek@fbmi.cvut.cz

A Noninvasive Method of Measuring Force-Frequency Relations to Evaluate Cardiac Contractile State of Patients during Exercise for Cardiac Rehabilitation

M. Tanaka[1], M. Sugawara[1], Y. Ogasawara[2], I. Suminoe[3], T. Izumi[4], K. Niki[5], and F. Kajiya[2]

[1] Faculty of Health Care Sciences / Himeji Dokkyo University, Himeji, Japan
[2] Department of Medical Engineering / Kawasaki University of Medical Welfare, Kurashiki, Japan
[3] Clinical Laboratory / Himeji Red Cross Hospital, Himeji, Japan
[4] Department of Physical Therapy / Health Sciences University of Hokkaido, Ishikari, Japan
[5] Biomedical Engineering Department / Tokyo City University, Tokyo, Japan

Abstract—Background Evaluation of the contractile state of the left ventricle during exercise is important for cardiac rehabilitation. As yet, no noninvasive methods for this purpose have been established. The force-frequency relation (FFR) during exercise has the potential for evaluating the contractile state noninvasively. Color Doppler- and echo tracking-derived carotid arterial wave intensity is a sensitive index of global left ventricular (LV) contractility.
Objectives We assessed the feasibility of measuring carotid arterial wave intensity and determining FFR's during exercise totally noninvasively.
Methods We enrolled 18 healthy men (age 20.6 ± 2.1 years). Using ultrasonic diagnostic equipment, we measured wave intensity in the carotid artery and heart rate (HR) before and during ergometer exercise. FFR's were constructed by plotting the maximum value of wave intensity (WD_1) against heart rate (HR).
Results WD_1 increased linearly with an increase in HR during exercise. The regression line of WD_1 on HR represented the FFR.
Conclusion Using WD_1 and gradual exercise test, we obtained FFR's noninvasively. These data should show the potential usefulness of FFR in practicing cardiac rehabilitation.

Keywords—Force-Frequency Relation, Echocardiography, Wave Intensity, Exercise.

I. INTRODUCTION

When we administer exercise training to a patient for cardiac rehabilitation, we should constantly monitor the patient's condition. In normal subjects, cardiac contractility increases with an increase in heart rate (HR). This phenomenon is called the Force-Frequency Relation (FFR). HR is an important determinant of myocardial performance, but increased HR does not necessarily increase cardiac contractility in patients with heart disease. The relation between cardiac contractility and HR in a diseased heart is different from that in healthy animals and humans(1). Conventionally, FFR's were obtained by measuring the maximum rate of LV pressure rise (Peak dP/dt) with a catheter-tipped micromanometer as an index of cardiac contractility, and using atrial pacing to change heart rate (Peak dP/dt - HR relation) (2). This is an invasive method and cannot be used repeatedly in the clinical setting.

It has been demonstrated that color Doppler- and echo tracking-derived carotid arterial wave intensity is a sensitive index of global left ventricular (LV) contractility (3). During exercise, HR increases with an increase in workload, therefore atrial pacing is not needed for changing HR. It has also been demonstrated that the Peak dP/dt-HR relation is markedly enhanced (the slope is increased) during exercise compared with during pacing in normal hearts, but the enhancement is limited in diseased hearts (2). Therefore, the FFR obtained by exercise may have higher power to discriminate cardiac contractile states than that obtained by pacing.

In this study, we assessed the usefulness of the measurement of wave intensity during exercise in obtaining FFR's, and evaluated the feasibility of an entirely noninvasive method for demonstration of the FFR.

II. METHODS

A. Noninvasive Measurements of Wave Intensity

Wave intensity (WI) is a hemodynamic index, which is defined as

$$WI = (dP/dt)(dU/dt), \qquad (1)$$

where dP/dt and dU/dt are the derivatives of blood pressure (P) and velocity (U) with respect to time, respectively. The maximum value of WI during a cardiac cycle (W_1) significantly correlates with Peak dP/dt (or max dP/dt) (4).

In our method of obtaining carotid arterial WI, carotid diameter-change waveform was used as a surrogate for carotid pressure waveform (3). Using measured diameter-change waveforms directly, we can also define a wave intensity (WD) as

$$WD = (1/D)(dD/dt)(dU/dt). \quad (2)$$

WD is obtained by measuring U and D without measuring upper arm pressure (Fig.1), which is easier to perform during exercise. The definition of the stiffness parameter (β) gives the relation

$$(1/D)(dD/dt) = (1/\beta P)(dP/dt). \quad (3)$$

Therefore,

$$WD = (1/\beta P) \, WI. \quad (4)$$

Hence, the maximum value of carotid arterial WD during a cardiac cycle (WD_1) correlates with the maximum value of WI (W_1) (Fig.2). Therefore WD_1 correlates with Peak dP/dt as W_1 does. The details of the method of measurements were described elsewhere (3).

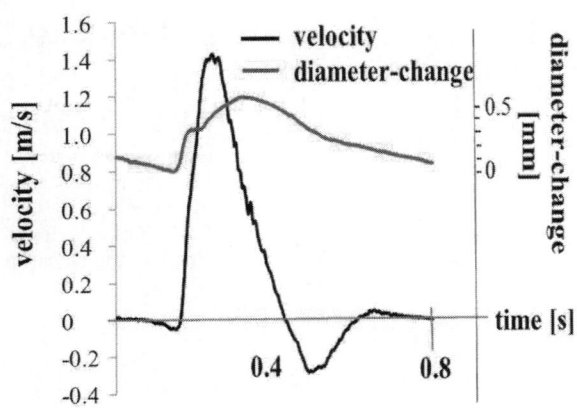

Fig. 2a.: Diameter-change and flow velocity waveforms is displayed.

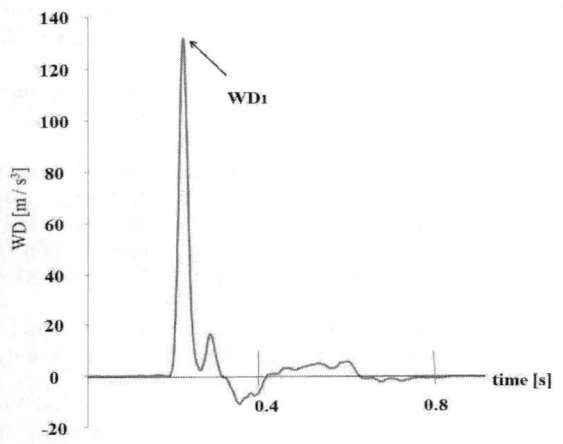

Fig. 2b Wave intensity (WD) is calculated as WD = (1/D)(dD/dt)(dU/dt).

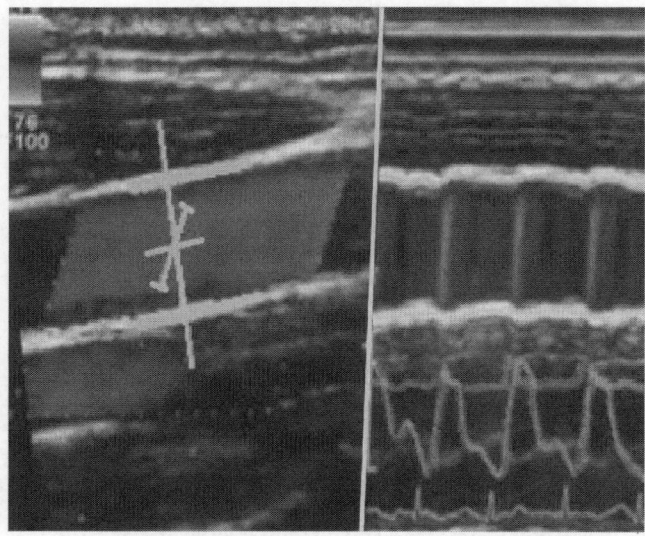

Fig. 1 Simultaneous measurements of diameter change wave form and blood flow velocity. View on the monitor during the measurements. *Left*: Color Doppler / B-mode long-axis view of the common carotid artery. Pink line and blue line indicate the ultrasonic beam for echo-tracking and for blood flow velocity measurement, respectively. By setting the tracking positions displayed as pink bars on the echo-tracking beam to arterial walls, echo-tracking automatically starts. *Right*: The diameter-change waveforms, which is calculated by subtracting the distance to the near wall from that to the far wall, and the velocity waveforms are displayed on the M-mode view.

B. Subjects

We studied 18 healthy male volunteers (mean age 20.6 ± 2.1 years, age range 19 – 22 years). We obtained informed consent from all the subjects. Approval from the Ethics Committee of Himeji Dokkyo University was obtained before study initiation.

C. Protocol

First, before the measurements of WD_1 and HR, the subjects lay down in the semi-supine position for 10 minutes on the strength ergometer. The location to be measured was the common carotid artery at about 2cm proximal to the carotid bulb. We used scanning in the long axis view, and obtained a B-mode image of a longitudinal section of the artery (Fig.1 left). With the B- and M-mode scans displayed simultaneously on a split screen, the echo-tracking system tracked the vessel wall movements to produce displacement waveforms of the anterior and posterior artery walls (Fig.1 right). This gave the diameter-change waveforms.

Next, after the measurements of WD_1 and HR at rest, gradual bicycle exercise was performed starting at an initial workload of 20 W and lasting for 2minutes; thereafter, the workload was increased stepwise by 20 W at 1-minute intervals. Electrocardiogram was continuously monitored. The criteria for the endpoint included increase of heart rate to [(220−age) × 0.8 (bpm)], and the impossibility of continuing exercise. We measured WD_1 and HR during the exercise.

D. Data Analysis

The obtained data are expressed as mean ± standard deviation. The scatter diagram of the points (HR, WD_1) for the data during exercise from each subject was analyzed by the linear regression method, and the regression line was regarded as the FFR.

III. RESULTS

All subjects at exercise were able to pedal the strength ergometer and effectively increased their HR (mean HR at rest vs. mean HR at peak exercise: 75 beats/ min vs. 154 beats/ min, $p < 0.0001$). WD_1 increased linearly with an increase in HR (Fig.3). The goodness-of-fit of the regression line of WD_1 on HR in each subject was very high ($r^2 = 0.67 \sim 0.94$, $p < 0.0001$, respectively). The slope of the WD_1-HR relation (FFR) varied with the individual ranging from 0.31 to 2.2 [m/s^3 (beat / min)] (Fig. 3).

IV. DISCUSSION

The basic property of the FFR to progressively enhance myocardial contractility as heart rate increases (frequency potentiation) has been observed in normal humans. This mechanism is augmented due to β-adrenergic stimulations induced by exercise in normal hearts. However, significant impairment of exercise-induced amplification of the FFR is observed in diseased hearts (2). Therefore, one would expect the diversity of responses of the FFR by exercise ensures increased sensitivity for detection of contractile impairment.

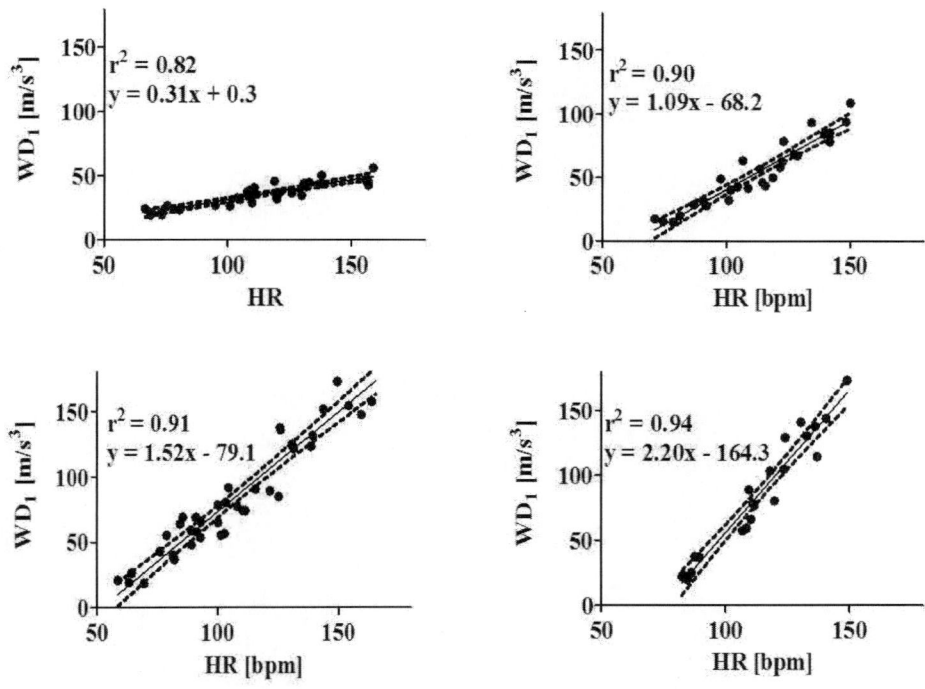

Fig. 3 Representative FFR's in the study.

Equation (4) shows that WD contains β. It is widely known that β increases with age. However, our study subjects only ranged from 19 to 22 in age. Therefore, β was considered to vary only slightly with the individual. In spite of this, the slope of the FFR varied relatively widely. This is considered to show that the FFR during exercise is determined mainly by the cardiac contractile state rather than arterial conditions.

Limitations

Our study subjects only ranged from 19 to 22 in age. We should enroll a greater number of subjects to be divided into age groups in future study. The final goal of our study is to apply the FFR in practicing cardiac rehabilitation. However, we did not enroll patients with heart disease in the present feasibility study.

V. CONCULUSION

Measurements of carotid arterial wave intensity by echocardiography during exercise give the FFR noninvasively. This will be useful for the evaluation of the contractile state of the left ventricle during exercise for cardiac rehabilitation.

REFERENCES

1. Hasenfuss G, Holubarsch C, Hermann HP, Astheimer K, Pieske B, Just H. Influence of the force-frequency relationship on haemodynamics and left ventricular function in patients with non-failing hearts and in patients with dilated cardiomyopathy. Eur Heart J. 1994;15(2):164-70. Epub 1994/02/01.
2. Ross J, Jr., Miura T, Kambayashi M, Eising GP, Ryu KH. Adrenergic control of the force-frequency relation. Circulation. 1995;92(8):2327-32. Epub 1995/10/15.
3. Sugawara M, Niki K, Ohte N, Okada T, Harada A. Clinical usefulness of wave intensity analysis. Medical & biological engineering & computing. 2009;47(2):197-206. Epub 2008/09/03.
4. Ohte N, Narita H, Sugawara M, Niki K, Okada T, Harada A, et al. Clinical usefulness of carotid arterial wave intensity in assessing left ventricular systolic and early diastolic performance. Heart and vessels. 2003;18(3):107-11. Epub 2003/09/05.

Author: Midori Tanaka,
Institute: Himeji Dokkyo University,
Street: Kamiohno7-2-1,
City: Himeji,
Country: Japan,
Email: mirakulu17.nikkori17@gmail.com

A Phase Lock Loop (PLL) System for Frequency Variation Tracking during General Anesthesia

C.A. Teixeira[1], B. Direito[1], A. Dourado[1], M.P. Santos[2], and M.C. Loureiro[2]

[1] Centre for Informatics and Systems, University of Coimbra, Coimbra, Portugal
[2] Anesthesiology Service, Centro Hospitalar e Universitário de Coimbra, Coimbra, Portugal

Abstract—We present a novel technique derived from the communication systems area, able to track frequency changes in electroencephalogram (EEGs) signals collected from patients subjected to general anesthesia. The technique is based on a phase lock loop (PLL) circuit, which is used for example for radio-FM demodulation.

The frequency variations extracted by the PLL circuit were compared with the widely used BIS trend, returned by the surgery room instrumentation. The results showed that the estimated PLL frequency variations are coherent with the BIS trend and also with the anesthesia induction, maintenance and recovering epochs. The developed system presents tolerance to EEG artifacts, due to several processing stages implemented, and due to the intrinsic capabilities of this type of circuits. It is envisaged that the system could bring improvements to the actual anesthesia deep monitors based on EEG. Other advantages are that the proposed system can be fully implemented in hardware, can work in real-time and is of low-cost.

Keywords—Anesthesia monitoring, EEG, Phase Lock Loop, BIS.

I. INTRODUCTION

General anesthesia (GA) can be defined as a combination of suppression of different functions of the central nervous system (CNS). This includes unconsciousness (amnesia and hypnoses), muscle relaxation analgesia and control of the sympathetic and parasympathetic reflexes [1]. The monitoring of the anesthetic procedure, i.e., the discrimination of the different depths of anesthesia, has been performed by subjective methods and also by quantitative approaches. On the one hand, the subjective, or conventional monitoring, includes the analysis of the autonomic nervous system (temperature, ventilation, pupil diameter, arterial blood pressure, and other hemodynamic parameters), and the patient response to surgery and external stimuli. On the other hand, quantitative monitoring involves the mathematical processing of signals collected from the patient.

The Electroencephalogram (EEG) is by far the most applied signal for anesthesia monitoring, since the brain is the body "central processing unit" and based on the assumption that several features from the EEG are dependent on the anesthetic depth. Different monitors are commercially available based on different EEG measures, such as Bispectral Index (BIS), Narcotrend, M-Entropy, Patient State Index, and aepEX [2]. BIS (Aspect Medical Systems) is the most widely applied one and was approved by the Food and Drugs Administration (FDA) in 1996. A 2009 report from the manufactured indicates that BIS is used in about 75% of the operating rooms in the US. The algorithm behind BIS is complex and partially unknown. The known information reports that it is based on a combination of time and frequency domain features of the EEG signal collected in the frontal scalp region. The time domain parameters intend to detect deeper levels of anesthesia by identifying isoelectric or quazi-isoelectric EEG epochs. The spectral features accounts for the anesthesia-dependent power transfer among different bands [3].

This paper aims to present an alternative, low-cost and open system to monitor changes in the frequency content of EEG signals. The new approach is inspired in the communication systems area and is based on the concept of FM demodulation by applying phase lock loops (PLLs). This principle was applied in the past for sleep spindle detection [4], but never for anesthesia monitoring, to the authors knowledge.

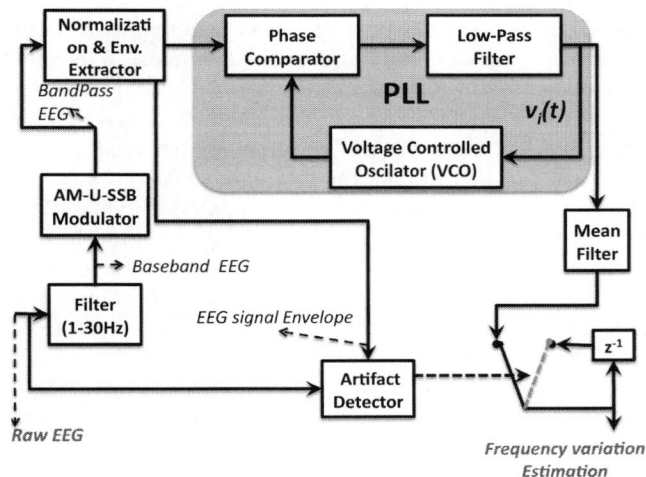

Fig. 1 Implemented system for frequency variation tracking.

II. DATA

Patients subjected to general anesthesia (GA) were monitored at the operating rooms of the Centro Hospitalar e Universitário de Coimbra (CHUC), Portugal. The CHUC ethical commission previously approved data collection, and the patient's consent was signed allowing us to analyze their data. Patients were monitored simultaneously by the standard operating room instrumentation and by additional EEG electrodes placed over the scalp, according to the 10-20 system [5]. The standard monitoring included respiratory parameters, hemodynamic parameters, and BIS module parameters (including the BIS index). The EEG electrodes are connected to an amplifier (SD-LTM32, Micromed, Italy) that sampled the analogue signals at 256 Hz.

III. METHODOLOGY

The system developed is presented in Fig. 1. First the raw EEG signal is filtered to confine its frequency range to the band 1-47Hz, which is the band considered by the BIS algorithm.

Because PLLs are circuits designed to synchronize with bandpass signals, the EEG signal is transferred to a high frequency band by modulating the filtered EEG using upper single side band amplitude modulation (AM-U-SSB). We decided to consider a carrier frequency of 100Hz, meaning that the modulated EEG signal will assume frequencies in the range 101-147 Hz. Thus for an adequate AM-U-SSB modulation, respecting the Nyquist frequency, the original EEG signal, sampled at 256 Hz, had to be up-sampled. We applied a linear interpolation to the EEG signal in order to obtain a 512 Hz sampled signal.

To minimize the effects of amplitude changes in the frequency estimation, we developed a system that extracts the envelope of the modulated signal and then makes a dynamic normalization to values of approximately ±1.

The EEG signal during surgery is contaminated by electrical interferences. The main source of interference is the electronic scalpel. In fact, the interference is so severe that creates high amplitude transients leading to saturation of the EEG signal, and also to the occurrence of null values. A reliable anesthesia system should account for these artifacts. The system developed detects the occurrence of high amplitude transients and also the presence of null values in the EEG. The system developed is based on a finite state machine, and enables the suppression of contaminated EEG epochs by maintaining the PLL system output at the level just observed before interference.

PLL is one of the most used circuits in communications, with applications ranging from modulation, demodulation, and frequency synthesization, among others. The aim behind PLL is to have a circuit that synchronizes (gets locked) itself with the instantaneous angle (i.e. both instantaneous frequency and phase) of an input bandpass signal [6]. The PLL has three basic sub-blocks, as represented in Fig. 1: a phase comparator, a low-pass filter, and a voltage controlled oscillator (VCO). The phase comparator produces an output that depends on the instantaneous angular difference between the input signal and the signal at the output of the VCO. The simplest, but yet efficient phase comparator is a signal multiplier. The low-pass filter defines the order of the PLL, eliminates high frequency components, and generates the voltage signal that will define the oscillation frequency of the VCO. To analyze and design this type of circuits an accepted procedure is to consider the PLL as a linear control system in the phase domain [6]. If a first order filter is assumed, then the PLL is a second-order system with two free parameters: the VCO gain (K_o) and the filter gain and pole (K_d), in (1):

$$\frac{\Phi_o(s)}{\Phi_i(s)} = \frac{K_o K_d}{s^2 + K_d s + K_o K_d}. \quad (1)$$

Where $\Phi_o(s)$ and $\Phi_i(s)$ are the phases in the Laplace domain corresponding to the signal at the VCO output and to the input signal, respectively. K_o and K_d can then be determined by assuming performance levels for the PLL. We consider the settling time (t_s) and the maximum overshoot time (t_{max}) as important performance specifications. t_s is the time required, after a step input, for the output oscillations of an underdamped second order system to decrease to a specified absolute percentage of the final value and thereafter to remain bounded by this value. We consider as a satisfactory operation when the system oscillations decrease and remain in the interval ±2% of the final value. t_{max} is the time of the first peak after a step input. We have (2) [7]:

$$t_s = \frac{4}{\xi \omega}; \quad t_{max} = \frac{\pi}{\omega_n \sqrt{1-\xi^2}} \quad (2)$$

Where ω_n is the natural undamped frequency and ξ is the damping ratio. By associating (1) with the generic second order transfer function, and considering the equations in (2) we get the solution for K_o and K_d as function of the performance parameters t_s and t_{max}, as in (4):

$$\begin{cases} \xi = \dfrac{4/t_s}{\sqrt{(\pi/t_{max})^2 + (4/t_s)^2}}; \quad \omega_n = \dfrac{4}{\xi t_s} \\ K_d = 2\xi\omega_n; \quad K_o = \dfrac{\omega_n^2}{K_d} \end{cases} \quad (4)$$

The definition of the VCO gain (K_o) and the filter gain and pole (K_d) affects the set of frequencies for which a locked PLL can maintain its synchrony. We found that the definition of t_s=3s and t_{max}=0.17s lead to K_o=128.7 and to K_d=2.66, correspondent to ω_n=18.5 and ξ=0.07. This result in a PLL that has can maintain synchrony over a range of approximately ± 55Hz and that can overcome sudden changes of ±9.5Hz. These values satisfied our problem, given that we aim a PLL that should follow gradual frequency changes in the range 0-47 Hz, and it is not expected a sudden change of more than 9.5 Hz.

The input signal to the VCO is considered as the output signal of the PLL ($v_i(t)$ in Fig. 1), because when the PLL is locked it is proportional to the variations in the signal frequency. The output of the PLL can present considerable oscillations, because the real EEG signal is a summation of several spectral components and noise, i.e., the PLL will never be synchronized with pure sinusoidal components. To extract visible frequency variations we apply a moving average filter to $v_i(t)$. The filter output is the average signal computed based on the past 100 seconds.

IV. RESULTS AND DISCUSSION

We describe in the following the frequency variation estimated by the proposed PLL system, as compared with the BIS descriptor obtained from the surgical room instrumentation. The PLL frequency estimation was obtained from the differential Fp1-Fpz combination. These electrodes were selected because they are also locations used by the surgery room instrumentation to compute the BIS index. The frequency variation estimated by the PLL system is obtained by multiplying the input voltage of the VCO by the VCO gain (K_o) expressed in Hz/V.

Illustrative results for two patients are shown in Fig. 2 where it can be observed that the proposed system is able to extract frequency variations. These variations are coherent with the computed BIS index and with the anesthesia induction (indicated by the red arrows in Fig. 2), maintenance and recovering epochs (with starting indicated by the green arrows in Fig. 2). The PLL system is able to minimize the effect of EEG artifacts (originated mainly by the electronic scalpel) because of the developed artifact suppression system and because of the dynamic normalization that discards the influence of amplitude variations. The epochs rejected are those that appear as flat plateaus in Fig. 2. Another advantage of the proposed system is that it is independent of the definition of the size and overlap of processing windows, because it works continuously over the fed EEG signal. This eliminates the problems related with signal delimitation by windowing, such as spectral leakage.

V. CONCLUSIONS

A novel system derived from the communication systems, based on a phase lock loop (PLL circuit), is reported. The system is characterized by presenting frequency variation estimations that are coherent with the anesthesia cycle and with the widely used BIS index. It is envisaged that the system could bring improvements to the actual anesthesia deep monitors based on EEG. Other advantages are that the proposed system can be fully implemented in hardware, can work in real-time and is of low-cost.

Future developments will face other PLL circuits with more robust capabilities, such as high order and active loop filters, and charge-pumps phase detectors. This will enable a more flexible design, improving the transient response.

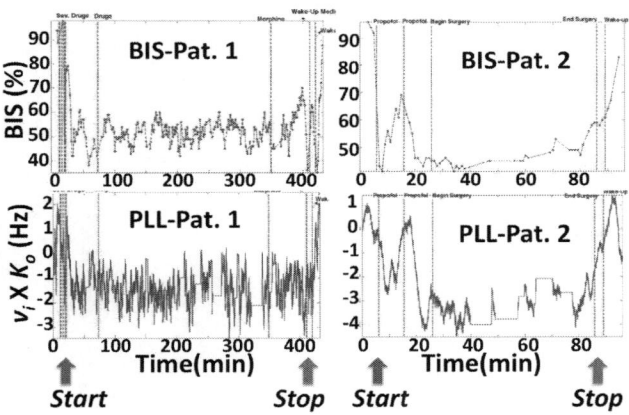

Fig. 2: Frequency variation estimation as compared with the BIS curve as computed by the operating room instrumentation. The red arrows indicate the beginning of anesthesia and the green arrows the end of the anesthesia maintenance.

ACKNOWLEDGEMENTS

The authors acknowledge the support of the project iCIS (CENTRO-07-0224-FEDER-002003). CT is supported by a research contract under the Ciência 2007 program of FCT. BD is supported by a grant from FCTUC.

REFERENCES

1. Woodbridge PD. Changing Concepts Concerning Depth of Anesthesia Anesthesiology. 1957;18:536-550.
2. Voss L, Sleigh J. Monitoring consciousness: the current status of EEG-based depth of anaesthesia mon- itors Best Pract Res Clin Anaesthesiol. 2007;21:313 - 325.
3. Rampil IJ. A Primer for EEG Signal Processing in Anesthesia Anesthesiology. 1998;89:980–1002.
4. Broughton R, Healey T, Maru J, Green D, Pagurek B. A phase locked loop device for automatic detection of sleep spindles and stage 2 Electroencephalogr Clin Neurophysiol. 1978;44:677 - 680.
5. Niedermeyer E, Lopes Da Silva FH. Electroencephalography: basic principles, clinical applications, and related fields. Lippincott Williams & Wilkins 2005.
6. Carlson AB. Communication Systems: An Introduction to Signals and Noise in Electrical Communication. McGRAW-HILL International Editions, Third ed. 1986.
7. D'Azzo JJ, Houpis CD. Linear Control System Analysis and Design: Conventional and Modern. McGraw- Hill Higher Education, 4th ed. 1995.

Author: César A. D. Teixeira
Institute: Centre for Informatics and Systems (CISUC)
Street: Polo II, Pinhal de Marrocos
City: Coimbra
Country: Portugal
Email: cteixei@dei.uc.pt

Evaluation of Arterial Properties through Acceleration Photoplethysmogram

R. Gonzalez[1], A. Manzo[1], E. Cardenas[1], J. Herrera[1], F. Martinez[1], J. Gomis[2], and J. Saiz[2]

[1] Instituto Tecnologico de Morelia/Departmento de Ing. Electronica, Morelia, Mexico
[2] Universidad Politecnica de Valencia/I3BH, Valencia, Spain

Abstract—This work aimes to provide, that acceleration photoplethysmogram produces useful information in assessing arterial properties. The development and progression of arterial vascular disease is a multifactorial process. Risk factors for cardiovascular disease mediate their effects by altering the structure, properties and function of wall and endothelial components of the arterial blood vessels that vary between different vascular beds. Monitoring arterial vascular walls as well as risk factors such as hypertension, hypercholesterolemia and other blood biochemical profiles can potentially help to identify individuals having an increased risk of developing cardiovascular disease in adulthood.

Pulse wave analysis has been shown to provide valuable information on aortic stiffness and elasticity, and it has been widely used to evaluate the vascular effects of aging, hypertension and atherosclerosis. The second derivative of the PPG or acceleration photoplethysmogram (APG) was developed as a method allowing more accurate recognition of the inflection points and easier interpretation of the original plethysmogram wave. Although obtained from the periphery of the circulation, APG provides information about both central and peripheral arterial properties.

A computer based PPG analyzer was developed. With the APG signal, four separate systole waves (named a-d) and a diastole wave (named e) were obtained. A study with 40 people, 20 healthy volunteers and 20 subjects with previously diagnosed cardiovascular disease (diabetes mellitus, atherosclerosis, hypertension) was carried out. A t–tested distribution between healthy volunteers and patients showed a significant differences in calculated parameters: b/a (-0.742 vs -0.361, $p < 0.0001$), d/a (-0.083 vs -0.482, $p < 0.0001$), e/a (0.182 vs 0.061, $p < 0.0001$), and for c/a (-0.039 vs -0.148, $p < 0.005$). In conclusion, APG signal has shown to be a noninvasive indicator for vascular assessments.

Keywords—Photoplethysmographic signal, acceleration photoplethysmogram, arterial properties.

I. INTRODUCTION

The complications of cardiovascular disease represent the leading cause of morbid and mortal events in Western society. The development and progression of arterial vascular disease is a multifactorial process. Risk factors for cardiovascular disease mediate their effects by altering the structure, properties and function of wall and endothelial components of the arterial blood vessels that vary between different vascular beds [1]. The ability to detect and monitor sub-clinical damage, representing the cumulative and integrated influence of risk factors in impairing arterial wall integrity, holds potential to further refine cardiovascular risk stratification and enable early intervention to prevent or attenuate disease progression [2].

Monitoring arterial vascular walls as well as risk factors such as hypertension, hypercholesterolemia and other blood biochemical profiles can potentially help to identify individuals having an increased risk of developing cardiovascular disease in adulthood. The importance of assessing arterial wall integrity has been highlighted by studies demonstrating that a reduction in the pulsatile function of large arteries represents an independent risk factor for future cardiovascular events. Accumulating evidence suggests that abnormalities in the pulsatile characteristics of arteries occur early in the disease processes associated with increased cardiovascular risk, and can be favourably modified by therapeutic interventions. Impaired pulsatile arterial function is recognized as an independent predictor of risk for vascular events with ageing and various disease states, including coronary heart disease, congestive heart failure, hypertension and diabetes mellitus [3–5].

The arterial pulse waveform is derived from the complex interaction of the left ventricular stroke volume, the physical properties of the arterial tree and the characteristics of the fluid in the system. Pulse wave analysis has been shown to provide valuable information on aortic stiffness and elasticity [6,7], and it has been widely used to evaluate the vascular effects of aging, hypertension and atherosclerosis [8,9].

Photoplethysmography (PPG) is an optical technique, which typically operates using infrared light, allowing the transcutaneous registration of venous and/or arterial blood volume changes in the skin vessels. The complex interaction between the heart and connective vasculature are the components of the mechanism that generates the PPG signal [10]. The PPG waveform comprises a pulsatile ('AC') physiological waveform attributed to cardiac synchronous changes in the blood volume with each heart beat, and is superimposed on a slowly varying ('DC') baseline with various lower frequency components attributed to the non-pulsatile arterial blood volume, the venous blood and skin, bone and tissue. See figure 1.

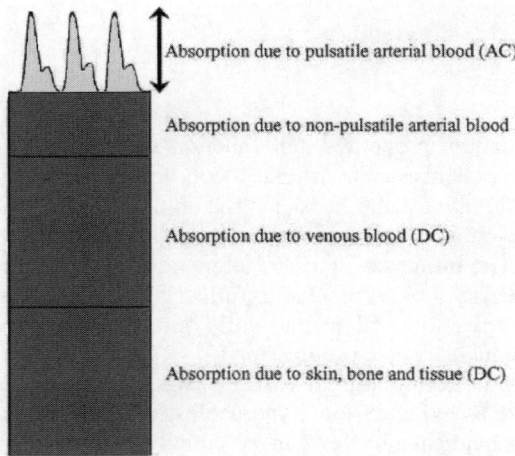

Fig 1 Photoplethysmographic signal components.

Photoplethysmography mainly reflects the pulsatile volume changes in the arteries/venous, has been recognized as a noninvasive method of measuring arterial pulse waves in relation to changes in wave amplitude. However, the wave contour itself has not been analysed because of the difficulty in detecting minute changes in the phase of the inflections. Accordingly, the second derivative of the PPG or acceleration photoplethysmogram (APG) was developed as a method allowing more accurate recognition of the inflection points and easier interpretation of the original plethysmogram wave. Although obtained from the periphery of the circulation, APG provides information about both central and peripheral arterial properties. It was shown that some APG quantifiers closely correlated with the ascending aortic augmentation index and distensibility of the common carotid artery, age and other atherosclerotic risk factors [11,12]. This work aimes to provide, that acceleration photoplethysmogram produces useful information in assessing arterial properties.

II. METHODS

The PPG waveform is formed due to wave transmission and reflection, which is illustrated in figure 2. The first part of the volume waveform in the finger is the result of pulse transmission along a direct path from the aortic root to the finger. The second part is formed by the pulse transmitted from the aortic root to the lower body, where it is reflected back along the aorta and subclavian artery to the finger as shown Fig. 2a. The APG waveform includes four systolic waves and one diastolic wave, namely a-wave (early systolic positive wave), b-wave (early systolic negative wave), c-wave (late systolic reincreasing wave), d-wave (late systolic redecreasing wave) and e-wave (early diastolic positive wave), as shown Fig. 2b.

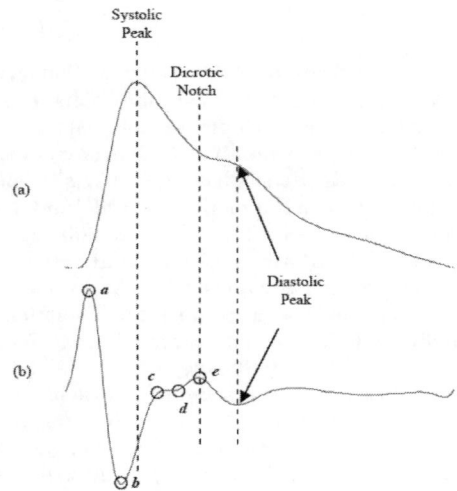

Fig. 2 (a) Original PPG waveform. (b) APG waveform.

A computer based PPG analyzer was developed. The signal was obtained by infrared light through the finger. It was conditioned and converted into digital domain by a signal processing circuitry, which contains amplifying and filtering steps, a microcontroller and an analog to digital converter. The acquired signal was displayed and analyzed.

With the APG signal, four separate systole waves (named a-d) and a diastole wave (named e) were obtained. The a and b waves are included in the early systolic phase of the PPG whereas the c and d waves are included in the late systolic phase. The height of each wave from the baseline was measured and their ratios b/a, c/a, d/a and e/a were calculated. A study with 40 people, 20 healthy volunteers and 20 subjects with previously diagnosed cardiovascular disease (diabetes mellitus, atherosclerosis, hypertension) was carried out.

III. RESULTS

The analysis of PPG signal is obtained as follows. One of acquired PPG signals from a healthy volunteer is chosen to be analyzed. In figure 3 is shown how are displayed, the original PPG signal and its APG signal.

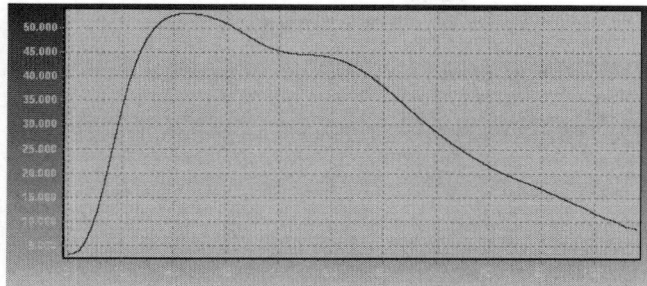

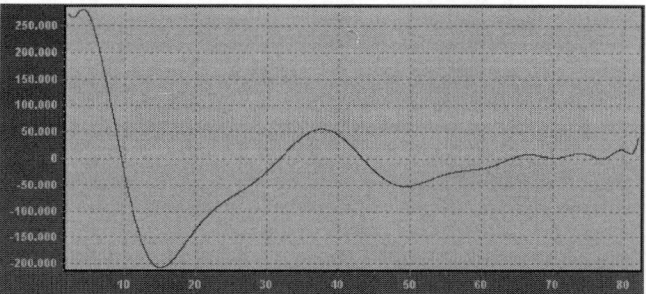

Fig 3 Original PPG waveform and APG waveform from a healthy volunteer.

Figure 4 shows a PPG signal and its APG signal from a second healthy volunteer.

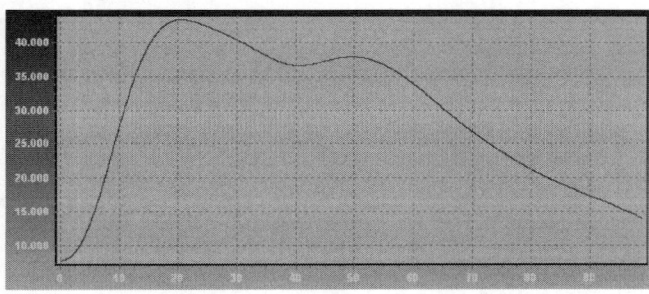

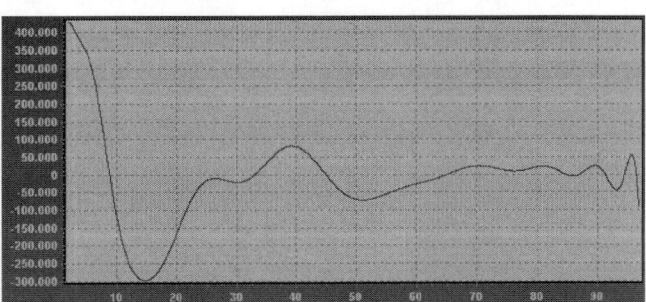

Fig 4 Original PPG waveform and APG waveform from a second healthy volunteer.

Analyzed PPG signal and its APG from a patient are shown in figure 5.

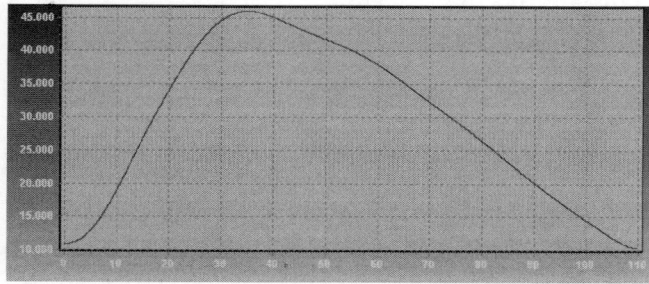

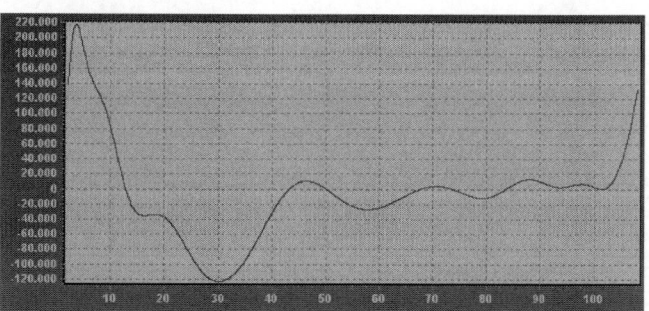

Fig. 5 Analized PPG waveform and APG waveform from a patient.

In figure 6 are shown the analyzed PPG signal and its APG signal from a second patient. For healthy subjects the dicrotic notch was predominantly seen in the PPG signal whereas in patients owing to the increase in arterial stiffness and a faster reflected wave augmenting the forward wave the pulse becomes rounded. It was noticed that the peripheral pulse has a steep rise and a notch on the falling slope in healthy volunteers whereas in patients a more gradual rise and fall and no pronounced dicrotic notch.

In subjects with previously diagnosed cardiovascular disease, the pulse can become smoothed, with changes in the blood pressure pulse producing less dramatic changes in the blood volume pulse at the periphery. The results show an overall elongation of the systolic rising edge, which could be explained on the basis of changes in resistance and compliance properties of arteries with cardiovascular diseases. The diminishing of the dicrotic notch in patients is mainly due to disease related increases in pulse wave velocity resulting in a faster reflected wave augmenting the forward wave.

A t–tested distribution between healthy volunteers and patients respectively showed a significant differences in calculated parameters: b/a (-0.742 vs -0.361, $p < 0.0001$), d/a (-0.083 vs -0.482, $p < 0.0001$), e/a (0.182 vs 0.061, $p < 0.0001$), and c/a (-0.039 vs -0.148, $p < 0.005$).

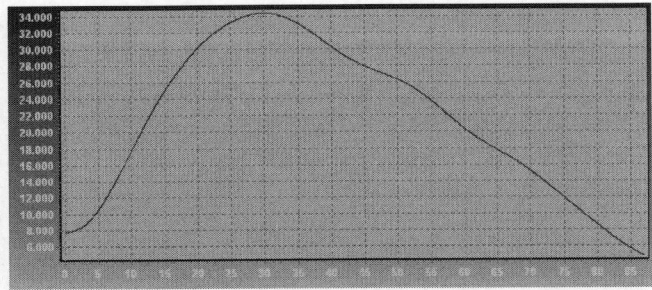

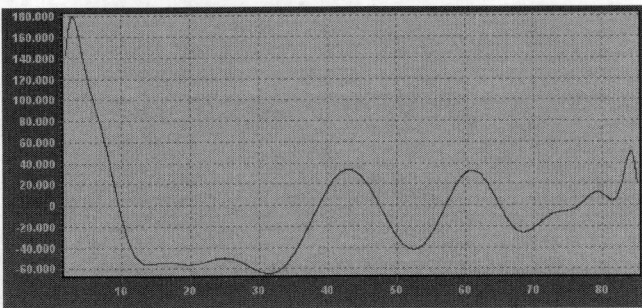

Fig 6 Analized PPG waveform and APG waveform from a second patient.

IV. CONCLUSIONS

The PPG analysis showed that such delicate changes in the waves were emphasized and easily quantified by double differentiating the original PPG signal with respect to time. The APG signal is used as a means to accentuate and locate inflection points.

The analysis of the APG waveform was carried out by a computer based photoplethysmographic analyzer. The calculated parameters of the APG signal: b/a, c/a, d/a, e/a have shown significant differences between healthy volunteers and patients. Takazawa et al [8] suggest that the b/a index might reflect large arterial stiffness and the d/a index indicates peripheral reflection.

In conclusion, the APG signal has shown to be a noninvasive indicator for vascular assessments. Using APG, inflection points identification of the photoplethysmographic signal is more accurate mainly in subjects with previously diagnosed cardiovascular disease.

ACKNOWLEDGMENT

This work was partially supported by DGEST Mexico grant (4426.11-PR).

REFERENCES

1. McVeigh GE, Hamilton PK, Morgan DR (2002) Evaluation of mechanical arterial properties: clinical, experimental and therapeutic aspects. Clinical Science 102:51–67
2. St John S, (2000) Aortic stiffness: a predictor of acute coronary events? Eur. Heart J 21:342-344
3. Benetos A, Rudnichi A, Safar M et al (1998) Pulse pressure and cardiovascular mortality in normotensive and hypertensive subjects. Hypertension 32:560–564
4. Vaccarino V, Holford TR, Krumholz HM (2000) Pulse pressure and risk for myocardial infarction and heart failure in the elderly. J. Am. Coll. Cardiol 36:130–138
5. Franklin SS, Khan SA, Wong ND et al (1999) Is pulse pressure useful in predicting risk for coronary heart disease? The Framingham Heart Study. Circulation 100:354-360
6. Kelly RP, Avolio A, O'Rourke MF (1989) Noninvasive determination of age-related changes in the human arterial pulse. Circulation 80:1652-1659
7. Elgendi M, (2012) On the analysis of fingertip photoplethysmogram signals. Curr. Cardio Rev 8:14-25
8. Takazawa K, Fujita M, Matsuoka O et al (1998) Assessment of vascular agents and vascular aging by the second derivative of photoplethysmogram wave. Hypertension 32:365-370
9. Bortolotto LA, Kondo T, Takazawa K et al (2000) Assessment of vascular aging and atherosclerosis in hypertensive subjects: second derivative of photoplethysmogram versus pulse wave velocity. Hypertension 13:165-171
10. Alnaeb ME, Alobaid N, Seifalian AM et al (2007) Optical Techniques in the Assessment of Peripheral Arterial Disease. Current Vascular Pharmacology 5:53-59
11. Takada H, Washino K, Harrel JS et al (1997) Acceleration Photoplethysmography to evaluate aging effect in cardiovascular system. Using new criteria of four wave patterns. Med Prog Technol, 21: 205-210
12. Millasseau SC, Kelly RP, Ritter JM et al (2002) Determination of age-related increases in large artery stiffness by digital pulse contour analysis. Clinical Science 103: 371–377

Author: Rodolfo Gonzalez
Institute: Instituto Tecnologico de Morelia
Street: Ave. Tecnologico 120
City: Morelia
Country: Mexico
Email: rodogon21@yahoo.com

Comparing over Ground Turning and Walking on Rotating Treadmill

A. Olenšek, J. Pavčič, and Z. Matjačić

University Rehabilitation Institute, Republic of Slovenia/Research Department, Linhartova 51, Ljubljana, Slovenia

Abstract—This paper presents rotating treadmill that combines forward walking of conventional treadmill and rotating motion of rotating platform. In this way we are able to some extent mimic similar turning maneuvers as people utilize when they turn during over ground walking. In a male subject with no known neurological or orthopedic disorders we compared pelvis and torso rotation during walking on rotating treadmill and during over ground walking in circles in four experimental conditions that combine selected levels of gait velocity and required time for completing one full circle. Results show that in all experimental conditions torso and pelvis orientation while the subject was walking on rotating treadmill closely follow torso and pelvis orientation when the subject was walking over ground in circles. Such setup has considerable potential for practical applications in gait rehabilitation when patients progress to a level when more challenging turning maneuvers should be included in rehabilitation program.

Keywords—Torso rotation, pelvis rotation, treadmill, rotating treadmill, over ground walking.

I. INTRODUCTION

Walking is elementary way of human movement. However after neurological damage or disease (stroke, Parkinson disease, multiple sclerosis...) or spinal cord injury gait is often severely constrained and accompanied by other long term disabilities such as muscle weakness, diminished dynamic stability, spasticity, osteoporosis and increased risk of falling due to physical inactivity[1]. Without immediate rehabilitation limited walking capabilities can persist for longer periods of time and postponing with rehabilitation cannot guaranty optimal recovery of walking function. It is expected that recovery will be faster in acute phase whereas in chronic phase the extent to which gait function could be restored is limited.

While therapist assisted gait training [2] is still the most dominant way of gait rehabilitation, robot supported gait training [3,4,5] combined with body weight support (e.g. Lokomat) and treadmill is successfully substituting manual gait rehabilitation in early phases. Robot supported gait training relieves the therapists from supporting the patient's body weight and provide assistance in restoring cyclical leg movement. Compared to therapist assisted gait training such approach often diminishes the need for at least one therapist being present. On the other hand, due to physical constraints that limit patient's gait robot assisted gait training is suitable only in early phases of gait rehabilitation, even more when combined with treadmill. Namely, limited number of degrees of freedom of rehabilitation robot confines the patient to sagittal movement but omits other aspects of human walking. Similar concerns apply in treadmill walking. On one hand treadmill gait exhibits similar gait kinematics as over ground walking [6,7], for which it is considered to be a good substitute for over ground walking and it additionally allows us to fairly easily access to lower extremities for manual manipulation if necessary. On the other hand treadmill like rehabilitation robots only allow forward movement whereas in everyday life maneuvering between two locations demands more challenging maneuvers like accelerating and decelerating as well as turning left and right. Such rehabilitation framework is therefore suitable only in early stages of gait rehabilitation. Eventually after improving gait function constraints should gradually decrease, ideally they should be completely eliminated, thus stimulating the patient to learn gait mechanisms associated with turning, accelerating and decelerating as fully as possible. From biomechanical perspective such maneuvers are more demanding and due to lack of proper equipment such training is exercised exclusively during over ground walking with usually more than one therapist assisting in proper completion of maneuvers and safety assurance.

To overcome this gap we designed an experimental setup that combines forward movement of treadmill and rotational degree of freedom in a way to impose turning capability. When combined with weight bearing system such setup has potential to establish fall safe training conditions not only in forward gait but also in maneuvers that to some extent resemble to turning in over ground walking. In this paper we investigate how torso and pelvis rotations under such experimental conditions relate to torso and pelvis rotations during over ground turning.

II. METHODS

A. Rotating Treadmill

Rotating treadmill is composed of three main components: base plate, rotating platform and conventional treadmill. Base plate creates ground surface and provides passive vertical axis of rotation in the center with mounting pads for fixating rotating platform thus confining the rotating platform to rotation about vertical axis. While fixating the rotating platform to mounting pads of vertical axis rotating platform is further supported by four small castor wheels at the corners that are positioned in a way to outline a circle when the platform is rotating and by two larger wheels with axes of rotations aligned with direction of walking on the treadmill. Only one of two larger wheels is motorized with speed of turning being controlled with controller in a feedback control loop. Finally a treadmill is bolted down to rotating platform making the treadmill to rotate about the same axis of rotation as rotating platform while retaining forward gait locally on the treadmill. Rotating treadmill is schematically presented in Fig. 1

B. Data Acquisition

One healthy male subject (age: 29, weight: 74kg, height: 180 cm) with no known neurological or orthopedic disorders participated in our preliminary study. Markers were placed on selected anatomical landmarks according to plug-in-gait model and Vicon motion capture system was used for recording three dimensional movement of pelvis and torso as well as rotating treadmill. In our preliminary study we focused on comparison between pelvis and torso orientation while walking on rotating treadmill and during over ground turning i.e. walking in circles respectively with selected speed of walking and time required for completing full circle. At least eight strides were averaged in each experimental condition.

C. Experimental Conditions

To be able to compare torso and pelvis orientation while walking on rotating treadmill and during over ground walking it was our demand that in both modes of walking the subject is walking with the same speed and is turning with the same angular velocity. This implied that while walking on rotating treadmill the speed of treadmill was set in a way to correspond to the speed of walking during over ground walking in circles. Similarly, angular velocity of rotating treadmill was selected in a way that the time needed for completing one turn on rotating treadmill was the same as the time needed to complete one full circle during over ground turning. In this sense we considered four experimental conditions of turning on rotating treadmill and during over ground walking in circles – we combined two levels of gait velocities (0.6 m/s and 1 m/s respectively) and two angular/turning velocities (30 deg/s and 40 deg/s during treadmill turning correspond to 12 s and 9 s respectively for completing one circle during over ground turning). While the speed of walking and angular velocity of rotating treadmill is enforced by feedback control, four circles with corresponding radii were outlined on the ground and stopwatch

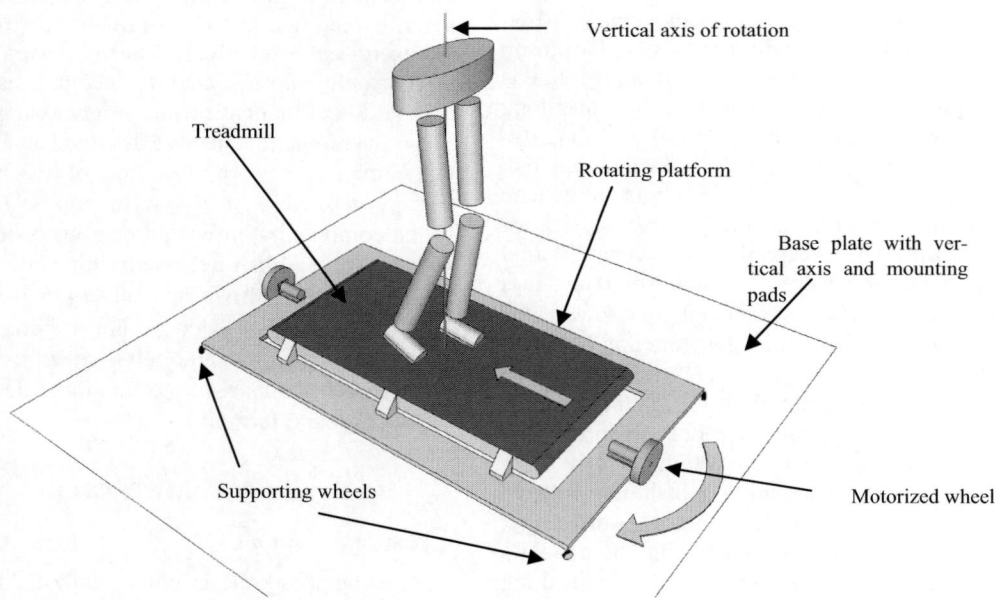

Fig. 1 Schematic representation of rotating treadmill.

Table 1 Experimental conditions

	Turning on rotating treadmill		Over ground turning		
	Speed of walking	Angular velocity	Speed of walking	Circle time	
RTT1	0.6 m/s	30 deg/s	0.6 m/s	12 s	OGT1
RTT2	1 m/s	30 deg/s	1 m/s	12 s	OGT2
RTT3	0.6 m/s	40 deg/s	0.6 m/s	9 s	OGT3
RTT4	1 m/s	40 deg/s	1 m/s	9 s	OGT4

was used in over ground turning to ensure required speed of walking according to selected experimental conditions. In all experimental conditions subject executed turning maneuvers first in clockwise direction followed by counter clockwise direction. Detailed description of experimental conditions is given in Table 1.

III. RESULTS

Fig. 2 compares torso and pelvis orientations in a single patient study according to experimental conditions listed in Table 1. Results show that in all experimental conditions torso and pelvis orientation while the subject was walking on rotating treadmill closely follow torso and pelvis orientation when the subject was walking over ground in circles. We recognize that the majority of torso rotation occurs approximately in the first half of gait cycle when the subject turns into direction of turning whereas in the second half the torso principally waits for stride completion. Pelvis on the other hand first follows the turning dynamics, then starts to lag behind the torso and catches the torso in the second half of gait cycle. There exist some deviations between over ground turning and walking on rotating treadmill, however they are in the range of a few degrees and do not distort the general kinematic pattern.

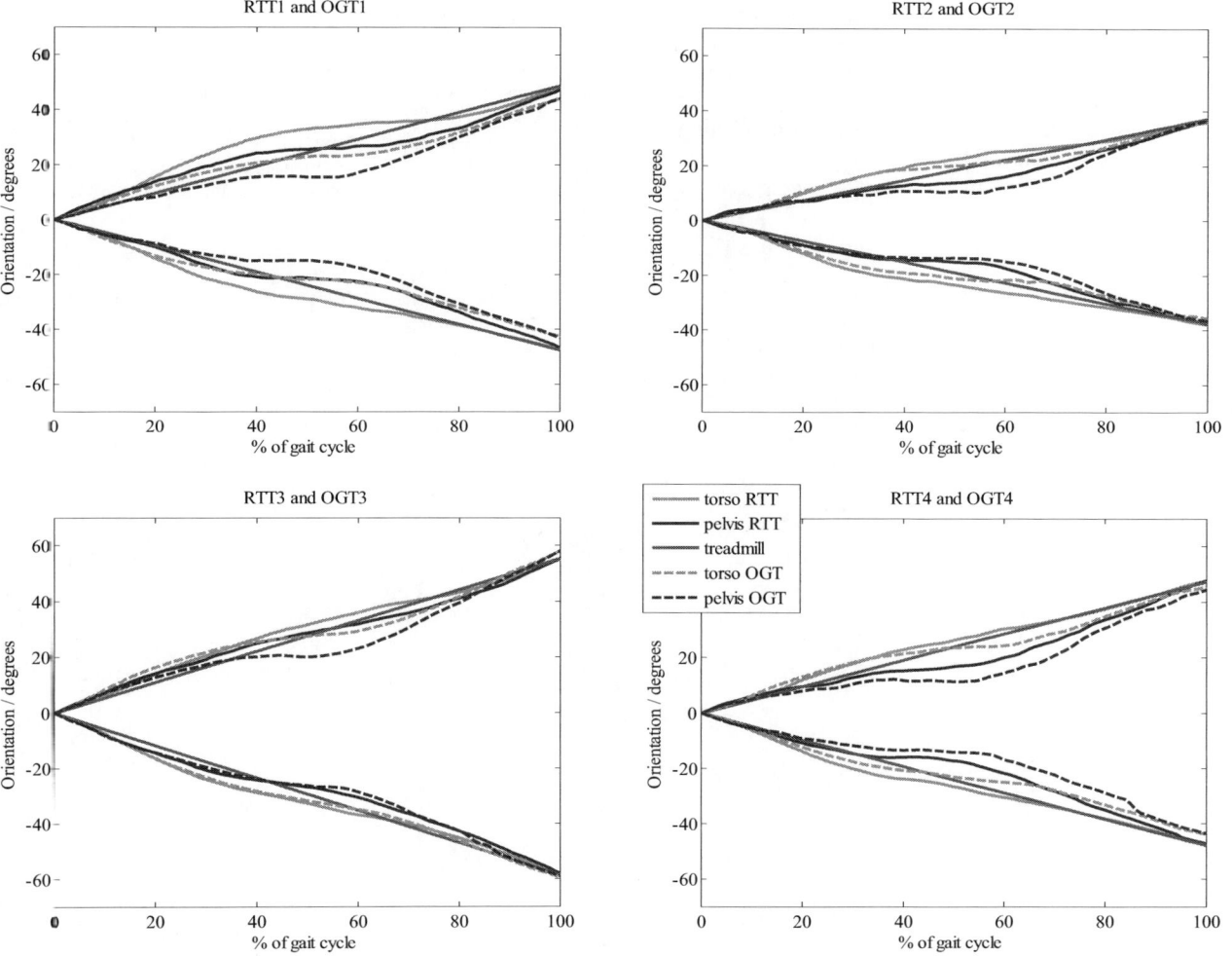

Fig. 2 Orientation of torso, pelvis and treadmill during clockwise and counter clockwise over ground turning and while walking on rotating treadmill.

IV. CONCLUSIONS

In both modes of walking two different approaches are used to impose turning maneuvers. In over ground walking it is the subject who initiates the turn with pelvis and the upper body as well as the opposite leg following to complete the turn. In walking on rotating treadmill however, the turning maneuver is initiated by rotating platform that rotates the stance leg but the subject must nonetheless utilize similar mechanisms as in over ground turning for the pelvis and upper body as well as opposite leg to follow the leading leg. Given that torso and pelvis orientation while walking on rotating treadmill correspond well to torso and pelvis orientation during over ground walking in circles this hypothesis is only partially supported. To further investigate our hypothesis studies should be continued in a larger group of subjects where kinematics as well as kinetics should be thoroughly investigated. Nevertheless such setup has considerable potential for practical applications. Presented experimental environment could be very easily upgraded with weight bearing system (to enhance safety and diminish the need for additional therapist), angular velocity of treadmill could be adjusted in a way to mimic speed of turning according to patient needs and capabilities or to motivate the patient during training the presented setup could be equipped with virtual environment.

ACKNOWLEDGMENT

This research was partially supported by the Slovenian Research Agency under research program number P2208 and by the European Commission 7th Framework Program as part of the CORBYS (Cognitive Control Framework for Robotic Systems) project under grant agreement number 270219.

REFERENCES

1. Latham NK., Jette DU, Slavin M (2005) Physical therapy during stroke rehabilitation for people with different walking abilities. Arch. Phys Med Rehabil 86:41-50
2. Latham NK, Jette D., Slavin (2005) Physical therapy during stroke rehabilitation for people with different walking abilities. Arch Phys Med Rehabil 86:41-50.
3. Hornby G, Campbell D, Zemon DH, Kahn JH (2005) Clinical and Quantitative Evaluation of Robotic-Assisted Treadmill Walking to Retrain Ambulation After Spinal Cord Injury. Topics in Spinal Cord Injury Rehabilitation 11: 1-17
4. Westlake KP, Patten C (2009) Pilot study of Lokomat versus manual-assisted treadmill training for locomotor recovery post-stroke. J Neuroeng Rehabil 6:1-11
5. Husemann B, Muller F, Krewew C, Heller S, Koenig E (2007) Effect of locomotor training with assistance of a robot-driven gait orthosis in hemiparetic patients after stroke. Stroke 38(2):349-354
6. Riley P, Paolini G, Della Croce U, Paylo KW, Kerrigan DC (2008) A kinematic and kinetic comparison of overground and treadmill walking in healthy subjects. Gait Posture 26:17-24
7. Lee JS, Hidler J (2008) Biomechanics of overground vs. treadmill walking in healthy individuals. J Appl Physiol 104:747-755

Author: Andrej Olenšek
Institute: University Rehabilitation Institute, Republic of Slovenia
Street: Linhartova 51
City: Ljubljana
Country: Slovenia
Email: andrej.olensek@mail.ir-rs.si

Displacement Measurement of a Medical Instrument Inside the Human Body

D.A. Fotiadis[1,2], A. Astaras[1], A. Kalfas[1], K. Papathanasiou[1], and P. Bamidis[1]

[1] Laboratory of Medical Informatics, School of Medicine, Aristotle University of Thessaloniki, Greece
[2] Technological Educational Institution (T.E.I.) of Western Macedonia,
Department of Informatics and Computer Technology, Kastoria Branch, Greece

Abstract—Several technologies have attempted to measure displacement of objects inside the human body; some of the leading challenges addressed by those technologies are measurement precision, the capacity to operate without clear Line-of-Sight (LOS), susceptibility to electromagnetic interference from other medical instruments, miniaturization, cost effectiveness and safety for both the patient and the medical personnel. The proposed novel method for measuring displacement and tracking the position of a medical instrument inside a human body achieves sub-millimeter accuracy, is characterized by a potentially low cost at high production volumes. It is simple to implement from the biomedical engineering point of view and has been developed to improve the function of medical devices used for invasive or minimally invasive surgery (MIS). This method is based on measuring phase shift displacement at an operating frequency in the gigahertz (GHz) range. The phase shift is detected from the signal emitted by a transmitter, which has been integrated into the medical instrument, with respect to a fixed receiver. It is subsequently translated into a very low frequency voltage, which carries all necessary information concerning transmitter displacement. Simulation results using Verilog-A models and the Spectre simulator provide a mathematical proof of concept model for our novel system design. These results were consequently validated using an experimental setup, which provided millimeter precision results, by the use of commercialy available discrete components and low cost measuring equipment.

Keywords—Phase shift, position tracking, medical instrument, Phase Locked Loop (PLL), Doppler Radar.

I. INTRODUCTION

Several techniques have been proposed to locate and track the position of a medical instrument inside a human body [1], [2]. With respect to invasive medical procedures and instruments, the use of X-ray imaging is the dominant method utilised for position tracking [3]. Though not as prevalent, electromagnetic [4]–[6] and mechanical tracking [7], [8] appear to have recently established a new trend in MIS.

The phase shift of a high frequency (sinusoidal) signal, generated by a transmitter integrated inside a medical instrument can be measured by receivers located at fixed positions outside the patient's body. When the signal frequencies involved are in the GHz range, the aforementioned phase shift can subsequently be interpreted into an exact displacement measurement accurate down to sub-millimeter levels [9].

The proposed system was implemented utilising a behavioural model developed using an analog and mixed-signal variety of the Verilog hardware description language. Once simulation results verified that the concept was sound, a proof of concept hardware model was set up and used to experimentally validate the entire precision measurement procedure.

Readily available doppler radar modules operating in the GHz range [10] were used in order to provide a low cost, real-time measurement experimental setup. A set of experiments were designed and conducted in which the transmitter is moving away from the receiver at a constant velocity. The only measurement device used was a low frequency digital oscilloscope, capable of extracting the desired signal while averaging provided noise filtering in the time domain.

In section II we introduce a basic description of the system. The materials used for the experimental setup and the methods for measuring the transmitter displacement are detailed in section III. In section IV we present the results obtained from our hardware prototype measurements. Finally in section V a path to a possible improvement of the system's accuracy is highlighted in conjunction with our conclusions.

II. SYSTEM DESCRIPTION

The system architecture is based on the phase shift method combined with a phase-lock approach. When a sinusoidal wave transmitter varies its distance with respect to a receiver with which it used to be synchronised in phase and frequency, its phase will change when compared to a reference signal. If the transmitter-receiver distance is R, a sinusoidal continuous wave (CW) signal is travelling with velocity v and the signal frequency is f, then the total number of wavelengths between the transmitter and receiver is R/λ, where λ is the sinusoidal signal wavelength. The sinusoidal signal wavelength λ is given by:

$$\lambda = \frac{v}{f} \quad (1)$$

The signal propagation speed during transmission within a specific medium is given by:

$$v = \frac{1}{\sqrt{\mu\varepsilon}} \quad (2)$$

In expression (2) μ is the transmission medium permeability and ε is the permittivity or dielectric constant of the medium (human body in our case).

Each wavelength displacement corresponds to a phase change of 2π radians. The total phase change or phase shift in the transmitter-receiver path can then be calculated by combining expressions (1) and (2):

$$\Delta\phi = 2\pi \times \frac{R}{\lambda} = \frac{2\pi R}{\lambda} = \frac{2\pi f R}{v} = 2\pi f R \sqrt{\mu\varepsilon} \quad (3)$$

The system basic architecture is depicted in Fig. 1.

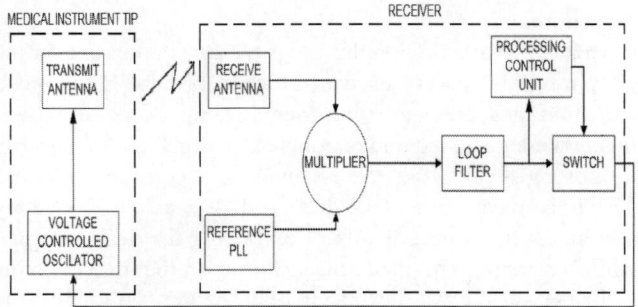

Fig. 1 Medical instrument displacement measurement architecture

Similar to phase locked loops (PLLs), our system consists of three core components. These are the phase detector (PD) or multiplier, the loop filter (LF) –a low pass filter (LPF) - and the voltage controlled oscillator (VCO). The multiplier, which also acts as a mixer, multiplies two sinusoidal signals, one from the VCO and another from a reference PLL. If we assume that a reference PLL produces a sinusoidal signal in the form of $x_1(t)=A_1\cos(2\pi ft)$ and the VCO transmits a signal in the form of $x_2(t)=A_2\sin(2\pi ft+\varphi(t))$, where $\varphi(t)$ denotes a phase shift, then the multiplier output is

$$x_1(t)\cdot x_2(t) = \frac{1}{2}A_1 A_2 \sin(\phi(t)) + \frac{1}{2}A_1 A_2 \sin(2\cdot 2\pi ft + \phi(t)) \quad (4)$$

The high frequency component is filtered by the LPF, resulting in a low frequency component around DC, which reflects the phase difference $\varphi(t)$ of the two multiplied signals. This voltage signal (V_{disp}) from the LPF provides information on the medical instrument displacement and is given by:

$$V_{disp} = x_1(t)\cdot x_2(t) = \frac{1}{2}A_1 A_2 \sin(\phi(t)) \quad (5)$$

The resultant voltage signal is of a low frequency nature as it is derived at the output of the LPF; it is a sinusoidal function of the phase shift $\varphi(t)$ between the VCO transmitted signal and the reference PLL signal. It is a down-converted signal which can be easily processed by common place signal acquisition topologies, in order to obtain a highly precise distance change measurement between the medical instrument tip and the receiver antenna.

Combining expressions (3) and (5) provide the following expression concerning displacement measurement:

$$R = \frac{2\cdot V_{disp}}{2\cdot \pi \cdot f \cdot A_1 \cdot A_2 \cdot \sqrt{\mu\cdot\varepsilon}} \quad (6)$$

Since V_{disp} is a sinusoidal voltage which is easily measured, parameters μ and ε can also be estimated [11] and thus displacement distance between transmitter and receiver can be calculated using expression (6). Having established a common reference point (e.g. x=0, y=0, z=0) during an initial calibration phase, it is straightforward to estimate the medical instrument tip's lateral displacement inside the human body. A recording and visualization device can therefore provide a full history of the movement of the instrument inside the human body, simply by recording and interpreting the associated output voltage.

In order to obtain a metric which will reflect the phase difference and therefore the nature and extent of movement, the proposed system needs to implement two discrete modes of operation. Initially the loop is always closed therefore the loop gain is utilized to correct any phase differences between transmitter and receiver, while the latter tracks the transmitted signal. In other words the feedback loop ensures that the VCO phase and frequency are continuously corrected by an error voltage, namely the output of the loop filter, which ensures that the frequency and the phase of the VCO are equal to the properties of the reference PLL and therefore the two track each other.

The closed PLL ensures a perfect synchronization between transmitter and receiver. On the other hand, the closed PLL loop does not permit displacement measurement, since the VCO would continuously correct its frequency and phase. A switch has therefore been placed in the circuit after the loop filter, allowing for the loop to open on demand. Since the PLL is locked the VCO transmits a sinusoidal signal which is synchronised in both phase and frequency with respect to the reference PLL. As the medical instrument moves the switch opens, while a sample and hold circuit maintains a stable voltage to control the VCO. As the transmitter starts deviating from its original position the resultant multiplier signal smoothed by the LF denotes an error signal, which represents the phase shift that occurs at the reception of the VCO signal and consequently the distance offset which has occurred between the transmitter and receiver antennas. This distance offset can subsequently be calculated measuring V_{disp} and using expression (6).

III. MATERIALS AND METHODS

Two commercially available X-Band doppler motion detection units were deployed for the experimental system implementation. Smooth and accurate displacement of the transmitter is achieved using a 4-phase step motor. Step motor function is controlled by an Intel 89C51 microcontroller. The motor's rotation displaces the transmitter pulling a string attached to the motor axis on one end and to a miniature roller vehicle carrying the transmitter on the other.

The MDU 1100T unit's basic principle of operation consists of detecting frequency or phase shift between a transmitted and a received signal, the latter is reflected back to a moving object, at the unit's field of view.

Instead of using two MDU 1100T units as doppler radar transceivers, they were modified so that one acts as a transmitter and another as a receiver. Both units have an operating frequency of 10.587 GHz. The MDU1100T schematic diagram can be seen in Fig. 2.

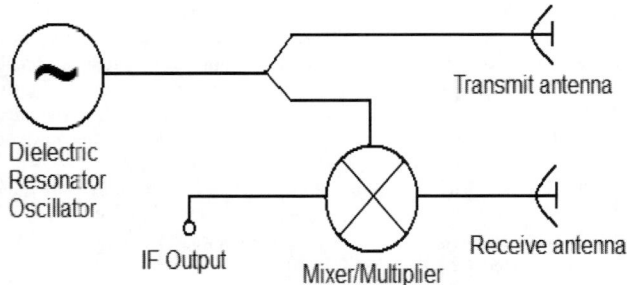

Fig. 2 MDU 1100T schematic diagram

A subset of the theoretical setup shown in Fig. 1 is implemented in Fig. 3. The MDU1100T units were utilized for our evaluation, because they were readily available operating at 10.587 GHz frequency, and offered good documentation of circuit implementation. In the mentioned board the transmitter contains an integrated Dielectric Resonator Oscillator (DRO) instead of a preferred VCO. Since we obtain location information by opening the switch of Fig. 1, the lack of a VCO circuit can be tolerated in our evaluation prototype.

The Intermediate Frequency (IF) output signal from receiver module is captured by a digital oscilloscope and is

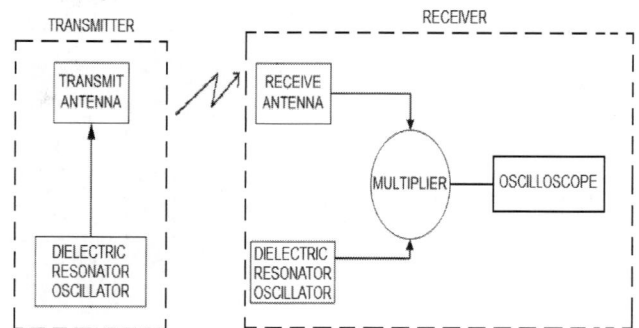

Fig. 3 Experimental system architecture

appropriately post-processed. The transmitter displacement in respect to the fixed receiver is 8cm and the movement is concluded in a 2s time interval.

IV. EXPERIMENTAL RESULTS

Ten measurements were obtained using a digital oscilloscope to capture the IF output in order to extract statistical results and derive system precision. A typical waveform is shown in Fig. 4.

It is clear that the resultant IF signal is of sinusoidal nature, which is in agreement with theoretical predictions. The phase shift method analysis of section II was therefore deemed valid.

All waveforms were curve fitted in order to accommodate subsequent noise or interference filtering. The time domain regression analysis method which was selected is based on Locally Weighted Scatterplot Smoothing (LOWESS) [12]. As a result the zero crossing of the best fit smoothed curve is observed. Unlike the standard Least Squared error method, this technique is nearly insensitive to outliers. The smoothing factor used is 15%, which means that the smoothing window used has a total width of 15% of the horizontal axis variable. In the waveform shown in Fig. 4, horizontal axis variable is time.

The standard deviation (σ) for ten repetitions is 0.383 mm and the standard error 0.12 mm. Based on the empirical 68% - 95% - 99.7% confidence intervals rule for sample data with a Gaussian-shaped distribution, as it is the case in our experiments, we can conclude that 99.7% of our transmitter displacement measurement values will lie within 3 standard deviations (3σ) from the mean value. This corresponds to a precision of 3 x 0.383 mm = ±1.149 mm.

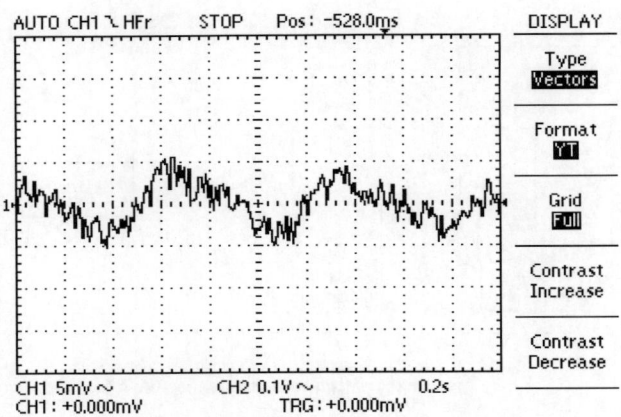

Fig. 4 Typical IF output waveform for 8cm transmitter displacement into 2s time period

V. CONCLUSIONS AND FURTHER WORK

This paper presented a novel method for tracking the position of medical instruments moving inside the human body. A phase shift method in conjunction to basic PLL topology provides a low frequency sinusoidal voltage signal as system output. Displacement of the transmitter on the medical instrument can subsequently be extrapolated from the voltage signal in a straight-forward fashion.

Controlled laboratory experiments were performed and the overall system was characterized with respect to its critical performance factors.

Regression analysis was used in order to filter out noise which exists in the recorded output signal. Furthermore, the IF of the output signal can be processed using spectral analysis, in order to obtain a more precise representation of the overall movement.

The methodology has been validated a 1-dimensional (1D) proof of concept setup. The displacement measurement error margins were calculated to be in the order of ±1.149 mm, using the Locally Weighted Scatterplot Smoothing (LOWESS) curve fitting method.

The highlighted experimental setup will be in the near future with more receivers, in order to be able to have a three-dimensional (3D) position tracking platform for future integration to the medical instrument.

Variable velocity transmitter movement will also provide more realistic experimental results, imitating a physician's hand movement as it displaces the medical instrument. This will also allow fully characterize the proposed system dynamic performance factors, such as spatial resolution, repeatability, precision and accuracy.

ACKNOWLEDGMENT

The authors would like to thank Microwave Solutions, Hertfordshire, U.K. for their kind offering of MDU 1100T units. Mr. Fotiadis, who is a lecturer at the Dept. of Informatics and Computer Technology, T.E.I. of Western Macedonia, Greece, is particularly appreciative of his institution's support during his sabbatical leave.

REFERENCES

[1] J. B. Hummel, M. R. Bax, M. L. Figl, Y. Kang, J. Calvin Maurer, W. W. Birkfellner, H. Bergmann, and R. Shahidi, 'Design and application of an assessment protocol for electromagnetic tracking systems', Medical Physics, vol. 32, no. 7, pp. 2371–2379, 2005.
[2] T. M. Peters and K. R. Cleary, Image-guided Interventions: Technology and Applications. Springer, 2008.
[3] S. Watson and K. Gorski, Invasive Cardiology: A Manual for Cath Lab Personnel: A Manual for Cath Lab Personnel. Jones & Bartlett Learning, 2010.
[4] B. J. Wood, H. Zhang, A. Durrani, N. Glossop, S. Ranjan, D. Lindisch, E. Levy, F. Banovac, J. Borgert, S. Krueger, J. Kruecker, A. Viswanathan, and K. Cleary, 'Navigation with electromagnetic tracking for interventional radiology procedures: a feasibility study', J Vasc Interv Radiol, vol. 16, no. 4, pp. 493–505, Apr. 2005.
[5] J. Krücker, S. Xu, N. Glossop, A. Viswanathan, J. Borgert, H. Schulz, and B. J. Wood, 'Electromagnetic Tracking for Thermal Ablation and Biopsy Guidance: Clinical Evaluation of Spatial Accuracy', J Vasc Interv Radiol, vol. 18, no. 9, pp. 1141–1150, Sep. 2007.
[6] J. M. Balter, J. N. Wright, L. J. Newell, B. Friemel, S. Dimmer, Y. Cheng, J. Wong, E. Vertatschitsch, and T. P. Mate, 'Accuracy of a wireless localization system for radiotherapy', Int. J. Radiat. Oncol. Biol. Phys., vol. 61, no. 3, pp. 933–937, Mar. 2005.
[7] E. Geist and K. Shimada, 'Position error reduction in a mechanical tracking linkage for arthroscopic hip surgery', Int J Comput Assist Radiol Surg, vol. 6, no. 5, pp. 693–698, Sep. 2011.
[8] R. M. S. M.D, K. S. Morgan, and H. B. Sieburg, Interactive Technology and the New Paradigm for Health Care. IOS Press, 1995.
[9] W. D. Boyer and W. D. Boyer, 'Continuous Wave Radar', U.S. Patent 315597203-Nov-1964.
[10] MICROWAVE SOLUTIONS LTD. UK, X-BAND DOPPLER MOTION DETECTOR UNITS, http://docs.microwave-solutions.com/createPdf.php?id=MDU1100T
[11] C. Gabriel, 'Compilation of the Dielectric Properties of Body Tissues at RF and Microwave Frequencies.', Report N.AL/OE-TR-1996-0037, Occupational and environmental health directorate, Radiofrequency Radiation Division, Brooks Air Force Base, Texas (USA), Jan. 1996.
[12] W. S. Cleveland, 'Robust Locally Weighted Regression and Smoothing Scatterplots', Journal of the American Statistical Association, vol. 74, no. 368, pp. 829–836, 1979.

Author: Dimitrios A. Fotiadis
Institute: School of Medicine, Aristotle University of Thessaloniki
Street: P.O. Box 323, 54124
City: Thessaloniki
Country: Greece
Email: dfotiadis@med.auth.gr

"Algorithmically Smart" Continuous Glucose Sensor Concept for Diabetes Monitoring

G. Sparacino, A. Facchinetti, C. Zecchin, and C. Cobelli, on behalf of the AP@home Consortium

Department of Information Engineering, University of Padova, Padova, Italy

Abstract—In the last 15 years, subcutaneous continuous glucose monitoring (CGM) sensors, portable devices able to measure glycemia almost continuously (1-5 min sampling period) for several days (up to 7), opened new frontiers in the treatment of diabetes, a chronic disease affecting about more than 300 million of people worldwide. However, glucose readings provided by current CGM devices still suffer of accuracy and precision problems. The present work illustrates technical details of algorithms and implementation issues of a new conceptual architecture to deal with those problems and render any commercial CGM sensor algorithmically smarter. In particular, three modules for denoising, enhancement and prediction are assessed on a dataset collected in 24 type 1 diabetic patients during the EU project AP@home.

Keywords—continuous glucose monitoring, sensor calibration, denoising, prediction, filtering, Diabetes.

I. INTRODUCTION

Diabetes is a chronic disease affecting more than 300 million individuals worldwide and is considered one of the socio-health emergencies of the 3rd millennium [1]. In diabetic patients, the pancreas does not secrete insulin (type 1 diabetes, T1D) or derangements in both insulin secretion and action occur (type 2 diabetes) and, as a consequence, glucose concentration in blood often exceeds the normal range (70-180 mg/dl). Hyperglycemia creates severe complications in the long-term, such as neuropathy, retinopathy and cardiovascular diseases, while hypoglycemia is a threat in the short-term, since it can degenerate into episodes (coma) particularly dangerous when the patient is asleep.

Several clinical studies demonstrated that diabetes complications can be reduced through a therapy based on a mix of diet, physical exercise, and drug delivery (including subcutaneous injections of exogenous insulin), tuned according to methods which in great part rely on the capability of correctly measuring glucose concentration levels in the blood. In the most commonly used approach, called self-monitoring blood glucose (SMBG), the diabetic patient self-monitors glycemia 3-4 times per day by using portable lancing sensor devices. Due to the sparseness of SMBG, however, glucose control is often suboptimal and it may happen that glucose exceeds the safe euglycaemic range without awareness of the patient. In the last 15 years, subcutaneous continuous glucose monitoring (CGM) sensors, portable devices able to measure glycemia in the interstitium in real time and almost continuously (1-5 min sampling period) for several days (up to 7) have been developed and assessed with encouraging results in clinical trials [2, 3]. Some of these devices exploit the glucose-oxidase enzyme using needles and are thus (minimally) invasive [4], while others, e..g. [5], take advantage of some superficial skin properties and, provided that some nontrivial mathematical models are developed, e.g. [6], result non-invasive. Rather comprehensive reviews of CGM sensors can be found in [7, 8]. CGM sensors stimulated investigations and applications previously hindered by the sparseness of SMBG. For instance, CGM data can be analyzed retrospectively to evaluate glucose variability [9] and used in real-time to generate alerts when glucose approaches, or exceeds, hypoglycemic or hyperglycemic thresholds [10, 11], with a potentially significant reduction of number and duration of hypoglycemic events [12]. Moreover, CGM sensors are a key element of artificial pancreas research prototypes for treating T1D [13, 14, 15].

In spite of exciting development perspectives, the performance of current state-of-art CGM sensors is still suboptimal and lower, in terms of accuracy and precision, than that of laboratory systems [16, 17, 18]. The present work illustrates algorithms to implement a new conceptual architecture aimed at rendering CGM sensors more accurate and reliable.

II. MATERIALS AND METHODS

A. Data Base

Data were collected, under the aegis of the EU 7th Framework Programme (FP7) project "AP@home" (2010-2014), at Academic Medical Center Amsterdam (Amsterdam, The Netherlands), Medical University of Graz (Graz, Austria), Profil Institute for Metabolic Research GmbH (Neuss, Germany), and Department of Clinical and Experimental Medicine, University of Padova (Padova, Italy). In 24 T1D patients, CGM was performed using the SEVEN Plus system (Dexcom Inc., San Diego, CA) while, in parallel, a time-series of BG references was measured using a laboratory apparatus (Yellow Springs Instruments, Yellow Spring, OH). More details can be found in [19]. Figure 1 presents a representative dataset (BG references, green dots, were linearly interpolated to improve their inspection). It is apparent that CGM is uncertain because of noise (see e.g. window 10:00-13:00), inaccurate (see e.g. delay in 20:00-23:00 and overestimation in 07:00-11:00) and it would allow generation of alerts only with some delay (e.g. compare hypoglycemic threshold crossing of BG and CGM at 23:00).

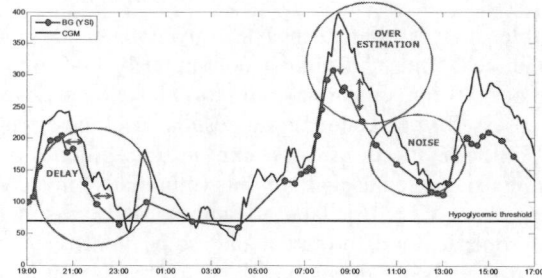

Fig. 1 Representative subject. CGM (blue line) and BG references (green dots, linearly interpolated by dashed line). Uncertainty, accuracy and delay problems are evidenced by red circles. Horizontal dashed line represents the 70 mg/dl hypoglycemic threshold.

B. Smart CGM Sensor Architecture and Algorithms

The smart CGM sensor architecture shown in Figure 2 consists of software modules placed in cascade to the output of a commercial CGM sensor, e.g.: (i) a denoising module to attenuate measurement noise; (ii) an enhancement module to improve accuracy by recalibrating CGM data against the 3-4 SMBG samples collected daily by the patient; (iii) a prediction module to forecast in real-time future glucose concentration after a certain prediction horizon (PH) in order to enable the possibility of generating timelier hypo-/hyper-glycaemic alerts. Several algorithms to implement the modules are available, see [20, 21] for recent reviews.

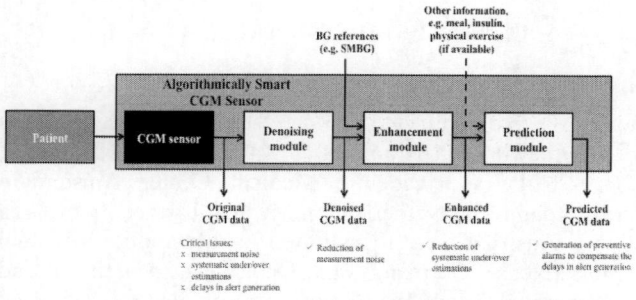

Fig. 2 The smart CGM sensor architecture, consisting of a CGM sensor and software modules placed in cascade e.g. for denoising, enhancement and prediction.

In denoising, the main problem to be faced is the inter-individual and intra-individual variability of the signal-to-noise ratio (SNR). The chosen denoising algorithm, described in detail in [22, 23], assumes that CGM data are modeled as

$$y(k) = u(k) + v(k) \quad (1)$$

where u(k) is the true (unknown) glycemic value and v(k) is the random measurement noise, uncorrelated from u(k) and with zero mean and unknown variance equal to $\sigma^2(k)$ (note the dependence on k). In its Bayesian setting, the algorithm takes advatage of a priori expectations on the smoothness of u(k), modeled as the double integration of a white noise:

$$u(k) = 2u(k-1) - u(k-2) + w(k) \quad (2)$$

where w(k) has (unknown) variance equal to $\lambda^2(k)$. A sliding window of N (usually some tens) CGM samples allows, at any k > N, the estimation of σ^2 and λ^2 according to a consistency criterion and thus automatically deals with the SNR variability. Then, every time a new noisy glycemic reading y(k) is produced, a denoised û(k) value is obtained (either using a matrix-vector or state-space representation) by exploiting a linear minimum variance estimator. At variance of other approaches with fixed parameters used in the literature, the above described algorithm is self tuneable, adaptive and, thanks to its Bayesian embedding, also provides a confidence interval of the denoised CGM values which can be useful for other modules which could be incorporated in the future in the architecture of Fig. 2, e.g. for alert generation or fault detection [24].

As far as the enhancement module is concerned, it is key to take into account that differences between BG and CGM (Figure 1) can be due to both plasma-to-interstitial fluid glucose kinetics and suboptimal (loss of) sensor calibration. The algorithm considered here is the one recently proposed in [25]. Briefly, the m-size vector **x** (m≥2) contains SMBG measurements, while the n-size vector **y** contains CGM samples, much more frequent than those in **x** but collected in the same temporal window. Exploiting a model of BG-to-IG kinetics [26] (with "population" time constant parameter) a n-size vector **û**, interpretable as the reconstruction of a nominal BG on the same grid of the CGM signal, is obtained by deconvolution [27]. Then, the vector **ψ** containing the m samples of **û** lying on the same grid of **x** is fitted against the former by linear least squares starting from

$$\mathbf{x} = a\,\mathbf{\psi} + b + \mathbf{\nu} \quad (3)$$

where **ν** is a m-size error vector and (using a Matlab-like notation) a and b are scalar parameters. The so-determined a and b are used to recalibrate the CGM data as

$$CGM_{enhanced}(k) = a\,CGM(k) + b \quad (4)$$

(no recalibration is performed in the ideal case in which a and b result equal to 1 and 0, respectively). With respect to similar literature algorithms, this enhancement method explicitly takes into account the influence of blood-to-interstitium glucose transport and has parameters that can be automatically and adaptively determined in real time.

As far as the prediction module is concerned, even if the architecture depicted in Figure 2 allows the incorporation of knowledge on meals, insulin administrations, and physical activity, as done e.g. by the literature prediction algorithms [28, 29], such an information is not available for our dataset. Therefore, here we implement a simple purely auto-regressive algorithm presented in [30]:

$$y(k) = \alpha(k) y(k-1) + \varepsilon(k) \quad (5)$$

where $y(k)$ is the CGM sample at time k, $\alpha(k)$ is a scalar parameter and $\varepsilon(k)$ is a random white noise process, with zero mean and variance $\rho^2(k)$. The estimation of α and ρ^2 at any k is performed in real time by using a recursive least square strategy, where past CGM data participate with different relative weights determined by a forgetting factor μ, i.e. at time k, the sample y(k-q) has weight μ^q. The value of μ can be determined on a training set by minimizing a prediction performance index [31]. The model is then used to predict glucose H steps ahead (H=PH/T_s, T_s being the sensor sampling rate) by iterating eq.(5) for j=k+1,..., k+H with $\varepsilon(j) \equiv 0$. Interestingly, the recursive implementation allows to maintain memory and processing requirements very low.

III. RESULTS

Figure 3 (top) illustrates a graphical example of the results obtained by the denoising module. The improvement in smoothness of CGM is evident by eye and can be quantified by the energy of second order differences (ESOD) index (the larger ESOD, the less smooth the time series), which is reduced from 3.8 to 1.9 for this particular subject and from 1.4 to 0.6 (p=0.001) in terms of median computed from all the 24 subjects. Noise attenuation can limit the number of false hypo/hyperglycemic alerts.

Figure 3 (middle) depicts original (blue) and enhanced (red) CGM, and BG reference data (green dots). The improvement in the accuracy of the sensor is evident (see e.g. the time interval 10:00–16:00 and the time interval around 01:00, where a false hypoglycemic alert is prevented). The improvement of accuracy can be quantified by the Mean Absolute Relative Difference (MARD) index, which decreases from 30.6% to 6.8% in this subject and from 13.1% to 9.6% (p=0.003) as median in the whole database. Accuracy increase gives obvious benefits to control actions (e.g. size of insulin administration).

Figure 3 (bottom) illustrates an example of short-term prediction (PH=30, μ=0.925). An hypoglycemic alert is generated at time 12:35 (blue arrow) on the basis of the original CGM, but using the output of the prediction module, the same alert would have been generated at 12:20 (red arrow), i.e. with a temporal gain of 15 min. On the 60 hypoglycemic episodes of the whole database, the prediction algorithm was able to forecast 55 hypo-events (91.6%) within the chosen PH, with a median amount of time gained of 15.1 min.

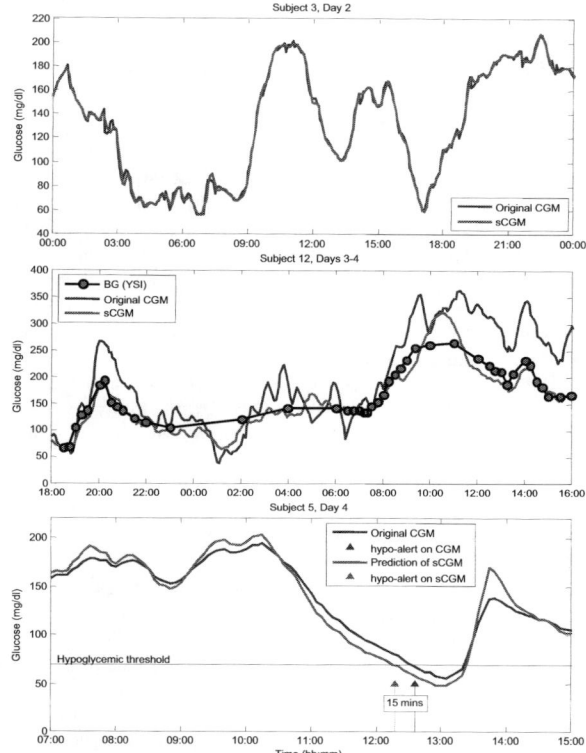

Fig. 3. Representative examples. Top. Original (blue) and denoised (red) CGM data. Middle. Original (blue) and enhanced (red) CGM, and BG references (green dots). Bottom. Original CGM data (blue) and glucose prediction (red). the vertical arrows (same color codes) indicating the alerts that could be generated at the crossing the hypoglycemic threshold.

An exhaustive report of the results and of their potential impact in diabetes treatment is documented in the clinical paper [19].

IV. CONCLUSIONS

The "algorithmically smart" sensor concept, implemented by suitable algorithms, can render CGM outcomes more accurate and reliable, with possible great benefit in several applications. While the architecture of Figure 2 is general, the algorithms specifically considered in the present work share some important features e.g. real-time functioning, adaptability, mutual independence. Further development of them is currently under investigation, for instance, the inclusion of additional information in the prediction algorithm, such as timing and composition of meals or level of physical activity measured in real-time [32].

ACKNOWLEDGMENTS

AP@home is supported by EU under FP7 (grant n. 247138).

REFERENCES

1. International Diabetes Federation at http://www.idf.org
2. Juvenile Diabetes Research Foundation Continuous Glucose Monitoring Study Group (2008) Continuous glucose monitoring and intensive treatment of type 1 diabetes. N Engl J Med 359:1464-1476
3. Bode BW, Battelino T (2010) Continuous glucose monitoring. Int J Clin Pract 166:11-15
4. McGarraugh G (2009) The chemistry of commercial continuous glucose monitors. Diabetes Technol Ther 11: S17-S24
5. Caduff A, Talary MS, Mueller M et al. (2009) Non-invasive glucose monitoring in patients with Type 1 diabetes: a Multisensor system combining sensors for dielectric and optical characterisation of skin. Biosens. Bioelectron 24: 2778-2784.
6. Zanon M, Sparacino G, Facchinetti A et al. (2012) Non-invasive continuous glucose monitoring: improved accuracy of point and trend estimates of the Multisensor system. Med. Biol. Eng. Comput 50:1047-1057
7. Cox M (2009) An overview of continuous glucose monitoring systems. J Pediatr Health Care 23:344-347
8. Sparacino G, Zanon M, Facchinetti A et al. (2012) Italian contributions to the development of continuous glucose monitoring sensors for diabetes management. Sensors (Basel) 12:13753-13780
9. Rodbard D (2011) Glycemic variability: measurement and utility in clinical medicine and research--one viewpoint. Diabetes Technol Ther. 13:1077-1080
10. McGarraugh G, Bergenstal R (2009) Detection of hypoglycemia with continuous interstitial and traditional blood glucose monitoring using the FreeStyle Navigator Continuous Glucose Monitoring System. Diabetes Technol Ther 11:145-150
11. Sparacino G, Facchinetti A, Maran A et al. (2008) Continuous glucose monitoring time series and hypo/hyperglycemia prevention: requirements, methods, open problems. Curr. Diabetes Rev. 4: 181-192
12. Zecchin C, Facchinetti A, Sparacino G et al (2013) Reduction of number and duration of hypoglycemic events by glucose prediction methods: a proof-of-concept in silico study. Diabetes Technol Ther 15:66-77
13. Weinzimer SA, Steil GM, Swan KL et al (2008) Fully automated closed-loop insulin delivery versus semiautomated hybrid control in pediatric patients with type 1 diabetes using an artificial pancreas. Diabetes Care 31:934-939
14. Hovorka R, Allen JM, Elleri D et al (2010) Manual closed-loop insulin delivery in children and adolescents with type 1 diabetes: a phase 2 randomised crossover trial. Lancet. 27: 375:743-751
15. Cobelli C, Renard E, Kovatchev BP et al (2012). Pilot studies of wearable outpatient artificial pancreas in type 1 diabetes. Diabetes Care. 35: 65-67
16. Kovatchev B, Anderson S, Heinemann L et al (2008) Comparison of the numerical and clinical accuracy of four continuous glucose monitors. Diabetes Care 31:1160-1164.
17. Langendam M, Luijf YM, Hooft L et al (2012) Continuous glucose monitoring systems for type 1 diabetes mellitus. Cochrane Database Syst Rev 1: CD008101
18. Damiano ER, El-Khatib FH, Zheng H et al (2013) A comparative effectiveness analysis of three continuous glucose monitors. Diabetes Care 36:251-259
19. Facchinetti A, Sparacino G, Guerra S et al on behalf of the AP@home Consortium (2013) Real-Time Improvement of Continuous Glucose-Monitoring Accuracy: The smart sensor concept. Diabetes Care 36:793-800
20. Sparacino G, Facchinetti A, Cobelli C (2010) "Smart" continuous glucose monitoring sensors: on-line signal processing issues. Sensors (Basel) 10: 6751-6772
21. Bequette BW (2010) Continuous glucose monitoring: real-time algorithms for calibration, filtering, and alarms. J Diabetes Sci Technol 4: 404-418
22. Facchinetti A, Sparacino G, Cobelli C (2010) An online self-tunable method to denoise CGM sensor data. IEEE Trans Biomed Eng 57: 634-641.
23. Facchinetti A, Sparacino G, Cobelli C (2011) Online denoising method to handle intraindividual variability of signal-to-noise ratio in continuous glucose monitoring. IEEE Trans Biomed Eng 58: 2664-2671
24. Facchinetti A, Del Favero S, Sparacino G et al (2013) An online failure detection method of the glucose sensor-insulin pump system: improved overnight safety of type-1 diabetic subjects. IEEE Trans Biomed Eng 60: 406-416.
25. Guerra S, Facchinetti A, Sparacino G et al (2012) Enhancing the accuracy of subcutaneous glucose sensors: a real-time deconvolution-based approach. IEEE Trans Biomed Eng 59: 1658-1669
26. Rebrin K, Steil GM, van Antwerp WP et al (1999) Subcutaneous glucose predicts plasma glucose independent of insulin: implications for continuous monitoring. Am J Physiol 277: E561-E571
27. Sparacino G, Cobelli C (1996) A stochastic deconvolution method to reconstruct insulin secretion rate after a glucose stimulus. IEEE Trans Biomed Eng 43:512-529.
28. Pappada SM, Cameron BD, Rosman PM et al (2011) Neural network-based real-time prediction of glucose in patients with insulin-dependent diabetes. Diabetes Technol Ther 13:135-141.
29. Zecchin C, Facchinetti A, Sparacino G et al (2012) Neural network incorporating meal information improves accuracy of short-time prediction of glucose concentration. IEEE Trans Biomed Eng 59:1550-1560.
30. Sparacino G, Zanderigo F, Corazza S et al (2007) Glucose concentration can be predicted ahead in time from continuous glucose monitoring sensor time-series. IEEE Trans Biomed Eng 54: 931-937
31. Facchinetti A, Sparacino G, Trifoglio E et al (2011) A new index to optimally design and compare continuous glucose monitoring glucose prediction algorithms. Diabetes Technol Ther 13:111-119.
32. Zecchin C, Facchinetti A, Sparacino G et al (2013) Physical activity measured by PAMS correlates with glucose trends reconstructed from CGM. Diabetes Technol Ther 15(10) (in press) DOI: 10.089/dia.2013.0105.

Piezoresistive Goniometer Network for Sensing Gloves

G. Dalle Mura[1,2], F. Lorussi[1,2], A. Tognetti[1,2], G. Anania[1], N. Carbonaro[1], M. Pacelli[3], R. Paradiso[3], and D. De Rossi[1,2]

[1] Research Center E.Piaggio, University of Pisa, Via Diotisalvi 2, Pisa, Italy
[2] Information Engineering Department, University of Pisa, Via Caruso 2, Pisa, Italy
[3] Smartex S.r.l., Via Giuntini 13L, 56023 Navacchio, Pisa, Italy

Abstract—This paper presents a kinesthetic glove realized with knitted piezoresistive fabric (KPF) sensor technology. The glove forefinger area is sensorized by two KPF goniometers obtained on the same piezoresistive substrate. The piezoresistive textile is used for the realization of both electrogoniometers and connections, thus avoiding mechanical constraints due to metallic wires. Sensors are characterized in comparison with commercial goniometers. The glove behavior is pointed out in terms of methacarpal-phalangeal and interphalangeal joint movement reconstruction.

Keywords—Knitted piezoresistive sensors, wearable goniometers, sensing glove.

I. INTRODUCTION

Monitoring joint angles through wearable systems allows reconstructing human posture and gesture for supporting physical rehabilitation both in clinics and at patient home. In the last years, the use of textile integrated deformation sensors made of conductive elastomer (CE) materials has allowed the realization of wearable and unobtrusive sensing garments for human motion detection and classification [1],[2], [3]. CE based sensing garments have shown good performance for slow and wide movements while the accuracy, transient time and hysteresis have limited their use in reconstructing fast and small movements [4]. Recently, knitted piezoresistive fabric (KPF) technology has been introduced for biomechanical and cardiopulmonary monitoring [5]. KPF based sensing garments have the same working principle of CE ones, but they present a consistent improvement in terms of limited transient time and hysteresis. In this paper, an innovative motion-sensing glove made of KPF goniometers is presented. Goniometers are obtained by arranging two identical KPF layers following the theoretical working principle and the derived electromechanical model previously proved for CE sensors [6], [7]. KPF goniometer capabilities in angle detection have been experimentally tested and the performances in angle reconstruction are reported. KPF goniometers have shown very good performances in terms of angle measurements both in quasi-static and dynamic working mode for velocities typical of human movement. By using KPF goniometers, a sensing glove has been realized. The glove is endowed by two goniometers on the metacarpal-phalangeal and proximal interphalangeal joints of the forefinger, which separately acquire the position of the two articulations. Two main innovations are introduced with the presented sensor configuration: (1) two goniometers are obtained on an unique sensing substrate and (2) the connections between the goniometers extremities and the acquisition electronics have been realized by KPF tracks in order to avoid mechanical constraints given by metallic inextensible wires. This device has been realized within the INTERACTION European project [8] (ICT, 7th framework), aimed at continuous daily-life monitoring of the functional activities of stroke survivors in their physical interaction with the environment. This kind of monitoring is essential for optimal guidance of rehabilitation therapy by medical professionals and coaching of the patient [9]. In the project, which foresees modular devices (shirt, trousers, glove, shoes) for kinesthetic and kinetic interaction with the environment, the kinesthetic glove is devoted to hand posture recognition and evaluation of the activity recover of long finger extensor or to verify the presence of compensatory movement of the wrist to perform daily life activity.

II. GONIOMETER WORKING PRINCIPLE

This section describes the working principle of double layer textile based goniometers starting from the electromechanical characteristics of single layer ones. Single layer (*SL*) sensors are made of piezoresistive textile attached to insulating textile. The following assumptions are considered: (i) the sensing layer is subjected to isovolumetric deformations, (ii) the resistivity (ρ) is constant and (iii) the fabric/flexible substrate is inextensible under bending. Under these assumptions, it was demonstrated that the layer electrical resistance RSL is function of the sensor total curvature through the following relation:

$$R_{SL} = l\frac{\rho}{d\,h} - \frac{\rho}{d}\Delta\alpha + O(\sup_{s\in(0,l)} k(s)^2) =$$

$$= l^2\frac{\rho}{V_0} - \frac{\rho}{d}\Delta\alpha + O(\sup_{s\in(0,l)} k(s)^2) \qquad (1)$$

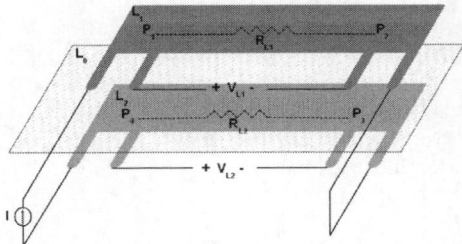

Fig. 1 A double layer KPF goniometer.

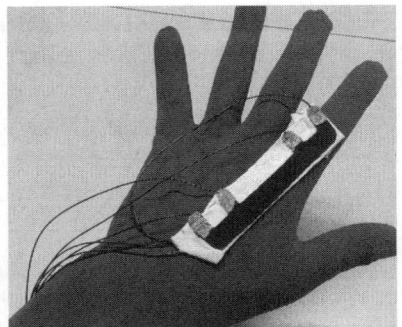

Fig. 2 A double layer KPF goniometer attached to a Lycra® glove.

where l is the length of the sensor, V_0 its volume, d its width, $\Delta\alpha$ is the angle between the tangent planes to the sensor extremities, h_0 is the specimen initial thickness, s is the arch length of the parametrized side of the sensor and $O(sup(k(s)^2))$ is a second order infinitesimal function which vanish if the curvature $k(s) \to 0$. Relationship (1) derives from the Corbino formula for the circumferential resistance of a conductive disk, expressed in Taylor expansion [1]. A bending in an opposite direction, with respect to the same arch length parameterization, is characterized by opposite sign in even position signs. Double layer (DL) angular sensors are based on the measurement of the resistance difference ($\Delta R_{DL} = R_{L1} - R_{L2}$) between two identical piezoresistive samples coupled via an insulating layer L_0. This device is shown in Fig 1, where L_1 and L_2 are the piezoresistive layers having the same rest thicknesses h_0. R_{L1} and R_{L2} are the electrical resistances of the two layers respectively. L_1 represents the insulating layer. In case of device deformation without bending, the electrical resistance difference between the two layers is negligible, since both the sensing layers undergo the same transformation. When the device is flexed it was proved that there is dependence between ΔR_{DL} and $\Delta\alpha$ (2).

$$\Delta R_{DL} = R_{L1} - R_{L2} = 2\frac{\rho}{d}\Delta\alpha + O(\sup_{s\in(0,l)} k(s)^3) \quad (2)$$

With this configuration, the error in angle estimation reduces to a third order infinitesimal function with the maximum of the curvature $k(s)$ [7] due to the alternate signs of the Taylor expansions of elecromechanical characteristics of the two layers. The hypothesis of inextensibility can be removed since in (2) there is no dependence on the sample actual length. By neglecting the third-order error term, the angle $\Delta\alpha$ can be estimated by:

$$\Delta\alpha = \frac{d}{2\rho}\Delta R_{DL} \quad (3)$$

Textile-based sensors were realized through knitted piezoresistive fabrics (KPF) which contain 75% electro-conductive yarn (Belltron®, produced by Kanebo Ltd) and 25% Lycra®, manufactured using a circular machine as described in previous works [3, 4]. The samples were realized according to the single and double layer design introduced in previous section, referred as SL and DL sensors respectively. Coupling one KPF rectangular sample with an elastic fabric through a double side adhesive membrane makes the SL structure. Adding another KPF layer to the back of the elastic fabric by using the same adhesive membrane produces the DL structure. Electrical connections, made of textile conductive yarns with PVC insulation (Bekinox®, produced by Bekaert), were fixed by using an ultrasonic welding device. In case of double layer sensors, the independence on the local curvature of the resistance difference of the two layers expressed by equation (3) makes the device work independent by the bending profile and consequently the physical properties of the joint to be monitored and the user body structure. Fig. 2 shows the KPF goniometer attached to a Lycra® glove on the metacarpal-phalangeal joint. Each piezoresistive layer has four semicircular pads specifically designed for sensor wiring. Each goniometer as a total of 8 connecting pads, which is a considerable number taking into account the user comfort requirements. The analog front-end for signal acquisition from the single electrogoniometer is showed in Fig. 3. Each piezoresistive layer can be represented as a series of three resistances. A constant current I is supplied through the external pads and the voltages ($V_{L1} = V_{P1} - V_{P2}$) and ($V_{L2} = V_{P3} - V_{P4}$) and the acquisition system is provided by a high input impedance stage realized by two instrumentation amplifiers ($A1$ and $A2$) to read this voltages (four point reading). These voltages can be related to resistances R_{L1} and R_{L2} (by dividing by I) and the resistance difference ΔR is related to the curvature angle (equation (3)).

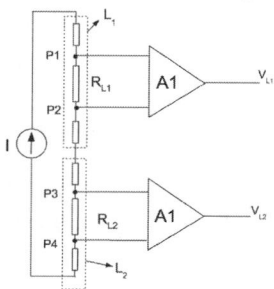

Fig. 3 Acquisition system for a double layer device (goniometer).

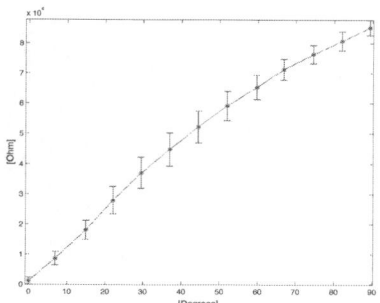

Fig. 4 Average ΔR^* vs. angle for flexion and extension (vertical bars represent two standard deviation units).

III. SENSOR CHARACTERIZATION

The double layer device was tested in stretching to confirm the output invariance with respect to elongation and in bending to verify the relationship between angles and ΔR. Both in elongation and shortening, it was demonstrated that ΔR is not dependent by the sensor length. Given this finding the ΔR value for the un-flexed device ($q_0 = 2.1 K\Omega$), mainly due to the inevitable dissimilarity between the two layers, can be subtracted from ΔR by measuring the initial offset of the sensor, obtaining:

$$\Delta R^* = \Delta R - q_0$$

Equation 3 has been experimentally verified to evaluate the error in angular measurement in comparison with SG110 by Biometrics ($\pm 2^o C$ accuracy) [10] during a bench static experiment. A representation of the data obtained in 15 trials for 13 different angles from $0°$ to $90°$, acquired both in increasing and in decreasing measurements, and characterized by:

$$\overline{\Delta R^*}_{\Delta \alpha_i} = \frac{1}{K} \sum_k \Delta R^*_{\Delta \alpha_i, k}$$

$$\sigma_{\Delta R^*_{\Delta \alpha_i}} = \sqrt{\sum_k \frac{1}{K} \left(\Delta R^*_{\Delta \alpha_i, k} - \overline{\Delta R^*}_{\Delta \alpha_i} \right)^2}$$

is shown in Fig. 4. $\Delta R^*_{\Delta \alpha_i}$ trend was approximated by a linear regression in last square sense and the angular sensitivity of the double layer sensor $S_{\Delta \alpha DL}$, is estimated as $955 \Omega/°$. The maximum standard deviation was evaluated in about 5 degrees. The double layer KPF sensor attached to the Lycra® glove of Fig. 2 was dynamically compared with the Biometrics electrogoniometer during metacarpal-phalangeal flexion/extension cycles. The purpose was to investigate how the static behaviour reflects the data obtained in dynamic conditions. The sensor shown a similar trend both in dynamic and in static conditions (Fig. 5) and the maximum error by using this linear approximation holds $5.7°$.

IV. KINESTHETIC GLOVE

An advanced kinesthetic glove has been realized for the independent monitoring of two degrees of freedom of the forefinger (metacarpal-phalangeal and proximal interphalangeal). This prototype introduces two main innovations with respect to the double layer device of Fig. 2. Firstly, two different goniometers are placed in series on the same sensing substrate, thus reducing the number of inter-connections. Secondly, the same piezoresistive and elastic material used for the sensing layer is employed to connect the goniometers with the electronic unit. In this way, flexible and elastic interconnections have been obtained, thus avoiding mechanical constraint due to the presence of metallic wires. As shown in Fig. 6-a, a conductive layer (black) is cut according to the desired topology (including sensors and connections), folded and applied on the two sides of an insulating layer by adhesive membranes. In terms of electronics the architecture of Fig. 3 is adapted to the sensor series. Having the two goniometers on the same substrate, just one current generator supplies both devices. Then, four high impedance stages acquire the voltage falls on the respective piezoresistive layers through the piezoresistive connection tracks. The high input impedances ensure a very

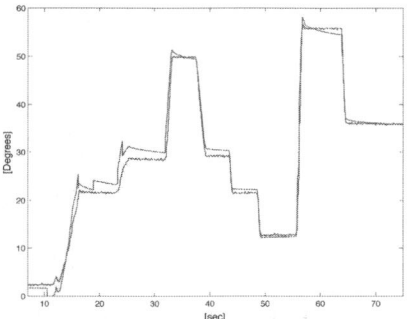

Fig. 5 Dynamic comparison between KPF sensor (blue line) and Biometrics electrogoniometer (red line).

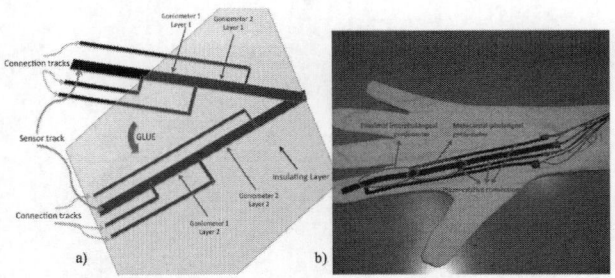

Fig. 6 a) Schematic realization of the goniometer series and b) advanced sensorized glove (dorsal view).

low current (amplifiers bias) in these piezoresistive tracks, making the relative voltage falls negligible (i.e. the voltage presented by tracks on the wrist is the same relieved at the goniometer extremities). This prototype (Fig. 6-b) has been realized by cutting (by hand) a pre-glued multilayer specimen (conductive layer adhesive membrane-insulating layer-adhesive membrane- conducting layer as shown in Fig. 6-b). In this case, the calibration procedure to determinate the characterizing coefficient of the two layers of each goniometer is crucial to avoid the cross-talk phenomenon among different goniometers. The prototype has performed good capabilities of relieving the considered joint movement as showed in Fig. 7. In particular, flexion and iper-extension of the metacarpal-phalangeral joint are detected (red graph) like so the proximal interphalangeal flexion (blue graph).

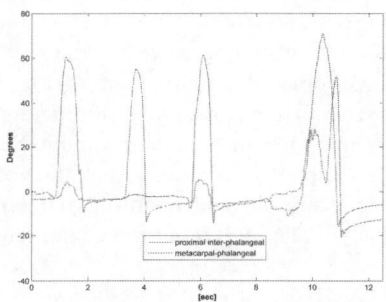

Fig. 7 Metacarpal-phalangeral and proximal inter-phalangeral angle reconstruction.

V. CONCLUSION

In this paper a novel wearable technology allowing the reconstruction of joint angles has been introduced. It consists in KPF goniometers linked by a connection topology realized by piezoresistive and elastic tracks. A sensorized glove has been developed and characterized in terms of reliability and its performance has resulted comparable with the ones provided by commercial devices. Movement of metacarpal-phalangeal and proximal phalangeal joints of the forefinger has been relieved by the sensing glove.

ACKNOWLEDGEMENTS

This research has been supported by the EU 7th framework project INTERACTION (FP7-ICT-2011-7-287351).

REFERENCES

1. Tognetti A, Carbonaro N, Zupone G, De Rossi D. Characterization of a novel data glove based on textile integrated sensors in *Proceedings of the IEEE International Conference of the Engineering in Medicine and Biology Society*:2510–2513 2006.
2. Tognetti A, Bartalesi R, Lorussi F, De Rossi D. Body segment position reconstruction and posture classification by smart textiles *Transactions of the Institute of Measurement and Control.* 2007;29:215–253.
3. Vanello Nicola, Hartwig Valentina, Tesconi Mario, et al. Sensing glove for brain studies: Design and assessment of its compatibility for fMRI with a robust test *IEEE/ASME Transactions on Mechatronics.* 2008;13:345–354.
4. Tognetti A, Lorussi F, Bartalesi R, et al. Wearable kinesthetic system for capturing and classifying upper limb gesture in post-stroke rehabilitation *Journal of NeuroEngineering and Rehabilitation.* 2005;2:8.
5. Pacelli M, Caldani L, Paradiso R. Textile piezoresistive sensors for biomechanical variables monitoring in *Proceedings of the IEEE International Conference of the Engineering in Medicine and Biology Society*:5358–5361 2006.
6. Lorussi F, Galatolo S, De Rossi D, Bartalesi R. Modeling and characterization of extensible wearable textile-based electrogoniometers *IEEE Sensors Journal.* 2011.
7. Lorussi F, Galatolo S, De Rossi D. Textile-based electrogoniometers for wearable posture and gesture capture systems *IEEE Sensors Journal.* 2009;9:1014–1024.
8. www.interaction4stroke.eu
9. Veltink et. al.. INTERACTION, Training and monitoring of daily-life physical interaction with the environment after stroke in *Proceedings of XII International Symposium on 3D Analysis of Human Movement*(Bologna, Italy) 2008.
10. www.biometricsltd.com/gonio.htm

Author: Alessandro Tognetti
Institute: Research Center E.Piaggio, University of Pisa
Street: Via Diotisalvi 2
City: Pisa
Country: Italy
Email: a.tognetti@centropiaggio.unipi.it

Preliminary Study of Pressure Distribution in Diabetic Subjects, in Early Stages

M.L. Zequera[1], L. Garavito[2], W. Sandham[3], Á. Rodríguez[1], J.A. Alvarado[4], and C.A. Wilches[1]

[1] Pontificia Universidad Javeriana/Department of Electronics - BASPI Research Group, Bogotá, Colombia
[2] Pontificia Universidad Javeriana - San Ignacio Hospital/School of Medicine, Bogotá, Colombia
[3] Scotsig, Glasgow, UK
[4] Pontificia Universidad Javeriana/Department of Industrial Engineering, Bogotá, Colombia

Abstract—Type 2 diabetes is considered a major public health challenge. Diabetic foot is a multifactorial disorder of vascular, nervous and mechanical origin that can seriously threaten the limb. Indeed, diabetes remains the leading cause of lower-limb amputations worldwide, and the use of imaging technologies can improve early diagnosis for prevention of ulcer development, but its costs can limit its implementation in clinical practice, for homecare monitoring and customized treatment. Previous studies have compared certain parameters between males and females, attributed to anatomical and physiological differences, which could be significant in the distribution of pressure on the foot. The aim of the present study was to evaluate the pressure platform EPS/R1 (LorAn Engineering) in subjects with and without diabetes, of both genders, for its future implementation in clinical practice. Fourteen volunteers, 6 diabetics (4 females, 2 males) and 8 non-diabetics (4 females, 4 males), aged 30 - 70 years, and body mass index below 35, were recruited. Pressure platform EPS/R1 (LorAn Engineering) was used for registering plantar pressure variables: lateral load distribution (LLD), antero-posterior load distribution (APLD), average pressure (AP), body barycenter (BB), foot barycenter (FB), and points of maximum pressure (PML). Variables BB and FB presented interactions with gender and condition. Variables LLD, PML and AP presented interactions with condition and marginal with gender. Variable APLD presented no interactions with gender or condition. In general, subjects without diabetes had similar measurements over the 3 sessions in both genders, while measurements of subjects with diabetes changed as trials progressed. Some gender differences were found. This study indicated that males have better balance compared to females. In particular, it was found that females, in search of balance, prioritized support in the hindfoot, making this area susceptible to ulcers.

Keywords— Diabetic foot, pressure distribution, Type 2 diabetes.

I. INTRODUCTION

Diabetes mellitus, is a metabolic disorder characterized by chronic hyperglycemia, which is a consequence of insulin deficiency, insulin resistance, or both. Uncontrolled chronic hyperglycemia causes microvascular and macrovascular pathologies, leading to a serious dysfunction of various organs [1]. Type 2 diabetes is considered one of the major public health challenges of today. There has been a dramatic increase in the number of cases of diabetes mellitus in the world, and it's estimated that people with type 2 diabetes doubled worldwide between 1995 and 2010. It's considered that this epidemic data are directly related to financial and lifestyle changes [1]. The diabetic foot is a multifactorial disorder [2], with vascular, nervous and mechanical origin that may threaten the limb [3]. Diabetes remains as the leading cause of lower limb amputations worldwide [2]. It's considered that patients with diabetes are at risk of amputation of a limb from 15 to 46 times greater than an individual without diabetes [4].

Several methods have been devised for measuring plantar pressure as insoles [5] and platforms [6], used in static studies [7] and dynamic plantar pressure [8], with the use of various footwear [9] as well as barefoot [5]. These imaging systems can improve early detection and prevention of ulcer formation, but their elevated costs limit their implementation in clinical practice or access to this technology to patients with diabetes, at risk of developing ulcers.

Some studies have compared parameters between men and women due to anatomical and physiological differences, which could make a difference in the distribution of pressure on the foot [10]. According to Periyasamy et al. there is a variation on plantar pressure distribution due to the larger foot contact area in men than women [7]. The study by Armstrong et al. showed higher prevalence of amputation (57%) in male subjects with diabetes, and an equal distribution by gender (50%) in subjects without diabetes [4].

The aim of the present study was to evaluate the pressure platform EPS/R1 (LorAn Engineering) in subjects with and without diabetes, of both genders, for its future implementation in clinical practice.

II. METHODS

A. Subjects

Fourteen volunteers were chosen for this study, 6 diabetic subjects (4 women, 2 men) and 8 non-diabetic subjects (4 women, 4 men), aged 30 - 70 years and body mass index below 35. Subjects were excluded if they were obese, or had a history of peripheral neuropathy, vasculopathy, or musculoskeletal injuries, which may affect measurements. Ethical approval was granted by the Ethics Committee for Medical Research of the Hospital San Ignacio. Written informed consent was signed by all subjects participating in the study.

B. Equipment

The measurement system installed at the Department of Electronics at Pontificia Universidad Javeriana (Bogotá, Colombia) has 3 components: pressure platform EPS/R1 (Loran Engineering SRL, Bologna, Italy), the software FootChecker 4.0, and a wooden frame.

The pressure platform EPS/R1, Loran Engineering SRL, consisted on 2304 resistive sensors, in a matrix of 48 × 48 sensors (1 sensor/cm^2), and sample rate of 30 Hz. Footchecker 4.0 was used to capture variables: lateral load distribution (LLD), antero-posterior load distribution (APLD), average pressure (AP), body barycenter (BB) foot barycenter (FB), and points of maximum pressure (PML).

The wooden frame controlled the relative position of the foot. A previous study showed that the wooden frame allows greater repeatability of measurements over a frame marked on the platform [11].

C. Protocol

At the beginning of each session, a doctor performed a comprehensive exam to the subjects involved, conducting an assessment of the state of his feet. Subjects were asked to be barefoot, with no objects in their pockets, jewelry or watches, and comfortable clothes to dress, to avoid changes in the measurement.

Each subject underwent three sessions of measurements, one every week. The method of positioning of the feet is a reference guide, a wooden guide placed on the platform, designed to control the position of the feet in an anatomical position and the position the body.

For the measurement process, subjects were on the platform for each measurement and retired from it after the measure was saved, performing 10 measurements per session. A total of 420 measurements were taken. This measurement process was supervised by a physician and a research assistant, who observed and guided the accommodation of the feet of the subject and body position.

D. Data Measured

The pressure platform EPS/R1 measures pressure distribution in feet, capturing the pressure at each sensor and calculating the different variables. The location coordinates of body barycenter BB (BBx,BBy) is calculated to determine the distribution; LLD indicates the percentage of load on each foot; APLD is percentage of load in rearfoot and forefoot; AP is the average pressure on each foot; the location coordinates of the foot barycenters FB (FBx,FBy) and points of maximum pressure PML (PMLx,PMLy) of each foot are given. Due to the instability of the subject during measurements, there are drastic changes in PML.

E. Statistical Analysis

It was performed a multivariate analysis of variance (MANOVA) for repeated measures, being the independent variables within-subjects: session (Ss), foot (Ft), condition (diabetic or non-diabetic, N-D), and position (anterior or posterior, pos). The dependent variables were the averages of 10 measurements for each of the 9 variables measured. Only were considered first and second level intra-subject interactions.

Table 1 presents the results of univariate tests of effects within-subject for each variable, using a level $\alpha = 1\%$ based on the Huynh-Feldt test. Significant values are marked in bold. This level is used instead of the traditional 5% because given the number of measurements taken, it would run the risk that only 1 in every 100 measurements come out significant by chance (1 to 2 false positives in total).

Table 1 MANOVA analysis and the univariate effects tests for averages

Variable	univariate effects tests for averages								
	APLD	LLD	AP	BBx	BBy	FBx	FBy	PMLx	PMLy
N-D	0.998	**<0,01**	0.730	**<0,01**	0.149	0.102	0.121	0.467	0.011
N-D * pos	0.012	1.000	1.000	1.000	1.000	1.000	1.000	1.000	1.000
Ss	1.000	0.022	**<0,01**	0.761	**<0,01**	0.090	**<0,01**	0.167	0.000
Ss * N-D	1.000	**<0,01**	**<0,01**	0.015	**<0,01**	0.588	**<0,01**	0.003	0.003
Ss * N-D * pos	**<0,01**	1.000	1.000	1.000	1.000	1.000	1.000	1.000	1.000
Ss * Ft	0.859	0.781	0.037	N/A	N/A	0.034	0.535	0.030	0.105
Ss * Ft * N-D	0.024	**<0,01**	**<0,01**	N/A	N/A	0.378	**<0,01**	0.031	0.001
Ss * Ft * N-D * pos	0.010	1.000	1.000	N/A	N/A	1.000	1.000	1.000	1.000
Ss * Ft * pos	0.918	1.000	1.000	N/A	N/A	1.000	1.000	1.000	1.000
Ss * pos	**<0,01**	1.000	1.000	1.000	1.000	1.000	1.000	1.000	1.000
Ft	**<0,01**	**<0,01**	**<0,01**	N/A	N/A	**<0,01**	**<0,01**	0.000	0.000
Ft * N-D	0.296	0.021	0.046	N/A	N/A	0.196	0.792	0.010	0.224
Ft * N-D * pos	0.542	1.000	1.000	N/A	N/A	1.000	1.000	1.000	1.000
Ft * pos	**<0,01**	1.000	1.000	N/A	N/A	1.000	1.000	1.000	1.000
pos	**<0,01**	1.000	1.000	1.000	1.000	1.000	1.000	1.000	0.681
tech	0.998	**<0,01**	0.075	0.468	0.304	0.027	0.272	0.179	0.222
tech * pos	0.099	1.000	1.000	1.000	1.000	1.000	1.000	1.000	1.000

III. RESULTS

Variable LLD presented differences in condition: it was higher in subjects without diabetes. It also showed an interaction of session with condition: subjects without diabetes remained the same while subjects with diabetes were increasing and approaching to the group without diabetes over time (Figure 1).

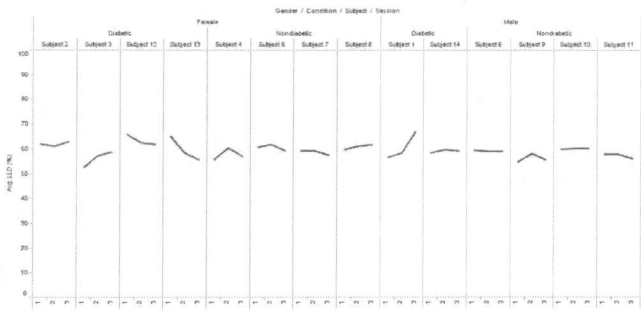

Fig. 1 Averages of Lateral Load Distribution in women and men

In the variable AP an effect was presented by condition, the average was higher in diabetics than in non-diabetics. There was an interaction of condition and session, the average decreased in the group with diabetes while the group without diabetes remained stable.

The variables BBx and FBx only presented a marginal effect on gender; average in men was higher than in women.

In variable BBy, the average presented an interaction of session with condition: average decreased in non-diabetic subjects, while diabetic subjects remained stable (Figure 2). By gender, men got higher average than women (Figure 3a).

Variable FBy presented a main effect in gender, average lower in women than in men. Session presented interactions with condition; subjects without diabetes decreased the average, while the average in subjects with diabetes remained the same; and a marginal interaction of gender with condition, the average for men was similar in both conditions, whereas women with diabetes had a greater average than without diabetes, but their average was lower than in men (Figure 3b).

Analysis of averages is not an appropriate for PML, since their locations are highly variable, so the average is not statistically significant. They were analyzed graphically through their dispersion (Figure 4).

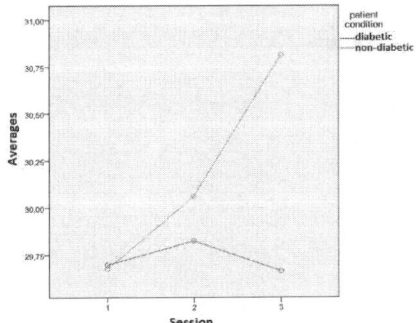

Fig. 2 Interaction of session and condition in averages of body barycenter

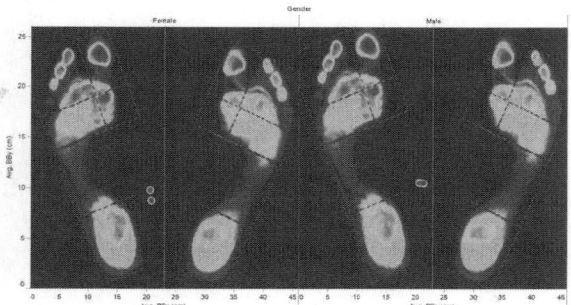

(a) Body Barycenters (BB)

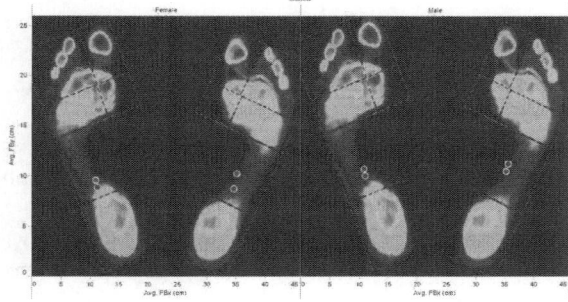

(b) Foot Barycenters (FB)

Fig. 3 Averages of BB and FB in women and men. Note: Background image is used only for guidance. Diabetics (blue), non-diabetics (red)

There is a greater concentration of points in the typical areas of support, i.e. metatarsals heads and the calcaneus. An effect was presented by gender, in which there was a greater dispersion of points, showing mainly in the calcaneus, in women. By condition, It can be distinguished a smaller dispersion in subjects without diabetes.

Variable APLD showed no gender effect or interaction of condition and session which were statistically significant.

IV. DISCUSSION

All measurements of BBx were displaced to the left of the ideal center, which may be due to laterality; however it presented higher averages in non-diabetic men. The interaction of gender with session shows that averages decreased in men and increased in women, over time. FBx variable, with only a marginal effect in gender, indicates an average higher in men than in women, but their difference is not statistically significant.

Variables LLD and AP presented differences of average in condition and an interaction of condition with session, showing a higher average in diabetics, indicating a difficulty of this group to maintain stability during the sessions, even when lateral movement is controlled. One factor that may have af-

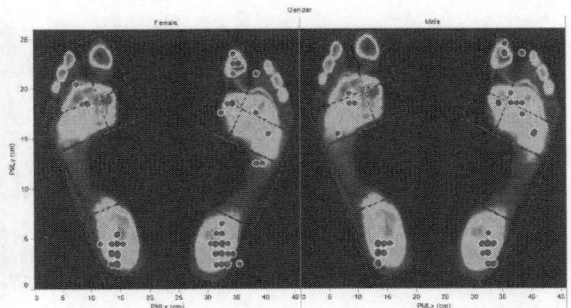

Fig. 4 Points of Maximum Pressure in women and men.

fected both the balance and the higher average pressure in diabetics is their BMI above 25, indicating overweight, and intensified by laterality.

BBy and FBy, presented similar results in the interaction of session with condition shows that in non-diabetic subjects the average decreased, while in diabetics remained stable, suggesting that over sessions, the non-diabetic group was increasing the support on the hindfoot, especially in women, who had lower averages than men.

Variables PMLx and PMLy, an effect of laterality can be seen by presenting greater dispersion of points of maximum pressure in the right foot, indicating greater stability with the left foot, consistent with the results of LLD. A greater dispersion of PML's in the calcaneus of women corroborates the result found in BBy, showing a greater support in the hindfoot, seeking greater stability with this region in this group of subjects.

In general, the non-diabetic group had similar measurements over the 3 sessions in both genders, while the diabetic group were more variable as the experiment progressed. This variation may be due to the physical condition of the subjects, a BMI over 25, which would make it difficult to maintain the standing position. Some gender differences were found, which would indicate that men have better balance than women. Women, in search of balance, prioritized support in the hindfoot, making this area vulnerable to ulcers.

The measurement protocol used in this study is very important to capture better plantar pressure measurements and make this system more viable to its future implementation in clinical practice and in a hospital network, for customized treatment and monitoring.

V. CONFLICT OF INTEREST

The authors declare that there is no conflict of interest associated with this work.

ACKNOWLEDGEMENTS

We sincerely acknowledge the efforts of the staff at the Orthopedics Department of San Ignacio Hospital in Bogotá, Colombia, for their help in recruiting suitable subjects for this study, and express our gratitude to all study participants. We are grateful to Javeriana University for having funded the study.

REFERENCES

1. Hussain A, Claussen B, Ramachandran A, Williams R. Prevention of type 2 diabetes: A review *Diabetes Research and Clinical Practice*. 2007;76:317–326.
2. Pataky Z, Vischer U. Diabetic foot disease in the elderly *Diabetes and Metabolism*. 2007;33:S56–S65.
3. Ikem R, Ikem I, Adebayo O, Soyoye D. An assessment of peripheral vascular disease in patients with diabetic foot ulcer *The Foot*. 2010;20:114–117.
4. Armstrong D G, Lavery L A, Houtum W H, Harkless L B. The impact of gender on amputation *The Journal of Foot and Ankle Surgery*. 1997;36:66–69.
5. Saito M, Nakajima K, Takano C, et al. An in-shoe device to measure plantar pressure during daily human activity *Medical Engineering and Physics*. 2011;33:638–645.
6. Burns J, Crosbie J, Hunt A, Ouvrier R. The effect of pes cavus on foot pain and plantar pressure *Clinical Biomechanics*. 2005;20:877–882.
7. Periyasamy R, Mishra A, Anand S, Ammini A C. Preliminary investigation of foot pressure distribution variation in men and women adults while standing *The Foot*. 2011;21:142–148.
8. Bus S A, Ulbrecht J S, Cavanagh P R. Pressure relief and load redistribution by custom-made insoles in diabetic patients with neuropathy and foot deformity *Clinical Biomechanics*. 2004;19:629–638.
9. Deleu P-A, Leemrijse T, Vandeleene B, Maldague P, Bevernage B D. Plantar pressure relief using a forefoot offloading shoe *Foot and Ankle Surgery*. 2010;16:178-182.
10. Putti A B, Arnold G P, Abboud R J. Foot pressure differences in men and women *Foot and Ankle Surgery*. 2010;16:21–24.
11. Zequera M, Garavito L, Sandham W, et al. Diabetic foot prevention: Repeatability of platform stabilometric parameters in bipedal standing for future home monitoring applications in diabetic patients *IFMBE Proceedings*. 2011;37:830-833.

Author: Martha Lucía Zequera Díaz
Institute: Pontificia Universidad Javeriana
Street: Carrera 7 No. 40 - 62
City: Bogotá D.C.
Country: Colombia
Email: mzequera@javeriana.edu.co

System for Precise Measurement of Head and Shoulders Position

J. Hejda[1], P. Kutílek[2], J. Hozman[1], and R. Černý[3]

[1] Czech Technical University in Prague, Faculty of Biomedical Engineering,
Department of Biomedical Technology, Kladno, Czech Republic
[2] Czech Technical University in Prague, Faculty of Biomedical Engineering,
Department of Natural Sciences, Kladno, Czech Republic
[3] Charles University in Prague, 2nd Medical Faculty, Medical School, Prague, Czech Republic

Abstract—The objective of our study is to develop a technique for precise head and shoulders posture measurement or, in other words, for measuring the native position of the head and shoulders in 3D space with accuracy to 1° in each direction. Until now, no cheap, widely applicable technique was presented.

Described measurement system and technique could have important applications, as there are many neurological disorders that affect the postural alignment position of the head and shoulders. In many cases, the abnormalities of the head position can be small and difficult to observe. Although an accurate method could contribute to the diagnosis of vestibular disorders and some other disorders, this issue has not been systematically studied. This paper describes improvements of measurement system based on experiences of clinical measurements and offers a new statistical method for evaluation of measured data.

Keywords—MOCAP, Head Position, Shoulders Position, Neurological Disorders

I. INTRODUCTION

A. The Importance of Accurate Head and Shoulders Position Measuring

Abnormal head posture (AHP) is an important clinical sign of disease in many medical specialties [1]. There is a large number of neurological disorders that affect the postural alignment position of the head and shoulders. These can be divided into three main groups:

- Cervical blockages and diseases of the cervical spine often cause a wide range of positional abnormalities.
- Dystonic "movement disorders". Abnormal body segment position is typical for dystonia.
- A paralysis of eye muscles also often causes a position that attempts to compensate the insufficient function.

The simultaneous measurement of head and shoulders position could contribute to a better definition of diseases affecting the vestibular system (labyrinthine) function in man [2], [3]. This also allows more precise diagnosis of disorders mentioned above.

Besides that, it can be used for physical medicine and rehabilitation, especially in the management and diagnosis of disorders affecting cervical spine. Habitual head anteflexion with chronic overload of cervical and upper thoracic spine and muscle imbalance is typical consequence of uncompensated sedentary way of life, starting already in school age.

B. Measurement Systems and Methods

Currently, an orthopedic goniometer is used as a standard way of simple and rapid measuring of angles in clinical practice. However, this method has some significant limitations, especially in the case of head and shoulder posture measurement.

Young [4] offered a new method for the study of head position using mirrors. This approach is based on the use of three mirrors and special head markers. Ferrario, V.F. et al., 1995, [5] developed a new method based on television technology that was faster than conventional analysis. Galardi, G. et al., 2003, [6] proposed an objective method for measuring posture by the use of an expensive *Fastrack* system. Mentioned systems and methods have common disadvantages, such as complex preparation, very large sensors or the inability to accurately define the anatomical coordinate system. Hozman, J. et al., 2005, [1], [7] proposed a new method based on the application of three digital cameras with stands and appropriate image processing software. The technique aimed at determining the differences between the anatomical coordinate system and the physical coordinate system with accuracy between one and two degrees for tilt and rotation. In our recent method [8], [9] two cameras are required for determining the head position.

The aim of our project is to simplify and completely optimize the technical equipment designed by Hozman, J. et al, 2005, [1], [7], and used for an otoneurological vestibular apparatus examination. Because of the insufficient accuracy of the current equipment (only two degrees in each measured angle), a new hardware and software is developed. Proposed system is able to determine tilt, rotation and inclination angles in physical coordinate system defined by the vector of the gravitational acceleration. Accuracy of the measurement should be less than one degree in each direction.

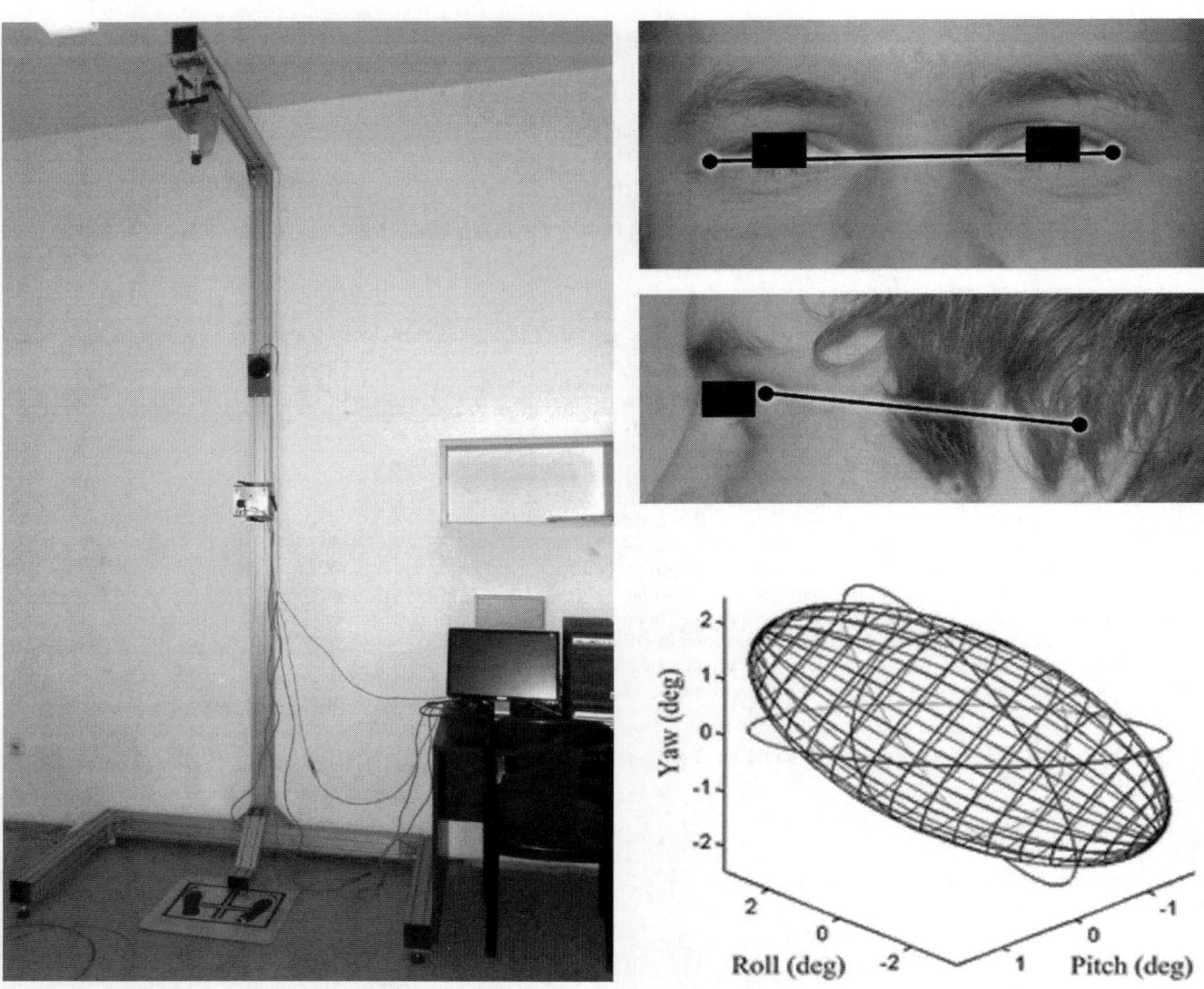

Fig. 1 (a) experimental setup for verification of measurement; (b, c) anatomical head axis; (d) example of a 3D confidence ellipsoid and three 2D confidence ellipses of the head angular displacement in the roll, yaw and pitch planes.

Main reference for sagittal plane is called *Frankfort horizontal* (line connecting meatus acusticus with the orbital floor or line connecting tragus with the outer eye canthus) [10], see Fig 1b, c. In most subjects this line is inclined forward bellow the space horizontal, in the extensor type of cervical positions it is reclined backwards, [9]. The shoulder axis is defined by two anatomical shoulders points, so called acromions, [11].

The technique of measurement [8] is aimed at determining differences between the anatomical coordinate system and the physical coordinate system with accuracy less than one degree for inclination, rotation and angles between the head axis and shoulder axis, [11].

II. MEASUREMENT SYSTEM AND METHODS

Our designed system is composed of a non-invasive examination device and specially developed software. The method of measurement is based on the combination of image recognition technique and capturing data from highly accurate motion tracking *inertial measurement units* (IMU) equipped by gyroscope, accelerometer and magnetometer. This approach should avoid the biggest disadvantage of IMUs based on the use of gyroscopes – low accuracy of rotation measurement. This drawback is eliminated by using a camera system for determining the mentioned angle.

A. Examination Device

The measuring device consists of two principal components – examination helmet and camera stand, see Fig 1a. The helmet is a component that is in contact with the examinee. The examiner must ensure correct positioning of the helmet on the patient's head. For this purpose, the helmet is equipped with calibration sliders, which must be placed according to the Frankfort horizontal. The essential part of helmet is an IMU *Trivisio Colibri* which is used for head flexion measurement. Among others the helmet is equipped with infrared LEDs, which are used as markers for tracking the head position by the camera system. Furthermore, the examiner's task is to attach LED markers on patients' shoulders, specifically on acromions.

Because it is necessary to capture an image from the top and front of the examinee, the universal camera stand was made. It holds the cameras on mentioned position and allows their adjustment. In addition, the stand is equipped by a platform marking the position of correct patient's stand.

The *Basler* industrial monochromatic cameras have been chosen for image capturing. They offer standard VGA (640x480px) resolution with frame rate 100fps. Because they are used for infrared (IR) LEDs capture, they are supplied by infrared light permeable optical filter *Kodak Wratten 2 No. 87* for visible optical spectrum suppression. The cameras are connected to Gigabit Ethernet switch which also provides their power supply. To ensure cameras calibration with the Earth (i.e. physical) coordinate system, the USB accelerometers *Phidget* are mounted to the cameras.

B. Software Equipment

Due to the specific needs of head position measurement, a completely new examination application has been created. This software is designed to offer the highest customization possible. In basic configuration it captures images from two cameras, collects data from accelerometers and reads data from IMU.

The ordinary PC equipped by Gigabit Ethernet card is used as a service computer. Due to the limited device driver support, the *Microsoft Windows* operating system was chosen. The examination software is built on *Microsoft .NET 4.0* platform and implemented in *C#* programming language.

As expected with the use of application in clinical practice, it is essential the program was easy to use. For this reason the program is widely configurable and thus adaptable to specific measurement requirements. Further the development was focused on high software modularity to allow flexible support for new sensors and use.

The application is divided into two principal components – *kernel* and *GUI*. The kernel is represented by application class library offering interface for the second mentioned component – the GUI which provides the interaction with user. Proposed software is very variable and it can be preconfigured by XML to make the simplest possible installation, maintenance and use.

The sensor data processing is performed by kernel and it is based on module pipelining. This approach brings great configuration options and multithreaded data processing ability. Kernel uses two types of modules – *DataSourceModule* intended for sensor data grabbing and *FilterModule* processing acquired data. Each module must implement a specified set of classes that depends on its type.

The key function of the measuring system lies in image recognition and it is ensured by *ImageToAngles* module. This module processes raw image data, performs distortion correction, markers detection and identification and it gives determined angles into the output. Principle of angle determination consists in finding the coordinates of markers and calculation of angle between the line connecting markers and the longer edge of sensor. The identification of markers in the picture is based on thresholding, distance transform [12] and pattern recognition technique which finds the marker's center.

C. Measurement

Numerous techniques were introduced to quantify trunk movements in both pitch (anterior–posterior) and roll (medial–lateral) directions during clinical stance and gait tests in freely moving individuals [13], [14]. In these studies, combinations of trunk pitch and roll sway measurements during specific balance and gait tasks identified differences in balance control of patients compared to that of normal subjects. It is well known that the sway of standing patients with a vestibular deficit is mainly side to side (in the roll plane) and they fall to the side of the deficit, [14]. Instability in other directions such as the fore-aft (pitch) plane are also seen, [15].

Traditional methods for assessing the postural instability are methods based on the envelope (convex hull), [16], or 2D *confidence ellipse*, [17]. These methods are based on the description of the behavior of only two variables (two angles in our case) in two planes/axis of human body. However, we need to model the distribution of the measured 3D data (pitch, roll, and yaw). We use a method based on statistical technique called *Principal Component Analysis* (PCA) [18]. PCA applied to a measured data forms a symmetric 3x3 covariance matrix. We can model the distribution of the measured data (pitch, roll, and yaw) by 3D ellipsoid. The volume of an ellipsoid is given by an ellipsoid matrix and the ellipsoid matrix is composed of entries from the covariance matrix, [19]. We use 95% *confidence ellipsoid* (CE), [20]. The covariance matrices were used to calculate the volumes of the 95% CEs. After calculating the volume of each 95% CE of each patient and healthy subject with eyes closed, the statistical analysis is performed. The measured data are compared to identify the normal and abnormal posture of the head of patients. According to the method described here, the volume of the 3D CE can inform about the postural stability and balance problems.

III. CONCLUSION

We have designed special equipment and implemented procedures for very accurate evaluation of head and shoulders positions in neurological practice. The new measured data evaluation methods have been described in this paper. The system is cheaper than sophisticated systems using accelerometers and magnetometers or expensive commercial camera system such as *Vicon* or *Optitrack*.

ACKNOWLEDGMENT

This work was supported by the Grant Agency of the Czech Technical University in Prague, grant No. SGS13/096/OHK4/1T/17.

REFERENCES

1. Cerny R, Strohm K, Hozman J, Stoklasa J, Sturm D (2006) Head in Space – Noninvasive Measurement of Head Posture. Bled
2. Murphy K, Preston C, Evans W (1991) The Development of Instrumentation for the Dynamic Measurement of Changing Head Posture. American Journal of Orthodontics and Dentofacial Orthopedics 99(6):520-6
3. Raine S, Twomey L (1997) Head and Shoulder Posture Variations in 160 Asymptomatic Women and Men. Archives of Physical Medicine and Rehabilitation 78:1215-23
4. Young J D (1988) Head Posture Measurement. Journal of Pediatric Ophthalmology and Strabismus 25(2):86-9
5. Ferrario V, Sforza C, Tartaglia G, Barbini E, Michileon G (1995) New Television Technique for Natural Head and Body Posture Analysis. Cranio 13:247-255
6. Galardi G, Micera S, Carpaneto J, Scolari S, Gambini M, Dario P (2003) Automated assessment of cervical dystonia. Movement Disorders 18(11): 1358-67
7. Hozman J, Sturm D, Stoklasa J, Cerny, R (2005) Measurement of Postural Head Alignment in Neurological Practice. EMBEC'05, Prague, Czech Republic
8. Kutilek P, Hozman J (2009) Non-contact method for measurement of head posture by two cameras and calibration means. The 8th Czech-Slovak Conference on Trends in Biomedical Engineering, Bratislava, Slovakia
9. Kutilek P, Hozman J, Cerny R, Hejda J (2012) Methods of Measurement and Evaluation of Eye, Head and Shoulders Position in Neurological Practice. Advanced Topics in Neurological Disorders. Rijeka: InTech, pp. 25-44
10. Ricketts R, Schulhof R, Bagha L (1976) Orientation-sella-nasion or Frankfort horizontal. American Journal of Orthodontics 69(6):648-54
11. Harrison A, Wojtowicz G (1996) Clinical Measurement of Head and Shoulder Posture Variables. JOSPT 23:353-61
12. Breu H, Gil J, Kirkpatrick D, Werman M (1995) Linear time Euclidean distance transform algorithms. IEEE Transactions on Pattern Analysis and Machine Intelligence, pp. 529-33
13. Allum J, Adkin A, Carpenter M, Held-Ziolkowska M, Honegger F, Pierchala K (2001) Trunk sway measures of postural stability during clinical balance tests: effects of a unilateral vestibular deficit. Gait Posture 14(3):227-37
14. Gill J, Allum J, Carpenter M, Held-Ziolkowska M, Adkin A, Honegger F, Pierchala K (2001) Trunk sway measures of postural stability during clinical balance tests: effects of age. The journals of gerontology 56(7):M438-47
15. Brandt T, Daroff R (1980) The multisensory physiological and pathological vertigo syndromes. Annals of neurology 7(3):195-203
16. Horlings C, Küng U, Bloem B, Honegger F, Van Alfen N, Van Engelen B, Allum J (2008) Identifying deficits in balance control following vestibular or proprioceptive loss using posturographic analysis of stance tasks. Clin Neurophysiol 119(10):2338-46
17. Prieto T, Myklebust J, Myklebust B (1992) Postural steadiness and ankle joint compliance in the elderly. Engineering in Medicine and Biology Magazine 11(4):25-7
18. Jolliffe I (2005) Principal Component Analysis. Encyclopedia of Statistics in Behavioral Science, John Wiley & Sons, Ltd.
19. Nash S G, Sofer A (1996) Linear and Nonlinear Programming. New York: McGraw-Hill Companies, Inc.
20. Rocchi M, Sisti D, Ditroilo M, Calavalle A, Panebianco R (2005) The misuse of the confidence ellipse in evaluating statokinesigram. Italian Journal of Sport Sciences 12(2):169-71

Author:	Jan Hejda
Institute:	Czech Technical University in Prague, Faculty of Biomedical Engineering
Street:	Nám. Sítná 3105
City:	Kladno
Country:	Czech Republic
Email:	jan.hejda@fbmi.cvut.cz

An Online Program for the Diagnosis and Rehabilitation of Patients with Cochlear Implants

T. Lopez-Soto and A. Castillo-Armero

Department of English Language, University of Seville, Seville, Spain

Abstract—This paper presents the general architecture and functionality of Co-Clear, an online program designed for the diagnosis and rehabilitation of patients with cochlear implants. The main features rely on the portability of the system, which can run multiplatform, allowing the patient to continue auditory rehabilitation outside the clinic. Co-Clear follows state-of-the-art techniques both in databases management and object organization. As for the perceptual model, it is based on the priming technique according to which the hypothesis for the rehabilitation method lies on the continuous presentation of synthetic stimuli. This paper presents a first version oriented towards the diagnosis of speech recognition and auditory rehabilitation of children. The easy-to-use GUI is also highlighted. The diagnostic model attempts to identify accuracy in the interpretation of frequency ranges characteristic of vowels. This approach complements traditional audiological tests based on pure tones. The rehabilitation model currently incorporated in the system allows for both discrimination and identification tasks and the same has been implemented on a multi-layered architecture made of a series of tasks chosen by the clinician which, allow for the selection of multiple curricula. Curricula here have to be understood as individualized and specific series of auditory tasks that need to be completed in different sessions (and days) by one patient. The system identifies the accuracy rate of each task and directs the patient towards the corresponding set of tasks. Both the tasks and the paths to complete the rehabilitation exercises can be managed at the clinic. This feature can guarantee a fast adaptation to the cochlear implant as the system focuses on the adaptation to natural spoken communicative situations.

Keywords—rehabilitation, cochlear implants, audiology, perception tests.

I. INTRODUCTION

Even though the appearance of cochlear implants has significantly improved the lives of many deaf people around the world, research has demonstrated that there is still a lot to be done in many ways. Big efforts are being taken in regards the design of these devices, which are currently going in two directions, namely the perception of music and how to cope with background noise.

Researchers agree that cochlear implant patients continue to achieve higher levels of performance, but one of the main technical limitations remains how to deal with background noise in real speech situations. The variability of noise is still a challenge for the signal processing accessories with which these devices are equipped. How to segregate speech/sounds and noise is of vital importance in children. Therapists need to guarantee that the adaptation of the implant is done effectively and fast so that the child can acquire communicative skills that allow him for a complete social and educational integration.

In a study carried out with 2,251 post-lingually deaf adults in 15 centers around the world, [1] show that the onset age of deafness and its duration are definite in predicting adaptation to cochlear implants. These results make us believe that the need to speed up the process for such an adaptation is even greater and probably more challenging in the case of pre-lingually deaf children.

This paper presents the general architecture and functionality of Co-Clear, an online program designed for the diagnosis and rehabilitation of patients with cochlear implants. The general features of the system are twofold: it runs online on multiplatform and it state-of-the-art techniques both in databases and object organization. From the point of view the audiological models, the system is inspired in current research on auditory testing and rehabilitation. The architecture of the system can be understood as a separation between the objects (sound stimuli and rules) and the interaction between them in the form of different tasks. At the moment, the system implements both discrimination and identification tasks. The philosophy behind is the widely accepted idea according to which perception can be enhanced by the continuous presentation of organized stimuli, which can be adequately structured to achieve one or various therapeutic objectives.

The main advantages of the system are probably its flexibility and portability. Its portability is demonstrated as being an online, multiplatform system. This has a clear advantage for it allows the patient to continue rehabilitation outside the clinic. As for its flexibility, we should mention that the incorporation of a well-structure stimuli database available online permits the clinician to organize and design different auditory tasks that can then be integrated in a general rehabilitation program. The system architecture combines the stimuli database with a multi-layered task architecture that incorporate different kinds of discrimination and identification auditory tasks. These tasks can be easily articulated by the clinician in the form of individual exercises or they can further be integrated into a wider framework of curricula. Curricula here have to be understood as individualized and specific series of auditory tasks that need to be completed in different sessions (and days) by one patient. The system identifies the accuracy rate of each task and directs the patient towards the corresponding set of tasks.

The first application of Co-Clear has been directed towards the rehabilitation of children (5-10 yrs old). The rehabilitation model has been focused on spoken language stimuli, namely synthesized Spanish vowels. The objective is to take advantage of the auditory tasks designed in the system in order to stimulate the early adaptation of patients so that their speech recognition skills can be enhanced. The hypothesis behind is that the repetition in the exposure to auditory stimuli reinforces the patient's perception skills [2].

II. SYSTEM ARCHITECTURE

One of the main objectives of this project was to make sure that the system was available at all times. The architecture had to be online and multi-platform so that patients could use it in different places, both in the clinic and at home. In order to facilitate these goals Co-Clear (www.co-clear.com) was implemented.

The functionality of the system make it necessary to implement an efficient database manager that can deal with large number of audio files in a fast and clean way.

This web application has been designed in order to compete with accessibility standards in the open-source environment for multiple web platforms. These standards include HTML5 and CSS3 as part of the front-end, which makes it possible for any patient to register to the system from any kind of computer.

At the back-end the system follows J2EE6 technologies in order to guarantee stability and scalability.

One of the main concerns prior to implementing the system was how to guarantee that data storing was clean and low-cost. We decided to use MySQL for being a widely accepted approach and the EJB and JPA standards in order to abstract the database layer. This decision has given the system a high degree of robustness and efficiency: on the one hand we need to guarantee that the presentation of audio stimuli is done in the correct temporal framework; on the other hand, the hierarchy of stimuli and tasks demands a highly efficient database management policy. Fig 1 shows the general architecture of the system.

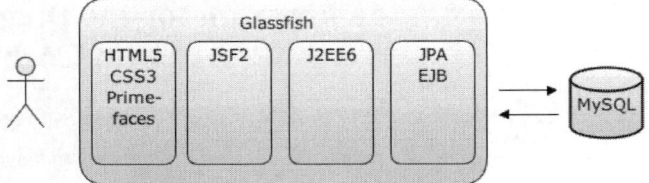

Fig. 1 General Architecture for Co-Clear

The presentation of stimuli for this first version of the system is oriented towards its use with children, so for discrimination and identification tasks we use easy-to-use pictures adapted in size and color to small children. All visual programming has been supported by JSF2 and component libraries such as Primefaces. Fig 2 shows a capture of a visual presentation in a discrimination task. In this task the child listens to two stimuli and has to click on one picture or another depending on the response chosen (same or different).

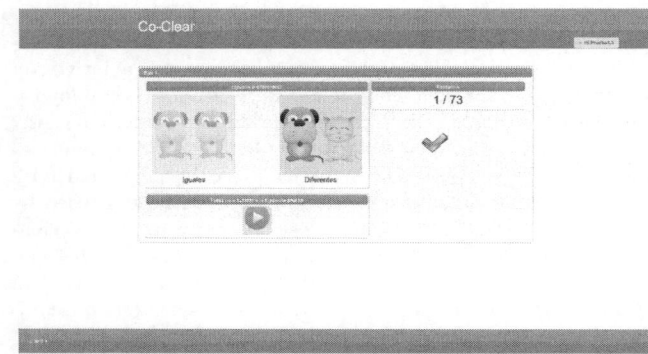

Fig. 2 Same/Different Exercise for Children

For dependency management we have chosen MAVEN2 while the application client container that the system uses is GlassFish3.

To make the system further flexible and applicable it offers API RESTful that can be used to integrate with future applications or external uses.

The functionality of Co-Clear is based on the management of audio stimuli, the creation of auditory tasks and the presentation of exercises. The system allows for discrimination and identification tasks that can be stored for future use. Each task contains an infinite number of stimuli that can also be stored and organized in families. Stimuli can be easily identified by a tagging system. Each patient can be assigned an infinite number of tasks, which can also be structured in the form of curricula. Curricula are itineraries of auditory exercises that can be set at different levels and according to different criteria. These levels and criteria can be easily set up by the clinician. The objective of curricula is to give the patient an organized number of exercises that are completed on the basis of previous results. So, according to the result obtained in one exercise, the patient will immediately to a similar or a different exercise that can be of a higher level or which measures different skills. Results are created on tabulated tables following the CSV format once an exercise is completed. These results are then treated statistically to assess progress and are used to set up the curricula.

III. SPEECH RECOGNITION

This section explains the approach taken for the rehabilitation model that is being used by a population of children with cochlear implants. The novelty of this model, when compared with traditional audiological approaches is that it put emphasis on spoken language sounds. These sounds are recorded by native speakers and then synthesized to accommodate for all variability in the language. This acoustic variability is achieved by changing the original sounds in terms of amplitude, time and frequency. The synthesis procedure is explained below.

The hypothesis behind the rehabilitation model adopted is the conviction that the repeated exposure to controlled audio stimuli can enhance the patient's speech recognition skills. This can facilitate a fast adaptation to the cochlear device and guarantee spoken skills too. The next two sections are devoted to review the benefits of intensive auditory training and the synthesis procedure.

A. Auditory Training

Traditional audiological and therapeutic techniques are based on the presentation of pure tones in order to diagnose and rehabilitate patients with hearing impairment. These tests are valid in as much as they can identify the frequency ranges which are poorly perceived. They can also delimit the amplitude levels of these patients and compare with normal hearing average values. However, a combination of them is firstly too simple for the design of rehabilitation exercises. Most of the time, therapists present isolated non-linguistic sounds to the patient or recordings of words and sentences. These tests can be considered tone patterns identification tests.

A selection of well-organized pieces of speech can guarantee better results. These pieces of speech can be considered to be vowels and consonants. But they need to be accordingly synthesized so we can control fixed values in terms of amplitude, time and frequency. Another reason why pure tones can poorly accomplish fast and steady adaptation to cochlear implant users is that spoken sounds are complex in terms of frequency. Let us take the vowels, for example, which show up to 5 different frequency ranges. The first formant of frequency (F0) is used to identify intonation, while F1 and F2 are used to distinguish sounds between them. For example, Spanish vowels /i/ and /u/ have similar F1 values but /u/ shows a very low F2 in comparison with /i/.

If we are capable of synthesizing vowels so that we create exemplars, we can then present them to the patient in a series of repetitive trials that, according to the evidence, can enhance speech recognition. Using spoken language material is decisive to promote the easy identification and discrimination of language, thus allowing the child to integrate themselves with their family and peers by the promotion of communicative skills that will eventually end up in spoken skills too.

Auditory training has successfully been applied to different populations, including second language learners, old people with cognitive deficiencies and hearing impaired patients [3].

B. Synthesis

All synthesis has been done using Klatt [4] and Praat [5]. The continuum and enveloping algorithms used guarantee the creation of synthesized vowels from a central point, which Praat takes from the frequency values associated to the natural stimuli. The stimuli used for the first version of Co-Clear were 24 Spanish consonants combined with 3 vowels (CV) (/i/, /a/, /u/). The selection of these vowels was made so that we could create a whole variety of spectrally different syllables, as can be seen in Fig. 3, Fig. 4 and Fig. 5.

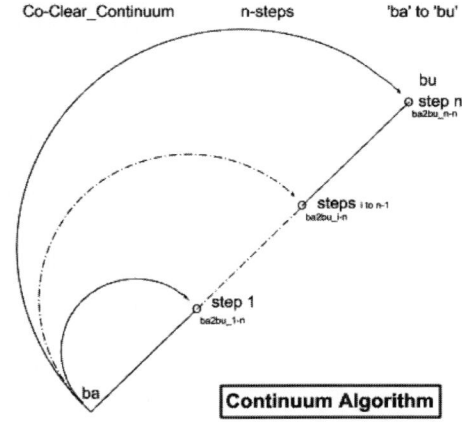

Fig. 3 Summary of the n-steps to be taken to amplify the spectral values from one steady frequency point to another

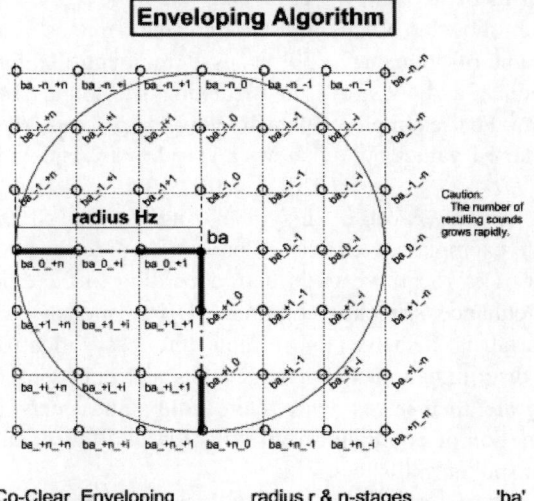

Fig. 4 Frequency grid for the expansion of synthesized vowels

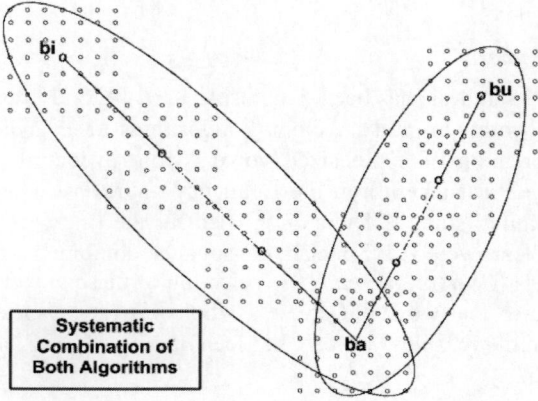

Fig. 5 The Systematic Combination of both the Continuum and the Enveloping Algorithm creates a whole set of exemplars for the Spanish vowels

IV. CONCLUSIONS

Co-Clear is an online platform that allows for the creation of auditory diagnosis and rehabilitation. At present, the system has been designed aiming at the therapy of patients with cochlear implants. It combines database and object management with visual integration that can be easily applied to pre-lingually deaf patients with a special emphasis on children. Its user-friendly interface makes it attractive for this type of population and being online guarantees that rehabilitation continues outside the clinic. This can enhance better adaptation to the device. The incorporation of task curricula makes it more efficient and useful for the clinician because progress can be anticipated and controlled statistically. All tools and techniques are open-source and follow web2.0 requirements.

REFERENCES

1. Holden LK, Finley CC, Firszt JB et al (2013) Factors affecting open-set word recognition in adults with cochlear implants. *Ear Hear.* 2013 May-Jun;34(3):342-60.
2. Holt, R (2011) Enhancing speech discrimination through stimulus repetition. *Speech Lang Hear.* 54(5):1431-1447.
3. Lively S, Pisoni DB, Yamada RA et al (1994) Training Japanese listeners to identify English /r/ and /l/. *J Acous Soc Am* 96(4): 2076-2087.
4. Klatt DH & Klatt LC (1990) Analysis, synthesis and perception of voice quality variations among male and female talkers. *J Acous Soc Am* 87:820-856.
5. Boersma P & Weenink D (2009) Praat: doing phonetics by computer (Version 5.1.05) [computer program] Retrieved May 10, 2013 from http://www.praat.org/

Analysis of Anomalies in Bioimpedance Models for the Estimation of Body Composition

D. Naranjo[1], L.M. Roa[1], L.J. Reina[2,1], M.A. Estudillo[1], N. Aresté[3,1], A. Lara[3,1], and J.A. Milán[3,1]

[1] Biomedical Engineering Group, CIBER-BBN and University of Seville, Seville, Spain
[2] Department of Signal Theory and Communications, University of Seville, Seville, Spain
[3] Service of Nephrology, Virgen Macarena University Hospital, Seville, Spain

Abstract—Bioimpedance analysis is a simple, safe and noninvasive method for Body Composition Estimation (BCE), which is of great interest for the monitoring of patients on renal replacement therapy. The most featured bioimpedance devices available are based on the bioimpedance spectroscopy technique, the extended Cole model and the Hanai Mixture theory. However, a set of anomalies using these methods has been found in this paper during the study of the evolution of body composition in patients on peritoneal dialysis. The main results obtained show that the estimates resulting from bioimpedance values that differ significantly from the single-dispersion Cole model have to be taken with some caution. This issue highlights the importance of medical assessment (technical or specialist) when interpreting any bioimpedance related data.

Keywords—Bioimpedance Spectroscopy, Body Composition Estimation, Cole Bioimpedance Model, Peritoneal Dialysis, Overhydration

I. INTRODUCTION

Uremic patients treated with both hemodialysis (HD) and peritoneal dialysis (PD) show alterations in the metabolism of water with continuous variations in hydration status [1–3]. After a hemodialysis session is quite common for patients to have a significant fluid excess or to be in an undesirable state of dehydration, which can cause or aggravate any cardiovascular disease [4].

It is very important for renal patients treated by dialysis to assess the amount of fluid excess to determine how much should be removed through ultrafiltration to achieve a desired state of normohydration [4]. Bioimpedance methods are common techniques to estimate body composition because they have not the restrictions of solution methods [5–7] and provide more accurate estimations than the anthropometric methods [7]. Thanks to bioimpedance techniques, it is possible to obtain an estimation of body fluids and body composition in both normal and disease states. Bioimpedance measurements also have many practical advantages that have led to their rapid development [5, 6].

After years of research about bioimpedance analysis for dialysis patients, this technique has significantly increased its clinical use [1, 8], both for estimating body volumes as to assess patient's nutritional status. The clinical utility of the body composition analysis through bioimpedance techniques for patients treated with PD has been demonstrated in numerous studies [6, 9, 10]. In this context, this paper describes a number of "anomalies" in the BCE using the bioimpedance analysis, which have been identified during an evolution study of a group of PD patients. These anomalies were not detected by a bioimpedance device that uses the models described below, requiring the analysis of nephrologist specialists in order to evaluate their importance for patient welfare assessment.

II. MATERIALS AND METHODS

A. Cole Bioimpedance Model

A simple model that describes the phenomenon of electric current conduction through the human body is represented by a parallel circuit in which a branch represents the current path through the extracellular medium and the other one the intracellular environment [1]. The extracellular path is modeled by a resistor (R_e, extracellular resistance) and the intracellular pathway by means of a resistor (R_i, intracellular resistance) in series with a capacity (C_m, membrane capacity) (see Fig. 1).

If the real part of the bioimpedance is plotted as a function of frequency versus the imaginary part in absolute values, the points obtained correspond to a semicircle (single-dispersion) in the first quadrant whose center is on the real axis (see Fig. 1). A more complete model of bioimpedance includes the effects of the variability of cell membranes, which are not perfect capacities due to different shape and size characteristics. This results in a shift pattern of the model in the center of the semicircle below the real axis, which can be expressed by the following equation (single-dispersion Cole model):

$$Z = R_\infty + \frac{R_0 - R_\infty}{1 + (j*\omega*(R_e+R_i)*C_m)^{1-\alpha}} \quad (1)$$

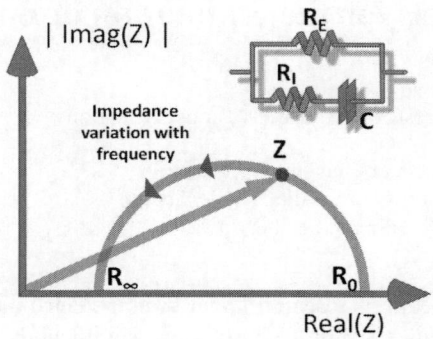

Fig. 1 Simple bioimpedance model and bioimpedance Cole diagram

R_0 is equivalent to R_e, R_∞ is the parallel of R_e and R_i, C_m is related to the characteristics of the cell membrane, $\omega = 2*\pi*frequency$ and α is a parameter ($0 \leq \alpha \leq 1$) related to the shift of the curve.

A more realistic model also includes the effects of delays in the signals caused by the electrodes, the wires and the hardware, which can be modeled by a delay in phase (T_d) which increases linearly with frequency (extended Cole model).

$$Z' = Z * e^{-j\omega T_d} \quad (2)$$

B. Bioimpedance Spectroscopy for BCE

The first equations for the estimation of body fluid volumes were based on linear regressions defined from $Height^2/Resistance$ at 50 kHz. The regression parameters differed depending on the population group with which they were obtained [11]. Subsequent equations included other components to improve the accuracy of the estimations, such as weight, age, gender, ethnicity or anthropometric measurements of trunk and/or limbs.

Thomasset became the first to use the Cole model in order to differentiate between the extracellular water and the body water [1]. In 1992, the bioimpedance spectroscopy technique was introduced. This tecnique uses a multifrequency bioimpedance based Cole model and Hanai's mixture theory in order to estimate the extracellular water and the body water, avoiding the population imbalances obtained by linear regression approaches [1, 4]. In this method the determination of body fluid volumes is based on the fact that low-frequency electric current does not penetrate cell membranes, so it flows only for the extracellular compartment, while high-frequency current flows through both the extracellular and intracellular compartments. Therefore, the resistance at low and high frequencies are related to the Extracellular Water and the Total Body Water, respectively. By applying the mixtures theory of Hanai, the human body is considered as a superposition of a conductive medium (water, electrolytes, soft tissue, etc.) and a non-conductive (bone, fat, air, etc.). These considerations improve estimations by introducing the effects of non-conductive substances in body water, eliminating the apparent population specificity found in the regression equations and improving its sensitivity to changes in body hydration status [1].

Bioimpedance spectroscopy have been broadly used to quantify body composition, estimate fluid volume and locate mass anatomy (muscles, fat, water) in certain parts of the body. This technique is used by devices with higher features/prices in the market for the monitoring of dialysis patients, such as the Body Composition Monitor of Fresenius Medical Care [4].

C. Description of the Study

The initial objective of the study (cross sectional, observational) was to analyze the evolution of the hydration and nutritional status in a representative sample of prevalent PD patients. In the center of the study, all prevalent patients on PD were chosen as possible candidates for their inclusion. Patients were excluded if they had a cardiac pacemaker or metallic implants, were amputees or were pregnant. The study lasted a year and a half and the measurements were repeated with the same group of patients during their clinical routine practice. BCEs were performed by a single PD physician or nurse, using a portable body bioimpedance spectroscopy device: the Body Composition Monitor (Fresenius Medical Care). This device and the bioimpedance spectroscopy technique have been intensively validated against gold-standard methods [4, 6].

Table 1 shows the characteristics of the patients under study:

Table 1 Patient characteristics

10 men/10 women	Min	Medium	Max	σ
Age (years)	31	61.8	86	15.6
Weight (kg)	45.8	71.1	107	17.3
Height (cm)	140	160.5	184	9.3
Body Mass Index	20.4	27.4	41.8	5.5

The evolution of the patients was analyzed using the following parameters derived from data provided by the device: hydration level, Fat percentage [12], Fat Mass Index [13] and Fat Free Mass Index [14].

III. RESULTS

Table 2 summarizes the average ratings of BCEs of the patients.

Table 2 BCE patients results

10 men/10 women	Min	Medium	Max	σ
Overhydration (l)	-3.7	1.1	6	1.9
Percentage of Fat	8.8	34.5	53.6	9.8
Fat Mass Index	1.9	9.4	17.8	3.5
Fat Free Mass Index	11.5	17.4	24	3

This seems to indicate that the patients usually present on average a slight over-hydration and a good nutritional status, but with a slight excess of fat. When analyzing the evolution of the patients, there were no significant trends or characteristics, at least during the time of the study.

However, the technical analysis of the bioimpedance data shown that the cases in which they did not fit the model of single-dispersion Cole, the nutritional and hydration status of the patient undergone significant changes in valuations not too temporarily far. This technical analysis was clinically confirmed with the help of the nephrology specialists. Below there are two examples that highlight the anomalies detected.

A. Example A

Figure 2 show bioimpedance values for two consecutive measurements on a Cole diagram (only two are shown for the sake of clarity). The dashed line represents the fitting for a single-dispersion Cole model. One of the measurements (number 4) show a relevant mismatch at high frequency which is corrected by the extended Cole model (solid line) and a major phase delay ($T_d = 5.27 nseg$).

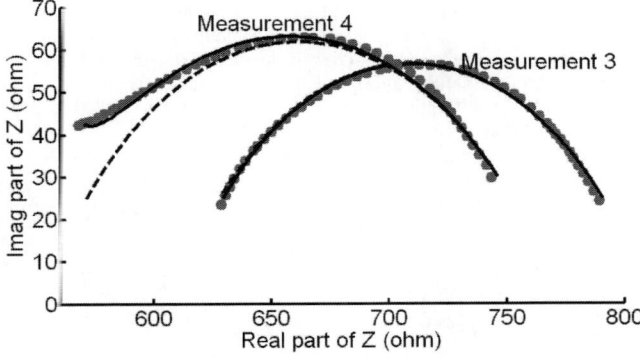

Fig. 2 Cole diagram of patient in example A

The body estimates in this case showed a significant variation with respect to other measures (see Fig. 3).

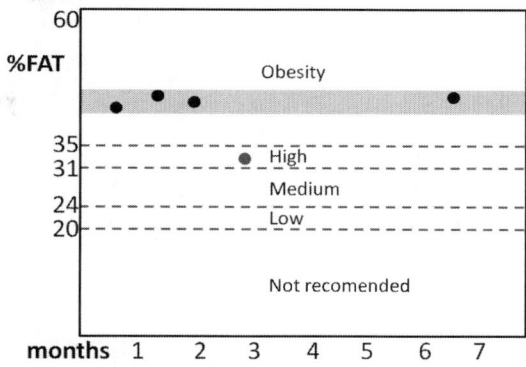

Fig. 3 Classification by percentage of FAT [12] of patient in example A

B. Example B

Figure 4 also shows bioimpedance values on two consecutive measurements on the Cole diagram. In this case the measurement 1 is corrected with a large phase lead ($T_d = -6.12 nseg$).

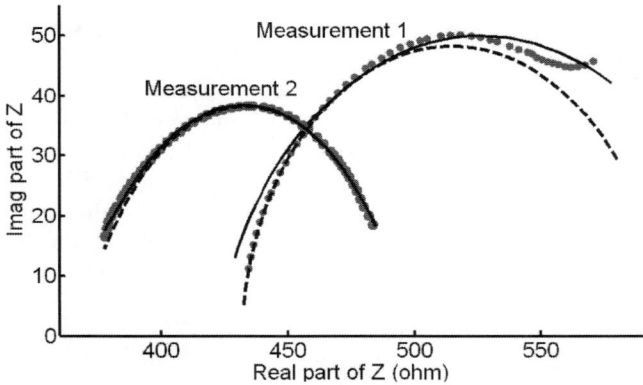

Fig. 4 Cole diagram of patient in example B

Assessments of BCE in this case also showed a significant variation with respect to the rest of measures (see Fig. 5).

IV. DISCUSSION AND CONCLUSIONS

The study described in this paper has analyzed the evolution of a group of peritoneal dialysis patients for over one year. BCEs were performed by a highly used device that employs the bioelectrical impedance spectroscopy technique, the extended Cole model and the Hanai Mixture theory. The results obtained have not shown relevant features to highlight regarding the clinical evolution of the patients. However, a number of anomalies were detected in those cases in which bioelectrical impedance values significantly differ compared

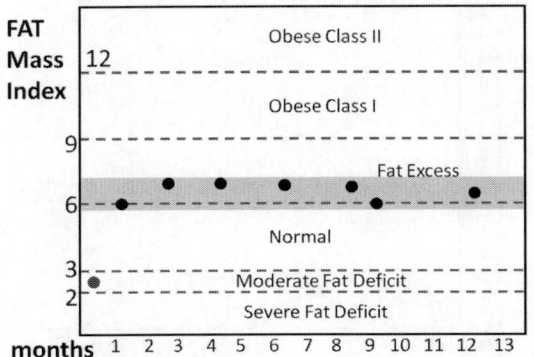

Fig. 5 Classification by Fat Mass Index [13] of patient in example B

to the single-dispersion Cole mole (the purpose of the extended Cole model is to correct and explain this effect). The anomalies were clinically evaluated in order to discard the related measurements for further studies.

BCEs were not compared with another reference method that could indicate a possible error in the measurement. However, in this paper the evolution of the patient has been analyzed, so that, assuming a progressive evolution of patient's condition, significant changes in the BCE on measures not too remote in time or regarding normal patient values in the patient may indicate a possible error in the estimation. If these estimates are considered to be mistaken, the possible sources of error can be multiple: defects in the electrodes or inappropriate placement on the body, the patient was not in the proper position, enough time was not spent so that the volume of fluid in the different compartments could be stabilized, etc.

According to the analysis carried out on the model parameters, the authors consider that those BCEs derived from the extended Cole model where the phase delay module T_d is greater than 5 nsec, must be taken with some caution. In these cases, it may be advisable to make a second estimate repeating the whole process of measurement (cleaning the skin, electrode placement, etc.) in order to compare the results obtained.

On the other hand, perhaps the peritoneal compartment of this particular type of patients has a greater influence and in certain situations it may be manifesting as a second dispersion. It is possible that more complex models (two dispersions maybe) could explain better these phenomena, and can provide more accurate measurements of BCEs for a higher number of situations.

Acknowledgements

All authors have participated actively in the development of this work. It was supported in part by the Consejería de Economía, Innovación y Ciencia, Government of Andalucía, under Grants P08-TIC-04069 and P10-TIC-6214, in part by the Fondo de Investigaciones Sanitarias, Instituto de Salud Carlos III, under Grants PI082023 and PI11/00111, and in part by CIBER-BBN under Grant PERSONA.

References

1. Roa L M, Naranjo D, Reina L J, et al. Applications of bioimpedance to end stage renal disease (ESRD). *Studies in Computational Intelligence.* 2013;404:689-769.
2. Liu L, Zhu F, G Raimann J, et al. Determination of fluid status in haemodialysis patients with whole body and calf bioimpedance techniques. *Nephrology.* 2012;17:131-140.
3. Chua H R, Xiang L, Chow P Y, et al. Quantifying acute changes in volume and nutritional status during haemodialysis using bioimpedance analysis. *Nephrology.* 2012;17:695-702.
4. Wabel P, Chamney P, Moissl U, Jirka T. Importance of whole-body bioimpedance spectroscopy for the management of fluid balance. *Blood Purification.* 2009;27:75-80.
5. Earthman C, Traughber D, Dobratz J, Howell W. Bioimpedance spectroscopy for clinical assessment of fluid distribution and Body cell mass. *Nutrition in Clinical Practice.* 2007;22:389-405.
6. Sipahi S, Hur E, Demirtas S, et al. Body composition monitor measurement technique for the detection of volume status in peritoneal dialysis patients: The effect of abdominal fullness. *International Urology and Nephrology.* 2011;43:1195-1199.
7. Moissl U M et al. Body fluid volume determination via body composition spectroscopy in health and disease. *Physiological Measurement.* 2006;27:921-933.
8. Laegreid I K, Bye A, Aasarod K, Jordhoy M. Nutritional problems, overhydration and the association with quality of life in elderly dialysis patients. *International urology and nephrology.* 2012;44:1885-1892.
9. Parmentier S P, Schirutschke H, Schmitt B, et al. Influence of peritoneal dialysis solution on measurements of fluid status by bioimpedance spectroscopy. *International Urology and Nephrology.* 2013;45:229-232.
10. Biesen W et al. Fluid status in peritoneal dialysis patients: The European body composition monitoring (EuroBCM) study cohort. *PLoS ONE.* 2011;6.
11. Kyle U G et al. Bioelectrical impedance analysis - Part I: Review of principles and methods. *Clinical Nutrition.* 2004;23:1226-1243.
12. Lohman Houtkooper L B Going S B. Body fat measurement goes hi-tech: not all are created equal. *ACSM's Health Fitness J.* 1997;1:30-35.
13. Kyle U G, Nicod L, Raguso C, Hans D, Pichard C. Prevalence of low fat-free mass index and high and very high body fat mass index following lung transplantation. *Acta Diabetologica.* 2003;40:S258-S260.
14. Coin A et al. Fat-free mass and fat mass reference values by dual-energy X-ray absorptiometry (DEXA) in a 20-80 year-old Italian population. *Clinical Nutrition.* 2008;27:87-94.

Author: David Naranjo Hernández (Biomedical Engineering Group)
Institute: CIBER-BBN and University of Seville
Street: Engineering School. Avd Descubrimientos s/n
City: Seville
Country: Spain
Email: dnaranjo@us.es

Risk Analysis and Measurement Uncertainty in the Manufacturing Process of Medical Devices

P.E. Baru[1], N.M. Roman[2], and C. Rusu[1]

[1] Technical University of Cluj-Napoca/ Electronics and Telecommunication Faculty, Cluj-Napoca, Romania
[2] Technical University of Cluj-Napoca/ Biomedical Engineering Department, Cluj-Napoca, Romania

Abstract—This paper presents a new risk analysis methodology coming from the measuring process of the medical device characteristics. Thus, one can determine the acceptance criteria for the functioning parameters of the medical device undergoing trials in order to be introduced to the market. We point out the fact that the proposed methodology is complementary to the medical device certification process and provides a technical support for its quality assessment.

Keywords—measurement uncertainty, quality, medical devices, clinical engineering

I. INTRODUCTION

Risk analysis is the activity through which decision makers are offered the information that allows for correct and optimal judgment about technical, economic and opportunity-related matters and that provides an estimation of future effects on variable terms.

For the decision making process to be efficient, a brief presentation of this issue is called for, in a hierarchical or even numerical form, as much as possible.

Measurement being the activity which allows for an objective hierarchization of events proves the importance of the process for risk analysis. Any measurement process is subject to statistical rules and is influenced by errors. Measurement uncertainty points to the error gravity associated with the measurement process.

II. METHODOLOGY

A. Descriptive Rules

The statistical processing of the data resulting from the measurement process involves at least the following stages:
- determining the measurement uncertainty specific to the laboratory
- conducting the actual product measurements
- determining the outliers and the values that do not fit the considered distribution type, respectively
- applying the statistical calculation rules on the value range remaining after removing the outliers

In the case of risk analysis this sequence is not appropriate, as the accident, being part of the risk itself, is a singular and abnormal phenomenon and as such, could be described precisely by the outliers. For this, we endeavor to set a few processing rules for outliers, in concrete cases. The processing outcome is represented by some acceptability criteria for each value. [1]

B. Definition of Measurement Uncertainty

If as a result of a measurement, the measurand is attributed value X, then the measurement result is expressed as: $X \pm u$, u being the uncertainty specified for a certain confidence level P (P=95% in general).

For instance, if while measuring a tension U, the value found is 20.5V and it is estimated that the measurement uncertainty is 0.1V, the result will be expressed as:
$U = 20.5 \pm 0.1$ V, meaning that the actual value of the tension is within the range of 20.4 V and 20.6 V having a 95% confidence level.

The main uncertainty sources are as follows: the object subject to measurement, the measurement means, the device-object interaction and outside influences.

Measurement uncertainty is composed of two elements and is given by eq1:

$$u(X_i) = \sqrt{A^2 + B^2} \quad (1)$$

Type A uncertainty is due to repeatability errors and in order to determine it, at least 10 measurements are made under the same technical conditions.

For type A uncertainty it is considered a normal distribution of the probability that the errors defined in equation 2 and 3 might appear:

$$u_A = \frac{s}{\sqrt{n}} \quad (2)$$

$$s = \sqrt{\frac{\sum (x_i - \overline{x_i})^2}{(n-1)}} \quad (3)$$

Where s is the standard deviation experimentally obtained, x_i is the value being measured, $\overline{x_i}$ is the average value, n represents the number of measurements made and u_A represents type A uncertainty.

Type B uncertainty is generally due to systematic errors and is considered a rectangular distribution of the systematic error probability, which is described in eq4:

$$u_B = \frac{a}{\sqrt{3}} \quad (4)$$

where a represents the range estimated for the systematic error value.

The expanded uncertainty is obtained from combined uncertainty multiplied by the amplification factor k, thus

$$U = k * u(X_i) \quad (5)$$

where k=2 for P=95% and n=5.
The expanded uncertainty is the value which shall be added to the final result of the measurement.

C. Estimating Uncertainty

In order to determine the measurement uncertainty, the following stages must be followed, for each measurement domain:

- identifying the uncertainty sources and determining the measurement equation
- conducting the necessary measurements, generally under the same technical conditions
- estimating type A and B uncertainties for each error source
- determining the correlation between the sources of the determined systematic errors. It is generally accepted that the errors are slightly correlated and as a consequence, type B uncertainty can also be written as follows

$$u_B = \sqrt{\sum_{i=1}^{n} u_i^2} \quad (6)$$

- calculating the correction which can be applied to the result of the measurement, for instance from the measurement device calibration certificate
- drafting the uncertainty budget, where the following error sources are taken into account: repeatability, those belonging to the measuring instrument, the ambient temperature, power supply voltage, manner of reading etc.
- calculating the combined uncertainty, is given by

$$u(X_i) = \sqrt{\sum_{i=1}^{n} c_i^2 * u_i^2} (\delta X_i) \quad (7)$$

where c_i is the sensitivity coefficient, u_i is Type A or type B standard uncertainty.

- calculating the expanded uncertainty by multiplying with k=2 for a P=95% confidence level.

D. Utilizing Measurement Uncertainty

The following prerequisites have been considered in the case of utilizing measurements uncertainty for risk analysis:

- the risks associated with a medical device can be separated into two categories, much the same as the values in operation are within or exceed the domain provided in the technical specification.
- the medical device acceptability criterion is considered as complied with when all the parameters are certainly within the domain provided in the technical specification
- the qualification criterion of the manufacturing process is complied only if the acceptability criterion is met in nine out of ten cases of measurements. If the elements involve the safety of human subjects all ten measurements must be complied by the acceptability criterion. [7]

Knowing that the outliers are either minimum or maximum within the measurement range, these values will be checked, and shall be referred to as min. and max. respectively. We have found that in practice the intervals containing the values of an element, are of two types, either closed or open, to the left or to the right. Using the abbreviation Val for the value provided in the technical specification, in the case of positive values, the acceptability criterion is met in the following cases:

min – U < Val   for the interval open to the left
max +U < Val   for the interval open to the right
max +U < Val   for the interval closed to the left
min – U < Val   for the interval closed to the right
For example:

- the elements defined on open intervals:
The value of an electric current generated by medical device, the voltage value, etc.

- the elements defined on closed intervals:
The value of the using power supply voltage, the using operating temperature domain, etc.

So as to avoid damaging the device, in the case of closed intervals, the attempts will be limited to the following values: max=Val–U, for the interval closed to the left; min =Val+U, for the interval closed to the right.

At one point, these values characterize the measuring and the manufacturing process.

The estimation process must be periodically repeated in order for the information to be relevant. [5]

III. RESULTS AND DISCUSSION

With the purpose of determining the measurement uncertainty, characteristic to the laboratory for final trials, we have used the algorithm described in II C section. Thus, as much as ten measurements have been made for each element, measuring domain and type of measuring device used. The results thus obtained have been processed and the expanded uncertainty values that have been obtained are given in Table 1.

Table 1 Values obtained for the expanded uncertainty

Trial type	Associated Uncertainty
DC voltage in 0.0 ÷ 3 mV domain	0.5 mV
DC voltage in 3.0 ÷ 50 mV domain	6.17 %
DC voltage in 50 ÷ 500 mV domain	0.96 %
DC voltage in 0.5 ÷ 5.000 mV domain	0.18 %
DC voltage in 5.00 ÷ 50.00 mV domain	0.19 %
DC current in 0.0 ÷ 10.0 µA domain	0.5 µA
DC current in 10.0 ÷ 500.0 µA domain	0.34 %
DC current in 500 ÷ 5000 µA domain	1.36 %
DC current in 5 ÷ 50.00 mA domain	0.35 %
DC current in 50 ÷ 500.0 mA domain	0.36 %
Measured period lower than 10 ns	1 ns
Measured period in ns	4.11 %
Measured period in µs	4.06 %
Measured period in ms	3.99 %
Measured period in tens of de ms	2.92 %
Triggering frequency in Hz	0.04 %
Triggering frequency in kHz	0.01 %
Triggering frequency in hundreds of kHz	0.01 %
Triggering frequency in MHz	0.29 %
Measured voltage lower than 10 mV	2 mV
Measured voltage in tens of mV	5.74 %
Measured voltage in hundreds of mV	4.79 %
Measured voltage in V	4.65 %

In order to check the acceptability and qualification criterion, we are considering ten of ATM11 medical devices. ATM11 is a medical device that generates low frequency monopolar-type currents which are used to stimulate muscles and nerves. The outcomes obtained during the trials conducted on the ten medical devices are in Table 2: [6]-[8]

Table 2 Font sizes and styles

Trial Type	Values [mA]		Result
	min	max	
Galvanic current 1mA +/- 0.5 mA	0.91	1.00	pass
Galvanic current 10mA +/- 10%	9.55	9.68	pass
Galvanic current 20mA +/- 10%	19.17	19.45	pass
Diadynamic current 1mA +/- 0.5 mA	0.94	1.11	pass
Diadynamic current 10mA +/- 10%	9.60	9.76	pass
Diadynamic current 60mA +/- 10%	57.61	58.81	pass
Trapezoidal current 1mA +/- 0.5 mA	0.92	0.98	pass
Trapezoidal current 10mA +/- 10%	9.61	9.68	pass
Trapezoidal current 60mA +/- 10%	57.21	58.01	pass

The outcome of the measurements is presented as histograms in Fig. 1a, 1b, 1c

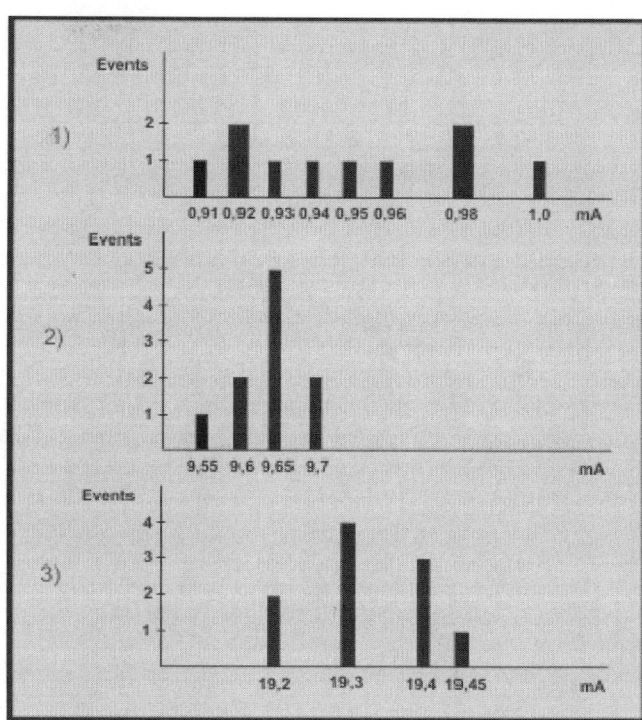

Fig. 1a Measured values for galvanic current

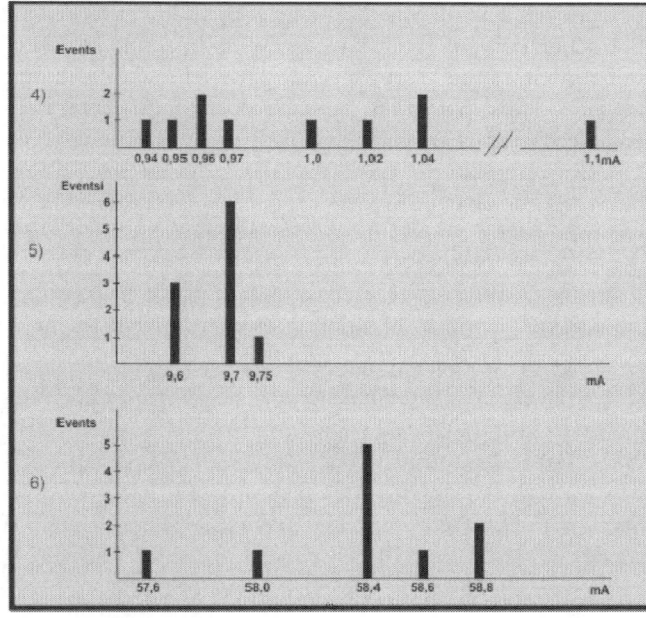

Fig. 2b Measured values for diadynamic current

IFMBE Proceedings Vol. 41

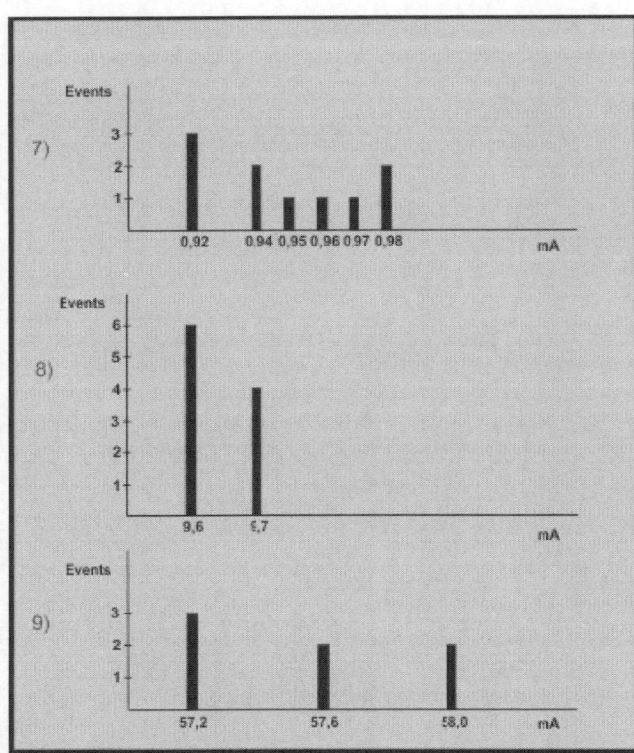

Fig. 3c Measured values for trapezoidal current

The values mentioned in Table 2 are defined for open intervals. It can be noticed that all 10 Type ATM 11 medical devices subject to trials, meet the acceptability criterion and the manufacturing process meets, in its turn, the qualification criterion. The trials concerning the safety of the human subjects have not been mentioned as they have been conduced in another specialized laboratory.

Following the analysis, it can be noticed that all the devices that have been subject to trials, comply with the technical specifications and thus the manufacturing process meets all the conditions required. [4]

If the manufacturing process fails to meet the required conditions, the following solutions can be applied:

- analysis of the required restrictions through the technical specifications using the price/performance criterion, with the purpose of modifying them
- slight change in the values of certain components with the purpose of aligning the intervals which contain the measured values with the interval accepted in the technical specification
- improving the technique used in the laboratory so as to minimize the measurement uncertainty
- replacing some components so as to minimize the dispersion of the measured values
- changing the medical device project

The conditions required before making any change concerning the medical device include: research, validation and acceptance of change.

These activities are conducted by individuals appointed through the quality management system, quality handbook or are appointed during the product certification process. The entire responsibility for complying with the rules and product quality lies with the manufacturer.

IV. CONCLUSIONS

This study renders a point of view regarding the mechanism of quality maintenance and the experience gathered along several years.

The risk analysis methodology suggested above has been discussed and accepted by the notified body, at the ATM11 medical device certification and also during the supervision activities.

The aspects mentioned in the study refer solely to involving the measuring process in maintaining quality for medical devices. [2]

ACKNOWLEDGMENT

We would like to thank Datronix Computer SRL for their support in our research activity during the ATM11 certification.

REFERENCES

1. Baru P, (2012) Risk analyse and measurement. PhD. The second reference, PhD
2. Roman N.M. (2001) Instrumentatie biomedicala ISBN 973-686-168-6 Cluj Napoca, Romania, 2001
3. Council Directive 93/42/EEC of 14 June 1993 concerning medical devices
4. EN ISO 14971:2009 Medical Devices. Application of risk management to medical devices
5. EN ISO 13485:2005 Medical Devices. Quality management systems. Requirements for regulatory purposes
6. EN 60601-1:2001 Medical electrical equipment. General requirements for safety. Collateral standard. Safety requirements for medical electrical systems
7. EN 60601-1:2006 Medical Electrical Equipment – Part 1: General requirements for basic safety and essential performance
8. EN 60601-2-10:2003 Medical electrical equipment - Part 2-10: Particular requirements for the safety of nerve and muscle stimulators

Author: Paul Emanuel Baru
Institute: Technical University of Cluj Napoca
Address: 26-28 Gh. Baritiu, Cluj Napoca, Romania
Email: paulbaru@yahoo.com

Development of an Equipment to Detect and Quantify Muscular Spasticity: Spastimed – A New Solution

V. Fernandes[1,4], I. Clemente[1], C. Quaresma[2,3], and P. Vieira[1,4]

[1] Departamento de física, Faculdade Ciências e Tecnologia,
Universidade Nova de Lisboa – Monte da Caparica, Portugal
[2] CEFITEC, Departamento de física da Faculdade de Ciências e Tecnologia – Universidade Nova de Lisboa – Monte da Caparica, Portugal
[3] Departamento de saúde, Instituto Politécnico de Beja – Campus do Instituo Politécnico de Beja, Beja, Portugal
[4] Centro de Física Atómica, Departamento de Física, Faculdade de Ciências e Tecnologia – Universidade Nova de Lisboa – Monte da Caparica, Portugal

Abstract—Spasticity is a known muscular tonus velocity dependent alteration which quantification in clinical practice is still based on subjective perception and scale grading through procedures that lack controlled protocols. In the research field, both physician's and engineer's researches have pointed the potentialities in the use of biomechanical magnitudes and their physiological meanings as much lesser subjective means of quantifying spasticity as well as its effects on patient daily life. Last, but not less important, this scientific and clinic urge is also justified by the high costs of treatments as well as the very tight relation they express between effectiveness and applied dose. As a consequence, this team of developers has been focused in creating a device to detect spasticity. During the validation of a first prototype with a small set of subjects, the obtained results were satisfyingly good as the device correctly detected 89% of the spastic subjects and 82% of the non-spastic subjects. Even so, the limitations found in the prototype concept itself led to a new development phase that resulted on a very different approach. SpastiMed, a motorized and electronically controlled device which is still on its validation phase but already showing an immense potential.

Keywords—**Biomedical Engineering, Medical device, Biomechanics, Muscular tonus, Spasticity assessment.**

I. INTRODUCTION

Spasticity consists on a muscular tonus alteration caused by a flawed neural central system [1] that fails to do its role in muscular inhibition which leads to a loss of the natural balance between activation and inhibition causing this hypertonic phenomenon [1]. Spasticity is normally perceptible by the rise of sensed "rigidity" during the passive mobilization of an affected limb [1]. This "rigidity" has been known to be velocity dependent [1] and can affect negatively the life of patients by turning daily simple tasks into hard or even impossible ones, greatly compromising the patient independency [2].

There are three main classes of methods used for muscular spasticity quantification: the *Likert* scales [3-8], electrophysiological measurements [4, 9] and biomechanical measurements [4, 6, 9]. The most used in clinical procedure are the *Likert* scales, especially the Modified Ashworth Scale (MAS) [4-8, 10] and Modified Tardieu Scale (MTS) [3, 7]. Both these scales quantify spasticity based on the perception of the muscular response trough the passive mobilization by an operator with no resource to any precise measuring tools [3-7]. In the research field, many equipment have been built based on the quantification of biomechanical magnitudes that have shown a direct relation with spasticity [4, 7, 11-16]. Most of these equipment had either unappropriated size for clinical usage, small inter and/or intra-patient result correlation, or a noticeable result dependence on the operator.

Like many other types of pathologies which assessment was or still is made by the use of subjective methods, spasticity assessment and quantification has been the target of many physicians and engineers with the goal of instrumenting it and turning it into an objective and precise process [4, 7, 11-16].

II. A FIRST STEP – "THE GLOVE"

Concept Design: The goal of this project was to develop a low-cost and easy-to-use prototype to detect the presence of muscular spasticity and, if possible, to do its quantification by grading it in a 6 levels custom scale (Inexistent, Very Slight to inexistent, Very Slight, Slight, Moderate and Severe) [12].

Through many studies, the few products and prototypes that showed potential for clinical use also expressed that one of their main problems was the need to create and apply a new protocol to attain a better spasticity measure aided by a tool in daily clinical procedures. For this reason, the design and conceptual idea was based on a main principle: "To build something that physicians could use without interfering with their usual protocol for spasticity assessment" [12]. As a result, the author [12] idealized a hand-held device that could be used trough the normal procedure. Based on its look, this prototype was nicknamed: "The Glove".

The final state of this prototype can be seen on Fig. 1 and can be swiftly described as a glove with a small water bag coupled with a pressure sensor connected to an acquisition system (MicroChip® PIC® based) in communication with a personal computer where a graphic interface developed in MATLAB® presents real time data to the user [12].

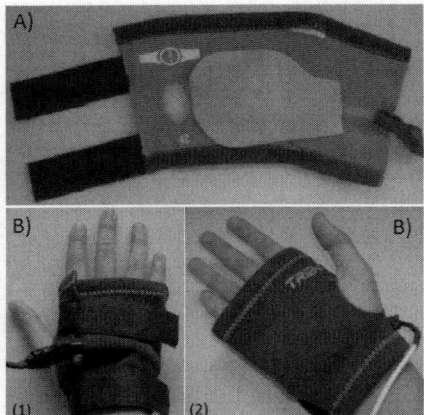

Fig. 1 Device developed by Inês Clemente [12] to assess muscular spasticity A) Inside view of "The Glove" (Small water bag inserted in between the glove tissue and the yellow cover tissue). B) Outside view of the Glove, (1) back of the hand (2) and palm view [12].

Results: To validate the equipment it was used to examine a small set of 29 subjects with ages comprised between 18 and 60 years old which expressed their ethic and informed consent on the procedure [12]. Among these, 4 had clinical record of muscular spasticity on flexor muscles and 25 subjects were deemed as having no muscular spasticity in

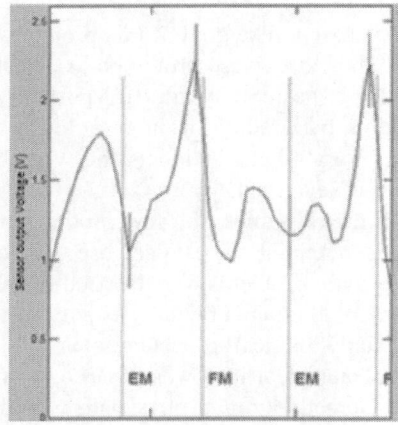

Fig. 2 Signal obtained from a healthy subject. The red markings represent the moments of max extension (EM) and max flection (FM) for better understanding of the signal and its peaks [12].

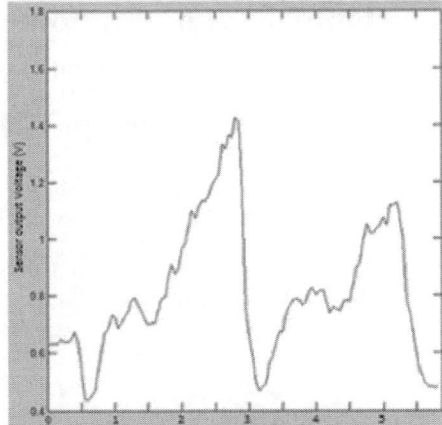

Fig. 3 Signal obtained from a subject with previously detected moderate upper limb flexor muscle spasticity [12].

the upper limb muscles [12]. Previously to the use of "The Glove", all of these 29 subjects were re-evaluated by a physician and their muscular spasticity was graded according to the before mentioned custom scale. In total, the team collected a pool of 83 (Physician + "The Glove") sets of data, 9 sets from the 4 spastic subjects and 74 sets from the 25 non-spastic subjects [12].

The evaluation procedure was done with the patient seated and starting with the limb on its max flexion parallel to the body [12]. During the data acquisition period the user, wearing "The Glove" should fully extend and fully flex the limb and repeat this cycle 5 times with the same time length trough cycles, if possible [12].

One characteristic signal resulting from the examination of a non-spastic subject can be seen in Fig. 2 and one in a spastic subject can be seen in Fig. 3. The signal analysis focused only on the peaks expressed during the flexion movements because of the spasticity presented by the previously mentioned evaluated subjects [12]. Those mentioned peaks were averaged and the resulting peak was parametrically adjusted to the first half-period of a sine wave function. The parameter that showed a good potential for the grading was the angular frequency (b_1). For that parameter and based on the physician classifications the author [12] b_1 gaps were defined for each degree of muscular spasticity and finally the results obtained with "The Glove" were given a grade. Those results are presented in Table 1.

Conclusions: Even if these results can't be considered statistically sound due to the small number of sets of data, from Table 1 we can conclude that "The Glove" was able to assess in accordance with the physician 74.6% of the "Inexistent" cases, 33.3% of the "Very Slight to Inexistent", 66.7% of the "Very Slight" cases as well as 100% of all the cases in the "Slight" and "Moderate" cases. As a global result, the equipment was able to detect correctly 89% of the spastic subjects and 82% of the healthy subjects.

During validation "The Glove" revealed itself prone to generate signals with considerably different shapes based on a simple change of user. This limitation was circumvented by doing this validation with just one operator [12] and relegating for later the development of a better software based on a more extensive signal analysis to fix this problem.

Table 1 Comparison of the results obtained by the equipment versus a trained physician in muscular spasticity assessment [adapted from 12].

Physician Evaluation Result		"The Glove" Evaluation Result		
Degree of Muscular Spasticity	no. of cases	Degree of Muscular Spasticity	Cases	
			no.	%
Inexistent	71	Inexistent	53	74.6
		Very Slight to Inexistent	6	8.5
		Very Slight	3	4.2
		Slight	9	12.7
Very Slight to Inexistent	3	Inexistent	1	33.3
		Very Slight to Inexistent	1	33.3
		Very Slight	1	33.3
Very Slight	3	Very Slight to Inexistent	1	33.3
		Very Slight	2	66.7
Slight	4	Slight	4	100
Moderate	2	Moderate	2	100

Another limitation was the before mentioned quantification scale which was adopted with the goal of being in accordance with the rating used in the *Hospital Curry Cabral* services of Physical Medicine and Rehabilitation where the instrument validation was made. In the end it revealed to have an excessive number of grades (four) from "Inexistent" to "Slight" which surely aggravated the flawed differentiation in between them which is a common problem even among all existing *Likert* scales and documented prototypes. Also, the fact that this scale it is not a standard scale, left an undeniable need for another study comparing this device results to a "gold standard" so it could be compared with other instruments as well as the most used *Likert* scales. Other identified limitations like the lack of velocity control and angular monitoring added to the above mentioned problems lead the team to a new development phase.

III. THE NEXT STEP – SPASTIMED

Following the limitations found on the conceptual idea behind "The Glove" as well as the improvable aspects in the prototype the final decision was to put this prototype development in standby and try a different approach. The team gathered all the crucial knowledge acquired previously and synthetized it to decide what the new approach should be based on. As a result, 5 main features were identified and "placed on the drawing table" as key-point characteristics that this new prototype should attain. These key-points were: being of easy and fast application, no need for a specialized operator, being portable, present a good result reproducibility and a good independency from the operator in the results produced.

After coming up with a first drawing of the prototype, the team applied for a short intern program lectured by physicians in *Centro de Medicina de Reabilitação de Alcoitão*, where they could contact with spastic patients as well as gather the physicians opinion on the prototype drawing and future perspective of using such tool. Their answers were highly considered and took a major role into the final design itself.

The whole prototype can be understood trough the block diagram presented in Fig. 4. To resume, the mechanical part (Fig. 5) consists on a DC motor and a set of gears attached on top of a metallic articulated arm responsible for making the patient limb move passively while a set of electronic circuits collect data like the motor consumed current (proportional to the torque output) and the metallic arm angle. All the data is collected by a MicroChip® PIC18F877A microcontroller and transmitted through USB communication to a control user interface developed in LabVIEW™ running on a personal computer where the muscular spasticity is assessed and quantified based on biomechanical velocity dependent information, based mainly in the incredible tool built to study spasticity made by *Ju MS et al.* [11], is extracted from the data.

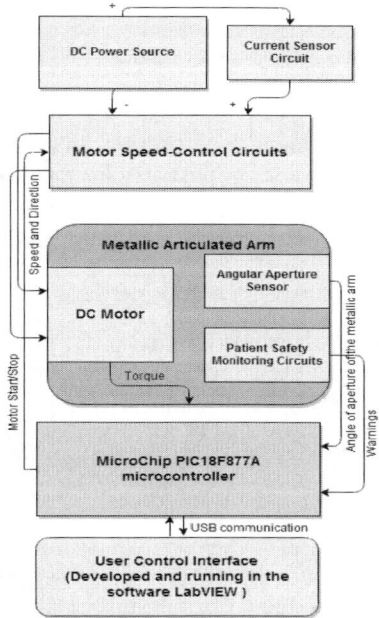

Fig. 4 Block Diagram of the SpastiMed device.

The development of this equipment was recently finished and it is now under its validation phase which is to be done by crossing the instrument results with results obtained with the clinically most used *Likert* scales: MAS and MTS.

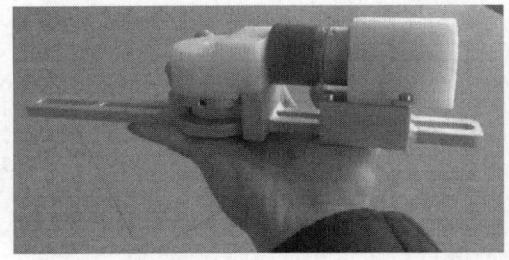

Fig. 5 Mechanical part of the SpastiMed equipment.

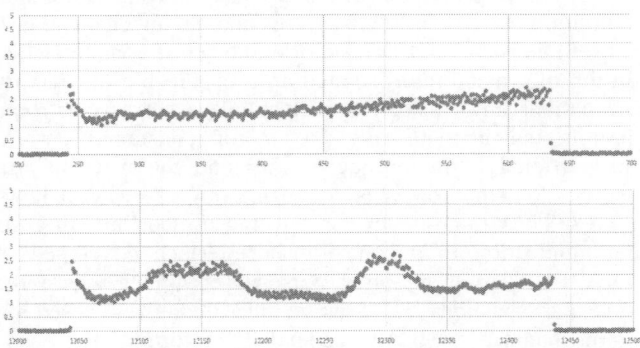

Fig. 6 Signals acquired from a healthy subject during an extension movement induced by SpastiMed. Relaxed arm (top). Small opposing force spikes (bottom). Axis labels: Y – current (A) X – time (cs)

In Fig.6, two different motor consumed current signals can be seen and the sensitivity of the device is evidenced. Last but not least, a safety protocol was developed into the software to ensure a painless and safe evaluation complemented by a set of push buttons that both the operator and the patient can click at any time to order the device to stop the motion.

IV. CONCLUSIONS

Even though the study on physically, physiologically and mathematically explained quantification methods for spasticity has started nearly half a century ago, not even a single device has been able to turn into a clinical new standard till now. With the development in the area of electronics and the decrease in the costs of components arose the possibility of producing smaller, smarter and more precise instruments to quantify spasticity and other pathologies from which scientific knowledge is still scarce. This last fact is mainly one of the causes that turns biomedical instrumentation into such an interesting area and with such relevance for both the improvement in quality of daily clinical procedure and in diluting barriers in scientific investigation towards the understanding of many biomechanical pathologies.

ACKNOWLEDGEMENTS

The authors would like to thank:

Centro de Medicina de Reabilitação de Alcoitão and all the personal involved in spastic evaluation and treatments: for the time and interest they presented in giving us such a valuable perspective as their perspective as physicians and possible future user of this kind of equipment.

NGNS - Ingenious solutions, LDA: for all the help, engineering knowledge, work space and tools provided that made this work possible.

REFERENCES

1. **Bhimani, RH., et al.** Clinical measurement of limb spasticity in adults: state of the science. *J Neurosci Nurs.* **2011** Apr;43(2):104-15.
2. **Adams, M., Hicks, A.** Spasticity after spinal cord injury. Spinal Cord. **2005**, 43(10): 577-586.
3. **Rekand, T.** Clinical assessment and management of spasticity: a review. *Acta Neurologica Scandinavica.* **2010**, 122(190): 62-66.
4. **Pandyan AD., et al.** Biomechanical examination of a commonly used measure of spasticity *Clin Biomech (Bristol, Avon).* **2001** Dec;16(10):859-65.
5. **Pizzi, A., et al.** Evaluation of Upper-Limb Spasticity After Stroke: A Clinical and Neurophysiologic Study. *Archives of Physical Medicine and Rehabilitation.* **2005**, 86(4): 410-415.
6. **Nordmark E. and Anderson G.** Wartenberg pendulum test: objective quantification of muscle tone in children with spastic diplegia undergoing selective dorsal rhizotomy. *Dev Med Child Neurol.* **2002** Jan;44(1):26-33.
7. **Supraja M, Singh U.** Study of Quantitative Assessment of Spasticity by Isokinetic Dynamometry. *IJPMR* 14, April **2003**; 15-18
8. **Ghotbi N., et al.** Measurement of lower-limb muscle spasticity: intrarater reliability of Modified Modified Ashworth Scale. *J Rehabil Res Dev.* **2011**;48(1):83-8.
9. **Pisano, F., et al.** Quantitative measures of spasticity in post-stroke patients. *Clinical Neurophysiology.* **2000**, 111(6): 1015-1022.
10. **Abolhasani H, et al.** Comparing the validity of the Modified Modified Ashworth Scale (MMAS) and the Modified Tardieu Scale (MTS) in the assessment of wrist flexor spasticity in patients with stroke: protocol for a neurophysiological study. *BMJ Open.* **2012** Nov 19;2(6).
11. **Ju MS., et al.** Time-course analysis of stretch reflexes in hemiparetic subjects using an on-line spasticity measurement system. *J Electromyogr Kinesiol.* **2000** Feb;10(1):1-14.
12. **Clemente I.,** Construção de um dispositivo de avaliação da espasticidade **2010**. Dissertação apresentada na FCT/UNL para obtenção do grau de Mestre em Engenharia Biomédica.
13. **Lehmann JF., et al.** Spasticity: quantitative measurements as a basis for assessing effectiveness of therapeutic intervention. *Arch Phys Med Rehabil.* **1989** Jan;70(1):6-15.
14. **Chen JJ.,** et al. The use of a portable muscle tone measurement device to measure the effects of botulinum toxin type a on elbow flexor spasticity. *Arch Phys Med Rehabil.* **2005** Aug;86(8):1655-60.
15. **Leonard CT., Mikhailenok EL., inventors.** Apparatus for measuring muscle tone. US patent 6,063,044. **2000 May 16**.
16. **Leonard CT., et al.** Assessing the spastic condition of individuals with upper motoneuron involvement: validity of the myotonometer. *Arch Phys Med Rehabil.* **2001** Oct;82(10): 1416-20.

Comparison between Two Exercise Systems for Rodents

T. Ödman[1], N. Ödman[2], E. Rabotchi[3], S. Åkervall[4], and M. Lindén[5]

[1,5] Mälardalen University, Box 883, 721 23, Västerås, Sweden
[2,4] Södra Älvsborgs sjukhus, Brämhultsvägen 53, SE-50182, Borås, Sweden
[3] Ericsson AB, Sandlidsgatan 3, SE-50111, Borås, Sweden

Abstract—This paper describes a comparison between two measurement systems for medical endurance test for rodents, a new running wheel system and an old system named ADEA. The ADEA test chamber system has been criticized because it is cumbersome and involves a lot of manual work which therefore limits the number of test cells that can be run simultaneously. Comparison between the systems has been done via measurements of dopamine levels and thereafter calculating correlation factor. The running wheel system has shown to provide at least as good physical strain as the well tested ADEA system.

Keywords—Dopamine, Infra-Red (IR), Locomotion, Running Wheel, Wireless.

I. INTRODUCTION

The goal of this study was to compare two different systems for medical endurance tests for rodents. The ADEA system has been widely used in about 300 research publications since 1986. The new system, a rodent exercise wheel, is a wireless system with possibilities to easily be expanded by increasing the amount of measurement stations, see figure 1. Further, the amount of manual work can be reduced and the measurements can be quantified and presented in the SI units meter and meter per second. The ADEA system, on the other hand presents the result in arbitrary units based on counting the number of IR beam interrupts, related to the numbers of revulsions.

Both systems are used at Uppsala University Hospital, department of Neurology.

Research at the department has been focused on the effects of hormones and on physical endurance of rodents.

The two systems were compared in a setting where dopamine depletion was measured against the degree of physical activity performed by the rodents.

After measurement of dopamine a correlation coefficient was calculated. These coefficients were used for comparison between the systems, since the running wheel system presents its result in the unit of meters and the ADEA system presents its result in arbitrary unit.

There are other systems for medical endurance test for rodent [1, 2]. These articles describe measurements of only one type of cage [1, 2]. Nevertheless, a variety of systems exist on the market that has been designed for multiple cages concurrently. Many of the activity wheel cages are connected to a wireless system linked to a management program on a personal computer (PC). Neither from the descriptions obtained from the articles nor from the market analysis performed, it was possible to identify an exercise measurement system consisting of mechanical running wheels, that could be enlarged to accommodate a dozen or even a hundred cages. (termed measurement stations).

Fig. 1 The test setup of the PC to left, under the PC on the lowest bench is the broadcasting unit.

II. SYSTEMS

ADEA, activity test chambers, see figure 2, is an automated device, consisting of polycarbonate (Makrolon) rodent test cages (400 x 250 x 150 mm) each placed within two series of infra-red beams (at two different heights, one low and one high, 20 and 80 mm, respectively, above the surface of the 10 mm deep sawdust). The distance between the infra-red beams was as follows: the low levels beams were 73 mm apart lengthwise and 58 mm apart breadthwise in relation to the test chamber; the high level beams, placed only along each long side of the test chamber were 28 mm apart. According to the procedures described previously [3-8], the parameter locomotion was measured by the low grid of infra-red beams. Counts were registered only when the mouse, in the horizontal plane, were ambulating around the test-cage.

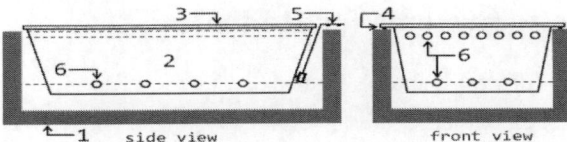

1. Base support for cages and sensors.
2. Transparent plastic cage.
3. Perforated aluminium lid.
4. Elastic rubber support.
5. Pick-up mounted on lever with counterweight.
6. Infrared (IR) detector.

Fig. 2. Components in the ADEA system.

The new wireless rodent exercise wheel has a test setup comprising a PC for the management software and database. The management software set up the test time, handles the database, and gets data from each test station, such as p if a test unit is connected and also checks the status of the test unit. A broadcasting unit (BCU) is connected the PC. It broadcasts the total test time to all test stations. The transmitter at the test station is named measurement unit (MU), see figure 3. It counts down the testing time to zero and also counts the number of interrupts caused by the broken IR beam. One turn of the rotating wheel is corresponding to one count. The number of turns is sent back to the BCU/PC and thereafter a calculation of the distance is done. The unit of the result is in meters.

The physical size of the rotation wheel has, in this setup, a width of up to 0.3 m.

Fig. 3. MU, the IR diode mounted on metallic arm.

III. MATERIAL

Male C57 Bl/6 mice were purchased from B&K, Sollentuna, Sweden, and were maintained, five-to-a-cage, in plastic cages in a room at temperature of $22 \pm 1°C$ and a 12/12 hours constant light/dark cycle (lights on between 06.00 and 18.00 hours). They were placed and maintained in groups of 4 to 6 animals in a room, maintained for male mice only, following arrival at the laboratory for about 2 weeks in order to acclimatize.

All mice had free access to food and water and were maintained throughout, except for the day previous to the initiation to wheel-running exercise which occurred at the end of the second week following arrival.

They were housed in groups of 6 animals, wheel-running exercised and activity chamber tested only during the hours of light (08.00-15.00 hours). All exercising and testing was performed in a normally lighted room.

Experiments were carried out in accordance with the European Communities Council Directive of 24th November 1986 (86/609/EEC) after approval from the local ethical committee (Uppsala University and Agricultural Research Council), and by the Swedish Committee for Ethical Experiments on Laboratory Animals (license S93/92 and S77/94, Stockholm, Sweden).

IV. EXPERIMENT

Ten mice were used in the ADEA system and compared to ten mice in the new system. The ADEA system measured the numbers of movements (locomotions) for 20 minutes and in the new system the mice run in a wheel for 20 minutes. The experiment was repeated five times.

The experiment consisted of five groups of mice with five different health statuses. The mice within each group had been regarded to have the same health status.

V. NEUROCHEMICAL

The mice were killed by cervical dislocation within two weeks of completion of behavioral testing. Determination of dopamine (DA) was performed using a high-performance liquid chromatograph with electrochemical detection (HPLC-EC).

Striatal regions were rapidly dissected out and stored at -80°C until neurochemical analysis. DA concentration was measured as follows: The frozen tissue samples were weighed and homogenized in 1 ml of 0.1 M perchloric acid, and alpha-methyl-5-hydroxytryptophan was added as an internal standard. After centrifugation (12 000 rpm, i.e. 18 600 g, 4°C, 10 min) and filtration, 20 µl of the supernatant was injected into the HPLC-EC to assay DA. The HPLC system consisted of a PM-48 pump (Bioanalytical Systems, BAS) with a CMA/240 autoinjector (injection volume: 20 µl), a precolumn (15 x 3.2 mm, RP-18 Newguard, 7 µm), a column (100 x 4.6 mm, SPHERI-5, RP-18, 5 µm), and an amperometric detector (LC-4B, BAS, equipped with an

Ag/AgCl reference electrode and a MF-2000 cell) operating at a potential of +0.85V. The mobile phase, pH 2.69, consisted of K2HPO4 and citric acid buffer (pH 2.5), 10% methanol, sodium octyl sulphate, 40 mg/l, and EDTA. The flow rate was 1 ml/min, and the temperature of the mobile phase was 35°C.

VI. RESULT

Table 1 shows four columns, groups, DA vs DIST, DA vs LOCOM and DIST vs LOCOM. The first column named "Groups" consists of five groups with ten mice in each group. The second column "DA v DIST" shows correlation coefficient of dopamine (DA) when the mice have been in the running wheel. Distance (DIST) is in unit of meter

The third column "DA vs LOCOM" shows correlation coefficient of dopamine when the mice have been in the ADEA test chamber. Locomotion (LOCOM) is measured in number of interrupted IR beams. The last column shows a comparison between the two systems. Figure 4 to 8 shows measured values of locomotion and distance in group 1 to 5. The range of dopamine has been from 1.08675 ng/ml to 7.99282 ng/ml.

Table 1. Correlation coefficient, (R) of dopamine and correlation coefficient between two measurement systems.

Groups	DA vs DIST (R)	DA vs LOCOM (R)	DIST vs LOCOM (R)
1	0.965	0.926	0.916
2	0.966	0.912	0.926
3	0.942	0.898	0.923
4	0.864	0.882	0.943
5	0.926	0.969	0.967

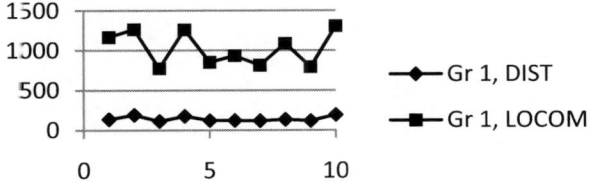

Fig 4. The diagram shows data of distance and locomotion on the Y axis related to the mouse on X axis in group 1.

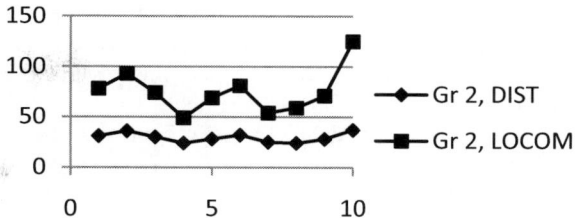

Fig 5. The diagram shows data of distance and locomotion on the Y axis related to the mouse on X axis in group 2.

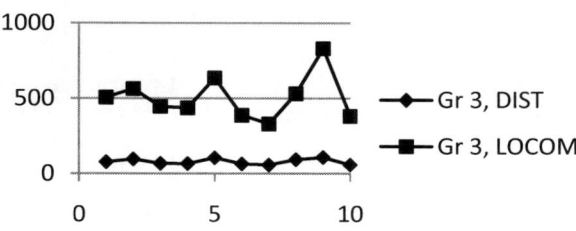

Fig 6. The diagram shows data of distance and locomotion on the Y axis related to the mouse on X axis in group 3.

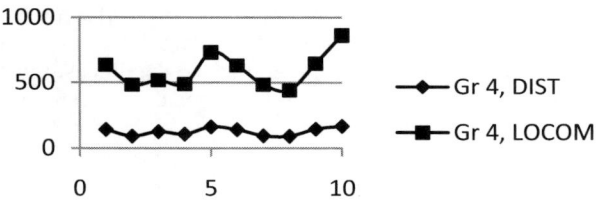

Fig 7. The diagram shows data of distance and locomotion on the Y axis related to the mouse on X axis in group 4.

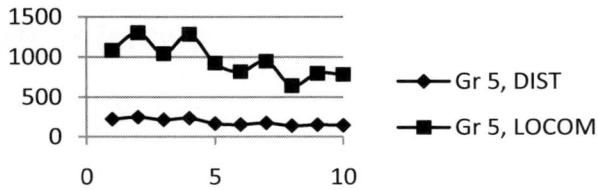

Fig 8. The diagram shows data of distance and locomotion on the Y axis related to the mouse on X axis in group 5.

VII. DISCUSSION

The correlation coefficients, presented above, all confirm that the rodent exercise running-wheel measurements were associated with measurements of locomotion counts in the activity test chambers to extremely marked extents, both within each group (N = 10 mice), and over the whole experimental population (N = 100 mice).

This result offers evidence for the reliability of the rodent exercise running-wheel system since the ADEA system has been used since 1986.

VIII. CONCLUSION

The correlation coefficient in the groups indirectly shows that the running wheel system provides at least as good physical strain compared to the well tested ADEA system.

However, physical performance can be better assessed with the new system, since both the distance and velocity can be quantified. These parameters are not available in the ADEA system.

ACKNOWLEDGMENT

Thanks to my daughters Amalia and Alina who tested interrupt generation of IR. Thanks also to Paul Joonas for valuable helping with technical questions about radio communications. Valuable thanks to MILMED AB, Anders Fredriksson (Uppsala Universitetssjukhus) and Trevor Archer (Göteborgs Universitet). They were fully responsible for the experiment and acquisition of the data measurements and gave information about neurochemical, material and the ADEAsystem in the article.

CONTRIBUTION TO AUTHORSHIP

S.Å, N.Ö and M.L have done correction of the text. M.L has also helped to guide us in this research. TÖ and ER have been designing the hardware, software system and communication protocols of the new running wheel system. TÖ and ER have also performed work of earlier systems and assisted in writing the article.

REFERENCES

[1] N.M. Bleehen, T.H.E. Bryant, R. Gallear, An Exercise cage designed to enable continuous electrical measurements to be made on small animals, Technical Note, (1966) Biological Medical Engineering Vol.4 (N.o.5) pages 505-506

[2] Oddist D. Murphree, Jack B. Johnson (1974) An Inexpensive Activity Measuring Device for Small Animals, Biological Medical Engineering integrative psychological and behavioral science Vol 9 (N.o.3) pages 169-171

[3] Patel JC, Rossignol E, Rice ME, Machold RP (2012) Opposing regulation of dopaminergic activity and exploratory motor behavior by forebrain and brainstem cholinergic circuits. Nat Commun. 2012;3:1172. doi: 10.1038/ncomms2144.

[4] Perlow MJ, Freed WJ, Hoffer BJ, Seiger A, Olson L, Wyatt RJ (1979) Brain grafts reduce motor abnormalities produced by destruction of nigrostriatal dopamine system. Science 204(4393), 643-7.

[5] Rousseaux MW, Marcogliese PC, Qu D, Hewitt SJ, Seang S, Kim RH, Slack RS, Schlossmacher MG, Lagace DC, Mak TW, Park DS (2012) Progressive dopaminergic cell loss with unilateral-to-bilateral progression in a genetic model of Parkinson disease. Proc Natl Acad Sci U S A. 2012 Sep 25;109(39):15918-23.

[6] Espadas I, Darmopil S, Vergaño-Vera E, Ortiz O, Oliva I, Vicario-Abejón C, Martín ED, Moratalla R (2012) L-DOPA-induced increase in TH-immunoreactive striatal neurons in parkinsonian mice: insights into regulation and function. Neurobiol Dis 48, 271-81. doi: 10.1016/j.nbd.2012.07.012.

[7] Plowman EK, Maling N, Rivera BJ, Larson K, Thomas NJ, Fowler SC, Manfredsson FP, Shrivastav R, Kleim JA (2013) Differential sensitivity of cranial and limb motor function to nigrostriatal dopamine depletion. Behav Brain Res. 2013 Jan 15;237:157-63. doi: 10.1016/j.bbr.2012.09.031.

[8] Archer T, Fredriksson A (2010) Physical exercise attenuates MPTP-induced deficits in mice. Neurotox Res 18, 313-327.

Tripolar Flexible Concentric Ring Electrode Printed with Inkjet Technology for ECG Recording

Y. Ye-Lin[1], E. Senent[1], G. Prats-Boluda[1], E. Garcia-Breijo[2], J.V. Lidon[2], and J. Garcia-Casado[1]

[1] Grupo de Bioelectrónica (I3BH, Universitat Politècnica de València), Valencia, Spain
[2] Centro de Reconocimiento Molecular y Desarrollo Tecnológico, Unidad Mixta UPV-UV, Valencia, Spain

Abstract—Laplacian potential on the body surface can be estimated by means of concentric ring electrodes. It has been proved that bioelectrical surface recordings performed with such concentric ring electrodes improve spatial resolution compared to unipolar or bipolar disc electrodes. However, concentric ring electrodes have commonly been implemented in rigid substrates. Therefore they do not adapt to the body surface curvature what provokes discomfort to the patient and a poor contact that affects the signal quality. The aim of this work was to develop a new active concentric ring electrode on a flexible substrate for obtaining high spatial resolution ECG recording. Specifically the active electrode is made up of a disposable tripolar concentric ring electrode printed onto polyester film using inkjet technology, and a reusable signal preconditioning circuit. Simultaneous laplacian ECG recordings with the developed electrode and reduced Lead-I ECG with conventional disc electrodes were conducted in 7 subjects. The results show that the flexible active sensor can be used for picking up laplacian ECG with a similar signal quality than standard Lead-I ECG. This sensor could be used for ECG monitoring with enhanced spatial resolution.

Keywords—Flexible electrode, concentric ring electrode, laplacian ECG, inkjet technology.

I. INTRODUCTION

Body surface cardiac electrical activation evaluation is currently of great interest for clinical diagnosis of a wide range of cardiac abnormalities including infarction, arrhythmia, etc. Although body surface potential mapping (BSPM) can provide additional information to that contained in the conventional electrocardiogram or vectorcardiogram [1], spatial resolution of BSPM is still limited due to the smearing effect caused by the torso volume conductor. In this respect, considerable efforts have been made to study the feasibility of body surface Laplacian electrocardiogram (LECG) to localize bioelectrical sources more accurately. The Laplacian potential is proportional to the derivative of the normal component of the current density at the body surface and can be interpreted as a filter that allocates more weight to the bioelectrical dipoles adjacent to the recording points [2]. Laplacian potential on the body surface can be estimated by means of concentric ring electrodes in bipolar configuration (BCR, one disc and one outer ring), tripolar configuration (TCR, one disc and two outer rings), and tripolar ring electrode in quasi-bipolar configuration (TCB, in which the outer ring and inner disc are electrically shorted); being the TCR the one that provides best spatial resolution [3], [4]. Nevertheless, these electrodes have commonly been developed on rigid substrates [3], [4], which limits electrode adaptability to the body contours and patient comfort, and provide signals of poorer quality [5].

In this context, inkjet printing has emerged as an alternative technique for manufacturing electronic devices that could be implemented on flexible substrates [6,7]. This technique consists in spraying ink through many tiny nozzles on the print head, placing the drops precisely where they are needed. Therefore, the aim of this work is to develop and test a new flexible active TCR electrode using inkjet technology, and to check its feasibility for LECG recording.

II. MATERIAL AND METHODS

A. TCR Electrode Design & Development

The active sensor is made up of two parts: a disposable sensing TCR electrode printed on flexible substrate, and a reusable battery powered signal conditioning circuit that filters and preamplifies the biosignals before transmission.

The sensing part consisted of two hook-shaped electrodes and an inner circular-shaped electrode (see Fig. 1) so as to be implemented using a monolayer design (no bias are needed). The outer diameter of the outer ring was set to 36 mm which is a compromise between signal amplitude and spatial resolution. Technical and physiological issues had been considered in the TCR electrode design [3]:

- The areas of the outer ring (A_{out}), the middle ring (A_{mid}) and the inner disc (A_{in}) should be equal so as to provide similar electrode impedance.
- The interelectrode distance (D) between the inner disc and middle ring should be the same as the distance between the middle ring and the outer ring to obtain a TCR Laplacian estimation, that is, $D = b-a = d-c$.

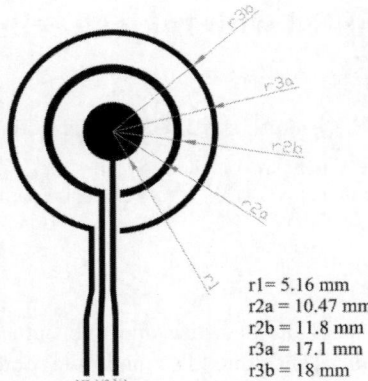

Fig. 1 Tripolar concentric electrode for monolayer implementation.

r1 = 5.16 mm
r2a = 10.47 mm
r2b = 11.8 mm
r3a = 17.1 mm
r3b = 18 mm

Other issues such as a minimum recording area of 50 mm² and a minimum ring width of 0.6 mm were also considered to ensure signal acquisition.

The flexible electrodes were implemented by inkjet printing technology. Firstly, the electrode pattern with 600 dpi resolution was obtained. Then nanoparticles Silver ink DGP40LT-15C (Advanced Nano Product, Korea) was printed onto polyester film (MelinexST506, Dupont) using Dimatix materials printer DMP-2800 (FujiFilm, USA). The curing temperature of conductive ink was set to 150°C. The printing process was repeated twice to achieve about 2 μm of ink thickness in order to guarantee electrode electrical continuity.

For the implemented TCR electrode, it was designed two ultra-high input impedance amplifiers [5] that performed the differential potential between the middle ring and inner disc between the middle ring and inner disc (V_2-V_1) and between the outer ring and inner disc (V_3-V_1). Each amplifier provide a gain G_T = 105 V/V and a bandwidth of 0.05-482.3 Hz. In this way, two bipolar laplacian potential and one tripolar laplacian potential can be estimated from each TCR electrode as follows [4]:

$$L_{BipA} \cong -\frac{1}{r^2}\cdot(V_2-V_1) \quad (1)$$

$$L_{BipB} \cong -\frac{1}{4r^2}\cdot(V_3-V_1) \quad (2)$$

$$L_{Trip} \cong -\frac{1}{3r^2}\left[16(V_2-V_1)-(V_3-V_1)\right] \quad (3)$$

Where V_1, V_2, V_3 are the biopotential picked up by the inner disc, middle ring and outer ring of the TCR electrode respectively, and r is the mean radius of the electrode.

In this work the scaling factors associated to the radius of the electrode are not considered so as to be able to directly compare the amplitudes of the recorded signals.

B. ECG Signal Acquisition

Seven recording sessions were carried out on healthy volunteers (2 women and 5 men) lying in supine position. While recording, the subjects were asked to relax and remain stationary to avoid the influence of fluctuation sof heart position on the body surface ECG and LECG. Skin was exfoliated and chest of male volunteers was shaved to reduce contact inmpedance. Reduced Lead-I ECG using commercial disposable Ag/AgCl wet electrodes and two LECG using dry TCR electrodes were simultaneously recorded. One TCR electrode was placed 6 cm above and 2 cm to the right of the left nipple (upper position, TCR1), and the second one 6 cm below the nipple and aligned with the first electrode (lower position, TCR2). The signals from TCR and disposable electrodes were amplified in the bandwidth 0.1-100 Hz, with gains of 50 and 500 respectively, by means of commercial amplifiers P511 (Grass Technologies, Warwick, USA) signals were simultaneously recorded using a sample rate of 1 kHz.

C. Data Processing

ECG is contaminated by different types of interferences such as baseline drift, power line interference and muscle noise. In order to compare tripolar LECG signal quality with Lead-I ECG with standard electrodes several parameters were worked out. Firstly, a fifth-order Butterworth high pass filter with a cut-off frequency at 0.5 Hz was applied to surface signals, giving rise to a filtered signal $x(t)$. ECG fiducial points were obtained by detecting the R-wave of the ECG signal with the algorithm proposed by Hamilton and Tompkins (1986). Then the averaged beat extending from 275 ms prior to the R-wave to 400 ms after it, was calculated in analysis windows of 60 s. Then the median values of the following parameters were obtained for each recording session:

- Signal amplitude of the averaged beat ($\overline{ECG}$).

- Signal-to-noise ratio (SNR). The noise was estimated as the difference between the filtered surface signal x(t), and the estimated target signal (ECG(t)), this last obtained by assigning the averaged beat to each detected beat,

$$SNR(dB) = 20\cdot\log_{10}\left(\frac{V_{rms}(\overline{ECG})}{V_{rms}(x(t)-ECG(t))}\right) \quad (4)$$

- Signal-to-baseline ratio (SBR). Similar to SNR but considering that the baseline drift is the components below 0.5 Hz.

III. RESULTS

Fig 2 shows the TCR electrode implemented using ink injection technology. The electrode flexibility that permits its adaptation to body surface curvature can be appreciated.

Fig. 3 shows 5 s of reduced Lead-I ECG simultaneously recorded with the signals derived from the TCR electrodes placed in upper position (TCR1) and in the lower position (TCR2) and its corresponding averaged beat performed on 60 s (right side). Firstly heartbeat fiducial points can be clearly distinguished in tripolar LECG recordings. In comparison to standard ECG, tripolar LECG morphology is sharper, as reported by others authors [8]. Concerning to the P wave, in all recording sessions it can be clearly identified in Lead-I ECG and in LECG from TCR1 electrode, which was placed in the upper position and therefore close to the atrial area. Nevertheless, TCR2 electrode, placed in the lower position, mainly picks up ventricular activity; being the P wave much more attenuated. It is also noticeable that, in this figure, signals from TCR2 electrode presented a stronger baseline drift than the rest of signals, which is later confirmed by the SBR results.

Mean and standard deviation of ECG parameters for Lead-I ECG, bipolar and tripolar LECG signal are shown in table I. Raw L_{Bip} ECG amplitude of TCR electrodes varied from 5.2 µV to 333 µV, which was considerably smaller in comparison to reduced Lead-I ECG (171.3 µV to 1039.2 µV). In general, both bipolar and tripolar LECG amplitudes were smaller in upper position than in lower position; and in L_{BipA} channel which corresponds to the biopotential between middle ring and inner disc than in L_{BipB} (15.3±7.6µV and 51.5±55.0 µV for L_{BipA}, vs 37.7±17.8 µV and 108±106.8 µV for L_{BipB}). Tripolar LECG amplitude varied from 79.1 uV to 2.38 mV, however it should be highlighted that this L_{Trip} is not a directly recorded signal, it is computed from equation (3) without considering the scaling factor. Regarding SNR, both bipolar and tripolar LECG in upper position presented similar values to that of standard Lead-I; being this parameter smaller for the signals obtained from TCR2 electrode in lower position. Respect to SBR, again signals from TCR2 provided worse values, showing they are more affected by baseline drifts. It is noteworthy that signals from TCR1 yield much higher values of SBR (about double) than reduced Lead-I ECG.

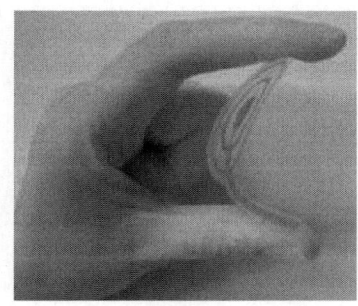

Fig. 2. TCR electrode implemented on a flexible substrate using inkjet technology.

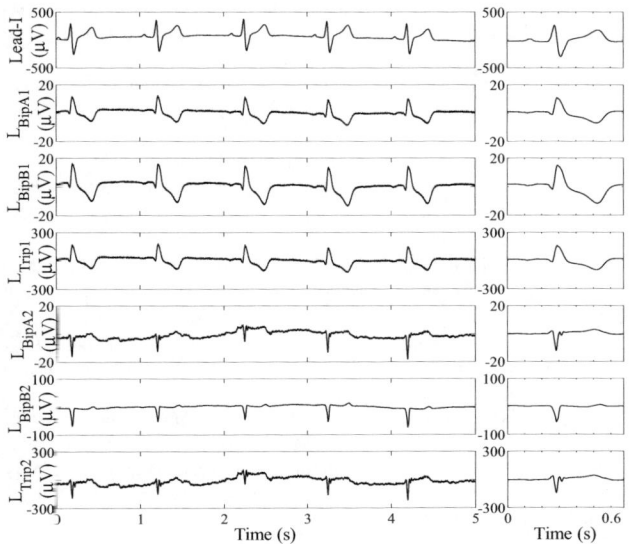

Fig. 3 Left: 10 seconds of reduced Lead-I simultaneously recorded with two TCRs one in upper position (TCR1) and the other in lower position (TCR2). For each TCR electrode, it was shown two raw bipolar laplacian recording (L_{BipA} y L_{BipB}) and one digitally estimated tripolar laplacian potential. Right: averaged beat corresponding to 1 minute of the chest surface signals shown on the left.

IV. DISCUSSION

In this work, active TCR electrodes were implemented on flexible substrates using single-layer inkjet printing technology for obtaining body surface LECG recording. The digital nature of this technique allows for direct CAD to board processing which permits to greatly reduce fabrication time of electrode prototypes. In addition, this technique can further improve electrode contour resolution since both the drops position and quantity of the drops to be placed can be precisely controlled during the manufacturing process[9]. The trade-off between electrode printing resolution and speed should be achieved for serial production. Note that it was necessary to repeat the printing process so as to increase the electrode thickness to guarantee the electrical continuity. Further analysis should be carried out for comparing this technique with other electronic circuit fabrication techniques such as screen printing [5], or gravure printing[10].

Table 1. Mean and standard deviation of signal amplitude, signal-to-noise ratio and signal-to-baseline ratio median value of ECG recording obtained from Lead-I and the two TCR electrodes in upper position (1) and in lower position (2). N=7 recording sessions.

	Lead-I	L_{BipA1}	L_{BipB1}	L_{Trip1}	L_{BipA2}	L_{BipB2}	L_{Trip2}
Amplitude	569.7±297.3 µV	15.3±7.6 µV	37.7±17.8 µV	208.8±105.0 µV	51.5±55.0 µV	108.0±106.8 µV	757.0±769.9 µV
SNR	16.4±3.5 dB	15.8±3.3 dB	18.3±2.9 dB	15.2±3.5 dB	8.1±6.0 dB	12.2±4.7 dB	7.7±6.2 dB
SBR	8.1±4.0 dB	16.2±5.3 dB	20.2±6.5 dB	15.6±5.2 dB	4.3±10.9 dB	8.8±9.1 dB	3.8±10.6 dB

Concerning to body surface LECG recording, signal amplitudes obtained using TCR electrode were significantly lower than that of standard Lead-I, which is mainly due to the smaller inter-electrode distance. As it was expected, the mean amplitude of L_{BipB} channel was twice that of L_{BipA} channel due to the double inter-electrode distance. Additionally, both bipolar and tripolar LECG amplitude and morphology is highly position-dependent, as reported in previous studies that used active TCR rigid electrodes for obtaining human body surface moment of activation isochronal mapping [8]. In our work the highest amplitudes are obtained in lower position, which agrees with other authors who found consistent and large amplitude LECG around the center right side of the sternal midline (subject's left 3rd and 4th rib) and subject left lateral false rib area [8]. Nevertheless, in our work, signal quality parameters of the developed TCR electrode were better for upper position. This contrasts with a previous similar study performed with flexible multilayer TCB electrodes of 24mm [5]. Despite the differences in configuration, size and manufacturing technology of the tested electrodes, further investigation with a more comprehensive database should be performed to clarify this issue. Nonetheless, it should be highlighted that the values of the signal quality parameters from the dry flexible TCR electrode in upper position were similar or even better than those obtained for Lead-I ECG picked up with conventional wet-disc electrodes. Future experiences using multiple units of the developed active flexible electrodes could be carried out to obtain body surface LECG maps, which can be very helpful to detect moment of activation so as to identify cardiac abnormalities.

V. CONCLUSIONS

It was developed a flexible active modular TCR electrode that combines the adaptability to body contours of conventional disposable electrodes with the enhanced spatial resolution of concentric ring electrodes. Experimental results showed that the proposed electrodes can be used for acquiring high quality body surface LECG which can suppose an advance in future high spatial resolution Laplacian-based monitoring systems.

ACKNOWLEDGMENT

This research was supported in part by the Ministerio de Ciencia y Tecnología de España (TEC 2010-16945) and by the Universidad Politècnica de València (SP20120469).

REFERENCES

1. See D. M. and Mirvis E. (1988) Body Surface Electrocardiographic Mapping Boston: Kluwer Academic.
2. He B. and Cohen R. J. (1992) Body surface Laplacian ECG mapping. IEEE Trans. Biomed. Eng, vol. 39, no. 11, pp. 1179-1191.
3. Lu C.-C. and Tarjan P. P. (2002) Pasteless, Active, Concentric Ring Sensors for Directly Obtained Laplacian Cardiac Electrograms. J. Med. Biol. Eng., vol. 22, pp. 199-203.
4. Besio W. G., Koka K., Aakula R., and Dai W. (2006) Tri-polar concentric ring electrode development for laplacian electroencephalography. IEEE Trans. Biomed. Eng, vol. 53, no. 5, pp. 926-933.
5. Prats-Boluda G., Ye-Lin Y., Garcia-Breijo E., Ibanez J., and Garcia-Casado J. (2012) Active flexible concentric ring electrode for non-invasive surface bioelectrical recordings. Measurement Science & Technology, vol. 23, no. 12
6. Singh M., Haverinen H. M., Dhagat P., and Jabbour G. E. (2010) Inkjet printing-process and its applications. Adv. Mater., vol. 22, no. 6, pp. 673-685.
7. Setti L., Fraleoni-Morgera A., Ballarin B., Filippini A., Frascaro D., and Piana C. (2005) An amperometric glucose biosensor prototype fabricated by thermal inkjet printing. Biosens. Bioelectron., vol. 20, no. 10, pp. 2019-2026.
8. Besio W. and Chen T. (2007) Tripolar Laplacian electrocardiogram and moment of activation isochronal mapping. Physiol Meas., vol. 28, no. 5, pp. 515-529.
9. Doraiswamy A., Dunaway T. M., Wilker J. J., and Narayan R. J. (2009) Inkjet printing of bioadhesives. J. Biomed. Mater. Res. B Appl. Biomater., vol. 89, no. 1, pp. 28-35.
10. Kang H., Kitsomboonloha R., Jang J., and Subramanian V. (2012) High-performance printed transistors realized using femtoliter gravure-printed sub-10 mum metallic nanoparticle patterns and highly uniform polymer dielectric and semiconductor layers. Adv. Mater., vol. 24, no. 22, pp. 3065-3069.

Author: Yiyao Ye-Lin
Institute: Grupo de Bioelectrónica, UPV
Street: Camino de Vera SN, Ed. 7F Acceso N
City: Valencia
Country: Spain
Email: yiye@gbio.i3bh.es

Measurement System for Pupil Size Variability Study

W. Nowak, A. Żarowska, E. Szul-Pietrzak, and A. Hachoł

Institute of Biomedical Engineering and Instrumentation, Wroclaw University of Technology, Wrocław, Poland

Abstract—This paper presents a measurement system for binocular pupil size variability study. The measurements can be taken at a rate up to 256 Hz with a linear accuracy better than 0.02 mm during both the pupil light reflex induced by computer-controlled light stimuli and the spontaneous pupil fluctuation. The unique system design and method of detecting a picture of each eye enables monitoring the right and left eye using a single camera with no overlap between each pictures. Using a synchronous signal this setup can cooperate with Neuron Spectrum system (41-channel multifunctional digital EEG system for neurophysiological studies) and Biopac system (complete systems for life science research and education). The metrological parameters and measurement abilities of the system can extend the scope of research using pupillometry.

Keywords—**pupil size variability, binocular dynamic pupillometer.**

I. INTRODUCTION

The extensive use of pupil size variability study has been used for years mainly to the evaluation of autonomic nervous system [1,2,3,4,5,6]. The standard approach to pupil monitoring is based on the specialized video camera which record monocular pupil picture in infrared. Unfortunately, the monocular operation is rather unsatisfactory, since on the one hand the pupil sizes may vary significantly with time as the effects of the higher centers of the brain activity which is not possible to control and on the other hand, in many cases, the diagnostic value has a comparison between both pupil responses. In addition, the higher frame rate (although the activity of the pupil is rather lower frequency) makes it possible to (1) determine time parameters with higher precision, and (2) examine the time-dependence between the pupil and the other biosignals. The higher resolution enables to study micro-fluctuation of the pupil size.

The dynamic pupillometry measurement system should allow for recording different types of pupil behavior (both spontaneous or induced by specific type of stimuli e.g. light or sound, which parameters can be set freely and which can be given either monocular or binocular). The possibility to synchronize the pupil measurement with other quantity measurement converters extends the application scope of such system. It is also important that the pupillometry measurement does not require the advanced level of cooperation from patient and they cannot be in any way stressful or inconvenient for the patient.

In the view of this necessities, we aim to develop a pupillometer that should meet the mentioned requirements.

II. MEASUREMENT SYSTEM DESCRIPTION

In the system presented in this paper, the left and right eyes are independently and alternately illuminated to generate alternating images which are successively transmitted by a simple optical system to a single camera. The PC stores the acquired images and operates offline, to get the pupil parameters, separate for the left and the right eyes. The system has a synchronous output and can cooperate with Neuron Spectrum system (41-channel multifunctional digital EEG system for neurophysiological studies) and Biopac system (complete systems for life science research and education). Fig.1. shows a schematic diagram presenting the system structure and principle of operation.

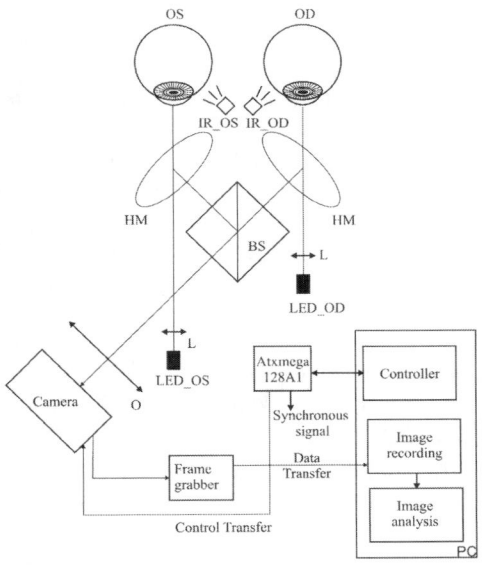

Fig. 1 The pupillometry system

The infra-red radiation (IR) is reflected by the corneal surface, transmitted by the optical tract and projected to high-speed high-resolution camera. The optical tract consists of two hot mirrors (HM), a single beam splitter (BS), and an objective (O). The camera (Photon Focus, MV-D102E) is a digital monochrome progressive scan camera with 1024x1024 pixels with each pixel size of 10.6 μm x 10.6 μm and a spectral response ranging from 400nm up to 1000 nm. The Matrix Solios type frame grabber allows real time storage of the camera output. The PC stores the acquired data as a sequence of gray-scale pupil images in bmp format.

The algorithm, developed by the authors under Vision Bülder performs analysis of recorded image sequence and determination, frame by frame, the measure of the pupil parameters, separate for the left and the right eyes. The algorithm includes four steps: (1) median filtering, (2) thresholding procedure, (3) pupil detection and best-circle and best-ellipse fitting, (4) average filtering. Median filtering is performed to smooth the image and remove any noise caused by changes in ambient lighting conditions. The thresholding procedure is used to remove potential artifacts (e.g. reflections) that arise during image acquisition. The procedure uses the average image intensity value which is calculated from the image histogram as a threshold to convert the image to a binary scale of black and white pixels. Additionally, during the thresholding process, the identified region of interest (i.e. pupillary region) is closed and filled. Once the pupil is isolated, object detection is performed to identify the boundary coordinates of the pupil. These coordinates are input into best-circle and best-ellipse fitting. The aim of the average filtering is to remove blinks and to reduce high-frequency noise. The obtained pupil waveforms, separately for the left and right eye, are written to a file and can be analyzed.

The computer-controlled light stimulator is composed of two LED sources, the one for the left eye (LED_OS) and the one for the right eye (LED_OD). A beam of stimuli light falls to an eye in the Maxwellian projection (i.e. to the pupil centre, and the spot of the light beam is smaller than the minimum pupil diameter). A retinal area with a diameter of about 20mm is stimulated. Each LED source is interchangeably and can be white, red, green or blue LED diode. The peak wavelengths of the three color LEDs were 470 nm, 520 nm and 625 nm with half-height bandwidths 5 nm, respectively. Output of each LED is controlled by μP controller using pulse-width modulation (PWM) technique. The light stimuli parameters can be adjusted by system operator utilizing the user interface as follow: wavelength, luminance level (from 1 cd/m^2 to 1000 cd/m^2), and pattern (any, e.g. a pseudorandom signal, a single pulse, a series of pulses (flicker), periodic signals (sinusoidal and rectangular), a positive/negative ramp). The start of stimuli light is synchronized with the beginning of recording. In the system the four classes of pupil response may be generated: both eyes response when the right or left eye is stimulated, or when both eyes are stimulated simultaneously or alternatively. Additionally, each eye can be stimulated using the same or different type of stimuli. The control and synchronization of the system is based on μP ATXMega128A1 which was chosen mainly because of the internal memory size and the ability to supply from USB (which is enough to power both the driver and LEDs connected).

III. TEST AND POSSIBILITIES OF THE MEASUREMENT SYSTEM

This section presents the experiments carried out to evaluate the accuracy and the measurement possibilities of the presented system. The experimental analysis of the system were carried out using set of phantoms and were divided into the examination of the system resolution and the evaluation how accurately and repeatable the system measure the pupil size. The set of 16, both circle and ellipse (major axis > minor axis and vice versa), black phantoms (grayscale bitmaps) with known size (similar to real pupil size) were prepared using a commercial graphics editing program (Corel Draw). Phantoms in the form of bmp files were prepared for software algorithm testing. The algorithm has provided correct results for the whole set of phantoms for best-ellipse fitting (differences between the size of the phantoms set in the graphic program and after using the analysis software are below 2%). The best circle fitting has provided correct results only for circle phantoms (the difference below 2%), the differences for ellipse phantoms are even more than 20%.

To study the measurement system, the circle phantoms were printed out using high level laser printer and then were measured with the microscope with accuracy better than 0.01 mm. During experiments each phantom was set on a tripod in a location corresponding to the eye position in a real-time measurements. For each phantom, the 1000 pictures sequence were recorded and repeated five times under the same measurement conditions. The phantom size in pixels were determined for each picture.

The system linear resolution was calculated as a ratio : phantom size in mm / phantom size in pixels. The ratio values were about 0.01 mm/pixel for all phantoms so the linear resolution of the proposed system is estimated to be better than 0.02 mm.

The system accuracy, calculated as difference between a measured quantity value and a true quantity value of a measured phantom was determined as below 0.5% for each phantom respectively. Standard deviation of the measurements was calculated as a measure of repeatability and it was below 4% for all phantoms.

To show the measurement possibilities of the system, some samples experiments were recorded for a woman of 30 years of age, who did not have any disease. The results (averaged for both eyes) are presented below. The measurement follows a 5-minute period of eye adaptation to darkness. During recording, the subject was asked to keep her eyes open and to look straight at the fixation point.

Pupillary reaction to a single light pulse consists of a single contraction followed by its dilation. An example of pupil light reflex (PLR) in response to single light pulse with 100cd/m^2 for red, green and blue light is presented in Fig.2.

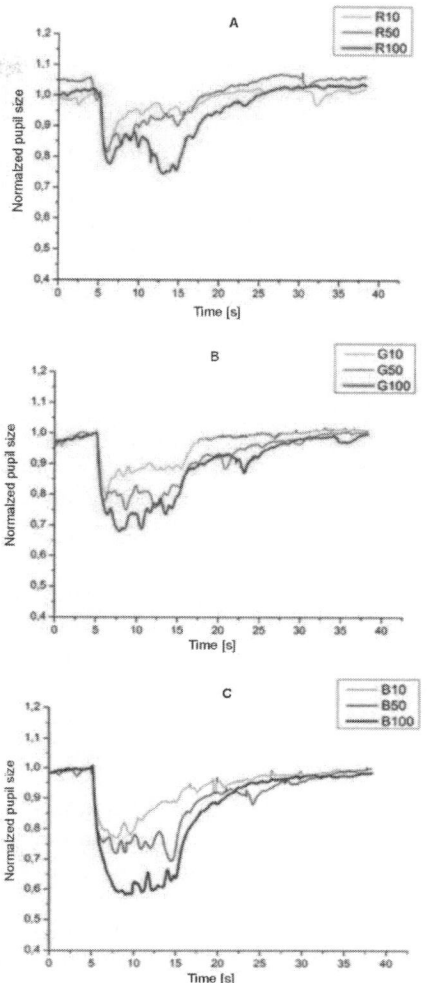

Fig. 3 Normalized PLR to chromatic light stimuli at a different luminance levels for red (A), green (C) and blue (B) lights.

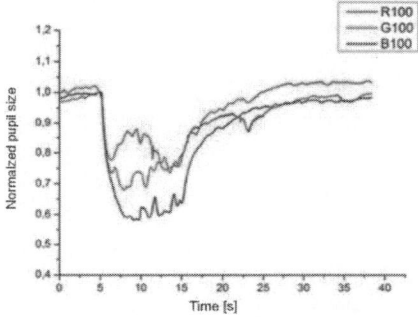

Fig.2. Normalized PLR to a single light pulse with 100 cd/m^2 for red (R), green (G) and blue (B) light.

In addition, Fig.3 presents PLR to chromatic light stimuli for different luminance levels (10, 50 and 100 cd/m^2). Possibility of testing the pupil response to chromatic stimuli is very important especially in context of the last discovery that a new photoreception process based on intrinsically photosensitive retinal ganglion cells (ipRGC), contributes to the pupillary light reflex in primates [7,8,9,10].

The example time-course of the spontaneous fluctuation of the pupil size is shown in Fig. 4. During the measurement, the patient looks at the fixation point (black cross on an illuminated white background) which is located at a distance of 4 meter.

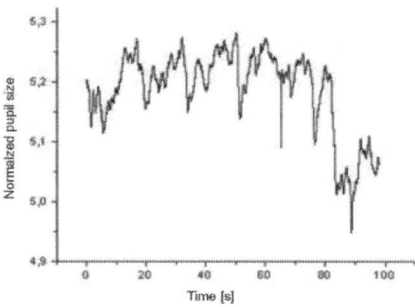

Fig. 4 Example of the spontaneous fluctuation in pupillary size.

IV. DISCUSSION

The article presents a binocular pupillometer enabling study the pupil size variability of the both eyes at a rate up to 256 Hz with a linear accuracy better than 0.02 mm. The system enables monitoring the right and left eye using a single camera with no overlap between each pupil pictures. The measurement method uses the principle of alternating the generation of the left and right pupil images, the unique design of the system which projects these images on the single camera and the original method developed to detect the pupil parameters from the recorded images. The pupil parameters are determined using both circle and ellipse shape approximation which gives the opportunity to test and compare the result. Results of the stimulation analysis performed showed that only best-ellipse approximation has provided correct results for both circle and ellipse phantoms.

The experimentally verified values of the system accuracy and repeatability are better than 0.5% and 4% respectively.

The system enables recording the spontaneous pupil fluctuation and the pupil light reflex induced by light stimuli which parameters and profile are computer-controlled independently for each eye. In addition, the system can cooperate with Neuron Spectrum system (41-channel multi-functional digital EEG system for neurophysiological studies) and Biopac system (complete systems for life science research and education). This possibility greatly expands research applications of the pupillometry.

The initial laboratory test have confirmed that a new system is a fast and precise device. It does not require a lot of operator experience, because it is a relatively simple and easy to use and most of all it is convenient to patients. The tests exemplify the system abilities and they have confirmed its comprehensive capabilities. Eventually, such a system could be an important step towards an pupillometer-based diagnostics systems.

ACKNOWLEDGMENT

This work was supported by the Polish Ministry of Science and Higher Education (Grant No. 405338)

REFERENCES

1. Ferrari G L, Marques J L B, Gandhi R A et al. (2010) Using dynamic pupillometry as a simple screening tool to detect autonomic neuropathy in patients with diabetes: a pilot study. Biomed Eng OnLine, 9:26
2. Pittasch D, Lobmann R, Behrens-Baumann W et al. (2002) Pupil signs of sympathetic autonomic neuropathy in patients with type 1 diabetes, Diabetes Care 25: 1545-1550
3. Fotiou F, Fountoulakis K N, Goulas A et al. (2000) Automated standardized pupillometry with optical method for purposes of clinical practice and research, Clin Physiol, 20: 336-47
4. McLaren J W, Hauri P J, Lin S C et al. (2002) Pupillometry in clinically sleepy patients. Sleep Med, 3:347-352
5. De Santis A, Iacoviello D.(2006) Optimal segmentation of pupillometric images for estimating pupil shape parameters. Comput Methods Programs Biomed, 84:174-187
6. Masi M G, Peretto L., Tinarelli R et al. (2009) Measurement of the Pupil Diameter under different light stimuli, I2MTC 2009, Int Instrum and Measur Technol Conference, Singapore, 2009,pp 1652-1656.
7. Chen K, Badea T C, Hattar S. (2011) Photoentrainment and pupillary light reflex are mediated by distinct populations of ipRGCs. Nature, 476:92-95
8. Young R S L, Kimura E. (2008) Pupillary correlates of light-evoked melanopsin activity in humans. Vision Res, 48:862-871.
9. Tsujimura S, Tokuda Y. (2011) Delayed response of human melanopsin retinal ganglion cells on the papillary light reflex, Ophthalmic Physiol Opt, 31: 469-479.
10. Tsujimura S, Ukai K, Ohama D et al. (2010) Contribution of human melanopsin retinal ganglion cells to steady-state pupil responses, Proc R Soc Lond B Biol Sci, 277: 2485-2492.

Author: Wioletta Nowak
Institute: Institute of Biomedical Engineering and Instrumentation
Street: Wybrzeze Wyspianskiego 27
City: Wroclaw
Country: Poland
Email: wioletta.nowak@pwr.wroc.pl

Patellar Reflex Measurement System with Tapping Force Controlled

Y. Salazar-Muñoz[1], L.A. Ruano-Calderón[2], J.A. Leyva[1], O.S. Martínez[1],
O. García-Cano[3], and B. García-Caballero[4]

[1] Instituto Tecnológico de Durango, Dpto. de Ing. Eléctrica-Electrónica, DGEST, Durango, México
[2] Hospital General de Durango, Servicios de Salud del Estado de Durango, México
[3] Instituto Tecnológico de Durango, Dpto. de Metal-Mecánica, DGEST, Durango, México
[4] Instituto Tecnológico de Durango, Dpto. de Ing. Química-Bioquímica, DGEST, Durango, México

Abstract—The clinical evaluation of muscular strength and stretching reflex are one of the most important neurological data to evaluate patients. Our system measure and capture variables generated by the patellar reflex with a non-invasive way and absolute control in the tapping force on the tendon. The digital saving data process is performed by a PC or Notebook with the Monitoring Reflex Software, where the measured variables are: angular position of the movement developed by the leg, angular rate, impact instant on the patellar tendon and the muscular distention in the quadriceps. All of them present changes according to the patellar reflex generated by the tapping force control.

Keywords—Patellar Reflex, accelerometer, muscular strength, device design.

I. INTRODUCTION

The clinical evaluation of muscular strength and stretching reflex are one of the most important neurological data to evaluate patients with system nervous affectation (1). Nowadays, the reflex data interpretation is very subjective and qualitative (2), (3), (4), it means that a doctor gives his diagnostic based on his own judgment and experience, causing several variation on multiple tests because there is no standard method for the patellar reflex evaluation and normalized data.

Besides, the patellar reflex exam is cheap and quick, and it is an elemental part for clinical diagnostics of the nervous system. There are many proposed projects like the biomechanical systems (5); (6); (7), electrophysiological (4) and (1) where the studies trying to define the behavior of the patellar reflex in a serious way but always in a special laboratory.

The patellar reflex research under isometric conditions (muscle stretch), developed by Safranov (8) explains a special method where in conditioned chair immobilize the patient with a solenoid and a group of measurements systems for capture the developed movement of the leg.

Another system capable of measuring the force generated time do by bending the leg of a patient, patient immobilization lacked a striker and automatic tendon since basically consisted of a LVDT frontally with the stem directed towards the leg, tibia level for mechanical motion of the leg directly impact the LVDT rod. This generated a voltage variation which was directed to a data acquisition card and whose information was displayed on the screen of a PC. Showing the interface according to the voltage generated by the LVDT classification impact force on a 4 grade scale (9). Therefore, it is necessary to define a more objective clinical classification of muscle stretch reflexes to standardize its measurement in monitoring patients, both in its initial approach and the follow-up therapy. And as part of the research of our group, biomedical instrumentation and neuromuscular diseases, has scheduled an electronic monitoring system for the knee jerk, which is composed of a mechanical part, one focused on data acquisition and the display of information. The system will maintain a constant impact force determined by the user, causing the knee jerk, to thereby measure the angle and speed of displacement of the tip, response time and the distension of the femoral quadriceps.

II. METODOLOGY

The design system was composed of two parts:

A. Mechanical Controlled Force System

The mechanical system consist of an aluminum pendulum rubber tip attached to a toothed gear angle with the hammer height adjustable, which allows you to select the impact force on the patellar tendon as a function of the elevation angle of the pendulum. The physician shall place the arm in the desired position and release it manually. The force applied will be the same for all test subjects to generating their own flexion.

B. Data Acquisition System

The data acquisition system is composed of the following elements:

Tapping Sensor. A piezoelectric sensor was used (LDT0-028K). It was connected to a charge amplifier circuit and an instrumentation amplifier to obtain a 5V pulse and detecting the instant of impact on the tendon and aligned the others measured variables.

Angular displacement and rate sensor. It was used an inertial measurement unit (IMU) SEN-11072 with 5 degrees of freedom. It contains a 2-axis gyroscope IDG500 model with sensitivity set to 2 mV / ° / s and a 3-axis ADXL335 accelerometer. The angular displacement is determined by combining the arc-tangent function of the Y and Z accelerometer.

$$\theta_{acc} = tan^{-1}(Y_{acc}/Z_{acc}) \quad (1)$$

Together with the numerical integration of the angular velocity of the gyroscope by the trapezoidal method.

$$\theta_{gyro} = \omega * dt \quad (2)$$

Both results are entered into a second order complementary filter resulting in an angular measurement generated by the quick action of the leg.

Calculating the angular velocity is determined directly with the signal generated by the gyroscope (previously referencing the voltage of the static moment)

$$\omega = (Gyro_{ADC})(ADC_{resolution}/Sensitivity_{gyro}) \quad (3)$$

Strain Sensor. Used a strain gauge (CH-8320) attached to the femoral quadriceps using an instrumentation amplifier for conditioning, thus allowing, recording and muscle distention indirectly monitor muscle contraction.

Control Unit. We used an Arduino development platform for the conversion and transmission of signals to the PC using the UART serial communication protocol.

Graphic User Interface (GUI). The variable storage and monitoring is done through a software platform developed in open source Processing programming. The program consists of an executable for the Windows operating system (with option on systems running Mac OS X and Linux). The graphical interface has the following indicators in real time (Fig. 1):

- Angular Displacement: A dial graduated in -180 ° to 180 ° and a marker allows knowing the number of degrees of flexing the leg.
- Angular Velocity: The speed developed by the flexing of the leg is monitored in the graph above.
- Impact and distension: Both signals are shown in the graph below, becoming apparent the instant of impact on the knee.

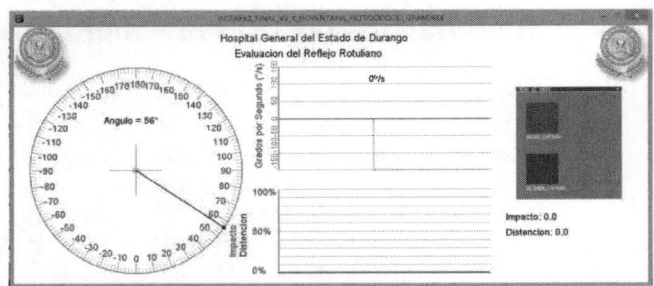

Fig. 1 Control and Monitor Window. On this screen we can see all the physical variables. A dial indicate the current position of the leg (having gravity axis as position Zero)

The multimedia control elements are formed by a virtual button that initiates storage of variables in an external text file at a frequency of 50 samples per second. A second closing button stops storing the text file with the name *.txt.

Results template: The outcomes template imports the data from the text file automatically and displays each of the parameters summarized in a sheet of calculus.

Experimental tests were performed on a group of healthy volunteers under the supervision of the medical team of the General Hospital of Durango. We evaluated a total of 12 people with an average age of 23 years. To each of them was applied controlled impact force of 450 grams force on the patellar tendon. Figure 2 shows the integration of the system and the location of the sensors used.

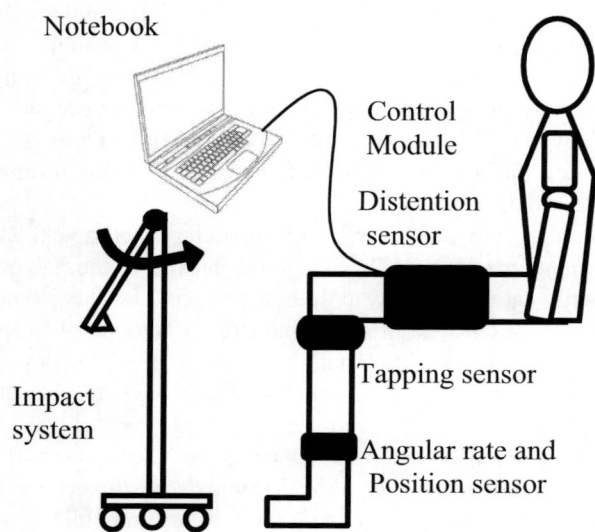

Fig. 2 System setup. All the sensors are connected in the control module allocated over the thigh.

III. RESULTS

Data obtained from tests conducted in healthy volunteers allowed to have a first evaluation of the behavior of variables and correlate with the observation of the physician based on your experience with the traditional method.

The figures represent the dynamic behavior of both changes in velocity as the change of elevation and muscle. Each set defines the speed, elevation and distension of the muscle in each patient.

The results obtained after the experimental analysis demonstrate the dynamic response for the patellar reflex. Figure 3 shows a representative subject with an angular velocity of 55°/s, in the first leg lift with a decrease greater than 65% in a second oscillation. The leg stops to move after 2.4s.

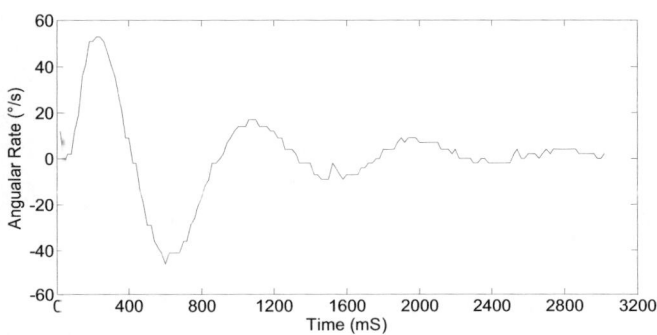

Fig. 3 Temporal change in angular velocity of the leg after the impact occurred at time 0.

The elevation angle of the leg for the same subject was 10° from the initial position which is normalized to zero. The second elevation decreases by 3° to then return to the initial position of zero degrees, (Figure 4).

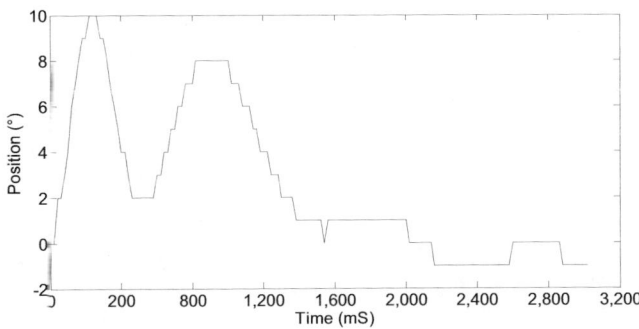

Fig. 4. Representative angular displacement of the leg.

Figure 5 shows the percentage of distension of the femoral quadriceps. This is a change corresponding to 80% with respect to its initial position as in the above variables each oscillation is coincident with the elevation of the subject's leg.

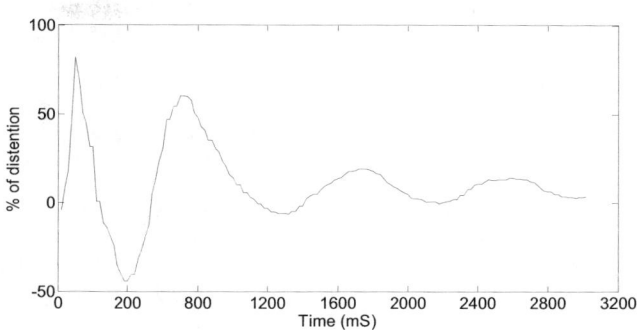

Fig. 5 The graphic show the relative change of the distention sensor with respect to the initial value measured.

IV. DISCUSSION

To quantitatively measure patellar tendon reflex was determined that the impact force were constant. Thus, in this first stage, the response analysis elevation and angular velocity changes would be only due to a change in the response of the subject itself and not by the force applied by the doctor over the patella to generate the reflex. Since there is an increase in the angular velocity due to an increase in impact strength according to the results obtained by (10).

Considering the time cero like the instant that impact is detected it was first measured the muscular distension resulting in elevation of the leg. Thus, in case of no show a significant increase could be a variable that determines an objective assessment of the patellar reflex response.

REFERENCES

1. **Voerman GE, Gregorič M and Hermens H.** 2005, *Neurophysiological methods for the assessment of spasticity: The Hoffmann reflex, the tendon reflex, and the stretch reflex.* Disability and Rehabilitation, Vol. 27, págs. 33-68. 10.1080/09638280400014600.
2. **Manschot S, L van Passel, E Buskens, A Algra, J van Gijn.** 1998, *Mayo and NINDS scales for assessment of tendon reflexes: between observer agreement and implications for communication.* J Neurol Neurosurg Psychiatry, Vol. 64, No. 2, págs. 253-255. doi:10.1136/jnnp.64.2.253.
3. **Tham LK, Abu Osman NA, Wan Abas WA, Lim KS.** 2013. *The validity and reliability of motion analysis in patellar tendon reflex assessment.* [ed.] University of Iowa Carver College of Medicine, United States of America Pedro Gonzalez-Alegre. 2, , PLoS ONE, Vol. 8, pág. e55702. doi:10.1371/journal.pone.0055702.
4. **Stam J, Tan KM.** 1987,*Tendon reflex variability and method of stimulation.* Electroencephalogr Clin Neurophysiol., Vol. 67, págs. 463-467.
5. **Simons DG, Lamonte RJ.** 1971. *Automated system for the measurement of reflex responses to patellar tendon tap in man.*, Am J Phys Med., Vol. 50, págs. 72-79.

6. **Zhang LQ, Huang H, Sliwa JA, Rymer WZ.** 1999. *System identification of tendon reflex dynamics.* IEEE Trans Rehabil Eng., Vol. 7, págs. 193-203.
7. **Wood DE, Burridge JH, van Wijck FM, McFadden C, Hitchcock RA, Pandyan AD, Haugh A, Salazar-Torres JJ, Swain ID.** 2005. *Biomechanical approaches applied to the lower and upper limb for the measurement of spasticity: a systematic review of the literature.* Disabil Rehabil, Vol. 27, págs. 19-32.
8. **Safronov, V.A** 2005. *Procedure for Investigating the Patellar Reflex under Isometric Conditions.* Human Physiology, Vol. 31, No. 6, págs. 724-727.
9. **Mohd Fauzulazim, M.Z. y Wan Suhaimizan, W.Z** 2008. *Development of patellar reflex measurement using Linear Variable Differential Transformer..* Malaysia : IEEE Conference Publications, Aug,. International Symposium on Information Technology. Vol. 4, págs. 26-28.
10. **Naotaka Mamizukaa, Masataka Sakanea, Koji Kaneokaa, Noriyuki Horib, Naoyuki Ochiaia.** 2007. *Kinematic quantitation of the patellar tendon reflex.*, Journal of Biomechanics, Vol. 40, págs. 2107–2111.

Author: Yolocuauhtli Salazar Muñoz
Institute: Instituto Tecnológico de Durango
Street: Blvd. Felipe Pescador 1830 ote.
City: Durango
Country: Mexico
Email: ysalazar@itdurango.edu.mx

A Bioelectric Model of pH in Wound Healing

M. Amparo Callejón[1], Laura M. Roa[1,2], and Javier Reina-Tosina[3,2]

[1] Biomedical Engineering Group, University of Seville, Seville, Spain
[2] CIBER de Bioingeniería, Biomateriales y Nanomedicina (CIBER-BBN), Spain
[3] Dept. of Signal Theory and Communications, University of Seville, Seville, Spain

Abstract—**pH has been suggested as a biomarker able to provide reliable information about wound healing. However, the underlying electrophysiological mechanisms responsible for pH changes observed in practice still remain unknown and a lack of bioelectric models capable of simulating these still exists. In this work, the authors have proposed a 2D rectangular wound model based on Nernst-Planck equations for the major ionic species involved in wound healing together with the implication of endogenous electric fields. In order to demonstrate the applicability of the model, both pH and electric potential results have been successfully compared with experimental data in the literature.**

Keywords—**Bioelectric model, endogenous electric fields, pH, wound healing.**

I. INTRODUCTION

Wound healing is a dynamic and complex regeneration process involving the interaction of many ionic species and biochemical reactions. From a clinical viewpoint, pH has proven to be a reliable indicator of wound status [1]. Although there are diverging pH data in the literature, it is widely accepted that intact skin presents a relatively acidic pH. However, upon injury, this acidic milieu is disturbed and pH changes can be observed over time. For instance, high pH values are usually associated with high rates of bacterial colonization and chronic wounds [2]. Recently, a 2D luminescence imaging technique for pH in vivo was proposed and applied to the study of pH in skin-graft wounds [3]. The results showed high pH values during the first few days with a progressive decrease of pH towards more acidic values as the epidermal barrier was reestablished. These outcomes evidence the potential of pH as a useful biomarker able to provide relevant clinical information about wound status as well as to help in the design of new wound therapies. However, the underlying electrochemical mechanisms by which these pH changes arise remain unknown and an important research effort is still needed in this field.

From a bioelectric point of view, pH is a measure of hydrogen ion concentration, the dynamics of which, along with the rest of ionic species and charged particles, is a consequence of a series of complex and interrelated electrochemical processes in the different phases of wound healing. It has been shown that when an epithelial layer, such as those forming the skin, is injured, endogenous electric fields (EFs) appear at the wound. It has also been demonstrated that they play an important role in cell migration [4]. Unbroken epithelium presents a transepithelial potential (TEP) of 20-50 mV, which is originated by Na^+, Cl^- and K^+ channels located at both apical and basolateral membranes [5]. When the skin is wounded, the high-resistive tight junctions between cells are disrupted, causing low-resistive current pathways to appear at the open wound site, which promotes diffusion and ionic current flow, and the eventual generation of the aforementioned EFs [6]. In this sense, experimental results reported in [7] showed that a negative electric potential is presented at the wound site with respect to the edges, which in turn is recovered as the wound closes.

In conclusion, it can be noticed that a great variety of experimental results in the literature confirm the essential role of EFs during wound healing as well as pH as a major indicator of wound status. However, there is still a lack of computational models able to reproduce such results. For instance, different modeling approaches have been reported [8–10], but only a few models have considered the bioelectric nature of tissue repair, and they have mainly focused on the electrochemical treatment of tumors [11]. The study of the bioelectric phenomenon of wound healing is an innovative topic that could help not only to explain the healing process related to pH dynamics but also to assess the design of new therapies promoting healing. For this reason, the modeling approach of this work is based on the bioelectric characteristics of wound healing, with the aim of explaining the pH results observed in the clinical practice not in isolation, but instead integrated within the rest of ion dynamics and the main electrochemical processes in wound healing. Recently, the authors introduced a 1D bioelectrical model that considered a constant electric field, considerably simplifying the solution of the equations for positive and negative carriers [12]. In this work, a more realistic 2D geometry of a wound is considered, and the electrical potential is not assumed constant but calculated from the model equations proposed.

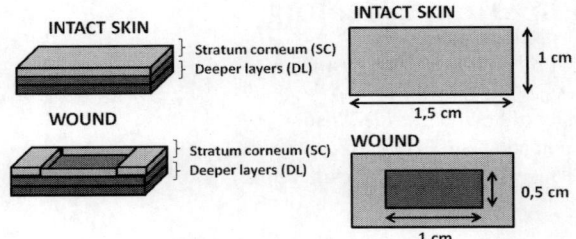

Fig. 1 The model geometry proposed represents an upper view of a 3D portion of the wounded skin.

II. METHODS

A. Wound Model Geometry

A simplified 2D rectangular geometry of the wound has been considered in this work. Therefore, the model can be defined as an upper view of a 3D portion of the skin. When the skin is intact, the model corresponds to a rectangular stratum corneum layer with a normal pH value of approximately 5.5. Conversely, when an injury occurs, the stratum corneum is removed and the deeper layers of the skin can be seen from above. The pH of such a broken area of the skin is no longer 5.5 but instead varies over time during the healing process. Fig. 1 shows a detailed picture of the geometry considered. For the sake of generalization, a specific wound area (1-cm width and 0.5-cm height) has been chosen, nevertheless the results can be extended to other values of wound area.

B. Modeling Hypothesis

In this model, wounded milieu has been taken to be as a diluted and electroneutral homogeneous aqueous solution. The major ionic species involved in wound healing have been considered: H^+ is shown to be vital in this process, not only according to experimental pH measurements in macroscopic wounds but also according to experimental evidence at molecular level. The Na^+/H^+ proton exchanger NHE1 plays an important role regulating the pH of stratum corneum and it is decisive for efficient cell migration [13]. In addition, it has been reported that cell lysis after injury implies consumption of adenosine triphosphate (ATP) and the production of H^+ ions in the wound [14]. On the other hand, the species Na^+, Cl^- and K^+ have also been considered in the model due to their role in the generation of the skin TEP and EFs [5].

We have tried to link the dynamics associated with hydrogen ions in the wound, supported by pH clinical measurements, with the rest of ionic species involved during wound healing, which are thought to generate the EFs that arise at wounds. With this end, our main modeling hypothesis relies on the assumption that, when an injury occurs, the tissue is disrupted and the repose ionic concentrations are altered. In this way, a concentration gradient between the intact and the injured site appears, thus leading to diffusion and migration currents in the wound. Therefore, both transport mechanisms (diffusion and migration in an electric field) are responsible for H^+ ions time dynamic, thus causing the pH changes observed in the practice. In the same way, such concentration gradients and the subsequent net movement of ions through the wound generates an endogenous electric potential, which tends to recover as the healing process is underway.

C. Model Equations

The proposed model considers diffusion, migration and reaction as the main transport mechanisms between two different domains: the intact skin and the injured skin. The equation system presents five unknown variables that account for four ionic species (H^+, Cl^-, Na^+ and K^+) and the electric potential, together with the electroneutrality condition:

$$\frac{\partial C_i}{\partial t} = -\nabla \cdot (D_i \nabla C_i + z_i u_{m,i} F C_i \nabla V) + R_i \quad (1)$$

$$u_{m,i} = \frac{D_i}{RT} \quad (2)$$

$$\sum z_i C_i = 0 \quad (3)$$

where C_i is the concentration of the specie i (mol/m³); t is the time; D_i denotes the diffusion coefficient (m²/s); z_i is the charge number, positive for cations and negative for anions; $u_{m,i}$ is the ionic mobility (mol·s/kg) satisfying the Nernst-Einstein relation in (2); F refers to Faraday's constant (A·s/mol), V is the electric potential (V); R_i accounts for the production term for the specie i (mol/m³·s); R is the universal gas constant (J/mol·K) and T is the temperature (K).

The current density is defined through Faraday's law, and the conservation of the electric charge is derived from the divergence of the current density, which in turn is used to compute the electric potential.

$$\mathbf{J} = F \sum z_i (-D_i \nabla C_i - z_i u_{m,i} F C_i \nabla V) \quad (4)$$

$$\nabla \cdot \mathbf{J} = F \sum z_i R_i \quad (5)$$

D. Solution Computation

Regarding Na^+ and K^+, we have assumed initial concentrations for intact and injured skin in the same physiological range as intracellular and interstitial cell concentrations (in the order of 1-100 mmol/liter) [15]. In the case of H^+, we have assumed an initial gradient concentration equivalent to an initial pH value in the injured skin of 8.3 and a normal pH value of 5.5 in the intact skin, according to [3]. A reaction term R_{H^+} has been added in order to model hydrogen release during cell lysis. Its value therefore represents the time constant for hydrogen ion generation during wound healing and was chosen according to the time intervals observed in pH clinical studies. The initial concentration for Cl^- was calculated through the electroneutrality condition in (3). Therefore, there is a concentration gradient of ionic species when the wound originates, thus producing an EF. In this way, as the wound closes, this initial gradient of charge carriers is compensated, and this eventually produces the observed recovery of the electric potential as the healing process continues [7]. An average healing period for acute wounds of 14 days has been considered. In this sense, the diffusion coefficients of the ionic species considered were adapted for a time dynamic in the range of days at macroscopic level. We focused on pH results on days 1, 6 and 14, which are related to inflammation, granulation and reephitelialization stages, respectively. We used a finite element method (FEM) software, Comsol Multiphysics, to solve the equation system (1)-(5) within a 2D geometry. All the input parameters of the proposed model are listed in Table 1. In this table, $C_{0,is}$ and $C_{0,iw}$ account for the initial concentration for the ionic specie i at intact and wounded skin, respectively.

III. RESULTS

As has previously been stated, a 14-day time interval has been considered in order to solve the model equations proposed in (1)-(5). The solutions obtained were the concentration for ionic species H^+, Na^+, K^+ and Cl^-, and the electric potential over time. Subsequently, pH dynamic was obtained by applying the well-known formula for H^+ molarity. The pH results for three different days corresponding to the major phase stages during wound healing are shown in Fig. 2. It can be seen that wounded skin presents a pH value of about 8.3 during the inflammation phase at day 1,

Table 1 Model parameters

Parameter	Value	Unit
D_H	9.31e-13	m²/s
D_{Cl}	2.03e-11	m²/s
D_{Na}	1.33e-11	m²/s
D_K	1.96e-11	m²/s
C_{0,H^+s}	3.1623e-3	mol/m³
C_{0,H^+w}	5.0119e-6	mol/m³
C_{0,Na^+s}	1	mol/m³
C_{0,Na^+w}	10	mol/m³
C_{0,K^+s}	100	mol/m³
C_{0,K^+w}	1	mol/m³
R_{H^+}	1e-10	mol/m³s

according to the initial condition for H^+ concentration considered, a value of about 7 during the granulation phase at day 6, and a more acidic value close to 6 during the reephitelialization stage at day 14. In contrast, the intact surrounding skin maintains a normal pH of 5.5, which was established as a boundary condition in the model. The simulated results for electric potential are presented in Fig. 3 for a cross-sectional cut of the wound. A negative value for electric potential can be observed in the center of the wound at the initial time with respect to the edges. It can also be noticed that the electric potential tends to zero as the wound healing process advances, which also matches the potential measurements reported in [7]. Fig. 4 shows pH and electric potential simulated results over time at the center of the wound, revealing that both of them tend to values that are close to those observed in the surrounding intact skin as the wound closes.

IV. DISCUSSION AND CONCLUSION

The simulated results are in the same range as the data reported in [3], [7], which demonstrates the applicability of the proposed model. In this work, an interrelated relationship

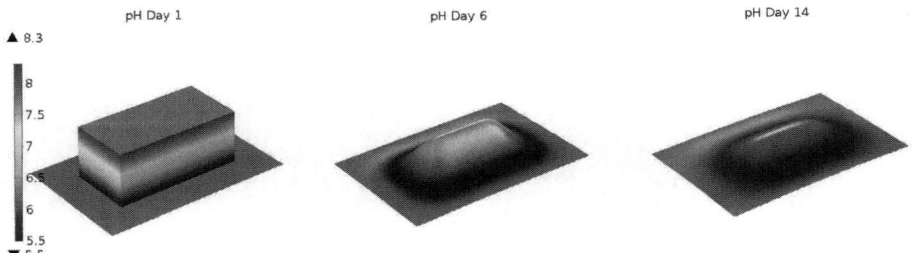

Fig. 2 Simulated pH evolution over days 1, 6 and 14, which are related to the inflammation, granulation and reephitelialization stages, respectively.

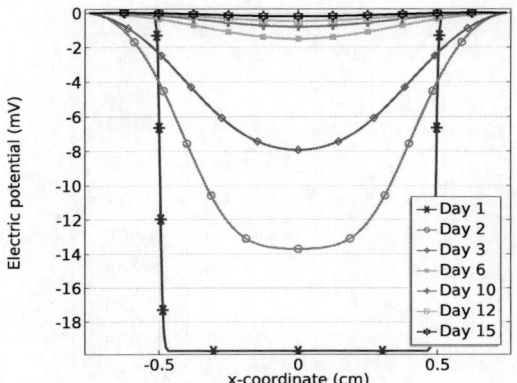

Fig. 3 Simulated electric potential for a cross-sectional cut of the wound.

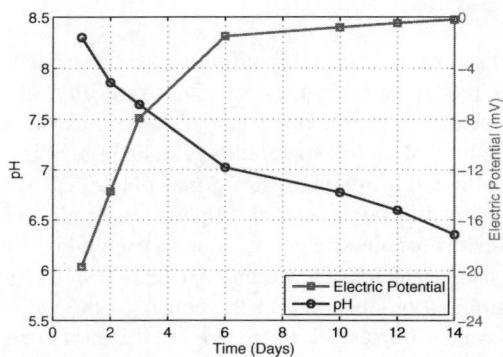

Fig. 4 Simulated pH and electric potential in the center of the wound.

between pH and electric potential has been derived. An initial concentration gradient between the center and the edges of the wound make the wounded tissue present an initial negative electric potential, which could be responsible for the appearance of other charged particles in the wound, thus influencing epidermal cell migration. In this way, a possible explanation for EFs in wounds is proposed in this model, as a consequence of a the appearance of a charge concentration gradient after injury. As a concentration balance is achieved, the healing process continues, and the electric potential finally tends to zero. Moreover, H^+ ions increase in the center of the wound during the healing process, due to diffusion from the edges, migration in an electric field and generation from cell lysis activity. This consequently leads to a decreasing pH value as the wound closes.

Acknowledgements

This work was supported in part by the Consejería de Economía, Innovación y Ciencia, Government of Andalucía, under Grants P08-TIC-04069 and P10-TIC-6214, the Fondo de Investigaciones Sanitarias, Instituto de Salud Carlos III, under Grant PI11/00111, and CIBER-BBN under Grants NANOPHOR and MICHORMON. CIBER-BBN is an initiative funded by the VI National Research and Development and Plan 2008-2011, Iniciativa Ingenio 2010, Consolider Program. CIBER Actions are funded by the Instituto de Salud Carlos III with assistance from the European Regional Development Fund.

References

1. Schneider LA, Korber A, Grabbe S, Dissemond J. Influence of pH on wound-healing: a new perspective for wound-therapy? *Arch Dermatol Res*. 2007;298:413-420.
2. Schreml S, Szeimies RM, Karrer S, Heinlin J, Landthaler, Babilas P. The impact of the pH value on skin integrity and cutaneous wound healing *J Eur Acad Dermatol Venereol*. 2010;24: 373-378.
3. Schreml S, Meier RJ, Wolfbeis OS, Landthaler M, Szeimies RM, Babilas P. 2D luminescence imaging of pH in vivo *PNAS* 2011;108 (6): 2432-2437.
4. Zhao M. Electrical fields in wound healing-An overriding signal that directs cell migration *Sem. Cell & Develop Biol*. 2009;20(6): 674-82.
5. Nuccitelli R. A role for endogenous electric fields in wound healing *Curr Top Dev Biol*. 2003;58:1-26.
6. Messerli MA, Graham DM. Extracellular electrical fields direct wound healing and regeneration *Biol Bull*. 2011;221(1):79-92.
7. Nuccitelli R. *Measuring Endogenous Electric Fields. The Physiology of Bioelectricity in Development, Tissue Regeneration and Cancer*. CRC Press 2011.
8. Tabatabai MA, Eby WM, Singh KP. Hyperbolastic modeling of wound healing *Mathematical and Computer Modelling* 2011;53(5-6):755-768.
9. Walker DC, Hill G, Wood SM, Smallwood RH, Southgate J. Agent-based computational modeling of wounded epithelial cell monolayers *IEEE Trans Nanobioscience*. 2004;3(3):153-63.
10. Azuaje F Computational discrete models of tissue growth and regeneration *Brief Bioinform*. 2011;12(1):64-77.
11. Turjanski P, Olaiz N, Abou-Adal P, Surez C, Risk M, Marshall G. pH front tracking in the electrochemical treatment (EChT) of tumors: Experiments and simulations *Electrochimica Acta* 2009; 54(26):6199-6206;
12. Callejón MA, Roa LM, Reina JL A First Approach to Bioelectric Modeling of Wound Healing in *IFMBE Proc, 5th Eur Conf. of the Int. Fed. for Med. and Biol. Eng.* 37: 263-266 2012.
13. Stock C, Schwab A. Role of the Na/H exchanger NHE1 in cell migration *Acta Physiol (Oxf)*. 2006;187(1-2):149-57.
14. O'Clock G. *Electrotherapeutic Devices: Principles, Design, and Applications*. Boston: Artech House 2007.
15. Guyton AC, Hall JE. Textbook of medical physiology. Elsevier 2006.

Author: M. Amparo Callejon
Institute: Biomedical Engineering Group
Street: C de los Descubrimientos s/n
City: Seville
Country: Spain
Email: amparocallejon@gmail.com

Part XI
Molecular, Cellular and Tissue Engineering and Biomaterials

Competitive Adsorption of Albumin, Fibronectin and Collagen Type I on Different Biomaterial Surfaces: A QCM-D Study

H. Felgueiras[1], V. Migonney[1], S. Sommerfeld[2], N.S. Murthy[2], and J. Kohn[2]

[1] Laboratory of Biomaterials and Specialty Polymers, LBPS-CSPBAT CNRS UMR 7244, Institut Galilée, Université Paris 13
93430 Villetaneuse, France
[2] New Jersey Center for Biomaterials, Rutgers University, 145 Bevier Road, Piscataway, New Jersey 08854, USA

Abstract—Adsorption of several proteins, individually and in a mixture, onto substrates was investigated. Five biomaterial surfaces were selected based on their functionality and/or hydrophilicity: Ti6Al4V, Ti6Al4V + poly(sodium styrene sulfonate) (poly(NaSS)), gold, poly(desamino tyrosyltyrosine ethyl ester carbonate) (poly(DTE)), polystyrene (PS). Their ability to interact with BSA, Fn and Col I was studied using a QCM-D technique.

The presence of sulfonate groups from the poly(NaSS) coating significantly increased protein adsorption. When two proteins were pooled, in the majority of the cases, the first to adsorb onto the surface inhibited the adsorption of the second. One exception is the combination of Col I with Fn, which could be attributed to probable changes in the protein conformation.

The advantage of these five surfaces in the biomedical field was confirmed by the ability of MC3T3-E1 osteoblastic cells to attach to the entire group.

Keywords—Fibronectin, Collagen I, Albumin, Biomaterials, Quartz Crystal Microbalance.

I. INTRODUCTION

In living systems, the blood is the first component to contact with biomaterials. Immediately after, rapid adsorption of plasma proteins occurs on the surface, mediating all the subsequent interactions. Adsorption of proteins onto biomaterial surfaces is a complex but fundamental process for a successful implantation.

There are several proteins in the extracellular matrix playing crucial role on this task. Albumin, the most abundant, is responsible for blocking nonspecific binding regions and works as media transport for various metabolites. Fibronectin and collagen type I that exhibit RGD (arginine-glycine-asparagine) sequences are known for inducing cell attachment [1–3]. Integrins bind to the RGD sequence, actin cytoskeletons are produced and the associated proteins form focal adhesion points at the surface. The cells that reach the surface establish contact with the protein coated substrate and attach to the extracellular matrix of those highly complex particles. Since there is no actual contact between the implant and the cells, the resultant biological activity will depend, among other factors, on the number, conformation and competition of the recruited proteins. Consequently, those factors will be a result of the surface physical and chemical properties influence [4].

There are numerous reports on the influence of the surface wettability, morphology, roughness, on the cell development [5, 6]. To further understand these effects, five different surfaces were selected from metals and polymers based on their potential to modulate albumin, fibronectin, and collagen I adsorption. These substrates were: Ti6Al4V, Ti6Al4V physisorbed with poly(sodium styrene sulfonate) (poly(NaSS)), poly(desamino tyrosyl-tyrosine ethyl ester carbonate) (poly(DTE)), polystyrene (PS) and gold.

In the last few years, there has been increased interest in understanding the mechanism of protein adsorption, and the influence they have on each other when present in a mixture with many studies being reported. However, there are still some issues that need further investigation. In this paper we study individual proteins and protein mixtures to test both the influence of the surface on the adsorption of specific proteins and the impact of one protein on the other. We also probe the ability of these materials to interact with osteoblastic cells

Different techniques have been used to quantify and characterize protein adsorption on solid surfaces [7]. In this study, we use a quartz crystal microbalance with dissipation (QCMD), a simple and high sensitive technique that detects the kinetics associated to the mass changes by following the frequency variations [8].

II. MATERIALS AND METHODS

Gold-coated QCM-D sensors with and without a 50 nm thick vapor deposited Ti6Al4V layer were purchased at QSense (Sweden). Half of the Ti6Al4V sensors were physisorbed (15 h) with 0.7 M poly(NaSS) in distilled water (Sigma), while the other half was kept untouched. The gold coated sensors were divided in three groups: the first was spin coated with 1% poly(DTE) carbonate) in tetrahydrofuran (OmniSolv), the second with 10% PS in toluene (Sigma) and the last was kept in the original conditions. The wettability of the entire group of crystals was determined using a static contact angle goniometer following the sessile drop method (2 μL, 8 seconds).

The proteins, bovine serum albumin (BSA, Sigma), human fibronectin (Fn, Sigma) and collagen Type I (Col I, Sigma), were used at different concentrations, mimicking their proportion (Fn/BSA) in the human plasma. BSA was used at 4 mg/mL in phosphate buffered saline solution (PBS, Sigma), Fn at 20 µg/mL in PBS and Col I at 10 µg/mL in acetate buffer (0.1 M, pH 5.6).

First part of this study consisted of individual protein adsorption. Each protein was injected into the QCM-D system (fundamental frequency of 5 MHz) at 25 µL/min and the flow was maintained until saturation was reached. In the second part, to analyze competition between the proteins, two proteins were injected on each surface. The second protein was introduced after the surface was saturated with the first protein. PBS was flushed into the system between protein injections to remove unattached fractions.

The attachment of MC3T3-E1 osteoblastic cells onto the five surfaces was tested in order to prove their ability to work as bone/tissue replacements. Before their introduction into the QCM-D system, the cells were expanded in Minimum Essential Eagle Medium – Alpha (MEM-α, Gibco) supplemented with 10% fetal bovine serum (FBS, Gibco) at 37°C and 5% CO_2 in air. Complete medium was primarily injected to establish a relative zero and just then the cells were introduced in a concentration of 5×10^4 cells/mL. They were continuously pumped for 2 hours.

The results were analyzed using the Sauerbrey equation ($\Delta f = -C.\Delta m$, $C = 17.7$ $ng/cm^2.s$). Data from the 3rd to the 11th overtone were analyzed.

III. RESULTS

Data in Table 1 show that polystyrene is the most hydrophobic, while Ti6Al4V is the least.

Table 1: Water contact angles of the five studied QCM-D crystals.

Crystals	Average (SD) (Degrees)
Ti6Al4V	30.9 (2.7)
Ti6Al4V + Poly(NaSS)	44.9 (2.5)
Gold	67.2 (4.8)
Gold + Poly(DTE)	77.3 (3.6)
Gold + PS	84.0 (1.9)

The morphology (scanning electron microscopy) and chemical composition (X-ray photoelectron spectroscopy) of these surfaces were also evaluated. However, since no differences between the crystals surfaces were detected, and the elemental composition was as expected, these results and analysis are not reported on this paper.

The results of the individual protein adsorption (Fig.1) show that Fn and Col I adsorb well to all surfaces. In general, the physisorbed crystals exhibited the highest adsorption. These three proteins possess a clear and significant preference for the sulfonated surface.

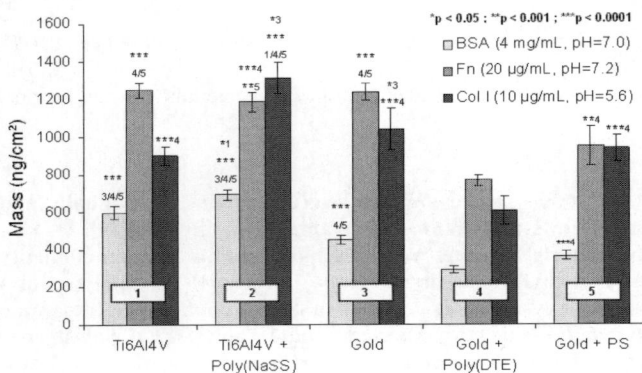

Fig. 1 Individual protein adsorption (BSA, Fn and Col I) on five different biomaterial surfaces.

In the majority of the cases, when two proteins were pooled, the first inhibited the second. For some (e.g., BSA after Col I, data not shown) this inhibition was up to 70%, for all the surfaces investigated.

In one case in which Col I followed Fn on gold + PS crystals, a substantial increase on the Fn adsorption levels was detected, reaching almost double the corresponding single protein values (Fig.2).

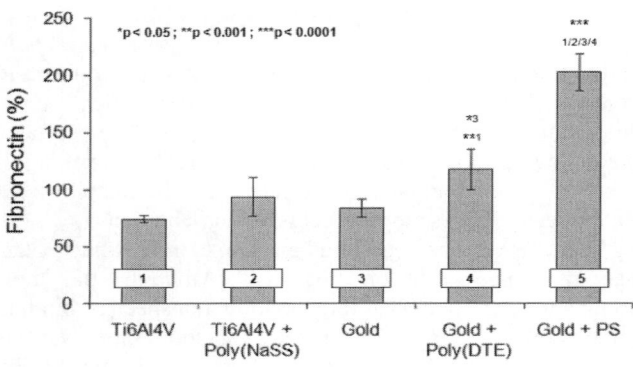

Fig. 2 Percentage of adsorbed Fn after pre-adsorption of Col I, onto five biomaterial surfaces. The percentage values were determined considering the 100% of adsorption when the protein in question is individually deposited on the surfaces.

Ability of the substrates to interact with osteoblastic cells was evidenced by a small but perceptible decrease on the frequency immediately after their injection (Fig.3). This observation was consistent for all the samples investigated in this study (data not shown).

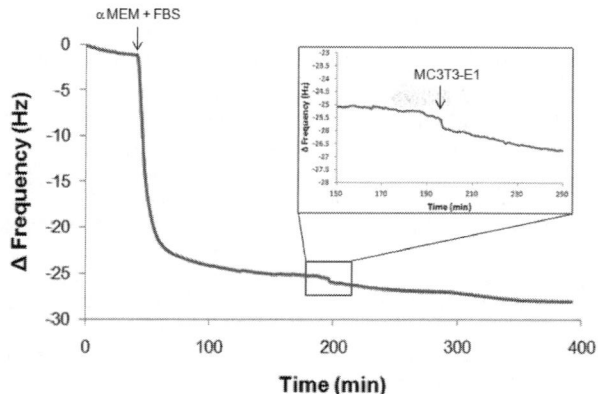

Fig. 3 Schematic representation of the MC3T3-E1 cells attachment on a Gold + DTE crystal for a 2 h period (n = 9, f = 5 MHz).

After a two-hour injection, the highest MC3T3-E1 attachment was seen on poly(DTE) carbonate coated crystals, followed by the titanium crystals (with and without physisorbed poly(NaSS)), while the smallest was on the PS surfaces (data not shown).

IV. DISCUSSION

The protein adsorption of namely Col I and BSA was significantly superior for the physisorbed crystals when compared to the control. The "bioactive" polymer poly(NaSS) is known to induce biological properties such as modulation of protein conformations and/or cell response (adhesion, proliferation, differentiation and signaling) [9,10]. The *in-vitro* results confirmed the ability of sulfonate groups from poly(NaSS) to stimulate the host response.

Protein adsorption onto biomaterials surfaces has been extensively studied and described on the kinetics and thermodynamic aspects. It is well schemed by a race for the surface which involves different parameters as protein affinity for the surface, concentrations of proteins and chemical nature of the surface [11]. As a general rule the first protein in contact with a surface will inhibit the adsorption of the second except if the affinity of the later is higher. Once a protein is adsorbed on a surface, the "free sites" for the adsorption of the second protein are reduced. Moreover the proteins weight/size, affinity and conformation will control this interaction. A less motile protein as well as a protein at a lower concentration but with a higher affinity replaces protein of a lower affinity already adsorbed on the substrate (Vrooman effect) [12]. It was observed that the larger proteins (Col and Fn) have a slight tendency to better adsorb onto hydrophilic surfaces, such as Ti6Al4V, when preceded by the adsorption of a smaller protein. That preference changed to more hydrophobic, like PS, when the opposite association happened (smaller proteins preceded by larger). The explanation resides on the proteins structure/class. Globular-like proteins, such as BSA (the smallest of the group), have a preference for more hydrophobic areas [13]. In addition, proteins tend to conserve more of their native structure on electrostatically neutral hydrophilic surfaces rather than on hydrophobic or charged surfaces [3]. This information taken together with the Vrooman effect theory shows that larger molecules will displace smaller on hydrophilic substrates in order to keep their original structure [3,13].

In contrast to previous investigations [1,2], in this QCM-D study, the kinetics of cell attachment was followed in a dynamic way (continuous flow) simulating *in vivo* conditions. The results showed a smaller attachment of osteoblastic cells compared to the former researches conducted under static conditions. This disparity was expected since in a "static environment" cells have more time to develop and consolidate interactions with the surfaces. We believe longer periods of cell injection will approach the observations.

Regarding the cellular attachment, it was expected the poly(NaSS) physisorbed crystals to show the highest cell attachment, since they exhibited a greater affinity to three of the proteins present in the cells culture medium (MEM-α supplemented with 10% FBS, a complex of proteins and nutrients). However, as discussed earlier, when different proteins are pooled together, as in the culture medium, their conformation can change affecting the cells interaction. As a consequence, the best cellular performance was seen on another surface, the poly(DTE) coated crystals. Unpublished data have confirmed its ability to accelerate cell attachment in early stages of interaction.

V. CONCLUSIONS

The presence of sulfonate groups from the poly(NaSS) polymer favors binding proteins adsorption and optimizes their conformation for cell response. As general rule, the first protein to contact with a surface tends to inhibit the adsorption of a second on the same material as far as its affinity for the surface is higher. The five surfaces that were studied were shown to be suitable for biomedical applications by successful selective proteins adsorption and osteoblastic interaction.

The influence of these proteins on the MC3T3-E1 osteoblastic cells behavior will be studied in a near future, using the QCM-D system.

ACKNOWLEDGMENT

We would like to thank the family Romano for funding the internship for H.F. at the New Jersey Center for

Biomaterials, as a part of the NJCBM International Exchange Program. H.F. would also like to thank the École Doctorale Galilée and the LBPS-CSPBAT lab for all the support and financial help.

REFERENCES

1. Tagaya M, et al. (2011) Detection of interfacial phenomena with osteoblast-like cell adhesion on hydroxyapatite and oxidized polystyrene by the quartz crystal microbalance with dissipation. Langmuir 27: 7635-7644.
2. Tagaya M, et al. (2011) Effect of Interfacial Proteins on Osteoblast-like cells adhesion to hydroxyapatite nanocrystals. Langmuir 27:7645-7653.
3. Gray JJ (2004) The interaction of proteins with solid surfaces. Curr. Opin. Struc. Biol. 14.
4. Moline PJ, et al. (2012) Fibronectin and bovine serum albumin adsorption and conformational dynamics on inherently conducting polymers: a QCM-D study. Langmuir 28:8433-8445.
5. Boyan BD, et al. (2001) Mechanisms involved in osteoblast response to implant surface morphology. Annu. Rev. Mater. Res 31:357–371.
6. Wei J, et al. (2009) Influence of surface wettability on competitive protein adsorption and initial attachment of osteoblasts. Biomed. Mater. 4:1-7.
7. Hlady V, et al. (1999) Methods for studying protein adsorption. Methods Enzymol 309:402-429.
8. Mecca VM, et al. (1996) Extensions of the quartz-crystal-microbalance technique. Sensors and Actuators A 53:371-378.
9. Michiardi A, et al. (2010) Bioactive polymer grafting onto titanium alloy surfaces. Acta Biomaterialia 6:667–675.
10. Mayingi J, et al (2008) Grafting of bioactive polymers onto titanium surfaces and human osteoblasts response. ITBM-RBM 29:1–6.
11. Horbett TA (1994) The role of adsorbed proteins in animal cell adhesion. Colloid Surf B: Biointerfaces 2:225-240.
12. Jung S-Y, et al. (2003) The Vroman effect: a molecular level description of fibrinogen displacement. J. Am. Chem. Soc. 42:12782–12786.
13. Hayens CA, W. Norde (1995) Structures and stabilities of adsorbed proteins. J Colloid Interface Sc 169:313-328.

Author: Helena Felgueiras
Institute: LBPS-CSPBAT CNRS UMR 7244, Institut Galilée, UP 13
Street: 99 av., Jean Baptiste Clément, 93430 Villetaneuse
City: Paris
Country: France
Email: felgueiras.helena@gmail.com

Biomimetic Poly(NaSS) Grafted on Ti6Al4V: Effect of Pre-adsorbed Selected Proteins on the MC3T3-E1 Osteoblastic Development

H. Felgueiras and V. Migonney

Laboratory of Biomaterials and Specialty Polymers, LBPS-CSPBAT UMR CNRS 7244, Université Sorbonne Paris Cité, Paris 13, 93430 Villetaneuse, France

Abstract—Titanium alloy Ti6Al4V has been extensively used in biomedical applications as implants and prostheses with successful results. In order to improve its long term performance in vivo, a bioactive polymer, poly(sodium styrene sulfonate) (poly(NaSS)), was chemically grafted onto its surface. The chemical characterization of poly(NaSS) grafted Ti6Al4V materials was performed by ATR-FTIR, XPS and toluidine blue colorimetric assay highlighting the presence of sulfonate groups. The poly(NaSS) grafting led to a significant increase in the hydrophilicity of the surface (contact angle and surface energy) as well as a strong improvement of the osteoblast cells response.

Proteins play a major role in the interactions between cells and surfaces since adsorbed proteins may modify the cell signaling process. The influence of three proteins – fibronectin (Fn), collagen I (Col I) and albumin (BSA) – involved in osteoblast cell response was studied on ungrafted and grafted surfaces. The osteoblastic cell behavior, differentiation and development/maturation was followed. Fn and Col I tend to facilitate cell adhesion and interactions with the surfaces, respectively in terms of quantity (proliferation) and cell spreading. This result was intensified in the presence of NaSS groups. The pre-adsorption of Col I on grafted surfaces was also associated with significant improvements on the osteoblastic early differentiation (higher alkaline phosphatase activity) compared to the rest of the tested groups.

Generally, it was seen that the poly(NaSS) grafting process on titanium based surfaces instigates osteoblastic development.

Keywords—Fibronectin, Collagen I, Albumin, grafted Ti6Al4V, osteoblasts.

I. INTRODUCTION

The first contact of one biomaterial implant with the living system usually starts by a rapid (within seconds) adsorption of plasma proteins onto the surface. These proteins will be responsible for medicating all the subsequent interactions. Fibronectin (Fn) and collagen type I (Col I), two of the most important, possess RGD (arginine-glycine-asparagine) sequences known for inducing cell attachment. On its turn, albumin, the most abundant, is responsible for blocking nonspecific binding regions [1-3].

The cells that reach the surface establish contact with the protein coated substrate and attach to the ECM of those highly complex particles. Since there is no actual contact between the implant and the cells, the resultant biological activity will depend, among other factors, on the number, conformation, orientation and competition of the recruited proteins [4,5]. Consequently, those factors will be a result of the surface physical and chemical properties influence [6].

Biocompatibility, corrosion resistance, low toxicity and good mechanical properties, namely high tensile strength and durability, high ductility and low density, are some of the requirements that a material to be used as a permanent implant must own [7]. Titanium and its alloys (Ti6Al4V) possess all of those attributes in abundance and for many years they have been successfully used in the biomedical field [7,8]. Nonetheless, the biomaterial-bone tissue interaction can introduce some challenges/difficulties to a long term success. In order to improve that relationship, along the years, many surface treatments have been proposed from simple modifications on the physical properties of the surface [9,10] to the immobilization of bioactive molecules, both from inorganic and organic origins (eg. hydroxyapatite [12] – precipitation of calcium and phosphate groups – or adhesion sequences such as RGD [13]).

The approach followed by our laboratory consists in the grafting of polymers bearing ionic sulfonate groups onto metallic surfaces, following a radical grafting method in two steps. The bioactive polymers are covalently bound to the metallic surface, which is created during the first step of the radical grafting process, allowing the biological properties to prevail [14,15]. Previous studies have pointed out that the distribution of these groups instigates proteins' adsorption and consequently influence the cell response [15,16]. Still, the individual impact of the specific proteins that are intimate related with the osteoblastic development has not yet been shown on such surfaces.

The aim of the current investigation was to confirm previous results by attesting the ability of grafted poly(sodium styrene sulfonate) (poly(NaSS)) to induce osteoblastic differentiation on titanium materials – this time using the Ti6Al4V alloy as substrate – and at the same time infer about the MC3T3-E1 osteoblastic development when biological molecules, such as Fn, Col I and bovine serum albumin (BSA), are individually pre-adsorbed on unmodified (control) and modified surfaces. The presence of the sulfonate groups onto the Ti6Al4V samples was confirmed by ATR/FTIR, XPS and the toluidine blue colorimetric method. The MC3T3-E1 maturation was followed up to 28 days of culture.

II. MATERIALS AND METHODS

Ti6Al4V disks (diameter = 13 mm) were kindly provided by Ceraver (France). Half of the samples were grafted with poly(NaSS) by following the process described by V. Migonney et al. [14-16]. To confirm the presence of the sulfonate groups on the grafted surfaces Fourier-Transformed InfraRed (FTIR) spectra, recorded in an attenuated total reflection (ATR), were obtained using a Nicolet Avatar 370 Spectrometer. X-ray photoelectron spectroscopy (XPS) measurements were also conducted. A K-Alpha XPS instrument (Thermo Scientific, USA) was used with a binding energy of 1350 eV and x-ray spot size of 400 μm. The poly(NaSS) density on the surfaces was acquired using the toluidine colorimetric method [14-16]. The wettability of both ungrafted and grafted Ti6Al4V was determined through dynamic contact angle measurements with four liquids: water (polar), formamide (polar, Sigma), ethylene glycol (polar, SDS) and diiodomethane (apolar, Reagent Plus®).

Prior to cell culture, all samples were washed with 1.5 M NaCl (sodium chloride, Fisher), 0.15 M NaCl, pure water and phosphate buffered saline solution (PBS, Gibco), three times each, and sterilized on each Ti6Al4V face with ultraviolet light (UV, 30 W) for 15 min. Then, they were incubated in a serum free medium (Dulbecco's modified eagle medium, DMEM, with 1% penicillin-streptomycin, 1% fungizone and 1% L-glutamine (Gibco)) for 1 night at 37 °C, to achieve pH equilibrium and antibiotic protection. The surfaces were coated with BSA, Fn and Col I (Sigma), 4 mg/mL in PBS, 20 μg/mL in PBS and 10 μg/mL in acetate buffer (0.1 M, pH 5.6), respectively for 1 h (37 °C). After, they were blocked in 1 % BSA solution in serum free medium for 30 min. The experiments where no specific proteins were pre-adsorbed, the samples were immersed in DMEM supplemented with 10% fetal bovine serum (FBS, Gibco), also known as complete medium (CM).

5×10^4 cells/mL were cultured on each surface until 28 days (37°C, 5% CO_2). DMEM with 10% FBS was the culture medium used. The osteoblastic morphology was evaluated by fluorescent microscopy (ZEISS Axiolab, Germany). The cells were stained with anti-vinculin (Sigma), phalloidin (FluoProbes$_{TM}$) and DAPI (Sigma) [6] and their size analyzed using the Image Pro Plus 5.0 software. Cellular attachment and proliferation were followed from 30 min to 14 days of culture (7 periods). After each period, the cells were detached using trypsin-EDTA (Gibco) and count in a Beckman Cell Coulter (Multisizer III). For alkaline phosphatase (ALP) and mineralizarion experiments, the media was supplemented with L-ascorbic acid (Sigma) and β-glycerophosphate (Sigma), to induce cell maturation [4]. The MC3T3-E1 development was followed for 28 days.

Statistical significance was determined by on-way analysis of variance (ANOVA) followed by the pos-hoc Bonferroni test, using the GraphPad Prism 5.0 software.

III. RESULTS

The ATR/FTIR spectra of the grafted surfaces revealed 4 picks characteristics attributed to the sulfonate groups (SO_3) from poly(NaSS) whereas those signals were not found on the ungrafted Ti6Al4V. The XPS analyses of grafted surfaces showed an increase in the carbon levels (almost 10 times) and the presence of sodium and sulfur confirming the success of the chemical grafting. The density of the grafted poly(NaSS) was found to be $\approx 1.63 \times 10^{-5}$ g/cm^2 ± 2.56×10^{-6} g/cm^2. Regarding the surfaces wettability measurements, it was seen that the polar liquids contact angle decreased with the grafting process while the inverse happened with the apolar liquid, putting in evidence the hydrophilic nature of the surfaces. This was also confirmed by the superior surface energy of grafted surfaces 57.3 mN/m compared to 29.3 mN/m for the ungrafted ones.

MC3T3-E1 cells revealed differences in the morphology according to the type of treated surface and pre-adsorbed protein. In the presence of CM the typical osteoblastic morphology was found. Cells exhibited a polygonal shape with cytoplasmatic protrusions extending in all directions. Here, no significant differences on the cells size were detected. This morphological predisposition prevailed on poly(NaSS) grafted Ti6Al4V pre-coated with BSA and Fn (on the Fn the cells were smaller but were found (visually) in a superior number, Fig 1A). In the absence of poly(NaSS) a more fibroblastic shape was observed. Significant differences on the MC3T3-E1 size were found between grafted and ungrafted surfaces (BSA p<0.001 and Fn p<0.0001, data not shown). The presence of Col I on Ti6Al4V materials conducted to significant expansion of the cells cytoplasm (Fig 1B).

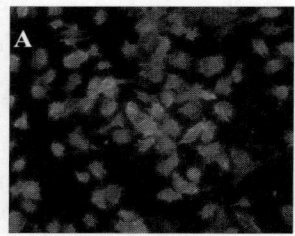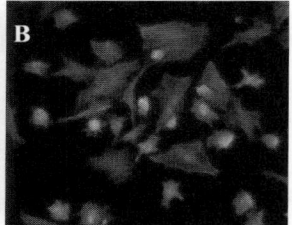

Fig. 1 Fluorescent microscopy images of MC3T3-E1 cells cultured after 4 h on Ti6Al4V grafted with poly(NaSS) and pre-adsorbed with (A) fibronectin and (B) collagen I.

The number of cells on the Ti6Al4V surfaces was determined from 30 min to 14 days, which includes both initial attachment and posterior proliferation. For the CM and BSA treated surfaces there were no significant differences in the kinetics of proliferation between the ungrafted and grafted samples. However when Fn and Col I were present, a slight but significant improvement on the cells number attached to the chemically modified materials was noticed at day 14.

The highest levels of ALP activity, early marker of osteoblastic differentiation, were registered at day 14, for the entire group of surfaces (Fig 2). The presence of poly(NaSS) on Ti6Al4V instigated significantly (p<0.0001) the production of this enzyme. The presence of Col I on the grafted substrates augmented significantly the ALP concentration, in contrast to the rest of the studied specimens.

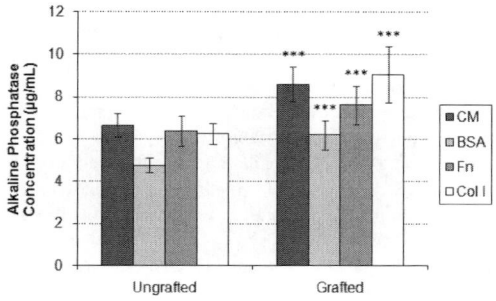

Fig. 2 Alkaline phosphatase produced by the MC3T3-E1 osteoblasts after 14 days of culture. Significant differences between ungrafted and grafted surfaces are indicated by * (*** p<0.0001).

Regarding the calcium and phosphate levels (mineralization), no big disparities were reported between proteins' coated substrates (data not shown). However, as before it was seen that the poly(NaSS) favors considerably more their production.

IV. DISCUSSION

The poly(NaSS) increases the hydrophilic character of the Ti6Al4V surfaces due to the presence of anionic charges along the macromolecular chains. It has been reported that more hydrophilic surfaces develop more focal contact points with cells by protein binding and, consequently, incite more intensively their spreading [16]. It has also been shown that rough surfaces increase cellular affinity [17]. Therefore, by associating these two properties, the grafting process created a sustainable environment for cell spreading, which was confirmed by the observation of the typical morphological characteristics of the osteoblastic cells on all groups of samples.

The proteins Fn and Col I possess bioactive motifs such as RGD cell-adhesive sequences known to be of extreme importance to promote cell attachment. The RGD oligopeptides support integrin-mediated adhesion, whose expression regulates, among other properties, the cells morphology, spreading, proliferation and differentiation [18]. A visual analysis showed that a superior number of cells can be found attached to the grafted surfaces pre-coated with Fn (this information was confirmed by the cell attachment numbers, however without significance). On its turn, the presence of Col I on the Ti6Al4V materials conducted to a significant expansion of the cells cytoplasm. They exhibited well defined nucleus and actin fibers. Aside from enhancing the cells attachment to the grafted surfaces compared to the ungrafted (p<0.05), the MC3T3-E1 cultured on Col I substrates displayed the biggest size (p<0.0001, compared to the rest of the treated surfaces).

Previous investigations have shown that the presence of poly(NaSS) alone has no influence on the cell proliferation rate [19]. The CM and BSA results are in accordance to this fact. However, in the presence of Fn and Col I, a significant improvement on the cells numbers attached to the grafted substrates was detected at day 14. This divergence to the general rule is most likely a result of the combined effect of the sulfonate groups with these particular proteins.

The osteoblasts ability to differentiate was confirmed by the ALP, calcium and phosphate production. While in activity (highest at the 14th day of culture), ALP is capable of releasing phosphate ions to the medium, which will interact with the pre-existent calcium ions and induce the precipitation of calcium phosphate, major component of the inorganic phase of bone. However, after the 14th day, the ALP levels decreased and the ECM maturation took place (superior calcium and phosphate concentration registered at day 28) [13]. It has been documented that the more the hydrophilic and rougher surfaces the more the osteoblastic differentiation occurs [20]. This study attests on this notion.

The presence of Col I on the grafted substrates augmented significantly the ALP concentration, in contrast to the rest of the studied pre-coated samples. The hypothesis concerns to the activation of specific signaling pathways. The effect of Col I on the expression of bone cell phenotypes associated to the ECM formation and maturation has been established [21]. Besides, the supplemented medium (β-glycerophosphate and L-ascorbic acid addition) is considered an instigator of osteogenic marker genes expression, which includes Col I. These supplements adopt an osteoblastic phenotype and secrete and organize the ECM. They trigger a series of molecular events that include the activation of signal transduction pathways that are known to affect osteoblast differentiation [22]. On the late differentiation (mineralization), the differences between pre-coated substrates were not as significant. However, it was once again seen that poly(NaSS) instigates the calcium and phosphate production.

V. CONCLUSIONS

In vitro tests using MC3T3-E1 osteoblastic cells attested on the Ti6Al4V grafted with poly(NaSS) ability for biomedical applications. Substrates pre-adsorbed with CM, BSA, Fn and Col I were studied. The morphological analyses showed that the presence of proteins with RGD adhesive-sequences facilitates the interaction between the cells and the surfaces. Fn was found to be a clear instigator of MC3T3-E1 proliferation when associated with poly(NaSS), right from the initial stages of development. On the other hand, the pre-adsorption of Col I on grafted surfaces was associated with significant improvements on the osteoblastic early differentiation (superior ALP activity). BSA did not reveal any particular effect. Generally, it was seen that the grafting process on titanium based surfaces promotes osteoblastic development, particularly its maturation.

ACKNOWLEDGMENT

Helena Felgueiras is a recipient from *Ministère de la Recherche*. The author would like to thank Dr. Joachim Kohn, from the New Jersey Center for Biomaterials (Rutgers University, USA), for the use of the XPS. The thank you is also extended to Dr. Pierre Ponthiaux from the *Laboratoire de Génie des Procédés et Matériaux* (École Centrale Paris, France) for the use of the Microtopographer.

REFERENCES

1. Tagaya M, et al. (2011) Detection of interfacial phenomena with osteoblast-like cell adhesion on hydroxyapatire and oxidized polystyrene by the quartz crystal microbalance with dissipation. Langmuir 27: 7635-7644.
2. Tagaya M, et al. (2011) Effect of Interfacial Proteins on Osteoblast-like cells adhesion to hydroxyapatite nanocrystals. Langmuir 27:7645-7653.
3. Gray JJ (2004) The interaction of proteins with solid surfaces. Curr. Opin. Struc. Biol. 14.
4. Moline PJ, et al. (2012) Fibronectin and bovine serum albumin adsorption and conformational dynamics on inherently conducting polymers: a QCM-D study. Langmuir 28:8433-8445.
5. Anderson JM (2001) Biological response to materials. Annu. Rev. Mater. Res 31:81-110.
6. Roach P, et al. (2007) Modern biomaterials: a review – bulk properties and implications of surface modifications. J Mater Sci: Mater Med 18:1263-1277.
7. Carew EO, et al. (2004) Chapter1 – Properties of Materials. Biomaterials Science. Elsevier Academic Press 23-65.
8. Long M, Rack HJ (1998) Titanium alloys in total joint replacement – A materials science perspective. Biomaterials 19:1621-1639.
9. Guéhennec LL, et al. (2007) Surface Treatments of titanium dental implants for rapid osseointegration. Dental Materials 23:844-854.
10. Liu X, et al. (2004) Surface modification of titanium, titanium alloys, and related materials for biomedical applications. Materials Science and Engineering 47:49–121.
11. Yang Y, et al. (2010) A review on calcium phosphate coatings produced using a sputtering process – an alternative to plasma spraying. Biomaterials 26:327–337.
12. Rammelt S, et al. (2006) Coating of titanium implants with collagen, RGD peptide and chondroitin sulfate. Biomaterials 27:5561–5571.
13. Hélary G, et al. (2009) A new approach to graft bioactive polymer on titanium implants: Improvement of MG63 cell differentiation onto this coating. Acta Biomaterialia 5:124-133.
14. Michiardi A, et al. (2010) Bioactive polymer grafting onto titanium alloy surfaces. Acta Biomaterialia 6:667–675.
15. Khadali FE, et al. (2002) Modulating fibroblast cell proliferation with functionalized poly(methyl methacrylate) based copolymers: chemical composition and monomer distribution effect. Biomacromolecules 3:51-56.
16. Vogler EA (1998) Structure and reactivity of water at biomaterial surfaces. Adv Colloid Interface Sci. 74:69-117.
17. Gabbi C, et al. (2005) Osteogenesis and bone integration: the effect of new titanium surface treatments. Ann. Fac. Medic. Vet. di Parma.
18. Garcia AJ, Boettiger D (1999) Integrin-fibronectin interactions at the cell-material interface: initial integrin binding and signaling. Biomaterials 20:2427-2433.
19. Oughlis S, et al. (2012) The osteogenic differentiation improvement of human mesenchymal stem cells on titanium grafted with poly(NaSS) bioactive polymer. J Biomedical Materials Research A 101A:582-589.
20. Advincula MC, et al. (2006) Osteoblast adhesion and matrix mineralization on sol-gel derived titanium oxide. Biomaterials 27:2201-2212.
21. Padial-Molina M, et al. (2011) Role of wettability and nanoroughness on interactions between osteoblast and modified silicon surfaces. Acta Biomater 7:771-8.
22. Higuchi C, et al. (2002) Continuous inhibition of MAPK signalling promotes the eraly osteoblastic differentiation and mineralization of the extracellular matrix. J Bone Miner Res. 17:1785-1794.

Author: Helena Felgueiras
Institute: LBPS-CSPBAT CNRS UMR 7244, Université Sorbonne Paris Cité, Université Paris 13
Street: 99 av., Jean-Baptiste Clément, 93430 Villetaneuse
City: Paris
Country: France
Email: felgueiras.helena@gmail.com

Bioactive Intraocular Lens – A Strategy to Control Secondary Cataract

Yi-Shiang Huang[1,2], Virginie Bertrand[1], Dimitriya Bozukova[3], Christophe Pagnoulle[3], Edwin De Pauw[1], Marie-Claire De Pauw-Gillet[1], and Marie-Christine Durrieu[2]

[1] Mass Spectrometry Laboratory, Department of Chemistry and Mammalian Cell Culture Laboratory, Department of Biomedical and Preclinical Sciences – GIGA R, University of Liège, Liège, Belgium
[2] CBMN UMR5248, Institut Européen de Chimie et Biologie, University of Bordeaux I, Pessac, France
[3] PhysIOL, Liège, Belgium

Abstract—Cataract is the opacity of the lens, causing impairment of vision or even blindness. Today, a surgery is still the only available treatment. The intraocular lens (IOL) is a polymer implant designed to replace the natural lens in the cataract surgery. However, the bioinert materials could not satisfy the unmet need in the secondary cataract control. Posterior capsular opacification (PCO, or Secondary Cataract), characterized by a thick and cloudy layer of lens epithelial cells (LECs), is the most common postoperative complication.

In our research, a bioactive molecule is immobilized onto the conventional acrylic hydrophilic polymer pHEMA (Poly(2-hydroxyethyl methacrylate)) using oxygen plasma treatment followed by deposition. The RGD peptide sequence, being well-known for its ability to promote cellular attachment by binding to integrin receptors, is designed to stimulate the adhesion of LECs on the IOL. Our data show the peptide immobilized biomaterial not only exhibits similar optical property, but also reveals enhanced biological properties in cell adhesion and cell morphology maintenance. By means of surface functionalization of IOL to stimulate LECs adhesion, the secondary cataract could be controlled.

Keywords—Intraocular Lens, Posterior Capsular Opacification, RGD Peptide, Lens Epithelial Cells, Secondary Cataract.

I. INTRODUCTION

Cataract is the opacity of the lens or capsule of the eye, causing impairment of vision or even blindness. The cataract surgery, with damaged native lens extraction and intraocular lens (IOL) implantation, is still the only currently available treatment. The materials for IOL require excellent optical properties for light transmission, mechanical properties for folding injection during surgery, and biological properties for preventing body rejection. The biocompatibility - or more specified: "bio-inert" - seems to be the prerequisite in selecting the materials. Nowadays, the conventional materials for IOLs include PMMA (Poly(methyl methacrylate)), silicone, hydrophobic acrylic and hydrophilic acrylic polymers [1]. The hydrophilic acrylic polymer, mainly composited by pHEMA (Poly(2-hydroxyethyl methacrylate)), has several superior characteristics. Surgeons benefit from its foldability and controlled unfolding behavior. Patients suffer less from glistening and glare phenomenon. For the manufacturers, the rigidity in dry state helps for easy machining. However, this material tends to induce secondary cataract [1].

Secondary cataract, or Posterior Capsular Opacification (PCO), is the most common postoperative complication of the cataract surgery [2]. PCO is raised from the cells response to the implant: the lens epithelial cells (LEC) undergo series of cellular responses to form a thick cloudy layer and enclose the intraocular lens, causing patients to lose vision again. Current treatment is using Nd:YAG laser capsulotomy. However, this also potentially creates other complications such as damage to the IOL, higher intraocular pressure, cystoid macular oedema, retinal detachment [1]. Ideally, to prevent the PCO is better than to treat PCO.

The "Sandwich Theory" model is proposed to elucidate the developmental process of PCO [3]. In this model, the residual LECs between the lens capsular bag and the IOL undergo proliferation, migration, as well as transdifferentiation and finally induce PCO if the affinity to the IOL material is low. The sandwich theory provides an idea to control PCO: LECs will remain attached, mitotically quiescent if the LEC can adhere to the surface of the IOL properly. The fact that hydrophobic acrylic material exhibits a higher LEC adhesion ratio and less incidence of PCO also gives a hint to prevent PCO.

Our strategy is to restore the normal cellular status by creating a bioactive surface of implant. In this study, we use the conventional 25% water content acrylic hydrophilic disks (AH25) as a starting material and improve its surface property by applying the well-known RGD peptide [4].

II. MATERIALS AND METHODS

Crosslinked AH25 polymer disks with thickness of 1 mm were obtained from PhysIOL (Belgium). Peptides with the sequence KRGDSPC was purchased from GeneCust (Luxembourg).

Peptide Immobilization

The AH25 disks were rinsed with deionized water before an overnight lyophilization. Dried disks were then subjected into the chamber of radio frequency glow discharge (RFGD) instrument (customized, Europlasma). The surface activation by plasma treatment was driven at 200 W for 10 minutes. The flow rate of oxygen gas was set at 15 Sccm, and the system pressure was maintained at 50 mTorr. After plasma treatment, the peptide deposition process was performed by incubating the disks in the peptide aqueous solution under vigorous shaking for 24 hours at room temperature. The peptide immobilized disks were washed with PBS solution for 24 hours.

X-ray Photoelectron Spectroscopy (XPS) Characterization

A VG Scientific 220 i-XL ESCALAB spectrometer was used for the surface analysis with a non-monochromatized MgKα source (hv=1253.6 eV) at 100 W (10 kV and 10 mA). A pressure of 10^{-7} Pa was maintained in the chamber during analysis. The analyzed area is about 150 μm in diameter. The full spectra (0-1150 eV) were obtained with a constant pass energy of 150 eV and high resolution spectra at a constant pass energy of 40 eV. Charge neutralization was required for all insulating samples. The peaks were referenced to C1s maximum shifted at 284.7 (±0.1) eV.

Cell Culture

Lens epithelial cells (LECs) were isolated from the lens crystalline capsule of pig eyes (Pietrain-Landrace pig; from Detry SA. Aubel, Belgium). The complete culture medium is composed of 87% Dulbecco's Modified Eagle's Medium (BE12-733, Lonza), 10% fetal bovine serum (10270-106, Gibco), 1% penicillin/streptomycin antibiotics (BE17-602, Lonza), 1% non-essential amino acids (NEAA) (BE13-114, Lonza), 1% sodium pyruvate (BE 13-115, Lonza), 1% Glutamax (35050 Gibco, Invitrogen), and 1% HEPES (17-737, Lonza). The cells were cultured under the condition of 5% CO_2 and 37°C. Trypsin-EDTA (Gibco, Invitrogen) was used for cell detachment after one rinse with PBS without calcium and magnesium (BE17-516, Lonza).

Cell Adhesion Assay

The peptide-immobilized polymer samples are cut into 14 mm diameter disks, washed and sterilized in PBS (BE17-513) at 1.5 bar, 120 °C, for 21 min. The polytetrafluoroethene (PFTE) cell culture (customized, SIRRIS) inserts were also sterilized by autoclave. Every disk was put into a well of a 12-well culture plate and fixed by an insert for seeding cells. The LECs cell concentration was adjusted into $5.3*10^4$ cells/ml. For each well, 750 μl of cell solution was added. The cells were cultured for 3 days.

Three days later, the culture medium was removed, and the sample was carefully washed with PBS (with Ca^{2+} and Mg^{2+}) in order to eliminate the non-adhering and dead cells. The adhering cells were fixed with 4% paraformaldehyde at 4 °C for 2 h, washed three times with PBS (with Ca^{2+} and Mg^{2+}), and kept in PBS (with Ca^{2+} and Mg^{2+}) at 4°C.

After hematoxylin (1.05174, Merck) and eosin (1.15935, Merck) coloration, the sample surface was observed with an optical inverted microscope Olympus IX81, equipped with UPlanFL objective Olympus (10x magnification) and a CCD camera (C-BUN-F-XC50). The images were visualized with the CellSens software (Olympus). Six independent photos were taken for each sample. The surface area occupied by the cells was determined with the ImagePro Analyzer management software (MediaCybernetics).

III. RESULTS

Characterization of Peptide Immobilization by XPS

XPS was employed to determine the surface chemical composition of surface modification. The XPS spectra of different status of chemical modifications were collected and their element compositions were calculated and shown in Table 1. The native polymer is a copolymer of 2-hydroxyethyl methacrylate (HEMA) (C/O=2) and ethoxyethyl methacrylate (EEMA) (C/O=2.66). From the experimental data, the virgin polymer surfaces exhibit only C and O elements as expected. The C/O experimental ratio on polymer surface is 2.21. The native polymer surface treated with oxygen plasma, followed by distilled water incubation shows a C/O ratio of 2.22. The samples after oxygen plasma treatment were incubated in 1 mM and 10 mM peptide solution. The nitrogen element, indicator of peptide, was observed and their nitrogen atomic ratios were 0.9% and 2.5%, respectively.

Table 1 Sample Element compositions obtained by XPS

Element	Polymer control	0 mM RGD	1 mM RGD	10 mM RGD
C %	68.9	68.9	72.1	67.6
O %	31.1	31.1	27.0	29.9
N %	-	-	0.9	2.5
C/O	2.21	2.22	2.67	2.26
C/N	-	-	80.16	27.32

Light Transmittance Assay

The light transmittances were measured of different concentrations of peptide during surface modification. (Fig. 1) The disks were homologous transparent and presented the light transmittance greater than 80% from green light to red light spectrum. The decreasing of light transmittance in blue light rises from the "Blue Filtering" design of the bulk polymer disk.

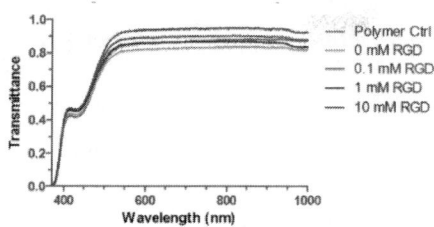

Fig. 1 Light transmittance assay

LEC Adhesion Assay

The evaluation of the bio-integration function of the RGD peptide was performed by culturing lens epithelial cell onto the polymer surface. The LECs were cultured onto the samples. After fixation and H&E staining, the images were taken and the cell coverage ratios were calculated. (Fig. 2) The polystyrene control stands for the conventional plastic surface optimized for cell culture and serves as a positive control for LECs proliferation and morphology. The cells adhere least onto the virgin AH25 surface. Along with RGD concentration increased, the adhesions of cells increase significantly. Moreover, in the sample with higher RGD concentration, the proliferated LECs clustered and spread similar to the polystyrene control, providing the evidence that the surface modification of RGD peptide facilitate the morphology maintenance as well as the adhesion of the porcine LEC.

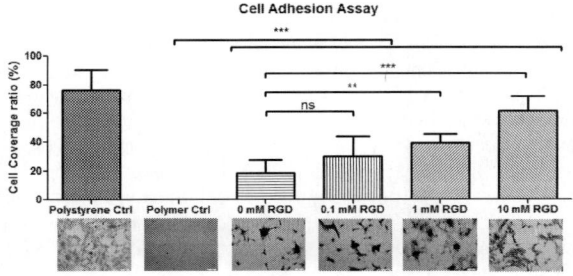

Fig. 2 Lens Epithelial Cell Adhesion Assay (ns nonsignificant, * $P<0.05$, ** $P<0.01$, *** $P<0.001$)

IV. DISCUSSION

Strategy Selection of Bioactive IOL to Control PCO

As a frequent complication of cataract surgery, scientists and engineers have designed a lot of methods to prevent PCO formation [2]. In geometry design, the square edged optic shape is considered better than round edged one, because the square edge can restrict the migration of LECs toward the space between the optic and the lens capsular bag, where the proliferation of LECs blocks the light and causes PCO [1]. In material selection, the hydrophobic acrylic polymer was reported less PCO incidence but causes other problems such as glistening and glare phenomenon. The hydrophobic polymer has lower water permeability and a higher protein adsorption rate [5]. When water molecules enter the hydrophobic optic, the condensation of water will change the direction of refracted light and induce the glistening which needs explanation [6]. The higher protein adsorption rate may lead the hydrophobic IOL optic surface coated by fibronectin and other proteins from the aqueous humor after the cataract surgery [7]. The coated protein layer will further provide a signal for the adhesion of cells [8]. Considering the sandwich theory [3], hydrophobic material provides a better cell adhesion, which seems to be a key factor for less PCO formation. However, the hydrophilic acrylic material has just the opposite properties. Therefore, in order to prevent PCO, as well as the glistening and the glare phenomenon in the same time, creating a bioactive surface which could attract LEC adhesion onto the hydrophilic acrylic polymer optic is a feasible strategy.

RGD peptide sequence was known for its recognition by the integrin on the cell surface [4]. While bound, integrins make transmembrane connections to the cytoskeleton and activate various intracellular signaling pathways, thus modulating many aspects of cell functions including adhesion, proliferation, polarity, spreading, motility, actin cytoskeleton organization and focal adhesion formation [9]. Nowadays, RGD peptide has been widely used in tissue fields including bone regeneration, tissue replacement, and drug targeting [10]. Comparing to the whole molecule of fibronectin, the RGD domain is the smallest domain for integrin recognition, less immune response, easy to prepare, and more stable to the immobilization chemistry [4]. In our case, the sequence KRGDSPC was applied because previous results showed that the sequence RGDSP provides a higher activity to integrins [4] and the flanking K and C is designed for the future work in grafting chemistry.

Considering the immobilization of the RGD peptide onto the hydrophilic acrylic material, the oxygen plasma, or radio frequency glow discharging (RFGD) method was applied. The plasma treatment can modify polymer surfaces in different ways: etching, cleaning, activation, cross-linking, and implantation, which specifically means incorporating oxygen in the chemical structure of the polymer [11]. The advantages of applying plasma treatment in the surface modification are reliable, reproducible, relatively inexpensive, and applicable to different sample geometries as well as different such as metals, polymers, ceramics, and composites [12]. In our case, oxygen plasma is designed to transform hydroxyl group into carboxyl group, and create more oxygen containing functional groups, which make the polymer more hydrophilic to favor LEC adhesion.

Immobilization of Peptide Not Altering Its Functions Required for IOL Implantation

As a candidate of new biomaterial in ophthalmic implant, the peptide immobilized polymer should provide appropriate features in optical properties, mechanical properties and biocompatibility. Since the starting material, AH25, is a conventional biomaterial used in IOL, it is appropriate to use it as a control to investigate the impact of the peptide surface functionalization process. From our data in light transmittance, the intensities of transmitted light in all the RGD immobilized samples are greater than 80% and spectrums resemble the virgin polymer control. In addition, the disks are clear and transparent, indicating the surface functionalization does not alter its optical property in light transmission.

LEC Adhesion Enhanced by Immobilization of Peptide and Its Potential of Controlling PCO

After cataract surgery, the remaining LECs undergo proliferation, migration, and epithelial-mesenchymal transition (EMT), which are all major causes of PCO [13]. Previous study shows the hydrophobic acrylic polymer surface favor protein adsorption as well as LEC adhesion [6]. Referring the Sandwich theory, the better LEC adhesion helps less EMT and induces less PCO. The surface functionalization by RGD peptide should provide a better cell adhesion and therefore control PCO formation. From our data, LECs responses to RGD modified surfaces with increased cell coverage (Fig. 2). In addition, the morphology of LECs in higher RGD concentration samples resembling the polystyrene control and suggesting a "normal" phenotype may imply a less EMT, which is a key process in PCO formation. The EMT process can be detected with molecular markers. For example, the alpha smooth muscle actin is an acquired marker and the E-cadherin is an attenuated marker during EMT [14]. In our further investigation, quantified information can be obtained by immunofluorescence technique with an internal standard of cytoskeleton protein.

V. CONCLUSIONS

The bioactive surface created by oxygen plasma implantation followed by RGD peptide immobilization exhibits improved LEC adhesion and cell morphology maintenance. The modification does not impede the light transmittance required for an ophthalmic implant material. The surface modification chemistry is clean, robust, and compatible for industrial production. It is also worthy to investigate the impact of LECs toward RGD grafted surfaces created by oxygen plasma implantation followed by crosslinker-assisted conjugation.

In order to confirm the PCO-controlling property of this new material, further researches including *in vitro* EMT test, *ex vivo* PCO model assay and *in vivo* implantation study are needed.

ACKNOWLEDGMENT

The authors thank the Walloon region, Belgium (Project LINOLA, 1117465) for financial support. We appreciate Dr. Christine Labrugère's (ICMCB, Pessac, France) assistance in XPS measurement.

REFERENCES

1. Bozukova, D., et al., Polymers in modern ophthalmic implants— Historical background and recent advances. Materials Science and Engineering: R: Reports, 2010. 69(6): p. 63-83.
2. Awasthi, N., S. Guo, and B.J. Wagner, Posterior capsular opacification- a problem reduced but not yet eradicated. Arch Ophthalmol, 2009. 127(4): p. 8.
3. Linnola, R., The Sandwich Theory 2001, Oulu: University of Oulu.
4. Hersel, U., C. Dahmen, and H. Kessler, RGD modified polymers: biomaterials for stimulated cell adhesion and beyond. Biomaterials, 2003. 24(24): p. 4385-4415.
5. Luensmann, D. and L. Jones, Albumin adsorption to contact lens materials: a review. Cont Lens Anterior Eye, 2008. 31(4): p. 179-87.
6. Pagnoulle, C., et al., Assessment of new-generation glistening-free hydrophobic acrylic intraocular lens material. J Cataract Refract Surg, 2012. 38(7): p. 1271-7.
7. Perez Olmedilla, M., et al., Response of human chondrocytes to a non-uniform distribution of hydrophilic domains on poly (ethyl acrylate-co-hydroxyethyl methacrylate) copolymers. Biomaterials, 2006. 27(7): p. 1003-12.
8. Anderson, J.M., A. Rodriguez, and D.T. Chang, Foreign body reaction to Biomaterials. Semin Immunol, 2008. 20(2): p. 15.
9. Lei, Y., et al., Peptide immobilization on polyethylene terephthalate surfaces to study specific endothelial cell adhesion, spreading and migration. J Mater Sci Mater Med, 2012. 23(11): p. 2761-72.
10. NIU, X., et al., Arg-Gly-Asp (RGD) Modified Biomimetic Polymeric Materials. J. Mater. Sci. Technol., 2005. 24(4): p. 6.
11. Hegemann, D., H. Brunner, and C. Oehr, Plasma treatment of polymers for surface and adhesion improvement. Nuclear Instruments and Methods in Physics Research Section B: Beam Interactions with Materials and Atoms, 2003. 208: p. 281-286.
12. Chu, P.K., et al., Plasma-surface modification of biomaterials. Materials Science and Engineering: R: Reports, 2002. 36: p. 64.
13. Liu, H., et al., The effects of rapamycin on lens epithelial cell proliferation, migration, and matrix formation: An in vitro study. Molecular Vision, 2010. 16: p. 8.
14. Zeisberg, M. and E.G. Neilson, Biomarkers for epithelial-mesenchymal transitions. J Clin Invest, 2009. 119(6): p. 1429-37.

Author: Yi-Shiang HUANG
Institute: Department of chemistry, University of Liège
Street: Institut de Chimie (B6c), Allée de la Chimie 3, Sart-Tilman
City: Liège
Country: Belgium
Email: yshuang@ulg.ac.be

Ultrathin Films by LbL Self-assembly for Biomimetic Coatings of Implants

M. Giulianelli, L. Pastorino, R. Ferretti, and C. Ruggiero

University of Genoa/Department of Informatics, Bioengineering, Robotics and Systems Engineering, Genoa, Italy

Abstract—The number of arthroplasties is rapidly increasing, but the ability of current biomaterials to mimic the natural environment of human tissues is very limited. It would be desirable to fabricate implant surface nanocoatings which recreate the main features of the biological interactions between the implant and the host tissue. Osseointegration is influenced by mechanical stability and biological stability and surface modification of implants at nanoscale level is a recognized method to favor it. In this respect the major problem is the incidence of implant infections on implant integration in the bone.

The work described here is focused on promoting osseointegration and reducing bacterial infections incidence on implanted biomaterials. We have used the Layer-by-Layer (LbL) self-assembly of Collagen I (Col) and Hyaluronic Acid (HA) to functionalize planar supports. LbL is an easy, versatile and affordable technique able to modify in a controlled way different kind of surfaces using opposite charged polyelectrolytes. Titanium substrates have been coated obtaining higher homogeneity and smoother surfaces with respect to pristine titanium. These coatings can be regarded as a first step for the development of more complex structures to modulate osteogenesis and avoid implant infections.

Keywords—Artificial implants, layer-by-layer, self - assembly, ultrathin film, Collagen I, Hyaluronic Acid.

I. INTRODUCTION

In the last decade the incidence of arthroplasty has increased not only in elder population but also in younger patients [1, 2]. Stable integration of orthopedic implants with host bone is crucial to avoid short term revisions [3-5].

In order to improve implant performances, a new and effective approach consists of applying bio-inspired nanocoatings on bone-implant interface to enhance bone tissue direct apposition rather than fibrous encapsulation [6, 7].

Layer-by-Layer self-assembly is a coating technique with which is possible to create thin film with nanoscale features to functionalize surfaces for biomedical applications [8]. Polyelectrolyte multilayer (PEM) thin film coatings on biomedical implants have the potential to ensure high longevity and excellent biocompatibility, guiding osteoblast adhesion, proliferation and differentiation at the implant-bone interface [9-12] and reducing the risks of bacterial infection, one of the primary cause of mechanical aseptic loosening in situ.

This technique is economic, highly reproducible, with designed architectures controllable at nanometer scale through fabrication parameters (such as pH, ionic strength, concentration of polyion solution). The constructive principle is quite simple and envision a proper alternation of positive and negative compounds. After every layer deposition, polyion resaturation is realized resulting in the alternation of the terminal charge. The coating process does not need surface of particular shape and size but it requires only charged supports [13, 14]. Extensive work has been carried out on the application of this technique to the fabrication of multilayer ultrathin films on surfaces of any shape and size, ranging from flat surfaces to round template capsules, using a wide variety of charged compounds [15-24].

Deposition of bioactive molecules as a coating thin film on titanium implants can significantly improve the implant-bone response. In vivo osseointegration and bone remodeling are influenced by several molecules and substances. The early healing phase is the most critical and affects the long term success of the arthroplasty[25]. The interaction between the surface and the tissue is a very complex phenomena in which, few seconds after the implantation, a cascade of adsorption events take place, involving not only water and ions but especially unspecific adsorption of plasma proteins. The entire process is influenced by the chemo-physical nature of the biomaterial surface (e.g. composition, energy, charge and the charge-transfer capabilities).

Main prerogatives of surface functionalization via PEM thin film are to avoid metallic ions release from the implant, to prevent unspecific adsorption of denatured proteins at the surface, to attract cells of the native tissue able to differentiate into the appropriate type and to provide biochemical signals to trigger healing mechanisms.

The approach described in our study aims to mimic the dimension, geometry, and arrangement of natural components of host tissue in order to create an ideal milieu for cellular adhesion, growth and differentiations, hampering bacterial adhesion [26]. The goal was realized immobilizing a PEM thin film of Collagen I and hyaluronic acid on titanium supports. Collagen I is the principal structural protein of the organic bone matrix and together with growth factors and adhesion proteins it affects cell-matrix interactions. Hyaluronic acid (HA), another integral part of ECM, is a biocompatible and biodegradable linear polysaccharide with bacterial inhibitory effect [27-29]. In order to define the better deposition protocol for each solution, a Quartz Crystal Microbalance (QCM) was used to monitor the ultrathin film fabrication. The protocol was then used to recreate this structure on titanium substrates and silicon substrates, further characterized with SEM and AFM respectively to assess the surface morphology and verify deposition.

II. MATERIALS AND METHODS

A. Materials

Cationic type I collagen from calf skin (COL, Sigma product number C8919), anionic Hyaluronic acid sodium salt from rooster comb (HA, Sigma product number H5388) and cationic poly (ethyleneimine) (PEI, average Mw_25.000, Aldrich product number 40,872-7) were used for the ultrathin film formation. PEI solution was prepared in pure water at a concentration of 5 mg/ml. COL was diluted in 0.1M acetic acid solution at a concentration of 1mg/mL, stirred for three hours in a becker using a magnetic bar, then diluted again in purified Milli-Q water at a final concentration of 0.2 mg/mL. HA was used as received and diluted in purified Milli-Q water at a final concentration of 0.5 mg/mL. All solutions were adjusted to a value of pH equal to 4 using HCl 0.1M. Water, used in the experiments for the solutions preparation and washing, was purified by Milli-Q system and had a resistance of 18.2 MΩcm.

B. Ultrathin Films Preparation

Study of the process of ultrathin films fabrication was performed on planar surfaces using a quartz crystal microbalance instrument working in liquid environment (QCM-Z500, KSV Instruments, Helsinki, Finland). The QCM-Z500 instrument measuring principle is based on the analysis of the quartz crystal impedance at multiple overtones. The obtained parameters are used to calculate the properties of adsorbed layers such as mass, density and thickness. PEI/(HA/COL)$_3$ multilayers were deposited on gold-coated 5 MHz AT-cut quartz crystals. Before adsorption, the quartz crystals were cleaned with H_2SO_4 at 150°C for 20 min followed by washing in purified Milli-Q water. After each use the quartz crystals were renewed. Considering that the quartz crystal surface is negatively charged, due to partial oxidation in air, a first layer of positive PEI was deposited to make easier the following deposition of the structure (HA/COL)$_3$. Specific solutions were alternatively introduced into the measurement chamber and were left in contact with the quartz crystal for 5 min for Milli-Q water (first calibration step), for 10 min for PEI adsorption, 20 min for HA deposition and 20 min for COL deposition. After each adsorption step, water at pH 4 was purred into the chamber and left in contact with the crystal for 5 min in order to remove the unabsorbed molecules. The data analysis and calculation of thickness were performed using the QCM Impedance Analysis software (KSV Instruments, version 3.11).

C. Characterization

A Scanning Electronic Microscope (SEM, Hitachi S-2500) at an acceleration voltage of 10kV observed the surface morphology of titanium substrates. Being the titanium samples conductive, there was no need to perform a substrate metallization. Samples were air dried at room temperature, and then analyzed at the AFM by using a custom build set-up driven by R9 advanced controller (RHK technology) in air at room temperature. Data acquisitions were carried out in tapping mode at scan rates between 0.4 and 0.7 Hz, using rectangular Si cantilevers (NCHR, Park Systems) having the radius of curvature less than 10 nm and with the nominal resonance frequency and force constant of 330 kHz and 42 N/m, respectively.

III. RESULTS AND DISCUSSION

The assembly of the multilayer ultrathin film on planar substrates of quartz having the architecture PEI/(HA/COL)$_3$ was characterized by QCM. The frequency shift due to the deposition of each successive layer of polyelectrolyte onto the quartz crystal revealed a gradual growth of the film with the number of deposition cycles (Fig. 1). The frequency shift of each assembly step is used to quantify the mass adsorption and thickness for each layer deposited thanks to Kanazawa-Gordon equation (Fig 2,3) [30-32].

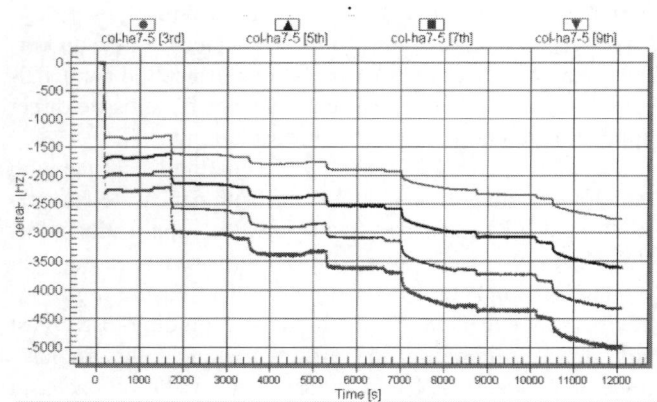

Fig. 1 Frequency shift versus time

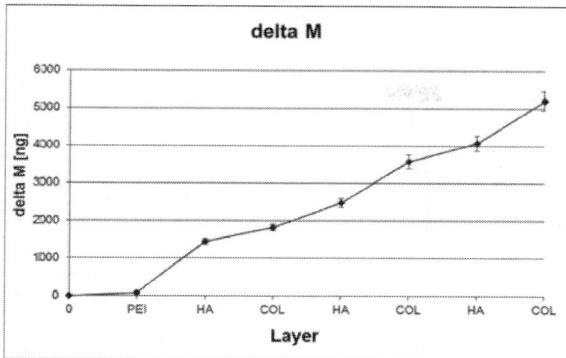

Fig. 2 Mass adsorption (in ng) for each layer deposited

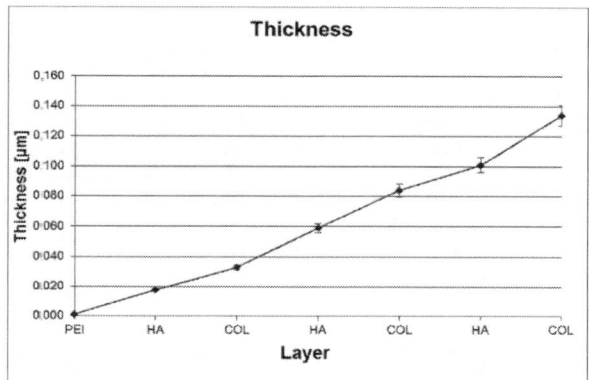

Fig. 3 Thickness (in µm) for each layer deposited

Polyelectrolyte multilayer coatings of implants provide an essential biological property at the implant–tissue interface, which leads to enhanced osseointegration. Surface biomaterial coatings with natural polyelectrolytes can improve adhesion, spreading and proliferation of cells [33].

In Fig. 4 non-treated titanium substrate (panel a) and functionalized titanium substrates with PEI/(HA/COL)$_3$ are shown. Titanium with an ultrathin film presents higher homogeneity and a smoother surface in respect to rough pristine titanium.

This confirm the success of the deposition of HA and COL that levelled the nanometric features of the titanium dioxide interface.

The same technique was performed on planar silicon to verify if the deposited fibers of collagen assumed a preferential orientation. PEI/(HA/COL) architecture was adsorbed on the silicon samples, then characterized via AFM. The superior spatial resolution of the AFM image (Fig. 5) reveals the presence of fibrillar aggregates and a random disposition of collagen fibers. LbL technique cannot

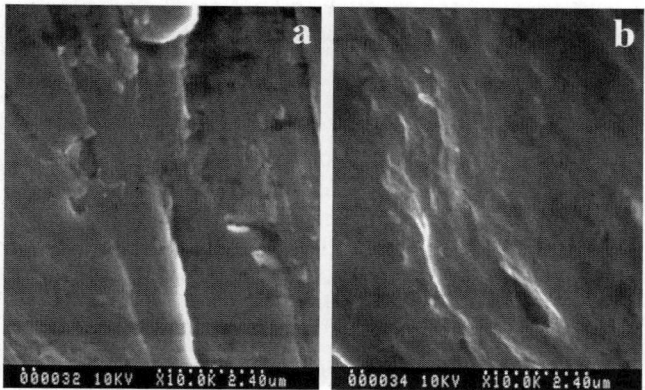

Fig. 4 SEM image of non-treated titanium substrate (a) and functionalized titanium substrates with PEI/(HA/COL)$_3$ (b)

clearly guide the assembly of fibers in larger ordered unit, but it is useful to modify the interface topography of biomaterials maximizing osteoconductive performances of orthopedic implants.

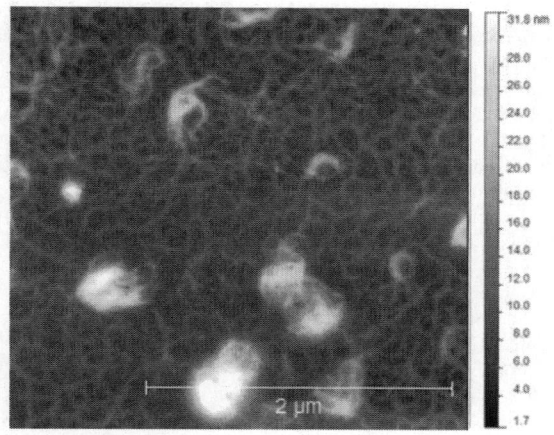

Fig. 5 AFM of silicon functionalized with PEI/(HA/COL) structure

IV. CONCLUSIONS

Even though the regenerative capabilities of bone tissue are not yet fully understood, it is recognized that future biomaterials need to be designed taking into account all complex phenomena that occur when a biomedical implant is inserted in the human tissue. Only an implant endowed with biomimetic potential can overcome main limitations of arthroplasties as micromotion at the interface, peri-implant infections, lack of personalized solutions (e.g. patients carrying metabolic disease as osteoporosis). Ultrathin films fabrication on the surface of biomedical implants can enhance functionality and biological efficacy of biomaterial used in current arthroplasties, especially if they are designed to avoid bacterial adhesion maximizing osteoconductive processes. Topography is able to affect cellular behavior on different levels and many efforts have to be done in order to define the key features of the osteoconductive phenomena. We can consider the development of these ultrathin films as a primary step to the realization of a smart coating capable of enhance osteoconductivity and reduce bacterial infections in biomaterials.

REFERENCES

1. Mäkelä, K.T., et al. (2008) Total Hip Arthroplasty for Primary Osteoarthritis in Patients Fifty-five Years of Age or OlderAn Analysis of the Finnish Arthroplasty Registry. The Journal of Bone & Joint Surgery. 90(10): p. 2160-2170
2. Jenkins, P., et al. (2013) Predicting the cost-effectiveness of total hip and knee replacement A health economic analysis. Bone & joint journal. 95(1): p. 115-121
3. Kurtz, S.M., et al. (2007) Future clinical and economic impact of revision total hip and knee arthroplasty. The Journal of Bone & Joint Surgery. 89(suppl_3): p. 144-151
4. Geetha, M., et al. (2009) Ti based biomaterials, the ultimate choice for orthopaedic implants–A review. Progress in Materials Science. 54(3): p. 397-425
5. Ulrich, S.D., et al. (2008) Total hip arthroplasties: What are the reasons for revision? International orthopaedics. 32(5): p. 597-604
6. Sun, J., et al. (2012) Fabrication of bio-inspired composite coatings for titanium implants using the micro-dispensing technique. Microsystem Technologies. 18(12): p. 2041-2051
7. Wilhelmi, M. and A. Haverich (2013) Functionalized medical implants in the era of personalized medicine. Clinical Practice. 10(2): p. 119-121
8. Goodman, S.B., et al. (2013) The future of biologic coatings for orthopaedic implants. Biomaterials. 34(13): p. 3174-3183
9. Samuel, R.E., et al. (2011) Osteoconductive protamine-based polyelectrolyte multilayer functionalized surfaces. Biomaterials. 32(30): p. 7491-7502
10. Klymov, A., et al. (2013) Understanding the role of nano-topography on the surface of a bone-implant. Biomaterials Science. 1(2): p. 135-151
11. Shah, N.J., et al. (2012) Osteophilic multilayer coatings for accelerated bone tissue growth. Advanced Materials. 24(11): p. 1445-1450
12. Caneva Soumetz, F., et al. (2010) Investigation of integrin expression on the surface of osteoblast-like cells by atomic force microscopy. Ultramicroscopy. 110(4): p. 330-338
13. Decher, G. (1997) Fuzzy nanoassemblies: toward layered polymeric multicomposites. science. 277(5330): p. 1232-1237
14. Iler, R. (1966) Multilayers of colloidal particles. Journal of Colloid and Interface Science. 21(6): p. 569-594
15. Lvov, Y.M. (2002) Polyion–protein nanocomposite films. Encyclopedia of Surface and Colloid Science. 3: p. 4162
16. Habibi, N., et al. (2012) Polyelectrolyte based molecular carriers: The role of self-assembled proteins in permeability properties. Journal of Biomaterials Applications
17. Soumetz, F.C., L. Pastorino, and C. Ruggiero (2008) Human osteoblast-like cells response to nanofunctionalized surfaces for tissue engineering. Journal of Biomedical Materials Research Part B: Applied Biomaterials. 84(1): p. 249-255
18. Habibi, N., et al. (2011) Nanoengineered polymeric S-layers based capsules with targeting activity. Colloids and Surfaces B: Biointerfaces. 88(1): p. 366-372
19. Pastorino, L., et al. (2009) Paclitaxel-containing nano-engineered polymeric capsules towards cancer therapy. Journal of nanoscience and nanotechnology. 9(11): p. 6753-6759
20. Pastorino, L., et al. (2011) Collagen containing microcapsules: Smart containers for disease controlled therapy. Journal of Colloid and Interface Science. 357(1): p. 56-62
21. Salerno, M., et al. (2013) Adhesion and proliferation of osteoblast-like cells on anodic porous alumina substrates with different morphology. IEEE transactions on nanobioscience. 12(2): p. 106-111
22. Soumetz, F.C., et al. (2010) Investigation of integrin expression on the surface of osteoblast-like cells by atomic force microscopy. Ultramicroscopy. 110(4): p. 330-338
23. Pastorino, L., et al. (2011) Permeability Variation Study in Collagen-Based Polymeric Capsules. BioNanoScience. 1(4): p. 192-197
24. Erokhina, S., et al. (2012) Release kinetics of gold nanoparticles from collagen microcapsules by total reflection X-ray fluorescence. Colloids and Surfaces A: Physicochemical and Engineering Aspects
25. Beutner, R., et al. (2010) Biological nano-functionalization of titanium-based biomaterial surfaces: a flexible toolbox. Journal of the Royal Society Interface. 7(Suppl 1): p. S93-S105
26. Chun, Y.W. and T.J. Webster (2009) The role of nanomedicine in growing tissues. Annals of biomedical engineering. 37(10): p. 2034-2047
27. Li, X., et al. (2012) The responses of preosteoblasts to collagen/hyaluronic acid polyelectrolyte multilayer coating on titanium. Polymers for Advanced Technologies. 23(4): p. 756-764
28. Burdick, J.A. and G.D. Prestwich (2011) Hyaluronic acid hydrogels for biomedical applications. Advanced Materials. 23(12): p. H41-H56
29. Pitt, W.G., et al. (2004) Attachment of hyaluronan to metallic surfaces. Journal of Biomedical Materials Research Part A. 68(1): p. 95-106
30. Keiji Kanazawa, K. and J.G. Gordon II (1985) The oscillation frequency of a quartz resonator in contact with liquid. Analytica Chimica Acta. 175: p. 99-105
31. Kanazawa, K.K. and J.G. Gordon (1985) Frequency of a quartz microbalance in contact with liquid. Analytical Chemistry. 57(8): p. 1770-1771
32. Melroy, O., et al. (1986) Direct determination of the mass of an underpotentially deposited monolayer of lead on gold. Langmuir. 2(6): p. 697-700
33. Zankovych, S., et al. (2012) The effect of polyelectrolyte multilayer coated titanium alloy surfaces on implant anchorage in rats. Acta Biomaterialia

Author: Carmelina Ruggiero
    Institute: University of Genoa
    Street: Via all'Opera Pia 13
    City: Genoa
    Country: Italy
    Email: carmel@dibris.unige.it

Nanohelical Shape and Periodicity Dictate Stem Cell Fate in a Synthetic Matrix

R.K. Das, O.F. Zouani, R. Oda, and M.-C. Durrieu

Institut Européen de la Chimie et Biologie, CBMN-UMR5248, Université de Bordeaux 1, Pessac, France

Abstract—Several aspects of the extracellular matrix (ECM) microenvironment, such as protein composition, topographical features and elasticity, have been reported to play critical roles in stem cell lineage specification. *In vivo*, the ECM is composed of nano-fibrillar networks of defined periodicity. Changes in the periodicity and shape of this nanofiber matrix have been implicated in the induction of several diseases. Despite this, till date, there has been no investigation of the effect of the nanoperiodicity of the ECM nanostructure on the fate of stem cells. In this report, we use helical organic nanoribbons based on self-assembled Gemini amphiphiles to access chiral silica nanoribbons with different shapes (twisted/helical) and periodicity. Glass substrates covalently grafted with 'RGD' functionalized helical nanoribbons of periodicity 63 nm induced specific hMSC adhesion through fibrillar focal contact formation, and commitment towards osteoblast lineage in absence of osteogenic-inducing media. In contrast, the substrate with twisted nanoribbons of periodicity 100 nm failed to induce osteogenic commitment. Inhibition of non-muscle myosin II with blebbistatin was sufficient to block osteogenic commitment on helical nanoribbon matrix, demonstrating that the stem cells interpreted nanohelical shape and periodicity environment in the same way they sense microenvironment elasticity. This study thus provides a promising tool to promote osteogenic capacity and may find applications in bone tissue engineering.

Keywords—silica nanostructures, nanohelical periodicity, stem cell microenvironment, cell differentiation.

I. INTRODUCTION

Construction of ECM mimetic synthetic systems is a topical area of stem cell research. Sophisticated design of nano-biomaterials is producing smart, responsive scaffolds that are increasingly capable of rivalling the *in vivo* ECM structural and functional complexity. In this context, the capability of self-assembly as a powerful bottom up approach to design artificial stem cell niche environments has been well demonstrated [1,2]. Mesenchymal stem cells (MSCs), a widely available autologous source of stem cells, have immense potential in regenerative medicine and tissue engineering. Precise control of MSCs lineage commitment is an important goal for their successful application in stem cell therapy. In the past decade, we have developed chiral self-assembled high aspect ratio nanostructures from non-chiral cationic bis-quaternary ammonium Gemini surfactants $[C_2H_4$-$1,2$-$((CH_3)_2N^+C_nH_{2n+1})_2]$, denoted as *n-2-n*, in presence of chiral tartrate counterions [3,4]. Sol-gel polycondensation with tetraethoxysilane (TEOS) enables access to chiral silica nanoribbons [5,6]. In this paper, we show that twisted and helical silica nanoribbons obtained by this method can be covalently immobilized on glass substrates, and investigate the effect of the periodicity and shape of this isotropically oriented nanoribbon matrix on hMSCs differentiation.

II. MATERIALS AND METHODS

A. Preparation of Silica Helical Nanoribbons and Twisted Nanoribbons

The 1 mM aqueous gels of *16-2-16* L-tartrate were aged for an appropriate time (24 h for transcription to twisted silica nanoribbons and 20 days for transcription to helical silica nanoribbons), and then were subjected to sol-gel polycondensation following reported procedure [6], finally to obtain helical silica (diameter 35±5 nm; periodicity 63±5 nm) and twisted silica (diameter 20±3 nm; periodicity 100±15 nm) nanoribbons.

B. Functionalization of Silica Nanoribbons with RGD Peptide

Silica nanoribbons were first reacted with (3-aminopropyl)triethoxysilane (APTES), following a reported procedure to introduce $-NH_2$ groups on nanoribbon surface [7]. 1 mL of amine modified silica dispersion in ethanol was then sonicated with 3-(maleimido)propionic acid (SMP; 3 mg in 1 mL ethanol) to introduce a reactive end group. Subsequent covalent attachment of this group to the cysteine end of the fluorescent peptide FITC-KRGDSPC in water (1 mL aqueous solution of 0.3 mM peptide) yielded the RGD peptide modified silica nanoribbons.

C. Activation of Glass Substrates

The glass substrates were first treated with APTES (2 mM hexane solution; 10 mM) for 2 h to functionalize the surface with $-NH_2$ [8]. Subsequent reaction with succinic anhydride in DMF (4 mg in 2 mL) for 3 h intruoduced a carboxylic acid function on the surface, which was converted to an activated ester by reaction with N-hydroxysuccinimide (104 mM) in 94 mM MES buffer (in presence of 172 mM EDC) for 20 h at 4 °C.

D. Covalent Grafting of RGD Modified Silica Nanoribbons on Activated Glass Substrates

The activated glass substrates were immersed in 2 mL of an aqueous dispersion of the RGD-functionalized silica nanoribbons (0.2 mg / mL) at room temperature for 24 h and then washed with water (1 h of sonication).

E. Characterization of Chemical Grafting

The functionalized substrates were characterized by X-ray photoelectron spectroscopy (nonmonochromatized Mg K 1253.6 eV source of 100 W), fluorescence microscopy (Leica microsystem DM5500B) and scanning electron microscopy (Hitachi S2500 at 10 kV).

F. Cell Culture and Immunofluorescence

Human mesenchymal stem cells (hMSCs) were seeded at a density of 10^4 cells/cm^2. After 4, 24 and 96 h of culture, hMSCs were fixed with paraformaldehyde. Fixed cells were permeabilized and blocked with 1% bovine serum albumin. Cells were then incubated with primary antibody and secondary antibody successively, then stained for F-actin using Alexa Fluor® 488 phalloidin. Cell staining was carried out for vinculin, osterix, OPN, β3 tubulin and MyoD1. Cell nuclei were counterstained with DAPI. After DAPI staining, samples were mounted with coverslips on microscope slides.

III. RESULTS

A 1 mM aqueous solution of *16-2-16* L-tartrate self-assembled to form helical nanoribbons after 1 day. Silica sol-gel polycondensation after only 1 day of aging leads to partial unfolding of these helices, eventually forming twisted silica nanoribbons after transcription [5]. In contrast, when the gel is aged for 20 days, silica transcription produces helical silica nanoribbons, preserving the native organic nanostructure. Utilizing this property of the self-assembly, helical and twisted silica nanoribbons of different periodicity were obtained (63 nm and 100 nm, respectively).The RGD-peptide functionalization scheme of the silica nanoribbons is illustrated in Fig. 1A. After APTES treatment, successful amine modification was confirmed by ninhydrin test (Fig. 1C). Citrate stabilized Au NPs were used as markers to confirm the homogeneous distribution of –NH$_2$ on the ribbon surface (Fig. 1B). Modification with SMP led to the disappearance of the absorption due to ninhydrin-amine complex (~ 600 nm). After grafing of the fluorescent peptide FITC-KRGDSPC,the aqueous dispersion of the silica nanoribbons showed the characteristic absorption of fluorescein moiety (Fig. 1D), confirming the success of functionalization.

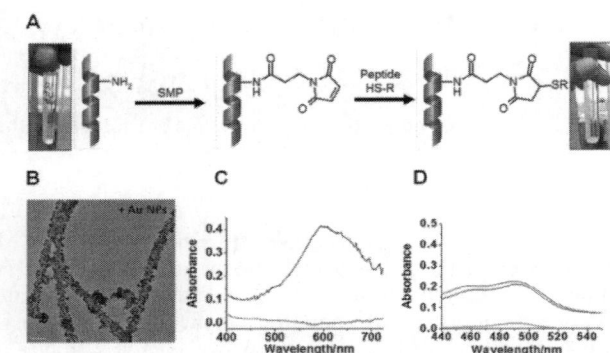

Fig. 1 (A) Schematic chemical representation of covalent immobilization of fluorescent RGD-peptide on silica helical nanoribbons. (B) TEM image of amine modified silica helical nanoribbons, after treatment with citrate-stabilized Au NPs. (C) Absorption spectra after treatment of silica nanoribbons with ninhydrin, with amine-modifed silica (black) and SMP-modified silica (red). (D) Absorption spectra of RGD-peptide grafted nanoribbons (black/blue) and the wash solvent (red).

These RGD-functionalized silica nanoribbons were then covalently grafted on activated glass substrates (Materials and Methods). Fluorescence microscopy image of these substrates showed the presence of fluorescent silica nanoribbons at high magnification (Fig. 2A and 2B). SEM analysis of these substrates confirmed this observation (Fig. 2C). A magnified SEM image also revealed that the nanoribbon structure was not affected after covalent grafting (Fig. 2D). Taken together, these observations confirmed the success of covalent grafting of RGD-functionalized peptides on activated glass substrates.

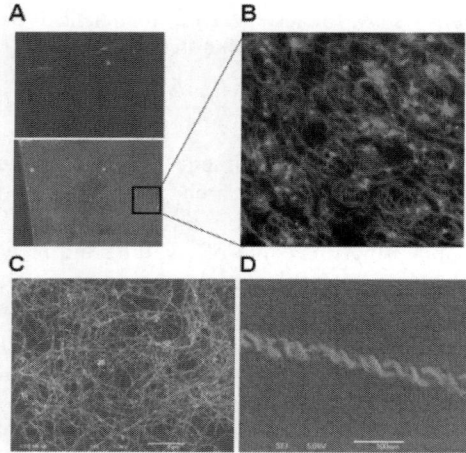

Fig. 2 Fluorescence microscopy images (A) (2.5X at excitation 488 nm) of control glass substrate (top) and functionalized glass substrate (bottom), after treatment with fluorescent RGD-peptide modified silica nanoribbons, and (B) at high magnification (40X) for functionalized glass substrate. (C) and (D) SEM images of helical silica nanoribbon modified glass substrates with scale bars 1 μm and 100 nm, respectively.

In order to study the influence of nanoribbon shape and periodicity on hMSCs fate, all cell adhesion experiments for the first 8 h were carried out in serum free condition. Unfunctionalized glass substrates and those homogeneously functionalized with the fluorescent peptide FITC-KRGDSPC were used as reference substrates. Within 4 h of cell culture, the presence of small filopodia extensions could be clearly observed on nanoribbon modified substrates (Fig. 3B and 1′), and not on the control substrates (Fig. 3A). After 24 h of hMSCs culture on silica nanoribbon functionalized glass substrates, hMSCs on the nanohelical ribbon (periodicity ~ 63 nm) modified substrate showed more stable cell adhesion and better cell spread compared to both control and twisted nanoribbon substrates. Besides, the nanohelical condition showed the presence of many fibrillar focal adhesions (FA), as shown in Fig. 4, left. FA size quantification revealed significantly higher size of FAs in nanohelical ribbon-RGD condition (Fig. 4, right).

ribbon-RGD condition and control substrate with only RGD (Fig. 5 right). OPN was expressed only on helical ribbon-RGD grafted substrate (data not shown).

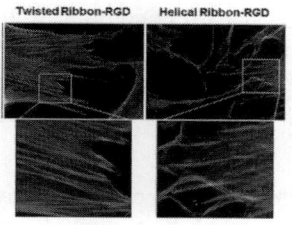
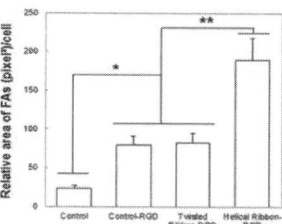

Fig. 4 Left : After 24 h of seeding, focal contact formation and stress fiber assembly are shown on different modified glass substrates as demonstrated by antivinculin staining (red), phalloidin staining (green) and DAPI (blue). Right : Relative area of focal adhesion contacts on different glass substrates (**p < 0.01, *p < 0.001).

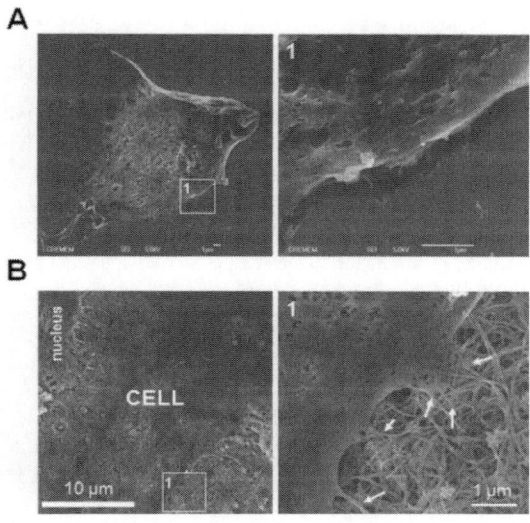

Fig. 3 (A) SEM images of hMSCs on control glass substrate and (B) on glass helical silica nanoribbon-RGD grafted glass substrate after 4 h of culture. (1 and 1′) Magnified SEM images of cell-substrate interactions. Filopodia-like structures are visible on helical ribbon-RGD modified glass substrate (white arrows in 1′).

hMSCs on grafted nanohelical-RGD substrate showed preferential commitment towards osteogenic lineage after 96 h of incubation (Fig. 5). This observation was confirmed by immunofluorescence staining of STRO-1 (stem cell marker) and two common protein osteogenic markers: osterix (Osx) and osteopontin (OPN). There was a clear decrease of STRO-1 expression in the helical ribbon-RGD condition (Fig. 5, left). Osx expression was significantly more in helical ribbon-RGD condition compared to twisted

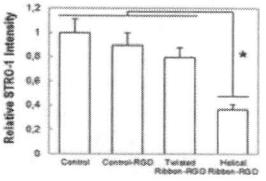
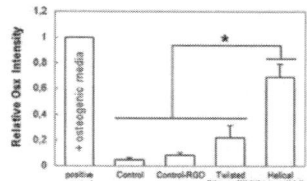

Fig. 5 Total cellular STRO-1 (left) and Osx immunofluorescence intensity (right) was quantified for hMSCs cultured after 96 h on different modified glass substrates (*p < 0.005).

IV. DISCUSSION

The results presented in this study clearly show that silica nanoribbons grafted on a glass substrate have a strong effect on the adhesion and differentiation behaviors of hMSCs. Our studies have established that stem cell fate is strongly linked to nanoperiodicity and dimensions. Indeed, it is demonstrated that it is possible to induce fibrillar focal adhesions and osteoblast differentiation solely by altering the periodicity and dimesion (shape and size) of an isotropically oriented silica nanoribbon matrix (twisted nanoribbons vs helical nanoribbons).

A critical question thus arises: how do the stem cells interpret the periodicity of the nanoribbons? In other words, which signalling pathway dictates the differentiation? It is well documented in literature that among the cytoskeletal motors, myosin II isoforms are implicated in actin contractility, and thus are involved in the process of the cell sensing the microenvironment elasticity of the ECM [9,10]. In our study, when non-muscle myosin II was inhibited by blebbistatin, the osteogenic commitment of hMSCs cultured on helical nanoribbon-RGD modified substrates was completely suppressed. These results suggest that hMSCs interpret the nanohelical periodicity through myosin II dependent pathway, similar to the mechanism by which they sense the microenvironment elasticity. This suggested mechanism is consistent with the general notion that when the stem cells are stressed and have mature focal contacts, they tend to differentiate into osteoblast lineage.

V. CONCLUSIONS

In this study, we demonstrated an important role of nanoribbon shape and periodicity in hMSCs commitment and showed that helical nanoribbons with a specific periodicity of ~ 60 nm induce significantly better cell adhesions through fibrillar focal contact formation and stronger commitment towards osteogenic lineage, compared to twisted nanoribbon with a periodicity of ~ 100 nm. Inhibition of non-muscle myosin II with blebbistatin was sufficient to block this osteogenic commitment, implying that the stem cells interpret the nanohelical shape and periodicity environment physically. This system closely models the ECM structural changes that are associated with the induction of several life-threatening disease conditions. The present study also provides a tool to promote osteogenic capacity of hMSCs, which may have important ramifications in bone tissue engineering.

ACKNOWLEDGEMENT

The authors thank Dimytro Dedovets for technical assistance and stimulating discussions. We also thank Michel Martineau at CREMEM, Université Bordeaux 1, for recording SEM images. This work was supported by the "Région Aquitaine", the GIS "Advanced Materials in Aquitaine" and the "Agence Nationale pour la Recherche" (ANR).

REFERENCES

1. Discher DE, Mooney DJ, Zandstra PW. (2009) Growth Factors, Matrices, and Forces Combine and Control Stem Cells. Science 324:1673-1677
2. Sargent TD, Guler MO, Oppenheimer SM et al. (2008) Hybrid Bone Implants : Self-Assembly of Peptide Amphiphile Nanofibers within Porous Titanium. Biomaterials 29:161-171
3. Oda R, Huc I, Schmutz M et al. (1999) Tuning Bilayer Twist Using Chiral Counterions. Nature 399:566-569
4. Berthier D, Buffeteau T, Leger JM et al. (2002) From Chiral Counterions to Twisted Membranes. J. Am. Chem. Soc. 124 :13486-13494
5. Oda R, Artzner F, Laguerre M et al. (2008) Molecular Structure of Self-Assembled Chiral Nanoribbons and Nanotubules Revealed in the Hydrated State. J. Am. Chem. Soc. 130 :14705-14712
6. Delclos T, Aime C, Pouget E et al. (2008) Individualized Silica Nanohelices and Nanotubes : Tuning Inorganic Nanostructures Using Lipidic Self-Assemblies. Nano Lett. 8 :1929-1935
7. Tamoto R, Lecomte S, Si S et al. (2012) Gold Nanoparticle Deposition on Silica Nanohelices: A New Controllable 3D Substrate in Aqueous Suspension for Optical Sensing. J. Phys. Chem. C 116:23143-23152
8. Porte-Durrieu MC, Labrugere C, Villars F et al. (1999) Development of RGD peptides Grafted onto Silica Surfaces: XPS Characterization and Human Endothelial Cell Interactions. J. Biomed. Mater. Res. 46:368-375
9. Engler AJ, Sen S, Sweeney HL et al. (2006) Matrix Elasticity Directs Stem Cell Lineage Specification. Cell 126: 677-689
10. Zouani OF, Kalisky J, Ibarboure E et al. (2013) Effect of BMP-2 from Matrices of Different Stiffnesses for the Modulation of Stem Cell Fate. Biomaterials 34: 2157-2166

Author: Rajat K. Das
Present Address:
Institute: Institute for Molecules and Materials, Radboud University
Street: Heyendaalseweg 135, 6525 AJ
City: Nijmegen
Country: Netherlands
Email: r.das@science.ru.nl

The Role of Adult Stem Cells on Microvascular Tube Stabilization

O.F. Zouani*, Y. Lei, and M.-C. Durrieu

Institut Européen de la Chimie et Biologie, CBMN-UMR5248, Université de Bordeaux 1, Pessac, France

Abstract—In the recent years, different studies have established a link between adult mesenchymal stem cells (MSCs) and perivascular cells like pericytes that envelop the vascular tubes. This link was validated mainly through shared marker signatures and raised the hypothesis that all MSCs are pericytes. A controversy however remains on whether there are functions specific only for one of the cell types. In this communication, we report the specific role of pericytes in vascular tube stabilization that is not shared with MSCs. An experimental model was introduced for the induction of endothelial cell (EC) tubulogenesis after 24 hours of incubation on micropatterned polymer surfaces. Pericytes or MSCs were added separately to this system in order to evaluate their effect the tubular stabilization. In the absence of additional cells or in the presence only of MSCs the tubular structures were lost after 36h. Adding only pericytes however stabilized the EC vasculogenic tubes. Furthermore, only pericytes, but not MSCs, were able to migrate through a mimetic basement membrane and could interact with the ECs to stabilize the lumen structures. We thus suggest that pericytes have specific cell functions of potential major importance in vascular and bone tissue regeneration. Their use in tissue engineering will thus become critical.

Keywords—surface micropatterning, tubulogenesis, pericyte, mesenchymal stem cell.

I. INTRODUCTION

Blood vessels are composed of two interacting cell types: endothelial cells (ECs), that form the inner lining of the vessel wall and perivascular cells, like pericytes, vascular smooth muscle cells (vSMCs) or mural cells, that envelop the surface of the vascular tubes [1]. Recent interesting studies have suggested a link between these perivascular cells (as in the case of pericytes) and different progenitor or multipotent stem cell types like adult human mesenchymal stem cells (hMSCs), [2] white adipocyte progenitors, [3] muscle stem cells, [4] and even neural stem cells. [5] This link was validated based on the analysis of different specific cell markers. It was indeed shown that, in certain conditions, the pericytes can express different hMSC markers like CD73, CD105, and CD90 [2]. Furthermore, cultured pericytes were shown to differentiate *in vitro* into osteoblasts, adipocytes, chondrocytes, vSMCs and skeletal muscle [4]. These observations have raised the hypothesis that all MSCs are pericytes [6]. A controversy however remains as the MSCs don't express several molecular markers specific for the pericytes such as PDGFRβ (platelet-derived growth factor receptor-beta), desmin, NG2 (chondroitin sulfate proteoglycan 4), and α-SMA (alpha-smooth muscle actin). This suggests the existence of cell functions specific only for the pericytes or for the MSCs, despite the shared markers between these cell types. Here, we report the specific role of pericytes in vascular tube stabilization that is not shared with MSCs.

II. MATERIALS AND METHODS

A. Fabrication and Preparation of Micropatterned Surfaces

Micropatterns on polymer surfaces were fabricated by photolithographic techniques as previously developed. [7] Briefly, the surfaces of materials were coated with S1818 photoresist and spun at 3000 rpm for 30 s to obtain a uniform photoresist layer with a thickness of approximately 2 μm. The surfaces were baked at 115 °C for 1 min for drying. The surfaces were then exposed to UV light (60 W) through a high-resolution Cr mask with predesigned pattern dimensions for 18s. Subsequently the surfaces were developed in Microposit Developer solution for 40s to dissolve the exposed photoresist, resulting in the desired pattern on material surfaces. Subsequently, the GDSVVYGLR peptides (Genecust, France) were functionalized onto the micropatterns by covalent immobilization as previously reported. [7] Finally, the photoresist surrounding the peptide micropatterns was removed by acetone, resulting in SVVYGLR peptide micropatterns on polymer surfaces.

B. Cell Culture

Human umbilical vein endothelial cells (HUVECs) and human pericytes were purchased from Promocell, France (C-12203 and C-12980, respectively). HUVECs are vWF positive, CD31 positive and smooth muscle alpha-actin (α-SMA) negative; pericytes are revealed to be CD146 positive. Human (Bone Marrow) Mesenchymal Stem Cells were obtained from LONZA, Switzland (PT-2501). And the cells are tested for purity by flow cytometry and for their ability to differentiate into osteogenic, chondrogenic and adipogenic lineages. Cells are positive for CD105, CD166, CD29, and CD44. Cells test negative for CD14, CD34 and CD45.

---

* Corresponding author.

HUVECs were grown in HUVEC culture medium (IMDM (Invitrogen, France) supplemented with 20% (v/v) fetal bovine serum (FBS) (PAA, France) and 0.4% (v/v) EC growth supplement/heparin kit (Promocell). [7, 8] Pericytes were maintained in pericyte growth medium (Promocell). Human MSCs were cultured in Minimum Essential Medium (Alpha-MEM, GIBCO) supplemented with 10% (v/v) FBS and 1% penicillin/streptomycin (GIBCO). [10, 11] These cells were incubated in a humidified atmosphere containing 5% (v/v) CO_2 at 37 °C. The cells were subcultured using trypsin/EDTA. HUVECs at passages 3 to 5, pericytes at passages 4 to 5 and hMSCs at passage 2 were used for experiments.

C. Cell Functions on Micropatterned Surfaces

In direct co-culture of HUVECs and pericytes, 25000 cells/cm2 HUVECs and 25000 cells/cm2 pericytes were mixed in EGM®-2 medium and seeded onto glass coverslips at 37 °C and 5% CO_2 for 24 h. Same conditions were employed for direct co-culture of HUVECs with hMSCs.

Subsequent co-culture of cells was developed on micropatterned surfaces in present study. In the first case, 50000 cells/cm2 HUVECs in EGM®-2 medium were seeded onto the micropatterned surfaces. After 24 h culture of ECs on the surfaces, medium was aspirated from the cells, and 50000 cells/cm2 hMSCs or pericytes were added onto the surfaces and further cultured for 16 h, respectively. In another case, basement membrane components were used to mimic the physiological condition of EC tubulogenesis. After 24 h culture of ECs on the micropatterned surfaces, medium on the surfaces was removed, and 500 µl of 4 °C Matrigel (1:4) (v/v) (BD Bioscience, France) or 500 µl of room temperature type IV collagen (Sigma, France) were added onto the surfaces and incubated for 1 h at 37 °C, respectively. Then the medium was removed, and 50000 cells/cm2 hMSCs or pericytes in EGM®-2 medium were added to the surfaces and further cultured for 16 h. In controlled condition, 50000 cells/cm2 HUVECs were seeded on the micropatterned surfaces in EGM®-2 medium and incubated for 24 h, 36 h, 48 h and 72 h, respectively.

D. Immunohistochemistry and Fluorescent Microscopy

In some cases, HUVECs were labeled with cell tracker green (CMFDA, Invitrogen) before cell seeding onto materials. Pericytes and hMSCs were both labeled with cell tracker red (CMTPX, Invitrogen) before cell seeding. Immunofluorescence staining was also performed to visualize the cells in different conditions. After cell culture, the samples were fixed with 4% (w/v) paraformaldehyde (PFA), permeabilized with 0.5 % triton X-100 in PBS (v/v) and stained with primary and secondary antibodies. The primary antibodies were: mouse anti-smooth muscle α-actin (α-SMA, Sigma), mouse anti-chondroitin sulfate proteoglycan (NG2, Millipore), rabbit anti-CD31 (R&D Systems), rabbit anti-von Willebrand Factor (Invitrogen), mouse anti-fibronectin (Santa Cruz). Cells were washed twice with PBS and incubated with Alexa Fluor® 647-, 568- or 488-conjugated secondary antibodies (Invitrogen) where appropriate. Cell nuclei were counterstained with DAPI. Then the samples were mounted with ProLong® Gold antifade reagent (Invitrogen) for observation. Imaging was performed using confocal microscopy (Leica SP5, Germany), and Imaris 7.0 software was used for three-dimensional (3D) reconstructions of confocal images of cells on micropatterned surfaces.

E. STATISTICS

All observation of morphology and proliferation was based on at least three populations for each condition, and experiments were performed in triplicates. Data were displayed as mean values ± standard deviation (SD). Statistical analysis was performed by one-way analysis of variance (ANOVA) (OriginPro 8). P-values less than 0.05 were considered statistically significant.

III. RESULTS

We have introduced an *in vitro* experimental model for tubulogenesis consisting of micropatterned polymer surfaces with 50 µm stripes of angiogenic SVVYGLR peptides [7]. These micropatterns were fabricated using a photolithographic technique as previously reported. [7, 8] The immobilized peptides were in fact fluorescently labeled (GDSVVYGLRK-FITC) in order to visualize the stripes. ECs were then seeded onto this 2D *in vitro* culture system and after 4h of culture, they began to align on the SVVYGLR peptide micropatterns. An almost complete alignment can be visualized after 24h of culture (Fig. 1A, top) with an increased elongation of the cell bodies as compared to unpatterned controls. Furthermore, after 24h of culture, the ECs also underwent morphogenesis with the formation of capillary tube-like structures on the peptide stripes. These structures can be visualized by vertical and horizontal cross sections of confocal images of the cells appearing as a negatively stained central lumen, extending along multiple cell lengths (Fig. 1A, middle and bottom). However, after 36h of culture, the vasculogenic tubular structures were lost with a rapid proliferation of the ECs forming a homogeneous monolayer of cells (Figure 1B, C, and D). This tendency was confirmed by the evaluation of the percentage of lumen structures in the cell culture over time (Figure 1E). Our *in vitro* experimental model of tubulogenesis is thus well adapted for the study of the effect of additional factors on the lumen stabilization. In fact, pericytes or hMSCs can be added to the culture in order to evaluate their effect on the vasculogenic tubular structure. Interestingly, on non-patterned glass coverslips, only direct co-cultures of ECs with pericytes but not with MSCs led to interactions between cells of the different types. In fact, hMSCs remained isolated from the ECs but the pericytes could clearly interact with them to form vascular-like structures.

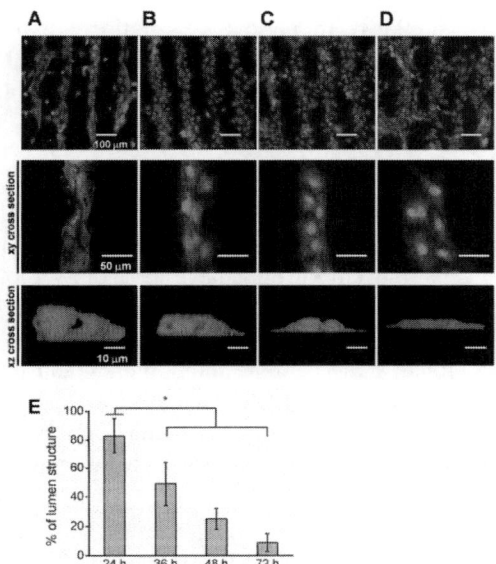

Fig. 1 Confocal images of ECs seeded on 50 μm SVVYGLR peptide micropatterned surfaces for (A) 24 h, (B) 36 h, (C) 48 h and (D) 72 h. (E) Percentage of lumen structures on micropatterned peptide stripes at different time point. ECs formed tubular structure at 24 h of culture. However, after 36 h of culture, ECs lost their lumen structures and formed a monolayer of cells. EC staining with Cell Tracker Green (CMFDA) and DAPI were represented in green and blue, respectively.

In order to evaluate their effect on vasculogenic stabilization, pericytes or hMSCs were added after 24h of incubation of ECs to our experimental model for 16h (Figure 2). We could observe that the vasculogenic tubes were stabilized only in the presence of pericytes but not hMSCs (Figure 2A, B). In fact, we also show that hMSCs did not affect the EC proliferation which was otherwise inhibited by the presence of perycites in our system (Figure 2C). This inhibition may indeed contribute to the blood vessel stabilization. Moreover, these pericytes can have different locations as seen from 3D reconstruction of the confocal images (Figure 2D). They are however more often located beneath the ECs (Figure 2E) and can thus act as "adhesion points" supporting the vasculogenic lumen stabilization in our 2D *in vitro* culture system.

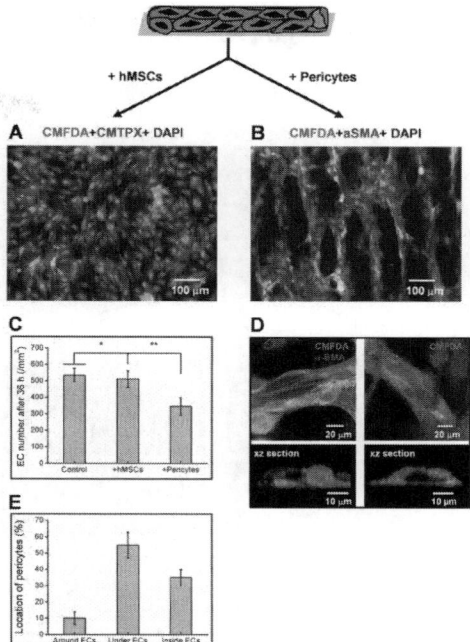

Fig. 2 Subsequent co-cultures of cells. ECs were cultured for 24 h on micropatterned surfaces, followed by the addition of (A) hMSCs or (B) pericytes for 16 h to the initial system. HUVECs were labeled with Cell Tracker Green (CMFDA), hMSCs were labeled with Cell Tracker Red (CMTPX), pericytes were stained with anti-αSMA (red). Cell nuclei were stained with DAPI (blue). (C) EC proliferation on Control (ECs seeded on peptide micropatterned surfaces for 36 h), on co-culture system of ECs with hMSCs or pericytes. * $P > 0.05$, ** $P < 0.01$. (D) Confocal images of location of ECs (green) and pericytes (red) in the co-culture system of ECs with pericytes for 16 h. Cell nuclei were stained in blue with DAPI. (E) Distribution of location of pericytes according to the EC tubular structures (%).

In order to further validate the distinct roles of pericytes and hMSCs, an additional component was added to our experimental model: collagen IV, used in this case to mimic the basement membrane as it is one of its major components. Pericytes or hMSCs were then added onto this mimicking matrix in our system. We show that only the pericytes could migrate through the type IV collagen and interact with ECs without however stabilizing the tubes which were not conserved. This can be seen after 3D reconstruction of the confocal images showing the lumen disappearance despite the presence of pericytes. In order to conserve this structure, the type IV collagen was replaced with Matrigel as it is known to intensively induce vasculogenic lumen formation. Interestingly, the appearance of vacuoles was observed after the addition of Matrigel to our experimental model (Figure 3A). Furthermore, like with collagen IV, we observed that only pericytes but not hMSCs migrated through the Matrigel. In this case however, the vasculogenic tubes were stabilized (Figure 3B, C).

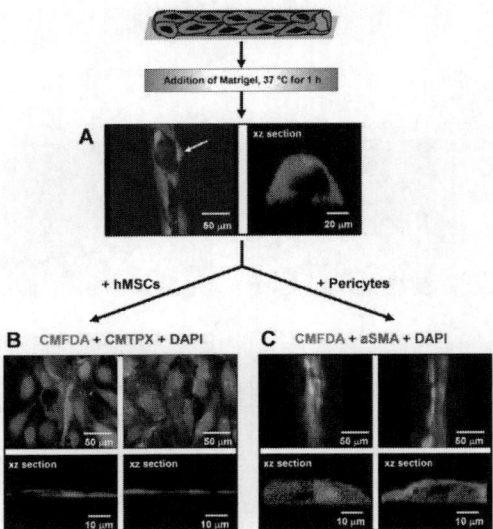

Fig. 3 Subsequent co-culture of cells with the recruitment of Matrigel. ECs were seeded for 24 h on micropatterned surfaces, added and incubated with Matrigel for 1 h (A), and followed by the addition of either (B) hMSCs or (C) pericytes for 16 h. HUVECs were labeled with Cell Tracker Green (CMFDA), hMSCs were labeled with Cell Tracker Red (CMTPX), pericytes were stained with anti-αSMA in red. Cell nuclei were stained with DAPI in blue.

The ensemble of our results suggests that hMSCs don't have any affinity for ECs. The migration of pericytes towards the ECs, on the other hand, is systematic regardless of the EC status. The function of the pericytes, however, is modified according to the EC status. In fact, it was recently shown [9] that the pericyte recruitment during vasculogenic tube assembly can stimulate the endothelial basement membrane matrix formation. In our experimental system, no formation of basement membrane components was observed after direct co-culture of hMSCs with ECs. However, in the case of pericytes/ECs co-culture, we could observe an increased expression of fibronectin, one of the major components of the basement membrane.

IV. DISCUSSION

A controversy exists regarding the precise identity of MSCs and pericytes. It has been indeed shown that both of the cell types share mesenchymal markers like CD44, CD90, CD105 and CD73 and that perivascular MSCs express the pericyte markers NG2 and PDGFR-β [2]. Moreover, it has been proposed that pericytes are stem cells themselves as they are multipotent for osteogenic, chondrogenic, adipogenic and myogenic lineages which are all hallmarks of the MSC identity. This pericyte multipotency has been observed in the central nervous system as well. A further complexity is added by the early in vivo studies as MSCs expressing the pericyte marker α-SMA were not capable of differentiation into pericytes and were not found at perivascular locations. Based on our results, we suggest that pericytes and MSCs are two separate cell types with a clear difference in their function despite the shared marker signatures. Moreover, MSCs may reside in the perivascular niches without having any functional similarity to pericytes. Additional experimenttation is thus required to establish whether MSCs participate in the regulation of the vascular neo-formation, sprouting or endothelial quiescence.

V. CONCLUSIONS

In summary, we show that only pericytes that are stem cell-like cells, but not hMSCs, have an affinity for ECs in order to stabilize the EC tubular structures. The pericyte expression of stem cell markers could be part of other functions. Finally, the pericyte multipotency and their role in the stabilization of the EC tubular structures suggest that they can play a major role in vascular and bone tissue regeneration. Their use in future tissue engineering will thus become critical.

ACKNOWLEDGEMENT

We sincerely thank Dr. Gocheva V. (Centre de Biochimie Structurale, Université Montpellier II, France) for useful discussions. This work was supported by the "Région Aquitaine" and the "Agence Nationale pour la Recherche" (ANR).

REFERENCES

1. Bergers, G.; Song, S. (2005) The role of pericytes in blood-vessel formation and maintenance. Neuro Oncol 7, (4), 452-64.
2. Crisan, M. et al. (2008) A perivascular origin for mesenchymal stem cells in multiple human organs.Cell Stem Cell 3, (3), 301-13.
3. Tang, W. et al. (2008) White fat progenitor cells reside in the adipose vasculature. Science 322, (5901), 583-6.
4. Dellavalle, A. et al. (2007) Pericytes of human skeletal muscle are myogenic precursors distinct from satellite cells. Nat Cell Biol 9, (3), 255-67.
5. Dore-Duffy, P.; Katychev, A.; Wang, X.; Van Buren, E. (2006) CNS microvascular pericytes exhibit multipotential stem cell activity. J Cereb Blood Flow Metab 26, (5), 613-24.
6. Caplan, A. I. (2008) All MSCs are pericytes? Cell stem cell 3, (3), 229-30.

7. Lei, Y.; Zouani, O. F.; Rémy, M.; Ayela, C.; Durrieu, M.-C. (2012) Geometrical microfeature cues for directing tubulogenesis of endothelial cells. PLoS ONE 7, (7), e41163.
8. Lei, Y.; Zouani, O. F.; Rami, L.; Chanseau, C.; Durrieu, M. C. (2012) Modulation of lumen formation by microgeometrical bioactive cues and migration mode of actin machinery. Small 8;9(7):1086-95.
9. Stratman, A. N.; Malotte, K. M.; Mahan, R. D.; Davis, M. J.; Davis, G. E. (2009) Pericyte recruitment during vasculogenic tube assembly stimulates endothelial basement membrane matrix formation. Blood 114, (24), 5091-101.
10. Zouani, O. F.; Chanseau, C.; Brouillaud, B.; Bareille, R.; Deliane, F.; Foulc, M.-P.; Mehdi, A.; Durrieu, M.-C. (2012) Altered nanofeature size dictates stem cell differentiation. Journal of Cell Science 125, 1-8.
11. Zouani, O. F.; Kalisky, J.; Ibarboure, E.; Durrieu, M. C. (2013) Effect of BMP-2 from matrices of different stiffnesses for the modulation of stem cell fate. Biomaterials 34(9):2157-66.

Author: Omar F. ZOUANI
Institute: Institut Européen de la Chimie et Biologie
Street: 2 Rue Escarpit
City: Pessac
Country: France
Email: omar.zouani@inserm.fr

Processes of Gamma Radiolysis of Soluble Collagen in the Aspect of the Scaffolds Preparation

K. Pietrucha

Department on Material and Commodity Sciences and Technology, Lodz University of Technology, Poland

Abstract—This paper presents the results of the impact that gamma radiation has on collagen derived from fish skin. The infrared spectroscopy (FTIR), differential scanning calorimetry (DSC) and UV-Vis methods were used to examine a triple helix structure of collagen. The mechanism of the occurring reactions and potential possibility to use the 3 D collagen network as scaffolds are discussed.

Keywords—**collagen, gamma irradiation, spectroscopic and DSC methods.**

I. INTRODUCTION

Collagen is a major component of the extra cellular matrix (ECM) and as a fibrillar protein plays a structural role supporting the resident cells. Reconstituted collagen from various sources primarily mammalian (porcine, bovine, equine) because of advantage of biocompatibility, biodegradability and low immunogenicity has been utilized in medical field. Highly organized collagenous structures are being increasingly used as scaffolds for the development of tissues and artificial organs.

Nowadays [1-7], the importance of fish collagen is growing and there is also growing number of possibilities of its usage including medicine applications. One of the major advantages of collagen from aquatic organisms sources, in comparison to land animals, is that they are not associated with the risk of outbreak of bovine spongiform encephalopathy (BSE). Besides, fish collagen is comparatively cheap and easily accessible protein.

Unfortunately, the dominant disadvantage of reconstituted collagen from fish is its fast biodegradability rate, the relatively low thermal stability and low mechanical strength. To overcome these undesirable features, the collagen before use to construction of scaffolds should be modified using a chemical or physical techniques. Production and evaluation of collagen from different kind of fish is actually not new, it has been prepared for many years. However, there is small amount of data about how to improve the functional properties of this protein. Only a few reports are devoted to the chemical modification of collagen produced from tilapia or salmon [8,9]. In this work the impact of gamma radiation ^{60}Co on the key structural parameters of collagen silver carp is investigated. This technique has not been proposed previously to modification of structure like this collagen.

II. MATERIALS AND METHODS

Collagen type I was prepared from fresh skin of silver carp and supplied by AAG Sp. z o.o., Poland. Hydroxyproline was purchased from Sigma-Aldrich, USA. All other chemicals of analytical grade were obtained from POCh-Gliwice, Poland and used as received.

A. Preparation of Samples

The experiments were carried out using collagen solutions or films. The concentration of collagen in the solution was established by estimating hydroxyproline (specific amino acid for collagen) using method recommended by the standard number PN-ISO 3496:2000. Solutions of collagen diluted with acetic acid at protein concentration 2.8 mg/ml and pH 3.5 were applied. The films were prepared from collagen solution by drying under a laminar air-low at room temperature. The moisture content of the test samples determined gravimetrically was about 17%.

B. Irradiation

Deaerated collagen films or aqueous solutions of collagen saturated with inert gas (Ar-medical grade) were irradiated with dose rate from 100 Gy to 6.0 kGy, at dose rate 1.3 kGy/h (source of gamma rays ^{60}Co). Some collagen solutions or films without any further treatment were taken as a reference material.

C. Dependence of Viscosity on Dose Irradiation

The changes in helical structure of collagen macromolecule in solution was determined from radiation dose induced intrinsic viscosity change using an Ubbelohde viscometer (type AVS 470, Schott, Germany).

D. Differential Scanning Calorimetry (DSC)

Thermal transition analysis was performed using a differential scanning calorimeter model NETZSCH DSC 204, and TG 204 analyzer. The instrument was calibrated with temperature and heat flow with indium. The samples of film at 5-6 mg were sealed in aluminium pans and heated at the constant rate of 5°C/min in nitrogen atmosphere with an empty aluminium pan as a reference probe. The endothermic peaks of the thermogram were monitored.

E. UV-Vis and FTIR-ATR Analysis

Ultraviolet spectra of irradiated collagen solution were recorded (220-350 nm) by UV-Vis spectrometer Jasco V-530, Japan. Spectra IR of the collagen foils were recorded at room temperature using Perkin Elmer 2000 FTIR spectrometer employing the technique-attenuated total refraction (ATR). Spectra were acquired at a resolution of 4 cm^{-1} and the measurement range was 4000-650 cm^{-1}.

III. RESULTS AND DISCUSSION

A. Effect of Gamma Radiation on Viscosity

Several strategies have been, so far, studied to enhance the functional properties of fish collagen. However, most often used methods, have a number of disadvantages [10]. In chemical methods, cytotoxicity and calcification of cross-linking agents have caused great concern. Dehydrothermal treatment in partial degradation and denaturation of collagen. Only a few studies on UV- irradiation on fish collagen have been published [11]. Besides, mechanism of ultraviolet light interaction with matter is different in comparison to gamma rays or electron beam. UV light allows to modify the surface properties without changing the characteristic of bulk material. To my knowledge, no studies have been undertaken towards exploring the impact of high-energy radiation on fish collagen. In this work the effect of gamma irradiation on the some pivotal properties of silver carp collagen macromolecules as a function of dose radiation is investigated.

The changes in the three-dimensional structure of collagen macromolecules are reflected by change of viscosity. The results of viscosity of collagen solution irradiated in absence of oxygen versus dose are collected in Fig.1.

The data in Fig.1 clearly shows that in only very small doses of up to 0.4 kGy there is a slight increase in viscosity. After crossing the 1.0 kGy dose followed substantial reduction in viscosity. The results indicated that the dose of gamma radiation of 100 to 400 Gy, leads to a slight cross-linking of collagen, but beyond about 1.0 kGy dose degradation of collagen occurs.

Summing up, it can be concluded that as a result of gamma irradiation of aqueous solutions of collagen reactive intermediates mainly free radicals of collagen and species radiolysis of water: hydrogen atom, hydroxyl radical and hydrated electron (H, $^{\bullet}$OH, e^{-}_{aq}) are formed [12]. The main radiolysis products of collagen associated with direct and indirect effects of irradiation are: 1) macro-radicals formed by abstraction of hydrogen atom from the α-carbon in protein chains, 2) radicals created from peptide bond scission, and 3) phenoxyl and orto- meta- and para- adducts of $^{\bullet}$OH with tyrosine residues. Mutual reactions between different products of transients are involved in the formation of intra- and intermolecular cross-linking. The relative contribution of the cross-linking effects to degradation depends on collagen source, type of collagen, water content in the microenvironment and conditions of the γ-ray (dose, dose rate, presence/absence of oxygen, etc.) to which the collagen is being exposed.

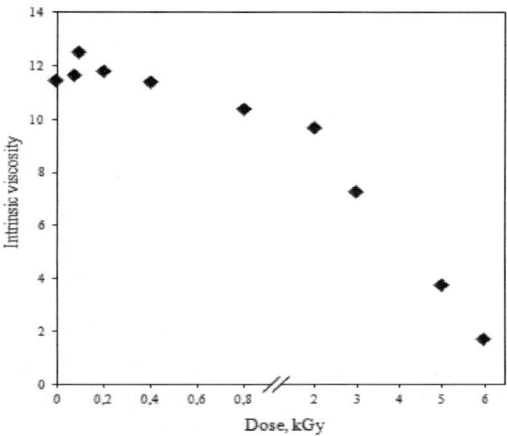

Fig. 1. Dependence of the intrinsic viscosity of acid soluble collagen of silver carp on the dose of gamma rays ^{60}Co.

B. UV-Vis Evaluation

Study using UV-Vis method provided additional evidence on the impact of radiation on aqueous solutions of collagen. In Fig.2 there are presented UV-Vis spectra irradiated collagen, and for comparison, UV-Vis spectra for phenylalanine and tyrosine in an insert are shown.

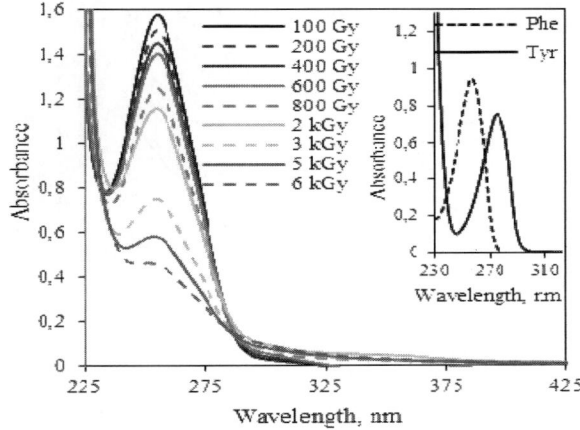

Fig. 2. UV-Vis spectra of collagen from silver carp after irradiation with different doses in absence of oxygen. Insert, UV-Vis spectra of phenylalanine and tyrosine.

The results show (Fig.2) that aqueous solution of collagen irradiated at different doses has a broad absorption band between 230 nm and 290 nm with maximum peak at 250 nm. This is in accordance with the characteristic absorption of phenylalanine and tyrosine with max peak at 250 nm and at 270 nm, respectively (see insert). Ii is known that the phenylalanine and tyrosine are the most radiosensitive among 19 different amino acid of collagen. As can be seen at Fig. 2 the phenylalanine and tyrosine sensitive chromophores have absorption bands in the range 250-290 nm, whereas the collagen has no evident absorption in this region. Early own studies and other experiences [12,13] have shown that collagen derived from various skins has a absorption spectrum of UV-Vis light between 190 nm and 240 nm. With increasing dose of gamma radiation overall absorption bands of collagen decreased. Such substantial changes in absorption decrement around 240-270 nm may indicate that deamination of tyrosine and phenylalanine residues occurs. This phenomenon influences destabilization of the secondary and tertiary structure of the collagen. However, based on my previous results [12] it can be concluded that as a result of multiple attack of OH radicals on various sites of collagen macromolecule, among others, the phenoxyl radicals and adducts of •OH with tyrosine residues of collagen are formed. These transients are involved in the formation of dityrosyl type-cross-links. Besides, in reaction of collagen in acid solutions at pH=3.5, interaction of e^-_{aq} atoms with the collagen macromolecules cannot be excluded [14].

C. Differential Scanning Calorimetry Studies

It is worth noticing that interesting phenomena can be observed after gamma irradiation of collagen films. The thermograms obtained by differential scanning calorimetry (DSC) and thermogravimetry (TG) are presented in Fig.3.

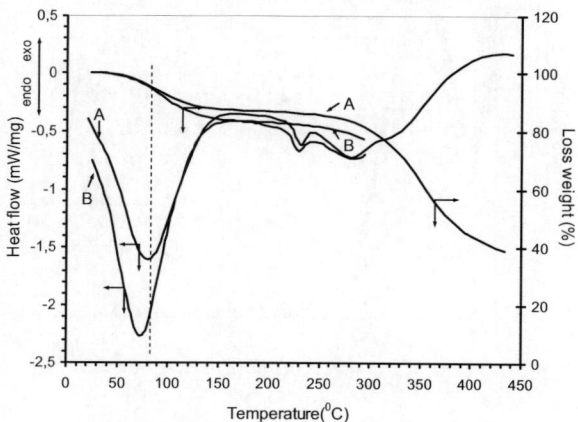

Fig. 3. DSC and TG thermograms of collagen films for: (A) after irradiation with dose 6 kGy and (B) before irradiation.

DSC thermograms of non-irradiated and irradiated collagen evidenced three different endothermal peaks: the first one related to temperature of thermal denaturation (T_d) of collagen and the other two were connected with a complex phenomenon of thermal modification, finally leading to the destruction of materials. These peaks relate to characteristic three-step thermogram TG, which suggests a three-stage decomposition of collagen. As can be seen in Fig.3 within the temperature range from 30° C to about 130° C, only broad endothermal peak appears in each sample. However, the maximum of the peaks for collagen irradiated (Fig.3A) compared to control sample (Fig.3 B) are significantly shifted to the region of higher temperatures. These peaks with maxima 70°C and 80°C are related to the transition from the triple helix to a randomly coiled conformation, taking place in the domains between the cross-links. It is known that intra- and intermolecular hydrogen bonds as well as hydrogen bound water are responsible for the stability of the triple helix conformation of collagen macromolecules [15]. Thereby, the obtained results indicate that partial changes in the secondary and tertiary structure of collagen macromolecule are preceded by the breakage of inter- and intra-molecular hydrogen bonds on heating and the release of loosely bound water. It is important that such a significant increase in T_d about 10^0 for collagen irradiated in comparison with unirradiated collagen indicates that the cross-linking processes occurred and thermal stability of irradiated samples increased. Recently, we have also reported that the T_d of silver carp collagen cross-linked using a carbodiimide increased about 13° as compared to unmodified collagen [2].

Generally, when collagen matrices are cross-linked, the thermal denaturation is hindered and T_d increases. Interesting is also a significant increase in an enthalpy change (ΔH) after irradiation of collagen (data not shown). This increase might be attributed to the gamma radiation induced cross-linking of fish collagen, which resulted in reinforcement of the network.

D. Infrared Spectroscopy

The representative infrared spectrum of irradiated collagen film together with the spectrum of non-irradiated collagen are shown in Fig.4.

The FTIR spectra presented in Fig.4 show typical features of proteins known as the amide I, amide II and amide III bands located at 1661, 1550 and 1451-1240 cm^{-1}, respectively. A comparison of the FTIR spectrum of irradiated collagen with that of collagen control shows no major differences since the infrared spectra are almost super impossible. Only slight differences can be seen around 2324 cm^{-1}, 1655 cm^{-1} and at 1532 cm^{-1}. The same sights shifts of the FTIR bands were also observed by others [11,16] They reported that this phenomenon might be possible due to the cross-linking induced by ultraviolet irradiation.

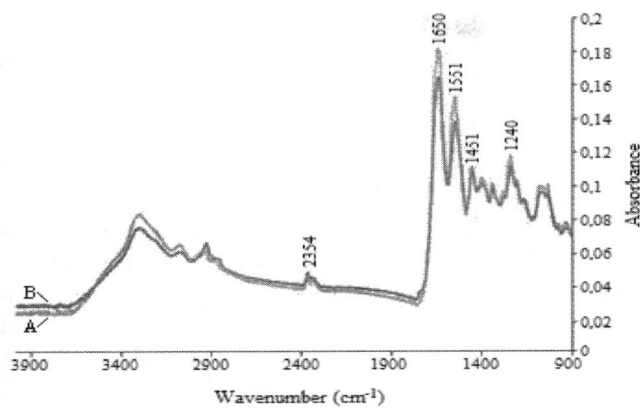

Fig. 4. Fourier transformation infrared spectrum (FTIR-ATR) of collagen (A) non-irradiated and (B) irradiated dose= 200 Gy, dose rate = 1.3kGy/h.

IV. CONCLUSIONS

1. The main γ-radiolysis products of aqueous solution of collagen are free radicals of collagen and species radiolysis of water H, $^{\bullet}$OH, e^-_{aq}. The formation of the intermediate products of collagen is associated with the three basic reactions: i/abstraction of hydrogen atom from the α-carbon in protein chains, ii/ radicals created from peptide bond scission, and iii/ creation of phenoxyl radicals and orto-, meta-, and para- adducts of $^{\bullet}$OH with tyrosine residues. Mutual reactions between different products of transients may lead to the formation of intra- and intermolecular cross-linking.
2. Collagenous thermostable matrix can be prepared with deoxygenated solutions of silver carp skin using γ-irradiation dose about 6 kGy.
3. The advance of this radiation manner is that the three-dimensional (3D) of collagen structure can be stabilized without any impurities in catalyst or chemical initiators.
4. It can be hypothesized that is possible to offer 3-D biodegradable and biocopatible collagen silver carp matrix as the suitable biomaterials or scaffolds for tissue engineering.
5. These findings suggest that silver carp skin can be used as the potential alternative for bovine or porcine, which have limitation for application in some religions.

ACKNOWLEDGMENT

The work is supported by National Science Centre via Grant No DEC-2011/03/B/ST8?05867. The technical assistance of PhD student M. Safandowska is gratefully acknowledged.

RERERENCES

1. Li Z, Wang B, Chi C et al. (2013) Isolation and characterization of acid soluble collagens and pepsin soluble collagens from the skin and bone of Spanmackerel (*Scomberomorous niphonius*). Food Hydrocolloid 31:103-113
2. Safandowska M, Pietrucha K (2013) Effect of fish collagen modification on its thermal and rheological properties. Int J Biol Macomol 53:32-37
3. Liu D, Liang L, Regenstein J M et al. (2012) Extraction and characterisation of pepsin-solubilised collagen from fins, scales, skins, bones and swim bladders of bighead carp (*Hypophthalmichthys nobilis*). Food Chem 133:1441–1448
4. Zeng S, Yin J, Yang S et al. (2012) Structure and characteristics of acid and pepsin-solubilized collagens from the skin of cobia (*Rachycentron canadum*). Food Chem 135:1975–1984
5. Pati F, Datta P, Adhikari B et al. (2012) Collagen scaffolds derived from fresh water fish origin and their biocompatibility. J Biomed Mater Res A 100A:1068–1079
6. Chen S, Hirota N, Okuda M et al. (2011) Microstructures and rheological properties of tilapia fish-scale collagen hydrogels with aligned fibrils fabricated under magnetic fields. Acta Biomaterials 7:644-652
7. Nalinanon S, Benjakul S, Kishimura H et al. (2011) Type I collagen from the skin of ornate threadfin bream (*Nemipterus hexodon*): Characteristics and effect of pepsin hydrolysis. Food Chem 125:500-507
8. Sugiura H, Yunoki S, Kondo E et al. (2009) *In Vivo* responses and bioresorption of tilapia scale collagen as a potential biomaterial. J Biomat Sci 20:1353-1368
9. Nagai N, Yunoki S, Suzuki T et al. (2004) Application of cross-linked salmon to the scaffold of human periodontal ligament cells. J Biosci Bioeng 97:389-394
10. Jiang B, Wu Z, Zhao H (2006) Electron beam irradiation modification of collagen membrane. Biomaterials 27:15-23
11. Bhat R, Karim A.A (2009) Ultraviolet irradiation improves gel strength of fish gelatin. Food Chem 113:1160-1164
12. Pietrucha K, Łubis M (1990) Some reactions of OH radicals with collagen and tyrosine in aqueous solutions. Radiat Phys Chem 36:155-160
13. Lin Y K, Liu D C (2006) Comparison of physical-chemical properties of type I collagen from different species. Food Chem 99:244-251
14. Pietrucha k, Góra L, Doillon CJ (1996) Reaction of hydrated electron with gentamycin and collagen-A pulse radiolysis study. Radiat Phys Chem 47:93-97
15. Miles C.A, Bailey A.J (1999) Thermal denaturation of collagen revisited. Proc Indian Acad Sci (Chem Sci) 111:71-80
16. Muyonga J H, Cole C C, Duodu K G (2004) Extraction and physicochemical characterization of Nile perch *(Lates niloticus)* skin and bone gelatin. Food Hydrocolloid 18:581-592

Building Up and Characterization of Multi-component Collagen-Based Scaffolds

K. Pietrucha

Department of Material and Commodity Sciences and Textile Metrology, Lodz University of Technology, Poland

Abstract—This report presents the effect of chemical synthesis of 3D hybrid collagen-based scaffolds on their functional and physicochemical properties. To achieve some data and information about conformational structure of superhelix, and conformational dynamic changes in collagen alone as well as in multi-components scaffolds, dynamic mechanical thermal analysis (DMTA) and differential scanning calorimetry (DSC) were included. The viscoelastic behavior and thermo-stability of scaffolds is manifested by a shift of the tan δ peak associated with the process of decomposition towards higher temperature.

Keywords—scaffolds, collagen, HA, CS, thermal stability.

I. INTRODUCTION

The tissue engineering strategies mostly utilize combination of various cells and biodegradable scaffolds to recapitulate natural process of tissue regeneration and development. Currently, collagen based scaffolds are the most important for tissue engineering and regenerative medicine [1]. Three decades of studies and clinical experience have shown that other natural or synthetic substitutes do not possess the biological subtleties of the natural collagen. In this study to synthesize biodegradable and biocompatible scaffolds, collagen alone or collagen enriched a minor proportion of hyaluronic acid (HA) and chondroitin sulfate (CS) was used. A major consideration in selecting the scaffold components are their complimentary properties such as:

- collagen - major component of the extracellular matrix (ECM) of many tissues has: a low mechanical strength, relatively rapid biodegradation, low antigenicity, excellent biocompatibility, good cell attachment, migration, proliferation and differentiation,
- HA - major intracellular component of soft connective tissues and synovial fluid is: non-immunogenic, high viscosity and water swelling ability, poor mechanical strength, non adhesive properties too many cell types but play an important role in cell differentiation and cell growth,
- CS present in all organisms from worms to humans and serves multiple important roles as vital structural component of some connective tissues, plays a special role as the enhancement of cell attachment, migration and division.

From the consideration of tissue engineering, none of the considered polymers on its own displays ideal properties. The indications are, however, that two or three-component compositions may prove to have the right balance of properties as carriers and as environment for cell culture.

Preparation of collagen matrix using glycos-aminoglycans (GAGs) is actually not new, it has been prepared for many years [2-7]. Earlier, I received very interesting results on the modification of collagen with HA [8]. However, practically no data is available on scaffolds consisting with multi-components system such as collagen-HA-CS. To my knowledge, only two reports are devoted to preparation of such material for dermal or cartilage tissue engineering [9,10]. It is known that the subtle changes in scaffold structure and physical properties can affect transcriptional events and associated cell phenotype and function [11]. Sterilization of scaffolds also leaves much to be desired [12]. In this study, particular attention has been paid to the preparation and characterization of cross-linked collagen-HA-CS matrices with potential for using in tissue engineering. These matrices were sterilized by radiation method. To characterize rheological properties of this scaffold, dynamic mechanical analysis as a function temperature (DMTA) and differential scanning calorimetry were carried out.

II. MATERIALS AND METHODS

A. Materials

Collagen type I was derived from purified calf skin by pepsin digestion and acetic acid dissolution to prepare 0.5 % w/v dispersion. High molecular weight (MW=1.6×10^6 Da) sodium hyaluronate from *Streptococcus equi*, chondroitin-6-sulfate, 1-ethyl-3-(3-dimethylamino-propyl) carbodiimide hydrochloride (EDC), N-hydroxy-sulfosuccinimide (NHS), and hydroxyproline were purchased from Sigma-Aldrich, USA. All other reagents and solvents of analytical grade were supplied by POCh Poland and used as received.

B. Preparation of Scaffolds

Dispersions of collagen (0.5% v/v) and subsequently Collagen-HA (6:1) and Collagen-CS (6:2) as well as Collagen-HA-CS (6:1:1) were prepared. A sponge-shape matrix was obtained by freezing the collagen dispersion or dispersion collagen with appropriate amounts of others components at -40° C. Then, they were lyophilized at -50° C (Labconco, USA). According to our previous studies [8] the sponges were subsequently cross-linked at room temperature by immersion in 80% ethanol-water solution containing 33mM EDC and 6 mM NHS. After reaction the matrices were washed in 0.1 M Na_2HPO_4 and then with deionized water. Finally, the cross-linked matrices were again lyophilized at -50° C. To sterilize, collagen alone or modified collagen scaffolds were placed in appropriate bags and irradiated by electron beam (Linear electron accelerator ELU-6e) with absorbed dose 20.6 kGy, dose rate 20 kGy/min.

C. Dynamical Mechanical Thermal Analysis (DMTA)

The dynamic mechanical analysis as a function temperature was performed with a DMA Q800 V20 B. Build 26 apparatus. Seventeen- millimeters square samples were subjected to mechanical testing. The measurements were carried out with a compression mode at a frequency of 1 Hz, a heating rate of 2° min^{-1} and temperature range from 25°C to 250°C. The set strain was 1.5% and the applied auto tension adjustment force was 50g. The changes in the elastic (E') and viscous (E'' = E' x tanδ) modules versus temperature were monitored.

D. Differential Scanning Calorimetry (DSC)

The thermal stability of above all collagen-based scaffolds were carried out with DSC Q2000 device (TA Instruments). The samples were weighed and introduced into aluminum pans. The pans were heated at the constant rate of 5°C min^{-1} in nitrogen atmosphere with an empty aluminum pan as the reference probe. The sample mass was in the range of 3-5 mg.

III. RESULTS AND DISCUSSION

Although collagen, HA and CS individually as well as in form of two-component systems as tissue-mimicking scaffolds have been constructed, an ideal scaffold exhibiting properties of native tissues is still not a reality. Here, I demonstrate preliminary results on the effects of synthesis of new 3D scaffolds composed with collagen, hyaluronic acid and chondroitin sulfate on their functional properties. To stabilize structure of these components water soluble 1-ethyl-3-(3-dimethyl aminopropyl) carbodiimide (EDC) with addition of N-hydroxysulfosuccinimide (NHS) in ethanol solution were used. These copolymers are supplemented by synthesis of two-component hybrid scaffolds consisting of collagen-HA and collagen-CS. To sterilize the scaffolds were exposed to electron beam radiation. Modifications in molecular organization of collagen-HA-CS scaffold induce variations in the material application properties, and more particularly in their thermal and mechanical properties. The typical DSC thermograms of collagen scaffolds are shown in Fig.1.

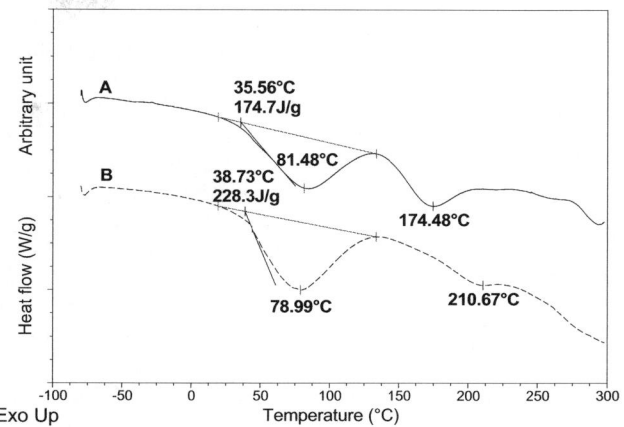

Fig. 1. DSC thermograms of collagen scaffold cross-linked by EDC with NHS: (A) before irradiation, (B) after irradiation absorbed dose 20,6 kGy, dose rate 20 kGy/min. Temperature denaturation (T_d) of collagen without any modification was 69° C (data not shown).

The thermograms obtained by DSC for both collagen samples non irradiated and irradiated, respectively (Fig.1) evidenced two different endothermal peaks: the one related to temperature of thermal denaturation (T_d) of collagen and the other around 200°C was connected with a complex phenomenon of thermal modification, finally leading to the destruction of materials. Previously, on the base thermogravimetric (TG) measurements, I showed that these peaks relate to the characteristic two-step thermogram TG, which suggests a two-stage decomposition of this matrix [8]. It is well known that temperature of thermal denaturation strongly depends on water content in collagen and its degree of cross-linking between chains. As it can be seen in Fig.1 over a temperature range from 30° C to about 130° C, in each sample only broad endothermal peak appears. However, the maximum peak for collagen modified by EDC/NHS and irradiated sterilizing dose compared to non irradiated collagen is slightly shifted to the region of lower temperature. These peaks with maximum at 81.5 °C and at 78.5° C result from transition of the triple helix to a randomly coiled conformation that takes place in the domains between the cross-links. It is known that intra- and intermolecular hydrogen bonds as well as hydrogen bound water are responsible for the stability of the triple helix conformation of collagen macromolecules. Thereby, the obtained results indicate that partial changes in the secondary structure of collagen macromolecule are preceded by the breakage of inter- and intra-molecular hydrogen bonds on heating and the release of loosely bound water. It is very important that such slightly decrease of T_d (3 °) of the cross-linked and irradiated scaffolds in comparison with the same non irradiated ones indicates that the radiation scission of collagen chains occurred in small extent. These results are different somewhat from those previously obtained by Möllers et al. [13]. They observed that micro-structured porcine collagen scaffold cross-linked using EDC after gamma irradiation with dose of 25 kGy had about 7° lower T_d compared to non-irradiated samples. These differences can be fully justified. It has been established that the relative contribution of the radiation effect cross-linking to degradation depends on collagen source, type of collagen, water content in the microenvironment and conditions of radiation (dose, dose rate, presence/absence of oxygen, etc.) [14]. Generally, when collagen scaffolds are cross-linked by EDC/NHS, the thermal denaturation is hindered and the T_d increases. Radiation sterilization of the cross-linked scaffolds only slightly lowered theirs thermal stability. The reaction of EDC and NHS with collagen results in the formation of NHS activated glutamic and aspartic acid residues, which react with ε-amino groups from lysine or hydroxylysine residues to form peptide bonds and release of NHS [8]. This way is preferable for cross-linking due to that chemical agent non-incorporation within structure of scaffolds and products exhibit no cytotoxicity.

Subsequent, hybrid tissue-mimicking scaffolds with micro-patterns of varying water content and viscoelasticity were synthesized from collagen, HA and CS. The efficiency of cross-linking of this copolymer was assessed by testing thermal stability. The thermograms obtained by differential scanning calorimetry for both collagen modified with HA and collagen modified by HA + CS, respectively are depicted in Fig.2.

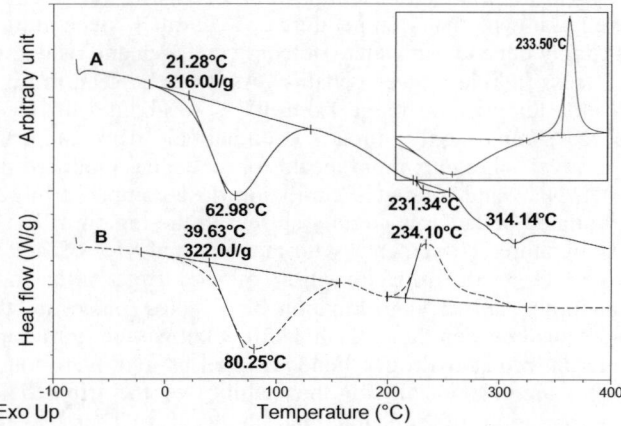

Fig. 2. DSC curves of multi-composite scaffolds both: (A) collagen + HA and (B) collagen + HA+ CS, respectively. All samples were cross-linked by EDC/NHS and irradiated with sterilizing dose equal to 20,6 kGy. Insert presents DSC curve for CS without any modification.

Thermally induced structural transitions in the collagenous network lead to changes of temperature denaturation. T_d of collagen modified with HA (Fig.2A) is about 17° lower than T_d of collagen without HA (Fig.1B).

This phenomenon may be explained by two facts: 1/HA is very sensitive to ionizing radiation, 2/HA has a highly hygroscopic nature.

It has been found that even low doses to 20 Gy of gamma radiation or electron beam resulted in a clear degradation of HA chains [15]. In addition, it is known that with the increase of water content in the sample decreases the temperature of denaturation [16]. Two weak endothermic peaks on the DSC curve with maximum around 200° C and 300° C for collagen-HA (Fig.2A) are associated with the evaporation of residual, strongly bound water and continued conformational changes of superhelix of collagen macromolecules. According to my and others authors previous studies [8,5], in result EDC/NHS cross-linked collagen in presence HA, carboxyl groups of HA reacted with amino group of collagen as well as with hydroxyl group of this protein or HA resulting in the formation of semi- and interpenetration network. However, very interesting is significant increase (about 88 J/g) in an enthalpy (ΔH) after irradiation of collagen modified by HA (see Fig. 1B and Fig.2A). This increase might be attributed to the induced cross-linking of scaffold, which resulted in reinforcement of the network.

It is worth noticing that interesting phenomena can be observed for tri-component scaffold consisting of collagen-HA-CS. Two broad peaks: one endothermic pick with max at 80 ° C and one exothermic peak with max at 234° C were recorded on DSC curve (Fig.2B). The first peak can be attributed to T_d for this hybrid scaffolds. On the contrary, the exothermic peak at max 234° C may be related to the total decomposition CS. We observed that in temperature 234 °C total destruction CS occurs (see insert). The reaction of EDC/NHS in such complex system collagen-HA-CS is still not fully understood and yet to be explored. Our findings suggest that similar reactions as collagen-HA may occur between collagen and CS, because the chemical structure of CS is the similar to other GAGs except that CS is sulfated disaccharide. Most likely at the system collagen-HA-CS there are other bonds. Tian et al. [17] demonstrate that the interactions between collagen and CS involve also electrostatic binding and other types of specific interaction such as hydrogen bond.

The molecular motions and viscoelastic nature of the scaffolds by applying a stress to the sample and monitoring its response studied with method dynamic mechanical thermal analysis (DMTA). Changes in tan δ equal energy lost to energy stored per cycle as a result of temperature dependence are plotted in Fig.3.

As it can be seen in Fig.3 the viscoelastic behavior and thermo-stability of all multi-components scaffolds is manifested by a shift of the tan δ peak associated with the process of decomposition towards higher temperature. Thus it can be concluded that the least sensitive to the changes in temperature are hybrid scaffolds consisting of Col-HA-CS.

Comparing even only a certain DSC and DMTA results indicate clearly that the trend in DSC is the same as the changes of rheological properties. Further, preliminary results show that the obtained hybrid scaffolds provide a good mechanical support for attachment and distribution of wide range of cell types including of neural cell.

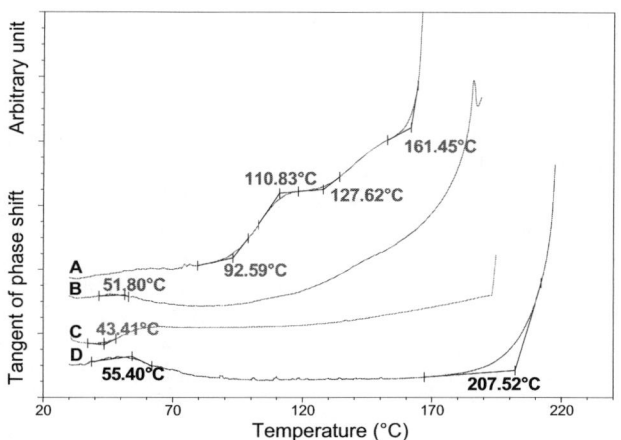

Fig. 3. Temperature relaxation of the collagen-based scaffolds cross-linked by EDC/NHS. Loss tan δ =E''/E' for sponges with the following composition: (A) Col non-sterile, (B) Col sterile, (C) Col-HA sterile, (D) Col-HA-CS sterile.

IV. CONCLUSION

Comparison of the thermal stability and rheological properties of unmodified collagen against modified collagen suggests that temperature denaturation and loss tangent depends more on the degree of hydratation and degree cross-linking of scaffolds than on the presence of HA or CS. The T_d for all scaffolds are very close around 80°C without collagen-HA. This may be explained by the fact that the hydroxyl group of HA may attach to the carboxyl group either on HA or collagen, resulting in aliphatic ester bond formation. With the consumption of this carboxyl groups, the amino groups on collagen may not have enough binding sites with carboxyl to form covalent linkage. Sterilisation of the cross-linked scaffold slightly lowered its thermal stability. These multi-components scaffolds could be useful as novel tissue engineering co-polymeric networks for regenerative medicine.

ACKNOWLEDGMENT

The work is supported by National Science Centre via Grant No DEC-2011/03/B/ST8?05867. The technical assistance of PhD M. Chrzanowski is gratefully acknowledged.

REFERENCES

1. Yannas IV (2013) Emerging rules for inducing organ regeneration. Biomaterials 34:321-330
2. Ellis DL, Yannas IV (1996) Recent advances in tissue synthesis in vivo by use of collagen-glycosaminoglycan copolymers. Biomaterials 17:291-299
3. Piper JS, Hafmans T, Veerkamp JH et al. (2000) Development of tailor made collagen glycosamino-glycan matrices: EDC/NHS cross-linking and ultrastructural aspects. Biomaterials 21:581-593
4. Wang TW, Spector M (2009) Development of hyaluronic acid-based scaffolds for brain tissue engineering. Acta Biomater 5:2371-2384
5. Suri S, Schmidt CE (2009) Photopatterned collagen-hyaluronic acid interpenetrating polymer network hydrogels. Acta Biomater 5:2385-2397
6. Haugh MG, Jaasma MJ, O'Brien FJ (2009)The effect of dehydrothermal treatment on the mechanical and structural properties of collagen-GAG scaffolds. J Biomed Mater Res 89A:363-369
7. Zhu J, Marchant RE (2011) Design properties of hydrogel tissue-engineering scaffolds. Expert Rev Ltd, ISSN 1743-4440, 10.1586/FRD.11.27
8. Pietrucha K (2005) Changes in denaturation and rheological properties of collagen-hyaluronic acid scaffolds as a result of temperature dependencies. Int J Biol Macrom 36:299-304
9. Wang W, Zhang M, Lu W et al. (2010) Cross-linked collagen-chondroitin sulfate-hyaluronic acid imitatin extracellular matrix as scaffold for dermal tissue engineering. Tissue Eng Part C 16:269-279
10. Zhang L, Li K, Xiao W, et al. (2011) Preparation of collagen - chondroitin sulfate - hyaluronic acid hybrid hydrogel scaffolds and cell compatibility in vitro. Carbohyd Polym 84:118-125
11. Kurniawan NA, Wong LH, Rajagopalan R (2012) Early stiffening and softening of collagen: Interplay of deformation mechanisms in biopolymer networks. Biomacromolecules 13:691-698
12. Odelius K, Plikk P, Albertsson A-C (2008) The influence of composition of porous copolymer scaffolds on reactions induced by irradiation sterilization. Biomaterials 29:129-140
13. Möllers S, Heschel I, Damink LO, et al. (2009) Cytocompatibility of a novel, longitudinally microstructured collagen scaffold intended for nerve tissue repair. Tissue Eng: Part A 15:461-472
14. Pietrucha K (1989) Effect of radiation on collagen solutions in relation to biomedical applications. Polym Med 19:3-18
15. Choi J, Kim J-K, Kim J-H, et al. (2010) Degradation of hyaluronic acid powder by electron beam irradiation, gamma ray irradiation, microwave irradiation and thermal treatment: A comparative study. Carbohyd Polym 79:1080–1085
16. Daar E, King L, Nisbet A, et al. (2010) Viscosity changes in hyaluronic acid: irradiation and rheological studies. Appl Radiat Isot 68:746-750
17. Tian H. Chen Y, Ding C et al (2012) Interaction study in homogeneous collagen/chondroitin sulfate blends by two-dimensional infrared spectroscopy. Carbohyd Polym 89:542-550

Fabrication of Gelatin/Bioactive Glass Hybrid Scaffolds for Bone Tissue-Engineering

S. Borrego-González[1], J. Becerra[2,3], and A. Díaz-Cuenca[1,3]

[1] Instituto de Ciencia de Materiales de Sevilla (ICMS), Centro Mixto CSIC-Universidad de Sevilla, Sevilla, España
[2] Departamento de Biología Celular, Genética y Fisiología, Facultad de Ciencias,
Universidad de Málaga, Málaga, España
[3] Networking Research Center on Bioengineering, Biomaterials and Nanomedicine (CIBER-BBN), Spain

Abstract—In this work, hybrid scaffold materials composed of gelatin and Bioactive Glass (BG) have been fabricated. BG particulates in the ternary system SiO_2-CaO-P_2O_5 have been synthesized initially using the sol-gel method. The gelatin/BG scaffolds were shaped by a freeze-drying process obtaining homogeneous and reproducible final macroporous structures. BG particulates precursor and the final hybrid macroporous structures (scaffolds) were characterized using Field Emission Scanning Electron Microscopy (FE-SEM), Energy Dispersive X-ray Analysis (EDX), X-ray diffraction (XRD) and thermal analysis. Finally, the scaffolds were essayed in Simulated Body Fluid (SBF) *in vitro* showing a bioactive response.

Keywords—Gelatin, Bioactive Glass, Bone Tissue Engineering, Macroporous scaffold.

I. INTRODUCTION

Population aging and enhance one's quality of life is demanding new approaches and therapies for bone reparation and regeneration. At present, bone substitutes are often required to replace defected tissue generated by disease, aging or trauma [1]. Tissue Engineering (TE) is a very promising option which has the potential to overcome the problem providing the development of constructs to be implanted for the repair or regeneration of injured or damaged tissues. As well as in natural tissues, cells, extracellular matrix (ECM) and signaling molecules are the basic elements of TE. Bone is a tissue whose special architecture has allowed a fast development of its associated TE. The fundamental constituents of bone are cells and the extracellular matrix (ECM). This ECM is a biphasic or hybrid matrix composed of organic and inorganic elements. The organic phase is mainly formed by collagen type I fibbers which correspond to the 90% of the total bone protein. The main component of the inorganic phase is calcium carbonate phosphate in the form of spindle or plate-shaped crystalline hydroxyapatite $(Ca_{18.3-0.7}(PO_4)_{4.3}(HPO_4, CO_3)_{1.7}(2OH, CO_3)_{0.15})$ [2]. It is this combination of the calcium phosphate mineral phase, collagen fibers and living cells which gives the bone its outstanding properties of hardness, toughness and self-renewal capacity [3].

The fabrication of synthetic bone extracellular matrix using calcium phosphates and natural polymers such as collagen or gelatin has leaded to the appearance of an entire field of biomaterials research. Gelatin (denatured form of collagen) has greater availability and it is easier and cheaper to obtain than collagen. Gelatin keeps collagen chemistry and amino acids polymeric architecture and it is readily reabsorbable by the body.

Calcium phosphates and bioceramic materials as tricalcium phosphate (TCP) or synthetic hydroxyapatite (HA) have a composition very close to the natural mineral of the bone. These materials are biocompatible and have been proved to enhance the bone growth [3,4]. However, studies *in vivo* have shown that bioactive glasses are the materials which are more efficiently and rapidly fixed in bone regeneration [5]. Bioactive glasses were discovered by Larry Hench's in 1969 when he found that a particular glass composition (45S5 Bioglass®) was able to form a chemical bond with bone [6]. Hench defined as bioactive a material which has the ability to bond and fix spontaneously in the body with the living tissue without the formation of a fibrous interface layer or capsule [7].

In bone TE the scaffold must act as a three-dimensional temporary template to guide the bone repair stimulating the natural regenerative mechanisms of the human body. The scaffolds should have a high porosity in the macropore range (100-500 μm) to facilitate cellular migration and vascularization, a proper mechanical integrity and biocompatibility, and their degradation rate should match the rate of neotissue formation. The scaffold should promote osteoconductivity (bone cell attach along the material surface from the bone–implant interface) and osteoinductivity (stimulates new bone formation) [8].

In this work, the fabrication of macroporous architectures based on gelatin/bioactive glass have been investigated using freeze-drying processing techniques. Different processing parameters as the gelatin hydrogel concentration, the processing cooling rate and the molding container configuration have been modified to study their influence in the final volume fraction porosity [10,11] and pore architecture [12-14] of the full gelatin (BG free) materials. From this preliminary study, selected parameters were used to prepare the gelatin/BG hybrid scaffolds. Final scaffolds were characterized in terms of composition, porosity, macrostructure and bioactivity *in vitro*.

II. MATERIALS AND METHODS

A. Bioactive Glass Synthesis

The Bioactive Glass (BG) particulate precursor was prepared using the reactants ratios as reported by Yan et al. [15]. Ethanol absolute (AnalaR Normapur) and 0,5 M HCl solution (Panreac, PA-ACS-ISO 37%) were mixed. The reactants $Ca(NO_3)_2 \cdot 4H_2O$ (Sigma-Aldrich), triethyl phosphate (TEP, $(C_2H_5O)_3PO$, Sigma-Aldrich) and tetraethyl orthosilicate (TEOS, $C_8H_{20}O_4Si$, Alfa Aesar) were added with a molar ratio of Si/Ca/P of 80:15:5 (or Si/Ca/P = 76:13:11 in weight ratio). The mixture (sol) was undergone an evaporation-induced self-assembly (EISA) process to form a gel. The gel was aged at 75°C and then heating at 700°C to obtain the final BG product which was ground to prepare the final particulate precursor.

B. Preparation of the Gelatin/BG Hybrid Scaffolds

The gelatin was dissolved in de-ionized distilled water at a concentration of 2.5% (w/w) and stirred for 1h at 40°C. Amounts of particulate BG were blended with gelatin water solution in 50% w/w, Fig.1a. The mixture was homogenized by stirring for 30 minutes. The resulting solution was poured into a Petri dish and kept at 4°C to produce a gel (Fig. 1b). A punch was used to obtain the scaffolds with 12mm diameter and 2mm height. The structures were cross-linked by immersion in 0.25%v/v glutaraldehyde (GA) aqueous solution (Fig. 1c) and freeze-dried with a freeze rate of 1°C/min to -20°C and held at this temperature under a vacuum of 0.04mbar for 24 hours (Fig. 1d).

C. Characterization of the BG particulates and the Gelatin/BG scaffolds

The particle size distribution (PSD) of the BG particulates was measured using the laser scattering method. A MalvernSizer Laser Diffraction (LD) instrument using active beam length of 2.4 mm and a 300-RF lens was used. X-ray diffraction (XRD) analysis was performed with a PA-Nanalytical X′Pert PRO diffractometer using $Cu\text{-}K_\alpha$ radiation ($\lambda = 0.154187$ nm). The diffractometer was operated at 45 kV and 40 mA. Wide-angle X-ray diffraction (WXRD) patterns were recorded using a step size of 0.02 ° and 800 s exposure time. BG particulate precursor and scaffolds were evaluated with an HITACHI S-4800 Field Emission Gun Scanning Electron Microscope (FEG-SEM). The compositional analysis of the BG was performed by X-Ray Fluorescence (XRF) using the Spectrometer Panalytical (AXIOS model). Thermo gravimetric (TG) and differential scanning calorimetry (DSC) analysis were carried out using an automatic thermal analyzer system (model SDT Q600). The total porosity of the hybrid scaffolds, P was calculated by the Eq. 1.

$$P = 1 - \frac{\rho_{scaffold}}{\rho_{solid}} = 1 - \rho_{relative} \quad (1)$$

The density of the scaffold ($\rho_{scaffold}$) was determined from the mass and dimensions of the final material pieces. The density of the solid (ρ_{solid}) was estimated using the quantitative analysis of the each component and the solid density values for the gelatin ($\rho_{gelatin} = 1.037$ g·cm^{-3}) [24] and BG ($\rho_{BG} = 2.321$ g·cm^{-3}, estimated using quantitative analysis). The assessment of the scaffolds bioactivity was carried out by soaking in Simulated Body Fluid (SBF) [16]. Each piece was immersed in 15 ml of SBF in polyethylene container for 3 days at 36.5°C. After the treatment, the scaffolds were washed thrice in distilled water and freeze-dried. Bioactivity was assessed by FE-SEM, EDX and XRD analysis.

III. RESULTS AND DISCUSSION

A. Characterization of the BG Particulates

Particle size distribution of the BG is shown in Fig. 2a. The plot shows a bimodal distribution (pink) which was deconvoluted using a Gaussian function by the Origin software. Deconvolution results (blue) show two symmetric

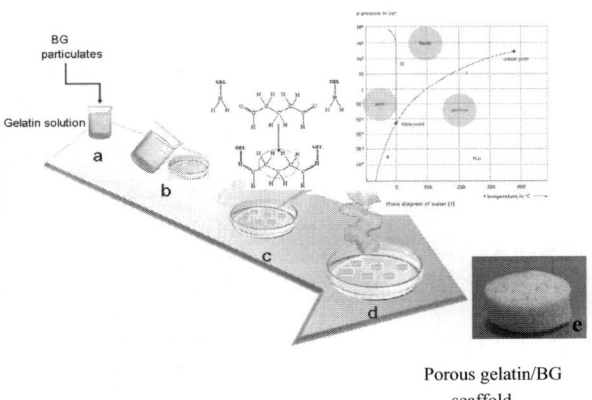

Fig. 1 Schematic diagram of the scaffolds fabrication. Amounts of particulate MBG were blended with gelatin water solution (a). Pouring into a Petri dish and shaping the scaffolds (b). Crosslinking (c). Freeze-drying (d). Final porous gelatin/BG scaffold (e)

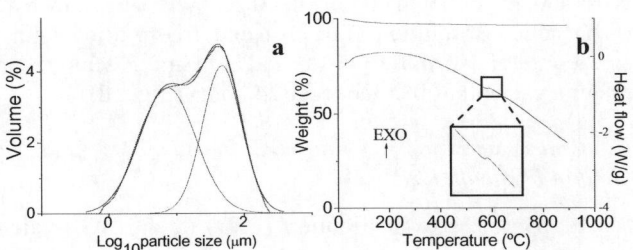

Fig. 2 Particle size distribution (a) and DTA curve (b) of the BG precursor particulates

distributions of 54 and 46 area %, with maximum at 7 and 47 μm respectively. The XRD analysis confirmed that the BG particulates were XRD amorphous as no diffraction peaks were observed apart from a broad band between 18 and 40° (2θ) as it has been reported previously [17,18]. The quantitative analysis of the BG indicates a final composition of $87SiO_2/10CaO/3P_2O_5$ wt. %. The elemental quantitative analysis indicates a lower percentage of P and Ca than could be expected from the reactants precursor ratio. The TGA (Fig. 2b) of the BG shows a mass loss of only 2% which would correspond to the removal of the moisture. The DSC (Fig. 2b) shows a typical endothermic event at 580°C characteristic of the glass transition temperature. Both, the DSC showing the glass transition temperature and the amorphous diffraction pattern confirm the obtention of a glass material.

B. Characterization of the Gelatin/BG Hybrid Scaffolds

The final macropore structure of the gelatin scaffolds (BG free) is presented in Fig. 3. The results show that the macropore size decreases with the increasing amount of gelatin concentration (Fig. 3a and b). Also, the pore size decreases with the increasing freezing rate (Fig 3a and c). An orientated pore alignment has been achieved using shaping moulds consisting on the combination of various materials which have different thermal conductivities [12] (Fig. 3a and d). From this preliminary work, a selection of parameters to prepare the gelatin/BG hybrid scaffolds was decided. These parameters were 2.5 wt.% gelatin concentration, a freezing rate of 1°C/min and the mould consisting on polystyrene (PS). The total porosity of the hybrid gelatin/BG scaffolds was measured as 95%±0.427. FE-SEM observations and EDX analysis of the total cross-section of these scaffolds are shown in Fig. 4. The FE-SEM images indicate an homogeneous macroporous structure in terms of size and macropore shape or morphology. The size of the macropore was measured in the range from 100 to 500 μm.

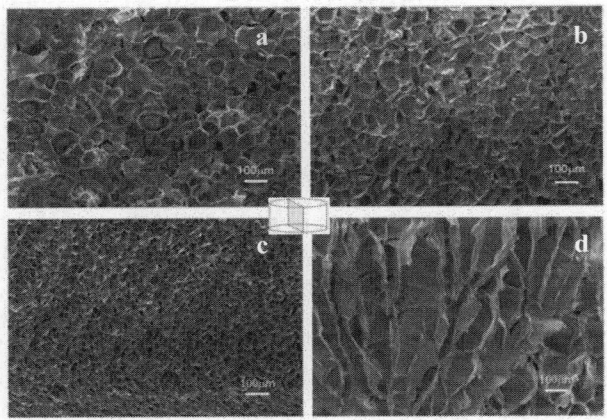

Fig. 3 FE-SEM images of full gelatin (BG free) systems study. 2.5% gelatin, cooling rate of 1°C/min and mould of polystyrene (PS) (a), 5% gelatin, 1°C/min and mould of PS (b), gelatin 2.5%, freezing rate of 2°C/min and mould of PS (c) and 2.5% gelatin, 1°C/min and mould of PS and aluminum (d)

Fig. 4 FE-SEM images of the gelatin/BG scaffold and EDX mapping of the scaffolds

Hence, both final total porosity and macropore size of the fabricated scaffolds were in the required values to allow tissue ingrowth generally agreed in the literature, of a porosity of 50% or higher and macropore size of 100 μm or higher. The EDX analysis of Ca, Si and P showed that the BG components are equally distributed within the scaffold cross-section.

C. Bioactivity Test In Vitro

Fig. 5a shows a FE-SEM image of the hybrid scaffold after 3 days in SBF treatment. As can be seen in the image, needle-shaped crystal aggregates characteristics of hydroxyapatite are formed on the surface of the scaffolds. The crystal size was measured by image analysis having values of about 200 nm. Also, the XRD pattern obtained after SBF (Fig. 5b) matched well with the hydroxyapatite phase

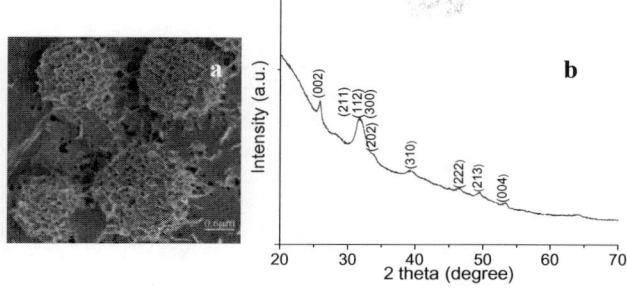

Fig. 5 FE-SEM images of the immersed scaffolds in SBF showing the hydroxyapatite crystals aggregates (a); and WXRD analysis (b)

according to the ICCD database (JCPDS 046-0905). The broad asymmetric peak with high intensity around 30-35 2θ may consists of four overlapping peaks, which could be assigned to HA characteristic triplet for the reflections (211), (112), and (300), with the (202) reflection appearing as a faint shoulder at 33.9 2θ [18, 20]. A control bioactivity experiment using a full gelatin scaffold (BG free) was performed. In this experiment, no hydroxyapatite formation was observed after 30 days in SBF treatment. This comparison between the positive bioactive response of the hybrid scaffolds and the negative response of the BG free full gelatin scaffolds experiment confirmed the incorporation of the BG particulates improves the bioactivity of the gelatin hydrogels. It is important to remark that although the obtained gelatin/BG structures have a high volume percentage of porosity and macropores in the 100-500 μm range size, there is another important requirement as the macropore cavities interconnectivity which must be improved. In their respect encouraging results have been obtained using a PS/Aluminium mould configuration (Fig. 3d).

IV. CONCLUSIONS

This work has shown the fabrication of homogeneous and reproducible gelatin/BG hybrid scaffolds. The porosity of the fabricated scaffolds was 95% and the macropore size was between 100-500μm. The results indicated that the introduction of the BG particulates in the full gelatin scaffold promotes a bioactive response *in vitro*.

ACKNOWLEDGEMENTS

We gratefully acknowledge the financial support provided by the Spanish Ministry of Science and Innovation (BIO2009-13903-C02-02), the Andalusian Ministry of Economy, Science and Innovation (Proyecto Excelencia CTS-6681), S. B-G is a fellow from the Andalusian Government Predoctoral Programme.

REFERENCES

1. Azami M, Moztarzadeh F, Tahriri M. (2010) Preparation, characterization and mechanical properties of controlled porous gelatin/hydroxyapatite nanocomposite through layer solvent casting combined with freeze-drying and lamination techniques. J Porous Mater 17:313-320
2. Fratzl P, Weinkamer R. (2007) Nature's hierarchical materials. Progress in Materials Science 57:1263-1334
3. Engel E, Castaño O, Salvagni E, Ginebra M, Planell (2009) J. Biomaterials for Tissue Engineering of Hard Tissues. Springer, Brighton
4. Chevalier J, Gremillard L. New Trends in Ceramics for Orthopedics.
5. Jones JR. (2013) Review of bioactive glass – from Hench to hybrids. Acta Biomaterialia 9:4457–4486
6. Hench L. (2006) The story of Bioglass. Journal of materials science. Materials in medicine 17:967-78.
7. Hench L. (1998) Bioceramics. J Am Ceram Soc 8:1705-1728.
8. Rahaman M, Day D, Bal BS, Fu Q, Jung S, Bonewald L, et al. (2011) Bioactive glass in tissue engineering. Acta Biomaterialia 7:2355-73.
9. Mozafari M, Moztarzadeh F, Rabiee M, Azami M, Maleknia S. (2010) Development of macroporous nanocomposite scaffolds of gelatin/bioactive glass prepared through layer solvent casting combined with lamination technique for bone tissue engineering. Ceram Int 36:2431-2439.
10. Van Vlierberghe S, Cnudde V, Dubruel P, Masschaele B, Cosijns A. (2007) Porous gelatin hydrogels: 1. Cryogenic formation and structure analysis. Biomacromolecules 8:331-337.
11. O'Brien F, Harley B, Yannas I, Gibson L. (2004) Influence of freezing rate on pore structure in freeze-dried collagen-GAG scaffolds. Biomaterials 25:1077-86.
12. Caliari S, Harley BAC. (2011) The effect of anisotropic collagen-GAG scaffolds and growth factor supplementation on tendon cell recruitment, alignment, and metabolic activity. Biomaterials 32:5330-5340
13. Davidenko N, Gibb T, Schuster C, Best SM, Campbell JJ, Watson CJ, et al. (2012) Biomimetic collagen scaffolds with anisotropic pore architecture. Acta Biomaterialia 8:667-76
14. Wu X, Liu Y, Li X, Wen P, Zhang Y, Long Y, et al. (2010) Preparation of aligned porous gelatin scaffolds by unidirectional freeze-drying method. Acta Biomaterialia 6:1167-77
15. Yan X, Yu C, Zhou X, Tang J, et al. (2004) Highly ordered mesoporous bioactive glasses with superior in vitro bone-forming bioactivities. Angew Chem Int Ed Engl. 43:5980-5984
16. Kokubo T, Takadama H. (2006) How useful is SBF in predicting in vivo bone bioactivity. Biomaterials 27:2907-2915
17. Yan X, Deng H, Huang X, Lu G, Qiao S. (2005) Mesoporous bioactive glasses. I. Synthesis and structural characterization. J Non Cryst Solids 351:3209-3217
18. Zhao S, Li Y, Li D. (2011) Synthesis of CaO-SiO2-P2O5 mesoporous bioactive glasses with high P2O5 content by evaporation induced self assembly process. J Mater Sci Mater Med. 22:201-208
19. HollingerJO, Brekke J, Gruskin E, Lee D. (1996) Role of Bone substitutes. Clin Orthop Relat Res 324:55-65.
20. Querido W, Abracado LG, Rossi AL, Campos APC, Abraçado LG, Farina M, et al. (2011) Ultrastructural and mineral phase characterization of the bone-like matrix assembled in F-OST osteoblast cultures. Calcif Tissue Int 89:358-71.

Collagen-Targeted BMPs for Bone Healing

P.M. Arrabal[1,2,3], R. Visser[1,2,3], L. Santos-Ruiz[3,2,1], J. Becerra[1,2,3], and M. Cifuentes[1,2,3]

[1] University of Malaga/Department of Cell Biology, Genetics and Physiology, Faculty of Science,
Campus de Teatinos s/n. 29071, Malaga. Spain
[2] BIONAND (Andalusian Center of Nanomedicine and Biotechnology) C/ Severo Ochoa,
35. PTA. 29590, Campanillas. Malaga. Spain
[3] CIBER-BBN(Networking Biomedical Research Center in Bioengineering,
Biomaterials and Nanomedicine)/Laboratory of Bioengineeing and Tissue Regeneration (LABRET-UMA),
University of Málaga. 29071. Malaga. Spain

Abstract—Among the members of the bone morphogenetic protein subfamily, only BMP-2 and BMP-7 are currently used for several applications in orthopaedic surgery. Due to the chemical and biological characteristics of BMPs, these growth factors are always used in combination with a carrier based on collagen type I.

Although the use of these growth factors is considered safe in short terms, the high doses needed make these treatments expensive and their safety uncertain at long term, since they are highly pleiotropic and have the capacity to induce ectopic ossification in the surrounding tissues.

Even though new bone formation can be achieved by direct application of BMPs alone, their use in combination with specific carriers improves their osteogenic abilities. The aim of the carrier is to retain the growth factors at the wound site, maintaining their local concentrations, since it has been demonstrated that the bone healing efficiency is correlated with the prolonged presence of BMPs.

There are several strategies to improve the currently used products in terms of efficiency, biosecurity and costs; our work is focused on modifying the biochemical features of the growth factors and complementing these systems with other growth factors or molecules, modified or not, toenhance or accelerate osteogenesis.These include the development of recombinant BMPs with additional features to increase their binding to the osteoconductive carrier; a recombinant human BMP-2 with improved collagen-binding properties has been produced.

Keywords—Tissue engineering, regenerative medicine, biomaterials, growth factors, recombinant proteins, BMPs.

I. INTRODUCTION

Reparation of bone defects and fractures is a major clinical and economic concern. From all cases registered in the hospitals of European countries, hip fracture is not only the most common, but also the most expensive injury to treat [1]. In the United States, over 7.9 million fractures are sustained each year, being trauma the second most expensive medical problem [2].

The expected time for a fracture to heal naturally is between six and twelve weeks, but there is a high rate of delayed unions. For the treatment of non-union fractures, autologous bone grafts are still considered the gold standard since they possess both important osteoconductive and osteoinductive properties. Although autologous bone grafting is generally successful, the limited amount of tissue that can be obtained from the donor and the high associated morbidity, are leading to the development of bone tissue engineering products for the restoration of damaged or lost bone [3, 4].

The bone morphogenetic proteins (BMPs) are a family of growth factors implicated in a variety of functions during development and in tissue regeneration [5]. Besides this fact, BMPs play a key role in the development and regeneration of the skeletal system [6], providing a solution for the treatment of bone fractures.

Recombinant BMPs have to be used in combination with a carrier to retain them at the wound site, and permitting a slow release into the extracellular milieu. For this purpose collagen is the only carrier approved to use in combination with rhBMP-2 (INFUSE®, Medtronic, Minneapolis,MN, USA). Since rhBMP-2 has low natural affinity to collagen, this approach requires the use of very high doses (milligrams) of BMPs [7, 8]. Besides the high costs of these treatments, the use of such amounts of BMPs can cause undesired side effects [9, 10].

A recombinant human BMP-2 with improved collagen-binding properties conferred by an additional domain at the N-terminus of the monomers, has been produced.

II. CLINICAL APPLICATION OF BONE MORPHOGENETIC PROTEINS AND COLLAGEN TYPE I

rhBMP-2, one of the only two approved BMPs for clinical use in combination with absorbable bovine type I collagen sponges (INFUSE®, Medtronic, Minneapolis), has shown excellent results for the treatment of open fractures in long bones and spinal fusions.

Due to its low natural affinity to collagen, the clinical studies carried out with rhBMP-2 concluded that the effective osteoinductive dose of this growth factor is 1.5 mg BMP-2 / mL [11, 12]. Nevertheless, concentrations in the order of just hundreds of nanograms per millilitre are sufficient to induce osteoblastic differentiation of mesenchymal cells *in vitro* while, in the human body, normal concentrations of BMPs are estimated at 2 ng/g of bone [13]. Thus, clinical application of BMPs implies raising their local concentration more than 10^6-fold over the physiological levels. Besides the high costs of these treatments, the use of such amounts of BMPs can cause undesired side effects [9, 10].

Although the use of BMP-2 is considered save, the longterm effects of the application of such amounts of these potent, highly pleiotropic growth factors are not well known. On the other hand, the immune mechanisms triggered upon BMP implantation are not well defined due to controversy in the literature. It seems that single applications of allogenic BMPs can activate a moderate production of anti-BMP antibodies [14].

Although it has been demonstrated that new bone formation can be achieved by direct application of BMPs alone [15, 16], application of the growth factors in combination with specific carriers can improve their osteogenic abilities [17].

Despite its poor biomechanical properties, collagen is the only carrier approved for clinical application of BMPs due to its high biocompatibility and biodegradability and low immunogenicity [18, 19]. Collagen is the main protein of connective tissue in animals, and is considered the most abundant protein in mammals. Clinical administration of BMPs for bone regeneration is done in combination with bovine type I absorbable collagen sponges (ACS), which are soaked with the growth factor before implantation [11]. It has been shown that this form of collagen allows proper cell infiltration during new bone formation [19].

Unfortunately, most growth factors have little natural affinity to collagen. Most of the problems associated with clinical application of growth factors could be palliated if these could be specifically retained at the wound site, with a slow and sustained liberation from their carrier. To achieve this, one strategy arising is the use of DNA technology to design and produce modified BMPs with special features.

III. COLLAGEN-TARGETED BMPs

In natural bone regeneration, the prolonged presence of BMPs at the local healing environment is provided by their interaction with components of the extracellular matrix, which is responsible for concentrating the active growth factors by avoiding their diffusion and modulating their local action [20]. Thus, BMPs are released as soluble active forms which are capable of diffusing away from the cell of origin or, due to natural mechanisms such as the presence of ECM-binding domains being tethered and concentrated within the ECM.

For their use in clinical applications, recombinant BMPs are combined with a carrier, not only with the aim of providing a support for bone ingrowth, but also to simulate the natural bone healing process in which BMPs are trapped in the extracellular matrix. The main problem of this approach is that the great majority of the available scaffolds do not have the ability to couple BMPs and provide a specific retention. Thus, osteogenic growth factors are commonly used with simple adsorption to the carrier by soak loading, what produces an initial burst release of them with a rapid decrease of biological activity, a fact that is considered inappropriate from a physiological point of view.

One of the strategies to accomplish the specific binding of the BMPs to their delivery material is to modify the growth factors with different matrix-binding domains, allowing a controlled slow release from their carrier and a protection from proteolytic degradation.

Since BMPs are structurally complex proteins, largescale *in vitro* production of these growth factors is not a simple task, especially when the coding region of the cloned gene is modified to obtain a non-native improved rhBMP. The active, mature form contains the typical cystine-knot motif that needs to be stabilized by three intracatenary and one intercatenary disulfide bonds. In addition, all BMPs have one or more putative N-glycosylation site in their mature domains, but the presence of N-moieties is not equally important for all the members of the BMP subfamily; whereas the lack of glycosylation does not affect the activity of rhBMP-2 [21, 22, 23], the binding of rhBMP-6 to its type I receptors seems to be strictly dependent on glycosylation [24, 23]. Other important factor is the low solubility of BMPs in aqueous solutions, what makes them prone to precipitate even at relatively low concentrations. Despite these facts, some members of the BMP subfamily have been produced to date as fusion proteins with several modifications in their sequence and additional domains which confer them specific affinity to several biomaterials or components of the extracellular matrix with no loss of their natural biological activity [23, 25-28].

Among all the BMP subfamily members, BMP-2 is one of the most studied, not only because of being involved in nearly all stages of the bone regeneration process, but also for its excellent results when used in clinical applications. In consequence, many researchers have concentrated their investigation on the modification of this protein to bind it to extracellular matrix components.

One of these modifications consists of the addition of collagen binding domains to the BMPs. Collagen-targeted rhBMP-2 are of special clinical interest, not only because direct administration of these molecules in soluble form could increase their local concentrations by direct binding to the collagen fibers at the site of injection, but when administered in combination with a collagenic carrier, the growth factors would be specifically retained, limiting their actions to the wound site. These approaches could reduce the doses of BMP needed to achieve bone regeneration when compared to the use of native molecules, improving the safety of the treatments and reducing their costs.

One of the strategies to achieve this goal consisted of the production of a fusion protein containing the mature rhBMP-2 sequence and an additional collagen type I-binding domain (CBD) derived from the bovine von Willebrand factor (vWF) fused to the N-terminal part of the molecule (Figure 1). This collagen-binding domain (CBD) has been identified as a decapeptide with the sequence Trp-Arg-Glu-Pro-Ser-Phe-Cys-Ala-Leu-Ser [29], and has demonstrated to be effective in enhancing the collagen binding properties of several proteins.

To avoid disulfide scrambling during the in vitro refolding procedure, the original Cys-7 of the decapeptide was replaced by a methionine [30]. An additional glycine acts as a linker between both sequences. Furthermore, this rhBMP2-CBD lacked any other additional sequences such as the commonly used 6xHis purification tag, and was purified by its natural affinity to heparin [23].

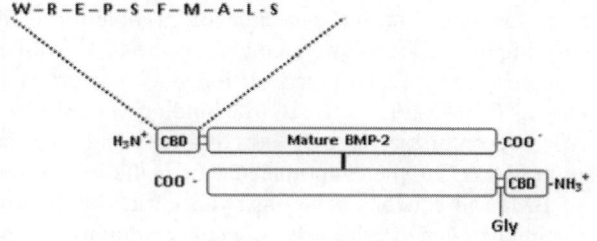

Fig. 1 Schematic representation of the genetically engineered rhBMP2-CBD fusion protein.

The resulting protein construct exhibited an enhanced specific affinity to absorbable bovine type I collagen sponges in a dose-dependent manner and this binding was demonstrated to be stable over time. In addition, when implanted in vivo together with ACSs which, unlike DBM, are free of any other endogenous growth factors, low concentrations of this rhBMP2-CBD was able to induce new bone formation in rats (Figure 2).

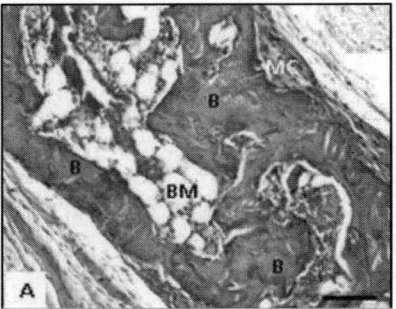

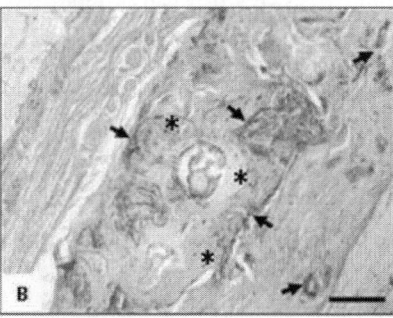

Fig. 2 In vivo osteogenic activity of rhBMP2-CBD. Ectopic bone formation when implanted in combination with ACS. (A) Hematoxylin-eosin staining of implants, showing the formation of a mature trabecular bone with medullar cavities. (B) Immunostaining with an anti-osteopontin antibody. Arrows: osteoblasts expressing osteopontin. Asterisks: osteocytes expressing osteopontin. B: mature bone trabecula. BM: bone marrow. MC: undifferentiated mesenchymal cells. Scale bars =100 μm

These and other studies suggest that the design and production of recombinant modified BMPs might be useful to improve the current results obtained with the clinical application of these growth factors.

In the future, a new strategy to complement and/or enhance BMP-induced osteogenesis is the use of other growth factors in the system, such as angiogenic signals or other molecules involved in bone induction.

Acknowledgment

We would like to thank Ms. Eva Jiménez-Enjuto for technical support. This work was partially supported by the Spanish Network on Cell Therapy (Red TerCel),BIO2009-13903-C01-01 (MICINN) and P07-CVI-2781 (Junta de Andalucía). CIBER-BBN isan initiative funded by the VI National R&D&I Plan 2008-2011, IniciativaIngenio 2010,Consolider Program, CIBER Actions and financed by the Instituto de Salud Carlos III withassistance from the European Regional Development Fund.

REFERENCES

1. Finnern HW, Sykes DP (2003) The hospital cost of vertebral fractures in the EU: estimates using national datasets. Osteoporos Int 14(5):429-36
2. Bishop GB, Einhorn TA (2007) Current and future clinical applications of bone morphogenetic proteins in orthopaedic trauma surgery. Int Orthop 31(6):721-7
3. Braddock M, Houston P et al. (2001) Born again bone: tissue engineering for bone repair. News Physiol Sci 16:208-13
4. Rush SM (2005) Bone graft substitutes: osteobiologics. Clin Podiatr Med Surg 22(4):619–30
5. Hogan BL (1996) Bone morphogenetic proteins in development. Curr Opin Genet Dev 6(4):432–8
6. Nakase T, Yoshikawa H (2006) Potential roles of bone morphogenetic proteins (BMPs) in skeletal repair and regeneration. J Bone Miner Metab 24(6):425–33
7. Geiger M, Li RH et al. (2003) Collagen sponges for bone regeneration with rh-BMP2. Adv Drug Deliv Rev 55(12):1613–29
8. Swiontkowski MF, Aro HT et al. (2006) Recombinant human bone morphogenetic protein-2 in open tibial fractures. A subgroup analysis of data combined from two prospective randomized studies. J Bone Joint Surg Am 88(6):1258–65
9. Itoh K, Udagawa N et al. (2001) Bone morphogenetic protein 2 stimulates osteoclast differentiation and survival supported by receptor activator of nuclear factor-kappa B ligand. Endocrinology 142(8):3656–62
10. Sciadini MF, Johnson KD (2000) Evaluation of recombinant human bone morphogenetic protein-2 as a bone-graft substitute in a canine segmental defect model. J Orthop Res 18(2):289–302
11. Valentin-Opran A, Wozney J et al. (2002) Clinical evaluation of recombinant human bone morphogenetic protein-2. Clin Orthop Relat Res 395:110-20
12. Govender S, Csimma C et al. (2002) BMP-2 Evaluation in Surgery for Tibial Trauma (BESTT) Study Group. Recombinant human bone morphogenetic protein-2 for treatment of open tibial fractures: a prospective, controlled, randomized study of four hundred and fifty patients. J Bone Joint Surg Am 84-A(12):2123-34
13. Rengachary SS (2002) Bone morphogenetic proteins: basic concepts. Neurosurg Focus 13(6):e2
14. Granjeiro JM, Oliveira RC et al. (2005) Bone morphogenetic proteins: from structure to clinical use. Braz J Med Biol Res 38(10):1463-73
15. Wozney JM, Rosen V et al. (1990) Growth factors influencing bone development. J Cell Sci Suppl 13:149-56
16. Einhorn TA, Majeska RJ et al. (2003) A single percutaneous injection of recombinant human bone morphogenetic protein-2 accelerates fracture repair. J Bone Joint Surg Am 85-A(8):1425-35
17. Peel SA, Hu ZM et al. (2003) In search of the ideal bone morphogenetic protein delivery system: in vitro studies on demineralized bone matrix, purified, and recombinant bone morphogenetic protein. J Craniofac Surg 14(3):284-91
18. Hubbell JA (1995) Biomaterials in tissue engineering. Biotechnology (N Y) 13(6):565-76
19. Friess W. Collagen--biomaterial for drug delivery (1998) Eur J Pharm Biopharm 45(2):113-36
20. Ruppert R, Hoffmann E et al. (1996) Human bone morphogenetic protein 2 contains a heparin-binding site which modifies its biological activity. Eur J Biochem 237(1):295-302
21. Vallejo LF, Brokelmann M et al. (2002) Renaturation and purification of bone morphogenetic protein-2 produced as inclusion bodies in high-cell-density cultures of recombinant Escherichia coli. J Biotechnol 94(2):185-94
22. Long S, Truong L et al. (2006) Expression, purification and renaturation of bone morphogenetic protein-2 from Escherichia coli. Prot Expr Purif 46(2):374-8
23. Visser R, Arrabal PM et al. (2009) The effect of an rhBMP-2 absorbable collagen sponge-targeted system on bone formation in vivo. Biomaterials 30(11):2032-7
24. Saremba S, Nickel J et al. (2008) Type I receptor binding of bone morphogenetic protein 6 is dependent on N-glycosylation of the ligand. FEBS J 275(1):172-83
25. Han B, Perelman N et al. (2002) Collagen-targeted BMP3 fusion proteins arrayed on collagen matrices or porous ceramics impregnated with Type I collagen enhance osteogenesis in a rat cranial defect model. J Orthop Res 20(4):747-55
26. Schmoekel HG, Weber FE et al. (2005) Bone repair with a form of BMP-2 engineered for incorporation into fibrin cell ingrowth matrices. Biotechnol Bioeng 89(3):253-62
27. Chen B, Lin H et al. (2007a) Homogeneous osteogenesis and bone regeneration by demineralized bone matrix loading with collagen-targeting bone morphogenetic protein-2. Biomaterials 28(6):1027-35
28. Chen B, Lin H et al. (2007b) Activation of demineralized bone matrix by genetically engineered human bone morphogenetic protein-2 with a collagen binding domain derived from von Willebrand factor propolypeptide. J Biomed Mater Res A 80(2):428-34
29. Takagi J, Asai H et al. (1992) A collagen/gelatin-binding decapeptide derived from bovine propolypeptide of von Willebrand factor. Biochemistry 31(36):8530-4
30. Tuan TL, Cheung DT et al. (1996) Engineering, expression and renaturation of targeted TGF-beta fusion proteins. Connect Tissue Res 34:1–9.

Author: Pilar M Arrabal
Institute: University of Malaga/Department of Cell Biology, Genetics and Physiology. Faculty of Science
Street: Campus de Teatinos s/n
City: 29071 Malaga
Country: Spain
Email: parrabalg@uma.es

Layer by Layer: Designing Scaffolds for Cardiovscular Tissues

B. Glasmacher[1], A. Repanas A., O. Gryshkov, F. AL Halabi, T. Rittinghaus, R. Kortlepel,
S. Wienecke, M. Müller, and H. Zernetsch

[1] Institute for Multiphase Processes, Leibniz Universität Hannover, Germany

Abstract—In the field of cardiovascular implant technology, synthetic implants, biological grafts and biohybrids are used for valves and vascular grafts. There is a great need for vascular grafts with small inner diameters and heart valve prostheses with growing potential for children. New opportunities for the design of artificial tissue structures via ice templating and electrospinning are described. Exemplarily, developments of vascular grafts, and heart valves will be presented..

Keywords—Cardiovascular tissue engineering, scaffold, ice templating, electrospinning, mimicking ECM, fibers, hemocompatibility.

I. INTRODUCTION

Since the implants that are available today still lack of some biocompatibility features, there is a need for improvement. Functional bioartificial constructs at the time of implantation remain a driving factor for further development. This is due to the active role of a functional endothelial cell layer in mediating blood compatibility and preventing adverse reactions. Thus designing a scaffold to present the correct mechanical and biochemical clues to cells and mechanically stimulating the construct after seeding are vital factors to support physiological tissue formation and cell activity.

Tissue engineering as a field of regenerative medicine is an interdisciplinary field that combines the principles of engineering and life sciences with the goal of restoration, repair or replacement of tissues and their functions. Its underlying principle is the combination of cells and scaffolds, a transitory extracellular matrix, to produce a new functional tissue. Bridging the gap between isolated cells and functioning tissue, the scaffolds become an instructive extracellular microenvironment that actively guides cells both locally and in time towards tissue formation and regeneration. This is achieved by the presentation of specific insoluble biochemical and structural cues to constituent cells in combination with the controlled release of soluble factors with time. In tissue engineering the combination of appropriate cells on a suitable biomaterial as a carrying structure poses a particular challenge. Thus, the cell type and source, the controlled cell seeding of the scaffolds and bioreactor systems for further cultivation and (stem cell) differentiation address further important challenges. Electrospinning and ice templating are versatile techniques in tissue engineering for the production of such scaffolds mimicking the extracellular matrix using different kinds of permanent or resorbable biomaterials with possible incorporation of specific drugs and different structural features.

II. MATERIALS AND METHODS

A. Electrospinning

Electrospinning is a technique for the production of ultrafine polymer fibers from polymer melts or polymer solutions through electrostatic interaction. The original technique of electrospinning was patented in the first half of the twentieth century by Anton Formhals and Richard Schreiber-Gastell. The filed assemblies and methods aimed at the production of silk-like threads from cellulose acetate. They used cellulose acetate solutions and their filed inventions already featured wheels, mandrels or a metal band to coil up and align the fibres. Formhals also proposed the use of precipitating baths and the possibility of needleless spinning assemblies [1]. Electrospinning as a technology for the production of micro- and nanostructured scaffold materials has gained widespread acceptance in the medical research community over the last decade. The process generates a non-woven fiber mat consisting of one continuous filament with diameters ranging from the micron to nanometer range. It is most often used as scaffold materials in tissue engineering applications due to its similarity to the filamentous microenvironment in native tissues. This similarity often promotes a more positive cell response to the generated fibers than to the bulk material alone. However, the reproducibility of the scaffold structure is often limited by the used electrospinning set-ups that so far do not fully utilize the available potential of the process technology. Here, we describe techniques for the production of aligned fiber structures, multilayered, multiscaled and multifiber scaffolds, fiber modification und functionalization and useful advances in process control.

B. Ice Templating

This technique enables to produce scaffolds with directional porous structures made out of suspensions such as collagen, chitosan, PEO (with alumina particles), fibrin. Examples are shown below in figure 1.

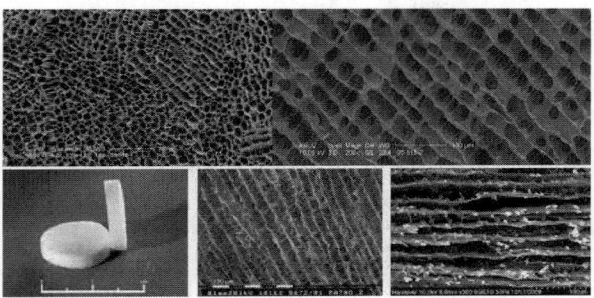

Fig. 1 Structures made by ice templating

C. Vascular Tissue Engineering

In the field of cardiovascular implant technology, synthetic implants, biological grafts and biohybrids are used for valves and vascular grafts. There is a great need for vascular grafts with small inner diameters and heart valve prostheses with growing potential for children. Since the implants that are available today still lack of some biocompatibility features, there is a need for improvement. Functional bioartificial constructs at the time of implantation remain a driving factor for further development. This is due to the active role of a functional endothelial cell layer in mediating blood compatibility and preventing adverse reactions. Thus designing a scaffold to present the correct mechanical and biochemical clues to cells and mechanically stimulating the construct after seeding are vital factors to support physiological tissue formation and cell activity.

Electrospinning offers a unique opportunity to process both synthetic and biological materials, either in pure form or as a combination, to generate nanofibrous mats as tissue engineering scaffolds. The generated fibrous microstructure closely resembles structures in the native extracellular matrix and can be oriented to provide tactile clues to influence cell migration and behaviour. Furthermore, processing different materials in parallel to finetune the internal microstructure allows a wide range in which to modulate mechanical and biochemical properties.

To support and guide neotissue formation, mechanical stimulation of the seeded construct up to the point of implantation is required. This has been recognized by the advent of specialized bioreactors to reproduce the mechanical stimuli in the cardiovascular environment. However, the effects of individual factors (e.g. shear rate and oxygenation level) are still not fully understood especially with regard to their temporal kinetics. Real time microscopic imaging of cells and their response to changes in their micromechanical stimulation may help to better define parameters for stimulation of tissue engineering constructs.

D. Hemocompatibility

Developing a synthetic vascular graft with a small inner diameter is one of the great challenges in biomedical engineering. Current prosthesis with an inner diameter less than 6 mm tend to occlude within the first months [2]. Tissue engineering, using cell-seeded scaffolds prepared by methods like salt leaching, directional solidification or electrospinning (fig. 2), is a promising way to overcome this limitation [3]. In order to screen such scaffolds for vascular tissue engineering, standard test methods were successfully adapted to highly porous structures. Tests on suture retention strength, longitudinal and circumferential stiffness, and compliance are performed at 37°C submerged in PBS solution. Hemocompatibility of the scaffold material is evaluated in a dynamic in vitro test setup with anticoagulated porcine whole blood. In a preliminary test run, electrospun polycaprolactone/polylactide (PCL/PLA)-Scaffolds (inner diameter of 4 mm) were shown to have excellent suture retention strength and tensile strength. Hemocompatibility testing resulted in a decrease in thrombocyte count of about 20 % over 1 h, which is comparable to the change caused by bare metal stents in a previous study. The set of test methods proved to be suitable for testing highly porous materials and will be used for screening of structures made from various materials produced by electrospinning or ice templating. [4] [5]

Improvements in hemocompatibility test setups for artificial vascular grafts are mandatory as a consequence of the accelerating need for blood vessel reconstructions. Nevertheless, standardized test systems for the evaluation of blood compatibility of biomaterials are still missing. This work is intended to define standardized procedures for hemocompatibility testing by the use of a modified CHANDLER-Loop system. Vascular grafts, consisting of thin polymer fibres (PCL/PLA), produced via electrospinning (72 kV/m) and electropolished stents were tested in a closed loop filled with citrated whole blood. Platelet count, hematocrit, and activation of coagulation were analyzed after dynamic incubation (20 cm/s, 1 h). In addition, platelet adhesion was investigated by light microscopy (LM) and SEM imaging. Furthermore, human umbilical vein endothelial cells were used for the endothelialization of grafts to examine differences in hemocompatibility analysis.

III. DISCUSSION AND CONCLUSION

Analysis of electropolished stents by our modified CHANDLER-loop system revealed only a weak activation of platelets and increase in coagulation parameters (FB, PT, TT). Moreover, electrospun fibers show a 12 % reduction in the platelet count, but no activation of the coagulation system. LM and SEM revealed isolated platelets and endothelial cells attached to the graft. In addition, endothelialization of grafts could be improved by a pretreatment of grafts with human whole blood. Test parameters for hemocompatibility analysis of artificial grafts should include analysis of platelet index and coagulation activation in association with qualitative methods such as LM and SEM analysis. In addition, our modified CHANDLER-loop system represents a suitable model for a dynamic test set up in hemocompatibility research. Furthermore, endothelialization of vascular grafts offers an interesting tool to improve the hemocompatibility foreign blood contacting surfaces.

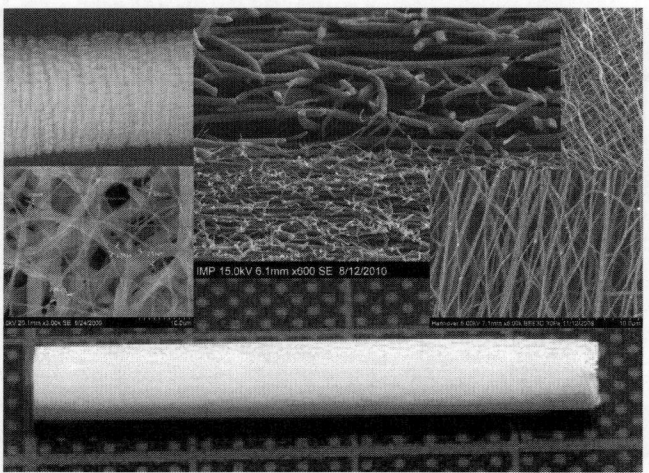

Fig. 2 2D and 3D structures via electrospinning

Scaffold design via electrospinng and ice templating resulted in versatile scaffolds made out of various synthetic respectively biological polymers. The scaffolds exhibited different macroscopic and microscopic structures, various pore sizes and porosities. Scaffolds reinforced with ceramic nanoparticles showed hard-elastic properties for bone replacement, aligned fibres for tendon and fiber mats for heart valve and small vascular graft replacement. Using biochemical cues, nerve guidance is under development. Kinetic drug release studies have been performed to assess the functionality of the scaffold design. Different cell types have been used. Collagen scaffolds made via ice templating were used in stem cell studies from the common marmoset monkey *(Callithrix jacchus)* to its genetic and physiological similarities to humans. Stem cells were used in cell seeding studies applying laser induced cell transfer to locally deposit the cells.

To guide tissue development, scaffolds must provide specific biochemical, structural and mechanical cues to cells and deliver them in a controlled manner over time. Electrospinning and ice templating have shown to be versatile techniques in tissue engineering for the production of scaffolds. Freeze-dried and electrospun scaffolds showed to address both controlled release and structural cues with adequate pore sizes and resulting porosity enabling cell proliferation and ingrowth. The generated fibrous microstructure closely resembles structures in the native extracellular matrix and can be oriented to provide tactile clues to influence cell migration and behaviour. Furthermore, processing different materials in parallel to finetune the internal microstructure allows a wide range in which to modulate mechanical and biochemical properties.

To support and guide neotissue formation, mechanical stimulation of the seeded construct up to the point of implantation is required. This has been recognized by the advent of specialized bioreactors to reproduce the mechanical stimuli in the cardiovascular environment. However, the effects of individual factors (e.g. shear rate and oxygenation level) are still not fully understood especially with regard to their temporal kinetics. [6] Real time microscopic imaging of cells and their response to changes in their micromechanical stimulation may help to better define parameters for stimulation of tissue engineering constructs.

ACKNOWLEDGEMANTS

The authors would like to thank the German Research Foundation (DFG) for the financial support of these studies within SFB 599, SFB/TR 37, and Exc 62/1.

REFERENCES

1. Formhals A, Schreiber-Gastel R: Process and apparatus for preparing artificial threads. US Pat. No: 1,975,504, 1934
2. Ratner BD. The catastrophe revisited: Blood compatibility in the 21st Century. Biomaterials 28,34 (2007), 5144–5147
3. Szentivanyi AL, Zernetsch H, Menzel H, Glasmacher B. A review of developments in electrospinning technology: New opportunities for the design of artificial tissue structures. Int J Artif Organs 34,10 (2011) 986-97
4. Szentivanyi AL, Chakradeo T, Zernetsch H, Glasmacher B. Electrospun Cellular Microenvironments: Understanding Controlled Release and Scaffold Structure. Adv Drug Deliv Rev.63,4-5(2011)209-20
5. Krolitzki B, Müller M, Glasmacher B, Validation of a test setup for hemocompatibility testing of small cardiovascular implants, Int J Artif Organs, 34, 8, (2011) 705
6. Dreyer L, Krolitzki B, Autschbach R, Vogt P, Welte T, Ngezahayo A, Glasmacher B: An Advanced Cone-And-Plate Reactor for the *in vitro*-Application of Shear Stress on Adherent Cells. Clin Hemorheol Microcirc. 2011 Jan 1;49(1):391-7.

Author: Birgit Glasmacher
Institute: Institute for Multiphase Processes
Street: Callinstrasse 36
City: Hannover
Country: Germany,
Email: glasmacher@imp.uni-hannover.de

Part XII
Neural and Rehabilitation Engineering

Voice Activity Detection from Electrocorticographic Signals

V.G. Kanas[1], I. Mporas[1], H.L. Benz[2], N. Huang[2], N.V. Thakor[3], K. Sgarbas[1],
A. Bezerianos[3], and N.E. Crone[4]

[1] Department of Electrical and Computer Engineering, University of Patras, Patras, Greece
[2] Department of Biomedical Engineering, Johns Hopkins University, Baltimore, MD 21205 USA
[3] Singapore Institute for Neurotechnology, National University of Singapore, Singapore
[4] Department of Neurology, Johns Hopkins University, Baltimore, MD 21205 USA

Abstract—The purpose of this study was to explore voice activity detection (VAD) in a subject with implanted electrocorticographic (ECoG) electrodes. Accurate VAD is an important preliminary step before decoding and reconstructing speech from ECoG. For this study we used ECoG signals recorded while a subject performed a picture naming task. We extracted time-domain features from the raw ECoG and spectral features from the ECoG high gamma band (70-110Hz). The RelieF algorithm was used for selecting a subset of features to use with seven machine learning algorithms for classification. With this approach we were able to detect voice activity from ECoG signals, achieving a high accuracy using the 100 best features from all electrodes (96%) or only 12 features from the two best electrodes (94%) using the support vector machines or a linear regression classifier. These findings may contribute to the development of ECoG-based brain machine interface (BMI) systems for rehabilitating individuals with communication impairments.

Keywords—Voice activity detection, electrocorticography, brain machine interface, machine learning.

I. INTRODUCTION

Brain-machine interface (BMI) systems attempt to rehabilitate paralyzed individuals and simultaneously allow direct communication between the human brain and an external machine [1]. In addition to BMIs for cursor control [2] and limb prosthetic control [3][4][5][6], there is growing interest in BMIs for restoring speech function, for example in patients with profound articulatory impairment [7][8].

Detecting and decoding speech from cortical activity is more complex than decoding movement. Language involves large-scale cortical networks that are dynamically engaged in phonological analysis, speech articulation and other processes [9][10]. In [11], the authors aimed to discriminate three different tasks from EEG recordings (imagined speech of vowels /a/ and /u/ and a no action state). They designed spatial filters using the common spatial pattern (CSP) method to extract features and the support vector machine (SVM) algorithm to discriminate the three tasks. More recently, in [12] the discrimination of vowels and consonants of overt and covert word production using ECoG recordings was proposed. They used spectral amplitudes in different frequency bands, estimated via an autoregressive (AR) model, and the local motor potential (LMP) as features and a Naïve Bayes model as classifier. ECoG recordings [13] have become recognized as an in vivo biocompatible option for a wireless chronically implantable BMI system. Such a system could be utilized as an assistive device to enable disabled individuals to produce speech through neural activity. This device could be feasible through speech reconstruction [14] from cortical brain activity. However, the first step before exploring decoding or reconstruction methodologies is to use neural activity to detect an individual's speech, i.e. to define the time intervals in which a subject speaks.

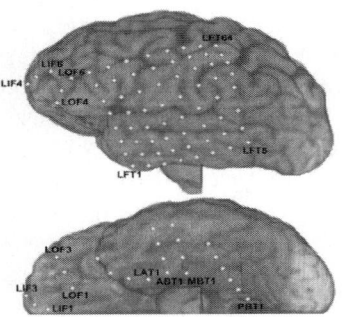

Fig. 1 The subject's electrode locations.

Voice activity detection (VAD) has been studied for more than ten years in the speech technology community [16]. The focus of VAD is to detect the time intervals in which vocal communication is occurring, especially when its acoustic spectrum overlaps that of other sources, such as music or noises. Current experimental protocols require human intervention to differentiate between such intervals; to be useful outside the laboratory, however, prosthetic systems will need to determine the epoch autonomously [15]. In this study, we propose a VAD framework using ECoG signals as a preliminary step before speech decoding methodologies. We intend to detect voice activity automatically by exploiting machine learning techniques, including feature selection and classification.

II. ECOG DATA COLLECTION

A. Subject and Data Collection

This experiment was conducted with one male patient with intractable epilepsy, who was temporarily implanted with subdural electrode arrays (Ad-Tech, Racine, Wisconsin; 2.3 mm exposed diameter, with 1 cm spacing between electrode centers) to localize his seizure focus for resection. The experimental protocol was approved by the Johns Hopkins Medicine Institutional Review Board, and the patient gave informed consent for this research. The electrode placement was chosen based on the clinical requirements of the patient, guided by estimation of the seizure focus prior to surgery. Localization of the ECoG electrodes after surgery was performed by co-registration of pre-implantation volumetric MRI with post-implantation volumetric CT using Bioimage [17].

One large eight-by-eight grid of electrodes was placed over the left hemisphere of the brain, covering portions of the frontal, temporal, and parietal lobes (LFT1-64). Additional electrode strips were placed on the frontal lobe and on the basal temporal surface (Fig. 1). Data was amplified and recorded through a NeuroPort System (Blackrock Microsystems, Salt Lake City, Utah) at a sampling rate of 10 kHz, low pass filtered with a cutoff frequency of 500 Hz, and then downsampled to 1 kHz. The patient's spoken responses were recorded by a Zoom H2 recorder (Samson Technologies, Hauppauge, New York), also at 10 kHz but without subsequent downsampling. Each dataset was visually inspected and all channels that did not contain clean ECoG signals were excluded. Channels 2, 12, 23, 44 and 53 were removed, which left 89 channels for our analyses.

B. Experimental Protocol

A picture naming task was performed by the patient during ECoG recording. The patient was seated in a hospital bed. During each trial, an image was presented to the patient for one second using E-Prime software (Psychology Software Tools, Inc., Sharpsburg, Pennsylvania). The patient was instructed to name the image as quickly as possible, or to say 'pass' if unable to name the image. Between trials, a fixation cross was presented for an average of seven seconds, with a jitter of up to a half second to avoid stimulus anticipation by the subject. Three hundred and fifty trials were conducted in one session lasting around one hour, with a short break in the middle. We used the open source Praat software [18] to segment and label the patient's spoken and ECoG responses as "silence", "speech" and "noise" to train our model. During our analysis the noisy epochs have been removed.

III. EXPERIMENTAL SETUP

A. Feature Extraction

Recorded data from each ECoG electrode were re-referenced by subtracting the common average (CAR) of electrodes in the same array, as defined by equation (1),

$$x[n]_{ch}^{CAR} = x[n]_{ch} - \frac{1}{N}\sum_{l=1}^{N} x[n]_l \quad (1)$$

where $x[n]_{ch}$ and $x[n]_{ch}^{CAR}$ are the ECoG and CAR referenced ECoG amplitudes on the ch-th channel out of a total of N recorded channels.

We aimed to characterize the temporal and frequency information in ECoG channels, so we chose hybrid features for investigation, which included channel-specific power spectra and autoregressive model coefficients. Each ECoG channel was segmented with a sliding Hamming window and two sets of features were extracted for each window. We used a window length of 256 samples and a step size of 128 samples.

First, the power spectra of these frames were calculated by applying a fast Fourier transform (FFT). Since the high gamma oscillations were previously reported to be highly correlated with speech-related cortical activation [9], we computed high gamma frequencies between 70 Hz and 110 Hz in 2 Hz bins. Power estimates in this frequency range were log-transformed to approximate normal distributions. We then averaged the power spectra in the above frequency range to obtain the final spectral features. We separately trained an autoregressive model of order 5 using the Yule-Walker method [19] for each channel and time window. The model order was found as a tradeoff between computational cost and model prediction accuracy. The AR model coefficients were used as temporal features. Consequently, a total of 534 features were used for our VAD analysis. In this paper, the k-th AR coefficient of n-th channel is denoted as channel(n) – AR(k) and the average power as channel(n) – PSD.

B. Feature Selection and Classification

Feature selection methods are typically applied to select a small set of effective features in order to improve generalization ability and classification performance. Consequently, feature selection was performed to reduce the dimensionality of the feature vector from $\mathbb{R}^{534}$ to a lower dimensional space $\mathbb{R}^N$, with N<534. The ReliefF algorithm was used [20] for the selection of the N most important features. The ReliefF algorithm evaluates the worth of a feature by iteratively sampling an instance and considering the value of the given feature for the nearest instance of the same and different class (here speech or silence). Using the feature ranking results, we evaluated the N-best features for N= {10, 50, 100, 150, 200, 300, 534} for the VAD task, i.e. the binary classification problem between speech/silence.

For classification we tested seven classifiers used in literature [21] to examine the robustness of our method: support vector machines (SVM), multilayer perceptron (MP), K- nearest neighbors (KNN), J48 tree, decision stump tree (DS), regression tree (RT) and logistic regression (LR). The evaluation of results was estimated using 10-fold cross validation, in order to avoid overlapping between training and test subsets.

For the SVM kernel we used the radial basis function (RBF), with parameters C=20.0 and γ=0.5, which were found as optimal values after a grid search at C= {1.0, 10.0, 20.0, 30.0} and γ= {0.001, 0.01, 0.1, 0.5, 1.0, 2.0, 5.0}. Additionally, for the KNN classifier K=20 was found as the optimal value after searching at K= {1, 10, 15, 20, 25, 30, 40, 50}.

IV. EXPERIMENTAL RESULTS

The ECoG features described above were evaluated for their appropriateness for the VAD task. The RelieF algorithm feature ranking results for the 10 most discriminative ECoG features are shown in Table 1. The most discriminative ECoG features are both spectral and temporal features. The most informative feature is the average high gamma power spectrum of channel 20, with a ranking score of 0.074.

Table 1 Ranking of the 10-best ECoG features for the Voice Activity Detection task as evaluated by the RelieF algorithm

Ranking	ECoG features	Ranking Score
1	channel(20)-PSD	0.074
2	channel(33)-AR(2)	0.063
3	channel(20)-AR(5)	0.061
4	channel(33)-PSD	0.058
5	channel(20)-AR(2)	0.049
6	channel(21)-PSD	0.047
7	channel(32)-PSD	0.046
8	channel(20)-AR(1)	0.041
9	channel(89)-AR(5)	0.035
10	channel(41)-AR(5)	0.034

We next evaluated subsets of ECoG features, in the order of their RelieF ranking outcome, to examine the optimal parametric subset. The VAD accuracy, in percentage, for the N-best ECoG features for the tested classifiers is shown in Table 2. As can be seen from Table 2, the optimal performance (average 96%) was achieved when using the 100 to 150 most discriminative ECoG features for SVM and LR classifiers. The use of fewer than 100 or more than 150 of the best ECoG features did not improve the speech/silence discrimination accuracy. This was due to the fact that the discriminative information of the best 100 ECoG features was not complementary to the information carried by the rest of the features.

Table 2 Voice Activity Detection accuracy (%) using seven classifiers for a varying number of ECoG features

Nf	SVM	KNN	LR	MP	DS	J48	RT
10	94.36	93.19	93.79	93.29	92.02	91.13	92.99
50	90.65	93.39	95.53	95.24	92.18	90.23	92.95
100	96.03	92.74	96.09	95.55	92.04	89.88	92.95
150	95.61	92.08	95.53	95.45	92.04	89.16	92.89
200	95.43	90.91	95.77	95.63	92.04	89.04	92.97
300	94.50	88.15	94.08	95.63	92.04	89.02	93.01
534	84.40	85.70	93.67	95.70	92.02	89.02	92.69

Nf: the number of N- best ECoG features

Since we are interested in the development of minimally invasive BMI systems, temporal and spectral features extracted from the "best" channels, as illustrated in Table 3, using the two "best" classifiers. Features extracted from the two "best" electrodes (channels 20 and 33) resulted in a 94% classification accuracy using SVM or LR classifier. Overall the feature selection and ranking showed that features extracted from channels 20 and 33 were more informative than features extracted from other channels. Additionally, the above step is crucial to investigate the channels' significance in relationship to their location on the brain. The two electrodes that were ranked highest by the RelieF algorithm, shown in Fig. 2 with enlarged circles, were located in cortical areas typically involved in speech and language processing. Channel 20 was located over the posterior superior temporal gyrus (STG), which contains auditory association cortex and is part of Wernicke's area, typically important for speech perception. Channel 33 was located over or near tongue motor cortex, which is important for articulation. Many of the other most highly ranked electrodes were located near these electrodes or in cortical areas also relevant to speech processing, including Broca's area.

V. CONCLUSIONS

While previous ECoG studies on speech decoding and reconstruction focused on phoneme or word level processing [12][14], they did not consider the problem of voice activity detection (VAD). Here, VAD from ECoG signals was studied. Temporal features were extracted from the raw ECoG signal, and spectral features were extracted from the high gamma (70 Hz – 110Hz) response. Then the machine learning RelieF algorithm was implemented to reduce the dimensionality of the feature space and several classification algorithms were tested. We achieved the highest average classification accuracy, 96%, using the 100-best ECoG features. Moreover, we found that channels 20 and 33, which were located in cortex typically important for speech production, were the most informative.

While results from the current study are encouraging, more extensive training using larger datasets is expected to further improve the generalization ability and increase the performance of our classification system. In the future, we aim to extract more complex features and investigate several machine learning techniques for feature selection and classification to detect voice activity, not only from overt but also from covert articulation. This may assist in fully automated natural speech BMIs, which will enable people to communicate silently using related brain activity.

Table 3 The 10-best ranked channels evaluated by the RelieF algorithm

Ranking	Channel	Ranking Score
1	channel(20)	0.041
2	channel(33)	0.034
3	channel(89)	0.024
4	channel(32)	0.019
5	channel(49)	0.016
6	channel(25)	0.016
7	channel(12)	0.015
8	channel(41)	0.013
9	channel(18)	0.012
10	channel(21)	0.011

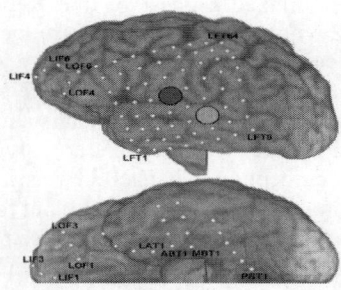

Fig. 2 The two best-ranked electrodes, channel 20 (blue) and channel 33 (red), were located in cortical areas relevant to the speech task.

ACKNOWLEDGMENT

This research was funded by the John S. Latsis Public Benefit Foundation, the European Social Fund, and the Operational Program "Education and Lifelong Learning" of the National Strategic Reference Framework, Research Funding Program THALES, and NINDS R01-40596 (NEC).

REFERENCES

1. Wolpaw J R, Birbaumer N et al. (2002) Brain-computer interfaces for communication and control. Clin Neurophysiol 113: 767–791
2. Schalk G, Kubanek J et al. (2007) Decoding two-dimensional movement trajectories using electrocorticographic signals in human. J Neural Eng 4: 264–275.
3. Benz H, Zhang H et al. (2012) Connectivity analysis as a novel approach to motor decoding for prosthesis control. IEEE Trans on Neural Systems and Rehabilitation Engineering 20:143-152.
4. Benz H L, Collard M et al. (2012) Directed Causality of the Human Electrocorticogram During Dexterous Movement. Engineering in Medicine and Biology Society (EMBC), 34th Annual International Conference of the IEEE, pp. 1872-1875.
5. Stavrinou M L, Moraru L et al. (2007) Evaluation of Cortical Connectivity During Real and Imagined Rhythmic Finger Tapping, Brain topography 19:137-145.
6. Fifer M S, Acharya S et al. (2012) Toward Electrocorticographic Control of a Dexterous Upper Limb Prosthesis: Building Brain-Machine Interfaces. IEEE Pulse 3:38-42.
7. Pei X, Hill J et al. (2012) Silent communication: Toward using brain signals. IEEE Pulse 3:43-46.
8. Guenther F H, Brumberg J S et al. (2009) A wireless brain–machine interface for real-time speech synthesis. PloS Biology 4: e8218.
9. Korzeniewska A, Franaszczuk P J et al. (2011) Dynamics of large-scale cortical interactions at high gamma frequencies during word production: Event related causality (ERC) analysis of human electrocorticography (ECoG). NeuroImage 56: 2218–2237.
10. Crone N E, Boatman D et al. (2001) Induced electrocorticographic gamma activity during auditory perception. Clinical Neurophysiol 112: 565-582.
11. DaSalla C S, Kambara H et al. (2009) Single-trial classification of vowel speech imagery using common spatial patterns. Neural Networks 22:1334-1339.
12. Pei X, Barbour D L et al (2011) Decoding vowels and consonants in spoken and imagined words using electrocorticographic signals in humans. J Neural Eng 8:046028
13. Schalk G, Leuthardt E C (2012) Brain-Computer Interfaces Using Electrocorticographic (ECoG) Signals. IEEE Reviews in Biomedical Engineering 4:140 – 154.
14. Pasley B N, David S V, et al. (2012) Reconstructing Speech from Human Auditory Cortex. PloS Biology 10:e1001251.
15. Linderman, M D, Santhanam G et al. (2008) Signal Processing Challenges for Neural Prostheses. IEEE Signal Processing Magazine 25:18-28.
16. Chang J, Kim N S et al. (2006) Voice Activity Detection Based on Multiple Statistical Models. IEEE Trans on Signal Processing 54: 1965-1976.
17. Duncan J S, Papademetris X et al. (2004) Geometric strategies for neuroanatomic analysis from MRI. Neuroimage 23, Suppl. 1: S34-45.
18. Boersma P, Weeninck D (2001) Praat, a system for doing phonetics by computer. Glot International 5:341-345.
19. Monson H (1996) Statistical Digital Signal Processing and Modeling. John Wiley & Sons.
20. Kononenko I (1994) Estimating Attributes: Analysis and Extensions of RELIEF. In Proc. of the European Conference on Machine Learning, pp. 171-182.
21. Bashashati A, Fatourechi M et al. (2007) A survey of signal processing algorithms in brain–computer interfaces based on electrical brain signals. J Neural Eng 4:32-57.

Author:	Vasileios G. Kanas
Institute:	University of Patras
Street:	Korinthou 1
City:	Patras
Country:	Greece
Email:	vaskanas@upatras.gr

Assessment of an Assistive P300–Based Brain Computer Interface by Users with Severe Disabilities

R. Corralejo, L.F. Nicolás-Alonso, D. Álvarez, and R. Hornero

Grupo de Ingeniería Biomédica, E. T. S. I. de Telecomunicación, Universidad de Valladolid, Valladolid, España

Abstract—The present study aims at assessing an assistive P300-based BCI tool for managing electronic devices at home. Fifteen subjects with motor and cognitive disabilities participated in the study. The assistive tool was designed to be simple and easy to interact with users. It allows managing 113 control commands from 8 different devices. Although most of the participants also showed cognitive impairments, nine out of the fifteen participants were able to properly manage the assistive BCI application with accuracy higher than 80%. Moreover, five out of them achieved accuracies higher than 95%. Maximum information transfer rate (ITR) values of 14.41 bits/min were reached. Hence, P300-based BCIs could be suitable for developing new control interfaces fulfilling the main needs of disabled people, such as comfort, communication and entertainment. Our results suggest that the degree of motor or cognitive disability is not a relevant issue in order to suitably operate the assistive BCI application.

Keywords—Brain Computer Interface (BCI), electroencephalogram (EEG), severe disabled people, home accessibility.

I. INTRODUCTION

A Brain Computer Interface (BCI) system monitors brain activity and translate specific signal features, which reflect the user's intent, into commands that operate a device [1]. The electroencephalogram (EEG) is generally used to monitor the brain activity since it is non-invasive, portable and it requires relatively simple and inexpensive equipment [1]. According to the nature of the input signals, BCIs can be classified into two groups. Endogenous BCIs depend on the user's control of endogenic electrophysiological activity, such as amplitude in a specific frequency band of EEG recorded over a specific cortical area [1]. BCIs based on sensorimotor rhythms or slow cortical potentials (SCP) are endogenous systems and often require extensive training. On the other hand, exogenous BCIs depend on exogenic electrophysiological activity evoked by specific stimuli and they do not require extensive training [1]. BCIs based on P300 potentials or visual evoked potentials (VEP) are exogenous systems.

P300-based BCIs allow selecting items displayed on a screen using the 'oddball' response: infrequent auditory, visual or somatosensory stimuli, when interspersed with frequent stimuli, evoke in the EEG a positive peak at about 300 ms over parietal cortex [1–4]. Some studies have verified the success of P300-based BCIs for disabled people [4–7].

The present study aims at assessing an assistive P300-based BCI tool for managing electronic devices at home in order to fulfill the main comfort, entertainment and communication needs. Participants will interact with the assistive BCI application in order to evaluate whether people with severe disabilities could use it to increase their personal autonomy in their usual environment. To that purpose, real end-users, i.e., people with motor and cognitive disabilities, will take part in the study.

II. MATERIALS

A. Subjects

Fifteen individuals (mean age: 50.3 ± 10.0 years; 7 males, 8 females) took part in the study. All of them were patients from the National Reference Center on Disability and Dependence (CRE-DyD), located in León (Spain). Participants showed motor impairments because of different pathologies. Thirteen out of them also presented cognitive impairments. All subjects gave their informed consent to participate in the study. Demographic and clinical data of all participants is summarized in Table 1.

B. EEG Signal Acquisition

EEG data was recorded using a g.USBamp biosignal amplifier (Guger Technologies OG, Graz, Austria). A total of 8 active electrodes were used: Fz, Cz, P3, Pz, P4, PO7, PO8 and Oz, according to the modified international 10-20 system [8]. Recordings were referenced to the right earlobe and grounded to the FPz electrode. EEG was sampled at 256 Hz, bandpass filtered at 0.1–60 Hz and Common Average Reference (CAR) was applied as spatial filter. In order to remove the main power interference signals were notch filtered at 50 Hz. Experimental design and data collection were controlled by the BCI2000 general-purpose system [9].

Table 1 Demographic and clinical data of all participants

Sex: male (M), female (F)
Motor/Cognitive impairment degree (MID, CID): absent (A), mild (m), moderate (M), severe (S), profound (P)
Sustained attention ability (SAA): very good (VG), good (G), moderate (M), poor (P), very poor (VP)

	Sex	Age	Diagnosis	MID	CID	SAA
P01	M	61	Arnold-Chiari malformation	m	A	VG
P02	F	44	Acquired brain injury	S	m	G
P03	F	36	Spastic cerebral palsy	S	m	M
P04	F	52	Extrapyramidal syndrome	M	m	G
P05	F	51	Acquired brain injury	M	M	G
P06	M	50	Spinal cord injury	S	A	VG
P07	M	57	Neurofibromatosis, severe kyphoscoliosis	S	m	G
P08	M	68	Spastic cerebral palsy	S	m	G
P09	M	65	Spastic cerebral palsy	S	m	G
P10	M	41	Acquired brain injury	M	M	M
P11	F	58	Multiple sclerosis	S	M	P
P12	F	35	Spastic cerebral palsy	m	m	VG
P13	M	47	Spastic cerebral palsy	S	m	G
P14	F	46	Acquired brain injury	S	M	G
P15	F	43	Spastic cerebral palsy	S	M	M

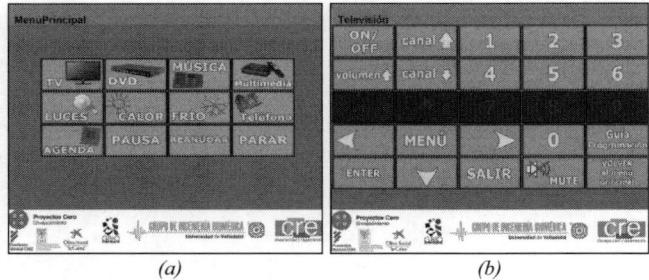

Fig. 1 (*a*) Main menu of the assistive P300-based BCI tool. (In Spanish). (*b*) Specific submenu for TV managing while the third row is dimmed. (In Spanish). From this submenu users can switch on/off the TV, select a specific TV channel, turn up/down or mute the sound, navigate through the TV menu and teletext service and go back to the main menu.

III. METHODS

A. BCI Assistive Application

A P300-based BCI tool for managing electronic devices at home was used in this study. This application allows users to control several devices related to comfort, communication and entertainment needs. Specifically, it manages the following devices: TV, DVD player, Hi-Fi system, multimedia hard drive, phone, heater, fan and lights. All devices are operated by means of an infrared (IR) emitter device (RedRat Ltd., Cambridge, UK). Thus, users are able to select up to 113 control commands of these devices switching on/off, switching TV channel, turning up/down the volume or making a phone call.

Firstly, the BCI application presents the main menu to the user. It consists of a 3 x 4 matrix of pictures, as shown in Fig. 1 (*a*). Each item of the matrix depicts one device: TV, DVD player, Hi-Fi system, multimedia drive, lights, heater, fan and phone. The last four items depict the address book and some control commands: pause, resume and stop. According to the typical P300 paradigm [3], 15 sequences of dimming stimuli are presented in order to select a single item. Each sequence contains one stimulus for each row and one for each column. Stimuli occur randomly every 187.5 ms: each stimulus dims for 62.5 ms and then the screen remains static for 125 ms. Users are asked to focus on a specific item from the matrix and silently count how many times it dims. Once the matrix finishes dimming, the selected command is performed: accessing a specific submenu or pause, resume or stop the system. Hence, from this main menu users can access to the submenus for managing a specific device. Submenus are implemented by means of variable size matrices, which consisted of images depicting the main functionalities of each device. Fig. 1 (*b*) shows the submenu for the TV set controlling. Similarly to the main menu, rows and columns of the matrix representing each submenu are randomly dimmed. Users are asked again to attend the item depicting the desired command and silently count how many times it dims. Once the system identifies the desired option, the appropriate control command is performed. Therefore, users can navigate through the application menus and manage electronic devices commonly present at their home.

B. Procedure

Participants were seated in a comfortable chair or in their own wheelchair facing a computer screen. Each subject performed three sessions. During the first session, data was collected in copy-spelling mode (Copy-Spelling Session, CSS) [10], [11]. The 5 x 5 TV submenu matrix was presented to the users. CSS was comprised of 10 runs. In each run, the user was asked to attend a specific item from a proposed task of 4-6 items. CSS approximately lasted one hour and each participant selected at least 40 items. Feedback was not provided to the users during this session. Participants who did not achieved minimum performance (minimum accuracy of 65%) during the first session repeated the CSS tasks in the next session.

The following two sessions were performed in online free mode (Free Mode Sessions, FMS) [12] and participants interacted with the assistive BCI tool. These sessions were comprised of 7 evaluation runs. In each run, participants were asked to select items across the different menus for completing a proposed task, e.g.: "access the TV submenu", "switch on", "select channel 8", "go back to the main menu", "access the DVD submenu" and "activate recording". During the last session, the number of sequences was decreased for each user depending on their performance. The less number of sequences needed to suitably detect the P300 peak, the faster users can navigate through menus.

C. EEG Signal Processing

Segments of 800 ms after each stimulus were extracted and low pass filtered for each EEG channel [12]. A Step-Wise Linear Discriminant Analysis (SWLDA) was used to compose the classifier. SWLDA performs feature space reduction by selecting suitable spatiotemporal features (i.e., the amplitude value at a particular channel location and time sample) to be included in a discriminant function based on the features with the greatest unique variance [4], [13]. The discriminant functions were obtained by using up to 60 spatiotemporal features from all the EEG channels [13]. The classifier built using CSS data was applied during the online running of FMS.

IV. RESULTS

Performance was measured in terms of Mean classification Accuracy (MA) and maximum Information Transfer Rate (ITR). The results achieved for each participant are summarized in Table 2. Mean accuracy (MA) of the CSS session was derived using the classifier built during the first 5 runs, which was applied over the subsequent runs. For the next sessions (FMS), mean accuracy was derived as the percentage of items accurately classified according to the previously proposed tasks.

Regarding CSS results, seven out of the fifteen subjects reached accuracy higher than 90% (four of them achieved 100% accuracy). Only two subjects (P11 and P13) needed to repeat CSS tasks during at least one additional session. In the particular case of P13, after three CSS sessions it was not possible to create a reliable classifier because the EEG recordings were excessively noisy due to frequent sudden muscle spasms. Participant P05 achieved poor accuracy during CSS tasks but the classifier performed well through the next session. Thus, this participant performed FMS tasks. In regard to FMS results, nine out of the fifteen participants achieved accuracy higher than 80% operating the assistive BCI application. Moreover, five of them reached accuracy higher than 95% and MaxITR up to 14.41 bit/min. The remaining participants were not able to manage the BCI assistive system (P05 and P13) or they achieved moderate accuracies ranging between 56-63% (P06, P07, P11 and P15). Regarding the ITR, subjects that achieved high accuracy were able to operate the assistive tool with fewer sequences of stimuli. ITR values range from 2.88 up to 14.41 bit/min. The lowest ITR values were usually related to the users with inferior sustained attention ability.

V. DISCUSSION

This study assesses the usefulness of P300-based BCIs to assist people with severe disabilities. In terms of the degree of motor disability, ten participants had severe impairments. Five of them achieved accuracy higher than 85% operating the assistive application, what is remarkable because they are the real end-users of these systems. Moreover, these five users showed mild or moderate cognitive impairments. Hence, the assistive tool seems to be simple and easy for most users. Regarding remaining users with severe motor impairments, four of them reached accuracies ranging 56-63% and the other one (P13) was not able to control the system due to the poor quality of recordings. Regarding the most severe patients (U11, U14 and U15), with both severe motor and moderate cognitive impairments, they achieved results ranging from moderate (63%) to excellent (89%). These results are promising considering that the population of this study presents motor impairments together with cognitive ones.

Many studies based on BCI applications are applied to healthy people. Nevertheless, some authors have studied the performance of BCIs for real end-users. Hoffman *et al* [5] applied a P300-based BCI to five subjects with different pathologies. Four out of the five participants achieved 100% accuracy. Nevertheless, this study was carried out using a

Table 2 Accuracy results for each participant

MA-Mean classification Accuracy for copy-spelling and free mode sessions
MaxITR-Maximum Information Transfer Rate for free mode sessions
*: Participant repeated the CSS tasks once
**: Participant repeated the CSS tasks twice
***: Participant achieved high accuracy during one FMS session but performance was not stable during all sessions

Participant	MA CSS (%)	MA FMS (%)	MaxITR FMS (bit/min)
P01	100.0	96.2	7.21
P02	100.0	95.6	14.41
P03	100.0	85.8	3.60
P04	100.0	95.5	14.41
P05***	35.7	50.9	-
P06	68.9	55.6	2.88
P07	78.6	62.3	2.88
P08	91.7	96.5	7.21
P09	96.7	95.3	7.21
P10	82.4	80.1	3.60
P11*	80.0	63.3	2.88
P12	87.5	92.3	14.41
P13**	37.5	-	-
P14	95.8	89.3	7.21
P15	67.9	62.5	2.88

quite different P300/BCI paradigm. Only one stimuli matrix, consisted of six images that flashed one by one, was used and only two sessions were performed, whereas our BCI application comprises 113 items from 10 menus and stimuli is presented over rows and columns. Nijboer et al [4] applied the typical 6 x 6 characters matrix [3] to eight subjects with ALS. Four out of the eight participants were able to control suitably the system. The mean accuracy ranged from 58% to 83%. Furthermore, this exhaustive study showed that the amplitude and latency of the P300 potential remained stable over 40 weeks. However, the number of participants, the percentage of them who managed properly the system and their mean accuracy is higher in our study.

Results show it could be possible to increase the autonomy of severe disabled people by means of assistive P300-based BCI applications. Nevertheless, this study has certain limitations. It would be suitable to increase the number of subjects in future works. Moreover, the quality of EEG signals of some patients who show sudden muscle spasms could be improved. Furthermore, new features and signal processing methods could be assessed in order to improve the P300 peaks detection. Besides, the system could be modified to add or remove devices adapting it to additional needs and requirements of end-users, decreasing their dependence from nurses, caregivers and relatives.

VI. CONCLUSIONS

In this study, an assistive BCI tool for managing electronic devices at home was assessed. Real end-users with motor and cognitive disabilities were involved in the study. The application allows users to operate electronic devices usually present at home, according to comfort, communication and entertainment needs. Hence, they can interact with their environment increasing their independence and improving their quality of life. The simple interface of the assistive tool allowed users to achieve promising accuracy along three sessions. Fifteen patients with motor disabilities participated in the study. Thirteen of them also showed mild or moderate cognitive impairment. Nevertheless, results of participants interacting with the assistive tool are encouraging. Nine out of the fifteen participants achieved accuracy higher than 80% managing the assistive BCI tool. Furthermore, five of them reached accuracy higher than 95%. Moreover, neither the degree of motor disability, nor the presence of cognitive impairments did affect the patient's performance. Thus, P300-based BCIs could be really proper to assist severe disabled people, covering their main needs and increasing their independence and personal autonomy.

ACKNOWLEDGMENT

This research was supported in part by the Project Cero 2011 on Ageing from *Fundación General CSIC*, *Obra Social La Caixa* and CSIC and by the *Ministerio de Economía y Competitividad* and FEDER under project TEC2011-22987. R. Corralejo was in receipt of a PIRTU grant from the Consejería de Educación (*Junta de Castilla y León*) and the European Social Fund. L.F. Nicolás-Alonso was in receipt of a PIF-UVa grant from University of Valladolid.

REFERENCES

1. Wolpaw JR, Birbaumer N, Heetderks WJ et al (2002) Brain-computer interface technology: A review of the first international meeting. IEEE Trans Rehab Eng 8:164–173
2. Donchin E, Smith DB (1970) The contingent negative variation and the late positive wave of the average evoked potential. Electroenceph Clin Neurophysiol 29:201–203
3. Donchin E, Spencer KM, Wijesinghe R (2000) The Mental Prosthesis: Assessing the Speed of a P300–Based Brain–Computer Interface. IEEE Trans Rehab Eng 8:174–179
4. Nijboer F, Sellers EW, Mellinger J et al (2008) A P300-based Brain–Computer Interface for People with Amyotrophic Lateral Sclerosis. Clin Neurophysiol 119:1909–1916
5. Hoffmann U, Vesin JM, Ebrahimi T et al (2008) An Efficient P300-based Brain–Computer Interface for Disabled Subjects. J Neurosci Methods 167:115–125
6. Sellers EW, Kubler A, Donchin E (2006) Brain–computer interface research at the University of South Florida Cognitive Psychophysiology Laboratory: the P300 speller. IEEE Trans Neural Syst Rehabil Eng 14(2):221–224
7. Corralejo R, Álvarez D, Hornero R (2012) A P300-based BCI Aimed at Managing Electronic Devices for People with Severe Disabilities, Int. Conf. NeuroRehab. Proc part I, Toledo, Spain, pp 641–645
8. Jasper HH (1958) The Ten Twenty Electrode System of the International Federation. Electroenceph Clin Neurophysiol 10:371–375
9. Schalk G, McFarland DJ, Hinterberger T et al (2004) BCI2000: a general-purpose brain–computer interface (BCI) system. IEEE Trans Biomed Eng 51(6):1034–1043
10. Krusienski DJ, Sellers EW, McFarland DJ et al (2008) Toward enhanced P300 speller performance. J Neurosci Methods 167:15–21
11. Sellers EW, Donchin E (2006) A P300-based brain–computer interface: initial tests by ALS patients. Clin Neurophysiol 117:538–48
12. Schalk G, Mellinger J (2010) A Practical Guide to Brain-Computer Interfacing with BCI2000. Springer-Verlag, London
13. Krusienski DJ, Sellers EW, Cabestaing F et al (2006) A comparison of classification techniques for the P300 Speller. J Neural Eng 3:299–305

Author: Rebeca Corralejo
Institute: Biomedical Engineering Group, Dpto. TSCIT, University of Valladolid
Street: Paseo Belén 15
City: Valladolid
Country: Spain
Email: rebeca.corralejo@gib.tel.uva.es

Single-Trial Detection of the Event-Related Desynchronization to Locate with Temporal Precision the Onset of Voluntary Movements in Stroke Patients

J. Ibáñez[1], M.D. del Castillo[1], J.I. Serrano[1], F. Molina Rueda[2,3], E. Monge Pereira[3,3], F.M. Rivas Montero[2,3], J.C. Miangolarra Page[2,3], and J.L. Pons[1]

[1] Bioengineering Group, CSIC, Arganda del Rey, Spain
[2] Departamento de Fisioterapia, Terapia Ocupacional y Medicina Fsica. Facultad de Ciencias de la Salud, Universidad Rey Juan Carlos, Madrid, Spain
[3] Lambecom, Facultad de Ciencias de la Salud, Universidad Rey Juan Carlos, Madrid, Spain

Abstract—Stroke patients may present motor impairments that in many cases require an intensive rehabilitation process with experts helping the patient to recover the functionality of the affected limb. A target during this rehabilitation process is to induce neural plasticity in brain regions associated with the motor control of the affected limb. Electrical stimulation tightly synchronized with the intention to perform a movement has proven to be an effective way of enhancing cortical excitability in healthy subjects. The electroencephalogram can help to detect voluntary movements online. We propose here an Electroencephalography-based system aimed to detect the instants at which stroke patients attempt to start voluntary movements with the affected upper-limb. To accomplish this, the analysis of the cortical rhythms and their variations are used. In the preliminary results obtained with 3 chronic stroke patients, $63 \pm 14\%$ of the movements were detected with a temporal precision in the detections of the onsets of the movements of -126 ± 313 ms.

Keywords—Stroke; Electroencephalography (EEG); Event-Related Desynchronization (ERD); Brain-Computer Interface (BCI).

I. INTRODUCTION

After stroke, the damage of neural networks in the brain may affect the ability to perform motor tasks with a part of the body. From that moment on, a successful recovery of the affected limb's functionality will depend mainly on two main factors: the characteristics (size and location) of the brain injury caused by the stroke, and the effectiveness of the rehabilitation therapy [1]. One third of stroke patients looses the functionality of the affected limb in a permanent way [2]. It is expected that the development of novel therapies successfully inducing neural rehabilitation may help them to improve their condition in the long term [2].

The measurement of the cortical activity by means of electroencephalographic (EEG) systems has been successfully used to characterize mental states associated to the execution of voluntary movements [3, 4]. It has been proposed that using this information to control brain computer interfaces (BCI) may help developing novel forms of inducing neural plasticity in brain regions targeting the affected limb. In this regard, it is expected that promoting the activation of cortical networks engaged in the generation of motor actions, and giving an appropriate proprioceptive feedback will improve the functionality of the affected limb [5]. Recent studies have provided evidence of increased cortical excitability after an intervention with peripheral electrical stimulation based on the EEG patterns associated to the execution of self-paced movements [6]. The achievement of a natural interface in terms of temporal association between the user's cortical commands and the proprioceptive feedback seems to be of special relevance in this kind of paradigms [6]. Under such conditions, the EEG becomes a valuable technology, given its capacity to detect voluntary movements online and with temporal precision, when they are to be started.

We present here the design of an EEG-based system aimed to detect the onset of voluntary movements with temporal precision. Niazi et al. achieved successful detection results with a system depending on the Bereitschaftpotential (slow variations of the cortical activity preceding voluntary movements [4]), with recall ratios above 80% on average with control subjects and of 55% with stroke patients. The latencies in their study were on of -66.6 ± 121 ms and -56.8 ± 139 ms with controls and patients respectively. Here we designed and experiment in which three stroke patients performed self-paced movements with their affected upper-limb in separated trials which were started only once the subject presented a basal EEG activity. Trials presented a resting period of around 3.5 s before the movements. The detector proposed relies on the event-related desynchronization, which refers to the reduction of the alpha and beta rhythms in the sensorimotor cortex, starting around 1.5 s before voluntary movements [3]. On average, 63 ± 14 of the movements were detected with an average delay with respect to the actual onset of the movements of -126 ± 313 ms. These results demonstrate the suitability of using the ERD to detect the initiation of voluntary movements and remark the importance of choosing an adequate experimental paradigm to achieve good performances of the EEG system.

II. METHODS

A. Patients

Three patients with chronic ischemic stroke (middle cerebral artery) took part in the experiments performed for this study. The patients' description is presented in Table 1. All patients signed an informed consent to participate in the study; the Ethical Committee at Universidad Rey Juan Carlos gave approval to the experimental protocol.

Table 1 Description of the patients participating in the present study

Patient code	Age	Years since accident	Gender	Affected hemisphere
P0	60	3	Male	left
P1	54	3	Male	right
P2	66	7	Male	left
Average	60±6	4.3±2.3	-	-

B. Experimental Protocol

One measurement session of one subject consisted of about 10 runs of 5 trials each (between runs, a resting period of a couple of minutes was given for the patient to relax). One trial consisted in a resting period followed by a self-initiated movement. Patients P01 and P02 performed a shoulder abduction given that they presented a strong difficulty in moving their affected arm. Patient P03 performed a reaching movement, with the arm starting in a resting position on a table. A screen was placed in front of the patients (about 1 m distant) while the tasks were performed. The screen was used to guide the patients throughout the experiment. Three cues were presented in the screen in each trial. First, the word "Rest" was shown. As soon as the patient reached baseline state (checked by an EEG expert), the screen message switched into "Whenever you want". At this time the patient had to remain still for a few seconds and start a movement. Once the patient started moving, the screen message switched to "Movement" until he returned to the initial position.

Trials free of artifacts and presenting resting periods of more than 3 s before the onset of the movement were kept to test the performance of the EEG-based detector.

C. Data Acquisition

EEG was acquired from 16 scalp positions over the motor area (F3, Fz, F4, FC3, FCz, FC4, C5, C3, C1, Cz, C2, C4, C6, CP3, CPz and CP4, all according to the international 10-20 system). Active Ag/AgCl electrodes were used to this end. The reference was set to the common potential of the two earlobes and Fz was used as ground. The amplifier (gUSBamp, g.Tecgmbh, Graz, Austria) was set to filter the signal between 0.1 and 60 Hz, and an additional 50 Hz notch filter was used. The data was acquired at 256 Hz.

The movements of the affected arm were analyzed with a surface electromyographic (EMG) amplifier (Zerowire Wireless EMG, Aurion, Milan, Italy). Electrodes were placed on the deltoids, biceps, triceps, wrist extensors and wrist flexors. The EMG data were acquired at 1000 Hz.

Synchronization between the two sources of information (EEG and EMG) was achieved by means of a common digital clock.

D. Detection of the Muscle Activation

EMG served to locate the onsets of the muscular activations associated to the performance of the voluntary movements. EMG from the deltoids was band-pass filtered (Butterworth, 4-th order, $0.1 \leq f \leq 4$ Hz) and the envelope of the resulting signal was extracted with the Hilbert transform. The onsets were set at the points where the amplitude of the processed EMG exceeded 15% of the maximum EMG found in the experiments with each patient.

E. Feature Selection, System Training and Classification

A leave one out methodology was used to test the performance of the proposed system. Therefore, to classify each trial, the rest of the trials of the same patient were used to train the system.

The core of the detector proposed here consists of a weighted naïve Bayes classifier [7]. Each of the features fed to the classifier corresponds to the spatially filtered logarithmic power values of the EEG signal within 2-Hz frequency subbands. The analyzed subbands are taken from 7 Hz to 22 Hz in steps of 1 Hz. The Common Spatial Patterns (CSP) method is applied with the training data of each frequency subband to obtain projections of the 16 channels maximizing the variance between the two states of the EEG signal, *i.e.* the desynchronized state (when the movement is about to be performed) and the basal state (periods of time preceding the planning and execution of the voluntary actions) [8]. To do so, first the EEG is introduced in a bank of filters (Butterworth, 4-th order, 2-Hz bandwidth). From each subband, epochs of 1 s are extracted from inactive states (from -3 s to -2 s with respect to the muscle activation points) and from the active states before the movement starts (from -1 s to 0 s with respect to the onset of the movement), and the data collected from these two classes are used to obtain the CSP projection matrix. The first column of the obtained matrix for each subband is kept for the signal spatial filtering. Therefore, a total of 16 CSPs (one per each analyzed frequency band) is obtained in this process. The weights applied to the features of the Bayesian classifier are obtained by computing the area under the ERD curve obtained for each frequency subband. The ERD is obtained following the methodology proposed in [9].

The raw EEG signal is spatially filtered with the obtained CSPs and the Power Spectral Density (PSD) is obtained (Welch's method, hamming windowing, 2 Hz resolution, 75% overlap) from segments of 1.5 s around the onset of all movements performed by each patient. The logarithmic power values at the central frequencies of the bank of filters are used to train the bayesian classifier.

The classification in each trial is performed in steps of 125 ms. Power values of the 16 projections of the EEG signal are extracted using the PSD of segments of 1.5 s. A threshold is selected following the criterion of maximizing the ratio between true positives (TP) and the sum of false positives (FP) and false negatives (FN), which are defined in subsection F.

The output probabilities over the threshold are considered detections or active outputs, whereas the rest of the classified segments are considered non-detections or inactive outputs. A Refractory Period (RP) is used to make the output of the classifier stable. The RP is configured to maintain each classifier's activation at least for 500 ms.

F. System Validation and Optimal Threshold Selection

Active outputs from the classifier starting in the interval of ±500 ms with respect to the muscle activations are considered TP. Activations observed more than 500 ms before the muscle activation are accounted as FP. Movements that are not detected or are detected latter than 500 ms after the muscle activation are considered FN.

III. RESULTS

Table 2 presents the results obtained with the proposed detector tested with each patient. The Recall results represent the percentage of trials in which the movement was detected with less than ±500 ms latency with respect to the actual onset of the movement. On average, 63±14 % of the movements were correctly detected and 1.70±0.59 false activations per minute were generated. The distances between detections and actual onsets of movements were on average -126±313ms.

The average of the CSPs across frequency bands obtained for each patient is showed in Fig. 1. The damage of cortical networks after a stroke generally forces cortical reorganization in these patients, which in turns results into a higher variability of cortical patterns when comparing with control subjects [10]. Here, channels with a higher relevance for

Table 2 Results of the EEG-based detector of the onset of the mucle activations

Pat. code	Nr. trials	Recall (%)	FP/min	Distance MO (ms)
P0	33	61	1.38	-6±338
P1	45	78	1.34	-287±245
P2	44	50	2.39	-86±356
Average	41±7	63±14	1.70±0.59	-126±313

the ERD detection (the darkest and brightest regions of the coloured scalp maps) vary between patients. P0 presents a bilateral desynchronization (activity of the C3 and C4 positions is summed), and P1 shows a significant predominance of the central Cz position, suggesting a cortical reorganization in both cases. As for P2, the average CSP reflects a maximization of the differences between the contralateral (right) and ipsilateral (left) hemispheres with respect to the moved arm, which is in line with what's known about the ERD in control subjects: it is first observed over the contralateral regions and it becomes bilateral once the movement starts [11]. Interestingly, P2 presented also higher functionality of the affected arm as compared to the other two patients. As for the weights assigned to the signals in the different frequency subbands (bottom part of Fig. 1), it is observed that both the alpha (7-12 Hz) and lower-beta (13-22 Hz) are found in regions of 1-2 Hz, and the specific frequencies of them vary between patients.

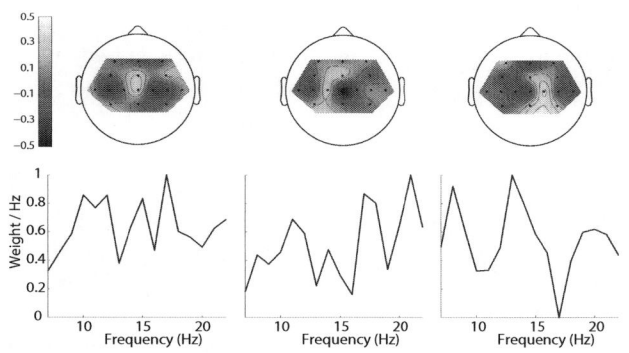

Fig. 1 Top: Average CSPs for each patient. Bottom: weights assigned to all frequency bands for the three patients. Left figures refer to P0, central figures to P1 and right figures to P2.

IV. DISCUSSION

The main contribution of this study is the validation with chronic stroke patients of an EEG-based detector of the onset of voluntary muscle activations relying on the single-trial detection of the ERD. One of the main difficulties of using the

ERD phenomenon to locate the time at which a movement starts is the variability of the anticipation of this pattern. To solve this problem, we have proposed a system combining the weighted information of a number of features covering both the alpha and lower-beta bands of the EEG. The preliminary results obtained demonstrate the ability of the system proposed to locate the onset of self-paced movements with a reduced number of false activations and an average latency of -126±313 ms with respect to the onset of the muscle activation. These results with stroke patients are comparable (and even better in terms of recall results) than those presented in previous experiments based on different patterns of the EEG signal. Nonetheless, the detector still needs to be validated on a larger group of patients and control subjects and with a higher number of trials. Additionally, it needs to be analyzed how adequate it may be to control with the proposed system an external device giving proprioceptive feedback to its user.

Finally, it has been demonstrated that the use of a protocol in which each trial does not begin until a baseline EEG is observed in the patient, allows achieving satisfactory results with the EEG. This sort of paradigm has demonstrated to be beneficial here in terms of how the EEG-based system benefits from the fact that all trials are starting from a baseline condition, where the classifier's output is expected to be inactive. Additionally, we suggest that the inclusion of this restriction in the protocol may provide an additional form of feedback to the patients, while doing the task. This way, they become aware about how to modulate their cortical rhythms, and they learn to focus on the task, which could also result in an increased activation of motor-related cortical regions.

V. CONCLUSION

We have provided evidence of the successful use of the ERD cortical pattern measured with EEG to locate with temporal precision and high recall ratios the times at which self-paced motor actions were initiated by chronic stroke patients. Results are comparable to similar systems based on other cortical patterns and represent an appealing alternative to control neural prosthesis to provide a natural feedback to patients during the rehabilitation.

ACKNOWLEDGEMENTS

This work has been partially funded by grant from the Spanish Ministry of Science and Innovation CONSOLIDER INGENIO, project HYPER (Hybrid NeuroProsthetic and Neuro-Robotic Devices for Functional Compensation and Rehabilitation of Motor Disorders, CSD2009-00067) and from Proyectos Cero of FGCSIC, Obra Social la Caixa, and CSIC.

REFERENCES

1. Murphy Timothy H, Corbett Dale. Plasticity during stroke recovery: from synapse to behaviour. *Nature reviews. Neuroscience.* 2009;10:861–72.
2. O'Dell Michael W, Lin Chi-Chang David, Harrison Victoria. Stroke rehabilitation: strategies to enhance motor recovery. *Annual review of medicine.* 2009;60:55–68.
3. Pfurtscheller G, Silva F H Lopes. Event-related EEG/EMG Synchronization and Desynchronization: Basic Principles *Clinical Neurophysiology.* 1999;110:1842–1857.
4. Shibasaki Hiroshi, Hallett Mark. What is the Bereitschaftspotential? *Clinical neurophysiology : official journal of the International Federation of Clinical Neurophysiology.* 2006;117:2341–56.
5. Daly Janis J, Wolpaw Jonathan R. Braincomputer interfaces in neurological rehabilitation *The Lancet Neurology.* 2008;7:1032–1043.
6. Mrachacz-Kersting Natalie, Kristensen Signe Rom, Niazi Imran Khan, Farina Dario. Precise temporal association between cortical potentials evoked by motor imagination and afference induces cortical plasticity. *The Journal of physiology.* 2012;590:1669–82.
7. Pomerol Jean-Charles, Barba-Romero Sergio. *Multicriterion Decision in Management - Principles and Practice.* Boston: Springer 2000.
8. Guger C., Ramoser H., Pfurtscheller G.. Real-time EEG analysis with subject-specific spatial patterns for a brain-computer interface (BCI) *IEEE Transactions on Rehabilitation Engineering.* 2000;8:447–456.
9. Graimann B, Huggins J E, Levine S P, Pfurtscheller G. Visualization of significant ERD/ERS patterns in multichannel EEG and ECoG data *Clinical Neurophysiology : official journal of the International Federation of Clinical Neurophysiology.* 2002;113:43–7.
10. Serrien Deborah J, Strens Lucy H A, Cassidy Michael J, Thompson Alan J, Brown Peter. Functional significance of the ipsilateral hemisphere during movement of the affected hand after stroke. *Experimental neurology.* 2004;190:425–32.
11. Bai O, Mari Z, Vorbach S, Hallett M. Asymmetric Spatiotemporal Patterns of Event-related Desynchronization preceding Voluntary Sequential Finger Movements: a High-resolution EEG Study *Clinical Neurophysiology.* 2005;116:1213–1221.

Hybrid Brain Computer Interface Based on Gaming Technology: An Approach with Emotiv EEG and Microsoft Kinect

Nuno André da Silva, Ricardo Maximiano, and Hugo Alexandre Ferreira

Institute of Biophysics and Biomedical Engineering of the Faculty of Sciences of the University of Lisbon, Portugal

Abstract—Brain computer interfaces (BCIs) can be used to improve quality-of-life of people with motor or communication impairments. However, these systems may not work properly for all users. The complex set-up, the intensive training required for BCI operation and the expensive devices often used, difficult the application to real-life scenarios, as well as, technology dissemination. Gaming technology devices may help to disseminate this technology and facilitate real-life validation due to lower costs and easier set-up. This paper presents the integration of gaming technology in a hybrid BCI based on Emotiv EEG and the Microsoft Kinect. This approach enables the collection of visual depth and RGB data from Microsoft Kinect, and electroencephalographic data from Emotiv EEG. Data combination from both devices and online labeling based on Kinect to avoid strict protocols are explored and system potentialities are discussed.

Keywords— Brain computer interface, hybrid brain computer interface, Emotiv EEG, Microsoft Kinect.

I. INTRODUCTION

Brain-computer interface (BCI) based on electroencephalography (EEG) is a technology with a large spectrum of applications, from rehabilitation and communication to entertainment (consumer electronics). However, an obstacle to practical use of BCI is the need for acquiring subject-specific extensive training data with strict protocols with many repetitions of the same task. This is a time-consuming procedure that may introduce fatigue in the user and deteriorate user performance [1]. Ideally, no intensive training would be necessary, and adaptation between BCI and user would be the most natural as possible. In order to achieve such natural adaptation and improve BCI performance, the combination of a specific EEG signal with other physiological signals, known as hybrid BCI (hBCI), has been proposed [2, 3]. However, these hBCI require more time to set-up and are more expensive when extra electrodes are used. The combination of gaming technologies, such as tracking systems based on visual monitoring (electrode-free) and EEG systems [4-7], is an unexplored and atractive topic for hBCI due to reduced cost and easy set-up, which may yield potential for daily applications.

This work proposes a hBCI data acquisition system based on two gaming technologies available to all users, the Emotiv EEG and the Microsoft Kinect. Moreover, it also explores the possibility of using the Kinect data as a real-time labeling as alternative to strict protocols

II. MATERIAL AND METHODS

A. Data Acquisition System

The proposed data acquisition system is comprised of two gaming technology devices: Emotiv EEG and Microsoft Kinect. The former is a high-resolution 14-channel EEG device, where the electrodes are located over international 10-20 system positions AF3, F7, F3, FC5, T7, P7, O1, O2, P8, FC6, F4, F8 and AF4 using two reference electrodes at the mastoids. The headset is a wireless device with a sampling frequency of 128Hz. It contains a rechargeable 12-hour lithium battery, gyroscopes (x and y directions), and has a band-pass filter (0.2-45Hz hardware built). The latter is a marker-free device for human-computer interaction, which enables the user to interact with a computer through natural body movements. It is comprised of a Red, Green, Blue (RGB) camera, a 3D depth sensor based on an infrared camera and a multi-array microphone. RGB and depth camera sampling frequency is up to 30Hz. OpenNI™ framework (OpenNI, USA) and the NiTE middleware (PrimeSense, Tel-Aviv, Israel) were used for subject recognition and skeleton tracking. In this work, the sampling frequency of 16Hz was used to acquire the signals.

Both devices were connected in a common MATLAB Simulink environment (Mathworks). For that, Epoc Simulink EEG Importer (Xcessity, Austria) and Simulink Support for Kinect (Mathworks) were used. Additionally, a stimuli presentation environment was developed for display of the acquisition protocol instructions. After data acquisition, feature extraction and classification were done.

B. Subjects

Three healthy volunteers (all male, 24 ± 2 years old, all right handed) participated in the experiment. All the subjects had no prior experience with BCI.

C. Experimental Paradigm

In order to evaluate the data combination from Emotiv headset and Kinect, two motor execution tasks were considered: hand grasping and arm movements (both actions used in strict protocols). While facing the display of the computer and at a distance of 2.0 meters away from the Kinect sensor, subjects were asked to follow the actions shown on the screen. When requested to move the right or left hand, subjects needed to grasp the respective hand energetically. When asked to move their arms, subjects needed to move the arm up and down along the body's coronal plane without moving the head. These actions were interleaved with baseline or resting periods, where subjects were asked to stay calm with hand palms open and arms down.

The acquisition protocol comprised 10 runs with 2 trials each, resulting in 20 trials for each subject. At the beginning of each run, there was also a 10-second period for calibrating Kinect's skeleton tracking. This period was ignored in the analysis step. Each trial was comprised of a baseline period of 9 seconds interleaved with a 5 seconds action period (left or right hand grasping and left or right arm movements), in a total trial duration of 56 seconds.

D. Data Pre-processing

In the case of EEG signals, the pre-processing steps consisted in normalization, band-pass filtering (Chebyshev filter between 5 – 40 Hz) and Laplacian spatial filtering [8, 9]. Due to the different electrode locations, the Laplacian filter was adapted depending on each channel location, where the 3 or 4 closest electrodes were considered. This ensures that the interference between the electrodes is reduced as much as possible, as well as other undesired artifacts such as eye blinking. In the case of Kinect, no pre-processing was performed.

E. Kinetic Based Labeling

In order to validate the possibility of using Kinect as an action labeler, four labels were considered: (i) right hand grasping, (ii) left hand grasping, (iii) right arm movement and (iv) left arm movement. In order to classify hand grasping, the variation of hand area was considered, and to classify arm movements, the wrist position was considered (described as follows).

E1. Hand Grasping and Arm Movements Identification

To identify hand grasping, a methodology based on three steps was proposed for each frame i: (i) hand identification, (ii) depth segmentation, and (iii) hand segmentation.

Firstly, a binary region-of-interest (ROI) close to each joint of interest k (right or left hand) was identified based on the 3D joint information J from the skeleton data. This region is named $ROIbm_i^k$. Secondly, on $ROIbm_i^k$, a depth segmentation was performed based on information from J_i^k. This new binary mask was applied to the RGB image. Lastly, this image was converted from RGB to YCbCr (Y-luma, Cb and Cr are the blue-difference and red-difference chroma components). In this color space the skin characteristics are well known and accurate hand segmentation can be performed based on the most representative range of chrominance skin values [10]. To improve segmentation, morphological operators were applied. After hand segmentation, the area of each hand (HA^k_i) was used as signal-of-interest for feature extraction in hand grasping classification.

To identify arm movements, the skeleton data was used. Wrist joints positions (HP^k_i) were monitored and used to classify the action.

E3. Features and Classification

To classify which of the four actions the user performed, our algorithm assumed that only one action could be performed at each observation window. In this work we assumed an observation window length of 1 second. The trend line of each signal was removed and the absolute value was considered. 1) Feature extraction. Several features based on time domain of HA^k_i and HP^k_i were used: mean absolute value (MAV), variance unbiased estimator (VAR) and wave length (WL) [11]. The above-mentioned features were normalized and used in classification individually and an exhaustive feature selection was done. 2) Classification. Three classifiers were considered: k-nearest neighbor (k-NN), linear classifier (Bayes' rule), and support vector machine (polynomial and radial kernel) (SVM) [12].

E4. Validation

To validate the action classification 10, 20, 30, 40 and 50% of the data from each subject were used to train the classifiers. The remaining data were used to access the classification performance. Note that, the labels from the stimuli presentation do not correspond to user actions due to reaction time. For that reason, RGB video analysis was performed to derive the true labels $TLabel_i$ based on user actions. For each training percentage, classification performance was evaluated based on the number of correctly classified observation windows based on Kinect ($KLabel_i$).

F. Feature Extraction and Classification Algorithms (EEG)

Initially, 32 features were extracted from each one of the 14 channels/sensors of the Emotiv system, meaning that a total of 448 features were obtained. After merging signals from all runs, EEG signals were split into observation windows with length of 1 second (also considered for kinetic data analysis), to simulate real-time online classification. Each one of those observation windows was associated with the respective label, which was used for training purposes. The extracted features used for this classification are described as follow: 1) Time domain features: mean, standard deviation, skewness, entropy and kurtosis; 2) Frequency domain features: 27 of the 32 features extracted from the Emotiv signals were related to their frequency domains. In this study, the Complex Morlet Wavelet Transform [13] was used to split the EEG signals into different frequency bands: Delta (0.2-4Hz), Theta (4-8Hz), Alpha (8-16Hz), Beta (16-32Hz) and Gamma (32-45Hz) bands. To estimate the power of these bands, periodogram and autoregressive models (Yule and Welch models) were used.

Since a relatively high number of features were obtained, floating forward selection was used to reduce the total number of features to 20 features [12]. This reduced number of features was determined by trial and error and taking into account the effects of the curse of dimensionality. These selected features derived from Emotiv signals were used for classification purposes. A multi-class classification and eight classifiers were used: k-NN with orders 1, 2 and 3; Nearest Mean Classifier (NMC); Parzen Classifier; Linear and Quadratic Discriminant Classifiers (LDC and QDC, respectively). The classifier with the lowest average error for each subject was selected for analysis of real online labeling using Kinect. Error rates are obtained using Leave-One-Out method [12].

III. RESULTS

A. Kinect Based Labeling Performance

The classification was done for the different features. The best results (smaller errors) were obtained using the feature combination MAV+VAR for whichever classifiers tested. Regarding the different classifiers, the best results were obtained using the support vector machine with radial kernel: an overall accuracy of 93±3% was achieved for all 4 actions using a training data set of 40%. No considerable improvement was found for training data set of 50%.

Figure 1 shows the displayed (stimuli protocol) labels and the labels determined using Kinect for one subject. A systematic delay in starting and in ending actions, after the corresponding stimulus presentation, was observed for all subjects. This delay is reduced when labels derived from Kinect are used.

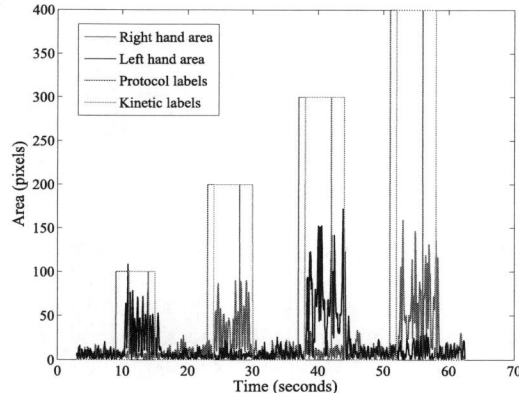

Fig. 1 Kinect based labeling and hand area variation during hand grasping (1st and 2nd actions) and arm movements (3rd and 4th actions). It is possible to observe that the user action is delayed from the stimuli presentation and does not have always the same duration. Note that the area of the hand might also change during the arm movements.

B. Emotiv and hCBI Performance

In order to understand the Emotiv's performance as an hBCI component, two different methods of evaluation were considered. Firstly, the protocol's block design was used to train the Emotiv's signals and its error rate was estimated. Secondly, the block design derived from Kinect labeling was used for training and classification steps, as well as, the estimation of the accuracy in percentage. These results are displayed in fig. 2.

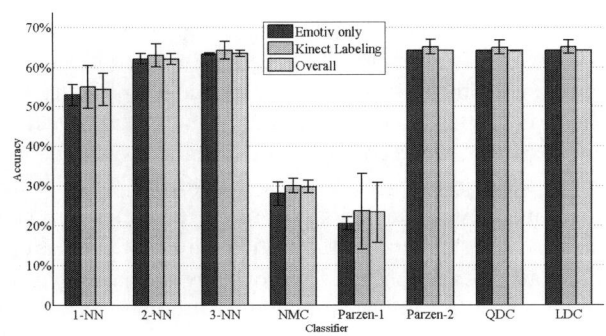

Fig. 2 Accuracy obtained when using protocol block design versus block design estimated by the Kinect device.

The classification accuracy of a standard BCI experiment for all 4 actions was always smaller than the obtained with the labels based on the kinect data. The best classification results were obtained with the LDC (64.25%), which were improved by 0.89% when labels based on Kinect were used (due to the delay of the response from the subject). The higher differences between standard and Kinect labels derived performances were observed in Parzen 1 Classifier (3.1%). Even though those numbers might be small, the classification performance is not deteriorated and Kinect has an advantage in making labeling much easier, not requiring the subject to follow a strict block design.

IV. DISCUSSION

The use of gaming technologies in BCI might boost technology dissemination and real-life validation. In this work an hBCI based on Emotiv EEG integrated with Kinect was studied and action labeling with Kinect was explored.

The proposed methodology showed that the Kinect enabled the identification of hand actions (grasping) and arm movements when integrated with Emotiv EEG in a hybrid measurement. The results obtained with our system show that Kinect can be used for online definition of action labels with an overall accuracy of >90%. When these labels were used instead of the labels imposed by the protocol at least the same classification performance of the EEG signals was obtained. This suggests that instead of a strict protocol, EEG labeling can be derived from users' natural actions monitored by Kinect. This might improve user performance by reducing the fatigue associated to repetitive protocols, while also increasing the motivation for BCI usage. Moreover, the proposed system also allows the validation of the user intentions by monitoring user movements. This can be used to reduce classification errors caused by the user's response delay, and may enable the development of new online user-BCI learning strategies.

The proposed integrated system uses a commercial EEG device with 14 channels. The location of those channels is non-optimum for motor imagery/execution. In [6] the electrodes were repositioned by exploiting the flexibility of the device arms. This reconfiguration is not easy to be performed by a regular user and for that reason was not performed in our work to simulate a real-life application. Nevertheless, the standard configuration of Emotiv electrodes enables the measurement of motor activity, namely ipsilateral activity [7]. Further validation will be performed with a larger number of subjects and against a medical EEG system in order to validate our results.

Regarding the cost, the Emotiv + Kinect approach is an attractive solution for home validation and dissemination of hBCI because of its low cost. The hardware costs under $500 while the medical hBCI (e.g. EEG+EMG) can cost over $12,000. While the proposed system needs 1-5 min to set-up, the medical hBCI needs at least 5 times more.

V. CONCLUSION

The presented paper proposes an hBCI based on gaming technology devices, Emotiv and Kinect. This combination allows the tracking of user actions without artificial markers and electrodes, and therefore BCI training without following a strict protocol. Moreover, it also yields potential for newer BCI-user learning strategies and technology dissemination.

ACKNOWLEDGMENT

The author's would like to thank the members of Brain Networks and Interfaces group from IBEB-FCUL for valuable discussions. Research supported by Fundação para a Ciência e Tecnologia (FCT) and Ministério da Ciência e Educação (MCE) Portugal (PIDDAC) under grant PEst-OE/SAU/UI0645/2011

REFERENCES

1. J. R. Wolpaw, N. Birbaumer, D. J. McFarland et al. (2002) Brain–computer interfaces for communication and control. Clin Neurophysiol, vol. 113, no. 6, p. 767–791
2. G. Pfurtscheller, B. Z. Allison, C. Brunner et al. (2010) The hybrid BCI. Front Neurosci., vol. 4, no. 30
3. R. Leeb, H. Sagha, R. Chavarriaga et al. (2011) A hybrid brain-computer interface based on the fusion of electroencephalographic and electromyographic activities. J Neural Eng., vol. 8, n.º 2
4. P. Ofner and G. R. Muller-Putz (2012) Decoding of velocities and positions of 3D arm movement from EEG. in IEEE EMBS
5. R. Scherer, J. Wagner, G. Moitzi et al. (2012) Kinect-based detection of self-paced hand movements: Enhancing functional brain mapping paradigms. in IEEE EMBS
6. F. Carrino, J. Dumoulin et al. (2012) A self-paced BCI system to control an electric wheelchair: Evaluation of a commercial, low-cost EEG device. in BRC, ISSNIP
7. S. Fok, R. Schwartz, M. Wronkiewicz et al. (2011) An EEG-based brain computer interface for rehabilitation and restoration of hand control following stroke using ipsilateral cortical physiology. in IEEE EMBS
8. B. Hjorth (1975) An on-line transformation of EEG scalp potentials into orthogonal source derivations. Electroencephalogr Clin Neurophysiol, vol. 39, pp. 526-30
9. T. Wang, J. Deng and B. He. (2004) Classifying EEG-based motor imagery tasks by means of time–frequency synthesized spatial patterns Clin Neurophysiol., vol. 115, no. 12, pp. 2744-2753
10. D. Chai and K. N. Ngan (1999) Face segmentation using skin-color map in videophone applications. Circuits and Systems for Video Technology, IEEE Transactions on, vol. 9, no. 4, pp. 551-564
11. M. Zecca, S. Micera, M. Carrozza et al. (2002) Control of multifunctional prosthetic hands by processing the electromyographic signal. Crit Rev Biomed Eng. vol. 30, no. 4-6, pp. 459-485
12. R. P. Duin, P. Juszczak, P. Paclik et al. (2007) PRTools4.1, A Matlab Toolbox for Pattern Recognition. Delft University of Technology
13. C. D'Avanzo, V. Tarantino, P. Bisiacchi et al. (2009) A wavelet Methodology for EEG Time-frequency Analysis in a Time Discrimination Task. Int J Bioelectromagnetism, Vols. 11-4, pp. 185-188

Rehabilitation Using a Brain Computer Interface Based on Movement Related Cortical Potentials – A Review

K. Dremstrup[1], IK. Niazi[1], M. Jochumsen[1], N. Jiang[2], N. Mrachacz-Kersting[1], and D. Farina[2]

[1] Center for Sensory-Motor Interaction, HST, Aalborg University, Denmark
[2] BCCN, University Medical Center Göttingen, Georg-August University, Göttingen, Germany

Abstract— This paper summarizes the research at Aalborg University within brain computer interfaces (BCI) used for rehabilitation done within the period 2006-2013. The work is based on movement related cortical potentials (MRCPs). MRCP's characterization for different task types was conducted in Nascimento et al [9] and showed that they potentially can be used for classification of planning and executions of lower limb movements. This combined with the novel protocol proposed in Mrachacz-Kersting et al. [8] showing that neural plasticity can be improved using imagination and afferents established the basis. To translate this synchronous protocol into an asynchronous BCI, a pseudo online detection of the movement intention was proposed in Niazi et al. [10]. The detector was then used to trigger peripheral electrical stimulation for inducing changes in excitability of the cortical projection of the target muscle in healthy subjects [11]. Development of this paradigm is described here. Furthermore, preliminary results from a pseudo online close loop BCI system are reported, where movements are detected and different movement types are classified.

Keywords— **MRCP, BCI, Plasticity, Rehabilitation**

I. INTRODUCTION

A BCI system can be used as a neuromodulatory system in the rehabilitation approach where neural functions are modulated by feedback triggered by the decoded brain activity. For such systems, the task of designing the feedback and its timing is very important in order to drive specific (rather than unspecific) cortical changes.

This review focuses on the technical challenges for the induction of plasticity by triggering peripheral electrical stimulation (PES) with a motor task decoded by a BCI system is needed. Prior to this focused effort of using BCI for rehabilitation and neuroplasticity was gone several years investigation different principles for communication using BCI e.g. using Steady State Visual Potentials and auditory and spatial navigation [1, 2].

In concordance with the increased focus on rehabilitation after stroke including the neural plasticity occurring in the brain it was natural to focus also on EEG-based BCI. Many different BCI-methods are used but many suffer from poor temporal resolution. Therefor focus was put on a time-domain method. The execution of a motor task in humans is preceded by a slow decrease in the EEG amplitude (by at least 500 ms) measured over the primary motor cortex, and this potential is termed Movement Related Cortical Potentials, MRCP. The MRCP consists of three events called: readiness potential, motor potential, and movement-monitoring potential, which are considered to reflect movement planning/preparation, execution, and control of performance, respectively [3, 9]. The MRCP has been further explored in normal subjects, in ALS-patients and in stroke-patients [3, 4, 6]. It has been verified that also imagined movements – ier. non-executed movements - generates MRCP's and this makes the MRCP useful for rehabilitation in patients hindered in movements but with the ability still to wish and imagine a movement, se Fig 1.

A novel technique based on this observation was presented using a conditioning protocol for inducing the changes in the excitability of the cortical projections to the tibialis anterior (TA) muscle [8]. The conditioning protocols consisted of a single electrical stimuli of the common peroneal nerve (CPN), delivered at motor threshold, paired with the MRCP to arrive during i) the preparation phase, ii) the movement execution phase or iii) the movement monitoring phase of the MRCP. A total of 50 pairings were applied. The mean peak-to-peak TA motor evoked potential (MEP) amplitude, using transcranial magnetic stimulation (TMS), measured prior to and following each intervention, was plotted against TMS intensities. It was demonstrated that a physiologically generated signal may be used to drive stimulation at the periphery leading to associative long term potentiation-like plasticity. The results demonstrate the importance of the timing of PES in relation to the different components of the MRCP, where only the intervention with

CPN stimulation in conjunction with the movement execution phase of MRCP led to significant excitability changes. The results also showed that afferent feedback from the periphery is necessary to induce the observed changes, as motor imagery and PES alone did not lead to significant changes in excitability (control experiments).

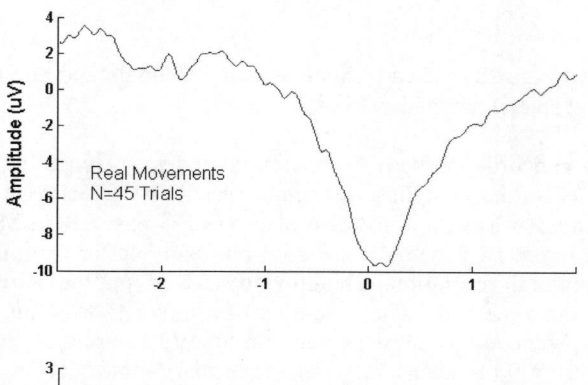
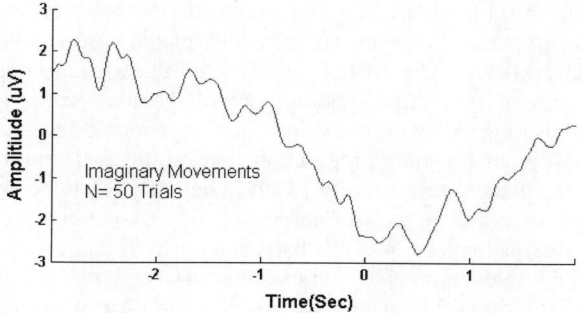

Fig 1. Grand average MRCP recorded over the primary motor cortex of a healthy subject during real (n=45 sweeps) and imaginary (n=50 sweeps) ballistic dorsiflextion. Task onset is at 0 sec.

II. AIMS AND METHODS

To translate the work of Mrachacz-Kersting et al. 2012 into a self-paced BCI paradigm, detection of the initial negative phase (INP) of the MRCP with short latency was required at a single-trial level. To accomplish this, the challenge to address was the signal-to-noise ratio (SNR). The spatial filtering approach was utilized to address the issue of the SNR. Commonly used methods for improving the SNR like Large Laplacian spatial filtering (LLSF) and common spatial pattern (CSP) filtering were compared with the proposed method of optimized spatial filtering (OSF) in both healthy subjects and stroke patients. The spatial filtering techniques were explored in a pseudo online manner [10] for motor execution (ME) and motor imagination (MI) tasks. In the next step a real-time online detector was built and tested. The detector was built from the template of the INP of the MRCP from a training set and a detection threshold based on the same data. The detector used a matched filter approach to detect movements in the testing data set. The online detection of the INP of the MRCP was used to trigger PES to observe the effects of such a system on the excitability of the cortical projections to TA in healthy subjects using MEPs through TMS [11].

In learning, task variability plays an important role [7] and to introduce task variability in the above mentioned system, a close loop BCI system is required which can detect a movement intention and classify the type of movement (in this case variations in force and speed). For investigating this, 6 subjects performed 3x50 cued dorsi-flexions of the right ankle: fast 20% maximum voluntary contraction (MVC), fast 60% MVC and slow 60% MVC. Fast: 0.5s to reach the force level, and slow: 3s to reach the force level. A pseudo online BCI was simulated from continuous EEG of nine channels. A matched-filter approach [10] was used to detect the movement intention and four temporal features were extracted and classified using a support vector machine.

The development of the methods for stroke rehabilitation, from analysis and definition of the MRCP's, over morphology investigations and design of the afferent feed-back protocol to the real-time implementation can be seen on Fig 2.

III. RESULTS

Detailed detector results based on INP of MRCP were presented in [10]. In summary the average true positive rate (TPR, %) of ME was 82.5±7.8 % (OSF), 68.7±14.9% (LLSF), 55.4±14.0 %(CSP), the average number of false positive detections per five minutes were 6.9±7.4 (OSF), 11.5±13.4 (LLSF), 57.3±17.8 (CSP) and the average detection latencies were -66.6±121ms (OSF), -79.7±92.8ms (LLSF), 153±148ms (CSP). The TPR based on OSF performed significantly (p<0.05) better than LLSF and CSP. For the MI task the average TPR (%) was 64.5±5.33% and the average number of false positive detections per five minutes was 15.8±9.42 when detecting MI using self-paced ME data as a training set. In stroke subjects, the TPR based on OSF was 55.0±12.0% with 8.63±7.32 false positive detections per five minutes and an average detection latency of -56.8±139ms.

Based on the above results from the pseudo online detection, a real online system was developed in [11]. In summary, the average results of the self-paced BCI system based on OSF across all subjects was 67.15±7.87 % for MI task with a false positive rate of 22.05±9.07%. On average the subjects had to perform the motor imagination task 75.38±9.01 times to get 50 true positives. MEP_{max}, in the post-TMS measurement of MEPs, increased significantly by 53±43% (P<0.05) compared to pre-TMS measurements of MEPs.

The system performance, defined as correct detection and classification, of the close loop BCI was 68±12% (variations in force) and 78±14% (variations in speed). While the number of movements that were correctly detected, but misclassified was 18±6% (variations in force) and 18±8% (variations in speed).

IV. CONCLUSIONS

In summary we can conclude it is possible to detect the initial negative phase (INP) of the MRCP with short latency (~100 ms) and an online system, based on such detector, can induce changes in the excitability of the cortical projections of the targeted muscle. The feasibility of a close loop BCI system was presented which can be used to deliver the task dependent afferent feedback in a similar fashion as that proposed in the protocol of [8].

Also described was the process of developing the real-time based stroke-rehabilitation method, from the early descriptive work to the implementation of the method.

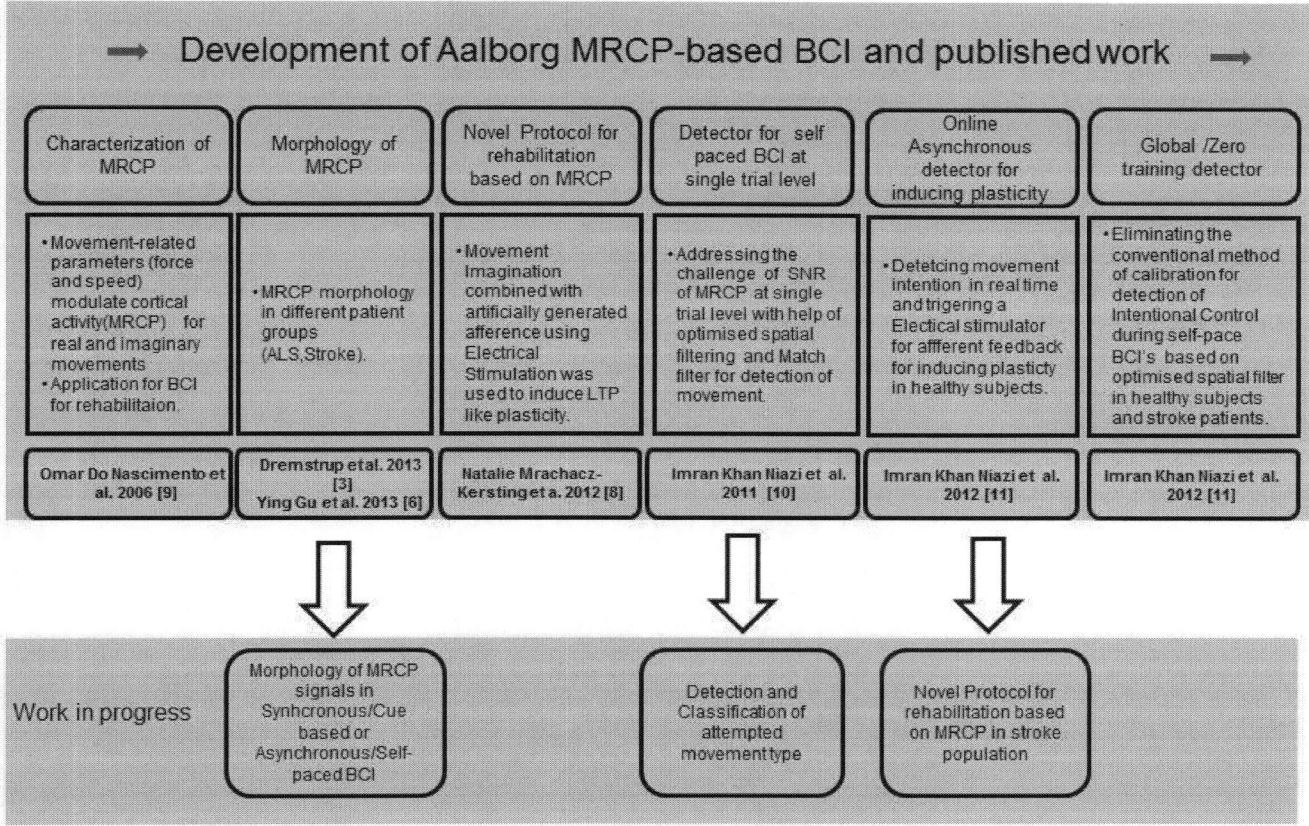

Fig 2. Development of MRCP-based BCI from earliest work in Aalborg with characterization of the MRCP, to real-time detection of movement intention to utilization in rehabilitation of stroke-patients.

REFERENCES

1. Cabrera, Alvaro Fuentes; Dremstrup, Kim. Steady-State Visual Evoked Potentials to drive a Brain Computer Interface, *HST-Report 2008-1. ISBN 978-87-90562-71-7 Department of Health Science and Technology, Aalborg University.* 2008

2. Cabrera, Alvaro Rodrigo; Dremstrup, Kim. Auditory and spatial navigation imagery in Brain-Computer Interface using optimized wavelets. *Journal of Neuroscience Methods*, Vol. 174, Nr. 1, 2009.

3. Dremstrup Kim, Gu Ying, Nascimento Omar Feix do, Farina Dario, Movement-related cortical potentials and their application in Brain Computer Interfacing. Chap 13 in *Introduction to Neural Engineering for Motor Rehabilitation*. John Wiley & Sons, Inc. 2013. In Print.

4. Gu, Ying; Farina, Dario; Murguialday, Ander Ramos; Dremstrup, Kim; Montoya, Pedro; Birbaumer, Niels. Offline identification of imagined speed of wrist movements in paralyzed ALS patients from single-trial EEG. *Frontiers in Neuroscience*, Vol. 3, 62, 2009.

5. Gu, Ying; Dremstrup, Kim; Farina, Dario. Single-trial discrimination of type and speed of wrist movements from EEG recordings. *Clinical Neurophysiology*, Vol. 120, Nr. 8, 2009.

6. Gu Y, Farina D, Murguialday AR, Dremstrup K and Birbaumer N. Comparison of movement related cortical potential in healthy people and amyotrophic lateral sclerosis patients. *Frontiers in Neuroscience.* **7**:65, 2013.

7. Krakauer JW. Motor Learning: its relevance to stroke recovery and neurorehabilitation. *Current opinion in neurology,* 19(1): 84-90, 2006.

8. Mrachacz-Kersting N, Kristensen SR, Niazi, IK, Farina D. Precise temporal association between cortical potentials evoked by motor imagination and afference induces cortical plasticity. *Journal of Physiology,* Vol. 590(7): 1669-1682, 2012.

9. Nascimento, Omar do; Nielsen, Kim Dremstrup; Voigt, Michael. Movement-related parameters modulate cortical activity during imaginary isometric plantar-flexions. *Experimental Brain Research*, Vol. 171, Nr. 1, 2006.

10. Niazi IK, Jiang N, Tiberghien O, Nielsen JF, Dremstrup K, Farina D. Detection of movement intention from single-trial movement-related cortical potentials. *Journal of Neural Engineering*, vol 8(6), pp. Article No. 066009, 2011.

11. Niazi IK, Mrachacz-Kersting N, Jiang N, Dremstrup K, Farina D. Peripheral electrical stimulation triggered by self-paced detection of motor intention enhances motor evoked potentials. *IEEE Transactions on Neural Systems and Rehabilitation Engineering*, 20(4): 595-604, 2012.

Author: Kim Dremstrup
Institute: Center for Sensory Motor Interaction, Department of Health Science and Technology, Aalborg Universty
Street: Fredrik Bajersvej 7 D
City: DK-9220 Aalborg
Country: Denmark
Email: kdn@hst.aau.dk

Numerical Simulation of Deep Transcranial Magnetic Stimulation by Multiple Circular Coils

M. Lu[1] and S. Ueno[2]

[1] Lanzhou Jiaotong University/Key Lab. of Opt-Electronic Technology and Intelligent Control of Ministry of Education, Lanzhou, 730070, P.R. China
[2] Kyushu University/Department of Applied Quantum Physics, Graduate School of Engineering, Fukuoka 812-8581, Japan

Abstract—Stimulation of deeper brain structures by tran-scranial magnetic stimulation (TMS) plays a role in the study of reward and motivation mechanisms, which may be beneficial in the treatment of several neurological and psychiatric disor-ders. This paper presents a numerical simulation of deep tran-scranial magnetic stimulation (dTMS) in a realistic head model by considering large circular coil surrounding the head combi-nation of smaller coils. Three-dimensional distributions of the induced electric fields in the head were calculated by employing impedance method. Simulation results show that multiple circu-lar coils provides a potential way to produce deeply penetrating electric fields in deep brain regions with less fields in superficial cortical tissues.

Keywords—Transcranial Magnetic stimulation, multiple cir-cular coils, impedance method, induced electric fields.

I. INTRODUCTION

Transcranial magnetic stimulation (TMS) is a technique for noninvasive stimulation of the human brain. Stimulation is produced by generating a brief, transient high intensity magnetic field by passing a brief, transient electric current through a magnetic coil placed upon the scalp. The induced electrical currents in the underlying cortical tissue produce a localized axonal depolarization [1][2][3]. As a noninvasive method to stimulate brain, TMS has attracted considerable interests as an important tool for studying the functional or-ganization of the human brain as well as a therapeutic tool to improve psychiatric disease.

Although TMS treatment for depression with a common target in the dorsolateral prefrontal cortex has improved over the last years [4], current TMS methodologies do not yet yield the desired results, especially for major depressive disorder which is a highly prevalent and disabling condition associ-ated with significant morbidity and mortality. Many studies indicate that reward circuit is the focus in the study of depres-sion. Medial prefrontal and orbitofrontal cortices and their connections to deeper brain sites are associated with reward processes and motivation [5][6]. The subgenual anterior cin-gulate cortex (SACC) is central to this network [7]. There is a reason to assume that activation of deeper prefrontal and limbic regions may increase the antidepressant effect. The promising targets for dTMS can be the SACC and the nu-cleus accumbens which lie at depths of approximately 6 and 7 cm, respectively, or the orbitofrontal, medial frontal cor-tices, and the frontal pole which lie at depths of 3 to 4 cm and have strong connectivity to anterior cingulate cortex.

To stimulate deeper neuronal regions such as reward-related pathways directly, much higher stimulation intensities are needed, as the electric field decreases rapidly as a func-tion of tissue depth. However, even if stimulation intensities could be highly increased at the source, the use of standard TMS coils (such as round or figure-of-eight coils) at such high stimulation intensities does not allow safe stimulation and can lead to undesirable side effects. These limitations have led to the development of novel coil designs suitable for dTMS, which allows direct stimulation of much larger and deeper brain regions by significant reduction of the de-cay rate. In the past decade, there are several coil configura-tions potentially suitable for dTMS, such as double cone and H-coils[8][9].

In the current study, we present numerical simulation of the induced electric fields in a realistic head model by consid-ering a large circular coil surrounding the head combination of one or two small circular coils. The purpose of the present study is to investigate the feasibility of multiple circular coils in deep transcranial magnetic stimulation.

II. SIMULATION METHODS

The realistic head model as shown in Fig. 1 was gener-ated from a man model developed by Virtual Family project, which consists of four anatomical resolution models, based on MRI data of two adults and two children [10]. The man model has been segmented into 77 tissues of which 36 tis-sues are involved in the present head model considered for the present simulation. The head model is composed of 10 million cubic voxels with resolution of 1mm. Some important brain subregions such as hippocampus, pineal-body, tha-lamus etc. have been included in the model.

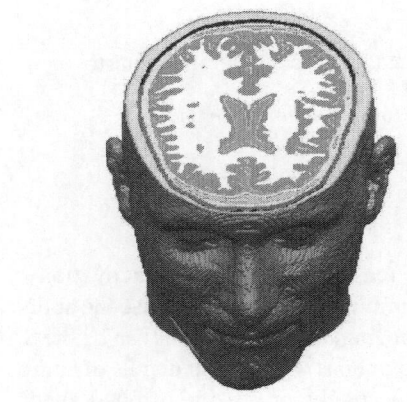

Fig. 1 Realistic head model in 3D.

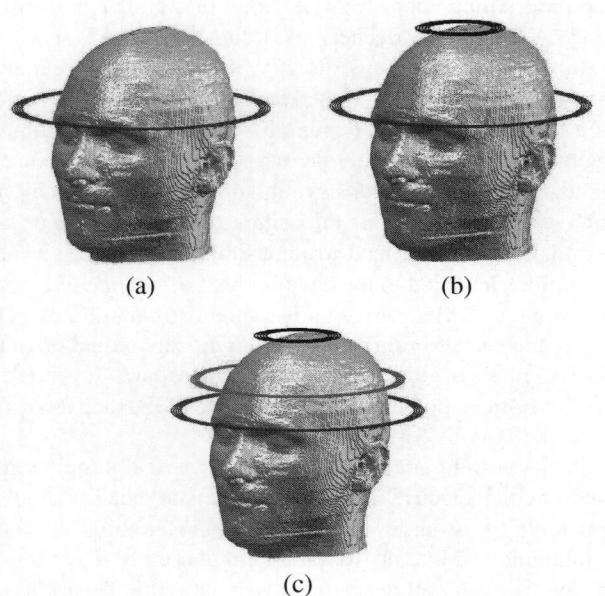

Fig. 2 3D head model with circular coils (a) one coil, (b) two coils, and (c) three coils.

A large circular coil surrounding the head originally defined as Halo coil [11] is shown in Fig. 2(a). The Halo coil with 5 turns has inner and outer radii of 138 mm and 150 mm, respectively. Fig. 2(b) shows the Halo coil operating si-multaneously with a smaller circular coil of mean diameter 90 mm and 14 turns located at the top of the head. Fig. 2(c) shows the Halo coil operating simultaneously with a two cir-cular coils. The inner and outer radii of the coil in the middle are 100 mm and 110 mm, respectively. The pulse currents with same working frequency of 2.38 kHz was fed into each of the coils. In both Figs. 2(b) and 2(c), the currents in neigh-boring coils was set in opposite directions.

The time variation of the applied magnetic field causes induced currents in the head through Faraday's induction mechanism. We have calculated these currents using the impedance method [12][13]. In this method, the human head model is described using a uniform 3D Cartesian grid and is composed of small cubic voxels. Assuming that, in each voxel, the electric conductivities are isotropic and constant in all directions. the model is represented as a 3-D network of impedances. The magnetic fields are calculated using Biot-Savart's law, the induced currents are calculated using the impedance method, and the induced electric fields are cal-culated using Ohm's Law.

Table 1 Tissue conductivities

Tissue	Conductivity $\sigma[S/m]$	Tissue	Conductivity $\sigma[S/m]$
Artery	7.00e-01	Hypothalamus	5.26e-01
Blood.Vessel	3.10e-01	Mandible	2.03e-02
Cartilage	1.75e-01	Marrow-red	2.44e-03
Cerebellum	1.24e-01	MO*3	4.65e-01
CSF	2.00e+00	Midbrain	4.65e-01
CA*1	6.44e-02	Mucosa	8.46e-04
CP*2	6.44e-02	Muscle	3.31e-01
Connective-tissue	2.04e-01	Nerve	3.04e-02
Ear-cartilage	1.75e-01	Pineal-body	5.26e-01
Ear-skin	2.00e-04	Pons	4.65e-01
Eye-cornea	4.25e-01	Skin	2.00e-04
Eye-lens	3.31e-01	Skull	2.03e-02
Eye-sclera	5.07e-01	Spinal-cord	3.04e-02
Eye-vitreous-humor	1.50e+00	Teeth	2.03e-02
FAT	2.32e-02	Thalamus	1.04e-01
Gray matter	1.04e-01	Tongue	2.76e-01
Hippocampus	1.04e-01	Vein	7.00e-01
Hypophysis	5.26e-01	White Matter	6.44e-02

*1 CA: Commissura-anterior; *2 CP: Commissura-posterior; *3 MO: Medulla-oblongata

The electrical properties are modeled using the 4-Cole-Cole model [14]. In this model, the biological tissues sub-ject to an electric field with angular frequency is modelled by relaxation theory and tissue properties can be obtained by fitting to experimental measurement. Since the number of tis-sue type in Virtual Family models is more than that in origi-nal Gabrial list [15], various tissues in Virtual Family models have been simulated with conductivities and permittivities of similar tissues (i.e. hippocampus and thalamus have the di-electric properties of brain grey matter from the Gabriel list). The tissue properties thus obtained and used in the present calculation are shown in Table 1.

III. RESULTS AND DISCUSSIONS

The field intensity of electric fields in brain tissues at dif-ferent depth from the top of the head (as shown in Fig. 3) in the lateral-medial direction are shown in Fig. 4. It is observed that for all three cases, the intensity of electric fields at depth of 3 cm are well below 100 V/m, the threshold of neural acti-vation. The Halo coil or the Halo coil combination of a small circular coil at the top of the head present large induced fields in deep brain regions at depth of 7 to 8 cm (Fig. 4(a)-(b)). The Halo coil working with two circular coils reduce the fields in deep brain regions greatly (Fig. 4(c)), which potentially provides the possibility of realization of deep TMS with im-proved focality.

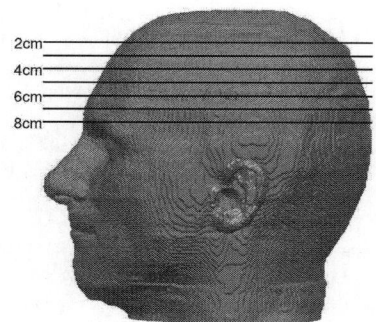

Fig. 3 Axial slice positions in head model

Fig. 5 shows the induced electric fields on the cross sec-tion of head model at coronal plane of y= 80 mm. In order to show the results more clearly, nonlinear colorbar was em-ployed in these figures. The color palette is linear up to stimu-lation threshold 100 V/m and then constant (dark red) above threshold. It is observed that both Halo coil and Halo coil combination of a small circular coil at the top of the head induce electric fields beyond 100 V/m over a large area in deep brain regions (Fig. 5(a)-(b)). While the effective stimu-lation area (electric fields exceeding 100 V/m) in deep brain regions is much less by Halo coil working with two circu-lar coils (Fig. 5(c)). Results shows that multiple circular coil with varied coil parameters provide a flexible way to stimu-late deep brain regions with improved focality.

IV. CONCLUSION

A large circular coil surrounding the head combination of smaller coils can produce deeply penetrating electric fields in deep brain regions with less fields in superficial cortical tis-sues. In the future, we intend to investigate the dependence of induced fields in the head on the coil positions and various pulse current protocols to realize focused stimulation in specific deep brain regions which are correlated to specific neural disorders.

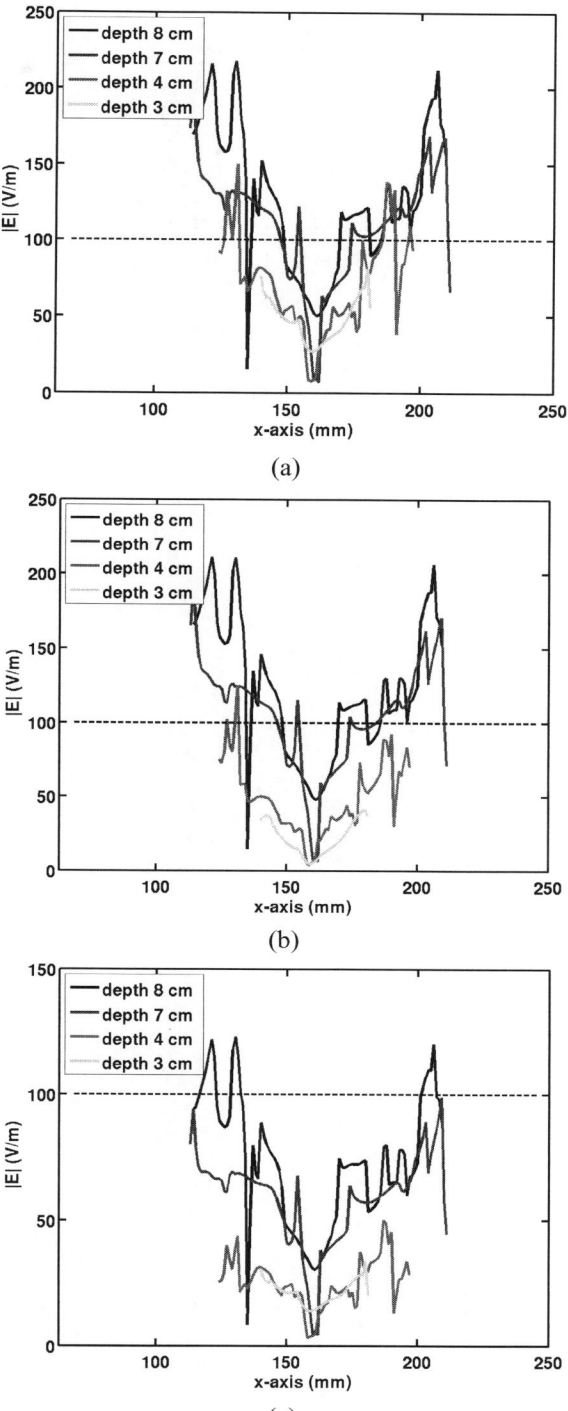

Fig. 4 Electric fields decay in brain tissues at different depth from the top of the head. (a) Halo coil, (b) Halo coil with a small circular coil at the top of the head, and (c) Halo coil with two circular coils. Dash line: threshold for neural activation.

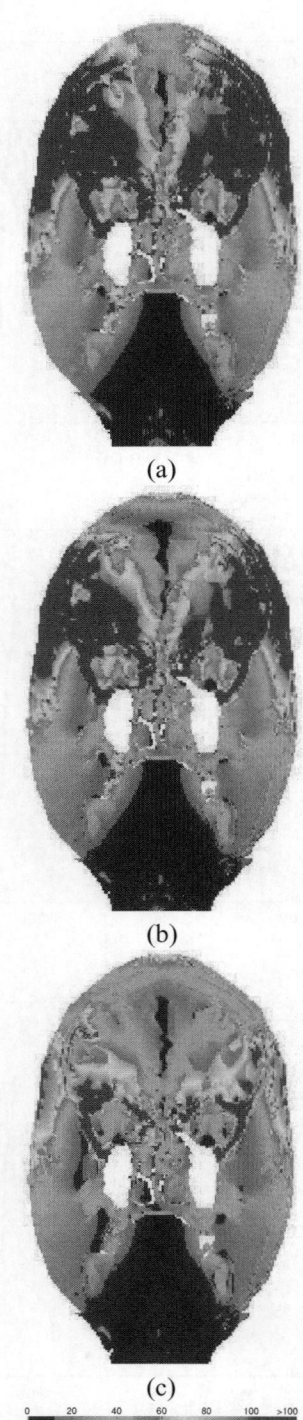

Fig. 5 Electric field distributions in the slice of 80 mm at coronal plane. (a) Halo coil, (b) Halo coil with a small circular coil at the top of the head, and (c) Halo coil with two circular coils.

ACKNOWLEDGMENTS

This work is supported by the National Nature Science Foundation of China under contract No. 51267010.

REFERENCES

1. Barker AT, Jalinous R and Freeston IL. Non-invasive magnetic stimula-tion of human motor cortex, Lancet. 1985;1:1106-1107.
2. Ueno S, Tashiro T, and Harada K, Localized stimulation of neural tissues in the brain by means of a paired configuration of time-varying magnetic fields, J. Appl. Phys. 1988;64:5862-5864.
3. Ueno S, Matsuda T and Fujiki M, Functional mapping of the human motor cortex obtained by focal and vectorial magnetic stimulation of the brain, IEEE Trans. Magn. 1990;26:1539-1544.
4. Gross M, Nakamura L, Pascual-Leone A, Fregni F. Has repetitive transcranial magnetic stimulation (rTMS) treatment for depression im-proved? A systematic review and meta-analysis comparing the recent vs. the earlier rTMS studies. Acta Psychiatr Scand. 2007;116:165-173.
5. Parkinson JA, Cardinal RN, Everitt BJ. Limbic cortical-ventral stri-atal systems underlying appetitive conditioning. Prog Brain Res. 2000;126:263-285.
6. Kalivas PW, Volkow ND. The neural basis of addiction: a pathology of motivation and choice. Am J Psychiatry. 2005;162:1403-1413.
7. Mayberg HS, Lozano AM, Voon V, McNeely HE, Seminowicz D, Hamani C, Schwalb JM, and Kennedy SH. Deep brain stimulation for treatment-resistant depression. Neuron. 2005;45:651-660.
8. Lontis ER, Voigt M, Struijk JJ. Focality assessment in transcranial magnetic stimulation with double and cone coils. J Clin Neurophysiol. 2006;23:462-471.
9. Zangen A, Roth Y, Voller B and Hallett M. Transcranial magnetic stim-ulation of deep brain regions: evidence for efficacy of the H-coil. Clin Neurophysiol. 2005;116:775-779.
10. Christ A, Kainz W, Hahn EG, Honegger K, Zefferer, M, Neufeld E, et al. The virtual family-development of surface-based anatomical models of two adults and two children for dosimetric simulations. Phys. Med. Biol. 2010,55:N23-N38.
11. Crowther LJ, Marketos P, et al. Transcranial magnetic stimulation: Improved coil design for deep brain investigation. J. Appl. Phys. 2011;109:07B314.
12. Orcutt N, Gandhi OP. A 3-D impedance method to calculate power de-position in biological bodies subjected to time varying magnetic fields, IEEE Trans. Biomed. Eng. 1988;35:577-583.
13. Lu M, Ueno S, Thorlin T and Persson M, Calculating the activating function in the human brain by transcranial magnetic stimulation. IEEE Trans. Magn. 2008;44:1438-1441.
14. Cole KS and Cole RH. Dispersion and absorption in dielectrics I. Alter-nating current characteristics. J. Chem Phys.. 1941;9:341-351.
15. Gabriel, S., Lau RW. The dielectric properties of biological tissues: III. Parametric models for the dielectric spectrum of tissues. Phys Med Biol. 1996;41:2271-2293.

Volitional Intention and Proprioceptive Feedback in Healthy and Stroke Subjects

M. Gandolla[1], S. Ferrante[1], F. Molteni[2], E. Guanziroli[2], T. Frattini[3], A. Martegani[3], G. Ferrigno[1], A. Pedrocchi[1,*] and N.S. Ward[4,*]

[1] Politecnico di Milano, Department of Electronics, Information and Bioengineering, NearLab, Milan, Italy
[2] Valduce Hospital, Villa Beretta, Rehabilitation Center, Costamasnaga, LC, Italy
[3] Valduce Hospital, Unità Operativa Complessa di Radiologia, Como, Italy
[4] Sobell Department of Movement Neuroscience, UCL Institute of Neurology, London, United Kingdom

Abstract—Increasing somatosensory input and concurrent volitional intention can enhance functionally relevant brain reorganization after stroke and is a potential mechanism of action of functional electrical stimulation (FES). Here we used fMRI to investigate this interaction in the brain during FES. FES is used as a method to augment proprioception during an active motor task. 17 controls and 11 chronic post-stroke patients were scanned during four conditions in a 2x2 factorial design: (1 & 2) repetitive unilateral active (moved by the subject) ankle dorsiflexion with and without concurrent electrical stimulation, (3 & 4) repetitive passive (moved by the experimenter) ankle dorsiflexion with and without concurrent electrical stimulation. In controls, all conditions elicited activity in regions known to be part of ankle motor control network. FES had an additional effect in primary sensorimotor areas (i.e. primary motor cortex – M1, primary somatosensory cortex – S1) during active compared to passive movement. In patients, all conditions elicited activity in a more widely distributed network, that included S1 and M1. In particular, FES had a greater effect during active compared to passive movement in secondary areas (i.e. ipsilesional postcentral and angular gyrus). Angular gyrus is a recipient of proprioceptive information encoded in the postcentral gyrus. These results suggest that the interaction takes place in a more widely distributed network of brain areas after stroke, and that patients may take advantage of secondary areas to support motor learning.

Keywords—fMRI, functional electrical stimulation (FES), stroke, augmented proprioception.

I. INTRODUCTION

After completing rehabilitation, approximately 50–60% of post-stroke patients still exhibit some degrees of motor impairment and require at least partial assistance in activities of day living [1]. Investigation of novel approaches to promote the recovery of motor impairments is therefore essential.

It has been suggested [2,3] that increase in somatosensory inputs and volitional intention to execute a movement could be an important strategy to enhance functionally relevant human brain reorganization after injury.

Functional Electrical Stimulation (FES) provides externally driven increased proprioceptive information that can be coupled or not with volitional intention to execute a movement. FES delivered on a mixed nerve trunk (i.e. nerve that contains both efferent motor and afferent sensory fibres) will synchronously depolarize motor and sensory axons that are bundled together, eliciting muscle contraction through a double path [4]: direct motorneurons depolarization and a proprioceptive signal artificially elicited by the sensory fibre stimulation that creates the impression that the muscle is extended (i.e. muscle spindles discharge), and leads to firing of the motoneurons in order to generate a contraction.

Moreover, FES is a recognized therapeutic procedure in rehabilitation [5]. Some hemiplegic patients treated with FES have shown a beneficial effect that outlasts the period of stimulation, known as 'carryover effect'.

Our aim was therefore to use FES during functional magnetic resonance imaging (fMRI) to explore (i) the effect of concurrent voluntary movement on brain regions receiving afferent proprioceptive signals in controls and post-stroke patients, and (ii) compare integration between augmented proprioception and volitional intention adopted by patients with respect to controls.

II. MATERIALS AND METHODS

A. Participants

11 patients were recruited (mean age 53.2, SD 14.5, 6 male and 5 female), who attended the Villa Beretta Rehabilitation center in Costamasnaga (Italy). All patients had suffered from first-ever stroke > 6 months previously, resulting in weakness of at least the tibialis anterior muscle (to <4+ on the Medical Research Council scale). Exclusion criteria consisted of impossibility to elicit an FES-induced ankle dorsiflexion; language or cognitive deficits; inability to walk even if assisted; incompatibility with MRI; high spasticity at ankle joint plantar flexor; and brain lesion

---
* Joint last co-authors.

including leg primary motor cortex area. An age matched control group composed by 17 healthy volunteers with no neurological impairment (8 male and 9 female) was also recruited. All subjects gave informed, written consent before the study, which was approved by the local research ethics committee.

B. Experimental Protocol

Experimental set-up - An integrated experimental set-up previously described and validated composed by a 1.5 T MRI scanner (CV/I), a motion capture system with three recording cameras and reflective markers (Smart μg -BTS, Italy); and a current-controlled electrical stimulator (RehaStim pro - HASOMED GmbH) was used [6,7].

Protocol design - A 2x2 event-related fMRI protocol with voluntary effort (V) and FES (F) as factors was performed using right ankle dorsiflexion (ADF). Given the 2x2 factorial design, 4 conditions were performed in a semi-randomized order during a single run lasting 10 minutes: FV = FES-induced ADF concurrently with voluntary movement by the subject; FP = FES-induced ADF, while the subject remains relaxed; V = voluntary ADF; P = passive dorsiflexion (by the experimenter) of the subject's ankle. ADFs were paced every 3.5 seconds (for 6 repetitions) with auditory cues. All subjects were free to choose the amplitude of their active movement to preclude fatigue. The experimenter moved the ankle to match the movements during volitional dorsiflexion.

Stimulation paradigm - FES was applied to the peroneal nerve through superficial electrodes, at 20 Hz fixed frequency. The current was set subject by subject; so as to reproduce the same movement amplitudes as during voluntary movements, within the tolerance threshold.

Kinematic measures - 3D trajectories of retro-reflective markers were acquired in order to measure the ankle angle during fMRI acquisitions and to determine the movement onsets and amplitude. Marker trajectories were analyzed with a custom algorithm running in Matlab (Matlab R2010b).

Table 1 Brain regions that resulted to be significantly active after small volume correction (p<0.05) in the different conditions (i.e. V, FV, FP, P) and in the interaction contrast (i.e. (FV-V)-(FP-P)) for C = controls; P = patients; P+ = patients > controls; C+ = controls > patients; and Cj = conjunction analysis. c = controlateral; i = ipsilateral; + = presence of at least one significant peak of activation in the selected regions; M1 = primary motor cortex; S1 = primary somatosensory cortex; SII = secondary somatosensory cortex; SMA = supplementary motor area; PM = premotor cortex.

Regions	V > Rest											FV > Rest											FP > Rest											P > Rest											Interaction										
	C		P		P+		C+		Cj		C		P		P+		C+		Cj		C		P		P+		C+		Cj		C		P		P+		C+		Cj		C		P		P+		C+		Cj	
	c	i	c	i	c	i	c	i	c	i	c	i	c	i	c	i	c	i	c	i	c	i	c	i	c	i	c	i	c	i	c	i	c	i	c	i	c	i	c	i	c	i	c	i	c	i	c	i	c	i
M1	+	+	+	+			+	+	+	+	+	+	+	+			+	+											+				+	+					+		+									
S1	+				+						+	+	+				+				+												+		+				+		+									
SII											+	+	+							+	+	+	+										+														+			
Postcentral gyrus (b5; b7)	+				+						+	+	+				+												+	+	+	+																		
SMA	+	+			+	+					+	+			+						+	+							+																					
PM	+	+			+	+					+	+																																						
Superior frontal gyrus (b10)																					+		+						+																					
Middle frontal gyrus															+						+		+						+													+	+						
Inferior frontal gyrus																																												+	+					
Rolandic operculum																	+				+																													
Middle cingulate cortex	+	+			+	+					+	+																	+											+										
Inferior parietal gyrus											+	+	+				+	+											+																					
Precuneus	+				+						+	+					+												+																					
Insula	+				+												+	+																																
Thalamus	+				+						+				+											+	+																							
Posterior putamen	+				+						+		+								+	+							+					+															
Caudate			+				+				+		+								+	+					+	+					+																	
Angular gyrus																					+	+	+						+											+		+							
Supramarginal gyrus																							+	+							+	+								+									

C. fMRI Images Analysis

fMRI Data Preprocessing – Images were analyzed using Statistical Parametric Mapping (SPM8, http://www.fil.ion.ucl.ac.uk/spm/) implemented in Matlab (MatlabR2010b). All fMRI volumes were realigned, unwarped, coregistered with the skull stripped structural image, normalized to the Montreal Neurological Institute reference brain, and smoothed with an isotropic 8 mm full-width half-maximum kernel. A threshold of 4 mm in translation and 5° in rotation was applied to realignment parameters [6].

Contrasts of Interest – Besides the main effect of each condition (i.e. FV>rest; V>rest; FP>rest; P>rest), thanks to the 2x2 factorial design a further contrasts of interest could be introduced: positive interaction defined as (FV-V)-(FP-P) that identifies regions where the FES augmented proprioception in the context of volitional intent (i.e. FV-V) produced a higher activation than FES augmented proprioception in absence of volitional movement (i.e. FP-P). The positive interaction contrast directly represents the sensorimotor integration during an active motor task, when in the brain we have the planning of the volitional movements coupled to augmented proprioceptive afference.

Group Analysis – The second stage of the analysis included three different aspects. First, to investigate the behavior of the fMRI response into the population from which our sample was drawn, i.e. group fMRI effects, contrast images for each subject were entered into a one sample t-test for each effect of interest, both for patients and controls separately. Second, to identify brain regions that were commonly activated between the two group of patients and controls a conjunction analysis was performed . Third, to investigate differences in the fMRI response between the two groups, the contrast images for all patients were entered into a two-sample t-test along with images of controls.

Regions of Interest - Statistical images were thresholded at $p<0.05$ corrected at voxel level for multiple comparisons within volumes of interest chosen a priori as spheres (radius in millimeters) centered on brain structures which are known to be part of the motor network [8,9,10,11]: M1 (10mm) centered in [±6; -28; 60]; S1 (10mm) in [±4;46;62]; SII (12mm) in [±58;-27;30]; Postcentral gyrus (12mm) in [±20;-36;72]; supplementary motor area (15mm) in [±8;-6;64]; premotor area (15mm) in [±20;-8;64]; superior frontal gyrus (12mm) in [±28;60;13]; middle frontal gyrus (12mm) in [±37;34;40]; inferior frontal gyrus (12mm) in [±46;31;12]; Rolandic operculum (12mm) in [±55;9;4]; middle cingulate cortex (12mm) in [±7;-12;44]; inferior parietal gyrus (12mm) in [±48;-24;20]; superior temporal gyrus (10mm) in [±54;15;-12]; precuneus (15mm) in [±8;-54;64]; insula (10mm) in [±44;12;-2]; thalamus (12mm) in [±13;-17;7]; posterior putamen (12mm) in [±30;0;9]; Caudate (12mm) in [±13;-13;13]; angular gyrus (12mm) in [±40;-58;50]; Supramarginal gyrus (12mm) in [±48;-48;34]. All coordinates are given with respect to the Montreal Neurological Institute reference brain.

III. RESULTS

fMRI Response in Controls and Patients - Realignment parameters were assessed for excessive motion, and one control subject and one patient violated our criteria and they were discarded from the group analysis. Both controls and patients show clear activation in motor and somatosensory areas known to be involved in ADF execution and previously reported [10] (Table 1).

A positive interaction ([FV-V] > [FP-P]) was seen in both M1 and S1 for control subjects (Figure 1). In other words, the effect of augmented proprioception is altered by volitional movement in both M1 and S1. The same positive interaction was seen in angular and postcentral gyri for patient group (Figure 1), where the effect of augmented proprioception depends on the presence of concurrent motor signals.

Conjunction Analysis - The conjunction analysis revealed a very clean pattern of activation in bilateral sensorimotor cortex, confirming that the motor task employed was valid (Table 1).

Controls versus Patients - Statistical comparison of patients and controls revealed that patients activated more than controls in several areas for different conditions, especially in the passive ones (i.e. P, FP) and in the interaction contrast. Rolandic opeculum was more active in controls than in patients for FP condition (Table 1).

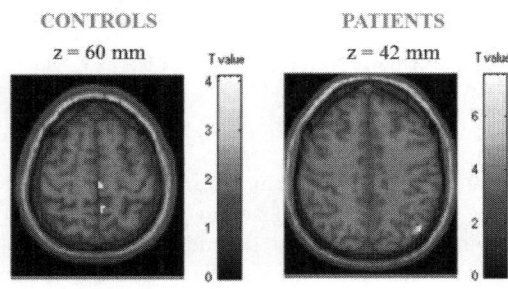

Fig. 1 Statistical parametric maps (thresholded at p<0.001, uncorrected) showing regions activated for the positive interaction contrast (i.e. (FV-V)-(FP-P)). The slice representation has been chosen for display purpose only. The clusters are located anatomically in M1 (anterior cluster) and S1 (posterior cluster) leg areas for controls, and in angular and Postcentral gyri for patients.

IV. DISCUSSION

The sensorimotor investigation in control subjects identified brain functional activation associated with ankle dorsiflexion task in all conditions (i.e., V, FV, P, FP), encompassing primary and secondary areas known in literature to be involved in movement processing and control of ankle movement [8,9,10,11,12]. Moreover, sensorimotor integration in chronic stroke patients was investigated, demonstrating the feasibility and the reliability of FES cortical correlates in patients, highlighting for the first time a clear network of activation related to FES-induced motor task in the context both of voluntary effort or passive movement. Patients significantly activated more than controls in a highly distributed pattern, which included regions that are activated when healthy controls perform complex or motor tasks [8,11]. It has been suggested that this finding might be indicative of higher demand and increased effort in patients to ensure adequate motor function despite existing deficits [13]. The conjunction analysis between patients and controls revealed clear pattern of activation in primary somatosensory cortex that has been reported to be principally involved during the same motor task execution. This result clearly shows that although patients use a quite widespread network in order to execute even a simple movement, they also share the principal regions devoted to motor control with healthy subjects.

The differential effect of action of FES in the context of active and passive movement (i.e. interaction between augmented proprioception and volitional intention) is shown in controls in primary sensorimotor areas (i.e. M1, S1) and is revealed in the contralateral angular and postcentral gyrus for patients. Postcentral gyrus has been described in literature to be active in healthy controls while executing complex or novel tasks [12], and angular gyrus has been suggested to be recipient of proprioceptive information encoded specifically in the postcentral gyrus [14].

V. CONCLUSIONS

The presented results may be important to disclose the central FES-induced mechanism that is yet not clear. Indeed, the localization in secondary motor areas of the interaction between augmented proprioception and volitional intention in patients could support the proposed mechanism of FES as a tool to promote improvement in motor function after central nervous system injury. This functional architecture may be important for future studies in patients during rehabilitation, by correlating changes in interaction contrast cortical correlates in a longitudinal design. Indeed, the brain correlates that has been shown to represent sensorimotor integration centers in the lesioned brain (i.e. interaction contrast) might be important in the investigation of the carryover effect brain signature.

REFERENCES

1. Bolognini N, Alvaro P, Felipe F (2009) Using Non-invasive Brain Stimulation to Augment Motor Training-induced Plasticity. J Neuroeng Rehabil 6: 8.
2. Ward NS, Cohen LG (2004) Mechanisms underlying recovery of motor function after stroke. Arch Neurol 61: 1844–1848.
3. Casellato C, Pedrocchi A, Zorzi G et al (2012) EMG-based visual-haptic biofeedback: a tool to improve motor control in children with primary dystonia. IEEE Trans Neural Syst Rehabil Eng. Oct 5.
4. Bergquist AJ, Clair JM, Lagerquist O et al. (2011) Neuromuscular electrical stimulation: implications of the electrically evoked sensory volley. Eur J Appl Physiol. 111: 2409–2426.
5. Ambrosini E, Ferrante S, Pedrocchi A et al (2011). Cycling induced by electrical stimulation improves motor recovery in postacute hemiparetic patients: a randomized controlled trial. Stroke 42: 1068-73.
6. Casellato C, Ferrante S, Gandolla M et al. (2010) Simultaneous measurements of kinematics and fMRI: compatibility assessment and case report on recovery evaluation of one stroke patient. J Neuroeng Rehabil 7:49.
7. Gandolla M, Ferrante S, Casellato C et al. (2011) fMRI brain mapping during motion capture and FES induced motor tasks: signal to noise ratio assessment. Med Eng Phys 33: 1027–1032.
8. Ciccarelli O, Toosy AT, Marsden JF et al. (2006) Functional response to active and passive ankle movements with clinical correlations in patients with primary progressive multiple sclerosis. J Neurol 253: 882–891.
9. Freund P, Weiskopf N, Ward Nick S et al. (2011) Disability, atrophy and cortical reorganization following spinal cord injury. Brain 134: 1610–1622.
10. Iftime-Nielsen SD, Christensen MS, Vingborg RJ et al. (2012) Interaction of electrical stimulation and voluntary hand movement in SII and the cerebellum during simulated therapeutic functional electrical stimulation in healthy adults. Hum Brain Mapp 33: 40–49.
11. Wegner C, Filippi M, Korteweg T et al. (2008) Relating functional changes during hand movement to clinical parameters in patients with multiple sclerosis in a multi-centre fMRI study. Eur J Neurol 15: 113–122.
12. Sahyoun C, Floyer-Lea A, Johansen-Berg H et al. (2004) Towards an understanding of gait control: brain activation during the anticipation, preparation and execution of foot movements. NeuroImage 21: 568–575.
13. Katschnig P, Schwingenschuh P, Jehna M et al. (2011) Altered functional organization of the motor system related to ankle movements in Parkinson's disease - Insights from functional MRI. Mov Disord 26: S245–S245.
14. Vesia M, Crawford JD (2012) Specialization of reach function in human posterior parietal cortex. Exp Brain Res 221: 1–18.

Corresponding author: Marta Gandolla – marta.gandolla@polimi.it

Politecnico di Milano, NearLab, Department of Electronics, Information and Bioengineering
via G. Colombo 40, 20133, Milano – Italy

Nonlinear Relationship between Perception of Deep Pain and Medial Prefrontal Cortex Response Is Related to Sympathovagal Balance

R. Sclocco[1,2], M.L. Loggia[2,3,4], R.G. Garcia[2,5], R. Edwards[3,4], J. Kim[2], S. Cerutti[1], A.M. Bianchi[1], V. Napadow[2,4], and R. Barbieri[6,7]

[1] Department of Electronics, Information and Bioengineering (DEIB), Politecnico di Milano, Milano, Italy
[2] MGH/MIT/HMS Athinoula A. Martinos Center for Biomedical Imaging, Charlestown, MA, USA
[3] Department of Anesthesiology, Perioperative and Pain Medicine, Brigham and Women's Hospital, Harvard Medical School, Boston (MA), USA
[4] Department of Psychiatry, Brigham and Women's Hospital, Harvard Medical School, Boston, MA, USA
[5] School of Medicine, Universidad de Santander (UDES), Bucaramanga, Colombia
[6] Department of Anesthesia, Critical Care and Pain Medicine, Massachusetts General Hospital, Boston, MA, USA
[7] Department of Brain and Cognitive Science, Massachusetts Institute of Technology, Cambridge, MA, USA

Abstract—Brain responses to evoked pain have been extensively investigated during the past years. Medial prefrontal cortex (MPFC) has been identified as a region showing a nonlinear, quadratic response to increasing evoked deep pain. Concurrently, MPFC plays an important role as an autonomic mediator in pain perception. Therefore, we hypothesize that MPFC response to pain is directly correlated with autonomic outflow. In order to test our hypothesis, we integrated autonomic modulation with functional neuroimaging data (BOLD fMRI) obtained during evoked deep tissue pain. We selected the LF/HF ratio, estimated instantaneously through a point process approach, as a measure of sympathovagal balance. Our findings reveal that mean LF/HF values show a nonlinear, U-shaped response to increasing pain, supporting our initial hypothesis. Furthermore, we found that subjects with higher LF/HF ratios, that is, higher shift towards sympathetic outflow, also show increased activation in the ventral portion of MPFC, thus providing further evidence in support of the relevant role of this region in sympathetic modulation.

Keywords—Deep pain, heart rate variability, fMRI, central autonomic network, medial prefrontal cortex.

I. INTRODUCTION

During the past years, brain responses to evoked pain have been extensively investigated through neuroimaging studies. These studies identified a widespread network of regions either activated (e.g., somatosensory cortices, anterior cingulate, insula, prefrontal cortices and thalamus (1)) or deactivated (e.g., medial prefrontal cortex, posterior cingulate and inferior parietal lobule (2)) by experimental pain. Moreover, our recent fMRI study showed the presence of both linear and nonlinear brain responses to different levels of evoked deep pain (3). A remarkable finding of this investigation is represented by the quadratic, U-shaped relationship revealed between medial prefrontal cortex (MPFC) and inferior parietal lobule (IPL) response to increasing levels of pain. Interestingly, medial prefrontal cortex has been shown to play an important role as an autonomic mediator for pain-like acupuncture stimuli (4) (5), and has been recently highlighted in a meta-analysis of autonomic fMRI studies (6).

In this study, we evaluated fMRI data from Loggia et al. (3) and concurrent ECG data, to investigate the relationship between pain-related brain activity and heart rate variability (HRV) response. More specifically, we hypothesized that, when using LF/HF ratio, an index of sympathovagal balance, as a regressor in fMRI data analysis, the resulting neural correlates would encompass aforementioned regions, thus confirming their connection with autonomic modulation by deep receptor pain. Finally, a closer linkage was investigated with a correlation analysis to investigate whether subjects with a higher shift towards sympathetic outflow also showed higher responses in the involved areas.

II. METHODS

A. Subjects and Experimental Protocol

Data from sixteen (16) healthy, right-handed subjects (11 male, mean age ± SD: 28.8 ± 9.7) were included in the study held at the Martinos Center for Biomedical Imaging at Massachusetts General Hospital in Boston, MA, USA. Deep pain sensation was evoked by a series of 14 seconds pressure stimuli (2-second ramp up, 10-second target pressure, 2-second ramp down), delivered to subjects' left calves by means of a 13.5 cm x 82.5 cm Velcro-adjusted pressure cuff (SC12D; Hokanson Inc, Bellevue, WA, USA) connected to a rapid cuff inflator (Hokanson E20 AG101, Hokanson Inc). Ten seconds after the end of each stimulus, subject rated pain intensity on a 0-100 scale (0 = "no pain", 100 = "the most intense pain tolerable") through a button box. A first training session was performed in order to obtain individual stimulus-response (S-R) curves, which were used to define 7 different pressure values ("p10" to "p70") corresponding to each subject's pain intensity levels (10 to 70); finally, the highest pressure value rated as nonpainful was selected as "p0".

During the imaging session, each stimulus was delivered 3 times, following a pseudorandom order, in 3 separate runs (8 stimuli per run, 24 stimuli in total); each painful stimulation was preceded by a 4-second visual cue and followed by 10 seconds of rest and by the intensity rating.

B. Data Acquisition and Analysis

fMRI data were acquired on a 3T Siemens TIM Trio MRI System (Siemens Medical, Erlangen, Germany), through a whole brain T2*-weighted gradient echo BOLD EPI pulse sequence (TR/TE = 2s/30ms, 32 slices, voxel sixe = 3.1 x 3.1 x 4 mm). High-resolution structural images were also collected using a multi-echo MPRAGE pulse sequence (TR/TE1/TE2/TE3/TE4 = 2530/1.64/3.5/5.36/7.22 ms, voxel size = 1 mm isotropic). fMRI data were analyzed using FEAT (part of FSL, FMRIB's Software Library). Pre-processing steps included motion correction, brain extraction, spatial smoothing (FWHM = 5 mm), grand-mean intensity normalization and high-pass temporal filtering. A first within-subject GLM analysis was carried out including in the design matrix the 24 stimuli, along with the 10-second rating periods, each as an independent block regressor convolved with a canonical double-gamma HRF. The resultant parameter estimates, as well as their variances, were then passed up to a second-level GLM analysis (see below).

Concurrently with fMRI acquisition, electrocardiogram (ECG) was recorded with an MRI-compatible Patient Monitoring system (Model 3150, InVivo Research Inc, Orlando, FL) using Chart Data Acquisition Software on a laptop equipped with Powerlab DAQ System (ADInstruments, Colorado Springs, CO). ECG traces were automatically annotated and R-R series were extracted. A point process model of heartbeat dynamics (7) was then applied to R-R intervals in order to compute instantaneous indices of HRV, further decomposed into low (LF) and high (HF) frequency components. Their ratio (LF/HF), considered as a measure of sympathovagal balance to the heart (8), was averaged over every subject for each pressure level, in order to inspect the shape of the stimulus-response curve to deep pain in terms of sympathovagal balance. In addition, each individual LF/HF ratio over each pressure level was used as a regressor in the second-level GLM analysis. In this case, only "p10" to "p70" levels were considered, since the present work aims to investigate the existence of a correlation between sympathovagal balance and cerebral response within the noxious continuum (3). Finally, individual maps resulting from these latter analyses were normalized to MNI space and passed up to group-level analysis carried out using FLAME (9).

III. RESULTS

A. Autonomic Response to Deep Pain

Autonomic outflow in response to different levels of deep pain, assessed by averaging LF/HF values across subjects for each level of applied pressure, showed the highest values in correspondence of the lowest and highest levels of pressure, while lower values were found in the range p20-p50 (with the exception of p30), as shown in Figure 1. The mean values of LF/HF, with their standard errors, are shown in Table 1.

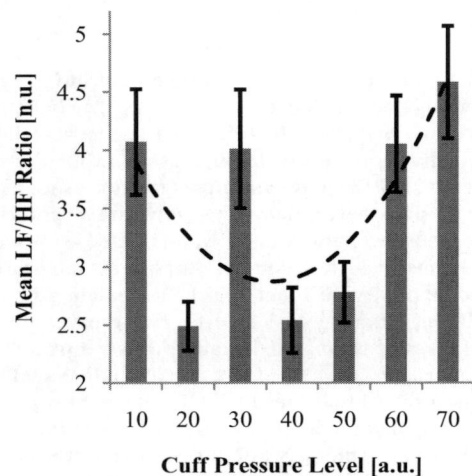

Fig. 1 LF/HF ratio values (mean ± SE) for each level of applied pressure.

Table 1 LF/HF ratios averaged across subjects for each cuff pressure level.

Cuff Pressure Level [a.u.]	LF/HF Ratio [mean ± SE, n.u.]
p10	4.07 ± 0.91
p20	2.49 ± 0.42
p30	4.01 ± 1.02
p40	2.54 ± 0.56
p50	2.78 ± 0.52
p60	4.05 ± 0.83
p70	4.58 ± 0.97

B. Brain Correlates of Autonomic Response to Deep Pain

According to our initial hypothesis, significant positive correlations were expected in brain areas that already showed a non-linear, quadratic response to pain in our previous study (3), in particular medial prefrontal cortices (MPFC) and inferior parietal lobule (IPL). Consistently with these previous results, the group map obtained in this analysis revealed two major clusters located in ventral MPFC and left IPL, both showing a positive correlation with LF/HF values. For visualization purposes, the aforementioned map was masked with the one from the previous study (Figure 2, Table 2), in order to highlight the existing overlap between cerebral correlates revealed by the two analyses.

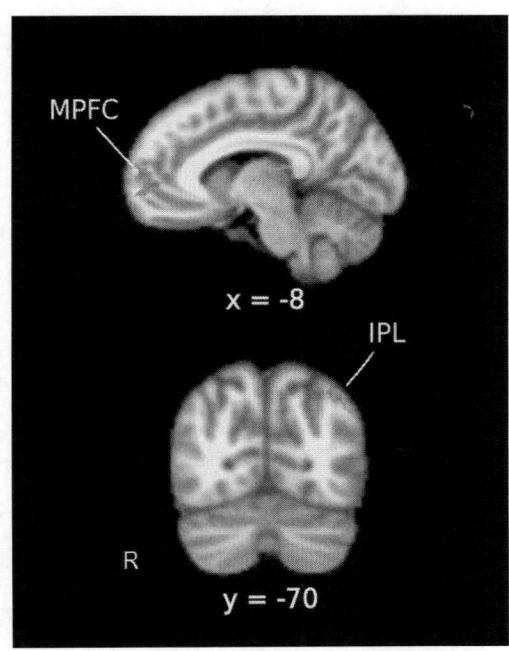

Fig. 2 Group map (Z>2.3, uncorrected) for LF/HF ratio correlation analysis, appropriately masked to reveal overlap with regions showing a quadratic stimulus-response relationship, as found in Loggia et al.(3).

Table 2 LF/HF correlated brain regions

Brain Region	Side	Location of maximum (MNI)			Z-score
		x	y	z	
MPFC	L	-8	52	2	2.56
IPL	L	-42	-70	44	2.42

IV. DISCUSSION

In the present work, we investigated the nonlinear, quadratic relationship previously found between MPFC and perception of deep pain. In particular, using LF/HF ratio as a regressor in BOLD fMRI data analysis, we showed how this interesting behavior is related to autonomic modulation during pain.

A possible interpretation for the relationship found between LF/HF values and increasing deep tissue pain is that such relationship describes two different phenomena in response to the increasing levels of cuff pressure. Initially, at lower pressure levels, there would be a negative relation, since each increment in cuff pressure would cause a higher venous return to the heart by acting as a mechanical pump, with a subsequent increase in the afferent input to cardiopulmonary baroreceptors and vagal activation; a similar phenomenon was observed in a previous study in healthy subjects exposed to lower-leg rhythmic cuff inflations (10). However, at some point the increasing deep pain resulting from higher pressure levels would counteract this hemodynamic phenomenon with a consequent increase in sympathetic activation. In any case, the activity of MPFC would be related with this autonomic modulation, irrespective of the cause.

The medial prefrontal cortex, along with anterior cingulate and insular cortices, is consistently associated with cardiovascular arousal (11). Previous studies observed a negative correlation between this region and heart rate (12) (13) (14), suggesting a primary executive control role for MPFC in parasympathetic modulation. However, in a recent meta-analysis Beissner et al. (6) showed that the ventral portion of MPFC is consistently associated with sympathetic modulation across studies. Our results showing nonlinear MPFC response and its relationship to sympathovagal balance in response to pain suggest an intriguing relationship between activity in this area and sympathovagal modulation by deep-receptor pain.

Our group analysis results showed a positive correlation between MPFC response and LF/HF ratio. This means that subjects showing higher LF/HF ratios, and therefore higher shifts towards sympathetic outflow, also show increased response in the aforementioned region. The correlation analysis between individual percent signal change in MPFC and LF/HF ratio is shown in Figure 3 (Spearman's $rho = 0.5$).

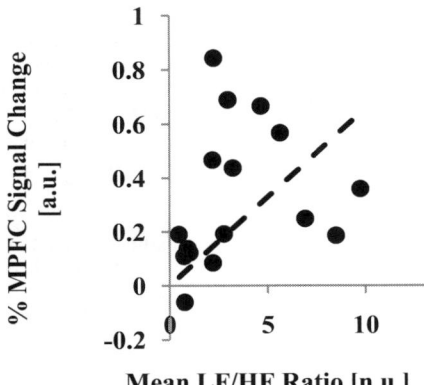

Fig. 3 Positive correlation between individual MPFC percent signal change and mean LF/HF ratio.

Interestingly, both MPFC and IPL have been shown to be involved in the Default Mode Network (15) (16), which is also closely linked with pain perception (17) and has been shown to be altered in chronic pain. A close link between the DMN and autonomic response may underlie the reports of autonomic dysregulation in some clinical pain populations.

V. CONCLUSION

The present investigation originates from a previous study, in which nonlinear relationships between cerebral activation in MPFC and perception of deep pain were found. Given the important role this region plays in central autonomic control, the present work demonstrates that a possible explanation for this behavior can be investigated by establishing a connection between this finding and the response of autonomic outflow to pain, as measured by an instantaneous characterization of the sympathovagal balance.

Future studies should confirm our results on a larger sample of subjects and in relation to different stimuli or tasks, as well as using different metrics for autonomic outflow estimation.

REFERENCES

1. Apkarian AV, Bushnell MC, Treede R, Zubieta J. Human brain mechanisms of pain perception and regulation in health and disease. European Journal of Pain. 2005;9(4):463-.
2. Kong J, Loggia ML, Zyloney C, Tu P, LaViolette P, Gollub RL. Exploring the brain in pain: Activations, deactivations and their relation. Pain. 2010;148(2):257-67.
3. Loggia ML, Edwards RR, Kim J, Vangel MG, Wasan AD, Gollub RL, et al. Disentangling linear and nonlinear brain responses to evoked deep tissue pain. Pain. 2012.
4. Beissner F, Deichmann R, Henke C, Bär K. Acupuncture—deep pain with an autonomic dimension? Neuroimage. 2012;60(1):653-60.
5. Napadow V, Lee J, Kim J, Cina S, Maeda Y, Barbieri R, et al. Brain correlates of phasic autonomic response to acupuncture stimulation: An event-related fMRI study. Hum Brain Mapp. 2012.
6. Beissner F, Meissner K, Bär K, Napadow V. The autonomic brain - an ALE meta-analysis for central processing of autonomic function. Journal of Neuroscience. in press.
7. Barbieri R, Matten EC, Alabi AA, Brown EN. A point-process model of human heartbeat intervals: New definitions of heart rate and heart rate variability. American Journal of Physiology-Heart and Circulatory Physiology. 2005;288(1):H424-35.
8. Montano N, Porta A, Cogliati C, Costantino G, Tobaldini E, Casali KR, et al. Heart rate variability explored in the frequency domain: A tool to investigate the link between heart and behavior. Neuroscience & Biobehavioral Reviews. 2009;33(2):71-80.
9. Beckmann CF, Jenkinson M, Smith SM. General multilevel linear modeling for group analysis in FMRI. Neuroimage. 2003;20(2):1052-63.
10. Niizeki K, Tominaga T, Saitoh T, Nishidate I, Takahashi T, Uchida K. Effects of lower-leg rhythmic cuff inflation on cardiovascular autonomic responses during quiet standing in healthy subjects. American Journal of Physiology-Heart and Circulatory Physiology. 2011;300(5):H1923-9.
11. Shoemaker JK, Wong SW, Cechetto DF. Cortical circuitry associated with reflex cardiovascular control in humans: Does the cortical autonomic network "speak" or "listen" during cardiovascular arousal. Anat Rec. 2012;295(9):1375-84.
12. Radhakrishnan R, Sluka KA. Deep tissue afferents, but not cutaneous afferents, mediate transcutaneous electrical nerve stimulation-induced antihyperalgesia. J Pain. 2005 Oct;6(10):673-80.
13. Lane RD, McRae K, Reiman EM, Chen K, Ahern GL, Thayer JF. Neural correlates of heart rate variability during emotion. Neuroimage. 2009;44(1):213-22.
14. Goswami R, Frances MF, Shoemaker JK. Representation of somatosensory inputs within the cortical autonomic network. Neuroimage. 2011;54(2):1211-20.
15. Raichle ME, MacLeod AM, Snyder AZ, Powers WJ, Gusnard DA, Shulman GL. A default mode of brain function. Proceedings of the National Academy of Sciences. 2001;98(2):676-82.
16. Greicius MD, Krasnow B, Reiss AL, Menon V. Functional connectivity in the resting brain: A network analysis of the default mode hypothesis. Proceedings of the National Academy of Sciences. 2003;100(1):253-8.
17. Seminowicz DA, Davis KD. Pain enhances functional connectivity of a brain network evoked by performance of a cognitive task. J Neurophysiol. 2007;97(5):3651-9.

Roberta Sclocco
Department of Electronics, Information and Bioengineering (DEIB),
Politecnico di Milano
Piazza Leonardo da Vinci, 32
20133, Milano
Italy
roberta.sclocco@polimi.it

Dynamics of Learning in the Open Loop VOR

P. Colagiorgio[1], G. Bertolini[2], C. Bockisch[2], D. Straumann[2], and S. Ramat[1]

[1] Department of Industrial and Information Engineering, University of Pavia, Pavia, Italy
[2] Department of Neurology, University Hospital Zurich, Zurich, Switzerland

Abstract—We present our preliminary results on motor adaptation of the angular vestibulo-ocular reflex (aVOR) in response to passive impulsive head rotations at constant acceleration (460 deg/s2). Human healthy subjects were repeatedly subjected to a 20 degrees yaw rotation (using a rotating chair) while they tried to maintain fixation on a visual target. We used an incremental velocity error signal in which target moved partially with the head, during rotations, causing adaptation of the initial eye velocity. We analyzed only the first 100 msec of the aVOR, i.e. the open loop portion of the response. In order to better understand the multiple-timescales dynamics of motor learning subjects were adapted to a aVOR gain reduction period followed by a shorter reverse adaptation block of trials, and a subsequent no-error feedback period which caused a recovery toward the initially adapted state. Adaptation mechanisms have been successfully described using a two hidden states model, in which a fast state learns quickly from motor error, but has poor retention, and a slow state that learns slowly but has a stronger retention. We modeled our data using such two-states model finding an underestimation of the spontaneous recovery trend.

Keywords—Vestibulo-ocular reflex, Adaptation, Motor-Learning, Time constants.

I. INTRODUCTION

Several studies on motor adaptation in response to external perturbations such as force-field, or saccadic target shift, showed the multiple time scales of motor memory [1–3]. The two state model proposed by Smith et al. [3] based on two hidden states, a fast one that learns quickly from motor error but has poor retention and a slow one that learns slowly but has stronger retention, can account for different adaptation phenomena, such as saving, i.e. a faster re-learning of the same task after the first adaptation, and spontaneous recovery, i.e. a rebound to the initially adapted state while the error is clamped to zero once an adaptation period is followed by a short reverse adaptation period. Despite several studies that account for adaptation in saccadic and in reaching movements with multiple time scales of memory, little is known about these processes in the adaptation of the aVOR.

The vestibular system provides our brain with information on the movement of the head to maintain balance and control posture, to program corrective responses, regulate other body rhythms and to drive the movement of the eyes in order to stabilize vision during head movements through the vestibulo-ocular reflex (VOR). The VOR is an open-loop control system, therefore an adaptive process of motor learning is required to keep the VOR calibrated. Thanks to its well studied neural circuitry, the VOR could be an excellent model system for studying neural mechanisms related to learning and memory in the CNS.

The majority of aVOR gain adaptation studies have been performed during passive, low-velocity, continuous sinusoidal head rotations. On the other hand recent studies explored the aVOR gain adaptation during self-generated and passive head-only impulsive rotations (high velocity horizontal head rotations) in humans, showing that aVOR can be efficiently adapted using a retinal velocity error (retinal slip) signal with a gradually increasing adaptive demand [4], [5]. Zhou and colleagues [6] showed the time adaptation of translational VOR in monkeys, in response to brief passive head translations.

In this work we present our preliminary results on adaptation of the human aVOR in response to passive impulsive head rotations during a gradually increasing velocity error demand. We demonstrate the existence of spontaneous recovery in aVOR motor memory by inducing a reduction in aVOR gain followed by a shorter reverse adaptation period, and a subsequent period of no error feedback trials, which reveal a recovery toward the initially adapted state.

Finally we tried to model the dynamics of adaptation with the two state model [3] and found that the model underestimates the entity of spontaneous recovery during the no error feedback trials. Such inability of the model to reproduce the experimental results may be due to the introduction of a gradual adaptation stimulus because, despite the effectiveness of the model has been shown for large sudden perturbations, although in other motor systems, some contradictions occur with the gradual adaptation condition [7].

II. MATERIALS AND METHODS

Subjects: data reported were collected from four human healthy subjects. Participation in this study was voluntary and all subjects consented to be a part of this project in accordance with a protocol approved by the Department of Neurology, University Hospital Zurich, Zurich, Switzerland.

Eye movement recording: horizontal and vertical right eye position was measured using the EyeSeeCam infrared video system (frequency 220 Hz).

Vestibular stimulation: rotational stimuli were delivered in complete darkness by a three-axis rotational stimulator driven by three servo-controlled motorized axes (Acutronic, Switzerland), controlled with Acutrol software and hardware, and interfaced with LabVIEW software. Subjects were comfortably seated in a chair, the center of the head was positioned at the center of rotations, in order to obtain purely rotational stimuli. Individually adjusted masks, made of a thermoplastic material, were molded to the contour of the head after warming, with openings in the mask made for the eyes, VOG system and mouth. The mask was attached to the back of the chair, and restricted head movements very effectively without causing discomfort. Rotational stimuli were 20 degrees yaw rotations at 460 deg/s^2 constant acceleration and 100 deg/s of peak velocity. We instructed subjects to hold their gaze on a projected laser dot.

Laser projection: We used a laser and real-time 2D mirror deflection system for displaying a visual target on a curved isovergence projection screen at 140 cm in front of the subject. Laser position was controlled using real-time horizontal position of the chair.

During the experiments each subject was exposed to several blocks of trials differing in terms of the behavior of the visual stimulus.

Space fixed (SF): the laser dot turns on 0.5 s before chair rotation in the straight ahead position and remains fixed in space for the entire rotation duration, asking for a normal aVOR gain of ~1).

Adaptation stimulus (AS): the laser dot turns on 0.5 s before chair rotation in straight ahead position and moves during the rotation with a percentage of the chair velocity depending on the planned adaptation requirement, producing retinal slip error. If the target moves in the direction of head rotation its fixation requires aVOR gain reduction, while a movement in opposite direction elicits aVOR gain enhancement.

Catch Trial stimulus (CT): the laser dot turns on 0.5 s before chair rotation in straight ahead position and turns off when the chair starts rotating. Thus, this condition provides no error feedback and the following fixation will be the SF after chair repositioning.

Incremental aVOR training: aVOR gain was adapted with a gradually and progressively changing stimulus, based on velocity error. Before adaptation aVOR gain was assessed with 20 SF trials (i.e. aVOR gain request of 1) and 20 CT trials. Then in 200 AD trials (4 blocks of 50 trials each) aVOR gain was gradually reduced by 40% (i.e. aVOR gain request progressively decreased from 1.0 to 0.6). After such first gain-down adaptation period subjects underwent a block of 75 AD trials of reversed (i.e. gain-up) adaptation (i.e. aVOR gain request from 1.0 to 1.2). A final block of 100 CT trials was then supplied to evaluate the spontaneous recovery of motor memory.

Data Analysis: Signals representing eye movements, target position, chair position and head velocity were sampled at 220 Hz with 16 bits resolution through the EyeSeeCam system and analyzed off-line using Matlab (Mathworks) software. Raw eye position data was calibrated, filtered with a 10 Hz low pass filter and differentiated to obtain eye velocity data. Onset of each head movement was found using an algorithm based on straight line and polynomial fit [8]. We have chosen to analyze only the first 100 msec of the aVOR response, as during such interval (the "open loop response"), the aVOR is uncontaminated by effects of external visual feedback [6]. Any trials in which the subject broke fixation or made a saccade within the first 100 msec of the onset of the head movement were rejected manually using a custom Matlab software. The effect of motor learning was evaluated by computing the mean of eye velocity (VM) between 65ms and 90 ms from head movement onset. The mean pre-adaptation velocity (*preS*) was assessed from the 20 SF trials. The course of adaptation was then computed for each trial as *(VM-preS)/preS* (Fig. 1B).

Mathematical Modeling: We modeled motor adaptation using the two state model proposed by Smith et al. [3]. Each state is thus modeled with a linear differential equation with a learning term and a forgetting term. For each trial *n* the motor error input *e* is determined by the difference between the external perturbation *f* and the motor output *y* as follows:

$$e(n) = f(n) - y(n).$$

The update equations for the motor output are giving by the following:

$$y(n) = x_f(n) + x_s(n)$$
$$x_f(n+1) = a_f * x_f(n) + b_f * e(n)$$
$$x_s(n+1) = a_s * x_s(n) + b_s * e(n).$$

The learning rates from the fast and the slow states are $1 > b_f > b_s > 0$ and their forgetting rates are $1 > a_s > a_f > 0$.

The model parameters were estimated using the Expectation Maximization (EM) algorithm [9] based on a maximum likelihood approach.

III. RESULTS

We induced aVOR adaptation during passive impulsive head rotations (at constant acceleration) on 4 human subjects, using an incremental error velocity stimulus. Fig. 1A shows the decrease of eye velocity for one subject during the backward adaptation period.

Dynamics of Learning in the Open Loop VOR

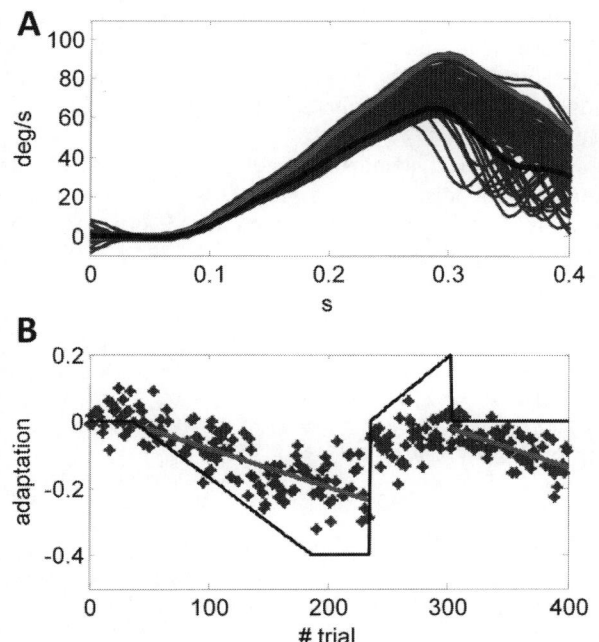

Fig. 1 A: decrease of eye velocity for one subject in response to constant passive head rotation, during gain decrease adaptation stimulus. In red the mean of the first ten movements, in black the mean of the last ten movements. B: eye velocity adaptation in response to the stimulus perturbation (black) for one subject. In red the linear fit of adaptation and catch trials.

Due to the retinal slip induced by the target moving in the same direction as the head, the CNS recalibrates the aVOR gain to stabilize the target on the retina. When the compensatory eye movement was not adequate, a covert catch up saccade was required to bring the eyes on target.

Considering spontaneous recovery of a previously learned aVOR motor memory as a fundamental proof of the existence of two (or more) memory processes, we induced a gain aVOR reduction followed by a shorter reversed adaptation period. Reversed gain adaptation was then followed by a block of catch trials to observe whether the aVOR gain returned toward the first adapted gain state. Fig. 1B shows the trend of the eye velocity adaptation for one representative subject together with the stimulus perturbation throughout the entire experiment. During the block of catch trials (trials 300 to 400) delivered after the block of forward adaptation, the eye velocity shows a *spontaneous recovery* toward the gain learned at the end of the backward adaptation trials.

Fig. 2 shows the evolution of the aVOR response before (SF) and after backward adaptation (ASend), after forward adaptation (deASend), and at the beginning (CTstart) and end (CTend) of the final catch trials for all subjects.

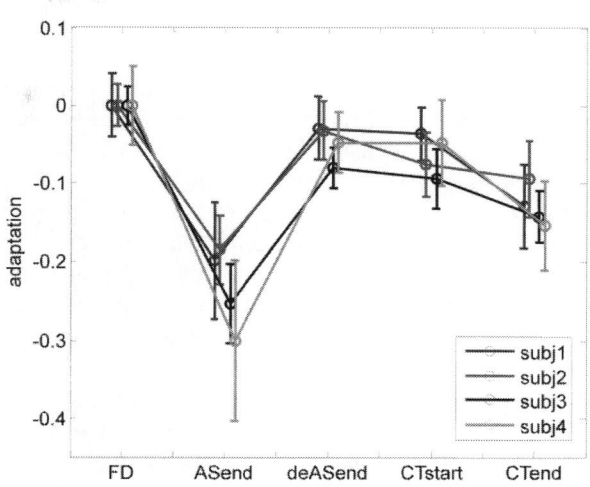

Fig. 2 Evolution of adaptation for all subjects. Dots represent mean and the bars the standard deviation of 10 movements.

Finally we modeled the dynamics of motor learning using the two state model [3]. The model parameters were estimated on the data obtained by averaging the data of all subjects at each trial. The values resulting from the estimation were: $a_s = 0.9999$, $a_f = 0.77$, $b_f = 0.0988$, $b_s = 0.071$.

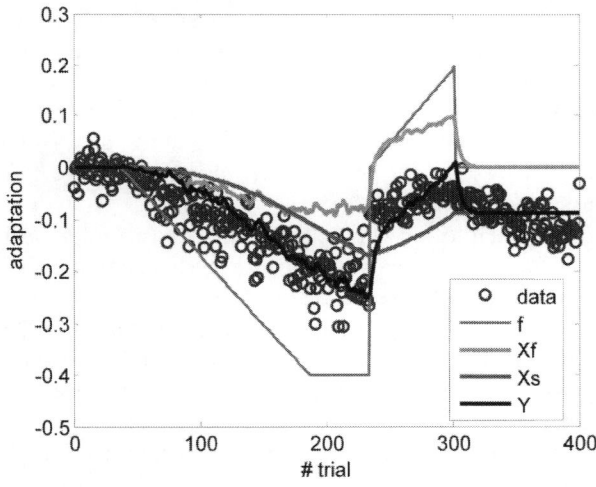

Fig. 3 Simulation of two state model on averaged data. The green line represents the fast state, the red line the slow state, the black line the model output.

As shown in Fig. 3 the model (black line) predicts the decrease of eye velocity during final set of catch trials (spontaneous recovery toward the slow state adapted in initial adaptation, red line) but underestimates its entity.

IV. DISCUSSION

We induced yaw aVOR gain adaptation in response to passive impulsive head rotations during a gradually increasing velocity error demand. A long backward adaptation training session was followed by a shorter set of reverse adaptation trials (reversed direction of learning errors). When, in the final set of catch trials the visual error feedback was abolished, the memory acquired during the first adaptation training appears to be re-expressed, producing a spontaneous recovery of the previously learned motor output.

Smith and colleagues [3] proposed a mathematical model in which the motor error affects two learning timescales, each one with its own internal state: one is highly sensitive to error but has poor retention, and the other is poorly responsive to error but has strong retention.

In our experiment, at the end of the backward adaptation period most of the motor output is due to the slow state (Fig.3 red line). In the reverse adaptation period, the prediction errors are very large, causing a rapid change in the fast state (Fig.3 green line). At the end of the reverse adaptation period the system appears to be near the baseline only because the fast and slow states are in competition, effectively canceling each other. The reverse adaptation block is followed by a block of catch trials, during which the fast state rapidly returns to zero, while the slow state slowly declines because there are no errors to learn from. In this condition, the system is not driven by an input and changes in the memory states are only due to the forgetting terms. The rapid decay of the fast state but the gradual decay of the slow state results in spontaneous recovery of the previously learned behavior.

The principles on which the model is based only partially account for our data, because the model shows an underestimation of the decrease of the motor output during the final block of catch trials. Wong and Shelhamer in 2011 compared saccadic adaptation performed with gradual and with sudden perturbations paradigms finding more robust adaptation as well as greater retention of learning (larger aftereffects) in response to a gradual adaptation stimulus, concluding that providing a large and consistent error signal is not as effective as a small but continuously varying error signal, in terms of driving a change in the gain of the motor system. They observed a conflict with error-based models of motor learning, which generally exhibit more robust adaptation in response to a large, consistent error signal.

Moreover Zarahn et colleagues [10] demonstrated that time invariant state-space models cannot explain the phenomenon of savings, finding that the data were better fit with a model whose parameters varied in each phase of the experiment.

Maybe the dynamics of motor learning needs more complex models than simple state-space formulations. However, we are collecting more data to formulate a more complete hypothesis.

REFERENCES

[1] Y. Kojima, Y. Iwamoto, and K. Yoshida, "Memory of learning facilitates saccadic adaptation in the monkey.," *The Journal of neuroscience: the official journal of the Society for Neuroscience*, vol. 24, no. 34, pp. 7531–9, Aug. 2004.

[2] V. Ethier, D. S. Zee, and R. Shadmehr, "Spontaneous recovery of motor memory during saccade adaptation.," *Journal of neurophysiology*, vol. 99, no. 5, pp. 2577–83, May 2008.

[3] M. a Smith, A. Ghazizadeh, and R. Shadmehr, "Interacting adaptive processes with different timescales underlie short-term motor learning.," *PLoS biology*, vol. 4, no. 6, p. e179, Jun. 2006.

[4] M. C. Schubert, C. C. Della Santina, and M. Shelhamer, "Incremental angular vestibulo-ocular reflex adaptation to active head rotation.," *Experimental brain research. Experimentelle Hirnforschung. Expérimentation cérébrale*, vol. 191, no. 4, pp. 435–46, Dec. 2008.

[5] A. a Migliaccio and M. C. Schubert, "Unilateral adaptation of the human angular vestibulo-ocular reflex.," *Journal of the Association for Research in Otolaryngology : JARO*, vol. 14, no. 1, pp. 29–36, Feb. 2013.

[6] W. Zhou, P. Weldon, B. Tang, and W. M. King, "Rapid motor learning in the translational vestibulo-ocular reflex.," *The Journal of neuroscience: the official journal of the Society for Neuroscience*, vol. 23, no. 10, pp. 4288–98, May 2003.

[7] A. L. Wong and M. Shelhamer, "Saccade adaptation improves in response to a gradually introduced stimulus perturbation.," *Neuroscience letters*, vol. 500, no. 3, pp. 207–11, Aug. 2011.

[8] S. Ramat and D. S. Zee, "Ocular motor responses to abrupt interaural head translation in normal humans.," *Journal of neurophysiology*, vol. 90, no. 2, pp. 887–902, Aug. 2003.

[9] S. Cheng and P. N. Sabes, "Modeling sensorimotor learning with linear dynamical systems.," *Neural computation*, vol. 18, no. 4, pp. 760–93, Apr. 2006.

[10] E. Zarahn, G. D. Weston, J. Liang, P. Mazzoni, and J. W. Krakauer, "Explaining savings for visuomotor adaptation: linear time-invariant state-space models are not sufficient.," *Journal of neurophysiology*, vol. 100, no. 5, pp. 2537–48, Nov. 2008.

Author: Paolo Colagiorgio
Institute: University of Pavia
Street: Via Ferrata 1
City: Pavia
Country: Italy
Email: paolo.colagiorgio@unipv

Numerical Modeling of Optical Radiation Propagation in a Realistic Model of Adult Human Head

N. Ortega-Quijano, F. Fanjul-Vélez, I. Salas-García, and J.L. Arce-Diego

Applied Optical Techniques Group, Electronics Technology, Systems and Automation Engineering Department,
University of Cantabria, Avenida de los Castros S/N, 39005 Santander, Cantabria, Spain
ortegan@unican.es, arcedj@unican.es

Abstract—In this work, optical propagation in human head is modeled by the Monte Carlo method in combination with a realistic high-resolution three-dimensional adult human head mesh which discriminates different brain layers. The method is applied to the analysis of optical radiation distribution in the brain using different types of sources. The results show the high spatial selectivity that can be achieved by optically irradiating human head, which entails a high potential for illuminating specific brain regions. The presented approach can be applied for predictive purposes in the emerging field of optical brain stimulation.

Keywords—Optical propagation, biological tissues, human brain, Monte Carlo, laser sources.

I. INTRODUCTION

Population ageing is a global trend in developed countries. It is estimated that the number of people older than 65 years will rise by 70% by 2050 in the EU [1]. Neurodegenerative diseases are strongly related to ageing, and constitute a major cause of dementia and personal mobility loss. There are multiple causes and factors that can provoke neurodegenerative diseases, but most of them entail neuronal synapses weakening and neurotransmitters malfunctions. It is well-known that nervous tissue stimulation can be achieved by several type of energy sources, which lead to electrical, magnetical, chemical, thermical, mechanical and optical stimulation methods [2]. The positive effects gene-rated by stimulation processes in nervous tissue, enables to use them as a therapeutic method.

This work is focused on the predictive analysis of light propagation in human head for optical brain stimulation. Infrared Neural Stimulation (INS) has emerged during the last years as a brand new technique which is achieving very promising results, with a high clinical potential [2,3]. We present the application of the Monte Carlo method to solve optical propagation in human head with different optical sources. We show several results that enable to observe the main characteristics of this type of stimulation. The analysis performed in this work constitute the first step towards the development of comprehensive predictive models that could enable to determine the optical radiation distribution in the brain, in order to appropriately control radiation parameters for enhancing and optimizing the stimulation process.

II. MONTE CARLO METHOD

Monte Carlo method is a widely used approach for modeling light propagation in biological tissues. It constitutes a stochastic method for statistically solving photon propagation in samples where scattering dominates over absorption (i.e. turbid media). It is computationally more efficient than the FDTD method at optical frequencies, and its accuracy and versatility makes it a widely used method in biomedical optics [4,5,6]. The method relies on the determination of several probability distributions based on a comprehensive model of light propagation in turbid media [4]. Due to the statistical nature of the method, a large number of photons is required to ensure the accuracy of the results. The flow diagram of this method is included in Fig. 1.

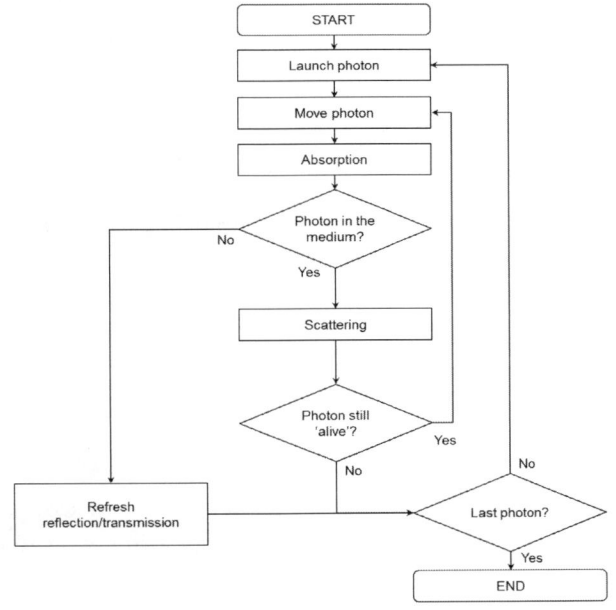

Fig. 1 Flow diagram of the Monte Carlo method.

The first step of the method is launching a packet of photons, which is initially given a weight $W = 1$, a position $\vec{r}$, and a propagation direction $\hat{n} = \begin{pmatrix} n_x & n_y & n_z \end{pmatrix}$. Then, the photon displacement Δs is calculated for each step by:

$$\Delta s = \frac{-\ln(\xi_1)}{\mu_e}, \qquad (1)$$

where μ_e is the extinction coefficient of the medium [6], and $\xi_1 \in (0,1]$ is a pseudo-random number. The new position of the photon packet is obtained directly from the displacement and the unitary propagation vector:

$$\vec{r}\,' = \vec{r} + \Delta s \cdot \hat{n}. \qquad (2)$$

The effect of absorption in each displacement step is modeled by the following reduction in the packet weight:

$$\Delta W = \frac{\mu_a}{\mu_e} W. \qquad (3)$$

The statistical sampling implemented in the displacement calculation implies that, if the photon continues in the medium, a scattering event takes place. The essential aspect in this step is to calculate the polar angle θ and the azimuthal angle ϕ of the scattered photon packet, that modify its propagation direction as a result of the elastic electromagnetic interaction of the photon packet with the scattering particle. The probability distribution function can be calculated by means of the Mie theory. However, in the Monte Carlo method it is usually modeled by the Henyey-Greenstein phase function:

$$p(\cos\theta) = \frac{1-g^2}{2(1+g^2-2\cos\theta)^{3/2}}, \qquad (4)$$

where g is the scattering anisotropy [6]. Therefore, the polar scattering angle is given by:

$$\theta = \begin{cases} \text{acos}\left\{ \dfrac{1}{2g}\left[1+g^2 - \left(\dfrac{1-g^2}{1-g+2g\xi_2}\right)^2 \right] \right\}, & g \neq 0 \\ \text{acos}(2\xi_2 - 1), & g = 0 \end{cases} \qquad (5)$$

In these equations, ξ_2 is a pseudo-random number with the same characteristics as ξ_1. Additionally, it is assumed that the azimuthal angle has a uniform probability distribution:

$$\phi = 2\pi\xi_3. \qquad (6)$$

According to both angles, the new direction vector of the photon packet is refreshed using the following equations:

$$n_x' = n_x \cos\theta + \frac{(n_x n_z \cos\phi - n_y \sin\phi)}{(1-n_z^2)} \sin\theta, \qquad (7)$$

$$n_y' = n_y \cos\theta + \frac{(n_y n_z \cos\phi + n_x \sin\phi)}{(1-n_z^2)} \sin\theta, \qquad (8)$$

$$n_z' = n_z \cos\theta - (1-n_z^2)\cos\phi \sin\theta. \qquad (9)$$

These steps are repeated till the photon packet is absorbed (according to the minimum weight threshold fixed a priori) or transmitted/backscattered. The process is performed for a high number of photon packets (usually around 10^6). It should be noted that the presence of anisotropy or multiple layers require additional equations for modeling such effects [6].

III. APPLICATION OF THE MONTE CARLO METHOD FOR OPTICAL PROPAGATION IN HUMAN HEAD

In the previous section, the generic Monte Carlo method has been described. In this section we apply a particular implementation to the study of optical propagation in human head. Specifically, we use the open-source Mesh-based Monte Carlo method recently developed by Fang [7]. The method uses a three-dimensional realistic adult head mesh publicly available (namely Colin27 adult brain atlas FEM mesh Version 2) [7].

Table 1 Optical properties of the brain layers at a wavelength of 632.8 nm

Layer	Refractive index	Absorption coefficient [mm^{-1}]	Scattering coefficient [mm^{-1}]	Scattering anisotropy
Skin and skull	1.37	0.019	7.8	0.89
Cerebrospinal fluid	1.37	0.004	0.009	0.89
Gray matter	1.37	0.02	9.0	0.89
White matter	1.37	0.08	40.9	0.84

Firstly, we consider a single optical collimated source with a wavelength of 632.8 nm (corresponding to a HeNe laser). Optical properties of the different head layers (skull, cerebrospinal fluid, grey matter and white matter) are specified for the source wavelength [7], and have been included in Table 1. The source is located above the upper rear part of the head, as shown in Fig. 2.

Numerical Modeling of Optical Radiation Propagation in a Realistic Model of Adult Human Head

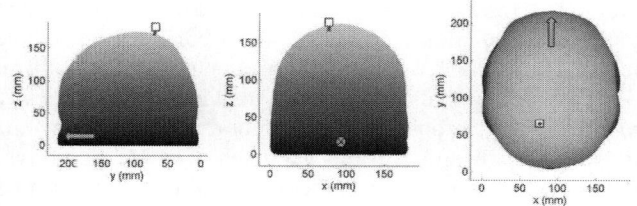

Fig. 2 Position of the HeNe laser used in the first Monte Carlo simulation. Arrows indicate the line along which the subject directs his sight.

The photons distribution calculated by the Mesh-based Monte Carlo method for a single optical source is depicted in Fig. 3, which represents the number of photons in each volumetric element of the cerebral cortex.

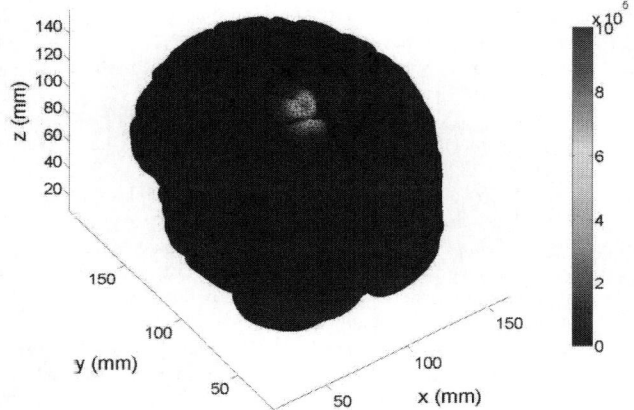

Fig. 3 Photons distribution in the cerebral cortex.

In order to better appreciate the characteristics of the photons distribution in the brain, the sagital plane at x=75.7 mm and the coronal plane at y=67 mm are shown in Figure 4. The different brain layers have also been included, where gray corresponds to the skin and skull, purple is associated with the cerebrospinal fluid, green represents gray matter, and blue is white matter. Firstly, it can be appreciated that optical illumination shows a high degree of spatial specificity, which enables to confine photons in a very narrow region of the head. Moreover, the results show that, although photons density rapidly decreases with depth, optical radiation is actually able to reach the cerebral cortex.

Next, we present a comparison of the radiation distribution obtained using different sources. Two different types of illumination at 632.8 nm are considered: a single collimated

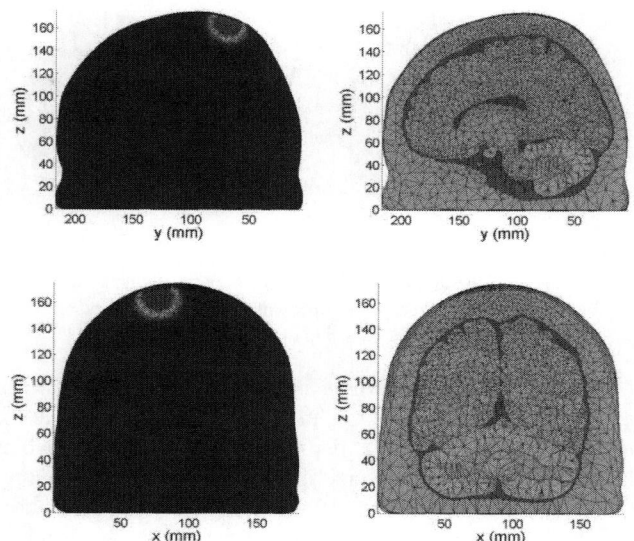

Fig. 4 Optical radiation distribution for the sagital plane at x=75.7 mm (upper left) and for the coronal plane at y=67 mm (bottom left). At right, brain layers for the corresponding planes have been depicted.

source located at (x, y, z) = (0, 155, 115) mm, and five collimated sources at the same wavelength located at a height of z = 115 mm and equally spaced 15 mm between each other from y = 125 mm to y = 185 mm (Fig. 5).

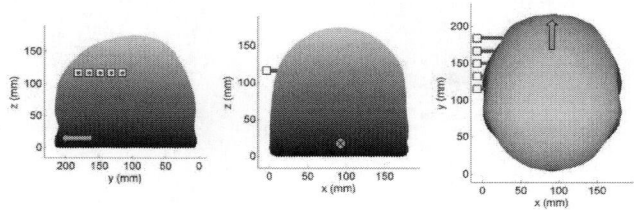

Fig. 5 Position of the five HeNe lasers. Arrows indicate the line along which the subject directs his sight.

The aim of using multiple light sources is to achieve the stimulation of a broader brain region. The optical radiation distribution in the cerebral cortex for the two options considered is shown in Fig. 6. This Figure enables to compare the difference between both illumination techniques. In fact, it can be appreciated that the region reached by optical radiation in the cortex region has been significantly broadened, achieving a reasonably uniform illumination over the region of interest.

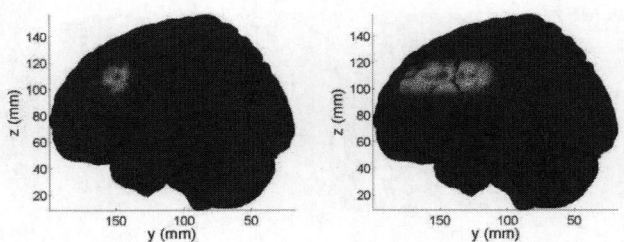

Fig. 6 Photons distribution in the cerebral cortex for illumination with a single optical source (left) and an array of five optical sources with an inter-source spacing of 15 mm (right).

The photons distribution in the transversal plane for z=115 mm (the position at which laser beams illuminate the modeled patient's head) is depicted in Fig. 7, which enables to better appreciate the versatility of using multiple sources for optical brain stimulation. These results illustrate how different illumination patterns can be achieved by irradiating with an array of optical sources, which can be very useful in practical situations.

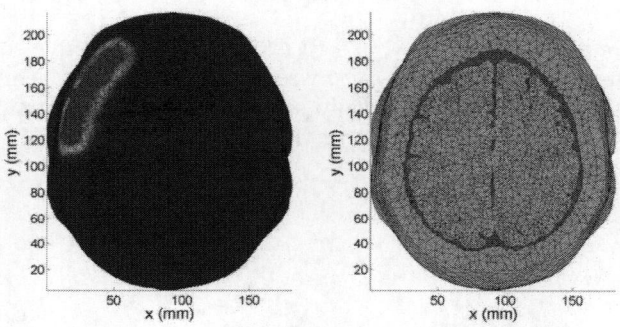

Fig. 7 Spatial distribution of the number of photons (left), and different layers of the head (right) for the transversal plane z = 115 mm.

IV. CONCLUSIONS

In this work, the Mesh-based Monte Carlo method has been applied to a three-dimensional realistic adult head mesh in order to analyze optical propagation in human brain. The results show that this method enables to predict the optical radiation distribution in different regions of the brain. The characteristics of the presented results confirm the potential of the Monte Carlo method for realistically modeling light propagation in cerebral tissue. Among the numerous applications of this approach, we would highlight the study of optical techniques for stimulating specific functional regions in the cerebral cortex.

ACKNOWLEDGMENT

This work has been partially supported by the San Cándido Foundation.

REFERENCES

1. European Technology Platform Photonics 21 (2010) Photonics 21 Strategic Research Agenda.
2. Wells J, Kao C, Mariappan K et al (2005) Optical stimulation of neural tissue in vivo, Optics Letters 30:504–506.
3. Groppa S et al (2012) A practical guide to diagnostic transcranial magnetic stimulation: Report of an IFCN committee, Clinical Neurophysiology 123:858–882.
4. Wang LV, Wu W (2007) Biomedical Optics. Principles and Imaging. Wiley, Hoboken.
5. Ortega-Quijano N, Fanjul-Vélez F, de Cos-Pérez J, Arce-Diego JL (2011) Analysis of the depolarizing properties of normal and adenomatous polyps in colon mucosa for the early diagnosis of precancerous lesions, Optics Communications 284:4852–4856.
6. Wang LH, Jacques SL, Zheng LQ (1995) Monte Carlo modeling of light transport in multi-layered tissues, Computer Methods and Programs in Biomedicine 47:131–146.
7. Fang Q (2010) Mesh-based Monte Carlo method using fast raytracing in Plucker coordinates, Biomed. Opt. Express 1:165–175.

Author: J. L. Arce-Diego
Institute: Applied Optical Techniques Group, Electronics Technology, Systems and Automation Engineering Department, University of Cantabria
Street: Avenida de los Castros S/N, 39005, Cantabria, Spain
City: Santander (Cantabria)
Country: Spain
Email: luis.arce@unican.es

Individual Noise Responses and Task Performance and Their Mutual Relationships

T. Niioka[1,2] and S. Ohnuki[2,3]

[1] Hokkaido University, Faculty of Environmental Earth Science, Sapporo 060-0810, Japan
[2] Hokkaido University, Graduate School of Environmental Science, Sapporo 060-0810, Japan
[3] National Institute of Technology and Evaluation, Tokyo 151-0066, Japan

Abstract—A computer-based modified Stroop color-word task was used as a cognitive task for 4 min. Priming effects were also examined. Alterations in cerebral oxygenated, deoxygenated, and total hemoglobin concentrations during two repetitive executions of the task, without noise and with a moderate/high level noise respectively, were measured in addition to the evaluation of cardiovascular-system response and personality traits. The results obtained in this study imply that individual differences in alterations in task performance under noise exposure are based on those in alterations by the noise condition in the cerebral hemodynamic and cardiovascular-system responses, which are considered to relate to personality traits.

Keywords—Cognitive task performance, individual difference, noise effect, near-infrared spectroscopy, cardiovascular-system response.

I. INTRODUCTION

Cerebral hemodynamic response is related to information processing in the brain [1-8]. We have so far reported relative changes in cerebral blood volume and oxygenation during a cognitive task by non-invasive measurements using a near-infrared spectroscopy system [9-12]. In the present study, we have investigated effects of noise on cognitive-task performance, hemoglobin concentrations in the prefrontal association cortex, and cardiovascular-system response, focusing on their individual differences and mutual relationships.

II. SUBJECTS AND METHODS

A. Subjects

Eighteen healthy, young, male volunteers with a mean age of 24.5 years participated in this study. All the subjects were right-handed normotensive nonsmokers. Informed consent was obtained from all the subjects before the participation in the experiment. Subjects were instructed to refrain from taking alcohol and caffeine the night before the experiment and on the day of the experiment. Subjects were encouraged to do their best in the task to earn money prizes awarded to them according to the number of correct answers.

B. Modified Stroop Color-Word Task

A five-color, computer-controlled version of a modified Stroop color-word task we developed was used as a cognitive task, and was described elsewhere (Patented, JP P4765059)[10]. Briefly, a subject is instructed to select one of five colored disks presented simultaneously with one color word on a computer screen, according to the instruction (i.e., "color" or "word"). The color of the presented word on the screen is discordant with the meaning of the word. The order of appearance of the instructions and color words is randomized. Subjects must overcome cognitive interference to respond properly. The original Stroop color-word task [13] was shown to actually activate the prefrontal cortex in a recent functional magnetic resonance imaging study [14].

C. Noise

As for a moderate/high level noise, intermittent pink noise with a sound level of about 84 dB (A), rise and decay times of 0.6 s, a repetition time of 10 s, and a duty ratio of about 0.5 was employed.

D. Cerebral Hemodynamic Response

Cerebral blood volume and oxygenation were measured using a tissue oximeter (HEO-200, OMRON, Tokyo, Japan), which provides relative changes in concentrations of oxygenated hemoglobin [Δoxy-Hb], deoxygenated hemoglobin [Δdeoxy-H], and total hemoglobin [Δtotal-Hb = Δoxy-Hb + Δdeoxy-Hb] based on continuous-wave near-infrared spectroscopy (NIRS). Optodes of the NIRS device with a 4-cm S-D distance were positioned high on the left side of the frontal region of a subject's head.

E. Cardiovascular-System Response

Cardiovascular-system response was evaluated based on systolic blood pressure (SBP) and diastolic blood pressure (DBP) as well as heart rate (HR).

F. Personality Traits

Scales of Maudsley Personality Inventory (MPI), Stress Self Rating Scale (SSRS), Manifest Anxiety Scale (MAS), and State-Trait Anxiety Inventory (STAI) were used for the evaluation of personality traits of the subjects.

G. Experimental Procedure

Two repetitive task-performing sessions (the first and the second) were administered as an experimental procedure. After a resting period, subjects participated in a 4-min performing session—the first session. After a 15-min break following the first task, subjects participated again in the second performing session—the second session. In the second session, subjects were exposed to the noise or not according to a randomized order of experiment participation. Each session consisted of the baseline (the last 2-min average of the resting period), the anticipation (2-min average), and the performance (4-min average). Alteration of the 4-min averaged value during the performing session from the baseline, in each cerebral hemoglobin concentration were analyzed in this study.

H. Data Reduction and Analysis

As for alterations in measurement values, the subtraction of an experimental value in the pre from that in the post was defined as "change". The subtraction of an experimental value in the "change" in the without-noise condition from that in the "change" in the with-noise condition was defined as "increased change".

III. RESULTS AND DISCUSSION

Deterioration by the noise in the task performance was obvious (Fig. 1 (a)). By contrast, there was no significant deterioration in simple action, which does not require color-disk selection time and needs only arm-movement time (Fig. 1 (b)).

On the other hand, there were no significant alterations by the noise in terms of mean values for the cerebral hemodynamic and cardiovascular-system responses.

Significant correlations were observed among several indexes of task performance, cerebral hemodynamic response, cardiovascular-system response, and personality trait. Hence multiple regression analysis was performed to analyze the data further.

Multiple regression analysis has revealed that a combination of personality traits, such as SSRS-PF (problem focused) and MPI-N (neuroticism), and a combination of alterations by the noise exposure in [Δdeoxy-Hb] and SBP can partially account for alterations by the noise exposure in reaction time (adjusted $R^2 = 0.369$, $P<0.05$, not shown) and a negative priming reaction time (adjusted $R^2 = 0.578$, $P<0.01$, Fig. 2), respectively, for instance. In addition, it was found that SSRS-PF accounted partially for the alteration by the noise in [Δdeoxy-Hb] (adjusted $R^2 = 0.312$, $P<0.01$).

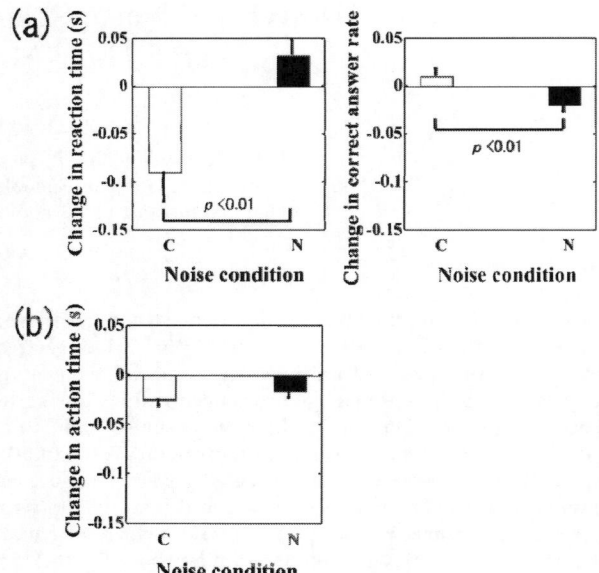

Fig. 1 Effects of noise on Stroop task performance (a) and simple action time without color-disk selection (b). C, calm; N, 84 dB (A) noise; n = 18; bar, S.E.M.

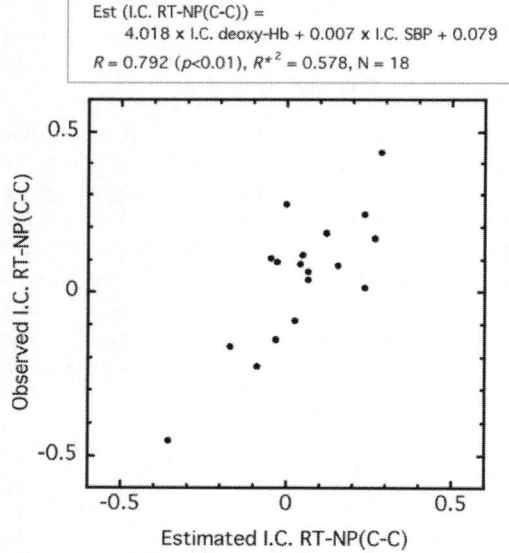

Fig. 2 Scatter plot of the multiple regression analysis on Stroop task performance. I.C. RT-NP (C-C), increased change in negative-priming reaction time to "color" to "color"; I.C. SBP, increased change in systolic blood pressure; R*2, adjusted coefficient of multiple determination.

The results obtained would indicate that cerebral hemodynamic response and cardiovascular-system response are influenced by noise in individuals and these changes cause alterations in Stroop task performance and so on, resulting in significant correlations among them. In addition, it is suggested that personality traits would play a key role in alterations in noise responses and in forming mutual relationships.

IV. CONCLUSIONS

The intermittent 84 dB (A) noise employed in the present study as a moderate/high level noise caused deterioration in the performance of a modified Stroop color-word task. It is suggested that a combination of personality traits as well as a combination of alterations by the noise exposure in cerebral hemodynamic and cardiovascular-system responses would account for those by the noise exposure in reaction time. Furthermore personality traits would also account for alteration by the noise in hemodynamic response. These analytical results obtained in the present study imply that individual differences in alterations in task performance under the noise condition are based on those in alterations by noise exposure in the cerebral hemodynamic and cardiovascular-system responses, which are considered to relate to personality traits.

ACKNOWLEDGMENT

This work was supported in part by Grant-in-Aid for Scientific Research (22570225).

REFERENCES

1. Jöbsis FF (1977) Noninvasive, infrared monitoring of cerebral and myocardial oxygen sufficiency and circulatory parameters. Science 198:1264-1267
2. Wyatt JS, Cope M, Delpy DT et al. (1986) Quantification of cerebral oxygenation and haemodynamics in sick newborn infants by near infrared spectrophotometry. Lancet ii:1063-1066
3. Chance B, Smith DS, Nioka S et al. (1989) Photon migration in muscle and brain. In: Chance B (eds.) Photon Migration in Tissues, pp. 121–135. Plenum, New York
4. Hoshi Y, Tamura M (1993) Detection of dynamic changes in cerebral oxygenation coupled to neuronal function during mental work in man. Neurosci Lett. 150:5–8
5. Kato T, Kamei A, Takashima S et al. (1993) Human visual cortical function during photic stimulation monitoring by means of nearinfrared spectroscopy. J Cereb Blood Flow Metab 13:516–520
6. Villringer A, Planck J, Hock C et al. (1993) Near infrared spectroscopy (NIRS): a new tool to study hemodynamic changes during activation of brain function in human adults. Neurosci Lett 154:101-104
7. Niioka T, Chance B (1997) Relative changes in optical path length of near infrared light during a mental task. In: Chance B, Tamura M, Zuo H et al. (eds.) Non-invasive optical diagnosis: basic science and its clinical application, pp. 3-6. Magazine Office of China-Japan Friendship Hospital, Beijing
8. Koizumi H, Yamamoto T, Maki A et al. (2003) Optical topography: practical problems and new applications. Appl Opt 42:3054-3062
9. Shiraiwa M, Hasegawa K, Niioka T (2000) Changes in the blood volume and oxygenation in the brain during a mental arithmetic task and their relationships to cardiovascular responses. Ther Res 21:1516-1519
10. Niioka T, Sasaki M (2003) Individual cerebral hemodynamic response to caffeine was related to performance on a newly developed Stroop color-word task. Opt Rev 10:607-608
11. Niioka T, Sasaki M (2004) Changes in cerebral hemodynamics during a cognitive task after caffeine ingestion. In: Nakagawa M, Hirata K, Koga Y et al. (eds.) Frontiers in Human Brain Topography, pp. 299-301. Elsevier B.V., Amsterdam
12. Niioka T, Ohnuki S, Miyazaki Y (2010) Individual differences in blood volume and oxygenation in the brain during a cognitive task based on time-resolved spectroscopic measurements. In: Takahashi E, Bruley DF (eds.) Advances in Experimental Medicine and Biology, pp. 251-255. Springer, New York
13. Stroop JR (1935) Studies of interference in serial verbal reactions. J Exp Psychol 18:643-662
14. Adleman NE, Menon V, Blasey CM et al. (2002) A developmental fMRI study of the Stroop Color-Word task. NeuroImage 16:61-75

Author: Dr. Tadashi Niioka
Institute: Hokkaido University, Faculty of Environmental Earth Science
Street: Kita10 Nishi 5, Kita-ku
City: Sapporo 060-0810
Country: Japan
Email: tniiw28@beige.plala.or.jp

Is the Software Package Embedding Dynamic Causal Modeling Robust?

P. Tayaranian Hosseini[1], S. Wang[2], S.L. Bell[1], J. Brinton[1], and D.M. Simpson[1]

[1] Institute of Sound and Vibration Research, University of Southampton, Southampton, UK
[2] Suzhou Institute of Biomedical Engineering and Technology, Chinese Academy of Sciences, Suzhou, China

Abstract—In recent years, Dynamic Causal Modeling (DCM) has become a very popular method in the field of effective connectivity measurement of the brain and it is being used widely for different purposes. The only software available that has embedded this algorithm is Statistical Parametric Mapping (SPM) which is a MATLAB-based toolbox. In this paper, we compare the results obtained from data analyses using different versions of the software. The dispersion of results suggest that researchers should approach the SPM toolbox with caution as the results of the analysis of the same set of data and the same defined model may vary depending on the version of SPM or MATLAB or even the operating system used.

Keywords—Dynamic Causal Modeling, Statistical Parametric Mapping, Effective Connectivity.

I. INTRODUCTION

In recent years, the idea of functional and effective connectivity measurement of the brain has attracted much attention. Our brain is an interconnected network. The interaction between brain regions is fundamental to brain function and this interaction can be quantified by brain connectivity measurement methods. Brain connectivity usually refers to three basic terms: structural, functional, and effective connectivity [1]. Structural connectivity describes the anatomical structure of the brain and the pattern of anatomical links inside the brain [1]. Functional and effective connectivity, on the other hand relate to signals passing between brain subsections. While functional connectivity refers to statistical dependencies of neural activities between distinct areas of the brain, effective connectivity measures the causality of brain interactions [1]. Different techniques have been introduced to measure both the functional and the effective connectivity [2, 3] of the brain using EEG signals, one of which is Dynamic Causal Modeling (DCM) [4]. DCM measures the influence that one area of the brain exerts over itself or another area. DCM has some advantages over other conventional connectivity methods, such as being a biophysically informed model, accounting for nonlinearities of the neuronal system, and having a known input to the model. Because of this, many researchers have become interested in this approach. The validity of this method has been reported for different sets of brain responses such as visual and auditory responses [3] and it has been used in many studies for varied purposes, such as understanding the neural interactions in psychological disorders [5] or the vegetative state [6]. Whilst DCM has potential advantages over other models, a possible weakness in the approach is the large number of parameters and initial assumptions which may cause instability in the algorithm.

The DCM algorithm has been implemented in the Statistical Parametric Mapping (SPM) software which is a MATLAB-based toolbox [7]. This software is used in this paper to model the effective connectivity of the brain using DCM in response to auditory stimulation.

In preparing to use this approach and the SPM toolbox, we tested the algorithm on different platforms and different versions of the toolbox and of Matlab. The current paper reports on inconsistent results obtained in this way and aims to alert to potentially misleading results arising with the use of this package. To the best of our knowledge, results of similar comparisons do not appear to have been published previously.

II. METHODS

DCM first considers a neuronal mass model for brain regions of interest [8]. Then, using the output response measured by EEG and some prior biological knowledge about model parameters and assuming an action-potential-like input to the system, DCM estimates the parameters so that the model fits the output. For this purpose, a Bayesian framework is employed in which the prior is the distribution of the parameters and the posterior is the probability distribution of the measured output. The prior hyperparameters are set according to knowledge about the architecture and behavioral characteristics of the brains neural networks [8]. The remaining parameters are identified iteratively by minimizing the free energy of the system, which can be regarded as the estimation error, using an expectation-maximization (EM) algorithm. In each iteration, first a posterior distribution is calculated according to the minimum free energy (E-step) and then a new set of parameters are computed according to the updated posterior distribution (M-step). The EM procedure iterates the E and M steps untill the decrease of the free energy stops. The important point is that various models (defined as different patterns of connectivity between brain regions) can be defined and compared according to the log-evidence (the probability of the output given a specific model) calculated from the model and the best fitted model is selected as the one with the highest log-evidence [8]. This best model is considered significantly better than other models if its log-evidence is at least 3 units larger than the log-evidence of other models.

III. MATERIALS

The DCM method was tested on data collected during auditory stimulation of normal-hearing subjects, in experiments that aimed to elucidate changes in brain connectivity during pure-tone and speech stimulation. Only results from the former protocol are reported here.

A. Stimulus Characteristics

Two pure tones were presented binaurally at 55dBHL approximately every 2 seconds at random intervals. The tones were 80ms-long 1kH (120 times) and 2 kHz (480 times) tones with 5ms-long rise and fall times.

B. Data Acquisition

The University of Southampton ethical approval was received for this study and the participant consented to take part in the experiment. A 66-channel EEG cap with equidistant electrode positions was placed over the head of a 30-year-old normal hearing male subject. The reference of the system was the nose tip and the ground electrode was placed on the line passing the nose tip and the brain vertex just above the forehead. The subject listened to the randomly played tones with eyes closed while sitting in a comfortable armchair.

C. EEG Pre-processing

EEG was filtered in the 0-30Hz band and epoched around the onset of the 2kHz stimulus (200ms pre- and 500ms post-stimulus). Data were visually checked and showed the expected evoked potential. Fifteen different sets of data were generated by randomly selecting 240 epochs from the 2kHz stimulus and averaging them. Electrode positions were co-registered to the template MRI map available in the SPM8 toolbox with nasion, right, and left auricular points being the fiducial points which were defined manually.

D. DCM Quantification

The GUI interface of SPM DCM was used and different models as in Figure 1 were defined with all forward, back-

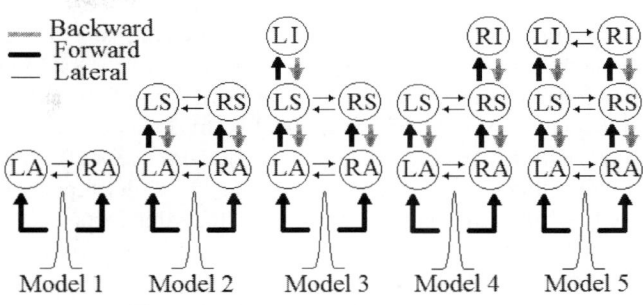

Fig. 1 Five different models used with DCM

ward, and lateral connections present [3]. In these models, Left and Right Primary Auditory Cortices (LA and RA), Left and Right Superior Temporal Gyri (LS and RS), and Left and Right Inferior Frontal Gyri (LI and RI) were used. These areas are shown to be related to sound perception in the brain [3]. Positions of these areas were taken from [9] and RI was assigned a symmetrical position to LI with respect to the sagittal line. Furthermore, the input of the system was defined to occur around 40ms after the stimulus onset and affect both LA and RA. The distributed spatial model was set to Equivalent Current Dipole (ECD) and other parameters of DCM GUI were left to default values.

DCM estimates model parameters and reports a probability value for each parameter being greater than zero. Parameters with probabilities higher than a set value can be considered responsible for the difference between the observed output and the baseline condition which is assumed zero output in this report. Here, the probability level is set to 90%.

E. Software Systems

Parameter estimations were performed using versions 4667 and 5236 of SPM8 (sv.4667 and sv.5236), two versions of MATLAB 64-bit (mv.2011a and mv.2012a), and three Personal Computers (PC) two with Windows 7 64-bit and one with Red Hat Enterprise Linux 64-bit as their Operating System (OS). Note that sv.4667 is older than sv.5236. Note that DCM GUI default values were the same in the two versions of SPM8 used.

IV. RESULTS

A. Reproducibility of DCM

Fifteen generated datasets were used to test the reproducibility of DCM in one subject. Models 1 to 5 of Figure 1 were defined and estimated for each dataset. In all except 2 datasets, model 5 was significantly better than other models.

The fact that 13 out of 15 datasets reported the same model as the best model speaks for the reproducibility of DCM but looking closer at estimated connection strengths of this model (model 5) showed that for each dataset, different connections were held responsible for the output. In the process of investigating this fact further, unexpectedly, some discrepancies were found in the estimation results of the same model when a new PC was used. This event motivated us to inspect the reliability of SPM before going any further with investigating the reproducibility of DCM.

B. Reliability of SPM

a) Data from the current research: To test the reliability of SPM, models 2 and 5 of Figure 1 were selected. The estimation of these two models were performed using two versions of SPM8 and MATLAB as explained in section III.E for one of the generated datasets. For each model, the results show very different connectivity patterns for this dataset when the version of SPM or MATLAB changes. This discrepancy was observed in both model 2 and model 5. Figure 2 shows examples of these estimates. To help interpretation of the results, only the connections with probabilities higher than 90% are plotted in this figure. For example in Figure 2.A.i, both lateral connections between LA and RA are responsible connections whereas in Figure 2.A.ii only the connection from RA to LA seems responsible and in Figure 2.A.iii no connection between RA and LA is reported responsible. As another example, Figure 2.B.i shows that the input enters both primary auditory cortices but Figure 2.B.ii presents that the input enters RA only. On the other hand, Figure 2.B.iii shows that for the same model, no input is responsible for the observed evoked response. It should be emphasised that all the pre-processing steps taken and all parameters entered the estimation algorithm for different versions of SPM and MATLAB, were the same. It should also be noted that when the same model was estimated more than once in the same combination of software versions, the results were found to be identical.

In another test, the versions of SPM and MATLAB were kept unchanged but the analyses were run on two different PCs with the same OS (Windows 7 64-bit). In this case the responsible connections did not vary for either of the two models (models 2 and 5). However, when the OS of one of the PCs changed from Windows 7 to Linux, even with the same SPM and MATLAB version, different results were obtained. The implementation of this condition for model 2 is presented in Figure 3.

It is worth mentioning that the log-evidence of the estimated models did not vary in a consistent way across software versions. For example, keeping the combination

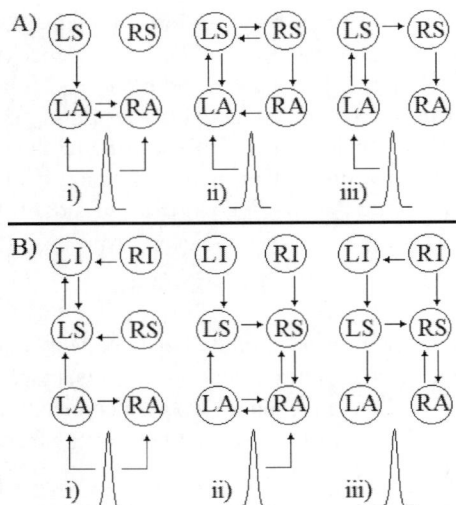

Fig. 2 Responsible connections in generating the evoked response in A) model 2 and B) model 5 for the same set of data. i) mv.2012a & sv.4667, ii) mv.2012a & sv.5236, iii) mv.2011a & sv.5236. The pulse acts as the input to the model.

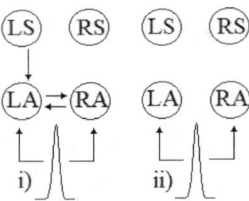

Fig. 3 Responsible connections in generating the evoked response in model 2. In both i and ii, sv.4667 and mv.2011a were used for the same set of data but OS was different: i) Windows 7 and ii) Linux. The pulse acts as the input to the model.

of mv.2011a and Windows 7 unchanged, the log-evidence of model 2 obtained from sv.5236 was significantly larger (greater than 3 units larger - see section II) than the one obtained from sv.4667 but the log-evidence for model 5 in sv.5236 was significantly smaller than the one in sv.4667. So, it cannot be said that the new changes to the SPM version have resulted in higher log-evidence as sometimes newer versions of SPM produce lower log-evidence values.

b) Data from the SPM website: Similar analyses were performed on a publicly available EEG dataset in SPM website [7]. This dataset is called the mismatch negativity (MMN) dataset and the EEG is recorded in response to similar stimuli as were used in the current research. More information about this dataset can be obtained from [3]. The data was already pre-processed so no further pre-processing was applied to the data. Model 4 of Figure 1 was used to initialize the DCM algorithm. All interactions were assumed connected in this model except for lateral connections between

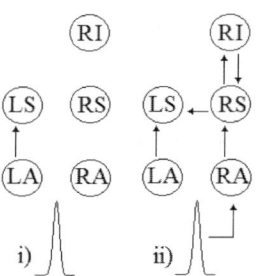

Fig. 4 Responsible connections in generating the evoked response in model 4 for MMN dataset available in SPM website. In both i and ii, sv.4667 and Windows 7 were used but the MATLAB version was different: i) mv.2011a and ii) mv.2012a. The pulse acts as the model input.

RA and LA to be consistent with [9]. The input pulse was defined to occur around 60ms after the onset. Other initialization steps were the same as described in section III.D except that the baseline condition was the averaged 1kHz response. To analyze the data with DCM, the SPM version and the OS were kept the same but the MATLAB version was changed. Once again, different responsible connections were obtained for the same model and the same set of data. The results of these analyses can be seen in Figure 4.

V. DISCUSSION

Possible reasons behind the variation in results could include different precision of the MATLAB version or the OS or different numbers of computation loops or slightly different estimation algorithms in various versions of SPM.

Another possible explanation for the differing results might be that different versions of the software use different random initialization values, but the same values are used in each repeat when using the same software setup. While this would be surprising, it does raise questions regarding the robustness of the algorithm. Furthermore, no mention appears to have been made in the documentation that random steps are used in the algorithm or could cause different results.

The current results clearly do not prove that connectivity measures derived from DCM are always unreliable. However, one example showing clear evidence of a lack of robustness raises the possibility of misleading results. To probe this a little further, a second example was tested, using the data made available by the developers of the toolbox. This also indicated large inconsistencies when different versions of the software were used.

It is not yet known what generates this variability in the results but whatever the reason is, caution should be employed in the interpretation of results of DCM using the SPM toolbox.

VI. CONCLUSION

It is shown for the first time in this paper that the results of estimating DCM parameters using SPM toolbox can vary greatly depending on the version of MATLAB or SPM, or the OS being used. This was observed in auditory evoked potentials and also test data provided on the SPM website: the responsible connections of DCM estimation algorithm may differ considerably if the version of MATLAB or SPM, or the operating system changes. It is thus suggested that the SPM toolbox should be used cautiously when implementing DCM.

ACKNOWLEDGEMENTS

This research is funded by the University of Southampton and its Auditory Implant Service.

REFERENCES

1. Daunizeau J., David O., Stephan K.E.. Dynamic causal modelling: A critical review of the biophysical and statistical foundations *NeuroImage.* 2011;58:312 - 322.
2. Sakkalis V.. Review of advanced techniques for the estimation of brain connectivity measured with EEG/MEG *Computers in Biology and Medicine.* 2011;41:1110 - 1117. ¡ce:title¿Special Issue on Techniques for Measuring Brain Connectivity¡/ce:title¿.
3. David O., Kiebel S.J., Harrison L.M., Mattout J., Kilner J.M., Friston K.J.. Dynamic causal modeling of evoked responses in EEG and MEG *NeuroImage.* 2006;30:1255 - 1272.
4. Friston K.J., Harrison L., Penny W.. Dynamic causal modelling *NeuroImage.* 2003;19:1273 - 1302.
5. Dima D., Roiser J.P., Dietrich D.E., et al. Understanding why patients with schizophrenia do not perceive the hollow-mask illusion using dynamic causal modelling *NeuroImage.* 2009;46:1180 - 1186.
6. Boly M., Garrido M.I., Gosseries O., et al. Preserved Feedforward But Impaired Top-Down Processes in the Vegetative State *Science.* 2011;332:858-862.
7. Statistical Parametric Mapping at http://www.fil.ion.ucl.ac.uk/spm/. Last accessed 24/04/2013
8. David O., Harrison L., Friston K.J.. Modelling event-related responses in the brain *NeuroImage.* 2005;25:756 - 770.
9. Garrido M.I., Kilner J.M., Kiebel S.J., Friston K.J.. Evoked brain responses are generated by feedback loops *Proceedings of the National Academy of Sciences.* 2007;104:20961-20966.

Author: Pegah Tayaranian Hosseini
Institute: ISVR, University of Southampton
Street: University Road
City: Southampton
Country: UK
Email: pth1v10@soton.ac.uk

The Focuses of Pathology of Electrical Activity of Brain in Assessment of Origin of Syncope

K. Peczalski[1,2], T. Palko[1], D. Wojciechowski[2,4], W. Jernajczyk[3], N. Golnik[1], and Z. Dunajski[1]

[1] Warsaw University of Technology, Warsaw
[2] Stopczyk Centre of Cardiology, Wolski, Hospital, Warsaw
[3] Institute of Psychiatry and Neurology, Warsaw
[4] IBBE PAS, Warsaw

Abstract—The origin and aetiology of some syncope attacks are still an unsolved issue. The hypothesis of focuses of origin of the syncope is presented in the paper. The results of our syncope study utilizing orthostatic and drug provocation showed: 1. EEG abnormalities prior to the syncope in the left occipital area are observed significantly more frequently than in the right occipital area, 2. Statistically significant differences of the blood flow and hemoglobin concentration in the brain present before and during syncope. The focuses of pathology of electrical activity in the brain common for patients with syncope history and healthy volunteers were found in our study. All published studies up to date were limited by the provoked mechanism of the syncope induction in contrast to the spontaneous syncope. We propose modeling the focuses of syncope origin for clinical verification with high resolution magneto encephalography. High resolution recording of the area of potential focus or focuses of patients with syncope history versus healthy volunteers are proposed as the first stage of model verification. The next step should consist of recording of the area of potential focus or focuses in patients undergoing provocation.

Keywords—EEG, syncope.

I. INTRODUCTION

The knowledge of symptoms and signs of patients with transient loss of consciousness (TLOC) is crucial for reliable clinical diagnosis of patients with syncope. The syncope, defined as a TLOC, is caused by a cerebral hypoperfusion [1]. The three main types of syncope are distinguished: cardiac syncope, neurally mediated syncope and syncope due to an autonomic failure (orthostatic hypotension).

The cardiac syncope can be evaluated by an invasive electrophysiological study. The head up tilt test (HUT) with the beat to beat measurements of heart rate and blood pressure of patients provides sufficient diagnosis of neurally mediated syncope and syncope caused by autonomic failure. For diagnostic purposes the standard classification of vasovagal reactions including cardiodepressive reaction (i.e. heart rate decrease) and vasodepressive reaction (i.e. arterial pressure decrease) consists of three types: I. mixed, II. cardiodepressive, III. vasodepressive.

The syncopes that are neurally mediated and caused by an autonomic failure develop hypotension and thus cerebral hypoperfusion. Therefore in the presented study we extended the commonly used set of measurements to include the electrical activity of brain examination [2]. Statistically significant differences of blood flow and hemoglobin concentration in the brain vessels during syncope induced by HUT were reported in our further study [3] in patients with induced lack of consciousness vs. patients with negative test results. The blood flow of the brain was assessed by an ultrasonic, transcranial doppler (TD) and the hemoglobin concentration changes (oxygenated and deoxygenated fractions) by near infrared spectroscopy (NIRS). Similar results were reported by other researchers. Our studies [4, 5] of electrical activity of the brain during HUT proved the appearance of pathology of EEG within tens of seconds (30 – 60) prior to a syncope attack. In our studies we did not determine the exact interval between observation of the pathology of EEG and the onset of syncope due to the lack of the strict criterion of onset. The time recognized as a start of syncope attack varies depending on the physician performing HUT.

The studies proved that the pathology of the electrical activity of brain is observed prior to the syncope attack in all patients with the positive passive HUT and in most patients with the pharmacological provocation in HUT. In other studies [5] we found the characteristic locations of the pathology of electrical activity of the brain and proposed the optimal lead system for the early detection of syncope in HUT.

II. OBJECTIVES

1. The localization of focuses of pathology of the electrical activity of brain in patients with a syncope history and healthy volunteers.
2. The model of focuses of the syncope origin.
3. The algorithm for clinical verification.

III. LOCALIZATION OF FOCUSES OF PATHOLOGY OF ELECTRICAL ACTIVITY OF BRAIN

The 43 patients (26 females, 17 males, medium age 36.23 +/- 21.30 years) randomly chosen from group of 103 patients (67 females, 36 males, medium age 37.73 +/- 20.67 years) with history of a syncope underwent the EEG examination during passive HUT. The passive test (without glycerin trinitrate provocation) was performed according to the protocol described in chapter 2. During the entire procedure the beat to beat of heart rate, blood pressure and electrical brain activity were recorded. The standard 10 - 20 configuration of 19 poles lead system and 16 bipolar channels of EEG were used for the electrical brain activity examination. The passive HUT performed as described above was also applied to 30 healthy volunteers (1 female, 29 males, medium age 33,46 +/- 6.24 years) in very good physical condition. The volunteers became the reference group. The volunteers examination was done with the permission given by Ethical Committee of Regional Medical Chamber of Warsaw dated April 03 2003.

The pathology of the the brain electrical activity observed in all of 43 patients was expressed by the burst of delta (1-3.9 Hz) or/and theta (4-7Hz) medium and high amplitude waves. The mean duration of pathology of EEG was 46.71 +/- 20.58 sec. The general pathology was found in 26 patients. The focal pathology was found in 17 patients. The general pathology switched into the focal pathology in one patient. The chain of changes: focal to general to focal to general to focal was observed in one patient. The EEG abnormalities in the left occipital area were found in 9 patients and in the right occipital or right temporo-occipitalis areas in 2 patients. Pathology of the brain electrical activity was observed more frequently in the left temporo-occipitalis area (Chi square test; $p \leq 0.05$). The EEG abnormalities in the left and right occipital areas were found in 6 patients. The left side domination was found in one patient and the domination of right side in another patient. In four remaining patients no domination was expressed. The similar results were presented in our previous publications and by other authors [6, 7].

The 2 of 30 volunteers developed syncope and the results of the passive test of 28 volunteers were negative. The pathology of the brain electrical activity was observed in 2 volunteers. The general pathology was found in a 34 years old volunteer. The focal pathology in the left and right occipital areas was found in a 40 years old volunteer. The pathology of the brain electrical activity in both volunteers was expressed by the burst of delta (1-3.9 Hz) or/and theta (4-7Hz) medium and high amplitude waves as in the group of above described patients.

The following lead systems were analyzed:

- system I; 16 leads (19 poles in 10-20 configuration): I: Fp2 - F8, II: F8 - T4, III: T4 - T6, IV: T6 - O2, V: Fp1 - F7, VI: F7 - T3, VII: T3 -T5,VIII: T5 - O1, IX: Fp2 - F4, X: F4 - C4, XI: C4 - P4, XII: P4 - O2, XIII: Fp1 - F3, XIV: F3 - C3, XV: C3 - P3, XVI: P3 - O1.
- system II; 10 leads: III: T4 - T6, IV: T6 - O2, V: Fp1 - F7, VI: F7 - T3, VII: T3 -T5, VIII: T5 - O1, IX: Fp2 - F4, X: F4 - C4, XI: C4 - P4, XII: P4 - O2, XIII: Fp1 - F3, XIV: F3 - C3,XV: C3 - P3, XVI: P3 - O1.
- system III; 3 leads: VIII: T5 - O1, XII: P4 - O2, XV: C3 -P3,

The clinical sensitivities of the detection of pathology of the brain electrical activity with different leads systems in group of 43 patients are presented in Table 1.

Table 1 Detection of EEG abnormalities.

Leads system	True positives	False negatives	Sensitivity
System I	43	0	1
System II	43	0	1
System III	41	2	0.95

The clinical sensitivities of the localization of pathology of the brain electrical activity with different leads systems in group of 43 patients are presented in Table 2.

Table 2 Localization of EEG abnormalities

Leads system	True positives	False negatives	Sensitivity
System I	43	0	1
System II	43	0	1
System III	40	3	0.93

The numbers of pathological waves of EEG detected in selected bipolar leads are presented in Table 3. The data was collected from the test records of the patients with focal EEG abnormalities. There were no pathological waves in remaining leads: I, II, V, VI, VII, IX, X, XI, XIII, XIV and XV.

Table 3 The distribution of pathology of electrical activity of brain.

Lead	III	IV	VIII	XII	XVI
	T4-T6	T6-O2	T5-O1	P4-O2	P3-O1
Pts with abnormalities of EEG	2	7	11	8	14

The results have showed the stable localization of focal EEG abnormalities in the left and right temporo-occipitalis areas with a significant prevalence of the left side location. The focal pathology of EEG was also observed in parallel in both sides in one patient. The general pathology was observed in 26 of 43 patients. The assessment of the clinical sensitivity of the tested systems of leads confirmed the high clinical effectiveness of all evaluated lead configurations. The clinical sensitivity calculated for combined group of patients and volunteers was different only for system III (0.95 for detection and 0.93 for localization of EEG abnormalities). The results presented in table 5 show that the substitution of lead XV by lead XVI gives the sensitivity of detection and localization for system III equal to one and provides the same sensitivity for two other systems. The data in table 5 also shows the high number of detected pathological waves in lead IV. Noting that only one additional pole/electrode is necessary for applying this lead we highly recommend such correction . In the system III the total number of electrodes needed for the realization of the proposed lead system is six. The system of four bipolar occipitalis leads (T6-O2, T5-O1, P4-O2, P3-O1) also provides the maximal clinical sensitivity for all examined by us patients.

We postulate that the focuses of the origin of pathology of electrical activity of brain are located in the area covered by occipitalis leads. The specific behavior of the electrical activity of brain in the occipital area prior to the syncope seems to be related to disturbances of vision and a neck ache radiating to the occipital region of the skull observed prior to the syncope reported by other authors [8, 9].

IV. THE MODEL OF FOCUSES OF SYNCOPE ORIGIN

The electrical activity in the brain is observed as a distribution of electric potentials on a scalp. The localization of the active sources within the brain based on the magnetography recordings is used as a highly sensitive diagnostic procedure. The observed spots of activity in the brain generate waves which can be recognized and located using signal analysis methods by dedicated programs. A computer program DIPOLE developed at the Institute of Metrology and Biomedical Engineering of Warsaw University of Technology for the mapping and localization of active sources in the brain was utilized. This program is based on the most popular method to locate an active source within the brain involving Levenberg-Marquardt algorithm.

The algorithm was chosen because of its efficiency and accuracy. In order to obtain the high localization accuracy the particular model of the head is tailored to a realistically shaped geometry and the detailed electrical properties of the tissues for each patient. The model used for this study consists of volume elements which can take into account an anisotropy and the inhomogeneities of the tissues. Therefore it can significantly increase the accuracy of the localization of the active sources within the brain. The localization of the active sources of the electrical potential inside the brain are performed using models of the head and of the generator. In the anticipation of further studies the program enables creation of the models with parameters which can be set by an operator for every individual case and which are conformable to a growing knowledge about the electrical properties of the head tissues.

V. THE CLINICAL VERIFICATION

A. The First Step Model Verification

The high resolution magnetoencephalography is performed in prone position. The spectrum of recordings from the areas of potential focus for the patients with a syncope history and healthy volunteers (as a reference group) are analyzed referring to the pathology of the brain electrical activity expressed by the burst of delta (1-3.9 Hz) or/and theta (4-7 Hz) waves.

The precise localization of focuses of pathology origin in both groups is the goal of the first stage of verification.

B. The Second Step Model Verification

The high resolution magnetoencephalography is performed in tilt position. The spectrum of recordings from the areas of the potential focus for the patients with the syncope history are analyzed referring to the pathology of the brain electrical activity expressed by the burst of delta (1-3.9 Hz) or/and theta (4-7Hz) waves.

The dynamic localization of the pathology of electrical activity of brain during the presyncope, syncope and recovery periods is the goal of the second stage of verification.

This stage of verification can be extended or substituted by a pharmacological provocation.

VI. CONCLUSIONS

1. The focus points of origin of the pathology of electrical activity are located in the area covered by occipitalis leads.
2. The proposed model appears to be a valuable tool for the localization of the sources of pathology of electrical activity of brain associated with syncopes.

REFERENCES

1. Wieling W, Thijs R, van Dijk N, Wilde A, Benditt D. (2009) Symptoms and signs of syncope: a review of the link between physiology and clinical clues. Brain: 132; 2630 – 2642.
2. Peczalski K, Wojciechowski D, Sionek P, Dunajski Z, Palko T. (2006). The system for synchronize registration of biological signals during Head up Tilt Test I. IFMBE Proceedings vol. 14, pp. 3561 – 3563.
3. Wojciechowski D, Peczalski K et al. (2002) Value of near infra red spectroscopy in assessment of patients during prolonged, passive standing. Abstract book of 67th ICB Seminar in Laser-Doppler Flowmetry and Near Infrared Spectroscopy in Medical Diagnosis Warsaw, pp. 16-18.
4. Peczalski K, Wojciechowski D, Jernajczyk W. (2002) The role of the brain bioelectrical activity examination in diagnosis of syncopic patients during presyncope or syncope induced by head up tilt test, Structures-Waves- Biomedical Engineering edited by Polish Acoustic Society in Krakow, Vol. XI, pp. 205-212.
5. Peczalski K, Jernajczyk W, Wojciechowski D. (2006) The system for synchronize registration of biological signals during Head up Tilt Test II - module EEG. IFMBE Proceedings vol. 14, pp. 963 – 966.
6. Abe H. Kohshi K, Kuroiwa K. (1998) Possible involvement of cerebral hypoperfusion as a trigger of neurally-mediated vasovagal syncope. Pacing and Clinical Electrophysiology, 21, 3, 613-616.
7. Ammirati F, Colivicchi F, Di Battista G, Garelli FF, Pandozi C, Santini M. (1998) Variable cerebral dysfunction during tilt induced vasovagal syncope. Pacing and Clinical Electrophysiology, 21, 11, II, 2420-2425.
8. C.J. Mathias, R. Mallipeddi, K. Bleasdale-Bar. (1999) Symptoms associated with orthostatic hypotension in pure autonomic failure and multiple system atrophy. Journal of Neurology, 246: 893-898.
9. R.W. Ross Russel, G. R. Page. (1983) Critical perfusion of brain and retina. Brain;106: 419-434.

Author: Kazimierz Pęczalski
Institute: Institute of Metrology and Biomedical Engineering Warsaw University of Technology
Street: Boboli 8
City: Warsaw
Country: Poland
Email: k.peczalski@mchtr.pw.edu.pl

Neuroanatomic-Based Detection Algorithm for Automatic Labeling of Brain Structures in Brain Injury

M. Luna[1], F. Gayá[1], A. García-Molina[4], L.M. González[1], C. Cáceres[1], M. Bernabeu[4], T. Roig[4], A. Pascual-Leone[3], J.M. Tormos[4], and E.J. Gómez[1]

[1] Bioengineering and Telemedicine Centre, ETSI Telemunicación, Universidad Politécnica de Madrid, Madrid, Spain
[2] Biomedical Research Networking Center in Bioengineering, Biomaterials and Nanomedicine (CIBER-BBN), Spain
[3] Berenson-Allen Center for Noninvasive Brain Stimulation, Beth Israel Deaconess Medical Center, Harvard Medical School, Boston, MA, USA
[4] Institut Guttmann, Institut Universitari de Neurorehabilitaci adscrit a la UAB (Universitat Autnoma de Barcelona), Spain

Abstract—The number and grade of injured neuroanatomic structures and the type of injury determine the degree of impairment after a brain injury event and the recovery options of the patient. However, the body of knowledge and clinical intervention guides are basically focused on functional disorder and they still do not take into account the location of injuries. The prognostic value of location information is not known in detail either. This paper proposes a feature-based detection algorithm, named Neuroanatomic-Based Detection Algorithm (NBDA), based on SURF (Speeded Up Robust Feature) to label anatomical brain structures on cortical and sub-cortical areas. The main goal is to register injured neuroanatomic structures to generate a database containing patient's structural impairment profile. This kind of information permits to establish a relation with functional disorders and the prognostic evolution during neurorehabilitation procedures.

Keywords—Neuroimaging, Descriptors, Landmarks, Magnetic Resonance Imaging (MRI), Neuroanatomic Structures.

I. INTRODUCTION

Brain injury can be defined as an acute event that causes damage to certain areas of the brain [1]. It may result in a significant impairment of an individual's physical, cognitive and psychosocial functioning. Traumatic brain injury (TBI), stroke (ischemic or hemorrhagic) and brain tumors are the main causes of brain injury. Brain injury is the most common cause of neurological disability accompanied by a long life expectancy. The cost of disability resulting from the sequelae of that pathology is high, including medical costs, lost salaries and low productivity. Neurorehabilitation aims to reduce the impact of disabling conditions, trying to improve the deficits caused by brain injury in order to reduce the functional limitations and increase the individual's ability to function in everyday life.

Neuroimaging is considered to be a promise for improving and personalizing medical care by providing objective information regarding patient's evolution and/or prognosis [2]. One of the main challenges of neuroimaging in brain injury is to develop robust automated image analysis methods to detect neuroanatomic features allowing the development of incremental databases to link such features with clinical outcome and functional impact of rehabilitation interventions.

Information related to anatomical structures, more precisely to injured structures, is contained on local intensity changes. Feature-based detection algorithms detect not only intensity changes but also store spatial information. The most known feature-based detection algorithms, also known as descriptors, are 'Scale Invariant Feature Transform' (SIFT) [3] and 'Speeded Up Robust Features' (SURF) [4].

The ultimate goal of this research is the creation of an image bank with labeled neurological injuries to extract knowledge of neurorehabilitation therapies. The automatic detection of landmarks to label anatomical brain structures is essential. In a previous research work recently submitted, SIFT and SURF algorithms were compared and evaluated. The results show that both algorithms obtain landmarks around the skull so only anatomical brain structures can be properly identified. However, SURF obtains better values of stability, efficiency and sample's representation. Consequently, this paper proposes an algorithm, named Neuroanatomic-Based Detection Algorithm (NBDA), based on the original SURF algorithm. The purpose of this algorithm is to label anatomical brain structures but not to divide the image into different tissues. High variability of brain morphology is one of the principal difficulties on neuroimaging processing; this variability tends to increase when any type of brain injury occurs. Therefore, it is necessary to validate algorithms on healthy subjects previously to use them on neuroanatomic imaging studies.

The remainder of this paper is organized as follows: section 2 describes the proposed algorithm and the data set used to validate it, section 3 presents the obtained results and finally, section 4 explains the conclusions and future works.

II. MATERIAL AND METHODS

As previously mentioned, the proposed algorithm in this paper is based on SURF algorithm [4]. SURF and NBDA algorithms take as input the cumulative distribution of image intensity values, also known as integral image. This imaging representation is directly related to the decrease of processing time in relation to other similar algorithms.

The main phases of a descriptor algorithm are: location of points of interest, orientation assignment and descriptor generation. At the first stage, the aim is to detect points featuring special characteristics, blobs in this case. A blob can be defined as the cross point where at least six direction gradient lines match [5]. Filters used to find them are structured in a pyramidal way, known as scale-space. At orientation assignment stage, maximum gradient direction of each landmark is obtained. Finally, information relative to location, orientation and gradient values is stored in a matrix, also known as descriptor.

Main differences and improvements between NBDA and original SURF to be applied on neuroanatomic images are described in the next methodological sections

A. Location of Points of Interest

Regarding blob detection, there are many algorithms aiming to detect these kind of imaging features such as Laplacian of Gaussian [6] or the Hessian-Laplace detector [7]. As in original SURF the determinant of the Hessian matrix is used to find blobs. Box filters have been used to approximate the Hessian matrix in the Cartesian coordinate x, y and xy direction.

In order to make our algorithm independent from local contrast changes, these filters are divided by the standard deviation (SD) of pixel values affected by them as in Eq. 1, where F is the filter used to detect blobs, FS is the size of the filter, SD is standard deviation and Hij is one of the three box filters (Hxx, Hyy and Hxy), used to obtain the Hessian matrix. If SD is zero, imaging points presenting very low contrast value are not considered as detected landmarks.

$$F(x,y,FS) = \begin{cases} \forall SD(x,y,FS) > 0 & \frac{H_{i,j}(x,y,FS)}{SD(x,y,FS)} \\ \forall SD(x,y,FS) = 0 & 0 \end{cases} \quad (1)$$

NBDA takes into account only intensity values of pixels affected by each filter and makes intensity dispersion independent regarding contrast. In order to calculate standard deviation in a fast way, an integral square representation of the image is used. It is obtained by using boxes and the same equation is used to calculate the integral image.

A pyramidal approximation to a Gaussian second order partial derivative representation based on box filters in x, y and xy direction is used to generate this scale-space. The scale-space is analyzed by up-scaling the filter size. It is divided into octaves and scales representing a series of response maps obtained by applying a convolution between the integral image and box filters of increasing size.

Finally, detected landmarks are obtained from the maximum of the determinant of the Hessian matrix by taking into account the size of each filter. Consequently, local intensity changes become independent from global ones. Original SURF considers the parameter w, which is the correction factor between the Gaussian kernels and the approximated Gaussian kernels, constant. Nevertheless, this paper proposes obtaining w, as in Eq. 2. It has been analytically obtained. Thus, the determinant of the Hessian matrix is obtained as in Eq. 3. The maximum of the determinant of the Hessian matrix is obtained by interpolating scale and octave.

$$w = \frac{119}{90} - 2 \cdot \Pi \cdot \left(\frac{2+LS}{LS^2}\right) \quad (2)$$

$$det(H) = H_{xx} \cdot H_{yy} - (w \cdot H_{xy})^2 \quad (3)$$

B. Orientation Assignment

For the purpose of calculating the orientation of each detected landmark, a Haar wavelet is obtained through x and y direction on a circular neighborhood of 6 times scale value where the landmarks were detected, around each detected landmark. The Haar wavelet is obtained by approximating box filters. Then, wavelet responses are weighted with a Gaussian centered at the detected landmark. The dominant orientation is estimated by calculating the sum of all responses within a sliding orientated window of size Π/3.

C. Descriptor Generation

In order to generate the descriptor, a square region is centered on each detected landmark. These regions are split up into smaller 4x4 square sub-regions. On each sub-region approximate Haar wavelets are computed at 5x5 spaced sample points. Obtained responses are weighted with a Gaussian to increase the robustness towards geometric deformations and are summed up to form a first set of entries in the matrix, named descriptor. Spatial information and intensity-related information is saved on the descriptor. Our descriptor contains spatial location and intensity information, owing to all this information will be useful to identify anatomical brain structures.

D. Data Acquisition and Image Processing Tools

Structural MRI data were obtained from a group of 42 healthy subjects. 21 women, age range 19-30, mean age 21.6 years, and 21 men, age range 17-28, mean age 20.7 years. Data was obtained using a 3.0 Tesla GE Medical Systems Signa. 42 image studies have been acquired with a TR=6 ms and TE=2 ms.

NBDA algorithm has been developed with Matlab and Windows 7 64 bits. The processor is an Intel Core i5-2430M with 6GB of RAM. A Matlab SURF implementation (available from Mathworks) has been used to compare and validate the proposed algorithm.

III. RESULTS AND DISCUSSION

This section presents the results obtained with NBDA algorithm and compares them with results obtained with SURF algorithm. The evaluation methodology followed has been described on [8]. Table 1 shows the seventeen selected anatomical brain structures. These structures have been selected due to their clinical relevance and several other structures related with them.

Homologous landmarks have been obtained by pairing descriptors obtained on each study, thus 21 pairs of descriptors have been compared. NBDA (1643 (1145-2153) homologous landmarks) obtains a higher average than SURF (1384 (1231-1494) homologous landmarks); this means that NBDA algorithm detects landmarks whose location and intensity values are repeatable among different slices. Repeatability is an important property owing to the fact that the identification of equivalent landmarks on different images permits to locate similar areas, anatomical brain structures in this case.

Regarding stability against imaging changes, the average number of homologous landmarks on original and rotated images has been obtained by applying NBDA and SURF algorithms. Original image has been rotated 2, 5, 10 and 20 degrees. The average number of homologous landmarks decreases when the rotation angle increases. Therefore, the more modified is the image, the less homologous landmarks are detected. However, NBDA obtains the highest average value in all the cases. Original SURF algorithm obtains 6%, 12% and 17% less homologous landmarks than when images are rotated only 2 degrees. NBDA experiments a reduction of 2%, 3% and 4% respect the rotation of 2 degrees.

In relation to sample's representation, Table 2 shows that SURF efficiency is below NBDA value. The proposed algorithm detects landmarks homogeneously on cortical and sub-cortical regions. Structures located around the skull are identified with similar average value of landmarks owing to the fact that NBDA and SURF algorithms obtain similar number of landmarks. However, structures located on sub-cortical regions such as lateral sulcus or superior frontal gyrus present notable differences between SURF and NBDA.

In summary, robust automatic location and identification of neuroanatomic structures is one of the main challenges of neuroimaging. The proposed approach consists of applying feature-based detection algorithms to identify anatomical brain structures. The goal of this paper is to validate the proposed algorithm (NBDA) with a set of healthy imaging studies in order to be applied in the next stage of our research work on a set of brain injury image studies. NBDA is a feature-based detection algorithm, also known as descriptor.

NBDA introduces important changes to obtain landmarks homogeneously distributed throughout the brain region. This algorithm is compared quantitatively and qualitatively. The main goal of this algorithm is to make intensity dispersion independent as regards contrast. Concerning the comparison among descriptor algorithms, NBDA obtains the highest number of landmarks in the processing time. However, the average number of homologous points gives more information about the repeatability of each descriptor algorithm. In relation to the stability of descriptors against imaging changes, the number of homologous landmarks decreases in relation to the rotation angle. In this case, NBDA

Table 1 Selected Brain Structures

Gyrus
Superior frontal gyrus (right and left)
Cingulate gyrus (right and left)
Sulcus
Lateral sulcus (rigth and left)
Parietoccipital sulcus (right and left)
Calcarine sulcus (right and left)
Sinus
Superior sagital sinus
Subcortical structures
Tapetum (right and left)
Frontal Horn (right and left)
Corpus Callosum (genu)
Cave of septum pellucidum
Anterior Horn of right/left lateral ventricle
Foramen of Monro
Third ventricle
Atrium and chroids plexus of lateral ventricle (right and left)
Internal capsule (right and left) (anterior limb)
Thalamus (right and left)
Head of Caudate nucleus (right and left)

Table 2 Descriptors Efficiency per Anatomical Brain Structure

	NBDA	SURF
Superior sagital sinus	11.8	8.2
Cingulate gyrus	14.4	11.9
Tapetum	13.4	10.3
Frontal Horn	61.7	57.7
Corpus Callosum	13.4	9.6
Cave of Septum Pellucidum	13.4	11.3
Anterior horn of lateral ventricle	16.3	14.4
Foramen of Monro	25.6	17.7
Third ventricle	14.7	12.3
Lateral sulcus	5.5	3.4
Atrium and Chroids plexus of lateral ventricle	12.1	11.0
Parietoccipital sulcus	58.0	53.0
Calcarine sulcus	17.2	15.1
Superior sagital sinus	21.2	19.9
Internal capsule (anterior limb)	35.6	31.8
Head of caudate nucleus	23.7	20.5
Thalamus	34.9	20.3

detects more pairs of homologous landmarks than the other two algorithms. In fact, the proposed algorithm detects a similar number of homologous landmarks with the highest value of rotation angle as SURF algorithm with the lowest angle.

Sample's representation per brain structure shows that the number of detected landmarks located around the skull is similar on the three algorithms, whereas NBDA obtains the best score of landmarks around anatomical structures located on sub-cortical regions. Therefore, NBDA is more efficient in terms of anatomical landmarks detection.

The obtained results confirm that NBDA obtains better results than SURF algorithm and it is more robust and efficient at the detection of brain anatomical structures. SURF algorithm detects landmarks which are mostly located around the skull. Therefore, many anatomical structures cannot be properly located and identified. Automatic location and identification of brain structures increase the information relative to neuro-anatomical structures and reduce the time spend by specialists. In future steps, an extension of this algorithm will be applied on imaging volumes to label anatomical structures based on the LPBA40 atlas.

IV. CONCLUSIONS

One of the main challenges of neuroimaging is to develop robust automated image analysis methods to detect brain injury features. The approach described in this paper consists of implementing a new feature-based detection algorithm, named NBDA. This algorithm is based on original SURF algorithm. NBDA introduces modifications to make it usable for neuroimage analysis. The obtained results confirm SURF algorithm detects landmarks which are mainly located close to the skull region. However, anatomical brain structures are not only located around skull, but also on sub-cortical areas. NBDA detects homogeneously landmarks over the brain area. Therefore, NBDA identifies anatomical structures located on cortical and sub-cortical regions.

This algorithm permits to asses the anatomical integrity and the spatial location of neuroanatomic structures, as well as structural disorders. It also permits to generate a database of dysfunctional profile of patients. It is known that the prognostic is not determined by the integrity of a neuroanatomic structure but by the set of injured ones. In order to improve the knowledge of the importance of neuroimaging in the medical care of brain injury is essential to obtain the dysfunctional profile of each patient.

ACKNOWLEDGMENT

This research has been partially founded by the Spanish Ministry of Economy and Finance (project TIN2012-38450, COGNITIO).

REFERENCES

1. S. Laxe U. Tschiesner R. Lpez-Blazquez J.M. Tormos M. Bernabeu. ICF use to identify common problems on a TBI neurorehabilitation unit in Spain *NeuroRehabilitation*. 2011;29:99-110.
2. Stinear C.M., Ward N.S.. How useful is imaging in predicting outcomes in stroke rehabilitation? *International Journal of Stroke*. 2013;8:33–37.
3. Lowe D.G.. Object recognition from local scale-invariant features in Computer Vision, 1999. *The Proceedings of the Seventh IEEE International Conference on*;2:1150-1157 vol.2 1999.
4. Bay H., Ess A., Tuytelaars T., Gool L. Van. Speeded-Up Robust Features (SURF) *Computer Vision and Image Understanding*. 2008;110:346 - 359.
5. Rosenfeld A., Sher Chiao-Yung. Detecting image primitives using feature pyramids *Information Sciences*. 1998;107:127 - 147.
6. Lowe D.G.. Distinctive Image Features from Scale-Invariant Keypoints *International Journal of Computer Vision*. 2004;60:91-110.
7. Mikolajczyk K., Schmid C.. Indexing based on scale invariant interest points in Computer Vision, 2001. ICCV 2001. *Proceedings. Eighth IEEE International Conference on*;1:525 -531 vol.1 2001.
8. Luna M., Gaya F., Caceres C., Tormos J.M., Gomez E. J.. Evaluation Methodology for Descriptors in Neuroimaging Studies in *Proceedings 8th International Conference on Computer Vision Theory and Applications*;2:114-117 2013.

Author: Marta Luna Serrano
Institute: Bioengineering and Telemedicine Centre,
ETSI Telecomunicación, Universidad Politécnica de Madrid
Street: Av. Complutense 30
City: Madrid
Country: Spain
Email: mluna@gbt.tfo.upm.es

Brain Activity Characterization Induced by Alcoholic Addiction: Spectral and Causality Analysis of Brain Areas Related to Control and Reinforcement of Impulsivity

J. Guerrero[1], A. Rosado[1], M. Bataller[1], J.V. Francés[1], T. Iakymchuk[1],
A. Luque-García[2], V. Teruel-Martí[2], and J. Martínez-Ricós[2]

[1] University of Valencia, Digital Signal Processing Group, Valencia, Spain
[2] University of Valencia, Neuronal Circuits Laboratory, Valencia, Spain

Abstract—Addiction to drugs generates modifications in the brain structure and its functions. In this work, an experimental model is described, using rats to characterize the brain activity induced by alcohol addiction. Four records were obtained using electrodes located in brain areas related to impulsivity control and reinforcement, i.e. the prelimbic (PL) and infralimbic (IL) cortex, together with the hippocampus (HPC). In the records, three main events related to the drinking action were selected: in the previous minute (T1), the first minute while drinking (T2) and the first minute after stopping drinking (T3).

In the frequency domain, the normalized energy and coherence were calculated for five bands: delta (2 - 5 Hz), theta (5 - 11 Hz), beta (13 – 30 Hz), slow gamma (35 – 45 Hz) and fast gamma (65 – 90 Hz). Granger causality in the time domain was also computed to obtain the causality relations among different brain areas.

Preliminary results show higher delta activity and higher beta and fast gamma coherence with higher causality values in the limbic channels compared to hippocampus. These results suggest a spectral and synchronization profile among different brain areas, which could lead to obtain parameters able measure the brain activity in case of alcohol addiction.

Keywords—addiction, alcohol, local field potential, coherence, G-causality.

I. INTRODUCTION

Regular psychostimulant drugs use is reported to alter the brain functions and its structure. Modifications occur at the cell and the neural system levels where different neuropsychiatric disorders are caused, together with a strong addiction mechanism [1][2].

Most of the studies related to the alcohol addiction and the alterations produced in the brain electrical response are based on electroencephalography (EEG) registers since it is a non-invasive method with a good temporal resolution [3]. Alternatively, Local Field Potentials (LFP) recordings are obtained by inserting electrodes using stereotaxic surgery. This method allows the measurement of the net current flow for wide neural areas as a function of time, with a high resolution.

This work studies the alterations in the electrical brain response induced by the addiction to alcohol. The study was done using an experimental rat model with liquid diet for alcohol ingestion, combined with abstinence periods.

The main aim in this work consists on the characterization of brain response depending on the analyzed region and their associated modifications related to addiction. In this work, the drinking action event is considered. Two main analyses were done, frequency domain analysis and its modifications, and causality relations. Causality relations show the existence of pairing or control actions between the neural activities of two different brain areas. In this study, spectral coherence is used, together with Granger causality in order to detect not only the causality but also the effective direction of causality [4][5][6].

II. MATERIAL AND METHODS

A. Experimental Procedure

For this experiment, four female Sprague-Dawley rats were used, weighting between 280 and 320 grams. All experimental protocols were performed according to the experimental animals care guidelines of the European Council (86/609/EEC) and approved by the Ethics Committee of the University of Valencia prior to the onset of the experiments.

During a period of 28 days, the animals are provided with a liquid diet containing alcohol as one of the ingredients. Thus, the rats develop an alcoholic addiction.

For data recording, intrabrain electrodes were implanted in the prelimbic and infralimbic cortex, (PL) and (IL) respectively, and the hippocampus (HPC). Special cages adapted to the free movement of rats where used, consisting on semi-spheres with a rotor in the topside to allow the movement without acquisition signal interruption. Simultaneously to signal recording, video recording to obtain rat's activity information is also performed.

Signal data acquisition was sampled at a 200Hz frequency with an approximate duration of 7-8 hours including sleeping period. During the recording period, rats were administered the same diet except alcohol to generate abstinence conditions.

B. Signal Processing

After analysis of the video records, the events where a rat is drinking are selected. Each drinking event was analysed in the previous minute before drinking (T1), the first minute of drinking (T2) and the first minute just after stopping drinking (T3). For each 20 second period of time, a 6 second segment was used for signal analysis and all three acquired channels were used (HPC, PL, IL).

Spectral analysis: To obtain the spectrum and the spectral coherence, the Welch method was used, with two averaged segments and 50% overlapping. The frequency bands were selected to exclude mains electricity frequency and its first harmonic. The frequency bands defined were the following: delta (2-5Hz), theta (5-11Hz), beta (13-30Hz), slow gamma (35-45Hz) and fast gamma (65-90Hz).

The computed parameters for each channel are:

- Spectral energy for each frequency band (ENB#).
- Spectral coherence normalized to the total energy, for each frequency band (CNB#).

Fig.1 shows an example of energy (EN) along recording time for each frequency band and the three acquired channels.

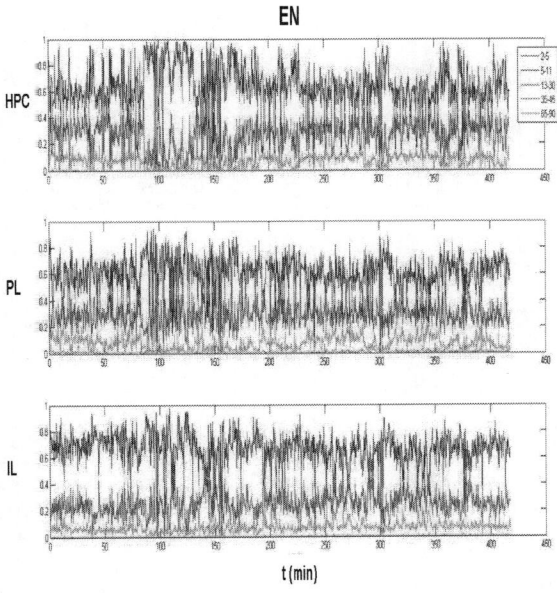

Fig. 1 Spectral normalized energy example for the five bands defined and three channels recorded.

Causality analysis: According to Granger's causality (G-causality) [4], a time series X_2 is 'causal' to X_1 when the addition of previous values in a regressive linear model of X_1 and X_2 reduces the predictions error compared to a model considering X_1 as the only source of data. In the general case of n time series signals $X_1...X_n$, G-causality must be done in pairs. In case of two signal time series, a bivariate autoregressive model is used according to [5]:

$$X_1(t) = \sum_{j=1}^{p} A_{11,j} X_1(t-j) + \sum_{j=1}^{p} A_{12,j} X_2(t-j) + \xi_1(t)$$

$$X_2(t) = \sum_{j=1}^{p} A_{21,j} X_1(t-j) + \sum_{j=1}^{p} A_{22,j} X_2(t-j) + \xi_2(t)$$

being p the order in the model, A the coefficient matrix, and ξ_1, ξ_2 the residues (prediction errors). In case of covariance stationary signals, the value of causality from X_2 to X_1 is given by:

$$C_{2 \rightarrow 1} = \ln \frac{\text{var}(\xi_{1R(12)})}{\text{var}(\xi_{1U})}$$

being $\xi_{1R(12)}$ the residue obtained by the model when the A_{12} is omitted, and ξ_{1U} the residue for the complete model.

Signal processing was done using Matlab® and the GCCA function library [7]. As signal pre-processing, linear tendencies and mean value are removed, and the signal is divided by its standard deviation. Secondly, the covariance stationarity condition is tested by the 'augmented Dickey-Fuller' (ADF) and Kwiatkowski, Phillips, Schmidt, and Shin (KPSS) tests.

In order to obtain the final model, its order is estimated by the Akaike information criterion (AIC). For the coefficients calculation, ordinary-least-squares method is used.

To verify that the obtained model is adequate for signal representation, the Durbin-Watson test was used to check the residues are non-correlated. Finally, the consistency of the model and the sum-square-error were verified.

For each 6-second segment, a multivariate model was used, corresponding to the HPC, PL and IL signals. The Granger's conditional causality in the time domain (CAT_C##) was computed for each case, obtaining the significant causality interactions (F-test with Bonferroni correction) for each pair of signals and each causality direction (C12 is for C1 causing C2, and C21 the other direction).

For all parameters, the Kolmogorov-Smirnov test was used to test the normality condition and the Levène test for variance equality. Average value comparisons were done using ANOVA as parametric method assuming normal distribution of data, and Friedmann test as non-parametric method. In all cases, $p<0.05$ statistical significance was used.

III. RESULTS

For the analysed parameters (EN, CA and CAT) as a function of frequency band and brain area, the following figures show the obtained results.

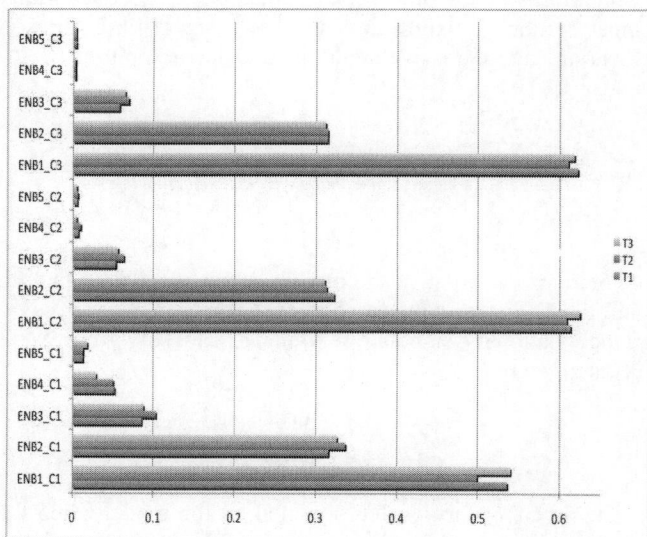

Fig. 2 Mean values of normalized spectral energy for all frequency bands defined (B1-5). Channels: C1: HPC, C2: PL, C3: IL. Segments: T1: previous to drinking, T2: drinking, T3: after drinking.

Fig. 2 shows the mean values of normalized spectral energy for all five bands defined. For a brain area, no differences were found regarding time moments. However, different brain areas provide significant differences for all frequency bands except theta band for each time event. Delta activity is significantly lower ($p<0.01$) in hippocampus (T1: 0.54±0,25; T2: 0.50±0,21; T3: 0.54±0,21) than PL (T1: 0.61±0,17; T2: 0.61±0,16; T1: 0.62±0,13) and IL (T1: 0.62±0,15; T2: 0.61±0,15; T3: 0.62±0,12). Beta activity in hippocampus (T1: 0.09±0,08; T2: 0.10±0,08; T3: 0.09±0,06) is slightly bigger than PL (T1: 0.05±0,03; T2: 0.06±0,03; T3: 0.06±0,04) and IL (T1: 0.06±0,03; T2: 0.07±0,03; T3: 0.06±0,03).

Spectral coherence results among brain areas are shown in Fig. 3. Mean values of spectral coherence are plotted. For the three brain areas, the maximum coherence values are observed in beta band (B3) (T1; HPC-PL: 0.26±0,07; HPC-IL: 0.29±0,07; PL-IL: 0.22±0,06; T2; HPC-PL: 0.28±0,07; HPC-IL: 0.29±0,08; PL-IL: 0.25±0,06; T3: HPC-PL: 0.28±0,07; HPC-IL: 0.27±0,07; PL-IL: 0.24±0,05; $p<0.01$) and fast gamma band (B5) (T1: HPC-PL: 0.35±0,09; HPC-IL: 0.31±0,09; PL-IL: 0.37±0,09; T2: HPC-PL: 0.34±0,10; HPC-IL: 0.31±0,08; PL-IL: 0.33±0,10; T3: HPC-PL: 0.34±0,09; HPC-IL: 0.32±0.08; PL-IL: 0.34±0.08; $p<0.01$). For this parameter, no significant differences were found in case of different time event (T1-T3).

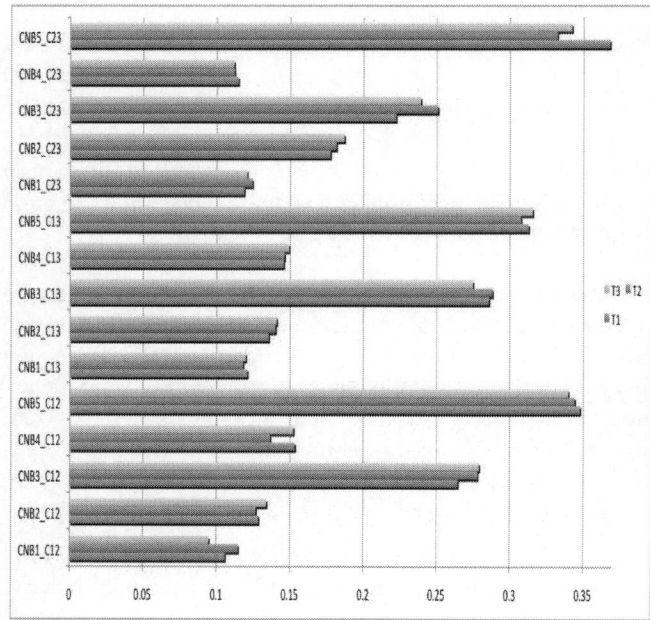

Fig. 3 Normalized spectral coherence mean values for the five frequency bands (B1-B5). Channels: C1: HPC, C2: PL, C3: IL. Segments: T1: previous to drinking, T2: drinking, T3: after drinking.

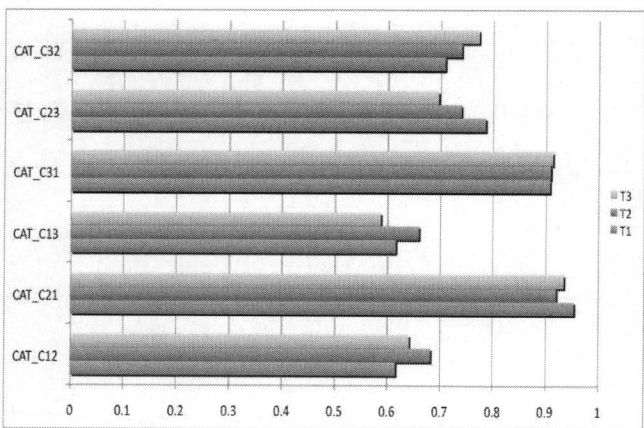

Fig. 4 Granger's causality mean values. The channel order shows causality direction: C12: HPC→PL, C13: HPC→IL, C23: PL→IL (and viceversa). Segments: T1: previous to drinking, T2: drinking, T3: after drinking.

Concerning temporal causality, Fig. 4 shows mean values for obtained results. Despite some bidirectional causality relations are observed, the most relevant is observed from the limbic channels to hippocampus (T1: PL→HPC:

0.95±0,21 vs. HPC→PL: 0.68±0,47; IL→HPC: 0.91±0.29 vs. HPC→IL: 0.62±0,49; T2: PL→HPC: 0.92±0,27 vs. HPC→PL: 0.62±0,49; IL→HPC: 0.91±0.29 vs. HPC→IL: 0.66±0,48; T3: PL→HPC: 0.93±0,25 vs. HPC→PL: 0.64±0,48; IL→HPC: 0.91±0.28 vs. HPC→IL: 0.59±0,50; $p<0.01$) for all temporal events (T1-T3). Mean values of causality between prelimbic and infralimbic areas are similar in both directions.

IV. CONCLUSIONS

In order to characterize neural activity under alcohol addiction, an experimental model for abstinence periods using alcoholic rats is used. Three brain areas are selected for signal recording, which are involved in the impulsivity control and reinforcement. Furthermore, three time events related to addiction are analysed (pre, during and post drinking activity).

An analysis about the frequency domains modifications (normalized spectral energy and coherence) was carried out, together with causality relations for the brain areas by calculating the Granger's temporal causality.

Despite the results are preliminary, higher delta activity is observed, associated to higher beta and fast gamma coherence as well as higher causality values in the limbic areas compared to hippocampus. These results suggest a different spectral and synchronization profile, which could lead to measure some parameters characterizing the brain activity under alcohol addiction. Further studies will complete the signal processing with a causality analysis in the frequency domain together with additional number of cases and situations (sleeping, type of activity, etc.), leading to a more precise characterization.

REFERENCES

1. Lapish C C, Chiang J, Wang J Z, Phillips A G. (2012) Oscillatory power and synchrony in the rat forebrain are altered by a sensitizing regime of D-Amphetamine. Neuroscience 203:108–121
2. Koob G F, Volkow N D. (2009) Neurocircuitry of Addiction. Neuropsychopharmacology 35:217-238.
3. Zhihua D, Ruifang F, Guangyu L, Tian L. (2010) Study on human brain after consuming alcohol based on EEG signal. Proceedings of the 3rd IEEE International Conference on Computer Science and Information Technology (ICCSIT), Chengdu, China, 2010 vol. 5, p. 406-409.
4. Granger C W (1969). Investigating causal relations by econometric models and cross-spectral methods, Econometrica 37:424-438.
5. Ding M, Chen Y, Bressler S L. (2006) Granger causality: Basic theory and application to neuroscience. In Handbook of Time Series Analysis (ed. S. Schelter, M. Winterhalder & J. Timmer), pp. 438-460. Wienheim: Wiley.
6. Seth, A K. (2007) Measuring autonomy via multivariate autoregressive modelling. Proceedings of the 9th European Conference on Artificial Life, Lisbon, Portugal, 2007 pp. 475-484
7. Seth A K. (2010) A MATLAB toolbox for Granger causal connectivity analysis. Journal of Neuroscience Methods 186:262–273

Author: Juan F. Guerrero
Institute: Dpt. Electronic Engineering. School of Engineering. University of Valencia
Street: Avda. Universitat s/n.
City: Burjassot. Valencia.
Country: SPAIN
Email: juan.guerrero@uv.es

A Novel RMS Method for Presenting the Difference between Evoked Responses in MEG/EEG– Theory and Simulation

I. Nemoto and M. Kawakatsu

Department of Information Environment, Tokyo Denki University, Inzai, Japan

Abstract—A novel method was previously proposed by one of the present authors for presenting a mismatch response in multichannel MEG (or EEG) measurements obtained with 'oddball' experiments. This paper briefly summarizes the theoretical analysis showing the advantage of this method and some simulation results with a simple source model in the auditory cortex in the brain model. The simulation results were in good agreement with the analysis.

Keywords—evoked response, mismatch response, MEG, RMS.

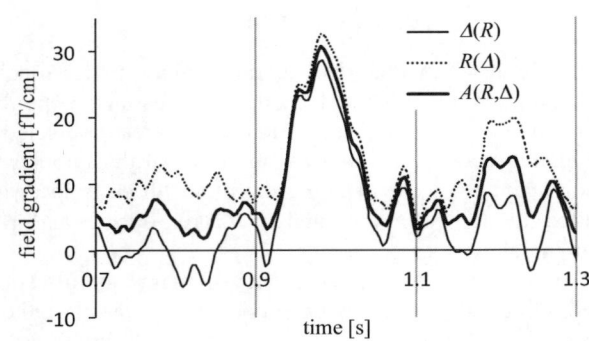

Fig. 1 Comparison of the three representations of the difference waveforms in an MEG 'oddball' experiment using the C major scale. See text.

I. INTRODUCTION

This paper deals with the 'N1 response' and the related 'mismatch response' observed in multichannel MEG (magnetoencephalography) or EEG (electroencephalography). When one hears a brief stimulus repeatedly (call it the 'standard' stimulus), N1 response is measured above the auditory cortex in response to each stimulus with a latency of about 100 ms. If the stimulus changes its feature (loudness, frequency, timber, etc.) occasionally, then the auditory cortex automatically responds to the change and elicits a 'mismatch response' which overlaps the N1 response and peaks around 130 - 200 ms [1]. The infrequently presented stimulus is called a 'deviant' stimulus. Mismatch response has become a widely used tool in investigating neural mechanisms in sensory systems non invasively. N1m and mismatch fields are small compared to the ongoing background processes in the brain and are usually obtained by averaging the measurements during many (e.g., 100) intervals ('epochs') corresponding to the stimuli. All the responses treated below are thus averaged responses.

Let the responses to the standard and deviant stimuli are denoted by the m−dimensional vectors $\mathbf{x}_S$ and $\mathbf{x}_D$, respectively where m is the number of the measurement channels. Let the L_2 norm of $\mathbf{x}$ be written $\|\mathbf{x}\| = \sqrt{\sum_{i=1}^{m} x_i^2}$. Then the RMS of $\mathbf{x} = m^{-1/2}\|\mathbf{x}\|$. In our previous paper, we compared

1. the difference of RMS: $\Delta(R) = m^{-1/2}(\|\mathbf{x}_D\| - \|\mathbf{x}_S\|)$
2. RMS of the difference: $R(\Delta) = m^{-1/2}\|\mathbf{x}_D - \mathbf{x}_S\|$
3. modified RMS: $A(R,\Delta) = (\Delta(R) + R(\Delta))/2$

and proposed the modified RMS, $A(R,\Delta)$ with some theoretical justification. Fig. 1 shows an example from our study of the musical scale with MEG [2]. The standard stimulus was the complete C major scale and the deviant lacked the tone G and the immediately preceding F tone was repeated. The mismatch field appeared at the repetition (0.8 – 1.0 s) of the tone F. $R(\Delta)$ shows the largest mismatch amplitude and positive DC shift outside the response period (before 0.9 s and after 1.1 s). $\Delta(R)$ fluctuates around the zero level outside the response period but there the variance looks large. The proposed $A(R,\Delta)$ seemed to have a smaller variance than the two outside the response period. The purpose of this paper is to give the results of the simulation study that we have conducted to underline our previous work of theoretical treatment.

II. REVIEW OF THE ANALYSIS

We briefly give below the results of the previous analysis [3]. We adopted a very simple model for the responses to the standard and deviant stimuli:

$$\mathbf{x}_S = \mathbf{s} + \mathbf{n}_S, \ \mathbf{x}_D = (1+k)\mathbf{s} + \mathbf{n}_D \quad (1)$$

where $k > 0$ and $\mathbf{n}_S$ and $\mathbf{n}_D$ are the residual noise (after averaging), independent from each other, and both follow the m−dimensional normal distribution with mean 0 and covariance matrix Σ. The noise is denoted simply $\mathbf{n}$ when $\mathbf{n}_S$ and $\mathbf{n}_D$ are not differentiated. We assume $\rho = \|\mathbf{n}\|/\|\mathbf{s}\| \ll 1$ and put

$r^2 = E[\rho^2]$ where E denotes ensemble average. We compared the behaviors of the three RMS representations in two situations, during the mismatch response period (around 1 s in Fig. 1) and outside the period. The mismatch period is modeled by eq. 1 with $k \gg \rho$ and the other situation can be simply modeled by putting $k = 0$. The terms of higher orders than ρ^2 and henceforth than r^2 were neglected. From the property of the norm,

$$\Delta(R) \leq A(R,\Delta) \leq R(\Delta). \quad (2)$$

A. The behaviors of $\Delta(R)$ and $R(\Delta)$

We first obtained for $k \gg \rho$,

$$E[\Delta(R)] \simeq m^{-1/2} k \|\mathbf{s}\| \left[1 - \frac{1}{2(k+1)} \left(r^2 - \frac{{}^t\mathbf{s}\Sigma\mathbf{s}}{\|\mathbf{s}\|^4} \right) \right] \quad (3)$$

$$E[R(\Delta)] = m^{-1/2} k \|\mathbf{s}\| \left[1 + \frac{1}{k^2} \left(r^2 - \frac{{}^t\mathbf{s}\Sigma\mathbf{s}}{\|\mathbf{s}\|^4} \right) \right] \quad (4)$$

Note that RMS of the true mismatch response in the model is $m^{-1/2} k \|\mathbf{s}\|$. From the property of the eigenvalues of Σ, we obtained $E[\Delta(R)] \leq m^{-1/2} k \|\mathbf{s}\| \leq E[R(\Delta)]$ indicating that $R(\Delta)$ is likely to overestimate and $\Delta(R)$ underestimate the RMS of the mismatch response. Next we saw that

$$V[R(\Delta)] = \simeq \frac{2\,{}^t\mathbf{s}\Sigma\mathbf{s}}{m\|\mathbf{x}\|^2} \simeq V[\Delta(R)] \quad (5)$$

Next we treated the case $k = 0$. Because $\mathbf{n}_S$ and $\mathbf{n}_D$ follow the same distribution,

$$E[\Delta(R)] = m^{-1/2}(E\|\mathbf{s} + \mathbf{n}_S\| - E\|\mathbf{s} + \mathbf{n}_D\|) = 0 \quad (6)$$

As it was difficult to obtain $E[R(\Delta)]$ in the general case, we assumed the simplest case where the components of $\mathbf{n}_S$ and $\mathbf{n}_D$ are all independent and have the same variance σ^2. In this case, each component of $\mathbf{n}_S - \mathbf{n}_D$ has variance $2\sigma^2$ and $\|\mathbf{n}_s - \mathbf{n}_D\|$ follows the χ distribution. From this we obtained

$$E[R(\Delta)] = 2m^{-1/2} \sigma g(m) \quad (7)$$

where $g(m) = \Gamma((m+1)/2)/\Gamma(m/2)$ and $\Gamma(\cdot)$ is the gamma function. It was shown that $2g(m)$ can be very closely approximated by $\sqrt{2m-1}$ for reasonably large m (> 6) and then

$$E[R(\Delta)] \simeq \sigma \sqrt{2 - m^{-1}} \quad (8)$$

$$V[\Delta(R)] \simeq \frac{2\,{}^t\mathbf{s}\Sigma\mathbf{s}}{m\|\mathbf{s}\|^2} \quad (9)$$

$V[R(\Delta)]$ was obtained in a similar manner as

$$V[R(\Delta)] \simeq \sigma^2/m. \quad (10)$$

Under the assumption $\Sigma = \sigma^2 I$ where I is the unit matrix,

$$V[R(\Delta)] \simeq \max V[\Delta(R)]/2 \quad (11)$$

follows.

B. The behavior of $A(R,\Delta)$

When $k \gg \rho$, from eqs. (6) and (7)

$$E[A(R,\Delta)] \simeq \frac{k\|\mathbf{s}\|}{\sqrt{m}} \left[1 - \frac{k^2 - 2k - 2}{4k^2(k+1)} \left(r^2 - \frac{{}^t\mathbf{s}\Sigma\mathbf{s}}{\|\mathbf{s}\|^4} \right) \right] \quad (12)$$

To compare the biases of the three RMS estimates, we define

$$\beta(\Delta) = 2^{-1}(k+1)^{-1}, \quad \beta(R) = k^{-2}$$
$$\beta(A) = |k^2 - 2k - 2|(2k)^{-1}(k+1)^{-1}$$

which indicates the dependency of the biases on k as Fig. 2 shows. Note that β's are positive by definition but the actual biases can be negative. When $k = 0$, the bias of $A(R,\Delta)$ is not 0 but half that of $R(\Delta)$.

When $k \gg \rho$, it was shown that

$$V[A(R,\Delta)] \simeq V[\Delta(R)] \simeq V[R(\Delta)] \quad (13)$$

For the case $k = 0$, $\text{cov}(\Delta(R), R(\Delta)) = 0$ can be easily seen and then

$$V[A(R,\Delta)] = (V[R(\Delta)] + V[\Delta(R)])/4. \quad (14)$$

C. Summary

When a mismatch response is reasonably large, $A(R,\Delta)$ has a smaller bias in the estimation of the mismatch response than $\Delta(R)$ and $R(\Delta)$ and a variance of the same degree as the other two. Outside the mismatch response ($k = 0$), $A(R,\Delta)$ has a positive bias but its variance is very likely to be smaller than the average of the variances of the other two estimates.

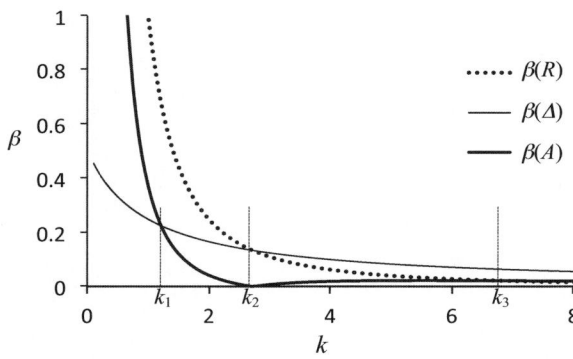

Fig. 2 Values of β determining the bias incurred by the three estimates. $k_1 = (1 + \sqrt{7})/3$, $k_2 = 1 + \sqrt{3}$, $k_3 = 3 + \sqrt{15}$. See text.

III. SIMULATION

A. Methods

Fig. 3 (a) shows the signal source used in the simulation. The thick line stands for the dipole current responding to the standard stimulus and the thin line to the deviant stimulus. The dotted line is the difference between the two signals and represents the mismatch response. The dipoles to the standard and deviant stimuli were both positioned at $(5,0,-1)$ [cm] in the right auditory cortex in the spherical brain model and was directed to $(0,0,-10)$. The deviant response dipole was linearly extended in time and amplitude and peaked at 0.14 s as shown. The dipole had jitter both in amplitude and the reflecting time points given by a normal noise with $\sigma = 0.02$. In (b) the value of k is shown at each time point.

The coil arrangement of the Neuromag 122 magnetometer was used for the forward calculation to obtain the sensor coil output at the 122 positions over the head. Here, we used the 12 sensor coil outputs which are normally the largest responding channels to sound stimuli in the right hemisphere. To each of their outputs, independent random noise following the normal distribution $N(0, \sigma^2)$ was added. The average RMS field (which turned out to be ~ 0.17 (arbitrary unit)) of the 12 outputs during the period 0 - 0.2 s was used as the signal intensity to determine σ to achieve a certain SN ratio.

B. Results

The data were all bandpass filtered between 1-40 Hz as in our usual procedure. Fig. 4(a) shows the result for $S/N = -10$dB. One hundred epochs for each of standard and deviant responses were averaged. In a usual experimental setting, the standard stimulus are presented 4 to 5 times more frequently than the deviant stimulus. Here the number of epochs was set to be common to the both stimuli in accordance with the assumption of the same variance for the two residual noises $\mathbf{n}_S$ and $\mathbf{n}_D$. The averaging improved the S/N to +10dB, achieving $r \sim 1/3.16$ although the S/N calculated for the simulation is not exactly the same as $1/r$ in the previous section. In the figure, (a) shows the standard and deviant responses measured in channel 10 and the difference waveform between the two. (b) shows the three RMS estimates of the mismatch response and the 'true' mismatch field shown by the fine dotted curve. The true mismatch field was calculated using the average data without noise contamination to the channels but with the jitter given to the signal source. Fig. 5 shows the result for $S/N = -20$dB in the raw data. Panels (a) and (b)

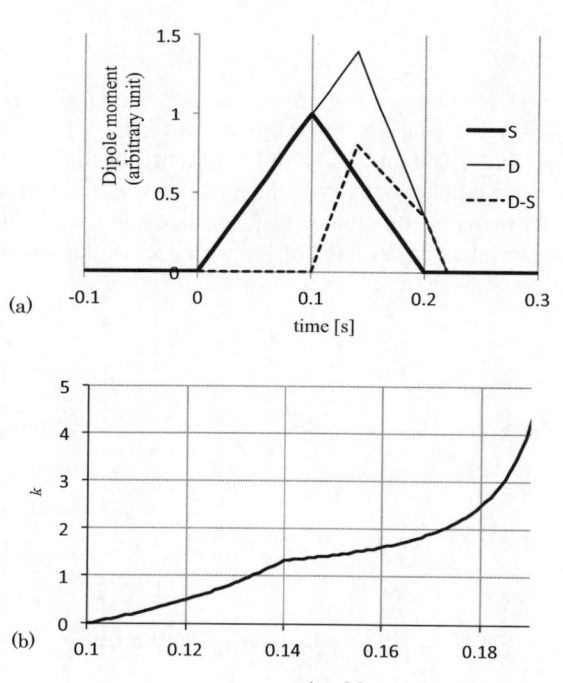

Fig. 3 (a) Source dipole model for the standard response (thick line) and the deviant response (thin line). The difference is shown by the dotted line. (b) The value of k defined in eq. (1) at each time point.

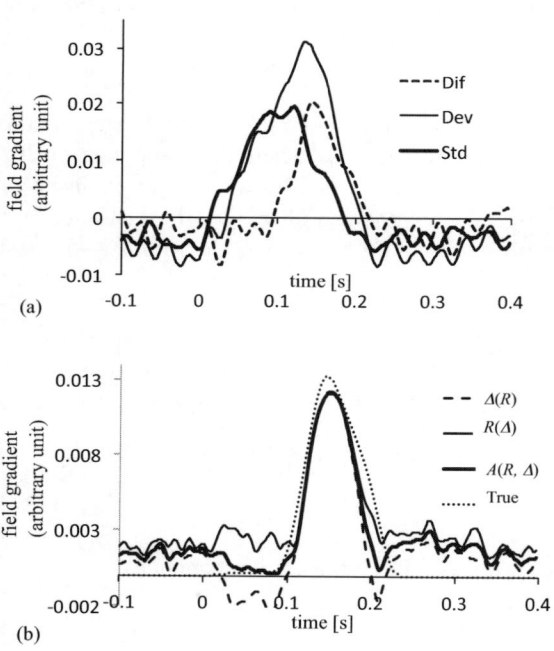

Fig. 4 Simulation result for $S/N = -10$dB in the raw data. (a)The average (over 100 epochs, each) standard and deviant responses and their difference at channel 10 (of Neuromag 122). (b) The three RMS representations of the mismatch response and the 'true' mismatch response.

IV. DISCUSSION AND CONCLUSION

The simulation results showed that the features of the proposed RMS presentation were well predicted by the theoretical analysis. In the case $S/N = -10$ dB, the peak of the mismatch field was estimated at almost the same value by the three methods (Fig.4), but the variance outside the mismatch response showed quite a large difference and $A(R, \delta)$ showed a cleaner baseline than the other two. The discrepancy with the 'true' response was probably due to the variance of the estimates. The case with more noise contamination (Fig. 5) showed further advantage of the proposed method. When the number of epochs for averaging was different for standard and deviant stimuli, the assumption of equal variance of noise was violated but the results were not far from the prediction as shown in (c).

ACKNOWLEDGEMENTS

The present work has been partly supported by Grant-in-Aid from the Ministry of Education, Culture, Sports, Science and Technology in Japan and partly by the Center for Research, Tokyo Denki University.

REFERENCES

1. Näätänen J, Picton T. The N1 wave of the human electric and magnetic response to sound: a review and an analysis of the component structure *Psychophysiol.* 1987;24:375-425.
2. I Nemoto, Evoked magnetoencephalographic responses to omission of a tone in a musical scale *J. Acoust. Soc. Am.* 2012;131:4770–4784.
3. I Nemoto, Discussion and proposal on RMS presentation of mismatch responses in MEG and EEG in *IFMBE Proc.* :37(Budapest, Hungary): 1161–1164, 2011.

Author: Iku Nemoto
Institute: Tokyo Denki University
Street: 2-1200 Muzai-gakuendai
City: Inzai
Country: Japan
Email: nemoto@mail.dendai.ac.jp

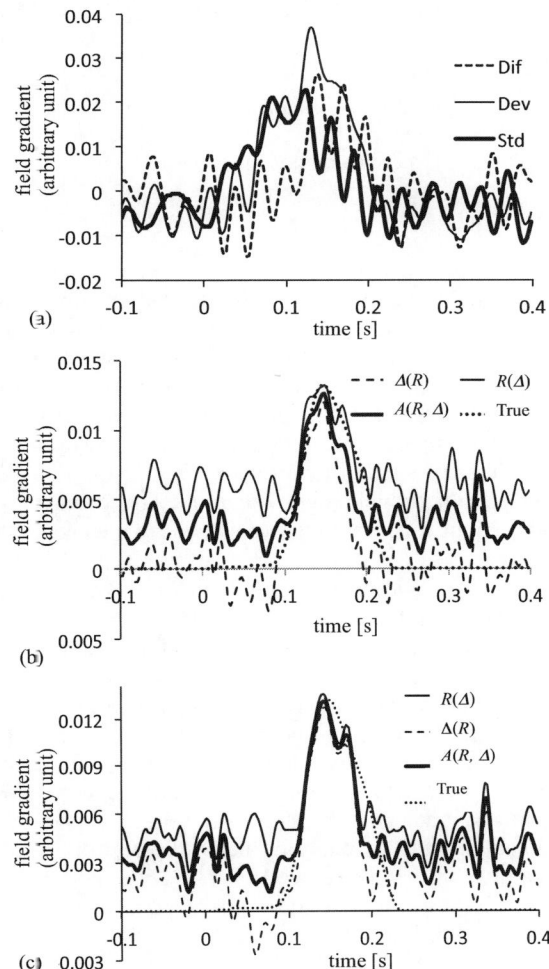

Fig. 5 Simulation result for $S/N = -20$ dB in the raw data. Panels (a) and (b) correspond to those in fig. 4. Panel (c) shows the result when the number of epochs of standard stimulus was increased to 400.

correspond to those in Fig. 4. Panel (c) shows the three presentations when the number of averaged epochs was increased to 400 for the standard stimulus as is usually done in the oddball experiments.

The Influence of Neuronal Density on Network Activity: A Methodological Study

E. Biffi[1], G. Regalia[1], A. Menegon[2], G. Ferrigno[1], and A. Pedrocchi[1]

[1] Politecnico di Milano, Department of Electronics, Information and Bioengineering,
Neuroengineering and Medical Robotics Laboratory, Milan, IT
[2] Advanced Light and Electron Microscopy Bio-Imaging Centre, Experimental Imaging Centre,
San Raffaele Scientific Institute, Milan, IT

Abstract—It is well known that cell density influences the maturation process of *in vitro* neuronal networks. Neuronal cultures plated with different cell densities differ in number of synapses per neuron and thus in single neuron synaptic transmission, which results in a density-dependent neuronal network activity. In this work, we present a methodological study to characterize the electrophysiological activity of neuronal cultures which were plated at different cell density and followed over maturation. We gathered data from 86 independent hippocampal cultures plated at six different neuronal densities, whose electrical activity was sampled by means of MicroElectrode Arrays (MEAs) from 4 to 32 days *in vitro* (DIV). Network activity was evaluated in terms of simple spiking, burst and network burst activities. We evaluated the distribution of the six density populations (ANOVA test) and we observed it was possible to merge our results in three categories (i.e. 900 cells/mm^2-like cultures - sparse, 1800 cells/mm^2-like cultures – medium, and 3600 cells/mm^2-like cultures- dense). We observed that electrical descriptors are characterized by a functional peak during the initial maturation, followed by a stable phase (for sparse and medium density cultures) or by a decrease (for high dense neuronal cultures). These results do constitute an helpful reference/benchmark for the planning of *in vitro* neuropharmacological and neurophysiological experiments with MEAs. Indeed, sparse cultures seem to be suitable for long lasting experiments (e.g. chronic modulatory effect of a drug treatment) while medium cultures are useful for experiments which require intense electrical activity (e.g. when the drug effect requires high firing rates, such as antiepileptics). Finally, cell dense cultures are more suitable for shorter experiments (e.g. rapid screening).

Keywords—micro electrode array, neuronal network, density, network development, electrophysiology.

I. INTRODUCTION

Neuronal signaling *in vitro* is mainly characterized by sequences of action potentials, whose amplitude and morphology depends on several factors [1]. Moreover, the characteristics of electrical activity of neuronal cultures change over time during neuronal network differentiation and maturation. Previous works reported the existence of a relationship between the age of the culture and its functionality, expressed as spiking activity [2-6]. Furthermore, it is already known that there is also a relationship between neuronal density and number of synapses per neuron, that may have implications in the functional characteristics of developing neuronal networks [7]. Therefore, neuronal cultures with different cell densities address the network maturation by modulating the number of synapses per neuron and thus the single neuron synaptic transmission which is reflected in neuronal network activity. This electrophysiological readout was previously analyzed by means of Micro Electrode Arrays (MEAs), matrices of planar metallic electrodes which provide new insights into the dynamics of *in vitro* networks and neuropathologies [8]. Thanks to MEAs, Wagenaar and coworkers observed that burst features changes when changing the neuronal density [6].

These studies provide evidence that neuronal density influences the number of synapses per neuron and the bursting behavior but they lack of an exhaustive and comprehensive evaluation of neuronal activity dependency. Accordingly, the goal of this study is to provide an experimental study of neuronal network electrophysiological features by varying cell density in cultures and analyzing a large pool of descriptors of network activity over maturation.

II. MATERIALS AND METHODS

A. Neuronal Culture Preparation

In this work we used standard 60-electrode MEA biochips (Multi Channel Systems, MCS GmbH). Substrates and neuronal cultures were prepared as previously described [9,10]. Cells were plated at 900, 1200, 1800, 2000, 2500 and 3600 cells/ mm^2 (see Table 1) and they were maintained in a humidified incubator up to 32 DIVs. The 30% of the total amount of medium was changed every 2 days before each electrophysiological experiment.

B. Electrophysiological Recordings

Extracellular recordings were carried out with a MEA1060 signal amplification and data acquisition system (MCS GmbH) at 37°C. Each recording started right after the stabilization of the electrical signals, lasted 10 minutes and was repeated every two days for each neuronal network, from 4 DIV to 32 DIV.

Table 1 Plating parameters of the cultures used

Density (cells/mm^2)	Number of cultures	Number of batches
900	5	3
1200	8	1
1800	38	5
2000	10	2
2500	14	2
3600	11	4

C. Data Analysis

After raw data recording, spikes were detected [11]. The off-line analyses were implemented in Matlab® and the descriptors of electrophysiological activity were extracted by means of a custom algorithm adapted from the literature [3,12,13]. We first evaluated the distribution of the six density populations with an ANOVA test. Then, we extracted (i) features describing the simple spiking activity, (ii) burst and (iii) network burst (NB) activity. Specifically, we extracted the median number of active channels (mean frequency>0.03 Hz) and the mean firing rate. Bursts (episodes of high frequency spiking) were detected when the minimum number of spikes was equal to 10 and the maximum inter spike interval was 100 ms. Network bursts (recurrent events of synchronized firing) were identified when the product of the number of active channels and the number of spikes was bigger than 9 and when the minimum inter network burst interval was 100 ms. Finally, we performed a statistical analysis of variance (ANOVA) and we chose the Tukey test as post hoc test (Statistica, StatSoft, Inc.). Data are given as mean and standard deviation. The significance level was established at $p<0.05$.

III. RESULTS

We observed 86 independent cell cultures, plated at six different cell densities, from 4 to 32 DIV. Results of ANOVA test suggested us to merge results related to cell cultures at 900 and 1200 cells/mm^2 (sparse cultures), 1800 and 2000 cells/mm^2 (medium cultures) and 2500 and 3600 cells/mm^2 (dense cultures).

A. Features of the Simple Spiking Activity

The number of active channels increases over cell maturation reaching its maximum at the end of the second week in vitro for all the cell densities. In cell cultures with 1800 cells/mm^2 it increases faster than in the others and also reaches higher values which are stable for almost one month in culture (Fig. 1). By contrast, cell cultures with density equal to 3600 cells/mm^2 show a reduction of the number of active channels after the end of the second week. Finally, cell cultures with 900 cells/mm^2 are characterized by a slower increase in the number of active channels which reaches its peak around 14 DIV and then it remains stable. The ANOVA test identified significant differences between cultures with 1800 cells/mm^2 (0.55±0.07, Tukey test $p<0.01$) with 900 cells/mm^2 (0.33±0.09) and 3600 cells/mm^2 (0.39±0.09).

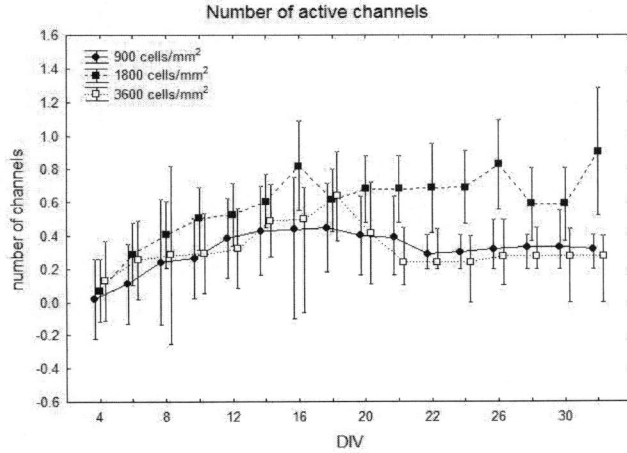

Fig. 1 Number of active channels changes during maturation for cell cultures with a density equal to 900 cells/mm^2 (black dot), 1800 cells/mm^2 (black square) and 3600 cells/mm^2 (black rhombus).

A similar behavior is shown by the mean frequency of neuronal activity which is higher in cell cultures with 1800 cells/mm^2 (2.13±0.61 Hz, Tukey test $p<0.01$) than in cell cultures with densities equal to 900 and 3600 cells/mm^2 (0.89±0.84 Hz and 0.67±0.83 Hz respectively).

B. Burst Features

The percentage of channels with burst activity is equal to 27±6%, 30±5% and 21±7% for neuronal cultures with cell density of 900, 1800 and 3600 cells/mm^2 respectively. This descriptor increases with cell maturation and has similar trends for the three densities. Concerning the burst duration (Fig. 2), it is roughly constant after the first week in vitro and it is equal to 0.17±0.04s for neuronal cultures at 900 cells/mm^2, to 0.21±0.03s ($p<0.05$ Tukey test) for 1800 cells/mm^2 and to 0.15±0.04s for 3600 cells/mm^2. The spiking rate inside the bursts is 115±10 Hz with no differences between the three densities. Finally the bursting rate increases during the first and the second week in vitro becoming stable for neuronal cultures at 900 and 1800 cells/mm^2. In contrast, cell cultures at 3600 cells/mm^2 show a fast decrease after the peak at 18 DIV. Bursting rate reaches the mean value of 2.73±1.14 burst per minute at 900 cells/mm², of 3.30±1.16 burst per minute at 3600 cells/mm² and of 4.34±1.13 burst per minute at 1800 cells/mm², which significantly differs from the other two densities (Tukey test p<0.05 for both couples) (Fig. 3).

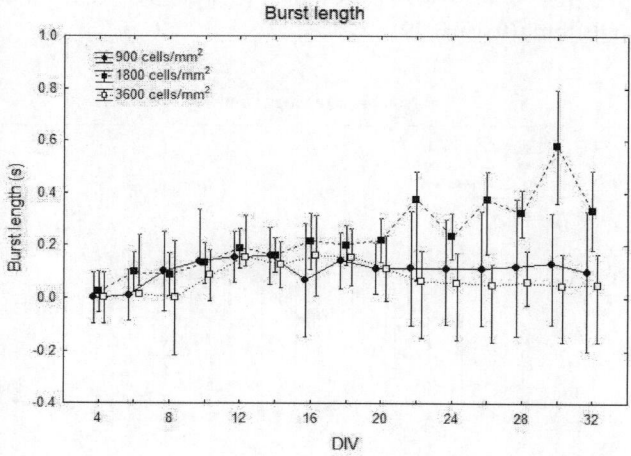

Fig. 2 Changes of burst duration (s) during maturation for cell cultures with a density equal to 900 cells/mm² (black dot), 1800 cells/mm² (black square) and 3600 cells/mm² (black rhombus).

C. Network Burst Features

Network burst length (Fig. 4) increases with neuronal network maturation and it is higher for cell cultures at 1800 cells/mm² (8.18s±3.23s) with respect to 900 cells/mm² (2.22±4.78s) and 3600 cells/mm² (1.92±6.24s) cell cultures. By contrast, the network bursting rate (Fig. 5) shows different trend: neuronal cultures at 900 cells/mm² have higher values (5.7±0.93 NB per minute) than 1800 cells/mm² (3.6±0.56 NB per minute) and 3600 cells/mm² (2.89±0.82 NB per minute). Concerning the intra network burst frequency, it is higher for cultures with 3600 cells/mm² (333Hz±90Hz, Tukey test p<0.05) than 900 cells/mm2 (205Hz±100Hz) and 1800 cells/mm² (211Hz±69Hz). Finally, the percentage of spikes involved in NB, in contrast to free spikes, is similar for all densities and it is equal to 31±5%.

IV. DISCUSSION

In this work we performed an experimental study to assess the influence of neuronal density and maturation on several comprehensive electrophysiological properties. Previous works have already characterized networks maturation on MEAs but only at one density [3] or by analyzing few parameters [6]. Here we take into account many descriptors making available a detailed picture of their behaviors.

We feel that this characterization is meaningful because it provides a reference on how to choose the right plating density, for MEA-based experiments, with different neurobiological purposes. The availability of established protocols and control data is important for the best planning of an experiment, according to the specific goals of the study. Despite the unavoidable culture to culture variation of the *in vitro* approach, neuronal networks grown using the protocol here described, show almost a reproducible behaviors. What we have described should represent a reference for the comparison of spontaneous behavior of neuronal networks at a certain cell density in terms of electrophysiological descriptors. Indeed, sparse cultures seem to be more adequate for long lasting experiments (e.g. chronic modulatory effect of a drug treatment), since control cultures display a prolonged and stable activity.

On the other hand, medium cultures seems to be more adequate for experiments that require intense electrical firing rates (e.g. screening of antiepileptics), since they show a spontaneous strong and intense electrical activity. Finally, dense cell cultures appears to be adequate for experiments in which time saving is more important than other aspects since they exhibit an earlier maturation with respect to the other cell densities (e.g. rapid screening).

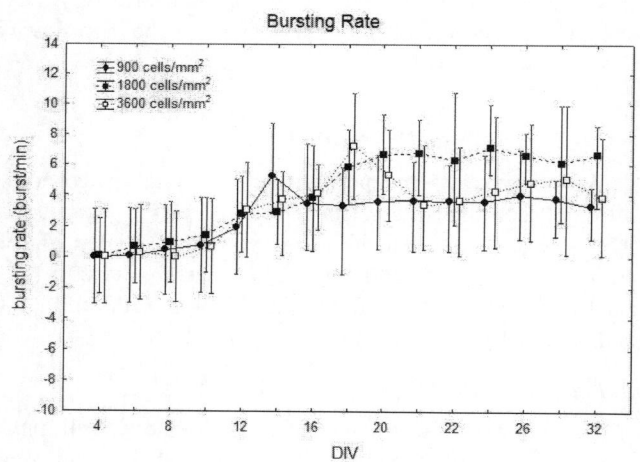

Fig. 3 Bursting rate (bursts/minutes) during maturation: cell cultures with a density equal to 900 cells/mm² (black dot), 1800 cells/mm² (black square) and 3600 cells/mm² (black rhombus).

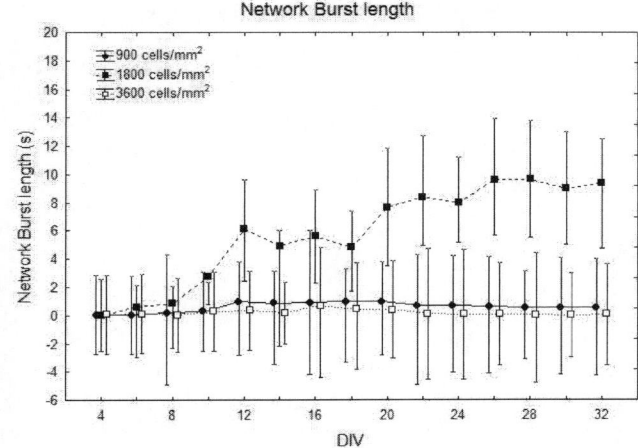

Fig. 4 Changes of mean duration of NBs (s) during maturation for cell cultures with a density equal to 900 cells/mm^2 (black dot), 1800 cells/mm^2 (black square) and 3600 cells/mm^2 (black rhombus).

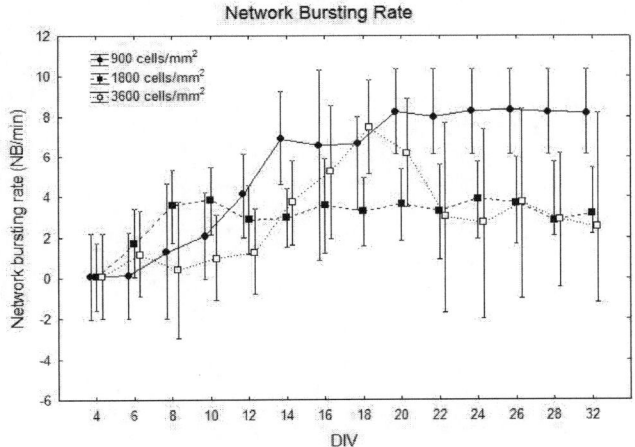

Fig. 5 Network bursting rate during maturation for cell cultures with a density equal to 900 cells/mm^2 (black dot), 1800 cells/mm^2 (black square) and 3600 cells/mm^2 (black rhombus).

V. CONCLUSIONS

This work is intended to provide a guide for electrophysiological studies of network activity by means of MEAs. Indeed, it provides some indications on how to choose the most adequate neuronal cell density conformed to the goal of the experiments.

ACKNOWLEDGMENT

Authors would like to thank people from the Alembic facility for their support and Dr. De Ceglia for the dissection of hippocampi.

REFERENCES

1. Morin F, Takamura Y and Tamiya E (2005) Investigating neuronal activity with planar microelectrode arrays: achievements and new perspectives. J Biosci Bioeng 100: 131-143.
2. Ben-Ari Y (2001) Developing networks play a similar melody. Trends Neurosci 24:353–360.
3. Chiappalone M, Bove M, Vato A et al. (2006) Dissociated cortical networks show spontaneously correlated activity patterns during in vitro development. Brain Res 1093: 41-53
4. Ichikawa M, Muramoto K, Kobayashi K, Kawahara M, Kuroda Y. 1993. Formation and maturation of synapses in primary cultures of rat cerebral cortical cells: An electron microscopic study. Neurosci Res 16:95–103.
5. Biffi E, Menegon A, Piraino F, et al. (2012) Validation of Long-Term Primary Neuronal Cultures and Network Activity Through the Integration of Reversibly Bonded Microbioreactors and MEA Substrates. Biotechnol Bioeng 109: 166-75.
6. Wagenaar DA, Pine J and Potter SM (2006) An extremely rich repertoire of bursting patterns during the development of cortical cultures. BMC Neurosci 7, 11 DOI:10.1186/1471-2202-7-11
7. Cullen DK, Gilory M, Irons HR et al. (2010) Synapse-to-neuron ratio is inversely related to neuronal density in mature neuronal cultures. Brain Res 1359: 44-55 DOI:10.1016/j.brainres.2010.08.058
8. Rossi S, Muzio L, De Chiara et al. (2011) Impaired striatal GABA transmission in experimental autoimmune encephalomyelitis. Brain Behav Immun 25:947-56
9. Biffi E, Regalia G, Ghezzi D et al. (2012) A novel environmental chamber for neuronal network multisite recordings. Biotechnol Bioeng 109:2553-2566
10. Ghezzi D, Menegon A, Pedrocchi A et al. (2008) A Micro-Electrode Array device coupled to a laser-based system for the local stimulation of neurons by optical release of glutamate. J Neurosci Methods 175:70-78
11. Biffi E, Ghezzi D, Pedrocchi A et al. (2010) Development and validation of a spike detection and classification algorithm aimed to be implemented on hardware devices. Comput Intell and Neurosci 659050 DOI 10.1155/2010/659050
12. Biffi E, Piraino F, Pedrocchi A et al. (2012) A microfluidic platform for controlled biochemical stimulation of twin neuronal networks. Biomicroflu 6:24106-2410610
13. Biffi E, Menegon A, Regalia G et al. (2011) A new cross-correlation algorithm for the analysis of "in vitro" neuronal network activity aimed at pharmacological studies. J Neurosci Methods 199:321-327

Corresponding author:

Author: Emilia Biffi
Institute: Politecnico di Milano, Department of Electronics, Information and Bioengineering, Neuroengineering and Medical Robotics Laboratory
Street: G. Colombo 40, 20133
City: Milan
Country: Italy
Email: emilia.biffi@mail.polimi.it

Effective Connectivity Patterns Associated with P300 Unmask Differences in the Level of Attention/Cognition between Normal and Disabled Subjects

S.I. Dimitriadis[1], Yu Sun[2], N.A. Laskaris[1], N. Thakor[2], and A. Bezerianos[2,*]

[1] Artificial Intelligence and Information Analysis Laboratory,
Department of Informatics, Aristotle University, Thessaloniki, 54124, Greece
stidimitriadis@gmail.com, laskaris@aiia.csd.auth.gr
[2] SINAPSE, National University of Singapore, 28 Medical Drive, 117456, Singapore
lsisu@nus.edu.sg, tassos.bezerianos@nus.edu.sg

Abstract—Recent findings indicated that both P300 and α-event-related desynchronization (a-ERD) were associated, and similarly involved in cognitive brain functioning, like attention, allocation and memory updating. In a very recent study, the causality between P300 and α-ERD (Event Related Desynchronization) generators was investigated in four sensory modalities, i.e., audition, vision, somatosensory, and pain. In the present study, we investigated the effective connectivity patterns related with P300 during a visual modality protocol with six different flashing images in a normal and a disabled subject (Cerebral Palsy). We constructed effective connectivity graphs (ECGs) with Partial Directed Coherence (PDC) focusing on target trials in the frequency range of 8 – 13 Hz. Adopting a tensorial classification scheme, we succeeded to differentiate the trials related with six different images with classification performance of 100% in both subjects. In addition, we estimated small-worldness as an indicator to quantify the difference in the level of attention/cognition between normal and disabled subject. Finally, small-world index can be used to measure how demanding a task is in terms of attention/cognition especially for disabled subjects.

Keywords—P300, alpha, attention, causality, small-world.

I. INTRODUCTION

P300 is an important event-related potential (ERP) component elicited by infrequent and task-relevant stimulus, and it reflects the processes of attention, stimulus classification, and memory updating [1-4]. Among many studies, a significant alpha-band (8–13 Hz in frequency) ERD (a-ERD) could be induced by both sensory stimulation (external event) across stimulus modalities [5] and cognitive processing (internal event) in various attention and memory tasks [6-8]. Previously, both P300 and a-ERD have been consistently triggered by the target stimuli in the oddball task paradigm, and P300 was showed to be functionally associated with the cognitive processing reflected by a-ERD [9, 10]. As revealed by time-varying effective connectivity, the cortical information was consistently flowed from a-ERD sources to P300 sources in the target condition for all four

---

* Corresponding author.

sensory modalities. All these findings showed that P300 in the target condition is modulated by the changes of a-ERD, which would be useful to explore neural mechanism of cognitive information processing in the human brain [11].

In the present work a six-choice P300 paradigm is tested using one normal and one disabled subject (Cerebral Palsy). Six different images were flashed in random order with a stimulus interval of 400 ms [12]. Figure 1 illustrates the flashed images used for evoking P300.

Fig. 1 The display used for evoking the P300. Images were flashed, one at a time, by changing the overall brightness of images.

In order to study the effective connectivity patterns related with P300 in both normal and disabled subjects, we adopted a well-known causality estimator called Partial Directed Coherence (PDC). Our analysis was based on α frequency range (8-13 Hz) as an attempt to study high cognitive activation and attention in terms of networks at the sensor level. Recent findings implied that P300 and a-ERD in the target conditions were independent of the stimulus modalities, and could mainly reflect the task-related high cognitive activation and attention while P300 in the target conditions was modulated by the changes of a-ERD, which may sub-serve the basic mechanism of high cognitive information processing in the human brain [11].

The principal scope of this work was to study the effective connectivity patterns related to a classical P300 protocol in both normal and disabled subjects. In our analysis, we used the complete electrode configuration with 32 sensors. At first, we adopted our methodological framework presented to an EEG Cognitive Workload paradigm [12] employing a subspace learning algorithm of Tensor Subspace Analysis (TSA) and applying algorithmic strategies that will treat the effective connectivity estimates directly (i.e. without computing network metrics), and, mainly, at the level of individual trials. In our approach, we treated the trial-dependent effective connectivity matrix as a tensor and used the TSA analysis to project it in a lower-dimensionality space where assessments of different visual stimulus related to P300 would be more efficient. After, quantifying the classification performance, we estimated the small-world index as an attempt to extract an indicator of how efficient is the communication in both normal and disabled subjects during a P300 evoked paradigm.

II. MATERIAL AND METHODS

A. Experimental Setup

Users were facing a laptop screen on which six images were displayed (see Fig. 1). The images showed a television, a telephone, a lamp, a door, a window and a radio. The images were selected according to an application scenario in which users can control electrical appliances via a BCI system [13]. The application scenario served however only as an example and was not pursued in further detail. The images were flashed in random sequences, one image at a time. Each flash of an image lasted for 100 ms and during the following 300 ms none of the images was flashed, i.e. the inter-stimulus-interval was 400 ms.

The EEG was recorded at 2048 Hz sampling rate from 32 electrodes placed at the standard positions of the 10-20 international system. A Biosemi Active Two amplifier was used for amplication and analog to digital conversion of the EEG signals.

B. Subjects

In our analysis, we employed a disabled and a healthy subject. The disabled subject (Cerebral Palsy) was wheelchair-bound but had varying communication and limb muscle control abilities. In addition, disabled subject was able to perform simple, slow movements with their arms and hands but were un-able to control other extremities. Spoken communication with disabled subject was possible, although he was suffered from mild dysarthria.

C. Experimental Schedule

Each subject completed four recording sessions. The first two sessions were performed on one day and the last two sessions on another day. For all subjects the time between the first and the last session was less than two weeks. Each of the sessions consisted of six runs, one run for each of the six images. For further details about the protocol followed on this experiment see the original paper related to this dataset [13].

D. Pre-processing

Signals were filtered within frequency range of 8 to 13 Hz. Biological artifacts were diminished by means of ICA [14,15] employing function *runica* from EEGLAB [16].

E. Effective Connectivity Measure

First, we used the ARFIT algorithm to fix an optimum model order for the time-varying MVAR model. ARFIT package estimates both the time-invariant parameters of the MVAR model and its optimum order [17]. The order estimation uses Schwarz's Bayesian Criterion (SBC) [18]. The estimated model order (p_{opt}) was then fixed for the remainder of the analysis. PDCs was computed based on the MVAR model fitted to the signal using an AR algorithm (ARFIT; [19]). A multivariate autoregressive (MVAR) model of order p is described by the following equation

$$\vec{Y}_t = A_1 x \vec{Y}_{t-1} + A_2 x \vec{Y}_{t-2} + ... A_p x \vec{Y}_{t-p} + \vec{X}_t \quad (1)$$

whereas $\vec{Y}_t$ is the observed EEG data at time t, $\vec{X}_t$ is the innovation process and A_k are the kth autoregressive parameters.

If we transform the MVAR model from the time-domain into the frequency-domain, the following transfer functions:

$$H(f) = A^{-1}(f) = Y(f) / X(f) \quad (2)$$

Partial Directed Coherence (PDC) [19] is frequency domain characterizations of causality. PDC is defined as:

$$PDC_{ij}(f) = \frac{A_{ij}(f)}{\sqrt{\sum_{i=1}^{M} |A_{ij}(f)|^2}} \quad (3)$$

PDC (Eq. 3) take values between zero and one where high values in a certain frequency band reflect a directionally linear influence from channel j to channel i in that band and is normalized by the sum of the influenced processes (j^{th} column of the matrix A).

F. Significant Links

The aforementioned procedures yielded a fully connected, weighted, asymmetric effective connectivity graph (EFG) representing influences among all cortical regions. The maximum number of possible directed connections in a network with k nodes is $N_{max} = k \times (k-1)$ and for $k = 32$, $N_{max} = 992$. Thus, the EFG is extremely dense. Therefore, this EFG must be filtered, so that the most significant connectivity patterns can emerge. We performed a ***Statistical filtering analysis***: A surrogate data method with 200 realizations was then used to select the most significant values of the measures at 99% confidence level. Surrogates were obtained by randomize signal to remove all causal relationships between them [20]. After surrogate analysis, we assigned a *p*-value to each of the N_{max} causal connections. To correct for multiple testing, the false discovery rate (FDR) method was adopted [21]. A threshold of significance was set such that the expected fraction of false positives was restricted to $q \leq 0.01$ [14,22].

G. Small-World Index

In the present study, we estimated Small-world index based on directed global (dGE) and local efficiency (dLE). A small-world network has a GE value that is less than a random network. Moreover, LE will have higher values when compared with a random network. The calculation of GE^{rand} and LE^{rand} was based on the procedure described here [24] which preserves the out-strength but not the in-strength distribution. We repeated this procedure 250 times. We then averaged across all random networks to obtain GE^{rand} and LE^{rand}. The small-world indices $\gamma = LE/LE^{rand}$ and $\lambda = GE/GE^{rand}$ were then calculated for the FCG under study, and the ratio $S = \gamma/\lambda$ was defined. This was greater than 1 for small-world networks [23][1].

III. RESULTS

A. Machine Learning Validation

The TSA algorithm, followed by a k-nearest-neighbor classifier (with $k=6$), was tested on trial-based connectivity data from all bands. The following results have been obtained through a cross-validation scheme that shuffle the trials and get 90% for training and 10% for testing. The following table summarizes the average classification rates derived after applying the above cross-validation scheme 100 times [12]. The classification performance was 100% for both subjects.

---

[1] Network analysis and machine learning algorithms were implemented in MATLAB and can be downloaded from the author's website: http://users.auth.gr/~stdimitr/software.html

B. Effective Connectivity Patterns

Fig. 2 illustrates the six effective connectivity graphs related to each one of the six images presented in Fig. 1. Hubs were defined as the nodes with the highest outgoing degree (outgoing degree = k-1=31) and stressed with red circles. The spatial distribution of hubs for the disabled subject is located mainly in bilateral frontal sites, left fronto-central, in P7 and in CP6. For the normal subject, two more sensor areas were detected as hubs: PO3 and O1. Visual inspection of EFGs uncovered greater connectivity strength for disabled subject compared to normal.

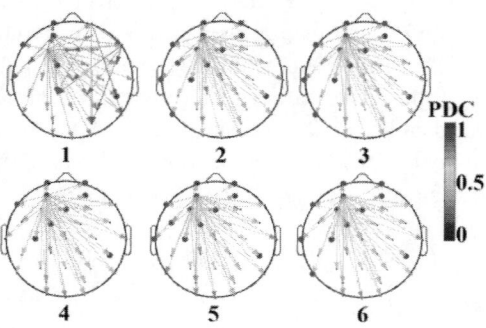

Fig. 2 (Disabled subject) Effective Connectivity Graphs related to P300 for each one of the six target images as presented in Figure 1. The color of the arrow (both body and head) is related with the strength of PDC. To improve visualization, we illustrated the 5% of the strongest connections. Red color stressed the hubs uncovered as the nodes that drive the rest of the network (outgoing degree = k-1=31).

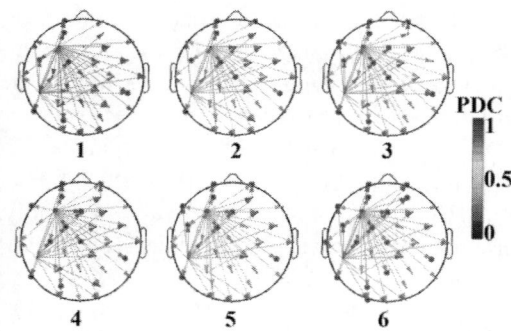

Fig. 3 (Normal subject) Effective Connectivity Graphs related to P300 for each one of the six target images as presented in Figure 1. The color of the arrow (both body and head) is related with the strength of PDC.

C. Small-World Index

EFGs of normal subject are closer to a small-world network compared to the EFGs of the normal subject for each of the flashing image (Fig.4).

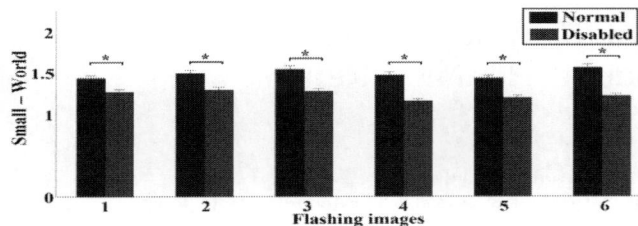

Fig. 4 Small-World index for each flashing image of Fig.1 in both subjects (* denotes significant difference $p < 0.001$ of SW between the two subjects across trials using Wilcoxon rank test; Bonferroni corrected $p' < p/6$).

IV. DISCUSSION

Based on our experiments, the tensorial treatment of ECGs succeeded to identical discriminate the six flashing images in normal and disabled subject. Our network analysis revealed important hubs areas consistent in both subjects located over frontal areas (bilaterally), left fronto-central and at sensors P7 and CP6. Additionally, in normal subject two more hubs were revealed: PO3 and O1. Disabled subject shall endeavor to successfully complete the task compared to normal. The above expected difference between the two subjects can be supported and quantified via network analysis applied to EFGs. By a visual inspection of the topographies based on the connection strength (Fig. 2, 3), it is obvious that hubs sensors in disabled subject entrusting the demands of the task related to high cognitive activation and attention [11]. Based on a recent source analysis in a multi-modal P300 task, the generators of α-ERD are located at the bilateral occipital cortices [11]. Even though our analysis based on sensor level, the lack of hubs located over left parieto-occipital sites in disabled subject can be related to visional issues related to cerebral palsy that balanced with more effort over frontal and fronto-central areas. Finally, small-world index uncovered a more well organized network in normal compared to disabled subject (Fig.4). It would be interesting in the near future to follow the same analysis in the source level as in [11].

ACKNOWLEDGEMENT

Dataset is downloaded from the Multimedia Signal Processing Group website:http://mmspg.epfl.ch/downloads.

REFERENCES

[1] Linden, D.E."The P300: where in the brain is it produced and what does it tell us? ," *Neuroscientist.*, vol.11, pp.563–576, Dec, 2005.

[2] Comercheroa, M.D., and Polich, J., "P3a and P3b from typical auditory and visual stimuli," *Clinical Neurophysiology.*, vol. 110, pp.24–30,1999.

[3] Polich, J., "Updating P300: an integrative theory of P3a and P3b," *Clinical neurophysiology*, vol.118, pp.2128–2148, Oct, 2007.

[4] Zaslansky, R., Sprecher, E., Tenke, C.E., et al., "The P300 in pain evoked potentials," *PAIN*, vol.66, pp.39–49, Jul, 1996.

[5] Pfurtscheller, G., Neuper, C., and Mohl, W., "Event-related desynchronization (ERD) during visual processing," *International Journal of Psychophysiology*, vol.16 pp.147–153, 1994.

[6] Basar, E., Basar-Eroglu, C., Karakas, S., et al., "Gamma, alpha, delta, and theta oscillations govern cognitive processes," *International Journal of Psychophysiology.*, vol.39, pp.241–248, 2001.

[7] Basar, E., Basar-Eroglu, C., Karakas, S., et al.," Brain oscillations in perception and memory," *International Journal of Psychophysiology*, vol.35, pp. 95–124, 2000.

[8] Klimesch, W.,"EEG-alpha rhythms and memory processes," *International Journal of Psychophysiology*, vol.26, pp.319–340, 1977.

[9] Yordanova, J., Kolev, V., Polich, J., "P300 and alpha event-related desynchronization (ERD)," *Psychophysiology*, vol.38, pp.143–152, 2001.

[10] Yordanova, J., and Kolev, V., Event-related alpha oscillations are functionally associated with P300 during information processing," *Neuro Report*, vol.9, pp.3159–3164, 1998.

[11] Peng, W., Hu, L., Zhang, Z., et al., "Causality in the Association between P300 and Alpha Event-Related Desynchronization," *PLoS ONE*, vol.7: e34163. doi:10.1371/journal.pone.0034163, 2012.

[12] Dimitriadis, S.I.,Yu, Sun., K, Kwok., et.al.," A Tensorial Approach to Access Cognitive Workload related to Mental Arithmetic from EEG Functional Connectivity Estimates," *35th Annual of the IEEE EMBC*, Osaka (Japan) 3 – 7 July 2013.

[13] Hoffmann, U., Vesin, J.M., Ebrahimi, T., et. al., "An efficient P300-based brain-computer interface for disabled subjects," *J Neurosci. Methods*, vol.167, pp.115-25, Jan 15, 2008.

[14] Dimitriadis, S.I., Laskaris, N.A., Tzelepi,A. et al., "Analyzingfunctional brain connectivity by means of commute times: a new approach and its application to track event-related dynamics," *IEEE Trans. Biomed. Eng.*, vol. 59, no. 5, pp. 1302-9, May, 2012.

[15] Delorme,A., Westerfield,M.,and Makeig,S., "Medial prefrontal theta bursts precede rapid motor responses during visual selective attention," *J. Neurosci.*, vol. 27, no. 44, pp. 11949-59, Oct 31, 2007

[16] Delorme,A., and Makeig, S., "EEGLAB: an open source toolbox for analysis of single-trial EEG dynamics including independent component analysis," *J. Neurosci. Methods*, vol. 134, no. 1, pp. 9-21, Mar 15, 2004.

[17] Tapio, S and Arnold,N.," Algorithm 808: ARfit - a matlab package for the estimation of parameters and eigenmodes of multivariate autoregressive models," *ACM Trans. Math. Softw.*, vol. 27, no. 1, pp. 58-65, 2001.

[18] Schwarz, G.," Estimating the Dimension of a Model," *The Annals of Statistics*, vol. 6, no. 2, pp. 461-464, 1978.

[19] Sameshima, K., and Baccala, L.A.,"Using partial directed coherence to describe neuronal ensemble interactions," *J. Neurosci. Methods*, vol.94, pp.93–103,1999.

[20] Hesse,W., Möller, E., Arnold, M., et al., "The use of time-variant EEG Granger causality for inspecting directed interdependencies of neural assemblies," *J. Neurosci. Methods*, vol. 124, no.1, pp. 27-44, 2003

[21] Benjamini, Y., and Hochberg, Y.,"Controlling the False Discovery Rate - a Practical and Powerful Approach to Multiple Testing," *J. R. Stat. Soc. Ser. B-Stat. Methodol.*, vol.**57**, pp.289-300,1995.

[22] Ioannides AA, Dimitriadis SI, Saridis G, et. al., "Source Space Analysis of Event-Related Dynamic Reorganization of Brain Networks. Computational and Mathematical Methods in Medicine Special Issue", Graph Theoretical Approaches in Brain Networks", vol. 2012, Article ID 452503, doi:10.1155/2012/452503.

[23] Dimitriadis, S.I., Laskaris, N.A., Tsirka, V., et al.," Tracking brain dynamics via time-dependent network analysis,**"** *J. Neurosci. Methods*, Volume 193, Issue 1, 30 October 2010, pp. 145-155,2010.

A General Purpose Approach to BCI Feature Computation Based on a Genetic Algorithm: Preliminary Results

S. Ramat[1,2] and N. Caramia[1,2]

[1] Dip. Ingegneria Industriale e dell'Informazione, Università di Pavia, Pavia, Italy
[2] Brain Connectivity Center, IRCCS Fondazione Istituto Neurologico Nazionale C. Mondino, Pavia, Italy

Abstract—BCI systems aim at developing man-machine communication channels independent of the intervention of muscles. This is accomplished by recognizing specific mental states and using their detection to trigger actions in a computer controlled environment. Brain activity is acquired, typically through EEG, and is then processed in order to compute features allowing the classification of the user's mental states being monitored. Several successful approaches to BCI based on different neural mechanism underlying the generation of the signal patterns to be recognized can be found in the literature. Yet, the signal processing leading to such meaningful features may be quite diverse between the different approaches due to the variability introduced by the different subjects, acquisition devices, experimental setup. Here we developed a new, general purpose approach to the computation of features allowing efficient trial classification based on a genetic algorithm. The algorithm was tested on three different datasets drawn from the BCI competition II and based on slow cortical potentials, motor imagery and self-paced movements.

Keywords—Brain-Computer Interfaces, EEG signal processing, BCI features computation, Genetic Algorithm.

I. INTRODUCTION

Brain Computer Interface systems process signals related to the neuronal activity in the brain in order to detect specific states or events, which are in turn associated with some form of action that is then activated by the BCI system. A variety of methods for monitoring brain activity may be considered, yet the most common approach is, by far, that based on EEG signals as it currently offers the best compromise of affordability, availability and low invasiveness.

In such scenario the problem faced by BCI systems is that of decoding the noisy EEG signals in order to recognize the occurrence of two or more different mental states, which can then be associated to some response through a *translation algorithm*.

Several neurophysiological principles responsible for producing specific patterns of neuronal activity have been investigated by the scientific community for the development of a relatively large range of BCI applications [1], [2].

These principles encompass changes in brain rhythms in various frequency bands (mu, alpha, beta and gamma), which result in event related desynchronization (ERD) and synchronization (ERS) [3], as well as movement-related potentials (MRP)[4] in tasks related to sensorimotor activity. Other neural mechanisms include slow cortical potentials (SCP), which a subject can learn to control and exploit for interaction and visual evoked potentials (VEP)[5], which are typically more evident on the occipital cortex, are evoked by visual stimuli and also constitute the basis for visually evoked steady state potentials (VSSEP)[6], and the P300 potential evoked by the appearance of infrequent or of especially significant somatosensory stimuli.

In spite of the relatively well-defined characteristics of these neural mechanisms, inter-individual differences introduce a large variability in time, space and frequency characteristics of the acquired EEG data. Variability is also introduced by the different recording equipment and experimental setup, so that the improvement of the signal-to-noise ratio of a BCI system is the main goal of all signal processing approaches. The selection of the appropriate EEG channels, of the spatial and temporal filtering approach and parameters is thus crucial to the performance of a BCI system, so that the development of the most effective solutions often requires the computation of extremely specific features. An example of such specificity is offered by the approached adopted by the winners of the various editions of the BCI competition, as detailed in the following section.

Thus the purpose of our work was to attempt at developing a general purpose approach for computing valuable features for EEG classification. To this goal we developed a standardized feature computation algorithm (Fig. 1) whose processing choices were coded in a binary string that was handled as a chromosome of a genetic algorithm (GA) searching the feature space and evaluating candidate solutions based on their performance in classifying the available trials.

II. MATERIALS AND METHODS

A. Data Sets

As stated, the purpose of our work was that of developing a widely applicable approach to the classification problem faced by BCI systems. We therefore chose three different datasets drawn from the BCI competition initiative and representative of a common approach used in BCI research: slow cortical potentials, motor imagery and self-paced movements.

DATA SET I: Self Regulation of Slow Cortical Potentials, Institute of Medical Psychology and Behavioral Neurobiology, University of Tübingen [8].

The EEG signal of a healthy subject was recorded during a BCI session on self-regulation of Slow Cortical Potentials (SCP). The subject was asked to move a cursor up or down on a computer screen using his SCP while receiving visual feedback. Positive cortical potentials lead to a downward movement of the cursor on the screen, while negative potentials produced an upward cursor movement.

The available training dataset includes 3.5 s of data for each of 268 trials recorded on two different days and mixed randomly. The signal was recorded with a 256 Hz sampling rate so that 896 samples per channel were available for every trial. EEG data was recorded from six electrodes in the following positions: A1-Cz, A2-Cz, 2 cm frontal of C3, 2 cm parietal of C3, 2 cm frontal of C4, 2 cm parietal of C4. The test set is composed of 293 trials collected during the second day.

The winner of the competition [9] used linear discriminant analysis on the dc potentials of the first two channels and on the value of high beta power (24-37 Hz) of channel 4 and channel 6 and obtained a classification error of 11.3%.

DATA SET II: Self Paced 1s, Fraunhofer-FIRST, Intelligent Data Analysis Group, and FreieUniversitätBerlin.

This dataset was recorded from a normal subject during a no feedback session. The subject sat in a normal chair, with relaxed arms resting on the table and fingers in the standard typing position at the computer keyboard. The task was to press, using either the index or the little finger of either the left or the right hand, one of four assigned keys in self-chosen order and timing ('self-paced key typing'). The experiment consisted of 3 session of 6 minutes each. All sessions were conducted on the same day with some minutes break in between. Typing was done at an average speed of 1 key per second [10]. The signal was recorded at 1 kHz and band pass filtered between 0.05 and 200Hz.

28 EEG channels were recorded at the following positions of the international 10/20-system (F3, F1, Fz, F2, F4, FC5, FC3, FC1, FCz, FC2, FC4, FC6, C5, C3, C1, Cz, C2, C4, C6, CP5, CP3, CP1, CPz, CP2, CP4, CP6, O1, O2).

The dataset consists of 426 epochs of 500ms each, ending 130ms before a key press. 316 epochs are labeled (0 for upcoming left hand movements and 1 for right hand movements), the remaining 100 epochs were not labeled, for competition purposes.

The competition winner [11] achieved a 16% classification error by using zero-phase 7 Hz low-pass filter and 10-33 Hz pass band filter, CSSD and Fisher Discriminant Analysis (FD) to obtain three features for each trial, which were used for training a perceptron neural network.

DATA SET III: Motor Imagery, Department of Medical Informatics, Institute for Biomedical Engineering, University of Technology Graz, Austria.

This data set was recorded from a healthy subject (female, 25 y.o.) during a feedback session. The subject sat in a relaxing chair with armrests. The task was to control a feedback bar by imagining left or right hand movements while directional cues were presented in random order [12].

The dataset contains 280 trials of 9s length. The first 2s were quiet, at t=2s an acoustic stimulus indicated the beginning of the trial and a cross "+" was displayed for 1s; then at t=3s, an arrow (left or right) was shown as cue indicating the subject to move a bar into the direction of the cue. Three bipolar EEG channels were recorded over C3, Cz and C4, sampled at 128Hz and filtered between 0.5 and 30Hz.

The competition winning team [13] achieved an error rate of 10.7%. They considered only signals from electrodes C3 and C4. The data were band-pass filtered using Morlet wavelets at the lower and higher frequency of mu-rhythm (10-22 Hz). Classification was accomplished by estimating a multivariate normal distribution for each class, using previous time instances weighed according to Bayes error.

B. Genetic Programming Algorithm

The idea that led to the development of our approach was that the features for EEG classifications may be considered as the result of a standardized computation algorithm, which may be conceived as a sequence of processing steps. The computation occurring at each step could in turn be coded as choices from predefined sets of operators. Thus, such choices could be coded as binary strings, one for each processing step, which represent the genes of a binary chromosome coding for the development of one classification feature. Such feature would then represent the phenotype of the chromosome, which is built through the execution of the standardized feature computation algorithm on the provided dataset. The coding approach therefore defines a feature search space in which each gene represents one dimension and which is suitable for GA-based exploration.

Thus, we developed a Genetic Algorithm working on a population of random chromosomes, each containing several binary genes arranged in a string. The GA evolves using elitism, tournament-k selection, single point crossover maintaining gene integrity, and mutation. The GA was set to evolve the population for a predefined number of generations.

Each binary chromosome is converted to its phenotype by applying the signal processing sequence it describes to a standardized data structure. The flow chart of the feature-processing algorithm is shown in figure 1.

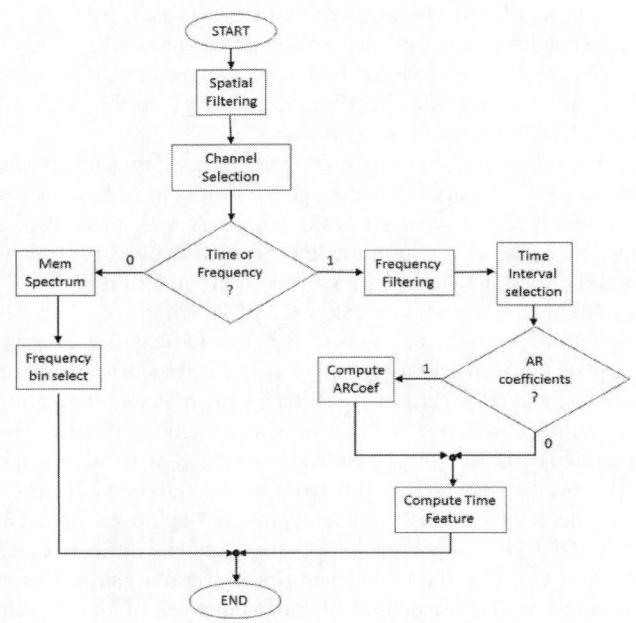

Fig. 1 Flow chart of the feature computation algorithm encoded by each chromosome.

Thus, the algorithm leading to the computation of one feature consists, for each trial, of a sequence of signal processing steps and of decisions following the choices coded in each chromosome. A corresponding fitness function for evaluating each chromosome was then computed in terms of correct classifications obtained by a support vector machine (SVM) with a linear kernel algorithm applied to the available dataset and exploiting the corresponding phenotype.

C. Feature Space and Operators

For each signal processing stage, the binary gene codes for one of the possibilities offered by the set of 'processing tools' available at that stage. The number of bits in a specific gene (ng) thus codes for 2^{ng} processing options, which may be easily customized and extended.

In our experiment we chose the following sets of signal processing functions at each stage.

Spatial Filtering: 2 bits

- no filter
- CAR, Common Average Reference filter.
- CSSD, Common Spatial Subspace Decomposition [14]
- ICA Independent component analysis

Channel Selection: uses a dataset-dependent number of bits and allows selecting either one of the recorded channels or some computed channels such. Among the latter an optimized spatial filter (OSF) channel, i.e. a computed channel maximizing the difference of variance between the samples acquired during the mental conditions to be recognized with respect to rest [15]. Moreover, depending on the available unused bit codes, we included the average of motor area channels and other combinations of existing channels.

Frequency Filtering: 4 bits, allowing no filtering and several combinations of low pass, high pass and band pass filters aimed at isolating alpha, mu and beta rhythms.

Time selection: 3 bits, allowing to select the entire trial, halves and quarters of the available time samples.

Time feature computation: 2 bits, allowing to compute the mean value, the variance, the slope and the difference between the mean value in the first 20% and last 20% of the available time samples.

Frequency bin selection: 4 bits selecting the power spectrum content, estimated using the maximum entropy method, in one of the 2Hz-wide bins covering the 0-32 Hz frequency range.

One bit was then used for each of the two decisions in Figure 1. The resulting feature-coding chromosome was then made out of 20, 21 and 19 bits for analyzing the described datasets I, II and III, respectively.

D. Experiments

The algorithm was run against each of the training datasets for finding three features for trial classification. Each feature is sought for incrementally: the algorithm initially searches for the single feature that best classifies the data. Then, for evaluating the fitness of the following features the algorithm combines previously estimated features with the new one as input to the classifier.

In evaluating the fitness of each chromosome, i.e. of the feature coded by the chromosome, a 5-fold leave-one-out approach was used, exploiting the same dataset partitioning for each chromosome being scored. This led to five scores for each individual, and the fitness of the chromosome was set to the mean of such scores.

The initial population was set to 100 chromosomes, and the algorithm ran for 100 generations for finding each feature. The entire GA was run for three times on each training dataset and the three resulting chromosomes, each coding for the three classifying features, were used without further training to classify the test data.

III. RESULTS

A. Data Set I (SCP)

The classification error achieved in the three runs of the GA on the training dataset of slow cortical potentials was of 10%, 9% and 9%. When these features were used for classifying the trials in the test dataset, while still using a linear kernel SVM, the performance of the classifiers was lower as they achieved classification errors of 20%, 19% and 19% respectively.

B. Data Set II (Self-Paced)

The three runs of the GA achieved a classification error of 19%, 18% and 17% on the training set. When these features were used for classifying the trials in the test set their performance degraded significantly and the classification error rose to 49%, 51% and 31%, respectively. Thus the best classifier in the training set resulted as the best performer also on the test set.

C. Data Set III (Motor Imagery)

The features resulting from the three runs of the GA on the training dataset allowed the linear kernel SVM classifier to achieve a classification error of 9.5%, 9.5% and 8.5%. The same features used on the provided test dataset achieved classification errors of 27%, 25% and 21%. Although the performance of the classifier decreased substantially between the training and the testing datasets, the best set of features found on the training set was the one achieving the best results also on the test set. Such features were:

1- Mean of AR coefficients computed on the 10-15Hz band-pass filtered third quarter of each trial taken from channel 3 after CSSD filtering.
2- Mean of AR coefficients computed on the 18Hz high-pass filtered third quarter of each trial taken from channel 1 without spatial filtering.
3- Slope of AR coefficients computed on 10-15Hz band-pass filtered first quarter of the data taken from channel 2 after CSSD spatial filtering.

With this dataset, the first feature was common to all three sets of results, the second feature was common to the first and second runs while its frequency filter was set to a 10Hz high-pass filter in the third run; the third feature differed in the three runs.

IV. CONCLUSIONS

We have developed a general purpose approach to automatic computation of features for classifying EEG signals for BCI applications based on a GA. The approach appears promising although the performance of the resulting classifiers on the test sets were lower than obtained by the best results reported in the literature. Such difference was especially evident in two of the three considered case studies, which showed a dramatic drop of performance between training and testing phases. Further refinements of our feature processing algorithm aimed at better grasping the time evolution of EEG signals, together with the extension of the sets of signal processing operators available at the different steps, are under way.

REFERENCES

[1] A. Bashashati, M. Fatourechi, R. K. Ward, and G. E. Birch (2007), A survey of signal processing algorithms in brain-computer interfaces based on electrical brain signals., *Journal of neural engineering*, 4, 2, R32–57.

[2] J. R. Wolpaw, N. Birbaumer, D. J. McFarland, G. Pfurtscheller, and T. M. Vaughan (2002), Brain-computer interfaces for communication and control., *Clinical neurophysiology: official journal of the International Federation of Clinical Neurophysiology*, 113, 6, 767–91.

[3] G. Pfurtscheller and A. Aranibar (1979), Evaluation of event-related desynchronization (ERD) preceding and following voluntary self-paced movement., *Electroencephalography and clinical neurophysiology*, 46, 2, 138–46.

[4] L. Deecke, B. Grözinger, and H. H. Kornhuber (1976), Voluntary finger movement in man: cerebral potentials and theory., *Biological cybernetics*, 23, 2, 99–119.

[5] J. J. Vidal (1973), Toward direct brain-computer communication., *Annual review of biophysics and bioengineering*, 2, 157–80.

[6] M. Middendorf, G. McMillan, G. Calhoun, and K. S. Jones (2000), Brain-computer interfaces based on the steady-state visual-evoked response., *IEEE transactions on rehabilitation engineering: a publication of the IEEE Engineering in Medicine and Biology Society*, 8, 2, 211–4.

[7] E. Donchin and D. B. Smith (1970), The contingent negative variation and the late positive wave of the average evoked potential., *Electroencephalography and clinical neurophysiology*, 29, 2, 201–3.

[8] N. Birbaumer, N. Ghanayim, T. Hinterberger, I. Iversen, B. Kotchoubey, A. Kübler, J. Perelmouter, E. Taub, and H. Flor (1999), A spelling device for the paralysed., *Nature*, 398, 6725. Nature Publishing Group, 297–298, 1999.

[9] B. D. Mensh, J. Werfel, and H. S. Seung (2004), BCI Competition 2003--Data set Ia: combining gamma-band power with slow cortical potentials to improve single-trial classification of electroencephalographic signals., *IEEE transactions on bio-medical engineering*, 51, 6, 1052–6.

[10] B. Blankertz, G. Curio, K. Müller, N. Group, and K. B. Franklin (2002), Classifying Single Trial EEG: Towards Brain Computer Interfacing, *Advances in neural information*, 1, c, 157–164.

[11] Y. Wang, Z. Zhang, Y. Li, X. Gao, S. Gao, and F. Yang (2004), BCI Competition 2003--Data set IV: an algorithm based on CSSD and FDA for classifying single-trial EEG., *IEEE transactions on bio-medical engineering*, 51, 6, 1081–6.

[12] A. Schloegl, K. Lugger, and G. Pfurtscheller (1997), *Using adaptive autoregressive parameters for a brain-computer-interface experiment*, 4, 1. Ieee, 1997, 1533–1535.

[13] S. Lemm, C. Schäfer, and G. Curio (2004), BCI Competition 2003--Data set III: probabilistic modeling of sensorimotor mu rhythms for classification of imaginary hand movements., *IEEE transactions on bio-medical engineering*, 51, 6, 1077–80.

[14] Y. Wang, P. Berg, and M. Scherg (1999), Common spatial subspace decomposition applied to analysis of brain responses under multiple task conditions: a simulation study., *Clinical neurophysiology: official journal of the International Federation of Clinical Neurophysiology*, 110, 4, 604–14.

[15] I. K. Niazi, N. Jiang, O. Tiberghien, J. F. Nielsen, K. Dremstrup, and D. Farina (2011), Detection of movement intention from single-trial movement-related cortical potentials., *Journal of neural engineering*, 8, 6, 066009.

Modular Control of Crouch Gait in Spastic Cerebral Palsy

D. Torricelli[1], M. Pajaro[1], S. Lerma[2], E. Marquez[2], I. Martinez[2], F. Barroso[1,3], and J.L. Pons[1]

[1] Grupo de Bioingeniería, Centro de Automática y Robótica, CSIC, Madrid, Spain
[2] Unidad de Neuroortopedia Servicio Ortopedia y Traumatología Infantil
Niño Jesús University Hospital, Madrid, Spain
[3] Adaptive System Behaviour Group, Industrial Electronics Department, University of Minho, Portugal

Abstract—The control of human movement is believed to be organized in sets of predefined muscular patterns usually referred to as motor modules or muscle synergies. Whether muscle synergies can be used to describe pathological control of movement is still under debate. In this paper, we present a preliminary description of the shapes, timing and weights of the synergistic patterns of three children with CP during crouch gait movements. The synergies and activations extracted from the activity of 8 lower limb muscles show strong differences with respect to healthy subjects and high similarities across patients.

Keywords—**Muscle synergies, Cerebral Palsy, Crouch Gait Pattern.**

I. INTRODUCTION

There is increasing evidence supporting that human movement control is organized in "functional units" or set of commands led by the Central Nervous System (CNS). These units are usually referred to as *motor modules* or *muscle synergies,* and currently, a growing number of scientists seek to decode the algorithms underlying these units of motion. A *motor module* consist of a specific pattern of temporal-spatial outputs of muscle activations [1]. The association of a number of *motor modules* produces complex movements across the limb joints that are required to perform motor tasks and activities of daily living such as walking [2]. During human adult locomotion, trunk and leg muscle activity is organized by a combination of four to five basic patterns coordinated to the phases of the gait cycle [3].

In children, there is good correlation between the developmental changes of the synergies and the parallel modifications in locomotion biomechanics [4]. For some authors, the determinants of a mature gait pattern (single-limb stance duration, walking velocity, cadence, step length and the ratio of pelvic span to ankle spread) are well established at the age of three years. According to other authors, maturation of gait occurs at 7 years of age [5]. After this age, only minor changes in cadence, velocity, and pelvic and joint rotation are expected [6]. Differences between typically developing (TD) children and adults gait patterns have been interpreted as evidence of the neuromuscular immaturity of children.

Muscle synergies extraction is based on the activity profile of the muscles measured by electromyography (EMG). Surface EMG (sEMG) is a reliable method frequently used to evaluate gait in adults and also in pediatric population [7]. Two basic modules present in newborns are retained between the four patterns extracted in toddlers. These patterns evolve over time, showing transitional forms during childhood until the acquisition of mature synergy patterns in the adulthood (4).

Cerebral Palsy (CP) is a permanent disorder of the movement and posture caused by a damage of the fetuses' brain, during birth or through the first years of life [8]. The damage of the CNS causes abnormal muscle tone and also poor selective motor control, lack of coordination between muscle agonist and antagonist muscles across joints, pathological muscle co-contractions, balance issues, weakness and pathological reflexes [9]. Spastic CP is the most frequent type and presents a velocity-dependent increase in the muscle tone resulting from the hyperexcitability of the stretch reflex. Children with CP usually walk later and present gait pattern abnormalities, low speed and high-energy costs, and many of these children remain dependent of braces and assistive devices [10]. Crouch gait pattern is one of the most common patterns in spastic CP, and is characterized by excessive knee flexion during the stance phase of walking. In addition, they frequently present excessive flexion on the hip, and increased ankle dorsiflexion during swing. Some authors have hypothesized that children with CP walk later because their locomotor networks are immature, and thereby their neural control patterns are similar to those extracted from younger TD children [11].

We hypothesize that *motor modules* in children with CP and crouch gait pattern are different from those described in matched TD children. The goal of this contribution is to give a preliminary description of the shape, timing and weight of the synergistic patterns of three children with CP during crouch gait movements.

II. MATERIALS AND METHODS

A. Subjects

Three children with diagnosis of bilateral spastic cerebral palsy and crouch gait pattern were recruited at Niño Jesús

University Hospital (Madrid, Spain). The approval to conduct this study was granted by the ethical committee of the hospital. The children and the parents gave the informed consent before the final inclusion of the subjects.

B. Experimental Protocol

Children were asked to walk overground on an 8-m straight pathway at a comfortable self-selected speed. All the experiments were performed in the Motion Analysis Laboratory of the Niño Jesús University Hospital (Madrid, Spain). Kinematics and EMG data have been collected during at least three walking cycles per leg. Limb and joint kinematics were analyze by means of a stereophotogrammetric system (BTS Bioengineering, Italy) based on retroreflective markers attached to the anatomical landmarks (Davis Protocol). Bilateral sEMG was recorded for 8 muscles of the legs: tibialis anterior (TA), gastrocnemius medialis (GAS), vastus medialis (VM), vastus lateralis (VL), rectus femoris (RF), adductor longus (AL), biceps femoris caput longus (BF), and semimembranosus (SEM). sEMG signals were recording by using a 16-channel wireless EMG system (BTS Bioengineering, Italy). SENIAM recommendations have been followed for electrodes placement.

C. Data Analysis

a) Kinematic Analysis

The ipsilateral heel strike event has been used to identify the beginning of each gait cycle. In our experiments, such event has been defined as the instant in which the height of the marker attached on the lateral malleolus reaches its minimum value. To check for its validity, this identification method has been compared to manual identification of heel strike events performed by an expert clinician. The difference between the two methods was negligible (<2% of the gait cycle).

b) Synergy Extraction

Muscle synergies have been extracted from EMG data, according to the following procedure. Each EMG signal has been pre-processed using band-pass filtering at 20-400Hz, demeaning, rectification, and low-pass filtering at 5 Hz. For each muscle, the resulting EMG envelope has been normalized by the average of its peak value across all the cycles [12], and then resampled at 1% of the gait cycle, resulting to 101 samples per cycle. All the normalized EMGs have been then combined into an m-by-t matrix, where m indicates the number of muscles (eight in this case) and t is the time base (n° of cycles multiplied by 101 samples). Muscle synergies (W) and activations in time (H) have been extracted by the non-negative matrix factorization (NMF) algorithm proposed by Lee et al. [13, 14]. The variability accounted for (VAF) coefficient [14] has been used to calculate the similarity between the original EMG and the EMG reconstructed by multiplying matrix W by matrix H. Being the number of synergies an input of the NMF algorithm, the overall procedure has been repeated four times, for 2, 3, 4, and 5 synergies respectively. The optimal number of synergies has been identified as the number by which the VAF coefficient either reached value greater than 90% or increases less than 5% with respect to the previous iteration. For each iteration the algorithm is repeated 10 times (12), to avoid local minima in the estimation of W and H.

III. RESULTS

Three children with bilateral spastic CP and crouch gait pattern participated in this study. Age, gender, diagnosis and the main spatial-temporal and kinematics parameters of the gait trial of the three subjects are described in Table 1.

Table 1 Clinical and kinematic data of the three children analyzed in this study. *Abbreviations*: GMFCS, Gross Motor Function Classification System; GMFM, Gross Function Measure; FMS, Function Mobility Scale; m, meter; SD, Standard Deviations; m, meters; Sag P, Sagittal Plan; Min, minimum; Max, maximum; R, Right; L, Left.

Subject		1		2		3	
Age (years)		15		14		14	
Gender		Male		Female		Male	
Diagnosis		Spastic Diplegia		Spastic Diplegia		Spastic Diplegia	
GMFCS		II		II		II	
GMFM							
Dimension D		87%		71%		66%	
Dimension E		73%		66%		68%	
FMS							
5 m		5		5		5	
50 m		5		5		1	
500 m		5		5		1	
Velocity (m/s)		0.63±0.06		0.68±0		0.85±0.05	
StrideLength (m)		0.76±0.03		0.86±0.02		0.85±0.03	
Kinematics of Stance (Means±SD) (Sag P)		R	L	R	L	R	L
Pelvis	Initial Contact	3.9±1.6	5.7±1.3	6.9±0.8	14.4±0.4	9.5±1.2	14.6±1.5
Hip	Initial Contact	38.8±2.1	37.7±0.8	27.3±0.4	40.8±1.1	50±0.7	46.9±1.2
	Min Flexion	11.5±2.5	6.9±1.7	2.6±0.7	2.6±0.7	20.8±1.1	9.3±1.2
Knee	Initial Contact	43.6±1.1	36.2±2.4	27.1±0.1	28.1±1.1	54.3±0.8	37.8±0.8
	Min Flexion	46.8±1.9	32.1±2	19.9±1	25.4±1.1	55±0.7	39.1±0.7
Ankle	Initial Contact	4.6±3.1	4.9±1.7	2.8±0.9	8.1±1.2	-3.8±1.8	6.2±0.2
	Max Dorsiflex	15.8±1.4	20.4±1.6	23±0.5	22.9±0.4	4.8±0.3	19.9±0.5

a) Kinematics

The registered kinematic data (Table 1) were in agreement with expected values according to the diagnosis and clinical characteristics of the children. Increased knee flexion at initial contact and diminished knee extension during stance were present in all the children.

b) EMG

Fig. 1 shows the EMG data from both legs of the patients in comparison with data of adult healthy subjects [15]. All the subjects present abnormal TA and GAS activity. TA does not present the peak of activity at the very beginning of the gate cycle (GC), while GAS has a prolonged activation during the first half of the GC, then decreases with no bell shape activation. In most cases, all the other muscles present a regular healthy shape, with the exception of VL, VM and RF in Subject 1 and 3 (no peak at the beginning of the GC), and BF and SEM in Subject 2 (very noisy).

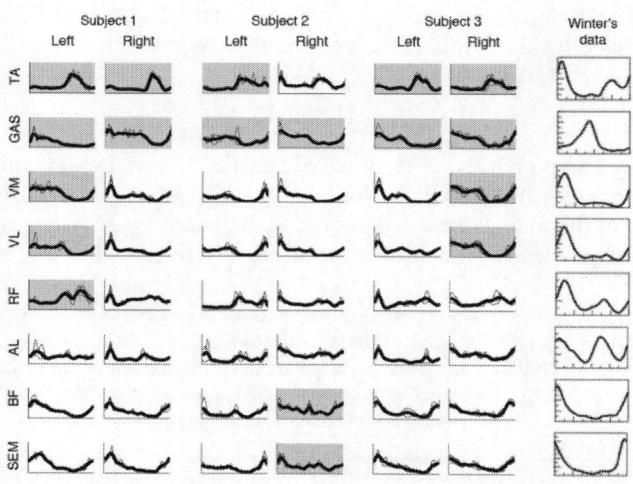

Fig. 1 EMG data of the three subjects. Highlighted in gray are those profiles that vary significantly from the healthy data (from Winter 1991).

c) Muscle Synergies

As shown in Table 2, all the subjects present a VAF greater than 90% at the first iteration. The second iteration (3 synergies) increased the VAF of 1-3%. During the next two iterations (4 and 5 synergies) produced a further global increase of 2 %. This means that only two synergies are sufficient to describe most of the variability of the EMG.

Table 2 Variability accounted for (VAF) as a function of the number of muscle synergies. In bold are the values chosen according to the decision rules established.

N° synergies	Subject 1		Subject 2		Subject 3	
	Left	Right	Left	Right	Left	Right
2	**93%**	**93%**	**90%**	**96%**	**93%**	**96%**
3	95%	96%	93%	97%	95%	97%
4	96%	98%	97%	98%	97%	98%
5	98%	98%	98%	98%	98%	99%

In Fig. 2 muscle synergies (W) and their corresponding time varying activations (H) are depicted. Synergy 1 is mostly associated with TA and RF, while the other muscles are prevalently included in synergy 2. The activations show quasi-sinusoidal time profiles, with a clearly observable phase shift of approximately 50% of the gait cycle between the two activations. Qualitatively, synergies and activations have similar shape across subjects and side.

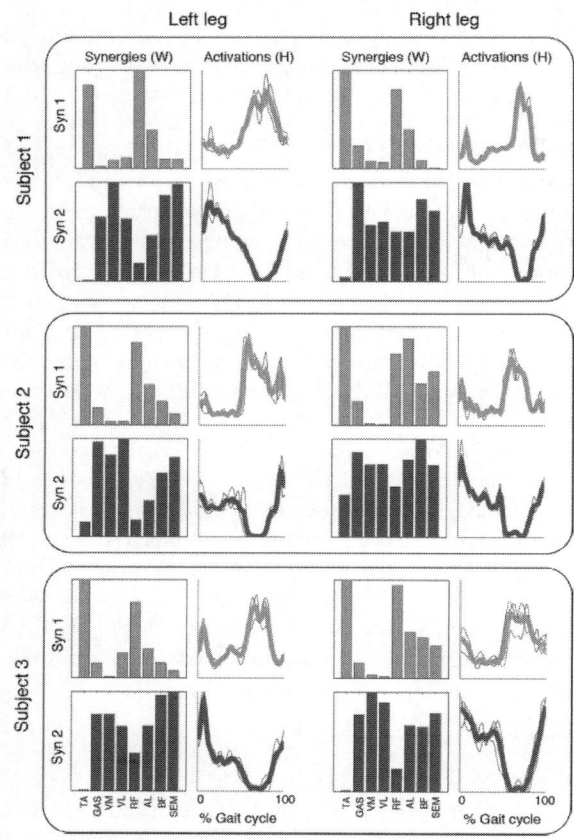

Fig. 2 Synergies (W) and activations (H) in right and left side.

IV. DISCUSSION

The results here presented show that the CP patients here analyzed showed evident synergistic control of the lower limb muscles, on both sides. Comparing these results with previous studies on walking [4, 14, 16], the first relevant difference concerns the number of synergies. The EMG recorded in our subjects can be well explained by only two synergies, while adult normal walking requires from three to five synergies. We found two possible explications for such evidence: a first possible cause is that, since only three gait cycles per person and per leg have been recorded, the low variability in the data permit to easily reconstruct the signal by a linear combination of a low-dimensional (two in this case) weights. The second explication is related to the hypothesis of this study, according to which children with CP would use very primitive control strategies, so that muscle activity result constrained by a very low number of control modules. In order to discriminate between these two possibilities, further experiments are needed, which should include more gait cycles and more numerous population.

Also, a statistical analysis of the results will be needed, which was not possible to perform in this study, due to the low number of trials and patients.

Another interesting result of our experiments is the high similarity of both synergies and activations across subjects. This similarity is maintained also when the EMG variability of individual muscles is consistently different (Fig. 1).

The last relevant point concerns the shape of synergies, as compared to synergies of healthy people. Synergy 1 resembles very well the synergy n° 3 of healthy subject, which is responsible for leg swing, triggering the activation of TA and RF. On contrary, synergy 2 presents no similarity to any other healthy synergies. Here, a possible mechanism that may be occurring is the 'merging' of different synergies in one. This mechanism has been already observed in stroke patients [14, 17] and may be present also in CP. According to this interpretation, we hypothesize that the healthy synergies 1, 2 and 4 may have merged in the second synergy. This fact may be supported by observing that, by summing the activation profiles of the three mentioned healthy synergies, we will obtain a time profile that is similar to the second activation.

V. CONCLUSION

With this work we have given preliminary results supporting the hypothesis that modular control of walking is affected in children with CP. The observed differences with respect to healthy subjects, together with the evident similarity across patients, support the idea that muscle synergies analysis has strong potential for objective evaluation of the pathological control of the movement. Nevertheless, a more extensive and quantitative analysis on wider population is needed to confirm these results.

REFERENCES

1. Roh J, Cheung VC, Bizzi E. Modules in the brain stem and spinal cord underlying motor behaviors. Journal of neurophysiology. 2011 Sep;106(3):1363-78. PubMed PMID: 21653716.
2. Bizzi E, Cheung VC, d'Avella A, Saltiel P, Tresch M. Combining modules for movement. Brain Res Rev. 2008 Jan;57(1):125-33.
3. Lacquaniti F, Ivanenko YP, Zago M. Patterned control of human locomotion. The Journal of physiology. 2012 May 1;590(Pt 10):2189-99.
4. Dominici N, Ivanenko YP, Cappellini G, d'Avella A, Mondi V, Cicchese M, et al. Locomotor primitives in newborn babies and their development. Science. 2011 Nov 18;334(6058):997-9.
5. Farmer SE. Key factors in the development of lower limb coordination: implication for the acquisition of walking in children with cerebral palsy. Disability and Rehabilitation. 2003;25(14):807-16.
6. Sutherland DH, Olshen R, Cooper L, Woo SLY. The development of mature gait. The Journal of Bone and Joint Surgery. 1980;62:336-53.
7. Granata KP, Padua DA, Abel MF. Repeatability of surface EMG during gait in children. Gait Posture. 2005 Dec;22(4):346-50. PubMed PMID: 16274917.
8. Rosenbaum P, Paneth N, Leviton A, Goldstein M, Bax M. A report: the definition and classification of cerebral palsy April 2006. Dev Med Child Neurol. 2007;49(Suppl.109):8-14.
9. Bache CE, Selber P, Graham HK. The management of spastic diplegia. Current Orthopaedics. 2003;17:88-104.
10. Glew GM, Bennett FC. Cerebral Palsy Grown Up. Journal of Developmental and Behavioral Pediatrics. 2011;32(6):469-75.
11. Lacquaniti F, Ivanenko YP, Zago M. Development of human locomotion. Current opinion in neurobiology. 2012 Oct;22(5):822-8.
12. Hug F, Turpin NA, Couturier A, Dorel S. Consistency of muscle synergies during pedaling across different mechanical constraints. Journal of neurophysiology. 2011;106:91-103.
13. Hug F. Can muscle coordination be precisely studied by surface electromyography? Journal of electromyography and kinesiology : official journal of the International Society of Electrophysiological Kinesiology. 2011 Feb;21(1):1-12. P
14. Clark DJ, Ting LH, Zajac FE, Neptune RR, Kautz SA. Merging of healthy motor modules predicts reduced locomotor performance and muscle coordination complexity post-stroke. Journal of neurophysiology. 2010 Feb;103(2):844-57.
15. Winter DA. The biomechanics and motor control of human gait: normal, elderly and pathological. Waterloo Biomechanics Press, Waterloo, Ontario. 1991.
16. Ivanenko YP, Poppele RE, Lacquaniti F. Five basic muscle activation patterns account for muscle activity during human locomotion. The Journal of physiology. 2004 Apr 1;556(Pt 1): 267-82.
17. Cheung VC, Turolla A, Agostini M, Silvoni S, Bennis C, Kasi P, et al. Muscle synergy patterns as physiological markers of motor cortical damage. Proceedings of the National Academy of Sciences of the United States of America. 2012 Sep 4;109(36):14652-6.

Corresponding Author

Author: Diego Torricelli
Institute: Spanish Research National Council (CSIC)
Street: Ctra. Campo Real, km 0.200
City: La Poveda, Madrid
Country: Spain
Email: diego.torricelli@csic.es

Time-Frequency Analysis of Error Related Activity in Anterior Cingulate Cortex

G. Mijatovic[1], T. Loncar Turukalo[1], E. Procyk[2], and D. Bajic[1]

[1] Department of Telecommunications and Signal Processing, University of Novi Sad, Serbia
[2] Inserm U846 Stem Cell and Brain Research Institute, University of Lyon, France

Abstract—Error related activity in local field potentials recorded in dorsal anterior cingulate cortex (dACC) of non-human primates is analyzed using ensemble empirical mode decomposition (EEMD). Neuronal activity in this region was recorded in a male rhesus monkey trained in the problem solving task (PS). It was hypothesized that activity in this region could reflect a modification of control just before execution errors, hence predicting the error about to be committed. The features obtained from time-frequency energy distribution, intrinsic mode functions (IMF) and higher order statistics calculated over IMFs were used to distinguish error related activity. The analysis has revealed the significant increase in energy of IMF4-IMF7, resulting in significant total energy increase immediately before and after execution errors. Higher order spectra analysis determined no difference in symmetry and shape of sample distribution when compared to control segments from successful trials. However, increased sample variance during errors combined with increase in energy indicates the unstable neural activation in this cortical region during execution errors resulting in less control over the current processing.

Keywords—ensemble empirical mode decomposition, higher order spectra, local field potentials, anterior cingulate cortex, execution errors.

I. INTRODUCTION

It is considered that dorsal anterior cingulate cortex (dACC) has an important role in problem solving, error recognition, adaptive response and emotional self-control [1]. dACC is engaged in both the novelty detection and response, providing for the fast environmental adaptations [2]. In primates, dACC is involved in detection and exploitation of errors, which is crucial for learning ability, adaptations of actions, movements and complex behaviors [3]. The invasive extracellular recordings which can be done in non-humane primates, can resolve among many hypothesis on dACC function. In a problem solving task alternating between exploration and exploitation periods, and where animals had to fixate eye position on central and peripheral targets, it was determined that dACC produced signals specific to controlling actions and monitoring feedback during the exploration period. Some signals were relevant for defining the shift between exploration and exploitation, other being selective of specific errors [4]. We have analyzed the error-related activity primarily focusing on the execution errors (breaks of fixation: BKF). Break of fixations are potentially arising from a drop in control of eye position, a drop in attention, or from impulsive responding.

There are several experimental studies analyzing error related activity in dACC of non-human primates, mainly restricted to single unit activities (SUA) [3]. Because dACC also produces anticipatory controlling or monitoring signals [4], it was hypothesized that activity in this region could reflect a modification of control just before BKF, hence predicting the error about to be committed. To do so single unit activity was used for identification of error related units and the activity of associated local field potentials (LFP) were analyzed in a short time span around the execution errors. We evaluated the possibility to identify changes in the LFP activity around the execution errors as compared to the LFP activity in the same time span of successful trials.

Empirical mode decomposition (EMD) was used for LFP analysis, for its adaptive, data driven decomposition basis, with no assumptions regarding linearity and stationarity of the signals [5]. The EMD enables analysis of time-varying data, giving the insight into mean frequency and the energy distribution among frequency bands in time. The signal decomposition onto intrinsic mode functions (IMF), the data driven decomposition base, might lead to certain problems, such as mode mixing. To overcome these problems ensemble EMD (EEMD) was used [6]

The higher order statistics calculated on IMF and the features of the IMFs and time-frequency energy distribution were used as descriptive statistics. The performance of these features is evaluated on the experimental signals obtained from dACC region in awake behaving macaque monkey.

II. MATERIALS AND METHODS

A. Experimental Procedures

All the experimental procedures followed the European Community Council Directive from 1986 and Départementale des Services Vétérinaires (Lyon, France) including: housing, surgical, electrophysiological, and histological procedures. The description of experimental procedures are given in [4] explaining surgical, experimental and task details. In brief a male rhesus monkey was trained in the problem solving task (PS). Monkey had to find by trial and error which target, presented in a set of four, was rewarded. A problem was composed of two periods: a search period that included all incorrect choices up to the first correct touch (exploration) and a repetition period wherein the animal was required to repeat the correct touch several times (exploitation).

At the start of a trial a lever is touched and central fixation spot illuminated. Subjects were required to fixate the fixation point during the delay period until targets onset (ON), and fixate the target once selected by eye. Any break in fixation requirements resulted in trial cessation. The animal then waits for a GO signal to touch the selected target. Afterwards, all targets switch off, and the feedback is given (no reward: negative; reward: positive, see Fig.1)

Neuronal activity was recorded using high impedance epoxy-coated tungsten electrodes, one to four in a set (1–4 MΩ at 1 kHz; FHC Inc, USA)[4].

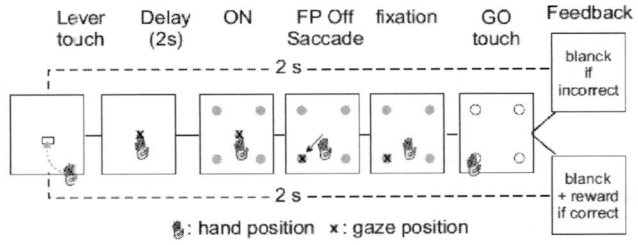

Fig. 1 The trial procedure

B. Preprocessing of Extracellular Recordings

Raw extracellular recordings were filtered using high pass two poles Butterworth filter with cutoff at 250Hz and 4 poles Butterworth low pass filter with cutoff 3KHz for extraction of multiunit activity (MUA) and 280Hz for LFP extraction. Sampling frequency for MUA was 12.5 kHz and for LFP 781.25 Hz. Identification of single unit activity was done using online spike sorting (MSD, AlphaOmega). Since the amplitude of waveforms depends on the electrode position with respect to the recorded cells, LFP data were normalized prior to analysis to zero mean and unit variance.

The complete data set has 487 recordings from dACC, yet in only 74 recordings BKF related activity was noticed. The length of recordings varied from 945s to 2205s. Among execution errors, the analysis was restricted to BKFs happened from all targets onset until the GO signal (see Fig.1). The total number of recordings containing relevant number of BKFs in this interval was 32 and the average number of BKF points per recording was 15.47±1.47. From both the LFP and SUA only period 1s before and 0.5s after BKF was analyzed.

C. Ensemble Empirical Mode Decomposition

The empirical mode decomposition (EMD), a non-linear method proposed by Huang [5], performs signal decomposition onto finite number of zero mean AM-FM components called intrinsic mode functions (IMF). EMD doesn't suffer from stationarity and linearity assumptions, thus being suitable for analysis of signals stemming from systems with nonlinear dynamics and signals with significant time variations.

The decomposition procedure is empirical and based on characteristic local time scales of the data resulting in IMFs with well-behaved Hilbert transform yielding instantaneous frequencies as functions of time. Energy-time-frequency distributions called Hilbert energy spectrum enables presentation of energy allocation in time and frequency domains.

The EMD decomposition, described in detail in [5], results in finite number K of IMFs and residual $r(n)$ allowing for the perfect reconstruction of the original signal x(n):

$$x(n) = \sum_{i=1}^{K} IMF_i(n) + r(n) \quad (1)$$

When implementing the basic EMD algorithms certain problems can be experienced. Mode mixing denotes the problem of the simultaneous presence of similar oscillations in several modes. Another problem might be encountered within a mode, if the amplitudes of the present oscillations are very disparate. To overcome these problems ensemble EMD (EEMD) is proposed [6] performing EMD over an ensemble of the signal with added Gaussian white noise of limited variance. The procedure can be simplified to the following steps:

1. Generation of ensemble $x_i(t)=x(t) + w_i(t)$, where $w_i(t)$ is the ith realization of Gaussian white noise, $i=1,...L$
2. EMD decomposition of $x_i(t)$ onto $IMF_i^k(t)$, $k=1,...K$
3. The kth IMF, $\overline{IMF^k}(t)$ is calculated as

$$\overline{IMF^k}(t) = \frac{1}{L}\sum_{i=1}^{L} IMF_i^k(t), k=1,...K$$

The IMF extraction is done on LFP segments of interest, 1.5s before and 1s after BKF point. The segments belonging to one recording were grouped representing a BKF ensemble. The comparison was done with an ensemble of control segments taken from all of the successful trials within the same recording. The reference time point for control segments (the point corresponding to BKF point) is determined as the mean value of the time instants of the BKF points within a trial.

Mean frequency and energy were calculated for extracted IMF. The energy of each IMF was estimated as:

$$E_{IMF}[k] = \sum_{n=1}^{N} IMF_k^2(n) \quad k=1,..K \quad (2)$$

Average Hilbert energy spectra (HES) were determined for all BKF LFP segments within the same recording. The time-course of energy concentrated in β [16-30], low γ [30-60], and high γ [60-140] bands was determined from average HES, as the special case of marginal spectra. The marginal energy spectrum over time was calculated from HES to determine the changes in overall energy around BKF points.

Additional descriptive statistics include variance, and higher order spectra statistics: skewness and kurtosis. The motivation for use of these statistics is tracking of eventual changes in distribution of samples, as they assess the asymmetry and concentration around the mean, respectively. These parameters were calculated in the time windows of 0.16s (128 samples) using sliding window procedure, to satisfy the stationarity assumptions needed for calculation of the basic statistics. As a result the time course of these parameters is obtained.

All results are presented as mean ± SEM. The statistical analysis was done in GraphPad Prism software. Wilcoxon matched pairs test was used to compare ensembles of BKF and the corresponding ensembles of control segments.

III. RESULTS

The LFP segments around BKF points and corresponding LFP control segments were decomposed using EEMD with addition of noise of variance 0.01 using ensemble of L=10 realizations. To account for edge effects LFP segments were cut 0.5s longer than specified on each side. The bottom panel of Fig. 2a presents the average LFP signal around BKF synchronized on the time of BKF (drawn at 0s) obtained from one of the recordings. Associated spiking activity in all BKF segments in the same recording is presented in the top panel, and peristimulus time histogram (PSTH) for BKFs is presented in the middle panel of Fig.2a. The activity during the corresponding control segments, cut from successful trials within the same recording site, is presented in Fig.2b.

In 68.75% of recordings spiking activity was significantly higher in BKF segments, while in 31,25% of recording sites spiking activity was significantly higher in control segments. The average spiking activity heavily depends on position of recording site in dACC, thus only matched observations result in the meaningful comparison.

The EEMD decomposition resulted consistently in 9 IMFs which average frequency matched for BKF segments and controls. The fastest component, IMF1, with average frequency of 196±5.89Hz captures noise and the spectral content of spiking activity which spreads partially in LFP band. Average frequency for other IMFs were: 98,18±2,71Hz (IMF2), 44,88±1,63Hz (IMF3), 20,06±0,99Hz (IMF4), 9,00±0,53Hz (IMF5), 4,18±0,30Hz (IMF6), 1,86±0,1 (IMF7), 0,90±0,05Hz (IMF8), 0,46±0,03Hz (IMF9).

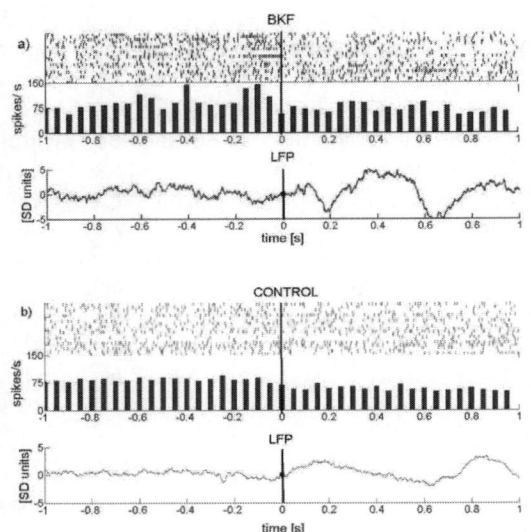

Fig. 2 Spiking activity, PSTH and average LFP segment for a)BKF and b)control segments from one recording site

The energies calculated from IMFs showed that the major energy contribution in LFP comes from IMF5, IMF6 and IMF7, i.e. IMFs in the frequency range of 2-12Hz. However, it could be noticed that IMF1 has significant impact on the overall energy, but without significant difference between BKF and controls. In Fig.3 the mean energy estimates over IMFs in BKF and control segments are presented. It could be noticed that IMF4, IMF5, IMF6 and IMF7 have significantly increased energy compared to controls.

Average HES for BKF and control segments of one of the recordings are presented in Fig.4, reflecting the typical changes in all recordings. The graphs indicate that significant LFP activity is as well present in β and low γ range. It could be noticed that during BKF, the increased high γ activity is present as well. The marginal energy in time over these three bands has showed that the mean value of energy in these bands is not significantly increased, but the variations of energy in these bands is more distinguished in BKF segments, especially in high γ band. The marginal HES spectra over time, presented in Fig.5, has revealed the significant increase in the total energy for BKF segments 50ms before and 0.5s after BKF points.

The time course of higher order spectra statistics, skewness and kurtosis, revealed no differences in sample distributions. However, the variances calculated over the small time windows of 0.16s, using sliding window approach, have revealed that the sample variance is significantly higher in BKF LFP segments, both before and after BKFs.

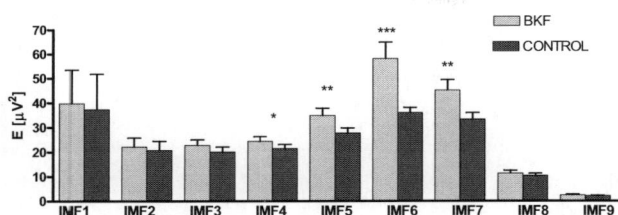

Fig. 3 Energy per IMFs in BKF and control LFP segments

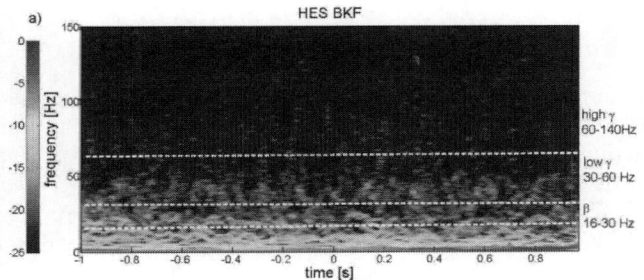

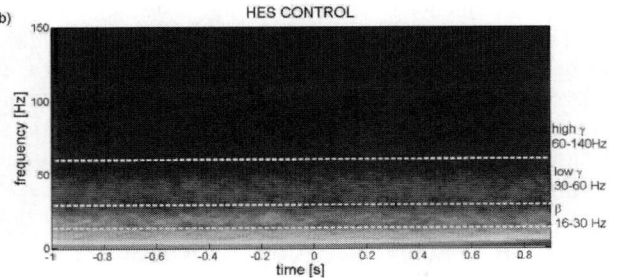

Fig. 4 Average HES for a)BKFs and b)control segments from one recording site

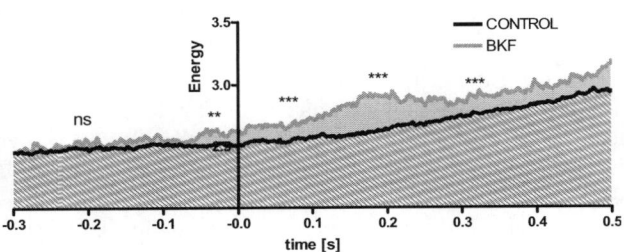

Fig. 5 Marginal average HES in time for control and BKF segments

IV. CONCLUSIONS

Time-frequency analysis of LFP segments during execution errors of breaking the fixation was done using EEMD. The resulting HES capture activities in all frequency bands of interest. The decomposition yields nine IMFs, whose energy changes in time could be tracked both separately and in total. The analysis has revealed the significant increase in energy of IMF4÷IMF7, resulting in significant total energy increase immediately before and after BKFs.

Combined with the increased variance of samples, these results might indicate that during execution errors the neural activation in this cortical region is unstable which might reflect less control over the current processing necessary to complete the trial properly. The energy increase after BKF might reflect the processing of the event (negative feedback) which was as well noticed at the level of unit activity [4].

Time-frequency analysis of single unit recordings and associated LFP can provide some features for prediction of BKF events. For more reliable prediction of BKF events analysis should include the simultaneous activity in surrounding ensemble of neurons.

ACKNOWLEDGMENT

This research is supported in part by grants TR32040 and III 43002 of Serbian Ministry of Science.

REFERENCES

1. Allman JM, Hakeem AA, Erwin JM et al. (2001) The anterior cingulate cortex. Annals of the New York Academy of Sciences 935: 107-117
2. Ranganath C, Rainer G (2003) Neural Mechanisms for Detecting and Remembering Novel Events. Nature Reviews Neuroscience 4:193-202
3. Amiez C, Joseph JP, Procyk E (2005) Anterior cingulate error-related activity is modulated by predicted reward. European Journal of Neuroscience 21: 3447–3452
4. Rothé M, Quilodran R, Sallet J, Procyk E (2011) Coordination of High Gamma Activity in Anterior Cingulate and Lateral Prefrontal Cortical Areas during Adaptation, The Journal of Neuroscience, 31(31):11110–11117
5. Huang NE et al.(1998) The empirical mode decomposition and the Hilbert spectrum for nonlinear and non-stationary time series analysis. Proc. R. Soc. Lond. A, vol. 454: 903–995, 1998.
6. Wu Z, Huang NE (2009) Ensemble empirical mode decomposition: A noise-assisted data analysis method. Advances in Adaptive Data Analysis, 1: 1–41

Author: Tatjana Loncar-Turukalo
Institute: Faculty of Technical Sciences, University of Novi Sad
Street: Trg Dositeja Obradovica 6
City: Novi Sad
Country: Serbia
Email: turukalo@uns.ac.rs

Depth-Sensitive Algorithm to Localize Sources Using Minimum Norm Estimations

B. Pinto[1,2], A.C. Sousa[2,3], and C. Quintão[1,3]

[1] Physics Department of the Faculty of Sciences and Technology of NOVA University of Lisbon,
2829-516 Caparica, Portugal
[2] Physics Department of the Faculty of Sciences and Technology of the University of Algarve, Faro, Portugal
[3] Institute of Biophysics and Biomedical Engineering of the Faculty of Sciences of the University of Lisbon,
Lisbon, Portugal

Abstract—The main objective of this paper is to apply and evaluate the neural source localisation accuracy of a new depth-sensitive algorithm using physiological data. This new algorithm is based on the behaviour of the dispersion of the minimum norm solutions (MNE), and was already tested with simulated data [1], yielding an accuracy of 2 to 4 mm in noise-free situations, and a mean accuracy of 10 mm for more disadvantageous situations. We estimate now the neural source depth in EEG recordings, namely in focal epileptic interictal paroxisms and in N100 auditory evoked potentials of normal volunteers. We show that the accuracy of the method is comparable with the commonly used dipolar localizations, when it was applied to those bio-signal recordings. It was revealed that, under adequate constraints, the algorithm is suitable for the estimation of the depth of one or two simultaneous neural generators, using a rather simple MNE approach. This study, demonstrating that MNE can handle spatial-limited sources successfully, opens the possibility to localize both quasi-punctual and extended neural generators using only the simplest MNE algorithm.

Keywords—Depth-Sensitive Miminum Norm Estimation; EEG Source Localization.

I. INTRODUCTION

The localization of the electric sources in the brain, which are responsible for the EEG signal, is the so-called inverse problem. This kind of problems is known to have an infinite number of solutions, and additional modelling criteria are required to choose among them. There are two main approaches to deal with this indeterminacy: the use of dipolar models, and the minimum norm based approaches [2]. In the former, the search for the solution is limited to the localization of a small number of dipolar sources. This model involves iterative procedures to reach the optimal solution, assumes that the sources are point-like [3], and needs *a priori* information about the number of such sources [4]. On the other hand, the minimum norm estimate (MNE), and its variants, are methods suitable for extended sources, which do not demand information of the number of active sources, since their output can be considered an image of current density [5]. Being this feature the main advantage of minimum norm methods when compared with the dipolar approach. However, these methods tend to spread the neural activity of point-like sources to an extended region of the cortex. In addition, MNE is extremely sensitive to white noise, and the solutions found are always superficial [6, 7].

We study, with physiological data, a new source localization method, DS-MNE (Depth-Sensitive Minimum Norm Estimates), which was introduced and tested on simulated data [1]. Despite the good results attained with simulations, in the current paper we verify if the behavior of the DS-MNE method is also reliable when it is applied to real data and how the results obtained through this method are in agreement with those obtained by the methods usually applied on the clinical procedures.

II. MATERIALS AND METHODS

A. Minimum Norm Estimate

The MNE is a mathematical method built around a lead field function. The lead field in electrode k, $\vec{L}_k(\vec{r})$, depends on the electric characteristics of the sources, or impressed currents, $\vec{J}^i$, and of its location $\vec{r}$.

In order to obtain the matrix of the lead fields it is necessary to calculate the potentials at the electrodes over the scalp, originated by unitary current dipoles along the three spatial co-ordinates arranged in a set of discretized points, wich fill the brain volume [8]. Knowing L, a system of equations, with J as the unknown vector should be solved. This system is clearly underdetermined, with an infinite number of possible solutions, since there are more unknowns (dipole components) than equations (number of electrodes). The Moore-Penrose pseudo-inverse procedure [9], is usually used to solve this kind of equation systems.

B. Dispersion Parameter

We studied if at certain depths the sources were more spatially confined and confirmed that when we were dealing with simulated punctual sources (dipoles). It was observed that the current density image with traditional MNE was less spread in the depth where the simulated dipole was placed (Fig. 1). This observation was guiding us in order to create a parameter, δ, which quantifies the extension of the source image created by the simple MNE at various depths. So, the parameter δ is an estimator of the dispersion of the solutions centered on their center of mass and is used to estimate the depth of the neural sources.

To calculate δ one needs to find the point on the discretized cortex associated to the maximum amplitude of the solution obtained by the usual MNE approach. After that, the reference frame should be rotated in order to make the z axis passing through this point. Then, the spherical co-ordinates of each grid point $(R, \vartheta_i, \varphi_i)$ were transformed in a radial, plain, reference frame, according to the equations:

$$X_i' = s_i \cos \vartheta_i = \varphi_i R \cos \vartheta_i \qquad Y_i' = s_i \sin \vartheta_i = \varphi_i R \sin \vartheta_i \quad (1)$$

where the variable s_i is the arc associated to i point, measured from the z axis. This procedure allows transforming the calculation of the dispersion parameter into a 2D problem. Then, a weighted center of mass of the source activity was defined as [6]:

$$X_{CM}' = \frac{\sum_{i=1}^{m} X_i' \left(J_{i,x}^2 + J_{i,y}^2 + J_{i,z}^2 \right)}{\sum_{i=1}^{m} \left(J_{i,x}^2 + J_{i,y}^2 + J_{i,z}^2 \right)} \quad (2)$$

where X_i' is the co-ordinate along the x axis of the ith point of the grid; $J_{i,x}$ is the component of the solution in the same point, along the x axis; and m is the number of the grid points.

After that, the co-ordinates of the center of mass (X_{CM}', Y_{CM}') calculated using equation (2) were used as the reference point in relation to which the dispersion of the solution, δ, was calculated:

$$\delta = \frac{\sum_{i=1}^{k} \left[(X_{CM}' - X_i')^2 + (Y_{CM}' - Y_i')^2 \right] J_i^2}{R \sum_{i=1}^{m} J_i^2} \quad (3)$$

where k is the number of the grid points and R is the radius of the grid. We decided to divide the parameter by R in order to obtain angular units for δ parameter, and so that we can compare the results between different depths.

C. The Evaluation of the δ Parameter

To find the depth where δ presents a minimum, we solved the inverse problem, by means the usual MNE procedure, using 46 grids with different radii, separately, and we calculated δ for all solutions. Then, we drew, using MatLab tools, a plot of the δ parameter against the depth of each grid, and we fitted a seven degree polynomial curve, in order to determine an unequivocal minimum, as it was illustrated in Fig. 2 [1].

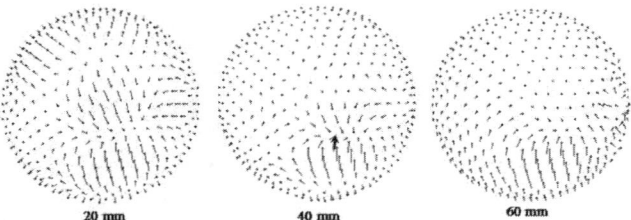

Fig. 1 Representation of three minimum norm solutions related to grids with different radii. The simulated dipole is represented in bold. Minimum spread of the solution at the depth of the simulated dipole is observed.

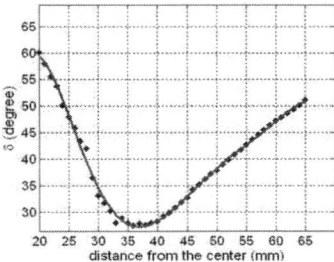

Fig. 2 Plot of the δ parameter associated to one dipole located 40 mm far way from the center of the sphere which models the head, against the radii of the grids. The line is the result of a seven degree polynomial.

D. Head Model and Inverse Problem

In the current paper, DS-MNE uses a four concentric spherical model to approximate the brain, the CSF, the skull and the scalp (constructed using a cover of 1280 similar triangles each). The direct problem, i.e. the search for the electric potentials generated at the scalp by a particular configuration of neuronal sources, was solved using an algorithmic implementation of Boundary Element Method (BEM), using C-language. The approach [10], constructs the lead field matrix allowing for the MNE to obtain the depth of all the sources within the four spheres. Furthermore, since the minimum norm localization techniques require discretization of the space where the sources are located (brain), we constructed a set of 46 spherical grids of 642 points each, with radii varying from 20 to 65 mm and an inter-grid distance of 1 mm.

To compare the results obtained by this method with the ones reached by dipolar sources we had to solve the inverse problem. The above mentioned C-program was used for the first data set and BESA software was used for the later. In both cases, we used a realistic head model, considering three head structures: the brain, the skull and the scalp.

E. Recorded Data

We apply the method in two distinct data sets. The first comprised paroxistic data, which is associated to focal epileptic activity and has its origin in a single neural generator, a group suitable for dipolar modelling. The second comprised N100 evoked potentials, originated by two well-known temporal sources. The former signals were recorded with 21 channel Biologic equipment. Our evaluation is made around the instant of maximum amplitude of a paroxysmal interictal signal of each patient [11].

The second data set was recorded using a 32 channel BioSemi. The record was made using a sampling frequency of 2048 Hz. The stimuli were 1000 Hz beeps during 60 ms, repeated every 1.3 s, and delivered to both ears. We have made a total of 120 stimuli, which was subsequently averaged. Each epoch had a total duration of 1000 ms, comprising a baseline of 250 ms prior to the stimuli. After averaging, we applied a high pass filter of 1.6 Hz, a low pass filter of 35 Hz and a notch filter of 50 Hz. The data was recording from 14 normal volunteers. The evaluation of our method was performed assuming that this response is generated by a set of two dipolar sources located in the primary auditory cortex. We selected 32 instants around the maximum amplitude of N100 evoked potential and imposed that the two sources, during this period, should stay at the same location, and be placed symmetrically to each other in relation to the inter-hemispherical commisure.

III. RESULTS

A. Application to Epileptic Patients

Concerning the epileptic interictal activity, we first localized an equivalent source using the dipolar model for all patients. These allow the generation of a suitable realistic solution. We determined the distance between the center of the sphere that better fits the head and the localization of the equivalent current. Then, we applied the MNE technique, plotted δ as a function of the depth of each grid and determined its minimum (Fig. 3). These results, and the comparisons between the two methods, are summarized in table 1.

Table 1 Summary of the results corresponding to the 5 patients with focal epilepsy: first row - distance between the localization of the current dipole obtained by BEM procedures and the centre of the sphere. Second row - distance between the localization obtained by DS-MNE and the centre of the sphere. Third row - differences between the two approaches.

Patients	1	2	3	4	5
Dipole model (mm)	60	38	26	43	42
DS-MNE (mm)	46	44	36	54	50
Difference (mm)	14	6	10	11	8

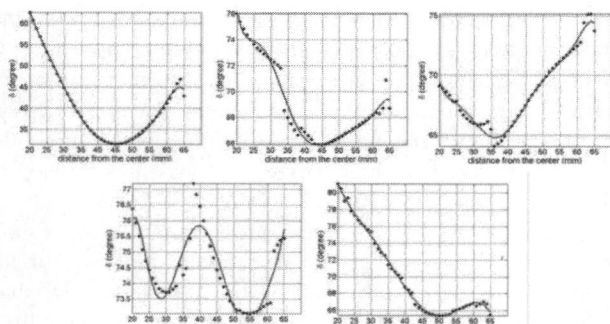

Fig. 3 Plots of the δ parameter against the radii of the grids, for the 5 cases of focal epilepsy. As in Fig 1, the line is the result of a seven degree polynomial fit and the minimum of each curve is presented in table 1.

B. Application to N100 Auditory Evoked Potential

Concerning the application of DS-MNE to N100 auditory evoked potential, it was assumed that there is more than one source active at the same time and to overcome this problem, we evaluated the behaviour of δ for each brain hemisphere separately, since we expect these sources to be symmetric (Fig 4). To obtain the localization of the two sources we proceeded as in the epileptic data set analysis. The two methods were then compared. The results are summarized in table 2.

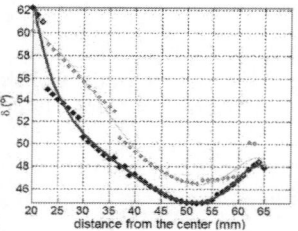

Fig. 4 Plots of the δ parameter against the radii of the grids, for a N100 auditory evoked potentials. As in Fig. 1, the lines are the result of a seven degree polynomial fit and a summary of the minima of both curves for the 14 subjects are presented in table 2. In these cases right and left cerebral hemispheres are studied separately.

Table 2 Summary of the results correspond to N100 potentials of 14 persons subjected to auditory stimuli. Minimum, maximum, mean and standard deviation of: first column - the distances between the localization of the current dipoles and the centre of the sphere; second column - distance between the localizations obtained by DS-MNE applied separately to each cerebral hemisphere and the centre of sphere; third column - differences between the two approaches.

	DP	MNE(R)	dist
Minimum (mm)	33	40	0
Maximum (mm)	59	61	23
Mean (mm)	47	51	7
Stand. Dev. (mm)	8	5	6

IV. CONCLUSIONS

Generally, it is very difficult, to find accurately the depth of neural sources using exclusively the minimum norm estimate approach [12]. Yet, solutions have been proposed, such as, LORETA (low resolution brain electromagnetic tomography[13]); FOCUSS (focal underdetermined system solution[14]); LAURA (local autoregressive average[12]); and EPIFOCUS, a linear inverse (quasi) solution [15] among others. Our approach involves less complex processing, is rather intuitive, might give better results than dipolar approach in the presence of extended sources, since it does not make any a priori assumptions about the geometry of the sources, and it seems that could be a new conception for dealing with MNE method. In addition, it is easy to be used by those who are familiarized by the common MNE approach.

In the current study, we evaluated the depth resolution of the sources and saw that the mean difference between the localizations obtained by the DS-MNE and those obtained by dipolar models is about 10 mm for epileptic discharges, and about 7 mm for N100 auditory evoked potentials. Such results show that using exclusively MNE procedures it is possible to find the depth of neural sources in physiological data with an attaining error bound of about 1 cm, a value clearly comparable to those acceptable in EEG source localization.

In summary, we can state that the major novelty of this work is the introduction of a new method, DS-MNE, based on the simple MNE approach, without using regularized matrix, which allows depth localization of the focal neural sources. In a previous publication we gave all the technical aspects of the method, and we evaluated its performance with simulated data. Now, we demonstrate that this method also succeed when it was applied to EEG data. The natural extension of this work is applying DS-MNE to different sources data sets, besides the pontual sources, which are not well modeled with dipolar methods.

ACKNOWLEDGMENT

The authors thank T. Oostendorp for providing the software of BEM, thank John Peter Foreid to help in the EEG recordings and selection of the relevant data from epileptic patients and Ricardo Vigário for his precious comments and suggestions about the manuscript.

REFERENCES

1. Pinto B, Quintão Silva C (2007) A simple method for calculating the depth of EEG sources using minimum norm estimates (MNE). Medical & Biological Engineering & Computing 45(7):643-652.
2. Koles ZJ (1998) Trends in EEG source localization. Electroenceph clin Neurophysiol 106:127-137.
3. Lopes da Silva F, Van Rotterdam A (1999) Biophysical aspects of EEG and magnetoencephalogram generation. In: Niedermayer E, Lopes da Silva F, editors. Electroencephalography, basic principles, clinical applications and related fields, Baltimore: William & Wilkins 93-109.
4. Knuth KH (1997) Difficulties applying recent blind source separation techniques to EEG and MEG. In Ericsson GJ., Rychert JT, Smith CR, editors, Maximum Entropy and Bayesian Methods, Dordrecht: Kluwer Academic.
5. Hämäläinen MS, Ilmoniemi RJ. (1994) Interpreting magnetic fields of the brain: minimum norm estimates. Med. & Biol. Eng. & Comput.;32:35-42.
6. Silva C, Maltez JC, Trindade E, Arriaga A, Ducla-Soares E (2004) Evaluation of L_2 and L_1 minimum norm performances on EEG localizations. Clinical Neurophysiol 115(7):1657-1668.
7. Quintão Silva C, Foreid JP (2006) Localization of EEG sources – an integrated approach. In: Chen FJ, editor, Trends in Brain Mapping Research, New York: Nova Science Publishers, Inc.
8. Sarvas J (1987) Basic mathematical and electromagnetic concepts of the biomagnetic inverse problem. Phys Med Biol 32(1): 11-22.
9. Penrose RA (1955) Generalized inverse for matrices. Proc Cambridge Phil Soc 51:406-13.
10. Oostendorp T, Van Oosterom A (1989) Source parameter estimation in inhomogeneous volume conductors of arbitrary shape. IEEE Trans Biom Engng 36:382-391.
11. Silva C, Almeida R, Oostendorp T, Ducla-Soares E, Foreid JP, Pimentel T. (1999) Interictal spike localization using a standard realistic head model: simulations and analysis of clinical data. Clin Neurophysiol 110:846-855.
12. Grave de Peralta-Menendez R, Gonzalez-Andino SL (1998) A critical analysis of linear inverse solutions to the neuroelectromagnetic inverse problem. IEEE Trans Biomed Eng 45:440-448.
13. Pascual-Marqui RD, Michel CM, Lehmann D (1994) Low resolution electromagnetic tomography: a new method for localizing electrical activity in the brain. Int. J Psychophysiol 18:49-65.
14. Gorodnitsky IF, George JS, Rao BD (1995) Neuromagnetic source imaging with FOCUSS: a recursive weighted minimum norm algorithm. Electroenceph clin Neurophysiol 95:231-251.
15. Grave de Peralta-Menendez R, Gonzalez-Andino SL (2002) Comparison of algorithms for the localization of focal sources: evaluation with simulated data and analysis of experimental data. International Journal of Bioelectromagnetism 4.

Author: Carla Maria Quintão Pereira
Institute: Institute of Biophysics and Biomedical Engineering
Street: Faculty of Sciences of the University of Lisbon, Campo Grande
City: 1749-016 Lisbon
Country: Portugal
Email: cmquintao@fct.unl.pt

Measurement of Gait Movements of a Hemiplegic Subject with Wireless Inertial Sensor System before and after Robotic-Assisted Gait Training in a Day

Takashi Watanabe[1] and Jun Shibasaki[2]

[1] Graduate School of Biomedical Engineering, Tohoku University, Sendai, Miyagi, Japan
[2] Minami Tohoku Hospital, Iwanuma, Miyagi, Japan

Abstract—This study aimed to test a possibility of using wireless inertial sensor system in evaluation of gait rehabilitation clinically. Gait movements of a hemiplegic subject were measured with the sensor system during 10 m walking before and after robotic-assisted gait training in a day. Segment inclination angles and joint angles of the lower limbs were calculated from measured angular velocities and acceleration signals based on Kalman filter. Gait event timings were also detected by foot angular velocities and shank acceleration signals. The measured data showed that gait movement changed after the training, which were considered to be improvements by the instruction of the therapist. The robotic-system could be effective for the training based on the instruction. The wireless inertial sensor system would be useful for measurement and evaluation of movements in rehabilitation clinically.

Keywords—gait, angle, gyroscope, accelerometer, lower limb, rehabilitation.

I. INTRODUCTION

Motor rehabilitation or motor training for health care become more important to patients with motor impairment and elderly persons. Evaluation of motor function is necessary to plan programs of rehabilitation and training and to give instruction in those trainings. Movements or range of motion (ROM) are measured to evaluate motor function. For these purposes, an optical 3D motion analysis system or electronic goniometers are effective for quantitative measurements. However, most of these systems are for laboratory use because of time-consuming process for setup and analysis. Therefore, therapists generally evaluate a level of motor function in rehabilitation by simple methods (e.g., watching movements, measurement of ROM with a manual goniometer, measurement of time for a task, and counting the number of steps in 10 m walking).

Inertial sensors such as accelerometers and gyroscopes can be useful in measurement and analysis of movements in rehabilitation because of its shrinking in size, easiness for settings and low cost. Therefore, applications of those inertial sensors have been studied on gait analysis: gait event detection [1-3], measurement of joint angle or segment tilt angle [4-8], and stride length estimation [9, 10]. Segment inclination angles and joint angles have important information for therapists and patients in evaluation of gait movements. Therefore, a motion measurement system using inertial sensors have to provide those angles calculating from angular velocities and/or acceleration signals.

In our previous study, aiming to realize a wearable gait evaluation system with inertial sensors for rehabilitation support, Kalman-filtering-based joint angle measurement method was tested in measurement of joint angles of lower limbs during gait with healthy subjects [11-13]. Joint angles in the sagittal plane were measured with root mean square error (RMSE) less than about 4 deg compared to results with camera-based 3D measurement system, which was considered to become practical since stick figure animation of gait of healthy subjects could be presented appropriately [13]. Based on the result, this study aimed to examine a possibility of using the wireless wearable inertial sensor system in gait rehabilitation clinically. In this paper, gait movements of a hemiplegic subject were measured in 10 m walking before and after robotic-assisted gait training in a day. Relationship between changes of gait movement after the training and instructions of a physical therapist during training were discussed.

II. METHODS

A. Measurement of Gait Movements

The wearable inertial sensor system developed in our research group measures inclination angles of trunk and lower limb segment angles and hip, knee and ankle joint angles based on Kalman filter [13]. The system consisted of seven wireless inertial sensors (WAA-010, Wireless Technologies) and a notebook computer (Fig. 1). Each sensor was attached on the body with a stretchable band with hook and loop fastener and a pocket for the sensor. The sensors are put inside of the pocket and attached with the bands on the feet, the shanks and the thighs of both legs, and lumbar region as shown in Fig. 1. The wireless sensor measures 3-axis components of angular velocity and acceleration. Acceleration and angular velocity signals of each sensor are sampled with a frequency of 100Hz, and transmitted to the

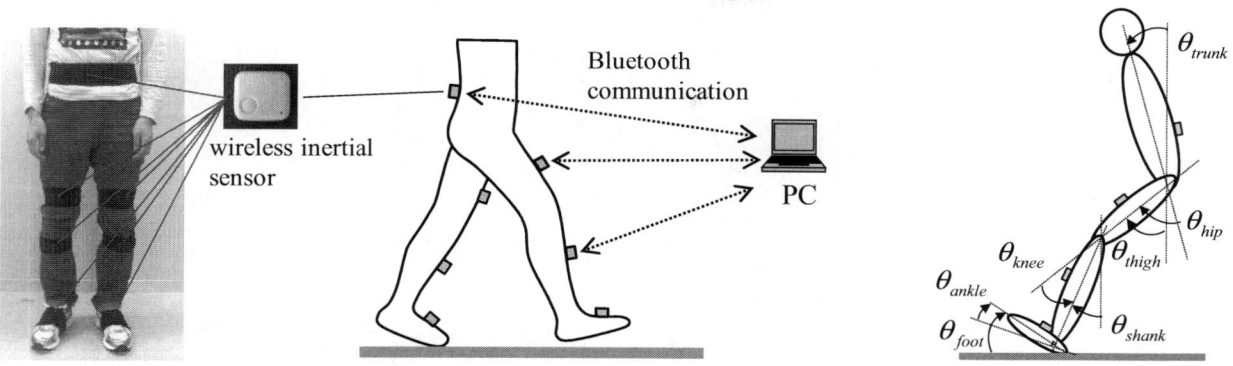

Fig. 1 The wireless wearable inertial sensor system for measurement of lower limb movements and definitions of segment inclination angles and joint angles.

PC via Bluetooth network and recorded.

Gait movements of a hemiplegic patient (80 y.o., male, C2 level spinal cord infarction, left side hemiplegia, chronic phase) were measured in 10 m walking with the wearable inertial sensor system before and after robotic-assisted rehabilitation training. The training was performed for about 30 min, in which he was instructed to increase the range of hip joint angle, to increase the time of loading during the stance phase and to reduce anterior inclination angle of the trunk using Hybrid Assistive Limbs (HAL) (CYBERDYNE Inc.) and manual instruction. In the measurement with the sensor system, sensors were attached on the body, while the subject wore his own wear and shoes. The subjects walked in 6 trials with a cane right side without prosthetic devices. Before the measurement, sitting position at 90 deg of the ankle joint angle, standing position with full knee extension and upright position were measured for angle calibration.

B. Analysis

Segment inclination angles and joint angles of the last 3 trials of 6 trials were analyzed after detecting individual stride. The measured data in a trial was divided into the performed strides by detecting the initial contact (IC) from acceleration signal of the sensor attached on the shank. Gait event timings of the foot flat (FF), the heel off (HO) and the toe off (TO) were measured from signals of gyroscope attached on the foot as shown in Fig. 2. First, quiet standing before walking was recognized by constant, very small angular velocity (almost 0 deg/s). Then, the HO was detected by increase of the angular velocity from the constant value. The TO was detected by zero cross point after the HO, the IC was detected by large peak acceleration, and the FF was detected as small angular velocity after the peak angular velocity near the IC. Threshold values to detect the gait event timings were determined through trial and

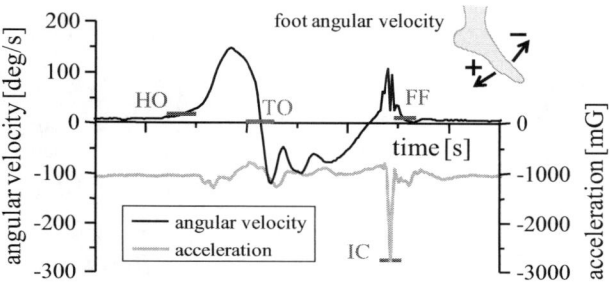

Fig. 2 Gait event detection from angular velocity and acceleration signals.

error based on measured data. The angles and the gait event timings for each stride were averaged after normalizing gait cycle by the time of 1 stride. Here, the strides that were not detected correctly were removed from the analysis. The first 2 strides and the last 2 strides in a trial were removed for analyzing steady state walking. The numbers of analyzed strides were 14 and 13 for the right side, 15 and 13 for the left side before and after the training, respectively.

III. RESULTS

Average angle patterns of inclination and joint angles of the subject are shown in Fig. 3. There were differences partly in measured angles between before and after the HAL training. As for the hip joint angle that was instructed to increase angle range in the training, the left hip joint extension angle (negative value) increased after the training. However, the hip extension angle of the right side decreased after the training. The movement range of the thigh inclination angle of the left side also increased, while there was no difference in the angle range of the right side. The foot inclination angle of both side showed increase of range of

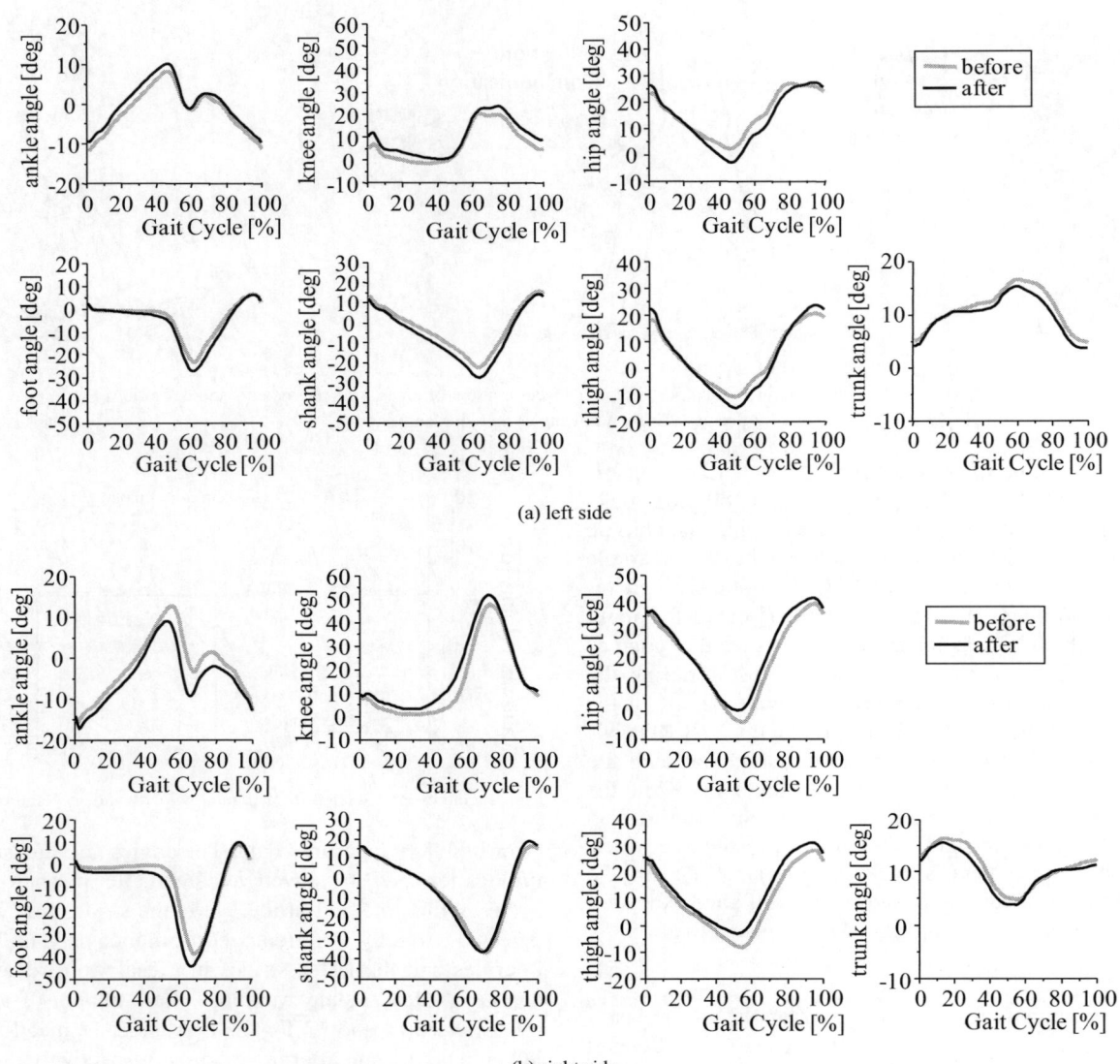

Fig. 3 Average angle patterns of inclination and joint angles measured before and after the robotic-assisted gait training. Since gait cycle was normalized by 1 stride time, 0% and 100% shows the IC.

motion during walking after the training, especially plantar flexion movement (negative value) increased at around the TO (60 – 65 %GC). The knee flexion angle also increased for both sides. The shank backward inclination angle (negative value) of the left side increased during the stride. The anterior inclination angle of the trunk (positive value) decreased a little after the training.

Average gait event timings of the patient are shown in Fig. 4. Although change of gait event timings was small, the FF timings of the both sides were delayed slightly after the training. The timings of the HO and the TO also became earlier slightly on the right side, decreasing the difference of the timings between the left and the right sides.

IV. DISCUSSIONS

The wearable inertial sensor system suggested that gait movements changed after the robotic-assisted gait training in a day. Left hip joint extension angle was suggested to increase after the training. Both foot inclination angles were also suggested to change after the training. Foot inclination angles mainly changed from around the HO to around the positive peak in the swing phase as seen in Fig. 3. It is considered that the subject pushed off more strongly than be-

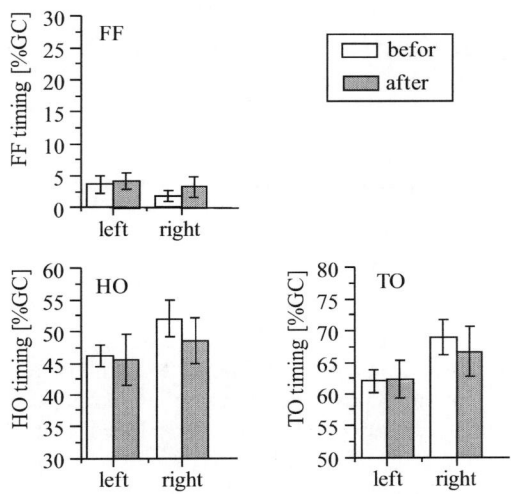

Fig. 4 Gait event timings before and after the training.

fore the training in order to increase hip joint movement. The strengthening of push off also increases thigh inclination angle (positive value) during the swing phase, and it leads increase of knee flexion angle in the swing phase.

Right hip joint angle changed mainly after the timing of the HO as seen in Fig. 3(b), in which hip flexion occurred earlier and maximum extension at around the TO decreased. Timings of the HO and the TO of the right side became earlier a little after the training as shown in Fig. 4. It is considered that the subject intended to move the right lower limb forward as early as possible increasing push off.

The movement changes caused by the training were considered to be improvements as the therapist instructed to the subject during the training. The HAL system could be effective to assist instruction by the therapist. Since the wearable inertial sensor system could measure the movement changes simply in a short time, it would be useful in clinical rehabilitation. However, it is necessary to evaluate movements from measured data in more detail.

V. CONCLUSIONS

From the gait movements of a hemiplegic subject measured with the wearable inertial sensor system, changes of gait movement after the robotic-assisted gait training were considered to be results of training instructed by the therapist. The robotic-assisted gait training could be effective for the rehabilitation under the instruction of the therapist. The inertial sensor-based wireless wearable sensor system would be useful for measurement of movements in gait rehabilitation clinically. It is necessary to evaluate movement changes from measured data in more detail.

ACKNOWLEDGMENT

This work was supported in part by the Ministry of Education, Culture, Sports, Science and Technology of Japan under a Grant-in-Aid for challenging Exploratory Research.

REFERENCES

1. Rueterbories J, Spaich E G, Larsen B, Andersen O K. Methods for gait event detection and analysis in ambulatory systems. Med Eng & Phys (2010) 32:545-552.
2. Lau H, Tong K The Reliability of Using Accel-erometer and Gyroscope for Gait Event Identification on Persons with Dropped Foot. Gait Posture (2008) 27:248-257
3. Jasiewicz J M, Allum J H, Middleton J W, Barriskill A, Condie P, Purcell B, Li R C. Gait Event Detection Using Linear Accelerometers or Angular Velocity Transducers in Able-bodied and Spinal-cord Injured Indi-viduals. Gait Posture (2006) 24:502-509
4. Tong K, Granat M H. A Practical Gait Analysis System Using Gyroscopes. Med Eng Phys (1999) 21:87-94
5. Dejnabadi H, Jolles B M, Aminian K. A New Ap-proach to Accurate Measurement of Uniaxial Joint Angles based on a Combination of Accelerometers and Gyroscopes. IEEE Trans Biomed Eng (2005) 52:1478-1484
6. Findlow A, Goulermas J Y, Nester C, Howard D, Kenney L P. Predicting Lower Limb Joint Kinematics Using Wearable Motion Sensors. Gait Posture (2008) 28:120-126
7. Cooper G, Sheret I, McMillian L, Siliverdis K, Sha N, Hodgins D, Kenney L, Howard D. Inertial sensor-based knee flexion/extension angle estimation. J Biomech (2009) 42:2678-285
8. Takeda R, Tadano S, Natorigawa A, Todoh M, Yoshinari S. Gait posture estimation using wearable accel-eration and gyro sensors. J Biomech (2009) 42:2486-2494
9. Alvarez J C, Gonzalez R C, Alvarez D, Lopez A M. Rodriguez-Uria J., Multisensor Approach to Walking Distance Estimation with Foot Inertial Sensing. Proc 29th IEEE EMBS (2007) 5719-5722
10. Bamberg S J, Benbasat A Y, Scarborough D M, Krebs D E, Para-diso J A. Gait Analysis Using a Shoe-integrated Wireless Sensor System. IEEE Trans Inf Technol Biomed (2008) 12:413-423
11. Saito H, Watanabe T. Kalman-filtering-based Joint Angle Measurement with Wireless Wearable Sensor System for Simplified Gait Analysis. IEICE Trans Inf & Syst (2011) E94-D:1716-1720
12. Watanabe T, Saito H, Koike E, Nitta K. A Preliminary Test of Measurement of Joint Angles and Stride Length with Wireless Inertial Sensors for Wearable Gait Evaluation System. Comput Intel Neurosci (2011) 2011:975193 doi: 10.1155/2011/975193.
13. Watanabe T, Saito H. Tests of Wireless Wearable Sensor System in Joint Angle Measurement of Lower Limbs. Proc 33rd IEEE EMBS (2011) 5469-5472 doi: 10.1109/IEMBS.2011.6091395.

Author: Takashi Watanabe
Institute: Graduate School of Biomedical Engineering, Tohoku University
Street: 901-7 Complex Build., 6-6-11 Aramaki-aza-Aoba
City: Sendai
Country: Japan
Email: nabet@bme.tohoku.ac.jp

Glenohumeral Kinetics in Gait with Crutches during Reciprocal and Swing-Through Gait: A Case Study*

E. Perez-Rizo[1], R. Casado-Lopez[2], V. Lozano-Berrio[1], M. Solis-Mozos[1],
M. Nieto-Diaz[3], S. Martin-Majarres[2], J.L. Pons[4], and A. Gil-Agudo[1]

[1] Biomechanics and Technical Aids Department, National Hospital for Spinal Cord Injury, Toledo, Spain
[2] Rehabilitation Department, National Hospital for Spinal Cord Injury, Toledo, Spain
[3] Experimental Neurology Unit, National Hospital for Spinal Cord Injury, Toledo, Spain
[4] Bioingeneering Group CSIC, Arganda del Rey, Madrid, Spain

Abstract—Many patients with incomplete spinal cord (SCI) injury and partial preservation of lower extremity function makes ambulation a mobility goal. Some of these patients have to use Lofstrand crutches to compensate insufficient strength in the lower limbs to permit ambulation. On the other hand, long-term crutch usage results in upper limb overuse pathologies. Here we present a kinetic model of right upper extremity based on rigid bodies to be applied to data generated by a movement capture system and an instrumented Lofstrand crutch with a six-degree-of-freedom load cell at the tip. This trial configuration may provide data for the upper extremity kinetic analysis. Data of forces and force moments of the right shoulder of a SCI patient walking with reciprocal and swing-through gait are shown. In both cases, the reaction force at the glenohumeral joint, during the support phase, was observed to be vertically upward, posterior and medial, whereas internal moments were mainly flexor, adductor and internal rotator.

Keywords—**Biomechanics, kinetic model, glenohumeral joint, crutch-assisted gait.**

I. INTRODUCTION

A SCI refers to any injury to the spinal cord that is caused by trauma or disease. Depending on where the spinal cord and nerve roots are damaged, the symptoms can vary widely, from pain to paralysis to incontinence.

For patients with incomplete SCI, partial preservation of lower-extremity function makes ambulation a realistic mobility goal. In addition, a considerable number of patients achieve ambulation, although the majority of them require lower-extremity orthoses to ambulate. Consequently, walking aids for the upper extremities are commonly prescribed to compensate for reduced strength in the lower limbs.

Literature shows that long-term crutch usage may result in overuse upper limb pathologies, such as destructive shoulder arthropathy, degenerative arthritis of the shoulder and wrist, or carpal tunnel syndrome [1].

Due to the large number of people with SCI who are dependent on crutches, we are interested in characterizing the moments and forces that occur during Lofstrand crutch-assisted ambulation. The information gained from this study may aid gait training.

In gait aided with assistive devices, studies based on biomechanical rigid body models of the lower limb provide valuable kinetic and kinematic information [2]. Furthermore, such data can provide information about device use-strategy and the influence of device position and load-transmission on a subject's gait parameters [3]. However there are few studies on the biomechanical behaviour of the upper limbs during walking with crutches in which a biomechanical model has been applied to data of a 3D movement capture system [4,5]. The goal of this work is to implement a kinetic model to analyse the upper limbs gait with crutches. Here we present data on forces and moments of force in the glenohumeral joint of patients walking with two types of gait patterns commonly used SCI subjects. On the one hand, two times reciprocal gait -progress alternates the right crutch and left foot with the left crutch and right foot-, and Swing-through gait -whose sequence occurs when one bears weight on both legs, then advances both crutches forward simultaneously, and then swings the body past the crutches-.

II. MATERIALS AND METHODS

A. Instrumentation

A Lofstrand crutch was instrumented with a six-degree-of-freedom load cell (AMTI, USA) at the distal end.

---

* This work was part of a project financed by FISCAM (Fundación para la Investigación Sanitaria de Castilla-La Mancha, Spain) which does not have any commercial interest in the results of this investigation. It was funded by FEDER. Ref no.: PI2010/50.

This research is part of the HYPER project funded by CONSOLIDER-INGENIO 2010 CSD2009-00067, Spanish Ministry for Science and Innovation. The institutions involved are: Spanish National Research Council (CSIC), Center for Electrochemical Technologies (CIDETEC), Visual Communication and Interaction Technologies Centre (VICOMTech), Fatronik-Tecnalia, University of Zaragoza, University of Rey Juan Carlos, University of Carlos III (UC3M), Bioengineering Institute of Cataluña (IBEC) and Hospital for Spinal Cord Injury of Toledo.

Fourteen active markers were placed on the thorax, the right upper extremity and the right crutch, whose trajectories were recorded by two scanners, placed at either side of the walkway and fitted to an active-marker motion-capture system (Charnwood Dynamics Ltd, England). Marker positions were recorded in synchronization with the signals of the load cell. The kinetic model was implemented in a biomechanical analysis software (C-Motion, Inc., USA).

B. Kinetic Model

Thorax, upper arm, forearm and hand were assumed to be rigid bodies segments with its mass uniformly distributed. Inverse dynamic Newton-Euler methodology was used to calculate the joint-kinetic values between two consecutive rigid bodies [6]. The thorax segment was deemed to be a cylinder with a radius of a half width of the subject's chest. The upper arm was considered to be a cylinder with a radius of half the distance from the lateral epicondyle marker (LE) to the medial epicondyle marker (ME) of the humerus. The forearm was deemed to be an elliptical cylinder with a proximal radius of half the distance from the LE to the ME and a distal radius of half the distance from the ulnar styloid process marker (US) to the radial stiloid process marker (RS). The hand was considered to be a sphere centered at the third metacarpal joint center (3MCJC) and with a radius estimated as the distance from 3MCJC to the lateral fifth metacarpal head marker (5MC). In the same way, Lofstrand crutch was divided into handle, shaft and load cell rigid body segments whose geometry corresponded to the crutch geometry. (Fig. 1)

The glenohumeral joint center (GHJC) was located 51.1mm in the caudal direction from right acromioclavicular joint marker (AC). The elbow joint center (EJC) was located at the midpoint between LE and ME; the wrist joint center (WJC) at the midpoint between US and RS; and the 3MCJC joint center at the point at which the ulnar styloid process of the third metacarpal head marker (M3) projected onto the plane formed by RS, US and 5MC [4]. (Fig. 1)

In line with International Society of Biomechanics (ISB) guidelines [7], the coordinate system of each every body segment were defined as follows: the X axis was defined as the anteroposterior axis, with the anterior direction being deemed positive; the Y axis was defined as the longitudinal axis, with the upward direction being considered deemed positive; and the Z axis was defined as the mediolateral axis, with the lateral direction being deemed positive.

The joint-reaction forces, i.e., the forces exerted by the lower segment on the upper segment of any given joint, were calculated and referenced to the upper segment coordinate system. The internal moments of each joint were calculated with respect to the proximal segment coordinate system, following the right-hand rule.

Due to the forces and moments reported in this study were referenced to the thorax coordinate system, this coordinate system is described: the Yt axis segment was parallel to the line connecting the midpoint between the xiphoid process marker (PX) and spinal process of the 8th thoracic vertebra marker (T8) and the midpoint between suprasternal notch marker (IJ) and spinal process of the 7th cervical vertebra marker (C7), with the anterior direction being considered positive; the Zt axis was perpendicular to the plane formed by IJ, C7, and the midpoint between PX and T8, with the right direction being deemed positive; and the Xt axis was perpendicular to the Yt and Zt axes, with the anterior direction being considered positive . (Fig. 1)

The forces and the moments measured by the load cell were deemed to be applied at the load cell's effective origin, with respect to its coordinate system.

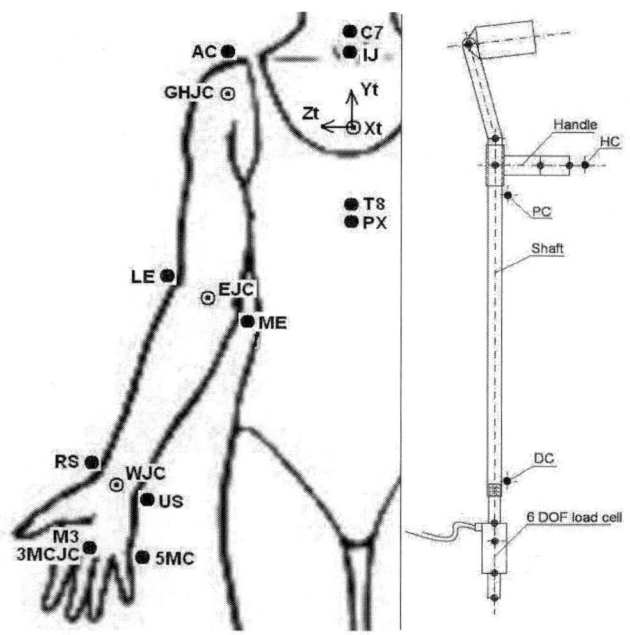

Fig. 1 Right upper limb and crutch marker set, joint landmarks and thorax segment coordinate systems [7]: C7, spinal process of the 7th cervical vertebra; IJ, suprasternal notch; T8, spinal process of the 8th thoracic vertebra; PX, xiphoid process; AC, acromioclavicular joint; LE, lateral epicondyle; ME, medial epicondyle; RS, radial stiloides process; US, ulnar styloid process; M3, third metacarpal head; 5MC, fifth metadarpal head; GHJC, glenohumeral joint; EJC, elbow; WJC, wrist; Xt, anteroposterior thorax axis; Yt, longitudinal thorax axis; Zt, mediolateral thorax axis. The crutch model was divided into handle, shaft, load cell, and tip segments, defined on the basis of the handle marker (HC), proximal marker (PC), and distal marker (DC) [2].

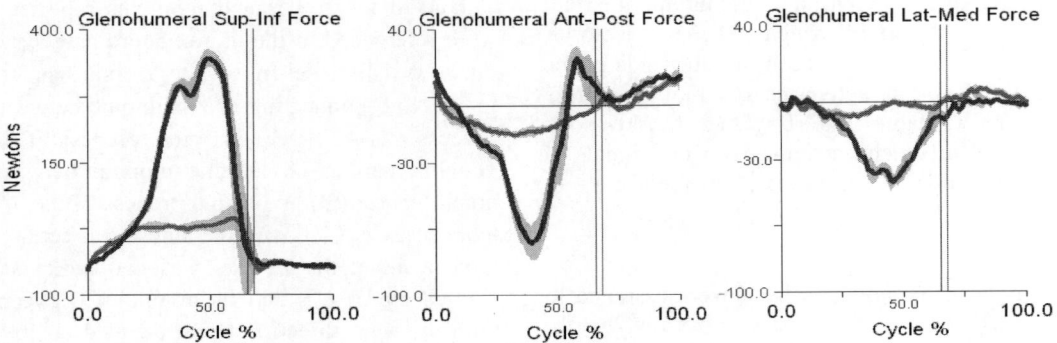

Fig. 2 Right glenohumeral reaction forces for one SCI subject, walking while using bilateral Lofstrand crutches. Mean (color lines) and standard deviation (light shadows) of five cycles are represented for both reciprocal gait (red) and swing-through gait (blue). Vertical line divides crutch cycle into stance and swing phases.

C. Test

A 45-years-old woman, with D12 spinal cord injury was instrumented. Her height and weight were 1.65m and 75kg, respectively. First, a static calibration recording was made. Then the marker positions and the load cell signals from a series of passes along a 8m-long walkway were recorded until 5 acceptable reads of each gait patter were obtained. Previously, informed consent was obtained from the subject.

D. Data Collection, Processing, and Analysis

Forces and moments measured by the crutch sensor were registered at 1000Hz and marker positions at 200Hz. 3D marker coordinates and kinetic data were filtered using a Butterworth low pass filter at 8Hz and 15Hz respectively. The crutch cycle and stance and the swing crutch period were defined according to the vertical force at the crutch. Crutch cycle time was normalized to 100%. Joint forces and moments were expressed as mean and standard deviation for each point of the crutch cycle.

Means and standard deviations of the forces and moments were calculated from the maximums and minimums of the five records for each gait pattern. In the same way, means and standard deviations of the instants in which those maximums and minimums occurred were calculated.

III. RESULTS

In the studied patient, the reaction force at the glenohumeral joint was observed to be mainly upward, posterior and medial in the support phase (Fig. 2). Greater mean peak superior, posterior and medial shoulder forces were observed during swing-through gait (Table 1). Lower mean cycle percent for the minimum posterior reaction force was observed at the reciprocal gait than at the swing-through gait (Table 1).

As expected, the maximum value of the vertical force at the swing-through gait corresponded approximately to a half of the subject's weight.

Table 1 Maximum (Max), minimum (Min) and instant (% of the cycle when Max and Min occur) of the glenohumeral joint reaction forces for reciprocal (RC) and swing-through (ST) gait. Values are given as mean (standard deviation).

Item	Max (N)	Max Cycle %	Min (N)	Min Cycle %
Sup-Inf , RC	50.0 (20.2)	48.0 (13.3)	-59.2 (3.0)	67.7 (0.5)
Sup-Inf , ST	357.7 (11.8)	52.45 (1.2)	-66.6 (2.01)	68.3 (0.3)
Ant-Post, RC	15.9 (3.6)	98.5 (3.0)	-17.3 (1.0)	30.0 (7.3)
Ant-Post, ST	29.17 (7.1)	48.6 (27.0)	-73.7 (9.4)	40.0 (2.7)
Lat-Med, RC	8.5 (1.0)	80.1 (2.1)	-8.7 (1.5)	25.1 (4.1)
Lat-Med, ST	3.5 (0.9)	7.9 (2.8)	-44.2 (4.2)	47.7 (1.9)

Table 2 Maximum (Max), minimum (Min) and instant (% of the cycle when Max and Min occur) of the glenohumeral joint internal moments for reciprocal (RC) and swing-through (ST) gait. Values are given as mean (standard deviation).

Item	Max (Nm)	Max Cycle %	Min (Nm)	Min Cycle %
Flex-Ext , RC	6.5 (1.2)	34.16 (30.8)	-1.3 (0.8)	90.2 (8.4)
Flex-Ext ,ST	21.0 (5.5)	54.3 (2.6)	-4.3 (0.8)	80.6 (8.6)
Add-Abd, RC	8.0 (1.9)	51.2 (5.6)	-8.0 (0.5)	0.4 (0.4)
Add-Abd, ST	18.2 (2.1)	55.4 (1.6)	-9.1 (0.7)	69.6 (1.3)
IRot-ERot, RC	2.4 (0.3)	24.6 (7.5)	-2.7 (1.2)	0.4 (0.5)
IRot-ERot, ST	5.1 (0.7)	38.3 (2.0)	-3.3 (0.3)	0.5 (0.3)

During the support phase, moments in the glenohumeral joint were mainly flexor, adductor and internal rotator both for the reciprocal gait and for the swing-through gait (Fig. 3). Greater flexor, adductor and internal rotator moments were observed at the swing-though gait (Table 2). We also observed a lower mean cycle percent for the maximum

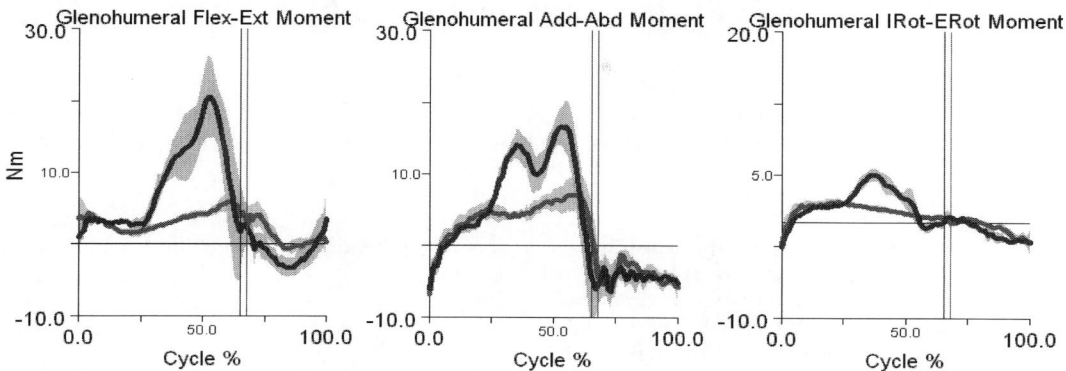

Fig. 3 Right glenohumeral internal force moments for one SCI subject, walking while using bilateral Lofstrand crutches. Mean (color lines) and standard deviation (light shadows) of five cycles are represented for both reciprocal gait (red) and swing-through gait (blue). Vertical line divides crutch cycle into stance and swing phase.

internal rotator moment at the reciprocal gait than at the swing-through gait (Table 2).

IV. DISCUSSION

In the present study we have developed a kinetic rigid bodies model of the right upper limb based on data generated from a 3D movement capture system. Application of this system to analyze gait with crutches during reciprocal and swing-through gait of a SCI patient shows that reaction forces at the glenohumeral joint are mainly upward, posterior and medial whereas internal moments are flexor, adductor and internal rotator.

Overall, the magnitudes of the forces and moments obtained in this study are comparable with the results from previous authors [4,5]. During the support phase, the three components of the reaction force in the shoulder are similar in form to the only previous three axes upper limb analysis of reciprocal and swing-through gait [5].

The predominance of the flexor moment observed in the shoulder during the support phase coincides with that found by other authors who have examined reciprocal gait [4,5] and swing-through gait [5].

V. CONCLUSIONS

The study reports the results of a pilot test which demonstrates both the functionality of the present trial configuration and the coherence of the results. This study may form the basis for future studies linking crutch kinetics to injury. Further applications of this system may offer valuable insight for crutch prescription, therapeutic walking strategies, and long-term usage effects.

REFERENCES

1. Lal S. Premature degenerative shoulder changes in spinal cord injury patients. Spinal Cord 1998;36:186–9.
2. Gil-Agudo A, Pérez-Nombela S, Pérez-Rizo E, et al. Comparative biomechanical analysis of gait in patients with central cord and Brown-Séquard syndrome. Disabil Rehabil. 2013 Apr 19. [Epub ahead of print] DOI: 10.3109/09638288.2013.766268.
3. Gil-Agudo A, Pérez-Rizo E, Del Ama-Espinosa A, et al. Comparative biomechanical gait analysis of patients with central cord syndrome walking with one crutch and two crutches. Clin Biomech (Bristol, Avon). 2009 Aug;24(7):551-7.
4. Requejo PS, Wahl DP, Bontrager EL, et al. Upper extremity kinetics during Lofstrand crutch-assisted gait. Med Eng Phys. 2005 Jan;27(1):19-29.
5. Slavens BA, Sturm PF, Harris GF. Upper extremity inverse dynamics model for crutch-assisted gait assessment. J Biomech. 2010 Jul 20;43(10):2026-31.
6. Zatsiorsky VM (2002) Kinetics of human motion. Human kinetics, Champaign, IL
7. Wu G, van der Helm FC, Veeger HE, et al.. ISB recommendation on definitions of joint coordinate systems of various joints for the reporting of human joint motion-Part II: shoulder, elbow, wrist and hand. J Biomech. 2005 May;38(5):981-992.

Corresponding author:

Author: Enrique Pérez Rizo
Institute: National Hospital for Spinal Cord Injury -Biomechanics-
Street: Finca la Peraleda, s/n
City: Toledo
Country: Spain
Email: enriquep@sescam.jccm.es

A Feasibility Study to Elicit Tactile Sensations by Electrical Stimulation

S.H. Hwang[1], J. Ara[1], T. Song[2], and G. Khang[1]

[1] Department of Biomedical Engineering, Kyung Hee University, Yongin-si, Republic of Korea
[2] Department of Biomedical Engineering, Jungwon University, Chungcheongbuk-do, Republic of Korea

Abstract—In this paper we investigated the feasibility of electrically-elicited tactile sensations. Eighteen subjects participated in the experiment and the constant current stimulation was applied to the subject's index finger pad. The cathodic monophasic rectangular pulse train was used throughout the experiment. The pulse amplitude was increased in every 2.5 seconds in such a way that the charge increment per pulse was kept constant at 70 nanoC. Four different pulse widths (200, 500, 700 and 1,000 us) and three different frequencies (20, 50 and 200 Hz) were selected to investigate the effect of the pulse parameters on the elicited tactile sensations.

The tickling sensation was elicited in all the subjects. Pressure, low and high frequency vibration were successfully evoked in 17, 13 and 12 out of 18 subjects. The activation threshold and the pain threshold (in mA) were low for large pulse widths. For each pulse width, each of the four sensations was perceived in almost the same number of subjects. We found that the sensation frequency was in accordance with the stimulation frequency; LH vibration and HF vibration were seen dominantly at low frequencies and high frequencies, respectively. The experimental results suggested that it is possible to elicit a group of tactile sensations by applying electrical stimulation to the skin.

Keywords—Mechanoreceptor, electrically-elicited tactile sensation

I. INTRODUCTION

The sensory feedback system has been drawing a great deal of attraction for a number of reasons. The major applications include the prosthetic arm, where it is highly important to enable the user to 'feel' the sensations as if the user were contacting objects with his/her healthy hands. That is, the contact information between the prosthetic hand and the object obtained by means of the transducers need to be transformed into a series of signals that can evoke the corresponding sensation to the user. Among the potential signals, we chose the electrical stimulation, and this paper discusses feasibility of the electrical stimulation together with the surface electrodes to be used for eliciting sensations such as pressure and vibration.

The ultimate goal of this study is to develop an electrical stimulation system that can selectively activate the mechanoreceptors responsible for most of the tactile sensations. The mechanoreceptors can be classified according to their adaptive properties and the size of their receptive field - the mechanoreceptive afferents are categorized as slowly-adapting (SA) nerve fibers or fast-adapting (FA) nerve fibers and these two are divided into type I and type II depending on their receptive field size. Each mechanoreceptive afferents responds to a specific mechanical stimulation. Meissner's corpuscles (FA I) and Pacinian corpuscles (FA II) are sensitive to tickling and vibration, and Merkel disks (SA I) and Ruffini endings (SA II) are responsible for sustained sensations like pressure and skin stretch[1].

There have been only a few researches to elicit tactile sensations by using electrical stimulation. A single mechanoreceptive unit was stimulated employing microelectrodes; 'flutter-vibration' sensation and 'sustained pressure' sensation were observed when stimulating the FA fibers and the SA I fibers, respectively[2]. They suggested that the electrical stimulation could generate the same sensation as the mechanical stimulation. Electrical stimulation of the SA II fibers, however, could rarely elicit a sensation[3].

Theoretical approaches has been also adopted to selectively stimulate the mechanoreceptive afferents by using a weighted array of electrodes and combining the electrical stimulation polarities based on the 'activation function' model[4][5][6]. Their approaches, however, could not be confirmed by the experimental results. We believe so many researches need to be done in the area before the electrical stimulation can be easily applied to the human skin to elicit tactile sensations including selection of the appropriate stimulus parameters, which was dealt with in this paper.

II. METHODS

A. Subjects

Eighteen healthy subjects at the age of 24.7 ± 3.17 participated in the experiments. They were fully informed of the objective, the experimental procedures, and any potential risks that could occur during the experiment, and provided written consents.

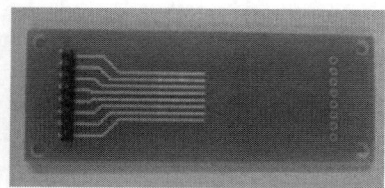

Fig. 1 Bar-type surface electrode designed and manufactured for this study

B. Equipment

A series of pulse trains were applied to the subject's index finger pad which is known to be the most sensitive in the human body. The waveform and the stimulus parameters (amplitude, width and frequency of the pulse) were created and controlled using LabVIEW® (National Instruments Co., Austin, Texas, USA), and NI 9263 (National Instruments Co., Austin, Texas, USA) generated the pulse train in voltage. The voltage was then converted into the current by the 8-channel stimulator developed in our laboratory for this research. The constant-current stimulation was employed to keep the stimulation intensity from changing due to the skin impedance variation during the experiment. Fig. 1 shows the bar-type electrodes (10 mm × 0.5 mm), where any two bar electrodes could be selected. Our previous study[7] led us to fix the inter-electrode distance at 5.5 mm. The distal electrode and the proximal one were used as the source and the reference, respectively.

C. Preparation

At the beginning of the experiment, the subject's index finger of the dominant hand was cleaned with an alcohol swap, and put in the room-temperature water for 10 seconds. Removing the water, we applied electrolytic gel to the finger pad and then the finger pad was placed on the electrodes with the center of the swirl located at the midpoint of the two electrodes.

D. Stimulation Parameters and Description of the Perceived Sensations

The cathodic monophasic rectangular pulse train was used throughout the experiment which was composed of three phases. In the first phase, the pulse amplitude modulation was employed to investigate any possibility to elicit tactile sensations such as pressure and vibration. The pulse width and the frequency were fixed at 200 us and 20 Hz, respectively, and the pulse amplitude was increased from 0 mA to the maximum value that caused any pain or uncomfortable sensation to the subject. The pulse amplitude was increased by 0.35 mA in every 2.5 seconds. The second phase of the experiment was to investigate the relationship between the electric charge per pulse and the resulting tactile sensation(s). Four different pulse widths were adopted at 200 us, 500 us, 700 us, and 1,000 us. The pulse amplitude increment was selected to be 0.35 mA, 0.14 mA, 0.1 mA, and 0.07 mA, respectively, for each pulse width so as to increase the same amount of the electric charge 70 nanoC. In the third phase of the experiment, we investigate how the frequency affected the perceived tactile sensation. The experiment described above was repeated to each subject at 3 different frequencies; 20 Hz, 50 Hz, and 200 Hz. The pulse width was fixed at 200 us, and the pulse amplitude was increased again by 0.35 mA in every 2.5 seconds.

Two variables, the activation threshold (AT) and the pain threshold (PT), were defined and monitored as the moments where the subject began to feel 'any' tactile sensation and felt any uncomfortable sensation, respectively. The AT and PT values were expressed either as the electric charge per pulse (CPP, nanoC) or as the pulse amplitude for a given pulse width. Between AT and PT, the subject described as much in detail as possible the sensation quality including the kind of sensation(s), the sensation intensity, the order of sensations, the sensation location, etc. The detailed inquiry form can be seen from our previous reports[8][9].

III. RESULTS

We were able to elicit, in the first phase of the experiment, up to 4 different tactile sensations by using a rectangular pulse train; tickling, pressure, low-frequency (LF) vibration, and high-frequency (HF) vibration. All 4 sensations were present in 7 (38.9%), 3 in 10 (55.6%), 2 only in only 1 (5.6%) out of 18 subjects. As the stimulation intensity increased, i.e. the pulse amplitude increased, the sensation intensity got stronger as well, and the sensation was switched into another. All the subjects felt tickling, the pressure sensation was reported in 17 (94.4%), the LF ibration and the HF vibration in 13 (72.2%) and 12 subjects (66.7%), as illustrated with solid bars in Fig. 3. The tickling sensation always appeared first among the 4 sensations. The second sensation was pressure in 11 (61.1%), and LF vibration in 9 (50%), where these numbers included 3 subjects (16.7%) who reported these two sensations were perceived simultaneously. Only 1 subject (5.6%) reported the tickling sensation was followed by HF vibration.

In the second phase of the experiment, the AT and PT values were measured and analyzed for different pulse widths. As shown in Fig. 2, both AT and PT (in mA) were low for large pulse widths, but, in terms of CPP (nanoC), the opposite.

Fig. 3 shows the perceived sensations for different pulse widths. The pulse width did not make a significant difference to the perception rate. The tickling sensation was elicited in almost all the subjects (95.8%), and pressure in more than 80% of the subjects regardless of the pulse width. LF and HF vibration were perceived in less subjects (70.8% and 65.3%, respectively) than tickling and pressure.

In the third phase of the experiment, where 3 different frequencies were selected, the AT values were not significantly different at different frequencies whereas the PT value was the lowest at the highest frequency 200 Hz (Fig. 4). Tickling and pressure were perceived in almost all the subjects regardless of the frequency as seen in Fig. 5. It

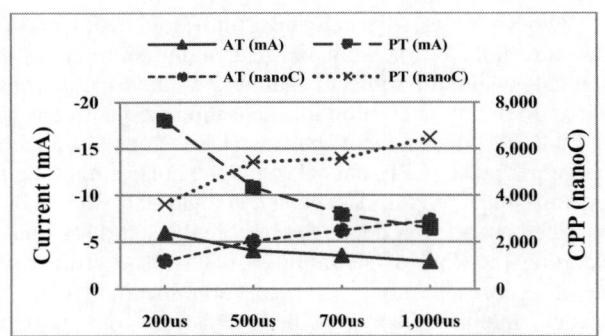

Fig. 2 AT and PT in mA and in nanoC (CPP) for different pulse widths

can be noted that the sensation frequency was in accordance with the stimulation frequency; LH vibration and HF vibration were seen dominantly at the low frequency and the high frequency, respectively. At low frequencies, e.g. 20 Hz, pressure was the dominant sensation as shown in Fig. 6.

IV. DISCUSSION

Generally speaking, the sensation order may well be related to the location of the mechanoreceptors and their nerve fibers, i.e. as the stimulation intensity increases, the superficial afferents can be activated earlier than the deep-rooted ones. Although not yet completely clear, most of the published literature agreed that a light touch such as tickling and LF vibration can be perceived through Meissner's corpuscle and that Merkel disk and Pacinian corpuscle are responsible primarily for pressure and HF vibration, respectively. It was observed in our study that tickling was, almost without exception, perceived earlier (i.e. with a low stimulation intensity) than any other sensation. Meissner's corpuscles are superficially located and therefore activated with a low stimulation intensity. On the other hand, Pacinian

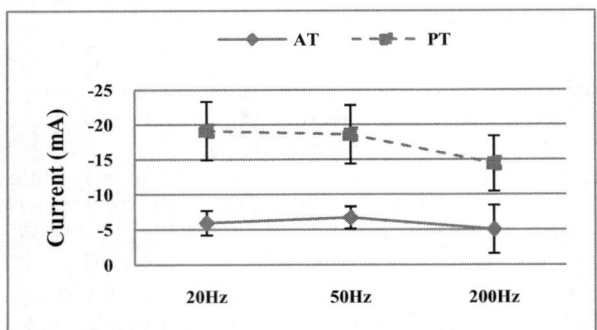

Fig. 4 AT and PT in mA at different frequencies

corpuscles are deeply located so that HF vibration appeared with a high stimulation intensity. Although Merkel disks are also superficially located, they were activated later than Meissner's corpuscles, which suggested that Merkel disks may have a higher activation threshold than Meissner's corpuscles. This result matches to a previously published report where a single unit of Meissner's corpuscle required a lower input current to get activated than that of Merkel disk[3].

Both the AT and PT values were low for large pulse widths. The stimulation intensity required to activate the sensory nerve fiber, as expected, needs to be determined not only by the current (i.e. the pulse amplitude in this study) but also the pulse width, implying that it may not be true that a specific sensory nerve fiber can be activated when a certain amount of electric charge is applied. Furthermore, most of the subjects reported that a large pulse width resulted in a less clear sensation, which needs further investigation regarding the underlying mechanism.

The AT value changed little when the frequency was increased from 20 Hz to 200 Hz, implying that AT may depend on the electric property of each pulse, not on the

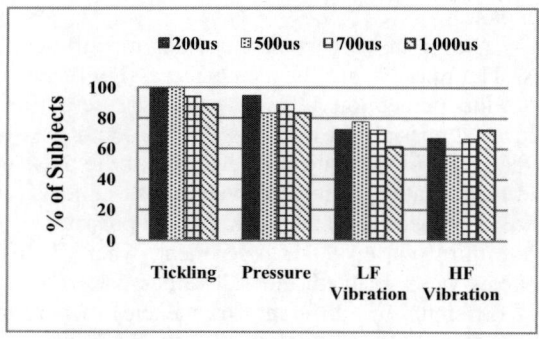

Fig. 3 Perception rate of each sensation for different pulse widths

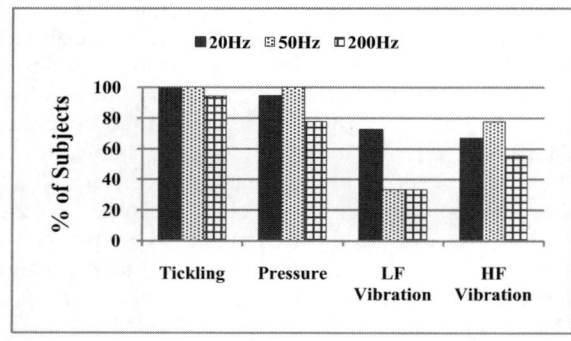

Fig. 5 Perception rate of each sensation at different frequencies

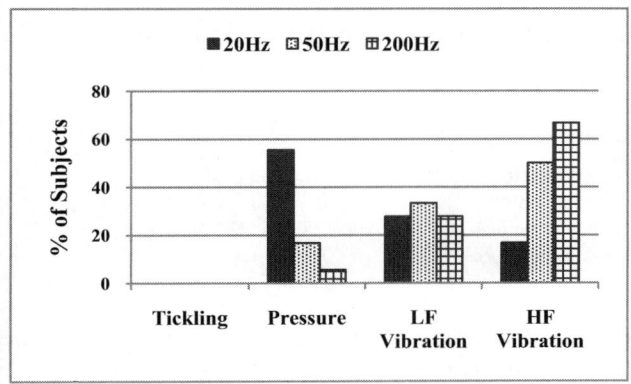

Fig. 6 Dominant sensations at different frequencies

number of pulses applied per second. It can be suggested that the effect of a single pulse is not accumulated, but instead the nerve fiber gets stimulated and activated by a pulse and then initialized before the next pulse arrives.

One of the hypotheses we had prior to this study was confirmed that the literature information and knowledge obtained from mechanical stimulation of the skin could be utilized to electrical stimulation. For instance, a force applied to the skin deforms the tissue medium beneath the skin and this deformation physically affects the mechanoreceptor. Then a series of action potentials are generated and propagated along the nerve fibers. Electrical stimulation, however, cannot 'deform' the mechanoreceptor but directly activate the nerve fibers. Despite such a difference, it was shown that the 'mechanical' frequency range to which a mechanoreceptor is the most sensitive was generally in accordance with the electrical stimulation frequency.

V. CONCLUSIONS

We have shown that it is possible to elicit a group of tactile sensations by applying electrical stimulation to the skin. Much more researches, however, are needed before each sensation can be selectively elicited. They include electrode design, study on the stimulus parameters, development of a generic multichannel stimulator, etc.

ACKNOWLEDGMENT

This research was supported by the Public welfare & Safety research program through the National Research Foundation of Korea(NRF) funded by the Ministry of Education, Science and Technology (No. 2010-0020447).

REFERENCES

1. Kruger L (1996) Pain and touch. Academic press, California
2. Macefield G, Gandevia S, Burke D (1990) Perceptual responses to microstimulation of single afferents innervating joints, muscles and skin of the human hand. J Physio 429:113-129
3. Vallbo , Olsson K, Westberg K (1984) Microstimulation of single tactile afferents from the human hand. Brain 107:727-749
4. Asamura N, Yokoyama N, Shinoda H (1998) Selective stimulating skin receptors for tactile display. IEEE Comp Graph App 18:32-37
5. Kajimoto H, Kawakami N, Tachi S (2002) Haptic Proc. vol. 1, Symposium on Hapt. Interf. Virt. Environ. & Tel. Sys., Japan, 2002, pp 303-310
6. Kajimoto H, Kawakami N, Meada T, Tachi S (1999) Tactile feeling display using functional electrical stimulation, ICAT proc. vol. 1., Artif. Real. & Telex., Tokyo, Japan, 1999, pp 107-114
7. Cordon S M G, Hwang S H, Song T, Khang G (2012) Current and frequency modulation for the characterization of electrically-elicited tactile sensations, Intn'l, J. Prec. Eng. & Manuf. 13:2051-2058
8. Ara J, Hwang S H, Song T et al. (2011) Effects of modulation type on electrically-elicited pressure sensation, KSPE Proc. vol. 2, Korean Society for Prec. Eng. Jeju, Republic of Korea, 2011, pp 1303-1304
9. Ara J, Hwang S H, Song T, Khang G (2012) Electrically-elicited tactile sensation for different modulation types polarities and waveforms of stimulation pulse trains, , Intn'l, J. Prec. Eng. & Manuf. 13:1911-1916

Author: Gon Khang
Institute: Kyung Hee University
Street: 1732 Deogyeong-daero
City: Youngin-si, Gyeonggi-do
Country: Republic of Korea
Email: gkhang@khu.ac.kr

Dystonia: Altered Sensorimotor Control and Vibro-tactile EMG-Based Biofeedback Effects

C. Casellato[1], S. Maggioni[1], F. Lunardini[1,2], M. Bertucco[2], A. Pedrocchi[1], and T.D. Sanger[2,3]

[1] Politecnico di Milano, Department of Electronics, Information and Bioengineering, NearLab,
Via G. Colombo 40, Milano, Italy
[2] University of Southern California, Department of Biomedical Engineering, Los Angeles, CA 90089, USA
[3] Children Hospital, Los Angeles, USA

Abstract—Dystonia has been associated with injury to the basal ganglia, in particular to the putamen and globus pallidus, even if there is evidence that other brain areas including cerebellum, brainstem, or sensory cortex can be causes of dystonia. Although dystonia is regarded as a movement disorders, several sensory phenomena occur. This suggests a dysfunction ascribable to a defective central processing of sensory inputs leading to a distorted afferent information causing a fixed input-output mismatch in specific motor programs. Still little is known about the altered sensorimotor integration in dystonia, therefore the need for a deeper investigation and for optimal complementary treatments is strong. We propose the use of the biofeedback technique as an exploratory tool and a possible future rehabilitation instrument. In this framework, the aim of our study is to investigate, in children with dystonia, the ability to voluntarily and selectively control the activation of a target muscle at different levels of the upper limb kinematic chain by means of a EMG-based vibro-tactile biofeedback. These tests should shed light on how an additional proprioceptive information affects the motor outcome in dystonia compared to the healthy behavior, both in kinematic and electromyographic quantitative terms.

Keywords—dystonia, sensorimotor integration, biofeedback, kinematics, EMG.

I. INTRODUCTION

Childhood dystonia is defined as "a movement disorder in which involuntary sustained or intermittent muscle contractions cause twisting and repetitive movements, abnormal postures, or both". Data from both functional imaging and electrophysiological experiments show functional abnormalities in premotor and primary sensorimotor cortical areas and in the cortico–striatal–thalamo–cortical motor circuits, suggesting an aberrant sensorimotor integration [1,2]. The abnormal or absent modulation of afferent and intracortical long-interval inhibition may indicate maladaptive plasticity that possibly contributes to the difficulty that dystonic patients have to learn a new sensorimotor task. Investigating these dystonic mechanisms in children is crucial since their brain and learning skills are still under development [3,4]. Biofeedback (BF) is a technique that consists in feeding back information about physiological processes, assisting the individual to increase awareness of these processes and to gain voluntary control over them.

Several evidences support the hypothesis that dystonic motor impairments involve a deficit of sensory processing. In this framework, the BF information is an additive re-afference to the CNS during movement execution that can effectively been integrated in the sensorimotor loop, thus leading to a better volitional motor control. For dystonic disorders, the most meaningful variable to be fed back is the electromyographic signal (EMG-based biofeedback). Promising results about the effectiveness of EMG-based biofeedback paradigms in dystonic children have been recently turned out. Young and colleagues [5] showed how a visual EMG-based biofeedback could be crucial in decreasing the excessive co-contraction on the upper-limb biceps and triceps muscles in tracking tasks. In the same framework, our lab developed an integrated set-up which provides the user with an EMG-based visual-haptic biofeedback during upper limb tracking tasks; tests targeting the brachioradialis on a group of primary dystonic children showed a muscle activity reduction when the BF was added into the loop, thus inducing a better muscle-specific voluntary motor control [6]. In addition, the effectiveness of a vibratory stimulus proportional to the ongoing EMG has been shown by Sanger's research group [7], which manufactured a portable EMG unit that includes a surface EMG sensor and amplifier, microcontroller-based nonlinear signal processing, and vibration feedback of muscle activity. Its usability during daily life activities allowed a successful preliminary long-term study on cerebral palsy children.

Our goal is to use the above device by integrating it in a set-up recording multi-muscles signals and upper limb kinematics, in order to accurately evaluate the role of this BF in the movement outcome. Data are recorded in three conditions: without BF ("No BF"), with BF on a distal upper-limb muscle ("Distal BF"), and with BF on a proximal upper-limb muscle ("Proximal BF").

II. METHODS

A. Participants

Inclusion criteria of this study were: I) primary or secondary dystonia; II) pediatric age (8-21 years); III) upper

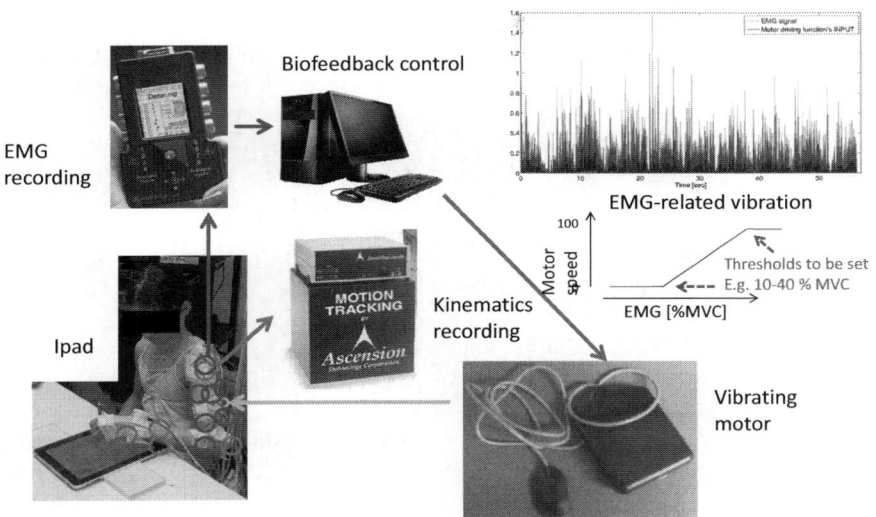

Fig.1 Set-up

limb control impairment compatible with the writing task execution; IV) no cognitive impairment that prevents from understanding the instructions. Participants consisted of a dystonic group of 10 children (13.4 ± 4.3 years) recruited from the Children's Hospital Los Angeles and diagnosed by a pediatric neurologist and of a control group of 9 neurologically healthy children (15.8 ± 4.0 years). The groups' age distributions were not statistically different. The University of Southern California Institutional Review Board approved the study. All parents gave informed written consent for participation and authorization for use of protected health information, and all children gave written assent.

B. Set-up and Protocol

The acquisition system integrated upper-limb kinematics (Flock of Birds (FOB), Ascension®, 100 Hz sample frequency), electromyography (DataLOG MWX8, Biometrics Ltd®, 1000 Hz), and 2D coordinates of the stylus pen tip on a tablet (iPad, Apple®, 60 Hz) (Fig 1).

FOB sensors were placed on elbow, dominant acromion, non-dominant acromion and anterior superior iliac spine (hip). Bipolar surface EMG electrodes were positioned on 8 muscles of the upper limb: Flexor Carpi Ulnaris (FCU), Extensor Carpi Radialis (ECR), Biceps (BIC), Triceps (TRIC), Anterior Deltoid (AD), Lateral Deltoid (LD), Posterior Deltoid (PD) and Supraspinatus (SS). An application had been ad-hoc developed in Cocoa (Apple®) to display the task interface and to record the 2D coordinates of the rubber pen tip. The subject was seated on an armless chair, at a distance from a height-adjustable table that allowed him to reach the furthest point of the tablet with the elbow at the 90% of its maximum extension. The child's trunk was fastened to the chair back with a Velcro® belt. The subjects were asked to draw, with their dominant arm, the figure "8" (15.7 x 7.8 cm) on the tablet at their natural speed, following a thin trace (0.3 cm thick) on the tablet. Participants performed three trials (consisting of ten figure-eight patterns executed in a row) for each condition: "No BF", "Distal BF", "Proximal BF". In the BF conditions the instruction was to reduce the level of contraction of the target muscle as much as possible, without neglecting the accuracy of the motor performance.

C. Data Analyses

Data analyses were executed with the software Matlab (Mathworks®). Representative parameters from kinematic and EMG data have been computed. Kinematic analyses focused on the task outcome and related end-effector features, on the whole upper body segments, with attention to joints motion size and variability. FOB data were processed with a low-pass Butterworth filter (5th order, 3 Hz). The cut-off frequency was determined with the Jackson "knee" method [8]. Each sequence was cut into single figure-eight, by thresholding the velocity profiles.

The joint movement size during the task was computed as the volume of the ellipsoids containing the center of the joint trajectories with a 95% probability [9].

In order to compute an index of the intra-subjects variability along the upper-limb kinematic chain, the variance explained by the first principal components was calculated (Principal Component Analysis). Lower PC1 explained variance means higher variability.

Each EMG signal is processed with a band-pass (5th order, 5-400 Hz) and a stop-band (5th order, 60 Hz) Butterworth filters. To extract the linear envelope the EMG signal was rectified and passed through a Butterworth low-pass filter (5th order, 15 Hz). In order to quantify the muscle activity, the Root Mean Square (RMS) value of the EMG envelope was computed.

III. RESULTS

Some deep analyses focused just on the "No BF" condition, comparing the control and dystonic groups. The frequency-domain analysis allows to exploit the strong coupling between kinematics and EMGs, to quantify the task-relevant and task-irrelevant components. The above distinction helps to separately investigate the motor planning phase from the execution phase, where noisy unwanted patterns should be strongly inhibited [*paper submitted*].

Here we mainly present differences due to the presence of BF and its localization along the involved upper-limb kinematic chain (after verifying that the tablet trace accuracy was constant across conditions).

Motion size is always higher in children with dystonia. In the "No BF" case, significant greater extent of movement was evident not just for elbow and acromion, involved in the focal motion, but also for non-dominant acromion and hip, which were supposed to be as still as possible. This result shows a poor stabilization in dystonia. An increase in the volume of each joint with respect to the "No BF" occurs in the "Distal BF" condition but not in the "Proximal BF" one (Fig 2).

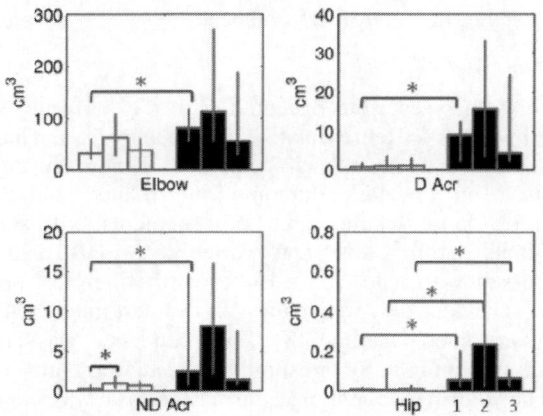

Fig. 2 Joint motion size for Elbow, Dominant Acromion (D Acr), Non dominant acromion (ND Acr) and Hip in the control (white) and pathological group (black for "NO BF" condition (1), "Distal BF" (2) and "Proximal BF" (3) conditions. Significant differences between populations are marked with a red asterisk (Mann-Whitney test, p<0.05). Significant differences between conditions within-group are marked with a blue asterisk (Wilcoxon test with Bonferroni correction, p<0.016).

The trend of the explained variance percentages is depicted in Fig. 3. First of all, an increasing tendency in variability is noticeable moving from distal trajectories (i.e. closer to movement final outcome) to proximal ones. This index shows lower values in the dystonic population in all conditions, except for the elbow trajectory which presents higher variability in the healthy controls than in children with dystonia for both BF conditions. The increase in variability induced by the BF is higher in healthy subjects than in children with dystonia.

Concerning the EMGs, the ΔRMS values represent a quantification of the BF-related goal achievement. With regard to distal target muscles, seven children with dystonia out of ten and eight control subjects out of nine managed to reach their goal (Fig. 4A). In general we can state that, even if the healthy group achieved a better performance, the control of distal muscles is easily executed by all subjects. On the other hand, only two children with dystonia out of nine managed to reduce the activity of the proximal target muscle. This task revealed to be tougher also for healthy controls, in fact only five out of nine succeeded in it (Fig. 4B).

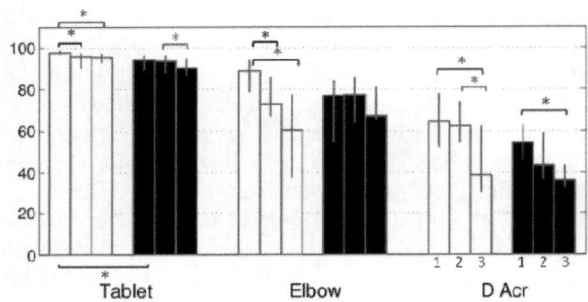

Fig. 3 Variability values for tablet, Dominant Acromion (D Acr) and Hip in the control (white) and pathological group (black) for group, "No BF" (1), "Distal BF" (2) and "Proximal BF" (3) conditions. Significant differences between populations are marked with a red asterisk (Mann-Whitney test, p<0.05). Significant differences between conditions within-group are marked with a blue asterisk for No BF vs Distal BF, a violet asterisk for No BF vs Proximal BF and a green asterisk for Distal BF vs Proximal BF (Wilcoxon test with Bonferroni correction, p<0.016).

IV. DISCUSSION

First of all, our preliminary results show that, if compared with healthy controls, the dystonic motor behavior is more variable [10-11] and characterized by wider joint motion sizes which are not fully functional to the required outcome. These findings support the hypothesis of a noisy CNS in dystonia.

Trial-to-trial variability increases starting from "No BF" condition, to "Distal BF" and finally to "Proximal BF". Since it is known that facing a new motor task leads to high

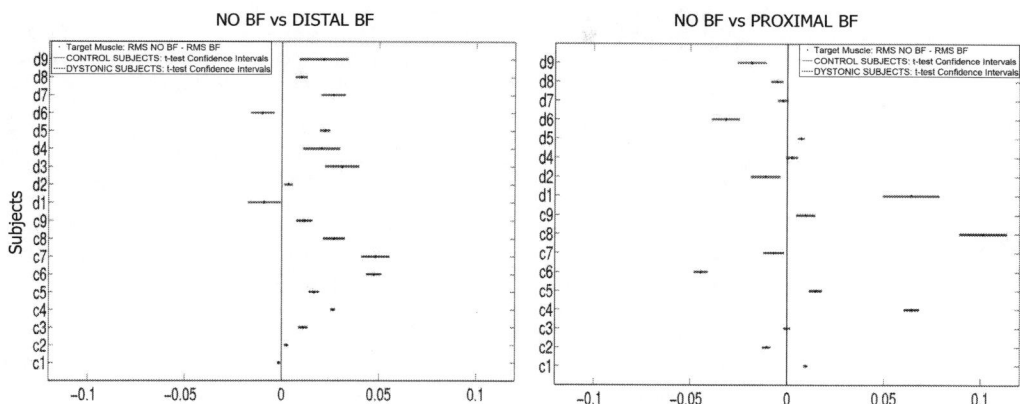

Fig. 4 Differences in the target muscle activation between the "BF" and the "No BF" conditions, for subjects with dystonia (red, d1-d9) and healthy controls (blue, c1-c9) (Left Panel: "Distal BF"; Right Panel: "Proximal BF"). Bars represent t-test confidence intervals.

trial-to-trial variability [12], this result was expected since the BF task pushed the subjects to adopt new motor strategies or to tune the standard ones. The increase in variability is however more marked for control subjects. Our hypothesis is that the strong increase in variability observed in healthy controls is useful to explore new motor solutions and to choose the optimal one. On the other hand, in children with dystonia, the abnormal variability at the baseline may cause the lack of a remarkable increment of variability when facing a new motor task, thus resulting in a failure of motor learning.

EMG results are in accordance with the hypothesis of poor motor adaptation in dystonia. In fact seeking a new motor solution to achieve a specific EMG modulation is harder for subject with dystonia, especially when the target is a proximal muscle. All the above results suggest that, in dystonia, the impairment is more evident in terms of proximal motor control. Under this hypothesis, distal dysfunctions can be seen as possible fall-outs of more proximal deficits. For this reason, in a rehabilitation paradigm in dystonia, BF should target proximal segments.

The haptic vibratory cue is proved to be a suitable channel to convey hints about the state of a physiological activity during an ongoing task [13]. Moreover, haptic training is different from the visual one since such training occurs in body centered coordinates as opposed to visuospatial coordinates. This may be especially helpful when learning motor tasks with complex kinematics, where a haptic presentation removes the need for complex sensorimotor transformations. In addition, vibro-tactile feedback proportional to the EMG activity is straightforward and easy to implement.

V. CONCLUSIONS

The use of BF provides patients affected by sensorimotor impairments with opportunities to regain the ability to better assess different physiological responses and to possibly learn self-control of those responses. Biofeedback paradigms can be customized in terms of physiological variable to be fed back (e.g. EMG, force, end-effector kinematics) and of sensory channels (e.g visual and/or haptic and/or auditory). Beside dystonia, this treatment can be effective in several movement disorders [14].

Future works will study the long-time use of EMG-based biofeedback and the sensorimotor integration underlying motor changes induced by this treatment, through movement-related fMRI protocols [15-16].

REFERENCES

1. Peterson DA, Sejnowski TJ, Poizner H (2010) Convergent evidence for abnormal striatal synaptic plasticity in dystonia. Neurobiol Dis 3:558–73.
2. Tinazzi M, Fiorio M, Fiaschi A et al (2009) Sensory functions in dystonia: insights from behavioral studies. Mov Disord 10:1427–36.
3. Casellato C, Pedrocchi A, Zorzi G et al (2012) Error-enhancing robot therapy to induce motor control improvement in childhood onset primary dystonia. J Neuroeng Rehabil 9:46.
4. Casellato C, Zorzi G, Pedrocchi A et al (2011) Reaching and writing movements: sensitive and reliable tools to measure genetic dystonia in children. J Child Neurol 26(7): 822-9.
5. Young SJ, van Doornik J, Sanger TD (2011) Visual feedback reduces co-contraction in children with dystonia. J Child Neurol 26:37-43.
6. Casellato C, Pedrocchi A, Zorzi G et al (2012) EMG-based visual-haptic biofeedback: a tool to improve motor control in children with primary dystonia. IEEE Trans Neural Syst Rehabil Eng. Oct 5.
7. Bloom R, Przekop A, Sanger TD (2010) Prolonged electromyogram biofeedback improves upper extremity function in children with cerebral palsy. J Child Neurol 25:1480-4.

8. Jackson KM (1979) Fitting of Mathematical Functions to Biomechanical Data. IEEE T Bio-Med Eng 26(2):122–4.
9. Oliveira LF, Simpson DM, Nadal J (1996) Calculation of area of stabilometric signals using principal component analysis. Physiol Meas 17(4):305–12.
10. Sanger TD, Kaiser J, Placek B (2005) Reaching Movements in Childhood Dystonia Contain Signal-Dependent Noise. J Child Neurol 20(6):489–96.
11. Berardelli A, Rothwell JC, Hallett M et al (1998) The pathophysiology of primary dystonia. Brain 121(7):1195–212.
12. Ölveczky BP, Andalman AS, Fee MS (2005) Vocal experimentation in the juvenile songbird requires a basal ganglia circuit. PLOS Biol 3(5):e153
13. Chatterjee A, Aggarwal V, Ramos A et al (2007) A brain-computer interface with vibrotactile biofeedback for haptic information. J Neuroeng Rehabil 4:40
14. Ambrosini E, Ferrante S, Ferrigno G, et al (2012) Cycling induced by electrical stimulation improves muscle activation and symmetry during pedaling in hemiparetic patients. IEEE Trans Neural Syst Rehabil Eng, 20(3):320-30
15. Casellato C, Ferrante S, Gandolla M et al (2010) Simultaneous Measurements of Kinematics and fMRI: Compatibility Assessment and Case Report on Recovery Evaluation of one Stroke Patient. J Neuroeng Rehabil 7:49
16. Gandolla M, Ferrante S, Casellato C et al (2011) fMRI brain mapping during motion capture and FES induced motor tasks: Signal to Noise Ratio assessment. Med Eng Phys 33:1027-1032.

Closed-Loop Modulation of a Notch-Filter Stimulation Strategy for Tremor Management with a Neuroprosthesis

J.A. Gallego[1], E. Rocon[1], J.M. Belda-Lois[2], and J.L. Pons[1]

[1] Bioengineering Group, Spanish National Research Council (CSIC), Madrid, Spain
[2] Instituto de Biomecánica de Valencia, Universitat Politècnica de València, Valencia, Spain

Abstract—Tremor is the most common movement disorder, and one of the major causes of functional disability. In spite of the existence of various treatments, tremor is not managed effectively in a large number of patients, which make it a major cause of loss of quality of life. Here we present a novel strategy for tremor suppression through neurostimulation that replicates an adaptive notch-filter at the frequency of the tremor. The controller, which adapts the neurostimulation to the ongoing amplitude, frequency and phase of the tremor, is implemented in a wearable neuroprosthesis for tremor management. Experimental results in one patient with severe essential tremor illustrate the interest of the approach.

Keywords—Tremor, Neurostimulation, Essential tremor, Neuroprosthesis, Electromechanical delay.

I. INTRODUCTION

Tremor is the most common movement disorder, and its prevalence is expected to increase in the next decades [1]. It appears due to a number of etiologically different disorders [2], none of which is fully understood [3]. Drugs constitute the standard treatment for the various types of tremor, and refractory patients—normally with severe symptoms—may undergo neurosurgery. In spite of the existence of these various treatments, a large proportion of patients exhibit significant disability when performing their daily chores [4, 5]. This motivates the research on novel forms to manage tremor.

Among the alternatives for tremor management that are currently under development, external devices that apply selective forces to attenuate the tremor have attracted great interest. These systems span devices fixed to a external frame and wearable orthoses, which are likely to provide the biggest functional benefit (see [6] for a brief review). Neuroprostheses (NPs) may be regarded as a special type of wearable orthoses, in which actuators are replaced by neural interfaces that stimulate the affected muscles. In [7], the authors proposed the first design of a table-mounted NP for tremor suppression, which provided a significant improvement [8] of the major types of tremor. In spite of the successful results it attained, a number of aspects should be improved to facilitate its translation into a product that improves the patients' quality of life, mainly: i) the NP should ideally be wearable system, and ii) the NP should adapt the stimulation it delivers to the ongoing characteristics of the tremor (mainly its severity). Therefore, in this work we present a novel controller for a tremor suppression NP that expands the originally proposed in [7] by implementing real-time adaptation to tremor amplitude, frequency and phase. The paper presents its concept and implementation, and provides experimental evidence in a essential tremor patient that demonstrate its potential. The strategy was implemented in a NP published elsewhere [9]. Further, we report a study of the electromechanical delay (EMD) of forearm muscles for this type of applications.

II. TREMOR SUPPRESSION STRATEGY

The tremor suppression strategy here presented relied on the same principle than that presented in [7]. There, the authors implemented a controller that, theoretically, stimulated a pair of antagonist muscles in counterphase to the ongoing tremor bursts. This resembled a notch filter at the selected frequency, which was selected in accordance to that expected for the different types of tremor [2, 7]. However, in [7], the assumed frequency was not corrected during the experiments, which might be problematic given that the characteristics of tremor vary depending on the conditions (as the authors mentioned in the report of the experimental validation [8]). A conceptually identical strategy, but with frequency adaptation was implemented and successfully validated in a recent work [10]. However, the authors of [10] developed a controller that delivered a constant stimulation level, which may have a number of drawbacks, e.g. it may facilitate the onset of muscle fatigue, it may cause that a current density higher than needed to compensate for the ongoing tremor is injected (possibly generating discomfort), and, importantly, it may impede the performance of concurrent voluntary movements.

Therefore, we developed a new notch-filter tremor suppression strategy, in which the stimulation delivered was adapted to the ongoing amplitude, frequency and phase of the tremor. Furthermore, it implemented a mechanism to compensate for the effects of the electromechanical delay (EMD, in this case understood as the time period between the delivery of neurostimulation and movement onset), and the possibility of producing brief periods of co-contraction when the direction of rotation switches due to the tremor oscillations. In more detail, the controller (summarized in Fig. 1) was implemented as follows. First, the raw motion, recorded with a pair of solid-state gyroscopes, was separated into concomitant voluntary movement and tremor, using a two-stage adaptive algorithm [11]. Next, to obtain accurate information about which muscle needed to be stimulated, the tremor obtained from the solid-state gyroscopes was differentiated, to compute the angular acceleration. Given that tremor is a rhythmical oscillatory movement, the angular acceleration has a phase of 180° relative to the angular displacement, which immediately provided the controller with information about the muscle in the antagonist pair that needed to be actuated. The estimated tremor also served to correct for the estimated tremor frequency and phase if a large discrepancy was detected (see Fig. 1). Next, the electromechanical delay was applied to the estimated tremor phase (see the right part of Fig. 1), which in turn served to choose whether stimulation needed to be delivered in co-contraction or in counterphase (see the switch in Fig. 1). The stimulation amplitude was computed using a proportional integral (PI) algorithm, and a saturation was applied to it, for comfort and safety reasons. Controller gains and the saturation values were identified during calibration. The last part of the controller (the blue rectangle in Fig. 1) was implemented independently for each muscle in the antagonist pair.

III. METHODS

A. Assessment of the Electromechanical Delay

We quantified the EMD in forearm muscles in order to investigate: *i*) the possible impact of this phenomenon on the performance of the NP, and *ii*) whether this parameter needed to be identified for each patient, and, in the case that this was not required, obtain a general estimate. Hence, we performed a experiment on 6 young volunteers (3 male and 3 female, age 28–31 years), who gave informed consent to participate.

The protocol aimed to quantify the EMD defined as the interval between stimulation onset and movement onset, as detected with the gyroscopes implemented in the NP [9]. Such definition was employed because the final goal was to integrate the EMD estimate in the controller of the NP. Our experimental design addressed the influence of inter- and intra-subject variability on the EMD, the latter understood as changes in EMD due to the stimulation with different amplitude. To this end, we applied 15 1-s stimulation bursts (separated by a 2-s pause) to the flexor carpi ulnaris and the extensor carpi radialis while the hand was supported against gravity, and fully supinated or pronated, respectively. The 15 stimulation bursts were applied at 3 different amplitudes (5 each), defined as low (L), medium (M) and high (H). These values were visually chosen after electrode setup. Stimulation frequency and pulse width were constant, and set to 30 Hz and 250 μs respectively. Trial order was randomized, and the experiment designed using an optimal algorithm [12]. Hand rotation was measured with a pair of solid state gyroscopes (Technaid S.L., Madrid, Spain) placed in differential configuration [11]. Neurostimulation was delivered with a multichannel monopolar stimulator that injected charge compensated pulses (UNA Systems, Belgrade, Serbia); the common electrode was placed at the dorsal side of the wrist. The data were stored for posterior analysis. Movement onset was determined manually based on the inspection of gyroscope data. The effect of stimulation amplitude and stimulation site (i.e. muscle) on each subject's data was assessed with a one-way ANOVA with repeated measures. Differences among subjects were also assessed with a one-way ANOVA with repeated measures. All results are reported as mean ± SD.

B. Pilot Testing of the Tremor Suppression Strategy

We present results for one patient with essential tremor (female) with severe postural and kinetic tremor. She was recruited from a outpatient clinic, informed beforehand, and gave written informed consent to participate. The ethical committee at Universidad Politécnica de Valencia, Valencia, Spain, approved the experimental protocol, and warranted its accordance to the declaration of Helsinki.

The experimental protocol replicated that presented in [9]. Briefly, the patient was asked to perform 12 repetitions of a task that triggered her tremor, which in this case was a postural task. The repetitions were split into two types of 30-s trials (referred to as ST and NO) which order was randomized. Each trial consisted of two 15-s sub-periods. In the ST trials, neurostimulation was delivered in the second sub-period, while in the NO trials neurostimulation was never delivered. The goal of this design was to avoid the influence of inter- and intra-trial tremor variations in the results. Details on the implementation of the NP are given in [9]. As mentioned above, the gains of the controller were identified manually at the beginning of the session: the proportional term was obtained as the best linear fit to the movement data obtained after applying a series of 5 1-s stimulation bursts, with increasing amplitude; the experimental setup replicated that employed to quantity the EMD. The integral gain was initially chosen as 1/2 the proportional gain. The saturation values were obtained as the maximum amplitude to elicit a movement with amplitude larger than that observed for the tremor, and perceived as comfortable by the patient. The EMD was set to 20 ms (according to the results presented below); the antagonist pair was co-contracted for a time period equivalent to 30 °.

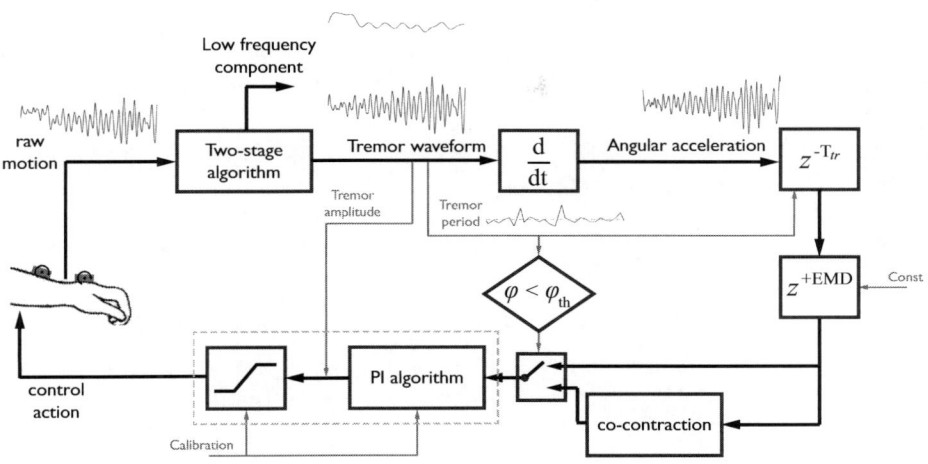

Fig. 1 Block diagram of the proposed tremor suppression strategy.

Tremor attenuation was computed as the ratio of the integral of the power spectral density (PSD) of the tremor in the second sub-period of the trial to the same variable calculated in the first sub-period (R_{att}) [13, 9], and reported as percentage. The pooled values of R_{att} for the ST and NO trials were compared using a Mann-Whitney test, to assess whether the control strategy attained a significant tremor attenuation.

IV. RESULTS

A. Electromechanical Delay

Fig. 2 shows a representative response of the response evoked by the stimulation of the extensor carpi radialis. Similar data were obtained for the flexor carpi ulnaris. The plot shows how the magnitude of the response increased as a higher current amplitude was injected to the muscle, and that, for the same stimulation amplitude, differences among repetitions were negligible. The delay observed for different stimulation amplitudes was very similar (average 19 ± 4 ms).

The pooled EMD data for the extensor carpi radialis (18 ± 8 ms) and flexor carpi radialis (22 ± 12 ms) did not present large discrepancies, and, for each subject, was statistically independent of the muscle ($P = 0.137$). Stimulation amplitude did not affect significantly the EMD either ($P = 0.743$ and $P = 0.541$ for the extensor carpi radials and the flexor carpi ulnaris respectively). Moreover, although we found a significant difference among subjects in the EMD for the same muscle ($P = 0.005$ and $P < 0.001$ for the extensor carpi radials and the flexor carpi ulnaris respectively) the maximum discrepancies were 9 ms for the extensor carpi radialis, and 16 ms for the flexor carpi radialis (3 ms in the latter if we remove subject 6 from the analysis, who could be considered as an outlier).

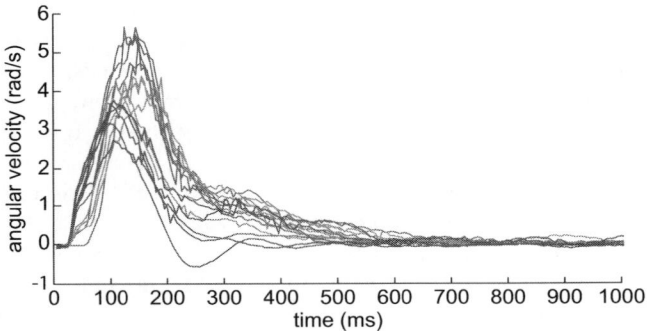

Fig. 2 Response evoked by the stimulation of the extensor carpi radialis with L (blue), M (green) and H (red) current amplitudes, in one subject.

B. Tremor Suppression Strategy

Fig. 3 shows a representative example of the performance of the controller, in which we observe that the control strategy proposed attained a fast and large reduction of the tremor (R_{att} = 11.55 %). The average reduction of tremor in all the ST trials was $R_{att} = 16.77 \pm 8.33$ %, ranging from 10.26 to 29.38 %. This tremor reduction was significantly larger ($P = 0.002$) than that measured in the NO trials (mean $R_{att} = 162.33 \pm 150.51$ %, ranging from 88.18 to 463.73 %), which indicates that the NP was the cause of the observed tremor attenuation.

V. DISCUSSION

This paper presented a new strategy for neuroprosthetic tremor management. The strategy relied on the traditional principle for tremor suppression through transcutaneous neurostimulation, but adapted, for the first time, the electrical

Fig. 3 Representative example of the performance of the controller, showing a large tremor attenuation. The two sub-periods of the trial are color coded (without stimulation in black, with stimulation in red).

charge delivered to the ongoing amplitude, frequency and phase of the tremor, which is expected to translate immediately in better functional outcome. Moreover, the proposed control strategy inherently compensated for the EMD, and implemented brief, adjustable co-contraction periods, which, in our opinion, could provide an important improvement of the controller. Results in one patient with severe tremor proved the feasibility and interest of the approach.

We have first reported a study to assess the EMD in forearm muscles, since we expected that due to the rationale of the controller, and the inherent relatively high frequency of some tremors (ET may be up to 10 Hz [2]), it could have a large influence on the performance of the controller. Our results showed that, for the same subject, the muscle and the stimulation amplitude did not cause a significant change in the EMD. Furthermore, although we obtained a statistically significant difference among subjects, such difference was very small, and given that the controller of the NP runs at 200 Hz, it may be neglected in the current context. We thus assumed that, for our application, the EMD may be considered to be constant, and equal to ~20 ms. This value is in agreement with the literature in the sense that the delay is shorter than for volitional contractions [14], although it cannot be directly compared given the different fashion in which it was obtained [14, 15]. As a matter of fact, much more precise measurements of EMD, e.g. using percutaneous stimulation combined with high frequency ultrasound recordings, found significant differences when the current density was changed [15]. Therefore, the results here presented are valid for applications such NPs for functional compensation of movement disorders, albeit they might not be completely generalizable.

The tremor suppression strategy provided systematic reduction of tremor amplitude, and interestingly, with a low stimulation amplitude once the tremor severity was very mild. This demonstrates the interest of adapting the amount of current injected to the ongoing severity of the tremor. Functional trials with a representative cohort of patients are needed to further demonstrate the interest of the approach.

ACKNOWLEDGMENT

This work has been partly supported by the EU Commission [NeuroTREMOR, grant number EU-FP7-2011-287739].

REFERENCES

1. Bach J.-P., Ziegler U., Deuschl G., et al . Projected numbers of people with movement disorders in the years 2030 and 2050. *Mov Disord.* 2011;26:2286–2290.
2. Deuschl G., Bain P., Brin M.. Consensus statement of the Movement Disorder Society on Tremor. Ad Hoc Scientific Committee. *Mov Disord.* 1998;13 Suppl 3:2–23.
3. Elble R.J.. Tremor: clinical features, pathophysiology, and treatment *Neurologic Clinics.* 2009;27:679–695.
4. Hariz G-M., Forsgren L.. Activities of daily living and quality of life in persons with newly diagnosed Parkinson's disease according to subtypeof disease, and in comparison to healthy controls. *Acta Neurol Scand.* 2011;123:20–27.
5. Louis E. D., Barnes L., Albert S. M., et al. Correlates of functional disability in essential tremor. *Mov Disord.* 2001;16:914–920.
6. Gallego J. A., Rocon E., Belda-Lois J. M., et al . Design and validation of a neuroprosthesis for the treatment of upper limb tremor in *Conf Proc IEEE Eng Med Biol Soc 2013* in press.
7. Prochazka A., Elek J., Javidan M.. Attenuation of pathological tremors by functional electrical stimulation. I: Method. *Ann Biomed Eng.* 1992;20:205–224.
8. Javidan M., Elek J., Prochazka A.. Attenuation of pathological tremors by functional electrical stimulation. II: Clinical evaluation. *Ann Biomed Eng.* 1992;20:225–236.
9. Gallego J. A., Rocon E., Belda-Lois J. M., Pons J. L.. A neuroprosthesis for tremor management through the control of muscle co-contraction. *J Neuroeng Rehabil.* 2013;10:36.
10. Popovic-Maneski Lana, Jorgovanovic Nikola, Ilic Vojin, et al. Electrical stimulation for the suppression of pathological tremor. *Med Biol Eng Comput.* 2011;49:1187–1193.
11. Gallego J. A., Rocon E., Roa J. O., et al . Real-Time Estimation of Pathological Tremor Parameters from Gyroscope Data *Sensors.* 2010;10:2129–2149.
12. Fedorov T. T.. *Theory of optimal experiments.* Academic Press 1972.
13. Rocon E., Belda-Lois J. M., Ruiz A. F., et al . Design and Validation of a Rehabilitation Robotic Exoskeleton for Tremor Assessment and Suppression *IEEE Tans Neural Syst Rehab Eng.* 2007;15:367–378.
14. Hopkins J.T., Feland J. B., Hunter I.. A comparison of voluntary and involuntary measures of electromechanical delay. *Int J Neurosci.* 2007;117:597–604.
15. Lacourpaille L., Nordez A., Hug F.. Influence of stimulus intensity on electromechanical delay and its mechanisms. *J Electromyogr Kinesiol.* 2013;23:51–55.

Author: J.A. Gallego
Institute: CAR - CSIC
Street: Ctra. Campo Real, km. 02, La Poveda
City: Arganda del Rey, Madrid
Country: Spain
Email: ja.gallego@csic.es

Objective Metrics for Functional Evaluation of Upper Limb during the ADL of Drinking: Application in SCI

A. de los Reyes-Guzmán[1], I. Dimbwadyo-Terrer[1], S. Pérez-Nombela[1], F. Trincado[1], D. Torricelli[2], and A. Gil-Agudo[1]

[1] Biomechanics and Technical Aids Department. National Hospital for Spinal Cord Injury (SESCAM), Toledo, Spain
[2] Centro de Automática y Robótica, Consejo Superior de Investigaciones Científicas, Arganda del Rey, Madrid, Spain

Abstract—Three-dimensional kinematic analysis provides quantitative assessment of upper limb motion and is used as an outcome measure to evaluate movement disorders. The aim of the present study is to present a set of kinematic metrics for quantifying characteristics of movement performance and the functional status of the subject during the execution of the activity of daily living (ADL) of drinking from a glass. Then, the objective is to apply these metrics in healthy people and a population with cervical spinal cord injury (SCI), and to analyze the metrics ability to discriminate between healthy and pathologic people. 19 people participated in the study: 7 subjects with metameric level C6 tetraplegia, 4 subjects with metameric level C7 tetraplegia and 8 healthy subjects. The movement was recorded with a photogrammetry system. The ADL of drinking was divided into a series of clearly identifiable phases to facilitate analysis. Metrics describing the time of the reaching phase, the range of motion of the joints analyzed, and characteristics of movement performance such as the efficiency, accuracy and smoothness of the distal segment and inter-joint coordination were obtained. The performance of the drinking task was more variable in people with SCI compared to the control group in relation to the metrics measured. Reaching time was longer in SCI groups. The proposed metrics showed capability to discriminate between healthy and pathologic people. Relative deficits in efficiency were larger in SCI people than in controls. These metrics can provide useful information in a clinical setting about the quality of the movement performed by healthy and SCI people during functional activities.

Keywords—Metrics, Kinematic, Assessment, Upper limb, Activities of Daily Living.

I. INTRODUCTION

The upper limb function is affected in more than 50% of spinal cord injuries [1].

Impairment and disability in clinical settings is generally assessed by ordinal scales such as the Fugl-Meyer Assessment, Frenchay Arm Test, Motor Assessment Scale, Box and Block Test, and the Nine-Hole Peg Test [2], but they may not be sensitive to small and more specific changes [3]. It's important that measures have the ability to detect a minimal clinically important change in subjects. At the same time, the measures must have the ability to give consistent responses on repeated assessments in the absence of change in the characteristic being studied [4]. Moreover, the use of these scales is not exempt from a degree of subjectivity. So more objective assessment methods are needed to evaluate and describe the upper limb function in detail.

On the other hand kinematic movement analysis can provide more specific information about movement component and strategies, although this requires special equipment. Kinematics describes movements of the body through space and time without reference to the forces involved [5]. It's of special importance to translate data from kinematic analysis in terms of objective metrics with a useful sense and easily understood in a clinical setting. These metrics can synthetize the main aspects of the movement that directly relate with functional ability.

The aim of this study is to present a set of objective metrics for the functional upper limb evaluation and for the measure characteristics of quality movement during functional activities. Then these metrics are applied to a population of healthy and cervical SCI people during the ADL of drinking from a glass. The last objective is to analyze the ability of these metrics as discriminative measures between healthy people and people with SCI.

II. MATERIALS AND METHODS

A. Population

Nineteen subjects participated in this study and were separated into three groups: a control group (CG) with 8 subjects; 7 people with C6 tetraplegia (C6 group) and 4 people with C7 tetraplegia (C7 group). The demographic and anthropometric characteristics are shown in Table 1. In the case of subjects with C6 or C7 tetraplegia, the etiology of the injury was trauma in every case. The patients screened had to fulfill the following criteria to be included in the study: age 16 to 65 years, injury of at least 6 months' duration and level of injury C6 or C7 classified according to the American Spinal Cord Injury Association (ASIA) [6] scales into grades A or B. The subjects were classified into

C6 or C7 tetraplegia by a physical examination. The upper limb Motor Index was obtained [6].

Table 1 Demographic characteristics of the sample analyzed (n=19).* Mean and standard deviation for continuous variables. †n y % for categorical variables

Item	CG	C6	C7
Sex (male)	3 (37.5)	4 (57.4)	4 (100)
Age (years)	26.0 (5.0)	34.0 (5.0)	30.5 (10.0)
Height (cm)	168.0 (20.0)	175.0 (10.0)	184.0 (10.0)
Weight (Kg)	65.0 (21.1)	90.2 (7.1)	79.0 (9.1)
Months since injury	-	8.5 (2.2)	7.5 (1.8)
ASIA (A)	-	3 (33.3)	2 (50)
ASIA (B)	-	4 (66.6)	2 (50)
Index Motor right arm	25.0 (0)	13.0 (3.0)	14.5 (2.0)

All patients signed an informed consent form before the study. The guidelines of the declaration of Helsinki were followed in every case and the study design was approved by the local ethics committee.

B. Experimental Set-Up and Data Processing

Three-dimensional movement capture was recorded with the photogrammetry equipment CodaMotion (Charnwood Dynamics, Ltd, UK) following the protocol in an earlier study [7].

All the subjects were right handed and performed the movement of the ADL of drinking with the right arm at a self-selected speed. Subjects with C6 or C7 tetraplegia sat in their own wheelchairs and the controls sat in a conventional wheelchair. The movement was performed in front of a table, from a standardized position. The complete experimental protocol was included in a previous study, in which the consistency and repeatability of the experimental setup was assessed. The recordings were processed with Visual 3D software (C-Motion, Inc., USA) [7].

The ADL of drinking from a glass was divided into a series of clearly identifiable phases to facilitate analysis. We used phases and events delimiting the phases that have been described previously: reaching, forward transport, drinking, back transport and return [7].

C. Kinematic Metrics Definition

A set of objective kinematic metrics was defined from experimental data preprocessed with Visual3D software. These metrics were separated into two groups and all of them were calculated in Matlab software:

a) Metrics in Relation to Spatio-temporal Variables from Preprocessed Data with Visual3D:

<u>Reaching time ($T_{Reaching}$)</u>: duration of the reaching phase normalized by the duration of the complete cycle (Tcycle) expressed in percentage.

$$T_{Reaching} = \left(\frac{T_{Reaching}}{T_{Cycle}}\right) 100 \quad (1)$$

<u>Range of motion (ROM)</u>: this metric is defined as the sum of the ROM in degrees for each of the six degrees of freedom (DoF), normalized by the sum of the total biologically allowed ROM for each DoF analyzed [8]. So the metric is obtained dimensionless expressed as a percentage. The ROM is defined here as the difference between the maximum and minimum angles recorded by a given joint over the course of the drinking task.

$$ROM_{Total} = \frac{\sum_{i=1}^{N}(\theta_{max} - \theta_{min})_{subject}}{\sum_{i=1}^{N}(\theta_{max} - \theta_{min})_{total}} * 100 \quad (2)$$

b) Metrics for the Evaluation of Functional Abilities and Skills of the Upper Limb

These parameters are defined from the kinematic of the distal segment, the hand, and they are measurements of the quality of the movement performed by the three groups of people. These metrics are:

<u>Efficiency (E)</u>: the movement efficiency is obtained by computing the path length (L_{path}) of the trajectory traveled by the subject to reach the target (in this study the glass used to perform the ADL of drinking from a glass) [9,10]. This parameter is obtained by summing the distances between two consecutive points of the subject's path; then, it's normalized to the straight line distance between the starting point of the task and the target (d_{sl}). A ratio between L_{path} and d_{sl} of 1 represents a direct movement to the target (ideal), and a reach path ratio slightly higher than 1 represents a healthy movement but a value proximal to 2 an abnormally curved movement. This ratio is calculated during the reaching phase of the task, and it is a measure of the error of movement efficiency, so a transformation function has been applied with the aim of higher values (maximum 100) correspond to a more efficient movement (equation 3).

$$E = 130 - 30 * \left(\frac{L_{path}}{d_{sl}}\right) \quad (3)$$

Accuracy (A): this metric is assessed by measuring the mean absolute value of the distance (d_{mean}) of each point of the path from the theorical path (the straight line between initial point and the target) [9] and the maximal distance (d_{max}). This result is inverted with the aim of higher values correspond to more accuracy movements.

$$A = 100 - (d_{max} + d_{mean}) \quad (4)$$

Smoothness (S): is defined from the number of peaks in the hand speed profile during the reaching and proximal transport phases [5]. To avoid signal noise effects, the time between two subsequent peaks had to be at least 150 milliseconds. Fewer peaks in speed represent fewer periods of acceleration and deceleration, making a smoother movement. For that reason, the smoothness metric has been defined as the inverse of speed peaks number (equation 5), so higher values of smoothness metric correspond to smoother movements and a better functional state.

$$S = 20 - npeaks \quad (5)$$

Coordination (C): the inter-joint coordination between the shoulder and elbow joint angles for the flexion-extension movement was characterized quantitatively by means of the Pearson correlation index [7], as expressed the following equation. This metric is obtained from covariance matrix and marginal deviations between both variables analyzed (Equation 6). A correlation index closer to 1.0 indicates stronger correlation and indicates that the joint motion of the 2 joints is tightly coupled. Then, this value is multiplied by a 100 factor for obtaining a coordination value between 0 and 100 points.

$$C = \frac{\frac{\sum F_S F_E}{N} - \overline{F_S F_E}}{S_{F_S} S_{F_E}} * 100 \quad (6)$$

D. Statistical Analysis

A descriptive analysis with SPSS for Windows (version 12.0) was made of the clinical and functional variables (Table 1). The Kruskal-Wallis test was used to find possible differences in each metric between the three groups analyzed. The Bonferroni correction was applied, which takes into account randomness due to multiple comparisons.

The repeatability of the protocol was evaluated in an earlier study [7].

III. RESULTS

There were no significant differences in demographic variables between participants with SCI and healthy people. The most relevant results are shown in Table 2.

The range of motion (ROM), expressed as percentage, is obtained by summing the ROM of each joint analyzed and then normalized by the total theoretically allowed.

In relation to the movement time, people with C6 tetraplegia had slower movement times than healthy people in reaching phase normalized by the duration of the complete cycle (p<0.01).

Table 2 Median and interquartile range of the metrics for the three groups analyzed.

METRICS	CG (n=8)	C6 (n=7)	C7 (n=4)
Range of motion (%)	24.52 (1.19)[a,b]	28.84 (4.80)[a]	30.18 (5.50)[b]
Reaching time (%)	14.10 (3.74)[c]	22.61 (15.52)[c]	14.58 (4.14)
Efficiency (%)	97.81 (2.04)[c,d]	77.27 (32.44)[c]	93.16 (8.83)[d]
Accuracy (points)	87.69 (5.86)[a]	80.01 (7.59)[a]	86.61 (4.12)
Smoothness (inverse peaks number)	18.00 (0.25)[c,d]	15.00 (2.50)[c]	16.50 (1.00)[d]
Coordination (%)	88.42 (6.98)	69.88 (22.89)	82.00 (6.15)

Same letters mean significant differences between groups
a,b: p<0.05 with Bonferroni correction
c,d: p<0.01 with Bonferroni correction

The metrics efficiency and accuracy were calculated from the hand path during the reaching phase within the ADL of drinking from a glass. The movement was accurate in C7 group as in the healthy people, but less efficient (p<0.01). The movement was less accurate (p<0.05) and efficient (p<0.01) in C6 group that in the controls (Figure 1) (Table 2).

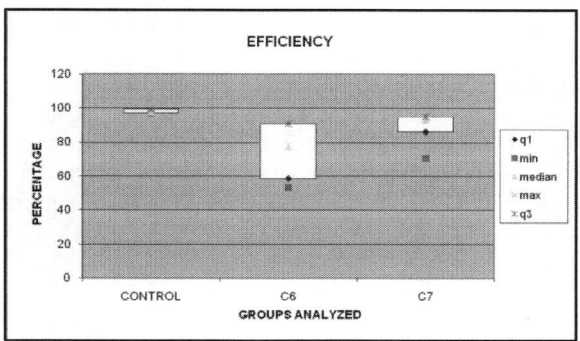

Fig. 1 Results of the efficiency metric for the three groups analyzed

Smoothness was obtained during the reaching and during the transport of the glass toward the mouth, so the minimum number of peaks in velocity profile is 2. Then the result of the three metrics is inverted for expressing it, as a greater value correspond to a better functional state. Statistically significant differences were found between healthy and C6 subjects for the metric efficiency and smoothness ($p<0.01$), and for the metric accuracy ($p<0.05$).

In relation to the inter-joint coordination, no statistically significant differences were found (Table 2).

In the C6 group, there was a great variability in the measured results for the efficiency (Figure 1) and coordination metrics (Table 2).

IV. Discussion

The metrics proposed can synthetize the main aspects of the movement that directly relate with functional capability. The main definition of the most of the metrics proposed in this study can be found in literature [9,10]. The inverse functions have been developed within this study with the aim of obtaining the results according to functional clinical tests.

It is worth noting that these parameters may provide useful information related to the quantitative data we can obtain, in situations in which, despite a high value of movement accuracy, the subject performs a longer path with the hand toward the target, and the consequence is a less efficient movement. The smoothness, efficiency and range of movement showed the strongest ability to discriminate movement quality between healthy and SCI people. This is confirmed, for smoothness metric, by Murphy et al. in a study in stroke people [5]. In this study, we calculated speed peaks during reaching and forward transport phases within the ADL of drinking because these two phases correspond to two different constraints of a purposeful movement.

The reaching phase, including grasping, it's the movement phase of the ADL of drinking from a glass with greater contribution to the total movement time [5,7].

On the other hand, kinematic analyses are increasingly being used in clinical research as evaluative measures [5]. It's necessary additional assessment to investigate the relation between kinematics and clinical scales. Moreover it is important to take into account that the analysis of a purposeful daily activity increases the validity of the study.

V. Conclusions

In the present study, the metrics proposed were applied to a population with SCI people, and their ability to discriminate between healthy and SCI people, has been shown.

For further studies the sample analyzed will be greater and the metrics will be applied to other neurological pathologies such as stroke, where the discriminative ability within this pathology will be investigated.

Acknowledgment

Fundings were obtained from the research project entitled "HYPER: Dispositivos Híbridos Neuroprotésicos y Neurorobóticos para Compensación Funcional y Rehabilitación de Trastornos del Movimiento". Referencia: CSD00C-09-61313. CONSOLIDER-INGENIO 2010. Ministerio de Ciencia e Innovación.

References

1. Wyndaele M, Wyndaele JJ (2006). Incidence, prevalence and epidemiology of spinal cord injury: what learns a worldwide literature survey?. Spinal Cord 21(8): 357-364
2. Wade DT (1992) Measurement in neurological rehabilitation. Oxford medical publications. Oxford University Press, 388
3. Mc Crea PH, Eng JJ, Hodgson AJ (2002) Biomechanics of reaching: clinical implications for individuals with acquired brain injury. Disabil Rehabil, 24: 534-541
4. Boyce W, Gowland C, Rosenbaum P et al. (1991) Measuring quality of movement in cerebral palsy: a review of instruments. Phys Ther. 71: 813-819
5. Murphy M, Willén C, Sunnerhagen K (2011) Kinematic variables quantifying upper-extremity performance after stroke during reaching and drinking from a glass. Neurorehabil Neural Repair 25:71 DOI 10.1177/1545968310370748
6. Maynard FM, Bracken MB, Creasey G et al. (1997) Standards for Neurological and Functional Classification of Spinal Cord Injury. American Spinal Injury Association. Spinal Cord, 35: 266-274
7. Ana de los Reyes-Guzmán, Angel Gil-Agudo, Benito Peñasco-Martín, Marta Solís-Mozos, Antonio del Ama-enEspinosa, Enrique Pérez-Rizo (2010) Kinematic analysis of the daily activity of drinking from a glass in a population with cervical spinal cord injury. http://www.jneuroengrehab.com/content/7/1/41
8. Kapandji AI (2012) Fisiología articular. Miembro superior. 6ª Edition. Editorial Medica Panamericana
9. Colombo R, Pisano F, Micera S et al (2008) Assessing mechanisms of recovery during robot-aided neurorehabilitation of the upper limb. Neurorehabil Neural Repair 22:50 DOI 10.1177/1545968307303401
10. Lang CE, Wagner JM, Bastian AJ et al (2005) Deficits in grasp versus reach during acute hemiparesis. Exp Brain Res 166: 126-136 DOI 10.1007/s00221-005-2350-6

Author: Ana de los Reyes-Guzmán
Institute: Biomechanics and Technical Aids Department. National Hospital for Spinal Cord Injury. SESCAM
Street: Finca La Peraleda,s/n
City: Toledo
Country: Spain
Email: adlos@sescam.jcmm.es

Efficacy of TtB-Based Visual Biofeedback in Upright Stance Trials

C. D'Anna, D. Bibbo, M. Goffredo, M. Schmid, and S. Conforto

Department of Engineering, University of Roma TRE, via Vito Volterra 62, 00146 Rome, Italy

Abstract—Several studies have shown the effect of visual biofeedback (VBF) on postural control with real-time presentation of centre of pressure (CoP). However, up to now no study has yet focussed on the effect that a predictive VBF could have on postural control. The aim of this study is thus to determine whether the Time-to-Boundary function (TtB) could be used as an efficacious VBF in static posturography.

The CoP coordinates were extracted from force plate data and elaborated to calculate TtB in real-time.

Two groups of six healthy young subjects executed the protocol in two different sequences composed of the following conditions: noVBF-VBF1 (real-time presentation of CoP) and noVBF-VBF2 (real-time TtB presentation). Each condition was repeated three times.

The effect of the two VBFs was studied by five parameters extracted directly from CoP coordinates (sway area, sway path, mean amplitude, frequency bandwidth that contains 95% of the power spectral density of antero-posterior and medio-lateral displacement) and two parameters by fraction Brownian motion model (exponential radial terms Hrs and Hrl corresponding to the short-term and long-term region Hurst exponents).

The comparison between the VBF conditions didn't show significant differences in the studied parameters. This evidence suggests that the participants react in similar way in both conditions, and it opens the possibility of using a predictive VBF as a tool to facilitate postural control in upright stance.

Keywords—time-to-boundary, centre of pressure, postural control, visual biofeedback.

I. INTRODUCTION

In spite of its seemingly simplicity, the maintenance of balance in upright stance is a rather complex task [1]: it involves the integration of data coming from multimodal sensory channels including vision and proprioception [2]. These data are processed by the nervous system to put in action the strategies needed to activate muscles [3], in order to maintain stability, whose impairment is a major risk factor for falling [4].

With the aim of evaluating the performance of the balance control system, different monitoring devices are generally used, including video-based systems [5,6,7], accelerometers [8], and force plates [9]. While the first ones are used to directly capture the movement of the body, and describe it as the result of the mechanical work, force plate data are directly linked to the reaction forces exerted by the ground and thus contain kinetic information.

As a consequence of this, the Centre of Pressure (CoP), which is usually extracted from force plates [10], is richer in information than its kinematic companion, the Centre of Mass (CoM), since it conveys both the effect of the forces acting to maintain stability and their causes.

Several parameters extracted from CoP data have been proposed in literature [11] and have been used in several studies to describe, interpret and assess the postural control [12] in different population samples [13].

Among these parameters the Time-to-Boundary function (TtB) was introduced in the framework of postural stability [14] by moving a concept defined in the theory of the time-to-collision to visual perception [15], and calculated in upright stance trials [16]. This variable takes position, velocity, and acceleration of the CoP trajectory into account, to estimate the temporal margin to the stability boundaries. When applied to experimental trials of upright stance, when subjects were asked to stand still, the TtB function shows a characteristic oscillating pattern. The temporal locations of particular points of the pattern (i.e. the minima/turning points) were interpreted as the time instances at which postural control intervened to escape the nest of instability [14].

The advantage of TtB is that the assessment of postural stability includes both spatial and temporal aspects of postural movements. Previous studies show that TtB may be perceived by the individual as predictive information of instability, as it provides information regarding the time needed to reverse a perturbation before loss of balance [17].

Since postural control is strongly affected by the perception and the integration of the sensory information, several protocols studied the effect of providing a sensory reinforcement to the subject. Thus, a number of studies have focused on the effect of real-time feedback [20,21]. Among these, recent papers have taken into account the effect of visual biofeedback (VBF) on motor tasks [22,23] and postural control, by highlighting how postural parameters [18] modify after showing to subjects standing on a force plate their own CoP displacements. Nevertheless, up to now no study has looked at the effect that a predictive VBF could have on postural control. The present study wants to see whether the TtB could thus be used as an efficacious VBF. This will be proven by comparing the effect of two different VBFs, based on CoP displacement and TtB, respectively.

II. MATERIALS AND METHODS

A. Participants and Procedure

12 healthy young subjects (age 29 ± 5 yrs, height 1.68 ± 0.1 m; weight 64.7 ± 16.3 kg), volunteered in the study. None of them reported neuropathies at the peripheral level, or vestibular pathologies; they had normal visual acuity and no colour blindness. They were instructed as for the experimental procedure that will be described in the following, and gave written informed consent according to the declaration of Helsinki.

The subjects stood barefoot, feet together, on a homemade force plate and were asked to maintain an upright posture with arms along their sides. Three different conditions, each consisting of three 40 s bouts, were considered:

- no VBF task (eyes open)
- VBF1 with real-time representation of CoP
- VBF2 with real-time representation of TtB

Six subjects executed the task in the sequence noVBF-VBF1, six subjects in the sequence noVBF-VBF2. The VBFs were presented on a 21" computer screen positioned at 1 m distance from the participants.

For the noVBF condition participants were asked to stand as still as possible looking in front of a fixed point.

The VBF1 task consisted of asking participants to maintain a spot on the screen within a red square. The spot is a real-time representation of CoP. At the onset of each trial, the spot was positioned at the centre of the square.

To present the VBF2, the CoP components were used to real-time compute the TtB function, according to the parabolic motion equation reported in [19]:

$$\begin{cases} x_i(\tau) = r_x(t_i) + \dot{r}_x(t_i)\tau + \ddot{r}_x(t_i)\tau^2/2 \\ y_i(\tau) = r_y(t_i) + \dot{r}_y(t_i)\tau + \ddot{r}_y(t_i)\tau^2/2 \end{cases} \quad (1)$$

At each time instant t_i, the corresponding TtB(t_i) sample is obtained as the τ value for which the parabolic motion expressed above crosses the boundary limits [24]. These are represented by an elliptical figure, whose axes were dimensioned on the basis of subjects' anthropometric features.

In the VBF2 condition, participants were asked to maintain the waveform of TtB over a horizontal red bar whose height was defined on the basis of subject-specific criteria.

C. Data Acquisition and Processing

CoP data in both antero-posterior and medio-lateral directions were obtained from signals coming from a home-made force plate, sampled at 100 samples/s (DAQCard-AI-16E-4, by National Instruments) and low-pass filtered at 10 Hz. CoP and TtB were calculated real-time by a Labview appplication (@ National Instruments ed. 2010 32-bit).

The CoP anteroposterior (AP) and mediolateral (ML) components were also stored for offline processing. This included mean value removal and digital low-pass filtering (cut-off frequency of 10 Hz, according to the recommendations reported in [12]). A subset of five summary measures were extracted directly from the CoP time series, following the classification reported in [11]: sway path (SP), mean amplitude (MA), sway area (SA), the frequency bandwidth that contains 95% of the power spectral density of AP and ML displacement (Fy95%, Fx95%). From the mathematical model called fraction Brownian motion [19], the exponential radial terms Hrs and Hrl, corresponding to the short-term and long-term region Hurst exponents, were also computed. Depending on these H values it can be inferred how much the CoP trajectory is controlled [18].

D. Statistical Analysis

Descriptive statistics were calculated for all the extracted parameters. All the parameters were also considered separately as dependent variables in a 2-way ANOVA test: repetition of task (RP) and type of task (TS) were the factors.

The test compares, for the two groups, the postural parameters in two different cases: considering all repetitions of the task and considering only the last two repetitions. Finally the test compares the parameters of VBF1 and VBF2. The level of significance was set at 0.05.

III. RESULTS

In general terms, both frequency parameters and Hurst exponents are in line with literature data and don't show significant differences due to the presence of VBF. In the next paragraphs differences on the remaining parameters are presented.

noVBF-VBF1- In the first group, the first ANOVA analysis (considering all the repetitions) shows no significant difference in RP for all parameters examined except SP, which shows a close-to-significant effect ($p<0.1$). In TS there is a significant difference for SP and MA parameter ($p<0.05$), and $p<0.1$ for SA. No significant difference for Fx95%, Fy95%, Hrs and Hrl. In RP*TS only SA shows a significant difference ($p<0.1$). The second ANOVA test (considering only the last two repetitions) no significant difference was present in RP and TS for all parameters. Numeric values are shown in Table 1.A and 2.A.

noVBF-VBF2 - In the second group, the ANOVA test (considering all repetitions) shows significant different in RP, TS and RP*TS for SP, SA, MA (p<0.05). No significant difference in Fx95%, Fy95%, Hrs and Hrl. Considering only the last two repetitions, no significant difference is showed in RP, TS, and RP*TS for all parameters except SA in TS, which shows a close-to-significant effect (p<0.1). Numeric values are shown in Table1.B and 2.B

Table 1A Mean±standard deviation of all parameters considering all repetitions in conditions noVBF-VBF1. Significance is reported as well †n.c, ~ p<0.1, * p<0.05

	noVBF	VBF1			
			RP	TS	RP*TS
SP(m/min)	1.73±0.22	2 ±0.5	~	*	~
SA(m²/s)	7.93E-05±3.5E-05	1.2E-04±9.08E-05	†	~	†
MA(m)	0.01±0.0036	0.01±0.006	†	*	†
Fx95%(Hz)	0.69±0.23	0.67±0.2	†	†	†
Fy95%(Hz)	0.89±0.16	0.8±0.32	†	†	†
Hrs	0.84±0.03	0.84±0.06	†	†	†
Hrl	0.3±0.12	0.23±0.13	†	†	†

Table 1B Mean±standard deviation of all parameters considering only the last two repetitions in condition noVBF-VBF1. Significance is reported as well †n.c, ~ p<0.1,* p<0.05

	noVBF	VBF1			
			RP	TS	RP*TS
SP(m/min)	1.73±0.53	1.9 ±0.7	†	†	†
SA(m²/s)	8.163E-05±4.25E-05	9.75E-05±5.42E-05	†	†	†
MA(m)	0.01±0.004	0.01±0.006	†	†	†

Table 2.A Mean±standard deviation of all parameters considering all repetitions in conditions noVBF-VBF2. Significance is reported as well †n.c, ~ p<0.1, * p<0.05

	noVBF	VBF2			
			RP	TS	RP*TS
SP(m/min)	1.69±0.34	2.76 ±2.18	*	*	*
SA(m²/s)	5.69E-05±2.36E-05	3.95E-04±0.6.E-04	*	*	*
MA(m)	0.008±0.002	0.018±0.018	*	*	*
Fx95%(Hz)	0.82±0.26	0.76±0.25	†	†	†
Fy95%(Hz)	1.07±0.26	1.01±0.18	†	†	†
Hrs	0.88±0.53	0.88±0.03	†	†	†
Hrl	0.1±0.08	0.08±0.17	†	†	†

Table 2B Mean±standard deviation of all parameters considering only the last two repetitions in condition noVBF-VBF2. Significance is reported as well †n.c, ~ p<0.1, * p<0.05

	noVBF	VBF2			
			RP	TS	RP*TS
SP(m/min)	1.63±0.34	1.77 ±0.3	†	†	†
SA(m²/s)	5.39E-05±2.11E-05	9.23E-05±5.45E-05	†	~	†
MA(m)	0.008±0.002	0.01±0.003	†	†	†

VBF1-VBF2 - The analysis compares two different VBFs considering all the repetitions first, and then only the last two repetitions. In the first case significant difference is shown in RP, TS and RP*TS for SP, SA and MA, no significant difference for Fx95%, Fy95%, Hrs and Hrl. In the second case no significant difference is shown for all parameters in RP, TS and RP*TS. Numeric values are shown in Table 3.A and 3.B.

Table 3A Mean±standard deviation of all parameters considering all repetitions in conditions VBF1-VBF2. Significance is reported as well †n.c, ~ p<0.1, * p<0.05

	VBF1	VBF2			
			RP	TS	RP*TS
SP(m/min)	2 ±0.5	2.76 ±2.18	*	*	*
SA(m²/s)	1.2E-04±9.08E-05	3.95E-04±0.6.E-04	*	*	*
MA(m)	0.014±0.006	0.018±0.018	*	*	*
Fx95%(Hz)	0.67±0.2	0.76±0.25	†	†	†
Fy95%(Hz)	0.8±0.32	1.01±0.18	†	†	†
Hrs	0.84±0.06	0.88±0.03	†	†	†
Hrl	0.23±0.13	0.08±0.17	†	†	†

Table 3B Mean±standard deviation of all parameters considering only the last two repetitions in condition noVBF1-VBF2. Significance is reported as well †n.c, ~ p<0.1, * p<0.05

	VBF1	VBF2			
			RP	TS	RP*TS
SP(m/min)	1.9 ±0.7	1.77 ±0.3	†	†	†
SA(m²/s)	9.75E-05±5.42E-05	9.23E-05±5.45E-05	†	~	†
MA(m)	0.01±0.006	0.01±0.003	†	†	†

IV. DISCUSSION AND CONCLUSIONS

The aim of this study was to evaluate whether predictive information could be used for a VBF.

The analysis of the results shows, on the one hand, the difference between two different conditions: maintaining stance without a VBF and with VBF (real-time CoP displacement and real-time TtB function). The analysis was done first by considering all the repetitions and then by considering only the last two repetitions. Moreover, the analysis compares two different biofeedbacks (VBF1 and VBF2) to evaluate the efficacy of predictive information on postural control.

The comparison noVBF-VBF1 shows an effect between the two tasks only for two parameters (SP and MA). This difference doesn't show if the first repetition of the trial was eliminated. The absence of postural improvement when the VBF was presented as compared to the eyes open condition is not in line with recent studies [21]. This effect is possibly depending on the type of instruction given to the subjects, so that they may maintain the postural performance in the preceding task.

The analysis of the second condition, noVBF-VBF2, shows an effect for SP, MA and SA. Also in this case, the analysis without the first repetition shows no difference, except SA. This latter result might be justified by a longer task learning time for TtB VBF as compared to CoP VBF: the subjects may need a period of training to learn the task and to apply the appropriate postural control strategies. When comparing the two VBFs, the presence of an effect for SP, SA and MA that disappears when the first repetition is eliminated might be associated with different learning rates. This evidence suggests that these subjects react in similar way in both conditions, and it opens the possibility of using predictive VBF to facilitate postural control in upright stance trials. Further studies will be needed to explore the most efficacious presentation of predictive VBF, by also capturing eye movements [25] to infer on subjects ability to follow the information contained in VBF.

REFERENCES

1. Balasubramaniam R, Wing AM (2002) The dynamics of standing balance. Trends Cogn Sci 6:531-536
2. Kiemel T, Oie KS, Jeka JJ(2002) Multisensory fusion and the stochastic structure of postural sway. Biol Cybern 87:262-277
3. Ting LH, McKay JL (2007) Neuromechanics of muscle synergies for posture and movement. Curr Opin Neurobiol 17:622-628
4. Piirtola M, Era P (2006) Force platform measurements as predictors of falls among older people - a review. Gerontology 52:1-16
5. Kejonen P, Kauranen K, Ahasan R, Vanharanta H(2002) Motion analysis measurements of body movements during standing: association with age and sex. Int J Rehabil Res 25:297-304
6. Goffredo M, Schmid M, Conforto S, D'Alessio T (2005)A Markerless sub-pixel motion estimation technique to reconstruct kinematics and estimate the centre of mass in posturography. Medical Engineering and Physics, 28: 719-726
7. Goffredo M, Schmid M, Conforto S, Carli M,. Neri A, D'Alessio T (2009)Markerless Human Motion Analysis in Gauss-Laguerre Transform Domain: An Application to Sit-To-Stand in Young and Elderly People. Information Technology in Biomedicine, IEEE Transactions on 13: 207-216
8. Kamen G, Patten C, Du CD, Sison S(1998) An accelerometry-based system for the assessment of balance and postural sway. Gerontology 44:40-45
9. Kapteyn TS, Bles W, Njiokiktjien CJ, Kodde L, Massen CH, Mol JM(1983) Standardization in platform stabilometry being a part of posturography. Agressologie 24:321-326
10. Goldie PA, Bach TM, Evans OM Force platform measures for evaluating postural control(1989) reliability and validity. Arch Phys Med Rehabil 70:510-517
11. Prieto TE, Myklebust JB, Hoffmann RG, Lovett EG, Myklebust BM (1996) Measures of postural steadiness: differences between healthy young and elderly adults. IEEE Trans Biomed Eng 43:956-966
12. Schmid M, Conforto S, Camomilla V, Cappozzo A, D'Alessio T(2000) The sensitivity of posturographic parameters to acquisition settings. Med Eng Phys 24:623-631
13. Schmid M, Conforto S, Lopez L, Renzi P, D'Alessio T (2005).The development of postural strategies in children: a factorial design study. Journal of NeuroEngineering and Rehabilitation 2: 29
14. Slobounov S M, Slobounova E S, Karl M N (1997). Virtual Time-to Collision and Human Postural Control. Journal of motor Behavior 29 : 263–281
15. Lee DN (1976) A theory of visual control of braking based on information about time-to-collision. Perception 5:437-459
16. Schmid M, Conforto S, Lopez L, D'Alessio T (2006) Cognitive Load affects Postural Control in Children. Experimental Brain Research, 179: 375-385
17. Riccio, G.E. (1993). Information in movement variability about the qualitative dynamics of posture and orientation. In K.M. Newell & D.M. Corcos (Eds)
18. Rougier P (1999) Influence of visual feedback on successive control mechanisms in upright quiet stance in humans assessed by fractional Brownian motion modeling. Neurosci Lett 266:157-160
19. Collins J J , De Luca C J (1993) Open-loop and closed-loop control of posture: a random-walk analysis of center-of-pressure trajectories. Exp Brain Res 95: 308-18
20. Boudrahem S, Rougier PR: Relation between postural control assessment with eyes open and centre of pressure visual feedback effects in healthy individuals (2009). Exp Brain Res 195:145-152
21. Dault MC, de Haart M, Geurts AC, Arts IM, Nienhuis B (2003) Effects of visual center of pressure feedback on postural control in young and elderly healthy adults and in stroke patients. Hum Mov Sci 22:221-236
22. Bibbo D, Conforto S, Bernabucci I, Carli M, Schmid M, D'Alessio T (2012) Analysis of different image-based biofeedback models for improving cycling performances . Proceedings of SPIE, 8295A: ,2012
23. De Marchis C, Schmid M, Bibbo D, Castronovo A M, D'Alessio T, Conforto S (2013) Feedback of mechanical effectiveness induces adaptations in motor modules during cycling. Frontiers in Computational Neuroscience 7:35
24. Schmid M, Conforto S (2007) Stability limits in the assessment of postural control through the Time-to-Boundary function. Proceedings of the 29th IEEE-EMBS Conference, Lyon, France: August 23-26
25. Gneo M, Schmid M, Conforto S, D'Alessio T (2012). A free geometry model-independent neural eye-gaze tracking system. Journal of NeuroEngineering and Rehabilitation 9:82

Author: C. D'Anna
Institute: Department of Engineering, University of Roma TRE
Street: via Vito Volterra 62
City: 00146 Rome
Country: Italy
Email: carmen.danna@uniroma3.it

A Data-Globe and Immersive Virtual Reality Environment for Upper Limb Rehabilitation after Spinal Cord Injury

Ana de los Reyes-Guzman[1], Iris Dimbwadyo-Terrer[1], Fernando Trincado-Alonso[1], Miguel A. Aznar[2], Cesar Alcubilla[2], Soraya Pérez-Nombela[1], Antonio del Ama-Espinosa[1], Begoña Polonio-López[3], and Ángel Gil-Agudo[1].

[1] Biomechanics and Technical Aids Department National Hospital for Spinal Cord Injury, Toledo, Spain
[2] AnswareTech, S.L., Madrid, Spain
[3] University of Castilla la Mancha, Talavera de la Reina, Spain

Abstract—While a number of virtual data-gloves have been used in stroke, there is little evidence about their use in spinal cord injury (SCI). A pilot clinical experience with nine SCI subjects was performed comparing two groups: one carried out a virtual rehabilitation training based on the use of a data glove, CyberTouch™, combined with traditional rehabilitation, during 30 minutes a day twice a week along two weeks; while the other made only conventional rehabilitation. Furthermore, two functional indexes were developed in order to assess the patient's performance of the sessions: normalized trajectory lengths and repeatability. While differences between groups were not statistically significant, the data-glove group seemed to obtain better results in the muscle balance and functional parameters, and in the dexterity, coordination and fine grip tests. Related to the indexes that we implemented, normalized trajectory lengths and repeatability, every patient showed an improvement in at least one of the indexes, either along Y-axis trajectory or Z-axis trajectory. This study might be a step in investigating new ways of treatments and objective measures in order to obtain more accurate data about the patient´s evolution, allowing the clinicians to develop rehabilitation treatments, adapted to the abilities and needs of the patients.

Keywords—Data-glove, rehabilitation, spinal cord injury, upper limbs, virtual reality.

I. INTRODUCTION

Recovery of function after spinal cord injury (SCI) largely depends on the preservation of some anatomic connections, and may also depends on the physiological reorganization of the brain and spinal cord [1].

In people with SCI, upper limbs are affected in more than 50% of cases [2]. In contrast with lower limbs, upper limbs have extensive functionality due to the mobility of numerous joints that can execute fine movements thanks to a complex neuromuscular control.

That's why considerable efforts have been directed towards the development of new upper limb function rehabilitation therapies based on robots, virtual reality, passive workstations, and functional electrical stimulation (FES) systems [3].

Use of virtual reality (VR) has emerged in an effort to promote task oriented and repetitive movement training of motor skills while using a variety of stimulating environments [4].

Comparing with conventional rehabilitation, VR technology increases the range of possible tasks, partly automating and quantifying therapy procedures, and improving patient motivation using real-time task evaluation and reward. Feedback can be provided either after a task, in the form of scores, or during the task using dynamic biofeedback, in the form of visual and auditory cues. There are systems that also provide physical assistance with movement and/or simulate haptic feedback [5]. Some studies affirm that afferent information may change cortical representations and/or improve motor performance in people with SCI. A possible underlying mechanism is that the somatosensory cortex has an important role in cortical reorganization after injury [6]. Therefore, afferent input may contribute to cortical reorganization and, ultimately, to functional recovery via increased communication between the cortex and the corticospinal tract in SCI subjects, which could contribute to improve the execution of functional movements.

While a number of virtual data-gloves have been used and have shown promised results in upper limb rehabilitation, in patients with stroke [7-8], as far we know, experiences with people with SCI are scarce.

Thus, the goal of this study was to test a data glove, CyberTouch™, combined with a virtual reality environment, for use in therapeutic training of upper limb movement after SCI. In addition, we wanted to provide objective data, which are obtained through the data glove, in order to adapt the rehabilitation treatments to the patient´s motor capacities; and to assess precisely the rehabilitation process.

II. MATERIALS AND METHODS

This was a pilot study of two groups: one intervention group (IG) who used the VR system as a complement to conventional therapy, and one control group (CG) who only was treated with traditional rehabilitations methods.

The motion capture system employed was a data-globe, CyberTouch™, which provides up to 22 high-accuracy joint-angle measurements and vibro-tactile actuators, one on each finger and the palm; and reproduces in real time and in the same orientation the hand movements of the patient on a 3D LCD monitor.

It has been observed that the first person view of a virtual representation of the hand induces stronger activation of primary and secondary motor areas associated with sensory motor control as opposed to only performing hand movements in the absence of such a representation [9].

The virtual scene developed for the current project consists of one room, two shelves and a trench (Figure 1).

Fig. 1 Virtual environment. Patient using CyberTouchTM glove in the virtual environment, that consists of one room, two shelves and a trench. The patient has to grasp and release or reach a set of virtual geometric elements.

There is also a reconstruction environment (Figure 2) to visualize the patients' recordings in order to analyze them.

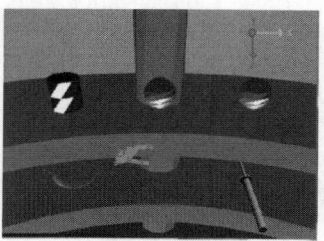

Fig. 2 Reconstruction environment. The figure shows the screen as it is seen by the user, and the direction of the 3 axes. This environment permits to visualize the patients' recordings in order to analyze them.

It was performed three virtual tasks to improve the patient's upper limb movements:

- GRASP AND RELEASE OBJECTS I: The task consisted of grip and drop three objects (ball, prism and cylinder) placed in front of the subject line. The user had to bend from waist while their trunk went forward and raise the arm to grab objects and drop them into a trench. The glove provides sensory feedback (vibration) by touching objects.

- GRASP AND RELEASE OBJECTS II: The task was similar to the previous one, but the virtual hand appeared only when the patient was very close to the object to be achieved, almost touching it, thus making the exercise less intuitive because there was less feedback.

- REACH OBJECTS: The reaching task consisted of five objects that appeared randomly and disappeared when the subject touched them. The glove provided sensory feedback when the user touched the object.

After we have tried the virtual glove with cervical SCI people and saw the great difficulty or not possibility to carry out the grasp and reach virtual reality tasks, we decided to do these virtual activities with high dorsal injuries patients, who have altered the trunk balance, necessary for body postural control and required for generating a stable base upper limb movement required to execute ADL.

Finally, six intervention subjects (5 males and 1 female; mean age 54.3±9.86 years, 5.83±2.99 months after injury) and 3 control subjects (2 males and one female, mean age 44.17±22.92 years, 5±1 months after injury) participated in the study. Subject's demographic and clinical characteristics are showed in the Table 1.

Table 1 Subject´s demographic and clinical characteristics (mean and standard deviation).

	CG	TG
Sex (female/male)	1/2	1/5
Age [years]	44.17±22.92	54.3±9.86
Dominance (right/left)	3/0	6/0
Level of injury (C5-C8)	D4 (2), D6 (1)	C4 (1), D4 (4)
ASIA (A-D)	A(3)	A(5), D(1)
Time since injury [months]	5±1	5.83±2.99
Etiology of damage (traumatic/postsurgical/vascular)	2/0/1	4/2/0

We did not include patients with: pacemaker or similar devices, joint injuries and/or upper limb muscles prior to provoke limitation of movement, cognitive condition and/or psychiatric pathologies, vision problems, technology addiction, epilepsy or pregnancy. Each subject gave informed consent voluntarily.

All subjects of the intervention group (IG) participated in the treatment two times per week (on alternate days) for two weeks, 30 minutes per session. There were two types of sessions: GRASP AND RELEASE SESSIONS, which consisted of doing three repetitions of the tasks Grasp and release objects I and II; and REACH SESSION, which involved three repetitions of the task Reach objects. At the same period, the users received conventional therapy rehabilitation. Patients assigned to the control group (CG) made only traditional rehabilitation.

Both groups were assessed pre-post treatment, with a battery of clinical scales to contrast the final results. We also evaluated the time to complete each session in the IG.

A descriptive analysis of the clinical variables was obtained by calculating the mean and standard deviation (SD) for the quantitative variables. Samples were analyzed with Kolmogorov-Smirnov test. After normal distribution was shown, data were analyzed by Mann-Whitney U test (p<0,05), to check differences among groups and between virtual reality sessions executed by IG. All statistical analyzes were performed using the program SPSS 17.0.

Additionally, two parameters were implemented in order to assess the patient's performance of the sessions:

-NORMALIZED TRAJECTORY LENGTH: The idea of this parameter is that, as long as the patient improves his performance, the trajectories followed during the reaching and grasping tasks should be shorter, thus indicating a more accurate movement.

$$L_{norm} = \frac{1}{N}\sum_{n=1}^{N}\sqrt{1+[x(n+1)-x(n)]^2}$$

x(n) is the trajectory along the axis of interest, and N is the length of x(n).

-REPEATABILITY: it computes the inverse of the area comprised between the upper and the lower envelope of the repetitions of the same movement during a session. An example can be seen in Figure 4 and in Figure 5. The idea is that, as long as the patient improves his performance, he should be able to repeat more accurately the same task, thus the area between the envelopes should decrease.

The aim of these parameters is to obtain an objective tool that could be used to evaluate the patients' functional abilities.

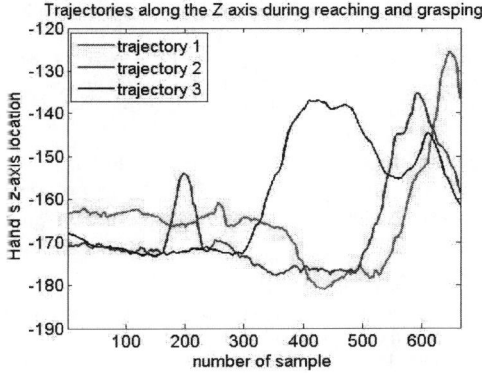

Fig. 3 Reaching and grasping trajectories. Trajectories of 3 repetitions of the same movement (reaching an object displayed on the right of the virtual shelf), all of them performed during the same session.

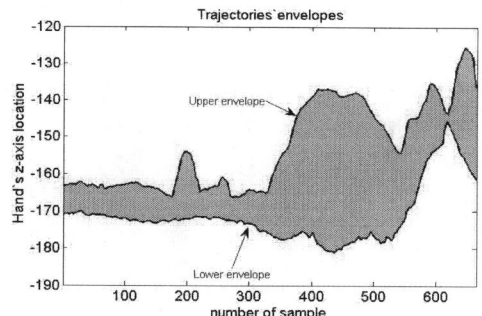

Fig. 4 Upper and lower envelopes. Upper and lower envelopes of the 3 trajectories represented in Figure 3, and the area comprised between them.

III. RESULTS

Comparing data obtained after treatment in both groups, no statistically significant differences were found in the functional independence and muscle balance, measured with the Spinal Cord Independence Measure (SCIM), Muscle Balance (MB) and Motor Index (MI). Nevertheless, there were better results in most items of the scales in the IG group with CyberTouch™.

We neither found significant differences in the coordination, dexterity nor grip function parameters, obtained in the scales Jebsen Taylor Hand Function (JTT) and Nine Hole Peg Test (NHPT). However, we observed that the time of complete 4 of the 7 test of the JTT and the test NHPT was significantly lower in the IG than in the CG after treatment.

We also observed that in the most of the patient (4 of 6), the third session was completed in less time than the first in both Grasp and release objects I-II sessions and Reach objects sessions (Figure 5).

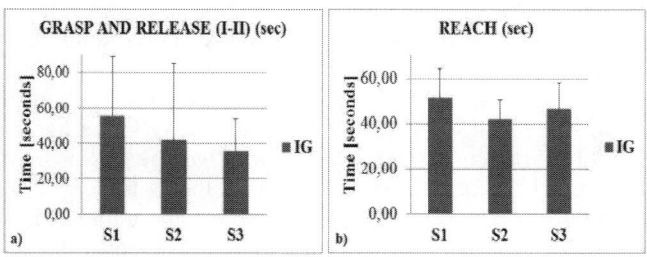

Fig. 5 Grasp and Release (I-II) and Reach sessions. Left figure shows the result of the Grasp and Release session (I-II) (a); Right figure shows the result of the Reach session (b).

We compared normalized trajectory lengths and repeatability indexes in the initial session and in the final session for every subject of the intervention group. We found that 4 out of 6 patients decreased their normalized trajectory lengths along the Z-axis (Figure 6), and 3 out of 6 patients increased their repeatability along the Y-axis (Figure 7), not showing significant differences in any of both indexes.

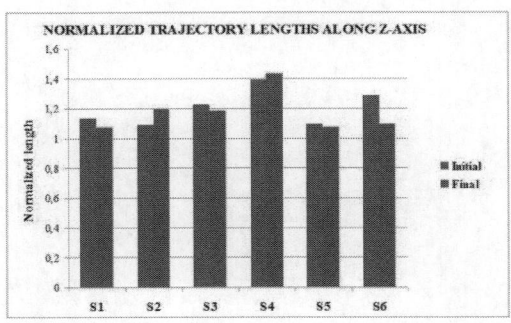

Fig. 6 Normalized trajectory lengths. Normalized trajectory lengths along the Z-axis of the 6 subjects of the intervention group, at the initial and at the final session.

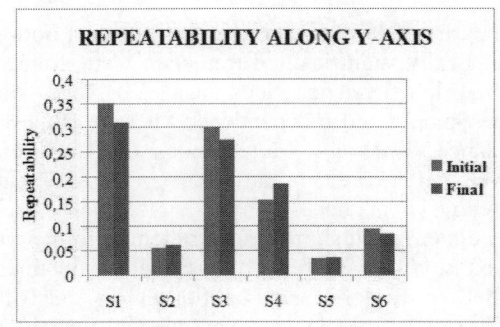

Fig. 7 Repeatability. Repeatability along the Y-axis of the 6 subjects of the intervention group, at the initial and at the final session.

IV. DISCUSSION

By comparing the data obtained after treatment in both groups, no statistically significant differences were found in any scale analyzed, but the IG improved more points than the CG when comparing between pre-post kinetic and functional evaluation. Similar studies in stroke patients showed improvements in functional activities of daily living and in the JTT and Fugl-Meyer scales after virtual upper limb exercises [7-8]. These results might be the consequence of an improvement in the trunk balance, which could cause a better stability of the upper limbs, and a more suitable hand function.

We could underline that the difference between pre and post evaluation in the IG (1.67 s) was bigger than in the CG (0.92 s) in the sub-test 7 of the JJT, which could be because of the similarity between the movements necessaries to perform this sub-test (picking up cans) and the virtual tasks.

In the normalized trajectory lengths and repeatability indexes, every patient showed an improvement in at least one of the indexes, either along Y-axis trajectory or along Z-axis trajectory, thus indicating the potential of this kind of objective measurements. They might be useful as a complement to the scales, allowing more accuracy in the assessment of rehabilitation progresses.

V. CONCLUSION

The present study is an attempt to provide knowledge about the application of virtual data-based gloves in patients with SCI. The use of this system, in addition with traditional rehabilitation treatments, suggests possible improvements in upper limb's kinetic and functional parameters. We have also developed two measure indexes: normalized trajectory lengths and repeatability, which allow knowing, objectively, about the patient's performance. Further research, with a larger sample and more number of sessions, is needed to improve the upper limb rehabilitation treatments supported by objective measures, which might allow the clinicians to assess and tailor their interventions.

REFERENCES

1. Green JB. Sora E. Bialy Y. Ricamato A. Thatcher RW. (1998) Cortical sensorimotor reorganization after spinal cord injury: an electroencephalographic study. Neurology 50:1115-21.
2. Wyndaele M. Wyndaele JJ. Incidence. (2006) Prevalence and epidemiology of spinal cord injury: what learns a worldwide literature survey? Spinal Cord 44(9):523-9.
3. Oess NP. Wanek J. Curt A. (2012) Design and evaluation of a low-cost instrumented glove for hand function assessment. J Neuroeng Rehabil 9:2.
4. Stewart JC, Yeh SC, Jung Y, et al. (2007) Intervention to enhance skilled arm and hand movements after stroke: A feasibility study using a new virtual reality system. J Neuroeng Rehabil. 23(4):21.
5. Eng K. Siekierka E. Pyk P et al. (2007) Interactive visuo-motor therapy system for stroke rehabilitation. Med Biol Eng Comput 45(9):901-7.
6. Green JB. Sora E. Bialy Y. Ricamato A. Thatcher RW. (1999) Cortical motor reorganization after paraplegia: an EEG study. Neurology 53:736-43.
7. Connelly L, Stoykov ME, Jia Y, Toro ML, Kenyon RV, Kamper DG. (2009) Use of a pneumatic glove for hand rehabilitation following stroke. Conf Proc IEEE Eng Med Biol Soc 2434-7.
8. Connelly L, Jia Y, Toro ML, Stoykov ME, Kenyon RV, Kamper DG. (2010) A pneumatic glove and immersive virtual reality environment for hand rehabilitative training after stroke. IEEE Trans Neural Syst Rehabil Eng 18(5):551-9.
9. Cameirão MS. Badia SB. Oller ED. Verschure PF. (2010) Neurorehabilitation using the virtual reality based Rehabilitation Gaming System: methodology. design. psychometrics. usability and validation. J Neuroeng Rehabil 22;7:48.

Author: Ana de los Reyes Guzmán.
Institute: Biomechanics and Technical Aids Department National Hospital for Spinal Cord Injury
Street: Finca la Peraleda s/n
City: Toledo
Country: Spain
Email: adlos@sescam.jccm.es

Wearable Navigation Aids for Visually Impaired People Based on Vibrotactile Skin Stimuli

M. Reyes Adame, K. Möller, and E. Seemann

Institute of Technical Medicine, Furtwangen University, Villingen-Schwenningen, Germany

Abstract—Blind and visually impaired people have a reduced capacity to perceive their environment which also affects their mobility and orientation. In order to improve this capacity, we introduce an approach based on vibrotactile stimulation of skin receptors using ordinary small vibration motors which are widely used in mobile phones.

Four different prototypes of vibrating belt systems, worn by blindfolded people and guided by wireless transmitted navigation instructions from a second person to the prototypes, have been constructed and tested in a course with obstacles. The aim of this study was to test and compare different vibrotactile encodings of commands to guide blind people. The comparison of the prototypes led to the assumption that simple kept systems could be a better solution for the vibrotactile navigation instructions than more complex systems.

Keywords—Visually impaired, navigation aid, vibrotactile signals, skin stimuli, collision avoidance.

I. INTRODUCTION

According to the World Health Organization (WHO), in the year 2012, 285 million people have been visually impaired worldwide. 246 million of them suffered from a vision disorder and 39 million from blindness [1].

These people have a reduced capacity to perceive their environment, thus for walking independently in an outdoor environment they mostly need some aids like the blindman's white cane or guide dog. These tools may have some disadvantages, e.g. a guide dog is very expensive and is not always available. With the white cane obstacles on the ground might be detected but obstacles at a height of the head or chest will not be detected in due time in advance in most of the cases [2].

Various approaches have been proposed to avoid these problems. Attempts which use the auditory system for information transfer could not find a wide acceptance among visually impaired people, because for spatial orientation, path finding and collision avoidance this sense is the most important for visually impaired. Additional acoustic signals can disturb this important sense and may complicate spatial orientation instead of improving the environmental perception [2].

Due to this fact, we focus in our work in another approach which utilizes vibrotactile stimulation of skin receptors to improve the spatial perception. In contrast to the auditory system, the majority of skin receptors are not active during spatial orientation. Besides the skin has a great surface area which can be used to partially compensate the missing sense. Furthermore skin receptors can represent information in two dimensions and also integrate signals over time [3].

Other research groups [3-11] also pursued the approach of transmitting environmental information via vibrotactile stimulation of skin receptors. Johnson and Higgins [6] for example used a system consisting of two belts around the waist, one containing 14 vibration motors with a spacing of 2 cm and another one with two affixed webcams for obstacles detection. Their system detects the obstacles using a computational stereo algorithm and transmits the information to the vibration motors.

Another group Zöllner et al. [8] presented a system based on a Microsoft Kinect and optical marker tracking to help blind people finding their way inside buildings. For point to point navigation, instructions were transmitted via short voice commands and for collision avoidance, commands were transmitted to six Lilypad Vibe boards on a belt located on the left, the center and right side of the waist.

In this work the main focus is set on the comparison of four vibrating systems with different encodings for navigation instructions in a trail with obstacles.

II. MARTERIALS AND METHODS

A. Basic Concept of the Prototypes

Four vibrating belt systems (s. Fig. 1) with different encodings of instructions for navigation were developed by four groups of engineering students. The students were supported during the developing phase by a professional employee at the Institute of Technical Medicine (ITeM).

The vibration motors (s. Fig. 2a, model number LA4-432A, Nidec Copal Company) utilized for the prototypes are mainly used in mobile phones. They were controlled wirelessly by an Arduino BT board (s. Fig. 2b) to stimulate the skin receptors. Due to their small size of 16 x 5 x 6 mm, these vibration motors are ideal for a wearable use without restricting the user's mobility. The rated voltage and current of these motors are 3 V and 66 A. The revolution speed is thereby

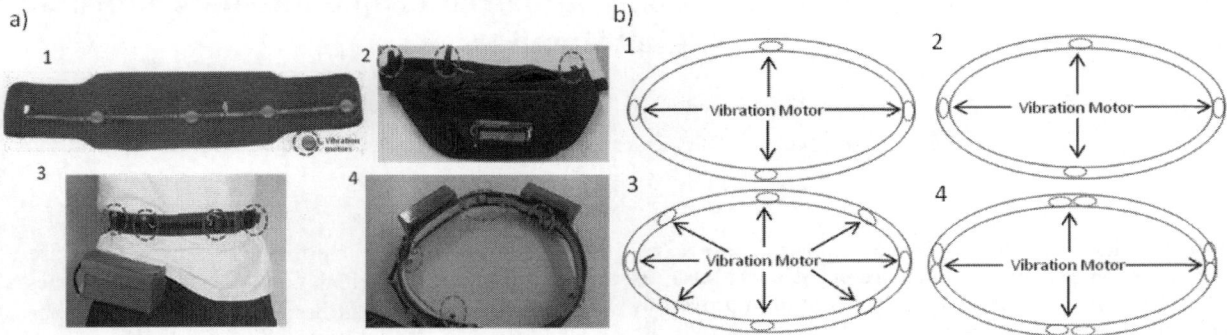

Fig. 1 Left: Vibrating belt systems for the reception of instructions for navigation via vibrotactile signals. Right: Sketch of the prototypes indicating the location of the attached vibration motors on the corresponding prototypes in the left image.

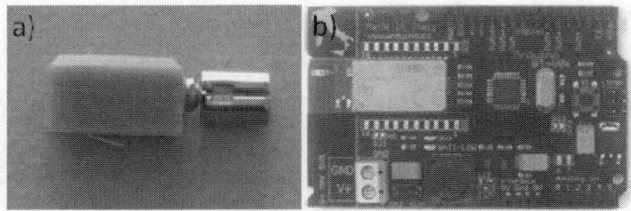

Fig. 2 Left: Vibration motor for the stimulation of skin receptors. Right: Arduino BT board to control the vibration motors wirelessly.

9500 rpm. Small plastic tubes were used to encapsulate the vibration motors to assure smooth revolutions. The Arduino BT board has an ATmega328 microcontroller and uses a BluegigaWT11 Bluetooth module to transmit the signals.

As a power source for the Arduino BT board, four 1.2 V AA batteries in a battery pack were used. The power supply for the vibration motors was realized with an additional battery pack containing three 1.2 V AA batteries. A dual full-bridge driver (L298 from the STMicroelectronics Company, Switzerland) was utilized to switch, by incoming digital signals from the Arduino BT to the required rated voltage of the vibration motors.

B. Obstacle Course

A course with obstacles was constructed to test the prototypes with one blindfolded member of each group wearing their constructed vibrating belt system. The instructions for navigation in order to pass the course with obstacles were transmitted wirelessly by commands from a Notebook, Game Controller or Tablet-PC from another person to the vibration motors on the belts. Fig. 3 shows a group completing the obstacle course and as can be seen from this picture, their vibrating belt system can be worn invisibly underneath the cloths.

Fig. 3 Navigation of a blindfolded person wearing a vibrating belt system through a course with obstacles by commands transmitted via Bluetooth to vibration motors on the belt.

Each group had to complete two runs. The course for the first run was about 30 m in length including diverse left / right curves of different curvature. At the end of the course, a small bump had to be crossed. After the first run the course was enlarged by additional 10 m for a second run and a crossing gate which had to be passed was added.

Table 1 Vibration motors activated on the vibrating belt prototypes by incoming navigational commands.

Command	Prototype 1 (4 Vibration Motors)	Prototype 2 (4 Vibration Motors)	Prototype 3 (8 Vibration Motors)	Prototype 4 (8 Vibration Motors)
Input Type	Xbox Game Controller	Touch screen of a Tablet-PC	Notebook Keyboard	Notebook Touchpad
Forward	Motor mounted at the front	Motor mounted at the front	Motor mounted at the front	Motors mounted at the front
Backward	Motor mounted at the back	Motor mounted at the back	Motor mounted at the back	Motors mounted at the back
Stop	Switching on the motor at the back for a short duration	Motor at the front, back, left and right at the same time for 0.5 s	Motors mounted at the front left/right and back left/right	Motors mounted at the back for 0.5 s
Left	*90° rotation to the left:* Motor at the left, *45° rotation to the left:* Motor at the front and left	*90° rotation to the left:* Motor at the left, *45° rotation to the left:* Motor at the front and left	*Step to the left:* motor at the left, *45° rotation to the left:* motor at the front and front left successively, *90° rotation to the left:* motor at the front, front left and left successively	*45° rotation to the left:* motors at the left for 0.5 s, *90° rotation to the left:* Twice enabling of the motors at the left for 0.5 s
Right	*90° rotation to the right:* Motor at the right, *45° rotation to the right:* Motor at the front and right	*90° rotation to the right:* Motor at the right, *45° rotation to the right:* Motor at the front and right	Same instructions as for the commands to the left, but enabling the motors mounted on the right side of the belt.	Same instructions as for the commands to the left, but enabling the motors mounted on the right side of the belt
Attention Bump	Motor at the front, back, left and right at the same time	Repetitive enabling of the front, back, left and right vibration motors at the same time	*Step up:* motors at the front left/right and motors at the back left/right successively, *Step down:* Vice versa to the step up instruction	Motors at the front, back, left and right at the same time

Differences between the Prototypes

Different approaches were pursued to give optimal navigation instructions through vibrotactile skin stimulation. In order to guide a blindfolded person safely and fast through a course with obstacles, a good coding of the stimulation signals is needed to be implemented. The encodings for the navigation commands of the presented prototypes are shown in Table 1. All vibrating systems were constructed as belts with attached vibration motors and can be worn around the waist.

For an enhanced guiding of blindfolded persons also an appropriate localization and number of vibration motors attached on the wearable systems have to be considered. A sketch of the prototypes indicating the position and number of the vibration motors on the belts is given in Fig. 1b.

Also different design ideas of the prototypes were pursued (s. Fig. 1a). The first prototype with four attached vibration motors was constructed using an elastic kidney belt. Small bags were sewed in to store the hardware parts. A waist bag was utilized to construct the second prototype, which also has four vibration motors affixed. The hardware components are kept in the bag. The third and fourth prototypes consist of ordinary belts with each eight vibration motors. The hardware for prototype 3 is mounted into a carry case attached to the belt and for prototype 4 two housings are used instead.

The communication between the PC and the prototypes was implemented in LabVIEW using the LabVIEW Toolkit for Arduino, except for prototype 2, where the communication was realized in Java.

III. RESULTS

To compare and evaluate the prototypes performance, two runs through an obstacle course with blindfolded persons wearing the vibrating belts and being navigated via these systems, were executed.

As a performance criterion, the measured time for finishing the two runs was chosen. Contacts with obstacles or the barrier tape were counted as a fault and were accompanied by a penalty of additional 5 s to the total time for each contact and were also considered as a performance criterion. In Table 2 an overview of results i.e. the required time for the runs, the number of penalties and the total time for the corresponding prototypes is depicted.

Table 2 Time needed to complete the runs in the course with obstacles and number of penalties.

Prototype	Time Run 1 [min]	Time Run 2 [min]	Penalties (+ 5)	Total Time [min]
1	1:44	2:05	5	4:14
2	1:38	2:16	4	4:14
3	1:56	2:18	2	4:24
4	1:51	3:14	4	5:25

IV. DISCUSSION

In the presented work we compared four different prototypes of vibrating belt systems, which were utilized to give blindfolded people navigation instructions by vibrotactile stimulation of the skin. Therefore a course with obstacles was constructed to evaluate the prototypes.

According to the results for the total time needed to complete the course and the number of contacts with obstacles presented in Table 2, the encodings for navigation instructions (s. Table 1) of prototype 1 and 2 seem to be the most efficient. It also demonstrates that a simple kept system which only uses four vibration motors on a belt can be very effectual for the guidance of blind people. In total the comparison of the prototypes give reason to presume that simple systems could be a better solution for the vibrotactile transmission of navigation instructions than more complex systems. A possible explanation for this finding is that the more encodings a system provide the more difficult is it for the user to distinguish between the single signals which could also lead in worst case to a misinterpretation. In order to prevent this, people using vibrating systems with more complex encodings would probably need more training before use. The test persons in our study had on average about two slight contacts with obstacles or the barrier tape per run.

One limitation of the present study is that only two runs in the course with obstacles were executed which makes it difficult to exactly determine the best encodings. Therefore more experiments are planned.

V. CONCLUSION

The stimulation of skin receptors in order to receive navigation instructions seems to be a promising method.

To improve the avoidance of collisions with obstacles and also to replace the second person which transmits the navigation commands, current efforts contain the integration of a 3D laser scanner for obstacles detection and acquisition of the environment.

ACKNOWLEDGMENT

The authors like to thank the engineering students Julian Keßler, Daniel Mantay, Janik Dorer, Andreas Rösch, Frederick Wursthorn, Claudius Ambs, Felix Jung, Jonas Scherer and Simon Krause from Furtwangen University for the construction of the prototypes.

This work was supported by the Bundesministerium für Bildung und Forschung (Grant 13EZ1129A and 01PL11008, TREFFER).

REFERENCES

[1] World Health Organization, "Visual impairment and blindness" - Fact Sheet N°282 (2012). Available online: http://www.who.int/mediacentre/factsheets/fs282/en/ Last access: January 2013.

[2] K. Möller, J. Möller, K. O. Arras, M. Bach, S. Schumann, and J. Guttmann, "Enhanced perception for visually impaired people evaluated in a real time setting", in *World Congress on Medical Physics and Biomedical Engineering*, O. Dössel and W. C. Schlegel, Eds., (Springer, Munich, Germany, 2009), vol. 25/4, pp. 283-6, 2009.

[3]]F. A. Geldard, "Some neglected possibilities of communication," *Science,* vol. 131, pp. 1583-1588, 1960.

[4] F. A. a. S. Geldard, C. E., "Multiple cutaneous stimulation: The discrimination of vibratory patterns," *The Journal of the Acoustical Society of America*, vol. 37, pp. 797-801, 1965.

[5] E. C. Lechelt, "Sensory-substitution systems for the sensorily impaired: the case for the use of tactile-vibratory stimulation," *Percept Mot Skills*, vol. 62 (2), pp. 356-8, 1986.

[6] L. A. Johnson and C. M. Higgins, "A navigation aid for the blind using tactile-visual sensory substitution," *Conf Proc IEEE Eng Med Biol Soc*, vol. 1, pp. 6289-92, 2006.

[7] R. Valazquez, Pissaloux, E. E., Guinot, J. C., and Maingreaud, F., "Walking using touch: Design and preliminary prototype of a noninvasive ETA for the visually impaired," presented at the Proceedings of the 2005 IEEE in Medicine and Biology 27th Annual Conference, 2005.

[8] M. Zöllner, S. Huber, H.-C. Jetter, and H. Reiterer, "NAVI: a proof-of-concept of a mobile navigational aid for visually impaired based on the microsoft kinect," presented at the Proceedings of the 13th IFIP TC 13 international conference on Human-computer interaction - Volume Part IV, Lisbon, Portugal, 2011.

[9] S. K. Nagel, C. Carl, T. Kringe, R. Martin, and P. Konig, "Beyond sensory substitution--learning the sixth sense," *J Neural Eng*, vol. 2 (4), p. 29, 2005.

[10] P. Bach-y-Rita, C. C. Collins, F. A. Saunders, B. White, and L. Scadden, "Vision substitution by tactile image projection," *Nature*, vol. 221 (5184), pp. 963-4, 1969.

[11] J. B. F. Van Erp, "Vibrotactile spatial acuity on the torso: effects of location and timing parameters", in Eurohaptics Conference, 2005 and Symposium on Haptic Interfaces for Virtual Environment and Teleoperator Systems, 2005. World Haptics 2005. First Joint, (2005), pp. 80-85, 2005.

Address of the corresponding author:

Author: Miguel Reyes Adame
Institute: Institute of Technical Medicine, Furtwangen University
Street: Jakob-Kienzle-Straße 17
City: Villingen-Schwenningen
Country: Germany
Email: rey@hs-furtwangen.de

Brain Injury MRI Simulator Based on Theoretical Models of Neuroanatomic Damage

L.M. González[1,2], M. Luna[1,2], A. García-Molina[3,4], C. Cáceres[1], J.M. Tormos[3,4], and E. J. Gómez[1,2]

[1] Bioengineering and Telemedicine Centre, ETSI Telecomunicación,
Universidad Politécnica de Madrid, Spain
[2] Biomedical Research Networking Center in Bioengineering,
Biomaterials and Nanomedicine (CIBER-BBN), Spain
[3] Institut Guttmann, Institut Universitari de Neurorehabilitació adscrit a la UAB, Badalona, Spain
[4] Universitat Autònoma de Barcelona, Spain

Abstract—In order to improve the body of knowledge about brain injury impairment is essential to develop image database with different types of injuries. This paper proposes a new methodology to model three types of brain injury: stroke, tumor and traumatic brain injury; and implements a system to navigate among simulated MRI studies. These studies can be used on research studies, to validate new processing methods and as an educational tool, to show different types of brain injury and how they affect to neuroanatomic structures.

Keywords—Brain Injury, Traumatic Brain Injury (TBI), Tumor, Stroke, Feature Point, Magnetic Resonance Imaging (MRI), Simulator.

I. INTRODUCTION

Acquired brain injury (ABI) refers to any brain damage occurring after birth [1]. It usually causes certain damage to portions of the brain and result in a significant impairment of an individual's physical, cognitive and/or psychosocial functioning. Traumatic brain injury (TBI), stroke and brain tumors are the main causes of an ABI among others like infections or hypoxia. The consequence of ABI is the loss of capacity to execute their daily activities that lead to a dramatic change in the individual's daily life. One of the main challenges of neuroimaging is the development of incremental databases to link brain injury features with clinical outcome.

Over the last years, different research groups have developed open databases. Some approaches are 'The Whole Brain Atlas'(WBA) [2], or 'Internet Brain Segmentation Repository'(IBSR) [3]. However, we do not detect injured neuroanatomic databases simulating different types of brain injury. In order to improve the body of knowledge about brain injury impairment is essential to develop image database with different types of injuries. This kind of database permits to simulate injuries on healthy images. These simulated imaging studies improve medical educational and research procedures.

As an educational tool permits to simulate different types of brain injuries and to study the alterations suffering from neuroanatomic structures.

As a research tool is a potential tool to obtain controlled imaging studies of diverse brain injuries. The validation of methods to label neuroanatomic structures in neuroimaging studies [4] requires a huge amount of MRI studies and clinical effort to obtain the clinical assessment.

This paper proposes a methodology for modeling three types of brain injury: stroke, tumor and traumatic brain injury; and it implements a system to simulate those different types of brain injuries.

The remainder of this paper is organized as follows: section 2 reviews the different types of brain injury, section 3 explains the injured model used on each case, section 4 explains the dataset used, section 5 describes obtained results and finally section 6 discusses the results and explains the conclusions.

II. TYPES OF BRAIN INJURY

ABI neuroimaging presents different morphologies depending on the physiological mechanisms causing the injury. This paper considers three types of brain injury: brain tumor, stroke and traumatic brain injury (TBI).These kinds of brain injury are the most common ones.

A. Brain Tumor

A brain tumor is defined as an intracranial solid neoplasm within the brain [5]. In T1-MRI, as shown in Fig. 1 a tumor can be considered as an amorphous mass of variable size and location. Mostly, brain tumors present evident intensity imaging characteristics. Depending on the acquisition sequence of the MR, the tumor will be a hiperintensity or a hipointensity area.

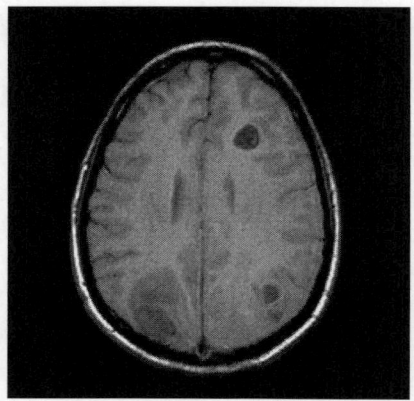

Fig. 1 Axial slice showing a sarcoma tumor T1-MRI

B. Stroke

A stroke or cerebrovascular accident is caused by an interruption of the blood flow in a region of the brain [6]. A stroke can be hemorrhagic or ischemic. A hemorrhagic stroke presents a diffuse area which is caused by bleeding or clotting (shown in Fig. 2). However, an ischemic stroke causes a necrosis on the infarction area.

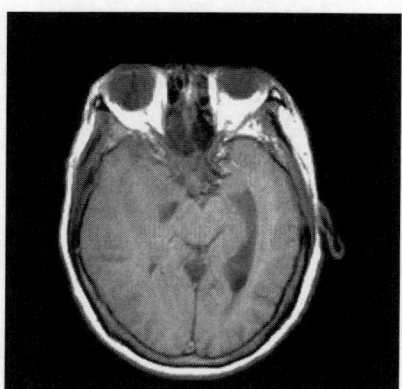

Fig. 2 Axial slice of a stroke

C. Traumatic Brain Injury

Traumatic Brain Injury (TBI) occurs when an external force traumatically injures the brain [7]. The trauma can be caused by a direct impact or by acceleration-deceleration process. MRI shows a TBI as a drift of the structures around the brain region affected by the impact (shown in Fig. 3) and on the opposite side of the brain, the brain area is smooth as a result of mass compression against the intracranial tissue.

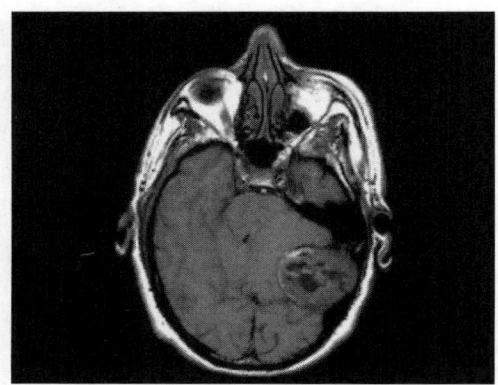

Fig. 3 Traumatic Brain Injury

III. METHODS

This section describes the methodology followed to model each type of brain injury. Firstly, the model of a brain tumor is described. Then, the model of a stroke is explained and finally, TBI model is characterized.

A. Brain Tumor

A brain tumor is considered to present a regular morphology. This paper considers a tumor as a spheroid that can appear in any location with random size. When a tumor is presented, not only an ellipsoid appears on the image but also neuroanatomic structures surrounding this area are moved.

Figure 4 shows a diagram of a simulated brain tumor imaging.

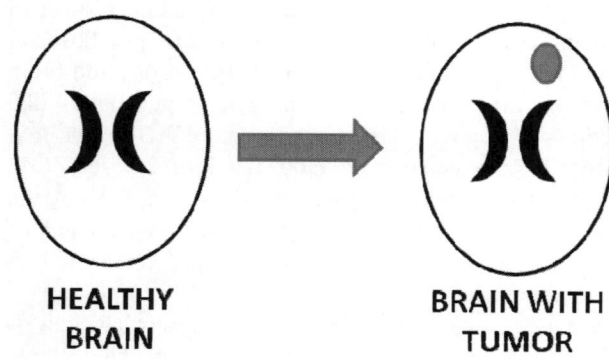

Fig. 4 Simulation of a brain tumor

As mentioned before, the proposed model generates a spheroid of variable size and random location. The volume defining the tumor is set as in Eq. 1(1).

$$S = \sqrt{\frac{(x-tumor_x)^2}{tumor_x} + \frac{(y-tumor_y)^2}{tumor_y} + \frac{(z-tumor_z)^2}{tumor_z}} \leq 1 \quad (1)$$

where $tumor_{xyz}$ is the radius of the tumor in the sagittal, coronal and axial plane. This parameter defines the location of the tumor within the brain volume. Finally, the image is smoothed by a Gaussian filter in order to achieve a non-uniform intensity area.

B. Stroke

In order to simulate a stroke brain a focus is defined. This focus represents where the blood supply is altered. As a consequence of the interruption of the blood supply a surrounding area around the focus is also affected. This area is also defined in the model, as shown in Fig. 5.

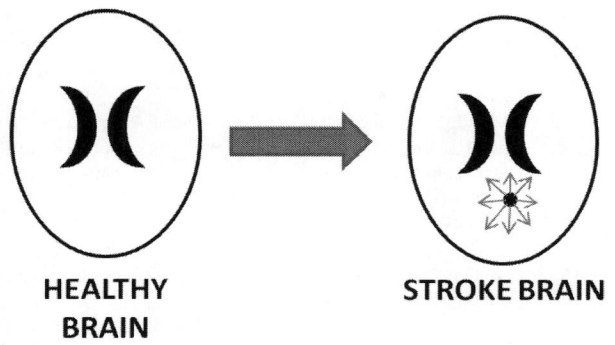

Fig. 5 Stroke

Then, this surrounding brain area is moved and intensity values are altered. The extension of the stroke and its flow direction are two parameters to be configured by the user.

C. Traumatic Brain Injury

A trauma is supposed to be originated by a hit on a side of the skull. This hit causes a neuroanatomic deformation. This deformation can be skull fracture or/and the displacement of brain tissues. The injured area is deformed due to different compression mechanism causing the disappearance of structures or/and a location change of them, as shown in Fig. 6.

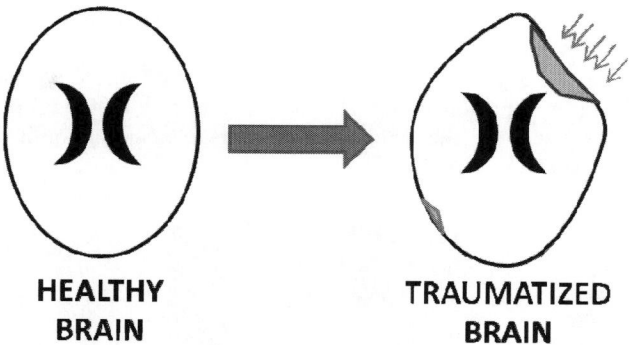

Fig. 6 Mechanism that produces a TBI

So as to simulate a TBI with a normal brain MRI. A mesh of points is defined and created over the brain region. This mesh computes the deformation on the selected brain region. The deformation is controlled with a local motion model by parameterized transformations based on B-splines [8]. The deformation is the product of 1D cubic B-splines (Eq. 2)

$$T(x,y,z) = \sum_{l=0}^{3} \sum_{m=0}^{3} \sum_{n=0}^{3} B_l(u) B_m(v) B_n(w) \phi_{i+l,j+m,k+n} \quad (2)$$

where $i = \lfloor x/n_x \rfloor - 1$, $j = \lfloor y/n_y \rfloor - 1$, $k = \lfloor z/n_z \rfloor - 1$, $u = x/n_x - \lfloor x/n_x \rfloor$, $v = y/n_y - \lfloor y/n_y \rfloor$, $w = z/n_z - \lfloor z/n_z \rfloor$ and B_l is the lth basis function of the B-spline.

A multiresolution approach is applying in order to increase the resolution of the control points of the mess. Control points are controlled by Eq. 3 (3):

$$\phi_{i,j,k}(s+1) = \phi_{i,j,k}(s) + \alpha * \sqrt{(i+j+k) - (c_i + c_j + c_k)} \quad (3)$$

The updating of the mess is a function of position of each control point in relation with the core of the injury $c(c_i, c_j, c_k)$ and a parameter α defining the intensity of the movement. The deformation stops when the difference between the original volume and the deformed volume reach a determinate threshold.

IV. MATERIALS

6 T1-MRI studies of healthy subjects have been used. These studies contain between 166 and 176 slices of 256x256 pixels.

Models and simulator system have been implemented on Matlab 2013a.

V. RESULTS

Figure 7 shows an axial slice where a dark circle within the slice, represents a brain tumor. The tumor is the section of a spheroid that moves determinate neuroanatomical structures.

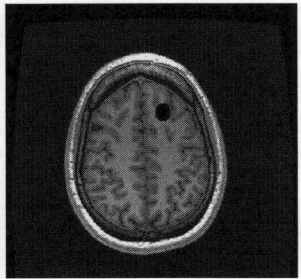

Fig. 7 Brain tumor model

Fig. 8 displays the difference between an original slice and the slice in a stroke case.

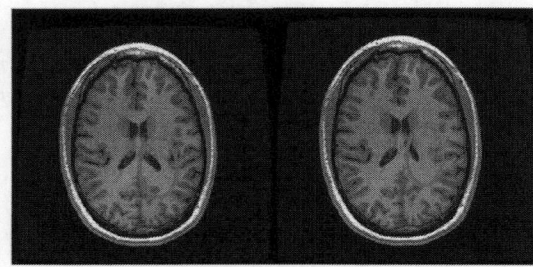

Fig. 8 Stroke simulation

The evolution of a TBI simulation is shown in Fig. 9. As observed, when increasing the force of deformation, more neuroanatomical structures are deformed.

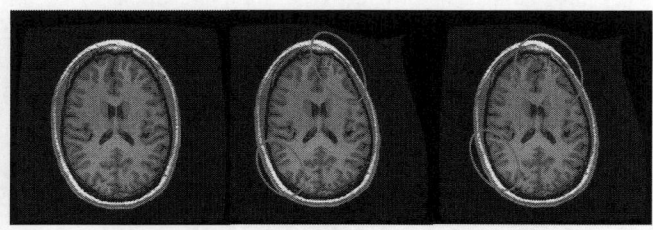

Fig. 9 TBI simulation

VI. CONCLUSIONS AND DISCUSSION

We propose a system to simulate different types of brain injury on T1-MRI studies. This paper proposes a new methodology to model alteration in the image increasing the variability and size of our database of images.

The results shown permit to modify the structural information in order to simulate tumors, strokes or traumatic brain injury by deformed or translated health MRI studies.

The three proposed models will be improved in future papers by including more geometries, mechanical properties of different brain tissues.

ACKNOWLEDGMENT

This research has been partially founded by the Spanish Ministry of Economy and Finance (project TIN2012-38450, COGNITIO).

REFERENCES

1. Laxe S, Zasler N, Tschiesner U, López-Blazquez R, Tormos JM, Bernabeu M. (2011) ICF use to identify common problems on a TBI neurorehabilitation unit in Spain. NeuroRehabilitation, vol. 29, n°1, pp. 99-110.
2. The Whole Brain Atlas(WBA) at http://www.med.harvard.edu/aanlib/
3. Internet Brain Segmentation Repository(IBSR) webpage at http://www.cma.mgh.harvard.edu/ibsr/
4. Luna M, Gayá F, Sánchez P, Cáceres C, Pascual-Leone A, Tormos JM, Gómez EJ. Brain structures identification based on feature descriptor algorithm for Traumatic Brain Injury. International Conference on recent advances in neurorehabilitation (ICRAN), vol. 1, pp.171-174, 2013
5. Herholz, Karl; Karl-Josef Langen, Christiaan Schiepers, James M. Mountz (2012). Brain Tumors. Semin Nucl Med. 42 (6): 356–70. doi: 10.1053/j.semnuclmed.2012.06.001.
6. Kidwell CS, Chalela JA, Saver JL et al. (2004) Comparison of MRI and CT for detection of acute intracerebral hemorrhage. *JAMA* 292 (15): 1823–30. doi:10.1001/jama.292.15.1823
7. Maas AI, Stocchetti N, Bullock R (2008) Moderate and severe traumatic brain injury in adults. Lancet Neurology 7 (8): 728–41. doi:10.1016/S1474-4422(08)70164-9.
8. Rueckert, D., Sonoda, L., Hayes, C. et al. (1999) Nonrigid Registration Using Free-Form Deformations: Application to Breast MR Images. IEEE Trans. Med. Imag. 18(8): 712–21

Author: Luis Miguel González Rivas
Institute: Bioengineering and Telemedicine Centre, ETSI Telecomunicación, Universidad Politécnica de Madrid
Street: Avenida Complutense n° 30, "Ciudad Universitaria"
City: Madrid
Country: Spain
Email: lmgonzalez@gbt.tfo.upm.es

Haptic Feedback Affects Movement Regularity of Upper Extremity Movements in Elderly Adults

M. Schmid[1], I. Bernabucci[1], S. Comani[2], S. Conforto[1], B. D'Elia[1], B. Fida[1], and T. D'Alessio[1]

[1] BioLab[3] - Engineering Department, Roma Tre University, Rome, Italy
[2] BIND - Behavioral Imaging and Neural Dynamics Center, University G. DAnnunzio of Chieti-Pescara, Italy

Abstract—Eight elderly adults were requested to perform circle movements with the hand through a commercial haptic platform, in two different conditions: with visual feedback, and with a facilitating force field produced by the machine. A measure of movement regularity (the mean square jerk in its normalized form) were captured to determine the effect of these feedbacks on hand kinematics. Regularity was higher when haptics feedback was given alone (MSJ_{ratio} 6.48 ± 0.15), as compared to combining it with visual feedback (MSJ_{ratio} 7.46 ± 0.18). We interpreted these differences as the ability to process visual information in trajectory tracking conditions as higher than the one to cope with external force fields, also when provided as a hypothetically facilitating one.

Keywords— Haptic feedback, smoothness measures, hand kinematics, motor skill learning.

I. INTRODUCTION

The use of haptics is widespread in a variety of application fields, from video-gaming [1], to simulation in surgery [2], to rehabilitation of the upper limb in post-stroke patients [3], and to representing the sense of touch when exploring 3D-models [4]. While there is a multiplicity of ways with which humans can interact with a computer, including the sight [5], [6], haptics can be handy in some specific fields: for instance, when combined with virtual reality systems, they play a key role in the interaction, as they provide the user with the sense of touch, which accompanies the provision of visual feedback [7]. In the case of people with visual impairments, haptics can be used as a facilitator, or even as a substitute to visual feedback. As of today, technology on haptic devices has been focussing on providing the user with vibration feedback, but if haptic systems are to be used to simulate the interaction with objects on everyday activities, vibration is uncommon, and haptics technology needs to be used to mimic force and pressure interaction [8]. Following this objective, a number of haptic devices able to provide these force fields are now at hand (see for instance the Phantom Desktop [9] and the Falcon [10].

While their use is getting popular, at present small is the number of studies aiming to determine whether the presence of haptics, when using these platforms, determines a change in kinematics associated with the finalized movements participants were requested to perform when using them. And while digital naïve people are accustomed to interaction through haptics, there may be differences when asking elderly people to cope with force fields that are generated externally. The present paper deals with this specific feature, by proposing a virtual reality system that is able to generate visual and haptic feedbacks, and by testing changes in movement kinematics when performing finalized movements with the help of assistive force fields. The focus of this paper will thus be on the presentation of a platform, based on a commercially available haptic device (Falcon© by Novint Technologies, inc.), able to provide the user with a visual and haptic feedback, and to gather trajectory data to assess movement smoothness in different conditions. The experiment that will be presented in the following has been planned to specifically investigate the influence of haptic feedback on movement smoothness.

II. SYSTEM DESIGN

The system used in this work is composed of a PC connected to the Falcon interface: the interface is characterized by 3 translational degrees of freedom allowing its handle to be moved inside a three-dimensional workspace (volume is approx. 1 dm^3). The handle can be moved describing trajectories similar to those drawn by the hand in everyday life activities to move objects or simple tools [11]. Referring to the haptic and force feedback, the interface can apply forces of about 10 N on the handle in any direction inside the workspace. As reported by different studies that focussed on the characterization of the Falcon [12], [13], forces are generated with an actuation frequency of 1000 samples/s. Figure 1 shows the Falcon interface.

The interface is connected to a PC by a USB 2.0 port, allowing the PC to control the handle and get data to record, such as the 3D position of the handle. A script file was done to log handle position data and to record them on a file.

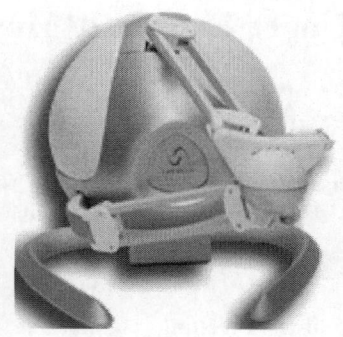

Fig. 1 Falcon interface, by Novint technologies

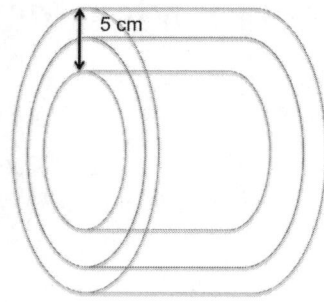

Fig. 2 Description of working volume and the corresponding force field

III. EXPERIMENTAL DATA

A. Participants and Procedure

For the aims of this study eight healthy elderly adults (age 65-77 years, all right-handed) were recruited, and informed about the procedure. Upon their assent, they were allowed to practice for about one minute to familiarize with the platform, and know the limits of the workspace: they were asked to avoid the workspace borders when doing the exercises. Then, they underwent the trials, and they were told to freely choose their own speed to draw the trajectories. In this study, they were asked to draw circular trajectories on a plane parallel to the coronal one, while sitting in front of a PC screen. These trajectories had to be executed by moving the handle of the haptic interface within the 3D-space. Each exercise was made of 5 quasi-2D circular trajectory repetitions under the following conditions: the use of right or left hand, the presence or absence of visual feedback, the presence or absence of haptic feedback, the direction of motor task (clockwise or counter-clockwise). No specific requests were made to the participants in terms of velocity with which circles had to be drawn, or of accuracy with which circles had to be tracked.

The visual feedback consisted in having a bright green circular target trajectory represented on a neutral background, with a bright point representing the current position of the projection of the handle on the screen, with respect to the target trajectory. The haptic feedback consisted of a radial force field that was null along the target trajectory and increases linearly up to 3 N at a distance of 2.5 cm, and directed towards the target circle, in such a way that the force would signal movements diverging from the target trajectory, and possibly facilitate the movement. The representation of the virtual space where the force field was present is drawn in Figure 2.

B. Data and Statistical Analysis

Handle 3-d position data were recorded into a file at a sampling frequency of 166 samples/s, and were processed by means of scripts and functions in MATLAB 8.5 (The Mathworks, inc.) to get kinematics (velocity, acceleration and then jerk as the time derivative of acceleration). Second-order Butterworth filtering, back and forward, with a cut-off frequency of 8.3 Hz, was applied to the set of extracted values of each variable. Jerk trajectories were then calculated by the processed position data as the third derivative of the position.

From the Jerk trajectory data, the mean squared jerk (MSJ) and the minimum value of mean squared jerk (MSJ_{min}) were calculated to extract the mean squared jerk ratio (MSJ_{ratio}): these parameters are considered capable of quantifying movement smoothness, and their values can be be related with the progress in rehabilitation and motor control recovery processes [14]. MSJ_{ratio} can be calculated as for rhythmic movements according to the following relation [15]:

$$MSJ_{ratio} = \frac{MSJ}{MSJ_{min}} = \frac{\frac{1}{d}\int_0^d J(t)^2 dt}{360 A^2/d^6} \quad (1)$$

where J(t) is the jerk to be integrated along the duration interval d, and A is the movement amplitude. The advantage of using MSJ_{ratio} as compared to other parameters of regularity relies in the fact that it is largely independent from the velocity with which an exercise is performed, and it is thus a robust parameter when setting up an ecological experimentation, where people are not requested to draw trajectories as fast (or as accurate) as possible, but just to draw them at their own will.

Once the kinematic variables were extracted, MSJ_{ratio} was calculated for each trial, a statistical analysis was done, since MSJ_{ratio} is considered one of the parameters more closely related with smoothness of movements [15]. More in particular, it is to be kept in mind that MSJ_{ratio} increases when movement regularity decreases, and in some way irrespective of the velocity with which the movement is performed. Descriptive statistics was thus calculated for each condition. As four where the different two-level factors that were tested (presence or absence of visual feedback, presence or absence of haptic feedback, exercises performed with the dominant or non-dominant hand, and the verse with which exercises were performed), the Analysis of Variance was performed with each of these factors as a way. First level interactions were checked for significance.

IV. RESULTS AND DISCUSSION

For the aim of this paper, 16 types of exercise, performed by each volunteer, were considered, 8 of them with haptic feedback condition. Group mean values of the MSJ_{ratio} as obtained for each different condition are reported in Fig. 2. The number of each type of exercise is reported on the x axis, while the y axis shows the common logarithm of MSJ_{ratio} (data averaged over 5 repetitions for each participant).

Table 1 Descriptive statistics for MSJ_{ratio}

Condition	Haptics ⊛⊛ ($\mu \pm \sigma$)	Haptics ⊙⊙ ($\mu \pm \sigma$)
Non dominant hand - ↺	7.19 ± 0.70	6.53 ± 0.85
Non dominant hand - ↻	7.47 ± 0.83	6.26 ± 0.79
Dominant hand - ↺	7.59 ± 1.07	6.50 ± 1.03
Dominant hand - ↻	7.56 ± 1.15	6.62 ± 0.93
Condition	No haptics ⊛⊛ ($\mu \pm \sigma$)	No haptics ⊙⊙ ($\mu \pm \sigma$)
Non dominant hand - ↺	6.75 ± 0.76	6.67 ± 0.74
Non dominant hand - ↻	6.77 ± 0.85	6.62 ± 0.75
Dominant hand - ↺	6.72 ± 0.83	6.65 ± 0.77
Dominant hand - ↻	6.73 ± 1.10	6.73 ± 0.74

All values are in common log.
⊛⊛ denotes eyes closed, ⊙⊙ denotes eyes open.

Table 2 ANOVA results for MSJ_{ratio}

Factor	Significance
Visual feedback	**
Haptic feedback	-
Hand dominance	-
Verse	-
Interaction	Significance
Visual x Haptic feedback	**

** $p < 0.005$, * $p < 0.05$, - $p > 0.05$
Only significant first-level interactions are reported.

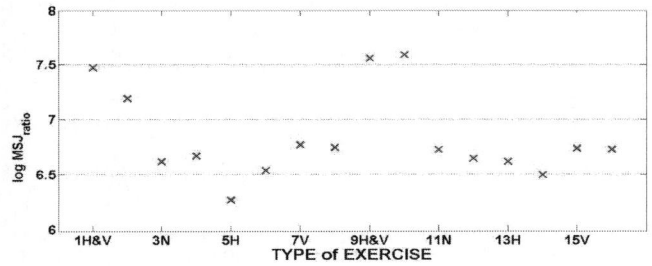

Fig. 3 MSJ_{ratio} values: H haptic feedback, N no feedback, V vision feedback. Non dominant (columns 1-8) vs. dominant hand (9-16); inward first (odd columns), outward first (even columns)

MSJ_{ratio} showed higher values for the exercises performed with the combination of haptic and visual feedback. Obtained values are in line with values obtained with different movements [16], and they resulted somewhat different from the ones that were obtained by asking young adults to perform the same exercises [17].

As it can be seen from table 1, the results of descriptive statistics for MSJ_{ratio} on 16 exercises show that hand dominance does not significantly affect movement regularity, and the same applies to the verse with which movements are performed. Inferential statistics confirms these results (see table 2), and it also highlights that, as an individual factor, vision clearly causes a change in movement smoothness, and this effect may be largely dependent on the combined presence of both haptic and visual feedback, as the significance of the first-level interaction (Visual Feedback x Haptic Feedback) seems to suggest.

There is an interesting debate as to which indicators calculated from kinematics associated with end-effector movements need to be used in order to quantify regularity, taking into account different durations, different amplitudes, and in general, different trajectories [18]. We chose the normalization of the mean-squared jerk ratio, as this parameter is considered able to evaluate smoothness or regularity of movements [15], where the increase in movement smoothness is agreed as a parameter associated with motor recovery [14]. This choice was also driven by the ecological setting of the protocol, as no specific instructions, in terms of velocity and accuracy of the movement were given to the participants.

Referring to the statistical analysis shown in the previous section, data collected on healthy elderly people can be used to draw some speculations about smoothness of basic hand movements. Regularity seems to diminish when both feedbacks are present, while this does not happen necessarily when only haptic feedback is present. This result may be interpreted by hypothesizing that healthy elderly people heavily rely on vision, and the presence of haptics, which, per se, may be considered as a facilitating factor, is in fact a disturbing one, as it brings to the necessity to merge both these inputs, a task that may be considered challenging. The reason of this reduced spread effect with respect to the age could be related with a reduced capacity to perceive the haptic stimulation and/or a reduced ability to control movements.

V. CONCLUSION

Its possible to conclude that haptic feedback, in the studied case a force field with direction transverse to motor task target, is a cause of disturbance (more than a facilitating factor) in the conditions that have been presented in this research. It can be speculated that people, trying to draw circular shape trajectories with increased accuracy in presence of the target, are more exposed to the disturbance of haptic force field so that smoothness is reduced and jerk is increased.

Haptic machines may be useful tools for different contexts, including robot-mediated therapy aimed to neuro-rehabilitation, and they could complement and integrate the activities of physical therapists. With the aim of targeting fine movements executed with the hand, in this research a haptic system was tested, by making healthy elderly perform exercises with the presence of different forms of feedback, including the presence of assistive force fields (i.e. haptics). The obtained results in the present research work may be also helpful in determining the best way to provide visually impaired users with haptic feedback when using such kind of interfaces when interacting with a computer: if haptics, when combined with vision, decreases regularity of movements, it may mean that different force field profiles need to be considered. In order to deepen this hypothesis, different force fields may need to be designed, and additional parameters associated with velocity and shape accuracy may need to be considered.

ACKNOWLEDGEMENTS

We thank all the volunteers for participating into the study. The present research has been partially funded by the Italian Ministry of Education and Research, under research grant no. 2009X3L8SW_004 (Interactive technologies and techniques for an ecology of movement).

REFERENCES

1. Jin SAA. Effects of 3D virtual haptics force feedback on brand personality perception: The mediating role of physical presence in advergames *CyberPsychology, Behavior, and Social Networking.* 2010;13:307–311.
2. Salkini MW, CR , Kiehl N, Broderick T, Donovan JF, Gaitonde K. The role of haptic feedback in laparoscopic training using the LapMentor II *Journal of Endourology.* 2010;24:99–102.
3. Broeren J, Rydmark M, Björkdahl A, Sunnerhagen KS. Assessment and training in a 3-dimensional virtual environment with haptics: a report on 5 cases of motor rehabilitation in the chronic stage after stroke *Neurorehabilitation and neural repair.* 2007;21:180–189.
4. Liu X, Dodds G, McCartney J, Hinds BK. Virtual DesignWorksdesigning 3D CAD models via haptic interaction *Computer-Aided Design.* 2004;36:1129–1140.
5. Torricelli D, Goffredo M, Conforto S, Schmid M. An adaptive blink detector to initialize and update a view-basedremote eye gaze tracking system in a natural scenario *Pattern Recognition Letters.* 2009;30:1144–1150.
6. Gneo M, Schmid M, Conforto S, D'Alessio T. A free geometry model-independent neural eye-gaze tracking system *Journal of neuroengineering and rehabilitation.* 2012;9:1–15.
7. Bibbo D, Conforto S, Bernabucci I, Carli M, Schmid M, D'Alessio T. Analysis of different image-based biofeedback models for improving cycling performances in *Proceedings of SPIE*;8295:829503 2012.
8. Yu W, Soma H, Gonzalez J. Analyzing upper limb reflexive responses for prosthetic applications *Journal of Mechanics in Medicine and Biology.* 2012;12.
9. Massie TH, Salisbury JK. The phantom haptic interface: A device for probing virtual objects in *Proceedings of the ASME winter annual meeting, symposium on haptic interfaces for virtual environment and teleoperator systems*;55:295–300Kluwer 1994.
10. Lehman AC, Wood NA, Farritor S, Goede MR, Oleynikov D. Dexterous miniature robot for advanced minimally invasive surgery *Surgical endoscopy.* 2011;25:119–123.
11. Muscillo R, Schmid M, Conforto S, D'Alessio T. Early recognition of upper limb motor tasks through accelerometers: real-time implementation of a DTW-based algorithm *Computers in biology and medicine.* 2011;41:164–172.
12. Martin S, Hillier N. Characterisation of the Novint Falcon haptic device for application as a robot manipulator in *Proc. Australasian Conf. Robotics and Automation* 2009.
13. Panarese A, Edin BB. A modified low-cost haptic interface as a tool for complex tactile stimulation *Medical engineering & physics.* 2011;33:386–390.
14. Rohrer B, Fasoli S, Krebs HI, et al. Movement smoothness changes during stroke recovery *The Journal of Neuroscience.* 2002;22:8297–8304.
15. Hogan N, Sternad D. On rhythmic and discrete movements: reflections, definitions and implications for motor control *Experimental Brain Research.* 2007;181:13–30.
16. Osu R, Ota K, Fujiwara T, Otaka Y, Kawato M, Liu M. Quantifying the quality of hand movement in stroke patients through three-dimensional curvature *Journal of NeuroEngineering and Rehabilitation.* 2011;8:1–14.
17. D'Elia B, Schmid M, Bernabucci I, Goffredo M, D'Alessio T. A Comparison between Influence of Visual and Haptic Feedback on Jerk Indicators in Hand Exercises in *Converging Clinical and Engineering Research on Neurorehabilitation*:317–320Springer 2013.
18. Hogan N, Sternad D. Sensitivity of smoothness measures to movement duration, amplitude, and arrests *Journal of motor behavior.* 2009;41:529–534.

Author: Maurizio Schmid
Institute: BioLab[3] - Engineering Department, Roma Tre University
Street: Via Vito Volterra, 62
City: Roma
Country: Italy
Email: maurizio.schmid@uniroma3.it

Dysfunctional Profile for Patients in Physical Neurorehabilitation of Upper Limb

M.A. Villán-Villán[1,2], R. Pérez-Rodríguez[1,2], C. Gómez[3], E. Opisso[3], J.M. Tormos[3],
J. Medina[3], and E.J. Gómez Aguilera[1,2]

[1] Bioengineering and Telemedicine Centre, ETSI Telecomunicación,
Universidad Politécnica de Madrid, Madrid, Spain
[2] Biomedical Research Networking Center in Bioengineering, Biomaterials and Nanomedicine (CIBER-BBN),
Zaragoza, Spain
[3] Guttmann University Institute for Neurorehabilitation affiliated to UAB, Barcelona, Spain

Abstract—This paper proposes a first approach to *Objective Motor Assessment* (OMA) methodology. Also, it introduces the Dysfunctional profile (DP) concept. DP consists of a data matrix characterizing the Upper Limb (UL) physical alterations of a patient with Acquired Brain Injury (ABI) during the rehabilitation process. This research is based on the comparison methology of UL movement between subjects with ABI and healthy subjects as part of OMA. The purpose of this comparison is to classify subjects according to their motor control and subsequently issue a functional assessment of the movement. For this purpose Artificial Neural Networks (ANN) have been used to classify patients. Different network structures are tested. The obtained classification accuracy was 95.65%. This result allows the use of ANNs as a viable option for dysfunctional assessment. This work can be considered a pilot study for further research to corroborate these results.

Keywords—Dysfunctional Profile, Classification, Physical Neurorehabilitation, Objective Assessment, Upper Limb.

I. INTRODUCTION

Acquired brain injury (ABI) is a brain structure lesion produced suddenly after birth. The cause of ABI can be either traumatic (road-traffic accidents, falls, etc.) or non-traumatic (strokes, brain tumors, infections, etc.). The most common ABIs are cerebrovascular disease (stroke) and traumatic brain injury (TBI) [1], [2]. These injuries, due to their physical, sensory, cognitive, emotional and socio-economic consequences, considerably change the life of both patients and families [3].

According to predictions by the World Health Organization, ABI will be among the ten most common causes of disability by the year 2020. This injury is the second leading cause of death and the eighth cause of severe disability in the elderly. Every year, nine million people suffer from stroke in the world [4] while 100.000 new cases in Spain, mostly people over 65 years old [5]. In USA, it is estimated that 5.3 million Americans, more than 2% of the country, are currently disabled due to TBI [6]. The annual incidence to TBI in Europe is estimated to be 235 per 100 000 population [7].

The ABI patients with functional alterations need treatments to reduce disability and improve their quality of life. Neurorehabilitation is a clinical process whose main objectives are restoring, minimizing or compensating the impairments in people with disabilities of neurological origin. The physical neurorehabilitation is intended to provide the patients with the capacity to perform specific activities of daily life (ADL) required for an independent life in terms of the handling of objects in the environments in which the upper limb (UL) is directly involved [8].

Assessment methods of UL motion depend on clinician experience and subjectivity. The UL functional evaluations are focused on clinical tests highly dependent on the examiner's criteria and require a clinician to score the performance of patients in some tasks on specific scales. Some of the most used clinical tests are: *Fugl Meyer* (FM) [9], *Action Research Arm Test* (ARAT) [10] and *Chedoke-Mc Master Stroke Assessment* [11].

The dysfunctional motion modeling of patients undergoing a rehabilitation treatment has been addressed by different methods based on kinematic analysis of UL movement [12-14]. Kinematic analysis provides objective data that allows making accurate descriptions of UL movements and activities executed by healthy subjects for establishing normal pattern which would be used in the comparison with pathological subjects [8]. Some methods focused on the comparison of movement between patients and healthy subjects [12]. Other methods developed and validated a 3D upper limb model on a motion assessment system that compares kinematics of the joint between the affected and unaffected arms [13,14].

Kinematic models of healthy subjects UL motion in the execution of ADL have been developed in scientific literature [15]. Due to lack dysfunctional Kinematic models, it is required to create new methods that allow making an accurate, precise and reliable assessment of UL motion, in order to provide objective data for the interpretation of the motor function of the patients. To solve the lack of dysfunctional kinematic models, the Dysfunctional Profile (DP) concept is proposed. DP consists of a data matrix characterizing the physical alterations of each patient. This profile is dynamic and can be updated during the rehabilitation process, allowing the tracking of the patients' evolution in a controlled fashion.

Previous authors' research is focused on the design of new methodologies to generate a change in the current model of UL physical neurorehabilitation, to improve the efficacy of treatment, personalize it and base it on evidence [3,8,16]. DP can be used in applications such as: 1. to allow robotic anticipatory assistance [16], and 2. to design an *Objective Motor Assessment* (OMA) methodology. The OMA methodology is conceived to provide an objective measurement of patient's DP. This assessment would assign an individual score for each patient that indicates the level of motor function the degree of similarity between the UL motion of the patient and the healthy subjects. OMA methodology consists of three consequential stages: 1. Calculation, 2.Classification and 3.Evaluation.

This paper is focuses on the stages 1 and 2 of the OMA methodology. The main goal of this research is to make a first approach to OMA based on classification of subjects according to their motor control in the execution of ADLs using the Artificial Neural Network (ANN).

II. MATERIALS AND METHODS

Fig.1. shows the proposed methodology diagram. First, kinematic analysis of UL movement is performed. Second, Principal Component Analysis (PCA) is applied in order to reduce the data dimension. Then, a Multilayer Perceptron (MLP) has been used for classifying subjects according to their dysfunctional deficits. The architecture of the network is established empirically.

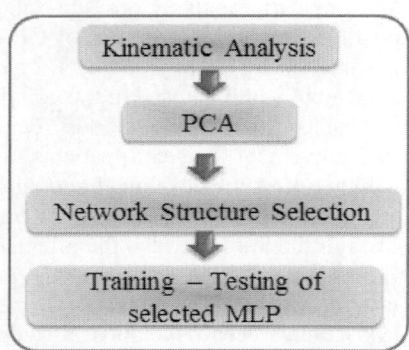

Fig.1 Proposed Methodology

A. Material

The BTS-SMART-D system [17] has been used to obtain all motion data of healthy subjects and ABI subjects. This device is a digital optoelectronic system with 6 infrared cameras with frequency of 140Hz and resolution of 1.4 Mp. Motion capture was performed with a model bimanual sixteen points [18]. The mathematical calculations of metrics and network training have been performed with the tool MATLAB® r2009b running on a computer with a 2.4 GHz Intel® Core™ Duo processor with a 4 GB RAM has been used. The PCA is performed with Statistic PASW v 18 (SPSS).

B. Biomechanical Model

A kinematic chain with three segments is used in this proposal. This biomechanical model consist of six degrees of freedom (DoF): three in the shoulder joint (flexion/extension, abduction/adduction and rotation), two in the elbow joint (flexion/extension and pronation/supination) and one in the wrist joint (flexion/extension) [8].

C. Calculation of Kinematic Variables

To make a more accurate kinematic analysis of the patients UL motion, an ADL modeling methodology [19] has been applied. Therefore, for each ADL that the patient executes, a state-chart diagram is obtained. Following this methodology, the activity gets partitioned into several transitions that are studied independently.

Fig. 2 shows the state diagram of the "picking up a bottle" ADL with three transitions. T1: Rest - bottle pick up, T2: bottle pick up - bottle on the table and T3: bottle on the table - rest.

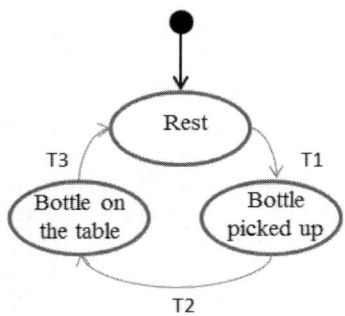

Fig. 2 State diagram of the "picking up a bottle" ADL

The temporal and spatial parameters, distributed in four metrics, were calculated in each transition of the ADL so that in each metric at least 3 Motion Parameters (MP), can be obtained. Below are the metrics and the MPs obtained in each transition:

- *Time* (s): time used to execute the ADL. 3 MPs.
- *Tangential velocity peak* (mm/s): maximum value of the rate of change of the position of the End Effector (EE) of human UL per second. 3 MPs.
- *Angular velocity peak* (deg/s): maximum value of the rate of change of angle in each joint DoF per second. 18 MPs.
- *Trajectory length index* (TLI): the ratio of the length of the actual path travelled by the EE in space (T) to the length of the straight line joining the initial and final end-point positions (D) [12]. 3 MPs.

$$TLI = \frac{T}{D} \quad (1)$$

D. Classification of Subjects

MLP is the most commonly used feedforward ANN [20]. In this study, MLPs used consisted of: *n* neurons in the Input layer (I) (one for each MP), a single neuron in Hidden layer (H), a neuron in Output layer (O) (category to which an individual belongs). Backpropagation learning [1], was used with a hyperbolic tangent as an activation function for neurons H and a lineal function for neurons in O. Therefore, when motor function of a subject is assessed, the algorithm can distinguish the category the subject belongs to. The categories according to the control of motor function are: Healthy subject's Category (HC) and Patients with ABI Category (PC).

III. EXPERIMENTAL WORK

For training and testing of the different MLPs, data from the "picking up a bottle" ADL have been used. Fig. 3 shows frontal view of this ADL: an empty plastic bottle with a capacity of 330ml is located on a shelf that is placed on a table. The subject is asked to put the bottle in the closest right corner of the table. The exact place is indicated by a solid dot. This ADL has been designed by therapists of the Guttmann Institute.

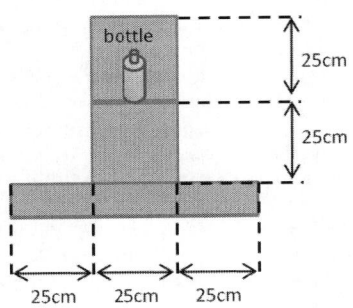

Fig. 3 Frontal view of the "picking up a bottle" ADL setup

The dataset used in this research consisted of 45 subjects divided into: 40 HCs and 5 PCs. To assess the proposed method, the network was trained with 48% of the data, of which 20 belonged to HCs and 2 belong to PC. The network was tested with 20 HC and 3 PC representing the 52% of data. Different structures and parameters were tested to determine the best network structure.

IV. RESULTS AND DISCUSSION

In the proposed method PCA of the input data was carried out (27 MP previously obtained) minimizing the loss of information. The number of initial MP was reduced to 10. These MP contain 90% of the original data information.

Table 1 shows the three best results with different network structures. Mean Squared Error (MSE) and Learning Rate (LR) are the parameters of network training, and accuracy in Classification (%C) is the result. The best result in classification accuracy was 95.65%. Results 2 and 3 obtained the same classification accuracy with similar architectures but different network parameters (MSE and LR). The difference between them and the reason for selecting the network structure 3 as the best result is because with this structure 10-12-1 (I-H-O) and network parameters 0.001 (MSE) and 0.7 (LR), the MLP classified 100% of HC. Additionally, with a small MSE the possibility that the network makes a mistake in classification is minimized because it approaches zero.

Table 1 The three best obtained results with different network structures and parameters.

Results	I	H	O	MSE	LR	% C
1	10	6	1	0.1	0.01	86.95
2	10	12	1	0.1	0.05	95.65
3	**10**	**12**	**1**	**0.001**	**0.7**	**95.65**

The results indicate that the use of a MLP can be useful tool for subjects classifying as part of OMA in physical neurorehabilitation of UL. Also, obtained results are in line with research claiming that the classification accuracy improves with ANNs when the PCA is used as method of feature extraction and selection [21].

Table 2 shows the confusion matrix of the best classification achieved by the selected MLP (result 3). The network classified 100% HC and 66.67% PC.

Table 2 Confusion matrix of the best classification achieved by the proposed method.

	HC	PC	Samples	Error
HC	20	0	20	0
PC	1	2	3	1

Although the results are promising, they are inconclusive, because one of the limitations of this research is the amount of movement data available for subjects with ABI, due to technical and clinical limitations necessarily present in this type of experimental studies. It is noted that with a greater amount of patient data, these results may vary, possibly increasing the classification accuracy. Therefore, this research can be considered as a first step in the development of the OMA methodology.

V. CONCLUSIONS

This research proposed a model based on a MLP for the classification of the level of dysfunction of ABI subjects and healthy subjects. The PCA was applied for the data dimensionality reduction (allowing for a choice of an appropriate data set for network training), to reduce the number of MP, increase classification accuracy, the lowest computational cost and the network training time reduction. The MLP enables assessment of motor function of subjects and the sorting of the category to which they belong (HC and PC). The results obtained by the MLP with the proposed method are promising. All healthy subjects are well classified and subjects with ABI are well classified in 2 of the 3 patients. This research can be considered as a first step in the development of a method that allows the objective assessment of UL motor control of patients undergoing a physical rehabilitation treatment.

Due to the fact that patients' motor function evolve over time, allowing the tracking of objective data during the rehabilitation treatment, DP of each patient could be continuously updated. Thereby, a useful database for medical research would be generated. Therefore, the implementation of OMA could optimize physical rehabilitation process. On the other hand, adding information to the database, such as patients' demographic data and tests results, it could be possible to create data exploitation and analysis system that focuses on extracting knowledge to improve the effectiveness of rehabilitation treatments, in a personalized way and based on evidence. In addition, new tools and applications for medical help and support platforms such as a management platform and therapeutic planning could be designed in the future.

REFERENCES

1. Güler I, Gökçil Z, & Gülbandilar E (2009) Evaluating of traumatic brain injuries using artificial neural network. Expert Syst. Appl. 36(7):10424–10427 DOI 10.1016/j.eswa.2009.01.036
2. Murray CJL, Lopez AD (1997) Alternative projections of mortality and disability by cause 1990-2020: Global Burden of Disease Study. Lancet, 349(9064):1498-1504 DOI 10.1016/S0140-6736(96)07492-2
3. Pérez R, Marcano A, Costa U, et al. (2012) Inverse kinematics of a 6 DoF human upper limb using ANFIS and ANN for anticipatory actuation in ADL-based physical Neurorehabilitation. Expert Syst. Appl. 39(10):9612–9622 DOI 10.1016/j.eswa.2012.02.143
4. World Health Organization, Burden of Disease Statistics, at http://www.who.org
5. Alvaro LC, Lopez-Arbolea P, Cozar R (2009) Hospitalizations for acute cerebrovascular accidents and transient ischemic attacks in Spain: temporal stability and spatial heterogeneity, 1998-2003. Revi. Calid. Asist. 24(1):16-23 DOI 10.1016/S1134-282X(09)70071-5
6. Brain Injury Association of America, at http://www.biausa.org
7. The Lancet Neurology (2010) Traumatic brain injury: time to end the silence. The Lancet Neurology 9(4):331 DOI 10.1016/S1474-4422(10)70069-7W
8. Pérez R, Costa Ú, Torrent M, et al. (2010) Upper Limb Portable Motion Analysis System Based on Inertial Technology for Neurorehabilitation Purposes. Sensors 10:10733–10751 DOI 10.3390/s101210733
9. FuglMeyer A, Jaasko L, Leyman I, et al. (1975). The post stroke hemiplegic patient. A method for evaluation of physical performance. Scand J Rehabil Med 7(1):13-31
10. McDonnell M (2008) Action Research Arm Test. Aust. J Physiother. 54(3):220-221.
11. Gowland C, Stratford P, Ward M, et al. (1993) Measuring physical impairment and disability with the Chedoke-McMaster Stroke Assessment. Stroke 24(1):58-63 DOI 10.1161/01.STR.24.1.58
12. Cirstea MC, Levin MF (2000) Compensatory strategies for reaching in stroke. Brain 123(5):940-953 DOI 10.1093/brain/123.5.940
13. Hingtgen BA, McGuire JR, Wang M, Harris GF (2006) An upper extremity kinematic model for evaluation of hemiparetic stroke. J Biomech 39(4):681-688 DOI 10.1016/j.jbiomech.2005.01.008
14. Hingtgen BA, McGuire, JR, Wang M, et al. (2004) Quantification of reaching during stroke rehabilitation using unique upper extremity kinematic model. In Proceedings of the 26th Annual International Conference of the IEEE EMBC, San Francisco, USA, 2004, vol. 7, pp 4916–4919
15. Costa U, Opisso E, Perez R, et al. (2010) 3D motion analisys of activities of daily living: implication in neurorehabilitation. In Proceedings of the International Gait and Clinical Movement Analysis Conference, Miami, USA, 2010
16. Peréz R, Costa Ú, Rodríguez C, et al. (2013) Assistance-as-needed robotic control algorithm for physical neurorehabilitation. In Proceedings of the International Conference on Recent Advances in Neurorrhabilitation, Valencia, Spain, 2013, pp 138-142
17. BTS Bioengineering, at http://www.btsbioengineering.com
18. Rab G, Petuskey K, Bagley A. (2002) A method for determination of upper extremity kinematics. Gait&Posture 15(2):113–119 DOI 10.1016/S0966-6362(01)00155-2
19. Pérez R, Costa Ú, Solana J, et al. (2009) Modelado de Actividades de la Vida Diaria para Neurorrehabilitación Funcional de miembro superior. In CASEIB 2009, In Proceedings of the XXVII Annual Congress of the Spanish Biomedical Engineering Society, Cádiz, Spain, 2009.
20. Haykin S. (1994) Neural networks, a comprehensive foundation book. Macmillan College Publishing Company, New York.
21. Burcu E, Tulay Y (2008) Improving classification performance of sonar targets by applying general regression neural network with PCA. Expert Syst. Appl. 5:472-475 DOI 10.1016/j.eswa.2007.07.021

Author: Mailin Adriana Villán Villán (M.A. Villán-Villán)
Institute: ETSI Telecomunicación, Universidad Politécnica de Madrid
Street: Avenida Complutense 30
City: Madrid
Country: Spain
Email: mvillan@gbt.tfo.upm.es

Video-Based Tasks for Emotional Processing Rehabilitation in Schizophrenia

R. Caballero-Hernández[1,2], A. Vila-Forcén[1,2], S. Fernandez-Gonzalo[3],
J.M. Martínez-Moreno[1,2], M. Turon[3], R. Sánchez-Carrión[4], and E.J. Gómez[1,2]

[1] Biomedical Engineering and Telemedicine Centre, ETSI Telecomunicación,
Universidad Politécnica de Madrid, Madrid, Spain
[2] Centro de Investigación Biomédica en Red en Bioingeniería, Biomateriales y Nanomedicina, Madrid, España
[3] Research Department, Fundació Parc Taulí. Universitary Institute, Universitat Autònoma de Barcelona,
Parc Taulí Sabadell, Universitary Hospital, Barcelona, España
[4] Guttmann University Institute for Neurorehabilitation affiliated to UAB, Barcelona, Spain

Abstract—Schizophrenia is a mental disorder characterized by a breakdown of cognitive processes and by a deficit of typical emotional responses. Effectiveness of computerized task has been demonstrated in the field of cognitive rehabilitation. However, current rehabilitation programs based on virtual environments normally focus on higher cognitive functions, not covering social cognition training. This paper presents a set of video-based tasks specifically designed for the rehabilitation of emotional processing deficits in patients in early stages of schizophrenia or schizoaffective disorders. These tasks are part of the Mental Health program of Guttmann NeuroPersonalTrainer® cognitive tele-rehabilitation platform, and entail innovation both from a clinical and technological perspective in relation with former traditional therapeutic contents.

Keywords—NeuroPersonalTrainer®, Schizophrenia, Emotional processing, Video, Tasks.

I. INTRODUCTION

Schizophrenia is a severe mental disorder, characterized by profound disruptions in thinking, affecting language, perception, and the sense of self. It often includes psychotic experiences, such as hearing voices or delusions. This mental disorder affects about 7 per thousand of the adult population, mostly in the age group 15-35 years [1].

Deficits in social functioning, including communicating with others, maintaining employment, and functioning in the community, are observed in many disorders but are a defining feature of schizophrenia [2]. Social cognition refers to the mental operations underlying social interactions, which include processes involved in perceiving, interpreting, and generating responses to the intentions, dispositions, and behaviors of others [3].

Most of the cognitive research in schizophrenia has focused on the following areas: emotional processing, theory of mind, social perception, social knowledge, and attributional bias. Emotional processing [3] refers broadly to aspects of perceiving and using emotion, and it relies on separate prerequisite abilities that are in the process of being identified.

Cognitive rehabilitation programs aim to restore normal functioning or to compensate cognitive deficits that involve higher cognitive functions (attention, memory, executive functioning and language). Besides pharmacological treatment and behavioral therapy, cognitive rehabilitation has been recognized as an important tool in the treatment of schizophrenia [4].

The combination of cognitive rehabilitation, virtual environments and collaborative tele-rehabilitation platforms is on the side of the efficacy of the treatment and maximizes the opportunity of the nervous system to reorganize those functions that will restore higher functionality in relation to Activities of Daily Living (ADL), improving patients' quality of life [5]. In addition, based on the principle that action observation activated the same neural structures responsible for the actual execution of those actions, several studies [6] have proved that ADL observation treatment using video is a good rehabilitative approach in patients with cognitive disorders.

Nowadays, several tele-rehabilitation platforms offer cognitive rehabilitation programs for schizophrenia patients: RehaCom [7], Gradior [8], PssCogReHab [9], BrainTrain [10], etc. However, available programs focus on main cognitive functions, not covering specific needs on social cognition training.

Guttmann NeuroPersonalTrainer® (NPT) [11][12] is a cognitive tele-rehabilitation system that allows the neuropsychologist to schedule a personalized rehabilitation program, each session consisted of specifically designed tasks, according to the patient's cognitive profile, offering an additional way of communication between therapists and patients. In addition, the platform offers a knowledge management module that allows the optimization of cognitive rehabilitation processes. NPT offers several modules of tasks for different neurological disorders, such as acquired brain injury, dementia, intellectual disability and mental illness, with specific modules for children.

Neuropsychologists from Parc Taulí Mental Health Hospital [13] have designed the Social Cognition Module

(SCM, from now onwards), as part of NPT's Mental Health program. The aim of this module is to cover the specific needs related with social cognition of patients with schizophrenia or schizoaffective disorders in early stages. SCM is compound of four different units: Emotional Processing, Cognitive Bias, Theory of Mind and Final Unit.

In this paper, we present a set of video-based tasks that compounds the SCM's Emotional Processing Unit, as part of NPT's therapeutic contents.

II. TASKS IMPLEMENTATION

This process entails a clinical-technological collaborative work, which results critical for the success of the whole project. Tasks implementation goes through four different phases, described below:

A. Requirements Analysis

The requirements analysis process entails the following steps: *Clinical problem*, studying schizophrenia and rehabilitation process to understand the specific needs of this kind of patients; *Technical study of NPT platform*, how it works and its software architecture; and *specifications analysis*. Neuropsychologists define the design of each interface element and the behavior of each task through a collaborative environment for tasks specification and validation [14] developed at the Bioengineering and Telemedicine Center (GBT-UPM).

B. Design

Every task has a common interface design, which aims to achieve the maximum usability, based on the special characteristics of schizophrenia patients. In general, a simple design is desired to make the patient feel in a well-known environment even though exercises change along the treatment. For improving usability, we have worked on patterns of consistency and coherence in order to maintain uniformity in screen design. We take care of the placement of the different elements. Likewise, the same color remains to the common elements of the different tasks for an easier and more intuitive identification. Screenshots of task interface are shown in Fig. 1 as an example.

Fig. 1 Examples of Emotional Processing Tasks

As for the execution, all the tasks follow a common sequence of well-defined actions. Tasks behavior can be modeled by a state diagram in which each state represents a screen. The transition between states can be determined automatically or by a user event, such a mouse click, or a temporal trigger. The detailed sequence between the different states is shown in Fig. 2.

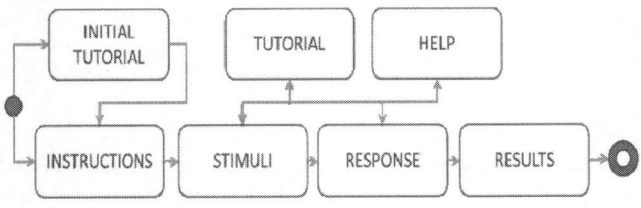

Fig. 2 State Diagram of tasks' execution

- **Initial Tutorial**

  Shows a test exercise and several tutorial screens if the neuropsychologist configures the task for the tutorial to be shown.

- **Instructions**

  Shows the directions for completing the task.

- **Stimuli**

  Generates task stimuli (images, texts, sounds, videos, relationships, etc.) and presents them to the patient for him to interact with the exercise.

- **Response**

  Gives feedback to the patient about his performance during the exercise.

- **Help**

  Displays a help message (usually instructions reminder) if the patient requires so.

- **Turorial**

  Displays the initial tutorial if the patient requires so.

- **Results**

  Represents patient's outcomes both numerically and graphically.

C. Software Implementation

Once the design process is completed, software implementation process begins. Java programming language is used for this. This language is chosen because NPT has been developed in this language, which means a restriction factor for selecting another one.

The programming scheme is based on a parent class, whose properties are inherited by all the classes that constitute a task. In addition, there is a set of classes that encapsulate a variety of common attributes and functions that are necessary for the development, representing entities such as the display or controlling the communication with the platform for launching the task.

In this work we introduce an innovation in stimuli representation in relation to traditional 2D-tasks [5]: the possibility of video visualization. For this, we use DJ Project, an open source project that allows playing a video in Java using VLC libraries as backend. To keep the integration with the NPT platform, it was necessary to create several classes to interact with the DJ Project library. One of the classes creates a panel and includes a video container while another class controls the video playback inside the task. A task can access to all functions of this new class and play a video when a screen is loaded or when the patient clicks a button, among other actions.

D. Technical Evaluation

Once a task is implemented, it is time to verify if it fulfills the initial specifications. As the specification, technical evaluation is carried out through the collaborative environment mentioned before [12], where engineers and neuropsychologists can access to a validation module (specific for each task) in which every aspect that must be checked is represented. A color code is used to indicate developers if a field has been checked with a positive or negative result (green/red), or to indicate neuropsychologists that it is waiting for validation (orange), acting like some kind of *semaphore*. This is a critical phase where a lot of information is exchanged between developers and neuropsychologists.

III. EMOTIONAL PROCESSING UNIT

As a result of this work, a set of 15 computerized tasks has been obtained. These tasks compound the NPT's Emotional Processing Unit, which is organized in blocks, as shown in Table 1.

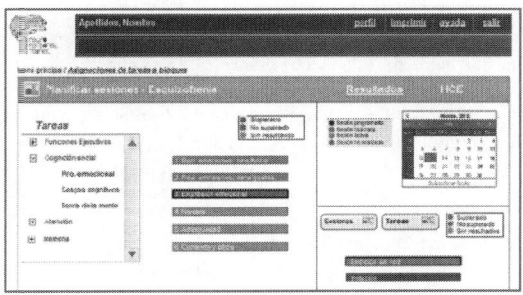

Fig. 3 NPT scheduler

Table 1 Emotional Processing Module

Block	Task
1. Emotion Recognition (Eyes/mouth)	Eyes & Mouths
	Discover the photograph
	Face puzzle I
	Same emotion
2. Emotion Recognition (Face/gesture/emotion/voice type)	Faces & Expressions
	Pleasant vs. Unpleasant
	The box
	Voice emotion
3. Emotion Levels	Order the emotions
	Face puzzle II
4. Ambiguity	Ask for more information I
	Ask for more information II
	End of the story
5. Context & Others	How will the character respond?
	What does the character feel?

Tasks in each block shares the same therapeutic goal:

1. **Emotion Recognition (Eyes/mouth)**

   To recognize basic emotions (happiness, sadness, anger, surprise, etc.) based on eyes and mouth expression.

2. **Emotion Recognition (Face/gesture/emotion/voice type)**

   To recognize basic and more complex emotions combining face expression with postural signs and gestures. Also, to identify whether an emotion is positive, negative or neutral.

3. **Emotion Levels**

   To identify both the type and degree (little, enough, and much) of an emotion, based on abilities trained in previous modules.

4. **Ambiguity**

   To be able to analyze and interpret an ambiguous situation, asking for more information when necessary.

5. **Context & Others**

   To work out how does a character feel or how he will react based on a particular context, previously presented.

Each block represents a more difficult challenge for the patient to achieve. Therefore, neuropsychologists have determined that planning tasks from one block must be restricted by the accomplishment of the previous ones. Consequently, NPT scheduler has been modified for SCM so that therapists can check whether the patient has completed a block or not by a color legend, as shown in Fig. 3.

Since May 2012, a randomized clinical trial (RCT) has been conducted at Parc Taulí Mental Healt Hospital [13] to test the effectiveness of this therapeutic program. Patients in the experimental group perform sessions of computerized tasks from SCM, planned by therapists, up to approximately 40 hours of treatment. Patients in the control group perform nonspecific office automation tasks and video viewing sessions for the same hours. Every subject is neuropsychologically assessed before and after the intervention, as well as three months after the end of it.

Preliminary results of 32 patients show an improvement in problem resolution, verbal memory, selective attention and emotional processing in the experimental group compared to the control one after the treatment. In spite of the small sample size of the RCT, results yielded by now suggest that SCM has a positive impact in the rehabilitation of patients in early stages of schizophrenia or schizoaffective disorder.

IV. DISCUSSION AND CONCLUSIONS

This paper presents a set of tasks that compound NPT's Emotional Processing unit.

From a technological perspective, video technology has been incorporated to the traditional 2D-tasks. For this, tasks framework has been modified and extended, providing a series of new classes and methods which open the door to the incorporation of another technologies in the near future, such as flash animations or interactive video [15].

Regarding economic aspects, it is obvious that video production is expensive. Professional actors have participated in the recording, which has been carried out by a film producer. Therefore, a video bank has been created so that clips can be reused in future tasks, resulting in a reduction in the development costs.

From a clinical point of view, this module focus on social cognition deficits, which remained uncovered by cognitive tele-rehabilitation programs up to now. Therefore, schizphrenia patients, who typically present this kind of disorders, can receive a complete rehabilitation program through NPT, combining both cognitive (attention, memory and executive functioning) and social cognitive tasks in their rehabilitation sessions. Preliminary results of the clinical trial suggest the efficacy of the intervention, expecting that it may lead in a clinical and functional impact in this group of patients.

It can therefore be concluded that tasks presented in this work entail innovation in NPT's therapeutic contents from both a clinical and a technological perspective.

ACKNOWLEDGMENT

This research has been partially founded by the Spanish Ministry of Economy and Finance (project IPT-300000-2010-30, NEUROCONTENT).

REFERENCES

1. World Health Organization (WHO) at http://www.who.int/
2. Couture S.M, Penn D.L, Roberts D.L. (2006) The Functional Significance of Social Cognition in Schizophrenia: A Review. Schizophrenia Bulletin 32:44–63 DOI 10.1093/schbul/sbl029
3. Green M.F, Olivier B, Crawley J.N et al. (2005) Social Cognition in Schizophrenia: Recommendations from the Measurement and Treatment Research to Improve Cognition in Schizophrenia. New Approaches Conference. Schizophrenia Bulletin 31:882–887 DOI 10.1093/schbul/sbi049
4. Krabbendam L, Aleman A. (2003) Cognitive rehabilitation in schizophrenia: a quantitative analysis of controlled studies. Psychopharmacology 169:376-382 DOI 10.1007/s00213-002-1326-5
5. Caballero-Hernández R, Martínez-Moreno J.M, García-Molina A et al. (2011) 2D-Tasks for Cognitive Rehabilitation. Proc. vol. 37, 5th European Conference of the International Federation for Medical and Biological Engineering (EMBEC 2011), Budapest, Hungary, 2011, pp 838-841
6. FranceschiniM, Agosti M, Cantagallo A, Sale P, Mancuso M, Buccino G. Mirror neurons: actions observation treatment as a tool in stroke rehabilitation. Eur J Phys Rehab Med 46 (2010), 517-523.
7. RehaCom at http://www.uam.es/personal_pdi/psicologia/pei/ pticexpo-0607/grupo8/software2.html.
8. Gradior at http://www.intras.es/index.php/que-hacemos/idi/gradior
9. PssCogRegHab at http://www.neuroscience.cnter.com/pss/psscr.html
10. BrainTrain at http://www.braintrain.com/
11. Solana J, García-Molina A et al. (2011) PREVIRNEC A new platform for cognitive tele-rehabilitation. The Third International Conference on Advanced Cognitive Technologies and Applications (COGNITIVE 2011), Roma, Italy; 2011. ISBN: 978-1-61208-155-7
12. Istitut Guttmann at http://www.guttmanninnova.com
13. ParcTaulí Mental Health Hospital at http://www.tauli.cat/
14. Orúe-Vega O, Caballero-Hernández R,García-Molina A, Martínez-Moreno JM,García-Rudolph A, Cáceres-Taladriz C,Tormos-Muñoz JM, Gómez-Aguilera EJ(2011) 'Entorno colaborativo de edición de tareas en neurorrehabilitación cognitiva', XXIX Congreso anual de la Sociedad Española de Ingeniería Biomédica, Cáceres. Libro de Actas CASEIB 2011, pp. 669-672.
15. J. M. Martínez Moreno, J. Solana Sánchez, R. Sánchez Carrión, S. González Palmero, P. Sánchez Gónzalez, C. Gómez Pérez, M. Morell Vilaseca, C. Cáceres Taladriz, T. Roig Rovira, J. M. Tormos Muñoz, E. J. Gómez Aguilera. Monitoring visual attention on a neurorehabilitation environment based on Interactive Video. International Conference on Recent Advances in Neurorehabilitation, Valencia. ICRAN Proceedings, 2013, pp. 182-185

Author: Ruth Caballero Hernández
Institute: Biomedical Engineering and Telemedicine Centre, ETSI Telecomunicación, Universidad Politécnica de Madrid
Street: Avda. Complutense 30
City: Madrid
Country: Spain
Email: rcaballero@gbt.tfo.upm.es

Preliminary Experiment with a Neglect Test

C. Lassfolk[1], M. Linnavuo[1], S. Talvitie[2], M. Hietanen[3], and R. Sepponen[1]

[1] Health Factory/Department of Electronics, Aalto University, Espoo, Finland
[2] ORTON Foundation / ORTON Rehabilitation Ltd, Helsinki, Finland
[3] Department of Neurology, Helsinki University Central Hospital, Helsinki, Finland

Abstract—This paper presents a limited (six healthy, right-handed participants) preliminary experiment with a test planned for diagnosing hemispatial neglect. Healthy participants were expected to show equal reaction times on both sides. Familiar participants were assumed to react faster than unfamiliar ones. The test included four separate vigilance experiments that were performed on a laptop computer. All participants showed slower reaction times to stimuli in the right observational field. Performance did not increase with familiarity with the experiments. The experiment did not confirm our hypotheses, but the consistency of the preliminary data serves as a good starting point for upcoming larger experiments with patients and healthy controls.

Keywords—Stroke rehabilitation, hemispatial neglect, vigilance test.

I. INTRODUCTION

Hemispatial neglect constitutes one of the ramifications of stroke [1-3]. Approximately one fourth of all surviving stroke patients suffer from neglect [4]. Neglect impairs the ability to attend to stimuli contralateral to a cerebral lesion [2, 3, 5, 6]. Its incidence and severity correlates with the rehabilitation predication. An accurate diagnosis of neglect forms the basis of an optimal rehabilitation, which also intends to restore the capabilities for activities of daily living. [1, 7]

Current practice mainly utilizes paper-and-pen tests to diagnose neglect [1]. Many ongoing research activities focus on computerizing the traditional tests or on moving to virtual realities in order to better mimic the three-dimensional reality and to obtain more accurate measurements of the deficiencies in spatial processing [8-10]. Our endeavors belong to the latter group.

We have been planning a computerized neglect test in which patients are assumed to miss more stimuli on the left side and to react slower to stimuli on the left side. Healthy controls, on the other hand, are presumed to show equal accuracy and reaction times on both sides. [2]

The experiment described in this paper tried to verify that the prototype vigilance test planned for diagnosing neglect performs reasonably well with healthy participants. Experiments including neglect patients will be performed later and reported separately. The present experiment was also expected to deliver early feedback for adjusting the test and the experiment design. We hypothesized, firstly that healthy participants show equal response patterns on both left and right sides, and secondly that persons familiar with the experiments might benefit and thus react faster than unfamiliar participants.

II. MATERIALS AND METHODS

A. Participants

The experiment was conducted on six healthy, right-handed volunteers (five male, one female). Ages varied between 29 and 59 (mean 45, SD 12). Two of the participants had been involved in developing the experiments, whereas the other four had no previous experience with the experiments.

B. Protocol and Apparatus

During the test the participants responded to four separate experiments. The three first experiments lasted two minutes each and the last one took three minutes. Responses were expected in 2000 milliseconds (ms) from the onset of the stimulus in experiments 1, 2 and 4, whereas in experiment number 3 the maximum reaction time was 1000 ms. Responses coming between these response windows were classified as false clicks. A failure to respond to a stimulus in the response window was categorized as a miss. In the three first experiments the participants were asked to direct their gaze on a varying digit displayed in the center of the task field. Brief oral instructions were given to the participants before each experiment. However, written instructions and training sessions were omitted in order to provide an insight into the clarity and intuitiveness of the experiments.

Visually the three first experiments looked identical (Figure 1). The center of the task field displayed a digit that changed every 800 ms. Each center digit (0-3) appeared an equal number of times (30). Occasional two-dimensional balls (diameter 10 mm) of different colors (red, green, blue, yellow and cyan) blinked in all parts of the task field for 400 ms. The balls (in total 100; 20 of each color) were evenly distributed over the entire task field. To achieve this, the task field had been segmented into ten invisible subfields by splitting it horizontally into a top and bottom part and vertically into five equal zones (far left, left, center, right and far right).

Fig. 1 Screen shot from the first three experiments show a varying digit in the center of the task field and occasional balls blinking

In the fourth experiment two-dimensional balls (diameter 10 mm) of different colors (red, green, blue, yellow and cyan) appeared at the top of the task field and then sank downwards until they disappeared after 2000 ms in the middle part of the screen. After a correct response the corresponding stimulus ball disappeared immediately. The task field was segmented into five horizontal segments like in the other experiments. An equal amount of balls (80) appeared in each segment, but the mid-segment contained no red balls.

Due to limitations in the prototype implementation the task field could not be shown in full screen mode. It covered approximately 80% of the screen and was always located in the bottom right part of the display. (Figure 1)

Experiment 1: The participants were instructed to press the mouse wheel always when a "2" appeared in the center. They were also told not to pay any attention to any other objects.

Experiment 2: The participants were instructed to press the mouse wheel always when a red ball blinked in any part of the task field. They were also told not to pay any attention to any other objects.

Experiment 3: The participants were instructed to press the mouse wheel always when a "2" appeared in the center or a red ball blinked anywhere in the task field.

Experiment 4: The participants were instructed to press the mouse wheel always when a red ball appeared. In response to the reaction the red ball would disappear.

The test was conducted on a Lenovo T410s laptop with an external Logitech wheel mouse attached. The task field measured 270 x 170 mm. The participants pressed the mouse wheel with the right hand. During the test reaction times were measured and erroneous false clicks were registered.

C. Experiment Design

Independent Variables: The experiment comprised three independent variables: experiment number, stimulus location and participant familiarity. The experiments had been numbered from one to four. Location took values left or right. Familiarity represented a dichotomous variable that took values familiar or unfamiliar.

Dependent Variables: The experiment comprised three dependent variables: the reaction time to the stimulus, reaction accuracy and false clicks. Reaction times were measured in milliseconds, reaction accuracy was computed as the ratio of correct responses to all expected responses and false clicks represented the sum of clicks occuring when none were expected.

Firstly, the setup generated a 2 x 4 between-subjects design for testing the effect of familiarity on reaction time. Secondly, the setup generated a 2 x 4 within-subjects design for testing reaction times in the left and right observational field. However, we also included a general reaction time and thus, in effect, ran a 3 x 4 within-subjects design.

First pair of Hypotheses: The hypothesis (H_1^1) proposed that healthy participants react equaly fast to stimuli in the left and right observational field. The null hypothesis (H_0^1) stated that reaction times to stimuli on the left and right differ.

Second pair of Hypotheses: The hypothesis (H_1^2) proposed that familiar participants react faster than unfamiliar participants. The null hypothesis (H_0^2) stated that familiar participants react as fast as or slower than unfamiliar participants.

After each experiment participants were asked to provide free form oral feedback about any observations they had made during the previous experiment and how they had experienced it.

For statistical analysis R version 2.14.2 was employed.

III. RESULTS

No outliers were detected in the data. The reaction times were normally distributed (Kolmogorov-Smirnov test, $p > .05$) and the homoscedasticity requirement would have been met (Levene's test for homogeneity of variance, $p > .05$), but the sample clearly consisted of too few subjects for allowing the application of parametric tests.

Reaction times (Table 1, Figure 2; left, Figure 3; left) to stimuli on the left and right differed (Friedman rank sum test, $\chi^2 (2) = 7.0$, $p < .05$). Nevertheless, a post hoc analysis indicated that the only significant difference appeared between the general reaction time and the reaction time to stimuli on the right side (Bonferroni adjusted Wilcoxon rank sum test, $p < .05$). Thus, H_0^1 could not be rejected.

Familiar participants reacted (Table 1, Figure 2; right) slower than unfamiliar ones (Wilcoxon rank sum test, U = 582, p < .01) and thus H_0^2 could not be rejected. The impact of familiarity proved to be reverse to our hypothesis. Nonetheless, familiar and unfamiliar participants showed similar reaction time patterns (Figure 3; right).

Table 1 Reaction time in hypothesis conditions

		a) Time type			b) Familiarity	
		General	Left	Right	Familiar	Unfamailiar
Reaction time [ms]	Mean	492.4	498.5	530.1	536.2	490.2
	SD	44.6	51.7	62.4	47.4	51.2

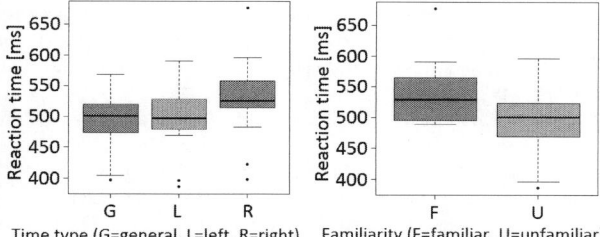

Fig. 2 Left: Correlation between the time type (general, left, right) and reaction time. Right: Correlation between participant familiarity (familiar, unfamiliar) and reaction time.

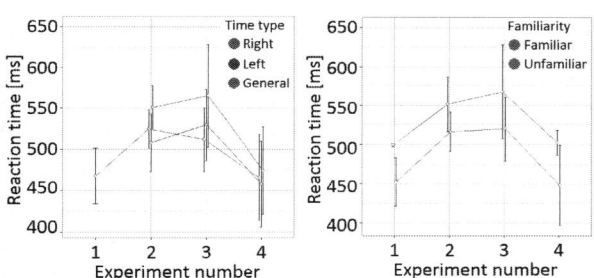

Fig. 3 Reaction time dependency on experiment number and time type (left) and participant familiarity (right). Both graphs display standard deviation confidence intervals.

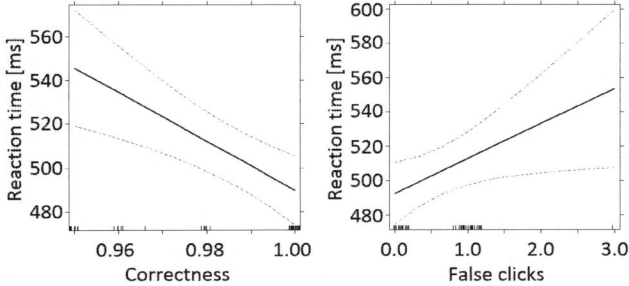

Fig. 4 Reaction time dependence on correctness (left) and false clicks (right).

Participants reacted faster the fewer errors they made (Figure 4; left) (Spearman, -0.41, p < .01). An opposite, however not significant (Spearman, p > .05), correlation between false clicks and reaction times seemed to emerge (Figure 4; right).

The qualitative feedback received revealed that most participants experienced the test as fairly stressful and hectic. Experiment 3 was considered most demanding. Everyone noticed their own misses. Several participants felt that clicking the mouse wheel did not represent the best possible response method, as the finger easily slipped. A few participants also reported that they did not manage to keep the gaze on the center of the task field throughout the first three experiments, as they either lost focus or the gaze unconsciously started to move about.

IV. DISCUSSION

The experiment reported in this paper indicates that participants reacted faster to stimuli in the left observational field than to stimuli in the right observational field, when using the right hand for responding. Familiar participants had slower reaction times than unfamiliar ones. We failed to reject both hypotheses, and thus the experiment actually proved more informative than expected.

Anzola et. al. [11] compared simple reactions of both hands and found that the reaction time was shorter for the right hand to stimuli on the right. However, they also concluded that in simple reaction time experiments elementary anatomical connectivity mainly affected reaction times, whilst spatial compatibility of stimulus and hand played a major role in selection type experiments. Lateral hand preference does neither seem to be based on faster reaction nor on most accurate reaction, but may correlate with the complexity of the task and the ability to perform a series of movements in an optimal way [12-14]. The conclusions of studies comparing reactions to stimuli in the left and the right observational field seem difficult to interpret as the results appear inconsistent. Perhaps the task field placement in the right bottom part of the screen affected the left-right outcome of this experiment. We did not control the relative position of the body of the participant and the computer.

The limited sample and the unfavorable age and gender distribution [15] in this experiment impeded verification of the impact of familiarity on reaction time. On the other hand the result could also indicate an advantage, as the test comes across as intuitive and easy to learn. The unfamiliar participants showed faster reaction times than the familiar ones, although written instructions and training session had been omitted.

The scroll wheel as a response button will be replaced by some other means later. A gaming oriented participant felt that a robust movable press button which the participant could keep in his hand while still being able to relax the arm in the lap would be optimal for quick responses.

The experiment was intended as a dry run of the experiments before starting clinical tests. With such a small sample we could not expect statistical significance. Anyhow, this experiment also permitted preliminary checking of normality and homoscedasticity and the data enables estimation of effect size in future experiments.

The fact that the participants perceived the test as fairly demanding can be viewed as an auspicious result, as lowering the degree of difficulty seems easier than raising it. However, the correct degree of difficulty can only be defined after more extensive experiments including neglect patients.

V. CONCLUSIONS

The preliminary finding of this study suggests that a reasonably difficult test sequence has been created, which seems to generate predictable results for healthy persons. The outcome looks promising, but we need a larger experiment with neglect patients and healthy controls.

AKNOWLEDGEMENTS

The Finnish Cultural Foundation and the Finnish Funding Agency for Technology and Innovation (Tekes) have supported this work. The authors are grateful to the testees for their participation and to Heini Heikkilä and Dima Smirnov for tutoring in data analysis and reporting practice.

REFERENCES

1. Ting D S J, Pollock A, Dutton G N, et al. (2011) Visual Neglect Following Stroke: Current Concepts and Future Focus. Surv. Ophthalmol. 56:114-134 DOI 10.1016/j.survophthal.2010.08.001
2. Rengachary J, He B J, Shulman G L, et al. (2011) A Behavioral Analysis of Spatial Neglect and its Recovery After Stroke. Frontiers in Human Neuroscience 5:29 DOI 10.3389/fnhum.2011.00029
3. Kerkhoff G (2001) Spatial hemineglect in humans. Prog. Neurobiol. 63:1-27 DOI 10.1016/S0301-0082(00)00028-9
4. Pedersen P M, Jørgensen H S, Nakayama H, et al. (1997) Hemineglect in Acute Stroke-incidence and Prognostic Implications: The Copenhagen Stroke Study1. American Journal of Physical Medicine & Rehabilitation 76:122-127
5. Smolyanskaya A and Born R T (2012) Neuroscience: Attention is more than meets the eye. Nature 489:371-372 DOI 10.1038/489371a
6. Graziano M S (2001) Neuroscience: Awareness of space. Nature 411:903-904 DOI 10.1038/35082182
7. Wolf T J (2011) Rehabilitation, Disability, and Participation Research: Are Occupational Therapy Researchers Addressing Cognitive Rehabilitation After Stroke? The American Journal of Occupational Therapy 65:46-59 DOI 10.5014/ajot.2011.002089
8. Dawson A M, Buxbaum L J and Rizzo A A (2008) The virtual reality lateralized attention test: Sensitivity and validity of a new clinical tool for assessing hemispatial neglect, Virtual Rehabilitation, 2008, 2008, pp 77-82 DOI 10.1109/ICVR.2008.4625140
9. Lee J H, Ku J, Cho W, et al. (2003) A virtual reality system for the assessment and rehabilitation of the activities of daily living. CyberPsychology & Behavior 6:383-388 DOI 10.1089/109493103322278763
10. Cameirão M S, Badia S B, Oller E D, et al. (2010) Neurorehabilitation using the virtual reality based Rehabilitation Gaming System: methodology, design, psychometrics, usability and validation. Journal of Neuroengineering and Rehabilitation 7:48 DOI 10.1186/1743-0003-7-48
11. Anzola G, Bertoloni G, Buchtel H, et al. (1977) Spatial compatibility and anatomical factors in simple and choice reaction time. Neuropsychologia 15:295-302 DOI 10.1016/0028-3932(77)90038-0
12. Nakamura R and Saito H (1974) Preferred hand and reaction time in different movement patterns. Percept. Mot. Skills 39:1275-1281 DOI 10.2466/pms.1974.39.3.1275
13. Kimura D and Vanderwolf C (1970) The relation between hand preference and the performance of individual finger movements by left and right hands. Brain 93:769-774 DOI 10.1093/brain/93.4.769
14. Kabbash P, MacKenzie I S and Buxton W (1993) Human performance using computer input devices in the preferred and non-preferred hands, Proceedings of the INTERACT'93 and CHI'93 Conference on Human Factors in Computing Systems, 1993, pp 474-481 DOI 10.1145/169059.169414
15. Silverman I W (2006) Sex differences in simple visual reaction time: A historical meta-analysis. Sex Roles 54:57-68 DOI 10.1007/s11199-006-8869-6

Address of the corresponding author:

Author: Christina Lassfolk
Institute: Aalto University
Street: Otakaari 7B
City: Espoo
Country: Finland
Email: christina.lassfolk(at)aalto.fi

A Subject-Driven Arm Exoskeleton to Support Daily Life Activities

E. Ambrosini[1], S. Ferrante[1], M. Rossini[2], F. Molteni[2], G. Ferrigno[1], and A. Pedrocchi[1]

[1] NearLab, Department of Electronics, Information, and Bioengineering, Politecnico di Milano, Milan, Italy
[2] Valduce Hospital, Villa Beretta Rehabilitation Center, Costa Masnaga, Lecco, Italy

Abstract—In the field of assistive technologies, the possibility to maximally exploit the residual capabilities of the user to increase the independence of persons with disabilities is a really promising approach. This prospective was the starting point of the European project MUNDUS that developed a customizable and modular system for recovering direct interaction capabilities of severely motor impaired people based on arm reaching and hand functions. The present work provides a quantitative evaluation of the simplest configuration of the MUNDUS system on three neurological patients. The apparatus consisted of a lightweight passive arm exoskeleton for weight relief, equipped with electromagnetic brakes for locking each degree of freedom. The user could autonomously activate and deactivate the brakes through a pre-defined contraction of a muscle of the contralateral arm. The subjects tested the system in a 3-day session. Each day, they were asked to execute four tasks: drinking, touching the left shoulder, touching the left hand, and pressing a button. A group of healthy volunteers were also involved in the trials to define normality ranges. Smoothness and straightness of the wrist trajectories were assessed. The usability of the system was also evaluated using the System Usability Scale (SUS). All subject were able to execute all the required tasks and to autonomously control the brakes from the first day, suggesting that no training was needed to learn how to use the system. In terms of performance over days, one subject's trajectories were more rectilinear and more smooth after the first day, one subject worsened some kinematic parameters on the third day and one subject did not show any significant differences. In terms of usability, all subjects showed a good satisfaction with the system (SUS > 90/100) and would like to use the system in their daily life.

Keywords—Assistive Technology, Exoskeleton, Upper limb, EMG signals, Kinematic analysis.

I. INTRODUCTION

Robotic devices have been increasingly used not only for rehabilitative purposes [1, 2] but also in the field of assistive technologies (AT) [3]. AT are technologies designed to assist and support people who would otherwise be unable to participate in social, cultural and economic activities due to impairments or disabilities. Robots can increase independence of persons with disabilities providing physical assistance. For this purpose, wheelchair-mounted general-purpose manipulators, such as the iARM produced by Exact Dynamics [4], are now commercially available. However, these solutions hardly surrogate the natural interaction with objects of daily life [3].

A novel approach might be offered by customizable systems able to exploit any residual motor capabilities of the user and to allow a natural motion of the arm and a natural interaction with objects of daily life. An example of such a system is the one developed during the European project "MUltimodal Neuroprosthesis for Daily Upper limb Support" (MUNDUS) [5]. MUNDUS consists of a modular platform where sensors, controllers, and actuators are adapted to the user's residual ability to support basic arm and hand functions. When the user preserves some residual muscle contractions, the movement can be driven by the user's own muscles supported by a passive spring-loaded arm exoskeleton for gravity compensation. When residual muscle contractions are too weak or completely absent neuromuscular electrical stimulation (NMES) can be used to support or restore arm motion [6, 7]. NMES as well as a robotic orthosis can be used to support hand functions for grasping and releasing of cylindrical objects. To allow the user to autonomously control the system, EMG signals, eye/hand motions, or brain signals can be alternatively used to detect the user's intention. The expected MUNDUS users are people affected by neurodegenerative and genetic neuromuscular diseases and high level spinal cord injuries (SCI).

The present work provides a quantitative evaluation of the simplest configuration of the MUNDUS system, consisting of the exoskeleton to support the arm movement and the acquisition of EMG signals to detect the user's intention. Three subjects were recruited for testing the system during daily life activities.

II. METHODS

A. Apparatus

The apparatus consisted of a lightweight passive arm exoskeleton (see Fig. 1) characterized by 3 degrees of freedom (DOF): shoulder elevation in the sagittal plane, shoulder rotation in the horizontal plane, and elbow flex-extension [8]. An easily adjustable weight support for the arm was integrated in the mechanical shoulder joint. A spring mechanism was implemented to compensate for gravity. Electromagnetic DC brakes could lock each DOF. Encoders to measure the angles at the three DOFs (sampling frequency of 50 Hz) were also included. The exo could be mounted on a wheelchair, as shown in Fig. 1, or used as a mobile system together with a body harness.

An EMG amplifier (Porti 32™, TMS International) was used to acquire a single channel EMG signal at 2048 Hz. Ag/AgCl pre-gelled self-adhesive electrodes were placed over a target muscle of the contralateral arm. For each subject, the muscle he/she could better control was selected. An ad-hoc real-time algorithm was developed to detect the user's intention: when a sequence of three consecutive contractions, lasting at least 1 s each, within a time interval of 10 s, was recognized by the algorithm, a trigger was sent to the exo and induced the activation or the deactivation of the brakes depending on their current status.

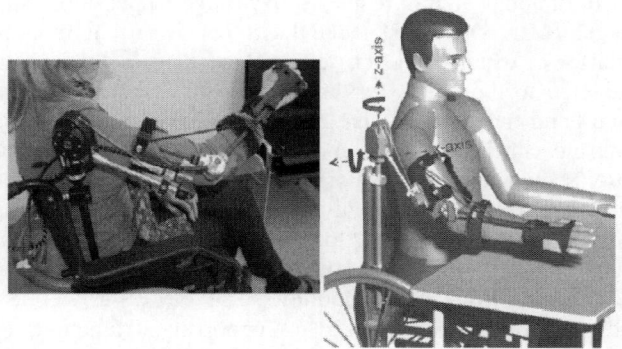

Fig. 1 The lightweight arm exoskeleton.

B. Participants

Three subjects (one incomplete SCI and two people affected by Friedreich's ataxia, FDRA) were involved in the study. Table 1 reports the clinical and demographic details of the participants. The trials were performed at the Villa Beretta Rehabilitation Center. The study was approved by the Ethical Committee of the Valduce Hospital and patients signed a written informed consent.

Five healthy subjects with no history of muscle weakness or neurological diseases (2 males and 3 females, mean age of 26.6 ± 10.4 years) were also involved in the trials to define the normality ranges for the kinematic parameters.

C. Experimental Protocol

Participants were asked to perform 3 experimental sessions over consecutive days. Each session consisted of the executions of four tasks: drinking, touching the left shoulder, touching the left hand, and pressing a button placed on the right side of the subject. For each task, 10 repetitions were performed. During each repetition, the subjects were asked to activate the brakes once reached the target (i.e., the mouth, the shoulder, the hand, or the button), wait for at least 10 s, deactivate the brakes and come back to the rest position. All the tasks were performed with the right arm since only a right-arm prototype of the exo was developed.

Healthy subjects were involved in a single day session, following the same protocol used for the patients.

Table 1 Clinical and demographic details of the participants (MI, Motricity Index; BI, Barthel Index, MRC, Medical Research Council).

Sub	Age	Sex	Pathology	MI right arm (0-100)	BI (0-20)	MRC right arm (0-5)	
						elbow flexion	elbow extension
S1	29	F	FDRA (1996)	76	4	3	3
S2	50	M	FDRA (1986)	71	7	4	3
S3	70	M	incomplete SCI C3-C4 (2012)	39	5	3	3

D. Data Analysis and Statistics

The execution of the movement was evaluated through a 3D kinematic analysis of the upper limb. The angles at the three DOFs of the exo were used to reconstruct the trajectory of the wrist. Each repetition was divided into two sub-actions: the forward movement, from rest to the brakes activation, and the backward movement, from the brakes deactivation to rest again. For each sub-action, the smoothness, as the ratio between the average speed and the maximal speed, and the straightness, as the distance between the initial and the final position divided by the trajectory length covered by the subject [9], were computed. To compare the performance of each subject over the 3 sessions, a Kruskal-Wallis test was used. In addition, a Mann-Whitney U test was used to compare the performance achieved by each patient during each experimental session with the performance obtained by the healthy subjects (control group).

The System Usability Scale (SUS), a ten-item attitude Likert scale ranging from 0 (no satisfaction) to 100 (extreme satisfaction), was also measured [10] at the end of the trials.

III. RESULTS

All the subjects were able to complete the repetitions required by the protocol with the support of the exo. No major problems in controlling the activation and deactivation of the brakes through the muscle contractions were highlighted for any subjects. In particular, the trapezius (for S1 and S3), and the brachioradialis muscles (for S2) were chosen for the detection of the user's intention.

Fig. 2 shows an example of a subject while performing one repetition of the "touching the left shoulder" task. In panel (A), the angular trajectories at the three DOFs are shown both for the forward (from rest to shoulder, 0-9 s) and the backward movement (from shoulder to rest, 17.5-20 s). Panel (B) represents the envelope of the EMG signal used to control the brakes and the trigger signal detected by the algorithm: the first trigger activated the brakes at the shoulder position, while the second trigger deactivated the brakes to come back to rest.

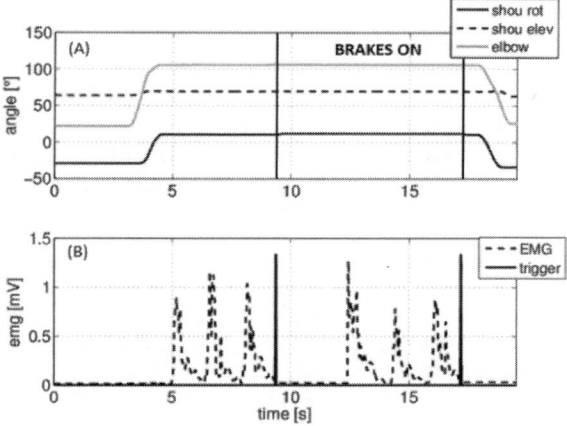

Fig 2 Example of a subject while performing the "touching the left shoulder" task. Panel (A) shows the angular data, while panel (B) depicts the envelope of the EMG signal and the trigger signal to activate/deactivate the brakes.

Due to page limit, we report only the results obtained for the "drinking" and the "touching the shoulder" tasks as representative of the whole performance. Fig. 3 and Fig. 4 show the results of the kinematic analysis for these two tasks. In each figure, panels (A-B) report the results achieved in the forward movement, while panels (C-D) report the results obtained in the backward movement. A worsening in the performance of S1 was observed comparing days 1/2 and day 3 in terms of both straightness and smoothness during the first sub-action of the drinking task. An improvement in the straightness from day 1 and day 2/3 was instead found for the second sub-action. Concerning the "touching the shoulder" task, S1 showed a slight improvement of the straightness and a worsening of the smoothness for both sub-actions. For S2, a general improvement in most of the kinematic parameters was observed for both tasks. Finally, S3 improved the straightness of the trajectory during the forward movement of the drinking task, but no other significant differences could be highlighted. Considering all subjects, more differences between patients and control group were found for the "drinking" task than for the "touching the shoulder" task. This result could be expected, being the drinking a more complex task, that requires to reach an object placed on a table in front of the subject, grasp the object, and bring it to the mouth.

In terms of usability, SUS scores of 95, 90, and 90 out of 100 were obtained for S1, S2, and S3, respectively. These results indicate a good satisfaction of all the subjects: they considered the system easy to use, non-cumbersome, and useful in supporting them during daily life activities.

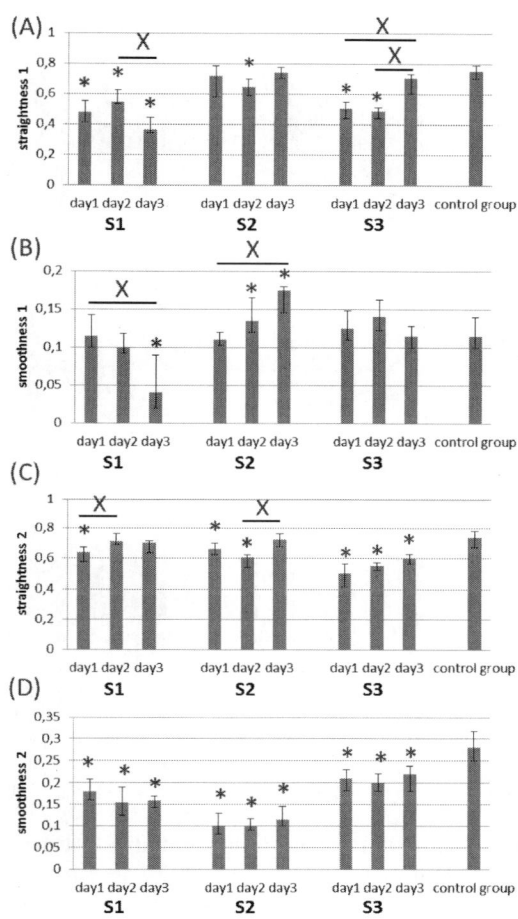

Fig. 3 Kinematic data obtained during the "drinking" task (forward movement, panels A-B; backward movement, panels C-D). Values: Median (interquartile range, IQR); X: Significant level of Kruskal-Wallis post-hoc analysis ($p<0.05$); *Significant level of Mann-Whitney U test ($p<0.05$).

IV. CONCLUSIONS

A subject-driven passive arm exoskeleton, light, non-cumbersome, and easy to use, has been developed and its feasibility has been assessed on three neurological patients.

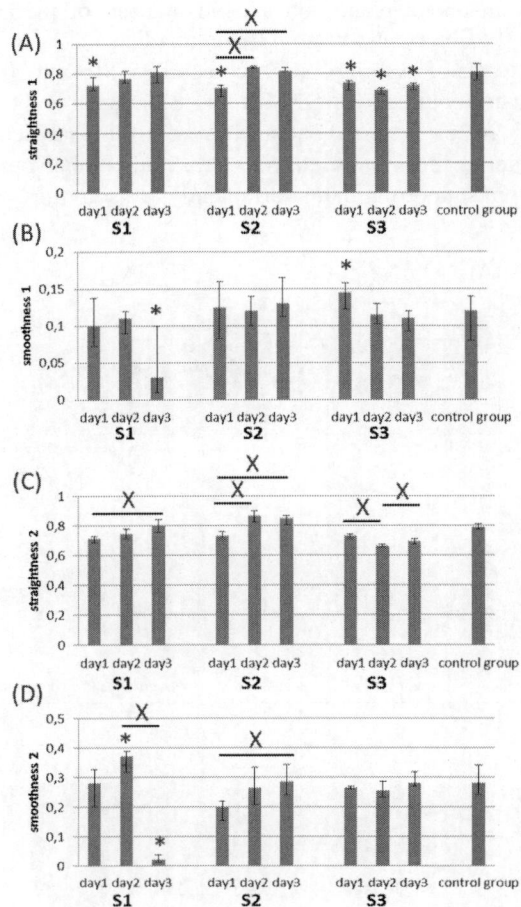

Fig. 4 Kinematic data obtained during the "touching the left shoulder" task (forward movement, panels A-B; backward movement, panels C-D). Values: Median (IQR): X: Significant level of Kruskal-Wallis post-hoc analysis ($p<0.05$); *Significant level of Mann-Whitney U test ($p<0.05$).

The subjects did not need any training to learn how to use the system and were able to complete all the required tasks from the first day. The possibility to activate and deactivate the brakes were considered very useful for all of them since it allowed to keep an anti-gravity position without any effort, thereby delaying the onset of muscular fatigue. Although the detection of the EMG signal to control the brakes resulted to be an acceptable solution for all the tested subjects, the possibility to use a USB-button activated by the contralateral hand was also provided. This solution can significantly reduce the time needed for controlling the brakes (from about 10 s to less than 1 s) for subjects who preserve some residual functions in the contralateral hand.

In terms of motor performance over time, the subjects were characterized by a different behavior: S2 showed a general improvement after the first day, suggesting a familiarization with the system; S1 worsened some of the kinematic parameters on the third day probably due to a worst general health status; finally, S3, who was the most impaired subject, showed no significant differences over time, suggesting that a longer training might be required for him to improve his performance. To conclude, all the subjects were really satisfied with the system and would like to use it at home during their daily life activities.

ACKNOWLEDGMENT

This work was supported by the Project MUNDUS (FP7 ICT 2009-4; grant agreement no.: 248326). We also acknowledge Andreas Jedlitschka from the Fraunhofer Institute (Kaiserslautern) for his help in the data analysis.

REFERENCES

1. Ferrante S, Pedrocchi A, Ferrigno G, et al. (2008) Cycling induced by functional electrical stimulation improves the muscular strength and the motor control of individuals with post-acute stroke. Eur J Phys Rehabil Med 44:159-67.
2. Casellato C, Pedrocchi A, Zorzi G, et al. (2012) Error-enhancing robot therapy to induce motor control improvement in childhood onset primary dystonia. J Neuroeng Rehabil; 9:46.
3. Brose SW, Weber DJ, Salatin BA, et al. (2010) The role of assistive robotics in the lives of persons with disability. Am J Phys Med Rehabil 89:509-521.
4. Driessen BJ, Evers HG, van Woerden JA. (2001) MANUS a wheelchair-mounted rehabilitation robot. Proc Inst Mech Eng H 215:285-90.
5. Pedrocchi A, Ferrante S, Ambrosini E, et al (2013) MUNDUS project: MUltimodal Neuroprosthesis for daily Upper limb Support, J. Neuroengineering Rehabil. 10:66, DOI: 10.1186/1743-0003-10-66.
6. Ambrosini E, Ferrante S, Tibiletti M, et al. (2011) An EMG-controlled neuroprosthesis for daily upper limb support: a preliminary study. IEEE Eng Med Biol Soc Proc, Boston, USA, pp 4259–4262.
7. Ambrosini E, Ferrante S, Schauer T, et al. (2010) Design of a symmetry controller for cycling induced by electrical stimulation - Preliminary results on post-acute stroke patients. Artif Organs 34:663-7.
8. Reichenfelser, W, Karner, J, Gfoehler M. (2012) Modular Instrumented Arm Orthosis with Weight Support for Application with NMES. Converging Clinical and Engineering Research on Neurorehabilitation, Springer, pp 1159–1163.
9. Gilliaux M, Lejeune T, Detrembleur C, et al. (2012) A robotic device as a sensitive quantitative tool to assess upper limb impairments in stroke patients: a preliminary prospective cohort study. J Rehabil Med 44:210-7
10. Brooke J. (1996). "SUS: a "quick and dirty" usability scale". In P. W. Jordan, B. Thomas, B. A. Weerdmeester, & A. L. McClelland. Usability Evaluation in Industry. London: Taylor and Francis.

Author: Emilia Ambrosini
Institute: NearLab, Department of Electronics, Information, and Bioengineering, Politecnico di Milano
Street: piazza Leonardo da Vinci 33
City: Milano
Country: Italy
Email: emilia.ambrosini@polimi.it

Evaluation of a Novel Modular Upper Limb Neuroprosthesis for Daily Life Support

S. Ferrante[1], E. Ambrosini[1], M. Rossini[2], F. Molteni[2], M. Bulgheroni[3], G. Ferrigno[1], and A. Pedrocchi[1]

[1] Politecnico di Milano, Department of Electronics, Information and Bioengineering, NearLab, Milan, Italy
[2] Valduce Hospital, Villa Beretta, Rehabilitation Center, Costamasnaga, LC, Italy
[3] Ab.Acus, Milano Italy

Abstract—The European Project "MUltimodal Neuroprosthesis for Daily Upper limb Support" (MUNDUS) aimed at the development of an assistive platform for recovering direct interaction capabilities of severely motor impaired people based on arm reaching and hand functions. Within this project the present study is focused on the evaluation of the MUNDUS prototype in terms of usability, user satisfaction and performance gained during daily life activities. Fourteen end-users were recruited: 8 individuals with Spinal Cord Injury, 1 with multiple sclerosis, 3 with Friedreich Ataxia, and 2 with Amyotrophic Lateral Sclerosis. Each subject used the MUNDUS system during a 3-day session performing 4 different tasks: drinking, touching the contralateral shoulder, and pressing a button. To assess the execution performance each task was divided into sub-actions and three experts evaluated over a 3-level score the functionality of the system for each sub-action. User satisfaction with the system was assessed using the Tele-healthcare Satisfaction Questionnaire – Wearable Technology (TSQ_WT) and the System Usability Scale (SUS). Furthermore, the Technology Acceptance Model questionnaire was adopted to assess the impact of the MUNDUS system on caregivers.

The system was always able to successfully support the end user in the accomplishment of the desired task. The MUNDUS system detected always perfectly the user intention during the tasks allowing a direct control of the system by the user. An overall high satisfaction was obtained: SUS median value was 85 (range 0-100) and TSQ-WT median value was 102 (range 0-120). All the caregivers agreed that usefulness of the MUNDUS system was very high. The general positive evaluation obtained by the MUNDUS prototype represents an important result for the potential future development of the system.

Keywords—Assistive technologies, neuromuscular electrical stimulation (NMES), upper limb, neurological disorders.

I. Introduction

People affected by severe diseases that led to a sudden or progressive loss of motor capabilities attribute a high value to the maintenance of a direct interaction with daily life objects. However, most of the assistive technologies (AT) currently available for people with severe motor impairments hardly surrogate the natural human interaction with daily life objects [1]. Restoring and augmenting human capabilities compensating for reduced motor functions and disabilities may be carried out by different approaches [1-4].

The European Project "MUltimodal Neuroprosthesis for Daily Upper limb Support" (MUNDUS) aimed at the development of an assistive platform for recovering direct interaction capabilities of severely motor impaired people based on arm reaching and hand functions (www.mundus-project.eu). The approach of the MUNDUS system is to maximally exploit any residual control of the end-users to allow them to directly control and drive the system during the selected tasks. MUNDUS offers a modular solution of sensors, actuators and controllers able to adapt to the level of severity or to the progression of the disease [5]. A passive exoskeleton provides gravity compensation; when needed, the movement execution can be also supported by neuromuscular electrical stimulation (NMES) applied both at the arm and hand levels [6-7]. The MUNDUS system was developed adopting a user-centered approach: the design process started with capturing the end-user needs through focus groups and interviews and the development and the optimization were continued until all the pre-defined requirements were fulfilled.

The aim of the present study was to evaluate the MUNDUS system in terms of performance gained during the selected interaction tasks, usability and user satisfaction.

II. Materials and Methods

A. Participants and Scenarios

MUNDUS target end-users were people affected by high level Spinal Cord Injury (SCI) and neurodegenerative diseases such as Amyotrophic Lateral Sclerosis (ALS), Friedreich Ataxia (FRDA), Multiple Sclerosis (MS).

The inclusion criteria were: age between 18 and 75 years; mini-mental state exam (MMSE) ≥ 23 to ensure that the subject was able to understand the interaction with the system and how to manage it; Modified Ashworth Scale ≤ 2 to ensure that there were no limitation to perform the task movement due to spasticity response of the muscles; absence of fixed contraction.

To assess eligibility the NMES responsiveness was also tested in order to assure that subjects were able to tolerate NMES and that functional movements could be induced.

Three different user scenarios were identified according to the residual ability of the subjects: in Scenario 1 subjects had a residual functional control of the arm and hand muscles but were too weak to accomplish repeatedly functional tasks; in Scenario 2 subjects had no residual functional activation of arm / hand muscles but could still control the head and gaze fixation; in Scenario 3 subjects could only interact through brain signals (the sight was preserved). An initial clinical evaluation of each subject based on the motricity index of the upper limb, and the Medical Research Council Scale (MRC) was performed to select the scenario and the MUNDUS configuration to use.

B. Experimental Setup

Fig. 1 shows the modularity of the MUNDUS system. To simplify the control of the movement and to optimize the interaction between the user and the system, each task was divided in sub-actions and the triggering of most of the sub-actions was directly controlled by the user. Different modules, depending on the end-user residual abilities, were used to detect the user intention and to allow him/her to directly drive the system during the selected task:

a. for Scenario 1: a USB button or an EMG signal coming from a muscle that the subject was able to control;
b. for Scenario 2: an Eye-tracking system;
c. for Scenario 3: Brain Computer Interface (BCI).

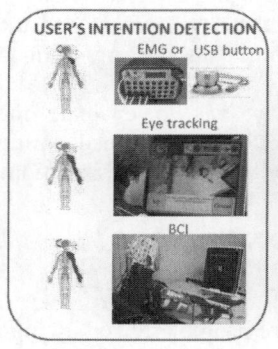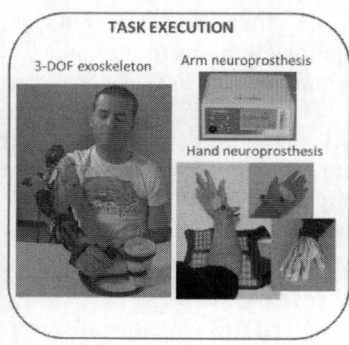

Fig. 1 The MUNDUS system modularity both for the user intention detection and for the task execution.

To support the arm and hand functions, the following modules were used:

a. for Scenario 1: The exoskeleton combined or not with myo-controlled NMES of arm muscles, and NMES to support hand functions if needed [6];
b. for Scenario 2 and 3: the exoskeleton with NMES of arm and hand muscles, depending on the user needs.

C. Experimental Procedure

Donning of the system and calibration - During this phase the following steps were performed:

a. configuration of the system (modules selection);
b. transfer of the subject in the wheelchair where the device was installed;
c. subject preparation, following the instruction of the graphical user interface (exo and sensors donning phase);
d. calibration of the required modules.

Movement execution - The following tasks were performed: drinking, touching the contralateral shoulder, touching the contralateral hand; pressing a button. Each movement was repeated 10 times, waiting about 30 s between consecutive repetitions. To assess the execution performance each task was divided into reaching and grasping sub-actions. The level of support provided by MUNDUS was evaluated for each sub-action with a 3-level score: 0, unsuccessful; 1, acceptable; 2, completely functional. The scores were agreed by 3 experts, one present at the experimental session and two analyzing the data acquired by the MUNDUS sensors and the videos.

De-briefing - To evaluate the user satisfaction with the system, two validated instruments were used: the "Tele-healthcare Satisfaction Questionnaire – Wearable Technology" (TSQ-WT; © RBMF 2010) and the System Usability Scale (SUS; © Digital Equipment Corporation) [8]. The TSQ-WT contains six subscales (benefit, usability, self-concept, privacy and loss of control, quality of life, and wearing comfort) that evaluate the satisfaction of the subject with the wearable part of the system. Each subscales includes 5 questions rated on a 5-point Likert scale between 0 (I strongly disagree with the statement) and 4 (I strongly agree with the statement). The final score, given separately for each subscale, ranges between 0 (no satisfaction) and 20 (extreme satisfaction). The SUS includes 10 questions rated by means of a five-point Likert scale, with 1 (I strongly disagree) to 5 (I strongly agree). The final score ranges between 0 (no satisfaction) and 100 (extreme satisfaction).

The Technology Acceptance Model (TAM) questionnaire was used to assess the impact of the system on people that can be involved in the preparation phase (i.e. formal and informal carers) [9]. The TAM questionnaire includes 44 items grouped in 16 subscales. A 7-point Likert scale ranging from 1 (I strongly disagree) to 7 (I strongly agree) is used to score each item. The results on each subscale are normalized between 0 (no satisfaction) and 100 (extreme satisfaction). The TAM was administered to 6 people involved as operators during the trials: 2 technicians (bioengineers), 2 therapists, and 2 caregivers.

Tests were performed at the Villa Beretta Rehabilitation Center and the study was approved by the Ethical Committee of the Valduce Hospital.

III. RESULTS AND DISCUSSION

Fourteen end-users were recruited: 8 SCI people, 1 affected by MS, 3 affected by FRDA, and 2 with ALS. 12 subjects resulted eligible and among them one dropped out before the end of the evaluation. The characteristics of the recruited subjects are reported in Table 1.

Table 1 End-users characteristics and clinical evaluations

Subject	sex	age	Pathology	MI right arm (0-100)	MRC (1-5)			
					Elbow extension	Elbow flexion	Finger extension	Finger flexion
PD001	M	50	SCI	30	2	2	3	3
VG002	F	29	FRDA	76	3	3	3	3
ES003	F	58	ALS	-	-	-	-	-
LR004	M	69	ALS	77	4	4	4	4
FL005	M	70	SCI	39	3	3	1	1
SB006	M	50	FRDA	71	4	3	3	3
GC007	M	33	SCI	40	2	2	1	1
BS008	M	71	SCI	60	3	3	2	2
GT009	M	23	FRDA	57	3	3	2	2
CR010	M	61	SCI	58	3	3	2	2
EL011	F	28	SCI	-	-	-	-	-
AN012	M	45	MS	77	4	4	4	4
GI013	M	57	SCI	40	3	4	2	0
MB014	M	27	SCI	84	4	4	4	4

Table 2 shows for the recruited subjects the scenario and the modules used in the experimental session.

Table 2 Scenarios and modules used by each end-user

Subject	Scenario	exo	environ-mental sensors	RFID	arm NMES EMG-based controller	arm NMES feedback controller	hand NMES	hand sensorized glove	intention detection module eye tracking	intention detection module BCI	intention detection module EMG/usb button
PD001	2	x	x	x			x	x	x		
VG002	1	x									x
ES003	not eligible										
LR004	2	x	x	x			x		x		
FL005	1	x			x						x
SB006	1	x									x
GC007	3	x	x	x			x			x	
BS008	1	x									x
GT009	drop out										
CR010	1	x									x
EL011	not eligible										
AN012	1	x			x						x
GI013	3	x	x	x			x			x	
MB014	1	x									x

The preparation time of the system strongly depended on the complexity of the configuration used. When Scenario 1 was tested using only the exoskeleton, 10 minutes were enough to don on and calibrate the system. Instead, when the most complex configuration was used (Scenario 3) a mean preparation time of about 40 minutes was required. The evaluation scores agreed by three experts and assigned for each sub-action of the performed interaction tasks are reported in Table 3. Considering all the evaluation obtained among subjects and sub-actions a median value of 2 (interquartile range of 0) was obtained. All users were perfectly able to control the system by means of the selected module for user intention detection (mean evaluation = 2).

Table 3 Evaluation of the MUNDUS functionality (-: Not Assisted by MUNDUS; 0: unsuccessful; 1: acceptable; 2: completely functional)

User	n. sessions	Task	Total n. of repetitions	Movement performance				User interaction performance
				reaching	grasping	coming back	releasing	
PD001	3	drinking	3	1	NA	2	NA	2
		reaching button	2	2	-	-	-	2
		reaching shoulder	3	2	-	2	-	2
		reaching other hand	0	-	-	-	-	-
VG002	3	drinking	10	2	2	2	2	2
		reaching button	10	2	2	2	2	2
		reaching shoulder	10	2	2	2	2	2
		reaching other hand	10	2	2	2	2	2
LR004	3	drinking	0	-	-	-	-	-
		reaching button	5	1	-	1	-	2
		reaching shoulder	2	1	-	1	-	2
		reaching other hand	2	1	-	1	-	2
FL005	3	drinking	10	2	2	2	2	2
		reaching button	10	2	2	2	2	2
		reaching shoulder	10	2	2	2	2	2
		reaching other hand	10	2	2	2	2	2
SB006	3	drinking	10	2	2	2	2	2
		reaching button	10	2	2	2	2	2
		reaching shoulder	10	2	2	2	2	2
		reaching other hand	10	2	2	2	2	2
GC007	3	drinking	0	-	-	-	-	-
		reaching button	4	2	-	2	-	2
		reaching shoulder	6	2	-	2	-	2
		reaching other hand	8	2	-	2	-	2
BS008	3	drinking	10	2	2	2	2	2
		reaching button	10	2	2	2	2	2
		reaching shoulder	10	2	2	2	2	2
		reaching other hand	10	2	2	2	2	2
CR010	3	drinking	10	2	2	2	2	2
		reaching button	10	2	2	2	2	2
		reaching shoulder	10	2	2	2	2	2
		reaching other hand	10	2	2	2	2	2
AN012	3	drinking	10	2	2	2	2	2
		reaching button	10	2	2	2	2	2
		reaching shoulder	10	2	2	2	2	2
		reaching other hand	10	2	2	2	2	2
GI013	1	drinking	1	1	1	1	1	2
		reaching button	1	2	-	1	-	2
		reaching shoulder	1	2	-	1	-	2
		reaching other hand	1	2	-	1	-	2
MB014	3	drinking	10	2	2	2	2	2
		reaching button	10	2	2	2	2	2
		reaching shoulder	10	2	2	2	2	2
		reaching other hand	10	2	2	2	2	2
N			319	41	29	40	29	41
median			10	2	2	2	2	2
iterquartile range			7	0	0	0	0	0

Fig. 2 shows the results of the TSQ-WT questionnaire in terms of median values and interquartile ranges. The satisfaction within most of the subscales was high. Only the "wearing comfort" performed poorly, with a satisfaction value of about 55%.

A median value of 85 (interquartile range of 20, min-max: 43-98) was found for the SUS score indicating a high level of usability of the MUNDUS system perceived by the end-users. Looking more specifically at the single questions of the SUS questionnaire we observed that the encumbrance of the system was not appreciated by the users, while the system usefulness and usability are recognized by most of the users.

Results of the TAM questionnaire are shown in Fig. 3. All the interviewed categories perceived that the MUNDUS system was useful but, except for the technicians, all the others perceived that easiness of use was moderate; this feedback perfectly correlated to their computer anxiety.

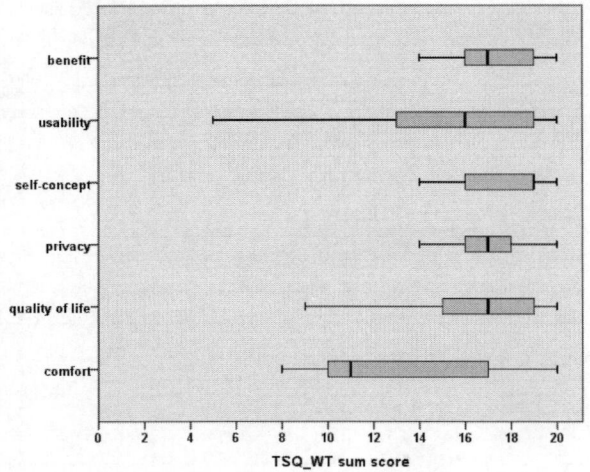

Fig. 2 - Results of the TSQ_WT questionnaire (0=worst, 20=best).

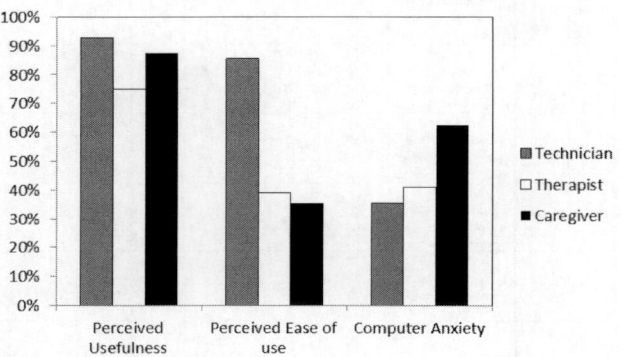

Fig. 3 - Results of the TAM questionnaire. Each column represents the mean of the 2 persons belonging to each group.

IV. CONCLUSION

The present work deals with the evaluation of a novel platform to support arm and hand functions during daily life activities. The evaluation was performed in terms of performance gained during the selected interaction tasks, usability and user satisfaction and showed encouraging results. The system effectively supported the users in performing daily life activities. For all the end-users tested, the use of MUNDUS was a renewed experience of using their arm to provide a function thanks to their own muscles and following their own intention. This experience was positive, and they did not mind about the slowness of the overall execution and the length of the preparation time. This feedback confirms that an increased autonomy in daily life activity is strongly required by the MUNDUS target population and very positively impacts over social inclusion, self-perception and independence. The general positive evaluation obtained by the MUNDUS prototype represents an important result for the future development of the system.

ACKNOWLEDGMENT

This work was supported by the European Project MUNDUS (FP7 ICT 2009-4; grant agreement no.: 248326).

REFERENCES

1. Brose SW, Weber DJ, Salatin BA, et al. (2010) The role of assistive robotics in the lives of persons with disability. Am J Phys Med Rehabil 89:509-521.
2. Casellato C, Pedrocchi A, Zorzi G et al (2012) Error-enhancing robot therapy to induce motor control improvement in childhood onset primary dystonia. J Neuroeng Rehabil 9:46.
3. Casellato C, Pedrocchi A, Zorzi G et al (2012) EMG-based visual haptic biofeedback: a tool to improve motor control in children with primary dystonia. IEEE T Neur Sys Reh. Oct 5.
4. S. Ferrante, E. Ambrosini, P. Ravelli, et al (2011) A biofeedback cycling training to improve locomotion: a case series study based on gait pattern classification of 153 chronic stroke patients, J. Neuroengineering Rehabil. 8: 47.
5. A. Pedrocchi, S. Ferrante, E. Ambrosini et al (2013) MUNDUS project: MUltimodal Neuroprosthesis for daily Upper limb Support, J. Neuroengineering Rehabil. 10:66, DOI: 10.1186/1743-0003-10-66.
6. Ambrosini E, Ferrante S, Tibiletti M, et al. (2011) An EMG-controlled neuroprosthesis for daily upper limb support: a preliminary study. IEEE Eng Med Biol Soc Proc, Boston, USA, 4259–4262
7. Ferrante S, Ambrosini E, Ferrigno G, et al (2012) Biomimetic NMES controller for arm movements supported by a passive exoskeleton. Conf Proc IEEE Eng Med Biol Soc. 1888-91
8. Brooke, J. (1996). SUS: a "quick and dirty" usability scale, In P. W. Jordan, B. Thomas, B. A. Weerdmeester, & A. L. McClelland. Usability Evaluation in Industry. London: Taylor and Francis.
9. Venkatesh V and Bala H (2008) Technology Acceptance Model 3 and a Research Agenda on Interventions, Decision Sciences 39:2,273-315.

Corresponding author:

Author: Simona Ferrante
Institute: Politecnico di Milano, Department of Electronics, Information and Bioengineering, Neuroengineering and Medical Robotics Laboratory
Street: via Giuseppe Colombo 40, 20133
City: Milano
Country: Italy
Email: simona.ferrante@polimi.it

Evaluating Spatial Characteristics of Upper-Limb Movements from EMG Signals

O. Urra[1,2], A. Casals[1,2], and R. Jané[1,2,3]

[1] Institut de Bioenginyeria de Catalunya (IBEC), Barcelona, Spain
[2] Universitat Politècnica de Catalunya. BarcelonaTech (UPC), Barcelona, Spain
[3] CIBER de Bioingeniería, Biomateriales y Nanomedicina (CIBER-BBN), Spain

Abstract—Stroke is a major cause of disability, usually causing hemiplegic damage on the motor abilities of the patient. Stroke rehabilitation seeks restoring normal motion on the affected limb. However, 'normality' of movements is usually assessed by clinical and functional tests, without considering how the motor system responds to therapy. We hypothesized that electromyographic (EMG) recordings could provide useful information for evaluating the outcome of rehabilitation from a neuromuscular perspective. Four healthy subjects were asked to perform 14 different functional movements simulating the action of reaching over a table. Each movement was defined according to the starting and target positions that the subject had to connect using linear trajectories. Bipolar recordings of EMG signals were taken from biceps and triceps muscles, and spectral and temporal characteristics were extracted for each movement. Using pattern recognition techniques we found that only two EMG channels were sufficient to accurately determine the spatial characteristics of motor activity: movement direction, length and execution zone. Our results suggest that muscles may fire in a patterned way depending on the specific characteristics of the movement and that EMG signals may codify such detailed information. These findings may be of great value to quantitatively assess post-stroke rehabilitation and to compare the neuromuscular activity of the affected and unaffected limbs, from a physiological perspective. Furthermore, disturbed movements could be characterized in terms of the muscle function to identify, which is the spatial characteristic that fails, e.g. movement direction, and guide personalized rehabilitation to enhance the training of such characteristic.

Keywords—EMG, movement spatial characteristics, pattern recognition, stroke rehabilitation, upper-limb.

I. INTRODUCTION

Stroke is considered a leading cause of disability [1], mostly related to upper limb impairment, since the recovery is slow and often limited [2], [3]. Quantitative evaluation of motor functions is essential to determine therapeutic efficiency and guide personalized rehabilitation. However, current available methods are subjective and do not reveal the behavior of the motor system during rehabilitation [4].

Movement execution entails symmetric processes of the two upper limbs [5]. Therefore, the impairment (or recovery) level of a patient could be assessed by comparing the motor function of the healthy and impaired limbs. Given that the EMG patterns of contractions producing different movements are distinct [6], we hypothesized that EMG signals may provide useful parameters to characterize motor improvement.

To prove such hypothesis we defined 14 different movements simulating reaching, since it is an important motor component of the daily life activities, and usually one of the basis of post-stroke rehabilitation [7]; and we classified them according to their spatial characteristics (direction, length and execution zone) using pattern recognition techniques with a set of EMG features.

II. METHODS

A. Data Acquisition

Four healthy subjects (three males, one female) participated in this study. All subjects were 27-35 years old and right handed. Electromyographic (EMG) signals were recorded using two pairs of circular disposable Ag-AgCl electrodes (1 cm in diameter, 1.5 cm inter-electrode distance; Foam electrode 50/PK – EL501, Biopac Systems Inc.) from the biceps brachii (C1) and triceps brachii (C2) of the right arm according to published guidelines [8]. A fifth reference electrode was placed at the wrist. EMG data were collected through the EMG 100C acquisition system (BIOPAC Systems, Inc.) at a sampling rate of 1000 Hz and a gain of 500.

B. Experimental Protocol

Subjects sat down in front of a table were a white grid indicated different target positions (Fig 1). The height of the chair was adjusted so that the elbow described an angle of 90º when placing the forearm on the table. The subjects were instructed to perform 14 different functional movements with their right hand by following linear trajectories. Each movement simulates the action of reaching and entails the proximal and distal upper-limb muscles, which are essential to perform daily life activities. Each subject repeated each movement 30 times. So at all, 420 EMG recordings were analyzed for each subject. All contractions began or ended with the subject's arm by the side in a comfortable neutral position. No constraints were imposed on the force or velocity of the contraction, except that the subject was asked not to drag the arm along the table and to be consistent in reproducing the linear trajectory specified in each case. EMG signals were manually segmented from movement onset (simultaneous with the auditory trigger) to the instant the limb-movement stopped after reaching the target position. Subjects were allowed to rest for 2-3 min between movements to avoid muscular and mental fatigue.

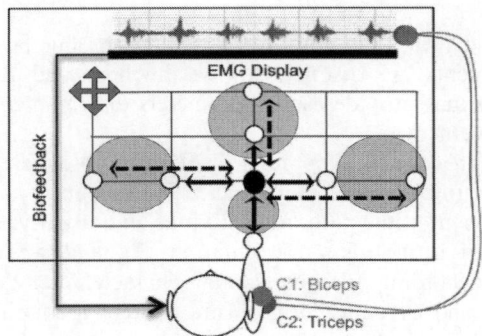

Fig. 1 **Experimental protocol.** Subjects were instructed to perform 14 functional movements with their right arm simulating reaching on a table. Movements were defined on a grid indicating the positions to reach (circles) by linear back and forth trajectories (arrows) starting and ending in the neutral position (black circle). EMG features from biceps and triceps were used to identify patterns of movement direction (red arrows), length (dotted vs continuous arrows) and execution zone (red circles).

C. Data Preprocessing and Feature Extraction

EMG signals were digitally filtered with fourth-order bi-directional Butterworth highpass (20 Hz) and lowpass (400 Hz) filters to remove movement artifacts and the high-frequency noise. A Notch filter was used to remove 50 Hz interference and mean was also subtracted. A third differential channel, C3, was offline computed by subtracting the signals of channels 1 and 2.

The feature set used to characterize the EMG data for movement classification was composed of the total power spectral density (PSD), mean PSD, mean and median frequency, root mean square (RMS), RMS ratio, signal variance and the feature set proposed by Hudgins *et al.* [6]. PSD was estimated using Burg's method; RMS ratio was defined as the ratio between the maximum absolute value and the RMS of the signal; the Hudgins' feature set was computed in the first 5 segments of 50 ms of each contraction. At all, 37 features were computed in each channel (C1: Biceps brachii, C2: Triceps brachii, C3: C1-C2). Additionally, estimations of Magnitude Squared Coherence (MSC) and cross-PSD (cPSD) between channel 1 (biceps brachii) and channel 2 (triceps brachii) were included. MSC indicates how well two signals correspond to each frequency.

D. Classification

Contractions were labeled according to their directions (ToUp, ToRight, ToDown, ToLeft); length (Long, Short); and the execution zone where they were performed in respect to the neutral position (Up, Right, Down, Left) (Fig 1). That is, each contraction was given three labels to allow applying pattern recognition techniques to the spatial characteristics of the 14 movement types separately.

Support vector machines (SVMs) were implemented using radial basis function (RBF) kernels [9] to classify contractions according to each of their three labels. The selection of gamma and C-parameters of the RBF was optimized to obtain maximum classification accuracy. Then, a 5-fold cross-validation was carried out with the optimized classifier. The accuracy of each classification was the percentage of correctly classified contractions. For each subject the performance was computed as the averaged accuracy of the five folds. The overall performance was evaluated as the mean accuracy across all subjects.

III. RESULTS

A. Qualitative Characterization of Muscle Activity Patterns

Fig. 2 shows the EMG recordings of the biceps brachii and triceps brachii muscles while performing the 7 pairs of back and forth movements. The EMG signal and activation pattern shape differs with each movement type. Similarly, the temporal coordination between the two muscles was different. For example, when performing movements upwards (plot 3), triceps brachii is activated prior to the recruitment of muscle fibers on biceps brachii. Conversely, for downward movements (plot 7) the activation of biceps muscle occurs before activity is detected on the triceps.

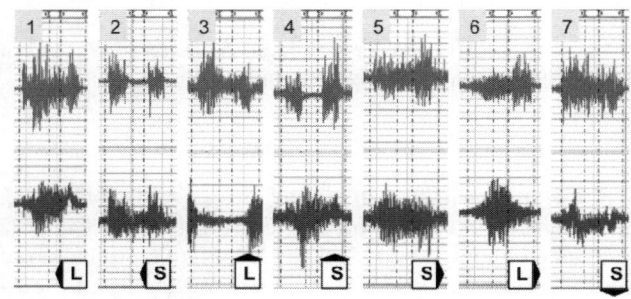

Fig. 2 Examples of the EMG signal for the seven pairs of movements (Subject 4). Red signals: biceps brachii, Blue signals: triceps brachii. L: Long Movements, S: Short Movements. Black arrows indicate the direction of the movement. Black dotted lines indicate movement onset, red dotted lines indicate movement end.

B. Model Selection

Akaike's Information Criterion (AIC) was used to select the most relevant EMG features determining a particular spatial characteristic of the movement (direction, length or execution zone). Feature selection was carried out iteratively, so that in the first step the AIC of the model built with the full set of features was calculated. In the next iteration a feature was left out and if the AIC of the reduced model diminished, then the feature was discarded. This process was repeated for every feature until the most informative (reduced) feature set was found for each movement characteristic. When comparing the three reduced feature sets we found that there was not a consensus on the features able to discriminate simultaneously the three spatial characteristics. Therefore, characterizing completely each contraction requires the complete feature set.

C. Detection of Movement Spatial Characteristics

Pattern recognition analysis revealed that the feature set and the proposed SVM classifier were able to accurately predict the direction, length and execution zone of the movement from EMG. Overall classification accuracies were 84.27%, 93.80% and 87.93%, respectively.

Table 1 Subject-specific classification accuracy of movement spatial characteristics.

Spatial Characteristics	Subject #			
	S1	S2	S3	S4
Movement Direction				
ToUp	0.96	0.89	0.90	0.95
ToRight	0.85	0.87	0.65	0.93
ToDown	0.84	0.81	0.68	0.95
ToLeft	0.85	0.87	0.71	0.94
Movement Length				
Short	0.98	0.96	0.86	0.98
Long	0.95	0.95	0.88	0.96
Execution Zone				
Up	0.89	0.92	0.82	0.98
Right	0.90	0.84	0.75	1.00
Down	0.97	0.92	0.81	0.95
Left	0.96	0.94	0.80	0.96

Table I contains subject-specific classification accuracies for the categories of the three movement characteristics. In general, we were able to classify the categories within each movement characteristic with accuracies above 85%, except for subject 3 for which classification performances were especially low comparing to the rest of the subjects. We did not find any category that was especially easy/tough to characterize: the patterns we found allowed recognizing specific categories with similar accuracy.

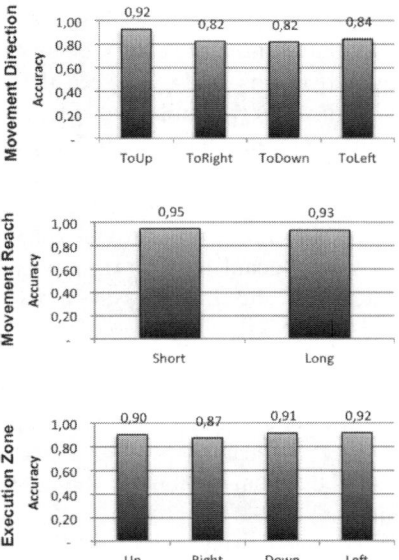

Fig. 3 Overall class-by-class classification accuracy of movement spatial characteristics from EMG

The accuracy with which upward movements (ToUp) were detected was significantly higher (92%) than other movement directions (82-84%) (Fig.3). In contrast, we identified different categories from movement length and execution zone with similar accuracy: Short and long movements were correctly characterized in 95% and 93% of the cases, while movements performed in the space located Up, Right, Down or Left of the neutral positions were in 90%, 87%, 91% and 92% of the cases respectively.

IV. DISCUSSION

This study demonstrates that EMG signal codifies spatial characteristics of movements that can be detected with high accuracy applying pattern recognition techniques. We have successfully applied SVMs to decode the direction, length and execution zone of 14 different planar movements that simulate reaching. At present, the outcome of rehabilitation is assessed using subjective and time consuming functional tests and kinematic analysis of movements [10], [11]. However a recent study showed that kinematic parameters do not reveal the behavior of the motor system, during rehabilitation [12]. Thus, this information may be very useful to assess rehabilitation from the perspective of the motor system and provide further understanding on the neuromuscular mechanisms behind functional recovery.

Quantitative evaluation of the rehabilitation is necessary to estimate therapeutic efficiency. Given that muscle activity is similar on both sides of upper limbs when performing the same movement [5], one of the objectives of our future research is confirming that the EMG features from right and left arms reflect such similarity. Thus, comparing the EMG features proposed in this study, between the impaired and healthy limb in hemiplegic stroke patients, could be used as a measure of "normal" motion. Besides, in the clinical practice the use of EMG is limited to analyze muscle activation patterns and evaluate muscle strength [13]. However, our study shows that EMG may provide much more information about spatial characteristics of movements, such as direction or length. Therefore, our system may enhance the potential of EMG and simplify the complementary instrumentation currently used to evaluate the kinematic aspects of movements. Furthermore, given that these parameters indicate specific motor skills, they may guide custom rehabilitation by indicating which aspects of motor recovery (e.g. movement direction) should be reinforced with rehabilitation.

Similar studies have been carried out using high density EMG (> 60 channels) [14]–[16]. Our approach is especially attractive because it achieves high accuracies using only 2 EMG channels, which makes the system easy and confortable to implement. Furthermore, unlike these studies, our system is able to discriminate the spatial characteristics from highly related movements that involve the same set of joints. Thus it is likely that the accuracy of the system will increase substantially when performing movements that combine joints and muscles differently. In addition, no averaging is needed for movement characterization, which allows assessing each contraction individually.

V. CONCLUSIONS

This study demonstrates that EMG codifies spatial characteristics of movements, such as direction, length or the execution zone. The set of EMG features proposed allows obtaining high accuracies in the spatial characterization of individual contractions corresponding to 14 different reaching movements by pattern recognition techniques. Thus, we provide a new framework that may be used to quantify the outcome of stroke rehabilitation directly from the perspective of motor system and guide personalized therapy to reinforce recovery on individual motor problems.

ACKNOWLEDGMENT

This work was supported by a grant of the Basque Government, (Programa de Formación de Investigadores del DEUI), the Spanish Ministry of Economy and Competitiveness under grant TEC2010-21703-C03-01 and developed in the frame of the HYPER project (grant CSD2009-00067 CONSOLIDER INGENIO 2010).

REFERENCES

1. Strong K, Mathers C, Bonita R. (2007) Preventing stroke: saving lives around the world. Lancet Neurol 6 (2):182–7
2. Nakayama H, Jørgensen HS, Raaschou HO et al. (1994) Recovery of upper extremity function in stroke patients: the Copenhagen Stroke Study., Arch Phys Med Rehabil 75 (4):394–8
3. Skilbeck CE, Wade DT, Hewer RL et al., (1983) Recovery after stroke. J Neurol Neurosurg Psychiatry 46(1):5–8
4. Hu XL, Tong KY, Song R et al. (2009) Quantitative evaluation of motor functional recovery process in chronic stroke patients during robot-assisted wrist training. J Electromyogr Kinesiol 19(4):639-50
5. Lum PS, Burgar CG, Kenney DE et al. (1999) Quantification of force abnormalities during passive and active-assisted upper-limb reaching movements in post-stroke hemiparesis. IEEE Trans Biomed Eng 46(6):652-62
6. Hudgins B, Parker P, Scott RN. (1993) A new strategy for multifunction myoelectric control. IEEE Trans Biomed Eng 40(1):82-94
7. Chang J, Yang Y, Guo L et al. (2008) Differences in Reaching Performance Between Normal Adults and Patients Post Stroke-A Kinematic Analysis. J Med Biol Eng 28(1):53-8
8. Stegeman DF, Hermens HJ. Standards for surface electromyography: the European project Surface EMG for non-invasive assessment of muscles (SENIAM). Available online: http://www.med.uni-jena.de/motorik/pdf/stegeman.pdf (accessed on 21 March 2013).
9. Chang C, Lin C. (2011) LIBSVM: a library for support vector machines. ACM Trans. Intell. Syst. Technol. 2(3)
10. Massie CL, Malcolm MP, Greene DP et al. (2013). Kinematic Motion Analysis and Muscle Activation Patterns of Continuous Reaching in Survivors of Stroke Continuous Reaching in Survivors of Stroke. J Mot Behav 44(3):213-22
11. Fugl-Meyer AR, Jääskö L, Leyman I et al. (1975)..The post-stroke hemiplegic patient. 1. a method for evaluation of physical performance. Scand J Rehabil Med 7(1):13-31
12. Rohrer B, Fasoli S, Krebs HI et al. (2002). Movement smoothness changes during stroke recovery. J Neurosci 22(18):8297–304
13. Wong L, Xie Q, Ji L. et al. (2011) Upper Limb Contralateral Physiological Characteristic Evaluation for Robot-Assisted Post Stroke Hemiplegic Rehabilitation. Universal Access in Human-Computer Interaction. Applications and Services Proc. vol.IV - 6th International Conference, UAHCI, Orlando, USA, 2011, pp. 458–63.
14. Rojas-Martínez M, Mañanas M, Alonso JF. (2012) High-density surface EMG maps from upper-arm and forearm muscles. J Neuroeng Rehab 9(1):85
15. Zhang X, Zhou P. (2012) High-Density Myoelectric Pattern Recognition. IEEE Trans Biomed Eng 59(6):1649-57
16. Zhou P, Suresh NL, Rymer WZ. (2011) Surface electromyogram analysis of the direction of isometric torque generation by the first dorsal interosseous muscle. J Neural Eng 8(3)

Author: Oiane Urra
Institute: Institut de Bioenginyeria de Catalunya d'(IBEC)
Street: Baldiri Reixac, 4, Torre I
City: Barcelona
Country: Spain
Email: ourra@ibecbarcelona.eu

An Adaptive Real-Time Algorithm to Detect Gait Events Using Inertial Sensors

N. Chia Bejarano[1], E. Ambrosini[1], A. Pedrocchi[1], G. Ferrigno[1], M. Monticone[2], and S. Ferrante[1]

[1] Politecnico di Milano, Department of Electronics, Information and Bioengineering, NearLab, Milan, Italy
[2] Physical Medicine and Rehabilitation Unit, Scientific Institute of Lissone, Salvatore Maugeri Foundation,
Institute of Care and Research, IRCCS, Lissone, Monza Brianza, Italy

Abstract—This study aimed at developing an adaptive algorithm to detect in real time temporal gait events, based on data acquired from inertial and magnetic measurement units.

Trials on 9 healthy subjects were performed to select the best body locations for the sensors out of 8 different possibilities, trying to optimize system portability, data inter-variability and real-time algorithm simplicity. Subjects walked over the GaitRite mat at different self-selected speeds: normal, fast, and slow. Results showed a significantly low variability ($p<0.05$) of the shank angular velocity in the sagittal plane, reducing the number of sensors required for the real-time algorithm to two (the ones placed on the shanks).

The detection of the Initial Contact (IC) and the End Contact (EC) was based on the shank angular velocity and flexion/extension angle. The gait events were identified as local minima on the sagittal-plane angular velocity. Features extracted from the signals of the previous steps were used to improve the events localization. These features were self-calibrated at the beginning of the trial and updated every step.

The algorithm was validated against the GaitRite system and was compared to two other real-time algorithms available in the literature to assess its reliability and performance. F1-scores of 0.9987 for IC and 0.9996 for EC were obtained. Our algorithm detected the gait events with a mean (SD) delay of 68.6 (15.1) ms for IC and 7.8 (23.6) ms for EC, with respect to the GaitRite, for the self-selected normal speed. These values were significantly lower than those obtained by other published algorithms.

Results indicated that the system is suitable for real-time gait monitoring, assessment and ambulatory rehabilitation, based on biofeedback or neuroprostheses.

Keywords—**Gait temporal parameters, Real-time processing, IMMS, Ambulatory system.**

I. Introduction

Gait and balance disorders are common in older adults and neurological patients. Thus, the detection of temporal gait parameters is crucial for monitoring or assessment purposes [1]. A real-time detection of gait temporal parameters, such as the initial contact (IC) and end contact (EC), is also needed to develop goal-oriented rehabilitation treatments based on biofeedback [2] or on Functional Electrical Stimulation (FES) [3,4].

Different sensors can be used to provide real-time information in ambulatory clinical settings. Body-mounted Inertial and Magnetic Measurement Systems (IMMS), combining data from tri-axial gyroscopes, accelerometers and magnetometers, have arisen as the optimal solution [1]. IMMS are small, light, easy to don and doff and, thanks to data fusion, they provide drift-free angles, angular velocities and linear accelerations with minimal latency. Accelerometers and gyroscopes have been widely used, but their output is strongly influenced by drift and, in the case of accelerometers, also by gravity and heel-strike vibrations [1]. Footswitches have also been used, especially to control FES. Their processing is very simple, but they do not provide as much gait information and they have been reported to cause discomfort, to be prone to mechanical failure and to be unreliable when used with patients with drop-foot or shuffling-feet [5].

Several algorithms have been proposed for gait assessment. Generally wavelet analysis [6] and low-pass filters have been applied to the signals to prepare it for derivatives, integration [7] or detection of peaks correlated to the desired event [6–8]. Most of the proposed offline algorithms have provided good reliability, but they're unsuitable for real-time applications. On the other hand, most of the proposed real-time algorithms are tested on a restricted sample, lack adaptability and introduce detection delays due to the processing.

The present study proposes a real-time, adaptive algorithm to provide accurate gait-event detection. To reach this aim, sensor placement was optimized, trying to minimize the number of sensors, the subject inter-variability and the software complexity. The developed algorithm was validated against the gold standard and compared to other real-time methods proposed in the literature.

II. Algorithm Design

A. Data Collection

Nine healthy subjects participated in the experiments (8 women and 1 man; age: 27.1 ± 4.4 years; height: 169.0 ± 8.8 cm; weight: 56.3 ± 10.3 kg). Eight IMMS (MTx sensors from Xsens technologies B.V., Netherlands) were placed over the sternum, S1 vertebra, mid-point of the external part of both thighs and calves, and on the insteps. They were fixed using Velcro and double-sided adhesive tape, minimizing motion artifacts. The sensors, sampled at 50 Hz, were synchronized with the GaitRite System (CIR Systems Inc., United States), sampled at 120 Hz.

Before starting the trials, the subjects were asked to keep an upright position in order to perform the initial coordinate alignment calibration of the Xsens system, according to [9]. Then, they were asked to walk over the GaitRite mat, at three different self-selected speeds: normal, fast and slow. Each condition was repeated 12 times.

B. Sensor Selection

The choice of the sensors used to design the algorithm was a trade-off between system portability, low data inter-variability and real-time algorithm simplicity.

To analyze the inter-variability, the acquired signals were separated into cycles using the IC provided by the GaitRite system, and then normalized in time and amplitude. For each sensor, only data related to the plane of the movement and the line of progression was considered, i.e. anterior-posterior and vertical acceleration, sagittal-plane angular velocity and flexion/extension angle. These four signals were analyzed in terms of correlation with gait events, and the computational load required to process them in real time. Thus, the accelerations were discarded due to the high number of oscillations, caused by noise and vibrations. For the two remaining signals, the root mean square error (RMSE) between each cycle and the average cycle of the rest of the subjects was computed. Data resulted not normally distributed, thus a non-parametric Kruskal-Wallis test ($p<0.05$) was performed to compare the RMSE obtained for the eight sensors and two signals. Dunn-Sidak post-hoc tests were performed ($p<0.05$) to determine which pairs of effects were significantly different. Median and interquartile ranges of the RMSE are shown on Fig. 1.

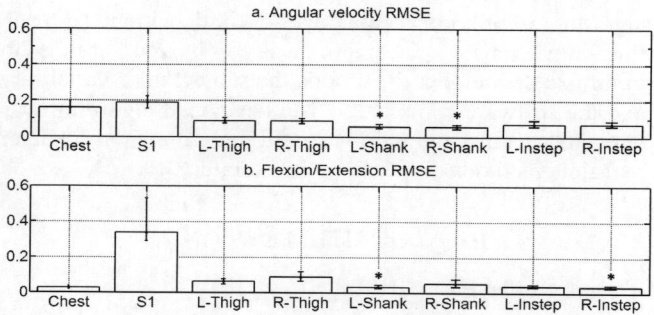

Fig. 1 Inter-variability of the angular velocity (a) and the Flexion/extension angle (b), in terms of normalized RMSE. L: Left; R: Right. * indicates that the sensor has a RMSE significantly lower ($p<0.05$) than all other sensors.

The sensors placed on the shanks were selected since they were characterized by the significantly lowest values of RMSE considering both the angular velocity and the flexion/extension angle. In addition, the shank angular velocity is highly correlated with the IC and EC [8], so the algorithm simplicity is also guaranteed. Thus, the algorithm was designed using, for each leg, the shank angular velocity and the shank flexion/extension angle.

C. Algorithm Description

An adaptive algorithm was designed to detect in real time the instants of the initial and end contact. As shown in Fig. 2, IC and EC were defined as two negative minima on the sagittal-plane angular velocity of the shank, as suggested by Lee et al. [8].

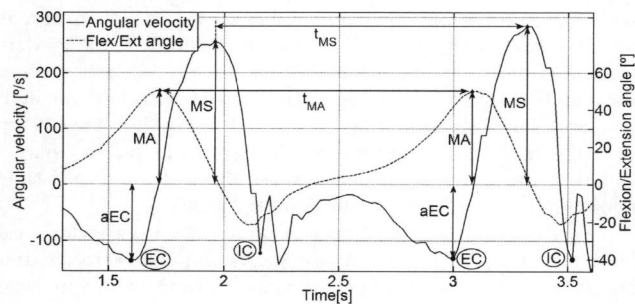

Fig. 2 Angular velocity and the flexion/extension angle of the shank used for event detection. Initial and End Contact as well as features extracted from the signals to optimize the localization of the events are highlighted.

The algorithm comprised four steps: signal conditioning, features initialization, peak detection and features update. The signal conditioning was a zero-delay, first-order FIR filter designed to reduce the noise.

The following features were defined (see Fig. 2): the peak of the flex/ext angle (Max-Angle, MA), the peak of the angular velocity (Mid-Swing, MS), their respective periods (t_{MA} and t_{MS}), and the value of the angular velocity at EC (aEC). These features were initialized during the first 3 steps of each leg and used to assure the robustness of the real-time event detection. IC was detected as the instant correspondent to the first negative minimum of the angular velocity after MS, within the 30% of t_{MS}. EC was detected as the instant correspondent to the minimum value of the angular velocity similar to aEC after the flex/ext angle reached the half of MA. After the event detection, the algorithm kept analyzing the signals, checking if the events were correctly detected. Thus, the features were updated every step with the true values, limiting the error propagation.

III. ALGORITHM VALIDATION

The algorithm performance was assessed using the GaitRite system as the goal standard.

A. Reliability Analysis

All true positives (TP), false positives (FP) and false negatives (FN) were counted; true negatives were omitted, due to the resulting unbalanced analysis dataset. TP, FP and FN were combined to compute the Precision (P), Recall (R) and F1-score metrics.

1193 IC events were correctly detected (TP), with 0 FP and 3 FN. In case of EC events, 1195 TP, 0 FP and 1 FN were obtained. This results in $P_{IC}=1$, $R_{IC}=0.9975$, $P_{EC}=1$ and $R_{EC}=0.9992$. The F1-scores were $F1_{IC}=0.9987$ and $F1_{EC}=0.9996$. All these values were above the 0.9 recommended by Rueterbories [1] for gait event detection reliability, in order to have a good system for ambulatory rehabilitation.

B. Agreement Analysis of Detection Timing

For all the TP, a Bland-Altman plot [10] was obtained for each speed condition to assess the agreement between the detection timing of the IC and EC events computed by the developed algorithm and the GaitRite (see figure 3). This agreement was evaluated as the difference between the GaitRite detection timings and the ones obtained by our algorithm. Thus, positive values corresponded to an early detection of our algorithm.

The mean values [95% Confidence Interval (CI)] of the difference between the detection timings were equal to -69.6 ms [-70.6, -68.6] for the slow self-selected speed, -68.6 ms [-69.5, -67.7] for the normal self-selected speed, and -71.0 ms [-72.0, -70.0] for the fast self-selected speed. For EC, the detection showed a mean difference [95% CI] of 3.3 ms [0.7, 5.9], -7.8 ms [-9.2, -6.4], and -7.8 ms [-9.1, -6.5] for slow, normal and fast self-selected speed, respectively. Limits of agreement are shown in Fig. 3.

The acceptable difference between both systems has to be defined taking into consideration the future application of the algorithm. Given that the sampling period is 20 ms, the mean difference in the detection timings are -3.49 samples in case of IC and -0.2 samples for EC. This makes the system suitable even for real-time applications.

Regardless of the detection timing variability, the proposed algorithm was always able to find the local minimum associated with IC. The differences in the detection timings came from the misalignment that sometimes happened between the local minimum of the angular velocity and the IC event detected by the GaitRite.

IV. COMPARISON WITH PREVIOUS PUBLISHED ALGORITHMS

The developed method was compared to two real-time algorithms previously published in literature [7,8]. The algorithm of Lee et al [8] used two shank-attached inertial sensors, detecting MS, IC and EC on a 3-Hz-filtered version of the raw signal. Its main drawback was the introduction of delays due to the filtering and the use of MS as reference for a previous event (EC). The algorithm of Gonzalez et al [7] used one sensor placed on the S1 vertebra, and located the IC and EC after a zero-cross on the 2-Hz-filtered version of the anterio-posterior acceleration. Both algorithms were assessed in terms of reliability and agreement analysis as explained in the section III.

For the reliability analysis, Lee's algorithm correctly detected all 1196 IC and EC events, but also extracted 3 false contacts (P=0.9975, R=1 and F1=0.9987). Gonzalez's algorithm worked perfectly (P=1, R=1 and F1=1).

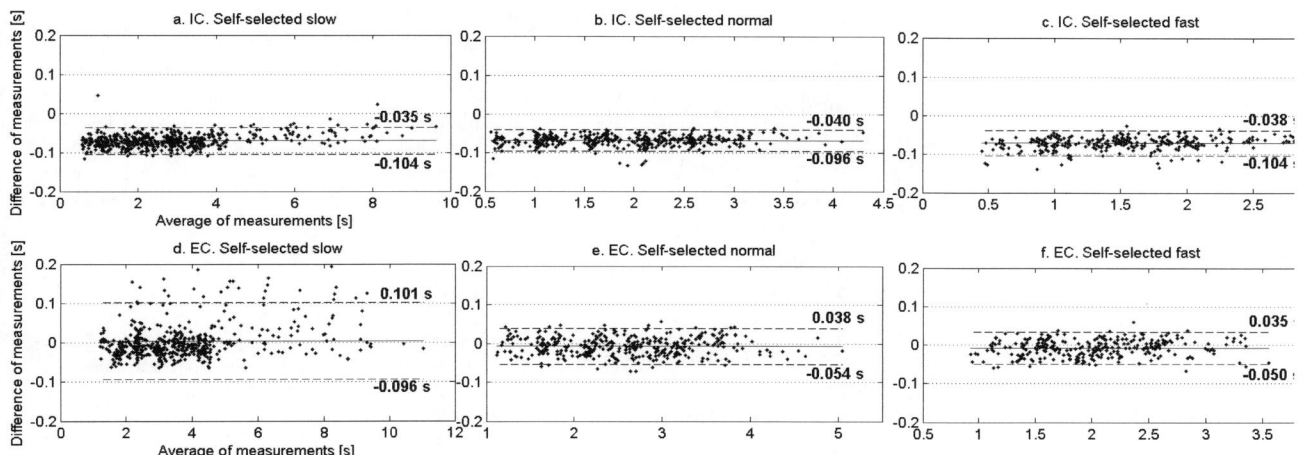

Fig. 3 Bland-Altman plots of IC and EC, to evaluate the agreement between the proposed algorithm and the GaitRite. Positive times correspond to an early detection of the proposed algorithm. The solid lines represent the mean difference of the detection times, while the dashed ones the limits of agreement (mean ± 1.96SD). Limits of agreement are reported in the figure.

For the timing agreement, a statistical analysis was done to compare the performance of the three algorithms in terms of detection timing. After verifying that all data was not normally distributed (Kolmogorov-Smirnov test), a non-parametric Kruskal-Wallis test (p<0.05) was performed. Six separate tests were used to analyze the two time events for the three speed conditions. Dunn-Sidak post-hoc tests were performed (p<0.05) to determine which pairs of effects were significantly different. Figure 4 shows the median and interquartile ranges of the detection timings obtained by the three algorithms, which were different for all speed conditions. The differences in the detection timings with respect to the GaitRite system obtained by the here proposed algorithm were always the lowest. Additionally, EC was sometimes detected in advance, which is tremendously useful for closed-loop rehabilitation treatments.

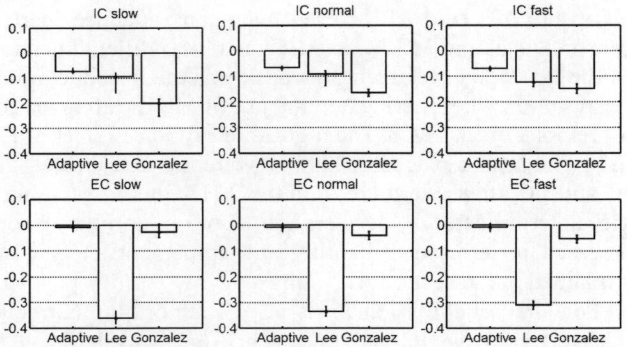

Fig. 4 Comparison between the 3 algorithms in terms of detection timing. Negative timings correspond to detections in delay with respect to the GaitRite.

V. CONCLUSION

This study presented a novel adaptive algorithm to detect gait events in real-time, using only two IMMS attached to the shanks. Its accuracy and reliability have been proved, indeed, from the comparison with the gold-standard system, where F1-scores of 0.9987 for IC and 0.9996 for EC were obtained. Our algorithm detected the gait events with a mean (SD) delay of 68.6 (15.1) ms for IC and 7.8 (23.6) ms for EC, with respect to the GaitRite, for the self-selected speed. These values were significantly lower than those obtained by other published algorithms. The obtained results suggested that the algorithm can be used to develop gait treatments based on biofeedback or neuroprostheses. The algorithm is adaptive and thus it can potentially be used for long-time applications. Additionally, since it used only information of the ipsilateral leg, it might be suitable for subjects with an asymmetrical gait. Further experiments are needed to validate the algorithm on a wider variety of ages and pathologies, such as elderly and post-stroke patients.

ACKNOWLEDGMENTS

This work was supported by the PRIN project (2010R277FT) funded by the MIUR.

REFERENCES

1. Rueterbories J, Spaich EG, Larsen B et al. (2010) Methods for gait event detection and analysis in ambulatory systems. Med. Eng. Phys 32:545–552.
2. Ferrante S, Ambrosini E, Ravelli P et al (2011) A biofeedback cycling training to improve locomotion: a case series study based on gait pattern classification of 153 chronic stroke patients, J. Neuroengineering Rehabil. 8:47.
3. Ambrosini E, Ferrante S, Pedrocchi A et al (2011) Cycling induced by electrical stimulation improves motor recovery in postacute hemiparetic patients: a randomized controlled trial, Stroke J. Cereb. Circ. 42:1068–1073
4. Ambrosini E, Ferrante S, Schauer T et al (2010) Design of a symmetry controller for cycling induced by electrical stimulation: preliminary results on post-acute stroke patients, Artif. Organs. 34:663–667
5. Monaghan C.C, van Riel W.J.B.M, Veltink P.H (2009) Control of triceps surae stimulation based on shank orientation using a uniaxial gyroscope during gait, Med. Biol. Eng. Comput. 47:1181–1188.
6. Aminian K, Najafi B, Büla C et al (2002) Spatio-temporal parameters of gait measured by an ambulatory system using miniature gyroscopes, J. Biomech. 35: 689–699.
7. González R.C, López A.M., Rodriguez-Uría J. et al (2010) Real-time gait event detection for normal subjects from lower trunk accelerations, Gait Posture. 31:322–325.
8. Lee J.K, Park E.J (2011) Quasi real-time gait event detection using shank-attached gyroscopes, Med. Biol. Eng. Comput. 49:707–712.
9. Lee J.K, Park E.J (2011) 3D spinal motion analysis during staircase walking using an ambulatory inertial and magnetic sensing system, Med. Biol. Eng. Comput. 49:755–764
10. Bland J.M, Altman D.G (1986) Statistical methods for assessing agreement between two methods of clinical measurement, Lancet. 1:307–310.

Corresponding author:

Author: Noelia Chia Bejarano
Institute: Politecnico di Milano, Department of Electronics, Information and Bioengineering, Neuroengineering and Medical Robotics Laboratory
Street: via Giuseppe Colombo 40, 20133
City: Milano
Country: Italy
Email: noelia.chia@polimi.it

A Graphical Tool for Designing Interactive Video Cognitive Rehabilitation Therapies

J.M. Martínez-Moreno[1,2], P. Sánchez-González[1,2], A. García[3], S. González[4], C. Cáceres[1,2], R. Sánchez-Carrión[3], T. Roig[3], J.M. Tormos[3], and E.J. Gómez[1,2]

[1] Biomedical Engineering and Telemedicine Centre, ETSI Telecomunicación, Universidad Politécnica de Madrid, Madrid, Spain
[2] Biomedical Research Networking Center in Bioengineering, Biomaterials and Nanomedicine, Madrid, Spain
[3] Institut Guttmann Neurorehabilitation Hospital, Badalona, Spain
[4] Lavinia Interactiva, Barcelona, Spain

Abstract—Acquired Brain Injury (ABI) has become one of the most common causes of neurological disability in developed countries. Cognitive disorders result in a loss of independence and therefore patients' quality of life. Cognitive rehabilitation aims to promote patients' skills to achieve their highest degree of personal autonomy. New technologies such as interactive video, whereby real situations of daily living are reproduced within a controlled virtual environment, enable the design of personalized therapies with a high level of generalization and a great ecological validity. This paper presents a graphical tool that allows neuropsychologists to design, modify, and configure interactive video therapeutic activities, through the combination of graphic and natural language. The tool has been validated creating several Activities of Daily Living and a preliminary usability evaluation has been performed showing a good clinical acceptance in the definition of complex interactive video therapies for cognitive rehabilitation.

Keywords—Interactive Video, Graphical Tool, Activities of Daily Living, Cognitive Rehabilitation, Acquired Brain Injury.

I. INTRODUCTION

Brain injury is a life-altering event that affects every area of a person's life including his/ her relationship with family members and close relatives. Acquired Brain Injury (ABI) refers to medical conditions that occur in the brain, altering its function. These conditions include stroke, traumatic brain injury, tumor and other diseases of the brain [1].

Cognitive processes determine individual performance in Activities of Daily Living (ADL) [2]. Thus, cognitive disorders after brain injury result in a loss of autonomy and independence, affecting the patient's quality of life.

Cognitive rehabilitation seeks to increase patients' autonomy and quality of life minimizing or compensating functional disorders showed by ABI patients. Moreover, rehabilitation programs aim to reduce the load, stress and isolation suffered by their family, who become an essential element in such patients' treatment.

Rehabilitation goals must focus on increasing ecological validity [3]. It is strongly recommended that therapeutic activities mimic as much as possible the patient's real environment and needs, merging the rehabilitation into his/her ADL routine. Ultimately, the rehabilitation environment must be adapted, reaching a compromise between realism and flexibility so that therapy can be tailored for any specific impairment profile. Moreover, to ensure maximum efficiency, successful rehabilitation should consider task reproducibility; accurate, effective formative and summative assessment, and the possibility of monitoring the rehabilitation process and the patient's performance [4].

The introduction of disruptive technological solutions in neurorehabilitation leads to higher intensity rehabilitation processes, extending the therapies in an economically sustainable way. The continuous technological development over the last few years on computer processing, graphics/image rendering, display systems, interface and tracking devices, etc.; has increased the application of virtual environment technologies for brain injury rehabilitation. A number of studies have emphasized the benefits of using virtual-based (virtual, augmented and mixed realities, serious games) rehabilitation methods that are both relevant to the patient's real-life context, and that can also be transferred to other ADLs [5][6].

A large sample of works based on serious games development can be found in the literature [7], among which the RehaCom system stands out [8]. Additionally, we can find augmented reality systems, which add virtual elements onto the real environments where the task is performed, such as GenVirtual [9] or ARVe [10]. Finally, Virtual Reality systems can reproduce everyday situations and environments as supermarkets [11], kitchen [12], town neighborhood [13] or even a city [14]. These environments simulate situations of the patients' ADLs encouraging him/her to deal with them as in his/her everyday life.

Interactive Video (IV) can be an appropriate technology to sustain personalized rehabilitation processes based on enhanced environments where real situations of daily living are reproduced [15]. IV refers to any video whose sequences and displayed information depend on the users' responses. Interactivity is provided by associating an interaction with any element that appears in the video scenes; the video flow is modified according to the way users interact with them.

IV effectiveness has been proven on the field of learning, where interactive dynamic visualization allows users to adapt the processes to their individual cognitive skills [16][17]. Moreover, some studies have proved that ADL observation treatment is a good rehabilitative approach in stroke patients [18]. The potential of IV lies in their capacity to combine videos portraying real life situations into guided tasks tailored to patient's cognitive profile and needs for achieving therapeutic observation. The combination of both aspects is aimed at reducing to the minimum cognitive requirements needed by the patient to interact with the ADL. Thereby, the neuropsychologist can define every single detail of the environment so that each session focuses only in the specific cognitive functions under rehabilitation. Using IV patient perceives a great quantity of information (cognitive stimulus), either presented as images, texts, audios, actions, etc. Proper structuring and sequencing of these multimedia elements enable the creation of controlled therapeutic situations for the patient to react to and interact with under a therapeutic goal.

Although IV offers many possibilities for the neuropsychologist to set up rehabilitation therapies, as the degrees of freedom increase, so does the complexity and effort required for designing an IV task. Such design complexity along with the inherent technological difficulty of IV management demands powerful tools to assist the neuropsychologist on the creation of IV therapies for cognitive rehabilitation. The main goal of this research work is to present and validate a graphical tool for the design of IV rehabilitation tasks based on ADLs, offering therapists the mean to easily edit the contents and control every aspect of the rehabilitation treatment. By mean of this tool, the therapist has to be able to set the stimuli to be presented, describe each possible interaction of the patient with them, either acceptable or pathological, select new stimuli for each one of them, many of them being "therapeutic" interactions that will be proven and validated, if the patient is able to successfully complete the exercise.

In addition to that, as soon as each interaction is the formal representation of a catalog of therapeutic interventions, each of them absolutely evaluated, it became unavoidable to face the challenge of buildup the system over a taxonomy oriented data base, allowing the emergence of a daily living rehabilitation ontology.

II. METHODS

In this research work, a graphical tool has been conceptualized for the therapist to be able to design ADL-based therapies taking clinical advantage of the potential of IV technology. This software allows clinicians to create a therapeutic script of the activity, designing and configuring the stimuli, interactions, assistance and feedback based on the therapeutic goals of the rehabilitation therapy.

The design of the graphical tool has been carried out following the principles of User Centered Design (UCD). The definition of its functional requirements was carried out by an expert panel of 6 neuropsychologists from Institut Guttmann Hospital by means of interviews and questionnaires.

The tool allows the design of the therapeutic script for each ADL, which are structured as a four-level hierarchical model, as shown in Fig. 1. This scheme is based on a therapeutic successive approximation model [3].

A designed ADL is divided into rehabilitation tasks, which in turn are divided into scenes, identifying situations for the generation of therapeutic interventions. Each scene is organized by means of 'pathways', which are composed of 'stimuli', 'actions' and 'responses'. The neuropsychologist can define normal, pathological or rehabilitative pathways.

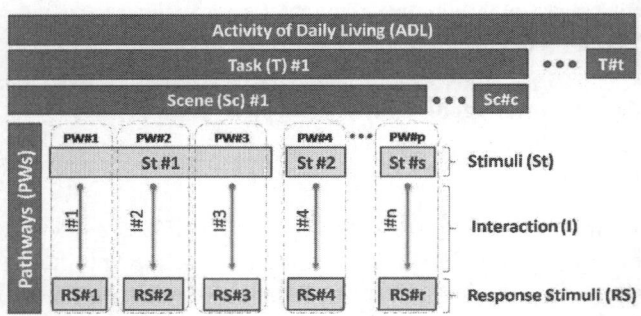

Fig. 1 ADL hierarchical structure

The main functionalities provided by the tool include the creation and edition of the therapeutic ADL script, interactive visualization, and generation of the descriptor file.

Graphical user interface design was performed generating wireframes by means of AxureRP Pro 5.6 software (Fig.2). This example shows the screen where user adds the different script elements. It represents the pathway definition where stimuli are configured and connected defining allowed interaction.

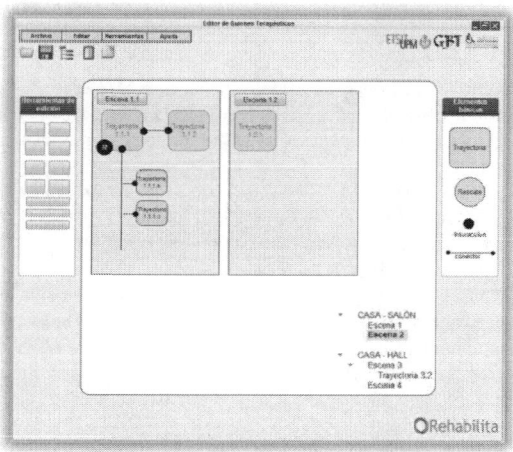

Fig. 2 Wireframe's screen example

The final implementation is based on a Model-View-Controller (MVC) framework (Symfony2) and ORM Doctrine in the server side. Extensive use of new functionalities provided by HTML5 combined with Javascript is used for interaction management in the client side.

A preliminary usability evaluation by likert-type questionnaires has been performed on the complete graphical tool in which five specialists from Guttmann hospital were involved. In order to validate the tool, an ADL 'buying bread' activity using IV technology was developed [15].

III. RESULTS

Figure 3 shows an example of the graphical user interface (GUI) in which the creation and visualization of an ADL exercise's task is displayed. Three main areas are shown on this screen: title task bar, menus bar and tools bar (top); structure task, it is a list of the task elements (right); and (center) the designing board, where boxes represent task's scenes and lines represent pathways.

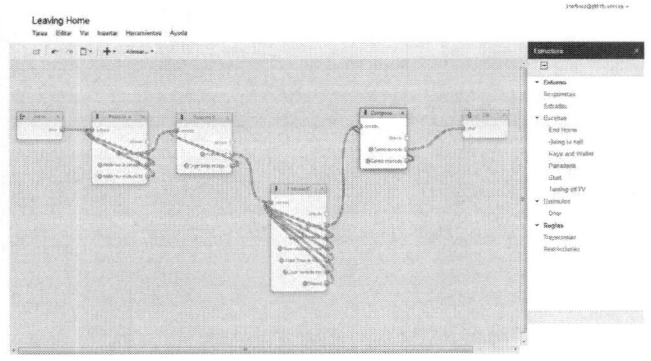

Fig. 3 Main Screen

The ADL 'buying bread' script was designed using the tool by Guttmann Hospital neuropsychologists and recorded with amateur actors in outdoor and indoor real settings (see Fig. 4). The ADL virtual environment allows patients to navigate through a series of scenes representing the different steps to follow in order to reach a final goal, from sitting on the couch at home (initial state) to buying bread at the bakery (end state) [15]. Throughout the task, patients are required to make decisions: choosing the next step, answering questions or interacting with other characters. Every single action in the task is under therapist control, who has previously adjusted the video scenes sequence to the cognitive capabilities required by the patient. Thus, all stimuli in the scenes may be preprogrammed by therapists in order to help the patient or even distract him/her from completing a task.

Fig. 4 Bakery task in ADL 'Buying Bread'.

Global results for the validation of the therapeutic designing graphical tool are shown in Fig 5. Ease of learning, ease of use, satisfaction, efficiency and efficacy have been analyzed, in many cases showing good preliminary results.

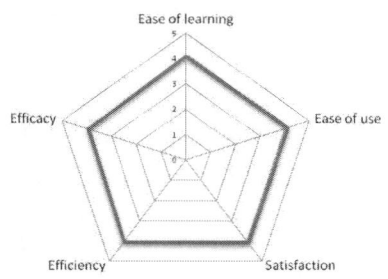

Fig. 5 Evaluation results. (1: Slightly satisfied; 5: Extremely satisfied)

IV. CONCLUSIONS

Cognitive neurorehabilitation therapies help brain injury patients enhance their ability to move through daily life by recovering or compensating for damaged cognitive functions. A restorative approach helps patients reestablish cognitive function, while compensatory approaches help the individual adapt to an ongoing impairment.

This paper has presented the design and development of a graphical tool to create cognitive rehabilitation ADL therapies base on interactive video technology. IV provides several advantages compared to traditional virtual environment technologies (virtual reality).Virtual environment technologies are offering many new possibilities to generate cognitive rehabilitation therapy interventions adapted to the patients' status and needed treatment.

Using real image provides more similarity between rehabilitation environment and patient's daily life. Organizing the video by pathways, neuropsychologist is allowed to adapt ADL to any cognitive profile. Thus, required compromise between realism and flexibility is reached. IV allows control of interactivity and screen movements, and reduces the risk of sickness symptoms (i.e. cybersickness).

The developed tool can be a valuable asset for the creation of cognitive rehabilitation therapies based on IV activities of daily living. It provides a powerful software to help neuropsychologists to conceptualize the therapeutic goals for brain injury patients and to create interactive videos supporting the rehabilitation processes.

Preliminary usability evaluation results of the therapies designing tool have shown that it is intuitive, ease to use and accessible to any clinician regardless of their experience with software developing tools. Further studies are required to confirm the clinical usage of the tool and the potentiality of interactive video environments for cognitive rehabilitation of brain injury patients.

Acknowledgment

This research has been partially funded by the project CENIT-E 'REHABILITA' CEN-20091043.

References

1. Woodward B. Understanding Brain Injury. A guide for the Family. 2008. MFMER.
2. Marcotte T, Cobb J, Kamat R. Neuropsychology and the Prediction of Everyday Functioning. In Neuropsychology of Everyday Functioning. (2010). pp. 5-39.
3. Wilson B, Herbert C, Shiel A. Behavioural Approaches in Neuropsychological Rehabilitation: Optimising Rehabilitation Procedures. (2003). ISBN 1-84169-183-6.
4. Lee JH, Ku J, Cho W et al. A virtual reality system for the assessment and rehabilitation of the activities of daily living. Cyberpsychol Behav.(2003) Aug;6(4). Pp 383-8.
5. Laver K, George S, Thomas S, Deutsch JE, Crotty M. Cochrane review: virtual reality for stroke rehabilitation. Eur J Phys Rehabil Med. 2012 Sep;48(3):523-30. Epub 2012 Jun 20.
6. Moreira MC, de Amorim Lima AM, Ferraz KM, Benedetti Rodrigues MA. Use of virtual reality in gait recovery among post stroke patients: a systematic literature review. DisabilRehabil Assist Technol. 2013 Apr 24.
7. Rego P, Moreira P.M., Reis L.P. Serious games for rehabilitation: A survey and a classification towards a taxonomy. Proceeding of the 5th Iberian Conference on Information Systems and Technologies (CISTI), 2010. ISBN 978-1-4244-7227-7.
8. Fernández E et al. Clinical impact of RehaCom Software for cognitive rehabilitation of patients with acquired brain injury. MEDICC rev. [online]. 2012, vol.14, n.4, pp. 32-35. ISSN 1555-7960.
9. Correa A, Grasielle D, et al. GenVirtual: An Augmented Reality Musical Game for Cognitive and Motor Rehabilitation. IEEE, 2007. Virtual Rehabilitation. pp. 1-6. ISBN: 978-1-4244-1204-4
10. Richard E, et al. Augmented Reality for Rehabilitation of Cognitive Disabled Children: A Preliminary Study. IEEE, 2007. Virtual Rehabilitation, 2007. pp. 102-108.
11. Rand D, Weiss PL, Katz N. Training Multitasking in a Virtual Supermarket: A Novel Intervention After Stroke. Occupational Therapy September/October 2009 vol. 63 no. 5 535-542.
12. Cao X, Douguet AS, Fuchs P, Klinger. Designing an ecological virtual task in the context of executive functions: a preliminary study. Proc. 8th Intl Conf. Disability, Virtual Reality & Associated Technologies. (2010). ISBN 978 07049 15022
13. E. Klinger E, Kadri A, Sorita E, Le Guiet JL et al. AGATHE: A tool for personalized rehabilitation of cognitive functions based on simulated activities of daily living. IRBM (2013) pp 113-118.
14. Dores AR, Carvalho IP, Mendes L. Virtual City: Neurocognitive rehabilitation of Acquired Brain Injury. Proceeding of the 7th Iberian Conference on Informtation Systems and Technologies (CISTI) (2012). ISBN: 978-1-4673-2843-2.
15. Martínez-Moreno JM, Solana J, Sánchez R, et al.Monitoring visual attention on a neurorehabilitation environment based on Interactive Video. Proceedings of the International Conference on Recent Advances in Neurorehabilitation (2013). pp 182-185
16. Schwan S,Riempp R. The cognitive benefits of interactive videos: learning to tie nautical knots. Learning and Instruction 14 (2004), 293-305.
17. Martin M, Weigand S, Heier A, Schwan S. Learning with videos vs. learning with print: The role of interactive features. Learning and Instruction 21 (2011), 687-704.
18. Franceschini M, Agosti M, Cantallo A, et al, Mirror neurons: actions observation treatment as a tool in stroke rehabilitation. European Journal of physical and rehabilitation medicine46 (2010), 517-523.

Author: José María Martínez Moreno
Institute: Biomedical Engineering and Telemedicine Centre
Street: Av. Complutense 30
City: Madrid
Country: Spain
Email: jmartinez@gbt.tfo.upm.es

Part XIII
Young Investigator Competition

Motion-Related VEPs Elicited by Dynamic Virtual Stimulation

P.J.G. Da Silva, B.P. Rosa, M. Cagy, and A.F.C. Infantosi

Biomedical Engineering Program / Coppe, Federal University of Rio de Janeiro, Rio de Janeiro, Brazil

Abstract—The coherent average was used for investigating the motion-related VEP (M-VEP). The electroencephalogram (EEG) signals of 29 healthy subjects were acquired during stabilometric protocol with distinct virtual scenes: static (SS) and dynamic (forward, DS_F, or backward motion, DS_B). The grand-averaged M-VEP for these visual conditions presented N2 peak dominance with higher values for parietal EEG derivation. The DS increased the N2 peak and latency, as well as the N2 — P3 amplitude, in comparison with SS. These components are not sensitive to motion orientation (t-test α = 0.05). However, the running *t*-test (α = 0.05) showed difference in the DS_F — DS_B waveforms 600 ms after the DS onset, with an increasing time delay from the occipital to central derivations. These findings suggest that the cognitive, planning and motor processing to maintain balance depends on the virtual scene direction.

Keywords—Coherent Average, Dynamic Visual Stimulation, EEG, Motion-Related VEP, Virtual Reality.

I. INTRODUCTION

Dynamic visual stimulation (DS) can also evoke cortical response in the visual and parietal areas [1-4]. Usually, this motion-related visual evoked potential (M-VEP) has been employed to investigate the cortical processing of the visual pathway and visual brain cortex. However, the M-VEPs components are associated only to the motion-onset [2,3].

According to Da_Silva et al. [1], by analyzing in the frequency domain the time evolution of the event-related desynchronization/synchronization index (ERD/ERS), it is possible to distinguish the time-locked effects of DS. In the time domain, the coherent average technique can indicate how this temporal information is encoded at the cortical areas [2-4], and could be useful in the study of M-VEP applied during stabilometric test.

This work aims at investigating the M-VEP elicited by distinct virtual stimulations: static and dynamic (forward or backward motion) scenes. The coherent average procedure was applied to the electroencephalogram (EEG) signals acquired during these visual conditions and it was statistically compared using the running *t*-test.

II. MOTION-RELATED VISUAL EVOKED POTENTIAL (M-VEP)

M-VEP provides an objective estimate of the time properties of the visual system by applying a dynamic stimulation. The M-VEP is typically composed of three main peaks (negative and positive with a latency of about 110 – 400 ms after movement onset), referred as P1, N2 and P3, which are dependent on the type of motion [2,3].

Since the evoked potential presents an amplitude at least ten times lower than the spontaneous EEG (sEEG), the VEP can only be highlighted by the use of a coherent average. Considering the additive linear model (Fig.1), in which the collected signal ($y[k]$) is composed by noise ($r[k]$: sEEG) and the stimulus-response ($s[k]$), and assuming that the sEEG is a zero-mean Gaussian noise, the effect of averaging is to increase the signal-to-noise ratio (SNR). This is expected since the cortical response is considered to be identical from stimulus-to-stimulus ($s_i[k] = s[k]$). Mathematically, the coherent average can be expressed as:

$$\hat{s}[k] = \frac{1}{M}\sum_{i=1}^{M} y_i[k] = \frac{1}{M}\sum_{i=1}^{M} s_i[k] + \frac{1}{M}\sum_{i=1}^{M} r_i[k] \quad (1)$$

where $\wedge$ denotes estimation and M is the number of EEG epochs. When $M \to \infty, \hat{s}[k] \to s[k]$.

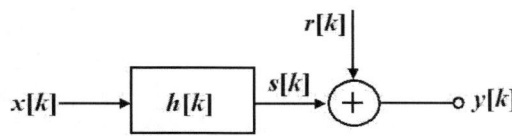

Fig. 1 Additive Linear Model for evoked potential generation: x[k] is the visual stimulation, h[k] is the transfer function (Cortex), s[k] is the stimulation response, r[k] is the spontaneous EEG and y[k] is the measured signal.

III. MATERIALS AND METHODS

A. Subjects

Signals from 29 healthy subjects (11 female), age ranging from 20 to 42 years, height of 172.7 ± 9.8 cm and mass of 73.3 ± 15.4 kg (mean ± standard deviation) were used in this study. All subjects present no history of neurological pathologies, osseous, muscles and joints diseases or equilibrium disorder. The anamnesis was carried out to obtain information about headache, illness, vertigo, eyestrain, and the use of medication that could compromise the balance. Nevertheless, subjects using glasses or corrective lens were included. All participants previously signed an informed consent form. The study was approved by the Research and Ethics Committee of IESC/UFRJ (Ref. 100/2011).

B. Experimental Protocols

The stabilometric tests with visual stimulation were performed in two sections within the same room and under controlled environmental condition (23°C, sound attenuated and light control). The sEEG was previously acquired for 5 min with the subject resting in a comfortable chair with eyes closed. During the sections, the EEG signals were acquired with the subject bare-footed in upright position, quite standing in a force platform. The feet position (angle: 30°; heels 2 cm apart) was previously demarked to maintain the same support base during the tests. The visual stimulation trials were performed with the subject observing a virtual scene (1.52 × 1.06 m) projected 1 m apart from the platform. This scene (Fig. 2), developed using IDE Delphi and OpenGL, consists of a room containing a chessboard-like floor (similar to pattern-reversal) with a table and chair placed in the center, and other objects in the periphery of the visual field. All subjects were instructed to keep the gaze on the chair.

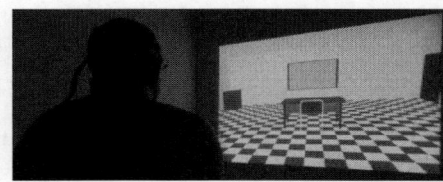

Fig. 2 The virtual scene with checkerboard floor and furniture

In the first section, the subject underwent a static visual stimulation with a virtual blue wall (denoted as BW) and the scene (denoted as SS) exposed in random order. For the second section, a dynamic visual stimulation (DS) was employed for inducing the perception of self-motion. In order to carry out the DS, the details of the virtual scene were randomly moved backward (DS_B) or forward (DS_F) with a velocity of 120 cm/s during 1 s, interspersed by 10 s of SS. The luminance of the virtual scene ranges from 33 to 39 cd/m². A set of 120 DS was performed into four blocks of 30 stimulations (15 of each DS_B and DS_F), with inter-block resting interval of 3 min (subject in sit position). The DS stimuli were codified by pulses with code number of 1200 (DS_F) and −1200 (DS_B) synchronized by the start of exhibiting scene. The sequence of pulses generates a trigger signal (Fig. 3a) to be used during signal processing.

C. Experimental Setup

The EEG signals were acquired according to the International 10/20 System (monopolar derivations, averaged bilateral ear-lobe reference and ground in FPz) using the ECI Electro-Cap Electrode (Electro-Cap International, Inc., EUA) and the BrainNet - BNT 36 biological amplifier (EMSA, Brazil). The electrode impedances were kept below 3 kΩ. The EEG signals were digitized at 600 Hz (resolution: 16 bits) with a digital notch filter in 60 Hz, and stored on a computer hard disk for offline analysis.

Only the EEG signals of the visual (occipital), the associative (parietal), the cognitive (frontal) and the motor (central) cortex were analyzed in this study. Figure 3b depicts 30 s of the O1 EEG derivation from subject #2 acquired during stabilometric test with dynamic virtual stimulation, showing a sequence of three DS (duration: 1 s, gray area).

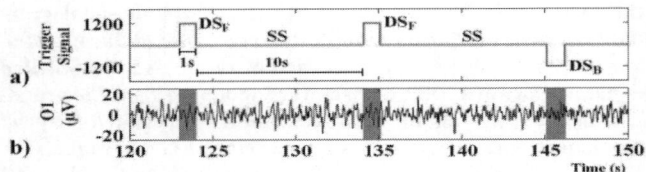

Fig. 3 The trigger signal (a) and EEG (O1 derivation, b) segment from subject #2 during stabilometric test with dynamic virtual stimulation. The gray area indicates the 1 s of DS_F (or DS_B) followed by 10 s of the static scene (SS), synchronized by the start of exhibiting DS.

D. EEG Signal Processing

The EEG signals were band-pass filtered by a 2nd order Butterworth (null phase) with cut-off frequency of 1-10 Hz. For the static virtual stimulation section, only the EEG signals of the transition between BW to SS were segmented into epochs of 2500 ms duration (−500 to 2000 ms, synchronized by the onset of the exhibiting SS, $t = 0$). Based on the sEEG, an algorithm for artifact rejection was applied to each signal epoch [1]. The M-VEP was then estimated using equation (1), i.e. coherently averaging the M epochs (free of artifact) relative to the exhibiting SS scene. The individual VEPs were low-pass filtered (cut-off frequency of 5 Hz, 2nd order Butterworth, null phase) and the M-VEP peaks were recognized from the zero-cross of the first approximate waveform derivative within the time-windows from 50 to 300 ms and 300 to 550 ms, respectively for N2 and P3. The resulting grand-averaged waveforms were obtained regarding all volunteers. The same procedure was used for the dynamic stimulation section, but each epoch was referred to the instant of applying the distinct DS_B (or DS_F).

E. The Running t-Test

The statistical differences between the N2 (or P3) M-VEP components (amplitude and latency) elicited by SS, DS_B and DS_F for all volunteers were obtained using the t-test ($\alpha = 0.05$). Moreover, using the differential waveform of DS_B and DS_F, the t-test ($\alpha = 0.05$) was also applied at each instant of time, hence called running t-test. In this procedure, the two-tails test comparing the sample-by-sample difference to zero within the time interval of interest, from −100 to 1100 ms. At a given instant of time, if the null hypothesis (H_0) of zero difference is rejected, one can assume that the resulting grand-averaged for DS_B is statistically more positive or less negative then that for DS_F.

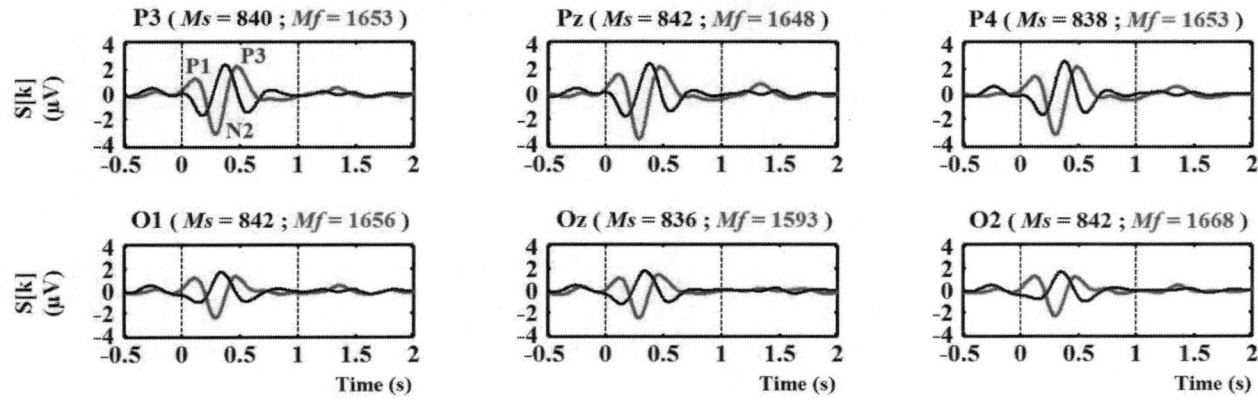

Fig. 4 The grand-averaged M-VEP (occipital and parietal EEG derivations) from virtual stimulation with static (SS, black) and dynamic forward scene (DS$_F$, red) estimated with Ms and Mf epochs, respectively. The vertical dashed line in $t = 0$ s indicates the onset of the exhibition SS or DS$_F$, while $t = 1$ s referrers to the DS$_F$ motion-offset presentation.

Table 1 N2 and P3 M-VEP components: latency (L), peak (P) and N2 — P3 amplitude (A) mean values ($\pm \sigma$) for occipital and parietal EEG derivations from all volunteers during virtual stimulation (VS) with static (SS), dynamic forward (DS$_F$) and backward (DS$_B$) scene. Values in bold indicate significant difference between SS and both DS conditions (t-test, $\alpha = 0.05$).

EEG	VS	N2 L (ms)	N2 P (µV)	P3 L (ms)	P3 P (µV)	N2 — P3 A (µV)
P4	SS	**146±7***	**-2.4±1.1***	**382±7***	3.3±1.4	**5.6±0.4***
	DS$_F$	302±4	-4.3±2.0	500±6	3.1±2.0	7.4±0.4
	DS$_B$	301±3	-4.5±2.2	489±5	3.5±1.8	8.1±0.3
Pz	SS	**141±6***	**-2.6±1.4***	**368±6***	3.3±1.4	**5.8±0.4***
	DS$_F$	293±4	-4.7±2.0	480±7	3.4±1.9	8.3±0.2
	DS$_B$	296±3	-5.0±2.5	472±5	3.9±1.8	8.9±0.3
P3	SS	**136±6***	**-2.4±1.8***	**369±10***	3.1±1.2	**5.4±0.5***
	DS$_F$	292±3	-4.0±1.8	465±6	3.1±1.9	7.2±0.4
	DS$_B$	298±3	-4.3±2.2	468±5	3.6±1.6	7.8±0.3
O2	SS	**125±7***	**-1.9±0.9***	**348±8***	2.2±1.3	**4.1±0.4***
	DS$_F$	299±4	-3.2±1.7	479±6	1.9±1.3	5.2±0.4
	DS$_B$	310±4	-3.2±1.9	472±6	2.0±1.3	5.1±0.3
Oz	SS	**122±6***	**-2.2±1.1***	**340±9***	2.8±1.4	**5.1±0.4***
	DS$_F$	298±4	-3.6±1.7	462±7	2.7±1.5	6.4±0.6
	DS$_B$	298±3	-3.5±2.2	469±6	2.6±1.5	6.2±0.4
O1	SS	**125±6***	**-1.9±0.9***	**323±8***	2.2±1.1	**4.0±0.4***
	DS$_F$	290±3	-3.2±1.7	458±6	2.1±1.3	5.1±0.4
	DS$_B$	299±4	-3.1±1.9	453±5	2.2±1.2	5.2±0.5

* indicates $p \ll 0.01$

IV. RESULTS

Figure 4 depicts the grand-averaged VEP (N2 peak dominance) for the occipital and parietal EEG derivations elicited by the static (SS, black) and dynamic forward stimuli (DS$_F$, red), synchronized by the onset of the exhibiting scene ($t = 0$). In all derivations, the N2 peak for the DS$_F$ condition occurs later, with higher values than that observed during SS ($p \ll 0.01$, indicated in bold in Table 1). The P3 peak also occurs later ($p \ll 0.01$), but with similar values when compared with SS one. For all stimulation conditions, the N2 peak and the N2 — P3 amplitude are higher for the parietal cortex. No difference was observed between N2 (or P3) elicited by DS$_F$ and DS$_B$, nor from N2 — P3 amplitude.

Figure 5 depicts the grand-averaged M-VEPs obtained from DS$_F$ (red) and DS$_B$ (blue) stimulation, including the frontal and central EEG derivation. Note that there is a pronounced positivity after P3 M-VEP, which is more evident in the frontal and central leads. The running t-test results pointed out two distinct time intervals when the waveforms from DS$_F$ differ from that of DS$_B$ ($p < 0.05$, gray area in Fig. 5). The first interval was observed approximately 600 ms after the onset of DS, and could be seen only in the occipital and parietal derivations (Table 2). The second one occurs in all derivations, but with a time delay among the leads, arranged in a spatial-temporal order of brain processing: occipital, parietal, frontal and central cortex.

V. DISCUSSION AND CONCLUSION

The grand-averaged VEP for all stimulation conditions presented N2 peak dominance with higher values for parietal EEG derivation, particularly in the right hemisphere cortex, as reported by [2,3]. These authors suggest that the

Table 2: Time interval (Δt) where the waveforms for dynamic forward (DS$_F$) and backward stimulation (DS$_B$) statistically differ (running t-test, $\alpha = 0.05$).

EEG	O1	Oz	O2	P3	Pz	P4	C3	Cz	C4	F3	Fz	F4
Δt (ms)	605:690	588:643	658:718	603:673	595:666	665:725	—	—	—	—	—	—
	835:978	836:948	865:966	891:1003	905:1076	906:1078	947:1020	945:1038	956:1066	921:1031	925:1051	923:1075

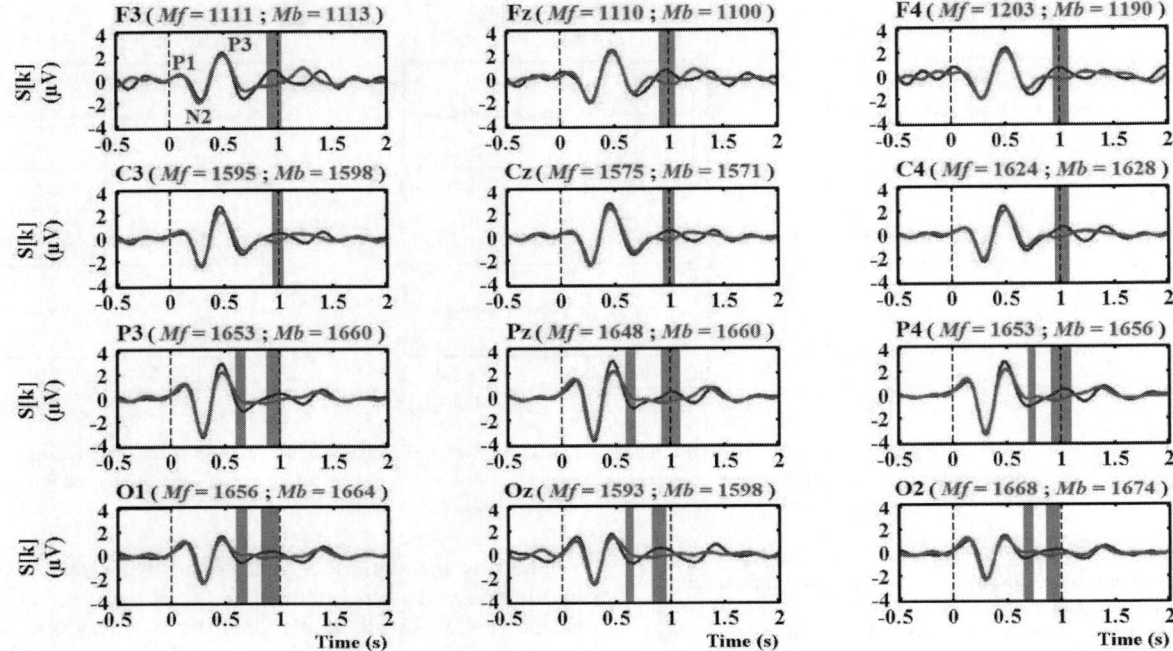

Fig. 5 The grand-averaged M-VEP (occipital, parietal, central and frontal EEG derivations) from virtual stimulation dynamic forward (DS_F, red) and backward (DS_B, blue) scene estimated with Mf and Mb epochs, respectively. The vertical dashed lines indicate the DS stimulation (1 s). In gray area, the running t-test results, where the waveforms from DS_F differ from that of DS_B ($p < 0.05$).

N2 component as being first generated from the occipital and associate parietal cortical areas with dominance related to time duration of the visual stimulus, at least 200 ms (in the study: 1 s) or by an unexpected visual motion onset.

The N2 VEP obtained during DS differs from that of SS. The application of DS increased the N2 peak and latency. Kuba et al. [2] observed that random radial motion (expansion / reduction of the scene) increased the N2 component when compared with translational motion (left – right). Moreover, the N2 can be attenuated by visual adaptation, as motion of the same orientation or as static landscape [3,4].

Although the P3 peak from SS and DS are similar, the N2 — P3 amplitude increases during DS, with large amplitude in parietal leads. However, no significant difference was found between DS_F and DS_B. This result is similar to that observed by [2,4], indicating that the P3 M-VEP component are not sensitive to motion orientation.

The running t-test showed difference in the DS_F — DS_B waveforms in the occipital and parietal derivations and in the pronounced positive peak after P3, with an increasing time delay from the occipital to central leads. These findings suggest that this difference perhaps represents a higher order of the cognitive, planning and motor processing to maintaining balance. Besides, the expansion of the scene evokes an illusion that something moves towards the subject, as pointed out by Da_Silva et al. [1]. Thus, it is in agreement with the physiological properties of the visual system, i.e. the sensitivity of the receptive fields to increasing size of the image at the retina and to motion velocity in the periphery of the visual field [2]. Therefore, the cortical response to maintain balance, as postural control strategies, depends on the virtual scene direction.

ACKNOWLEDGMENT

This work received the financial support from FAPERJ and the Brazilian Research Council (CNPq).

REFERENCES

1. Da_Silva P J G, Nadal J, Infantosi A F C (2013) Time evolution of the Event-related synchronization/desynchronization for investigating cortical response detection induced by dynamic visual stimuli. IFMBE Proc. vol. 33, Latin America Congress on Biomedical Engineering, Habana, Cuba, 2011, pp 1198–1201 DOI: 10.1007/978-3-642-21198-0_304
2. Kuba M, Kubová Z, Kremlácek J, Langrová J (2007) Motion-onset VEPs: Characteristics, methods, and diagnostic use. Vision Research 47:189–202 DOI: 10.1016/j.visres.2006.09.020
3. Kremlácek J, Hulan M, Kuba M, Kubová Z, Langrová J, Vít F, Szanyi J (2012) Role of latency jittering correction in motion-onset VEP amplitude decay during prolonged visual stimulation. Doc Ophthalmol 124:211-223 DOI: 10.1007/s10633-012-9321-6
4. Mercier M, Schwartz S, Michel C M, Blanke O (2009) Motion direction tuning in human visual cortex. European Journal of Neuroscience 29:424-434 DOI: 10.1111/j.1460-9568.2008.06583.x

Author: Antonio Fernando Catelli Infantosi
Institute: Biomedical Engineering Program (COPPE/UFRJ)
Street: P. O. Box 68.510
City: Rio de Janeiro
Country: Brazil
Email: afci@peb.ufrj.br

A Distributed Middleware for the Assistance on the Prevention of Peritonitis in CKD

M.A. Estudillo-Valderrama[1,2], A. Talaminos-Barroso[2], L.M. Roa[2,1], D. Naranjo-Hernández[1,2], and L.J. Reina-Tosina[3,2]

[1] CIBER de Bioingeniería, Biomateriales y Nanomedicina, Zaragoza, Spain
[2] Biomedical Engineering Group (GIB), University of Seville, Seville, Spain
[3] Department of Signal Theory and Communications, University of Seville, Spain

Abstract— This paper presents the feasibility study of using a distributed middleware for the management of infections from Chronic Kidney Disease (CKD) patients. The middleware follows the publish-subscribe pattern, and supports the OMG DDS (Data Distribution Service) standard, which facilitates the real-time monitoring of the distributed information, as well as the scalability and interoperability of the solution developed regarding the different stakeholders and resources involved. The proof of concept studied shows DDS viability for the activation of emergency protocols in real-time to manage alarms among the stakeholders subscribed to the information published by the infection detection system at patient's home.

Keywords— CKD, peritonitis, distributed middleware, real-time, publish-subscribe pattern.

I. INTRODUCTION

The application of Information and Communication Technologies (ICT) has revolutionized the domain of health care in many of its areas of practice. For instance, regarding the Chronic Kidney Disease (CKD) pathology, ICT can provide advanced capabilities for remote control sessions of hemodialysis and peritoneal dialysis, allowing data and alarms to be transmitted from home to the health organizations [1-3]. However, these developments are still not widespread implemented in everyday practice because there are several obstacles that hinder an effective implementation of ICT. Among them, it stands out the diversity of procedures, actors and stakeholders involved in the real-time exchange of information: patients, doctors, nephrology services at hospitals, ICUs, among others [4]. To this, it is added the heterogeneity and distribution of the agents and the different ICT infrastructures deployed, whose services and systems may be distant and non-interoperable.

In order to address the challenge described, this paper proposes the use of the publish-subscribe pattern as a basis for the exchange of information, where each agent plays the role of information producer, consumer or both at the same time. This communication paradigm enhances the autonomy among the components involved in a distributed architecture [5], as the telecare of chronic patients and in particular those suffering from CKD, might demand. From a technological point of view, the Data Distribution Service (DDS) [6] facilitates the efficient distribution of data in a distributed system for the communications based on this kind of paradigm. Among its most important advantages, DDS excels as platform independent from other messaging-oriented middlewares like JMS, and object-oriented as CORBA. In addition, the interoperability between software provided from different DDS vendors is also guaranteed thanks to the Interoperability Wire Protocol specification (DDSI). Moreover, the performance of DDS over competitors has also been analyzed [7] using benchmarking environments, which showed a significantly better performance compared to the publish-subscribe paradigms referred before (JMS and CORBA Notification Service). In conclusion, DDS seems to be a perfect candidate to fill the gap and satisfy the previously described requirement.

Thus, the main objective of this paper is the viability study of the application of the DDS specification for real-time clinical settings, which is performed through a particularly critical issue for CKD patients: the appearance of infection episodes called peritonitis. This study is developed in the context of a Research Project being carried out by the authors [8]. For this purpose, the next section briefly describes the distributed scenario required for the e-Nefro project, as well as the main features of the DDS approach to be considered from a methodological perspective. Later, in Section 3, the main variables to be handled by the DDS middleware are presented to prevent the development of peritonitis in real-time. Finally, in Section 4 the main results are analyzed to be followed by the conclusions.

II. METHODS AND MATERIALS

A. e-Nefro Distributed Scenario

The e-Nefro project (Adaptive Modular Architecture for the comprehensive telecare of the renal patient) is a multi-center effort between the University of Seville and several Spanish hospitals, whose goal is to establish an extensible and adaptable architecture for the remote care of pre-dialysis and peritoneal dialysis patients.

The Nephrology Services of the hospital partners in e-Nefro, and consequently the patients being monitored within the project, are geographically dispersed along Spain (Madrid, Canary Islands, and Andalusia autonomous communities). Thus, the aim is to provide these Services with a distributed architecture, as shown in Figure 1, able to assess differences and improve the procedures that might arise when dealing with information from patients at different locations. To this, it is added the need of developing a tool that could ease the real-time communication between the different stakeholders involved.

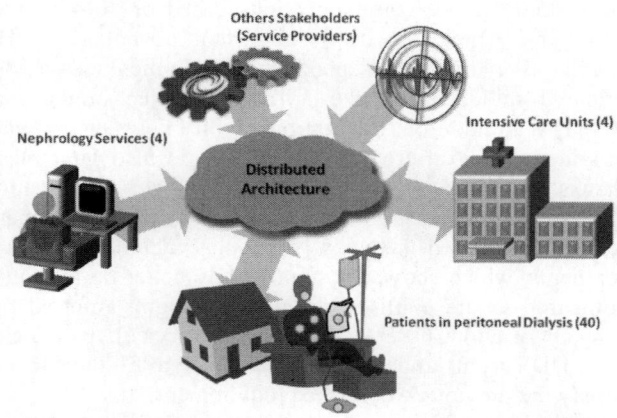

Fig. 1 e-Nefro distributed scenario. Yellow circles represent the approximate location of the sensors that compose the infection detection system at home. Numbers in parenthesis represent potential publishers/subscribers.

In particular, this paper studies the problematic of managing potential episodes of peritonitis coming from the 40 patients under study, which are controlled by the nephrology services from the different regions considered in the project, and how they can be optimally addressed regarding available resources. For this purpose, up to three different actor roles are defined that have access to the real-time data within the distributed architecture, in an effort to cover the most common stakeholders located in different sites: CKD patients in treatment with peritoneal dialysis (PD) (40), nephrologists (4), and intensive care units (4).

B. DDS Publish-Subscribe Pattern

Data Distribution Service (DDS) is a specification adopted by the OMG (Object Management Group) to standardize data-based communications based on the publish-subscribe pattern. The DDS communication is carried out within the so-called DDS domains, which represent isolated logical contexts where the publishers and subscribers coexist. The communication channel to exchange data between applications is called Topic, and is characterized by a name and a data structure that can vary dynamically at runtime [9].

Recently, DDS has been fully deployed in heterogeneous scenarios, ranging from financial and military environments up to other critical systems [10]. This great disparity is possible due to the wide spectrum of temporal requirements for real-time systems and high parameterization of communication of DDS-based applications.

DDS was originally designed to be used exclusively in controlled environments like isolated Local Area Networks (LAN). However, some scenarios require that different geographically dispersed and distributed applications must work together to achieve a single aim. In this regard, safety requirements concerning information and communication between DDS domains separated by a WAN (Wide Area Network) were not initially covered in the specification.

To address this problem, important efforts to extend the functionality of DDS are being carried out today. For example, the DDS-IS Routing Engine [11] enhances the need to improve interoperability of DDS-based heterogeneous applications that are located in different and possibly remote DDS domains. Other security-related aspects are also being addressed by the OMG for future standardization [12].

From the viewpoint of the portability of applications, commercial software vendors of DDS are working to transfer the technology to other environments where devices efforts are allowing DDS to keep growing continuously, adapting to the new technological advances and increasing integration, such as mobile communications [13] or cloud computing [14]. All technology developed by these manufacturers can communicate with each other thanks to the DDSI protocol, also adopted by the OMG, this way channeling all efforts towards a single direction and enhancing opportunities for interoperability and scalability.

III. RESULTS

A. Selection of Parameters

The nephrologists that participate in the e-Nefro project agreed to consider still valid the definitions of the ACCP/SCCM Consensus Conference [15]. Thus, physiological variables such as temperature, blood pressure, CO_2 partial pressure (pCO_2), heart rate and respiratory rate may be indicators of infection. In the case of use of infection detection under study, these thresholds are set as follows:

- 36°C < Temperature > 38°C
- 40mmHg < Systolic Blood Pressure < 90mmHg
- paCO2 Sensor < 32mmHg
- Respiratory Rate > 20breaths/min
- Heart Rate > 90 beats/min

Besides, the inclusion of a biosensor to detect the presence of bacteria in biological samples from the catheter, could be added (see Figure 1, bottom part).

B. DDS Application

The e-Nefro architecture must facilitate the communication of variable measurements from the system of infection detection. Besides it must automate, as far as possible, the detection and management of notifications that alert anomalous situations related to possible signs of peritonitis of the 40 patients.

In order to check the validity of DDS for the communication of data obtained for the CKD telecare, and in order to optimize the management of notifications resulting from this telecare, a proof of concept has been developed in which three types of alarm notification are defined:

1. Alarm Type 1: when one of the sensors plus the bacteria sensor, exceed the thresholds.
2. Alarm Type 2: when two or more sensors exceed the thresholds in an elapsed time of 1 hour.
3. Alarm Type 3: when no data is sent in a time period longer than 3 days.

This proof of concept requires the participation of the following actors:

- Agents at home, in this case, the infection detection system at home of PD patients, which collect information and publish it within the architecture,
- Health center agents (subscribed to topics), the ICUs, which will implement the most appropriate protocol for each event, and
- Supervisors agents, nephrologists, who are subscribed to the information so as to identify anomalous situations and trends.

Alarms will be activated based on the processing of the information published on the distributed architecture. There is a topic called CKD aimed at collecting all the information of a patient, shown below. Its structure is composed of the variables defined in subsection III.A, including an identification field for the patient, and two other fields to specify the dates on which the current and previous sample were collected.

Struct CKD {
  long id;
  string date;
  string previous_date;
  long temperature;
  long systolic_bp;
  long paCO2;
  long respiratory_rate;
  long heart_rate;
  boolean bacteria_sensor;
};

Figure 2 shows an example of this proof of concept, where all defined actors are within the same DDS domain.

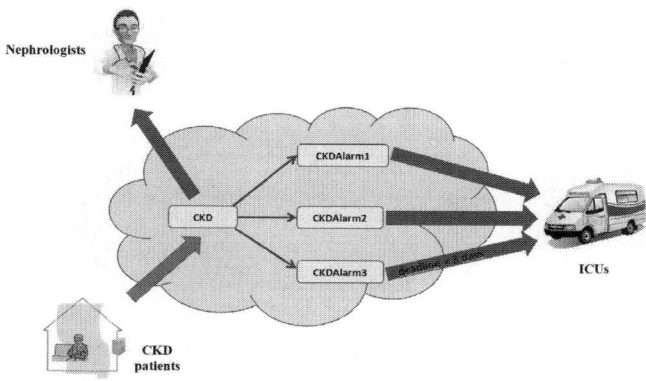

Fig. 2 Example of DDS application for eNefro

On the other hand, the different alarms considered have been managed through DDS as the following:

1. Alarm 1: Subscribers interested in this alarm perform a content-based subscription based on the CKD topic. For this, it is necessary to create a ContentFilteredTopic called CKDAlarm1. This topic is associated with CKD and specifies, using an SQL statement, the samples whose values exceed preset thresholds in subsection III.A. This way, actors subscribed to CKDAlarm1 only receive samples when the conditions established in the following statement are fulfilled:
   SELECT CKD where (bacteria_sensor = true) and ((temperature < 36) || (temperature > 38) || (systolic_bp < 40) || (systolic_bp > 90) || (paCO2 < 32) || (respiratory_rate > 20) || (heart_rate > 90))
2. Alarm 2: in this case, the interested subscribers are subscribed to the topic CKDAlarm2 which is also ContentFilteredTopic type. The SQL statement for this alarm incorporates an extra operator to compare the date of the current sample with the previous. Subscribers of this alarm only receive samples when this condition is satisfied.
3. Alarm 3: publishers and subscribers specify a deadline period of 3 days within the corresponding QoS policy. If the deadline is expired for a given patient and it does not send a sample during this period, the subscriber of this alarm will be automatically notified to handle it.

The implemented scenario uses a university license of RTI Connext software v5.0 [16]. This DDS implementation was chosen because of its robustness, high penetration in the industry and support for the extension of DDS Dynamic Topic Types. The implementation employs solely standardized characteristics of DDS specification, easing that future developments shall also be integrated using DDS implementations from other vendors.

IV. CONCLUSIONS

This paper studies the feasibility of a distributed middleware for managing alarms related to the occurrence of episodes of infection in CKD patients, based on the publisher-subscriber model and developed according to the DDS specification. The study demonstrates the potential of DDS as a communication infrastructure for real-time distributed systems strongly decoupled. In addition, other benefits of message-oriented intermediation software are: automatic discovery of information without knowing source's location, filtering of information through content-based subscriptions, independence of the platform or the programming language used granted by the DDS standard, the real-time support, and rich QoS policies that enhances the predictability in the delivery of information.

The proof of concept under study defines 40 participants (patients) pertaining to a DDS domain. However, the future increase of patients to be handled by e-Nefro platform does not substantially affect its performance. DDS allows thousands of participants without a significant performance impact [17], so the main restrictions come from network and machine constraints. Besides, the e-Nefro project will develop more complex scenarios that include not only the variables selected in the project for the detection of peritoneal infections, but any other information relevant to improving the quality of life of patients. Therefore, the work presented here is the first step in the development of the architecture. Besides, the opportunities offered by DDS and the publish-subscribe pattern in the healthcare environment have not yet been sufficiently exploited. Thus, the benefits of DDS in terms of independence of technologies and programming languages will lead to future developments likely to be published.

ACKNOWLEDGMENT

This work was supported in part by the Fondo de Investigaciones Sanitarias under Grant PI11/00111, in part by the Dirección General de Investigación, Tecnología y Empresa, Government of Andalucía, under Grants TIC-6214 and P08-TIC-04069, and in part by the CIBER de Bioingeniería, Biomateriales y Nanomedicina (CIBER-BBN) under intramural project PERSONA. CIBER-BBN is an initiative funded by the VI National R&D&i Plan 2008-2011, Iniciativa Ingenio 2010, Consolider Program, CIBER Actions and financed by the Instituto de Salud Carlos III with assistance from the European Regional Development Fund.

REFERENCES

1. Kaldoudi E, Vargemezis V (2010) Renal telemedicine and telehealth - Where do we stand? IFMBE Proceedings; 29: 920-3
2. Musso C, et al. (2012) Informatic nephrology. Int Urol Nephrol; 1-6
3. M.H. Rosner, C. Ronco (2012) Peritoneal Dialysis – State-of-the-Art. Contrib Nephrol, 178:68–73
4. Yee J, Faber MD (2012) Chronic Kidney Disease: Changing the Mean by Changing the Mien. Great Health Care, pp 143-157
5. Nageba E, Rubel P, Fayn J (2013) Towards an intelligent exploitation of heterogeneous and distributed resources in cooperative environments of eHealth. IRBM, 34(1): 79-85
6. OMG (2006) Data-Distribution Service for Real-Time Systems (DDS). v1.2. Tech. Rep., at http://www.omg.org/cgi-bin/doc?formal/07-01-01.pdf
7. Joshi R (2006) A Comparison and Mapping of Data Distribution Service (DDS) and Java Message Service (JMS), PhD, 2006
8. PI11/00111 - Adaptive Modular Architecture for the comprehensive telecare of the kidney patient. FIS, Instituto de Salud Carlos III, 2012-2014
9. OMG (2011) Extensible And Dynamic Topic Types For DDS (DDS-XTypes) 1.0 (FTF 2) - Beta 2. Tech. Rep
10. Bovenzi A, et al. (2013) JFIT: an Automatic Tool for Assessing Robustness of DDS-Compliant Middleware, Innovative Technologies for Dependable OTS-Based Critical Systems, pp. 69-81
11. Lopez-Vega JM, et al. (2013) A content-aware bridging service for publish/subscribe environments, J Syst Softw, 86(1):108–124
12. OMG Document (2011) DDS Security Extensions RFP Proposal.
13. Twin Oaks (2012) What can DDS do for Android?
14. An K, Pradhan S, Caglar F, Gokhale A (2012) A publish/subscribe middleware for dependable and real-time resource monitoring in the cloud», Procs of the Workshop on Secure and Dependable Middleware for Cloud Monitoring and Management, pp. 3:1–3:6
15. Bone RC, Balk RA, et al. (1992) The ACCP/SCCM Consensus Conference Committee. Chest, 101(6): 1644-1655
16. RTI (2013) Real-Time Innovations (RTI) DDS Data Distribution Service. http://www.rti.com/
17. RTI (2013) Ultra Low Latency Messaging for Financial Services, at https://www.rti.com/docs/RTI_DDS_FS.pdf

Author: Miguel Ángel Estudillo Valderrama
Institute: Biomedical Engineering Group, University of Seville
Street: Av. Descubrimientos s/n (41092)
City: Seville
Country: Spain
Email: mestudillo@us.es

Bioactive Nanoimprint Lithography: A Study of Human Mesenchymal Stem Cell Behavior and Fate

Z.A. Cheng[1,2], O.F. Zouani[1], K. Glinel[2], A.M. Jonas[2], and M.-C. Durrieu[1]

[1] Institut Européen de la Chimie et Biologie, CBMN-UMR5248, Université de Bordeaux 1, Pessac, France
[2] Institut de la matière condensée et des nanosciences, Université catholique de Louvain, Louvain-la-Neuve, Belgium

Abstract—Biomaterials aim to mimic *in vivo* extracellular matrices where cell interactions occur on the nanoscale. Thus, incorporation of nanosized components is interesting in the preparation of bioactive surfaces. We present a technique using nanoimprint lithography to create chemical nanopatterns on silicon surfaces functionalized with bioactive motifs. Due to high throughput and versatility, a wide range of geometries and dimensions can be efficiently patterned. In our study, we prepared and characterized two types of bioactive nanodots (150 nm diameter with 350 nm spacing, and 80 nm diameter with 110 nm spacing) functionalized with cell adhesion-promoting RGD peptides. We examined mesenchymal stem cell adhesion and commitment on these modified material surfaces with respect to homogeneous RGD and non-functionalized surfaces. We report that bioactive nanostructures induce fibrillar adhesions on human mesenchymal stem cells with an impact on their behavior and dynamics specifically in terms of cell spreading, cell-material contact, and cell differentiation.

Keywords—nanoimprint lithography, mesenchymal stem cell, surface functionalization focal adhesion, differentiation.

I. INTRODUCTION

In biomaterials design, an effective product combines the right type of cells, an appropriate scaffold, and a smart choice of signaling molecules to be incorporated in the system [1]. Human mesenchymal stem cells (hMSCs) are widely used in regenerative medicine as their multi-potent capabilities show promise in repairing damaged tissues and organs [2]. In addition, nanopatterned scaffolds have gained interest in tissue engineering due to the unique potential of mimicking an *in vivo* extracellular matrix (ECM) [3]. It is important to understand the mechanisms by which stem cells differentiate in the nanoscale environment, particularly in relations to cell dynamics such as responses in cell morphology and focal adhesions (FAs) to bioactive nanostructures. We develop a method of preparing and characterizing nanostructured RGD motifs using nanoimprint lithography (NIL) and surface functionalization [4,5]. We demonstrate that bioactive surface nanostructures have an impact on hMSC behavior by inducing fibrillar FA formation, increasing cell-material contact, and affecting hMSC commitment.

II. MATERIALS AND METHODS

A. Materials

Silicon wafers were purchased from Active Business Company GmbH, Germany. Poly(methyl methacrylate) (PMMA) was purchased from Agilent Technologies, Belgium. 3-aminopropyldimethylethoxysilane (APDMS) and 2-[Methoxypoly(ethyleneoxy)propyl]trimethoxysilane (PEO silane) were purchased from ABCR GmbH, Germany. Dry dimethylformamide (DMF) and 3-Succinimidyl-3-MaleimidoPropionate (SMP) were purchased from Sigma-Aldrich, France. Dry toluene was purchased from Fisher Scientific, Belgium. Customized GRGDSPC peptides were synthesized by Genecust, Luxembourg.

B. Nanoimprint Lithography

PMMA was spin-coated onto Si to create a polymer mask. A Si mold was pressed onto the polymer mask using an Obducat nanoimprinter. The sample was heated at 170 °C for 3 minutes, then the pressure was increased to 60 bars and left for 3 minutes to perform the imprint. The system was then cooled down to 70 °C and the mold was detached from the sample. Samples were subjected to a descum process in O_2 plasma to remove residual PMMA from the patterned regions.

C. Surface Functionalization and Characterization

Imprinted samples were placed in a Schlenk reactor injected with APDMS. The reaction was run overnight at 80°C. Samples were then washed in acetone using a Soxhlet apparatus, and immersed overnight in a solution of non-adhesive PEO silane in dry toluene. Passivated samples were immersed in a SMP solution in dry DMF for 2 hours at room temperature. Samples were then rinsed with Milli-Q water and immersed in an RGD peptide solution. The reaction was run for 4 hours at room temperature under gentle agitation. After surface preparation, atomic force microscopy (AFM) was performed in contact mode on functionalized nanopatterned samples to characterize surface topography and roughness.

D. Cell Culture and Immunofluorescence

Human mesenchymal stem cells (hMSCs) were seeded at a density of 10^4 cells/cm^2. After 24 hours or 4 weeks in culture (for adhesion and differentiation studies, respectively), hMSCs were fixed with paraformaldehyde. Fixed cells were permeabilized and blocked with 1% bovine serum albumin. Cells were then incubated with primary antibody and secondary antibody successively, then stained for F-actin using Alexa Fluor® 488 phalloidin. Cell nuclei were counterstained with DAPI. After DAPI staining, samples were mounted with coverslips on microscope slides.

E. Image Acquisition and Analysis

A Leica DM5500B epifluorescence microscope was used to image hMSCs. Cells with morphologies representative of each condition were imaged. Fluorescent images of at least 50 cells at each surface condition were taken for quantitative analysis, both for adhesion and differentiation studies. Quantification of FA count, FA area, projected cell area, and STRO-1 expression was carried out using ImageJ software.

III. RESULTS

A. Nanopattern Characterization

Two types of ordered nanodots ($D_{150}S_{350}$ and $D_{80}S_{110}$, with D and S denoting the nanodot diameter and inter-dot spacing in nanometer, respectively) were prepared. AFM was performed in contact mode on both types of surfaces to show the chemical contrast between the PEO background and the nanopatterned regions (Figures 1A and 1B for $D_{150}S_{350}$ and $D_{80}S_{110}$ respectively). Data was acquired in height mode to obtain chemical topography, with 3D rendering shown in Figures 1C and 1D.

B. Stress Fiber Organization and Focal Adhesion Formation

To evaluate cell behavior, hMSCs were cultured on RGD-grafted $D_{150}S_{350}$ and $D_{80}S_{110}$ surfaces as well as bare, polished silicon surfaces (Si poli) and homogeneous peptide-grafted silicon surfaces (RGD H) as controls. Cell morphology was observed at each condition (Figure 2). Adherent hMSCs on bare Si samples are smaller and lacked defined cytoskeletal organization, whereas on RGD H samples, cells are larger with a more organized cytoskeletal structure as shown by the arrangement of the F-actin stress fibers. $D_{150}S_{350}$ and $D_{80}S_{110}$ show a mixture of cell shapes and sizes, but cytoskeletal arrangement remains organized with defined stress fibers.

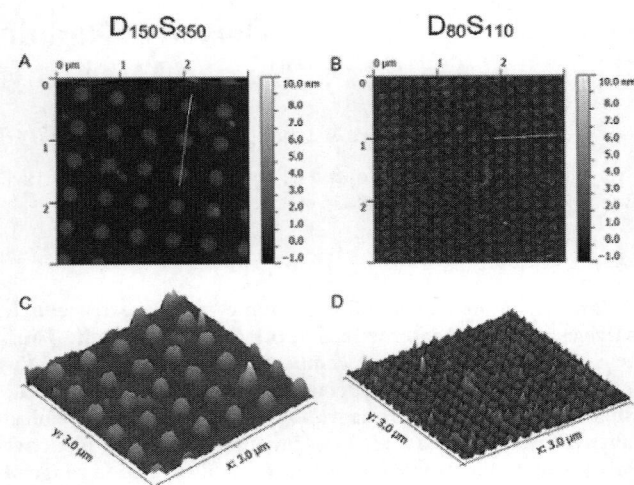

Fig. 1 Contact mode AFM is performed on (A) $D_{150}S_{350}$ and (B) $D_{80}S_{110}$ surfaces after the grafting of SMP (hetero-bifunctional cross-linker) in the nanopatterns. The difference in intensity between the background and the nanodots is indicative of topographical and chemical contrast between the SMP and the PEO silane. 3D rendering of the (C) $D_{150}S_{350}$ and (D) $D_{80}S_{110}$ surfaces was performed to illustrate chemical topography.

To observe the formation of FAs, we stained for vinculin, an important protein at the site of integrin-mediated FAs. Figure 2 highlights the typical appearance of FAs found on each type of surface with magnified views of selected regions shown in Figures 2A to 2D. FAs on Si poli were scarce, while RGD H induced thin, sparse clusters of vinculin both around the periphery and around the nucleus of the cell. In contrast, FAs were concentrated exclusively around the cell periphery on $D_{150}S_{350}$ and $D_{80}S_{110}$, with thicker and more pronounced fibrillar contacts, representing locally concentrated integrin clustering. Notably, FAs on RGD H are arranged in a random fashion, whereas on $D_{150}S_{350}$ and $D_{80}S_{110}$, the elongated FAs are aligned along the orientation of the stress fibers.

C. Focal Adhesion Area

To implicate the role of FA clustering, we quantified the area of each FA. For each type of material, the individual FAs are classified based on their area: > 25 um^2, $10 - 25$ um^2, $5 - 10$ um^2, and < 5 um^2. Figure 3A is a hMSC on a $D_{150}S_{350}$ surface with FAs in the first three classes. Magnified views of FAs are shown in Figures 3B, 3C, and 3D for FA areas of 25 um^2, 10 um^2, and 5 um^2 respectively. Figure 3E expresses the number of FAs in each class as a percentage of the total number of FAs for each material. FAs on both $D_{150}S_{350}$ and $D_{80}S_{110}$ were more abundant in each class compared with homogeneous surfaces, whether bare Si or RGD-grafted.

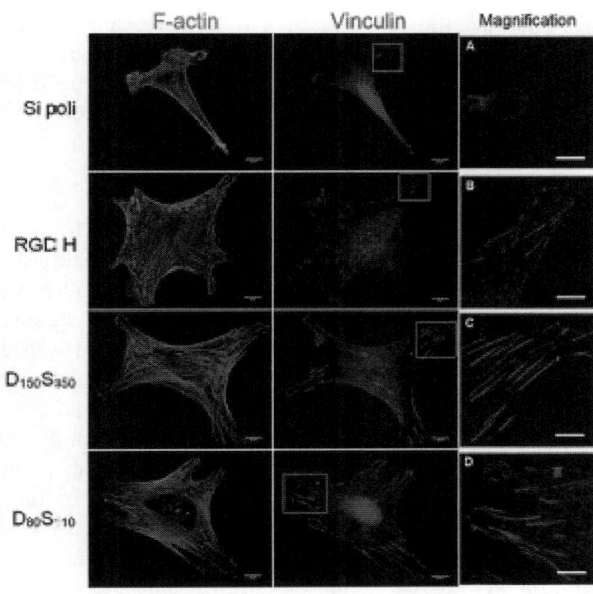

Fig. 2 Immunofluorescent staining showing typical morphologies of hMSCs grown on various substrates, with F-actin in green and vinculin in red, scale bar = 20 μm. Cells are larger on RGD-grafted substrates, with a more organized cytoskeletal structure and more distinct points of focal adhesion on nanopatterns as represented by vinculin clustering. Magnifications of focal adhesions are shown in (A) – (D), scale bar = 10 μm.

D. STRO-1 Expression

To investigate whether FA fibrillar shape formation induced by nanopatterns changes hMSC commitment, we cultured hMSCs for 4 weeks and stained for STRO-1, a mesenchymal stem cell-specific marker (Figures 4A to 4D). The STRO-1 fluorescence signal was quantified by measuring the mean signal density of each cell (Figure 4E). We noted a significant decrease in STRO-1 expression on RGD H, and again a significant decrease on nanopatterned RGD-grafted surfaces. A lower STRO-1 activity implies that stem cell population has decreased over 4 weeks on nanopatterns relative to homogeneous controls. While a part of the population is still STRO-1 positive (hence retaining its "stemness"), the decrease indicates that more cells have differentiated on nanopatterns than homogeneous controls.

IV. DISCUSSION

Stem cell differentiation is affected by molecular composition and physical forces present in their ECM environment [6]. Nanoscale topographies have a noticeable effect on stem cell behavior and fate [7]. The study of hMSCs on bioactive nanopatterned surfaces offers interesting insights. We prepared materials with chemical

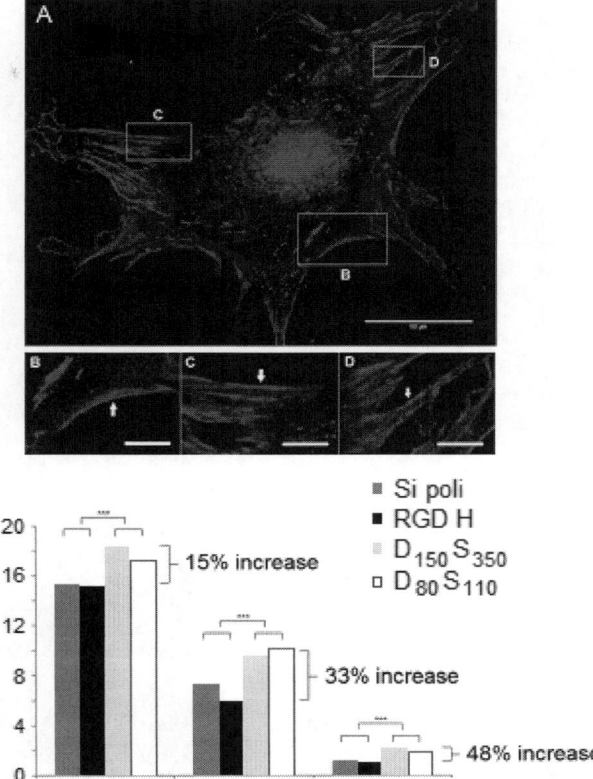

Fig. 3 A hMSC expressing vinculin is shown in (A), scale bar = 50 μm. FAs with sizes (B) 25 μm^2, (C) 10 μm^2, and (D) 5 μm^2 are magnified, scale bar = 10 μm. FAs are sorted into three classes. The number of FAs (vinculin clusters) in each class is expressed as a percentage of the total number of FAs (E). Large FAs are more abundant in nanopatterns compared with control surfaces, with a 48% increase for FA areas > 25 um^2, 33% increase for FA areas between 10 and 25 um^2, and a 15% increase for FA areas between 5 and 10 um^2. Two sample t-test was used to compare significant difference between the percentages. *** represents a *p-value* of less than 0.05.

nanopatterns, providing a bioactive stage for specific cell response. The onset of efficient cell adhesion requires the clustering of integrins and their receptors which, when mature, establish FAs and enhance adhesive strength, in turn initiating a host of related signal transductions [8]. We note the formation of fibrillar FAs on nanopatterned surfaces (Figure 2), which are absent on bare Si and RGD H surfaces. Moreover, these fibrillar FAs are aligned in the direction of stress fibre elongation, serving as anchors for that maintain fiber tension. On the other hand, FAs on RGD H orient in a random and disordered way, without the fibrillar shape and cytoskeletal alignment observed on nanopatterned surfaces. Since cytoskeletal contractility is implicated in cellular signal transduction, the alignment of fibrillar FAs with stress fibers

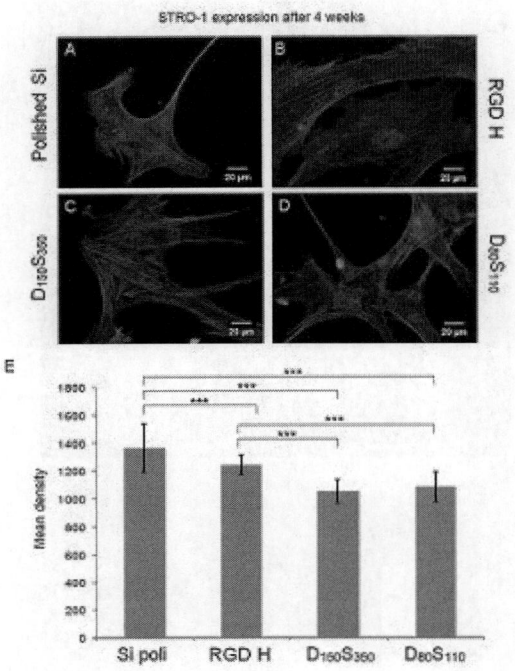

Fig. 4 Commitment studies of hMSCs after 4 weeks. (A) – (D) STRO-1, a hMSC marker, is immunofluorescently stained and shown in red, with F-actin stained in green and cell nucleus in blue, scale bar = 20 μm. (E) The amount of STRO-1 present in the cell is expressed as mean fluorescent density. Compared with bare Si controls, STRO-1 activity is lower on RGD H surfaces and still lower on patterned RGD surfaces, both $D_{150}S_{350}$ and $D_{80}S_{110}$, indicating that the cells have lost some "stemness" and differentiated into mature lineages. *** represents a p-value of less than 0.05.

may upregulate this contractility through tension caused by the pulling action of FAs, inducing stem cell behaviors such as differentiation. The amount of mature fibrillar FAs is also significantly greater on surfaces with nanopatterned bioactivity (Figure 3), confirming the impact of nanopatterns on FA configuration.

As FAs have a direct effect on cell mechanotransduction and signaling pathways, we attempted to establish a direct link between FA activity and hMSC commitment and performed differentiation-specific immunofluorescence staining on hMSCs cultured on various surfaces for 4 weeks. STRO-1 is expressed when cell stemness is present (Figure 4). The decrease in STRO-1 expression on $D_{150}S_{350}$ and $D_{80}S_{110}$ relative to RGD H is a sign that cells are less "stem" on nanopatterns than RGD H surfaces after 4 weeks in culture, indicating that they have differentiated into a mature cell lineage. We can attribute this change in commitment behavior to the way the FAs form stable contacts between the cells and the substrate, causing changes in the cytoskeletal contractility and altering the cell mechanism and chemical signal pathways. hMSC multi-potency allows them to differentiate into any of a variety of mature lineages, and it is currently unknown whether our nanopatterns favor a particular route for hMSC differentiation. Future comprehensive differentiation studies will address this interest.

V. CONCLUSIONS

The patterning technique we have developed in this study allows the deposition of biomolecules on nanopatterned surfaces for the study of stem cells on spatially organized bioactivity. NIL is a comprehensive technique of surface fabrication as it is versatile in terms of the geometries that can be patterned and the types of biomolecules that can be grafted. As we have demonstrated in our study that nanopatterned RGD peptides can induce a noticeable effect on the specific fibrillar adhesion and differentiation behavior of hMSCs, the same approach can be applied on other types of biomolecules to examine the lineage commitment and differentiation of hMSCs in future studies.

ACKNOWLEDGEMENT

The authors thank André Crahay, Cédric Burhin, Sylvie Derclaye, Bernard Nysten (UCL, Belgium), Yifeng Lei (IECB, Bordeaux) for technical assistance, and the Région Aquitaine and Wallonie-Bruxelles International for financial support.

REFERENCES

1. Liao S, Chan CK, Ramakrishna S. (2008) Stem cells and biomimetic materials strategies for tissue engineering. Mater Sci Eng C 28:1189-1202
2. Caplan AI. (2007) Adult Mesenchymal Stem Cells for Tissue Engineering Versus Regenerative Medicine. J Cell Physio 213:341-347
3. Anselme K, Davidson P, Popa AM et al. (2010) The interaction of cells and bacteria with surfaces structured at the nanometre scale. Acta Biomater 6:3824-3846
4. Pallandre A, Glinel K, Jonas AM et al. Binary Nanopatterned Surfaces Prepared from Silane Monolayers. Nano Lett 4:365-371
5. Porté-Durrieu MC, Guillemot F, Pallu S et al. (2004) Cyclo-(DfKRG) peptide grafting onto Ti–6Al–4V: physical characterization and interest towards human osteoprogenitor cells adhesion. Biomaterials 25:4837-4846
6. Discher DE, Mooney DJ, Zandstra PW. (2009) Growth Factors, Matrices, and Forces Combine and Control Stem Cells. Science 324:1673-1677
7. Dalby MJ, Gadegaard N, Tare R et al. (2007) The control of human mesenchymal cell differentiation using nanoscale symmetry and disorder. Nat Mater 6:997-1003
8. Adams JC. (2001) Cell-matrix contact structures. Cell Mol Life Sci 58:371-392

Author: Zhe A. CHENG
Institute: Institut Européen de la Chimie et Biologie
Street: 2 Rue Escarpit
City: Pessac
Country: France
Email: za.cheng@iecb.u-bordeaux.fr

Automated Normalized Cut Segmentation of Aortic Root in CT Angiography

Mustafa Elattar[1], Esther Wiegerinck[2], Nils Planken[3], Ed vanbavel[1],
Hans van Assen[4], Jan Baan Jr.[2], and Henk Marquering[1,2]

[1] Department of Biomedical Engineering and Physics
[2] Deparment of Cardiology
[3] Department of Radiology, Academic Medical Center, University of Amsterdam, 1105AZ Amsterdam, Netherlands
[4] Department of Electrical Engineering, Eindhoven University of Technology, 5600 MB Eindhoven, Netherlands

Abstract—Transcatheter Aortic Valve Implantation (TAVI) is a new minimal-invasive intervention for implanting prosthetic valves in patients with aortic stenosis. This procedure is associated with adverse effects like paravalvular leakage, stroke, and coronary obstruction. Accurate automated sizing for planning and patient selection is expected to reduce these adverse effects. Segmentation of the aortic root in CTA is pivotal to enable automated sizing and planning. We present a fully automated segmentation algorithm to extract the aortic root from CTA images consisting of a number of steps: first, ascending aorta and aortic root centerline were extracted. Subsequently, high intensities due to calcifications are masked to improve segmentation. Next, the aortic root is represented in cylindrical coordinates. Finally, the aortic root is segmented using 3D normalized cuts. We validated the method against manual delineations by calculating Dice coefficients and average distances. The method successfully segmented the aortic root in all 20 image datasets. The mean Dice coefficient was 0.945±0.03 and mean radial absolute error was 0.74 ± 0.39 mm. The proposed algorithm showed accurate results compared to manual segmentations.

Keywords—Aortic Root, Medical Image Segmentation, Normalized Cut, TAVI, CT Angiography.

I. INTRODUCTION

Aortic stenosis is the most common heart valve disease. Approximately one third of all patients with severe symptomatic aortic stenosis are not eligible for surgery, mainly because of high age, left ventricular dysfunction or other co-morbidities [1]. Transcatheter aortic valve implantation (TAVI) has recently been introduced as an alternative treatment for these high-risk patients. TAVI provides sustained clinical and hemodynamic benefits in selected high-risk patients declined for conventional aortic valve replacement [2, 3]. However, TAVI is associated with a number of potential adverse effects, such as paravalvular leakage, coronary obstruction, and conduction disorders [4]. These potential adverse effects may be reduced with improved patient selection, intervention planning, and aortic sizing with the assistance of imaging and image analysis. For better sizing, automated image analysis is required that supports preoperative planning and allows alignment of pre-operative CT data with intra-operative imaging, providing additional 3D information during the procedure.

A large number of studies have focused on artery segmentation, including the aortic segmentation in particular [5–7]. To our knowledge four studies have presented methods for the segmentation of the aortic root in CT-based images. Zheng et al. have introduced a fully automatic segmentation of the aortic root in peri-procedural planning C-arm images using marginal space learning [8]. Lavi et al. have proposed a 2-D watershed based algorithm to detect the aortic root in CT axial images. This proposed technique was semiautomatic and showed inadequate accuracy in low quality volumes [9]. In [10], all heart chambers and the aortic structure were extracted in CTA images using a model based segmentation technique. Also Grbic et al. have proposed the usage of a new constrained multi-linear shape model conditioned by anatomical measurements to segment the heart valves in 4D Cardiac CT [11]. The studies [8], [9] and [11] did not address the effect of calcifications, which are quite common in our patient population, on their performance. All pre-mentioned studies did not perform an inter-observer analysis.

In this paper, we introduce an algorithm for localizing and segmenting the aortic root in 3D CTA images. We localized the aortic root structure using fuzzy classification. In this aortic root, the centerline was determined, which was used to represent the volume of interest in cylindrical coordinates. The segmentation was performed using a 3D normalized cut method in the cylindrical coordinates. We evaluated our proposed technique by comparing the automated segmentation with expert manual delineations for 20 cases using radial variations, Dice-metric, and accuracy measures.

II. METHODS

A. Aortic Root Localization and Centerline Estimation

To segment the aortic root and ascending aorta in 3D volume, it is important to have initial estimation of the position and its centerline to guide the segmentation process. This centerline is also used to resample the volume in cylindrical coordinates. We propose a detection algorithm based on morphological operators, connected component analysis and fuzzy classification. Figure 1 shows a schematic overview of the proposed algorithm. We down-sampled the volume to reduce computation time and smoothed the volume using a 3D Gaussian filter to reduce the noise level. We applied thresholding based on Hounsfield units and eroded the volume to remove connectivity between artery structures. Connected component analysis was used to identify separated objects based on 6-voxels connectivity. Image volumes without the presence of an aortic arch were detected by calculating the area of the ascending and descending aorta in the superior slices of the volume. Based on the arch existence, a cost function was used to calculate a value for each connected structure. This cost was used to identify the aorta structure. We included seven features in the cost function: (1) center of mass; (2) the most superior position; (3) ratio of the principal components of the axial projection of object voxels; (4) volume; (5) anterior-posterior range; (6) height; (7) spread in caudal-cranial direction. To separate the ascending aorta and aortic root from the whole aorta structure, we calculated the maximum area of the aorta in each slice. We aligned a predefined reference curve defining the start and the end of the ascending aorta. Given the final object of interest, we calculated the centerline and extended it in the direction of the aortic root to cover the region of the Left Ventricle Outflow Tract (LVOT) using cubic-spline.

B. Preprocessing

Using the aortic root centerline, the aortic root was represented in cylindrical coordinates, resulting in three new dimensions; radius r, angle θ, and length along the root axis z (figure 2). Reformatting was performed using nearest neighbor interpolation. To cover the whole diameter of the aortic root, which ranges between 20 and 40 mm [12], we sampled a radius of 30 mm with 70 steps. The angle was sampled with 64 points, and the aortic root axis was sampled with 100 samples with a step size of approximately 0.4 mm. The high intensity of voxels with calcium was adjusted by assigning contrast-enhanced blood like intensities.

To reduce the noise, a 3D Gaussian filtering was performed with a standard deviation of 0.63 mm in the z- and r-direction and of 8.44° degrees along θ. The filter size was 3mm*3mm*25°.

C. Normalized Cut Segmentation

The normalized cut is a graph based segmentation technique which measures the total dissimilarity between different groups as well as the total similarity within the group [16]. A graph g is composed of nodes v and edges ε as shown in equation (1). Nodes v represent the cylindrical volume voxels. Edges ε are connecting neighboring voxels N. The normalized cut segmentation can be performed by selecting a minimal cost cut through the graph, which is associated with the separation between the structure of interest and other non-interesting structures.

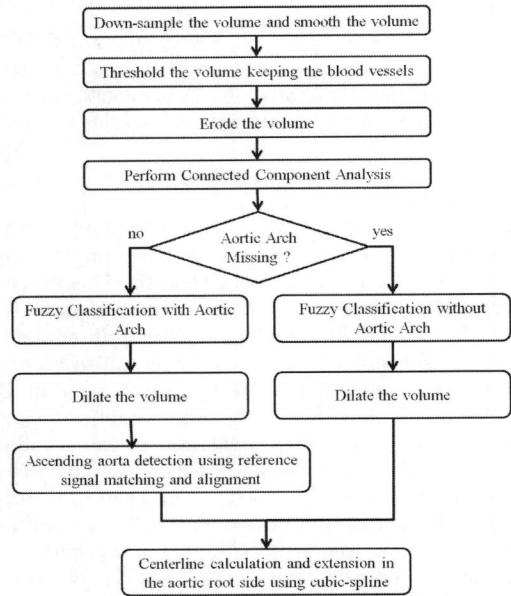

Fig. 1 Schematic overview of the proposed algorithm

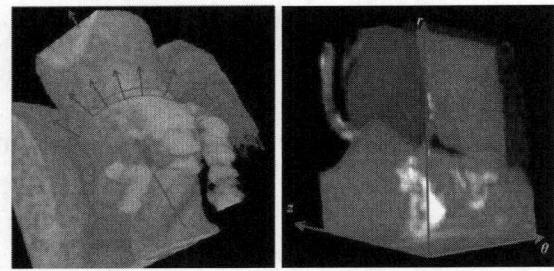

Fig. 2 (Left) 3D aortic root and 3 colored axes showing the sampling dimensions, the radial direction (red arrow), angular direction (green circles) and the root axis (blue axis). (Right) The same aortic root represented in 3D cylindrical coordinates.

The cost of the cut C is defined as the sum of the costs of its associated edges e, where ω_e is the cost of the single cut along edge e. These weights ω_e were calculated using the gradient magnitude $G(p,q)$ of the intensity between these voxels. The weights were scaled by a normalization factor σ^2 to steer towards large-scale segmentations.

$$\text{Cost}(C) = \sum_{e \in C} \omega_e \quad (1)$$

$$\omega_e = \omega_{(p,q)} = \omega_{(q,p)} = \begin{cases} 0 & Dist(p,q) > 1 \\ e^{\left(-\frac{G(p,q)}{\sigma^2}\right)} & Dist(p,q) \leq 1 \end{cases} \quad (2)$$

Normalized cut has been proposed by Shi et al. [13] introducing a solution for the graph cut technique drawbacks, e.g. the difficulty of cutting between nodes that should remain connected [14].

$$NCut(A,B) = \frac{C(A,B)}{assoc(A,\epsilon)} + \frac{C(A,B)}{assoc(B,\epsilon)}, \text{ where}$$

$$assoc(A,\epsilon) = \sum_{e \in A} \omega_e, \quad (3)$$

where $C(A,B)$ represents a cut separating groups A and B. Edges were calculated only for 6-connected neighborhood voxels (figure 3). The gradients were calculated in the radial direction only, with a scale of 1.26 mm to reduce the effect of vertical gradients caused by calcifications $G = \frac{\partial I}{\partial r}$, where I is the intensity.

This technique is computationally demanding for large volumes. The use of a sparse matrix and Eigen vectors is strongly contributing to the complexity of the calculations, which is $O(n^3)$ [13], where n is the number of nodes. Our volume of interest consists of up to 100 slices. To keep the computation time limited, we applied the normalized cut on separated patches of smaller volumes of 10 slices per patch or less. As a post processing step, a 3D averaging filter was used to smooth the segmented surface. Thereby, we reduced any discontinuities between segmentation results of the various patches.

D. Validation

We assessed the accuracy of the localization and segmentation. The localization and centerline detection were evaluated visually. The accuracy of the segmentation was assessed by comparing the segmentation results with manual delineations of the aortic root by two experienced observers. The observers delineated the aortic root at three oblique planes: at the annular level, at the level of the maximum bulging of the aortic sinuses, and at the level of the ascending aorta. We employed two accuracy measures: (1) average distance between two contours, which is calculated from the center to the contour and averaged over the angle; (2) the Dice metric (figure 3). The accuracy was compared to the inter-observer variability. We compared the performance of our proposed algorithm with other literature methods.

III. RESULTS

The algorithm was applied to 20 datasets for 20 consecutive candidates for TAVI, after optimizing the algorithm on a training set of 10 patients. The aortic root detection algorithm was perfect by detecting all 20 cases. Figure 4 shows some examples of segmentation results and manual delineations. The Bland Altman plot in figure 5 shows the error of the normalized cut segmentation versus manual delineations. Results of interobserver analysis and the proposed technique evaluation are shown in table 1. The Dice coefficient of the automated technique and observer 1 was 0.945±0.03. The inter-observer Dice coefficient was 0.951±0.01. The performance of this and previous published methods is shown in table 2.

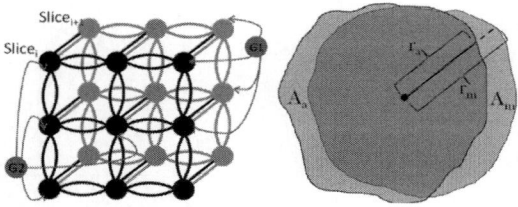

Fig. 3. (Left) A schematic representation of a graph of two successive slices and edge connections between the 6-connected nodes. (Right) Radial distance and area measures. Subscripts m and a stand for manual and automatic segmentation respectively.

IV. DISCUSSION & CONCLUSIONS

We have introduced a fully automatic aortic root segmentation that uses connected component analysis, fuzzy classification, normalized cuts and additional pre-processing and post-processing. The proposed technique was evaluated by comparison with manual delineation and showed good accuracy in terms of Dice coefficient and radial distance error. The proposed algorithm proves its high performance in comparison with work done in [8], [9] and [11]. Although [10] has better performance in terms of mean error, The complexity of this method was not reported.

Table 1 The performance of proposed algorithm compared with observer 1 and the inter-observer variability.

	Dice Coefficient [Range]	Distance Error in mm [Range]
3D Normalized Cut vs. Observer1	0.945 ± 0.03 [0.85-0.988]	0.74 ± 0.39 [0.21: 1.98]
Inter-observer Variability	0.951 ± 0.03 [0.85 : 0.986]	0.68 ± 0.34 [0.21: 1.81]

Table 2 The performance of proposed algorithm compared with other literature methods.

	Modality	Subjects	mm/pixel	Automatic	Speed	Mean Mesh Error
Grbic et al.[11]	4D CT	640	0.28-1.00	Yes	N/A	1.22 mm
Lavi et al.[9]	CTA	34	N/A	No	N/A	N/A
Waechter et al.[10]	CT	20	N/A	Yes	N/A	0.5 mm
Zheng et al.[8]	C-arm CT	276	0.70-0.84	Yes	0.8 sec	1.08 mm
Proposed Algorithm	CTA	20	0.39-0.45	Yes	80 sec	0.74 mm
Interobserver Variability	CTA	20	0.39-0.45	No		0.68 mm

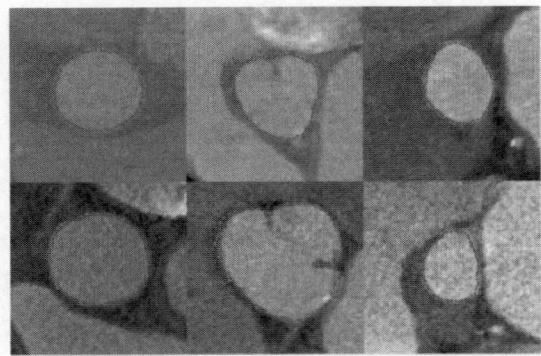

Fig. 4 Six images for three different planes showing the automatic segmentation in red and the manual delineation in green.

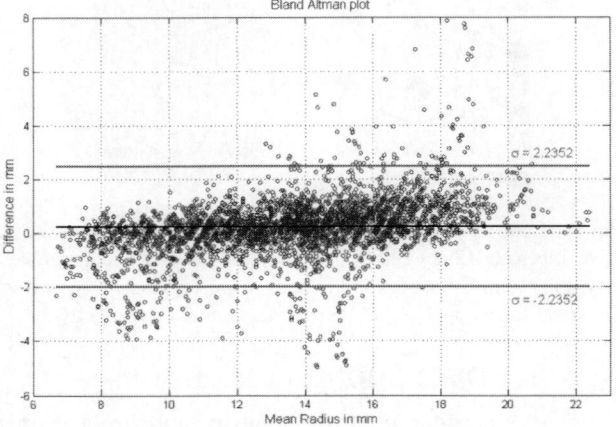

Fig. 5 Bland Altman plot shows the error for different radiuses.

A number of commercial solutions for aortic root segmentations have been introduced (e.g. Delgado et al.) [15]. However, these segmentation methods and their accuracy have not been addressed in detail in the literature. This method has the potential to be important for better planning and patient selection. We have now a solid foundation for aortic root segmentation, which enables us to segment the aortic root landmarks correctly.

ACKNOWLEDGMENT

The authors wish to thank for the support from the Technology Foundation STW, The Netherlands under Grant 11630.

REFERENCES

1. Iung, B.: A prospective survey of patients with valvular heart disease in Europe: The Euro Heart Survey on Valvular Heart Disease. European Heart Journal. 24, 1231–1243 (2003).
2. Vahanian, A., Alfieri, O., Al-Attar, N., Antunes, M., Bax, J., Cormier, B., Cribier, A., De Jaegere, P., Fournial, G., Kappetein, A.P., Kovac, J., Ludgate, S., Maisano, F., Moat, N., Mohr, F., Nataf, P., Piérard, L., Pomar, J.L., Schofer, J., Tornos, P., Tuzcu, M., Van Hout, B., Von Segesser, L.K., Walther, T.: Transcatheter valve implantation for patients with aortic stenosis: a position statement from the European Association of Cardio-Thoracic Surgery (EACTS) and the European Society of Cardiology (ESC), in collaboration with the European Association of Percu. European heart journal. 29, 1463–70 (2008).
3. Ye, J., Cheung, A., Lichtenstein, S. V, Nietlispach, F., Albugami, S., Masson, J.-B., Thompson, C.R., Munt, B., Moss, R., Carere, R.G., Jamieson, W.R.E., Webb, J.G.: Transapical transcatheter aortic valve implantation: follow-up to 3 years. The Journal of thoracic and cardiovascular surgery. 139, 1107–13, 1113.e1 (2010).
4. Baan, J., Yong, Z.Y., Koch, K.T., Henriques, J.P.S., Bouma, B.J., Vis, M.M., Cocchieri, R., Piek, J.J., De Mol, B. a J.M.: Factors associated with cardiac conduction disorders and permanent pacemaker implantation after percutaneous aortic valve implantation with the CoreValve prosthesis. American heart journal. 159, 497–503 (2010).
5. Duquette, A.A., Jodoin, P.-M., Bouchot, O., Lalande, A.: 3D segmentation of abdominal aorta from CT-scan and MR images. Computerized medical imaging and graphics: the official journal of the Computerized Medical Imaging Society. 36, 294–303 (2012).
6. Uday Kurkure , Olga C . Avila-Montes , Ioannis A . Kakadiaris.Automated Segmentation Of Thoracic Aorta In Non-Contrast Ct Images 29–32 (2008).
7. Isgum, I., Staring, M., Rutten, A., Prokop, M., Viergever, M. a, Van Ginneken, B.: Multi-atlas-based segmentation with local decision fusion--application to cardiac and aortic segmentation in CT scans. IEEE transactions on medical imaging. 28, 1000–10 (2009).
8. Zheng, Y., John, M., Liao, R., Nöttling, A., Boese, J., Kempfert, J., Walther, T., Brockmann, G., Comaniciu, D.: Automatic aorta segmentation and valve landmark detection in C-arm CT for transcatheter aortic valve implantation. IEEE transactions on medical imaging. 31, 2307–21 (2012).
9. Lavi, G., Lessick, J., Johnson, P.C., Khullar, D.: Single-Seeded Coronary Artery Tracking in CT Angiography. 00, 3308–3311 (2004).
10. Waechter, I., Kneser, R., Korosoglou, G., Peters, J., Bakker, N.H., Van der Boomen, R., Weese, J.: Patient specific models for planning and guidance of minimally invasive aortic valve implantation. Medical image computing and computer-assisted intervention: MICCAI ... International Conference on Medical Image Computing and Computer-Assisted Intervention. 13, 526–33 (2010).
11. Grbic, S., Ionasec, R., Vitanovski, D., Voigt, I., Wang, Y., Georgescu, B., Navab, N., Comaniciu, D.: Complete valvular heart apparatus model from 4D cardiac CT. Medical image analysis. 16, 1003–14 (2012).
12. Erbel, R., Eggebrecht, H.: Aortic dimensions and the risk of dissection. Heart (British Cardiac Society). 92, 137–42 (2006).
13. Malik, J.: Normalized cuts and image segmentation. IEEE Transactions on Pattern Analysis and Machine Intelligence. 22, 888–905 (2000).
14. Vicente, S., Kolmogorov, V., Rother, C.: Graph cut based image segmentation with connectivity priors. 2008 IEEE Conference on Computer Vision and Pattern Recognition. 1, 1–8 (2008).
15. Delgado, V., Ng, A.C.T., Schuijf, J.D., Van der Kley, F., Shanks, M., Tops, L.F., Van de Veire, N.R.L., De Roos, A., Kroft, L.J.M., Schalij, M.J., Bax, J.J.: Automated assessment of the aortic root dimensions with multidetector row computed tomography. The Annals of thoracic surgery. 91, 716–23 (2011).

Author: Mustafa A. Elattar
Institute: Academic Medical Center – University of Amsterdam
Street: Meibergdreef 9,1105AZ
City: Amsterdam
Country: The Netherlands
Email: M.Elattar@amc.nl

Classification Methods from Heart Rate Variability to Assist in SAHS Diagnosis

J. Gómez-Pilar[1], G.C. Gutiérrez-Tobal[1], D. Álvarez[1], F. del Campo[2], and R. Hornero[1]

[1] Grupo de Ingeniería Biomédica, ETSI Telecomunicación, Universidad de Valladolid, Valladolid, España
[2] Servicio de Neumología, Hospital Río Hortega, Valladolid, España

Abstract—The aim of this study is to analyze different feature classification methods applied to heart rate variability (HRV) signals in order to help in sleep apnea-hypopnea syndrome (SAHS) diagnosis. A total of 240 recordings from patients suspected of suffering from SAHS were available. This initial dataset was divided into training set (96 subjects) and test set (144 subjects). For this study, spectral and nonlinear features have been extracted. Spectral characteristics were obtained from the power spectral density (*PSD*) from HRV records. On the other hand, the nonlinear features were obtained from HRV records in the time domain. Afterwards, some features were selected automatically by forward stepwise logistic regression (FSLR). We constructed two classifiers based on logistic regression (LR) and support vector machines (SVMs) with the selected features. Our results suggest that there are significant differences in various spectral and nonlinear parameters between SAHS positive and SAHS negative groups. The highest sensitivity, specificity and accuracy values were reached by the SVMs classifier: 70.8%, 79.2% and 73.6%, respectively. Results showed that feature selection of optimum characteristics from HRV signals could be useful to assist in SAHS diagnosis.

Keywords—Sleep apnea hypopnea syndrome, HRV, stepwise feature selection, logistic regression, support vector machines.

I. INTRODUCTION

Sleep apnea-hypopnea syndrome (SAHS) is characterized by repetitive pharyngeal collapse during sleep, causing intermittent cessations of breathing (apnea) or marked reduction (hypopnea) in airflow [1]. These airflow interruptions are related to hypoxemia and variations in heart rate [2]. SAHS has been usually related to major cardiovascular diseases [3], occupational accidents, and motor-vehicle collisions [4]. Epidemiological studies estimate the prevalence of SAHS up to 5% of adult men in western countries [2].

Nocturnal polysomnography (PSG) is the gold standard in SAHS diagnosis. However, it presents some drawbacks. A high number of physiological signals and data are acquired in each polysomnographic test. The specialist conducts an inspection of the signals to derive the apnea-hypopnea index (AHI), which is used to determine SAHS severity. Hence, PSG is complex, costly and time-consuming [5].

In recent years, due to limitations of the PSG, alternative diagnostic approaches have emerged. These alternatives have focused on automatic methods based on a reduced number of biomedical signals. One of them is the analysis of heart rate variability (HRV) derived from electrocardiogram (ECG) [6, 7, 8].

In this study, we performed a feature extraction stage where different spectral and nonlinear features were computed. Then, forward stepwise logistic regression (FSLR) were applied to automatically select optimum features. Finally, these features were classified by two methods: logistic regression (LR) and support vector machines (SVMs).

Relative power (*RP*), peak amplitude (*PA*), spectral entropy (*SE*) and median frequency (*MF*) were extracted from the power spectral density (*PSD*) of HRV recordings. Three nonlinear characteristics were also computed: central tendency measure (*CTM*), Lempel-Ziv complexity (*LZC*) and sample entropy (*CTM*).

Our initial hypothesis was that these features could contain complementary information that could be helpful in the diagnosis of SAHS. This study was aimed at assessing the usefulness of the proposed techniques to obtain relevant information from HRV recordings in SAHS diagnosis.

II. DATA SET

The study population used in this work was 240 subjects. All had typical symptomatology of suffering from SAHS. ECG recordings were obtained through a polygraph Alice 5 of Philips Healthcare-Respironics as part of the PSG. The signal acquisition was performed with 200 Hz of sampling frequency. AHI was the average calculated from the number of apneic events detected in PSG. Note that medical specialists have considered positive diagnosis of SHAS provided an AHI $\geq$ 10 events / hour.

A positive diagnosis of SAHS was confirmed in 160 subjects. There were no differences in age and body mass index (BMI) between SAHS positives and negatives. Table 1 summarizes the clinical characteristics of both groups, including age, sex, BMI and AHI (mean $\pm$ standard deviation).

HRV signals were obtained from ECG recordings. Firstly, QRS detection algorithm was applied [9]. Next, physiologically impossible beats no fitting the following criteria were removed: *i)* $0.33s <$ interval between beats $< 1.5s$ and *ii)* the maximum difference between consecutive intervals were 0.66 s [2]. In order to perform spectral analysis, the HRV signal was interpolated by a linear interpolation at a frequency of 3.41 Hz [2].

III. METHODOLOGY

A training set with 96 records (40%) and a test set with 144 records (60%) were randomly derived from the initial population to developed and evaluated our algorithm. Accordingly, the training set was used to perform feature extraction and feature selection processes. Once LR and SVMs classifiers were built, the test set was used to evaluate them.

A. Feature Extraction

PSD was calculated using the non-parametric Welch method. This is suitable for the analysis of non-stationary signals [10]. Hamming window of 2^{11} samples (10 minutes), along with a 50% overlap and a 2^{15} points *PSD* was used. Then, each *PSD* was normalized dividing the amplitude by its total power.

Nonlinear characteristics were obtained from HRV recordings in the time domain. Each signal was divided into epochs of 500 samples corresponding to 5 minutes signal because, during an episode of apnea, cyclical heartbeat variations usually appears periodically between 25 and 100 seconds [11]. Hence, the selection of several of this cycles is ensured. We calculate the average of the nonlinear features extracted in all periods of 5 minutes to obtain a single value for each feature.

A1 Spectral Parameters

Firstly, every single frequency of the *PSD* was analyzed to obtain the spectral bands that showed significant differences between SAHS positive and SAHS negative groups (*p*-value < 0.01). We used Mann-Whitney U test in the training set. Two spectral bands were obtained: $B_1 \in (0.0242 - 0.274)$ Hz. and $B_2 \in (0.0309 - 0.0341)$ Hz. Figure 1 shows the *p*-value for each frequency in the training set.

Next, four spectral features were computed from spectral bands B_1 and B_2:

1. *RP*, which is the ratio of the area enclosed under the *PSD* in the frequency band to the total area under the *PSD*.
2. *PA* in the frequency band, which is the local maximum of the spectral content in the apnea frequency range.
3. *SE*, which is a disorder quantifier related to the flatness of the spectrum [12]:

Table 1 Clinical data of the study population

Features	All	SAHS negative	SAHS positive
Nº subjects	240	80	160
Age (years)	52.3 ± 13.7	47.2 ± 12.2	54.8 ± 13.8
Male (%)	77.5	65.0	83.8
BMI ($Kg./m^2$)	29.8 ± 4.4	27.8 ± 3.7	30.8 ± 4.3
AHI (events/hour)	-	3.9 ± 2.4	36.6 ± 25.7

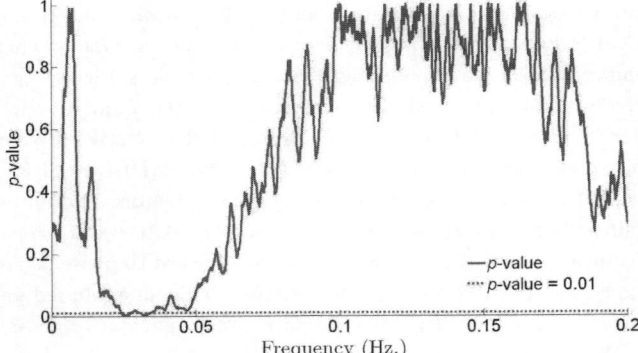

Fig. 1 *p*-value for each frequency in the training set

$$SE = -\sum_j p_j \cdot ln(p_j), \quad (1)$$

where p_j is the normalized value of the *PSD*.

4. *MF*, which is defined as the spectral component which comprises 50% of the total signal power [12].

A2 Nonlinear Parameters

Three nonlinear characteristics were computed:

1. *CTM*, which provides a variability measure from second order difference plots, assigning larger values to lower variability [13]:

$$CTM = \frac{1}{N-2} \sum_{i=1}^{N-2} \delta(d_i), \quad (2)$$

$$\delta(d_i) = \begin{cases} 1 & \text{if } [(x(i+2)-x(i+1))^2 + (x(i+1)-x(i))^2]^{1/2} < \rho \\ 0 & \text{otherwise} \end{cases}, \quad (3)$$

where, in our implementation, $\rho = 0.54$.

2. *LZC*, which is a nonparametric measure of complexity linked with the rate of new subsequences and their repetition along the original sequence [14]. $c(n)$ is increased every time a new subsequence is encountered:

$$LZC(n) = \frac{c(n)}{b(n)} \quad (4)$$

$$b(n) = \lim_{n \to \infty} c(n) = \frac{n}{\log_2(n)}. \quad (5)$$

3. *SampEn*, which quantifies irregularity in time series, with larger values corresponding to more irregular data [15]:

$$SampEn(m,r,N) = -\ln\left[\frac{A^m(r)}{B^m(r)}\right], \quad (6)$$

where A^m and B^m are the average number of (m)-length and $(m+1)$-length segments $X_m(i) (1 \leq i \leq N - m + 1)$ with $d[X_m(i), X_m(j)] \leq r (1 \leq j \leq N - m, j \neq i)$. According to [16], in our algorithm we use $m = 3$ and $r = 0.25$ because all SampEn values were different from $log(0)$.

B. Feature Selection

Feature selection was performed using the algorithm proposed by Hosmer and Lemeshow based on automatic step forward feature selection [17].

This algorithm chooses the strongest variables in a data set. The likelihood ratio test was used to assess statistical differences (*p*-value) between nested LR models differing in one degree of freedom. Iterative LR processes were applied to describe the relationship between a dependent variable and the independent variables. At each iteration, the stepwise method performs a test for backward elimination followed by a forward selection procedure [17]. Features that contribute with significant information are added to the model, whereas no significant features are removed.

If the likelihood *p*-value of the likelihood ratio is less than a threshold input α_E, FSLR model includes the feature. Also excludes features when the likelihood ratio is greater than a threshold output $\alpha_S > \alpha_E$.

C. Feature Classification

In the classification stage, two classification methodologies were assessed:

1. LR relates a set of input features with a categorical dependent variable. The input patterns are classified into two mutually exclusive categories. The probability density function for the response variable can be modelled by a Bernoulli distribution. LR classifier assigns an input vector to the class with the maximum a posteriori probability value. The maximum likelihood criterion is used to optimize the coefficients of the independent input features in the model.
2. SVMs map the input data into a much higher dimensional space. The goal is finding an optimal separating hyperplane between outputs belonging to two classes. SVMs attempt to maximize the separation between the two classes and minimize training error. This is controlled by means of the regularization parameter C. In this study a linear SVM kernel was used. C was obtained by leave-one-out cross validation in the training set. A value of $C = 10^{-8}$ was selected.

D. Statistical Analysis

Statistical differences between groups were evaluated by means of the Mann-Whitney U test. The LR and SVM classifiers were assessed on the test set. Sensitivity (*Se*), specificity (*Sp*), positive predictive value (*PPV*), negative predictive value (*NPV*) and accuracy (*Ac*) were computed to quantify classification performance.

IV. RESULTS

Table 2 shows the average value (mean ± standard deviation) and the *p*-value for each feature under study in the training set. *RP*, *PA* and *SE* achieved statistical significant differences between SAHS negative and SAHS positive groups in B_1 and B_2.

Table 3 summarizes the diagnostic results of the two classification algorithms in the test set. It can be seen that the SVM classifier achieved higher diagnostic accuracy than the LR model. Furthermore, the results demonstrate the utility of performing feature selection.

V. DISCUSSION AND CONCLUSIONS

In the present study, were assessed in order to help in SAHS diagnosis two classifiers, based on LR and SVM. Independent training and test sets were used to compose and assess each classifier. A total of 11 spectral and nonlinear features composed the initial feature set. After the feature selection stage, 3 optimum variables were selected: PA_{B_2}, *LZC* and *CTM*. The spectral characteristics were obtained in two very low frequency bands, which showed significant differences.

Our results are consistent with previous studies that evaluated the relationship between changes in HRV (tachycardia, bradycardia) and physiological changes due to periodic hypoxia [6, 7, 8].

The optimal characteristics were extracted from the spectral parameters and the nonlinear parameters, which shows that the information is complementary. The model using SVM performed better than LR, reaching a diagnostic accuracy of 73.6%.

The study has some limitations that must be noted. First, it is clear that it would be appropriate to extend the total number of subjects and the rate of negative SAHS subjects. It would

Table 2 Average (mean ± standard deviation) for each feature under study

Features	SAHS negative	SAHS positive	p-value
RP_{B_1}	0.0128 ± 0.0045	0.0205 ± 0.0158	< 0.01
PA_{B_1}	0.0153 ± 0.0100	0.0340 ± 0.0454	< 0.01
SE_{B_1}	0.0104 ± 0.0033	0.0153 ± 0.0097	< 0.01
MF_{B_1}	0.0257 ± 0.0001	0.0258 ± 0.0001	> 0.01
RP_{B_2}	0.0103 ± 0.0041	0.0161 ± 0.0117	< 0.01
PA_{B_2}	0.0113 ± 0.0065	0.0255 ± 0.0286	< 0.01
SE_{B_2}	0.0086 ± 0.0031	0.0125 ± 0.0076	< 0.01
MF_{B_2}	0.0325 ± 0.0001	0.0325 ± 0.0001	> 0.01
CTM	0.6613 ± 0.2117	0.6923 ± 0.1746	> 0.01
LZC	0.3727 ± 0.0625	0.3546 ± 0.0684	> 0.01
$SampEn$	0.4121 ± 0.0867	0.4053 ± 0.1034	> 0.01

Table 3 Diagnostic evaluation of the classifiers (test set)

Model	$Se(\%)$	$Sp(\%)$	$PPV(\%)$	$NPV(\%)$	$Ac(\%)$
LR (PA_{B_2}, LZC, CTM)	76.0	58.3	78.5	54.9	70.1
SVM (PA_{B_2}, LZC, CTM)	70.8	79.2	87.0	56.7	73.6
LR (all features)	70.8	47.9	74.7	47.2	63.2
SVM (all features)	51.6	71.9	77.5	41.1	58.3

also be desirable to use other feature selection methods, such as principal component analysis (PCA). Thus, it is possible that other features also provide useful information.

In conclusion, the results reported in this study suggest that the joint analysis of spectral and nonlinear features from HRV recordings obtained from ECG, automatically selected by means of FSLR, could provide additional useful information to help in SAHS detection.

Acknowledgements

This research was supported in part by the Proyecto Cero 2011 on Ageing from Fundación General CSIC, Obra Social La Caixa and CSIC, the "Ministerio de Economía y Competitividad" and FEDER under project TEC2011-22987, and project VA111A11-2 from Consejería de Educación (Junta de Castilla y León).

References

1. Malhotra A, White DP. Obstructive sleep apnoea. *Lancet*. 2002;360:237–245.
2. Penzel P, Kantelhardt JW, Grote L, Peter JH, Bunde A. Comparison of Detrended Fluctuation Analysis and Spectral Analysis for Heart Rate Variability in Sleep and Sleep Apnea. *IEEE Transactions on Biomedical Engineering*. 2003;50:1143–1151.
3. Lopez-Jiménez F, Kuniyoshi FHS, Gami A, Somers VK. Obstructive Sleep Apnea. *Chest*. 2008;133:793–804.
4. Sassani A, Findley LJ, Kryger M, Goldlust E, George C, Davidson TM. Reducing Motor-Vehicle Colissions, Costs, and Fatalities by Treating Obstructive Sleep Apnea Syndrome. *Sleep*. 2003;27:453–458.
5. Bennet JA, Kinnear WJM. Sleep on the cheap: the role of overnight oximetry in the diagnosis of sleep apnoea hypopnoea syndrome. *Thorax*. 1999;54:958–959.
6. Penzel T, McNames J, Chazal P De, Raymond B. Systematic comparison of different algorithms for apnoea detection based on electrocardiogram recordings. *Medical and Biological Engineering and Computing*. 2002;40:402–407.
7. Drinnan MJ, Allen J, Langley P, Murray A. Detection of Sleep Apnoea from Frequency Analysis of Heart Rate Variability. *IEEE Computers in Cardiology*. 2000;27:259–262.
8. Chazal P De, Heneghan C, Sheridan E, Reilly E, Nolan P, O'malley M. Automatic processing of the single-lead electrocardiogram, for the detection of obstructive sleep apnea. *IEEE Transactions on Biomedical Engineering*. 2003;50:686–696.
9. Benitez D, Gaydecki PA, Zaidi A, Fitzpatrick AP. The use of the Hilbert transform in ECG signal analysis. *Computers in Biology and Medicine*. 2001;31:399–406.
10. Welch PD. The Use of Fast Fourier Transform for the Estimation of Power Spectra: A Method Based on time Averaging Over Short, Modified Periodograms. *IEEE Transactions on Audio Electroacoustics*. 1967;15:70–73.
11. Canisius S, Ploch T, Gross V, Jerrentrup A, Penzel T, Kesper K. Detection of Sleep Disordered Breathing by Automated ECG Analysis. in *30th Annual International IEEE EMBS Conference*;1(Vancouver, Canada):2602–2605 2008.
12. Poza J, Hornero R, Abásolo D, Fernández A, García M. Extraction of spectral based measures from MEG background oscillations in Alzheimer's disease. *Med. Eng. Phys.*. 2007;29:1073–1083.
13. Álvarez D, Hornero R, García M, Campo F., Zamarrón C. Improving diagnostic ability of blood oxygen saturation from overnight pulse oximetry in obstructive sleep apnea detection by means of central tendency measure. *Artif. Intell. Med.*. 2007;41:13–24.
14. Álvarez D, Hornero R, Abásolo D, Campo F., Zamarrón C. Nonlinear characteristics of blood oxygen saturation from nocturnal oximetry for obstructive sleep apnoea detection. *Physiol. Meas.*. 2006;27:399–412.
15. Richman JS, Moorman JR. Nonlinear characteristics of blood oxygen saturation from nocturnal oximetry for obstructive sleep apnoea detection. *Physiological time series analysis using approximate entropy and sample entropy*. 2000;278:2039–2049.
16. Al-Angari HM, Sahakian AV. Use of Sample Entropy Approach to Study Heart Rate Variability in Obstructive Sleep Apnea Syndrome. *IEEE Trans. Biomed. Eng.*. 2007;54:1900–1904.
17. Hosmer DW, Lemeshow S. *Applied Logistic Regression*. New York, USA: John Wiley and Sons 2000.

Author: Javier Gómez-Pilar
Institute: Biomedical Engineering Group
Street: Paseo Belén, 15
City: Valladolid
Country: Spain
Emll: jgompil@gmail.com

AdaBoost Classification to Detect Sleep Apnea from Airflow Recordings

G.C. Gutiérrez-Tobal[1], D. Álvarez[1], J. Gómez-Pilar[1], F. del Campo[1,2], and R. Hornero[1]

[1] Grupo de Ingeniería Biomédica, ETSI Telecomunicación, Universidad de Valladolid, Valladolid, España
[2] Hospital Universitario Rio Hortega, Valladolid, España

Abstract—**In this paper, we focus on the automatic detection of sleep apnea-hypopnea syndrome (SAHS) from single-channel airflow (AF) recordings. Spectral data from a very low frequency band of AF is used to feed classifiers based on linear discriminant analysis (LDA). These are iteratively obtained through the *AdaBoost.M1* (ABM1) algorithm, which combines their performance in order to reach a higher diagnostic ability. We built an ABM1-LDA model, using a training set, which showed generalization ability as well as high diagnostic statistics in an independent test set (94.1% sensitivity, 85.7% specificity, and 92.7% accuracy). These results outperform those from recent studies focused on scoring apneas and hipopneas. Hence, the utility of our approach to assist in SAHS diagnosis is showed.**

Keywords—**sleep apnea hypopnea syndrome, airflow, spectral analysis, linear discriminant analysis, boosting.**

I. INTRODUCTION

The sleep apnea-hypopnea syndrome (SAHS) is a prevalent illness that affects both health and life quality of diseased [1]. Patients suffering from SAHS experiment recurrent episodes of complete cessation (apneas) and significant reduction (hypopneas) of airflow (AF) during sleep [2]. Apneas and hypopneas lead to oxygen desaturations and arousals [3], which avoid resting while sleeping. Inadequate rest derives in poor life quality due to daytime symptoms such as hypersomnolence, cognitive impairment, and depression [1]. Some of them have been related to occupational accidents and motor vehicle collisions [4], [5]. Moreover, SAHS is usually associated with major cardiovascular diseases such as stroke, myocardial infarction, cardiac failure, and hypertension [3]. Recently, it has been also related to an increase in cancer incidence [6].

The "gold standard" for SAHS diagnosis is overnight polysomnography (PSG) [7]. Despite its effectiveness, the PSG is a complex test since it requires monitoring and recording multiple physiological signals from subjects during sleep, such as electroencephalogram (EEG), electrocardiogram (ECG), oxygen saturation of hemoglobin (SpO_2) or airflow (AF) [7]. It is also costly due to the expensive equipments and specialized workforce required for the acquisition of the signals and the patients' care, respectively [2]. Furthermore, physicians must perform an offline inspection of the recordings in order to derive the apnea-hypopnea index (AHI), the parameter used to establish SAHS severity. Hence, PSG is also time-consuming [2]. These drawbacks, in turn, lead to increased time to reach diagnosis and treatment due to large waiting lists [2].

The search for diagnostic alternatives has mainly relied on studying a reduced set of signals from PSG [2]. The automatic analysis of a single signal has been proposed in order to minimize complexity, cost, and time of the test [2]. This approach facilitates the implementation of diagnostic portable devices. Since AF is directly modified by apneas and hypopneas [8], its investigation is a natural way of dealing with the problem of SAHS detection. A number of studies aimed at diagnosing SAHS in time domain from single-channel AF by automatic scoring of apneas and hypopneas [9]-[11]. Alternatively, due to the recurrence of apneic events, we propose a global analysis of AF recordings in the frequency domain.

Previous studies showed the usefulness of data from the very low frequency band of AF spectrum to help in SAHS diagnosis [12], [13]. Thus, our first step is to characterize this band by the extraction of several spectral features. Then, we propose the use of *AdaBoost.M1* (ABM1) along with linear discriminant analysis (LDA) in order to classify the spectral data. ABM1 is a boosting algorithm commonly used to combine the diagnostic ability of several weak classifiers to reach generalized models [14]. LDA, which acts as the weak classifier, has been already used to detect SAHS from oximetry recordings [15].

In this study, the ability of the proposed methodology to help in SAHS diagnosis is assessed. Our hypothesis is that the information contained in the very low frequency band of single-channel AF can be used along with the generalization ability of ABM1 to accurately detect SAHS.

II. POPULATION AND SIGNAL UNDER STUDY

This study involved overnight AF recordings from 104 subjects: 86 diseased (SAHS-positive) and 18 non-diseased (SAHS-negative). The recordings were acquired during the PSG, which was performed using a polygraph (E-Series, Compumedics) in the sleep unit of the Hospital Universitario Río Hortega (Valladolid, Spain). The sensor used to obtain AF was a nasal prong pressure (NPP) and the sample rate was 128 Hz. All the subjects were suspected of suffering from

SAHS before undergoing PSG due to common symptoms such as daytime sleepiness, loud snoring, nocturnal choking and awakenings, and referring apneic events. Physicians scored apneas and hypopneas according to the American Academy of Sleep Medicine (AASM) rules [8]. They established AHI = 10 events per hour (e/h) as the threshold for a positive diagnosis. Subjects were randomly divided into a training set (60%) and a test set (40%). All of them gave their informed consent to participate in the study and the Review Board on Human Studies accepted the protocol. Table 1 shows demographic and clinical data of the subjects, such as body mass index (BMI) or age (mean ± standard deviation), from the entire set and each group.

III. Methods

Our methodology uses data from the very low frequency components of AF power spectral density (PSD) to feed several LDA classifiers. These were obtained iteratively, through ABM1, in order to combine the diagnostic ability of all classifiers and improve the overall performance.

A. Power Spectral Density and Feature Extraction

A PSD for each AF recording was estimated by the Welch periodogram, which is suitable for non-stationary signals [16]. We used 50% overlap, a Hamming window with 2^{15} samples, and a discrete Fourier transform (DFT) of 2^{16} points. PSDs were normalized (PSDn) by dividing the amplitude at each frequency by the total spectral power of the signal. Figure 1 shows the averaged PSDns for the SAHS-positive and SAHS-negative groups in the training set. As reported in previous studies [12], [13], there exists a PSD increase in the very low frequency components of the SAHS-positive subjects. Since apneic events last 10 seconds or more [8], their corresponding frequency components are located in the range 0.01-0.10 Hz. Hence, we characterized it by extracting five common features:

- First-to-fourth statistical moments in frequency domain (Mf_1-Mf_4), corresponding to mean, standard deviation, skewness, and kurtosis in the band 0.01-0.10 Hz.
- Peak amplitude (PA), which corresponds to the local maximum in the band 0.01-0.10 Hz.

Mf_1-Mf_4 and PA values from each subject were stored into an associated vector, $\mathbf{x}_i \in \mathbf{x}$, with $i = 1, 2, \ldots, N$, where N is the total number of subjects, and $\mathbf{x}$ is the whole dataset.

B. Linear Discriminant Analysis

LDA is a supervised classifier which assigns data, $\mathbf{x}$, into one out of k classes, C_k. It relies on the assumption that the conditional class density function of each class, $p(\mathbf{x}|C_k)$, follows a multivariate normal distribution (normality), with

Table 1 Demographic and clinical data

	All	SAHS-negative	SAHS-positive
All subjects			
# Subjects	104	18	86
Age (years)	52.1 ± 15.0	42.3 ± 12.8	54.1 ± 14.7
Men (%)	73 (70.2)	12 (66.7)	61 (70.9)
BMI (kg/m^2)	31.2 ± 6.2	29.3 ± 6.1	31.6 ± 6.1
AHI (e/h)	-	6.2 ± 2.3	44.6 ± 26.9
Training set			
# Subjects	63	11	52
Age (years)	51.5 ± 15.6	42.2 ± 13.1	53.6 ± 15.4
Men (%)	44 (69.8)	8 (72.7)	36 (69.2)
BMI (kg/m^2)	30.3 ± 5.9	28.8 ± 6.4	30.7 ± 5.8
AHI (e/h)	-	6.7 ± 1.3	43.6 ± 27.5
Test set			
# Subjects	41	7	34
Age (years)	52.9 ± 14.2	43.6 ± 14.1	54.9 ± 13.7
Men (%)	29 (70.7)	4 (57.1)	25 (73.5)
BMI (kg/m^2)	32.5 ± 6.4	30.5 ± 6.4	32.8 ± 6.4
AHI (e/h)	-	5.2 ± 3.1	46.0 ± 26.3

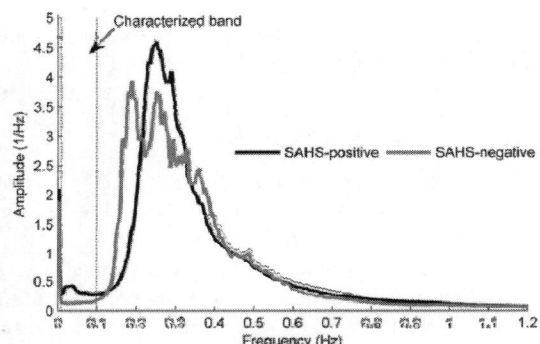

Fig. 1 Averaged PSDn for SAHS-positive and SAHS-negative subjects in the training set

identical covariance matrices, Σ, for all the classes (homocedasticity) [17]. A discriminant score $y_k(\mathbf{x})$ is computed for each class following [15]:

$$y_k(\mathbf{x}) = \boldsymbol{\mu}_k^T \Sigma^{-1} \mathbf{x} - \frac{1}{2} \boldsymbol{\mu}_k^T \Sigma^{-1} \boldsymbol{\mu}_k + \ln P(C_k), \quad (1)$$

where $\boldsymbol{\mu}_k$ is the mean vector for class C_k and $P(C_k)$ its corresponding prior probability, i.e., the initial proportion of vectors $\mathbf{x}_i$ belonging to class C_k. Since we only consider two classes (SAHS-positive and SAHS negative), the classification task is performed by the decision rule, "assign a new vector $\mathbf{x}_i$ to the class C_k if $y_k(\mathbf{x}_i) = \max_{k=1,2} y_k(\mathbf{x}_i)$". In our study, $P(C_k)$, $\boldsymbol{\mu}_k$, and Σ were computed from the training set, since they form the LDA model.

C. AdaBoost.M1

Boosting procedures are iterative algorithms designed to combine models that complement one another [14]. These techniques combine models of the same type using a weighted vote of the prediction from each one [14], [17]. *AdaBoost.M1* (adaptive boosting, ABM1) is a widely used method to perform boosting, originally developed by Freund and Schapire [18]. ABM1 can be used along with any classifier [14]. However, applying ABM1 to too-complex ones could lead to a poor performance when classifying new data [14]. Hence, simple procedures known as weak classifiers, such as LDA, are preferable [14].

ABM1 starts assigning the same weight, w_i, to each instance or vector $\mathbf{x}_i$ in the training set. Typically, $1/N_{tr}$ is used [17], being N_{tr} the number of training instances. The iterative process begins by assessing the performance of the first classifier when assigning weighted instances into the right class. Thus, an error, ε, is computed by dividing the sum of the weights of the misclassified instances by the total weights of all instances [14]. This error is used to determine a weight, α_m, for the current *m*-classifier following:

$$\alpha_m = \ln \frac{1-\varepsilon_m}{\varepsilon_m} \quad (2)$$

Then, the weights of the misclassified instances are updated by the expression:

$$w_i^{m+1} = w_i^m \cdot \frac{1-\varepsilon_m}{\varepsilon_m} \quad (3)$$

Next, all the weights w_i^{m+1} are normalized in order to sum the same as the previous ones, w_i^m, and an additional classifier is assessed using the updated weighted instances [14]. The iterative process automatically ends when $\varepsilon_m = 0$ or $\varepsilon_m \geq 0.5$. The updating of w_i provides higher values to those instances misclassified during the previous iteration [14]. Hence, the new classifier provides these instances with more relevance, being more likely to classify them rightly [14]. The final classification task is performed by returning the class C_k with the highest sum of the votes along all classifiers, weighted by the corresponding α_m value. Thus, those classifiers with smaller ε_m contribute more to the final decision. It has been proved that using the given expressions for ε, w_i, and α_m is equivalent to a sequential minimization of the exponential error function [17].

D. Statistical Analysis

The diagnostic ability of each single spectral features, a conventional LDA classifier, and ABM1-LDA were assessed in terms of sensitivity (Se, proportion of diseased subjects rightly classified), specificity (Sp, proportion of non-diseased subjects rightly classified), and accuracy (Acc, proportion of all subjects rightly classified). To find an optimum threshold, u_o, for the assessment of each single feature, a receiver operating-characteristics (ROC) analysis was done. For each feature, u_o was selected in the training set according to the minimum Euclidean distance between the pair (Se, 1-Sp) and the point (1, 0).

IV. RESULTS

ABM1 iteratively formed five LDA models until satisfying the stopping criterion ($\varepsilon_m = 0$ or $\varepsilon_m \geq 0.5$). These were obtained from the instances in the training set. Table 2 shows the performance of each one, its corresponding ε and α values, and the performance of ABM1. The LDA with the lowest ε value was reached at iteration # 1. Hence, its contribution to the final voting, α, was the highest. None of the single classifiers outperformed ABM1-LDA in terms of accuracy (84.1%). The five LDA models were subsequently applied to the test instances. Their predictions were considered according to each α, in order to perform the final classification. Table 3 summarizes the diagnostic performance of ABM1-LDA, every single feature, and a conventional LDA in the test set. It also shows the optimum thresholds u_o to classify the subjects by the use of single features. As expected, ABM1-LDA generalizes better since it widely outperformed each single feature and LDA. It improved the performance of ABM1-LDA in the training set as well. ABM1-LDA not only reached the highest accuracy (92.7%), but also achieved a balanced sensitivity/specificity pair (94.1% / 85.7%, respectively).

V. DISCUSSION AND CONCLUSIONS

We obtained an ABM1 model to detect SAHS. It was composed of five LDA classifiers and was developed using features from a very low spectral band in AF (0.01-0.10 Hz). Our model showed high generalization ability, reaching high diagnostic performance when applied to independent test data (94.1% Se, 85.7% Sp, and 92.7% Acc).

Recent studies addressed the automatic diagnosis of SAHS through single-channel AF [9]-[11]. The common aim was to detect and score respiratory events in AF time series. These studies involved from 59 to 131 subjects and reported Se, Sp, and Acc values ranging 80.4-91.5 %, 82.3-87.5 %, and 81.2-89.3 %, respectively [9]-[11]. When using single-channel AF, the event-by-event approach scores AF reductions which are not truly hypopneas, since they should be accompanied by a 3% or more decrease in SpO_2 [8]. Alternatively, the very low spectral band that we used was previously related to desaturations [12]. This could be a reason for our higher results. However, further investigation is required to address this issue.

Table 2 Performance of each LDA classifier and ABM1-LDA (Training set)

		ε	α	Se(%)	Sp(%)	Acc(%)
LDA iteration		0.175	1.553	98.1	1.0	82.5
	#2	0.355	0.598	63.5	72.7	65.1
	#3	0.277	0.961	75.0	81.8	76.2
	#4	0.407	0.378	76.9	36.4	69.8
	#5	0.372	0.523	69.2	63.6	68.2
ABM1-LDA		-	-	90.4	54.5	84.1

Table 3 Diagnostic ability of single features, LDA, and ABM1-LDA (Test set)

	u_0	Se(%)	Sp(%)	Acc(%)
PA	0.269	76.5	85.7	82.9
Mf_1	0.184	70.6	100.0	75.6
Mf_2	0.040	70.6	85.7	73.2
Mf_3	0.549	82.3	71.4	80.5
Mf_4	2.816	76.5	71.4	75.6
LDA	-	100.0	14.3	85.3
ABM1-LDA	-	94.1	85.7	92.7

In spite of the effectiveness showed by our methodology, some limitations need to be addressed. First, more subjects are required to increase the statistical power of our results. Accordingly, the number of SAHS-negative subjects should be higher. However, our sample reflects a realistic proportion of diseased and non-diseased subjects who undergo PSG. Finally, several weak classifiers could be used along with ABM1 and newer versions of *AdaBoost*. The assessment of an optimum combination of classifiers and boosting algorithms to help in SAHS diagnosis is a future goal.

Summarizing, the usefulness of AF spectral data from very low frequencies was showed. We outperformed recent studies focused on a common event-by-event scoring approach. We also obtained an ABM1-LDA model which achieved generalization ability as well as high accuracy, showing its utility to help in SAHS detection.

Acknowledgment

This research was supported in part by the Proyecto Cero 2011 on Ageing from Fundación General CSIC, Obra Social La Caixa and CSIC, the "Consejería de Educación (Junta de Castilla y León)" under project VA111A11-2, and project TEC2011-22987 from Ministerio de Economía y Competitividad and FEDER. G. C. Gutiérrez-Tobal was in receipt of a PIRTU grant from the Consejería de Educación de la Junta de Castilla y León and the European Social Fund (ESF).

References

1. Young T, Peppard PE, Gottlieb DJ (2002) Epidemiology of Obstructive Sleep Apnea: A Population Health Perspective. Am J Respir Crit Care Med 165:1217-39
2. Flemmons WW, Littner MR, Rowley et al. (2003) Home diagnosis of sleep apnea: a systematic review of the literature. Chest 124:1543-79
3. López-Jiménez F, Kuniyoshi FHS, Gami A, et al. (2008) Obstructive sleep apnea: implications for cardiac and vascular disease. Chest 133:793-804
4. Lindberg E, Carter N, Gislason T, et al. (2001) Role of snoring and daytime sleepiness in occupational accidents. Am J Respir Crit Care Med 164:2031-35
5. Sassani A, Findley LJ, Kryger M, et al. (2004) Reducing motor-vehicle collisions, cost, and fatalities by treating obstructive sleep apnea syndrome. Sleep 27:453-58
6. Campos-Rodríguez F, Martínez-García MA, Martínez M, et al. (2013) Association between obstructive sleep apnea and cancer incidence in a large multicenter Spanish cohort. Am J Respir Crit Care Med 187:99-105
7. Patil SP, Schneider H, Schwartz AR, et al. (2007) Adult obstructive sleep apnea: pathophysiology and diagnosis. Chest 132:325-337
8. Iber C, Ancoli-Israel S, Chesson A, et al. (2007) The AASM manual for the scoring of sleep and associated events: rules, terminology and technical specifications (Westchester, IL, USA: American Academy of Sleep Medicine)
9. Erman MK, Stewart D, Einhorn D et al. (2007) Validation of the ApneaLinkTM for the screening of sleep apnea: a novel and simple single-channel recording device. J Clin Sleep Med 3:387-392
10. Ragette R, Wang Y, Weinreich G et al. (2010) Diagnostic performance of single airflow channel recording (ApneaLink) in home diagnosis of sleep apnea. Sleep Breath 14:109-14
11. Nigro CA, Dibur E, Aimaretti S et al. (2011) Comparison of the automatic analysis versus the manual scoring from ApneaLink™ device for the diagnosis of obstructive sleep apnoea syndrome. Sleep Breath 15:679-86
12. Álvarez D, Gutiérrez GC, Marcos JV, et al. (2010) Spectral Analysis of Single-Channel Airflow and Oxygen Saturation Recordings in Obstructive Sleep Apnea Detection, Proc. of the 32nd Ann. Inter. Conf. of the IEEE EMBS, Buenos Aires, Argentina, 2010, pp 847–850
13. Gutiérrez-Tobal GC, Hornero R, Álvarez D, et al. (2012) Linear and nonlinear analysis of airflow recordings to help in sleep apnoea-hypopnoea syndrome diagnosys. Physiol Meas 33:1261-75
14. Witten IH, Frank E, Hall MA (2011) Data Mining: Practical Machine Learning Tools and Techniques. Morgan Kaufmann/Elsevier, Burlington, MA
15. Marcos JV, Hornero R, Álvarez D, et al.(2009) Assessment of four statistical pattern recognition techniques to assist in obstructive sleep apnoea diagnosis from nocturnal oximetry, Med Eng Phys 31:971–78
16. Welch PD (1967) The use of fast Fourier transform of the estimation of power spectra: a method based on time averaging over short, modified periodograms. IEEE Trans Aud Elec Acoust AU-15 70–3
17. Bishop CM (2006) Pattern Recognition and Machine Learning. Springer, New York: NY.
18. Freund Y, and Schapire RE (1996) Experiments with a new boosting algorithm. In L. Saitta (Ed.), Thirteenth International Conference on MachineLearning, pp. 148–156. Morgan Kaufmann.

Author: Daniel Álvarez González
Institute: ETSI Telecomunicación, Universidad de Valladolid
Street: Paseo Belén, 15
City: Valladolid
Country: Spain
Email: dalvgon@gmail.com

Effect of Electric Field and Temperature in E.Coli Viability

A.M. Oliva[1,2,3], A. Homs[1,2,3], E. Torrents[4], A. Juarez[5], and J. Samitier[1,2,3]

[1] Nanobioengineering group, Institute for Bioengineering of Catalonia (IBEC), Spain
[2] Centro de Investigación Biomédica en Red en Bioingeniería, Biomateriales y Nanomedicina (CIBER-BBN), Spain
[3] Department of Electronics, University of Barcelona, Spain
[4] Bacterial infections: antimicrobial therapies group, Institute for Bioengineering of Catalonia (IBEC), Spain
[5] Microbial biotechnology and host-pathogen interaction group, Institute for Bioengineering of Catalonia (IBEC), Spain

Abstract—Electromagnetic Fields are increasingly used to manipulate bacteria. However, there is no systematic and definitive study on how the different electric parameters change bacteria viability. Here we present preliminary data on the effect of electric field intensity and temperature application. E. Coli colonies have been exposed to different voltages at 1MHz during 5 minutes by means of a custom-made microfluidic device. Results show that E.Coli survival rate is already reduced by applying field intensities as low as 220V/cm during 5 minutes. The use of stronger fields resulted in death rates increase also. Viability of survived bacteria was maintained. On the other hand, temperature has shown a synergistic effect with voltage. When temperature is increased results seem to indicate stronger sensitivity of cells to the electric field. It is necessary to continue studying the contribution of other parameters as intensity, time, frequency or concentration, to study further synergies.

Keywords—E. Coli, Electromagnetic Field, Viability, Temperature.

I. INTRODUCTION

Living organisms are under forces and stresses from different sources, including electromagnetic fields (EMF). Biological effects of EMF have been seen as definitive and crucial to life origins, and they can modify cellular behavior [1]. Interacting EMF can be originated externally (exogenous) or internally (endogenous) to the biological entities. Our work has been focused on the study of the effect of exogenous fields on E. Coli. Due to the existing controversies on the roles of endogenous EMF effects we cannot disregard possible interactions of these two types of fields in this study [2].

Relatively recent reviews published data about the effects on cellular biology related with electrical, magnetic and electromagnetic fields. Nevertheless, new researches are being published in which the effects of electromagnetic interactions are being considered in signaling transduction processes. Tiny fields or currents, far too weak to power any cellular activity or to produce an effect due to the heating, can trigger a change at a regulation level, and that will lead to a physiological response. Different components of the regulation cascades, including receptors, channels and enzyme processes, are sensible to the EMF [3].

Electrical fields are usually used to manipulate cells in Microsystems. In the last 30 years a huge number of devices have been developed for sorting and trapping cells. The impact of these fields on the cells is not fully understood. However, it is known that electrical fields induce local temperature increments around the cells, and that frequently can create reactive species in the electrode-electrolyte interface. These reactions can become interactions between field and cell in the cellular membrane, and can make changes in the phenotype [4].

Membrane is considered the place in which interaction between electric fields and cells occur. It is also regarded as the only active element in cells. It is generally accepted in scientific literature that the field required to exceed the transmembrane potential is about 1000V/cm [5], but no frequency or wave form is associated at this electric field. Under this EMF no apparent biological effect is found and mortality rates remain unchanged.

Researches show stress evidence when EMF in the KHz-MHz range are applied [4] and enzymatic changes [6]. Changes in E.Coli bacteria membrane structure and morphology have also been reported while thermal effects were not present [7], although they seem to remain viable and able to use their biomolecular machinery to produce proteins under pDEP and nDEP exposure [8].

There are different approaches to the study of the effect of EMF on microorganisms, but there is no standardization neither in the parameters (voltage, field intensity, electrical intensity, power, time, concentration) nor the methodology they apply. This and the fact that some studies contain limited information make comparing the results a difficult task. In our work we have designed an experimental setup to perform a systematic study on these parameters. We present data on the effects of electric field and temperature on cell viability and metabolism.

II. MATERIALS AND METHODS

A. Bacterial Samples Preparation

E.Coli MG1655 wild type were cultured overnight, reinoculated and grown until a concentration of 1×10^{11} cfu/ml was achieved. The sample was rinsed and finally, the pellet was resuspended in a final volume of 2.5mL sterile deionised water.

The conductivity of the sample was adjusted to 10 μS/cm by addition of microliter volumes of 0.01M NaCl solution. This was done in order to reproduce the experiments with the same conditions.

B. Electrode Chamber Design and Fabrication

A microfluidic chamber with embedded electrodes was designed for the study. The design was verified by using finite element analysis software (Comsol Multiphysics 3.5, Comsol) in order to guarantee that the electric field distribution was homogeneous.

The chip has two glass slides with gold electrodes, and a double side adhesive tape defining the chamber.

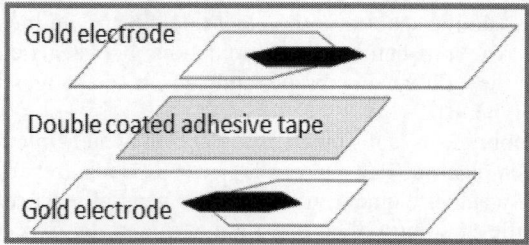

Fig 1 Schematic description of chip structure

Electrodes have 2cm x 1cm. Production of electrodes has been by standard photolithographic techniques. This consists in initial preparation and cleaning of glass slides with acetone, isopropanol and ethanol, and in plasma cleaning of 15 minutes at medium power. The entire slide was coated with a layer of negative photoresist AZ1512 and exposed to the photolitography. The photoresist was developed with AZ 726MF. The metal were deposited with e-beam technique, with 20nm Ti and 80nm Au. After deposition, the lift-off was made with the AZ100 remover, by sonicating during 5 minutes. All AZ products were acquired from MicroChemicals.

The chamber was fabricated using a double side adhesive tape (SCAPA 43023 - TTSL).

Before each experiment, the microfluidic device was filled and analyzed with electrical impedance spectroscopy (Agilent 4294A). The anomalous devices were discarded.

C. Protocols

The irradiation protocol is the following: Once the experimental set up is ready, bacterial solution is mixed for 5 seconds with a vortex. Cells are introduced in the chip. In control sample, cells are maintained in the chip during 5 minutes without voltage. In irradiated samples, a voltage is applied during the same time. In both cases, cells are harvested and resuspended in LB. After that they are incubated during 5minutes. In every set of experiments there is always at least one control in the beginning and another control at the end of the experiment, in order to be sure that cells are still alive at the end of the experiment, and the hypothetical death is not due to the time during experimentation.

In order to study changes in viability, two different methods have been followed:

1. Seeding 100μL of the samples in an agar Petri dish after 5minutes incubation and then continue incubating the samples and seeding every 30 minutes. In this way it is possible to observe the *evolution of the number of viable cells*. Every sample is seeded at each dilution in at least 3 different petri dishes to have a statistical measurement. The number of irradiate viable cells is then compared to the number of control CFU.

2. Measuring the optical density (OD) with a spectrophotometer at 550nm every 30 minutes. In this way it's possible to observe the *growing rate*.

D. Set Up Control

In order to be sure about there are not (macroscopic) thermal effects, microfluidic devices were studied under a infrared camera (FLIR Systems A40M), applying different voltages at different conductivity values. Temperature measured with the camera is only in the surface, so temperature inside the chips was estimated mathematically taking into account materials, thickness and thermal conductivity of each material. Results showed that for voltages as applied, (17Vp), conductivity must be under 20μS/cm in order to assure a final temperature lower than 37ºC. In case of using a thermal bath at 25ºC it seems that using higher conductivities is not a problem.

Temperature was monitored during the experiments, by introducing a thermal probe (THERMOWORKS THS-307-159) in the microdevice.

Conductivity of bacterial sample has been measured (Corning 441) before each experiment.

III. EXPERIMENTS AND RESULTS

Cells were exposed to different voltages at 1MHz during 5 minutes. For voltages upper than 6Vp (equivalent to 220V/cm) a decrease in the number of viable cells was observed. As expected, results showed that the higher the voltage, the higher the mortality.

Nevertheless, no changes have been found in the slope of the curve representing the viable counting (Fig 2). The slopes at different electric fields have been evaluated by visual inspection by a senior biologist researcher.

The maximum values of the optical density achieved after an overnight culture didn't show differences.

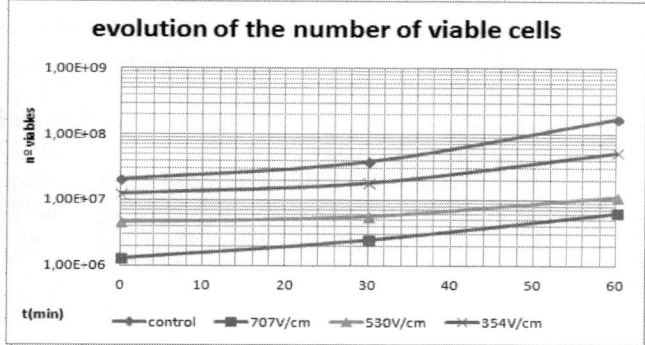

Fig 2 Growth curves: there is no change in evolution of viable cells along time

A new culture was exposed to 40°C (without electric field), which was the higher temperature measured with the thermal probe. Growth curves were overlapped. Results (Fig. 3) showed that only temperature don't produce the observed effects.

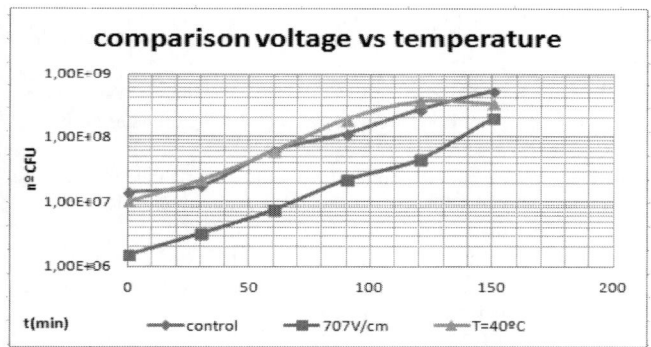

Fig 3 Only temperature do not inactivate bacteria

When the growth curve is represented, the effect of irradiated cells can be compared to a dilution effect, as the Figure 4 shows.

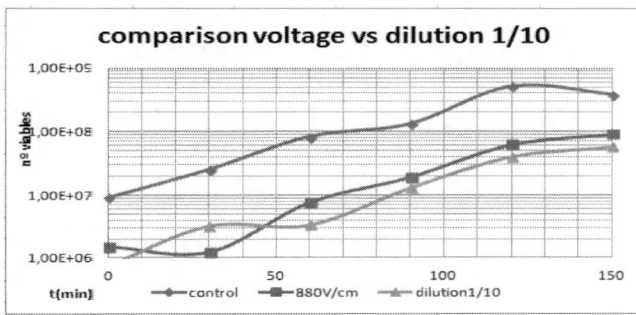

Fig 4 Irradiated cells behave as a dilution

To calculate the survival rate, and to compare between different experiments, number of CFU is normalized, by dividing the number of irradiated CFU by the number of control CFU. In Fig 5 we can see the results of experiments of the effect of Electric Field Intensity in viability:

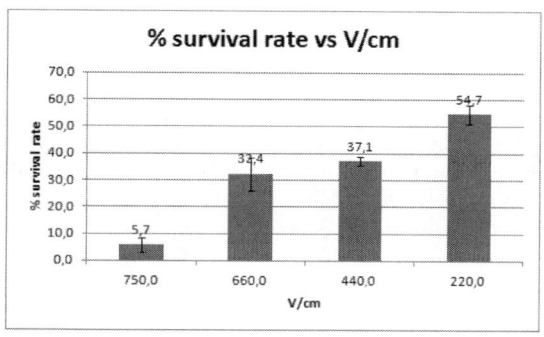

Fig 5 Summary of the effect of electric field intensity in viability

Another set of experiments was done to analyze the synergistic effect between temperature and voltage. Results are summarized in the following graph:

In it, we can see the survival rate in different situations.

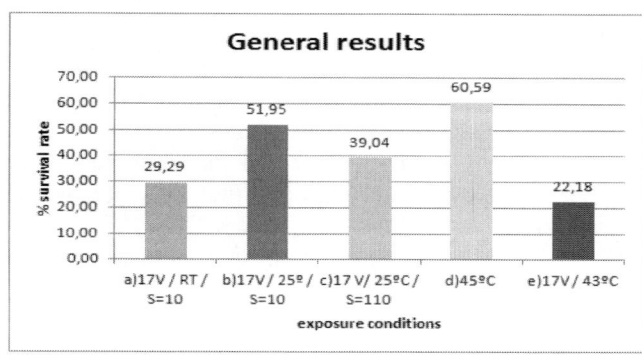

Fig 6 Synergistic effect of voltage and temperature

The first one (a) corresponds to "normal" exposure at 17V peak, conductivity of 10µS/cm, 1MHz and at room temperature. The final temperature measured with the infrared camera along the experiment was less than 27°C. If the same conditions are applied in a thermal bath, with temperature fixed at 25°C (b), the survival rate increases a 40%, even when the temperature increment seems to be insignificant.

After that, conductivity was increased in order to have a higher electrical current. It is not possible to avoid thermal effects at such a big conductivity if a thermal bath is not used. That is the reason why there is no data about this situation without thermal bath. When cells (c) at high conductivity are exposed at the same conditions as (b), there is a decrease in the survival rate.

Case (d) represents the survival rate when cells are only exposed to a temperature of 43°C, without any electricity. And (e) represents the case of cells exposed to a temperature of 43°C and also an electric field of 17V, conductivity 10μS/cm.

IV. DISCUSSION

Results obtained regarding electric field intensity showed that the higher the voltage, the higher the mortality. Results also suggest that there is an initial mortality and after that, survival cells grow at the same rate. A possible explanation to this result could be that surviving cells adapt themselves to new conditions, so they conserve original viability after exposure.

Temperature showed to have a synergistic effect with electricity. When a voltage is applied at room temperature, mortality is higher than if a thermal bath is used, even when the maximum temperature reached without a thermal bath is the same than with the bath (at least, global measured temperature). It may suggest that it is not possible to ignore local microheating due to Joule effect that occurs in all electric setup.

When an electric field is applied with the thermal bath at 43°C cells were a little bit more sensitive to EMF, although this change is only less than a 10%. In any case, it is very important to use a thermal bath to be sure that results are due to electricity and not to heat. It is important to remember that we can only control "global" (not local) parameters.

V. CONCLUSIONS

We have performed a set of experiments in order to analyze the specific contribution of voltage, field intensity and temperature to the inactivation of bacteria E.Coli.

Field intensities lower than 1000V/cm have produced already changes in the survival rate. Cells mortality higher than 50% was observed at values starting at 220V/cm while an increase of one order of magnitude was observed at 750V/cm. That means that is possible to kill bacteria at lower fields which is consistent with previous researches [7]. Analysis from growth curves and viable counting show that there seem to be an initial "shock" that represents a decrease in the number of viable cells. There are not observable changes in the curve slope, as it has been reported in literature [9].

Results on temperature showed that the effect of EMF is highly influenced by thermal conditions, which determine the viability of bacteria.

It is necessary to continue studying the contribution of other parameters, as intensity, frequency, time, or initial concentration. It is probable to find relations between them, and perhaps may be a combination of conditions, highly influenced by thermal conditions, what determine the viability of bacteria.

REFERENCES

[1] Zhou, S.-A. and M. Uesaka (2006). "Bioelectrodynamics in living organisms." International Journal of Engineering Science 44: 67-92.
[2] Gowrishankar, T. R. and J. C. Weaver (2003). "An approach to electrical modeling of single and multiple cells." Proceedings of the National Academy of Sciences 100(6): 3203-3208.
[3] Funk, R. H. W., T. Monsees, et al. (2009). "Electromagnetic effects - From cell biology to medicine." Progress in Histochemistry and Cytochemistry 43(4): 177-264.
[4] Desai, S. P. and J. Voldman (2011)"Cell-based sensors for quantifying the physiological impact of microsystems." Integrative Biology 3(1): 48-56.
[5] Pethig, R. (2010"Review Article---Dielectrophoresis: Status of the theory, technology, and applications." Biomicrofluidics 4(2): 022811.
[6] Nawarathna, D., J. H. Miller, Jr., et al. (2005). "Harmonic Response of Cellular Membrane Pumps to Low Frequency Electric Fields." Physical Review Letters 95(15): 158103.
[7] Machado, L. s. F., R. N. Pereira, et al.(2010) "Moderate electric fields can inactivate Escherichia coli at room temperature." Journal of Food Engineering 96(4): 520-527.
[8] Donato, S. S., V. Chu, et al. (2013). "Metabolic viability of Escherichia coli trapped by dielectrophoresis in microfluidics." Electrophoresis 34(4): 575-82.
[9] Yang, L., P.P.Banada, et al (2008). "Effects of Dielectrophoresis on Growth, Viability and Immuno-reactivity of Listeria monocytogenes." J Biol Eng 2(6):1754-1611.

Author:	Ana Mª Oliva
Institute:	IBEC
Street:	Baldiri Reixac 10-12
City:	Barcelona
Country:	Spain
Email:	daya13@gmail.com

3D Shape Landmark Correspondence by Minimum Description Length and Local Linear Regularization

M. Valenti[2,1], C. Chen[2], E. De Momi[1], G. Ferrigno[1], and G. Zheng[2]

[1] Dept. of Electronics, Information and Bioengineering, Politecnico di Milano, Italy
[2] Institute for Surgical Technologies and Biomechanics, Universität Bern, Switzerland

Abstract—**Statistical Shape Models (SSMs) are currently used in orthopaedic surgery to allow accurate position of prosthetic components through bone morphing and to assess the correct post-operative follow up by virtually reconstructing the surgical site. Focusing on computer assisted Total Knee Arthroplasty (TKA) applications, in this paper we propose a new approach for establishing landmark correspondence of 3D shapes for building SSMs of anatomical structures around the knee joint. Our method is based on the landmark correspondence method by Minimum Description Length (MDL) and enforces local geometric similarity. Our new constraint, which is in the form of local linear regularization, ensures that the local shape geometry of corresponding landmarks on different shapes is similar. We tested our method on building SSMs of three anatomical structures from 24 MRI images of pathological knees, namely femur, patella and tibia. Compared with the original method using only the MDL criterion, our method shows significant improvement in two out of the three structures.**

Keywords—**Computer Assisted Surgery, Image Processing, Statistical Shape Model, Minimum Description Length, Landmark Correspondence.**

I. INTRODUCTION

In the field of computer vision and medical image processing, statistical shape analysis [2] is an important research tool. Different types of SSMs have been proposed, allowing accurate modelling of shape structure and variation. Most SSMs treat a shape instance as a vector which is built from landmarks. Therefore, to make different shape vectors comparable and to construct a meaningful SSM, it is crucial that the landmarks identified on different shape instances are well corresponded.

There has been a considerable work on automatic landmark correspondence in literature. An earlier work of Brett and Taylor [1] tackled this problem by ICP algorithm. In [9], the landmark sliding algorithm was proposed which features an objective function which encodes both global shape deformation and local shape topology. Xie and Heng [10] developed an algorithm where the shape correspondence is first established by the shape skeleton features and then refined via point matching by a assignment problem.

Recently, methods based on MDL criterion have shown promising results. The MDL criterion was first employed for landmark correspondence in [3], and was shown to generate superior results in [7]. In [4], Heimann et al. proposed a new procedure based on the MDL criterion which is more computationally inexpensive and easier to implement.

In this paper, we propose a new extension to the existing landmark corresponding method. Our method is based on the framework as in [4]. Motivated by the fact that MDL criterion pays more attention on *global* consistency, we introduce a new constraint that enforces the *local* shape similarity. Our new constraint, which is based on the local linear regularization, enforces that the local shape geometry is similar on the corresponded landmark on different shapes. By combining the standard MDL criterion with our new constraint, we end up with an objective function which enforces the correspondence from both global and local points of view.

We tested our new algorithm using 24 MRI images of pathological knees, entailing femur, patella and tibia. We use the bipartite matching difference and the Wald-Wolfowitz test [6] as performance measures. We compared our result with the original MDL method and observe improvement of our method in the case of constructing femur and patella SSMs.

II. THE PROPOSED METHOD

A. Problem Formulation

Considering a set of M training shapes, each of which is a triangulated mesh $\{S_m = (V_m, E_m)\}_{m=1...M}$, V_m and E_m are the sets of vertices and edges of the mth training shape. Each shape is a Genus Zero Surface, a closed surface that can be parametrized onto the unit sphere. Let us denote $\Omega_m(S_m)$ as the spherical parametrization of the mth training shape [8]. For any vertex $v \in V_m$, $\Omega_m(v) \in \mathbb{R}^3$, where $|\Omega_m(v)| = 1$, specifies the coordinate of v on the unit sphere.

A set of N landmarks, $\{(\tilde{\theta}_n, \tilde{\phi}_n)\}_{n=1...N}$ is also defined on the sphere, where $(\tilde{\theta}_n, \tilde{\phi}_n)$ is the spherical coordinates of the nth landmark. To calculate the actual position of the nth landmark on the mth training shape, we build a ray from the origin to $(\tilde{\theta}_n, \tilde{\phi}_n)$ on the parametrization sphere, and calculate the intersection of the ray with the mth shape mesh. We denote a_m^n

as the position of the nth landmark on the mth training shape.

Our goal is to establish landmark correspondence over the training shapes. The landmarks are defined on the parametrization sphere (i.e. $\{(\tilde{\theta}_n, \tilde{\phi}_n)\}_{n=1...N}$ is fixed for all shapes). For every training shape S_m, its parametrization Ω_m solely determines $\{a_m^n\}_{n=1...N}$, the actual position of each landmarks on that shape. Therefore, our goal is searching for the optimal parametrizations $\{\Omega_m\}_{m=1...M}$ under which the landmark positions over the training shapes have optimal correspondences.

B. Miminum Description Length Cost Function

Given the set of all landmark positions on all training shapes $\{a_m^n\}_{m=1...M, n=1...N}$, in [4], the quality of landmark correspondence is defined as:

$$F_{MDL} = \sum_k \mathscr{L}_m, \text{where} \quad (1)$$

$$\mathscr{L}_m = \begin{cases} 1 + \log(\lambda_m/\lambda_{cut}), & \text{if } \lambda_m \geq \lambda_{cut} \\ \lambda_m/\lambda_{cut}, & \text{if } \lambda_m < \lambda_{cut} \end{cases} \quad (2)$$

where λ_{cut} is a parameter which represents the expected noise in the training data and λ_m is the mth eigenvalue of the distribution. In [4], the parametrization $\{\Omega_m\}_{m=1...M}$ is optimized so that the corresponding landmark positions $\{a_m^n\}_{m=1...M, n=1...N}$ generates the minimum F_{MDL}.

C. Local Linear Regularization Cost Function

In this paper we extend the original MDL-based approach with a new objective function considering the local linear similarity. The idea is to enforce the geometric consistency in the local neighbourhoods of each landmark over different shapes. To do this, we first create a neighbourhood system $\mathscr{N}$ of the landmarks. Specifically, for the nth landmark, $\mathscr{N}(n) = \{\mathscr{N}(n)_k\}_{k=1...K_n}$ is the set of K_n landmark indices that are within its local neighbourhood. Each neighbour shares with the selected landmark one edge of the landmark shape. Then, considering a_m^n, which is the position of the nth landmark on the mth training shape, it should be reasonably reconstructed using its neighbouring landmarks on the same shape:

$$a_m^n = w_{m,n}^1 a_m^{\mathscr{N}(n)_1} + ... + w_{m,n}^{K_n} a_m^{\mathscr{N}(n)_{K_n}} = A_m^n W_{m,n} \quad (3)$$

where $A_m^n = \left[a_m^{\mathscr{N}(n)_1}, ..., a_m^{\mathscr{N}(n)_{K_n}}\right]$ is the matrix of neighbouring landmarks of a_m^n, and $W_{m,n} = \left[w_{m,n}^1, ..., w_{m,n}^{K_n}\right]^\top \in \mathbb{R}^{K_n}$ is the reconstruction coefficient vector. Note that Eq. (3) only considers reconstructing a single landmark on a single shape. In a usual non-degenerate case, Eq. (3) is underdetermined as long as $K_n > 3$. However, if we consider the nth landmark on *every* training shape, it is natural to require that the same reconstruction weight is used to reconstruct the same landmark on all shapes. That is, $W_{m,n}$ should be independent of m. We thus drop the subscript m, and denote the reconstruction weight as W_n, and Eq. (3) becomes a system of equations defined on all shapes:

$$\forall m = 1,...M : a_m^n \approx A_m^n W_n \quad (4)$$

Note that since in our case the number of training shapes is larger than the number of neighbours, Eq. (4) becomes overdetermined and W_n can be solved by Least Mean Squares (LMS) method, given that the landmark positions a_m^n and A_m^n are known. In this way we can compute $\{W_n\}$ for $n = 1...N$, and then the quality of landmark correspondence (F_{LLR}) can be expressed as the summation of reconstruction errors of all landmarks over all shapes:

$$F_{LLR} = \sum_{n=1}^{N} \sum_{m=1}^{M} \|a_m^n - A_m^n W_n\|^2 \quad (5)$$

Eq.(5) is our objective in terms of the local linear regularization. We add it to the original objective based on MDL, and get the final objective function (F):

$$F = F_{MDL} + \alpha F_{LLR} \quad (6)$$

with α being a positive weighing parameter controlling the relative importance of the new term.

D. Optimization Process

To find the parametrizations $\{\Omega_m\}$ that optimize the objective function, we adopt a similar optimization strategy as in [4] and adapt it to our objective function with the new term.

Initialization. For each training shape S_m, we initialize Ω_m as a conformal parametrization as in [8].

Iterative Optimization. We iteratively optimize the set of parametrizations $\{\Omega_m\}$. In each iteration, for each shape, we locally update the parametrization using an update function $\Omega' = \Phi(\Omega)$ which is parametrized as $\Omega' = \Phi_{c,\sigma,\Delta\theta,\Delta\phi}(\Omega)$, where c and σ are the centre and bandwidth of the update kernel, and $\Delta\theta$ and $\Delta\phi$ specify the update direction. We use a Gaussian envelope kernel:

$$\rho(x) = \begin{cases} \exp\left(\frac{-\|x-c\|^2}{2\sigma^2}\right) - \exp\left(\frac{-(3\sigma)^2}{2\sigma^2}\right) & \text{for } \|x-c\| < 3\sigma \\ 0 & \text{for } \|x-c\| \geq 3\sigma \end{cases} \quad (7)$$

Eq. (7) gives the magnitude of change at any point x on the mesh. Combined with the direction of update, we actually change the spherical coordinate of x by $\rho(x) \times (\Delta\theta, \Delta\phi)$.

The update direction $(\Delta\theta, \Delta\phi)$ in each iteration is determined through the gradient of the objective function with regard to $(\Delta\theta, \Delta\phi)$. Since the influence of $(\Delta\theta, \Delta\phi)$ to the objective F is via the landmark positions we have:

$$\frac{\partial F}{\partial(\Delta\theta, \Delta\phi)} = \frac{\partial F}{\partial a_i^j} \frac{\partial a_i^j}{\partial(\Delta\theta, \Delta\phi)} \qquad (8)$$

where $\frac{\partial a_i^j}{\partial(\Delta\theta, \Delta\phi)}$ is calculated by finite difference method. $\frac{\partial F}{\partial a_i^j}$ is calculated analytically. From Eq. (6), we have:

$$\frac{\partial F}{\partial a_i^j} = \frac{\partial F_{MDL}}{\partial a_i^j} + \alpha \frac{\partial F_{LLR}}{\partial a_i^j} \qquad (9)$$

where $\frac{\partial F_{MDL}}{\partial a_i^j}$ is calculated as in [4]. For $\frac{\partial F_{LLR}}{\partial a_i^j}$, note from Eq.(5) that in each component of summation $\|a_m^n - A_m^n W_n\|^2$, depending on the relation of (m,n) and (i,j), a_i^j might not appear, or might appear in a_m^n or A_m^n (but not both). Therefore:

$$\frac{\partial F_{LLR}}{\partial a_i^j} = \sum_{m=1}^{M} \sum_{n=1}^{N} f'_{(m,n)}(i,j), \text{ where} \qquad (10)$$

$$f'_{(m,n)}(i,j) = \begin{cases} 2a_i^j - 2W_j A_i^j & \text{if } (m,n) = (i,j) \\ 2W_n^d(W_n A_i^n - a_i^n) & \text{if } m = i \text{ and } j = \mathcal{N}(n)_d \\ 0 & \text{otherwise} \end{cases} \qquad (11)$$

III. EXPERIMENTS

In order to evaluate the performances of our new method, we used a dataset of 24 MRI images of pathological knees. The images were manually segmented using Amira® (VSG3D, France), then reduced and rigidly aligned. We then used both methods to register three different groups of anatomical structures, namely femurs, patellas and tibias. Table 1 shows the number of points of each mesh and the number of landmarks used for each group.

Table 1 Schema of the three different experiments done

Group	Anatomical part	# points	# landmarks
1	Femur	5002	2562
2	Patella	5002	642
3	Tibia	5002	2562

For each group, we ran the algorithm with α equal to 0 (for the original algorithm) and with $\alpha = 0.5$ representing our new implementation (cfr. Eq.(6)). To evaluate the performance of the two algorithms we used the coefficient described in [6] which states that the well known measurements of compactness, specificity and generalization could have some limitation in the evaluation of a statistical shape model. In [6] Munsell et al. describe a new benchmark for the evaluation of 2D shape-space based on a given ground truth. We extended this method to 3D volume-space, and used the original shapes as ground truth. Formally, the evaluation of the shape correspondence follows these steps:

- each shape resulting from the two algorithms ($\{S_i^r\}_{i=1}^M$) is rigidly realigned with its original shape ($\{S_i^o\}_{i=1}^M$)
- each shape, including the ground-truth shapes, is then voxelized with a grid of $0.5 \times 0.5 \times 0.5$ mm.

At this point we need to introduce the Jaccard coefficient. This is defined as

$$\Delta(S_1, S_2) = 1 - \frac{|R(S_1) \cap R(S_2)|}{|R(S_1) \cup R(S_2)|} \qquad (12)$$

where S_1, S_2 are the two shapes considered and $|R|$ computes the volume enclosed in the surface.

A. Bipartite Matching Difference Measure

The first measure we define is based on the bipartite-matching difference between $\{S_i^o\}_{i=1}^M$ and $\{S_j^r\}_{j=1}^M$. We build up the graph that has $2M$ vertices for the shapes $\{S_i^o\}_{i=1}^M$ and $\{S_j^r\}_{j=1}^M$. The weight of the link between two different shapes is given by the Jaccard coefficient between the two shapes linked. Then, applying the bipartite matching algorithm (with the Hungarian method [5]), we can match each ground truth shape with each result shape in order to minimize the sum of the weights. The bipartite matching difference measure is defined as

$$\Delta_b = \frac{\sum_{i=1}^{M} \Delta\left(S_i^o, S_{b(i)}^r\right)}{M} \qquad (13)$$

where $\Delta\left(S_i^o, S_{b(i)}^r\right)$ is the Jaccard difference of the identified corresponding shapes. Thanks to the normalization, Δ_b is always a value in $[0,1]$; $\Delta_b = 0$ implies that the two shape space compared come from the same distribution, while if $\Delta_b = 1$ they describe two completely different shape spaces.

B. Wald-Wolfowitz Test

The second measure we take into consideration is the Wald-Wolfowitz generalized test, based on the minimum spanning tree (MST) algorithm. For this algorithm, we build a fully connected undirected graph with 2M vertices, that represent both $\{S_i^o\}_{i=1}^M$ and $\{S_i^r\}_{i=1}^M$. Then we define the weight of each edge connecting two shapes (both inter and intra the two spaces) as the Jaccard coefficient between the two shapes. We then find the MST of the constructed graph, that is the spanning tree with the minimum total edge weight. On this tree, we count the number of edges that connect two shapes from the same space, either inside $\{S_i^o\}_{i=1}^M$ or $\{S_i^r\}_{i=1}^M$. We can call this number W. Normalizing W over $2M-2$ we finally get the Wald-Wolfowitz difference measure (Δ_w).

The Δ_w value is thus always included in $[0,1]$. In particular, a smaller value of Δ_w indicates that the two distributions most likely come from the same shape space.

C. Results

We evaluate the performances of the two algorithms both with the two methods described above and with a visual/qualitative comparison.

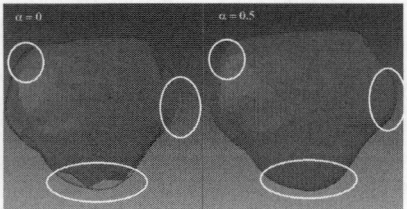

Fig. 1 Same patella mesh processed with two different values of α

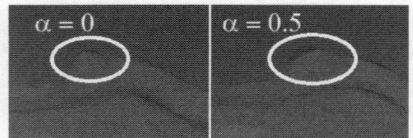

Fig. 2 Same tibia mesh processed with two different values of α

Qualitative differences between the two models are highlighted in Figure 1 and 2. For the quantitative results, Table 2 describes the two index achieved with $\alpha = 0$ (as in [4]) and $\alpha = 0.5$ (the present algorithm) for the investigated bones.

D. Discussion

Quantitative results show that our method performs better in case of femur and patella, while no improvement is made in case of tibia. Evaluating visually the performances, we can see that tibia has some sharp contours that cannot be rightly approximated by our algorithm, based on the similarity of neighbouring points. However, such low levels of Δ_b

Table 2 Schema of the results achieved

Anatomical part	index	$\alpha = 0$	$\alpha = 0.5$
Femur	Δ_b	0.0226	0.0204
	Δ_w	0.4348	0.3913
Patella	Δ_b	0.0786	0.0449
	Δ_w	0.500	0.4783
Tibia	Δ_b	0.0108	0.0117
	Δ_w	0.4565	0.500

achieved for tibia with both algorithms, indicates that the two shape space contain similar meshes and also that a shape has the same probability density in these two shape spaces.

IV. CONCLUSIONS

We describe a new algorithm to improve landmark correspondences on different shapes for statistical shape analysis. We evaluate our new method with two different quantitative measures and with a qualitative overview of the results.

While our method is better in two cases (femur and patella) out of three, we cannot state that for every shape we can achieve better results. Further investigation will be done on other bones, in order to asses the improvement given by our landmark correspondences optimization method.

REFERENCES

1. A.D. Brett, C.J. Taylor. A method of automated landmark generation for automated 3D PDM construction. Image and Vision Computing 18 (2000) 739-748.
2. T.F. Cootes, C.J. Taylor. Active shape models - smart snakes. In: BMVC (1992).
3. R.H. Davies, C.J. Twining, T.F. Cootes, J.C. Waterton, C.J. Taylor. 3D statistical shape models using direct optimisation of description length. In: ECCV (2002).
4. T. Heimann, I. Wolf, T. Williams, H.-P. Meinzer. 3D active shape models using gradient descent optimization of description length. In: IPMI (2005).
5. H.W. Kuhn. The Hungarian method for the assignment problem. Naval Research Logistics (1955).
6. B.C. Munsell, P. Dalal, S.Wang. Evaluating Shape Correspondence for Statistical Shape Analysis: A Benchmark Study. In: IEEE Pattern Analysis and Machine Intelligence (2008)
7. M. Styner, K.T. Rajamani, L.P. Nolte, G. Zsemlye, G. Szekely, C.J. Taylor, R.H. Davies. Evaluation of 3D correspondence methods for model building. In: IPMI (2003).
8. H.H. Thodberg. Minimum description length shape and appearance models. In: IPMI (2003).
9. S. Wang, T. Kubota, T. Richardson. Shape correspondence through landmark sliding. In: CVPR (2004).
10. J. Xie, P. Heng. Shape modeling using automatic landmarking. In: MICCAI (2005).

β-Band Peak in Local Field Potentials as a Marker of Clinical Improvement in Parkinson's Disease after Deep Brain Stimulation

P.D. Frangou[1], K.P. Michmizos[2], P. Stathis[3], D. Sakas[3], and K.S. Nikita[1]

[1] National Technical University of Athens, Athens, Greece
[2] Massachusetts Institute of Technology, Cambridge, MA, USA
[3] Evangelismos Hospital, National and Kapodistrian University of Athens, Greece

Abstract—Although locating the stimulation contact in Deep Brain Stimulation (DBS) requires a sub-mm-precision, it remains a trial-and-error, patient-specific procedure that is usually the main cause of post-operational side-effects. In this work, we used microelectrode recordings from Parkinson's disease (PD) patients, acquired at the Neurosurgery Clinic, Evangelismos Hospital, Athens, Greece, to relate the β-band peak, a known neurophysiological signature of the sensorimotor pathways with the clinical outcome of DBS, as assessed by an expert neurologist after a follow-up of at least 1 year. By combining recordings from 5 microelectrodes, we estimated a summed β-band amplitude peak, per recording depth. We suggest that the maximum aggregate β-band peak is related to the stimulation target. We verified our method in 6 patients that responded well in a bilateral DBS treatment (average increase of Unified Parkinson's Disease Rating scale by 32.6 ± 5.4). In 7 out of 12 hemispheres, the distance between the stimulation depth and that of the maximum β-band peak was 0 and for the rest cases that distance was smaller than 2 mm which is a typical effective radius of a stimulation point. Our method needs to be further supported by data acquired from patients with good and poor clinical responses after DBS.

Keywords—Deep Brain Stimulation, Subthalamic Nucleus, microelectrode recordings, Parkinson's disease.

I. INTRODUCTION

Parkinson's disease (PD) is estimated to affect 7 to 10 million people worldwide [1]. Although a combination of drugs such as Levodopa and Carbidopa is usually effective in alleviating most of the motor symptoms in PD, pharmaceutical treatment over a prolonged period of time becomes gradually less effective and other treatments need to be considered. Deep Brain Stimulation (DBS), initially developed in 1987 [2], has been used as an alternative invasive therapeutic approach in PD patients since 2002 [3]. The surgical procedure can briefly be described as the implantation of electrodes in the patient's brain, along with a pacemaker that regulates the stimulation, usually placed below the clavicle. DBS, when successful, moderates the need for pharmaceutical treatment. In some cases, DBS has no improvement of PD motor symptoms and sometimes is even related to side effects, such as psychiatric and speech disorders.

DBS outcome, which is largely based on the clinician's experience, entails some level of uncertainty. Clinical results are generally believed to be related to the accurate placement of the stimulation contact inside the dorsolateral area of the subthalamic nucleus (STN), where sensorimotor neurons are believed to predominate [4]. Selecting the stimulation contact is patient-specific and includes trial-and-error procedures that are discomforting for the patient and sometimes may cause side-effects. An expert neurologist initially locates the stimulation target combining preoperative magnetic resonance imaging (MRI) scans with microelectrode recordings (MERs) of the neuronal activity [5]. After the target is empirically confirmed, the final stimulation macroelectrode replaces the microelectrodes. The stimulation parameters are adjusted to avoid side-effects, while maximizing the improvement of motor symptoms, based on the sole judgement of the expert neurologist intraoperatively and during follow-up. Following a trial-and-error process, the neurologist decides on the stimulation contacts based on the amelioration of the motor symptoms and the absence of both long and short-term side-effects.

In this paper, we seek to develop a MER-driven clinical decision support system to guide the placement of the stimulation contact at the optimal, in terms of clinical results, point. Our ability to record the STN's neural activity as close to its generator as possible promises maximal spatial resolution and accuracy for the localization of the stimulation contact. MERs inside the STN have been used before to predict the spike activity from the local field potentials (LFPs) [6] and even the STN detection, per se [7]. We now seek to use the increased power and coherence observed in the β-band of the STN- LFPs in PD patients that are in «off-state» (known as the reduced medication efficacy state) [8,9]. In addition, previous studies rely on features acquired by a single microelectrode to determine the location of the STN [10]. In this study we propose the combination of features acquired from multiple microelectrodes.

Table 1 Motor evaluation of the patients in terms of UPDRS (III) score, before and after the operation in «off state» (reduced medication efficacy) (mean UPDRS(III) score improvement: 32.67 ± 5.4).

Patient ID	Age	UPDRS (III) preoperative "off state"	UPDRS (III) postoperative "off state"
101	62	75	38
105	65	60	35
109	50	66	28
110	62	70	41
113	53	61	24
122	64	78	48

II. MATERIALS AND METHODS

A. Patient Recordings

Six male PD patients with average age of 59 ± 6 years old were included in this study. Patients underwent a bilateral DBS implantation procedure according to the CAPSIT-PD protocol [11], in the Neurosurgery Clinic of Evangelismos Hospital, Athens, Greece. Following the operation and for a period of at least 1-year, the patients presented an average UPDRS (Unified Parkinson's Disease Rating Scale) (III) scale improvement of 32.67 (± 5.4). The motor evaluation for all patients is presented in Table I. No mental disorders were reported in follow-up psychiatric evaluations.

DBS implantation was guided by an expert neurologist (co-author of this study, P.S.) by listening to MERs sent to an audio scope. MERs were acquired on spontaneous STN activity, defined as the neuronal activity acquired during periods in which the PD patient lied down immobile in the operational table. No electrical stimulation was performed during recordings. Neither active nor passive movements were executed during analyzed MERs.

Analysis of MERs was conducted separately for each hemisphere. Both hemispheres were considered to contribute equally to the patient's clinical responses. Therefore, each clinical result was mapped to 2 STN, one at each side of a patient's brain. MERs were acquired using a Ben Gun formation, consisting of five microelectrodes in a cross formation, namely central, anterior, posterior, medial, lateral. The distance between the central and the surrounding electrodes was 2 mm and the signals recorded from each electrode lasted 10s each [12].

The permanent DBS lead (Medtronic®) had 4 contacts that were 0.5mm apart and had a diameter of 1.5mm. Stimulation was either monopolar or bimonopolar (i.e. two contacts with the same negative polarity).

Stimulation parameters (contact, pulse amplitude, width and frequency) were also chosen by neurologist (P.S.) for optimal clinical benefit intraoperatively and during follow-up. In both cases, the neurologist had no access to the β-peak information acquired from MER data. Hence, contact point and stimulation parameters were only based on clinical outcome.

B. Identifying the Stimulation Target

Each of the 5 electrodes entered and exited the STN at different depths. The STN entrance and exit for each trajectory was determined off-line by P.S., after visual inspection of MERs. Signals outside the STN were excluded from this study.

Power spectral density (PSD) for each MER, normalized by its electrode impedance, was estimated using the Welch's modified periodogram with a data window length of 0.68 s and 50% overlap. The β-band [12 – 30 Hz] is considered to be a neurophysiological signature of the location of sensorimotor neurons in non-moving patients [13]. That is why we isolated the amplitude peaks in that range.

The stimulation seems to be effective within a spherical area of 3mm radius around the stimulation contact [14]. Knowing the final position of the stimulation contact, we estimated the 3mm spherical area where the DBS signal had an effect.

We calculated the β-band amplitude peaks and compared the maximum aggregate β-band peak with the 3mm sphere. Our hypothesis was that the depth where the maximum β-band amplitude peak was present coincided with the optimal stimulation target [15]. We tested our hypothesis by estimating the distance, in mm, between stimulation point and maximum aggregate β-peak in vertical direction, as shown in Figure 1. For each STN we identified the suggested stimulation target as the depth where the sum of the five β-band amplitude peaks was maximum.

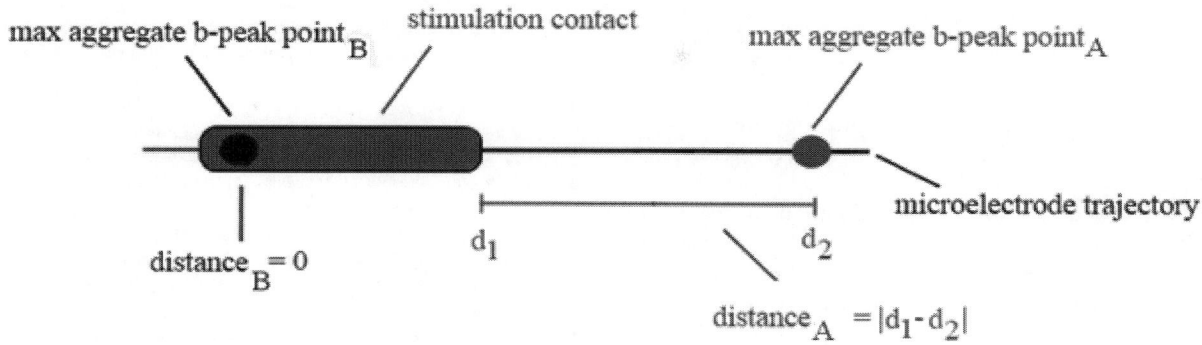

Fig 1 In case A, the distance between the stimulation contact and the maximum aggregate β-peak in vertical direction is calculated as the absolute value of the difference between the depth of the maximum aggregate β-peak (point A) and the depth of the contact's tip that is closest to point A. In case B, the maximum aggregate β-peak lies inside the clinically selected stimulation contact, hence the distance is 0.

III. RESULTS

The measured distances for the 12 cases are presented in Figure 2. In 7 cases the distance was 0, meaning that the maximum aggregate β-band peak lied inside the clinically selected stimulation contact. In the other 5 cases, that distance ranged from 1 to 2mm, as measured from the closest tip of the stimulation contact. In all 12 STN from patients with good clinical response, the distance was always smaller than 3mm. In other words, the suggested stimulation depth was within the spherical area of effective stimulation. Overall, the distance between the suggested stimulation depth and the actual stimulation contact had an average value of 0.67 mm (± 0.86).

IV. CONCLUSIONS

In this paper, we introduced a method that combines the β-band peak from 5 electrodes to propose a location for the optimal stimulation depth inside the STN of PD patients. Considering the β-band amplitude peaks as a neurophysiological signature of the location of sensorimotor neurons in non-moving patients, we found that the stimulation at the depth where the sum of β-band amplitude peaks was maximum was related to a good clinical response, in terms of UPDRS(III) score improvement and psychiatric evaluation, in all 6 patients.

These encouraging results call for an extended study on more patients, in order to verify our hypothesis. Specifical-

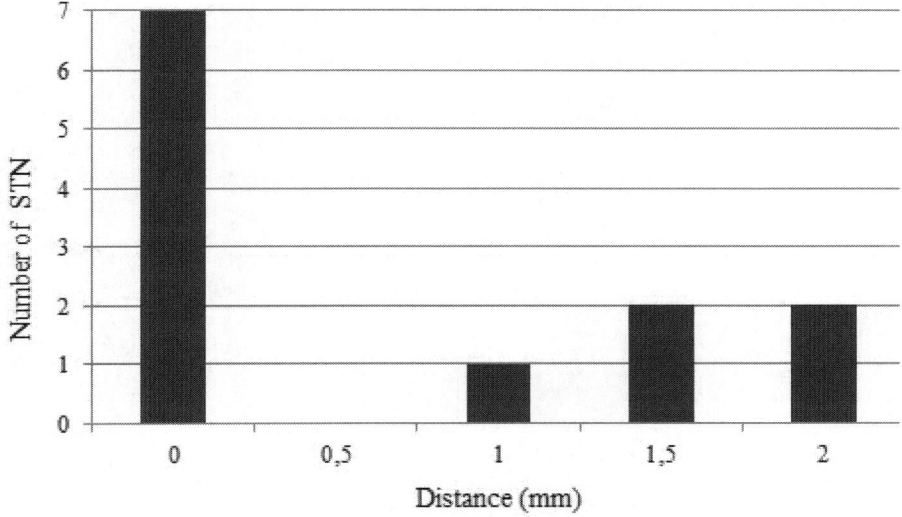

Fig 2 Distribution of the distance values in the 12 STN.

ly, we aim to verify our hypothesis on patients that developed side-effects after DBS implantation. We seek to relate the poor outcome of the DBS procedure to stimulation of an STN area that is farther than 3 mm away from the maximum β-peak. Another interesting application of this work would be to test how the stimulation of each hemisphere contributes separately to the patient's clinical response, under the scope of the hemispheric preponderance of Parkinson's disease. This way, we can possibly determine whether unilateral stimulation determines the final clinical response of the patient or not.

REFERENCES

1. Parkinson's Disease Foundation, Inc. at http://www.pdf.org/
2. McIntyre C, Savasta M, Kerkerian-Le Goff L, Vitek J, (2004) Uncovering the mechanism(s) of action of deep brain stimulation: activation, inhibition, or both, Clinical Neurophysiology, 115(6):1239-48.
3. U.S. Food and Drug Administration, U.S. Department of Health & Human Services at www.fda.gov
4. Romanelli P, Esposito V, Schaal D, Heit G, (2005) Somatotopy in the basal ganglia: experimental and clinical evidence for segregated sensorimotor channels, Brain Research Brain Research Reviews, 48(1):112-128
5. Hutchison W. D, Allan R. J, Opitz H, Levy R, Dostrovsky J. O, Lang A. E, Lozano A.M, (1998) Neurophysiological identification of the subthalamic nucleus in surgery for Parkinson's disease, Ann Neurol, vol. 44, pp. 622–628.
6. Michmizos K, Sakas D, and Nikita K, (2012) Prediction of the Timing and the Rhythm of the Parkinsonian Subthalamic Nucleus Neural Spikes using the Local Field Potentials, IEEE Transactions of Information Technology in Biomedicine
7. Chen C.C, Brown P et al, (2006) Intra-operative recordings of local field potentials can help localize the subthalamic nucleus in Parkinson's disease surgery, Experimental Neurology, vol. 198 (1): 214-221
8. Kuhn A, Williams D, Kupsch A, Limousin P, Hariz M, Schneider G, Yarrow K, Brown P, (2004) Event-related beta desynchronization in human subthalamic nucleus correlates with motor performance, Brain, vol. 127, pp. 735–746.
9. Levy R, Hutchinson W, Lozano A, Dostrovsky J, (2002) Synchronized neuronal discharge in the basal ganglia of parkinsonian patients is limited to oscillatory activity, Journal of Neuroscience, vol. 22, pp. 2855–2861.
10. Novak P, Przybyszewski A et al, (2011) Localization of the subthalamic nucleus in Parkinson disease using multiunit activity, Journal of the Neurological Sciences, vol. 310, pp. 44-49
11. Defer G. L, Widner H, Marie R. M, Remy P, Levivier M, (1999) Core Assessment Program for Surgical Interventional Therapies in Parkinson's Disease (CAPSIT-PD), Mov Disord, vol. 14, pp. 572-584.
12. Michmizos K, Sakas D, Nikita K, (2011) Towards relating the subthalamic nucleus spiking activity with the local field potentials acquired intranuclearly, IOP Measurement Science and Technology
13. Kuhn A, Trottenberg T, Kivi A, Kupsch A, Schneider G, Brown P, (2005) The relationship between local field potential and neuronal discharge in the subthalamic nucleus of patients with Parkinson's disease, Experimental Neurology, 194
14. Ranck J.B, (1975) Which elements are excited in electrical stimulation of mammalian central nervous system? A review, Brain Res 98:417–440.
15. Michmizos K, Frangou P, Stathis P, Sakas D, Nikita K, (2013) β-band peak localizes the optimal Deep Brain Stimulation contact in Parkinson's disease patients, International DBS Conference, Düsseldorf, Germany

Polytimi D. Frangou

National Technical University of Athens
Iroon Polytechniou 9
Athens, Greece
pfrangou@biosim.ntua.gr

Ergonomics during the Use of LESS Instruments in Basic Tasks: 2 Articulated vs. 1 Straight and 1 Articulated Graspers

M. Lucas-Hernández, F.J. Pérez-Duarte, A.M. Matos-Azevedo,
J.B. Pagador, and F.M. Sánchez-Margallo

Jesús Usón Minimally Invasive Surgery Centre (JUMISC), Cáceres, Spain

Abstract—Laparoendoscopic single site surgery (LESS) is now consolidated as a real alternative to conventional laparoscopy. The introduction of advanced instruments (articulated forceps) and technical modifications will eliminate/reduce the need to use several trocars to triangulate inside the cavity. In this study, we compare surgeons' muscle activity during the use of two common configurations of surgical instruments (C1: two articulated instruments, and C2: one straight instrument on the right hand and one articulated instrument on the left hand) in LESS surgery, during the performance of two basic tasks, cutting and coordination, on physical simulator. As conclusion, the coordination exercise and the use of the C1 configuration generated lower muscle activity, representing the less physically demanding scenarios for the surgeon.

Keywords—Laparoscopy, ergonomics, electromyography, LESS.

I. INTRODUCTION

Laparoendoscopic single site surgery (LESS) is now consolidated as a real alternative to conventional laparoscopic surgery, with numerous studies sustaining its feasibility and therapeutic safety [1-2]. It consists of a new development/evolution in minimally invasive surgery which uses a skin incision of 2-3 cm as the only access to the surgical field, and through which all necessary laparoscopic instrumentation and optics are introduced. The surgical principles do not differ much from those of laparoscopy. Nevertheless, the physical, spatial and mechanical restrictions inherent to LESS are important influences on the use of traditional laparoscopic instrumentation, as its design is unsuitable for working in the new conditions of this new approach.

There are still many challenges in terms of intraoperative ergonomics and laparoscopic skill requirements for surgical procedures performed by LESS. New, specially designed, instruments and trocars may represent a solution to solve the ergonomic problems that arise during this surgery [3]. The introduction of advanced instruments (articulated forceps) and technical modifications has reduced the need to use several trocars to triangulate inside the abdominal cavity, rendering LESS surgery feasible.

Other ergonomics studies in LESS [4-5] supported the surgeons claim of discomfort, musculoskeletal disorders and physical fatigue due to static and forced postures during the surgical process.

This study aims to compare surgeons' muscle activity during the use of two common LESS configurations of surgical instruments in the performance of two basic tasks, cutting and coordination, hands-on physical simulator.

II. MATERIALS AND METHODS

24 surgeons, all right-handed, were included, after being adequately informed of the protocol and voluntarily accepting to participate in this study. According to their experience level the sample was divided in three groups:

- Novice (n=10), without any experience in minimally invasive surgery.
- Intermediate (n=10), with over 50 procedures carried out by laparoscopy.
- Experienced (n=4), with more than 100 procedures carried out by laparoscopy and over 30 performed through LESS.

Executed tasks were:

- Coordination (Fig. 4): the subject had to place small rough and irregular objects in the wells of a coordination plate, following a predefined pattern and alternating between left and right instrument.s Electromyography (EMG) signals were obtained during a total of 10 minutes.
- Cutting (Fig. 5): the subject had to cut following a predefined pattern printed on a training plate. EMG was measured during total exercise length.

All tasks were accomplished on the SIMULAP® box-trainer (JUMISC, Cáceres, Spain) with a 10 mm 0 degrees rigid laparoscope (Karl Storz GmbH & Co. KG, Tuttlingen, Germany) as vision system. The structure of a SILS™ Port (Covidien, Mansfield, MA, USA) was used due to its high flexibility which promotes maneuverability (Fig. 1), without inserting the rigid cannulae included in the market package. The simulator was placed on a surgical table, which was later adjusted in height by each surgeon. Furthermore, surgeons were allowed to define ideal monitor positioning and height, measured at its central point. These were later registered for posterior analysis.

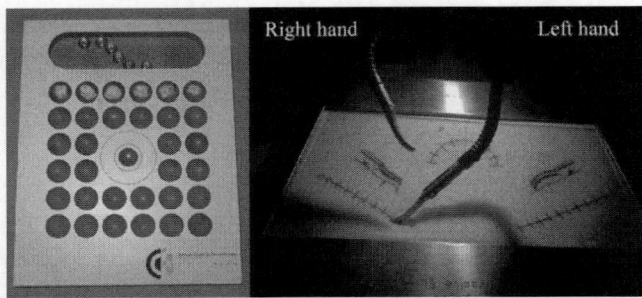

Fig. 1 Tasks: a) Coordination, and b) Cutting

Each task was performed with two different instrument configurations:

- C1 / two articulated (Fig. 2a):
  a. Coordination: articulated laparoscopic forceps SILS™ DISSECTOR (Covidien, Mansfield, MA, USA) on both hands.
  b. Cutting: Laparoscopic scissors SILS™ SHEARS (Covidien, Mansfield, MA, USA) on right hand and articulated laparoscopic forceps SILS™ DISSECTOR (Covidien, Mansfield, MA, USA) on left hand.
- C2 / one straight and one articulated (Fig.2b):
  a. Coordination: Laparoscopic dissection forceps ENDO DISSECT™ (Covidien, Mansfield, MA, USA) on right hand and articulated laparoscopic forceps SILS™ DISSECTOR (Covidien, Mansfield, MA, USA) on left hand.
  b. Cutting: Laparoscopic scissors ENDO SHEARS™ (Covidien, Mansfield, MA, USA) on right hand and articulated laparoscopic forceps SILS™ DISSECTOR (Covidien, Mansfield, MA, USA) on left hand.

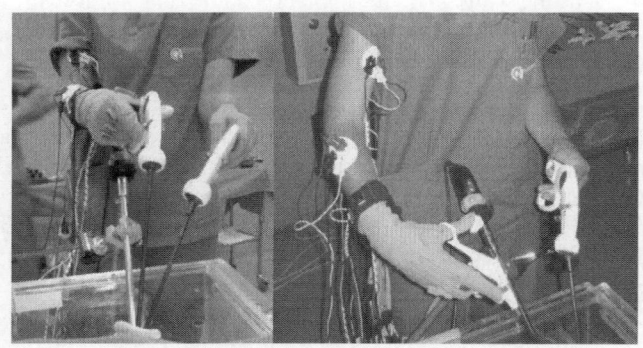

Fig. 2 Instruments configurations: a) C1, and b) C2

The surgeons were instrumented with surface electrodes in the biceps brachii, triceps brachii, flexor carpi ulnaris, extensor digitorum communis and trapezius muscles on right arm. EMG signals were transmitted at a frequency of 1,000 Hz to EMG MP100 Biopac® (BIOPAC Systems, Inc., Goleta, CA, USA). For registry the software used was AcqKnowledge 3.8.1 (BIOPAC Systems, Inc., Goleta, CA, USA) along with Matlab 7.0 MathWorks, Natick, MA, USA) used for posterior analysis, in which a trimmed mean was applied to 5%.

Before starting the trial, maximum voluntary contraction (MVC) measures were obtained. Final muscle contraction values are normalized as muscle strain in percentage of MVC.

Descriptive statistics for each variable was obtained, by calculation of characteristic parameters: average and standard deviation. For every case a Kolmogorov-Smirnov test was used to determine normal distribution of variables. In the cases in which this condition was verified, a T-Student test followed for comparison between groups. All variables determined no parametric, were further analyzed with Wilcoxon signed-rank test. Significance level was set at $p<0.05$ for all tests.

All statistical analyses were performed with statistic software SPSS 15.0 for Windows OS (SpSS Inc., Chicago, Illinois).

III. RESULTS

A. Coordination Task

Muscle activity was lower during the use of C2 configuration instruments when compared with C1 in all muscles and surgeon levels. Furthermore, triceps had the lower values and extensor the highest in all cases.

In novice surgeons (Table 1), there were significant differences during the use of C1 configuration in the muscle activity of the biceps (C1 9.02±3.84 vs. C2 7.82±3.71; $p<0.05$), and the extensor muscles (C1 16.18±9.37 vs. C2 13.25±6.20; $p<0.05$).

Table 1 Muscle strain in percentage of MVC. Surgeon level: novice. Task: Coordination. Sig. (bil): T-Student test bilateral significance level *$p<0.05$, **$p<0.01$, ***$p<0.001$

Muscle	C1	C2	Sig. (bil)
Biceps	9.02±3.84	7.82±3.71	0.022*
Triceps	3.37±1.47	3.23±1.22	0.314
Trapezius	10.33±9.60	10.11±6.72	0.754
Flexor	9.38±5.70	8.49±5.57	0.243
Extensor	16.18±9.37	13.25±6.20	0.047*

On other hand, extensor muscles had the greatest activity during the use of both configurations (C1 16.18±9.37 and C2 13.25±6.20).

Respecting intermediate surgeons (Table 2), we found significant differences in the extensor muscles during the use of C1 configuration (C1 23.35±9.37 vs. C2 18.99±6.13; p<0.05).

In the same manner as novices surgeons, extensor muscles had the greatest activity during the use of both configurations (C1 23.35±9.37 and C2 18.99±6.20).

Table 2 Muscle strain in percentage of MVC. Surgeon level: intermediate. Task: Coordination. Sig. (bil): T-Student test bilateral significance level *p<0.05, **p<0.01, ***p<0.001

Muscle	C1	C2	Sig. (bil)
Biceps	10.27±5.34	8.78±3.73	0.296
Triceps	4.27±2.68	4.12±2.82	0.784
Trapezius	10.48±12.29	8.81±8.56	0.292
Flexor	10.33±4.67	8.26±3.62	0.082
Extensor	23.35±9.37	18.99±6.13	0.014*

Regarding experienced surgeons, there were no significant differences between muscle activities (Table 3). In this case, extensor muscles presented the highest activity values during the use of C1 configuration (C1 17.51±6.66), whilst with C2 greatest values were obtained for the trapezius muscle (C1 13.12±3.59).

Table 3 Muscle strain in percentage of MVC. Surgeon level: experienced. Task: Coordination. Sig. (bil): T-Student test bilateral significance level *p<0.05, **p<0.01, ***p<0.001

Muscle	C1	C2	Sig. (bil)
Biceps	8.07±5.72	6.23±2.92	0.368
Triceps	3.95±4.45	1.82±0.46	0.873
Trapezius	16.24±13.87	10.33±7.80	0.511
Flexor	8.49±9.73	3.08±1.41	0.512
Extensor	17.51±6.66	13.12±3.59	0.282

With regard to the highest values obtained for each muscle during C1 configuration, we present them as follows:

- Biceps: intermediate surgeon.
- Triceps: intermediate surgeon.
- Trapezius: experienced surgeon.
- Flexor: intermediate surgeon.
- Extensor: intermediate surgeon.

B. Cutting Task

Muscle activity was lower during the use of C2 configuration instruments when compared with C1 in all muscles and surgeon levels. Triceps had the lower values in all cases.

There were significant differences in the muscle activity of the biceps muscle during the use of C1 configuration in novice surgeons (C1 9.43±4.76 vs. C2 7.73±3.95; p<0.001), as shown on Table 4.

On other hand, trapezius muscle had the greatest activity during the use of C1 configuration (C1 18.91±9.10), while during the use of C2 this was observed on the extensor muscles (C1 18.57±9.22).

Table 4 Muscle strain in percentage of MVC. Surgeon level: novice. Task: Cutting. Sig. (bil): T-Student test bilateral significance level *p<0.05, **p<0.01, ***p<0.001

Muscle	C1	C2	Sig. (bil)
Biceps	9.43±4.76	7.73±3.95	0.005**
Triceps	4.11±1.29	3.88±1.25	0.142
Trapezius	18.91±9.10	14.53±7.39	0.114
Flexor	11.85±8.48	10.58±6.75	0.396
Extensor	18.57±10.47	18.57±9.22	0.999

Intermediate surgeons group (Table 5) presented significant differences in the flexor muscles during the use of C1 configuration (C1 12.40±6.83 vs. C2 10.32±4.82; p<0.05). In this task, extensor muscle had the greatest activity during the use of both instrument configurations (C1 24.85±9.82 and C2 22.95±9.60).

Table 5 Muscle strain in percentage of MVC. Surgeon level: intermediate. Task: Cutting. Sig. (bil): T-Student test bilateral significance level *p<0.05, **p<0.01, ***p<0.001

Muscle	C1	C2	Sig. (bil)
Biceps	10.21±4.48	9.09±4.06	0.195
Triceps	5.28±3.07	4.85±3.82	0.477
Trapezius	13.71±7.09	10.75±7.49	0.303
Flexor	12.40±6.83	10.32±4.82	0.048*
Extensor	24.85±9.82	22.95±9.60	0.051

Regarding experienced surgeons, there were no significant differences between muscle activities (Table 6). Similarly to novice surgeons, trapezius muscle presented the greatest activity during the use of C1 configuration (C1 26.58±19.47), while during the use of C2 it was the extensor muscles that emitted the highest muscle contraction signal (C1 21.46±5.44).

Table 6 Muscle strain in percentage of MVC. Surgeon level: experienced. Task: Cutting. Sig. (bil): T-Student test bilateral significance level *p<0.05, **p<0.01, ***p<0.001

Muscle	C1	C2	Sig. (bil)
Biceps	9.57±6.70	5.72±2.44	0.214
Triceps	5.93±7.63	5.70±6.90	0.640
Trapezius	26.58±19.47	13.60±9.06	0.397
Flexor	10.81±12.27	4.62±1.98	0.398
Extensor	23.50±8.94	21.46±5.44	0.193

Considering the highest levels of muscle activity during C1 configuration, we can list each muscle according to the group in which this was verified as follows:

- Biceps: intermediate surgeon.
- Triceps: experienced surgeon.
- Trapezius: experienced surgeon.
- Flexor: intermediate surgeon.
- Extensor: intermediate surgeon.

IV. CONCLUSIONS

The use of one straight instrument on the right hand and one articulated instrument on the left hand proved to be the most adequate configuration for the performance of LESS basic maneuvers on physical simulator as it showed the lowest EMG muscle activity values. Novice surgeons showed lower levels of muscle activity, which was also observed in other studies [6] where new instruments were tested. In contrast, intermediate surgeons obtained the highest values in most of the muscles during the execution of both tasks, with the extensor muscles representing the most active muscle group. Trapezius muscle constituted the second more active muscle, in which experienced surgeons reached the highest values.

Furthermore, it is noteworthy that in both tasks, the expert surgeons reach much higher values of muscle activity in the trapezius compared to the intermediate surgeons. In cutting task, trapezius muscle activity of expert surgeons doubles with respect to intermediate one. This issue is interesting and needs to be analyzed in future studies.

Regarding the different tasks, novice and intermediate surgeons obtained similar values during both exercises and with both configurations, in biceps and triceps muscles, and lower values in coordination task. For experienced surgeons we also registered very similar values during both exercises and configurations in the biceps brachii, and lower values in coordination task.

Thus, as expected after gathering surgeons' subjective impression at the end of the trials, the coordination exercise and the use of one straight instrument on the right hand and one articulated instrument on the left hand were the less physically demanding scenarios during the performance of LESS basic tasks on simulator.

Articulated instruments in both hands is a configuration that produces more muscle activity because this instrument is for single use [7], and therefore there are certain mechanical deficiencies such as backlash and loss of sensation in the distal tool. This involves greater muscular effort for the surgeon.

ACKNOWLEDGMENT

This work has been partially funded by the project PI12/01467 from the Institute of Health Carlos III - FIS of Spain. Authors are especially grateful to JUMISC´s personnel for their technical assistance and participation.

REFERENCES

1. Raman JD, Bagrodia A, Cadeddu JA. (2009) Single-incision, umbilical laparoscopic versus conventional laparoscopic nephrectomy: a comparison of perioperative outcomes and short-term measures of convalescence. Eur Urol 55(5):1198-1204.
2. Park HS, Kim TJ, Song T, Kim MK, Lee YY, Choi CH, et al (2011) Single-port access (SPA) laparoscopic surgery in gynecology: a surgeon's experience with an initial 200 cases. Eur J Obstet Gynecol Reprod Biol 154(1):81-84.
3. Stolzenburg JU, Kallidonis P, Oh MA, Ghulam N, Do M, Haefner T et al EN (2010) Comparative assessment of laparoscopic single-site surgery instruments to conventional laparoscopic in laboratory setting. J Endourol 24(2):239-245.
4. Morandeira-Rivas A, Millán-Casas L, Moreno-Sanz C, Herrero-Bogajo ML, Tenías-Burillo JM, Giménez-Salillas L (2012) Ergonomics in Laparoendoscopic Single-Site Surgery: Survey Results. J Gastrointest Surg 16(11):2151-2159.
5. Montero PN, Acker CE, Heniford BT, Stefanidis D (2011) Single incision laparoscopic surgery (SILS) is associated with poorer performance and increased surgeon workload compared with standard laparoscopy. Am Surg 77(1):73-77.
6. Lucas Hernández M, Matos Azevedo AM, Pérez Duarte FJ, Pagador Carrasco JB, Sánchez Margallo FM (2012) Entrenamiento en cirugía laparoscópica: el uso de instrumental en la actividad muscular del cirujano. Libro de Actas XXX CASEIB 2012.
7. Raman JD, Cadeddu JA, Rao P, Rane A (2008) Single-incision laparoscopic surgery: initial urological experience and comparison with natural-orifice transluminal endoscopic surgery. BJU Int., 101(12):1493-1496.

Author: Marcos Lucas Hernández
Institute: Jesús Usón Minimally Invasive Surgery Centre (JUMISC)
Street: Ctra. N-521, km 41.8.
City: Cáceres
Country: Spain
Email: mlucas@ccmijesususon.com

Network-Based Modular Markers of Aging across Different Tissues

Aristidis G. Vrahatis[1], Konstantina Dimitrakopoulou[2], Georgios N. Dimitrakopoulos[3],
Kyriakos N. Sgarbas[3], Athanasios K. Tsakalidis[1], and Anastasios Bezerianos[2,4]

[1] Department of Computer Engineering and Informatics, University of Patras, Patras, 26500, GR
[2] School of Medicine, University of Patras, Patras, 26500, GR
[3] Department of Electrical and Computer Engineering, University of Patras, Patras, 26500, GR
[4] SINAPSE Institute, Center of Life Sciences, National University of Singapore, Singapore 117456

Abstract—Aging is a highly complex biological process and a risk factor for many diseases. Motivated by the high availability of diverse high-throughput data in the mouse model organism, we provide a systemic view of the age-related mechanisms. In particular, we present a robust network-based integrative approach that provides, based on protein interaction and microarray data, reliable modules that alter significantly in terms of expression during aging. Our modular meta-analysis provides novel information about the involvement of several established as well as recently reported longevity-associated pathways across different tissues.

Keywords—aging, module, protein interaction, microarray.

I. INTRODUCTION

Aging is a highly complicated process characterized by progressive decline and destabilization of the system at cellular and organismal level. In addition, aging is a significant factor in many complex diseases like neurodegenerative disorders, cancer, diabetes and others. In recent decades, the release of whole-genome scale data derived from high-throughput technologies, like microarrays, ChIP-chip, ChIP-seq, protein microarrays and others, enabled the identification of genes and pathways governing aging mechanisms. More recently, systems biology approaches re-addressed the study of aging on two levels. First, they integrated the diverse information of various levels: chromosomal, RNA, protein and metabolite level and second, they revealed the crosstalk among pathways, offering so a more holistic view of longevity mechanisms and aging-associated phenotypes.

In this context, several studies employed network theory to offer a more comprehensive view of the underlying age-related processes [1]. Topological attributes of networks such as the existence of hubs (nodes with the most links) or modules (dense sub-networks with distinct functional role) have assisted significantly in revealing the regulatory circuitry. For example, the work of [2] showed in yeast protein-protein interaction (PPI) network that the regulators of known age-reported genes are linked through shortest paths. The work of [3] on human PPI network found that aging regulators, homologs of regulators as predicted by invertebrates and their first-order neighbors tend to be hubs nodes. Moving further, studies like [4] examined the modular structure of PPI networks and showed that aging refers to a small number of modules and the genes linking different modules are highly related to aging. Finally, the work of Wang *et al.* [5] studied the relationships between aging and genetic disease genes and found that disease genes close to known longevity genes are more likely to have a hub position on the PPI topology.

In this work, we present a novel methodological framework that combines diverse 'omics' data with network-based approaches, with the scope to define modular markers of aging. Based on a large cohort of diverse tissue datasets in mouse model, we successfully identified well-established as well as recently implicated age-related pathways. Meta-analysis of the identified modules managed to designate them as tissue-specific or cross-tissue markers.

II. METHODS

Our large scale integrative framework (Fig. 1) that identifies module-based markers discriminative of two states (young/old) is summarized in four steps: (a) the successful integration of heterogeneous data (PPI, multiple tissue microarray), (b) the nomination of differentiated interacting proteins/genes in the transition from young to old state, (c) the employment of a module detecting algorithm that efficiently identifies dense interacting groups of proteins that simultaneously alter in terms of expression during aging, and (d) the meta-analysis of derived modules at the level of tissue information so as to enable the revelation of tissue-specific and cross-tissue longevity mechanisms.

A. Datasets

We downloaded the Z-transformed mouse AGEMAP expression data, whose raw form is also available in NCBI Gene Expression Omnibus (accession GSE9909). Our microarray collection contains the male data across 15 different tissues aged at 1 and 24 months of age. For all probes on the array corresponding to the same Unigene ID, we averaged the Z-scores from each of the probes together. The samples at 1 month were characterized as 'young', whereas at 24 months as 'old'.

We downloaded all mouse protein-protein interaction (PPI) data from BioGRID, IntAct, MINT, DIP and InnateDB databases. In particular, we isolated for analysis only the PPIs mapped in the Unigene IDs present in expression datasets and ended up with 3,020 Unigene IDs and 6,742 interactions (after removing self-loops) among them. The final PPI network consists of a large component of 2,889 proteins and 73 smaller components with 1-9 members.

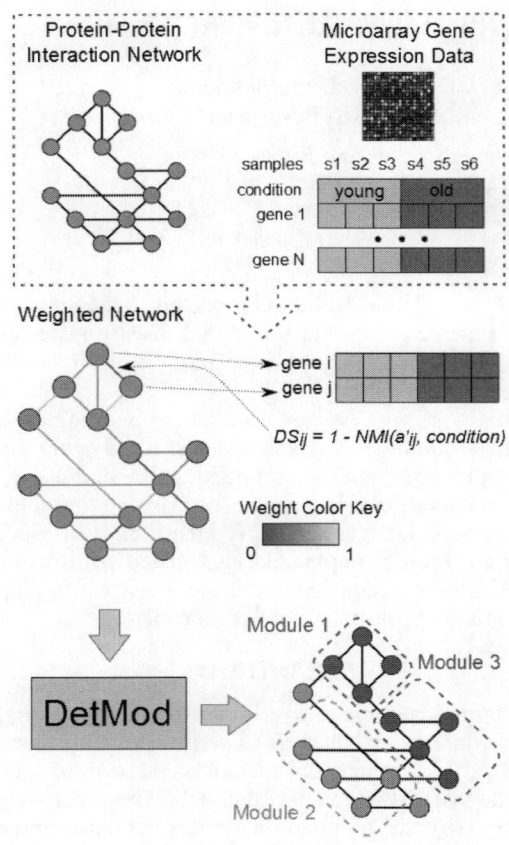

Fig. 1 Overview of our integrative module detecting approach.

B. Data Integration

We constructed a weighted graph for each tissue, which captures the significant expression changes during the transition from young to old state. To accomplish that, we assigned the weights onto the PPI graph with the use of the discriminative score, described in [6], which was adapted in order to weigh each interacting pair of proteins. Initially, for each gene the expression profiles in both classes (young/old) were reshaped in a vector followed by another vector representing the corresponding class (c) labels. Next, we calculate the activity score (a) of each interacting pair of genes i and j as follows:

$$a_{ij} = (g_i + g_j)/\sqrt{2},$$

where g_i and g_j are the Z-transformed expression values. The activity score is discretized (a'_{ij}) into $\lfloor \log 2(\# of\ samples) +1 \rfloor$ equally spaced bins as described in [7]. Our adapted discriminative score (DS_{ij}) derives from the normalized mutual information (NMI) between a' and c.

$$DS_{ij} = 1 - NMI(a'_{ij}, c)$$

The DS, which serves as the weight onto the PPI graph, ranges in [0,1]. Smaller values indicate that the interacting nodes discriminate significantly in terms of expression between the two states (young and old). The resulting weighted graph is used as input in the module-detecting DetMod algorithm in order to define highly confident discriminative modules that serve as signature of aging.

C. DetMod Algorithm

DetMod is a module-detecting algorithm, which has provided robust results when applied to yeast microarray and PPI data [8]. The main advantage of DetMod is its ability to capture the inter-module cross-talk mediated by a controlled degree of overlap among the identified modules. It operates in two phases. Initially, it iteratively applies Detect Module from Seed Protein (DMSP) algorithm [9] to every node of the graph, which is regarded as 'seed' protein upon which a module is created. DMSP firstly accepts one 'seed' protein and selects a subset of its best neighbors and then expands this initial kernel with more proteins. This expansion considers the number of neighbors for the specific protein as well as the weights of these connections. At this point DetMod checks each newly constructed module in terms of overlap with the rest of the modules that have been previously created. If this overlapping degree is above a certain threshold then the module is discarded. In the second phase of the DetMod, the derived modules are further examined in order to determine if merging is plausible. The parameters of DetMod were set as $p_1=0.4$, $p_2=0.95$, L=3 and overlapping degree threshold equal to 0.7, so as to acquire small-sized (5-20 members) and highly confident modules (strict thresholds in module expansion).

D. Modular Meta-analysis

DetMod was applied on the weighted graph for each tissue separately and the number of detected modules ranged in {17-24}. The small number of identified modules per tissue is explained by the weight distributions of edges. As shown in Fig. 2A, a small fraction of edges (9%) had weights lower than 0.4, which indicates that only a small fraction of interacting proteins changed, in terms of expression, substantially during aging.

The final pool of modules across all tissues, which was further analyzed, consisted of 191 modules. Next, we filtered this pool after searching for modules with highly over-represented KEGG pathway terms (Fisher's exact P-value ≤ E-2) and ended up with 104 modules (Fig. 2B).

We chose to categorize the modules into four categories based on their tissue occurrence and high node overlap ratio (NOR) ≥ 0.8. The node overlap ratio was calculated [10] as:

$$NOR(M_i, M_j) = 2 * \frac{M_i \cap M_j}{M_i \cup M_j},$$

where M_i and M_j represent the compared modules.

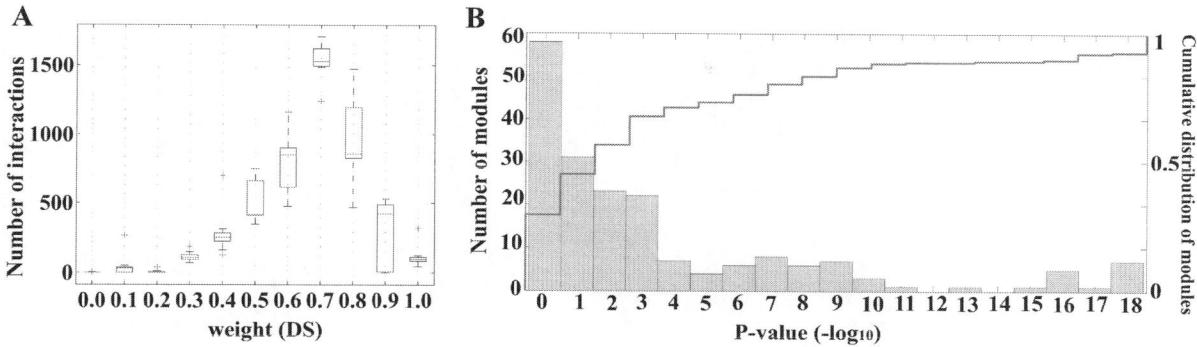

Fig. 2 (A) Boxplot summarizing the weight distribution of PPI interactions in all tissues. As shown, a small fraction of interactions had weight lower than 0.4, i.e. only a fraction of interacting proteins changed significantly between the two states (young/old). (B) P-value distribution in terms of KEGG pathways of the 191 modules identified across all tissues. In order to detect over-represented pathways related to aging, we isolated the modules with P < E-2.

The categories are: (A) super tissue-specific (found only in one tissue), (B) tissue-specific (found in 2-4 tissues), (C) middle tissue-specific (found in 5-9 tissues), (D) cross-tissue (found in 10-14 tissues) and (E) super cross-tissue (found in all tissues). After examining the set of 104 modules based on the above criteria, the 104 modules were grouped into 27 subsets; each subset is represented by a module derived from the intersection of modules among the involved tissues. The 27 modules incorporate 276 unique proteins (9% of the total proteome included in our study).

III. RESULTS

In Table 1, we report the most indicative over-represented pathways of the 27 examined modules along with the Fisher Exact P-values as detected by DAVID tool [11]. As evident, no module is reported in the super cross-tissue category. This finding does not necessarily imply that no single pathway serves as cross-tissue marker of aging. Probably, the explanation lies in the fact that certain tissues like liver, striatum and bone marrow miss multiple array experiments (i.e. they had less than four biological repeats in a single age group) and thus module construction procedure was severely affected.

Interestingly, the majority of modules in the rest categories are highly enriched in pathways reported as age-related. The nomination of cell cycle, ubiquitin mediated proteolysis and proteasome as cross-tissue longevity markers accords well with the so far reported aging mechanisms [12]. The implication of adherens junction and Wnt signaling in longevity has been recently reported [13-14] (Fig. 3A). Also, we comment on NOD-like receptor pathway, whose role in inflammaging (low-grade pro-inflammatory phenotype accompanying aging) has been lately elucidated. NOD-like receptors can be activated by many danger signals, which in turn provoke cellular stress and aging process [15]. Also, our results corroborate to the findings of [16] about the modifications of cysteine and methionine metabolism in aging. With regard to super tissue-specific category, we report the module enriched in MAPK and insulin signaling pathway which was identified in hippocampus tissue (Fig. 3B). Our findings corroborate to two recent experimental studies; the first showed that MAPK signaling has differentiated function during aging in terms of gene expression in hippocampus tissue, which may be implicated in anxiety and long term depression [17]. The second study showed that impairments of insulin signaling in the brain are prominent factors for age-dependent cognitive decline and Alzheimer's disease [18]. Another interesting case is the

Table 1 KEGG pathways associated to aging across tissues

Pathways	Members*	%**	P-value (log10)
Cross-tissue			
Cell cycle†	12	91.62	-16
Cell cycle†	18	72.20	-18
Adherens junction/Wnt signaling	11	45.00	-7
Ubiquitin mediated proteolysis	11	45.43	-6
NOD-like receptor signaling	7	42.81	-3
Cysteine and methionine metabolism	6	50.00	-3
Middle tissue-specific			
Proteasome	11	81.81	-16
TGF-beta signaling	8	37.50	-3
p53 signaling	8	50.00	-5
Tissue-specific			
Jak-STAT signaling	9	66.66	-6
ErbB signaling	11	54.54	-7
Pathways in cancer	13	46.15	-4
mTOR signaling	11	27.27	-3
MAPK signaling	7	42.85	-3
Neurotrophin signaling	20	25.00	-5
Super tissue-specific			
Insulin signaling/MAPK signaling (Hippocampus)	6	83.33	-6
T cell receptor signaling (Eye)	7	57.14	-4
Chronic myeloid leukemia (Bone Marrow)	9	44.44	-4
Oxidative phosphorylation (Kidney)	17	29.41	-4

* Module size ** Percentage of pathway related proteins relative to the module size †These modules had NOR < 0.2.

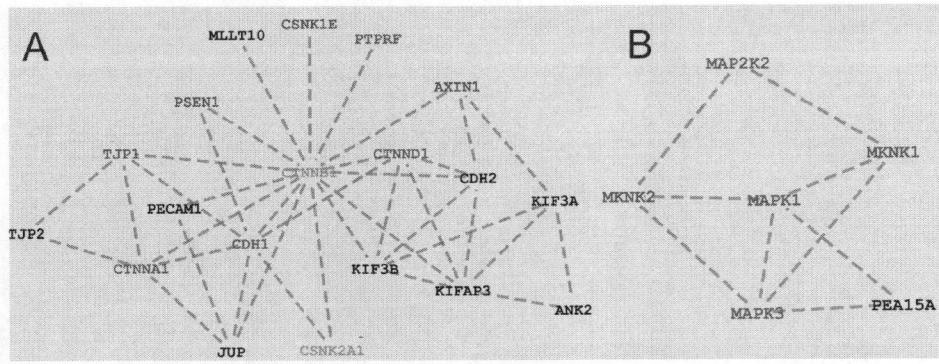

Fig. 3 Network visualization of two modules characterized as (A) cross-tissue and (B) super tissue-specific. The (A) module is highly enriched (P ≤ E-7) in the adherens junction (pink nodes) and Wnt signaling pathways (green nodes). The orange nodes denote common members between pathways. The (B) module is enriched (P ≤ E-6) in insulin and MAPK signaling pathways. The pink nodes are common between the two pathways.

super tissue-specific module enriched in the oxidation phosphorylation pathway, which was detected in kidney tissue. Generally, oxidative phosphorylation is a widely known pathway contributing to aging and impaired cellular metabolism [19]. However, our study revealed its role only in kidney tissue. Evidence of recent study showed that the renal oxidative phosphorylation of mitochondria declined in elderly animals due to decreased cytochrome content [20].

IV. CONCLUSIONS

Our large scale integrative methodological framework provided robust modular aging markers that could serve as targets for anti-aging intervention. Finally, our approach is applicable in several complex biological processes and diseases where massive amount of multiple tissue 'omics' data is available.

ACKNOWLEDGMENT

This research has been co-financed by the European Union (European Social Fund ± ESF) and Greek national funds through the Operational Program "Education and Lifelong Learning" of the National Strategic Reference Framework (NSRF) - Research Funding Program: Thalis. Investing in knowledge society through the European Social Fund.

REFERENCES

1. Hou L, Huang J, Green CD et al. (2012) Systems biology in aging: linking the old and the young. Curr Genomics 13:558-65
2. Managbanag JR, Witten TM, Bonchev D et al. (2008) Shortest-path network analysis is a useful approach toward identifying genetic determinants of longevity. PLoS One 3:e3802
3. Bell R, Hubbard A, Chettier R et al. (2009) A human protein interaction network shows conservation of aging processes between human and invertebrate species. PLoS Genet 5:e1000414
4. Xue H, Xian B, Dong D et al. (2007) A modular network model of aging. Mol Syst Biol 3:147
5. Wang J, Zhang S, Wang Y et al. (2009) Disease-aging network reveals significant roles of aging genes in connecting genetic diseases. PLoS Comput Biol 5:e1000521
6. Chuang HY, Lee E, Liu YT et al. (2007) Network-based classification of breast cancer metastasis. Mol Syst Biol 3:140
7. Tourassi GD, Frederick ED, Markey MK et al. (2001) Application of the mutual information criterion for feature selection in computer-aided diagnosis. Med Phys 28:2394–2402
8. Maraziotis IA, Dimitrakopoulou K, Bezerianos A (2008) An in silico method for detecting overlapping functional modules from composite biological networks. BMC Syst Biol 2:93
9. Maraziotis IA, Dimitrakopoulou K, Bezerianos A (2007) Growing functional modules from a seed protein via integration of protein interaction and gene expression data. BMC Bioinformatics 8:408
10. Wang X, Wang Z, Ye Z (2011) HKC: An Algorithm to Predict Protein Complexes in Protein-Protein Interaction Networks. Journal of Biomedicine and Biotechnology 2011:480294
11. Huang DW, Sherman BT, Lempicki RA (2009) Systematic and integrative analysis of large gene lists using DAVID Bioinformatics Resources. Nature Protoc 4:44-57
12. Hoppe T (2010) Life and destruction: ubiquitin-mediated proteolysis in aging and longevity. F1000 Biol Rep 2:79
13. Wolfson M, Budovsky A, Tacutu R et al. (2009) The signaling hubs at the crossroad of longevity and age-related disease networks. Int J Biochem Cell Biol 41:516-520
14. Naito AT, Sumida T, Nomura S et al. (2012) Complement C1q activates canonical Wnt signaling and promotes aging-related phenotypes. Cell 149:1298-313
15. Salminen A, Kaarniranta K, Kauppinen A et al. (2012) Inflammaging: disturbed interplay between autophagy and inflammasomes. Aging (Albany NY) 4:166-75
16. Mercier S, Breuillé D, Buffière C et al. (2006) Methionine kinetics are altered in the elderly both in the basal state and after vaccination. Am J Clin Nutr 83:291-8
17. Wefers B, Hitz C, Hölter SM et al. (2012) MAPK signaling determines anxiety in the juvenile mouse brain but depression-like behavior in adults. PLoS One 7:e35035
18. Yin F, Jiang T, Cadenas E (2013) Metabolic triad in brain aging: mitochondria, insulin/IGF-1 signalling and JNK signalling. Biochem-Soc Trans 41:101-5
19. Lesnefsky EJ, Hoppel CL (2006) Oxidative phosphorylation and aging. Ageing Res Rev 5:402-33
20. O'Toole JF, Patel HV, Naples CJ et al. (2010) Decreased cytochrome c mediates an age-related decline of oxidative phosphorylation in rat kidney mitochondria. Biochem J 427:105-12

Author: Anastasios Bezerianos
Institute: SINAPSE Singapore Institute for Neurotechnology, National University of Singapore
Street: 28 Medical Drive, 117456
City: Singapore
Country: Singapore
Email: tassos.bezerianos@nus.edu.sg

Experimental Characterization of Active Antennas for Body Sensor Networks

D. Naranjo[1], L.M. Roa[1], L.J. Reina[2,1], G. Barbarov[1], and A. Callejón[1]

[1] Biomedical Engineering Group, CIBER-BBN and University of Seville, Seville, Spain
[2] Department of Signal Theory and Communications, University of Seville, Seville, Spain

Abstract—This paper addresses methods for experimental characterization of active antennas in the context of Body Sensor Networks (BSN). In a previous work, the authors proposed an active antenna design to be used in a biomedical monitoring network, following specifications such as consumption, size and cost minimization. In this work, such an antenna has been integrated with the transceiver in the same device and characterized experimentally using active measurement methods. First, two sets of measurements were performed both with and without the presence of a human body in order to analyze its influence on the antenna performance. Subsequently, the active antenna was characterized inside an anechoic chamber in order to obtain more comprehensive parameters, thus presenting a method for obtaining the radiation patterns of antennas integrated into battery-powered smart sensors.

Keywords—Active antenna, anechoic chamber, Body Sensor Networks, radiation pattern.

I. INTRODUCTION

The antenna is a key element within Body Sensor Networks (BSN) [1]. Important advances have been achieved in the development of miniaturized portable antennas over previous years [2]. However, the strict requirements of size, consumption, and cost, together with other design challenges, such as the well-known dilemma between antenna efficiency and size [3], are the main reasons for which research in portable technologies is still needed. It must also be noticed that the environment where the antennas are located in BSN is the human body itself, which considerably affects the antenna efficiency [4]. On the other hand, in the BSN scenario, the characterization of the antenna has usually been performed without considering the transceiver to which it is connected, and hence effects such as the existence of a finite ground plane or the characteristics of the transceiver may considerably affect the antenna characteristics [5], Consequently, this highlights the fact that the characterization of active antennas in BSN still poses technical challenges which need to be researched.

The passive antenna measurement in an anechoic chamber is the most commonly used method for antenna characterization [6]. However, such a scheme may not be the most suitable when the purpose is to analyze an active antenna in a self-powered device, due to the possibility of the hardware elements significantly affecting the radiation characteristics [7–9]. Furthermore, the antenna may be adapted to an impedance of a different value than that of the measurement equipment (typically 50 ohms). It is necessary in this case the use of baluns which may affect the antenna properties [6]. An standardized methodology for the characterization of active antennas including the transceiver is yet to be proposed in the field of BSN.

In this work, a first approach into the active characterization of an antenna for BSN is addressed. The antenna under test was previously selected from a set of low-profile antennas specifically designed and optimized in order to comply with BSN specifications [10]. Active measurement methods in the laboratory have been performed both with and without the presence of a human body, thus allowing us to analyze its influence. Moreover, in order to obtain a more comprehensive characterization of the antenna, a second set of experiments was performed inside an anechoic chamber. The method described solves the issues raised during the development of the experiments.

II. MATERIALS AND METHODS

A. Description of the Biomedical Application, Technology Selection and Antenna Design

The biomedical application is based on an accelerometer sensor called SoM (Sensor of Movements), which is is used in a fall detection system including capabilities for activity monitoring [11, 12]. SoM communicates with a second device, called DAD (Decision-Analysis Device), which carries out a further processing of sensorial information, and if necessary, communicates with a remote medical-care center. Due to the fact that a low data rate is required, the IEEE 802.15.4 standard was selected with the aim of minimizing energy consumption [13]. Regarding the transceiver, Texas Instruments CC2430/CC2431 was selected given its low cost and consumption [14]. Regarding the antenna design, an orthogonal folded dipole antenna (OFD) (see Fig. 1) was selected because of its robustness and adaptability to the design specifications from an initial set of eighteen antennas, some of which were proposed by the transceiver manufacturer. An electromagnetic finite-element-method (FEM) simulation software, Ansys HFSS, was used in order to design the antenna and study the radiation characteristics (e.g. return loss, radiation diagram, gain, etc.).

B. Experimental Protocol for Measurements in Laboratory

A prototype of the SoM device with the OFD designed antenna was implemented, and a experimental validation was carried out by establishing a continuous transmission mode in the transceiver. In order to calculate the received power, an FSL8 spectrum analyzer of Rodhe & Schwarz was tuned to the transmission frequency of 2.4425 GHz and an omnidirectional monopole receiving antenna was used. Two different experimental cases were considered: a transmitting antenna without the presence of a human body, and a transmitting antenna attached to a human body (on-body antenna), in order to evaluate the influence of the body on the performance of the antenna. Furthermore, in both cases, measurements were repeated 10 times at different instants to provide a better statistical significance. In the first case, without the presence of a human body, the antenna was placed in a non-conductive and insulating 10-cm high empty cylinder. The cylinder was 1 m away from the omnidirectional monopole receiver antenna and four different azimuth angles were considered: 0°, 90°, 180° and 270°, as shown in Fig. 1. In addition, measurements were taken in both horizontal and vertical positions. In the case of the on-body antenna measurements, SoM was placed on the back close to the sacrum, since this was shown to be the most suitable position for the particular monitoring system according to the results of a previous research of the authors [11, 12]. The volunteer was 1m distance away from the receiving antenna and four different orientations of the subject along with four different positions of the antenna were tested, as can be seen in Fig. 1.

C. Experimental Protocol for Measurements in Anechoic Chamber

In the anechoic chamber used in the present work, both transmitter and receiver antennas are located on metallic supports allowing configurable orientation. For this reason, an insulating support made with PVC pipes was developed in order to avoid the effects derived from the presence of such conductive elements (PVC presents dielectric properties comparable to air) as much as posible . This support separated the antenna from the metallic support. Therefore, the gap between them was subsequently covered with prisms of absorbent material. The rest of metallic elements were also covered with absorbent material at the direct reflection points. As can be seen in Fig. 2, the antenna was then placed in the

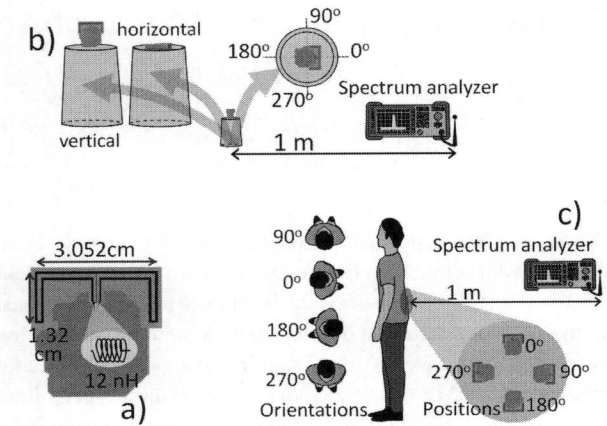

Fig. 1 a) Antenna design. b) Experimental setup without a human body. c) On-body experimental setup.

center of the outer plane of the insulating support following three different positions.

The measurement equipment used in this work presented a 50-ohmio impedance and worked with unipolar measurements. Therefore, since the OFD antenna is differential and is adapted to the characteristic impedance of the transceiver (60+164j), an impedance matching network was necessary but difficult to adjust (a passive antenna implementation with the balun described in [15] did not show satisfactory results). In order to solve these issues, a configuration in which the antenna was fed by the sensor device, with its own power supply, was used. Such scheme also avoided the use of long coaxial cables on the side of the transceiver and allowed to include in the study the effects deriving from the characteristics of the sensor (size, shape, arrangement of components, metallic components, dielectric properties of the printed circuit board, etc.).

The device was configured to transmit data on a pseudo-continuous infinite loop transmission mode with random data frames at a frequency of 2.4448 GHz. In this scheme, the network analyzer of the anechoic chamber was unable to synchronize with the received signal, probably due to the fact that the transmitted signal was not a pure carrier. It must be noticed that the transceiver uses the physical layer of IEEE 802.15.4 standard, in which logic '1' corresponds to half a positive sine and logic '0' to half a negative sine. However, since the data are codified in 4-bit groups for 16 pseudo-random sequences, none of which correspond to the desired transitions of a pure carrier, continuous 180° phase changes were produced, thus avoiding the network analyzer to detect the received signal.

This problem was resolved using a spectrum analyzer, which only measures gain. Fig. 3 shows a scheme of the arrangement of the transmitter (Tx) and receiver (Rx) devices

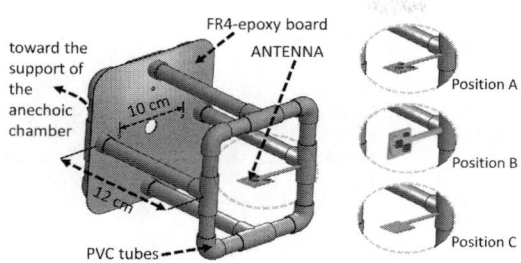

Fig. 2 Insulating support for the antenna in the anechoic chamber.

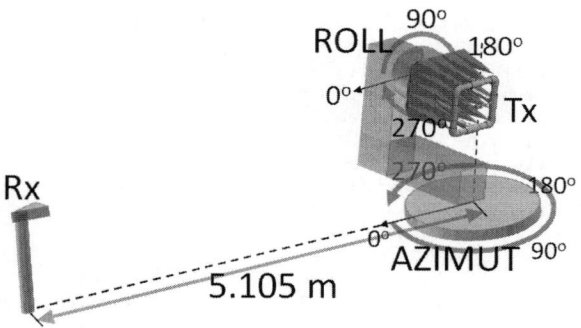

Fig. 3 Scheme of the arrangement of the transmitter (Tx) and receiver (Rx) devices inside the anechoic chamber.

Table 1 Received mean power (dBm) in laboratory experiments.

Without body	Azimut			
	0°	90°	180°	270°
Horizontal	−45.7	−57.4	−44.3	−44.7
Vertical	−45.3	−48.1	−47.7	−60.5
On-body	Orientations			
Positions	0°	90°	180°	270°
0°	−60.4	−62.1	−46	−46.4
90°	−55.6	−53.8	−45.3	−45.5
180°	−55	−56.2	−40.7	−55.6
270°	−57.4	−52.2	−40.7	−49.5

inside the anechoic chamber. In order to obtain the radiation pattern in the H plane, ROLL was set to 0°, the antenna was in position A, and AZIMUT was varied in increments of 5°. The value of received power was registered for each position over three different time instants, with the average value being taken. When the metallic support was in the way between the Tx and Rx, AZIMUT, ROLL and antenna position were modified to obtain an equivalent position, but with the metallic support just on the opposite side of the AZIMUT rotation axis. In order to obtain the radiation pattern in the E plane, AZIMUT was fixed to 0°, the antenna was in position C and ROLL was varied in increments of 5°. In the same way, three measures were taken for each orientation.

III. RESULTS

A. Experimental Results in Laboratory

Once the proposed OFD antenna was implemented, a set of measurements was performed without the presence of a human body, according to the experimental protocol explained in a previous section. The average received power was calculated for all the configurations considered (see Table 1). In the case of the on-body antenna, the mean received power was measured for all positions and configurations proposed, but in this case, the antenna was attached to the human body. The results show that the antenna performance is affected by the human body influence. The best position was 90° (see Table 1).

B. Experimental Results in Anechoic Chamber

The spectrum analyzer used within the anechoic chamber provided an estimation of the average received power. Since the distance between transmitter and receiver (5.105 m), the gain of the receiver antenna (8.575 dB at 2.4448 GHz, LB-20245 Broadband Horn Antenna of Chengdu AINFO Inc.), and the transmitted power (0 dBm, configured in the transceiver CC2430) are known parameters, it is possible to obtain an approximation to the gain of the transmitter antenna by using Friss Transmission Equation. Fig. 4 shows the experimental radiation patterns obtained within the anechoic chamber for both planes E and H and the simulated radiation patters for the sake of comparison.

The differences obtained between the experimental and the simulated results can be due to some losses that have not been considered in Friss equation, such as those originated at the receiver front-end (receiver antenna, coaxial cable and spectrum analyzer), mismatching impedance in connectors, etc. Nevertheless, such differences can also be related to effects derived from the hardware implementation of the sensor device and not taken into account during the design and antenna simulation stages (electronic components, metallic planes, printer circuit pads, etc.). Therefore, the results obtained evidence the necessity of incorporating an experimental analysis during the antenna design stage that allows these effects to be considered in order to optimize the antenna performance.

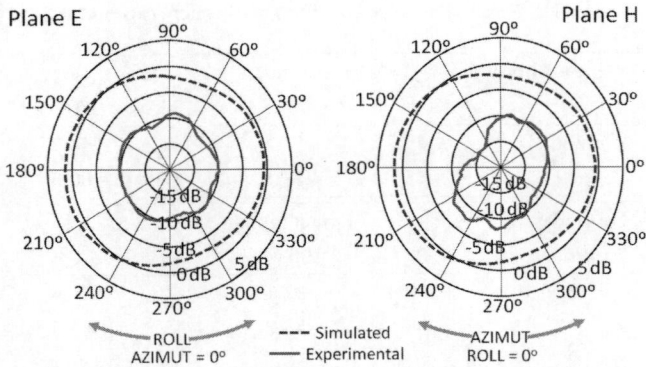

Fig. 4 Simulated and experimental radiation pattern in E and H planes.

IV. CONCLUSIONS

An experimental characterization of an active antenna has been performed with the implementation of a biomedical smart sensor and both the antenna and transceiver integrated into a self-powered device. Faced with the common methods of characterization of passive antennas, this procedure allows to analyze the effects of the transceiver, the size and shape of the ground plane, the distortions caused by the presence of electronic components, metallic elements and dielectric materials. A first set of laboratory experiments highlights the fact that the human body itself can modify the antenna performance. The results evidenced the necessity for personalization and adaptation of the antennas not only to the particular application but also to the presence of the human body. The issues raised during the experimental characterization inside an anechoic chamber (metallic elements in the support of the anechoic chamber, 50-ohmio adaptation, unipolar to differential transformation, synchronization of the network analyzer to the received signal, interposition of the metal support in the direct path between antennas, estimate of the antenna gain) were analyzed and resolved, which in turn allowed us to conclude a method for obtaining the radiation patterns of antennas integrated into battery-powered smart sensors.

ACKNOWLEDGEMENTS

The authors would like to thank Verónica Roldán and Prof. Francisco Medina and Vicente Losada for their helpful support in this work. This work was supported in part by the Consejería de Economía, Innovación y Ciencia, Government of Andalucía, under Grants P08-TIC-04069 and P10-TIC-6214, in part by the Fondo de Investigaciones Sanitarias, Instituto de Salud Carlos III, under Grants PI082023 and PI11/00111, and in part by CIBER-BBN under Grant PERSONA.

REFERENCES

1. Wang S, Ji L, Li A, Wu J. Body Sensor Networks for Ubiquitous Healthcare *J Control Theory Appl.* 2011;9:3-9.
2. Zhang Lanlin, Wang Zheyu, Volakis J L. Textile Antennas and Sensors for Body-Worn Applications *IEEE Antennas and Wireless Propagation Letters.* 2012;11:1690-1693.
3. Caimi FM. Antenna design challenges for 4G *IEEE Wireless Communications.* 2011;18:4-5.
4. Munoz MO, Foster R, Hao Yang. On-Body Channel Measurement Using Wireless Sensors *IEEE Transactions on Antennas and Propagation.* 2012;60:3397-3406.
5. Cheng S, Hallbjorner P, Rydberg A, Vanotterdijk D, Engen P. T-matched dipole antenna integrated in electrically small body-worn wireless sensor node *Microwaves, Antennas Propagation, IET.* 2009;3:774-781.
6. Baker DE, Du Toit JB. A compact 1 to 18 GHz planar spiral antenna for interferometer and other direction finding applications in *IEEE-APS Topical Conference on Antennas and Propagation in Wireless Communications (APWC) 2012*:1016-1019 2012.
7. Buckley J, Aherne K, O'Flynn B, Barton John, Murphy A, O'Mathuna C. Antenna performance measurements using wireless sensor networks in *Electronic Components and Technology Conference, 2006. Proceedings. 56th*:1652-1657 2006.
8. Okano Yoshiki, Cho Keizo. Antenna Measurement System for Mobile Terminals *NTT DoCoMo Technical Journal.* 2007;9:43-50.
9. Salas MA, Martinez R, Haro L. Procedure for Measurement, Characterization, and Calibration of Active Antenna Arrays *Instrumentation and Measurement, IEEE Transactions on.* 2013;62:377-391.
10. Naranjo D, Roldán V, Callejón A, Roa LM, Reina LJ. Evaluación experimental de antenas energéticamente eficientes para redes de sensores corporales in *XXX Congreso Anual de la Sociedad Española de Ingeniería Biomédica (CASEIB 2012)*:1-4 2012.
11. Naranjo D, Roa LM, Reina J, Estudillo MA. Personalization and Adaptation to the Medium and Context in a Fall Detection System *IEEE Transactions on Information Technology in Biomedicine.* 2012;16:264-271.
12. Naranjo D, Roa LM, Reina J, Estudillo MA. SoM: A Smart Sensor for Human Activity Monitoring and Assisted Healthy Ageing *IEEE Transactions on Biomedical Engineering.* 2012;59:3177-3184.
13. Choi Jin Soo, Zhou MengChu. Design issues in ZigBee-based sensor network for healthcare applications in *9th IEEE International Conference on Networking, Sensing and Control (ICNSC 2012)*:238-243 2012.
14. Instruments Texas. System-on-Chip for 2.4 GHz ZigBee(TM)/IEEE 802.15.4 with Location Engine (Rev. B). http://www.ti.com/lit/ds/symlink/cc2431.pdf 2013.
15. Instruments Texas. Anaren 0404 (BD2425N50200A00) balun optimized for Texas Instruments CC2430 Transceiver. http://www.ti.com/lit/an/swra156a/swra156a.pdf 2013.

Author: David Naranjo Hernández (Biomedical Engineering Group)
Institute: CIBER-BBN and University of Seville
Street: Engineering School. Avd Descubrimientos s/n
City: Seville
Country: Spain
Email: dnaranjo@us.es

Part XIV
Special Sessions

Models of Arrhythmogenesis in Myocardial Infarction*

Natalia A. Trayanova, Senior Member

Department of Biomedical Engineering, Johns Hopkins University, Baltimore, MD 21218, USA
ntrayanova@jhu.edu

Abstract—**Ventricular tachycardia, a life-threatening regular and repetitive fast heart rhythm, frequently occurs in the setting of myocardial infarction. Recently, the peri-infarct zones surrounding the necrotic scar (termed gray zones) have been shown to correlate with ventricular tachycardia inducibility. However, it remains unknown how the latter is determined by gray zone distribution and size. The goal of this study is to examine how tachycardia circuits are maintained in the infarcted heart and to explore the relationship between the tachycardia organizing centers and the infarct gray zone size and degree of heterogeneity. To achieve the goals of the study, we employ a sophisticated high-resolution electrophysiological model of the infarcted canine ventricles reconstructed from imaging data, representing both scar and gray zone. The baseline canine ventricular model was also used to generate additional ventricular models with different gray zone sizes, as well as models in which the gray zone was represented as different heterogeneous combinations of viable tissue and necrotic scar. The results of the tachycardia induction simulations with a number of high-resolution canine ventricular models (22 altogether) demonstrated that the gray zone was the critical factor resulting in arrhythmia induction and maintenance. The findings of this study have important implications for the advancement of improved criteria for stratifying arrhythmia risk in post-infarction patients and for the development of new approaches for determining the ablation targets of infarct-related tachycardia.**

I. Introduction

Ventricular arrhythmia, the pathogenesis of which results from abnormal impulse propagation in the heart, is a leading cause of death in the industrialized world. Ventricular tachycardia (VT), a life-threatening regular and repetitive fast heart rhythm, frequently occurs in the setting of myocardial infarction (MI). Implantation of a cardioverter-defibrillator (ICD) is the most effective measure for preventing lethal arrhythmias in post-MI patients. Despite this critical survival benefit, ICD therapy is costly and can be associated with procedural complications, infections, device malfunctions and diminished quality of life. Current clinical criteria for identifying ICD candidates for the primary prevention of sudden cardiac death rely almost exclusively on a nonspecific reduction in global left ventricular function (ejection fraction35%). Only 5% of patients who meet this criterion and thus undergo device implantation receive life-saving appropriate defibrillation shocks. Development of patient selection criteria of higher specificity for arrhythmia risk is currently hindered by the lack of understanding of the relationship between the complex structural remodeling in MI and infarct-related VTs. It has been shown that magnetic resonance imaging (MRI) with late gadolinium enhancement (LGE) can be used to detect infarct location and distribution in animal models[1, 2] and humans [3, 4]. Recently, the peri-infarct (border) zones surrounding the necrotic scar, also known as gray zones (GZ) based on their appearance as regions of intermediate intensity in the LGE-MRI scans, have been shown to correlate with post-MI mortality [5], clinical VT [6], and VT inducibility [7]. Histological studies have shown these GZ regions to be a heterogeneous mix of viable myocardium and necrotic scar [8]. Animal experimental evidence has implicated the GZ as the arrhythmogenic substrate in MI[9]. However, it remains unclear how infarct-related VTs relate to the specific GZ distribution and size. Addressing this question would provide an impetus to the development of improved criteria for stratifying arrhythmia risk in post-MI patients. The present study aims to address this question. We employ a sophisticated high-resolution MRI-based model of the infarcted canine ventricles to explore the infarct-related VT circuits as well as the relationship between the locations of the VT organizing centers and GZ distribution. The approach allows us to determine how the specific region of infarct (GZ and/or scar) maintains the VT circuits, and whether VT inducibility depends on GZ size.

II. Methods

A. Baseline Post-MI Canine Ventricles Model: Representation of Infarct Morphology

To understand comprehensively how VT is sustained in the post-MI heart, we used a biophysically-detailed model of an individual canine heart post-MI, reconstructed from high-resolution ex-vivo MRI and diffusion tensor (DT)-MRI scans (Fig. 1A, leftmost image). Detail regarding the image acquisition and partial description of the model

---

* This project was supported by NIH grants R01-HL094610, R01-HL103428, and R01-HL105216.

reconstruction has been published previously [10]. Briefly, a canine heart 4 weeks post-infarction was scanned (ex-vivo MRI and DTMRI) at a resolution of 350x350x800m3 and interpolated, using cubic splines, to a resolution of 200x200x200m3, to ensure accurate segmentation. A level set method was applied to the MR image stack to separate the myocardium from the surrounding suspension media. Once an accurate reconstruction of the myocardium was obtained, the ventricles were delineated from the atria. To perform this step, a closed spline curve was fitted through landmark points placed around the ventricles and along the atrio-ventricular border; all voxels that belonged to the tissue inside the closed curve were marked as ventricular. The identification of landmark points was performed manually for a number of slices that were evenly distributed in the image stack. The landmarks for the remaining slices were obtained by linearly interpolating the manually identified points. After the separation of the ventricles from the atria, infarct tissue was labeled. Since the high-resolution ex-vivo MRI acquisition protocol did not involve the use of contrast agents such as gadolinium, we developed a two-step methodology that combined information from both the DTMRI and the structural MRI scans to segment out the two infarct zones, the scar and the electrically-remodeled gray zone (GZ). First, the effective diffusion tensor (Deff) at each pixel was calculated from the DTMRI dataset. Next, fractional anisotropy (FA) values were calculated from Deff to quantify water diffusivity at each pixel in the ventricles. FA values range from 0 to 1, with 0 indicating isotropic diffusion and 1 indicating diffusion restricted along one axis only. Remodeling within the infarct (both in scar and in GZ) is characterized with increased presence of cells such as necrotic tissue, collagen, macrophages, and fibroblasts, results in a lower FA value compared to the normal myocardium. Thus, level-set thresholding of the FA images was used to discriminate infarcted tissue from the normal myocardium. In the second step, the infarct region was further subdivided into inactive scar and GZ by thresholding the structural MR image based on the intensity values. Scar tissue appears in the MRI scans as central dark regions surrounded by hyperenhanced regions. Thus, pixels with high ($>75\%$) and low ($<25\%$) gray-level intensities within the low-FA region were designated as electrically-inert scar, while pixels with intermediate intensities (25% and 75%) were designated as GZ. A finite element computational mesh was constructed directly from the segmented images; it preserved the fine geometric detail of the ventricles and of the different zones within the infarct. The mesh of the infarcted canine ventricles consisted of 3,177,732 nodes (2,185,112 nodes in the ventricles) and 3,981,196 (2,603,624 in the ventricles) mixed-type elements with an average element edge length of $396 \mu m$. Finally, fiber orientations were assigned to each element by calculating the primary eigenvector from the corresponding DT-MRI. A similar approach for ex-vivo MRI-based heart reconstruction has been used in our recent studies[11, 12, 10].

B. Baseline Post-MI Canine Ventricles Model: Representation of Electrophysiological Properties

Mathematical description of cardiac tissue was based on the monodomain representation. The scar was modeled as an insulator (collagen). Anisotropic conductivities in the normal myocardium were assigned to match canine conduction velocities. Within the GZ, the transverse conductivity was decreased by 90% to match reported conduction velocities, thus reflecting connexin 43 (Cx43) downregulation and lateralization. Membrane behavior in the canine ventricular model was represented by the Luo-Rudy dynamic model (LRd). The same membrane model was used, with modifications based on experimental data, to represent the electrophysiology of GZ cells. Specifically, patch clamp studies using cells harvested from the epicardial border zone of infarcted canine hearts have reported a reduction in peak sodium current to 38% of the normal value; in peak L-type calcium current to 31% of normal; and in peak potassium currents IKr and IKs to 30% and 20% of the normal, respectively. These modifications were implemented in the LRd model to obtain a GZ action potential; the latter was characterized, consistent with experimental recordings, with 25% longer action potential duration, 31% smaller upstroke velocity, and 32% lower peak action potential amplitude compared to that of the normal myocardium.

C. Additional Post-MI Ventricular Models to Explore the Role of GZ Size and GZ Level of Heterogeneity in VT Morphology

To explore the role of GZ size in VT dynamics, the baseline infarcted canine model described above was modified to generate infarcted canine ventricular models with decreased GZ volumes. The rationale for this step was that canine hearts have an extensive epicardial GZ [38], while human hearts have typically intramural infarcts with smaller size GZs. The decrease in GZ volume thus allowed us to obtain insight into human infarct-related VT. We used the software ImageJ (http://imagej.nih.gov/ij/) to perform the operation 3D morphological erosion, which decreases the object volume without changing the 3D structure [39]. The operation was performed such that only the position of the boundary between GZ and normal tissue was relocated, while the boundary between scar and GZ remained unchanged. We decreased GZ volume to reach values found in arrhythmogenic human hearts [12]. The operation was performed 3 times to obtain 3 models with GZ volumes that were 64%, 37%, and 15% of the baseline model. In the baseline model and the models above, the GZ was represented as a homogeneous region of averaged remodeled electrophysiological properties. Presence of patches of scar in the GZ was not explicitly represented, but rather via the changes in conduction velocity.

D. Simulation Protocol and Analysis

To examine the arrhythmogenic propensity of the infarcted ventricular models, programmed electrical stimulation (PES), similar to protocols used for clinical evaluation of post-MI patients, was simulated. The models were paced from an endocardial location for 6 beats (S1) at a cycle length of 300 ms followed by a premature stimulus (S2) initially given at 90% of S1 cycle length. The timing between S1 and S2 was progressively shortened until VT was induced. If VT was not induced, a second premature stimulus (S3) was delivered after S2. PES was performed from 27 different endocardial sites to ensure that all possible VT morphologies arising from the infarct geometry will be evaluated.

III. RESULTS

A. Location of the 3D Organizing Centers of Infarct-Related VT

VTs were induced in the baseline canine post-MI ventricular model following PES from 8 out of the 27 pacing sites. All VTs persisted for the entire 2s of simulated time interval. For all VTs induced, reentry initiation took place within the GZ. Figure1A (three middle images) presents activation maps depicting the events leading to reentry initiation for PES from an endocardial site near the LV apex. The GZ exhibited slowed conduction and longer recovery time compared to the surrounding healthy tissue. This resulted in conduction block, wavebreak, and the formation of reentry. For all PES sites resulting in VT induction, the reentrant circuit manifested itself as a figure-of-eight pattern on the epicardium and breakthrough(s) on the endocardium (Fig.1A).

The VT morphologies induced from the 8 pacing sites were not all unique. Comparison of pseudo-ECGs demonstrated two distinct VT morphologies. The first VT morphology resulted from PES at two sites, both on RV, and had an average cycle length of 19014 ms. The reentrant circuit was a figure-of-eight pattern on the epicardium and RV endocardium. For this VT morphology, the reentry revolved around two I-type filaments with endpoints at the epicardium and RV endocardium. The filaments were fully contained within the GZ and the endpoints remained in the same locations for the duration of the VT.

The second VT morphology resulted from PES at six LV endocardial sites. The average cycle length, 22217 ms, was longer than that of the first VT morphology. The figure-of-eight reentry on the epicardium had a direction of rotation (chirality) opposite to that of the first VT morphology, and was manifested as breakthroughs on the LV and RV endocardial surfaces. This was due to the reentrant activity being organized around two I-type filaments with endpoints at the epicardium and the infarct scar (Fig.1A, rightmost image, pink lines). Since the filaments did not extend to the endocardium, no rotational activity was observed there. Both filaments were stably located within the GZ throughout the duration of the simulation.

For all induced VTs in these models, the reentrant waves propagated through the small viable tissue channels in the scar, sometimes with the appearance of an apparent reentry. However, in such instances, the behavior was simply wave propagation around an obstacle (the latter part of the scar) and was not sustained. Thus, the reentrant activity underlying the monomorphic VT was driven by reentry filaments that were always located, in their entirety, within the bulk GZ.

B. Role of GZ Size in VT Morphology

As stated in the Methods, since dog hearts exhibit a more extensive epicardial GZ compared to humans, we created additional ventricular models (example in Fig.1B, leftmost image) by decreasing GZ volume to represent values reported in arrhythmogenic human hearts. All decreases in GZ volume implemented here (see Methods) resulted in the GZ becoming intramural and no longer extending to the epicardium as in the original canine ventricular model. In the model with GZ at 64% of the original volume (Fig.1B, GZ=3.23 cm^3), PES from the same 27 endocardial sites induced 9 VTs (average cycle length 22723 ms). For both VT morphologies, the VT manifested itself as a breakthrough on both endo- and epicardium (Fig.1B, second image from left), with a figure-of-eight intramural pattern. In both cases, the reentrant activity was organized around a single U-type filament attached with both ends to the scar and fully contained within the GZ (Fig.1B, middle image). However, the position of the U-filament and the locations of the breakthrough sites were different for the two distinct VTs. Further reduction of GZ to 37% of the original volume (1.88 cm3) resulted in VT induction by PES from 7 sites with an average VT cycle length of 1967 ms; all VTs had the same morphology. VT was similarly organized around a U-type filament located in its entirety within the GZ, which remained stable for the duration of the simulation. Consistent with spiral wave behavior organized around a U-type filament attached to an intramural boundary, reentry was again intramural with breakthroughs on both epi- and endocardial surfaces. Further morphological erosion of GZ resulting in critical GZ volume of 15% of the original (0.76 cm^3, Fig.1C) resulted in inability to induce VT from any pacing site. In this case, the GZ volume was not large enough to support the formation of stable filaments. No sustained VT could be induced at any values of GZ volume below this critical value. These results indicate that there is a minimum GZ volume necessary in order to support the formation of reentry filaments.

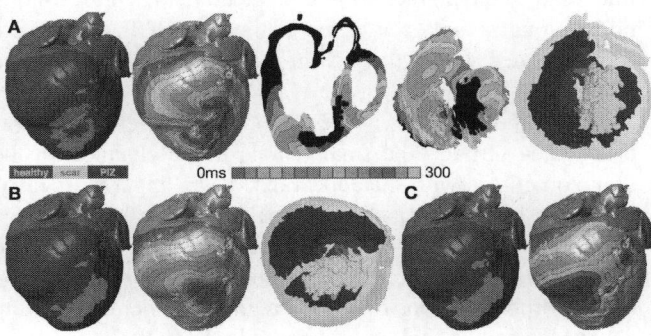

Fig. 1 VT induction in the canine heart for a pacing site near RV apex. **A, left to right**: model geometry with original GZ extending on the epicardium (bright green); epi-, transmural and endocardial activation maps; two I-type filaments (pink) sustain the reentry. **B, left to right:** geometry with PIZ volume 64% of original (i.e. intramural GZ); epicardial activation maps; U-type filament. **C, left to right**: geometry with GZ volume 15% of original; activation during pacing.

IV. DISCUSSION

In this work, we successfully developed a highly detailed computational model of the infarcted canine ventricles. Previous efforts in modelling of infarcted hearts have been limited to 2D representations or lacked full characterization of the entire ventricles including specific fiber orientations. The individualized infarcted canine ventricular model used in this study incorporated accurate geometry, infarct distribution, and fiber orientation obtained from high resolution ex-vivo MRI and DTMRI. Simulations with this computational model, where VTs were induced following a clinical PES protocol from numerous pacing sites provided novel mechanistic insight into the relationship between infarct-related VTs and the GZ in the post-MI heart. The baseline canine ventricular model was also used to generate additional ventricular models with different GZ sizes, as well as models in which the GZ was represented as different heterogeneous combinations of viable tissue and necrotic scar. To determine whether the GZ is the perpetrator in maintaining infarct-related VT, the organizing centers of VT, the scroll-wave filaments, were calculated and their locations with respect to the infarction zone components (scar and GZ) were determined. We also determined whether changing the size of the GZ while maintaining its shape altered VT morphology as well as the location, number and type of the scroll-wave filaments, and whether similar changes took place when GZ heterogeneity was progressively increased. The results of the VT induction simulations with a number of high-resolution canine ventricular models demonstrated that the GZ was the critical factor resulting in VT induction and maintenance. In all inducible models, the VT scroll-wave filaments were contained entirely within the GZ, regardless of GZ size or level of heterogeneity of its composition. GZ was thus the arrhythmogenic substrate that promoted wavebreak and reentry formation. While the necrotic scar played a role in determining the reentrant pathway, GZ always contained the VT organizing centers; all induced VTs were thus of both structural and functional nature. Our simulations also showed that there is a minimum volume of GZ required to render post-MI hearts arrhythmogenic. The critical GZ size obtained in our simulations is comparable with those determined in experiments[9]. Our results further demonstrate that large GZ volumes were able to support a larger number of stable filaments, resulting in multiple VT morphologies arising from the same infarct geometry. Intermediate GZ volumes were able to support typically a single filament, giving rise to the same VT morphology regardless of the PES site, while GZ volumes below the critical value resulted in VT non-inducibility due to insufficient amount of electrically remodelled tissue to support reentrant activity. The shape of the filaments transitioned from I-type to U-type (attached to the endocardium) to U-type (attached to the scar) as GZ size progressively decreased, with GZ becoming fully intramural. Our finding that GZ contains all the VT scroll-wave filaments has an important clinical significance. It supports promising new use of MRI to evaluate the arrhythmia risk of patients with coronary artery disease, assessing the GZ extent from the clinical scans. The potential advantages of such an approach are that it is a noninvasive technique, it is applicable to a wide range of patients, and the reproducibility of the image data is high. Furthermore, one would envision an even more targeted patient-specific approach to the assessment of infarct-related arrhythmia risk, in which computer simulations with in-vivo clinical-MRI-based computer model of the infarcted patient heart (with reconstructions of both scar and GZ, similar to the model and simulations presented in this study) would be used to determine noninvasively whether VT is inducible in the patient heart, the latter then warranting an ICD implantation.

REFERENCES

[1] D. S. Fieno, R. J. Kim, E.-L. Chen, J. W. Lomasney, F. J. Klocke, and R. M. Judd, "Contrast-enhanced magnetic resonance imaging of myocardium at riskdistinction between reversible and irreversible injury throughout infarct healing," *jacc*, vol. 36, no. 6, pp. 1985–1991, 2000.

[2] R. J. Kim, D. S. Fieno, T. B. Parrish, K. Harris, E.-L. Chen, O. Simonetti, J. Bundy, J. P. Finn, F. J. Klocke, and R. M. Judd, "Relationship of mri delayed contrast enhancement to irreversible injury, infarct age, and contractile function," *Circulation*, vol. 100, no. 19, pp. 1992–2002, 1999.

[3] E. Wu, R. M. Judd, J. D. Vargas, F. J. Klocke, R. O. Bonow, and R. J. Kim, "Visualisation of presence, location, and transmural extent of healed q-wave and non-q-wave myocardial infarction," *The Lancet*, vol. 357, no. 9249, pp. 21–28, 2001.

[4] O. P. Simonetti, R. J. Kim, D. S. Fieno, H. B. Hillenbrand, E. Wu, J. M. Bundy, J. P. Finn, and R. M. Judd, "An improved mr imaging technique for the visualization of myocardial infarction1," *Radiology*, vol. 218, no. 1, pp. 215–223, 2001.

[5] A. T. Yan, A. J. Shayne, K. A. Brown, S. N. Gupta, C. W. Chan, T. M. Luu, M. F. Di Carli, H. G. Reynolds, W. G. Stevenson, and R. Y. Kwong, "Characterization of the peri-infarct zone by contrast-enhanced cardiac magnetic resonance imaging is a powerful predictor of post-myocardial infarction mortality," *Circulation*, vol. 114, no. 1, pp. 32–39, Jul 2006. [Online]. Available: http://www.hubmed.org/display.cgi?uids=16801462

[6] S. D. Roes, C. J. W. Borleffs, R. J. van der Geest, J. J. Westenberg, N. A. Marsan, T. A. Kaandorp, J. H. Reiber, K. Zeppenfeld, H. J. Lamb, A. de Roos, et al., "Infarct tissue heterogeneity assessed with contrast-enhanced mri predicts spontaneous ventricular arrhythmia in patients with ischemic cardiomyopathy and implantable cardioverterdefibrillatorclinical perspective," *Circulation: Cardiovascular Imaging*, vol. 2, no. 3, pp. 183–190, 2009.

[7] A. Schmidt, C. F. Azevedo, A. Cheng, S. N. Gupta, D. A. Bluemke, T. K. Foo, G. Gerstenblith, R. G. Weiss, E. Marban, G. F. Tomaselli, J. A. Lima, and K. C. Wu, "Infarct tissue heterogeneity by magnetic resonance imaging identifies enhanced cardiac arrhythmia susceptibility in patients with left ventricular dysfunction," *Circulation*, vol. 115, no. 15, pp. 2006–2014, Apr 2007. [Online]. Available: http://www.hubmed.org/display.cgi?uids=17389270

[8] H. Arheden, M. Saeed, C. B. Higgins, D.-W. Gao, P. C. Ursell, J. Bremerich, R. Wyttenbach, M. W. Dae, and M. F. Wendland, "Reperfused rat myocardium subjected to various durations of ischemia: Estimation of the distribution volume of contrast material with echo-planar mr imaging1," *Radiology*, vol. 215, no. 2, pp. 520–528, 2000.

[9] H. Estner, M. Zviman, D. Herzka, F. Miller, V. Castro, S. Nazarian, H. Ashikaga, Y. Dori, R. Berger, H. Calkins, A. Lardo, and H. Halperin, "The critical isthmus sites of ischemic ventricular tachycardia are in zones of tissue heterogeneity, visualized by magnetic resonance imaging," *Heart Rhythm*, vol. 8, pp. 1942–9, 2011

[10] F. Vadakkumpadan, H. Arevalo, A. Prassl, J. Chen, F. Kickinger, P. Kohl, G. Plank, and N. Trayanova, "Image-based models of cardiac structure in health and disease," *Wiley Interdisciplinary Reviews: Systems Biology and Medicine*, vol. 2, pp. 489–506, 2010.

[11] F. Vadakkumpadan, L. Rantner, B. Tice, P. Boyle, A. Prassl, E. Vigmond, G. Plank, and N. Trayanova, "Image-based models of cardiac structure with applications in arrhythmia and defibrillation," *Journal of Electrocardiology*, vol. 42, no. 2, pp. 157 e1–e10, 2009.

[12] E. Vigmond, F. Vadakkumpadan, V. Gurev, H. Arevalo, M. Deo, G. Plank, and N. Trayanova, "Towards predictive modelling of the electrophysiology of the heart," *Experimental Physiology*, vol. 94, no. 5, pp. 563–577, 2009

Multiscale-Multiphysics Models of Ventricular Electromechanics – Computational Modeling, Parametrization and Experimental Validation

G. Plank, A.J. Prassl, R. Arnold, Y. Rezk, T.E. Fastl, E. Hofer, and C.M. Augustin

Medical University of Graz/Institute of Biophysics, Graz, Austria

Abstract—Computational modeling of ventricular electromechanics is considered to be among the most promising approaches to gain novel insight into cardiac function in health and disease. Such models allow to integrate the wealth of available experimental data into a mechanistic framework which allows to study complex cause-effect relationships across several scales of biological organization – ranging from subcellular processes such as cellular force generation up to the organ scale. However, due to the enormous challenges involved in constructing such models, the number of available multiscale models is very limited. Their implementation resorts to very crude approximations which ignores many aspects of known biophysical details and/or uses extremely coarse spatio-temporal discretizations to reduce computational complexity.

In this study we focus on a subset of the parametrization and validation issues in models of cardiac electromechanics. An experimental setup for measuring passive stress-strain relationships as well as isometric force transients in thin preparations of ventricular trabeculae or papillary muscles was developed along with a matching computer model to parametrize and validate passive and active mechanical model responses at the tissue scale.

Keywords—Cardiac electrophysiology, Cardiac electromechanics, Bidomain model.

I. Introduction

Computational modeling of ventricular electromechanics is considered to be among the most promising approaches to gain novel insight into cardiac function in health and disease. Such models allow to integrate the wealth of available experimental data into a mechanistic framework which facilitates studying complex cause-effect relationships across several scales of biological organization – ranging from subcellular processes up to the organ scale. However, due to the enormous multi-faceted challenges involved in constructing such models, the number of available multiscale-multiphysics frameworks is limited. Moreover, most implementations resort to rather crude approximations which ignore many known biophysical details and employ fairly coarse spatio-temporal discretizations to reduce computational complexity.

Conceptually, the problem of building multiscale organ models of cardiac function is tackled by assembling basic building blocks which characterize different functional aspects and length scales. A fundamental component are models of cellular dynamics which describe electrophysiological (EP) behavior. Numerous models specialized for different species, tissues and pathologies have been reported in the literature over the past few decades. Most of them predict intracellular calcium transients, the main mediator signal of cellular force generation, thus allowing to couple the electrical excitation and recovery process to mechanical contraction and relaxation, a phenomenon referred to as excitation contraction coupling (ECC). Various models describing cellular force generation based on sliding filament theory exist [1] which can be coupled to EP models.

At the tissue and organ scale, the cardiac bidomain equations are the most relevant model for describing tissue responses to stimuli and the spread of activation and repolarization throughout the organ. The subject-specific parametrization of the bidomain equations remains to be a challenging and unresolved problem. Only a limited number of experimental reports on key parameters such as tissue conductivities exist which are afflicted with a large margin of error [2]. Further, there is a significant inter-individual variation in the electrical activation sequence, as evidenced by variations in electrocardiogram waveform recorded from the body surface of individuals, due to factors such as topological variability of the specialized conduction system [3], variations in spatial heterogeneity of tissue structure [4], and EP heterogeneities which profoundly influence repolarization sequences [5]. During the contraction phase active forces have to act against the passive mechanical stiffness of myocytes which are embedded in the extracellular matrix and reinforced by the collagen fiber fabric at the tissue scale. These passive mechanical properties are investigated by multi-axial stretch experiments to parametrize constitutive relations [6] which are incorporated in the continuum mechanics framework. Finally, at the organ scale additional mechanical loads are imposed by time-varying pressures acting in the cavities against the endocardium. These pressures are typically accounted for by lumped parameter models of the circulatory system which are coupled as inhomogeneous Neumann type boundary conditions.

While all these models are carefully validated at the individual length scales and experimental conditions they were designed for, their integration into a comprehensive organ model leads to inconsistencies due to the numerous differences in terms of species, temperature, experimental protocol, etc. In general, one cannot expect models based on the direct assembly of these sub-models to yield a sufficiently close match with experimental observations at the organ scale. Rather, significant deviations at the organ scale are common, requiring a complex reparametrization of basic model parameters to achieve a better match with the organ's global response. Unlike in recent large-scale studies where it was attempted to parametrize *in silico* models of ventricular electromechanics in a patient-specific fashion at the organ scale, in this study we report on our efforts aimed at building an experimental setup suitable for measuring passive strain-stress relationships and active force transients under isometric conditions in thin strands of ventricular tissue, such as trabeculae or small papillary muscles. The limited geometric complexity of such preparations facilitates a quick iteration between experimental recordings and computer simulations to achieve a very close match between experimental observation and model prediction.

II. METHODS

Modeling of cardiac electromechanics relies upon discretizations of the bidomain equations to characterize electrophysiology at the tissue scale, and the finite deformation elasticity equations to describe cardiac deformation under mechanical loads and intrinsic active force generation and relaxation. In a strongly coupled multi-physics scenario both sets of equations have to be solved together, necessitating the bidirectional exchange of data between the two physics. Hence, the two physics are bidirectionally coupled. Electrical activation triggers ECC – generated stresses influence the state of deformation – and mechanical deformation influences upon the electrical tissue state via mechanoelectric feedback (MEF) – mechanical stresses or strains influence upon cellular physiology via different mechanisms such as length dependent tension, calcium sensitivity or stretch-activated channels.

A. Experimental Model

From an experimental point of view, a rigorous validation of an electromechanical model of a specific preparation across several scales of biological organization – ranging from the subcellular to the organ scale – is not feasible. This is due to the large number of variables in such models which cannot be measured at a sufficiently high spatio-temporal resolution at the same time. Thus, to assess the validity of computational models of ventricular electromechanics a reduction of complexity is indispensable. For this sake, a simplified experimental setup for measuring force transients in preparations of ventricular trabeculae was developed. Analog to pressure-volume relationships such as measured in passive inflation experiments at the ventricular level, this setup allows to measure passive stress-strain relationships to parametrize passive constitutive equations as well as active force transients under isometric conditions, as elicited by electrical stimulation.

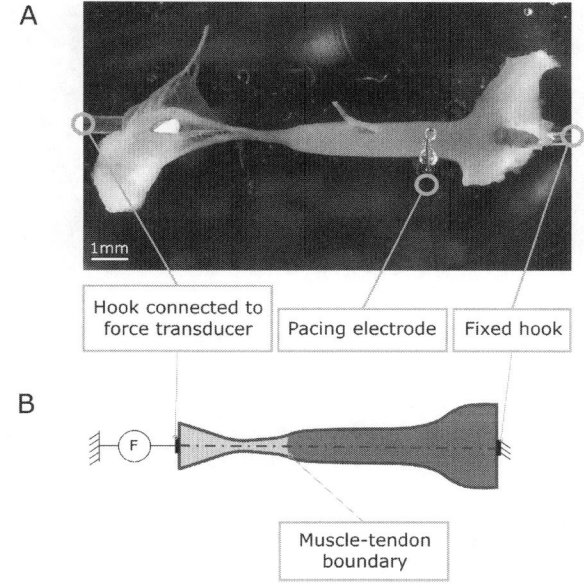

Fig. 1 A: Experimental setup for measuring force transients in small muscle preparations. B: Cross section of a computer model built to match the geometry of the preparation shown in A.

The experimental setup is shown in Fig. 1A. Rabbit ventricular trabeculae and papillary muscles were excised and stored in cold cardioplegic Tyrode's solution containing (in mM): Na$^+$ 104, KCl 5.4, NaHC0$_3$ 24, MgCl$_2$ 1.15, NaH$_2$PO$_4$ 0.42, CaCl$_2$ 0.2, glucose 5.6, C$_8$H$_{18}$N$_2$O$_4$S (HEPES) 10, and 2,3-butanedione-monoxime (BDM) 30, equilibrated with 95% O$_2$ and 5% C0$_2$ to a pH of 7.4.

The muscle stripes were transferred to a temperature controlled organ bath at 36°C and superfused with oxygenated Tyrode's solution containing (in mM): Na$^+$ 132, KCl 5.4, NaHC0$_3$ 24, MgCl$_2$ 1.15, NaH$_2$PO$_4$ 0.42, CaCl$_2$ 2.5, glucose 5.6, C$_8$H$_{18}$N$_2$O$_4$S (HEPES) 10, pH 7.4. Preparations were paced at a constant cycle length of 1000 ms. Length changes ΔL were applied in steps of $50-200\,\mu$m, increasing the stretch ratio $\lambda = (L_0 + \Delta L)/L_0$ from 1 to 1.2. Length L was kept constant for $15-40$ min after each step change until active force transients reached steady state. Maximum magnitudes of force transients and diastolic forces, measured as the baseline force in between active force twitches, were recorded to facilitate the construction of force-stretch ratio diagrams (Fig. 2).

B. Governing Equations

The set of equations governing cardiac electromechanics is given by

$$\nabla \cdot (\boldsymbol{\sigma}_i + \boldsymbol{\sigma}_e)\mathbf{C}^{-1}\nabla \phi_e = -\nabla \cdot \boldsymbol{\sigma}_i \mathbf{C}^{-1} \nabla V_m, \quad (1)$$

$$\nabla \cdot \boldsymbol{\sigma}_i \mathbf{C}^{-1} \nabla V_m = -\nabla \cdot \boldsymbol{\sigma}_i \mathbf{C}^{-1} \nabla \phi_e + \beta I_m, \quad$$

$$I_m = C_m \frac{\partial V_m}{\partial t} + I_{ion}(V_m, \boldsymbol{\eta}), \quad (2)$$

$$V_m = \phi_i - \phi_e, \quad (3)$$

$$\frac{\partial \boldsymbol{\eta}}{\partial t} = f(\boldsymbol{\eta}, V_m, \sigma_a), \quad (4)$$

$$-\operatorname{div}\boldsymbol{\sigma}(\mathbf{u}) = \mathbf{0}, \quad (5)$$

$$\boldsymbol{\sigma} = \boldsymbol{\sigma}_p + \boldsymbol{\sigma}_a, \quad (6)$$

$$\boldsymbol{\sigma}_p = 2J^{-1}\mathbf{F}\frac{\partial \Psi(\mathbf{C})}{\partial \mathbf{C}}\mathbf{F}^T, \quad (7)$$

$$\boldsymbol{\sigma}_a = \sigma_a \left(\hat{\mathbf{f}} \otimes \hat{\mathbf{f}}\right), \quad (8)$$

$$\sigma_a = h(V_m, \boldsymbol{\eta}, \lambda, \dot{\lambda}). \quad (9)$$

In the bidomain equations (1)-(4) describing electrophysiology, ϕ_i and ϕ_e are the intracellular and extracellular potentials, respectively, $V_m = \phi_i - \phi_e$ is the transmembrane voltage, $\boldsymbol{\sigma}_i$ and $\boldsymbol{\sigma}_e$ are the intracellular and extracellular conductivity tensors, respectively, β is the membrane surface to volume ratio, I_m is the transmembrane current density, C_m is the membrane capacitance per unit area, and I_{ion} is the membrane ionic current density which depends on V_m and a set of state variables, $\boldsymbol{\eta}$. Deformation is governed by (5)-(8) with (8) linking electrics and mechanics.

The deformation is governed by the equilibrium equation (5) with unknown displacement field $\mathbf{u}$. The stress tensor $\boldsymbol{\sigma}$ consists of a passive (7) and an active contribution (8), while the latter depends on the normalized myocyte orientation $\hat{\mathbf{f}}$. Here, $\mathbf{F}$ is the deformation gradient, $\mathbf{C} = \mathbf{F}^T\mathbf{F}$ is the right Cauchy-Green tensor, and Ψ is a function which describes the strain energy density of the specific material. Using a particular cell model, indicated by the function $h(\bullet)$, the scalar-valued active stress term σ_a is computed (9), thus providing the link for ECC. To specify prescribed displacements or tractions, e.g., pressure, additional boundary conditions are incorporated.

C. Discretization and Numerical Solution

Contours of experimental preparations were manually extracted and discretized. Assuming rotational symmetry virtual 3D image stacks were generated. Image stacks were fed into the image-based unstructured mesh generation software Tarantula (CAE Software Solutions, Eggenburg, Austria) which builds fully unstructured, boundary fitted, locally refined, hex-dominant hybrid tessellations [7]. Subsequently, the hybrid mesh was tessellated into tetrahedral elements and each finite element was tagged either as electrically and mechanically active myocardium or as electrically and mechanically passive connective tissue (Fig. 1B).

Spatial discretization relied upon the finite element (FE) method where linear tetrahedral elements were used for both electrical and mechanical grid to discretize a strand model of a thin papillary muscle (diameter ≈ 0.9 mm). Homogeneous Dirichlet boundary conditions were applied at the root of the muscle as well as at the distal end of the muscle where tendons, modeled as a passive nonlinear material, were also fixed, impeding displacement there in any direction (Fig. 1B). The same FE grid was used for both problems, thus facilitating an easier exchange of data in the multiphysics problem. The Mahajan-Shiferaw model of a rabbit ventricular myocyte [8] coupled to an active stress model [9] was employed to describe ECC. MEF remained unaccounted for at this preliminary stage of the model development. Data between electrical grid where quantities are represented at the nodes spanning the elements and the integration points of the mechanics grid were transferred by interpolation using FE shape functions. An electrical stimulus was applied at the proximal end of the papillary muscle to initiate propagation and mechanical contraction. Under the given boundary conditions contractions were isometric, however, inner shortening of sarcomeres was possible due to the compliance of the tendons and the dynamics of the contraction process which led to prestretching of later activating regions of the muscle. Standard iterative solvers were used for solving the bidomain equations [10]. A custom-tailored unsmoothed aggregation algebraic multigrid preconditioner for an iterative conjugate gradient solver was used to solve the equilibrium equation (5).

III. Results

A set of 8 experiments on rabbit trabeculae and papillary muscles was performed. Both baseline passive forces as well as magnitudes of active force transients were recorded to construct force-stretch ratio diagrams. A sample result of one preparation is shown in Fig. 2. Further experimental results are in the process of being analyzed to provide a reference against which computer simulations can be gauged.

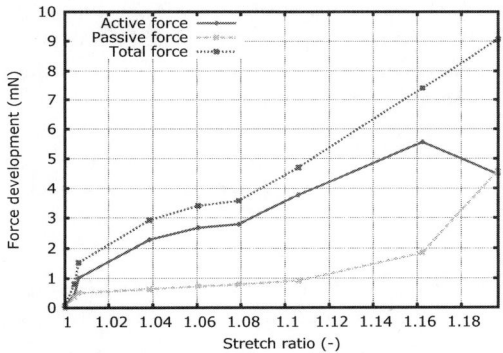

Fig. 2 Example of measured force magnitudes as a function of stretch ratio.

IV. Discussion

In this study we describe the development of a basic experimental setup for measuring passive forces and active force transients under isometric conditions in thin strand-like ventricular tissue preparations. So far only a small number of experiments has been performed to provide a reference dataset against which a matching computer model can be validated. More experiments have to be conducted to better quantify the variability in both passive and active mechanical properties. A computer model matching one particular setup has been built using generic model parameters taken from the literature. The Mahajan-Shiferaw model was weakly coupled to the Rice active stress model in a coupled electromechanical tissue scale simulation setup. In a next step the computer model will be modified to allow the replication of the experimental protocol. All available datasets will be analyzed to quantify the range of variability in these experiments. These data will serve to steer the further development of the computer model to achieve a close match between experimental observation and model prediction.

Acknowledgments

This research is supported by a grant of the Austrian Science Fund FWF (F3210-N18).

References

1. Trayanova NA, Rice JJ. Cardiac electromechanical models: from cell to organ. *Front Physiol.* 2011;2:43.
2. Roth BJ. Electrical conductivity values used with the bidomain model of cardiac tissue. *IEEE Trans Biomed Eng.* 1997;44:326–328.
3. Dobrzynski H, Anderson RH, Atkinson A, et al. Structure, function and clinical relevance of the cardiac conduction system, including the atrioventricular ring and out ow tract tissues. *Pharmacol Ther.* 2013. in press
4. Bayer JD, Blake RC, Plank G, Trayanova NA. A novel rule-based algorithm for assigning myocardial fiber orientation to computational heart models. *Ann Biomed Eng.* 2012;40:2243–2254.
5. Keller DU, Weiss DL, Dossel O, Seemann G. Influence of I(Ks) heterogeneities on the genesis of the T-wave: a computational evaluation. *IEEE Trans Biomed Eng.* 2012;59:311–322.
6. Holzapfel GA, Ogden RW. Constitutive modelling of passive myocardium: a structurally based framework for material characterization. *Philos Trans A Math Phys Eng Sci.* 2009;367:3445–3475.
7. Prassl AJ, Kickinger F, Ahammer H, et al. Automatically generated, anatomically accurate meshes for cardiac electrophysiology problems. *IEEE Trans Biomed Eng.* 2009;56:1318–1330.
8. Mahajan A, Shiferaw Y, Sato D, et al. A rabbit ventricular action potential model replicating cardiac dynamics at rapid heart rates. *Biophys J.* 2008;94:392–410.
9. Rice JJ, Wang F, Bers DM, de Tombe PP. Approximate model of cooperative activation and crossbridge cycling in cardiac muscle using ordinary differential equations. *Biophys J.* 2008;95:2368–2390.
10. Plank G, Liebmann M, Weber dos Santos R, Vigmond EJ, Haase G. Algebraic multigrid preconditioner for the cardiac bidomain model. *IEEE Trans Biomed Eng.* 2007;54:585–596.

Effect of Purkinje-Myocyte Junctions on Transmural Action Potential Duration Profiles

R. Walton[1], O. Bernus[1], and E.J. Vigmond[2]

[1] Université Bordeaux 2, IHU LIRYC, Pessac, France
[2] Université Bordeaux 1, IHU LIRYC, Pessac, France

Abstract—Action potential characteristics change across the wall of the ventricles. It is generally accepted that epicardial action potential duration (APD) is shorter than that on the endocardium, although this difference diminishes with pacing frequency. Some studies have reported midmyocardial islands of increased APD and suggested a third cell type to account for this. Furthermore, human studies report this profile changes during heart failure. We postulate Purkinje system coupling may account for increased subendocardial APD prolongation. This study used computer simulations to model the effect of a purkinje-myocyte junction (PMJ) on the cut surface of a ventricular wedge preparation measuring $0.5 \times 0.5 \times 1$ cm. Transmurally, the 1 cm was divided into epicardium (3 mm) and endocardium (7 mm) and a human ventricular ionic model with appropriate adjustments for each zone was used. A single 3 mm long Purkinje strand was inserted 2 mm into the wedge. To model Purkinje cells, the ventricular ionic model was modified to increase resting level and APD. Coupling parameters (resistance, junction size, conduction asymmetry) were varied to assess their effects. Pacing was performed by transmural current stimulation to a region on the endocardium. The PMJ significantly affected APD as measured on the cut surface. This affect was more pronounced the closer the PMJ was to the cut surface and the larger the PMJ was. We conclude that a third type of cell is not needed to produce islands of increased APD. Intrinsically longer APDs of the Purkinje system increase local myocardial APD in the vicinity of the junction. This complicates interpretation of myocardial APD data as the density and proximity of PMJs to the recording surface will affect surface APD.

Keywords—transmural heterogeneity, heart failure, computer modeling, wedge preparation, action potential duration.

I. INTRODUCTION

Action potential morphology displays a gradient across the wall of the ventricles[1]. Epicardial action potentials are shorter as well as have a more marked notch. At fast pacing rates, the difference in APD diminishes. Recent studies of human heart wedge preparations[2] have also revealed islands of prolonged action potential duration (APD) located subendocardially. The profile was also shown to change under heart failure (HF). The cause of this prolongation remains unknown. One explanation for this is that cells in this region could have intrinsically longer APD due to protein expression. Interesting, this region is also where the Purkinje System (PS) forms Purkinje-myocyte junctions (PMJs) with the myocardium. Since Purkinje cells have intrinsically longer APDs, they could electrotonically increase the APDs of the myocytes to which they are coupled.

This paper uses a computer model of the ventricles with PS to explore how electrotonic interactions near the PMJ alter local APD. Parameters affecting behavior were analyzed included PMJ size, tissue-PS resistance. Furthermore, conditions of HF were also implemented. The factors necessary to reproduce experimentally recorded APD transmural profiles were ascertained. PMJ coupling could account for the local regions of increased APD observed.

II. METHODS

A. Geometry

Simulations were performed in a slab measuring $0.5 \times 0.5 \times 1$ cm was modelled with 1 cm being the transmural thickness. This represented a wedge of tissue as used in experiment, with one of the "cut" faces defined as the imaging face from which surface measurements were taken. The first 3 mm in the transmural direction was considered as epicardium while the remaining 7 mm was treated as endocardium. A single 3 mm long straight PS fibre oriented along the transmural direction was inserted 2 mm into the tissue. Simulations were performed with the PS fibre in the middle of the tissue as well as offset to closer to the imaging face. See Fig.1. The Purkinje fibre was isolated from the myocardium except at the tip of the fibre, i.e. the PMJ.

Myocyte orientation was incorporated into the model. A laminar structure was assumed with layers perpendicular to the transmural axis and 120 degree rotation from endocardium to epicardium.

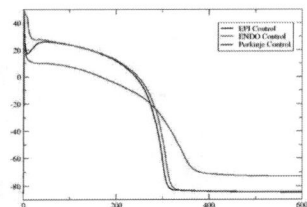 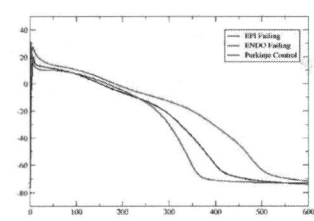 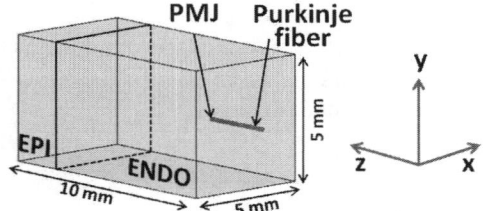

Fig. 1 Action potentials and model. **Left**: Control single cell action potentials for epicardial (black), endocardial (red) and Purkinje (blue) cells. **Middle**: Heart failure single cell action potentials for epicardial (black), endocardial (red) and Purkinje (blue) cells. **Right**: Geometry of the model with the transmural direction as z.

B. Electrical System

The propagation of electrical activity was described by the monodomain equation

$$\nabla \cdot \overline{\sigma}_m \nabla V_m = \beta I_m \quad (1)$$

where $\overline{\sigma}_m$ is the harmonic mean of the homogenized conductivity values of the intra- and extracellular media, β is the ratio of membrane surface to tissue volume, V_m is the transmembrane voltage and I_m is the total transmembrane current composed of a capacitive component and the currents through all the membrane proteins. Electrical propagation was initiated by a small transmembrane current stimulus to the center of the endocardial face.

C. Ionic Model

The Ten Tusscher human ventricular ionic model[3] was used with model parameters as published in the paper for epicardial and endocardial cells used as starting points. Some additional changes were made as indicated in Table 1. For the Purkinje cell action potential, parameters were modified in the same ionic model as explained elsewhere[4].

Table 1 Changes to published ionic model parameters

Parameter	Purkinje	ENDO	EPI - HF	ENDO - HF
Gto	x0.5	x0.2	x0.636	x0.0
Gks	x0.8	x0.9		x0.81
Gkr	x0.3	x0.9		x0.81
GNa	x2.94		x0.372	x0.372
GpK	x6		x1.2	x1.2
GCaL	x0.75			
xr2_off	-8			
knak	x0.3		x0.75	x0.75
GbCa	x13.2		x4.6	x4.5
GbNa	x0.5			
Vrel	x0.6		x0.56	x0.56
Vmaxup	x0.6		x0.83	x0.83
GK1	x0.7		x0.856	x0.856
knaca	x0.3		x1.57	x1.57
GpCa	x11			
D_CaL_off	-5		-5	-5
Vleak			x4	x4

III. RESULTS

Activation of the wedges are seen in Fig.2. In the failing wedge, it takes much longer to activate, primarily because of the downregulation of sodium current. Also, the anisotropy is more apparent. In failing tissue, the wavefront had trouble exciting the PMJ, and activity progressed around it, before fully depolarizing the PMJ, in contrast to non-failing tissue.

APD distributions are displayed in Fig. 3. For both failing and non-failing tissue, when the PMJ was in the center of the tissue, its effect on the imaged surface was negligible. For the non-failing case, the PMJ was clearly visible when it was

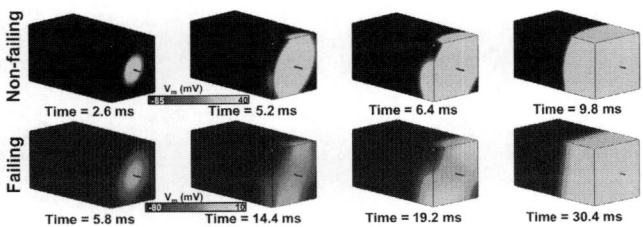

Fig. 2 Activation maps. **Top Row**: Control tissue. **Bottom Row**: Control tissue. Time since stimulation is given below each frame.

moved closer to the imaged surface. An island of prolonged APD appears which is quite distinct and about 10 ms longer than the surrounding tissue. For the failing case, the effect is attenuated and not as dramatic.

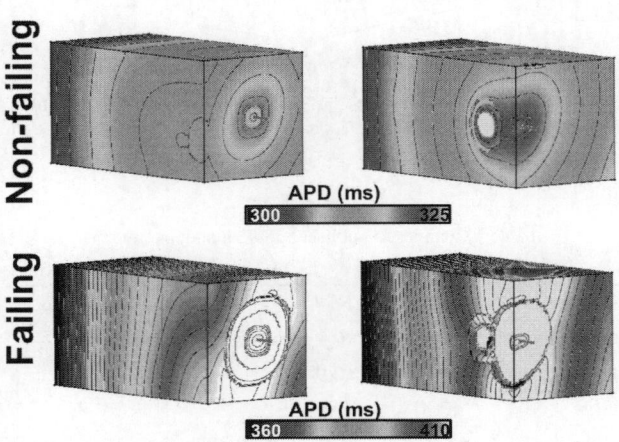

Fig. 3 APD distribution. **Left**: PMJ depth of 2.5 mm. **Right**: PMJ depth of 0.25 mm.

IV. DISCUSSION

Our study indicates that islands of locally prolonged APD may be the result of the interaction of the PS and myocytes. The intrinisically much longer APD of the PS prolongs the APD of the myocytes through electrotonic interaction, but only in the immediate vicinity of the PMJ. In HF, many changes occur. Consistent with literature[2], we observe smaller and less pronouned isalnds of APD prolongation. This effect is made greater by larger region of PMJ, as well as the distance of the PMJ to the imaged surface. The latter factor depends on exactly how the wedge was excised from the tissue. Experimentalist do not know exactly where PMJs are, nor have the requisite precision in cutting, so this distance will vary from experiment to experiment and have a probabilistic distribution based on PMJ density.

Our model uses a continuum approach. Thus, we treat the PMJ as a black box and adjust parameters to match known conduction properties. The area affected by a single Purkinje fibre still remains unelucidated as does the role and number of any transitional cells, if present.

V. ACKNOWLEDGEMENT

This work received funding from the French Government through the Agence National de la Recherche and Investissements d'Avenir ANR-10-IAHU-04. Richard Walton is supported by a Marie Curie Fellowship.

REFERENCES

1. Antzelevitch C., Fish J.. Electrical heterogeneity within the ventricular wall. *Basic Res Cardiol.* 2001;96:517–527.
2. Glukhov Alexey V., Fedorov Vadim V., Lou Qing, et al. Transmural dispersion of repolarization in failing and nonfailing human ventricle. *Circ Res.* 2010;106:981–991.
3. ten Tusscher K H W J., Panfilov A. V.. Alternans and spiral breakup in a human ventricular tissue model. *Am J Physiol Heart Circ Physiol.* 2006;291:H1088–H1100.
4. Tusscher K H W J Ten, Panfilov A. V.. Modelling of the ventricular conduction system. *Prog Biophys Mol Biol.* 2008;96:152–170.

Morphometry and Characterization of Electrograms in the Cavotricuspid Isthmus in Rabbit Hearts during Autonomic Sinus Rhythm

E. Hofer[1], D. Sanchez-Quintana[2], and R. Arnold[1]

[1] Institute of Biophysics, Medical University of Graz, Graz, Austria
[2] Departamento de Anatomia y Biologia Cellular, Universidad de Extremadura, Badajoz, Spain

Abstract—The cavotricuspid isthmus (CTI) in the right atrium is the origin of the most common type of atrial flutter (AFL) and therefore focus for arrhythmia research in anatomy and clinical electrophysiology. The impact of complexities and discontinuities of macro- and microstructure on the intercaval conduction in the subregions Crista Terminalis, Pectinate muscles and Vestibulum is little known in detail. Discontinuities of structure result in signals of non-uniform conduction with double or multiple deflections of signals indicating a fractionation of the depolarization wave. From experiments with 13 Rabbit hearts we analyzed topology and morphology of the CTI as well as the corresponding signal parameters taken at 1290 recording sites during sinus rhythm. A representative branching topology of the network of Pectinate muscles is described here with 4 proximal, 7 central and 4 sections. We found the largest portion of non-fractionated signals and of large signal amplitudes in Pectinate muscles, the highest portion of complex fractionated signals in the Vestibulum. The results allow to reconstruct a map of signal characteristic in detailed subregions of the CTI and to estimate the impact of macro- and microstructure on the activation sequence during sinus rhythm.

Keywords—Rabbit atrium, cavotricuspid isthmus, morphometry, electro-anatomical characterization, signal analysis.

I. INTRODUCTION

The cavotricuspid isthmus (CTI) in the right atrium is the origin of the most common type of atrial flutter (AFL) and therefore focus for arrhythmia research in anatomy and electrophysiology [1, 2]. This region of interest (ROI) comprises the Crista Terminalis (CT), the vestibule area (VB) along the tricuspid valve and a comb shaped network of Pectinate muscles (PectM) connecting the CT with the VB electrically. From the rather diffuse sinus node area activation enters the CT, spreads out over the network of PectMs and ends in the VB. Pathways for cavostricuspidal activation sequence are determined by the macro-anatomy of the network of cable like PectMs. Branching and merging sites within this network represent discontinuities of conduction due to the abrupt change of current source and current load.

In animal experiments PectM have been identified as anatomical substrates for reentrant wave-fronts during AFL [3]. The role of complexities and discontinuities of the microstructure in the intercaval conduction is widely unknown. They might alter the activation patterns expected due to the macrostructure and generate additional complexities of excitation spread even at sites were smooth and continuous conduction is expected. Such microstructural effects become visible in extracellular electrograms Φ with lower amplitudes, with double or multiple deflections of $d\Phi/dt$, and in complex vector loops of the local gradient of Φ, the cardiac near field $\mathbf{E}$ [4, 5]. The most relevant microstructural elements for discontinuous conduction are connective tissue layers. They separate adjacent myocytes electrically and allow the development of separated individual wavelets with delays to each other. Since these structures, representing electrical obstacles for conduction, are not visible directly during in-vitro experiments, the changes of electrical signal waveforms associated with micro-discontinuities could be used to characterize the local microstructure at a given site. In this work we describe and analyze the macro-anatomy of the CTI in Rabbit hearts with corresponding parameters of local conduction during sinus rhythm. Such electro-anatomical maps related to sinus rhythm might serve as a basis to study arrhythmias induced by rapid pacing or by drugs as well as for the development of rule-based computer models of the CTI. In addition a representative and characteristic branching topology of the ROI can be obtained from morphometric data.

II. METHODS

A. Experimental Setup

Hearts were rapidly excised from 13 anesthetized Rabbits. The procedure was carried out in accordance to the national ethic guidelines. The right atria including the sinus node region were dissected and mounted with the epicardial surface downside on transparent silicon carrier and placed in a tissue bath streamed with oxygenated Tyrodes solution at 36.4 °C. A custom designed trans-illumination unit in LED technology at the bottom of the bath allowed to identify thin fibers with diameters in the range of just some tens of micrometers.

B. Morphometry

During the experiment a digital camera-system (Canon EOS 5D Mark II, lens EF 180 mm/3.5 Macro, Canon Inc., Japan) produced high-resolution digital images of the preparation and of the recording positions (21 Megapixel with 10.5 μm pixel size). Post experimentum the atria were fixed in 10% formaldehyde for later histological analysis. The ROI was documented with ultra-high-resolution (1.28 μm per pixel) with images obtained from zoomed and stitched segments taken with a special macro lens (Canon MP-E 65mm/2.8, 1-5x, Canon Inc., Japan).

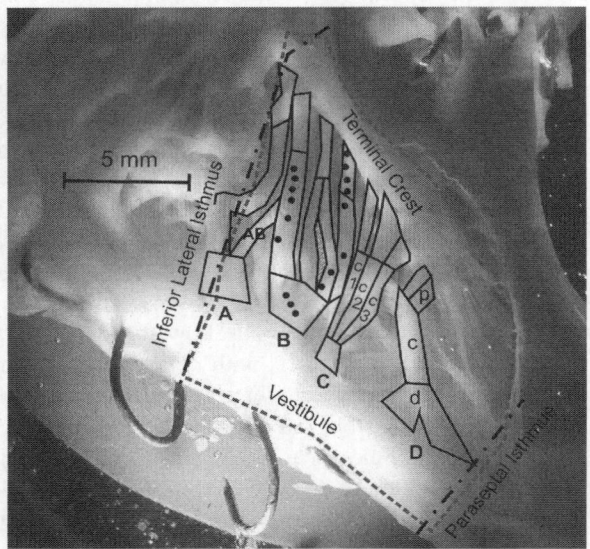

Fig. 1 Right atrium with the region and subregions of interest. PectMs are indexed by the sequence of the distant roots from A to D. Proximal p, central c and distant sections d were used to classify subregions of the PectM. Recording positions are indicated with full circles.

The ROI depicted in Figure 1 is bounded by the paraseptal and inferolateral isthmus, the CT and the VB and is characterized by the network of PectMs separated by crevices. The space between PectMs, termed non pectinate tissue (NPT), is formed by very thin layers of myocytes and connective tissue and constitutes the smooth epicardium. These parts of the tissue are transparent for the LED light source placed at the bottom of the tissue bath. Morphometry was executed after importing the digital images into CorelDRAW X5 (Corel Inc., Mountain View, CA, USA) and applying freely available plug-in macros for calculating areas, lengths and diameters of tissue parts and fibers. For a systematical description of branching geometry we introduced a sub-classification of the PectMs into three sections: a the proximal section describing the part from the ramification from the CT to the first branching site, a central section with a higher number of parallel strands due to the branching, and distal part from the merging sites to the junction with the VB. A corresponding classification is shown in Figure 1. Areas of the ROI as well as subareas of the PectMs and NPT were measured to estimate the size of active tissue and the ratio of PectM to NPT. Numbers of PectMs in their proximal, central and distal subsections were evaluated to get a measure for branching topology. Diameters of CT and VB were obtained by averaging 3 measurements taken at the paraseptal, at the central, and at the inferolateral side. Diameters of PectM were determined at the middle of the proximal, the central and the distal section. The diameter of muscle fibers is important to estimate long term viability of tissue during the experiment [6] and the contribution to the endocardial signal Φ. The network topology of the ROI thus is described systematically.

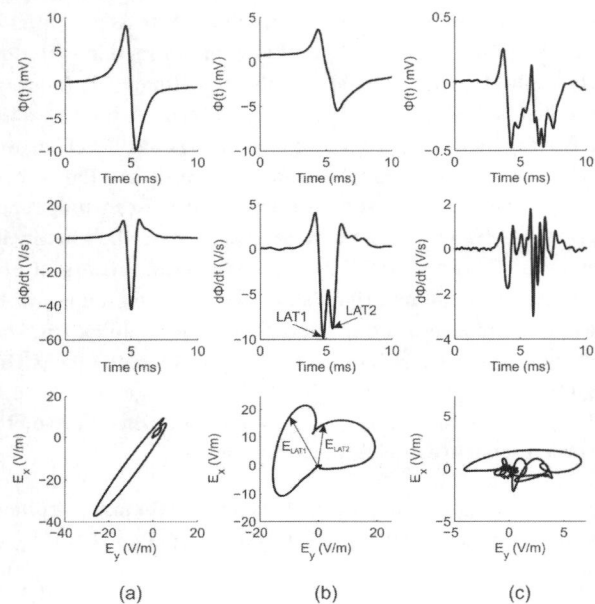

Fig. 2 Representative waveforms of Φ, $d\Phi/dt$, and of the vector loop of the cardiac near field $\mathbf{E}$ during uniform conduction (a), showing slight fractionation with $FI = 2$ (b), and severe fractionation with $FI > 3$ (c).

C. Electrical Signal Recordings and Analysis

Electrical signals were recorded with a custom designed 20-channel-amplifier and data acquisition system, amplified (gain 100), low-pass filtered (4th-order Bessel, -3 dB point at 20 kHz) and digitized with 16 bit at a sampling rate of 100 kHz per channel (NI PXI-6123, National Instruments, Austin, Texas). Data acquisition, analysis and monitoring during the experiment were done with custom designed software written in MATLAB (Mathworks, Natick, MA, USA) and in LabVIEW (National Instruments, Austin, Texas, USA). Surface potentials at the endocardium were recorded with non-contact electrode-arrays (four electrodes arranged at square corners with 50μm spacing) fabricated in thin film technique [4]. The propagating wave of depolarization produces single or in case of microstructural disturbances double or multiple extracellular deflections of $d\Phi/dt$. $d\Phi/dt$ and the cardiac near field $\mathbf{E}$ were computed from each set of four electrograms of Φ [7]. Local activation time (LAT) was precisely determined by taking the instant of peak negative derivative $d\Phi/dt$, fractionation index FI was computed using a signal model based method [8]. Peak-to-peak amplitude Φ_{PP}, the peak negative deflection $d\Phi/dt_{min}$ and the magnitude $\mathbf{E}_{max}$ were evaluated at all recording sites to characterize local activation. Magnitude of the local conduction velocity (LCV) was determined from delay components in LAT obtained from four orthogonally arranged electrode signals. A symmetry-factor a was determined derived from the waveform of Φ to indicate if the character of local propagation is rather free running ($a=1$), starting ($a=-1$) or ending ($a=+1$) in the cable like muscles [9].

D. Electro-anatomical Maps

Since the position of recording sensors can be determined precisely from the digital images taken during the experiment, maps for each signal parameter of excitation spread could be reconstructed. Each recording was assigned to its corresponding region or sub-region in order to cluster and to categorize the signal data. By this we obtained a representative profile of signal parameters for each subcategory of tissue appropriate for electro-anatomical mapping.

III. RESULTS

A. Morphometry

Morphometric data are given as *mean ± std*. From 13 experiments we evaluated the area of the CTI (including CT and VB) with $114.2 \pm 35.7\,mm^2$. PectM covered $40.8 \pm 15.1\,mm^2$ and NPT tissue $32.2 \pm 14.0\,mm^2$. The length of the inferolateral and paraseptal isthmus was $14.8 \pm 2.2\,mm$ and $7.3 \pm 1.8\,mm$ respectively. Diameters of CT, PectM and VB were 1.7 ± 0.4, 0.7 ± 0.2, and $2.4 \pm 0.7\,mm$. Due to bifurcations in the central part the number of PectMs in the proximal, central and distal section was different indicating the branching character of the fiber network. We counted in sum 66 proximal, 96 central and 54 distal sections. A representative topology (calculated as median and range of PectM) gives 4 (3-9) proximal, 7 (4-12) central, and 4 (3-6) distal sections.

B. Distribution of Signal Parameters

The signals from 1290 recording sites in the ROI were analyzed, results were clustered in recording areas from CT (205), PectM (837) and VB (248). We found uniform ($FI = 1$) as well as fractionated signals ($FI > 1$) in all types of tissue, but with very different distribution of FI (see Figure 3a) in each category of tissue. Undisturbed electrograms ($FI = 1$) were found in 49% of signals from PectMs whereas in the CT and the VB this value was only 21% and 44%. The portion of high degree of fractionation ($FI > 9$) was 2% in CT, 1% in PectM and 10% in VB. Quantitative signal parameters like Φ_{PP}, $d\Phi/dt_{min}$, and $\mathbf{E}_{max}$ showed large variations with no normal distribution. Therefore parameters are described by median and interquartile ranges. As can be seen in Figure 3b there were statistically significant differences between signals of CT, PectM and VB (see asterisk in Figure 3b). Largest values of Φ_{PP}, $d\Phi/dt_{min}$, and $\mathbf{E}_{max}$ were obtained in PectM.

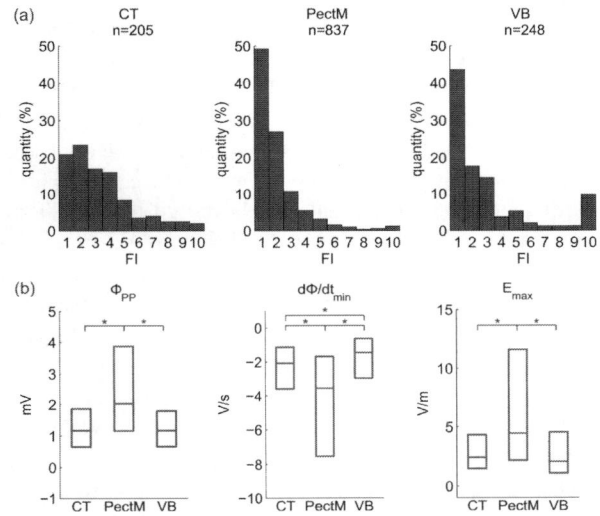

Fig. 3 Signal evaluation of 1290 recordings in the sub-regions CT (205), PectM (837), and VB (248). (a) Distribution of FI given in %. (b). Amplitude of Φ, $d\Phi/dt$, and magnitude of $\mathbf{E}$. Median (red) and 25/75% quartiles (blue) describe the distribution within CT, PectM, and VB. *, $\alpha \leq 0.01$.

The median values of LCV did not differ substantially between CT ($1.37\,m/s$), PectM ($1.03\,m/s$) and VB ($1.05\,m/s$) and interquartile ranges were large in all recordings ($0.80\,m/s$–$1.76\,m/s$). The symmetry factor a of the signals Φ_e was categorized into 3 intervals. -1 to -0.3, -0.3 to $+0.3$ and $+0.3$ to $+1$ to give a quantitative measure for starting, free running, or ending propagation. Fraction of free running signals was largest in PectM (77%), of starting potentials in CT (25%), and of ending and colliding potentials in VB (26%).

IV. CONCLUSION

There was a distinct variability in size, diameters and length of ROI as expected. This might be caused (i) by the fact that we did not select Rabbits in terms of a small range of weight and (ii) due to inherent variability in morphology found between different hearts. This holds also for branching structures and topologies within the network of PectMs. In all atria we found a typical increase in the number of PectMs in the central section doubling its value in average compared to the proximal or distal section. We also emphasize that the diameter of PectM is small enough to provide the network of muscles with sufficient oxygen to keep normal conduction for hours just by superfusion. Similar to morphological parameters a large variability of electrical signal parameters was found. Nevertheless, differences between the sub-regions could be detected. The largest portion of non-fractionated electrograms ($FI = 1$) was found in PectM, followed by a large value of $FI = 2$ and 3. This can be explained by the microstructure we found histographs from PectMs. Muscle fiber orientation was strongly parallel with just a few aligned connective tissue layers leading to fractionation of electrograms in thicker segments close to merging areas of PectM. Not synchronized wavelets there can propagate with small delays along electrically separated parts of the PectM. It should also be noted that the largest portion of high fractionation indices was found in the VB. This might be due to the fact, that this subarea represents the end-points of cavotricuspidal conduction representing a complex network of crossing fibers. Amplitudes of Φ, $d\Phi/dt$, and of $\mathbf{E}$ differed considerably in median and interquartile range (Figure 3b). This can be explained by the circumstance that the direction of the propagating wave-front of depolarization related to the fiber axis is different in CT, PectM and VB (longitudinal vs. transverse conduction). In addition to these directional effects desynchronization of local wavelets caused by the presence of connective tissue structures might influence amplitude as well as waveform of Φ (fractionation index). By this, despite the larger diameter of CT and VB, amplitudes of Φ are smaller in average compared to those of PectM. The decreased values of signal symmetry in CT and VB compared to PectM let us conclude that in the network of PectM the largest portion of free running activation waves can be expected whereas CT represents rather an area of activations starting from the CT into the network of PectM and signals in the VB represent rather the signals of waves arriving at the end of cable-like structures. This work describes the distribution of electrical signal parameters of normal intercaval conduction during sinus rhythm. The description of the signals at critical sites of macro- and micro-conduction under arrhythmogenic conditions like rapid pacing or increased heart rate induced by drugs remains open. The results shown here represent normal conduction during sinus rhythm and could be used for construction of rule-based computer models of the CTI.

ACKNOWLEDGEMENTS

This work has been supported by the Austrian Science Fund (FWF) grant P19993-N15 and grant F3210-N18.

REFERENCES

1. Gami Apoor S., Edwards William D., Lachman Nirusha, et al. Electrophysiological anatomy of typical atrial flutter: the posterior boundary and causes for difficulty with ablation. *J Cardiovasc Electrophysiol.* 2010;21:144–149.
2. Sánchez-Quintana Damián, López-Mínguez José Ramón, Pizarro Gonzalo, Murillo Margarita, Cabrera José Angel. Triggers and anatomical substrates in the genesis and perpetuation of atrial fibrillation. *Curr Cardiol Rev.* 2012;8:310–326.
3. Wu T. J., Kim Y. H., Yashima M., et al. Progressive action potential duration shortening and the conversion from atrial flutter to atrial fibrillation in the isolated canine right atrium. *J Am Coll Cardiol.* 2001;38:1757–1765.
4. Hofer E., Keplinger F., Thurner T., et al. A new floating sensor array to detect electric near fields of beating heart preparations. *Biosens Bioelectron.* 2006;21:2232–2239.
5. Plank G., Vigmond E., Leon L. J., Hofer E.. Cardiac near-field morphology during conduction around a microscopic obstacle–a computer simulation study. *Ann Biomed Eng.* 2003;31:1206–1212.
6. Campos Fernando O., Prassl Anton J., Seemann Gunnar, Weber dos Santos Rodrigo, Plank Gernot, Hofer Ernst. Influence of ischemic core muscle fibers on surface depolarization potentials in superfused cardiac tissue preparations: a simulation study. *Med Biol Eng Comput.* 2012;50:461–472.
7. Plank G., Hofer E.. Model study of vector-loop morphology during electrical mapping of microscopic conduction in cardiac tissue. *Ann Biomed Eng.* 2000;28:1244–1252.
8. Wiener Thomas, Campos Fernando O., Plank Gernot, Hofer Ernst. Decomposition of fractionated local electrograms using an analytic signal model based on sigmoid functions. *Biomed Tech (Berl).* 2012;0:1–12.
9. Jacquemet Vincent, Henriquez Craig S.. Genesis of complex fractionated atrial electrograms in zones of slow conduction: a computer model of microfibrosis. *Heart Rhythm.* 2009;6:803–810.

Author:	Ernst Hofer
Institute:	Institute of Biophysics, Medical University of Graz, Graz, Austria
Street:	Harrachgasse 21
City:	A8010 Graz
Country:	Austria
Email:	ernst.hofer@medunigraz.at

Cavotricuspid Isthmus: Anatomy and Electrophysiology Features: Its Evaluation before Radiofrequency Ablation

D. Sánchez-Quintana[1] and J.A. Cabrera[2]

[1] Departamento de Anatomia y Biologia Cellular, Universidad de Extremadura, Badajoz, Spain
[2] Hospital Quirón Madrid. Universidad Europea de Madrid, Madrid, Spain

Abstract—The cavotricuspid isthmus (CTI) in the lower pan of the right atrium, between the inferior caval vein and the tricuspid valve, is considered crucial in producing a conduction delay and. hence, favoring the perpetuation of a reentrant circuit. Non-uniform wall thickness, muscle fiber orientation and the marked variability in muscular architecture in the CTI should be taken into consideration from the perspective of anisotropic conduction, thus producing an electrophysiologic isthmus. The purpose of this article is to review the anatomy and electrophysiology of the CTI in human hearts to provide useful information to plan CTI radio frequency ablation for the patients with atrial flutter.

Keywords—cavotricuspid isthmus, wall thickness, muscular architecture, anisotropic conduction.

I. INTRODUCTION

Creation of a complete bidirectional conduction block across the inferior right atrial cavo-tricuspid isthmus (CTI) is the accepted marker for long-term success in patients with isthmus-dependent atrial flutter [1]. Histopathologic findings have demonstrated that transmural ablation of the atrial wall is a prerequisite for success [2]. The CTI is limited posteriorly by the Eustachian valve/ridge and anteriorly by the annular insertion of the septal leaflet of the triscuspid valve. Post-mortem and in vivo imaging examination in normal hearts and in patients with atrial flutter have pointed to the anatomic variability of the dimensions, endocardial geometry, and muscular architecture across the anatomic landmarks of the inferior right atrium and its impact as anatomic determinant for catheter ablation [3][4]. These observations suggest that in some patients the length of the isthmus that needs to be ablated, i.e. the functional isthmus, can be considerably narrower than the area bounded by the anatomic barriers of the tricuspid valve anteriorly and the inferior caval vein posteriorly. In this work we describe and analyze the anatomy and electrophysiology features of the CTI in human hearts to provide useful information to plan CTI radiofrequency ablation.

II. METHODS

We examined 50 formalin-fixed human hearts. The isthmus area was prepared for histology. Deparaffinized sections were stained with Masson´s trichrome.

III. RESULTS

A. Gross Morphological Features of the Human Cavotricuspid Isthmus

The right atrium as displayed in attitudinally orientation, the CTI shows on its endocardial surface an irregular quadrilateral shape (Fig. 1). The superior border of the quadrilateral is the paraseptal isthmus (so-called septal isthmus) extending between the septal insertion of the coronary sinus, and the most inferior paraseptal insertion of the tricuspid valve. The length of the paraseptal isthmus was 24±4mm (range 14-33 mm). The inferolateral border of the quadrilateral area (so-called inferolateral isthmus) contained pectinate muscles as a continuation of the final ramifications of the terminal crest. The length of the inferolateral isthmus was 30±3mm (range 18-36 mm). The middle zone of the isthmus, between the paraseptal and the inferolateral levels, is the central isthmus. The central isthmus is the shortest distance between the orifices of the inferior caval vein (its length was 19±4 mm, range 13-26mm), and the tricuspid valve and traverses through a pouch-like recess, the subeustachian sinus. A posterior pouch-like recess deeper than 5 mm can be seen in 20% of patients and it usually causes the most difficulty in achieving a complete line of block during isthmus ablation [5]. When pouches occur, they tend to be closer to the coronary sinus orifice than the free wall of the right atrium and can vary significantly in the depth as well as the anteroposterior dimension. There also appears to be an association, possibly developmental, between prominence of subeustachian pouches and a prominent Thebesian valve guarding the opening of the coronary sinus [6]. Thus, it is unusual to find a large subeustachian pouch without significant evidence of a Thebesian valve.

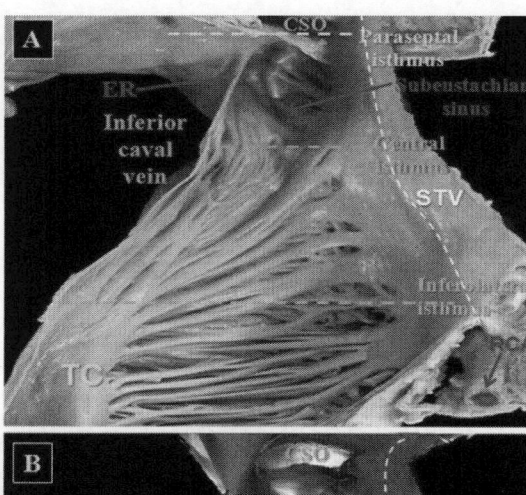
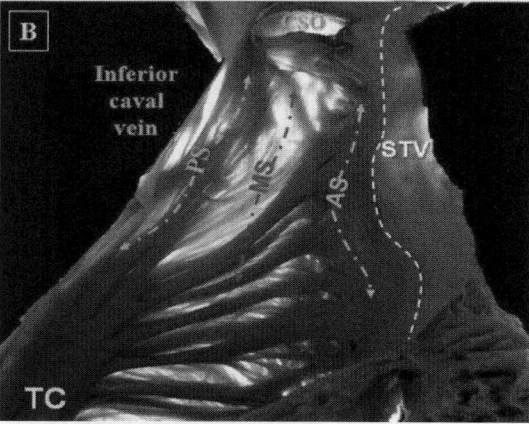

Fig.1. Normal (**A**) and transillumination (**B**) the endocardial surface of the human right atrial isthmus is displayed to show the three levels (inferolateral, central, and paraseptal) and the three sectors: posterior (PS), middle (MS) and anterior (AS). Note the pouch (subeustachian sinus) at the central isthmus and the distal branching of the terminal crest (TC) that feed into the inferolateral isthmus. CSO: Coronary sinus orifice. ER: Eustachian ridge/valve. STV: Septal leaflet of the tricuspid valve.

In the normal heart, the topography of the CTI is complex and not flat (Fig. 1). The Eustachian ridge (ER) occurs as an elevation on the CTI and divides this isthmus into two portions: the first between the tricuspid valve and the ER and then from the crest of the ER sloping downward to the inferior caval vein. In addition, a prominent and muscular Eustachian ridge (such as found in 26% of specimens) may require tricky angling of the catheter for good contact [7]. The ridge itself varies in terms of prominence and can be very well developed in some adult hearts. The ratio of myocardium to fibrous tissue in the ridge is also variable and in part, determines the conduction properties of this structure. It should be emphasized, however, that in the majority of hearts, myocardium is present in the ER, and thus, ablation purely between the tricuspid valve and the ER cannot be expected to be completely efficacious. When a prominent ER is present, the ridge acts as a fulcrum, and clockwise torque will result in the ablation electrode pointing away from the septum. This can make it difficult and necessitate the use of a guiding sheath that is placed distal to the ER [6]. Usually the ER is less prominent more laterally, and when a particularly prominent ridge is present, inferolateral CTI ablation may be beneficial.

Angiography demonstrated larger dimensions of the right atrium and the CTI in patients with isthmus-dependent atrial flutter than in normal controls. Furthermore, studies have reported that both longer isthmus 35 mm and morphological characteristics such as concave deformation of the isthmus and deep pouch-like recesses were associated with a significantly increased procedure duration and number of RF applications during atrial flutter ablation [4]. The subeustachian pouch may be a source of difficulty when ablating the CTI for several reasons: i)catheter contact may not occur within the depths of the pouch resulting in incomplete transaction of the CTI. ii)when catheter contact does occur, power delivery is often limited by impedance rise and coagulation formation as a result of poor blood flow within the pouch. Thus, larger electrode surface area or irrigation may be required when ablating in the depths of the pouch. However, care to avoid injury to the right coronary artery or perforation is required. To avoid possible complications, the ablation line can be done more laterally (8 o'clock to 9 o'clock on the left anterior oblique projection), making use of the anatomic fact that the pouches are deepest and most likely to occur close to the coronary sinus ostium [6].

Recent studies have shown that 3D electroanatomic mapping systems provide a good reconstruction of the inferior right atrial isthmus with a good correlation with angiography for the isthmus anatomy and dimensions. Imaging evaluation of the isthmus characteristics has also been reported using multi-slice computed tomography (CT) and cardiac magnetic resonance. These studies demonstrated its feasibility to visualize the morphological variants of isthmus morphology and its impact on catheter ablation [8].

B. Architectural Factors of the Isthmus: Myocardial Fiber Orientation and Fibro-fatty Tissues Thickness

The pectinate muscles all originated from the terminal crest are most prominent laterally and gradually decrease in thickness as they traverse across the CTI towards the ostium of the coronary sinus. At all three sectors of the CTI, the anterior part was consistently muscular, whereas the posterior part was composed of mainly fibro-fatty tissue (Fig. 2). Therefore, there is a non-uniform muscular content and thickness in the isthmus of evident relevance to create transmural lesion of the atrial wall during atrial flutter ablation [5]. In the majority of the specimen (64%), the wall of the sector immediately anterior to the orifice of the inferior vena cava, dubbed the posterior sector, is composed mainly of fibrous and fatty tissue, with minimal muscular fibers coursing through it. The clinical implication is that this part of the isthmus needs the least radiofrequency energy. The middle part of the quadrilateral is, in the majority of specimens (64%), made up of muscular trabeculations separated by delicate membranes (Figs. 1and 2). The trabeculations originated either from the wall of the coronary sinus or else were the continuations of the muscular bundles of the terminal crest. The anatomy of these pectinate extensions has several implications for the interventional electrophysiologist: i)When ablating on a pectinate muscle, energy delivery may not be sufficient to create a transmural lesion leaving a gap at this site. High amplitude electrograms prior to ablation delivery should suggest the presence of a pectinate muscle and energy delivery maximized at that site or an alternative position for the line considered. ii)Since the pectinates are most prominent laterally, when their presence is recognized (large amplitude electrograms or ultrasound imaging), the electrophysiologist can place the ablation line more septally close to the coronary sinus ostium. Care, however, should be taken when doing a very septal line that the catheter tip is not inadvertently within the middle cardiac vein since damage to the adjacent arterial vasculature (right coronary artery or AV nodal artery) may occur [6].

iii)When pectinates are traversing posterior to the inferior caval vein beyond a posterior ridge, they may serve as a conduit for electrical conduction that may be utilized in a flutter circuit around the inferior caval vein (lower loop reentry). More commonly, however, these pectinate muscles may allow passive activation of the free wall of the right atrium, giving the appearance of continued medial to lateral conduction despite adequate ablation of the CTI [6].

The third or anterior sector of the quadrilateral isthmus, adjacent to the hinge of the tricuspid valve, is smooth, showing no evidence of trabeculations in any heart. Histological sections confirmed that in this region, the myocardium and fibro-fatty tissues formed the full thickness of the atrial wall (Fig. 2). In 26% of the heart specimens, the pectinate muscles radiated from the terminal crest in almost parallel fashion without an appreciable cross-over. In 74% of the heart specimens, the muscular arrangement showed abundant cross-over and interlacing trabeculae, particularly in three zones: a)the zone immediately inferior to the coronary sinus ostium. In this area, trabeculae to the right- and left-hand sides of the ostium are interconnected along the inferior rim. b)the zone between the end of the pectinate muscles and the vestibule of the tricuspid valve. c)the zone between the individual parallel pectinate muscles has abundant crossovers with different orientation [9][5]. On this structural basis (Fig. 2) one can easily speculate that nonuniform anisotropic conduction occurs, and thus, may lead to a conduction delay.

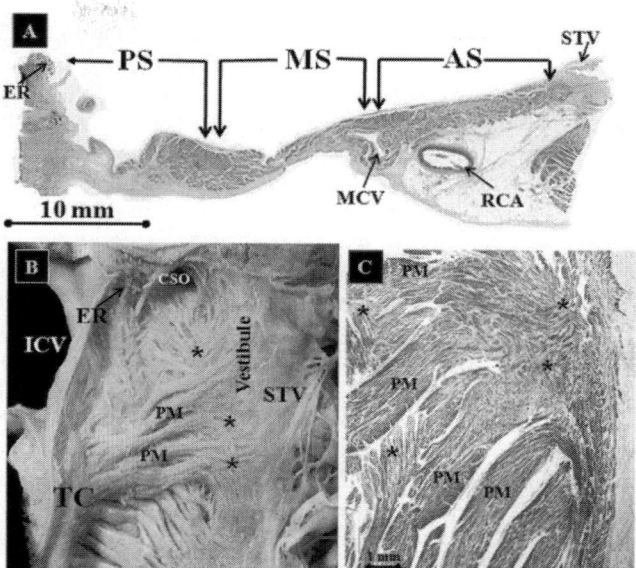

Fig. 2 A (sagittal) and C (frontal) histological sections (Masson's trichrome stain) at the level of the central isthmus illustrating abundant crossovers with different fiber orientation (asterisks) and fibro-fatty tissues thickness. Note in B the fibers change directions at sharp angles or cross in different directions (asterisks). PS: Posterior sector. MS: Middle sector. AS: Anterior sector: MCV: Middle cardiac vein. RCA: Right coronary artery. STV: Septal leaflet of the tricuspid valve. ER: Eustachian ridge/valve. CSO: Coronary sinus orifice. ICV: Inferior caval vein. TC: Terminal crest. PM: Pectinate muscle.

C. Electrophysiological Mechanisms of Atrial Flutter

Typical atrial flutter circuit, is after atrial fibrillation the most common among atrial tachycardias, and has anisotropic conduction at the terminal crest and vestibule of the tricuspid annulus as essential parts of its mechanism [10]. Furthermore, has been demonstrated that during sinus rhythm, incremental pacing from the low lateral right atrium and coronary sinus ostium could produce rate-dependent conduction delays [10], culminating in unidirectional block in the low right atrial isthmus, and induction of counterclockwise or clockwise atrial flutter in patients with or without clinical atrial flutter. These findings were confirmed by other authors and suggested that slow conduction in the low right atrial isthmus may be mechanistically important for the development of human typical atrial flutter [11]. The mechanism of slow conduction in the isthmus was not clear. It has been suggested that conduction velocity of atrial impulses is faster parallel to the long axis of myocyte fibers and slower along the plane transverse to myocyte fiber orientation [12]. This phenomenon was explained by higher axial resistance due to scant cell-to-cell coupling encountered when impulses propagated perpendicular to the long axis of muscle fibers. With aging or atrial dilatation, intercellular fibrosis can change the density of gap junctions and produce nonuniform anisotropic conduction through the trabeculations of the low right atrial isthmus [13]. But the role of anisotropy in the flutter circuit may not end just here. Relatively slow conduction has been described in the isthmus between the anterior rim of the inferior caval vein and the inferior rim of the tricuspid ring, and this may be important both for initiation and maintenance of re- entry. Fine studies of fibre orientation in the isthmus [3][9] show terminal ramifications of the crest running perpendicular to insert in the tricuspid ring, thus making again a perfect anatomic set-up for local conduction slowing during circular activation. This hypothesis is supported by a recent anatomic study of the low right atrial isthmus in humans [5]. Particularly interesting is the complex relation between the tightly packed, parallel myocardium of the crest itself and the fibres making the pectinate muscles, connected to the crest in varying angles that should favour block during rapid activation. It has been proposed that in human flutter transverse block may occur close to, but not precisely at, the terminal crest level [14], and experimental work in dogs showed how block can occur preferentially at the junction of the terminal crest with the pectinate muscles region [151]. Furthermore, other observations in dogs with natural and evoked atrial flutter suggest that thinning of atrial myocardium with intervening spaces may predispose to both slow and nonuniform conduction.

IV. CONCLUSIONS

Typical atrial flutter is a reentrant arrhythmia and needs anatomic or functional barriers to maintain its activation. Typical atrial flutter rotates around the tricuspid annulus with the crista terminalis and tricuspid annulus as barriers. Radiofrequency ablation of the isthmus between the boundaries can cure this arrhythmia.

ACKNOWLEDGMENT

This work has been supported by the Ministerio de Economia y Competitividad grant TIN2012-37546-C03-02

REFERENCES

1. Feld GK, Fleck RP, Chen PS, Boyce K, Bahnson TD, Stein JB et al. (1992) Radiofrequency catheter ablation for the treatment of human type 1 atrial flutter: identification of a critical zone in the reentrant circuit by endocardial mapping techniques. Circulation 86(4):1233–1240
2. Kohno I, Ishihara T, Umetani K, Sawanobori T, Ijiri H, Komori S et al. (2000) Pathological findings of the isthmus between the inferior vena cava and tricuspid annulus ablated by radiofrequency application. PACE 23(5):921–923
3. Cabrera JA, Sánchez-Quintana D, Ho SY, Medina A, Anderson RH. (1998) The architecture of the atrial musculature between the orifice of the inferior caval vein and the tricuspid valve. J Cardiovasc Electrophysiol 9(11):1186–1195
4. Da Costa A, Romeyer-Bouchard C, Dauphinot V, Lipp D, Abdellaouri L, Messier M et al. (2006) Cavotricuspid isthmus angiography predicts atrial flutter ablation efficacy in 281 patients randomized between 8 mm- and externally irrigated-tip catheter. Eur Heart J. 27(15):1833–1840
5. Cabrera JA, Sánchez-Quintana D, Farré J, Rubio JM, Ho SY. (2005) The inferior right atrial isthmus revisited: further architectural insights for current and coming ablation technologies. J Cardiovasc Electrophysiol 16(4):402–8
6. Sehar N, Mears J, Bisco S, Patel S, Lachman N, Asirvatham SJ. (2010) Anatomic guidance for ablation: atrial flutter, fibrillation, and outflow tract ventricular tachycardia. Indian Pacing Electrophysiol J. 10(8):339-356.
7. Asirvatham SJ. (2009) Correlative anatomy and electrophysiology for the interventional electrophysiologist: right atrial flutter. J Cardiovasc Electrophysiol. 20(1):113-122
8. Saremi F, Pourzand L, Subramanian K, Ashikyan O, Gurudevan SV, Narula J et al. (2008) Right atrial cavotricuspid isthmus: anatomic characterization with multidetector row CT. Radiology 247(3):658–668.
9. Waki K, Saito T, Becker AE. (2000) Right atrial flutter isthmus revisited: normal anatomy favors nonuniform anisotropic conduction. J Cardiovasc Electrophysiol 11(1):90–94
10. Tai CT, Chen SA, Chiang CE, Lee SH, Ueng KC, Wen ZC, Huang JL, et al. (1997) Characterization of low right atrial isthmus as the slow conduction zone and pharmacological target in typical atrial flutter. Circulation 96(8):2601–2611
11. Feld GK, Mollerus M, Birgersdotter-Green U, Fujimura O, Bahnson TD, Boyce K, Rahme M. (1997) Conduction velocity in the tricuspid valve-inferior vena cava isthmus is slower in patients with type I atrial flutter compared to those without a history of atrial flutter. J Cardiovasc Electrophysiol 8(12):1338–1348
12. Spach MS, Dolber PC, Heidlage JF. (1988) Influence of the passive anisotropic properties on directional differences in propagation following modification of the sodium conductance in human atrial muscle: a model of reentry based on anisotropic discontinuous propagation. Circ Res 62(4):811–832
13. Spach MS, Dolber PC. (1986) Relating extracellular potentials and their derivatives to anisotropic propagation at a microscopic level in human cardiac muscle: evidence for electrical uncoupling of side-to-side fiber connections with increasing age. Circ Res. 58(3):356–371
14. Friedman PA, Luria D, Fenton AM, Munger TM, Jahangir A, Shen WK, et al. (2000) Global right atrial mapping of human atrial flutter: the presence of posteromedial (sinus venosa region) functional block and double potentials: a study in biplane fluoroscopy and intracardiac echocardiography. Circulation 101(13): 1568–1577
15. Becker R, Bauer A, Metz S, Kinscherf R, Senges JC, Schreiner KD. et al. (2001) Intercaval block in normal canine hearts: role of the terminal crest. Circulation 103(20):2521–2526.

Author: Damián Sánchez-Quintana
Institute: Facultad de Medicina. Departamento de Anatomía
Street: Avenida de Elvas s/n. 06071
City: Badajoz
Country: Spain
Email: sanchezquintana55@gmail.com

Preoperative Prognosis of Atrial Fibrillation Concomitant Surgery Outcome after the Blanking Period

A. Hernández[1], R. Alcaraz[2], F. Hornero[3], and J.J. Rieta[1]

[1] Biomedical Synergy, Electronic Engineering Department, Universidad Politécnica de Valencia, Spain
[2] Innovation in Bioengineering Research Group, University of Castilla-La Mancha, Cuenca, Spain
[3] Cardiac Surgery Department, General University Hospital Consortium, Valencia, Spain

Abstract—Despite recent advances in the treatment of atrial fibrillation (AF), Cox-Maze surgery still is the therapy with the highest success rate. Long-term outcome prediction of this procedure has been widely studied in previous works. However, only few studies addressed the short-time prediction issue. Moreover, they all presented a very limited predictive ability. In the present work, preoperative information from the surface electrocardiogram (ECG) is analyzed to predict the patient's rhythm after the blanking period, which lasts 3 months after the surgery. Three aspects have been studied: the dominant atrial frequency (DAF), the pattern repetitiveness in the atrial activity via sample entropy (SampEn) and the fibrillatory waves mean power (fWP). They all revealed a considerably high prognosis ability of 68.97%, 72.42% and 75.86%, respectively. Additionally, the combination of these parameters through a decision tree improved substantially the prognosis accuracy up to 82.76%, with sensitivity of 81.25% and specificity of 84.62%. Hence, the preoperative ECG can yield clinically relevant information on the patient's rhythm after the blanking period and could be helpful in the development of tailored postoperative therapies.

Keywords—ECG, atrial fibrillation, Cox-Maze surgery, organization indices, fibrillatory waves amplitude.

I. INTRODUCTION

Atrial fibrillation (AF) is the most common sustained cardiac arrhythmia, increasing its prevalence with age [1]. Given the significantly increased morbidity and mortality associated with AF, there have been aggressive attempts to find a cure over the last decades. An effective and widely extended surgical therapy for restoring sinus rhythm (SR) in AF is the Cox-Maze procedure [2]. Furthermore, with the introduction of ablation devices, less invasive techniques have been proposed and the cut-and-sew model was replaced with ablation technologies [3]. These techniques range from changes in the Maze procedure (Maze IV, mini-Maze, etc.) to minimally invasive approaches such as catheter ablation [3]. However, studies comparing different techniques are heterogeneous but they all point out that Cox-Maze III yields better long-term outcomes [3].

After surgery, patients are routinely evaluated at three, six and twelve months; and then, every year. Moreover, they are habitually treated with oral anticoagulants and antiarrhythmic drugs at the moment of discharge, independently of the presented rhythm: SR or AF [4]. Then, if AF persists after the blanking period, i.e., 3 months after the surgery, electrical cardioversion (ECV) is applied to restore SR [4]. In contrast, antiarrhythmic drugs are withdrawn if a stable sinus rhythm is recorded. Hence, preoperative information about patient's rhythm after the blanking period could be clinically relevant. For patients with high risk of AF, ECV could be planned before surgery without the need of waiting for a routine evaluation. In this way, clinical resources could be optimally managed and risks for the patient could be minimized by reducing postoperative AF duration and, therefore, the arrhythmia perpetuation probability [5]. Contrarily, an aggressive treatment with antiarrhythmic drugs could be avoided for those patients with a low risk of maintaining AF after the three first postoperative months. Thus, clinical costs could be reduced and the patient's quality of life could be improved.

Although a wide variety of works dealing with the long-term prediction of Cox-Maze outcome can be found in the literature [2], there are few studies addressing the prediction at 3 months post-surgery [6]. Moreover, they have revealed a notably limited diagnostic ability [6]. Therefore, the present work focuses on defining a robust method able to predict preoperatively the patient's rhythm after the post-surgery blanking period. For this purpose, the atrial activity (AA) extracted from surface electrocardiographic (ECG) recordings will be characterized through the use of signal processing tools.

II. METHODS

A. Study Population and Data Preprocessing

The database consisted of 29 patients (age 68.5 ± 12.7 years) in permanent AF for more than 3 months. The Cox-Maze III surgery was applied concomitantly to another heart surgery. After 3 months of follow-up, 13 patients relapsed to AF (45%) and the remaining 16 (55%) maintained SR.

A preoperative 12-lead ECG recording was acquired for each patient with a sampling rate of 1 kHz and an amplitude resolution of 0.4 μV. A 20 second-length segment from lead V1 was analyzed for each patient, because the atrial signal is larger in this recording. To improve later analysis, the ECG segment was first preprocessed using forward/backward highpass filtering (0.5 Hz cut-off frequency) to remove baseline wander, lowpass filtering (70 Hz cut-off frequency) to reduce high frequency noise and notch adaptive filtering at 50 Hz to remove powerline interference. Moreover, a wavelet denoising was also applied to reduce muscle noise [7]. Finally, the AA signal was extracted by making use of an adaptive QRST cancelation method [8].

B. Dominant Atrial Frequency

Previous studies have demonstrated that the atrial component during AF is typically an oscillation with a main frequency in the 3-9 Hz range [9]. Hence, the dominant atrial frequency (DAF) has been defined as the frequency with the highest amplitude within this range in the AA power spectrum, its inverse being directly related to the atrial cycle length [9]. In order to obtain this frequency, the Power Spectral Density (PSD) of the AA signal was computed using the Welch's overlapped segmented average method. A Hamming window of 4096 points in length, 50% overlapping between adjacent windowed sections and a 8192-points fast Fourier transform (FFT) were used as computational parameters.

C. Organization in the AA Signal

Pattern repetitiveness in the AA signal and, thus, its organization degree, has been successfully quantified making use of a nonlinear regularity index, such as sample entropy (SampEn) [10]. It examines time series for similar epochs and assigns a non-negative number to the sequence, with larger values corresponding to more irregularity in the data [11]. SampEn is defined as the negative natural logarithm of the conditional probability that two sequences similar for m points remain similar at the next point, where self-matches are not included in calculating the probability [11]. Given that SampEn is sensitive to noise and ventricular residua, it was computed over the main atrial wave (MAW) to reduce the influence of these nuisance signals [10]. The MAW was extracted by applying a selective filtering to the AA signal centered on the DAF. As in previous works, a linear phase Chebyshev FIR filter with a 3 Hz bandwidth and 40 decibels of relative sidelobe attenuation was used for this purpose [10].

D. Amplitude of the AA Signal

In previous works, the amplitude of the fibrillatory (f) waves characterizing the arrhythmia from the surface ECG has been estimated by their mean power (fWP) [12]. This index can be considered as a robust estimation of the energy carried by the AA signal within the interval under analysis. As a consequence, it can also be considered as a reliable indicator of the AA signal amplitude [12]. Thereby, the fWP was obtained by computing the root mean square value of the AA [12]. Before extracting the AA signal, each analyzed ECG segment was normalized to its maximum R peak amplitude. This operation avoided all the effects that can modify the ECG amplitude as a function of the different gain factors during recording, electrodes impedance, skin conductivity, etc [12]. In any case, this intra-patient normalization did not affect the computation of DAF and SampEn. In the first case, because the spectral distribution of a signal is independent of its amplitude. In the second, because the tolerance r was normalized to the standard deviation of the data.

E. Statistical Analysis

Statistical differences between patients in SR and AF after the blanking period were assessed with a Student's t-test for normal and homoscedastic variables and with a U Mann-Whitney test for not normal and homoscedastic ones. In both cases, a statistical significance (p) lower than 0.05 was considered as significant. To assess normality and homoscedasticity in the analyzed parameters, the Shapiro-Wilk and Levene tests were used, respectively.

On the other hand, the value of each parameter providing maximum discrimination between patient groups, that is, the optimum threshold, was obtained by means of a receiver operating characteristic (ROC) curve. It is a graphical representation of the trade-offs between sensitivity and specificity. Sensitivity (i.e., the true positive rate) was the proportion of patients maintaining SR 3 month after surgery correctly identified, whereas specificity (i.e., the true negative rate) represented the percentage of patients in AF precisely classified. Other related parameter is accuracy, which quantifies the total number of patients properly predicted. The value providing the highest accuracy was selected as optimum threshold.

Finally, a decision tree was performed to investigate non-monotonic relationships among parameters and, thus, group classification. The used stopping criterion for the tree growth was that each node contained only observations of one class or fewer than 5 observations. Moreover, the impurity-based Gini index was used to look for the best parameter and its threshold for the splitting of each node [13].

F. Nonlinear Analysis Applicability

As SampEn is a nonlinear metric, the nonlinearity in the AA signal was assessed through the surrogate data test to confirm the suitability of its use in the analysis. The test consists of obtaining a surrogate dataset from the original data. Then, a nonlinear metric that quantifies some aspect of the series, called discriminating statistic, has to be computed over the original and surrogate data. When the original series discriminating statistic is significantly different than the surrogate data values, nonlinearity can be assumed [14]. In the present work 40 surrogate data were generated for each AA signal and the analyzed statistic was SampEn. The Wilcoxon T test for paired data was used to evaluate statistical differences between original and surrogate data.

III. RESULTS

Statistically significant differences among original and surrogate data were noticed with the Wilcoxon T test. Therefore, nonlinearity in the AA signal can be assumed, thus confirming the suitability of its analysis with SampEn.

On the other hand, Table 1 shows sensitivity, specificity and accuracy values for each analyzed single parameter. As can be observed, SampEn and fWP presented higher sensitivity than the DAF. Nonetheless, the fWP showed the highest specificity and accuracy values. Although the discriminant ability reported for the three metrics was high, around or higher than 70%, only slight statistically significant differences between groups were observed for SampEn. The notable overlapping among the values obtained for DAF, SampEn and fWP from the two patient groups, shown in Figs. 1, 2 and 3, could explain this result.

Finally, the obtained decision tree is presented in Fig. 4. It is noteworthy that this algorithm outperformed the discriminant ability of each single parameter, reaching a sensitivity of 81.25%, a specificity of 84.62% and an accuracy of 82.76%. Hence, its accuracy outperformed by a 7% the results yielded by the best single predictor, i.e., the fWP.

IV. DISCUSSION AND CONCLUSSIONS

To the best of our knowledge, this work presents for the first time an attempt to predict the Cox-Maze surgery outcome after a 3 months blanking period of follow-up using only preoperative parameters extracted from the surface ECG. In view of the obtained results, the fWP seems to be the best single predictor. It was able to identify appropriately the patient's rhythm in more than 75% of the analyzed cases. Moreover, in agreement with previous works where a longer

Table 1 Classification results for DAF, SampEn, fWP together with the statistical significance obtained for each parameter.

	Sensitivity	Specificity	Accuracy	p
DAF	62.50%	76.92%	68.97%	0.115
SampEn	68.75%	76.92%	72.42%	0.048
fWP	68.75%	84.62%	75.86%	0.156

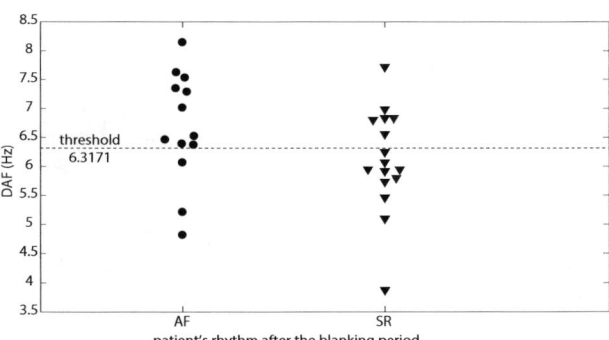

Fig. 1 Prediction of patient's rhythm into AF and SR after the Cox-Maze surgery blanking period obtained with the DAF.

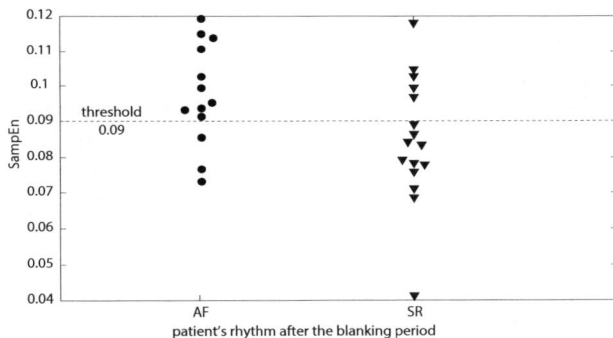

Fig. 2 Prognosis of patient's rhythm into AF and SR after the Cox-Maze surgery blanking period obtained with SampEn.

follow-up was considered [15], patients with postoperative AF showed lower fWP values than those in SR. Given that f waves amplitude has been previously associated with structural parameters of the atria [15], this result suggests that the higher the preoperative atrial structural remodeling, the higher the probability of AF recurrence after the surgery.

On the other hand, previous works have reported that DAF and SampEn can estimate qualitatively the number of propagating wavefronts throughout the atria [9, 10]. Therefore, both parameters could be considered as indirect estimators of AF organization and, thus, of the atrial electrical remodeling. From this point of view, results suggest a higher probability of AF recurrence after surgery when the preoperative

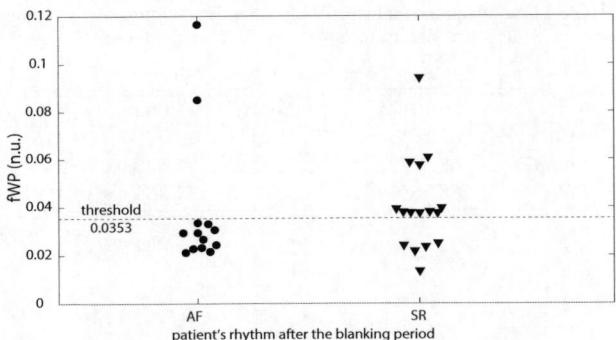

Fig. 3 Classification of patient's rhythm into AF and SR after the Cox-Maze surgery blanking period obtained with the fWP.

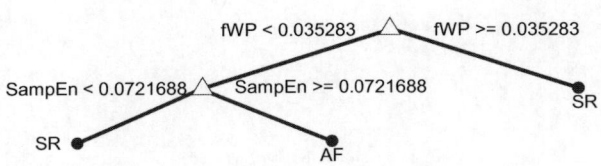

Fig. 4 Decision tree based on the impurity Gini index generated from the computed DAF, SampEn and fWP values.

electrical remodeling is in a more advanced state. In fact, higher DAF and SampEn values were appreciated for patients who maintained AF after the Cox-Maze procedure. Nonetheless, these indices reported lower discriminant ability between patient groups than the fWP. This finding could be justified by the fact that atrial lesions produced during surgery block reentrant wavefront paths [2], thus producing a notable impact on the atrial electrical remodeling.

It is also interesting to note that the analyzed parameters combination, via a decision tree, allowed to improve the patient's rhythm prognosis after surgery. Thus, the obtained results suggest that patients with higher f waves amplitude, and therefore, fewer structural remodeling are more likely to recover SR. On the other hand, in patients with small f waves amplitude, the electrical remodeling could play a key role in the Cox-Maze surgery outcome. Thus, small SampEn values elucidated SR maintenance, whereas patients in AF presented high entropy values. As a consequence, the information extracted from the preoperative ECG by means of parameters as the fWP, SampEn and DAF showed auspicious results and could be clinically relevant to plan in advance postoperative treatment of patients undergoing concomitant surgery of AF. Nonetheless, further analyses with a wider database are required to corroborate repeatability and robustness of the obtained results. In this way, more statistically significant differences between patient groups could also be yielded.

Acknowledgements

Work funded by TEC2010–20633 from the Spanish Ministry of Science and Innovation and PPII11–0194–8121 from Junta de Comunidades de Castilla-La Mancha.

References

1. Kannel W B, Abbott R D, Savage D D, McNamara P M. Epidemiologic features of chronic atrial fibrillation: The Framingham Study. *N Engl J Med*. 1982;306:1018–1022.
2. Ad N. The Cox-Maze procedure: history, results, and predictors for failure. *J Interv Card Electrophysiol*. 2007;20:65–71.
3. Harling L, Athanasiou T, Ashrafian H, Nowell J, Kourliouros A. Strategies in the surgical management of atrial fibrillation. *Cardiol Res Pract*. 2011;2011:439312.
4. Hornero F, Montero J A, Cánovas S, Bueno M. Biatrial radiofrequency ablation for atrial fibrillation: epicardial and endocardial surgical approach. *Interact Cardiovasc Thorac Surg*. 2002;1:72–77.
5. Wijffels M C, Kirchhof C J, Dorland R, Allessie M A. Atrial fibrillation begets atrial fibrillation. A study in awake chronically instrumented goats *Circulation*. 1995;92:1954-68.
6. Damiano R J, Schwartz F H, Bailey M S, et al. The Cox-Maze IV procedure: predictors of late recurrence. *J Thorac Cardiovasc Surg*. 2011;141:113–121.
7. Sörnmo L, Laguna P. *Bioelectrical signal processing in cardiac and neurological applications*. Amsterdam: Elsevier Academic Press 2005.
8. Alcaraz R, Rieta J J. Adaptive singular value cancelation of ventricular activity in single-lead atrial fibrillation electrocardiograms. *Physiol Meas*. 2008;29:1351–1369.
9. Holm M, Pehrson S, Ingemansson M, et al. Non-invasive assessment of the atrial cycle length during atrial fibrillation in man: introducing, validating and illustrating a new ECG method *Cardiovasc Res*. 1998;38:69-81.
10. Alcaraz R, Rieta J J. A review on sample entropy applications for the non-invasive analysis of atrial fibrillation electrocardiograms *Biomedical Signal Processing and Control*. 2010;5:1 - 14.
11. Richman J S, Moorman J R. Physiological time-series analysis using approximate entropy and sample entropy. *Am J Physiol Heart Circ Physiol*. 2000;278:H2039–H2049.
12. Alcaraz R, Rieta J J. Time and frequency recurrence analysis of persistent atrial fibrillation after electrical cardioversion. *Physiol Meas*. 2009;30:479–489.
13. Breiman L, Friedman J, Stone C J, Olshen R A. *Classification and Regression Trees*. Chapman & Hall/CRC 1984.
14. Palus M, Hoyer D. Detecting nonlinearity and phase synchronization with surrogate data. *IEEE Eng Med Biol Mag*. 1998;17:40–45.
15. Kamata J, Kawazoe K, Izumoto H, et al. Predictors of sinus rhythm restoration after Cox maze procedure concomitant with other cardiac operations. *Ann Thorac Surg*. 1997;64:394–398.

Corresponding Author: Antonio Hernández Alonso
Institute: Biomedical Synergy
Address: Universidad Politécnica de Valencia,
        Camino de Vera s/n. Building 7F, 5th Floor
City: Valencia
Country: Spain
Email: anheral@upv.es

A Decision Support System for Operating Theatre in Hospitals: Improving Safety, Efficiency and Clinical Continuity

R. Miniati[1], G. Cecconi[1], F. Frosini[1], F. Dori[1], G. Biffi Gentili[1], F. Petrucci[2], S. Franchi[2], and R. Gusinu[2]

[1] Information Engineering Department, School of Engineering, University of Florence, Florence, Italy
[2] AOU Careggi Teaching Hospital, Florence, Italy

Abstract—Surgical area is one of the most critical components within the hospital system as it represents the core activity in terms of health services' production, patients' infections and technology management.

Guaranteeing continuity of a safe and productive Operating Theatre (OT) is essential for hospitals. Post-surgery infections represent a serious issue in healthcare, where monitoring engineering parameters, related to the air system installation, is considered a proper indicator for the risk evaluation. Moreover, OT clinical activity must be continuously ensured not depending on technology failures and that activity should be planned for the highest usage in order to amortize and reduce the OT management costs (including fixed plus variable cost).

Hence, this paper aims to design a Decision Support System (DSS) for a sustainable web-based application which is able to support decision makers with fast and usable online reports on patient safety, productivity and technology dotation in OT.

Keywords—DSS, Operating Theatre, BCM, Patient Safety.

I. INTRODUCTION

The Operating Theatre (OT) is a key component to the general performance of health facilities as its correct management and usage is directly related to important aspects in healthcare such as patient safety, clinical continuity and cost reduction.

Given the high OT complexity, which merges different scientific areas (eg. engineers, physicians, physicists, biologists, analysts, etc.), there is real need for decision support systems which must organize and report few important and reliable information for helping decision makers to sort out situations such as safety, OT cost reduction and continuity. Most of the times these problems are strongly connected: for instance, the Italian health system estimates annual post-surgical infection rate of 5-10% on the whole hospital inpatients, with a subsequent increase of general costs of about 1 billion euro/year.

The developed OT-Decision Support System (OT-DSS) is composed of three modules which are as follows:

Safety – OT activity should be carried out by ensuring proper patient safety by complying minimum technical standards, related to technology and basic systems, for guaranteeing effective and safe performances, and protect patients from post-surgery infections [1]. For instance, a continuous monitoring system for OT is required to reduce the airborne infection risk [2] and properly define improvement plan and actions based on continuous analysis of technical parameters (eg. Air Change per Hour 'ACH', air pressure and velocity, filtration, etc.).

Clinical continuity - the technological research in the biomedical sector saw in the OT application the main promising area by becoming responsible for both higher clinical improvements and increase of systemic complexity. A technology failure can badly influence, or even interrupt, the surgical treatment of patients. It is therefore essential to medical facilities, guarantee, foresee and manage technological continuity as basic condition for clinical continuity [3]. This goal must be included as part of decisional support to hospitals' decision makers.

Cost reduction – technology progress is also responsible for recent higher acquisition and maintenance costs in OT. One way to reduce costs in health structures remains the increase of "clinical production" including modifications on number and type of interventions. Hence, a "productivity dashboard" regarding OT interventions divided per specialty and temporal scale can provide useful information for optimizing clinical activity and medical reimbursement, as in case of the DRG system (diagnosis related group)[4].

Finally, from a technical view, the decision system was designed by following these requirements: usability, technically easy to improve and update, applicability to any hospital or medical facility, and large accessibility to decision makers, not depending on geographic position (at office, home or train) nor to specific PC/Tablet and operating system.

Fig. 1 OT-DSS log-in interface.

II. DESIGNING THE OT-DSS

The OT-DSS can be seen as a feedback system, where information regarding the three modules (safety, continuity and cost reduction) are entered, and the system outputs useful information for decisional support. As reported in figure 2, the structure of the system is composed of three action/levels which are differently defined according to the specific aspect to be considered:

- Key Performance Indicator (KPI) definition and evaluation;
- Data collection;
- Information reporting.

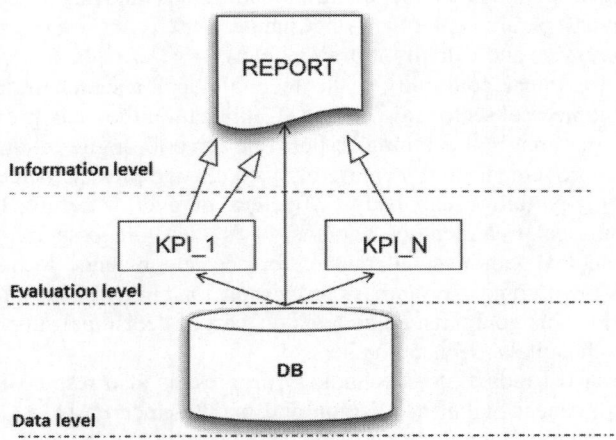

Fig. 2 OT-DSS structure [3].

A more detailed description of the system, according to the specific module (safety, continuity and cost reduction), will be reported as follows.

A. Patient Safety

KPI definition - Technical parameters belonging to the OT systems (air, electric and plumbing system plus technology) were taken into consideration as indirect KPI for monitoring patient safety, especially regarding OT infections. They include safety tests, maintenance verification and performance indicators as indicated in the current national and international state of art. The main areas of interest are: risk evaluation process, environmental monitoring of anesthetic gases and microclimatic conditions, airborne particles, ACH, microbiology controls, technology safety and performance tests planning (which also includes preventive maintenance), planning of safety and maintenance test for medical gas and electric system, noise, lightning and plumbing quality control.

Data collection - Data can be periodically collected through interviews, specific requests or directly connecting to other hospital DB.

Information reporting - The last step of the methodology deals with the development of the reporting system through the evaluation of KPI and information which are considered useful in providing decisional support to decision makers and developing the graphic forms of the reports themselves. In general the information report three main points: current situation, temporal trend analysis and scenario simulations according to specific corrective actions.

As reported in figure 3, the reporting system follows a user oriented-approach by reporting only the essential information according to the specific type of user: all the info can be accessible to the highest level (directorate), only the ones regarding the specific department at the 2nd level and only the specific information on the single surgical room at the ward/room level.

Moreover, on the server side of the system, data entry exclusively involves the ward/room level while maintenance/update actions are carried out to every level of the system.

When the parameters listed above are out of the valid range (obtained by norms/guidelines) in a specific room, this warning/alarm affects the ward, then the correspondent department and finally the whole hospital.

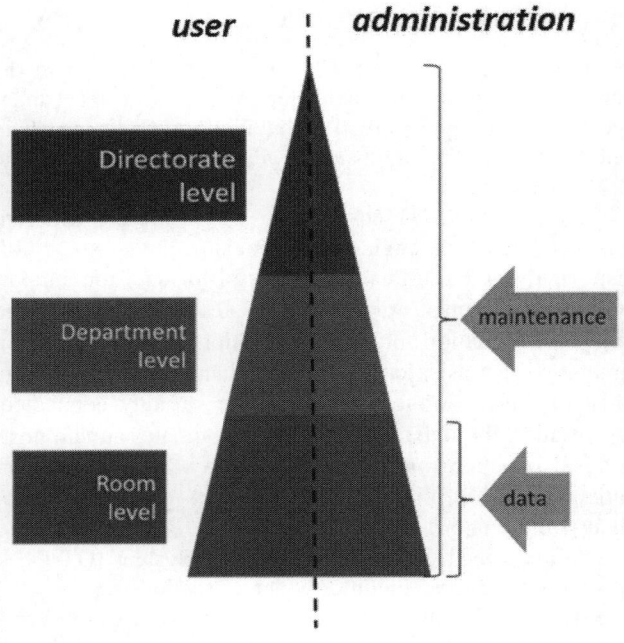

Fig. 3 OT-DSS structure.

B. Clinical Continuity

KPI definition - Both technical aspects and clinical area of use of medical equipment were considered. Technical aspects include technological renewal and technical maintenance, while for the area of use, KPI on technology safety are integrated with the ones including OT process analysis, intrinsic criticality [5], usability and user-safety [6-10].

Data collection – This includes data from the technical DB (equipment position within the hospital and maintenance status) and other data obtained from other hospital databases (OT specialty, urgency/emergency availability, etc) or from interviews and inspections on the field (back-up device availability, device importance for the completion of clinical activities, etc).

Information reporting – The system keeps the general features presented in previous module regarding the patient safety module, with the addition of the "substitution device searching tool", see figure 4.

Fig. 4 OT-BCM device searching tool.

The system is designed for providing operative support in real time on searching for an alternative device to the one which is malfunctioning [3].

The interface enables to search for devices by the following two criteria: the nearest device or the specific model (nearest); while the output lists those devices satisfying the search criteria inserted by the healthcare operator and providing further info/actions: availability, name, brand, hospital area, booking option and download of the operational withdrawn forms.

Once the device is used, in terms of returning to the original hospital area, the notification system works by "negative logic": the operator to whom the device is not returned notifies the system and the system then, by one or more message systems (e-mail, sms, etc), sends out the invite to return the device (to the operator who was using it)[3]. For safety reason, the search algorithm needs at least two devices to make one available.

C. Cost Reduction

KPI definition – Clinical and organizational aspects coming from the OT activity are taken into account. As reported in figure 5, they include all the different phases in the OT process, from the surgical room being "ready-to-operate" to the patient returning to the specific ward or to the recovery room.

Aim of the system is supporting decision makers in reducing the lost times between the different phases as they cost to the medical facility the same money as they cost when surgical room is actively working.

Data collection – This includes data from the discharge system and from the OT check-list and management software.

Information reporting – Most of the information reports activity efficiency according to the specific OT within the hospital or according to the specific surgical procedure (eg. laparoscopy versus laparotomy versus robotic).

Also in this case, the overall organization of the reporting system follows the structure reported in figure 3.

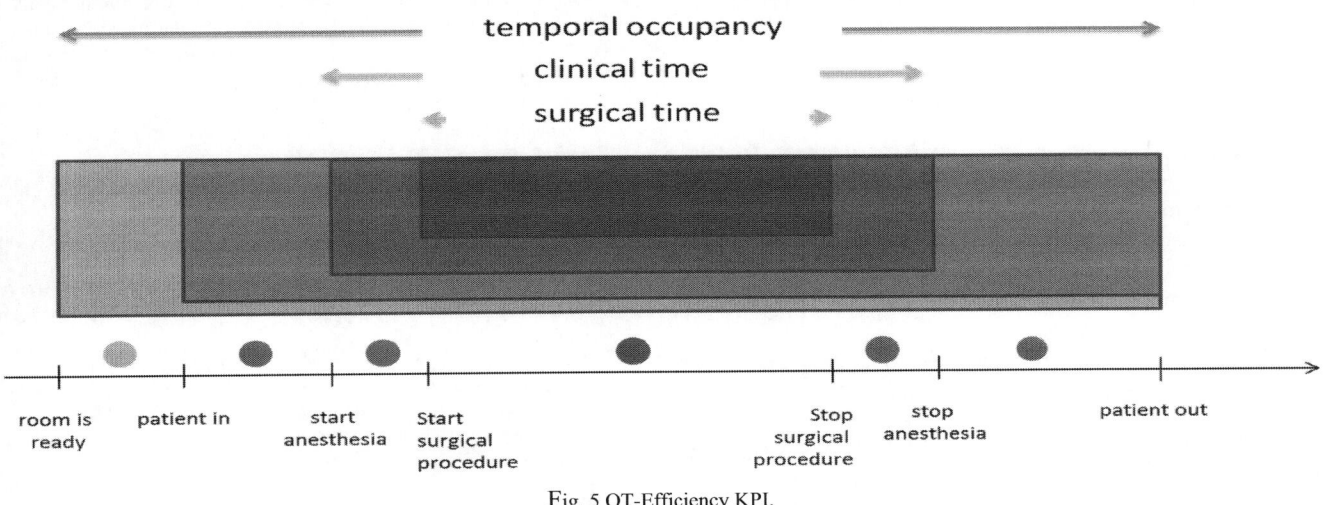

Fig. 5 OT-Efficiency KPI.

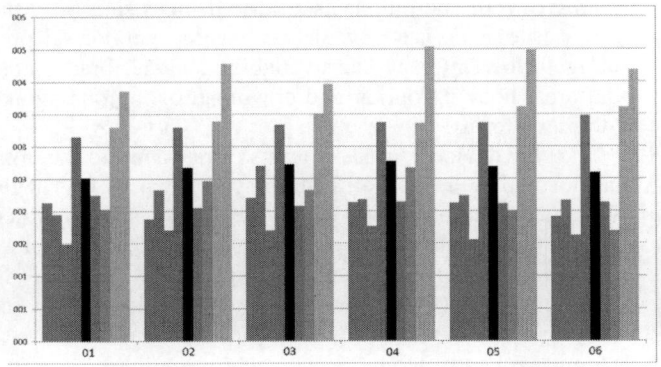
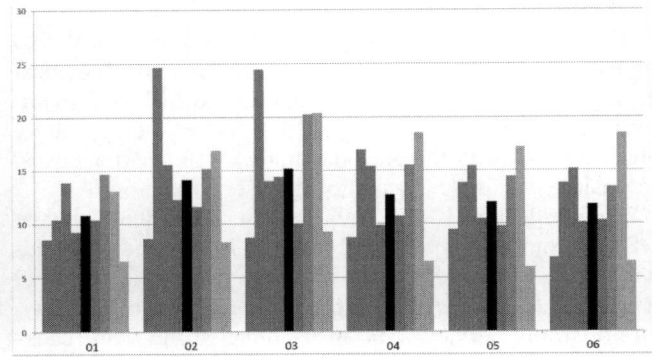

Fig. 6 Simulation results: (a) number of interventions per OT and per month; (b) Patient in – Room ready delay (minutes) per OT and per month.

III. RESULTS

From a simulation of a hospital with OT activity comparable to the AOU Careggi, the following OT specialties were analyzed by considering a total of around 30 surgery rooms: orthopaedics, neurology, general, thoracic, urology, maternity delivery and gynecology, ophthalmology and laryngology.

Example of supporting reports are given in figure 6 concerning, as representative example, the efficiency analysis through the graphic representation of monthly usage rate (number of interventions per OT, per month) and monthly lost time rate coming from organizational inefficiencies and estimated as total minutes counted between the phase "patient-in" and "room ready" per OT, per month. Many differences were found between different OTs (corresponding to different columns in the figure) and different specialties/rooms within the same OT.

Similar representations and report organization are used for the module "OT-safety" as well, while for the module business continuity the searching tool provided with useful and reliable support, see [3] for more details.

IV. CONCLUSIONS

The designed system is beneficial for decision makers since it provides useful information for patient safety, continuity implementation and efficiency of OT in hospitals. Applications with simulated scenarios showed a capacity to solve 100% of the problems related to the continuity during the night and public holidays, and 80% during regular hours (Monday to Friday, 8am to 5pm).

For patient safety module, the OT-DSS outputs could be the base for defining the priority of technical interventions within the OT structures.

Higher efficiency in OT activity is a key point for cost reduction in hospitals. The OT-DSS simulations can be essential for an efficiency-based OT-planning through the individuation of the most critical points in the whole surgical process as opportunities spots where, if possible, applying corrective actions for mitigating the "lost times" and increasing the OT-usage rate.

REFERENCES

1. H. Humphreys a, b, *, J.E. Coia c, A. Stacey d, M. Thomas e, A.-M. Belli f, P. Hoffman g, P. Jenks h, C.A. Mackintosh, "Guidelines on the facilities required for minor surgical procedures and minimal access interventions". Journal of Hospital Infection, 2012.
2. T.T. Chow, X.Y. Yang "Ventilation performance in operating theatres against airborne infection: review of research activities and practical guidance", Journal of Hospital Infection, 2004Smith J, Jones M Jr, Houghton L et al. (1999) Future of health insurance. N Engl J Med 965:325–329
3. Roberto Miniati, Fabrizio Dori, Giulio Cecconi, Roberto Gusinu, Fabrizio Niccolini and Guido Biffi Gentili. HTA decision support system for sustainable business continuity management in hospitals. The case of surgical activity at the University Hospital in Florence. Technology and Health Care 21 (2013) 49–61.
4. Diagnosis Related Groups (DRGs) and the Medicare Program – Implications for Medical Technology. A 1983 document found in the "CyberCemetery: OTA Legacy" section of University of North Texas Libraries Government Documents department.
5. Yadin David, Thomas M. Judd. Management and Assessment of Medical Technology In: The Biomedical Engineering Handbook, Joseph D. Bronzino. 2005; 2507-2516.
6. European Norm EN 62366:2008-01. Medical devices – Application of usability engineering to medical devices.
7. Gosbee J. Human factors engineering and patient safety. Qual. Saf. Health Care. 2002; 11: 352-354.
8. EN 60601-1:2006-10. Medical electrical equipment – Part 1: General requirements for basic safety and essential performance. European Standard.

Author: Roberto Miniati
Institute: Information Engineering Dept., University of Florence.
Street: Via Santa Marta 3, 50139
City: Florence
Country: Italy
Email: roberto.miniati@unifi.it

LICENSE: Web Application for Monitoring and Controlling Hospitals' Status with Respect to Legislative Standards

E. Iadanza[1], L. Ottaviani[1], G. Guidi[1], A. Luschi[1], and F. Terzaghi[2]

[1] Department of Information Engineering, University of Florence, Florence, Italy
[2] Technical Department, Azienda Ospedaliero Universitaria Careggi, Florence, Italy

Abstract—This article describes LICENSE, an expert system that is able to automatically verify the compliance with legislative requirements, both structural and technological. These requirements have to be observed by a health facility, public or private, in order to be accredited and to work on behalf of the Regional Healthcare Service. The system consists of a web application that can be used by the hospital top management as well as by the clinical engineer or by the quality management office, so as to evaluate and monitor the hospital's status. The evaluation is performed on the basis of the intended use of the premises, the surface, the number of beds or technical beds, and the class of the biomedical technologies available. The software's structure has been made adaptable to the legislative changes through the creation of an associated database.

Keywords—License, qualify, expert system, monitoring.

I. INTRODUCTION

The current regulations establish that a facility, both public and private, must have certain structural, plant, organizational and technological requirements, in order to deliver health care services. The principle is the need, for public and private entities interested in performing on account of the Regional Healthcare Service, to define a compulsory regulatory instrument: at the national and regional level, the necessary requirements and standards are defined, the level of adherence to them is verified and, according to this, the business and the possibility to access to public financing is authorized. Each requirement section has associated checklists of self-assessment with which the structure declares the possession of the required standards.

The developed software, called LICENSE, provides an automatic estimation of the structural and technological requirements, by establishing which standard is satisfied and which is not and leaving to the user the final decision regarding the compliance to the requirements. Thus, it computes the percentage of fulfilled standards and stores the value as accreditation coefficient. So, the software is an expert system that supplies a decisional support to the evaluation of the status of a complex building, as the hospital, but it can be useful also for the management of those processes for which the clinical engineering is responsible, such as: the management of requirements' inadequacy situations, the scheduling of an order of priority for the interventions, and the risk analysis in order to reduce, with the available resources, the major risks.

II. MATERIALS AND METHODS

LICENSE has been developed using Microsoft Visual Studio 2008 Professional Edition. The employed technology is ASP.NET and the programming language is VB. SQL Server 2008 has been used as DBMS, while the report implementation worked out through Crystal Reports 9.2.

III. DISCUSSIONS

The proposed system can be applied to every healthcare facility that makes use of IT for the gathering and storage of the structure and equipments' data; for the purpose of our analysis, it has been applied to the hospital of Careggi, in Florence, Italy.

The standards, relating to the structural, organizational, technological and plant conditions, are established, in Tuscany, by the regional law n. 51/2009 and a series of implementing provisions (regulation n.61/2010 and subsequent changes and integrations). The operating requirements, articulated into sections, are divided into:

• general: they must be possessed by every type of building and are the same for everyone;
• specific: they are related to the type of facility (that can be ambulatory, hospitalization for severe illness or post-acute phase).

The acknowledgment of standards' achievement has to be presented by every structure, that must demonstrate to have at least the 100% of satisfied requirements. [1]

The general requirements must be owned by every type of building, while the specific ones differ depending on the performance delivered (and so, they refer to the activity undertaken in the facility). [2] [3] [4] [5]

Regarding the structural and plant equipment, the hospital of Careggi is currently using a software called SACS. SACS is a custom software that drives Autocad. It provides a mapping of every single hospital room in terms of beds, square meters, intended use, functional area and many other details. Each room is considered as a part of a ward introducing the concept of Main Destination of Use (MDU). System reports can be used via web by the hospital top management as well as by the clinical engineer, who can assess parameters like the maintenance status of an equipment or its obsolescence, being aware of the hospital context in which this equipment has been used. [6]

For the classification and inventory of health technologies, the information system utilized by Careggi uses a custom software which resorts to a database created through an Oracle platform. The information contained in the database concerns the identification and sorting of the equipment, its age and functionality, the hospital location, the area of activity to which is assigned and several other details. Therefore, the system will interface with the other abovementioned systems.

In order to make a software the most flexible as possible to any regulation changes (or to make it available also in other regions, in which there are different requirements) and adaptable to different companies, LICENSE is equipped with its own database. The database is realized from a process of standardization of legislative requirements, through the identification of objective parameters useful to make a comparison with the parameters available in the hospital's IT systems. In this way, in case of changes, the building and the software operation will remain unchanged, because only the database population will change.

In the matter of structural standards, the identified parameters for the comparison are the intended use of the premises, its surface, the number of beds or technical beds, and the presence of certain rooms that must exist in every floor, if the activity is undertaken on more than one floor.

In terms of technological requirements, the assessment of the compliance to legislative regulations is based on the presence of a particular technological resource. This check is carried out searching for a tool (among those available and assigned to the analyzed activity) through the code of the CIVAB class. The CIVAB encoding is a national (Italian) system that allows unique identification of biomedical technologies through a code of eight alphanumeric characters, through which it can be individuated:

- the class of technology (first three characters) (e.g.: ECT= Ultrasound Tomograph);
- the manufacturer (second tern of characters);
- the specific model of technology of that class and that manufacturer (last two characters).

Hence, the research is based on the first tern of characters.

An example of how a requirement is standardized in the database can be made. The requirement S.02 of the list C1.3 (hospitalization area), has an ID equal to 952. It demands the presence of a minimum surface for the multiple hospitalization room of 9 m² for the first bed and of 7 m² for each further bed. This requirement refers to the hospitalization rooms, which have intended use (column DU) with ID equal to 18. Since the requirement is referred to the multiple hospitalization rooms, then it will be also present the entry condition regarding the beds (column COND_PL), that will obviously be ">1". The equation will evaluate if the hospitalization room's size is 9 m² for the first bed and 7 m² for the remaining ones (column ESPRESSIONE). Note that [TB_QUERY.PPLL] in the equation will be substituted in a dynamical way by the system, that will obtain, for every room, the number of available beds and will utilize it for the computation. The value of the column GRUPPO will be True (1), since the requirement has to be verified for each multiple hospitalization room associated to that activity area.

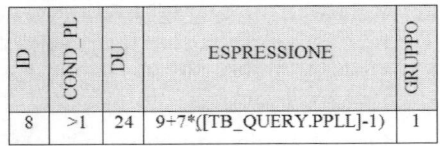

ID	COND_PL	DU	ESPRESSIONE	GRUPPO
8	>1	24	9+7*([TB_QUERY.PPLL]-1)	1

Fig. 1 Standardized requirement in the database

The software will extrapolate the information regarding the environments and the technologies assigned to the activity in analysis and compare them with the legislative standards (formalized in the database) expected for that task; then, it will establish which requirements are satisfied and which are not, providing a percentage estimation on the total number of questions. The system has been realized as a web application so as to meet the needs of heterogeneous utilities, such as the clinical engineering services, the management, the technical department and the quality management office, each with different needs and purposes.

In the following figure, it is shown the software use cases' diagram, which describes the functions or services offered by a system, from the user's point of view.

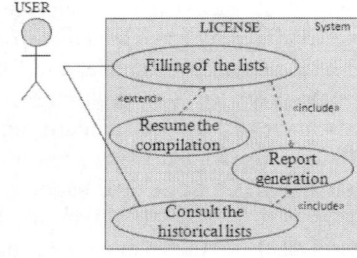

Fig. 2 Use cases diagram

On the first page, there is a TreeView containing the list of Careggi facilities and of the activities made inside each one of them, on each floor. The user, expanding the node relative to the structure he wants to analyze, will be provided with the index of the hospital floors, with the list of the activities carried out within every facility in that floor. On the second page, the user can choose to consult the historical lists compiled, or to proceed with the filling of the lists associated with that activity. It is possible to access to this page directly from SACS, through the combination with every single polyline and hyper-textual link. In this case, the data regarding the building and the selected activity (the unique identifiers) will be recovered through the query-string included in the URL.

If the user chooses the first option, it will be shown a GridView containing the list of the compilations saved in chronological order, from the most recent to the oldest.

If, on the other hand, the user's choice is the second one, it will be displayed, through a TreeView, the register of the lists related to the selected activity's area. Once the user has chosen the list, the system checks if a saved compilation for the same list already exists or if it is a first compilation; in the first case, it asks the user if he wants to resume the compilation from the previously saved version or start a new compilation from the beginning. The system selects the rooms and technologies assigned to the chosen activity and the requirements associated to the desired list, and, for each automatically analyzable standard, recalls the necessary functions to compare (in terms of intended use, surface, number of beds, technologies and other identified parameters) the available data and the database stored ones. At the end of this process, it shows the final page of the software (Fig. 2).

On the page, there is a table including the different requirements that make up the list of exercise in analysis; there are reported the chapter and the requirement code (which uniquely identify the list's question), the question text, a column reporting the automatic responses of the system and a column for a note's insertion.

The "system's responses" column contains, for every standard, a checkbox corresponding to those questions that can be evaluated automatically, while it is empty in correspondence of the questions that must necessarily be answered by the user. The checkboxes are "checked" if the requirement is satisfied, whilst they are "unchecked" if the standard is not fulfilled (indeed, the system can only provide an answer of the yes/no type). The traffic light icon next to the column heading shows if the number of requirements positively and automatically evaluated reaches the 100%. in the positive case, the traffic light will have the green light (the level of legislative adherence is respected), otherwise the light will be red. Notice that this first assessment only makes a consideration of the system responses, which could a priori not represent the 100% of the requirements included in the list in analysis.

In the "requirement" column, it is reported the standard's text joined to a widget that could be a checkbox, a dropdown list, a label or a radiobutton according to the answers that might be expected from the user.

The choice of the widget to load is performed by reading a database field of LICENSE. The control is loaded with a default value equal to that of the response of the system if present, conversely set to NO. The answers that the user can give are YES, NO, P (i.e. "action plans" to be expected in the case of deviations from the requirements) and, in some cases, NP (not pertinent). If the user modifies the answer suggested by the system, then this dissimilarity will be highlighted, placing the text of the requirement in red and in bold, and the auto-reply checkbox will be marked with an asterisk.

The "note" column shows an icon that can be enabled if the user's answer is P (a note is expected only for the insertion of intervention plans). If the icon is enabled and the user clicks on it, he/she will see a JavaScript editor that replaces the traditional text box, providing functions such as formatting and alignment of the text. In the system it has been integrated TinyMCE, one of the most common editor JavaScript. It is a free product, open source licensed under LGPL by Moxiecode Systems AB. This editor has the ability to convert HTML TEXTAREA fields or other HTML elements to editor instances, and it is a WYSIWYG editor that resembles that of the toolbars of Microsoft Word. For data saving, TinyMCE uses the HTML format: everything that is written in the text-box, the objects and formatting applied are made through HTML tags.

Finally, on the left of the table, is a div in the "fixed" position (always visible where it was placed, ignoring any movements of the scrollbar) that contains an image that represents the percentage level of compliance with the standards: if less than 100%, the image shown will be that of a "rejected" list, while, if greater than 100%, the image will be that of an "approved" list. The image will update automatically every time the user changes a response; so, it may vary during the compilation.

At the end of the page are always visible the two buttons "save" and "report". The save button is always enabled and, when pressed, saves the automatic responses of the system and those of the compiler user in the LICENSE database. If the saving is carried out correctly, the user is presented with a window box where he/she is asked if he wants to enter the compilation just made in the historic, closing it permanently (no longer editable).

The button for the report generation is instead enabled only if there is at least a compilation saved. Thus, if the user is compiling a list for the first time, he/she must, before choosing how to generate reports, perform a saving. The user can choose in which format, among those available (.xls and .pdf), the report will be displayed. The PDF report consists of a summary of the general information (list of self-assessment, building, business, accreditation coefficient), a pie graph for the percentage representation of the requirements to be satisfied and a table containing all the requirements, their responses and any notes formatted. The XLS report, instead, has the same structure as the self-assessment lists provided by the region.

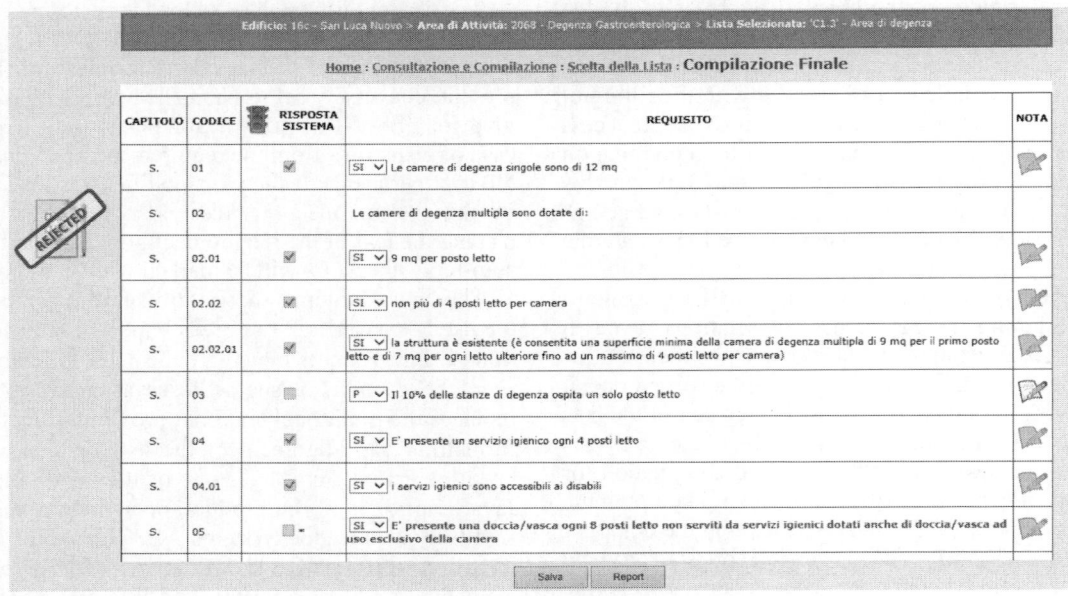

Fig. 3 Final page of LICENSE

IV. CONCLUSIONS AND FUTURE WORKS

The developed software, LICENSE, tries to meet as fully as possible the need for a simple and effective tool for the evaluation and monitoring of the hospital's business data. Given the nature of the software and its scope, it is not possible to think of a "final" version, because over time the standards required may change. For this reason, the basic structure of the program was made easily adaptable to new legislative changes, by modifying the associated database.

The requirements were satisfied during testing phase alpha. The software is in its beta version; it will be up to users, with the use of the application, to assess its real clarity and efficiency, and possibly open the door for future developments both in terms of improvement of the existing functions and in terms of implementation of new functions.

LIST OF ABBREVIATIONS

SACS the Italian acronym for "system for the analysis of the hospital equipment"
VB (Microsoft) Visual Basic
DBMS Database Management System
URL Uniform Resource Locator
WYSIWYG acronym for "what you see is what you get"
HTML HyperText Markup Language

REFERENCES

1. Regional law n. 51/2009 (Tuscany)
2. DPGR n.61/2010 (Tuscany)
3. DGR n. 1623/2011 (Tuscany)
4. DGR n. 1409/2011 (Tuscany)
5. DGR n. 2128/2012 (Tuscany)
6. Iadanza E (2009) An Unconventional Approach to Healthcare (Geographic) Information Systems using a Custom VB Interface to AutoCAD. International workshop on Mobilizing health information to support healthcare-related knowledge work – MobiHealthInf 2009
7. Iadanza E, Dori F, Biffi Gentili G (2006) "The role of bioengineer in hospital upkeep and development". IFMBE Proceedings Aug, Vol. 14, 2006, Seoul, Korea. ISBN: 3 540 36839 6
8. Iadanza E, Marzi L, Dori F, Biffi Gentili G, Torricelli MC (2006) Hospital health care offer. A monitoring multidisciplinar approach. IFMBE Proceedings WC 2006 "World Congress on Medical Physics and Biomedical Engineering", Vol. 14, 2006, Seoul, Korea. ISBN: 3 540 36839 6

Author: Ernesto Iadanza
Institute: Department of Information Engineering, Florence
Street: via di Santa Marta, 3
City: Florence
Country: Italy
Email: ernesto.iadanza@unifi.it

Relation among Breathing Pattern, Sleep Posture and BMI during Sleep Detected by Body Motion Wave

Jun`ya Wada[1], Tadashi Yajima[2], Takenori Imamatsu[3], and Hiroaki Okawai[4]

[1] Department of Civil and Environmental Engineering, Iwate University, Morioka, Japan
t2512010@iwate-u.ac.jp
[2] Department of Civil and Environmental Engineering, Iwate University, Morioka, Japan
[3] Department of Civil and Environmental Engineering, Iwate University, Morioka, Japan
[4] Department of Civil and Environmental Engineering, Iwate University, Morioka, Japan
hokawai@iwate-u.ac.jp

Abstract—Sleep is an important theme in order to know and keep health. In this study, we investigated the relationship between breathing pattern and sleep posture by using the dynamic air-pressure sensor and a pressure sensor array. In addition, the value of BMI was considered to describe the physical characteristics of the subjects. As a result, we found breathing patterns and sleep postures associated with the physical characteristics such as BMI, and further study would suggest a new index of biological information.

Keywords—sleep, Body Motion Wave (BMW), breathing pattern, sleep posture, Body Mass Index (BMI),

I. INTRODUCTION

Recently, health problems due to overwork, apparent suicide and lifestyle diseases produced by the progress of the social stress have frequently been reported. From this, it has become of interest to avoid these problems and to keep the health.

Sleep would be one of the significant themes to study how to check and to promote health precisely. During sleep, the information of physical condition would appear because of predominant activity of autonomic nerve system under the state of unconsciousness. In this point of view, the polysomnography has traditionally been adopted to detect biological information and promoted sleep study. However, it requires high cost because only qualified staff in medical or research facilities can take such a data [1].

In late years, we have studied sleep in daily living by adopting a pressure sensor named "dynamic air-pressure sensor" and a pressure sensor array [2] [3]. Thus, the biological information such as rates of pulse and respiration were calculated from pressure wave named "Body Motion Wave" detected by the above dynamic air-pressure sensor, and sleep posture and distribution of body pressure were obtained by a pressure sensor array. In our previous study, we have found some characteristics in the wave pattern of respiration, percentage of duration of each body posture through the night and the effects of body posture changes on the rates of pulse and respiration. However, relations between the two in detailed time zone have not yet found. Now, at the present study, we focused on respiratory information detected by the above sensors and unconscious behavior. In addition, we included the value of BMI (Body Mass Index) in order to consider the physical characteristics of a subject.

II. METHOD

A. Measurement System

Here, two types of pressure sensor will be described.

At first, the dynamic air-pressure sensor (M.I.Labs) was adopted to detect information of unconscious physical motions during sleep. The motions obtained here have biological information such as pulse and respiration, and so on. A pressure sensor, fabricating a non-restraint and non-attachment measurement system, can detect the above motions through the night without any psychological or physiological stress. As shown in Fig.1, the sensor was set between the bed and the subject's body lying on a bed to measure air-pressure arisen from subject's body. The pressure variation thus obtained was converted to electric signals, "Body Motion Wave" as mentioned later in detail, sampled at the rate 400 Hz, 16 bit and stored in a personal computer. Signal was processed with Chart ver5.0 (AD Instrument).

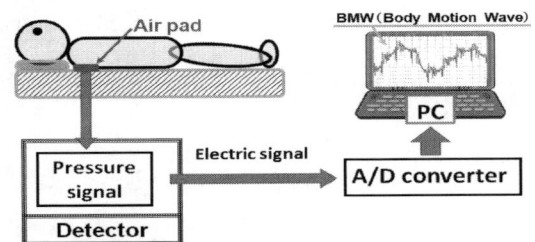

Fig. 1 Measurement system for the dynamic air-pressure sensor

Secondly, the pressure sensor array, pressure distribution measurement system BIG-MAT (Nitta Co., Ltd.) was adopted in order to determine the changes of sleep postures and distribution of body pressure through the night and also fabricate a non-restraint and non-attachment measurement system. As shown in Fig.2, it was set under the bed. This sensor array sheet has 2288 sensor cells to detect and describe a distribution of body pressure in two dimensions. The signals thus obtained were sampled at the rate 1 Hz, 16 bit and stored in a personal computer.

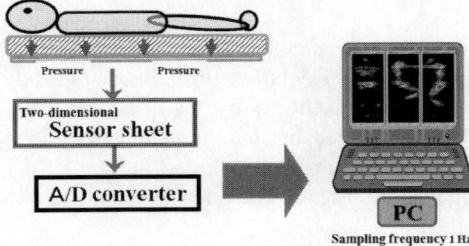

Fig. 2 Measurement system for the pressure sensor array

B. Body Motion Wave

As previously mentioned, Body Motion Wave (BMW) is a pressure wave detected as a result of the physical motions associated with the unconscious behavior, such as pulse and respiration, emanating from the subject's body during sleep by the dynamic air-pressure sensor. Thus, the pressure waves were neither electrocardiogram origin nor respiration gas origin. Therefore, we named this wave "Body Motion Wave, BMW", as reported by Okawai et al [3].

An example of BMW was shown in Fig.3. The top wave, original wave, was filtered into two waves as traces in the lower two. The one is "respiration-origin BMW, R-BMW" and the other is "pulse-origin BMW, P-BMW". Waveform with a period of approximately five sec. indicates the breathing. Ascending and falling phases show inspiratory phase and expiratory phase, respectively. In addition, the wave lasting for approximately one sec. indicates the pulse. In previous study, these waveforms were consistent with that of ECG and respiratory belt transducer [4].

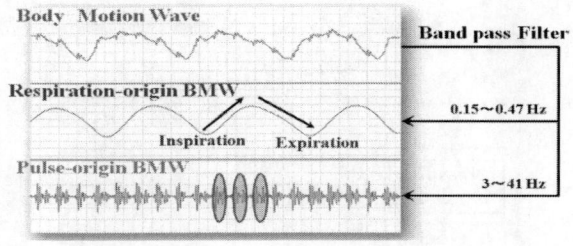

Fig. 3 Body Motion Wave (BMW) can be filtered into components of respiration-origin and pulse-origin BMWs.

C. Calculation of the Rates of Pulse and Respiration during Sleep

To calculate the rates of pulse and respiration, we made a program by using a program software VEE Pro ver6.0 (Agilent Technologies). Thus, R-BMW and P-BMW were basically processed respectively as follows; (1) detection of peak position, as shown in Fig.4, (2)peak interval detection, (3)peak interval measurement, and (4)Calculation of the average per minute of each peak interval. An example of the respiration rate thus calculated was shown in Fig.5.

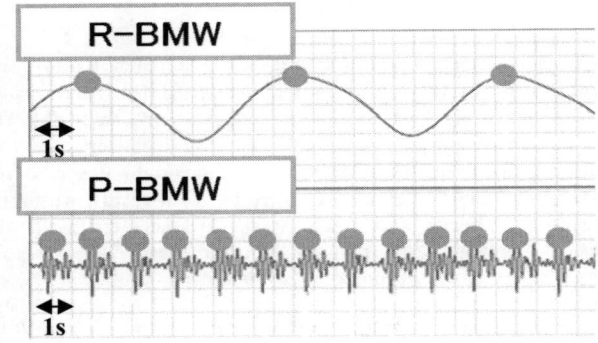

Fig. 4 Detected peak positions for R-BMW and P-BMW

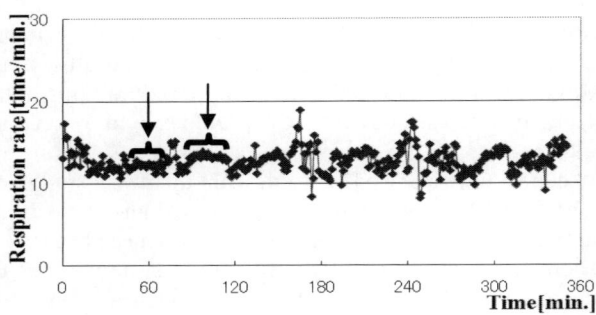

Fig. 5 An example of the respiration rate through the night

D. Classification of Breathing Pattern

To characterize R-BMW detected by the dynamic air-pressure sensor, we classified breathing patterns into seven types; ①Convex upward, ②Convex downward, as

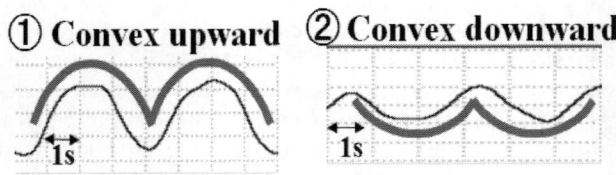

Fig. 6 Examples of breathing pattern

described in Fig. 6, ③Sin, ④Long inspiration phase, ⑤Long expiration phase, ⑥Complex(Composite of ① ~ ⑤), and ⑦ Indistinguishable.

E. Classification of Sleep Posture

The sleep postures obtained by the body pressure distribution measurement system were classified into six types as shown in Fig.7; (1)Supine, (2) Right Lateral, (3) Left Lateral, (4) Prone, (5) Right Sims, and (6)Left Sims.

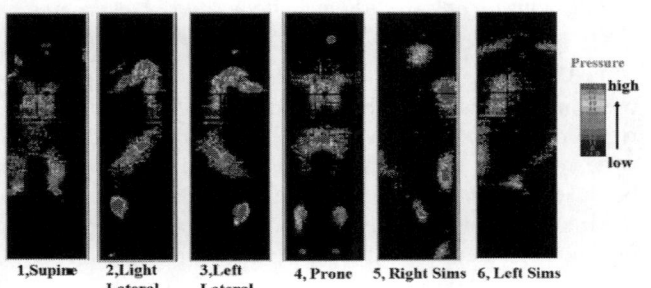

Fig. 7 Six patterns of sleep posture

F. Selection of the Time Zone of Interest

The rates of pulse and respiration obtained by the procedure illustrated in II-C were made up of three parts roughly; "stable", "rising" and "falling". In this study, we adopted time zone for five minutes that the transitions were stable by visual inspection. As shown in Fig 8, "stable" in the time zone means that the variation in the rates of pulse and respiration was much smaller than others. Totally, we selected 56 time zones.

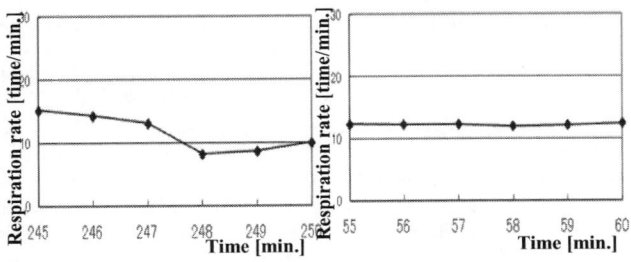

Fig. 8 An example of non-stable rate (left) and stable rate (right) of respiration

III. EXPERIMENT

Eighteen healthy young adults (20`s male; 15, 20`s female; 3) participated in this study. A subject stayed at Health care experiment house in Iwate University for four nights. We performed sleep experiments by using two types of pressure sensor as described in II-A simultaneously. We classified subjects into two groups by the value of BMI; group A (BMI<25.0) and group B (BMI>25.0). For bedding material, we adopted ones of mainly made of polyester and cotton on sale. An air pad of the dynamic air-pressure sensor was placed under the scapula.

IV. RESULT

A. Breathing Pattern during the Stable Rates of Pulse and Respiration

The time percentage for each breathing pattern as shown in II-D in 56 time zones of 18 subjects during the stable rates of pulse and respiration as described in II-F was shown in Fig.9. We visually confirmed the breathing pattern in the time zone as described in the II-F. As a result, it was noted that the percentage of two types, ⑤Long expiratory phase (73%) and ③Sin (19%), was greater than the others.

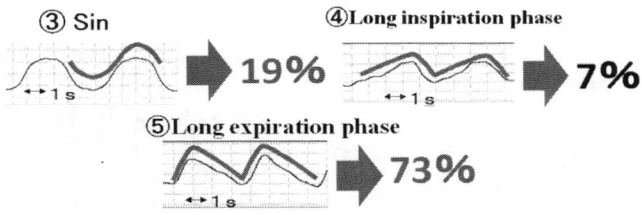

Fig. 9 Percentages of each breathing pattern

B. Characteristic of Breathing Pattern by BMI

We paid attention to type ⑤Long expiratory phase because of greater proportion in the IV-A to characterize breathing pattern. To take into account the physical characteristics, male subjects were divided into two groups depending on BMI. As a result, eleven male subjects corresponded: seven of group A and four of group B. Waveforms in Fig.10 show characteristics of group A and group B. Here, in all breathing patterns through five minutes, we defined a waveform in the appearing proportion at more than 50% as a characteristic waveform for that group. From this, in the waveform for group B, we got a feature that waveform would be flat at the middle in expiratory phase as shown in Fig.10. However, the proportion of that waveform was lower in group A.

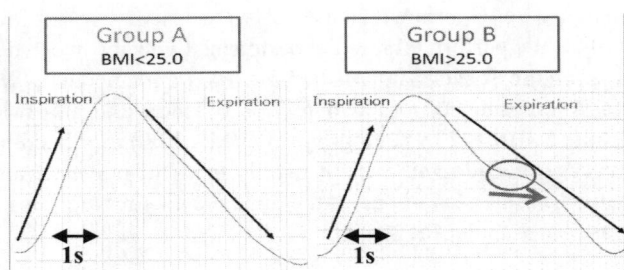

Fig. 10 Characteristics of breathing pattern for group A (left) and group B (right).

C. Relationship between Sleep Posture and BMI

The time percentage of each body posture as mentioned in II-E in 56 time zones of 18 subjects during the stable rates of pulse and respiration as described in II-F was shown in Fig.11. As a result, it was noted that the percentages of Supine (35%) and Lateral of Right and Left (31%) were greater than the others. In addition, we compared these sleep positions among the groups of BMI as well as IV-B for fifteen male subjects; ten of group A (BMI <25.0) and five of group B (BMI> 25.0). As a result, the group A had a large percentage of the Supine, while the group B had a large percentage of the lateral of Right and Left.

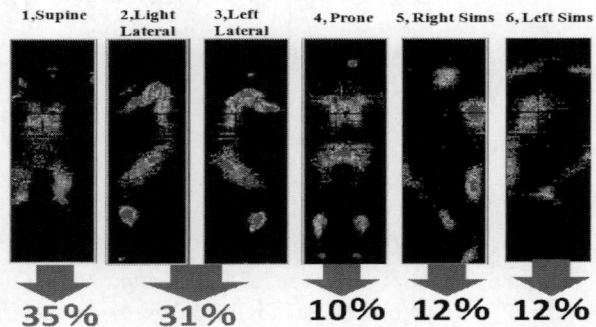

Fig. 11 Percentages of each sleep posture

D. Relations among Breathing Pattern, Body Posture and BMI

From the above results, the Long expiratory phase breathing pattern was seen with the Supine posture for group A (BMI <25.0), while Lateral of Right and Left was seen with the characteristic Long expiratory phase as shown in Fig.10 for group B (BMI> 25.0).

V. DISCUSSION

A. Variation in the Rates of Pulse and Respiration

At the present study, from the point of view of the time span for several minutes, we adopted the time zone that the variation in the rates of pulse and respiration were stable. It was considered that stable time zone at relatively low rate as shown by arrows in Fig.5 in the rates of pulse and respiration was related to relax during sleep.

B. Relations among Breathing Pattern, Body Posture and BMI

In IV-A and B from the point of view of one period of breathing, it was found that the duration of expiration phase was longer than that of inspiration. It can be considered that wider duration of expiration indicated relax during sleep, because the diaphragm and the related muscles else did not work anymore at that phase, i.e., all of muscles related to breathing relaxed, so that characteristics of expiration phase itself appeared honestly.

From IV-D, we found that long expiratory breathing pattern has a relationship between the supine posture for group A, and characteristic waveform in long expiration bears relevance to Lateral of Right and Left posture for group B. This result implied one example of relationship of producing relaxing of autonomic nervous system. Moreover, since it depends on the subject's physical characteristics, the results of this study may suggest the possibility that a set of breathing patterns and sleep postures will be a new type of biological information.

REFERENCES

1. Tero Myllymaki, Heikki Kyrolainen, Katri avolainen et al;Effects of vigorous late-night exercise on sleep quality and cardiac autonomic activity. J. Sleep Res.20, pp146~153,2011
2. H. Okawai, S. Ichisawa, K. Numata; Detection of influence of stimuli or services on the physical condition and satisfaction with unconscious response reflecting activities of autonomic nervoussystem., N.A.Abu Osman et al.(Eds.); IOMED2011,IFMBE Proceedings35,pp.420-423,2011
3. H. Okawai, T. Yajima, T. Imamatsu; Transition of rates of respiration and pulse, and sleeping posture during night detected by body motion wave., Conference on Biomedical Engineering & Sciences(IECBES 2012), 2012
4. H.Kuno, M.Takashima, H. Okawai: Measurement of Respiration, Heart Beat and Body Movement on a Bed Using Dynamic Air-Pressure Sensor. ,IEEJ Trans. EIS, Vol.124, No.4, 2004 (in Japanese)

Sophisticated Rate Control of Respiration and Pulse during Sleep Studied by Body Motion Wave

H. Okawai, T. Yajima, T. Imamatsu, and J. Wada

Graduate School of Engineering, Iwate University, Morioka 020-8551 Japan

Abstract—The purpose of the present study is to consider how respiration and pulse are controlled by autonomic nervous system during sleep, i.e., unconscious state. In addition, how and why sleeping posture change occurs also under unconscious state. For the methodology for the above question, a dynamic air pressure sensor was adopted to detect body motion wave (BMW) to produce information of respiration, pulse and unconscious actions. As a result, it was found that i) transition of rates of respiration and pulse had three slope patterns with some difference between the two, ii) both transitions had periodicity but not same between the two, and iii) the change of sleeping posture occurred to significantly vary both rates under the control by autonomic nervous system. Also for the change of sleeping posture at unconscious state, it can be considered to vary above two rates under some direct or indirect control of autonomic nervous system.

Keywords—Respiration, pulse, autonomic nervous system, body motion wave (BMW).

I. INTRODUCTION

For the biological phenomena it is well known that respiration is performed by motions of the diaphragm and related muscles (Respiratory system), and pulse, or heartbeat, is done by heart muscles (Cardiovascular system) with related to autonomic nervous system. In addition, the respiratory system is also performed by motions of musculoskeletal system with related to cerebrum [1]. These two activities of autonomic nervous system through involuntary muscles and cerebrum through voluntary muscles are (I) unconscious activity and (II) conscious activity, respectively.

Then, during sleep, in the state of relaxed, it is understood that both respiration and pulse are controlled by autonomic nervous system and the rates of those two are becomes less [1][2].

However, detailed description of phenomena during sleep is not yet well known. Are both respiration and pulse controlled in the same way? If not, we are interested in what and how differences are. In addition to that, what makes sleeping posture change? For the change of sleeping posture, striated muscles must work under unconscious state nevertheless these muscles work voluntarily during awake in principle except respiration and pulse related.

In this respect, it was reported that autonomic nervous system works on the systems of respiration and cardiovascular in some different manner [3]. Now, we studied more detailed phenomena by means of the body motion wave (BMW) [3]-[5], to consider above some questions.

II. METHOD

A. Instrumentation

A pressure sensor, named "dynamic air pressure sensor" (M.I.Lab) was adopted in order to fabricate a non-restraint, non-attachment measurement system. As shown in Fig. 1, in principle, it was set on a bed to measure dynamic air pressure arises between the sensor and a subject's body at lying. The pressure variation detected with the sensor was converted to electric signals, "body motion wave" as mentioned later, sampled at the rate 400 Hz, 16 bit and stored in a personal computer. The signal was processed with Chart v4.2.2 (AD Instrument) and programmable software VEE Pro ver6.0 (Agilent Technologies).

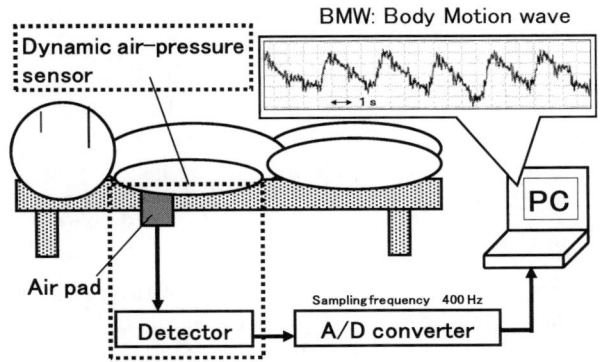

Fig. 1 Measurement system for body motion waves

B. Body Motion Waves Detected

As reported by Okawai et al[3]-[6], Here, in subject's body during sleep some continuous motions are generated resulting in respiration and pulse, so that thus motions can be detected as pressure waves named "body motion wave (BMW)", "respiration-origin BMW(R-BMW)" and "pulse-origin BMW (P-BMW)". In addition during sleep, some frequent extra motions are generated resulting in unconscious actions et al, these can be detected as pressure waves also. For these extra waves, approximately two types of waves appear. The one was named Tremble-origin BMW (T-BMW) having a small magnitude and short duration of wave due to a slight action of a portion of a body. The other was named Action-origin BMW (A-BMW) having a large magnitude and wide duration of wave. The latter usually means the change of sleeping posture. Such changing posture was made sure with a sensor array as described in next sub-section C. Accuracy of the present method for detecting rates of respiration and pulse was already confirmed by comparing with the data taken from a thermistor and an electrocardiogram [6].

C. Detection of Sleeping Posture during Sleep

An instrument with using a pressure sensor array, on the market, for detecting sleeping position was adopted. This is for detecting rather static signs revealing an image of in two dimensional pressure distribution to show body postures at lying.

III. RESULT

Experiments were carried out for more than 70 in the number of subjects with age range 20 to 80. A subject usually stayed at Healthcare Experimental House in Iwate University for four successive nights. In the first and second days physical reproducibility was checked, and in the third and fourth days an experimental condition was changed with aim at a theme. Thus, data were stored more than 800 in the total number of days up to now. At the present study, the data were analyzed in detail for 70 nights data obtained from 20 healthy subjects in 20's. All of the subjects agreed to be a subject with our privacy policy. A subject spent a day, went to bed as usual and awoke naturally in these experiment days.

A. Rate Transitions of Respiration and Pulse

Examples of the rate transitions of respiration and pulse are shown in Fig.2 and Fig.3, respectively. In Fig.2, a considerable number of small deletions traced like pillars getting to zero, by a handmade signal processing, shows count error because of T-BMW or A-BMW. Such deletions were interpolated to see in another viewpoint in Fig.3. As a result, it was found as follows.

(1) Transition patterns of the rates were roughly classified into almost three: i) it moved downward with time as shown in the upper graphs of Fig.2 and Fig.3, ii) it remained constant as shown in the lower graphs, respectively, and iii) it moved a little bit upward though not shown here.

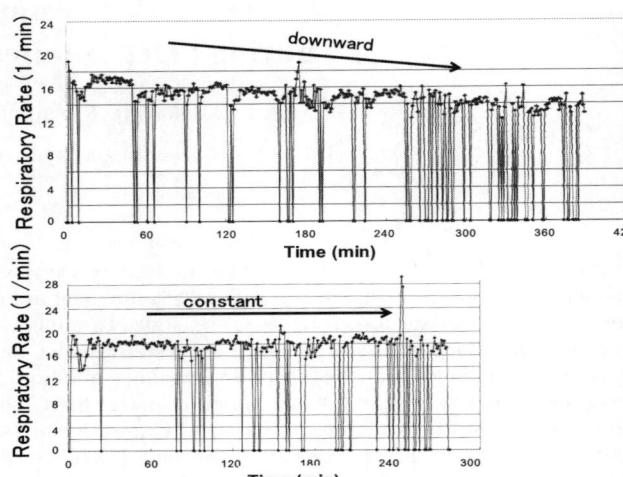

Fig.2 Transitions of rates of respiration for two subjects

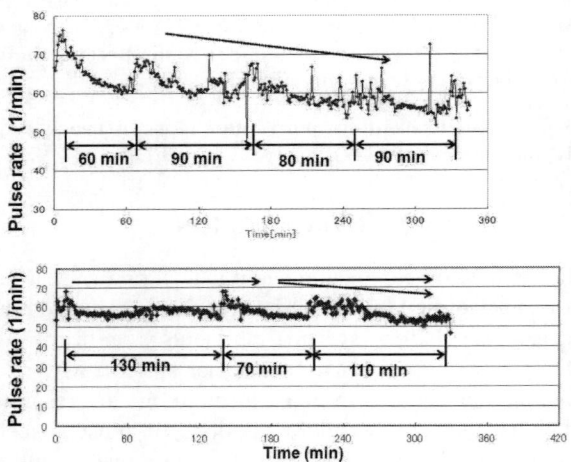

Fig.3 Transitions of rates of pulse for two subjects

Table 1 Transition patterns of two rates of respiration and pulse classified

Transition pattern	Respiration (%)	Pulse (%)
downward	20	60
constant	70	30
upward	10	10

The data were summarized to be, in percentage, roughly 20, 70, 10 for the respiration rate and 60, 30, 10 for the pulse rate, up to now, as shown in Table 1. Thus the transition patterns between respiration and pulse were not always same for even one subject.

(2) Periodicity was recognized in the transitions both of respiration and pulse. The period varied through a night for

both respiration and pulse for even a subject with the range approximately 30 to 130 minute.

(3) Periods between transitions of the rates of respiration and pulse did not always agree as shown in Fig.4. This figure also shows that roughly constant pattern for respiration, on the contrary, downward pattern for pulse. In addition, the period varied through night for both transitions.

B. Transitions of Action and Posture

Transition of action during sleep was obtained from both T-BMW and A-BMW as previously mentioned by significant number of deletions in Fig.2. Concrete number of the rate of action through one night was, on average, in the range approximately one or two times for every minute [3]. While the time duration of action per minute was in the range from approximately 5 s to 30 s. Here, the action had a tendency of not appearing in time range 50 to 150 minute and of occurring more after rather three hours from getting to bed.

For the change of sleeping posture, it was detected through A-BMW and made sure by the method with a pressure sensor array as mentioned in section II C. In fact, A-BMW had a feature in its numerical data itself that all of subjects of healthy 20s changed their postures at least one time per one hour in average through one night. In concretely, more than four times per hour for 20% of the subjects, two or three times for 70 % and one time for 10 %. Here, it is worthy of special attention to note that there was no one who did not change his/her posture through one night sleep [3]. In the next, more of rolls of changing sleeping posture were studied. There are significant numbers of deletions along the transition of pulse rate in Fig.5 as well as Fig.2. In Fig.5 the points in time axis changing posture occurred were added by thick lines above the pulse rate.

In order to consider more of posture easily these the thick lines were overlaid with two rates as shown in Fig.6 by sophisticating Fig.5. As a result, these lines implied that they related to occurrence of significant scale of fluctuation for the rates of respiration and pulse as marked by quadrilaterals in Fig. 6. Note that these lines, i.e., actions, related to occurring fluctuation but sometimes not related simultaneously both of respiration and pulse.

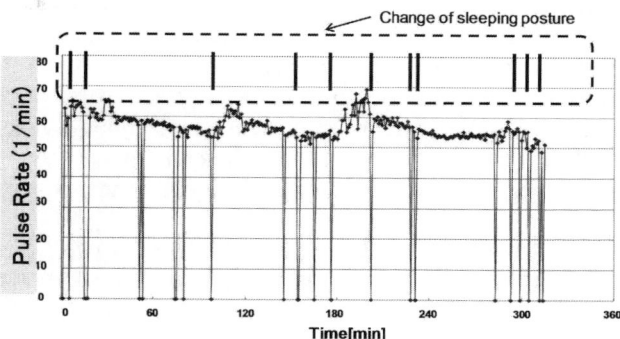

Fig.5 transition of all actions described by deletions, as mentioned in Fig.2, along transition of pulse rate as same as in Fig.4 and the transition of change of sleeping posture was added by thick lines above the pulse rate

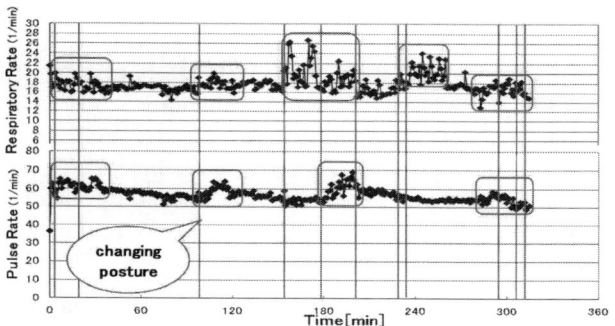

Fig.6 Transitions of changing posture added by solid lines on the graphs of rates of respiration and pulse as same as in Fig.4

IV. DISCUSSIONS

A. Slope and Periodicity in the Rate Transitions of Respiration and Pulse

As shown in Figs.2-5, a transition pattern showed a slope or a flat through entire night and periodicity confirmed by a combination of increase and decrease in the range 30 to 130 minute. It means that mainly two frequency components of reciprocal of several hours and of 30 to 130 minute appeared.

For the periodicity, in the traditional sleep research by e polysomnography adopting mainly electric signals such as Electroencephalogram, an explanation for sleep depth has been well recognized. Then the sleep depth increases and

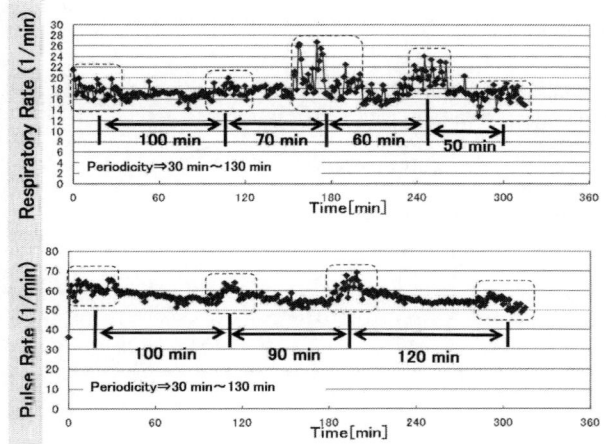

Fig.4 Transitions of rates of respiration and pulse for one subject to demonstrate different periodicity

decreases periodically in approximately 80 to 100, and typically 90 minute under the control of autonomic nervous system [2].

While at the present study using dynamics of pressure, the periodicity was confirmed. So that, it is sure that the autonomic nervous system makes muscles related to respiration and pulse work. In addition, however as shown in Fig4, the periods in two rate transitions were not same and pulse of that was clearer. So far it can be considered that autonomic nervous system controls priority to pulse rate compared with respiratory rate under the condition that each tidal volume is unknown. Thus, autonomic nervous system controls respiration and pulse in a different manner.

B. Roll of Body Actions and Change of Sleeping Posture

For body actions during sleep confirmed at the present study there were two types: the one was small in motion scale and short in time duration, i.e., T-BMW, the other was large in motion and wide in time duration, A-BMW. InFig.2, there are more than 60 of pillars and real number of T-BMW and A-BMW gets to more large up to several hundred through a night.

Some numbers of pillars were A-BMWs, in Fig.5, introducing the change of postures as shown by vertical solid lines in Fig.6. Here, thus A-BMWs introducing the change of postures occurred at the same time when the rate of both or one of respiration or pulse varied significantly or showed fluctuation, especially for respiration rate.

Therefore, it can be considered that the change of sleeping posture occurred to remarkably vary both or either of respiration or pulse, particularly respiration.

C. Another Control of the Rates of Respiration and Pulse by Autonomic Nervous System

Skeletal muscles, i.e., voluntary muscles, excluding ones related to respiration and pulse, produce body actions under control of somatic nervous system during awake.

While it was also found the change of sleeping posture occurred to make two rate transitions introduce remarkable variation or fluctuation. It is due to skeletal muscles that such actions occurred at the timing not in a random order but in a needed order. From this, it can be explained that an extra certain direct or indirect control by autonomic nervous system also must have introduced changing posture through skeletal muscles.

The roll of the change of sleeping posture is well recognized to result in adjusting body pressure distribution and varying circulations of blood and lymph fluid in the body. The present study must have suggested one more idea for the roll of changing posture.

ACKNOWLEDGMENT

Authors would like to thank subjects for data producing and Toyo Feather Industry Co. Ltd. for some instrumentation.

REFERENCES

1. Kaniusas E, (2012)"Sleep." Biomedical signals and Sensors1, Springer, pp 270-282.
2. Hori T (2008) Sleep psychology. Kitaoji Shobou, Kyoto, Japan(in Japanese)
3. Okawai H, Yajima T, Imamatsu T(2012)Transition of Rates of Respiration and Pulse, and Sleeping Posture During Night Detected by Body Motion Wave. 2012 IEEE international Conference on Biomedical Engineering and Sciences, MALAYSIA, 17th-19th DEC2012.
4. Okawai H, Ichisawa H, Numata K(2011) Detection of influence of stimuli or services on the physical condition and satisfaction with unconscious response reflecting activities of autonomic nervous system. N.A.Abu Osman et al. (Eds.); BIOMED2011, IFMBE Proceedings35, MALAYSIA, 2011, pp 420-423.
5. Okawai H, Kato K, Baya D (2012) Entrusting the reply of satisfaction or physical condition for services to unconscious responses reflecting activities of autonomic nervous system. 4th International Conference on Applied Human Factors and Ergonomics (AHFE), San Francisco, 21-25 July 2012, pp 911-920
6. Kuno H, Takashima M, Okawai H (2004) Measurement of Respiration, heart beat and body movement on a bed using dynamic airpressure sensor. IEEJ Trans. EIS.vol.124,No.4, pp 935-940 (in Japanease).

Author: Hiroaki Okawai
Institute: Iwate University
Street: 4-3-5 Ueda
City: Morioka, Iwate 020-8551
Country: Japan
Email: hokawai@iwate-u.ac.jp

A Tool for Patient Data Recovering Aimed to Machine Learning Supervised Training

G. Guidi[1], M.C. Pettenati[2], M. Milli[3], and E. Iadanza[1]

[1] University of Florence, Dept. of Information Engineering, Italy
[2] ICON Foundation, Italy
[3] Azienda Sanitaria Fiorentina Dept. of Cardiology, Italy

Abstract—In this paper we present a method and a tool to acquire data of outpatients suffering from Heart Failure and to populate a database so that it is suitable for the supervised training of machine learning techniques. In our studies we had to train an artificial intelligence-based system to recognize different severity and to predict worsening of heart failure patients, using as input various parameters acquirable during outpatient visits. We have therefore developed a tool that would allow the cardiologist to populate a "supervised database" suitable for machine learning during his regular outpatient consultations. The idea comes from the fact that in literature there are few databases of this type and they are not scalable in our case. The tool includes a management part for the patient demographics, a part for the data acquisition, a part for displaying the follow-ups of a patient, a part of artificial intelligence to provide the smart output, and a part called "score based prognosis" in which we have computerized some prognostic models known in the literature as the SHFM, the CHARM, the EFFECT, and ADHERE.

Keywords—Tool, supervised training, machine learning, heart failure.

I. INTRODUCTION

A. Scope

In this paper we present a tool that allows cardiologists to populate, during their regular outpatient consultations, a database suitable for supervised training of machine learning techniques. This has the aim to realize a special-purpose Clinical Decision Support System (CDSS) for Heart Failure (HF) patients, accurately described in [1–3]. Some of the functions provided in the final system are already inserted in the tool herein described, as well as a management structure for the patient registry. There are two approaches to provide decision support: an expert system based approach and a machine learning based approach. In expert systems, the knowledge is inserted into the system by an human expert through a specific interface; in machine learning techniques, knowledge is automatically acquired, extracting information from the raw data (data mining). This means that, unlike the expert systems, in machine learning-based methods it is not necessary a previous knowledge of the phenomenon we are dealing with and how the outputs are related to the inputs: the linking model between inputs and outputs, which is often not known a priori but it is "hidden" among data, it is automatically created. In developing the CDSS we chose to use these machine learning techniques because there were no well-defined rules in literature for the outputs that we wanted the system to provide. To achieve some of these outputs it was necessary to have a structured database in order to train in supervised manner the artificial intelligences. Unfortunately, this database does not exist in the literature, so we have developed a tool that would allow our cardiologist partners to generate it.

B. Supervised Training

In this section we briefly describe what is meant by supervised training. The supervised training consists of providing input to a machine learning technique to a series of "examples" of the phenomenon we are observing, each linked by its "Desired output" which is the output we have when as input we provide that particular parameters pattern. Then the input of a supervised training process is a series of n pairs of inputs / desired output as in Fig. 1. As a result of the training process we get a reusable model that, during the use phase, will respond to user inputs according to what it learned during the training phase. Note that the learned phenomenon should not necessarily be known, in fact the system learns from the evidence of the data and it independently discovers the rules that binds the inputs to the desired output. As shown in Fig 1, in the train phase, some input vectors coupled with the respective desired outputs are supplied to the machine. The machine will organize its internal parameters to reproduce these outputs in the use phase, thus producing the model. In the use phase, when the user enters as input the vector 15,34,20,62 the model provides the output that was associated with the input pattern, learned during the train phase, more similar to the one entered by the user (input 14,35,23,64 - output 3).

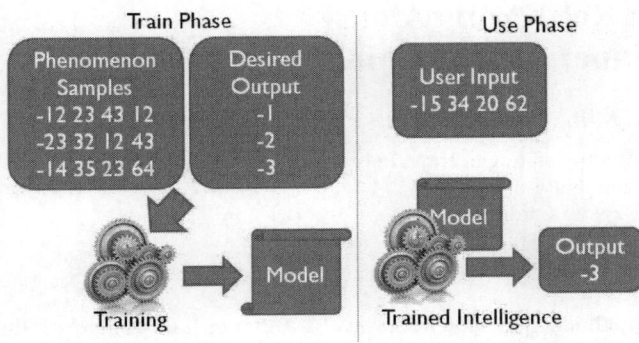

Fig. 1 Supervised training schema

C. Heart Failure System Outputs and Relative Required Training

According to our specialist cardiologist partners we decided what were the outputs that the Heart Failure CDSS system should provide. These outputs will require different applicative scenarios for the system and therefore different trainings. The outputs that system should provide are:

1. Severity assessment: a three level evaluation of actual HF severity (Mild, Moderate, Severe)
2. Score-based Prognosis: a risk stratification that provides percentage of mortality according with models from literature
3. Chronological follow-up comparison: a graphic view of parameters of the selected patient in the various dates of follow-up. An indication of whether the patient status has worsened, improved or remained stable compared to the last follow-ups
4. Long Term Predictivity: an indication of the fact that the patient, in the next follow-up, will get worse, will remain stable or will improve, basing on the past trend of follow-ups. The follow-ups are temporally spaced by 1-6 months
5. Short Term Predictivity: an indication of the occurrence of an impending acute episode. This output requires a scenario in which patient is in a daily telemonitoring. The system is trained by analyzing the measured parameters which preceded an acute event.

The proposed tool described in this paper provides the first 4 outputs as well as a support in the management of the patient registry.

II. TOOL DESCRIPTION

A. Tool Design

This tool has been designed involving physicians in order to satisfy any practical requirement (intended as both content and usability) that they have during their work. Indeed, given that this software is designed to be used during routine outpatient visits, it was necessary to design it so that it would impact as little as possible on the outpatient visits workflow. It includes also some practical features such as the Modification of Diet in Renal Disease calculation (MDRD), smart discovery of drug molecules to prescribe basing on dose, and also some score based models. The physician can take real advantage from the use of the tool in its cardiology outpatient visits, instead of using it only to get the data for machine learning training that is an advantage not immediately perceivable in his eyes. In this section we will describe in detail the tool in all its parts which are: parameters acquisition, patient's follow-up displaying, severity assessment basing on artificial intelligence, module for asynchronous compiling, score based prognosis, enrollment calculation. All these parts are supported by a management module for the patient demographics.

B. Patient Management

Through this section it is possible to select patients already included in its database or add new ones. In addition to standard biographical data one can also insert and then visualize the data related to patient's GP (General Practitioner). As shown in Fig. 2 this interface allows access to parameter input interface, follow-ups display, and calculation of score-based prognosis.

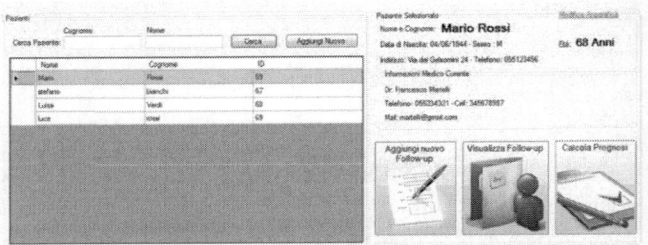

Fig. 2 Patient Management Interface

C. Enrollment Score Calculation

Using this function it is possible to calculate a customizable score to assess whether the patient is at high risk of re-hospitalization. Usually this score is calculated at discharge time. In this case it can be entered manually, otherwise by using this tool it can be automatically calculated by entering the physiological parameters at discharge.

D. Score-Based Prognosis

In order to provide a complete tool, useful also in the physician's clinical routine we have computerized 4 prognostic models known in the literature and included them in the tool[4–7]. In this way the cardiologist, rather than have to manually calculate the prognostic indices or having to re-enter all the values in other software, can easily obtain such indexes in this tool, recovering much of the data needed for the models directly from the database.

E. Parameters Acquisition

This is the most important section and the one that the cardiologist uses during the visit. With reference to figure 4 we notice that in section 1 are shown the patient's personal details. In section 2 one can specify whether the patient is today classified as a stable patient, or if in the past has developed rare or frequent exacerbations; this is useful in training the system to provide the output denominated Long term prediction. In the part of Figure 4 marked with the number "3" is possible to input the patient's parameters. As discussed above there are parameters that need a "frequent update" and others - such as the BNP (Brain Natriuretic Peptide) or EF (Ejection Fraction) - that, in case of a close follow-up, are not to be re-entered but are retrieved automatically from previous follow-up record. On these numerical form there are various controls that prevent from entering non-numeric values or numbers out of range. The user must then enter other parameters related to the ECG (for example, the presence of bundle branch block or pacemaker), the etiology and comorbidities. Note that, as mentioned earlier, in case of renal failure as comorbidity it is possible to calculate the MDRD by entering the race and creatinine (age and sex are retrieved from the database) using an abbreviated formula [8]. There is also a whole part of therapy management, marked with the number "4", in which the physician can enter the therapy that is assigned to the patient. For some categories of drugs (ACE inhibitors, angiotensin receptor blockers, beta blockers, diuretics) by filling the dose in milligrams the system, on the basis of pre-set thresholds, automatically recognizes the active ingredient and if the dose for that drug is considered as high, medium or low. Section 5 contains an indication that the physician must provide in order to have a supervised train of the system. With reference to fig. 1 these are the "Desired Output". All various input parameters entered by this acquisition mask will then be associated with these desired outputs that are mild-moderate-severe and improvement-stable-worsening, useful for outputs 1 and 4 described above in Part C of Introduction section. Using buttons marked with the number 6 it is possible to save the follow-up or analyze it. If you press on save, the system will highlight in red possible blank fields before adding the follow-up to the database. If you click on "analyze button" you can choose whether to use artificial intelligence trained with your own database (if it's consistent) or with a default database embedded in the system that is the one for which they are guaranteed the performance published in [3].

F. Graphical View Follow-Ups

Using this interface, shown in Fig. 5, you can choose a follow-up of the selected patient "1", view its numerical values "3" and a summary report of comorbidity, etiology and treatment "4". It is also possible to have a graphical view of all the patient's follow-ups, being able to choose if displaying or not some parameters and to select one out of three different types of chart "2". This satisfies the output 3

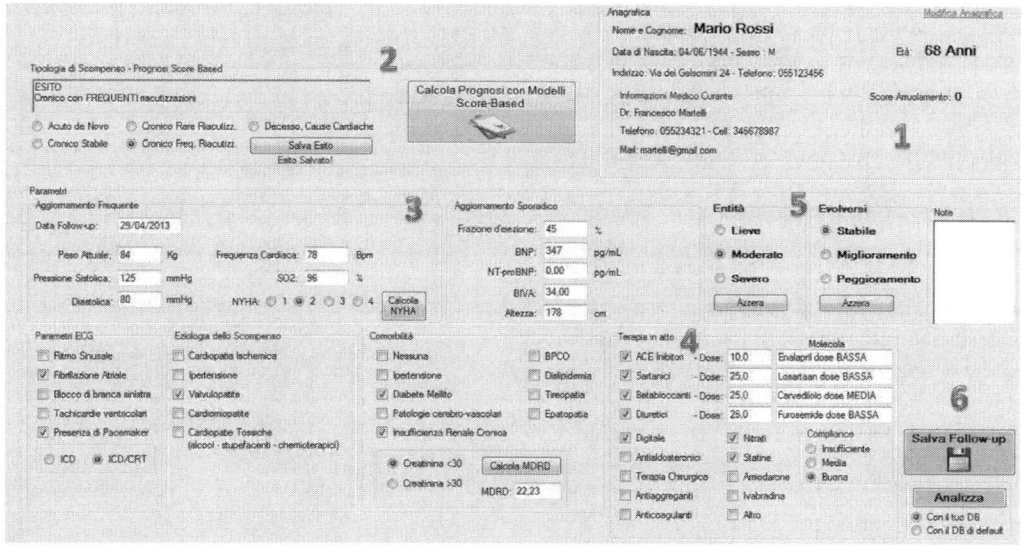

Fig. 3 Input Mask

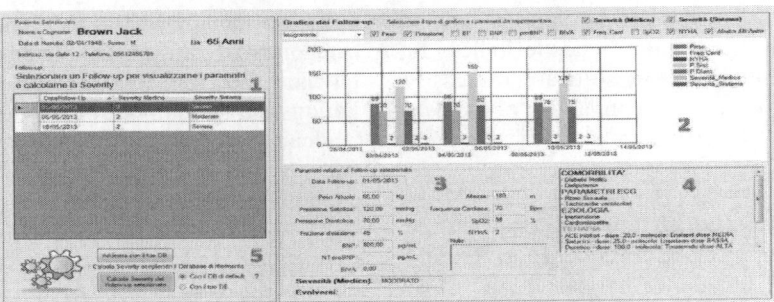

Fig. 4 Graphical view of follow-ups

of Section C of the Introduction. If you have not already done, you can also analyze a follow-up by using artificial intelligence "5".

G. Asynchronous Doctor / Nurse Compiling

From the "clinical practice" of outpatient visits of our case study in the Santa Maria Nuova Hospital, it was found that filling the input forms during the visit was made by a nurse, while the cardiologist performs operations on patients (ultrasound, ECG, etc...). The "Desired Output" (mild-moderate-severe), however, needed to be inserted by the cardiologist, who preferred to insert it at the same time for all patients after finishing all visits. It was therefore necessary to create a special interface that highlighted the patients for whom there were some follow-up with some empty fields (including the Desired output).

III. DISCUSSION

Our partners cardiologists were very satisfied in using the software, because, in addition to actively contribute to the training of a smart system which they may later use, they take advantage of immediate benefits such as the score-based prognosis and the graphical view of follow-ups. The need for a specific interface for asynchronous doctor / nurse compiling is an example of the importance to develop clinical software side by side with doctors who will have to use it. Through this tool we are acquiring data with a rate of 10 patients per week. A problem still to solve is that, as the average follow-up for this type of patients is 6 months, in order to obtain a structured series of follow-ups for each patient so that we can get the output "Long Term predicticity" it takes at least 2 years of use of the tool.

REFERENCES

1. G. Guidi, E. Iadanza, M. C. Pettenati, M. Milli, F. Pavone, and G. Biffi Gentili, "Heart Failure Artificial Intelligence-based Computer Aided Diagnosis Telecare System," *ICOST 2012, Lecture Notes in Computer Science*, vol. 7251, pp. 278–281, 2012.
2. G. Guidi, E. Iadanza, and M. C. Pettenati, "Heart Failure Artificial Computer Aided Diagnosis Telecare System Using Various Artificial Intelligence Techniques". In: CONGRESSO NAZIONALE DI BIOINGEGNERIA 2012. ATTI. Roma, 2012, BOLOGNA: Pàtron, ISSN: 978-88-555. .
3. G. Guidi, M. C. Pettenati, R. Miniati, and E. Iadanza, "Heart Failure analysis Dashboard for patient's remote monitoring combining multiple artificial intelligence technologies," in *2012 Annual International Conference of the IEEE Engineering in Medicine and Biology Society*, 2012, pp. 2210–2213.
4. W. C. Levy, D. Mozaffarian, D. T. Linker, S. C. Sutradhar, S. D. Anker, A. B. Cropp, I. Anand, A. Maggioni, P. Burton, M. D. Sullivan, B. Pitt, P. a Poole-Wilson, D. L. Mann, and M. Packer, "The Seattle Heart Failure Model: prediction of survival in heart failure.," *Circulation*, vol. 113, no. 11, pp. 1424–33, Mar. 2006.
5. D. S. Lee, P. C. Austin, J. L. Rouleau, P. P. Liu, and D. Naimark, "Predicting Mortality Among Patients Hospitalized for Heart Failure Derivation and Validation of a Clinical Model (EFFECT)," *Hospitals*, vol. 290, no. 19, pp. 2581–2587, 2003.
6. G. C. Fonarow, K. F. Adams, W. T. Abraham, C. W. Yancy, and W. J. Boscardin, "Risk stratification for in-hospital mortality in acutely decompensated heart failure: classification and regression tree analysis. (ADHERE)," *JAMA*, vol. 293, no. 5, pp. 572–80, Feb. 2005.
7. S. J. Pocock, D. Wang, M. a Pfeffer, S. Yusuf, J. J. V. McMurray, K. B. Swedberg, J. Ostergren, E. L. Michelson, K. S. Pieper, and C. B. Granger, "Predictors of mortality and morbidity in patients with chronic heart failure.(CHARM)," *European heart journal*, vol. 27, no. 1, pp. 65–75, Jan. 2006.
8. A. S. Levey, J. Coresh, E. Balk, A. T. Kausz, A. Levin, M. W. Steffes, R. J. Hogg, R. D. Perrone, J. Lau, and G. Eknoyan, "National Kidney Foundation Practice Guidelines for Chronic Kidney Disease: Evaluation, Classification, and Stratification," *Annals of Internal Medicine*, vol. 139, no. 2, pp. 137–147, 2003.

The Potential of Machine-to-Machine Communications for Developing the Next Generation of Mobile Device Monitoring Systems

R.S.H. Istepanian[1], B. Woodward[2], D.J. Mulvaney[2], S. Datta[2], P. Harvey[2], A.L. Vyas[3], and O. Farooq[4]

[1] Medical Information and Network Technologies Research Centre, Kingston University, London, UK
[2] Loughborough University/School of Electronic, Electrical and Systems Engineering, Loughborough, UK
[3] Instrument Design and Development Centre, Indian Institute of Technology, Delhi, India
[4] Department of Electronics Engineering, Aligarh Muslim University, Aligarh, India

Abstract—This paper describes an m-health system developed for monitoring patients with heart disease and diabetes in India. It has been developed to allow patients in remote rural regions to access health centers in large cities. Wireless medical sensor readings and personal data acquired by health workers are combined into a set of patient monitoring parameters. Mobile device applications running on a web browser are then used to upload the data over the communications network infrastructure to a central server located at a main hospital for storage and clinical diagnosis. The system has great potential for M2M (machine-to-machine) communications.

Keywords—Cardiac monitoring, m-health, communications, wireless sensors.

I. INTRODUCTION

Communications technology using mobile devices for medical applications, referred to as m-health, is well-established[1]. It is recognized as being particularly use in the prevention and management of chronic conditions, such as diabetes, hypertension and cardiovascular disease. The remit of the project outlined here, which is under the auspices of The British Council's UK-India Education and Research Initiative (UKIERI), is to improve the monitoring of these and other medical conditions in India by taking advantage of the existing mobile telecommunications infrastructure there. Technical tests have been carried out and initial trials have shown promising first results.

UKIERI, which was instigated at a meeting of the Prime Ministers of the UK and India, has enabled close collaboration between partners from two UK universities, two Indian universities and an Indian hospital. All partners were allotted work packages to complete that exploited their interests and strengths. Web browser application and database systems were developed at Loughborough University, diabetes monitoring systems at Kingston University, patient sensor networks and telemetry protocols at the Indian Institute of Technology Delhi, a cardiac spreadsheet for data analysis at the All-India Institute of Medical Sciences Delhi, and data security and trend analysis at Aligarh Muslim University [2,3].

II. METHODS

The components and features of the patient monitoring system in India are shown in Fig. 1. The three main sections comprise a patient and local clinician sub-system; a communications network; and a remote clinician or hospital sub-system.

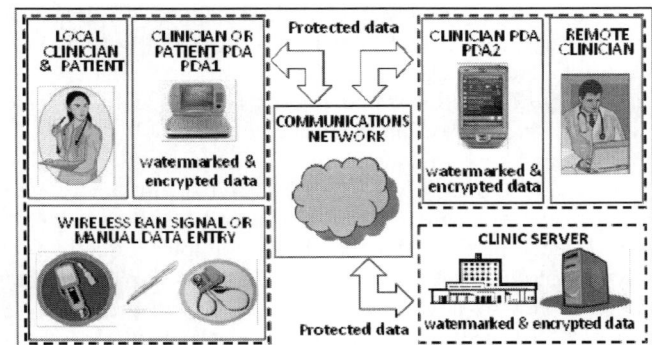

Fig. 1 m-health system implemented for patient monitoring in India

The main component at the patient end is a body area network (BAN) of wireless sensor nodes to record the patient's vital signs. Connectivity between nodes is achieved by a low power, high data rate protocol, which is used as an alternative to ZigBee and Bluetooth. The initial concept was to use a basic mobile phone, which has been demonstrated successfully, but much greater system flexibility is possible with a more advanced device, such as a personal digital assistant (PDA), as used here, although this is clearly a more expensive approach for a developing country. To allow access using web browsers, including standard mobile web browsers, obviates the need to install a bespoke application. By using features available in the latest browsers, the system's web pages can provide all the facilities required for successful deployment, many of which, until recently, would only have been available in a dedicated application [4].

Watermarking and encryption provide security of the transmitted patient data; this is achieved by embedding identification information to detect any illegal tampering [5]. The protected data are then transmitted over the communications network using a standard such as GSM or GPRS. In India, a fibre-optic network and satellite communications are also available, both of which are normally used transparently.

At the receiving end the data may be stored in a server for later access or, unusually, viewed in real time by a clinician. Any response needed can then be transmitted back to the patient end.

A. System Overview

The body area network worn by the patient is arranged as multiple slaves and one master. The slave nodes communicate with the master node via wireless transceivers. The master node is a similar device but also has a Bluetooth transceiver to link to the PDA. There are several options to link the wireless nodes, including Bluetooth, ZigBee and proprietary protocols such as the Nordic 2.4 GHz transceiver used here because of its low-power, long battery life capability. This allows one master node to communicate with a star network of up to six slave nodes.

Although such measurements as temperature, respiration and ECG lend themselves to be acquired electronically with no invasive procedure, the measurement of blood sugar is still largely performed using blood samples directly [6]. Some instruments to do this have wireless capability, but most do not so it is apposite to include the provision of wireless-free instruments in the patient's network, which allows the patient to read the instrument's display and record the value. A wireless-free LifeScan OneTouch Ultra 2 glucose meter was used during system evaluation and clinical trials.

The patient PDA has Bluetooth for the node communication and also 3G/EDGE/GPRS mobile phone and Wi-Fi capabilities for Internet access. Its Windows Mobile operating system has similar capabilities as a Java-based OS on other comparable devices. The PDA application is used to acquire, store, upload, download and display medical readings. Using Bluetooth, it can configure and receive telemetry from the master node to be stored locally to the PDA for uploading later. For standard instruments, their display values are entered into the application's spreadsheet forms, which are designed for the purpose of monitoring specific parameters associated with cardiac and diabetic conditions.

The server and database platform used to store the patient measurements is Apache, PHP and MySQL. The server and database are situated in England and can readily be accessed by the project partners in India. To increase the scope of participants to view the data, any clinician should be able to access the data without having to download a specific software application. By taking into account the lesser capabilities of a mobile device web browser compared to a PC browser, the system allows a browser on a PC or mobile to be used to search and view patient data as text or images with a similar capability to the PDA application.

B. Wireless Sensor Nodes

The prototype wireless slave and master nodes share a common design with power supply, signal acquisition, microcontroller and wireless transceivers. Each slave node worn by the patient is connected to a sensor with a wired link. To contribute to a prolonged node battery life, the node's microcontroller unit (MCU) is the Texas Instruments MSP430F2618E mixed signal device. This is a 16 MHz 16-bit RISC ultra-low power MCU including 12-bit ADC, four universal serial communication interfaces (USCI) and direct memory access (DMA) controller. Three slave nodes measure ECG, temperature and respiration respectively. A fourth node is available for monitoring any other parameter. The master node is also worn by the patient and has the same design as the slaves but is also populated with a Bluetooth module to communicate with the patient PDA.

The Nordic transceiver is in receive mode and 'listens' for incoming data from each slave in turn as it is polled by the master. The incoming data is read from the transceiver by the MCU on a USCI port and sent to the Bluetooth module on a second USCI port. The Bluetooth module is a KC Wirefree KC-21 device that supports the Bluetooth serial port profile (SPP) and is driven by AT commands from the MCU. Once the master node is Bluetooth-paired with the patient PDA, the PDA initiates a connection by opening its virtual serial port to the node and communications can begin.

C. Node Nordic Transceiver

Two considerations for the selection of the Nordic transceivers were the size of the nodes and their battery life. The wireless transceivers on the patient nodes use the international ISM 2.4 GHz band since this is a common option for BANs and multiple devices are available. It also allows the antenna to be small and placed discretely within the transceiver module, which allows the node to be made small enough for placing around the body. A consequence of a wearable node is that the antenna is very close to the patient's body, which may produce RF losses due to absorption. To mitigate these losses, the nodes can increase their output power, with the penalty of reduced battery life and increased localized radiation absorption. Further power saving is achievable by adopting the nodes' medium access control (MAC) protocol so that the transceiver is only powered when necessary.

D. Mobile Device Application

The application on the patient-end PDA is written in C# for the Windows Mobile.net platform to acquire, store, upload, download and display medical readings. The data acquisition can come from the master node via Bluetooth and also by spreadsheet user entry for standard instruments. The data is stored on a small database on the mobile device to give secure and searchable access for the different data files. An Internet connection to the server allows a file to be searched for and uploaded or downloaded. Incoming master node telemetry, together with stored or downloaded spreadsheet, image or waveform files, can also be displayed.

To allow users to enter readings from standard instruments (i.e. without communications facility) and also regarding the patient's general health, spreadsheets were devised for recording specific conditions helpful in cardiac and diabetic monitoring. Specifically, these values are entered, typically by a health worker, selecting from the discrete value options of multiple-list boxes. Discrete selections allow some commonality between patient responses and trends can be monitored more easily between monitoring sessions. List boxes also allow the user to point and select, rather than typing explicitly on a small or virtual keypad. Color-coding on general health responses also give a visual indication of the general condition of the patient. Text areas are also available to type notes if required. The drawback of viewing many parameters across multiple pages on a small screen is self-evident, so further work is necessary to distil the information into a more concise representation of the patient's condition.

Multiple measurement files from different sessions and patients are stored on the PDA using Microsoft SQL Server Compact 3.5, which is a secure and searchable database for desktop and mobile devices. From search fields, a particular file can be selected for uploading to the server or displaying on the PDA. Similarly, a clinician can search the MySQL database at the server to download a file over an Internet connection to the patient end. The database search criteria used to distinguish the files are patient identification number, timestamp (date and time) and signal type (e.g. ECG, cardiac spreadsheet) that populate the search list boxes. These boxes are actively updated with database results as the search is narrowed and show options such as all measurements for a particular patient and on a particular date and time. An example search is shown in Fig. 2.

III. RESULTS

In the evaluation and technical trials, the system was able to acquire, transmit and display all the information as required in our clinical needs study. We have demonstrated that medical data from a remote patient can be transmitted and that the displayed information can be dynamically changed. An example electrocardiograph image gathered using the sensor nodes and transmitted to a mobile device is shown in Fig. 3. Now, with the system operational, further work is required on data presentation, which is all that the end-users, namely doctors, nurses and paramedics, would normally see.

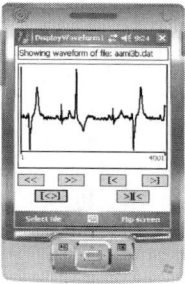

Fig. 3 Electrocardiograph image displayed on a mobile device

Although it remains to be seen which of the many parameters being monitored will prove to be the most informative in practice in India, initial studies with a sample of around 25 patients have provided some early indicators. The cardiac data gathered from these patients has been processed using frequency and wavelet transforms and a suitably small number of features have been selected. The system has shown a success rate of over 75% in being able to classify abnormal cardiac pulses as emanating from patients with a variety of medical issues, including gastric and prostate problems. The results of the study will be particularly useful in the development of a system for the initial screening of patients. In the longer term, trend analysis using the measurements recorded would need to be carried out to determine which are the principal parameters required and how they can be best incorporated in a health profile. There is clearly scope for adopting M2M concepts to automate some of the system functionality, although this has not yet been attempted in this project.

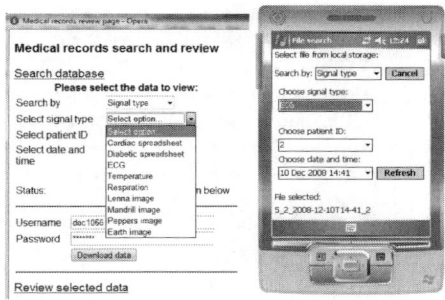

Fig. 2 Example of database search on browser (left) and PDA (right)

The existing prototype system is suitable for demonstration under conditions controlled by the development team. The plan for the next stage of the project is to design a robust miniaturized version of the monitoring system to facilitate proof of operation in rural regions of India. As the system is intended for portable use, consideration will also be given to the potential power consumption savings that result from burst and low-power transmission operation modes. In addition, a series of tests will be carried out to ensure that the transmission of data from the PDA to a remote server can be achieved in a sufficiently secure manner for commercial use.

IV. Discussion

It is envisaged that in India a local health worker would normally be present to supervise the monitoring of a patient in a remote area, where poorly educated patients may be unable to cope alone. As well as the vital signs from body sensors, such as an ECG, readings from standard instruments recording a patient's health status, such as blood pressure or blood glucose, can be transmitted with the data.

A future aim will be to study the potential for M2M communications, i.e. machine-to-machine, mobile-to-machine, or machine-to-mobile [7]. M2M generally requires low-bandwidth and is therefore applicable for use with GSM and GPRS protocols that allow very wide coverage; also it does not need new infrastructure and is easy to install. Further, it can permit an "always on" capability, thereby allowing automatic data transfer at any time.

The concept is that the communication is not initiated by a person, but by a device, such a PDA or iPhone. A scenario that the UKIERI team has envisaged is to have a tamper-proof 'intelligent black box' sited in a clinic in a remote village, such that any patient wearing 'intelligent clothing' triggers the box to interrogate the body area network in the clothing and transmit the data. At the receiving end, a corresponding intelligent black box processes the data, transfers it to a server, or responds to it accordingly. This would be an entirely automatic process that, although futuristic in the health domain, is at least technically feasible.

M2M is most appropriate for identifying specific data that may provide indicators for trend analysis or early warnings of the onset of specific medical conditions.

Along with advances in connectivity and increasing data rates, other important concepts, such as Internet of Things (IoT) and Internet of Services (IoS), will also be more readily adopted, especially for 4G protocol applications [8].

V. Conclusions

In this paper we have presented an m-health system whose potential for adoption of M2M technologies to improve healthcare services in India is substantial, with major opportunities to increase disease awareness and implement better management.

Several key issues have been identified and these are summarized as follows.

- The availability of communications and networking infrastructures in India can assist the rapid deployment of these applications and systems on a larger scale.
- The inclusion of all the relevant participants (clinicians, patients, telecommunications providers and operators) in India is vital to the larger adoption and successful deployment of m-health services.
- An initial clinical trial has indicated that the acceptability of innovative health delivery mechanisms and services by patients and clinicians in India is encouraging and positive. This is an important factor in any successful, larger-scale adoption for patient and clinician-led m-health services.
- There is a need for a strategic plan for m-health in India to deliver not only a patient-led approach but also a cost effective mechanism for m-health services.
- Future technical developments must include consideration of M2M aspects to 'automate' certain components of the system that are 'always on' and cost-effective.

References

1. Istepanian RSH, Jovanov E, Zhang YT (2004) Beyond seamless mobility and global wireless health-care connectivity. Guest editorial on m-health: IEEE Trans Inf Technol in Biomed, 8: 405–414
2. Mulvaney D, Woodward B, Datta S, Harvey P, Vyas A, Thakker B, Farooq O (2010) Mobile communications for monitoring heart disease and diabetes. Proc 32nd Ann IEEE EMBS Int. Conf., Buenos Aires, Argentina, pp 2208-2210
3. Harvey P, Woodward B, Datta S, Mulvaney D (2011) Data acquisition in a wireless diabetic and cardiac monitoring system. Proc. 33rd Ann IEEE EMBS Int. Conf., Boston, MAS, USA, pp 3154-3157
4. Mulvaney D J, Woodward B, Datta S, Harvey P D, Vyas A L, Thakkar B, Farooq O, Istepanian R S H(2012), Monitoring heart disease and diabetes with mobile internet communications, Int J Telemed & Appl, 2012:ID195970, DOI 10.1155/2012/195970
5. Farooq O, Mulvaney D, Vyas A, Datta S, Woodward B, Harvey P, Istepanian R (2012), Chaotic encryption of physiological signals using multi-level non-uniform quantizer, Proc. 4th Int. Conf. on Computer Technol. & Development, Bangkok, Thailand, pp 9-15
6. Harper R, Nicholl P, McTear M, Wallace J, Black L-A, Kearney P (2008) Automated phone capture of diabetes patients readings with consultant monitoring via the web. Proc 15th Ann IEEE Int. Conf. & Workshop on Engrg of Computer Based Systems, pp 219-226
7. Christaldi L, Faifer M, Grande F, Ottoboni R (2005), An improved M2M platform for multi-sensors agent application, Sensors for Industry Conf, Houston, Texas, USA, pp 79-83
8. Istepanian R (2012), Guest Editorial: 4G health – the long-term evolution of m-health, IEEE Trans Info Technol in Biomed 16: 1-5

A Study on Perception of Managing Diabetes Mellitus through Social Networking in the Kingdom of Saudi Arabia

T.M. Alanzi, R.S.H. Istepanian, N. Philip, and A. Sungoor

Medical Information and Network Technologies Research Centre, Kingston University, London, UK

Abstract—The Kingdom of Saudi Arabia (KSA) has the seventh highest prevalence of diabetes in the world with estimates of 20% of the populations diagnosed with diabetes. Furthermore, there is a lack of good educational and management programs of diabetes in the kingdom. In parallel to this major health problem, there is an increasing trend of smart mobile phone usage and access to social networking in the Kingdom, especially in the younger population. In this paper, we conducted a preliminary study on the perception of managing diabetes mellitus through mobile technologies and social networking in the kingdom of Saudi Arabia. A mixed-method design with interviews and a survey were used to gather data. Most of the participants were younger users aged between (10-30 years). The key outcomes of this study indicate the high percentage acceptance of using smart phone technologies and social networking within the participation.

Keywords—m-health, mobile diabetes management, social networking for healthcare, diabetes mellitus, telemedicine, e-health, Kingdom of Saudi Arabia.

I. INTRODUCTION

The global prevalence of diabetes is alarming with approximately 366 million individuals are living with this long term condition. According to the recent International Diabetes Federation (IDF) report, The Kingdom of Saudi Arabia (KSA) has the seventh highest prevalence of diabetes in the world with estimates of 20% of the populations diagnosed with diabetes [1].

Diabetes mellitus has two main types: Type 1 and Type 2. Type1 occurs when the cells in the pancreas are damaged. Globally, the Type 1 diabetes affects nearly 10 % of all people with diabetes as it hits young people under 25 years old. Type1 diabetes is also called Insulin Dependent Diabetes (IDD). Type 2 occurs when the cells in the pancreas are not working effectively. Type 2 diabetes is known as noninsulin dependent diabetes (NIDD). There are many different causes and risk factors behind both types of diabetes including environmental factors, family history, high blood pressure, unhealthy awareness of diabetes , alcohol intake, smoking, sedentary lifestyle, and obesity. Extreme fatigue, increased thirst, weight loss and blurred vision are the main symptoms of diabetes [2].

There has been a revolution in the use of smart mobile phones and applications in the Kingdom of Saudi Arabia.

According to a recent United Nations Conference on Trade and Development (UNCTAD) report, the Kingdom of Saudi Arabia has the largest percentage of smart mobile phone users globally [3]. In parallel with this revolution in the use of smart mobile phone technologies, there is an increase in the use of social networking in the kingdom, especially among educated and younger citizens. Thus, there is an urgent need for an innovative strategy for the KSA to deliver health care services and medical education to diabetic patients through these technologies, and for enhancing diabetes management by using a mobile management system. There is also a need to provide emotional support and health education via social networking.

In recent years, social networking has become an important tool for exchanging health care information between users and patients [4]. An example is the social site PatientsLikeMe, which aims to create a community-web environment for patients, nurses, and society to provide medical information and education, and empower patients [5]. Patients can share experiences, explore their medical conditions, symptoms and routines, and support one another.

A recent survey of existing social networks in health care was presented in [6]. This study illustrates the influence of social networking on health care outcome models. It classified network services into three categories: i) health care social networking, ii) consumer personalized medicine, and iii) quantified self-tracking with the potential of four major health care services offered to the clients through the social networks: 1) clinical trial access, 2) emotional support and information sharing, 3) quantified self-tacking, 4) Q&A with a professional physician. Other social networking health care sites such as CureTogether, MedHelp, and mCare provide different health services supported by their own delivery models [7, 8, 9].

To date, there is no study that addresses this concept with a focus on social networking tailored specifically for Saudi diabetic patients. Furthermore, there has not been any study conducted to date on the impact and evaluation of mobile diabetes management systems integrated with social networking in the kingdom. In order to achieve this innovative strategy and tailor it to diabetic patients in the KSA, the life cycle of this research is divided into four parts, 1) a study on perception of managing diabetes mellitus through social networking in the KSA, 2) proposing and designing general models based on the results of this phase, 3) implementation of the system, 4) evaluation of the system on sample patients population in the KSA.

This paper presents a preliminary study of the first phase on patients' perception of using social networking for diabetes management in the KSA. The study is based on two key research instruments: a tailored questionnaire for KSA diabetic patients, followed by a detailed interview with Saudi and UK clinicians. The general outcomes of this study indicate that the structure and functionality of the proposed system architecture should include the following key models: 1) a mobile diabetes management tailored for Saudi patients, 2) a specific social networking, and 3) a coaching strategy and implementation of those key elements which constitutes a successful strategy for deployment and usage of the system in the KSA. The remaining of the paper is organized as follows: Section II describes the methodology, Section III provides the result of this study and Section IV provides discussion, conclusions and future work.

II. METHODOLOGY

A. Study Setting and Participants

In order to explore clinical requirements and needs, a preliminary study on social networking was carried out both in the KSA and the UK. The study was divided into two phases: (1) Initial interview with diabetic specialist clinician from the KSA and the UK to understand the need for smart mobile management and social networking for diabetes management, and (2) a follow-up questionnaire designed to understand and identify the key social-networking elements required from a diabetic patient's perspective in the KSA. Two senior clinicians from the UK and the KSA participated in this initial phase that included a total of 30 patients with both Type 1 and Type 2 diabetes. Patients were recruited by clinical staff during an office visit or by sending an SMS message.

B. Research Instruments

There were two research instruments used in this study.

1) Interviews:

An interview program contained 10 in-depth questions about the need for social networking tools, diabetes management tools, and the main concern of the clinician.

2) Study Survey:

A follow-up questionnaire based on the mobile diabetes management system of an earlier work was used [10]. The questionnaire was chosen because it matches the areas addressed in this work both diabetes management and social networking issues. The questionnaire used in this study is divided into four sections: basic information, diabetes management, social networking with social support, and a Saudi Arabia social network for diabetics. It was also translated into Arabic. The questionnaire structure as shown in Fig. 1 is a sample snapshot of the online questionnaire used in this study. The design of questionnaire based on multiple choices and open questions.

The questionnaire was then published and posted online to a Saudi diabetic-patient group in the Dammam region using Google forms. All of the diabetic patients were informed that the information given would be kept confidential. A summary of the project is given at the top of the questionnaire, followed by a description of its importance and a notification that information gathered will be used for research purposes. Study participants completed a total of 30 questionnaires.

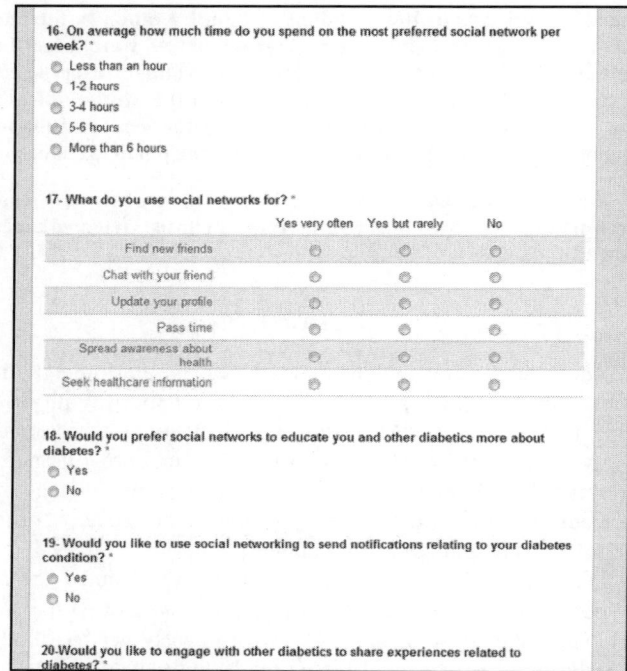

Fig. 1 A sample snapshot of the online questionnaire used in this study.

III. RESULTS

A. Patient Demographics

Table 1 shows an overview of the demographics of patients used in this study. Participating patients (N=30) were predominantly male (n=21, 70%). The majority of the participants were between 21-30 years (n = 17, 57%). Most had been diagnosed with diabetes Type 2 (n = 19, 63%). Almost all participants (n = 23, 77%) consulted a diabetes specialist at least once a month.

We also measured the way of accessing internet (n = 29, 97%) of respondents who had access the internet using PCs, laptops, and smart phones. All participants were engaged with online social network such as Facebook and Twitter (n = 30, 100%). The majority of the participants spent more than 6 hours weekly (n = 24, 80%) on social networks, 63% of mobile phone users use the iPhone as the most popular smart phone followed by HTC mobile type at 23%. Most of participants preferred to engage with a private social network tailored to Saudi diabetic patients (n = 24, 80%).

Table 1 Patients' demographics

General characteristics		N	%
Gender	Male	21	(70)
	Female	9	(30)
Diabetes type	Type 1	11	(36)
	Type 2	19	(63)
Age group	10-20	5	(17)
	21-30	17	(57)
	31-40	8	(27)
Diabetes specialist consultation	More than once	23	(77)
	Once a month	5	(17)
	Other	2	(7)
Computerized characteristics			
Internet access	PC/Labtop	29	(97)
	Smartphone	29	(97)
Smart phone type	iPhone	19	(63)
	Blackberry	4	(13)
	HTC	7	(23)
	Nokia	0	(0)
	Other	0	(0)
Engaged with web social network	Yes	30	(100)
	No	0	(0)
Average time spending on social network	Less than an hour	0	(0)
	1-2 hours	0	(0)
	3-4 hours	0	(0)
	5-6 hours	6	(20)
	More than 6 hours	24	(80)
Prefer to use private social network for Saudi diabetics	Yes	24	(80)
	No	6	(20)

B. Patients Requirements and Needs

The following results were the most significant outcomes and proposed design decisions were made based upon them:

i. A total number of diabetic patient participating (n=30) prefer the following social networking functionalities to be incorporated in the proposed system which are: (i) ask a doctor (n = 30, 100%), (ii) messaging (n =30, 100%), (iii) blogs (n = 27, 90%), (iv) video education tutorials on diabetes (n =20, 67%). Fig. 2 shows a graphical chart presented those functionalities.

ii. A total of (n=30,100%) and (n=23, 77%) of respondents prefer to receive real-time feedback by SMS and E-mail respectively.

iii. A total of (n=22, 73%) of respondents prefer to manage their diabetic condition by using a social networking system.

iv. A total of (n=27, 87%) of respondents would like to share experiences of their diabetic condition with other diabetics friends and relatives by using social networking system.

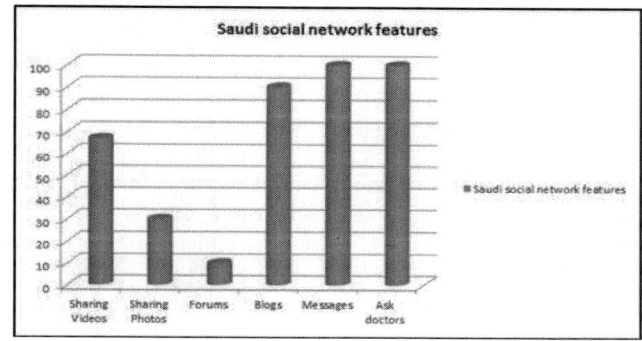

Fig. 2 Patients preference for social-networking functionalities in the proposed system

C. Clinician Requirements and Recomnedations

The summary of the outcomes from the UK and the KSA clinician surveyed in this study indicated the following requirements:

i. Diabetes self-management is a new and challenging issue for KSA clinicians.

ii. The system should include an emphasis on privacy and limited information exchange due to the social and religious issues in the KSA.

iii. The system must include social-networking elements and diabetes self-management elements for enhanced diabetes management in the KSA.

iv. The system must contain a reminder element delivered either via SMS, e-mail, or a private message.
v. The system must have a simple mechanism for interactivity between patients and clinicians.
vi. The system must include coaching strategies based on behavioral change theory, such as goal-setting.

IV. DISCUSSION AND CONCLUSION

The results of this preliminary study can conclude that the acceptance of Saudi patients for using social networking as a tool for better managements of their diabetes is relatively high. This acceptance is specifically high in younger population (10 – 30 years), whom preferred the use of a private Saudi social networking tool for managing their condition. Another important finding was that the preferred social-networking functionalities —such as ask a doctor, messaging, blogs and video tutorials, had the highest percentages of suggested functionalities. It is interesting to note that most of the patients prefer to have a real-time feedback via SMS services. In addition, the most interesting finding was that all of the Saudi patients engaged with some web social networking, which is consistent with other research outcomes and statistics [3].

Furthermore, the results of this study also indicate that support of the KSA specialist for such new concept, provided that specific and key restrictions apply to any proposed network structure. In particular, emphasis was on privacy and limited information exchange. Another interesting finding was that e-coaching strategies need to be included in the proposed system to assist patients in changing their behavior as part of the management cycle.

In conclusion, there is substantial support for the adoption of social networking in enhancing healthcare awareness in the KSA. The research work is currently ongoing with focus on the design and development of an integrated architecture combining social-networking system with mobile diabetes and e-coaching strategies to be further evaluated in the KSA healthcare sectors.

ACKNOWLEDGEMENTS

The authors would like to acknowledge the advice and assistant of Dr. Donatella Casiglia, Consultant Diabetologist from Bromley Healthcare, NHS, UK, Dr. Waled Albakr, Consultant Diabetologist from King Fahd University Hospital – Dammam University, KSA and Dr.Naif Alsuwaidan, Family doctor, King Fahd Hospital, Dammam, KSA.

REFERENCES

[1] 5th Edition of the Diabetes Atlas released on World Diabetes Day. International Diabetes Federation. Available at: http://www.idf.org/diabetesatlas/news/fifth-edition-release.
[2] Understanding Diabetes – audio information. Diabetes UK, Available at: http://www.diabetes.org.uk/Guide-to-diabetes/Introduction-to-diabetes/Understanding-Diabetes-audio-information/.
[3] Statistics Available at: http://unctad.org/en/Pages/Statistics.aspx.
[4] J.F. Pearson, C.A. Brownstein, and J.S. Brownstein,(2011) "Potential for electronic health records and online social networking to redefine medical research," Clin.Chem, vol 57, no. 2, pp.196-204.
[5] "Patients Like You," Available at: http://www.patientslikeme.com/.
[6] M. Swan, (2009) "Emerging patient-driven health care models: An examination of health social networks, consumer personalized medicine and quantified self-tracking," International journal of environmental research and public health, vol 6, no. 2, pp. 492-525.
[7] MedHelp, Available at: http://www.medhelp.org/.
[8] CureTogether.com, Available at: http://curetogether.com/.
[9] D. Yu Weider and Asim Siddiqui,(2009) "Toward a wireless mobile social network system design in health care," Multimedia and ubiquitous Engineering, MUE '09, Third International Conference on IEEE.
[10] D.J. Jacobs,(2011) "E-health diabetes management system," Honours Project Report, Unveristy of Cap Twon, Saouth Africe.

Author: Turki Alanzi
Institute: Kingston University
City: London
Country: UK
Email:k1148892@kingston.ac.uk

Author: R.S.H.Istepanian
Institute: Kingston University
City: London
Country: UK
Email: R.Istepanian@kingston.ac.uk

Validity of Smartphone Accelerometers for Assessing Energy Expenditure during Fast Running

C. Easton[3], N. Philip[1], A. Aleksandravicius[1], J. Pawlak[2],
D.J. Muggeridge[3], P.A. Domene[4], and R.S.H. Istepanian[1]

[1] Medical Information and Network Technologies Research Centre (MINT), Faculty of Science,
Engineering and Computing, Kingston University, Kingston upon Thames, UK
[2] Department of Movement Science, Grand Valley State University, Grand Rapids, MI, USA
[3] Institute for Clinical Exercise and Health Science, School of Science,
University of the West of Scotland, Hamilton, UK
[4] School of Life Sciences, Faculty of Science, Engineering and Computing,
Kingston University, Kingston upon Thames, UK

Abstract—Recent advances in smartphone technology have facilitated the generation of mobile applications to monitor exercise performance. The aim of the present study was to investigate the validity and reliability of the triaxial accelerometers in the HTC and Samsung platforms for assessing energy expenditure (EE) during walking and fast running.

Eleven endurance-trained participants (nine males and two females, mean ± s.d.: Age 35 ± 10 years; Maximal aerobic capacity 58.5 ± 7.2 ml·min^{-1}·kg^{-1}) completed two identical discontinuous incremental exercise tests on a treadmill. The tests consisted of walking (4, 5, and 6 km·h^{-1}) and running (8, 10, 12, 14, 16, 18, and 20 km·h^{-1}, or until volitional exhaustion) for 3 min at each speed, followed by 3 min of recovery for all speeds faster than 10 km·hr^{-1}. Acceleration was recorded in all three axes of motion using a novel application on each device and expressed as vector magnitude (VM). Respiratory variables used to assess EE were measured using indirect calorimetry. The reliability of EE and acceleration measured between tests 1 and 2 were assessed using intraclass correlation coefficients (ICC). The validity of the accelerometers to assess EE was determined using Pearson's correlation coefficient.

Measurements of EE and VM recorded using both devices were highly reproducible between tests 1 and 2 (ICC: EE = 0.976; HTC VM = 0.984; Samsung VM = 0.971). All accelerometer outputs rose linearly with speed during walking and running up to and including 20 km·h^{-1}. EE was significantly correlated with both HTC VM ($R = 0.98$, $P < 0.001$) and Samsung VM ($R = 0.99$, $P < 0.001$).

Data from the present study suggests that the inbuilt triaxial accelerometers in HTC and Samsung smartphone devices offer a valid and reliable method for determining the change in EE during fast running.

Keywords—HTC, Samsung, Exercise, Mobile applications

I. INTRODUCTION

Recent developments in smartphone technology has facilitated the measurement of physical activity and exercise intensity via inclusion of inbuilt sensing techniques, such as accelerometers, gyroscopes, and Global Positioning Systems (GPS). The capacity of mobile phones to monitor physical activity has aided development of numerous mobile applications (apps) aimed at enhancing exercise training, performance and weight loss. The apps typically exploit measurements of movement (acceleration and GPS data) to provide indirect estimates of exercise intensity. These apps are popular on all platforms although the reliability and validity (through comparison with gold standard measurements) has not always been established.

Heo et al. [1] reported that a smartphone accelerometer was a valid technique to assess the activity classification during sitting, standing and walking. Furthermore, Nolan et al. [2] established that an Apple iPhone® can predict aspects of locomotion during walking and running up to 11.3 km · h^{-1} with accuracy similar to other accelerometer-based tools. However, while these data suggest smartphones can provide reliable measurements of typical free-living physical activity, the accuracy during high-intensity exercise such as fast running is currently unknown. Indeed, we have previously demonstrated [3] that some commercial models of accelerometer were incapable of distinguishing between activity classifications at faster running speeds due to an alteration in running biomechanics.

To our knowledge, no study has yet investigated the validity of smartphone accelerometers for assessing exercise intensity at faster running speeds. Therefore, the aim of this study was to assess the reliability and validity of device accelerometers in the HTC and Samsung platforms for assessing changes in energy expenditure (EE) during walking and fast running.

II. MATERIALS AND METHODS

A. Participants

Eleven well-trained middle and long distance runners (nine males and two females, mean ± s.d.: Age 35 ± 10 years; Stature 178 ± 8 cm; Body mass 76.2 ± 14.0 kg; Maximal aerobic capacity (VO_{2max}) 58.5 ± 7.2 ml·min⁻¹·kg⁻¹) all gave their written informed consent and completed a Physical Activity Readiness Questionnaire prior to taking part in the study. The study was approved by the Faculty Ethics Committee at Kingston University and all procedures were conducted in accordance with the Declaration of Helsinki.

B. Procedures

Each participant was required to report to the Exercise Physiology Laboratory, Kingston University at the same time of day on two separate occasions separated by at least one week to perform an exercise test. The preparation and conduct of each exercise test was identical to allow reliability of the assessment techniques between different testing days to be established. Participants arrived at the laboratory having refrained from alcohol and exercise in the previous 24 hours prior to each test and from food and fluid with the exception of water in the previous three hours.

The exercise tests required running at progressively faster speeds on a motorized treadmill (H/P/Cosmos Venus, H/P/Cosmos, Nussdorf-Traunstein, Germany) until volitional exhaustion. The initial phase of each test was continuous, beginning with walking at speeds of 4, 5, and 6 km · h⁻¹ and then running at speeds of 8 and 10 km · h⁻¹ at 0% gradient (Fig. 1). Participants completed 3 min of exercise at each speed and the velocity was increased at a constant rate between each bout. The subsequent discontinuous phase of the test consisted of running for 3 min at 12, 14, 16, 18 and 20 km · h⁻¹ (or until volitional exhaustion) with each stage separated by 3 min of active recovery (walking at 4 km · h⁻¹). The protocol was designed to allow participants to run at the higher speeds for a sufficient period of time to collect steady state respiratory data [3]. Some participants did not complete the full stages at 18 km · h⁻¹ (n = 3) and 20 km · h⁻¹ (n = 6).

Respiratory variables (oxygen consumption (VO_2) and carbon dioxide production (VCO_2)) were measured continuously during exercise using an indirect calorimeter (Oxycon Pro, VIASYS GmbH, Eric Jaeger, Hoechberg, Germany). The final 30 s of each exercise stage were used for analysis. The Oxycon Pro was calibrated immediately prior to each test according to the manufacturer's guidelines. Participants were fitted with a facemask (Hans-Rudolf Inc., Kansas City, MO) which was secured using an adjustable nylon head strap to allow unrestricted collection of variables during the test. The EE during each stage of exercise was calculated using the Weir equation [4] as follows:

$$EE\ (kcal/min) = (VO_2 \times 3.941) + (VCO_2 \times 1.1)$$

C. Smartphone Accelerometers

In this work a native app was developed based on the Android platform. The app was designed to achieve the following three functions:

- Data Acquisition
- Data processing and analyzing
- Local data storage

For the data acquisition, accelerometer data was collected from the embedded triaxial accelerometers in the HTC and Samsung platforms. The triaxial accelerometer type used in these phones is of type BMA 150 accelerometer from Bosh with resolution of 0.153, max range of 39.24 and power of 0.2 mA. The data acquisition frequency was in the range of 15-20 Hz as this was dependent on the smartphone resources available at the time. The received raw acceleration data in all vectors from the accelerometers then was processed and average vector magnitude (VM) for the last 30 s of each stage of exercise was calculated as follows:

$$VM = \sqrt{(axis\ 1)^2 + (axis\ 2)^2 + (axis\ 3)^2}$$

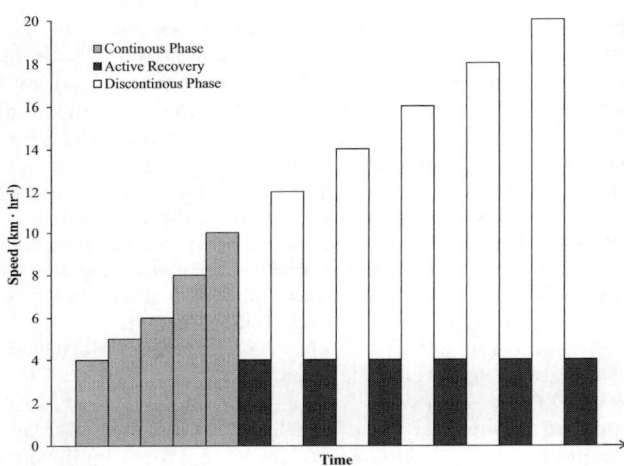

Fig. 1 Exercise protocol consisting of a continuous walking and running phase (4–10 km· h⁻¹) with 3 min at each speed and a discontinuous running phase (12–20 km· h⁻¹ or until voluntary exhaustion) with 3 min at each speed (3 min of active recovery between bouts).

D. Data Analysis

Data are expressed as mean ± standard deviation. The reliability of EE and acceleration measured between tests 1 and 2 were assessed using intraclass correlation coefficients (ICC). The relationship between EE and VM acceleration measured by both devices was determined using Pearson's correlation coefficient. Statistical significance was set at $P < 0.05$.

III. RESULTS

A. Reliability

The measurements of EE and VM acceleration measured using both smartphones were highly reproducible between tests 1 and 2 (ICC: EE = 0.976; HTC VM = 0.984; Samsung VM = 0.971) (Fig. 2 and Fig. 3).

B. Validity

All accelerometer outputs rose linearly with speed during walking and running up to and including 20 km · h-1 (Fig. 4). EE was significantly correlated with both HTC VM ($R = 0.98$, $P < 0.001$) and Samsung VM ($R = 0.99$, $P < 0.001$).

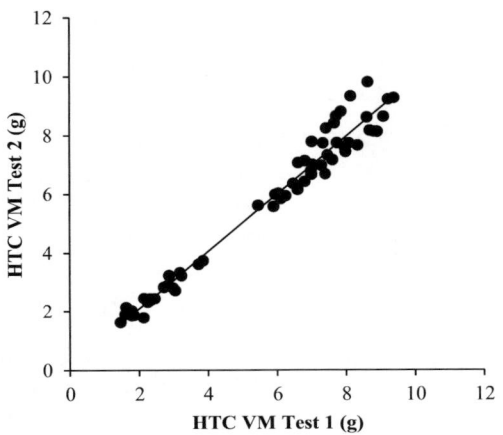

Fig. 2 Correlation of HTC vector magnitude measured in Tests 1 and 2.

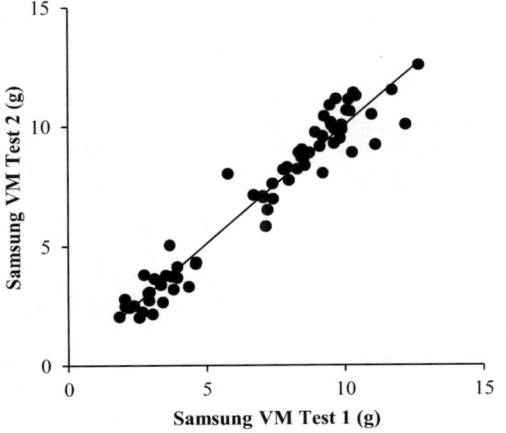

Fig. 3 Correlation of Samsung vector magnitude measured in Tests 1 and 2

IV. DISCUSSION

This investigation was the first to assess the efficacy of smartphone accelerometers for assessing changes in locomotion and EE during fast running. The present data suggest that inbuilt triaxial accelerometers in both HTC and Samsung devices offer reliable and valid means to assess EE in trained runners up to speeds of 20 km · h^{-1}. While further research in this area is unquestionably required, there is evidence to suggest that custom apps could be designed to provide accurate real time and summary estimates of EE and other physiological parameters during exercise.

The data in the present study is consistent with our previous work [3] demonstrating that standalone research models of triaxial accelerometers are capable of detecting changes in locomotion during fast running speeds. The capacity of smartphone accelerometers to offer a similar level of accuracy is less well established. Indeed, it was recently reported [5] that a selected Apple iPhone® pedometer app did not provide valid estimates of physical activity during treadmill walking. Contrastingly, Nolan et al. [2] established that a custom designed app for the iPhone® was capable of classifying the type of physical activity the participant was performing with 99% accuracy and could provide accurate predictions of both running speed and EE. Nevertheless, these authors only had participants running to a maximum speed of 11.3 km · h^{-1} which is slower than the typical training and competition speeds of trained middle and long-distance runners.

The importance of assessing the accuracy of accelerometer detected changes in locomotion at faster running speeds is highlighted by previous work [6] establishing that despite running from 7 to 32 km · h^{-1} vertical power is almost constant, whereas horizontal power increases by a factor of > 10. Fudge et al. [3] demonstrated that a reduction in vertical acceleration at approximately 14–16 km · h^{-1} was compensated for by concomitant increases in acceleration in the anterior–posterior and medial–lateral directions. Therefore, accelerometers that are capable of detecting subtle changes in acceleration in all three axes of motion are required if EE is to be adequately estimated during running. The data from the present study suggests that smartphone accelerometers, at least in the HTC and Samsung platforms are compatible with such requirements.

One possible limitation of using accelerometery alone to assess EE is that the biomechanical characteristics of locomotion such as vertical oscillation may vary between individuals. Consequently, the metabolic cost of exercise may be over- or underestimated for certain individuals.

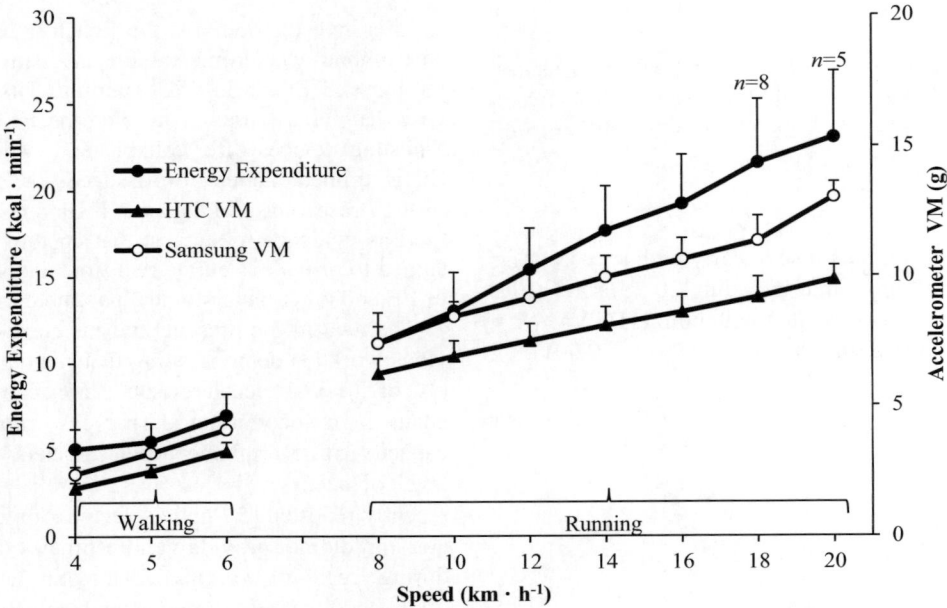

Fig. 4 Relationships between walking and running speed and energy expenditure and vector magnitude from both devices. N = 11 unless otherwise stated.

Furthermore, the increased physiological demand of carrying a load or uphill running will not be recognized by the devices. One alternative is to incorporate a physiological measurement into the calculation which may strengthen the estimation of metabolic rate. For example, previous research indicates that combining accelerometry with measurement of heart rate provides a more accurate prediction of EE when compared to either measure alone [3,7]. Indeed the technology has already been developed for acquisition of heart rate using smartphone technology [8] although this has not yet been validated during high-intensity exercise.

V. Conclusions

The present study has demonstrated that inbuilt device accelerometers in HTC and Samsung platforms offer a valid and reliable method to assess EE during fast running on a level treadmill. Further research is required to determine the accuracy of EE predictions using this accelerometry data during fast running outdoors and whether the addition of a physiological parameter such as heart rate would strengthen the predictive power.

Acknowledgments

This project was partly funded by a Summer Studentship Award from the Faculty of SEC at Kingston University.

References

1. Heo S, Kang K, Bae C. (2012) Activity classification using a single tri-axial accelerometer of smartphone. LNEE 182:269–275
2. Nolan M, Mitchell JR, Doyle-Baker PK. (2013) Validity of the Apple
3. Fudge BW, Wilson, J, Easton C et al. (2007) Estimation of oxygen uptake during fast running using accelerometry and heart rate. Med Sci Sports Exerc 39:192–198
4. Weir J. (1949) New methods for calculating metabolic rate with special reference to protein metabolism. J Physiol 100:1–9
5. Bergman RJ, Spellman JW, Hall ME et al. (2012) Is there a valid app for that? Validity of a free pedometer iPhone application. J Phys Act Health. 9:670–676
6. Cavagna GA, Thys H, Zamboni A. (1976) The sources of external work in level walking and running. J Physiol. 262:639–657
7. Crouter SE, Churilla JR, Bassett DR. (2007) Accuracy of the Actiheart for assessment of energy expenditure in adults. Eur J Clin Nutr 62:704–711
8. Gregoski MJ, Mueller M, Vertegel A et al. (2012) Development and validation of a smartphone acquisition application for health promotion and wellness telehealth applications. Int J Telemed Appl. doi:10.1155/2012/696324

Medical Quality of Service Analysis of Ultrasound Video Streaming over LTE Networks

Ali Alinejad[1], R.S.H. Istepanian[2], Senior Member, IEEE, and N. Philip[2], Member, IEEE

[1] Department of Computer, Islamshahr Branch, Islamic Azad University, Islamshahr, Iran
a.alinejad@iiau.ac.ir
[2] Medical Information and Network Technologies Centre, Kingston University, London, KT1 2EE, U.K
{r.istepanian,n.philip}@kingston.ac.uk

Abstract—It is well known that 4G systems aim to provide such high speed, high capacity, and IP based personalized services for nomadic and mobile wireless environment. From the mobile health perspective, there are some restrictions to existing broadband wireless technologies and their strategies for different care services that can utilize the mobile broadband capabilities of 4G networks. In particular, the medical Quality of Service (m-QoS) issues and their guarantee to provide robust m-healthcare services with clinically acceptable quality.

In this paper, we address the relevant medical quality of service requirements for real-time ultrasound video streaming in LTE networks and the corresponding parameters that are specified from the doctors and clinical end perspective. The relevant performance analysis of medical video streaming model is validated using LTE OPNET® modeler.

Keywords—LTE advanced, QoS, medical QoS, video streaming.

I. INTRODUCTION

The evolution of m-health system towards providing mobile broadband services on the move is closely correlated with the evolution of 4G networks [1]. A decade ago, wireless broadband systems were introduced to substitute wired broadband systems with high speed wireless data transfer. These days, 4G networks are deployed in near future which can be a significant growth in m-health area [11]. It is well known that 4G systems aims to provide such high speed, high capacity, low cost, IP based services for nomadic and mobile wireless environment [12]. In general, mobile multimedia service, mobile ubiquitous network, and AAA (Anytime, Anywhere, and Always on) access are main targets of beyond 3G (3rd Generation) networks. The performance of LTE (long term evolution) networks for delivering ultrasound video is studied in this work.

There have been major research initiatives on m-health systems using current broadband wireless access technologies [2]. However, there are some restrictions to existing broadband wireless technologies and their strategies for health care services which hindered the wider applications of m-health technologies across healthcare systems. The Quality of Service (QoS) issues and their guarantee to provide robust m-healthcare services with clinically acceptable quality. In order to future 4G-based m-health systems to be clinically acceptable. These services must provide the end users (patients, healthcare provider) with an acceptable medical quality of service (m-QoS) from the doctors and clinical end perspective.

The main research focus of this paper is to investigate particularly the LTE performance and QoS challenges to assign the proper value to QoS class parameters.

The paper is organized as follows. In Section II, we review LTE technology and architecture. In Section III, we explain the LTE quality of service parameters. In Section IV, we explain the network simulation test bed. In Section V, we present the results and analysis. Finally, Section VI concludes the paper with recommendations for future work in this area.

II. LTE ARCHITECTURE

LTE is a significant broadband standard that was introduced by 3G partnership project (3GPP). This technology is a suitable choice for different m-health applications and scenarios. With the introducing of HSDPA, the significant aspects of the management of the air interface have been moved from the radio network controller (RNC) down into the base station or Node B in 3G terminology. Moving this functionality closer to the air interface means that the 3G system can react more quickly to changes in the quality of the wireless link between the user and the Node B and to the user's data requirements. In turn, this efficient and reactive control allows for higher data throughput and greater cell capacity. LTE, are formally known as E-UTRA (Evolved UMTS terrestrial radio access), represents the latest 3GPP Release 8 technology and aims to provide optimized packet data support in the uplink and downlink. LTE provides higher peak data rate, better capacity, low latency, and better coverage [5]. LTE deploys e-NodeB

Figure 1 shows the evolution line of Universal Mobile Telecommunications System (UMTS) networks. HSUPA is deployed on the top of the WCDMA network either on the same carrier or using another carrier [6]. The LTE technology is based on 3GPP standard that includes a considerable improvement on HSPA networks. LTE supports both frequency and time Duplexing mode (FDD and TDD). LTE network is also called E-UTRAN (Evolved UMTS Terrestrial Radio Access Network).

Figure 2 compares the 3G and LTE network architectures. The LTE uses e-NodeB instead of NodeB in 3G network which includes radio network controller (RNC) functionality.

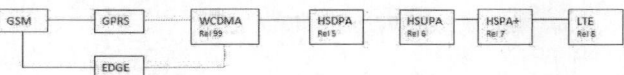

Fig. 1 LTE migration path [79]

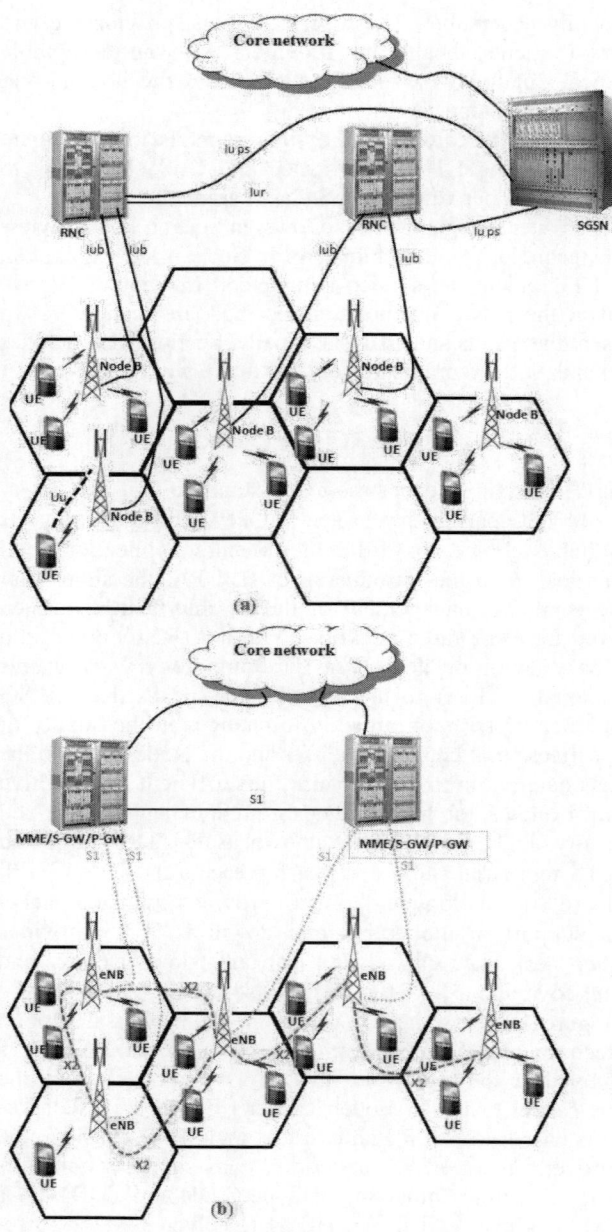

Fig. 2 Network architecture (a) 3G, (b) LTE [6]

The UMTS core network consists of two domains: the Circuit Switched (CS) service domain and the Packet Switched (PS) service domain. These two domains are responsible for providing appropriate services to the circuit switched traffic such as voice and the packet switched traffic such as web and other Internet Protocol related applications, respectively.

The NodeB/eNB provides wireless coverage for UEs, the interface between the UE and the Node B is known as Uu. One or more Node Bs are controlled by the RNC through Iub interface, and finally RNCs communicate with core network through the SGSN with Iu-ps interface [6]. A set of RNCs and Node Bs make up a Universal Terrestrial Radio Access Network (UTRAN). The Iu-cs and Iu-ps interfaces are used to connect the UTRAN to the circuit switched service domain and the packet switched service domain respectively. LTE offers enhanced data rates, fast packet retransmission mechanisms, and reduced packet latencies.

eNB can communicate with other eNBs through X2 interface to provide handover mechanism and mesh network functionality. The uplink data rate for LTE is increased up to a theoretical maximum of 75 Mb/s. One of the techniques used to achieve this data rate is adaptive channel coding which adjusts the amount of error correction according to load and channel conditions. And Hybrid ARQ (HARQ) packet transmission techniques and the 2-ms TTI (transmission time interval) also are copied from HSDPA.

Table 1 compares LTE, HSPA, and HSxPA in terms of data rate and round trip time (RTT). In this paper, we consider an m-health scenario including ultrasound video streaming node (Figure 3).

Table 1 3G/4G technology comparison [r2]

Technology	LTE	HSPA+	HSxPA
Data rate (Downlink/ Uplink)	150 Mbps/75 Mbps	28 Mbps / 11 Mbps	14 Mbps/ 5.7 Mbps
RTT	~10 ms	< 50 ms	<100 ms

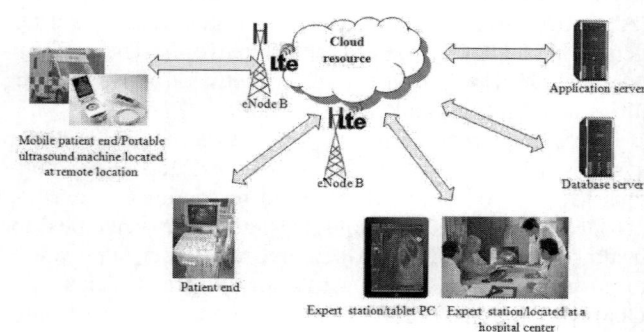

Fig. 3 LTE m-health scenario

III. LTE QOS

The QoS in LTE technology is similar to UMTS, it is implemented through EPS-Bearer (Evolved Packet System). The bearers set-up a data flow between UE and packet data network gateway. Each UE (User Equipment) uses up to 8 EPS bearer. LTE architecture supports two types of Evolved Packet System bearers (EPS-B) as follows [8]:

GBR: admission control allocates required resources permanently.

Non-GBR: there is no admission control policy. This type is used as default bearer in initial connection stage [7].

QCI (QoS class identifier) includes a set of parameters such as: priority, packet delay, packet error rate, and data rate.

The GBR service type use uplink/downlink guaranteed bit rate and uplink /downlink maximum bit rate to implement scheduling algorithm. Two other kind of bearer, default and dedicated bearer, are also used, the default bearer is a non-GBR service for unassigned traffic. And dedicated bearer is either GBR or non-GBR associated service with a specific TFT (traffic flow template) packet filter.

Table 2 describes the QoS parameters. QCI value specifies the QoS class definition. The QCI range is between 1 to 9 which QCI 9 is usually used for as a default QoS bearer.

Admission and pre-emption function block uses ARP (Admission retention policy) to control service admission policy and pre-emption of existing services. However this parameter is usually deployed in GBR service type.

Table 2 LTE QoS parameters

QCI	Quality of service classification			Admission and pre-emption	Typical services
	Service Type	Packet latency	Packet loss	ARP	
1	GBR	100 ms	10^{-2}	2	Conversational voice
2		150 ms	10^{-3}	4	Conversation video
3		50 ms	10^{-3}	3	Real-time gaming
4		300 ms	10^{-6}	5	Non-conversational video
5	Non-GBR	100 ms	10^{-6}	1	IMS signaling
6		300 ms	10^{-6}	6	Video, browsing, ftp, file sharing,…
7		100 ms	10^{-3}	7	Voice, live video, interactive gaming
8		300 ms	10^{-6}	8	Video, browsing, ftp, file sharing,…
9		300 ms	10^{-6}	9	Video, browsing, ftp, file sharing,…

IV. SIMULATION SET-UP AND RESULTS

The m-health LTE simulation set-up is shown in Figure 4. This figure illustrates both the ultrasound traffic and cross traffic paths. The main traffic is used to simulate ultrasound video streaming between the patient side (node UE) and the expert side (hospital); and the cross traffic is deployed to emulate the real environmental traffic between the nodes. The simulation scenario includes 1 eNB and 8 UEs, where the UE is generating ultrasound video streaming data traffic presented in [4]. And the rest of UEs are used to generate cross traffic. The medical expert's station is connected to the eNB through a LAN as shown in the simulation set-up (Figure 4). The following evaluation metrics are used as performance outcomes: Application response time for different HARQ retransmission number 0 and 3, Uplink Packet dropped rate. Based on different simulation results against different wireless SNR, the LTE QoS parameter is set to QCI 4 that means packet latency less than 300 ms and packet loss probability is less than 10^{-6}.

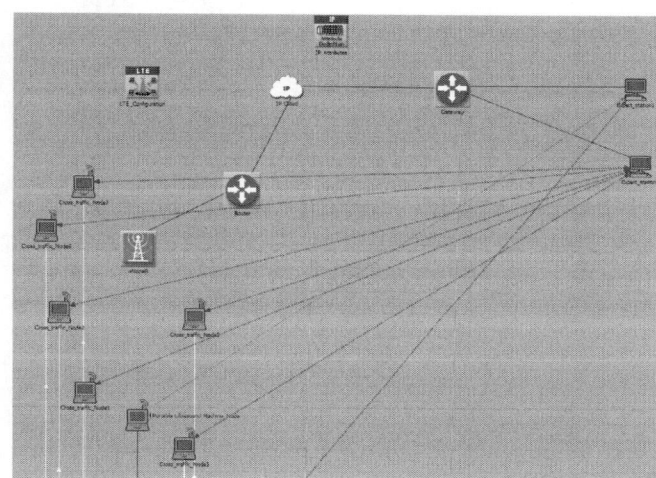

Fig. 4 Network simulation topology

Figure 5 shows a sample of the simulation results. Figure 5(a) shows the application response time against SNR, and figure 5(b) shows uplink packet loss rate against SNR for different HAR retransmission numbers. It is clear that a proper modulation and coding should be chosen. The simulation results show that HARQ number in medical video streaming with m-QoS considerations [4] can have different HARQ number to overcome packet loss issue. In the mean time it is acceptable in terms of delay index.

The network parameters are set as a hard code. However, a cross layer optimiser can be implemented to configure the modulation and coding rate and QCI priority to have better network throughput.

From a technical point of view, the LTE network can provide an acceptable m-QoS. However, the QoS mapping and optimize modulation and coding increase system performance and throughput in terms of network resource allocation and m-health diagnostics criteria [9].

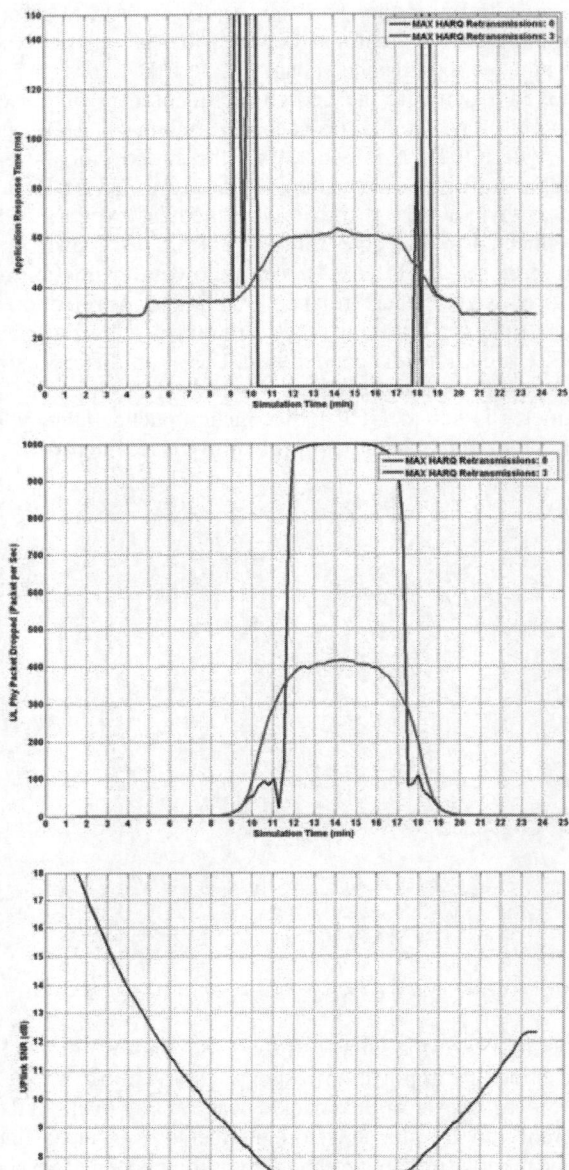

Fig. 5 HARQ Retransmission analysis. (a) Application response time. (b) Uplink packetloss

V. CONCLUSION AND FUTURE WORKS

In this paper, we described the LTE technology and its QoS parameters, and we assigned the proper QoS parameters to LTE QoS class's parameters to provide acceptable medical QoS. Also a typical ultrasound video streaming has been simulated over LTE using OPNET® modeler to investigate the network performance and QoS issues for medical applications. The paper investigated HARQ retransmission effect on medical video streaming. A Cross layer implementation can guarantee an optimal HARQ number against wireless network status.

Future studies such as QCI mapping to m-QoS and cross layer HARQ and modulation in m-health applications are also ongoing as part of this research study.

REFERENCES

[1] Istepanian, R. S. H. and Zhang Y.T., "4G health: The Long Term evolution of m-health- Editorial," *IEEE Transactions on Information Technology in Biomedicine*, vol. 16, no. 1, pp. 1-5, Jan. 2012.

[2] Panayides, A., Pattichis, M.S., Pattichis, C.S. and Pitsillides, A. (2011) 'A Tutorial for Emerging Wireless Medical Video Transmission Systems [Wireless Corner]', *IEEE Antennas and Propagation Magazine*, vol. 53, no. 2, pp. 202-213

[3] Kaur, B. (2010) 'Factors Influencing Implementation of 4G with Mobile Ad-hoc Networks In m-Governance Environment', *International Journal of Computer Applications*, vol. 3, pp. 141-146

[4] Alinejad, A. ,Philip, N.Y., and Istepanian, R.S.H., "Cross-Layer Ultrasound Video Streaming Over Mobile WiMAX and HSUPA Networks," *IEEE Transactions on Information Technology in Biomedicine*, vol. 16, no. 1, pp. 31-39, Jan. 2012.

[5] Stefan Parkvall, anderS furuSkÄr, erik dahlman, "Next generation LTE, LTE-Advanced," *Ericsson review*, 2010.

[6] Johnson, C. (2008). Radio Access Networks for UMTS: Principles and Practice. Wiley-Blackwell.

[7] S. M. Chadchan and C. B. Akki, "Priority-Scaled Preemption of Radio Resources for 3GPP LTE Networks", International Journal of Computer Theory and Engineering, Vol. 3, No. 6, December 2011

[8] 3GPP TS 23.203 V8.12.0 (2011-06), *3rd Generation Partnership Project;Technical Specification Group Services and System Aspects; Policy and charging control architecture (Release 8)*, 2011.

[9] Alinejad, A., Istepanian, R.S.H., and Philip, N.Y., "Dynamic Subframe Allocation for Mobile Broadband m-health using IEEE 802.16j Mobile Multihop Relay Networks," *IEEE EMBC12*, Sep. 2012.

Smart Social Robotics for 4G-Health Applications

R.S.H. Istepanian[1], A. Good[2], and N. Philip[1]

[1] Faculty of Science, Engineering and Computing, Kingston University, London, UK
[2] School of Computing, University of Portsmouth, Portsmouth, Hampshire, UK

Abstract—The concept of 4G health was recently introduced and defined as *'The evolution of m-health towards targeted personalized medical systems with adaptable functionalities and compatibility with the future 4G networks'*. One of the primary applications of 4G health is in the social robotics domain. A key challenge in this domain will be the full understanding of the interaction between humans and social robots. In particular in understanding the issues of QoS and QoE in such environments, which contribute to the quality of interaction.

In this paper we highlight some of the key challenges and opportunities of human robotics interaction using 4G-healthtechnologies and their relevant QoS and QoE issues.

Keywords—Social robotics, m-health, 4G health, QoE, HRI.

I. INTRODUCTION

The concept of m-health was first introduced and defined as "mobile computing, medical sensor, and communications technologies for healthcare" [1]. Since then, it has become one of the main pillars within the ICT for healthcare domain bringing together major academic research and industry disciplines worldwide. The introduction of 4G technologies and networks in this decade will bring new services and consumer usage models that will be compatible with these emerging mobile network architectures. More recently the concept of 4G health was introduced and defined as "The evolution of m-health towards targeted personalized medical systems with adaptable functionalities and compatibility with the future 4G networks." [2].

Fig. 1 shows the general concept of 4G-health. One of the emerging applications within 4G –Health will be in the social robotics domain.

This emerging domain is becoming increasingly used to provide better interaction between users and their daily activities in general [3]. In the healthcare domain especially in the assisted living and chronic disease management, the potential of social robotics to sustain people's engagement as well as to motivate, coach, educate, facilitate communication, monitor performance, improve adherence to health regimen, and provide social support to people are being increasingly investigated in recent years [4].

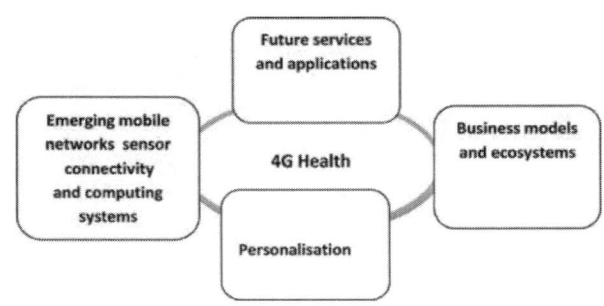

Fig. 1 General concept of 4G - Health

The recent advancement in smart social robots for mobile health applications implies greater emphasis in fully understanding the interactions between humans and robots. In this paper we address and discuss the potential challenges relevant to the quality of human experience when interacting with social robotics particularly in mobile health environments. In such environment the smart social robotic can help the patient in managing ubiquitously their user network and their wireless medical devices and sensor connectivity and the communications with their carers. Fig 2 shows a conceptual diagram of a future 4G health using social robotic technology to support Type 1 Diabetes (T1D) children in their daily management and behavioral change issues. In this system the social robotic system acts as the intelligent conduit between the patient and his daily medical sensors and behavioral patterns from one end and his/her healthcare provider and family members from the other end. This paper is organized as follows; in section two we introduce the relevant medical quality of experience issues. .In section three we discuss the Human Robotic Interaction (HRI) in such environment. In section four we present the relevant medical Quality of Experience (m-QOE) in the context of (HRI).. Finally, section five concludes the paper with future research directions in this area.

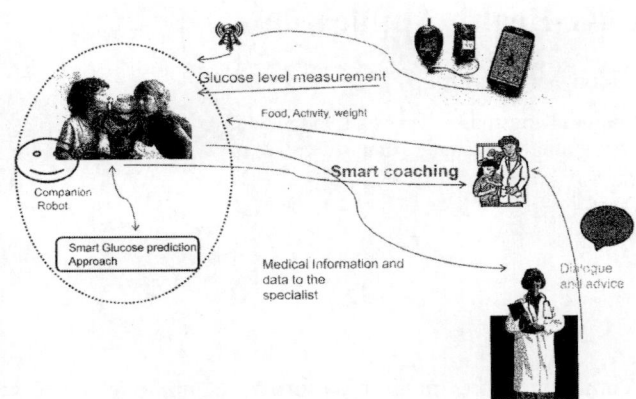

Fig. 2 Conceptual model of 4G health-robotic coaching system for T1D management.

II. QUALITY OF EXPERIENCE FOR 4G HEALTH

More recently, the focus of different network service providers is switching from network quality of service to user quality of experience that describes the overall performance of network from the user's perspective.

Implementing QoE management in real systems is a complex process. The complexity of QoE is mainly due to the difficulty of its modeling, evaluation, and translation to quality of services (QoS). Whereas QoS can easily be objectively measured, monitored, and controlled at both the networking and application layers, and at both the end system and network ends. However the relevant QoE issues are still quite complex to manage especially from the m-health perspective. Indeed, the variables that affect QoE are many and changeable, and span several disciplinary areas including not only multiple technologies but also human-computer interaction, psychological, and sociological factors. These mechanisms constitute major challenges on future 4G health systems in assuring both end-to-end QoE and QoS for next generation networks.

As mentioned earlier, the QoS/QoE indices are dependent on lower network layers and their QoS requirements. In order to match the required resources with the available network conditions, there must be a tradeoff between these QoS requirements in order to satisfy the diagnostic (clinical) requirements and the required bounds of relevant QoS metrics. The concept of m-QoS was introduced as a sub category of QoS from the m-health perspective and defined as: ' An augmented requirements of critical mobile health care applications and the traditional wireless Quality of Service requirements' [5].

From the 4G health perspective, the medical quality of experience (m-QoE) can be defined 'as the overall acceptability of m-health application as perceived subjectively by the end users of these applications' [6]. In general, the m-QoE considers all the patients and stakeholders acting as healthcare providers (nurses, doctors) and their perception of the underlying next generation network services and their performance from the medical perspective. In general, QoE has many contributing factors, among which some are subjective and not controllable, while others are objective and can be controlled, some specific subjective factors include factors such as healthcare provider/patient emotion, experience, and expectation. These issues become more clearly evident in areas of social robotic for mobile health applications that warrant further studies and work in this area.

III. HUMAN ROBOTIC INTERACTION

The recent advancement in smart social robots for mobile health applications implies greater emphasis in fully understanding the interactions between humans and robots. Primarily, the success of the robots lies not only in their utility, but also the enablement of a natural and intuitive interaction. Research into the quality of effective HRI is increasingly showing that people respond better when robots are able to respond in more natural and human-like ways [7]. The robot needs to appear to be socially intelligent, even though it is essentially a machine. This however requires significant understanding into the myriad of human behaviors that relate to interaction and how that can be modeled. Particularly, how robots can respond to individual personalities, in as human-like manner as is possible.

The quality of interaction between human are dependent upon the effectiveness of the robot in managing users' health problems. However, to ensure high quality interaction, there are numerous factors to consider, including: acceptance; adaptation; empowerment; embodiment; Natural Language capabilities, the extent of autonomy and many others. Understanding and modeling this interaction between humans and robots is a complex procedure and involves a multidisciplinary approach, with knowledge and application of multiple theories and practices.

When designing robots to deliver healthcare, one of the most important factors to consider is acceptance. Clearly, if the user has difficulties or an unwillingness to accept the robot, this will impact upon the robot's effectiveness in providing health care. Acceptance can be affected by a number of factors, including the naturalness of interaction as well as naturalness of the design of the robot [8] We already know that people are more likely to engage with an embodied as opposed to a virtual agent [9]. Indeed the design of the robot itself presents major challenges, further complicated by the fact that different types of users are likely to respond to the degree of anthropomorphic design [10]

People's mental model of robots is a crucial factor impacting upon the quality of interaction. To some extent this is interrelated to the design of the robot. People who have anthropomorphic mental models of robots tend to evaluate the social awareness of these robots in terms of personality traits that are relevant to humans [11].

In this paper however, we focus more upon interaction and the challenges of acceptance as opposed to the specific design of the robot.

Many studies have addressed the issue of 'social acceptance' [12] given that the social capability of the robot can potentially impact upon user acceptance and subsequently, the quality of the interaction. Social interaction is an important consideration to the design of health care robots for children [13]. For example, how socially aware is the robot? Can it detect emotional expressions and other non verbal cues? How well does it perceive and respond to emotional cues? Social and expressive gestures are crucial components of both human-human and human-robot interaction. Furthermore, Breazeal 's extensive research in this area also shows that robots need to display behaviour such as attention and reactivity [14,10] Clumsy robots with jerky movements can potentially impact upon acceptance and so body pose and movement are also important sources of information for social interaction [15]. The quality of social interaction is an important factor because it dictates the extent to which the user will respond to the robot, and in turn, the effectiveness of the health care.

Adaptivity ensures the robot adapts to users' needs, expectations (partly based upon mental models) and history. With children in particular, engagement is likely to be much higher at the beginning, however, the robot needs to be able to adapt to the child's 'boredom' once the novelty has worn off. This would of course be particularly important with healthcare robots. To enable the robot to respond to users' personality is far more complex than speech or gestures. In a recent work in this area, the importance of robot adaptivity and its role in long term of adoption of health care robots in children has been highlighted [13].

Patient empowerment is a priority. If robots can teach children to take responsibility for their health care in important area such as diabetes and from a young age, they are less likely to neglect it as they reach puberty. These social and care robots will be designed to support patients but at the same time, can be empowering to the user in much the same way as domestic robots support and empower elderly in independent living conditions[13, 16]. Furthermore, the challenges of behavioral change remain open for further work and in particular how these robotic systems can be used to assist in the adaptation of such change. This becomes more important in T1D for younger patients and their long term disease management issues.

IV. M-QOE IN SOCIAL ROBOTIC ENVIROMENTS

In this section, we describe briefly the concept of m-QoE presented earlier in the context of social robotics within the 4G health environments. In m-heath applications providing the necessary QoS and QoE in uncertain wireless environments is a challenging problem that requires further research work.

Furthermore, there are a number of factors that determine the quality of human robot interaction and their effect on the QOE issues. In general, HRI metrics are likely to include effectiveness, efficiency and performance, in relation to system performance, in much the same way as one would expect from any human machine interaction. In looking specifically at QoE, we can see there are many psychological and sociological factors that contribute to the complexity of designing such modeling interaction. Metrics are then extended to include trust; engagement; persuasiveness and enjoyment.

In particular, for social robotics environment, we have identified the following as the main QoE metrics in ensuring a quality interaction:

Acceptance - This is one of the driving factors in ensuring the robot is effective in delivering healthcare. One of the main issues here is the naturalness of the interaction.

Engagement – how well the robot captures attention, which will be dependent upon many social characteristics. This is as important in human robot interaction as it is with human to human.

Trust – this is impacted by both design and interaction. One of the key factors in ensuring reliance on a system is based precisely upon trust.

Enjoyment – the degree of fun and enjoyment from the interaction can significantly affect the appeal, particularly relevant with children.

These factors are important in the relevant estimations of m-QOE issues for mobile health applications.

V. CONCLUSIONS

In this paper, we discussed the key challenges and opportunities of the integration of social robotics within the 4G health domain. In particular, we highlighted some of the key categories of medical quality of experience issues from the social robotics perspective and also the key human robotic interactions relating to acceptance. The promise of these technologies will depend on how to augment key 4G wireless and short range communication technologies with these robotic systems that can provide new mobile healthcare services ranging from chronic disease management to assistive living and aging support. These need to address not only the important communication and HRI challenges that have been highlighted here, but also the business and relevant ecosystems models that will be both cost effective and can be scalable and acceptable to wider and diverse users.

1. This is a growing and emerging field of research that continues to require further work in order to fully understand the possibilities and research challenges including:
2. How can a new generation of 4G Health based social robotics be designed and tailored for key mobile health applications and services?
3. How can we address ethical and security issues so that these robots are more acceptable to clinicians and medical practices?
4. How can these robotic systems be compatible with the current wireless and network standards and compatible with their interoperability issues?
5. How can we design for the highest quality of interaction, based upon the identified QoE: acceptance, engagement, trust and enjoyment?

Ongoing work is currently underway to design and implement such system with the focus on T1D patients. The aim is to study the relevant challenges both from the clinical, technological and psychological perspective.

In conclusion this area will be one of the key enabling research domains that will integrate different medical, social and engineering disciplines for new and innovative mobile healthcare solutions and services.

REFERENCES

1. Istepanian RSH, Jovanov E, Zhang YT (2004) Beyond seamless mobility and global wireless health-care connectivity. Guest editorial on m-health: IEEE Trans Inf Technol in Biomed, 8: 405–414.
2. Istepanian, R. (2012), Guest Editorial: 4G health – the long-term evolution of m-health, IEEE Trans Info Technol in Biomed 16: 1-5 .
3. Bickmore, T. and Picard, R. (2005). Establishing and maintaining long-term human-computer relationships. ACM Transactions on Computer-Human Interaction (TOCHI), 12(2):293–327.
4. Okamura, A. M., Mataric, M. J. and Christensen, H. I. (2010), Medical and healthcare robotics: Achievements and opportunities. IEEE Robotics and Automation Magazine vol 17, issue 3: 26-37.
5. Istepanian, R., Philip, N., & Martini, M. (2009). Medical QoS Provision Based on Reinforcement Learning in Ultrasound Streaming over 3.5G Wireless Systems. Selected Areas in Communications, IEEE Journal on , 27 (4), pp. 566-574 .
6. Istepanian, R. S. H., Alinejad, A. and N. Philip (2012) Medical Quality of Service (m-QoS) and Quality of Experience (m-QoE) for 4G-Health Systems. In: Reuben A. Farrugia and Carl J. Debono (eds.) Medical Multimedia Networking and Coding. IGI Global. pp. 359-377. ISBN13: 9781466626607.
7. Scheutz, et al. (2007). First Steps toward Natural Human-Like HRI. Autonomous Robots 22(4):411–423.
8. Scholtz, J. and Bahrami, S. (2003). Human-Robot Interaction: Development of an Evaluation Methodology for the Bystander Role of Interaction. *Proceedings of the IEEE Conference on System, Man, and Cybernetics, 2003*. Washington, DC. Oct. 2003.
9. Kidd C., and Breazeal C. (2004) Effect of a robot on user perceptions. In IEEE/RSJ International Conference on Intelligent Robots and Systems, pages 3559–3564, Sendai, Japan.
10. Lee, J. and Breazeal, C. (2010) Human social response toward humanoid robot's head and facial features. CHI Extended Abstracts: 4237-4242.
11. Kiesler, S., and Goetz, J. (2002) Mental Models and Cooperation with Robotic Assistants, Proceedings of the ACM CHI'02 Conference on Human Factors in Computing Systems, pp. 576–577.
12. Beer, J., Smarr, C., Chen, T., Mitzner, T., Kemp, C. and Rogers, W. (2012). The domesticated robot: Design Guidelines for assisting older adults to age in place. Proceedings of the 7th IEEE/ACM International Conference on Human-Robot Interaction (HRI'12), Boston, MA.
13. Baxter, P., Wood, R., Baroni, I., Kennedy, J., Nalin, M. & Belpaeme, T (2013). Emergence of Turn-taking in Unstructured Child-Robot Interactions. In Human Robot Interaction (HRI) 2013, Japan.
14. Breazeal, C., Gray, J. and Berlin, M. (2009) An Embodied Cognition Approach to Mindreading Skills for Socially Intelligent Robots. I. J. Robotic Res. 28(5): 656-680
15. Rizzolatti G, and Arbib M A (1998). "Language Within Our Grasp." Trends Neurosci. 21, 188–194. doi: 10.1016/S0166-2236(98)01260-0.
16. Prescott T, Epton, T, Evers V, McKee K, Hawley M, Webb T and the Robot Companions for Citizens Society Community Working Group (2013), Robot Companions For Citizens: Roadmapping The potential for future robots In empowering older people, BRAID (Bridging Research in Ageing and ICT Development) Final Conference, Prague, May 29th, 2012. Springer-Verlag.

Regional Cooperation in the Development of Biomedical Engineering for Developing Countries

K.I. Nkuma-Udah[1], G.I. Ndubuka[1], and E.E.C. Agoha[2]

[1] Nigerian Institute for Biomedical Engineering and Department of Biomedical Technology,
Federal University of Technology, Owerri, Nigeria
[2] Nigerian Institute for Biomedical Engineering and Department of Food Science and Technology,
Abia State University, Uturu, Nigeria

Abstract—Specific challenges have been identified in previous IFMBE conferences and other global biomedical engineering events where it was noted that many of the developing (less developed) countries, especially those of the sub-Saharan Africa are not usually represented due mainly to those factors that made them really developing.

The Council of Societies (CoS) of the International Federation for Medical and Biological Engineering (IFMBE) has recently subdivided the biomedical engineering world into CoS Regional Groups, which are expected to meet independently in the respective Regions. Therefore the CoS through the Regional Groups serves as the main communication and feedback channel between the International Federation for Medical and Biological Engineering (IFMBE) and its Constituent Societies.

The groupings are Asian-Pacific Rim, North-America, Latin-America and Europe-Africa. Here, the mapping of Africa with Europe in same group should be seen as strategic. For one, IFMBE has provided a forum for a part of developed biomedical engineering world to enable the development of a less developed biomedical engineering world.

Therefore, Africa and other developing countries so mapped are expected to grab this opportunity with its attendant benefits. These benefits doubtlessly include the opportunities of joint conferences, exchange academic and professional programmes, tapping and exposure to the global industrial world.

Keywords—Regional, Cooperation, Developing, Biomedical, Engineering, Countries.

I. INTRODUCTION

Regional Cooperation can be said to be a frame work in which the containing entities agree to work together. The goal of a regional cooperation is usually to promote mutual existence and to inspire development in the region to the benefit of the component entities of the region.

Biomedical engineering (BME) defined as a discipline that advances knowledge in engineering, biology and medicine and improves human health through cross-disciplinary activities that integrate the engineering sciences with the biomedical science and clinical practice, studies the human body and other living systems from the perspective of engineering. It uses engineering concepts to appreciate how living system works and applies them for the development of devices and processes that improve biology, medicine and healthcare.

Biomedical engineering is responsible for design and development of the technology and devices that are at the heart of the far-reaching improvements in human health. It has grown as a result of the integration of various disciplines and so is performing a wide and vital role of integrating the entire sciences as one by the opportunities of inter – and cross- disciplinary activities it offers.

Now, developing countries, sometimes called less-developed countries, are nations with a low level of material well-being [1]. Though, the levels of development may vary widely within the developing countries group, with some having high average standards of living.

From the World Bank's perspectives, countries are classified into four income groups as (a) low income countries having GNI per capita of US$1,005 or less, (b) lower middle income countries having GNI per capita between US$1,006 and US$3,975, (c) upper middle income countries having GNI per capita between US$3,976 and US$12,275, (d) high income countries having GNI above US$12,276 [2]. The World Bank classifies all low- and middle-income countries as developing but notes that, "Classification by income does not necessarily reflect development status."

In the context of the biomedical engineering community, the factor of *developing region* includes a region in which biomedical engineers and technicians do not have sufficient resources to function effectively, whether individually or collectively. This may be due to limited logistical and financial resources, limited or lack of access to professional information or the lack of opportunities to meet with professional colleagues. As a result, there is poor representation of the developing countries in the global community of biomedical engineering.

The poverty level in developing countries have resulted in health care per capita spending that is almost two orders of magnitude lower than in developed countries [3]. Again it is noted that low-income countries spend US$ 32 per capita on health care, and high-income countries spend US$ 3,724 per capita. So, low-income countries health expenditures fall far short of the US$ 60 per capita that the WHO posits is necessary for an adequately functioning health system [4].

As developed countries delve into the realms of advanced medical devices, nano-biotechnology and decoding the human genome, many developing countries are still struggling to give the conventional plain x-rays machines to those who need them, to provide the sick with non-counterfeit drugs, and to accurately diagnose common ailments such as malaria, typhoid fever and HIV in the field. So despite all the research and progress that has gone into medical technologies in developed countries, developing countries, which comprise a majority of the world's population, have minimal, if any, access to many of these innovations [5].

II. BACKGROUND

A major reason why there should be region cooperation between the developed and the developing countries to enhance the biomedical engineering of developing countries is that most of the major health problems on this planet, which the biomedical technologies are expected to solve are not in the developed countries. For instance, life expectancies in developed countries are now 75-80 years of age, but in many developing countries, life expectancies are about half that. In many developing countries, 10 to 15 percent of babies born do not live to their first birthday.

The Human Development Report of the United Nations Development Programme (UNDP) in 2005 indicated that most developing countries spend US $100 per capita or less on their health ministries/departments while the US spend $5274 per capita. This translates to a lower percentage allocation to medical (or health) technology and to eventually a meager percentage to manpower development in the health technology sub-sector. Therefore many developing countries have limited access to healthcare technologies [6].

The factor of poverty is a serious issue for many countries in the developing countries regions. This affects their technology development in general and the development of biomedical technology in particular. The World Health Organisation has estimated that 40- 60% of medical equipment in the developing world is unserviceable and that there is a serious shortage of resources and trained personnel.

Aside from the limited access to healthcare technologies, the potential of the technologies that are available is often not realised. One reason for this is the lack of appropriate maintenance management because the healthcare technology management (HTM) of such country is not developed. Where they are present, the HTM practitioners and other decision-makers often cannot find the basic technology-related information. Sometimes the HTM practitioners are not trained and often lack the logistical and financial resources to meet their peers from other countries and regions.

Many developing countries depend on medical equipment donation. When this happens, some of these countries still find it difficult to foot the bills of clearing, transportation, installation, manpower training and maintenance. Besides this, most donated equipment are out dated and refurbished, so the recipient countries do not usually enjoy the benefit of modern and advanced medical technology until some 20-30yrs of existence of such technologies.

There is paucity of biomedical engineering professionals in the developing countries. Where they exist they are mainly technicians. Very few are engineers and these are seen in the public health system. There are very few training institutes so most of the technical personnel are foreign-trained or came from externally supported programmes.

There has been a lot of concern about technology transfer from the developed to the developing world. This concern have not really been addressed neither has the need been met. One reason for this unmet need is the issue of appropriateness aside from the affordability of these technologies. Yet the need for a robust system of engineering and innovation within developing countries is what is needed more than these desire for superfluity. This is true because, foreign technology can only take a resource-limited country so far; in order to be truly sustainable, innovation must be indigenous – that is, it must come from within.

Specific challenges have been identified in previous IFMBE conferences and other global biomedical engineering events where it was noted that many of the developing (less developed) countries, especially those of the sub-Saharan Africa are not usually represented due mainly to those factors that made them really developing.

III. THE IFMBE STRATEGIC PLAN

IFMBE is the International Federation for Medical and Biological Engineering. It is primarily a federation of national and transnational professional societies, which represent interests in medical and biological engineering. The IFMBE is also a Non-Governmental Organization (NGO) for the United Nations and the World Health Organization (WHO), where it is uniquely positioned to influence the delivery of health care to the world through Biomedical and Clinical Engineering.

The IFMBE's objectives are scientific and technological as well as educational and literary. Within the field of medical, biological and clinical engineering IFMBE's aims are to encourage research and application of knowledge, and to disseminate information and promote collaboration through organizing World Congresses and Regional Conferences, publishing its flagship journal Medical & Biological Engineering & Computing (MBEC), web-based newsletter – IFMBE News, Congress and Conference Proceedings, and books. The ways in which IFMBE promote collaborations is through networking programs, workshops and partnerships with other professional groups, e.g., Engineering World Health.

In IFMBE's Strategic Plan, the Clinical, Medical and Biological Engineering professions are seen as integrating objects for physical, mathematical and life sciences with engineering principles for the study of biology, medicine and health systems and for the application of technology to improving health and quality of life.

In line with this, the mission of the IFMBE is to encourage, support, represent and unify the world-wide Medical and Biological Engineering community in order to promote health and quality of life through advancement of research, development, application and management of technology.

In order to realize this mission, IFMBE developed six achievable Goals. Goal 4 is specifically: To advance collaboration between national and transnational societies, industry, government and non-governmental organizations engaged in health care and in biomedical research and its applications; to achieve this, IFMBE is meant to:

- Establish effective communication between constituent societies.
- Establish an Industrial Forum. Collaborate and work with industry at the national, regional and international levels to promote "better health for all humanity".
- Assist in addressing special needs of developing countries and regions.

IV. REGIONAL COOPERATION IN BIOMEDICAL ENGINEERING

The Council of Societies (CoS) of the International Federation for Medical and Biological Engineering (IFMBE) in its charter subdivided the biomedical engineering world into CoS Regional Groups, which are expected to meet independently in the respective Regions. Therefore the CoS through the Regional Groups serves as the main communication and feedback channel between the International Federation for Medical and Biological Engineering (IFMBE) and its Constituent Societies.

An IFMBE Region is defined by a justifiable distribution of Constituent Societies in a Region and an appropriate number of members in the Constituent Societies. The discussion on the Region of Constituent Societies and their members required for formation of a Region is initiated by the Members of CoS, and a recommendation is made by the Executive Board of the CoS to IFMBE Administrative Council (AC). Redefined and/or Additional Regions are formed based on approval by AC in response to justifiable requests from associated Constituent Societies through the Executive Board of the CoS. The Constituent Societies have to be in good standing, in order for the Members of the CoS and RG Representatives to vote.

The IFMBE Regional Groupings are Asian-Pacific Rim, North-America, Latin-America and Europe-Africa. Here, the mapping of Africa with Europe in same group should be seen as strategic. For one, IFMBE has provided a forum for a part of developed biomedical engineering world to enable the development of a less developed biomedical engineering world.

Therefore, Africa and other developing countries so mapped are expected to grab this opportunity with its attendant benefits. These benefits doubtlessly include the opportunities of joint conferences, exchange academic and professional programmes, tapping and exposure to the global industrial world.

V. CONCLUSIONS

Regional cooperation in biomedical engineering will entail organisation of of joint conferences, exchange academic and professional programmes, tapping and exposure to the global industrial world. Then developing biomedical engineering in the developing countries entails developing resources for biomedical engineering in those countries including both material and manpower resources. This should be a matter of concern for the international biomedical engineering community considering the population of these group and the global resources waiting to be tapped.

Therefore, IFMBE should encourage more regional conferences and other programmes within the intervals between the world congresses. In addition to this, IFMBE should facilitate exchange of academic and professional programmes

One way the regional cooperation can be encouraged for the sake of developing biomedical engineering in developing countries is by encouraging the hosting of some of the regional conferences in the developing countries themselves and formation of national or a more manageable transnational societies where they do not exist.

In line with the IFMBE's goal to "assist in addressing special needs of developing countries and regions", part of these special needs of developing countries include undertaking projects like establishing international programmes in biomedical engineering in the developing countries and the need for appropriate national and regional biomedical engineering societies where these are not present. One way IFMBE can strengthen and contribute to sustainable development of biomedical engineering in the developing countries is by providing regional training courses, supporting and sponsoring the organisation of regional conferences in these countries and supporting their participation in the world congresses.

REFERENCES

1. Wikipaedia the Free Encyclopaedia. "Developing Country". *http://en.wikipedia.org/wiki/Developing_country*. Retrieved 7th January, 2012.
2. The World Bank "How we Classify Countries". *http://dataworldbank.org/about/country-classifications*. Retrieved 7th January, 2012.
3. World Bank: International Development Goals: Strengthening Commitments and Measuring Progress. Washington DC: World Bank; 2001.
4. World Bank: *World Development Indicators*. London: World Bank; 2007.
5. World Health Organization Department of Essential Health Technologies, "Medical devices: managing the mismatch - an outcome of the Priority Medical Devices project," WHO, Geneva, Switzerland, 2010 [Online]. Available: http://whqlibdoc.who.int/publications/2010/9789241564045 eng. pdf
6. Nkuma-Udah K.I. et al, "Developing Biomedical Engineering in Africa: A Case for Nigeria", IFMBE Proceedings, Volume 14, Springer.

Address of the corresponding author:

Author: Dr Kenneth I. Nkuma-Udah
Institute: Federal University of Technology
Street: Department of Biomedical Technology
City: Owerri
Country: Nigeria
Email: drnkumaudah@yahoo.com

Author Index

A

Aadamsoo, K.   694
Abascal, J.   491
Abascal, J.F.P.J.   202, 214, 233, 415
Abásolo, D.   575, 698, 706, 795
Abdulla, T.   1225
Abella, M.   245, 257, 491
Abernethy, Laurence   309
Abreu, Miguel Henriques   1366
Abreu, Pedro Henriques   1366
Abtahi, F.   551
Accardo, A.   194, 579, 587, 591, 702
Accornero, N.   639
Acha, Begoña   352, 356, 360
Acosta, C.D.   811
Adame, M. Reyes   1763
Adamek, T.   939
Afonso, Noémia   1366
Afsharipour, B.   293, 674
Aggelinos, George   1306
Agnelo, José   419
Agnoletti, V.   1111, 1310
Agoha, E.E.C.   1923
Aguilera, E.J. Gómez   1775
Aguirre, E.   1159
Aguirre-Echeverry, C.A.   658
Ahanathapillai, Vijayalakshmi   1201, 1515
Ahmed, K.S.   1245
Ahn, H.C.   1413
Aiolli, Fabio   261
Aja-Fernández, S.   253
Ajcevic, M.   587, 591
Akan, A.   631
Akay, A.   1394
Åkervall, S.   1575
Alam, N. Riyahi   206
Alam, S. Riyahi   206
Alamedine, D.   779
Al-ani, T.   599
Alanzi, T.M.   1907
Alberola-López, C.   253, 989
Alcalde, M.A.   1481
Alcañiz, M.   273
Alcaraz, R.   973, 1005, 1009, 1013, 1025, 1879
Alcubilla, Cesar   1759

Aleksandravicius, A.   1911
Alemán-Gómez, Y.   218
AL Halabi, F.   1638
Ali, Zulfiqur   888, 892
Alinejad, Ali   1915
Almeida, R.H.   1182, 1186, 1489
Alonso, J.B.   571
Alvarado, J.A.   1551
Álvarez, A.   919
Álvarez, D.   762, 969, 1647, 1825, 1829
Á lvarez, Mauricio A.   182
Alves, José Nuno   157
Alves, Sergio S.   157
Amaro, Hugo   1366
Amat, Josep   109
Ambrosini, E.   1787, 1791, 1799
Amo, C.   627
Amor, James D.   1201, 1515
Anania, G.   93, 1547
Andersen, P. Berg   1358
Andersson, B.M.   1124
Ando, Joji   19
Andreadis, I.I.   305
Angeleri, M.   1079
Angulo, Jesús   317
Antoniadis, A.   376
Antoniou, Ioannis   783
Ara, J.   1738
Arana, S.   852, 864, 880
Aranda, J.   85
Arce-Diego, J.L.   427, 615, 1679
Arcentales, A.   1021
Arcuri, Giovanni   1468
Aresté, N.   1563
Argelagós, A.   962
Armas, J.A. Hernández   190
Armentano, R.L.   1049
Armitage, Roger   1477
Arnold, R.   1864, 1871
Arnoult, C.   1250
Arrabal, P.M.   1634
Arredondo, Maria Teresa   1468
Arrigo, P.   662
Arruda, M.B.   1237
Arvanitis, Theodoros N.   309, 734
Assecondi, S.   734
Astaras, A.   65, 1539
Athanasiou, A.   65

Atienza, F.   340, 423
Attanasio, M.   1132
Auer, Dorothee P.   309
Auger, V.   892
Augustin, C.M.   1864
Aviles, A.I.   372
Avrahami, I.   29, 33
Aydin, N.   819
Ayres-de-Campos, D.   555
Aza, I. Velázquez   1254
Azevedo, F.M.   666, 770, 1464
Aznar, Miguel A.   1759
Azpeitia, J.M. Álvarez   856
Azpilicueta, L.   1159

B

Baan Jr., Jan   1821
Babic, A.   1358, 1390, 1413, 1417
Bacchereti, M.   93
Bachiller, A.   583, 799
Bachmann, M.   694
Badnjevic, Almir   911
Bagshaw, A.P.   734
Bailón, Raquel   710
Bajelani, K.   509
Bajic, D.   1722
Bąk, M.I.   515
Bakas, A.   455
Bakola, Iliana   1213
Balay, G.   1049
Balestra, G.   1059
Balodis, G.   1326
Balogiannis, G.   407, 439
Balsa-Lozano, I.   344
Balsells, M.   1370
Bamidis, Panagiotis D.   65, 403, 783, 787, 1539
Bamparopoulos, Georgios   783
Barbancho, J.   923
Barbarov, G.   1853
Barbic, F.   535
Barbieri, C.   1233
Barbieri, R.   1671
Barca, Carlos Cavero   845, 1278, 1334
Barea, R.   627
Bargellini, S.   1166

Bari, V.   531, 535
Baroni, M.   269
Barrón-González, H.G.   1459
Barros, J.   1447
Barroso, F.   738, 1718
Barroso, N.   571
Baru, P.E.   1567
Bassani, T.   531, 535
Bataller, M.   527, 1698
Battista, C.E.   1079
Bauzano, E.   105
Becerra, J.   1630, 1634
Becerro, S.   880
Beda, A.   678
Beenen, L.F.M.   336
Bejarano, N. Chia   1799
Belacortu, Yaiza   845
Belda-Lois, J.M.   1747
Bell, S.L.   1686
Bellos, Christos   1213
Belo, J.   1489
Benavente-Babace, A.   864
Benito, S.   1021
Benz, H.L.   1643
Benz, T.   475
Berges, E.   113
Berkhemer, Olvert A.   237
Bermúdez, L.   714
Bernabeu, M.   884, 1694
Bernabucci, I.   639, 1771
Bernardes, J.   555
Bernardes, Rui   419
Bernus, O.   1868
Bertoldo, A.   225, 241
Bertolini, G.   1675
Bertollo, M.   137, 141, 149, 639
Bertrand, Virginie   1605
Bertucco, M.   1742
Berzuini, A.   1079
Bescos, C.   1147
Bezerianos, Anastasios   1643, 1710, 1849
Bezerra, F.N.   1182, 1186
Bianchessi, M.A.   872
Bianchi, Anna M.   690, 803, 1338, 1671
Bibas, Athanasios   1213
Bibbo, D.   1755
Biffi, E.   876, 1706
Biffi Gentili, G.   1883
Bifulco, P.   643, 647, 651, 655, 1096
Bilbao, S. de Miguel   190
Blanco, R.   627
Blinowska, K.J.   791
Bliznakov, Z.   1174

Bliznakova, K.   479, 483, 495
Bocanegra, David Escot   325
Bockisch, C.   1675
Bolea, Juan   710
Bollensdorff, C.   340, 423
Bolotin, G.   29
Bona, A.A.   1115
Bond, Raymond   1439
Boom, L.   1421
Boos, C.F.   770
Boquete, L.   627
Bordone, Maurizio   1334
Borges, F.S.   1182, 1186
Borrego-González, S.   1630
Borromeo, Susana   815, 1398
Bortoli, L.   137, 141
Boscá, Diego   1258, 1266
Boschetto, D.   225
Bosnjak, A.   321
Bottenberg, P.   229
Boyd, Kyle   1439
Bozukova, Dimitriya   1605
Bragós, Ramon   900
Brand, M.   29, 33
Brignol, A.   599
Brindeiro, R.M.   1237
Brink, P.A.   531
Brinton, J.   1686
Brites, I.S.G.   1029
Brun, F.   194
Bruña, Ricardo   815
Buccioli, M.   1111, 1310
Buda, A.   435
Bueno, G.   750
Bulgheroni, M.   1791

C

Caballero, E.   1409
Caballero-Hernández, R.   1779
Caballero-Ruiz, E.   1370
Cabiddu, R.   690
Cabrera, J.A.   1875
Cáceres, C.   1694, 1767, 1803
Cáceres, J.   1302
Cagy, M.   607, 1809
Cakrt, O.   1519
Calabrese, M.   241
Calderón, C. Parra   45
Calil, S.J.   1178, 1182, 1186
Callejón, A.   1853
Callejón, M. Amparo   1591
Calligaris, W.   702
Calvillo, J.   1286, 1290

Calvillo, L.   531
Calvo, A.   1021
Calvo, F.   37
Caminal, P.   623
Campos, Jordi   109
Campos Pavón, Jaime   325
Canepa, Silvia   1314
Canesi, M.   535
Cánovas-Paradell, R.   1100
Capdessus, C.   726
Cappa, P.   69
Caramia, N.   1714
Caravaca, J.   527
Carbonaro, N.   93, 1547
Cardenas, E.   1531
Cardona, Hernán Darío Vargas   182, 742
Cardona-Cuervo, G.P.   955
Cardozo, G.   1386
Carmona, M.   1302
Carniel, S.   69
Carreras, J.L.   447
Carrero, A. Muñoz   1254
Carretero, J.   257
Carrino, F.   758
Casado-Lopez, R.   1734
Casals, Alicia   85, 109, 113, 372, 1795
Casaseca-de-la-Higuera, P.   989
Casellato, C.   1742
Castellanos-Domínguez, G.   811, 947, 955, 658
Castellaro, M.   225
Castelli, E.   69
Castillo-Armero, A.   1559
Castro, A.L.   1302
Castronovo, A.M.   137, 149
Català, P.Ll. Miribel   856
Cataldi, P.   194
Cecconi, G.   1144, 1883
Cereda, M.   872
Černý, R.   1555, 1519
Cerutti, S.   531, 690, 1233, 1671
Cervinka, T.   289
Cesarelli, M.   81, 643, 647, 651, 655, 1096
Ceschia, P.   579
Cevik, A.   265
Chamorro, J.   491
Chamorro-Servent, J.   399, 415
Chang, Ching-Yi   849
Chang, Walter H.   849
Chapman, A.M.   1151
Charalambous, H.P.   1053
Charisis, V.S.   297

Chatzistergos, S.E. 305
Chatzizisis, Y.S. 376
Chavarrías, C. 202, 214, 399
Cheimariotis, G. 376
Chen, C. 1837
Chen, Y. 575
Cheng, Z.A. 1817
Chikalanov, A. 1221
Chil, R. 1398
Chkeir, A. 1127, 1503
Chlup, H. 939, 1041
Chondros, Petros 1278
Chouvarda, I. 376, 1147
Chrissoulidis, D.P. 837
Cía, T. Gómez 45
Cifrek, Mario 911
Cifuentes, M. 1634
Cirelli, C. 706
Ciriza, G. Gómez 45
Clariá, F. 623
Clark, C.A. 309
Claudel, J. 833, 841
Cleland, Ian 1140, 1477
Clemente, F. 651, 655, 1096
Clemente, I. 1571
Climent, A.M. 340, 423
Cobelli, C. 539, 686, 1543
Cocci, A. 872
Cohen, M.E. 1402
Colagiorgio, P. 1675
Colantuoni, A. 647
Colier, W.N.J.M. 467
Collier, G. 145
Colvin, C. 1163
Comani, S. 137, 141, 149, 639, 1771
Condemi, F. 915
Conesa, José Luis 845
Conforto, S. 69, 137, 141, 149, 639, 1755, 1771
Conti, Claudio 1468
Corbera, A. Homs 856
Corino, V.D.A. 670
Corralejo, R. 762, 1647
Cosentino, F.B. 1464
Cosmi, Erich 249
Costa, A.D.T. 872
Costa, D.C. 133
Cot, A. 443
Crespo, L.F. 603, 1455
Crespo-Ruiz, B. 153
Cristóbal, R. Somolinos 1254
Crone, N.E. 1643
Crotti, L. 531

Cruz, I. 218, 447
Cuesta-Vargas, A.I. 101, 125, 1075
Cuevas-Rodriguez, M. 105
Cunha, M.J. 1386
Cuniberti, G. 860
Cusenza, M. 702
Cussó, L. 392, 399
Cvetković, Božidara 1334
Cybulski, G. 766, 993
Cymberknop, L. 1049
Czabański, R. 563, 754

D

da Cunha, R.F.M. 61
D'Addio, G. 81, 643, 647, 651
D'Alessio, T. 69, 149, 639, 1771
Dal Molin, Renzo 845
D'Anna, C. 1755
Das, R.K. 1613
da Silva, Nuno André 1655
Da Silva, P.J.G. 1809
da Silva, R. Morgado 61, 706
Daskalaki, A. 479, 483
Datta, S. 1903
David, S.L. 455
Davies, Nigel P. 309
Daza, Genaro 182, 742
de Aldecoa, J.C. Fernández 190
de Azevedo, F. 1386
De Blas, M. 186
de Canete, J. Fernandez 923
De Falco, I. 1378
De Fazio, M. 872
De Jong, H. 336
De La Cuerda, R. Cano 738
del Ama-Espinosa, Antonio 153, 1759
de la Peña, Pedro 1334
de la Torre-Díez, I. 1425, 1430
del Campo, F. 969, 1825, 1829
del Castillo, M.D. 1651
D'Elia, B. 1771
de los Reyes-Guzmán, Ana 1751, 1759
del Portillo, H.A. 884
del Pozo, Francisco 815, 868
de Madariaga, R. Sánchez 1254
De Marchis, C. 69, 141, 149
de Miguel-Bilbao, S. 1155, 1159, 1205
Demirci, S. 277
de Molina, C. 245, 257
De Momi, E. 1837
Demongeot, J. 722, 1250
de Niz, M. 884

Déniz, O. 750
de Pablo Gómez de Liaño, Lucía 325
De Pauw, Edwin 1605
De Pauw-Gillet, Marie-Claire 1605
De Pietro, G. 1378
Dermitzakis, A. 495
De Rosario, H. 965
De Rossi, D. 93, 1547
de Sá, A.M.F.L. Miranda 678
de Santiago, Luis 325, 627
Desco, M. 37, 202, 214, 218, 233, 245, 257, 301, 344, 392, 399, 415, 447, 491, 1398
Deutscher, L. 860
de Villoria, J. Guzmán 447
Diab, A. 718
Dias, J.M. 1374
Díaz, M.A. Martín 190
Díaz-Cuenca, A. 1630
Diazzi, Chiara 1468
Dickinson, R.J. 1322
Diego, I.M. Alguacil 738
Díez, A. 799
Díez, R. 1447
Di Fronso, S. 137, 141
Dimbwadyo-Terrer, Iris 1751, 1759
Dimitrakopoulos, Georgios N. 1849
Dimitrakopoulou, Konstantina 1849
Dimitriadis, S.I. 1710
Di Nardo, F. 611, 619
Di Prima, P. 225
Direito, B. 519, 1527
Dogrusoz, Y. Serinagaoglu 505
Domene, P.A. 1911
Domingues, José P. 419
Donati, G. 93
Donini, Michele 261
Donnelly, Mark 1439
Dopido, I. 368
Dordoni, Elena 1045
Dori, F. 1144, 1883
Dorronzoro-Zubiete, E. 1382
Doswald, A. 758
Doulaverakis, C. 376
Douma, S. 330
Dourado, António 519, 1366, 1527
Dovrolis, N. 1362
Doyle, John 509
Dragomir, A. 1394
Draisci, A. 702
Drechsler, K. 348
Dremstrup, K. 1659
Dubini, Gabriele 1045

Duchêne, J.   1127, 1503
Dudek, H.   993
Dunajski, Z.   1690
Duque, L.   811
Duque-Muñoz, L.   658
Durrieu, Marie-Christine   1605, 1613, 1617, 1817
Duun-Henriksen, J.   539, 686

E

Easton, C.   1911
Easty, A.C.   1163
Ecay-Torres, M.   571
Echevarría, C.   1447
Edwards, R.   1671
Eftestøl, T.   165
Egiraun, H.   571
Einav, Shmuel   27
Elattar, Mustafa   1821
Elena, M.   1447
Elias, G.A.   1178
Elizalde, A.   884
Elle, O.J.   57
Elmazria, O.   841
El-Metwally, S.M.   1245
Elwahed, M. Abd   1059
Endres, J.   475
Engan, K.   165
Eqlimi, E.   206
Erlandsson, B.E.   1394
Eroğlu, H.H.   198,161
Escalona, M.J.   1282
Escobar, G.   1447
Escobar, N.   1421
Escoffier, J.   1250
Eshaghi, A.   206
Esteban, O.   443
Estrada, L.   977
Estudillo, M.A.   1563
Estudillo-Valderrama, M.A.   1813
Eyüboğlu, B.M.   161, 198,265

F

Fabris, C.   539, 686
Facchin, S.   435
Facchinetti, A.   1543
Fahmi, F.   336
Faiella, G.   647, 651, 1096
Fakih-Gomez, N.   77
Falcone, F.   1159
Fanelli, A.   823, 1435
Fanjul-Vélez, F.   427, 615, 1679

Farina, D.   1659
Farooq, O.   1903
Farrarons, J. Colomer   856
Fastl, T.E.   210, 1864
Faundez-Zanuy, Marcos   567
Favaretto, A.   241
Felgueiras, H.   1597, 1601
Feng, Z.   907
Fenili, A.M.   1083
Fernandes, V.   1571
Fernández, A.   583
Fernandez, Luis   845
Fernández, T.   868
Fernández-Aldecoa, J.C.   1100
Fernández-Avilés, F.   340, 423
Fernandez-Becerra, C.   884
Fernández-Carrobles, M.M.   750
Fernández-Chimeno, M.   962, 965
Fernandez-Gallardo, T. Martin   178
Fernandez-Gonzalo, S.   1779
Fernández-Luque, L.   1382
Fernández-Rodríguez, A.J.   627
Fernández-Santos, M.E.   340, 423
Ferrante, S.   1667, 1787, 1791, 1799
Ferrario, M.   1233
Ferreira, A.E.K.   61
Ferreira, B.C.   1374
Ferreira, F.B.   1104
Ferreira, Hugo Alexandre   1655
Ferrer, A.   714
Ferrero, J.M.   927, 931
Ferretti, L.   93
Ferretti, R.   1609
Ferrigno, G.   876, 1667, 1706, 1787, 1791, 1799, 1837
Ferriol, P.   1481
Fico, Giuseppe   1468
Fida, B.   1771
Filho, E.   137, 141
Filos, D.   1147
Finlay, Dewar D.   1262, 1477
Fioravanti, Alessio   1468
Fioretti, S.   611, 619
Firouznia, K.   206
Fiz, J.A.   981, 1000
Floor-Westerdijk, M.J.   467
Florian, Z.   89
Foltyński, P.   515, 1071
Fondón, Irene   356
Fonseca, D.S.   678
Fontanilla-Arranz, P.   57
Fontenelle, H.   57
Foresti, Francesco   1334

Fornasa, E.   587, 591
Fortunato, P.   269
Fotiadis, D.A.   1539
Fotiadis, Dimitrios I.   1045, 1213
Fountos, G.P.   455, 459, 463, 471
Fraccaro, Paolo   1274
Fragomeni, G.   915
Fragua, J.A.   1302
Fraile, L.M.   1398
Francés, J.V.   1698
Franchi, S.   1144, 1883
Frangi, A.F.   443
Frangou, P.D.   1841
Fratini, A.   655, 1096
Frattini, T.   1667
Freire, M.A. Landeira   97
Frosini, F.   1144, 1883
Fuertes, J.J.   281
Fuis, V.   89
Furlan, R.   535

G

Gabriel, A.   1088
Gabriel, P.   1190
Gallego, J.A.   1747
Gallego-Pérez, D.   864
Gallinaro, J.V.   666
Gallo, P.   241
Gambale, G.   1111, 1310
Gambus, P.L.   623
Gandolla, M.   1667
Garavito, L.   1551
García, A.   1803
García, G.   186
García, J.   257, 1155, 1159, 1205
García, M.   583, 795
García, N.   714
Garcia, R.G.   1104, 1671
García, R. López   45
Garcia-Breijo, E.   1579
García-Caballero, B.   1587
García-Cano, O.   1587
Garcia-Casado, J.   977, 1579
García-Castro, J.   392, 399
García-González, M.A.   962, 965
García-Laencina, P.J.   746
García-Linares, A.   285
García-Molina, A.   1694, 1767
García-Novoa, J.   52, 1443
García-Rojo, M.   750
García-Sáez, G.   1370, 1409
García-Sagredo, P.   1205

Author Index

García-Vázquez, V.   37, 392, 399, 447
Garré-Olmo, Josep   567
Gatti, E.   1233
Gatzounis, G.   495
Gavgani, A. Mazloumi   505
Gayá, F.   1694
Gay-Fernández, J.A.   1473
Gazzarata, Roberta   1274, 1294, 1314
Gemelli, M.   1067
Genna, M.   579
Gentili, G. Biffi   1144, 1166, 1170
Gewehr, P.M.   1063
Ghanaati, H.   206
Ghetti, G.   619
Giacomini, Mauro   1274, 1294, 1314
Gialelis, John   1278
Giannakopoulos, A.E.   1053
Giannoglou, G.   376
Giannopoulos, A.   376
Gibson, Lorraine   1439
Gil, O. Moreno   1254
Gil-Agudo, Ángel   153, 1734, 1751, 1759
Gili, Gabriel   109
Giraldo, B.F.   1021
Girardengo, G.   531
Girling, A.J.   1151
Giulianelli, M.   1609
Giussani, P.   1079
Glasmacher, B.   1638
Glinel, K.   1817
Godoy, E.   714, 931
Goffredo, M.   1755
Göksu, C.   161, 198
Goljahani, A.   539, 686
Golnik, N.   1108, 1690
Gómez, C.   73, 583, 799, 1775
Gómez, E.J.   41, 52, 57, 73, 1217, 1370, 1409, 1443, 1694, 1767, 1779, 1803
Gomez, E.J.   210
Gomez-Aranzadi, M.   852
Gómez-Cía, Tomás   356
Gomez-deGabriel, J.   105
Gomez-Foix, Anna M.   900
Gómez-Pilar, J.   969, 1825, 1829
Gomis, J.   1531
Gonçalves, H.   555
González, J.C. Ramos   97
González, L.M.   575, 1346, 1694, 1767
Gonzalez, R.   1531
González, S.   1803
González, V. Ramos   190

Gonzalez-Perez, L.M.   77
González-Sánchez, M.   125, 1075
Good, A.   1919
Gordo, M.A. López   1455
Graña, M.   186
Granero, M.A. Fernández   603, 1455
Graziani, A.C.   872
Gripp, P.A.G.   1115
Grisan, Enrico   225, 241, 249, 261, 380, 435
Grundy, Richard   309
Gryshkov, O.   1638
Guanziroli, E.   1667
Guerrero, Guillermina   803
Guerrero, J.F.   527, 1698
Guidi, G.   1354, 1887, 1899
Guijarro, Enrique   182, 742
Guillem, M.S.   340, 423
Guillén, Elena López   325
Guitart, Maria   900
Gülan, U.   23
Gulcur, H.O.   819
Gurgen, F.   730
Gusinu, R.   1144, 1883
Gutiérrez-Tobal, G.C.   969, 1825, 1829
Guzmán-De-Villoria, J.   218

H

Hachol, A.   807
Hachoł, A.   1583
Hadjileontiadis, L.J.   297
Haj-Ali, Rami   27
Haj-Hosseini, N.   49
Halak, M.   33
Hallberg, J.   1262
Hamashima, S.   1033
Hamie, A.   722
Han, M.H.   1037
Hanak, J.   996
Hännikäinen, J.   951
Hansen, O.   722
Hansford, D.J.   852, 864
Hardegger, M.   1481
Hargrave, Darren   309
Haritopoulos, M.   726
Harvey, P.   1903
Hassan, M.   718
Hassani, K.   509
Hatem, G.   435
Hattori, Takumu   1406
Hazgui, H.   1250

Hejda, J.   117, 1519, 1555
Hemm, S.   49
Henriques, J.   690
Hernández-Armas, J.   1005, 1100, 1879
Hernández-Tamames, Juan Antonio   815, 1270
Hernando, M.E.   1217, 1302, 1370, 1409
Herrera, J.   1531
Hewson, D.J.   1127, 1503
Hietanen, M.   1783
Hinrikus, H.   694
Hodasova, T.   996
Hofer, E.   1864, 1871
Hogrel, J.-Y.   1127
Holčík, J.   1241
Holcik, J.   996
Holguin, German A   742
Holguin, Mauricio   742
Holzner, M.   23
Homs-Corbera, A.   884, 1833
Honkonen, J.   13
Hornero, F.   1005, 1009, 1879
Hornero, R.   583, 762, 795, 799, 969, 1647, 1825, 1829
Horný, L.   1041
Horny, L   939
Horoba, K.   559, 754
Hosseini, P. Tayaranian   1686
Hozman, J.   1555
Huang, N.   1643
Huang, Sheng-Cheng.   635
Huang, Yi-Shiang   1605
Hudson, D.L.   1402
Huh, S.J.   1485
Hwang, S.H.   1738
Hyttinen, J.   289

I

Iadanza, E.   1132, 1166, 1170, 1342, 1350, 1354, 1887, 1899
Iakymchuk, T.   1698
Ibáñez, J.   1651
Ibrahim, M.   841
Ichioka, S.   1033
Iglesias, M.   1182, 1186
Ignesti, C.   1170
Iitomi, Y.   523
Illana, C.   37

Imamatsu, Takenori   1891, 1895
Infantosi, A.F.C.   607, 1809
Insa, L.   1266
Ioannides, A.A.   403
Isaila, F.   257
Isernia, P.   1079
Islam, Meez   888
Istepanian, R.S.H.   1903, 1907, 1911, 1915, 1919
Iucksch, D.D.   61
Iuppariello, L.   81, 647
Izumi, T.   1523

J

Jaber, R.   1127, 1503
Jackson, Duncan E.   1140
Jacquet, W.   229
Jagomägi, K.   958
Jahn, P.   996
James, Christopher J.   1201, 1515
Jan, Hao-Yu   635
Jané, R.   977, 981, 1000, 1017, 1021, 1795
Janicek, P.   89
Janssen, N.Y.   336
Jara, I.   321
Jaspan, Tim   309
Jedrzejczak, W.W.   791
Jensen, E.W.   623
Jentoft, A. Cruz   1205
Jeong, W.   1037
Jernajczyk, W.   1690
Jesus, A.P.   1489
Jezewski, J.   559, 754
Jezewski, M.   559
Jiang, N.   1659
Jimenez-Díaz, J.F.   153
Joao, Filipa   157
Jochumsen, M.   1659
Jonas, A.M.   1817
Joo, S.K.   1485
Jospin, M.   623
Joutsen, A.   951
Juárez, A.   856, 1833
Juez, Alejandro   845
Juhl, C.B.   539, 686
Julián, M.   973, 1013

K

Kajiya, F.   1523
Kaldoudi, E.   1362
Kalfas, A.   1539

Kalivas, N.I.   455
Kamiya, Akira   19, 1033
Kanas, V. G   1643
Kandarakis, I.S.   455, 459, 463, 471
Kanza, G.   1417
Karadimas, Dimitris   1278
Karamaounas, P.   330
Karami, E.   206
Karanasiou, G.S.   1045
Karimata, T.   907
Karlsson, B.   718
Karnafel, W.   515
Kartsidis, P.   65
Kassif, Yigal   27
Kataoka, S.   384
Katashev, A.   129
Katsikas, Sokratis   1306
Kawakatsu, M.   1702
Kawiak, J.   515
Keene, Kevin   1278
Khalil, M.   779
Khang, G.   1738
Kim, J.   1671
Kimira, Yutaka   1406
Kimura, Kazushige   364
Kimura, Yuichi   1406
Kinzelbach, W.   23
Kirsner, R.L.G.   9
Kitney, R.I.   1322
Kivastik, J.   958
Kivikunnas, S.   1197
Klados, Manousos A.   783, 787
Knaut, L.A.M.   61
Kobayashi, Takakazu   364
Kohn, J.   1597
Kompatsiaris, I.   376
Kontomaris, S.V.   407
Kortelainen, Juha M.   803
Kortlepel, R.   1638
Koruga, Dragan   911
Koukal,   89
Koukou, V.   459, 463
Kounadi, E.   471
Kourkoutas, K.   455
Kourtiche, D.   833, 841
Koutkias, V.   376
Koutny, Tomas   1229
Koutsouris, Dimitrios   1213
Kovatcheva, E.   1221
Kowarschik, M.   475
Kramer, M.R.   935
Krieger, M.A.   872
Krishnan, S.M.   1119
Kristensen, T.   774

Kronek, J.   939
Krueger, E.   121
Krzymień, J.   515
Kupka, T.   559
Kursun, O.   730
Kusunoki, M.   907
Kutílek, P.   1555
Kutilek, P.   117, 1519
Kwaskiewicz, K.   791

L

Lacalle-Aurioles, M.   218, 447
Lacković, I.   388, 682
Ładyżyński, P.   515, 1071
Laforêt, J.   718
Lago, P.   1079, 1083
Laguna, M.A.   1270
Laguna, Pablo   710
Lagunar, Carlos Marcos   845
Lai, Wen-Fu.   849
Lambrecht, S.   1497
Lansdown, J.   344
Lanzola, G.   1435
Lapi, D.   647
Laporte, Enric   109
Lara, A.   1563
Larrey-Ruiz, Jorge   317
Larsson, E.   194
Laskaris, N.A.   1710
Lass, J.   694
Lassfolk, C.   1783
La Torre, A.   269
Laura, C. Oyarzun   348
Lauznis, J.   1451
Leach, Martin O.   309
Leal, G.   827
Led, S.   1459
Ledesma-Carbayo, M.J.   443
Lee, P.   340, 423
Legaz-Aparicio, Álvar-Ginés   317
Lei, Y.   1617
Lekkala, J.   1511
León, A   603, 1455
Lerma, S.   1718
Leyva, J.A.   1587
Liberzon, A.   23
Lidon, J.V.   1579
Lin, Cheng-An J.   849
Lin, Geng-Hong   635
Lin, Kang-Ping   635
Lin, Wen-Chen   635
Lindahl, O.A.   1124
Lindecrantz, K.   551

Author Index

Lindén, M. 1575
Linguraru, Marius 249
Linnavuo, M. 1783
Liria, E. 257
Löfgren, N. 551
Loggia, M.L. 1671
Lombardi, C. 1079
Lomeña, F. 443
Lönn, U. 1413
Lopes, G. 431
Lopes, M.C. 1374
López, F. 1205
López, J. 714
López, L. 1421
López, M. 969
López, R. Nocelo 1473
López-Coronado, M. 1425, 1430
Lopez-de-Ipiña, Karmele 567, 571
López-Guerra, J.L. 360
López-Mir, F. 273, 281
López-Rodríguez, D. 285
López-Sánchez, L. 399, 392
Lopez-Soto, T. 1559
Lorussi, F. 1547
Loureiro, M.C. 519, 1527
Lozano, M. 981
Lozano-Berrio, V. 1734
Lu, M. 1663
Lucangelo, U. 591
Lucas, M. 41
Lucas-Hernández, M. 1845
Lucena, R. 1489
Lullo, F. 81
Luna, M. 1694, 1767
Lunardini, F. 1742
Lundström, R. 1124
Luque, J. 923
Luque-García, A. 1698
Lurbe, E.F. 714
Luschi, A. 1350, 1887
Luštrek, Mitja 1334
Lüthi, B. 23
Lwoga, E.T. 1298

M

Machado, Penousal 1366
MacPherson, Lesley 309
Maday, P. 475
Madjarova, M. 1221
Maeno, Soichi 1406
Magenes, G. 823, 1435
Maggi, N. 662
Maggioni, S. 1742

Magjarevic, R. 313
Maglaveras, N. 376, 1147
Mahdavi, H. 1507
Mainardi, L.T. 221, 670
Maiora, J. 186
Majoie, Charles B.L. 237, 336
Makikawa, Masaaki 595
Maldaner, M. 1136
Maldonado, José Alberto 1258, 1266
Malliori, A. 479
Malmivuo, J.A. 13
Malpica, N. 178, 1270
Mañas, Alejandro 1258
Mandarino, C. 579
Manstad-Hulaas, F. 277
Manzanedo, Eva 815
Manzo, A. 1531
Maramis, C. 1147
Marcan, M. 313
Marceglia, S. 1338
Marchi, A. 531, 535
Marco, L. 1266
Maretić, I. Silva 388
Margotti, A.E. 1104
Mari, F. 1233
Mariani, S. 690
Marín, C. 281
Marinetto, E. 37, 233, 344
Marino-Neto, J. 1464
Marjanović, T. 682
Markovica, I. 1326, 1451
Markovics, Z. 1326, 1451
Marozas, V. 985
Marque, C. 718, 779
Marquering, Henk A. 237, 336, 1821
Marques, C.M.G. 666
Marquez, E. 1718
Martegani, A. 1667
Marti, B. 443
Martin, Alexandra 845
Martínez, B. 1021, 1425
Martinez, F. 1531
Martinez, I. 1718
Martínez, J.D. 811
Martínez, O.S. 1587
Martínez-Espronceda, M. 1459
Martínez-García, A. 1282
Martinez-Lage, P. 571
Martínez-Moreno, J.M. 1443, 1779, 1803
Martínez-Pérez, B. 1430
Martínez-Ricós, J. 1698
Martinez-Tabares, F.J. 947, 955

Martín-Fernández, M. 253, 989
Martini, N. 459, 463
Martin-Jaular, L. 884
Martin-Majarres, S. 1734
Martín-Martín, J. 101, 1075
Martín-Martínez, D. 989
Martino, M.E. 218, 447
Martín-Ruiz, D. 427
Martins, L. 1489
Martins, R.V. 1182, 1186
Marzi, L. 1350
Masinghe, W.D. 145
Mateos-Pérez, J.M. 218, 301
Matisone, D. 1326
Matjačić, Z. 1535
Matonia, A. 563, 754
Matos-Azevedo, A.M. 1845
Matsuoka, T. 523
Maximiano, Ricardo 1655
Mazzitelli, R. 915
McClean, Sally I. 1140
McDonald, H.A. 1262
McKenzie, J.S. 9
Mediavilla, César 845
Medina, J. 73, 1775
Medvedec, M. 1092
Meirelles, D. 1136
Meirson, T. 29
Mejías, C. Suárez 45
Melanda, A.G. 61
Melià, M. 1481
Melia, U. 623
Melillo, P. 543, 1132, 1342, 1354
Melkova, P. 996
Mell, N 935
Melnik, O.V. 501
Menegon, A. 876, 1706
Meneses, Cristina 157
Mengarelli, A. 619
Merilahti, J. 1197
Merletti, R. 293, 674
Mert, A. 631
Messina, A. 221
Michail, C.M. 455, 459, 463, 471
Michalis, L.K. 1045
Micheloyannis, Sifis 783
Michmizos, K.P. 1841
Migliavacca, Francesco 1045
Migonney, V. 1597, 1601
Miguel-Jiménez, J.M. 627
Mijatovic, G. 1722
Mikheev, A.A. 501
Miklavcic, D. 313, 896

Milán, J.A. 1563
Milanesi, B. 1079
Milli, M. 1354, 1899
Milosevic, B. 1481
Minato, Kotaro 1406
Miniati, R. 1144, 1350, 1883
Mirones, I. 392, 399
Molina, Elena 815
Molina, F. 73
Molina, V. 799
Möller, K. 1763
Molteni, F. 1667, 1787, 1791
Moneda, A.P. 837
Moner, David 1258, 1266
Montero, B. 1205
Montero, F.M. Rivas 738, 1651
Montesinos, P. 202, 214, 233, 399
Monti, F. 702
Monticone, M. 1799
Montilla, G. 321
Montin, E. 221
Moore, G. 1262
Morales, S. 273, 281
Morales-Sánchez, Juan 317
Morán, Roberto Román 325
Moreno, Aitor 1334
Moreno, J. 41
Moreno, O. 1302
Moreno, P.A. 1217
Moreno-Perez, O. 395
Moreno-Sánchez, P. 1443
Morgado, António Miguel 419
Morillo, Daniel S 603, 1455
Morillo, M. 1370
Moustakas, N. 65
Moyano-Cuevas, J.L. 368
Mporas, I. 1643
Mrachacz-Kersting, N. 1659
Muggeridge, D.J. 1911
Mujika, M. 852, 864
Müller, M. 1638
Mulvaney, D.J. 1903
Muñoz, A. 1302
Muñoz, V.F. 105, 923
Muñoz-Macho, A.A. 1382
Mura, G. Dalle 1547
Murray, Alan 3
Murthy, N.S. 1597
Muuraiskangas, S. 1197

N

Nadi, M. 833, 841
Naftali, S. 935

Nakamura, H. 943
Nakamura, T. 907
Nanayakkara, T. 145
Nandi, A.K. 726
Napadow, V. 1671
Naranjo, A. 1447
Naranjo, D. 1563, 1853
Naranjo, V. 165, 273, 281
Naranjo-Hernández, D. 1813
Nardi, A. 33
Navab, N. 277, 475
Navarro, Augusto 710
Navarro, Rafael Barea 325
Navea, A. 273
Navidbakhsh, M. 509
Ndubuka, G.I. 1923
Negreira, C. 1049
Nemoto, I. 1702
Neokleous, K.C. 1318
Neves, C. 827
Neves, E.B. 61, 121, 1063, 1115
Niazi, I.K. 1659
Nicolás-Alonso, L.F. 762, 1647
Niessen, Wiro J. 237
Nieto-Diaz, M. 1734
Niewiadomski, W. 766
Niioka, T. 1683
Niki, K. 1523
Nikiforidis, G. 459, 463
Nikita, K.S. 305, 1841
Nikolov, R. 1221
Niñerola, A. 443
Nishina, A. 907
Nkuma-Udah, K.I. 1923
Nobre, F.F. 1237
Noseworthy, Michael 157
Nowak, W. 807, 1583
N. Philip, N. 1907, 1911, 1915, 1919
Nugent, Chris D. 1140, 1262, 1439, 1477
Nyssen, E. 229

O

Oda, R. 1613
Ödman, N. 1575
Ödman, T. 1575
Ørn, S. 165
Ogasawara, Y. 1523
O'Hare, Liam 888
Ohnuki, S. 1683
Oizumi, Tatsuya 364
Okada, Shima 595
Okawai, Hiroaki 1891, 1895

Oki, N. 1029
Olazarán, J. 218, 447
Olenšek, A. 1535
Oliva, A.M. 1833
Oliveira, M.C.N. 121
Opisso, E. 73, 1775
Opri, E. 1338
Opršalová, K. 1241
Orion, E. 29
Oropesa, I. 41, 52, 1443
Orozco, Álvaro 182,742
Orrico, A. 1132, 1342
Ortega-Quijano, N. 427, 615, 1679
Ortiz, Miguel 325, 627
Otsuki, M. 547
Ottaviani, L. 1887

P

Pacelli, M. 1547
Pacifici, E. 1132, 1342
Padovani, E. 1111, 1310
Pagador, J.B. 368, 1845
Page, J.C. Miangolarra 1651
Pagkalos, I. 1330
Pagnoulle, Christophe 1605
Pajaro, M. 1718
Palazis, K. 1318
Pałko, T. 766, 1108, 1690
Pallikarakis, N. 479, 483, 1174
Palovuori, K. 951
Panayiotakis, G.S. 455
Pander, T. 563
Pantarotto, E. 1182, 1186
Papadelis, C. 403
Papathanasiou, K. 1539
Papathanasiou, Nikolaos 783
Pappone, N. 81
Paradiso, R. 1547
Paredes, J. 880
Parejo, M. 1447
Parejo-Bellido, M. 1382
Parra, C.L. 1447
Parra-Calderón, C.L. 1282
Parrini, G. 93
Pascau, J. 37, 233, 245, 344, 399
Pascual, M. 1205, 1302
Pascual-Leone, A. 1694
Pasirayi, Godfrey 888
Pastorino, L. 1609
Patané, F. 69
Pauleto, A.C. 61
Pauws, S. 1147
Pavčič, J. 1535

Author Index

Pavia, J. 443
Pavliha, D. 313
Pawlak, J. 1911
Payne, Geoffrey S. 309
Pecchia, L. 1132, 1342
Pęczalski, K. 1108, 1690
Pedrocchi, A. 876, 1667, 1706, 1742, 1787, 1791, 1799
Peet, Andrew C. 309
Peleg, M. 1409
Pellicci, G. 93
Peltokangas, Mikko 1493, 1511
Peñas, Carlos 109
Peña-Zalbidea, S. 392, 399
Penhaker, M. 1190
Perälä, J. 1197
Pereira, Ana Rita 157
Pereira, E. Monge 1651
Pérez-Carrasco, J.A. 352, 360
Pérez-Duarte, F.J. 1845
Pérez-Lorenzo, E. 864
Pérez-Martín, C. 1100
Pérez-Nombela, Soraya 1751, 1759
Pérez-Rizo, E. 153, 1734
Pérez-Rodríguez, R. 73, 1775
Perez-Somarriba, B. Gonzalez 77
Perger, P. 1111, 1310
Perrone, I. 579
Peterzen, B. 1413
Petrakis, E.G.M. 1045
Petrarca, M. 69
Petrou, L. 1330
Petrucci, F. 1144, 1883
Pettenati, M.C. 1354, 1899
Petti, A. 1059
Piazza, S. 738
Picado, E. 1398
Pierce, D.M. 210
Pietrucha, K. 1622, 1626
Pinkney, S. 1163, 1182, 1186
Pino, A.V. 133
Pinto, B. 1726
Pinto, P. 555
Pirini, Giampiero 1468
Pirola, S. 1338
Piskulak, P. 766
Pizer, Barry L. 309
Planes, X. 443
Plank, G. 1864
Planken, Nils 1821
Platero, C. 174
Plaza, A. 368

Pol, S. 121
Poletti, Enea 261, 380, 411, 435
Poli, F. 587
Politis, C. 229
Politopoulos, K. 439
Pollazzi, L. 269
Polonio-López, Begoña 1759
Pons, B. 1370
Pons, J.L. 738, 1497, 1651, 1718, 1734, 1747
Porta, A. 531, 535
Poza, J. 583, 795, 799
Prassl, A.J. 1864
Prats-Boluda, G. 1579
Prisco, L. 702
Procyk, E. 1722
Przybyła, T. 559, 563
Pucihar, G. 896
Puddu, Paolo Emilio 1278, 1334
Pupella, Valeria 1274, 1294, 1314
Puppato, D. 1059

Q

Quaresma, C. 1088, 1489, 1571
Questier, F. 1298
Quintão, C. 1726
Quintero, Ana María 845

R

Raamat, R. 958
Rabotchi, E. 1575
Raffaele, L. 1079
Rajasekaran, V. 85
Ramat, S. 1675, 1714
Ramos, J.A. 868
Ramos, V. 1155, 1159
Ramos-Castro, J. 962, 965
Ramos-Llordén, G. 253
Rampazzo, R.C.P. 872
Ramser, K. 1124
Raposo, L.M. 1237
Rashid, M. 892
Ratan, M. 33
Ratnovsky, A. 935
Redel, T. 475
Regalia, G. 876, 1706
Reina-Tosina, L. Javier 919, 1563, 1813, 1591, 1853
Remartínez, Jose Maráa 710
Rems, L. 896

Remvig, L.S. 539, 686
Renner, L.D. 860
Renzulli, A. 915
Repanas A., A. 1638
Rezk, Y. 1864
Rhee, K. 1037
Ribeiro, B. 519
Riederer, Stephen J. 169
Rieta, J.J. 973, 1005, 1009, 1013, 1025, 1879
Riga, M. 376
Rigat-Brugarolas, LG. 884
Rigla, M. 1370, 1409
Ringeval, F. 758
Riordan, A. 336
Ripka, W.L. 121, 1063
Ripoll, J. 415
Rittinghaus, T. 1638
Rivas-Blanco, I. 105
Rivolta, M.W. 670
Roa, Laura M. 919, 1286, 1290, 1563, 1591, 1813, 1853
Robazza, C. 137, 141, 149, 639
Robles, Montserrat 1258, 1266
Roca-Dorda, J. 746
Rocha, H. 1374
Rocon, E. 1747
Rodríguez, Á 1551
Rodríguez, C. 73
Rodriguez, Daniel 309
Rodriguez, J. 395
Rodríguez, Juan Mario 1278, 1334
Rodríguez-Ascariz, J.M. 627
Rodríguez-Bermúdez, G. 746
Rodríguez-Ibáñez, N. 965
Rodriguez-Moreno, A. 1398
Rodriguez-Sanchez, M.C. 1398
Rodriguez-Vila, B. 57, 210
Roggen, D. 1481
Roggerone, Sabrina 1314
Roig, T. 1694, 1803
Roj, D. 563
Román, I. 1286, 1290
Roman, N.M. 1567
Romaneli, E.F.R. 1067
Romano, M. 81, 643, 651, 1096
Romero, L. 927, 931
Roncalés, Miguel 845
Ros, D. 443
Rosa, B.P. 1809
Rosado, A. 527, 1698
Rosário, B.R. 121

Rosell-Ferrer, Javier 900, 1507
Rossini, M. 915, 1787, 1791
Roure-Alcobé, Josep 567
Roussel, J. 726
Roussis, P.C. 1053
Rovira, Eva 710
Ruales, C. 1421
Ruano-Calderón, L.A. 1587
Rubio, D. 1155
Rueda, F. Molina 738, 1651
Ruggeri, A. 411
Ruggiero, C. 662, 1609
Rugnone, Alberto 1334
Ruiz, D. 1346
Ruiz-Fernández, D. 1209
Ruiz-Muñoz, M. 1075
Rusu, C. 1567
Rutoli, G. 1096

S

Sabalińska, S. 515, 1071
Sabater-Navarro, J. 395
Saccà, F. 1144
Sahraian, M.A. 206
Saito, H. 943
Sáiz, J. 714, 927, 1531
Sakar, B.E. 819
Sakar, C.O. 730
Sakas, D. 1841
Sakaue, Yusuke 595
Sakellarios, A.I. 1045
Salameh, E. 1318
Salas-García, I. 427, 615, 1679
Salazar-Muñoz, Y. 1587
Saleh, N. 1059
Salo, A. 1511
Salpavaara, Timo 1493, 1511
Salvador, C.H. 1205, 1302
Samitier, J. 856, 884, 1833
Sánchez, C. 868
Sánchez, L.F. 368
Sánchez, M.G. 1473
Sánchez, P.L. 340, 423
Sánchez, Tomás García 900
Sánchez-Carrión, R. 1779, 1803
Sánchez-de-Madariaga, R. 1302
Sánchez-González, P. 52, 1443, 1803
Sánchez-Laguna, F.J. 1382
Sánchez-Margallo, F.M. 41, 52, 368, 1845

Sánchez-Margallo, J.A. 41, 52, 368
Sánchez-Morla, E.M. 627
Sánchez-Quintana, D. 1875, 1871
Sandham, W. 1551
Sanger, T.D. 1742
Sanguino, J. 174
Sannino, G. 1378
Sansone, M. 655
Santamarta, D. 795
Santana, V.M. Febles 190
Santini, R. 1079
Santoro, D. 543
Santos, A. 443
Santos, Emilie M.M. 237
Santos, F.A. 1104
Santos, J.P. 431
Santos, M.P. 519, 1527
Santos-Miranda, J.A. 37, 344
Santos-Ruiz, L. 1634
Sanz-Ruiz, R. 340, 423
Sartori, R. 587
Sarturi, M.F. 1136
Sassi, R. 670
Sato, D. 907
Scarpellini, S. 1435
Schiza, E.C. 1318
Schizas, C.N. 1318
Schleich, J.-M. 1225
Schmid, M. 69, 137, 141, 149, 639, 1755, 1771
Schonenberg, H. 1147
Schuchter, A. 165
Schwartz, P.J. 531
Sclocco, R. 1671
Scolaro, G.R. 770
Scott, Simon M. 888
Secca, Mario Forjaz 157, 1088
Secci, I. 1435
Seemann, E. 1763
Seferis, I.E. 455, 471
Sejling, A.S. 539, 686
Senent, E. 1579
Seoane, F. 551
Sepponen, R. 1783
Sepúlveda, L.M. 811
Serbes, G. 819
Sergiadis, G.D. 297
Serraino, G.F. 915
Serrano, A. Castro 1254
Serrano, Carmen 352, 356, 360

Serrano, J.I. 1651
Serrano, J.J. 868
Serrano, L. 1459
Serrano-López, A.J. 527
Sesa-Nogueras, Enric 567
Sgarbas, Kyriakos N. 1643, 1849
Shalev, Omer 27
Shalom, E. 1409
Sharawi, A. 1059
Shibasaki, Jun 1730
Shibata, M. 1033
Shimokawa, E. 384
Shin, D.I. 1485
Shirzadfar, H. 833, 841
Shishlova, E. 129
Sievänen, H. 289
Signorini, M.G. 823, 1233, 1435
Silva, Daniel Castro 1366
Silva, Susana F. 419
Silverberg, D. 33
Similä, H. 1197
Simons, S. 698, 706
Simpson, D.M. 678, 1686
Sisniega, A. 491
Sørheim, H. 1390
Solana, Ana Beatriz 815
Solano, F. 1155
Solé-Casals, Jordi 567, 571
Solís-Mozos, M. 153, 1734
Solosenko, A. 985
Sommerfeld, S. 1597
Somolinos, R. 1302
Song, J.H. 1485
Song, T. 1738
Soriano, A. 1346
Soria-Olivas, E. 527
Sotiropoulou, P. 459, 463
Soudah, Eduardo 1334
Sousa, A.C. 1726
Souza, M.N. 133
Sparacino, G. 539, 686, 1543
Stadnik, A.M.W. 1115, 1136
Stathis, P. 1841
Stavrianou, K. 1174
Stefanadis, C. 376
Steger, T. 487
Sterritt, Roy 1439
Straumann, D. 1675
Streekstra, G.J. 336
Stromatia, F. 471

Author Index

Strurniolo, G.C. 435
Styliadis, Charalampos 403, 787
Stylianou, A. 407
Suárez O.J. 1155
Suárez-Bagnasco, D. 1049
Suárez-Mejías, C. 360, 1447
Suazo, V. 799
Such-Belenguer, L. 527
Sugawara, M. 1523
Sugiyama, Katsumi 364
Suhhova, A. 694
Suminoe, I. 1523
Summers, R. 1225
Sun, Yu 1710
Sungoor, A. 1907
Suzuki, D.O.H. 666
Suzuki, M. 384
Svoboda, Z. 117
Szul-Pietrzak, E. 807, 1583

T

Tacchino, Giulia 803
Tadeusiak, M. 1515
Tagliati, Alessandro 1294
Talaminos, A. 919
Talaminos-Barroso, A. 1813
Talts, J. 958
Talvitie, S. 1783
Tamburini, Elena 1334
Tamura, Toshiyo 1406
Tanaka, M. 1523
Tanaka, Y. 547
Tantisatirapong, Suchada 309
Tapia, A. 186
Tapia, E. Sánchez 97
Tarjuelo-Gutiérrez, J. 57, 210
Tassin, S. 535
Tavares, Nuno Jalles 157
Taylor, C.A. 1151
Teixeira, C.A. 519, 1527
Tello, J. Cáceres 1254
Tendeiro, D. 431
Teruel-Martí, V. 1698
Terzaghi, F. 1887
Thakor, N.V. 1643, 1710
Titapiccolo, J. Ion 1233
Tobar, M.C. 174
Tognetti, A. 93, 1547
Tononi, G. 706

Tormos, J.M. 73, 1694, 1767, 1775, 1803
Torrado-Carvajal, A. 178
Torrents, E. 1833
Torres, A. 977, 1000
Torricelli, D. 738, 1718, 1751
Tous, F. 1481
Toutouzas, K. 376
Toyoshima, H. 547
Travieso, C.M. 571
Trayanova, Natalia A. 1859
Trbovich, P.L. 1163, 1182, 1186
Trenor, B. 927
Triantafyllou, A. 330
Trincado, F. 1751
Trincado-Alonso, Fernando 1759
Tripoliti, Evanthia E. 1045, 1213
Tromba, G. 194
Tsakalidis, Athanasios K. 1849
Tsampoulatidis, I. 376
Tsinober, A. 23
Tsukahara, V.H.B. 1182, 1186
Turon, M. 1779
Turukalo, T. Loncar 1722

U

Udías, J.M. 1398
Ueno, S. 1663
Ulbricht, L. 1063, 1067, 1136
Ullah, K. 293, 674
Urra, O. 1017, 1795
Uvdal, K. 194

V

Vadursi, M. 543
Valais, I.G. 455, 459, 463, 471
Valdivieso, M. 37, 344
Valencia, J.F. 623
Valenti, M. 1837
Vallone, I. 1083
Vallverdú, M. 623
Valverde, I. 919
van Assen, Hans 1821
van Bavel, E. 336, 1821
van der Burght, R.J.M. 467
van der Lugt, Aad 237
van der Zwam, Wim 237
Vande Vannet, B. 229
Vaquero, J.J. 214, 233, 301, 415, 491, 1398

Vázquez, Luciano Boquete 325
Vecchia, L. Dalla 535
Vegas-Sánchez-Ferrero, G. 253
Vehkaoja, Antti 1493, 1511
Vela, M. 1209
Velasco, O. 174
Veloso, Antonio P 157
Vendina, V. 129
Verdú-Monedero, Rafael 317
Verho, Jarmo 1493, 1511
Verlinden, Jan-Marc 1278
Veronese, Elisa 241, 249, 435
Veselý, J. 1041
Vicent, C. García 714
Vieira, P. 431, 827, 1088, 1489, 1571
Vigmond, E.J. 1868
Vila-Forcén, A. 1779
Villanueva, E. 281
Villán-Villán, M.A. 1775
Villegas, R. 321
Visconti, J.V. 927
Visentin, Silvia 249
Visser, R. 1634
Vitetta, N. 93
Võhma, Ü. 694
Vrahatis, Aristidis G. 1849
Vyas, A.L. 1903
Vyazovskiy, V.V. 706

W

Wada, Junýa 1891, 1895
Wakatsuki, T. 907
Wald, S 935
Walton, R. 1868
Wang, S. 1686
Ward, N.S. 1667
Wårdell, K. 49
Watanabe, Takashi 1730
Weiss, Dar 27
Wendel, K.E. 13
Wesarg, S. 348, 487
Westerteicher, C. 1147
Wiegerinck, Esther 1821
Wienecke, S. 1638
Wilches, C.A. 1551
Woie, L. 165
Wójcicki, J.M. 515, 1071
Wojciechowski, D. 1690
Woodward, B. 1903
Wróbel, J. 563, 559, 754

Y

Yajima, Tadashi 1891, 1895
Yamada, S. 833
Yamamoto, K. 19
Yamanoi, T. 547
Yamazaki, T. 547
Ye-Lin, Y. 977, 1579
Ylikauppila, M. 1197
Yoshida, M. 943
Yova, D. 407, 439

Z

Zabulis, X. 330
Zakeri, Z. 734
Zakrzewski, M. 951
Zamboulis, C. 330
Zamora, B. del Moral 856
Zarco, M.J. 1447
Żarowska, A. 1583
Zecchin, C. 1543
Zelarain, F. 571

Zequera, M.L. 1551
Zernetsch, H. 1638
Zervakis, M.E. 1045
Zhang, Shuai 1140
Zheng, G. 1837
Zielińska, A. 993
Žitný, R. 939, 1041
Zorraquino, C. 451
Zouani, O.F. 1613, 1617, 1817
Zurbarán, M. 1382

Keyword Index

111-Indium 392
2D-3D 277
3D 368
3D reconstruction 439
3D-PTV 23
4G health 1919
8 bit 1464

A

abdominal 1115
abdominal aortic aneurysm 210
Abdominal effort and Inattention 965
Absolute Efficiency 455
Acceleration 947
acceleration photoplethysmogram 1531
Accelerometer 1485, 1515, 1587, 1730
accelerometry 1477
Acquired brain damage 1447
action potential duration 1868
action potential duration restitution 927
Active antenna 1853
Activities of Daily Living 73, 1386, 1515, 1751, 1803
Activity 1485
Activity classification 1515
activity monitoring 1201
Activity recognition 1477
AD 447
ADA Boost 380
Adaptation 762, 1675, 1679
adaptive approximations 791
Adaptive Canceller 1000
addiction 1698
adherent cells 900
Adhesion 1225
adiposity 1115
Adoption 1439
Adverse events 1174, 1166
Affective Computing 787
AFM 407
Africa 1298
Agent Based 1225
Aging 1473, 1849
airflow 1829
Airway 935

Airway clearance 587
Airway Obstruction 935
Albumin 1597, 1601
alcohol 1698
alcohol consumption 655
Alertness state 965
alpha 1710
Alzheimer disease diagnosis 567, 571
Alzheimer's Disease 218, 698
ambient assisted living 1217
Ambient Intelligence 1213
Ambulatory monitoring 766, 1201
Ambulatory system 1799
AMELIE 1443
amplitude map 7941
Amplitude Maps639
amplitude-based characterization 958
integral-based characterization 958
Anastomosis 1053
anatomy 229
anechoic chamber 1853
Anesthesia 519
Anesthesia monitoring 1527
Aneurysm 186, 1037
angiography 269
angle 1730
anisotropic conduction 1875
anisotropic diffusion filtering 479
Anisotropy 253, 939
Anistropic Diffusion Filters 505
ankle flexor muscles 619
Annotation 261
anonymisation
anterior cingulate cortex 1722
Anthropometry 121
anti-smoking 1136
Aorta 915
Aortic Arch Aneurism 33
aortic bifurcation 939
Aortic Root 1821
application 1358
Approximate Entropy 686
approximation 1229
apps 1430
AR model 996
Archetype 845, 1258, 1266, 1278
ARMA 989

arousal 403
artefact detection 599
artefact removing 599
arterial bifurcation 1049
arterial properties 1531
arterial stiffness 23
Artificial implants 1609
Artificial Neural Network 269, 639, 770
ARV 293
ARX model 996
ascending aorta 23
Ascorbic Acid 892
Assessment 52, 101, 1751
assistance-asneeded 73
Assistive technologies 1791
Assistive Technology 1787
asthma 981
asymmetric functions dictionaries 791
asymmetry 117
Atherosclerosis 939, 1045
atlas-based segmentation 174
Atrial Fibrillation 340, 670, 973, 1005, 1009, 1013, 1025, 1879
atrial flutter 670
attention 1710
audiology 1559
augmented proprioception 1667
auscultation 509
authoring tool 1443
Autism 774
Automated segmentation 165, 237, 376, 435
Automation 1079, 1088
Autonomic Nervous System 531, 535, 827, 1895

B

Balance Disorders 1213
baroreflex 678
Basal nuclei 321
BCI 1659
BCI features computation 1714
BCM 1883
Beam Angle Optimization 1374
best practices 1147

Bidomain model 1864
Bifurcation angles 273
big data 1417
Bimetallic stimulation 13
binocular dynamic pupillometer 1583
Bioactive Glass 1630
Bioelectric model 1591
biofeedback 1742
biofilm 880
Bioimpedance Spectroscopy 1563
biological cells 841
Biomaterials 1597, 1634
Biomechanics 145, 153 157, 1571, 1734
Biomedical 1923
Biomedical Engineering 1124, 1571
Biomedical signal processing 623
biomimetic 93
Biosensor 841, 845, 852
Biosignal 947
Biosignal Pattern Analysis 787
Biosignal processing 1435
biosignals 1338
biotemplating 860
Bipolar current supply 161
bipolar disorder 1358
BIS 1527
Blood pressure 1406
Blood pressure variability 1021
BME Cooperative Education 1119
B-mode ultrasound 249
BMPs 1634
Body Composition Estimation 1563
Body Mass Index (BMI) 1891
Body Motion Wave (BMW) 1891, 1895
Body Sensor Networks 1853
Body-machine interface 65
BOLD 202
bone assessment 463
Bone strength 289
Bone Tissue Engineering 1630
boosting 1829
Brain 261
brain activity 547
Brain Computer Interface (BCI) 65, 762, 1647, 1651, 1655, 1714
Brain Imaging 495
Brain Injury 1767
brain machine interface 1643
Brain maturation 583
brain tumor segmentation 265
Breast Cancer 1425
breast tomosynthesis 479
Breathing 962, 1891

breathing training 943
bronchodilator response 981
bronchoscopy 487
Brugada Sindrome 423
burn 356
Business development 1124
Business Intelligence 1101

C

Ca/P mass ratio 463
CABG 29
CAD 356
CAGB 1041
calibration 245
calvarium diploe 194
camera system 1519
cancer 395
cancer registries 1417
Cancer treatment response 221
capacitive electrode 1511
cardiac action potential 931
cardiac autonomic nervous system 943
cardiac cine MRI 214
Cardiac electromechanics 1864
Cardiac electrophysiology 1864
Cardiac monitoring 1903
Cardiac output 993
Cardiac tissue engineering 907
Cardio belt 1221
Cardiomyocytes 907
Cardiotocography 555, 823
cardiovascular events 1233
Cardiovascular System 923
Cardiovascular tissue engineering 1638
cardiovascular system response 1683
care coordination 1147
Carers 1439
C-arm 245
C-arm pose estimation 487
CART 1354
Case Based Reasoning 1413
Catechol 892
Catheterization 57
causal estimates 678
causality 1710
caveolae 19
cavotricuspid isthmus 1871, 1875
CBCT 229, 491
CDMAM phantom 467
CDSS 1382
CE marking 1170

cell 1817
Cell Concentrator 856
cell differentiation 1613
cell migration 666
Cellular Potts 1225
central autonomic network 1671
centre of pressure 1755
cerebral aneurysm 475
Cerebral blood flow 218
Cerebral circulation 915
Cerebral Palsy 121, 1718
certification 1092
CFD 29, 33
CFD modeling 1049
characteristic ratio 958
chemoprophylaxis 1382
Chest physiotherapy 587
Child 1063, 1067
Children 121
chimney vs. hybrid techniques 33
chronic illness 1147
circle detection 487
Circulatory hemodynamics 915
CIS (clinical information system) 1322
CKD 1813
Class E amplifier 856
Classification 762, 1366, 1370, 1485, 1775
classification of signals 1029
clinical care continuity 1254
clinical dashboard 1322
Clinical Decision Support System 1334
Clinical Document Architecture 1274
Clinical Engineering 1083, 1092, 1104, 1567
clinical guidelines 1409
Clinical information model 1258
clinical pathways 1322
Clinical Trial 1270
Clusgter Clinical evaluation 1406
clustering coefficient 799
CMOS 471
Cobalt-chromium alloys for surgical implants 77
cochlear implants 1559
cochlear models experiment 523
co-culture 864
Cognitive 1447
Cognitive Rehabilitation, Acquired Brain Injury 1803
Cognitive robotics 113

Keyword Index

Cognitive task performance 1683
coherence 799, 1698
Coherent Average 1809
Cole Bioimpedance Model 1563
Collagen I 407,, 1597, 1601, 1609, 1622, 1626
collision avoidance 1763
Colony counting 388
color 356
color and monochrome map 364
color models 750
Colorectal Cancer 1425
Coma 702
combined therapy 1115
Common Terminology Services 2 (CTS2) framework 1314
communications 1903
comorbidity management 1362
compartmental modeling 301
competence 52
complex medical devices 1163
complex tone 523
complexity 535, 623
Compressed-sensing 202, 214, 225, 233, 415, 491
Computational fluid dynamics (CFD) 1037
Computational Intelligence 285
Computational method 1245
computational modeling 919
computed microtomography
Computer aid 237
Computer aided diagnosis
Computer Aided System 336
Computer applications 915
Computer Assisted Surgery 1837
computer modeling 1868
computer performance 372
computer vision 1209
Computerised cardiotocography 651
concentric ring electrode 1579
conditional entropy 535
confidence ellipsoid 1519
confocal laser endomicroscopy 435
conforming hexahedral meshes 210
Congestive Heart Failure (CHF) 1132, 1334, 1342
constitutive model 1041
consulting 1451
Context Aware 947
context awareness 1481
context-aware applications 1217
continuous adventitious sound 981

continuous glucose monitoring 515, 1543
continuous maxflow 360
Continuous measurements 1507
contrast agent 849
contrast kinetics 225
convex optimization 360
Cooperation 1923
COPD 603,911, 1455
Corneal Endothelium 411
Correlation 1009
cortical bone 289
Cost Accounting 1111
Cost Control 1111
Countries 1923
Cox-Maze surgery 1005, 1879
cranial bones 194
Crosssample entropy 1025
Crouch Gait Pattern 1718
crutch-assisted gait 1734
crypts 435
CS 1626
CT Angiography 237, 1821
CT Brain Perfusion 336
CUDA 257
current mirror 161
current steering 198
cyclic coherence 726
Cyclic RNAs inhibition 1250
cycling 137
cyclogram 117
cyclostationarity 726
Cyprus 1318
Cystic fibrosis 587

D

daily living activity 1205
DARTEL 447
data acquisition 214
data mining 1417
Data Provenance 1270
data visualization 1417
Databank 1338
database 1029
Data-glove 1759
Data-Mining 1394
D-band 407
decision support 1370
Decision Support Rules 1262
decision support system 1326, 1409
decision trees 1370
Deep Brain Stimulation 1841

Deep Brain Stimulation (DBS) 49, 182
Deep pain 1671
Deformation model 372
denoising 1543
density 1706
dentate gyrus 575
Dentin impedance 682
dependent people 1209
Depression 694
Depth-Sensitive Minimum Norm Estimation 1726
Derivative-free Optimization 1374
Dermatology 352
Descriptors 1694
Design 1, 1186
Design methodology 113
design research 1390
Detailed Clinical Models 1266
detection 348
detrended fluctuation analysis 714
Developing 1923
Development 1, 1225
device design 1587
DEXA 121
DFA 694
Diabetes 249, 539, 686, 1370, 1543
Diabetes mellitus 515, 1907
Diabetic foot 1551
diabetic foot syndrome 1071
diabetic ulceration 1071
diagnosis 911
diagnostic 872
diagnostics 1326, 1451
diagnostic-therapeutic process 1166
Diaphragm muscle 977, 1000
diastolic period 509
diastolic time intervals 1021
dielectric properties of dentin 682
Dielectrophoresis 856
differentiation 1817
Diffusion 221
diffusion and conventional MR images 309
Diffusion Tensor Images (DTI) 182
digital diary 1390
digital healthcare 1201
Digital imaging 471
digital mammography 467
Digital Signal Processing 662
digital subtraction angiography 475
digital tablets 579
digital to analog converter 161
digital tomosynthesis 483

Digital Volume Pulse s 635
Dimensionality Reduction 819
Disability Evaluation 1386
Discrete Complex Wavelet 819
discrete wavelet 1029
Discriminant Function 770
discriminative CCA 730
disease models 1402
disequilibrium 583
disocclusion 277
Distance Bivariate Analysis 698
distributed middleware 1813
Dopamine 892, 1575
dopaminergic system 443
Doppler Radar 1539
Dorkin's pulmonary model 591
dorsal flexion 1075
Dosimeter 1190
Double inversion recovery 241
Drilling Parameters 97
Drug Resistance 1237
DSC-MRI 225
DSS 1883
Dual Energy 491, 495
dual energy x-ray 463
dual-energy mammography 459
DWI 221
dyadic Green's function 837
Dynalets 722
Dynamic CausalModeling 1686
Dynamic Features 742
dynamic imaging 301
Dynamic Posturography 69
dynamic pupillometry 807
dynamic susceptibility 225
Dynamic Visual Stimulation 1809
dynamometry 1075
Dysfunctional Profile 1775
Dystocia 555, 1742

E

E. Coli 1833
early detection 1455
Early Diagnosis 285
Early economic 1151
ECG 722, 955, 973, 1013, 1005, 1879
ECG lead 962
ECG reconstruction 989
Echocardiography 1523
Edge Detection 1346
EEG 519, 539, 639, 686, 694, 702, 734, 762, 815, 141, 1527, 1690, 1809

EEG coherence 137
EEG denoising 631
EEG signal 770
EEG signal processing 1714
EEG Source Localization 1726
Effect and Criticality Analysis (FMECA) 1096
Effective Connectivity 1686
Efficiency 1310
e-health 1451, 1326, 1907
eHealth 1459
EHR 1278, 1286,1334
Eigenvalues 674
elastography 395
Elderly 1205, 1485, 1503
e-learning 1443
electric field 1155, 1159
Electrical Stimulation 133
electrically-elicited tactile sensation 1738
electro-anatomical characterization 1871
Electrocardiogram 973, 1009, 1013, 1025, 947
Electrocardiography 655
electrochemical detection 880
electrochemotherapy 313
electrocorticography 1643
Electrodoctor 947, 955
Electroencephalogram 623, 698
Electroence-phalogram (EEG) 1647
Electroencephalograms 547
electroencephalograph 595
Electroencephalography(EEG) 1651
electrofusion 896
Electromagnetic Field 1833
electromagnetic fields 190
Electromechanical delay 1747
electromyography 1075, 1845
electronic health care record (EHR) 1302
Electronic Health Record 845, 1254, 1258, 1266, 1294, 1318
electronics 856
Electrophysiology 340, 1706
Electroporation 313, 896, 900
electrospinning 1638
Embedded space 563
Embolic Signals 819
emergency 1451
EMFi 1493
EMFi pulse wave sensor 1493

EMG 69, 125, 1000, 1742, 1795
EMG Amplitude 293
EMG signals 1787
Emotional processing 1779
Emotional Recognition 787
Emotions 787
Emotiv EEG 1655
EMT 1225
EN/ISO 12967 1286
EN 13606 1278
endogenous electric fields 1591
Endothelial cell 19
endovascular repair 33
end-to-end 1053
Energy substrate 907
Engineering 1923
ensemble empirical mode decomposition 1722
Ensemble Methods 1366
ensembles of classifiers 186
Enterprise Resource Planning 1101
entropy 555, 583
environmental risk 1178
epicardical border zone 927
epileptiform activity 575
Epileptiform Pattern 770
equivalent current dipole source localization 547
ERD/ERS 141
Ergonomics 1178, 1845
Error 1079
ESD 29
Essential tremor 1747
estimation 595
Euclidean distance 815
EVA 52
Evaluation 376, 1088
Evaluation of visibility 364
evaluation, Headroom 1151
EVAR 186
event related potentials 547
Event-Related Desynchronization (ERD) 1651
Evoked Potentials 607
evoked response 1702
Evolution model 989
exacerbation 1455
exacerbations 603
excitation 575
execution errors 1722
Exercise 1523, 1911
Exoskeleton 85, 1787

Keyword Index

Expectation-maximization algorithm 388
Experiential Learning Models 1119
Expert system 1386, 1887
Exponentially damped sinusoids 658
exposure threshold 190, 1155, 1159
extension package 1459
External uterine contractions 555
Extrasystoles 985
Extreme Learning Machine 746
eye movement 364

F

Facebook 1425, 1439
faculty 1298
Failure Mode 1096
fall recording 1378
Fast 253
Fast Fourier Transform 419
Fatigue 125, 965, 977
Fatty acid composition 907
FDG-PET 447
feature extraction 178, 297, 730, 779
Feature Point 1767
feature selection 779, 1233
feature sets 758
FECG 543
feedforward selection 309
FEM 29
fetal electrocardiogram 559
fetal heart rate monitoring 543
Fetal Monitoring 559, 754, 823
Fetal ultrasound 249
FHR 651
FHR signal 563
fibers 1638
fibrillatory waves amplitude 1005, 1879
Fibronectin 1597, 1601
figure-eight elastic ankle wrap 117
Filter 253
Filter learning 380
filtering 1543
finite element analysis 210
Finite element method 1045
finite element model 896
first heart sound 509
flagella 860
Flecainide 423
Flexible electrode 1579
flow diversion 475
Fluid-Repellent 955

fluorescence 415, 888
fluorescent protein 849
fluoroscopy 487
FMEA 1166, 1170, 1667
fMRI 202, 206, 1671
foetal electrocardiogram 726
foot drop 117
force control 85
force myography 93
Force-Frequency Relation 1523
Fourier Transform 647, 722
fractal 714
Fractal Dimensions 567, 571, 702
frequential descriptors 750
frog muscle 13
Function 157
functional and effective connectivity 718
Functional Connectivity 783
functional electrical stimulation (FES) 1667
Fuzzy 1354
fuzzy logic 911
fuzzy system 754

G

Gabor 352
gadolinium 849
GAGG:Ce 455
gait 1730
gait analysis 611
gait assistance 85
gait deviation 61
Gait Profile Score 61
Gait temporal parameters 1799
Gait velocity 1503
gamma irradiation 1622
Gamma Rays 1190
gamma-spectroscopy 1398
gastrocnemius lateralis 611
Gaussian Mixture Model 388
G-causality 1698
Gelatin 1630
gene electrotransfer 900
gene target prediction 662
Genetic 1354
Genetic Algorithm 1714
Genetic regulatory networks 1250
geriatric assessment 1205
Gestational diabetes 249
GFPUV 888
GIMIAS 443

glenohumeral joint 1734
Global Optimization 1374
glucose level 1229
GMR Sensor 833
gold 849
Gold nanorods 868
gradient 352
grafted 1601
Graph 1394
graph network 799
Graph Theory 206, 783
Graphical Tool 1803
Graphical User Interface 845, 479
gray level distance 165
gray matter/white matter segmentation 265
Grip strength 1127
growth factors 1634
guided migration 864
gyroscope 1730

H

HA 1626
Haemodynamics 993
hand 101
hand kinematics 1771
hand prosthesis 93
Handwriting 579
Haptic feedback 1771
HCI design 1417
head orientation 1519
Head Position 1555
Health alarm systems 1290
Health care 1124, 1350
Health emergencies 1209
Health informatics 1286
Health Technology Assessment 1104, 1151, 1132
Health technology assessment and Patient-centred care 1140
Health Technology Management 1104
health worker 1092
Health care Technology Management 1059
Heart 919
heart auscultation 1029
Heart Failure 690, 1354, 1413, 1868, 1899
Heart Rate Variability (HRV) 531, 535, 655, 678, 710, 714, 823, 962, 996, 1342, 1671
Heart sounds 1029

Heart valve 915
Helmholtz coil 833
Hemiplegia 69
hemispatial neglect 1783
hemo-compatibility 1638
Hemodialysis 1233
hemodynamic simulator 1049
Hemodynamics 1037
Hemoglobin A1c 515
Hessian Matrix 674
Heuristic Analysis 1182, 1186
HFPV 591
Hidden Markov Chain (HMC) 742
High frequency 587
high resolution ultrasonography 1049
High Speed Data Acquisition 419
higher order spectra 1722
hippocampus segmentation 174
HISA 1286
HIV 1237
HL-1 cells 340
HL7 1274, 1282, 1294
HL7 CDA 1258
Hodgkin-Huxley 931
Holter 955
home accessibility 1647
Home Care 559, 1096, 1421
Home monitoring 1132, 1205, 1473
homeML and XML 1262
homing 392
Hospital Based Assessment 1144
Hospital Based-Health Technology Assessment 1083
Hospital Information Systems 1306
Hospital Internship 1119
Hospital Management 1111
Hospital-Based HTA 1079
Hough transform 165
HPLC spectra 774
HRI 1919
HRV 690, 1825
HTC 1911
human 710
Human artery 1045
human brain 595
Human factors 1096, 1178
Human Factors Engineering 1182, 1186
Human retina 431
human walking 117
Hurst Exponents 973, 1013
Hyaluronic Acid 1609

hybrid brain computer interface 1655
Hydrocephalus 795
Hypertension 273
Hypnogram 1017
Hypoglycaemia 539, 686

I

ICA 734
ice templating 1638
IFĚTHEN rules 1378
image analysis 194
Image Classification 178, 411
image enhancement 305
Image Guided Surgery 321
Image Management 1270
Image Processing 23, 194, 257, 376 1837
Image quality 471
Image reconstruction 233
Image registration 221
Image segmentation 178, 281, 321 388, 411
image segmentation accuracy 313
Image-guided 41
IMMS 1799
Immunohematology and Transfusion 1079
Impedance cardiography 766, 993
impedance method 1663
impedance microbiology 880
impedance spectroscopy 841
Independent living 1201
Indicators 1310
individual difference 1683
individual healthcare 1402
induced electric fields 1663
induced transmembrane voltage 896
inertial sensing 93
Inertial sensors 101, 1497
infarction 927
Inferior Alveolar Nerve (IAN) 229
inflation test 1041
information modeling 1362
information system development 1358
Information systems infrastructure 1306
Information Technology 1111
Information theory 623, 631
Infra-Red (IR) 1575
Infusion Pump 1182, 1186
initialization 348

inkjet technology 1579
innovation 1124
integrated software suite 911
intelligent sensing 1398
Intensity normalization 218
Intensity-modulated Radiation Therapy 1374
interaction 403
Interaction design 1390
Interactive Video 1803
interdigitated microelectrode 880
Internet of things (IoT) 1221
Interoperability 1294, 1302, 1459
interpolation 1229
interpolation methods 372
intervention 41
intervertebral kinematics 643
Intima-media thickness 249
Intracoronary Optical Coherence Tomography (OCT) 376
intracranial pressure 795
intraluminal thrombus 210
Intraocular Lens 1605
Intrapulmonary Percussive Ventilation 587
Inverse problem of electrocardiography 505
Investment decisions 1151
IOL 1346
ion channel remodeling 927
ionic current 931
IORT 37
Ischemic Stroke 237, 336, 993
ischemic zone 927
ISO EN 13606 standard 1254
ISO/EN 13606 1302
ISO/IEEE 11073 1459
Isolated Langendorff 423
isometric contractions 149
Isophotes 439

J

Joint ECG-PPG modeling 989
joint motion fitting 643
Joint torque 145
K-edge filtering 459
K-edge techniques 463
Kinect sensor 1209
kinematic 1075, 1751
Kinematic analysis 1787
kinematics 61, 81, 101, 1742

Keyword Index

kinetic analysis 301
kinetic model 1734
Kingdom of Saudi Arabia 1907
k-means 706
K-nearest neighbours 563
Knitted piezoresistive sensors 1547
Knowledge extraction 1378
Knowledge management 1266
KSB models 1221
k-space method 427

L

Label fusion 174
lab-on-a-chip (LOC) 856
Laboratory LOINC based coding system 1314
Lagrangian flow field 23
Landmark Correspondence 1837
Landmarks 1694
Laparoscopic surgery 105
Laparoscopic tool tracking 41
Laparoscopy 1845
laplacian ECG 1579
Laplacian electrode 977
laser Doppler flowmetry 647
Laser Doppler perfusion monitoring 49
laser modulation 868
layer-by-layer 1609
Lean 1083
learning 579
legislation 1092
Lempel-Ziv complexity 575, 698
Lens Epithelial Cells 1605
Lesion segmentation 241
LESS 1845
LGE-CMR 165
License 1887
life charting 1358
light emitting diode 888
limb 1730
line detection 487
linear discriminant analysis 1342, 1829
lipid order 19
lipofilling 45
Liver 368
loading rate 1041
local field potential 706, 1698
local field potentials 1722
Locomotion 1575
locus 364

Logistic Models 1237
logistic regression 1825
LOINC 1294
long QT syndrome 531
long-term monitor 943
long-term recordings 876
low cost 856
lower 1730
LTE advanced 1915
Lumen-Endothelium Border 376
Lysinibacillus sphaericus 860

M

machine learning 1354, 1643, 1899
Maching Factor 455
Macroporous scaffold 1630
Magnetic flux density 833
Magnetic induction 1507
Magnetic Resonance Imaging (MRI) 49, 210, 261, 1694, 1767
magnetoencephalogram 583
magnetoencephalography (MEG) 403
malaria 884
mammogram 305
mammography 483
Management 1310, 1350
management dashboard 1322
mandible 229
MAP model 137, 141
Markov 931
mass lesion 305
Matched Filters 411
matching pursuit 791
Mathematical Cognition 783
Mathematical model 666
Mathematical modeling 515
mathematical morphology 281
Mathematics 783
measurement uncertainty 1567
measurements 229
mechanical misalignments 245
mechanical properties 939
Mechanoreceptor 1738
medial prefrontal cortex 1671
median frequency 815
Medical Device 1, 1088, 1571, 1174, 1567
Medical Diagnosis 746
Medical education 1443
Medical Equipment 1059
Medical Equipment life-cycle 1104
Medical image processing 265

medical image registration 277
Medical Image Segmentation 313, 1821
medical instrument 1539
medical ontologies 1362
medical physics 1108
medical QoS 1915
medical technology biomedical engineering 1108
medium frequency current 1115
MEG 1702
melanoma 666
MELFrequency Cepstral Coefficients (MFCC) 742
membrane fluidity 19
meningococcal disease 1382
mental disorder 694
MER signals 742
mesenchymal stem 1817
Mesenchymal stem cellS 392, 1617
Metamodel 1282
Metrics 1751
mfVEP 627
m-health 1409, 1430, 1903, 1907, 1919
micro electrode array 1706
microarray 1849
microbioreactor 888
microcalcifications 459
microcontroller 161, 198
Microelectrode arrays 876
microelectrodes 900, 1841
Microfluidics 841, 852, 864, 884
MicroRNAs 1250
Microscopy 411
Microsoft Kinect 1655
microstructures 852
Microwave 1170
Mild Cognitive Impairment 285
mimicking ECM 1638
miniature robots 105
Minimum Description Length 1837
MIS 52
mismatch response 1702
Mixture models 249
MLP 774
mobile 1451
Mobile applications 1911
mobile diabetes management 1907
mobile wireless sensor network 1398
MOCAP 1555
Model Driven Engineering 1282
Model Interchange Format 1282

MODELICATM Simulation Language 923
modelling 958
Modified electrode 892
Modular System 1318
Module 1849
monitoring 1350, 1464, 1887
monoximetry 1136
Monte Carlo method 431, 427
Monte Carlo simulation 483
morphology 165
morphometry 1871
Motion Analysis 41, 384
motion artifacts 766
motion capture 1497
motion estimation 233
Motion-Related VEP 1809
motivation 1197
motor control theories 719
motor imagery 762
motor skill learning 1771
Motor-Learning 1675, 1679
mouse 399
movement spatial characteristics 1795
MPI 257
MRCP 1659
MRI 157, 218, 221, 225, 368
mRMR 309
MS 206
multiclass 762
multidimensional decision making 1402
multidimensional independent component analysis 726
multidimensional scaling 356
multimodal imaging 451
multimodality 392, 399
multiple circular coils 1663
multiple orientation estimation 317
Multiple sclerosis 241, 627
multiscale analysis 297
Multiscale Entropy 686
murmurs 509
Muscle synergies 719, 1718
muscular architecture 1875
muscular strength 1587
Muscular tonus 1571
Music Therapy 384
Mutual Information 631, 783

N

Nanoelectronic 845
nanohelical periodicity 1613
nanoimprint lithography 1817
Nanoindentation 407
nanoparticle 837
nanowire 860
navigation 57
navigation aid 1763
Naïve Bayes Classifier 985
NcRNA 662
near field exposure 1155, 1159
Nearest Neighbor algorithm 1413
near-infrared spectroscopy 1683
Necrosis 97
network development 1706
neural implementation of CCA 730
Neural networks 352, 911, 1370, 603 1354
Neuro Personal Trainerő 1779
neuro rehabilitation 73
Neuroanatomic Structures 1694
Neuroimage 285
Neuroimaging 1694
Neurological Disorders 1555, 1791
neurological rehabilitation 81
neuromuscular electrical stimulation (NMES) 1791
Neuromuscular fatigue 149
neuronal cultures 876
neuronal network 1706
Neuro-Physiology 827
Neuroprosthesis 1747
neurorobot 85
Neurostimulation 1747
newborn outcome 754
next-generation network 1217
NGO 1136
Night-time heart rate 1511
NMO 206
nodule 395
Noise 766
noise effect 1683
non melanoma skin cancer 615
Non-linear analysis 690, 698
Nonlinear Speech Processing 567, 571
non-obtrusive system 803
non-pharmacological 1136
non-rigid registration 57
non-spherical 837
non-stationary 762
normalized average 509
Normalized Cut 1821
Number of Degrees of Freedom 607
Numerical Simulation 935

O

Object recognition 388
Objective Assessment 1775
Objective Response Detection 607
Object-Oriented Modeling 923
occupation 1092
occupational risk prevention 190
OCT 419
ODK 1382
olfaction 815
One-class classifier 571
Online Social Networks 1439
ontology 1217
Opacification 1346
open access 1298
openEHR 1278
Operating Room 1310
Operating Theatre 1883
Optic neuritis 627
optical diagnosis 427
Optical Flow 384, 475
Optical light 431
Optical Mapping 340, 423
Optical pose tracker 41
Optical Tracking 37
Optimal IED 293
Organization 1009
organization indices 1005, 1879
organ-on-a-chip 884
orientation tracker 1519
Oscillometric blood pressure measurement 958
osteoblasts 1601
Outcomes 1310
Outliers 651
over ground walking 1535
Overall Survival Prediction 1366
overfitting and overlearning in covariates 730
Overhydration 1563
oximetry 969

Keyword Index

P

P300 1710
Parallel architectures 257
parametric identification 591
Parametric methods 1241
Parkinson's disease (PD) 182, 443, 535, 742, 746, 1841
particle swarm optimization 779
Patellar Reflex 1587
path length 799
Patient monitoring 1473
Patient safety 1163, 1883
Patient satisfaction 1140
Pattern 1891
pattern recognition 1795
PCA 309, 989
PDMS 852
pediatric brain tumors 309
pelvis rotation 1535
Pentaho-CE 1101
perception 523
perception tests 1559
performance 137, 141
Perfusion-weighted imaging 218
pericyte 1617
Peritoneal Dialysis 1563
Peritonitis 1813
personal dosimetry SiPM 1398
Personal health records 1402
personalized care 1409
Person-centric 1334
PET 301, 451
pH 1591
Phase Imaging 407
Phase Locked Loop (PLL) 1527, 1539
Phase Rectified Signal Average 823
Phase shift 1539
phase space 599
phenotypes 603
phonocardiography 509
Photoacoustic Tomography 427
Photodynamic therapy 615
Photogrammetry 129, 1063, 1067
Photoplethysmographic signal 1531
photoplethysmography 985
photosensitizer fluorescence 615
physical environment 1178
physical exercising 1197
Physical Neurorehabilitation 1775
physiological monitoring 1511
Physiotherapy 1386

Planification tool 73
planning 45
plasma membrane 19
Plasticity 1659
point of care 872
Point-of-care systems 1435
Polysomnography 1017
portable device 856
position tracking 1539
Positron emission tomography 281
Posterior Capsular Opacification 1605
Posterior Capsule 1346
postural control 719, 1755
Postural Response 69
Postural transitions 1515
Posture 129
Posture Classification 1489
Posture Correction 1489
posture processing 1481
power electronics converter 161
power spectral density 815
Power spectrum 702
PPG waveform 635
PPI 1245
pQCT 289
preclinical research 1108
prediction 1543
predictive model 615
pregnancy 710
Pre-processing 651, 734
Pressure Bed Sensor (PBS) 803
Pressure Control 923
pressure distribution 1551
pressure waves 868
Pressure-distribution sensors 1489
preterm labor 779
prevalent health conditions 1430
Preventive Maintenance 1059
prioritization 1059
Product development 113
Product liability 113
Programmable current source 198
Prony method 627
prospective studies 710
Protein function 1245
protein interaction 1849
prototyping 1421
Pseudomonas fluorescens 860
psychiatry 1358
publish-subscribe pattern 1813
PUK kernel 758
pulley system 153

pulmonary circulation 919
Pulse 1895
Pulse Decomposition Analysis 635
pulse transition time (PTT) 1493
pulsewave recording 1493
pupil size variability 1583
Pupillometry 827

Q

QFD 1059
QoE 1919
QoS 1915
QT variability 531
quadriceps 149
qualify 1887
quality 1567
quality control 467
Quantification 443
Quartz Crystal Microbalance 1597
questionnaire 1455

R

R 301
Rabbit atrium 1871
radial basis functions 372
Radiation 1190
Radiation level monitoring 1398
radiation pattern 1853
radio coverage 1473
Random Forest 1354
rapid prototyping 45
RBF 1455
Reactivity index 539
Real time PCR 872
real-time 57, 1813
real-time monitoring system 1378
Real-time processing 1799
Recalls 1174
recombinant proteins 1634
Reconstruction 257
reconstruction algorithms 479
rectal cancer 178
Reference region 218
regenerative medicine 1634
Regional 1923
regional evaluation 1147
Registration 348, 399, 1559, 1659, 1730, 1759
Relevance rhythm diagram 658

Remote monitoring 1473
reproducibility 81
resection 348
resonance magnetic imaging 190
Respiration 1895
Respiratory Analysis 1063, 1067
respiratory mechanics 591
respiratory sound intensity 981
respiratory sounds analysis 603
Resting State Functional Connectivity 206
retina 269
Retinal Fundus 317, 380
Retinal Imaging 419
Retinal vascular tree 273
Retinal vessels 273
Retinopathy of Prematurity 380
retroperitoneal tumors 360
Reuse of Clinical Information 1274
RFId 1166
RGD Peptide 1605
Risk assessment 1334
Risk Management 1096, 1166, 1170
risk of falls 1503
RM-ODP 1286
RMS 293, 1702
robot - mediated - therapy 81
Robot Assisted Surgery 109
Robot Knotting 109
Robot Surgery 97
robotic 65
Robotic Platform 69
Robotic surgery 113, 1144
Roentgen 1190
Roentgen Rays 1190
root canal length measurement 682
rotating treadmill 1535
Rounded shoulder 129
Running Wheelv 1575
Rural/Frontier Countries 1473

S

Safety 113, 1310
SAHS 1017
Samsung 1911
SAP-Business Objects 1101
saphenous vein 1041
saphenous vein 29
SAP-Net Weaver 1101
sarcoma 427
scaffold 1638

scaffolds 1626
scaling exponents 714
Schizophrenia 1390, 1779, 799
science centre 1124
screening 1326, 1451
searching distance and time 364
second heart sound 509
Secondary Cataract 1605
secondary research development 1254
secondary use 1302
Security 1318
Segmentation 186, 249, 289, 360, 368, 487, 509
self - assembly 1609
self-monitoring 1197
Self-Organizing Map 1233, 1394
Self-reporting 1358
semantic interoperability 1254, 1258, 1266, 1314
Semantic technologies 1290
Semantic Web 1330
sEMG classification 758
sEMG image 674
sEMG, biomechanics 149
Sensing chair 1489
sensing glove 1547
Sensitivity 833, 852
sensor calibration 1543
Sensor data fusion 951
sensor placement 1477
sensorimotor integration 1742
Sensors 1330
sepsis 714
serious gaming 1481
severe disabled people 1647
Shannon energy 509
Shannon entropy 815
Shape-from-Shading 439
shear stress 19
Shewanella oneidensis 860
shooting 141
Short Term Variability (STV) 651
Short Time Fourier Transform 647
Shoulder index 129
Shoulders Position 1555
signal analysis 1871
Signal Processing 599, 722, 1338
signal quality 766
signal source 595
silica nanostructures 1613
silicon chip 872
silicone gel implant 483

SILS 105
silver 849
Simulation 73, 931
Simulator 1767
Single-beam Metrics 1374
Single-cell 864
singular value analysis 415
sinus rhythm 670
Sitting posture 1489
skin blood flow 647
Skin optics 439
skin stimuli 1763
skinfold thickness 121
Sleep 690, 706, 1891
sleep apnea hypopnea syndrome 969, 1825, 1829
sleep posture 1891
Sleep Structure 1017
Sleep Transitions 1017
Sleep-Disordered Breathing (SDB) 803
small-world 1710
smart environments 1213
Smart Home Environments 1262
smartphones' stores 1430
smoothing parameter 643
smoothness measures 1771
SNR 459
social alarm devices 1155, 1159
social exchange theory 1298
Social Isolation 1439
Social Media 1394
Social Network 1330, 1425
social networking for healthcare 1907
Social robotics 1919
Social Sensor 1330
software 45, 1421
Software failures 1174
FDA Enforcement reports 1174
sonomyography 1075
Spasticity assessment 1571
Spatial aliasing 293
Spatial normalization 447
spatiotemporal filtering 543
special data types 1402
Speckle 253
SPECT 443
SPECT/MRI 392, 399
Spectral Analysis 555, 639, 1829
spectral features 803
spectral turbulence 795

Keyword Index

spectroscopic and DSC methods 1622
spermatozoa morphogenesis regulation 1250
spinal anaesthesia 710
spinal cord injury 153, 1759
Spine 1088
spiral 900
spleen 884
spline interpolation 643
Split Bregman 214, 233, 415
spontaneous fluctuations of pupil size 807
Spontaneous Speech 567, 571
Sports performance 157
Stacked learning 261
Standard Interface 1274
Standard Language 1294
Standards 1266
statistical complexity 583, 815
statistical descriptors 750
statistical gait analysis 619
Statistical methods 1140
Statistical Parametric Mapping 1686
Statistical Shape Model 1837
stem cell microenvironment 1613
stenosis 1049
Stent 277, 935, 1037, 1045
step-up DC-DC convertor 198
Stepwise feature selection 1825
Stereotaxy 49
Stress 935
stress test 996
Stroke 1651, 1667, 1767
Stroke rehabilitation 1783, 1795
structural mathematical model 919
structured scintillators 471
subcortical 403
Subthalamic Nucleus(STN) 182, 1841
Superhydrophobic 955
superimposed technique 133
Supervised Classification 380
supervised training 1899
support 1350
Support Vector Machine (SVM) 368, 321, 819, 969, 1825
surface electromyography 611, 977
surface EMG 619
surface functionalization focal adhesion 1817
surface micropatterning 1617
surgery 45
surgery planning 37
Surgical Path 1310

surrogate data 706
Survey 1306
Survival analysis 1241
Sustainability 1310
sustainable design 1390
Suture 109
suture-artery interaction 1053
SVM 774, 1354
Swept Source 419
Symbolic analysis 531
Symmetry 237
synaptic inhibition 575
synchronization 1025
Synchrony analysis 527
Synchrosqueezing 965
syncope 1690
Synthetic atrial electrogram 670
systematic effects compensation 543
systemic circulation 919
systolic blood pressure 678
systolic period 509
systolic time intervals 993, 1021

T

Takens theory 563
Tanzania 1298
Tasks 1779
TAVI 1821
Taylor series 674
teaching methods 579
technology acceptance model 1298
technology design 1163
telehealth 1147
telehealth, symptoms 1455
Telemedicine 1079, 1132, 1221, 1326, 1421, 1451, 1464, 1907
telemonitoring 1132
tele-monitoring 1278, 1455
Tele-monitoring, e-Health 1435
Telerehabilitation 1443, 1447, 1481
Temperature 1833
temperature sensor 880
temporal-spatial parameters 61
temporomandibular joint 77
Tennis serve 145
tensor image 305
texture features 186, 750
Textures 261
TF-IDF 1394
the missing fundamental 523
thermal ablation 1170
thermal expansion 868

thermal stability 1626
Theta band 539
thoracic surgery 1413
Thresholding 1346
Thyroid 395
Ti6Al4V 1601
tibialis anterior 611, 1075
Time Analysis 639
Time constants 1675, 1679
Time synchronization 951
Time variant autorregressive model 658
time-frequency analysis 807
time-frequency distributions 791
time-to-boundary 1755
Time-Variant Autoregressive Model (TVAM) 803
Tissue engineering 1634
titanium alloys for surgical implants 77
TMA (Tissue Microarray) 750
tomography 245, 415
Tone-Entropy analysis 943
Tool 1899
Topographic maps 519
Torque 125
Torso rotation 1535
tortuosity 435
total joint replacement 77
Total Laboratory Automation 1083
tracking 269
Training 1, 1439
trajectory planning 85
Transcranial Magnetic stimulation 1663
Transformed distributions 1241
transmural heterogeneity 1868
transtibial amputees 61
Traumatic Brain Injury (TBI) 1767
Traumatic Brain Injury and Cerebral Vascular Accident 1447
treadmill 153, 611, 1535
treatment 1136
treatment planning 313
Tremor 1747
TrEndo 52
triage 1421
Trypanosoma cruzi 872
tubulogenesis 1617
Tumor 1767
Twitter 1425
Type 2 diabetes 1551

U

u-health 1464
ulcer 297
Ulcerative colitis 435
ultra-high molecular weight polyethylene for surgical implants 77
ultrasonic sensors 868
ultrasonography 395
Ultrasound 125, 253, 1115
ultrathin film 1609
UML 1282
under sampling 202
under sampling pattern 214
Unipolar Cardiac Mapping 527
Unobtrusive monitoring 1507
unsupervised image segmentation 265
Upper limb 73, 81, 145, 1497, 1751, 1759, 1775, 1787, 1791, 1795
Usability 1182, 1186
Usability Problem 1182
use 1
user assessment 65
user model 1217
User preferences 1290
user training 1163

V

Uterine EMG 718, 779
valence 403
Validation 285
variable ranking 969
Ventricular Fibrillation 527
vertical jump height 133
vessel 435
vessel bifurcations 317
Vessel caliber 273
Vessel Segmentation 269, 380
Vestibulo-ocular reflex 1675, 1679
Viability 1833
vibrotactile signals 1763
Video 1779
video streaming 1915
vigilance test 1783
virtual coach 1197
Virtual Reality 45, 57, 1809, 1759
visual biofeedback 1755
visual feedback 1127
Visually impaired 1763
vital signs 1464, 1507
Voice activity detection 1643
voltage divider 595
voltage to current convertor 198
Volume of Tissue Activated (VTA) 182
voluntary contractions 133
Voxel-wise analysis 447

W

walking step 1406
wall thickness 1875
watershed transformation 281
Wave Intensity 1523
Waveform model 989
Wavelet 352
Wavelet Analysis 827
wavelet decomposition 509
wavelet reconstruction 509
wavelet transform 807
Wavelets 372, 722
wavelets analysis 647
Wearable 955
wearable computing 1481
wearable devices 1201
wearable goniometers 1547
wearable sensors 1378
Wearable technologies 1435
Web 2.0, Social Media 1439
web- based system 1413
wedge preparation 1868
wheelchair propulsion 153
wheezes 981
White noise reduction 631
Wireless 1575
wireless body sensor network 1493
wireless sensors 1903
wireless transmission 1451
Women Breast Cancer Dataset 1366
word recognition 547
workplace 1350
World Health Organization 1430
wound area measurement 1071
wound healing 1591
wrapper 1233
Wrist 1485
Wrist-worn 1515

X

X-ray 245
X-ray imaging 455
X-Ray Imaging Simulation 495

Z

Zero crossing 702

Printed by Publishers' Graphics LLC